FIFTH EDITION

MECHANICAL AND ELECTRICAL SYSTEMS
in Architecture, Engineering, and Construction

JOSEPH B. WUJEK
Advanced Building Consultants, LLC

FRANK R. DAGOSTINO

Prentice Hall
Upper Saddle River, New Jersey
Columbus, Ohio

Library of Congress Cataloging-in-Publication Data

Wujek, Joseph B.
  Mechanical and electrical systems in architecture, engineering, and construction / Joseph B. Wujek, Frank Dagostino. — 5th ed.
    p. cm.
  Rev. ed. of Mechanical and electrical systems in construction and architecture / Frank Dagostino, Joseph B. Wujek.
  ISBN-13: 978-0-13-500004-5
  ISBN-10: 0-13-500004-1
  1. Buildings—Mechanical equipment.   2. Buildings—Electric equipment.   I. Dagostino, Frank R.   II. Dagostino, Frank R. Mechanical and electrical systems in construction and architecture.   III. Title.
  TH6010.D33 2010
  696—dc22   2008053804

Vice President and Executive Publisher: Vernon R. Anthony
Acquisitions Editor: Eric Krassow
Editorial Assistant: Sonya Kottcamp
Production Manager: Wanda Rockwell
Creative Director: Jayne Conte
Cover Designer: Bruce Kenselaar
Cover photo: Corbis
Director of Marketing: David Gesell
Marketing Manager: Derril Trakalo
Senior Marketing Coordinator: Alicia Wozniak

This book was set in Times by Aptara®, Inc. and was printed and bound by R.R. Donnelley. The cover was printed by DPC.

Pearson Prentice Hall™ is a trademark of Pearson Education, Inc.
Pearson® is a registered trademark of Pearson plc
Prentice Hall® is a registered trademark of Pearson Education, Inc.

Pearson Education Ltd., London
Pearson Education Singapore Pte. Ltd.
Pearson Education Canada, Inc.
Pearson Education—Japan

Pearson Education Australia Pty. Limited
Pearson Education North Asia Ltd., Hong Kong
Pearson Educación de Mexico, S.A. de C.V.
Pearson Education Malaysia Pte. Ltd.

Prentice Hall
is an imprint of

www.pearsonhighered.com

10 9 8 7 6 5 4 3 2
ISBN-13: 978-0-13-500004-5
ISBN-10: 0-13-500004-1

ACKNOWLEDGMENTS

This book is dedicated to my mother and father, Sophie and Joseph Wujek, Sr., and my family. The constant encouragement I received from my parents causes me to believe that anything is possible through hard work. I thankfully acknowledge my best friend and wife, Shauna, for her patience, assistance, and guidance. Mostly, I thank her for her love and companionship. I also recognize my sons, Blaze and Bryce. They gave Dad the free time needed for the completion of a project of this size. These commitments were all necessary in making this undertaking possible.

I am grateful to those students, faculty, professionals, colleagues, and others who have contributed to this work, either through direct contributions or through feedback. The many professional associations and governmental entities that supplied technical information in this book must be acknowledged. Their cooperation and support are greatly appreciated. I thank the following reviewers for their input: Daphene Cyr Koch, *Purdue University*; Bruce W. Smith, *Auburn University*; and Russell Walters, *University of Florida*.

The original author, the late Frank Dagostino, should be recognized because his insight and effort over many decades served as a foundation for a fine book.

I am thankful to Sonya Kottcamp, Editorial Assistant, Eric Krassow, Acquisitions Editor, and Wanda Rockwell, Production Manager with Pearson Prentice Hall, who all worked patiently and prudently to keep me on schedule (as best they could). Last, Evelyn Perricone, my copy editor, and Nitin Agarwal, my Project Manager, should also be recognized for their hard work in turning my roughly written manuscript into its present professional form.

Joseph Wujek

PREFACE

Mechanical and Electrical Systems in Architecture, Engineering, and Construction, fifth edition, is for those who must have a broad understanding of building mechanical and electrical materials, equipment, and systems to successfully envision, design, draw, construct, evaluate or operate a building or building project. It is written specifically for those interested in building heating, ventilating, and air conditioning (HVAC), plumbing and piping (water supply and sanitary drainage), storm drainage, illumination, electrical power distribution, building telecommunications, acoustics and acoustical control, vertical/horizontal transportation and conveying, fire protection and suppression, and building renewable energy and energy conservation systems.

This book is intended to provide a broad-scope introduction to building mechanical and electrical materials, equipment, systems, design concepts, and engineering principles. It presents material that can provide the future architect, architectural engineer, and architectural engineering technician with a basic working-level knowledge of principles and practices. The audience of this text will likely be undergraduate college and university students in architecture, architectural technology, architectural engineering technology, construction engineering technology, construction management, and elementary architectural engineering.

The fifth edition of this book has been fully reformatted and updated from the previous edition. This transformation is intended to better accommodate its use in introductory college and university courses. Nine chapters have been rewritten to reflect changes in the industry and the other chapters have been updated. Chapters on occupant transportation/conveying systems and emerging sustainability technologies have been added to expand coverage. The new chapter, "Emerging Sustainable Technologies," addresses emerging building mechanical and electrical technologies that are being incorporated into the whole system that makes up an advanced building.

Elementary engineering concepts and design principles are introduced in a straightforward manner. Over 150 new photographs and 40 new figures have been added to help improve the reader's understanding of these subjects. Topics are presented on an intermediate mathematics level, requiring that the student have a working knowledge of college algebra. Homework exercises and design problems are written with the intent of introducing basic principles with, in most instances, real-world connections.

A text like this is needed because those in the architecture, engineering, and construction (AEC) industry must have an understanding of whole building design. Building mechanical and electrical technologies are integral elements that make up the whole building system—that is, they are really elemental systems within a single system, each of which must function and interact effectively with the other systems. Successful integration during design and construction depends on knowledge and teamwork of all involved—the architectural designer, the mechanical and electrical engineer, the technician, the draftsperson, the construction manager, the general contractor, the subcontractors, workers in the trades, and facilities managers and personnel. All involved should be familiar with basic design strategies, construction procedures, system characteristics, space requirements, and the time frame and progression at which such work must be done on the job. Simply put, everyone involved in building design, construction, operation, maintenance, and deconstruction should be familiar with building mechanical and electrical systems.

Joseph Wujek

A general introduction to common building industry practices and trends ensures that the reader has a basic understanding of the industry. Such an understanding is beneficial because it validates the need for all building industry professionals to understand the subjects presented in this text: building mechanical and electrical materials, components, equipment, and systems. This introduction is particularly helpful to the reader who has little or no experience in the building industry.

THE BUILDING INDUSTRY

The global architecture, engineering, and construction (AEC) industry accounts for about 10% of the world's gross domestic product, 7% of all employment, and approximately half of all resource use, including about 40% of all energy consumption. In the United States, the AEC industry is over a trillion dollar business ($1069 trillion for construction alone in 2005). In 2005, the U.S. construction industry directly employed 7.3 million people and another 1.3 million people in architecture and engineering. The AEC industry is big business in the United States and worldwide.

In the AEC industry, *architects* and their support staff design buildings, while *engineers* and their support staff design the engineering systems within these buildings. *Constructors,* serving as contractors, and their employees and subcontractors build buildings. *Construction managers* supervise the construction project. *Facilities managers* and staff operate and maintain buildings. All players must effectively work together as the building design, construction, and operation team.

THE BUILDING MECHANICAL AND ELECTRICAL SYSTEMS

Well-designed, modern buildings are made up of many components and pieces of equipment that are integrated so that, when they are operated and maintained properly, they mutually perform as a single system. Simply put, an efficient building system is made up of many elemental systems. In buildings, mechanical and electrical technologies are among the most expensive and labor-intensive of these elemental systems. These mechanical and electrical technologies are used for heating, ventilating, and air conditioning (HVAC), illumination, electrical power distribution, plumbing and piping (water supply and sanitary drainage), storm drainage, building telecommunications, acoustics and acoustical control, vertical/horizontal transportation and conveying, fire protection and suppression, renewable energy sources, heat recovery, and energy conservation.

Mechanical and electrical systems in the building construction industry fit within classifications known as *mechanical/electrical/plumbing (MEP)* or *electrical/mechanical/plumbing/fire protection (E/M/P/FP)* systems. MEP systems influence occupant health, comfort, and productivity, and greatly affect costs, including the first cost and operating (energy use and maintenance) costs. MEP systems are the heart and nervous system of a building.

ENVIRONMENTAL IMPACT OF BUILDINGS

The earth's natural resources are limited and world population continues to increase. With the passing of each day, there is a greater and greater reliance on natural resources and more degradation of the environment. Buildings account for a large amount of resource (energy and water) consumption, land use, atmospheric greenhouse gas emissions, and generation of environmental waste and pollution.

With about 4.5% of the world's population, the United States consumes nearly 23% of the total global energy. This means that the U.S. consumes energy over 5 times the world per capita average and over 100 times more per capita than many undeveloped countries. The United States is not alone in its energy-use intensity. Countries like Qatar, Kuwait, Norway, and Canada use energy at a higher per capita rate. From a global perspective, more developed, industrialized countries (e.g., countries in Europe, North America, Australia, New Zealand, and Japan) make up only about 17% of the world's population but use about three-quarters of the world's energy resources.

About 40% of the energy consumed in the United States is used in buildings. Thus, U.S. buildings use between 9 and 10% of the energy consumed worldwide. Of the energy consumed in U.S. buildings, about 40% is for space (comfort) heating, cooling, and ventilation; about 18% is used for lighting; and almost 20% is used for domestic water heating. (See Tables P.1 and P.2). About two-thirds of the electrical power produced is consumed in buildings. In addition, U.S. buildings generate about 40% of the atmospheric emissions that make up greenhouse gases. Comparable magnitudes are used in most developed European and Asian countries. As a result, MEP systems have a significant influence on global resource consumption and associated waste and pollution.

Developed and developing countries are totally dependent on natural (material and energy) resources. Many less developed countries (e.g., countries in Africa, Asia [excluding Japan], regions of Melanesia, Micronesia, and Polynesia, Latin America,

TABLE P.1   RESIDENTIAL BUILDING ENERGY USE,
BY PERCENTAGE.

End Use	Single Family	Multifamily
Space heating	52%	33%
Appliances	22%	23%
Water heating	17%	32%
Refrigerators	4%	6%
Space cooling	4%	5%
Dishwashers	1%	1%

Computed from data provided in U.S. DOE Commercial Buildings Energy
Consumption Survey (CBECS) 2003.

and the Caribbean) are striving to become more industrialized. Many Asian countries (e.g., China, India, Taiwan, and South Korea) and some Middle Eastern countries (e.g., Dubai, Qatar, and United Arab Emirates) are examples of countries experiencing rapid growth. These countries are becoming more resource-use intensive at a time when their rate of population growth is substantial. As developing countries move toward industrialization, resource use in these countries increases substantially, so that limited global resources are taxed more and will be exhausted sooner. A growing global population coupled with ever increasing reliance on natural resources combines to create an outcome that is alarming. This concern makes a strong case for integration of sustainable design practices in building MEP systems.

SUSTAINABLE BUILDINGS

Sustainability is our ability to meet current needs without harming the environmental, economic, and societal systems on which future generations will rely for meeting their needs. It simply means using resources wisely. A *sustainable* or *green building* is designed to lessen the overall impact of a building on the en-

vironment and human health by efficiently using resources (i.e., energy, materials, and water), enhancing occupant health and employee productivity, and eliminating or reducing waste and pollution. Reducing the amount of natural resources required in constructing and operating buildings and the amount of pollution generated by buildings is crucial for future sustainability. In buildings, this can be accomplished by effectively using materials, increasing efficiency, and, developing and using new and renewable energy technologies.

MEP DESIGN AND LAYOUT

MEP components and equipment influence building design and layout. Dedicated building spaces or rooms must be reserved for MEP components and equipment and serve as the nucleus of these technologies. This can include, but is not limited to, central utility plants, boiler and chiller rooms, fuel rooms, electrical switchboard rooms, transformer vaults, and metering and communications closets. These spaces can make up a significant portion of a building floor area. Large commercial buildings have a single mechanical room of considerable size and often require additional rooms throughout the building. Skyscrapers may have mechanical spaces that occupy one or more complete floors. In contrast, a small commercial building or single-family residence may only have a small utility room.

The size of MEP rooms is typically tied to building occupancy type and is usually proportional to the building size; that is, hospitals and medical centers require more MEP space than schools, offices, and residences. For example, in offices, department stores, and schools, the MEP floor area is typically in the range of about 3 to 8% of the gross floor area; in hospitals, it is about 7 to 15%; and in residences, it is typically less than 3%. Allowances must be made by the building designer to locate spaces near the habitable spaces, especially those spaces with the largest demand for heating, cooling, power, and water (i.e., kitchens, restrooms, bathrooms, and so forth).

TABLE P.2   COMMERCIAL BUILDING ENERGY USE, BY PERCENTAGE.

End Use	Office	Health Care	Retail	K–12 Schools	Colleges/ Universities	Governmental	Lodging
Space heating	25%	23%	30%	45%	32%	36%	16%
Space cooling	9%	4%	10%	6%	5%	5%	6%
Ventilation	5%	3%	4%	2%	2%	3%	1%
Water heating	9%	28%	6%	19%	24%	17%	41%
Lighting	29%	16%	37%	19%	22%	21%	20%
Cooking	1%	5%	3%	2%	1%	2%	4%
Office equipment	16%	6%	4%	2%	2%	6%	3%
Refrigeration	Negl.	2%	1%	1%	1%	2%	2%
Miscellaneous	5%	13%	5%	2%	11%	7%	6%

Computed from data provided in U.S. DOE Commercial Buildings Energy Consumption Survey (CBECS) 2003.

TABLE P.3   CHARACTERISTIC PERCENTAGE OF CONSTRUCTION AND COST BREAKDOWN OF A 100 000 FT² OFFICE BUILDING PROJECT.

Division of Work	Percentage of Construction	Cost Breakdown	Description
General conditions	3%	$ 750 000	General procedures, superintendent, trailer, fence, traffic control, insurance
Site work	4%	$ 1 000 000	Excavation/backfill, roads, walks, landscaping
Concrete	8%	$ 2 000 000	Concrete foundations, framing, slabs
Masonry	7%	$ 2 000 000	Concrete masonry, brick, stone, reinforcement, mortar, grout
Metals	11%	$ 2 750 000	Structural steel framing, light-gauge framing
Woods and plastics	4%	$ 1 000 000	Wood framing, millwork, cabinetry
Thermal/moisture protection	10%	$ 2 500 000	Insulation, roof coverings, caulking, cladding
Doors and windows	8%	$ 2 000 000	Doors, windows, glass, and glazing
Finishes	9%	$ 2 250 000	Drywall, plaster, floor and ceiling coverings, paint
Specialties	4%	$ 1 000 000	Signs, flagpoles, restroom accessories
Equipment	5%	$ 1 250 000	Kitchen equipment, laboratory casework
Furnishings	4%	$ 1 000 000	Artwork, furniture, room partitions
Special construction	3%	$ 750 000	Computer rooms, clean rooms
Conveying systems	4%	$ 1 000 000	Elevators, escalators, moving ramps, walkways
Mechanical	10%	$ 2 500 000	HVAC, plumbing, fire protection
Electrical	6%	$ 1 500 000	Power distribution, lighting, telecommunication
Total	100%	$25 000 000	Complete project construction costs

MEP SYSTEM COSTS

Design and construction costs of MEP components and equipment are significant in buildings. Commercial and institutional buildings and large residences necessitate that an engineer design the MEP systems, whereas, in small residences, design is done by the mechanical and electrical trades. Design fees for MEP systems in commercial buildings typically range from 20 to 40% of the overall design costs, depending on building occupancy type and size.

Actual MEP construction costs for buildings vary by building occupancy type and construction method. In residences and retail stores, the range is generally between 10 and 20% of the construction costs; for kindergarten through high schools (K–12), it ranges from 15 to 30%; for office and university classroom buildings, it ranges between 20 and 30%; and for hospitals and medical centers, it is typically between 25 and 50%. A characteristic percentage of construction and cost breakdown of a commercial office building project is provided in Table P.3. In this example, mechanical and electrical systems (including conveying systems) account for 20% of the overall project construction costs.

MEP ASSOCIATIONS, SOCIETIES, AND AGENCIES

Many professional associations and trade organizations support the various MEP fields. Professional and trade associations are membership organizations, usually nonprofit, that serve the interests of members who share a common field and promote professional and technical competence within the sustaining industry. Professional organizations, frequently called *societies*, consist of individuals of a common profession, whereas *trade associations* consist of companies in a particular industry. However, the distinction is not uniform; some professional associations also accept certain corporate members, and conversely, trade associations may allow individual members. The activities of both trade and professional associations are similar and the ultimate goal is to promote, through cooperation, the economic activities of the members while maintaining ethical practices. Additionally, professional associations have the objectives of expanding the knowledge or skills of its members and writing professional standards. Many *governmental agencies* also exist that support the work of these industries. Examples of professional, trade, and governmental entities are provided in Table P.4.

CONSTRUCTION STANDARDS AND BUILDING CODES

In the AEC industry, a *standard* is a set of specifications and design/construction techniques written by a standards writing organization (see Table P.4) or group of industry professionals that seek to standardize materials, components, equipment, or methods of construction and operation. In the United States, a *building code* is a law adopted by a state or is an ordinance (a local law) approved by a local authority (a municipality or county) that establishes the minimum requirements for design, construction, use, renovation, alteration, and demolition of a building and its systems. The intent of a building code is to

TABLE P.4   EXAMPLES OF PROFESSIONAL SOCIETIES, TRADE ASSOCIATIONS, GOVERNMENTAL
AGENCIES, AND STANDARDS AND CODE-WRITING ENTITIES.

Professional Societies

- American Institute of Architects
- American Society of Heating, Refrigerating and Air-Conditioning Engineers
- American Society of Mechanical Engineers
- American Society of Plumbing Engineers
- American Society of Sanitary Engineering
- Architectural Engineering Institute of the American Society of Civil Engineers
- Association of Energy Engineers
- Illuminating Engineering Society of North America
- National Council of Acoustical Consultants
- National Society of Professional Engineers
- Refrigeration Service Engineers Society
- Society of Fire Protection Engineers
- Society of Women Engineers

Standards and Code-Writing Entities

- American National Standards Institute
- American Society of Testing and Materials
- International Association of Plumbing and Mechanical Officials
- International Code Council
- International Fire Code Institute
- National Fire Protection Association
- Underwriters Laboratories

Governmental Agencies

- National Institute of Building Sciences
- U.S. Department of Energy

- U.S. Environmental Protection Association
- National Renewable Energy Laboratory

Trade Associations

- Air Conditioning Contractors of America
- Air Movement and Control Association
- Air-Conditioning and Refrigeration Institute
- American Boiler Manufacturers Association
- American Gas Association
- American Water Works Association
- Construction Specifications Institute
- Gas Appliance Manufacturers Association
- Heating, Air Conditioning & Refrigeration Distributors International
- Hydronic Heating Association
- International Telecommunications Union
- Mechanical Contractors Association of America
- National Association of Electrical Distributors
- National Association of Home Builders
- National Association of Lighting Management Companies
- National Electrical Contractors Association
- National Electrical Manufacturers Association
- National Fire Protection Association
- National Fire Sprinkler Association
- Plumbing-Heating-Cooling Contractors Association
- Sheet Metal and Air Conditioning Contractors' National Association
- U.S. Green Building Council

ensure health, safety, and welfare of the building occupants. Building codes began as fire regulations written and enacted by several large cities during the 19th century, and have evolved into a code that contains standards and specifications for materials, construction methods, structural strength, fire resistance, accessibility, egress (exiting), ventilation, illumination, energy conservation, and other considerations.

A *model building code* (i.e., International Building Code, National Electrical Code, and International Mechanical Code) is a standardized document written by a standards writing organization (a group of professionals) and made available for adoption by state and local jurisdictions. A municipality, county, or state may write its own building code, but typically it relies on adoption of model codes as the base of its building code, mainly because it is easier. Amendments are usually made to the text of a model code to address local issues. Some states adopt a uniform statewide building code while others legally assign code adoption to local authorities (counties and municipalities). Technically, a model building code is not a code (a law) until it is formally adopted. Model codes are periodically revised, usually every 3 to 5 years, to remain current with advancements and new practices in industry. Each time a model code is revised, it needs to be reviewed and adopted into law by the governmental authority having jurisdiction (control). As a result, different code editions may be in effect in neighboring municipalities at a specific time, which can cause confusion that can lead to design/construction errors. Professionals in the AEC industry must become familiar with and maintain a working-level understanding of current codes and standards, and must work hard to keep abreast of revisions in each edition of the code.

CONTENTS

THERMAL, ENVIRONMENTAL, AND COMFORT CONCEPTS

1.1 THERMAL CONCEPTS

Heat

Heat is the agitation or motion of atoms and molecules. It is thermal energy in motion. Heat always flows from a substance at a higher temperature to the substance at a lower temperature, raising the temperature of the lower temperature substance and lowering the temperature of the higher temperature substance.

Quantity of heat (Q) is measured in British thermal units and joules. In scientific terms, the *British thermal unit (Btu)* is defined as the amount of heat required to raise the temperature of 1 lb (0.45 kg) of water from 59.5°F (15.3°C) to 60.5°F (15.8°C) at constant pressure of standard atmosphere pressure. In broad terms, one Btu is about the amount of heat given off by the combustion of one wooden match. The Btu is equivalent to 1055 *joules (J)*. A Btu is equivalent to 0.293 watt-hour (W-hr). Relationships are as follows:

$$1 \text{ Btu} = 1055 \text{ joules (J)}$$
$$= 0.293 \text{ watt-hour (W-hr)}$$
$$= 252 \text{ calories (c)}$$
$$= 0.252 \text{ kilocalories (C)}$$
$$1 \text{ Joule} = 0.00095 \text{ Btu}$$
$$= 0.239 \text{ calories (c)}$$
$$1 \text{ watt-hour (W-hr)} = 3.413 \text{ Btu}$$
$$1 \text{ kilocalorie (C)} = 1000 \text{ calories}$$

In the construction industry, heat is customarily expressed in much larger quantities, such as one thousand Btu (called the MBtu or kBtu) or one million Btu (MMBtu). M is the Roman numeral that represents 1000, so MM is one million (1000 times 1000). A *therm* is used to express the energy content of fuels and equal to 100 000 Btu.

$$1000 \text{ Btu} = 1 \text{ MBtu} = 1 \text{ kBtu}$$
$$1\,000\,000 \text{ Btu} = \text{MMBtu} = 1000 \text{ kBtu}$$
$$1 \text{ therm (natural gas)} = 100\,000 \text{ Btu}$$
$$1000 \text{ joules (J)} = \text{kilojoules (kJ)}$$

The *heating (calorific) value* of a fuel, including food, is the quantity of heat produced by its combustion under specified conditions. It is the energy released as heat when it undergoes complete combustion with oxygen. Some heating values of common fuels are shown in Table 1.1. Magnitudes of energy consumption are provided in Table 1.2.

A typical home in a heating climate will consume about 100 million Btu of heating energy over the heating season.

Temperature

Temperature (T) is the measure of the average kinetic energy associated with the chaotic microscopic motion of atoms and molecules within a substance. It is the measure of the intensity of the heat. The temperature of an object determines the sensation of warmth or coldness felt from contact with it. The same temperature relates to the same average kinetic energy in a substance.

A *thermometer* is an instrument that measures the temperature of a body or substance in a quantitative way. A thermometer can be read on any of a number of different scales, as

TABLE 1.1   HEATING (CALORIFIC) VALUES OF COMMON FUELS.

Fuel	Heating (Calorific) Value[a]
Coal (bituminous)	18 100 000 Btu/ton
Fuel oil number 2	140 000 Btu/gal
Natural gas	100 000 Btu/CCF[b]
Propane	91 000 Btu/gal
Wood (Lodgepole Pine)	21 000 000 Btu/cord[c]
Wood (oak)	30 700 000 Btu/cord[c]

[a]The heating value varies with quality of fuel.
[b]A hundred cubic feet (CCF) is 100 ft^3. Value of gas varies with altitude. Value of natural gas is typically about 83 000 CCF in Denver, CO, at 5280 ft (1600 m) above sea level.
[c]A cord is a tightly stacked $4' \times 8' \times 8'$ pile of wood, that is, 128 ft^3 of wood and space.

TABLE 1.2   MAGNITUDES OF ENERGY CONSUMPTION.

Use	Energy Consumption
Heat house for year in moderate heating climate	100 000 000 Btu
Apollo 17 to the moon	5 600 000 000 Btu
Hiroshima atomic bomb	80 000 000 000 Btu
Annual use—30 African countries	1 000 000 000 000 000 Btu
Annual use—United States	100 000 000 000 000 000 Btu
Annual use—world	470 000 000 000 000 000 Btu

TABLE 1.3   THE RELATIONSHIP BETWEEN TEMPERATURE SCALES AT SELECTED TEMPERATURES.

Fahrenheit (°F)	Rankine (R)	Celsius (°C)	Kelvin (K)	Associated Temperature
−459.67	0	−273.15	0	Absolute zero
0	460	−18	255	
32	492	0	273	Water freezes
72	532	22	295	Typical target room air temperature (winter)
78	538	26	299	Typical target room air temperature (summer)
98.6	558	37	310	Average body temperature
200	660	93	366	Boiling point of water at 6500 ft (1987 m) above sea level
212	672	100	373	Boiling point of water at standard conditions

described in the paragraphs that follow. Table 1.3 shows the relationship between temperature scales at selected temperatures.

Common temperature scales are *Fahrenheit (°F)* and *Celsius (°C)*. They are defined by using the point at which ice melts and water boils at the standard atmospheric pressure. On the Celsius scale, the interval between the ice point and boiling point of water are divided into 100 equal parts. The Celsius ice point is at 0 and the boiling point is at 100. (Actually, on the Celsius scale the boiling point of water at standard atmospheric pressure is 99.975°C in contrast to the 100 degrees defined by the Centigrade scale.) The Fahrenheit ice point is at 32 and the boiling point is at 212. Conversion from one form to the other is as follows:

$$°C = (°F − 32)/1.8$$
$$°F = (1.8 × °C) + 32$$

For example, 20°C is equal to 68°F [°F = (1.8 × 20°C) + 32]. In other words, 20°C is approximately equal to normal room temperature.

In engineering and scientific computations, it is often necessary to express temperature on an *absolute temperature* or *thermodynamic temperature* scale. At absolute zero, molecular motion is at rest and the substance contains no internal energy. Absolute zero is the lowest temperature possible. For all substances this occurs at about −459.67°F (−273.15°C).

The *Kelvin (K)* scale has its zero point at −273.15°C and the *Rankine (R)* temperature scale has its zero point at −459.67°F. Therefore, the ice point of water occurs at 273 K and 460 R. The degree sign (°) is not customarily used when expressing temperature on the Kelvin and Rankine scales. Some engineering fields in the U.S. express thermodynamic temperature using the Rankine scale, but in scientific fields the Kelvin scale is used exclusively. Conversion is found by the following expressions, where °F is degrees Fahrenheit, R equals degrees Rankine, °C is degrees Celsius, and K equals degrees Kelvin:

$$R = °F + 459.67$$
$$K = °C + 273.15$$

The difference between heat and temperature is often confused. Consider two containers of water: a small container holding one cup of water at 150°F and the other holding ten cups of water at the same temperature. The water in both containers is at the same temperature. However, the larger container holds ten times the amount of heat (in Btu or J).

Density

Mass density (ρ) of a substance is the mass per unit volume. Density is expressed in units of pounds per cubic foot (lb/ft³) or kilograms per cubic meter (kg/m³). Densities of common materials are provided in Table 1.4.

Specific Heat

In defining the fundamental definition of the Btu, it was found that one pound of liquid water increases in temperature by one °F when one Btu of heat is added. Samples of other substances react differently to the addition or removal of a given amount of heat. This reaction is dependent on the molecular composition of a substance. *Specific heat (c)* is defined scientifically as the amount of heat that must be added or removed from one pound of substance to change its temperature by one degree. Specific heats of common materials are provided in Table 1.4. The unit of specific heat in the engineering system is Btu per pound per degree temperature change (Btu/lb · °F) or Joules per pound per degree temperature change (J/kg · °C). In equation form:

$$ΔT = Q/cM$$
$$Q = cM\,ΔT$$

where ΔT equals the change in temperature, M equals the mass of the substance in lb or kg, Q equals the amount of heat removed or added in Btu or J, and c equals specific heat of the substance in Btu/lb · °F or J/kg · °C.

For example, the specific heat of standard concrete is 0.22 Btu/lb · °F. It takes the addition or removal of 0.22 Btu to change the temperature of one pound of concrete one °F. The

TABLE 1.4   SELECTED PHYSICAL AND THERMAL PROPERTIES OF COMMON MATERIALS. THERMAL PROPERTIES ARE AT A TEMPERATURE OF 68°F (20°C), UNLESS SPECIFIED OTHERWISE. DATA WERE COMPILED FROM A VARIETY OF SOURCES.

Product	Mass Density		Specific Heat		Specific Heat Capacity	
	kg/m^3	lb/ft^3	$kJ/kg \cdot K$	$Btu/lb \cdot °F$	$kJ/m^3 \cdot K$	$Btu/ft^3 \cdot °F$
Water						
Water (distilled) at 39°F (4°C)	1000	62.4	4.19	1.00	4190	62.4
Water (distilled) at 68°F (20°C)	99.8	62.3	4.15	0.99	414	61.7
Water (seawater) at 39°F (4°C)	1030	64	3.94	0.94	4058	60.2
Ice at 32°F (0°C)	910	57	2.11	0.504	1920	28.7
Liquids						
Alcohol	790	49	2.93	0.70	2315	34.3
Ammonia	610	38	0.47	0.11	287	4.2
Ethylene glycol	1104	69	2.38	0.57	2628	39.4
Glycerin	1270	79	2.41	0.58	3061	45.8
Paraffin	810	50	2.14	0.51	1733	25.5
Gasoline	700 to 750	44 to 47	2.09	0.50		
Turpentine	870	54	1.98	0.33	1723	17.8
Solid Metals and Alloys						
Aluminum	2690	168	0.912	0.218	2453	36.6
Brass	8100	505	0.377	0.090	3054	45.5
Bronze	8450	529	0.435	0.104	3676	55.0
Copper	8650	551	0.389	0.093	3365	51.2
Gold	19 200	1200	0.130	0.031	2496	37.2
Iron (cast)	7480	467	0.460	0.110	3441	51.4
Iron (wrought)	7850	486	0.460	0.110	3611	53.5
Lead	11 340	705	0.130	0.031	1474	21.9
Nickel	8830	551	0.452	0.108	3991	59.5
Platinum	21 450	1340	0.134	0.032	2874	42.9
Silver	10 500	655	0.234	0.056	2457	36.7
Steel	7824	489	0.460	0.120	3599	58.7
Tin	7280	455	0.230	0.055	1674	25.0
Zinc	7200	444	0.393	0.094	2830	41.7
Solids						
Asphalt	1650	103	0.80	0.19	1320	19.6
Brick	1000 to 2000	62 to 134	0.92	0.22	920–1840	13.6–29.5
Ceramic porcelain	2300	143	1.07	0.26	2461	37.2
Concrete	2240	140	1.13	0.27	2531	37.8
Glass	2640	164	0.84	0.20	2218	32.8
Granite	2130	133	0.75	0.18	1598	23.9
Gypsum	1249	78	1.09	0.26	1364	20.3
Limestone	3170	198	0.84	0.20	2663	39.6
Marble	2650	165	0.88	0.21	2332	34.7
Paraffin	897	56	2.89	0.69	2592	38.6
Plaster	1180	73	0.84	0.20	991	14.6
Rubber	920	67	1.1 to 4.1	0.27 to 0.98	1012 to 3772	18.1 to 65.7
Sand/gravel	1441 to 2000	90 to 125	0.82	0.19	1182 to 1640	17.1 to 23.8
Wood	700 to 900	44 to 56	2.3 to 2.7	0.55 to 0.65	1610 to 2430	24.2 to 36.4

addition of one Btu would raise the temperature of one pound of standard concrete 4.55°F.

Specific Heat Capacity

The property of specific heat capacity is similar to specific heat, except that it is expressed with volume rather than weight.

Specific heat capacity (C) is defined as the amount of heat required to change the temperature of a specific volume of substance one degree. The property of specific heat capacity of a substance is measured in Btu per cubic foot per degree temperature change ($Btu/ft^3 \cdot °F$). Specific heat capacities of common materials are provided in Table 1.4.

Specific heat capacity (C) of a substance in units of Btu/ft^3 · °F (J/m^3 · °C) is found by the following expression where specific heat (c) is expressed in Btu/lb · °F (J/kg · °C) and density (ρ) is expressed in lb/ft^3 (kg/m^3):

$$C = c\rho$$

Example 1.1

The specific heat of water is 1.0 Btu/lb · °F and its density is 62.4 lb/ft^3. Determine the specific heat capacity of water.

$$C = c\rho = 1.0 \text{ Btu/lb} \cdot °F \cdot 62.4 \text{ lb/ft}^3 = 62.4 \text{ Btu/ft}^3 \cdot °F$$

Example 1.2

The specific heat of concrete is 0.27 Btu/lb · °F and its density is 140.0 lb/ft^3. Determine the specific heat capacity of concrete.

$$C = c\rho = 0.27 \text{ Btu/lb} \cdot °F \cdot 140.0 \text{ lb/ft}^3 = 37.8 \text{ Btu/ft}^3 \cdot °F$$

Sensible and Latent Heating

Sensible heat is the heat associated with change in temperature of a substance. It appears that the addition or removal of heat from a substance changes its temperature in proportion to the heat added or removed; that is, as heat is added to a substance its temperature typically increases, and when heat is removed the temperature of the substance typically decreases. This phenomenon is referred to as *sensible heating*. It is "sensible" because it can be observed by the sense of touch.

Sensible heating does not occur when the substance undergoes a change in state from a solid to a liquid, a liquid to a gas, or visa versa. Materials exist in three states: as solids, liquids, and gases. When heat is added to a substance, it may change its state from a solid to a liquid or from a liquid to a gas. It may change from a gas to a liquid or from a liquid to a solid when heat is removed. A change of state is commonly referred to as a *phase change*. Heat is stored or released whenever a substance changes phase.

Latent heat is the release or storage of heat associated with change in phase of a substance, without a change in the substance's temperature. It is the "hidden" quantity of heat absorbed or released when a material changes phase without a change in temperature. For example, when a container of water is heated, the temperature of the water will increase until it reaches its boiling point. With additional heating, the water begins to go through a phase change from liquid water to water vapor. As heat continues to be added, more of the liquid water is changed to water vapor. There is no increase in temperature. The water remains at the boiling temperature. Heat is required to cause the liquid water to undergo a phase change from a liquid to a gas. Heat absorbed by the water that causes this phase change is latent heat.

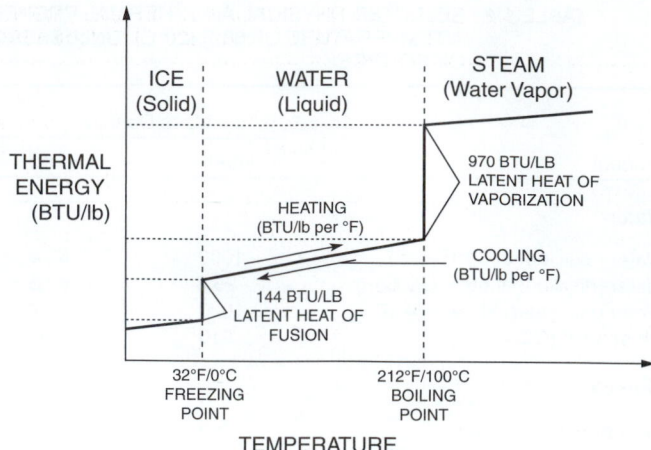

FIGURE 1.1 A graph of changes in enthalpy (thermal energy) of water with temperature change. Drastic changes in thermal energy (added or released) occur when water makes a phase change between ice and water (latent heat of fusion) and when it makes a phase change between water and steam (latent heat of vaporization).

Latent heat of vaporization is required when a substance changes from a liquid to a gaseous state. It is released when the substance changes from a gaseous to a liquid state. The latent heat of vaporization of water is about 970 Btu/lb. As water vaporizes from a liquid to a gas, it absorbs 970 Btu/lb. Conversely, when water condenses from a gas to a liquid, it releases 970 Btu/lb of heat.

Latent heat of fusion occurs when a substance is converted from a liquid to a solid state. For water, the latent heat of fusion is 144 Btu/lb. As one pound of water is changed to ice, 144 Btu are released. When ice changes back to a liquid, an equivalent amount of heat is absorbed.

Figure 1.1 is a graph of changes in thermal energy (enthalpy) in water with temperature change. Drastic changes occur when water makes a phase change between ice and water (latent heat of fusion) and when it makes a phase change between water and steam (latent heat of vaporization). Different materials behave differently when there temperatures change; they have sensible heats and latent heats. Latent heat of fusion and latent heat of vaporization for selected materials are provided in Table 1.5.

The temperature at which a change of phase occurs is dependent on the chemical composition of the substance and the pressure at which it is stored. As a substance changes phase from a solid to liquid, or liquid to gas, energy is absorbed. Energy is released when a substance changes from a gas to liquid or liquid to solid. Energy is store when a substance changes from a liquid gas to liquid or a solid to liquid.

Power

Power is defined as the measure of energy consumed over a period of time. When energy is extracted, converted into a useful form, and used in an application, power is the result. The

TABLE 1.5 LATENT HEAT OF FUSION AND LATENT HEAT OF VAPORIZATION FOR SELECTED MATERIALS. DATA WERE COMPILED FROM AN ASSORTMENT OF SOURCES.

Product	Freezing Point Temperature		Boiling Point Temperature at Standard Conditions		Latent Heat of Fusion		Latent Heat of Vaporization	
	°F	°C	°F	°C	Btu/lb	kJ/kg	Btu/lb	kJ/kg
Water	32	0	212	100	143.5	334	970.3	2257
Other Materials (for Comparison)								
Alcohol, ethyl	−179	−117	173	78	46.4	98	368	790
Ammonia	−108	−78	−28	−33	48.8	103	589	1248
Ethylene glycol	13	−11	388	198	77.9	165	344	729
Common Refrigerants								
HCFC-12 (R-12)	−252	−158	−21.6	−29.8	—	—	71.0	150
HCFC-22 (R-22)	−256	−160	−41.4	−40.8	—	—	100.5	213
HCFC-123 (R-123)	−161	−107	82.2	27.9	—	—	73.3	155
HFC-134A (R-134A)	−142	−97	−15.1	−26.2	—	—	93.2	197
HFC-410A (R-410A)				−51.7	—	—		136

amount of energy delivered by a furnace firing at maximum capacity is its output power rating. A furnace power rating depends upon how fast the energy is converted, which means time must be considered.

The unit used to define power in the customary system is Btu per hour (Btu/hr). For brevity, it is sometimes written as Btu/hr (or MBH for thousand Btu per hour and MMBH for million Btu per hour). The amount of energy used to fuel a furnace at its maximum firing capacity (usually for a period of one hour) is its input power rating (e.g., the power rating of an appliance that can deliver 100 000 Btu in an hour is 100 MBH). Examples of power consumption are provided in Table 1.6.

When referring to electricity, the watt (W) is used to define power. Oftentimes the watt is written with a prefix K that is equal to a kilowatt (kW) or one thousand watts. In addition,

TABLE 1.6 EXAMPLES OF POWER CONSUMPTION.

Use	Power Capacity
Bring 1 gallon of tap water to boil	9500 Btu/hr
Large room air conditioner (1 ton)	12 000 Btu/hr
House air conditioner (5 ton)	60 000 Btu/hr
Home furnace capacity	100 000 Btu/hr
Automobile on highway (25 mpg)	375 000 Btu/hr
Home furnace	100 000 Btu/hr
High school chiller capacity (100 ton)	1 200 000 Btu/hr
High school boiler capacity	10 000 000 Btu/hr
Space shuttle takeoff	37 500 000 000 Btu/hr
Average hourly energy use— United States	11 000 000 000 000 Btu/hr
Average annual use— United States	100 000 000 000 000 000 Btu/year

horsepower (hp) may be used as a unit of power. The watt, Btu/hr, and horsepower are interrelated because they refer to power:

$$1\ W = 3.413\ Btu/hr$$
$$1\ hp = 745.7\ W$$
$$1\ hp = 2544.3\ Btu/hr$$

For example, A 100 W lamp emits 341.3 Btu/hr. An electric baseboard heater rated at 1000 W (or 1 kW) will release 3413 Btu/hr if it operates at full capacity for one hour; at half capacity it releases half its rated capacity, 1707 Btu/hr.

1.2 PSYCHROMETRICS

Psychrometrics is the study of the thermodynamic properties of gas–water vapor mixtures under varying temperatures and pressures. In heating, ventilating, and air conditioning system design, to analyze conditions and processes involving moist air, including indoor air in buildings and ventilation (outdoor) air is introduced to building interior. An understanding of air and water vapor in air is necessary to the study of psychrometrics.

Air

Atmospheric air, air found in the atmosphere, is a mixture of several constituent gases, water vapor, and contaminants such as dust, pollen, and smoke. It consists of nitrogen, oxygen, and other trace gases. See Table 1.7. In contrast, the average composition of constituent gases in the exhaled air of a seated person by volume is as follows: oxygen, 16.5%; carbon dioxide, 4.0%; nitrogen and argon, 79.5%; and other trace gases. Note

TABLE 1.7   GASEOUS CONSTITUENTS OF DRY AIR
BY PERCENTAGE OF VOLUME AND MASS.

Gaseous Constituent	Chemical Symbol	Ratio Compared to Dry Air (%)	
		By Volume	By Mass
Nitrogen	N_2	78.09	75.47
Oxygen	O_2	20.95	23.20
Argon	Ar	0.933	1.28
Carbon dioxide	CO_2	0.0300	0.046
Neon	Ne	0.0018	0.0012
Helium	He	0.0005	0.00007
Krypton	Kr	0.0001	0.0003
Hydrogen	H_2	0.00005	~0
Xenon	Xe	0.000009	0.00004

the decrease in oxygen (about 6% drop) and the increase in carbon dioxide (about 4%). This illustrates the need for introducing outdoor air as ventilation air in buildings.

By definition, *dry air* is air that contains no water vapor. Dry air is used as the reference in psychrometrics. *Moist air* is air that contains water vapor. All atmospheric air contains some water vapor. *Saturated air* is air that contains the maximum amount of moisture it can hold at that temperature. Water vapor in air is superheated steam at low pressure and temperature. Air containing superheated vapor is clear, except under certain conditions where small water droplets are suspended in air, causing foggy conditions. Because of the variability of atmospheric air, the terms *dry air* and *moist air* are used in psychrometrics. For practical purposes, moist air and atmospheric air can be considered equal under the range of conditions normally encountered in building systems.

The mass of air per unit volume differs with air pressure (see Figure 1.2), so the properties of air are typically expressed based on unit weight (lb or kg). Air is heavier than most people realize. One cubic yard of air at sea-level pressure and temperature of 70°F weighs almost exactly 2 lb. The air in a 12 ft × 14 ft (3.7 m × 4.3 m) room with an 8 ft (2.4 m) high ceiling weighs almost exactly 100 lb (45 kg).

Standard air is generally used to rate heating, ventilating, and air conditioning equipment. The properties of standard air are provided in Table 1.8. The subscript, s, denotes properties at

TABLE 1.8   THE PROPERTIES OF STANDARD AIR.
THE SUBSCRIPT, s, DENOTES PROPERTIES
AT STANDARD CONDITIONS.

Property	Customary Units	SI (Metric) Units
Absolute pressure (P_s)	14.696 psi	101.325 kPa
Temperature (T_s)	70°F	21.1°C
Mass density (ρ_s)	0.075 lb/ft^3	1.189 kg/m^3
Specific heat (c_s)	0.24 Btu/lb · °F	0.000 0573 J/kg · °C
Specific heat capacity (C_s)	0.018 Btu/ft^3 · °F	1.195 kJ/m^3 · °C

standard conditions. At nonstandard conditions, air temperature and pressure have a significant influence on air density:

- With a decrease in air pressure, density of air decreases. Similarly, with an increase in air pressure, density increases.

- As the temperature of air decreases, density of air decreases. Likewise, as the temperature of air increases, air density increases.

A decrease in the density of dry air results in a proportional decrease in specific heat and specific heat capacity. This influence makes it necessary for the designer to adjust for actual conditions. This is particularly true of buildings.

Psychrometric Variables

Psychrometric variables are described below. Units of measurement used in psychrometrics are summarized in Table 1.9.

Dry bulb temperature (DBT) is the temperature of the mixture of air and moisture at rest. It is measured with a common thermometer and is not affected by the amount of water vapor in the air or thermal radiation from the surroundings. Dry bulb temperature is expressed in units of °F (°C). A thermometer that is freely exposed to the air but is shielded from radiation and moisture measures the dry bulb temperature of air.

Wet bulb temperature (WBT) is the temperature measured by a thermometer with a wetted cloth sock covering its

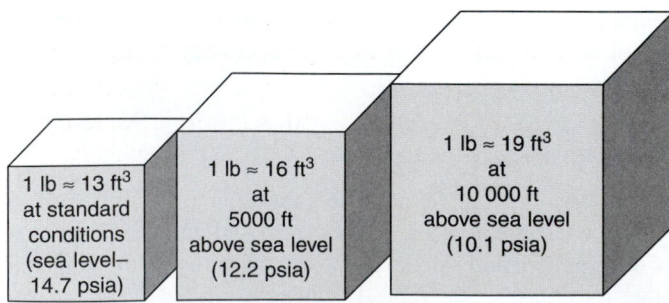

FIGURE 1.2 The mass of air per unit volume differs with air pressure, so the properties of air are typically expressed based upon unit weight (lb or kg).

TABLE 1.9   UNITS OF MEASUREMENT USED
IN PSYCHROMETRICS.

Property	Customary (U.S.) System	Metric (SI) System
Dry bulb temperature	°F	°C
Wet bulb temperature	°F	°C
Dew point temperature	°F	°C
Relative humidity	%	%
Specific humidity	lb of vapor/lb of dry air or grains/lb[a] of dry air	g/kg or kg/kg of dry air
Enthalpy	Btu/lb of dry air	kJ/kg of dry air
Specific volume	ft^3/lb of dry air	m^3/kg of dry air

[a]7000 grains = 1 lb of water vapor.

bulb as fast-moving air passes across it. It is expressed in °F (°C). Moisture in the wet sock evaporates, typically causing the wet bulb temperature to stay lower than dry bulb temperature. Wet bulb and dry bulb temperature readings are equal only when air is fully saturated (100% relative humidity) because evaporation from the wetted sock cannot occur. Wet bulb temperature is the lowest temperature that air can be cooled to by the evaporation of water.

Dew point (or saturation) temperature (DPT) is the temperature of the air/water-vapor mixture when it is fully saturated (100% relative humidity). It is expressed in units of °F (°C). It is the temperature at which water vapor from the air begins to form water droplets (condenses) on surfaces that are colder than the dew point of the air. The more moisture the air contains, the higher the dew point temperature. When the dew point temperature falls below freezing, it is called the *frost point temperature*, as the water vapor no longer creates dew but instead creates frost.

Relative humidity (RH) can be thought of as a ratio of moisture present in the air to the maximum amount of moisture it could hold at the same temperature and pressure (when fully saturated). Relative humidity is expressed as a percentage. The amount is based upon the mole fraction of water vapor, but mass of water vapor is an accepted (close) approximation.

Water vapor exerts a pressure just like the atmospheric air. The amount of water vapor present in air can be expressed by its *partial pressure*, the pressure that the water vapor contributes to the total pressure of the mixture. Technically, it is defined as a ratio of the *partial pressure of the water vapor (P_p)* present in the gas mixture and the *saturation pressure (P_s)* at the temperature of the gas.

$$RH = (\text{moisture present/moisture when fully saturated}) \, 100 = 100 \cdot P_p/P_s$$

Relative humidity can be measured by an instrument called a *hygrometer* or indirectly by using wet bulb and dry bulb thermometers. Relative humidity is a temperature-dependent variable because as air temperature changes, the maximum amount of moisture the air can hold changes. As dry bulb temperature increases, relative humidity of air decreases as long as water vapor is neither added nor removed; in contrast, as dry bulb temperature decreases, relative humidity increases.

Humidity ratio (W) is the ratio of the mass of water vapor (m_w) to the mass of dry air (m_a) contained in a sample of air: $W = m_w/m_a$. It is expressed with respect to the weight of dry air. It is expressed in grains of moisture per pound of dry air (gr/lb), pounds of moisture per pound of dry air (lb/lb), or grams of moisture per kilogram of dry air (gr/kg). A grain is a small amount of moisture: a grain

is equal to 1/7000 of a pound (7000 grains equal 1 lb). Sometimes humidity ratio is expressed as "lbm$_w$/lbm$_a$," to differentiate between the mass of water vapor and dry air. *Specific humidity* is the ratio of water vapor to the total mass of moist air (including water vapor *and* dry air). *Absolute humidity* (or water vapor density) is the ratio of the mass of water vapor to the total volume of the sample.

Specific volume (v) is a ratio of unit volume of dry air per unit weight of dry air, expressed in cubic feet of air per pound of dry air (ft^3/lb) or cubic meters of air per kilogram of dry air (m^3/kg). It is the inverse of density.

Enthalpy (h) is the quantity of heat contained in air under specific psychrometric conditions (e.g., a specific dry bulb temperature and relative humidity). It is the sum of the sensible heat contained in dry air and the sensible and latent heat contained in moisture in the air at a specific temperature and pressure. Enthalpy is usually measured from 0°F (−18°C) for air, and expressed in Btu per pound of dry air (Btu/lb) or kilojoules per kilogram of dry air (kJ/kg).

Psychrometric Chart

A *psychrometric chart* is a graphical representation of the thermodynamic properties of moist air. It describes the relationships between psychrometric variables. A point on the chart can be identified by any two variables and associated psychrometric variables can then be determined. The chart may also be used to trace a process undergone by an air/water-vapor mixture. The design of the psychrometric chart is such that a straight line between two points represents a change in the condition of air. Air is assumed to be at constant pressure on a psychrometric chart, so charts are related to a specific atmospheric pressure and, thus, these charts are altitude dependent. A skeleton psychrometric chart for normal temperatures and standard pressure is shown in Figure 1.3.

ASHRAE (American Society of Heating, Refrigerating and Air-Conditioning Engineers, Inc., Atlanta, Georgia 30329) and other sources make various psychrometric charts available. Most charts are based on standard atmospheric pressure at sea level. For higher altitude locations, charts based on lower barometric pressure should be used. ASHRAE makes five charts available:

Chart 1:	Normal temperatures	32 to 100°F	Sea level
Chart 2:	Low temperatures	−40 to 50°F	Sea level
Chart 3:	High temperatures	50 to 250°F	Sea level
Chart 4:	Normal temperatures	32 to 100°F	5000 ft above sea level
Chart 5:	Normal temperatures	32 to 100°F	7500 ft above sea level

Several vendors offer software packages that find psychrometric variables or create psychrometric charts for various

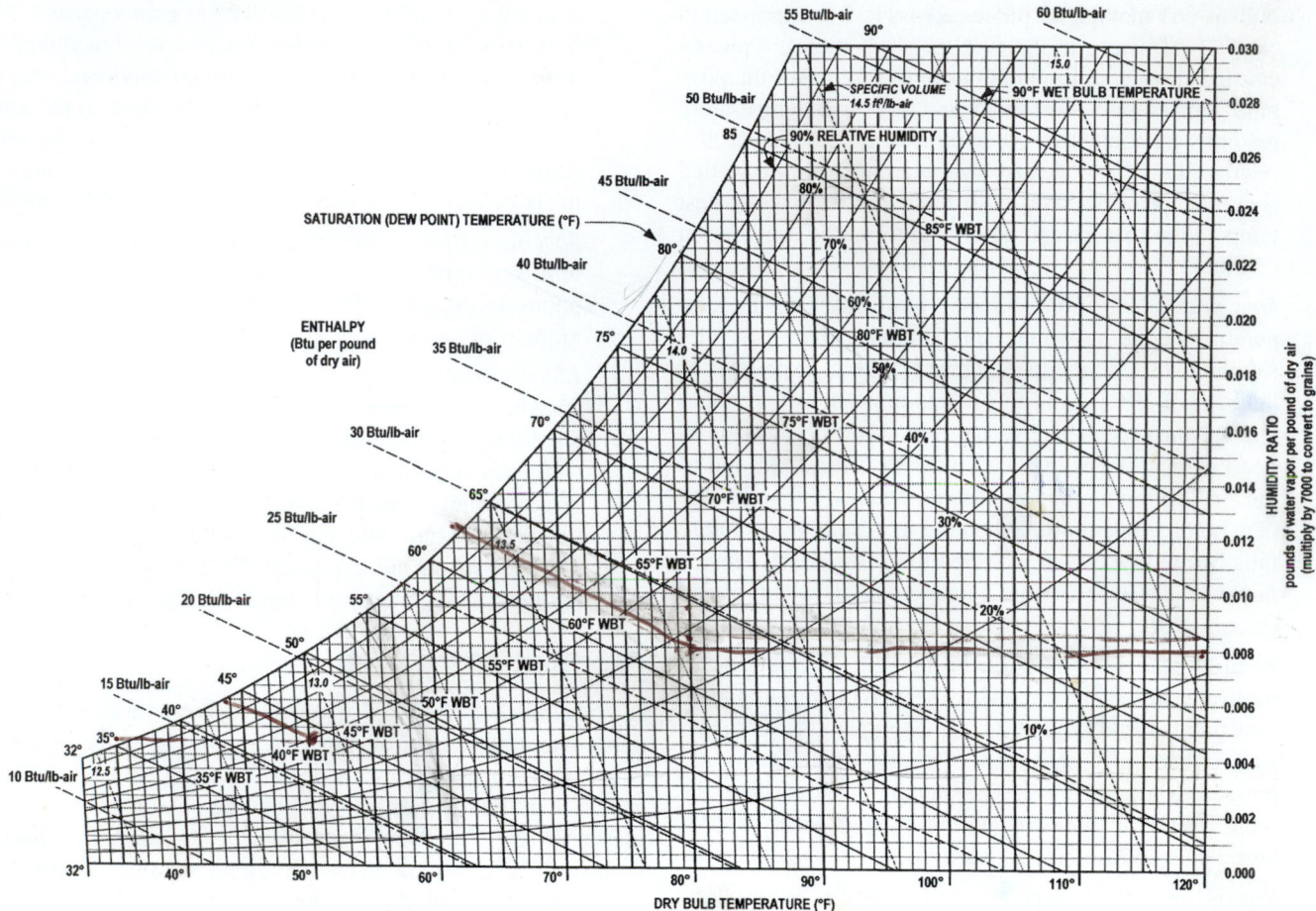

FIGURE 1.3 A psychrometric chart for normal temperatures at standard conditions (sea level). A larger chart is provided at the end of the chapter. See Figure 1.9.

temperature ranges and altitudes. Many are available over the Internet free of charge. Such software packages allow the user to perform some basic calculations that depend on psychrometrics.

A psychrometric chart is a mix of lines that represent psychrometric variables. A review of lines identifying variables on the psychrometric chart is necessary in understanding how to use the chart. (See, for instance, Figure 1.4.)

The *dry bulb temperature scale* is located at the base (bottom) of the chart. Vertical lines indicate constant dry bulb temperature.

The *wet bulb temperature scale* is located along the upper curved portion of the chart. Lines sloping to the right indicate equal wet bulb temperatures.

The *dew point temperature scale* is located along the same curved portion of the chart as the wet bulb temperature scale. Horizontal lines extending from the upper curved portion of the chart indicate equal dew point temperatures.

The *relative humidity scale* is represented by curved lines that sweep from the lower left to the upper right of the

psychrometric chart. The 100% relative humidity (saturation) line follows the upper curved area of the chart and corresponds to the wet bulb and the dew point temperature scale line. The line for 0% relative humidity lies along the base (bottom) of the chart. A 50% relative humidity line is determined by connecting points at various dry bulb temperatures that represent the humidity ratio when the air contains one-half of its maximum water content; that is, the 50% relative humidity points are halfway between the 100% and 0% lines at a specific dry bulb temperature.

The *humidity ratio* or *specific humidity* is found on the right side of the chart. Lines of constant moisture extend horizontally.

The *specific volume scale* is represented by diagonal lines that run from left to right.

The *enthalpy scale* is typically located near the upper curved portion of the chart. Lines of enthalpy slope to the right in a manner similar to wet bulb temperature lines. They do not, however, correspond exactly with the wet bulb temperature lines.

USE OF PSYCHROMETRIC CHART (Sea Level)

Psychrometric Properties
1. dry-bulb temperature (80°F)
2. wet-bulb temperature (62°F)
3. dew point temperature (50°F)
4. relative humidity (36%)
5. humidity ratio (0.0077 lb_water/lb_air)
6. specific volume (13.8 ft³/lb)
7. enthalpy (27.7 Btu/lb_air)

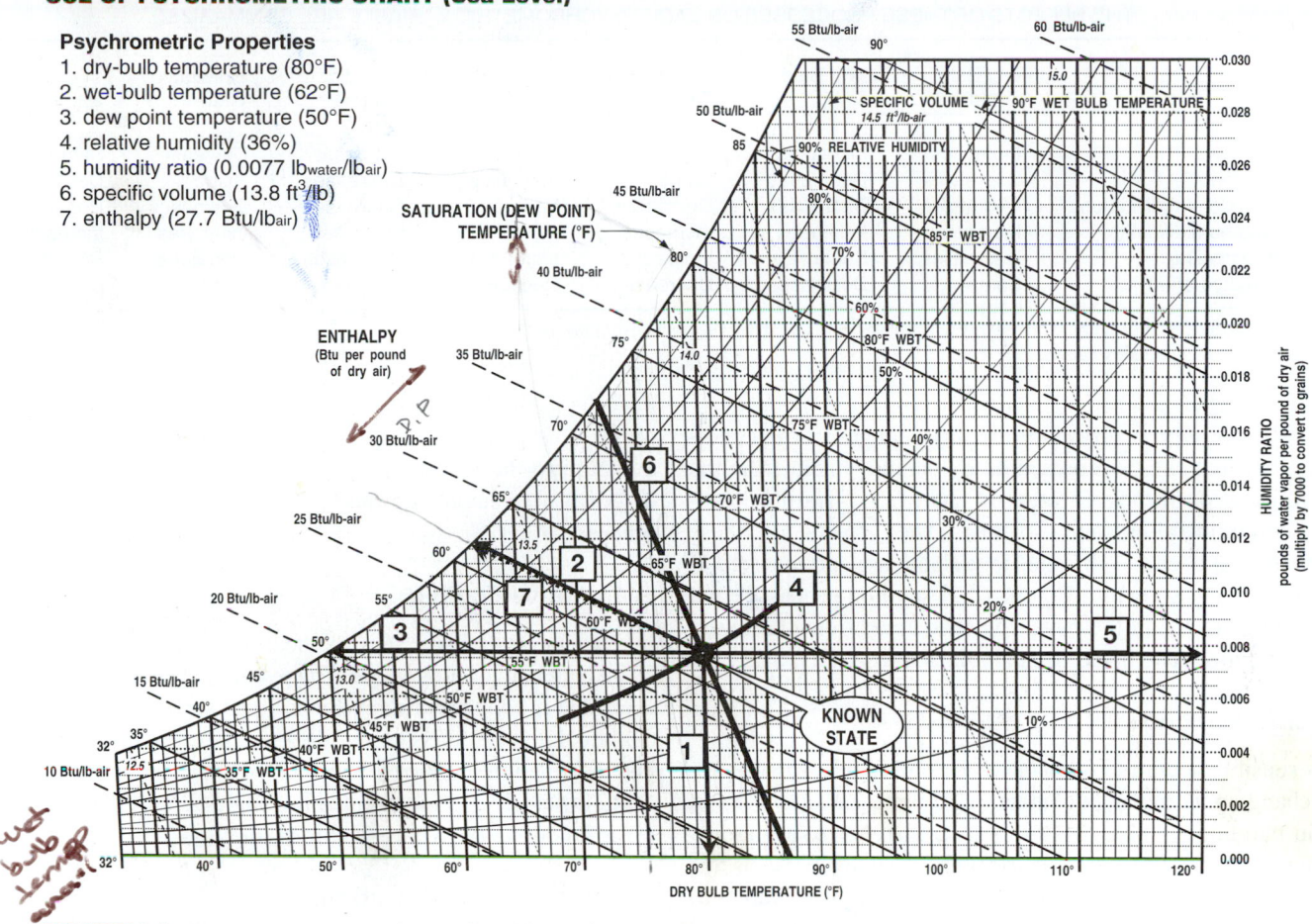

FIGURE 1.4 Psychrometric properties are found from a known state.

Example 1.3

Moist air (at sea level) is at a dry bulb temperature of 80°F and a wet bulb temperature of 67°F. Find the relative humidity and the dew point temperature.

On the psychrometric chart provided, locate the intersection of the 80°F dry bulb temperature line and the 67°F wet bulb temperature line. From this intersecting point, follow the relative humidity and dew point temperature lines to the correct scales and find that:

Relative humidity: 50%.

Dew point temperature: 59°F

Example 1.4

The wet bulb thermometer reads 60°F and a dry bulb thermometer reads 72°F. Find the dew point temperature, relative humidity and humidity ratio.

On the psychrometric chart provided, locate the intersection of the 60°F dry bulb temperature line and the 72°F wet bulb temperature line. From this intersecting point, follow the dew point temperature, relative humidity, and humidity ratio lines to the correct scales and find that:

Dew point temperature: 52°F

Relative humidity: 50%

Humidity ratio: 0.0084 lb/lb

Using the Psychrometric Chart

Heat gain or loss in moist air may be described by two terms: sensible heating and latent heating. Sensible heating and cooling occur by a change dry bulb temperature alone; that is, moisture is not added or removed. Humidification and dehumidification involve adding or removing moisture and therefore involve latent heat. Latent heat is the energy required to change the phase of a substance, in this case of moist air, water vapor. Evaporative cooling involves an increase in latent heat of the air (evaporation of water) for sensible heat decrease (decrease in the dry bulb temperature of air). Wet bulb temperature is the lowest temperature that air can be theoretically cooled by the evaporation of water.

TABLE 1.10   PSYCHROMETRIC PROCESSES CHANGE THE THERMODYNAMIC AND PHYSICAL PROPERTIES OF AIR. THE EFFECTS OF THESE PROCESSES ON EACH PSYCHROMETRIC VARIABLE ARE SUMMARIZED.

Psychrometric Process	Dry Bulb Temp	Wet Bulb Temp	Dew Point Temp	Relative Humidity	Specific Volume	Humidity Ratio	Enthalpy
Sensible heating	Increases	Increases	Same	Decreases	Increases	Decreases	Same
Sensible cooling	Decreases	Decreases	Same	Increases	Decreases	Same	Decreases
Cooling and dehumidification	Decreases	Decreases	Decreases	Varies[a]	Decreases	Decreases	Decreases
Heating and humidification	Increases	Increases	Increases	Varies[a]	Increases	Increases	Increases
Evaporative cooling	Decreases	Same	Increases	Increases	Decreases	Increases	Same[b]
Chemical dehydration	Same	Decreases	Decreases	Decreases	Increases	Decreases	Decreases

[a]Determined by the relative amounts of energy and water vapor added to or removed from the air.
[b]A slight increase in enthalpy occurs.

Psychrometric processes are any changes in the condition of air that can be represented by the movement from the initial point on the psychrometric chart (e.g., heating, cooling, dehumidifying, and so on). The effects of these processes on each psychrometric variable are summarized in Table 1.10 and shown in Figure 1.5. Common processes and relate movement on the psychrometric chart from the initial point include those listed in the table above.

Sensible Heating

This sensible heating process involves adding heat to air without changing its moisture content (e.g., winter heating or room air in buildings). On a psychrometric chart, horizontal movement to the right along the specific humidity line represents this process. The result is increases in dry bulb and wet bulb temperatures, and enthalpy, and a decrease in specific volume and relative humidity results. Because moisture is neither added nor removed, no change occurs in humidity ratio and dew point temperature.

Sensible Cooling

The sensible cooling process is removing heat from air without changing its humidity ratio (e.g., cooling room air in buildings). The result is decreases in dry bulb and wet bulb temperatures, and enthalpy. An increase in specific volume and relative humidity also results. Because moisture is neither added nor removed, no change occurs in humidity ratio and dew point temperature. On a psychrometric chart, horizontal movement to the left along the specific humidity line represents this process.

Cooling and Dehumidification

Cooling and dehumidifying is the process of lowering both the dry bulb temperature and the humidity ratio of the moist air. On a psychrometric chart, this process is represented by diagonal movement down and to the left.

Heating and Humidification

Heating and humidifying is the process of simultaneously increasing both the dry bulb temperature and humidity ratio of the air (e.g., human respiration adds both sensible heat and moisture to the surrounding air). On a psychrometric chart, diagonal movement up and to the right represents this process.

Evaporative Cooling

In evaporative cooling, sensible heat in the air is converted to latent heat in the added vapor. The result of evaporative cooling is air that is cooler and more humid, and surrounding surfaces that are dryer. For this reason, the process of evaporative cooling is sometimes referred to as *drying*. On a psychrometric chart, this process is represented by diagonal movement up and to the left along the wet bulb temperature line.

Desiccant Dehumidification

Desiccant dehumidification involves the use of chemical or physical absorption of water vapor to dehumidify air. On a psychrometric chart, this process is represented by diagonal movement down and to the left.

Example 1.5

A wet bulb thermometer reads 60°F and a dry bulb thermometer reads 72°F. If the air is heated in a furnace to a dry bulb temperature of 120°F, what will be its relative humidity?

No moisture is added as air is heated in a furnace so sensible heating is occurring in this psychrometric process. Sensible heating is represented by moving horizontally to the right from the initial temperature conditions along the specific humidity line. The horizontal specific humidity line intersects the vertical 120°F dry bulb temperature line at about 12% relative humidity.

As air is sensibly heated, the amount of moisture in it remains constant. In all cases, as the dry bulb temperature of the air increases without adding or removing moisture (sensible heating), the relative humidity of the air decreases.

Basic Psychrometric Processes
Includes sensible heating, sensible cooling, humidifying and dehumidifying, or a combination.

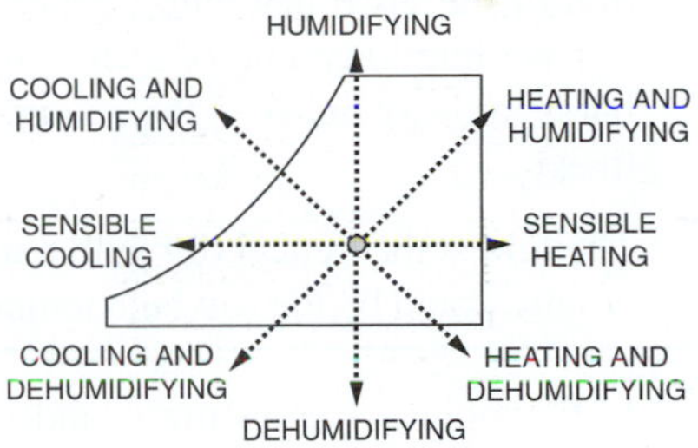

Sensible Heating
Adding heat to air without changing humidity ratio, e.g., space heating a room or building.

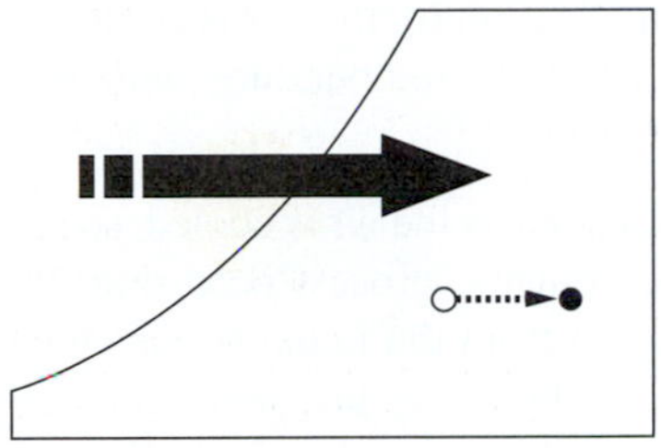

Sensible Cooling
Removing heat from air without changing its humidity ratio.

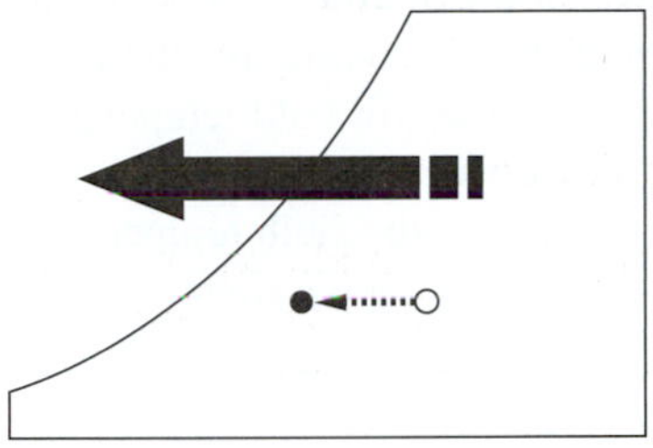

Humidifying
Adding moisture to air without changing its dry bulb temperature.

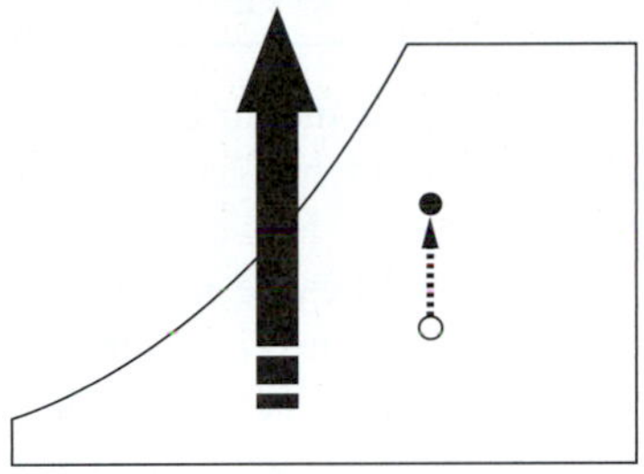

Dehumidifying
Removing moisture to air without changing its dry bulb temperature.

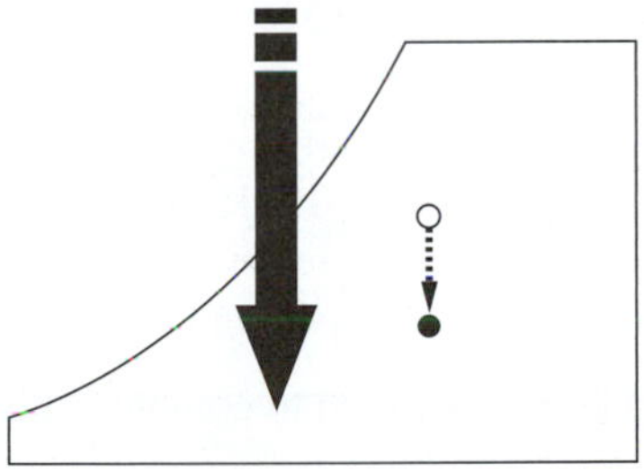

Evaporative Cooling
Sensible heat in air is converted to latent heat (added vapor) following the WB temperature line.

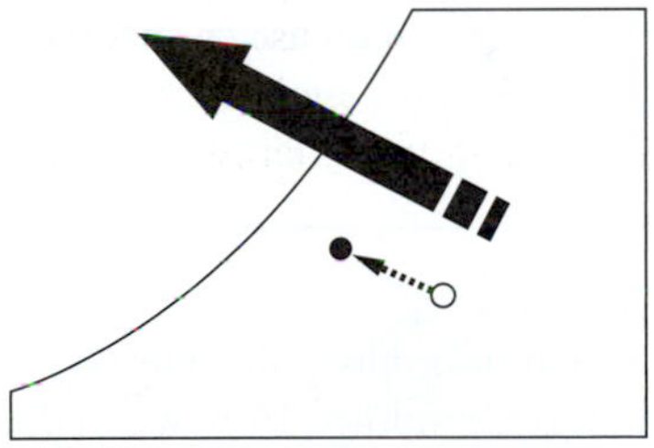

Heating and Humidifying
Simultaneously increasing both dry-bulb temperature and humidity ratio of the air.

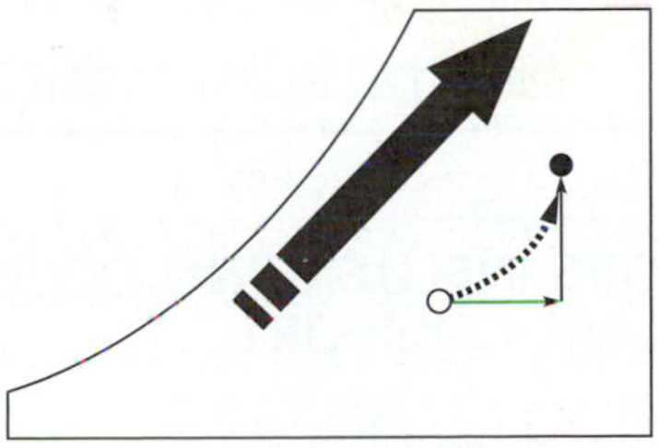

Cooling and Dehumidifying
Lowering both the dry-bulb temperature and the humidity ratio of the moist air.

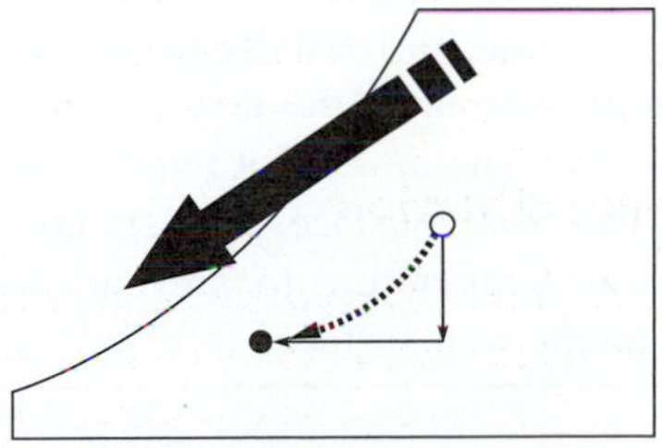

FIGURE 1.5 Psychrometric processes are any changes in the condition of air that can be represented by the movement from the initial point on the psychrometric chart (e.g., heating, cooling, dehumidifying, and so forth).

Example 1.6

A wet bulb thermometer reads 60°F and a dry bulb thermometer reads 72°F in outdoor air (similar readings as the previous problem). On a calm, clear night, this air is cooled by night sky radiation. At what dry bulb temperature will the cooled air begin to condense moisture?

No moisture is added as the air is cooled, so sensible cooling is the psychrometric process occurring. Moving horizontally to the left from the initial temperature conditions along the specific humidity line represents sensible cooling. The horizontal specific humidity line intersects the dew point temperature line at 52°F (and 100% relative humidity). If the air is cooled below 52°F, it will begin to condense some of the water vapor contained in the air (causing dew to fall). The air will have less moisture content.

In Examples 1.5 and 1.6, if the air cooled to 52°F (and 100% relative humidity), it would still have the same relative humidity after heating to 120°F. However, if the air were cooled to less than 52°F, it would cause moisture to condense from the air and the air would have less moisture content. Now, if this cooled air were heated to 120°F, it would have less than 12% relative humidity.

As the dry bulb temperature of the air decreases without adding or removing moisture (sensible cooling), the relative humidity of the air increases up to a relative humidity of 100%. If cooling of the saturated air continues, moisture is extracted from the air in the psychrometric process known as *cooling and dehumidification*. As the dry bulb temperature drops in the cooling and dehumidification process, the air follows the 100% relative humidity line, moving to the left until cooling ends.

Example 1.7

Air (at sea level) in a building has a dry bulb temperature of 80°F and is at 70% relative humidity. How warm does the wall surface have to be to prevent condensation?

In this case, the dew point temperature is needed. On the psychrometric chart provided, locate the intersection of the 80°F dry bulb temperature line and the 70% relative humidity line. Proceed horizontally to the left along the specific humidity line until the dew point temperature scale is intersected. The dew point temperature is found to be 69°F. The wall surface temperatures must be warmer than this temperature to prevent condensation from forming on the wall surface.

Example 1.8

A building owner is considering installing an evaporative cooling system. What is the lowest temperature that can theoretically be obtained from a air coming off the moist cooling pads

of the evaporative cooling system if the outside air has a 90°F dry bulb temperature and 35% relative humidity?

On the psychrometric chart provided, locate the intersection of the 90°F dry bulb temperature line and the 35% relative humidity line (which is not shown on the chart, but is midway between the 30 and 40% relative humidity lines).

The lowest theoretical dry bulb temperature for these conditions would be the wet bulb temperature of 69°F.

In Example 1.8, moisture is added to the air as it passes through the wet pads of an evaporative cooling system. This psychrometric process, called *evaporative cooling*, causes the air to be cooled. The lowest possible dry bulb temperature for these conditions would be the wet bulb temperature of 69°F. However, because of the ineffectiveness of evaporative coolers, the temperature of the air coming off the cooling pads will be about 3 to 5°F above the wet bulb temperature. This evaporative cooler can provide air at a temperature of about 73°F under these conditions. Air at a dry bulb temperature of 73°F would have a relative humidity of approximately 85%.

Example 1.9

1000 ft³ of moist air (at sea level) with a dry bulb temperature of 50°F and a relative humidity of 60% is sensibly heated to a dry bulb temperature 80°F.

a. What will be the relative humidity of the air once this heat is added?

On the psychrometric chart provided, begin at initial conditions (50°F DBT and 50% RH) and move horizontally along the constant moisture line to the 80°F DBT line. A relative humidity of 22% is found.

b. How much heat is required to complete this process?

On the psychrometric chart provided, an enthalpy of 17.2 Btu/lb and a specific volume of 12.95 ft³/lb is found at initial conditions (50°F DBT and 50% RH). After heating of the air, an enthalpy of 24.6 Btu/lb is found. Energy required to heat 1000 ft³ of moist air is found by:

$(1000 \text{ ft}^3) \cdot (24.6 \text{ Btu/lb} - 17.2 \text{ Btu/lb})/12.95 \text{ ft}^3/\text{lb} = 571.4 \text{ Btu}$

1.3 FACTORS INFLUENCING OCCUPANT THERMAL COMFORT

Body Temperature

The average core temperature of the typical human body must be maintained at a constant temperature of about 98.6°F (37°C), although temperatures in the range of 97.5° to 99°F (36.4° to 37.2°C) are within normal limits. Body temperature will fluctuate slightly depending upon time of day. A change in average core temperature of more than one degree points toward illness.

Heat is generated within the human body by combustion of food. The body is continuously attempting to maintain a heat balance through an intricate control system. This system balances the rate of heat generation with the rate of heat loss to the surrounding environment. If the environment is maintained at suitable conditions such that an energy balance can be maintained, a sensation of thermal comfort will result. When environmental conditions transfer heat away from the body too rapidly or too slowly, thermal discomfort will result.

The two basic mechanisms that the body uses to involuntarily adjust body temperature are change of metabolism and change in the rate of blood circulation near the surface of the skin. A change in the metabolic rate is the increase or decrease in internal energy production as body temperature rises or falls. The body begins to shiver when it is cold. An increase in circulation of blood near the surface of the skin increases heat transfer to the environment. These mechanisms control heat in the body by compensating for heat lost from the body.

The modes by which the body loses heat and the relationship between these modes can be expressed in a simple equation:

$$H = S + E + R + C$$

H: rate of internal heat production
S: rate of heat storage in the body
E: rate of loss by evaporation (perspiration and respiration)
R: rate of radiant energy exchange with surroundings
C: rate of loss by convection

As revealed in the equation, internal heat production must be in balance with the heat production and loss. Some heat may be stored, or drawn from storage within the body, with a corresponding change in body temperature over short periods. Most of the heat, however, must be rejected from the body surfaces through convection losses to the surrounding air, by radiative energy exchanges with surrounding surfaces, and by the evaporation of perspiration from the skin when required.

Heat loss from the body is transferred to the surroundings through convection, radiation, and evaporation. Typically, heat is lost from the human body by:

- Conduction and convection about 25%
- Radiation about 43%
- Evaporation and moisture about 30%
- Exhaled air about 2%

Convection removes heat by the movement of air around the body because of the difference between skin and air temperatures. It is dependent on exposed skin area and air movement affecting the body. Evaporation of moisture from the body is affected by the relative humidity of the surrounding air, air velocity, and the skin wetness. Perspiration affects evaporative heat loss. A portion of evaporative loss is also caused by respiration. Evaporation is greatest at high ambient temperatures, whereas conduction and convection are greatest at low ambient temperatures.

The effect of radiation is dependent on the mean radiant temperature of the surrounding surfaces and the exposed areas of the body. Mean radiant temperature is discussed later in this chapter. The remaining 40% of the heat lost from the body is from radiation.

Body Metabolism

The human body is constantly generating heat through metabolic activity. The average adult, when sleeping, generates about 250 Btu/hr (73 W) and about 400 Btu/hr (120 W) when awake but sedentary. While walking at a slow pace (2 miles per hour), the body produces about 750 Btu/hr (220 W). The total heat loss from an adult is approximately 400 Btu/hr (118 W) at indoor temperatures between 66°F and 93°F (19°C and 34°C).

By comparison, a seated adult generates about the same amount of heat as a 100 W light bulb. An athlete at peak exertion generates over 1000 W, about the output of a small portable space heater or hair dryer. As activity increases, the rate of oxidation of food increases.

The energy intake of humans in the form of food is often expressed in dietary calories (kilocalories). The average U.S. diet is about 2000 calories (kilocalories) per day. The daily intake of food energy can be expressed in other units:

1 calorie (kilocalorie) = 3.97 Btu = 4186 Joules

By comparison a king-size, fast-food hamburger with supersize fries and soda contains about 1600 calories. It potentially generates about 6400 Btu (6.7 MJ) of heat through metabolic activity.

Conditions of Thermal Comfort

Thermal comfort means different things to different people because individuals have different thermal preferences and sensitivities. Variations in feeling thermally comfortable are from differing physiological factors (e.g., weight, age, gender, metabolic rate, and the ability to adjust to the conditions of the surrounding environment), but are ultimately determined by individual preference; that is, some people simply prefer to be warm and others cool. As result, it is impractical to ensure that greater than 80% of the room occupants feel an acceptable thermal sensation.

By definition, *thermal comfort* is a condition of mind that expresses satisfaction with the thermal environment; it requires subjective evaluation. ASHRAE Standard 55–2004, *Thermal Environmental Conditions for Human Occupancy*, specifies combinations of indoor space environmental and personal factors that will produce thermal environmental conditions where 80% of the occupants do not express dissatisfaction with the room conditions; that is, the given condition is not acceptable to 20% of the occupants. Occupants are evaluated on their *thermal sensation*—a conscious feeling commonly graded on a thermal sensation scale: cold, cool, slightly cool, neutral, slightly warm, warm, and hot. A predicted mean vote (based on the thermal sensation scale) of a large population of people exposed to a certain environment is used to determine the predicted percentage of dissatisfied people. The definition of thermal comfort gives insight into the intricate problem of comfortably satisfying as many occupants as possible with one set of thermal conditions.

Overall thermal comfort will still vary from person to person, as well as with activity level, level of clothing, temperature and relative humidity of the room air, air velocity, types of heating and cooling systems, and types and materials of construction. Factors affecting occupant comfort relate to the thermal environment in the space and personally controlled parameters.

Environmental Factors Affecting Thermal Comfort

Besides the physiological adjustments made by the body, heat transfer from the body is influenced by several environmental factors. These factors affect an occupant's temperature sensation in a space. They can be reasonably controlled by the designer of the space and usually play a larger role in body comfort than the physiological adjustments made by the body. Environmental factors affecting occupant comfort are typically grouped into four classifications.

Air Temperature

Dry bulb temperature refers to the ambient air temperature. For design of heating systems, the lower temperature in the winter design temperature range is 68°F (20°C). In the summer, the upper limit for the cooling and air conditioning design temperature is 79°F (26°C). Differences in design temperature limits reflect that people adjust clothing levels to the season. The body is relatively insensitive to dry bulb temperature variations in the 68° to 79°F (20° to 26°C) range, when adjustments in clothing levels are made.

With typical heating systems, temperatures near the floor are lower than those at shoulder level because of heat stratification in the air. Such differences are not a concern if floor temperatures are maintained above 65°F (18°C) and if the maximum temperature gradient from ankle to the neck is no more than about 4°F (2°C). During cooling, which is the predominant mode in most occupied spaces, local air temperature differentials generally are not more than 1° to 2°F from ankle to neck region with properly designed air distribution systems. However, if the temperature falls outside these ranges, comfort is likely to be determined by the temperature required to keep the feet warm.

Relative Humidity

Indoor air is mixed with water vapor, in proportions that vary with climate, seasonal and daily weather conditions, and indoor occupancies. Humidity in air affects the rate of evaporative heat loss from the body. A high relative humidity will cause the surrounding air to evaporate less perspiration from the body, making the occupant feel warmer. A low relative humidity allows the air to absorb greater quantities of moisture, making the occupant feel cooler.

The human body is relatively insensitive to moderate variations (less than about 10%) in indoor relative humidity levels within the range of about 30 to 70%. Relative humidity levels below about 30% cause dryness of skin, cracking of nasal membranes, and sinus, asthmatic, and allergic reactions in humans. Low humidity levels result in the generation of static electricity, which is damaging to electronic equipment. At relative humidity levels below 20%, these problems increase drastically. In contrast, relative humidity levels above about 60% are favorable to the survival and proliferation of bacteria, fungi, and viruses. A relative humidity of 30 to 60% is optimal.

ASHRAE standards have specified that the humidity of a zone occupied by people doing light or primarily sedentary activity shall fall approximately within a range between 30 and 60% relative humidity. In regions of the United States requiring space heating and in arid climates, most buildings typically have indoor relative humidity levels below 30% during the heating season unless moisture is supplied to the air through mechanical humidification. The most recent standard (2007) sets an upper limit of 65% humidity for occupied spaces.

At comfortable room air temperatures and at relative humidity levels within the range of about 30 to 70%, an increase in relative humidity increases the temperature sensation slightly, probably about 1°F (0.5°) per 10 to 15% change in relative humidity. Within the acceptable relative humidity range, about a 3 to 4°F (1 to 2°C) difference in temperature may be experienced. This difference in warmth is usually not significant enough to warrant mechanical humidification, because it is generally less costly to heat air than it does to introduce moisture into the conditioned space.

Air Movement

The rate of air movement has a significant bearing on the sensation of freshness and thermal comfort. The effect of air movement becomes evident when considering the effect of a fan during hot weather: a person will sense a lower temperature sensation standing in front of a fan without there actually being a decrease in the room air temperature. In the summer, increased air motion increases the evaporation rate of heat from the body to help keep it cool. This is why buildings without cooling systems often employ fans to increase the movement of the air on hot days. On the other hand, in the winter, a slower rate of air motion is more desirable, or the occupants will feel too cool from evaporative heat loss. But even in the winter it is desirable to have some air movement to keep the air from becoming stagnant.

Air movement in building interiors is commonly expressed in feet per minute (fpm) and, in SI (metric) units, in meters per second (m/s). Air flow should normally be in the range of 20 to 160 fpm (0.1 to 0.8 m/s). (By comparison, a one-mile per hour wind has a velocity of 88 fpm). At velocities below 20 fpm (0.1 m/s), the air feels stagnant because little evaporation and convection takes place. Above about 160 fpm (0.8 m/s) air movement is uncomfortable and disturbing (e.g., hair blowing, paper moving, and so forth). To maintain the 80% comfort level, however, the local air velocities should not exceed about 50 fpm (0.1 m/s).

Research shows that air movement lowers the thermal sensation of the body by about 1°F (0.6°C) for an increase in air movement of about 15 fpm (0.075 m/s). Adverse effects from movement of air are easily minimized. With proper design, a greater effect on cooling the body during the summer season will result.

The term *wind chill factor* is used to describe the effect of the combination of low temperature and air movement in outdoor environments. Cooling or freezing of exposed flesh increases rapidly as wind velocity increases. The wind chill effect is significant even at low air velocities. Drafts may be caused by natural convection at cold exterior surfaces, such as windows and uninsulated walls. Likewise, mechanically induced air movement (fans and blowers) may impact comfort if high velocity air is directed at the occupant.

The temperature sensation of occupants in a space will be increased in the winter by directing the flow of warm air over the colder surfaces. This is why most space heating distribution equipment (e.g., registers, baseboard heaters) is located below a window or at an exterior wall.

Mean Radiant Temperature

The *mean radiant temperature (MRT)* of a space is a measure of the combined effects of the temperatures of surfaces within that space; it is a weighted average of the various radiant influences in a space. The larger the surface area and the closer an occupant is to a surface, the more effect the surface temperature of that surface has on the occupant. As an occupant moves in a room, the MRT changes with the new location. The closer an occupant is to a warm or cold surface, the more effect that surface has and the greater or smaller the MRT is at that location.

MRT can be more instrumental than air temperature in affecting the thermal sensation of a body. The surface temperatures in the room may vary widely. Windows and exterior walls are typically cooler in the winter than interior walls and furniture. The temperatures of the floors and ceilings will depend on the air temperature on the other side of the construction assembly.

MRT may be *approximated* by the following formula, where T is the temperature of a surface, θ is the horizontal angle of wall exposure relative to the position of the occupant, and MRT is the mean radiant temperature:

$$MRT = [(T_1\theta_1) + (T_2\theta_2) + (T_3\theta_3) + (T_4\theta_4) + \cdots]/360$$

MRT is affected by all surfaces of a space (i.e., ceiling, walls, floor, furniture, occupants, and so on). This equation neglects the influence of radiative effects of other surfaces in a space. It is provided to demonstrate the effects of MRT in a simplified manner. A more complex mathematical definition is beyond the scope of this text.

The MRT in a space affects the radiant heat loss from the body. An increase in MRT reduces radiant heat loss, thereby making the occupant feel warmer. In fact, the MRT of surrounding surfaces has a greater effect on thermal comfort than surrounding air temperature. See Table 1.11.

The MRT of a space has about 40% more of an influence on thermal comfort than does the DBT. Thus, a change in MRT by 1.0° will require a 1.4° opposite change in air temperature to maintain the same thermal condition. For example, if the MRT of a room were 80°F (26.7°C) and the air temperature were

TABLE 1.11   THE RELATIONSHIP BETWEEN VARYING MEAN RADIANT TEMPERATURES (MRT) OF SURROUNDING SURFACES IN A SPACE AND REQUIRED DRY BULB TEMPERATURE (DBT) TO MAINTAIN A TEMPERATURE SENSATION OF 70°F (21°C).

Customary (U.S.) Units		Metric (SI) Units	
MRT of Surrounding Surfaces	DBT of Surrounding Air Required to Achieve a 70°F Temperature Sensation	MRT of Surrounding Surfaces	DBT of Surrounding Air Required to Achieve a 21°C Temperature Sensation
°F	°F	°C	°C
58	86.8	15	29.4
60	84.0	16	28.0
62	81.2	17	26.6
64	78.4	18	25.2
66	75.6	19	23.8
68	72.8	20	22.4
70	**70.0**	**21**	**21.0**
72	67.2	22	19.6
74	64.4	23	18.2
76	61.6	24	16.8
78	58.8	25	15.4
80	56.0	26	14.0

56°F (13.3°C), the occupant would feel a temperature sensation of 70°F (21°C). However, if the MRT were 56°F (13.3°C), the air temperature would need to be maintained at almost 90°F to achieve the same thermal condition of 70°F (21°C).

Example 1.10

The 12 ft by 14 ft room illustrated in Figure 1.6 has an exterior wall at a surface temperature of 60°F, a window surface temperature of 40°F (double glazing on a cold day). The two interior wall surfaces are a temperature of 72°F.

a. Determine the MRT sensation of a person standing in the center of the room (see Figure 1.6a).

$$MRT = [(T_1\theta_1) + (T_2\theta_2) + (T_3\theta_3) + (T_4\theta_4)]/360$$

$$MRT = [(60°F \cdot 82°) + (40°F \cdot 98°) + (72°F \cdot 82°F)$$
$$+ (72°F \cdot 98°)]/360$$

$$MRT = 60.5°F$$

b. Determine the required indoor air temperature for the occupant to sense a comfortable thermal sensation of 72°F.

$$T = [(72 - 60.5°F) \cdot (1.4°F - DBT/1°F - MRT)]$$
$$+ 72°F = 88.1°F$$

c. Determine the MRT sensation of a person standing 3 ft from the center of the window (see Figure 1.6b).

$$MRT = [(60°F \cdot 75°) + (40°F \cdot 134°) + (72°F \cdot 75°)$$
$$+ (72°F \cdot 76°)]/360$$

$$MRT = 57.6°F$$

d. Determine the required indoor air temperature for the occupant to sense a comfortable thermal sensation of 72°F.

$$T = [(72 - 57.6°F) \cdot (1.4°F - DBT/1°F - MRT)]$$
$$+ 72°F = 92.2°F$$

e. Heavy drapes are used in front of window with a surface temperature of 60°F. Determine the MRT of the person standing 3 ft from the window (see Figure 1.6c).

$$MRT = [(60°F \cdot 75°) + (60°F \cdot 134°) + (72°F \cdot 75°)$$
$$+ (72°F \cdot 76°)]/360$$

$$MRT = 65.0°F$$

f. Determine the required indoor air temperature for the occupant to sense a comfortable thermal sensation of 72°F.

$$T = [(72 - 65.0°F) \cdot (1.4°F - DBT/1°F - MRT)]$$
$$+ 72° = 81.8°F$$

Upon review of Example 1.10, it is evident that MRT greatly influences occupant comfort. Surface areas of exterior walls with little insulation have low surface temperatures during cold weather. Windows and poorly insulated walls, ceilings, and floors are prime examples. A low MRT requires a

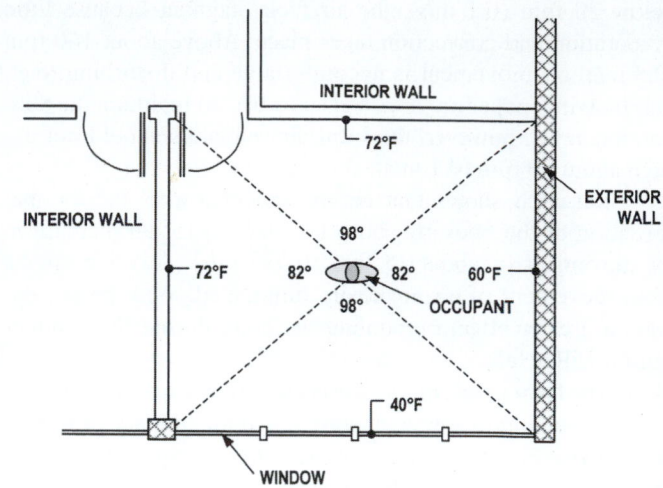

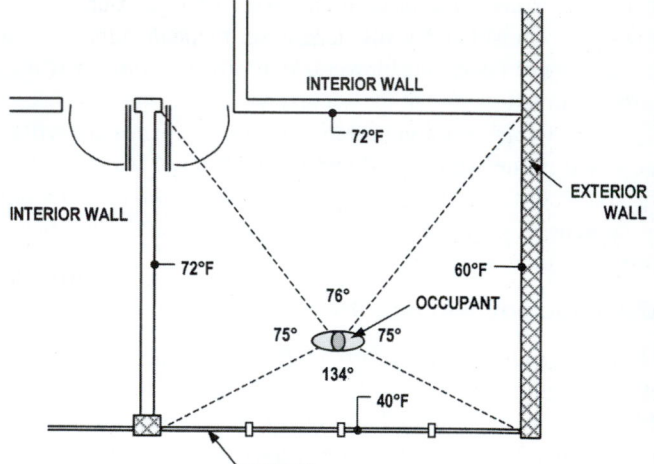

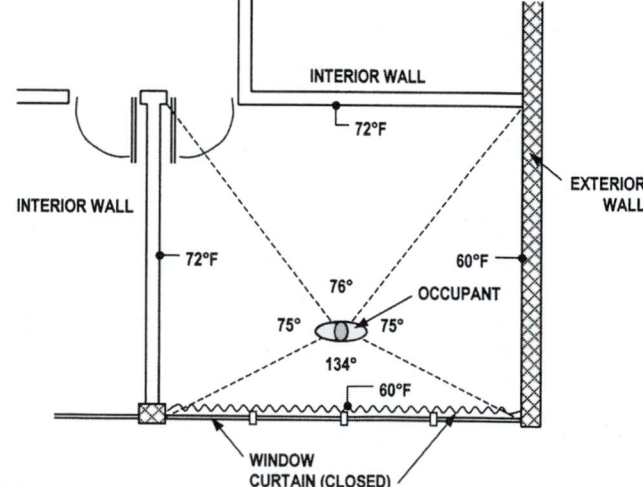

FIGURE 1.6 A 12 ft by 14 ft room and the position of the occupant and surface temperatures for Example 6.10.

higher space air temperature to maintain thermal comfort, thereby significantly increasing heating costs.

During winter months, the low surface temperatures of uninsulated walls and large windows reduce the MRT. To maintain thermal comfort of occupants, a much higher dry bulb temperature must be provided to achieve a thermally comfortable condition. Double or triple glazing not only helps reduce heat loss, but leads to higher glass surface temperatures and a higher MRT in the space.

All environmental factors interact with one another to create a comfortable or uncomfortable thermal sensation. If they blend in the correct proportionate mix, the body will experience thermal comfort.

There are, however, acceptable limits to adjustment of these variables. The tolerable range of MRT, clothing, and activity levels are normally considered to be between 60 and 90°F dry bulb temperature (15.5 to 32°C). Although air temperature and relative humidity have an acceptable range of about 70 to 90°F (15.5 to 32°C) dry bulb temperature, the designer may directly control only some of these factors.

Occupant-Related Factors Affecting Thermal Comfort

Besides the environmental factors described above and the body's ability to make slight physiological adjustments, body heat generation and heat transfer is influenced by two additional factors that are not related to the environment but rather to the occupant. These include:

Activity Level

Occupant activity level, as it increases, will generate additional body heat because of an increase in metabolic rate. The metabolic rate of the body increases in proportion to the exercise level of the physical activity. The *met* is the unit used to describe the activity level of a typical person. One met is equivalent to the metabolic rate of a quietly seated, average male, which corresponds to about 350 Btu/hr (about 370 kJ/hr).

$$1 \text{ met} = 58.2 \text{ W/m}^2 \text{ (SI units)} = 18.4 \text{ Btu/hr} \cdot \text{ft}^2$$

The surface area of an average person is about 19 ft^2 (1.8 m^2).

Sleeping has an activity level equal to about 0.7 met. High activity, such as playing basketball or racquetball, will have an activity level equivalent to between 5.0 and 7.6 met, which is 5 to 7.6 times the "resting" metabolic rate. An average healthy male 20 years of age will have a maximum energy capacity of 12 met, and he can maintain about half of this on a continuous basis. Women will normally have maximum levels about 30% lower than men. Metabolic rates (met) for selected activities are provided in Table 1.12.

At higher activity levels, air temperature must be lowered for the body to maintain a comfortable thermal sensation. In most building spaces, activity levels cannot be controlled by the designer: usually a specific activity will occur in a space (e.g.,

TABLE 1.12 APPROXIMATE METABOLIC RATES (MET) FOR SELECTED ACTIVITIES. VARIES BY PERSON AND RIGOR OF ACTIVITY. DATA WERE COMPILED FROM GOVERNMENTAL SOURCES.

Activity	Approximate met
Climbing stairs	10.0+
Sleeping	0.7
Seated, quiet	1.0
Seated, light work	1.2
Typing	1.4
Driving	1.5
Standing, light activity	1.6
Cooking	1.8
Walking leisurely on level ground [at a 2 mph or 3 ft/s (0.9 m/s) pace]	2.0
Walking briskly on level ground [at a 3 mph or 4.5 ft/s (1.35 m/s) pace]	2.3
Power walking on level ground [at a 4.2 mph or 6 ft/s (1.8 m/s) pace]	3.5
Dancing	3.5
Calisthenics	4.0
Basketball	7.0
Wrestling	8.0

sitting in a classroom, about 1.0 met; high activity in a gymnasium, about 5.0 met; and so forth). The designer must accommodate these activities in the design of the space.

Clothing Level

Because of its insulating properties, clothing plays an important role in body heat loss and comfort. Clothing may be described in terms of its clo value. The *clo* is an index of clothing insulation that describes the resistance to sensible heat transfer provided by a clothing ensemble (an assembly of more than one clothing garment) from the skin to the outer surface of the clothing ensemble, not including the insulation provided by the air layer around the clothed body. One clo is equal to a thermal conductivity of about 1.13 Btu/hr · °F · ft^2 (6.41 W/m^2 · °C) or a thermal resistance or R-value of about 0.88 hr · °F · ft^2/Btu (0.156 m^2 · °C/W). The European unit of clothing insulation is the *tog* (1 tog = 0.645 clo).

The clo index ranges from a naked body (0.0 clo) to artic clothing (about 4.0 clo). A clothing insulation level of 1.0 clo will keep an average sedentary man (1.0 met) thermally comfortable when surrounding air conditions are at a dry bulb temperature of 70°F (21°), 50% relative humidity, and in nearly still air.

Tables 1.13 and 1.14 provide typical clo values for individual garments and clothing ensembles. Heavier clothing generally has a higher insulating value in comparison to lighter clothing and thus a higher clo value. With no wind penetration or body movements to move air around, clothing insulation equals 0.15 clo per lb of clothes (e.g., 10 lb clothes equal 1.5 clo). In addition, occupants seated in a chair derive some insulating effect from the chair, so chairs have clo values ranging from 0.1 to 0.3.

TABLE 1.13 CLOTHING INSULATION VALUES (CLO) OF SELECTED INDIVIDUAL ARTICLES OF CLOTHING. DATA WERE COMPILED FROM AN ASSORTMENT OF GOVERNMENTAL SOURCES.

Article of Clothing	Clothing Insulation Values (clo)	Thermal Resistance ft² · °F · hr/Btu	m² · °C/W
Underwear			
Undershorts	0.04	0.035	0.006
T-shirt	0.09	0.079	0.014
Shirt with long sleeves	0.12	0.106	0.019
Socks	0.02	0.018	0.003
Thick, long socks	0.10	0.088	0.016
Shirts			
Short sleeve	0.09	0.079	0.029
Light shirt, long sleeves	0.20	0.176	0.031
Normal shirt, long sleeves	0.25	0.220	0.039
Flannel shirt, long sleeves	0.30	0.264	0.047
Pants			
Shorts	0.06	0.053	0.009
Normal pants	0.25	0.220	0.039
Flannel pants	0.28	0.246	0.043
Dresses			
Light dress, below knee	0.18	0.158	0.028
Light dress, sleeveless	0.25	0.220	0.039
Winter dress, long sleeves	0.40	0.352	0.062
Sweaters and Coats			
Thin sweater	0.20	0.176	0.031
Long sleeves, turtleneck (thin)	0.26	0.229	0.040
Thick sweater	0.35	0.308	0.054
Long sleeves, turtleneck (thick)	0.37	0.326	0.057
Light summer jacket	0.25	0.220	0.039
Jacket	0.35	0.308	0.054
Coat	0.60	0.528	0.093
Parka	0.70	0.616	0.109
Shoes			
Street shoes	0.02	0.018	0.003
Work shoes or hiking boots	0.04	0.035	0.006

The insulating value of clothing worn can be influenced by season and outdoor weather conditions. Appropriate winter attire, such as a warm sweater or jacket along with street clothes, will have a value of about 1.2 clo. In contrast, lightweight clothing typically worn in summer months will have insulating values ranging from 0.35 to 0.6 clo (e.g., a skirt, blouse, underwear, and no socks has a value of 0.5 clo). Clothing levels may be adjusted to produce a substantial difference in the body's thermal sensation. A 5 to 8°F (3 to 4.5°C) difference in temperature sensation exists between normal winter attire and lightweight summer clothing.

Trends in the business world show that women tend to adjust their attire by season more than men. In contrast, women are less likely to wear heavy socks, so they tend to be more sensitive to floor temperatures than men. These traditional differences between men and women's clothing often create conditions where a single interior dry bulb temperature will fully satisfy neither. Additionally, as heating and air conditioning technologies have improved, people tend to wear less clothing in the winter.

Occupant-related factors interact with environmental factors to create a comfortable or uncomfortable thermal sensation. They all play a large role in affecting thermal comfort because the body can only maintain a heat balance within a small range of surrounding conditions.

1.4 THERMAL COMFORT

Measuring Temperature and Relative Humidity

Many different techniques are used to measure temperature and relative humidity. These are described below:

Measuring Dry Bulb Temperature

A *thermometer* is an instrument that measures the dry bulb temperature of air. An ordinary thermometer consists of a uniform-diameter glass tube that opens into a bulb at one end. Liquid mercury, alcohol, ether, or some other liquid fills the bulb. The glass assembly is sealed to preserve a partial vacuum in the tube. As air temperature increases, the liquid expands and rises in the tube. The temperature is read on an adjacent scale. Accurate measurement of temperature can also be made electronically by use of a thermocouple, a thermoelectric device composed of two dissimilar wires that generates a small voltage difference when heated.

Dry bulb temperature of air in a room should be taken at a central location at about face level. Bright sunlight or other radiant heat sources (e.g., motors, lights, external walls, and people) should be avoided. This can be done by placing the thermometer where it cannot "see" the warm object or by protecting it with a radiant heat shield assembly.

Measuring Wet Bulb Temperature

A *wet bulb thermometer* is an ordinary thermometer with a cloth wick covering its bulb. The cloth wick is moistened and air is rapidly moved across the wet bulb. As fast-moving air passes across the moistened sock, the sock is cooled by evaporation of the water, thereby cooling the bulb. The amount of evaporation and consequent cooling of the thermometer is dictated by the moisture in the air: the drier the air, the faster the water evaporates. Once the temperature reaches equilibrium with the air, the wet bulb temperature is read on an adjacent scale much like a common thermometer. Rate of evaporation

TABLE 1.14 TYPICAL CLOTHING INSULATION VALUES (CLO) FOR SELECTED CLOTHING ENSEMBLES. DATA WERE COMPILED FROM AN ASSORTMENT OF GOVERNMENTAL SOURCES.

Clothing Attire	Description of Clothing Ensemble	Approximate Clothing Insulation Values (clo)	Thermal Resistance	
			ft² · °F · hr/Btu	m² · °C/W
Naked body	Minimum clo level	0.0	0	0
Light summer clothes	Panties; normal, short-sleeve shirt; A-line, knee-length skirt; panty hose; thongs/sandals	0.5	0.44	0.08
Normal working clothes	Panties; full slip; normal, short-sleeve shirt; light skirt; long-sleeve sweater; pantyhose; thin-soled shoes	0.6	0.53	0.09
	Undershorts; normal, long-sleeve shirt; long-fitted pants; belt; socks; dress shoes	0.7	0.62	0.11
Formal indoor business clothes	Undershorts; normal, long-sleeved shirt; single-breasted suit jacket; tie; straight, long-fitted pants; calf-length socks; thin-soled shoes	1.0	0.88	0.16
Heavy indoor winter clothes	Undershorts; normal, long-sleeved shirt; single-breasted suit jacket; tie; straight, long-fitted pants; calf-length socks; thick-soled shoes; cotton topcoat (or heavy wool suit without topcoat)	1.5	1.32	0.23
Winter outdoor work clothing	Long underwear; heavy long-sleeved flannel shirt; thick sweater; flannel pants; thick, long socks; thick-soled shoes; vest; parka; gloves	2.5	2.20	0.39
Artic outdoor clothing	(full Eskimo outdoor attire) maximum clo level	4.0	3.52	0.62

from the wetted wick is tied to air speed. A minimum air speed of about 500 feet per minute is required for accurate readings.

Measuring Mean Radiant Temperature

The instrument used to measure radiant temperature is a *black globe thermometer*, which consists of a thin-walled copper sphere, typically of 1½ to 6 in (40 to 150 mm) diameter, painted flat black that contains a thermometer. A globe thermometer suspended with its bulb at the center of the sphere is allowed to reach thermal equilibrium with its surroundings (usually 20 minutes). See Figure 1.7.

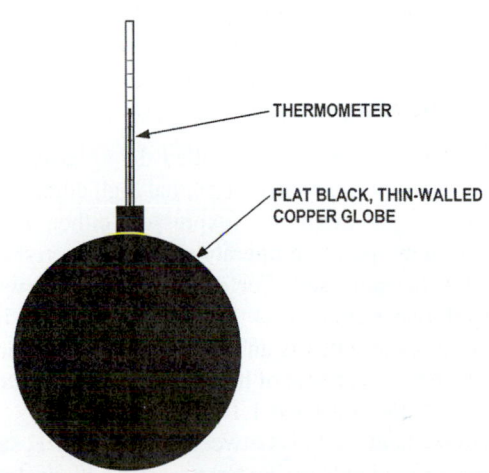

THERMOMETER

FLAT BLACK, THIN-WALLED COPPER GLOBE

FIGURE 1.7 A sketch of a black globe thermometer used to measure mean radiant temperature.

The temperature recorded by a globe thermometer is the mean radiant temperature of the space *only* if there are no air currents. Typically, local convective air currents cause the globe temperature to be between the air temperature and the actual mean radiant temperature. The faster the air moves over the globe thermometer, the closer globe temperature approaches air temperature. As a result, corrections need to be made to the globe temperature to ascertain the actual mean radiant temperature.

Measuring Relative Humidity

A *hygrometer* is an instrument that measures relative humidity in air. The approximate humidity inside a building can be found by putting a hygrometer in the space; it will show the percentage of humidity just as a thermometer shows temperature. A psychrometer can also be used to measure dry bulb and wet bulb temperature to ascertain relative humidity on a psychrometric chart.

The Psychrometer

A *psychrometer* is an instrument that contains a dry bulb thermometer and a wet bulb thermometer and that is used to determine dry bulb and wet bulb temperatures of air. The two thermometers are mounted side by side; the dry bulb thermometer has its bulb fully exposed to the atmosphere; the wet bulb thermometer, wrapped in a cloth sock material, is then moistened with water and serves as a wick. Air is forced across the thermometers either in a device with a fan that blows air across the thermometers or in a sling psychrometer. A *sling psychrometer* is an instrument that contains a dry bulb thermometer and a wet bulb thermometer and that is whirled in the air. See Figure 1.8, as well as Photos 6.1 and 6.2.

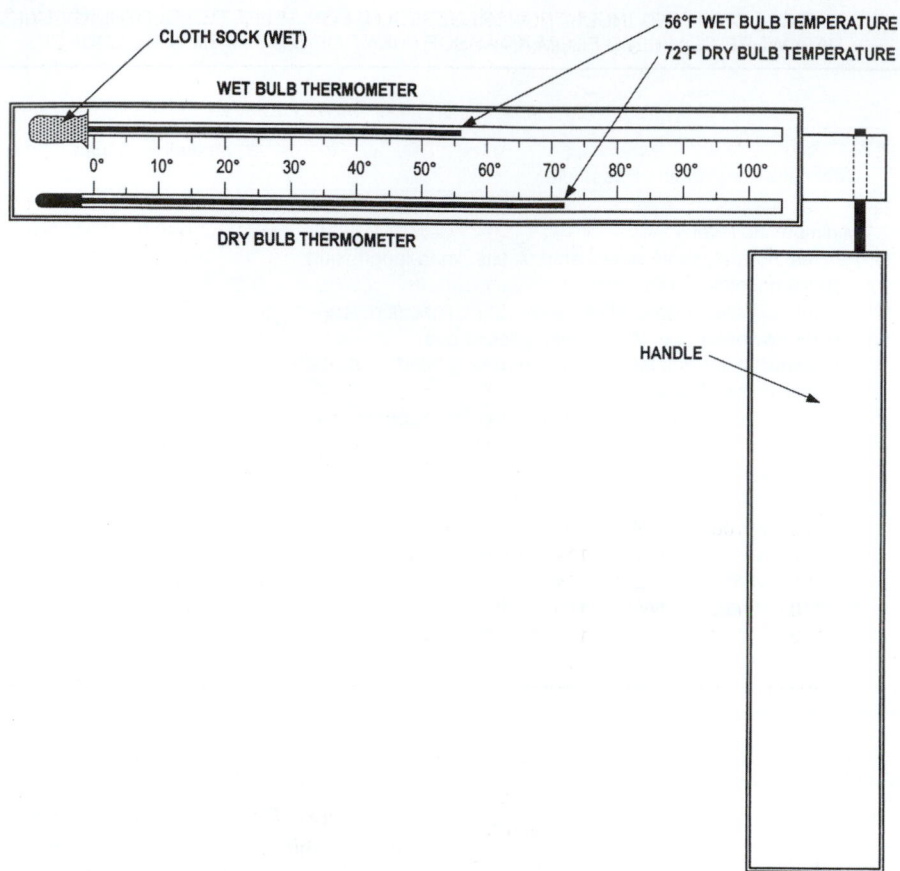

FIGURE 1.8 A sketch of a sling psychrometer used to measure the wet bulb temperature and dry bulb temperature of air.

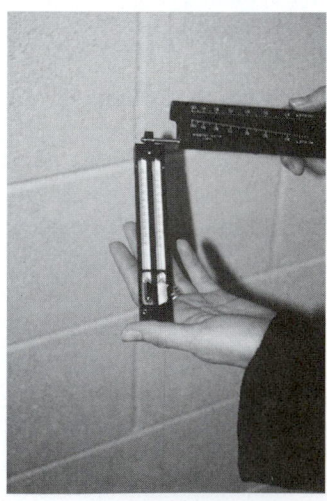

PHOTO 1.1 The thermometers of a sling psychrometer. Note the cloth sock on the wet bulb thermometer. *(Used with permission of ABC)*

PHOTO 1.2 A sling psychrometer is whirled in the air until the wet bulb temperature and dry bulb temperature achieve equilibrium. *(Used with permission of ABC)*

When the temperatures read in a psychrometer reach equilibrium with the air (e.g., temperatures stabilize), the dry and wet bulb temperatures are determined. A table accompanying the instrument provides the relative humidity in terms of the readings of the wet bulb and dry bulb thermometers. In another approach, the dry bulb and wet bulb temperatures can be plotted on a psychrometric chart to establish the related conditions.

The Heat Index

The *heat index (HI)*, sometimes called the *apparent temperature*, is a measure of the contribution that high ambient air temperature and high humidity (expressed either as relative humidity or dew point temperature) make in reducing the body's ability to cool itself. For example, with a dew point of 79°F and relative humidity of 70%, the heat index is 106°F. Essentially, the heat index is an accurate measure of how hot it really feels when the effects of humidity are added to high temperature. See Tables 1.15 and 1.16.

When the heat index is between 90° and 104°F, sunstroke, heat cramps, or heat exhaustion are possible with prolonged exposure and physical activity. When the index is between 105° and 129°F, sunstroke, heat cramps, or heat exhaustion is likely

TABLE 1.15 HEAT INDEX BASED ON DRY BULB TEMPERATURE AND DEW POINT TEMPERATURE. EXPOSURE TO FULL SUNSHINE CAN INCREASE HEAT INDEX VALUES BY UP TO 15°F.

Dew Point Temperature (°F)	Ambient Dry Bulb Temperature (°F)															
	90	91	92	93	94	95	96	97	98	99	100	101	102	103	104	105
65	94	95	96	97	98	100	101	102	103	104	106	107	108	109	110	112
66	94	95	97	98	99	100	101	103	104	105	106	108	109	110	111	112
67	95	96	97	98	100	101	102	103	105	106	107	108	110	111	112	113
68	95	97	98	99	100	102	103	104	105	107	108	109	110	112	113	114
69	96	97	99	100	101	103	104	105	106	108	109	110	111	113	114	115
70	97	98	99	101	102	103	105	106	107	109	110	111	112	114	115	116
71	98	99	100	102	103	104	106	107	108	109	111	112	113	115	116	117
72	98	100	101	103	104	105	107	108	109	111	112	113	114	116	117	118
73	99	101	102	103	105	106	108	109	110	112	113	114	116	117	118	119
74	100	102	103	104	106	107	109	110	111	113	114	115	117	118	119	121
75	101	103	104	106	107	108	110	111	113	114	115	117	118	119	121	122
76	102	104	105	107	108	110	111	112	114	115	117	118	119	121	122	123
77	103	105	106	108	109	111	112	114	115	117	118	119	121	122	124	125
78	105	106	108	109	111	112	114	115	117	118	119	121	122	124	125	126
79	106	107	109	111	112	114	115	117	118	120	121	122	124	125	127	128
80	107	109	110	112	114	115	117	118	120	121	123	124	126	127	128	130
81	109	110	112	114	115	117	118	120	121	123	124	126	127	129	130	132
82	110	112	114	115	117	118	120	122	123	125	126	128	129	131	132	133

and heatstroke is possible. Heat indices of 130° or higher will result in heatstroke or sunstroke quickly. To ensure health and safety of personnel, employers institute work policies based on the heat index that establish maximum times of exposure to conditions of high temperature and high humidity.

The heat index is not used in the actual design of a building. It is, however, a measure related to thermal comfort and the potential for environmental heat stress that can affect worker health and safety on the construction work site. It is introduced here because it helps the student understand the relationship between temperature sensation, dry bulb temperature and relative humidity and wet bulb temperature.

Wet Bulb Globe Temperature

The *wet bulb globe temperature (WBGT)* is an index of various environmental factors (air temperature, humidity, radiant heat, and ventilation) that can be easily measured in the workplace. It is commonly used as a measure related to environmental heat stress to prevent heatstroke during physical exercise or while at work.

The WBGT is not directly used in design of a building. It is considered, however, an important measure related to health and safety of personnel employed in the construction and maintenance of buildings. (As a result, it is introduced here.) With

TABLE 1.16 HEAT INDEX BASED ON DRY BULB TEMPERATURE AND RELATIVE HUMIDITY. EXPOSURE TO FULL SUNSHINE CAN INCREASE HEAT INDEX VALUES BY UP TO 15°F.

Relative Humidity (%)	Ambient Dry Bulb Temperature (°F)															
	90	91	92	93	94	95	96	97	98	99	100	101	102	103	104	105
90	119	123	128	132	137	141	146	152	157	163	168	174	180	186	193	199
85	115	119	123	127	132	136	141	145	150	155	161	166	172	178	184	190
80	112	115	119	123	127	131	135	140	144	149	154	159	164	169	175	180
75	109	112	115	119	122	126	130	134	138	143	147	152	156	161	166	171
70	106	109	112	115	118	122	125	129	133	137	141	145	149	154	158	163
65	103	106	108	111	114	117	121	124	127	131	135	139	143	147	151	155
60	100	103	105	108	111	114	116	120	123	126	129	133	136	140	144	148
55	98	100	103	105	107	110	113	115	118	121	124	127	131	134	137	141
50	96	98	100	102	104	107	109	112	114	117	119	122	125	128	131	135
45	94	96	98	100	102	104	106	108	110	113	115	118	120	123	126	129
40	92	94	96	97	99	101	103	105	107	109	111	113	116	118	121	123
35	91	92	94	95	97	98	100	102	104	106	107	109	112	114	116	118
30	89	90	92	93	95	96	98	99	101	102	104	106	108	110	112	114

TABLE 1.17   THE WET BULB GLOBE TEMPERATURE (WBGT) IS AN INDEX OF VARIOUS ENVIRONMENTAL FACTORS (AIR TEMPERATURE, HUMIDITY, RADIANT HEAT, AND VENTILATION) THAT CAN BE EASILY MEASURED IN THE WORKPLACE. IT IS COMMONLY USED AS A MEASURE RELATED TO ENVIRONMENTAL HEAT STRESS TO PREVENT HEATSTROKE DURING PHYSICAL EXERCISE OR WHILE AT WORK. STANDARDS ARE EXTRACTED FROM A PUBLIC COLLEGE'S POLICY ON EMPLOYEE SAFETY.

WBGT		
°F	°C	Types of Warning and Required Rest
Below 70	Below 21	Safe (needs occasional supply of water)
70 to 77	21 to 25	Safe (needs immediate supply of water)
77 to 82	25 to 28	Caution (needs to take rest)
82 to 88	28 to 31	Warning (stop active physical exercise)
Above 88	Above 31	Dangerous (cancel physical exercise)

high WBGT readings at the work site, employees are provided with more work breaks and removed from the extreme environment. Refer to Table 1.17.

WBGT is a measurement method to determine heat stress given to human body at work in thermally harsh environment. The WBGT Index is computed from temperatures recorded from three types of thermometers: a stationary wet bulb thermometer exposed to the sun and to the prevailing wind, and measured in wet bulb temperature (WBT); a six-inch black globe thermometer similarly exposed, expressed in temperature (BGT); and a dry bulb thermometer shielded from the direct rays of the sun, measured in dry bulb temperature (DBT).

All WBGT readings are taken at a location representative of the conditions to which individuals are exposed. The wet bulb and globe thermometers are suspended in the sun at a height of 4 ft above ground. A period of at least 20 min after deploying the apparatus should elapse before readings are taken. For outdoors with a solar load, WBGT is calculated as:

$$WBGT = 0.7\,WBT + 0.2\,BGT + 0.1\,DBT$$

For indoor and outdoor conditions with no solar load, WBGT is calculated as:

$$WBGT = 0.7\,WBT + 0.3\,BGT$$

Wet bulb temperature, and thus humidity, has the greatest influence on this index.

Example 1.11

Conditions within an unconditioned warehouse are recorded at a WBT of 71°F, a BGT of 88°F, and a DBT of 93°F. Determine the WBGT.

$$WBGT = 0.7\,WBT + 0.2\,BGT + 0.1\,DBT = 0.7\,(71°F)$$
$$+ 0.2\,(88°F) + 0.1\,(93°F) = 76.6°F$$

In the Example 1.11, the WBGT was found to be 76.6°F (24.8°C). Upon reviewing Table 1.17, this temperature is deemed safe for working conditions if an immediate supply of water is available.

Effective and Operative Temperatures

Several parameters have been devised to measure the cooling effect of the air on a human body.

ASHRAE Standard 55-1992, *Thermal Environmental Conditions for Human Occupancy*, defines *effective temperature (ET)* as an arbitrary, experimentally determined index used to rate the various combinations of dry bulb temperature, relative humidity, radiant conditions, and air movement that create the same thermal sensation. ET of a specific space may be defined as the dry bulb temperature of an environment at 50% relative humidity and a specific uniform radiation condition; heat exchange of the environment is based upon 0.6 clo, still air (40 fpm or less), 1 hr exposure time, and sedentary activity level (about 1 met). It is a reliable indicator of comfort with the thermal environment. For example, a space having an ET of 68°F (20°C) will induce a temperature sensation equivalent to a condition described by 68°F (20°C) at 50% relative humidity in nearly still air and at a metabolic rate of 1 met.

ASHRAE Standard 55-2004 explains the *operative temperature* as being based on an arbitrary set of indexes that combine into a single number the effects of dry bulb temperature, radiant temperature, and air motion on the sensation of warmth or cold felt by the human body. The operative temperature is equal to the temperature at which a specified hypothetical environment would support the same heat loss from an unclothed, reclining human body as the actual environment.

Comfort Zones

ASHRAE Standard 55-2004 specifies a comfort zone representing the optimal range and combinations of thermal factors (air temperature, radiant temperature, air velocity, humidity) and personal factors (clothing and activity level) with which at least 80% of the building occupants are expected to express satisfaction. The standard views thermal comfort as a specific combination of thermal conditions that will elicit the desired physiological state of "comfortable" (thermal comfort temperature). Generally, temperature and humidity should be maintained within the comfort zone of 68 to 79°F (20° to 26°C) and 30 to 60% relative humidity, depending on the season.

Overall Thermal Comfort in a Space

The combined effects of environmental and individual factors can be represented on a psychrometric chart. The comfort zone is the combination of temperature and humidity where people report comfort. If conditions are to the right of the comfort zone, then there is a need to increase cooling (e.g., by decreasing dry bulb temperature or increasing airflow). If conditions are to the left of the comfort zone, then there is a need to increase heating (e.g., by increasing dry bulb temperature). There are different

TABLE 1.18 EXAMPLES OF ACCEPTABLE OPERATIVE
TEMPERATURE RANGES BASED ON COMFORT
ZONE DIAGRAMS IN ASHRAE STANDARD-55-2004.

Conditions	Acceptable Operative Temperatures	
	°F	°C
Summer (clothing insulation = 0.5 clo)		
Relative humidity 30%	76 to 82	24.5 to 28
Relative humidity 60%	74 to 78	23 to 25.5
Winter (clothing insulation = 1.0 clo)		
Relative humidity 30%	69 to 78	20.5 to 25.5
Relative humidity 60%	68 to 75	20 to 24

Based on activity levels of 1.0 to 1.3 met; air speed of <0.2 m/s (40 fpm); humidity ratio (W) <0.012 lbmw/lbma (pounds mass of water vapor per pounds mass of dry air).

charts for different levels of activity and different comfort zones for different clo values and different air velocities.

Women will generally prefer higher effective temperatures than men. Similarly, older people and people who live in southern climates prefer higher temperatures. Of course, whether a person will be comfortable or not also depends on the level of activity and the type of clothing the individual wears. A person wearing a sweater who is cleaning the house will need a lower effective temperature than someone who is wearing a cotton T-shirt or blouse and is watching television.

The suggested inside winter dry bulb temperatures for a variety of spaces are shown in Table 1.18. Those spaces, in which people will be relatively inactive, such as classrooms and hospital rooms, will require higher temperatures than spaces in which people will be more active.

Thermal comfort is greatly affected by the relationship between indoor and outdoor dry bulb temperatures. This factor is especially true in spaces that people occupy for short periods of time. People who are in residences or offices for a long period of time acclimate to a specific dry bulb temperature and are comfortable in it. However, people occupying a space for short durations (up to 1 hr) will be most comfortable if there is just a 10° to 15°F (6° to 9°C) difference between the outside and the inside temperatures. So, if it is 95°F (35°C) outside, it should be about 80°F inside for a person to feel comfortable. The common practice in stores, shops, grocery markets, and so on, of setting their cooling at a 68° to 70°F (20° to 21.1°C) inside temperature does not make for comfortable conditions in the summer. This is obvious when you see the number of people who must take sweaters into grocery stores, fast food restaurants, and other stores and shops during the summer. Yet, it is comfortable to most employees because they have taken time to adjust to the indoor dry bulb temperature.

The mean radiant temperature of a space also influences comfort. No matter how high the dry bulb temperature is in a space, it is very difficult for an occupant to feel comfortable when standing or sitting next to a very cold surface, such as a window. The radiant heat loss on the side of the body nearest the cool surface will be significant. In one case, the dry bulb temperature in a person's office was turned up to 78°F (23°C) yet the occupant still felt uncomfortable. By rearranging the office so that there was an interior wall at the occupant's back instead of a window, the occupant felt comfortable at 72°F (22°C). An alternative would be to install a thermally reflective, insulating glass with a higher R-value. High performing glass raises the interior surface temperature of the window surface and improves the MRT, thereby improving occupant comfort.

STUDY QUESTIONS

1-1. Distinguish between heat and temperature.

1-2. Define the properties of:

a. Density

b. Specific heat

c. Specific heat capacity

1-3. Distinguish between sensible heat and latent heat.

1-4. Describe psychrometrics.

1-5. Describe the psychrometric variables.

1-6. Describe the psychrometric processes.

1-7. Briefly describe the three natural processes by which body heat is transferred and how they affect comfort.

1-8. Describe the occupant-related factors that influence thermal comfort.

1-9. Describe the environmental factors that influence thermal comfort.

1-10. In what ways does the surrounding air temperature affect the amount of body heat given off?

1-11. What effect does humidity have on the comfort of a body?

1-12. What is meant by effective temperature in relation to comfort?

1-13. What is meant by operative temperature in relation to comfort?

Design Exercises

1-14. A boiler used to heat a school building is rated at 8 000 000 Btu/hr. Determine its rating in:

a. MBH   b. MBH   c. kW

1-15. A furnace used to heat a home is rated at 30 kW. Determine its rating in:

a. Btu/hr   b. MBH

1-16. Convert a temperature of 80°F to:

a. °C   b. R   c. K

1-17. Convert a temperature of 22°C to:

a. °F   b. R   c. K

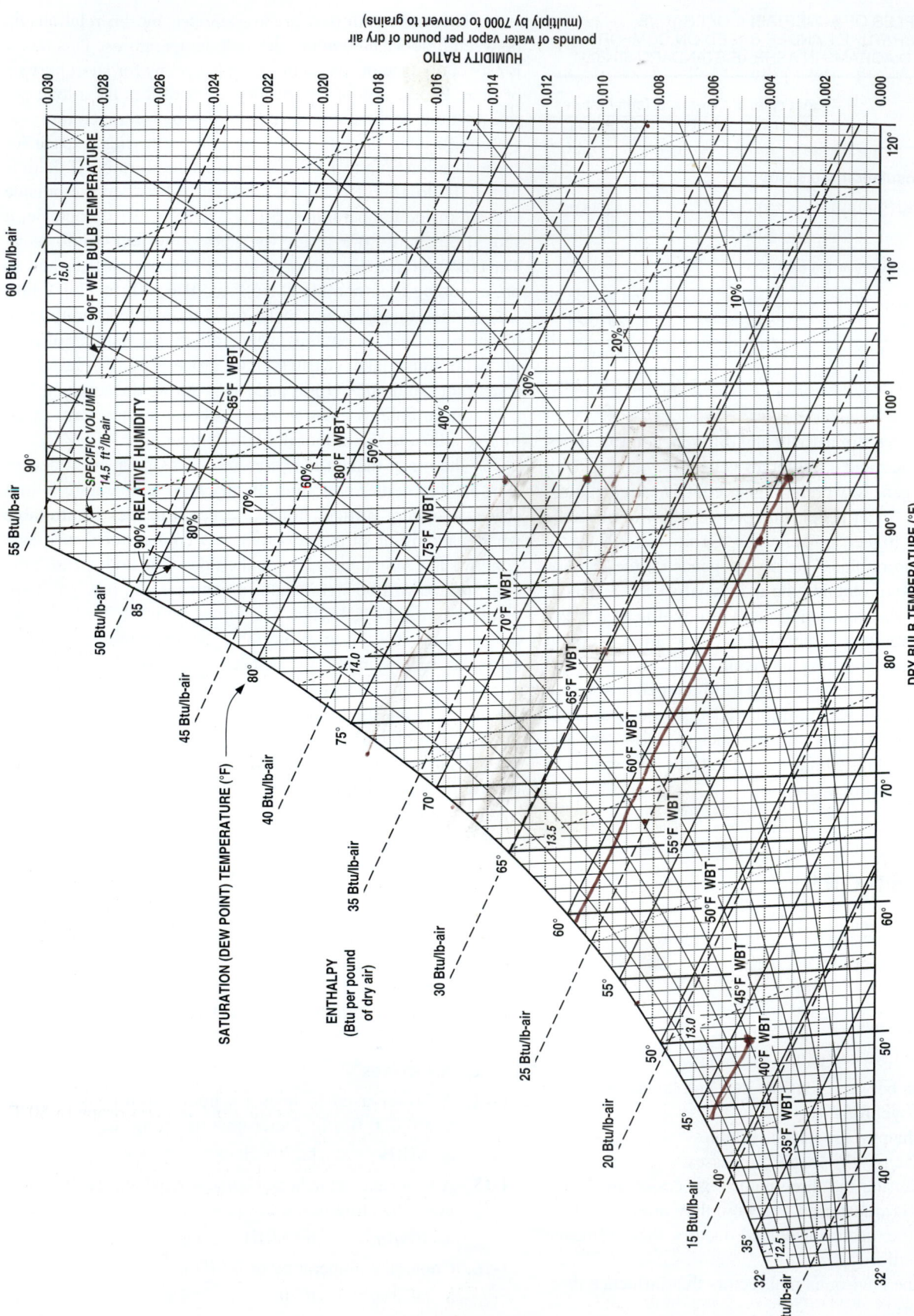

FIGURE 1.9 A full-size psychrometric chart for normal temperatures at standard conditions (sea level).

24

1-18. Ten pounds of liquid water is at a temperature of 180°F at standard atmospheric conditions. Determine the energy required to convert this to water vapor (steam) at a temperature of 212°F.

1-19. Moist air at standard conditions is at a dry bulb temperature of 93°F and a wet bulb temperature of 59°F. Use the psychrometric chart to find:

a. The relative humidity

b. The dew point temperature

c. The humidity ratio

d. The specific volume

e. The enthalpy

1-20. Moist air at standard conditions is at a dry bulb temperature of 93°F and a wet bulb temperature of 70°F. Use the psychrometric chart to find:

a. The relative humidity

b. The dew point temperature

c. The specific volume

d. The humidity ratio

e. The enthalpy

1-21. Moist air at standard conditions is at a dry bulb temperature of 50°F and a relative humidity of 70%. It is sensibly heated (constant moisture) to 72°F. Use the psychrometric chart to determine the following properties of the air after it is heated:

a. The relative humidity

b. The dew point temperature

c. The specific volume

d. The humidity ratio

e. The enthalpy

1-22. A sling psychrometer is used to measure conditions in an indoor space. The dry bulb temperature is 76°F and the wet bulb temperature is 58°F. Use this data and the psychrometric chart to determine the following properties of the indoor air:

a. The relative humidity

b. The dew point temperature

c. The specific volume

d. The humidity ratio

e. The enthalpy

1-23. The following table contains outdoor cooling design temperatures for various cities. Use a psychrometric chart to find the following:

a. The dew point temperature at each city under these design temperatures

b. The relative humidity at each city under these design temperatures

c. The humidity ratio at each city under these design temperatures

d. The six cities with the highest amount of moisture content in air under these design conditions

e. The six cities with the lowest amount of moisture content in air under the following design conditions:

City	Dry Bulb Temperature, °F	Wet Bulb Temperature, °F
Anchorage, AK	68	57
Atlanta, GA	91	74
Boston, MA	87	71
Denver, CO	90	59
Honolulu, HI	88	73
Houston, TX	94	77
Los Angeles, CA	81	64
Miami, FL	90	77
New York, NY	88	72
Newark, NJ	90	73
Phoenix, AZ	108	70
Seattle, WA	81	64

1-24. Moist air at standard conditions is at a dry bulb temperature of 80°F and a relative humidity of 50%. It is sensibly cooled (constant moisture) to 72°F. Use the psychrometric chart to determine its saturation temperature. Draw a simple sketch showing this process on a psychrometric chart.

1-25. Moist air at standard conditions in an indoor building space has a dry bulb temperature of 72°F and relative humidity of 50%. At what dry bulb temperature will condensation begin to occur on wall and window surfaces?

1-26. A building owner is considering replacing single-glazed windows with double-glazed windows to improve efficiency. It will also eliminate condensation moisture problems during cold weather. Moist air at standard conditions in the indoor building space reaches a maximum dry bulb temperature of 72°F and relative humidity of 40%.

a. At what dry bulb temperature will condensation begin to occur on the window surfaces?

b. Will the double-glazed windows prevent condensation under these indoor space conditions if the interior surface temperature of the glass drops to 50°F in cold weather?

1-27. Moist air at standard conditions is at a dry bulb temperature of 50°F and a relative humidity of 70%. It is sensibly heated (constant moisture) to 72°F. Use the psychrometric chart to determine the energy added to the air after it is heated:

a. In Btu/lb of dry air

b. In Btu/ft^3 of dry air

1-28. Draw a simple sketch showing the process outlined in the previous example on a psychrometric chart.

1-29. Moist air at standard conditions is at a dry bulb temperature of 80°F and a relative humidity of 50%. It is sensibly cooled (constant moisture) to 72°F. Use the psychrometric chart to determine the energy removed from the air after it is cooled:

   a. In Btu/lb of dry air

   b. In Btu/ft³ of dry air

1-30. Draw a simple sketch showing the process outlined in the previous example on a psychrometric chart.

1-31. Moist air at standard conditions is at a dry bulb temperature of 80°F and a relative humidity of 50%. It is cooled to 50°F. Use the psychrometric chart to determine:

   a. The energy removed from the air after it is cooled, in Btu/lb of dry air

   b. The energy removed from the air after it is cooled, in Btu/ft³ of dry air

   c. The amount of moisture removed from the air after it is cooled, in lb/lb of dry air

   d. The amount of moisture removed from the air after it is cooled, in lb/ft³ of dry air

1-32. Draw a simple sketch showing the process outlined in the previous example on a psychrometric chart.

1-33. Moist air at standard conditions is at a dry bulb temperature of 93°F and a wet bulb temperature of 59°F. It is cooled in an evaporative (swamp) cooler by adding water along the wet bulb temperature line. Use the psychrometric chart to find:

   a. The temperature of the air at a relative humidity of 60%

   b. The amount of moisture added to the air after it is cooled, in lb/lb of dry air

   c. The amount of moisture added to the air after it is cooled, in lb/ft³ of dry air

   d. The amount of water that must be added to the air per hour at an volumetric flow rate of 250 ft³/min

   e. Draw a simple sketch showing the process on a psychrometric chart

1-34. A 12 ft by 12 ft square room has one exterior wall with a surface temperature of 58°F and three interior wall surfaces with a temperature of 72°F.

   a. Approximate the mean radiant temperature sensation of a person standing in the center of the room. Neglect the effects of the floor and ceiling.

   b. Approximate the mean radiant temperature sensation of a person standing 2 ft from the exterior wall. Neglect the effects of the floor and ceiling.

1-35. A 3 m by 5 m rectangular room has one 4-m exterior framed wall with a surface temperature of 16°C. The three interior wall surfaces are at a temperature of 22°C.

   a. Approximate the mean radiant temperature sensation of a person standing in the center of the room. Neglect the effects of the floor and ceiling.

   b. Approximate the mean radiant temperature sensation of a person standing 1 m from the exterior wall. Neglect the effects of the floor and ceiling.

1-36. A 12 ft by 12 ft square corner room has an exterior wall with a surface temperature of 58°F and a glass window surface covering one wall at a surface temperature of 48°F. The two interior wall surfaces are adjacent to one another and are at a temperature of 72°F.

   a. Approximate the mean radiant temperature sensation of a person standing in the center of the room. Neglect the effects of the floor and ceiling.

   b. Approximate the mean radiant temperature sensation of a person standing 2 ft away from the center of the window. Neglect the effects of the floor and ceiling.

   c. Approximate the mean radiant temperature sensation of a person standing 2 ft away from the center of the window. Assume that the window curtains are closed and the temperature of the interior surface of the curtains is 62°F. Neglect the effects of the floor and ceiling.

1-37. A 3 m by 4 m rectangular corner room has one 4-m exterior framed wall with a surface temperature of 16°C, and one window glass surface covering on one 4-m wall that is at a surface temperature of 7°C. The two interior wall surfaces are adjacent to one another and are at a temperature of 22°C.

   a. Approximate the mean radiant temperature sensation of a person standing in the center of the room.

   b. Approximate the mean radiant temperature sensation of a person standing 1 m away from the center of the window.

1-38. Conditions at a construction site are recorded as a wet bulb temperature (WBT) of 70°F, a black globe temperature (BGT) of 89°F, and a dry bulb temperature (DBT) of 94°F.

   a. Determine the wet bulb globe temperature (WBGT).

   b. Is this temperature deemed safe for working conditions if an immediate supply of water is available?

1-39. Conditions at a construction site are recorded as a wet bulb temperature (WBT) of 18°C, a black globe temperature (BGT) of 32°C, and a dry bulb temperature (DBT) of 36°C.

   a. Determine the wet bulb globe temperature (WBGT).

   b. Is this temperature deemed safe for working conditions if an immediate supply of water is available?

FUNDAMENTALS OF HEAT TRANSFER

2.1 MODES OF HEAT TRANSFER

Introduction

In cold weather, heat is lost as it moves from the building interior to the exterior. A building's heating system (e.g., furnace, boiler) introduces heat into the building to replace the lost heat so that comfortable indoor conditions can be maintained. Heat transfer theory is used to estimate the heat loss from the building's interior. As a result, an understanding of the basic principles of heat transfer is valuable to the designer.

Basic Theory

Heat can be transferred through molecular motion or radiative transport by the three fundamental modes of heat transfer: conduction, convection, and radiation. Conduction and convection heat transfer are tied to molecular motion. These types of heat transfer occur when molecules of a substance in contact with or close to a heat source vibrate more vigorously as they are heated, and this additional molecular kinetic energy (vibration) spreads to neighboring molecules. Radiation heat transfer is by electromagnetic waves that do not require a medium (e.g., air, water) through which to travel. The surface of a warm temperature emits radiative energy to surrounding surfaces that are at a lower temperature. In all cases, transfer of heat only occurs when a temperature difference exists and will always be from a higher temperature region to a lower temperature region. Thus, the driving force behind heat transfer is temperature difference.

The basic theory and computations associated with the three modes of heat transfer are described in the sections that follow. This explanation is based on simple, steady state, one-dimensional flow of heat, which means that heat is driven by a temperature difference that does not fluctuate so that heat flow is always in one direction. Several properties of materials affect heat transfer. Examples include thermal conductivities, specific heats, material densities, fluid velocities, fluid viscosities, and surface reflectance and emittance characteristics. The effects of these properties make the solution of many heat transfer problems an involved process. An explanation of more complex (actual) dynamic, three-dimensional flow is beyond the scope of a text of this nature.

2.2 CONDUCTION

The movement of heat through a substance or between two substances in contact with each other is called *conduction*. Heat moves from one molecule to another molecule in contact with it. See Figure 2.1. For example, conduction heat transfer occurs when heat moves along a metal fireplace poker when one end is placed in and heated by a fire. The result is an increase in temperature as heat moves from the hot poker end to the handle of the poker. On a molecular scale, regions with greater molecular kinetic energy pass their thermal energy to regions with less molecular energy through direct molecular collisions. Conduction is the most efficient mode of heat transfer.

Joseph Fourier (1768–1830), a French mathematician, developed the basic heat transfer relationship for conduction. It is known as *Fourier's law of conduction*:

$$q = (k \cdot A \cdot \Delta T)/L$$

where k is the thermal conductivity of the material in Btu $\cdot$ in/hr $\cdot$ °F $\cdot$ ft^2 (W/m $\cdot$ °C), A is the area in ft^2 (m^2), ΔT is the temperature difference, L is the thickness of the material in inches (m), and q is the rate of heat transfer in Btu/hr (W). The reader is cautioned that at times k is expressed in units of Btu/hr $\cdot$ °F $\cdot$ ft, which is Btu $\cdot$ ft/hr $\cdot$ °F $\cdot$ ft^2 simplified and is based upon the conductivity of a material with a thickness of 1 ft (rather than 1 in).

The thermal conductivities of common materials are provided in Table 2.1. Thermal conductivity varies with temperature of the body or medium through which heat passes. The rate of conduction heat flow through the material is dependent on the capability of the molecules to send or receive heat. A lower temperature means slower movement of molecules in the body or medium, and thus a lower rate of heat transfer. Within small temperature ranges, thermal conductivity is relatively constant. Reported values of k are average values for a common temperature range for buildings and building systems.

The rate of conduction heat flow through the material is proportional to the temperature difference between each surface and inversely proportional to the thickness of the material. From this equation, it can be observed that if the thickness of a homogenous material is increased, the rate of heat transfer is reduced. The reduction in heat transfer is proportional to thickness—that is, double the thickness results in half the heat

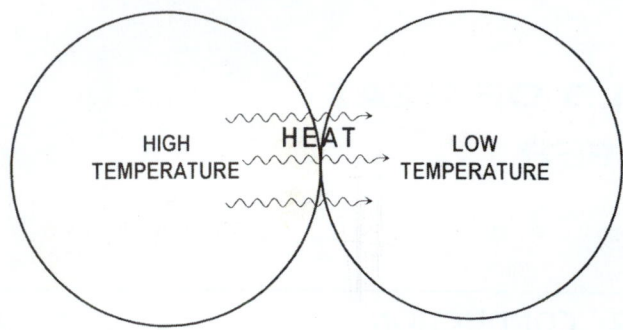

FIGURE 2.1 The movement of heat through a substance or between two substances in contact with each other is called *conduction*.

TABLE 2.1   THE THERMAL CONDUCTIVITY (k) OF COMMON MATERIALS AT A TEMPERATURE OF 68°F (20°C), UNLESS SPECIFIED OTHERWISE. DATA WERE COMPILED FROM A VARIETY OF SOURCES.

Material	Thermal Conductivity (k)	
	Btu · in/hr · °F · ft^2	W/m · °C
Adobe/clay (dry)	2.6	0.38
Aluminum	1430	205
Air	0.18	0.026
Steel (framing)	324	46.0
Brick (building)	2.8	0.40
Brick (fire brick)	4.0	0.58
Concrete (heavy)	6.9	1.00
Concrete (light)	3.0	0.43
Concrete (normal)	3.7	0.54
Glass (ordinary)	6.5	0.93
Gypsum wallboard	0.7	0.10
Insulation (cellulose)	0.27	0.04
Plaster	2.9	0.42
Sand (dry)	0.4	0.05
Sand (saturated)	2.7	0.39
Water (liquid)	4.1	0.60
Water (ice)	14.6	2.1
Wood (softwood)	1.00	0.14

loss, triple the thickness results in one-third the heat loss, and so on. A larger temperature difference between the surfaces of substances causes a greater rate of heat flow.

There are wide variations in the ability of different materials to conduct heat, so there are significant differences in the thermal conductivity of materials. Metals exhibit the greatest ability to conduct heat, because their molecular structure allows electrons to move around freely instead of being bound to particular

atoms. Ceramics (e.g., concrete, glass) and some plastics are also good thermal conductors. Gases are poor conductors because their molecules are relatively far apart. The molecules of liquids and nonmetals are closer together, offering somewhat higher rates of conduction heat transfer.

Example 2.1

Determine the rate of conduction heat transfer through the following materials. Assume the surface temperatures of the materials are 0° and 68°F (or −20°C and 20°C for SI units).

a. For 100 ft^2 of a one inch thickness of building brick, where k is from Table 2.1:

$$q = [(2.8 \text{ Btu} \cdot \text{in/hr} \cdot °\text{F} \cdot \text{ft}^2) \cdot (100 \text{ ft}^2) \cdot (68 - 0)°\text{F}]/1 \text{ in}$$
$$= 19\,040 \text{ Btu/hr}$$

b. For 100 ft^2 of a one inch thickness of steel, where k is from Table 2.1:

$$q = [(324 \text{ Btu} \cdot \text{in/hr} \cdot °\text{F} \cdot \text{ft}^2) \cdot (100 \text{ ft}^2) \cdot (68 - 0)°\text{F}]/1 \text{ in}$$
$$= 2\,203\,200 \text{ Btu/hr}$$

c. For 100 ft^2 of a one inch thickness of normal weight concrete, where k is from Table 2.1:

$$q = [(3.7 \text{ Btu} \cdot \text{in/hr} \cdot °\text{F} \cdot \text{ft}^2) \cdot (100 \text{ ft}^2) \cdot (68 - 0)°\text{F}]/1 \text{ in}$$
$$= 25\,160 \text{ Btu/hr}$$

d. For 100 ft^2 of a one inch thickness of cellulose insulation, where k is from Table 2.1:

$$q = [(0.27 \text{ Btu} \cdot \text{in/hr} \cdot °\text{F} \cdot \text{ft}^2) \cdot (100 \text{ ft}^2) \cdot (68 - 0)°\text{F}]/1 \text{ in}$$
$$= 1836 \text{ Btu/hr}$$

e. For 10 m^2 of a 25 mm (0.025 m) thickness of steel, where k is from Table 2.1:

$$q = [(46 \text{ W}/°\text{C} \cdot \text{m}) \cdot (10 \text{ m}^2) \cdot (20 - (-20))°\text{C}]/0.025 \text{ m}$$
$$= 736\,000 \text{ W}$$

f. For 10 m^2 of a 25 mm (0.025 m) thickness of cellulose insulation, where k is from Table 2.1:

$$q = [(0.04 \text{ W}/°\text{C} \cdot \text{m}) \cdot (10 \text{ m}^2) \cdot (20 - (-20))°\text{C}]/0.025 \text{ m}$$
$$= 640 \text{ W}$$

Materials that resist heat flow (e.g., cellulose, foam) are called *thermal insulators*. As shown in Example 2.1, cellulose insulation is a good thermal insulator because it conducts heat slowly in comparison to materials like steel, concrete, and brick.

Example 2.2

Determine the conduction heat transfer through the glass of a 3 ft by 4 ft single-glazed window with ⅛ in thick (double-strength thickness) glass and a single-glazed window with 3⁄32 in thick (single-strength thickness) glass. Assume the temperatures of the glass surfaces are 0°F and 70°F.

For ⅛ in thick glass, where k is from Table 2.1:

$$q = [(6.5 \text{ Btu} \cdot \text{in/hr} \cdot °\text{F} \cdot \text{ft}^2) \cdot (3 \text{ ft} \cdot 4 \text{ ft}) \cdot (70 - 0)°\text{F}]/⅛ \text{ in}$$
$$= 43\ 680 \text{ Btu/hr}$$

For 3/32 in thick glass:

$$q = [(6.5 \text{ Btu} \cdot \text{in/hr} \cdot °\text{F} \cdot \text{ft}^2) \cdot (3 \text{ ft} \cdot 4 \text{ ft}) \cdot (70 - 0)°\text{F}]/3/32 \text{ in}$$
$$= 58\ 240 \text{ Btu/hr}$$

2.3   CONVECTION

Heat transfer by the motion of a heated or cooled mass is called *convection* heat transfer. It is transfer of heat between a surface and a moving fluid (a gas or a liquid), or heat transfer by the movement of molecules from one region in a fluid to another region because of movement of the heated fluid. Convection involves conduction heat transfer because heat is transferred by contact of molecules. Natural or forced movement of a fluid is involved, so convection heat transfer is less efficient than conduction heat transfer.

Natural Convection

When the convection process occurs naturally, it is known as *natural convection* or *free convection*. As heat conducts into a motionless region of fluid, an increased agitation of the molecules leads to volumetric expansion of the fluid; that is, the molecules collide with one another forcing them farther apart and causing the fluid to expand. The expanded fluid is less dense, becomes more buoyant because it is lighter, rises, and displaces a cooler, more dense region of fluid, thereby transporting heat by fluid motion; in other words, a warm fluid rises and cold fluid falls because of the buoyancy forces acting upon them. Natural movement is much like the buoyant behavior of a hot air balloon when the air contained in the balloon is heated or cooled: Less dense warm air rises above the higher density cool air.

The natural convection process also works in reverse. As a warm fluid comes in contact with a cooler surface, the warmer molecules transfer some of their molecular energy to the cooler surface. The cooled fluid becomes heavier (more dense) and sinks. Molecules of warmer fluid replace it. An example of natural convection is the free-flowing downward draft of cool air at the base of a window during the winter.

Free convection is an important factor of heat lost through a window, or by which heat is lost through the glazing of a solar collector. See Figure 2.2. Circulating boundary air films that form on both sides of a window (or a wall for that matter) act to reduce heat transfer because it creates an insulating layer.

Forced Convection

When a cool liquid is forced across a warm object by mechanical means (e.g., a fan or pump), it tends to cool the object much faster than if convection were allowed to occur naturally. This

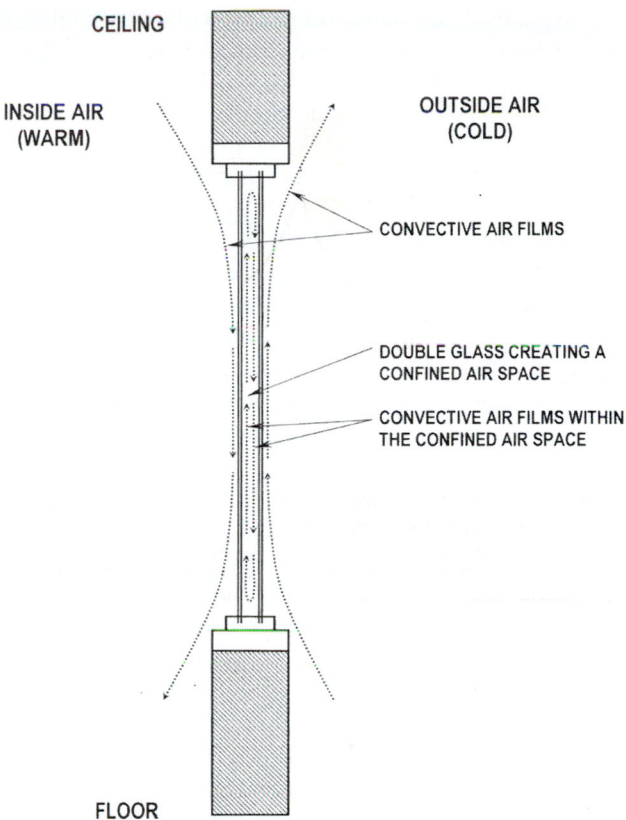

FIGURE 2.2   A section of a window illustrating natural convection. The buoyancy effects of heated or cooled air cause natural convection.

process is referred to as *forced convection*. Forced convection is a more efficient mode of heat transfer in comparison to free convection because molecules are forced to come in contact with other molecules. An example of forced convection heat transfer occurs when air moves across a hot surface and is heated (e.g., air flowing across the hot metal surfaces of the heat exchanger in a forced-air furnace).

The rate heat transfer from convection is proportional to the temperature difference, and is similar to the equation for conduction:

$$q = f \cdot A \cdot \Delta T$$

where f is the individual film coefficient (like C for solids) in Btu/hr · °F · ft² (W/m · °C), A is the area in ft² (m²), ΔT is the temperature difference in °F (°C), and q is the heat transfer rate in Btu/hr (W).

Example 2.3

Assume that the 3 ft by 4 ft window from Example 2.2 has a convection coefficient of 1.0 Btu/hr · °F · ft². Determine the convection heat loss if the outdoor ambient temperature is 0°F and the indoor temperature is 70°F.

$$q = 1.0 \text{ Btu/hr} \cdot °\text{F} \cdot \text{ft}^2 \cdot (3 \text{ ft} \times 4 \text{ ft}) \cdot (70 - 0)°\text{F}$$
$$= 840 \text{ Btu/hr}$$

The individual film coefficient (f) will vary with the conditions of flow (e.g., natural versus forced flow) and the thermal properties of the fluid (e.g., air, water). With forced convection, a fluid is blown across a surface and the rate of heat transfer is increased because more molecules of air are contacting the surface. The faster the fluid is forced across the surface, the faster the rate of heat transfer.

In buildings, the velocity air greatly affects the rate of forced convection heat transfer: a greater wind velocity leads to a greater rate of heat flow. Because air movement is greater on the outside of a building (from wind), the convection air film on the exterior side of the wall will have a greater effect on heat flow in comparison to the air film on the interior surface of the wall. These air films create a thermal obstruction that actually resists heat flow.

When warm air is used to heat (or cold air is used to cool) a building, the heat transfer rate is also by a form of forced convection. For example, in the winter outside air is heated to room temperature and forced through the volume of the structure. Such heat flow is described in equation form as:

$$q = m \cdot c \cdot V \cdot \Delta T = C \cdot V \cdot \Delta T$$

where m is the mass flow rate of the fluid in lb/hr or lb/min (kg/min or kg/s), c is the specific heat of the fluid in Btu/lb $\cdot$ °F (kJ/kg $\cdot$ °C), C is the specific heat capacity of the fluid in Btu/ft^3 $\cdot$ °F (kJ/m^3 $\cdot$ °C), ΔT is the temperature difference in °F (°C), and q is the heat transfer rate in Btu/hr (W). Under standard conditions, the specific heat capacity of air is 0.018 Btu/ft^3 $\cdot$ °F (1.195 kJ/m^3 $\cdot$ °C).

Example 2.4

A quantity of air equivalent to the volume of air in a 12 ft by 12 ft by 8 ft room is replaced with outdoor air once per hour from air leakage into and out of the room. The specific heat capacity of air is 0.018 Btu/ft^3 $\cdot$ °F. Determine the rate of convection heat loss if the outdoor ambient temperature is 0°F and the indoor temperature is 70°F.

$$q = C \cdot V \cdot \Delta T = 0.018 \text{ Btu/ft}^3 \cdot °F \cdot (12 \text{ ft} \cdot 12 \text{ ft} \cdot 8 \text{ ft})$$
$$\cdot (70 - 0)°F = 1452 \text{ Btu/hr}$$

Example 2.5

A bathroom fan in a residence exhausts heated indoor air at a rate of 75 ft^3/min. The specific heat capacity of air is 0.018 Btu/ft^3 $\cdot$ °F. Determine the rate of convection heat loss in Btu/hr if the outdoor ambient temperature is 0°F and the indoor temperature is 70°F.

$$q = C \cdot V \cdot \Delta T = 0.018 \text{ Btu/ft}^3 \cdot °F \cdot [(75 \text{ ft}^3/\text{min})$$
$$\cdot (60 \text{ min/hr})] \cdot (70 - 0)°F = 5670 \text{ Btu/hr}$$

2.4   RADIATION

Radiation heat transfer involves movement of energy by electromagnetic waves. All substances above absolute zero (above −460°F/−273°C) can emit radiant energy. Unimpeded, radiative energy travels at about the speed of light (186 000 miles per second or 300 000 000 meters per second) in a straight-line path unless influenced by a gravitational, magnetic, or some other force. When two surfaces of different temperatures are in view of each other, there will be an exchange of radiative energy from the hot body to the cold body. See Figure 2.3.

Radiant heat transfer is the mode by which the thermal energy of the sun is transferred to earth. The warmth one feels facing a campfire on a cool evening is from radiant energy transmitted by the fire. Because radiant energy moves in a straight path, only the body surface area facing the fire feels the fire's warmth. The body surface area facing away from the fire is cooled by the low temperature of the surrounding air (through convection).

Radiant heat transfer does not require a molecular medium (e.g., fluid or solid) for transfer to occur. Radiative energy travels through a vacuum better than it does through a gas, because, when passing through a gaseous substance, a portion of the radiative energy is absorbed and reflected by the gas molecules. For example, the sun's radiative energy travels the 93 million miles to earth through space with little change, in comparison to the decrease caused as it passes through the earth's atmosphere (a mixture of gases).

The transfer of radiant energy to and from building components is a study that is extremely complex and even challenging to simulate with a computer. An understanding of some basic principles of how radiative energy is transferred is helpful.

Radiative energy is classified by length of wavelength. *Wavelength* (λ) is measured as the distance from the peak of one wave to the peak of the next wave. The units of wavelengths of thermal radiative energy are expressed in millionths of a meter, which is a *micron (μm)* or *nanometer (nm)*, where 1 μm = 1000 nm. Shorter wavelengths tend to be more powerful than longer wavelengths.

Gamma waves, x-rays, ultraviolet waves, visible light waves, infrared waves, radio waves, and TV waves are all electromagnetic waves. Only ultraviolet, visible light, and infrared radiative energy are important in our study of thermal radiation. These are introduced in the sections that follow.

Ultraviolet Radiation

Ultraviolet radiation (UV) or short-wavelength radiative energy is not visible to the human eye. These are the rays that produce sunburn and account for the degradation of plastics, and discoloration of certain fabrics. UV wavelengths are below about 0.38 μm (380 nm).

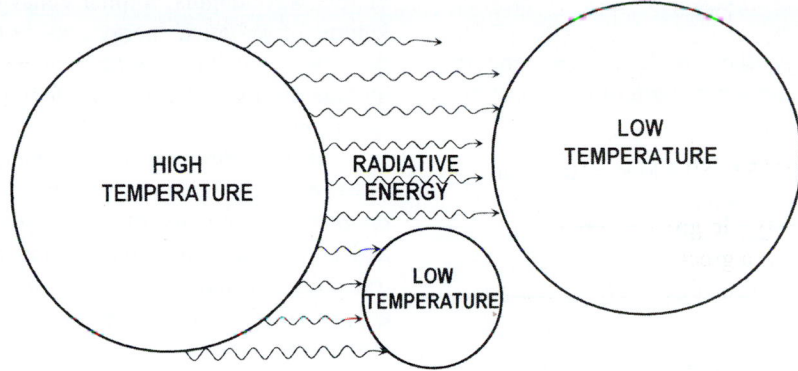

FIGURE 2.3 Radiation heat transfer involves movement of radiative energy by electromagnetic waves. Radiative energy travels in a straight-line path until it contacts a body. It does not require a molecular medium (e.g., fluid or solid) for transfer to occur. When two surfaces of different temperatures are in view of each other, there will be an exchange of radiative energy from the hot body to the cold body.

Visible Light Radiation

The middle wavelengths are referred to as *visible light (VL),* because the human eye can perceive them. The wavelengths of VL range from about 0.38 μm (380 nm) to about 0.75 μm (750 nm). When light passes through a prism, the varying speeds that visible wavelengths travel cause VL to be broken up into colors that vary from violet and blues, through greens and yellows, to reds.

Infrared Radiation

Long wavelengths of thermal radiation are referred to as *infrared radiation (IR)*. The wavelengths of IR range from about 0.75 μm (750 nm) to about 1000 μm (1 000 000 nm). IR is often subdivided into *near-infrared* (NIR, 0.7 to 5 μm in wavelength), *mid-* or *intermediate-infrared* (MIR or IIR, 5 to 30 μm), and *far-infrared* (FIR, 30 to 1000 μm). The sun and other very high temperature bodies emit near-infrared radiative energy. Bodies with temperatures close to room temperature emit intermediate-infrared radiative energy.

Radiative energy released (emitted) by a surface is dependent on surface temperature and surface characteristics. For natural sources, the radiative energy emitted by a body is in wide range of wavelengths. For example, radiative energy emitted from the sun consists of radiation of varying wavelengths known as the *solar spectrum*. Visible light accounts for about 46% of the energy emitted, even though it is a relatively small segment of the wavelengths emitted. About 49% is infrared and slightly less than 5% is ultraviolet. As a metal object is heated above a temperature of about 1000°F (800 K), it begins to glow a dull red. (See Photo 2.1.) If it is heated to a higher temperature, the predominant wavelength of radiative energy released shifts to different colors (as perceived by the human eye):

Dull red glow ~1000°F (800 K)
Orange glow ~2500°F (1650 K)
Bright yellow glow ~5000°F (3000 K)
White glow ~8500°F (5000 K)

The predominant wavelength at which radiative energy is emitted varies inversely with the temperature of the emitting surface; that is, higher temperature bodies emit smaller, more powerful wavelengths. The wavelength at which maximum radiative power occurs is based on *Wein's displacement law:*

$$\lambda_{max} = 5215.6/T$$

where T is the absolute temperature in R, and λ_{max} is the dominant wavelength in microns (μm).

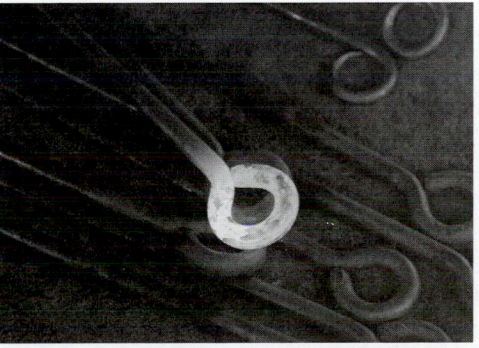

PHOTO 2.1 Glow is the visible part of the thermal radiative energy emitted from the high temperature of the metal object.

Example 2.6

The sun emits radiant energy like a body at a temperature of about 10 800°R. Determine the wavelength of maximum radiative power.

$$\lambda_{max} = 5215.6/T = 5215.6/10\ 800\ R = 0.48\ \mu m$$
$$(\text{within the VL range})$$

Example 2.7

The exterior building siding emits radiant energy at a temperature of about 500 R (about 40°F/4°C). Determine the wavelength of maximum radiative power.

$$\lambda_{max} = 5215.6/T = 5215.6/500\ R = 10.4\ \mu m$$
$$(\text{within the IR range})$$

The maximum radiative power of the sun occurs at a wavelength of 0.48 μm. Because this dominant wavelength is within the visible light range, a large fraction of its power is radiated as visible light. In contrast, a typical building structure radiates energy at a temperature of between about 500 to 550 R (about 40° to 90°F, or 4° to 32°C). The dominant wavelengths occur at wavelengths of between 9.5 and 10.4 μm, which means that maximum power is not visible because it is within the mid/intermediate-infrared range.

As radiant energy in different wavelengths strike a surface, they are reflected, absorbed, or transmitted. With visible light, the color seen is the reflection of a visible wavelength or combination of wavelengths. Black is the absence of reflected visible wavelengths and pure white is the reflection of all

visible wavelengths. Similar behaviors take place with wavelengths of radiative energy outside the visible light spectrum; the outcome just cannot be seen by the naked eye. Similar behaviors are exhibited by wavelengths outside the visible light spectrum.

If the surface the radiant energy is striking is *reflective,* most of the radiative energy will be reflected. If the surface is *opaque,* at least some of the radiative energy will be absorbed and converted to heat energy, but no radiant energy will be transmitted. If the surface is *translucent,* at least some radiative energy will be transmitted through it. When radiative energy strikes a translucent surface, part will be absorbed, part will be reflected, and part will be transmitted through the receiving body:

$$\alpha + \rho + \tau = 1.0$$

This is based on the fraction of radiative energy absorbed or the *absorptivity* (α) of the surface, the fraction of radiative energy reflected or *reflectivity* (ρ) of the surface, and the fraction of radiative energy transmitted or *transmissivity* (τ) of the surface. Average reflectivities and absorptivities for surfaces of various building materials are provided in Table 2.2.

Most materials used in the building industry behave in an opaque manner (glass is an exception)—that is, they absorb and reflect all radiant energy and do not transmit radiant energy. In the case of an opaque surface, transmissivity (τ) is omitted and the equation is reduced to:

$$\alpha + \rho = 1.0$$

Transmissivity of glass varies with the angle at which the radiative energy strikes the glass and the size of the wavelength striking the glass. Ordinary window and skylight glasses are nearly transparent with solar radiation; about 85% of the direct

TABLE 2.2   AVERAGE REFLECTIVITIES, EMISSIVITIES, AND ABSORPTIVITIES FOR SURFACES OF VARIOUS BUILDING MATERIALS.

Surface	Reflectivity	Emissivity or Thermal Absorptivity at 50° to 100°F (10° to 38°C)	Absorptivity for Solar Radiation
Window glass (clear)	0.06 to 0.10	0.90 to 0.95	0.06 to 0.10
Black nonmetallic surfaces	0.10 to 0.20	0.90 to 0.98	0.85 to 0.98
Red brick, concrete, and stone, dark paints	0.30 to 0.50	0.85 to 0.95	0.65 to 0.80
Yellow brick and stone	0.40 to 0.60	0.85 to 0.95	0.95 to 0.70
White brick, tile, paint, whitewash	0.80 to 0.90	0.85 to 0.95	0.30 to 0.50
Bright aluminum paint	0.40 to 0.50	0.40 to 0.60	0.30 to 0.50
Dull copper, aluminum, galvanized steel	0.50 to 0.70	0.20 to 0.30	0.40 to 0.65
Polished copper	0.60 to 0.90	0.02 to 0.05	0.30 to 0.50
Highly polished aluminum	0.90 to 0.95	0.02 to 0.04	0.10 to 0.40

solar radiation striking the glass is transmitted through the glass. The remaining 15% of the solar radiation is absorbed or reflected. However, at steep angles, more radiative energy glances off the glass surface, causing a decrease in transmissivity and an increase in reflectivity.

With translucent materials, the capability of the surface of the receiving body to transmit, absorb, and reflect radiative energy varies with the size of wavelength striking that surface. For example, the change in the properties of glass with various wavelengths is significant. Ordinary window glass has a transmissivity of about 0.85 between wavelengths of 0.2 and 2.7 μm. Beyond wavelengths of about 2.7 μm, glass's transmissivity falls to about 0.1. At these wavelengths, glass becomes opaque and is a reasonably good reflector.

Much of the sun's radiative energy will be transmitted through glass. Materials inside the building that receive sunlight absorb the transmitted radiant energy and heat up (increase in temperature). With an increase in surface temperature, energy is emitted by these surfaces, but in much larger (infrared) wavelengths. Because glass is nearly opaque to infrared energy, only about 10% of the infrared energy is transmitted back through the glass, and a large portion is reflected back. This shift in the properties of glass is desirable in capturing the sun's radiant energy during the winter but is undesirable because it increases a building's cooling load in the summer.

A *black body* is a hypothetical globe-shaped body that absorbs all of the radiative energy striking it (absorptivity = 1.0) and emits all of this radiative energy (emissivity = 1.0). Of course, a black body is theoretical and does not exist. Real-life materials are not perfect emitters or absorbers of radiative energy. A *gray body* emits only a fraction of the thermal energy emitted by an equivalent black body. By definition, a gray body has a surface emissivity less than 1.0 and a surface reflectivity greater than 0.0.

Emissive power or *emittance* is the radiative energy emitted by the surface of a body. The *emissivity (ε)* of a surface is the ratio of radiant energy emitted by a surface and radiant energy emitted by a black body at the same temperature. According to *Kirchhoff's Law,* a body capable of absorbing radiative energy at a specific wavelength is equally capable of emitting radiative energy at that wavelength. In other words, at a particular wavelength, the emissivity of a surface is equivalent to its surface absorptivity.

Surface emissivity and absorptivity are affected by several variables; the most important are geometric shape of the surface, surface temperature, surface emissivity, and wavelength dependence of the surface. For example, surface color is a good indicator of the ability of a surface to absorb solar radiation; dark colors absorb more sunlight than light colors because a large portion of the radiative energy of sunlight is in the visible light band. However, surface color is a poor indicator of the ability to absorb infrared radiative energy.

The value of emissivity/absorptivity rarely remains constant over extended wavelengths. For example, white paint appears perfectly white in the visible light spectrum, but becomes gray at about 2 μm, and beyond 3 μm it is almost black (fully opaque). This shift in performance is useful in thermal design. This performance does not violate Kirchoff's Law, because it is wavelength dependent. Selective surfaces and coatings use this change in performance to control emittance and absorptance of radiant energy.

Selective surfaces/coatings (e.g., black chrome, black nickel, or black crystal) are specially designed surface coatings that have a high absorptivity and high emissivity for visible light and short wavelength infrared radiative energy, and a low absorptivity and low emissivity for long wavelength infrared radiative energy. Unlike flat black painted surfaces, selective coatings can restrict long wavelength radiative energy loss from surfaces at or near room temperature. The end result is a surface that absorbs visible light and solar radiative energy well, but does not radiate thermal energy very well. Use of these surfaces in solar collectors reduces heat loss from radiative energy by as much as ten times.

Low emissivity coatings, also called *low emittance coatings* or simply *low-e coatings,* are used on window glass to reduce infrared radiative energy heat transfer by up to 20 times. They do this by reflecting infrared wavelengths while transmitting most of the visible wavelengths in the solar radiation. To the naked eye, visible light transmission makes it appear identical to an untreated sheet of glass, but the invisible infrared rays are not transmitted. Infrared wavelengths are reflected back inside in the winter and back outside in the summer.

High emittance surfaces easily release most (80 to 90%) of their radiant energy to adjacent surfaces. A rougher surface has a higher emittance. Dull coatings or surfaces (e.g., most building materials, unpolished metals, painted surfaces, mass-type insulations, fibers, nonmetalized plastics, wood) will typically have an emittance of between 0.8 and 0.95. *Low emittance surfaces* release very little (<10%) radiant energy to adjacent surfaces. A shiny metal surface (e.g., polished aluminum, silver, gold) can have an emittance of less than 0.1. Reflective insulations can have values as low as 0.03.

The theoretical rate of heat transfer by thermal radiative energy (q) from a surface to an infinitely large black body may be calculated as follows in customary (U.S.) units:

$$q = 1.714 \times 10^{-9} \cdot \varepsilon \cdot A \cdot (T_h^4 - T_c^4)$$

where q is the theoretical heat transfer rate by radiative energy in Btu/hr, ε is the emissivity of the surface, T_h is the temperature of the emitting surface in R, T_c is the temperature of the receiving surface in R, and A is the area of the surface in ft^2. The constant of $1.714 \ 10^{-9}$ Btu/hr $\cdot$ ft^2 $\cdot$ R (e.g., 0.000 000 001 714) is known as the *Stefan-Boltzmann constant*. In metric (SI) units, the Stefan-Boltzmann constant is 5.669×10^{-8} W/m^2 $\cdot$ K^4.

Note that temperatures used in radiation heat transfer computations (e.g., T_h and T_c) must be expressed in absolute temperature units (e.g., R or K).

In calculating the theoretical rate of heat transfer by radiation, it assumed that the receiving surface is a black body (absorptivity = 1.0). This is not practical in building applications where receiving surfaces are not fully absorptive. Even the atmosphere is not a perfect absorber because of the gases it contains. Therefore, it is necessary to account for the actual absorptivity and emissivity of the interacting surfaces by substituting the effective emissivity (E). The following equation is used to determine the *effective thermal emissivity (E) between two parallel surfaces* with surface areas that are large in comparison to the space between them:

$$E = 1/[1/\varepsilon_1 + 1/\varepsilon_2 - 1)]$$

where ε_1 and ε_2 are the emissivities of the interacting surfaces and E is the effective emissivity of these surfaces.

Effective emissivity (ε) of the interacting surfaces is substituted for emissivity (ε) of a surface to compute the rate of heat transfer by radiation (q) with the Stefan-Bolzmann constant:

$$q = 1.714\ 10^{-9} \cdot E \cdot A \cdot (T_h^4 - T_c^4)$$

Example 2.8

Approximate the radiation component of heat transfer between the following closely spaced interior surfaces of a wall cavity. Assume surface temperatures of 60° and 20°F and an area of 1.0 ft².

$$T_h = 60° + 460° = 520\ R$$
$$T_c = 20° + 460° = 480\ R$$

a. An uninsulated wall cavity with a gypsum wallboard surface ($\varepsilon = 0.9$) and a rough plywood surface ($\varepsilon = 0.9$) on the sheathing enclosing the wall cavity.

$$E = 1/[1/\varepsilon_1 + 1/\varepsilon_2 - 1)] = 1/[1/0.9 + 1/0.9 - 1)] = 0.818$$

$$\begin{aligned} q &= 0.000\ 000\ 001\ 714 \cdot E \cdot A \cdot (T_h^4 - T_c^4) \\ &= 0.000\ 000\ 001\ 714 \cdot 0.818 \cdot 1.0\ ft^2 \cdot (520\ R^4 - 480\ R^4) \\ &= 28.1\ Btu/hr \end{aligned}$$

b. An uninsulated wall cavity with a gypsum wallboard surface ($\varepsilon = 0.9$) and a highly polished aluminum surface ($\varepsilon = 0.03$) on the sheathing enclosing the wall cavity.

$$E = 1/[1/\varepsilon_1 + 1/\varepsilon_2 - 1)] = 1/[1/0.9 + 1/0.03 - 1)] = 0.030$$

$$\begin{aligned} q &= 0.000\ 000\ 001\ 714 \cdot E \cdot A \cdot (T_h^4 - T_c^4) \\ &= 0.000\ 000\ 001\ 714 \cdot 0.030 \cdot 1.0\ ft^2 \cdot (520\ R^4 - 480\ R^4) \\ &= 1.03\ Btu/hr \end{aligned}$$

c. An uninsulated wall cavity with highly polished aluminum surfaces ($\varepsilon = 0.03$) on the sheathing and on wall finish surface enclosing the wall cavity.

$$E = 1/[1/\varepsilon_1 + 1/\varepsilon_2 - 1)] = 1/[1/0.03 + 1/0.03 - 1)] = 0.0152$$

$$\begin{aligned} q &= 0.000\ 000\ 001\ 714 \cdot E \cdot A \cdot (T_h^4 - T_c^4) \\ &= 0.000\ 000\ 001\ 714 \cdot 0.0152 \cdot 1.0\ ft^2 \cdot (520\ R^4 - 480\ R^4) \\ &= 0.52\ Btu/hr \end{aligned}$$

In Example 2.8, observe the large difference in rates of radiation heat transfer between surfaces having different effective emissivities. It is for this reason that the polished aluminum foil placed on the surfaces of various commercially available insulation products (e.g., polystyrene or polyisocyanurate insulation board) will greatly reduce transfer of heat.

These basic radiative energy calculations provide the reader with a basic insight on some of the characteristics of radiant energy exchange and radiation heat transfer. Other surface configurations and characteristics will change effective emittance (E) considerably. An overall discussion of radiation heat transfer would account for the myriad of physical relationships of the interacting materials and surfaces. Such analysis is very complex and beyond the scope of this text.

2.5 TRANSMISSION HEAT LOSS

Thermal Conductivity, Conductance, and Resistance

The three basic modes of heat transfer (e.g., conduction, convection, and radiation) are important in building heat loss and gain. The quantity of heat transfer may be calculated through each material and for each mode (e.g., conduction, convection, and radiation). The heat transfer characteristics of the many different construction assemblies (e.g., walls, windows, doors, ceilings) that make up the building envelope must be determined. Because heat passes through different materials at different rates, each material is considered to have a specific resistance to the flow of heat. Heat passing through materials in a building envelope (e.g., glass) or through an assembly of materials (e.g., composite walls, roofs, floors, and the like) is called *transmission heat loss*. Several terms are used to describe transmission heat loss, including those in the paragraphs that follow.

Thermal Conductivity

Thermal conductivity (k) describes a homogenous material's ability to transfer heat. A homogeneous material is uniform in composition throughout (such as a solid concrete wall). Thermal conductivity is expressed in Btu · in/hr · °F · ft² or Btu/hr · °F · ft (W · m/°C · m² or W/°C · m). The thermal conductivity of a material is inversely proportional to the insulating value of that material. The reader is cautioned that sometimes the k-value is described in units of Btu · ft/hr · °F · ft²,

---

Symbols Used in Thermal Science

In thermal science, the symbol C is commonly used for both specific heat capacity and thermal conductance. In addition, the symbol R is used for thermal resistance and thermal resistivity. These symbols remain unchanged in this text because this is an industry standard. The reader is cautioned that this likeness is a confusing convention.

---

where the thickness of the material is expressed in feet rather than inches.

Thermal Conductance

Thermal conductance (C) is a heterogeneous or composite material's ability to transfer heat. A heterogeneous material is a composite of different materials such as a concrete masonry unit, hollow brick, or insulated panel. These masonry units are made with air spaces (e.g., cells or cores) in the concrete or clay unit. The number and size of these air voids, and thus the size of these masonry units, will affect rate of heat transfer. Thermal conductance is expressed in units of Btu/hr $\cdot$ °F $\cdot$ ft^2 (W/°C $\cdot$ m^2). It differs from thermal conductivity in that a thickness for the material must be specified, such as 0.90 Btu/hr $\cdot$ °F $\cdot$ ft^2 for an 8 in thick standard weight CMU (concrete masonry unit or concrete block).

Thermal Resistance

Thermal resistance (R) is expressed in units of hr $\cdot$ °F $\cdot$ ft^2/Btu (°C $\cdot$ m^2/W). It is a measure of the ability of a material to resist heat transfer. It is a measure of the effectiveness of insulation: a material with a high R-value is a good insulator. In Canada, the metric (SI) equivalent is referred to as the RSI. The relationships between R and RSI are:

$$1.0 \text{ hr} \cdot \text{°F} \cdot \text{ft}^2/\text{Btu} = 0.1761\text{°C} \cdot \text{m}^2/\text{W}$$

$$R = 0.1761 \text{ RSI}$$

$$\text{RSI} = 5.679 \text{ R}$$

The value of thermal resistance is the same R-value used in technical literature to rate commercially available insulation. Assuming a 1°F temperature difference between its outer surfaces, it will take 11 hr for one Btu to pass through one square foot of R-11 insulation. With a 10°F tem~~~~~~~~, it will take one-tenth the time (1~~~~~~~~~~~~~~~~ss through.

R is the reciprocal of transmission values of k a~~~. It can be associated with 1/C, in which case it will b~ expressed in terms of a specified thickness; for example, an 8 in thick standard weight concrete block has an R-value of 1.11 hr $\cdot$ °F $\cdot$ ft^2/Btu. It can also be associated with 1/k and expressed in terms of unit thickness (e.g., one inch); for example, fiberglass batt insulation has an R-value of about 3.14 hr $\cdot$ °F $\cdot$ ft^2/Btu per inch of thickness. In the case of expressing it per unit thickness (e.g., per inch, per mm), the R-value is called *thermal resistivity,* which is also referred to by the symbol R.

In general, the denser a material, the less resistance it has to the flow of heat and the lower its resistance (R) value; the lighter (less dense) the material, the higher its R-value and the better its insulating value.

A summary of terms used to describe heat transmission is provided in Table 2.3. Properties of heat transmission for common building materials are provided in Table 2.4.

TABLE 2.3   A SUMMARY OF TERMS USED TO DESCRIBE THE RATE OF HEAT TRANSMISSION.

Symbol	Thermal Property	Brief Definition	Units Customary (U.S.)	Units Metric (SI)	Relationship Between Units
k	Thermal conductivity	Homogenous material's ability to transfer heat	$\dfrac{\text{Btu} \cdot \text{in}}{\text{hr} \cdot \text{°F} \cdot \text{ft}^2}$	$\dfrac{\text{W} \cdot \text{m}}{\text{°C} \cdot \text{m}^2}$	k = 1/R
C	Thermal conductance	Composite material's ability to transfer heat	$\dfrac{\text{Btu}}{\text{hr} \cdot \text{°F} \cdot \text{ft}^2}$	$\dfrac{\text{W}}{\text{°C} \cdot \text{m}^2}$	C = 1/R
R	Thermal resistivity	Insulating ability of a material	$\dfrac{\text{hr} \cdot \text{°F} \cdot \text{ft}^2}{\text{Btu}}$	$\dfrac{\text{°C} \cdot \text{m}^2}{\text{W}}$	R = 1/C = x/k = x(1/k)
R_t	Total thermal resistance	Ability of a construction assembly to insulate heat, including air films	$\dfrac{\text{hr} \cdot \text{°F} \cdot \text{ft}^2}{\text{Btu}}$	$\dfrac{\text{°C} \cdot \text{m}^2}{\text{W}}$	$R_t = 1/U = R_1 + R_2 + R_3 + \cdots$
U	Overall coefficient of heat transmission	Ability of a construction assembly to transfer heat, including air films	$\dfrac{\text{Btu}}{\text{hr} \cdot \text{°F} \cdot \text{ft}^2}$	$\dfrac{\text{W}}{\text{°C} \cdot \text{m}^2}$	$U = 1/R_t$

x = Material thickness (in or mm).

TABLE 2.4 PROPERTIES OF HEAT TRANSMISSION FOR COMMON BUILDING MATERIALS.

Material	Thermal Conductivity (k) Btu·in / hr·°F·ft²	W / °C·m	Thermal Conductance (C) Btu / hr·°F·ft²	W / °C·m²	Thermal Resistivity (R) for Thickness — in hr·°F·ft² / Btu·in	mm °C·m² / W·mm	Thermal Resistance (R) hr·°F·ft² / Btu	°C·m² / W
Construction Materials								
Brick, common	5.00	0.72			0.20	0.0014		
Brick, common (4 in/100 mm)			1.25	0.220			0.80	0.141
Brick, face	9.09	1.31			0.11	0.0008		
Brick, face (4 in/100 mm)			2.27	0.400			0.44	0.077
Cedar logs and lumber	0.75	0.11			1.33	0.0092		
Concrete block (4 in/100 mm), standard weight			1.25	0.220			0.80	0.141
Concrete block (6 in/150 mm), standard weight			1.06	0.187			0.94	0.166
Concrete block (8 in/200 mm), standard weight			0.90	0.159			1.11	0.195
Concrete block (10 in/250 mm), standard weight			0.83	0.146			1.21	0.213
Concrete block (12 in/300 mm), standard weight			0.78	0.138			1.28	0.225
Concrete block (4 in/100 mm), lightweight			0.84	0.148			1.19	0.210
Concrete block (6 in/150 mm), lightweight			0.80	0.141			1.25	0.220
Concrete block (8 in/200 mm), lightweight			0.69	0.121			1.45	0.255
Concrete block (10 in/250 mm), lightweight			0.65	0.114			1.55	0.273
Concrete block (12 in/300 mm), lightweight			0.61	0.107			1.65	0.291
Concrete block (6 in/150 mm), filled with perlite			0.25	0.045			3.95	0.696
Concrete block (8 in/200 mm), filled with perlite			0.22	0.038			4.65	0.819
Concrete block (10 in/250 mm), filled with perlite			0.18	0.031			5.65	0.995
Concrete block (12 in/300 mm), filled with perlite			0.14	0.025			7.05	1.242
Concrete, autoclaved aerated	0.26	0.04			3.90	0.0270		
Concrete, cast-in-place, sand and gravel	12.50	1.80			0.08	0.0006		
Lumber, softwood	0.80	0.12			1.25	0.0087		
Lumber, softwood—2 in nominal (1½ in/38 mm)			0.53	0.094			1.88	0.331
Lumber, softwood—2 × 4 (3½ in/89 mm)			0.23	0.040			4.38	0.771
Lumber, softwood—2 × 6 (5½ in/140 mm)			0.15	0.026			6.88	1.212
Sheathing Materials								
Oriented strand board (OSB) or plywood (softwood)	0.80	0.12			1.25	0.0087		
OSB or plywood (¼ in/6 mm)			3.23	0.568			0.31	0.055
OSB or plywood (⅜ in/9 mm)			2.13	0.375			0.47	0.083
OSB or plywood (½ in/13 mm)			1.60	0.280			0.62	0.111
OSB or plywood (⅝ in/16 mm)			1.30	0.229			0.77	0.136
OSB or plywood (¾ in/19 mm)			1.06	0.187			0.94	0.166
Fiberboard	0.38	0.05			2.64	0.0183		
Fiberboard (½ in/13 mm)			0.76	0.133			1.32	0.232
Fiberboard (²⁵⁄₃₂ in/20 mm)			0.49	0.085			2.06	0.363
Fiberglass insul board (¾ in/19 mm)			0.33	0.059			3.0	0.528
Fiberglass insul board (1 in/25 mm)			0.25	0.044			4.0	0.704
Fiberglass insul board (1½ in/38 mm)			0.17	0.029			6.0	1.057
Extruded polystyrene insul board (¾ in/19 mm)			0.27	0.047			3.75	0.660
Extruded polystyrene insul board (1 in/25 mm)			0.20	0.035			5.0	0.881
Extruded polystyrene insul board (1½ in/38 mm)			0.13	0.023			7.5	1.321
Polyisocyanurate insul board, foil-faced two sides (¾ in/19 mm)			0.19	0.033			5.4	0.951

TABLE 2.4 PROPERTIES OF HEAT TRANSMISSION FOR COMMON BUILDING MATERIALS. (CONTINUED)

Material	Thermal Conductivity (k)		Thermal Conductance (C)		Thermal Resistivity (R) for Thickness		Thermal Resistance (R)	
					in	mm		
	Btu · in / hr · °F · ft²	W / °C · m	Btu / hr · °F · ft²	W / °C · m²	hr · °F · ft² / Btu · in	°C · m² / W · mm	hr · °F · ft² / Btu	°C · m² / W
Sheathing Materials (continued)								
Polyisocyanurate insul board, foil-faced two sides (1 in/25 mm)			0.14	0.024			7.2	1.268
Polyisocyanurate insul board, foil-faced two sides (1½ in/38 mm)			0.09	0.016			10.8	1.902
Vapor Diffusion Materials								
Building (felt) paper			16.67	2.40			0.06	0.0004
Plastic film vapor diffusion retarder			0.0	0.0			0.0	0.0
Siding and Cladding Materials								
Aluminum, steel, vinyl (hollow back)			1.64	0.289			0.61	0.107
Aluminum, steel, vinyl (½ in/13 mm insul board)			0.56	0.098			1.80	0.317
Brick veneer (4 in/100 mm)			2.27	0.400			0.44	0.077
Hardboard lapped (7/16 in/11 mm)			2.94	0.518			0.34	0.060
OSB or plywood (¾ in/19 mm)			1.08	0.189			0.93	0.164
OSB or plywood (⅝ in/16 mm)			1.30	0.229			0.77	0.136
Stucco (½ in/13 mm)			10.0	1.761			0.10	0.018
Wood bevel lapped (½ in/13 mm)			1.24	0.220			0.81	0.141
Wood shingles	1.15	0.166			0.87	0.0060		
Interior Finish Materials								
Gypsum wallboard (drywall ½ in/13 mm)			2.22	0.391			0.45	0.079
Gypsum wallboard (drywall ⅝ in/16 mm)			1.79	0.314			0.56	0.099
Paneling (⅜ in/6 mm)			2.13	0.375			0.47	0.083
Flooring Materials								
Carpet (fibrous pad)			0.48	0.085			2.08	0.366
Carpet (rubber pad)			0.81	0.143			1.23	0.217
Hardwood flooring	1.10	0.158			0.91	0.0063		0.000
Hardwood flooring (25/32 in/20 mm)			1.47	0.259			0.68	0.120
Particleboard (underlayment)	0.76	0.110			1.31	0.0091		0.000
Particleboard (⅝ in/16 mm)			1.22	0.215			0.82	0.144
OSB or plywood	0.80	0.115			1.25	0.0087		0.000
OSB or plywood (¾ in/19 mm)			1.08	0.189			0.93	0.164
Tile, vinyl			20.0	3.5			0.05	0.009
Roofing Materials								
Asphalt shingles			2.27	0.328			0.44	0.0031
Wood shingles/wood shakes			1.03	0.149			0.97	0.0067
Built-up roof (⅜ in/6 mm), no topping	4.17	0.601			0.24	0.0017		
Built-up roof (⅜ in/6 mm), aggregate topping	3.03	0.437			0.33	0.0023		
Single-ply membrane (without insulation)			16.67	2.40			0.06	0.0004

TABLE 2.4   PROPERTIES OF HEAT TRANSMISSION FOR COMMON BUILDING MATERIALS. (CONTINUED)

Material	Thermal Conductivity (k)		Thermal Conductance (C)		Thermal Resistivity (R) for Thickness		Thermal Resistance (R)	
					in	mm		
	$\frac{Btu \cdot in}{hr \cdot °F \cdot ft^2}$	$\frac{W}{°C \cdot m}$	$\frac{Btu}{hr \cdot °F \cdot ft^2}$	$\frac{W}{°C \cdot m^2}$	$\frac{hr \cdot °F \cdot ft^2}{Btu \cdot in}$	$\frac{°C \cdot m^2}{W \cdot mm}$	$\frac{hr \cdot °F \cdot ft^2}{Btu}$	$\frac{°C \cdot m^2}{W}$
Insulation Materials								
Cellulose blown (attic)	0.32	0.046			3.13	0.0217		
Cellulose blown (wall)	0.27	0.039			3.70	0.0257		
Cotton (loose and blown)	0.30	0.043			3.33	0.0231		
Polystyrene expanded (unfaced)	0.25	0.036			4.00	0.0277		
Polystyrene extruded (unfaced)	0.20	0.029			5.00	0.0347		
Fiberglass batt	0.32	0.046			3.14	0.0218		
Fiberglass blown (attic)	0.45	0.066			2.20	0.0153		
Fiberglass blown (wall)	0.31	0.045			3.20	0.0222		
Perlite, expanded	0.37	0.053			2.70	0.0187		
Polyisocyanurate (foil-faced two sides)	0.14	0.020			7.20	0.0499		
Polyurethane (foamed-in-place)	0.16	0.023			6.25	0.0433		
Rigid fiberglass (less than 4 lb/ft³ density)	0.25	0.036			4.00	0.0277		
Rock wool batt	0.32	0.046			3.14	0.0218		
Rock wool blown (attic)	0.32	0.047			3.10	0.0215		
Rock wool blown (wall)	0.33	0.048			3.03	0.0210		
Vermiculite	0.47	0.068			2.13	0.0148		

TABLE 2.5   THERMAL RESISTANCE (R) OF SELECTED CONVECTIVE AIR FILMS FOR STILL (INSIDE) AIR BASED ON REFLECTIVITY OF THE SURFACE.

Orientation of the Surface in Contact with the Air Film	Direction of Heat Flow	Probable Location of Surface	Nonreflective Surfaces ($\varepsilon = 0.9$)		Reflective Surfaces ($\varepsilon = 0.03$)	
			$\frac{hr \cdot °F \cdot ft^2}{Btu}$	$\frac{°C \cdot m^2}{W}$	$\frac{hr \cdot °F \cdot ft^2}{Btu}$	$\frac{°C \cdot m^2}{W}$
Horizontal	Up	Interior ceiling surface in winter, interior floor surface in summer	0.61	0.11	1.32	0.23
	Down	Interior ceiling surface in summer, interior floor surface in winter	0.92	0.16	4.55	0.80
Vertical	Horizontal	Interior wall surface	0.68	0.12	1.70	0.30

TABLE 2.6   THERMAL RESISTANCE (R) OF CONVECTIVE AIR FILMS FOR MOVING (OUTSIDE) AIR.

7½ mph (12 km/hr) wind		15 mph (24 km/hr) wind		Probable Location of Surface
$\frac{hr \cdot °F \cdot ft^2}{Btu}$	$\frac{°C \cdot m^2}{W}$	$\frac{hr \cdot °F \cdot ft^2}{Btu}$	$\frac{°C \cdot m^2}{W}$	
0.25	0.04	0.17	0.03	All exterior (outside) surfaces

TABLE 2.7   THERMAL RESISTANCE (R) OF SELECTED CONFINING AIR SPACES (CAVITIES) BASED ON REFLECTIVITY OF THE SURFACES CONFINING THE AIR SPACE AND COLD AND WARM CAVITY TEMPERATURE CONDITIONS.

Cavity Air Temperature	Cavity Surfaces Orientation	Direction of Heat Flow	Probable Location of Cavity	Cavity Width	Nonreflective Surfaces ($\varepsilon = 0.9$)		Reflective Surfaces ($\varepsilon = 0.03$)	
					$\frac{hr \cdot °F \cdot ft^2}{Btu}$	$\frac{°C \cdot m^2}{W}$	$\frac{hr \cdot °F \cdot ft^2}{Btu}$	$\frac{°C \cdot m^2}{W}$
Cold (winter)	Horizontal	Up	Ceiling or flat roof over space	½ in (13 mm)	0.84	0.148	2.05	0.361
				¾ in (19 mm)	0.87	0.153	2.20	0.387
				1½ in (38 mm)	0.89	0.157	2.40	0.423
				3½ in (89 mm)	0.93	0.164	2.66	0.468
	Vertical	Horizontal	Exterior wall	½ in (13 mm)	0.91	0.160	2.54	0.447
				¾ in (19 mm)	1.01	0.178	2.46	0.433
				1½ in (38 mm)	1.02	0.180	3.55	0.625
				3½ in (89 mm)	1.01	0.178	3.40	0.599
	Horizontal	Down	Floor over unheated space	½ in (13 mm)	0.92	0.162	2.55	0.449
				¾ in (19 mm)	1.02	0.180	3.59	0.632
				1½ in (38 mm)	1.15	0.203	5.90	1.039
				3½ in (89 mm)	1.24	0.218	9.27	1.633
Warm (summer)	Horizontal	Up	Floor over uncooled space	½ in (13 mm)	0.73	0.129	2.13	0.375
				¾ in (19 mm)	0.75	0.132	2.34	0.412
				1½ in (38 mm)	0.77	0.136	2.55	0.449
				3½ in (89 mm)	0.80	0.141	2.84	0.500
	Vertical	Horizontal	Exterior wall	½ in (13 mm)	0.77	0.136	2.47	0.435
				¾ in (19 mm)	0.84	0.148	3.50	0.616
				1½ in (38 mm)	0.87	0.153	3.99	0.703
				3½ in (89 mm)	0.85	0.150	3.69	0.650
	Horizontal	Down	Ceiling or flat roof over space	½ in (13 mm)	0.77	0.136	2.48	0.437
				¾ in (19 mm)	0.85	0.150	3.55	0.625
				1½ in (38 mm)	0.94	0.166	6.09	1.072
				3½ in (89 mm)	1.00	0.176	10.07	1.773

Example 2.9

Determine the thermal resistance (R) values of the following materials:

a. ½ in thick plywood (softwood).

From Table 2.4, the thermal resistance (R) is read directly as 0.62 hr · °F · ft²/Btu.

b. A plastic film vapor retarder.

From Table 2.4, the thermal resistance (R) is negligible (R = 0.00).

c. 12 in of cellulose insulation used in attic space.

From Table 2.4, the thermal resistance (R) is found as 3.13 hr · °F · ft²/Btu · in:

R = x/k = x(1/k) = 12 in · 3.13 hr · °F · ft²/Btu per inch
= 37.6 hr · °F · ft²/Btu

d. 8 in thick, standard weight concrete block.

From Table 2.4, the thermal resistance (R) is read directly as a value of 1.11 hr · °F · ft²/Btu.

e. 8 in thick, ordinary cast-in-place concrete.

From Table 2.4, the thermal resistance (R) is found as 0.08 hr · °F · ft²/Btu · in:

R = x/k = x(1/k) = 8 in · 0.08 hr · °F · ft²/Btu per inch
= 0.64 hr · °F · ft²/Btu

f. 8 in thick autoclaved aerated concrete.

From Table 2.4, the thermal resistance (R) is read directly as a value of 3.90 hr · °F · ft²/Btu · in.

R = x/k = x(1/k) = 8 in · 3.90 hr · °F · ft²/Btu per inch
= 31.2 hr · °F · ft²/Btu

g. 7/16 in thick hardboard siding.

From Table 2.4, the thermal resistance (R) is read directly as 0.34 hr · °F · ft²/Btu.

Thermal Resistance of Air Films and Air Spaces

As a surface heats up or cools down, its temperature affects the temperature of the air (or another fluid) adjacent to the surface. With natural convection, air rises or falls depending on whether it is hotter or colder. The influence of these convective air films is an increase in the resistance of the material to the flow of heat. So, in addition to the resistance to heat flow through solid materials, air films that result from convection currents at the surface of a material create a resistant to heat flow—that is, there is an added thermal resistance associated with air films and confined air spaces.

Thermal resistance (R) values for convective air films are based on orientation of the surface in contact with the air film, direction of heat flow, air movement (e.g., still air, moving air), and emissivity of the surface. A condensed set of values of thermal resistance (R) for convective air films for still (inside) air is provided in Table 2.5. A more complete set of data is found in the *ASHRAE Handbook—Fundamentals*. From Table 2.5, a thermal resistance (R) value for a convective air film is selected based upon orientation of the surface adjacent to the air film and direction of heat flow. A condensed set of values of thermal resistance (R) for convective air films for moving air is provided in Table 2.6. By convention, values associated with a 15 mph (24 km/hr) wind are used for winter conditions and values associated with a 7½ mph (12 km/hr) wind are used for summer conditions, unless there are reasons to opt otherwise.

Air (or another fluid) confined within a cavity will also create a resistance to flow of heat. The thermal resistance (R) values for confining air spaces (cavities) are based upon orientation of the surface adjacent to the cavity, direction of heat flow, temperature of the air in the cavity, temperature difference between the cavity surfaces, and reflectivity of the surfaces. A condensed set of values of thermal resistance (R) for common confining air spaces (cavities) is provided in Table 2.7. A more detailed set of data is found in the *ASHRAE Handbook—Fundamentals*. From Table 2.7, a thermal resistance (R) value is selected based on cavity air temperature (generalized for winter or summer), orientation of the cavity surfaces, direction of heat flow, and cavity width or thickness. Confining air space (cavity) widths provided are for common construction cavity widths. For widths beyond the 3½ in (89 mm) width noted, the value for 3½ in (89 mm) should be used.

Example 2.10

Determine the thermal resistance (R) values of the following air films and air spaces based upon winter temperatures:

a. A convective air film on an inside wall covered with gypsum wallboard.

From Table 2.5, the thermal resistance (R) is 0.68 hr · °F · ft²/Btu for still air based on winter temperatures, vertical surface orientation, horizontal direction of heat flow, and a nonreflective surface (ε = 0.9).

b. A convective air film on an outside wall.

From Table 2.6, the thermal resistance (R) is 0.17 hr · °F · ft²/Btu based on winter conditions [15 mph (24 km/hr)] wind.

c. 3½ in wide cavity between gypsum wallboard and plywood sheathing in an uninsulated wall.

From Table 2.7, the thermal resistance (R) is 1.01 hr · °F · ft²/Btu based on winter temperatures, vertical cavity orientation, horizontal direction of heat flow, a 3½ in cavity width, and nonreflective surfaces (ε = 0.9).

d. 3½ in wide cavity between foil-covered (reflective, ε = 0.03) sheathings in an uninsulated wall.

From Table 2.7, the thermal resistance (R) is 3.40 hr · °F · ft²/Btu based on winter temperatures, vertical cavity orientation, horizontal direction of heat flow, a 3½ in cavity width, and reflective surfaces (ε = 0.03).

Example 2.11

Determine the thermal resistance (R) values of the following confined air spaces in a wall with gypsum wallboard and plywood sheathing with nonreflective surfaces (ε = 0.9) based on winter temperatures:

a. ½ in wide cavity between nonreflective surfaces (ε = 0.9):

From Table 2.7, the thermal resistance (R) is 0.91 hr · °F · ft²/Btu.

b. ¾ in wide cavity between nonreflective surfaces (ε = 0.9):

From Table 2.7, the thermal resistance (R) is 1.01 hr · °F · ft²/Btu.

c. 1½ in wide cavity between nonreflective surfaces (ε = 0.9):

From Table 2.7, the thermal resistance (R) is 1.02 hr · °F · ft²/Btu.

d. 3½ in wide cavity between nonreflective surfaces (ε = 0.9):

From Table 2.7, the thermal resistance (R) is 1.01 hr · °F · ft²/Btu.

Example 2.11 demonstrates that the thermal resistance of a confined air space does not increase proportionately with increase in cavity width. Heat flow across a confined air space is mainly affected by radiative and convective heat flow. The thermal resistance of a narrow air space tends to be affected mostly by radiative energy transfer because the surfaces are close, allowing efficient transfer by radiation and because development of convective air currents is subdued so heat transfer by convection is restricted. On the other hand, a wider confined air space tends to allow development of convective air currents, but radiative heat transfer is restricted by the mass of air that radiation must travel through.

As cavity width increases, the thermal resistance of a confined air space reaches a point where increases in convection heat transfer balance out with decreases in radiative heat transfer. As a result, under most conditions found in building construction, the thermal resistance of a confined air space between nonreflective surfaces is very nearly 1.0 hr · °F · ft²/Btu (0.176°C · m²/W). The exception is for cases with very narrow confined air spaces and spaces where one or both confining surfaces is highly reflective.

Total Thermal Resistance

A building structure is composed of different materials that make a construction assembly. For example, a simple residential wall assembly may be constructed of siding, sheathing, studs, insulation, vapor diffusion retarder, and gypsum wallboard. With respect to heat transfer, a construction assembly is considered to act either in series or parallel. When materials are assembled one on top of the other, they are constructed in series. When a construction assembly is constructed in parallel, the thermal conductance values for each material are added and the total energy flow is increased for a given temperature difference. (The concept of parallel flow is introduced later in this chapter.)

The *total thermal resistance* (R_t) is the insulating ability of a construction assembly of materials (e.g., wall, ceiling, floor, roof) or fenestration (e.g., doors, windows), including air films. The total thermal resistance of a construction assembly of materials in series is determined by adding the thermal resistances of each component, including materials, confined air spaces, and surface air films:

$$R_t = R_1 + R_2 + R_3 + \cdots$$

The thermal resistances (R) of a specific material, air space, or boundary air film for each of the various components of a building section are obtained, and then added together to compute the total R-value for the specific construction assembly. Note that k- and C-values for each of the components may not be added. These values must first be converted to an R-value before being added with the R-values of the other components.

Overall Coefficient of Heat Transmission

The *overall coefficient of heat transmission (U)* is the ability of a construction assembly to transfer heat, from exterior to exterior, including air films. The U-factor is the reciprocal of the total thermal resistance of the construction assembly:

$$U = 1/R_t$$

One way to lower the amount of heat transfer through the building envelope is to increase R_t (or decrease U). The only practical means of doing this is to add insulation. Materials used as insulation significantly retard the flow of heat because of their high R-values (or low k-values). Selection of the appropriate R_t is based on temperature difference, cost of insulation, and cost of fuel.

TABLE 2.8 THE RELATIONSHIP BETWEEN TOTAL THERMAL RESISTANCE (R_t) AND THE OVERALL COEFFICIENT OF HEAT TRANSMISSION (U).

Total Thermal R (R_t) $\dfrac{\text{hr} \cdot °F \cdot ft^2}{\text{Btu}}$	Overall Coefficient of Heat Transfer (U) $\dfrac{\text{Btu}}{\text{hr} \cdot °F \cdot ft^2}$	Percentage of Heat Transfer over an R-Value of 1.0
1.0	1.000	100.0%
1.1	0.909	90.9%
1.2	0.833	83.3%
1.3	0.769	76.9%
1.4	0.714	71.4%
1.5	0.667	66.7%
2.0	0.500	50.0%
2.5	0.400	40.0%
3.0	0.333	33.3%
4.0	0.250	25.0%
5.0	0.200	20.0%
6.0	0.167	16.7%
7.0	0.143	14.3%
8.0	0.125	12.5%
9.0	0.111	11.1%
10.0	0.100	10.0%
12.0	0.083	8.3%
15.0	0.067	6.7%
20.0	0.050	5.0%
25.0	0.040	4.0%
30.0	0.033	3.3%
33.0	0.030	3.0%
40.0	0.025	2.5%
50.0	0.020	2.0%
60.0	0.017	1.7%
75.0	0.013	1.3%
100.0	0.010	1.0%

The relationship between values of total thermal resistance (R_t) and the overall coefficient of heat transfer (U) is shown in Table 2.8. A graph of the relationship between total thermal resistance (R_t) and the overall coefficient of heat transfer (U) is provided in Figure 2.4.

Example 2.12

An insulated wall is composed of ½ in wood bevel siding, ½ in plywood sheathing, 3½ in fiberglass insulation, a vapor retarder (plastic film), and ½ in gypsum board. See Figure 2.5. Determine the R_t-value and U-factor for this wall under winter conditions.

The following R-values are found:

Outside air film, 15 mph wind (from Table 2.6)	0.17 hr · °F · ft²/Btu
½ in wood bevel siding (from Table 2.4)	0.81 hr · °F · ft²/Btu
½ in plywood sheathing (from Table 2.4)	0.62 hr · °F · ft²/Btu

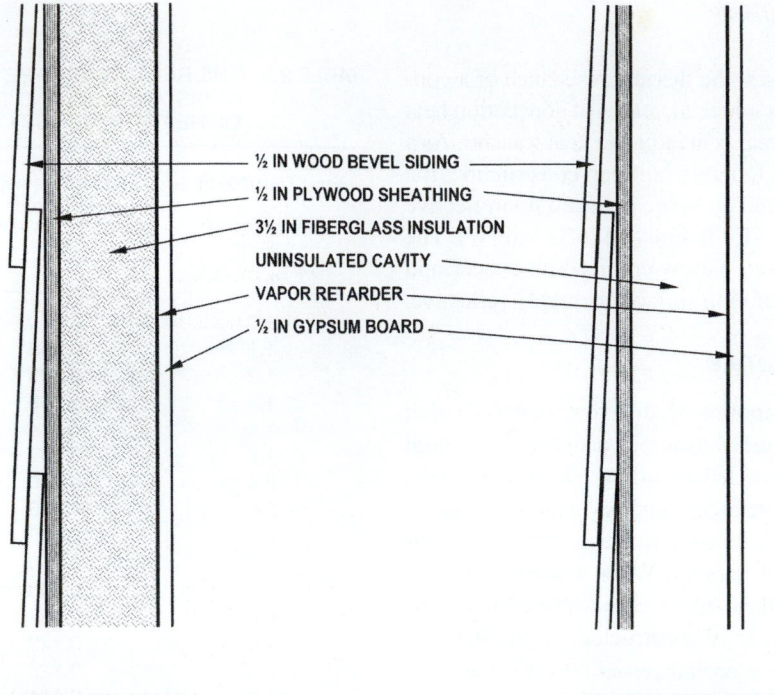

½ IN WOOD BEVEL SIDING
½ IN PLYWOOD SHEATHING
3½ IN FIBERGLASS INSULATION
UNINSULATED CAVITY
VAPOR RETARDER
½ IN GYPSUM BOARD

INSULATED WALL   **UNINSULATED WALL**

FIGURE 2.4   Detail of insulated and uninsulated wall assemblies for Examples 2.12 and 2.13.

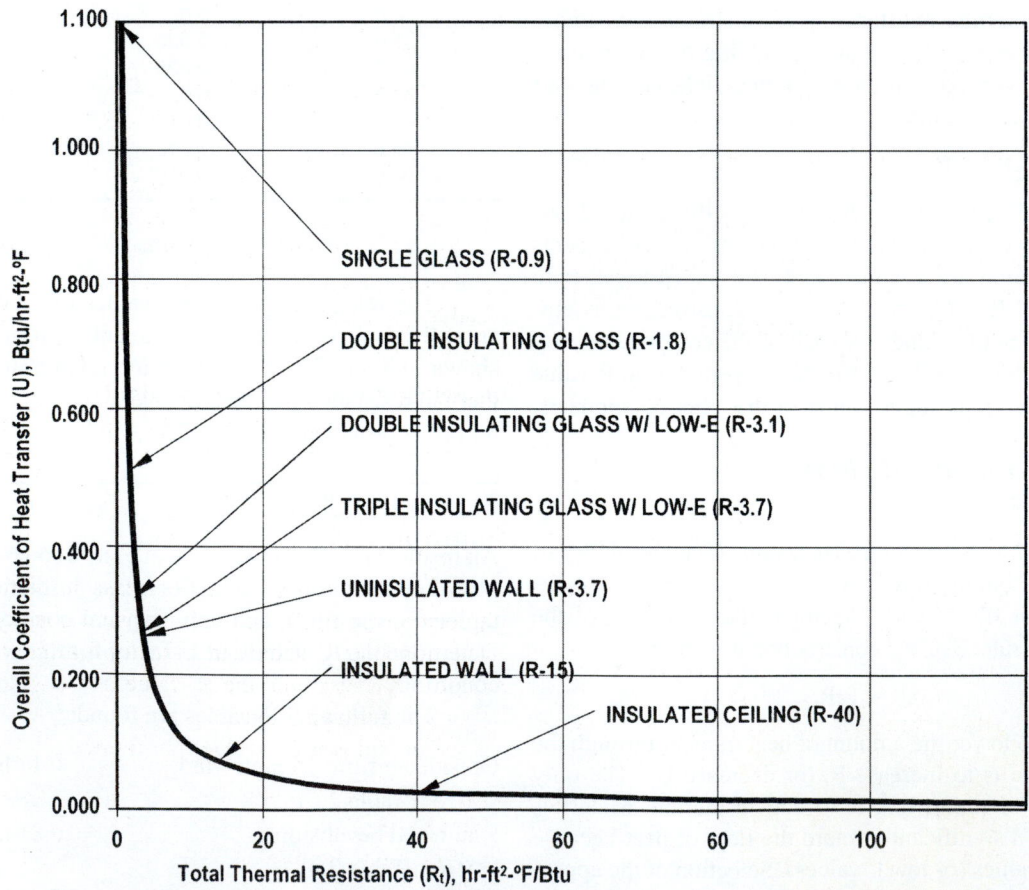

SINGLE GLASS (R-0.9)

DOUBLE INSULATING GLASS (R-1.8)

DOUBLE INSULATING GLASS W/ LOW-E (R-3.1)

TRIPLE INSULATING GLASS W/ LOW-E (R-3.7)

UNINSULATED WALL (R-3.7)

INSULATED WALL (R-15)

INSULATED CEILING (R-40)

Overall Coefficient of Heat Transfer (U), Btu/hr-ft²-°F

Total Thermal Resistance (R_t), hr-ft²-°F/Btu

FIGURE 2.5   A graph of the relationship between total thermal resistance (R_t) and the overall coefficient of heat transfer (U).

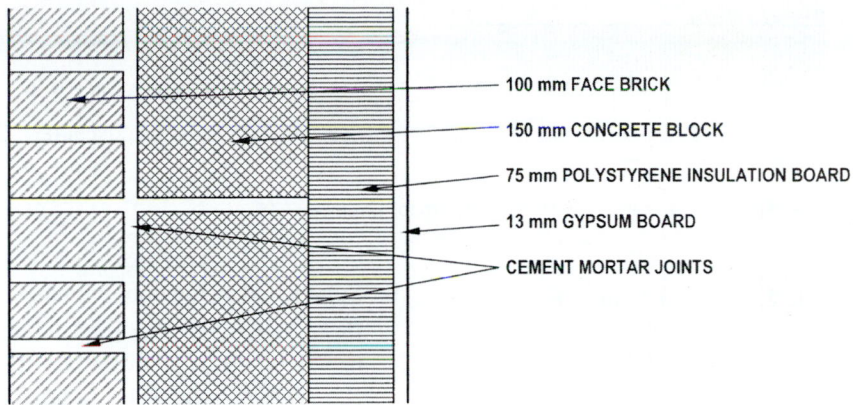

FIGURE 2.6 A solid load-bearing masonry wall for Example 2.14.

3½ in fiberglass batt insulation (from   11.00 hr · °F · ft²/Btu
   Table 2.4: 3.5 in · 3.14 hr · °F · ft²/Btu)

vapor retarder, plastic film   negligible
   (from Table 2.4)

½ in gypsum board (from Table 2.4)   0.45 hr · °F · ft²/Btu

Inside air film (from Table 2.5 for   0.68 hr · °F · ft²/Btu
   vertical surface, horizontal flow)

Total thermal resistance   13.73 hr · °F · ft²/Btu
   $(R_t) = R_1 + R_2 + R_3 + \cdots =$

Overall coefficient of heat transmission
   $(U) = 1/R_t = 1/13.73$ hr · °F · ft²/Btu $= 0.073$ Btu/hr · °F · ft²

Example 2.13

An uninsulated wall is composed of ½ in wood bevel siding, ½ in plywood sheathing, a vapor retarder (plastic film), and ½ in gypsum board. See Figure 2.5. Determine the R_t-value and U-factor for this wall under winter conditions.

   The following R-values are found:

Outside air film, 15 mph wind   0.17 hr · °F · ft²/Btu
   (from Table 2.6)

½ in wood bevel siding   0.81 hr · °F · ft²/Btu
   (from Table 2.4)

½ in plywood bevel siding   0.62 hr · °F · ft²/Btu
   (from Table 2.4)

3½ in cavity (from Table 2.7, cold,   1.01 hr · °F · ft²/Btu
   vertical orientation, horizontal flow)

vapor retarder, plastic film   negligible
   (from Table 2.4)

½ in gypsum board (from Table 2.4)   0.45 hr · °F · ft²/Btu

Inside air film (from Table 2.5 for   0.68 hr · °F · ft²/Btu
   vertical surface, horizontal flow)

Total thermal resistance   = 3.74 hr · °F · ft²/Btu
   $(R_t) = R_1 + R_2 + R_3 + \cdots$

Overall coefficient of heat transmission
   $(U) = 1/R_t = 1/3.74$ hr · °F · ft²/Btu $= 0.267$ Btu/hr · °F · ft²

Example 2.14

A solid load-bearing masonry wall is composed of 100 mm face brick, 150 mm concrete block, 75 mm expanded polystyrene insulation board (unfaced), and 13 mm gypsum board. See Figure 2.6. Determine the R_t-value and U-factor for this wall under winter conditions.

   The following R-values are found:

Outside air film, 24 km/hr wind   0.03°C · m²/W
   (from Table 2.6)

100 mm face brick (from Table 2.4)   0.077°C · m²/W

150 mm concrete block (from Table 2.4)   0.166°C · m²/W

75 mm polystyrene insulation board   2.08°C · m²/W
   (from Table 2.4: 0.0277 · 75 mm)

13 mm gypsum board (from Table 2.4)   0.079°C · m²/W

Inside air film (from Table 2.5 for   0.120°C · m²/W
   vertical surface, horizontal flow)

Total thermal resistance   2.55°C · m²/W
   $(R_t) = R_1 + R_2 + R_3 + \cdots =$

Overall coefficient of heat transmission
      $U = 1/R_t = 1/2.55°C · m²/W = 0.392$ W/°C · m²

Example 2.15

A roof system is constructed of a built-up roof covering with aggregate topping, 4 in thick extruded polystyrene (unfaced) insulation board, and 6 in cast-in-place concrete deck. See Figure 2.7. Determine the R_t-value and U-factor for this wall under summer conditions.

   The following R-values are found:

Outside air film, 7½ mph wind   0.25 hr · °F · ft²/Btu
   (from Table 2.6)

Built-up roof covering   0.33 hr · °F · ft²/Btu
   (from Table 2.4)

4 in thick extruded polystyrene (from   20.00 hr · °F · ft²/Btu
   Table 2.4: 5.00 hr · °F · ft²/Btu · 4 in)

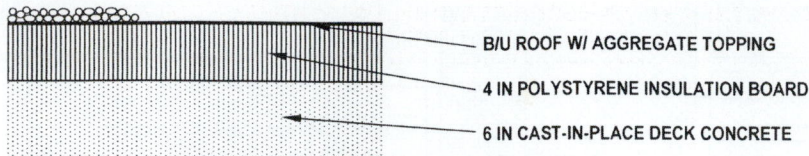

FIGURE 2.7   A built-up (B/U) roof system for Example 7.15.

6 in concrete deck (from Table 2.4: 0.08 hr · °F · ft²/Btu · 6 in)	0.48 hr · °F · ft²/Btu
Inside air film (from Table 2.5 for horizontal surface, vertical flow)	0.92 hr · °F · ft²/Btu
Total thermal resistance (R_t) = $R_1 + R_2 + R_3 + \cdots$ =	21.98 hr · °F · ft²/Btu

Overall coefficient of heat transmission

$(U) = 1/R_t = 1/21.98$ hr · °F · ft²/Btu

$$= 0.045 \text{ Btu/hr} \cdot °F \cdot ft^2$$

As evident in the preceding examples, the R_t-value and U-factor of the various construction assemblies varies considerably. A general comparison of thermal properties of conventional and alternative wall construction techniques is provided in Table 2.9. For some building components, particularly windows and doors, the total thermal resistance (R_t) and overall coefficient of heat transfer (U) are found directly from tabular information.

The overall or whole window U-factor of a window or skylight depends on the type of glazing, frame materials and size, glazing coatings, and type of gas between the panes. Different combinations of window or skylight frame style, frame material, and glazing change the actual thermal performance of a window unit. Window frames and spacers, not glazing type, account for a large fraction of the heat transfer in advanced glazings. The National Fenestration Rating Council (NFRC) provides reliable ratings on window, door, and skylight products through the NFRC Energy Performance Label program.

Table 2.10 lists general total thermal resistance (R_t) and overall coefficient of heat transmission (U) values for windows, skylights, and doors. Data provided are overall values that include air films. Actual values vary by manufacturer because of differences in fabrication and the thermal properties of the frame. They also vary slightly by season.

Example 2.16

Find the total thermal resistance (R_t) and overall coefficient of heat transmission (U) of the following windows and doors:

a. Double insulating glass with a ½ in air space.
   From Table 2.10, the total thermal resistance (R_t) of 2.0 hr · °F · ft²/Btu and overall coefficient of heat transmission (U) of 0.49 Btu/hr · °F · ft² are read directly.

b. Solid wood swing door with metal storm door (1¾ in thick)
   From Table 2.10, the total thermal resistance (R_t) of 3.2 hr · °F · ft²/Btu and overall coefficient of heat transmission (U) of 0.31 Btu/hr · °F · ft² are read directly.

2.6 THERMAL BRIDGING

A *thermal bridge* in a building envelope assembly or component is a penetration of the insulation layer by a highly conductive or noninsulating material (e.g., concrete or steel). Those areas of the building envelope assembly with high thermal conductivity will increase heat flow, thereby lowering the average U-factor of the building envelope assembly. In commercial construction, steel or concrete members incorporated in exterior wall or roof construction often form thermal bridges. Metal ties in cavity walls are another type of thermal bridge commonly found in masonry construction. A common example in residential construction is the wood framing of an exterior frame wall where insulation is installed between studs and plates in the wall cavity. The wood framing has a larger thermal conductivity than the insulation, which provides an enhanced path for heat to flow through the wall assembly. See Figure 2.8.

Thermal bridges created by highly conducting construction materials can significantly reduce the R-value of the construction assembly. For example, the light-gauge steel framing (from 14 to 25 gauge sheet metal) that is frequently used in commercial buildings presents more of a thermal insulating challenge than wood because of its much higher thermal conductivity. Light-gauge steel framing does have a much smaller cross-sectional area than standard 2 × 4 wood framing and thus a much narrower heat transfer path. However, the higher thermal conductivity of steel over wood (324 Btu · in/hr · °F · ft² for steel and about 1.0 Btu · in/ hr · °F · ft² for wood, from Table 2.1) overshadows this size advantage. The product (kA) of thermal conductivity (k) and area (A) of the heat flow path can be used in evaluating heat flow. The following values for kA are computed based on heavy (16 gauge) steel framing (0.054 in thick) and 2 × 4 wood framing (1½ in thick):

16 gauge steel stud:
kA = 324 Btu · in/hr · °F · ft² · [(0.054 in/12 in per ft) · 1 ft]
   = 1.458 Btu · in/hr · °F

2 × 4 wood stud:
kA = 1 Btu · in/hr · °F · ft² · [(1.5 in/12 in per ft) · 1 ft]
   = 0.125 Btu · in/hr · °F

TABLE 2.9   A COMPARISON OF THERMAL PROPERTIES OF CONVENTIONAL AND ALTERNATIVE WALL CONSTRUCTION TECHNIQUES.

Wall Type	R-Value		U-Factor	
	$\frac{hr \cdot ft^2 \cdot °F}{Btu}$	$\frac{°C \cdot m^2}{W}$	$\frac{Btu}{hr \cdot ft^2 \cdot °F}$	$\frac{W}{°C \cdot m^2}$
Conventional Construction				
Wood Frame				
2 × 4 (38 mm × 88 mm) wall w/ R-11 (RSI-1.9) insulation	10.2	1.8	0.098	0.557
2 × 6 (38 mm × 140 mm) wall w/ R-19 (RSI-3.3) insulation	15.4	2.7	0.065	0.369
Light-Gauge Steel Frame				
2 × 4 (38 mm × 88 mm) wall w/ R-11 (RSI-1.9) insulation	7.5	1.3	0.133	0.761
2 × 6 (38 mm × 140 mm) wall w/ R-19 (RSI-3.3) insulation	9.1	1.6	0.110	0.627
Concrete				
8 in (200 mm) standard weight uninsulated wall	1.5	0.3	0.667	3.786
12 in (200 mm) standard weight uninsulated wall	1.6	0.3	0.625	3.549
Masonry				
8 in (200 mm) solid brick uninsulated wall	2.4	0.4	0.417	2.366
12 in (200 mm) solid brick uninsulated wall	3.2	0.6	0.313	1.775
8 in (200 mm) CMU uninsulated wall	2.0	0.4	0.500	2.839
12 in (200 mm) CMU uninsulated wall	2.1	0.4	0.476	2.704
8 in (200 mm) CMU wall, perlite insulated cores	5.5	1.0	0.182	1.032
12 in (200 mm) CMU wall, perlite insulated cores	7.9	1.4	0.127	0.719
Alternative Construction				
Adobe[a]				
10 in (250 mm) uninsulated	3.8	0.7	0.284	1.494
10 in (250 mm) insulated[b]	12.2	2.1	0.084	0.465
14 in (350 mm) uninsulated	5.2	0.9	0.204	1.092
14 in (350 mm) insulated[b]	13.6	2.4	0.075	0.418
18 in (350 mm) uninsulated	6.6	1.2	0.159	0.860
18 in (350 mm) insulated[b]	15.0	2.6	0.068	0.379
24 in (600 mm) uninsulated [c]	7.1	1.3	0.147	0.800
24 in (600 mm) insulated[b,c]	15.4	2.7	0.066	0.369
Log/Cordwood				
8 in (200 mm) solid	10.0	1.8	0.100	0.568
12 in (300 mm) solid	15.0	2.6	0.067	0.379
16 in (400 mm) solid	20.0	3.5	0.050	0.284
24 in (600 mm) composite (8 in/200 mm insulation)	40.0	7.0	0.025	0.142
Papercrete				
12 in (300 mm) (R-2.4/in or 0.017 RSI/mm)	28.8	5.1	0.035	0.197
Pumicecrete				
12 in (300 mm) (R-1.5/in or 0.01 RSI/mm)	18.0	3.2	0.056	0.315
Rammed/Compressed Earth				
12 in (300 mm) uninsulated	3.0	0.5	0.333	1.893
12 in (300 mm) insulated[d]	11.4	2.0	0.088	0.498
Straw Bale				
23 in (575 mm) bale (R-2.4/inch or 0.017 RSI/mm)[e]	56.5	9.9	0.018	0.101
23 in (575 mm) bale (R-1.45/inch or 0.01 RSI/mm)[f]	33.4	5.9	0.030	0.170
Structural Insulated Panels (SIP)				
SIP wall with EPS core (3½ in/88 mm)	14	2.5	0.071	0.406
Agriboard® straw panels (4⅜ in/110 mm)	13	2.3	0.077	0.437
Agriboard® straw panels (7⅞ in/200 mm)	25	4.4	0.040	0.227
Ttire Bale				
60 in (1.5 m) tire bale (R-0.69/in or 0.005 RSI/mm)	41.4	7.3	0.024	0.137

[a] Walls finished with ¾ in (19 mm) stucco on exterior surface and ½ (13 mm) gypsum plaster on interior surface.
[b] Walls with an additional 2 in (50 mm) of polystyrene added to exterior surface.
[c] Wall is two 10 in (250 mm) wythes (layers) with 4 inch void.
[d] Walls finished with ¾ in (19 mm) stucco on exterior surface and ½ (13 mm) gypsum wallboard (drywall) on interior surface.
[e] Based on measurements done in a 1993 study by Joseph McCabe at the University of Arizona.
[f] Based on measurements done in a 1998 study at Oak Ridge National Laboratory (ORNL).

TABLE 2.10   APPROXIMATE TOTAL THERMAL RESISTANCE (R_t) AND OVERALL COEFFICIENT OF HEAT TRANSMISSION (U) VALUES FOR SELECTED WINDOWS AND DOORS DURING WINTER MONTHS. ACTUAL VALUES VARY BECAUSE OF DIFFERENCES IN THE THERMAL PROPERTIES OF THE FRAME. VALUES PROVIDED INCLUDE AIR FILMS.

Material	Customary (U.S.)		Metric (SI)	
	Total Thermal Resistance (R_t)	Overall Coefficient of Heat Transmission (U)	Total Thermal Resistance (R_t)	Overall Coefficient of Heat Transmission (U)
	$\dfrac{\text{hr} \cdot {}^\circ\text{F} \cdot \text{ft}^2}{\text{Btu}}$	$\dfrac{\text{Btu}}{\text{hr} \cdot {}^\circ\text{F} \cdot \text{ft}^2}$	$\dfrac{{}^\circ\text{C} \cdot \text{m}^2}{\text{W}}$	$\dfrac{\text{W}}{{}^\circ\text{C} \cdot \text{m}^2}$
Windows				
Single glass	0.9	1.10	0.16	6.2
Double insulating glass				
³⁄₁₆ in (4.7 mm) air space	1.6	0.62	0.28	3.5
¼ in (6.4 mm) air space	1.7	0.58	0.30	3.3
½ in (13 mm) air space	2.0	0.49	0.36	2.8
Double insulating low-e ($\varepsilon = 0.1$) glass	3.1	0.32	0.55	1.8
Triple insulating glass				
¼ in (6.4 mm) air space	3.2	0.31	0.57	1.8
½ in (13 mm) air space	3.8	0.26	0.68	1.5
Triple insulating low-e ($\varepsilon = 0.15$) glass	3.7	0.27	0.65	1.5
Aerogel (theoretical R-20)	20.0	0.05	3.52	0.3
Skylight				
Single glass	0.8	1.23	0.14	7.0
Double insulating glass				
³⁄₁₆ in (4.7 mm) air space	1.4	0.70	0.25	4.0
¼ in (6.4 mm) air space	1.5	0.65	0.27	3.7
½ in (13 mm) air space	1.7	0.59	0.30	3.4
Double insulating low-e ($\varepsilon = 0.1$) glass	2.5	0.40	0.44	2.3
Doors				
Solid wood swing door				
1⅜ in (35 mm) thick	1.9	0.52	0.34	3.0
1¾ in (45 mm) thick	2.2	0.46	0.38	2.6
2 in (50 mm) thick	2.3	0.43	0.41	2.4
Solid wood swing door with metal storm door				
1⅜ in (35 mm) thick	2.8	0.36	0.49	2.0
1¾ in (45 mm) thick	3.2	0.31	0.57	1.8
2 in (50 mm) thick	3.4	0.29	0.61	1.6
Steel swing door with mineral fiber core (1¾ in/45 mm thick)	1.7	0.59	0.30	3.4
Aluminum frame swing door with double insulating glass	2.0	0.50	0.35	2.8
Steel swing door with XPS foam core (1¾ in/45 mm thick)	2.1	0.47	0.37	2.7
Steel swing door with urethane foam core (1¾ in/45 mm thick)	5.3	0.19	0.93	1.1
Insulated steel overhead door (1½ in/38 mm XPS)	5.6	0.18	1.60	1.0
Insulated steel overhead door (2 in/50 mm XPS)	9.1	0.11	0.99	0.6

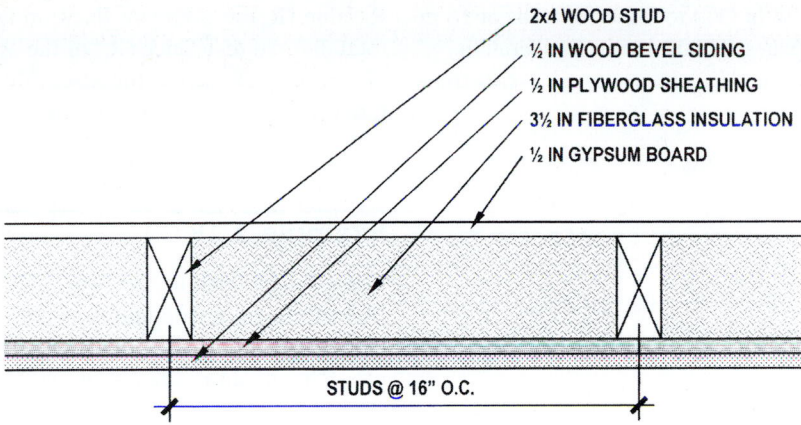

FIGURE 2.8 A composite construction assembly. Framing members such as studs create a thermal bridge, which reduces the overall insulating ability of the wall.

The rate of heat flow through 16 gauge steel framing is almost 12 times $(1.458/0.125 = 11.7)$ the rate through 2×4 wood framing for the same temperature difference. This difference is less for lighter-gauge framing (e.g., 25 gauge) and greater for heavier-gauge framing (14 gauge). Normally, 14 to 20 gauge steel framing is used in load-bearing applications and on exterior walls and 20 to 25 gauge steel is used for furring and to frame partitions.

Thermal bridges can seriously interfere with the performance of a building envelope. The temperature of the inside surface over a thermal bridge is lower than that of the adjacent construction during the heating season and is higher in the cooling season. This can significantly increasing winter heat loss and summer heat gain. Additionally, the formation of condensation on thermal bridging elements can result in mold and mildew growth, staining of surfaces, and serious moisture damage to building components. The difference in the temperature between the thermal bridge and adjacent construction can cause thermal stressing that may result in structural damage.

Many thermal bridges that occur in present-day construction can be avoided, or their effects minimized, if they are recognized in the early stages of design. Thermal bridging can also be avoided by separating conducting paths with insulation. Problems can be overcome with insulation placed over the entire exterior of a building, enclosing all construction elements. This practice is particularly important with steel framing where exterior insulating sheathing should be installed in moderate to cold climates to lessen the problems of thermal bridging. However, screws attached to steel framing can reduce the insulating value of foam sheathing by up to 40%, making fasteners themselves a thermal bridging concern.

2.7 TRANSMISSION HEAT LOSS IN PARALLEL CONSTRUCTION ASSEMBLIES

In many construction assemblies, framing members such as studs and joists create a thermal bridge, which reduces the overall insulating ability of the wall. These thermal bridges can significantly reduce local R-values and change dynamic response for building envelope components. For some conventional wall systems, the whole-wall R-value can be as much as 40% less than what is measured for the clear-wall section.

There are several methods used to express the thermal resistance (R) properties of a construction assembly:

The *center-of-cavity R-value* is an R-value estimation at a point in the assembly's cross-section containing the most insulation.

The *clear-wall R-value* is an R-value estimation for the exterior wall area containing only insulation and necessary framing materials for a clear section with no fenestration (e.g., windows and doors), corners, or connections between other envelope elements such as roofs, foundations, and other walls.

The *whole-wall R-value* is an R-value estimation for the whole opaque wall including the thermal performance of the clear wall area (with insulation and construction elements) and envelope interface details (e.g., wall corners, wall/roof, wall/floor, wall/door, and wall/window intersections and connections).

Accurate heat loss computations are based upon the clear-wall R-value. Before proceeding, it is necessary to compute the R-value of the assembly through the framing (R_{af}), between the framing (R_{bf}) and the fraction of wall area covered by the framing. The adjusted R-value of the wall (R_{adj}) may be calculated as follows:

$$R_{adj} = 1/[S/R_{af} + ((1 - S)/R_{bf})]$$

where R_{af} is the R-value for the area backed by framing, R_{bf} is the R-value for the area between framing members, and S is the fraction of wall area backed by framing over the gross wall area. Standard stud spacing is 16 in on center (OC) and standard studs are 1½ in thick. Thus, on average, S is equal to 1.5 in/16 in or 0.094 of the gross wall area. In whole-wall analysis, S will be about

0.15 (for wall with studs at 24 in OC) to 0.20 (for walls at 16 in OC) when beams, sills, and plates are taken into consideration.

The whole-wall U-factor (U_{adj}) for a complex construction assembly is found by:

$$U_{adj} = 1/R_{adj}$$

Example 2.17

Calculate the whole-wall U-factor for the following standard frame wall with 2 × 4 studs at 16 in OC based on 20% of the wall area backed by framing.

	R_{bf}	R_{af}
outside air film, 15 mph wind	0.17	0.17
Wood bevel siding	0.81	0.81
½ in plywood sheathing	0.62	0.62
2 × 4 wood stud @ 16 in OC	—	4.38
3½ in fiberglass insulation	11.0	—
Vapor retarder, plastic film	negligible	negligible
½ in gypsum board	0.45	0.45
Inside air film	0.68	0.68
Total thermal resistance (R_t)	13.73 hr · °F · ft²/Btu	7.11 hr · °F · ft²/Btu

$$R_{adj} = 1/[(S/R_{af}) + ((1 - S)/R_{bf})] = 1/[(0.20/7.11)$$
$$+ ((1 - 0.20)/13.73)] = 11.6 \text{ hr} \cdot °F \cdot ft^2/Btu$$

$$U_{adj} = 1/R_{adj} = 1/11.6 \text{ hr} \cdot °F \cdot ft^2/Btu$$
$$= 0.086 \text{ Btu/hr} \cdot °F \cdot ft^2$$

Note the significant difference in the corrected value of heat loss—an 18% decrease from the center-of-cavity R-value of 13.73 hr · °F · ft²/Btu (from Example 2.12) to the whole-wall R-value of 11.6 hr · °F · ft²/Btu (from Example 2.17. Results for the whole-wall R-value more accurately represent steady state heat loss properties in comparison to the center-of-cavity R-value.

2.8   TEMPERATURE GRADIENT IN CONSTRUCTION ASSEMBLIES

A *temperature gradient* exists when there is a change in temperature over some distance across a material. When a construction assembly (e.g., wall, ceiling, floor) is exposed to different but steady temperatures, a difference in temperature develops across the thickness of the assembly. In a construction assembly, a temperature gradient exists between the two exterior surfaces and the materials within the assembly during design conditions.

Under steady state conditions for a homogenous assembly, the temperature gradient is linear between the two surfaces. For composite construction assemblies, the slope of the temperature gradient is proportional to the resistances of individual layers. The temperature gradient across composite construction assemblies can be computed by multiplying the ratio of the

R-value (R_x) to a known location (x) in the assembly and the total thermal resistance (R_t) of the assembly by the temperature difference (ΔT) across the assembly, where ΔT_x is the temperature difference at location x from a known temperature:

$$\Delta T_x = (R_x/R_t) \cdot \Delta T$$

Example 2.18

The insulated wall in Example 2.12 has the following R-values. Determine the temperatures at the surfaces of each material in the construction assembly based on an outside air temperature of −2°F and an inside air temperature of 68°F (a ΔT of 70°F).

outside air film, 15 mph wind	0.17 hr · °F · ft²/Btu
½ in wood bevel siding	0.81 hr · °F · ft²/Btu
½ in plywood sheathing	0.62 hr · °F · ft²/Btu
3½ in fiberglass insulation	11.00 hr · °F · ft²/Btu
vapor retarder, plastic film	negligible
½ in gypsum board	0.45 hr · °F · ft²/Btu
Inside air film	0.68 hr · °F · ft²/Btu
Total thermal resistance (R_t) =	13.73 hr · °F · ft²/Btu

a. Temperature between outside air film and wood bevel siding: $R_x = 0.17$.
$$\Delta T_x = (R_x/R_t) \cdot \Delta T = (0.17/13.73) \cdot 70°F = 0.9°F$$
$$T_x = -2°F + 0.9°F = -1.1°F$$

b. Temperature between wood bevel siding and plywood sheathing: $R_x = 0.17 + 0.81 = 0.98$.
$$\Delta T_x = (R_x/R_t) \cdot \Delta T = (0.98/13.73) \cdot 70°F = 5.0°F$$
$$T_x = -2°F + 5.0°F = 3.0°F$$

c. Temperature between plywood sheathing and fiberglass insulation: $R_x = 0.17 + 0.81 + 0.62 = 1.60$.
$$\Delta T_x = (R_x/R_t) \cdot \Delta T = (1.60/13.73) \cdot 70°F = 8.2°F$$
$$T_x = -2°F + 8.2°F = 6.2°F$$

d. Temperature between fiberglass insulation and vapor retarder: $R_x = 0.17 + 0.81 + 0.62 + 0 = 1.60$.
$$\Delta T_x = (R_x/R_t) \cdot \Delta T = (1.60/13.73) \cdot 70°F = 8.2°F$$
$$T_x = -2°F + 8.2°F = 6.2°F$$

e. Temperature between vapor retarder and gypsum board: $R_x = 0.17 + 0.81 + 0.62 + 0 + 11.00 = 12.60$.
$$\Delta T_x = (R_x/R_t) \cdot \Delta T = (12.60/13.73) \cdot 70°F = 64.2°F$$
$$T_x = -2°F + 64.2°F = 62.2°F$$

f. Temperature between gypsum board and inside air film: $R_x = 0.17 + 0.81 + 0.62 + 0 + 11.00 + 0.45 = 13.05$.
$$\Delta T_x = (R_x/R_t) \cdot \Delta T = (13.05/13.73) \cdot 70°F = 66.5°F$$
$$T_x = -2°F + 66.5°F = 64.5°F$$

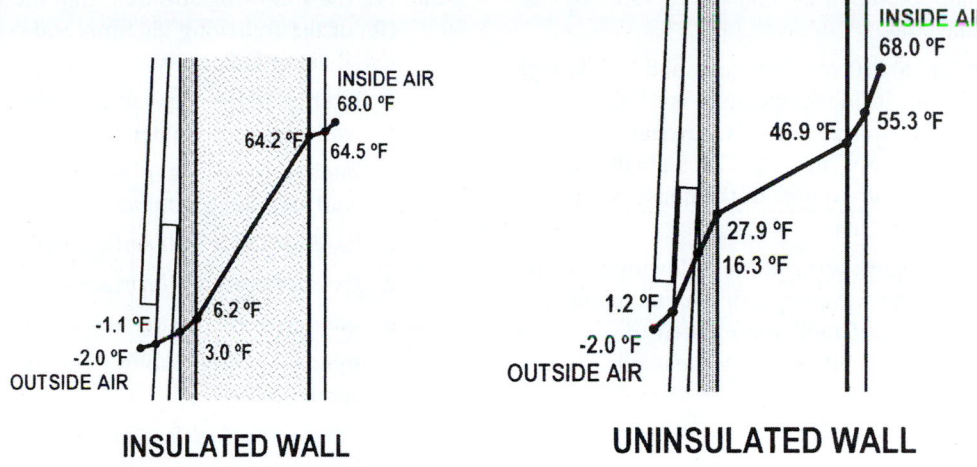

FIGURE 2.9 Temperature gradients across insulated and uninsulated wall assemblies.

Temperature gradients across insulated and uninsulated wall assemblies are shown in Figure 2.9. The temperature gradient for the insulated wall was computed in Example 2.18.

Example 2.19

Determine the temperature at the midpoint of the insulation of the wall described in Example 2.18.

At the midpoint of the insulation: $R_x = 0.17 + 0.81 + 0.62 + 0 + 11.00/2 = 7.10$.

$$\Delta T_x = (R_x/R_t) \cdot \Delta T = (7.10/13.73) \cdot 70° F = 36.2°F$$
$$T_x = -2°F + 36.2°F = 34.2°F$$

STUDY QUESTIONS

2-1. Describe the three fundamental modes of heat transfer.

2-2. Describe the difference between natural convection and forced convection.

2-3. With respect to radiation heat transfer, describe opaque and translucent surfaces.

2-4. With respect to radiation heat transfer, describe the terms *absorptivity, reflectivity,* and *transmissivity*.

2-5. With respect to radiation heat transfer, describe the term *emissivity*.

2-6. Identify and describe types of coatings and surfaces used on buildings that intentionally affect radiant heat transfer.

2-7. With respect to buildings, describe the terms *heat loss* and *heating load*.

2-8. With respect to buildings, identify and describe the factors that contribute to heat loss in a building.

2-9. Describe the difference between thermal conductivity (k) and thermal conductance (C).

2-10. Describe the relationships between R-, C-, k-, and U-values.

2-11. What is transmission heat loss, and where does it occur in a building?

2-12. What is a thermal bridge in a building envelope assembly, and how does it affect heat loss in a building?

2-13. Describe the differences between the center-of-wall R-value, clear-wall R-value, and whole-wall R-value.

Design Exercises

2-14. Determine the rate of conduction heat transfer through 1 ft² of the following materials. Assume the surface temperatures of the materials are 30° and 72°F.

   a. 3½ in brick

   b. 8 in normal weight concrete

   c. ¼ in thick glass

2-15. Determine the rate of conduction heat transfer through 1 m² of the following materials. Assume the surface temperatures of the materials are −1° and 22°C.

   a. 87.5 mm brick

   b. 200 mm normal weight concrete

   c. 6 mm thick glass

2-16. A residence has a 2000 ft² floor area and 8 ft high ceilings. Inside air inside temperature is 70°F and outside ambient temperature is 10°F. Assume the heat capacity of air is 0.018 Btu/ft³ · °F. Calculate the heat removed if the entire air volume of the house is replaced by outdoor air in one hour.

2-17. A residence has a 210 m² floor area and 2.5 m high ceilings. Inside air inside temperature is 21°C and outside ambient temperature is −5°C. Assume the heat capacity of air is 0.35 W/m³ · °C. Calculate the heat removed if the entire air volume of the house is replaced by outdoor air in one hour.

2-18. Surfaces emit radiant energy at the following temperatures. Determine the wavelength of maximum radiative power and the type of radiation emitted (e.g., visible light, UV, IR) by the predominant wavelength.

   a. 0°F
   b. 80°F
   c. 200°F
   d. 1000°F
   e. 10 000°F

2-19. Surfaces emit radiant energy at the following temperatures. Determine the wavelength of maximum radiative power and the type of radiation emitted (e.g., visible light, UV, IR) by the predominant wavelength.

   a. 0°C
   b. 25°C
   c. 100°C
   d. 500°C
   e. 5000°C

2-20. Find the thermal resistance (R) values of the following materials:

   a. 8 in thick lightweight concrete block
   b. ⅜ in thick plywood
   c. 2 × 8 (7¼ in actual thickness) softwood lumber
   d. ¾ in fiberglass insulation board
   e. ⅝ in thick gypsum wallboard
   f. Asphalt shingles (⅛ in thick)
   g. 12 in thick cellulose insulation blown into an attic space

2-21. Find the thermal resistance (R) values of the following materials:

   a. 200 mm thick lightweight concrete block
   b. 9 mm thick plywood
   c. 89 mm thick softwood lumber
   d. 25 mm thick fiberglass insulation board
   e. 16 mm thick gypsum wallboard
   f. Asphalt shingles (3 mm thick)
   g. 300 mm thick cellulose insulation blown into an attic space

2-22. For the following surfaces, find the thermal resistance (R) of the following air films and confining air spaces (cavities) in U.S. units:

   a. Interior painted gypsum wallboard (nonreflective) wall surface in winter
   b. Interior painted gypsum wallboard (nonreflective) wall surface in summer
   c. Exterior brick wall surface in winter
   d. Exterior brick wall surface in summer
   e. Interior wood (nonreflective) floor surface in winter
   f. Interior wood (nonreflective) floor surface in summer
   g. Exterior roof surface in winter
   h. Exterior roof surface in summer
   i. ½ in wide wall cavity between nonreflective surfaces (ε = 0.9) based on winter temperatures
   j. ½ in wide wall cavity between nonreflective surfaces (ε = 0.9) based on summer temperatures
   k. 3½ in wide cavity between nonreflective surfaces (ε = 0.9) based on winter temperatures
   l. 3½ in wide cavity between nonreflective surfaces (ε = 0.9) based on summer temperatures

2-23. For the following surfaces, find the thermal resistance (R) of the following air films and confining air spaces (cavities) in metric (SI) units:

   a. Interior painted gypsum wallboard (nonreflective) wall surface in winter
   b. Interior painted gypsum wallboard (nonreflective) wall surface in summer
   c. Exterior brick wall surface in winter
   d. Exterior brick wall surface in summer
   e. Interior wood (nonreflective) floor surface in winter
   f. Interior wood (nonreflective) floor surface in summer
   g. Exterior roof surface in winter
   h. Exterior roof surface in summer
   i. 13 mm wide wall cavity between nonreflective surfaces (ε = 0.9) based on winter temperatures
   j. 13 mm wide wall cavity between nonreflective surfaces (ε = 0.9) based on summer temperatures
   k. 89 mm wide wall cavity between nonreflective surfaces (ε = 0.9) based on winter temperatures
   l. 89 mm wide wall cavity between nonreflective surfaces (ε = 0.9) based upon summer temperatures

2-24. From the tables provided in this text, find the total thermal resistance (R_t) and overall coefficient of heat transmission (U) of the following windows and doors in U.S. units:

a. Double insulating glass (¼ in air space)

b. Triple insulating glass (¼ in air space)

c. Double insulating glass (low-e)

d. Triple insulating glass (low-e)

e. Solid wood swing door with metal storm door (1¾ in thick)

2-25. From the tables provided in this text, find the total thermal resistance (R_t) and overall coefficient of heat transmission (U) of the following windows and doors in metric (SI) units:

a. Double insulating glass (6.4 mm air space)

b. Triple insulating glass (6.4 mm air space)

c. Double insulating glass (low-e)

d. Triple insulating glass (low-e)

e. Solid wood swing door with metal storm door (45 mm thick)

2-26. An uninsulated solid load-bearing masonry wall is constructed of a ½ in stucco exterior finish on 8 in standard weight concrete block (CMU) and ⅝ in gypsum wallboard.

a. Determine the R_t-value and U-factor for this construction assembly under winter conditions.

b. Determine the R_t-value and U-factor for this construction assembly under summer conditions.

2-27. A solid load-bearing masonry wall is constructed of 4 in face brick, 8 in standard weight concrete block, 2 in expanded polystyrene insulation board (unfaced), and ⅝ in gypsum wallboard.

a. Determine the R_t-value and U-factor for this construction assembly under winter conditions.

b. Determine the R_t-value and U-factor for this construction assembly under summer conditions.

2-28. A solid load-bearing masonry wall is constructed of 4 in face brick, 4 in standard weight concrete block, 3½ in fiberglass batt insulation, and ⅝ in gypsum wallboard.

a. Determine the R_t-value and U-factor for this construction assembly under winter conditions.

b. Determine the R_t-value and U-factor for this construction assembly under summer conditions.

2-29. A cavity masonry wall is constructed of 4 in face brick, 1½ in hollow (air) cavity, 4 in standard weight concrete block, 3½ in fiberglass batt insulation, and ⅝ in gypsum wallboard.

a. Determine the R_t-value and U-factor for this construction assembly under winter conditions.

b. Determine the R_t-value and U-factor for this construction assembly under summer conditions.

2-30. An insulated wall is constructed of 11 mm hardboard lapped siding, 19 mm extruded polystyrene insulation board sheathing, 89 mm fiberglass batt insulation, a vapor retarder (plastic film), and 13 mm gypsum wallboard.

a. Determine the R_t-value and U-factor for this construction assembly under winter conditions.

b. Determine the R_t-value and U-factor for this construction assembly under summer conditions.

2-31. An insulated wall is constructed of ⁷⁄₁₆ in hardboard lapped siding, ¾ in foil-faced (both sides) polyisocyanurate insulation board sheathing, 3½ in cellulose insulation, a vapor retarder (plastic film), and ½ in gypsum board.

a. Determine the R_t-value and U-factor for this construction assembly under winter conditions.

b. Determine the R_t-value and U-factor for this construction assembly under summer conditions.

2-32. A commercial roof is constructed of a single-ply roof membrane, 4 in fiberglass insulation sheathing board, and 6 in cast-in-place concrete deck.

a. Determine the R_t-value and U-factor for this construction assembly under winter conditions.

b. Determine the R_t-value and U-factor for this construction assembly under summer conditions.

2-33. An insulated wall is constructed of ⁷⁄₁₆ in hardboard lapped siding, ½ in OSB sheathing, 5½ in fiberglass batt insulation, a vapor retarder (plastic film), and ½ in gypsum board. The wall is framed with 2 × 6 solid softwood lumber at 16 in O.C. Calculate the whole-wall U-factor for this construction assembly under winter conditions.

2-34. An insulated wall is constructed of ⁷⁄₁₆ in hardboard lapped siding, ½ in OSB sheathing, 5½ in fiberglass batt insulation, a vapor retarder (plastic film), and ½ in gypsum board. The wall is framed with 2 × 6 solid softwood lumber at 24 in OC. Calculate the whole-wall U-factor for this construction assembly under winter conditions.

2-35. A solid load-bearing masonry wall is constructed of 4 in face brick, 8 in standard weight concrete block, 2 in expanded polystyrene insulation board (unfaced), and ⅝ in gypsum wallboard. Determine the temperatures at the surfaces of each material in the construction assembly based on an outside air temperature of 10°F and an inside air temperature of 72°F (winter conditions).

2-36. An insulated wall is constructed of 11 mm hardboard lapped siding, 19 mm extruded polystyrene insulation board sheathing, 89 mm fiberglass batt insulation, a vapor retarder (plastic film), and 13 mm gypsum board. Determine the temperatures at the surfaces of each material in the construction assembly based on an outside air temperature of $-10°C$ and an inside air temperature of 21°C (winter conditions).

2-37. An insulated wall is constructed of $\frac{7}{16}$ in hardboard lapped siding, ¾ in foil-faced (both sides) polyisocyanurate insulation board sheathing, 3½ in cellulose insulation, a vapor retarder (plastic film), and ½ in gypsum board. Determine the temperatures at the surfaces of each material in the construction assembly based on an outside air temperature of 10°F and an inside air temperature of 72°F.

CONCEPTS IN BUILDING SCIENCE

3.1 BUILDING SCIENCE

The disruptions in Middle East oil supplies that resulted in the energy crises in 1973 and 1979 coupled with the recognition that U.S. oil production had peaked (about 1970) abruptly caused modifications in building construction methods, such as tighter construction standards, increased insulation, reduced ventilation regulation, and more efficient systems. These changes resulted in complaints about poor indoor air quality and sick buildings, and thermal discomfort linked to problems in performance and durability. These issues sparked an increased effort of scientific study in building performance that, decades later, involves recognizing the relationships among building components, equipment, and building systems during design, construction, and operation of the building. In the new millennium, climate change, limited resources, volatility in energy markets, and a need for sustainable practices have emerged as issues that further drive scientific study of buildings.

Building science is the study of building dynamics and the functional relationships between a building's components, equipment, and systems, and the effects associated with occupancy and operation, and the outdoor environment to understand and prevent problems related to building design, construction, and operation. Through application of information coming from many disciplines (i.e., construction technology, material science, physics, meteorology, engineering, and architectural design), the building science field works to understand the physical performance of building parts, buildings, and the built environment, and translates it into proper design and construction methods. The goal is to maximize occupant comfort, health, and safety; and optimize performance through energy efficiency and structural durability. Building science plays a vital role in recognizing how changes and advancements in components, equipment, and systems affect the role of the various building materials and systems.

In this chapter, selected concepts related to building science are introduced. These topics include a discussion of thermal characteristics of the building envelope, indoor air, moisture dynamics, ventilation, and thermal insulation. They are introduced because of their connection to building electrical and mechanical systems.

3.2 THE BUILDING

The Building Envelope

A building consists of the superstructure and the substructure. The *superstructure* is that part of the building structure that is above the top of the foundation walls. The *substructure* is the building's foundation system. The *building envelope* or building shell includes the elements that enclose the conditioned (heated and cooled) space and interface the indoor spaces with the outdoor environment. It includes the wall, roof, and opening assemblies (e.g., walls, windows, doors, roof, floor, and foundation), insulation, air barriers, vapor retarders, weather stripping, and caulking. The building envelope provides shelter from severe weather. It also plays a major role in sustaining the indoor environment and, as a result, influences the heating, ventilating and air conditioning (HVAC) systems in a building.

Historically, structural integrity and durability were the prime characteristics of a well-designed building envelope. Today, the design and construction of a building envelope must incorporate other characteristics that contribute to high performance, such as airtightness, thermal insulation, heating and cooling load reduction, and amount of embodied energy. Optimal design of the building envelope increases construction and design costs. However, the additional cost for a high-performance envelope must be paid for through cost avoidance achieved by a need for smaller heating, cooling, ventilating, and lighting equipment and in reduction of operating and maintenance costs.

Embodied Energy

One important factor that contributes to overall performance of a building is the energy it uses, with one aspect being the energy consumed in constructing it. *Embodied energy* is the sum of all energy used to extract, process, refine, fabricate, transport, and install a product or material. It includes all energy consumed from removing natural resources (e.g., mining iron ore, cutting timber) through material or product installation. It is a concept that attempts to measure the energy expense of an item. Both direct and indirect energy consumed are included in determining a material's embodied energy. For example, consider a brick: embodied energy of the brick includes the energy used to extract clay from the pit, transport the raw material to the manufacturing plant, refine and mix the clay, extrude and cut the brick, fire the brick in a kiln (oven), stack the brick on a crate, transport and deliver it to the building site, and eventually lay the brick in a wall. It also includes the indirect energy required, including energy required to fabricate the equipment and materials needed to manufacture a brick (e.g., trucks, kilns, and mining equipment).

Computation of embodied energy for a specific material is an inexact science because it is affected by many factors. The manufacturing technology used to refine a material or fabricate a product can be different with different manufacturers. Factors such as distance of the building site from manufacturer and the distance trades people must travel to and from the site during construction are also all part of the embodied energy calculation. Furthermore, embodied energy can be estimated for the material or product before it is installed, after it is installed, or after it is installed and including maintenance, repair, demolition, and disposal of the building structure. As a result, the published values of embodied energy are not universally fixed.

A relationship exists among the embodied energy of materials, the complexity of processing the material, and pollution. Generally, the fewer and simpler the steps involved in a material's extraction, production, transportation, and installation, the lower its embodied energy. As a result, the embodied energy of a material is often reflected in its cost. Release of pollutants is also linked with embodied energy. For example, on average, about 1000 lb (450 kg) of CO_2 are produced per million Btu of embodied energy.

The unit measure for embodied energy is typically the Btu/lb (MJ/kg or GJ/kg). Table 3.1 provides embodied energy values for common materials for comparison purposes. Generally, plastics, virgin (not recycled) metals, concrete, and bitumen/asphalt products are high in embodied energy. Engineered wood products [e.g., medium-density fiberboard (MDF), oriented strand board (OSB), and glued-laminated timber products] and recycled metals and glass have a moderate amount of

embodied energy. Earthen (e.g., brick, adobe), and sawn timber products, if they are harvested and produced nearby, are low in embodied energy.

Embodied energy is a significant contributor to the life cycle cost of a building material, product, or system. Total embodied energy of a building ranges from about 400 000 to 500 000 Btu/ft^2 (4.5 to 5.5 GJ/m^2) of finished floor area depending on floor size and type, cladding material, and number of stories. Embodied energy can be the equivalent of several years of operational energy consumption (e.g., from lighting, heating, cooling), ranging from approximately 10 years (for typical houses) to over 30 years (for commercial buildings). As buildings become more energy efficient in their operation, total embodied energy can approach half the lifetime energy consumption.

Building Classifications

There are many ways to classify a building. For example, model codes classify a building based on occupancy (e.g., R for residential, A for assembly, E for educational, and so on) and construction type (e.g., Type I, Type II, Type III, and so forth). When taking into account energy use, buildings can be classified by thermal load and thermal mass.

Heating/Cooling Load Classifications

Energy is consumed in buildings in many different ways. A heating energy profile of a typical American residence is shown in Table 3.2. Although this is based upon an average, space heating and cooling generally account for the largest fraction of energy consumed in most homes. As a result, transmission and infiltration (air leakage) through the building envelope play an important role in energy use in residences.

Cooling load profiles of a typical residence and a typical office building are shown in Table 3.3. Notice the differences in factors that contribute to the typical loads. In a residence,

TABLE 3.1   RANGE OF REPORTED VALUES OF EMBODIED ENERGY FOR COMMON BUILDING MATERIALS. THESE VALUES ARE FOR COMPARISON ONLY. EMBODIED ENERGY IS NOT A UNIVERSALLY FIXED VALUE; IT VARIES FROM ONE LOCATION TO ANOTHER.

Material	Customary (U.S.) Units				Metric (SI) Units			
	Btu/lb		Btu/in³		MJ/kg		GJ/m³	
	Low	High	Low	High	Low	High	Low	High
Steel (structural)	10 320	25 360	2950	7150	24	59	190	460
Glass	5590	13 330	530	1260	13	31	34	81
Cement	1850	3355	100	180	4.3	7.8	6.5	11.7
Plaster	475	2880	20	125	1.1	6.7	1.3	8.0
Bricks	430	4040	26	250	1.0	9.4	1.7	16
Timber (sawn softwood)	225	3050	4	55	0.52	7.1	0.26	3.6
Aggregates	13	52	0.8	14.5	0.03	0.12	0.05	0.93

TABLE 3.2   ENERGY PROFILE OF A TYPICAL RESIDENCE.

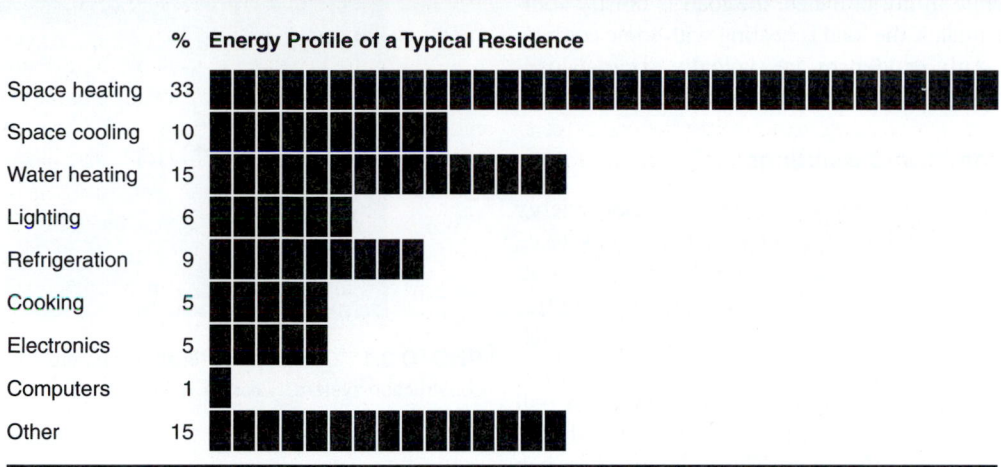

% Energy Profile of a Typical Residence

	%
Space heating	33
Space cooling	10
Water heating	15
Lighting	6
Refrigeration	9
Cooking	5
Electronics	5
Computers	1
Other	15

From: U.S. Department of Energy, *Energy Efficiency and Renewable Energy,* 2000 data.

transmission, ventilation/infiltration, and solar gains dominate the loads. However, in the typical office building, heat from solar gains, occupants, lights, and equipment affect the cooling load much more and transmission and ventilation/infiltration affect the cooling load much less. As a result, the performance of the building envelope is less of a concern in office buildings in contrast to residential buildings.

From the perspective of energy use, buildings can be classified by how they use energy for space heating and cooling.

External-Load-Dominated Buildings

External-load-dominated buildings have the largest part of their energy consumption associated with the thermal characteristics of the building envelope; that is, heating and cooling loads are tied mainly to characteristics like thermal insulation and glazing type, orientation, and area. Such buildings are referred to as *skin-load-dominated* or *envelope-load-dominated* buildings. Climatic conditions also greatly affect the thermal performance of these buildings. Therefore, they can be referred

TABLE 3.3   COOLING LOAD PROFILES OF A TYPICAL RESIDENCE AND OFFICE BUILDING.

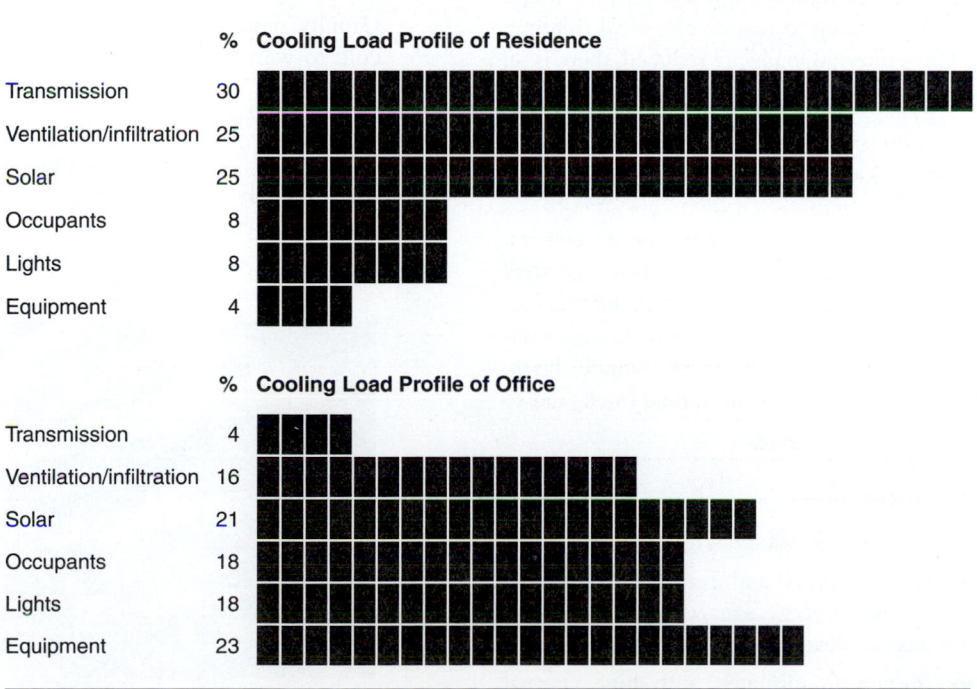

% Cooling Load Profile of Residence

	%
Transmission	30
Ventilation/infiltration	25
Solar	25
Occupants	8
Lights	8
Equipment	4

% Cooling Load Profile of Office

	%
Transmission	4
Ventilation/infiltration	16
Solar	21
Occupants	18
Lights	18
Equipment	23

From: U.S. Department of Energy, *Energy Efficiency and Renewable Energy.*

to as *climate-dominated* buildings. In cold climates, the load is mostly heating, while in hot climates, the load is mostly cooling. In temperate climates, the load is heating with some cooling. Single and multifamily residences are typically external-load-dominated buildings.

Internal-Load-Dominated Buildings

Internal-load-dominated buildings have most of their energy consumption associated with large internal heat gains (e.g., from occupants, lighting, equipment); that is, heating and cooling loads are largely generated within the building itself. Office buildings, schools, hospitals, and factories are typically internal-load-dominated buildings.

In building design, it is helpful to know how loads will dominate a building, because this will direct the focus of design strategies related to energy efficiency. The difference between classifications influences the significance of insulation standards and glazing area because of the desirability of solar heat gain. In exterior-load-dominated buildings in temperate to cold climates, strategies include high levels of insulation, minimizing infiltration (air leakage) through sealing the envelope, high-performance windows (e.g., double or triple low-e glazing), and passive solar incorporated into the design that controls the amount of radiant heating loads in the winter and summer with window orientation and size and location and amount of heat absorbing thermal mass. In moderate and hot climates, solar control is crucial, because of the need to keep cooling loads under control.

Specific strategies to reduce energy use in internal-load-dominated buildings include improvement of energy efficiency of equipment, appliances, and lighting and use of natural and climatic factors (e.g., daylighting to reduce electrical lighting, natural ventilation). Once internal loads are reduced, there is an increase in the importance of the thermal properties of the building envelope.

Building Mass Classifications

Buildings can also be classified by their mass content. *Lightweight* construction uses light timber or light-gauge steel framing as the structural support system for nonstructural cladding (e.g., fiber cement, plywood, and steel). *Heavyweight* construction systems are usually masonry and include brick, concrete, concrete block, tiles, rammed earth, mud brick, and so on. See Photos 3.1 and 3.2.

Heavyweight Construction

- Generally has higher embodied energy.
- Improves thermal comfort and reduces operational (heating and cooling) energy use, when used in conjunction with passive design and good insulation.
- Is most appropriate in climates with high diurnal (day–night) temperature ranges and significant heating and cooling requirements.

PHOTO 3.1   Cast-in-place (in situ) concrete is a heavyweight construction system. (Courtesy of DOE/NREL)

- Requires more substantial foundation systems and causes greater site impact and disturbance.
- Should be avoided on remote sites where there is a high transport component.

Lightweight Construction

- Generally has lower embodied energy.
- Can yield lower total life cycle energy use, particularly where the diurnal range is low.
- Responds rapidly to temperature changes and can provide significant benefits in warmer climates by cooling rapidly at night.
- Is preferred on remote sites with a high materials transportation component.
- Usually requires more heating and cooling energy in cold to warm climates (where solar access is achievable) when compared to heavyweight construction with similar levels of insulation and passive design.
- Can have low production impact (e.g., sustainably sourced timber) or high impact (unsustainably sourced timber or metal frame).

PHOTO 3.2   Lightweight construction typically consists of light timber frame. (Courtesy of DOE/NREL)

3.3 THERMAL INSULATION

Whenever there is a temperature difference, heat flows naturally from a warmer space to a cooler space. Heat moves through the building envelope by radiation, conduction, and convection, but the main mode of heat transfer for solid, opaque building elements is by conduction and radiation. When installed correctly, thermal insulation reduces heat transfer through the building envelope. It makes sense to use thermal insulation to reduce energy consumption while increasing comfort and saving money. Thermal insulation is an important component of a high-performance building.

Characteristics of Thermal Insulation

As introduced in Chapter 2, the "R-value" used to rate insulation products is *thermal resistance (R)*. It is a measure of the ability of a material to resist heat transfer. In the United States, it is expressed in units of hr · °F · ft^2/Btu (°C · m^2/W). The higher the R-value, the better the insulation will be at keeping the heat in (or out). In Canada and other countries, the metric equivalent referred to as RSI, is used. The RSI is about 1/6th the R-value. The exact relationships between R and RSI are:

$$1.0 \text{ hr} \cdot °F \cdot ft^2/Btu = 0.1761 °C \cdot m^2/W$$

$$R = 0.1761 \text{ RSI}$$

$$\text{RSI} = 5.679 \text{ R}$$

Insulation is a material, such as fiberglass, cellulose, mineral wool, foam, or fiberboard, that is installed in the building envelope to significantly reduce heat loss or heat gain. Insulation functions by trapping gases (e.g., air or another gas), which reduces conduction and convection heat transfer through the material. The thermal conductivity of most gases is very low, but gases tend to transfer heat well if natural convection currents are allowed to develop. By trapping gasses in extremely small pockets in the insulation, the formation of convective currents is curtailed.

Because air is a frequently used gas in insulation, most commercially available insulations approach the thermal performance of air. As a result, air-based insulations have similar theoretical thermal properties regardless of their basic composition—that is, thermal conductivities of no less than about 0.24 Btu-in/hr · °F · ft^2 (an R of about 4.2 hr · ft^2 · °F/Btu per inch of thickness or 0.03 m^2 · °C/W per mm). Because of the different materials used to trap air and limitations of manufacture, R-values will be somewhat lower for commercial grades of insulation, usually about 3.0 hr · ft^2 · °F/Btu per inch of thickness (0.021 m^2 · °C/W per mm).

Other types of insulations, particularly closed-cell polyisocyanurate and polyurethane foams that trap a gas with a lower thermal conductivity than air, offer slightly better insulating capability. The closed-cell structure of the foam retains the gas and yields a thermal conductivity of about 0.17 Btu-in/hr · °F · ft^2 (an R of about 6.0 hr · ft^2 · °F/Btu per inch of thickness or 0.042 m^2 · °C/W per mm). Manufacturers frequently produce rigid foam insulation boards with a reflective foil adhered to one or both surfaces, which improves the R-value of the insulation board by about 0.5 hr · ft^2 · °F/Btu per reflective surface (0.0035 m^2 · °C/W per mm per reflective surface). Insulations constructed of multiple layers of reflective films with air voids in between the films are also very effective. These offer a thermal conductivity of about 0.13 Btu-in/hr · °F · ft^2 (an R of about 8.0 hr · ft^2 · °F/Btu per inch of thickness or 0.056 m^2 · °C/W per mm). A graphic comparison of average R-value per unit of thickness by insulation type is provided in Table 3.4.

TABLE 3.4 A GRAPHIC COMPARISON OF AVERAGE R-VALUE PER UNIT OF THICKNESS BY INSULATION TYPE.

Customary (U.S.) Units	hr · ft^2 · °F/Btu per inch									
Type of Insulation	1	2	3	4	5	6	7	8	9	10
Fiberglass (loose and blown fill)										
Cellulose (loose and blown fill)										
Fiberglass (blanket and batt)										
Rock wool (loose and blown fill)										
Polystyrene, expanded (rigid board)										
Polystyrene, extruded (rigid board)										
Polyurethane (rigid board)										
Polyisocyanurate (rigid board)										
Layered reflective films w/air voids										
Type of Insulation	0.01		0.02		0.03		0.04		0.05		0.06		0.07
SI (metric) units	m^2 · °C/W per mm										

The average temperature of the material affects the thermal properties of insulation. A slight increase in thermal conductivity and decrease in R-value occurs when the insulation temperature increases is typical for most types of insulation. This increase is from the increased agitation of the gas molecules as temperature rises. Because building insulations are used within a small temperature range, heat transfer values are assumed to remain constant.

Density also affects the thermal properties of insulation. The resistance to heat flow is dependent on gas trapped between the fibers of the insulation. As insulation is compressed, more fibrous materials fill the same thickness. This compression creates smaller air spaces and offers more resistance to heat transfer. However, at a specific density, part of the air spaces close and the fibrous material conducts heat more rapidly. Thus, stuffing R-19 insulation into a space only large enough to contain R-11 may actually reduce the effectiveness of the R-19 insulation to below R-11. The optimum density of "loose" insulation is typically achieved when the insulation can support its own weight without settling. Examples of insulation installations are provided in Photos 3.3 through 3.12.

PHOTO 3.5   Loose-fill fiberglass insulation being scrapped to create a flat surface to accept gypsum wallboard finish. Excess insulation is vacuum collected for reuse. (Used with permission of ABC)

PHOTO 3.3   Loose-fill fiberglass being blown into an attic space. (Used with permission of ABC)

PHOTO 3.6   Loose-fill insulation is blown through a hose connected to a blower in a truck. Insulation is manually loaded into a hopper connected to the blower. This is one method of installing loose fill insulation. (Used with permission of ABC)

PHOTO 3.4   Loose-fill fiberglass insulation being sprayed into wall cavities. (Used with permission of ABC)

PHOTO 3.7   Blanket insulation installed between the cavities of wall studs or joists. (Used with permission of ABC)

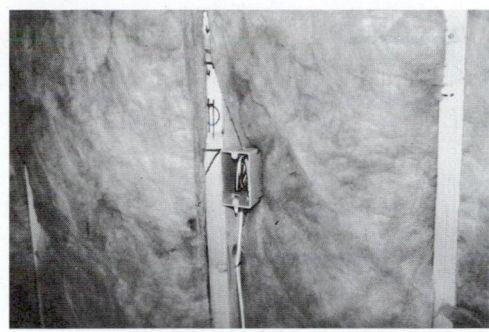

PHOTO 3.8 Care must be exercised in placement of blanket insulation. When it is compressed significantly, such as at electrical outlets, it looses its thermal effectiveness. (Used with permission of ABC)

PHOTO 3.9 Icynene® spray foam insulation and air-sealing system installed in ceiling and wall cavities. (Used with permission of ABC)

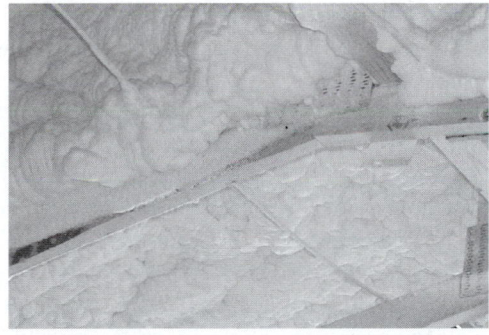

PHOTO 3.10 Close-up view of Icynene® spray foam insulation. (Used with permission of ABC)

Types of Thermal Insulation

There are many different types of insulation. A description and comparison of types of insulation follows:

Blankets, Batts, or Rolls

Rolls and batts (or blankets) are flexible products made from mineral fibers (e.g., fiberglass, rock wool). They are available

PHOTO 3.11 Foil-faced rigid foam insulation installed on basement foundation wall. (Used with permission of ABC)

PHOTO 3.12 Foil-faced sheathing improves the thermal performance of insulation by reducing radiant heat transfer in the winter and summer. (Used with permission of ABC)

in widths suited to standard spacings of wall studs, attic or floor joists, and roof rafters. They are available with or without vapor retarder facings. Batts with a special flame-resistant facing are available in various widths for basement walls where the insulation will be left exposed. Continuous rolls can be hand cut and trimmed to fit, but this must be done precisely. Poor-fitting insulation diminishes expected thermal performance because of gaps in insulation coverage and random air movement within the insulation.

Loose-Fill/Spray Applied

Loose-fill insulation is made from a blend of virgin and recycled waste materials (e.g., fiberglass, rock wool, or cellulose) cut into shreds, granules, or nodules. These small particles are blown into spaces using special pneumatic equipment. Loose-fill insulation is best suited for places where it is difficult to install other types of insulation because the blown-in material easily fills and conforms to building cavities in wall studs and attic or floor joists. Additional resistance to air infiltration (air leakage) can be provided if the insulation is sufficiently dense or thick. The performance of loose-fill insulation is strongly affected by its installation.

Materials used for loose-fill insulation (e.g., fiberglass, rock wool, or cellulose) can also be sprayed into building cavities. Water is generally added to activate an adhesive in the insulation, which causes the insulation to adhere within building cavities as it is sprayed. Sprayed insulation products completely fill voids in areas where it is difficult to get roll insulation types to fit properly. Because sprayed insulation tightly conforms and adheres to the building cavities, it reduces air infiltration (air leakage) and increases the insulation's effectiveness. It forms a uniform covering throughout the cavity and forms a good air seal between electrical wiring, pipes, framing members, and anything else inside the building cavity. Spayed insulation tends not to settle as the insulation fibers adhere to each other and the sides of the building cavity.

Liquid foam insulation can be sprayed into building cavities as a liquid or in larger quantities as a pressure-sprayed product (foamed-in-place). It will completely conform to a building cavity, sealing it thoroughly. Expanding liquid foams can seal small cavities but allowances for expansion must be made to prevent blowout of an assembly.

Rigid Insulation

Rigid insulation is typically made with foam, including molded expanded polystyrene (EPS or bead board), extruded expanded polystyrene (XPS), polyurethane, polyisocyanurate, or other related chemical mixtures. Rigid insulation is typically more expensive than fiber insulation (e.g., fiberglass, rock wool, or cellulose), but it is very effective in buildings with space limitations and where higher insulating values are needed. Rigid boards may be faced with a reflective foil that reduces heat flow when next to an air space. Rigid foam boards are lightweight and can provide structural support (i.e., sheathing) but must be covered with a fire barrier (i.e., gypsum wallboard).

Reflective Insulation and Reflective Barriers

A reflective insulation is formed with layers of aluminum foil (or another type of low emittance material) and enclosed air spaces that provide highly reflective or low emittance cavities. Some reflective insulation systems also use other layers of materials such as paper or plastic to form additional enclosed air spaces. The performance of the system is based on the emittance (e.g., 0.1 or less) of the material(s). The smaller the air space, the less heat will transfer by convection. Therefore, to lessen heat flow by convection, a reflective insulation, with its multiple layers of aluminum and enclosed air space, is positioned in a building envelope cavity (e.g., stud wall, furred-out masonry wall, floor joist, ceiling joist) to divide the larger cavity into smaller air spaces. These smaller trapped air spaces reduce convective heat flow. Like other types of insulation, reflective insulation is labeled with R-values that provide a measure of thermal performance.

Radiant barriers (i.e., polished aluminum foils) can be installed directly under roof rafters or on the attic floor (above insulation) to reduce heat gain from the sun by reflecting energy back to the roof. They are very effective in reducing a building's cooling load but not at reducing the heating load. As a result, radiant barriers are more effective in hot climates. Radiant barriers must have a low emittance (e.g., 0.1 or less) and high reflectance (e.g., 0.9 or more). Unlike other insulation products, there is no standard method for equating the performance of a radiant barrier to standard insulations. Manufacturers may use the term equivalent R-value but this really has no scientific meaning for a radiant barrier, and often reflects optimum conditions and not necessarily climate conditions. A radiant barrier used in the attic floor installation must allow water vapor to pass through it.

Insulating Techniques

The primary method used to reduce transmission heat transfer in a building is insulating or improving the insulation of the building envelope. Typically, insulation is situated as close to the heated space as possible. For example, in an open unheated attic above a sloped roof, it is best to lay the insulation on the attic floor (on the ceiling of floor below) rather than along the sloped roof rafters. Figure 3.1 provides examples of where to insulate a residence.

Typically, loose fill, blanket, or batt insulation made from fiberglass, cellulose, or mineral wool is used to fill cavities and voids in construction assemblies such as frame walls, floors, and attic spaces. Insulation is also installed against a foundation wall or below the perimeter area of a concrete slab. Loose-fill insulation may be used to fill hollow voids in masonry, but this is less effective than covering the surface of the masonry wall.

Depth and rating (R-value) of insulation is dependent on construction technique, climate, and energy and construction costs. In an attic floor, insulation thickness can generally be increased without added cost (excluding cost of additional insulation). In a wall, however, the wall thickness limits maximum insulation thickness to the depth of the cavity. Additional insulation will require an increase in wall framing thickness and thus wall framing costs.

For a specific set of energy and insulation costs, it is no longer cost-effective to insulate beyond a specific insulation thickness. Simply, energy cost savings from added insulation becomes so small that it is no longer cost-effective to invest in the additional insulation. Table 3.5 shows a comparison of heat loss and heating cost per square foot of construction assembly for a typical heating season in Denver. The approximate heating cost per square foot of construction assembly with an R-value of 1.0 hr · ft^2 · °F/Btu for a typical Denver heating season are $1.00 for natural gas heating, $4.00 for electric resistance heating, and $2.00 for heating with a geothermal heat pump. Costs vary with type of heating source. This means that one square

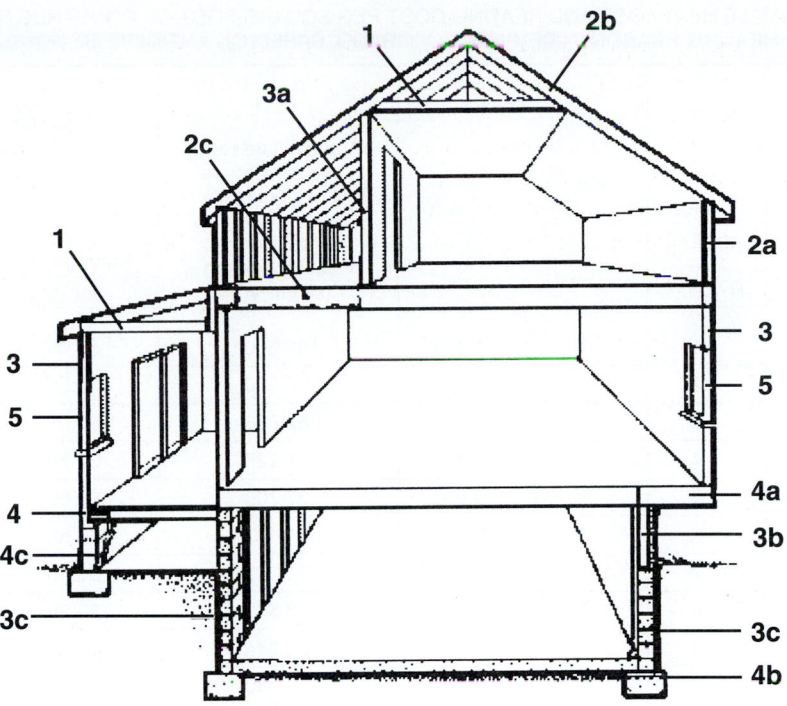

1. Unfinished attic spaces, insulate between the ceiling joists to seal off living spaces below.*

2. In finished attic rooms with or without dormers, insulate:
   a Between the studs of "knee" walls
   b. Between the studs and rafters of exterior walls
   c. Ceiling with cold spaces above

3. All exterior walls, including:
   a. Wall between living spaces and unheated garages or storage areas
   b. Foundation walls above ground level
   c. Foundation walls in heated basements (foundation can be insulated or inside or outside of wall

4. Floors above cold spaces, such as vented crawl spaces and unheated garages, and:
   a. Any portion of the floor in a room that is cantilevered beyond the exterior wall below
   b. Slab floors built directly on the ground**
   c. Foundation walls of crawl spaces and perimeter plates
   d. Storm windows as recommended

*Well-insulated attics, crawl spaces, storage areas, and other closed cavities should be adequately ventilated to prevent excessive moisture buildup.

**Slab on grade is almost always insulated, in accordance with building codes, when the house is constructed.

FIGURE 3.1 Examples of where to insulate a residence (U.S. Department of Energy).

TABLE 3.5   APPROXIMATE HEAT LOSS AND HEATING COST PER SQUARE FOOT OF CONSTRUCTION ASSEMBLY FOR A TYPICAL HEATING SEASON IN DENVER, COLORADO, BASED ON ANTICIPATED (2010) COSTS.

Construction Assembly	R-Value	U = 1/R		Approximate Heat Loss per Square Foot of Construction Assembly for a Typical Denver Heating Season (Btu/ft²-yr)[a]	Approximate Heating Cost per Square Foot of Construction Assembly for a Typical Denver Heating Season		
						Electricity	
					Natural Gas[b] $/ft²-yr	Resistance Heating[c] $/ft²-yr	Geothermal Heat Pump[d] $/ft²-yr
Single glass (R-0.9)[a]	1	1/1	1.000	188 500	$3.00	$4.00	$2.00
Double glass (R-2.0)[a]	2	1/2	0.500	94 250	$1.50	$2.00	$1.00
Triple glass (R-3.2)[a]	3	1/3	0.333	62 833	$1.00	$1.33	$0.67
Uninsulated wall (R-4)	4	1/4	0.250	47 125	$0.75	$1.00	$0.50
	5	1/5	0.200	37 700	$0.60	$0.80	$0.40
	6	1/6	0.167	31 417	$0.50	$0.67	$0.33
	7	1/7	0.143	26 929	$0.43	$0.57	$0.29
	8	1/8	0.125	23 563	$0.38	$0.50	$0.25
	9	1/9	0.111	20 944	$0.33	$0.44	$0.22
Insulated wall (R-13)	10	1/10	0.100	18 850	$0.30	$0.40	$0.20
	15	1/15	0.067	12 567	$0.20	$0.27	$0.13
Well-insulated wall (R-20)	20	1/20	0.050	9425	$0.15	$0.20	$0.10
	30	1/30	0.033	6283	$0.10	$0.13	$0.07
insulated ceiling (R-40)	40	1/40	0.025	4713	$0.08	$0.10	$0.05
	50	1/50	0.020	3770	$0.06	$0.08	$0.04
Superinsulated ceiling (R-60)	60	1/60	0.017	3142	$0.05	$0.07	$0.03
	70	1/70	0.014	2693	$0.04	$0.06	$0.03
	80	1/80	0.013	2356	$0.04	$0.05	$0.03
	90	1/90	0.011	2094	$0.03	$0.04	$0.02
	100	1/100	0.010	1885	$0.03	$0.04	$0.02

[a] Excludes solar gain.
[b] Based on $1.32/CCF (hundred cubic feet), 83 000 Btu/CCF, 80% furnace efficiency, normal heating season in Denver, CO.
[c] Based on $0.106/kWh, 3413 Btu/kWh, 98% heater efficiency, normal heating season in Denver, CO.
[d] Based on $0.106/kWh, 3413 Btu/kWh, COP (coefficient of performance) = 2.0 (conservative average), normal heating season in Denver, CO.

foot of an R-1 assembly in a Denver building will contribute this amount to the annual heating costs. In reviewing this table, it is evident that an assembly rated at R-2 results in half the cost; an assembly rated at R-3, a third of the cost; and an assembly rated at R-10, one-tenth the cost. Increasing insulation decreases heating load and cooling load costs; however, this decrease is not linear.

Table 3.5 can be used to analyze the cost-effectiveness of insulation levels. For example, natural gas cost per square foot of construction assembly for a typical Denver heating season is $0.15/ft²-yr for an assembly rated at R-20 and $0.08/ft²-yr for an assembly rated at R-40. An additional 6 in of insulation (from R-20 to R-40) in an attic space would result in cost avoidance (savings) of $0.07/ft² per typical heating

season with constant natural gas costs. If the home were heated with electricity, a $0.10/ft² savings would result: $0.20/ft² − $0.10/ft² = $0.10/ft². Depending on the costs of insulation and heating fuel used, it may be cost-effective to invest in the additional 6 in of attic insulation needed to go from R-20 to R-40. In the case of natural gas heating, adding an additional R-20 insulation (from R-40 to R-60) yields a savings of only $0.03/ft²-yr: $0.08/ft² − $0.05/ft² = $0.03/ft². This is less than half the savings achieved when going from R-20 to R-40. This means that the first inch of insulation saves more than the second inch, which saves more than the third inch, and so on. At some point it is no longer cost-effective to add insulation; this point is referred to as the *point of diminishing return*. It no longer makes economic sense when more is spent on the insulation than the savings it generates over its life.

Tables 3.6 and 3.7 provide recommended total R-values for new house construction in the United States and Canada. Table 3.8 indicates the thickness of insulation required to obtain commonly used R-values. R-values typically refer to the nominal cavity insulation R-value. Actually, thermal performance will yield lower R-values once the framing members are taken into account. Table 3.9 indicates the thickness of insulation required to obtain commonly used R-values.

Superinsulation

Superinsulation in buildings is the use of vast amounts of insulation, coupled with airtight construction, and controlled ventilation without sacrificing comfort, health, or aesthetics. The extra insulation in the building envelope is well beyond what is considered the current standard for insulation levels. Superinsulation is generally used in cold and moderate climates, but it is finding its way into warm climate design.

Superinsulation in cold climates is associated with design features with *average* U-values less than 0.033 Btu/ft^2-°F (0.2 W/m^2-°C) for chief opaque (nontransparent) structural assemblies. Superinsulation typically results in R-24 (RSI-4.3) or more insulation in exterior walls and over R-60 (RSI-10.6) insulation in the roof. As a result of these elevated insulation values, typical thicknesses of insulation materials are likely to be at least 8 in (200 mm) in walls and 20 in (500 mm) or more in roofs. Limits on insulation thicknesses used are often defined by the allowable cavity widths in cavity wall and attic construction—that is, of how much insulation fully fills the cavity.

Thermal Storage

Walls and floors with high heat capacities can store more energy, will have a larger thermal lag, and thus will generally be more effective for thermal storage and peak load shifting. The thermal diffusivity of wood is about 75 times less than steel, so steel conducts energy through it about 75 times faster than wood. Materials with low thermal diffusivities, such as concrete and masonry, have a slow rate of heat transfer relative to the amount of heat storage. These materials are effective for thermal storage and peak load shifting of heating and cooling loads.

In buildings, thermal diffusivity corresponds to the interaction between the effectiveness of thermal insulation (R-value) and the heat capacity of the materials. Both factors should be large to obtain the maximum thermal lag. Insulation is most effective for this purpose when it is installed on the exterior of concrete masonry walls. This keeps the mass in direct contact

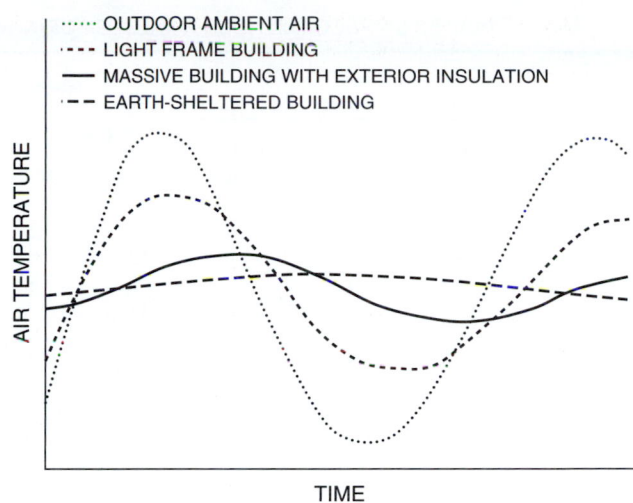

FIGURE 3.2 Daily interior temperature swings vary with thermal mass of the structure.

with the interior air for maximum heat transfer efficiency. Exterior insulation minimizes the effect of outdoor temperature swings on the temperature of the mass. For example, rigid board insulation is frequently used on the exterior of concrete masonry walls, and then covered with a mesh or fabric and a weather-resistant finish. Interior finishes on exterior walls should be minimal to prevent insulating the mass from the interior air. See Figure 3.2.

Mass and Effective Thermal Resistance

In certain climates, construction of a heavyweight building envelope (e.g., concrete, earth, and insulating concrete forms) can be an effective way of reducing building heating and cooling loads. Several comparative studies have shown that heating and cooling energy demands in buildings containing massive walls can be lower than those in similar buildings constructed using lightweight wall technologies. This better performance results because the thermal mass integrated in the structure reduces temperature swings and absorbs energy excess both from solar gains and from heat produced by internal energy sources such as lighting, computers, and appliances.

Massive walls delay and moderate thermal fluctuations caused by daily exterior temperature swings. The steady state R-value introduced in Chapter 2 and traditionally used to measure the thermal performance of a wall does not accurately reflect the dynamic thermal performance of heavyweight building envelope systems. For example, manufacturers of insulating concrete form (ICF) walls suggest the effective R-value of an 8 in (200 mm) ICF wall is between 26 and 30 hr · °F · ft^2/Btu (148 and 170°C · m^2/W).

TABLE 3.6   RECOMMENDED R-VALUES FOR NEW HOUSE CONSTRUCTION IN SIX INSULATION ZONES.

Zone	Heating Source				Recommended Total R-Values for New House Construction (hr · ft^2 · °F/Btu)							
	Gas	Heat Pump	Fuel Oil	Electric Furnace	Ceiling		Wall[a]	Floor	Crawl Space[b]	Slab Edge	Basement	
					Attic	Cathedral					Interior	Exterior
1	x	x	x		49	38	18	25	19	8	11	10
1				x	49	60	28	25	19	8	19	15
2	x	x	x		49	38	18	25	19	8	11	10
2				x	49	38	22	25	19	8	19	15
3	x	x	x	x	49	38	18	25	19	8	11	10
4	x	x	x		38	38	13	13	19	4	11	4
4				x	49	38	18	25	19	8	11	10
5	x				38	30	13	11	13	4	11	4
5		x	x		38	38	13	13	19	4	11	4
5				x	49	38	18	25	19	8	11	10
6	x				22	22	11	11	11	c	11	4
6		x	x		38	30	13	11	13	4	11	4
6				x	49	38	18	25	19	8	11	10

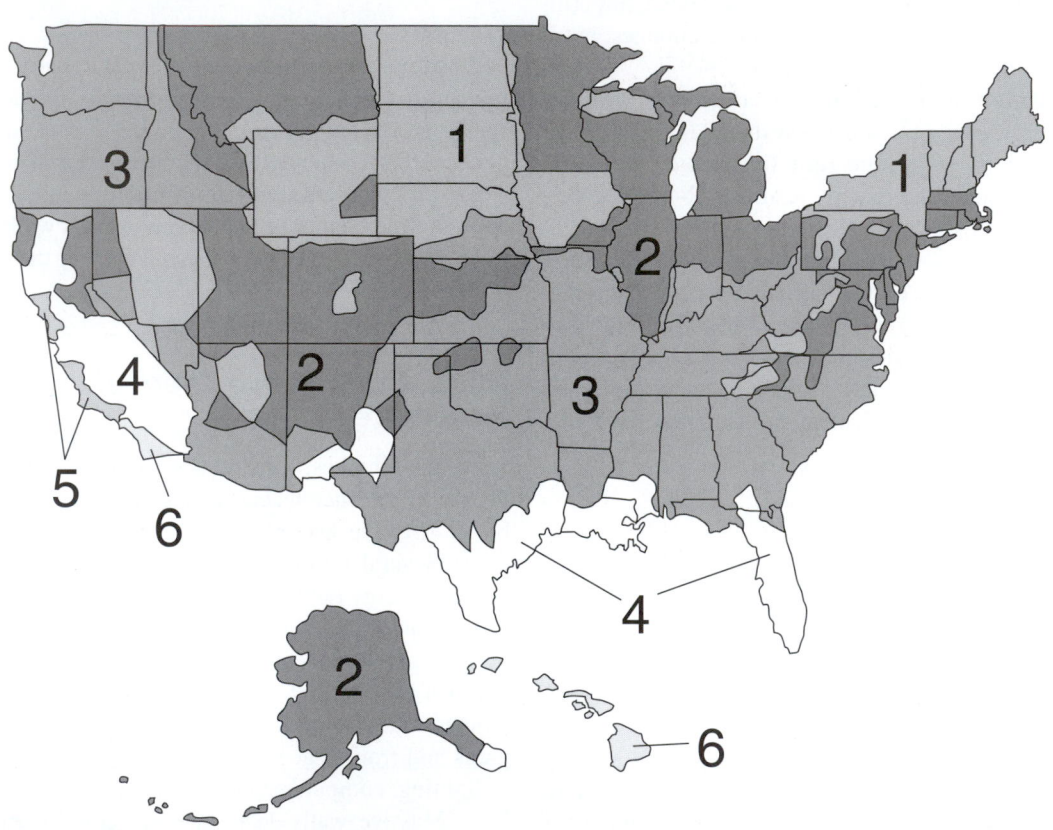

[a] R-18, R-22, and R-28 exterior wall systems can be achieved by either cavity insulation or cavity insulation with insulating sheathing. For 2 × 4 in walls, use either 3½ in thick R-15 or 3½ in thick R-13 fiberglass insulation with insulating sheathing. For 2 × 6 in walls, use either 5½ in thick R-21 or 6¼ in thick R-19 fiberglass insulation.

[b] Insulate crawl space walls only if the crawl space is dry all year, the floor above is not insulated, and all ventilation to the crawl space is blocked. A vapor retarder (e.g., 4- or 6-mil polyethylene film) should be installed on the ground to reduce moisture migration into the crawl space.

[c] No slab edge insulation is recommended.

TABLE 3.7 RECOMMENDED R-VALUES FOR NEW HOUSE CONSTRUCTION IN CANADA BASED ON HEATING DEGREE-DAYS (°C-DAYS).[a]

Zone	Zone A		Zone B		Zone C	
Heating Degree Days	Up to 5000°C-days		5000 to 6500°C-days		Over 6500°C-days	
Region	Vancouver Island Lower mainland South interior North coast		North interior Center interior		Far north	
City	Argentia, Nfld Charlottetown, PEI Corner Brook, Nfld Fredericton, NB Halifax, NS Kamloops, BC Kitchener, Ont Lethbridge, Alb London, Ont Moncton, NB Montreal, Que Ottawa, Ont Prince Rupert, BC Quebec, Que St. John, NB St. John's, Nfld Summerside, PEI Sydney, NS Toronto, Ont Vancouver, BC Victoria, BC Yarmouth, NS		Arvida, Que Brandon, Man Calgary, Alb Cochrane, Ont Edmonton, Alb Fort William, Ont Gander, Nfld Kapuskasing, Ont North Bay, Ont Prince Albert, Sask Prince George, BC Regina, Sask Saskatoon, Sask Winnipeg, Man		Aklavik, NWT Churchill, Man Dawson, Sask Fort Norman, NWT Goose, Nfld Mayo Landing, Sask Resolution Island, NWT The Pas, Man	
R-Value	hr · ft² · °F/Btu	m² · °C/W	hr · ft² · °F/Btu	m² · °C/W	hr · ft² · °F/Btu	m² · °C/W
Attic/ceiling	40	7.0	48	8.5	56	9.9
Basement walls	20	3.5	20	3.5	20	3.5
Cathedral ceiling and solid roof deck	28	4.9	40	7.0	40	7.0
Exterior walls	20	3.5	24	4.2	26	4.6
Floor (over unheated spaces)	28	4.9	40	7.0	40	7.0

[a] Abbreviations: Nfld = Newfoundland; PEI = Prince Edward Island; NB = New Brunswick; NS = Nova Scotia; BC = British Columbia; Ont = Ontario; Alb = Alberta; Man = Manitoba; Sask = Saskatchewan; NWT = Northwest Territories.

TABLE 3.8 THICKNESS OF INSULATION REQUIRED TO OBTAIN COMMONLY USED R-VALUES.

R-Value		Loose-Fill Insulation						Blanket and Batt Insulation	
		Fiberglass		Rock Wool		Cellulose			
hr · ft² · °F/Btu	m² · °C/W	in	mm	in	mm	in	mm	in	mm
11	1.94	4 to 5¼	102 to 133	3½	89	3¾	95	3 to 4	76 to 102
13	2.29	5¼ to 6	133 to 152	6¼	159	4½	114	3½	89
19	3.35	7 to 8¾	178 to 222	6¼	159	6½	165	5½ to 6½	140 to 165
30	5.28	11 to 14	279 to 356	9¾	248	10½	267	8¼ to 10	210 to 254
38	6.69	14 to 17¾	356 to 451	12¼	311	13	330	10 to 13	254 to 330
49	8.63	18 to 23	457 to 584	16	406	17	432	13 to 17	330 to 483

TABLE 3.9   RANGE OF DIFFERENCES IN R-VALUE ESTIMATIONS FOR CENTER-OF-WALL AND WHOLE-WALL R-VALUES.

Wall Framing Material	Ratio Between Thermal Conductivities of Wall Framing and Cavity Insulation Materials	Difference in R-Value Estimations
Wood framing	3	1.4 to 1.8%
Concrete framing	40	17.9 to 27.5%
Steel framing	1332	28.0 to 44.4%

Courtesy of Jan Kosny and Jeffrey E. Christian, Oak Ridge National Laboratory.

3.4   WINDOWS

Windows can have a significant effect on performance of the building envelope (e.g., heat transfer, natural ventilation, passive solar heating and cooling), and, as a result, are addressed here. Windows are constructed of the glazing (glass) and frame. *Glazings* are translucent or transparent materials (i.e., glass and some plastics that allow light to pass through the building envelope). The type of glazing used in a window or skylight can have a dramatic effect on energy performance. Window frames, sash, and mullions are available in a variety of materials including aluminum, wood, vinyl, and fiberglass. Frames may be composed of one material or may be a combination of different materials (i.e., wood clad with vinyl or aluminum-clad wood). Window frame and sash assemblies comprise only 10 to 25% of the window area, but can account for up to half of the window heat loss and can be the principal location for the formation of condensation. The thermal weak point is the edge of the glass and the window frame. To improve performance, manufacturers use thermal breaks in metal frames, increase the use of wood and clad wood sash and frames, and increase the use of frame materials with lower thermal conductivity, such as vinyl.

Performance Measures

The *National Fenestration Rating Council (NFRC),* a non-profit, public/private organization created by the window, door, and skylight industry, provides consistent ratings on window, door, and skylight products. Builders and consumers use the *NFRC Energy Performance Label* to reliably compare one product with another, and make informed decisions about the windows, doors, and skylights they buy. An example of this label is shown in Photo 3.13. The energy performance label lists the manufacturer, describes the product, provides a source for additional information, and includes ratings for one or more energy performance characteristics. NFRC rates all products in two standard residential and commercial sizes (i.e., "Res" and "Non-Res"). A description of the principal measures follows.

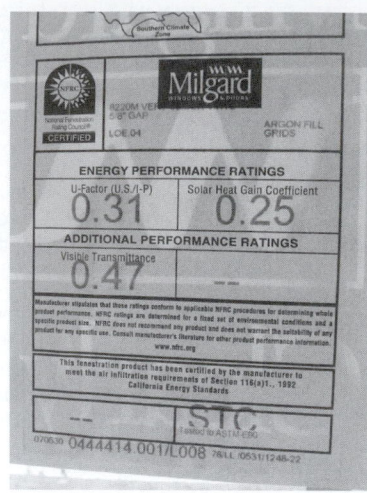

PHOTO 3.13   The NFRC Energy Performance Label that includes ratings of energy performance characteristics on a window. (Used with permission of ABC)

Overall Coefficient of Heat Transfer (U) and Thermal Resistance (R)

The *overall coefficient of heat transfer (U)* is a measure of how easily heat travels through an assembly of materials: the lower the U-factor, the lower the rate of heat transfer through the glazing and the more efficient the glazing. The U-factor has units of Btu/hr · ft^2 · °F (W/m · °C). Thermal insulating ability is also measured by the *thermal resistance (R):* a higher R-value indicates a better insulating performance. The R-value has units of hr · ft^2 · °F/Btu (m · °C/W). The R-value is the inverse of the U-factor (e.g., R = 1/U and U = 1/R; a U-factor of 0.5 Btu/hr · ft^2 · °F is the same as an R-value of 2.0 hr · ft^2 · °F/Btu). These values are discussed in Chapter 2.

The *overall* or *whole-window* U-factor of a window or skylight depends on the type of glazing, frame materials and size, glazing coatings, and type of gas between the panes. The overall U-factor should be used because energy-efficient glazings can be compromised with poor frame designs. R-values for common glazing materials range from 0.9 to 3.0 (U-factors from 1.1 to 0.3), but some highly energy-efficient exceptions also exist. Experimental super window glazings have a center-of-glass R-value of 8 to 10 (U-factor of about 0.1), but have an overall window R-value of only about 4 to 5 (U-factor of about 0.25 to 0.2), because of edge and frame losses.

Solar Heat Gain Coefficient

The *solar heat gain coefficient (SHGC)* is the fraction of solar heat that is transmitted through the glazing and ultimately becomes heat. This includes both directly transmitted and absorbed solar radiation. The lower the SHGC, the less solar heat is transmitted through the glazing, and the greater its shading ability.

Solar transmission through windows and skylights can provide free heating during the heating season, but it can cause an external-load-dominated building to overheat during the cooling season and an internal-load-dominated building to require cooling during much of the year. Solar-induced cooling needs are generally greater than heating benefits in most regions of the United States. In some climates, solar transmission through windows and skylights may account for 30% or more of the cooling requirements in a residence.

In general, south-facing windows in buildings designed for passive solar heating should have windows with a high SHGC to allow in beneficial solar heat gain in the winter. East- and west-facing windows that receive undesirable direct sunlight in mornings and afternoons should have lower SHGC assemblies.

Visible Transmittance

Visible transmittance (VT) is the percentage of visible light (light in the 380–720 nm range) that is transmitted through the glazing. When daylight in a space is desirable, glazing is a logical choice. However, low VT glazing such as bronze, gray, or reflective-film windows are more logical for office buildings or where reducing interior glare is desirable. A typical clear, single-pane window has a VT of about 0.88, meaning it transmits 88% of the visible light.

Types of Window Glazings

Common and innovative types of glazings available for use in windows and skylights include:

Single-Pane Glazings

Single-pane glazings are a single layer (pane) of glazing. They transmit about 88% of the solar light that strikes them. In the cooling season they are a significant source of heat gain and are also often a source of glare. Single-pane glazings should only be used where space heating and cooling are not needed in the building, such as a warehouse in a mild climate.

Insulated Glazings

Insulated glazings trap air or an inert gas such as argon or krypton between two or more panes (layers) of glass, creating a glazing product with improved resistance to heat transmission. Some typical U-factor ranges for common glass assemblies are:

Double glazed:

$0.43–0.57$ Btu/hr $\cdot$ ft^2 $\cdot$ °F (2.44–3.24 W/m $\cdot$ °C)

Triple glazed:

$0.15–0.33$ Btu/hr $\cdot$ ft^2 $\cdot$ °F (0.85–1.87 W/m $\cdot$ °C)

Historically, insulating windows were produced with a space between two glass layers (panes) that was filled with dehydrated (dry) air or flushed with dry nitrogen immediately before sealing the glass layers. The insulating effect of the confined air space between the two panes of glazing made a significant improvement in heat transmission. The optimal spacing for an air-filled glazing unit is between ½ and ¾ in (12 to 19 mm). At this spacing, radiant and natural convection heat transfer between the glazing surfaces is the best possible. Additional confined air spaces created by adding panes improve this insulating effect.

Convective air currents develop very easily with air and these currents carry heat to the top of the confined air space and settle into cold pools at the bottom of the space. An improvement to the thermal performance of insulating glazing units can be made by use of a less conductive, more viscous slow-moving gas that minimizes the convection currents within the confined air space. This "heavier" gas reduces conduction so that the overall transfer of heat is reduced. Manufacturers typically use argon and krypton gas fills to improve thermal performance. Argon is inexpensive, nontoxic, inert, clear, and odorless. The optimal spacing for an argon-filled unit is about ½ in (11–13 mm). Krypton offers better thermal performance but is more expensive to produce. Krypton is particularly useful when the confined air space must be thinner than normally desired (e.g., ¼ in, or 6 mm). The optimum gap width for krypton is ⅜ in (9.5 mm). A mixture of krypton and argon gases can also be used as a compromise between thermal performance and cost.

Spacers, sometimes called *edge-spacers*, are used to separate and hold apart multiple panes of glass within insulating windows. Because of its excellent structural properties, window manufacturers began using aluminum spacers. However, aluminum is an excellent conductor of heat and the aluminum spacer used in most standard edge systems represented a significant thermal bridge at the edge of the insulating glass unit. During cold weather, the thermal resistance around the edge of a window is lower than that in the center of the glass; thus, heat can escape, and condensation can occur along the edges. To address the edge problem, window manufacturers have developed a series of innovative edge systems. *Warm-edge spacers* either incorporate a thermal break in the spacer assembly or are constructed from a low-conductivity material. If warm-edge technology is not used with low-e windows (see below), much of the benefit of these technologies is lost because heat is conducted along the glass and across the edge spacer.

Tinted (Heat-Absorbing) Glazings

Tinted glazings, also known as *heat-absorbing glass,* block solar transmission by absorbing heat in the glass itself. Unfortunately, this causes the glass temperature to rise, increasing the radiation coming off the glazing into the indoor space. Thus, tinting by itself yields only a modest shading coefficient, with SHGCs in the range of 0.5 to 0.8. With glazings having SHGCs below about 0.5, indoor vegetation grows more slowly or could die from the reduced sunlight. The most common colors for tinted glass (e.g., bronze and gray) block visible light and solar heat in equal proportions, that is, they are not spectrally selective. Green- and blue-tinted glazings are more spectrally selective; they offer greater transmission of visible light and slightly reduced heat transfer compared with other colors of tinted glass. Black-tinted (gray) glass should be avoided because it absorbs much more visible energy than near-infrared radiation.

Reflective Glazings

Reflective glazings have a semitransparent metallic coating applied to the surface(s) of clear or tinted glass. They have better shading coefficients because they reflect rather than absorb infrared energy. However, most reflective glazings block daylight (visible light) more than solar heat. Reflective glass has worked well in hot climate applications, where a high level of solar control is necessary. However, reflective glass reduces cooling loads at the expense of daylight transmittance, so the reduction is offset somewhat by the heat created by the additional electric lighting required. Reflective coatings are available for single-pane applications, while some coatings must be sealed inside multiple-glass units.

Low-e Coatings

Low-emittance (low-e) coatings are microscopically thin, virtually invisible, metal or metallic oxide layers deposited directly on one or more surfaces of glass or on plastic films between two or more glazings to suppress radiative heat flow and reduce the U-factor. *Emittance (e)* is the ratio of emitted radiation to that of a black body at the same temperature. It is a measure of the amount of radiant heat transfer between two surfaces with a given roughness at a different temperature. A low-e coating rated at 0.20 reduces infrared radiation heat transfer by 5 times. It is based upon the emissivity and physical configuration of the surfaces. (See Chapter 2.)

The principal mechanism of heat transfer in an argon or krypton gas-filled insulating glass unit is thermal radiation from a warm pane of glass to a cooler pane. Coating a glass surface with a low-e material and facing that coating into the confined space between the glass panes impedes radiant heat transfer significantly, thus lowering the total heat flow through the window. When applied inside a multiple-pane glazing, a low-e coating will reflect heat back into the habitable space during the heating season. This same coating will reflect heat back outdoors during the cooling season, slightly reducing heat gain.

Low-e coatings are transparent to visible light but not to infrared radiation. Some low-e glazings reduce solar gain, which can be undesirable in winter months. Different types of low-e coatings have been designed to allow for high solar gain, moderate solar gain, or low solar gain.

Spectrally Selective Glazings

Spectrally selective glazings have optical coatings that filter out much (about 40 to 70%) of the invisible solar heat gain that is normally transmitted through clear glass, yet allow the full amount of daylight to be transmitted. These coatings can be combined with tinted glazings to produce special customized selective coatings that can maximize or minimize solar gain to achieve the desired aesthetic effects and still balance seasonal heating and cooling requirements.

Suspended coated film (SCF) is a low-e system, which insulates better than typical low-e. Using the sputtering process, a wavelength-selective low-e coating is applied to thin plastic film. The film is then suspended between two plates of glass, which creates a unit that, as far as convection and conduction are concerned, is essentially a triple-pane unit. Additionally, SCF provides a wavelength-selective coating on the suspended film. The coating blocks 99% of the ultraviolet wavelengths and varying amounts of infrared wavelengths, without blocking significant portions of visible light. The result is a glazing system with a low shading coefficient and high visible light transmission.

Super Glazings

Super glazings are innovations in glazing technology. They can achieve high thermal resistance (an R-value as high as 10) by blending multiple low-e coatings, low-conductance gas fills, barriers between panes that reduce convective circulation of the gas fill, and insulating frames and edge-spacers.

Under development are *chromogenic* (optical switching) glazings that will adapt to the frequent changes in the lighting and heating or cooling requirements of buildings. Passive glazings will be capable of varying their light transmission characteristics according to changes in sunlight (*photochromic*) and their heat transmittance characteristics according to ambient temperature swings (*thermochromic*). Active (*electrochromic*) glazings use a small electric current to alter their transmission properties.

Aerogels

An *aerogel* is a low-density, highly porous material similar to a translucent, super lightweight foam that is formed by extracting liquid from a gel-like substance. It contains extremely small air spaces or pores that are only a few hundred times larger than atoms. Aerogels are the lightest existing solid material, having a density of only about three times that of air. They are so lightweight and translucent that they are often referred to a "solid smoke." Aerogels have extraordinary thermal insulating properties and have near glass-like optical characteristics.

Silica aerogels are produced of pure silicon dioxide and sand, much like glass. Some silica aerogels are over 99.5% air, making them 1000 times less dense than glass. The best aerogel can provide 39 times more insulating ability than the best fiberglass insulation. One manufacturer (Aspen) produces silica aerogels that have thermal insulation values between R-13 and 16 hr $\cdot$ ft^2 $\cdot$ °F/Btu per inch (thermal conductivity values as low as 0.011W/m $\cdot$ °C) at 100°F (38°C) and standard atmospheric pressure.

Aerogels have pore sizes ranging from 1 to 100 nm, and typically average between 3 and 5 nm. This is below the wavelength size of visible light. The pores act as particles that scatter white light and make the aerogel appear blue. Commercially available aerogels are slightly less transparent than window glass (about 90%) and have a slight bluish tinge.

Aerogels can be easily molded into different shapes (e.g., cylinders, cubes, plates of varying thickness), making them suitable for use as building materials. They have extraordinary thermal insulation values that are much higher than conventional thermal insulations used in buildings and appliances. One distinctive potential application is as a composite with glass that could be used as a high-performance glazing product in windows. Present commercially available aerogels, with their slight hazy appearance and bluish tint, would be deemed unsatisfactory for most building occupants. Much research is being done in this area.

Whole-Window Performance

Different combinations of window or skylight frame style, frame material, and glazing change the actual thermal performance of a window unit. Commercially available insulating glazing systems can achieve a center-of-glass R-value of 10 (U-factor of 0.1) by sandwiching translucent glass fiber or foam insulation between plastic or composite glazings. There is a compromise between heat transmission and light transmission. Window frames and spacers account for a large fraction of the heat transfer in advanced glazings. The result is a reduction in overall thermal performance for the window as a whole. For example, a wood-frame window with a center-of-glass thermal resistance of 8 hr · ft^2 · °F/Btu has a whole-window thermal resistance of about 4.5 hr · ft^2 · °F/Btu from the frame.

Air infiltration affects performance of the entire window unit. Manufacturers take steps to reduce air infiltration through the operable window sash by adding advanced weatherstripping that limits air leakage between the window sash and frame. During construction, flashing is installed around a window or door unit to reduce air and moisture infiltration. See Photo 3.14.

PHOTO 3.14 Self-adhered flashing installed to reduce air and moisture infiltration around window and door. (Used with permission of ABC)

3.5 THERMAL TESTING

Thermographic Scanning

Thermography or *infrared scanning* uses specially designed infrared video or still cameras to produce images called thermograms, which show surface temperature variations. Thermographic equipment can be used as a tool to help detect heat losses and air leakage in building envelopes. Infrared scanning allows energy auditors to check the effectiveness of insulation in a building's construction. The resulting thermograms help auditors determine whether a building needs insulation, and where in the building it should go. Because wet insulation conducts heat faster than dry insulation, thermographic scans of roofs can often detect roof leaks.

Several types of infrared sensing devices can be used on an on-site inspection. A *spot radiometer* (also called a *point radiometer*) measures radiation one spot at a time, with a simple meter reading showing the temperature of a given spot. The auditor pans the area with the device and notes the differences in temperature. A *thermal line scanner* shows radiant temperature viewed along a line. A *thermogram* shows the line scan superimposed over a picture of the panned area. This process shows temperature variations along the line.

Thermographic scans can be done inside or outside a structure. Exterior scans, while more convenient for the homeowner, have a number of drawbacks. Warm air escaping from a building does not always move through the walls in a straight line. Heat loss detected in one area of an outside wall might originate at some other hard-to-find location inside the wall. Air movement also affects the thermal image. On windy days, it is harder to detect temperature differences on the outside surface of the building. The reduced air movement and ease of locating air leaks often make interior thermographic scans more effective.

The most accurate thermographic images usually occur when there is a large temperature difference [at least 20°F (14°C)] between inside and outside air temperatures. In northern states, thermographic scans are generally done in the winter. In southern states, however, scans are usually conducted during warm weather with the air conditioner running.

Blower Door Testing

Infiltration is airflow through unintentional openings in the building envelope that is caused by wind and stack pressure differences between indoor and outdoor air. Energy is required to heat or cool unconditioned air that has leaked into the structure. The rate of infiltration in a building depends on weather conditions, equipment operation, and occupant activities. The characteristics of infiltration airflow may be determined by measuring the air leakage of the building envelope, which describes the relative tightness.

A *blower door* is a large assembly with a powerful fan that is placed in an exterior door of an existing building to measure infiltration and determine the location of air leaks in the building envelope. It creates a strong draft that depressurizes the building interior by drawing indoor air out of the building. Energy contractors use a blower door to determine how much air leaks through the building envelope. In a residence, it takes about 15 min to set up the blower door, and about 5 min to run a simple blower door test.

The typical test pressure is about 0.20 in of water (50 Pa), about equivalent to a 20 mile/hr (32 km/hr) wind. Internal and external temperatures and the external barometric pressure are measured during the test to make corrections for changes in the air volume flow rate. All exterior doors and windows must be closed and ventilation openings sealed during the test to provide a realistic measurement of the actual building envelope leakage. Once the building is depressurized, the contractor can identify leakage through the external envelope. Identification of air leakage paths is pinpointed by feeling with the hand or with a hand-held smoke generator.

In blower door testing, airflow measured in ft^3/min (or cfm) that creates a change in building pressure of 50 Pa (CFM50) is a commonly used estimate of building airtightness. This measure will vary by building size. A range for a small residence is 1200 CFM50 (tight) to 3000 CFM50 (very leaky). Another commonly used measure is *air changes per hour at 50 Pa* (ACH50), which is the number of complete air changes that will occur in one hour with a 50 Pa pressure being applied uniformly across the building envelope. Typical ACH50 rates are from 5 to 10 air changes per hour. The ACH50 rate method is useful in comparing leakage rate by size (volume) of building. The natural infiltration rate can be approximated by dividing the measured ACH50 by 20; for example, for a 10 ACH50: 10 ACH50/20 = 0.5 ACH natural infiltration rate. In another method, infiltration is expressed as the *effective leakage area (ELA),* the area of a special nozzle-shaped hole (in a blower door) that would leak the same amount of air as the building does at a pressure of 10 Pa. The LEED Green Building Rating System has set an airtightness standard for multifamily dwelling units of 1.25 square inches of leakage area per 100 ft^2 (0.868 cm^2/m^2) of enclosure area.

3.6 QUALITY OF INDOOR AIR

Indoor Air

It is estimated that most people spend as much as 90% of their time indoors (about 70% in our home). U.S. Environmental Protection Agency (EPA) studies of human exposure to air pollutants suggest that pollutants in indoor air may be 2 to 5 times greater, and occasionally more than 100 times higher, than outdoor levels. Indoor air quality affects both comfort and health. U.S. EPA is a federal agency created in 1970 to permit coordinated governmental action for protection of the environment by systematic abatement and control of pollution through integration of research, monitoring, standards setting, and enforcement activities.

According to the U.S. EPA, poor quality of indoor air is the third leading cause of death and claims an estimated 335 000 lives/year. Additionally, deaths attributed to poor indoor air quality have risen faster than most other major diseases in the last decade. Worldwide, according to the World Health Organization, 1 out of 3 workers are in a workplace that is making them sick. Thus, the quality of indoor air has become a primary concern in all types of buildings. As a result, in addition to tending to the thermal comfort of occupants in the building, the designer must address the quality of indoor air.

In normal breathing, a typical human breathes in about 0.25 ft^3 (7 L) per min; the equivalent of about 360 ft^3 (10 000 L) per day. One ft^3 (about 4 L) of indoor air contains about 400 million particles, which equates to a typical human inhaling about 40 billion particles a day. Some of these particles adversely affect comfort and health.

Indoor Air Quality

Indoor air quality (IAQ) refers to the physical, chemical, and biological characteristics of air in the interior spaces of a building or facility. The quality of indoor air can be influenced by many factors, even though a building does not have the industrial processes and operations found in factories and plants. A building heating, ventilating, and air conditioning system that is properly designed, installed, operated, and maintained can promote indoor air quality. When proper procedures are not followed, indoor air problems may result. Indoor air must satisfy three basic requirements, including having normal concentrations of respiratory gases, adequately diluted airborne contaminants, and be thermally satisfactory.

As a result of efforts to save energy and lower utility costs, buildings are being built tighter than ever, with thicker walls, more insulation, and better windows and doors. Making structures more energy efficient does, however, exact an unexpected price: Clean outdoor air stays out and indoor air pollutants such as smoke particles, dust, bacteria, fungus, mold spores, and chemical vapors become trapped inside. This leaves the potential for pollutants to build up to harmful levels.

Poor IAQ can lead to serious health, productivity, and financial consequences. Pollutants found in the indoor air have been shown to heighten many of the health conditions from which people suffer. People exposed to indoor air pollutants over a long period of time may become more vulnerable to outdoor air pollutants, especially groups that include the young, the elderly, and the chronically ill who suffer from respiratory and cardiovascular diseases. Additionally, research indicates that poor air quality results in several negative effects on the productivity of employees in office buildings.

TABLE 3.10   SUBSTANCES THAT POTENTIALLY CAUSE INDOOR AIR POLLUTION IN BUILDINGS. COMPILED FROM VARIOUS SOURCES.

Source	Substances
Cleaning and maintenance	Cleaning chemicals Pesticides
Equipment and work activities	Copy machines (ozone) Office supplies such as glues and correction fluids Printing machines Laboratory use of chemicals
Building materials and furnishings	Damaged asbestos insulation, fireproofing, and flooring lead from paint Formaldehyde from furniture, curtains, and carpeting
Human activities	Smoking Cosmetics, soaps, and lotions Preparing food
Outside pollution	Exhaust from vehicles Industrial pollution Dumpsters and other unsanitary debris near air intakes Leakage from underground fuel tanks and landfills

Indoor Air Contaminants

Several indoor air contaminants exist in buildings that fall into roughly three categories:

- biologically active organisms (i.e., bacteria, viruses, mold, and spores)
- gases and other odor-causing compounds (products of combustion, radon)
- fine particulate matter (i.e., asbestos, dust)

Substances that potentially cause indoor air pollution in buildings are listed in Table 3.10. Indoor air contaminants include the following:

Biological Contaminants

Biological contaminants, also referred to as *microbials,* are a living organism; was living; or was a product of something living. A *bioaerosol* is a biological contaminant that is airborne and causes indoor air problems. Some release spores into the air and present the biggest health concern. Examples of biological contaminants are the following:

Bacteria

Bacteria are simple, one-celled microscopic organisms of which less than 1% are infectious and cause disease in humans. When infectious, bacteria enter the body and cause illness by invading "good" cells, rapidly reproducing, and producing toxins, which are powerful chemicals that damage specific cells in the tissue. Bacteria can flourish in dust, HVAC

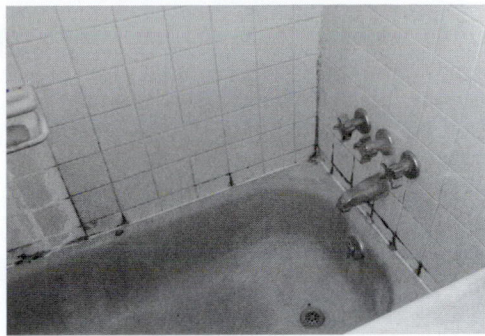

PHOTO 3.15   Evidence of mold growth in joints of bathtub surround tile. (Used with permission of ABC)

systems, swimming pools, physical therapy equipment, showers, whirlpool spas, and potting soil.

Viruses

Viruses are tiny microscopic capsules that contain genetic material (DNA or RNA). They can be inhaled, absorbed through the skin, or ingested and need a suitable host to reproduce.

Types of viruses include the common cold (rhinovirus + 200 more), influenza (flu), viral hepatitis, avian influenza (bird flu), mononucleosis (mono), Ebola, and human immunodeficiency virus (HIV). Viruses may be spread around an office via the mechanical ventilation system.

Molds

Molds are microscopic fungi. Fungi grow everywhere. To thrive, mold requires air, moisture (40 to 60% relative humidity), and food (paper, wood, drywall, insulation, and natural fibers). It manifests itself in a variety of appearances, such as black, grey-brown, grey-green, white with orange spots, or pink or purple splotches. All homes and buildings have molds, with about 160 species commonplace. Evidence of mold growth is shown in Photos 3.15 through 3.17.

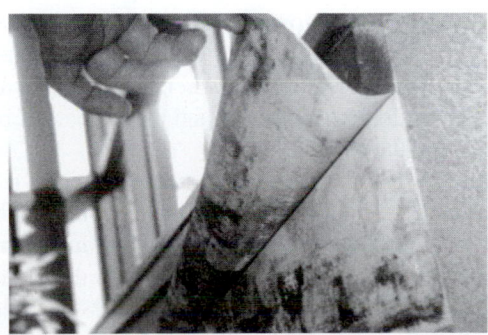

PHOTO 3.16   Evidence of mold growth under wallpaper; the cause was leakage in the return ducts. (Courtesy of DOE/NREL)

PHOTO 3.17 Evidence of mold growth caused by small leaks and inadequate airflow. (Courtesy of DOE/NREL)

PHOTO 3.18 Grains of pollen (magnified). (Courtesy of DOE/NREL)

Mold spores vary in shape and size (2 to 100 μm). They become airborne and may travel in several ways. They may be passively moved (by a breeze or water drop), mechanically disturbed (by a person or animal), or actively discharged by the mold (usually under moist conditions or high humidity). Spores can be inhaled, absorbed through the skin, or ingested with our food. People with allergies, immune suppression, or underlying lung disease are more sensitive and susceptible to infections. Toxic black mold and other fungi produce volatile organic compounds (VOCs) that irritate mucous membranes and the central nervous system.

Allergens

Many indoor sources produce allergenic particles including the feces of dust mites, which are minute creatures that feed on the dead skin that humans shed every day; the excrement of cockroaches; the dander and hair from animals (dogs, cats, rats, and mice); and pollens from flowering plants. Magnified grains of pollen are shown in Photo 3.18.

Biological contaminants can cause or exacerbate many undesirable health effects. Allergic reactions are the most common health problem. Symptoms often include watery eyes, runny nose and sneezing, nasal congestion, itching, coughing, wheezing and difficulty breathing, headache, dizziness, and fatigue. Respiratory disorders (e.g., runny nose, cough, nasal congestion, and aggravation of asthma), hypersensitivity diseases, and infectious diseases can also result. There have been many reports linking health effects in office workers to offices contaminated with moldy surfaces and in residents of buildings contaminated with fungal growth.

Many IAQ concerns tied to microbials are related to moisture problems such as water intrusion from plumbing and roof leaks, and floods, or condensation buildup from excessive humidity within a space. The need for energy conservation in buildings over the past several decades has introduced new construction materials and techniques. It has also introduced concerns with molds. The building does not breathe as easily, trapping moisture vapors inside the building envelope

cavities. Additionally, leaky roofs, windows, and plumbing from poor construction or lack of timely repairs result in the growth of mold and mildew and release of spores. The microbial spores become airborne, spreading into the air and settling inside wall cavities, behind cabinets and wallpaper, and throughout ductwork.

When moisture is present, explosive growth of microbials can occur in a surprisingly short period of time (e.g., 12 to 48 hr). Mold, mildew, and other fungi and bacteria need only oxygen (air), moisture, food, and an ideal temperature to thrive. Nourishment comes from organic building materials such as gypsum wallboard, ceiling tile, cardboard, wood, carpets, furniture, and clothing. Liquid water is not necessary because most species propagate well in conditions of 60 to 70% relative humidity. An ideal temperature is in the range of 68° to 86°F (20° to 30°C).

One of the first examples of a biological contaminant is *Legionnaires' disease*. This term was coined in 1976 after a respiratory disease affected many delegates attending a convention in Philadelphia held by the American Legion of Pennsylvania. Eventually, 182 people became ill and 32 died. Symptoms begin with a headache, pain in the muscles, and general flu-like symptoms. These symptoms are followed by high fever (some topping 107.4°F) and shaking chills. Nausea, vomiting, and diarrhea may occur. This can lead to dry coughing, breathing difficulties, and chest pain. Mental changes, such as confusion, disorientation, hallucination, and loss of memory, can occur to an extent that seems out of proportion to the seriousness of fever. Complete recovery can take several weeks. About 10% of known cases of Legionnaires' disease have been fatal. Legionella bacteria have been found in hot water tanks, hot water propelled from showerheads and faucets, and in whirlpool spas. The use of hot water with production of aerosols allows Legionella, if present in the water, to get into the lungs.

Carbon Dioxide

Carbon dioxide (CO$_2$) is a naturally occurring gas that is produced by combustion and is a by-product of the natural metabolism of living organisms. A small amount of CO_2 exists in indoor air at all times and is acceptable. At moderately high concentrations, however, CO_2 causes discomfort. It raises people's breathing rate and may cause minor eye irritation, particularly in people who wear contact lenses. Problems associated with high CO_2 levels are drowsiness, fatigue, and the sick building syndrome.

HVAC engineers often use indoor CO_2 concentrations to estimate air exchange with the outdoors. Controls used for this purpose mostly reflect comfort issues, not health concerns. CO_2 concentrations should be maintained at levels below 1000 ppm (parts per million).

Environmental Tobacco Smoke

Environmental tobacco smoke (ETS), commonly termed as "second hand smoking" or "passive smoking," is one of the major sources of indoor air pollution. Environmental tobacco smoke is a complex mixture of more than 4000 chemicals that exist in vapor and particle states. Many of these chemicals are known toxic or carcinogenic agents. Nonsmoker exposure to ETS-related toxicants would occur in indoor spaces where there is smoking. The U.S. EPA has classified ETS as a known human carcinogen and estimates that it is responsible for approximately 3000 lung cancer deaths every year in the United States. Children's lungs are even more susceptible to harmful effects caused by ETS. There is also strong evidence of increased middle ear effusion, reduced lung function, and reduced lung growth in children exposed to ETS. Although the trend in the United States is toward elimination of smoking, which reduces the concern, the rate of tobacco smoking remains high in many parts of the world. In Europe almost one-half of the men and one-quarter of the women smoke, and in the western Pacific area of Asia two in five men smoke.

Combustion Pollutants

Combustion pollutants from stoves, space heaters, furnaces, and fireplaces that may be present at harmful levels in the home or workplace stem chiefly from malfunctioning heating devices, or inappropriate, inefficient use of such devices. Another source can be motor vehicle emissions from, for example, proximity to a garage or a main road. A variety of particulates acting as irritants or, in some cases, carcinogens may also be released in the course of combustion. Possible sources of contaminants include gas ranges that malfunction or are used as heat sources, improperly ventilated fireplaces, furnaces, wood or coal stoves, gas water heaters and gas clothes dryers, and improperly used kerosene or gas space heaters. The gaseous pollutants from combustion sources within a building follow.

Carbon monoxide (CO) is an odorless, colorless gas that can cause asphyxiation. At high concentrations, it can kill people in minutes; at lower concentrations, it can worsen the symptoms of heart disease, or cause headache, dizzi-

TABLE 3.11 PHYSIOLOGICAL RESPONSE TO VARIOUS CONCENTRATIONS OF CARBON MONOXIDE (CO). EXTRACTED FROM GOVERNMENTAL SOURCES.

Physiological Response	Parts of CO per Million Parts of Air
Threshold limit value	50
Concentration that can be inhaled for 1 hr without appreciable effect	400 to 500
Concentration causing unpleasant symptoms after 1 hr of exposure	1000 to 2000
Dangerous concentration for exposure of 1 hr	1500 to 2000
Concentrations that are fatal in exposure of less than 1 hr	4000 and above

ness, nausea, fatigue, and vomiting. Low-level CO poisoning is often mistaken for the flu. Susceptible groups who are especially vulnerable to low-level CO effects include pregnant women and developing fetuses, and people with cardiovascular or cerebral vascular illnesses. The physiological reaction to various concentrations of carbon monoxide is summarized in Table 3.11.

Each year, over 200 people in the United States die from CO produced by fuel-burning appliances (furnaces, ranges, water heaters, room heaters). Others die from CO produced by cars left running in attached garages. Several thousand people go to hospital emergency rooms for treatment for CO poisoning.

Nitrogen dioxide (NO$_2$) is a corrosive oxidant gas that can injure lung tissues at high concentrations. At lower concentrations, there is evidence that NO_2 can hamper the body's immune defenses, creating increased susceptibility to respiratory infections. Researchers also report indications that the gas reduces lung function and increases the allergic response of some asthma sufferers.

Sulfur dioxide (SO$_2$) is a colorless gas that acts mainly as an irritant, affecting the mucosa of the eyes, nose, throat, and respiratory tract. Continued exposure to high SO_2 levels can contribute to the development of acute or chronic bronchitis.

Additional ventilation air is not the solution for eliminating high levels of combustion contaminants in a space. Instead, eliminating or repairing the source of combustion pollutants is the best approach. Measures to prevent high levels of combustion contaminants in a space include ensuring that fuel-fired appliances are installed according to manufacturer's instructions, and local building codes and make certain they are inspected and serviced regularly.

Ozone

Ozone (O$_3$) is a type of oxygen molecule that has three atoms per molecule instead of the usual two (O_2). It is produced from oxygen by UV radiation and is naturally present in air. It can be produced indoors by electrical discharges from electrical equipment such as photocopiers and electrostatic precipitators. Symptoms include irritation to mucous membranes in eyes, nose, and throat, headaches, dizziness, and severe fatigue. Long-term effects are lung damage at higher exposures, potential genetic damage, and premature death in people with heart/lung disease. It mimics effects of ionizing radiation (x-rays and gamma rays).

Volatile Organic Compounds (VOCs)

VOCs are emitted as gases from certain solids or liquids, including building materials, fabric furnishings, carpet, adhesives, fresh paint, new paneling, pesticides, solvents, and cleaning agents. VOCs are consistently found at higher levels indoors than outdoors. Products used in home, office, school, and arts/crafts and hobby activities emit a wide array of VOCs including scents, hair sprays, rug, oven cleaners, dry-cleaning fluids, home furnishings, and office material like copiers, certain printers, correction fluids, graphics, and craft materials.

Methylene chloride, which is found in some common household products like paint strippers, can be metabolized to form carbon monoxide in the blood. Formaldehyde, another industrial product, can off-gas from materials made with it, such as foam, insulation, and engineered wood products (e.g., plywood, OSB, and so on) and has been classified as a probable human carcinogen. Pesticides sold for household use are technically classified as semivolatile organic compounds. Symptoms like nose and throat irritation, headache, allergic skin reaction, and nausea may indicate the presence of these VOCs.

Airborne Lead

Airborne lead affects children in the form of cognitive and developmental deficits, which are often cumulative and subtle. Lead poisoning via ingestion has been the most widely publicized, stressing the roles played by nibbling of flaking paint by toddlers and by the use of lead containing foodware like glass and soldered metal-ceramic ware by adults. Among children, long-term lead poisoning is linked to irreversible learning disabilities, mental retardation, and delayed neurological and physical development. Lead toxicity may alternatively manifest itself as acute illness. Signs and symptoms in children may include irritability, abdominal pain, emesis, marked ataxia, and seizures or loss of consciousness.

The U.S. Department of Housing and Urban Development (HUD) reports that 65 million homes built before 1978, especially those built before 1950, contain some lead paint. HUD also estimates that over 1.7 million children nationwide have elevated blood lead levels.

During the middle of the 1900s, paint containing 30 to 40% lead pigment was used on interior and exterior surfaces of buildings because of its colorfastness and hiding power. Lead was also used in pipes and soldered pipe joints for carrying drinking water in older buildings. In 1955, the paint industry adopted a voluntary standard of no more than 1% lead by weight in interior paints. The U.S. Consumer Product Safety Commission lowered the maximum allowable lead content in paint to 0.5% in 1973, to 0.06% in 1977, and to zero in 1978. However, the presence of lead-containing soil, dust, and paint chips can continue to create potential health hazards.

Radon

Radon is a naturally occurring gas produced by radioactive decay of radium. According to the U.S. Surgeon General and EPA, when propagated in air and inhaled into the lungs, the low-level radiation emitted by the products of decay of radon (polonium) damages lung tissue and increases the risk of lung cancer over the long-term. The U.S. Surgeon General has warned that radon is the second leading cause of cancer. This adverse health effect is significantly compounded if you smoke. U.S. EPA literature advises that radon causes between 7000 and 30 000 deaths per year in the United States alone. The U.S. EPA reported risks of living with different levels of radon gas over a lifetime are summarized in Table 3.12.

TABLE 3.12   REPORTED RISKS OF LIVING WITH DIFFERENT LEVELS OF RADON GAS OVER A LIFETIME. FROM U.S. EPA LITERATURE.

Average Radon Level (pCi/L)	If You've Never Smoked Number of people out of 1000 that EPA suggests would get lung cancer if exposed to this level of radon gas over a lifetime	If You Smoke Number of people out of 1000 that EPA suggests would get lung cancer if exposed to this level of radon gas over a lifetime	EPA Recommendation If this radon level is found in the home
20	8	135	Fix the home
10	4	71	Fix the home
8	3	57	Fix the home
4	2	29	Fix the home
2	1	15	Consider fixing the home
1.3	Less than 1	9	Difficult to improve
0.4	Less than 1	3	Difficult to improve

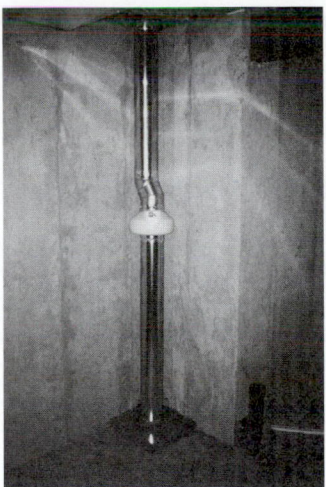

PHOTO 3.19 A subfloor depressurization system used to reduce radon levels by ventilating a crawl space below a structural basement floor. (Used with permission of ABC)

PHOTO 3.20 An inline blower draws soil gasses, including radon, and exhausts them to the outdoors before they enter the living space of the home. (Used with permission of ABC)

Because radon is a gas, it can move freely through soil pores and rock fractures and eventually into air. Movement into a building is driven by a difference in soil gas pressure and air pressure; the soil gas attempts to equalize the lower pressure of indoor air in the building. Radon easily migrates through foundation cracks, utility (pipe and wiring) entries, seams between foundation walls and basement floor slabs, sumps, drains, and uncovered soil in crawl spaces.

Radon levels in indoor air can vary considerably from floor to floor and from day to day. Variations in levels in a building are from fluctuations in the radon's ability to freely migrate through the soil and into the building. These fluctuations are caused by changes in occupancy patterns and weather conditions that influence a building's indoor air pressure. Variations in test measurements between neighboring buildings are generally because of differences in the physical features of the buildings and geological variations at the building sites. Differences in radon quantities from region to region are generally from differences in the abundance of radium found in the underlying soils and bedrock below the building.

Several methods of mitigation for high levels of radon are used. Most involve sealing and ventilating the building envelope. Shown in Photos 3.19 and 3.20 is a subfloor depressurization system that ventilates a crawl space below a structural basement floor. A blower built into the system draws in gases from the soil, including radon, and exhausts them to the outdoors before they enter the living space of the home.

Suspended Particulate Matter

Suspended particulate matter (SPM) includes airborne particles from smoke and dust. Indoor particulate pollution may come from sources within the building or infiltrate from out-

doors. In most buildings, particulate pollution is a major problem because most of the air-handling units use synthetic filters that only remove particles of sizes larger than 20 μ. The smaller sized particles enter these buildings freely, and are more dangerous because these are respirable.

Asbestos

Asbestos is the name of a group of naturally occurring materials that separate into strong, very fine fibers. Asbestos fiber masses tend to break easily into a dust composed of tiny particles about 1/1200 the size of a human hair. Magnified asbestos fibers are shown in Photo 3.21. Any material that has

PHOTO 3.21 Asbestos fibers (magnified). (Courtesy of DOE/NREL)

asbestos as an ingredient is known as an *asbestos-containing material (ACM)*. ACMs were used in older buildings as pipe, duct, furnace and boiler insulation, linings, gaskets or wrappings; sprayed- or toweled-on surface finishes on ceilings and walls; wall and roof cement boards and shingles; flooring and ceiling tiles and sheets; and components of cooking appliances and heaters.

ACMs are of concern when the fibers are *friable*, that is, they can be easily crumbled or crushed and become airborne. Stable, well-bonded ACMs are a concern only when they begin to deteriorate, can be easily disturbed, or are disturbed by drilling, sawing, sanding, or removal that causes release of asbestos fibers into the air. Released fibers remain suspended in air and can easily penetrate body tissues as they are inhaled or ingested. The durability of these fibers causes them to remain in the body for many years.

Long-term exposure to asbestos may increase the risk of several serious diseases: lung cancer; asbestosis, a chronic lung ailment that can produce shortness of breath and permanent lung damage and increase the risk of dangerous lung infections; and mesothelioma, a relatively rare cancer of the thin membranes that line the chest and abdomen. Symptoms of these diseases did not typically manifest themselves for decades. Since 1979, use of asbestos in building materials has been reduced substantially.

Sick Building Syndrome

The term *sick building syndrome (SBS)* was first coined in the late 1970s. SBS describes a situation in which reported symptoms among a population of building occupants can be temporarily associated with their presence in that building. Today, SBS is also known as the *tight building syndrome*. Typical complaints from occupants include lethargy, headache, dizziness, nausea, eye irritation, nasal congestion, and inability to concentrate. The cause of SBS is frequently tied to poor design, maintenance, and operation of the building's ventilation system. Other contributing elements may include poor lighting and poor ergonomic conditions, temperature extremes, noise, and psychological stresses that may have both individual and interpersonal impact. *Building-related illness (BRI)* is the general term for a medically diagnosable illness, which is caused by or related to occupancy of a building.

SBS should be suspected when a substantial proportion of occupants who spend extended times in a building report or experience acute on-site discomfort. SBS is the condition of a building in which more than 20% of the occupants are suffering from adverse health effects, but with no clinically diagnosable disease present. It is the condition of the building—not of the occupants. Examples include Legionnaire's disease if caught from a building's cooling tower, CO poisoning from a malfunctioning water heater, and so forth.

IAQ problems that cause building related illnesses and/or SBS include the following:

Lack of Fresh Air

If insufficient fresh air is introduced into occupied spaces, the air becomes stagnant and odors and contaminants accumulate. Lack of fresh air in occupied areas is the number one cause of SBS.

Inadequately Maintained or Operated Ventilation Systems

Mechanical ventilation systems must be properly maintained and operated based on original design or prescribed procedures. If systems are neglected, their capability to provide adequate IAQ decreases. For example, missing or overloaded filters can cause high levels of dust, pollen, and cigarette smoke to enter occupied spaces. Clogged condensate drain pans and drain lines in HVAC systems allow water to accumulate, which can lead to microbial contamination and BRI.

Disruption of Ventilation Air

The quantity of air depends on the effectiveness of air distribution. File cabinets, bookshelves, stored boxes, dropped ceiling tiles, added office walls, cubicles, and partitions can block or divert the supply of air in occupied spaces. If air circulation is disrupted, blocked, or otherwise not introduced to occupied areas, room air can become stagnant, causing increased levels of contaminants.

Poorly Regulated Temperature and Relative Humidity Levels

If the temperature and/or relative humidity levels are too high or too low, employees may experience discomfort, loss of concentration, eye and throat irritation, dry skin, sinus headaches, nosebleeds, and the inability to wear contact lenses. If relative humidity levels are too high, microbial contamination can build up and can cause BRIs.

Sources of Contamination

Chemical emissions can contribute to building related illnesses and SBS. Chemical contaminants in a building environment either originate from indoor sources or are introduced from outdoor sources. Common sources include emissions from office machinery or photocopiers, cigarette smoke, insulation, pesticides, wood products, synthetic plastics, newly installed carpets, glues and adhesives, new furnishings, cleaning fluids, paints, solvents, boiler emissions, vehicle exhaust, roof renovations, and contaminated air from exhaust stacks. Contaminants found in indoor environments can include radon, ozone, formaldehyde, VOCs, ammonia, CO, particulates, nitrogen and sulfur oxides, and asbestos.

Methods for Improving IAQ

Methods for improving IAQ fall into four categories: eliminating the source, ventilation, design, and operations and maintenance procedures. Practical measures for IAQ control are listed as follows:

Eliminating the Source

- Elimination: complete removal of biological agent, toxic substance, hazardous condition, and/or contamination source.
- Substitution: the intentional use of less hazardous materials wherever possible.
- Isolation: such as by encapsulation, shielding, sealing, and the use of distance.
- Housekeeping and dust suppression: keep surfaces clean of contaminants, prevent their redispersion, and/or eliminate personal contact.

Ventilation

- Ventilation: increase outside air or exhaust for dilution.
- Filtering and purification: filtering system and its proper maintenance.

Design

- New construction: design steps to prevent problems from occurring; procedures to commission a new building (such as air purging).
- Renovation design: avoid changes that restrict original airflow and that increase contaminant load beyond HVAC capabilities.

Operation and Maintenance

- Maintenance and work practices: specification for the proper work procedures to reduce or control contaminant releases.
- Replacement: insulation, carpeting, wall coverings, and so on, which when wet can serve as breeding grounds for microorganisms, need to be checked regularly and replaced when damaged.
- Education, training, labeling, and warning procedures for building occupants and maintenance staff.
- Sanitary procedures and personal protective devices for building maintenance staff.
- Proper storage and disposal of some contaminants: care and good control practices.

Control of pollutants at the source is the most effective strategy for maintaining clean indoor air. Control or mitigation of all sources, however, is not always possible or practical. Ventilation, either natural or mechanical, is the second most effective approach to providing acceptable indoor air.

3.7 VENTILATION

Basic Types

Buildings must be ventilated to maintain a healthy and comfortable environment. It is necessary to introduce outside air into a space to replenish oxygen, remove unpleasant odors, and dilute concentrations of water vapor, CO_2, and other indoor air pollutants. *Ventilation* is a combination of processes that results in the supply and removal of air from inside a building. These processes typically include bringing in outdoor air, exhausting some indoor air to the outdoors, conditioning and mixing outdoor air with indoor air, and distributing the mixed air throughout the building. Inadequate ventilation in a densely occupied building space can cause the level of CO_2 to increase, leading to occupant sleepiness and reduced efficiency at school and work; with other pollutants it can result it can adversely influence comfort and short- and long-term health.

Natural ventilation occurs when the air in a space is changed with outdoor air without the use mechanical systems, such as a fan. This type of ventilation occurs naturally through infiltration of outdoor air—that is, by air leakage through small cracks in the building's envelope, especially around windows and doors. It can also be achieved through temperature and pressure differences between spaces. Such random and uncontrolled ventilation may be sufficient for some buildings (such as some residences), but is not acceptable for most buildings, especially those with large numbers of occupants such as offices, schools, and theaters. In industrial plants, ventilation systems must also remove hazardous airborne contaminants from the workplace. Restaurants require extra ventilation to make up exhaust air needed to remove fumes in the cooking area.

Mechanical ventilation, sometimes simple referred to simply as ventilation in the HVAC industry (i.e., the *V* in HVAC), is the intentional movement of air from outside a building to the inside. A ventilation system typically works in combination with the heating and/or cooling system, but can work separately. Simple ventilation devices include fans or blowers that are arranged either to exhaust the stale indoor air from the building or to force outdoor air into the building, or both. Ventilating systems may be combined with heaters, filters, humidity controls, or cooling devices. Modern HVAC systems and separate ventilation systems include heat exchangers to improve effectiveness. Heat exchangers use outgoing exhaust air to heat or cool incoming outdoor air, thereby increasing the efficiency of the system by reducing the amount of energy needed to operate it.

In its recent standard, ASHRAE addresses recirculating reconditioned air in specific spaces. Air within individual building spaces is now classified into one of four classes (Class 1–4). The higher the potential for the air in a space to have odors or contaminants, the higher the numerical classification of the space. For example, a break room that is expected to have relatively few odors or contaminants is a Class 1 space; a private toilet would be a Class 2 space; a room for storage of soiled laundry would be a Class 3 space; and a chemical storage room would be a Class 4 space. There are of course exceptions, but the general concept is that recirculating air from numerically higher class spaces into lower class spaces is not acceptable. Nor is it acceptable to recondition and recirculate Class 4 air back into the space from which it came.

ASHRAE Standard 62

The amount of ventilation air required is established by building codes and industry standards. ASHRAE Standard 62 is a standard that outlines ventilation requirements for acceptable indoor air quality that has existed for three decades. Standard 62 has undergone some key changes over the years, reflecting improvements in knowledge, experience, and research related to ventilation and air quality. Although the purpose of the standard has remained consistent—to specify minimum ventilation rates and other measures intended to provide indoor air quality that is acceptable to human occupants and that minimizes adverse health effects—the methods of achieving this goal have evolved. The latest editions of ASHRAE Standard 62 include:

- Standard 62.1-2007 Ventilation for Acceptable Indoor Air Quality
- Standard 62.2-2007 Ventilation and Acceptable Indoor Air Quality in Low-Rise Residential Buildings

Most building codes reference these standards either in part or in entirety, as the minimum requirement for ventilation system design.

Standard 62 specifies minimum ventilation rates and IAQ that will be acceptable to occupants in most types of facilities. According to Standard 62, the specified rates at which outdoor air must be supplied to each room is within the range of 15 to 60 cubic feet per minute (ft^3/min or cfm)/person (7.5 to 30 L/s per person), depending on the activities that normally occur in that room. The need for this amount of ventilation generally dictates mechanical devices such as fans and blowers to supplement the natural flow of air. Tables 3.13 and 3.14 provide an example of outdoor (ventilation) air requirements for selected occupancies.

3.8   MOISTURE IN BUILDINGS

Moisture Generated in Buildings

A considerable amount of moisture is produced in a building and introduced into indoor air as water vapor. Sources of moisture include occupant respiration, food preparation, body cleaning (e.g., bathing), and maintenance (e.g., mopping floors). Table 3.15 provides sources and quantities of moisture generated in a typical building. Although some moisture in a building

is acceptable because it can maintain good relative humidity levels, too much moisture creates problems. Moisture (water) in various forms is a principal cause of building component deterioration, so it must receive considerable attention in building design.

Moisture Problems in Buildings

Over a half century ago, buildings in the United States and Canada were built with a fairly loose building envelope with respect to flow of airborne moisture. When building cavities got wet, air moved freely through and out of the building envelope, carrying excess water vapor with it. The introduction of thermal insulation in the 1940s and increased use of insulation in the building envelope in response to the energy crisis in the 1970s limited the ability of these cavities to breathe and expel moisture. Simply, water vapor movement inside a building envelope was not important until the introduction of thermal insulation. It is a problem because when insulation is added, the temperature of the water vapor can drop very quickly as it is being isolated from the heat of the building (in the winter) or from the outdoors in the summer if the building is being cooled. With changes in construction techniques over the last decade, the number of buildings experiencing moisture-related problems has risen sharply.

Most building materials are adversely affected by moisture. Wood expands with increases in moisture content and contracts with a decrease. Irregular levels of humidity can cause parts of a single wood member to swell and contract in different ways, which causes warping of the member (e.g., one edge may get wet and swell while the other edge remains at a constant moisture level). High humidity levels (above about 20% moisture content) can cause wood to begin to decay and rotting problems to develop. Exposure to water induces corrosion of irons and steels. Recurring freezing and thawing of water-saturated concrete and clay masonry products can lead to deterioration (i.e., cracking and spalling). When soluble salts are present in or in contact with concrete and masonry, moisture caused by condensation may contribute to efflorescence, a white, crystalline deposit that is aesthetically unpleasing. The performance of insulating and finish materials is greatly reduced by the absorption of water. Additionally, expansion from freezing water may also create structural problems.

Causes of moisture problems in buildings can be summarized in three areas:

- Highly insulated and sealed cavities that reduce air exchange in the assembly (i.e., wall, attic, and floor).
- Lighter construction materials that have less mass to store water and organic materials that are more susceptible to moisture deterioration.
- Improper use of vapor diffusion retarders and air barriers that trap moisture in by limiting moisture transmission.

TABLE 3.13   OUTDOOR (VENTILATION) AIR REQUIREMENTS FOR SELECTED OCCUPANCIES.

Occupied Space

Facility/Application	Space Type	cfm per Person	cfm/ft² of Floor Area	L/s per Person	L/s /m² of Floor Area
Commercial					
Food/beverage	Bar, cocktail lounge	30		14	
	Dining room, cafeteria, fast food	20		9	
	Kitchen	15		7	
Hotel, motel, dorm	Baths		0.35		0.015
	Bedroom, living room		0.30		0.013
	Gambling casinos	30		14	
	Conference rooms	20		9	
	Lobbies, assembly rooms, dorm sleep areas	15		7	
Office	Office spaces, conference rooms	20		9	
	Reception areas	15		7	
Public spaces	Elevators		1.00		0.044
	Locker rooms, dressing rooms		0.50		0.022
	Corridors, utility rooms		0.05		0.002
	Smoking lounge	60		27	
	Public restrooms	50		23	
Retail store, showroom	Basement and street		0.30		0.013
	Upper floors, dressing rooms		0.20		0.009
	Malls, arcades		0.20		0.009
	Storage rooms, shipping and receiving		0.15		0.007
	Warehouse spaces		0.05		0.002
	Ticket booths, lobbies	20		9	
	Auditorium, stages	15		7	
Institutional					
Education	Corridors		0.10		0.004
	Laboratories, training shops	20		9	
	Classrooms, music rooms	15		7	
	Libraries, auditoriums	15		7	
	Autopsy rooms		0.50		0.022
	Operating rooms	30		14	
	Patient rooms	25		12	
	Medical procedure, recovery, intensive care unit, therapy	15		7	
Residential					
Private dwellings	Living areas	15		7	
	Kitchens	25 cfm per room		12 L/s per room	
	Baths, toilet rooms	20 cfm per room		9 L/s per room	

TABLE 3.14   REQUIRED MINIMUM VENTILATION RATE IN CFM PER PERSON FOR THE 2006 INTERNATIONAL MECHANICAL CODE AND ASHRAE STANDARD 62.1-2007. METRIC (SI) UNITS ARE APPROXIMATE.

Occupancy Category	International Mechanical Code		ASHRAE Standard 62.1	
	cfm/person	L/s per Person	cfm/person	L/s per Person
Educational classroom (students ages 5–8)	15	7	15	7
Educational classroom (students age 9+)	15	7	13	6
General or office conference room	20	9	6	3
Hotel, motel, resort, and dormitory lobbies	15	7	10	5
Office building office space	20	9	17	8
Public assembly space or theater auditorium	15	7	5	2

TABLE 3.15   MOISTURE PRODUCED BY SELECTED TASKS IN A TYPICAL HOME. DATA WAS COLLECTED FROM A VARIETY OF GOVERNMENTAL SOURCES.

Sources of Moisture	Quantity	
	gal	L
Clothes drying[a] (unvented)	3.2	12.0
Clothes washing[a] (unvented)	0.5	2.0
Cooking with electric range/oven (unvented)[a]	0.2	0.9
Cooking with gas range/oven (unvented)[a]	0.3	1.2
Dishwashing[a]	0.1	0.5
Respiration, occupant per hour (average)	0.05	0.2
Respiration, occupant per day (24 hour average)	1.1	4.3
Activity		
Washing kitchen floor (150 ft^2 (14 m^2))	0.56	2.1
Average shower	0.06	0.24

[a] Based on a family of 4.

Modes of Moisture Movement

The movement of moisture into and through building assemblies generally takes place by any of four modes:

- *Liquid flow* is the movement of water under the influence of a driving force (such as gravity, or suction caused by air pressure differences).

- *Capillary flow* is the movement of liquid water in porous materials resulting from surface tension forces. Capillarity, or capillary suction, can also occur in the small space created between two materials.

- *Diffusion* is the movement of water vapor resulting from a vapor pressure difference.

- *Moisture-laden air movement* is to the movement of water vapor in air resulting from airflow through spaces and materials.

Liquid flow and capillary flow into the building envelope occur primarily with exterior source moisture (e.g., rainwater and groundwater) while movement of moisture into the building envelope by air movement and diffusion can occur with interior or exterior source moisture. Of the four modes of moisture movement, liquid flow and capillary flow are the most common and significant. As a result, builders and designers place their primary emphasis on control of rain and groundwater penetration into the building. Although less significant, air movement and vapor diffusion can also cause moisture problems.

Types of Moisture Problems

Moisture problems found in buildings can be caused by three factors: water intrusion, airborne water vapor infiltration, and water vapor diffusion. Water intrusion can be from penetration of water through the above-ground building envelope (the superstructure) or through the building foundation. Airborne water vapor infiltration and water vapor diffusion are directly associated with the formation of condensation in cavities of the building envelope. These condensation problems are tied to development of mold and mildew and structural or material deterioration and decay.

Water Intrusion

Water intrusion is leakage of moisture into the building envelope. It can cause degradation of underlying materials inside wall, ceiling, and floor assemblies. Intrusion can occur through and/or around building components such as windows, doors, gable vents, penetrations, flashings, and construction details. It can also take place through foundation walls in basements and crawl spaces. Additionally, water from leaking pipes, plumbing fixtures, and bathtub and shower surrounds can leak unnoticed into wall and floor cavities. Water can enter behind cladding or roofing materials, become trapped, and dampen unprotected substrate (building sheathing) and in some case the wood structural members. Depending on construction of the building assembly, trapped moisture may not easily escape, so materials may not readily dry out. If concealed water intrusion continues to occur, it can accumulate to levels substantial enough to cause moisture damage (e.g., mold, rot, and so forth).

Airborne Water Vapor Infiltration

Water vapor in moist air leaks into wall, attic, or floor cavity openings with infiltrating air currents and by heat transfer. Pressure differences from wind and mechanical system operation can drive airflow. Additionally, pressure differences from temperature gradients inside the building envelope cause air to move through a heat transfer mode called *natural convection;* that is, warm air is less dense and thus more buoyant than cool air, so warm air rises. In the heating season, then, warm air exfiltrates through the upper area of the building and infiltrates through the lower area.

Air travels by the easiest path possible, which is generally through any available opening in the building envelope. Moist air can also be mechanically introduced through the improper discharge of bathroom exhaust fans and clothes dryer vents that incorrectly discharge directly into building envelope cavities, crawl spaces, and attics.

Water Vapor Diffusion

Water vapor can pass through (not around) a building material and become trapped in a wall, floor, or attic cavity. Some building materials are fairly permeable to water vapor and others are nearly impermeable. The rate at which water vapor is transmitted through a material is based on the properties of the material and the pressure difference that drives vapor movement. These factors are discussed in the following:

Water Vapor Pressure

Vapor pressure in air is based upon the amount of water vapor that exists in the air. The constituent gases that make up a mixture of gases each exert a *partial pressure* that contributes to the total pressure exerted by the gas mixture. For example, atmospheric air consists of about 75% nitrogen, by weight, so 75% of the total pressure exerted by the air is due to the nitrogen constituent. Thus under standard conditions (14.696 psia), the partial pressure of nitrogen is 11 psia, 75% of the total pressure. The partial pressure of a constituent gas in a mixture of gases equals the pressure it would exert if it occupied the same volume alone at the same temperature.

Atmospheric air contains some water vapor. Like all constituent gases in air, water vapor exerts a pressure, which is known as vapor pressure. *Water vapor pressure* is the contribution of water vapor to the total pressure exerted by a gas. In buildings, this gas is atmospheric air. Water vapor pressures at standard pressure at various relative humidities are shown in Table 3.16.

Vapor pressure difference is the principal driving force behind water vapor diffusion. When a water vapor pressure difference exists, water vapor will attempt to equalize the pressure with water vapor moving from a high vapor pressure region to one of low vapor pressure. In comparing the vapor pressures in Table 3.16, one can find that *typically* during the heating season water vapor is driven out of a building, and, during the cooling season, water vapor is driven into a building.

It is not only the vapor pressure difference in surrounding air that drives water vapor diffusion. Recent studies indicate that solar heating of wet cladding materials (e.g., siding, brick veneer) will create a vapor pressure force that drives water vapor inward into the building envelope.

Moisture Transmission Rate

Water vapor *permeability* is the ability of a material to permit the passage of water vapor, the rate of water vapor transmission between two surfaces of a homogeneous material under a specified vapor pressure difference. Simply, it is the measure of the ease with which water vapor passes through a material. Units of permeability are gr-in/hr · ft^2 · inHg (ng-m/Pa · s · m^2). The lower the permeability, the less water vapor will come through the material. Permeability is not dependent on the materials' thickness.

Water vapor *permeance* is the rate at which water vapor passes through a material with parallel surfaces when induced by a vapor pressure difference, per unit thickness. It is the ratio of water vapor flow to the differences of the vapor pressures on the opposite surfaces of a material or an assembly of materials with parallel surfaces. The unit of permeance is called the *perm,* which is defined as one grain of water passing through one square foot in one hour under the action of a vapor pressure differential of one inch of mercury. In U.S. Customary (I-P) units, a perm (gr/hr · ft^2 · inHg) is equal to the transfer of one grain (1/7000th of a lb) of water per square foot of material per hour under a pressure differential of one inch of mercury

TABLE 3.16   WATER VAPOR PRESSURES AT STANDARD PRESSURE AT VARIOUS RELATIVE HUMIDITIES AND TEMPERATURES.

Temperature	Vapor Saturation Pressure	Water Vapor Pressure (psi) at Various Relative Humidities at Standard Atmospheric Pressure (14.696 psia)									
°F	psi	10%	20%	30%	40%	50%	60%	70%	80%	90%	100%
32	0.09	0.01	0.02	0.03	0.04	0.04	0.05	0.06	0.07	0.08	0.09
40	0.12	0.01	0.02	0.04	0.05	0.06	0.07	0.09	0.10	0.11	0.12
50	0.18	0.02	0.04	0.05	0.07	0.09	0.11	0.12	0.14	0.16	0.18
60	0.26	0.03	0.05	0.08	0.10	0.13	0.15	0.18	0.20	0.23	0.26
70	0.36	0.04	0.07	0.11	0.15	0.18	0.22	0.25	0.29	0.33	0.36
80	0.51	0.05	0.10	0.15	0.20	0.25	0.30	0.35	0.41	0.46	0.51
90	0.70	0.07	0.14	0.21	0.28	0.35	0.42	0.49	0.56	0.63	0.70
100	0.95	0.09	0.19	0.28	0.38	0.47	0.57	0.66	0.76	0.85	0.95

Temperature	Vapor Saturation Pressure	Water Vapor Pressure (kPa) at Various Relative Humidities at Atmospheric Standard Pressure (101 325 Pa)									
°C	kPa	10%	20%	30%	40%	50%	60%	70%	80%	90%	100%
0	0.60	0.06	0.12	0.18	0.24	0.30	0.36	0.42	0.48	0.54	0.60
4.4	0.84	0.084	0.17	0.25	0.34	0.42	0.50	0.59	0.67	0.76	0.84
10.0	1.23	0.123	0.25	0.37	0.49	0.62	0.74	0.86	0.98	1.11	1.23
15.6	1.77	0.177	0.35	0.53	0.71	0.89	1.06	1.24	1.42	1.59	1.77
21.1	2.50	0.25	0.50	0.75	1.00	1.25	1.50	1.75	2.00	2.25	2.50
26.7	3.49	0.349	0.70	1.05	1.40	1.75	2.09	2.44	2.79	3.14	3.49
32.2	4.81	0.481	0.96	1.44	1.92	2.41	2.89	3.37	3.85	4.33	4.81
37.8	6.54	0.654	1.31	1.96	2.62	3.27	3.92	4.58	5.23	5.89	6.54

(1.134 ft of water). In metric units (SI), a perm (ng/Pa · s · m²) is equal to the transfer of one nanogram of water per square meter of material per second under a pressure difference of one Pascal. The corresponding unit of permeability is the "perm-inch," the permeance of unit thickness.

A material's permeance is dependent on thickness, much like the R-value in heat transmission. If a material has a perm rating of 1.0, then one grain of water vapor will pass through one square foot of the material when the vapor pressure difference between the one side and the other side of the material is equal to one inch of mercury. One grain of water is equal to 1/7000th of a pound (about a droplet of water). For homogenous materials, doubling material thickness generally cuts water vapor transmission in half; that is, if 1 inch of a material has a perm rating of 2.0, then for 2 inches of the same material, the perm rating would be 1.0. With paints, however, adding a second coat more than halves the water vapor transmission.

Permeance is used to compare various products in regard to moisture transmission resistance: the lower the perm rating of a material, the more effectively it will retard diffusion and thus the more effective the material will be in controlling moisture vapor transmission. Reported perm ratings are based upon standardized dry and wet tests at a temperature of 73.4°F (23°C). Under the *desiccant (dry-cup) test,* one side of the material is maintained at 0% relative humidity and the other side is maintained at 50% relative humidity. Under *water (wet-cup) test,* one side of the material is maintained at 50% relative humidity and the other side is maintained at 100% relative humidity. The water test method is preferred when testing materials that are likely to be exposed to high humidity in service.

Reporting permeance of generic types of materials is difficult because these properties are sensitive to small changes in material composition and the moisture content of the material being tested. Hydroscopic materials adsorb (i.e., gather and store a substance on their surface) moisture and lose moisture with changes in ambient relative humidity and temperature. Water vapor permeance of hydroscopic materials (e.g., wood and paper products) increases greatly with increase in relative humidity. For example, the perm rating of bitumen-impregnated kraft paper varies by a factor of about 10 between the desiccant and water permeance tests. Data provided in Table 3.17 is presented to permit comparisons of perm ratings of various building materials. Exact values for permeance should be obtained from the manufacturer of the materials under consideration or secured as a result of laboratory tests.

Condensation from Water Vapor Infiltration and Diffusion

In cold climates, components of the building envelope that are outside the insulation (e.g., roof and wall sheathings) are commonly cooler than the dewpoint temperature of air in a conditioned space. Under certain conditions, the dew point temperature occurs within the insulation itself. If water vapor from diffusion or moisture-laden infiltration air comes into contact with a component at or below the dew point temperature of the air, the vapor will condense to a liquid (water droplets). If the temperature is below freezing, condensation will form as frost (ice crystals). Frost can collect in the cavity through the heating season and melt during the first thaw of the spring season.

Condensation can form within a wall, attic, or floor cavity, where it remains concealed. It can accumulate unnoticed, where it can cause rot of framing members and sheathing, to a point where structural collapse is possible; deterioration of interior room finishes, from gravity flow of water; corrosion and failure of metal truss plates and clips; and decrease of thermal resistance or deterioration of insulation. Thus, it is very important that unintended paths that infiltrating moist air may follow be permanently sealed. It is also important that an impediment exist that limits diffusion of moisture into the building envelope cavity because reducing the amount of water vapor entering a cavity will reduce the quantity of condensation in the cavity.

Moisture Control

Moisture control in building envelope cavities is generally accomplished by eliminating (or minimizing) water intrusion, ventilation of indoor spaces where moisture is generated, and construction techniques that properly retard water vapor.

Eliminating Water Intrusion

Designing and constructing proper details and using appropriate materials will prevent the likelihood of water intrusion. After construction, intrusion can still take place if maintenance of these details and other critical areas (e.g., caulk joints) is neglected. In addition, even slight structural movement as a building settles during its early life can create openings that were not there initially.

Damage can be significant if moisture intrusion goes undetected, so early detection of water intrusion is crucial to minimizing or preventing such damage. The location of water entry is often difficult to detect (e.g., damage to substrate and structural members behind the exterior wall cladding frequently cannot be detected by visual inspection). In cases where intrusion is suspected, testing must be conducted. A *moisture meter* is used to detect elevated levels of water in building materials. Two types of moisture meters should be used: a noninvasive moisture meter that scans through the assembly for the presence of moisture without penetrating and damaging the surface; and an invasive, probe-type moisture meter that penetrates the assembly surface and gives moisture readings of materials in contact with the probes.

In cases where water intrusion is discovered, repairs should be made promptly. The primary objective of any repair

TABLE 3.17 CHARACTERISTIC WATER VAPOR PERMEANCE (PERM) RATINGS FOR SELECTED BUILDING MATERIALS. DATA IS PRESENTED TO PERMIT COMPARISONS OF PERM RATINGS. EXACT VALUES FOR PERMEANCE SHOULD BE OBTAINED FROM THE MANUFACTURER OF THE MATERIALS UNDER CONSIDERATION OR SECURED AS A RESULT OF LABORATORY TESTS. DATA WAS COMPILED FROM AN ASSORTMENT OF MANUFACTURER AND GOVERNMENTAL SOURCES.

Material	Material Thickness	Water Vapor Permeance	
		U.S. Customary grain/ft$^2 \cdot$ hr $\cdot$ inHg	Metric (SI) ng/Pa $\cdot$ m$^2 \cdot$ s
Vapor Retarders			
Aluminum foil	1 mil (0.03 mm)	0.00	0.0
	0.35 mil (0.01 mm)	0.05	2.9
STEGO WRAP® Underslab polyolefin geomembrane	15 mil (0.38 mm)	0.01	0.6
Polyethylene film	6 mil (0.15 mm)	0.06	3.4
	4 mil (0.10 mm)	0.08	4.6
	2 mil (0.05 mm)	0.16	9.1
Asphalt-impregnated kraft facing	varies	0.30[a]	17.0[a]
Building felt/tar paper (15#)	37.4 mil (0.95 mm)	4.00	230
Air Barriers			
Prime Wrap®	5 mil (0.12 mm)	5 to 15	290 to 860
Tyvek® CommercialWrap®	9 mil (0.22 mm)	28	1680
Tyvek® HomeWrap®	6 mil (0.15 mm)	58	3330
Insulation			
Polystyrene (extruded)	1 in (25 mm)	0.40 to 1.60	23 to 92
Polystyrene (expanded)	1 in (25 mm)	2.0 to 5.8	115 to 333
Polyurethane	1 in (25 mm)	1.20	69
Cellulose	1 in (25 mm)	29	1666
Fiberglass	1 in (25 mm)	29	1666
Building Materials			
Built-up roofing	varies	0.01	0.6
Glazed structural (clay) tile masonry	4 in (100 mm)	0.10 to 0.16	6 to 9
Wood (solid board)	¾ in (19 mm)	0.30 to 4.03	17 to 232
Oriented strand board (OSB)	½ in (13 mm)	0.70	40
Plywood, Douglas fir (exterior glue)	⅜ in (8 mm)	0.78	44.6
Plywood, southern pine (exterior glue)	⅜ in (8 mm)	1.43	82
Brick masonry	4 in (100 mm)	0.80	46
Concrete	4 in (100 mm)	1.25	72
Gypsum wallboard (painted)	½₃ in (13 mm)	2.0 to 3.0	114 to 172
Concrete masonry	8 in (200 mm)	2.38	137
Building felt/tar paper (15#)	37.4 mil (0.95 mm)	4.00	230
Hardboard (tempered)	⅛ in (3 mm)	5.0	287
Hardboard (standard)	⅛ in (3 mm)	11.0	630
Finishes			
Gypsum wallboard—knockdown texture, semigloss paint	½ in (13 mm)	13.3	761
Gypsum wallboard, 2 coats latex flat paint	½ in (13 mm)	19.5	1115
Gypsum wallboard—knockdown texture, flat paint	½ in (13 mm)	21.5	1230
Gypsum wallboard (unpainted)	½ in (13 mm)	50.0	2860
Plaster on wood lath	¾ in (19 mm)	11.0	630
Plaster on metal lath	¾ in (19 mm)	15.0	860
Plaster on gypsum board lath	¾ in (19 mm)	20.0	1150
Paint			
Asphalt paint on plywood	2 coats	0.40	23
Latex VDR paint or primer	1 coat	0.45	26
Oil-based paint on plaster	2 coats	1.6 to 3	91 to 172

[a] 5 to 10 grain/ft$^2 \cdot$ hr $\cdot$ in Hg (85 to 170 ng/Pa $\cdot$ m$^2 \cdot$ s) when wet.

is to first eliminate water intrusion. Areas of elevated moisture with an absence of damage or decay may require no more than eliminating the source of water intrusion. It has been discovered that undamaged but wet substrate can dry out over time once the source of the water intrusion has been eliminated. Where structural damage has occurred, replacement of decayed lumber is required.

One of the chief sources of water intrusion into a building interior is basement or under-floor crawl-space moisture. Water vapor from a wet or damp basement or crawl space travels into the indoor spaces and ultimately migrates into the building interior and envelope. Improper grading around the base of the exterior of the building foundation directs storm water towards the house. Not extending downspouts so they discharge away from the foundation also contributes to the problem. Solutions to basement or under-floor crawl-space moisture are improving the slope of grade around the foundation and adding downspout extensions, which are extended as far away from the foundation as practical.

PHOTO 3.22   Installation of a polyethylene VDR on the "warm" side of a framed wall in a building in a heating climate. (Used with permission of ABC)

Retarding Vapor Diffusion

A *vapor diffusion retarder (VDR)* is a material or system that adequately impedes the transmission of water vapor through the building envelope. A VDR has a rating of 1.0 perms or less. The main reason to limit or stop transmission of water vapor is to prevent it from condensing to liquid water in the building envelope. VDR membranes commonly used in the building construction industry include thin, flexible aluminum foils, plastic (e.g., polyrthylene) films, and bitumen-impregnated kraft paper. Other products (i.e., CertainTeed's MemBrain™ SMART Vapor Retarder) are designed specifically for use as a VDR.

Many insulation batts, blankets, and boards come with a VDR attached to one face. A common example of this is aluminum- or kraft-paper-faced fiberglass insulation blankets. Polyethylene, a thermoplastic sheet material, is the most commonly used VDR in cold climates. It is used over unfaced, vapor-permeable insulation (i.e., fiberglass and mineral wool) in wall and ceiling cavities. Coatings may be semifluid, mastic, or paint type, which are applied on a substrate such as gypsum wallboard. Photographs of VDRs during construction are shown in Photos 3.22 and 3.23.

According to industry standards, a *vapor barrier* is essentially impermeable. It has a rating of 0.01 perm (0.60 metric perms) or less. A material with a perm rating of 1.0 perm (60 metric perms) or less is considered to be a *nonpermeable VDR,* which includes most polyethylene films, elastomeric coatings, oil-based paint, aluminum foils, solid glass, metals, and vinyl wall coverings. Materials rated at 10 perms (60 metric perms) or less are *semipermeable VDRs,* including solid and engineered wood products (e.g., plywood, OSB, hardboard), unfaced expanded polystyrene insulation board, latex paints, and bitumen-impregnated kraft-paper facing on insulation. Those materials rated above 10 perms (60 metric perms) are permeable and are not really vapor retarders at all, such as unpainted gypsum

PHOTO 3.23   Installation of a blanket of insulation. Attached to the bottom face of the blanket is a bitumen-impregnated kraft paper facing (installed by manufacturer) that will serve as a VDR. (Used with permission of ABC)

wallboard, cellulose and unfaced fiberglass insulation, bitumen-impregnated cellulose sheathing boards, and building air barriers and house wraps.

Current building code regulations require placement of a VDR rated at 1.0 perm (60 metric perms) or less on the heated side of the building envelope, except in hot, humid climates. Placement of a VDR in a construction assembly (e.g., a wall) is climate dependent. See Figure 3.3. The rationale used in determining VDR placement was that in cold climates water vapor pressure is typically higher inside the building, so placement of the VDR should be on the interior side of the wall insulation. Having a single VDR on the side of the wall with the highest vapor pressure retards wetting and encourages drying of the cavity. Such placement makes sense in extremely cold climates. However, this flawed rationale does not account for hot and/or humid weather and the use of air conditioning for cooling.

Under summer conditions, vapor pressure is typically higher outside the building, causing vapor to move to the inside of the building envelope and condense on the surface of the VDR, which is cooled by air conditioning. Because of such inconsistencies in climate, type and placement of a VDR cannot be treated with a single comprehensive specification that covers all conditions in all climates.

The assembly of the building envelope plays a significant role in vapor diffusion control requirements in terms of the order in which layers of different permeance materials are arranged. A single VDR is typically installed that limits the rate of vapor diffusion so moisture can move through the other components of the building envelope assembly more rapidly than it passes through the VDR. For example, in a cold climate a low-permeance VDR (e.g., a polyethylene film or bitumen-impregnated kraft facing) is installed on the interior side of a wall to reduce the transmission of water vapor through the building envelope. In a second example, an low-permeance layer (e.g., a roofing

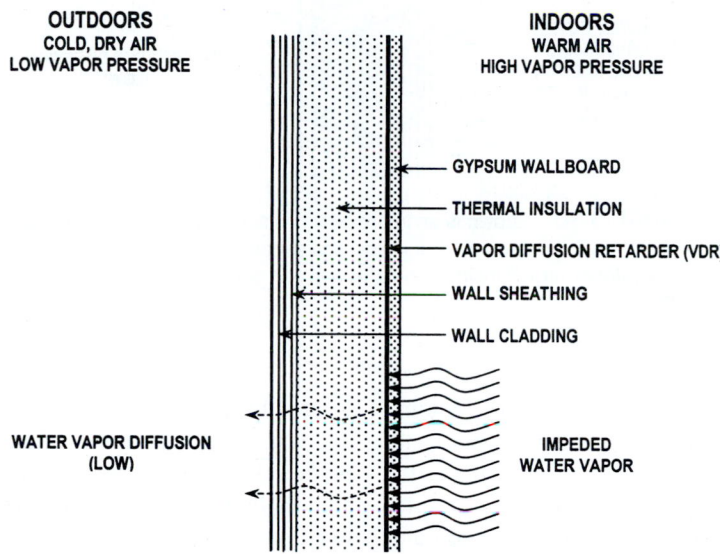

FIGURE 3.3A   A vapor diffusion retarder (VDR) is a material of system that adequately impedes the transmission of water vapor through the building envelope. Shown is a properly positioned cold climate VDR.

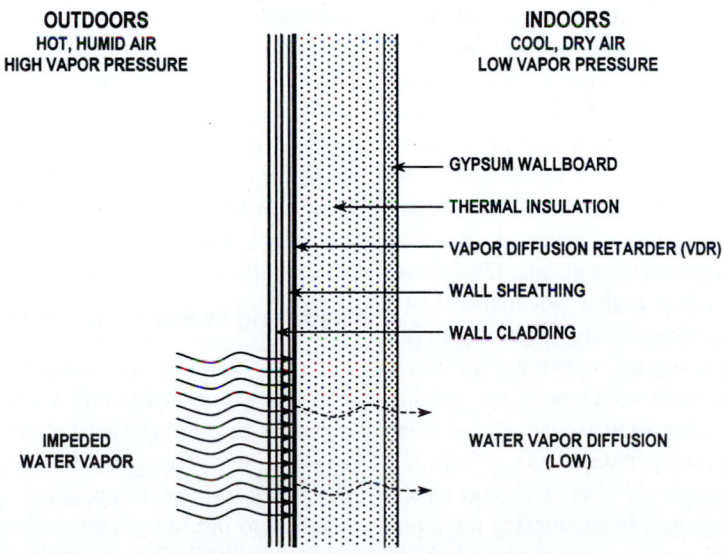

FIGURE 3.3B   A Properly positioned warm climate VDR.

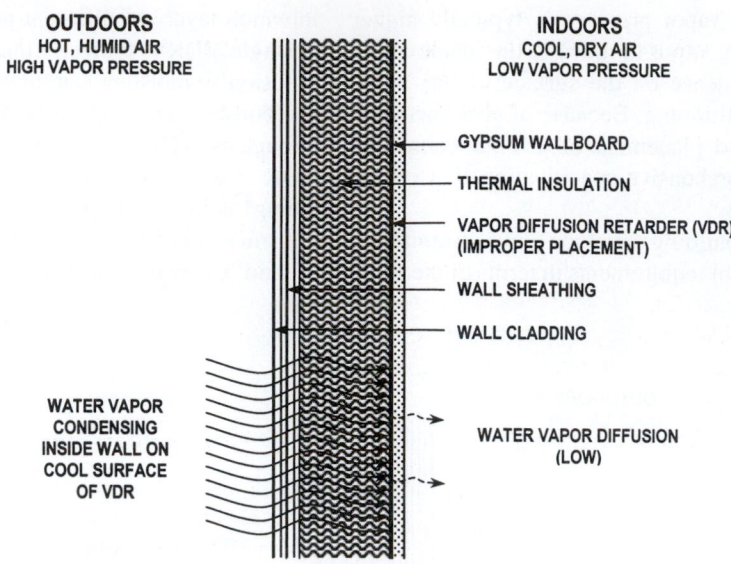

FIGURE 3.3C Shown is an improperly positioned VDR for a warm climate With the VDR located near the inside wall surface, water vapor contained in moist outdoor air will infiltrate into the wall and condense on the cooler inner wall surface.

membrane) is installed on the outside surface of the building envelope in a cold climate. This outer layer will prevent water vapor from escaping to the exterior, which slows drying to the outside of the envelope. In this case, the permeance of interior layers must be considerably less than the permeance of outer layers. To prevent trapping moisture in a cavity, the cold-side material's perm rating should be at least five times larger than the value of the warm side (various rules place the ratio at 3:1 to as much as 10:1).

Local climatic conditions and heating and cooling needs of a building dictate whether, what type, and where a VDR should be installed in a building envelope. In most climatic regions, the building envelope assemblies must be designed to promote drying to both the interior and exterior, so a semipermeable VDR is recommended on the heated side of the building envelope. The varying permeance of the bitumen-impregnated kraft facing on batt and blanket fiberglass insulation is low under dry winter conditions but its permeance rises significantly under very humid conditions, which allows water vapor to flow out of the cavity. In hot, humid regions (Florida and Gulf Coast) where vapor pressure is more often higher outside, use of a VDR is advised on the outside of the building envelope. In hot, dry regions, a VDR is typically not needed. In extremely cold climates, a nonpermeable VDR on the heated side of the building envelope is needed. A nonpermeable VDR is also advisable behind high moisture areas such as shower surrounds.

Basements and under-floor crawl spaces should have a vapor retarder to prevent soil moisture from entering the building interior. In under-floor crawl spaces, a nonpermeable VDR should be installed on top of the earthen floor. Concrete is per-

meated by microscopic voids that allow the passage of water and water vapor so, in basements with a concrete slab, a nonpermeable underslab VDR should be placed just below the concrete at building sites with moderate to heavy annual rainfall, low-lying building sites, or those sites where soil testing reveals sufficient hydrostatic pressure. Foundation walls should also be treated with a VDR coating or membrane.

To be effective, an under-slab VDR must possess a number of key performance characteristics, including low permeance [usually less than 0.3 perm (17 metric perms)], high tensile strength, high puncture resistance, and resistance to chemical or environmental attack. Under-slab VDRs are classified by the industry as Class A, B, and C, with Class A being the best. The key distinction between the classes of materials is their differing ability to withstand the abuse of installation without being punctured or torn—that is, being capable of handling the rigors of job-site installation. Although 4 or 6 mil (1 mil is a thickness of 1/1000 in or 0.0254 mm) polyethylene film is commonly used below concrete slabs, it typically does not meet the classification standard.

Reducing Water Vapor Infiltration

Airborne water vapor infiltration accounts for the largest part of water vapor movement into building envelope cavities, yet it typically receives the least attention in design and construction. In cold temperatures, moisture-laden indoor air can pass through construction openings (e.g., electrical boxes, unsealed joints, and plumbing penetrations) where moisture easily condenses in the building envelope cavity. In hot, humid conditions, damp outdoor air travels in the opposite direction through similar openings. Openings in the building envelope do little to reduce airborne water vapor infiltration and reduce, if not eliminate, the effect of a vapor retarder. Eliminating these paths is very important. Simply put, an effective VDR needs to be a continuous barrier, free of any holes or open seams. The VDR should be continuously lapped over all joints and should be carefully cut to size. The key to making this method work effectively is to carefully and permanently seal all openings, seams, and penetrations, including around windows, doors, electrical outlets, plumbing stacks, and vent fans.

Air barriers are breathable water vapor permeable materials intended to impede airflow into the building envelope. Air barriers are effective in controlling moisture from airborne water vapor infiltration by obstructing much of the air movement into a building envelope, yet still allowing water vapor that does collect in the cavity to diffuse back through the barrier material. Most air barriers are also effective in reducing water intrusion into cavities by providing an additional layer beyond the cladding. They have microscopic pores, small enough to resist flow of liquid water and air molecules, but large enough to allow smaller vapor molecules to pass through the barrier material.

Common materials used for this purpose are house wrap, plywood, drywall (gypsum) board, and foam board. See Photo 3.24. Many of these materials serve a dual purpose in that they are used for insulation, structural purposes, and finished surfaces. The most common air barrier materials in use today are fibrous spun polyolefin plastic, matted into sheets (e.g., Tyvek HomeWrap® or CommercialWrap®), and woven and coated polyethylene fabric with microscopic-size perforations (Prime Wrap®). These air barriers are sheets that are wrapped around the exterior of a building's sheathing before the siding or cladding is installed. Sealing all of the joints with a special tape improves the wrap's performance significantly.

Use of an air barrier is appropriate in all climates, but especially in hot, humid climates where keeping moist outdoor air from entering the building cavities is important during the extended cooling season. However, some house wraps react poorly to certain kinds of wood siding (e.g., redwood, cedar) when they get wet, such as in rainy or humid climates. Natural substances in wood (lignin) act as a detergent when moistened, which makes it easier for liquid water to pass through the barrier material. Additionally, most house wraps are susceptible to ultraviolet degradation, so they must be protected from prolonged exposure to sunlight.

Alternative Approaches to Reducing Water Vapor Infiltration

There are alternative approaches to the use of VDR and air barrier in controlling moisture in the building envelope. These include:

Airtight Drywall Approach

The *airtight drywall approach (ADA)* uses drywall installed over gaskets adhered to building framing members and ordinary paint to control air and vapor movement.

In this approach, drywall must be carefully joined to provide continuity between drywall sheets. Gaskets installed between wall plates, rim joist, subfloor, and other members create an air barrier. Two coats of latex paint on gypsum wallboard behave in a manner similar to bitumen-impregnated kraft facing and provide a sufficient semipermeable vapor retarder.

Attention to detail is necessary during installation of interior walls, window openings, electrical boxes, plumbing and electrical penetrations, around bathtubs, at split levels and stairs on outside walls. The barrier created reduces air leakage and moisture penetration into cavities in the building envelope.

Dynamic Buffer Zone Ventilation Systems

A *dynamic buffer zone (DBZ) ventilation system* consists of an exterior wall or roof of a building coupled with an air-handling system that is arranged so that the cavities in the building envelope are mechanically ventilated with preheated dry air during the winter to prevent and control of cavity condensation. A DBZ system typically includes supply and exhaust fans, temperature, relative humidity and pressure sensors and controllers, sealed cladding components, and a sealed interior plane of air tightness. There are two types of DBZ systems: the ventilated cavity DBZ system and the pressure-controlled cavity DBZ system.

In the *ventilated cavity* DBZ ventilation system, the cavities in the building envelope are ventilated with dry outdoor air and pressure relieved or controlled through a return and exhaust system. In some cases, the cavity air is recirculated if the dew point temperature of the air is low and the cavity air requires supplementary heat. In this system, it is the ventilation air that dilutes and evacuates any cavity humidity.

In the *pressurized cavity* DBZ ventilation system, cavities in the building envelope are pressurized slightly above indoor air pressure with preheated outdoor air but without a pressure relief or return air system. In this system, it is the

PHOTO 3.24   An installed air barrier wrapping the exterior sheathing before installation of the cladding (siding and masonry veneer). (Used with permission of ABC)

cavity pressure generated by the system fans, which prevent further contamination of the cavity air with humid indoor air. These systems are less expensive to design and build and more efficient at controlling construction cavity moisture conditions.

A DBZ system is used when efforts to fully seal an existing building envelope to prevent condensation have not been successful. It is best suited to restored buildings that are upgraded to higher indoor humidity levels.

Need for Indoor Ventilation

Excessive water vapor levels in air must be removed by ventilation of indoor air. In restrooms, bathrooms, kitchens, laundries, indoor swimming pools, and other moisture-producing areas, exhaust vents are normally necessary and called for by most building codes. All vents must terminate and discharge to the exterior of the building.

In tight structures with heavy moisture loads, mechanical dehumidification may be required. Humidification during dry winter months and heavy use of evaporative (swamp) cooling are not recommended.

Need for Attic Ventilation

In cold and extremely cold climates, attic ventilation has been effective in controlling moisture problems where the objective is to maintain cold attic temperatures in winter to avoid ice dams created by melting snow and to vent moisture that moves from the conditioned space into the attic. *Ice dams* develop in conditions when air in the attic space is at above-freezing temperature (if insulation is insufficient or warm building air enters the attic) and melting snow or ice on the roof surface backs up at the roof overhang and seeps through the shingles into the roof and damaging materials.

Sufficient attic ventilation is usually defined as having a net free ventilating area equal to 1/150 of the attic floor area. When an attic vapor retarder is used, ventilation requirements are cut in half with a net free vent area can be 1/300 of the attic floor area. The net free ventilating area of a roof vent is usually least than 50% of the actual size of the vent area. The vent area should be divided evenly between the top and bottom of the attic space to effectively promote airflow by natural convection.

Attic ventilation may cause moisture problems in other climates: In hot, humid climates, moist outdoor air that comes in contact with cold surfaces in the attic will condense, particularly if low interior temperatures are maintained with summer air conditioning. Recent research suggests that in hot, humid areas, the best approach to avoiding moisture condensation in attics may be to keep the moisture out of the attic completely by sealing the attic from the outdoors. A continuous nonpermeable VDR can prevent condensation problems in the attic by preventing moisture diffusion through the ceiling and infiltration into the attic cavity. It must serve as a barrier to airborne water vapor infiltration and vapor diffusion.

Roof systems without attic spaces (e.g., low-slope roof structures, cathedral ceilings) are characteristically not vented and are protected by roof coverings with very low vapor permeance. The roof covering does not diffuse moisture well. In this case, moisture control in the roof cavity is extremely difficult. A continuous nonpermeable VDR can prevent condensation problems in cathedral ceilings by obstructing flow of water vapor. The VDR must be completely sealed to prevent moisture diffusion through the ceiling and infiltration into the roof cavity. It must serve as a barrier to vapor diffusion and airborne water vapor infiltration. Particular attention is necessary around skylights, paddle fan outlets, recessed light fixtures, and other potential openings in the barrier.

STUDY QUESTIONS

3-1. Define building science.

3-2. Define embodied energy.

3-3. With respect to energy use, describe the difference between external-load-dominated and internal-load-dominated buildings.

3-4. Describe the difference between lightweight and heavyweight building construction, as associated with:

a. Embodied energy

b. Energy use

3-5. Explain how thermal insulation restricts the flow of heat.

3-6. List some of the ways in which the heat loss of a building can be controlled.

3-7. Why is limiting of window areas important in keeping the heat loss low?

3-8. Based upon governmental recommendations, identify the recommended R-values for new house construction in the geographical location at which you reside.

3-9. Climates with high daily temperature swings (i.e., Arizona, New Mexico, and Colorado) benefit from the thermal mass effect for space heating while climates with low daily temperature swings experience little benefit. Why?

3-10. What is indoor air quality?

3-11. Describe the three categories of indoor air contaminants.

3-12. Describe the four types of biological contaminants.

3-13. Identify indoor air contaminants that are produced by combustion.

3-14. At what concentrations is carbon monoxide (CO) fatal?

3-15. What does the term *sick building syndrome* mean?

3-16. Radon levels in a building interior are recorded at 8 pCi/L. What are the U.S. Environmental Protection Agency recommendations for this level?

3-17. Identify problems related to building related illness.

3-18. Identify methods used to improve indoor air quality.

3-19. Describe two methods of controlling moisture in a building.

3-20. Describe the purpose of ventilating a building.

3-21. Describe the two basic types of ventilating a building.

3-22. Describe the three basic categories of moisture problems in buildings.

3-23. Identify and describe the four modes of moisture movement into and through building assemblies.

3-24. Describe water vapor permeability and permeance.

3-25. Describe the purpose of a vapor diffusion retarder (VDR).

3-26. Identify materials that can serve as a nonpermeable VDR.

3-27. Identify materials that can serve as a semipermeable VDR.

3-28. What is the difference between a nonpermeable VDR and a semipermeable VDR?

3-29. Why does bitumen-impregnated kraft paper work effectively as a VDR?

3-30. How can moisture problems associated with water intrusion be reduced?

3-31. How can moisture problems associated with vapor diffusion be reduced?

3-32. How can moisture problems associated with airborne water vapor infiltration be reduced?

3-33. What is a building air barrier, and where and why is it used in a construction assembly?

3-34. Describe the airtight drywall alternative to a VDR.

3-35. What is an ice dam and how can it be prevented?

3-36. Describe a dynamic buffer zone (DBZ) ventilation system.

Design Exercises

3-37. For single-family dwellings in Denver, Colorado, the International Energy Conservation Code (IECC) offers the following minimum R-value ranges for the construction assemblies shown (depending on option). Convert these R-values to the metric equivalent thermal resistance (RSI).

   a. Ceiling: 38 to 49 hr $\cdot$ °F $\cdot$ ft^2/Btu

   b. Wall cavity: 17 to 19 hr $\cdot$ °F $\cdot$ ft^2/Btu

   c. Basement wall: 9 to 13 hr $\cdot$ °F $\cdot$ ft^2/Btu

   d. Floor (above crawl space): 21 to 30 hr $\cdot$ °F $\cdot$ ft^2/Btu

   e. Slab-on grade: 5 to 13 hr $\cdot$ °F $\cdot$ ft^2/Btu

3-38. For multifamily dwellings in Denver, Colorado, the IECC requires the following minimum R-values for the construction assemblies shown. Convert these R-values to the metric equivalent thermal resistance (RSI).

   a. Ceiling: 38 hr $\cdot$ °F $\cdot$ ft^2/Btu

   b. Wall cavity: 17 hr $\cdot$ °F $\cdot$ ft^2/Btu

   c. Basement wall: 9 hr $\cdot$ °F $\cdot$ ft^2/Btu

   d. Floor (above crawl space): 21 hr $\cdot$ °F $\cdot$ ft^2/Btu

   e. Slab-on grade: 5 hr $\cdot$ °F $\cdot$ ft^2/Btu

3-39. For a three-story, 100 000 ft^2 office building in Denver, Colorado, the IECC requires the following minimum R-values for the construction assemblies. Convert these R-values to the metric equivalent thermal resistance (RSI).

   a. Ceiling: 25 hr $\cdot$ °F $\cdot$ ft^2/Btu

   b. Wall cavity: 13 hr $\cdot$ °F $\cdot$ ft^2/Btu

   c. Basement wall: 9 hr $\cdot$ °F $\cdot$ ft^2/Btu

   d. Floor (above crawl space): 17 hr $\cdot$ °F $\cdot$ ft^2/Btu

3-40. For single-family dwellings in Denver, Colorado, the IECC requires window assemblies to have minimum U-factors in the range of 0.35 to 0.45 Btu/hr $\cdot$ °F $\cdot$ ft^2. Convert these U-factors to:

   a. The thermal resistance (R-values), in units of hr $\cdot$ °F $\cdot$ ft^2/Btu

   b. The metric equivalent thermal resistance (RSI), in units of °C $\cdot$ m^2/W

   c. The metric equivalent U-factor, in units of W/°C $\cdot$ m^2

3-41. For multifamily dwellings in Denver, Colorado, the IECC requires window assemblies to have a minimum U-factor of 0.45 Btu/hr $\cdot$ °F $\cdot$ ft^2. Convert this U-factor to:

   a. The thermal resistance (R-values), in units of hr $\cdot$ °F $\cdot$ ft^2/Btu

   b. The metric equivalent thermal resistance (RSI), in units of °C $\cdot$ m^2/W

   c. The metric equivalent U-factor, in units of W/°C $\cdot$ m^2

3-42. For a three-story, 100 000 ft^2 office building in Denver, Colorado, the (IECC) requires window assemblies to have a minimum U-factor of 0.70 Btu/hr $\cdot$ °F $\cdot$ ft^2 (for IECC climate zone 5). Convert this U-factor to:

   a. The thermal resistance (R-values), in units of hr $\cdot$ °F $\cdot$ ft^2/Btu

   b. The metric equivalent thermal resistance (RSI), in units of °C $\cdot$ m^2/W

   c. The metric equivalent U-factor, in units of W/°C $\cdot$ m^2

3-43. A conference room is designed with seating for 20 occupants. Determine the minimum outdoor air (ventilation) rate for this space, in cfm. A classroom is designed with seating for 32 occupants. Determine the minimum outdoor air (ventilation) rate for this space, in cfm, based on:

a. The International Mechanical Code

b. ASHRAE Standard 62.1

3-44. An auditorium-like lecture room is designed with seating for 300 occupants. Determine the minimum outdoor air (ventilation) rate for this space, in cfm, based on:

a. International Mechanical Code

b. ASHRAE Standard 62.1

3-45. A 64 000 ft^2 warehouse is designed to store electronic equipment in a big box retail store. Determine the minimum outdoor air (ventilation) rate for this space, in cfm.

3-46. A 7000 m^2 warehouse is designed to store electronic equipment in a big box retail store. Determine the minimum outdoor air (ventilation) rate for this space, in liters per second.

3-47. A restaurant has an 800 ft^2 cocktail lounge and a 3800 ft^2 dining room, which are served by a single rooftop HVAC unit. Determine the outdoor (ventilation) air requirements for this unit based upon the selected occupancies.

3-48. A restaurant has a 90 m^2 cocktail lounge and a 420 m^2 dining room, which are served by a single rooftop HVAC unit. Determine the outdoor (ventilation) air requirements for this unit based on the selected occupancies.

3-49. A building has indoor air conditions of 70°F dry bulb temperature and 40% relative humidity. Determine the direction of water vapor flow under the following outdoor air conditions that represent a dry, heating climate:

a. 32°F dry bulb temperature and 40% relative humidity (winter conditions)

b. 90°F dry bulb temperature and 20% relative humidity (summer conditions)

3-50. For the conditions in the previous exercise, describe a good strategy for use of a vapor diffusion retarder.

3-51. A building has indoor air conditions of 70°F dry bulb temperature and 50% relative humidity. Determine the direction of water vapor flow under the following outdoor air conditions that represent a cooling climate that is warm and humid:

a. 60°F dry bulb temperature and 60% relative humidity (winter conditions)

b. 90°F dry bulb temperature and 70% relative humidity (summer conditions)

3-52. For the conditions in the previous exercise, describe a good strategy for use of a VDR.

3-53. A building has indoor air conditions of 70°F dry bulb temperature and 40% relative humidity. Determine the direction of water vapor flow under the following outdoor air conditions that represent a climate that is warm and humid in the summer and cold and dry in the winter:

a. 32°F dry bulb temperature and 30% relative humidity (winter conditions)

b. 90°F dry bulb temperature and 70% relative humidity (summer conditions)

3-54. For the conditions in the previous exercise, describe a good strategy for use of a VDR.

CHAPTER FOUR

HEATING LOAD COMPUTATIONS FOR BUILDINGS

4.1 HEAT LOSS IN BUILDINGS

Building Heat Load

The building envelope continuously interacts with the outside environment, and its thermal performance has a substantial effect on the inside environment and occupant comfort. In cold weather, buildings are heated to maintain a comfortable inside temperature. Because a difference in temperature exists between inside and outside, heat will move through walls, windows, ceilings, floors, and doors at a rate related to temperature difference and the ability of the structure's materials to transfer heat.

Heat loss is the amount of heat that a building or building space loses from heat transfer during cold weather. It is dependent on the size, type, and quality of construction, weather conditions, and the climate in the geographical area where the building is constructed. The rate of heat loss from the building will determine the size of the heating plant (e.g., furnace or boiler), the size of terminal heating units (radiators, ducts, and so on), and ultimately, the heating cost. The *heating load* is the heat that the HVAC equipment must generate and introduce into the building to maintain comfortable conditions in the building interior. A designer needs to know the peak heating load so that a space heating system can be designed.

Heat losses that contribute to the heating load in a building occur generally through transmission, infiltration, and ventilation. See Figure 4.1. The rate of individual heat losses and the total heat loss will vary by building construction, occupancy type, and climate.

Transmission

Transmission heat losses are the result of heat passing through a material in the building envelope (e.g., glass) or through an assembly of materials (e.g., walls, ceilings, floors, and so forth). The quantity of heat transfer may be calculated through each material and for each mode (e.g., conduction, convection, and radiation as described in Chapter 2). Instead of using this lengthy and cumbersome process, methods are used that account for the combined effects. This concept includes use of thermal resistance and coefficient of transmission.

Infiltration

Infiltration heat losses relate to air leakage through the building envelope and the energy required to heat unconditioned air that has leaked into the structure. This air passes around or through the building envelope. In every building, no matter how well constructed, there is a certain amount of cold air that leaks into the building, referred to as *infiltration heat loss,* and an equal amount of hot air that leaks out. Most commonly, infiltration will occur around doors and windows. The tight construction of the building will save the building owner a considerable amount of money over the life of the building.

FIGURE 4.1   Heat losses from a building consists of transmission heat loss ($q_{transmission}$), infiltration heat loss ($q_{infiltration}$), and ventilation heat loss ($q_{ventilation}$).

Ventilation

Ventilation is the introduction of outdoor air into the building, or parts of the building, at a controlled rate with the intent to maintain or improve indoor air quality. Ventilation heat losses are tied to the energy required to condition (heat, cool, humidify, or dehumidify) outside air that is intentionally introduced into the building. Many times the air used to ventilate the building is heated before it is introduced into the building, and this must be considered in the design of the system.

Units of Heat Loss

In the heating, ventilating, and air conditioning industry, metric (SI) and U.S. customary units, typically called *inch-pound (I-P) units,* are used. In the heating sector, *quantity of heat loss (Q)* is commonly expressed in units of measure of Btu or W-hr and *rate of heat loss (q)* is expressed in Btu/hr or W. Metric (SI) units make use of standard prefixes such as k for 1000 (e.g., kWh or KW). On the other hand, Roman-based numerical prefixes are often used with I-P units such as C for 100, M for 1000, MM for 1 000 000 (e.g., CCF, MBtu, MMBtu, MMBtu/hr). Additionally, in many cases, Btu/hr is expressed as Btu/h, Btuh, and BTUH, or in larger units of MBH or MMBH. The designer must grow familiar with the different methods of expressing I-P units because practices are frequently not consistent from one manufacturer to the next.

4.2 TRANSMISSION HEAT LOSS COMPUTATIONS

Transmission Heat Loss in Construction Assemblies

Transmission heat loss is heat lost through the building envelope. Heat lost through a building assembly is calculated by multiplying the coefficient of heat transmission (U-factor), by the area of the assembly, the difference between inside and outside (ambient) temperature, and a time period. In equation form, *transmission heat loss (Q)* and *rate of transmission heat loss (q) are* found by:

$$Q = U \cdot A \cdot \Delta T \cdot t$$
$$q = U \cdot A \cdot \Delta T$$

where U is the overall heat transmission coefficient in Btu/hr · °F · ft^2 (W/m^2 · °C), A is the area in ft^2 (m^2), ΔT is the temperature difference in °F (°C), t is the time period in hours, Q is the quantity of heat loss in Btu (W-hr), and q is the rate of heat transfer in Btu/hr (W). The area, A, is of the construction assembly (e.g., wall, ceiling, floor) that is between the heated and unheated spaces. Note that ΔT is found by subtracting the outside air temperature from the air temperature of the heated space ($\Delta T = T_{outside} - T_{inside}$). Typically, these are design temperatures.

The various portions of the building envelope may differ in the amounts of heat conducted through them because they have different U-factors (R-values). Such differences will depend upon the conductive qualities of the materials used in each section. The concrete used in the foundation, wood in the walls, glass of the windows, and shingles of the roof, all have different conductive characteristics. Heat transfer analysis is required for each of the different construction assemblies of a building.

Example 4.1

A house has 1000 ft^2 of net exterior wall area (after subtracting for windows, doors, and so on). The outside air temperature is 0°F and the inside temperature is 68°F.

a. Compute the rate of heat loss through an insulated wall (U = 0.073 Btu/hr · °F · ft^2).

$$q = U \cdot A \cdot \Delta T = 0.073 \text{ Btu/hr} \cdot °F \cdot ft^2 \cdot 1000 \text{ ft}^2$$
$$\cdot (68 - 0)°F = 4964 \text{ Btu/hr}$$

b. Compute the rate of heat loss through an uninsulated wall (U = 0. 267 Btu/hr · °F · ft^2).

$$q = U \cdot A \cdot \Delta T = 0.267 \text{ Btu/hr} \cdot °F \cdot ft^2 \cdot 1000 \text{ ft}^2$$
$$\cdot (68 - 0)°F = 18 \ 156 \text{ Btu/hr}$$

c. Compute the transmission heat loss through an insulated (U = 0.073 Btu/hr · °F · ft^2) over a period of one day (24 hr).

$$Q = U \cdot A \cdot \Delta T \cdot t = 0.073 \text{ Btu/hr} \cdot °F \cdot ft^2 \cdot 1000 \text{ ft}^2$$
$$(68 - 0)°F \cdot 24 \text{ hr} = 119 \ 136 \text{ Btu}$$

Upon review of Example 4.1, it is clearly evident that the uninsulated wall assembly has a rate of heat loss much greater than the insulated wall.

Transmission Heat Loss in Construction Assemblies Below Grade

Earth is an excellent conductor of heat; one foot of dry earth has an R-value equal to about one. However, earth has the ability to absorb huge quantities of heat over a long period of time and then release the stored heat over an equally long period of time. In essence, the heat capacity of earth causes it to act as an excellent thermal moderator.

Ground temperature at 10 ft below grade is about the same as water temperature at depths of 20 to 60 ft. In the United States, average ground temperature varies from 50 to 80°F, depending upon location. Because of these constant temperatures, soil above a 10 ft depth changes temperature much slower than does the rapidly changing air temperature. This slow changing characteristic is known as thermal lag. Thermal lag not only reduces the daily temperature swing below grade, but also forces a lag in weekly and monthly average soil temperatures. For example, at a 10 ft depth, soil temperatures lag average air temperatures by about three months. Normally there is a lag period of 1 to 1.5 weeks for each foot of depth below grade (down to about 10 ft, where the temperature swing begins to level off).

In addition to moderation of the outside temperature, heat loss below grade is affected by heat gain from the structure to the earth. This results in an increase in the temperature of the soil surrounding the structure and a further reduction in heat loss. It is important to note that because the average soil temperature remains below a comfortable inside temperature, insulating construction assemblies below grade is necessary in cold climates.

The rate of heat flow through a wall assembly decreases with the depth of the soil, because the total thermal moderation provided by the soil depends upon depth below grade. A precise calculation is extremely difficult, and involves such variables as moisture content and type of soil. Snow cover also greatly affects temperature moderation, because of its insulat-ing ability above grade. An approximate overall coefficient of heat transfer based upon depth below grade and insulation levels of the construction assembly is provided in the Appendix. These values are based upon depth below grade and R-value of the wall. These values can be multiplied by the area of the assembly at a certain depth below grade and the temperature difference:

$$Q = U_s \cdot A \cdot \Delta T \cdot t$$
$$q = U_s \cdot A \cdot \Delta T$$

where U_s is the overall coefficient of heat transfer based upon soil depth and insulation level, A is the area of the construction assembly at a specific depth below grade, ΔT is the temperature difference, and t is the time period in hours. Values for U_s are listed in Table 4.1. These values are based upon depth below grade. See Figure 4.2.

Selection of an accurate temperature difference (ΔT) for temperatures below grade is complicated because earth temperatures are different than air temperature at winter design conditions. ASHRAE suggests subtracting a ground-temperature fluctuation of 18° to 25°F from the average air temperature.

TABLE 4.1   OVERALL COEFFICIENT OF HEAT TRANSFER (U_S) FOR UNINSULATED AND INSULATED WALLS BELOW GRADE.

Depth Below Grade	Overall Coefficient of Heat Transmission (U) in Btu/hr · °F · ft² With R-Value of Wall Insulation in hr · °F · ft²/Btu			
ft	Uninsulated	R = 4.2	R = 8.3	R = 12.5
0 to 1	0.410	0.152	0.093	0.067
1 to 2	0.222	0.116	0.079	0.059
2 to 3	0.155	0.094	0.068	0.053
3 to 4	0.119	0.079	0.060	0.048
4 to 5	0.096	0.069	0.053	0.044
5 to 6	0.079	0.060	0.048	0.040
6 to 7	0.069	0.054	0.044	0.037
7 to 8	0.061	0.049	0.041	0.035

Depth Below Grade	Overall Coefficient of Heat Transmission (U) in W/°C · m² With R-Value of Wall Insulation in °C · m²/W			
m	Uninsulated	R = 0.7	R = 1.5	R = 2.2
0 to 0.3	2.33	0.86	0.53	0.38
0.3 to 0.6	1.26	0.66	0.45	0.36
0.6 to 0.9	0.88	0.53	0.38	0.30
0.9 to 1.2	0.67	0.45	0.34	0.27
1.2 to 1.5	0.54	0.39	0.30	0.25
1.5 to 1.8	0.45	0.34	0.27	0.23
1.8 to 2.1	0.39	0.30	0.25	0.21
2.1 to 2.4	0.35	0.27	0.24	0.20

Extracted from 2005 ASHRAE *Handbook—Fundamentals.* Used with permission from American Society of Heating, Refrigerating and Air-Conditioning Engineers, Inc.

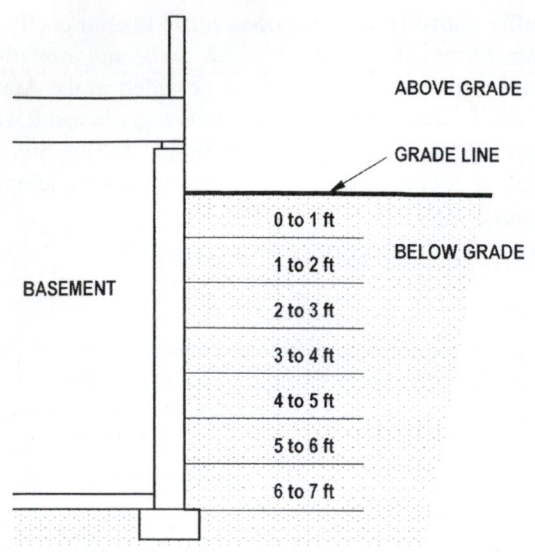

FIGURE 4.2 The overall coefficient of heat transfer (U_s) is based on soil depth below grade.

Example 4.2

A house has an 8 ft deep basement (foundation) wall that is in a 24 ft by 40 ft rectangular shape based on interior dimensions. On average, 7 ft of the wall is below grade. Assume the outside air temperature is 0°F and the inside temperature is 68°F; there is an average ground temperature of 40°F.

a. Determine the rate of heat loss through the wall if it is insulated with an R-value of about 12 hr · °F · ft²/Btu.

The house has 128 ft of basement wall: 2(24 ft + 40 ft) = 128 ft

The 8 in thick insulated wall has a U-factor of: 1/12 hr · °F · ft²/Btu = 0.083 Btu/hr · °F · ft²

Above grade: q = U · A · ΔT
From 0 to 1 ft: 0.083 Btu/hr · °F · ft² · 128 ft² · (68 − 0)°F = 722 Btu/hr

Below grade: q = U_s · A · ΔT
From 0 to 1 ft: 0.067 Btu/hr · °F · ft² · 128 ft² · (68 − 40)°F = 240 Btu/hr
From 1 to 2 ft: 0.059 Btu/hr · °F · ft² · 128 ft² · (68 − 40)°F = 211 Btu/hr
From 2 to 3 ft: 0.053 Btu/hr · °F · ft² · 128 ft² · (68 − 40)°F = 190 Btu/hr
From 3 to 4 ft: 0.048 Btu/hr · °F · ft² · 128 ft² · (68 − 40)°F = 172 Btu/hr
From 4 to 5 ft: 0.044 Btu/hr · °F · ft² · 128 ft² · (68 − 40)°F = 158 Btu/hr
From 5 to 6 ft: 0.040 Btu/hr · °F · ft² · 128 ft² · (68 − 40)°F = 143 Btu/hr
From 6 to 7 ft: 0.037 Btu/hr · °F · ft² · 128 ft² · (68 − 40)°F = 133 Btu/hr
                                                        q = 1969 Btu/hr

b. Determine the rate of heat loss through the wall if it is uninsulated. Assume the 8 in thick concrete wall has an R_t-value of 1.49 hr · °F · ft²/Btu.

The house has 128 ft of basement wall: 2(24 ft + 40 ft) = 128 ft

The 8 in thick concrete wall has an R-value of: 1/1.49 hr · °F · ft²/Btu = 0.671Btu/hr · °F · ft²

Above grade: q = U · A · ΔT
From 0 to 1 ft: 0.671 Btu/hr · °F · ft² · 128 ft² · (68 − 0)°F = 5840 Btu/hr

Below grade: q = U_s · A · ΔT
From 0 to 1 ft: 0.410 Btu/hr · °F · ft² · 128 ft² · (68 − 40)°F = 1469 Btu/hr
From 1 to 2 ft: 0.222 Btu/hr · °F · ft² · 128 ft² · (68 − 40)°F = 796 Btu/hr
From 2 to 3 ft: 0.155 Btu/hr · °F · ft² · 128 ft² · (68 − 40)°F = 556 Btu/hr
From 3 to 4 ft: 0.119 Btu/hr · °F · ft² · 128 ft² · (68 − 40)°F = 426 Btu/hr
From 4 to 5 ft: 0.096 Btu/hr · °F · ft² · 128 ft² · (68 − 40)°F = 344 Btu/hr
From 5 to 6 ft: 0.079 Btu/hr · °F · ft² · 128 ft² · (68 − 40)°F = 283 Btu/hr
From 6 to 7 ft: 0.069 Btu/hr · °F · ft² · 128 ft² · (68 − 40)°F = 247 Btu/hr
                                                        q = 9961 Btu/hr

It is clearly evident that the uninsulated basement wall assembly in Example 4.2 has a rate of heat loss much greater than the insulated wall. These results make a strong economic case for insulating walls even when they are below grade.

Transmission Heat Loss in Floors on Grade

For design purposes, the heat loss of a slab on or near grade is primarily from energy passing from the slab edge at the exterior building perimeter. Although the slab area away from exterior walls contributes somewhat to heat loss, it only accounts for a very small fraction (usually less than a few percent with perimeter insulation). Thus, transmission heat loss in floors on grade is based upon length of exposed perimeter edge rather than the area of the slab itself.

Heat loss through the exterior edge of a concrete slab on grade or a slab within 3 ft (1 m) of grade is based on:

$$Q = U_f \cdot L \cdot t$$
$$q = U_f \cdot L$$

where rate of heat loss per unit length (U_f) is expressed in Btu/hr-ft (W/m), L is lineal feet (ft) or meters (m) of edge of the slab exposed to the outside conditions, t is the time period in

TABLE 4.2   RATE OF HEAT LOSS PER UNIT LENGTH (U_f) THROUGH THE EXTERIOR EDGE OF CONCRETE SLABS ON GRADE BASED ON WINTER (OUTSIDE) DESIGN TEMPERATURE.

Winter Design Temperature (Outside)		Rate of Heat Loss (U_f)					
		Btu/hr per ft of Edge			W/hr per m of Edge		
			Perimeter Insulation			Perimeter Insulation	
°F	°C	No Insulation	1 in thick R = 2.5	2 in thick R = 5	No Insulation	100 mm R = 0.44	200 mm R = 0.88
20 to 10	−7 to −12	50	40	30	48	38	29
10 to 0	−12 to −18	55	45	35	62	43	34
0 to −10	−18 to −23	60	50	40	58	48	38
−10 to −20	−23 to −29	65	55	45	53	53	43
−20 to −30	−29 to −34	75	60	50	48	58	48

hours, and Q is the quantity of heat loss in Btu or rate of heat loss (q). Values for U_f are listed in Table 4.2. In colder climates, it is common practice to place insulation around the exterior perimeter of a building.

Example 4.3

A house is built on a slab on grade foundation that is a 25 ft-4 in by 41 ft-4 in rectangular shape. Assume the outside design temperature is −5°F.

a. Determine the rate of heat loss through the exterior edge of a concrete slab if the slab edge perimeter is insulated with 2 in of perimeter insulation (an R-value of about 5).

The house has 133.33 ft of exterior edge of a concrete slab: 2(25.33 ft + 41.33 ft) = 133.33 ft

$q = U_f \cdot L$ 40 Btu/hr per ft of edge · 133.33 ft
  = 5333 Btu/hr

b. Determine the rate of heat loss through the exterior edge of a concrete slab if the slab edge is not insulated.

The house has 133.33 ft of exterior edge of a concrete slab: 2(25.33 ft + 41.33 ft) = 133.33 ft

$q = U_f \cdot L$ 60 Btu/hr per ft of edge · 133.33 ft
  = 8000 Btu/hr

4.3   INFILTRATION HEAT LOSS COMPUTATIONS

Infiltration

Infiltration occurs when unconditioned outside air leaks into the building through the building envelope, is conditioned, and then leaks out (exfiltrates) from the structure. Infiltration air enters through openings in walls, cracks around doors and windows, electrical outlet holes, and a variety of other openings. The quantity of heat lost through infiltration depends on many factors, including inside–outside air temperature difference, wind velocity, occupancy patterns, furnace operation, and the physical condition of the structure. Warm inside air is more buoyant and rises to the top of the structure interior (chimney effect). In the winter, inside air leaks out through openings in the building envelope and is replaced by colder outside air that infiltrates through the building envelope. Wind creates a difference in pressure within the building, thereby increasing the rate of infiltration. As wind velocity increases, the amount of air leakage also increases. The proximity of large trees, fences, and other buildings will affect the amount of infiltration caused by wind.

Infiltration is the exchange of conditioned air with outside air, so it is necessary to know the heat capacity of air. The heat capacity (C) of air is the amount of heat necessary to change the temperature of one cubic foot (or cubic meter) of air one degree. It is expressed in units of Btu/ft³ · °F (W/m³ · °C). Heat capacity is dependent on the density of air, which varies with temperature and atmospheric pressure. The heat capacity of air will generally range from 0.014 to 0.018 Btu/ft³ · °F (0.26 to 0.35 W/m³ · °C); the lower value applies to higher altitudes (e.g., Denver). It is common practice to use 0.018 Btu/ft³ · °F (0.35 W/m³ · °C) as the heat capacity of air in heating load calculations.

Infiltration Heat Loss

The quantity of heat lost through infiltration is difficult to calculate with precision. However, a good approximation is important, because infiltration heat loss is usually a significant fraction of the total energy consumed in space heating or cooling a building. There are two methods used to calculate heat loss from infiltration. The *crack-estimation method* is calculated on the basis of physical size and length of known cracks, such as around doors and windows. The *air-change method* is based on an approximation of the number of air exchanges per hour based upon the volume of the structure or room under consideration. Both methods require engineering judgment regarding quality of construction. The air-change method is normally used in design of modern buildings, because all known cracks are sealed and infiltration is already reduced.

An *air change per hour (ACH)* is a way of expressing the rate at which air in an interior space is replaced by outside air caused by infiltration. It is expressed in terms of the exchange rate of the equivalent of one volume of the space or room under consideration; that is, 1.0 ACH relates to an infiltration rate equivalent to the volume of the space or room in one hour, a 2.0 ACH means that twice the volume of the space or room has been replaced by infiltration, and so on.

Instantaneous rates of infiltration vary significantly from seasonal (average) infiltration rates. Instantaneous infiltration rates as low as 0.1 ACH and as high as 5.0 ACH (or more) have been recorded in buildings. As a rule of thumb, seasonal (average) infiltration rates of 1.0 to 2.0 ACH are encountered in an older building; and 0.5 to 1.0 ACH may be anticipated in new residences, office buildings, and multifamily dwellings (e.g., condominiums, town homes, and apartments).

Establishing the ACH for heating load computations requires engineering judgment on the quality of construction of the building envelope because the only reliable way to determine the ACH is to measure it after the building is built. One method of estimating infiltration rates is reported in the 2001 *ASHRAE Handbook—Fundamentals*. It is based on outside design temperature and construction quality:

For tight, energy-efficient construction, the ACH is between 0.41 and 0.59.

For medium construction, the ACH is between 0.69 and 1.05.

For loose construction, the ACH is between 1.11 and 1.47.

These air exchange rates are based on outside design temperatures ranging from −40 to 50°F (−40 to 10°C); the higher ACH value in each construction category is for the highest outside design temperature (50°F/10°C). It is important to note that an ACH of at least 0.5 is desirable to maintain air quality over the long-term.

Once an estimate of air exchange rate has been established, *infiltration heating load (Q_{infil})* and *rate of infiltration loss (q_{infil})* may be computed using the following equation:

$$Q_{infil} = C \cdot ACH \cdot V \cdot \Delta T \cdot t$$
$$q_{infil} = C \cdot ACH \cdot V \cdot \Delta T$$

where C is the heat capacity of air (e.g., 0.014 to 0.018 Btu/ft³ · °F or 0.27 to 0.35 W/m³ · °C), ACH is the number of air changes per hour, V is the volume of the space in ft³ (m³), and ΔT is the inside–outside temperature difference. As with transmission heat losses, infiltration heat losses are proportional to the temperature difference.

Example 4.4

Calculate the rate of infiltration heat loss of a room with a 200 ft² floor area and 8 ft high ceilings. Use an inside temperature of 68°F, an outside ambient temperature of 0°F, and an hourly ACH of 0.5. Assume the heat capacity of air is 0.018 Btu/ft³ · °F.

$$V = 200 \text{ ft}^2 \cdot 8 \text{ ft} = 1600 \text{ ft}^3$$

$$q_{infil} = C \cdot ACH \cdot V \cdot \Delta T$$
$$q_{infil} = 0.018 \text{ Btu/ft}^3 \cdot °F \cdot 0.5 \text{ air changes/hr}$$
$$\qquad \cdot 1600 \text{ ft}^3 \cdot (68°F - 0°F)$$
$$q_{infil} = 979 \text{ Btu/hr}$$

4.4   HEAT LOSS FROM VENTILATION

Ventilation air is unconditioned outdoor air that is mechanically introduced into the building interior that, as a result, contributes to the building heating load. *For residences,* ventilation heat loss is generally ignored in heating load computations because it is typically not a code requirement and is generally small. The introduction of ventilation air is considered in combination with the infiltrated air and is handled effectively with the infiltration component of the computation described above. *For nonresidential buildings,* ventilation heat loss is a significant consideration in heating load computations because ventilation air is required in these buildings for occupant comfort and health. This is particularly true of high-occupancy buildings. Ventilation requirements and rates were discussed in Chapter 1.

In large buildings, ventilation heat loss is treated as a separate component of the heating load. For example, ASHRAE Standard 62 specified rates at which outside air must be supplied to a room, which is generally within the range of 15 to 60 cfm/person (7.5 to 30 L/s per person), depending on the activities that normally occur in that room. (A cfm is a cubic foot per minute of airflow). For example, assume a classroom space is designed for an occupancy of 30 persons. The ASHRAE Standard calls for a minimum outside airflow rate for classrooms of 15 cfm per person. Therefore, the classroom requires 15 cfm/ person · 30 persons, or 450 cfm.

Sensible heat loss from ventilation air is based on the quantity and conditions of outdoor air. The sensible heating load from ventilation can be computed as follows, based on the outside design temperature and inside design temperature difference (ΔT), in °F, and the volumetric airflow rate ($Q_{airflow}$), in cfm. The constant 1.1 is based on the heat capacity of air under standard conditions multiplied by 60 min in an hour: 0.018 Btu/ft³ · °F · 60 min/hr = 1.08 (rounded to 1.1).

$$q_{ventilation} = 1.1 \cdot Q_{airflow} \cdot \Delta T$$

In metric (SI) units, the sensible heating load from ventilation can be computed as follows, based on the outside design temperature and inside design temperature difference (ΔT), in °F, and the volumetric airflow rate ($Q_{airflow}$), in L/s.

$$q_{ventilation} = 1.2 \cdot Q_{airflow} \cdot \Delta T$$

Example 4.5

A college classroom is designed for an occupancy of 30 persons. The ASHRAE Standard calls for a minimum outside airflow rate for classrooms of 15 cfm per person. The target inside design temperature is 72°F and the outside design temperature is 3°F. Determine the sensible heating load from ventilation.

The classroom ventilation rate is found by:

$$15 \text{ cfm/person} \cdot 30 \text{ persons} = 450 \text{ cfm}$$

$$q_{\text{ventilation}} = 1.1 \cdot Q_{\text{airflow}} \cdot \Delta T = 1.1 \cdot 450 \text{ cfm} \cdot (72°F - 3°F)$$
$$= 34\,155 \text{ Btu/hr}$$

As shown in Example 4.5, the ventilation component of the heating load of a building with high occupancy can be significant.

4.5 HEATING LOAD ANALYSIS

The primary requirement for thermal comfort of building occupants during cold weather is a comfortable inside air dry bulb temperature. Heat introduced by heating equipment is constantly lost through the building envelope. Calculation of the heating load consists of estimating the maximum probable heat loss of each room or space to be heated while maintaining a selected inside air temperature during design outside weather conditions.

Heating Load Computation Methods

Heating load calculations are made to size an HVAC heating system (e.g., furnace, boiler) and its components (e.g., coils, ductwork, piping) based on heat loss and building performance at near extreme conditions. A building experiences a vast range of total heat losses, which vary by hour, day, month or year, and range from zero (no heating required) to the peak (maximum) heating load. There are several methods used to compute the heating load of a building from use of a rule-of-thumb method (e.g., 25 000 to 75 000 Btu/hr per 1000 ft^2 of floor area, based on geographic location) to detailed sophisticated computer modeling.

One of the most common problems encountered in HVAC heating system design is improper sizing of equipment. This is particularly true of residential systems where design normally involves use of unsophisticated rapid rule-of-thumb methods that lead to improper equipment size. For example, a study of design, construction, and performance of residential heating systems installed between 1994 and 1999 in Fort Collins, Colorado, showed that 70% of heating systems were notably oversized by an average of 158% of the design heating requirements. Oversized heating equipment leads to equipment short cycling, higher operating costs, and premature wear.

For the heating load to be accurately computed, the areas of the building envelope where heat transfer occurs and the heat transfer properties of the many different construction assemblies (e.g., walls, windows, doors, ceilings) must be identified. Heating load is computed on the basis of these characteristics. Except in very small buildings with a single zone, load calculations should be made on a room-by-room or zone-by-zone basis because an HVAC system must properly deliver heat to specific rooms or spaces. By knowing the individual room or zone load, the designer can determine duct or pipe size required to deliver the necessary heated air or water to the room or space.

Procedures that result in precise heating load calculation results should be used to compute the heating load of a building. Unsophisticated rule-of-thumb methods should only be used as approximations for preliminary design and should never be used for actual system sizing. ASHRAE publishes procedures for accurately computing the design heating and cooling loads in the *ASHRAE Handbook—Fundamentals*. The Air Conditioning Contractors of America, Inc. (ACCA), and the Air-Conditioning and Refrigeration Institute, Inc. (ARI), jointly developed *Manual J—Residential Load Calculation*, another widely accepted method of calculating the heating and cooling loads under design conditions for residential structures. ACCA's *Manual N—Commercial Load Calculation* can be used to estimate heating and cooling loads for small- and mid-sized commercial structures.

Design Conditions

The designer uses *winter design conditions* as the basis for computing the peak operating conditions under which HVAC systems and equipment must operate during the heating season. Inside dry bulb temperature and outside dry bulb temperature are used as the basis for design conditions.

Inside Design Conditions

The *inside design conditions* are the target indoor environmental conditions intended to achieve thermal comfort. As introduced in Chapter 1, thermal comfort is associated with many variables including air temperature, radiant temperature, air movement, gender, clothing, activity level, age, and adaptation to local climate. There is no single set of inside design temperature/humidity levels recommended to achieve acceptable comfort.

The typical inside design temperature range for the heating season in most U.S. homes is usually in the range of 68 to 72°F (20 to 22°C). In commercial buildings, indoor temperature and humidity levels should be in accordance with the comfort criteria established in ASHRAE Standard 55-2004, or the *ASHRAE Handbook—Fundamentals*. Local code requirements will vary. For example, one model code requires that inside design temperature shall be 70°F for heating.

Suggested indoor design dry bulb temperature ranges for selected occupancies are provided in Table 4.3. Inside design

TABLE 4.3   SUGGESTED INDOOR DESIGN (DRY BULB) TEMPERATURE AND RELATIVE HUMIDITY RANGES FOR SELECTED OCCUPANCIES. LOWER TEMPERATURES TYPICALLY APPLY TO THE HEATING SEASON. CODE REQUIREMENTS MAY VARY BY JURISDICTION (E.G., ONE CODE REQUIRES THAT INDOOR DESIGN TEMPERATURE SHALL BE 70°F FOR HEATING AND 78°F FOR COOLING).

Occupancy	Function of Space	Minimum Temperature		Maximum Temperature		Relative Humidity
		°F	°C	°F	°C	%
Residences	Habitable spaces	68	20	79	26	30 to 50
Schools	Classroom	68	20	75	24	30 to 50
	Administrative offices	70	21	75	24	30 to 50
	Service areas	70	21	80	26	30 to 60
	Gymnasium	66	19	72	22	30 to 50
	Swimming pool	75	24	82	28	30 to 70
Offices	Office, conference rooms	68	20	75	24	30 to 50
Hospitals	Operating rooms	68	20	75	24	50
	Patient rooms	75	24	75	24	30 to 50
	Intensive care	75	24	75	24	40
	Administrative offices	70	21	75	24	30 to 50
	Service areas	70	21	80	26	30 to 60
Libraries and museums	Viewing rooms	70	21	73	23	40 to 50
	Rare manuscript and storage vaults	70	21	72	22	45
	Art storage areas	65	18	72	22	50
Corridors, stairwells	Foot traffic	65	18	80	26	30 to 60

temperatures for heating tend to be on the low end of the scale. These recommendations apply only to inactive occupants wearing typical levels of indoor winter clothing. It may be necessary to adjust inside conditions based on other anticipated clothing levels in the space (e.g., higher temperatures for swimming pools, doctors' examination rooms), occupant age (e.g., higher temperatures for retirement homes), and activity level (e.g., lower temperatures for gymnasiums).

Outside Design Conditions

Outside dry bulb temperature based on climatic conditions at a specific geographic location are used as the basis for outside design conditions. Climate conditions used are from a long-term historical database and do not represent an actual year, but are statistical representations of average long-term conditions. The heating load is estimated for a winter design temperature occurring during the early morning hours. It is not economical to choose the minimum recorded outdoor air temperature as the basis for design because it leads to equipment oversizing to meet demands that seldom occur. Instead, design values near these extreme conditions are used.

Recommended outside winter design conditions are published in sources such as the *ASHRAE Handbook—Fundamentals*. ASHRAE tables containing winter design conditions for geographical locations in the United States and Canada are provided in Table 4.6. (Because of the length of these tables, they are placed at the end of the chapter, beginning on page 110.) These winter design conditions are reported as 99.6% and 99% design values that represent the dry bulb temperatures that are exceeded 99.6% and 99% of the time. In other words, of the

8760 total hours in a year, the outside dry bulb temperature at the specific geographical location historically fell below the 99.6% design temperature for 0.4% of the time in a year or 35 hr over an average year; and the 99% design temperature, for 88 hr. These design value percentages reflect the annual cumulative frequency of occurrence of outdoor climate conditions *during an average year*. For comparison, the new 99.6 and 99% winter conditions are slightly colder than the obsolete 99 and 97.5% conditions used in the past.

In Denver, Colorado (from Table 4.6), the 99.6% dry bulb temperature for heating is −3°F. The 99.0% dry bulb temperature for heating is 3°F. The 99.6% condition represents the most extreme design conditions and the 99.0% the least extreme. The designer must choose which set of design temperature conditions applies to a particular project. Traditionally, the 99.6% design values are used for computations involving light mass residential buildings (e.g., house, apartment, townhouse, manufactured home, dwelling, condominium, convent, monastery, hotel, and so forth). For massive, masonry structures that have little glass, the 99.0% design values should be used.

Design Temperature Difference (ΔT)

In heating load computations, the *design temperature difference (ΔT)* is found by subtracting outside temperature from inside temperature. For example, in Denver an inside temperature of 72°F minus the 99.6% outside design temperature of −3°F yields a design temperature difference of 75°F. The 75°F is substituted for ΔT in heat transfer equations to find the heating load at design conditions.

Temperatures in Unheated Spaces

To determine the heating load from a heated space to an unheated space (e.g., from a house to an unheated garage), it is first necessary to determine the temperature in the unheated space. The temperature in an unheated space will lie somewhere in the range between the inside and outside temperatures.

Temperatures in Unventilated Spaces

Temperatures in unventilated spaces will vary depending on the transmission characteristics and quantity of surface area adjacent to heated rooms compared to the transmission characteristics and quantity of surface area adjacent to the exterior. In theory, this temperature can be estimated using the following equation:

$$T_u = T_i(A_1U_1 + A_2U_2 + \cdots) + T_o(A_aU_a + A_bU_b + \cdots)/$$
$$(A_1U_1 + A_2U_2 + \cdots) + (A_aU_a + A_bU_b + \cdots)$$

where

T_u = temperature in unheated space (°F or °C)
T_i = inside design temperature of heated room (°F or °C)
T_o = outside design temperature (°F or °C)
$A_1, A_2, A_3, \ldots$ = areas of the construction assemblies that are exposed to the heated space (ft^2 or m^2)
$A_a, A_b, A_c, \ldots$ = areas of the construction assemblies that are exposed to the exterior (ft^2 or m^2)
$U_1, U_2, U_3, \ldots$ = coefficients of heat transmission for the construction assemblies in $A_1, A_2, A_3, \ldots$
$U_a, U_b, U_c, \ldots$ = coefficients of heat transmission for the construction assemblies in $A_a, A_b, A_c \ldots$

Temperatures in Unheated Attics, Basements, and Crawl Spaces

To accurately calculate the heating load, it is necessary to estimate the temperature in unheated attics, basements, and crawl spaces adjacent to the conditioned spaces. These temperatures cannot be easily calculated, especially when considering natural ventilation and earth temperatures. The designer can use the following as guidelines for estimating the air temperatures in unheated spaces:

- Unheated basements and crawl spaces located almost entirely below ground will normally have a temperature roughly midway between the inside and outside design temperatures being used.

- If the basement has open windows or the crawl space or attic is vented, the temperature will be lower than if it does not have windows or vents. For crawlspace or attic spaces separated with high levels of insulation, an indoor air temperature very near outside air temperature should be used, usually within a range of 2° to 10°F (1° to 5°C) above outside air temperature.

- If an operating heating unit (e.g., boiler or furnace) and uninsulated piping or ductwork is located in the

basement, crawl space, or attic space, the indoor temperature will be higher than if the heater were located elsewhere.

In summary, the designer must use judgment in approximating the temperature to use in the heating load calculations for unheated attics, basements, and crawl spaces adjacent to conditioned spaces.

Internal Heat Gain

Building heat gain is the sensible and latent heat emitted within an indoor space by the occupants, lighting, electric motors, and electronic equipment and solar heat gains from windows. For example, an occupant gives off about 250 Btu/hr, the heat given off by a 75 W lamp. A kitchen range may generate heat of 7000 Btu/hr (about 2000 W) or more when cooking a family meal. These and other sources account for a considerable heat gain, and it is for this reason that a building interior maintains a comfortable temperature even when the outside temperature falls well below inside temperatures. Additionally, buildings, especially those with heavy thermal mass, have the ability to coast past the coolest part of the day.

As a result of heat gains and effects of thermal mass, residential structures may require space heating only when the outside temperature drops about 15°F (9°C) below room temperature. Well-insulated buildings or buildings with heavy thermal mass may only need space heating at much lower outside temperatures. Buildings with high occupancies (e.g., churches, schools, and theaters), and those with significant internal heat gains from equipment and lights, may not have a space heating load until outside air temperatures drop substantially (i.e., below freezing). These buildings may only need a heating system for periods of low occupancy (e.g., vacation periods). Additionally, some buildings may rely upon heat rejected from equipment to provide space heating (e.g., refrigeration equipment at grocery stores).

Generally, heating load computations are made with the assumption that internal heat gains do not exist. Any heat gains during severe winter conditions serve as a factor of safety in design.

General Procedure

The process of estimating the design heating load involves computation of heat losses in each of the individual spaces at design conditions. Heating load calculations are carried out for each space (or zone) within the structure. The process of computing the heating load on a room-by-room basis makes the sizing of equipment (e.g., ducts, fan-coil units) serving these individual spaces easier. The process is outlined in Example 4.6, which follows. Designers use forms similar to those provided in Figure 4.3 to assemble data and aid in computing the repetitive computations. The general procedure is as follows:

1. Determine the outside design temperature.
2. Select the inside temperature to be maintained in the heated space during design conditions.

HEATING LOAD ANALYSIS WORKSHEET

ADVANCED BUILDING CONSULTANTS, LLC
12450 WEST BELMONT AVENUE
LITTLETON, CO 80127-1515
303.933.1300
advbldg1@aol.com

Project:		Job No.:
Address:		
Designer:		Date:
Checked by:		Date:
Room/Space Description:		Room No.:
		Page No.

1	**TRANSMISSION - ABOVE GRADE**					
2	**Description**	**U-factor**	**Area**	**ΔT**	**Subtotal**	Notes:
3		Btu/hr-ft^2-°F	ft^2	°F		
4	Glass					
5						
6						
7	Walls - above grade					
8						
9						
10						
11	Roof					
12						
13						

14	**TRANSMISSION - BELOW GRADE**						
15	**Wall Description**	**Depth**	**U-factor**	**Area**	**ΔT**	**Subtotal**	Notes:
16		ft	Btu/hr-ft^2-°F	ft^2	°F		
17	Walls - below grade	0 to 1					
18		1 to 2					
19		2 to 3					
20		3 to 4					
21		4 to 5					
22		5 to 6					
23		6 to 7					
24		7 to 8					
25	**Floor Description**		**U-factor**	**Area**	**ΔT**	**Subtotal**	Notes:
26			ft^2	Btu/hr-ft^2-°F	°F		
27	Basement Floors	>3ft below grade					
28	**Slab Description**		**Length**	**Heat Transfer Value**		**Subtotal**	Notes:
29			ft	Btu/hr-ft			
30	Exposed Slab Edge	<3ft below grade					

31	**INFILTRATION**						
32	Infiltration/Exfiltration	**Heat Capacity**	**Air change/hr**	**Volume**	**ΔT**	**Subtotal**	Notes:
33		Btu/ft^3-°F	ACH	ft^3	°F		
34							
35							
36							

37	**VENTILATION (non residential)**						
38	**Description**	**Occupant Load**	**Outdoor Air Requirements**	**Heat Flow Rate**	**ΔT**	**Subtotal**	Notes:
39		No. of occupants	cfm/occupant	Btu/cfm-°F	°F		
40							
41							
42							
43							
44				**TOTAL HEATING LOAD (Btu/hr)**			

FIGURE 4.3   A sample worksheet for heating load analysis.

3. Estimate air temperatures in adjacent unheated spaces.

4. Determine the overall coefficient of heat transfer (U-factor) for all construction assemblies that are between heated spaces and outside air or the heated space and unheated space.

5. Calculate the net area of all construction assemblies surrounding the heated spaces and outside or the heated space and an unheated space. Use inside dimensions.

6. Compute heat transmission losses for appropriate construction assemblies by using methods previously described.

7. Compute losses associated with infiltration.

8. If mechanical ventilation is used to maintain inside air quality (very tightly sealed structures), determine the rate of air exchange and calculate ventilation loads based upon design conditions.

9. Sum transmission and infiltration losses (and ventilation losses). This represents the design heating load of the space at design conditions (DL_S).

Example 4.6

A home in Denver, Colorado, has a 12 ft by 14 ft family room as shown in Figure 4.4. Temperature in the space is to be maintained at 70°F. It is exposed to the outside on two adjacent sides with three windows and a swing door on the south exterior wall, and a window on the east exterior wall. The room is above a finished (heated) basement. The space has the following characteristics:

- Ceiling/roof is well insulated (R = 45.2 hr · ft² · °F/Btu, U = 0.022 Btu/hr · °F · ft²) with 9 ft ceiling height.

- Exterior walls are well insulated (R = 18.2 hr · ft² · °F/Btu, U = 0.055 Btu/hr · °F · ft²).

- South windows are 3 ft by 4 ft with low-e double glass (R = 3.1 hr · ft² · °F/Btu, U = 0.32 Btu/hr · °F · ft²).

- East window is 2 ft-6 in by 4 ft with low-e double glass (R = 3.1 hr · ft² · °F/Btu, U = 0.32 Btu/hr · °F · ft²).

- Door is steel with urethane foam core that is 1¾ in thick (R = 5.3 hr · ft² · °F/Btu, U = 0.19 Btu/hr · °F · ft²).

- The infiltration rate is 0.5 ACH. There is no mechanical ventilation.

Find the heating load at design conditions for the room described.

In Denver, Colorado (from Table 4.6 at the end of the chapter), the 99.6% dry bulb temperature for heating is −3°F, so the design temperature difference is ΔT = 70°F − (−3°F) = 73°F.

Areas of the different construction assemblies are found by:

Area of ceiling:	12 ft · 14 ft = 168 ft²
Area of windows:	3 (3 ft · 4 ft) + 1(2.5 ft · 4 ft) = 46 ft²
Area of door:	2.67 ft · 6.67 ft = 17.8 ft²
Area of walls (net):	[(12 ft + 14 ft) · 9 ft] − 46 ft − 17.8 ft = 170.2 ft²

Transmission losses are found by:

$$\begin{array}{llll}
 & U & \cdot A & \cdot \Delta T = q \\
q_{ceiling} & = 0.022 \text{ Btu/hr}\cdot°F\cdot ft^2\cdot & 168 \text{ ft}^2 & \cdot 73°F = 270 \text{ Btu/hr} \\
q_{windows} & = 0.32 \text{ Btu/hr}\cdot°F\cdot ft^2\cdot & 46 \text{ ft}^2 & \cdot 73°F = 1075 \text{ Btu/hr} \\
q_{door} & = 0.19 \text{ Btu/hr}\cdot°F\cdot ft^2\cdot & 17.8 \text{ ft}^2 & \cdot 73°F = 247 \text{ Btu/hr} \\
q_{walls} & = 0.055 \text{ Btu/hr}\cdot°F\cdot ft^2\cdot & 170.2 \text{ ft}^2 & \cdot 73°F = 683 \text{ Btu/hr}
\end{array}$$

Infiltration losses are found by:

Volume of space: 12 ft · 14 ft · 9 ft = 1512 ft³

$$\begin{array}{lllll}
 & C & \cdot ACH \cdot & V & \cdot \Delta T = q \\
q_{infil} & = 0.018 \text{ Btu°F}\cdot ft^3\cdot & 0.5 & \cdot 1512 \text{ ft}^3\cdot & 73°F = 993 \text{ Btu/hr}
\end{array}$$

Heating load from space at design conditions (DL_S) = 3268 Btu/hr

Upon review of Example 4.6, it is evident that heat losses from infiltration represent a significant portion of the heating load. In an older home (ACH = 1.0), infiltration would be significantly greater. This makes a strong case for attention to sealing the structure (e.g., caulking and weather stripping) to minimize air leakage through the building envelope. In addition, when one considers the small area of windows, it is apparent that they contribute significantly to the overall heating load. Heat loss through windows could be reduced if high-performance windows were specified.

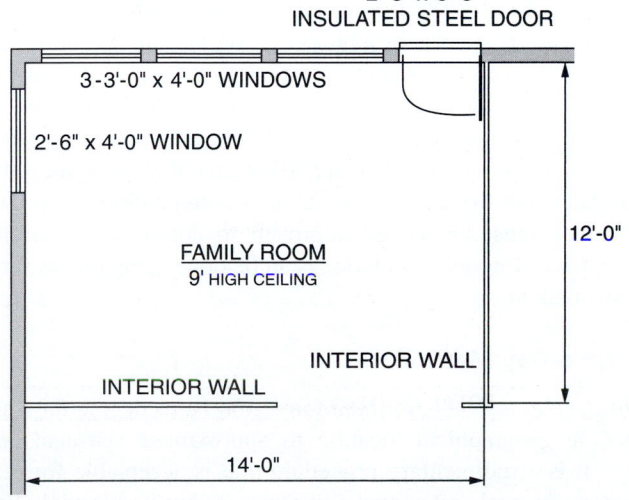

FIGURE 4.4 Floor plan for Example 4.6.

Heating load computations are not carried out to several decimal places, or even to single digits. Unknown factors associated with quality of materials and construction techniques affect the precision of these calculations to some extent. Moderate "rounding" in heating load calculations is acceptable in the building design industry. For example, it would be acceptable to say that the heating load in the family room from Example 4.6 is 3270, 3275, or even 3300 Btu/hr.

Total Heating Load

In small buildings with a single heating zone (e.g., a residence with a single central unit, a multifamily dwelling with apartments with individual heating units), the total heating load is determined by summing the loads of the individual spaces, where the *total heating load at design conditions (DL$_T$)* and the *individual space heating load at design conditions (DL$_S$)* are expressed in Btu/hr (W):

$$DL_T = DL_{S1} + DL_{S2} + DL_{S3} + DL_{S4} + \cdots$$

Example 4.7

With the following spaces and heating loads of each space at design conditions, calculate the total heating load.

Space	Heating Load (Btu/hr)
Bedroom #1	4800
Bedroom #2	5900
Master bedroom	6150
Bath #1	2800
Bath #2	3200
Living room	9700
Dining room	6800
Kitchen	4700
Foyer/hallways	2700
Total heating load at design conditions (DL$_T$)	46 750 Btu/hr

In buildings with heating systems divided into multiple, separately controlled zones (e.g., residences with multiple heating systems, multifamily dwellings served by a single heating system with fan-coil units in each apartment), *zone loads* or *block loads* are calculated with consideration only to the peak heat losses for that zone. Equipment serving a zone is sized to meet the peak block load for that zone. The total heating load at design conditions is calculated by taking into consideration variations in block loads in all zones. Simply put, the whole-building peak loads may not occur at the same instant as the peak load of any of its zones. Accurate determination of the time of occurrence of the total whole-building load requires either extensive hand calculations or, more realistically, an hourly computer simulation.

Future and Pick-Up Loads

Once total heating loads have been determined from the procedures outlined above, the capacity of the heating system may be increased by 10 to 20% to allow for future added loads or unexpected loads that may arise as space usage changes. Additionally, when HVAC systems are operated intermittently (e.g., night setback), the system must be capable of heating up the inside temperature, which has been allowed to coast to temperatures below comfort ranges. These additional loads are called *pick-up allowances*. The need for pickup of heating generally occurs during the early morning hours when the inside–outside temperature difference is at its greatest. A total heating load is generally increased in the range of 20 to 40% to provide an allowance for pickup.

Most HVAC system components operate less efficiently at part load than they do at full load. Oversizing results in equipment cycling, which causes premature equipment wear and adversely affects efficiency and performance. Design methodology results in equipment sized for peak conditions, so equipment is already oversized when operating at nonpeak, part-load duty the majority of the time. The routine practice of adding safety factors of 50 or 100% is unacceptable for nearly all applications. The tendency to significantly oversize HVAC equipment must be restrained.

4.6 HEATING ENERGY ESTIMATION

Energy Estimation Methods

As a part of the design of HVAC systems, it is frequently necessary to estimate annual heating energy consumption and operating costs. This analysis is useful in budget planning and in the evaluation of alternative system selection. There are both simple, approximate methods and complex, (more) accurate methods. Energy consumption and operational costs of HVAC systems will depend on weather and usage patterns, which are difficult to predict.

Although the inside temperature of a heated building remains relatively constant, outside temperatures fluctuate considerably. As a result of these dynamic conditions, it is difficult to estimate energy use by a building over the entire heating season, so all methods used are merely estimates. For a commercial building, this type of calculation can rarely, if ever, be performed manually to obtain a reliable estimate. There are a few simplified methods that are still used today to obtain rough estimates, but these should be treated with caution given their extremely unsophisticated approach to commercial building energy use. Energy calculations can be done using one of three basic methods.

Degree-Day Method

The *degree-day energy estimation method* uses degree-days at a specific geographical location to approximate seasonal heat loss. It is a rudimentary procedure that is acceptable for residential seasonal and annual computations but is extremely limited for commercial buildings.

Bin Method

The *bin energy estimation method* uses hourly occurrences of dry bulb temperatures that are collected and compiled in temperature ranges (usually 5°F/3°C), and then characterized by the midpoint temperature of the range of the individual temperature bin. The bin method represents a good way to obtain estimates for annual energy consumption because it accounts for the changing outside temperature and partial load conditions. It can be automated easily in a spreadsheet program and can be used to treat multiple-zone buildings. Hand calculations of simple, residential-type buildings are possible but drawn out and tedious. Commercial buildings would be more difficult to treat using hand calculations.

Hour-By-Hour Simulation Method

The *hour-by-hour simulation method* is the most advanced technique for performing annual energy use calculations because it involves computer simulation. Many commercial programs are available that perform an hourly simulation of the building to predict energy use and cost for all types of buildings based on data input provided by the user on building layout and construction methods, building systems (e.g., lighting, HVAC), and utility rates in combination with weather data. Although potentially very accurate, validity of results will depend to a great extent on accuracy of data input.

The degree-day energy estimation method is presented here to introduce the concept of estimating energy and costs. It offers a rough estimate of seasonal heat loss that is generally within 15% of actual performance. In presenting this method, it is first necessary to introduce the concept of degree-days.

Degree-Days

Degree-days are a measure of weather severity that make use of recorded daily dry bulb temperature extremes and an established inside dry bulb temperature called the *base temperature.* Heating and cooling degree-days are introduced in the following.

Heating Degree-Days

Heating degree-days are an index used to estimate the amount of energy required for heating during the cool season. The *heating degree-day (HDD)* is defined as the difference between a selected base temperature (T_{base}) and the daily average temperature (T_{ave}). In the United States, the base temperature is 65°F and in Canada it is 18°C. When the mean temperature is above 65°F (18°C), the heating degree-day total is 0. If the mean temperature is below the base temperature, the heating degree-day amount is the difference between the base temperature and the mean temperature.

$$HDD = T_{base} - T_{ave}$$

Cooling Degree-Days

Cooling degree-days are used as an index to approximate the amount of energy required for cooling during the warm season. When the temperature rises above 65°F, many buildings use air conditioning to maintain a comfortable inside temperature. The *cooling degree-day (CDD)* is defined as the difference between a selected base temperature (T_{base}), commonly 65°F (18°C), and the daily average temperature (T_{ave}). The daily mean temperature is found by adding together the high and low temperature for the day and dividing by 2. When the mean temperature is below 65°F (18°C), the cooling degree-day total is 0. If the mean temperature is above the base temperature, the cooling degree-day is the difference between the mean temperature and the base temperature.

$$CDD = T_{ave} - T_{base}$$

Example 4.8

The average daily temperature on a particular day was 30°F. With a 65°F base temperature, determine the HDD accrued.

Because the average daily temperature is below 65°F, heating degree-days are accrued:

$$HDD = T_{base} - T_{ave} = 65°F - 30°F = 35°F \cdot days$$

In the *simple degree-day method,* the *heating degree-day* and *cooling degree-day* is found as difference between a selected base temperature (T_{base}), commonly 65°F (19°C), and the median of the high (T_{high}) and low (T_{low}) extreme temperatures:

$$HDD = T_{base} - ((T_{high} + T_{low})/2) = T_{base} - T_{high/low\ ave}$$
$$CDD = ((T_{high} + T_{low})/2) - T_{base} = T_{high/low\ ave} - T_{base}$$

Example 4.9

The high temperature for a particular day was 33°F and the low temperature was 25°F. With a 65°F base temperature, determine the heating degree-days accrued.

The temperature median for that day was:

$$(33°F + 25°F)/2 = 29°F$$

Because the result is below 65°F:

$$HDD = 65°F - 29°F = 36°F \cdot days (HDD)$$

Example 4.10

The high temperature for a particular day was 90°F and the low temperature was 66°F. With a 65°F base temperature, determine the cooling degree-days accrued.

The temperature median for that day was:

$$(90°F + 66°F)/2 = 78°F$$

Because the result is above 65°F:

$$CDD = 78°F - 65°F = 13°F \cdot days \ (CDD)$$

The relationships between the Fahrenheit degree-day (°F·days) and the Celsius degree-day (°C·days) are:

$$°C \cdot day = 5/9 \ (°F \cdot days)$$
$$°F \cdot day = 9/5 \ (°C \cdot days)$$

Degree-days only accrue when they are positive; that is, on a warm day, a 65°F base temperature minus a 70°F average temperature produces 0 degree-days. During the winter months, heating degree-days accrue almost every day, except in warm climates. Summer months have few or no heating degree-days, except in cold climates.

Typical degree-day totals that are used for design purposes are based on historical monthly and yearly over many years. Historical degree-day information is normally available through the weather bureau, local utility company, and most fuel suppliers. Table 4.7 lists the 30-year seasonal normal heating degree-days for locations throughout the United States. Table 4.8 lists seasonal normal heating degree-days for selected cities in Canada. (Because of the length of these tables, they are placed at the end of the chapter, beginning on page 110).

The total number of degree-days over an entire heating season is a measure of the temperature severity of the winter. In a severe winter, the number of degree-days may be 10 or 15% above average, while in a mild winter, the number will be below average. Even in a winter when the number of degree-days is near average, individual months may be exceedingly abnormal. Because of these variations, the heating demand will vary from year to year. By monitoring degree-day totals and energy usage over a period of time, a building's energy consumption per degree-day can be calculated and used for monitoring fuel consumption, in energy efficiency evaluation and future fuel supply estimates.

The degree-day concept was established several decades ago by utilities and fuel companies to correlate energy demand with weather severity. The 65°F base temperature was based on an assumption that the average building had an indoor temperature of 75°F, with a balance point temperature of 65°F. It was assumed that the 10°F temperature difference was usually handled by internal heat gains so that below the 65°F base temperature heating would be required. Heating load calculations continue to be computed using this basis, despite better insulation, a large increase in internal heat gain, and lower temperature settings. Persistent use is primarily from availability of data.

Monthly and Seasonal Calculations

In using the degree-day concept, the equation for transmission heat loss over an extended period may be simplified. Because degree-days are the product of temperature difference (ΔT) and time (t) in days instead of hours, it is possible to substitute degree-days for ΔT and t, provided days are converted to hours (i.e., 24 hours per day):

$$Q = U \cdot A \cdot 24 \cdot HDD$$

where U is the overall coefficient of heat transfer in Btu/hr · °F · ft² (W/m² · °C), A is the area of the assembly in ft² (m²), 24 is an hours per day conversion, HDD is degree-days in °F · day (°C · day), and Q is the quantity seasonal heat loss through the construction assembly in Btu (J).

Similarly, the equation for air infiltration may be substituted with 24 hr/day and degree-days (°F · days) to estimate the seasonal heat loss from infiltration:

$$Q = C \cdot ACH \cdot V \cdot 24 \cdot HDD$$

where C is the heat capacity of air in Btu/ft³ · °F (W/m³ · °C), ACH is the number of air changes per hour, V is the volume of the heated space in ft³ (m³), 24 is hours per day, HDD is degree-days in °F · day (°C · day), and Q is the quantity of seasonal heat loss from the structure from infiltration in Btu (J).

Example 4.11

A house in Denver, Colorado, has insulated walls (U = 0.073 Btu/hr · °F · ft²). Assume the house has 886 ft² of net wall area. Calculate the annual heat loss through the insulated walls over a typical heating season.

From Table 4.7, annual heating degree-days in Denver, Colorado, are 6128°F · day/year

$$Q = U \cdot A \cdot 24 \cdot HDD$$

$$Q_{walls} = 0.073 \ Btu/hr \cdot °F \cdot ft² \cdot 886 \ ft² \cdot 24 \ hr/day \cdot 6128°F \cdot day/year = 9.5 \ MMBtu/year$$

Example 4.12

The house described in Example 4.11 is located in Atlanta, Georgia. Calculate the annual heat loss through the insulated walls over a typical heating season.

From Table 4.7, annual heating degree-days in Atlanta, Georgia, are 2827°F · day/year:

$$Q = U \cdot A \cdot 24 \cdot HDD$$

$$Q_{walls} = 0.073 \ Btu/hr \cdot °F \cdot ft² \cdot 886 \ ft² \cdot 24 \ hr/day \cdot 2827°F \cdot day/year = 4.4 \ MMBtu/year$$

Example 4.13

A house in Denver, Colorado, has a floor area of 1400 ft^2 and ceilings that are 8 ft high. Assume the house has an ACH of 0.6. Calculate the annual infiltration heat loss.

$$Q = C \cdot ACH \cdot V \cdot 24 \cdot HDD$$

$$Q_{infil} = 0.018 \text{ Btu/°F} \cdot \text{ft}^3 \cdot 0.6 \text{ ACH} \cdot (1400 \text{ ft}^2 \cdot 8 \text{ ft})$$
$$\cdot 24 \text{ hr/day} \cdot 6128°F \cdot \text{day/year} = 17.8 \text{ MMBtu}$$

Frequently, a building's balance point temperature is well below the base temperature of the degree-days used in the seasonal heat loss computation. A more accurate method of determining quantity of seasonal heat loss from the structure is based on a base temperature equivalent to the balance point temperature of the building—that is, the temperature at which the building first requires heating.

Approximating Heating Fuel Costs

The following method can be used to approximate the fuel consumed over the heating season based on the total design load (DL$_T$), heating degree-days, seasonal efficiency of the heating system or appliance (η_{hs}), the design temperature difference (ΔT_{design}), and the heating value or heat of combustion value of the fuel (HV):

$$\text{Fuel consumed} = (DL_T \cdot HDD \cdot 24)/(\eta_{hs} \cdot \Delta T_{design} \cdot HV)$$

Seasonal efficiency varies with types of heating system or appliance. The seasonal efficiencies of various types of heating systems and appliances are provided in Table 4.4. Approximate heating values of common heating fuels are listed in Table 4.5. These values are general references for residential heating applications only. Commercial and industrial users should obtain more precise values from their fuel vendors.

TABLE 4.4   PERFORMANCE OF DIFFERENT TYPES OF RESIDENTIAL SPACE HEATING SYSTEMS. COMPILED FROM MANUFACTURERS' INFORMATION.

Heating System	Typical Seasonal Efficiency
Gas- and Oil-Fired Boilers and Furnaces	
Old converted coal burner (converted to natural gas or oil), poorly maintained	35
Old converted coal burner (converted to natural gas or oil), well maintained	45
Standard low-efficiency furnace or boiler (pre-1985), poorly maintained	55
Low-efficiency furnace or boiler (pre-1985) with pilot light, well maintained	65
Low-efficiency furnace or boiler (pre-1985) with pilot light and vent damper, well maintained	70
Low-efficiency furnace or boiler (pre-1985) with electronic ignition and vent damper, well maintained	77
Medium-efficiency, induced draft furnace or boiler (post-1985), new or well maintained	80
High-efficiency, condensing furnace or boiler (post-1985), new or well maintained	92
High-efficiency, pulse furnace or boiler (post-1985), new or well maintained	96
Electric Units	
Electric resistance baseboard heater	100
Electric resistance furnace or boiler	97
Wood/Pellet/Coal-Burning Units	
Standard fireplace without glass doors	0
Standard fireplace with glass doors and tight damper	10
Standard fireplace with metal liner or tubular grate and tight damper	20
Standard fireplace with metal liner or tubular grate, tight damper, and glass doors	30
Standard updraft wood stove (e.g., Franklin stove)	30
Standard airtight wood or pellet stove, well maintained	50
Premium airtight wood or pellet stove, well maintained	60
Wood/coal furnace, well maintained	60
Wood/coal furnace with catalytic converter, well maintained	70
Heat Pumps	
Air source heat pump with electric resistance backup, well maintained	150–250
Geothermal (air or water) heat pump with electric resistance backup, well maintained	300–400

[a] For a moderate heating climate (5000°F · days). Varies significantly with climate.

TABLE 4.5 APPROXIMATE HEATING VALUES OF COMMON HEATING FUELS. THESE VALUES ARE GENERAL REFERENCES FOR RESIDENTIAL HEATING APPLICATIONS ONLY. COMMERCIAL AND INDUSTRIAL USERS SHOULD OBTAIN MORE PRECISE VALUES FROM THEIR FUEL VENDORS.

Commercial Fuel	Average Heating Values[a]	
	Customary (U.S.) Units	Metric (SI) Units
Coal (anthracite)	14 200 Btu/lb 28 400 000 Btu/ton	33 000 kJ/kg
Coal (bituminous)	9000 Btu/lb 18 100 000 Btu/ton	21 000 kJ/kg
Fuel oil number 1	136 000 Btu/gal	16 340 kJ/L
Fuel oil number 2	140 000 Btu/gal	16 820 kJ/L
Fuel oil number 4	149 000 Btu/gal	17 900 kJ/L
Fuel oil number 5	150 000 Btu/gal	18 020 kJ/L
Fuel oil number 6	154 000 Btu/gal	18 500 kJ/L
Natural gas	103 000 Btu/CCF[b] 1 030 000 Btu/MCF[b] 100 000 Btu/therm	37.2 kJ/m^3
Liquid petroleum gas (LPG)/propane	91 000 Btu/gal	11 000 kJ/L
Wood (lodge pole pine)	6000 Btu/lb 21 000 000 Btu/cord	14 000 kJ/kg
Wood (oak)	8700 Btu/lb 30 700 000 Btu/cord	20 000 kJ/kg
For Comparison		
Electricity	3 413 Btu/kWh	3 413 Btu/kWh
Gasoline	124 000 Btu/gal	14 900 kJ/L
Diesel fuel	139 000 Btu/gal	16 700 kJ/L

[a] The heating value varies to some extent with quality of fuel.
[b] A CCF is 100 ft^3. An MCF is 1000 ft^3. The heating value of natural gas varies significantly at high altitude. For example, the heating value of natural gas is typically about 83 000 Btu/CCF in Denver, CO, 5280 ft (1600 m) above sea level. Natural gases used at higher altitudes may be enriched with LPG.

Example 4.14

The home in Example 4.7 has a heat loss of 46 750 Btu/hr. Assuming a design temperature difference of 73°F, a climate with 6000°F · days, and a natural gas-fired, medium-efficiency, induced-draft furnace (at sea level).

a. Determine the fuel consumed over the heating season.

Fuel consumed = $(DL_T \cdot HDD \cdot 24)/(\eta_{hs} \cdot \Delta T_{design} \cdot HV)$
(46 750 Btu/hr · 6000°F · days
· 24 hr/day)/(0.80 · 73°F
· 103 000 Btu/CCF)
Fuel consumed = 1119 CCF

b. At $0.90/CCF, approximate the fuel cost over the heating season.

Fuel cost = 1119 CCF · $0.90/CCF = $1007

Limitations

This method of energy estimating is rudimentary and provides only rough approximations because it is based on the degree-day method and does not account fully for system efficiencies. Studies indicate that buildings coast through a portion of the heating season, resulting in a load that is less than that computed. In a moderate climate (5000°F · days), the seasonal load will be 20 to 30% less than the computed load using the degree-day method of energy estimating. On the other hand, additional system losses (e.g., piping losses, duct losses) have the effect of increasing the load because some of the energy consumed is waste heat. These efficiency losses can be significant, particularly when the additional part load efficiency losses are considered. These factors are opposite in effect and have the outcome of roughly canceling one another, which results in an estimate that more closely (but still roughly) represents actual performance.

STUDY QUESTIONS

4-1. What is transmission heat loss, and where does it occur in a building?

4-2. What is infiltration heat loss, and where does it occur in a building?

4-3. Define and describe the difference between building heat loss and building heat load.

4-4. The transmission heat load is computed by the equation, UA(ΔT). Describe each term in this equation.

4-5. With respect to a heating load, what is a pick-up allowance?

4-6. Air infiltration rates (i.e., air change per hour) vary considerably. Describe typical infiltration rates in new and older buildings.

4-7. How is the outside design temperature determined?

4-8. When are the following winter design conditions used?

   a. 99.6% dry bulb temperature

   b. 99.0% dry bulb temperature

4-9. Find the outside design temperature for the geographical location at which you reside.

4-10. Describe the degree-day, and how it will affect the amount of fuel used per year.

4-11. How does seasonal efficiency of the heating system affect heating costs?

4-12. List some of the ways in which the heat loss of a building can be controlled.

4-13. Why is limiting of window area important in keeping the heat loss low?

4-14. Describe three methods used to estimate energy consumption.

Design Exercises

4-15. Calculate the number of simple degree-days in a day if the high temperature were 40°F and the low temperature were 0°F.

   a. For a base temperature of 65°F.

   b. For a base temperature of 60°F.

   c. For a base temperature of 55°F.

4-16. Calculate the number of simple degree-days in a day if the high temperature were 10°C and the low temperature were 0°C.

   a. For a base temperature of 18°C.

   b. For a base temperature of 15°C.

   c. For a base temperature of 12°C.

4-17. Calculate the number of simple degree-days in a day if the high temperature were 70°F and the low temperature were 50°F.

   a. For a base temperature of 65°F.

   b. For a base temperature of 60°F.

   c. For a base temperature of 55°F.

4-18. Calculate the number of degree-days in a 30-day month if every day had an average daily temperature of 30°F.

   a. For a base temperature of 65°F.

   b. For a base temperature of 60°F.

   c. For a base temperature of 55°F.

4-19. Calculate the number of degree-days in a 30-day month if every day had an average daily temperature of 10°C.

   a. For a base temperature of 18°C.

   b. For a base temperature of 15°C.

   c. For a base temperature of 12°C.

4-20. Calculate the annual heat loss (in U.S. units) through the following wall assemblies using U-factors from Table 2.9 (in Chapter 2) and the annual heating degree-days over a typical heating season at the geographical location where you reside. Base your analysis on 1000 ft² of wall area.

   a. 2 × 4 wood-framed wall with R-11 insulation

   b. 2 × 6 wood-framed wall with R-19 insulation

   c. 8 in concrete uninsulated wall

   d. 8 in solid brick uninsulated wall

   e. 8 in concrete masonry unit (CMU) wall, with perlite-insulated cores

   f. 14 in thick uninsulated adobe wall

   g. 16 in solid log wall

   h. 23 in straw bale (R-1.45/inch) wall

   i. 3½ in structural insulated panel (SIP) with EPS core wall

4-21. Calculate the annual heat loss [in metric (SI) units] through the following wall assemblies using U-factors from Table 2.9 (in Chapter 2) and the annual heating degree-days over a typical heating season at the geographical location where you reside. Base your analysis on 100 m² of wall area.

   a. 38 mm × 88 mm wood-framed wall with RSI-1.9 insulation

   b. 38 mm × 140 mm wood-framed wall with RSI-3.3 insulation

   c. 200 mm standard-weight concrete uninsulated wall

   d. 200 mm solid brick uninsulated wall

   e. 200 mm CMU wall, with perlite insulated cores

   f. 350 mm thick uninsulated adobe wall

g. 400 mm solid log wall

h. 575 mm straw bale (RSI-0.01/mm) wall

i. 88 mm SIP with EPS core wall

4-22. A building has a 9 ft deep basement (foundation) wall that is 160 ft long. The wall has no windows. On average, 8 ft of the wall is below grade (underground). Assume an outside air temperature of 10°F, an inside temperature of 70°F, and an average ground temperature of 45°F.

  a. Determine the rate of heat loss through the wall below grade if it is uninsulated.

  b. Determine the rate of heat loss through the wall below grade if it is insulated with an R-value of about 12 hr · °F · ft^2/Btu.

4-23. A building has a 2.4 m deep basement (foundation) wall that is 100 m long. The wall has no windows. On average, 2.1 m of the wall is below grade (underground). Assume the outside air temperature is −10°C, the inside temperature is 20°C, and the average ground temperature is 10°C.

  a. Determine the rate of heat loss through the wall below grade if it is uninsulated.

  b. Determine the rate of heat loss through the wall below grade if it is insulated with an R-value of about 2.2°C · m^2/W.

4-24. A commercial building is built on a slab on grade foundation that is a 120 ft by 48 ft rectangular shape. Assume the outside design temperature is −5°F. Determine the rate of heat loss through the exterior edge of a concrete slab.

  a. For an uninsulated slab edge perimeter.

  b. For a slab edge perimeter insulated with 1 in of perimeter insulation (an R-value of about 2.5).

  c. For a slab edge perimeter insulated with 2 in of perimeter insulation (an R-value of about 5).

4-25. A commercial building is built on a slab on grade foundation that is a 40 m by 18 m rectangular shape. Assume the outside design temperature is −15°C. Determine the rate of heat loss through the exterior edge of a concrete slab.

  a. For an uninsulated slab edge perimeter.

  b. For a slab edge perimeter insulated with 25 mm of perimeter insulation (an RSI-value of about 0.44).

  c. For a slab edge perimeter insulated with 50 mm of perimeter insulation (an RSI-value of about 0.88).

4-26. Calculate the rate of infiltration heat loss of a room with a 380 ft^2 floor area and 10 ft high ceilings. Use an inside temperature of 72°F, an outside ambient temperature

of −5°F. Assume the heat capacity of air is 0.018 Btu/ft^3°F. Base the analysis on the following hourly air exchange rates:

  a. tight, energy efficient construction (ACH = 0.5)

  b. medium construction (ACH = 1.0)

  c. loose construction (ACH = 1.5)

4-27. Calculate the rate of infiltration heat loss of a room with a 42 m^2 floor area and 4 m high ceilings. Use an inside temperature of 21°C and an outside ambient temperature of −15°C. Assume the heat capacity of air is 0.35 W/m^3 · °C. Base the analysis on the following hourly air exchange rates:

  a. tight, energy efficient construction (ACH = 0.5)

  b. medium construction (ACH = 1.0)

  c. loose construction (ACH = 1.5)

4-28. A college lecture auditorium is designed for an occupancy of 300 persons. The ASHRAE Standard calls for a minimum outside airflow rate for classrooms of 15 cfm per person. The target inside design temperature is 74°F and the outside design temperature is −5°F. Determine the sensible heating load from ventilation.

4-29. A college lecture auditorium is designed for an occupancy of 300 persons. The ASHRAE Standard calls for a minimum outside airflow rate for classrooms of 2 L/s per person. The target inside design temperature is 22°C and the outside design temperature is −15°C. Determine the sensible heating load from ventilation.

4-30. A building has a heat loss of 100 MBtu/hr at design conditions at the geographical location where you reside. Calculate fuel consumed over the heating season at the geographical location where you reside. Base your analysis on the following fuels and efficiencies:

  a. Natural gas (80% efficiency)

  b. Natural gas (92% efficiency)

  c. Liquid petroleum gas/propane (80% efficiency)

  d. Fuel oil number 2 (78% efficiency)

  e. Electricity, resistance heating (100% efficiency)

  f. Wood (pine) (50% efficiency)

  g. Coal (60% efficiency)

4-31. For the previous exercise: Approximate the fuel cost over the heating season, based on current energy costs. (It will be necessary for you to contact local suppliers to get local costs.)

4-32. A building has a heat loss of 100 MJ/hr at design conditions at the geographical location where you reside. Calculate fuel consumed over the heating season at the geographical location where you reside. Base your analysis on the following fuels and efficiencies:

a. Natural gas (80% efficiency)

b. Natural gas (92% efficiency)

c. Liquid petroleum gas/propane (80% efficiency)

d. Fuel oil number 2 (78% efficiency)

e. Electricity, resistance heating (100% efficiency)

f. Wood (pine) (50% efficiency)

g. Coal (60% efficiency)

4-33. For the previous exercise: Approximate the fuel cost over the heating season, based on current energy costs. (It will be necessary for you to contact local suppliers to get local costs.)

INFORMATION FOR DESIGN EXERCISES 4.34 AND 4.35. HISTORICAL ENERGY CONSUMPTION DATA FOR AN OFFICE BUILDING WITH 182 400 ft^2 OF FINISHED FLOOR AREA IS PROVIDED IN THE FOLLOWING TABLE.

			Natural Gas				Electricity			
	Degree-Days (°F · days)		Consumption (MCF)[a]		Costs ($)		Electric Consumption (kWh)		Costs ($)	
Month	2006	2007	2006	2007	2006	2007	2006	2007	2006	2007
Jan	1007	1263	709	1200	$6 921.06	$12 752.01	250 160	223 721	$24 663.76	$21 947.61
Feb	878	928	610	872	$5 956.04	$9 264.05	229 481	197 946	$22 853.97	$19 886.37
Mar	816	1030	661	978	$6 448.92	$10 393.48	223 133	203 674	$22 336.70	$20 257.57
Apr	494	570	316	319	$3 087.82	$3 394.96	250 241	205 424	$25 145.54	$20 831.28
May	111	189	83	75	$811.30	$799.91	242 138	196 739	$24 957.63	$21 129.08
Jun	22	30	52	50	$507.52	$531.50	245 209	223 050	$25 369.99	$23 654.68
Jul	0	0	48	47	$468.48	$499.61	270 159	249 103	$27 172.34	$25 439.93
Aug	0	2	47	51	$458.72	$542.13	241 649	234 817	$25 030.63	$21 592.35
Sep	112	91	62	58	$605.12	$616.54	231 217	216 259	$24 037.24	$19 867.58
Oct	414	420	225	316	$2 194.78	$3 360.41	202 332	190 205	$21 576.98	$17 867.22
Nov	669	569	626	583	$6 112.20	$6 198.62	179 429	176 098	$18 457.49	$16 925.50
Dec	906	987	621	654	$6 064.62	$6 953.35	182 334	181 048	$18 710.29	$16 476.64
Total	5430	6079	4061	5203	$39 636.58	$55 306.56	2 747 480	2 498 085	$280 312.57	$245 875.81

[a]An MCF is 1000 ft^3 of natural gas with an approximate heating value of 1 030 000 Btu.

4-34. For the office building in the preceding table, compute the following for 2006:

a. Monthly natural gas consumption based on Btu/ft^2 of floor area.

b. Annual natural gas consumption based on Btu/ft^2 of floor area.

c. Monthly natural gas consumption normalized for weather conditions (degree-days), based on Btu/ft^2 · (°F · day).

d. Annual natural gas consumption normalized for weather conditions (degree-days), based on Btu/ft^2 · (°F · day).

4-35. For the office building in the preceding table, compute the following for 2007:

a. Monthly natural gas consumption based on Btu/ft^2 of floor area.

b. Annual natural gas consumption based on Btu/ft^2 of floor area.

c. Monthly natural gas consumption normalized for weather conditions (degree-days), based on Btu/ft^2 · (°F · day).

d. Annual natural gas consumption normalized for weather conditions (degree-days), based on Btu/ft^2 · (°F · day).

4-36. Calculate the heat loss for a top-floor apartment and an apartment that is below the top floor, using the apartment building in Appendix A. Assume the apartment will be built at the geographical location where you reside.

4-37. Calculate the heat loss for the residence in Appendix D on a room-by-room basis. Assume the residence will be built at the geographical location where you reside.

TABLE 4.6   WINTER DESIGN CONDITIONS FOR THE UNITED STATES AND CANADA.

United States

Station	WMO#	Lat.	Long.	Elev., ft	StdP, psia	Dates	Heating Dry Bulb 99.6%	99%	Extreme Wind Speed, mph 1%	2.5%	5%	Coldest Month 0.4% WS	MDB	1% WS	MDB	MWS/PWD to DB 99.6% MWS	PWD	0.4% MWS	PWD	Extr. Annual Daily Mean DB Max.	Min.	StdD DB Max.	Min.
1a	1b	1c	1d	1e	1f	1g	2a	2b	3a	3b	3c	4a	4b	4c	4d	5a	5b	5c	5d	6a	6b	6c	6d
ALABAMA																							
Anniston	722287	33.58	85.85	610	14.374	8293	19	24	16	14	13	18	47	15	46	6	300	7	240	98	10	3.2	7.4
Birmingham	722280	33.57	86.75	630	14.364	6193	18	23	19	17	15	20	41	18	42	7	340	9	320	98	9	3.3	6.4
Dothan	722268	31.32	85.45	400	14.484	8293	28	32	18	17	15	19	45	17	47	9	320	8	320	99	16	1.6	7.2
Huntsville	723230	34.65	86.77	643	14.357	6193	15	20	23	20	18	23	40	21	40	10	340	10	270	97	7	3.0	7.5
Mobile	722230	30.68	88.25	220	14.579	6193	26	30	22	19	17	23	48	21	48	10	360	9	320	97	18	1.9	6.3
Montgomery	722260	32.30	86.40	203	14.588	6193	24	27	20	17	15	20	45	18	45	7	360	8	270	98	15	2.9	6.3
Muscle Shoals/Florence	723235	34.75	87.62	551	14.405	8293	16	21	18	16	14	19	42	17	42	9	360	7	290	98	7	3.1	9.2
Ozark, Ft. Rucker	722269	31.28	85.72	299	14.538	8293	28	31	16	13	12	17	49	15	47	5	340	5	300	99	18	2.3	5.9
Tuscaloosa	722286	33.22	87.62	171	14.605	8293	20	24	17	14	13	18	47	16	51	5	360	7	240	99	11	1.8	6.8
ALASKA																							
Adak, NAS	704540	51.88	176.65	13	14.688	8293	19	23	34	30	27	40	34	34	35	4	210	10	170	67	11	3.4	2.9
Anchorage, Elemendorf AFB	702720	61.25	149.80	213	14.583	8293	−13	−8	17	14	12	18	26	15	26	3	50	7	260	77	−18	3.2	6.5
Anchorage, Ft. Richardson	702700	61.27	149.65	377	14.496	8293	−19	−13	19	14	11	20	35	15	36	3	50	5	270	80	−23	2.2	6.3
Anchorage, Intl Airport	702730	61.17	150.02	131	14.626	6193	−14	−9	22	19	17	23	18	19	18	4	10	8	290	77	−18	2.9	7.2
Annette	703980	55.03	131.57	112	14.636	6193	13	17	31	27	23	31	41	28	40	10	40	8	320	81	10	3.8	5.4
Barrow	700260	71.30	156.78	13	14.688	6193	−41	−36	28	25	22	30	3	26	−1	7	140	12	90	65	−45	4.7	4.4
Bethel	702190	60.78	161.80	151	14.615	6193	−28	−24	31	27	24	34	8	30	5	13	20	12	360	78	−32	3.3	6.6
Bettles	701740	66.92	151.52	643	14.357	6193	−49	−44	18	16	14	19	11	16	7	2	340	8	190	85	−55	4.0	5.8
Big Delta, Ft. Greely	702670	64.00	145.73	1283	14.027	6193	−45	−39	34	29	25	38	0	33	3	3	180	9	180	84	−48	3.3	7.5
Cold Bay	703160	55.20	162.73	102	14.642	6193	6	10	38	34	30	46	34	40	34	15	340	16	140	67	2	4.0	5.3
Cordova	702960	60.50	145.50	43	14.673	8293	−4	1	22	19	16	24	40	22	38	1	340	8	240	79	−9	5.0	5.4
Deadhorse	700637	70.20	148.47	56	14.666	8293	−36	−34	32	28	25	34	−1	30	−7	12	240	12	60	78	−51	14.2	5.2
Dillingham	703210	59.05	158.52	95	14.645	8293	−20	−13	25	22	20	28	20	24	21	5	40	10	180	74	−27	3.1	9.4
Fairbanks, Eielson AFB	702650	64.67	147.10	548	14.407	8293	−33	−31	14	12	10	14	21	11	16	0	150	5	290	87	−46	3.8	7.7
Fairbanks, Intl Airport	702610	64.82	147.87	453	14.457	6193	−47	−41	18	15	13	16	11	12	11	2	10	8	220	87	−48	3.8	7.8
Galena	702220	64.73	156.93	151	14.615	8293	−33	−31	18	15	13	19	14	16	15	0	270	5	320	84	−50	2.5	10.4
Gulkana	702710	62.15	145.45	1578	13.877	6193	−44	−39	26	24	21	22	17	19	18	3	360	7	180	82	−46	3.2	7.4
Homer	703410	59.63	151.50	72	14.657	8293	0	4	22	20	18	23	24	21	27	9	30	10	270	70	−5	4.0	6.8
Juneau	703810	58.37	134.58	23	14.683	8293	4	7	27	23	20	29	39	25	38	5	360	9	230	81	−1	2.5	4.9
Kenai	702590	60.57	151.25	95	14.645	8293	−22	−14	23	20	18	25	25	22	24	2	30	9	270	75	−27	3.4	7.4
Ketchikan	703950	55.35	131.70	95	14.645	8293	13	20	25	22	19	29	42	24	42	5	280	11	320	78	7	1.8	5.2
King Salmon	703260	58.68	156.65	49	14.669	6193	−24	−19	32	28	24	33	36	28	36	7	360	12	300	78	−31	3.5	7.2
Kodiak, State USCG Base	703500	57.75	152.50	112	14.636	6193	7	12	34	30	26	34	28	30	30	18	300	11	320	76	1	3.6	6.1
Kotzebue	701330	66.87	162.63	16	14.687	6193	−36	−31	35	31	28	38	14	32	14	7	70	12	300	75	−39	4.8	6.5
McGrath	702310	62.97	155.62	338	14.517	6193	−47	−42	18	16	14	18	23	14	12	1	310	7	340	83	−52	3.3	7.0
Middleton Island	703430	59.43	146.33	46	14.671	8293	18	21	40	34	30	42	35	37	36	18	330	8	260	66	15	4.9	6.8
Nenana	702600	64.55	149.08	361	14.505	8293	−51	−44	16	14	12	18	10	15	8	2	250	7	60	87	−52	4.1	7.2
Nome	702000	64.50	165.43	23	14.683	6193	−31	−26	30	26	23	31	17	28	18	4	20	12	260	76	−35	4.2	6.3
Northway	702910	62.97	141.93	1722	13.803	8293	−34	−32	15	13	12	14	−13	12	−6	0	300	7	290	83	−54	2.7	5.9
Port Heiden	703330	56.95	158.62	95	14.645	8293	−6	−2	38	32	28	38	36	32	35	17	60	15	160	74	−11	4.3	7.4
Saint Paul Island	703080	57.15	170.22	30	14.680	6193	−2	3	41	37	33	47	24	41	21	19	350	14	240	58	−3	5.2	6.9
Sitka	703710	57.07	135.35	66	14.661	8293	16	21	23	21	19	24	40	22	41	8	70	9	230	76	11	6.1	5.0
Talkeetna	702510	62.30	150.10	358	14.507	8293	−28	−21	17	16	14	19	13	17	15	4	50	8	200	82	−35	2.8	8.0
Valdez	702750	61.13	146.35	33	14.678	8293	4	7	24	19	16	28	13	22	15	15	70	10	240	76	1	3.6	6.1
Yakutat	703610	59.52	139.67	30	14.680	6193	−3	2	24	19	16	25	36	21	33	2	100	9	320	75	−8	4.0	7.0
ARIZONA																							
Flagstaff	723755	35.13	111.67	7011	11.335	6193	1	8	21	18	17	21	29	18	30	3	20	9	220	89	−10	2.5	7.3
Kingman	723700	35.27	113.95	3389	12.983	8293	22	27	26	23	20	24	46	21	43	5	90	13	240	103	15	1.8	6.8
Page	723710	36.93	111.45	4278	12.561	8293	20	24	19	16	13	16	42	12	40	4	300	7	360	104	8	3.6	12.2
Phoenix, Intl Airport	722780	33.43	112.02	1106	14.118	6193	34	37	19	16	14	17	59	14	58	5	90	9	270	114	30	2.2	4.6
Phoenix, Luke AFB	722785	33.53	112.38	1089	14.126	8293	35	38	19	15	13	16	58	13	55	4	340	9	210	115	30	2.2	3.8
Prescott	723723	34.65	112.42	5043	12.208	6193	15	20	22	19	17	21	42	18	42	6	190	11	230	98	7	2.2	6.2
Safford, Agri Center	722747	32.82	109.68	3117	13.114	8293	21	26	17	14	12	15	50	13	48	4	110	7	310	106	11	3.8	11.5
Tucson	722740	32.12	110.93	2556	13.388	6193	31	34	24	21	18	24	56	21	56	7	140	12	300	108	25	2.8	4.0
Winslow	723740	35.02	110.73	4882	12.281	8293	10	14	26	22	19	24	46	19	45	5	140	9	250	100	3	4.9	6.3
Yuma	722800	32.65	114.60	207	14.586	8293	40	44	19	17	15	20	59	17	58	4	30	7	280	116	29	1.8	11.9
ARKANSAS																							
Blytheville, Eaker AFB	723408	35.97	89.95	256	14.560	8293	12	18	22	19	17	23	36	21	38	10	10	6	240	99	6	5.0	9.0
Fayetteville	723445	36.00	94.17	1250	14.044	8293	6	13	21	19	18	21	44	19	44	9	350	10	190	100	−1	3.2	9.2
Ft. Smith	723440	35.33	94.37	463	14.451	6193	13	19	20	18	16	21	46	18	41	9	320	9	270	102	6	3.9	6.6
Little Rock, AFB	723405	34.92	92.15	312	14.531	6193	16	21	20	18	16	20	42	18	42	9	360	9	200	101	10	3.8	6.2
Texarkana	723418	33.45	93.98	390	14.489	8293	20	25	19	17	15	20	47	18	48	9	50	9	190	101	13	3.1	7.6
CALIFORNIA																							
Alameda, NAS	745060	37.78	122.32	13	14.688	8293	40	42	21	18	16	20	51	17	52	6	120	8	300	93	25	4.9	14.2
Arcata/Eureka	725945	40.98	124.10	217	14.581	6193	30	32	21	19	17	21	53	18	51	5	90	10	320	82	26	4.5	3.2
Bakersfield	723840	35.43	119.05	492	14.436	6193	32	35	19	16	14	19	56	14	54	5	90	12	310	108	28	2.3	3.4
Barstow/Daggett	723815	34.85	116.78	1926	13.701	6193	28	32	30	27	23	30	58	25	54	6	270	12	290	111	22	2.5	4.8
Blue Canyon	725845	39.28	120.72	5285	12.097	8293	21	24	15	12	11	16	35	14	35	5	70	6	290	89	11	3.8	18.4
Burbank/Glendale	722880	34.20	118.35	774	14.289	8293	39	41	18	14	12	20	56	17	57	2	330	8	180	106	33	3.1	4.5

WMO# = World Meteorological Organization number   AAF = Army Air Field   Lat. = North latitude °   StdP = standard pressure at station elevation, psia
AFB = Air Force Base   NAS = Naval Air Station   MCAS = Marine Corps Air Station   Long. = West longitude °   WS = wind speed, mph
WSO = Weather Service Office   DB = dry-bulb temperature, °F   Elev. = elevation, ft   PWD = prevailing wind direction, ° True

United States

Station	WMO#	Lat.	Long.	Elev., ft	StdP, psia	Dates	Heating Dry Bulb 99.6%	99%	Extreme Wind Speed, mph 1%	2.5%	5%	Coldest Month 0.4% WS	MDB	1% WS	MDB	MWS/PWD to DB 99.6% MWS	PWD	0.4% MWS	PWD	Extr. Annual Daily Mean DB Max.	Min.	StdD DB Max.	Min.
1a	1b	1c	1d	1e	1f	1g	2a	2b	3a	3b	3c	4a	4b	4c	4d	5a	5b	5c	5d	6a	6b	6c	6d
Fairfield, Travis AFB	745160	38.27	121.93	62	14.662	8293	31	34	28	24	22	26	53	22	52	4	20	9	240	105	26	3.4	4.0
Fresno	723890	36.77	119.72	328	14.522	6193	30	32	17	15	13	17	53	14	52	4	90	9	290	107	26	2.2	3.8
Lancaster/Palmdale	723816	34.73	118.22	2346	13.492	8293	22	24	30	28	25	29	48	26	49	2	260	14	240	107	15	2.0	5.9
Lemoore, Reeves NAS	747020	36.33	119.95	236	14.570	8293	30	32	19	16	14	20	49	16	51	4	150	7	360	110	17	4.1	10.4
Long Beach	722970	33.82	118.15	39	14.675	6193	40	43	19	16	14	19	58	16	58	4	300	10	270	102	35	4.5	2.8
Los Angeles	722950	33.93	118.40	105	14.640	6193	43	45	21	18	16	20	56	17	56	6	70	10	250	97	38	5.1	3.0
Marysville, Beale AFB	724837	39.13	121.43	112	14.636	8293	31	34	20	17	14	23	53	19	53	3	20	5	200	106	26	3.2	4.1
Merced, Castle AFB	724810	37.38	120.57	187	14.596	8293	30	32	18	15	12	21	51	17	49	2	110	9	320	104	26	2.7	3.6
Mount Shasta	725957	41.32	122.32	3543	12.909	8293	16	21	14	12	10	14	36	12	37	4	60	4	180	95	10	2.7	6.7
Mountain View, Moffet NAS	745090	37.42	122.05	39	14.675	8293	36	39	19	17	15	19	54	16	52	1	140	9	330	98	23	2.5	12.2
Ontario	722865	34.05	117.60	942	14.202	8293	35	38	22	19	17	28	62	21	57	4	10	13	240	108	29	3.4	2.5
Oxnard, Pt. Mugu NAWS	723910	34.12	119.12	7	14.692	8293	39	41	22	19	16	25	57	21	58	5	20	12	50	93	24	5.0	10.6
Paso Robles	723965	35.67	120.63	837	14.257	8293	26	29	22	20	18	21	52	18	51	3	110	11	300	108	21	2.2	4.9
Red Bluff	725910	40.15	122.25	354	14.508	8293	29	32	23	21	19	26	53	23	50	6	340	9	160	111	25	3.2	3.8
Riverside, March AFB	722860	33.88	117.27	1539	13.896	8293	34	36	18	15	13	22	51	18	55	1	210	9	300	107	29	2.3	3.2
Sacramento, Mather Field	724835	38.55	121.30	95	14.645	8293	30	32	20	17	14	24	53	20	51	2	120	6	310	105	26	5.0	4.3
Sacramento, McClellan AFB	724836	38.67	121.40	75	14.655	8293	31	34	20	16	14	23	53	19	52	2	340	5	220	107	27	2.5	4.9
Sacramento, Metro	724839	38.70	121.58	23	14.683	6193	31	33	22	19	17	23	51	20	50	3	340	8	220	107	27	6.6	3.0
Salinas	724917	36.67	121.60	85	14.650	8293	33	35	21	19	18	23	51	21	51	6	130	11	310	95	29	4.7	2.2
San Bernardino, Norton AFB	722866	34.10	117.23	1158	14.091	8293	34	36	17	13	11	21	56	16	55	2	50	8	250	109	29	2.5	2.7
San Diego, Intl Airport	722900	32.73	117.17	30	14.680	6193	44	46	18	16	15	20	59	16	60	3	70	10	310	95	39	6.4	4.4
San Diego, Miramar NAS	722930	32.85	117.12	420	14.474	8293	39	42	13	11	9	15	59	12	59	3	90	6	310	102	27	4.0	13.7
San Francisco	724940	37.62	122.38	16	14.687	6193	37	39	29	26	23	27	53	22	52	5	160	13	300	94	33	4.3	3.0
San Jose Intl Airport	724945	37.37	121.93	56	14.666	8293	35	38	20	18	17	20	56	17	56	1	160	10	320	101	27	3.1	9.0
Santa Barbara	723925	34.43	119.83	10	14.690	8293	34	37	20	17	14	19	58	16	58	1	40	10	260	97	28	6.7	6.5
Santa Maria	723940	34.90	120.45	240	14.569	6193	32	35	23	21	19	21	59	18	59	4	110	11	300	95	27	5.0	2.8
Stockton	724920	37.90	121.25	26	14.681	8293	30	32	22	19	17	24	52	21	49	4	110	11	280	106	26	3.1	3.4
Victorville, George AFB	723825	34.58	117.38	2874	13.232	8293	27	30	22	19	16	22	49	18	47	3	160	9	180	106	21	3.1	5.6
COLORADO																							
Alamosa	724620	37.45	105.87	7543	11.108	6193	−17	−11	26	23	21	23	33	20	30	3	190	12	240	88	−27	2.0	7.9
Colorado Springs	724660	38.82	104.72	6171	11.701	6193	−2	4	29	25	21	28	35	23	33	7	20	12	160	95	−9	2.0	6.9
Craig	725700	40.50	107.53	6283	11.652	8293	−20	−12	26	20	17	22	33	17	27	2	270	9	250	93	−31	2.0	10.6
Denver	724699	39.75	104.87	5331	12.076	8293	−3	3	24	21	18	25	39	21	40	6	180	9	160	97	−11	2.3	7.0
Eagle	724675	39.65	106.92	6539	11.539	6193	−13	−7	22	19	17	20	33	18	32	3	90	11	230	93	−23	3.2	7.8
Grand Junction	724760	39.12	108.53	4839	12.301	6193	2	7	22	19	17	17	33	14	30	5	70	11	290	100	−3	2.0	8.5
Limon	724665	39.27	103.67	5364	12.062	8293	−6	1	27	23	21	27	29	22	25	9	160	12	200	96	−13	2.2	4.5
Pueblo	724640	38.28	104.52	4721	12.355	6193	−1	5	32	27	24	30	44	26	43	5	270	12	140	102	−12	1.9	7.7
Trinidad	724645	37.27	104.33	5761	11.883	8293	−2	6	25	22	19	24	41	21	42	5	290	10	210	98	−10	2.0	6.8
CONNECTICUT																							
Bridgeport	725040	41.17	73.13	16	14.687	6193	8	12	27	23	21	34	29	30	29	14	320	14	230	93	2	2.8	4.9
Hartford, Brainard Field	725087	41.73	72.65	20	14.685	6193	2	6	23	20	18	23	25	20	26	7	320	11	250	97	−6	2.4	5.7
Windsor Locks, Bradley Fld	725080	41.93	72.68	180	14.600	8293	3	8	21	19	17	22	30	20	29	7	360	11	240	97	−5	2.0	5.8
DELAWARE																							
Dover, AFB	724088	39.13	75.47	30	14.680	8293	14	18	22	19	17	23	36	21	35	8	340	9	240	97	6	3.2	6.1
Wilmington	724089	39.68	75.60	79	14.654	6193	10	14	25	22	19	27	29	23	30	11	290	11	240	96	3	2.7	6.8
FLORIDA																							
Apalachicola	722200	29.73	85.03	20	14.685	8293	31	35	19	17	15	19	51	17	51	6	360	9	220	93	23	6.7	7.4
Cape Canaveral, NASA	747946	28.62	80.72	10	14.690	8293	38	42	19	17	15	21	60	19	60	8	320	8	220	96	29	1.4	6.1
Daytona Beach	722056	29.18	81.05	36	14.676	6193	34	37	21	19	17	22	61	19	61	7	310	11	240	96	27	1.9	4.4
Ft. Lauderdale/Hollywood	722025	26.07	80.15	23	14.683	8293	46	50	22	20	18	22	69	20	71	9	330	11	120	97	39	1.1	6.1
Ft. Myers	722106	26.58	81.87	16	14.687	8293	42	47	19	18	16	20	64	18	66	6	30	9	70	97	34	1.3	4.7
Gainesville	722146	29.68	82.27	151	14.635	8293	30	33	19	17	14	19	65	17	62	4	300	9	270	97	21	1.8	7.2
Homestead, AFB	722026	25.48	80.38	7	14.692	8293	48	52	17	15	13	17	70	15	70	6	360	7	120	95	41	2.2	5.6
Jacksonville, Cecil Fld NAS	722067	30.22	81.88	82	14.652	8293	31	34	18	16	14	19	62	17	62	3	290	7	270	100	20	2.0	8.8
Jacksonville, Intl Airport	722060	30.50	81.70	30	14.680	6193	29	32	21	18	17	21	54	19	55	6	310	9	230	98	22	2.1	5.1
Jacksonville, Mayport Naval	722066	30.40	81.42	16	14.687	8293	34	39	19	17	14	21	54	18	55	6	310	7	270	99	20	2.2	13.1
Key West	722010	24.55	81.75	20	14.685	6193	55	58	22	20	18	24	65	21	66	12	50	9	140	91	51	1.2	4.5
Melbourne	722040	28.10	80.65	36	14.676	8293	38	43	21	19	18	22	62	20	62	9	320	11	120	97	30	1.8	6.7
Miami, Intl Airport	722020	25.82	80.28	13	14.688	6193	46	50	23	20	18	22	68	20	69	10	340	11	150	94	39	2.1	5.1
Miami, New Tamiami A	722029	25.65	80.43	10	14.690	8293	45	49	21	19	18	21	72	19	72	8	360	11	130	95	39	2.0	6.3
Milton, Whiting Field NAS	722226	30.72	87.02	200	14.589	8293	28	31	18	16	14	19	50	17	52	6	340	6	330	99	17	1.8	8.6
Orlando	722050	28.43	81.32	105	14.640	8293	37	42	20	18	16	21	66	19	65	8	330	9	290	96	29	1.6	6.5
Panama City, Tyndall AFB	747750	30.07	85.58	16	14.687	8293	33	37	18	16	14	19	52	17	52	9	360	7	240	94	24	2.3	6.3
Pensacola, Sherman AFB	722225	30.35	87.32	30	14.680	8293	28	32	23	20	18	25	43	22	48	9	360	10	200	100	15	6.3	7.6
Saint Petersburg	722116	27.92	82.68	10	14.690	8293	43	47	21	19	17	22	65	20	63	11	10	10	230	97	35	1.8	4.7
Sarasota/Bradenton	722115	27.40	82.55	30	14.680	8293	39	43	22	19	17	23	67	20	67	5	40	9	270	97	29	1.6	12.1
Tallahassee	722140	30.38	84.37	69	14.654	6193	25	28	18	16	14	19	52	17	54	3	350	8	300	98	17	2.0	4.6
Tampa, Intl Airport	722110	27.97	82.53	10	14.690	6193	36	40	19	17	15	21	59	19	59	8	20	10	270	95	29	1.2	4.8
Valparaiso, Eglin AFB	722210	30.48	86.53	85	14.650	8293	30	33	19	16	14	18	49	16	51	6	360	7	210	97	19	2.0	6.1
Vero Beach	722045	27.65	80.42	26	14.681	8293	39	43	20	19	17	21	67	19	67	8	310	11	240	96	31	2.0	6.5

WMO# = World Meteorological Organization number
AFB = Air Force Base   NAS = Naval Air Station
WSO = Weather Service Office

AAF = Army Air Field
MCAS = Marine Corps Air Station
DB = dry-bulb temperature, °F

Lat. = North latitude °
Long. = West longitude °
Elev. = elevation, ft

StdP = standard pressure at station elevation, psia
WS = wind speed, mph
PWD = prevailing wind direction, ° True

United States

Station	WMO#	Lat.	Long.	Elev., ft	StdP, psia	Dates	Heating Dry Bulb 99.6%	99%	Extreme Wind Speed, mph 1%	2.5%	5%	Coldest Month 0.4% WS	MDB	1% WS	MDB	MWS/PWD to DB 99.6% MWS	PWD	0.4% MWS	PWD	Extr. Annual Daily Mean DB Max.	Min.	StdD DB Max.	Min.
1a	1b	1c	1d	1e	1f	1g	2a	2b	3a	3b	3c	4a	4b	4c	4d	5a	5b	5c	5d	6a	6b	6c	6d
West Palm Beach	722030	26.68	80.12	20	14.685	6193	43	47	24	21	19	24	69	21	70	9	320	12	110	94	35	2.0	5.0
GEORGIA																							
Albany	722160	31.53	84.18	194	14.593	8293	27	30	19	17	15	19	50	18	50	4	360	9	250	100	17	2.2	7.2
Athens	723110	33.95	83.32	810	14.270	6193	20	25	19	17	15	20	40	18	40	10	290	9	270	98	11	3.5	6.6
Atlanta	722190	33.65	84.42	1033	14.155	6193	18	23	22	19	17	23	37	21	36	12	320	9	300	96	9	3.5	7.3
Augusta	722180	33.37	81.97	148	14.617	6193	21	25	20	18	15	21	45	19	46	5	290	9	250	100	13	3.7	5.6
Brunswick	722137	31.15	81.38	20	14.685	8293	30	34	18	17	16	19	49	18	49	8	350	10	250	98	22	2.5	7.7
Columbus, Ft. Benning	722250	32.33	85.00	233	14.572	8293	23	27	16	13	11	17	46	15	46	3	320	5	240	100	14	2.9	6.7
Columbus, Metro Airport	722255	32.52	84.93	397	14.486	6193	23	27	17	15	14	18	44	16	46	7	310	9	310	99	14	2.3	6.1
Macon	722170	32.70	83.65	361	14.505	6193	23	27	19	17	15	20	46	18	45	7	320	9	270	100	14	2.7	6.4
Marietta, Dobbins AFB	722270	33.92	84.52	1070	14.136	8293	21	26	18	16	13	20	35	18	38	9	340	6	300	97	12	3.6	6.7
Rome	723200	34.35	85.17	643	14.357	8293	15	21	14	12	10	14	42	13	42	5	340	6	270	98	4	3.8	7.0
Savannah	722070	32.13	81.20	49	14.669	6193	26	29	20	17	15	21	49	19	49	7	270	9	270	98	18	3.0	5.4
Valdosta, Moody AFB	747810	30.97	83.20	233	14.572	8293	30	34	15	13	12	16	53	14	52	4	360	5	300	99	21	2.5	7.6
Valdosta, Regional Airport	722166	30.78	83.28	203	14.588	8293	28	31	17	15	14	18	55	16	56	4	340	8	300	99	17	3.2	7.7
Waycross	722130	31.25	82.40	151	14.615	8293	29	32	16	14	12	16	52	14	52	4	250	7	240	98	21	7.0	7.6
HAWAII																							
Ewa, Barbers Point NAS	911780	21.32	158.07	49	14.669	8293	59	61	20	18	16	22	73	19	75	5	40	11	60	93	35	1.6	21.4
Hilo	912850	19.72	155.07	36	14.676	6193	61	63	19	16	14	21	76	18	76	7	230	12	110	88	58	1.6	1.8
Honolulu	911820	21.35	157.93	16	14.687	6193	61	63	23	21	20	23	74	21	75	5	320	15	60	91	58	1.9	2.2
Kahului	911900	20.90	156.43	66	14.661	6193	59	61	27	25	24	32	76	28	76	6	160	19	50	92	54	1.5	4.4
Kaneohe, MCAS	911760	21.45	157.77	10	14.690	8293	67	68	20	18	17	21	74	19	74	7	190	10	70	88	40	1.4	29.0
Lihue	911650	21.98	159.35	148	14.617	6193	60	62	26	24	21	25	73	23	73	8	270	14	60	87	57	1.4	3.0
Molokai	911860	21.15	157.10	449	14.458	8293	60	61	24	22	21	22	74	21	74	4	70	13	60	92	43	4.0	22.0
IDAHO																							
Boise	726810	43.57	116.22	2867	13.235	6193	2	9	24	21	18	22	37	19	37	6	130	11	320	103	−4	2.7	9.1
Burley	725867	42.55	113.77	4150	12.621	8293	−5	2	23	21	19	23	30	22	28	7	60	8	280	98	−11	4.0	8.5
Idaho Falls	725785	43.52	112.07	4741	12.346	8293	−12	−6	27	23	21	28	32	23	29	7	360	12	180	96	−20	3.6	9.0
Lewiston	727830	46.38	117.02	1437	13.948	8293	6	15	20	17	14	24	38	20	40	5	280	7	310	103	3	2.7	9.9
Mountain Home, AFB	726815	43.05	115.87	2995	13.173	8293	0	5	23	21	18	23	33	21	31	2	90	8	350	105	−6	3.2	8.5
Mullan	727836	47.47	115.80	3317	13.017	8293	−1	7	10	10	9	11	18	9	21	2	10	4	10	92	−7	2.0	7.9
Pocatello	725780	42.92	112.60	4478	12.468	6193	−7	0	29	25	23	30	36	27	36	6	50	11	250	98	−15	2.3	9.1
ILLINOIS																							
Belleville, Scott AFB	724338	38.55	89.85	453	14.457	8293	3	10	21	18	15	23	32	20	31	7	360	7	190	100	−3	3.1	7.2
Chicago, Meigs Field	725340	41.78	87.75	623	14.367	8293	−4	3	23	22	19	26	17	23	30	12	240	13	220	97	−10	3.2	8.1
Chicago, O'Hare Intl A	725300	41.98	87.90	673	14.342	6193	−6	−1	26	23	21	27	24	23	23	10	270	12	230	96	−12	2.8	6.5
Decatur	725316	39.83	88.87	682	14.337	8293	−2	3	24	22	20	27	24	24	27	13	310	12	210	99	−10	5.8	7.2
Glenview, NAS	725306	42.08	87.82	653	14.352	8293	−3	4	22	19	17	23	17	20	25	11	250	10	240	98	−10	3.1	7.7
Marseilles	744600	41.37	88.68	738	14.308	8293	−5	1	26	22	20	28	18	25	21	12	290	10	250	96	−11	4.0	5.9
Moline/Davenport IA	725440	41.45	90.52	594	14.383	6193	−8	−3	26	23	20	28	16	24	18	9	290	12	200	97	−14	2.7	6.0
Peoria	725320	40.67	89.68	663	14.347	6193	−6	−1	25	22	20	26	16	23	19	9	290	11	180	96	−12	3.3	6.1
Quincy	724396	39.95	91.20	768	14.292	8293	−4	2	26	23	20	28	23	24	22	12	330	12	210	97	−10	3.6	8.1
Rockford	725430	42.20	89.10	741	14.306	6193	−10	−4	26	23	21	26	18	23	20	9	290	13	200	95	−16	3.1	5.5
Springfield	724390	39.85	89.67	614	14.373	6193	−4	2	25	23	21	27	25	24	27	10	270	12	230	97	−11	2.8	5.5
West Chicago	725305	41.92	88.25	758	14.297	8293	−7	0	23	21	19	25	13	23	20	11	290	11	240	96	−14	3.2	7.7
INDIANA																							
Evansville	724320	38.05	87.53	387	14.491	6193	3	9	22	19	17	22	33	20	34	7	320	9	240	97	−4	2.7	8.5
Ft. Wayne	725330	41.00	85.20	827	14.262	6193	−4	2	25	23	20	27	19	24	22	10	250	12	230	95	−11	3.6	5.2
Indianapolis	724380	39.73	86.27	807	14.272	6193	−3	3	24	21	19	25	26	22	27	8	230	11	230	94	−10	2.8	6.8
Lafayette, Purdue Univ	724386	40.42	86.93	607	14.376	8293	−5	1	22	20	18	24	26	22	27	9	270	12	220	97	−11	3.8	7.7
Peru, Grissom AFB	725335	40.65	86.15	810	14.270	8293	−3	4	24	21	18	29	20	24	22	11	270	9	210	96	−8	3.8	7.4
South Bend	725350	41.70	86.32	774	14.289	6193	−2	3	25	23	20	26	22	23	23	13	230	12	230	95	−10	3.3	5.8
Terre Haute	724373	39.45	87.32	584	14.388	8293	−3	5	23	20	18	23	31	21	32	8	150	11	230	96	−10	3.2	7.9
IOWA																							
Burlington	725455	40.78	91.13	699	14.328	8293	−4	1	21	19	17	24	12	21	18	9	310	11	200	98	−10	4.0	6.8
Cedar Rapids	725450	41.88	91.70	869	14.240	8293	−11	−5	25	22	20	29	12	26	14	10	300	11	180	96	−15	3.6	5.4
Des Moines	725460	41.53	93.65	965	14.190	6193	−9	−4	27	24	21	28	14	24	19	11	320	12	180	98	−15	3.4	5.1
Ft. Dodge	725490	42.55	94.18	1165	14.087	8293	−13	−7	27	23	21	29	10	26	10	11	340	11	190	96	−17	4.9	4.9
Lamoni	725466	40.62	93.95	1122	14.109	8293	−6	0	19	17	15	21	23	19	20	7	320	9	210	99	−12	4.3	6.8
Mason City	725485	43.15	93.33	1214	14.062	6193	−15	−10	27	23	22	30	9	27	12	12	300	14	200	97	−23	3.6	11.4
Ottumwa	725465	41.10	92.45	866	14.251	8293	−5	0	29	26	23	31	20	28	24	13	320	15	200	98	−12	4.0	4.8
Sioux City	725570	42.40	96.38	1102	14.119	6193	−11	−6	29	25	22	31	14	28	16	11	320	14	180	99	−18	3.6	4.7
Spencer	726500	43.17	95.15	1339	13.998	6193	−16	−11	24	22	20	25	13	23	13	10	300	12	180	99	−20	6.3	4.0
Waterloo	725480	42.55	92.40	879	14.234	6193	−14	−9	27	24	22	29	10	25	13	9	300	13	180	96	−20	3.5	5.9
KANSAS																							
Concordia	724580	39.55	97.65	1483	13.925	8293	−4	3	28	25	22	28	32	25	32	13	360	16	10	104	−8	4.0	9.4
Dodge City	724510	37.77	99.97	2592	13.370	6193	0	6	30	27	24	31	31	27	32	13	10	17	200	104	−6	2.8	5.6
Ft Riley, Marshall AAF	724550	39.05	96.77	1066	14.138	8293	−2	5	21	18	16	20	39	18	37	5	350	9	180	104	−5	3.1	9.0
Garden City	724515	37.93	100.72	2890	13.224	8293	−3	4	30	26	23	29	32	25	34	12	360	16	190	104	−4	3.9	6.5
Goodland	724650	39.37	101.70	3688	12.840	6193	−3	2	32	28	24	31	27	27	30	12	270	13	180	102	−11	2.9	6.6

WMO# = World Meteorological Organization number   AAF = Army Air Field   Lat. = North latitude °   StdP = standard pressure at station elevation, psia
AFB = Air Force Base   NAS = Naval Air Station   MCAS = Marine Corps Air Station   Long. = West longitude °   WS = wind speed, mph
WSO = Weather Service Office   DB = dry-bulb temperature, °F   Elev. = elevation, ft   PWD = prevailing wind direction, ° True

United States

Station	WMO#	Lat.	Long.	Elev., ft	StdP, psia	Dates	Heating Dry Bulb 99.6%	99%	Extreme Wind Speed, mph 1%	2.5%	5%	Coldest Month 0.4% WS	MDB	1% WS	MDB	MWS/PWD to DB 99.6% MWS	PWD	0.4% MWS	PWD	Extr. Annual Daily Mean DB Max.	Min.	StdD DB Max.	Min.
1a	1b	1c	1d	1e	1f	1g	2a	2b	3a	3b	3c	4a	4b	4c	4d	5a	5b	5c	5d	6a	6b	6c	6d
Russell	724585	38.87	98.82	1864	13.732	8293	−4	3	29	26	23	29	33	25	35	11	10	16	190	105	−8	3.6	8.5
Salina	724586	38.80	97.65	1273	14.032	8293	−3	4	27	23	22	28	33	24	34	11	360	15	180	106	−7	2.3	9.9
Topeka	724560	39.07	95.62	886	14.231	6193	−2	4	25	22	20	25	28	22	29	9	320	12	180	100	−8	3.8	7.4
Wichita, Airport	724500	37.65	97.43	1339	13.998	6193	2	8	29	25	23	28	30	26	31	13	360	16	200	105	−4	2.9	6.3
Wichita, McConnell AFB	724505	37.62	97.27	1371	13.982	8293	2	10	25	23	20	25	38	23	36	11	360	12	190	105	−1	2.7	7.7
KENTUCKY																							
Bowling Green	746716	36.97	86.42	548	14.407	8293	7	14	20	19	17	21	40	19	40	6	220	9	230	97	−2	3.2	10.3
Covington/Cincinnati A	724210	39.05	84.67	876	14.236	6193	1	7	22	20	18	25	30	22	33	9	250	10	230	95	−7	3.1	8.5
Ft. Campbell, AAF	746710	36.67	87.50	571	14.395	8293	9	15	19	16	14	20	40	17	43	4	330	6	240	98	0	3.1	9.5
Ft. Knox, Godman AAF	724240	37.90	85.97	755	14.299	8293	9	15	17	15	13	18	42	16	39	4	290	6	270	97	0	4.1	7.9
Jackson	724236	37.60	83.32	1381	13.977	8293	7	14	17	14	13	18	44	16	40	7	230	6	230	94	−3	2.7	9.4
Lexington	724220	38.03	84.60	988	14.179	6193	4	10	21	19	17	23	38	20	38	8	270	9	240	94	−4	3.5	8.3
Louisville	724230	38.18	85.73	489	14.438	6193	6	12	22	19	17	22	40	20	34	10	290	10	250	96	−1	3.1	7.9
Paducah	724350	37.07	88.77	413	14.477	8293	7	13	22	19	17	22	45	19	42	8	40	9	180	98	−1	2.9	9.4
LOUISIANA																							
Alexandria, England AFB	747540	31.33	92.55	89	14.648	8293	27	30	16	13	12	17	53	15	49	7	360	3	180	98	20	2.2	6.3
Baton Rouge	722317	30.53	91.15	69	14.659	6193	27	30	20	18	16	21	48	19	49	8	360	8	270	97	20	2.2	5.4
Bossier City, Barksdale AFB	722485	32.50	93.67	167	14.607	8293	22	27	18	16	14	19	49	16	51	7	360	5	180	99	15	2.3	6.7
Lafayette	722405	30.20	91.98	43	14.673	8293	28	32	21	18	16	21	54	19	53	9	10	8	200	97	19	1.6	8.1
Lake Charles	722400	30.12	93.22	33	14.678	6193	29	32	22	19	17	24	50	21	49	10	20	8	230	96	23	2.3	4.7
Leesville, Ft. Polk	722390	31.05	93.20	328	14.522	8293	27	30	16	13	12	16	51	14	52	4	20	4	180	98	20	2.0	5.9
Monroe	722486	32.52	92.03	79	14.654	8293	22	27	19	17	15	20	50	18	47	9	10	7	230	99	17	1.8	8.5
New Orleans, Intl Airport	722310	29.98	90.25	30	14.680	6193	30	34	21	19	17	21	48	19	49	7	340	8	360	96	23	2.0	5.3
New Orleans, Lakefront A	722315	30.05	90.03	10	14.690	8293	35	39	22	19	18	21	49	20	50	14	360	9	300	94	21	8.1	12.4
Shreveport	722480	32.47	93.82	259	14.558	6193	22	26	20	18	16	22	46	19	48	9	360	8	180	99	16	3.1	5.6
MAINE																							
Augusta	726185	44.32	69.80	351	14.510	8293	−3	1	23	21	19	25	20	22	22	10	320	11	210	93	−10	3.1	3.4
Bangor	726088	44.80	68.83	194	14.593	8293	−7	−2	22	19	17	24	18	21	20	6	300	10	240	94	−16	2.9	5.9
Brunswick, NAS	743920	43.88	69.93	75	14.655	8293	−2	2	20	17	15	21	27	19	25	4	340	9	190	96	−12	7.9	6.1
Caribou	727120	46.87	68.02	623	14.367	6193	−14	−10	28	24	22	30	13	27	11	10	270	13	250	90	−23	2.8	4.5
Limestone, Loring AFB	727125	46.95	67.88	745	14.304	8293	−13	−9	23	20	18	25	12	22	11	7	300	9	260	91	−20	2.3	2.9
Portland	726060	43.65	70.32	62	14.662	6193	−3	2	24	21	18	24	26	21	25	7	320	12	270	93	−13	3.6	5.5
MARYLAND																							
Camp Springs, Andrews AFB	745940	38.82	76.87	282	14.546	8293	13	18	21	18	16	23	30	21	32	7	350	9	230	98	4	2.9	6.7
Baltimore, BWI Airport	724060	39.18	76.67	154	14.614	6193	11	15	24	21	19	25	31	22	31	10	290	11	280	97	4	2.9	5.8
Lex Park, Patuxent River NAS	724040	38.28	76.40	39	14.675	8293	16	21	20	17	15	22	30	19	35	9	340	9	270	98	8	2.3	6.1
Salisbury	724045	38.33	75.52	52	14.668	8293	13	18	20	18	16	20	35	19	37	6	10	9	240	97	4	2.7	5.8
MASSACHUSETTS																							
Boston	725090	42.37	71.03	30	14.680	6193	7	12	29	25	23	30	30	27	28	17	320	14	270	96	0	2.7	4.7
East Falmouth, Otis ANGB	725060	41.65	70.52	131	14.626	8293	11	14	26	22	20	26	34	23	33	9	300	10	240	90	5	2.5	3.8
Weymouth, S Weymth NAS	725097	42.15	70.93	161	14.610	8293	6	11	19	16	14	18	29	16	29	7	320	9	260	97	−2	3.8	3.8
Worcester	725095	42.27	71.88	1010	14.167	6193	0	5	27	23	20	29	22	26	21	14	270	10	270	90	−6	1.9	4.1
MICHIGAN																							
Alpena	726390	45.07	83.57	692	14.332	6193	−7	−1	21	19	17	22	20	19	20	5	270	11	240	93	−17	3.4	5.9
Detroit, Metro	725370	42.23	83.33	663	14.347	6193	0	5	27	23	21	28	28	24	27	11	240	13	230	95	−7	3.0	5.4
Flint	726370	42.97	83.75	764	14.294	6193	−2	3	25	22	20	27	24	23	23	8	230	13	230	93	−10	3.1	5.0
Grand Rapids	726350	42.88	85.52	804	14.274	6193	0	5	25	22	20	26	25	23	24	8	180	13	240	93	−9	2.1	5.3
Hancock	727440	47.17	88.50	1079	14.131	6193	−9	−4	21	19	18	23	18	20	16	8	270	10	250	90	−16	2.9	5.6
Harbor Beach	725386	44.02	82.80	600	14.379	8293	9	12	26	22	19	26	27	23	27	10	220	9	230	94	2	2.9	4.1
Jackson	725395	42.27	84.47	1001	14.172	8293	−3	4	20	19	17	23	22	20	23	9	240	11	210	93	−11	2.5	5.6
Lansing	725390	42.77	84.60	873	14.238	6193	−3	2	26	23	20	28	23	25	24	8	290	13	250	94	−13	2.8	5.9
Marquette, Sawyer AFB	727435	46.35	87.40	1220	14.059	8293	−11	−6	24	21	18	26	18	23	17	6	280	10	210	91	−18	4.7	4.7
Marquette/Ishpeming A	727430	46.53	87.55	1424	13.955	8293	−13	−8	22	19	18	22	20	20	16	8	270	11	230	90	−22	4.5	4.5
Mount Clemens, ANGB	725377	42.62	82.83	581	14.390	8293	3	7	21	18	16	25	21	21	24	7	280	9	230	95	−3	4.0	2.7
Muskegon	726360	43.17	86.25	633	14.362	6193	3	7	27	24	22	28	25	25	26	10	290	12	200	90	−5	2.7	5.0
Oscoda, Wurtsmith AFB	726395	44.45	83.40	633	14.362	8293	0	3	21	19	17	23	26	21	24	6	220	11	200	95	−7	4.1	4.7
Pellston	727347	45.57	84.80	719	14.318	8293	−9	−3	26	23	20	28	22	24	22	4	300	14	250	92	−21	3.1	4.9
Saginaw	726379	43.53	84.08	669	14.343	8293	0	4	23	21	19	25	22	22	23	10	260	13	240	96	−6	5.8	4.5
Sault Ste. Marie	727340	46.47	84.37	725	14.314	6193	−12	−7	23	20	18	24	19	21	18	7	90	10	230	89	−22	3.5	5.4
Seul Choix Point	726379	45.92	85.92	591	14.385	8293	0	4	28	24	22	30	27	26	27	9	300	8	200	82	−5	2.3	6.3
Traverse City	726387	44.73	85.58	623	14.367	6193	−3	2	21	19	18	23	23	21	23	7	180	13	230	94	−13	2.8	7.3
MINNESOTA																							
Alexandria	726557	45.87	95.40	1424	13.955	8293	−20	−15	25	22	20	28	12	24	8	10	300	14	180	96	−26	3.6	4.5
Brainerd, Pequot Lakes	727500	46.60	94.32	1280	14.029	8293	−24	−17	11	10	9	11	8	10	11	3	320	5	190	95	−30	7.9	6.8
Duluth	727450	46.83	92.18	1417	13.958	6193	−21	−16	25	22	20	25	12	22	11	10	310	12	230	90	−28	2.8	4.7
Hibbing	727455	47.38	92.83	1352	13.992	8293	−25	−20	20	19	17	20	13	19	13	6	330	11	200	92	−34	2.5	4.7
International Falls	727470	48.57	93.38	1184	14.077	6193	−29	−23	22	20	18	22	10	20	8	6	270	11	180	92	−37	3.4	3.8
Minneapolis-St. Paul	726580	44.88	93.22	837	14.257	6193	−16	−11	25	22	20	25	12	22	14	9	300	14	180	97	−22	3.5	5.4
Redwood Falls	726556	44.55	95.08	1024	14.160	8293	−17	−12	26	22	20	28	14	24	15	11	280	14	180	99	−22	4.1	5.2
Rochester	726440	43.92	92.50	1319	14.008	6193	−17	−12	29	26	24	32	12	28	12	13	300	15	200	94	−23	3.7	5.2

WMO# = World Meteorological Organization number
AFB = Air Force Base   NAS = Naval Air Station
WSO = Weather Service Office

AAF = Army Air Field
MCAS = Marine Corps Air Station
DB = dry-bulb temperature, °F

Lat. = North latitude °
Long. = West longitude °
Elev. = elevation, ft

StdP = standard pressure at station elevation, psia
WS = wind speed, mph
PWD = prevailing wind direction, ° True

United States

Station	WMO#	Lat.	Long.	Elev., ft	StdP, psia	Dates	Heating Dry Bulb 99.6%	99%	Extreme Wind Speed, mph 1%	2.5%	5%	Coldest Month 0.4% WS	MDB	1% WS	MDB	MWS/PWD to DB 99.6% MWS	PWD	0.4% MWS	PWD	Extr. Annual Daily Mean DB Max.	Min.	StdD DB Max.	Min.
1a	1b	1c	1d	1e	1f	1g	2a	2b	3a	3b	3c	4a	4b	4c	4d	5a	5b	5c	5d	6a	6b	6c	6d
Saint Cloud	726550	45.55	94.07	1024	14.160	6193	−20	−14	22	20	18	23	11	20	10	8	300	12	200	95	−27	3.0	5.6
Tofte	727554	47.58	90.83	791	14.280	8293	−10	−6	24	20	17	25	16	22	18	8	260	8	330	86	−19	4.5	4.9
MISSISSIPPI																							
Biloxi, Keesler AFB	747686	30.42	88.92	33	14.678	8293	31	35	17	14	13	18	49	16	50	8	360	7	210	97	23	2.0	7.4
Columbus, AFB	723306	33.65	88.45	220	14.579	8293	20	25	18	15	13	19	43	16	46	6	360	6	240	100	12	2.7	6.8
Greenwood	722359	33.50	90.08	154	14.614	8293	20	24	19	17	14	19	46	18	47	6	360	6	180	99	13	2.3	7.6
Jackson	722350	32.32	90.08	331	14.520	6193	21	25	20	18	16	21	45	19	46	7	340	8	270	98	14	2.7	5.8
McComb	722358	31.18	90.47	413	14.477	8293	23	28	17	14	13	17	49	15	49	6	350	7	230	98	15	2.0	7.2
Meridian	722340	32.33	88.75	308	14.532	6193	21	25	19	17	15	19	43	17	46	6	360	8	360	99	13	2.9	5.9
Tupelo	723320	34.27	88.77	361	14.505	8293	18	22	19	17	15	20	44	17	44	7	10	7	260	99	10	2.9	8.5
MISSOURI																							
Cape Girardeau	723489	37.23	89.57	341	14.515	8293	6	13	21	19	18	22	35	20	36	9	360	10	200	100	−1	2.9	9.2
Columbia	724450	38.82	92.22	899	14.224	6193	−1	5	25	22	20	25	27	22	28	11	310	11	200	99	−8	4.4	6.2
Joplin	723495	37.15	94.50	981	14.182	8293	3	11	23	21	19	24	50	21	47	10	10	11	220	100	−2	4.0	9.4
Kansas City	724460	39.32	94.72	1024	14.160	6193	−1	4	26	23	20	26	34	23	33	10	320	13	190	100	−7	4.1	6.6
Poplar Bluff	723300	36.77	90.47	479	14.443	8293	8	13	18	15	13	17	40	15	38	7	360	7	200	101	2	6.8	9.4
Spickard/Trenton	725400	40.25	93.72	886	14.231	8293	1	6	23	20	18	25	29	22	31	8	360	11	200	100	−5	5.2	7.4
Springfield	724400	37.23	93.38	1270	14.034	6193	3	9	24	21	19	23	35	21	35	10	340	10	230	98	−4	3.5	6.4
St. Louis, Intl Airport	724340	38.75	90.37	564	14.398	6193	2	8	26	23	20	26	26	23	27	12	290	11	240	99	−5	3.5	6.2
Warrensburg, Whiteman AFB	724467	38.73	93.55	869	14.240	8293	1	7	22	19	17	23	34	21	34	9	360	9	190	101	−5	4.0	7.7
MONTANA																							
Billings	726770	45.80	108.53	3570	12.896	6193	−13	−7	28	24	22	30	25	27	30	10	230	10	240	99	−19	2.8	6.2
Bozeman	726797	45.78	111.15	4475	12.469	8293	−20	−12	21	18	15	20	36	17	34	4	140	9	360	96	−29	2.9	7.7
Butte	726785	45.95	112.50	5545	11.980	8293	−22	−14	23	21	18	21	29	19	30	4	150	13	120	92	−34	2.5	7.9
Cut Bank	727796	48.60	112.37	3855	12.760	6193	−21	−16	34	30	27	40	36	34	36	7	320	13	200	93	−28	4.0	5.7
Glasgow	727680	48.22	106.62	2297	13.516	6193	−22	−17	29	26	23	28	18	25	15	8	330	13	160	99	−29	3.2	6.6
Great Falls, Intl Airport	727750	47.48	111.37	3658	12.854	6193	−19	−13	33	29	26	34	38	31	38	7	240	12	230	98	−25	3.2	7.4
Great Falls, Malmstrom AFB	727755	47.50	111.18	3527	12.917	8293	−17	−11	28	24	21	33	38	29	38	4	240	8	260	99	−22	3.2	7.9
Havre	727770	48.55	109.77	2598	13.367	8293	−25	−19	24	21	19	26	35	23	33	6	240	9	270	102	−33	5.0	8.1
Helena	727720	46.60	112.00	3898	12.740	6193	−18	−10	25	22	19	25	40	22	35	5	290	12	280	96	−24	3.3	7.2
Kalispell	727790	48.30	114.27	2972	13.184	6193	−12	−3	24	20	17	25	12	21	18	7	20	9	170	95	−19	2.9	8.6
Lewistown	726776	47.05	109.47	4167	12.613	6193	−18	−12	26	23	20	29	35	25	35	7	250	11	90	95	−25	3.5	7.3
Miles City	742300	46.43	105.87	2628	13.352	6193	−19	−13	27	23	20	28	25	23	27	8	290	11	140	102	−25	2.7	6.5
Missoula	727730	46.92	114.08	3189	13.079	6193	−9	−1	22	19	17	22	17	19	21	7	120	10	290	97	−15	2.9	8.2
NEBRASKA																							
Bellevue, Offutt AFB	725540	41.12	95.92	1047	14.148	8293	−5	1	22	19	17	26	23	22	23	8	330	10	190	100	−9	4.5	6.3
Grand Island	725520	40.97	98.32	1857	13.736	6193	−8	−2	30	26	23	29	21	26	19	11	270	15	180	102	−14	3.2	5.2
Lincoln	725510	40.85	96.75	1188	14.076	8293	−7	−2	27	23	21	28	25	24	27	9	350	15	180	103	−11	6.5	8.1
Norfolk	725560	41.98	97.43	1552	13.890	6193	−11	−5	29	25	22	33	20	28	21	11	340	15	190	101	−18	3.0	5.4
North Platte	725620	41.13	100.68	2785	13.275	6193	−10	−4	29	25	22	28	24	24	26	7	320	12	180	101	−16	2.9	6.6
Omaha, Eppley Airfield	725500	41.30	95.90	981	14.182	6193	−7	−2	26	23	20	27	21	23	17	10	340	12	180	100	−14	3.3	4.8
Omaha, WSO	725530	41.37	96.02	1332	14.002	8293	−8	−2	22	20	18	25	23	22	25	10	310	11	170	98	−14	4.0	6.5
Scottsbluff	725660	41.87	103.60	3957	12.712	6193	−11	−3	30	26	23	32	35	27	35	8	300	11	300	101	−19	2.9	8.0
Sidney	725610	41.10	102.98	4304	12.549	8293	−8	−1	29	24	22	31	32	26	35	9	290	12	160	101	−18	4.5	8.6
Valentine	725670	42.87	100.55	2598	13.367	8293	−16	−8	27	23	21	26	25	23	28	9	250	15	180	104	−22	4.3	8.5
NEVADA																							
Elko	725825	40.83	115.78	5135	12.166	6193	−5	1	21	18	16	20	36	16	37	4	70	10	230	98	−13	3.2	8.0
Ely	724860	39.28	114.85	6263	11.660	6193	−6	0	28	24	21	26	33	22	30	11	190	13	230	93	−15	2.3	7.3
Las Vegas, Intl Airport	723860	36.08	115.17	2178	13.575	6193	27	30	30	26	23	25	48	22	49	7	250	12	230	111	21	2.2	4.7
Mercury	723870	36.62	116.02	3310	13.021	8293	24	28	25	22	19	25	44	21	42	8	50	12	230	102	19	17.6	6.3
N. Las Vegas, Nellis AFB	723865	36.23	115.03	1870	13.729	8293	28	31	24	21	18	23	52	19	49	2	20	9	210	112	21	2.0	4.5
Reno	724880	39.50	119.78	4400	12.505	6193	8	13	26	22	19	26	46	21	44	3	160	10	290	99	1	2.2	8.4
Tonopah	724855	38.05	117.08	5427	12.033	6193	7	13	25	22	20	24	37	22	36	9	340	12	180	98	1	2.0	6.5
Winnemucca	725830	40.90	117.80	4314	12.544	6193	1	7	23	19	17	21	39	18	38	5	160	11	250	101	−9	2.3	10.2
NEW HAMPSHIRE																							
Concord	726050	43.20	71.50	344	14.513	6193	−8	−2	23	20	17	23	20	20	21	4	320	10	230	95	−18	2.9	5.5
Lebanon	726116	43.63	72.30	597	14.381	8293	−7	−3	18	15	14	18	25	16	26	2	360	9	220	94	−17	2.0	5.2
Mount Washington	726130	44.27	71.30	6266	11.659	8293	−23	−19	88	81	73	99	−14	92	−15	73	280	21	270	65	−33	2.3	4.1
Portsmouth, Pease AFB	726055	43.08	70.82	102	14.642	8293	4	9	21	18	16	22	26	20	27	8	280	8	270	94	−2	2.2	3.2
NEW JERSEY																							
Atlantic City	724070	39.45	74.57	66	14.661	6193	8	13	27	23	20	29	36	25	34	9	310	11	250	96	0	2.9	5.7
Millville	724075	39.37	75.07	82	14.652	8293	10	15	19	18	17	20	35	19	35	7	300	11	240	96	0	2.3	7.4
Newark	725020	40.70	74.17	30	14.680	6193	10	14	26	23	20	27	28	23	29	13	260	13	230	98	4	2.6	4.8
Teterboro	725025	40.85	74.07	10	14.690	8293	10	14	21	19	17	21	29	19	30	11	280	12	240	97	2	2.5	5.6
Trenton, McGuire AFB	724096	40.02	74.60	135	14.624	8293	11	15	22	19	17	22	31	21	31	8	330	8	240	97	2	2.2	5.2
NEW MEXICO																							
Alamogordo, Holloman AFB	747320	32.85	106.10	4094	12.647	8293	20	23	20	17	14	18	50	15	48	3	10	8	250	102	13	2.9	3.6
Albuquerque	723650	35.05	106.62	5315	12.084	6193	13	18	29	25	22	26	34	22	37	8	360	10	240	100	6	2.6	7.3
Carlsbad	722687	32.33	104.27	3294	13.028	8293	19	23	25	22	19	25	57	21	54	8	340	12	150	104	9	3.6	7.0
Clayton	723600	36.45	103.15	4970	12.241	8293	1	9	30	27	24	30	40	26	39	10	40	13	200	98	−5	2.5	7.4

WMO# = World Meteorological Organization number
AFB = Air Force Base   NAS = Naval Air Station
WSO = Weather Service Office

AAF = Army Air Field
MCAS = Marine Corps Air Station
DB = dry-bulb temperature, °F

Lat. = North latitude °
Long. = West longitude °
Elev. = elevation, ft

StdP = standard pressure at station elevation, psia
WS = wind speed, mph
PWD = prevailing wind direction, ° True

United States

Station	WMO#	Lat.	Long.	Elev., ft	StdP, psia	Dates	Heating Dry Bulb 99.6%	99%	Extreme Wind Speed, mph 1%	2.5%	5%	Coldest Month 0.4% WS	MDB	1% WS	MDB	MWS/PWD to DB 99.6% MWS	PWD	0.4% MWS	PWD	Extr. Annual Daily Mean DB Max.	Min.	StdD DB Max.	Min.
1a	1b	1c	1d	1e	1f	1g	2a	2b	3a	3b	3c	4a	4b	4c	4d	5a	5b	5c	5d	6a	6b	6c	6d
Clovis, Cannon AFB	722686	34.38	103.32	4295	12.554	8293	10	15	26	23	20	26	40	23	39	8	50	11	220	101	5	2.3	4.0
Farmington	723658	36.75	108.23	5502	11.999	8293	8	13	23	21	18	22	35	19	34	6	60	10	240	99	−1	3.8	7.2
Gallup	723627	35.52	108.78	6470	11.570	8293	−1	5	23	20	18	19	39	18	37	1	140	11	270	94	−12	2.3	7.9
Roswell	722680	33.30	104.53	3668	12.849	8293	14	20	22	19	17	20	51	18	48	8	360	11	140	105	6	4.7	6.5
Truth Or Consequences	722710	33.23	107.27	4859	12.292	8293	22	26	25	21	18	24	43	21	41	8	350	10	170	102	6	2.9	23.8
Tucumcari	723676	35.18	103.60	4065	12.661	6193	9	15	25	22	20	28	50	23	45	8	50	12	230	102	1	3.1	7.2
NEW YORK																							
Albany	725180	42.75	73.80	292	14.541	6193	−7	−2	24	22	19	23	20	20	22	5	300	10	230	95	−18	3.0	6.3
Binghamton	725150	42.22	75.98	1631	13.850	6193	−2	2	24	21	19	24	20	22	19	13	270	11	220	89	−9	3.1	4.4
Buffalo	725280	42.93	78.73	705	14.325	6193	2	5	29	26	23	34	25	30	24	12	270	13	240	91	−6	2.3	5.3
Central Islip	725035	40.80	73.10	98	14.643	6193	11	15	23	32	21	31	10	340	11	210	94	2	3.2	5.4
Elmira/Corning	725156	42.17	76.90	955	14.195	8293	−2	3	21	19	17	23	20	20	27	5	240	11	210	95	−10	3.6	6.1
Glens Falls	725185	43.33	73.62	328	14.522	8293	−10	−4	18	16	14	19	22	17	22	2	350	10	190	93	−20	2.7	4.7
Massena	726223	44.93	74.85	213	14.583	6193	−15	−10	21	18	17	23	22	21	22	4	270	10	230	92	−27	3.0	6.0
New York, JFK Airport	744860	40.65	73.78	23	14.683	6193	11	15	27	24	21	30	29	27	28	17	320	13	230	96	6	2.6	4.6
New York, La Guardia A	725030	40.77	73.90	30	14.680	8293	13	17	28	25	22	30	29	27	28	18	310	12	280	97	6	2.2	4.3
Newburgh	725038	41.50	74.10	492	14.436	8293	6	10	23	20	18	26	17	23	26	8	260	10	230	92	−4	3.1	6.5
Niagara Falls	725287	43.10	78.95	591	14.385	8293	4	7	26	22	20	30	24	27	23	11	240	13	230	91	−4	3.2	6.1
Plattsburgh, AFB	726225	44.65	73.47	236	14.570	8293	−9	−4	21	18	16	22	27	19	24	2	350	8	260	93	−17	2.7	4.7
Poughkeepsie	725036	41.63	73.88	167	14.607	8293	2	6	18	16	14	19	25	17	25	3	250	9	250	96	−8	3.1	5.9
Rochester	725290	43.12	77.67	554	14.403	6193	1	5	27	23	21	29	22	26	21	10	230	12	250	93	−7	2.8	4.9
Rome, Griffiss AFB	725196	43.23	75.40	505	14.429	8293	−5	1	22	19	16	23	22	20	22	3	330	8	260	93	−15	2.5	3.8
Syracuse	725190	43.12	76.12	407	14.481	6193	−3	2	26	22	20	28	20	25	21	7	90	11	250	93	−13	3.0	6.0
Watertown	726227	44.00	76.02	325	14.524	8293	−12	−6	21	19	18	24	24	21	25	5	80	11	240	90	−25	2.9	7.0
White Plains	725037	41.07	73.70	440	14.463	8293	7	12	19	17	15	19	29	18	29	13	310	9	260	95	0	2.9	4.5
NORTH CAROLINA																							
Asheville	723150	35.43	82.55	2169	13.580	6193	11	16	25	22	19	26	26	23	28	11	340	9	340	91	3	2.6	6.8
Cape Hatteras	723040	35.27	75.55	10	14.690	6193	26	29	26	22	20	27	47	23	47	11	340	11	230	91	20	2.0	4.9
Charlotte	723140	35.22	80.93	768	14.292	6193	18	23	20	17	15	20	44	18	45	6	50	9	240	97	10	2.9	6.0
Cherry Point, MCAS	723090	34.90	76.88	30	14.680	8293	24	28	19	16	14	19	43	17	48	5	10	7	240	100	12	2.5	8.5
Fayetteville, Ft. Bragg	746930	35.13	78.93	243	14.567	8293	22	27	17	14	12	19	42	16	44	4	10	6	240	100	15	3.8	6.8
Goldsboro, Johnson AFB	723066	35.33	77.97	108	14.638	8293	22	27	17	14	12	18	46	15	44	4	270	8	260	100	14	3.1	7.4
Greensboro	723170	36.08	79.95	886	14.231	6193	15	19	19	17	15	20	40	18	40	7	290	8	230	96	7	2.7	5.0
Hickory	723145	35.73	81.38	1188	14.076	8293	18	23	17	15	13	18	41	16	41	4	320	9	240	97	8	3.2	6.8
Jacksonville, New River MCAF	723096	34.72	77.45	26	14.681	8293	23	27	18	16	14	19	49	17	47	5	350	7	240	99	13	2.0	8.8
New Bern	723095	35.07	77.05	20	14.685	8293	22	27	18	16	14	19	49	17	47	6	10	8	240	99	13	1.4	8.3
Raleigh/Durham	723060	35.87	78.78	440	14.463	6193	16	20	21	18	16	21	42	19	43	8	360	9	240	96	9	2.9	5.3
Wilmington	723013	34.27	77.90	33	14.678	6193	23	27	21	19	17	22	51	20	48	7	320	10	220	97	17	2.2	5.7
Winston-Salem	723193	36.13	80.22	971	14.187	8293	18	23	19	17	15	21	38	19	38	7	290	8	240	96	8	2.7	5.8
NORTH DAKOTA																							
Bismarck	727640	46.77	100.75	1660	13.835	6193	−21	−16	29	25	22	29	13	25	16	7	290	13	180	100	−30	3.6	6.4
Devils Lake	727580	48.10	98.87	1453	13.940	8293	−23	−19	26	22	20	27	12	24	10	9	300	11	10	98	−27	7.0	5.0
Fargo	727530	46.90	96.80	899	14.224	6193	−22	−17	31	27	24	32	7	28	7	8	180	14	160	98	−27	3.4	4.4
Grand Forks, AFB	727575	47.97	97.40	912	14.217	8293	−20	−16	27	24	21	30	9	26	13	7	290	13	180	98	−25	4.7	4.9
Minot, AFB	727675	48.42	101.35	1667	13.832	8293	−21	−16	28	24	21	30	18	27	16	10	310	12	150	101	−25	3.8	7.4
Minot, Intl Airport	727676	48.27	101.28	1716	13.807	6193	−20	−16	28	24	21	30	14	27	14	12	290	13	200	98	−25	3.0	4.8
Williston	727670	48.18	103.63	1906	13.711	8293	−24	−18	27	23	21	28	25	24	20	8	220	14	150	101	−30	4.5	8.3
OHIO																							
Akron/Canton	725210	40.92	81.43	1237	14.050	6193	0	5	24	21	19	25	26	22	26	11	270	10	230	92	−7	2.9	7.0
Cincinnati, Lunken Field	724297	39.10	84.42	482	14.441	8293	5	12	21	19	17	22	35	19	33	9	260	10	210	96	−3	3.2	9.4
Cleveland	725240	41.42	81.87	804	14.274	6193	1	6	26	23	20	27	28	24	28	12	230	12	230	93	−6	2.8	6.3
Columbus, Intl Airport	724280	40.00	82.88	817	14.267	6193	1	6	23	20	18	24	30	21	25	9	270	11	270	94	−6	2.6	7.1
Columbus, Rickenbckr AFB	724285	39.82	82.93	745	14.304	8293	3	10	21	18	16	23	26	20	27	7	210	8	270	96	−4	4.5	8.6
Dayton, Intl Airport	724290	39.90	84.20	1004	14.170	6193	−1	5	24	21	19	25	26	22	28	11	270	11	240	95	−8	2.9	7.0
Dayton, Wright-Patterson	745700	39.83	84.05	823	14.263	8293	1	8	21	18	16	23	28	21	30	7	270	9	240	96	−7	3.2	7.7
Findlay	725366	41.02	83.67	810	14.270	8293	−2	4	23	20	19	25	34	22	29	11	250	12	210	94	−9	3.8	7.9
Mansfield	725246	40.82	82.52	1296	14.020	6193	−1	4	25	22	20	28	28	25	26	13	240	12	240	91	−8	2.8	6.0
Toledo	725360	41.60	83.80	692	14.332	6193	−2	3	23	20	18	25	25	22	21	10	250	11	230	95	−10	3.0	5.4
Youngstown	725250	41.27	80.67	1184	14.077	6193	−1	4	23	21	19	24	22	22	21	10	230	10	230	91	−8	2.5	5.8
Zanesville	724286	39.95	81.90	899	14.224	8293	2	9	19	18	16	21	32	19	31	7	240	9	220	94	−7	3.6	8.5
OKLAHOMA																							
Altus, AFB	723520	34.67	99.27	1378	13.978	8293	13	19	23	21	19	24	40	21	42	9	20	10	190	107	7	3.4	7.7
Enid, Vance AFB	723535	36.33	97.92	1306	14.015	8293	5	12	26	23	20	27	38	23	38	12	10	11	190	105	1	3.6	6.7
Lawton, Ft. Sill/Post Field	723550	34.65	98.40	1188	14.076	8293	12	19	24	21	19	26	35	22	36	11	10	11	170	103	8	2.5	7.4
McAlester	723566	34.88	95.78	771	14.291	8293	10	17	20	18	16	21	47	19	45	9	360	9	190	102	4	4.0	8.3
Oklahoma City, Tinker AFB	723540	35.42	97.38	1293	14.022	8293	10	17	24	22	19	25	42	22	42	10	10	11	190	103	6	3.2	6.1
Oklahoma City, W. Rogers A	723540	35.40	97.60	1302	14.017	6193	9	15	27	25	23	29	33	26	37	15	360	13	180	103	4	3.4	4.9
Tulsa	723560	36.20	95.90	676	14.340	6193	9	14	25	23	21	24	46	22	40	11	360	12	180	103	3	3.6	5.6
OREGON																							
Astoria	727910	46.15	123.88	23	14.683	6193	25	29	25	22	19	29	51	24	49	8	90	12	320	87	20	4.5	6.1

WMO# = World Meteorological Organization number   AAF = Army Air Field   Lat. = North latitude °   StdP = standard pressure at station elevation, psia
AFB = Air Force Base   NAS = Naval Air Station   MCAS = Marine Corps Air Station   Long. = West longitude °   WS = wind speed, mph
WSO = Weather Service Office   DB = dry-bulb temperature, °F   Elev. = elevation, ft   PWD = prevailing wind direction, ° True

United States

Station	WMO#	Lat.	Long.	Elev., ft	StdP, psia	Dates	Heating Dry Bulb 99.6%	99%	Extreme Wind Speed, mph 1%	2.5%	5%	Coldest Month 0.4% WS	MDB	1% WS	MDB	MWS/PWD to DB 99.6% MWS	PWD	0.4% MWS	PWD	Extr. Annual Daily Mean DB Max.	Min.	StdD DB Max.	Min.
1a	1b	1c	1d	1e	1f	1g	2a	2b	3a	3b	3c	4a	4b	4c	4d	5a	5b	5c	5d	6a	6b	6c	6d
Eugene	726930	44.12	123.22	374	14.498	6193	21	26	20	18	16	22	46	19	45	8	360	12	360	99	16	3.7	7.9
Hillsboro	726986	45.53	122.95	203	14.588	8293	19	24	19	17	15	23	26	19	34	8	60	9	360	100	15	3.8	6.3
Klamath Falls	725895	42.15	121.73	4091	12.649	8293	4	10	25	22	19	28	39	23	33	6	320	9	320	97	−4	4.1	8.6
Meacham	726885	45.52	118.40	4055	12.666	8293	−9	0	12	10	9	13	33	11	33	1	130	5	360	93	−21	4.7	12.2
Medford	725970	42.37	122.87	1329	14.003	6193	21	24	19	16	13	20	51	15	50	3	130	9	290	104	15	3.4	6.4
North Bend	726917	43.42	124.25	13	14.688	6193	30	32	25	23	20	23	51	20	50	7	140	14	340	82	24	4.0	5.5
Pendleton	726880	45.68	118.85	1496	13.918	6193	3	11	28	24	20	27	44	23	42	6	140	9	310	102	−1	3.6	11.2
Portland	726980	45.60	122.60	39	14.675	6193	22	27	25	21	18	28	37	25	39	13	120	11	340	99	18	4.4	6.0
Redmond	726835	44.25	121.15	3077	13.133	6193	1	9	20	17	16	20	42	18	41	6	320	11	340	98	−7	3.2	10.2
Salem	726940	44.92	123.00	200	14.589	6193	20	25	23	19	17	25	46	22	46	5	350	10	340	100	14	3.3	6.6
Sexton Summit	725975	42.62	123.37	3842	12.767	8293	21	24	24	22	19	27	37	24	38	9	120	6	340	89	16	4.9	8.5
PENNSYLVANIA																							
Allentown	725170	40.65	75.43	384	14.493	6193	5	10	27	23	21	28	26	25	24	9	270	11	240	95	−2	2.8	5.2
Altoona	725126	40.30	78.32	1503	13.915	8293	5	10	20	18	17	23	20	20	22	9	270	8	250	92	−5	3.6	7.9
Bradford	725266	41.80	78.63	2142	13.593	6193	−6	−1	19	18	16	22	22	19	21	7	270	9	240	87	−15	2.8	5.0
Du Bois	725125	41.18	78.90	1818	13.756	8293	0	5	21	19	17	23	20	21	20	11	280	10	270	90	−9	3.1	7.0
Erie	725260	42.08	80.18	738	14.308	6193	2	7	27	24	22	29	28	26	28	14	200	12	250	90	−4	3.1	6.4
Harrisburg	725115	40.20	76.77	308	14.532	6193	9	13	22	20	18	24	29	22	29	8	270	10	250	97	2	3.3	5.7
Philadelphia, Intl Airport	724080	39.88	75.25	30	14.680	6193	11	15	24	21	19	26	31	23	30	12	290	11	230	96	5	2.8	5.6
Philadelphia, Northeast A	724085	40.08	75.02	121	14.631	8293	11	15	21	19	17	22	30	19	29	10	300	10	260	97	3	2.5	6.1
Philadelphia, Willow Gr NAS	724086	40.20	75.15	361	14.505	8293	10	14	18	15	13	19	30	16	30	5	300	6	250	99	2	5.4	5.8
Pittsburgh, Allegheny Co. A	725205	40.35	79.93	1253	14.042	8293	4	11	21	19	17	23	24	21	24	11	250	11	240	94	−4	3.1	9.4
Pittsburgh, Intl Airport	725200	40.50	80.22	1224	14.057	6193	2	7	25	21	19	26	24	23	25	10	260	11	230	93	−6	3.1	6.7
Wilkes-Barre/Scranton	725130	41.33	75.73	948	14.199	6193	2	7	20	18	16	21	26	19	25	8	230	11	220	92	−5	2.8	4.9
Williamsport	725140	41.25	76.92	525	14.419	6193	2	7	23	20	18	24	23	21	25	8	270	10	250	94	−6	3.1	5.9
RHODE ISLAND																							
Providence	725070	41.73	71.43	62	14.662	6193	5	10	27	23	21	27	31	23	32	12	340	13	230	95	−2	3.7	5.0
SOUTH CAROLINA																							
Beaufort, MCAS	722085	32.48	80.72	39	14.675	8293	28	31	18	15	13	19	46	17	45	4	300	7	270	101	13	2.9	8.3
Charleston	722080	32.90	80.03	49	14.669	6193	25	28	22	19	17	22	52	19	51	7	20	10	230	98	18	2.3	5.6
Columbia	723100	33.95	81.12	226	14.576	6193	21	24	20	17	15	20	48	18	49	5	220	9	240	100	13	3.1	5.6
Florence	723106	34.18	79.72	148	14.617	8293	23	27	19	17	15	20	51	18	50	7	360	10	240	100	14	2.7	7.6
Greer/Greenville	723120	34.90	82.22	971	14.187	6193	19	23	20	18	16	21	45	18	44	6	50	9	230	97	11	2.6	5.5
Myrtle Beach, AFB	747910	33.68	78.93	26	14.681	8293	25	29	18	15	13	18	49	15	47	4	360	7	290	98	17	2.9	7.4
Sumter, Shaw AFB	747900	33.97	80.47	243	14.567	8293	24	29	18	16	14	19	48	17	48	5	10	8	240	100	17	3.1	6.1
SOUTH DAKOTA																							
Chamberlain	726530	43.80	99.32	1739	13.795	8293	−13	−7	27	24	21	28	18	25	20	11	270	13	190	106	−12	8.1	18.4
Huron	726540	44.38	98.22	1289	14.024	6193	−17	−12	29	25	22	29	14	25	15	9	290	14	180	102	−25	4.6	5.9
Pierre	726686	44.38	100.28	1742	13.794	6193	−14	−9	29	25	22	32	15	27	20	11	320	14	180	106	−20	3.8	5.7
Rapid City	726620	44.05	103.07	3169	13.089	6193	−11	−5	36	31	27	37	26	32	26	9	350	13	160	102	−17	3.4	5.4
Sioux Falls	726510	43.58	96.73	1427	13.953	6193	−16	−11	28	25	22	30	15	26	17	8	310	15	180	100	−23	4.1	4.9
TENNESSEE																							
Bristol	723183	36.48	82.40	1519	13.906	6193	9	14	20	17	15	21	35	19	36	6	270	8	250	92	−1	3.0	7.5
Chattanooga	723240	35.03	85.20	689	14.333	6193	15	20	19	17	15	20	37	18	38	7	360	8	280	97	7	3.6	7.0
Crossville	723265	35.95	85.08	1880	13.724	8293	7	15	16	14	13	18	33	16	36	4	310	8	270	93	−3	4.0	8.6
Jackson	723346	35.60	88.92	433	14.467	8293	12	18	20	18	16	21	46	19	44	9	360	8	240	98	4	2.3	8.8
Knoxville	723260	35.82	83.98	981	14.182	6193	13	19	21	18	15	21	48	19	45	7	50	8	250	95	4	3.0	7.8
Memphis	723340	35.05	90.00	285	14.544	6193	16	21	22	19	17	22	42	20	42	10	20	9	240	99	9	2.8	7.2
Nashville	723270	36.13	86.68	591	14.385	6193	10	16	22	19	17	22	46	20	42	8	340	9	230	97	1	3.3	7.8
TEXAS																							
Abilene	722660	32.42	99.68	1791	13.769	6193	16	22	27	24	22	26	48	23	46	12	0	11	140	102	10	2.8	5.9
Amarillo	723630	35.23	101.70	3606	12.879	6193	6	12	30	27	24	30	40	27	38	14	20	15	200	100	−1	2.8	5.5
Austin	722540	30.30	97.70	620	14.369	6193	25	30	23	20	18	25	41	22	43	12	10	10	180	101	20	2.4	5.9
Beaumont/Port Arthur	722410	29.95	94.02	23	14.683	6193	29	32	22	20	18	23	51	21	51	10	340	9	200	97	22	2.5	4.5
Beeville, Chase Field NAS	722556	28.37	97.67	190	14.595	8293	28	33	22	20	18	23	58	20	53	13	350	9	150	104	22	2.5	8.3
Brownsville	722500	25.90	97.43	20	14.685	6193	36	40	27	24	22	26	64	23	62	13	330	16	160	98	31	2.4	5.2
College Station/Bryan	722445	30.58	96.37	322	14.525	8293	22	29	21	19	17	21	47	19	49	12	350	9	170	101	17	2.3	8.3
Corpus Christi	722510	27.77	97.50	43	14.673	6193	32	36	28	25	23	27	59	24	58	13	360	15	140	98	25	1.9	5.2
Dallas/Ft. Worth, Intl A	722590	32.90	97.03	597	14.381	8293	17	24	26	23	21	26	46	24	47	13	350	10	170	103	14	3.1	8.3
Del Rio, Laughlin AFB	722615	29.37	100.78	1083	14.130	8293	28	32	22	19	17	22	47	19	50	7	10	9	140	105	22	3.1	5.8
El Paso	722700	31.80	106.40	3917	12.731	6193	21	25	25	21	18	24	51	21	49	5	20	8	180	104	14	3.3	6.1
Ft. Worth, Carswell AFB	722595	32.77	97.45	650	14.354	8293	18	24	22	20	18	23	43	20	45	11	10	8	10	103	15	2.2	8.1
Ft. Worth, Meacham Field	722596	32.82	97.37	709	14.323	6193	19	24	27	24	21	27	40	24	44	13	350	10	180	103	14	3.0	5.8
Guadalupe Pass	722620	31.83	104.80	5453	12.022	8293	13	19	51	45	41	50	39	46	37	19	70	13	250	98	10	2.9	6.8
Houston, Hobby Airport	722435	29.65	95.28	46	14.671	8293	29	34	22	20	18	23	52	21	52	13	350	7	190	98	24	2.0	8.1
Houston, Inter Airport	722430	29.97	95.35	108	14.638	6193	27	31	20	18	16	22	47	20	52	8	340	10	180	98	22	3.1	5.4
Junction	747400	30.50	99.77	1713	13.808	8293	19	23	19	16	15	19	53	16	53	6	360	9	150	104	12	2.3	6.7
Killeen, Ft. Hood	722576	31.07	97.83	1014	14.165	8293	20	27	22	19	17	22	48	19	53	11	360	9	160	102	15	2.0	8.6
Kingsville, NAS	722516	27.50	97.82	49	14.669	8293	31	36	23	21	19	22	61	20	60	11	350	12	140	102	18	2.0	10.1
Laredo	722520	27.55	99.47	509	14.427	8293	32	36	24	22	20	22	59	20	62	9	320	13	140	106	28	2.2	6.7

WMO# = World Meteorological Organization number   AAF = Army Air Field   Lat. = North latitude °   StdP = standard pressure at station elevation, psia
AFB = Air Force Base   NAS = Naval Air Station   MCAS = Marine Corps Air Station   Long. = West longitude °   WS = wind speed, mph
WSO = Weather Service Office   DB = dry-bulb temperature, °F   Elev. = elevation, ft   PWD = prevailing wind direction, ° True

United States

Station	WMO#	Lat.	Long.	Elev., ft	StdP, psia	Dates	Heating Dry Bulb 99.6%	99%	Extreme Wind Speed, mph 1%	2.5%	5%	Coldest Month 0.4% WS	MDB	Coldest Month 1% WS	MDB	MWS/PWD to DB 99.6% MWS	PWD	MWS/PWD to DB 0.4% MWS	PWD	Extr. Annual Daily Mean DB Max.	Min.	StdD DB Max.	Min.
1a	1b	1c	1d	1e	1f	1g	2a	2b	3a	3b	3c	4a	4b	4c	4d	5a	5b	5c	5d	6a	6b	6c	6d
Lubbock, Intl Airport	722670	33.65	101.82	3241	13.054	6193	11	17	30	26	23	30	43	27	44	12	0	14	160	102	4	2.6	5.6
Lubbock, Reese AFB	722675	33.60	102.05	3337	13.008	8293	11	18	25	22	19	25	48	22	44	10	20	11	170	102	6	3.1	4.9
Lufkin	722446	31.23	94.75	289	14.543	6193	23	27	18	16	14	18	44	17	46	6	330	8	230	99	17	3.2	5.3
Marfa	722640	30.37	104.02	4859	12.292	8293	15	19	24	21	18	25	44	22	45	5	360	9	220	97	5	2.3	5.0
McAllen	722506	26.18	98.23	108	14.638	8293	34	40	24	22	20	23	68	21	68	11	350	14	130	106	27	4.3	8.1
Midland/Odessa	722650	31.95	102.18	2861	13.238	6193	17	22	28	25	22	27	50	23	48	9	20	13	180	103	9	2.6	6.9
San Angelo	722630	31.37	100.50	1909	13.709	6193	20	24	26	23	21	25	52	22	51	10	20	11	160	103	13	2.8	6.1
San Antonio, Intl Airport	722530	29.53	98.47	794	14.279	6193	26	30	22	19	17	23	43	20	45	10	350	10	160	100	19	2.9	5.2
San Antonio, Kelly AFB	722535	29.38	98.58	689	14.333	8293	27	32	19	17	15	21	51	18	52	8	360	8	160	103	22	2.9	6.5
San Antonio, Randolph AFB	722536	29.53	98.28	761	14.296	8293	27	31	19	17	15	20	45	17	48	7	340	7	150	101	20	2.2	6.7
Sanderson	747300	30.17	102.42	2838	13.250	8293	23	28	19	16	13	20	44	17	48	6	360	7	120	102	9	2.9	8.3
Victoria	722550	28.85	96.92	118	14.633	6193	29	33	26	23	21	26	50	23	51	12	360	12	180	99	23	2.5	5.2
Waco	722560	31.62	97.22	509	14.427	6193	22	26	26	23	21	29	38	25	42	13	360	12	180	104	16	2.8	6.4
Wichita Falls, Sheppard AFB	723510	33.98	98.50	1030	14.157	6193	14	19	29	25	23	28	42	25	43	12	360	13	180	107	7	3.4	6.6
UTAH																							
Cedar City	724755	37.70	113.10	5623	11.945	6193	2	8	26	22	20	24	38	21	39	4	140	12	200	97	−6	2.3	8.3
Ogden, Hill AFB	725755	41.12	111.97	4787	12.325	8293	6	11	22	19	17	22	27	19	28	9	110	6	190	96	1	2.9	6.3
Salt Lake City	725720	40.78	111.97	4226	12.586	6193	6	11	27	23	20	27	42	22	40	7	160	11	340	100	−3	1.9	6.7
VERMONT																							
Burlington	726170	44.47	73.15	341	14.515	6193	−11	−6	23	21	18	24	30	21	27	6	70	11	180	93	−19	2.7	5.6
Montpelier/Barre	726145	44.20	72.57	1165	14.087	8293	−10	−6	21	19	17	22	20	20	20	4	320	9	220	91	−18	3.6	5.9
VIRGINIA																							
Ft. Belvoir	724037	38.72	77.18	69	14.659	8293	12	18	18	14	12	19	35	17	34	2	320	6	160	100	2	2.3	7.6
Hampton, Langley AFB	745980	37.08	76.37	10	14.690	8293	21	24	22	19	17	22	41	20	40	10	330	9	240	97	13	3.2	6.1
Lynchburg	724100	37.33	79.20	938	14.204	6193	12	17	19	17	15	21	35	18	35	8	360	9	230	95	5	2.9	5.8
Newport News	723086	37.13	76.50	43	14.673	8293	18	22	19	18	16	20	40	18	41	8	350	10	220	99	11	2.3	4.7
Norfolk	723080	36.90	76.20	30	14.680	6193	20	24	25	22	20	26	40	23	40	12	340	12	230	97	14	2.8	5.4
Oceana, NAS	723075	36.82	76.03	23	14.683	8293	22	25	21	19	17	21	42	19	42	8	310	9	220	98	14	1.8	6.8
Quantico, MCAS	724035	38.50	77.30	13	14.688	8293	16	21	17	14	12	19	36	15	38	6	340	5	230	100	8	3.6	5.9
Richmond	724010	37.50	77.33	177	14.602	6193	14	18	20	18	16	21	40	18	39	7	340	10	230	98	6	2.6	5.8
Roanoke	724110	37.32	79.97	1175	14.082	6193	12	17	23	20	17	27	31	23	32	10	320	10	290	96	4	3.3	5.6
Sterling	724030	38.95	77.45	322	14.525	6193	9	14	22	19	16	25	32	21	31	7	340	10	250	97	−1	3.3	7.0
Washington, R. Reagan A	724050	38.85	77.03	66	14.661	8293	15	20	23	20	18	24	34	21	35	11	340	11	170	99	8	2.5	6.8
WASHINGTON																							
Bellingham	727976	48.80	122.53	157	14.612	8293	15	21	23	20	18	28	33	23	34	17	40	9	290	87	11	3.1	7.4
Hanford	727840	46.57	119.60	732	14.311	8293	5	12	25	21	18	24	44	19	44	6	20	8	20	105	2	3.1	9.0
Olympia	727920	46.97	122.90	200	14.589	6193	18	23	21	18	16	21	45	19	45	5	180	8	50	94	10	4.0	8.1
Quillayute	727970	47.95	124.55	203	14.588	6193	23	27	33	27	21	41	45	35	45	7	60	9	240	87	19	8.4	6.4
Seattle, Intl Airport	727930	47.45	122.30	449	14.458	6193	23	28	22	19	17	24	44	21	44	10	10	10	350	92	19	3.6	6.8
Spokane, Fairchild AFB	727855	47.62	117.65	2461	13.435	6193	1	7	27	23	20	28	39	25	38	7	50	9	240	98	−7	3.2	8.7
Stampede Pass	727815	47.28	121.33	3967	12.708	8293	3	10	21	19	16	27	19	22	25	13	90	7	100	84	2	3.2	7.2
Tacoma, McChord AFB	742060	47.13	122.48	322	14.525	8293	18	24	18	15	13	22	45	18	46	2	180	7	20	94	12	2.7	6.8
Walla Walla	727846	46.10	118.28	1204	14.067	8293	4	12	22	19	17	24	49	22	47	6	180	9	300	105	1	3.2	11.7
Wenatchee	727825	47.40	120.20	1243	14.047	8293	3	9	22	19	17	17	36	12	31	3	100	9	280	101	−2	2.5	7.2
Yakima	727810	46.57	120.53	1066	14.138	6193	4	11	24	20	17	23	47	19	43	7	250	7	90	101	−2	3.2	8.5
WEST VIRGINIA																							
Bluefield	724125	37.30	81.20	2858	13.240	8293	5	12	15	13	12	18	34	15	33	6	270	6	290	88	−6	4.0	8.5
Charleston	724140	38.37	81.60	981	14.182	6193	6	11	18	16	14	20	38	18	34	7	250	8	240	94	−2	2.8	6.7
Elkins	724170	38.88	79.85	1998	13.665	6193	−2	5	20	18	16	22	30	19	30	4	280	8	290	88	−12	2.8	5.4
Huntington	724250	38.37	82.55	837	14.257	6193	6	11	19	16	14	20	32	17	32	8	270	8	270	94	−2	5.0	7.6
Martinsburg	724177	39.40	77.98	558	14.402	8293	8	14	21	18	16	23	33	20	34	7	270	9	290	99	−3	4.0	8.3
Morgantown	724176	39.65	79.92	1247	14.045	8293	4	11	18	15	13	19	32	17	33	6	210	8	240	93	−4	3.6	8.6
Parkersburg	724273	39.35	81.43	860	14.245	8293	4	11	18	16	14	20	32	18	29	7	240	8	270	95	−4	3.1	9.2
WISCONSIN																							
Eau Claire	726435	44.87	91.48	906	14.221	6193	−18	−13	22	19	17	21	14	20	13	7	250	13	220	95	−25	3.2	5.7
Green Bay	726450	44.48	88.13	702	14.326	6193	−13	−8	25	22	20	25	19	22	18	10	270	12	200	93	−19	2.8	5.6
La Crosse	726430	43.87	91.25	663	14.347	6193	−14	−8	23	20	18	23	13	21	13	7	310	12	180	97	−21	3.2	6.2
Madison	726410	43.13	89.33	866	14.241	6193	−11	−6	24	21	19	25	16	22	17	8	300	12	230	94	−18	3.2	6.0
Milwaukee	726400	42.95	87.90	692	14.332	6193	−7	−2	28	24	22	28	19	24	20	13	290	15	220	95	−12	3.2	6.7
Wausau	726463	44.93	89.63	1201	14.069	8293	−15	−9	19	17	15	19	17	17	17	7	300	10	200	93	−22	3.1	4.7
WYOMING																							
Big Piney	726710	42.57	110.10	6969	11.353	8293	−22	−15	24	20	17	22	25	19	21	3	60	11	260	87	−33	2.7	8.5
Casper	725690	42.92	106.47	5289	12.096	6193	−13	−5	34	30	27	35	35	32	32	9	260	13	240	97	−22	2.2	8.4
Cheyenne, Warren AFB	725640	41.15	104.82	6142	11.714	6193	−7	0	34	29	26	38	36	33	34	10	290	13	290	92	−15	2.2	7.5
Cody	726700	44.52	109.02	5095	12.184	8293	−14	−7	34	28	23	35	35	30	35	6	40	11	70	95	−20	4.1	9.4
Gillette	726650	44.35	105.53	4035	12.675	8293	−16	−7	28	25	22	30	34	27	33	8	260	11	140	101	−20	5.9	10.1
Lander	725760	42.82	108.73	5558	11.974	6193	−14	−7	23	19	16	25	38	19	37	3	120	10	270	95	−20	2.5	7.8
Rock Springs	725744	41.60	109.07	6759	11.444	6193	−9	−2	28	25	23	32	25	29	24	7	70	13	280	90	−17	2.0	8.0
Sheridan	726660	44.77	106.97	3967	12.708	6193	−14	−8	28	24	20	29	32	23	27	5	280	9	120	99	−22	3.0	6.4
Worland	726665	43.97	107.95	4245	12.577	8293	−22	−13	22	19	16	20	28	17	28	3	210	9	220	103	−30	2.2	10.4

WMO# = World Meteorological Organization number
AFB = Air Force Base NAS = Naval Air Station
WSO = Weather Service Office

AAF = Army Air Field
MCAS = Marine Corps Air Station
DB = dry-bulb temperature, °F

Lat. = North latitude °
Long. = West longitude °
Elev. = elevation, ft

StdP = standard pressure at station elevation, psia
WS = wind speed, mph
PWD = prevailing wind direction, ° True

Canada

Station	WMO#	Lat.	Long.	Elev., ft	StdP, psia	Dates	Heating Dry Bulb 99.6%	99%	Extreme Wind Speed, mph 1%	2.5%	5%	Coldest Month 0.4% WS	MDB	1% WS	MDB	MWS/PWD to DB 99.6% MWS	PWD	0.4% MWS	PWD	Extr. Annual Daily Mean DB Max.	Min.	StdD DB Max.	Min.
1a	1b	1c	1d	1e	1f	1g	2a	2b	3a	3b	3c	4a	4b	4c	4d	5a	5b	5c	5d	6a	6b	6c	6d
ALBERTA																							
Calgary Intl A	718770	51.12	114.02	3556	12.902	6193	-22	-17	28	24	21	32	29	28	27	7	0	11	160	89	-28	2.7	5.6
Cold Lake A	711200	54.42	110.28	1785	13.772	6193	-31	-26	21	18	16	21	18	18	11	3	270	9	180	88	-40	3.2	6.3
Coronation	718730	52.07	111.45	2595	13.369	6193	-27	-23	25	21	19	28	12	23	13	9	320	11	160	92	-35	3.1	6.3
Edmonton Intl A	711230	53.30	113.58	2372	13.479	6193	-28	-23	24	21	18	24	12	21	11	6	180	9	180	87	-36	3.1	8.1
Fort McMurray A	719320	56.65	111.22	1211	14.064	6193	-32	-29	17	15	13	18	16	15	11	3	90	9	250	90	-42	3.6	4.9
Grande Prairie A	719400	55.18	118.88	2195	13.567	6193	-32	-27	27	23	19	29	32	24	28	3	320	8	270	87	-41	2.7	6.7
Lethbridge A	718740	49.63	112.80	3048	13.147	6193	-22	-16	36	32	28	45	39	39	38	5	250	13	270	94	-30	3.2	6.5
Medicine Hat A	718720	50.02	110.72	2349	13.490	6193	-24	-19	26	22	20	29	36	25	33	5	230	11	220	97	-32	3.6	7.2
Peace River A	710680	56.23	117.43	1873	13.727	6193	-32	-27	21	18	17	22	30	19	24	4	0	9	270	87	-42	2.9	6.7
Red Deer A	718780	52.18	113.90	2969	13.186	6193	-27	-21	22	19	17	27	13	22	13	6	200	10	180	88	-35	3.1	6.5
Rocky Mtn. House	719280	52.43	114.92	3245	13.052	6193	-25	-20	19	16	13	19	26	16	20	3	340	8	160	87	-36	2.7	5.0
Vermilion A		53.35	110.83	2028	13.650	6193	-30	-25	22	19	17	21	13	19	11	3	270	11	180	90	-43	3.6	6.7
Whitecourt A	719300	54.15	115.78	2566	13.383	6193	-30	-24	17	15	14	19	22	17	18	4	270	7	90	87	-41	2.0	5.4
BRITISH COLUMBIA																							
Abbotsford A	711080	49.03	122.37	190	14.595	6193	15	20	20	17	15	29	33	25	34	12	90	7	220	92	10	4.0	6.7
Cape St. James	710310	51.93	131.02	302	14.536	6193	25	29	50	46	40	60	40	54	42	22	50	11	300	69	22	3.4	5.6
Castlegar A	718840	49.30	117.63	1624	13.853	6693	5	9	18	15	14	21	18	19	21	8	0	7	180	98	-3	2.9	7.0
Comox A	718930	49.72	124.90	79	14.654	6193	21	25	29	25	21	31	43	28	42	7	290	7	340	87	17	3.8	5.2
Cranbrook A	718800	49.60	115.78	3081	13.131	7093	-15	-8	20	18	16	20	33	18	33	2	200	10	210	94	-22	2.7	6.7
Fort Nelson A	719450	58.83	122.58	1253	14.042	6193	-33	-30	16	14	12	15	8	12	2	1	220	5	120	88	-42	3.2	6.3
Fort St. John A	719430	56.23	120.73	2280	13.524	6193	-30	-25	25	22	19	29	23	25	22	7	0	9	230	86	-36	2.9	6.3
Kamloops A	718870	50.70	120.45	1135	14.103	6693	-8	-1	23	20	18	25	26	22	27	4	90	8	270	98	-14	2.7	8.8
Penticton A	718890	49.47	119.60	1129	14.106	6193	5	10	23	20	17	28	34	25	35	8	340	9	180	96	1	2.7	7.2
Port Hardy A	711090	50.68	127.37	72	14.657	6193	22	26	28	24	21	33	38	29	39	8	110	9	340	76	18	3.4	4.9
Prince George A	718960	53.88	122.68	2267	13.531	6193	-25	-18	21	18	15	27	32	23	24	2	0	6	180	85	-37	9.2	5.9
Prince Rupert A	718980	54.30	130.43	112	14.636	6393	7	13	28	23	20	30	44	26	43	6	70	8	270	75	2	4.5	6.8
Quesnel A	711030	53.03	122.52	1788	13.770	6193	-22	-14	17	15	14	19	18	17	20	1	340	5	340	92	-31	4.0	8.3
Sandspit A	711010	53.25	131.82	20	14.685	6193	21	25	38	32	27	42	44	37	42	18	320	9	270	72	18	3.4	5.0
Smithers A	719500	54.82	127.18	1716	13.807	6193	-19	-12	17	15	13	18	23	16	19	3	140	6	320	88	-26	4.0	7.0
Spring Island	714790	50.12	127.93	322	14.525	6193	29	31	41	35	29	44	46	40	45	6	50	6	320	78	25	6.1	4.5
Terrace A	719510	54.47	128.58	712	14.321	6193	-2	2	26	23	20	32	10	29	14	19	0	8	270	89	-5	4.0	5.8
Tofino A	711060	49.08	125.77	79	14.654	6193	25	29	24	20	18	29	46	24	45	5	70	7	290	81	21	4.0	5.6
Vancouver Intl A	718920	49.18	123.17	7	14.692	6193	18	24	22	19	16	25	41	21	42	6	90	7	290	82	14	2.9	6.3
Victoria Intl A	717990	48.65	123.43	62	14.662	6193	23	26	20	16	14	24	37	20	38	10	50	6	90	87	18	3.2	5.8
Williams Lake A	711040	52.18	122.05	3084	13.130	6193	-20	-14	22	19	17	24	29	21	30	3	320	6	140	88	-29	4.0	7.9
MANITOBA																							
Brandon A	711400	49.92	99.95	1342	13.997	6193	-29	-24	27	23	20	28	2	24	2	9	270	12	160	94	-36	3.1	4.5
Churchill A	719130	58.75	94.07	95	14.645	6193	-36	-33	34	30	26	36	-11	30	-14	15	270	13	230	86	-41	4.5	4.0
Dauphin A	718550	51.10	100.05	1001	14.172	6193	-28	-23	28	25	22	31	2	28	5	9	250	13	200	93	-36	3.4	4.3
Portage La Prairie A	718510	49.90	98.27	883	14.233	6193	-25	-21	26	23	20	29	1	25	1	9	250	12	180	95	-31	3.4	4.1
The Pas A	718670	53.97	101.10	889	14.229	6193	-32	-28	24	21	19	25	-6	22	-2	6	290	11	160	89	-40	3.4	4.1
Thompson A	710790	55.80	97.87	715	14.320	6893	-38	-34	20	18	16	19	-5	17	-7	3	270	10	180	89	-48	3.8	4.3
Winnipeg Int'l A	718520	49.90	97.23	784	14.284	6193	-27	-23	29	25	23	30	5	26	5	7	320	13	180	94	-33	3.4	4.7
NEW BRUNSWICK																							
Charlo A	717110	47.98	66.33	125	14.629	6793	-14	-10	24	21	19	27	3	24	8	11	250	11	250	89	-21	2.7	4.5
Chatham A	717170	47.00	65.45	102	14.642	6193	-12	-7	24	21	18	27	16	24	16	7	270	11	230	93	-20	2.2	4.3
Fredericton A	717000	45.87	66.53	66	14.661	6193	-12	-7	23	20	17	25	17	22	18	5	270	11	230	92	-21	2.5	5.4
Moncton A	717050	46.12	64.68	233	14.572	6193	-10	-5	26	23	20	30	19	26	19	13	270	13	250	89	-17	2.0	4.5
Saint John A	716090	45.32	65.88	358	14.507	6193	-9	-4	26	23	20	32	24	28	23	9	340	11	230	84	-18	4.0	4.7
NEWFOUNDLAND																							
Battle Harbour	718170	52.30	55.83	26	14.681	6193	-14	-10	40	35	32	48	17	42	15	18	270	17	230	78	-18	4.7	5.9
Bonavista	711960	48.67	53.12	89	14.648	6193	3	7	43	38	34	48	21	42	23	24	280	17	230	81	0	2.7	5.8
Cartwright	718180	53.70	57.03	46	14.671	6493	-18	-15	36	30	27	40	18	35	19	12	220	12	210	84	-23	4.1	4.5
Daniels Harbour	711850	50.23	57.58	62	14.662	6693	-7	-3	40	35	31	45	16	39	21	12	270	15	230	75	-12	3.2	6.5
Deer Lake A	718090	49.22	57.40	72	14.657	6693	-13	-7	24	22	18	26	21	22	19	3	240	14	220	86	-23	2.3	6.7
Gander Intl A	718030	48.95	54.57	495	14.434	6193	-4	0	32	28	25	37	20	32	22	16	270	13	230	85	-8	2.9	6.3
Goose A	718160	53.32	60.37	160	14.362	6193	-23	-20	26	22	20	30	3	26	5	11	250	12	250	89	-29	4.0	3.6
Hopedale	719000	55.45	60.23	26	14.681	6493	-21	-18	36	30	27	40	13	35	11	12	250	13	250	79	-25	4.5	5.8
St. John's A	718010	47.62	52.73	459	14.453	6193	3	7	37	33	29	41	24	37	25	17	290	17	250	82	-2	2.5	4.9
Stephenville A	718150	48.53	58.55	85	14.650	6193	-2	3	33	28	23	36	21	30	21	11	50	9	250	79	-7	2.5	6.3
Wabush Lake A	718250	52.93	66.87	1808	13.760	6193	-33	-30	23	20	17	25	-5	21	-4	5	270	12	240	82	-43	3.8	4.3
NORTHWEST TERRITORIES																							
Cape Parry A	719480	70.17	124.68	56	14.666	6193	-34	-33	31	28	25	34	-12	29	-12	7	270	9	110	66	-42	4.9	4.3
Fort Smith A	719340	60.02	111.97	666	14.345	6193	-34	-32	18	17	15	20	-3	18	-5	3	150	10	180	88	-48	3.4	5.6
Inuvik UA	719570	68.30	133.48	223	14.577	7393	-43	-40	17	15	14	19	-7	17	-7	1	70	8	180	84	-52	2.5	5.0
Norman Wells A	710430	65.28	126.80	243	14.567	6193	-40	-36	24	21	18	29	-4	24	-7	2	170	8	140	87	-49	3.1	5.4
Yellowknife A	719360	62.47	114.45	676	14.340	6193	-39	-36	22	19	17	23	-8	20	-9	5	50	10	160	82	-47	3.6	4.5
NUNAVUT																							
Baker Lake A	719260	64.30	96.08	59	14.664	6393	-41	-39	36	32	28	42	-26	37	-26	12	0	10	270	77	-50	4.9	4.3
Cambridge Bay A	719250	69.10	105.12	89	14.648	6193	-38	-35	35	31	27	36	-19	31	-19	9	320	11	140	68	-50	4.9	3.8
Chesterfield	719164	63.33	90.72	36	14.676	6193	-35	-34	33	29	26	35	-26	31	-26	14	320	13	320	73	-49	13.9	4.9
Coral Harbour A	719150	64.20	83.37	210	14.584	6193	-40	-38	37	32	27	39	-5	32	-7	9	340	13	270	71	-49	3.8	5.2
Hall Beach A	710810	68.78	81.25	26	14.681	6193	-42	-38	33	29	25	34	-20	30	-21	10	320	11	180	64	-53	5.2	5.4

WMO# = World Meteorological Organization number   Elev. = elevation, ft   PWD = prevailing wind direction, ° True   DB = dry-bulb temperature, °F
Lat. = North latitude, °   Long. = West longitude, °   StdP = standard pressure at station elevation, psia   WS = wind speed, mph

Canada

Station	WMO#	Lat.	Long.	Elev., ft	StdP, psia	Dates	Heating Dry Bulb 99.6%	99%	Extreme Wind Speed, mph 1%	2.5%	5%	Coldest Month 0.4% WS	MDB	1% WS	MDB	MWS/PWD to DB 99.6% MWS	PWD	0.4% MWS	PWD	Extr. Annual Daily Mean DB Max.	Min.	StdD DB Max.	Min.
1a	1b	1c	1d	1e	1f	1g	2a	2b	3a	3b	3c	4a	4b	4c	4d	5a	5b	5c	5d	6a	6b	6c	6d
Iqaluit A (Frobisher)	719090	63.75	68.55	108	14.638	6193	−39	−36	32	28	25	40	−12	34	−12	4	320	12	320	68	−43	4.3	4.7
Resolute A	719240	74.72	94.98	220	14.579	6393	−42	−40	39	34	30	42	−15	37	−18	9	320	12	110	55	−49	3.8	4.9
NOVA SCOTIA																							
Greenwood A	713970	44.98	64.92	92	14.647	6193	−2	3	29	25	22	35	25	30	23	7	300	15	250	89	−11	2.3	5.4
Halifax Intl A	713950	44.88	63.50	476	14.445	6993	−2	2	27	23	20	30	26	27	26	11	320	12	200	87	−8	2.7	4.3
Sable Island	716000	43.93	60.02	13	14.688	6193	14	17	38	34	23	43	30	39	29	24	290	13	200	73	10	2.5	4.0
Shearwater A	716010	44.63	63.50	167	14.607	6193	1	5	29	25	22	33	24	28	24	12	340	10	230	85	−5	3.2	4.1
Sydney A	717070	46.17	60.05	203	14.588	6193	−1	3	30	26	23	35	23	30	24	13	270	14	230	87	−7	2.5	4.5
Truro	713980	45.37	63.27	131	14.626	6193	−9	−4	24	21	18	30	25	26	25	5	0	11	270	86	−16	3.2	5.8
Yarmouth A	716030	43.83	66.08	141	14.621	6193	7	10	28	25	22	30	26	28	27	12	320	10	190	79	1	2.7	3.6
ONTARIO																							
Armstrong A	718410	50.30	89.03	1152	14.094	6193	−33	−30	22	19	17	21	−2	18	−3	3	270	12	0	88	−49	3.2	4.9
Atikokan	717480	48.75	91.62	1289	14.024	6793	−31	−27	16	14	12	16	6	14	6	1	270	8	230	89	−42	3.6	3.8
Big Trout Lake	718480	53.83	89.87	735	14.309	6793	−32	−30	23	21	18	24	−1	21	−2	6	290	9	200	86	−42	3.2	4.9
Earlton A	717350	47.70	79.85	797	14.277	6193	−27	−21	21	19	17	24	14	21	12	4	320	12	200	91	−38	3.8	5.0
Geraldton A	718340	49.78	86.93	1152	14.094	6893	−32	−28	21	18	16	22	3	19	4	2	270	11	0	87	−45	4.1	6.7
Gore Bay A	717330	45.88	82.57	633	14.362	6193	−12	−7	27	23	21	29	23	26	22	7	0	11	180	86	−21	4.5	5.9
Kapuskasing A	718310	49.42	82.47	745	14.304	6193	−30	−25	20	17	16	21	5	19	5	4	270	10	230	90	−39	2.7	5.0
Kenora A	718500	49.78	94.37	1348	13.993	6193	−27	−22	21	18	17	21	7	19	6	8	320	11	180	89	−32	3.4	5.0
London A	716230	43.03	81.15	912	14.217	6193	−3	2	25	22	20	30	21	26	21	9	250	11	250	90	−11	3.4	5.6
Mount Forest	716310	43.98	80.75	1362	13.987	6293	−7	−3	25	21	19	28	19	25	18	6	90	10	250	87	−16	2.2	4.5
Muskoka A	716300	44.97	79.30	925	14.211	6193	−17	−11	21	19	18	23	21	20	21	7	320	9	270	87	−30	2.2	5.6
North Bay A	717310	46.35	79.43	1217	14.060	6193	−18	−13	20	17	16	23	16	20	16	6	0	10	230	86	−26	3.2	5.4
Ottawa Int'l A	716280	45.32	75.67	374	14.498	6193	−13	−8	23	20	17	27	17	23	15	9	290	10	250	91	−19	2.7	5.0
Sault Ste. Marie A	712600	46.48	84.52	630	14.364	6293	−13	−8	27	23	20	28	18	24	18	4	90	9	220	88	−25	3.2	5.9
Simcoe	715270	42.85	80.27	791	14.280	6293	−2	3	24	20	18	28	23	24	24	10	270	11	230	91	−10	2.9	4.7
Sioux Lookout A	718420	50.12	91.90	1280	14.029	6193	−30	−25	17	15	14	19	2	17	2	4	270	9	200	89	−39	3.2	4.9
Sudbury A	717300	46.62	80.80	1142	14.099	6193	−19	−14	30	26	23	30	13	27	13	10	0	14	230	89	−27	4.0	5.2
Thunder Bay A	717490	48.37	89.32	653	14.352	6193	−22	−18	24	21	18	25	9	22	6	9	250	11	200	90	−31	3.8	4.7
Timmins A	717390	48.57	81.37	968	14.189	6193	−28	−23	21	19	17	22	6	20	7	5	180	10	250	91	−39	3.6	4.7
Toronto Int'l A	716240	43.67	79.63	568	14.397	6593	−4	1	26	22	20	29	22	26	23	9	340	12	270	92	−11	2.9	5.6
Trenton A	716210	44.12	77.53	282	14.546	6193	−8	−3	26	22	19	30	23	26	24	6	50	12	230	89	−16	3.1	5.6
Wiarton A	716330	44.75	81.10	728	14.313	6193	−5	0	25	22	19	27	27	24	25	7	340	12	230	88	−15	2.3	7.0
Windsor A	715380	42.27	82.97	623	14.367	6193	2	6	28	24	21	30	22	26	22	12	230	12	250	94	−5	2.7	5.4
PRINCE EDWARD ISLAND																							
Charlottetown A	717060	46.28	63.13	177	14.602	6193	−6	−2	24	22	20	35	17	29	18	13	270	12	230	85	−12	2.3	4.7
Summerside A	717020	46.43	63.83	79	14.654	6193	−5	−1	32	28	25	40	18	35	18	15	270	13	200	85	−11	2.7	4.1
QUEBEC																							
Bagotville A	717270	48.33	71.00	522	14.421	6193	−23	−19	26	23	20	30	3	26	5	6	270	10	270	91	−30	2.5	5.0
Baie Comeau A	711870	49.13	68.20	72	14.657	6593	−19	−14	27	24	21	30	14	25	14	11	270	13	230	82	−28	3.2	6.8
Grindstone Island		47.38	61.87	194	14.593	6193	−1	3	49	43	39	55	20	49	21	24	290	18	250	79	−5	2.0	5.6
Kuujjuarapik A	719050	55.28	77.77	39	14.675	6193	−33	−30	29	25	22	26	1	24	−2	8	120	13	180	85	−43	4.0	4.7
Kuujuaq A	719060	58.10	68.42	121	14.631	6193	−34	−31	29	25	21	31	2	27	−2	5	230	11	180	82	−41	3.4	3.6
La Grande Riviere A	718270	53.63	77.70	640	14.359	7793	−33	−30	22	20	17	23	0	20	−2	6	270	13	240	85	−39	3.6	3.4
Lake Eon A	714210	51.87	63.28	1841	13.744	6193	−31	−27	23	21	18	23	4	21	0	5	270	10	230	80	−42	2.5	4.0
Mont Joli A	717180	48.60	68.22	171	14.605	6193	−12	−8	28	25	22	35	8	30	9	15	290	14	230	87	−18	3.1	4.5
Montreal Intl A	716270	45.47	73.75	118	14.633	6193	−12	−7	23	20	18	30	19	26	18	7	250	11	230	90	−19	2.3	4.5
Montreal Mirabel A	716278	45.68	74.03	269	14.553	7693	−16	−11	21	18	16	25	13	21	11	6	240	9	240	88	−25	1.8	4.3
Nitchequon		53.20	70.90	1759	13.785	6193	−33	−31	25	22	20	29	−6	25	−5	6	270	11	230	78	−46	3.6	4.7
Quebec A	717140	46.80	71.38	240	14.569	6193	−16	−11	24	21	19	30	13	26	11	10	250	12	250	89	−23	2.3	4.9
Riviere Du Loup	717150	47.80	69.55	486	14.439	6693	−13	−10	21	19	17	24	14	21	13	10	180	11	230	85	−18	2.0	4.7
Roberval A	717280	48.52	72.27	587	14.386	6193	−23	−19	23	20	18	27	8	23	10	7	270	12	220	90	−30	3.1	4.3
Schefferville A	718280	54.80	66.82	1709	13.810	6193	−33	−31	26	23	21	30	−10	26	−9	7	320	12	270	81	−43	4.0	5.0
Sept–Iles A	718110	50.22	66.27	180	14.600	6893	−20	−15	27	23	20	30	16	25	13	8	300	11	220	82	−26	4.3	4.3
Sherbrooke A	716100	45.43	71.68	791	14.280	6393	−20	−14	19	17	15	22	15	19	14	5	110	9	250	88	−30	2.2	5.0
St. Hubert A	713710	45.52	73.42	89	14.648	6193	−12	−7	27	23	20	30	18	27	19	7	20	13	250	91	−19	2.5	4.9
Ste. Agathe Des Monts	717200	46.05	74.28	1296	14.020	6693	−19	−15	19	17	15	22	12	19	11	4	290	9	270	86	−28	2.9	4.0
Val d'Or A	717250	48.07	77.78	1106	14.118	6193	−27	−22	21	18	16	22	12	20	10	5	310	11	230	88	−37	2.9	4.9
SASKATCHEWAN																							
Broadview	718610	50.38	102.68	1975	13.676	6693	−30	−25	26	23	20	28	9	25	9	8	290	12	160	94	−37	3.6	5.8
Estevan A	718620	49.22	102.97	1906	13.711	6193	−25	−21	30	26	23	32	12	28	12	10	290	14	180	97	−32	3.2	5.4
Moose Jaw A	718640	50.33	105.55	1893	13.718	6193	−26	−21	30	26	23	35	21	30	19	9	290	13	180	97	−33	3.2	5.4
North Battleford A	718760	52.77	108.25	1798	13.765	6193	−31	−26	25	22	20	26	8	22	7	4	320	11	140	93	−38	3.1	5.4
Prince Albert A	718690	53.22	105.68	1404	13.965	6193	−34	−29	23	20	18	24	2	21	3	3	270	11	180	91	−43	3.8	5.9
Regina A	718630	50.43	104.67	1893	13.718	6193	−29	−24	30	27	23	34	8	29	10	9	270	14	180	96	−36	3.6	5.0
Saskatoon	718660	52.17	106.68	1654	13.838	6193	−31	−26	27	23	20	27	12	23	6	7	290	13	180	94	−37	3.8	5.9
Swift Current A	718700	50.28	107.68	2684	13.325	6193	−25	−21	33	28	25	36	17	30	14	14	270	13	180	95	−33	3.1	5.8
Uranium City A	710760	59.57	108.48	1043	14.150	6393	−38	−35	20	18	16	18	−7	16	−9	4	70	7	230	86	−49	4.0	3.4
Wynyard	718650	51.77	104.20	1841	13.744	6593	−29	−25	26	23	21	27	13	24	11	8	270	13	180	92	−35	3.4	6.1
Yorkton A	711380	51.27	102.47	1634	13.848	6193	−30	−26	25	22	20	27	6	23	2	7	290	12	180	93	−37	3.2	5.6
YUKON TERRITORY																							
Burwash A	719670	61.37	139.05	2644	13.344	6793	−34	−32	28	24	21	29	31	25	31	0	290	9	110	79	−55	3.4	6.5
Whitehorse A	719640	60.72	135.07	2306	13.511	6193	−34	−31	23	20	18	24	15	21	14	2	340	8	140	83	−47	4.0	5.9

WMO# = World Meteorological Organization number   Elev. = elevation, ft   PWD = prevailing wind direction, ° True   DB = dry-bulb temperature, °F
Lat. = North latitude, °   Long. = West longitude, °   StdP = standard pressure at station elevation, psia   WS = wind speed, mph

TABLE 4.7   NORMAL HEATING DEGREE-DAYS (°F · DAYS), 65°F BASE TEMPERATURE, FOR SELECTED CITIES IN THE UNITED STATES FROM 1971 TO 2000.

City	Years	Jul	Aug	Sep	Oct	Nov	Dec	Jan	Feb	Mar	Apr	May	Jun	Total
BIRMINGHAM AP, AL	30	0	0	11	133	359	590	691	514	339	154	31	1	2823
HUNTSVILLE, AL	30	0	0	18	165	417	669	780	587	404	180	41	1	3262
MOBILE, AL	30	0	0	2	51	204	387	455	326	182	57	3	0	1667
MONTGOMERY, AL	30	0	0	3	84	278	487	568	415	250	98	9	0	2192
ANCHORAGE, AK	30	206	268	505	957	1297	1472	1526	1295	1212	861	560	311	10 470
ANNETTE, AK	30	215	206	337	572	760	885	928	781	791	637	484	321	6917
BARROW, AK	30	763	815	1016	1564	1978	2346	2440	2267	2443	1967	1391	903	19 893
BETHEL, AK	30	280	355	587	1085	1428	1724	1813	1608	1566	1175	738	410	12 769
BETTLES, AK	30	170	366	721	1437	1975	2245	2365	2041	1888	1280	642	227	15 357
BIG DELTA, AK	30	144	308	620	1270	1761	2016	2097	1759	1576	987	536	228	13 302
COLD BAY, AK	30	447	409	518	774	915	1054	1142	1048	1085	945	780	573	9690
FAIRBANKS, AK	30	121	283	615	1287	1882	2199	2315	1926	1670	999	504	179	13 980
GULKANA, AK	30	250	370	658	1199	1786	2064	2163	1733	1543	1018	655	358	13 797
HOMER, AK	30	338	349	514	844	1069	1215	1290	1125	1103	860	663	451	9821
JUNEAU, AK	30	257	288	453	704	953	1125	1219	1010	973	728	529	335	8574
KING SALMON, AK	30	290	317	521	984	1254	1481	1538	1384	1286	957	667	425	11 104
KODIAK, AK	30	339	310	468	766	931	1067	1096	983	1007	833	667	474	8941
KOTZEBUE, AK	30	327	407	696	1297	1703	2022	2092	1918	2023	1606	1037	607	15 735
MCGRATH, AK	30	174	318	611	1233	1779	2132	2223	1847	1653	1078	583	256	13 887
NOME, AK	30	387	446	664	1134	1444	1756	1836	1663	1727	1361	867	533	13 818
ST. PAUL ISLAND, AK	30	569	518	603	828	957	1122	1220	1174	1267	1097	911	693	10 959
TALKEETNA, AK	30	193	294	563	1043	1425	1613	1676	1391	1317	923	596	293	11 327
UNALAKLEET, AK	30	297	373	644	1192	1572	1826	1915	1703	1685	1271	791	481	13 750
VALDEZ, AK	30	306	353	537	832	1101	1251	1336	1126	1091	821	596	383	9733
YAKUTAT, AK	30	353	364	506	742	980	1129	1216	1027	1040	837	664	460	9318
FLAGSTAFF, AZ	30	33	56	224	554	850	1085	1099	930	880	668	446	174	6999
PHOENIX, AZ	30	0	0	0	11	117	305	304	174	99	29	1	0	1040
TUCSON, AZ	30	0	0	0	33	195	397	401	275	194	76	7	0	1578
WINSLOW, AZ	30	0	1	31	290	649	965	961	713	580	357	133	12	4692
YUMA, AZ	30	0	0	0	5	90	246	228	114	73	23	3	0	782
FORT SMITH, AR	30	0	0	23	145	448	745	854	619	410	172	34	1	3451
LITTLE ROCK, AR	30	0	0	13	124	400	666	775	563	369	150	24	0	3084
NORTH LITTLE ROCK, AR	30	0	0	8	99	383	669	770	549	343	128	31	0	2980
BAKERSFIELD, CA	30	0	0	2	51	283	534	521	324	236	119	31	3	2104
BISHOP, CA	30	1	1	46	276	609	847	843	643	545	344	138	21	4314
EUREKA, CA	30	216	198	232	326	423	532	530	451	491	430	354	263	4446
FRESNO, CA	30	0	0	3	70	344	597	578	377	283	140	37	4	2433
LONG BEACH, CA	30	0	0	1	16	128	265	268	205	186	99	39	5	1212
LOS ANGELES AP, CA	30	1	0	2	21	121	234	252	205	212	141	78	19	1286
LOS ANGELES C.O., CA	30	0	0	1	11	91	201	206	149	144	83	36	5	927
MOUNT SHASTA, CA	30	66	65	196	451	753	936	921	753	738	563	374	175	5991
REDDING, CA	30	0	0	13	131	420	611	606	445	390	239	99	7	2961
SACRAMENTO, CA	30	0	0	11	84	359	595	580	387	335	208	97	10	2666
SAN DIEGO, CA	30	0	0	1	12	109	231	227	176	160	90	47	10	1063
SAN FRANCISCO AP, CA	30	77	56	62	131	298	476	482	354	339	266	201	120	2862
SAN FRANCISCO C.O., CA	30	133	107	95	100	232	383	396	283	288	233	214	150	2614
SANTA BARBARA, CA	30	22	23	54	92	234	368	369	277	263	193	151	75	2121
SANTA MARIA, CA	30	68	49	70	141	288	422	419	337	350	291	230	135	2800
STOCKTON, CA	30	0	0	5	76	348	609	592	391	313	169	54	6	2563

TABLE 4.7   NORMAL HEATING DEGREE-DAYS (°F · DAYS), 65°F BASE TEMPERATURE, FOR SELECTED CITIES IN THE
United STATES FROM 1971 TO 2000. (CONTINUED)

City	Years	Jul	Aug	Sep	Oct	Nov	Dec	Jan	Feb	Mar	Apr	May	Jun	Total
ALAMOSA, CO	30	47	91	302	675	1082	1475	1551	1189	983	719	451	169	8734
COLORADO SPRINGS, CO	30	11	20	163	471	827	1082	1114	915	816	568	306	76	6369
DENVER, CO	30	1	9	136	436	826	1078	1111	892	788	524	267	60	6128
GRAND JUNCTION, CO	30	1	1	59	342	766	1118	1194	860	643	397	151	20	5552
PUEBLO, CO	30	1	3	88	381	779	1058	1092	843	694	431	172	24	5566
BRIDGEPORT, CT	30	2	4	68	320	591	918	1089	944	803	489	207	32	5467
HARTFORD, CT	30	3	12	120	413	697	1054	1218	1024	844	486	195	38	6104
WILMINGTON, DE	30	1	2	49	297	564	871	1029	864	687	376	132	15	4887
WASHINGTON DULLES AP, D.C.	30	1	4	60	323	589	882	1025	847	670	362	139	21	4923
WASHINGTON NAT'L AP, D.C.	30	0	1	19	202	467	755	906	741	562	269	72	5	3999
APALACHICOLA, FL	30	0	0	0	32	152	328	408	285	169	42	3	0	1419
DAYTONA BEACH, FL	30	0	0	0	6	67	185	245	183	99	29	1	0	815
FORT MYERS, FL	30	0	0	0	1	16	76	103	75	28	3	0	0	302
GAINESVILLE, FL	30	0	0	0	21	121	268	321	208	123	40	2	0	1104
JACKSONVILLE, FL	30	0	0	0	30	148	314	374	272	155	55	5	0	1353
KEY WEST, FL	30	0	0	0	0	0	14	26	18	6	0	0	0	64
MIAMI, FL	30	0	0	0	0	4	38	58	39	15	1	0	0	155
ORLANDO, FL	30	0	0	0	2	40	142	220	128	57	9	0	0	598
PENSACOLA, FL	30	0	0	1	35	175	352	416	299	171	48	1	0	1498
TALLAHASSEE, FL	30	0	0	1	49	193	369	428	315	185	71	5	0	1616
TAMPA, FL	30	0	0	0	4	46	144	187	136	63	13	0	0	593
VERO BEACH, FL	30	0	0	0	0	18	101	166	114	48	8	0	0	455
WEST PALM BEACH, FL	30	0	0	0	0	10	58	83	60	27	4	0	0	242
ATHENS, GA	30	0	0	10	133	353	597	687	522	349	152	28	1	2832
ATLANTA, GA	30	0	0	11	126	352	600	692	523	346	150	26	1	2827
AUGUSTA,GA	30	0	0	5	112	313	540	617	469	301	129	21	1	2508
COLUMBUS, GA	30	0	0	3	78	263	481	559	415	252	94	8	0	2153
MACON, GA	30	0	0	4	91	277	494	570	427	262	104	12	0	2241
SAVANNAH, GA	30	0	0	1	56	204	403	472	350	202	72	6	0	1766
HILO, HI	30	0	0	0	0	0	0	0	0	0	0	0	0	0
HONOLULU, HI	30	0	0	0	0	0	0	0	0	0	0	0	0	0
KAHULUI, HI	30	0	0	0	0	0	0	0	0	0	0	0	0	0
LIHUE, HI	30	0	0	0	0	0	0	0	0	0	0	0	0	0
BOISE, ID	30	12	16	126	408	769	1088	1102	819	675	460	254	80	5809
LEWISTON, ID	30	10	10	102	401	717	951	962	742	616	411	218	71	5211
POCATELLO, ID	30	21	26	201	536	907	1240	1274	1003	842	584	353	129	7116
CHICAGO, IL	30	5	9	112	401	759	1147	1333	1075	858	513	232	49	6493
MOLINE, IL	30	3	8	108	394	782	1191	1374	1090	831	450	172	19	6422
PEORIA, IL	30	2	7	94	368	738	1136	1316	1045	788	423	159	19	6095
ROCKFORD, IL	30	5	14	136	444	832	1244	1430	1150	912	522	215	36	6940
SPRINGFIELD, IL	30	1	4	77	319	674	1060	1239	980	726	376	126	14	5596
EVANSVILLE, IN	30	0	1	45	262	565	891	1047	825	591	295	85	5	4612
FORT WAYNE, IN	30	3	11	105	394	722	1094	1275	1063	835	479	188	29	6198
INDIANAPOLIS, IN	30	2	4	77	335	659	1020	1192	957	724	394	141	16	5521
SOUTH BEND, IN	30	6	13	114	392	721	1090	1270	1055	844	498	213	41	6257
DES MOINES, IA	30	1	6	103	386	804	1223	1385	1090	826	439	153	16	6432
DUBUQUE, IA	30	8	21	158	465	879	1307	1492	1192	949	536	226	40	7273
SIOUX CITY, IA	30	3	10	128	434	888	1308	1439	1131	885	473	172	25	6896
WATERLOO, IA	30	7	20	155	478	903	1343	1538	1221	948	528	205	29	7375

TABLE 4.7   NORMAL HEATING DEGREE-DAYS (°F · DAYS), 65°F BASE TEMPERATURE, FOR SELECTED CITIES IN THE UNITED STATES FROM 1971 TO 2000. (CONTINUED)

City	Years	Jul	Aug	Sep	Oct	Nov	Dec	Jan	Feb	Mar	Apr	May	Jun	Total
CONCORDIA, KS	30	1	2	76	307	722	1068	1195	927	702	380	131	13	5524
DODGE CITY, KS	30	1	2	65	273	674	978	1087	826	647	351	121	12	5037
GOODLAND, KS	30	5	10	117	407	807	1079	1147	916	776	490	224	35	6013
TOPEKA, KS	30	1	1	73	287	665	1030	1174	898	647	336	106	7	5225
WICHITA, KS	30	0	0	49	235	620	965	1087	819	594	302	89	5	4765
JACKSON, KY	30	0	4	44	263	522	830	966	761	557	273	128	10	4358
LEXINGTON, KY	30	1	2	53	284	574	877	1026	819	616	332	119	13	4716
LOUISVILLE, KY	30	0	1	36	240	527	838	992	779	569	280	84	6	4352
PADUCAH, KY	30	0	0	38	229	516	833	978	750	529	250	67	2	4192
BATON ROUGE, LA	30	0	0	2	49	211	386	456	325	179	55	2	0	1665
LAKE CHARLES, LA	30	0	0	1	38	191	363	434	304	163	47	1	0	1542
NEW ORLEANS, LA	30	0	0	0	30	169	332	403	288	150	44	1	0	1417
SHREVEPORT, LA	30	0	0	6	78	296	522	597	416	250	89	8	0	2262
CARIBOU, ME	30	58	103	344	691	1039	1505	1719	1466	1254	805	417	159	9560
PORTLAND, ME	30	19	37	199	523	790	1152	1346	1145	988	649	361	116	7325
BALTIMORE, MD	30	0	1	41	279	549	839	986	816	647	345	119	12	4634
BLUE HILL, MA	30	9	22	138	422	698	1040	1207	1034	894	562	271	74	6371
BOSTON, MA	30	4	8	84	344	604	932	1104	951	815	503	233	48	5630
WORCESTER, MA	30	9	20	158	478	764	1119	1284	1094	952	601	278	74	6831
ALPENA, MI	30	46	78	260	583	894	1244	1447	1285	1133	723	394	150	8237
DETROIT, MI	30	5	12	121	429	742	1099	1270	1084	894	527	221	45	6449
FLINT, MI	30	13	28	163	474	781	1149	1329	1147	957	577	267	66	6951
GRAND RAPIDS, MI	30	10	24	159	471	793	1147	1317	1135	956	571	255	58	6896
HOUGHTON LAKE, MI	30	41	74	250	577	901	1271	1468	1283	1115	685	338	117	8120
LANSING, MI	30	17	35	179	493	805	1167	1341	1160	970	585	277	69	7098
MARQUETTE, MI	30	92	134	348	700	1083	1484	1659	1405	1280	859	468	200	9712
MUSKEGON, MI	30	15	27	168	476	784	1117	1288	1124	968	602	296	78	6943
SAULT STE. MARIE, MI	30	91	107	320	642	979	1383	1606	1405	1253	798	434	212	9230
DULUTH, MN	30	69	106	331	682	1124	1587	1772	1435	1248	787	421	180	9742
INTERNATIONAL FALLS, MN	30	55	102	360	723	1217	1744	1946	1551	1304	775	378	140	10 295
MINNEAPOLIS-ST.PAUL, MN	30	7	20	178	516	978	1428	1616	1279	1034	560	222	44	7882
ROCHESTER, MN	30	23	50	208	558	1014	1479	1650	1305	1066	609	281	65	8308
SAINT CLOUD, MN	30	19	49	253	604	1077	1551	1742	1381	1135	637	285	79	8812
JACKSON, MS	30	0	0	7	100	305	516	607	440	272	110	11	0	2368
MERIDIAN, MS	30	0	0	6	106	303	506	598	435	274	111	14	0	2353
TUPELO, MS	30	0	0	14	150	400	639	750	559	368	160	29	1	3070
COLUMBIA, MO	30	1	2	72	291	642	1004	1153	891	656	336	115	10	5173
KANSAS CITY, MO	30	0	7	58	269	668	1047	1182	897	658	331	124	8	5249
ST. LOUIS, MO	30	0	1	46	246	583	943	1097	845	613	294	83	6	4757
SPRINGFIELD, MO	30	1	1	62	248	578	899	1034	790	581	300	100	8	4602
BILLINGS, MT	30	20	25	205	516	909	1195	1280	1001	876	575	312	90	7004
GLASGOW, MT	30	22	38	253	609	1088	1506	1671	1290	1055	610	308	91	8541
GREAT FALLS, MT	30	48	64	275	579	952	1237	1323	1063	948	639	389	158	7675
HAVRE, MT	30	33	49	270	622	1067	1410	1546	1201	999	613	324	113	8247
HELENA, MT	30	42	56	283	631	1018	1348	1397	1093	932	634	384	157	7975
KALISPELL, MT	30	85	94	341	705	1014	1297	1359	1079	933	640	411	213	8171
MISSOULA, MT	30	55	62	276	637	985	1287	1291	1019	852	596	384	178	7622
GRAND ISLAND, NE	30	3	7	114	401	835	1192	1310	1031	819	452	175	23	6362
LINCOLN, NE	30	1	5	100	377	806	1188	1328	1043	799	425	154	16	6242

TABLE 4.7   NORMAL HEATING DEGREE-DAYS (°F · DAYS), 65°F BASE TEMPERATURE, FOR SELECTED CITIES IN THE UNITED STATES FROM 1971 TO 2000. (CONTINUED)

City	Years	Jul	Aug	Sep	Oct	Nov	Dec	Jan	Feb	Mar	Apr	May	Jun	Total
NORFOLK, NE	30	4	11	130	432	879	1266	1388	1099	872	478	180	28	6767
NORTH PLATTE, NE	30	6	14	158	481	902	1222	1316	1026	853	519	240	46	6783
OMAHA EPPLEY AP, NE	30	1	6	105	384	806	1211	1349	1053	805	424	151	17	6312
OMAHA (NORTH), NE	30	4	15	83	349	800	1204	1323	1022	783	400	154	16	6153
SCOTTSBLUFF, NE	30	7	11	157	489	885	1183	1233	969	837	544	257	53	6625
VALENTINE, NE	30	11	19	175	516	952	1285	1386	1101	932	571	260	57	7265
ELKO, NV	30	53	57	237	569	916	1208	1222	943	820	612	383	161	7181
ELY, NV	30	26	48	258	605	938	1220	1240	996	903	690	459	178	7561
LAS VEGAS, NV	30	0	0	1	57	320	581	583	380	247	90	16	1	2276
RENO, NV	30	12	22	130	416	732	987	984	757	683	502	285	91	5601
WINNEMUCCA, NV	30	12	23	173	509	829	1101	1088	820	742	554	315	105	6271
CONCORD, NH	30	22	44	212	548	835	1220	1402	1188	999	623	302	90	7485
MT. WASHINGTON, NH	30	504	542	740	1079	1333	1702	1857	1639	1594	1262	914	619	13 785
ATLANTIC CITY AP, NJ	30	1	6	69	323	573	868	1019	873	725	437	187	32	5113
ATLANTIC CITY C.O., NJ	30	0	0	20	228	481	772	924	787	674	409	167	18	4480
NEWARK, NJ	30	1	2	41	264	541	863	1030	869	697	371	120	13	4812
ALBUQUERQUE, NM	30	0	0	29	248	614	898	914	670	525	294	85	4	4281
CLAYTON, NM	30	1	7	85	325	685	937	964	767	661	407	179	27	5045
ROSWELL, NM	30	0	0	18	144	485	755	775	542	378	182	51	2	3332
ALBANY, NY	30	10	26	168	484	772	1142	1330	1136	938	553	240	62	6861
BINGHAMTON, NY	30	22	43	200	514	812	1160	1331	1156	997	617	292	90	7234
BUFFALO, NY	30	8	21	149	442	737	1081	1256	1111	961	594	268	65	6693
ISLIP, NY	30	0	1	49	339	604	909	1060	913	782	479	197	24	5357
NEW YORK CENTRAL PARK, NY	30	1	2	43	261	524	841	1009	853	695	372	127	16	4744
NEW YORK (JFK AP), NY	30	1	2	42	264	532	838	1007	868	723	420	166	19	4882
NEW YORK (LAGUARDIA AP), NY	30	1	1	40	249	524	836	1008	861	713	392	136	16	4777
ROCHESTER, NY	30	10	24	154	447	741	1085	1263	1117	958	582	266	66	6713
SYRACUSE, NY	30	10	25	158	460	748	1108	1294	1131	959	572	254	66	6785
ASHEVILLE, NC	30	1	2	52	285	531	769	872	708	550	302	116	15	4203
CAPE HATTERAS, NC	30	0	0	2	72	244	464	587	518	400	187	44	3	2521
CHARLOTTE, NC	30	0	0	16	165	404	655	747	585	409	180	44	3	3208
GREENSBORO-WNSTN-SALM-HGHPT, NC	30	0	1	32	232	480	742	851	679	501	245	77	8	3848
RALEIGH, NC	30	0	1	20	194	425	679	783	627	456	214	61	5	3465
WILMINGTON, NC	30	0	0	3	95	277	497	589	474	331	134	28	1	2429
BISMARCK, ND	30	19	44	256	625	1112	1539	1711	1335	1110	660	305	93	8809
FARGO, ND	30	17	37	245	614	1137	1610	1808	1446	1185	652	271	73	9095
GRAND FORKS, ND	30	27	53	276	655	1186	1660	1860	1484	1233	689	294	88	9505
WILLISTON, ND	30	22	43	274	645	1146	1567	1734	1336	1104	646	315	90	8922
AKRON, OH	30	7	16	120	412	704	1040	1220	1026	836	498	219	50	6148
CLEVELAND, OH	30	7	13	110	385	677	1023	1205	1025	847	516	235	54	6097
COLUMBUS, OH	30	3	7	81	354	656	982	1154	954	742	421	165	27	5546
DAYTON, OH	30	2	7	90	358	669	1016	1185	973	760	427	167	24	5678
GREATER CINCINNATI AP, OH	30	1	3	68	326	626	953	1110	899	684	373	138	19	5200
MANSFIELD, OH	30	8	19	122	407	714	1066	1236	1045	852	509	227	53	6258
TOLEDO, OH	30	6	18	129	431	745	1107	1281	1087	878	517	224	45	6468
YOUNGSTOWN, OH	30	15	26	148	439	723	1063	1243	1057	879	530	252	71	6446
OKLAHOMA CITY, OK	30	0	0	30	152	482	780	884	648	446	197	43	1	3663
TULSA, OK	30	0	0	29	152	468	781	898	658	437	179	38	1	3641
ASTORIA, OR	30	151	130	197	386	542	688	695	583	585	492	375	244	5068

TABLE 4.7   NORMAL HEATING DEGREE-DAYS (°F · DAYS), 65°F BASE TEMPERATURE, FOR SELECTED CITIES IN THE UNITED STATES FROM 1971 TO 2000. (CONTINUED)

City	Years	Jul	Aug	Sep	Oct	Nov	Dec	Jan	Feb	Mar	Apr	May	Jun	Total
BURNS, OR	30	89	96	313	638	968	1245	1259	982	869	661	439	226	7785
EUGENE, OR	30	40	31	115	370	594	780	769	615	564	443	308	152	4781
MEDFORD, OR	30	10	7	69	316	632	837	804	610	550	402	233	69	4539
PENDLETON, OR	30	14	15	115	400	711	962	971	747	623	433	247	83	5321
PORTLAND, OR	30	21	15	78	318	558	756	765	605	529	393	234	94	4366
SALEM, OR	30	39	34	116	376	592	771	765	623	574	452	301	141	4784
SEXTON SUMMIT, OR	30	127	101	211	440	737	860	853	741	788	648	470	267	6243
ALLENTOWN, PA	30	3	8	95	374	657	985	1147	966	784	450	176	26	5671
ERIE, PA	30	13	18	123	398	684	1016	1209	1074	926	585	288	75	6409
HARRISBURG, PA	30	0	1	52	338	621	937	1076	901	723	390	148	14	5201
MIDDLETOWN/HARRISBURG INTL APT	30	0	1	52	338	621	937	1076	901	723	390	148	14	5201
PHILADELPHIA, PA	30	1	2	39	269	545	857	1020	858	681	362	113	12	4759
PITTSBURGH, PA	30	6	13	105	397	677	996	1163	979	788	462	200	43	5829
AVOCA, PA	30	9	18	138	431	711	1047	1214	1040	866	512	219	53	6258
WILLIAMSPORT, PA	30	5	12	116	417	708	1033	1201	1014	824	471	196	38	6035
PROVIDENCE, RI	30	3	9	101	377	637	961	1125	965	817	494	221	44	5754
CHARLESTON AP, SC	30	0	0	2	68	222	428	510	394	242	95	11	1	1973
CHARLESTON C.O., SC	30	0	0	0	47	183	390	489	362	224	57	3	0	1755
COLUMBIA, SC	30	0	0	8	121	325	552	628	485	321	131	23	1	2595
GREENVILLE-SPARTANBURG AP, SC	30	0	0	19	178	417	655	750	586	420	197	47	3	3272
ABERDEEN, SD	30	11	27	206	569	1066	1506	1678	1318	1072	591	251	59	8354
HURON, SD	30	8	21	180	530	996	1417	1572	1242	1004	567	242	49	7828
RAPID CITY, SD	30	16	21	190	521	934	1233	1314	1061	925	595	313	88	7211
SIOUX FALLS, SD	30	7	20	175	519	986	1417	1563	1236	988	558	231	46	7746
BRISTOL-JHNSN CTY-KNGSPRT, TN	30	1	2	44	279	541	810	919	744	556	303	108	11	4318
CHATTANOOGA, TN	30	0	0	16	180	442	697	797	618	432	195	48	2	3427
KNOXVILLE, TN	30	0	0	22	210	470	732	841	652	467	223	65	3	3685
MEMPHIS, TN	30	0	0	13	121	381	651	770	565	366	144	22	0	3033
NASHVILLE, TN	30	0	0	24	189	457	730	858	664	462	217	56	1	3658
OAK RIDGE,TN	30	0	0	30	230	518	787	882	696	510	254	80	6	3993
ABILENE, TX	30	0	0	18	93	353	608	678	477	299	113	20	0	2659
AMARILLO, TX	30	1	1	56	239	594	874	920	706	542	291	94	7	4325
AUSTIN/CITY, TX	30	0	0	2	32	205	406	475	319	163	44	2	0	1648
AUSTIN/BERGSTROM, TX	30	0	0	1	38	248	480	532	360	180	45	5	0	1889
BROWNSVILLE, TX	30	0	0	0	6	64	170	206	123	45	8	0	0	622
CORPUS CHRISTI, TX	30	0	0	0	12	103	246	299	191	77	19	0	0	947
DALLAS-FORT WORTH, TX	30	0	0	2	52	312	571	650	448	248	74	13	0	2370
DALLAS-LOVE FIELD, TX	30	0	0	7	62	281	527	605	415	238	75	9	0	2219
DEL RIO, TX	30	0	0	2	24	183	384	423	262	111	27	1	0	1417
EL PASO, TX	30	0	0	9	96	388	626	641	435	285	113	11	0	2604
GALVESTON, TX	30	0	0	0	6	112	245	316	220	94	15	0	0	1008
HOUSTON, TX	30	0	0	1	37	189	367	427	298	156	48	2	0	1525
LUBBOCK, TX	30	0	0	33	158	472	744	800	588	409	178	38	2	3422
MIDLAND-ODESSA, TX	30	0	0	19	103	380	622	680	472	302	120	18	0	2716
PORT ARTHUR, TX	30	0	0	1	34	181	349	411	286	143	41	1	0	1447
SAN ANGELO, TX	30	0	0	12	82	325	558	617	427	258	92	13	0	2384
SAN ANTONIO, TX	30	0	0	2	33	197	390	455	303	149	42	1	0	1572
VICTORIA, TX	30	0	0	1	22	145	314	372	249	113	28	1	0	1245
WACO, TX	30	0	0	6	58	271	512	589	409	235	77	7	0	2164

TABLE 4.7   NORMAL HEATING DEGREE-DAYS (°F · DAYS), 65°F BASE TEMPERATURE, FOR SELECTED CITIES IN THE UNITED STATES FROM 1971 TO 2000. (CONTINUED)

City	Years	Jul	Aug	Sep	Oct	Nov	Dec	Jan	Feb	Mar	Apr	May	Jun	Total
WICHITA FALLS, TX	30	0	0	18	106	395	676	762	550	354	140	23	0	3024
MILFORD, UT	30	1	4	112	451	821	1129	1144	879	726	503	280	66	6116
SALT LAKE CITY, UT	30	2	3	87	370	737	1067	1108	857	665	448	215	48	5607
BURLINGTON, VT	30	17	38	203	538	834	1240	1457	1273	1063	642	283	77	7665
LYNCHBURG, VA	30	1	1	43	268	526	798	918	749	572	288	102	13	4279
NORFOLK, VA	30	0	0	8	150	368	623	758	638	487	240	66	4	3342
RICHMOND, VA	30	0	1	27	224	464	736	863	705	526	250	75	7	3878
ROANOKE, VA	30	1	2	48	276	528	798	911	745	569	290	107	13	4288
OLYMPIA, WA	30	90	78	195	465	666	829	819	678	644	511	347	201	5523
QUILLAYUTE, WA	30	193	174	250	460	626	762	758	646	656	548	421	296	5790
SEATTLE C.O., WA	30	52	50	139	362	571	735	729	593	564	423	266	131	4615
SEATTLE SEA-TAC AP, WA	30	55	45	138	383	592	754	747	613	582	447	291	150	4797
SPOKANE, WA	30	44	42	196	554	897	1168	1169	916	790	557	338	149	6820
WALLA WALLA, WA	30	9	9	99	323	659	928	939	709	574	373	191	69	4882
YAKIMA, WA	30	28	28	155	487	813	1100	1090	821	671	463	258	98	6012
SAN JUAN, PR	30	0	0	0	0	0	0	0	0	0	0	0	0	0
BECKLEY, WV	30	10	16	115	388	647	926	1068	882	714	414	199	48	5427
CHARLESTON, WV	30	1	3	56	300	558	837	977	794	604	319	122	18	4589
ELKINS, WV	30	15	23	134	456	719	996	1133	959	792	498	251	70	6046
HUNTINGTON, WV	30	1	2	55	292	557	843	991	796	596	308	116	14	4571
GREEN BAY, WI	30	19	38	208	540	925	1350	1537	1270	1065	638	301	85	7976
LA CROSSE, WI	30	6	17	152	467	893	1347	1545	1225	975	526	204	38	7395
MADISON, WI	30	12	33	183	504	892	1298	1490	1209	978	576	261	63	7499
MILWAUKEE, WI	30	13	18	134	443	808	1200	1384	1132	949	611	318	86	7096
CASPER, WY	30	15	24	229	581	961	1252	1309	1073	921	661	393	115	7534
CHEYENNE, WY	30	28	39	238	574	914	1145	1187	1000	928	686	414	136	7289
LANDER, WY	30	16	24	220	580	1027	1355	1397	1122	921	643	373	116	7794
SHERIDAN, WY	30	27	33	238	581	979	1293	1351	1078	925	628	374	129	7636

Source: U.S. National Oceanic and Atmospheric Administration (http://ols.nndc.noaa.gov/plolstore/plsql/olstore.prodspecific?prodnum=C00095-PUB-A0001).

TABLE 4.8   NORMAL HEATING DEGREE-DAYS (°C · DAYS), 18°C BASE TEMPERATURE, FOR SELECTED CITIES IN CANADA.

City	°C · days	°F · days	City	°C · days	°F · days
Abbotsford	2981	5366	Ottawa-Gatineau	4602	8284
Alma	5821	10 478	Owen Sound	4148	7466
Alma	5821	10 478	Pembroke	5178	9320
Baie-Comeau	6014	10 825	Penticton	3431	6176
Barrie	4379	7882	Peterborough	4537	8167
Bathurst	5056	9101	Port Alberni	3173	5711
Belleville	4024	7243	Portage la Prairie	5624	10 123
Brandon	5951	10 712	Prince Albert	6277	11 299
Brantford	3924	7063	Prince George	5132	9238
Brockville	4217	7591	Prince Rupert	3967	7141
Calgary	5108	9194	Québec	5202	9364
Campbell River	3462	6232	Red Deer	5696	10 253
Campbellton	5469	9844	Regina	5661	10 190
Charlottetown	4715	8487	Rimouski	5217	9391
Chicoutimi-Jonquière	5793	10 427	Rivière-du-Loup	5449	9808
Chilliwack	2832	5098	Rouyn-Noranda	6304	11 347
Cold Lake	5970	10 746	Saint John	4755	8559
Corner Brook	4766	8579	Saint-Hyacinthe	4571	8228
Cornwall	4234	7621	Saint-Jean-sur-Richelieu	4483	8069
Courtenay	3083	5549	Sarnia	3882	6988
Cranbrook	4576	8237	Saskatoon	5852	10 534
Drummondville	4621	8318	Sault Ste. Marie	5057	9103
Duncan	3179	5722	Sept-Iles	6277	11 299
Edmonton	5189	9340	Shawinigan	5046	9083
Edmundston	5358	9644	Sherbrooke	5151	9272
Estevan	5361	9650	Sorel-Tracy	4654	8377
Fort McMurray	6346	11 423	St. Catharines-Niagara	3659	6586
Fort St. John	5847	10 525	St. John's	4881	8786
Fredericton	4751	8552	Stratford	4210	7578
Gander	5198	9356	Sudbury	5343	9617
Grande Prairie	5888	10 598	Summerside	4631	8336
Guelph	4352	7834	Swift Current	5251	9452
Halifax	4193	7547	Sydney	4618	8312
Hamilton	4012	7222	Thompson	7743	13 937
Joliette	4734	8521	Thunder Bay	5718	10 292
Kamloops	3571	6428	Timmins	6149	11 068
Kelowna	3869	6964	Toronto	4066	7319
Kenora	5749	10 348	Trois-Rivières	4929	8872
Kentville	4166	7499	Truro	4518	8132
Kingston	4289	7720	Val-d'Or	6213	11 183
Kitchener-Waterloo	4288	7718	Vancouver	2926	5267
Lethbridge	4600	8280	Vernon	3991	7184
London	4058	7304	Victoria	3041	5474
Medicine Hat	4632	8338	Whitehorse	6811	12 260
Midland	4301	7742	Williams Lake	5073	9131
Moncton	4806	8651	Windsor	3525	6345
Montréal	4575	8235	Winnipeg	5778	10 400
Moose Jaw	5276	9497	Woodstock	4047	7285
Nanaimo	3056	5501	Yellowknife	8256	14 861
North Bay	5295	9531	Yorkton	6066	10 919
Oshawa	3918	7052			

COOLING LOAD COMPUTATIONS
FOR BUILDINGS

5.1 HEAT GAIN IN BUILDINGS

Introduction

Heat gain is the amount of heat that a building or building space accumulates from external sources (e.g., high outside temperature and solar radiation) and internal sources (e.g., occupants, lights, equipment, and appliances). The *cooling load* is the heat the HVAC equipment must remove to maintain the building interior at inside design conditions. Accurate computation of the cooling loads in a building is an important step in sizing cooling equipment.

Many of the considerations for heat gains in buildings are fundamentally the same as those used in heat loss. Concepts of heat transfer and heat loss from transmission and infiltration were introduced in Chapter 4. As a review, factors for heat gain that are common with heat loss include the following:

Transmission

Transmission heat gains, mainly from conduction and convection heat transfer, are the result of heat passing through a material in the building envelope (e.g., glass) or through an assembly of materials (e.g., walls, ceilings, floors, and so on). The quantity of heat transfer may be calculated through each assembly by the method involving use of thermal resistance (R and R_t) and the overall coefficient of heat transmission (U). The concepts of the R-value and U-factor were introduced in Chapter 2.

Infiltration

Infiltration heat gains relate to the energy required to cool and dehumidify unconditioned air that has leaked into the building space through the building envelope (i.e., through cracks and openings in and around walls, doors and windows, electrical outlet holes, and a variety of other openings).

Ventilation

Ventilation heat gains are tied to the energy required to condition (e.g., cool and dehumidify) outside air that is mechanically introduced into the building to maintain occupant comfort and avoid adverse health effects. Although outdoor air can sometimes be used to cool the building interior, most often air used to ventilate a building must be cooled and dehumidified before it is introduced into the building.

Other factors that were ignored in heat loss analysis must be considered in cooling load computations. These include the following:

Solar Radiation

Solar heat gain is the energy that a building absorbs from solar energy striking its exterior and conducting to the interior or energy transmitted through windows and being absorbed by materials in the building interior. Heat gains from windows can be a huge factor in cooling loads. Exterior surfaces of the building envelope are heated by solar radiation creating a larger temperature difference than what exists between inside air and outside air temperatures. As a result, it creates a greater thermal driving force of transmission.

Internal Heat

Occupants, lights, equipment, computers, and appliances reject heat into the occupied space. In fact, in some climates, internal loads may be enough to completely heat the building most of the year, leaving only a few hours of temperature pickup in the morning on the coldest days for the heating system to perform. This heat can contribute significantly to the cooling loads of a building.

Sensible and Latent Gains

Occupants, appliances, and equipment generate moisture. Heating load analysis was based on sensible loads only because moisture was only a negligible concern. With cooling, latent heat gain can have a significant influence on the total load. Sensible and latent heat gains are comprised of the following:

Sensible Heat Gains

Sensible heat is heat that is associated with a change in temperature of a substance. It is heat that a substance absorbs, and while its temperature goes up, the substance does not change its state. In cooling load computations, sensible heating loads are associated with:

- Heat transmission through opaque envelope assemblies (e.g., roofs, walls, floors, ceilings)
- Solar heat gain through transparent or translucent envelope assemblies (e.g., skylights, windows, glazed openings)
- Leakage of outside air through the building envelope (infiltration)

- Air introduced for ventilation
- Occupants
- Lights, equipment, and appliances

Latent Heat Gains

Latent heat is the energy required to change the state of a substance from solid to liquid (latent heat of fusion) or a liquid to gas (latent heat of vaporization), without a change in temperature. The latent load is the energy associated with the addition (or loss) of moisture in the conditioned space. This load comes into play only under cooling conditions, when the cooling systems (e.g., air conditioners, heat pumps) are actually removing some of that moisture to lower levels of relative humidity and improve comfort. In cooling load computations, latent heating loads are related to:

- Moisture-laden outside air from infiltration and ventilation
- Occupant respiration and activities (e.g., bathing, cleaning)
- Moisture from equipment and appliances

Together, sensible and latent heat gains make up the total heat gain of a building. Because of the differences in loads, sensible and latent heat gains are computed separately in cooling load computations.

Heat Gain and the Cooling Load

Heat that enters a building or that is generated in a building interior can affect the cooling load instantly or may be temporarily stored in materials or furnishings and released to affect the cooling load at a later time. Convective and latent heat flows are instantaneous; gain equals load. In contrast, radiant gains (e.g., from solar radiation, lights, occupants, and so forth) tend to be partly stored in building materials (the fraction depends on a number of factors) and released at a later time. Furthermore, the varying effect of solar radiation caused by movement of the sun across the sky produces a variation in sensible heat gain. In contrast, latent loads tend to immediately affect the cooling load. So, because of the time-delay effects of radiation, and thermal mass and thermal lag, a building's heat gain and cooling load are not equal:

$$\text{heat gain} \neq \text{cooling load}$$

Additionally, the cooling load in one room might not occur when the total cooling load for the building occurs. For example, a room with east-facing windows might have its cooling load peak in the morning, when the building's total peak load occurs later in the day. These factors combine to make cooling load computations very complex.

Units of Heat Gain and Cooling

In the cooling sector of the HVAC industry, metric (SI) and inch-pound (I-P) units are used. *Quantity of heat gain (Q)* is universally expressed in Btu or W-hr and *rate of heat gain (q)*

in Btu/hr or W. Metric (SI) units make use of standard prefixes such as k for 1000 (e.g., kW or kWh).

In the United States, a unique unit of measure is used for the amount of heat removed (the cooling capacity of an air conditioning system). It is known as the *ton of refrigeration* or simply the *ton*. A ton of refrigeration is based on the cooling effect of one ton of ice at 32°F melting in a day. As the latent heat of fusion of ice is 144 Btu/lb, one ton of ice is 2000 lb $\cdot$ 144 Btu/lb = 288 000 Btu. Consequently, one ton of refrigeration is the removal of 288 000 Btu/day. This rate of cooling is equivalent to:

$$\text{one ton of refrigeration} = 12\ 000\ \text{Btu/hr} = 3.516\ \text{kWh}$$

A one ton air conditioner is rated at 12 000 Btu/hr, a 2 ton unit at 24 000 Btu/hr, and so on. Typically, residential central systems provide from 2 to 5 tons of cooling. Commercial rooftop units are typically 3 to 20 tons each and central chillers can range from 5 tons up 1500 tons.

5.2   COOLING LOAD COMPUTATIONS

General Approach

Cooling load calculations are made to size an HVAC cooling system and its components based upon the effects of heat gains and building performance at extreme conditions, usually midday during the summer. A building experiences a vast range of total heat gains, which vary by hour, day, month, or year, and range from zero (no cooling required) to the peak (maximum) cooling load. As discussed previously, heat gain at a particular instant is not the same as the cooling load handled by the HVAC system. Because building heat and moisture transfer are based on weather conditions, occupant behavior, and equipment operation, they are difficult to predict.

Variations in heat gains and building performance must be accounted for in cooling load computations. Typically, building occupancy is assumed to be at full design capacity and equipment and appliances are considered to be operating at typical capacity. Lights are assumed to be operating as anticipated for a typical day of design occupancy. Heat flow is analyzed assuming dynamic conditions, such that heat storage in the building envelope and in interior materials is considered. Latent and sensible loads are taken into account, but are computed separately.

Except in small buildings with a single zone, load calculations should be made on a room-by-room or zone-by-zone basis because an HVAC system must properly deliver heating/cooling to the specific rooms or spaces. By knowing the individual room or zone load, the designer can determine duct or pipe size required to deliver the necessary conditioned air or water to the room or space. Additionally, HVAC equipment serving a set of rooms or spaces must have the capacity to meet the calculated loads.

One of the most common problems encountered in HVAC cooling system design is improper sizing of equipment. This is particularly true of residential systems where design normally involves use of rapid rule-of-thumb analysis that leads to improper size. For example, a study of design, construction, and performance of residential cooling systems built between 1994 and 1999 in Fort Collins, Colorado, showed that cooling systems were significantly oversized—ranging from 143% to 322% of the design cooling requirements. This validates that residential cooling equipment tends to be oversized by substantial margins, at least in this part of the country. In dry climates, oversized equipment leads to short cycling, higher operating costs, and premature wear. In geographical locations with high heat and humidity, similar studies show that residential cooling systems tend to be improperly sized. In these cases, oversized equipment leads to short cycling, higher operating costs, high electrical demand, poor humidity control, and premature wear.

Cooling Load Computation Methods

There are several methods and ranges of methods used to compute the cooling load of a building. Analysis of measured and observed data and computer simulations have created a range of computation techniques from fairly simple hand calculations through sophisticated computer modeling methods. The most basic and least accurate of these methods is use of a rule-of-thumb method (e.g., 400 to 600 ft^2 of floor area per ton of cooling required based upon geographic location). A typical commercial interior space will have an internal heat gain from lights, people, and equipment of about 10 to 25 Btu/hr · ft^2 of floor area (without solar or transmission heat gain from the outdoors). Another method treats the building with static sets of conditions (like heat loss computations). Although these methods are easy, they are undependable because of variations in building heat gains, even with similar building types. Quick rule-of-thumb methods should only be used as approximations during preliminary design and should never be used for actual system design. Procedures that result in precise cooling load calculation results should be used.

ASHRAE publishes procedures for the determination of design cooling loads in the *ASHRAE Handbook—Fundamentals.* The ACCA and the ARI jointly developed *Manual J—Residential Load Calculation,* another widely accepted method of calculating the sensible and latent cooling (and heating) loads under design conditions for residential structures. ACCA's *Manual N—Commercial Load Calculation* can be used to estimate heating and cooling loads for small- and mid-sized commercial structures.

Methods used to compute a cooling load in a commercial building include the following: The *heat balance (HB) method* indicates the balance of all heat transfer in a room to determine the cooling load. It is the most reliable and rigorous theoretical approach to estimating the cooling load. It must be calculated using computer programs. The *radiant time series (RTS) method* uses an array of repetitive equations to produce accurate results and is less dependent on approximated data. It is suitable for small and large commercial buildings, but because of its complexity, hand calculations are somewhat impractical. Both the

heat balance and radiant time series methods are preferred but require the use of a computer program or advanced spreadsheet.

The *transfer function method (TFM)* is a derivative of the heat balance method. One of the older methods, it is complex and requires the use of a computer program or advanced spreadsheet. The *cooling load temperature difference/solar cooling load factor/cooling load factor (CLTD/SLF/CLF) method* is derived from the transfer function method and uses tabulated data to simplify the calculation process. This method has some limitations because of the use of tabulated data, but it can be hand computed. The *total equivalent temperature differential/time-averaging (TETD/TA)* was the preferred method for hand or simple spreadsheet calculation before the introduction of the CLTD/SLF/CLF method. The *admittance method,* developed in the United Kingdom, allows the calculation of overheating temperature for the room, or peak cooling load calculation for a constant (internal) environmental temperature.

Each cooling load computation method offers different benefits and limitations. Accuracy and computation simplicity tend to be conflicting undertakings. The CLTD/SLF/CLF method is introduced in this text because it is a suitable hand computation method used in residences and simple commercial buildings, and it is fairly easy to understand. Furthermore, there are some commonalities between the CLTD method and other methods.

Design Conditions

Summer design conditions are used by the designer to determine the peak operating conditions under which HVAC systems and equipment (e.g., equipment for cooling, ventilating, dehumidifying, condensing) must operate during the summer. Climate conditions used are from a long-term historical database. Climatic data do not represent an actual year, but are a statistical representation of average long-term conditions at the specific geographic location. In design of building cooling systems, inside dry bulb temperature, outside dry bulb temperature, wet bulb temperature, and range of dry bulb temperature are used as the basis for design conditions.

Inside Design Conditions

As introduced in Chapter 4, the *inside design conditions* are the target indoor environmental conditions intended to achieve thermal comfort. The sensation of thermal comfort is related to many variables including air temperature, radiant temperature, air movement, gender, clothing, activity level, age, and adaptation to local climate. These were introduced in Chapter 1. There is no single set of inside design temperature/humidity levels recommended to achieve acceptable comfort. Suggested indoor design temperature and relative humidity ranges for selected occupancies and recommended indoor temperature ranges were provided in Chapter 1. Code requirements will vary (e.g., one model code requires that indoor design temperature shall be 70°F for heating and 78°F for cooling).

The typical inside design temperature range for cooling in most U.S. homes is usually from 74° to 78°F (23° to 25°C).

Practical experience suggests that the comfort range for cooling in commercial applications may actually be much lower than is prescribed in ASHRAE Standard 55; typically, dry bulb temperatures are between 73° and to 76°F (23° and 26°C), depending on relative humidity and indoor relative humidity between 30% and 60%. These recommendations apply only to inactive occupants wearing typical levels of clothing. It may be necessary to adjust inside conditions to expected clothing levels in the space (e.g., swimming pools, doctors' examination rooms), occupant age (e.g., retirement homes), and activity level (e.g., gymnasiums, athletic clubs).

Outside Design Conditions

The cooling load is estimated for summer design conditions that typically occur at midday. It is not economical to choose the maximum recorded outdoor air temperature as the basis for design because it leads to equipment oversizing to meet demands that seldom occur. Design values near these extreme conditions are used. Additionally, temperature and humidity levels must be considered. As a result, outside summer design conditions are typically described by dry bulb (DB) temperature coupled with a *mean coincident wet bulb (MWB)* temperature.

Recommended outside summer design conditions are published in sources such as the *ASHRAE Handbook—Fundamentals*. ASHRAE tables containing cooling (summer) design conditions for geographical locations in the United States and Canada are provided in Table 5.12. Because of the length of these tables, they are placed at the end of the chapter, beginning on page 152.

The cooling (summer) design conditions provided by ASHRAE are reported as 0.4%, 1.0%, and 2.0% design values that represent the dry bulb temperatures with mean coincident wet bulb temperature that are exceeded 0.4%, 1.0%, and 2.0% of the time. These design value percentages reflect the annual cumulative frequency of occurrence of outdoor climate conditions *during an average year*. In other words, of the 8760 total hr in a year, the outside dry bulb temperature climbed above the 0.4% design temperature for 0.4% of the time or 35 hr over the year; the 1.0% design temperature, for 88 hr; and, the 2.0% design temperature, for 175 hr at the specific geographical location. (For historical comparison, the 0.4% annual value is about the same as the obsolete 1% design temperature, and the 1.0% value is about 1°F lower than the outdated 2.5% design temperature.)

In Denver, Colorado (from Table 5.12), the 0.4% outdoor design conditions for cooling are 93°F DB/60°F MWB. The 1.0% design conditions for cooling are 90°F DB/59°F MWB and, the less extreme, 2.0% design conditions for cooling are 87°F DB/59°F MWB. The 0.4% condition represents the most extreme design conditions. Actual weather conditions are sometimes even more severe than design conditions. For example, the dry bulb temperature in Denver does exceed 100°F occasionally, but this is rare and only for a few hours a year.

The designer must choose which set of design conditions applies to a particular project. Traditionally, the 0.4% design

values are used for computations involving light mass residential buildings (e.g., house, apartment, townhouse, manufactured home, dwelling, condominium, convent, monastery, hotel, and so forth). For massive, masonry or concrete building structures that have little glass, the 2.0% design values should be used. The 1.0% value is used with heavy mass structures with a large amount of glass area.

Range of Dry Bulb Temperature

Another design measure is the range of dry bulb temperature. The *range of dry bulb temperature* is related to the mean temperature swing on the design day—that is, the average difference between the maximum and minimum dry bulb temperatures during the hottest month. This value is used to account for the effects of thermal mass (thermal lag) in cooling load computations. A greater range of dry bulb temperature value results in a smaller cooling load because the building's mass can be cooled during the evening hours and coast through hot midday temperatures.

5.3 RESIDENTIAL COOLING LOAD COMPUTATIONS

A simplified method prescribed by the ASHRAE for use in computing residential (single detached and multifamily) cooling loads is outlined in the following text to introduce the reader to the process of computing cooling loads. This can be used to compute the heating load of a house, apartment, townhouse, manufactured home, dwelling, condominium, convent, monastery, or hotel. Other buildings require a more comprehensive approach that is broadly introduced later in this chapter.

Cooling Load from Transmission through Exterior Construction Assemblies

The cooling load resulting from heat transmission through opaque (nontranslucent) components of above-grade exterior envelope assemblies (e.g., walls, roofs, ceilings, floors, and doors) occur when heat from the outside flows through the external building envelope. Heat transfer occurs almost entirely by conduction; however, some convection and radiation may occur within cavities. Opaque elements receive sunlight during the day, which heats up exposed surfaces, causing an increase in the rate of heat flow through the building envelope. Transmission of heat through walls, roofs, ceilings, floors, and doors depends largely on the area and U-factor.

The cooling load resulting from heat transmission is calculated in a manner similar to the approach used to compute transmission heat loss. The process of computing the total thermal resistance (R_t) was introduced in Chapter 2. It first involves establishing the total thermal resistance (R_t) of the construction assembly, and then converting R_t to the overall coefficient of heat transmission (U): $U = 1/R_t$. Note that although the thermal resistance of solid materials in an assembly will not change from those used in heating load computations, the thermal resistance of air films and air spaces will be different (slightly). Envelope performance parameters (e.g., U-factors) must be consistent with the values used to show compliance with the local energy code.

The equation used to compute the transmission component of the cooling load for opaque surfaces (i.e., walls, roof, doors, and so on) is based on the overall coefficient of heat transmission (U), the net exposed area (A) of the assembly being considered (e.g., area of wall, roof, floor, door, or ceiling), and a hypothetical temperature difference called the *cooling load temperature difference (CLTD)*:

$$q_{transmission} = U \cdot A \cdot CLTD$$

The CLTD is a theoretical temperature difference (similar to but not the same as the ΔT used in the heating load transmission computations in Chapter 4). The CLTD accounts for the combined effects of inside and outside air temperature difference, daily temperature range, solar radiation, heat storage in the construction assembly, and radiation storage in building mass. In theory, a temperature difference (ΔT) equivalent to the CLTD would generate the same heat flow through a shaded assembly. The CLTD is affected by orientation, tilt, month, day, hour, latitude, solar absorptivity of the exterior surface, and construction (mass).

Values of CLTD for single-family detached and multifamily residences are provided in Tables 5.1 and 5.2, as extracted from the 2001 *ASHRAE Handbook—Fundamentals*. A CLTD

TABLE 5.1   COOLING LOAD TEMPERATURE DIFFERENCE (CLTD) VALUES FOR SINGLE-FAMILY RESIDENCES.[a]

Daily Temperature Range[b]	85 L	85 M	90 L	90 M	90 H	95 L	95 M	95 H	100 M	100 H	105 M	110 H
All walls and doors												
North	8	3	13	8	3	18	13	8	18	13	18	23
NE and NW	14	9	19	14	9	24	19	14	24	19	24	29
East and West	18	13	23	18	13	28	23	18	28	23	28	33
SE and SW	16	11	21	16	11	26	21	16	26	21	26	31
South	11	6	16	11	6	21	16	11	21	16	21	26
Roofs and ceilings												
Attic or flat built-up	42	37	47	42	37	51	47	42	51	47	51	56
Floors and ceilings												
Under conditioned space, over unconditioned room, or over crawl space	9	4	12	9	4	14	12	9	14	12	14	19
Partitions												
Inside or shaded	9	4	12	9	4	14	12	9	14	12	14	19

[a]Cooling load temperature differences (CLTDs) for single-family detached houses, duplexes, or multifamily, with both east and west exposed walls or only north and south exposed walls, °F.
[b]L denotes low daily range, less than 16 °F; M denotes medium daily range, 16 to 25 °F; and H denotes high daily range, greater than 25 °F.

Extracted from 2001 *ASHRAE Handbook—Fundamentals*. Used with permission from American Society of Heating, Refrigerating and Air-Conditioning Engineers, Inc.

TABLE 5.2   COOLING LOAD TEMPERATURE DIFFERENCE (CLTD) VALUES FOR MULTIFAMILY RESIDENCES.[a]

Daily Temperature Range[b]		85 L	85 M	90 L	90 M	90 H	95 L	95 M	95 H	100 M	100 H	105 M	110 H
Walls and doors[c]													
N	Light	14	11	19	16	12	24	21	17	26	22	27	32
N	Medium	13	10	18	15	11	23	20	16	25	21	26	31
N	Heavy	9	6	15	11	7	20	16	12	21	17	22	27
NE	Light	23	17	28	22	17	33	27	22	32	26	31	36
NE	Medium	20	15	25	20	16	30	25	21	29	25	29	34
NE	Heavy	16	12	21	17	13	26	22	18	26	22	26	31
E	Light	32	27	37	32	27	43	38	32	42	37	42	47
E	Medium	30	24	34	29	24	40	34	29	39	33	39	44
E	Heavy	23	18	28	23	18	34	29	23	33	28	33	38
SE	Light	31	27	35	31	26	41	37	31	42	37	42	47
SE	Medium	28	22	32	27	22	37	32	27	37	33	38	43
SE	Heavy	21	16	26	22	17	32	27	22	31	27	32	37
S	Light	25	22	29	26	22	35	31	26	36	32	37	43
S	Medium	22	18	26	22	18	31	26	22	31	27	32	38
S	Heavy	16	11	20	16	12	26	21	17	26	21	27	33
SW	Light	39	36	44	40	35	50	46	40	51	47	52	58
SW	Medium	33	29	37	34	29	44	40	35	45	40	46	52
SW	Heavy	23	18	28	24	19	36	31	25	35	30	36	42
W	Light	44	41	48	45	40	54	51	46	56	52	57	63
W	Medium	37	33	41	38	33	46	42	38	48	43	49	55
W	Heavy	26	22	31	27	23	37	32	27	37	32	38	44
NW	Light	33	30	37	34	30	43	39	34	44	40	45	50
NW	Medium	28	25	32	29	24	37	33	29	39	35	40	45
NW	Heavy	20	16	25	20	16	31	26	21	31	26	32	37
Roof and ceiling													
Attic or flat built-up	Light	58	53	65	60	55	70	65	60	70	65	72	77
Flat built-up	Medium or heavy	21	18	23	21	18	25	23	21	25	23	25	28
Floors and ceiling													
Under or over unconditioned space, crawl space		9	4	12	9	4	14	12	9	14	12	14	19
Partitions													
Inside or shaded		9	4	12	9	4	14	12	9	14	12	14	19

[a]Cooling load temperature differences (CLTDs) for multifamily low-rise or single-family detached if zoned with separate temperature control for each zone, °F.
[b]L denotes low daily range, less than 16 °F; M denotes medium daily range, 16 to 25 °F; and H denotes high daily range, greater than 25 °F.
[c]Light denotes lightweight; medium denotes medium-weight; and heavy denotes heavyweight construction.

Extracted from 2001 *ASHRAE Handbook—Fundamentals*. Used with permission from American Society of Heating, Refrigerating and Air-Conditioning Engineers, Inc.

value is selected based on orientation of the surface, outside design temperature, daily design temperature range, and the mass of construction. The design daily temperature range is related to the temperature swing on the design day based on low, L (less than 16°F range); medium, M (16° to 25°F range); and high, H (greater than 25°F range). In multifamily construction, the mass of construction is for light (e.g., wood frame), medium (e.g., masonry veneer), and heavy (e.g., solid masonry) construction.

Example 5.1

A wood-framed single-family detached residence in Denver, Colorado, has a corner room with an east-facing wall with a net exposed area of 98 ft^2. The wall has an R$_t$ of 20.8 hr · ft^2 · °F/Btu. Determine the sensible heat transmission component of the cooling load for the east-facing wall.

$$U = 1/R_t = 1/20.8 \text{ hr} \cdot \text{ft}^2 \cdot °F/\text{Btu} = 0.048 \text{ Btu/hr} \cdot \text{ft}^2 \cdot °F$$

From Table 5.12 (at the end of this chapter), the outside air design temperature is 93°F and the daily (temperature) range is 26.9°F. The daily (temperature) range is considered high because it is greater than 25°F.

From Table 5.1, the CLTD must be interpolated because the outside air design temperature is 93°F. For an east-facing wall, with a 90°F outside design temperature and a high (H) daily range, the CLTD is 13°F. It is 18°F for a 95°F outside design temperature. Through interpolation, the CLTD is 16°F for an east-facing wall.

$$q_{transmission} = U \cdot A \cdot CLTD = 0.048 \text{ Btu/hr} \cdot \text{ft}^2 \cdot °F \cdot 98 \text{ ft}^2 \cdot 16°F = 75 \text{ Btu/hr}$$

For construction assemblies that separate a conditioned space from unconditioned attics, floors, and ceilings, the CLTD values provided in Tables 5.1 and 5.2 are used in the transmission equation. Construction assemblies (e.g., interior walls) that separate a conditioned space and an unconditioned space (e.g., a wall separating the home interior and an attached garage or an office and a warehouse) also contribute to the cooling load.

When CLTD values provided in Tables 5.1 and 5.2 are not applicable, the equation used to compute the transmission component of the cooling load is the standard equation used for heating load computations, based on the overall coefficient of heat transmission (U), the net exposed area (A) of the assembly (e.g., area of wall, roof, floor, door, or ceiling), and the temperature difference (ΔT):

$$q_{transmission} = U \cdot A \cdot \Delta T$$

Example 5.2

A residence in Denver, Colorado, has a wall separating the unconditioned garage and the home interior. It has a net exposed area of 192 ft^2. The wall has an R$_t$ of 13.9 hr · ft^2 · °F/Btu. The indoor design temperature is 75°F. Assume the garage door is left open, exposing the garage interior to design conditions. Determine the sensible heat transmission component of the cooling load for the wall.

$$U = 1/R_t = 1/13.9 \text{ hr} \cdot \text{ft}^2 \cdot °F/\text{Btu} = 0.072 \text{ Btu/hr} \cdot \text{ft}^2 \cdot °F$$

From Table 5.12, the outside air design temperature is 93°F.

$$q_{transmission} = U \cdot A \cdot \Delta T = 0.072 \text{ Btu/hr} \cdot \text{ft}^2 \cdot °F \cdot 192 \text{ ft}^2 \cdot (93 - 75)°F = 249 \text{ Btu/hr}$$

Cooling Load from Glass/Windows

Heat gain through glass, windows, and other transparent/translucent assemblies must take heat gains from solar radiation (both direct and reflected) and conduction into consideration. The choice of window type and type of shade control requires careful analysis because it has a large influence on the cooling load.

Before the recent technological advancements that led to high-performance glazings (i.e., use of tints, films, and coatings), a typical window with one or two layers of glass allowed about 75% to 85% of the solar energy to enter a building. Historically, this heat gain had a considerable effect on comfort and cooling costs, especially in hot climates. In recent years, use of high-performance glazings has resulted in a major reduction in window heat gain without significant loss of visible light. However, heat gain through glass can still be a large component of the cooling load, especially for a space located at the perimeter of a building.

Solar radiation transmitted through glass is greatly affected by window orientation. In the northern hemisphere, south-facing windows generally have a smaller heat gains at midsummer than east or west windows of the same size, but this advantage diminishes as the summer moves to fall. In contrast, in the winter, solar gain through south-facing glass is high, so in buildings with low internal heat gains (like residences) this helps offset the high rate of heat loss from the low outside air temperature. Because the noon sun is high in the summer sky and low in the winter sky, horizontal or near-horizontal glass (e.g., skylights) has high heat gains during the summer and low heat gains during the winter. Additionally, because the summer sun is relatively low in the morning and afternoon sky, heat gains from east- and west-facing glass tends to be higher than from south-facing glass, regardless of latitude.

The cooling load through glass and windows can be computed as follows, based on the *glass load factor (GLF)*, in units of Btu/hr · ft^2, and glass area (A), in ft^2:

$$q_{glass} = GLF \cdot A$$

The glass load factor, in units of Btu/hr · ft^2, accounts for the effects solar radiation (both direct and reflected) and conduction. It will vary according to window orientation, type of glass, type of interior shading, and outside design temperature. The solar radiation component of the window load is dependent on orientation. Values of the glass load factor for single-family detached and multifamily residences are provided in Tables 5.3 and 5.4, as extracted from the 2001 *ASHRAE Handbook—Fundamentals*. GLF values are for sites at a latitude of 40°. For other latitudes, the south, southeast, and southwest GLF values must be corrected. ASHRAE suggests adding 30% to the glass load factor value for sites at a latitude of 48° and subtraction of 30% for sites at a latitude of 32°. Linear interpolation from the latitude of 40° can be used between latitudes of 32° and 48°.

TABLE 5.3 WINDOW GLASS LOAD FACTORS (GLFS) FOR SINGLE-FAMILY DETACHED RESIDENCES.

Design Temperature, °F	Regular Single Glass						Regular Double Glass						Heat-Absorbing Double Glass						Clear Triple Glass		
	85	90	95	100	105	110	85	90	95	100	105	110	85	90	95	100	105	110	85	90	95
No inside shading																					
North	34	36	41	47	48	50	30	30	34	37	38	41	20	20	23	25	26	28	27	27	30
NE and NW	63	65	70	75	77	83	55	56	59	62	63	66	36	37	39	42	44	44	50	50	53
East and West	88	90	95	100	102	107	77	78	81	84	85	88	51	51	54	56	59	59	70	70	73
SE and SW[b]	79	81	86	91	92	98	69	70	73	76	77	80	45	46	49	51	54	54	62	63	65
South[b]	53	55	60	65	67	72	46	47	50	53	54	57	31	31	34	36	39	39	42	42	45
Horizontal skylight	156	156	161	166	167	171	137	138	140	143	144	147	90	91	93	95	96	98	124	125	127
Draperies, venetian blinds, translucent roller shades, fully drawn																					
North	18	19	23	27	29	33	16	16	19	22	23	26	13	14	16	18	19	21	15	16	18
NE and NW	32	33	38	42	43	47	29	30	32	35	36	39	24	24	27	29	29	32	28	28	30
East and West	45	46	50	54	55	59	40	41	44	46	47	50	33	33	36	38	38	41	39	39	41
SE and SW[b]	40	41	46	49	51	55	36	37	39	42	43	46	29	30	32	34	35	37	35	36	38
South[b]	27	28	33	37	38	42	24	25	28	31	31	34	20	21	23	25	26	28	23	24	26
Horizontal skylight	78	79	83	86	87	90	71	71	74	76	77	79	58	59	61	63	63	65	69	69	71
Opaque roller shades, fully drawn																					
North	14	15	20	23	25	29	13	14	17	19	20	23	12	12	15	17	17	20	13	13	15
NE and NW	25	26	31	34	36	40	23	24	27	30	30	33	21	22	24	26	27	29	23	23	26
East and West	34	36	40	44	45	49	32	33	36	38	39	42	29	30	32	34	35	37	32	32	35
SE and SW[b]	31	32	36	40	42	46	29	30	33	35	36	39	26	27	29	31	32	34	29	29	31
South[b]	21	22	27	30	32	36	20	20	23	26	27	30	18	19	21	23	24	26	19	20	22
Horizontal skylight	60	61	64	68	69	72	57	57	60	62	63	65	52	52	55	57	57	59	56	57	59

[a]Glass load factors (GLFs) for single-family detached houses, duplexes, or multifamily residences, with both east and west exposed walls or only north and south exposed walls, Btu/h·ft².
[b]Correct by +30% for latitude of 48° and by −30% for latitude of 32°. Use linear interpolation for latitude from 40 to 48 and from 40 to 32°.

To obtain GLF for other combinations of glass and/or inside shading: $GLF_a = (SC_a/SC_t)(GLF_t − U_tD_t) + U_aD_t$, where the subscripts a and t refer to the alternate and table values, respectively. SC_t and U_t are given in Table 5. $D_t = (t_a − 75)$, where $t_a = t_o − (DR/2)$; t_o is the outdoor design temperature and DR is the daily range.

Extracted from 2001 *ASHRAE Handbook—Fundamentals.* Used with permission from American Society of Heating, Refrigerating and Air-Conditioning Engineers, Inc.

TABLE 5.4 WINDOW GLASS LOAD FACTORS (GLFS) FOR MULTIFAMILY RESIDENCES.[a]

Design Temperature, °F	Regular Single Glass						Regular Double Glass						Heat-Absorbing Double Glass						Clear Triple Glass		
	85	90	95	100	105	110	85	90	95	100	105	110	85	90	95	100	105	110	85	90	95
No inside shading																					
North	40	44	49	54	58	64	34	36	39	42	44	47	23	24	26	29	30	33	30	32	34
NE	88	89	91	95	97	100	78	79	80	83	84	85	52	52	53	55	55	57	71	71	73
East	136	137	139	142	144	147	120	121	122	125	126	127	79	79	81	83	83	84	109	109	111
SE	129	130	134	139	141	144	109	113	116	119	120	122	72	75	77	79	79	81	99	103	105
South[b]	88	91	96	101	105	110	76	78	81	84	86	89	50	52	54	56	58	60	68	70	72
SW	154	159	164	169	174	179	134	137	140	143	145	148	89	91	93	95	97	99	121	123	125
West	174	178	183	188	192	197	151	154	157	160	162	165	100	102	104	106	108	110	137	139	141
NW	123	127	132	137	141	147	107	109	112	115	117	121	71	72	75	77	79	81	96	98	100
Horizontal	249	252	256	261	264	268	218	220	223	226	228	230	144	146	148	150	152	154	198	200	202
Draperies, venetian blinds, translucent roller shades, fully drawn																					
North	21	25	29	33	36	40	18	21	23	26	28	31	15	17	19	21	23	25	17	19	21
NE	43	44	46	50	51	52	39	40	41	44	45	46	33	33	34	36	36	37	39	39	40
East	67	68	70	74	75	76	61	62	63	65	66	67	50	50	51	54	54	55	60	60	61
SE	64	65	69	73	74	77	58	59	61	63	64	66	48	48	50	52	52	54	57	57	59
South[b]	45	48	52	56	59	63	40	42	44	47	49	52	33	34	36	39	40	42	38	40	42
SW	79	83	87	91	94	98	70	72	75	78	80	83	57	59	62	64	66	68	68	69	71
West	89	92	96	100	103	107	79	81	84	86	88	91	65	66	69	71	72	75	76	78	80
NW	63	66	70	74	77	81	56	58	61	63	66	68	46	48	50	52	54	56	54	55	57
Horizontal	126	128	132	135	137	141	113	115	117	120	121	124	93	94	96	98	100	102	110	111	113
Opaque roller shades, fully drawn																					
North	17	21	25	29	32	36	15	17	20	23	25	28	14	15	18	20	22	24	15	16	18
NE	33	34	35	39	40	42	31	32	33	36	35	37	29	28	30	32	32	34	32	31	33
East	51	52	53	57	61	65	48	49	50	53	52	55	45	45	46	48	48	49	49	49	50
SE	49	50	53	57	58	61	46	47	49	52	52	55	42	43	45	47	47	49	46	46	48
South[b]	35	38	42	46	49	53	32	34	37	40	42	42	29	31	33	35	37	39	32	33	35
SW	61	65	69	73	77	81	57	59	62	65	67	70	52	54	56	58	60	62	56	58	60
West	68	71	75	80	83	87	64	66	68	71	73	76	58	60	62	64	66	68	63	64	66
NW	49	52	56	60	63	67	45	47	50	53	55	58	41	43	45	47	49	51	45	46	48
Horizontal	97	99	102	106	108	111	91	93	95	97	99	102	83	85	87	89	90	92	90	92	93

[a]Glass load factors (GLFs) for multifamily low-rise or single-family detached residences if zoned with separate temperature control for each zone, Btu/h·ft².
[b]Correct by +30% for latitude of 48° and by −30% for latitude of 32°. Use linear interpolation for latitude from 40 to 48 and from 40 to 32°.

To obtain GLF for other combinations of glass and/or inside shading: $GLF_a = (SC_a/SC_t)(GLF_t − U_tD_t) + U_aD_t$, where the subscripts a and t refer to the alternate and table values, respectively. SC_t and U_t are given in Table 5. $D_t = (t_a − 75)$, where $t_a = t_o − (DR/2)$; t_o is the outdoor design temperature and DR is the daily range.

Extracted from 2001 *ASHRAE Handbook—Fundamentals.* Used with permission from American Society of Heating, Refrigerating and Air-Conditioning Engineers, Inc.

TABLE 5.5   SHADE LINE FACTORS (SLFs).

Direction Window Faces	Latitude, Degrees N						
	24	32	36	40	44	48	52
East	0.8	0.8	0.8	0.8	0.8	0.8	0.8
SE	1.8	1.6	1.4	1.3	1.1	1.0	0.9
South	9.2	5.0	3.4	2.6	2.1	1.8	1.5
SW	1.8	1.6	1.4	1.3	1.1	1.0	0.9
West	0.8	0.8	0.8	0.8	0.8	0.8	0.8

Note: Shadow length below the overhang equals the shade line factor times the overhang width. Values are averages for the 5 h of greatest solar intensity on August 1.

Extracted from 2001 *ASHRAE Handbook—Fundamentals*. Used with permission from American Society of Heating, Refrigerating and Air-Conditioning Engineers, Inc.

Use of external window shading devices (e.g., awnings, roof overhangs, shutters, and solar screens) and internal shading devices (e.g., curtains and blinds) also affects the entry of solar radiation. The use of internal shading devices should typically be assumed because occupants should be expected to control solar gain on summer days. The designer must treat all glass in direct sunlight using the appropriate glass load factor. Glass areas that are *fully* shaded by awnings, overhangs, or other "solid" shading devices are treated as north-facing glass. However, not all of the glass area protected by an awning or overhang may be in the shade. The *shade line factor (SLF)* is the ratio of the distance a shadow falls beneath the horizontal projection of an overhang to the width of the overhang. Values of the shade line factor based on latitude and window orientation are found in Table 5.5, as extracted from the 2001 *ASHRAE Handbook—Fundamentals*. The shadow depth beneath the horizontal projection of overhang (L_{shadow}) is found by:

$$L_{shadow} = SLF \cdot \text{overhang width}$$

Example 5.3

A wood-framed single-family detached residence in Denver, Colorado (40° north latitude), has a corner room with east-facing windows. The fixed windows are regular double glass with a total glass area of 25.5 ft². Determine the cooling load attributed to the windows.

a. For windows with no inside shading.

From Table 5.3, for this type of window, based on a 90°F outside design temperature, the GLF is 78. It is 81 with a 95°F outside design temperature. Through interpolation, the GLF is 79 Btu/hr · ft² for an east-facing window with no inside shading, with a 91°F outside design temperature.

$$q_{glass} = GLF \cdot A = 79 \text{ Btu/hr} \cdot \text{ft}^2 \cdot 25.5 \text{ ft}^2 = 2015 \text{ Btu/hr}$$

b. For windows with fully drawn venetian blinds.

From Table 5.3, for this type of window, based on a 90°F outside design temperature, the GLF is 41. It is 44 with a 95°F outside design temperature. Through interpolation, the GLF is 42 Btu/hr · ft² for an east-facing window with

fully drawn venetian blinds, with a 91°F outside design temperature.

$$q_{glass} = GLF \cdot A = 42 \text{ Btu/hr} \cdot \text{ft}^2 \cdot 25.5 \text{ ft}^2 = 1071 \text{ Btu/hr}$$

c. For windows that are fully shaded by an exterior overhang and with no inside shading.

The GLF for a window that is fully shaded by an exterior overhang is treated as north facing. From Table 5.3, based on a 90°F outside design temperature, the GLF is 30. It is 34 with a 95°F outside design temperature. Through interpolation, the GLF is 31 Btu/hr · ft² for an east-facing window that is fully shaded by an exterior overhang and with no inside shading, with a 91°F outside design temperature.

$$q_{glass} = GLF \cdot A = 31 \text{ Btu/hr} \cdot \text{ft}^2 \cdot 25.5 \text{ ft}^2 = 791 \text{ Btu/hr}$$

d. For two fixed windows with no inside shading and a 3 ft wide exterior overhang. The window glass is 3 ft wide by 4.25 ft high. The top of the glass is 1 ft below the overhang.

From Table 5.5, the SLF for east-facing windows and a site at 40° north latitude is 0.8. The shadow depth beneath the horizontal projection of overhang (L_{shadow}) is found by:

$$L_{shadow} = SLF \cdot \text{overhang width} = 0.8 \cdot 3 \text{ ft} = 2.4 \text{ ft}$$

The tops of the windows are 1 ft below the overhang, so the top 1.4 ft of the windows are in the shade.

The area of shaded glass = 2 windows · 1.4 ft · 3 ft wide = 8.4 ft²
The area of unshaded glass = 2 windows · (4.25 ft – 1.4 ft) · 3 ft wide = 17.1 ft²

From c and a above, the GLF for fully shaded glass is 31 and the GLF for unshaded glass is 79:

For fully shaded glass: $q_{glass} = GLF \cdot A = 31 \text{ Btu/hr} \cdot \text{ft}^2 \cdot 8.4 \cdot \text{ft}^2 = 260 \text{ Btu/hr}$

For unshaded glass: $q_{glass} = GLF \cdot A = 79 \text{ Btu/hr} \cdot \text{ft}^2 \cdot 17.1 \text{ ft}^2 = 1351 \text{ Btu/hr}$

The total load is found by:

$$260 \text{ Btu/hr} + 1351 \text{ Btu/hr} = 1611 \text{ Btu/hr}$$

Sensible Cooling Load from Infiltration/Ventilation

In residential cooling load computations, the heat gain from infiltrated air is included with the heat gain from outdoor ventilation air mainly because there are no code requirements for ventilating residences and because the load associated with ventilation air is comparatively small in comparison to ventilation requirements for commercial buildings. The heat gain from infiltration of outside air is difficult to estimate precisely because it depends on the inside–outside air temperature difference, wind velocity, and airflow patterns around the building, occupancy patterns, and the airtightness of the building envelope.

Sensible heat gain resulting from infiltration is typically determined with the air-change technique using a volume of air equivalent to the number of air exchanges in the structure or room per hour, which is expressed as the *air exchange rate (ACH)*. One *air change* (ACH = 1.0) is equal to the volume of the room or space under consideration being exchanged in one hour. Establishing the ACH requires engineering judgment on the quality of construction of the building envelope because the only reliable way to determine the ACH is to measure it after the building is built. One method of estimating infiltration rates is reported in the 2001 *ASHRAE Handbook—Fundamentals*. It is based on outside design temperature and construction quality:

For tight, energy-efficient construction, the ACH is between 0.33 and 0.38

For medium construction, the ACH is between 0.46 and 0.56

For loose construction, the ACH is between 0.68 and 0.78

These air exchange rates are based on outside design temperatures ranging from 85° to 110°F (29° to 43°C); the higher ACH value in each construction category is for the highest outside design temperature (110°F/43°C). It is important to note that an ACH of at least 0.5 is desirable to maintain air quality over the long-term.

The sensible cooling load resulting from infiltration airflows can be approximated based on the outside design temperature and inside design temperature difference (ΔT), in °F; the air exchange rate (ACH); and the volume of the room or space (V), in ft³:

$$q_{infil} = 1.1 \cdot ACH \cdot V/60 \cdot \Delta T$$

The constant 1.1 is based on the heat capacity of air under standard conditions multiplied by 60 minutes in an hour: 0.018 Btu/ft³ · °F · 60 min/hr = 1.08 (rounded to 1.1).

In metric (SI) units, the sensible cooling load resulting from infiltration airflows can be approximated based on the outside design temperature and inside design temperature difference (ΔT), in °C; the air exchange rate (ACH); and the volume of the room or space (V), in m³:

$$q_{infil} = 0.33 \cdot ACH \cdot V \cdot \Delta T$$

Example 5.4

A well-built, wood-framed, single-family detached residence in Denver, Colorado, has a 16 ft by 13 ft room with 8 ft high ceilings. The target inside design temperature is 78°F. Determine the sensible cooling load attributed to infiltration. Use an air exchange rate of 0.35.

The volume of the room is found by:

$$V = 16 \text{ ft} \cdot 13 \text{ ft} \cdot 8 \text{ ft} = 1664 \text{ ft}^3$$

From Table 5.12 (at the end of the chapter), the outside air design temperature is 93°F. The sensible cooling load resulting from infiltration air is found by:

$$q_{infil} = 1.1 \cdot ACH \cdot V/60 \cdot \Delta T = 1.1 \cdot 0.35 \cdot 1664 \text{ ft}^3/60$$
$$\cdot (93°F - 75°F) = 192 \text{ Btu/hr}$$

Sensible Cooling Load from Internal Heat Gains

Sensible heat gains from occupants, appliances, and lights causes the temperature of the air in the room to rise. A small part of this sensible heat gain is stored in the mass of the structure as an increase in temperature and then released after the occupants leave. In buildings with a low occupancy density (e.g., residences, apartments), these thermal lag effects are usually small and can be neglected.

Occupants release both sensible heat and latent heat into an occupied conditioned space. The sensible heat production of an occupant will vary based on activity level and body type (e.g., weight, gender, and so on). Heat released from an average adult female has been found to be about 85% of an average adult male, and heat released from an average child is 75% of an average adult male. In a residence, it is about 230 Btu/hr (67 W) per person. It is difficult to project the number of occupants in a residence. ASHRAE recommends that if the number of occupants is not known, then an assumption of two occupants for the first bedroom and one additional person per additional bedroom should be made.

Electric lighting, appliances, and equipment create either sensible or latent loads, or sometimes both. For example, an operating coffee maker generates both sensible and latent loads. Although heat gain from these devices is typically a major component for commercial buildings, it is generally comparatively small in residential buildings. In residential computations, the appliance and lighting load is typically assumed to be 1600 Btu/hr (470 W) for single-family detached residences and 1200 Btu/hr (350 W) for multifamily dwelling units. The appliance and lighting load should be divided evenly between the kitchen, laundry area, and adjacent spaces.

Example 5.5

Approximate the sensible cooling load attributed to lighting, appliances, and equipment in a four-bedroom, single-family detached residence.

With two occupants for the first bedroom and one additional person for each of the three additional bedrooms, the assumption of five occupants is made.

Occupant load:
230 Btu/hr (67 W) per person · 5 persons = 1150 Btu/hr (335 W)

Appliance and lighting load: 1600 Btu/hr (470 W)
Total cooling load from internal heat: 2750 Btu/hr (805 W)

Cooling Loads from Moisture (Latent Loads)

Moisture is introduced into a structure through people, equipment and appliances, ventilation air, and air infiltration. Occupants give off water vapor through respiration and through daily activities (e.g., bathing, food preparation, laundry, maintenance, and so forth). Another major component of cooling load is associated with latent heat from infiltration and ventilation air. Much of the moisture generated in a building is exhausted through exhaust fans in the bathroom, kitchen, and laundry areas.

Traditionally, the latent cooling load of a residence is computed as 30% or less of the total sensible load. ASHRAE, in its residential heating load computation technique, suggests a slightly more complex approach that involves computation of the *latent load (LF) multiplier*. Homes in North America generally have an LF of less than 1.3. The LF for moderate construction is based on a design humidity ratio (W), in lb of vapor/lb of dry air:

$$LF = 0.58 + 42W$$

For poor (loose) construction, the LF is 3% to 7% higher than computed for moderate construction. For good (tight) construction, the LF is 2% to 4% lower than computed for moderate construction. The lower percentage variation is related to dry climates and the higher value is tied to humid climates.

In residential computations, the total load is based on the latent load (LF) multiplier, which accounts for the addition of moisture to the air. In this technique, the total load is found by:

$$\text{total load} = \text{sensible load} \cdot LF$$

Total Cooling Load

The total cooling load is the sum of the sensible loads and the latent loads:

$$\text{total cooling load} = \text{sensible load} + \text{latent load}$$

With the total and sensible loads known, the latent load can be found by:

$$\text{latent load} = \text{total cooling load} - \text{sensible load}$$

Sensible Heat Ratio

The *sensible heat ratio (SHR)* is the ratio between the sensible heating load and the total heat load:

$$SHR = \text{sensible load}/\text{total cooling load}$$

Dehumidification of air containing water vapor will result in an extra cooling load (latent heat load) on the evaporator compared to the cooling load from changing the temperature of the air (sensible heat load). A sensible heat ratio value of 100% corresponds to a cooling load on the evaporator only from cooling of the air (sensible load). A sensible heat ratio value of 80% indicates that 80% of the load on the evaporator origins from cooling the air (sensible load) and 20% origins from dehumidification of the air (latent load).

Cooling equipment at design conditions operates at a specific equipment sensible heat ratio, which must be capable of achieving the sensible heat ratio of the space or building. Commercially available direct expansion (DX) air conditioning units typically have an equipment sensible heat ratio in the range of 0.65 to 0.80. Systems in schools, businesses, and most other buildings typically have an equipment sensible heat ratio of about 0.65. This means 65% of the cooling is dedicated to lowering temperature, and 35% is used to lower humidity. Computer center environments require a 0.80 to 0.90 sensible heat ratio for effective and efficient data center cooling. This is fairly easily accomplished because, except for occupants, loads are sensible as long as outdoor ventilation air is not too humid. In many cases in hot and humid climates, the use of standard DX air conditioning units having a constant-speed compressor and relying on on–off cycling often leads to either an uncontrolled indoor humidity level or space overcooling, because of high latent cooling load resulted from the hot, humid climate. In hot, humid climates, dehumidification or heat recovery is needed to maintain an acceptable SHR.

Example 5.6

A single-family detached residence of moderate construction has a total sensible load of 27 150 Btu/hr. The design humidity ratio is 0.013 lb of vapor/lb of dry air.

a. Determine the total load.

For moderate construction:

$$LF = 0.58 + 42W = 0.58 + 42(0.013) = 0.126$$
$$\begin{aligned}\text{total cooling load} &= \text{sensible load} \cdot LF \\ &= 27\,150 \text{ Btu/hr} \cdot 1.126 \\ &= 30\,571 \text{ Btu/hr}\end{aligned}$$

b. Determine the total load.

$$\begin{aligned}\text{latent load} &= \text{total cooling load} - \text{sensible load} \\ &= 30\,571 \text{ Btu/hr} - 27\,150 \text{ Btu/hr} \\ &= 3421 \text{ Btu/hr}\end{aligned}$$

c. Determine the sensible heat ratio.

$$\begin{aligned}SHR &= \text{sensible load}/\text{total cooling load} \\ &= 27\,150 \text{ Btu/hr}/30\,571 \text{ Btu/hr} = 0.89\end{aligned}$$

Applying the Total Cooling Load

Cooling (DX refrigeration) systems operate less efficiently at part load than they do at full load. Oversizing a cooling system adversely influences efficiency and performance. Routinely adding safety factors of 50% or 100% is unacceptable for nearly all applications. Oversizing results in equipment cycling and typically a warmer average coil temperature. It also may result in too much airflow, which increases the coil temperature and decreases the system's ability to remove moisture. Given that outdoor design conditions occur only a few hours each year, the system will operate at part load for most of the cooling season. Design methodology results in equipment already sized for peak conditions, so equipment is oversized when operating at nonpeak, part-load duty the majority of the time. The propensity to oversize equipment must be restrained. In fact, a slight undersizing (by a few percent) of a DX refrigeration system can be efficient.

Example 5.7

A single-family detached residence of moderate construction has a total sensible load of 27 150 Btu/hr. From Example 5.6, the latent load is 3421 Btu/hr. Determine the total load.

$$\text{total cooling load} = \text{sensible load} + \text{latent load}$$
$$= 27\ 150\ \text{Btu/hr} + 3421\ \text{Btu/hr}$$
$$= 30\ 571\ \text{Btu/hr}$$

General Procedure for Computing Residential Cooling Loads

The general process of estimating the design cooling load involves computation of heat gains in each of the individual spaces at design conditions. As with heating load computations, cooling load calculations are carried out for each space (or zone) within the structure. The process of computing the heating load on a room-by-room basis makes the sizing of equipment (e.g., ducts, fan-coil units) serving these individual spaces easier. The process is demonstrated in Example 5.8. Designers use forms similar to those provided in Figure 5.1 to assemble data and aid in computing the repetitive computations. The general procedure is as follows:

1. Determine outside design conditions.

2. Select the inside temperature to be maintained in the heated space during design conditions. Estimate air temperatures in adjacent unheated spaces.

3. Determine the overall coefficient of heat transfer (U-factor) for all construction assemblies that are between heated spaces and outside air or the heated space and unheated space.

4. Calculate the net area of all construction assemblies surrounding the heated spaces and outside or the heated space and an unheated space. Use inside dimensions.

5. Compute the sensible cooling load associated with heat transmission through exterior construction assemblies.

6. Compute the sensible cooling load associated with heat transmission to unconditioned spaces.

7. Compute the sensible cooling load associated with solar gains from glass and windows.

8. Compute the sensible cooling load associated with infiltration and ventilation gains.

9. Compute the sensible cooling load associated with internal heat gains.

10. Sum all sensible cooling loads.

11. Multiply the total sensible cooling load by the latent load (LF) multiplier. This represents the total design cooling load of the space at design conditions (DL_S).

Example 5.8

A tightly constructed home in Denver, Colorado (about 40° north latitude), has a 12 ft by 14 ft family room as shown in Figure 5.2. Temperature in the space is to be maintained at 74°F. It is exposed to the outside on two adjacent sides with three windows and a swing door on the south exterior wall, and a window on the east exterior wall. The room is above a finished basement that is not cooled. The ceiling has an attic space above. Assume exterior construction assemblies are unshaded (i.e., no roof overhang or other exterior shading) and there is no inside window shading. Assume an occupant load of 3 persons and an additional 400 Btu/hr appliance (from adjacent kitchen) and lighting heat gain. The space has the following characteristics:

- Ceiling/roof is well insulated (R = 45.2 hr · ft^2 · °F/Btu, U = 0.022 Btu/hr · °F · ft^2) with 9 ft ceiling height.

- Exterior walls are well insulated (R = 18.2 hr · ft^2 · °F/Btu, U = 0.055 Btu/hr · °F · ft^2).

- Floor above (unconditioned) basement is insulated (R = 13.2 hr · ft^2 · °F/Btu, U = 0.076 Btu/hr · °F · ft^2).

- South windows are 3 ft by 4 ft with low-e double glass (R = 3.1 hr · ft^2 · °F/Btu, U = 0.32 Btu/hr · °F · ft^2).

- East window is 2 ft-6 in by 4 ft with low-e double glass (R = 3.1 hr · ft^2 · °F/Btu, U = 0.32 Btu/hr · °F · ft^2).

- Door is steel with urethane foam core that is 1¾ in thick (R = 5.3 hr · ft^2 · °F/Btu, U = 0.190 Btu/hr · °F · ft^2).

- The infiltration rate is 0.35 ACH. There is no mechanical ventilation.

Find the sensible cooling load at design conditions for the room described.

In Denver, Colorado, from Table 5.12 (at the end of the chapter), the 0.4% design conditions for cooling are 93°F DB/60°F MWB. The 0.4% design conditions are used because the structure is light frame construction. The design daily temperature range from Table 5.12 (at the end of the chapter) is 26.9°F. The daily (temperature) range is considered high (H) because it is greater than 25°F. The design temperature difference is:

$$\Delta T = 93°F - 74°F = 19°F$$

COOLING LOAD ANALYSIS WORKSHEET (RESIDENTIAL)

ADVANCED BUILDING CONSULTANTS, LLC
12450 WEST BELMONT AVENUE
LITTLETON, CO 80127-1515
303-933-1300
advbldg1@aol.com

Project:		Job No.:
Address:		
Designer:		Date:
Checked by:		Date:
Room/Space Description:		Room No:
		Page No:

1				DESIGN CONDITIONS						
2	Case 1	DB	MWB	% RH	W	Case 2	DB	MWB	% RH	W
3	outside design					outside design				
4	inside design					inside design				
5	difference (Δ)					difference (Δ)				

6	SENSIBLE LOADS

7	Transmission

8	type of assembly (walls, floors, ceilings)	direction	mass	daily range	U-factor	area	CLTD/ΔT	subtotal	Notes:
9				°F	Btu/hr-ft^2·°F	ft^2	°F		
10									
11									
12									
13									
14									
15									
16									
17									
18									
19									
20									
21									
22									
23									

24	Windows/Glass

25	type of glass	type of shading	direction	GLF	area (ft^2)	subtotal	Notes:
26							
27							
28							
29							
30							
31							
32							
33							

34	Infiltration/Ventilation

35	volume (V) of space or room	constant	rate of air exchange	room volume: (V)/60	ΔT	subtotal	Notes:
36			ACH	(ft^3)	°F		
37		1.1		/60			

38	Internal (Sensible) Heat Gain
39	Notes:
40	TOTAL SENSIBLE LOAD

41	LATENT LOADS		
41	total sensible load	latent load factor (LF) = 0.58 + 42W (or specified)	Notes:
42		LF:_____ -1	TOTAL LATENT LOAD
43	TOTAL COOLING LOAD (Btu/hr)		

FIGURE 5.1 A sample worksheet for residential cooling load computations.

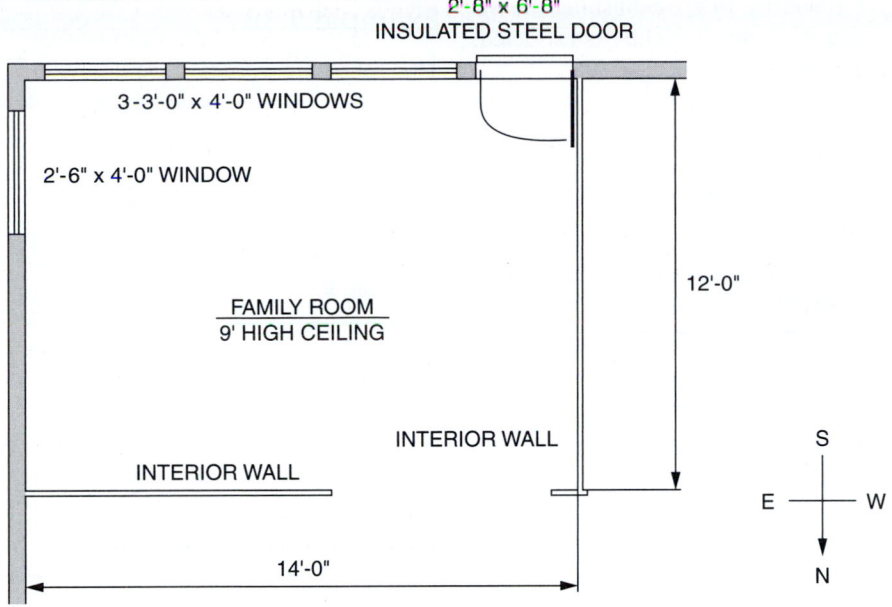

2'-8" x 6'-8"
INSULATED STEEL DOOR

3-3'-0" x 4'-0" WINDOWS

2'-6" x 4'-0" WINDOW

FAMILY ROOM
9' HIGH CEILING

INTERIOR WALL

INTERIOR WALL

12'-0"

14'-0"

S

E — W

N

FIGURE 5.2 Floor plan for Example 5.8.

Areas of the different construction assemblies based on orientation are found by:

Area of ceiling/roof:	$12 \text{ ft} \cdot 14 \text{ ft} = 168 \text{ ft}^2$
Area of south windows:	$3(3 \text{ ft} \cdot 4 \text{ ft}) = 36 \text{ ft}^2$
Area of east window:	$1(2.5 \text{ ft} \cdot 4 \text{ ft}) = 10 \text{ ft}^2$
Area of south door:	$2.67 \text{ ft} \cdot 6.67 \text{ ft} = 17.8 \text{ ft}^2$
Area of south wall (net):	$(14 \text{ ft} \cdot 9 \text{ ft}) - 36 \text{ ft} - 17.8 \text{ ft} = 72.2 \text{ ft}^2$
Area of east wall (net):	$(12 \text{ ft} \cdot 9 \text{ ft}) - 10 \text{ ft} = 98 \text{ ft}^2$
Area of floor:	$12 \text{ ft} \cdot 14 \text{ ft} = 168 \text{ ft}^2$

Transmission computations require determination of the CLTD based on a design dry bulb temperature of 93°F and a high (H) daily temperature range. From Table 5.1, the CLTD must be interpolated between 90° and 95°F outside design temperature data. The CLTD for a south-facing wall, with a 90°F design temperature and a high (H) daily range is 6°F. The CLTD is 11°F for a 95°F design temperature. So, through interpolation, the CLTD is 9°F for a south-facing wall. Interpolated CLTD values are:

CLTD from Table 5.1 for south-facing wall, through interpolation of 6° and 11°F:	9°F
CLTD from Table 5.1 for east-facing wall, through interpolation of 13° and 18°F:	16°F
CLTD from Table 5.1 for ceiling/attic, through interpolation of 37° and 42°F:	40°F
CLTD from Table 5.1 for unconditioned floor, through interpolation of 4° and 9°F:	7°F

Transmission load computations are:

$$q_{\text{south wall}} = U \cdot A \cdot CLTD = 0.055 \text{ Btu/hr} \cdot °F \cdot ft^2 \cdot 72.2 \text{ ft}^2 \cdot 9°F = 36 \text{ Btu/hr}$$

$$q_{\text{south door}} = U \cdot A \cdot CLTD = 0.190 \text{ Btu/hr} \cdot °F \cdot ft^2 \cdot 17.8 \text{ ft}^2 \cdot 9°F = 30 \text{ Btu/hr}$$

$$q_{\text{east wall}} = U \cdot A \cdot CLTD = 0.055 \text{ Btu/hr} \cdot °F \cdot ft^2 \cdot 98 \text{ ft}^2 \cdot 16°F = 86 \text{ Btu/hr}$$

$$q_{\text{ceiling}} = U \cdot A \cdot CLTD = 0.022 \text{ Btu/hr} \cdot °F \cdot ft^2 \cdot 168 \text{ ft}^2 \cdot 40°F = 148 \text{ Btu/hr}$$

$$q_{\text{floor}} = U \cdot A \cdot CLTD = 0.076 \text{ Btu/hr} \cdot °F \cdot ft^2 \cdot 168 \text{ ft}^2 \cdot 7°F = 89 \text{ Btu/hr}$$

Window load computations are based on window GLFs from Table 5.3. Again, values must be interpolated between 90° and 95°F outside design temperature data. From Table 5.3, for south-facing, heat-absorbing double glass, based on a 90°F design temperature, the GLF is 31. It is 34 with a 95°F design temperature. Through interpolation, the GLF is 33 Btu/hr · ft^2 for a south-facing window. For the east-facing, heat-absorbing double glass, the GLF interpolated to be 53 Btu/hr · ft^2. Values do not need to be corrected for latitude because Denver is very near 40° north latitude.

Window load computations are:

$$q_{\text{south glass}} = GLF \cdot A = 33 \text{ Btu/hr} \cdot ft^2 \cdot 36 \text{ ft}^2 = 1188 \text{ Btu/hr}$$

$$q_{\text{east glass}} = GLF \cdot A = 53 \text{ Btu/hr} \cdot ft^2 \cdot 10 \text{ ft}^2 = 530 \text{ Btu/hr}$$

Load computations from internal heat gains are based on the assumption of three occupants and an appliance and lighting load of 400 Btu/hr:

$$q_{\text{south glass}} = (3 \text{ persons} \cdot 230 \text{ Btu/hr per person}) + 400 \text{ Btu/hr} = 1090 \text{ Btu/hr}$$

Infiltration gains are found by first establishing the air exchange rate. Because the home is tightly constructed, an ACH of 0.35 shall be used.

The infiltration load computation is:

$$q_{\text{infil}} = 1.1 \cdot \text{ACH} \cdot V/60 \cdot \Delta T$$
$$= 1.1 \cdot 0.35 \cdot (12 \text{ ft} \cdot 14 \text{ ft} \cdot 9 \text{ ft})/60 \cdot 19°F$$
$$= 184 \text{ Btu/hr}$$

The sensible cooling load at design conditions (the sum of all sensible loads) = 3381 Btu/hr.

Example 5.9

The room in Example 5.8 has a sensible cooling load at design conditions of 3381 Btu/hr. Based on a latent load (LF) multiplier of 1.15,

a. Determine the total load for this space.

$$\text{total load} = \text{sensible load} \cdot \text{LF}$$
$$= 3381 \text{ Btu/hr} \cdot 1.15$$
$$= 3888 \text{ Btu/hr}$$

b. Determine the latent load for this space.

$$\text{latent load} = \text{total load} - \text{sensible load}$$
$$= 3888 \text{ Btu/hr} - 3381 \text{ Btu/hr}$$
$$= 507 \text{ Btu/hr}$$

Computing the Total Cooling Load

In small buildings with a single cooling zone (e.g., a residence with a single central unit, a multifamily dwelling with apartments with individual cooling units), the total cooling load is generally determined by summing the sensible and latent loads of the individual spaces, where the total cooling load at design conditions (DL_T) and the individual space cooling load at design conditions (DL_S) are expressed in Btu/hr (W):

$$DL_T = DL_{S1} + DL_{S2} + DL_{S3} + DL_{S4} + \cdots$$

Example 5.10

The following spaces have individual (sensible and latent) cooling loads at design conditions as shown. Calculate the total cooling load.

Space	Cooling Load (Btu/hr)
Bedroom #1	3200
Bedroom #2	4300
Master bedroom	5100
Bath #1	1260
Bath #2	1380
Living room	6220
Dining room	4300
Kitchen	4550
Foyer/hallways	1100
Total cooling load at design conditions (DL_T)	31 410 Btu/hr

In buildings with cooling systems divided into multiple, separately controlled zones (e.g., residences with multiple cooling systems, multifamily dwellings served by a single cooling system with fan-coil units in each apartment), zone loads or block loads are calculated with consideration only to the peak heat gains for that zone. Equipment serving a zone is sized to meet the peak block load for that zone. The total cooling load at design conditions is calculated, taking into consideration variations in block loads in all zones. Simply put, the whole-building peak load may not occur at the same instant as the peak load of any of its zones. This concept is explained in Table 5.6.

Table 5.6 shows hourly cooling loads by room in a classroom wing served by a single cooling zone. Peak hourly loads in each room (in bold) range from 650 Btu/hr (in offices) to 17 800 Btu/hr at 3 pm in Classroom #4. The peak hourly load for the classroom wing is 84 780 Btu/hr, about 7 tons. Only one room (Classroom #6) has its peak load at the same time as the peak block load for the classroom wing. All other rooms have their peak load at different times of the school day. The sum of the peak room loads is 104 300 Btu/hr, almost 9 tons.

TABLE 5.6   HOURLY COOLING LOADS BY ROOM IN A CLASSROOM WING, IN BTU/HR. PEAK HOURLY LOADS IN EACH ROOM ARE IN BOLD. PEAK TOTAL LOAD IS IN ITALICS.

| Period | Classrooms | | | | | | Restrooms | Offices | | Conf. Room | Block Load |
	1	2	3	4	5	6		1	2		
8 am	4200	4200	4200	3200	3200	3200	600	**650**	**650**	1100	25 200
9 am	11 200	4200	10 400	11 600	10 400	11 600	1100	300	300	480	61 580
10 am	4200	16 100	12 400	16 800	11 700	11 700	1100	300	300	480	75 080
11 am	**17 000**	**16 800**	12 400	3200	3200	3200	**2600**	**650**	300	**1600**	60 950
12 pm	4200	4200	4200	12 760	11 440	12 760	2500	300	**650**	**1600**	54 610
1 pm	12 800	12 800	4200	17 200	16 200	**18 200**	2300	300	300	480	*84 780*
2 pm	12 800	12 800	**12 800**	4200	**16 200**	17 300	2000	300	300	480	79 180
3 pm	12 400	12 400	12 400	**17 800**	5200	16 800	2000	**650**	**650**	480	80 780
4 pm	3200	3200	3200	6400	6400	6400	1300	**650**	**650**	480	31 880

Equipment serving this classroom wing should be sized to meet the peak block load for that zone (7 tons). Using the sum of the peak room loads would result in equipment oversizing, which would lead to part load inefficiencies and premature equipment wear caused by excessive equipment (on–off) cycling.

Precise determination of the time of occurrence and the actual total whole-building load requires either extensive hand calculations or, more pragmatically, an hourly computer simulation. As a result, computer modeling is used in all but simple buildings.

5.4 COOLING ENERGY ESTIMATION

Energy Estimation Methods

Estimating the annual energy consumption and operating costs of a building's cooling system is useful in budget planning and in the assessment of alternative systems. As with heating system analysis, energy consumption and operational costs of a cooling system will depend on weather and usage patterns, which are difficult to predict. There are a few simplified methods that are still used today to obtain rough estimates of energy consumption in a residence, but these should be used with caution given their extremely unsophisticated approach. For a commercial building, this type of calculation can rarely, if ever, be performed manually to obtain a dependable estimate.

As with estimation of heating energy consumption and operating costs, calculations can be done using one of three basic methods: the *degree-day energy estimation, bin energy estimation,* and *hour-by-hour simulation.* These methods were covered in Chapter 4. Normal cooling degree-days (°F · days), 65°F base temperature, for selected cities in the United States are provided in Table 5.13. Because of the length of this table, it is located at the end of this chapter.

Simplified Estimation Method for Cooling

A simplified method used to approximate annual energy consumption and operating costs associated with cooling is based on cooling load hours accrued over the cooling season and several system efficiency measures (e.g., seasonal energy efficiency ratio, energy efficiency ratio, kW/ton). This method is an unsophisticated rule-of-thumb technique and should only be used as an approximation for preliminary design.

Cooling load hours is a time quantity developed by the HVAC industry to specify the number of hours that a cooling system would typically operate at a specific geographical location. A higher number of cooling load hours indicates that the cooling system will operate longer, resulting in higher energy consumption. Cooling load hours for selected cities in the United States are provided in Table 5.14. Because of the length of this table, it is located at the end of this chapter.

Cooling load hours can be adjusted to correct for differences in operation patterns and house construction: add 200 cooling hours if the air conditioner is left on all summer; subtract 200 cooling hours if the house is unoccupied during the day and the thermostat is set back to cool just during the afternoon; add 300 if the home has extra west-facing glass; and subtract 100 cooling hours if the house is well shaded or has no west-facing glass.

Residential air conditioning systems are rated by the *seasonal energy efficiency ratio (SEER)* or *energy efficiency ratio (EER).* (See Chapter 6). The EER is the energy of cooling (Btu or W) delivered by the compressor divided by the watt-hours consumed. The SEER is the energy (Btu or W) of cooling divided by the watt-hours consumed, including fan energy, but it is averaged over a typical cooling season and takes into account seasonal inefficiencies such as part loading. The SEER is a more realistic rating to use for estimating actual energy usage. Commercial cooling systems are rated in kW/ton. The *kW/ton rating* is a performance measure based on the actual power input to the compressor motor divided by tons of cooling produced, or kilowatts per ton (kW/ton). A lower kW/ton rating indicates higher efficiency. By comparison, an EER rating of 12.0 is equivalent to 1 kW/ton.

Characteristic efficiency ratings for residential and commercial cooling systems are provided in Table 5.7. Only a few decades ago, the SEER of residential air conditioners was in the range of 5 to 7. To be sold in the United States today, a residential air conditioner is required to have a SEER of at least 13. Higher efficiency models have a SEER of between 14 and 15. Twenty years ago, chillers for commercial buildings had efficiencies of approximately 2.4 kW/ton but are now less than 0.7 kW/ton.

The equation used to provide an approximation of the annual energy consumption (in kWh) of a residential air conditioner is found in the next example, where the total cooling load is expressed in Btu/hr, SEER is seasonal energy efficiency ratio, and cooling load hours is for the specific geographical location:

$$\text{energy consumption} = (\text{cooling load/SEER}) \\ \cdot (\text{cooling load hours/1000 W})$$

Example 5.11

Approximate the annual energy consumption and annual cost of operation of an air conditioner with an SEER of 11.3 that is serving a cooling load of 31 410 Btu/hr (from Example 5.10). Use an energy cost of $0.12/kWh.

a. For a home in Denver, Colorado.

From Table 5.14, the cooling load hours for Denver, Colorado, are 450 hr.

$$\begin{aligned}\text{energy consumption} &= (\text{total cooling load/SEER}) \cdot (\text{cooling} \\ &\quad \text{load hours/1000 W}) \\ &= (31\ 410\ \text{Btu/hr/11.3}) \cdot (450\ \text{hr/1000 W}) \\ &= 1251\ \text{kWh}\end{aligned}$$

$$\text{cost of operation} = 1251\ \text{kWh} \cdot \$0.12/\text{kWh} = \$150$$

b. For a home in San Antonio, Texas.

From Table 5.14, the cooling load hours for San Antonio, Texas, are 1750 hr.

TABLE 5.7   CHARACTERISTIC EFFICIENCY RATINGS FOR RESIDENTIAL AND COMMERCIAL COOLING SYSTEMS. COMPILED FROM VARIOUS GOVERNMENTAL AND INDUSTRY SOURCES.

Equipment	Efficiency		Compressor kW/ton		Auxiliaries kW/ton	Total kW/ton	
Residential Split Systems	Minimum	Maximum	Minimum	Maximum		Minimum	Maximum
Air conditioners	12 SEER	15 SEER	0.9	1.1	0.2	1.1	1.3
Air-cooled heat pumps	11 SEER	12 SEER	1.1	1.2	0.2	1.3	1.5
Water-source heat pumps	12 EER	15.8 EER	0.75	1.1	0.2	0.95	1.3
Ground-source heat pumps	15.8 EER	22 EER	0.55	0.75	0.2	0.75	0.95
Historical Systems (for Comparison)							
20 years old—not well maintained	6.0 SEER		2.2		0.2	2.4	
20 years old—well maintained	6.5 SEER		2.1		0.2	2.3	
10 to 20 years old—not well maintained	7.0 SEER		1.9		0.2	2.1	
10 to 20 years old—well maintained	8.0 SEER		1.7		0.2	1.9	
5 to 10 years old—not well maintained	7.5 SEER		1.8		0.2	2.0	
5 to 10 years old—well maintained	8.5 SEER		1.6		0.2	1.8	
less than 5 years old	9.0 SEER		1.5		0.2	1.7	
Commercial Rooftop and Package AC Units							
Air cooled, 5 to 20 ton			1.2		0.2	1.4	
Air cooled, 20 ton and greater			1.2		0.2	1.4	
Air-cooled condensing units (all sizes)			1.1		0.2	1.3	
Packaged terminal units (all sizes)			1.2		0.2	1.4	
Commercial Central Chillers							
Air cooled (all types)			1.1	1.2	0.2	1.3	1.4
Reciprocating (water cooled)			0.8	0.9	0.2	1.0	1.1
Scroll/screw (water cooled)			0.6	0.8	0.2	0.8	1.0
Centrifugal (water cooled)			0.5	0.8	0.2	0.7	1.0

energy consumption = (total cooling load/SEER) · (cooling
load hours/1000 W)
= (31 410 Btu/hr/11.3) · (1750 hr/1000 W)
= 4864 kWh

cost of operation = 4864 kWh · $0.12/kWh = $584

c. For a home in San Antonio, Texas, with a 20-year-old air conditioner (SEER = 6.0).

From Table 5.14, the cooling load hours for San Antonio, Texas, are 1750 hr.

energy consumption = (total cooling load/SEER) · (cooling
load hours/1000 W)
= (31 410 Btu/hr/11.3) · (1750 hr/1000 W)
= 9161 kWh

cost of operation = 9161 kWh · $0.12/kWh = $1100

The equation used to approximate the annual energy consumption (in kWh) of a commercial cooling system is found in Example 5.12, where the total cooling load is expressed in tons, kW/ton is the performance rating for the cooling system, and cooling load hours is for the specific geographical location:

energy consumption = cooling load
· kW/ton · cooling load hours

Example 5.12

Approximate the annual energy consumption and annual cost of operation of a commercial centrifugal (water cooled) cooling system rated at 0.73 kW/ton. It is serving a total cooling load of 2 150 000 Btu/hr, including auxiliaries. Use an energy cost of $0.10/kWh.

The cooling load of 2 150 000 Btu/hr is converted to tons of refrigeration:

(2 150 000 Btu/hr)/(12 000 Btu/hr per ton) = 179 tons

a. For a building in Denver, Colorado.

From Table 5.14, the cooling load hours for Denver, Colorado, are 450 hr.

energy consumption = cooling load · kW/ton · cooling load hours
= 179 tons · 0.73 kW/ton · 450 hr
= 58 800 kWh

cost of operation = 58 800 kWh · $0.10/kWh
= $5880 (about $6000)

b. For a building in San Antonio, Texas.

From Table 5.14, the cooling load hours for San Antonio, Texas, are 1750 hr.

energy consumption = cooling load · kW/ton · cooling load hours
= 179 tons · 0.73 kW/ton · 1750 hr
= 228 670 kWh

cost of operation = 228 670 kWh · $0.10/kWh
= $22 867 (about $23 000)

5.5 COOLING LOAD COMPUTATIONS FOR COMMERCIAL BUILDINGS

Cooling Load Computation Approach

Computation of cooling loads for commercial buildings (e.g., offices, schools, hospitals, and so on) is complex because loads in a room or zone vary throughout the hour, day, or season because of varying heat gains caused by changing occupancy and operation patterns in the individual spaces. The peak load of an HVAC system will occur when the sum of the loads in the individual rooms or zones being served is the greatest, which is typically less than the sum of the peak loads in the individual rooms or zones being served. For example, the peak load for an office building may occur in early afternoon even though the peak load in the east wing of the building occurs in midmorning and the peak load in the west wing take place in late afternoon.

In performing cooling load calculations for commercial buildings, the spaces need to be divided into zones that have similar load parameters. This approach allows different thermal conditions in the different spaces to be served with separate equipment using independent temperature controls. There are two principal zones: *Perimeter* or *external zones* include rooms and spaces with at least one exterior wall. In these spaces, loads vary over a broad range from heating to cooling and are affected by both interior and exterior (e.g., solar gain, outdoor temperature, and so on) factors. *Interior* or *central zones* include rooms and spaces that are core spaces that have no exterior walls. Heat generated by lights, occupants, and office equipment dominates the load, which in most climates usually provides a continuous cooling demand.

Zoning can be accomplished by having separate equipment or a separate sensor/control arrangement in each space. For large building floor areas, a minimum of five zones per floor (one zone for each exposure: north, south, east, west, and an interior zone) is recommended. Any single zone should be limited to floor areas with perimeter walls not exceeding 40 ft (12 m).

Cooling Load Computations

There are many considerations in CLTD/SLF/CLF cooling load computations. Some considerations are similar to the methodology outlined previously for CLTD residential load computations. In commercial load computations, several factors that were ignored in residential computations are considered. A detailed explanation of this method is beyond the scope of this text. Some considerations are addressed in general terms in the following sections.

Transmission

In most commercial buildings, the cooling load resulting from heat transmission through opaque (nontranslucent) components of exterior envelope assemblies (e.g., walls, roofs, ceilings, floors, and doors) is a smaller fraction of the total load, especially those buildings with high occupancy, lighting, and equipment loads because they have high internal heat gains. The cooling load resulting from heat transmission is calculated in a manner similar to the method outlined earlier.

The cooling load resulting from heat transmission is calculated based on the customary formula of $q = UA \Delta T$, but indirect solar gains heat the opaque surfaces of a building, making transmission computations slightly more complex. The incident solar radiation increases the external surface temperature, causing more heat to flow from outside to inside the building envelope. The *sol–air temperature* is the equivalent outdoor temperature that would cause the same rate of heat flow at the surface and the same temperature distribution through the material as the current outdoor air temperature with the solar gains on the surface and the net radiation exchange between the surface and its environment. When the sol–air data is combined with the inside design temperature, a CLTD is obtained. ASHRAE tables provide hourly CLTD values for one typical set of conditions (i.e., outdoor maximum temperature of 95°F, mean temperature of 85°C, and daily range of 21°F); the equation is adjusted further with correction factors related to surface color, average indoor–outdoor, outdoor temperatures, latitude, and month.

The equation used to compute the transmission component of the cooling load ($q_{transmission}$) for opaque surfaces (i.e., walls, roof, doors, and so forth) is based on the overall coefficient of heat transmission (U), the net exposed area (A) of the assembly being considered (e.g., area of wall, roof, floor, door, or ceiling) and the corrected cooling load temperature difference ($CLTD_{corrected}$):

$$q_{transmission} = U \cdot A \cdot CLTD_{corrected}$$

In cases where transmission heat gain through interior partitions, ceilings, and floors is calculated, the temperature difference (ΔT) is used in lieu of CLTD:

$$q_{transmission} = U \cdot A \cdot \Delta T$$

Example 5.13

A steel roof above a cooling zone in a commercial building has a U-factor of 0.12 Btu/hr $\cdot$ ft^2 $\cdot$ °F and an area of 1240 ft^2. Compute the transmission component of the cooling load at the following hours:

a. A corrected CLTD of 34 (at 9 am).

$$
\begin{aligned}
q_{transmission} &= U \cdot A \cdot CLTD_{corrected} \\
&= 0.12 \text{ Btu/hr} \cdot 1\text{ft}^2 \cdot 1°F \cdot 1240 \text{ ft}^2 \cdot 34 \\
&= 5059 \text{ Btu/hr}
\end{aligned}
$$

b. A corrected CLTD of 71 (at noon).

$$
\begin{aligned}
q_{transmission} &= U \cdot A \cdot CLTD_{corrected} \\
&= 0.12 \text{ Btu/hr} \cdot \text{ft}^2 \cdot °F \cdot 1240 \text{ ft}^2 \cdot 71°F \\
&= 10\,565 \text{ Btu/hr}
\end{aligned}
$$

c. A corrected CLTD of 79 (at 2 pm).

$$
\begin{aligned}
q_{transmission} &= U \cdot A \cdot CLTD_{corrected} \\
&= 0.12 \text{ Btu/hr} \cdot \text{ft}^2 \cdot °F \cdot 1240 \text{ ft}^2 \cdot 79°F \\
&= 11\,755 \text{ Btu/hr}
\end{aligned}
$$

Loads Through Transparent/ Translucent (Glass) Elements

In commercial buildings, window/glazing selection involves maximizing daylighting and keeping summer heat out. As a result, solar heat gains vary with quantity of fenestration, window treatments (e.g., tinted and reflective glazing, shading and draperies), and thermal mass of the building structure. Computations involving solar gains account for the delay from the time the room floor and furnishings absorb solar radiation until it convects to the room air and become a cooling load some time later. The solar load through glass has two components: thermal transmission ($q_{transmission}$) and solar transmission (q_{solar}) for transparent/translucent elements:

$$q_{transmission} = U \cdot A \cdot CLTD$$

$$q_{solar}^* = A \cdot (SC \text{ or } SHGC) \cdot SHGF \cdot CLF$$

---

*Cooling load from solar transmission (q_{solar}) for transparent/translucent elements may also be alternatively calculated by $q_{solar} = A \cdot SC \cdot SCL$, where *solar cooling load (SCL)* values were introduced because of occasional problems with the CLTD/CLF method.

For transparent/translucent elements (i.e., for glass/glazing), the CLTD is approximately equal to the indoor–outdoor temperature difference (ΔT). The *solar heat gain factor (SHGF)* is the maximum clear-day solar radiation that will pass through a double-strength sheet glass at a specific latitude for a specific orientation (i.e., north vertical, horizontal) on a specified month, day, and hour, expressed in units of Btu/hr $\cdot$ ft^2 (W/m^2). At high elevations and very clear days, the actual maximum SGHF may be 15% higher, and in very dirty industrial areas, SGHF may be 20% to 30% lower. *Area (A)* is the net exposed area of the glass/glazing under consideration.

The *shading coefficient (SC)* is a value represented by the ratio of solar heat gain through a glazing system under specific conditions to the solar gain through reference glazing (⅛ in clear single glazing) under the same conditions. The *solar heat gain coefficient (SHGC)* is the fraction of external solar radiation that is admitted through a window or skylight, both directly transmitted, and absorbed and released inward. SHGC is a contemporary value that is determined in a laboratory, while SC is the long-established value. The relationship between SHGC and SC is: SHGC = SC $\cdot$ 0.87. A SHGC of 0.3 will allow 30% of the sun's heat to pass through. Clear single glass has SHGC values from 0.82 (commercial) to 0.86 (residential) and double glass values are from 0.70 (commercial) to 0.76 (residential). Reflective double glass has values from 0.17 to 0.35, which significantly reduces solar heat gains.

In the case of solar transmission (q_{solar}) for transparent/translucent elements, the *cooling load factor (CLF)* relates to the fraction of radiant energy that is stored in or discharged from the interior thermal mass of the building. CLF is a dimensionless value that may be less than or greater than 1.0. It is based on affected by time of day and type of thermal mass.

ASHRAE tables provide values for SHGF, SC, SHGC, and CLF. Typical center-of-glass performance measures for selected types of glazings are found in Table 5.8.

Example 5.14

East-facing windows in a space of a commercial building have an area of 150 ft^2. The double-glazed, low-e (e = 0.10) windows have performance ratings of: U-factor of 0.32 Btu/hr $\cdot$ ft^2 $\cdot$ °F; a cooling load factor is 0.83; and a solar heat gain coefficient of 0.60. At design conditions (July at 48° south latitude), the solar heat gain factor is 214 Btu/hr $\cdot$ ft^2. The indoor–outdoor temperature difference is 19°F. Determine the following:

a. Thermal transmission through the windows.

$$
\begin{aligned}
q_{transmission} &= U \cdot A \cdot CLTD = U \cdot A \cdot \Delta T \\
&= 0.32 \text{ Btu/hr} \cdot \text{ft}^2 \cdot °F \cdot 150 \text{ ft}^2 \cdot 19°F \\
&= 912 \text{ Btu/hr}
\end{aligned}
$$

b. Solar transmission through the windows.

$$
\begin{aligned}
q_{solar} &= A \cdot SHGC \cdot SHGF \cdot CLF \\
&= 150 \text{ ft}^2 \cdot 214 \text{ Btu/hr} \cdot \text{ft}^2 \cdot 0.60 \cdot 0.83 \\
&= 15\,986 \text{ Btu/hr}
\end{aligned}
$$

TABLE 5.8 TYPICAL CENTER-O-GLASS PERFORMANCE MEASURES FOR SELECTED TYPES OF GLAZINGS. TOTAL WINDOW VALUES ARE SIGNIFICANTLY DIFFERENT. COMPILED FROM VARIOUS INDUSTRY SOURCES.

Window and Glazing Types	Center-of-Glass Thermal Properties		SHGC	
	R-Value	U-Factor	Total Window	Center of Glass
Single glazed, clear	0.91	1.10	0.79	0.86
Double glazed, clear	2.04	0.49	0.58	0.76
Double glazed, bronze	2.04	0.49	0.48	0.62
Double glazed, spectrally selective	2.04	0.49	0.31	0.41
Double glazed, low-e (e = 0.10)	3.13	0.32	0.26	0.32
Triple glazed, low-e (e = 0.15)	3.70	0.27	0.37	0.49

c. Total loads through the windows.

912 Btu/hr + 15 986 Btu/hr = 16 898 Btu/hr

d. Solar transmission through spectrally selective (solar control) glass with SHGF = 0.32.

$$q_{solar} = A \cdot SHGC \cdot SHGF \cdot CLF$$
$$= 150 \text{ ft}^2 \cdot 214 \text{ Btu/hr} \cdot \text{ft}^2 \cdot 0.32 \cdot 0.83$$
$$= 8526 \text{ Btu/hr}$$

In Example 5.14, it is evident that use of a high-performance, spectrally selective (solar control) glass significantly reduces the solar load over standard double-glazed, low-e glass. High-performance glazings, although more costly, can have a significant influence on cooling equipment and operating costs. Usually these savings offset added costs.

Example 5.15

Each wall of a single room commercial building has a window with an area of 200 ft^2. Walls are oriented toward the true north, south, east and west. The double-glazed, low-e (e = 0.10) windows have performance ratings of the following: the U-factor is 0.32 Btu/hr · ft^2 · °F; the cooling load factor is 0.83; and the solar heat gain coefficient is 0.60. Determine the solar transmission through the windows for each direction for July at 48° south latitude.

a. North, with the solar heat gain factor is 37 Btu/hr · ft^2.

$$q_{solar} = A \cdot SHGC \cdot SHGF \cdot CLF$$
$$= 200 \text{ ft}^2 \cdot 37 \text{ Btu/hr} \cdot \text{ft}^2 \cdot 0.60 \cdot 0.83$$
$$= 3685 \text{ Btu/hr}$$

b. South, with the solar heat gain factor is 146 Btu/hr · ft^2.

$$q_{solar} = A \cdot SHGC \cdot SHGF \cdot CLF$$
$$= 200 \text{ ft}^2 \cdot 146 \text{ Btu/hr} \cdot \text{ft}^2 \cdot 0.60 \cdot 0.83$$
$$= 14 542 \text{ Btu/hr}$$

c. East and West, with the solar heat gain factor is 214 Btu/hr · ft^2.

$$q_{solar} = A \cdot SHGC \cdot SHGF \cdot CLF$$
$$= 200 \text{ ft}^2 \cdot 214 \text{ Btu/hr} \cdot \text{ft}^2 \cdot 0.60 \cdot 0.83$$
$$= 21 314 \text{ Btu/hr}$$

Lighting Load

The cooling load from the heat gain from electric lights is a large component of the total cooling load for commercial buildings, especially in buildings having a large interior zone. Electric lights contribute to sensible load only. Approximate heat gains from selected luminaires (including ballast, if needed) are provided in Table 5.9. The lighting load ($q_{lighting}$) is computed by the following formula:

$$q_{lighting} = 3.41 \cdot W \cdot F_{ballast} \cdot F_{usage} \cdot CLF$$

The constant 3.41 is a conversion factor that when multiplied by the *installed lamp wattage (W)* converts from wattage to Btu/hr; that is, one watt equals 3.41 Btu/hr. When working in SI units, the 3.41 conversion factor is neglected.

The full wattage of a light fixture and lamp(s) must be considered, including any ballast. A ballast is a voltage transformer and current limiting device designed to start and properly control the flow of power to discharge light sources [i.e., fluorescent and high intensity discharge (HID) lamps], which consumes more power than the lamp(s) alone. A ballast is not required on incandescent lamps (i.e., ordinary "light bulbs," tungsten lamps, and so on). It is addressed in Chapter 20. A *ballast factor ($F_{ballast}$)* is an adjustment factor that is used when a lamp in a fixture requires a ballast. Ballasts introduce about 15% to 30% more heat into the space, so the ballast factor is typically from 1.15 to 1.30. In instances when the full wattage of the luminaire [light fixture and lamp(s) and ballast] is known, the ballast factor is neglected.

The *lighting usage factor (F_{sa})* is defined as the ratio of wattage in use at design condition to the installation condition. It is an estimated adjustment factor that accounts for any lamps that are installed but that would not be operated under conditions

TABLE 5.9   APPROXIMATE HEAT GAINS FROM SELECTED LIGHTS, APPLIANCES, AND EQUIPMENT.

Description	Sensible Heat Gain		Latent Heat Gain		Total Heat Gain	
	Btu/hr	W	Btu/hr	W	Btu/hr	W
Computer with monitor	425	125	—	—	425	125
Computer with monitor (power saver mode)	85	25	—	—	85	25
Copier (office size)	3400	1000	—	—	3400	1000
Copier (power saver mode)	850	250	—	—	850	250
Printer (large desktop laser)	850	250	—	—	850	250
Printer (power saver mode)	135	40	—	—	135	40
Coffee maker (10 cup)	5000	1470	3500	1030	5000	1470
Microwave oven (small)	2000	600	—	—	2000	600
Refrigerator (kitchen type)	1500	440	—	—	1500	440
Vending machine (cold beverage)	3250	950	—	—	3250	950
Vending machine (hot beverage)	2900	850	—	—	2900	850
Vending machine (snack)	940	275	—	—	940	275
100 W incandescent light fixture	340	100	—	—	340	100
Fluorescent light fixture with two 32 W lamps	250	75	—	—	250	75
250 W metal halide light fixture	950	280	—	—	950	280

assumed for load calculations. Some luminaires will not be operated. In installations provided with local switching (or occupancy sensing), lights in unoccupied areas may not be on. For example, if only 80% of the lights in a space or zone will be used at design conditions (e.g., lights in storage and closet rooms are off), the usage factor will be 0.80.

CLFs for lighting are used to account for the time lag of the cooling load caused by lighting. CLF data related to lighting is related to the number of hours that electric lights are switched on and the mass of building construction and furnishings. The CLF is 1.0 with continuous (24 hr) occupancy, and/or when the cooling system is shut down for extended periods (i.e., for energy savings at night or during weekends).

Sensible heat released from electric lights is caused by convective heat transfer directly from the lamp and fixtures, and by radiation absorbed by the walls, floors, and furniture, and then transferred into the space by convection of air after a time lag. Some of the heat from electrical power used by the lights appears immediately as a part of the cooling load but some does not. The difference between the energy input to the lights and the cooling load is heat energy stored by the structure and furnishings. This stored heat energy is released after the lights are turned off. For this reason, the cooling load from lighting continues after the lights are switched off. The cooling load increases continuously when the lights are on, but even after 10 hr it reaches only 80% of the input power. Increasing the mass of the building, particularly the mass of the floor slab, increases the heat storage capacity of the structure and causes the cooling load from lights to be spread more evenly over the entire 24 hr.

Example 5.16

A lobby in an office building has 16 fluorescent luminaires, each containing three 32 W fluorescent lamps. The luminaires

have ballasts with a ballast factor of 1.15. All lights remain lit and operate continuously, 24 hours a day, 365 days a year. The cooling system is shut down during weekend hours. Determine the lighting load.

Because all lights are lit, $F_{usage} = 1.0$ (100%). Because cooling is shut down during weekend hours, CLF = 1.0.

$$\begin{aligned} Q_{lighting} &= 3.41 \cdot W \cdot F_{ballast} \cdot F_{usage} \cdot CLF \\ &= 3.41 \cdot (16 \cdot 3 \cdot 32W) \cdot 1.15 \cdot 1.00 \cdot 1.0 \\ &= 6023 \ Btu/hr \end{aligned}$$

Example 5.17

As in the previous example, a lobby in an office building has 16 fluorescent luminaires, each containing three 32 W fluorescent lamps. The luminaires are rated at 110 W (including ballast). All lights remain lit and operate continuously, 24 hours a day, 365 days a year. The cooling system is shut down during weekend hours. Determine the lighting load.

Because the luminaire rating is provided, the ballast factor is neglected. Because lights are shut down during weekend hours, CLF = 1.0.

$$\begin{aligned} Q_{lighting} &= 3.41 \cdot W \cdot F_{usage} \cdot CLF \\ &= 3.41 \cdot (16 \cdot 110W) \cdot 1.00 \cdot 1.0 \\ &= 6002 \ Btu/hr \ (rounding \ error) \end{aligned}$$

Appliance and Power Equipment Loads

In estimating the cooling load, heat gains from equipment (e.g., computers, printers, copiers) and appliances (e.g., ranges, refrigerators, coffee makers) must be considered. Heat gains for some appliances (e.g., refrigerators, freezers, phones, fax machines) are independent of occupant activities, while operation of other appliances (e.g., computers, televisions, elevators) depends on occupant activity.

Appliance and power equipment loads may contribute to either sensible or latent loads, and sometimes both. Latent heat gains can be significant, especially when they come from appliances (coffee pot), indoor plantings, fountains, pools, and Jacuzzis. Approximate heat gains from selected appliances and equipment are listed in Table 5.9. Appliance and equipment loads ($q_{equipment}$) are computed by the following formula:

$$q_{equipment\ (sensible)} = N \cdot HG_{sensible} \cdot F_{usage} \cdot CLF$$
$$q_{equipment\ (latent)} = N \cdot HG_{latent} \cdot F_{usage}$$

The *number (N)* of appliances or pieces of equipment affects the cooling load proportionately. *Sensible heat gain (HG$_{sensible}$)* by the appliance or piece of equipment is based on the appliance wattage · 3.413 Btu/hr per watt. *Latent heat gain (HG$_{latent}$)* is directly related to the latent heat output by the appliance or piece of equipment, usually provided from tabular data. See Table 5.9.

The *usage factor (F$_{sa}$)* is related to the percentage of appliances or pieces of equipment in use. It is an estimated adjustment factor that accounts for appliances that are installed but switched off at design conditions; for example, if 20% of the computers are switched off, the usage factor will be 0.80. Motors usually only operate at a fraction of their nameplate rating (i.e., the mechanical load on the rotating shaft is less than rated). This could be from operating variations in load or to conservative equipment selection by the designer. Some loads are cyclic in nature (e.g., sump pumps, sewerage ejector pumps, air compressors, lifts, and so on). The cyclic nature of these loads creates a probability that not all will be simultaneously operating. Some loads rarely operate except in unusual circumstances (e.g., fire sprinkler booster pumps). Additionally, the electrical energy consumed by an electric motor that is not converted to power is instantaneously dissipated as heat. Depending on the motor type, size, and service, this heat gain can be 1 to 30% of the energy used by the motor.

For power driven appliances and equipment, CLFs already are used in the same manner for other components to account for the radiant part of the sensible heat gain that is delayed in contributing to the instantaneous cooling load. The CLF is 1.0 with high density, continuous (24 hr) occupancy, and/or when the cooling system is shut down for extended periods (i.e., for energy savings at night or during weekends). The latent load from equipment and appliances ($q_{equipment\ (latent)}$) is assumed to be instantaneous because it is not greatly affected by storage of water vapor in the building mass or furnishings, so there is no CLF multiplier.

Equipment loads are frequently oversized when they are based on the nameplate or connected load of equipment (i.e., computers, copiers, printers, and so forth) and on the assumption that operation of this equipment occurs simultaneously.

However, most equipment operates at a fraction of the nameplate value, and rarely operates at the same time. Many HVAC designs are based on plug load assumptions of about 5 W/ft^2 in office spaces. According to a recent study, plug loads in commercial office buildings had equipment load densities from 0.4 and 1.2 W/ft^2.

Example 5.18

A commercial laundromat contains 16 clothes washers (500 Btu/hr sensible and 85 Btu/hr latent loads) and 9 large clothes dryers (7600 Btu/hr sensible and 4200 Btu/hr latent loads). Assume that at design conditions, 70% of these machines are being operated. The cooling system is shut down at night (CLF = 1.0).

a. Determine the sensible load associated with these machines.

Washers: $q_{equipment\ (sensible)} = N \cdot HG_{sensible} \cdot F_{usage} \cdot CLF$
$= 16 \cdot 500 \text{ Btu/hr} \cdot 0.70 \cdot 1.0$
$= 5600 \text{ Btu/hr}$

Dryers: $q_{equipment\ (sensible)} = N \cdot HG_{sensible} \cdot F_{usage} \cdot CLF$
$= 9 \cdot 7600 \text{ Btu/hr} \cdot 0.70 \cdot 1.0$
$= 47\,880 \text{ Btu/hr}$

b. Determine the latent load associated with these machines.

Washers: $q_{equipment\ (latent)} = N \cdot HG_{latent} \cdot F_{usage}$
$= 16 \cdot 85 \text{ Btu/hr} \cdot 0.70$
$= 952 \text{ Btu/hr}$

Dryers: $q_{equipment\ (latent)} = N \cdot HG_{latent} \cdot F_{usage}$
$= 9 \cdot 4200 \text{ Btu/hr} \cdot 0.70$
$= 26\,460 \text{ Btu/hr}$

Occupant Load

Occupants release both sensible heat and latent heat to the conditioned space. The quantity of heat released is related to gender and activity. Approximate heat gains from occupants for various activities are listed in Table 5.10. Approximate heat gains from occupants for various occupancies are listed in Table 5.11. The cooling load from occupants is significant in spaces with high occupancy (e.g., classrooms, assembly areas, theaters, churches, and so on). Occupant loads (q_{people}) are computed by:

$$q_{people-sensible} = N \cdot HG_{sensible} \cdot CLF$$
$$q_{people-latent} = N \cdot HG_{latent}$$

The *number (N)* of occupants affects the cooling load proportionately. Approximate *sensible* and *latent heat gains (HG$_{sensible}$ and HG$_{latent}$)* from occupants for various activities are listed in Table 5.10. Like lighting, the sensible load component from occupants ($q_{people-sensible}$) is related to the time people spend in the conditioned space and time elapsed since first entering the space. The appropriate CLF is selected from tables

TABLE 5.10 APPROXIMATE HEAT GAINS FROM OCCUPANTS FOR VARIOUS ACTIVITIES.

Activity	Sensible Heat Gain		Latent Heat Gain		Total Heat Gain	
	Btu/hr	W	Btu/hr	W	Btu/hr	W
Seated at rest	205	60	135	40	340	100
Seated, light work	255	75	255	75	510	150
Slow walking	300	90	330	95	630	185
Briskly walking	345	100	380	110	725	210
Light work	340	100	700	205	1040	305
Hard work	560	165	1040	305	1600	470
Athletic exercise	640	185	1160	340	1800	525

TABLE 5.11 APPROXIMATE HEAT GAINS FROM OCCUPANTS BASED ON BUILDING TYPE.

Occupancy Type	Sensible Heat		Latent Heat		Total Heat	
	Btu/hr	W	Btu/hr	W	Btu/hr	W
Bank	220	64	280	82	500	146
Office	215	63	185	54	400	117
Apartment	215	63	185	54	400	117
Restaurant	240	70	310	91	550	161
Retail store	220	64	230	67	450	132
Theater, church sanctuary	215	63	135	40	350	103

according to zone type, occupancy period, and number of hours after entering. If the temperature in a space is not maintained constant (e.g., because of setback, or weekend setback), all of the heat gained affects the cooling load. In such cases, the sensible load component from occupants ($q_{people\text{-}sensible}$) and the CLF are equivalent to 1.0. The latent load from occupants ($q_{people\text{-}latent}$) is assumed to be instantaneous because it is not greatly affected by storage of water vapor in the building mass or furnishings, so there is no CLF multiplier.

Example 5.19

A large convention center conference room seats 800 occupants. All conference attendees are seated at rest. The cooling system is shut down during off days.

a. Determine the sensible load associated with the people seated in this space.

Because cooling is shut down during weekend hours, CLF = 1.0. $HG_{sensible}$ is from Table 5.11.

$$q_{people\text{-}sensible} = N \cdot HG_{sensible} \cdot CLF$$
$$= 800 \text{ occupants} \cdot 205 \text{ Btu/hr} \cdot 1.0$$
$$= 164\,000 \text{ Btu/hr}$$

b. Determine the latent load associated with the people seated in this space.

$$q_{people\text{-}latent} = N \cdot HG_{latent} = 800 \text{ occupants} \cdot 135 \text{ Btu/hr}$$
$$= 108\,000 \text{ Btu/hr}$$

c. Determine the total load associated with the people seated in this space.

$$164\,000 \text{ Btu/hr} + 108\,000 \text{ Btu/hr} = 272\,000 \text{ Btu/hr}$$

Cooling Load from Ventilation

Ventilation is outdoor air that is mechanically introduced into the building interior to ensure occupant comfort and health. Heat gain from ventilation in cooling load computations for residences is generally ignored because it is typically not a code requirement and is handled adequately with the infiltration component of the computation. Ventilation is an important consideration in cooling load computations for nonresidential buildings, mainly because ventilation is required in these buildings. This is particularly true of high-occupancy buildings. With respect to cooling load computations, ventilation loads consist of both sensible and latent cooling loads.

Sensible heat gain from ventilation air is based on the quantity and conditions of outdoor air. The sensible cooling load from ventilation can be computed as follows based on the outside design temperature and inside design temperature difference (ΔT), in °F, and the volumetric airflow rate ($Q_{airflow}$), in cfm. (A cfm is a cubic foot per minute of airflow).

$$q_{ventilation} = 1.1 \cdot Q_{airflow} \cdot \Delta T$$

In metric (SI) units, the sensible cooling load from ventilation can be computed as follows, based on the outside design temperature and inside design temperature difference (ΔT), in °F, and the volumetric airflow rate ($Q_{airflow}$), in L/s.

$$q_{ventilation} = 1.2 \cdot Q_{airflow} \cdot \Delta T$$

In large buildings, ventilation requirements are larger and more complex, so they are treated as a separate component of the cooling load. Ventilation requirements and rates were discussed in Chapter 3. For example, ASHRAE Standard 62 specified rates at which outside air must be supplied to a room, which is generally within the range of 15 to 60 cfm/person (7.5 to 30 L/s per person), depending on the activities that normally occur in that room. (A cfm is a cubic foot per minute of airflow). For example, assume a classroom space is designed for an occupancy of 30 persons. The ASHRAE Standard calls for a minimum outside airflow rate for classrooms of 15 cfm per person. Therefore, the classroom requires 15 cfm/person · 30 persons, or 450 cfm.

The inside–outside design temperature difference (ΔT) is computed by the designer. The outside design temperature can be selected from Table 5.12 (located at end of this chapter). This table provides the summer design dry bulb temperature for locations in the United States and Canada. The inside design temperature depends on the type of occupancy; again, a temperature of between 73° to 76°F (23° to 26°C) is typically used.

Example 5.20

A college classroom in Denver, Colorado, is designed for an occupancy of 30 persons. The ASHRAE Standard calls for a minimum outside airflow rate for classrooms of 15 cfm per person. The target inside design temperature is 74°F. Determine the sensible cooling load from ventilation.

From Table 5.12 (located at end of chapter), the 1% outside air design temperature is 90°F.

The classroom ventilation rate is found by:

$$15 \text{ cfm/person} \cdot 30 \text{ persons} = 450 \text{ cfm}$$

$$\begin{aligned}q_{ventilation} &= 1.1 \cdot Q_{airflow} \cdot \Delta T \\ &= 1.1 \cdot 450 \text{ cfm} \cdot (90°F - 74°F) = 7920 \text{ Btu/hr}\end{aligned}$$

Latent heat gain from infiltration or ventilation air is based on the quantity of airflow and conditions of outdoor air. The latent cooling load from infiltration or ventilation can be computed as follows, based on the difference in absolute humidity between indoor and outdoor air (ΔW), in lb of water vapor/lb of dry air (extracted from a psychrometric chart), and the volumetric airflow rate ($Q_{airflow}$), in cfm. (A cfm is a cubic foot per minute of airflow):

$$q_{latent} = 4840 \cdot Q_{airflow} \cdot \Delta W$$

The constant 4840 is a unit conversion factor based on: (60 min/hr · Btu/hr)/(ft^3 · lb of water). It applies to customary (U.S.) units only.

Example 5.21

A college classroom in Lihue, Hawaii, is designed for an occupancy of 30 persons. The ASHRAE Standard calls for a minimum outside airflow rate for classrooms of 15 cfm per person. The 1% outside design conditions are 85°F DB and 74°F WB. The inside design conditions are 75°F dry bulb temperature and 60% relative humidity. Determine the latent cooling load from ventilation for this set of design conditions.

From a psychrometric chart (see Chapter 1), for the outside design conditions, W = 0.0155 lb of water/lb of dry air and W = 0.0111 lb of water/lb of dry air for the inside design conditions.

The classroom ventilation rate is found by:

$$15 \text{ cfm/person} \cdot 30 \text{ persons} = 450 \text{ cfm}$$

$$\begin{aligned}q_{latent} &= 4840 \cdot Q_{airflow} \cdot \Delta W \\ &= 4840 \cdot 450 \text{ cfm} \cdot (0.0155 \text{ lb/lb} - 0.0111 \text{ lb/lb}) \\ &= 9583 \text{ Btu/hr}\end{aligned}$$

A more detailed computation technique for commercial (nonresidential) buildings can be found in the *ASHRAE Handbook—Fundamentals*. The method is more comprehensive than the computation procedure used for residential loads. The Air Conditioning Contractors of America, Inc. (ACCA), developed *Manual N—Commercial Load Calculation*. It can be used to estimate heating and cooling loads for small- and mid-sized commercial structures.

Because cooling load computations for commercial buildings are complex and cumbersome, a number of software programs are available to help designers compute and evaluate loads. Most computer-modeling programs rely upon the ASHRAE method or similar methods and can be very accurate with proper assumptions (e.g., input data on building operations, occupancy patterns, and so on).

STUDY QUESTIONS

5-1. Describe the factors that must be considered when determining heat gain in buildings.

5-2. Define sensible and latent heat.

5-3. Describe a ton of refrigeration.

5-4. Describe why oversizing of cooling equipment is a problem.

5-5. Identify the design conditions at the geographical location where you reside.

5-6. Define the CLTD.

5-7. Define the GLF.

5-8. How does the orientation of the windows in the building affect the heat gain?

5-9. Why must the number of occupants and the type of activity be considered in a heat gain calculation?

5-10. How is infiltration calculated for heat gain, and how does this compare with infiltration heat load calculations?

Design Exercises

5-11. Find the CLTD for a wood-framed single-family detached residence at the geographical location where you reside, for exterior walls with the following orientations:

   a. Facing south

   b. Facing north

   c. Facing east

   d. Facing west

5-12. Find the CLTD for a wood-framed single-family detached residence in Houston, Texas, for exterior walls with the following orientations:

   a. Facing south

   b. Facing north

   c. Facing east

   d. Facing west

5-13. Find the CLTD for a wood-framed single-family detached residence in Newark, New Jersey, for exterior walls with the following orientations:

   a. Facing south

   b. Facing north

   c. Facing east

   d. Facing west

5-14. Find the CLTD for a wood-framed single-family detached residence in Phoenix, Arizona, for exterior walls with the following orientations:

   a. Facing south

   b. Facing north

   c. Facing east

   d. Facing west

5-15. A wood-framed single-family detached residence in Houston, Texas, has a south-facing wall with a net exposed area of 100 ft^2. The wall has an R_t of 15.1 hr · ft^2 · °F/Btu. Determine the sensible heat transmission component of the cooling load for this wall.

5-16. A wood-framed single-family detached residence in Austin, Texas, has a south-facing wall with a net exposed area of 100 ft^2. The wall has an R_t of 15.1 hr · ft^2 · °F/Btu. Determine the sensible heat transmission component of the cooling load for this wall.

5-17. A wood-framed multifamily residence in Columbus, Ohio (about 40° north latitude), has a room with south-facing windows. The fixed windows are regular double glass with a total glass area of 100 ft^2. Determine the cooling load attributed to the windows.

   a. For windows with no inside shading

   b. For windows with fully drawn venetian blinds

   c. For windows that are fully shaded by an exterior overhang and with no inside shading

5-18. A wood-framed multifamily residence in Savannah, Georgia (about 32° north latitude), has a room with south-facing windows. The fixed windows are regular double glass with a total glass area of 100 ft^2. Determine the cooling load attributed to the windows.

   a. For windows with no inside shading

   b. For windows with fully drawn venetian blinds

   c. For windows that are fully shaded by an exterior overhang and with no inside shading

5-19. A well-built, wood-framed, single-family detached residence in Savannah, Georgia, has a 12 ft by 20 ft family room with 10 ft high ceilings. The target inside design temperature is 75°F. Determine the sensible cooling load attributed to infiltration. Use an air exchange rate of 0.4.

5-20. A single-family detached residence of moderate construction has a total sensible load of 53 300 Btu/hr. The design humidity ratio is 0.021 lb of vapor/lb of dry air.

   a. Determine the latent load.

   b. Determine the total cooling load.

5-21. Approximate the sensible cooling load attributed to lighting, appliances and equipment in a two bedroom, dwelling unit in an apartment.

5-22. Approximate the annual energy consumption and annual cost of operation of an air conditioner with an SEER of 13 that is serving a cooling load of 36 000 Btu/hr. Use an energy cost of $0.10/kWh.

   a. For a home in San Francisco, California

   b. For a home in Miami, Florida

   c. For a home in Columbia, Missouri

   d. For a home in Raleigh, North Carolina

5-23. Approximate the annual energy consumption and annual cost of operation of an air conditioner with an SEER of 13 that is serving a cooling load of 36 000 Btu/hr. Use current electricity costs. (It will be necessary for you to contact your local utility to get energy costs.)

5-24. A church sanctuary is designed for an occupancy of 300 persons. The outside airflow rate is 15 cfm per person. The inside design temperature is 74°F and the outside air design temperature is 89°F. Approximate internal heat gains per occupant are 215 Btu/hr for sensible heat and 135 Btu/hr for latent heat.

a. Determine the sensible cooling load component from ventilation.

b. Determine the sensible cooling load component from occupants.

c. Determine the latent cooling load component from occupants.

d. Determine the total cooling load component from occupants.

5-25. Calculate the cooling load for the residence in Appendix D. Assume the residence will be built at the geographical location where you reside.

5-26. Calculate the cooling load for a top-floor apartment and an apartment that is below the top floor, using the apartment building in Appendix A. Assume the residence will be built at the geographical location where you reside.

TABLE 5.12   COOLING DESIGN CONDITIONS FOR THE UNITED STATES AND CANADA.

United States

Station	Cooling DB/MWB 0.4% DB	MWB	1% DB	MWB	2% DB	MWB	Evaporation WB/MDB 0.4% WB	MDB	1% WB	MDB	2% WB	MDB	Dehumidification DP/MDB and HR 0.4% DP	HR	MDB	1% DP	HR	MDB	2% DP	HR	MDB	Range of DB
1	2a	2b	2c	2d	2e	2f	3a	3b	3c	3d	3e	3f	4a	4b	4c	4d	4e	4f	4g	4h	4i	5
ALABAMA																						
Anniston	95	76	93	76	90	75	79	90	78	88	77	86	77	143	84	76	137	82	75	133	81	19.6
Birmingham	94	75	92	75	90	74	78	89	77	88	76	87	75	135	83	74	131	82	73	127	81	18.7
Dothan	95	76	93	76	92	76	80	90	79	88	78	87	77	144	83	76	139	82	76	136	82	17.5
Huntsville	94	75	92	74	90	74	78	89	77	88	76	86	75	135	83	74	130	82	73	126	81	18.5
Mobile	94	77	92	76	91	76	79	89	79	88	78	87	77	142	83	76	139	83	76	135	82	16.5
Montgomery	95	76	93	76	91	76	79	91	78	89	78	88	76	139	85	75	134	84	75	130	83	18.7
Muscle Shoals/Florence	96	76	94	75	92	74	78	90	78	89	77	87	76	137	82	75	133	82	74	130	81	20.0
Ozark, Ft. Rucker	95	77	94	77	92	76	81	90	79	89	78	88	78	146	85	77	142	84	76	138	83	18.0
Tuscaloosa	95	77	94	77	92	76	80	90	79	89	78	88	77	142	84	76	137	83	75	134	82	19.6
ALASKA																						
Adak, NAS	59	55	57	53	55	51	55	59	53	57	51	54	53	59	58	51	55	56	49	51	53	9.7
Anchorage, Elemendorf AFB	71	58	69	57	66	56	60	69	58	66	57	64	57	69	62	55	65	61	53	61	60	12.6
Anchorage, Ft. Richardson	74	60	71	58	68	57	61	72	59	69	58	66	56	69	64	54	63	62	53	61	61	15.5
Anchorage, Intl Airport	71	59	68	57	65	56	60	69	58	66	57	63	56	68	62	55	64	61	53	61	60	12.6
Annette	74	61	70	59	66	57	62	72	60	68	59	65	58	71	65	56	68	63	55	65	61	10.5
Barrow	57	51	52	49	48	46	52	56	49	52	46	48	49	53	54	46	46	51	44	42	48	10.6
Bethel	72	59	68	57	64	55	60	69	58	66	56	63	56	68	62	55	64	60	53	60	58	13.4
Bettles	79	61	75	59	72	58	63	76	61	73	59	70	58	72	66	56	67	64	54	63	63	19.4
Big Delta, Ft. Greely	78	59	75	58	71	56	61	74	59	72	58	69	56	70	65	54	65	63	52	61	61	17.3
Cold Bay	60	54	57	53	55	52	55	59	54	56	53	55	54	62	56	53	59	55	51	56	54	7.4
Cordova	70	59	67	57	63	56	60	69	58	65	56	62	56	67	63	54	62	60	53	60	59	13.5
Deadhorse	66	57	61	54	58	53	58	64	55	62	53	58	54	61	62	51	56	59	49	51	56	13.7
Dillingham	69	57	65	56	62	54	59	67	57	64	55	61	56	67	62	53	61	59	52	57	57	13.1
Fairbanks, Eielson AFB	81	61	78	60	75	59	64	78	62	75	60	72	58	74	66	56	69	64	54	63	66	19.4
Fairbanks, Intl Airport	81	61	77	59	74	58	63	77	61	74	59	71	58	72	65	56	68	64	54	64	63	18.6
Galena	78	61	74	59	71	58	63	74	61	71	59	69	58	73	66	56	69	65	54	63	64	15.3
Gulkana	77	58	73	56	69	55	59	73	57	70	56	67	53	63	62	51	60	60	50	55	59	20.3
Homer	65	56	62	55	60	54	57	63	56	61	55	59	54	64	59	53	60	58	52	57	57	11.9
Juneau	74	60	69	58	66	57	61	71	59	68	58	64	57	70	63	56	67	61	55	64	60	13.9
Kenai	68	56	65	55	62	54	58	65	56	63	55	61	55	64	59	53	60	58	52	58	57	13.3
Ketchikan	71	60	68	59	66	58	62	69	60	67	59	64	59	74	64	57	71	62	56	68	61	10.3
King Salmon	71	58	67	56	64	55	59	68	57	65	56	62	56	66	61	54	62	60	52	58	58	15.5
Kodiak, State USCG Base	68	58	65	56	62	55	59	66	57	63	56	61	56	67	61	55	64	59	53	61	57	11.2
Kotzebue	68	59	64	58	61	56	60	67	58	64	56	61	57	70	64	55	65	61	54	62	59	8.8
McGrath	77	60	73	58	70	56	61	74	59	70	58	67	57	69	63	55	65	62	54	62	61	17.4
Middleton Island	62	54	60	51	59	51	55	61	54	59	53	57	52	57	56	51	56	56	50	54	55	5.8
Nenana	80	60	76	59	73	57	62	75	60	73	59	70	57	69	65	55	65	65	53	60	63	21.2
Nome	69	57	65	55	61	54	58	66	56	63	55	60	55	64	61	53	60	59	51	56	57	10.9
Northway	78	58	74	57	71	56	60	76	58	71	57	69	54	66	62	53	62	61	51	60	59	20.0
Port Heiden	64	54	61	52	59	51	56	62	54	60	52	58	51	57	59	50	54	57	49	51	55	9.7
Saint Paul Island	54	51	52	50	51	49	51	53	50	52	49	50	50	55	52	49	52	51	48	50	50	5.4
Sitka	66	59	64	58	61	57	60	65	59	62	58	60	58	74	62	57	71	60	56	68	59	9.2
Talkeetna	77	60	73	58	70	57	62	74	60	70	58	67	57	71	64	56	67	62	54	64	61	16.4
Valdez	69	56	65	55	62	54	58	67	56	64	55	61	53	60	59	53	59	57	52	57	56	12.2
Yakutat	66	56	63	55	60	54	58	62	57	60	56	59	56	67	58	55	64	57	54	62	57	12.0
ARIZONA																						
Flagstaff	85	56	83	55	80	55	61	74	60	73	59	72	58	93	65	56	88	64	55	83	63	27.6
Kingman	99	64	97	63	95	62	71	82	67	85	66	86	67	112	77	62	92	75	59	85	76	24.8
Page	99	62	97	62	95	61	66	85	65	86	64	86	60	92	74	58	85	74	56	80	74	23.8
Phoenix, Intl Airport	110	70	108	70	106	70	76	97	75	96	74	95	71	118	82	69	111	84	67	104	85	23.0
Phoenix, Luke AFB	110	71	107	71	105	71	78	97	76	97	75	96	74	130	85	71	118	85	69	111	86	25.2
Prescott	94	60	91	60	89	60	67	81	66	80	64	79	63	104	71	61	98	71	60	93	70	25.4
Safford, Agri Center	102	66	99	66	97	65	71	89	71	89	69	88	67	111	77	66	106	76	64	102	77	34.7
Tucson	104	65	102	65	100	65	72	88	71	87	70	86	69	116	76	67	111	76	66	106	77	29.4
Winslow	95	60	93	60	91	59	65	80	64	81	63	80	61	95	71	59	91	69	58	85	69	27.4
Yuma	111	72	109	72	106	72	80	96	78	95	77	95	76	136	87	74	127	88	71	117	89	23.8
ARKANSAS																						
Blytheville, Eaker AFB	97	78	95	77	93	77	82	92	80	91	78	89	78	149	88	77	142	86	76	135	85	18.7
Fayetteville	95	75	93	75	90	74	78	90	77	89	76	87	75	136	85	74	132	84	72	124	81	21.4
Ft. Smith	99	76	96	76	93	75	79	92	78	91	77	90	75	134	85	74	130	84	73	126	83	21.5
Little Rock, AFB	97	77	95	77	92	76	80	92	79	91	78	89	77	141	86	76	137	85	75	133	84	19.5
Texarkana	97	77	95	77	93	76	80	91	79	90	78	89	77	143	85	76	139	85	75	135	84	20.5
CALIFORNIA																						
Alameda, NAS	83	65	79	64	76	63	67	79	65	76	64	73	62	85	70	61	80	69	60	78	67	14.8
Arcata/Eureka	70	60	67	59	65	58	62	67	61	65	60	64	60	78	64	59	75	63	58	71	62	15.5
Bakersfield	104	70	101	69	99	69	73	98	71	96	70	95	64	92	84	62	85	83	60	79	83	26.5
Barstow/Daggett	107	68	105	67	102	67	72	95	71	95	69	95	66	103	81	63	91	85	59	81	85	27.8
Blue Canyon	84	59	81	57	79	56	62	80	60	78	58	75	54	74	70	52	69	70	50	64	68	16.6
Burbank/Glendale	98	69	95	69	92	68	74	90	72	89	71	86	69	108	80	67	103	79	66	98	77	23.4

DP = dew-point temperature, °F
MDB = mean coincident dry-bulb temp., °F
StdD = standard deviation, °F

MWB = mean coincident wet-bulb temp., °F
MWS = mean coincident wind speed, mph
HR = humidity ratio, grains of moisture per lb of dry air

A = airport
ANGB = Air National Guard Base
MCAF = Marine Corps Air Facility

NAF = Naval Air Facility
NAWS = Naval Air Weapons Station
RAF = Royal Air Force

United States

Station	Cooling DB/MWB 0.4% DB	MWB	1% DB	MWB	2% DB	MWB	Evaporation WB/MDB 0.4% WB	MDB	1% WB	MDB	2% WB	MDB	Dehumidification DP/MDB and HR 0.4% DP	HR	MDB	1% DP	HR	MDB	2% DP	HR	MDB	Range of DB
1	2a	2b	2c	2d	2e	2f	3a	3b	3c	3d	3e	3f	4a	4b	4c	4d	4e	4f	4g	4h	4i	5
Fairfield, Travis AFB	98	67	94	67	91	66	70	92	69	90	67	88	62	85	76	61	79	74	59	75	73	29.0
Fresno	103	71	101	70	98	69	73	98	71	96	70	94	64	92	85	62	85	84	61	80	82	30.9
Lancaster/Palmdale	101	66	98	65	96	64	70	94	69	91	67	90	62	92	80	60	84	81	58	78	81	27.9
Lemoore, Reeves NAS	103	72	101	71	98	70	75	97	73	96	71	94	67	101	89	65	94	87	62	85	86	32.9
Long Beach	92	67	88	67	84	66	71	85	70	82	69	80	67	101	76	66	96	75	65	92	75	16.7
Los Angeles	85	64	81	64	78	64	70	78	69	76	68	75	67	99	75	66	95	73	65	92	72	10.9
Marysville, Beale AFB	101	70	98	69	95	68	72	97	71	95	69	92	63	86	85	61	80	82	60	76	81	29.9
Merced, Castle AFB	99	69	97	69	94	68	72	96	71	93	69	92	64	90	81	62	82	84	60	78	81	30.2
Mount Shasta	91	62	88	61	85	60	64	87	63	84	61	82	56	76	74	53	69	73	51	64	71	32.0
Mountain View, Moffet NAS	88	65	84	65	80	64	68	82	67	80	65	78	62	83	74	61	80	73	60	76	72	18.0
Ontario	102	71	98	70	95	69	75	94	73	92	72	90	70	113	80	68	106	80	66	101	78	27.7
Oxnard, Pt. Mugu NAWS	83	62	79	64	77	64	70	77	69	75	67	74	68	103	74	66	97	73	65	93	72	14.6
Paso Robles	102	68	98	67	95	65	70	97	68	94	67	91	61	81	76	58	75	73	57	71	71	37.8
Red Bluff	105	70	102	69	98	67	72	98	71	95	69	93	65	94	82	62	85	80	61	80	78	29.5
Riverside, March AFB	101	68	98	68	95	67	72	92	71	91	70	90	67	104	79	65	97	80	62	90	79	29.0
Sacramento, Mather Field	101	69	97	68	95	67	71	97	69	94	68	92	61	80	79	60	77	77	58	74	79	33.7
Sacramento, McClellan AFB	102	70	98	69	95	68	72	97	71	95	69	92	63	85	84	61	80	81	60	77	79	29.7
Sacramento, Metro	100	69	97	69	94	68	72	96	70	94	69	91	62	84	82	61	79	80	59	76	78	33.3
Salinas	83	63	78	62	75	61	66	78	65	75	63	72	62	82	69	60	78	68	59	76	67	18.7
San Bernardino, Norton AFB	103	70	101	70	97	69	74	94	73	94	71	92	68	107	83	66	101	83	65	95	82	31.5
San Diego, Intl Airport	85	67	81	67	79	67	73	79	71	78	70	77	70	111	77	68	104	76	67	99	74	8.9
San Diego, Miramar NAS	92	69	88	67	85	67	72	85	71	83	69	81	68	104	78	67	99	77	65	96	75	17.5
San Francisco	83	63	78	62	74	61	64	79	63	75	62	72	59	76	67	58	73	66	57	71	65	16.7
San Jose Intl Airport	93	67	89	66	86	65	70	88	68	85	67	83	63	85	77	61	81	76	60	78	74	22.3
Santa Barbara	83	64	80	64	77	64	69	77	67	76	66	74	66	96	74	65	91	71	63	85	70	18.0
Santa Maria	86	63	82	62	78	61	66	81	65	78	64	75	61	80	70	60	77	69	59	74	68	19.4
Stockton	100	69	97	68	94	67	71	96	70	94	68	92	62	83	78	60	78	78	59	75	77	30.4
Victorville, George AFB	101	65	98	65	96	64	69	88	68	88	67	88	65	102	78	61	90	79	59	83	78	28.3
COLORADO																						
Alamosa	84	55	82	55	80	54	60	75	59	74	58	73	55	87	62	54	81	62	52	77	62	31.2
Colorado Springs	90	58	87	58	84	58	63	78	62	77	61	76	59	92	66	57	88	66	56	83	65	24.9
Craig	88	57	85	56	83	55	60	79	59	78	57	77	53	77	66	52	72	65	50	68	64	36.4
Denver	93	60	90	59	87	59	65	81	63	80	62	79	60	96	69	58	90	68	57	85	68	26.9
Eagle	88	58	86	57	83	57	62	80	60	78	59	76	57	88	64	55	82	65	53	76	65	36.1
Grand Junction	96	61	94	60	92	60	65	84	64	83	63	82	60	93	70	58	87	71	56	79	71	26.6
Limon	90	60	88	60	85	59	64	79	63	78	62	77	60	96	67	59	92	66	58	88	66	26.8
Pueblo	97	62	94	62	92	62	67	84	66	83	65	83	63	104	71	62	98	71	60	92	71	29.4
Trinidad	93	61	90	60	87	60	65	84	64	83	63	81	60	96	71	58	91	69	57	86	69	28.3
CONNECTICUT																						
Bridgeport	86	73	84	72	82	71	76	83	74	81	73	79	74	126	79	72	120	78	71	115	77	14.1
Hartford, Brainard Field	91	73	88	72	85	70	76	87	74	84	73	82	73	121	81	71	116	79	70	110	78	20.9
Windsor Locks, Bradley Fld	92	73	88	71	85	70	76	87	74	84	72	82	72	119	81	71	114	79	69	109	77	20.9
DELAWARE																						
Dover, AFB	93	76	89	75	87	74	79	88	78	86	76	84	77	141	84	76	135	82	74	129	81	16.2
Wilmington	91	75	89	74	86	73	78	87	76	85	75	83	75	132	82	74	125	81	72	120	80	17.0
FLORIDA																						
Apalachicola	92	79	90	78	89	78	81	88	80	87	79	87	79	148	85	78	145	84	77	141	84	13.3
Cape Canaveral, NASA	92	78	90	78	89	77	80	88	79	87	79	87	78	145	84	77	141	84	76	138	83	16.0
Daytona Beach	92	77	90	77	88	77	79	88	79	87	78	86	77	141	84	76	137	84	76	134	83	15.4
Ft. Lauderdale/Hollywood	92	78	90	78	89	78	81	88	80	87	79	87	78	147	85	78	145	84	77	141	84	11.3
Ft. Myers	94	77	93	77	92	77	80	88	80	88	79	87	78	147	84	78	144	83	77	140	83	16.9
Gainesville	94	77	92	77	90	76	80	89	79	88	78	87	77	143	84	76	139	83	76	136	82	18.7
Homestead, AFB	92	79	90	79	89	78	81	89	80	88	80	87	79	150	87	78	145	86	77	141	85	11.7
Jacksonville, Cecil Fld NAS	96	76	95	76	93	76	79	91	78	90	77	89	76	138	84	75	134	83	75	130	82	20.0
Jacksonville, Intl Airport	94	77	93	77	91	77	80	90	79	89	78	88	77	142	85	76	138	84	76	134	83	17.8
Jacksonville, Mayport Naval	95	78	92	78	90	77	81	89	80	89	79	88	78	147	86	77	142	85	77	139	85	15.3
Key West	90	79	89	79	89	79	81	87	80	87	80	87	79	149	85	78	146	85	77	143	85	8.1
Melbourne	93	79	91	79	89	79	82	89	81	88	80	87	80	155	86	79	150	85	78	146	85	15.3
Miami, Intl Airport	91	77	90	77	89	77	80	87	79	87	78	86	78	144	83	77	141	83	76	138	82	11.4
Miami, New Tamiami A	92	78	91	78	90	77	80	89	79	88	79	87	78	145	83	77	141	83	76	138	83	15.5
Milton, Whiting Field NAS	95	78	93	77	92	76	81	90	80	89	79	88	78	148	86	77	143	85	76	138	84	18.5
Orlando	94	76	93	76	92	76	79	88	79	88	78	87	77	142	83	76	139	82	76	136	81	16.6
Panama City, Tyndall AFB	91	79	89	79	88	79	83	88	82	87	81	86	81	160	86	80	154	85	79	150	84	12.2
Pensacola, Sherman AFB	93	78	92	78	90	78	81	89	80	88	79	88	79	150	85	78	144	85	76	138	85	15.3
Saint Petersburg	94	80	93	79	92	79	82	90	82	89	81	88	80	156	86	80	153	85	79	150	85	13.5
Sarasota/Bradenton	93	80	92	79	90	79	82	90	81	89	80	88	79	153	87	79	148	86	78	146	86	15.8
Tallahassee	95	77	93	76	91	76	80	89	79	88	78	87	77	142	83	76	138	82	76	135	82	18.5
Tampa, Intl Airport	92	77	91	77	90	77	80	88	79	88	78	87	78	144	85	77	140	84	76	137	83	15.0
Valparaiso, Eglin AFB	92	78	90	78	89	77	81	88	80	87	79	86	79	149	85	78	144	84	77	141	83	13.9
Vero Beach	92	77	90	78	89	77	80	88	79	88	79	87	77	141	85	77	139	84	76	137	84	15.7

DP = dew-point temperature, °F
MDB = mean coincident dry-bulb temp., °F
StdD = standard deviation, °F

MWB = mean coincident wet-bulb temp., °F
MWS = mean coincident wind speed, mph
HR = humidity ratio, grains of moisture per lb of dry air

A = airport
ANGB = Air National Guard Base
MCAF = Marine Corps Air Facility

NAF = Naval Air Facility
NAWS = Naval Air Weapons Station
RAF = Royal Air Force

United States

Station	Cooling DB/MWB 0.4% DB	MWB	1% DB	MWB	2% DB	MWB	Evaporation WB/MDB 0.4% WB	MDB	1% WB	MDB	2% WB	MDB	Dehumidification DP/MDB and HR 0.4% DP	HR	MDB	1% DP	HR	MDB	2% DP	HR	MDB	Range of DB
1	2a	2b	2c	2d	2e	2f	3a	3b	3c	3d	3e	3f	4a	4b	4c	4d	4e	4f	4g	4h	4i	5
West Palm Beach	91	78	90	78	89	77	80	88	79	88	78	87	77	143	84	77	139	84	76	137	83	13.1
GEORGIA																						
Albany	96	76	95	76	93	75	79	90	78	89	78	88	77	141	83	76	136	82	75	133	81	19.8
Athens	94	75	92	75	90	74	78	89	77	87	76	86	75	133	82	74	129	81	73	125	80	18.4
Atlanta	93	75	91	74	88	73	77	88	76	87	75	85	74	133	82	73	128	81	72	124	80	17.3
Augusta	96	76	94	76	92	75	79	91	78	89	77	88	76	135	84	75	130	83	74	127	82	20.2
Brunswick	93	78	91	79	88	78	81	89	80	88	79	87	78	147	86	78	144	85	77	141	84	14.4
Columbus, Ft. Benning	97	76	94	76	92	76	80	91	79	89	78	88	77	142	85	76	136	83	75	133	82	20.5
Columbus, Metro Airport	95	76	93	75	91	75	79	90	78	88	77	87	76	139	82	75	134	82	74	130	81	18.0
Macon	96	76	94	75	92	75	79	91	78	89	77	88	76	136	83	75	132	82	74	129	82	19.3
Marietta, Dobbins AFB	94	74	91	74	89	74	77	88	76	87	75	86	74	134	82	73	130	81	72	123	79	17.1
Rome	96	74	94	74	91	74	78	90	77	89	76	88	75	134	83	74	130	83	73	127	83	20.7
Savannah	95	77	93	76	91	76	79	90	78	89	78	87	77	139	84	76	135	83	75	132	82	17.5
Valdosta, Moody AFB	95	77	94	77	92	76	80	91	79	89	78	88	77	142	85	76	139	84	76	135	83	17.8
Valdosta, Regional Airport	95	77	94	76	92	76	80	90	79	89	78	88	77	144	83	76	139	82	76	136	82	19.4
Waycross	96	76	94	76	93	75	78	91	78	90	77	89	75	134	84	75	130	83	74	127	83	20.3
HAWAII																						
Ewa, Barbers Point NAS	92	73	90	72	89	72	76	86	75	86	75	85	74	126	83	72	118	82	71	113	82	15.8
Hilo	85	74	84	74	83	73	76	82	76	81	75	81	75	130	79	74	127	79	73	123	78	13.3
Honolulu	89	73	88	73	87	73	76	84	75	84	74	84	74	125	80	72	120	80	71	116	79	12.2
Kahului	89	74	88	74	87	73	76	85	76	85	75	84	74	127	80	73	122	80	72	118	80	15.6
Kaneohe, MCAS	86	75	85	74	84	74	78	82	77	82	76	82	76	138	81	75	133	81	74	128	80	7.4
Lihue	85	75	85	74	84	74	77	83	76	82	75	82	75	132	80	74	128	80	73	125	79	9.6
Molokai	88	73	87	73	86	72	76	85	75	83	74	83	74	128	80	73	124	79	71	118	79	13.3
IDAHO																						
Boise	96	63	94	63	91	62	66	90	64	89	63	87	58	79	72	55	72	71	53	67	71	30.3
Burley	94	63	90	62	87	61	67	86	65	84	63	83	60	90	75	58	84	72	56	78	72	29.0
Idaho Falls	92	61	89	60	86	60	64	84	63	82	61	81	58	88	71	56	81	69	54	73	68	34.0
Lewiston	97	65	93	64	90	63	67	91	65	89	64	87	58	76	72	56	71	71	54	65	71	26.5
Mountain Home, AFB	99	63	96	62	93	61	66	91	64	91	63	89	58	79	71	54	70	69	52	64	71	32.8
Mullan	87	62	84	61	80	60	65	81	63	79	62	77	60	86	69	58	80	68	56	75	66	28.1
Pocatello	93	61	90	60	87	59	64	84	62	83	61	82	57	83	70	55	76	70	53	70	69	32.1
ILLINOIS																						
Belleville, Scott AFB	95	78	93	77	90	76	80	92	78	90	77	88	77	141	87	76	136	85	74	131	84	19.8
Chicago, Meigs Field	92	74	89	73	86	71	77	88	76	85	74	83	74	132	84	72	121	80	71	115	80	16.0
Chicago, O'Hare Intl A	91	74	88	73	86	71	77	88	75	85	74	83	74	130	84	72	123	82	71	115	80	19.6
Decatur	94	76	91	75	88	74	79	90	78	89	76	86	76	140	86	75	133	84	73	127	83	20.0
Glenview, NAS	93	75	89	73	87	71	78	90	76	87	74	84	74	130	85	72	120	82	70	113	81	17.6
Marseilles	93	74	89	73	86	71	78	89	76	86	74	84	75	135	85	73	126	82	71	117	81	19.4
Moline/Davenport IA	93	76	90	74	87	73	78	90	77	87	75	85	75	134	85	73	127	83	72	120	82	20.0
Peoria	92	76	89	74	86	73	78	89	77	86	75	84	75	137	85	74	130	83	72	123	81	19.5
Quincy	94	76	91	75	88	74	78	89	77	88	76	85	76	138	84	74	132	82	73	126	82	18.9
Rockford	91	74	88	73	85	71	77	87	75	85	74	82	74	132	84	73	124	81	71	116	79	19.8
Springfield	93	76	91	75	88	74	79	89	77	88	76	85	76	139	86	75	132	84	73	125	82	19.4
West Chicago	91	75	88	74	86	72	78	88	76	85	74	83	76	138	85	74	130	83	71	119	80	19.8
INDIANA																						
Evansville	94	77	92	76	90	75	79	90	78	89	77	87	76	137	86	75	132	84	73	126	83	19.8
Ft. Wayne	90	74	88	73	85	71	77	86	75	84	74	82	74	131	83	72	124	81	71	117	79	19.9
Indianapolis	91	75	88	74	86	73	78	88	77	87	75	83	75	137	84	74	131	82	73	125	81	18.9
Lafayette, Purdue Univ	93	75	90	75	88	73	79	89	77	86	75	84	76	139	85	74	132	83	73	125	82	20.9
Peru, Grissom AFB	93	75	89	75	87	73	79	90	77	86	75	83	76	142	85	75	134	83	73	127	81	18.5
South Bend	90	73	87	72	85	71	77	86	75	84	73	81	74	130	83	72	123	80	71	116	78	18.6
Terre Haute	93	76	90	76	88	75	80	89	78	87	76	85	77	144	86	76	136	84	74	131	82	19.6
IOWA																						
Burlington	94	76	91	76	88	73	78	89	77	88	76	85	75	136	85	74	131	83	72	124	82	18.7
Cedar Rapids	93	75	89	74	86	72	78	89	76	86	74	84	75	136	84	74	129	83	71	120	80	20.0
Des Moines	93	76	90	74	87	73	78	89	76	87	75	85	74	133	85	73	126	83	71	120	81	18.5
Ft. Dodge	92	75	88	73	86	71	77	88	75	86	74	83	74	133	84	72	123	82	70	116	79	18.9
Lamoni	96	74	92	74	89	72	77	89	76	87	75	85	74	134	83	73	127	82	71	120	80	18.9
Mason City	91	74	88	73	85	71	77	87	75	85	74	82	75	135	84	73	126	82	71	118	80	20.8
Ottumwa	95	75	92	75	88	73	78	90	76	89	75	86	75	136	84	74	130	83	72	121	81	18.7
Sioux City	94	75	90	74	88	72	78	89	76	87	75	85	75	135	86	73	127	84	71	120	82	20.4
Spencer	91	75	88	73	85	71	77	88	75	86	73	82	74	134	84	72	123	82	70	117	79	20.2
Waterloo	91	75	88	73	85	71	77	87	75	85	74	83	74	132	84	72	124	82	71	117	80	20.0
KANSAS																						
Concordia	100	73	96	72	93	72	77	90	76	89	74	88	74	133	84	72	123	82	70	118	81	22.5
Dodge City	100	70	97	70	94	69	74	90	73	89	71	88	70	120	79	68	114	78	67	109	77	24.3
Ft Riley, Marshall AAF	99	75	96	74	93	74	78	90	77	90	75	88	75	136	86	73	130	83	71	120	82	22.7
Garden City	100	69	97	69	94	69	73	89	72	89	71	88	69	118	79	67	113	78	66	108	77	27.5
Goodland	97	66	94	66	91	65	70	86	69	84	68	84	66	111	74	65	106	73	63	100	73	26.5

DP = dew-point temperature, °F   MWB = mean coincident wet-bulb temp., °F   A = airport   NAF = Naval Air Facility
MDB = mean coincident dry-bulb temp., °F   MWS = mean coincident wind speed, mph   ANGB = Air National Guard Base   NAWS = Naval Air Weapons Station
StdD = standard deviation, °F   HR = humidity ratio, grains of moisture per lb of dry air   MCAF = Marine Corps Air Facility   RAF = Royal Air Force

United States

Station	Cooling DB/MWB						Evaporation WB/MDB						Dehumidification DP/MDB and HR									Range of DB
	0.4%		1%		2%		0.4%		1%		2%		0.4%			1%			2%			
	DB	MWB	DB	MWB	DB	MWB	WB	MDB	WB	MDB	WB	MDB	DP	HR	MDB	DP	HR	MDB	DP	HR	MDB	
1	2a	2b	2c	2d	2e	2f	3a	3b	3c	3d	3e	3f	4a	4b	4c	4d	4e	4f	4g	4h	4i	5
Russell	100	72	96	72	94	72	76	91	75	90	73	88	72	126	83	71	120	82	69	116	80	24.1
Salina	101	74	97	73	94	73	77	92	76	90	75	89	74	132	85	72	123	83	71	118	82	23.0
Topeka	96	75	93	75	90	75	79	90	78	89	76	88	76	139	87	74	132	85	73	126	83	20.3
Wichita, Airport	100	73	97	73	94	73	77	91	76	90	74	89	73	129	83	72	123	82	71	118	81	22.2
Wichita, McConnell AFB	100	73	97	73	94	73	77	92	76	90	75	89	74	133	84	72	124	83	71	119	82	21.8
KENTUCKY																						
Bowling Green	94	76	91	75	88	74	78	89	77	87	76	86	76	136	84	75	132	82	74	127	81	20.0
Covington/Cincinnati A	91	74	89	73	86	72	77	87	76	86	74	83	74	132	84	73	126	81	72	120	80	18.9
Ft. Campbell, AAF	95	77	93	76	90	76	80	90	78	89	77	87	77	143	85	76	136	84	74	132	83	19.4
Ft. Knox, Godman AAF	94	76	92	74	89	74	78	90	77	88	76	86	76	138	85	74	132	83	73	126	82	19.4
Jackson	90	74	87	73	85	72	77	87	76	85	74	83	74	135	83	73	130	81	71	122	79	18.2
Lexington	91	74	89	73	86	72	77	87	75	86	74	83	74	130	83	72	124	81	71	120	80	18.4
Louisville	93	76	90	75	88	74	78	90	77	88	76	86	75	134	85	74	129	84	73	125	82	18.2
Paducah	96	77	93	76	92	75	80	91	79	90	78	88	77	143	86	76	138	85	75	132	83	20.2
LOUISIANA																						
Alexandria, England AFB	95	78	94	78	92	77	81	90	80	90	79	89	78	147	86	77	142	85	76	138	85	18.4
Baton Rouge	94	78	92	77	91	77	80	89	79	88	78	87	78	145	84	77	141	84	76	137	83	16.7
Bossier City, Barksdale AFB	96	77	94	77	93	77	80	90	79	90	78	89	77	144	84	76	139	83	76	134	83	20.0
Lafayette	94	78	93	78	91	77	80	89	80	89	79	88	78	146	84	77	143	83	77	140	83	17.1
Lake Charles	93	78	91	78	90	77	80	88	80	88	79	87	78	148	84	78	145	84	77	141	84	16.2
Leesville, Ft. Polk	95	77	94	76	92	76	79	89	79	88	78	87	77	144	83	76	140	82	76	136	82	18.2
Monroe	96	78	94	78	93	77	81	91	80	90	79	89	78	147	86	77	143	85	77	139	84	19.3
New Orleans, Intl Airport	93	79	92	78	90	78	81	90	80	88	80	87	79	151	86	78	146	85	77	142	84	15.5
New Orleans, Lakefront A	93	78	92	78	90	77	81	88	80	87	79	87	79	150	85	78	145	84	77	141	83	11.9
Shreveport	97	77	95	77	93	76	79	91	79	90	78	89	76	139	84	76	135	83	75	132	83	19.1
MAINE																						
Augusta	87	71	84	69	80	67	73	83	71	80	69	77	70	113	77	68	106	75	67	100	74	18.4
Bangor	87	71	84	69	81	67	73	83	71	81	69	77	70	111	78	68	104	75	67	99	73	20.5
Brunswick, NAS	87	71	84	69	80	67	73	83	71	80	70	77	70	111	78	69	105	76	67	100	74	19.1
Caribou	85	69	82	67	79	66	72	81	70	77	68	76	70	112	76	68	104	75	66	97	72	19.5
Limestone, Loring AFB	84	68	80	66	78	64	71	79	69	76	67	74	68	107	75	67	101	72	65	94	71	18.7
Portland	86	71	83	70	80	68	74	83	72	80	70	77	71	114	79	69	107	76	67	101	74	18.7
MARYLAND																						
Camp Springs, Andrews AFB	94	75	91	74	88	73	78	88	77	87	75	85	75	134	83	74	129	82	73	124	80	18.7
Baltimore, BWI Airport	93	75	91	74	88	73	78	88	76	86	75	85	75	132	83	74	125	81	72	120	80	18.8
Lex Park, Patuxent River NAS	93	76	90	75	87	74	79	88	77	87	76	85	76	136	84	75	131	83	74	125	82	15.8
Salisbury	93	77	90	76	88	75	80	88	78	86	77	85	78	144	84	76	137	82	75	132	81	18.7
MASSACHUSETTS																						
Boston	91	73	87	71	84	70	75	87	74	83	72	81	72	119	80	71	113	79	69	108	78	15.3
East Falmouth, Otis ANGB	85	72	82	72	79	69	75	81	74	78	72	76	74	125	78	72	118	76	71	113	75	14.6
S. Weymouth NAS	92	73	87	72	85	71	77	87	75	84	73	81	74	129	82	72	118	79	70	111	78	19.6
Worcester	85	71	83	69	80	68	74	82	72	80	70	77	71	119	78	69	112	76	68	105	75	16.6
MICHIGAN																						
Alpena	87	71	84	69	81	67	74	83	72	81	70	78	71	116	79	69	107	76	67	100	74	22.9
Detroit, Metro	90	73	87	72	84	70	76	86	74	84	73	81	73	125	83	71	118	80	70	111	78	20.4
Flint	88	73	86	71	83	70	75	84	74	82	72	80	73	125	81	71	116	78	69	110	77	20.6
Grand Rapids	89	73	86	71	84	70	76	85	74	83	72	81	73	126	81	71	118	79	70	112	77	20.7
Hancock	86	71	83	69	80	67	73	82	71	80	70	77	70	116	79	69	109	76	67	103	74	20.6
Harbor Beach	90	71	86	69	83	68	74	86	72	83	70	80	70	113	82	68	106	80	67	100	78	14.4
Jackson	88	74	86	73	84	71	77	86	75	83	73	81	74	134	83	72	123	81	71	117	78	20.3
Lansing	89	73	86	72	84	70	76	85	74	83	73	81	73	127	81	72	120	79	70	114	78	21.7
Marquette, Sawyer AFB	86	69	83	68	79	65	72	83	70	79	68	75	69	113	77	67	106	74	66	99	73	22.1
Marquette/Ishpeming, A	85	69	82	67	78	65	72	82	70	78	68	75	69	111	77	67	104	75	65	98	72	22.1
Mount Clemens, ANGB	90	74	87	72	84	71	77	87	75	83	73	80	74	131	83	72	120	81	70	113	78	19.6
Muskegon	85	71	83	70	81	69	75	82	73	80	71	78	72	122	80	70	115	77	69	109	76	18.1
Oscoda, Wurtsmith AFB	89	72	86	71	83	69	75	86	73	83	71	79	72	120	80	70	112	79	68	106	77	21.4
Pellston	87	71	85	69	81	68	74	83	72	81	70	78	70	115	78	69	108	76	67	103	75	23.9
Saginaw	90	74	87	72	84	70	77	86	75	84	73	81	74	132	83	72	120	80	70	112	78	21.2
Sault Ste. Marie	83	69	80	68	77	66	72	80	70	77	68	74	69	111	76	67	103	74	65	95	72	21.9
Seul Choix Point	78	66	76	65	74	64	70	76	68	72	66	71	68	106	74	67	101	72	65	94	70	13.9
Traverse City	89	71	86	70	83	68	74	84	72	82	70	80	71	117	80	69	109	78	67	103	76	22.0
MINNESOTA																						
Alexandria	89	72	86	70	83	69	75	86	73	82	71	80	72	123	82	70	116	79	68	109	77	19.3
Brainerd, Pequot Lakes	88	70	85	68	81	66	72	85	70	82	68	78	68	108	81	66	102	77	65	96	75	21.6
Duluth	84	69	81	67	78	65	72	81	69	78	67	75	68	110	77	66	102	75	64	94	72	20.2
Hibbing	85	70	81	68	78	66	73	82	71	78	68	75	70	116	78	68	108	76	66	101	73	23.2
International Falls	86	69	83	67	80	66	72	82	70	79	68	77	69	112	78	67	103	75	65	96	73	21.8
Minneapolis-St. Paul	91	73	88	71	85	70	76	88	74	84	72	82	73	124	83	71	116	81	69	109	79	19.1
Redwood Falls	92	74	88	72	86	70	77	89	75	85	73	82	75	135	83	72	123	81	70	116	80	20.7
Rochester	88	72	85	71	82	70	76	85	74	82	72	80	73	128	81	71	120	79	69	111	77	19.7

DP = dew-point temperature, °F
MDB = mean coincident dry-bulb temp., °F
StdD = standard deviation, °F
MWB = mean coincident wet-bulb temp., °F
MWS = mean coincident wind speed, mph
HR = humidity ratio, grains of moisture per lb of dry air
A = airport
ANGB = Air National Guard Base
MCAF = Marine Corps Air Facility
NAF = Naval Air Facility
NAWS = Naval Air Weapons Station
RAF = Royal Air Force

United States

Station	Cooling DB/MWB						Evaporation WB/MDB						Dehumidification DP/MDB and HR									Range of DB
	0.4%		1%		2%		0.4%		1%		2%		0.4%			1%			2%			
	DB	MWB	DB	MWB	DB	MWB	WB	MDB	WB	MDB	WB	MDB	DP	HR	MDB	DP	HR	MDB	DP	HR	MDB	
1	2a	2b	2c	2d	2e	2f	3a	3b	3c	3d	3e	3f	4a	4b	4c	4d	4e	4f	4g	4h	4i	5
Saint Cloud	91	72	88	71	85	70	76	87	74	84	72	82	72	125	83	70	116	80	68	109	78	21.5
Tofte	79	64	75	62	71	61	66	74	64	72	63	70	64	92	70	61	83	69	59	77	68	13.0
MISSISSIPPI																						
Biloxi, Keesler AFB	92	79	91	78	89	78	81	89	80	88	80	87	79	151	86	78	147	85	78	144	84	13.0
Columbus, AFB	96	77	94	76	92	76	80	91	78	89	78	88	77	141	85	76	136	83	75	132	82	19.3
Greenwood	96	78	94	78	93	77	81	91	80	90	79	89	78	148	86	77	143	85	76	139	84	19.1
Jackson	95	77	93	76	92	76	80	90	79	89	78	88	77	142	84	76	138	83	75	134	82	19.2
McComb	94	76	92	76	91	76	79	89	78	88	78	87	77	141	83	76	138	82	75	135	81	19.8
Meridian	96	77	94	76	92	76	79	91	78	90	77	88	76	139	84	75	134	83	74	130	83	20.3
Tupelo	96	76	94	76	92	75	79	89	78	89	77	88	76	137	83	75	134	83	74	131	82	18.9
MISSOURI																						
Cape Girardeau	96	77	94	77	91	76	80	92	78	90	78	88	77	141	86	76	136	85	75	132	83	19.8
Columbia	95	75	92	75	89	74	78	89	77	88	75	86	75	137	85	74	130	83	72	124	82	20.3
Joplin	96	75	94	75	91	74	78	90	77	89	76	88	75	137	85	74	132	85	72	125	83	20.0
Kansas City	96	75	93	75	90	74	78	90	77	89	76	87	75	137	85	74	130	84	73	125	83	18.8
Poplar Bluff	95	77	92	76	90	76	80	90	78	88	77	87	77	144	85	76	138	83	75	133	82	20.0
Spickard/Trenton	96	74	93	73	89	72	78	88	76	88	75	86	76	139	83	73	128	83	71	118	81	19.6
Springfield	95	74	92	74	89	74	78	89	76	88	75	86	74	134	84	73	128	83	72	124	81	20.8
St. Louis, Intl Airport	95	76	93	76	90	74	79	90	78	88	76	87	76	138	85	75	132	83	73	127	82	18.3
Warrensburg, Whiteman AFB	96	76	93	76	90	75	79	91	78	90	76	88	76	139	86	75	134	85	73	128	83	19.3
MONTANA																						
Billings	93	63	90	62	87	61	65	86	64	84	62	83	59	83	71	57	78	71	55	74	70	25.8
Bozeman	91	61	87	60	85	59	64	83	62	82	61	81	58	83	69	56	78	67	53	71	66	31.7
Butte	86	57	84	56	80	55	60	76	58	76	57	76	54	76	61	52	70	63	50	66	62	31.5
Cut Bank	87	60	84	59	80	58	62	81	60	79	59	77	56	77	67	54	70	65	51	65	63	26.1
Glasgow	94	64	90	63	86	62	68	85	66	83	64	82	62	91	74	60	83	71	58	77	69	25.3
Great Falls, Intl Airport	92	61	88	60	85	59	64	84	62	82	61	81	57	81	69	55	74	67	53	69	66	27.2
Great Falls, Malmstrom AFB	93	62	89	61	86	60	65	85	63	83	62	81	59	84	71	57	78	69	54	71	68	26.3
Havre	94	63	90	62	86	61	66	87	64	84	62	82	60	84	72	58	78	69	56	74	68	27.9
Helena	90	60	87	59	84	59	63	82	61	81	60	80	57	80	68	55	73	66	52	68	66	28.0
Kalispell	89	62	86	61	82	60	65	83	63	81	61	79	59	82	69	57	76	67	55	71	67	29.9
Lewistown	89	61	86	60	82	59	64	81	63	80	61	78	58	85	71	56	79	69	54	74	67	28.3
Miles City	97	66	93	65	90	64	69	89	67	86	66	84	63	95	76	61	88	75	59	82	73	25.9
Missoula	91	62	88	61	85	60	65	84	63	82	62	81	58	82	68	56	76	68	55	71	66	31.3
NEBRASKA																						
Bellevue, Offutt AFB	95	76	91	75	88	74	79	89	77	88	76	86	76	141	85	74	134	83	73	127	82	18.4
Grand Island	97	72	93	72	90	70	76	89	74	88	73	86	72	127	82	70	120	81	69	113	79	22.4
Lincoln	97	74	94	74	91	73	78	90	76	89	75	87	75	136	84	73	130	83	71	121	82	22.3
Norfolk	95	74	92	72	89	72	76	90	75	88	73	86	73	129	83	71	121	82	70	115	81	20.8
North Platte	95	69	92	69	89	68	73	87	72	86	70	85	69	118	80	67	111	78	66	105	77	25.5
Omaha, Eppley Airfield	95	75	92	75	89	73	78	90	77	88	75	86	75	136	85	73	128	84	72	121	82	19.9
Omaha, WSO	94	75	90	75	87	73	77	89	76	87	74	85	74	134	84	72	126	83	71	120	82	17.6
Scottsbluff	95	65	92	64	89	64	69	87	68	85	66	84	64	102	76	62	97	74	60	91	73	28.9
Sidney	95	63	92	63	89	63	67	84	66	84	64	84	62	97	73	60	91	72	58	86	71	27.9
Valentine	97	68	94	67	90	67	72	90	71	89	69	87	67	110	79	65	103	78	63	94	77	26.5
NEVADA																						
Elko	95	60	92	59	90	59	63	85	61	84	60	84	57	84	68	54	75	66	51	67	67	38.4
Ely	89	56	87	56	85	55	60	78	59	78	58	78	55	82	64	53	75	64	50	68	65	34.6
Las Vegas, Intl Airport	108	66	106	66	103	65	71	95	70	93	69	93	65	102	79	63	92	81	60	84	85	24.8
Mercury	102	65	100	64	98	63	69	88	67	89	66	89	64	102	72	60	89	77	58	80	80	25.9
N. Las Vegas, Nellis AFB	108	68	106	67	104	66	72	94	71	94	70	94	67	106	79	64	97	82	61	86	84	26.3
Reno	95	61	92	60	90	59	63	87	62	86	60	85	56	77	69	53	69	69	50	63	68	37.3
Tonopah	94	58	92	57	89	57	62	83	61	82	60	81	56	83	67	53	74	68	50	67	69	31.1
Winnemucca	97	61	94	60	92	59	63	88	62	87	60	86	56	79	68	53	69	67	50	62	68	37.4
NEW HAMPSHIRE																						
Concord	90	71	87	70	84	68	74	85	73	82	71	79	71	118	79	70	111	77	68	105	76	24.1
Lebanon	88	71	86	69	83	68	74	84	72	82	70	79	70	113	79	69	108	77	67	103	75	23.0
Mount Washington	60	56	58	54	56	54	58	59	56	57	54	56	58	90	58	56	84	57	54	78	55	8.5
Portsmouth, Pease AFB	89	72	85	70	83	70	75	84	73	82	72	79	73	123	85	71	113	77	69	106	76	18.2
NEW JERSEY																						
Atlantic City	91	74	88	73	86	72	77	87	76	84	75	82	75	131	81	74	125	80	72	120	79	18.1
Millville	92	75	89	74	87	73	78	87	76	86	75	83	75	134	81	74	129	80	73	125	80	18.7
Newark	93	74	90	73	87	71	77	88	76	85	74	83	74	127	81	73	121	80	71	116	80	15.9
Teterboro	92	74	89	74	87	73	78	88	77	87	75	83	76	134	84	74	128	82	72	119	81	18.4
Trenton, McGuire AFB	93	75	90	74	87	73	78	89	76	87	75	84	75	132	83	74	127	82	72	118	80	18.9
NEW MEXICO																						
Alamogordo, Holloman AFB	98	63	96	63	93	63	68	87	67	85	67	85	65	106	72	62	98	72	61	92	73	30.2
Albuquerque	96	60	93	60	91	60	65	83	64	82	64	81	61	98	68	60	93	69	58	89	69	25.4
Carlsbad	101	65	98	66	96	66	72	88	71	87	70	85	69	121	76	68	116	76	67	111	75	25.4
Clayton	94	62	91	62	88	62	67	84	65	84	65	82	61	98	72	60	94	71	59	90	70	26.1

DP = dew-point temperature, °F   MWB = mean coincident wet-bulb temp., °F   A = airport   NAF = Naval Air Facility
MDB = mean coincident dry-bulb temp., °F   MWS = mean coincident wind speed, mph   ANGB = Air National Guard Base   NAWS = Naval Air Weapons Station
StdD = standard deviation, °F   HR = humidity ratio, grains of moisture per lb of dry air   MCAF = Marine Corps Air Facility   RAF = Royal Air Force

TABLE 5.12   COOLING DESIGN CONDITIONS FOR THE UNITED STATES AND CANADA. (CONTINUED)

United States

Station	Cooling DB/MWB 0.4%		1%		2%		Evaporation WB/MDB 0.4%		1%		2%		Dehumidification DP/MDB and HR 0.4%			1%			2%			Range of DB
	DB	MWB	DB	MWB	DB	MWB	WB	MDB	WB	MDB	WB	MDB	DP	HR	MDB	DP	HR	MDB	DP	HR	MDB	
1	2a	2b	2c	2d	2e	2f	3a	3b	3c	3d	3e	3f	4a	4b	4c	4d	4e	4f	4g	4h	4i	5
Clovis, Cannon AFB	96	64	93	64	91	65	70	84	69	83	68	83	66	114	75	65	109	74	64	105	73	24.5
Farmington	94	60	92	60	89	60	65	83	64	83	63	82	60	94	69	58	90	68	57	85	69	28.8
Gallup	89	57	87	56	85	56	62	76	61	76	60	75	59	94	65	57	90	64	56	85	64	30.6
Roswell	98	65	96	65	94	65	70	87	69	86	68	85	66	111	73	65	108	73	64	104	73	24.8
Truth Or Consequences	97	61	95	61	93	61	66	85	65	85	64	84	60	94	71	59	90	71	58	87	72	25.0
Tucumcari	98	64	95	65	93	64	69	87	68	85	67	84	65	109	73	64	104	73	63	100	72	24.9
NEW YORK																						
Albany	90	71	86	70	84	69	74	85	73	82	71	79	72	118	79	70	111	77	68	106	76	23.7
Binghamton	85	70	82	69	80	67	73	81	71	79	70	77	70	118	77	69	111	75	67	106	74	17.5
Buffalo	86	70	84	69	81	68	74	82	72	80	71	78	71	118	78	70	113	77	68	106	75	17.7
Central Islip	88	73	85	72	83	70	76	83	75	81	74	79	74	129	79	73	124	78	71	116	77	15.1
Elmira/Corning	90	72	87	71	84	69	75	86	73	82	72	80	72	122	81	70	116	78	69	110	76	24.1
Glens Falls	88	73	85	71	83	70	76	85	74	82	72	80	74	127	81	71	116	79	69	108	76	22.1
Massena	87	72	84	71	82	69	75	84	73	81	71	79	72	118	80	70	111	78	68	105	76	21.8
New York, JFK Airport	91	74	88	72	85	71	76	86	75	84	74	82	74	125	80	72	120	80	71	114	79	13.9
New York, La Guardia A	92	74	89	73	86	72	77	87	76	85	74	83	74	128	81	73	125	80	71	116	80	14.6
Newburgh	88	74	85	72	83	70	76	85	74	83	73	80	74	130	82	72	119	80	70	111	78	17.1
Niagara Falls	87	72	85	71	83	69	75	84	74	81	72	79	73	125	81	71	116	78	69	111	76	18.9
Plattsburgh, AFB	86	71	83	69	80	68	74	82	72	80	70	78	71	115	79	69	108	76	67	102	75	19.6
Poughkeepsie	92	75	88	72	85	71	76	87	75	85	73	82	74	126	82	71	116	80	70	111	78	23.0
Rochester	89	73	86	71	83	70	75	85	74	82	72	80	73	123	81	71	116	79	69	109	77	20.1
Rome, Griffiss AFB	88	71	86	70	83	69	74	84	73	82	71	79	71	117	80	70	111	78	68	105	76	22.9
Syracuse	88	72	85	71	83	70	75	85	73	82	72	80	72	120	80	70	113	78	69	107	77	20.3
Watertown	85	71	83	70	80	69	74	82	72	80	71	77	71	118	78	70	111	77	69	106	75	20.5
White Plains	89	74	87	72	84	71	76	86	75	83	73	80	74	128	80	72	120	80	71	114	78	18.0
NORTH CAROLINA																						
Asheville	88	72	85	71	83	70	75	84	73	82	72	80	72	128	79	71	123	78	70	118	76	19.4
Cape Hatteras	88	78	86	77	85	77	80	86	79	84	78	83	78	147	83	77	142	83	76	138	82	11.4
Charlotte	94	74	91	74	89	73	77	88	76	87	75	86	74	130	82	73	125	80	72	122	80	17.8
Cherry Point, MCAS	95	79	92	78	90	77	81	91	80	90	79	88	78	146	87	77	141	86	76	136	85	16.6
Fayetteville, Ft. Bragg	96	77	94	76	92	75	79	91	78	89	77	88	76	139	84	76	135	83	75	131	83	18.2
Goldsboro, Johnson AFB	96	77	94	76	91	76	80	91	78	89	77	87	76	139	84	76	135	83	75	132	82	18.4
Greensboro	92	75	90	74	88	73	77	88	76	86	75	85	74	132	82	73	127	81	72	123	80	18.5
Hickory	94	73	91	72	88	72	76	87	75	85	74	84	74	133	80	73	128	80	71	120	78	19.6
Jacksonville, New River MCAF	94	79	92	78	89	77	81	90	79	89	78	87	78	145	86	77	140	85	76	136	84	17.1
New Bern	94	78	92	78	90	76	81	91	79	89	78	87	78	144	86	77	139	84	76	134	83	17.1
Raleigh/Durham	93	76	90	75	88	74	78	88	77	87	76	85	75	134	82	74	130	81	73	125	80	18.8
Wilmington	93	79	91	78	89	77	80	89	79	88	78	86	78	146	85	77	141	83	76	137	83	15.7
Winston-Salem	92	74	89	74	87	73	77	86	76	86	75	85	74	134	81	73	129	80	72	121	79	17.6
NORTH DAKOTA																						
Bismarck	93	68	90	67	86	66	72	86	70	84	68	82	68	109	79	65	100	77	63	92	75	26.5
Devils Lake	91	69	87	67	84	66	72	86	70	83	68	80	68	108	78	66	100	77	63	90	75	21.1
Fargo	91	71	88	70	85	69	75	86	73	84	71	81	72	122	82	69	112	80	67	104	77	22.3
Grand Forks, AFB	91	71	88	69	85	68	75	86	72	83	70	80	71	118	81	69	109	78	67	101	76	22.9
Minot, AFB	94	68	90	67	86	66	72	87	70	85	68	82	68	109	80	65	100	78	63	90	75	24.7
Minot, Intl Airport	92	67	88	66	84	65	71	85	69	83	67	80	67	106	78	64	96	75	62	89	73	22.9
Williston	96	67	92	66	87	65	71	87	69	86	67	83	66	103	78	63	92	76	61	85	73	25.7
OHIO																						
Akron/Canton	88	72	85	71	83	70	75	84	73	82	72	80	72	125	80	71	118	78	69	113	77	18.8
Cincinnati, Lunken Field	93	74	90	75	88	73	77	89	76	87	75	84	75	132	82	74	128	81	72	120	80	20.0
Cleveland	89	73	86	72	84	71	76	85	74	83	72	81	73	125	82	71	118	80	70	112	78	18.6
Columbus, Intl Airport	90	74	88	73	86	71	77	87	75	85	74	82	73	128	82	72	123	81	71	117	79	19.3
Columbus, Rickenbckr AFB	92	74	89	73	87	72	77	88	75	86	74	84	74	130	83	72	120	82	70	115	79	19.8
Dayton, Intl Airport	90	74	88	73	86	71	76	87	75	85	74	82	73	129	82	72	123	80	71	117	79	19.2
Dayton, Wright-Paterson	92	74	89	74	87	73	78	88	76	86	74	84	75	136	84	73	127	83	71	119	81	19.8
Findlay	90	74	87	72	85	71	76	86	75	83	73	81	74	132	81	72	121	80	71	116	78	18.9
Mansfield	88	73	85	72	83	71	76	85	74	83	73	80	73	128	81	71	122	79	70	116	78	17.8
Toledo	90	73	87	72	85	71	77	86	75	84	73	81	74	129	82	72	122	80	70	115	78	20.9
Youngstown	88	72	85	70	83	69	74	84	73	82	71	79	72	122	80	70	116	78	69	110	76	20.6
Zanesville	90	74	88	73	86	71	76	87	75	85	74	82	74	130	82	72	120	80	71	116	78	20.7
OKLAHOMA																						
Altus, AFB	102	73	100	73	97	73	77	93	76	92	75	91	74	132	84	72	124	83	71	119	82	23.6
Enid, Vance AFB	101	74	98	74	95	73	77	92	76	91	75	90	73	130	85	71	121	83	70	116	82	21.8
Lawton, Ft. Sill/Post Field	99	73	97	73	95	73	77	90	76	90	75	89	74	135	83	73	129	82	71	121	81	20.7
McAlester	98	76	96	76	93	76	79	92	78	91	77	89	76	141	85	75	137	83	74	133	83	21.8
Oklahoma City, Tinker AFB	98	74	96	75	94	74	78	92	77	91	76	89	75	138	87	74	132	85	72	123	83	19.4
Oklahoma City, W. Rogers A	99	74	96	74	94	73	77	91	76	90	75	89	73	129	83	72	125	82	71	120	81	21.0
Tulsa	100	76	97	76	94	75	79	92	78	92	77	90	76	137	87	74	132	85	73	127	84	19.5
OREGON																						
Astoria	76	64	72	62	69	61	65	75	63	71	62	68	61	81	69	60	76	66	59	74	65	14.2

DP = dew-point temperature, °F
MDB = mean coincident dry-bulb temp., °F
StdD = standard deviation, °F

MWB = mean coincident wet-bulb temp., °F
MWS = mean coincident wind speed, mph
HR = humidity ratio, grains of moisture per lb of dry air

A = airport
ANGB = Air National Guard Base
MCAF = Marine Corps Air Facility

NAF = Naval Air Facility
NAWS = Naval Air Weapons Station
RAF = Royal Air Force

United States

Station	Cooling DB/MWB 0.4%		1%		2%		Evaporation WB/MDB 0.4%		1%		2%		Dehumidification DP/MDB and HR 0.4%			1%			2%			Range of DB
	DB	MWB	DB	MWB	DB	MWB	WB	MDB	WB	MDB	WB	MDB	DP	HR	MDB	DP	HR	MDB	DP	HR	MDB	
1	2a	2b	2c	2d	2e	2f	3a	3b	3c	3d	3e	3f	4a	4b	4c	4d	4e	4f	4g	4h	4i	5
Eugene	91	67	87	65	83	64	69	87	67	84	65	81	62	83	74	60	78	73	59	74	71	27.6
Hillsboro	92	69	88	67	84	65	71	89	68	86	66	82	64	90	79	61	82	75	60	78	72	26.6
Klamath Falls	91	64	87	62	85	61	66	87	64	84	63	81	58	85	74	57	80	73	55	75	71	34.2
Meacham	87	59	84	58	80	57	61	82	59	80	58	78	52	67	66	50	63	64	49	59	64	37.1
Medford	98	67	95	66	91	65	69	94	67	91	66	88	60	81	75	58	76	74	56	71	73	33.7
North Bend	71	60	69	60	67	59	62	69	61	67	60	66	60	76	65	58	73	64	57	70	63	12.8
Pendleton	97	64	93	63	90	62	66	92	64	90	63	87	57	74	71	55	68	69	53	62	70	27.2
Portland	90	67	86	66	83	64	69	87	67	84	65	80	62	83	75	60	78	72	59	75	71	21.6
Redmond	93	62	89	61	86	59	63	88	62	86	60	83	55	71	68	52	65	67	50	60	66	35.0
Salem	92	67	87	66	83	64	68	89	67	85	65	81	61	81	75	59	76	73	58	72	71	27.9
Sexton Summit	83	60	80	59	77	58	62	80	61	77	59	74	55	76	70	53	69	68	52	66	66	18.9
PENNSYLVANIA																						
Allentown	90	73	88	72	85	71	76	86	74	84	73	82	73	123	81	71	117	79	70	111	78	19.4
Altoona	89	72	86	70	83	69	74	85	72	83	71	80	71	119	79	69	113	77	68	109	76	19.4
Bradford	83	69	80	68	78	66	72	79	70	77	68	75	69	116	75	68	111	73	66	105	72	21.2
Du Bois	86	70	84	69	81	67	72	81	71	79	70	78	70	116	76	69	112	74	67	108	73	19.4
Erie	85	72	83	70	80	70	74	82	73	80	71	78	72	122	79	70	115	77	69	109	76	15.6
Harrisburg	92	74	89	73	86	72	77	87	76	85	74	83	74	130	82	73	123	80	72	118	79	18.8
Philadelphia, Intl Airport	92	75	89	74	87	73	78	88	77	86	75	84	75	132	83	74	126	81	73	121	80	17.7
Philadelphia, Northeast A	93	76	90	74	88	73	78	88	77	87	75	84	76	135	83	74	129	82	72	121	82	19.1
Philadelphia, Willow Gr NAS	93	75	90	74	88	72	78	89	76	87	75	85	74	131	83	73	125	82	71	116	81	19.4
Pittsburgh, Allegheny Co. A	90	72	87	71	85	70	75	85	74	84	72	81	71	122	79	70	117	78	69	113	77	18.0
Pittsburgh, Intl Airport	89	72	86	70	84	69	74	85	73	82	71	80	71	121	80	70	115	78	68	109	77	19.5
Wilkes-Barre/Scranton	88	71	85	70	83	69	74	83	73	81	71	79	71	120	79	70	115	77	69	109	76	18.8
Williamsport	90	73	87	71	84	70	76	85	74	83	73	80	73	125	82	72	118	78	70	113	77	20.3
RHODE ISLAND																						
Providence	89	73	86	71	83	70	76	85	74	82	73	80	73	124	80	72	118	78	70	112	77	17.4
SOUTH CAROLINA																						
Beaufort, MCAS	95	78	93	78	92	77	80	90	80	89	79	88	78	145	85	77	141	85	76	137	84	16.7
Charleston	94	78	92	77	90	77	80	90	79	88	78	87	78	145	84	77	139	83	76	134	83	16.2
Columbia	96	76	94	75	92	74	78	90	77	89	77	87	75	134	82	75	130	81	74	127	81	19.9
Florence	96	76	94	76	92	76	80	90	78	89	78	88	77	142	85	76	136	83	75	132	82	19.8
Greer/Greenville	93	74	91	74	88	73	77	88	76	87	75	85	74	130	81	73	126	80	72	122	80	18.2
Myrtle Beach, AFB	93	79	90	78	88	78	81	89	80	88	79	87	79	150	87	78	144	86	77	140	84	14.4
Sumter, Shaw AFB	95	76	93	75	90	75	78	89	77	88	76	86	76	136	83	75	132	82	74	129	81	18.5
SOUTH DAKOTA																						
Chamberlain	98	72	94	71	90	70	76	91	74	89	72	87	71	124	84	70	116	82	68	109	80	23.8
Huron	95	72	91	71	88	70	76	89	74	87	72	84	72	126	84	70	117	81	69	110	79	24.1
Pierre	99	70	95	69	91	68	74	90	72	89	71	86	70	116	81	68	109	80	66	102	78	25.6
Rapid City	95	65	91	65	88	64	70	85	68	84	67	82	65	104	76	63	98	75	61	92	73	25.3
Sioux Falls	94	73	90	72	87	71	76	89	75	87	73	84	73	127	84	71	119	82	69	112	80	22.1
TENNESSEE																						
Bristol	89	72	87	72	85	71	75	85	74	84	73	82	72	125	81	71	120	79	70	116	77	19.2
Chattanooga	94	75	92	75	89	74	78	89	77	88	76	86	75	134	82	74	130	82	73	125	81	19.5
Crossville	89	73	87	72	85	72	76	85	74	83	73	82	74	134	80	72	125	79	71	121	78	19.8
Jackson	95	77	93	76	91	76	80	91	78	90	78	88	76	140	85	75	135	85	75	132	84	19.8
Knoxville	92	74	90	74	87	73	77	88	76	86	75	85	74	131	82	73	127	81	72	123	80	18.1
Memphis	96	78	94	77	92	77	80	92	79	91	78	89	77	143	87	76	137	86	75	133	84	16.8
Nashville	94	76	92	75	90	74	78	89	77	88	76	86	75	134	83	74	130	82	73	126	81	19.1
TEXAS																						
Abilene	99	71	97	71	95	71	75	89	74	89	73	88	71	123	81	70	119	80	69	115	79	20.5
Amarillo	96	67	94	66	92	66	71	86	70	86	69	85	67	112	76	65	107	75	64	104	74	23.3
Austin	98	74	96	74	94	74	78	89	77	88	76	87	76	137	81	75	134	80	74	130	80	20.1
Beaumont/Port Arthur	94	79	92	79	91	78	81	90	81	89	80	88	79	152	86	79	148	85	78	145	84	15.9
Beeville, Chase Field NAS	101	77	98	77	96	77	82	91	81	91	80	90	80	155	86	78	148	85	78	144	84	21.6
Brownsville	95	78	94	77	93	77	80	89	79	88	79	88	78	146	83	77	142	83	77	140	82	16.5
College Station/Bryan	98	75	96	75	94	75	79	89	78	89	78	88	77	141	82	76	138	81	75	134	81	21.4
Corpus Christi	95	78	94	78	92	78	81	90	80	89	79	88	79	148	84	78	146	83	77	143	83	16.5
Dallas/Ft. Worth, Intl A	100	74	98	74	96	74	78	92	77	91	76	91	75	132	82	74	130	82	73	126	81	20.3
Del Rio, Laughlin AFB	101	72	98	73	96	72	78	91	77	90	76	89	75	136	82	74	131	82	72	124	81	20.9
El Paso	101	64	98	64	96	64	70	85	69	84	68	84	67	114	73	65	109	73	64	103	74	28.0
Ft. Worth, Carswell AFB	100	75	97	75	96	75	79	92	78	91	77	90	76	141	85	75	135	84	74	130	84	19.3
Ft. Worth, Meacham Field	100	75	98	74	96	74	78	91	77	90	76	89	75	135	83	74	131	82	73	127	82	20.0
Guadalupe Pass	92	61	89	60	87	60	66	82	65	80	64	79	62	102	71	60	96	71	59	91	71	20.9
Houston, Hobby Airport	94	77	93	77	92	77	80	89	80	88	79	87	78	147	84	78	144	83	77	141	82	16.6
Houston, Inter Airport	96	77	94	77	92	77	80	90	79	89	79	88	78	144	83	77	141	83	76	137	83	18.2
Junction	100	72	98	71	96	71	76	89	75	88	74	87	73	130	80	71	121	79	70	118	79	24.8
Killeen, Ft. Hood	98	74	96	73	95	74	78	90	77	89	76	88	75	137	81	74	132	81	73	128	80	21.4
Kingsville, NAS	97	77	96	78	95	78	81	91	80	91	80	90	79	149	85	78	144	84	77	141	84	19.8
Laredo	102	73	101	74	98	74	79	92	78	91	77	89	76	138	82	75	136	81	75	132	81	21.2

DP = dew-point temperature, °F MWB = mean coincident wet-bulb temp., °F A = airport NAF = Naval Air Facility
MDB = mean coincident dry-bulb temp., °F MWS = mean coincident wind speed, mph ANGB = Air National Guard Base NAWS = Naval Air Weapons Station
StdD = standard deviation, °F HR = humidity ratio, grains of moisture per lb of dry air MCAF = Marine Corps Air Facility RAF = Royal Air Force

United States

Station	Cooling DB/MWB 0.4%		Cooling DB/MWB 1%		Cooling DB/MWB 2%		Evaporation WB/MDB 0.4%		Evaporation WB/MDB 1%		Evaporation WB/MDB 2%		Dehumidification DP/MDB and HR 0.4%			Dehumidification DP/MDB and HR 1%			Dehumidification DP/MDB and HR 2%			Range of DB
	DB	MWB	DB	MWB	DB	MWB	WB	MDB	WB	MDB	WB	MDB	DP	HR	MDB	DP	HR	MDB	DP	HR	MDB	
1	2a	2b	2c	2d	2e	2f	3a	3b	3c	3d	3e	3f	4a	4b	4c	4d	4e	4f	4g	4h	4i	5
Lubbock, Intl Airport	97	67	95	67	93	67	73	87	72	86	71	85	69	120	77	68	115	76	67	111	76	22.1
Lubbock, Reese AFB	98	67	95	67	93	67	73	87	72	86	71	85	69	122	78	68	115	77	66	110	77	23.8
Lufkin	97	76	95	77	93	76	79	90	79	89	78	89	77	143	83	76	139	83	75	134	82	20.9
Marfa	94	62	92	61	89	62	68	82	67	81	66	80	65	110	72	63	103	71	62	98	71	31.3
McAllen	100	76	98	76	97	76	80	91	80	90	79	89	78	146	83	77	143	82	77	140	82	20.7
Midland/Odessa	99	67	97	67	95	67	73	87	72	86	71	86	69	120	76	68	115	75	67	111	75	23.7
San Angelo	100	70	97	70	95	70	75	90	74	89	73	88	71	123	80	70	118	79	69	116	78	22.3
San Antonio, Intl Airport	98	73	96	73	94	74	78	87	77	87	76	86	76	139	81	75	135	81	74	132	80	19.1
San Antonio, Kelly AFB	99	74	97	74	96	74	79	89	78	88	77	88	77	145	83	76	140	82	75	136	81	20.5
San Antonio, Randolph AFB	98	74	96	74	94	74	78	90	77	89	76	88	76	138	82	75	134	81	74	132	81	22.3
Sanderson	97	67	95	68	94	68	74	86	73	86	72	86	70	123	79	69	119	78	68	114	77	20.7
Victoria	95	76	94	76	92	77	80	88	79	88	78	87	78	145	83	77	141	82	76	139	82	17.4
Waco	101	75	99	75	97	75	79	93	78	92	77	91	75	135	83	74	131	82	74	127	82	21.6
Wichita Falls, Sheppard AFB	103	74	100	73	98	73	77	93	76	92	75	91	73	129	82	72	124	82	71	120	81	23.9
UTAH																						
Cedar City	93	59	91	59	88	58	64	80	62	80	61	79	59	93	68	57	85	68	55	78	68	28.5
Ogden, Hill AFB	93	61	90	60	87	60	65	83	64	81	62	81	60	91	72	57	83	73	55	77	73	22.0
Salt Lake City	96	62	94	62	92	61	66	85	65	85	64	85	60	92	73	58	84	73	56	77	73	27.7
VERMONT																						
Burlington	87	71	84	69	82	68	74	83	72	81	70	78	71	115	79	69	109	77	67	102	75	20.4
Montpelier/Barre	85	70	83	68	80	67	72	82	70	80	69	77	69	111	78	67	106	75	66	99	73	21.1
VIRGINIA																						
Ft. Belvoir	95	78	93	76	89	75	80	92	78	89	77	87	77	139	86	75	133	85	74	127	83	20.9
Hampton, Langley AFB	94	78	91	77	88	76	80	90	79	89	78	86	77	141	85	76	136	84	75	132	83	14.9
Lynchburg	93	74	90	74	88	73	77	88	76	87	75	85	74	129	81	73	125	80	72	120	79	18.2
Newport News	95	78	92	77	89	76	80	91	78	89	77	87	77	139	84	76	135	83	75	132	82	18.2
Norfolk	93	77	91	76	88	75	79	89	77	89	77	85	76	135	83	75	130	82	74	126	81	15.3
Oceana, NAS	94	77	91	76	88	75	79	89	78	87	77	86	77	139	85	76	134	83	74	129	82	15.7
Quantico, MCAS	94	77	92	76	89	74	79	91	78	89	76	87	76	136	87	75	130	85	73	125	83	18.5
Richmond	94	76	92	75	89	74	79	90	78	88	76	86	76	137	84	75	131	82	74	126	81	19.1
Roanoke	92	73	89	72	87	71	75	88	74	86	73	84	72	123	80	71	118	79	70	115	78	19.6
Sterling	93	75	90	74	88	73	77	88	76	87	75	85	74	130	83	73	125	81	72	120	80	21.0
Washington, National A	95	76	92	76	89	74	79	89	78	88	76	86	76	137	83	75	132	83	74	127	81	16.6
WASHINGTON																						
Bellingham	79	65	76	64	74	62	67	78	65	75	63	72	61	81	73	60	78	70	59	74	67	16.7
Hanford	100	67	96	65	93	64	68	96	66	94	65	90	58	73	72	56	68	75	53	62	74	26.5
Olympia	87	67	83	65	79	64	68	85	66	81	64	78	61	81	73	60	76	71	58	73	69	25.2
Quillayute	80	62	74	61	70	59	64	76	62	72	60	67	60	76	65	58	74	63	57	71	62	15.4
Seattle, Intl Airport	85	65	81	64	78	62	66	83	65	79	63	76	60	78	71	59	74	69	57	71	68	18.3
Spokane, Fairchild AFB	92	62	89	61	85	60	65	86	63	84	61	82	57	77	68	55	71	68	53	67	67	26.1
Stampede Pass	78	57	74	56	71	54	59	74	57	71	56	69	53	70	63	51	65	61	50	62	58	16.0
Tacoma, McChord AFB	86	65	82	63	78	62	67	83	65	80	63	76	60	79	71	59	76	70	58	72	68	22.5
Walla Walla	98	66	95	65	92	64	68	92	67	91	65	88	60	82	74	58	76	72	57	71	72	27.0
Wenatchee	95	67	92	65	88	63	67	91	66	89	64	85	59	78	75	57	73	75	55	68	74	25.2
Yakima	95	65	92	64	88	63	67	90	66	89	64	86	59	78	75	57	71	74	55	67	72	31.1
WEST VIRGINIA																						
Bluefield	85	69	83	69	80	67	72	81	71	79	70	77	69	120	75	68	116	75	67	111	73	16.4
Charleston	91	73	88	73	86	71	76	86	75	85	74	82	73	129	81	72	123	80	71	118	78	19.1
Elkins	85	71	83	70	81	69	73	82	72	80	71	78	71	121	78	69	116	77	68	111	75	21.1
Huntington	91	74	89	73	86	72	77	87	76	85	74	83	74	132	82	73	127	81	72	121	79	19.1
Martinsburg	94	74	91	73	88	72	77	87	75	86	74	85	74	130	81	72	120	80	71	116	79	21.8
Morgantown	89	72	87	71	85	70	75	85	74	83	73	82	72	124	79	71	119	78	70	115	76	20.3
Parkersburg	91	74	88	72	86	72	76	87	75	85	74	82	74	132	82	72	122	80	71	118	78	19.6
WISCONSIN																						
Eau Claire	90	73	87	71	84	70	76	86	74	83	72	81	73	125	82	71	116	80	69	109	78	20.6
Green Bay	88	73	85	72	82	70	76	85	74	82	72	80	73	124	82	71	116	79	69	109	77	20.7
La Crosse	91	74	88	73	85	71	77	87	75	84	74	82	75	132	83	73	125	81	71	117	78	20.1
Madison	90	73	87	72	84	70	76	86	74	84	72	82	73	126	83	71	118	80	69	111	78	21.9
Milwaukee	89	74	86	72	83	70	76	86	74	83	72	81	73	127	83	71	119	80	70	111	78	16.6
Wausau	88	71	85	70	82	69	74	83	72	82	71	78	71	120	79	69	113	77	68	108	75	19.6
WYOMING																						
Big Piney	83	54	80	53	78	52	56	75	55	74	53	74	50	69	60	48	64	60	45	57	59	32.8
Casper	92	59	89	58	86	58	62	81	61	80	60	79	57	85	66	55	78	66	53	73	65	30.4
Cheyenne, Warren AFB	87	58	85	57	82	57	62	77	61	76	60	75	58	90	66	56	85	65	55	80	64	25.7
Cody	91	59	87	58	84	57	61	83	60	81	58	80	54	76	70	52	69	66	50	64	65	25.4
Gillette	94	61	91	61	87	60	65	84	63	83	62	82	59	88	73	57	80	69	54	73	68	28.6
Lander	90	59	87	58	85	57	62	81	61	80	59	80	56	81	69	53	74	68	51	69	67	26.7
Rock Springs	86	54	84	54	82	53	58	75	57	74	56	74	54	78	62	51	71	61	49	66	61	27.7
Sheridan	93	62	90	61	86	61	66	85	64	83	63	81	60	88	71	58	82	71	56	76	69	29.1
Worland	96	63	93	63	90	62	67	88	66	86	64	84	61	94	75	59	86	75	57	80	73	31.0

DP = dew-point temperature, °F MWB = mean coincident wet-bulb temp., °F A = airport NAF = Naval Air Facility
MDB = mean coincident dry-bulb temp., °F MWS = mean coincident wind speed, mph ANGB = Air National Guard Base NAWS = Naval Air Weapons Station
StdD = standard deviation, °F HR = humidity ratio, grains of moisture per lb of dry air MCAF = Marine Corps Air Facility RAF = Royal Air Force

Canada

Station	Cooling DB/MWB 0.4%		Cooling DB/MWB 1%		Cooling DB/MWB 2%		Evaporation WB/MDB 0.4%		Evaporation WB/MDB 1%		Evaporation WB/MDB 2%		Dehumidification DP/MDB and HR 0.4%			Dehumidification 1%			Dehumidification 2%			Range of DB
	DB	MWB	DB	MWB	DB	MWB	WB	MDB	WB	MDB	WB	MDB	DP	HR	MDB	DP	HR	MDB	DP	HR	MDB	
1	2a	2b	2c	2d	2e	2f	3a	3b	3c	3d	3e	3f	4a	4b	4c	4d	4e	4f	4g	4h	4i	5
ALBERTA																						
Calgary Intl A	83	60	80	59	77	57	62	78	61	75	59	73	57	79	67	55	74	65	53	69	64	22.0
Cold Lake A	82	64	78	62	75	60	66	78	64	76	62	72	62	88	71	60	82	69	58	76	67	20.0
Coronation	85	62	82	60	78	59	65	80	63	78	61	75	60	84	69	58	78	67	56	73	66	22.1
Edmonton Intl A	82	63	78	62	75	60	66	77	64	75	62	73	62	89	73	59	83	69	57	77	66	21.8
Fort McMurray A	84	64	80	62	76	60	66	79	64	76	62	73	61	84	71	59	78	68	57	74	66	22.0
Grande Prairie A	81	62	78	60	75	58	64	77	62	74	60	71	59	81	68	57	76	65	55	71	63	20.9
Lethbridge A	88	61	84	61	81	60	65	81	63	79	61	77	59	84	70	57	78	69	55	73	67	24.8
Medicine Hat A	90	63	87	62	84	61	66	84	64	83	62	80	60	83	70	58	78	69	56	72	68	25.0
Peace River A	81	62	78	60	75	59	65	77	62	75	60	71	60	82	69	58	76	67	56	71	65	21.4
Red Deer A	82	62	79	61	76	59	65	78	63	75	61	73	60	85	71	58	80	69	56	74	66	22.9
Rocky Mtn. House	80	62	78	61	75	59	64	78	63	75	61	72	60	87	70	58	81	68	56	76	66	22.5
Vermilion A	83	64	80	62	77	61	66	78	64	77	62	74	62	88	72	60	83	69	58	77	67	22.0
Whitecourt	80	61	77	60	74	59	65	77	62	74	60	71	60	86	69	58	80	67	57	75	64	23.4
BRITISH COLUMBIA																						
Abbotsford A	85	67	80	66	77	64	69	83	66	79	64	76	62	85	77	61	80	73	59	76	70	21.4
Cape St. James	64	59	62	58	60	57	60	63	59	61	58	60	59	74	61	58	71	59	57	69	59	7.6
Castlegar A	92	64	88	63	84	62	67	85	65	83	63	81	60	83	71	59	78	70	57	74	68	27.9
Comox A	80	63	76	62	73	61	65	76	63	74	62	71	60	78	68	59	75	66	58	71	65	16.4
Cranbrook A	88	61	85	60	81	59	63	83	62	80	60	78	57	77	66	55	72	66	53	67	65	24.8
Fort Nelson A	82	62	78	60	75	59	64	77	62	74	60	72	60	80	68	58	75	66	56	70	64	21.1
Fort St. John A	79	61	76	59	73	58	63	76	61	72	59	70	59	80	67	57	74	65	55	69	63	18.7
Kamloops A	93	65	88	63	85	62	66	88	65	85	63	81	59	78	70	57	74	69	56	69	68	24.7
Penticton A	90	65	87	64	83	63	67	85	65	83	64	81	60	81	73	59	76	72	57	72	71	26.5
Port Hardy A	68	59	65	58	63	57	60	66	59	64	58	62	58	72	62	57	69	61	56	67	60	12.4
Prince George A	81	60	78	59	74	58	63	78	61	74	59	71	58	77	66	56	73	64	54	67	62	23.2
Prince Rupert A	66	58	63	57	61	56	60	64	58	62	57	60	58	71	61	57	69	60	56	66	59	10.4
Quesnel A	85	62	81	60	77	59	64	80	62	77	61	74	59	81	67	57	75	65	56	71	63	25.4
Sandspit A	68	60	65	59	63	58	61	66	60	64	59	62	59	75	62	58	71	61	57	69	60	8.6
Smithers A	81	61	77	59	73	58	62	78	61	74	59	71	57	74	65	56	70	64	54	65	62	22.0
Spring Island	68	60	66	59	63	58	61	66	60	64	59	62	59	76	62	58	74	61	57	71	60	8.8
Terrace A	83	62	78	60	74	59	64	79	62	76	60	72	58	73	66	56	69	64	55	65	64	17.1
Tofino A	72	62	68	60	66	58	62	70	60	67	59	64	59	76	63	58	73	62	57	71	60	12.2
Vancouver Intl A	76	65	74	64	71	62	66	75	64	72	63	70	62	83	71	61	79	70	60	76	68	14.0
Victoria Intl A	79	63	75	62	72	61	64	77	63	74	61	71	59	75	69	58	71	67	57	69	65	18.4
Williams Lake A	83	59	79	57	75	56	60	77	59	75	57	72	55	73	62	53	67	61	51	62	60	22.0
MANITOBA																						
Brandon A	87	67	84	66	80	65	71	82	69	80	66	77	67	105	73	65	97	74	62	88	72	23.6
Churchill A	77	62	72	60	67	58	64	74	61	70	58	66	59	76	68	56	68	65	53	60	62	16.7
Dauphin A	87	67	84	66	80	64	70	82	68	80	66	77	67	102	77	64	94	74	62	86	72	22.1
Portage La Prairie A	88	68	85	67	81	65	72	83	69	80	68	78	68	108	78	66	99	75	64	92	73	20.5
The Pas A	83	66	79	64	76	62	68	79	66	76	64	73	64	93	73	62	85	70	60	80	68	18.4
Thompson A	83	64	79	62	76	60	66	78	64	75	62	73	62	85	71	60	78	69	58	73	66	23.0
Winnipeg Int'l A	87	68	84	67	81	66	72	82	70	80	68	78	68	107	78	66	99	75	64	92	73	20.5
NEW BRUNSWICK																						
Charlo A	83	68	79	66	75	65	70	78	68	76	66	73	68	102	74	66	97	72	64	90	70	18.4
Chatham A	86	69	83	67	79	65	71	81	69	78	68	75	68	104	75	67	98	73	65	92	71	20.3
Fredericton A	86	69	83	68	79	66	72	82	70	79	68	76	68	104	77	67	99	75	65	93	72	20.7
Moncton A	83	68	80	67	77	65	71	79	69	77	68	74	68	104	75	67	99	73	65	93	71	19.4
Saint John A	78	65	75	64	72	62	68	75	66	72	64	69	66	96	71	64	90	69	62	84	66	16.9
NEWFOUNDLAND																						
Battle Harbour	65	58	60	55	58	53	59	63	56	60	54	57	57	69	61	54	62	58	52	57	55	10.4
Bonavista	74	65	71	63	68	62	67	72	65	69	63	67	65	92	70	63	85	68	61	80	66	11.7
Cartwright	75	62	70	59	67	58	63	72	61	68	59	66	59	76	67	57	70	65	55	64	63	17.5
Daniels Harbour	69	63	66	62	65	61	65	67	63	66	62	64	64	90	67	62	83	65	60	78	63	9.7
Deer Lake A	81	66	77	64	74	62	68	77	66	74	65	71	66	95	73	64	88	71	61	81	69	21.4
Gander Intl A	79	65	76	63	72	62	68	75	66	72	64	70	65	95	71	63	88	70	61	83	68	17.8
Goose A	81	63	77	61	73	60	66	77	63	73	61	71	61	82	70	59	76	68	57	71	65	18.2
Hopedale	70	59	65	57	61	55	60	68	58	64	55	61	57	69	64	54	62	61	52	57	59	12.6
St. John's A	76	65	73	64	70	63	68	73	66	71	64	68	66	98	71	64	92	69	62	85	67	15.7
Stephenville A	74	64	71	64	69	63	67	71	65	69	64	68	66	95	70	64	88	68	62	83	66	12.4
Wabush Lake A	76	60	72	58	68	57	63	71	61	69	59	66	60	83	66	58	76	63	56	70	62	16.9
NORTHWEST TERRITORIES																						
Cape Parry A	58	53	54	50	50	47	53	58	50	53	47	50	50	53	56	47	48	52	45	43	50	9.7
Fort Smith A	82	63	78	61	75	60	65	78	63	75	61	72	61	81	69	59	76	67	57	71	66	21.4
Inuvik UA	78	60	75	59	71	57	62	75	60	72	58	69	56	67	67	53	61	65	52	57	64	18.4
Norman Wells A	80	62	77	60	74	59	64	77	62	74	60	71	59	75	68	57	71	66	56	67	65	18.5
Yellowknife A	77	60	74	59	70	57	62	73	60	71	59	68	58	73	66	56	68	65	53	62	63	14.4
NUNAVUT																						
Baker Lake	69	57	65	55	61	53	59	67	56	63	54	60	55	63	64	52	57	59	50	53	57	16.4
Cambridge Bay A	60	53	57	51	53	48	54	59	51	56	49	53	50	55	56	48	50	53	46	46	51	12.4
Chesterfield	66	54	60	52	56	50	55	65	52	60	50	56	50	53	60	48	49	55	46	46	52	13.7
Coral Harbour A	64	53	60	51	56	50	55	63	52	59	50	56	50	54	58	48	50	55	46	46	52	14.8
Hall Beach A	56	50	52	47	49	45	50	55	48	52	45	49	47	48	53	44	42	50	42	39	47	9.9

MDB = mean coincident dry-bulb temperature, °F
MWB = mean coincident wet-bulb temperature, °F
MWS = mean coincident wind speed, mph
StdD = standard deviation, °F
HR = humidity ratio, grains of moisture per lb of dry air
A = airport
DP = dew-point temperature, °F

TABLE 5.12   COOLING DESIGN CONDITIONS FOR THE UNITED STATES AND CANADA. (CONTINUED)

Canada

Station	Cooling DB/MWB 0.4% DB	MWB	1% DB	MWB	2% DB	MWB	Evaporation WB/MDB 0.4% WB	MDB	1% WB	MDB	2% WB	MDB	Dehumidification DP/MDB and HR 0.4% DP	HR	MDB	1% DP	HR	MDB	2% DP	HR	MDB	Range of DB
1	2a	2b	2c	2d	2e	2f	3a	3b	3c	3d	3e	3f	4a	4b	4c	4d	4e	4f	4g	4h	4i	5
Iqaluit A (Frobisher)	60	50	57	48	53	46	51	59	49	55	47	52	46	46	53	44	43	52	42	40	50	12.4
Resolute A	50	45	48	43	44	41	46	50	43	47	41	44	42	39	47	40	36	45	38	34	43	8.5
NOVA SCOTIA																						
Greenwood A	84	69	80	67	78	66	72	79	70	77	69	75	69	108	76	68	102	74	66	96	72	20.0
Halifax Intl A	80	68	78	66	75	64	70	77	69	74	67	72	68	106	73	67	99	71	65	95	69	16.7
Sable Island	70	67	69	66	67	65	68	69	67	68	66	67	68	101	69	66	97	68	65	92	66	8.3
Shearwater A	78	66	75	64	72	63	69	74	67	72	66	69	68	102	71	66	96	69	65	92	68	13.1
Sydney A	81	68	78	67	75	65	71	78	69	75	67	72	68	105	74	67	99	72	65	93	70	17.3
Truro	79	69	77	67	75	66	71	77	69	75	68	72	69	106	75	68	102	73	66	97	70	19.3
Yarmouth A	73	66	71	64	69	63	68	71	66	69	64	67	67	98	69	65	93	68	63	86	66	13.1
ONTARIO																						
Armstrong A	81	66	78	65	75	63	69	78	67	75	65	73	67	102	73	64	94	71	61	85	69	24.7
Atikokan	84	67	80	66	77	64	71	80	69	77	67	74	68	109	75	66	102	73	64	95	71	23.0
Big Trout Lake	79	64	75	62	72	61	67	75	65	72	63	70	64	91	70	62	84	68	60	78	67	16.4
Earlton A	85	69	81	67	78	65	71	80	69	77	67	75	69	108	76	67	101	74	65	95	72	21.6
Geraldton	81	66	78	65	75	63	69	77	67	75	65	73	67	103	74	65	97	72	62	87	69	22.1
Gore Bay A	80	68	78	67	75	65	71	77	69	75	68	73	69	108	74	67	102	72	66	97	71	16.4
Kapuskasing A	84	67	80	66	77	64	70	79	68	77	66	74	67	102	75	65	95	72	62	87	70	22.5
Kenora A	84	67	81	65	78	64	70	79	68	77	66	75	68	106	75	66	99	73	63	92	72	16.4
London A	85	71	83	70	80	69	74	82	72	80	71	77	71	119	79	70	113	77	68	108	75	19.8
Mount Forest	83	70	80	68	77	67	72	80	70	77	69	75	70	115	77	68	109	74	67	103	73	20.3
Muskoka A	84	69	80	68	78	66	72	80	70	77	68	75	70	113	76	68	106	74	66	100	72	20.7
North Bay A	81	67	78	66	76	64	70	77	68	75	67	73	68	108	74	66	102	72	65	96	70	17.1
Ottawa Int'l A	86	70	83	69	80	67	73	82	71	80	69	78	70	112	78	68	106	76	67	99	75	18.5
Sault Ste. Marie A	83	69	79	67	76	65	71	79	69	76	67	74	69	108	76	67	102	73	65	95	71	20.9
Simcoe	85	72	83	70	80	69	74	83	72	80	71	77	71	118	79	70	112	77	68	107	75	19.3
Sioux Lookout A	84	67	80	65	78	63	70	80	68	76	66	74	67	104	74	65	97	71	63	90	70	18.9
Sudbury A	84	67	81	66	78	64	70	80	68	77	66	74	68	105	74	66	99	72	64	92	71	19.1
Thunder Bay A	84	68	80	66	77	64	70	80	68	77	66	73	67	101	76	65	93	72	62	86	70	21.8
Timmins A	84	67	81	65	78	64	70	80	68	77	66	74	67	102	74	65	95	72	63	88	70	23.0
Toronto Int'l A	87	71	84	70	81	68	74	83	72	80	70	78	71	116	79	69	110	77	68	104	75	20.2
Trenton A	84	71	81	70	78	68	74	81	72	78	70	76	71	116	78	70	110	76	68	104	74	18.0
Wiarton A	82	70	80	69	77	67	72	80	70	77	69	75	70	113	77	68	106	75	67	101	73	18.0
Windsor A	89	73	86	71	83	70	76	85	74	83	73	80	73	125	82	71	118	79	70	112	77	17.5
PRINCE EDWARD ISLAND																						
Charlottetown A	79	69	77	67	74	65	71	77	69	74	67	72	68	105	75	67	99	72	65	94	71	15.1
Summerside A	79	68	77	66	74	65	71	76	69	74	67	72	68	104	74	67	99	72	65	94	70	14.4
QUEBEC																						
Bagotville A	84	67	80	65	77	63	69	79	68	76	66	74	66	99	74	65	93	72	62	85	70	19.8
Baie Comeau A	75	63	71	61	69	60	65	71	63	69	61	67	62	85	68	61	79	66	59	75	64	17.1
Grindstone Island	73	66	70	65	68	64	68	71	67	69	65	68	67	100	70	66	95	69	64	90	68	8.6
Kuujjuarapik A	75	61	70	59	66	57	63	72	61	68	58	65	60	76	67	57	69	64	54	62	62	15.8
Kuujjuaq A	74	60	69	57	65	55	62	69	59	66	56	64	58	72	65	55	65	63	52	58	60	18.7
La Grande Riviere A	78	62	75	60	71	58	65	73	62	70	60	68	62	83	68	59	77	65	57	70	63	21.2
Lake Eon A	74	60	70	58	67	57	63	70	61	67	59	64	60	83	65	58	78	63	57	74	62	16.4
Mont Joli A	80	67	76	65	74	64	69	77	67	75	65	72	66	97	75	64	90	72	62	83	70	16.4
Montreal Intl A	85	71	83	70	80	68	74	82	72	80	70	77	70	112	79	69	106	77	67	101	75	17.6
Montreal Mirabel A	84	71	81	69	78	67	73	81	71	78	69	76	70	110	78	68	103	75	66	98	73	20.2
Nitchequon	72	60	69	58	66	57	63	68	61	65	59	64	61	85	65	59	79	63	57	74	62	14.0
Quebec A	84	70	80	68	78	66	73	80	70	77	68	75	70	111	77	68	103	75	66	95	73	19.1
Riviere Du Loup	79	68	76	65	74	65	70	76	69	74	67	72	68	106	74	67	99	72	65	93	70	15.5
Roberval A	83	68	79	66	76	65	71	79	69	77	67	74	68	105	76	66	98	74	64	92	72	17.8
Schefferville A	74	58	69	57	66	55	61	69	59	66	57	64	58	76	64	56	71	62	53	64	60	16.0
Sept-Iles A	72	60	69	59	67	58	64	69	62	67	60	65	61	82	66	60	77	64	58	73	62	13.9
Sherbrooke A	84	70	80	68	78	66	72	81	70	78	68	75	69	109	77	67	103	75	66	97	73	22.3
St. Hubert A	86	71	83	69	80	68	74	83	72	80	70	78	71	113	78	69	106	77	67	101	75	19.1
Ste. Agathe Des Monts	81	68	78	66	76	65	71	77	69	75	67	73	69	110	75	67	103	72	65	97	70	19.4
Val d'Or A	83	67	80	65	77	63	69	79	68	76	66	73	67	102	73	65	95	72	62	88	70	20.7
SASKATCHEWAN																						
Broadview	87	65	83	64	79	63	69	81	66	79	64	76	64	97	74	62	89	71	60	83	70	23.8
Estevan A	90	66	86	65	83	64	70	83	68	81	66	79	66	103	76	63	92	73	61	86	71	23.8
Moose Jaw A	90	64	87	64	83	62	68	83	66	81	64	79	63	92	73	61	85	71	59	79	69	23.8
North Battleford A	86	64	82	63	78	61	67	80	65	78	63	76	62	88	71	60	83	69	58	77	68	21.1
Prince Albert A	84	65	81	64	78	62	68	80	65	77	64	75	63	90	74	61	84	71	59	78	69	21.8
Regina A	89	64	85	64	82	62	68	82	66	80	64	78	63	92	75	61	85	71	59	79	69	23.6
Saskatoon A	87	64	84	63	80	62	67	81	65	79	63	77	62	88	73	60	82	70	58	76	68	22.7
Swift Current A	88	63	84	62	80	61	66	81	64	79	62	77	61	88	71	59	82	68	57	76	67	23.0
Uranium City A	79	62	76	60	72	58	64	76	62	72	60	69	59	78	68	58	74	66	56	69	64	17.1
Wynyard	85	64	81	63	78	62	68	79	65	77	63	75	64	94	74	61	85	70	59	81	68	20.7
Yorkton A	86	65	82	64	79	62	68	80	66	78	64	75	64	96	74	62	87	71	60	81	69	22.0
YUKON TERRITORY																						
Burwash A	73	57	69	55	66	53	57	71	56	68	54	64	52	62	60	50	58	58	48	55	57	21.4
Whitehorse A	77	57	73	55	69	53	58	73	56	70	55	67	52	62	61	50	58	59	48	55	58	20.9

MDB = mean coincident dry-bulb temperature, °F
MWB = mean coincident wet-bulb temperature, °F
MWS = mean coincident wind speed, mph
StdD = standard deviation, °F
HR = humidity ratio, grains of moisture per lb of dry air
A = airport
DP = dew-point temperature, °F

Extracted from 2001 *ASHRAE Handbook—Fundamentals*. Used with permission from American Society of Heating, Refrigerating and Air-Conditioning Engineers, Inc.

TABLE 5.13   NORMAL COOLING DEGREE DAYS (°F · DAYS), 65°F BASE TEMPERATURE, FOR SELECTED CITIES IN THE UNITED STATES FROM 1971 TO 2000.

City	Years	Jan	Feb	Mar	Apr	May	Jun	Jul	Aug	Sep	Oct	Nov	Dec	Total
BIRMINGHAM AP, AL	30	1	3	16	51	167	351	476	455	280	69	9	3	1881
HUNTSVILLE, AL	30	0	1	8	40	142	326	446	417	238	47	5	1	1671
MOBILE, AL	30	9	11	45	108	282	450	529	520	384	151	41	18	2548
MONTGOMERY, AL	30	2	4	24	73	225	415	519	502	350	106	23	9	2252
ANCHORAGE, AK	30	0	0	0	0	0	0	3	0	0	0	0	0	3
ANNETTE, AK	30	0	0	0	0	0	0	5	8	0	0	0	0	13
BARROW, AK	30	0	0	0	0	0	0	0	0	0	0	0	0	0
BETHEL, AK	30	0	0	0	0	0	0	0	1	0	0	0	0	1
BETTLES, AK	30	0	0	0	0	0	11	20	7	0	0	0	0	38
BIG DELTA, AK	30	0	0	0	0	0	1	12	12	2	0	0	0	27
COLD BAY, AK	30	0	0	0	0	0	0	0	0	0	0	0	0	0
FAIRBANKS, AK	30	0	0	0	0	0	20	42	11	1	0	0	0	74
GULKANA, AK	30	0	0	0	0	0	0	0	0	0	0	0	0	0
HOMER, AK	30	0	0	0	0	0	0	0	0	0	0	0	0	0
JUNEAU, AK	30	0	0	0	0	0	0	0	0	0	0	0	0	0
KING SALMON, AK	30	0	0	0	0	0	0	0	0	0	0	0	0	0
KODIAK, AK	30	0	0	0	0	0	0	0	0	0	0	0	0	0
KOTZEBUE, AK	30	0	0	0	0	0	0	8	4	0	0	0	0	12
MCGRATH, AK	30	0	0	0	0	0	4	11	6	0	0	0	0	21
NOME, AK	30	0	0	0	0	0	0	2	0	0	0	0	0	2
ST. PAUL ISLAND, AK	30	0	0	0	0	0	0	0	0	0	0	0	0	0
TALKEETNA, AK	30	0	0	0	0	0	1	4	0	0	0	0	0	5
UNALAKLEET, AK	30	0	0	0	0	0	0	1	2	0	0	0	0	3
VALDEZ, AK	30	0	0	0	0	0	0	0	0	0	0	0	0	0
YAKUTAT, AK	30	0	0	0	0	0	0	0	0	0	0	0	0	0
FLAGSTAFF, AZ	30	0	0	0	0	0	23	64	36	3	0	0	0	126
PHOENIX, AZ	30	2	15	71	223	473	744	900	859	664	350	53	1	4355
TUCSON, AZ	30	0	3	25	107	300	577	672	625	477	211	20	0	3017
WINSLOW, AZ	30	0	0	0	3	39	221	384	327	124	6	0	0	1104
YUMA, AZ	30	12	29	118	251	466	714	900	882	694	381	82	11	4540
FORT SMITH, AR	30	0	0	8	43	155	364	520	491	275	59	5	1	1921
LITTLE ROCK, AR	30	1	1	14	52	188	408	542	502	296	72	8	2	2086
NORTH LITTLE ROCK, AR	30	0	6	10	66	212	412	564	528	307	83	8	0	2196
BAKERSFIELD, CA	30	0	1	7	56	205	392	580	541	374	144	4	0	2304
BISHOP, CA	30	0	0	0	3	45	191	357	293	106	8	0	0	1003
BLUE CANYON, CA	30	0	0	0	1	15	41	129	125	84	28	0	0	423
EUREKA, CA.	30	0	0	0	0	0	0	0	2	5	0	0	0	7
FRESNO, CA	30	0	0	3	40	170	351	530	483	316	97	1	0	1991
LONG BEACH, CA	30	3	5	10	28	55	135	260	302	244	119	20	2	1183
LOS ANGELES AP, CA	30	6	7	6	15	19	58	135	175	154	81	22	4	682
LOS ANGELES C.O., CA	30	15	23	26	58	84	178	295	325	281	164	44	13	1506
MOUNT SHASTA, CA	30	0	0	0	0	7	31	100	66	29	2	0	0	235
REDDING, CA	30	0	0	3	22	133	310	504	430	263	74	2	0	1741
SACRAMENTO, CA	30	0	0	6	24	110	204	320	303	210	66	5	0	1248
SAN DIEGO, CA	30	2	4	5	17	32	81	183	230	199	97	15	1	866
SAN FRANCISCO AP, CA	30	0	0	0	4	10	21	23	26	38	20	0	0	142
SAN FRANCISCO C.O., CA	30	0	2	4	3	9	14	19	26	56	22	5	0	160
SANTA BARBARA, CA	30	0	0	5	8	23	50	84	133	125	45	9	0	482
SANTA MARIA, CA	30	0	1	1	4	5	10	23	25	33	18	3	0	123

TABLE 5.13   NORMAL COOLING DEGREE DAYS (°F · DAYS), 65°F BASE TEMPERATURE, FOR SELECTED CITIES IN THE UNITED STATES FROM 1971 TO 2000. (CONTINUED)

City	Years	Jan	Feb	Mar	Apr	May	Jun	Jul	Aug	Sep	Oct	Nov	Dec	Total
STOCKTON, CA	30	0	0	0	18	111	254	390	363	247	73	0	0	1456
ALAMOSA, CO	30	0	0	0	0	0	7	27	10	0	0	0	0	44
COLORADO SPRINGS, CO	30	0	0	0	1	5	86	184	131	35	1	0	0	443
DENVER, CO	30	0	0	0	2	23	135	261	217	57	0	0	0	695
GRAND JUNCTION, CO	30	0	0	0	3	45	247	414	350	119	4	0	0	1182
PUEBLO, CO	30	0	0	0	2	31	191	343	276	91	2	0	0	936
BRIDGEPORT, CT	30	0	0	0	1	21	125	286	258	91	7	0	0	789
HARTFORD, CT	30	0	0	1	5	38	144	277	220	68	5	1	0	759
WILMINGTON, DE	30	0	0	2	9	62	215	368	317	135	16	1	0	1125
WASHINGTON DULLES AP, D.C.	30	0	0	4	11	60	203	345	302	132	15	3	0	1075
WASHINGTON NAT'L AP, D.C.	30	0	0	4	21	108	307	464	410	210	32	4	0	1560
APALACHICOLA, FL	30	25	14	35	95	284	448	522	517	423	193	63	23	2642
DAYTONA BEACH, FL	30	36	40	86	150	306	441	513	502	436	277	122	52	2961
FORT MYERS, FL	30	97	99	174	260	423	516	564	568	518	394	222	122	3957
GAINESVILLE, FL	30	17	17	70	131	308	439	507	498	408	207	74	26	2702
JACKSONVILLE, FL	30	15	21	58	116	277	437	535	509	400	183	64	21	2636
KEY WEST, FL	30	189	183	277	361	485	552	604	600	550	473	340	233	4847
MIAMI, FL	30	155	154	236	315	442	510	568	568	517	433	291	194	4383
ORLANDO, FL	30	91	60	129	201	373	485	539	543	484	319	154	79	3457
PENSACOLA, FL	30	11	11	42	107	296	466	541	529	412	174	48	19	2656
TALLAHASSEE, FL	30	9	12	44	96	273	440	515	509	400	162	49	16	2525
TAMPA, FL	30	56	59	124	204	393	501	550	549	489	323	157	76	3481
VERO BEACH, FL	30	104	83	132	202	347	462	517	513	471	354	183	92	3460
WEST PALM BEACH, FL	30	122	121	195	266	408	485	544	549	499	408	255	160	4012
ATHENS, GA	30	0	1	8	46	164	351	471	430	257	54	7	1	1790
ATLANTA, GA	30	0	1	11	52	170	354	463	430	262	58	8	1	1810
AUGUSTA, GA	30	1	2	15	52	191	385	511	468	296	77	15	3	2016
COLUMBUS, GA	30	1	4	25	77	234	429	533	511	349	107	21	6	2297
MACON, GA	30	1	3	23	73	227	418	528	495	325	95	19	6	2213
SAVANNAH, GA	30	6	11	39	101	267	435	547	506	367	141	40	11	2471
HILO, HI	30	198	180	215	223	268	303	336	350	336	328	268	223	3228
HONOLULU, HI	30	249	225	288	319	379	436	490	521	497	472	382	303	4561
KAHULUI, HI	30	210	195	250	277	330	379	429	448	422	407	329	260	3936
LIHUE, HI	30	207	188	239	266	323	382	433	455	435	408	327	257	3920
BOISE, ID	30	0	0	0	3	30	116	281	260	75	4	0	0	769
LEWISTON, ID	30	0	0	0	3	29	105	283	285	84	3	0	0	792
POCATELLO, ID	30	0	0	0	0	3	51	167	143	23	0	0	0	387
CHICAGO, IL	30	0	0	1	9	48	159	283	234	91	10	0	0	835
MOLINE, IL	30	0	0	1	12	62	205	322	254	101	12	0	0	969
PEORIA, IL	30	0	0	1	11	64	210	325	263	112	12	0	0	998
ROCKFORD, IL	30	0	0	1	7	49	162	263	205	74	7	0	0	768
SPRINGFIELD, IL	30	0	0	2	17	84	249	358	291	140	23	1	0	1165
EVANSVILLE, IN	30	0	0	4	23	108	304	425	356	173	27	2	0	1422
FORT WAYNE, IN	30	0	0	1	7	53	183	278	212	88	8	0	0	830
INDIANAPOLIS, IN	30	0	0	2	10	69	221	331	272	122	14	1	0	1042
SOUTH BEND, IN	30	0	0	1	10	53	172	268	214	86	9	0	0	813
DES MOINES, IA	30	0	0	1	12	60	219	353	285	110	12	0	0	1052
DUBUQUE, IA	30	0	0	0	6	37	138	233	175	61	6	0	0	656
SIOUX CITY, IA	30	0	0	0	11	53	198	311	246	87	8	0	0	914

TABLE 5.13 NORMAL COOLING DEGREE DAYS (°F · DAYS), 65°F BASE TEMPERATURE, FOR SELECTED CITIES
IN THE UNITED STATES FROM 1971 TO 2000. (CONTINUED)

City	Years	Jan	Feb	Mar	Apr	May	Jun	Jul	Aug	Sep	Oct	Nov	Dec	Total
WATERLOO, IA	30	0	0	0	7	47	168	261	198	70	7	0	0	758
CONCORDIA, KS	30	0	0	2	15	63	265	436	366	163	22	1	0	1333
DODGE CITY, KS	30	0	0	2	18	79	291	462	407	193	28	1	0	1481
GOODLAND, KS	30	0	0	0	4	26	173	320	266	99	6	0	0	894
TOPEKA, KS	30	0	0	3	22	85	278	419	357	166	26	1	0	1357
WICHITA, KS	30	0	0	2	19	93	330	503	454	221	35	1	0	1658
JACKSON, KY	30	0	0	0	11	100	201	310	277	130	29	1	0	1059
LEXINGTON, KY	30	0	0	3	16	80	228	350	307	147	21	2	0	1154
LOUISVILLE, KY	30	0	0	6	24	109	287	421	374	189	29	3	1	1443
PADUCAH KY	30	0	0	6	33	122	320	444	377	191	37	3	0	1533
BATON ROUGE, LA	30	12	15	55	128	303	462	538	523	389	157	48	22	2652
LAKE CHARLES, LA	30	9	13	49	126	312	467	544	534	399	178	54	20	2705
NEW ORLEANS, LA	30	15	19	62	136	320	466	538	534	413	182	62	29	2776
SHREVEPORT, LA	30	5	7	31	87	242	436	553	532	353	119	24	7	2396
CARIBOU, ME	30	0	0	0	0	7	39	80	56	9	0	0	0	191
PORTLAND, ME	30	0	0	0	0	7	51	144	120	24	1	0	0	347
BALTIMORE, MD	30	0	0	4	13	71	236	390	332	153	19	2	0	1220
BLUE HILL, MA	30	0	0	1	3	21	93	215	173	49	3	0	0	558
BOSTON, MA	30	0	0	1	4	32	139	282	235	76	7	1	0	777
WORCESTER, MA	30	0	0	0	0	7	64	166	122	12	0	0	0	371
ALPENA, MI	30	0	0	0	3	13	54	117	83	22	1	0	0	293
DETROIT, MI	30	0	0	0	6	41	144	252	204	75	5	0	0	727
FLINT, MI	30	0	0	1	5	33	110	199	151	52	4	0	0	555
GRAND RAPIDS, MI	30	0	0	1	6	38	124	218	165	56	5	0	0	613
HOUGHTON LAKE, MI	30	0	0	0	3	20	64	127	89	24	1	0	0	328
LANSING, MI	30	0	0	1	6	34	113	195	151	53	5	0	0	558
MARQUETTE, MI	30	0	0	0	0	12	30	72	50	3	0	0	0	167
MUSKEGON, MI	30	0	0	0	4	24	86	181	145	44	3	0	0	487
SAULT STE. MARIE, MI	30	0	0	0	1	6	20	56	50	12	0	0	0	145
DULUTH, MN	30	0	0	0	0	7	28	82	60	12	0	0	0	189
INTERNATIONAL FALLS, MN	30	0	0	0	1	17	47	91	67	10	0	0	0	233
MINNEAPOLIS-ST.PAUL, MN	30	0	0	0	4	41	146	259	190	56	3	0	0	699
ROCHESTER, MN	30	0	0	0	1	30	99	181	135	26	1	0	0	473
SAINT CLOUD, MN	30	0	0	0	2	26	90	172	121	31	1	0	0	443
JACKSON, MS	30	5	8	32	83	232	424	525	505	340	101	25	10	2290
MERIDIAN, MS	30	4	6	26	70	213	400	509	495	331	91	20	8	2173
TUPELO, MS	30	0	1	14	50	171	364	488	453	274	60	8	2	1885
COLUMBIA, MO	30	0	0	3	18	70	245	396	341	152	20	1	0	1246
KANSAS CITY, MO	30	0	0	0	12	101	264	418	367	151	12	0	0	1325
ST. LOUIS, MO	30	0	0	7	32	114	316	461	396	196	36	3	0	1561
SPRINGFIELD, MO	30	0	0	3	20	83	258	415	379	179	27	1	0	1365
BILLINGS, MT	30	0	0	0	2	13	90	227	204	44	3	0	0	583
GLASGOW, MT	30	0	0	0	1	17	80	185	182	28	1	0	0	494
GREAT FALLS, MT	30	0	0	0	1	7	47	126	121	22	2	0	0	326
HAVRE, MT	30	0	0	0	1	10	59	146	141	19	1	0	0	377
HELENA, MT	30	0	0	0	0	3	39	122	100	13	0	0	0	277
KALISPELL, MT	30	0	0	0	0	3	17	62	63	4	0	0	0	149
MISSOULA, MT	30	0	0	0	0	3	33	111	99	10	0	0	0	256
GRAND ISLAND, NE	30	0	0	1	11	48	218	349	285	107	8	0	0	1027

City	Years	Jan	Feb	Mar	Apr	May	Jun	Jul	Aug	Sep	Oct	Nov	Dec	Total
LINCOLN, NE	30	0	0	1	13	56	244	390	315	123	12	0	0	1154
NORFOLK, NE	30	0	0	0	11	48	202	324	261	93	7	0	0	946
NORTH PLATTE, NE	30	0	0	0	4	22	139	279	230	70	2	0	0	746
OMAHA EPPLEY AP, NE	30	0	0	1	14	60	233	365	296	114	12	0	0	1095
OMAHA (NORTH), NE	30	0	0	0	8	71	209	330	295	101	5	0	0	1019
SCOTTSBLUFF, NE	30	0	0	0	2	19	138	279	225	63	1	0	0	727
VALENTINE, NE	30	0	0	0	5	27	141	286	242	75	3	0	0	779
ELKO, NV	30	0	0	0	0	2	62	181	135	31	1	0	0	412
ELY, NV	30	0	0	0	0	0	22	98	69	7	0	0	0	196
LAS VEGAS, NV	30	0	1	17	93	310	597	792	734	470	150	4	0	3168
RENO, NV	30	0	0	0	0	11	72	204	164	41	1	0	0	493
WINNEMUCCA, NV	30	0	0	0	0	11	81	232	174	28	0	0	0	526
CONCORD, NH	30	0	0	0	2	18	82	173	133	33	1	0	0	442
MT. WASHINGTON, NH	30	0	0	0	0	0	0	0	0	0	0	0	0	0
ATLANTIC CITY AP, NJ	30	0	0	1	5	44	168	322	269	110	15	1	0	935
ATLANTIC CITY C.O., NJ	30	0	0	0	0	25	147	316	302	136	25	0	0	951
NEWARK, NJ	30	0	0	2	10	70	240	403	350	146	19	2	0	1242
ALBUQUERQUE, NM	30	0	0	0	6	70	297	417	343	148	9	0	0	1290
CLAYTON, NM	30	0	0	0	6	37	173	274	228	77	2	0	0	797
ROSWELL, NM	30	0	0	1	47	194	391	488	431	228	32	2	0	1814
ALBANY, NY	30	0	0	1	3	27	102	206	157	46	2	0	0	544
BINGHAMTON, NY	30	0	0	1	4	23	74	158	115	32	2	0	0	409
BUFFALO, NY	30	0	0	0	4	28	101	203	158	50	4	0	0	548
ISLIP, NY	30	0	0	0	0	15	129	296	251	72	7	0	0	770
NEW YORK CENTRAL PARK, NY	30	0	0	2	10	63	214	379	335	138	17	2	0	1160
NEW YORK (JFK AP), NY	30	0	0	0	2	31	162	335	306	125	13	1	0	975
NEW YORK (LAGUARDIA AP), NY	30	0	0	1	6	54	209	377	336	141	17	1	0	1142
ROCHESTER, NY	30	0	0	1	5	32	109	210	162	54	4	0	0	577
SYRACUSE, NY	30	0	0	1	4	29	105	203	158	48	3	0	0	551
ASHEVILLE, NC	30	0	0	0	6	47	165	278	243	104	8	0	0	851
CAPE HATTERAS, NC	30	1	1	5	29	122	297	440	422	297	96	24	4	1738
CHARLOTTE, NC	30	0	1	7	40	142	323	451	405	226	43	5	1	1644
GREENSBORO-WNSTN-SALM-HGHPT,NC	30	0	0	4	25	97	263	398	345	172	24	3	1	1332
RALEIGH, NC	30	0	1	9	38	119	293	429	379	206	39	6	2	1521
WILMINGTON, NC	30	3	4	17	65	187	361	501	455	304	90	25	5	2017
BISMARCK, ND	30	0	0	0	2	18	80	180	161	30	0	0	0	471
FARGO, ND	30	0	0	0	3	33	104	191	162	38	2	0	0	533
GRAND FORKS, ND	30	0	0	0	2	30	85	148	127	27	1	0	0	420
WILLISTON, ND	30	0	0	0	1	20	82	177	166	24	1	0	0	471
AKRON, OH	30	0	0	1	7	41	136	232	189	69	4	0	0	679
CLEVELAND, OH	30	0	0	2	7	40	140	239	195	80	8	1	0	712
GREATER CINCINNATI AP, OH	30	0	0	3	13	71	209	334	280	126	16	1	0	1053
COLUMBUS, OH	30	0	0	2	9	61	188	296	251	106	11	1	0	925
DAYTON, OH	30	0	0	2	9	62	194	305	246	105	11	1	0	935
MANSFIELD, OH	30	0	0	1	7	39	136	228	181	73	7	0	0	672
TOLEDO, OH	30	0	0	1	7	42	148	248	190	73	6	0	0	715
YOUNGSTOWN, OH	30	0	0	1	8	33	112	190	154	57	5	1	0	561
OKLAHOMA CITY, OK	30	0	1	7	38	145	360	527	497	271	58	3	0	1907
TULSA, OK	30	0	1	10	50	163	385	568	524	277	64	6	1	2049

TABLE 5.13   NORMAL COOLING DEGREE DAYS (°F · DAYS), 65°F BASE TEMPERATURE, FOR SELECTED CITIES
IN THE UNITED STATES FROM 1971 TO 2000. (CONTINUED)

City	Years	Jan	Feb	Mar	Apr	May	Jun	Jul	Aug	Sep	Oct	Nov	Dec	Total
ASTORIA, OR	30	0	0	0	0	1	2	4	7	7	1	0	0	22
BURNS, OR	30	0	0	0	0	0	19	117	70	12	0	0	0	218
EUGENE, OR	30	0	0	0	0	5	21	95	93	32	1	0	0	247
MEDFORD, OR	30	0	0	0	2	24	90	253	240	95	7	0	0	711
PENDLETON, OR	30	0	0	0	2	23	86	243	224	63	3	0	0	644
PORTLAND, OR	30	0	0	0	1	15	44	138	145	53	2	0	0	398
SALEM, OR	30	0	0	0	0	7	25	95	98	31	1	0	0	257
SEXTON SUMMIT, OR	30	0	0	0	0	0	11	84	76	62	15	0	0	248
GUAM, PC	30	390	350	414	435	473	463	458	449	439	449	439	431	5190
JOHNSTON ISLAND, PC	30	389	354	399	409	457	485	526	545	523	522	448	412	5469
KOROR, PC	30	507	454	518	518	545	503	509	515	509	527	524	526	6155
KWAJALEIN, MARSHALL IS., PC	30	512	470	536	515	538	515	529	534	518	537	513	522	6239
MAJURO, MARSHALL IS., PC	30	487	451	502	482	504	485	498	507	496	510	490	493	5905
PAGO PAGO, AMER SAMOA, PC	30	511	469	526	498	493	460	454	457	460	486	485	515	5814
POHNPEI, CAROLINE IS., PC	30	490	449	506	485	501	480	486	487	471	486	477	493	5811
CHUUK, E. CAROLINE IS., PC	30	510	456	513	501	522	497	506	492	488	508	511	511	6015
WAKE ISLAND, PC	30	396	344	409	427	488	527	567	567	556	546	480	443	5750
YAP, W CAROLINE IS., PC	30	467	428	488	493	518	479	484	478	466	486	476	484	5747
ALLENTOWN, PA	30	0	0	1	6	45	164	292	235	83	7	0	0	833
ERIE, PA	30	0	0	1	5	30	105	197	166	63	6	0	0	573
HARRISBURG, PA	30	0	0	0	1	54	186	337	279	87	11	0	0	955
MIDDLETOWN/HARRISBURG INTL APT	30	0	0	0	1	54	186	337	279	87	11	0	0	955
PHILADELPHIA, PA	30	0	0	2	10	70	234	395	351	152	19	2	0	1235
PITTSBURGH, PA	30	0	0	2	8	41	143	244	203	78	6	1	0	726
AVOCA, PA	30	0	0	1	5	35	113	220	174	57	4	0	0	609
WILLIAMSPORT, PA	30	0	0	0	6	39	135	251	206	68	4	0	0	709
BLOCK IS., RI	30	0	0	0	0	1	51	184	177	48	5	0	0	466
PROVIDENCE, RI	30	0	0	0	3	25	122	265	223	71	5	0	0	714
CHARLESTON AP, SC	30	3	7	29	85	242	408	532	494	348	122	35	8	2313
CHARLESTON C.O., SC	30	16	7	27	83	266	431	551	514	377	156	47	11	2486
COLUMBIA, SC	30	2	4	20	69	206	388	515	467	296	76	15	5	2063
GREENVILLE-SPARTANBURG AP, SC	30	0	0	5	30	127	304	430	384	207	35	3	1	1526
ABERDEEN, SD	30	0	0	0	3	29	112	235	196	49	2	0	0	626
HURON, SD	30	0	0	0	4	29	138	273	228	66	3	0	0	741
RAPID CITY, SD	30	0	0	0	2	13	86	227	208	59	3	0	0	598
SIOUX FALLS, SD	30	0	0	0	5	37	151	278	217	65	4	0	0	757
BRISTOL-JHNSN CTY-KNGSPRT, TN	30	0	0	1	10	61	200	309	274	128	11	1	0	995
CHATTANOOGA, TN	30	0	0	5	32	124	312	450	418	229	35	2	1	1608
KNOXVILLE, TN	30	0	1	5	27	110	282	408	381	205	28	3	0	1450
MEMPHIS, TN	30	1	2	15	72	210	428	554	504	307	84	11	2	2190
NASHVILLE, TN	30	0	0	9	37	136	321	455	416	230	46	5	1	1656
OAK RIDGE, TN	30	0	0	2	19	95	254	380	347	180	23	1	0	1301
ABILENE, TX	30	0	5	28	94	253	442	568	535	327	118	15	1	2386
AMARILLO, TX	30	0	0	2	18	90	285	405	345	173	26	0	0	1344
AUSTIN/CITY, TX	30	7	18	59	147	323	495	605	610	439	207	51	13	2974
AUSTIN/BERGSTROM, TX	30	8	11	23	107	307	453	551	532	377	148	32	8	2557
BROWNSVILLE, TX	30	60	76	180	292	463	551	607	608	497	338	170	82	3924
CORPUS CHRISTI, TX	30	32	43	121	229	402	520	594	598	485	300	123	51	3498
DALLAS-FORT WORTH, TX	30	2	11	10	72	265	478	621	601	376	118	15	2	2571

TABLE 5.13   NORMAL COOLING DEGREE DAYS (°F · DAYS), 65°F BASE TEMPERATURE, FOR SELECTED CITIES
IN THE UNITED STATES FROM 1971 TO 2000. (CONTINUED)

City	Years	Jan	Feb	Mar	Apr	May	Jun	Jul	Aug	Sep	Oct	Nov	Dec	Total
DALLAS-LOVE FIELD, TX	30	2	9	39	110	290	511	659	646	417	162	28	5	2878
DEL RIO, TX	30	2	18	86	207	401	545	630	622	453	217	41	4	3226
EL PASO, TX	30	0	1	8	65	238	481	535	473	293	69	2	0	2165
GALVESTON, TX	30	30	23	65	163	367	515	596	601	482	286	122	30	3280
HOUSTON, TX	30	15	21	63	147	328	485	573	563	412	196	65	25	2893
LUBBOCK, TX	30	0	0	7	48	180	382	472	413	225	49	1	0	1777
MIDLAND-ODESSA, TX	30	0	2	15	77	254	438	512	473	281	83	4	0	2139
PORT ARTHUR, TX	30	11	16	55	140	324	480	553	546	417	195	63	23	2823
SAN ANGELO, TX	30	1	4	30	107	277	446	554	519	322	113	16	1	2390
SAN ANTONIO, TX	30	7	19	68	161	344	505	607	601	439	215	57	15	3038
VICTORIA, TX	30	18	26	84	181	368	514	601	597	454	248	83	29	3203
WACO, TX	30	2	8	39	111	292	497	637	628	416	170	34	6	2840
WICHITA FALLS, TX	30	0	2	19	66	220	448	618	574	339	99	10	1	2396
MILFORD, UT	30	0	0	0	1	18	115	286	239	54	2	0	0	715
SALT LAKE CITY, UT	30	0	0	0	4	34	184	395	355	111	6	0	0	1089
BURLINGTON, VT	30	0	0	0	3	23	96	192	139	35	1	0	0	489
LYNCHBURG, VA	30	0	0	3	20	72	218	348	308	141	19	2	0	1131
NORFOLK, VA	30	1	2	8	35	119	303	456	403	240	50	11	2	1630
RICHMOND, VA	30	0	1	8	33	107	282	428	374	193	33	6	1	1466
ROANOKE, VA	30	0	0	4	20	74	217	355	309	136	17	2	0	1134
OLYMPIA, WA	30	0	0	0	0	2	10	38	40	7	0	0	0	97
QUILLAYUTE, WA	30	0	0	0	0	1	3	7	8	4	0	0	0	23
SEATTLE C.O., WA	30	0	0	0	0	4	15	67	79	27	0	0	0	192
SEATTLE SEA-TAC AP, WA	30	0	0	0	0	5	19	65	65	19	0	0	0	173
SPOKANE, WA	30	0	0	0	1	11	46	155	154	26	1	0	0	394
WALLA WALLA, WA	30	0	0	0	4	29	138	329	323	131	3	0	0	957
YAKIMA, WA	30	0	0	0	1	18	68	187	163	28	0	0	0	465
SAN JUAN, PR	30	360	332	388	421	484	513	533	539	515	513	436	392	5426
BECKLEY, WV	30	0	0	1	8	29	99	183	149	56	4	0	0	529
CHARLESTON, WV	30	0	0	7	25	77	204	324	280	125	18	3	1	1064
ELKINS, WV	30	0	0	0	2	16	76	153	126	41	2	0	0	416
HUNTINGTON, WV	30	0	0	8	25	82	218	341	300	135	19	3	1	1132
GREEN BAY, WI	30	0	0	0	3	24	95	177	126	36	2	0	0	463
LA CROSSE, WI	30	0	0	0	8	49	162	272	208	70	6	0	0	775
MADISON, WI	30	0	0	0	6	33	123	214	154	48	4	0	0	582
MILWAUKEE, WI	30	0	0	0	5	27	114	222	180	63	5	0	0	616
CASPER, WY	30	0	0	0	0	3	64	186	154	28	0	0	0	435
CHEYENNE, WY	30	0	0	0	0	1	41	126	92	20	0	0	0	280
LANDER, WY	30	0	0	0	0	3	70	190	153	29	0	0	0	445
SHERIDAN, WY	30	0	0	0	1	4	52	165	154	30	1	0	0	407

Source: U.S. National Oceanic and Atmospheric Administration (http://ols.nndc.noaa.gov/plolstore/plsql/olstore.prodspecific?prodnum=C00095-PUB-A0001).

TABLE 5.14   COOLING LOAD HOURS FOR SELECTED CITIES IN THE UNITED STATES.

City	State	Cooling Load Hours	City	State	Cooling Load Hours
Birmingham	AL	1000	Fort Myers	FL	1750
Huntsville	AL	1000	Gainesville	FL	1250
Mobile	AL	1500	Jacksonville	FL	1250
Montgomery	AL	1250	Key West	FL	2250
Anchorage	AK	0	Miami	FL	2000
Annette	AK	0	Orlando	FL	1500
Barrow	AK	0	Pensacola	FL	1500
Bethel	AK	0	Tallahassee	FL	1250
Bettles	AK	0	Tampa	FL	1750
Big Delta	AK	0	Vero Beach	FL	1750
Cold Bay	AK	0	West Palm Beach	FL	2000
Fairbanks	AK	0	Athens	GA	750
Gulkana	AK	0	Atlanta	GA	750
Homer	AK	0	Augusta	GA	1000
Juneau	AK	0	Columbus	GA	1000
King Salmon	AK	0	Macon	GA	1000
Kodiak	AK	0	Savannah	GA	1000
Kotzebue	AK	0	Hilo	HI	2500
Mcgrath	AK	0	Honolulu	HI	2500
Nome	AK	0	Kahului	HI	2500
St. Paul Island	AK	0	Lihue	HI	2500
Talkeetna	AK	0	Boise	ID	550
Unalakleet	AK	0	Lewiston	ID	450
Valdez	AK	0	Pocatello	ID	450
Yakutat	AK	0	Chicago	IL	450
Flagstaff	AZ	450	Moline	IL	550
Phoenix	AZ	1500	Peoria	IL	550
Tucson	AZ	1750	Rockford	IL	450
Winslow	AZ	750	Springfield	IL	750
Yuma	AZ	2000	Evansville	IN	750
Fort Smith	AR	1000	Fort Wayne	IN	450
Little Rock	AR	1000	Indianapolis	IN	550
North Little Rock	AR	1000	South Bend	IN	450
Bakersfield	CA	500	Des Moines	IA	550
Bishop	CA	1000	Dubuque	IA	450
Eureka	CA.	250	Sioux City	IA	450
Fresno	CA	750	Waterloo	IA	450
Long Beach	CA	125	Concordia	KS	750
Los Angeles	CA	125	Dodge City	KS	750
Mount Shasta	CA	450	Goodland	KS	750
Redding	CA	250	Topeka	KS	750
Sacramento	CA	750	Wichita	KS	750
San Diego	CA	125	Jackson	KY	750
San Francisco AP	CA	250	Lexington	KY	750
Santa Barbara	CA	250	Louisville	KY	750
Santa Maria	CA	250	Paducah	KY	750
Stockton	CA	750	Baton Rouge	LA	1250
Alamosa	CO	550	Lake Charles	LA	1250
Colorado Springs	CO	550	New Orleans	LA	1250
Denver	CO	450	Shreveport	LA	1250
Durango	CO	750	Caribou	ME	150
Grand Junction	CO	750	Portland	ME	150
Pueblo	CO	550	Baltimore	MD	350
Bridgeport	CT	350	Blue Hill	MA	350
Hartford	CT	350	Boston	MA	350
Wilmington	DE	350	Worcester	MA	350
Washington	D.C.	350	Alpena	MI	250
Apalachicola	FL	1500	Detroit	MI	350
Daytona Beach	FL	1500	Flint	MI	250

TABLE 5.14   COOLING LOAD HOURS FOR SELECTED CITIES IN THE UNITED STATES. (CONTINUED)

City	State	Cooling Load Hours	City	State	Cooling Load Hours
Grand Rapids	MI	250	Grand Forks	ND	250
Houghton Lake	MI	250	Williston	ND	450
Lansing	MI	350	Akron	OH	350
Marquette	MI	150	Cleveland	OH	350
Muskegon	MI	250	Columbus	OH	550
Sault Ste. Marie	MI	150	Dayton	OH	550
Duluth	MN	150	Mansfield	OH	450
International Falls	MN	150	Toledo	OH	450
Minneapolis-St. Paul	MN	350	Youngstown	OH	350
Rochester	MN	350	Oklahoma City	OK	1000
Saint Cloud	MN	250	Tulsa	OK	1000
Jackson	MS	1250	Astoria	OR	250
Meridian	MS	1000	Burns	OR	350
Tupelo	MS	1000	Eugene	OR	350
Columbia	MO	750	Medford	OR	450
Kansas City	MO	750	Pendleton	OR	450
St. Louis	MO	750	Portland	OR	250
Springfield	MO	750	Salem	OR	250
Billings	MT	350	Sexton Summit	OR	350
Glasgow	MT	350	Allentown	PA	450
Great Falls	MT	250	Erie	PA.	350
Havre	MT	250	Harrisburg	PA	450
Helena	MT	250	Philadelphia	PA	450
Kalispell	MT	250	Pittsburgh	PA	350
Missoula	MT	250	Avoca	PA	350
Grand Island	NE	550	Williamsport	PA	450
Lincoln	NE	550	Providence	RI	350
Norfolk	NE	550	Charleston Co.	SC	1000
North Platte	NE	550	Columbia	SC	1000
Omaha	NE	550	Greenville	SC	750
Scottsbluff	NE	450	Aberdeen	SD	450
Valentine	NE	450	Huron	SD	450
Elko	NV	550	Rapid City	SD	450
Ely	NV	550	Sioux Falls	SD	450
Las Vegas	NV	1250	Bristol-Johnson City-Kingsport	TN	750
Reno	NV	250	Chattanooga	TN	750
Winnemucca	NV	750	Knoxville	TN	750
Concord	NH	250	Memphis	TN	1000
Mt. Washington	NH	250	Nashville	TN	750
Atlantic City	NJ	250	Oak Ridge	TN	750
Newark	NJ	250	Abilene	TX	1250
Albuquerque	NM	750	Amarillo	TX	750
Clayton	NM	550	Austin/City	TX	1250
Roswell	NM	1000	Austin/Bergstrom	TX	1250
Albany	NY	350	Brownsville	TX	2000
Binghamton	NY	250	Corpus Christi	TX	1750
Buffalo	NY	250	Dallas-Fort Worth	TX	1250
Islip	NY	350	Del Rio	TX	1750
New York	NY	150	El Paso	TX	1250
Rochester	NY	250	Galveston	TX	1250
Syracuse	NY	350	Houston	TX	1250
Asheville	NC	750	Lubbock	TX	1000
Cape Hatteras	NC	750	Midland-Odessa	TX	1250
Charlotte	NC	750	Port Arthur	TX	1250
Greensboro-Winston	NC	750	San Angelo	TX	1250
Raleigh	NC	750	San Antonio	TX	1750
Wilmington	NC	1000	Victoria	TX	1500
Bismarck	ND	350	Waco	TX	1500
Fargo	ND	350	Wichita Falls	TX	1250

TABLE 5.14   COOLING LOAD HOURS FOR SELECTED CITIES IN THE UNITED STATES. (CONTINUED)

City	State	Cooling Load Hours	City	State	Cooling Load Hours
Milford	UT	750	Beckley	WV	550
Salt Lake City	UT	750	Charleston	WV	550
Burlington	VT	250	Elkins	WV	450
Lynchburg	VA	350	Huntington	WV	550
Norfolk	VA	750	Green Bay	WI	250
Richmond	VA	750	La Crosse	WI	350
Roanoke	VA	750	Madison	WI	350
Olympia	WA	150	Milwaukee	WI	350
Quillayute	WA	150	Casper	WY	450
Seattle	WA	150	Cheyenne	WY	350
Spokane	WA	350	Lander	WY	250
Walla Walla	WA	450	Sheridan	WY	350
Yakima	WA	350			

HVAC EQUIPMENT

6.1 HEATING, VENTILATING, AND AIR CONDITIONING

History of Heating and Cooling For Human Comfort

The need for heating and cooling for human comfort was likely first recognized when early humans wandered into a cave to escape summer heat and winter cold. The need for ventilation was likely recognized when early humans desired to air out a smoky cave. Since then, humans have continually searched for ways to keep warm in the winter and cool in the summer and breathe fresh air year round.

History of Space Heating

There is archeological evidence that as early as 1.5 million years ago, prehistoric humans began using campfires. At some point thereafter, the campfire was brought inside caves and huts for cooking and space heating. The central fire and a central roof opening for smoke to escape was the oldest arrangement. Later the fire was moved to different parts of a building.

Early central heating systems date as far back as 2500 B.C. in Greece and 1300 B.C. in Turkey with the use of hearths and fireplaces, stoves, and under-floor systems. About 2000 years ago, Roman systems heated building floors, which heated the building interior through radiant energy. Later, systems in upper-class Roman houses were designed as warm-air heating systems, which introduced heated air into the building space through openings in the floor. Crude fireplace heating was used in Europe as early as about 800 C.E. In about 1200 C.E., the Luneberg City Hall in Germany had a central warm-air system with furnaces and a heating chamber connected to rooms above with round ducts. Fireplaces and stoves made from clay bricks were prevalent in Europe by about 1300 C.E.

Important advancements in space heating technology were the invention of the chimney and the "smokeless stove." The large chimneys of that time discharged combustion products above the building and could be easily cleaned, but were quite drafty. The earliest metal stoves, made of cast iron hearth appeared in about 1400 C.E. In North America and Europe, the cast iron stove was a well-established method of comfort heating by the 1850s.

The Industrial Revolution served as the catalyst for more advanced warm-air systems in buildings. William Strutt in England invented a warm-air furnace that consisted of a riveted, wrought iron air chamber encased in brick about 1805. Ducts connected to the furnace fed heated air into rooms and the room openings were fitted with dampers to regulate airflow. Warm-air systems were introduced in the United States in larger institutional buildings before 1820. The first U.S. building to be centrally heated was most likely the Massachusetts Medical College in 1816, which used a gravity warm-air system with a basement furnace and ducts extending to the individual rooms.

Steam heating evolved in England during the late 1700s, but was only used to heat industries. A number of steam heating systems were installed in the United States after about 1810. These steam systems used coils or rows of pipe to heat rooms. Residential steam systems were introduced in 1854. Steam heating was used in large buildings but never became popular for home heating because of its complexity, noise, and the fear of explosions. The choice system for residences and small commercial buildings became hot water systems.

The earliest recorded use of hot water heating in buildings can be tied to a monastery in Greenland that used hot spring water about 1400 C.E. In the late 1700s, an actual hot water heating system using a boiler was introduced in France. Early hot water systems used the thermosiphon principle for circulation and used very large pipes because it was thought this was necessary to ensure adequate circulation. Radiators were patented in 1863, and by the 1880s, cast iron sectional radiators became popular. The concept of a circulator (pump) was not introduced until the early 1900s.

Early developments in forced air systems began in the mid-1800s with use in large buildings and factories, where fans were cranked by hand and later were driven by steam engines. They moved only outside air. The concept of recirculated air was not introduced until almost 1900. In 1846, a fan like a paddle wheel was used in the United States to move heated air. In 1855, parts of the U.S. Capitol were heated and ventilated with large centrifugal fans powered by steam engines.

Before about 1925, buildings had hand-fired fireplaces, stoves, furnaces, or boilers. Occupants had to shovel fuel (coal or wood) and clear ash on a daily basis, fully stoking fires at night and waking up early to stoke the fire with fresh fuel. Guesswork and experience dictated how much fuel to use. The introduction of oil, gas, and electricity coupled with the automatic thermostat has led to the modern automated space heating systems we use today. Such systems allow more complex and energy-efficient buildings that are extremely comfortable.

History of Ventilation

Natural ventilation through openings (e.g., operable windows and doors) was the only means of ventilating buildings prior to the development of the steam engine and electric power industry in the 1800s. As late as 1925, placement and geometry of open windows and room exhausts, which used the buoyancy effect of hot air to exhaust room air, were still used in building design. The development of mechanical ventilation and air conditioning in the 1930s caused the reliance on natural ventilation to gradually become out of date.

Mechanical ventilation was used in the early 1800s. Air was moved through ducts by fans driven by steam engines. Ventilation rates of 30 cfm (15 L/s) per person were common practice for school and public buildings in the 1900s. The oil embargo of 1974 brought attention to energy use. As a result of energy conservation efforts, minimum acceptable ventilation rates were dropped to as low as 5 cfm (2.5 L/s) per person in the 1980s. Building-related illnesses caused by poor ventilation resulted. Demand for better indoor air quality gradually led to improvements in ventilation. Today, ventilation systems are designed and operated to achieve a proper balance involving thermal comfort, air quality, and energy consumption. Currently, more effective ventilation rates of 20 cfm (10 L/s) per person or more are common to ensure health and comfort of occupants.

History of Air Conditioning

It has only been in the last 100 years that mechanical refrigeration (air conditioning) has been in use. Before then, the only means of cooling a building was by natural ventilation, ventilating with fans, or blowing air across large chunks of ice before introducing the cooled air into the building space.

Refrigeration was originally used in controlling humidity in manufacturing plants, such as textile mills and printing plants, to preserve meat and perishable foods, or to chill beer. One of the first uses of air conditioning for occupant comfort was in 1902 when the New York Stock Exchange's new building was equipped with a central cooling and heating system. Alfred Wolff, an engineer from Hoboken, New Jersey, helped design the new system, transferring air conditioning technology from industrial plants to commercial buildings.

Willis Carrier, a mechanical engineer, was an early pioneer of "controlled air." Starting in 1902, he designed a spray-type temperature and humidity controlled system. His induction system for office buildings, hotels, apartments, and hospitals was just one of his air-related inventions. Many industry historians consider him the father of air conditioning.

Early on, industries and some office buildings and schools were the primary focus of air conditioning in buildings. From 1911 to 1930, many movie theaters were air conditioned, providing theater occupants with entertainment and an escape from a hot and humid environment. With these advancements, air conditioning technology was providing the means to control humidity and temperature that improved personal comfort. Following World War II, use of air conditioning in warm climates found its way into commercial and residential buildings.

Modern HVAC Systems

A *heating, ventilation, and air conditioning (HVAC)* system refers to the system or systems that condition air in a building. It includes heating, cooling, humidifying, dehumidifying, and cleaning (filtering) air in building spaces. *Air conditioning* is a process that involves heating, cooling, humidifying, dehumidifying, and cleaning air. However, whenever the term *air conditioning* is used, most everyone thinks of cooling air only. It is important to recognize that while manufacturers will continue to sell air coolers as air conditioners and when clients say they want (or do not want) air conditioning, they mean cooling, and the technician and the engineer must be aware of the real meaning of the term.

Space heating equipment is used for heating air in building spaces. It can be a heat pump, furnace, boiler, packaged-heating unit, individual space heater, and district steam or hot water piped in from outside the building. *Space cooling equipment* is used to cool air in building spaces. It involves the use of refrigeration (e.g., room air conditioner or heat pump), an evaporative cooling unit (e.g., evaporative cooler), or a central cooling or district cooling system that circulates chilled water. A *humidifier* is a piece of equipment that adds water vapor to the air in a building to increase relative humidity. A *dehumidifier* removes moisture from air by chemical or physical methods.

Chilled water systems cool water at a central point, in a condenser, and then distribute chilled water throughout the building, either to coils in an air distribution system or through pipes embedded in the floor or ceiling. They are used for cooling in many larger buildings.

Heating and cooling *distribution equipment* is that part of a heating or cooling system that distributes conditioned water and/or air throughout a building by means of pipes, ducts, or fans. It includes radiators, baseboards, individual space heaters, unit ventilators, and fan-coil units. Air distribution equipment also provides ventilation air into the building spaces.

Selection of a type of heating and cooling system and the associated equipment will depend on local climate, heating and cooling load, level of desired occupant comfort, client's budget, and energy costs. Climate and fuel or energy costs tend to vary significantly by location. Types of HVAC equipment are introduced in this chapter. Design of basic HVAC systems is introduced in Chapters 8 through 10.

6.2 SPACE HEATING EQUIPMENT

In buildings, heating equipment is used to generate heat for space heating. Two types of equipment are typically used: a boiler, which generates hot water or steam, and a furnace, which produces hot air.

Boilers

A *boiler* is piece of equipment consisting of a vessel or tank-like container where heat produced from the combustion of fuels such as natural gas, fuel oil, or coal or electrical resistance heating is used to produce hot water or steam. In a basic analogy, a boiler heats water in a manner similar to how a kettle of water is heated on a stove. In buildings, a boiler can be used to generate hot water or steam for space heating and domestic hot water. In industrial settings, boilers are used to produce hot water and steam for manufacturing processes and steam to generate electrical power.

A boiler consists of two main components: the furnace, which produces heat by burning a fuel, and the closed vessel, a container where hot water or steam is generated. Fuel and air fed to a burner in the furnace is converted to hot combustion gases. The heating surfaces between the furnace and the vessel place the hot combustion gases in indirect contact with the heat transfer medium (water). Heat from the combustion gases heats the water, creating hot water or steam. A pressure/temperature safety valve is an important device that is used to prevent explosions by releasing steam if the pressure in the boiler becomes too great. The American Society of Mechanical Engineers' *Boiler and Pressure Vessel Code* governs the safe construction of boilers in the United States. See Photo 6.1.

PHOTO 6.1 A commercial-size, hot water fire tube boiler. (Used with permission of ABC)

Types of Boilers

Three common types of boilers are used to generate hot water or steam in buildings: fire tube, water tube, and electric boilers.

Fire Tube Boilers

Fire tube boilers are constructed of a vessel that a hold water and has hollow passages (tubes) within the vessel that are surrounded by the water. Hot combustion gases from the burning fuel pass through the inside of the tubes, driving heat to the water. The tubes may be arranged horizontally or vertically above the combustion area of the furnace. Sometimes baffles or turbulators are inserted in the tubes to control gas velocity and improve heat transfer by forcing the hot gases into more intimate contact with the tube walls. A cross-section of a boiler is shown in Figure 6.1.

Fire tube boilers have a fairly large amount of contained water so that there is a considerable amount of stored heat energy in the boiler. This characteristic allows for load swings where large amounts of steam or hot water are required in a relatively short period of time, as often happens in the morning wake-up of the space heating system and in process applications at a manufacturing plant.

Cast Iron Sectional Boilers

Cast iron sectional boilers are a type of fire tube boiler. In most original boiler rooms, the building was built around or on top of the boiler room. When it came time to replace many of these boilers, there was no convenient way to bring in a new boiler. The cast iron sectional boiler provided a solution to this problem. Cast iron sections are taken individually into the boiler room and field assembled at the job site. Although the initial cost is low, much on-the-job labor is required to make the boiler fully operational. Cast iron boilers are used for closed, low-pressure heating systems only. These boilers also have small steaming areas so that they are very sensitive to changes in demand or water levels.

Water Tube Boilers

Water tube boilers hold water in tubes and allow hot combustion gases from the burning fuel to pass over the tubes in which the boiler water is contained. Metal or refractory baffles direct the gas flow to improve heat transfer. Water tube boilers are characterized by smaller-diameter tubes, which are longer, in bent, curved, or straight configurations. Many of these boilers have tubes configured in a serpentine or spiral. These tubes are often aligned in parallel to make up a complete wall or panel of heat-absorbing water tubes. For this reason, the heat absorption rate per unit of furnace volume or furnace wall area is relatively high as compared to fire tube boilers. Most water tube boilers contain a small amount of boiler water so they are very sensitive to changes in demand and water level. They require close attention and more sophisticated control systems, and demand good consistent maintenance and water treatment.

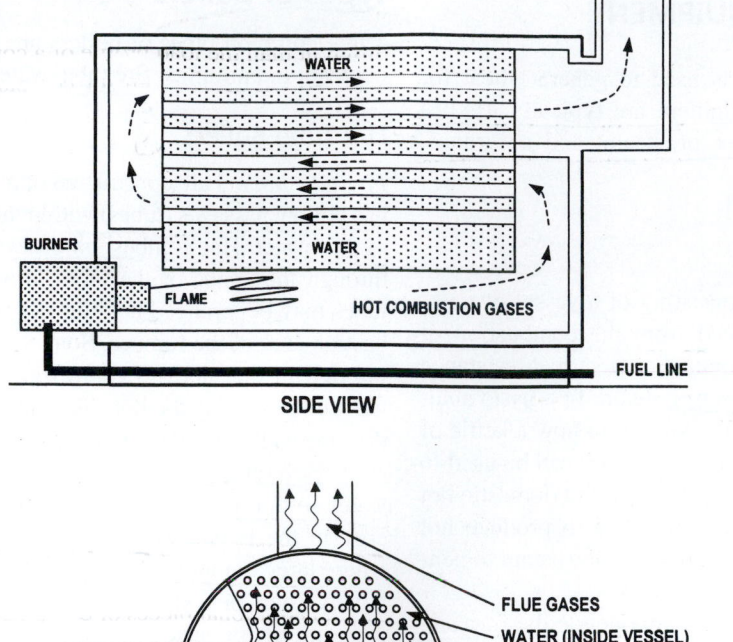

FIGURE 6.1 A cross-section and front cut-away view of a commercial-size, three-pass, fire tube boiler.

Electric Boilers

Electric boilers are in a separate classification because they are not heated with combustion of a fuel and no flue openings are needed. Instead, in these boilers an electrode or electrical resistance heating heats water.

Fire tube and water tuber boilers are called "one-pass," "two-pass," "three-pass," or "four-pass" according to the number of times the heat released from combustion gases is conducted through the boiler before the gases exit the vent. The furnace in which the combustion takes place is counted as one "pass." If the tubes are arranged so that the combustion gases must make a 180° turn to enter them, then each group of flue passages requiring a reversal of gas direction is referred to as a separate "pass." For example, a three-pass boiler consists of a furnace (first pass), from which the gases exit with a 180° turn into a course of flue tubes (second pass), from which they exit with a 180° turn into another course of flue tubes (third pass), before exiting into the vent. If the tubes are arranged so that they are actually an extension of the combustion chamber, as in a vertical tube unit, then they are not considered as a separate pass and the unit is then called a "one-pass" boiler.

In recent years several features have been added to boiler design to improve the efficiency of a boiler heating system, including what follows: electric ignition to eliminate the need for a standing pilot light; motorized dampers that shut the exhaust vent when the boiler cycles off to avert the loss of heated room air between boiler cycles; and fan-assisted combustion systems (forced draft, power burner, pressure fired, induced draft, power vent) that do not require a large natural draft chimney and can be vented through a smaller-size vent directly through a wall.

Boiler Ratings

Commercial boiler ratings are related to the amount of heat that will be produced when the boiler is fired at specified fuel input under specified conditions. The basic factor governing boiler output is the rate of heat release that can be maintained in the furnace of the boiler as determined by the type of firing equipment used. The *heat release rate* of a boiler is an expression of the rate at which the combustion of the fuel liberates heat energy based on the fuel input in Btu per hour. Heat release rate may be defined as either Btu per hour per cubic foot of furnace volume or as Btu per hour per square foot of radiant heating surface.

Gross Output Rating

Gross output rating is the total amount of heat available at the boiler outlet for a specified fuel input. Gross output is usually expressed in thousands of Btu per hour (MBH), boiler horsepower (BHP), or pounds of steam per hour. Electric boilers are rated in kilowatts (kW). The gross output rating in SI units is the kilowatt (kW).

Boiler Horsepower

Boiler horsepower is defined as the evaporation of 34.5 pounds of water into steam at 212°F at standard atmospheric pressure. One boiler horsepower is equal to 33 475 Btu/hr (9.81 kW). Boiler horsepower may also be related to the heat transfer rate of the boiler expressed by the square feet of heat transfer surface per horsepower. A rating based on 5 square feet is equivalent to a heat transfer rate of approximately 6.7 MBH per square foot of heating surface. (Similarly, at 4 square feet per horsepower, the heat transfer rate is 8.370 MBH/ft^2.)

Net Rating

Net ratings represent the net connected design load that can be supplied with heat by a boiler of a given output, allowing for normal piping losses and pickup from a cold start. Net ratings are usually stated both in thousands of Btu per hour (MBH), kilowatts (kW), and in square feet of radiation. The Hydronic Institute of Boiler and Radiator Manufacturers (I=B=R or IBR) rating for boilers is based on an allowance of 1.15 (additional 15%) for piping and pick-up losses.

Boiler Capacities

The range of capacities of manufactured fuel-fired boilers on the market is:

Residential: 40 to 300 MBH (12 to 88 kW)
Commercial: 300 MBH to 30 000 MBH (88 to 8800 kW)
Industrial: 30 000 to 100 000 MBH (8800 to 29 300 kW)
Utility: 100 000 MBH and up (29 300 kW and up)

Electric boilers are available up to 12 000 kW for generation of steam and hot water.

Boiler Efficiency

The *overall efficiency* of a boiler is the gross heat output, the heat in the hot water or steam (in Btu), divided by the energy input (Btu of fuel or electricity). For example, a boiler that delivers 40 MMBH (millions of Btu per hour) of hot water and that consumes 50 MMBH of natural gas has an overall efficiency of 80%. All boilers are rated by their overall efficiency.

Small (up to 300 MBH) gas- and oil-fired boilers are rated according to their *annual fuel utilization efficiency (AFUE)*. AFUE accounts for the effect of part-load efficiency and cyclic losses; a single AFUE number represents performance under a specific set of conditions. The conditions, chosen to represent operation in an average climate with a certain usage pattern, include flue and infiltration losses during on and off cycles. AFUE is based on a residential load profile, which may be quite different than the load profile of a commercial building.

AFUE serves well for comparing two boilers under the same test conditions, but it is less useful for predicting annual fuel use in the field, where local conditions may not match the AFUE test conditions and calculation assumptions. Since 1992, the National Appliance Energy Conservation Act has required that small gas boilers maintain combustion efficiencies at a minimum of 80% for gas fired and 83% for oil fired models. The Energy Star Program, which is run by the U.S. Environmental Protection Agency and Department of Energy, awards an Energy Star label to boilers with an AFUE of 85% or better. The most efficient boilers on the market have an AFUE of about 97%. See Tables 6.1 through Table 6.4.

Boilers typically operate most efficiently at full load. As they cycle on and off, their efficiency can fall off by as much 20%.

Ancillary Boiler Equipment

Several additional pieces of equipment are attached to and used by a boiler in a building space heating system. These are shown in Figure 6.2. These include:

Expansion Tank

The expansion tank, also called a compression tank, allows for the expansion of the water in the system as it is heated. It is located above the boiler.

Automatic Filler Valve

When the pressure in the system drops below 12 psi (82.7 kPa), this valve opens, allowing more water into the system; then the check valve automatically closes as the pressure increases. This maintains a minimum water level in the system.

Circulating Pump

Hot water is circulated from the boiler through the pipes and back to the boiler by a pump. The pump provides fast distribution of hot water through the system, thus delivering heat as quickly as possible. The circulating pump size is based on the delivery of water required, in gpm (L/s), and the amount of friction head allowed. Circulating pumps are not used on gravity systems.

Flow Control Valve

The flow control valve closes to stop the flow of hot water when the pump stops, so that the hot water will not flow through the system by gravity. Any gravity flow of the water would cause the temperature in the room to continue to rise.

Boiler Relief Valve

When the pressure in a low-pressure boiler system exceeds a pressure of 30 psi (207 kPa), a spring-loaded valve opens, allowing some of the water to bleed out of the system and the pressure to

TABLE 6.1 PERFORMANCE SPECIFICATIONS FOR HIGH-EFFICIENCY RESIDENTIAL HOT WATER BOILERS. COMPILED FROM MANUFACTURERS' INFORMATION.

Gas Fired				
Input		Output		
Btu/hr	kW	Btu/hr	kW	AFUE
26 000	7.6	21 970	6.4	84.5
35 520	10.4	33 424	9.8	94.1
38 000	11.1	32 794	9.6	86.3
38 000	11.1	32 414	9.5	85.3
42 000	12.3	35 448	10.4	84.4
45 000	13.2	40 500	11.9	90.0
47 000	13.8	46 530	13.6	99.0
53 400	15.6	50 249	14.7	94.1
55 000	16.1	48 400	14.2	88.0
68 000	19.9	61 200	17.9	90.0
70 000	20.5	68 600	20.1	98.0
74 000	21.7	68 080	19.9	92.0
79 920	23.4	74 006	21.7	92.6
81 000	23.7	76 302	22.4	94.2
90 000	26.4	81 360	23.8	90.4
90 000	26.4	81 000	23.7	90.0
94 000	27.5	92 120	27.0	98.0
95 000	27.8	90 250	26.4	95.0
98 000	28.7	83 790	24.6	85.5
100 000	29.3	90 000	26.4	90.0
112 000	32.8	105 392	30.9	94.1
115 400	33.8	106 860	31.3	92.6
117 000	34.3	114 660	33.6	98.0
129 000	37.8	118 680	34.8	92.0
140 000	41.0	135 800	39.8	97.0

Oil Fired				
Input		Output		
Btu/hr	kW	Btu/hr	kW	AFUE
62 000	18.2	54 312	15.9	87.6
74 000	21.7	64 750	19.0	87.5
76,000	22.3	68 476	20.1	90.1
80 000	23.4	70 400	20.6	88.0
83 000	24.3	72 957	21.4	87.9
90 000	26.4	78 930	23.1	87.7
91 000	26.7	81 081	23.8	89.1
92 000	27.0	81 236	23.8	88.3
93 000	27.2	81 375	23.8	87.5
94 000	27.5	83 942	24.6	89.3
97 000	28.4	85 942	25.2	88.6
98 000	28.7	87 122	25.5	88.9
100 000	29.3	88 000	25.8	88.0
105 000	30.8	93 135	27.3	88.7
111 000	32.5	97 791	28.7	88.1
112 000	32.8	99 344	29.1	88.7
127 000	37.2	111 506	32.7	87.8
130 000	38.1	113 880	33.4	87.6
132 000	38.7	118 800	34.8	90.0
147 000	43.1	128 772	37.7	87.6
156 000	45.7	139 152	40.8	89.2
167 000	48.9	146 292	42.9	87.6

TABLE 6.2 PERFORMANCE SPECIFICATIONS FOR SELECTED COMMERCIAL GAS-FIRED HOT WATER BOILERS. (MBH = 1000 BTU/HR). COMPILED FROM MANUFACTURERS' INFORMATION.

Input		Output		I=B=R Net Rating	
MBH	kW	MBH	kW	MBH	kW
360	105	298	87	258	76
420	123	342	100	297	87
500	146	412	121	358	105
600	176	480	141	417	122
750	220	618	181	537	157
794	233	610	179	530	155
900	264	720	211	626	183
1000	293	824	241	717	210
1191	349	915	268	796	233
1200	352	960	281	835	245
1500	439	1200	352	1043	306
1588	465	1220	357	1061	311
1800	527	1440	422	1252	367
1985	582	1525	447	1326	389
2100	615	1680	492	1461	428
2382	698	1830	536	1591	466
2400	703	1920	563	1675	491
2779	814	2135	626	1857	544
3176	931	2440	715	2122	622
3573	1047	2745	804	2387	699
3970	1163	3050	894	2652	777
4367	1280	3355	983	2917	855
4764	1396	3660	1072	3183	933
5161	1512	3965	1162	3448	1010
5558	1628	4270	1251	3714	1088
5955	1745	4575	1340	3978	1166
6352	1861	4880	1430	4243	1243
6749	1977	5185	1519	4509	1321
7146	2094	5490	1609	4774	1399

drop. The valve should be located where the discharge will not cause any damage. On high-pressure systems, the pressure limit is increased.

Heating Plant

In a typical commercial building, a heating plant consists of one or several boilers that produce hot water or steam. This hot water or steam is pumped to one or more air handlers, where the heat of the hot water or steam is transferred to indoor air. The heated air is distributed around the building through a network of ducts. The ducts run to terminal units that control the flow of air to diffusers.

Use of Multiple Small Boilers

An installation of multiple small boilers is typically more efficient and effective than an installation of a single large boiler. A small boiler is normally more efficient than a comparable

TABLE 6.3   PERFORMANCE SPECIFICATIONS FOR SELECTED COMMERCIAL OIL-FIRED HOT WATER BOILERS (MBH = 1000 BTU/HR). COMPILED FROM MANUFACTURERS' INFORMATION.

Input		Output		I=B=R Net Rating	
MBH	kW	MBH	kW	MBH	kW
420	1433	338	1154	295	1007
560	1911	454	1550	395	1348
660	2253	532	1816	463	1580
770	2628	612	2089	532	1816
840	2867	676	2307	588	2007
990	3379	798	2724	694	2369
1155	3942	918	3133	798	2724
1320	4505	1064	3631	925	3157
1540	5256	1224	4178	1064	3631
1925	6570	1530	5222	1330	4539
2310	7884	1836	6266	1596	5447
2695	9198	2142	7311	1862	6355
3080	10 512	2448	8355	2129	7266
3465	11 826	2754	9399	2395	8174
3850	13 140	3060	10 444	2661	9082
4235	14 454	3366	11 488	2927	9990
4620	15 768	3672	12 533	3193	10 898
5005	17 082	3978	13 577	3459	11 806
5390	18 396	4284	14 621	3725	12 713
5775	19 710	4590	15 666	3991	13 621
6160	21 024	4896	16 710	4257	14 529
6545	22 338	5202	17 754	4523	15 437
6930	23 652	5508	18 799	4790	16 348

TABLE 6.4   PERFORMANCE SPECIFICATIONS FOR SELECTED COMMERCIAL DUAL-FUEL (GAS- AND OIL-FIRED) HOT WATER BOILERS (MBH = 1000 BTU/HR). COMPILED FROM MANUFACTURERS' INFORMATION.

Input		Output		I=B=R Net Rating	
MBH	kW	MBH	kW	MBH	kW
770	2628	612	2089	532	1816
1155	3942	918	3133	798	2724
1540	5256	1224	4178	1064	3631
1925	6570	1530	5222	1330	4539
2310	7884	1836	6266	1596	5447
2695	9198	2142	7311	1862	6355
3080	10 512	2448	8355	2129	7266
3573	12 195	2754	9399	2395	8174
3970	13 550	3060	10 444	2661	9082
4367	14 905	3366	11 488	2927	9990
4764	16 260	3672	12 533	3193	10 898
5161	17 614	3978	13 577	3459	11 806
5558	18 969	4284	14 621	3724	12 710
5955	20 324	4590	15 666	3991	13 621
6352	21 679	4896	16 710	4257	14 529
6749	23 034	5202	17 754	4523	15 437
7146	24 389	5508	18 799	4790	16 348

large boiler. A boiler is most efficient when it operates at or near its rated capacity. Therefore, if building loads are highly variable, as is often the case in commercial and institutional buildings, multiple small boilers can operate more efficiently than a single large boiler because each boiler can operate more often at or close to its full load.

Multiple small boilers also provide redundancy, which can reduce system downtime if one of the boilers is down for repairs or maintenance. Small boilers can also reduce installation

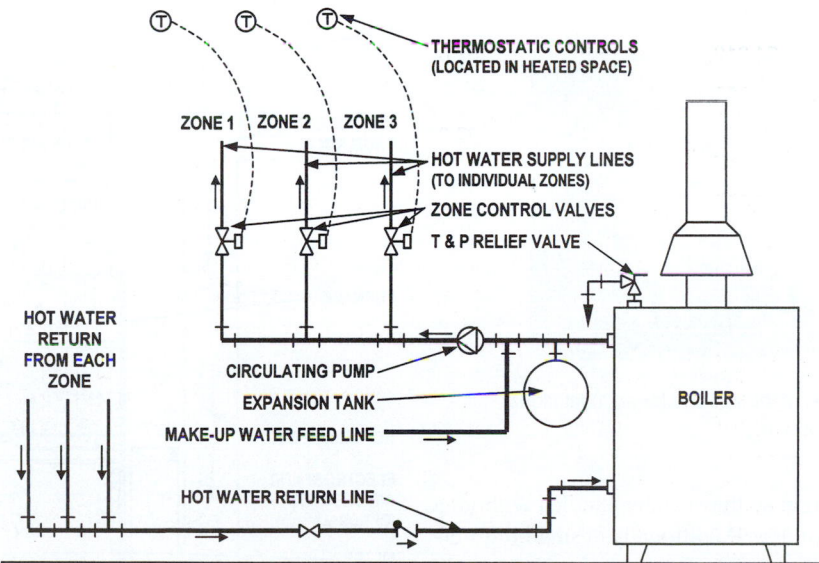

FIGURE 6.2   A drawing of the key components of a hydronic space heating system with a boiler.

PHOTO 6.2   A close-up of the vent for the boilers in the next photograph. (Used with permission of ABC)

PHOTO 6.3   Two boilers used to provide hot water for heating an institutional building. (Used with permission of ABC)

PHOTO 6.4   The main power plant facilities for an institutional building. (Used with permission of ABC)

costs because of lighter weight so they can be handled without a crane (unlike many large boilers). Additionally, a small high-efficiency boiler can also be teamed with a large, inefficient, old boiler to improve overall efficiency. Applied in this manner, the small unit is used whenever there is a heating load, but the old unit is operated only during periods of high load.

Water Treatment of Boilers

Consistent maintenance and careful water treatment go a long way towards ensuring the long life of a boiler. Untreated water is never pure; it contains varying amounts of gases, solids, and pollutants. Although most suspended solids can been removed, gases and salts still remain. These manifest themselves in a boiler system as scale or corrosion. All boiler systems require water treatment.

Warm-Air Furnaces

A *warm-air furnace* is an appliance with an enclosed chamber where fuel is burned or electrical resistance heating is used to heat air directly. See Figure 6.3. The heated air is then distributed throughout the building, typically by air ducts. A furnace with a fan contained within the unit is called a *forced-air furnace*. A forced-air furnace operates by drawing air into a heat exchanger, where the air is warmed with a flame of gas or oil or with heated electric coils. As the air is heated, the fan drives the warmed air into supply ducts that are attached to the furnace. Air is distributed by these ducts throughout the building and then enters the space through registers, diffusers, or grills in the ceiling, floor, or wall. Indoor air is circulated continuously through the system as room air is drawn back to the furnace for reheating.

Furnace Design

The basic components of a gas-fired forced-air furnace are a gas valve, burner assembly, heat exchanger, fan, and cabinet. A gas-fired forced-air furnace combusts natural gas or liquefied petroleum gas (LPG) in an open or sealed chamber to generate heat. Hot combustion gases heat one side of a heat exchanger. Room air, driven by a fan, is blown across the other side of the heat exchanger where the air contacts the hot surface of the heat

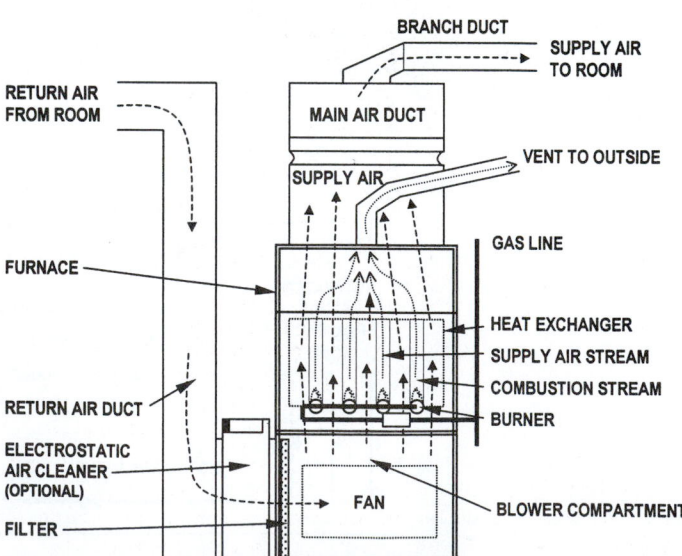

FIGURE 6.3   A sectional view of an upflow warm-air furnace system with optional electronic air cleaner.

exchanger. Heat from the exchanger is transferred to the air. The heated air is then introduced into ductwork, which delivers it to the space. The gas valve controls flow of gas. Manufacturers offer a variety of proprietary combustion chamber and heat exchanger designs.

An oil-fired forced-air furnace uses fuel oil burned in a sealed chamber to heat a heat exchanger and ultimately the room air. The basic components of an oil-fired forced-air furnace are the burner, heat exchanger, firebox, fan, and cabinet, sometimes known as the jacket.

An electric forced-air furnace uses resistance elements to create heat directly in the airstream. Inside the jacket or cabinet will be controls, a fan, and the circuit breakers for the heating elements. Some furnaces have the breakers accessible from the outside of the cabinet. The basic components of an electric force-air furnace are the cabinet or jacket, a fan, heating elements, and controls.

Commercially available forced-air furnaces are classified according to the direction of the airflow for the supply air. They are available in three configurations:

Upflow Furnaces An *upflow furnace* takes in return air from the bottom of the cabinet and discharges heated air through an opening in the top of the cabinet. It is best suited when situated in basement locations. The supply air duct system is attached to the furnace plenum, a sheet metal chamber anchored to the top of the furnace with sheet metal screws. The air conditioning evaporator coil is mounted on top of the furnace. Return air enters the bottom of the furnace. The blower serves for both heating and air conditioning mode, but airflow direction remains the same. See Photo 6.5.

PHOTO 6.5 An upflow, warm-air furnace. The supply air plenum (above furnace) carries heated air to the rectangular main duct or trunk suspended from the ceiling. Return air recirculates back to the furnace in the rectangular return air duct also suspended from the ceiling (in foreground). Main ducts are isolated from the furnace with a flexible duct connection that allows movement and isolates furnace vibration. Round ducts on the side of furnace introduce air for combustion. (Used with permission of ABC)

PHOTO 6.6 A high-efficiency, horizontal furnace being used as a duct heater. (Used with permission of ABC)

Downflow Furnaces A *downflow furnace* takes in return air from the top of the cabinet and discharges heated air through an opening in the bottom of the cabinet. This type of furnace is best suited for applications where ductwork is positioned below the furnace such as on an upper story closet or in an attic space. The air conditioning evaporator coil is mounted on the bottom of the furnace, inside the plenum. Supply air travels from the plenum through ductwork to the living spaces. Return air enters the top of the furnace. The blower serves for both heating and air conditioning.

Horizontal Furnaces A *counterflow* or *horizontal furnace* lays parallel with floor or ceiling members. It takes in return air from the side of the cabinet and discharges heated air through an opening in the opposite side. This type of furnace works best when located in a crawl space or attic. Supply air exits at one end and return air enters through the other end. See Photo 6.6.

There are many different types of noncondensing furnaces, which extract heat from combustion without extracting the latent heat from the water vapor that is generated during combustion. A *natural-draft furnace* is a noncondensing furnace in which the natural flow of air from around the furnace provides the air to support combustion. Natural-draft furnaces are fairly simple with few moving parts and thus easy to maintain. An *induced-draft furnace* is a noncondensing furnace in which a motor-driven fan draws air from the surrounding area or from outdoors to support combustion. These furnaces are also called fan-assisted combustion, forced-draft, power-burner, power-combustion and pressure-fired furnace. The efficiency of standard forced-draft furnaces can be increased by adding extra passes to the combustion heat exchanger. The number of passes and increased efficiency are limited by the amount of heat that can be extracted. It is critical not to remove so much heat as to condense the flue gases. *Direct-vent furnaces,* also called sealed combustion furnaces, draw combustion air directly from outside through a venting system. Sealed combustion is a way to prevent furnaces from inducing infiltration into a building and to more carefully control the combustion process.

Condensing furnaces differ from conventional furnaces in that they recover waste heat from their exhaust gases that would otherwise be lost to the atmosphere. This is achieved through the use of an enlarged heat exchanger surface, which extracts sensible heat and, under certain conditions, extracts the latent heat from the water vapor generated during combustion. The principal component of natural gas is methane (CH_4). The difference between methane's gross caloric value of 23 875 Btu/lb and its net caloric value of 21 495 Btu/lb is the 2380 Btu/lb (10%) of contained in the latent heat of vaporization. This latent heat content is not released unless the combustion gas is condensed. Thus, condensing furnaces are approximately 10% more efficient than the efficient noncondensing furnaces noted earlier. Condensing furnaces have an acidic condensate (pH of ~3.8) that requires heat exchanger materials to be noncorrosive.

The *pulse combustion furnace* is one form of condensing furnace. It is unlike all other types of furnaces in that during firing the combustion is not a continuous equilibrium reaction. Instead, it burns with discrete "charges" of gas/air mixtures in rapid succession, similar to the firing process in an automobile engine. The combustion of the fuel/air mixture creates pressure waves at sonic velocity, which drives the flow of combustion gases out of the combustion chamber and across the heat transfer surfaces. The turbulent flow in pulse furnaces produces high heat transfer rates compared to furnaces.

Residential furnaces are available in heat only models and heat–cool models. The *heat-only furnaces* are designed to generate hot air only. Heat–cool furnaces produce hot air and also serve as the air-handling unit of a split air conditioning system in which the furnace fan moves air but the furnace does not generate heat.

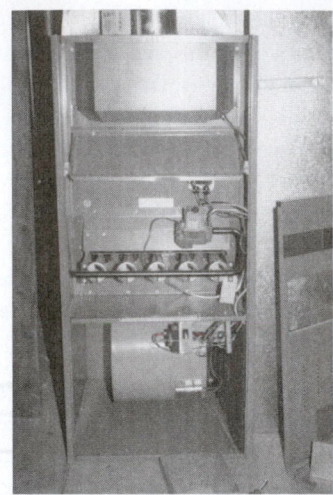

PHOTO 6.9   A standard, natural-draft, upflow warm-air furnace with access panels removed. The fan is evident in the lower blower compartment just below the inlets for five gas burners. The gas regulator is just above the burners. The regulator has not yet been connected to the natural gas line. (Used with permission of ABC)

PHOTO 6.7   Two high-efficiency furnaces used to heat a residence. The two furnaces provide zone control for separate (two) zones. (Used with permission of ABC)

PHOTO 6.8   Sidewall, thermoplastic vents (under construction) to the two high-efficiency furnaces. (Used with permission of ABC)

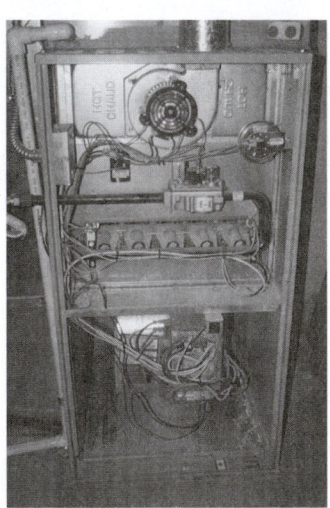

PHOTO 6.10   A medium-efficiency, induced-draft, upflow warm-air furnace with access panels removed. The fan is evident in the lower blower compartment just below the inlets for five gas burners. The gas regulator is just above the burners and has been connected to the natural gas line. An induced-draft fan is evident in the top of the unit. This blower controls the exhaust of combustion products and forces combustion gases up through the vent to prevent condensation in the vent pipe. (Used with permission of ABC)

PHOTO 6.11   A flexible connection isolates furnace vibration and noise from the main supply and return air ducts. (Used with permission of ABC)

PHOTO 6.12   A furnace roof vent before roof-covering materials are installed. Note the metal flashing at the roof line. (Used with permission of ABC)

Furnace Controls and Accessories

A *thermostat* located in the conditioned space typically controls a furnace. A valve or burner controls flow of fuel, allowing fuel flow when the thermostat calls for heat. A *fan switch* controls operation of the fan. The switch turns on the fan once the heat exchanger is adequately heated and switches off the fan once most of the heat is extracted. A *limit switch* prevents overheating by directing the fuel valve to close in the event the heat exchanger overheats.

Accessories that can be installed inside or outside of a forced-air furnace include a fan center (usually a low-voltage relay that can turn the fan on independent of the heat cycle), an air filter or high-performance media filter, electronic air cleaner, and humidifier. When a split air-conditioning system is installed, an evaporator coil is added in the ductwork of the furnace installation. With air conditioning, a condensate pump may be installed to remove the condensate (moisture) discharged in the dehumidification action of the air conditioning system. A zone control panel and motor-actuated dampers can be attached to the furnace in the ductwork if the system is zoned.

Furnace Ratings

Ratings for fuel-fired furnaces are related to the amount of heat that is input and that which will be produced when the furnace is fired under specific conditions. Ratings are determined in accordance with U.S. Department of Energy test procedures with an isolated combustion system. They are expressed as *heating input* and *heating output* of the furnace in Btu per hour (Btu/hr) or thousands of Btu per hour (MBH). For example, a furnace serving a single-family dwelling may be rated at 120 000 Btu/hr heating input and 100 000 Btu/hr heating output. All-electric furnaces are rated in kilowatts (kW). The ratings expressed in SI units are in kilowatt (kW).

Furnace Capacities

The range of capacities of commercially available furnaces is:

Residential:	35 MBH to 175 MBH input	(about 10 to 50 kW)
Commercial:	320 MBH to 2 MMBH input	(about 50 to 586 kW)

Residential furnaces are manufactured in much larger quantities than commercial furnaces, making them relatively inexpensive and readily available with a variety of options. The most efficient residential furnaces are far more efficient than their commercial market counterparts.

Furnace Efficiency

Efficiency of a gas furnace corresponds to the type of furnace:

A *low-efficiency* or *natural aspirating furnace* operates at efficiencies of only 60 to 70%. This type of furnace has low efficiencies because it has a large air requirement in order for exhaust gases to vent naturally. Federal law has required that all new residential furnaces built after January 1992 operate with an AFUE efficiency of 78% or higher. Many low-efficiency furnaces are still in use.

A *mid-efficiency furnace* achieves efficiencies of 78 to 85%. It does so by incorporating an induced-draft fan. Some units also have a motorized damper that closes when the unit is off so that heated air is not lost up the chimney. This type of furnace is the most popular type used in residential construction.

A *high-efficiency* or *condensing furnace* achieves efficiencies of 90 to 97%. Increased efficiency is caused by a secondary heat exchanger that extracts heat from the exhaust gases that would normally flow up the vent. Instead of a flue, the unit incorporates an induced-draft fan or power-vent fan and plastic piping to vent the cooled gases through a wall in the building envelope. Condensate drainage piping carries the condensate from the cooled gases to a waste pipe. In some units, a pump discharges the condensate to the exterior. Units with a sealed combustion system also have an intake air duct so that outside air is used for combustion. A pulse-combustion produces about 60 to 70 tiny explosions of air/gas mixture per second in the combustion chamber instead of a continuous flow of fuel as with most common high-efficiency units. Because this system is noisy, a vibration isolator (canvas fabric) is installed between the furnace and plenum.

Furnaces typically operate most efficiently at full load. As they cycle on and off, their efficiency can fall off by as much 20%.

The furnace industry uses three measures of efficiency to rate furnace performance: steady state efficiency, utilization efficiency, and annual fuel utilization efficiency. Each efficiency rating is determined using a prescribed testing procedure.

Steady State Efficiency   *Steady state efficiency* is computed by measuring energy input less exhaust (vent and condensate) losses divided by energy input. For example, a furnace that delivers 100 MBH of heat and that consumes 120 MBH of natural gas has an overall efficiency of 80%. *Utilization efficiency* is computed by deducting cyclic (on–off), leakage, and pilot light losses in addition to the exhaust losses, so this rating provides an efficiency value that is less than steady state efficiency.

Annual Fuel Utilization Efficiency   Small (up to 300 MBH) gas- and oil-fired furnaces are rated according to their AFUE. AFUE accounts for the effect of part-load efficiency and cyclic losses; a single AFUE number represents performance under a specific set of conditions. The conditions, chosen to represent operation in an average climate with a certain usage pattern, include vent and infiltration losses during on and off cycles. AFUE is based on a residential load profile, which may be quite different than the load profile of a commercial building.

AFUE serves well for comparing two furnaces under the same test conditions, but it is less useful for predicting annual fuel use in the field, where local conditions may not match the AFUE test conditions and calculation assumptions. Since 1992, the National Appliance Energy Conservation Act has required that small gas furnaces maintain combustion efficiencies at a minimum of 78%. The most efficient furnaces have an AFUE of about 97%. See Table 6.5.

Use of Multiple Furnaces

In medium to large residences and some small commercial buildings, an installation of multiple furnaces is typically more effective than an installation of a single large furnace. Use of multiple furnaces offers better air distribution (e.g., less ductwork) and zoning. Furnaces are available in relatively small sizes so they can accommodate nearly any zoning scheme. For example, a common solution to heating a large home is to heat the first floor and basement with an upflow furnace and the second story with a counterflow furnace in the attic. It is necessary to provide power and fuel to each furnace location, but these costs are generally offset by the need for less ductwork.

Individual/Room Space Heaters

An *individual space heater,* often called a *room heater,* is a space heating device that is a freestanding or a self-contained

TABLE 6.5   PERFORMANCE SPECIFICATIONS FOR HIGH-EFFICIENCY RESIDENTIAL WARM-AIR FURNACES. COMPILED FROM MANUFACTURERS' INFORMATION.

Gas Fired				
Capacity				
Input		Output		
Btu/hr		Btu/hr		AFUE
25 000	7.3	24 150	7.1	96.6
38 000	11.1	36 708	10.8	96.6
45 000	13.2	43 200	12.7	96.0
47 000	13.8	44 415	13.0	94.5
57 000	16.7	53 865	15.8	94.5
67 000	19.6	63 985	18.7	95.5
71 000	20.8	67 095	19.7	94.5
75 000	22.0	70 575	20.7	94.1
76 000	22.3	71 820	21.0	94.5
80 000	23.4	75 280	22.1	94.1
90 000	26.4	86 130	25.2	95.7
95 000	27.8	89 775	26.3	94.5
100 000	29.3	94 500	27.7	94.5
109 000	31.9	104 422	30.6	95.8
113 000	33.1	106 333	31.2	94.1
118 000	34.6	111 156	32.6	94.2

Oil Fired				
Capacity				
Input		Output		
Btu/hr		Btu/hr		AFUE
66 500	19.5	63 175	18.5	95.0
71 000	20.8	60 066	17.6	84.6
83 000	24.3	71 380	20.9	86.0
85 500	25.1	81 225	23.8	95.0
91 000	26.7	78 260	22.9	86.0
99 750	29.2	94 763	27.8	95.0
107 000	31.4	90 950	26.6	85.0
116 000	34.0	98 600	28.9	85.0
116 000	34.0	97 672	28.6	84.2
116 000	34.0	97 440	28.5	84.0
145 000	42.5	124 555	36.5	85.9

unit that produces and delivers heat to a single space or local zone within the building. The heater may be permanently mounted in a wall or floor, or may be portable. Examples of individual space heaters include electric baseboards, electric radiant or quartz heaters, heating panels, unit heaters, wood stoves, and infrared radiant heaters. These heaters are characterized by a lack of pipes or ductwork for distributing hot water, steam, or warm air through the building.

Unit Heaters

Unit heaters can be a type of individual forced-air heater that is suspended from the ceiling or hung from the wall. A fan built into a unit heater throws air directly into the space in which the heater is situated. They operate much like a furnace but without the ductwork. Unit heaters are normally gas-fired or electric re-

PHOTO 6.13 A gas-fired unit heater suspended from a warehouse ceiling before natural gas piping and venting are installed. (Used with permission of ABC)

PHOTO 6.14 A gas-fired infrared heating unit. Combustion products fill the tube extending from the unit. (Used with permission of ABC)

sistant heating units. Another type of unit heater is fed by steam or hot water from a boiler when it is available. Unit heaters are commonly used to heat warehouses, attic spaces, mechanical rooms, storage rooms, and garages. See Photo 6.13.

Electric Resistance Heating

Electric baseboard heaters have electric resistance coils mounted behind shallow panels along baseboards. Electric resistant heating consists of one or more baseboard heaters. Heat is transferred from the unit to the room air through convection and, to a lesser degree, radiation. Room air warmed by the heating unit rises to the top of the room, and cooler room air is drawn into the bottom of the heater. Each unit has a separate thermostat to allow for different temperatures in each space.

Electric resistance heating was popular during the 1940s and 1950s because of low initial and operating (electricity) costs. They were again used in the mid-1970s because of energy shortages and a moratorium on installation of natural gas appliances. Today, they tend to be used only in areas where other heating fuels are not readily available.

Infrared Heaters

Infrared heaters consist of a hot surface that radiates heat energy in a specific direction. In large space heating designs, adjustable reflectors that are part of the heating unit control distribution patterns. Radiated energy warms floors and other objects in the space. Little energy is lost because air is a poor absorber. Heat is released from these objects by convection. See Photos 6.14 through 6.16.

The most common gas- and oil-fired infrared heaters are hung from the ceiling. They operate by burning fuel, usually a gas, in a tube or a grate. In the tube system, the inner surface of the tube is heated by contact with the flame or hot combustion gases. The outer surface heats up and radiates heat toward the floor with aid from the reflector. Electric infrared heaters generate heat by electrical current flowing through a high-resistant wire, a graphite ribbon, or a film element. An infrared heat lamp is an example of the simplest electric infrared heater.

PHOTO 6.15 The heated tube and reflector of a gas-fired infrared heater are mounted at the ceiling–wall line. The heated tube emits radiant thermal energy to the space without directly heating the air. The reflector directs thermal radiation to the space, which improves efficiency of the unit. (Used with permission of ABC)

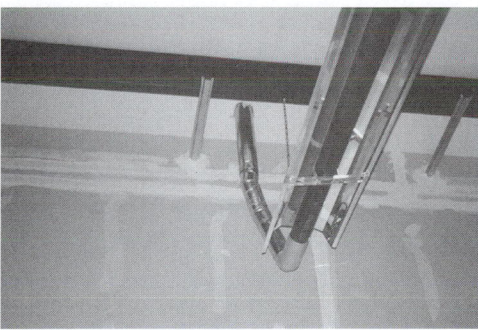

PHOTO 6.16 Combustion products from the infrared heater tube are exhausted through a roof vent. (Used with permission of ABC)

Infrared heaters work well as spot heaters at workbenches in unheated environments or in exterior applications (e.g., sitting areas at ice hockey rinks or outdoor cafes). They should not be used in spaces containing explosive gases, vapors, or dust. See Figure 6.4.

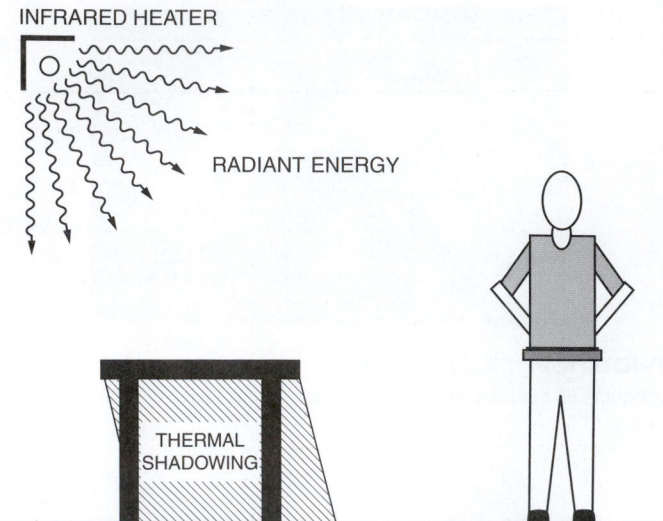

INFRARED HEATER

RADIANT ENERGY

THERMAL SHADOWING

FIGURE 6.4   A schematic sketch of an infrared heater.

Heat Pumps

A *heat pump* is a type of heating and/or cooling device that draws heat into a building from outside during the heating season. If designed with cooling capability also, the heat pump expels heat from the building to the outside during the cooling season. Heat pumps are direct expansion systems whose indoor–outdoor coils are used reversibly as condensers or evaporators, depending on the need for heating or cooling. (The direct expansion refrigeration process is described in this chapter under Space Cooling Equipment.)

Heat pumps can be a factory-assembled packaged unit that includes a fan, compressor, inside coil, and outside coil. Most common is a split system with the air handler (inside coil, fan, and supplementary electric heat strip) installed inside the building and the condensing unit (outside coil and compressor) installed outside. Heat pumps that provide heating and cooling capabilities are like air conditioners except that they include a reversing valve with a different refrigerant circuit design for the refrigerant metering device.

Like a refrigerator, *air-source heat pumps* use electricity to move heat from a cool space into a warm place. Outside air or water is used as a heat source or heat sink, depending on whether the system is heating or cooling. During the heating season, heat pumps move heat from the cool outdoors into the heated building and during the cooling season; heat pumps move heat from the building into the warm outdoors. *Water-source heat pumps* use water in a pond or lake as the heat source and heat sink.

Geothermal heat pumps offer a renewable energy feature that uses the natural heat storage ability of the earth and/or the earth's groundwater to heat and/or cool the building. The earth has the ability to absorb and store heat energy from the sun. To use that stored energy, heat is extracted from the earth through a liquid medium (i.e., groundwater, antifreeze solution, or a refrigerant) and is pumped to the heat pump or heat exchanger. This heat is used to heat the building. In the summer, the process is reversed and indoor heat is extracted from the building and transferred to the earth through the liquid. The geothermal heat pump is more efficient than an air-source heat pump. Geothermal heat pumps are covered in greater detail in Chapter 25.

Heat pumps are normally powered by electricity, although gas-fired units are available. For heating, they transfer heat from the outside coil to the inside coil. If there is not enough heat for the outside coil because the outdoor air temperature is too low, the system or unit turns on a supplementary heat strip. High-efficiency heat pumps also dehumidify better than standard central air conditioners, resulting in less energy usage and more cooling comfort in summer months.

The *heating seasonal performance factor (HSPF)* is a measure of the seasonal efficiency of an electric heat pump using a standard heating load and outdoor climate profile over a standard heating season. It is the total heating output of a heat pump (in Btu) during its normal usage period for heating divided by the total electrical energy input (in watt-hour) during the same period. The higher the HSPF, the more efficient the unit. The resultant value for HSPF has units of Btu/W-hr. Most new heat pumps have HSPF ratings from 7.0 to 9.4.

The efficiency of gas-fired air-source heat pumps is indicated by their *coefficient of performance (COP)*. COP is the ratio of either heat removed (for cooling) or heat provided (for heating) in Btu/hr per Btu/hr of energy input. The higher the COP, the more efficient the unit. Efficient heat pumps should have a COP of 1.2 or greater and a cooling efficiency of 1.25 COP or greater. Performance specifications for high-efficiency heat pumps are provided in Table 6.6.

Packaged Heating Units

A *packaged unit* is a type of heating and/or cooling equipment that is assembled at a factory and installed as a self-contained unit. Packaged units are in contrast to engineer-specified units built up from individual components for use in a given building. A *roof top unit (RTU)* is a packaged heating and/or cooling unit that conditions a structure. An RTU is mounted on the roof and requires adequate support by the roof structure.

6.3   SPACE COOLING (REFRIGERATION) EQUIPMENT

Vapor-Compression Refrigeration Cycle

The *vapor-compression refrigeration* process also known as the *direct expansion (DX) refrigeration cycle* is a method of transferring heat from a low-temperature region to a high-temperature region, essentially driving heat in a direction that is

TABLE 6.6   PERFORMANCE SPECIFICATIONS FOR HIGH-EFFICIENCY HEAT PUMPS.

Type	Capacity	Minimum Efficiency	High Efficiency
Air source (split system)	<65 000 Btu/hr	HSPF V = 6.7 SEER = 13.0	HSPF V = 7.1 (ES) SEER = 14.0 HSPF V = 7.4 (CEE) SEER = 14.0
Air source (single package)	<65 000 Btu/hr	HSPF V = 6.7 SEER = 13.0	HSPF V = 7.0 (ES) SEER = 14.0
	≥65 000 to <135 000 Btu/hr	COP @ 8.3°C = 3.2 COP @ −8.3°C = 2.1 EER = 10.1	
	≥135 000 to <250 000 Btu/hr	COP @ 8.3°C = 3.1 COP @ −8.3°C = 2.0 EER = 9.3	
	≥250 000 Btu/hr	EER = 9.0 (no COP requirement)	
Air source (through-the-wall split system)	<65 000 Btu/hr	HSPF V = 6.2 HSPF V = 6.4 > Jan. 23/2010 SEER = 10.9 SEER = 12 > Jan. 23/2010	
Air source (through-the-wall single package)	<65 000 Btu/hr	HSPF V = 6.1 HSPF V = 6.4 > Jan. 23/2010 SEER = 10.6 SEER = 12 > Jan. 23/2010	
Packaged terminal	All	COP = 3.2 − (0.026 × Cap/1000) (new construction) COP = 2.9 − (0.026 × Cap/1000) (replacement market) EER = 12.3 − .213 × Cap/1000) (new construction) EER = 10.8 − .213 × Cap/1000) (replacements)	
Water source	<17 000 Btu/hr >17 000 to 135 000 Btu/hr	COP = 4.2, EER = 11.2 COP = 4.2, EER = 12.0	
Ground-source	<135 000 Btu/hr	COP @ 0°C = 3.1 ER @ 15°C =16.2	COP @ 0°C = 3.3 (est.)

Note: ES = Energy Star Program; CEE = Consortium for Energy Efficiency; HSPF V = heating seasonal performance factor; Cap = rated cooling capacity, in Btu/hr.

opposite to the direction it naturally flows. In this process, the refrigeration or cooling segment of a building air conditioning system absorbs heat from within the building and transfers it to the outside. The process is used in a kitchen refrigerator that removes heat from the icebox and discharges it outside of the refrigerator cabinet. Vapor-compression refrigeration equipment is used in most buildings that have space-cooling capabilities.

The vapor-compression refrigeration process is used in most building air conditioning systems. Room air conditioners and central air conditioning systems found in homes and light commercial buildings cool using this process. Large commercial buildings use chillers and water towers that rely upon the vapor-compression refrigeration process to generate chilled water and cooling coils and air-handling units to distribute the cooling effect to the building spaces.

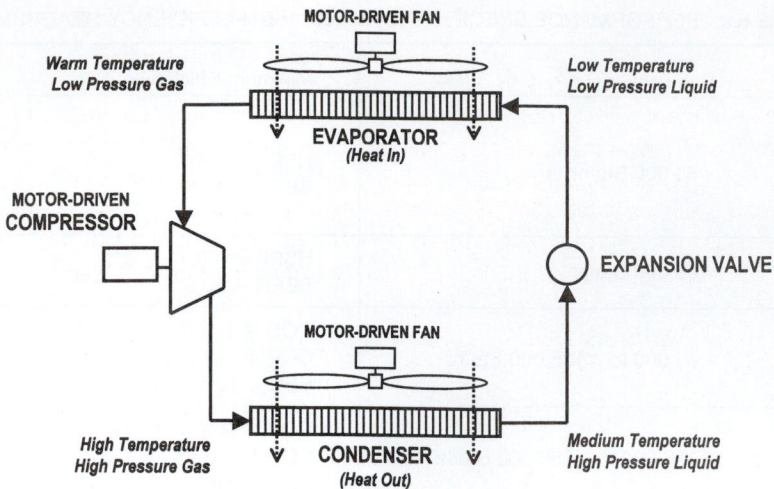

FIGURE 6.5 A diagram of the components and process of the vapor-compression refrigeration cycle.

A vapor-compression refrigeration system has the following components (refer to Figure 6.5).

Refrigerants

A *refrigerant* is a substance that produces a refrigerating effect while expanding or vaporizing as it circulates as a gas or liquid through a closed vapor-compression refrigeration system. It picks up heat by evaporating at a low temperature and pressure and gives up heat by condensing at a higher temperature and pressure. A good refrigerant must absorb and release heat readily (high latent heat of vaporization), be easily converted from a gas to a liquid, be relatively nontoxic and have good lubricating qualities. Properties of refrigerants are provided in Table 6.7.

Hydrochlorofluorocarbon (HCFC) refrigerants such as HCFC-22 (R-22), HCFC-123 (R-123), and hydrofluorocarbon (HFC) refrigerants such as HFC-134a (R-134a) are commonly used in building systems. They should not be confused with those refrigerants with ozone-destroying chlorofluorocarbons (CFCs) such as CFC-12 (R-12), which were common in older chillers but are not used today. The signing of the Montreal Protocol 1987 by 191 countries banned the production of CFCs. In the United States, this began in 1996. When retiring a chiller that contains CFCs or HCFCs, the Clean Air Act requires that a certified technician recover the refrigerant on-site.

HCFC-22 is used as a refrigerant in a wide range of HVAC cooling equipment from room air conditioners to large chillers for more than four decades. It is subject to phase out in 2030 under the international Montreal Protocol. The U.S. Clean Air Act will prohibit the use of this refrigerant in new equipment in 2010, along with prohibiting new production of the refrigerant after 2020. Refrigeration equipment presently manufactured for HCFC-22 is being altered to HFC-134a, R-410A, or R-407C (R-410A and R407C are blends of HFC refrigerants). The new refrigerants help combat the growing

TABLE 6.7   SELECTED THERMAL AND ENVIRONMENTAL PROPERTIES OF OZONE-DEPLETING REFRIGERANTS.

Refrigerant	Heat of Vaporization (Btu/lbm)	Global Warming Potential (GWP)	Ozone Depletion Potential
CFC-11 (abandoned)	81	4000	1
CFC-12 (abandoned)	65	7100	1
HCFC-22	86	1700	0.055
HCFC-123	77	93	0.016
HFC-134a	83	1300	0
R-407C	95	1600	0
R-410A (Prozone® or Puron®)	95	1890	0
R-290 (propane)	159	~20	0
R-600a (isobutane)	151	~20	0
R- 717 (ammonia)	536	<1	0
R-744 (carbon dioxide)	94	1	0
R-718 (water)	1070	0	0

ozone depletion in the earth's atmosphere, as they contain no chlorine.

Ammonia is a low-cost refrigerant, which has good heat transfer properties and thus good efficiency. It has been used extensively for industrial refrigeration applications and is being developed for commercial air conditioning, mainly in Europe. Ammonia is highly corrosive of the copper-based alloys used in conventional air conditioning systems. More expensive stainless steel tubing or aluminum is needed. Ammonia is moderately flammable and toxic. Ammonia systems must use open-drive compressors, which result in some refrigerant leakage. Therefore, because of safety and liability issues, ammonia is not being considered seriously for air conditioning applications in the United States.

Global warming potential (GWP) is a measure of how much a gas is estimated to contribute to global warming. It is a relative scale that compares the gas in question to that of the same mass of carbon dioxide (CO_2). For example, the obsolete refrigerant called Freon (CFC-12) has a GWP of 7100 (7100 times the adverse global warming effect of CO_2), whereas HCFC-22 has a GWP of 1700 (1700 times the adverse global warming effect of CO_2), and R-717 ammonia has a GWP of less than 1. *Ozone depletion potential (ODP)* is a measure that refers to a refrigerant's ozone depletion. It is a relative scale that compares the gas in question to that of the same mass of CFC-11, so the ODP of CFC-11 is defined to be 1.0. Other CFCs and HCFCs have ODPs that range from 0.01 to 1.0. The ODP and GWP values of ozone-depleting refrigerants are listed in Table 6.7. These values are used to ascertain the environmental friendliness of a refrigerant.

Evaporator

The evaporator is a network of tubes filled with refrigerant located inside the building that absorbs heat from air or liquid and moves it outside the refrigerated area by means of a refrigerant.

In the split air conditioning system typically used in light commercial and residential applications, the evaporator is a coil (network of tubes) located in the plenum (ductwork) near the furnace. The furnace serves as an air-handling unit, blowing air across the coil where it is cooled and delivered through ductwork to the indoor rooms. It is also known as a cooling coil, fan coil, or indoor coil. In large commercial installations, the evaporator is part of the chiller, where it cools chilled water that is piped to cooling coils located in ductwork.

Condenser

The condenser is a network of tubes that discharges unwanted heat from the refrigerant to a medium (e.g., air, water, or a combination of air and water) that absorbs the heat. As the heat is discharged, the refrigerant condenses and becomes liquid again.

There are three types of condensers: air-cooled, water-cooled, and evaporative condensers. In the split air conditioning system typically used in light commercial and residential applications, the condenser is air cooled. It is a coil (network of tubes) located outdoors. A fan blows outdoor air directly across the condensing coil, cooling the hot refrigerant and releasing heat to the outdoors. In large commercial installations, the water-cooled condenser is part of the chiller, where it cools water that is piped to a water tower that discharges heat to the outdoors through evaporation. The evaporative condenser uses a combination of air and water as its condensing medium.

Compressor

A pump-like device that drives the refrigerant from the indoor evaporator to the outdoor condenser and back to the evaporator again. The compressor increases the pressure (and thus increases temperature) of the refrigerant that circulates through the closed system. It is often referred to as the heart of the system because it circulates the refrigerant through the refrigeration loop. In the split air-conditioning system typically used in light commercial and residential applications, the compressor is located outdoors as part of the condensing unit. In large commercial installations, the compressor is part of the chiller. On most systems, the compressor is driven by an electric motor. On large units, the compressor may be driven by a gas internal combustion engine or gas turbine.

Expansion Valve

By sensing the temperature of the evaporator, the expansion valve controls liquid refrigerant passing through a very small orifice, which causes the refrigerant to expand to a low-pressure (and thus low-temperature) gas.

This vapor-compression refrigeration process relies on three physical principles:

- Heat moves from a high-temperature region to a low-temperature region.
- A change in temperature of a gas is proportional to a change in pressure in the gas—that is, a pressure increase results in a temperature increase in the gas and a pressure decrease results in a temperature decrease.
- A change in phase results in the absorption or release of latent heat: evaporation of a liquid results in energy being absorbed by the substance and condensation results in the substance releasing energy.

Operation of a vapor-compression refrigeration system occurs as follows: The system contains liquid refrigerant that is forced through an expansion valve, causing it to change into a low-temperature liquid. The low-temperature liquid flows through the evaporator coil, cooling the coils and extracting heat directly from recirculated room air or indirectly from a recirculated liquid medium (chilled water) that is used to cool room air. As heat is extracted, the refrigerant absorbs energy, increasing its temperature and converting it to a gas. Then the warm gas is drawn into the compressor where its pressure is increased, causing its temperature to rise. The hot refrigerant gas then enters the condenser where coils dissipate heat directly to the outside air or indirectly to a liquid medium (water) that carries the heat outdoors. As the gaseous refrigerant passes through the condenser, it condenses into a liquid, releasing latent heat. The liquid refrigerant travels to the expansion valve where the process begins again.

Cooling/Refrigeration Capacity

Cooling and refrigeration systems are rated in *tons* of cooling: One ton of cooling is the amount of cooling obtained by melting one ton of ice in one day. One ton of cooling is equivalent to 12 000 Btu/hr or 3.516 thermal kW. This unusual unit is from the days when ice was actually used for cooling. Ice has a latent heat of fusion of 144 Btu/lb (288 000 Btu per ton). Melting a ton of ice over a 24-hr period (one day) produced a cooling effect of 288 000 Btu/day or 12 000 Btu/hr.

Direct Expansion Equipment

Direct expansion (DX) is a generic term that is used to identify vapor-compression refrigeration equipment having two or more components, one usually positioned externally and the other usually positioned within the building. Site-installed refrigeration piping that is charged with a refrigerant connects the indoor and outdoor components.

The indoor components consist of fan coil units or air handling units that are located in the room being air condi-

tioned or located remotely in a mechanical room. Some manufacturers produce external units that may be located internally and in the case of these units ductwork is connected to outdoor atmosphere to reject or extract heat. DX refrigeration systems are in direct contrast to chilled water systems in which cooling is achieved by circulating chilled water.

Central Air Conditioners

A *central air conditioner* is a DX refrigeration cooling system. It has four basic parts: a condensing unit (containing the compressor, condenser, and expansion valve), a cooling coil (an evaporator in the air-handling unit or ductwork near the furnace), ductwork (to distribute the cooled air), and a control mechanism such as a thermostat.

The most common type of central air conditioner is a split system. A *split system* has the condensing unit (containing the compressor, condensing coil and fan, and expansion valve) located outside and a cooling coil (an evaporator in an air-handling unit or in the ductwork near the furnace) located inside. The outdoor condensing unit is connected to the indoor evaporator coils by two copper pipes. The smaller pipe is the liquid line, which carries the high-pressure liquid refrigerant from the condenser to the expansion valve. The larger suction line, which should be insulated, carries low-pressure gas from the evaporator coils to the compressor. See Figures 6.6 and 6.7, as well as Photos 6.17 through 6.23. Table 6.8 provides performance data for residential central air conditioners.

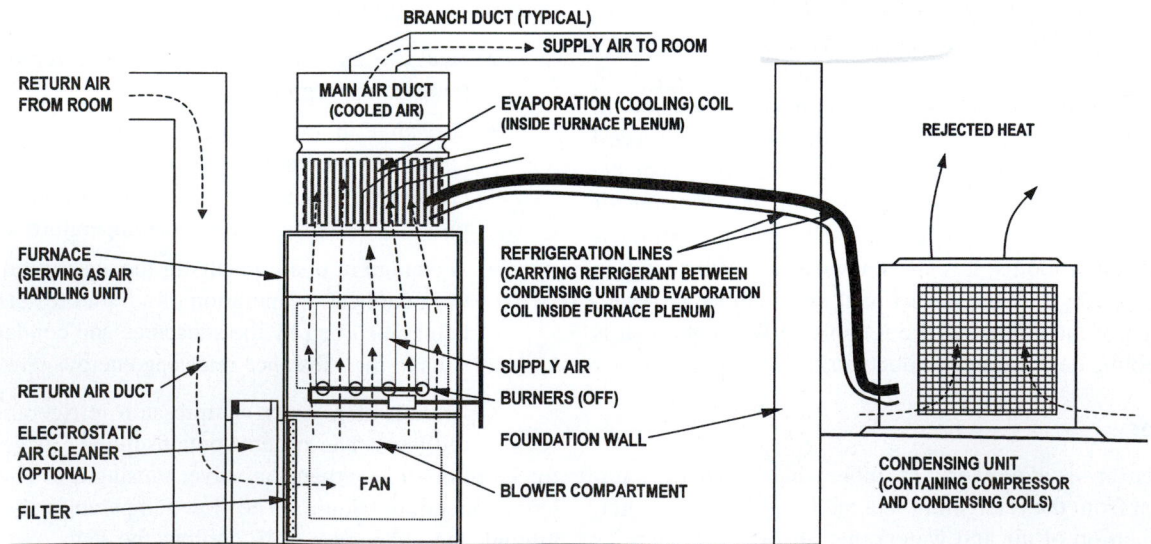

FIGURE 6.6 A split system has the condensing unit (containing the compressor, condensing coil and fan, and expansion valve) located outside and a cooling coil (an evaporator in an air-handling unit or in the ductwork near the furnace) located inside. The outdoor condensing unit is connected to the indoor evaporator coils by two pipes that carry refrigerant between the two units.

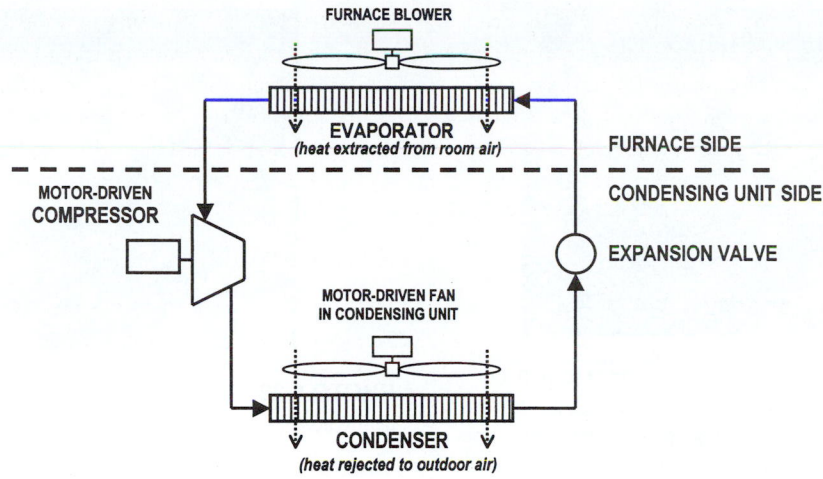

FIGURE 6.7 Splitting of the vapor-compression refrigeration components in a system like that shown in Figure 6.6.

PHOTO 6.17 An air-cooled condensing unit on a split system. (Used with permission of ABC)

PHOTO 6.18 Four small air-cooled condensing units as part of an individual split system for four condominiums. (Used with permission of ABC)

PHOTO 6.19 A 10-ton rooftop PTAC. (Used with permission of ABC)

PHOTO 6.20 Rooftop units providing cooling and heating for an elementary school. (Used with permission of ABC)

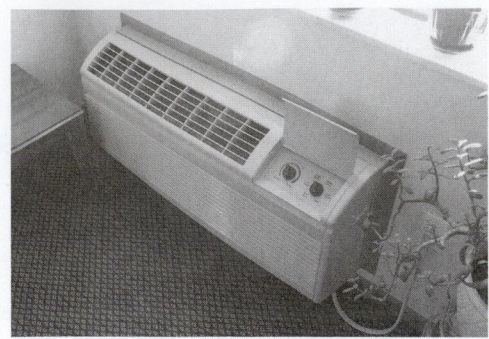

PHOTO 6.21 A through-the-wall PTAC unit rated at 9200/9100 Btu/hr for heating and 12 000/9000 Btu/hr for cooling. Unit is powered by electricity on individual 120V/208V single phase circuits split from a 208Y/120V, 3-phase, 4-wire building voltage panelboard. (Used with permission of ABC)

PHOTO 6.22 A packaged, air-cooled condensing unit. (Used with permission of ABC)

Packaged-Terminal Air-Conditioning Unit

A *packaged-terminal air-conditioning (PTAC) unit* is a self-contained, air-cooled DX cooling unit (e.g., with integral compressor, evaporator, and condenser), an optional heating unit, and an optional minimum ventilation intake. A PTAC unit contains all four components encased in one unit and is usually located in a utility closet or an exterior wall. A wall-mounted PTAC unit is interchangeable; it fits in a sleeve that extends through the wall so outdoor air can be accessed. PTACs are

TABLE 6.8   PERFORMANCE SPECIFICATIONS FOR HIGH-EFFICIENCY RESIDENTIAL CENTRAL AIR CONDITIONERS. COMPILED FROM MANUFACTURERS' INFORMATION.

Capacity		SEER Range		EER Range	
Tons-Cooling	Btu/hr	Minimum	Maximum	Minimum	Maximum
2.0	24 000	14.60	16.50	12.85	14.55
		15.00	16.20	13.00	14.25
		13.00	15.60	11.40	13.50
2.5	30 000	13.50	16.00	12.00	14.44
		13.50	16.00	12.70	13.55
		13.00	16.00	11.40	13.95
		13.50	15.75	11.70	14.00
		13.50	15.75	11.70	14.00
3.0	36 000	15.85	18.00	11.35	13.20
		15.50	18.00	11.45	13.15
		15.20	17.20	11.00	11.00
		15.00	17.10	10.05	12.00
		14.50	16.50	10.65	12.85
		13.60	16.50	11.95	11.95
		15.00	16.00	13.35	14.40
3.5	42 000	14.00	16.00	12.10	14.00
		13.25	15.50	11.30	13.20
4.0	48 000	15.15	16.50	10.75	12.30
		15.10	16.40	10.15	11.60
		14.00	16.00	12.20	14.50
		11.20	15.60	9.90	13.00
		14.00	15.50	10.15	11.85
5.0	60 000	14.35	16.40	9.95	10.90
		14.85	15.80	9.80	10.35
		13.50	15.50	12.00	14.00
		13.00	15.25	11.60	13.70

PHOTO 6.23 A rooftop central hot/chilled water plant that produces both hot water and chilled water (in lieu of a boiler and chiller). (Courtesy of NREL/DOE)

used to heat or cool a single living space quietly and efficiently using only electricity. PTACs are commonly used in hotels and apartments, as shown in Photo 6.21.

A *roof top unit (RTU)* is a type of PTAC unit that is mounted on the roof and requires adequate support by the roof structure. It consists of cooling equipment, air-handling fans,

and frequently gas or electric heating equipment that are self-contained in a single unit. Rooftop units can be installed at ground level. Rooftop units are available in sizes ranging from 1 ton to more than 100 tons (12 000 Btu/hr or about 3.5 kW) of cooling capacity. More than half of all North American commercial floor space is cooled by self-contained, packaged air conditioning units, most of which sit on rooftops. Refer to Table 6.9 for a sample of RTU performance specifications.

Individual/Room Air Conditioners

An *individual air conditioner,* frequently called a *room air conditioner,* is a cooling device that is installed in either walls or windows and exposed to the outdoor air. These self-contained units are characterized by a lack of pipes or ductwork for distributing the cool air. Through-the-wall models are installed in an outside wall, usually during construction or remodeling. Individual air conditioners condition air only in the room or areas where they are located. They typically range in output capacity from 4000 to 36 000 Btu/hr (1.2 to 10.5 kW). Energy efficiency ratios range from 8.0 to 12. See Table 6.10.

TABLE 6.9 PERFORMANCE SPECIFICATIONS FOR SELECTED COMMERCIAL PACKAGED ROOFTOP UNITS. COMPILED FROM MANUFACTURERS' INFORMATION.

			Gas–Electric, Heating and Cooling Units					
Nominal Tons of Cooling	Net Cooling (Btu/hr)	Number of Cooling Stages	EER	IPLV	Heating Input (Btu/hr)	Heating Output (Btu/hr)	Steady State Heating Efficiency	Number of Heating Stages
15	176 000	2	9.5	9.5	230 000	186 000	81	2
15	176 000	2	9.7	9.5	300 000	243 000	81	2
18	202 000	2	9.7	10.2	275 000	223 000	81	2
18	202 000	2	9.7	10.1	360 000	292 000	81	2
20	236 000	2	9.5	10.1	275 000	223 000	81	2
20	236 000	2	9.5	10.0	360 000	292 000	81	2
25	278 000	2	9.7	10.4	275 000	223 000	81	2
25	277 000	2	9.5	10.0	360 000	292 000	81	2

			Electric–Electric Heating and Capacity Cooling Units					
Nominal Tons of Cooling	Net Cooling (Btu/hr)	Number of Cooling Stages	EER	IPLV	Heating Capacity (kW)	Heating Output (Btu/hr)	Steady State Heating Efficiency	Number of Heating Stages
15	178 000	2	9.7	9.9	30 / 50 / 73	102 390 / 170 650 / 249 149	100	3
18	204 000	2	10.0	10.5	30 / 50 / 73	102 390 / 170 650 / 249 149	100	3
20	236 000	2	9.7	10.4	30 / 50 / 73	102 390 / 170 650 / 249 149	100	3
25	278 000	2	9.8	10.5	30 / 50 / 73	102 390 / 170 650 / 249 149	100	3

TABLE 6.10   PERFORMANCE SPECIFICATIONS FOR HIGH-
EFFICIENCY ROOM AIR CONDITIONERS. COMPILED
FROM MANUFACTURERS' INFORMATION.

Cooling Capacity				Power Consumption (kW)
Tons-Cooling	Btu/hr	KW (of Cooling)	EER	
0.48	5700	1.67	11.2	0.15
0.58	7000	2.05	11.7	0.18
0.77	9200	2.70	11.5	0.23
0.85	10 200	3.00	11.4	0.26
1.25	15 000	4.39	11.5	0.38
1.50	18 000	5.27	10.8	0.49

Performance Ratings for Air Conditioners and Heat Pumps

Air conditioners and heat pumps of the same capacity are usually available with a wide range of efficiencies. Measures used to rate the efficiency of an air conditioner are:

Energy Efficiency Ratio (EER):   The EER is measure of the instantaneous energy efficiency of cooling equipment. It is the ratio of the cooling capacity of the air conditioner or heat pump in (in Btu/hr) to the total electrical input (in watts) under industry-specified test conditions. The higher the EER, the more efficient the unit. It applies to air conditioners with cooling capacities equal to or greater than 65 000 Btu per hour Btu/hr (equal to or greater than 5.4 tons).

Seasonal Energy Efficiency Ratio (SEER):   This is a measure of the seasonal cooling efficiency of an electric air conditioner or heat pump using a standard cooling load and outdoor climate profile over a standard cooling season. It is a rating that denotes the total cooling of a heat pump (in Btu) during its normal annual usage period for heating divided by the total electrical energy input (in watt-hours) during the same period. The higher the SEER, the more efficient the unit. It applies only to air conditioners with a cooling capacity of less than 65 000 Btu/hr (less than 5.4 tons). New units have SEER ratings from 10 to 17 Btu per watt.

Integrated Part-Load Value (IPLV):   The IPLV is a seasonal efficiency rating method based on representative annual commercial loads. It applies to air conditioners with cooling capacities equal to or greater than 65 000 Btu per hour Btu/hr (equal to or greater than 5.4 tons).

EER is the preferred rating when determining which air conditioner or heat pump will operate most efficiently during full-load cooling conditions. SEER and IPLV are better indicators of which air conditioner or heat pump will use less energy over the course of an entire cooling season.

Central Chilled Water Systems

Central chilled water systems are used in large buildings and are bigger and more complex than standard split or packaged DX systems. These systems typically consist of a central chiller to cool water. The *chilled water* is then distributed by the appropriate fan coil or air handling system.

Chillers

A *chiller* is a large piece of cooling equipment that produces chilled water, which is used to remove heat from the building. Chillers generally use the vapor-compression refrigeration cycle to extract heat from water and reject it to either outdoor air or water. Absorption chillers, another type of chiller, use a thermal or chemical process to produce the refrigeration effect necessary to provide chilled water. Selection of a chiller depends upon availability and cost of fuel/electricity and the size of the cooling load. See Photos 6.24 and 6.25.

PHOTO 6.24   A commercial-size chiller is a water-cooled condensing unit. The slender tubes are fluid-to-fluid heat exchangers. (Used with permission of ABC)

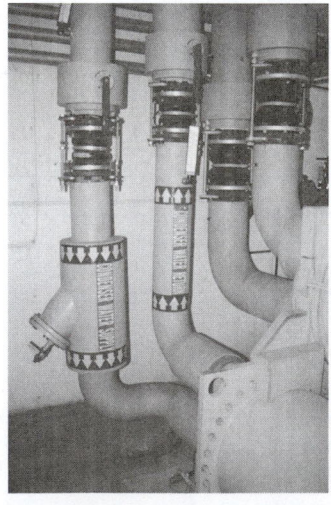

PHOTO 6.25   Condenser water supply and return and chilled water supply and return lines on the rear of the chiller in the previous photograph. Chilled water circulates to and from the chiller and cooling coils in air-handling units, where it cools air. Condenser water circulates to and from the chiller and a cooling tower, where heat is rejected outdoors. (Used with permission of ABC)

Mechanical Vapor-Compression Refrigeration Chillers

Most chillers operate using the vapor-compression refrigeration process. They consist of a compressor, evaporator, condenser, and an electric motor, a steam or gas turbine or internal combustion engine to drive the compressor. These chillers use a compressor as part of the vapor-compression refrigeration process and are classified by the type of compressor they use:

Centrifugal Chillers *Centrifugal chillers* have a compressor that acts like a centrifugal fan, compressing the vapor flowing through it by spinning it from the center of an impeller wheel radially outward so that centrifugal forces compress the vapor. Some machines use multiple impellers to compress the refrigerant in stages. They have very few moving parts and are the workhorses of the space-cooling industry.

Screw Chillers *Screw chillers* have a compressor that is a positive displacement device in which the refrigerant chamber is actively compressed to a smaller volume. The rotary screw chiller operates by the twisting motion of two meshing rotors. The single-screw compressor consists of a cylindrical main rotor positioned between identical gate rotors. The screw or double-helical screw compressor consists of two mating twin-grooved rotors. Rotary screw chillers are smaller and lighter than centrifugal chillers. They have fewer parts than a reciprocating compressor, are generally more reliable and usually found in systems rated over 2.5 tons. The screw compressor is a relative newcomer to the space cooling industry, although it has long been used for air compressors and low-temperature refrigeration. They lead the market for small- to mid-size chillers.

Scroll Chillers *Scroll chillers* employ a rotating twisted rod as the screw compressor. Twisting of the scroll compresses the refrigerant, increasing its pressure. They have fewer parts than a reciprocating compressor, are generally more reliable.

Reciprocating Chillers *Reciprocating chillers* are used in small installations. These chillers use a reciprocating compressor, a type of positive displacement compressor that relies on the back-and-forth motion of a piston in a cylinder to compress the refrigerant.

An *electric chiller* has the compressor driven by an electric motor. *Engine-driven* and *turbine-driven chillers* use the same refrigeration cycle as electric chillers, but have compressors driven by a gas internal combustion engine or gas turbine. Packaged engine-driven chillers can utilize reciprocating, screw, or centrifugal compressors.

Absorption Chillers

Absorption chillers differ from mechanical vapor-compression refrigeration chillers in that they utilize a thermal or chemical process to produce the refrigeration effect necessary to provide chilled water. There is no mechanical compression of the refrigerant taking place within the machine as occurs within more traditional vapor-compression refrigeration type chillers.

The thermochemical process used in absorption chillers takes advantage of a chemical property called *affinity,* the ability of some chemicals to easily dissolve into other chemicals. An absorption cycle uses two fluids: a refrigerant and an absorbent. Two common refrigerant/absorbent combinations are water and lithium bromide (a salt) and ammonia and water. The refrigerant in an absorption chiller dissolves into an absorbent solution for which it has a high affinity. An electric pump pushes the absorbent solution into a generator section, where heat is applied to drive the refrigerant vapor out of the solution and into the evaporator. Substituting thermal energy for mechanical compression means that absorption chillers use much less electricity than mechanical compressor chillers.

To produce the cooling effect necessary to make approximately 45°F (7°C) chilled water, the shell side of the chiller must be maintained at an extremely low pressure (a vacuum) to allow the refrigerant (water) to boil at approximately 40°F (4.5°C). The lithium bromide solution absorbs the vaporized refrigerant, diluting it before it is pumped to the generator section of the chiller where heat is added to reconcentrate the dilute solution. The water vapor boiled off in the generator is then condensed, returning to the evaporator as liquid. The reconcentrated lithium bromide returns to the absorber section as strong solution to begin the cycle again. Chilled water temperature leaving the evaporator is typically between 40° and 60°F.

Many absorption chillers use two- or three-stage cycles. In multistage equipment, generators at different temperatures, and often heated by different sources, are used to improve the COP.

Direct-fired absorption chillers contain a burner that runs on natural gas or another fuel to produce the heat required for the absorption process. *Indirect-fired absorption chillers* use steam or hot water produced externally by a boiler or cogeneration system. A system of piping and heat exchangers transfers the heat to the chiller.

Absorption chillers are cost-effective when the thermal energy they consume is less expensive than the electricity that is displaced. Cost-effectiveness depends on the relative costs of gas and electricity, the relative efficiencies of the two types of equipment, the cooling loads, and the operating hours. This type of cooling is also cost-effective when a source of heat between 212°F (100°C) and 390°F (200°C) is available. For example, if an industrial process results in a steady supply of waste steam or hot water, absorption cooling may be able to provide very cost-effective cooling. Similarly, the waste heat from electricity generation can be used cost-effectively for absorption cooling in summer and space heating in winter. Absorption cooling can also be cost-effective when peak electricity charges are high.

Chiller Systems

In a typical commercial building installation, a *central chilled water plant* consists of one or several chillers that produce chilled water. This chilled water is pumped to one or more air handlers, where it cools the indoor air. The cool air is then distributed within the building through a network of ducts. The ducts run to terminal units that control the flow of air to diffusers. The chilled water plant also requires several additional devices, known as *auxiliaries*, to move chilled water between the chilled water plant and the air handlers. In addition, the waste heat from the chilled water plant must be rejected to the outside air using pumps and a cooling tower. Chillers are generally located in, or just outside, the building they serve. See Figure 6.8.

Hybrid Chiller Systems

Hybrid chiller systems use electric and gas chillers in the same plant. This can help reduce first costs and operating costs and can minimize the loss of cooling during electricity outages. Chiller operation in hybrid systems can be alternated so that, at any given moment, the chillers that are operating are powered by the less-expensive energy source. For example, in an electric and natural gas hybrid chiller system, the electric chillers would only operate when inexpensive off-peak electric rates were available. When expensive on-peak electric rates applied, the gas-fired equipment would operate. Both electric and gas chillers may be operated simultaneously to meet peak cooling loads.

Chiller Ratings

Centrifugal chillers are available in sizes ranging from 70 to 2500 tons factory assembled and up to 10 000 tons field assembled. They use HCFC-123, HFC-134a, or HCFC-22 refrigerants. Chillers that use HCFC-123a currently have the highest efficiencies, down to 0.49 kilowatts per ton (kW/ton) at full load and Air Conditioning and Refrigeration Institute (ARI) conditions.

Rotary screw chillers are available in sizes ranging from 100 to 1100 tons. They are most commonly used in applications of 300 tons or less. Rotary screw chillers use HCFC-22 or HFC-134a refrigerant. Water-cooled rotary screw chiller efficiencies range from 0.62 to 0.72 kilowatts per ton (kW/ton) at full load. Air-cooled screw chillers range from slightly more than 1.0 to 1.2 kW/ton. Some screw compressors operate well at partial loads and are stable down to about 10% capacity.

Engine-driven chillers are available with a variety of compressors: reciprocating (up to about 700 tons), screw (about 100 to 1000 tons), or centrifugal (about 350 to 5000 tons). The most common configuration in use today is a reciprocating engine powered by natural gas and driving a screw or centrifugal chiller.

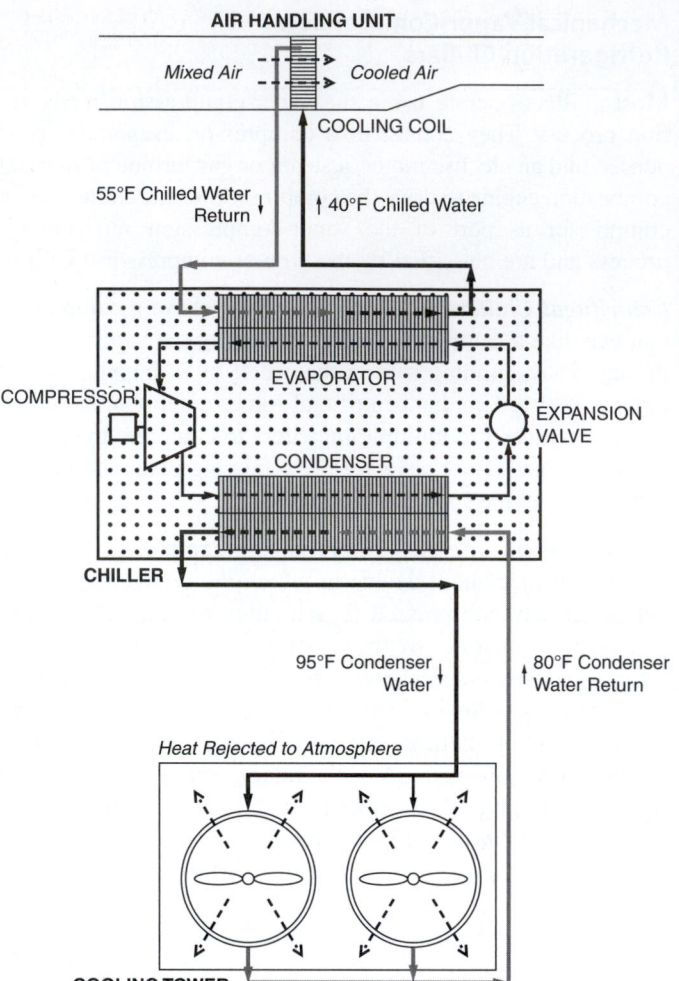

FIGURE 6.8 A diagram of the components and processes of a central chilled water system. A chiller (composed of an evaporator, compressor, condenser, and expansion valve) has a refrigerant flowing through its components as part of the vapor-compression refrigeration cycle. The chiller feeds chilled water from its evaporator to a coil in an air-handling unit, which cools air. Chilled water is heated as it passes through the coil so it is returned back to the chiller to be re-cooled in the evaporator. On the condenser side of the chiller, warmed water is pumped from the chiller's condenser to a cooling tower where heat is rejected from the condenser water to the atmosphere. Cooled condenser water is then returned back to the chiller's condenser.

Chiller Performance

The performance of a chiller is rated in the following terms:

Coefficient of Performance (COP) This is the cooling output (in Btu) divided by the energy input (electric power in Btu): the higher the COP, the more efficient the unit.

kW/ton Rating Commonly referred to as efficiency, but actually is power input to the compressor motor divided by tons of cooling produced, or kilowatts per ton (kW/ton). Lower kW/ton indicates higher efficiency.

TABLE 6.11   PERFORMANCE SPECIFICATIONS FOR SELECTED LARGE CENTRIFUGAL CHILLERS. COMPILED FROM MANUFACTURERS' INFORMATION.

Nominal Cooling Capacity	
Tons	Btu/hr
200	2 400 000
225	2 700 000
250	3 000 000
300	3 600 000
350	4 200 000
450	5 400 000
500	6 000 000
550	6 600 000
600	7 200 000
700	8 400 000
800	9 600 000
1000	12 000 000
1200	14 400 000
1300	15 600 000
1500	18 000 000

Energy Efficiency Ratio (EER)   The EER is calculated by dividing a chiller's cooling capacity (in Btu/hr) by its power input (in watts) at full-load conditions. Performance of smaller chillers is frequently measured using the EER. A higher EER indicates higher efficiency.

Integrated Part-Load Value (IPLV)   A measure that strives to attain a more representative average chiller efficiency over a representative operating range. It is the efficiency of the chiller, measured in kW/ton, averaged over four operating points, according to a standard industry formula. IPLV is preferred for more variable loads, a situation much more common in air-cooled chiller applications.

Because chiller efficiency varies depending on the load under which it operates, determining annual energy performance can be complicated.

Standard reference conditions, defined by the ARI, at which chiller performance is measured are: 44°F water leaving the chiller and, for water entering the condenser, 85°F at 100% load and 60°F at zero% load. ARI conditions are based on an Atlanta, Georgia, climate with a design wet bulb temperature of 78°F. See Table 6.11.

Water Treatment of Chillers

Consistent maintenance and careful water treatment ensure the long life of a chiller. Untreated water is never pure; it contains varying amounts of gases, solids, and pollutants. Although most suspended solids can been removed, gases and salts still remain. These manifest themselves in a chiller system as scale or corrosion. Scale and corrosion may be controlled by bleed off. Bleed off will not control biological contaminants. All chiller systems require water treatment.

Use of Multiple Small Chillers

A chiller is most efficient when it operates at or near its rated capacity. Therefore, if building loads are highly variable, as is often the case in commercial and institutional buildings, multiple small chillers can operate more efficiently than a single large chiller because each chiller can operate more often at or close to its full load. Common methods of sizing smaller chillers are based upon a percentage of total cooling load, such as 40%-40%-40% or 50%-50%-20%.

Air-Cooled Condensers

PTAC units, small central air conditioning systems, individual/room air conditioners, individual heat pumps, and small chillers rely upon air to cool the refrigerant in the condenser. *Air-cooled condensers* pass outdoor air over a dry surface coil to condense refrigerant contained inside the coil. This results in a higher condensing temperature and lower performance under peak conditions in comparison to water-cooled condensers. An air-cooled condenser is preferred, however, on the systems noted previously because of their simplicity and low maintenance requirements. An air-cooled condenser should be avoided with larger central chilled water systems because chiller efficiency is reduced significantly.

Evaporative Condensers

Evaporative condensers pass air over coils sprayed with water, taking advantage of the latent heat of vaporization to reduce the condenser temperature.

Cooling Towers

A *cooling tower* expels heat from a chilled water system when water is cooled by directly or indirectly contacting the atmosphere. Cooling towers are used as the condenser on large-scale air conditioning installations because they are most efficient. Cooling tower systems, however, increase water use, have extensive maintenance requirements, require freeze protection, and close control of water treatment to function successfully. See Figure 6.9, as well as Photos 6.26 and 6.33.

In a cooling tower, secondary chilled water carrying heat from the chiller moves through a series of louvers and baffles. Outdoor air passing through the louvers and baffles is pushed by a natural draft or a mechanical draft (fan) system. Mechanical draft towers are most common because they do not depend on wind to function properly. By bringing the cooling water into contact with the outdoor airstream, an exchange of latent heat and sensible heat occurs that lowers the temperature of the cooling water.

Cooling Tower Operation

All cooling towers operate on the principle of removing heat from water by evaporating a small portion of the water that is

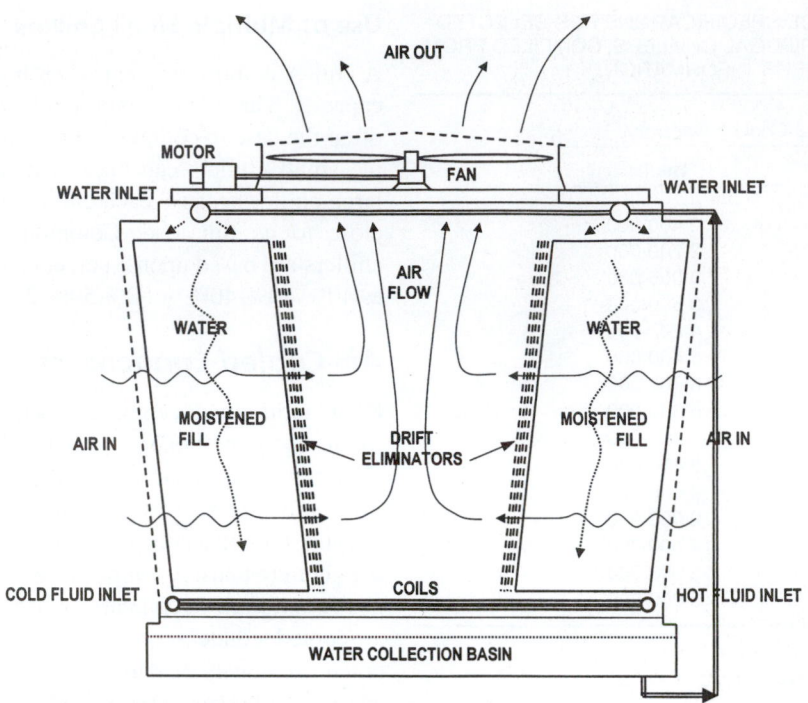

FIGURE 6.9 A cross-section of a cooling tower.

PHOTO 6.26 A roof-mounted cooling tower. (Used with permission of ABC)

PHOTO 6.28 Condenser water is piped to the top of the cooling tower, where it is sprayed over the cooling tower fill. (Used with permission of ABC)

PHOTO 6.27 A side view of a cooling tower. (Used with permission of ABC)

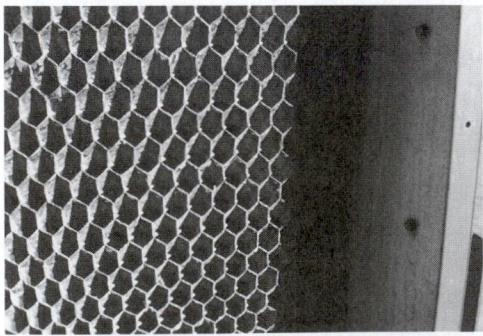

PHOTO 6.29 A close-up view of the fill in a cooling tower. (No water is flowing.) During cooling operations, water trickles down through the fill. The majority of heat removed from the water is by evaporation. (Used with permission of ABC)

196

PHOTO 6.30 A fan draws outdoor air through the wet fill on the sides of the cooling tower to increase evaporation and cooling. Air is discharged above the tower. (Used with permission of ABC)

PHOTO 6.31 A side view of the cooling tower. Evident on the tower side is the fill. At the tower bottom is piping carrying cooled condenser water that is piped back to the chiller. (Used with permission of ABC)

PHOTO 6.32 Pumps used to pump chilled water to several air-handling units located in different zones in the building. (Used with permission of ABC)

PHOTO 6.33 Labeled chilled water (from chiller) and hot water (from boiler) pipes on their way to and from several air-handling units located in different zones in the building. (Used with permission of ABC)

recirculated through the unit. The heat that is removed is called the *latent heat of vaporization,* about 970 Btu/lb evaporated.

There are two primary heat transfer mechanisms by which water is cooled inside a cooling tower: sensible heat and latent heat transfer. Sensible heat transfer takes place when the temperature of the circulating water is higher than the incoming outdoor air temperature, thereby causing heat from the circulating water to be absorbed by the colder air. Most of the cooling that takes place inside a cooling tower (usually greater than 80%) is by evaporation of some of the circulating water. Evaporation requires latent heat transfer, so when water is evaporated in a cooling tower, heat is removed with the water vapor and leaves in the exiting airstream from the top of the tower. The result is that the remaining water is cooled significantly, even to temperatures below the actual ambient temperature. Each pound of water that is evaporated removes approximately 970 Btu in the form of latent heat.

If sensible heat transfer were the only cooling mode that took place inside a cooling tower, the ambient (outdoor) air temperature would limit the circulating water temperature. Instead, because evaporation accounts for most of the cooling effect, wet bulb temperature of the ambient air more greatly affects the circulating water temperature.

The *entering wet bulb temperature (EWBT)* is a measure of the level of humidity in the ambient air entering the cooling tower. Wet bulb temperature is the lowest temperature that water theoretically can reach by evaporation. In general, the higher the wet bulb temperature, the more moisture exists in the air. The wet bulb temperature is a key design parameter in the designing/sizing of a cooling tower, as it determines the rate at which water can be evaporated.

Cooling towers provide a warm, wet environment that is very conducive to both corrosion and biological growth. As a result, they are made of wood, plastic, ceramics, galvanized sheet metal, and concrete.

Cooling towers must be located so that the discharge air diffuses freely without recirculating through the tower and so that air intakes are not restricted. Cooling towers should be located as near as possible to the refrigeration systems they serve, but should never be located below them so as to allow the condenser water to drain out of the system through the tower basin when the system is shut down.

Types of Cooling Towers

Direct or *open cooling towers* expose the cooling water directly to the outdoor air. The warm cooling water is sprayed over a fill in the cooling tower to increase the contact area, and outdoor air is blown through the fill. The majority of heat removed from the cooling water is by evaporation. A small fraction of the water (about 1 to 5%) is carried away by the humidified outdoor airstream. The remaining cooling water drops into a collection basin in the water tower and is circulated to the chiller evaporator.

An *indirect* or *closed cooling tower* circulates the cooling water through the tubes of a coil bundle in the tower. A separate external circuit sprays water over the cooling tubes. The spray evaporates, cooling the circulating water in the coils.

Airflow in either class of cooling towers may be crossflow or counterflow with respect to the falling water. Most HVAC systems use a *counterflow* design, with the air moving up through the tower and the water cascading down through the airstream. Some systems are forced draft (FD), with the fan pushing the air into the tower; others are induced draft (ID), with the fan drawing the air out of the tower. Another type in use is the *crossflow* design, where outdoor air moves through horizontally and the water cascades down. These are generally found in smaller sizes.

Forced-draft cooling towers employ an air distribution system that forces air into the tower. Fans positioned on the side of the cooling tower characterize them. *Induced-draft cooling towers* operate by pulling air through the tower and use fans located on top of the cooling tower.

Capacities of Cooling Towers

Packaged cooling tower systems are manufactured in ranges from 10 to 250 tons in a single unit. Systems up to 1000 tons are typical.

Performance of Cooling Towers

Cooling towers operating in high wet bulb temperatures require a proportionately larger tower than those found in lower wet bulb regions of the country (e.g., Houston versus Phoenix). The performance of a cooling tower degrades when the efficiency of the heat transfer process declines. Some of the common causes

of this degradation include buildup of scale deposits, clogging of spray nozzles, and poor airflow.

In every cooling tower, there is a loss of water to the environment in the form of pure water, which results from the evaporative cooling process. This evaporated water leaves the tower in a pure vapor state, and thus presents no threat to the environment.

Drift consists of small unevaporated droplets of water that become entrained in the exhaust airstream of a cooling tower. Drift is the undesirable loss of liquid water to the environment. These water droplets carry with them minerals, debris, microorganisms, and water treatment chemicals from the circulating water, which potentially impacts the surrounding environment.

Minimizing drift losses in a cooling tower reduces the risk of impacting the environment with potentially corrosive water treatment chemicals. Drift is usually responsible for damage to property near the cooling tower yard (e.g., buildings, cars, and so forth). Water use and chemical consumption are also reduced as more remains in the circulating flow, thus generating savings in operating costs. Excessive drift losses also pose serious health risks, not only because of the chemicals released, but also because of microorganisms that can be transmitted through drift such as the microbial that causes Legionnaires' disease.

Evaporative Coolers

An *evaporative cooler* is a type of cooling device that turns incoming dry, hot outdoor air into moist, cool air by saturating the air with water vapor. Outside air is cooled by about 30° to 40°F (17° to 22°C) as it is drawn through the wet pads and blown into the building. An evaporative cooler does not cool air by use of refrigeration.

There are three types of evaporative coolers: single-stage, direct coolers; indirect evaporative cooling; and, two-stage, direct/indirect coolers.

Single-Stage, Direct-Type Coolers

Single-stage, direct-type evaporative coolers, sometimes called a *swamp cooler,* are the most common type used. They are available in three pad types:

Fixed Fiber Pad Coolers   Fixed fiber pad evaporative coolers have pads made of shredded aspen wood or synthetic fibers packaged in plastic netting. A small pump moves water from a catch sump at the bottom of the cooler up to the top of the pads. The pads are wetted by water dripping onto them from the top. Natural fiber pads require replacement every year or two if the cooler is used frequently. See Figure 6.10.

Rotating Pad Coolers   Rotating pad evaporative coolers have pads made of a rough polyester material sewn into a wide belt shape. The belts are rotated by pad motors through a trough of

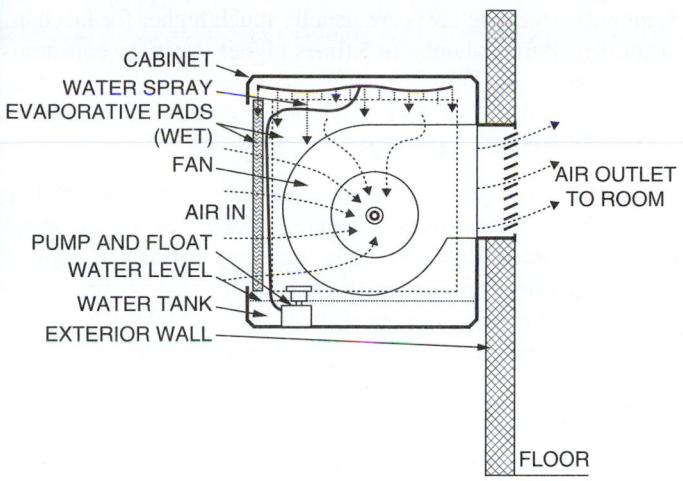

FIGURE 6.10 A cross-section of a fixed fiber pad evaporative cooler discharging cooled air through a wall.

PHOTO 6.34 A commercial-scale direct evaporative cooler built into an air-handling unit and partially hidden from view by the hot water and hot water return lines (vertical pipes). The evaporative cooler is fed by a single pipe off of the hot water supply line (left pipe). (Used with permission of ABC)

PHOTO 6.35 A commercial-scale indirect evaporative cooler. (Used with permission of ABC)

water at the bottom of the cooler. No pump is required in this type of cooler, and the belts tend to last longer than aspen pads.

Rigid-Sheet Pad Coolers *Rigid-sheet pad evaporative coolers* have a stack of corrugated sheet material that allows air to flow through it like a pad. The pads are wetted by a pump distributing water to the top of the pads, similar to the fiber pad type. This type of pad is more expensive than fiber pads, but can last much longer if water quality is maintained.

Indirect Evaporative Coolers

Indirect evaporative cooling is similar to direct evaporative cooling, but uses a heat exchanger that separates indoor and outdoor airstreams and prevents moisture from being added to the indoor airstream. The cooled moist air never comes in direct contact with the conditioned environment, so this kind of cooler does not produce humidity levels as high as that produced by traditional single-stage evaporative coolers. Indirect evaporative cooling systems tend to be less efficient than direct systems, however, because the heat exchange is never complete and because powering two airstreams requires twice the fan energy.

Two-Stage, Direct/Indirect Coolers

Two-stage, direct/indirect coolers use an air-to-water heat exchanger called a *precooler* that reduces the incoming air temperature without raising the relative humidity, then puts the incoming air through a direct evaporation stage, further reducing its temperature. Because of their expense, direct/indirect units are typically only used where daytime temperatures consistently exceed 100°F (38°C).

A basic direct evaporative cooler consists of a sheet metal or plastic cabinet mounted on the roof or besides the building. Inside the cabinet, water is sprayed on sponge-like pads on the sides of the cabinet. Then hot outside air is drawn through the pads by a fan. As the air flows through the pads, the water evaporates and cools the air by exchanging sensible heat in the air with latent heat in the water. The cool, moist air is then blown into the building, forcing hot stale air through open windows.

An evaporative cooler moves significant quantities of air through the building interior, so it not only cools occupants by dropping the dry bulb temperature of indoor air but also cools by moving air over the body surfaces of building occupants. It is not abnormal to achieve lower indoor air temperatures with evaporative cooling than with vapor-compression refrigeration during periods of very low humidity because of the lowered performance of mechanical refrigeration equipment in these conditions.

The wetted pads on an evaporative cooler are fairly efficient air filters, trapping particles on their wet surfaces. The continuous wetting of the pads flushes the trapped particulates into the sump, where they are contained. Regular inspection of these units is necessary for health (e.g., mold, mildew, and so on) reasons. A seasonal shutdown is also required in climates prone to freezing temperatures.

Packaged evaporative coolers are commercially available in sizes ranging from 2000 to 80 000 ft/min. Smaller, stand-alone units are available in sizes as small as 50 ft/min. Water consumption in residential and units can be from 3 to 15 gal/day (11 to 57 L/day) depending upon climatic conditions. In large commercial units, water consumption can be 50 to 80 gal/hr (190 to 300 L/hr) at extreme design conditions.

Evaporative cooling is well suited for warm to temperate, dry climates, such as those that exist in the western region of the United States, excluding coastal California. In the desert southwest region of the United States, it is a common practice to use both evaporative and vapor-compression refrigeration cooling in buildings. Direct evaporative cooling increases humidity, which, in dry climates, may improve thermal comfort. In humid regions, however, evaporative coolers do not work well for human comfort cooling because the air is nearly saturated, so adding moisture will not significantly lower the temperature of the air. See Table 6.12 for an example of performance specifications for evaporative coolers.

Acquisition cost of vapor-compression refrigeration systems (e.g., air conditioning systems) is usually about 3 times that of evaporative cooling system for a similar structure. Costs vary widely because of type of structure, climate, and other factors. Upkeep and maintenance costs are somewhat lower with evaporative cooling partially because of the technical expertise required. Operating costs are usually much higher for mechanical refrigeration, about 3 to 5 times higher in energy consumption alone.

Cool Thermal Storage

A *cool thermal storage* system produces and stores chilled water or generates a phase change in water (ice) or a eutectic salt/water mix. It stores "cooling" capacity during low cooling demand periods for use during peak cooling load periods. The types of cool thermal storage systems used in buildings are outlined in Table 6.13, and are described in the following.

Ice Thermal Storage

Ice thermal storage systems use the latent heat of fusion of water (144 Btu/lb · °F) to store cooling capacity. Basically, ice is stored, and when cooling is needed water is circulated through the ice storage area and then distributed at about 34°F (1.7°C) to provide space cooling (although lower temperatures can be used with appropriate fluids). To store energy at the temperature of ice requires refrigeration equipment that provides charging fluids at temperatures below the normal operating range of conventional air conditioning equipment. Special ice-making equipment or standard chillers modified for low-temperature service are used. See Figures 6.11 and 6.12, as well as Photo 6.36.

TABLE 6.12   PERFORMANCE SPECIFICATIONS FOR EVAPORATIVE COOLERS. COMPILED FROM MANUFACTURERS' INFORMATION.

Flow Rate Range (ft³/min)	Typical Blower Motor Horsepower Range	Blower Size (in)	Pump Size (Horsepower)
1000 to 3000	0.5 to 5	12	$\frac{1}{40}$
3000 to 6000	0.75 to 10	15	$\frac{1}{40}$
6000 to 9000	1.5 to 15	18	$\frac{1}{40}$
9000 to 13 000	2 to 20	20	$\frac{1}{40}$
13 000 to 18 000	5 to 25	22	$\frac{1}{30}$
18 000 to 23 000	5 to 30	25	$\frac{1}{3}$
23 000 to 31 000	7.5 to 40	28	$\frac{1}{3}$
31 000 to 38 000	15 to 50	32	$\frac{1}{3}$
38 000 to 49 000	15 to 75	36	$\frac{1}{2}$
49 000 to 60 000	20 to 100	40	$\frac{1}{2}$

TABLE 6.13   PERFORMANCE CHARACTERISTICS OF COMMON TYPES OF COOL THERMAL STORAGE SYSTEMS.

Storage Medium	Volume		Storage Temperature		Discharge Temperature		Strengths
	ft³/ton-hr	m³/kW	°F	°C	°F	°C	
Ice	2.4 to 3.3	0.77 to 1.09	32	0	34 to 36	1.1 to 2.2	High discharge rates; potential for low temperature air system
Chilled water	10.7 to 21	3.5 to 7.0	39 to 44	3.9 to 6.7	41 to 46	5.0 to 7.8	Can use existing chillers; water in storage tank can do double duty for fire protection
Eutectic salts	6	1.97	47	8.3	48 to 50	8.9 to 10	Can use existing chillers

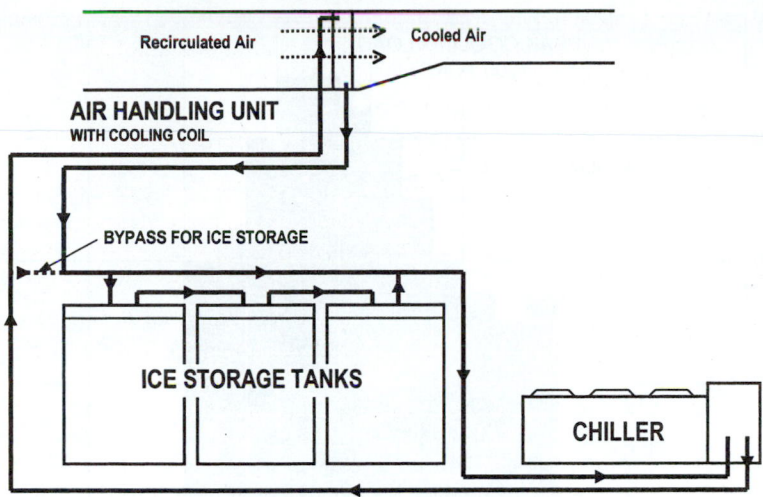

FIGURE 6.11   A diagram of an ice thermal storage system. A chiller makes ice during off-peak hours, which is stored in ice storage tanks. The ice is melted during off-peak hours and used to make chilled water.

Chilled Water Thermal Storage

Chilled water thermal storage systems use the sensible heat capacity of water (1 Btu/lb · °F) to store cooling capacity. They operate at temperature ranges compatible with standard chiller systems and are most economical for systems greater than 2000 ton-hours in capacity.

Eutectic Salt Thermal Storage

Eutectic salt thermal storage systems use a combination of inorganic salts, water, and other elements to create a mixture that freezes at a desired temperature. The material is encapsulated in plastic containers that are stacked in a storage tank through which water is circulated. A commonly used mixture for thermal storage freezes at 47°F (8°C), which allows the use of standard chilling equipment to charge storage, but leads to higher discharge temperatures.

Chilled water, ice, or frozen salt–water mix is produced by a chiller system and it is stored until needed. Steel and concrete are the most commonly used types of tanks for chilled water storage. Most ice systems use site-built concrete. Concrete tanks with polyurethane liners are common for eutectic salt systems.

Two common strategies are used for charging and discharging storage to meet cooling demand during peak hours. These are full storage and partial storage.

Full-Storage Strategy   The *full-storage strategy* shifts the entire on-peak cooling load to off-peak hours. The system is typically designed to operate at full capacity during all non-peak hours to charge storage on the hottest anticipated days. This strategy is most attractive where on-peak demand charges are high or the on-peak period is short.

Partial-Storage Strategy   The *partial-storage strategy* runs the chiller to meet part of the peak period cooling load, and the remainder is met by drawing from storage. The chiller is sized at a smaller capacity than the design load. Partial-storage systems may be run as load-leveling or demand-limiting operations.

In a *load-leveling* approach, the chiller is sized to run at its full capacity for 24 hr on the hottest days. The strategy is most effective where the peak cooling load is much higher than the average load. In a *demand-limiting* approach, the chiller runs at reduced capacity during on-peak hours and is often controlled to limit the facility's peak demand charge. Demand savings and equipment costs are higher than they would be for a load-leveling system, and lower than for a full-storage system.

Most electric utilities have rate structures that have high demand charges, ratchet charges, or a high differential between on- and off-peak energy rates. So, a cool thermal storage system offers the potential for substantial operating cost savings by using off-peak electricity to produce chilled water or ice for

PHOTO 6.36   A small-scale ice storage system uses off-peak energy at night to chill ice used for cooling the house on hot days. (Courtesy of NREL/DOE)

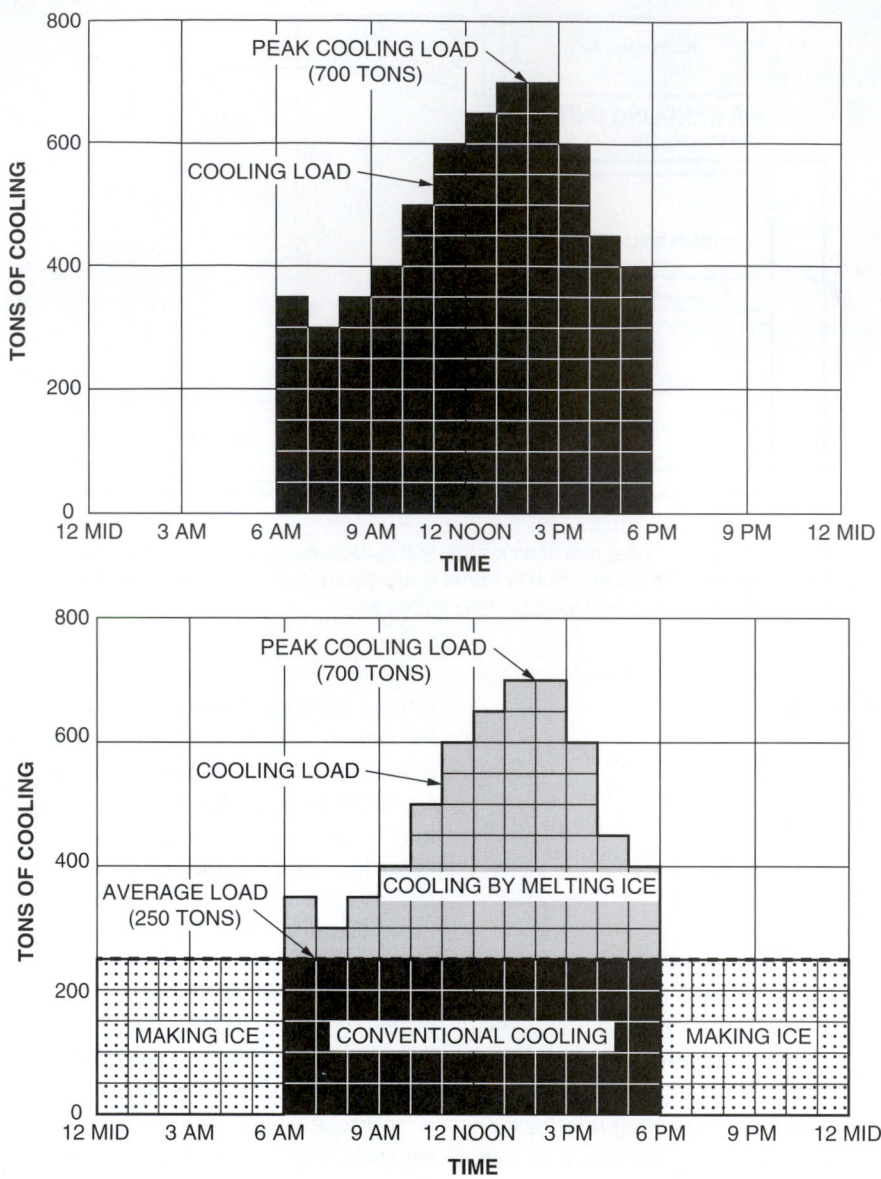

FIGURE 6.12 A graph of potential savings with an ice thermal storage system. In this hypothetical example, smaller chilling equipment (about 250 tons-cooling) can meet the 700-ton peak demand.

use in cooling during peak hours. Furthermore, operating a chiller at night when outdoor air temperatures are better suited for cooling results in additional operating cost savings.

6.4 VENTILATION EQUIPMENT

Natural Ventilation

Natural ventilation was commonplace in most buildings through about 1950, prior to the commercial availability and acceptance of mechanical air conditioning. Reliance on natural ventilation has become less common in recent times. According to most codes, natural ventilation is still an option to mechanical air conditioning.

Natural ventilation works well when the outdoor air temperature and humidity levels are below the indoor air conditions. Outdoor air can flow through windows, creating an air exchange with warm and moist, indoor air. Wind and the buoyancy effect (hot air rises) drive flow of air. Natural ventilation is especially viable for most of the year in dry, moderate climates.

Additionally, the capability of having operable windows for natural ventilation does have other potentially important advantages. For example, in some instances it may be desirable to extensively ventilate a space (e.g., during cleaning or painting activities or when a complete purge of the room air is needed). Also, operable windows do provide some potential for cooling and ventilating during times when the air conditioning system is inoperable or undergoing repair or replacement.

Ceiling Fans

A *paddle-blade ceiling fan* can be used to provide air motion, improving comfort and allowing a higher thermostat set point during the cooling season. Research suggests that air motion permits comfort at temperatures from 2° to 6°F higher than would be possible without ceiling fans. This results in a moderate reduction in energy consumption. However, ceiling fans may create problems with air motion on papers and the potential distraction from moving shadows caused by fan blades moving below recessed ceiling luminaires.

Whole-House Fans

A *whole-house fan* is located in the ceiling separating the attic from the occupied space and is operated when outdoor air temperature and humidity levels are below the indoor air conditions. Cooler outdoor air is drawn into the building through open windows and exhausted through the fan into the attic space at an air exchange rate of several times per hour. Even when outdoor air conditions are near room air conditions, air movement in the space permits comfort at indoor air temperatures higher than would be possible without air movement. Airflow in the attic also reduces the attic temperature and thus heat gain into the building.

In climates and seasons having significant daily temperature swings from day to night [e.g., 90°F (32°C) during the day and 55°F (13°C) at night], the whole-house fan can be operated during the cooler evening hours. In a light frame (residential) building, nighttime operation can result in a drop the indoor air temperature to within 10°F (6°C) of the early morning outdoor air temperature.

By closing windows and doors before the heat of the day (e.g., midmorning), effectively shutting down the exchange of warm outdoor air, a light frame building can coast through the heat of the day. By beginning to ventilate in the early evening hours, when the day's heat has migrated through the building envelope, a comfortable temperature can be maintained by drawing outdoor air into the space.

Economizers

An air-side *economizer* is designed into a building's air handling system or into a terminal device (i.e., a unit ventilator). It can save energy in buildings by using cool outside air as a means of cooling the indoor space. When the enthalpy of the outside air is less than the enthalpy of the recirculated air, conditioning the outside air is more energy efficient than conditioning recirculated air. When the outside air is sufficiently cool, no additional conditioning of it is needed; this portion of the air-side economizer control scheme is called *free cooling*. Air-side economizers can reduce HVAC energy costs in cold and temperate climates while increasing ventilation and improving indoor air quality. They are not well suited to warm humid climates because the moisture content of outdoor air is typically too high.

On a large all-air system, air-side economizers can be a separate and additional, dampered cabinet with a collection of

PHOTO 6.37 Outdoor air and exhaust vents on a packaged central air system. In cold weather, using an economizer cycle, a large volume of low-temperature outdoor air can be brought into the building for cooling without having to mechanically cool the outdoor air with refrigeration equipment. (Courtesy of NREL/DOE)

louvers, dampers with actuators, and fans. They can also be built into a conventional air-handling system that opens the outdoor air damper that draws air from the outside when possible. In both systems, sensors and logic devices monitor indoor and outdoor air conditions and decide how much outside air to bring into the building. See Photo 6.37.

6.5 AIR HUMIDIFICATION EQUIPMENT

Air Humidification

Air is mixed with water vapor, in proportions that vary with climate, season, and daily weather conditions. As discussed in Chapter 1, relative humidity levels below about 40% begin to cause dryness of skin, cracking of nasal membranes, and asthmatic and allergic reactions in humans. Low humidity levels result in the generation of static electricity, which is damaging to electronic equipment. At relative humidity levels below 20%, these problems increase drastically. In contrast, relative humidity levels above about 60% are favorable to the survival and proliferation of bacteria, fungi, and viruses. A relative humidity (RH) of 30 to 60% is optimal. Yet, in heating climate regions, buildings typically have indoor relative humidity levels well below 30% during the heating season unless moisture is supplied to the air. In many cases, indoor relative humidity levels fall well below 20%.

To maintain a good air quality in a building interior, water vapor must be added to indoor air in a process called *air humidification*. There are three methods of air humidification.

Atomization

Atomization reduces water to very small particles, which are dispersed in the airstream where they evaporate. That leads to lowering of room temperature.

Vaporization

In the *vaporization method* water is heated to its boiling point and the steam produced is emitted to the surrounding air. This is quite a simple way of humidification but requires substantial energy consumption for heating and vaporization of water.

Evaporation

In the evaporation process, room air is blown over a large water surface or through a wetted media by a fan. The air flowing over the water absorbs water vapor. Modern evaporating units cannot only humidify, but also efficiently reduce pollution, such as dust particles, noxious gases, and disagreeable odors.

A *humidification system* introduces moisture to building air to maintain a healthy environment for building occupants and allow proper operation of equipment (e.g., electronic equipment).

Residential Humidifiers

Residential humidifiers are the evaporation or atomization type. They rely upon heated air flowing across a water surface, air circulating over or through a wetted media, or small particles of water introduced into the airstream. The fan drives air on a furnace. Types of residential humidifiers include the following.

Pan-Type Humidifiers

Pan-type humidifiers are installed in the ductwork directly above a furnace (e.g., in the furnace plenum). Heated air exiting the furnace passes across a shallow pan containing water. As air moves across the water surface, water is evaporated into the airstream, where it is distributed to the indoor spaces. *Electrically heated humidifiers* heat the water, increasing its temperature and improving the rate of heat transfer. *Water-absorbent wicking plates* can be added that extend from the pan into the airstream. These plates wick water from the pan and increase the wetted area, which improves the rate of evaporation.

Wetted Media Humidifiers

Wetted media humidifiers rely upon a heated airstream flowing across a fixed pad that is moistened by trickling water or a spray, or through a paddle, wheel, or drum that is wetted by rotating it through a water reservoir. A small electric motor drives rotation of the paddle, wheel, or drum-type media. In the *fan-type humidifier,* air is drawn from the furnace plenum into the humidifier by a fan; it passes through a fixed or moving pad and then is returned to the plenum. *Bypass humidifiers* are installed between the return air duct and the supply air duct and rely upon the pressure difference to drive air across a wetted pad. *Duct-mounted humidifiers* are installed in the supply air duct, where heated air passes across a fixed or moving pad.

Atomizing Humidifiers

Atomizing humidifiers introduce small droplets of water into the airstream by using specially designed spray nozzles; by dropping water on a spinning disc, which slings the water into the airstream; or by ultrasonically vibrating water to break it up into small particles.

Portable Humidifiers

Portable humidifiers are packaged, stand-alone units located directly in the space or room requiring humidification. They rely upon the humidification principles described in the humidifiers previously. Most common is the fan type with a fixed pad or rotating wheel setting in or passing through a reservoir of water. Water in the reservoir is typically filled manually. Some units are equipped with an automatic fill requiring a float valve, shut-off valve, and tubing for connection to the water supply.

Commercial Humidification Systems

Commercial systems tend to be more complex than residential humidifiers. Types of humidification systems used in large buildings include the following:

Spray Atomization Systems

In a *high-pressure spray atomization system,* water is forced through small orifices in spray-atomizing nozzles at pressures up to 1000 psi (6900 kPa). Turbulence created by the high-pressure flow breaks the water into droplets, approximately 15 to 20 microns in size. The relatively large droplet size makes this system difficult to apply to many buildings because the droplets do not readily evaporate in the air-handling unit and must be taken out of the airstream by mist eliminators. If water gets past the mist eliminator into the ductwork, unsafe microbial growth may result. As a result, this system must be used with a mist eliminator treated with an antibacterial agent. These systems are well suited for direct dispersion into large, open spaces.

The *compressed air spray atomization system* uses compressed air and specially designed nozzles to aid in atomization of water. Compressed air causes additional turbulence in the nozzle that breaks the droplets into smaller sizes. The smaller size droplets (about 0.3 to 10 microns) evaporate better into the airstream. Although mist eliminators should still be used with this system, it is much safer than water spray nozzles in avoiding water in ductwork. Nozzles are arranged in arrays inside supply air ductwork or are available as packaged units for direct dispersion into a space.

Steam Injection Systems

Direct steam injection is used where a large humidification system is needed. In this system, steam is evaporated into an airstream by steam injection manifolds that are mounted in ductwork. Steam may be provided by the building's space or water heating boiler or by a dedicated boiler. Steam from a space heating boiler typically contains chemical additives intended to extend the life of the boiler, piping, and related equipment. Chemicals used should be safe for human inhalation/exposure and safe for equipment in the space. A better option is to provide a dedicated boiler for the steam injection system.

A *steam-to-steam heat exchanger injection system* is similar to the direct steam injection system, except that steam comes from a steam-to-steam heat exchanger that separates boiler water and chemicals from the injected steam. Chemically treated steam from the space heating boiler is used to make pure steam that is injected into the ducted airstream. This system can also be used where a large humidification system is needed.

The *gas-fired steam injection system* blends direct steam injection and a special boiler in a packaged system. The boiler is built to withstand the thermal shock that results from high levels of water entering the boiler. This is typically a small-size system in comparison to other injection systems.

Disposable Electrode Steam Generators

A *disposable electrode steam generator* is a self-contained canister unit that makes steam with electrodes immersed directly in water. The steam may either be injected into an air supply duct or supplied directly to the humidified space. The electrodes and water are contained in a disposable canister that is discarded periodically along with the built-up minerals left behind when the water is evaporated away. Units are powered with electricity or natural gas, with natural gas-fired units typically the most cost-effective. They can be a packaged gas-fired humidifier, dedicated boiler system, or a steam-to-steam heat exchanger system.

Electric Resistance Steam Generators

An *electric resistance steam generator* is similar to the self-contained electrode unit except that it uses an electric resistance element to boil the water. In addition, the unit does not have a disposable canister system. Instead, it relies on an automatic flushing system to remove minerals from the boiling chamber.

Ultrasonic Vaporization System

Ultrasonic vaporization takes place when a small piezoelectric element is vibrated at radio frequencies in a water bath. As the element contracts, the inertia of the water prevents it from filling in the void and the water boils from the local low pressure. Ultrasonic systems are small in capacity, reaching their limit at about 50 lb/hr. This makes them good candidates for computer rooms and one-room laboratories, but too small for entire buildings.

Performance

Air humidification equipment must be installed in an airstream that will not fall below its dew point temperature, otherwise condensation and problems with moisture result. The relative humidity in a building must not rise above levels where condensation begins to develop on surfaces of the building envelope (e.g., windows) or in concealed materials (e.g., within insulation).

The output rates of air humidification equipment are expressed in lb/hr or lb/day (kg/hr or kg/day). Outputs range from several lb/day in residential systems to 3000 lb/hr (1360 kg) or more in commercial steam injection humidification systems.

6.6 AIR DEHUMIDIFICATION EQUIPMENT

Conventional Dehumidification Systems

Conventional HVAC systems remove humidity by passing the supply air across a cooling coil maintained at a temperature below the dew point temperature of the air so that water vapor will condense on the coils as liquid. The condensate (excess moisture) is drained away from the coil. Generally, DX units are more effective at removing moisture than chilled water systems.

Reducing supply airflow will decrease air velocity at the cooling coil and result in increased moisture removal. It will also, however, reduce system relative efficiency somewhat and may affect air filtration and the adequacy of ventilation air supplied to conditioned spaces. Lowering the chilled water supply temperature will increase the relative moisture removal, although it will also reduce system chiller efficiency. Both approaches may require increased need for supply air reheat.

In dry climates, dry outside air can be drawn into a central air-handling unit and mixed with return air to lower relative humidity in the supply air prior to its introduction into the building's interior.

Desiccant Dehumidification Systems

Desiccant dehumidification involve the use of chemical or physical absorption of water vapor to dehumidify air and reduce the latent cooling load in a building HVAC system. Reducing moisture (and relative humidity) levels in the air-handling system and the building spaces during the cooling season will improve indoor air quality by preventing condensation in equipment and reducing the propagation of microbiologicals that can contaminate indoor air. See Photo 6.38.

A *desiccant* is a drying substance or agent. Desiccant dehumidification uses a desiccant to absorb moisture from ventilation air and thus reduce the latent cooling load. With less moisture in the indoor air, air conditioning equipment can be used to meet the demands of sensible cooling, so smaller air

PHOTO 6.38 A rooftop desiccant unit designed to handle the make-up air requirements. (Courtesy of NREL/DOE)

conditioning capacity is required. There are two types of desiccant systems: liquid (sorbent) and dry.

> *Liquid desiccant systems* commonly use two chambers with air/liquid contact surfaces such as sprayed coils. In the conditioning chamber, ventilation air is dehumidified as the concentrated desiccant absorbs moisture from the air. In the regeneration chamber, building exhaust air is humidified as moisture is transferred from the dilute desiccant to the exhaust air. The exhaust air and/or desiccant are usually heated to promote desiccant regeneration. A desiccant pump, level controls, and heat exchanger are typically included in the system.

> *Dry desiccant systems* operate in a manner similar to liquid desiccant systems, but use a desiccant coating on a rotary enthalpy heat exchanger. Dry desiccant systems do not require energy for desiccant regeneration.

Liquid desiccant systems remove more moisture from ventilation air than do dry desiccant systems; the air produced by dry desiccant systems, however, is warmer than the air produced by dry desiccant systems.

Desiccant cooling systems are best suited to for hospital buildings, where healthy indoor air is critical, and industrial facilities with humidity-sensitive processes (e.g., microelectronics, photography, and printing operations). Desiccant cooling may be practical for office buildings in humid climates that rely on a central cooling system. Dehumidification can offset the cost of other system components. These systems may also be advantageous when increased ventilation requirements will exceed the capacity of the existing chiller.

6.7   FUEL AND ENERGY SOURCES

The most commonly used sources of building heat are the electricity, gas, oil, and coal. Use of the sun as an energy source is increasing and is discussed in Chapter 11. The decision on which fuel to use is based on site and commercial availability and cost of operation. In this section, advantages and disadvantages of each fuel and the method used to determine the cost of operation.

Electricity

Electricity is used as a fuel for a variety of heating systems, including baseboard radiant heat and electric coils in the ceiling and/or walls (refer to Chapter 10). In addition, there are electric furnaces for forced hot air systems, and electricity is used with heat pumps both to operate the system and to provide supplemental electric resistance heat for the system.

Electricity has as its advantage its simplicity. It requires no chimney to remove toxic combustion gases, and when baseboard strips and ceilings and wall coils are used, the system has individual room controls, providing a high degree of flexibility and comfort. Electric systems cost significantly less to install than other systems (including electric furnaces for forced hot

air), and all electric systems (except heat pumps) require much less upkeep and maintenance than those using the other fuels.

The primary disadvantage of electric heat is its annual cost of operation when compared with those of other fuels. Electricity rates vary tremendously throughout the country, but electricity costs tend to be 2 to 3 times higher than other heating fuels. It is necessary to determine the rates in the geographical area of construction by doing a cost analysis.

Electric heat is quite popular in remote regions where other fuels are not available. It is frequently economical, in the southern regions because winter is shorter and heating bills are generally lower. Electric heat is also used in apartments, offices, and similar buildings where the developer is primarily interested in building the units as inexpensively as possible and where the tenant renting the space usually pays the cost of heating. Its minimal need for maintenance and repairs is also a factor in such construction.

Gas

Gas, also a popular fuel for heating, is used to heat the water in hot water systems and air in forced-air systems. Gas fuels available include natural gas, which is piped to the building, and propane gas, which is delivered in pressurized cylinders in trucks and tanks and stored in tanks at or near the building. Because natural gas is available as needed, with no storage or individual delivery required, it is considered simpler to use.

Natural gas is extracted from the earth and delivered to the building through a pipeline. Gas supplied to the building is measured by a meter. See Photo 6.39. The primary advantages of natural gas have been its relatively low cost and its simple and clean burning, which reduces the maintenance required on the heating unit. As with all of the fuels, as costs go up in the future, it is difficult to say which will be the most economical.

Liquid petroleum gas (LPG), also known as *propane gas,* is equally as clean as natural gas. Its availability is not limited

PHOTO 6.39   A natural gas meter. (Used with permission of ABC)

to areas where supply pipes have been installed. The cost of propane gas has generally been higher than that of natural gas.

Natural gas is not available in all areas: the more rural the area, the less likely it is that natural gas will be available. The designer must first determine if natural gas is available at the building site. If it is not available, the designer should investigate the possibility and cost of making it available. If natural gas is unavailable, use of propane should be investigated.

In the past, there have been occasional short- and long-term shortages of natural gas. Although these shortages have been mostly alleviated, some building sites may not be permitted new natural gas hookups. In periods of shortages, residential customers can be reasonably assured that they will have sufficient natural gas for use, but industrial and commercial customers cannot be so assured. Because of an extremely cold winter and shortages in the winter of 1975, hundreds of businesses faced the option of converting to another fuel or closing, and many did close for the winter months. Large consumers of natural gas are offered a reduced *interruptible supply* rate if they agree in advance to switch to another fuel or shut down during a period of shortages.

Heating Oil

Fuel oil is a derivative of the petroleum fractional distillation process, which separates crude oil into fractions. When used for space heating buildings, it is called *heating oil.* Fuel oil is available in various weights having different heating values and different costs; it is numbered 1 through 6. Generally, the lower the number, the more refined the oil is and the higher the cost. Number 2 fuel oil is commonly used in residences and most commercial buildings, while Number 4 and sometimes Number 6 are used in some large commercial and industrial projects.

The demand for heating oil has decreased owing to the widespread use of natural gas. Because of its availability, it remains a popular heating fuel in the northeastern United States, where nearly half of all buildings use it for space heating. It requires delivery by trucks that fill storage tanks located in or near the building. Because of added costs associated with tighter environmental regulations, tanks are generally not located underground. Oil-heating units generally require more maintenance than gas-fired heating units. The selection of oil as a heating fuel has diminished somewhat since the oil embargos in the 1970s: the cost of oil has risen dramatically since that time.

Coal

Coal is rarely used for heating in new residential construction, and its use in industrial and commercial construction fluctuates. Its primary advantage is that it is available, and there are ample supplies so that a shortage seems unlikely. Generally, its cost is competitive with those of other fuels. Its disadvantages lie in the amount of space required for storage and the fact that it does not burn as completely as oil or gas, thus producing more pollution. Government regulations for clean air have also limited the use of coals to industrial purposes.

TABLE 6.14   HEATING VALUES OF COMMON COMMERCIAL FUELS.

Commercial Fuel	Average Calorific Heating Values[a]	
	Customary (U.S.) Units	Metric (SI) Units
Coal (anthracite)	14 200 Btu/lb 28 400 000 Btu/ton	33 000 kJ/kg
Coal (bituminous)	9000 Btu/lb 18 100 000 Btu/ton	21 000 kJ/kg
Fuel oil number 1	136 000 Btu/gal	16 340 kJ/L
Fuel oil number 2	140 000 Btu/gal	16 820 kJ/L
Fuel oil number 4	149 000 Btu/gal	17 900 kJ/L
Fuel oil number 5	150 000 Btu/gal	18 020 kJ/L
Fuel oil number 6	154 000 Btu/gal	18 500 kJ/L
Natural gas	103 000 Btu/CCF[b]	37 500 kJ/m^3
Propane	91 000 Btu/gal	11 000 kJ/L
Wood (lodgepole pine)	6000 Btu/lb 21 000 000 Btu/cord	14 000 kJ/kg
Wood (oak)	8700 Btu/lb 30 700 000 Btu/cord	20 000 kJ/kg

[a] The calorific value varies to some extent with quality of fuel.
[b] A CCF is 100 ft^3. Calorific heating value of natural gas varies with altitude. For example, the calorific heating value of natural gas is typically about 830 000 Btu/ft^3 in Denver, CO, 5280 ft (1600 m) above sea level.

Coal's decreased use for space heating in buildings is because of the handling required in delivery and the inconvenience of having a coal bin in the basement. In addition, originally the coal had to be shoveled into the furnace or boiler by a building occupant; a sufficient reason to change to oil, gas, or electricity. Automatic stokers that require no hand shoveling feed most coal-heating units. Because coal is used so little as a residential and small commercial heating fuel, it will not be considered further.

Combustion Air Requirements

All fuel-fired equipment needs oxygen for the combustion reaction to take place. Oxygen for combustion is obtained from the atmosphere, which is about 21% oxygen by volume or 23% by weight. Most of the 79% of air that is not oxygen is nitrogen, with traces of other elements. Nitrogen is inert at ordinary flame temperature and forms few compounds as the result of combustion. Nitrogen is an unwanted constituent that must be accepted in order to obtain the oxygen. It contributes nothing to combustion reaction, yet it increases the volume of combustion products to be vented and it robs heat from the reaction.

Air required in combustion is classified as primary air, secondary air, and excess air. *Primary air* controls the rate of combustion, which determines the amount of fuel that can be burned. *Secondary air* controls combustion efficiency by controlling how completely the fuel is burned. *Excess air* is air supplied to the burner that exceeds the theoretical amount needed to burn the fuel.

Combustion air requirements are based on the composition of the fuel used. Commonly used fuels contain nitrogen, ash, oxygen, sulfur, carbon, and hydrogen. When a fuel has a large volume of nitrogen that must be accepted along with the desired oxygen, more excess air must be provided. That excess air has a chilling effect on the flame. Some fuel particles fail to combine with oxygen and pass out of the vent unburned, so excess air is required to ensure adequate combustion and reduce output of deadly carbon monoxide.

Water vapor is a by-product of burning hydrogen. It also removes heat from the flame and becomes steam at flue gas temperature, passing out of the vent as vapor mixed with the combustion products. Natural gas contains more hydrogen and less carbon per unit of heat content than oil and consequently its combustion produces a great deal more water vapor, which withdraws a greater amount of heat from the flame. Therefore, gas efficiency is always slightly less than oil efficiency.

Air requirements for combustion are generally expressed in cubic feet of air per gallon of oil or per cubic foot of gas for convenience because fans, ducts, and other air-moving devices are rated in cubic feet per minute or cubic feet per hour. The *fuel/air ratio* for combustion is actually a weight ratio based on the required weight of oxygen for a given weight of fuel. About 2000 cubic feet of air is required to burn one gallon of fuel oil at 80% efficiency at sea level. About 15 cubic feet of air is required to burn one cubic foot of natural gas at 75% at sea level. For example, a 100 HP (3.35 MMBH, 981 kW) boiler requires 75 000 ft^3 (2145 m^3) of fresh air per hour for combustion to take place.

Fuel-fired (gas and oil) boilers and furnaces need air for combustion. When considering air requirements, two roles must be considered: air must be present for combustion to take place and air is required to remove the products of combustion. This requires *draft,* which is simply air movement through the boiler or furnace. Because the inner structure of a boiler or furnace offers resistance to combustion gas movement (commonly called "draft loss") the draft must be strong enough to overcome this resistance.

With a *forced draft,* air is delivered to the combustion zone by the burner blower at sufficient pressure to not only provide the combustion air but also to expel the combustion products from the vent against the draft resistance of the boiler. The combustion zone is then under a positive pressure (pressure higher than atmospheric pressure).

With *natural draft,* the burner blower delivers air required for combustion, but the draft required to move the combustion products through the boiler is created by a stack or chimney. A stack or chimney creates draft because the column of air it encloses is under negative pressure (pressure less than atmospheric) so the combustion zone then is under negative pressure.

If the stack or chimney provides inadequate negative pressure to vent the products of combustion, an *induced-draft fan* may be used to compensate for the deficiency of the stack. In that case, the burner still operates with negative draft in the combustion zone.

Table 6.15 shows typical seasonal efficiencies for different types of residential space heating systems. Table 6.16 provides the percentage of heating energy saved when a comparing high-efficiency system to a low-efficiency system.

TABLE 6.15   PERFORMANCE OF DIFFERENT TYPES OF RESIDENTIAL SPACE HEATING SYSTEMS. COMPILED FROM MANUFACTURERS' INFORMATION.

Heating System	Typical Seasonal Efficiency
Gas- and Oil-Fired Boilers and Furnaces	
Old converted coal burner (converted to natural gas or oil), poorly maintained	35
Old converted coal burner (converted to natural gas or oil), well maintained	45
Standard low-efficiency furnace or boiler (pre-1985), poorly maintained	55
Low-efficiency furnace or boiler (pre-1985) with pilot light, well maintained	65
Low-efficiency furnace or boiler (pre-1985) with pilot light and vent damper, well maintained	70
Low-efficiency furnace or boiler (pre-1985) with electronic ignition and vent damper, well maintained	77
Medium-efficiency, induced-draft furnace or boiler (post-1985), new or well maintained	80
High-efficiency, condensing furnace or boiler (post-1985), new or well maintained	92
High-efficiency, pulse furnace or boiler (post-1985), new or well maintained	96
Electric Units	
Electric resistance baseboard heater	100
Electric resistance furnace or boiler	97
Wood/Pellet/Coal-Burning Units	
Standard fireplace without glass doors	0
Standard fireplace with glass doors and tight damper	10
Standard fireplace with metal liner or tubular grate and tight damper	20
Standard fireplace with metal liner or tubular grate, tight damper, and glass doors	30
Standard updraft wood stove (e.g., Franklin stove)	30
Standard airtight wood or pellet stove, well maintained	50
Premium airtight wood or pellet stove, well maintained	60
Wood/coal furnace, well maintained	60
Wood/coal furnace with catalytic converter, well maintained	70
Heat Pumps	
Air-source heat pump with electric resistance backup, well maintained	150 to 200[a]
Geothermal (air or water) heat pump with electric resistance backup, well maintained	250 to 300[a]

[a] For a moderate heating climate (6000°F · days). Varies significantly with climate.

6.8   DISTRICT HEATING AND COOLING

District Heating

A district heating system generates hot water or steam produced in a central plant outside of the building and piped into the building as an energy source for heating. The heat can be provided from a variety of sources, including geothermal, cogeneration plants, waste heat from industry, and purpose-built heating

TABLE 6.16   PERCENTAGE OF HEATING ENERGY SAVED IN COMPARING HIGH-EFFICIENCY SYSTEM TO LOW-EFFICIENCY SYSTEM.

Seasonal Efficiency of Low-Efficiency Heating System	Seasonal Efficiency of High-Efficiency Heating System								
	55%	60%	65%	70%	75%	80%	85%	90%	95%
Percentage of Heating Energy Saved									
50%	9	16	23	38	33	37	41	44	47
55%		8	5	21	26	31	35	38	42
60%			7	14	20	25	29	33	37
65%				7	13	18	23	27	32
70%					6	12	17	22	26
75%						6	11	16	21
80%							5	11	16
85%								5	11

plants. Steam may be purchased from a utility or provided by a central physical plant in a separate building that is part of the same multibuilding facility (e.g., a hospital complex or university campus). A steam district heating system has been in use at the U.S. Naval Academy since 1853. District heating has various advantages compared to individual heating systems; most significant is that it is more energy efficient because of cogeneration capabilities, the simultaneous production of heat and electricity in combined heat and power generation plants.

District heating systems consists of feed and return lines, with the heat transfer medium being water or steam. Usually the pipes are installed underground in tunnels that run to the buildings being served. Central system distribution pipes are connected to heat exchangers (heat substations) at each building's central heating system. The water or steam used in the district heating system is not mixed with the water of the central heating system of the building. A district heating distribution system, boiler stations, and cogeneration plants require a high initial capital expenditure but normally are a wise long-term investment.

District Cooling

A *district cooling* system distributes chilled water or other media to multiple buildings for air conditioning or other uses. The cooling (actually heat rejection) is usually provided from a dedicated cooling plant. Chilled water may be purchased from a utility or provided by a central physical plant in a separate building that is part of the same multibuilding facility (e.g., a hospital complex or university campus). The first district cooling operation began operating in Denver, Colorado, in late 1889, as the Automatic Refrigerator Company. District cooling is now widely used in downtown business districts and institutional settings such as college campuses. District cooling systems can displace peak electric power demand with storage using ice or chilled water.

Deep water source cooling (DWSC) refers to the renewable use of a large body of naturally cold water as a heat sink for process and comfort space cooling. Water at a constant 40° to 50°F (4° to 10°C) or less is withdrawn from deep areas within lakes, oceans, aquifers, and rivers and is pumped through the

primary side of a heat exchanger. On the secondary side of a heat exchanger, clean chilled water is produced with one tenth the average energy required by conventional, chiller-based systems. Coincident with significant energy and operating cost savings, DWSC offers reductions in airborne pollutants and the release of environmentally harmful refrigerants.

6.9   ENERGY RECOVERY

Energy recovery is the reuse of heat or cooling energy that would otherwise be lost. Recovered energy can be used for heating and/or cooling/dehumidifying outdoor air brought into a building for ventilation, space heating, and water heating. In commercial and institutional buildings, waste heat is recovered from liquids or gases at temperatures below 450°F (232°C). It results in savings in energy consumption and cost in buildings.

Applications for recovering heat depend on the temperature of the gases or liquids containing the waste heat. High-temperature recoverable heat at temperatures above 1200°F (650°C) from manufacturing processes can also be recovered for reuse. Such high-temperature processes are only available in heat-intensive industries (e.g., cement and metals manufacturing).

Some common applications for energy recovery in commercial and institutional buildings are discussed the in following.

Heat Recovery Ventilation

The need for good indoor air quality to maintain healthy and comfortable conditions for building occupants coupled with increasing energy costs has created an increased requirement for ventilation air (fresh outdoor air brought in to replace exhausted air) in all types of commercial and institutional buildings. Energy recovery of heat in exhaust ventilation is growing in availability and application throughout the country.

The three most common *heat recovery ventilation (HRV)* devices are: flat plate air-to-air heat/energy exchangers; rotary heat and energy (enthalpy) wheels; and heat pipes. All require adjacent and parallel outdoor air intake and exhaust ducts. When heat needs to be transferred from a remote location, a run-around loop system can be installed. This requires the installation of heat transfer coils in both the intake and exhaust ducts (or other source of heat). A pump circulates the heat transfer fluid, moving heat to where it is needed. Air-to-air heat pumps are occasionally used to extract heat from warm exhaust air. Higher installation and operating costs limit this application.

When an HRV system is integrated with HVAC systems, the resulting conditioning and/or heat recovery allow installation of smaller HVAC systems. This provides savings in both capital investments and operating costs.

Heat Recovery for Space Heating

Heat generated in one area or system of a building (e.g., kitchens, laundries, printing plants, computer rooms, refrigerated compressor rooms in supermarkets, and condensers of air conditioners and chillers) can be recovered and delivered to another space in a building for space heating. Use of heat pumps is a suitable method of removing heat from a heat source (hot air, water, process fluid, and so on) and delivering it to another point of need.

Lights in commercial buildings also produce large quantities of heat. Normally, only a fraction of this heat directly contributes to space heating. One method of heat recovery in lighting systems uses aluminum reflector housings that contain integral water channels. Water pumped through those channels absorbs heat produced by the lights. This warm water then circulates through a heat pump or a heat exchanger to provide space heating or to warm supply air. Other methods involve using the light fixture itself as a return grille to a return air duct. Heated air from the lights can then be distributed to cooler perimeter zones or recirculated through the central HVAC system.

STUDY QUESTIONS

6-1. What is a boiler and what does it produce?

6-2. Describe the types of boilers.

6-3. Identify and explain types of boiler ratings.

6-4. What is a warm-air furnace and what does it produce?

6-5. Describe the types of warm-air furnace classifications based on the direction of airflow for the supply air.

6-6. Describe the types of furnace efficiency classifications.

6-7. How do boilers and furnaces differ?

6-8. Describe the difference between an air-source heat pump and a geothermal heat pump.

6-9. Describe the function of an infrared heater and identify where it can be used effectively.

6-10. What is a chiller and what does it produce?

6-11. What is the vapor-compression refrigeration cycle?

6-12. In a vapor-compression refrigeration system, what is the function of the following components?
   a. Evaporator
   b. Condenser
   c. Compressor
   d. Expansion valve

6-13. What is a roof top unit?

6-14. What type of system is most commonly used for residential cooling?

6-15. When might a chilled water system be used?

6-16. What is a cooling tower and how does it function?

6-17. Explain the performance ratings used for air conditioners and heat pumps.

6-18. Explain the difference between a mechanical (vapor-compression refrigeration) chiller and an absorption chiller.

6-19. What is an evaporative cooler and how does it function?

6-20. Describe ice thermal storage system.

6-21. Describe the types of ventilation equipment.

6-22. Describe humidification and why is it needed in buildings.

6-23. Describe dehumidification and why is it needed in buildings.

6-24. Describe heat recovery ventilation.

6-25. Describe how cogeneration can be used in an HVAC system.

6-26. Describe district heating and district cooling.

Design Exercises

6-27. The heating load of a building at winter design conditions is 1 510 000 Btu/hr (with pick-up allowance). From the performance specifications for commercial gas-fired hot water boilers provided in Table 6.2, select a boiler that can efficiently meet this load based on the
   a. MBH output rating
   b. I=B=R net rating

6-28. The heating load of a building at winter design conditions is 3 210 000 Btu/hr (with pick-up allowance). From the performance specifications for commercial oil-fired hot water boilers provided in Table 6.3, select a boiler that can efficiently meet this load based on the
   a. MBH output rating
   b. I=B=R net rating

6-29. The heating load of a building at winter design conditions is 3 600 000 Btu/hr (with pick-up allowance). From the performance specifications for commercial dual fuel (gas- and oil-fired) hot water boilers provided in Table 6.4, select a boiler that can efficiently meet this load based on the
   a. MBH output rating
   b. I=B=R net rating

6-30. The heating load of a residence at winter design conditions is 88 400 Btu/hr (with pick-up allowance). From the performance specifications for high-efficiency, gas-fired residential warm-air furnaces provided in Table 6.5, select a furnace that can efficiently meet this load based on the MBH output rating.

6-31. The heating load of a residence at winter design conditions is 88 400 Btu/hr (with pick-up allowance). From the performance specifications for high-efficiency, oil-fired residential warm-air furnaces provided in Table 6.5, select a furnace that can efficiently meet this load based on the MBH output rating.

6-32. The cooling load of a residence at summer design conditions is 46 800 Btu/hr. From the performance specifications for high-efficiency residential central air conditioners provided in Table 6.8, select an air conditioner that can efficiently meet this load in tons-cooling.

6-33. In a commercial building, the heating load at winter design conditions is 211 400 Btu/hr (with pick-up allowance) and the cooling load at summer design conditions is 232 800 Btu/hr. From the performance specifications for commercial, gas-electric, heating and cooling, packaged rooftop units provided in Table 6.9, select an RTU that can efficiently meet these loads.

Information for Exercises 6-34 through 6-43

Selecting the best-suited HVAC system depends on factors such as cost and availability of the energy source; appliance or system efficiency; cost to purchase, install, and maintain the appliance or system; and environmental impacts associated with the fuel. Energy prices tend to vary somewhat by geographic location because of availability of fuel type and supply, market pressures, production capacity, weather, season, and politics. A common way to compare heating energy costs is by determining the delivered heating cost, expressed in consistent units (i.e., $/MMBtu or $/GJ), based on the cost of the fuel ($/unit), heating value of the fuel (HV), and seasonal efficiency of the system or appliance (η_{hs}) expressed in decimal form (i.e., 80% = 0.80):

Delivered Heating Cost ($/MMBtu)
$$= (\$/\text{unit} \cdot 1\ 000\ 000\ \text{Btu/MMBtu})/(\eta_{hs} \cdot \text{HV})$$
Delivered Heating Cost ($/GJ)
$$= (\$/\text{unit} \cdot 1\ 000\ 000\ \text{kJ/GJ})/(\eta_{hs} \cdot \text{HV})$$

6-34. Compute the delivered heating cost ($/MMBtu) for a natural gas fired heating unit (i.e., furnace, boiler) based on a heating value of 103 000 Btu/CCF and a cost of $1.50/CCF.

  a. For an older, low-efficiency unit with a seasonal efficiency of 65%.

  b. For a modern, middle-efficiency unit with a seasonal efficiency of 80%.

  c. For a modern, high-efficiency unit with a seasonal efficiency of 96%.

6-35. Compute the delivered heating cost ($/GJ) for a natural gas fired heating unit (i.e., furnace, boiler) based on a heating value of 37 500 kJ/m³ and a cost of $.50/m³.

  a. For an older, low-efficiency unit with a seasonal efficiency of 65%.

  b. For a modern, middle-efficiency unit with a seasonal efficiency of 80%.

  c. For a modern, high-efficiency unit with a seasonal efficiency of 96%.

6-36. Compute the delivered heating cost ($/MMBtu) for a fuel oil fired heating unit (i.e., furnace, boiler) based on a heating value of 140 000 Btu/gal and a cost of $3.50/gal.

  a. For an older, low-efficiency unit with a seasonal efficiency of 65%.

  b. For a modern, middle-efficiency unit with a seasonal efficiency of 78%.

  c. For a modern, high-efficiency unit with a seasonal efficiency of 92%.

6-36. Compute the delivered heating cost ($/GJ) for a fuel oil fired heating unit (i.e., furnace, boiler) based on a heating value of 16 820 kJ/L and a cost of $1.00/L.

  a. For an older, low-efficiency unit with a seasonal efficiency of 65%.

  b. For a modern, middle-efficiency unit with a seasonal efficiency of 78%.

  c. For a modern, high-efficiency unit with a seasonal efficiency of 92%.

6-38. Compute the delivered heating cost ($/MMBtu) for a propane-fired heating unit (i.e., furnace, boiler) based on a heating value of 91 000 Btu/gal and a cost of $2.50/gal.

  a. For an older, low-efficiency unit with a seasonal efficiency of 65%.

  b. For a modern, middle-efficiency unit with a seasonal efficiency of 80%.

  c. For a modern, high-efficiency unit with a seasonal efficiency of 96%.

6-39. Compute the delivered heating cost ($/GJ) for a propane-fired heating unit (i.e., furnace, boiler) based on a heating value of 11 000 kJ/L and a cost of $0.75/gal.

  a. For an older, low-efficiency unit with a seasonal efficiency of 65%.

  b. For a modern, middle-efficiency unit with a seasonal efficiency of 80%.

  c. For a modern, high-efficiency unit with a seasonal efficiency of 96%.

6-40. Compute the delivered heating cost ($/MMBtu) for the following electricity-powered heating units (i.e., furnace, boiler) based on a heating value of 3413 Btu/kWh and a cost of $0.15/kWh.

a. An electric resistance heating unit (i.e., furnace, baseboard heater) with a seasonal efficiency of 100%.

b. An air source heat pump with a seasonal efficiency* of 150%.

c. An electricity-powered geothermal heat pump with a seasonal efficiency* of 350%.

6-41. Compute the delivered heating cost ($/MMBtu) for coal-fired heating units (i.e., furnace, boiler) based upon a heating value of 18 100 000 Btu/ton and a cost of $150/ton.

a. For an older, low-efficiency unit with a seasonal efficiency of 65%.

b. For a modern, middle-efficiency unit with a seasonal efficiency of 80%.

6-42. Compute the delivered heating cost ($/GJ) for coal-fired heating units (i.e., furnace, boiler) based upon a heating value of 21 000 kJ/kg and a cost of $0.205/kg.

a. For an older, low-efficiency unit with a seasonal efficiency of 65%.

b. For a modern, middle-efficiency unit with a seasonal efficiency of 80%.

6-43. Compute the delivered heating cost ($/MMBtu) for a wood-fired heating units based upon a heating value of 21 000 000 Btu/cord and a cost of $225/cord.

a. For a fireplace with glass doors with a seasonal efficiency of 15%.

b. For a modern, middle-efficiency unit with a seasonal efficiency of 75%.

6-44. A client is considering heating systems for a 2400 ft^2 residence anticipated to have an annual heating load of 90 MMBtu/year. Compute the annual cost of heating the building with these systems:

a. A modern, high-efficiency furnace with a seasonal efficiency of 96% that is fired with natural gas with a heating value of 103 000 Btu/CCF and a cost of $1.50/CCF.

b. A modern, high-efficiency furnace with a seasonal efficiency of 92% that is fired with fuel oil with a heating value of 140 000 Btu/gal and a cost of $3.50/gal.

c. An electric-resistance furnace with a seasonal efficiency of 100% that is electrically powered with a heating value of 3413 Btu/kWh and a cost of $0.15/kWh.

d. An air source heat pump with a seasonal efficiency* of 150% that is electrically powered with a

heating value of 3413 Btu/kWh and a cost of $0.15/kWh.

e. A geothermal heat pump with a seasonal efficiency* of 350% that is electrically powered with a heating value of 3413 Btu/kWh and a cost of $0.15/kWh.

6-45. A school district is considering heating systems for a high school anticipated to have an annual heating load of 6200 MMBtu/year. Compute the annual cost of heating the building with these systems:

a. A modern, high-efficiency boiler with a seasonal efficiency of 96% that is fired with natural gas with a heating value of 103 000 Btu/CCF and a cost of $1.50/CCF.

b. A modern, high-efficiency boiler with a seasonal efficiency of 92% that is fired with fuel oil with a heating value of 140 000 Btu/gal and a cost of $3.50/gal.

c. An electric-resistance boiler with a seasonal efficiency of 100% that is electrically powered with a heating value of 3413 Btu/kWh and a cost of $0.15/kWh.

d. An air source heat pump system with a seasonal efficiency* of 150% that is electrically powered with a heating value of 3413 Btu/kWh and a cost of $0.15/kWh.

e. A geothermal heat pump system with a seasonal efficiency* of 350% that is electrically powered with a heating value of 3413 Btu/kWh and a cost of $0.15/kWh.

6-46. A school district is considering heating systems for a high school anticipated to have an annual heating load of 7200 GJ/year. Compute the annual cost of heating the building with these systems:

a. A modern, high-efficiency boiler with a seasonal efficiency of 96% that is fired with natural gas with a heating value of 37 500 kJ/m^3 and a cost of $.50/m^3.

b. A modern, high-efficiency furnace with a seasonal efficiency of 92% that is fired with fuel oil with a heating value of 140 000 Btu/gal and a cost of $3.50/gal.

c. An electric-resistance furnace with a seasonal efficiency of 100% that is electrically powered with a heating value of 3413 Btu/kWh and a cost of $0.15/kWh.

d. An air source heat pump with a seasonal efficiency* of 150% that is electrically powered with a heating value of 3413 Btu/kWh and a cost of $0.15/kWh.

e. A geothermal heat pump with a seasonal efficiency* of 350% that is electrically powered with a heating value of 3413 Btu/kWh and a cost of $0.15/kWh.

*Technically, the "efficiency" of a heat pump is indicated by the coefficient of performance (COP), which is the ratio of heat provided (for heating) in Btu/hr per Btu/hr of energy input. The COP varies significantly by temperature and thus with climate. Values provided are comparable "efficiencies" for a moderate heating climate (approximately 5000°F · days).

HVAC DISTRIBUTION COMPONENTS AND SYSTEMS

7.1 DISTRIBUTION SYSTEMS AND EQUIPMENT

Basic HVAC Systems

In large buildings, boilers, packaged heating units, and heat pump systems are used extensively for space heating; central chilled water systems, and in some climates, large direct and indirect evaporative coolers are used for cooling. In light commercial and residential buildings, furnaces, small boilers, individual/room space heaters, and heat pumps are the source of space heating. Packaged split-system, individual/room air conditioners, and in some climates, direct (swamp) evaporative coolers, are common small cooling installations. Types and operation of these pieces of equipment were introduced in the previous chapter.

Although these units generate space heating and/or cooling, additional equipment and components are needed to distribute the heating/cooling throughout the building spaces. *Distribution equipment* consists of that part of the heating and cooling system that conveys and delivers conditioned water and/or air to interior spaces in the building by means of pipes, pumps, ducts, or fans. Frequently, the distribution equipment provides both heating and cooling.

Cooling equipment has the cooling effect it produces distributed primarily by the use of one or combinations of three types of distribution systems:

- Cold air produced by packaged units (e.g., rooftop units) and split-system air conditioners (e.g., with furnaces) distributed primarily by ducts.
- Chilled water generated by chilled water systems transported to and passed through terminal fan-coil units, terminal unit ventilators, or air-handling units with cooling coils.
- Individual air conditioning units cooling the surrounding areas directly without a separate distribution system.

Heating equipment has the heat it produces distributed primarily by the use of one or combinations of four types of distribution systems:

- Warm air produced by packaged units (e.g., rooftop unit, air-handling units, and furnaces) that is distributed primarily by ducts.

- Hot water or steam generated by boilers transported to convectors or radiators to heat air by natural heat transfer.
- Hot water or steam generated by boilers transported to and passed through terminal fan-coil units, terminal unit ventilators, or air-handling units with heating coils.
- Individual space heaters giving off heat directly to surrounding areas without a separate distribution system.

HVAC Distribution Components

HVAC distribution components can be divided into an air side and a water side. Components are categorized as follows:

Air-Side Components

- Ductwork (ducts, fittings, and dampers)
- Fans
- Air-handling units
- Terminal units
- Diffusers, registers, and grilles

Water-Side Components

- Piping
- Valves
- Pumps
- Expansion tanks
- Heat exchangers
- Terminal units

The advantages and disadvantages of common types of HVAC systems are compared in Table 7.1. These systems and their system components are introduced in much of the remainder of this chapter.

7.2 AIR DELIVERY SYSTEM COMPONENTS

Ductwork

A *duct* is a hollow pipe or conduit that conveys and transfers air in a heating, ventilating, and cooling system. A *fitting* is a coupling, tee, elbow, or similar duct section that changes the direction of airflow or changes the size or shape of the duct

TABLE 7.1   ADVANTAGES AND DISADVANTAGES OF
TYPES OF HVAC DISTRIBUTION SYSTEMS.

All-Air Systems

Advantages	Disadvantages
• Mechanical equipment can be isolated. • No drain pipes, electrical wiring, and filters at the conditioned space. • Seasonal changeover very simple and easy to automate. • Good flexibility because heating and cooling can occur in different zones simultaneously.	• Additional duct clearance required, which can reduce usable floor space and increase building height. • Air balancing is complicated and requires periodic care.

All-Water Systems

Advantages	Disadvantages
• Central air-handling units and ductwork are not required. • Unused zones can be shut down from central control. • Individual room control with little or no chance of cross-contamination of recirculated air from one space to another. • Easier to retrofit to existing buildings.	• More maintenance required. Often, the maintenance is done in occupied areas. • Units that have condensate and drip pans must be flushed and cleaned. • Filters are small, low in efficiency, and require frequent replacement. • Ventilation is unsophisticated—usually accomplished by opening windows or by installing outside wall vents.

Air–Water Systems

Advantages	Disadvantages
• Rooms can be individually controlled at relatively low cost. • Size of the central air-handling unit can be reduced.	• Controls tend to be complex. • Secondary air can cause induction coils to foul. • Filtering can be done at the terminal unit; however this requires frequent in-room maintenance. • A low chilled-water temperature is needed to accurately control humidity levels.

(called a transition). A duct system, frequently called *ductwork,* is a network of ducts that confines and channels air and serves as the distribution system in a heating, ventilating, and cooling air system. Examples of ductwork are shown in Photos 7.1 through 7.9).

Ducts and duct fittings are made of rigid steel and aluminum sheet metal, a fiberglass composite (fiberglass with a thin metal facing), or thermoplastics such as polyvinylchloride (PVC), polyethylene (PE), polypropylene (PP), and acrylonitrile-butadiene-styrene copolymer (ABS) resin. Fabric ducts are made

PHOTO 7.1   Ducts suspended from the ceiling of a gymnasium under construction. An air-handling unit (AHU) (not shown) behind the masonry wall blows conditioned air into the rectangular main duct (left side). Air then branches off into round branch ducts, where it is eventually introduced to the space through diffusers attached to the branch ducts. (Used with permission of ABC)

PHOTO 7.2   A section of lined rectangular duct. (Used with permission of ABC)

PHOTO 7.3   Unlined round ducts. (Used with permission of ABC)

PHOTO 7.4   A section of unlined oval duct. (Used with permission of ABC)

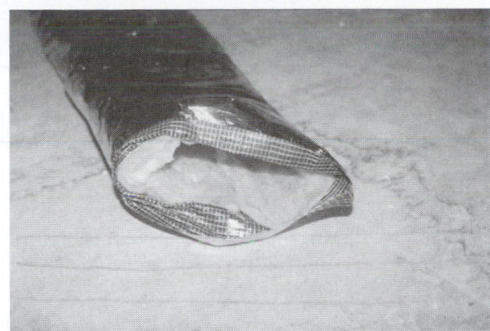

PHOTO 7.5   Lined flexible duct. (Used with permission of ABC)

PHOTO 7.6   A round, 90° elbow fitting. (Used with permission of ABC)

PHOTO 7.7   Turning vanes inside a lined rectangular bend. Turning vanes reduce pressure drop in a bend. (Used with permission of ABC)

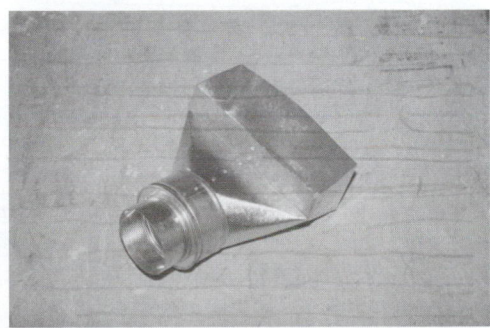

PHOTO 7.8   A register box fitting. (Used with permission of ABC)

PHOTO 7.9   A register box fitting installed as part of a branch duct. This fitting introduces air through an opening in the subfloor to the floor above. (Used with permission of ABC)

of a permeable material such as polyester. Galvanized steel ducts are normally 26 to 30 gauge thick, and aluminum ducts are typically 23 to 26 gauge.

Underground ducts can be made of coated steel, plastic, concrete, and reinforced fiberglass. Ducts and duct fittings can be round, oval, or rectangular in cross-section. Round ducts channel air most efficiently while rectangular ducts save headroom because they can be wider than they are deep. Flat-oval ducts are a compromise between these characteristics. *Flexible ducts* are typically used to connect diffusers, mixing boxes, and other terminals.

Ductwork will have a designation depending on the function of the duct:

- *Outdoor air* ducts supply outdoor air to the air-handling unit. Outdoor air is used for ventilating the occupied building spaces.

- *Return air* ductwork removes air from the conditioned building spaces and returns the air to the air-handling unit, which reconditions the air. In some cases, part of the return air in this ductwork is exhausted to the building exterior.

- *Exhaust (relief) air* ductwork carries and discharges air to the outdoors.

- *Mixed air* ductwork mixes air from the outdoor air and the return air ductwork and supplies this mixed air to the air-handling unit.

- *Supply air* ductwork supplies conditioned air from the air-handling unit to diffusers and registers that introduce the occupied space.

- *Supply air inlets* (diffusers and registers) allow the conditioned air from the supply air ductwork to enter the occupied space.

- *Return air outlets* (e.g., return air grilles) allow air from the occupied space to enter the return air ductwork.

Common duct sizes used in small buildings are shown in Table 7.2.

TABLE 7.2   TYPICAL DUCT, REGISTER, AND GRILLE SIZES FOR SMALL INSTALLATIONS (E.G., RESIDENTIAL, LIGHT COMMERCIAL), BASED ON DIFFERENT LOW-VELOCITY FLOW RATES. COMPILED FROM MANUFACTURERS' INFORMATION.

Flow Rate		Minimum Duct Size				Minimum Supply Air Floor Register Size		Minimum Wall Return Air Grille Size	
		Round		Rectangular					
ft³/min	m³/s	in	mm	in	mm	in	mm	in	mm
60	27	5	125	2¼ × 10	56 × 250	4 × 10	100 × 250	4 × 12	100 × 300
				3¼ × 8	81 × 200	2¼ × 12	56 × 250		
				4 × 6	100 × 150				
100	45	6	150	2¼ × 12	56 × 300	4 × 10	100 × 250	6 × 12	150 × 300
				3¼ × 10	81 × 250	4 × 12	100 × 300		
				4 × 10	100 × 250				
150	68	7	175	3¼ × 14	56 × 350	4 × 14	100 × 350	8 × 12	200 × 300
				6 × 8	150 × 200				
200	90	8	200	4 × 14	100 × 350	6 × 14	150 × 350	8 × 14	200 × 350
				8 × 8	200 × 200				
300	135	9	225	8 × 8	200 × 200	8 × 14	200 × 350	10 × 14	250 × 350
400	180	10	250	8 × 10	200 × 250	Multiple required		14 × 14	350 × 350
500	225	12	300	8 × 12	200 × 300	Multiple required		20 × 14	500 × 350
600	270	12	300	8 × 14	200 × 350	Multiple required		20 × 14	500 × 350
700	315	14	350	8 × 16	200 × 400	Multiple required		24 × 14	600 × 350
800	360	14	350	8 × 18	200 × 450	Multiple required		24 × 14	600 × 350
900	405	14	350	8 × 20	200 × 500	Multiple required		30 × 12	750 × 300
1000	450	14	350	8 × 22	200 × 550	Multiple required		30 × 16	750 × 400
1200	540	16	400	8 × 24	200 × 600	Multiple required		Multiple required	
1400	630	16	400	10 × 22	250 × 550	Multiple required		Multiple required	
1600	720	18	450	10 × 24	250 × 600	Multiple required		Multiple required	
2000	900	18	450	10 × 30	250 × 750	Multiple required		Multiple required	

Duct Insulation

Supply and return air ductwork and air plenums are frequently lined with acoustical materials to reduce noise caused by airflow. Supply and return ducts in unconditioned spaces must be thermally insulated. Exhaust ductwork need not be insulated because insulation will have no impact on building energy usage. Caution should be exercised in cold-air ductwork because condensation can occur if insulation is improperly compressed or if there are openings or gaps in the vapor-retarding envelope of the insulation.

As a rule, all supply and return air ducts and plenums should be insulated with a minimum of R-5 (RSI-0.88°C · m²/W) insulation when located in unconditioned spaces (e.g., attics, crawl spaces, unheated basements, unheated garages) and with a minimum of R-8 (RSI-1.41°C · m²/W) when located outside the building envelope. When located within a building envelope assembly, the duct or plenum shall be separated from the building exterior or unconditioned spaces by a minimum of R-5 (RSI-0.88°C · m²/W) insulation. Insulation is generally not required for:

- Ducts located within equipment
- When the design temperature difference between the interior and exterior of the duct or plenum does not exceed 15°F (8°C)

- Exhaust (relief) air ducts
- Outside air supply ducts located outside of the building envelope

Sealants and Tape

Proper sealing of plenums, air-handling units, and ducts is indispensable in eliminating leaks in a duct system. Sealing openings is extremely important in maintaining indoor air quality and operating an efficient system.

All joints, longitudinal and transverse seams, and connections must be securely fastened and sealed. Duct registers, grilles, and diffusers must be sealed to the gypsum board or other interior finish. Penetrations into the supply or return plenum (taps, takeoffs, and starting collars) and any structural cavities used for air plenums or ducts must also be sealed. However, sealants should not be used as a substitute for mechanical fastening of duct system components.

Duct mastic is a thick adhesive that is applied moist to fill gaps and dries to a soft solid. It is a flexible sealant that can move with the expansion, contraction, and vibration of the duct system components. See Photo 7.10. Duct mastic with embedded fabric is duct mastic that is strengthened with fiberglass strands for increased strength. Aerosol sealant is a sticky vinyl polymer that is applied to the leaks internally. It is pumped through the duct system, where it spans leaks and dries. Duct

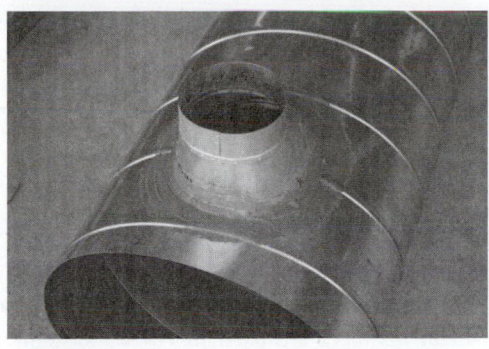

PHOTO 7.10 Duct mastic used to seal a tap-in connection of a branch duct into the main duct. (Used with permission of ABC)

tape is cloth backed and has a rubber-based adhesive. Duct mastics and aerosol sealants are the preferred duct sealing materials. Duct tape is no longer acceptable because its adhesive tends to dry out, rendering it ineffective.

Dampers

A *damper* is an airflow control device used in ductwork to vary the volume of air passing through an air outlet, air inlet, or duct.

A damper opens and closes to control airflow. Air dampers are placed in ductwork and allow the quantity of air in different parts of the ventilation ductwork to be adjusted. The door-like mechanism of a damper manually or automatically opens and closes, thereby controlling flow of air. Dampers are the part of an air distribution system that controls the system's return, outside, supply, exhaust, and relief air. Dampers are used in zoning to regulate airflow to specific zones or spaces. See Figure 7.1 and Photos 7.11 and 7.12.

There are basically two types of dampers: manual and motorized. A *manual damper* generally consists of a sheet metal (or similar material) flap, shaped to fit the inside of a round or rectangular duct. By rotating a handle located outside of the duct, a technician can adjust air flow to match the needs of a particular zone or room. A *motorized damper* is generally used in a zoned system (see Thermal Zoning section) to automatically deliver conditioned air to specific rooms or zones.

Volume control dampers are designed to control air flow in a HVAC system. These dampers require operation by either manual, electric, or pneumatic actuators. *Manually adjusted dampers* are typically set to control airflow in ducts at the time the system is balanced during initial startup of the system. Once balanced, dampers provide the correct amount of air to a

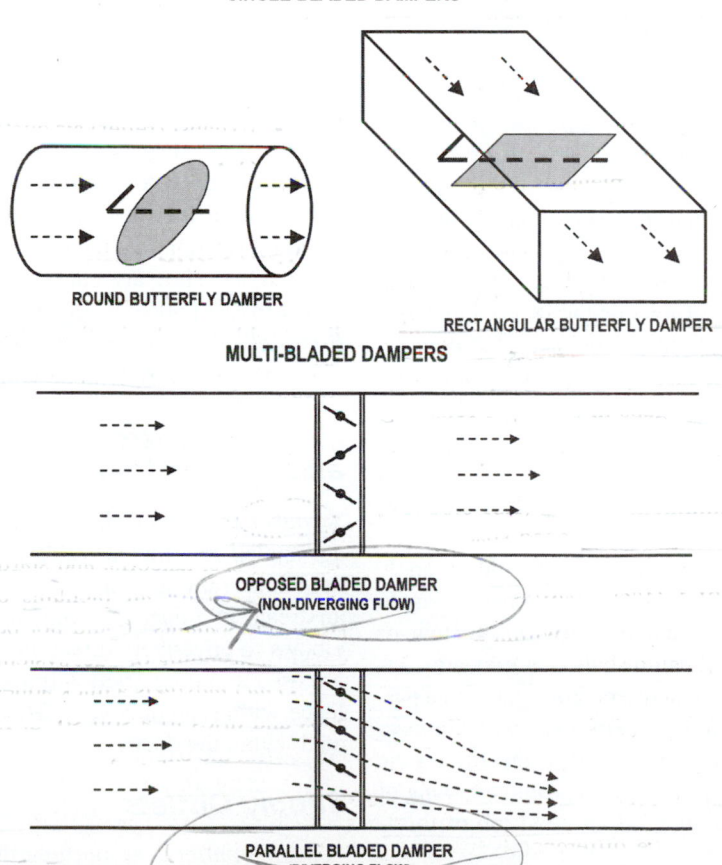

FIGURE 7.1 Dampers are used to regulate flow of air.

PHOTO 7.11   A butterfly damper in a round tap-in connection. Note use of duct mastic sealant. (Used with permission of ABC)

PHOTO 7.12   A parallel blade damper. (Used with permission of ABC)

given space or zone. They are not intended to be used in applications as a positive shutoff or for automatic control. *Automatic dampers* can be motorized or self-acting, responding to temperature or pressure changes, or can be remotely controlled by a central computer system to coordinate their operation. An electric or pneumatic actuator powers a motorized damper. An *actuator* is a mechanism that causes a device such as a damper or valve to be adjusted or turned on or off.

Dampers used in HVAC systems may be either single blade (leaf) or multiblade. Single-blade dampers have a single shutter that is positioned to be fully closed to shut off flow, or partially closed to throttle flow. *Multiblade dampers* are constructed of a number of blades (or shutters) configured in a parallel- or opposed-blade configuration. *Parallel-blade dampers* impart a directional characteristic to the airflow that directs the air streams into each other. They are used to promote mixing of air streams. *Opposed-blade dampers* do a good job of throttling airflow.

Automatic dampers in some systems are computer controlled and are automatically adjusted to balance the system.

For example, buildings typically use outdoor air instead of cooling equipment whenever possible to provide cooling inside the building (e.g., when outdoor temperatures are substantially lower than indoor temperature). Computerized systems automatically adjust the quantity of outside air brought into the building as cooling needs change inside the occupied space. As the amount of outdoor air supplied to the building interior increases, the amount of air exhausted must also increase. Several air dampers are adjusted to instantaneously to control airflow in a computerized system.

Backdraft dampers are used in HVAC systems to allow airflow in one direction and prevent airflow in the opposite direction. For example, exhaust dampers (in the exhaust ducts) allow airflow out of buildings while preventing airflow into the building.

A common way for fire and smoke to spread from one compartment to another is through the HVAC ductwork. *Fire dampers* automatically close to obstruct smoke and fire from a building blaze. Fire dampers are installed in the plane of the firewall to protect these openings. Upon detection of heat, the fusible link available in 165°F (74°C), 212°F (100°C), and 285°F (141°C) melts, closing the fire damper blades and blocking the flame from penetrating the partition into the adjoining compartment. Fire dampers are required by all building codes to maintain the required fire resistance ratings of walls, partitions, and floors when they are penetrated by air ducts or other ventilation openings.

Smoke dampers close upon detection of smoke, preventing the circulation of air and smoke through a duct or a ventilation opening. They can be part of an engineered smoke control system designed to control smoke migration using walls and floors as barriers to create pressure differences. Pressurizing the areas surrounding the fire prevents the spread of smoke into other areas. They are controlled by a smoke or heat detector signal that is part of a fire alarm control system. *Combination fire/smoke dampers* are designed for applications where building codes require both a fire and smoke damper.

Louvers, Splitters, and Turning Vanes

A *louver* is a shutter-like device with fixed slats or vanes that are angled to admit or release air, to provide privacy, and to keep out rain and noise. *Splitters* are used to direct the airstream in a duct. *Turning vanes* are used in duct bends and elbows to effectively direct the airflow through the turn. They are used to reduce friction (pressure) losses by reducing turbulence within the ductwork. They will also reduce noise generated within the ducts.

Supply Outlets

A *supply outlet* is an opening through which air is introduced into the space. Diffusers, grilles, and registers are supply outlet devices. The proper selection of diffusers, grilles, and registers

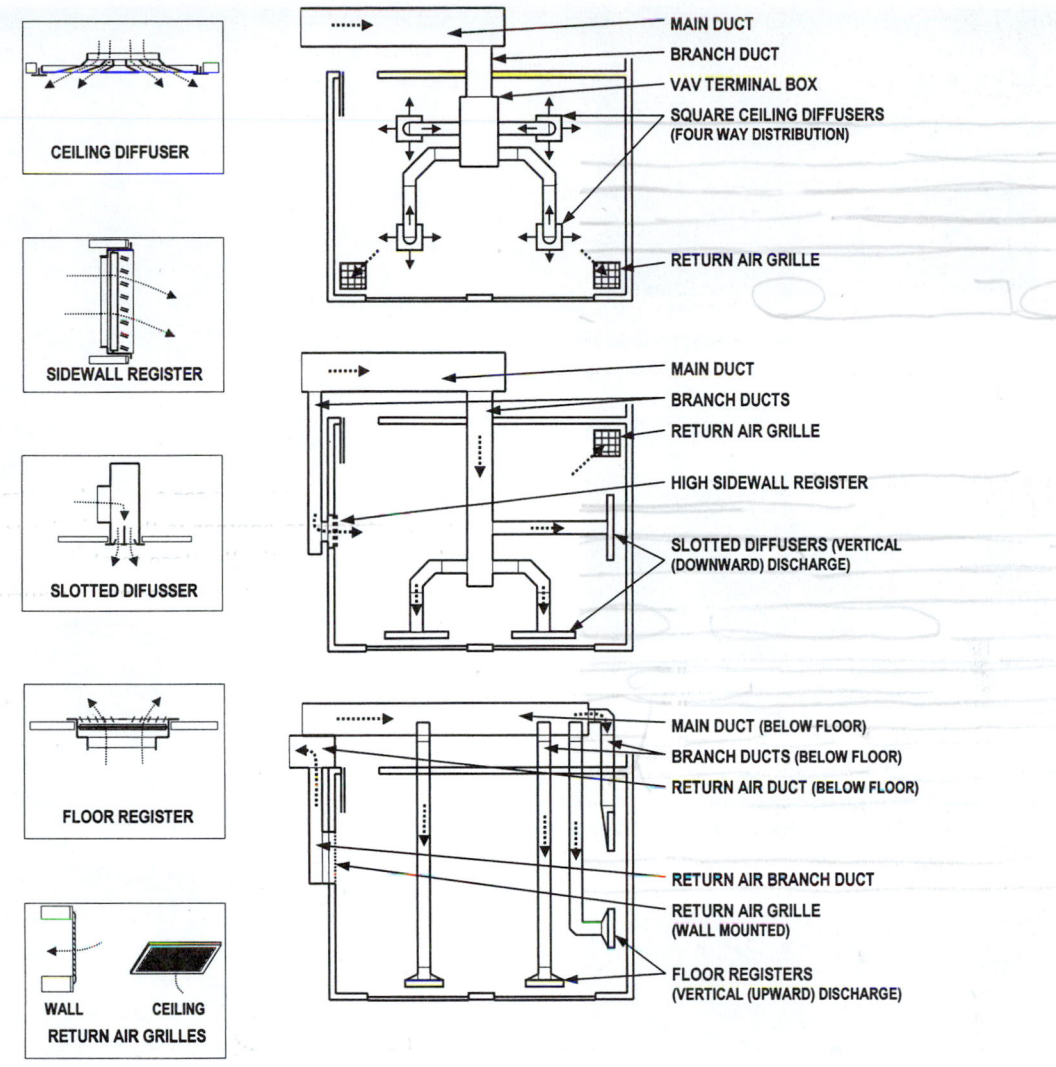

FIGURE 7.2   Diffusers, registers, and grilles direct the flow of air.

is necessary to ensure that both occupant comfort and adequate ventilation mixing are provided. See Figure 7.2.

Diffusers are the supply outlets of an air distribution (duct) system with built-in louvers or deflection devices that throw and disperse air to promote good air circulation and temperature equalization in the conditioned space. Types of diffusers include round, square, perforated, linear slot, linear louvered, and linear bar. Circular floor swirl diffusers introduce air at very low velocities to cause minimal mixing of the room air for displacement ventilation. Photos 7.13 through 7.19 show different types of diffusers. A *fire-rated diffuser* has a damper-like device that closes to isolate the space under fire conditions.

Grilles have sets of vertical and horizontal vanes that deflect the air stream vertically and horizontally. The vanes on return air grilles are typically fixed. A *register* is a grille equipped with a built-in damper. Typical register and grille sizes for small (e.g., residential, light commercial) installations are shown in Table 7.2. Selected diffuser sizes for small (e.g., residential, light commercial) installations and acceptable airflow rates are provided in Table 7.3.

Fans

A *fan* is a mechanical device that uses rotating airfoil blades or vanes to continuously move air. In HVAC systems, fans move air through a system of ductwork. Two basic types of fans are used in HVAC applications, classified according to the direction of the airflow through the impeller: axial and centrifugal.

Axial-Flow Fans

Axial-flow fans are propeller-type fans that use a propeller to draw air into the fan and discharge it in the same axial direction; airflow, at all times, from entry to exit is predominately

PHOTO 7.13 A ceiling diffuser placed in a suspended ceiling. (Used with permission of ABC)

PHOTO 7.14 A ceiling diffuser connected to a flexible branch duct, which is tapped in to a main duct. Use of a flexible duct makes for easier placement of the diffuser once the suspended ceiling grid is installed. (Used with permission of ABC)

PHOTO 7.15 A large, round ceiling diffuser. (Used with permission of ABC)

PHOTO 7.16 Round, high sidewall diffusers. (Used with permission of ABC)

PHOTO 7.17 A wall register. (Used with permission of ABC)

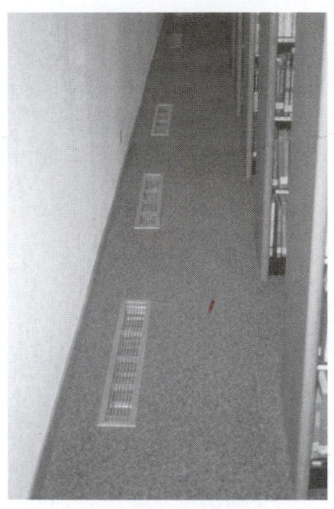

PHOTO 7.18 Floor registers located along an exterior wall. (Used with permission of ABC)

PHOTO 7.19 High wall registers used for cooling in a residential installation. (Courtesy of NREL/DOE)

TABLE 7.3 SELECTED DIFFUSER SIZES FOR SMALL INSTALLATIONS (E.G., RESIDENTIAL, LIGHT COMMERCIAL). COMPILED FROM MANUFACTURERS' INFORMATION.

Diffuser Inlet (Duct) Size		Diffuser Frame Size		Diffuser Frame Type	Diffuser Airflow at NC 30[a]	
					ft³/min	m³/s
Square—4-Way Discharge Pattern						
in	mm	in	mm			
6	150	12 × 12	300 × 300	Suspended ceiling lay-in	195	88
6	150	12 × 12	300 × 300	Surface mounted	225	101
6	150	24 × 24	600 × 600	Suspended ceiling lay-in	175	79
6	150	24 × 24	600 × 600	Suspended ceiling lay-in	250	113
8	200	12 × 12	300 × 300	Suspended ceiling lay-in	270	122
8	200	12 × 12	300 × 300	Surface mounted	355	160
8	200	24 × 24	600 × 600	Suspended ceiling lay-in	365	164
8	200	24 × 24	600 × 600	Suspended ceiling lay-in	405	182
10	250	24 × 24	600 × 600	Suspended ceiling lay-in	535	241
10	250	24 × 24	600 × 600	Suspended ceiling lay-in	560	252
12	300	24 × 24	600 × 600	Suspended ceiling lay-in	730	329
12	300	24 × 24	600 × 600	Suspended ceiling lay-in	775	349
14	350	24 × 24	600 × 600	Suspended ceiling lay-in	910	410
14	350	24 × 24	600 × 600	Suspended ceiling lay-in	920	414
15	375	24 × 24	600 × 600	Suspended ceiling lay-in	1040	468
15	375	24 × 24	600 × 600	Suspended ceiling lay-in	1060	477
Perforated Ceiling, Adjustable—1-, 2-, 3-, or 4-Way Discharge Pattern						
6	150	24 × 24	600 × 600	Suspended ceiling lay-in	175	79
8	200	24 × 24	600 × 600	Suspended ceiling lay-in	265	119
10	250	24 × 24	600 × 600	Suspended ceiling lay-in	350	158
12	300	24 × 24	600 × 600	Suspended ceiling lay-in	470	212
14	350	24 × 24	600 × 600	Suspended ceiling lay-in	575	259
Round—Full Discharge						
6	150	12	300	Surface mounted	190	86
8	200	16	400	Surface mounted	280	126
10	250	16	400	Surface mounted	420	189
12	300	18	450	Surface mounted	630	284
14	350	24	600	Surface mounted	800	360
Slotted—Two Slots						
8	200	24 in long	600 mm long	Recessed ceiling discharge	160	72
10	250	48 in long	1200 mm long	Recessed ceiling discharge	270	122

[a]Noise criteria (NC) is a continuous noise rating for HVAC equipment. An NC rating for classrooms, private offices and conference rooms is usually NC-30. See Chapter 23.

parallel to the axis of rotation. Axial-flow fans resemble most residential fans that get plugged into the wall for space cooling; the air is passed straight through. Axial fans are often directly connected to their motors, avoiding losses associated with a drive belt. They also have a central hub that allows the motor to fit neatly behind the fan with little penalty in efficiency. The weight distribution of their blades allows for low starting torque.

Axial fans can be subdivided into three categories: *propeller fans* (used to move high air volume against low or no static pressure), *tube-axial fans* (fans that encase the propeller in a duct section), and *vane-axial fans* (fans that use straightening fins to convert circular, twisting air to get the fan moving). Vane-axial fans tend to be the most efficient fans available for HVAC air-handling units largely because the direction of the airflow is little changed as it passes through the fan.

Centrifugal Fans

A *centrifugal fan*, also known as a *squirrel cage fan*, is a fan in which the air is turned from parallel to the axis of rotation on entry to a direction tangential to the arc described by the tips of the rotating blades or vanes. The air, instead of passing straight through, makes a 90° angle turn as it travels from the inlet to the outlet, and it is thrown from the blade tips. Centrifugal fans have more mass farther from the axle, which requires more starting torque, but they are generally quieter than axial fans. Despite their lower efficiencies, centrifugal fans greatly outnumber axial fans.

There are several arrangements of fan blades for centrifugal impellers. Airfoil blades are curved backward but have an airfoil shape, while backward-curved blades are of a single thickness of metal. Straight radial fan blades are used mostly in industrial applications. The main advantage of radial blades is that they permit the passage of foreign objects in the airstream such as sawdust, metal filings, and other debris. They have no advantages for HVAC use and should not be used for air distribution in buildings. Forward-curved fan blades have low efficiency and are typically used to move high volume against low pressure in applications such as window air conditioners and hotel unitary packages.

HVAC fans are typically driven by an electric motor called a *drive*. Historically, HVAC fan drives are designed to meet the full heating or cooling load. They were preset to maintain a constant rotational speed so the fan would deliver air a constant flow rate. Today, a *variable-frequency drive (VFD) controller*, also known as a *variable-speed drive* or *variable-frequency inverter*, adjusts the rotational speed of the electric motor so that it closely corresponds to the varying heating or cooling load requirements. By controlling (and reducing) fan drive speed, VFD technology saves energy and reduces airflow noise. As a result, VFDs are used on almost all fans used in commercial HVAC installations.

Heating/Cooling Coils

A *coil*, also known as a heat exchanger coil, consists of a network of tubes in which a hot or cold fluid such as water is circulated. Air is blown across the exterior surface of the coil by a fan to promote heat transfer from the fluid to the air. Fins that extend from the tubes in the coil increase heat transfer effectiveness. Coils resemble a car radiator and have metal fins attached to metal tubes. See Figure 7.3 and Photos 7.20 and 7.21.

Hot water or steam is circulated through a *heating coil* so it heats the airstream before the air is distributed for space heating. Heating coils are normally located downstream of the air filtra-

PHOTO 7.20 A heating (or cooling) coil. (Courtesy of NREL/DOE)

PHOTO 7.21 A reheat coil installed in a duct. (Used with permission of ABC)

tion system. Hot liquids are pumped through the metal tubes with the attached metal fins. The airstream passes through the coil and picks up heat from the metal tubes in the coil.

A *cooling coil* has chilled water circulating through it so the airstream is cooled, decreasing the temperature and humidity level of the airstream. During the cooling-dehumidification process, the airstream temperature will drop below the dew point temperature of the air, so water will condense onto the cooling coils and drip into a drip pan located at the base of the cooling coils. Cooling coils resemble heating coils and are normally located immediately downstream of the heating coils. Cooling coils can also carry refrigerant. These coils are referred to as *direct expansion (DX) coils*.

Reheat coils are part of some air conditioning systems. These coils are located in air ducts near the space being conditioned and are used to raise the temperature of circulated air after it was overcooled to remove moisture. Some buildings use reheat coils as their sole heating source. Reheat coils can resemble heating coils and are normally located immediately downstream of the heating coils.

Coil selection is based on the amount of sensible or latent heat that must be transferred to the airstream; the design conditions of air entering and leaving the coil; coil construction (e.g., size and spacing of fins and tubing, number of rows of tubing, and so forth); water or refrigerant velocity; and face velocity of the airstream.

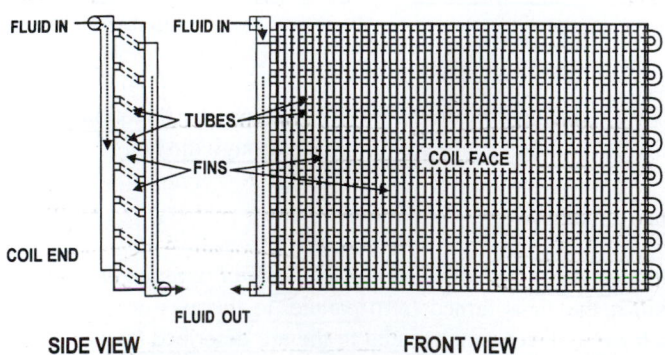

FIGURE 7.3 A coil is a liquid-to-air or air-to-liquid heat exchanging device that consists of a network of tubes in which a hot or cold fluid such as water is circulated. Air is blown across the exterior surface of the coil by a fan to promote heat transfer from the fluid to the air. Fins that extend from the tubes in the coil increase heat transfer effectiveness.

Air Cleaners and Filters

The air filtration system cleans the air of undesirable elements. *Air cleaners* are classified by the method used to remove particles of various sizes from the air. There are three general types of commercially available air cleaners: mechanical filters, electronic air cleaners, and ion generators. Because they may reduce some pollutants present in indoor air through condensation, absorption, and other mechanisms, devices such as air conditioners, humidifiers, and dehumidifiers may technically be considered air cleaners.

A *mechanical filter* is a fibrous or fabric medium that traps particulates and/or contaminants in an indoor airstream. It consists of a low packing density of coarse glass fibers, animal hair, vegetable fibers, or synthetic fibers often coated with a viscous substance (e.g., oil) to act as an adhesive for particulate material, or slit and expanded aluminum. Mechanical filters may be installed in ductwork near a furnace or air-handling unit or that serves as part of a fan filter unit, or they may be used in portable devices that contain a fan to force air through the filter.

Flat or *panel filters* slip into the receptacle and can be replaced manually. *Pleated* or *extended surface filters* generally attain greater efficiency for capture of respirable size particles than flat filters. Their greater surface area allows the use of smaller fibers and an increase in packing density of the filter without a large drop in airflow rate. Automatic and manual roll-type filters come in rolls of filter media and scroll through the airstream. As the filter media becomes congested, new filter media is unrolled from a continuous roll and the old media is scrolled onto a waste roll on the other side of the airstream. See Photo 7.22.

Flat filters may efficiently collect large particles, but remove only a small percentage of respirable size particles. A basic filter prevents the largest particulates, such as lint, hair and large dust, from getting into the air system. These filters operate at a cleaning efficiency of about 5 to 10% (They remove about 5 to 10% of all the particulates in the air stream).

High-efficiency particulate arresting (HEPA) filters remove 99.97% of particles as small as 0.3 microns (1/25 000 of an inch). HEPA filters are used in hospitals, manufacturing

PHOTO 7.22 A set of flat-panel, mechanical air filters in an air-handling unit (AHU). (Courtesy of NREL/DOE)

PHOTO 7.23 An in-duct electrostatic air cleaner. (Used with permission of ABC)

clean rooms and wherever else particle-free air is essential. *Sound-attenuating filters* reduce the level of sound moving through the ductwork. In many cases, a filter is relied on to mix two airstreams before they contact a coil. Mixing return air with cold outdoor air will prevent freezing of the heating coil.

An *electronic air cleaner* uses an electrical field to trap charged particles. It is an electronic device located in ductwork near a furnace or air-handling unit or that serves as part of a fan filter unit that filters large particles and contaminants out of indoor air. See Photo 7.23.

Electronic air cleaners are usually electrostatic precipitators or charged-media filters. In *electrostatic precipitators,* particles are collected on a series of flat plates. In *charged-media filters,* the particles are collected on fibers in a filter. In most electrostatic precipitators and some charged-media filters, the particles are deliberately ionized (charged) before the collection process, which results in a higher collection efficiency.

Ion generators use static charges to remove particles from indoor air. These devices come in portable units only. They act by charging the particles in a room, so they are attracted to walls, floors, tabletops, draperies, occupants, and so on. In some cases, these devices contain a collector to attract the charged particles back to the unit.

In addition to particle removal devices, air cleaners may also contain adsorbents and/or reactive materials to facilitate removal of gaseous materials from indoor air. Air cleaners that do not contain these types of materials will not remove gaseous pollutants.

The effectiveness of air cleaners in removing pollutants from the air depends on both the efficiency of the device itself (e.g., the percentage of the pollutant removed as it goes through the device) and the amount of air handled by the device. For example, a filter may remove 99% of the pollutant in the air that passes through it, but if the airflow rate is only 10 cubic feet per minute (ft³/min, which is frequently expressed as "cfm" in the HVAC industry), it will take a long time to process the air in a typical room of 1000 cubic feet.

The *weight arrestance test,* described in the ASHRAE Standard 52, is generally used to evaluate *low-efficiency filters* designed to remove the largest and heaviest particles; these filters are commonly used in residential furnaces and/or air conditioning systems or as upstream filters for other air-cleaning devices. For the test, a standard synthetic dust is fed into the air cleaner and the proportion (by weight) of the dust trapped on the filter is determined. Because the particles in the standard dust are relatively large, the weight arrestance test is of limited value in assessing the removal of smaller, respirable-size particles from indoor air.

The *atmospheric dust spot test,* also described in ASHRAE Standard 52-76, is used to rate *medium-efficiency air cleaners* (both filters and electronic air cleaners). The removal rate is based on the cleaner's ability to reduce soiling of a clean paper target, an ability dependent on the cleaner removing very fine particles from the air.

ASHRAE Standard 52-76 addresses the overall efficiency of removal of a complex mixture of dust. However, removal efficiencies for different size particles may vary widely. Recent studies by U.S. EPA, comparing ASHRAE ratings to filter efficiencies for particles by size, have shown that efficiencies for particles in the size range of 0.1 to 1 micron are much lower than the ASHRAE rating.

Air-Handling Units

An *air-handling unit (AHU),* or *air handler,* is a preassembled or site-built equipment device containing coils and fans that is used to condition and circulate air as part of an HVAC system. An AHU is a large sheet metal box-like cabinet containing a fan, cooling and/or heating coils, filter racks or chambers, sound attenuators, and dampers. A humidifier may be required to add moisture to the air under certain conditions. See Photos 7.24 and 7.25.

PHOTO 7.24 A medium-sized air-handling unit (AHU) in a mechanical room. This unit contains fans (blowers), a bank of filters, and heating and cooling coils. (Used with permission of ABC)

PHOTO 7.25 A close-up view of chilled water and hot water supply and return pipes connected to the air-handling unit (AHU) in the previous photograph. (Used with permission of ABC)

Large commercial AHUs have coils that circulate hot water or steam for heating, and chilled water for cooling, and transfer the conditioning effect to the airstream. The hot water or steam is provided by a central boiler, and the chilled water is provided by a central chiller. Small air handlers may contain a fuel-burning heater or a refrigeration evaporator, placed directly in the air stream. Electric resistance elements and heat pumps may be used, also. Evaporative cooling is possible in dry climates, in which case a large AHU may have an arrangement of direct evaporative pads, and indirect evaporative and DX cooling coils.

Most AHUs connect to ductwork that distributes the conditioned air through the building and returns it to the AHU. Some AHUs discharge (supply) and draw in (return) air directly to and from the space served, without ductwork. A furnace serves as an AHU for a split-system air conditioning system.

Small AHUs have a single blower (fan) that drives airflow. The blower may operate at a single speed, a variety of preset speeds, or be driven by a variable frequency drive (VFD), which allows a wide range of airflow rates and reduces power requirements. The AHU flow rate will also be controlled by inlet vanes or outlet dampers on the fan. In large commercial AHUs, multiple blowers may be used. Supply fans are placed at the outflow end of the AHU and the beginning of the supply ductwork, and return fans in the return air duct push the air into the AHU.

A *fan filter unit (FFU)* combines a filter(s) and a fan(s) in an air-handling unit. Controls can be local or, in the case of a large-scale ventilation network, they can be remotely controlled to coordinate their operation.

A *makeup air unit (MAU)* is an air handling unit that conditions 100% outside air. MAUs are typically used in industrial or commercial settings, or in *once-through* (blower sections that only blow air one way into the building), *low-flow* (air handling systems that blow air at a low flow rate), or *primary-secondary* (air-handling systems that have an air handler or rooftop unit connected to an add-on make-up unit or hood) commercial HVAC systems. They are necessary to prevent undesirable or unhealthy negative pressures from developing because of large amounts of exhaust in spaces with large commercial kitchen vent hoods, laboratory draft hoods, large open-hearth fireplaces, and appliances without sealed combustion.

Small AHUs deliver up to 10 000 ft³/min (4500 L/s) while very large units deliver over 50 000 ft³/min (22,500 L/s). Because they use VFD-driven fans, output varies. AHUs are configured as either packaged or built-up units. Packaged units can deliver up to 100 000 ft³/min (45 000 L/s). Built-up (custom designed and built) units are used for larger systems or when special conditions are encountered. Furnaces used as AHUs in residential installations deliver up to about 1500 ft³/min (675 L/s).

Terminal Boxes

A *terminal box* is a factory-made assembly that is part of the air distribution (duct) system that manually or automatically controls the velocity, pressure, humidity, and/or temperature of the supply air by controlling mixing of air and/or the airflow rate. A terminal box will generally consist of some of these parts: a casing (cabinet), air-mixing section, damper, a damper actuator, heat exchanger, airflow rate controller, air induction section, and sound attenuation baffles. See Photos 7.26 and 7.27.

Constant Volume Terminal Box

A *constant volume (CV) terminal box* is a factory-made assembly that uses special design and construction features that provide unusually quiet operation. The primary air section is enclosed in a sound-attenuating chamber. Instead of the usual primary air butterfly damper, there is a special design damper assembly mounted in the primary air section enclosure. This air valve reduces noise-producing turbulence. Other quiet performance features are a more rigid casing, special baffling, and a fan specially selected for low noise levels. Terminals of this type are used in broadcast studios, libraries, and other applications where a minimum-noise, premium-quality terminal is required.

PHOTO 7.26 A terminal box branching off from an insulated main duct. (Used with permission of ABC)

PHOTO 7.27 A damper motor used to control the position of a set of dampers (not visible) inside the duct. (Used with permission of ABC)

Reheat Terminal Box

A *reheat terminal box* (RH) is a factory-made assembly that has hot water or steam reheat coils or electric-resistant heaters within the box. This type of device can increase the temperature of the airstream by reheating it before it enters a space that does not need the desired cooling. A reheat configuration tends to trade energy consumption for comfort because the device reheats air that was cooled in the central cooling system.

Variable Air Volume Terminal Box

A *variable air volume (VAV) terminal box* is a factory-made assembly with a fan added to recirculate plenum air, for heating only. Figure 7.4 shows a diagram of a basic VAV terminal box. The heating cycle occurs generally when the primary air is off or at minimum flow. Heat is picked up as the recirculated air is drawn from the ceiling space and the fan motor. Additional heating capability can be provided by a hot water or electric coil in the terminal. During the cooling cycle, the fan is off and cool primary air is supplied from the central system. A backdraft damper prevents reverse flow through the fan. The flow of the primary air is regulated by VAV controls. In another VAV design, the fan runs continuously, fed by a mixture of primary and plenum air. The more primary air is forced in, the less plenum air is drawn in. The result is variable volume from the central system, and constant volume (and sound) to the room. Because the central system need only deliver air as far as the fan, the inlet static pressure can be lower than in the parallel flow terminal. The fan, however, is sized to handle the total airflow.

Engineers designing air systems try to match the airflow capacity of fan-powered terminals to the needs of the space. Exact matches are uncommon. Although design may not allow an exact match, a product other than the one that is the subject of the design might be selected, or system balancing might require a different airflow to meet field conditions. Mechanical

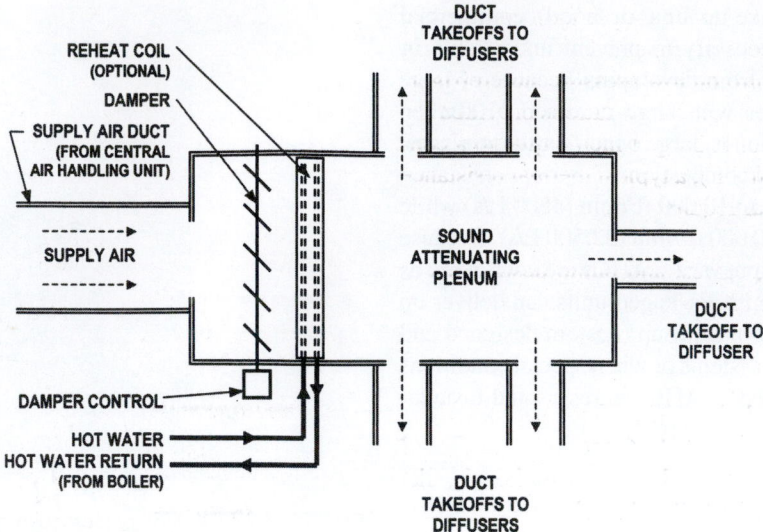

FIGURE 7.4 A variable air volume terminal box. Figure 7.2 shows a VAV box location in an air system.

trimming involves the use of a mechanical device, such as a damper, to adjust the fan airflow to meet the design requirements. Voltage adjustment of fan-powered terminals typically involves the use of a *VFD*. The revolutions of the motor is reduced by the VFD, lowering the speed of the fan.

7.3 WATER (HYDRONIC) DISTRIBUTION SYSTEM COMPONENTS

Piping, Fittings, Valves, and Pumps

Pipe is a tube used to transport liquids or gases from one point to the next. In a building HVAC system, pipes transport hot and chilled water, steam and refrigerant. Pipe materials most commonly used in building plumbing systems include copper, steel, plastic, and composite.

A variety of *fittings* must be used to connect pipe lengths and make all the pipe turns, branch lines, couplings that join the straight runs, and stops at the end of the runs. Fittings include couplings, elbows, tees, and unions. *Valves* are used to control the flow of the water throughout the system. Proper location of valves simplifies repairs to the system, fixtures, or equipment being serviced.

Pipes and fittings are joined through a number of techniques: threaded joints, insert fittings with crimped or clamped connections, and flared (metal-to-metal) joints are accepted mechanical joining techniques. Soldering, brazing, and welding are ways of joining metal surfaces. Solvent cementing and fusion welding can join some plastic pipe materials.

HVAC piping systems can be divided into two subsystems: *equipment piping* in the main mechanical equipment room (e.g., fuel lines, refrigerant lines, water connections, and so on),

and *distribution piping* that runs to the air handling systems or terminal units (e.g., terminal boxes, fan-coil units, and so on).

A *pump* is a mechanical device used to compress a liquid or drive flow in a fluid. In building space heating and cooling systems, the fluid is usually water. HVAC pumps move hot water (for hydronic systems), boiler feed water, primary and secondary chilled water, and condenser water. See Photo 7.28. The most commonly used pump in HVAC systems is a centrifugal pump. Identical pumps are frequently used in a parallel arrangement in case one pump must go down for maintenance or pump failure.

Pumps are typically driven by an electric motor that is linked to the pump. Like fans, HVAC pumps were traditionally designed to operate at a constant rotational speed so the pump would deliver water at a constant flow rate that matched the peak load. Currently, VFD controllers are used on pump drives to control the rotational speed of the electric motor so pump output can be adjusted to varying loads.

PHOTO 7.28 Pumps used to deliver hot water (supply) to several AHUs located in different zones in the building. (Used with permission of ABC)

Pipe insulation requirements for simple systems are: 1 in (25 mm) on hot water space heat piping, 2 in (50 mm) on steam piping, and ¾ in (18 mm) on low pressure side refrigerant piping. Insulation thickness recommendations provided are based on an insulation value of R-3.7 per inch (equal to a conductivity of 0.27 Btu · in/hr · ft² · F), a typical thermal resistance for pipe insulation. Additional requirements apply to complex piping systems.

Pipe materials, fittings, valves, and pump design theory will be introduced in Chapter 13.

Convectors

A *convector,* sometimes called a *convector heater,* is a heating distribution device that is constructed of one or more finned tubes that carry steam or hot water and that transfer heat by circulation of air (natural convection). *Baseboard convectors* have a finned tube hidden behind a decorative sheet metal panel that is mounted at the intersection of the wall and floor (along the baseboard). *Cabinet convectors* are constructed of finned tubes that carry steam or hot water and that are hidden behind a sheet metal cabinet. Baseboard and cabinet convectors rely on natural convection to distribute heated air in the space. See Figures 7.5 through 7.7 and Photos 7.29 through 7.31.

Convector output is based upon the size and configuration of the coil and the temperature of the water or steam passing through the coil. A typical output for a convector that is used in residential and small commercial installations is in the range of 500 to 2000 Btu/hr per linear ft (45 to 180 W/m) of convector length. Output can be controlled by opening/closing a valve or by throttling flow of water through the unit. The unit may have thermostatic controls that are self-contained or may have an actuator-operated valve that is controlled by an externally-located thermostat.

Radiators

A *radiator* is a heating distribution device that transfers heat from steam or hot water by radiation to objects within visible

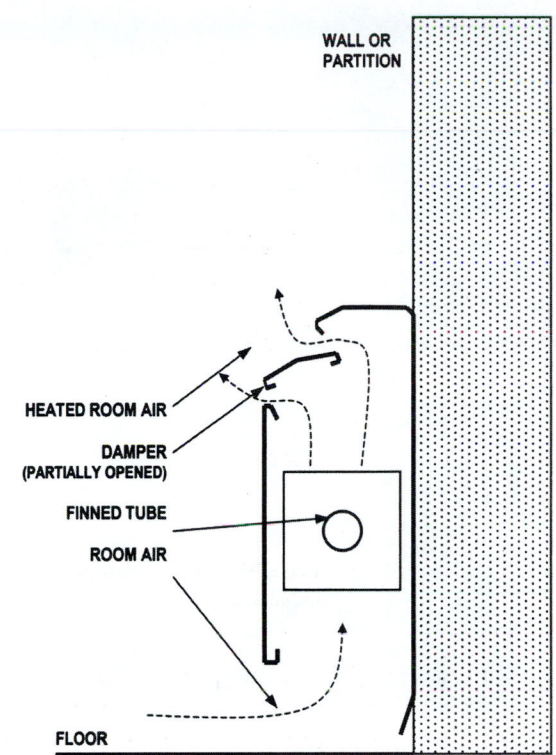

FIGURE 7.6 A baseboard convector. Room air comes in contact with the fins, is heated, rises, and is replaced by more air.

range and by conduction to the surrounding air that, in turn, is circulated by natural convention. It is a freestanding fixture typically made of formed steel or cast iron. It is usually visibly exposed within the room or space to be heated.

Radiator output is based on the surface area of the radiation surface and the temperature of the water or steam passing through the unit. Output can be controlled by cycling a valve on and off. A radiator may have a self-contained thermostatic control or may have an actuator-operated valve that is connected to a thermostat located in the space.

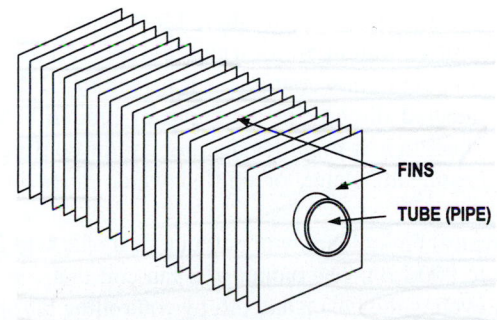

FIGURE 7.5 A finned tube convector. Hot water flows through the tube, heating the tube and fins. In the natural convection heating process, room air comes in contact with the fins, is heated, rises, and is replaced by more air.

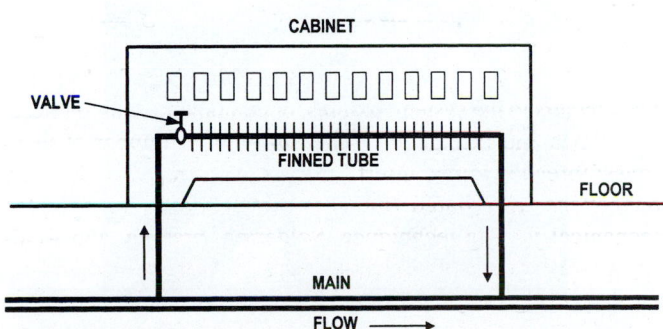

FIGURE 7.7 A finned tube cabinet heater. Hot water from an underfloor main is diverted to the unit.

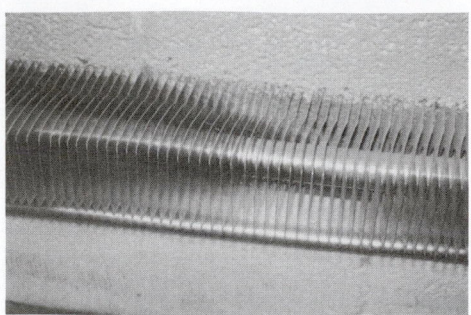

PHOTO 7.29   A finned tube (exposed). (Used with permission of ABC)

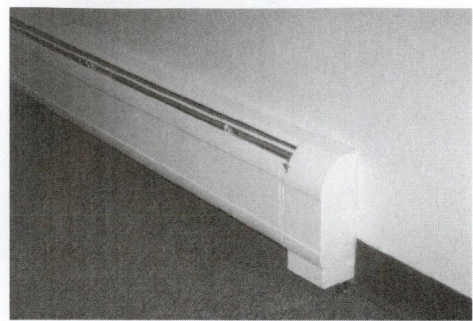

PHOTO 7.30   A finned tube, baseboard convector used for residential installations. (Used with permission of ABC)

PHOTO 7.31   A finned tube, baseboard convector, typically used in commercial installations. (Used with permission of ABC)

PHOTO 7.32   A convector radiator is placed along a curtain window wall to counter the downward convective drafts and radiant temperature effects. (Used with permission of ABC)

PHOTO 7.33   A close-up view of the front of a convector radiator. (Used with permission of ABC)

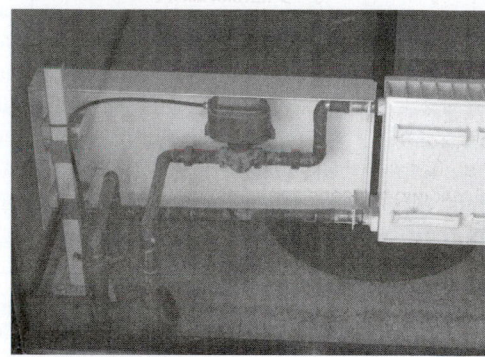

PHOTO 7.34   A backside view of the control valve and piping configuration serving a convector radiator. (Used with permission of ABC)

Convector Radiators

A *convector radiator* is a heating distribution device that transfers heat from steam or hot water by radiation and convection. In this device, steam or hot water runs through a pipe core, heating metal plates or fins attached to it at short intervals. See Photos 7.32 through 7.34. Output and control of a convector radiator is similar to a convector.

Fan-Coil Units

A *fan-coil unit* consists of a filter, fan(s), and separate hot water and chilled water coils or a single water coil that may be used for both functions contained in a factory-made cabinet assembly located directly in the room or hidden behind the wall or in a ceiling. It provides heating and/or cooling by circulating steam, hot water, or chilled water piped in from a central plant.

A typical fan-coil unit will output a total of 500 to 2000 ft³/min (225 to 900 L/s). The output of a fan-coil unit can be controlled by cycling the fan on and off, by controlling fan speed, by throttling water flow into the coil, or by switching electric coils on and off. The unit may have a thermostatic control that is self-contained or may have an actuator-operated valve that is connected to a thermostat located in the conditioned space. Some

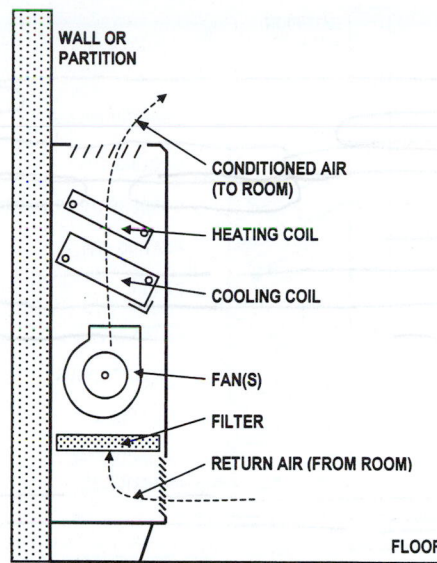

FIGURE 7.8 A fan-coil unit uses a fan (or fans) to force air across the coil (either cooling or heating) to induce heat transfer. Air is then discharged back into the conditioned space.

units have control panels to allow occupants to select heating or cooling, and to select fan speed.

When cooling, condensation frequently forms on the cooling coil, so a condensate drain must be provided. Built-in fans draw air from the room into the unit cabinet where the air is forced across a heating or cooling coil and then introduced as conditioned air back into the space. A fan-coil unit generally is a small unit located in the conditioned environment. It may also be located above a ceiling and connected to a very small amount of ductwork. See Figure 7.8 and Photos 7.35 through 7.40.

Unit Ventilators

A *unit ventilator* is a type of fan-coil unit that has the capacity to introduce ventilation (outdoor) air into a room in addition to

PHOTO 7.35 A ceiling-mounted fan-coil unit. (Used with permission of ABC)

PHOTO 7.36 A close-up, inside view of a fan-coil unit. Evident are part of the filters (left side), four fans, and the drip pan below the cooling coil. (Used with permission of ABC)

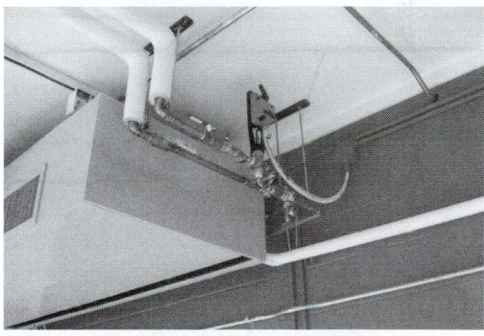

PHOTO 7.37 Pipes connected to the coil of a fan-coil unit. Note the automatic control valve. (Used with permission of ABC)

PHOTO 7.38 A cabinet heater. (Used with permission of ABC)

being able to provide space cooling and/or heating. It generally consists of a filter, fan, cooling and/or heating coil, and a set of dampers that control room air and outdoor ventilation air that enters the unit. See Figure 7.9. When cooling, condensation is typically produced, so a condensate drain must be provided.

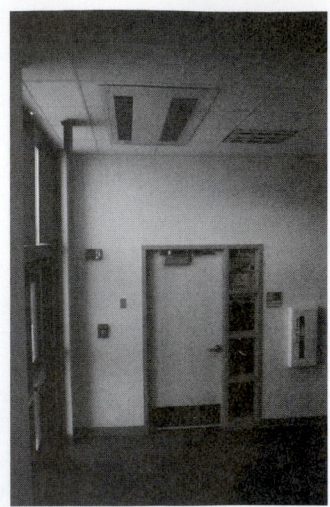

PHOTO 7.39 A flush-mounted ceiling fan-coil unit. (Used with permission of ABC)

PHOTO 7.40 A suspended, fan-coil unit heater (under construction).

Fans in the unit ventilator cabinet draw recirculated air from the room and ventilation air from outdoors through louvers on an exterior wall. The air is mixed by a filter and then pushed across a coil containing hot water, steam, or chilled water, where it is heated or cooled before it is introduced back into the space.

A unit ventilator may have thermostatic controls that are entirely self-contained or may have actuators that are connected to a thermostat located in the space. The actuators control the valves and dampers. Some units have control panels to allow occupants to select heating or cooling, to select fan speed, and to control outside air ventilation. The output of a unit ventilator can be controlled by cycling the fan on and off, by controlling fan speed, by throttling water rate to the coil, or by switching electric coils on and off. Additionally, an economizer feature is typically designed into unit ventilator controls so it can cool with outside air instead of relying on a chilled

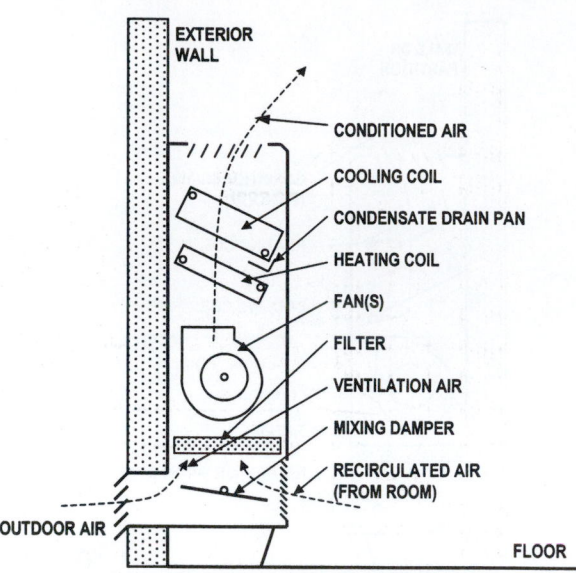

FIGURE 7.9 A unit ventilator is a type of fan-coil unit. It mixes outdoor air with recirculated room air before blowing the mixed air across the coil (either cooling or heating) to induce heat transfer. Air is then discharged back into the conditioned space.

water coil to cool air. In extremely cold weather, it is often necessary to close the outside damper to prevent freezing in the coil and rely on infiltration for ventilation.

A unit ventilator is generally a small unit located against an exterior wall in the room being conditioned. It may also be located above a false ceiling and connected to a small amount of ductwork. A typical unit ventilator will circulate a total of 750 to 2000 ft^3/min (338 to 900 L/s), a portion of which is outdoor air used for ventilation.

Induction Units

An *induction unit* is a type of fan-coil unit that receives primary air from a central air-handling unit, which includes ventilation (outdoor) air. It consists of a filter, cooling and/or heating coil, and inlet for primary air. No fan is required. See Figure 7.10. Primary air enters a small plenum, where it is forced out through small nozzles at high velocity. The low pressure created by high-velocity primary airflow induces secondary (recirculated room) air into the unit, where it passes through a cooling and/or heating coil and mixes with the primary air.

Chilled Beams

Chilled beams are modern terminal devices designed to be mounted near or in the ceiling and have room air circulate through coils fed by chilled or hot water. They have been used successfully in Europe and Australia and are slowly being introduced in the United States and Canada. There are two basic types of chilled beam technologies: passive and active.

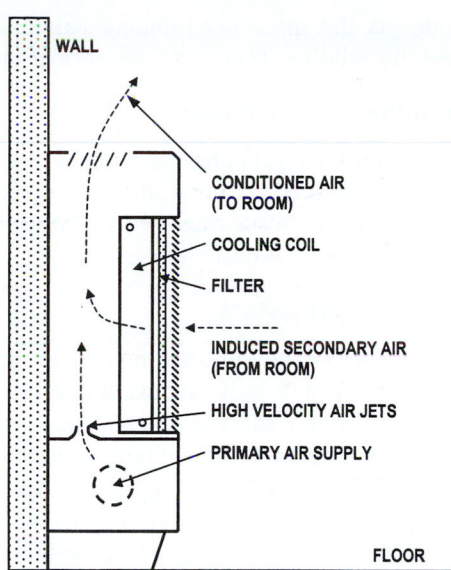

FIGURE 7.10 An air induction unit is a type of unit ventilator that does not require fans to move air. Instead, high-velocity air passes through air jets, which induces flow of secondary air into the unit. Primary air and secondary air mix and are introduced back into the conditioned space. Primary air carries ventilation air into the unit. A coil cools (or heats) the secondary air.

Passive chilled beams use a type of all-water terminal unit that consists of a finned coil contained in a formed metal casing with openings that allow room air to flow into and out of the unit. These units are recessed in the ceiling or suspended to hang below the ceiling in the building zone being conditioned. Warm room air naturally rises to the ceiling level and enters the top of the beam. As the room air enters the coil, it is cooled by contacting the coil surface. The cooled air falls through the chilled beam unit being driven by natural convection. The air exits through outlet slots on the underside of the beam and descends into the room. Air movement occurs naturally without the use of fans and with minimal air movement, so noise and air velocity are minimized. These units are completely passive because natural convection transfers heat.

Active chilled beams, also called *induction diffusers,* are a type of air–water terminal unit that is integrated into the room's primary air supply ducting. Conditioned ventilation air supplied from conventional ductwork flows through slots or nozzles that pushes air across the coil. In a manner similar to an air induction unit, high-velocity airflow induces flow of warm room air into the chilled beam unit and through the cooling coil. Supply and chilled room air mix and then enter the room through outlet slots on the underside of the beam. Active chilled beams can also be used for space heating, by reheating room air. Active chilled beam technology offers greater cooling capacity over passive units because airflow is induced so heat transfer is more effective.

7.4 THERMAL ZONING

Different sections of a building frequently have different air conditioning requirements (differing heating and cooling loads), depending on building orientation, direction of the sun, differences in occupancy patterns, and so on. For example, because of solar gains from the eastern position of the early-morning sun, spaces on the east side of a building may need more cooling than spaces on the west side during morning hours. Delivering an equal amount of cooled air to all building spaces will provide needed cooling to spaces on the east side of a building, but the west side spaces (those not exposed to the early-morning sun) become too cold as a result of the cooling provided.

Thermal zoning, or simply *zoning,* is a way of dividing a building into independent areas with similar heating and cooling needs. A *zone* is an area of a building, such as a single room (e.g., motel room or classroom) or set of rooms, that has thermal and/or humidity requirements requiring separate control. Elements to consider when establishing zones are different occupancy loads, different operating schedules, different solar radiation exposures (an east-facing space versus a west-facing space, or rooms with differences in window areas).

A thermostat or sensor in each zone controls the temperature and/or humidity level in that zone. This way the different rooms in a building can be set at different operating conditions. Humidity sensors, indoor and outdoor temperature sensors, and adjustable heat/cool dead band controls can be included in the control scheme to more efficiently control conditions in a zone. Buildings and spaces that typically need zone control include:

- Large homes (generally with a floor area over 3000 ft²/ 300 m²).
- Large commercial buildings or small commercial buildings with different occupancy types.
- Rooms with heating and cooling load changes based on the sun, seasons, or occupancy such as classrooms, offices, and conference rooms.
- Buildings that need independently controlled rooms such as dormitories, inns, hotels, and motels.
- Buildings that require immediate heating or cooling such as hospitals, care centers, and nursing homes.

In residential buildings, zones may be assigned based upon activity in a room or group of rooms. Bedrooms may be placed in a separate zone from living areas (e.g., family room, living room, kitchen, and so on).

In commercial buildings, there are typically two principal zones. *Perimeter zones* include rooms and spaces with at least one exterior wall. In these spaces, loads vary over a broad range from heating to cooling and are affected by both interior and exterior (e.g., solar gain, outdoor temperature, and so on) factors. *Interior* or *central zones* include rooms and spaces that are core spaces that have no exterior walls. Heat generated by lights, occupants, and office equipment dominates the load, which in most climates usually provides a continuous cooling demand. Zoning can be accomplished by having separate equipment or a separate sensor/control arrangement in each space.

One zoning method is to have each zone served by separately controlled pieces of equipment. For example, a large residence may be served by two zones created with furnaces: an upflow furnace in the basement serving the first floor and basement areas as a single zone, and a horizontal furnace in the attic serving the upstairs bedroom areas as a second zone. In an office building, packaged-terminal air conditioning units mounted on the roof, called rooftop units (RTU) can operate independently. Each RTU can serve 1000 to 2000 ft^2 (100 to 200 m^2) of office floor area, with each group of offices being separately controlled. A sensor and controller in each zone controls the RTU serving only that zone.

A second method is to provide zone control on a single HVAC system by dividing the building into zones that are served by a single HVAC distribution system. Each zone has its own sensors and controls. In an air system, the temperature-controlling device acts on an actuator-controlled damper that is placed in a duct or part of a terminal box and is located at key points in ductwork. In a water system, the controlling device acts on a motor-driven control valve that opens or closes the flow of hot or chilled water through a coil. By varying the degree to which each damper or valve is open or by simply opening or closing each damper or valve, air or water flow to a specific zone, and thus temperature and/or humidity in that zone, is adjusted independently of the others. The heating or cooling equipment is brought on when the sensors call for heating or cooling. When all zone sensors are satisfied, the heating or cooling system equipment turns off and all zone dampers return to the open position and control valves close to allow for continuous air circulation.

In laying out zones, the designer must establish those rooms or spaces that need to be individually controlled. Solar gains, seasonal changes, occupancy types, and occupancy patterns must be considered. Often it is necessary to locate similar occupancy types on the same zone, so it is essential to physically locate these rooms together. Thermal zones must be established very early in the building design process. A zoning scheme established while the floor plane of the building is under development is imperative. It is difficult to zone a building after it is designed.

7.5   CONTROLS

Thermostatic Controls

In buildings, a *thermostat* is a temperature-sensitive switch that regulates the temperature of a space conditioning unit or system, such as a furnace or air conditioner. When the indoor temperature drops below or rises above the thermostat set point, the thermostat directs the space conditioning unit or system to warm or cool the building. There are two types of thermostats:

Nonprogrammable Thermostat

A *nonprogrammable thermostat* has the ability to regulate room temperature to one desired set point, the desired temperature. The set point on the thermostat must be manually changed if a new temperature set point is desired.

Programmable Thermostats

Programmable thermostats allow flexibility to the building occupant because they can be programmed to match temperature set points with occupancy patterns in a building space. A programmable thermostat changes the set point according to a programmed schedule: lowering the set point when the building is scheduled to be unoccupied and raising it when occupants are scheduled to be in the building space. There are five basic types of programmable thermostats: electromechanical, digital, hybrid, occupancy, and light sensing.

Electromechanical (EM) thermostats have manual controls such as movable tabs to set a rotary timer and sliding levers for night and day temperature set points. These thermostats work with most conventional heating and cooling systems, except heat pumps. EM controls have limited flexibility and can store only the same set points for each day, although at least one manufacturer has a model with separate set points for each day of the week.

Digital thermostats have digital readouts and data entry pads or buttons. They offer a wide range of features and flexibility, and can be used with most heating and cooling systems. They provide precise temperature control and permit custom scheduling.

Hybrid thermostats combine the technology of digital controls with manual slides and knobs to simplify use and maintain flexibility. Hybrid models are available for most systems, including heat pumps.

Occupancy thermostats maintain the setback temperature until someone presses a button to call for heating or cooling. They do not rely on the time of day. The ensuing preset comfort period lasts from 30 min to 12 hr, depending on how the thermostat was set. Then, the temperature returns to the setback level. These thermostats are best suited for spaces that remain unoccupied for long periods of time on an unscheduled basis.

Light-sensing thermostats rely on a preset lighting level to activate heating systems. When lighting is reduced, a photocell inside the thermostat senses unoccupied conditions and allows space temperatures to fall 10° below the occupied temperature set point. When lighting levels increase to normal, temperatures automatically adjust to comfort conditions. These thermostats are designed primarily for stores and offices where occupancy determines lighting requirements, and therefore heating requirements.

PHOTO 7.41   An electromechanical, clock-type thermostat. (Used with permission of ABC)

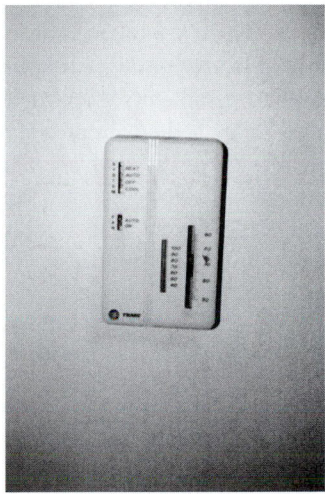

PHOTO 7.42   A programmable, electronic thermostat. (Used with permission of ABC)

PHOTO 7.43   A nonprogrammable, thermostatic sensor. The unit is controlled by a remote energy management system. (Used with permission of ABC)

Thermostat Operation

A thermostat has a dead band temperature differential of a degree or two to prevent the thermostat from causing short-cycling the space conditioning unit or system. Typically, this temperature differential is about 2°F (1°C). For example, if the set point for heating is 70°F (21°C), the space conditioning unit or system will not be energized until the thermostat senses a temperature of 68°F (20°C), and the unit or system will continue heating until the thermostat senses a temperature of 71°F (22°C). This differential may be adjusted on a small calibration set point inside the thermostat housing. The less the differential, the more comfortable the space will feel. But it also means that the system will be turned on and off more frequently (to maintain a close tolerance of temperature), which will lower system efficiency and cause premature wear.

Thermostats used for hot water systems are usually either the single set point thermostat or a day–night thermostat. The single set point thermostat is the one most commonly used. The desired temperature is set and the thermostat will operate the cycles of the heating system. The temperature may be changed at any time by simply adjusting the temperature set point on the thermostat.

A day–night thermostat permits the setting of one temperature for the daytime and another (usually lower) for nighttime. With this thermostat, a clock is set for a given time (say, 8 am), and at that time the temperature will go to the daytime set point (perhaps 72°F or 22°C); then at the other time set (say, 5 pm) the temperature will go to the nighttime set point (perhaps 60°F or 16°C). The times and temperatures are set on the thermostat and may be changed as desired.

The programmable thermostat is used in residences, stores, offices, apartments, institutions, and commercial and industrial projects. In a residence, the temperature differential may be 5° or 10°F (3° or 6°C), set to begin to coast to a lower temperature just before bedtime and heat up as the occupants arise. In apartments where the heat is furnished and paid for by the apartment owner and not the tenants, this type of thermostat is commonly used to control heat costs for the owner. In this case, the thermostat used would have two parts: one located in the apartments to sense the temperature and the second located in a remote location that the tenants cannot enter. It controls the temperature desired and the times of operation.

Because many stores and offices are open only limited hours, it would be foolish to maintain the temperature needed during hours of operation through the hours in which the space is not in use. Rather than depend on the occupants to turn down the temperature before they leave (which they may forget to do) and to turn up the temperature in the morning (no one will begin work until it warms up), the programmable thermostat takes care of these functions automatically.

In locations where the possibility of occupants tampering with the thermostats exist, the unit may have a locking cover or may be designed so that the temperature set points can be changed only with a special keying device.

Thermostats are usually placed on a wall, about 5 ft (1.5 m) off the floor line. The location should be carefully checked to be certain that the thermostat would provide a true representation of the temperature in the spaces it serves. Guidelines used in thermostat location include the following:

- Always mount it on an inside wall (on an outside wall, the cold air outside will affect the readings).
- Keep it away from the cold air drafts (such as near a door or window), away from any possible warm air drafts (near a radiation unit, fireplace, or stove), and away from any direct sunlight.

Carbon Dioxide (CO₂) Sensors

Carbon dioxide (CO₂) sensors can be a very effective method of providing for occupancy-based ventilation control. This is often useful in portions of a facility characterized by high occupancy density for relatively short periods of time (e.g., meeting rooms, auditoriums, gymnasiums, cafeterias, and classrooms). Under CO_2 sensor control, these areas receive the design ventilation rate during periods of occupancy but receive reduced outside air during unoccupied periods, which provides a substantial savings in energy.

Relative Humidity Sensors

Relative humidity sensors are used extensively in HVAC applications since they have a direct impact on occupant comfort and indoor air quality. Relative humidity sensors are typically used to control humidification and/or dehumidification equipment in environments such as retail stores, restaurants, schools, hospitals, office buildings, data processing centers, warehouses, and food storage buildings; and in critical environmental and industrial applications because relative humidity effects electrostatic discharge and moisture content of materials (e.g., semiconductor fabrication, laboratories, clean rooms, museums and agricultural processes).

Residential Control Systems

In residential installations, the programmable thermostat is typically used. It offers the capability to be programmed for an *automatic setback,* a scheduled change to a lower set point that drops the temperature in the space during periods when occupants are scheduled to be sleeping or not occupying the space. An *automatic wakeup* is a scheduled increase in the set point scheduled for the beginning of periods when occupants are scheduled to be actively occupying the space.

Commercial Building Control Systems

In commercial buildings, control systems typically fall into one of three categories:

Pneumatic Control Systems

Pneumatic control systems use compressed air to control equipment. A pneumatic thermostat regulates air pressure in an air line that is attached to a damper or valve motor. Changes in air pressure cause the damper motor to open or close the damper or valve.

Stand-Alone Controls

Stand-alone controls are built into individual pieces of large equipment, such as packaged rooftop units and chillers, to maintain comfort conditions and manage energy consumption. Some of these stand-alone controllers are quite sophisticated and are capable of implementing many control strategies.

Direct Digital Control Systems

Direct digital control (DDC) systems use a microprocessor-based controller that directly controls equipment based on sensor inputs and set point parameters. Electronic signals are sent to a computer to be processed. A preprogrammed control sequence determines the output to and operation of the equipment. DDC systems allow building control and information to be centralized at a single location so building operators can readily observe and control all building systems from a single computer terminal instead of having to keep track of a variety of different control locations throughout the building.

DDC control systems are more expensive than pneumatic systems, hybrids, or stand-alone controls, but they offer many more benefits. Mainly, DDC systems offer precise control, making it possible to use accurate sensors to save energy and provide occupant comfort.

Building Energy Management Systems

A *building energy management system (BEMS),* also known as a *building automation system (BAS),* controls energy-consuming equipment in a building to make it operate more efficiently, while maintaining a comfortable environment. A BEMS may include other features such as maintenance planning, fire- and physical-safety functions, and security services as shown in Figure 7.11. A BEMS controller is shown in Photo 7.44. The most common functions that energy management systems employ to use energy more efficiently:

> *Scheduling functions* turn equipment on or off depending on time of day, day of the week, day type, or other variables such as outside air conditions. Improving equipment schedules is one of the most common and effective measures for saving energy in commercial buildings.
>
> *Lock-out functions* make certain that equipment does not come on unless it is needed. They protect against nuances in the programming of the control system that may inadvertently cause equipment to turn on. For example, a chiller and its associated pumps can be locked out according to

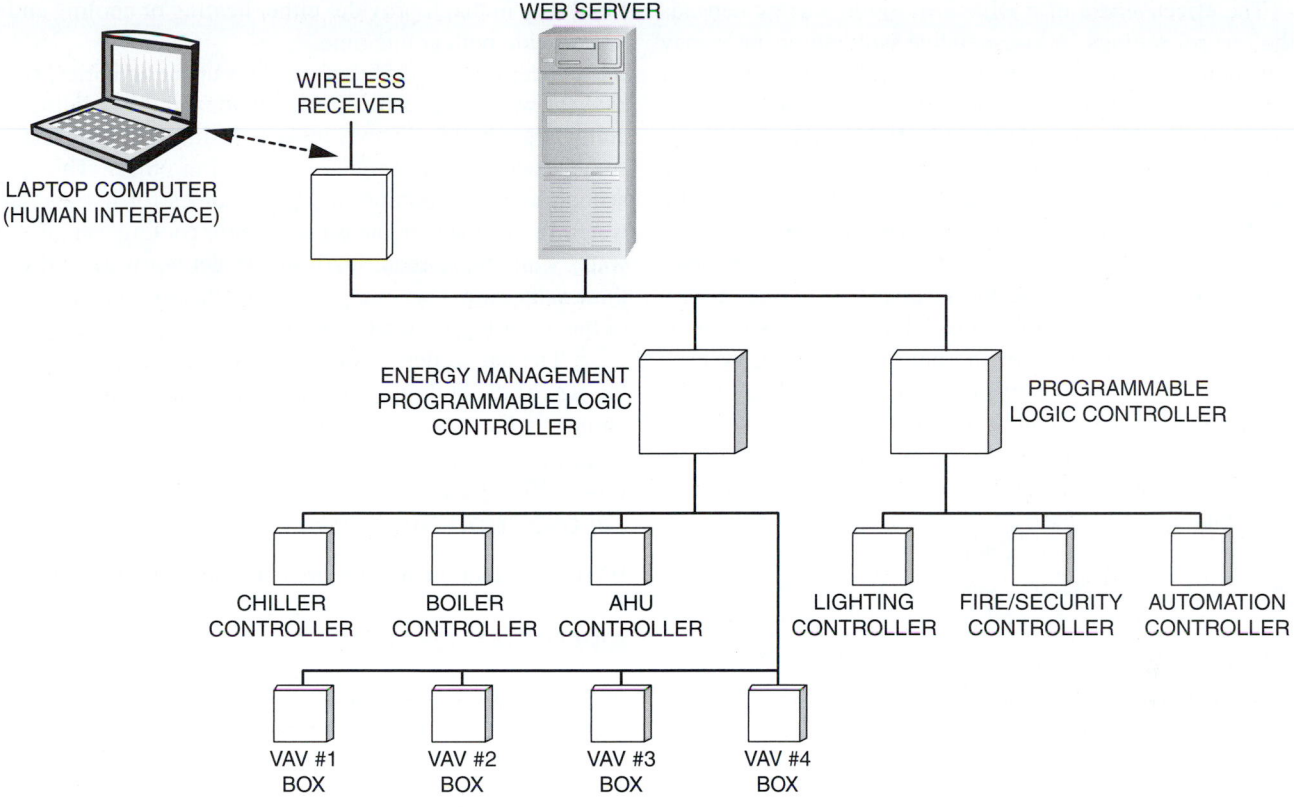

FIGURE 7.11   A building energy management system (BEMS) may control HVAC, lighting, fire, safety, and security functions.

calendar date, when the outside air falls below a certain temperature, or when building cooling requirements are below a minimum.

Reset functions keep equipment from operating at greater capacity than necessary to meet building loads. A BEMS can allow equipment to operate at the minimum capacity required by resetting operating parameters. Traditional

PHOTO 7.44   Staff member checking an energy management system. (Courtesy of NREL/DOE)

design practice is to use a proportional reset schedule based on outdoor temperature. Although that method works reasonably well, a more effective method is to base resets directly on building loads. Examples of building control parameters that can be reset include supply air and discharge air temperatures for fan systems that use terminal reheat, hot-deck and cold-deck temperatures for multizone HVAC systems, and heating-water supply temperature.

Demand-limiting or *load-shedding functions* reduce electrical demand charges tied to peak demand for electrical power. Demand charges can make up 40% or more of a utility bill. For example, when the demand on a building meter or piece of equipment, such as a chiller, approaches a predetermined set point, the BEMS does not allow the equipment to load up any further. Another way to minimize peak demand is to program time delays between the startup of major pieces of electrical load equipment so that several pieces of equipment do not all start up at once.

Diagnostic functions monitor information such as temperatures, flows, pressures, and actuator positions and use that data to determine whether equipment is operating incorrectly or inefficiently and to troubleshoot problems. A building diagnostic arrangement typically requires the BEMS monitor more points than those used to control building systems.

The effectiveness of a BEMS in saving energy depends on the control settings. A conservative building operator may set the control system to never shut off equipment to avoid the risk of tenant complaints. This is hardly an effective strategy. Whenever the building is unoccupied, HVAC equipment (heating plant, cooling plant, outdoor and exhaust dampers, and HVAC fans and pumps) should be shut off and the temperature allowed to setback (in winter) or setup (in summer). Temperature setbacks of 5° to 10°F (3° to 6°C) are typical. If the temperature setback is too large, the heating system may take a long time to recover in the morning and there is a risk of freezing piping in isolated parts of the building. Cooling systems can be shut down completely at night. The cooling system should be able to return the building to comfort conditions fairly easily because cooling loads are small in the early morning. An economizer can provide cooling at night or when the outdoor air temperature is below the indoor air temperature set point.

Energy management systems can save an average of about 10% of overall building energy consumption and cut 20 to 30% off of electricity demand costs. In addition to saving energy and reducing demand costs, these systems typically reduce the costs of building maintenance. Typically buildings with more than 25 000 ft^2 (2300 m^2) of floor area or those with complex HVAC systems are best controlled by a BEMS.

7.6 HEATING AND COOLING SYSTEM ARRANGEMENTS

A myriad of heating and cooling system arrangements are used in buildings. It is not unusual for industrial and commercial buildings to use different types of systems to heat and cool different areas of a building. The systems may vary in use of hot water heat, with finned tube units being used in offices and convectors units being used in storage or warehouse areas. Similarly, completely different systems may be used, with a hot water heating system in an office area and gas-fired unit heaters in the warehouse areas. Any combination may be used, and the selection may vary, depending on the type of heat required, the type of fuels available, and to a great extent, how much heat is required, how often it is required, and what method of supplying the heat will provide the best results in the designer's opinion.

Residential designs may use combinations of systems such as hot water for heat and forced air for cooling, hot water or forced air heat with electric supplements (particularly in bathrooms or kitchens), and finned tube units with supplemental hot water unit heaters in areas such as a kitchen or basement.

Chilled water may be used to provide cooling in a building by running it through the same pipes used for hot water heating. This means that once the system is changed over from one function to the other (e.g., from heating to cooling as summer approaches) it is not easily reversed if cold weather comes for several days. This type of system has little flexibility in that it provides either heating or cooling and cannot provide both at one time.

Additionally, different uses, occupant activities, types of exterior wall construction, and locations in the building (especially larger buildings) may make it desirable to have heating in some spaces while cooling is required in others. This may be accomplished by putting in separate pipes for hot and chilled water. In this manner, areas that require cooling can get chilled water while those requiring heat can get hot water. Of course, there is the extra cost of piping and controls to regulate the flow of the water to the desired location.

The actual design of any particular system will involve the size of the building, amount of insulation, doors, windows, climate in the area, and fuel to be used.

Classification of HVAC Distribution Systems

HVAC distribution systems are classified into three categories:

All-Air HVAC Systems

All-air HVAC systems heat, cool, and ventilate by providing air only to the space being conditioned. As shown in Figure 7.12, centrally located air conditioning equipment heats and cools air as it passes through coils and circulates the conditioned air to and from the rooms through long runs of ductwork. Coils in the central handling unit contain circulating hot water from a boiler or chilled water from a chiller. The mechanical room containing the air conditioning equipment is remote from the spaces being conditioned. All-air systems are the most commonly used central HVAC systems in the United States.

Temperature control in an all-air system is by varying the temperature of the airstream. This system provides excellent ventilation, heating, cooling, humidification, and dehumidification, and it offers use of economizer capabilities when outdoor air temperature is cool. All-air systems require that most of the air supplied to a space be returned to the air-handling unit or exhausted from the building. This return air may be conveyed in a return air duct system or through plenums formed by building elements, such as a corridor or the ceiling plenum consisting of the suspended ceiling and the underside of the floor structure.

Maintenance of these systems occurs in the mechanical room away from building occupants because most of the equipment is away from the occupied spaces.

All-Water HVAC Systems

All-water HVAC systems heat and cool by circulating hot water from a boiler or chilled water from a chiller through a finned coil in a fan-coil unit or unit ventilator in the space. Ventilation is provided through a unit ventilator, a separate ventilation system, or by infiltration.

Temperature control in the all-water system is by varying the temperature of the water and by regulating flow of water

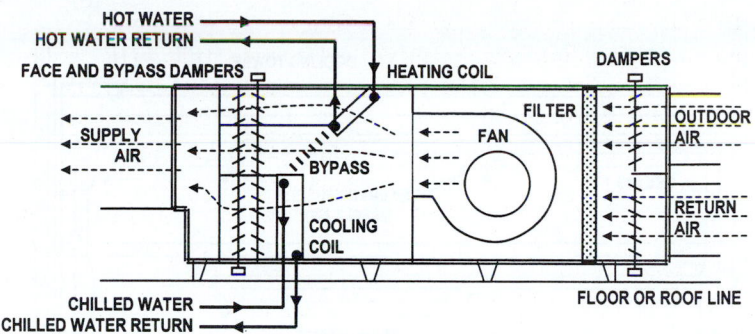

FIGURE 7.12 Central air conditioning equipment.

and air through the coils. Ductwork is not required in this system, so the system is extremely compact. It is also effective, because pumping horsepower is considerably less than fan horsepower required to move air. Because much of the equipment is in occupied spaces, disadvantages of this system are that much of the maintenance activity must occur in occupied spaces and that occupants have access to the equipment.

Air–Water HVAC Systems

Air–water HVAC systems consist of an air side and a water side, which contribute to heating, cooling, and ventilating. The *water side* consists of a pump, distribution piping network, and coils in terminal boxes or fan-coil units. Usually, the majority of space load is carried by conditioned water on the water side. The *air side* consists of the central air conditioning equipment, distribution ductwork, and terminal units, or can be simple fan-coil units. Just enough supply air to meet ventilation demands is supplied through the ductwork (called primary air) or air is mixed with recirculated room air in an induction or fan-coil unit before it is introduced into the space.

Temperature control in the air–water system is by varying the temperature of the airstream or the temperature of the water, and by regulating flow of water through the coils. These capabilities offer excellent control of temperature in the individual spaces. The air–water system is more compact than the all-air system because ductwork does not need to be as large. Some of the maintenance activities on this type of system must occur in occupied spaces because some of the equipment is in these spaces.

Air Distribution System Arrangements

There are two primary configurations of air distribution systems (excluding packaged units): constant air volume (CAV or CV) systems and variable air volume (VAV) systems. Within these configurations, single-duct systems tend to be less costly and less expensive than dual-duct and multizone systems. VAV systems are generally more efficient. The common types of air distribution systems are introduced as follows.

CAV Distribution Systems

A *CAV distribution system* is an air-handling configuration that provides a steady airflow into a room or group of rooms while varying the temperature of the air to meet heating and cooling demands. Temperature control of a CAV system works in a manner similar to regulating temperature in an automobile air conditioner by adjusting the temperature control and leaving fan speed constant. As a result of the constant volume of flow and varying temperature of air, these systems are often referred to as *constant volume variable temperature (CVVT)* systems. CAV systems are used on small to large projects to supply regulated outdoor air to each zone or floor as well as to extract return air from the building.

Ducting of a CAV distribution system is identified as a single-duct or dual-duct arrangement:

Single-duct CAV systems have central supply and return fans and one duct that carries either hot air in the range of 100° to 120°F (38° to 49°C) or cold air in the range of 50° to 55°F (10° to 13°C) to diffusers where the air is introduced into the conditioned space.

Dual-duct CAV systems have central supply and return fans and two sets of ducts, one with a cooling coil (called the cold deck) and the other with a heating coil (called the hot deck). Two duct networks distribute heated and cooled air throughout the building. In each zone, the hot and cool air streams are mixed as required to meet the zone heating and cooling load. The airflow to each zone is constant. At maximum cooling demand, only air from the cold deck is supplied to the zone. As cooling demand drops, more air from the hot deck is added. Similarly, during peak heating needs, only air from the hot deck is added.

Dual-duct systems are less efficient because of leakage of warm air from the hot duct, which provides additional load to space, which must be cooled. They are not widely used, because of the extra installation cost (i.e., additional ductwork) and the space required to route the second duct.

Single-zone CAV systems introduce conditioned air to the entire building or a single zone within a large building.

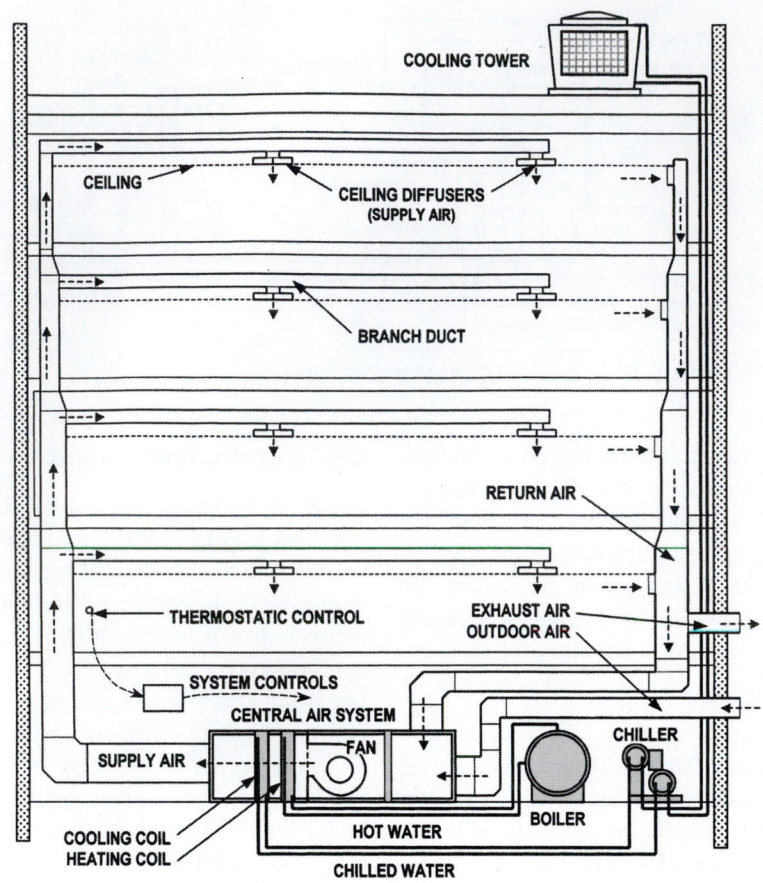

FIGURE 7.13   An all-air, single-zone, constant air volume (CAV) distribution system.

See Figure 7.13. A single thermostat controls operation of a CAV system so the area being served is a single zone. Buildings with rooms with different loads will require a single-zone CAV system in each zone. A residential furnace arrangement is a basic type of single-zone CAV system.

Multizone CAV systems have a single fan and mixing boxes located at the central air-handling unit and distribute supply air through multiple ducts, with one duct per zone. The hot deck carries preconditioned ventilation air at a constant supply flow. Ventilation air temperature can then be modulated to maintain the temperature in the most critical space.

CAV terminal reheat (CVTR) systems are a multizone arrangement that cools a constant flow of mixed (return and outdoor) air to the minimum required cold air supply temperature and then reheats it to the desired room temperature. Electric or water heating coils in terminal reheat boxes near air diffusers in each zone provide zone temperature control by reheating the cold supply air. When heating is required in the water heating coils, a valve opens to allow hot water to flow through the coil and heat

the air. Opening or closing the valve as opposed to modulating airflow controls the amount of heat delivered.

This system is simple and inexpensive to install and offers excellent temperature control. It is expensive to operate because of the excessive cooling of supply air and because a lot of heating energy is required for reheating the cold supply air. Proper operation of this system requires that heating and cooling equipment be energized simultaneously for all but the most extreme seasons. As a result of these inefficiencies, this system is limited by most codes.

VAV Distribution Systems

A *VAV (or VV) distribution system* is an air-handling configuration that holds supply air to a constant temperature and controls temperature within a space by varying the quantity of supply air. See Figure 7.14. A VAV system works in a manner similar to regulating the temperature in an automobile air conditioner by adjusting fan speed control and not adjusting the temperature control. As a result of the varying volume of flow and constant temperature of air, these systems are often referred to as *variable volume constant temperature (VVCT)* systems.

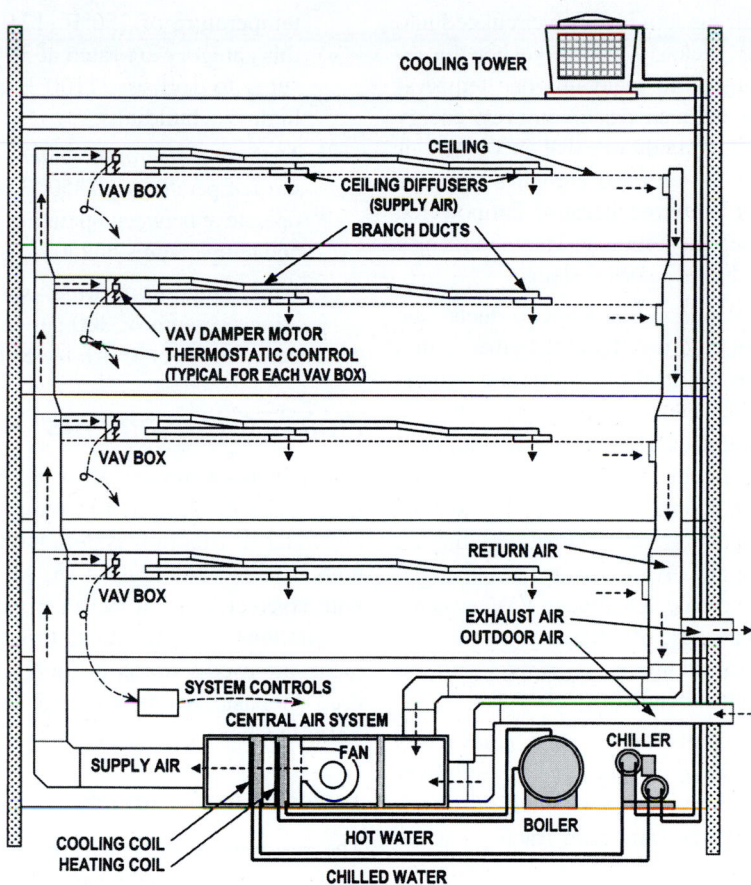

FIGURE 7.14　A four-zone, variable air volume (VAV) distribution system.

In a more sophisticated VAV system, the temperature of air supplied by the air-handling unit may be varied occasionally to adapt to building-wide changes in loads, but day-to-day control of each zone is achieved through varying supply airflow rate. These systems are often referred to as *variable volume variable temperature (VVVT)* systems.

VAV systems adjust the flow of cooled air to each building zone. Airflow is typically cooled to 50° to 55°F (10° to 22°C) by a central air-handling unit. A VAV box (with a motorized damper) located in each building zone opens and closes in response to the zone air temperature. Total system airflow is modulated to maintain a constant pressure in the ductwork. As dampers in more VAV boxes close, the pressure in the ductwork rises. A control system senses this pressure increase and reduces the fan speed using a variable speed drive and lowers the total system flow rate.

At maximum cooling demand, the central system generates and supplies the maximum volume of cold supply air to the zones. As demand for cooling in a zone decreases, airflow to the zone is reduced to a minimum fraction, which is typically between 15 and 30% of the full flow rate. The minimum fraction is the lowest acceptable airflow to provide required ventilation air and to limit maximum humidity levels. Reheat of the airstream occurs only when this point is reached and the airstream is too cold. Heating can be provided in three ways in VAV systems:

- By placing a heating coil in, or adjacent to, each VAV box. When heating is required, a valve opens to allow hot water to flow through the coil and heat the air. Opening or closing the valve as opposed to modulating airflow controls the amount of heat delivered. When in heating mode, the flow of air to the zone is held constant. In cold climates, this flow is usually above the minimum flow required to meet the peak heating load. This configuration is limited by most codes.

- Using a fan-powered induction unit instead of a VAV terminal box. Fan-powered induction units are similar to VAV boxes, except that they include a fan and a duct opening to recirculate some of the zone air. With this system, VAV flow can be set to the minimum and recirculated air can be used to meet the peak-heating load. As system air volume is decreased to the minimum stop at lower cooling loads, a fan between the VAV box and the zone air inlet (the power induction unit) starts adding return plenum air to the supply air. Heat from light fixtures rejected to the return air plenum is thus recirculated into the zone. Conventional reheat is added only when this re-circulated air is inadequate to maintain zone temperature. However, plenum return air systems are very energy inefficient and can lead to conditions that encourage air quality problems. This system also complies with the most codes but is not preferred because fan-powered boxes typically have inefficient motors. They should also be avoided in favor of ducted return systems.

- Using a *dual-duct VAV system* that has two ducts, one with a cooling coil (the cold deck) and the other with a heating coil (the hot deck). It also has variable volume central fans and variable volume mixing boxes at each zone. At maximum cooling demand, this system supplies the maximum volume of cold air to the zone. As cooling demand diminishes, the volume is reduced. As the zone begins to need heating, the hot deck starts supplying hot air in a second duct at gradually increasing rates up to the maximum design air volume. Dual-duct VAV systems are not as widely used because of efficiency losses, extra installation cost (i.e., for additional ductwork), and the space required to route the second duct.

- Using a separate perimeter heating system. Convectors or wall- or ceiling-mounted radiant panels can deliver heat. In this approach, the VAV box supplies only conditioned ventilation air (the minimum fraction) when heat is required. This system also complies with codes but requires the expense of a separate heating system.

A VAV all-air system offers the most energy-efficient method of providing fresh air and high cooling/heating load requirements for medium- to large-sized buildings. It provides excellent temperature control that rivals the CAV terminal reheat system.

Potential problems with VAV systems include the following: reduced airflows at part loads that can provide low air turnover rates and poor filtration for the space; reduced dehumidification from low airflow; and use of fan-powered induction boxes, which can lack effective filtration equipment that can load the ductwork downstream with dust and dirt, leading to potential indoor air quality problems.

Compared to the CAV reheat system, a VAV systems save energy in three ways: reheat energy consumption is minimized; fan energy consumption is decreased at low volumes; cooling coil consumption decreases significantly as less volume of air passes across the coil.

Water Temperature Classifications

All-water systems may be classified according to temperature and pressure limitations:

Low-temperature water systems have a maximum operating pressure of 160 psi (1100 kPa) and a maximum

temperature of 250°F (121°C). Most HVAC systems in this category are rated at 30 psi (207 kPa), although pressures to 160 psi (1100 kPa) are available and used in high-rise buildings.

Medium-temperature water systems have maximum operating temperatures of 350°F (177°C) or less. Most systems operate at between operating pressure of 150 psi (1000 kPa) and temperatures of 250° to 325°F (121° to 163°C).

High-temperature water systems have a maximum operating pressure of 300 psi (2100 kPa) and operating temperatures over 350°F (over 171°C).

Piping Arrangements

Piping in air–water and all-water systems in commercial installations can be designed in many different configurations. An air–water distribution system with unit ventilators and fan-coil units is shown in Figure 7.15. An all-water distribution system with convectors is shown in Figure 7.16. In most commercial installations, the two-pipe, three-pipe, or four-pipe arrangements are used. The series loop, one-pipe water arrangement is used in basic residential and light commercial installations (refer to Chapter 9).

Two-Pipe Arrangement

The *two-pipe arrangement* has a supply pipe to the coil in an AHU or distribution device and a return pipe back to the chiller or boiler. It is the simplest commercial arrangement. Simultaneous heating and cooling is not possible with this configuration. Seasonal changeover (e.g., system change from cooling to heating in fall) is required in most climates, which presents overheating or overcooling potential for some spaces between seasons. A more complex zoning configuration is also required in buildings where a sun-exposed zone may require cooling when the remainder of the building requires heating.

Three-Pipe Arrangement

The *three-pipe arrangement* has separate supply pipes for hot water and chilled water to separate coils in an AHU or distribution device and a single return pipe. It allows hot and cold water to be circulated year round, depending on loads within the building spaces. Return water is mixed in a common return line, which is extremely inefficient. As a result, the three-pipe system is seldom used.

Four-Pipe Arrangement

The *four-pipe arrangement* is the most complex arrangement with the highest initial cost. It has two separate supply pipes (hot supply and cold supply) and two separate return pipes, each set carrying hot water or chilled water to separate coils in the AHU or distribution device. This configuration is like two separate two-pipe arrangements, one for heating and one for

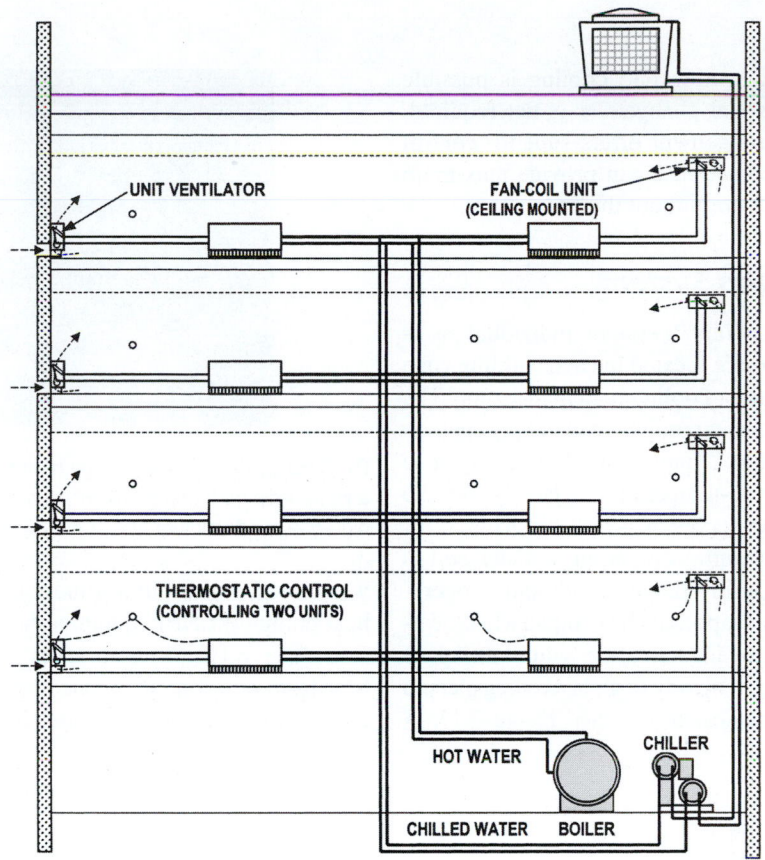

FIGURE 7.15   An air–water distribution system with unit ventilators and fan-coil units.

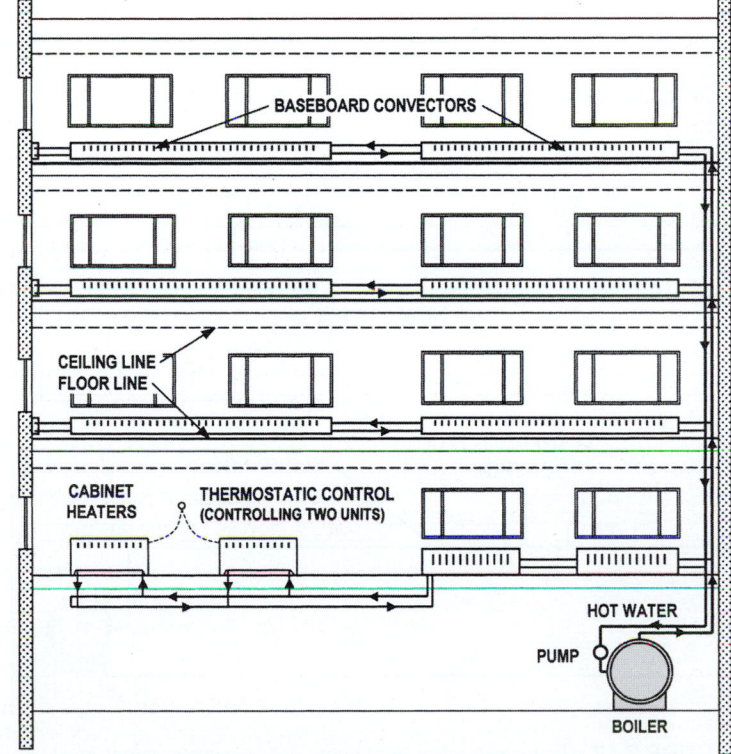

FIGURE 7.16   An all-water distribution system with convectors. Ventilation is required. Addition of a separate air system is needed to introduce ventilation air into the building.

cooling. Thus, simultaneous heating and cooling is possible with this configuration. Seasonal changeover is not required. As a result, the four-pipe arrangement offers superior control because the terminal unit in any space can provide maximum heating or cooling at any time throughout the year.

Water-Loop Heat Pump System

In the *water-loop heat pump (WLHP) system,* individual packaged water-source heat pumps are located in each building zone and operate to meet the heating or cooling load in that zone. The heating/cooling source for the heat pumps is a water-piping loop that supplies all building zones. Water in the loop is piped to each heat pump. The neutral-temperature (usually 60° to 90°F/16° to 32°C) water loop serves as a heat sink/source for the heat pumps. During heat pump operation, a water coil is used for condenser cooling during the air conditioning operation of the zone, and for the evaporator heat input when heating. The heat pump units can be located either within the space (e.g., low, along outside wall) or remotely (e.g., in a ceiling plenum or in a separate nearby mechanical room). See Figure 7.17. A

PHOTO 7.45   A water-source heat pump (under construction) for a water-loop heat pump (WLHP) system, which will heat and cool a high-rise condominium unit. (Courtesy of NREL/DOE)

water-source heat pump (under construction) for a water-loop heat pump (WLHP) system is shown in Photo 7.45.

The WHLP system saves energy by transferring excess heat from one zone to a zone requiring heat. For example, during the winter, the heat generated in the interior of a large building

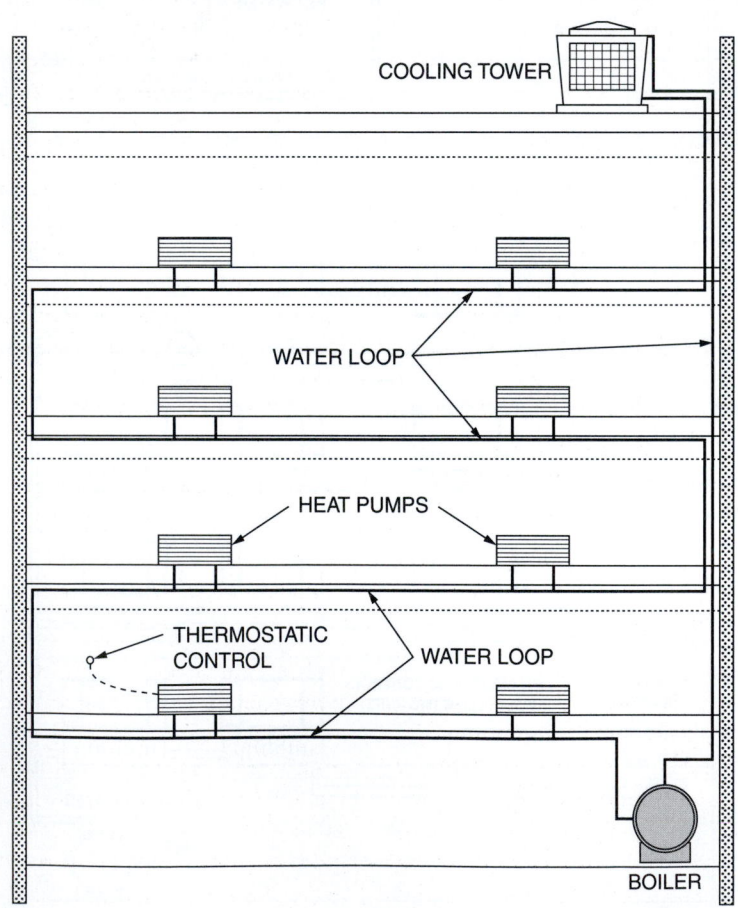

FIGURE 7.17   In the water-loop heat pump (WLHP) system, individual packaged water-source heat pumps are located in each building zone and operate to meet the heating or cooling load in that zone. The water loop serves as a heat sink/source for the heat pumps.

(e.g., building core zone) can be transferred to the perimeter zone for heating, thus reducing the net building heating load. As required depending on the net building heating or cooling load, water circulated in the building's water loop is heated with a boiler or cooled in a cooling tower. The ventilation issues of all-water systems apply to a WHLP system.

Chilled Beam System

A *chilled beam system* is an all-water or air–water system arrangement that consists of chilled (or hot) water piping feeding terminal units called *chilled beams*. Passive or active chilled beans are situated near or in the ceiling in each building zone. Most systems remove most of the heat gain instead of it being carried away by a primary air system. In unique installations, chilled beams can actually be used to provide heating.

In a chilled beam system, a large fraction of space cooling is performed using chilled water instead of cold supply air. Because water has a much higher cooling capacity, the air distribution equipment (i.e., ductwork, AHUs, and exhaust fans) normally used to provide cooling for a space can be scaled back. The reduced ductwork requirement results in a more compact ceiling cavity, which results in a reduction in the building's floor-to-floor dimension and envelope cladding. These reductions offset the added cost of the chilled beam units and the required additional piping. Chilled beams have lower capacities than standard air diffusers and take more ceiling space.

The basic design strategy is for the chilled beams to provide the required sensible cooling and for the central system to circulate only the amount of air needed for ventilation and latent load purposes. This results in a substantial reduction in the air circulated by the central system, which saves on fan power requirements. By removing a large fraction of the cooling load from the less efficient air distribution system (fans and ductwork), and shifting it to the more efficient water distribution system (pumps and piping), energy consumption and operating costs are reduced. Additionally, cooling occurs with minimal air movement, more comfortable cooling temperatures, and more uniform cooling.

Water is typically chilled to a temperature in the range of 55° to 60°F (13° to 16°C), which is higher than a conventional chilled water system. The higher-than-normal chilled water temperature is desirable because of the potential for condensation to form on the chilled beam coil when contacted by very moist room air. If the dew point temperature of room air is above the temperature of the water being supplied to the chilled beam, condensation will develop. As a result, the water temperature entering the chilled beam must be above the dew point temperature of the room air. In moist climates, an open window in proximity of a chilled beam can cause condensation to form on the coil. Some chilled beams have drip pans integrated into the unit to capture unwanted condensation.

Chilled water can be produced by a chiller system or extracted from a geothermal source such as a deep water source cooling (i.e., water from lakes, oceans, aquifers, and rivers). The concept of deep water source cooling (DWSC) was introduced in Chapter 6. In many climates, this temperature range is well suited to a free cooling, which eliminates the need for mechanical chiller operation because chilled water can be produced directly by a cooling tower. Chilled beams can also make it possible to eliminate reheat and reduce the HVAC energy for the building.

Active chilled beams can also be used for space heating because room air is induced into the unit and there is no reliance on natural convection. Hot water for heating with chilled beams is typically at a temperature in the range of 85° to 140°F (30° to 45°C). Higher temperatures result in an inability of the room air to mix properly with primary air, causing a significant (and uncomfortable) temperature gradient to develop in the room air.

Chilled beam systems are not stand-alone systems as they do not introduce ventilation air and remove only sensible heat gains (they do not remove moisture). It is necessary to have a separate central air system to deliver ventilation air and to remove latent heat (moisture). This is typically accomplished with a conventional air distribution system through a network of ductwork and diffusers in the ceiling or walls.

The chilled beam system arrangement has been used successfully in Europe and Australia, mainly to supplement an all-air system. This system and the related technologies are gradually being introduced in the United States. It is best suited for environments with high sensible cooling and low latent (humidity) loads. In certain cases it can be used in spaces that require high ventilation rates (more air changes per hour), so it has the greatest potential for use in laboratory and health care environments. It has been effectively used in offices, schools, retail stores, and computer centers. Advantages of this system include noise reduction, significant energy savings, and increased occupant comfort.

7.7 VENTILATION SYSTEMS

Mechanical Ventilation

A *mechanical ventilation system* is a network of ducts and fans that introduce outdoor air into building. The goal of a ventilation system is to ensure good indoor air quality in the breathing zone. The *breathing zone* is the area from which the occupant draws air. The definition of the breathing zone from American National Standards Institute (ANSI)/ASHRAE Standard 62.1-2004 *Ventilation for Acceptable Indoor Air Quality* is that "region within an occupied space between planes 3 and 72 in (75 and 1800 mm) above the floor and more than 2 ft (600 mm) from the walls or fixed air-conditioning equipment."

There are several basic types of mechanical ventilation systems. A *balanced system* draws outdoor air into the building and expels an approximately equal amount of indoor air to the

PHOTO 7.46 A small-scale (residential), supply-only air handler for outside air ventilation. (Courtesy of NREL/DOE)

PHOTO 7.47 An air-to-air heat exchanger for ventilation air in a commercial application. (Courtesy of NREL/DOE)

outdoors. A *supply-only system* relies on a supply fan to bring outdoor air into a building, raising the air pressure inside the building spaces and increasing the outward flow through cracks and openings in the building envelope. An *exhaust-only system* works in the opposite way. It uses an exhaust fan to expel indoor air to the outdoors, thereby lowering the air pressure inside the building so that outdoor air is drawn in through cracks and openings in the building envelope.

Buildings with a supply-only system are more susceptible to condensation problems than buildings with either of the other systems. The exfiltration of warm, humid air from inside these building spaces occurs mainly through unintentional leakage paths in the building envelope. As a result, excess moisture can condense and be absorbed by the building materials at the lower temperatures encountered outside the vapor retarder.

Generally, a balanced system is preferred. It is suited to buildings with high occupancies, with fuel-fired heating appliances, or where radon gas or other contaminants may collect in the building structure. An exhaust-only system is quite satisfactory in small buildings with low occupancies, unless the building is too tight.

In the *conventional mixed-air ventilation* method, the air in the entire room is fully mixed. Diffusers are used as the air outlets to enhance the mixing. Most often, the air outlets and inlets of the system are placed in the ceiling. Air is supplied at a relatively high velocity and at a temperature difference of about 20°F below/above the desired room temperature. Because the supply air fully mixes with the room air, the contaminant concentration is fairly uniform throughout the entire room and temperature variations are small. The full mixing results in a relatively low ventilation efficiency.

Displacement ventilation is an alternative ventilation method that introduces air at very low velocities to cause minimal mixing of the room air. Supply air is introduced at or near the floor level at a temperature only slightly below the desired room temperature. The cooler supply air displaces the stratified warmer room air, creating a zone of fresh cool air at the occupied level. Heat and contaminants produced by activities in the space rise to the ceiling level from the natural buoyancy of warm air. The air is exhausted from the space. In contrast to the conventional mixed-air ventilation method, displacement ventilation provides improved indoor air quality and comfort. An added advantage is that the low-velocity air supply produces less drafts and noise than a typical mixed-air supply diffuser. In contrast, displacement ventilation may not be appropriate in spaces with low ceilings because of inadequate stratification, nor may it be adequate in extremely warm climates because it requires an uncomfortably cold air supply to make up the high cooling load. As a result, displacement ventilation is generally appropriate for classrooms, lecture halls, and theaters with high ceilings in moderate to cold climates.

Demand-controlled ventilation (DCV) involves using CO_2 sensors and a CO_2 set point to dictate and control the amount of ventilation for the actual number of occupants. This makes it possible to maintain proper ventilation and improve air quality while saving energy. During design occupancy, a unit with the DCV system will deliver the same amount of outdoor air as specified by the ventilation standard but will result in savings whenever the space is occupied below the design level.

Economizers

An air-side *economizer* is designed into a building's air-handling system to cool with outside air instead of using air conditioning equipment to cool indoor air. It can be an additional, dampered cabinet opening that draws air from the outside when outside air is cooler than the temperature inside the building, thereby providing "free" cooling. On large all-air systems, it can be a control sequence that opens the outdoor air damper that draws air from the outside when possible.

Commercial buildings typically have a large cooling load even when the outdoor air temperature is significantly below the desired room air temperature. For example, the large internal heat gain in a classroom (from students, computers, lights, and incoming sunlight) may mandate cooling even though the temperature outdoors is 30°F (−1°C). When the outdoor temperature and relative humidity are suitable, economizers save energy by cooling buildings with outside air instead of using air conditioning equipment to cool recirculated air.

Economizer dampers can be controlled by either temperature or enthalpy (energy content of air). Temperature controls are recommended in dry and mild climates because they are less expensive, require little maintenance, and are typically more reliable. The simplest control is a fixed dry bulb high-limit controller. When the outside air temperature is below the thermostat set point, the economizer calls for maximum outside air. When the outside air temperature is above the thermostat set point, the economizer is disabled and the outside air damper closes to its minimum position. Greater energy savings can be achieved through the use of a differential dry bulb controller, which calls for maximum outside air whenever the outside air temperature is lower than the *return* air temperature.

Night Precooling

Night ventilation precooling or *night flush ventilation* involves operating the building mechanical cooling system to circulate outdoor air into the space during the naturally cooler nighttime hours. Except for any fan power requirement needed to circulate the outdoor air through the space, this can be considered a passive technique. The night ventilation precooling system improves indoor air quality through a cleansing effect of introducing more ventilation air. In a method called *mechanical precooling,* the building mechanical cooling system is operated during the nighttime hours to precool the building space to a set point that is typically lower than the normal daytime temperature. By cooling the building mass at night, the peak cooling load can be reduced as the building coasts past the daily peak temperature. The electric utility rate structure for peak and off-peak loads can mean significant off-peak savings. Because of high moisture levels in outdoor air in humid climates, the concept of night precooling is better suited for drier climates. Additionally, reliance on this cooling technique will reduce the required refrigeration capacity because of smaller peak loads.

7.8 COMPENSATION FOR NONSTANDARD ALTITUDE AND TEMPERATURE

An increase in altitude results in decreasing atmospheric pressure and thus a decrease in air density. At altitudes up to 2000 ft (610 m) above sea level, air density does not vary sufficiently to create heat transfer problems. At higher altitudes, lower air density has to be taken into consideration. For example, at 5000 ft (1640 m) above sea level, air has approximately 4/5 of its weight at sea level; therefore, about 20% more air by volume must be introduced to obtain the required oxygen for combustion. Air density change also affects heat transfer. Lower air

density results in less heat rejection. High altitude and nonstandard temperature affects performance related to:

- Airflow (e.g., ductwork, fans, and so on)
- Atmospheric combustion equipment (e.g., boilers, furnaces, and so on)
- Air-cooled equipment (e.g., condensers, chillers, motors, and other electrical and electronic equipment)
- Equipment involving evaporation (e.g., evaporative coolers, water towers, and so on)

This necessitates a compensation for altitude and temperature in HVAC system design. Examples of adjustments for high altitude and temperature are shown as follows:

Compensation for Ducts and Fans

Fans are rated on the basis of standard air having a dry bulb temperature of 70°F (21.1°C), an absolute pressure of 14.696 psi (101.325 kPa), and a mass density 0.075 lb/ft^3 (1.189 kg/m^3). When a fan must operate at some other condition, its ability to move air and develop pressure and the horsepower needed to do so will change.

A fan is similar to buckets on a waterwheel: they will hold and move the same volume of air regardless of what the air weighs. A fan will move less air by weight at higher temperatures for the reason that air is less dense at a higher temperature. A fan will move less air by weight at higher altitudes because air is less dense at a higher altitude. Correction factors for altitude (absolute pressure) and temperature are provided in Tables 7.4 and 7.5.

Example 7.1

A fan is needed to move 5000 standard cfm (ft^3/min) warm air at a temperature of 140°F. The fan will be installed in a building at an elevation of 5500 ft above sea level. Approximate the required fan rating in standard cfm at this elevation and temperature.

Altitude correction factor from Table 7.4 = 1.22
Temperature correction factor from Table 7.5 = 1.13
5000 standard cfm · 1.22 · 1.13 = 6893 cfm

A fan rated at approximately 7000 cfm is needed.

Friction loss from air moving through a duct is affected by density and viscosity of the air flowing through the duct. Changes in viscosity are relatively negligible with change in altitude (less than about 5% at typical altitudes). Friction loss with change in air density occurs in a manner related to the altitude correction factor. See Table 7.4.

Compensation for Combustion Equipment

Natural gas and liquid petroleum gas (LPG) obey the same physical laws as air. Thus, at high altitudes the energy content

TABLE 7.4   PROPERTIES AND CORRECTION FACTORS OF DRY AIR AT DIFFERENT ALTITUDES ABOVE MEAN SEA LEVEL.

Altitude Above Mean Sea Level		Density Correction Factor	Mass Density		Specific Heat Capacity		Altitude Correction Factor
ft	m		lb/ft³	kg/m³	Btu/ft³ · °F	kJ/m³ · °C	
0	0	1.00	0.075	1.189	0.0180	1.195	1.00
500	152	0.98	0.074	1.168	0.0177	1.174	1.02
1000	305	0.96	0.072	1.147	0.0174	1.153	1.04
1500	457	0.95	0.071	1.126	0.0171	1.132	1.06
2000	610	0.93	0.070	1.106	0.0167	1.112	1.07
2500	762	0.91	0.069	1.086	0.0164	1.092	1.09
3000	915	0.90	0.067	1.067	0.0162	1.072	1.11
3500	1067	0.88	0.066	1.048	0.0159	1.053	1.13
4000	1220	0.87	0.065	1.029	0.0156	1.034	1.16
4500	1372	0.85	0.064	1.011	0.0153	1.016	1.18
5000	1524	0.83	0.063	0.993	0.0150	0.998	1.20
5500	1677	0.82	0.061	0.975	0.0148	0.980	1.22
6000	1829	0.81	0.060	0.957	0.0145	0.962	1.24
6500	1982	0.79	0.059	0.940	0.0142	0.945	1.26
7000	2134	0.78	0.058	0.923	0.0140	0.928	1.29
7500	2287	0.76	0.057	0.907	0.0137	0.912	1.31
8000	2439	0.75	0.056	0.891	0.0135	0.895	1.33
8500	2591	0.74	0.055	0.875	0.0132	0.879	1.36
9000	2744	0.72	0.054	0.859	0.0130	0.864	1.38
9500	2896	0.71	0.053	0.844	0.0128	0.848	1.41
10 000	3049	0.70	0.052	0.829	0.0125	0.833	1.43
10 500	3201	0.68	0.051	0.814	0.0123	0.818	1.46
11 000	3354	0.67	0.050	0.799	0.0121	0.803	1.49
11 500	3506	0.66	0.050	0.785	0.0119	0.789	1.51
12 000	3659	0.65	0.049	0.771	0.0117	0.775	1.54
12 500	3811	0.64	0.048	0.757	0.0115	0.761	1.57

of gas fuels is less than at elevations near sea level. For example, at 5000 ft above sea level, the energy content of natural gas is about 830 Btu/ft³ in comparison to about 1000 Btu/ft³ at sea level. In contrast, fuel oil and coal, as a liquid or solid, have the same weight at any altitude and no compensation is necessary.

Industry standards recommend that an adjustment for altitude is required at 2000 ft above sea level and beyond at a reduction in heating capacity of 4% for every 1000 feet above sea level. This leads to a reduction in rated capacity as follows:

80% of rated capacity at 5000 ft above sea level

76% of rated capacity at 6000 ft above sea level

60% of rated capacity at 10 000 ft above sea level

TABLE 7.5   PROPERTIES AND CORRECTION FACTORS OF DRY AIR AT DIFFERENT TEMPERATURES.

Air Temperature		Density Correction Factor	Mass Density		Temperature Correction Factor	Air Temperature		Density Correction Factor	Mass Density		Temperature Correction Factor
°F	°C		lb/ft³	kg/m³		°F	°C		lb/ft³	kg/m³	
−50	−46	1.30	0.097	1.560	0.77	180	82	0.83	0.062	0.996	1.21
−25	−32	1.22	0.092	1.469	0.82	200	93	0.80	0.060	0.966	1.25
0	−18	1.16	0.087	1.388	0.87	225	107	0.78	0.058	0.930	1.29
20	−7	1.11	0.083	1.330	0.91	250	121	0.75	0.056	0.897	1.34
32	0	1.08	0.081	1.296	0.93	275	135	0.72	0.054	0.866	1.39
40	4	1.06	0.080	1.277	0.94	300	149	0.70	0.052	0.837	1.43
60	16	1.02	0.076	1.224	0.98	325	163	0.68	0.051	0.810	1.48
70	21	1.00	0.075	1.203	1.00	350	177	0.65	0.049	0.785	1.53
80	27	0.98	0.074	1.179	1.02	375	191	0.63	0.048	0.761	1.58
100	38	0.95	0.071	1.137	1.06	400	204	0.62	0.046	0.741	1.62
120	49	0.92	0.069	1.098	1.09	450	232	0.58	0.044	0.699	1.72
140	60	0.88	0.066	1.062	1.13	500	260	0.55	0.041	0.663	1.81
160	71	0.86	0.064	1.028	1.17						

TABLE 7.6 EXAMPLES OF COMPENSATION FACTORS FOR EFFECT OF HIGH ALTITUDE. CHECK WITH MANUFACTURER FOR COMPENSATION FACTORS FOR SPECIFIC EQUIPMENT.

Type	Elevation Above Sea Level					
	5000 ft	1500 m	6000 ft	1800 m	10 000 ft	3000 m
Copper baseboard or finned tube	0.85		0.82		0.71	
Steel finned tube or cast iron baseboard	0.91		0.89		0.83	
Radiant baseboard	0.96		0.96		0.93	
Air-cooled condenser	0.89		0.86		0.74	
Water-cooled condenser	0.97		0.94		0.81	

Example 7.2

A furnace is rated at 125 000 Btu/hr heating output and will be installed at 7500 ft above sea level. Approximate its output at this elevation.

$$7500 \text{ ft} \cdot 0.04/1000 \text{ ft} = 0.30 \text{ (a 30\% reduction)}$$
$$100\% - 30\% = 70\% \text{ or } 0.70$$
$$125\,000 \text{ Btu/hr} \cdot 0.70 = 87\,500 \text{ Btu/hr adjusted output}$$

Compensation for Heat Transfer Equipment

Air-cooled chillers and condensers require adjustment for altitude. Terminal units (e.g., radiators, convectors, baseboard units, finned tube units, coils, and so forth) also require compensation for effect of altitude based on elevation above sea level. See Table 7.6.

Example 7.3

An air-cooled condenser rated at 50 tons will be installed at an elevation of 5000 ft (1640 m) above sea level. The manufacturer advises a correction factor of 0.89 at an altitude of 5000 ft (1640 m). Determine the condenser rating at this elevation.

$$\text{Altitude correction factor} = 0.89$$
$$50 \text{ tons} \cdot 0.89 = 44.5 \text{ tons adjusted output}$$

Example 7.4

A copper baseboard unit is designed to output 600 Btu/hr per ft at design conditions. Estimate its output at 6000 ft above sea level.

$$\text{Altitude correction factor} = 0.82$$
$$600 \text{ Btu/hr} \cdot 0.82 = 492 \text{ Btu/hr adjusted output}$$

Compensation for altitude and temperature is too complex to fully address in this text. It necessitates following specified corrections provided by the manufacturer. However, as demonstrated in Examples 7.1 through 7.4, neglecting to adjust for altitude can lead to severe consequences associated with under design of equipment.

7.9 MECHANICAL ROOMS

About 4 to 6% of a building's gross floor area should be available at each floor for air-handling equipment. In commercial buildings, usually about 1 to 2% of the building's gross floor area is needed for the mechanical room for the central heating and cooling plant. In large buildings, main mechanical equipment rooms generally shall have clear ceiling heights of not less than 12 ft (3.6 m). Where maintenance requires the lifting of heavy parts [100 pounds (45 kg) or more], hoists and hatchways should be installed.

Space requirements of mechanical and electrical equipment rooms shall be based on the layout of required equipment. Clear circulation aisles and adequate access to all equipment should be provided. The arrangement should take into consideration the future removal and replacement of equipment. Adequate doorways or areaways and staging areas should be provided to permit the replacement and removal of equipment without the need to demolish walls or relocate other equipment.

Adequate access is needed for chillers, boilers, heat exchangers, cooling towers, reheat coils, VAV boxes, pumps, hot water heaters, and all devices that have maintenance service requirements. Chillers should be positioned to permit pulling of tubes from all units. Clearance required should equal the length of the tubes plus 3 feet (900 mm). AHUs typically require a minimum clearance of 2.5 ft (750 mm) on all sides.

Access doors or panels should be provided in ventilation equipment, ductwork, and plenums as required for in-site inspection and cleaning. Additionally, the replacement of major equipment is necessary over the life of a building. Large central equipment shall be located to make replacement possible, including removal and replacement, without damage to the structure.

7.10 CODES AND STANDARDS

Codes

A *mechanical code* is a set of standards adopted into law by local governmental entities to regulate mechanical systems and equipment including HVAC, exhaust systems, chimneys and vents, ducts, appliances, boilers, water heaters, refrigeration, hydronic piping, and solar systems. A *model code* is written and published by a code-writing organization with the intent that local governments adopt it into law to ensure public and occupant safety. A model code does not become law (the code) until the local government having jurisdiction formally adopts it.

Historically, there were several model codes related to building mechanical systems. These include the *Uniform Mechanical Code* published by the International Association of Plumbing and Mechanical Officials, Walnut, California; the *Standard Mechanical Code* by the Southern Building Code Congress International, Inc., Birmingham, Alabama; and the *National Mechanical Code* by the Building Officials and Code Administrators International, Inc., Country Club Hills, Illinois.

The *International Mechanical Code* is the newest model code written by the International Code Council (ICC), Whittier, California. The above-mentioned model code-writing organizations in the United States created the ICC with the intent to develop a single set of model codes (e.g., building codes, plumbing codes, mechanical codes, and so on) to replace the corresponding model code of each organization. These efforts resulted in a single set of national model codes referred to as the International Codes.

The *Model Energy Code* (MEC) published by the Council of American Building Officials (CABO), was first published in 1983, with full editions published every three years through 1995. It is the most commonly used residential building energy. More than half of new residences constructed to code in the United States today are required to be built to codes that meet or exceed the MEC.

The MEC establishes minimum requirements for energy-related features of new residential and commercial buildings and some additions to existing buildings. It covers low-rise buildings (e.g., buildings with three stories or less) and one- and two-family residences. It does not apply to existing buildings, including those being renovated, unless there is a change in use that increases the building's energy use. It is not applicable to historic structures specifically designated as historically significant by the state or local governing body, listed in the National Register of Historic Places, or eligible for listing.

The MEC states detailed criteria that the designer and builder must satisfy. Tables are provided that specify the following: the thermal resistance required for each part of the building envelope (e.g., walls, roof/ceiling, and floors), considering the climate (as expressed in degree days); required performance criteria are specified for various types of heating and cooling equipment, and illumination level standards are established for the lighting system.

The MEC was replaced by the *International Energy Conservation Code (IECC)* in 1998. The ICC developed the IECC as part of its family of codes. The IECC covers energy efficiency provisions for residential and commercial buildings, prescriptive- and performance-based approaches to energy-efficient design, and building envelope requirements for thermal performance and air leakage. The U.S. Energy Policy of Act of 1992 requires all states must certify that they have energy codes in place by July 15, 2004.

American Society of Heating, Refrigerating and Air-Conditioning Engineers

The American Society of Heating, Refrigerating and Air-Conditioning Engineers, Inc. (ASHRAE, is a professional organization of engineers in the heating, ventilating, air conditioning, and refrigeration (HVAC&R) industry. Research and education are key elements in the mission of ASHRAE. The ASHRAE handbooks are accepted as the authoritative source for information in the HVAC&R industry. ASHRAE standards serve as the basis for federal, state, and local building codes. ASHRAE also develops standards for energy efficiency in buildings.

ASHRAE 90.1-1989, *Energy Standard for Buildings Except Low-Rise Residential Buildings,* is a standard that sets energy-efficiency requirements for commercial buildings related to exterior and interior lighting, thermal envelope (insulation levels and maximum window areas), heating and cooling systems, and service water heating. These standards are incorporated in the IECC.

HVAC Construction Drawings

A complete set of construction drawings and specifications of the HVAC system is needed to convey design information to the contractor. Construction drawings show the layout and design of a HVAC system installation. The following construction drawings and specifications are generally required:

1. Heating equipment data, including the following information:

    a. Equipment capacity
    b. Controls
    c. Appliance layouts showing location, access, and clearances
    d. Disconnect switches
    e. Indoor and outdoor design temperatures

2. Ventilation data, ductwork, and equipment, including the following:

    a. Ventilation schedule indicating the amount of outside air (in cfm or L/s supplied to each room or space
    b. Layout showing outside air intakes
    c. Construction of ducts, including support and sheet metal thickness
    d. Duct lining and insulation materials with flame spread and smoke-developed ratings
    e. Exhaust fan ductwork layout and termination to the outside
    f. Size of louvers and grilles for attic ventilation

3. Boiler and water heater equipment and piping details, including safety controls and distribution piping layout

4. Gas and fuel oil piping layout, material, sizes, and valves

5. Combustion air intake quantities and details

HVAC ACRONYMS AND ABBREVIATIONS

AC	Air conditioning		HP	High pressure
ACH	Air changes per hour		HPS	High pressure steam
ACM	Asbestos containing material		HR	Heat recovery
ASHRAE	American Society of Heating, Refrigerating and Air Conditioning Engineers, Inc		HRU	Heat recovery unit
			HRV	Heat recovery ventilator
AFUE	Annual fuel efficiency ratio		HSPF	Heating seasonal performance factor
AHU	Air handling unit		HVAC	Heating ventilating and air conditioning
ATC	Automatic temperature control		HWG	Hot water generator
AVB	Atmospheric vacuum breaker		HWP	Hot water pump
B	Boiler		HWR	Hot water return
BAS	Building automation system		HWS	Hot water supply
BEP	Best efficiency point		HX	Heat exchanger
BOD	Bottom of duct		IAQ	Indoor air quality
BOM	Bill of material		IBR	Institute of Boiler and Radiator Manufacturers
C	Chiller		IR	Infrared
CAV	Constant air volume		LP	Liquefied petroleum
CD	Cold deck		LPS	Low pressure steam
CDD	Cooling degree days		LRA	Locked rotor amps
CFC	Chlorofluorocarbon		LWBT	Leaving wet bulb temperature
CHWP	Chilled water pump		LWCO	Low water cut off
CHWR	Chilled water return		LWT	Leaving water temperature
CHWS	Chilled water supply		MAU	Make-up air unit
CLF	Cooling load factor		MBU	Thousand Btu per hour
CLTD	Cooling load temperature difference		MMBU	Million Btu per hour
COP	Coefficient of performance		MCC	Motor control center
CT	Cooling tower		MOV	Motor operated valve
CTH	Constant temperature and humidity		NEC	National Electrical Code
CV	Control valve		NC	Noise criteria
CWP	Chilled water pump		NEMA	National Electrical Manufacturers Association
CWP	Condenser water pump		NPSH	Net positive suction head
CWR	Chilled water return		OA	Outdoor air
CWR	Condenser water return		OBD	Opposed blade damper
CHWS	Chilled water supply		PT	Pressure & temperature
CWS	Condenser water supply		PTAC	Packaged terminal air conditioner
DA	Discharge air		R	Return
DB	Dry bulb		RA	Return air
DD	Degree day		RAG	Return air grille
DDC	Direct digital control		RF	Return fan
DH	Duct heater		RH	Relative humidity
DP	Dew point		RTU	Rooftop unit
DX	Direct expansion		RV	Reversing valve
EAT	Entering air temperature		S	Supply
EER	Energy efficiency ratio		SA	Supply air
EWT	Entering water temperature		SC	Shading coefficient
EDH	Electric duct heater		SCFM	Standard cubic feet per minute
EF	Exhaust fan		SD	Smoke damper
EMS	Energy management system		SEER	Seasonal energy efficiency ratio
ERV	Energy recovery ventilator		SF	Service factor
FCU	Fan coil unit		SF	Supply fan
FLA	Full load amps		SHFG	Solar heat gain factor
FOB	Flat on bottom		SHR	Sensible heat ratio
FOT	Flat on bottom		SP	Static pressure
FV	Face velocity		SWP	Steam working pressure
FTU	Fan terminal unit		T	Temperature
FW	Feed water		TD	Temperature difference
G	Gas		TOD	Top of duct
HEPA	High efficiency particulate arresting		TXV	Thermostatic expansion valve
HCFC	Hydrochlorofluorocarbon		UV	Unit ventilator
HD	Hot deck		VAV	Variable air volume
HDD	Heating degree days		VD	Volume damper
HFC	Hydrofluorocarbon		VFD	Variable frequency drive
HHWP	Heating hot water pump		VP	Velocity pressure
HHWR	Heating hot water return		WB	Wet bulb
HHWS	Heating hot water supply		WC	Water column
HI	Hydronics Institute		ZD	Zone damper
HL	High limit			

FIGURE 7.18 Common HVAC abbreviations, acronyms, and symbols. Symbols will vary with sets of drawings.

6. Commercial kitchen exhaust equipment details including hood and fan drawings

7. Chimney and chimney connector or vent and vent connector details and connector gauges and clearances

8. Mechanical refrigeration equipment data and details

9. Solid fuel burning appliance details, including incinerator and fireplace drawings and details

10. Energy conservation equipment data and details

Common abbreviations and symbols used on HVAC drawings are provided in Figure 7.18.

Building Commissioning

A new building project or major retrofit to an existing building with highly sophisticated or innovative systems generally requires additional attention to be sure that it operates as designed. *Building commissioning* is the process of ensuring that all systems

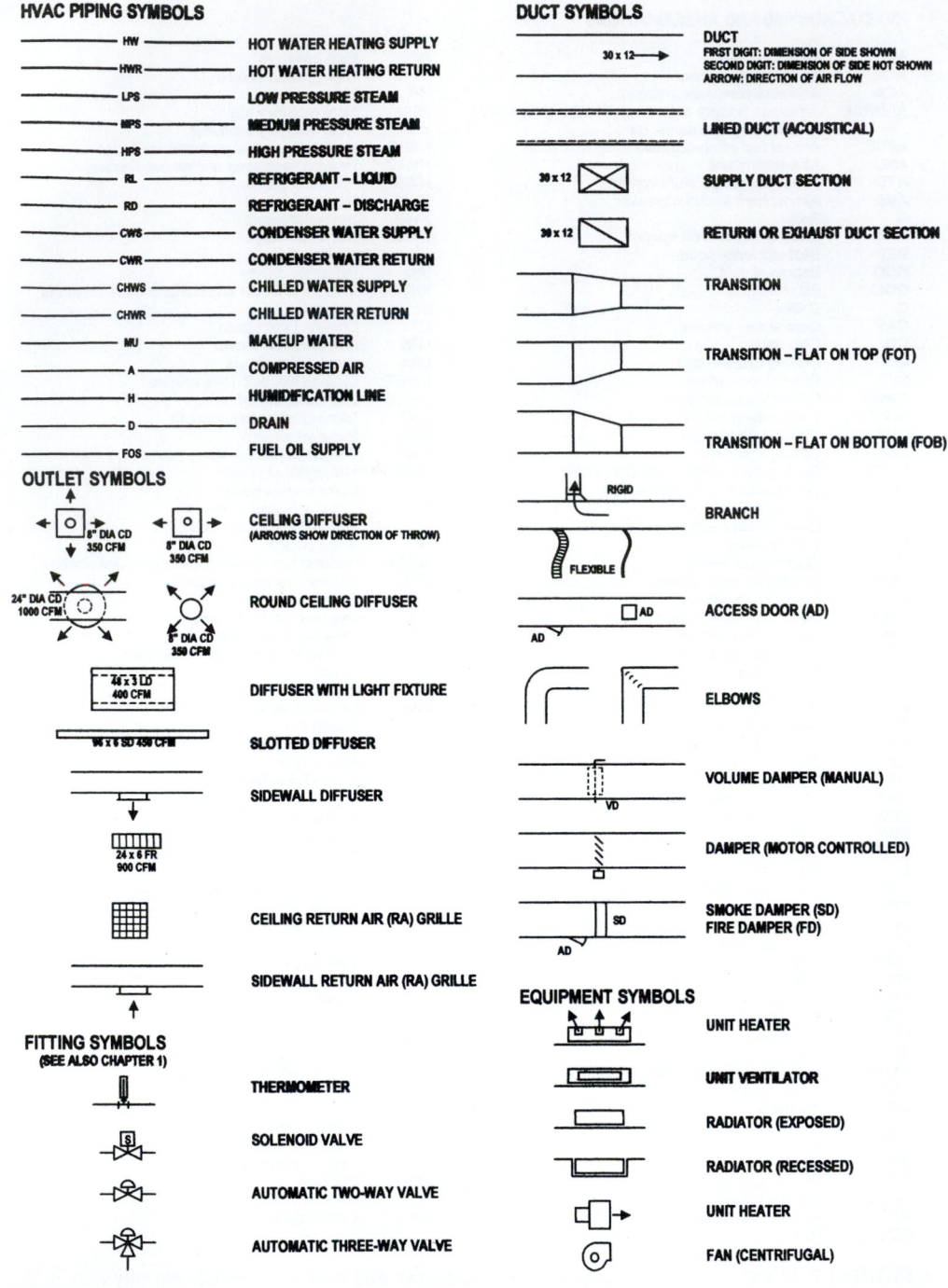

FIGURE 7.18   *(Continued)*

are built and function as intended. It usually covers HVAC, electrical, communications, security, and fire management systems and their controls. On HVAC systems, commissioning is a fairly new practice that partially includes what was formerly referred to as *testing, adjusting, and balancing (TAB)*. Commissioning has been found to be very valuable, because it maximizes energy savings and maintains a quality indoor environment.

Commissioning begins by recording design objectives from drawings and specifications, developed during the design process. During building construction, individual system components are inspected and tested when they arrive at the job site to make sure they meet design specifications. After installation, components and systems are tested, adjusted, and balanced to establish specified air and water flows. On an HVAC system, this involves testing, adjusting, and balancing fans, coils, dampers and ducts, and pumps, chillers, coils, valves, and piping. Testing, adjusting, and balancing the system will minimize temperature control problems that can occur within an indoor space (e.g., hot and cold spots and drafts). In some spaces, such as in medical facilities and laboratories, pressure differentials are required within the building to protect the building occupants (e.g., negative pressure isolation rooms for patients suffering from tuberculosis). A well-balanced system is environmentally safe and much more efficient to operate.

When systems are balanced and functioning properly but before the building is in full operation, equipment operation and maintenance manuals and training for operating staff are provided. A final commissioning report is submitted to the owner that contains all records of the commissioning procedures, testing results, deficiency notices, and records of satisfactory corrections of deficiencies. In some cases, commissioning may also extend to testing the building and systems several months or a year after occupancy.

A consultant with special experience and training is employed for commissioning a building and its systems. Preferably, this consultant is independent of the design professional and contractor and is retained by and responsible directly to the building owner. A special commissioning consultant may be responsible for whole-building commissioning for very sophisticated projects such as shopping malls, office buildings, health care facilities, and airports. The project architect or engineer may oversee the completion of commissioning.

Commissioning is needed in buildings with sophisticated HVAC systems, cogeneration or renewable energy equipment, on-site water treatment systems, or other sophisticated or atypical technologies. Commissioning is usually not cost-effective for projects with very little mechanical or electrical intricacy, such as most residential projects.

STUDY QUESTIONS

7-1. Describe the following air distribution components:

    a. Ductwork

    b. Duct insulation

    c. Damper

    d. Diffusers

    e. Grilles

    f. Fans

    g. Coils

7-2. What is an air-handling unit? Describe its chief components. Where is it used?

7-3. What is the difference between a roof-mounted air-handling unit (AHU) and an rooftop unit (RTU)?

7-4. Describe the purpose of a fan filter unit (FFU) and where it may be used.

7-5. Describe the purpose of a makeup air unit (MAU) and where it may be used.

7-6. Describe the function of the following types of ducts:

    a. Supply air

    b. Outdoor air

    c. Return air

    d. Mixed air

7-7. Describe the types of multiblade dampers.

7-8. Describe the purpose of installing smoke and fire dampers.

7-9. How are duct joints sealed?

7-10. Why are filters needed in an air distribution system?

7-11. Describe the function of:

    a. Constant volume (CV) terminal box

    b. Reheat terminal box

    c. Variable air volume (VAV) terminal box

7-12. Describe the following hydronic distribution components:

    a. Convection heaters

    b. Radiator

    c. Fan-coil unit

    d. Unit ventilator

7-13. Describe the following components of hydronic systems:

    a. Pipe

    b. Fittings

    c. Valves

    d. Pump

    e. Insulation

7-14. Describe the following components of hydronic systems:

    a. Convector

    b. Radiator

    c. Convector radiator

    d. Fan-coil unit

    e. Unit ventilator

    f. Induction unit

7-15. Describe the difference between a fan-coil unit and a unit ventilator.

7-16. What is zoning?

7-17. What are the types of zones and where are they used?

7-18. Describe the types of thermostats.

7-19. Describe thermostat operation.

7-20. Describe the types of sensors used in HVAC control applications.

7-21. Describe the operation of a direct digital control (DDC) system.

7-22. Describe the functions of a building energy management system (BEMS).

7-23. Describe the types of HVAC distribution system classifications.

7-24. Describe the types of HVAC air distribution system arrangements.

7-25. Describe the types of pipe arrangements used in all-water and air-water HVAC distribution systems.

7-26. Explain the components and operation of a water-loop heat pump (WLHP) system.

7-27. Explain the operation of a conventional mixed-air ventilation system.

7-28. Explain the operation of a displacement ventilation system.

7-29. Explain the operation of a demand controlled ventilation (DCV) system.

7-30. Explain the components and operation of an air-side economizer.

7-31. Describe the operation of night ventilation precooling (night flush ventilation).

7-32. How does a higher altitude affect performance of a HVAC system? What is affected?

Design Exercises

7-33. A fan in a residential furnace is needed to move 1700 standard cfm (ft³/min) of warm air at a temperature of 120°F. The fan will be installed in a building at an elevation of 6000 ft above sea level. Approximate the required fan rating in standard cfm at this elevation and temperature.

7-34. A fan in an AHU is needed to move 20 000 standard cfm (ft³/min) of warm air at a temperature of 130°F. The fan will be installed in a building at an elevation of 5000 ft above sea level. Approximate the required fan rating in standard cfm at this elevation and temperature.

7-35. At a mountain ski resort, a fan in an AHU is needed to move 15 000 standard cfm (ft³/min) of warm air at a temperature of 120°F. The fan will be installed in a building at an elevation of 5000 ft above sea level. Approximate the required fan rating in standard cfm at this elevation and temperature.

7-36. A large boiler is rated at 8000 MMBH heating output and will be installed at 5000 ft above sea level. Approximate its actual Btu/hr output at this elevation.

7-37. A furnace is rated at 100 000 Btu/hr heating output and will be installed at 6000 ft above sea level. Approximate its actual Btu/hr output at this elevation.

7-38. A furnace at a home in a mountain ski resort is rated at 150 000 Btu/hr heating output and will be installed at 10 000 ft above sea level. Approximate its actual Btu/hr output at this elevation.

7-39. An air-cooled condenser is rated at 40 tons will be installed at an elevation of 5000 ft (1640 m) above sea level. The manufacturer advises a correction factor of 0.89 at an altitude of 5000 ft (1640 m). Determine the condenser rating at this elevation.

7-40. A 10 ft long copper baseboard unit is designed to output 550 Btu/hr per ft at standard design conditions. Estimate its output at 5000 ft above sea level.

7-41. Make a visit to a residence. Examine the HVAC system. Make a sketch of the system and identify chief components of the system. Write out control strategies used.

7-42. Make a visit to a commercial building. Examine the HVAC system. Identify chief components of the system. Describe control and energy management strategies used.

7-43. Review a set of construction drawings of a residence. Examine drawings of the HVAC system. Make a sketch of the system and identify chief components of the system.

7-44. Review a set of construction drawings of a commercial building. Examine drawings of the HVAC system. Make a sketch of the system and identify chief components of the system.

HVAC AIR DISTRIBUTION SYSTEMS

8.1 AIRFLOW FUNDAMENTALS

Air distribution systems distribute conditioned air through ductwork to the spaces in a building for space heating, cooling, and ventilating. Design of an air distribution system requires an understanding of the properties of air and characteristics of airflow. These are introduced in the sections that follow.

Standard Air

Standard air is defined as dry air at standard atmospheric pressure (14.696 lb/in^2, 101 325 Pa, 29.92 inches of mercury) and room temperature (70°F, 21.1°C); its mass density is normally 0.075 lb/ft^3 (1.189 kg/m^3). One cubic yard ($\sim$1 m^3) of air weights approximately 2 lb ($\sim$1.2 kg). As air temperature increases, its density decreases; air becomes lighter when it is heated and heavier when it becomes colder. Air at higher altitudes is less dense than air at sea level.

Air Velocity

Velocity (v) is the average rate at which air flows past a specific point in the system. It can be computed as the volume of air that passes a specific location per unit of time (ft/min, m/s) divided by the flow cross-section area (A) in ft^2 (m^2) of the duct:

$$v = Q/A$$

The velocity of an airstream is not uniform across the cross-section of a duct. Molecules of air are moving in all directions—some are moving against the flow but most are moving in the direction of average airflow. Friction slows air moving close to and contacting the duct walls, so the velocity of air is greatest in the center of the duct. An HVAC technician must traverse an airstream and take several measurements to ascertain the average velocity in a duct.

Volumetric Flow Rate

The *volumetric flow rate (Q)* is defined as the quantity of air that passes a specific location per unit of time (ft^3/min, L/s). It is related to the average velocity (v) of airflow and the flow cross-section area (A) in ft^2 (m^2) of the duct:

$$Q = vA$$

In the HVAC industry, the term *volume* is used to describe volumetric flow rate. Units of cubic feet per minute are often expressed as "CFM" or "cfm" in lieu of ft^3/min. Actual cubic feet per minute (ACFM or acfm) is the volume of air flowing anywhere in a system independent of its density.

The properties of air (e.g., density, specific heat, heat capacity, and so on) change as the air is heated or cooled and, because of changes in atmospheric pressure, these properties are different at various elevations from sea level. Standard cubic feet per minute (SCFM or scfm) is the volumetric flow rate of air corrected to "standardized" conditions of temperature, pressure, and relative humidity, thus representing a precise mass flow rate. A variation in temperature can result in a significant volumetric variation for the same mass flow rate. For example, at sea level, a mass flow rate of 1000 lb/hr of air at standard pressure is 206 SCFM when defined at 32°F (0°C) but 218 SCFM when defined at 60°F (15.6°C).

Air Pressure

The concept of pressure is central to the study of fluids. The definition of pressure is force (F) per unit area (A). Air pressure in an HVAC system is comparatively small ($<$1 lb/in^2). As a result, air pressure is traditionally expressed in water pressure in inches of water column (in w.c.) or Pascal (Pa).

$$1 \text{ pound per square inch (psi or lb/in}^2) = 6895 \text{ Pa}$$
$$1 \text{ psi or lb/in}^2 = 2.31 \text{ ft of water column}$$
$$1 \text{ foot water column (ft w.c.)} = 0.433 \text{ lb/in}^2 \text{ (psi)}$$
$$1 \text{ psi or lb/in}^2 = 27.7 \text{ in water column}$$
$$1 \text{ inch water column (in w.c.)} = 0.0361 \text{ lb/in}^2 \text{ (psi)}$$
$$1 \text{ inch water column (in w.c.)} = 248.8 \text{ Pa}$$
$$1 \text{ inch water column (in w.c.)} = 0.0739 \text{ in Hg (mercury)}$$

There are three types of air pressures of concern in ducts:

Static Pressure

Static pressure is the pressure exerted against the side walls of the duct no differently than atmospheric pressure presses against your body's surface. It is the difference between the air pressure on the inside of the duct and the air pressure outside the duct. Static pressure can be positive or negative. In pressure or supply systems, static pressure will be positive on the discharge side of the fan and negative on the inlet side of the fan.

Velocity Pressure

Velocity pressure (or dynamic pressure) is the pressure created by the flow of air down the duct. It is calculated by subtracting the static pressure from the total pressure. Velocity pressure can be thought of as the "added" pressure you feel against your body when you stand in front of a moving airstream (i.e., air being blown by a fan).

Total Pressure

The *total pressure* is the combined effect of both static and velocity pressure. As air flows through a duct system from a fan to the terminal the total pressure will always decrease. Velocity pressure and static pressure will either increase or decrease because of changes in velocity. When a fitting changes the direction or velocity of airflow, there is a friction loss.

Air Flow

Flow of air is caused by a pressure difference (ΔP). Flow will originate from a region of high pressure and proceed to an area of lower pressure. A high pressure difference results in more flow. Airflow can be turbulent or laminar: *Laminar flow,* sometimes known as streamline flow, occurs when a fluid flows in parallel layers, with no disruption between the layers. *Turbulent flow* occurs when there is a disruption between the laminar layers within the airflow from the formation of eddies and chaotic motion. This turbulence increases the resistance to airflow significantly. *Transitional flow* is a mixture of laminar and turbulent flow, with turbulence in the center of the duct, and laminar flow near the edges. Most HVAC applications fall in the transitional range.

Air in a duct moves according to the fundamental laws of physics including laws of conservation of mass, energy, and momentum.

Conservation of mass states that the amount of air mass remains constant—that is, mass is neither created nor destroyed. The amount of air mass entering a section of duct is equal to the amount of air mass leaving that section of duct. At a branch, the sum of the two flow rates that enter the fitting must be equivalent to the total leaving the fitting. At differing temperatures and pressures, the mass airflow rate will differ for the same volumetric flow rate because air is compressible. In many cases the air in an HVAC system is assumed to be incompressible, an assumption that overlooks the change of air density that occurs as a result of pressure loss and flow in the ductwork. (Incompressibility can be assumed if the total system pressure does not exceed 20 in w.c. [5 kPa].) With the assumption that air is incompressible, a duct size can be recalculated for a new air velocity using the simple equation, based on velocity (v) and area (A):

$$v_2 = (v_1 \cdot A_1)/A_2$$

According to the *law of energy conservation,* the amount of energy in a system remains constant—that is, energy is neither created nor destroyed. Energy can be converted from one form to another (e.g., potential energy can be converted to kinetic energy) but the total energy within the domain remains fixed. The Bernoulli's equation in its simple form shows that,

for an elemental flow stream, the difference in total pressures (P) between any two points in a duct is equal to the pressure loss (ΔP) between these points:

$$\Delta P = P_{\text{Point 1}} - P_{\text{point 2}}$$

Momentum is defined to be the mass of an object multiplied by the velocity of the object. *Conservation of momentum* is based on Newton's law that a body will maintain its state of rest or uniform motion unless compelled by another force to change that state—that is, the amount of momentum remains constant. This law is useful to explain flow behavior in a duct system's fitting.

Noise and Airflow

Still air is silent, while moving air generates noise. The faster the air moves, the louder the noise. When airflow is forced to change direction, the noise level is increased. Noise is related to air molecules banging and clanging into the walls of the duct and fluctuations in air pressure caused by turbulence.

8.2   ALL-AIR SYSTEMS

Large All-Air Systems

HVAC air systems in medium to large commercial buildings use a duct system to circulate cooled and/or heated air to all the conditioned spaces in a building. This type of system is illustrated in Figure 8.1.

In the system shown in this figure, *supply air (SA)* is the conditioned air delivered to the building spaces. When supply air is delivered into a room, an equal amount of air must be removed from the room. This *return air (RA)* is transported back to the central air system for reconditioning. Because there is a need for ventilation, only a fraction of the stale RA can be reused. *Exhaust air (EA),* sometimes called *relief air,* is discharged to the outdoors and replaced by fresh air that is brought into the system as *outside air (OA).* The EA, OA, and RA ducts have dampers and controls that all operate in conjunction with one another. The automatic control system operates the damper motors to maintain the proper mix of air. The OA and the EA dampers open together and close together to balance the air entering and leaving the system. After the air is mixed in the proper proportions, the mixed air is drawn through a filter before it enters the fan and returns to the conditioned spaces.

The air-handling unit (AHU) typically consists of a mixing box, filter bank, heating coil, cooling coil, and fan, which is a packaged unit or custom built. It conditions supply air by heating, cooling, dehumidifying, humidifying, and/or filtering it. It is normally a large piece of equipment designed to move about 2000 ft³/min (944 L/s) in residences to about 50 000 ft³/min (23 500 L/s) in commercial installations, and in exceptionally large systems, even more. Smaller units facilitate flexible zone control, particularly for spaces that involve off-hour or high-load operating conditions.

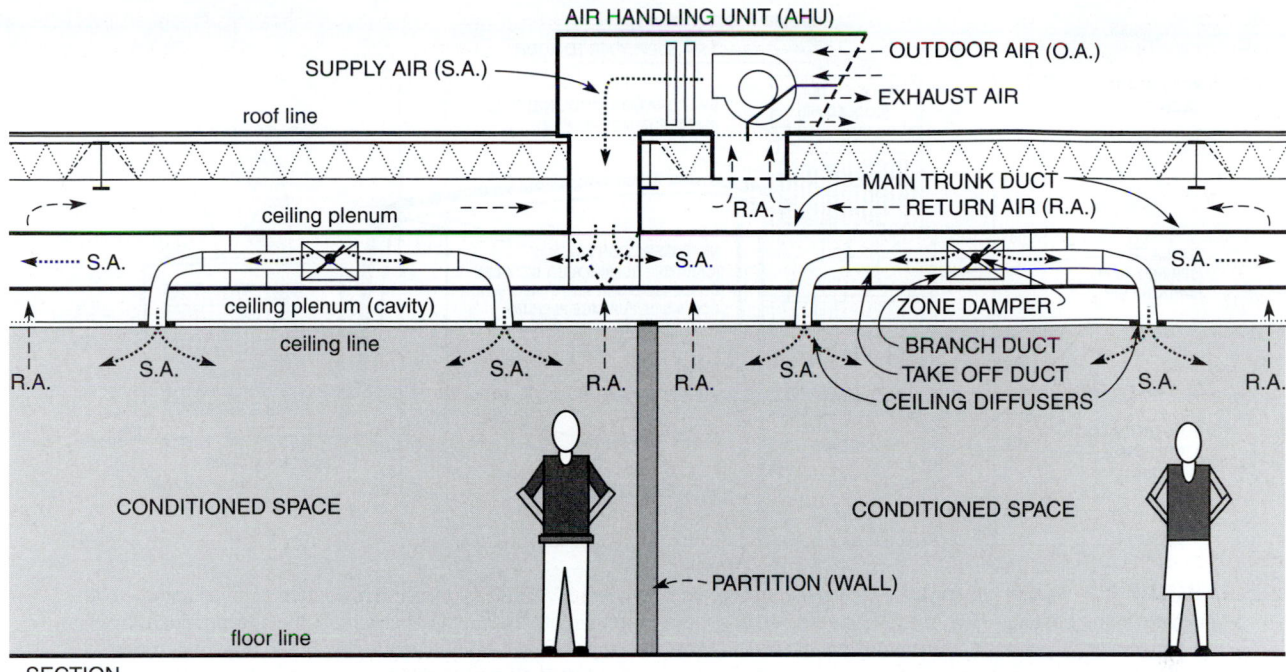

FIGURE 8.1   A HVAC air system in a commercial building.

A chilled water coil in the AHU cools the SA as chilled water from a chiller flows through the tubes in the coil and air passes across the coil. Heat removed from the airstream in the process of cooling air is conducted to the coil fins, to the tubes carrying the chilled water. This has the effect of heating the chilled water, which is pumped back as chilled water returns water to the chiller where it is chilled again. A hot water coil can be located in the AHU and works similar to the chilled water coil. Heat is conducted from the hot water supply water to the outside of the coil. Air passing across the coil tubes and fins is heated, which has the effect of cooling the supply hot water. This water is pumped back to the boiler where it is reheated and circulated back to the AHU heating coil. Filters in the AHU airstream keep the fan and motor clean and keep dust and dirt off the heating and cooling coils.

The AHU shown in Figure 8.1 is located on the roof and may be exposed as a weather-tight unit or may be enclosed in a penthouse. An AHU can be mounted on the roof or in a mechanical room located in the building, at floor level, or in a basement, as shown in Figures 7.13 and 7.14 in the previous chapter. AHUs take up a significant portion of the mechanical room. Rooftop equipment does not require a mechanical room, which improves the usable floor area to total floor area ratio. It can detract, however, from the aesthetics of the building exterior. In all cases, the OA and EA louvers (openings) must be far enough apart to avoid recirculation.

In a typical system, the AHU drives (blows) air into supply ducts that hang from the underside of the roof structure. SA ducts distribute air to the thermal zone where it is introduced (dis-

charged) into a space through diffusers or registers. A diffuser is an air outlet that discharges supply air in a spreading pattern. It is typically located on a wall. A grille is a louvered covering for an opening through which air passes. A register is a grille equipped with a damper and directs air in a nonspreading pattern.

The introduction of SA into the space counters the heating or cooling load and moisture levels in the space—that is, cool SA is introduced when the space needs to be cooled; warm SA is introduced when the space needs to be heated; dry air is introduced when air is too moist and needs to be dehumidified, and so on.

RA is removed from the space and flows through RA ductwork or through a plenum created by the ceiling and the roof or floor assembly above. It travels back to the AHU where some air is exhausted outdoors (the EA) and some is mixed with OA for ventilation. Mixing produces air with psychrometric conditions somewhere between that of the RA and OA. The mixed air is then conditioned (i.e., heated, cooled, dehumidified, or humidified) as it passes through the AHU. Finally, it is conveyed back through SA ducts and introduced as SA into the conditioned space. AHUs often have an economizer or inlet damper that allows for a small amount of OA or make-up air to be pulled in through the air handler. A fan or set of fans in the AHU drive airflow and dampers built into the AHU control airflow.

Small All-Air Systems

In residential and small commercial buildings, HVAC central air systems often consist of a packaged rooftop unit (RTU) with direct expansion (DX) cooling and/or heating capabilities,

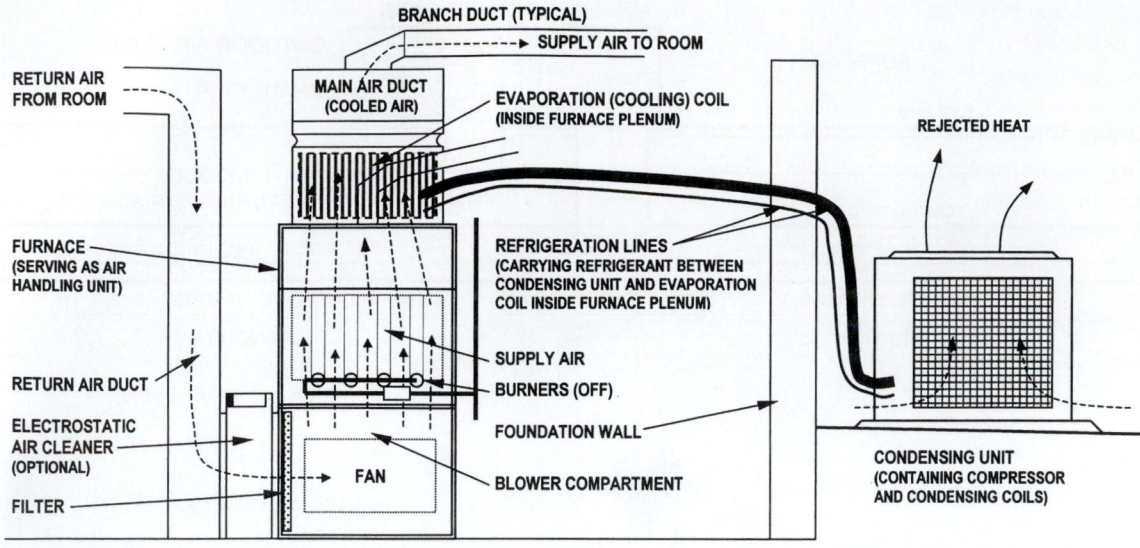

FIGURE 8.2 A split heating/air conditioning unit can function as a HVAC air system in a residence or small commercial building. In the heating mode, the furnace creates and blows warm air into the supply ductwork. In the cooling mode, a split air conditioning system has a condensing unit on the exterior to reject heat, and an evaporation coil inside the furnace plenum to absorb heat from the indoor air stream. The furnace fan blows air across the evaporation coil, to cool the air, and into the supply ductwork. Return air comes back into the blower compartment of the furnace to be reconditioned.

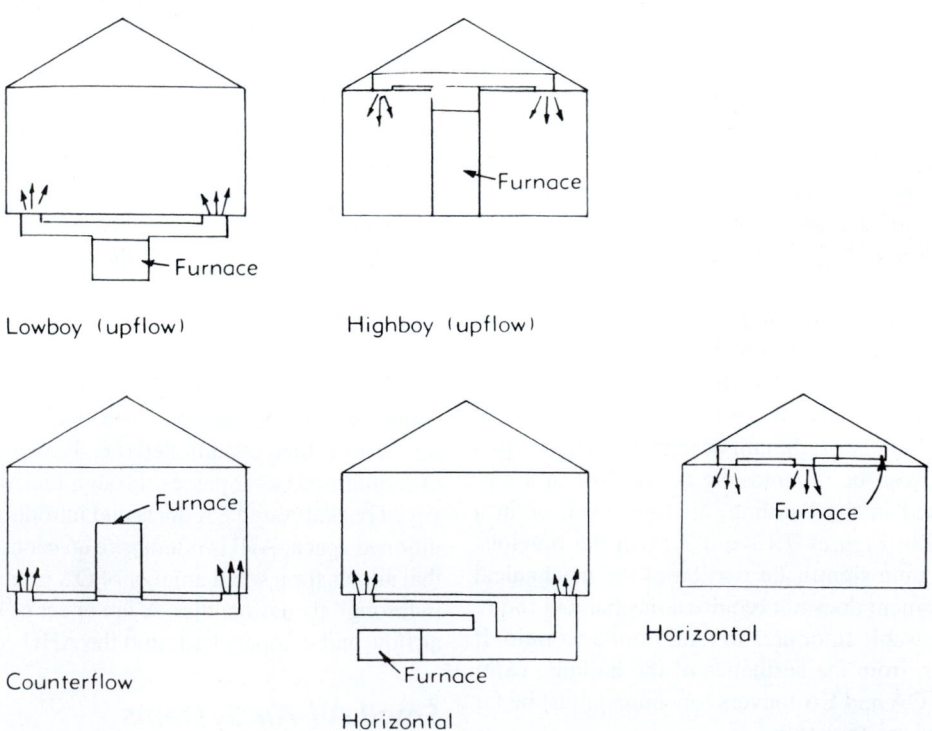

FIGURE 8.3 Furnace types and locations for heating and/or cooling (split heating/air conditioning unit). A furnace can heat and serve as an air-handling unit for cooling in a residence or small commercial building.

forced-air furnace and/or split air conditioning system, or a heat pump unit for heating and/or cooling. In light commercial installations, these systems are normally an electrical air conditioner/compressor system for cooling and/or a gas-fired heating system. Where gas is not available, an electric resistance heating system or heat pump is used. Packaged RTUs with DX cooling and heat pumps are found in a large fraction of the new light commercial installations. The most popular packaged DX system sizes are 5 and 10 tons, but units between 2 and 25 tons of cooling are used. Most residential installations include a forced-air furnace and/or split air conditioning system. These systems are typically 2 to 5 tons and 150 000 Btu/hr and less.

HVAC central air systems are typically located in the center of the building or the thermal zone they serve. In commercial buildings, they are frequently located on the roof, alongside the building, or in specifically designed mechanical rooms on each floor. In residences, this space may be the basement, crawlspace, attic, utility room, or garage.

In most small central air system installations, a ducted, fan-powered furnace, RTU, or heat pump serves as the AHU. The unit has a motor-driven fan that circulates filtered, conditioned air through supply ducts to each of the spaces (rooms) being conditioned. As the SA is delivered through the ducts and into the spaces through the supply outlets, stale air from the spaces is returned through RA grilles, into RA ducts, and back through the furnace, RTU, or heat pump to be reconditioned. It is then conveyed back to the space. Photographs of a residential forced air heating system are shown in Photos 8.1 through 8.18.

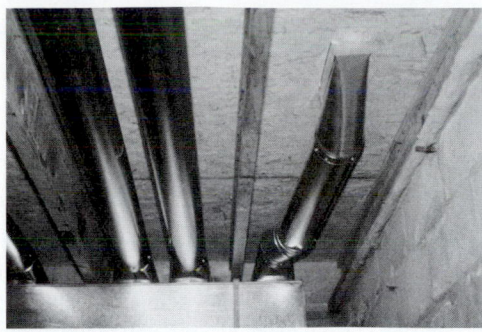

PHOTO 8.2 Branch ducts extend off the main duct to convey air to the heated spaces. (Used with permission of ABC)

PHOTO 8.3 A fitting called a floor diffuser box or floor register box conveys air to a floor register. (Used with permission of ABC)

PHOTO 8.4 The register box fitting conveys heated air through the subfloor to the heated space. (Used with permission of ABC)

PHOTO 8.1 An upflow furnace installation. The supply air plenum (above furnace) carries heated air to the rectangular main duct or trunk suspended from the ceiling. Return air recirculates back to the furnace in the rectangular return air duct, also suspended from the ceiling (in foreground). Main ducts are isolated from the furnace with a flexible duct connection that allows movement and isolates furnace vibration. Round ducts on the side of furnace introduce air for combustion. (Used with permission of ABC)

PHOTO 8.5 A floor register fits in the opening created for the register box fitting. Louvers and a damper of the register control airflow. (Used with permission of ABC)

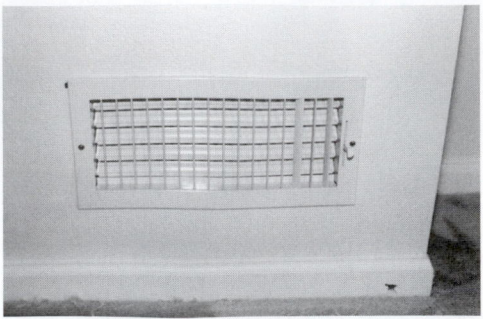

PHOTO 8.6 Conditioned air can be delivered into a room through a wall register. Louvers and a damper that are part of the register control airflow. (Used with permission of ABC)

PHOTO 8.7 A branch duct can deliver heated air to a ceiling register. (Used with permission of ABC)

PHOTO 8.8 A ceiling register. (Used with permission of ABC)

PHOTO 8.9 Ductwork must pass around obstacles. The design shown in this photograph is not effective because it has significant pressure loss. (Used with permission of ABC)

PHOTO 8.10 Room air returns back to the furnace through a return air grille. (Used with permission of ABC)

PHOTO 8.11 Return air can be conveyed through wall cavities created by construction materials. (Used with permission of ABC)

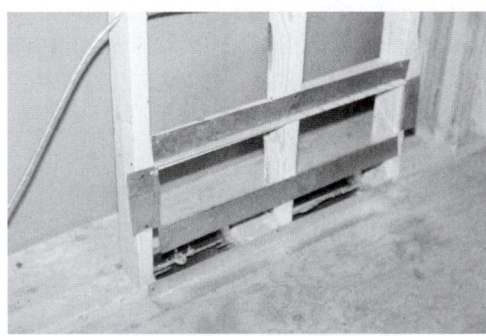

PHOTO 8.12 Openings in the subfloor allow return air to be conveyed to the main return air duct. (Used with permission of ABC)

PHOTO 8.13 The opening in the subfloor through which return air flows. Note that the floor decking has been removed. (Used with permission of ABC)

PHOTO 8.14   Sheet metal creates a return air duct between floor joist spacing. (Used with permission of ABC)

PHOTO 8.15   A return air duct created by sheet metal nailed to the bottom side of the floor joists. The cavity between the joists carries return air that then enters through openings cut in the top side of the return air duct. (Used with permission of ABC)

PHOTO 8.16   An individual supply duct system. (Courtesy of NREL/DOE)

PHOTO 8.17   A furnace serves as an air-handling unit in many split-system air conditioning installations. In this installation, the absorption (cooling) coil is mounted in a plenum above the furnace. Refrigeration lines extend out of the furnace plenum (just above and to the right of the furnace) and are connected to the condensing unit located outdoors. (Used with permission of ABC)

PHOTO 8.18   A pad-mounted condensing unit. Note the refrigeration lines connecting to the back side of the unit and the electrical box feeding power to the unit. (Used with permission of ABC)

8.3   DUCTWORK

Ducts

As introduced in the previous chapter, a *duct* is a sheet metal (galvanized steel, aluminum, or stainless steel) or plastic tube through which air is moved from one location to another. In both large and small air systems, air is distributed in a building through connected individual duct sections that make up the duct system. Selection of duct material is based on price, performance, and installation requirements. Galvanized steel sheet metal is the most common duct material and can be used on most all supply and return duct applications (for plenums, trunks, branches, and runouts). Sheet metal ducts have a smooth interior surface that offers the least resistance to air flow. Integrated duct designs that use the building structure or framing (e.g., building cavities, closets, raised-floor air-handler plenums, platform returns, wall stud spaces, panned floor joists) as supply or return ducts are also common, but are difficult to seal or control. When located outdoors or in an unconditioned space, ducts must be insulated.

Ducts are traditionally fabricated in circular and rectangular shapes. Round ducts are available in actual diameters, usually in 2 in (50 mm) increments. Square and rectangular ductwork is typically custom built to a specific size, usually in 2 in (50 mm) increments. Flat oval ducts are not truly ellipse shaped, but are much like a rectangular shape with semicircular sides. Flat oval ducts combine the advantages of both round ducts and rectangular ducts; mainly they can fit in spaces where there is not enough vertical space for round ducts. As a result, they are fast becoming a substitute for rectangular ducts. One disadvantage is that fittings for flat oval ducts are difficult for a contractor to fabricate or modify in the field. Most round and flat oval ducting is spirally wound sheet metal (i.e., a narrow strip or metal is formed into an airtight spiral). Flexible ducting, with a flexible sheath around a wire helix, is common in the range of 4 to 16 in (100 to 400 mm) diameter. In some cases, hollow construction cavities and plenums between ceiling and floor are used as a duct.

Fittings, such as tees, crosses, or transitions, connect duct sections. A variety of fittings may be used to make all of the reductions, branch takeoffs, turns, and bends required in many duct systems. Figure 8.4 shows different types of duct fittings. When more than one type of fitting can be used, the fitting with a lower pressure loss should be used.

Duct type and size depends on the amount of air that must flow through them to maintain the desired temperature of the room. Round ducts are generally the most cost-effective in comparison to rectangular ducts. They use less material and have a smaller pressure loss per unit area. Round spiral ducts can be made to almost any length. In contrast, rectangular ducts allow more headroom, saving significantly on the floor-to-floor distance and the cost of extra height in the building (for example, a 26 in × 8 in rectangular duct can be substituted for a 16 in diameter round duct but will save one 8 in high course of concrete masonry and 2 or 3 courses of brick). Rectangular ducts are more easily fabricated and easily shipped when bro-

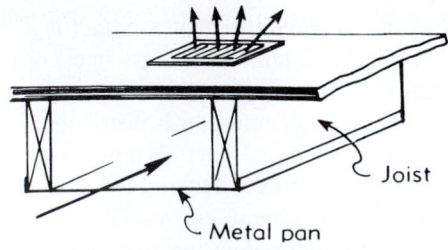

FIGURE 8.4 Joist cavities used as ducts.

ken down into half sections that fit into one another. The flat surfaces of rectangular duct make for easier tap-in connections. Disadvantages to using rectangular ducts over other duct shapes are as follows: they create a higher pressure loss; they use more material for the same airflow rate; length is limited to the sheet metal widths stocked by the contractor; and joints tend to be more difficult to seal.

Duct Layout

The basic conventional duct arrangement can be compared to a tree: the AHU or central air system at the base of the tree, a main duct or trunk duct, and duct branches connected to terminal units, usually diffusers and registers. The roots of the tree are the return air ducts. The AHU/central air system is in between the supply and return/outside air parts of the system. Basically, there is a supply side of the system and a return side, with the AHU/central air system in between.

Downstream of the AHU, the supply air *trunk duct* will commonly fork, providing air to many individual air outlets such as diffusers, grilles, and registers. When the system is designed with a main duct branching into many subsidiary branch ducts, fittings called *takeoffs* allow a small portion of the flow in the main duct to be diverted into each branch duct. Takeoffs may be fitted into round or rectangular openings cut into the wall of the main duct. The takeoff commonly has many small metal tabs that are then bent to retain the takeoff on the main duct; round versions are called *spin-in* fittings. The outlet of the takeoff then connects to the rectangular, oval, or round branch duct.

A variety of duct arrangements, or basic designs, may be used; several of the most common are shown in Figure 8.5. These systems are discussed in the following sections:

Extended Plenum Arrangement

In the *extended plenum* or *trunk and branch* arrangement, a main supply *trunk* is connected directly to the air handler or its supply plenum and serves as a supply plenum or an extension to the supply plenum. Smaller *branch* ducts and runouts are connected to the trunk. It provides airflows that are easily balanced and can be easily designed to be located inside the conditioned space of the building.

There are several variations of the trunk and branch system. An extended plenum system uses a main supply trunk that is one size and is the simplest and most popular design. Based on AHU fan size, the length of the trunk is usually limited because the velocity of the air in the trunk gets too low and airflow into branches and runouts close to the AHU becomes poor. This design is most popular in residential installations, where a large supply trunk and slightly smaller return air trunk are placed side by side. Branches and runouts are connected to the trunk.

Reducing Plenum Arrangement

A *reducing plenum (trunk)* arrangement is a modified extended plenum system that uses intermittent trunk reductions at intervals

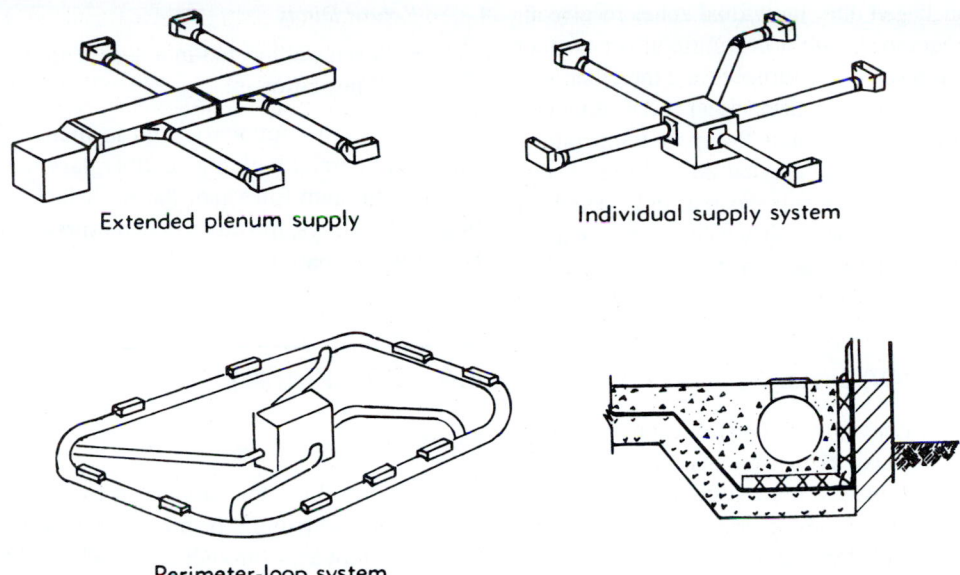

Extended plenum supply

Individual supply system

Perimeter-loop system

FIGURE 8.5 Typical duct systems.

in the trunk—that is, the trunk cross-section gets smaller as the trunk gets farther away from the AHU. It is also known as a *reducing trunk and branch system*. This design maintains a more uniform pressure and air velocity in the trunk and improves airflow in branches and runouts closer to the AHU. Because the trunk of a reducing trunk system decreases in cross-sectional area after every branch duct or runout, it is a more complex system to design. It consists of sections of straight duct connected to reducing transitions that are placed between the straight sections at intervals of about every 20 to 30 ft. A branch duct or runout is tapped into the trunk as necessary. As a result, this system is used in larger systems.

Radial Arrangement

In a *radial system* or *individual supply* arrangement, there is no main supply trunk; branch ducts or runouts that deliver conditioned air to individual supply outlets are essentially connected directly to the AHU, usually using a small supply plenum. The short, direct duct runs maximize airflow. The radial system is most adaptable to single-story buildings. Traditionally, this system is associated with an AHU that is centrally located so that ducts are arranged in a radial pattern. Symmetry is not necessarily mandatory, and designs using parallel runouts can be created so that duct runs remain in the conditioned space (i.e., installed above a suspended ceiling).

Perimeter Loop Arrangement

A *perimeter loop* arrangement uses a perimeter duct fed from a central supply plenum with several feeder ducts. This system is typically limited to buildings constructed on slab in cold climates.

Ideally, the best system layout has the fewest fittings and simplest duct configuration, because fittings restrict airflow and increases the pressure loss in the system. The pressure loss of the various fittings used must be included in the duct size calculations.

Basic Systems

Although there are many system options, what follows are the basic types:

Single-Zone, Constant Air Volume Systems

Single-zone, constant air volume systems are the most commonly installed air distribution system. A single thermostat located in a central location controls the conditioned air delivered to the zone; a fan delivers a constant volume of heated or cooled air (not a mixture of the two) to every space when a thermostat calls for conditioned air. Some constant-volume, single-zone systems (i.e., packaged RTUs) can easily integrate air-side economizers to take advantage of "free" cooling, but usually there is no reheat in these systems for temperature control. Fan energy for most constant volume, single-zone systems is generally high. These systems can cause some discomfort under some common conditions. These are the usual systems installed in residential and light commercial construction.

Constant Air Volume, Multizone Systems

Constant air volume, multizone systems are designed to provide simultaneous heating and cooling to multiple temperature zones from a single packaged unit. Individual zones receive air from two ducts, one equipped with a heating coil (or another heating method, such as a small gas furnace) and the other with a cooling coil. The thermostat energizes either the heating coil or the cooling coil as needed to maintain the space temperature. These systems, which were widely applied 20 to 30 years ago, are very inefficient because they mix heated and cooled air streams for space temperature control. They can be very expensive because of the need for two duct systems, each sized to meet the total load.

Variable Air Volume (VAV) Systems

Variable air volume (VAV) systems can be very efficient when well designed and carefully operated. These systems vary the amount of air supplied to a space. Typically, these systems are more efficient for the following reasons: (1) less energy is used for cooling because the volume of air cooled is reduced; (2) less energy is required for heating because less air needs to be reheated; and (3) fan energy is reduced because the amount of air the fan needs to move is reduced. These systems generally employ economizers. VAV, variable temperature systems have been applied to packaged equipment to overcome the constant-volume, single-zone system's lack of zone control. A specific area is served by a single unit but the area is divided into temperature control zones and each zone has a thermostat. Each thermostat operates an automatic control damper located in the branch duct that serves the zone. In the cooling mode, if the zone is too cold, the damper is closed. If the zone is too warm, the damper is opened. Just the opposite occurs in the heating mode. All the thermostats are connected to a central control panel that determines which zone demands the most heating or cooling. All zones must have comparable loads, so some are not calling for heating while others require cooling.

Duct Insulation and Sealing

Recent studies of existing ductwork suggest that sealing and insulating ducts can improve the efficiency of building heating and cooling systems by up to 30%. Duct sealant (mastic) or metal-backed (foil) tape is used to seal the seams and connections of ducts. Sealing is typically done before the duct is insulated.

Ducts that run through enclosed, unconditioned spaces or in spaces that are exposed to outdoor temperatures should be insulated. Ducts that are located in conditioned spaces do not need to be insulated. Although glass fiber ducts are made of an insulating material (the glass fiber), sheet metal ducts must be wrapped in insulation. Thicknesses of insulation recommended follow.

Supply Ducts
- When located in enclosed, unheated spaces, 1 in (25 mm) of insulation.
- When located in a space exposed to outdoor temperatures, 2 in (50 mm) of insulation.

Return Ducts
- When not located in a heated space, use 1 in (25 mm) of insulation.

The most commonly used insulation on residential and small commercial buildings is fiberglass with a facing of reinforced aluminum foil vapor-barrier, which goes to the outside. In large buildings, the ducts may be sprayed with an insulating/fire-resistant coating.

8.4 EQUIPMENT

Fans

A fan is a machine for moving air by means of a rotating impeller using centrifugal or propeller action. Fan energy has to be used to move air through ductwork. Two basic types of fans are used in HVAC applications: axial and centrifugal. These are classified according to the direction of the airflow through the impeller. Types of fans were introduced in Chapter 7.

Some HVAC systems will have separate supply air fans and return air fans while other systems will use a single fan to both supply and return air to an occupied space. The supply fan is normally located downstream of the cooling coils; however it can be located anywhere inside the AHU. Fans should be isolated from ductwork to prevent vibration noise. This is accomplished by connecting the fan inlets and outlets to the ductwork with a flexible material.

With respect to airflow in a duct, friction causes static pressure loss. Static pressure is a measure of a fan's ability to push air against the friction created in the duct system. The total pressure produced by a fan is made up of the static pressure (the useful working pressure available for overcoming the resistance of a ventilating system) and velocity pressure (the pressure from the speed of the air). Fan pressures are of a very low order, so using small units (i.e., inches of water column and Pascals) is warranted.

A fan is selected on these two pressures. But for low-velocity systems (velocity of air below 2000 fpm) the fan is selected on only static pressure, while for high-velocity systems (velocity of air greater than 2000 fpm) the fan will be selected on total pressure. Additionally, some fans (i.e., centrifugal fan) are constant volume devices. This means that for a constant fan speed, the fan will move a constant volume of air, which is not the same as moving a constant mass of air. The air velocity and the volumetric flow rate in a system is the same even though mass flow rate through the fan is not. This design needs to compensate for this effect.

Fans in an AHU can create substantial noise and vibration. The duct system will transmit this noise and vibration to the spaces in the building, which can irritate occupants. To avoid this condition, *vibration isolators* (rubberized, canvas-like, flexible sections) are usually located in the ductwork immediately before and after the AHU. Vibration isolators allow the AHU to vibrate without transmitting much vibration to the attached ducts.

Basic Equipment

The heating and cooling equipment should be selected based on availability of different fuels (e.g., natural gas, electricity), installation costs, occupant preferences, operating costs, and the general type of air distribution system (supply and return duct systems). The general designs and duct materials for the supply and return duct systems should be selected after considering the type of equipment selected, its location in the building, local climate, architectural and structural features of the building, zoning requirements, and installation and operating costs.

Dampers

Ducting systems must have a method of adjusting the volume of airflow to various parts of the system. Balancing and flow-control dampers provide this function. Dampers are normally fitted within the ducts themselves and can be manual or automatic. Supply ducts should be equipped with a locking-type, *adjustable damper* (Figure 8.6) so that the air volume can be controlled and the system balanced. The damper should be located in an accessible location in the branch duct as far from the supply outlet as possible. This allows a measure of control over the flow of the air and also allows a branch to be shut off when desired. There are

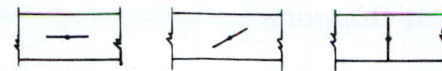

FIGURE 8.8 Squeeze dampers.

two common types of control dampers. *Splitter dampers* (Figure 8.7) are used to direct part of the air into the branch where it is taken off the trunk. They do not give precise volume control. *Squeeze dampers* (Figure 8.8) are placed in a duct to provide a means to control the air volume in the duct.

Air Terminals and Terminal Units

Air terminals are supply air outlets and return or exhaust air inlets. For the supply, diffusers are most common, but grilles, and for very small HVAC systems such as in residences, registers are also used widely. Return or exhaust grilles are used primarily for appearance reasons. Single-zone constant-air volume systems typically use them. Other types of air distribution systems often use terminal units in the branch ducts. Usually there is one terminal unit per thermal zone. Some types of terminal units are VAV boxes of either single- or dual-duct, fan-powered mixing boxes of either parallel or series arrangement, and induction terminal units. Terminal units may also include either a heating or cooling coil, or both.

8.5 BASIC SYSTEM HYDRAULICS

Aspect Ratio

Aspect ratio is the relationship between the width (w) and height (h) of a duct, expressed as a ratio: w/h. A square duct has an aspect ratio of 1 to 1, or simply 1.0; a rectangular duct has an aspect ratio greater than 1.0. For example, a 24 in × 8 in duct has an aspect ratio of 3 to 1, or 3.0. The aspect ratio is a way to classify the cost and airflow effectiveness of the shape and perimeter of a rectangular duct.

Effect of Duct Shape

The duct shape that is the most efficient (offers the least resistance) in conveying moving air is a round duct, because it has the greatest cross-sectional opening area and a minimum contact surface. Rectangular and square duct sections have a greater contact surface that allows air turbulence in the corners, which creates greater resistance to flow. A rectangular duct section with an aspect ratio close to 1 yields the most efficient rectangular duct shape in terms of conveying air. A duct with an aspect ratio above 4 is much less efficient in use of material and experiences great pressure losses. Aspect ratios of 2 to 3 are ideal in trading off added duct cost of material and fan energy for headroom savings.

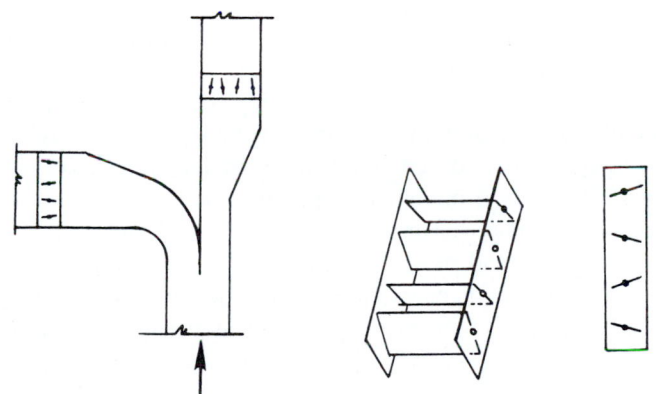

FIGURE 8.6 Adjustable dampers.

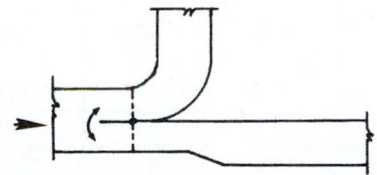

FIGURE 8.7 Splitter damper.

Equivalent Diameter

The *equivalent diameter (D_e)* of a rectangular duct is the diameter of a round duct that has the same pressure loss per unit length with other airflow parameters being the same (i.e., velocity, wall surface roughness, air density). For any rectangular duct there is a round duct that will have the same amount of friction loss. Round ducts are more efficient in conveying air, in contrast to rectangular ducts. For example, for a specific airflow, the friction loss in a 30 in diameter round duct is less than a 35 × 20 rectangular duct, even though the rectangular duct has approximately the same cross-sectional area (area of 700 in²) as the 30 in round duct (707 in²). As a result, the cross-sectional area of a rectangular duct is always larger than a round duct with the same equivalent diameter.

The equivalent diameter of a rectangular duct is found by the following formula, where w and h are the rectangular width (w) and height (h) of duct dimensions.

Rectangular ducts: $D_e = 1.30 (wh)^{0.625}/(w + h)^{0.250}$

The equivalent diameter of a flat oval duct is found by the following formula, where A is the cross-sectional area of the flat oval cross section, and P is the flat oval perimeter.

Flat oval ducts: $D_e = 1.55A^{0.625}/P^{0.250}$

Rectangular ducts with the same equivalent diameter will have different cross-section areas.

System Velocities

HVAC systems are defined as *low velocity* (low pressure) with operating velocities up to 2000 ft/min (10 m/s), *medium velocity* (medium pressure) 2000 to 4000 ft/min (10 to 20 m/s), and *high velocity* (high pressure), which use velocities between 4000 and 8000 ft/min (20 to 40 m/s). High velocities relate to high-velocity pressures and higher pressure losses through fittings. Low-velocity systems are used in most HVAC installations because the low pressures used allow lightweight duct construction, lower fan power, higher air leakage rates, and lower noise levels. High-velocity systems tend to be used on very large projects where the increased cost

TABLE 8.1   TYPICAL FACE VELOCITIES OF EQUIPMENT AND COMPONENTS.

Installation	Velocities	
	ft/min	m/s
Air filter	500	2.5
Air-handling unit	500	2.5
Cooling coil	700	3.6
Heating coil	1800	9.2

of noise reduction (use of noise silencers) and fan power are offset by the reduction in duct dimensions and duct space with their added costs.

Traditional design for office, school, and similar buildings bases the size of the AHU on a face velocity of 500 ft/min (2.5 m/s) at the coil face, which is based on a balance between the first cost and the lifetime energy cost of the equipment. Other building occupancies will find a lower face velocity more cost-effective. Typical face velocities of equipment and components are provided in Table 8.1.

Duct Velocities

Maximum air velocity in the ducts should be kept below certain limits to maintain a reasonable pressure loss. Velocities in main trunks and major duct branches in most air-handling systems are in the 1000 to 3000 ft/min range. The velocities in the smaller distribution ducts serving small terminal equipment and diffusers are often below about 800 fpm. Table 8.2 provides recommended maximum duct velocities.

The sound caused by air flowing through a duct is directly proportional to the velocity of the air passing through it. Velocity can be reduced significantly by increasing duct size (see Table 8.3). If minimizing system noise is important (e.g., television studio, theatre complex), velocities should be reduced by 15% and 20%.

Automatic dampers must be designed to operate so they close gradually. A sudden damper closure can generate air

TABLE 8.2   RECOMMENDED MAXIMUM DUCT VELOCITIES.

Occupancy	Main Ducts				Branch Ducts			
	Supply		Return		Supply		Return	
	ft/min	m/s	ft/min	m/s	ft/min	m/s	ft/min	m/s
Residences	1000	4.0	800	3.2	600	2.4	600	2.4
Multifamily dwellings	1500	6.0	1300	5.2	1200	4.8	1000	4.0
Businesses (offices, stores, banks)	2000	8.0	1500	6.0	1600	6.4	1200	4.8
Manufacturing plants	3000	12.0	1800	7.2	2200	8.8	1500	6.0

TABLE 8.3   AIR VELOCITY (FT/MIN) IN SELECTED SIZES OF ROUND DUCT SIZES FOR DIFFERENT VOLUMETRIC FLOW RATES.

Air Flow Rate (ft³/min)	Round Duct Size (inches)						
	4	5	6	8	10	12	16
100	1146	733	509	286	183		
200		1467	1019	573	367	255	
400			2037	1146	733	509	286
800					1467	1019	573

hammer effects that can explode or implode ductwork. For example, a typical commercial AHU can move approximately 20 000 ft/min (about 1600 lb/min) of air. In a commercial duct system, this air is moving at a velocity of 1500 to 2000 fpm or more, which correlates to a speed of about 20 mph. A spring-loaded fire damper or pneumatically actuated smoke damper closes quickly (often in seconds). Such a sudden stoppage of the moving air can generate huge pressures, which can damage the damper and ductwork. Systems need to be designed to minimize this potential for damage.

Pressure Losses in Fittings

Air flowing through fittings, and in particular elbows or turns, can become very turbulent because of the chaotic change of direction. This can result in large pressure loss and undesirable vibration and noise. The airstream undergoes change of either direction or velocity, as velocity pressure loss occurs. Unlike friction losses in straight duct, fitting losses are from internal turbulence rather than skin friction. Duct fitting loss is a function of the velocity pressure in the duct, and that duct velocity pressure is a function of the square of the flow rate. In practical terms, it means that the pressure losses associated with a poor fitting in a low velocity duct system will be much less than the losses associated with the same fitting in a higher velocity duct system.

Turning vanes are added to large elbows and turns to cut down on chaotic flow, direct the flow of air smoothly around a corner or a bend, and thus reduce static pressure loss. See Figure 8.9. Transition fittings (i.e., fittings with an increase or decrease in cross section) should gradually change in cross-section to minimize pressure loss. Converging transition fittings should have a taper that does not exceed a 20° deviation from the duct side. Diverging transition pieces (i.e., those that increase in cross-section) that have a taper that exceeds 10° should have splitters fitted within the duct to reduce the expansion angle to less than 10°.

Improperly applied and supported flexible duct can cause numerous operating problems. The sagging duct between supports that are spaced too far apart in effect adds close-coupled elbows to the system. Sagging aggravates the loss coefficient associated with the flexible duct itself, which is often high relative to an equivalent diameter sheet metal duct because of the rougher interior surface conditions. It is also easy to make a very sharp turn with the flexible duct, which may solve a space problem but can result in a very high pressure loss at the turn. If it is used, the duct should be supported to minimize sagging and any turns should have a gradual radius to minimize bend losses. Where long runs or sharp turns are unavoidable, increasing duct size is effective in reducing pressure loss. Generally, flexible duct runs should not exceed 10 ft (3 m) nor contain more than two bends.

Supply Registers and Diffusers

From a performance design perspective, a supply register or diffuser is sized based on airflow rate (ft³/min or L/s), face velocity, throw, and pressure loss. The *face velocity* (ft/min or m/s) of air emerging from the outlet is measured in the plane of the outlet in fpm (m/s). *Throw* is the distance (ft or m) that the airstream travels from the outlet to a point where the terminal velocity occurs. The *terminal velocity* is the measured velocity at the point at which the discharged air from the outlet decreases to 50 ft/min. *Pressure loss* (in w.c. or Pa) is the amount of total pressure required to move air through the diffuser. The *core area* (ft² or m²) is the actual "open" area measured perpendicular to the fins of a diffuser, grille, or register through which air passes. Design data for round ceiling diffusers, floor registers, and return air grilles are provided in Tables 8.4 through 8.6. Other design criteria include size limitations, noise criteria, aesthetics, and cost. The topic of noise criteria (NC) is addressed in Chapter 23.

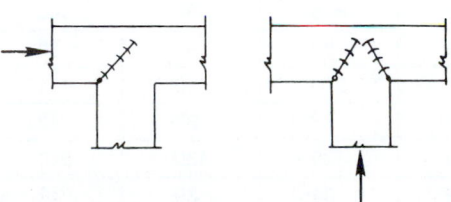

FIGURE 8.9   Turning vanes.

TABLE 8.4   DESIGN DATA FOR ROUND CEILING DIFFUSERS.

Face Velocity, ft/min		300	400	500	600	700	800	900	1000
Pressure Loss, in w.c.		0.006	0.010	0.016	0.022	0.031	0.040	0.050	0.062
6 in diameter	ft³/min	—	55	65	80	95	105	120	135
Area: 0.135 ft²	Throw	—	2.5	3.0	3.5	4.0	4.5	5.0	5.5
8 in diameter	ft³/min	70	90	115	135	160	180	200	225
Area: 0.225 ft²	Throw	2.0	3.0	3.5	4.5	5.0	5.5	6.5	7.0
10 in diameter	ft³/min	105	140	175	210	240	275	310	345
Area: 0.345 ft²	Throw	2.5	3.5	4.5	5.0	6.0	7.0	8.0	8.5
12 in diameter	ft³/min	150	200	250	300	350	400	450	500
area: 0.500 ft²	Throw	3.0	4.0	5.0	6.0	7.5	8.5	9.0	10.5
14 in diameter	ft³/min	190	250	315	375	440	500	565	625
Area: 0.625 ft²	Throw	3.5	4.5	5.5	6.5	8.0	9.0	10.0	11
18 in diameter	ft³/min	310	415	520	625	730	830	935	1040
Area: 1.04 ft²	Throw	4.5	6.0	7.0	8.5	10.0	11.5	13.0	14.5

Throw is the radial throw, ft. Terminal velocity of 50 ft/min.

TABLE 8.5   DESIGN DATA FOR A FLOOR REGISTER.

Face Velocity, ft/min		300	400	500	600	700	800	900	1000
Pressure Loss, in w.c.		0.006	0.010	0.016	0.022	0.031	0.040	0.050	0.062
4 × 10	ft³/min	60	80	100	120	140	160	180	200
Area: 0.20 ft²	throw	8	11	13	16	19	22	24	26
4 × 12 and 6 × 8	ft³/min	75	100	125	150	175	200	225	250
Area: 0.25 ft²	throw	9	12	15	18	21	25	28	31
4 × 14	ft³/min	85	110	140	170	195	225	250	280
Area: 0.28 ft²	throw	6	10	15	19	22	25	28	32
6 × 10 and 8 × 8	ft³/min	95	130	160	190	225	255	290	320
Area: 0.32 ft²	throw	10	14	17	21	24	27	31	34
6 × 12	ft³/min	115	150	190	230	265	305	340	380
Area: 0.38 ft²	throw	12	15	19	23	26	30	34	38
8 × 10	ft³/min	125	170	210	250	295	335	380	420
Area: 0.42 ft²	throw	12	15	18	24	28	32	37	41
6 × 14	ft³/min	130	175	220	265	310	350	395	440
Area: 0.44 ft²	throw	13	16	20	24	28	32	36	41
6 × 16	ft³/min	145	195	245	295	345	390	440	490
Area: 0.49 ft²	throw	11	16	21	25	30	35	39	44
8 × 12 and 10 × 10	ft³/min	160	210	265	320	370	425	475	530
Area: 0.53 ft²	throw	12	17	21	26	32	37	43	48
9 × 12 and 8 × 14	ft³/min	180	240	300	360	420	480	540	600
Area: 0.60 ft²	throw	14	20	24	29	34	39	44	50

Throw is the throw, ft. Terminal velocity of 50 ft/min.

TABLE 8.6   DESIGN DATA FOR A RETURN AIR GRILLE.

Application	Maximum Recommended Velocity	
	ft/min	m/s
Return air grille for low noise level	250	1.25
Return air grille at low level	300	1.50
Fresh air grilles	300	1.50
Discharge louvers	300	1.50
Return air grille at high level	350	1.75
Wall registers	500	2.50

Example 8.1

A round ceiling diffuser must deliver 250 ft³/min. Identify diffusers that can deliver this airflow.

From Table 8.4, the following diffusers that can deliver 250 ft³/min:

- 14 in diameter (face velocity of 400 ft/min, 4.5 ft throw, 0.0010 in w.c. pressure loss).
- 12 in diameter (face velocity of 500 ft/min, 5.0 ft throw, 0.0016 in w.c. pressure loss).
- 10 in diameter (face velocity of 729 ft/min, 6.3 ft throw, 0.0034 in w.c. pressure loss).*

Any of the diffusers selected in Example 8.1 would be suitable if throw requirements are acceptable. In some "quiet" spaces, a lower velocity is recommended to minimize airflow noise (see Table 8.7). The 10 in diameter diffuser has a face velocity of 729 ft/min, which may be considered high in some spaces (e.g., churches, residences, apartments, television studios) because of concerns with noise. The sound caused by a supply register or diffuser in operation is directly proportional to the velocity of the air passing through it. By selecting outlets of proper sizes, air velocities can be controlled within safe sound limits. Additionally, flexible ducts may be used for low-pressure ductwork downstream of the terminal box or duct outlet, which serves to attenuate duct noise.

8.6   DESIGNING A DUCT SYSTEM

The main goal in designing HVAC ductwork is to select the smallest (and lowest cost) duct sizes that can be used without producing a large total pressure loss necessitating a powerful (and noisy) fan and without generating excessive noise from high velocities. Additionally, the duct system should be routed to minimize the individual lengths of the various duct runs. This design approach minimizes material costs and creates a naturally balanced system where the static pressure and air velocities available at each diffuser are as similar as possible.

Pressure balancing is an important component in the design of a duct system. The system pressure is balanced when the fan pressure is equal to the sum of the pressure losses through each section of a branch. This is true for each system branch. A different interpretation of pressure balancing is that pressure losses need to be balanced at each junction. If the sum of the pressure losses in a branch does not equal the fan pressure, the duct system will automatically redistribute air, which will result in airflows different from those designed. Designing a duct system involves sizing the ducts and selecting the fittings.

Most HVAC systems are designed to maintain a slightly higher pressure than outdoor atmospheric conditions, which creates a positive pressure in indoor spaces. The higher indoor pressure reduces infiltration and prevents unconditioned and untreated outdoor air from entering the space directly. In indoor spaces where toxic or hazardous contaminants exist or objectionable gases are produced (i.e., restrooms, swimming pools, laboratories, and hospital [quarantine] rooms), they must be contained within the space so they do not escape into neighboring spaces. In these spaces a slightly lower pressure than the surrounding indoor spaces (a negative pressure) should be maintained to prevent diffusion of these contaminants to the surrounding area. In some of these spaces, this can be accomplished by a separate exhaust system (e.g., an exhaust fan in a restroom or draft hood in a laboratory). In other systems, this is accomplished through precise balancing of the air distribution system.

Straight Duct Pressure Losses

Pressure losses in straight ducts are related to volumetric flow rate (ft³/min or L/s), average air velocity (ft/min or m/s), duct size and shape, duct surface roughness, and duct length. Testing has led to a variety of pressure loss charts for ducts of many different materials. Table 8.8 shows pressure loss in selected sizes of round duct sizes for different volumetric flow rates. Note that for a specific volumetric flow rate, pressure loss decreases as duct size increases.

TABLE 8.7   RECOMMENDED MAXIMUM DIFFUSER AND REGISTER FACE VELOCITIES TO CONTROL NOISE.

Occupancy	Face Velocity	
	ft/min	m/s
Television studios	500	2.5
Residences, multifamily dwellings, churches, hotel bedrooms	500 to 750	2.5 to 3.8
Offices (private)	500 to 1000	2.5 to 3.8
Theaters, auditoriums	1000 to 1250	5.1 to 6.4
Offices (public), stores	1200 to 1500	6.1 to 7.6
Manufacturing plants	1500 to 2000	7.6 to 10.2

*Based on 250 ft³/min. Found through interpolation between 240 and 275 ft³/min.

TABLE 8.8   PRESSURE LOSS (IN W.C.) PER 100 FEET OF DUCT IN SELECTED SIZES OF
ROUND DUCT SIZES FOR DIFFERENT VOLUMETRIC FLOW RATES.

| Air Flow Rate (ft³/min) | Pressure Loss (in w.c. per 100 feet of duct) | | | | | | |
| | Round Duct Size (in) | | | | | | |
	4	5	6	8	10	12	16
100	0.65	0.21	0.09	0.02	0.01		
200		0.8	0.32	0.08	0.02	0.01	
400			1.19	0.28	0.09	0.04	0.01
800					0.34	0.14	0.03

Pressure loss charts are provided in Figures 8.16 (at the end of the chapter on pages 309–310). A review of the pressure loss chart in Figure 8.16 for air in a smooth (i.e., galvanized steel) duct shows pressure loss related to flow rate, velocity, and equivalent diameter of the duct. The volumetric flow rate or simply "volume" in ft³/min (L/s) is plotted horizontally along the bottom of the chart. Friction loss in inches of water (in w.c.) per 100 feet of straight duct is plotted vertically along left side of chart. Duct sizes, expressed as equivalent diameters (in or mm), are shown on the diagonal lines sloping upward from left to right. Air velocity in ft/min is shown along the diagonal lines sloping downward from left to right. Table 8.41 (at the back of this chapter–see pages 300–303) can be used to convert an equivalent diameter to rectangular duct sizes. Pressure loss with change in air density occurs in a manner related to the altitude correction factor. See Chapter 7.

Example 8.2

A 12 in diameter galvanized steel (smooth) straight duct section is 22 ft long and carries air at 700 ft³/min.

a. Determine the pressure loss.

From Figure 8.16 (at the end of the chapter on pages 309–310), the pressure loss is found as 0.1 in w.c. per 100 ft of duct. Pressure loss for a 22 ft long duct section is found by:

0.1 in w.c. per 100 ft · (22 ft/100 ft) = 0.0022 in w.c.

b. Determine the average velocity of air flowing through the duct.

From Figure 8.16, velocity is 900 ft/min

c. Find duct section with a depth of 8 in.

From Table 8.41 (at end of chapter on pp. 300–303), for an equivalent diameter of 12 in, a 14 × 8 rectangular duct is selected.

Table 8.41 shows the rectangular duct size for an equivalent diameter of round duct for a specific flow rate, based on

a pressure loss of 0.1 in w.c. Duct size increases with an increase in volumetric flow rate. The designer has many options of rectangular duct size for a specific equivalent diameter.

Fitting Pressure Losses

Fitting pressure losses can be expressed in many ways: as an equivalent length of straight duct (e.g., in ft or m); or as a fraction of velocity pressure; or directly in inches of water column (in w.c.). To simplify our discussion, the equivalent length of straight duct method will be used.

Example 8.3

Select the friction loss as equivalent length of straight duct (ft) for the following fittings:

a. A 90° elbow of round duct.

A 90° elbow of round duct is listed in Figure 8.15 (pp. 305–308) as E-10
Friction loss of the fitting is the equivalent length of 10 ft (3.0 m) of straight duct.

b. Four 90° elbows of round duct connected together.

From a. above, the friction loss of one 90° elbow of round duct is the equivalent length of 10 ft (3.0 m) of straight duct. For four elbows it is found by:

4 · 10 ft = 40 ft equivalent length of straight duct

c. Boot Fitting A (the piece used to connect the branch duct to the supply register).

The friction loss of Boot Fitting A is listed in Figure 8.15 (pp. 305–308) as A-30.
Friction loss of the fitting is the equivalent length of 30 ft (9.1 m) of straight duct.

Sizing Ductwork

There are a number of accepted methods used in duct design (i.e., static regain, equal friction, or constant velocity methods). In

TABLE 8.9 RECTANGULAR DUCT SIZE FOR AN EQUIVALENT DIAMETER OF ROUND DUCT FOR A SPECIFIC FLOW RATE, BASED ON A PRESSURE LOSS OF 0.1 IN W.C.

Volumetric Flow Rate (ft³/min)	Equivalent Diameter (in)	Rectangular Duct Height (in)				
		4	6	8	10	12
50	5	6 × 4				
75	6	6 × 4				
100	6	8 × 4	6 × 6			
125	7	10 × 4	6 × 6			
150	7	10 × 4	8 × 6			
175	8	12 × 4	8 × 6			
200	8	14 × 4	8 × 6			
225	8	16 × 4	10 × 6			
250	9	16 × 4	10 × 6			
275	9		12 × 6	8 × 8		
300	9		12 × 6	8 × 8		
400	10		14 × 6	10 × 8		
500	11		18 × 6	12 × 8	10 × 10	
600	12		20 × 6	14 × 8	12 × 10	
700	12		24 × 6	16 × 8	12 × 10	
800	13		26 × 6	18 × 8	14 × 10	12 × 12
900	14		30 × 6	20 × 8	16 × 10	12 × 12
1000	14			22 × 8	16 × 10	14 × 12
1100	15			24 × 8	18 × 10	16 × 12
1200	15			26 × 8	20 × 10	16 × 12
1300	16			28 × 8	20 × 10	18 × 12
1400	16			30 × 8	22 × 10	18 × 12
1500	16				24 × 10	20 × 12
1600	17				24 × 10	20 × 12
1700	17				26 × 10	22 × 12
1800	18				28 × 10	22 × 12
1900	18				30 × 10	22 × 12
2000	18					24 × 12

most cases, duct systems in small buildings are sized using the *equal friction method* or a modified equal friction method. In this method, a duct size is selected using the design parameters of friction loss (in w.c. per 100 ft.) and air velocity (for noise control).

In most cases in low-velocity (low-pressure) duct systems when ducts are carrying 40 000 ft³/min and less, the duct diameter is selected with pressure loss in the range of 0.05 to 0.2 in w.c. of ductwork. A traditional pressure loss target used for duct sizing in schools, offices, and other similar buildings is 0.1 in. w.c. per 100 ft of ductwork. Occasionally, higher pressure losses are used on smaller systems or on systems where the known fan performance can handle the additional pressure loss. However, decreasing this design parameter to 0.05 in. w.c. per 100 ft is becoming a contemporary target in large installations because it can significantly cut fan energy consumption (~15% to 20%) as there is a decrease in pressure loss in the duct system.

The equal friction approach does not automatically balance flow through parallel branches of ducts. Thus, balancing dampers are required to achieve the desired flow rate in each zone. In practice this is not a significant problem, because balancing dampers are typically part of actual duct systems. The balancing dampers are adjusted during building commissioning to achieve design flow rates.

A design approach using the equal friction method of duct design is explained in the following.

1. Draw duct layout. The duct system should be designed at the same time the building layout is being developed. Specify distances and discharge volume flow rates

(ft³/min). Number all locations where ducts split, turn, or discharge air as nodes. Node 1 should be the exit of the supply air fan.

2. Determine design pressure loss (i.e., 0.1 in w.c./100 ft).

3. Determine the diameter of each duct from the duct pressure loss chart (provided in Figures 8.16, at the end of the chapter on pages 309–310) using the design pressure loss and volume flow rate as input values.

4. Determine the actual pressure loss per length of duct from the duct pressure loss chart using the duct diameter determined. Calculate the total pressure loss through each duct as the product of the pressure loss per length of duct and the length of the duct.

5. Use the duct diameters to determine the diameters of all elbows, branches, and reductions.

6. Determine the pressure loss for each fitting from Figure 8.15 (at the end of the chapter on pages 305–308) using the fitting diameter.

7. Sum the pressure losses calculated for each branch from the fan to an air discharge point. The branch with the largest pressure loss sets the system pressure drop. Adding balancing dampers is required for all other branches to create the required pressure loss and the desired flow rate.

The designer must account for pressure losses in ductwork, components, and pieces of equipment in the system. Table 8.10

TABLE 8.10   TYPICAL PRESSURE LOSSES IN DUCTS AND EQUIPMENT USED IN RESIDENTIAL OR LIGHT COMMERCIAL INSTALLATIONS.

Component	Typical Pressure Losses	
	in w.c.	Pa
Duct (100 ft length)	0.05 to 0.20	12.5 to 50
Cooling and heating coils	0.15 to 0.50	37 to 125
Elbows (with turning vanes)	0.01 to 0.10	2.5 to 25
Manual dampers	0.03 to 0.06	7.5 to 15
Return air grilles	0.02 to 0.15	5 to 37
Supply air diffusers	0.02 to 0.15	5 to 37
Transitions, boots	0.05 to 0.35	12.5 to 87
Disposable filters	0.05 to 0.30	12.5 to 75
Pleated filters	0.10 to 0.45	25 to 112
Electrostatic filters	0.20 to 0.80	50 to 200
Heat recovery ventilator (HRV)	0.35 to 1.0	87 to 250
Total duct system (supply and return)	0.5 to 5.0	125 to 1250
Duct noise control (silencers)	0.1 to 1.0	25 to 250
VAV terminal unit	0.1 to 0.60	25 to 150
Terminal unit with reheat coil	0.2 to 0.60	50 to .60

indicates ranges of typical pressure losses in ducts and equipment used in residential or light commercial installations.

8.7   SIZING THE EQUIPMENT

HVAC central air equipment (i.e., AHU, packaged RTU, forced-air furnace, or heat pump unit) must have a heating and/or cooling capacity that meets the heating and cooling loads. In small to medium installations, the designer selects the equipment from the manufacturers' catalog information. Table 8.11 is an example of specifications for a RTU. In large installations, individual pieces of equipment (i.e., heating coil, cooling coil, filter, fans, and so on) are specified on an individual basis such that each piece meets specific design requirements.

Central air equipment is designed to heat and or cool a specific amount of air (as measured in ft³/min or L/s) at a designed static pressure or pressure loss (in lb/in² or Pa). The amount of motor horsepower (HP) required to move this amount of air at the designed static load is then applied. Static pressure or pressure loss is the greatest single factor affecting the performance of a fan. For example, a 3-ton heating/cooling system may be designed to move up to 1200 ft³/min at a static pressure of 0.5 in w.c. with a ½ HP motor. This pressure loss includes the air resistance in the supply air ductwork, return air ductwork, the heating coil or burner system, the cooling coil or air conditioning coil, and the filter.

A fan of an AHU must be selected to deliver a specific volumetric flow rate (standard ft³/min, scfm) and a specific *external static pressure*. Basically, the external static pressure generated by the fan must overcome the pressure loss of the duct system (both the supply and return system) while delivering the required volume of air. This pressure loss from the ductwork and other components in the system is known as *system resistance*.

Fan volumetric flow rate is based on the external static pressure at different pressure intervals to account for interference/resistance created by ductwork, fan placement, and so forth. Increases in system resistance results in an increased pressure differential across the fan and slightly reduced airflow. So, at a specific system resistance, the fan will deliver a specific volumetric flow rate. If system resistance is decreased, the fan will deliver more air; if it is increased, the volumetric flow rate will be less. In an ideally designed system, the fan's external static pressure should equal the system resistance—that is, the fan selected has the capacity to deliver the design airflow under normal operating conditions.

An example of fan performance data for a small AHU is provided in Table 8.12. In this table, there are four AHU models (A500, A1000, A1500, and A2000). Each unit has three fan speed settings (high, medium, and low) that offer different volumetric flow rates at the same external static pressure. For example, the A1000 model delivers 1320 ft³/min at the high setting, 1240 ft³/min on medium, and 1060 ft³/min on low.

TABLE 8.11   AN EXAMPLE OF SPECIFICATIONS FOR A ROOF TOP UNIT.

U.S. Customary (IP) units

Model Number	Cooling Data					Heating Data							Air Flow Data			Physical Data	
	Nominal Cooling Capacity	Gross Capacity	ARI Net Rated Capacity	Full Load EER	Part Load SEER	Gross Heating Input			Heating Output			Efficiency	Volumetric Air Flow		Static Pressure	Dimensions	Shipping Weight
						low	medium	high	low	medium	high		min	max		H × W × D	
	Tons	MBH	MBH	—	—	MBH	MBH	MBH	MBH	MBH	MBH	—	cfm	cfm	in w.c.	inches	lb
JB36GY	3	36	35	11.4	13.4	78	—	—	62	—	—	80	500	1660	0.20–1.8	37 × 45 × 86	785
JB48GY	4	51	49.5	11.5	13.5	78	—	125	62	—	100	80	660	2210	0.20–1.8	37 × 45 × 86	850
JB60GY	5	63	61	10.8	12.8	78	—	125	62	—	100	80	840	2760	0.20–1.8	37 × 45 × 86	860
JB72GY	6	75	72	10.4	12.2	78	—	125	62	—	100	80	1020	3310	0.20–1.8	37 × 45 × 86	890
JB90GY	7.5	93	90	11.1	12.3	130	180	240	104	144	192	80	1260	4140	0.20–1.8	50 × 58 × 100	1400
JB102GY	8.5	101	99	11.1	12.2	130	180	240	104	144	192	80	1440	4690	0.20–2.6	50 × 58 × 100	1400
JB120GY	10	128	122	10.3	11.3	130	180	240	104	144	192	80	1680	5520	0.20–2.6	50 × 58 × 100	1450
JB150GY	12.5	148	140	9.6	10.2	130	180	240	104	144	192	80	2100	6900	0.20–2.6	50 × 58 × 100	1480
JB156GY	13	159	154	11.6	13.3	169	180	240	135	144	192	80	2180	7170	0.20–2.6	55 × 91 × 133	2550
JB180GY	15	190	184	11.4	12.8	1697	360	480	135	288	384	80	2520	8280	0.20–2.6	55 × 91 × 133	2550
JB210GY	17.5	212	204	11.2	11.6	169	360	480	135	288	384	80	2940	9660	0.20–2.6	55 × 91 × 133	2700
JB240GY	20	254	244	10.5	11.5	260	360	480	208	288	384	80	3360	11040	0.20–2.6	55 × 91 × 133	2750
JB252GY	21	257	248	11.7	12.7	260	360	480	208	288	384	80	3540	11590	0.20–2.6	65 × 91 × 145	3350
JB300GY	25	311	300	11.0	11.8	260	360	480	208	288	384	80	4200	13800	0.20–2.6	65 × 91 × 145	3350
JB360GY	30	359	344	10.1	11.2	260	360	480	208	288	384	80	5040	16560	0.20–2.6	65 × 91 × 145	3350

Metric (SI) units

Model Number	Cooling Data					Heating Data							Air Flow Data			Physical Data	
	Nominal Cooling Capacity	Gross Capacity	ARI Net Rated Capacity	Full Load EER	Part Load SEER	Gross Heating Input			Heating Output			Efficiency	Volumetric Air Flow		Static Pressure	Dimensions	Shipping Weight
						low	medium	high	low	medium	high		min	max		H × W × D	
	Tons	kW	kW	—	—	kW	kW	kW	kW	kW	kW	—	L/s	L/s	Pa	m	kg
JB36GY	3	10.5	10.3	11.4	13.4	22.9	—	—	18.2	—	—	80	236	784	50–450	0.9 × 1.1 × 2.2	356
JB48GY	4	14.9	14.5	11.5	13.5	22.9	—	36.6	18.2	—	29.3	80	312	1043	50–450	0.9 × 1.1 × 2.2	385
JB60GY	5	18.5	17.9	10.8	12.8	22.9	—	36.6	18.2	—	29.3	80	396	1303	50–450	0.9 × 1.1 × 2.2	390
JB72GY	6	22.0	21.1	10.4	12.2	22.9	—	36.6	18.2	—	29.3	80	481	1562	50–450	0.9 × 1.1 × 2.2	403
JB90GY	7.5	27.2	26.4	11.1	12.3	38.1	52.7	70.3	30.5	42.2	56.3	80	595	1954	50–450	1.3 × 1.5 × 2.5	635
JB102GY	8.5	29.6	29.0	11.1	12.2	38.1	52.7	70.3	30.5	42.2	56.3	80	680	2214	50–650	1.3 × 1.5 × 2.5	635
JB120GY	10	37.5	35.7	10.3	11.3	38.1	52.7	70.3	30.5	42.2	56.3	80	793	2605	50–650	1.3 × 1.5 × 2.5	657
JB150GY	12.5	43.4	41.0	9.6	10.2	38.1	52.7	70.3	30.5	42.2	56.3	80	991	3257	50–650	1.3 × 1.5 × 2.5	671
JB156GY	13	46.6	45.1	11.6	13.3	49.5	52.7	70.3	39.6	42.2	56.3	80	1029	3384	50–650	1.4 × 2.3 × 3.4	1156
JB180GY	15	55.7	53.9	11.4	12.8	49.5	105.5	140.6	39.6	84.4	112.5	80	1189	3908	50–650	1.4 × 2.3 × 3.4	1156
JB210GY	17.5	62.1	59.8	11.2	11.6	49.5	105.5	140.6	39.6	84.4	112.5	80	1388	4560	50–650	1.4 × 2.3 × 3.4	1224
JB240GY	20	74.4	71.5	10.5	11.5	76.2	105.5	140.6	60.9	84.4	112.5	80	1586	5211	50–650	1.4 × 2.3 × 3.4	1247
JB252GY	21	75.3	72.7	11.7	12.7	76.2	105.5	140.6	60.9	84.4	112.5	80	1671	5470	50–650	1.7 × 2.3 × 3.7	1519
JB300GY	25	91.1	87.9	11	11.8	76.2	105.5	140.6	60.9	84.4	112.5	80	1982	6514	50–650	1.7 × 2.3 × 3.7	1519
JB360GY	30	105.2	100.8	10.1	11.2	76.2	105.5	140.6	60.9	84.4	112.5	80	2379	7816	50–650	1.7 × 2.3 × 3.7	1519

TABLE 8.12   AN EXAMPLE OF FAN PERFORMANCE DATA FOR A SMALL AIR-HANDLING UNIT: VOLUMETRIC FLOW RATE (STANDARD FT³/MIN, SCFM) FOR A SPECIFIC EXTERNAL STATIC PRESSURE AT A SPECIFIC FAN SPEED SETTING.

Fan Speed Setting	External Static Pressure (in w.c.)	Air-Handling Unit Model No.			
		A500	A1000	A1500	A2000
High	0.1	1280	1540	1840	2200
	0.2	1210	1500	1780	2120
	0.3	1160	1400	1720	2030
	0.4	1100	1320	1650	1930
	0.5	1020	1240	1560	1830
	0.6	940	1150	1470	1730
Medium	0.1	940	1420	1580	1770
	0.2	890	1360	1530	1710
	0.3	830	1310	1490	1630
	0.4	780	1240	1440	1560
	0.5	720	1180	1390	1460
	0.6	645	1080	1320	1360
Low	0.1	690	1210	1170	1560
	0.2	630	1170	1140	1490
	0.3	580	1120	1110	1400
	0.4	540	1060	1070	1330
	0.5	470	990	1030	1240
	0.6	460	980	1010	1220

8.8   DUCT DESIGN EXAMPLE (HEATING)

1. Determine the heat loss of each room and list them.

   The heating load is computed using the method outlined in Chapter 4 (see Table 8.13).

2. Determine the location and number of supply outlets (in this case, located on outside walls under windows). Sketch the proposed locations on the floor plan as shown in Figure 8.10. As a general rule, in residential design, no one supply outlet should supply more than 8000 Btu/hr (2340 W).

TABLE 8.13   ROOM HEAT LOSS.

	BTUH	Watts
Entry	4061	1287
Dining	3056	898
Kitchen	4160	1205
Living	9327	2690
Bedroom 3	3483	1013
Bath	1092	322
Bedroom 1	4326	1267
Bedroom 2	2772	807
Int. Bath	203	57
	32,480	9546

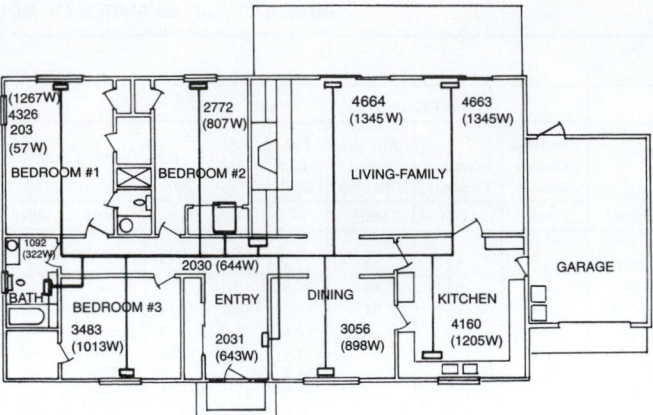

FIGURE 8.10   Duct schematic floor plan for example.

Figure 8.10 has a proposed layout for the ductwork. The ducts are shown as single lines and the outlets as rectangles, and the Btu/hr for each outlet is listed (the watts for each room are in parentheses).

Table 8.14 has a tabulation of each supply outlet and the Btu/hr.

3. Note on the sketch the types of fittings that will be used and the actual length of each duct. The fittings may be selected from Figure 8.15, which also gives the equivalent length for each type of fitting. The plan in Table 8.15 shows the fittings and their equivalent lengths.

   Now, on the sketch, note the actual length of each run. Then add the equivalent and actual lengths for each run to obtain the total equivalent length for each run from the bonnet to the supply outlet (Figure 8.11).

   The total equivalent length of each run is tabulated in Figure 8.11.

TABLE 8.14   HEAT LOSS FOR EACH OUTLET.

Supply Outlet	Heat Loss (BTUH)	Watts
Bedroom #1	4529	1324
Bedroom #2	2772	807
Living	4664	1345
Living	4663	1345
Kitchen	4160	1205
Dining	3056	898
Entry	2031	643
Hall	2030	644
Bedroom #3	3483	1013
Bath	1092	322
Total	32,480	9546

Note: Bedroom 1 includes interior bath.

TABLE 8.15   EQUIVALENT LENGTHS.

Outlet	Equivalent length	Actual length	Total equivalent length
Bedroom (1)	F 35 B 15 G 30 — 80	36	116
Bedroom (2)	F 35 + 25 B 15 G 30 — 105	26	131
Living 1	F 35 B 15 + 25 G 30 — 105	33	138
2	F 35 B 15 G 30 — 80	55	135
Kitchen	F 35 B 15 G 30 — 80	46	126
Dining	F 35 + 25 B 15 G 30 — 105	27	132
Entry	F 35 + 25 B 15 E 10 G 30 — 115	16	131
Hall	F 35 B 15 G 30 — 80	10	90
Bedroom (3)	F 35 + 25 B 15 G 30 — 105	30	135
Bath	F 35 B 15 E 10 G 30 — 90	25	115

SI UNITS

Outlet	Equivalent length	Actual length	Total equivalent length
Bedroom (1)	F 10.7 B 4.6 G 9.1 — 24.4	11	35.4
Bedroom (2)	F 10.7 + 7.6 B 4.6 G 9.1 — 32	7.9	39.9
Living 1	F 10.7 B 4.6 + 7.6 G 9.1 — 32	10	42
2	F 10.7 B 4.6 G 9.1 — 24.4	16.8	41.2
Kitchen	F 10.7 B 4.6 G 9.1 — 24.4	14	38.4
Dining	F 10.7 + 7.6 B 4.6 G 9.1 — 32	8.2	40.2
Entry	F 10.7 + 7.6 B 4.6 E 3.1 G 9.1 — 35.1	5	40.1
Hall	F 10.7 B 4.6 G 9.1 — 24.4	3.1	27.5
Bedroom (3)	F 10.7 + 7.6 B 4.6 G 9.1 — 32	3.1	35.1
Bath	F 10.7 B 4.6 E 3.1 G 9.1 — 27.5	7.6	35.1

4. A tentative blower size is selected next. For the most efficient operation and uniform heating, it is recommended that the total ft³/min (L/s) should not be less than 400 ft³/min (190 L/s) per 12 000 Btu/hr (3510 W).

$$\frac{32\,480}{12\,000} \cdot 400 = 1083 \text{ ft}^3/\text{min (use 1200)}$$

In metric (SI) units:

$$\frac{9546}{3510} \cdot 190 \text{ L/s} = 518 \text{ L/s (use 550 L/s)}$$

5. Calculate the ft³/min required from each of the supply outlets by dividing heat loss of each supply outlet by the total heat loss and multiplying by the total airflow.

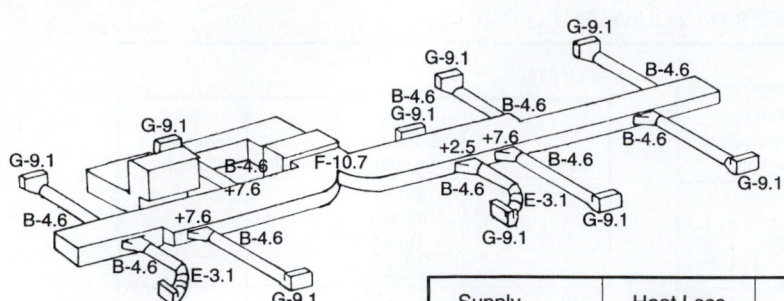

Supply Outlet	Heat Loss (BTUH)	Actual Length (ft)	Equivalent Length (ft)	Total Equiv. Length (ft)
Bedroom #1	4529	36	80	116
Bedroom #2	2772	26	105	131
Living	4664	33	105	138
Living	4663	55	80	135
Kitchen	4160	46	80	126
Dining	3056	27	105	132
Entry	2031	16	115	131
Hall	2030	10	80	90
Bedroom #3	3483	30	105	135
Bath	1092	25	90	115
Totals				

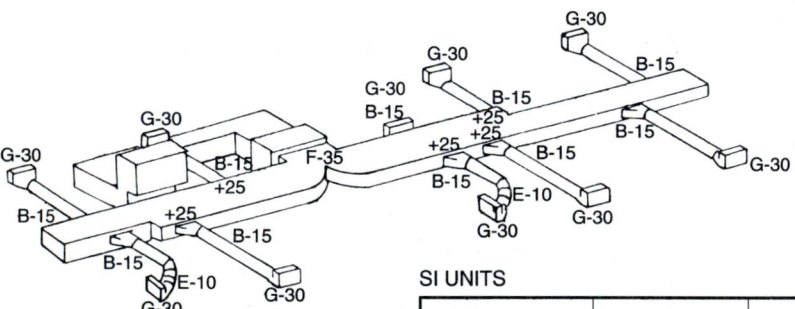

SI UNITS

Supply Outlet	Heat Loss (Watts)	Actual Length (m)	Equivalent Length	Total Equiv. Length
Bedroom #1	1,324	11	24.4	35.4
Bedroom #2	807	7.9	32	39.9
Living	1,345	10	32	42
Living	1,345	16.8	24.4	41.2
Kitchen	1,194	14	24.4	38.4
Dining	898	8.2	32	40.2
Entry	643	5.0	35.1	40.1
Hall	644	3.1	24.4	27.5
Bedroom #3	1,013	3.1	32	35.1
Bath	322	7.6	7.5	35.1
Totals	9,535			

FIGURE 8.11 Total equivalent lengths for example.

In this design, for Bedroom 1:

$$\frac{4529}{32\,480} \cdot 1200 = 167$$

In metric (SI) units:

$$\frac{1324}{9546} \cdot 550 = 76 \text{ L/s}$$

Following this example, calculate the required ft³/min for each room and tabulate as shown (Table 8.16).

6. Bonnet pressure.

   a. The bonnet pressure is selected next. The approximate bonnet pressure required for a trunk (main duct) to carry the maximum volume of air (ft³/min) is found in Table 8.42 (p. 304).

   In this design, the total cfm required is 1200 (the total of all the supply outlets). From Table 8.42, from a quiet operation in a residence (Type A), the suggested bonnet pressure is a minimum of 0.10 in (24.8 Pa) of water.

   b. If a furnace has been selected, the rated capacity of the unit must be checked and its pressure used. In this design, assume a furnace with 0.10 in (24.8 Pa) of water was selected.

7. The available bonnet pressure is divided proportionally by length between the supply and return runs. In perimeter heating (Figure 8.5), the supply runs are longer than

TABLE 8.16   HEATING CFM REQUIRED.

Supply Outlet	Heat Loss (BTUH)	Actual Length (ft)	Equivalent Length (ft)	Total Equiv. Length (ft)	Heating CFM
Bedroom #1	4529	36	80	116	167
Bedroom #2	2772	26	105	131	103
Living	4664	33	105	138	173
Living	4663	55	80	135	173
Kitchen	4160	46	80	126	154
Dining	3056	27	105	132	113
Entry	2031	16	115	131	75
Hall	2030	10	80	90	75
Bedroom #3	3483	30	105	135	129
Bath	1092	25	90	115	40

SI UNITS

Supply Outlet	Heat Loss (Watts)	Actual Length (m)	Equivalent Length (m)	Total Equiv. Length (m)	Heating (L/s)
Bedroom #1	1324	11	24.4	35.4	77
Bedroom #2	807	7.9	32	39.9	47
Living	1345	10	32	42	78
Living	1345	16.8	24.4	41.2	78
Kitchen	1205	14	24.4	38.4	69
Dining	898	8.2	32	40.2	52
Entry	643	5	35.1	40.1	37
Hall	644	3.1	24.4	27.5	37
Bedroom #3	1013	3.1	32	35.1	59
Bath	322	7.6	27.5	35.1	19

the returns, and a 0.10 in (24.8 Pa) pressure might be divided with 70% on the supply and 30% on the return. For other systems, the sketch plan layout shows the approximate proportion of supply to return runs.

The sketch plan layout in Figure 8.10 shows that there are longer supply than return runs, so a division of 0.07 in (17.36 Pa) for supply and 0.03 in (7.44 Pa) for return is selected and tabulated (Table 8.17).

Note: If later during the design the proportion selected does not provide satisfactory results, it can be reapportioned between supply and return.

8. Next, the supply outlet size and its pressure loss are selected from the manufacturer's engineering data. See Table 8.17. Typically, the pressure loss will range from

0.01 to 0.02 in, and many designers simply allow 0.02 in (4.8 Pa).

In this design, the pressure loss has been tabulated in Table 8.18, p. 277.

9. Next, the actual pressure that is available for duct loss is obtained by subtracting the supply outlet loss (Step 8) from the total pressure available for supply runs (Table 8.18).

In this design, the pressure for supply runs is 0.07 in. Beginning with the Bedroom 1 supply run, from Step 8 the supply outlet loss is 0.02 in, so for this run the pressure available for actual duct loss is 0.07 in − 0.02 in = 0.05 in. Calculate the actual pressure available for duct loss for each of the supply runs,

TABLE 8.17 SUPPLY AIR PRESSURE.

Supply Outlet	Heat Loss (BTUH)	Actual Length (Ft)	Equivalent Length (Ft)	Total Equiv. Length (Ft)	Heating CFM	Supply Air Pressure
Bedroom #1	4529	36	80	116	167	0.07
Bedroom #2	2772	26	105	131	103	0.07
Living	4664	33	105	138	173	0.07
Living	4663	55	80	135	173	0.07
Kitchen	4160	46	80	126	154	0.07
Dining	3056	27	105	132	113	0.07
Entry	2031	16	115	131	75	0.07
Hall	2030	10	80	90	75	0.07
Bedroom #3	3483	30	105	135	129	0.07
Bath	1092	25	90	115	40	0.07

SI UNITS

Supply Outlet	Heat Loss (Watts)	Actual Length (m)	Equivalent Length (m)	Total Equiv. Length (m)	Heating (L/s)	Supply Air Pressure
Bedroom #1	1,324	11	24.4	35.4	77	17.36
Bedroom #2	807	7.9	32	39.9	47	17.36
Living	1,345	10	32	42	78	17.36
Living	1,345	16.8	24.4	41.2	78	17.36
Kitchen	1,205	14	24.4	38.4	69	17.36
Dining	898	8.2	32	40.2	52	17.36
Entry	643	5	35.1	40.1	37	17.36
Hall	644	3.1	24.4	27.5	37	17.36
Bedroom #3	1,013	3.1	32	35.1	59	17.36
Bath	322	7.6	27.5	35.1	19	17.36

TABLE 8.18   SUPPLY OUTLET PRESSURE.

Supply Outlet	Heat Loss (BTUH)	Actual Length (ft)	Equivalent Length (ft)	Total Equiv. Length (ft)	Heating CFM	Supply Air Pressure	Supply Outlet Pressure Loss
Bedroom #1	4529	36	80	116	167	0.07	0.02
Bedroom #2	2772	26	105	131	103	0.07	0.02
Living	4664	33	105	138	173	0.07	0.02
Living	4663	55	80	135	173	0.07	0.02
Kitchen	4160	46	80	126	154	0.07	0.02
Dining	3056	27	105	132	113	0.07	0.02
Entry	2031	16	115	131	75	0.07	0.02
Hall	2030	10	80	90	75	0.07	0.02
Bedroom #3	3483	30	105	135	129	0.07	0.02
Bath	1092	25	90	115	40	0.07	0.02

SI UNITS

Supply Outlet	Heat Loss (Watts)	Actual Length (m)	Equivalent Length (m)	Total Equiv. Length (m)	Heating (L/s)	Supply Air Pressure	Supply Outlet Pressure Loss
Bedroom #1	1,324	11	24.4	35.4	77	17.36	4.8
Bedroom #2	807	7.9	32	39.9	47	17.36	4.8
Living	1,345	10	32	42	78	17.36	4.8
Living	1,345	16.8	24.4	41.2	78	17.36	4.8
Kitchen	1,205	14	24.4	38.4	69	17.36	4.8
Dining	898	8.2	32	40.2	52	17.36	4.8
Entry	643	5	35.1	40.1	37	17.36	4.8
Hall	644	3.1	24.4	27.5	37	17.36	4.8
Bedroom #3	1,013	3.1	32	35.1	59	17.36	4.8
Bath	322	7.6	27.5	35.1	19	17.36	4.8

and tabulate the information as shown (Table 8.19, p. 278).

In this design in metric (SI) units, the pressure for supply runs is 16.8 Pa. Beginning with the Bedroom 1 supply run, from Step 8 the supply outlet loss is 4.8 Pa, so for this run the pressure available for actual duct loss is 16.8 − 4.8 = 12 Pa. Calculate the actual pressure available for each of the supply runs, and tabulate the information as shown in Table 8.19.

10. To make a workable table of duct sizes, the pressure drop for duct loss must be calculated as the pressure drop per *100 ft* (per meter) of the duct. Because the various supply runs are all different lengths, it is necessary to find the allowable *pressure drop per 100 ft* (per meter) based on the *allowable duct loss* and the *total equivalent length* of the supply run. This allowable pressure drop is calculated by using the equation:

$$\text{Allowable pressure drop per 100 ft} = \frac{\text{Allowable duct loss} \cdot 100}{\text{Total equivalent length}}$$

In this design, the Bedroom 1 supply run has an allowable duct loss of .05 in (Step 9) and an equivalent length of 116 ft (Step 3). Use the formula to determine the allowable pressure drop per 100 ft.

In metric (SI) units:

$$\text{Allowable pressure drop per 100 ft} = \frac{0.05 \text{ in} \cdot 100 \text{ ft}}{116 \text{ ft}} = 0.43 \text{ in}$$

TABLE 8.19 AVAILABLE DUCT PRESSURE.

Supply Outlet	Heat Loss (BTUH)	Actual Length (ft)	Equivalent Length (ft)	Total Equiv. Length (ft)	Heating CFM	Supply Air Pressure	Supply Outlet Pressure Loss	Pressure Avail. for Duct Loss
Bedroom #1	4529	36	80	116	167	0.07	0.02	0.05
Bedroom #2	2772	26	105	131	103	0.07	0.02	0.05
Living	4664	33	105	138	173	0.07	0.02	0.05
Living	4663	55	80	135	173	0.07	0.02	0.05
Kitchen	4160	46	80	126	154	0.07	0.02	0.05
Dining	3056	27	105	132	113	0.07	0.02	0.05
Entry	2031	16	115	131	75	0.07	0.02	0.05
Hall	2030	10	80	90	75	0.07	0.02	0.05
Bedroom #3	3483	30	105	135	129	0.07	0.02	0.05
Bath	1092	25	90	115	40	0.07	0.02	0.05

SI UNITS

Supply Outlet	Heat Loss (Watts)	Actual Length (m)	Equivalent Length (m)	Total Equiv. Length (m)	Heating (L/s)	Supply Air Pressure	Supply Outlet Pressure Loss	Pressure Avail. for Duct Loss
Bedroom #1	1,324	11	24.4	35.4	77	17.36	4.8	12.56
Bedroom #2	807	7.9	32	39.9	47	17.36	4.8	12.56
Living	1,345	10	32	42	78	17.36	4.8	12.56
Living	1,345	16.8	24.4	41.2	78	17.36	4.8	12.56
Kitchen	1,205	14	24.4	38.4	69	17.36	4.8	12.56
Dining	898	8.2	32	40.2	52	17.36	4.8	12.56
Entry	643	5	35.1	40.1	37	17.36	4.8	12.56
Hall	644	3.1	24.4	27.5	37	17.36	4.8	12.56
Bedroom #3	1,013	3.1	32	35.1	59	17.36	4.8	12.56
Bath	322	7.6	27.5	35.1	19	17.36	4.8	12.56

The allowable pressure drop is calculated by using the equation:

$$\text{Allowable pressure drop} = \frac{\text{Supply static pressure available}}{\text{Total equivalent length of branch supply run}}$$

In this design, the Bedroom 1 supply run has an allowable duct loss of 12.0 Pa and an equivalent length of 35.4 m (Step 3). Use the formula to determine the allowable pressure drop:

$$\text{Allowable pressure drop} = \frac{12}{35.4} = 0.34$$

Calculate the allowable pressure drop for each supply run, and tabulate the information as shown (Table 8.20).

11. Figure 8.16 (at the end of the chapter on pages 309–310) is used to size the round duct based on the ft³/min required and the allowable pressure drop per 100 ft (Step 12). The size of the round duct may be converted to an equivalent rectangular duct size, which will handle the required cfm within the allowable pressure drop.

In this design, the Bedroom 1 supply run has an allowable duct loss of 0.043 in per 100 ft (Step 12) and a required ft³/min of 170 (Step 5). Using the table in Figure 8.16, a 9.0 in round duct would be selected. Table 8.21 shows the use of the table. Begin by finding the pressure drop per 100 ft on the left; then the required cfm along the bottom; then move horizontally to the right and vertically from the bottom to read the round duct size required.

TABLE 8.20   PRESSURE AVAILABLE.

Supply Outlet	Heat Loss (BTUH)	Actual Length (ft)	Equivalent Length (ft)	Total Equiv. Length (ft)	Heating CFM	Supply Air Pressure	Supply Outlet Pressure Loss	Pressure Avail. for Duct Loss	Pressure Avail. Per 100 Ft
Bedroom #1	4529	36	80	116	167	0.07	0.02	0.05	0.043
Bedroom #2	2772	26	105	131	103	0.07	0.02	0.05	0.038
Living	4664	33	105	138	173	0.07	0.02	0.05	0.036
Living	4663	55	80	135	173	0.07	0.02	0.05	0.037
Kitchen	4160	46	80	126	154	0.07	0.02	0.05	0.040
Dining	3056	27	105	132	113	0.07	0.02	0.05	0.038
Entry	2031	16	115	131	75	0.07	0.02	0.05	0.038
Hall	2030	10	80	90	75	0.07	0.02	0.05	0.056
Bedroom #3	3483	30	105	135	129	0.07	0.02	0.05	0.037
Bath	1092	25	90	115	40	0.07	0.02	0.05	0.043

SI UNITS

Supply Outlet	Heat Loss (Watts)	Actual Length (m)	Equivalent Length (m)	Total Equiv. Length (m)	Heating (L/s)	Supply Air Pressure	Supply Outlet Pressure Loss	Pressure Avail. for Duct Loss	Pressure Avail. Per Branch
Bedroom #1	1,324	11	24.4	35.4	77	17.36	4.8	12.56	0.35
Bedroom #2	807	7.9	32	39.9	47	17.36	4.8	12.56	0.31
Living	1,345	10	32	42	78	17.36	4.8	12.56	0.30
Living	1,345	16.8	24.4	41.2	78	17.36	4.8	12.56	0.30
Kitchen	1,205	14	24.4	38.4	69	17.36	4.8	12.56	0.33
Dining	898	8.2	32	40.2	52	17.36	4.8	12.56	0.31
Entry	643	5.0	35.1	40.1	37	17.36	4.8	12.56	0.31
Hall	644	3.1	24.4	27.5	37	17.36	4.8	12.56	0.46
Bedroom #3	1,013	3.1	32	35.1	59	17.36	4.8	12.56	0.36
Bath	322	7.6	27.5	35.1	19	17.36	4.8	12.56	0.36

From Figure 8.16, determine the round duct sizes for each of the supply runs and tabulate the information as shown in Table 8.21.

In this design in metric (SI) units, the Bedroom 1 supply run has an allowable duct loss of 0.35 (Step 12) and a required L/s of 77 (Step 7). Using the table in Figure 8.16, a 250 mm round duct would be selected. Begin by finding the pressure drop on the left and the L/s along the bottom; then find where they intersect and read the duct size to the right of the intersection.

12. The round duct sizes selected in Step 11 may be changed to equivalent rectangular duct sizes, carrying the same ft³/min (L/s) required while maintaining the allowable pressure drops. Using Table 8.25 (p. 284), the 9-in round duct for Bedroom 1 can be changed to a rectangular duct of 8 in · 8 in, 11 in · 7 in, or 14 in · 6 in and others.

From Table 8.41 (at end of chapter on pp. 300–303):

The round duct sizes are found along the side of the table. The rectangular sizes are to the right.

The actual size selected depends to a large extent on the space available for installation of the duct.

From Table 8.41, determine a rectangular duct size for each supply run that could be used instead of the round duct previously selected. Tabulate the rectangular duct sizes as shown (Table 8.22).

13. Add the air volumes of all of the branch supply runs from each trunk duct. When there is more than one trunk duct

TABLE 8.21   TABULATED ROUND DUCT SIZES.

Supply Outlet	Heat Loss (BTUH)	Actual Length (ft)	Equivalent Length (ft)	Total Equiv. Length (Ft)	Heating CFM	Supply Air Pressure	Supply Outlet Pressure Loss	Pressure Avail. for Duct Loss	Pressure Avail. Per 100 Ft	Round Duct Size (In)
Bedroom #1	4529	36	80	116	167	0.07	0.02	0.05	0.043	9.0
Bedroom #2	2772	26	105	131	103	0.07	0.02	0.05	0.038	7.0
Living	4664	33	105	138	173	0.07	0.02	0.05	0.036	9.0
Living	4663	55	80	135	173	0.07	0.02	0.05	0.037	9.0
Kitchen	4160	46	80	126	154	0.07	0.02	0.05	0.040	8.0
Dining	3056	27	105	132	113	0.07	0.02	0.05	0.038	8.0
Entry	2031	16	115	131	75	0.07	0.02	0.05	0.038	7.0
Hall	2030	10	80	90	75	0.07	0.02	0.05	0.056	6.0
Bedroom #3	3483	30	105	135	129	0.07	0.02	0.05	0.037	8.0
Bath	1029	25	90	115	40	0.07	0.02	0.05	0.043	6.0

SI UNITS

Supply Outlet	Heat Loss (Watts)	Actual Length (m)	Equivalent Length (m)	Total Equiv. Length (m)	Heating (L/s)	Supply Air Pressure	Supply Outlet Pressure Loss	Pressure Avail. for Duct Loss	Pressure Avail. Per Branch	Round Duct Size (mm)
Bedroom #1	1,324	11	24.4	35.4	77	17.36	4.8	12.56	0.35	250
Bedroom #2	807	7.9	32	39.9	47	17.36	4.8	12.56	0.31	200
Living	1,345	10	32	42	78	17.36	4.8	12.56	0.30	250
Living	1,345	16.8	24.4	41.2	78	17.36	4.8	12.56	0.30	250
Kitchen	1,205	14	24.4	38.4	69	17.36	4.8	12.56	0.33	200
Dining	898	8.2	32	40.2	52	17.36	4.8	12.56	0.31	200
Entry	643	5	35.1	40.1	37	17.36	4.8	12.56	0.31	160
Hall	644	3.1	24.4	27.5	37	17.36	4.8	12.56	0.46	160
Bedroom #3	1,013	3.1	32	35.1	59	17.36	4.8	12.56	0.36	200
Bath	322	7.6	27.5	35.1	19	17.36	4.8	12.56	0.36	160

(such as in this design), keep the totals for each trunk duct separate.

In this design, there are two trunk ducts (one toward the living room and kitchen, the other toward the bedrooms). Tabulate and total the air volume for each trunk duct as shown (Table 8.23).

14. For best airflow distribution through the ducts, it is important that the friction losses per 100 ft be approximately equal in both trunk ducts.

Using Figure 8.16, select the trunk sizes required. When there is more than one trunk duct, size each separately using the required air volume (ft^3/min) and the allowable duct friction (Step 10).

The bedroom branch trunk must handle 439 ft^3/min with an allowable pressure drop of 0.37 in per 100 ft (0.05 times 100 divided by 135 ft). Using Figure 8.16, the bedroom trunk duct will be a 14 in round duct.

The living space trunk duct has 763 ft^3/min and an allowable 0.040 in per 100 ft pressure drop. Using

TABLE 8.22   TABULATED RECTANGULAR DUCT SIZES.

Supply Outlet	Heat loss (BTUH)	Actual Length (Ft)	Equivalent Length (Ft)	Total Equiv. Length (Ft)	Heating CFM	Supply Air Pressure	Supply Outlet Pressure Loss	Pressure Avail. for Duct Loss	Pressure Avail. Per 100 Ft	Round Duct Size (In.)	Rectangular Duct Size (In.)
Bedroom #1	4,529	36	80	116	167	0.07	0.02	0.05	0.060	9.0	6 × 12
Bedroom #2	2,772	26	105	131	103	0.07	0.02	0.05	0.054	7.0	6 × 8
Living	4,664	33	105	138	173	0.07	0.02	0.05	0.051	9.0	6 × 12
Living	4,663	55	80	135	173	0.07	0.02	0.05	0.052	9.0	6 × 12
Kitchen	4,160	46	80	126	154	0.07	0.02	0.05	0.056	8.0	6 × 9
Dining	3,056	27	105	132	113	0.07	0.02	0.05	0.053	8.0	6 × 9
Entry	2,031	16	115	131	75	0.07	0.02	0.05	0.054	7.0	6 × 7
Hall	2,030	10	80	90	75	0.07	0.02	0.05	0.078	6.0	6 × 5
Bedroom #3	3,483	30	105	135	129	0.07	0.02	0.05	0.052	8.0	5 × 9
Bath	1,092	25	90	115	40	0.07	0.02	0.05	0.061	0.6	6 × 5

SI UNITS

Supply Outlet	Heat loss (Watts)	Actual Length (m)	Equivalent Length (m)	Total Equiv. Length (m)	Heating (L/s)	Supply Air Pressure	Supply Outlet Pressure Loss	Pressure Avail. for Duct Loss	Pressure Avail. Per Branch	Round Duct Size (mm)	Rectangular Duct Size (mm)
Bedroom #1	1,324	11	24.4	35.4	77	17.36	4.8	12.56	0.35	250	150 × 400
Bedroom #2	807	7.9	32	39.9	47	17.36	4.8	12.56	0.31	200	150 × 225
Living	1,345	10	32	42	78	17.36	4.8	12.56	0.30	250	150 × 400
Living	1,345	16.8	24.4	41.2	78	17.36	4.8	12.56	0.30	250	150 × 400
Kitchen	1,205	14	24.4	38.4	69	17.36	4.8	12.56	0.33	200	150 × 225
Dining	898	8.2	32	40.2	52	17.36	4.8	12.56	0.31	200	150 × 225
Entry	643	5.0	35.1	40.1	37	17.36	4.8	12.56	0.31	160	100 × 200
Hall	644	3.1	24.4	27.5	37	17.36	4.8	12.56	0.46	160	100 × 200
Bedroom #3	1,013	3.1	32	35.1	59	17.36	4.8	12.56	0.36	200	150 × 225
Bath	322	7.6	27.5	35.1	19	17.36	4.8	12.56	0.36	160	100 × 200

Figure 8.16, the living space trunk duct will be a 16 in round duct.

In metric (SI) units, the bedroom trunk must handle 202 L/s with an allowable pressure drop of 0.31 Pa. Using Figure 8.16, the bedroom trunk will be a 315 mm round duct.

The living space trunk duct has 351 L/s with an allowable pressure of 0.30 Pa. Using Figure 8.16, the living space trunk will be a 400 mm round duct.

Because the trunk ducts have been sized separately based on airflow and allowable pressure drop, they are balanced.

The round duct sizes may be converted to rectangular duct sizes using Table 8.41 (at end of chapter on pp. 300–303). Once again, the space available for installation will have a large influence on the shape of the rectangular duct selected. Tabulate the trunk duct sizes as shown in Table 8.24.

TABLE 8.23   TRUNK DUCT VOLUMES.

Duct		CFM
Main		
	A-B	1,200
Bedroom		
Branch	B-C	439
	C-D	336
	D-E	207
	E-F	167
Living		
Branch	B-G	763
	G-H	688
	H-I	613
	I-J	500
	J-K	327
	K-L	173

SI UNITS

Duct		L/s
Main		
	A-B	550
Bedroom		
Branch	B-C	202
	C-D	155
	D-E	96
	E-F	77
Living		
Branch	B-G	351
	G-H	314
	H-I	277
	I-J	225
	J-K	147
	K-L	78

TABLE 8.24   TABULATED TRUNK SIZES.

Duct		CFM	Supply Static Pressure*	Total Equiv. Length (Ft)	Supply Pressure Avail. Per 100 Ft	Round Duct Size (In.)	Rectangular Duct Size (In.)
Main							
	A-B	1,200	0.05	138	0.049	18.0	12 × 24
Bedroom							
Branch	B-C	439	0.05	135	0.037	14.0	8 × 22
	C-D	336				12.0	8 × 14
	D-E	207				10.0	7 × 11
	E-F	167				8.0	7 × 9
Living							
Branch	B-G	763	0.05	138	0.048	16.0	8 × 28
	G-H	688				16.0	8 × 28
	H-I	613				14.0	8 × 22
	I-J	500				14.0	8 × 22
	J-K	327				10.0	8 × 12
	K-L	173				9.0	8 × 8

*Allows 0.02 for supply outlet

SI UNITS

Duct		L/s	Supply Static Pressure* (Pa)	Total Total Equiv. Length (m)	Supply Pressure Avail. (Pa)	Round Duct Size (mm)	Rectangular Duct Size (mm)
Main							
	A-B	550	12.56	42.1	0.30	500	200 × 1300
Bedroom							
Branch	B-C	202	12.56	40	0.31	315	200 × 450
	C-D	155				315	200 × 450
	D-E	96				250	200 × 275
	E-F	77				250	200 × 275
Living							
Branch	B-G	351	12.56	42.1	0.30	400	200 × 750
	G-H	314				400	200 × 750
	H-I	277				400	200 × 750
	I-J	225				400	200 × 750
	J-K	147				315	200 × 450
	K-L	78				250	200 × 275

*Allows 4.8 Pa for supply outlet.
Note: Rounding off causes some variations in L/s.

Duct	Cfm	Actual Length	Equivalent Length	Total Equivalent Length	Supply Pressure Avail. per 100 ft	Round Duct Size (in.)	Rectangular Duct Size (in.)
Return	1,200	12	70	82	0.036	18.0	12 × 24

SI UNITS

Duct	L/s	Actual Length	Equivalent Length	Total Equivalent Length	Supply Pressure Avail. (Pa)	Round Duct Size	Rectangular Duct Size
Return	550	3.7	21.4	25.1	0.30	500	200 × 1300

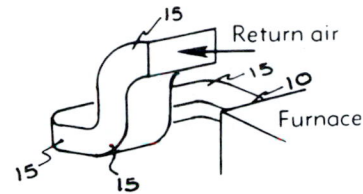

FIGURE 8.12 Return duct sizes for the example.

15. Review the supply trunk duct to determine if it is desirable to reduce the size of the duct as each duct leaves the trunk. Generally, such a reduction is suggested so that the velocity of air through the duct will not drop too low. Using the remaining air volume in the trunk duct, after each branch, and the allowable pressure drop of duct (Step 10), select the reduced trunk sizes from Figure 8.16.

    Using tables in Figure 8.16 and Table 8.41 (at end of chapter on pp. 300–310), determine the reduced trunk duct sizes for both the bedroom and the living space ducts, and tabulate the sizes as shown in Figure 8.12.

16. Next, the design turns to the return ducts. The first step is to select an allowable return air pressure drop. This was previously decided (Step 10), but it is reviewed at this point as the designer reviews the supply trunk and branch duct sizes to see if he or she might want to increase the duct sizes (reducing the allowable air pressure drop) or perhaps to reduce the duct sizes (increasing the allowable air pressure drop for the supply runs). Remember that a change in allowable air pressure drop for the supply will affect the allowable air pressure drop for the return runs.

    A review of the supply duct sizes is made, and it is decided that they will not be revised. Therefore, the original decision, in Step 7, to use 0.03 in as the allowable pressure drop for the return, is unchanged, and the return design will be based on it.

17. The return trunk duct size is determined by the air volume it must handle and the allowable pressure drop. This allowable pressure drop must be converted to allowable pressure drop per 100 ft or Pa/m, just as was done for the supply runs. Using the air volume and the allowable pressure drop, the round return duct size may be selected from Figure 8.16.

The rectangular duct of equivalent size may be found in Table 8.41 (at end of chapter on pp. 300–303). The return trunk duct will have to handle an air volume equal to what the supply trunk duct handled: from Step 5, this was 1200 ft³/min. The allowable pressure drop selected for the return was 0.03 in (Step 7). First, the allowable pressure drop must be converted into the allowable pressure drop *per* 100 ft: (0.03 in · 100 ft) ÷ 82 ft = 0.036 in per 100 ft allowable pressure drop.

Using this information (1200 ft³/min, 0.036 in per 100 ft), select the round return duct size from the table in Figure 8.16 and its equivalent rectangular size from Table 8.41. Tabulate the information as shown.

In metric (SI) units, the return duct will have to handle an air volume equal to what the supply trunk handled: from Step 5, this was 550 L/s. The allowable pressure drop selected was 7.44 Pa (Step 7). First the allowable drop must be converted into the allowable pressure drop available for this length of run:

$$(7.44 \text{ Pa} \div 25.1 \text{ m}) = 0.30 \text{ Pa}$$

Using this information (550 L/s, 0.30 Pa), select the round duct size from Figure 8.16 and its equivalent rectangular duct size from Table 8.41. Tabulate the information as shown in Figure 8.12.

If the return trunk sizes are too large, they may be reduced in size by:

a. Reapportioning the pressure drop available so there is less drop allowed for supply runs and more pressure allowed for return runs. This may require a revision of supply branch and trunk duct sizes.

b. Checking to see if the return air grille can be located closer to the furnace, which would reduce the actual length of duct. This increases the pressure drop, resulting in a smaller duct size.

18. The size of the blower on the furnace is determined from the total air volume (the total cfm to be delivered) and the total static pressure requirements.

The total static pressure (bonnet pressure) of a furnace-blower combination unit is the total of the actual pressure drops in the supply and return ducts. Most residential and small commercial buildings have furnace-blower combination units.

When the blowers are selected separately from the furnace, the total static pressure is the sum of the actual pressure losses in the supply and return ducts, the filter loss, the casing loss, and losses through any devices that are put on the system, such as air washers and purifiers.

In this design, the pressure drop, on which the designs of ducts were based, has been selected to provide a standard blower-furnace combination. The actual drop was not calculated because each duct selected is large enough to handle the air volume within the allowable pressure drop.

8.9   DUCT DESIGN EXAMPLE (COOLING)

1. Determine the heating load and heat gain of each room, and tabulate them.

The cooling load (Chapter 5) should be computed following the methods previously outlined (see Table 8.25).

2. The cooling unit size is selected based on the heat gain calculated. The unit selected should be as close as possible to the computed heat gain.

When the heat gain calculations are more than the capacity of an available unit, but the next available unit would provide far too much capacity, the designer may want to review the calculations and suggest changes (perhaps in insulation, type of glass, or sunshields) that will reduce the load.

From Step 1, the heat gain calculations are 23 150 Btu/hr (about 2 tons). A 2½ ton unit is selected to provide enough capacity to allow for a pick-up allowance and for cooling lost in the ductwork.

3. Determine the location and number of supply outlets and return air intakes. The layout should allow:

Heat loss: no more than 8000 Btu/hr (2340 watts) per outlet.

Heat gain: no more than 4000 Btu/hr (1170 W) per outlet.

Extra outlets may be desirable in some rooms to provide the best air distribution. This is particularly true in large rooms.

TABLE 8.25   TABULATE HEAT LOSS AND GAIN.

SI UNITS

Supply Outlet	Heat Gain (BTUH)	Heat Loss (BTUH)	Supply Outlet	Heat Gain (Watts)	Heat Loss (Watts)
Bedroom #1	2,887	4,529	Bedroom #1	849	1324
Bedroom #2	1,330	2,772	Bedroom #2	403	807
Living	3,562	4,664	Living	981	1345
Living	3,562	4,663	Living	980	1345
Kitchen	2,341	2,080	Kitchen	706	603
Kitchen	2,341	2,080	Kitchen	706	602
Dining	2,175	3,056	Dining	628	898
Entry	1,038	2,031	Entry	313	644
Hall	1,037	2,030	Hall	312	643
Bedroom #3	2,180	3,483	Bedroom #3	666	1013
Bath	697	1,092	Bath	220	322
Totals	23,150	32,480	Totals	6764	9546

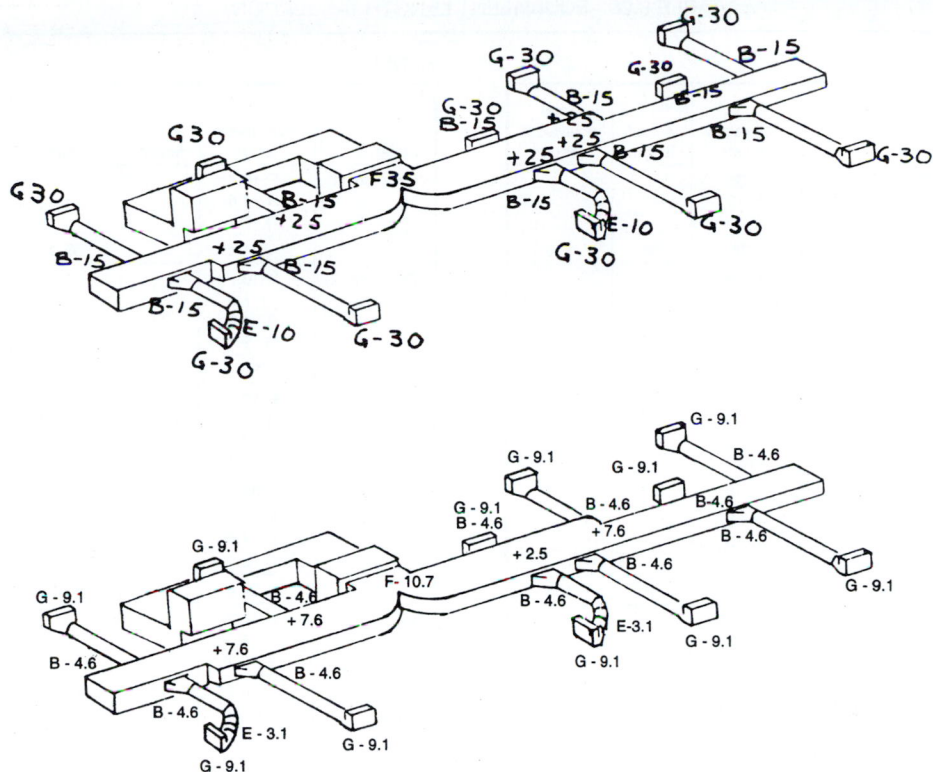

FIGURE 8.13 Duct layout for example.

4. Note on the sketch the types of fittings that will be used and the actual length of each run from the furnace to the outlet. The fittings are the same as those used for heating and are also included at the end of this chapter. Fittings may be selected from Figure 8.15 (pp. 305–308), and the equivalent lengths are given for each type of fitting shown in the illustrations. Note the equivalent length of each fitting on the sketch.

The sketch in Figure 8.13 shows the actual length of duct, the type of fitting, and each fitting's equivalent length.

5. Determine the total equivalent length of each run from bonnet to outlet by adding the actual length and the equivalent length of each fitting in the run, and tabulate the totals.

The tabulated total equivalent length of each run is calculated in Tables 8.26 and 8.27.

6. In metric (SI) units, determine the total air volume required for the system. First determine the air volume required in ft³/min per ton or L/s per 3510 W from Table 8.28. Using this information, the air volume for the system can be found as shown here.

In this design, the ft³/min for cooling is

$$400 \cdot \frac{23\,150}{12\,000} = 772 \text{ ft}^3/\text{min}$$

The ft³/min for heating is

$$400 \cdot \frac{32\,480}{12\,000} = 1089 \text{ ft}^3/\text{min (use 1200 ft}^3/\text{min)}$$

In this design in metric (SI) units, the L/s for cooling is

$$190 \text{ L/s} \cdot \frac{6764}{3510} = 366 \text{ L/s}$$

The L/s for heating is

$$190 \text{ L/s} \cdot \frac{9546}{3510} = 518 \text{ L/s (use 550)}$$

7. The heating requirement for each supply outlet is found by dividing the number of outlets in each room into the heat loss of each room.

In this problem, the living room and kitchen have two outlets; they are noted twice in Table 8.27.

8. Calculate the air volume that will be delivered through each supply outlet. The air volume is calculated by dividing the Btu/hr for the outlet by the total Btu/hr or heat gain and then multiplying it times the total ft³/min.

In Bedroom 1, the heat gain is 2887. With a total heat gain of 23 313, the system is 1200 ft³/min.

$$\frac{2887}{23\,150} \cdot 1200 = 150 \text{ ft}^3/\text{min}$$

TABLE 8.26   EQUIVALENT LENGTH TABULATION.

Outlet	Equivalent length	Actual length	Total equivalent length
BEDROOM (1)	F 35		
	B 15		
	G 30		
	80	36	116
BEDROOM (2)	F 35		
	+ 25		
	B 15		
	G 30		
	105	26	131
LIVING 1	F 35		
	B 15		
	+ 25		
	G 30		
	105	33	138
2	F 35		
	B 15		
	G 30		
	80	55	135
KITCHEN 1	F 35		
	B 15		
	G 30		
	80	46	126
KITCHEN 2	F 35		
	B 15		
	G 30		
	80	40	120
DINING	F 35		
	+ 25		
	B 15		
	G 30		
	105	27	132
ENTRY	F 35		
	+ 25		
	B 15		
	E 10		
	G 30		
	115	16	131
HALL	F 35		
	B 15		
	G 30		
	80	10	90
BEDROOM (3)	F 35		
	+ 25		
	B 15		
	G 30		
	105	30	135
BATH	F 35		
	B 15		
	E 10		
	G 30		
	90	25	115

SI UNITS

Outlet	Equivalent length	Actual length	Total equivalent length
Bedroom (1)	F 10.7		
	B 4.6		
	G 9.1		
	24.4	11	35.4
Bedroom (2)	F 10.7		
	+ 7.6		
	B 4.6		
	G 9.1		
	32	7.9	39.9
Living 1	F 10.7		
	B 4.6		
	+ 7.6		
	G 9.1		
	32	10	42
2	F 10.7		
	B 4.6		
	G 9.1		
	24.4	16.8	41.2
Kitchen 1	F 10.7		
	B 4.6		
	G 9.1		
	24.4	14	38.4
Kitchen 2	F 10.7		
	B 4.6		
	G 9.1		
	24.4	12	36.4
Dining	F 10.7		
	+7.6		
	B 4.6		
	G 9.1		
	32	8.2	40.2
Entry	F 10.7		
	+ 7.6		
	B 4.6		
	E 3.1		
	G 9.1		
	35.1	5	40.1
Hall	F 10.7		
	B 4.6		
	G 9.1		
	24.4	3.1	27.5
Bedroom (3)	F 10.7		
	+ 7.6		
	B 4.6		
	G 9.1		
	32	3.1	35.1
Bath	F 10.7		
	B 4.6		
	E 3.1		
	G 9.1		
	27.5	7.6	35.1

In metric (SI) units, in Bedroom 1, the heat gain is 849 W. With a total heat gain of 6764, the system is 550 L/s.

$$\frac{849}{6764} \cdot 550 = 69 \text{ L/s}$$

This is repeated for each outlet, for both heat gain and heat loss, and tabulated as shown in Table 8.29.

9. The bonnet pressure is selected next. The approximate bonnet pressure required for a trunk (main duct) to carry the maximum volume of air (either heating or cooling ft³/min, whichever is larger) is found in Figure 8.15 (pp. 305–308).

In this design, the total ft³/min (L/s) required is 1200 ft³/min (550 L/s). From Figure 8.15 for quiet operation in a residence (Type A), the suggested bonnet pressure

TABLE 8.27   DUCT LENGTHS.

Supply Outlet	Heat Gain (BTUH)	Heat Loss (BTUH)	Actual Length (ft)	Equivalent Length (ft)	Total Equiv. Length (ft)
Bedroom #1	2,887	4,529	36	80	116
Bedroom #2	1,330	2,772	26	105	131
Living	3,562	4,664	33	105	138
Living	3,562	4,663	55	80	135
Kitchen	2,341	2,080	46	80	126
Kitchen	2,341	2,080	40	80	120
Dining	2,175	3,056	27	105	132
Entry	1,038	2,031	16	115	131
Hall	1,037	2,030	10	80	90
Bedroom #3	2,180	3,483	30	105	135
Bath	697	1,092	25	90	115

SI UNITS

Supply Outlet	Heat Gain (Watts)	Heat Loss (Watts)	Actual Length	Equivalent Length	Total Equiv. Length
Bedroom #1	849	1324	11	24.4	35.4
Bedroom #2	403	807	7.9	32	39.9
Living	981	1345	10	32	42
Living	980	1345	16.8	24.4	41.2
Kitchen	706	603	12	24.4	36.4
Kitchen	706	602	14	24.4	38.4
Dining	628	898	8.2	32	40.2
Entry	313	644	5	35.1	40.1
Hall	312	643	3.1	24.4	27.5
Bedroom #3	666	1013	3.1	32	35.1
Bath	220	322	7.6	27.5	35.1

TABLE 8.28   AIR FLOW VOLUME.

Air Flow	cfm/ton	L/s per 3510 W
Typical*	400	190
Some Heat Pumps	450	215
Humid Areas	360	170
Dry Areas	429	204

*For normal residential cooling (70% sensible and 30% latent heat)
Reprinted with permission from ASHRAE, *HVAC Handbook*, 1987.
Note: Metric added by author

is about 0.10 in (24.8 Pa) of water; assume a system with 0.10 (24.8) is selected.

10. Next, the supply outlet size and its pressure loss are selected from the manufacturer's engineering data. A typical example of manufacturer's data is shown in Figure 8.14 (on p. 288). The exact pressure loss will depend on variables such as the duct velocity and the angles at which the register blades are set. Typically, the pressure loss will range from 0.01 to 0.02 in (2.4 to 4.8 L/s); many designers assume a 0.02 in (4.8 L/s) loss and make

TABLE 8.29 BRANCH AIR FLOW.

Supply Outlet	Heat Gain (BTUH)	Heat Loss (BTUH)	Actual Length (ft)	Equivalent Length (ft)	Total Equiv. Length (ft)	Cooling cfm	Heating cfm
Bedroom #1	2,887	4,529	36	80	116	150	167
Bedroom #2	1,330	2,772	26	105	131	69	103
Living	3,562	4,662	33	105	138	185	173
Living	3,562	4,663	55	80	135	185	173
Kitchen	2,341	2,080	46	80	126	121	77
Kitchen	2,341	2,080	40	80	120	121	77
Dining	2,175	3,056	27	105	132	113	113
Entry	1,038	2,031	16	115	131	54	75
Hall	1,037	2,030	10	80	90	54	75
Bedroom #3	2,180	3,483	30	105	135	113	129
Bath	697	1,092	25	90	115	36	40

SI UNITS

Supply Outlet	Heat Gain (Watts)	Heat Loss (Watts)	Actual Length	Equivalent Length	Total Equiv. Length	Cooling (L/s)	Heating (L/s)
Bedroom #1	849	1314	11	24.4	35.4	69	77
Bedroom #2	403	807	7.9	32	39.9	33	49
Living	981	1345	10	32	42	80	78
Living	980	1345	16.8	24.4	41.2	80	78
Kitchen	706	603	12	24.4	36.4	58	35
Kitchen	706	602	14	24.4	38.4	58	34
Dining	628	898	8.2	32	40.2	51	52
Entry	313	644	5	35.1	40.1	25	37
Hall	312	643	3.1	24.4	27.5	25	37
Bedroom #3	666	1013	3.1	32	35.1	54	59
Bath	220	322	7.6	27.5	35.1	18	19

LISTED WIDTH

6	8	10	12	14	16	18	20	22	24	26	28	30	32	34	36
6	4														
8	6	5		4											
10	8		6		5		4								
12	10		8			6			5						4

V = Duct Vel.	300			400			500			600		
Blade Set °	0	22½	45	0	22½	45	0	22½	45	0	22½	45
Pₜ	.01	.012	.027	.014	.021	.050	.023	.034	.082	.034	.051	.120
CFM	75			100			125			150		
T PWL-NC	8	L		10	L		13	L		16	L	
CFM	130			180			220			260		
T PWL-NC	9	L		13	L		17	L		20	L	
CFM	210			280			350			400		
T PWL-NC	13	L		17	L		21	L		24	28	
CFM	300			400			500			600		
T PWL-NC	15	L		20	L		25	26		29	32	

SYMBOLS:

V = Duct velocity in fpm.
CFM = Quantity of air in cubic ft./min.
NC = Noise criteria (8 db room attenuation).
D = Drop in feet.

P_t = Total pressure inches H_2O
T = Throw in Feet
PWL-NC INDEX = A single number which expresses the PWL (sound power level) in relation to NC (noise criteria) curves.

FIGURE 8.14 Supply outlet data for example.

TABLE 8.30   SUGGESTED BONNET AND RETURN PRESSURES (INCHES OF WATER).

CFM	Type A	Type B	Type C
1000	0.10	0.15	0.20
1500	.12	.17	.22
2000	.13	.18	.24
2500	.14	.20	.27
3000	.16	.22	.29
3500	.17	.24	.31
4000	.18	.26	.33
4500	0.20	0.28	0.36
5000	.21	.29	.38
5500	.22	.31	.40
6000	.24	.33	.42
6500	.25	.34	.45
7000	.27	.36	.47
7500	.28	.38	.49
8000	0.29	0.39	0.51
8500	.30	.41	.53
9000	.32	.43	.56
9500	.34	.45	.58
10,000	.34	.47	.60
10,500	.37	.48	.60
11,000	.39	.50	.60
11,500	0.40	0.50	0.60
12,000	.40	.50	.60
12,500	.40	.50	.60
13,000	.40	.50	.60
14,000	.40	.50	.60

CFM = Maximum cfm in any one duct.
Type A = Systems for quiet operation in residences, churches, concert halls, broadcasting studios, funeral homes, etc.
Type B = Systems for schools, theaters, public buildings, etc.
Type C = Systems for industrial buildings.

a final selection later. Suggested bonnet and return pressures are shown in Table 8.30.

The selected supply outlet pressure losses are tabulated in Table 8.31.

11. The available bonnet pressure is divided between the supply and the return runs. It should be divided in proportion to the number of supply and return runs on the project. A review of the sketch plan layout shows the approximate proportions of supply to return runs.

A review of the sketch plan layout (Figure 8.13) shows that there are considerably more supply than return runs, so a division of 0.07 in (17.36 Pa) for supply and 0.03 in (7.44 Pa) for return is selected.

Note: If later during the design the proportion selected does not provide satisfactory results, the available bonnet pressure can be reapportioned between supply and return.

12. Determine the pressure available for duct loss by subtracting the supply outlet loss (Step 10) from the total pressure available for supply runs (Step 11).

In this design, the pressure allowed for supply runs is 0.07 in (17.36 Pa) (see Step 11). Beginning with the supply run in Bedroom 1, the supply outlet loss is 0.02 in (4.8 Pa) (see Step 11), and the pressure available for duct drop is 0.05 in (12.56 Pa). Calculate the pressure available for duct loss for each of the supply runs, and tabulate the information as shown in Table 8.31.

13. To make a workable table of duct sizes, the pressure drop for duct loss must be calculated. Because the various supply runs are all of different lengths, it is necessary to find the *allowable pressure drop per 100 ft* based on the *allowable duct loss* and the *total equivalent length* of the supply run. This allowable pressure drop may be calculated by using the equation:

$$\text{Allowable pressure drop per 100 ft} = \frac{\text{Allowable duct loss} \cdot 100}{\text{Total equivalent length}}$$

In this design, in Bedroom 1, the supply run has an allowable duct loss of 0.05 in (see Step 12) and an equivalent length of 116 ft.

TABLE 8.31   PRESSURE TABULATION.

Supply Outlet	Heat Gain (BTUH)	Heat Loss (BTUH)	Actual Length (ft)	Equivalent Length (ft)	Total Equiv. Length (ft)	Cooling cfm	Heating cfm	Supply Air Pressure	Supply Outlet Pressure Loss	Pressure Avail. for Duct Loss
Bedroom #1	2,887	4,529	36	80	116	150	167	0.07	0.02	0.05
Bedroom #2	1,330	2,772	26	105	131	69	103	0.07	0.02	0.05
Living	3,562	4,662	33	105	138	185	173	0.07	0.02	0.05
Living	3,562	4,663	55	80	135	185	173	0.07	0.02	0.05
Kitchen	2,341	2,080	46	80	126	121	77	0.07	0.02	0.05
Kitchen	2,341	2,080	40	80	120	121	77	0.07	0.02	0.05
Dining	2,175	3,056	27	105	132	113	113	0.07	0.02	0.05
Entry	1,038	2,031	16	115	131	54	75	0.07	0.02	0.05
Hall	1,037	2,030	10	80	90	54	75	0.07	0.02	0.05
Bedroom #3	2,180	3,483	30	105	135	113	129	0.07	0.02	0.05
Bath	697	1,092	25	90	115	36	40	0.07	0.02	0.05

SI UNITS

Supply Outlet	Heat Gain (Watts)	Heat Loss (Watts)	Actual Length	Equivalent Length	Total Equiv. Length	Cooling (L/s)	Heating (L/s)	Supply Air Pressure (Pa)	Supply Outlet Pressure Loss (Pa)	Pressure Avail. for Duct Loss
Bedroom #1	849	1314	11	24.4	35.4	69	77	17.36	4.8	12.56
Bedroom #2	403	807	7.9	32	39.9	33	49	17.36	4.8	12.56
Living	981	1345	10	32	42	80	78	17.36	4.8	12.56
Living	980	1345	16.8	24.4	41.2	80	78	17.36	4.8	12.56
Kitchen	706	603	12	24.4	36.4	58	35	17.36	4.8	12.56
Kitchen	706	602	14	24.4	38.4	58	34	17.36	4.8	12.56
Dining	628	898	8.2	32	40.2	51	52	17.36	4.8	12.56
Entry	313	644	5	35.1	40.1	25	37	17.36	4.8	12.56
Hall	312	643	3.1	24.4	27.5	25	37	17.36	4.8	12.56
Bedroom #3	666	1013	3.1	32	35.1	54	59	17.36	4.8	12.56
Bath	220	322	7.6	27.5	35.1	18	19	17.36	4.8	12.56

Using the formula, the allowable pressure drop per 100 ft is:

Allowable pressure drop per 100 ft

$$= \frac{0.05 \text{ in } (100)}{116 \text{ ft}} = 0.0431 \text{ in}$$

Calculate the allowable pressure drop per 100 ft for each supply run, and tabulate the information as shown in Table 8.32.

In metric (SI) units:

Allowable pressure loss

$$= \frac{\text{Allowable duct loss}}{\text{Total equivalent length}}$$

In this design, in Bedroom 1, the supply run has an allowable duct loss of 12.56 Pa (see Step 12) and an equivalent length of 35.4 m. Using the formula, the

TABLE 8.32   AVAILABLE PRESSURE TABULATION.

Supply Outlet	Heat Gain (BTUH)	Heat Loss (BTUH)	Actual Length (ft)	Equivalent Length (ft)	Total Equiv. Length (ft)	Cooling cfm	Heating cfm	Supply Air Pressure	Supply Outlet Pressure Loss	Pressure Avail. for Duct Loss	Pressure Avail. Per 100 ft
Bedroom #1	2,887	4,529	36	80	116	150	167	0.07	0.02	0.05	0.043
Bedroom #2	1,330	2,772	26	105	131	69	103	0.07	0.02	0.05	0.038
Living	3,562	4,662	33	105	138	185	173	0.07	0.02	0.05	0.036
Living	3,562	4,663	55	80	135	185	173	0.07	0.02	0.05	0.037
Kitchen	2,341	2,080	46	80	126	121	77	0.07	0.02	0.05	0.040
Kitchen	2,341	2,080	40	80	120	121	77	0.07	0.02	0.05	0.042
Dining	2,175	3,056	27	105	132	113	113	0.07	0.02	0.05	0.038
Entry	1,038	2,031	16	115	131	54	75	0.07	0.02	0.05	0.038
Hall	1,037	2,030	10	80	90	54	75	0.07	0.02	0.05	0.056
Bedroom #3	2,180	3,483	30	105	135	113	129	0.07	0.02	0.05	0.037
Bath	697	1,092	25	90	115	36	40	0.07	0.02	0.05	0.043

SI UNITS

Supply Outlet	Heat Gain (Watts)	Heat Loss (Watts)	Actual Length	Equivalent Length	Total Equiv. Length	Cooling (L/s)	Heating (L/s)	Supply Air Pressure (Pa)	Supply Outlet Pressure Loss (Pa)	Pressure Avail. for Duct Loss	Pressure Avail. (Pa)
Bedroom #1	849	1314	11	24.4	35.4	69	77	17.36	4.8	12.56	0.35
Bedroom #2	403	807	7.9	32	39.9	33	49	17.36	4.8	12.56	0.31
Living	981	1345	10	32	42	80	78	17.36	4.8	12.56	0.30
Living	980	1345	16.8	24.4	41.2	80	78	17.36	4.8	12.56	0.30
Kitchen	706	603	12	24.4	36.4	58	35	17.36	4.8	12.56	0.35
Kitchen	706	602	14	24.4	38.4	58	34	17.36	4.8	12.56	0.31
Dining	628	898	8.2	32	40.2	51	52	17.36	4.8	12.56	0.31
Entry	313	644	5	35.1	40.1	25	37	17.36	4.8	12.56	0.31
Hall	312	643	3.1	24.4	27.5	25	37	17.36	4.8	12.56	0.46
Bedroom #3	666	1013	3.1	32	35.1	54	59	17.36	4.8	12.56	0.36
Bath	220	322	7.6	27.5	35.1	18	19	17.36	4.8	12.56	0.36

pressure drop per meter is:

$$\text{Allowable pressure loss} = \frac{12.56}{35.4} = 0.3548 \text{ (use 0.35)}$$

Calculate the allowable pressure drop for each of the supply runs and tabulate the information as shown in Table 8.32.

14. Round branch duct sizes may be determined by using Figure 8.15 or by using an air duct calculator if one is available. Figure 8.15 gives the size of round ducts based on the larger ft^3/min required (either heating or cooling) and the allowable pressure drop (Step 13). The size of the round duct may be converted to an equivalent rectangular duct size (Table 8.41, pp. 300–303) that will

handle the required ft^3/min within the allowable pressure drop.

In this design, Bedroom 1 has an allowable duct loss of 0.043 in per 100 ft (Step 13). Using the table in Figure 8.16, a 9.0 in round duct is selected.

In this design, in metric (SI) units, Bedroom 1 has an allowable duct loss of 0.35 Pa per meter. Using the table in Figure 8.16, a 250 mm round duct is selected.

a. Find the allowable pressure drop (along the bottom) (friction loss).

b. Air quantity, ft^3/min, is found along the bottom.

c. Move horizontally from the friction loss and vertically from the air quantity until the two lines meet.

d. This intersection will occur on or near the diagonal lines that will give the duct diameters required (sizes on the line). If it falls right on the line, using that size duct will use the amount of friction loss available. If it falls between two lines, using the larger duct will use slightly less friction loss (the smaller duct would use more).

From Figure 8.15, determine the round duct sizes for each of the supply runs and tabulate the information as shown in Table 8.33.

15. Round duct sizes selected in Step 14 may be changed to equivalent rectangular duct sizes, carrying the same air volume required while maintaining the allowable pressure drops, by using Figure 8.15 or by using an air duct calculator.

The round duct for Bedroom 1 can be changed to a comparable rectangular duct by using Table 8.41. To use the table, the sizes of the rectangular duct are found with one dimension along the top and the other along the side. Their circular equivalent is the number in the middle of the page where the two sizes intersect.

To convert circular ducts to rectangular, find the circular equivalent in the center of the table and get the rectangular equivalents by reading up and to the side. There are many different sizes that can be used. The actual size selected depends to a large extent on the space available for installation of the duct.

From Table 8.41, determine a rectangular size for each supply run that could be used instead of the round duct previously selected. Tabulate the rectangular duct size as shown in Table 8.34.

TABLE 8.33   ROUND DUCT SIZE TABULATION.

Supply Outlet	Heat Gain (BTUH)	Heat Loss (BTUH)	Actual Length (ft)	Equivalent Length (ft)	Total Equiv. Length (ft)	Cooling cfm	Heating cfm	Supply Air Pressure	Supply Outlet Pressure Loss	Pressure Avail. for Duct Loss	Pressure Avail. Per 100 ft	Round Duct Size (in.)
Bedroom #1	2,887	4,529	36	80	116	150	167	0.07	0.02	0.05	0.043	9.0
Bedroom #2	1,330	2,772	26	105	131	69	103	0.07	0.02	0.05	0.038	8.0
Living	3,562	4,662	33	105	138	185	173	0.07	0.02	0.05	0.036	9.0
Living	3,562	4,663	55	80	135	185	173	0.07	0.02	0.05	0.037	9.0
Kitchen	2,341	2,080	46	80	126	121	77	0.07	0.02	0.05	0.040	8.0
Kitchen	2,341	2,080	40	80	120	121	77	0.07	0.02	0.05	0.042	8.0
Dining	2,175	3,056	27	105	132	113	113	0.07	0.02	0.05	0.038	8.0
Entry	1,038	2,031	16	115	131	54	75	0.07	0.02	0.05	0.038	7.0
Hall	1,037	2,030	10	80	90	54	75	0.07	0.02	0.05	0.056	6.0
Bedroom #3	2,180	3,483	30	105	135	113	129	0.07	0.02	0.05	0.037	8.0
Bath	697	1,092	25	90	115	36	40	0.07	0.02	0.05	0.043	6.0

SI UNITS

Supply Outlet	Heat Gain (Watts)	Heat Loss (Watts)	Actual Length	Equivalent Length	Total Equiv. Length	Cooling (L/s)	Heating (L/s)	Supply Air Pressure (Pa)	Supply Outlet Pressure Loss (Pa)	Pressure Avail. for Duct Loss	Pressure Avail. (Pa)	Round Duct Size (mm)
Bedroom #1	849	1314	11	24.4	35.4	69	77	17.36	4.8	12.56	0.35	250
Bedroom #2	403	807	7.9	32	39.9	33	49	17.36	4.8	12.56	0.31	200
Living	981	1345	10	32	42	80	78	17.36	4.8	12.56	0.30	250
Living	980	1345	16.8	24.4	41.2	80	78	17.36	4.8	12.56	0.30	250
Kitchen	706	603	9	24.4	36.4	58	35	17.36	4.8	12.56	0.35	200
Kitchen	706	602	14	24.4	38.4	58	34	17.36	4.8	12.56	0.31	200
Dining	628	898	8.2	32	40.2	51	52	17.36	4.8	12.56	0.31	200
Entry	313	644	5	35.1	40.1	25	37	17.36	4.8	12.56	0.31	160
Hall	312	643	3.1	24.4	27.5	25	37	17.36	4.8	12.56	0.46	160
Bedroom #3	666	1013	3.1	32	35.1	54	59	17.36	4.8	12.56	0.36	200
Bath	220	322	7.6	27.5	35.1	18	19	17.36	4.8	12.56	0.36	160

TABLE 8.34   RECTANGULAR DUCT SIZE TABULATION.

Supply Outlet	Heat Gain (BTUH)	Heat Loss (BTUH)	Actual Length (ft)	Equivalent Length (ft)	Total Equiv. Length (ft)	Cooling cfm	Heating cfm	Supply Air Pressure	Supply Outlet Pressure Loss	Pressure Avail. for Duct Loss	Pressure Avail. Per 100 ft	Round Duct Size (in.)	Rectangular Duct Size (in.)
Bedroom #1	2,887	4,529	36	80	116	150	167	0.07	0.02	0.05	0.043	9.0	7 × 10
Bedroom #2	1,330	2,772	26	105	131	69	103	0.07	0.02	0.05	0.038	8.0	6 × 9
Living	3,562	4,662	33	105	138	185	173	0.07	0.02	0.05	0.036	9.0	7 × 10
Living	3,562	4,663	55	80	135	185	173	0.07	0.02	0.05	0.037	9.0	7 × 10
Kitchen	2,341	2,080	46	80	126	121	77	0.07	0.02	0.05	0.040	8.0	6 × 9
Kitchen	2,341	2,080	40	80	120	121	77	0.07	0.02	0.05	0.042	8.0	6 × 9
Dining	2,175	3,056	27	105	132	113	113	0.07	0.02	0.05	0.038	8.0	6 × 9
Entry	1,038	2,031	16	115	131	54	75	0.07	0.02	0.05	0.038	7.0	6 × 7
Hall	1,037	2,030	10	80	90	54	75	0.07	0.02	0.05	0.056	6.0	4.5 × 7
Bedroom #3	2,180	3,483	30	105	135	113	129	0.07	0.02	0.05	0.037	8.0	6 × 9
Bath	697	1,092	25	90	115	36	40	0.07	0.02	0.05	0.043	6.0	4.5 × 7

SI UNITS

Supply Outlet	Heat Gain (Watts)	Heat Loss (Watts)	Actual Length	Equivalent Length	Total Equiv. Length	Cooling (L/s)	Heating (L/s)	Supply Air Pressure (Pa)	Supply Outlet Pressure Loss (Pa)	Pressure Avail. for Duct Loss	Pressure Avail. (Pa)	Round Duct Size (mm)	Rectangular Duct Size (mm)
Bedroom #1	883	1314	11	24.4	35.4	70	77	17.36	4.8	12.56	0.35	250	150 × 400
Bedroom #2	410	807	7.9	32	39.9	33	49	17.36	4.8	12.56	0.31	200	150 × 225
Living	1027	1345	10	32	42	82	78	17.36	4.8	12.56	0.30	250	150 × 400
Living	1027	1345	16.8	24.4	41.2	82	78	17.36	4.8	12.56	0.30	250	150 × 400
Kitchen	706	597	12	24.4	36.4	56	35	17.36	4.8	12.56	0.35	200	150 × 225
Kitchen	706	597	14	24.4	38.4	56	34	17.36	4.8	12.56	0.31	200	150 × 225
Dining	628	898	8.2	32	40.2	50	52	17.36	4.8	12.56	0.31	200	150 × 225
Entry	313	644	5	35.1	40.1	25	37	17.36	4.8	12.56	0.31	160	100 × 200
Hall	312	643	3.1	24.4	27.5	25	37	17.36	4.8	12.56	0.46	160	100 × 200
Bedroom #3	666	1013	3.1	32	35.1	53	59	17.36	4.8	12.56	0.36	200	150 × 225
Bath	229	322	7.6	27.5	35.1	18	19	17.36	4.8	12.56	0.36	160	100 × 200

16. Add the air volume of all the branch supply runs from each trunk duct. When there is more than one trunk duct (such as in Figure 8.13), keep the totals for each trunk duct separate.

In this design, there are two trunk ducts, one toward the living room and kitchen, the other toward the bedrooms. Tabulate and total the air volume for each trunk duct.

The totals for both heating and cooling must be tabulated for each trunk and the largest value used. The trunk serving the living areas will have a higher cooling air volume requirement, while the trunk serving the bedroom areas will have a higher heating air volume requirement. Each trunk must be sized for the maximum air volume it will receive.

17. For best airflow distribution through the ducts, it is important that the friction losses be approximately equal in both trunk ducts.

Using the table in Figure 8.15, select the trunk sizes required. When there is more than one trunk duct, size each separately using the required air volume (ft^3/min) and the allowable duct friction (Step 12).

The bedroom trunk must handle 443 ft^3/min with an allowable pressure drop of 0.038 in per 100 ft. Using Figure 8.15, the bedroom trunk duct will be a 14 in round duct.

The living space trunk duct has 757 ft^3/min and an allowable 0.036 in pressure drop per 100 ft. Using Figure 8.15, the living space trunk duct will be a 16.0 in round duct.

Because the trunk ducts have been sized separately, based on the ft^3/min and the allowable pressure drop, they are balanced.

The round duct sizes may be converted to rectangular duct sizes by use of Table 8.41. Once again, the space available for installation will have a large influence on the shape of the rectangular duct selected.

The bedroom trunk must handle 202 L/s with an allowable pressure drop of 0.31 Pa. Using Figure 8.16, the bedroom trunk will be a 315 mm round duct.

The living room trunk duct has 376 L/s and an allowable pressure drop of 0.30 Pa. Using Figure 8.16, the living space trunk duct will be a 400 mm round duct.

18. Review the supply trunk duct to determine if it is desirable to reduce its size as each branch duct leaves the trunk. Generally, such a reduction is suggested so that the velocity of air through the duct will not drop too low. Using the remaining air volume in the trunk duct, after each branch and the allowable pressure drop (Step 9), select the reduced trunk sizes from Figure 8.15 and Table 8.41 (pp. 300–308).

Using the information in Table 8.36 and the tables in Figure 8.15 and Table 8.41, determine the reduced trunk duct sizes for both the bedroom and the living space ducts, and tabulate the sizes as shown in Table 8.37.

19. Next, the design turns to the return ducts. The first step is to select an allowable RA pressure drop. This was previously decided (Step 11), but it is reviewed at this point as the designer reviews the supply trunk and branch duct sizes to see if it is desirable to increase the duct sizes (reducing the allowable air pressure drop) or perhaps to reduce the duct sizes (increasing the allowable air pressure drop for the supply runs). Remember that a change in allowable air pressure drop for the supply will affect the allowable air pressure drop for the return runs (Step 11).

A review of the supply duct sizes is made, and it is decided that they will not be revised, as duct sizes selected so far seem reasonable. Therefore, the original decision, in Step 11, to use 0.03 in (7.44 Pa) as the allowable pressure drop for the return, is unchanged and the return design will be based on it.

20. The return trunk duct size is determined by the air volume it must handle and the allowable pressure drop. This allowable pressure drop must be converted to allowable pressure drop, just as was done for the supply runs. Using the air volume and the allowable pressure drop, the round return duct size may be selected from Figure 8.5. The rectangular duct of equivalent size may be found in Table 8.41.

The return trunk duct will have to handle an air volume equal to what the supply trunk duct handles; from Table 8.35, this is 1200 ft³/min. The allowable pressure drop selected for the return in 0.03 in (Steps

TABLE 8.35   TABULATED MAIN DUCT SIZES.

Duct	cfm	Supply Static Pressure*	Total Equiv. Length (ft)	Supply Pressure Avail. Per 100 ft	Round Duct Size (in.)	Rectangular Duct Size (in.)
Main Branch	1,200	0.05	138	0.036	18.0	10 × 28
Bedroom Branch	439**	0.05	131	0.038	14.0	10 × 18
Living Branch	863***	0.05	138	0.036	16.0	10 × 20

*Allows 0.02 for supply outlet
**Based on heating requirements
***Based on cooling requirements

SI UNITS

Duct	L/s	Supply Static Pressure* (Pa)	Total Equiv. Length (m)	Supply Pressure Avail. Per branch	Round Duct Size (mm)	Rectangular Duct Size (mm)
Main Branch	550	12.56	42.1	0.30	500	200 × 1300
Bedroom Branch	202**	12.56	40	0.31	315	200 × 450
Living Branch	377***	12.56	42.1	0.30	400	200 × 750

*Allows 0.02 for supply outlet
**Based on heating requirement
***Based on cooling requirement

TABLE 8.36   REDUCED TRUNK SIZES.

Duct		cfm	Supply Static Pressure	Total Equiv. Length (ft)	Supply Pressure Avail. Per 100 ft	Round Duct Size (in.)	Rectangular Duct Size (in.)
Main Branch	A-B	1,200	0.05	138	0.036	18.0	10 × 28
Bedroom Branch	*B-C	439	0.05	131	0.038	14.0	10 × 18
	C-D	336					
	D-E	207					
	E-F	167					
Living Branch	*B-G	863	0.05	138	0.036	16.0	10 × 20
	G-H	788					
	H-I	713					
	I-J	600					
	J-K	427					
	K-L	306					
	L-M	185					

*Does not total the main because the maximum need of heating or air conditioning must be used for each length.

SI UNITS

Duct		L/s	Supply Static Pressure*	Total Equiv. Length	Supply Pressure Avail.	Round Duct Size	Rectangular Duct Size
Main	A-B	550	12.56	42.1	0.30	500	200 × 1300
Bedroom Branch	*B-C	204	12.56	40	0.31	315	200 × 450
	C-D	155					
	D-E	96					
	E-F	77					
Living Branch	*B-G	377	12.56	42.1	0.30	400	200 × 750
	G-H	297					
	H-I	239					
	I-J	181					
	J-K	101					
	K-L	50					
	L-M	25					

*Does not total the main because the maximum need of heating or air conditioning must be used for each length.

11 and 12). First, the allowable pressure drop must be converted into the allowable pressure drop *per* 100 ft (Step 13). The equivalent length is shown in Table 8.38.

$$(0.03 \text{ in} \cdot 100) \, 4 \div 82 \text{ ft} = 0.036 \text{ in}$$
allowable pressure drop per 100 ft

In metric (SI) units, the return duct will have to handle the air volume equal to that handled by the supply duct. From Table 8.35, this is 550 L/s. The allowable pressure drop calculated for the return is 0.30 (using the information from Steps 11 and 12 and the total equivalent length).

Using the allowable pressure drop, select the round return duct size from Figure 8.15 and its equivalent rectangular size from Table 8.41. Tabulate the information as shown in Table 8.38.

If the return trunk sizes are too large, they may be reduced in size by the following:

a. Reapportioning the pressure drop available so there is less drop allowed for the supply runs and more pressure allowed for the return runs. This may require a revision of supply branch and trunk duct sizes.

TABLE 8.37   REDUCED TRUNK SIZES.

Duct		cfm	Supply Static Pressure	Total Equiv. Length (ft)	Supply Pressure Avail. Per 100 ft	Round Duct Size (in.)	Rectangular Duct Size (in.)
Main	A-B	1,200	0.05	138	0.036	18.0	10 × 28
Bedroom Branch	B-C	439	0.05	131	0.038	14.0	10 × 18
	C-D	336				12.0	10 × 13
	D-E	207				9.0	10.0 × 11.7
	E-F	167				6.0	9.0 × 8.8
Living Branch	B-G	863	0.05	138	0.036	16.0	10 × 20
	G-H	788				14.0	10 × 18
	H-I	713				14.0	10 × 18
	I-J	600				14.0	10 × 18
	J-K	427				12.0	10 × 13
	K-L	306				10.0	10 × 8
	L-M	185					

SI UNITS

Duct		L/s	Supply Static Pressure	Total Equiv. Length	Supply Pressure Avail.	Round Duct Size	Rectangular Duct Size
Main	A-B	550	12.56	42.1	0.30	500	200 × 1300
Bedroom Branch	B-C	204	12.56	40	0.31	315	200 × 450
	C-D	155				315	200 × 450
	D-E	96				250	200 × 275
	E-F	77				250	200 × 275
Living Branch	B-G	377	12.56	42.1	0.30	400	200 × 750
	G-H	297				400	200 × 750
	H-I	239				400	200 × 750
	I-J	161				315	200 × 450
	J-K	101				250	200 × 275
	K-L	50				200	200 × 200
	L-M	25				160	200 × 160

TABLE 8.38   RETURN DUCT SIZES.

Duct	Cfm	Actual length	Equivalent length	Total equivalent length	Loss per 100′	Round duct size (in)	Rectangular duct size (in)
Return	1200	12	70	82	0.036	18.0	12 × 24

Duct	L/s	Actual length	Equivalent length	Total equivalent length	Return pressure avail.	Round duct size	Rectangular duct size
Return	550	3.7	21.4	25.1	0.30	500	200 × 1300

b. Checking to see if the return air grilles can be located closer to the furnace, which would reduce the actual length of duct. This reduces the allowable pressure drop, resulting in a smaller duct size.

21. The size of the blower on the furnace is determined by the total air volume (the total ft^3/min to be delivered) and the total static pressure requirements.

The total static pressure for furnace-blower combination units is the total of the actual pressure drops in the supply and return ducts. Most residential and small commercial buildings have furnace-blower combination units.

When the blowers are selected separately from the furnace, the total static pressure is the sum of the actual pressure losses in the supply and return ducts, filter loss, casing loss, and losses through any devices that are put on the system, such as air washers and purifiers.

In this design, the 0.10 in pressure drop, on which the designs of the ducts are based, is selected for a standard blower-furnace combination. Although the actual pressure drop is slightly less than 0.10 in, the actual drop is not calculated because, as each duct is selected, it is large enough to handle the air volume within the allowable pressure drop.

8.10   SYSTEM INSTALLATION

Air distribution systems need space to run ductwork throughout the building. In wood frame construction, the ductwork may be run in a crawl space, basement, or in an attic space. In truss construction, the ducts may be run in the open spaces between the webs. The duct may either be placed in the slab or the attic or ceiling area. In commercial construction, ductwork is typically integrated within the structure such as the space between the ceiling and floor.

The heating and/or cooling unit may be placed in a crawl space, basement, or attic. Small residential and commercial units have also been placed in centrally-located mechanical closets. In commercial installations, equipment is located in a large dedicated mechanical room. The mechanical equipment may be found on the roof, either exposed or contained in a penthouse.

Large ductwork that must run vertically in the building will need to have a space to run. Residential ducts are typically located in a furred-out area of a closet. In commercial installations, vertical ductwork is installed in a *chase* that runs from floor to floor. This chase must be constructed of fire-resistant materials and must be sealed to prevent fire from traveling outside a chase. This is addressed in Chapter 21.

A cooling unit will usually need a condensate drain to dispose of condensed moisture. This is sometimes done by means of a pipe to the exterior or by draining to a floor drain or into a small sump pump that pumps the water into a drain line.

8.11   HEAT PUMPS

The use of refrigeration to provide both heating and cooling for a space is accomplished with a device called a *heat pump*. Heat pumps are cost-effective when space heating is needed and electricity is the only energy source available. Cooling is also available with a heat pump.

With basic cooling principles and a valve control, it is possible to reverse the refrigeration cooling cycle, causing the refrigerant to absorb heat from an outside surrounding medium (usually air, but water is sometimes used) and to release this heat inside the building (in the air or water being used to heat the building).

The reversing valve control allows refrigerant to be used to provide cooling or heating, depending on the direction of flow after it leaves the compressor. If the refrigerant flows to the condenser, it provides cooling; if it flows toward the evaporator section, it provides heating.

The most commonly used surrounding mediums are the OA and the forced air in the ductwork.

OA is the surrounding medium most likely to be used to draw heat from in the winter for heating and to release heat to in the summer for cooling. The temperature of this OA affects the efficiency of the unit in providing heating and cooling. Typically, in the winter as OA temperature decreases, the amount of heat that the heat pump will produce decreases. In effect, the more heat needed in the space, the less heat output the unit will provide.

At this point, it is important to realize that a heat pump is not electric. Actually, it is a vapor compression refrigeration system that draws heat from the OA. It requires electrical power to run the equipment, but it is more efficient than electric resistant heating. The efficiency of the heat pump can be determined from the manufacturer's data. Although there are many sizes and capacities of units available, data for a 3 ton (36 000 Btu/hr) cooling capacity unit will be reviewed.

Note in Table 8.39 that at a 47°F (8°C) outdoor temperature, the unit will provide 36 600 Btu/hr for heating. As the temperature drops, the unit will supply less heat, and yet more heat is required inside the space. When it gets cold (say, 17°F [−8°C]), the unit will provide 19 600 Btu/hr. It is obvious that as the temperature decreases, so does the ability of the heat pump to produce heat. Table 8.40 shows performance data for 2-ton and 4-ton heat pumps with similar output variations.

It is when the heat pump alone is producing the heat that the unit is economical to operate. The coefficient of performance is not always provided by the manufacturer, but it has been calculated for the 3-ton unit in Table 8.39. This coefficient of performance offers a comparison of the heat pump and electric resistance heat. Table 8.39 shows a COP of 2.95 at 47°F (8°C), dropping to 1.80 at 17°F (−8°C). Because of this fluctuation, it

TABLE 8.39   TYPICAL THREE-TON HEAT PUMP.

Typical 3-ton (36,000 Btuh) heat pump
Cooling capacity, 62° outside wet bulb,
85° air temperature entering evaporator.

Outside air temperature		Cooling		Operating	
db°F	db°C	Btuh	Watts	Watts	SEER
85	29	38,600	11,300	3750	10.3
95	35	36,400	10,655	3950	9.2
105	41	32,000	9,370	4210	7.6

Heating capacity

Outside air temperature		Heating		Operating	
		Btuh	Watts	Watts	COP
47°F/8°C		36,600	10,540	3640	2.95
17°F/−8°C		19,600	5,730	3200	1.80

TABLE 8.40   FOUR-TON AND TWO-TON HEAT PUMPS.

4-ton heat pump, heating capacity

Outside air temperature		Heating		Operating
db°F	db°C	Btuh	Watts	Watts
47	8	50,000	10,540	5293
17	−8	29,000	8,490	4315

2-ton heat pump, heating capacity

Outside air temperature		Heating		Operating
db°F	db°C	Btuh	Watts	Watts
47	8	24,600	7,220	3250
17	−8	14,000	4,100	2665

is difficult to give an actual comparison with other fuels and systems. It depends very much on the geographic area it is being used in and how many hours it operates at the different temperatures.

By comparison, electric heat has a COP of 1.0. Electric resistant heating tends to cost about twice as much as oil or natural gas. So, for the heat pump to be competitive, it needs to operate at a COP of about 2.0 and at 40°F (4°C). But at lower outdoor temperatures, it becomes less and less efficient.

Because the heat pump puts out 19 600 Btu/hr (5730 W) at 17°F (−8°C), the remainder of the heat is supplied by electric resistance heating elements, which are mounted inside the evaporator blower discharge area; or by electric duct heaters, which are placed in the ductwork. These elements or heaters are used to supplement the heat pump output during cold weather. These supplemental units are powered by electricity

and provide 3413 Btu/hr per 1000 W (1 kW). These electric resistance units are generally available in increments of 3, 5, 10, 15, and 20 kW, depending on the manufacturer. Many of the smaller heat pump units have space to install only the smaller sizes, perhaps to a limit of 10 kW, and the manufacturer's information must be reviewed. The large units, such as the 3-ton unit being discussed here, will take the 15 kW unit.

The 15-kW unit would be made up of three 5-kW elements that would operate at three different stages set to provide the required heat to the space. The first stage is controlled by the room thermostat, and typically operates on a 2°F (1°C) differential from the setting on the room thermostat. This means that if the room thermostat is set at 70°F (21°C), as the thermostat calls for heat, the heat pump goes on and begins to send heat to the space. If the temperature of the room falls below 68°F (20°C), then the first 5-kW unit would come on and begin to provide additional heat to supplement the heat pump. The second and third stages of electrical resistance heat are activated by outdoor thermostat settings, which are adjustable (from 50° to 0°F or 10°C to −18°C).

This differential of 2°F (1°C) between the thermostat setting and the room temperature, which activates the first stage, may mean that in cold weather the room temperature may stabilize at 68°F (20°C). This is because 68°F (20°C) is the temperature at which the first stage shuts off; to get a 70°F (21°C) reading on those days, it may be necessary to raise the thermostat reading to about 72°F (22°C).

The cooling efficiency of the heat pump should also be reviewed. Many times, less expensive units have much lower efficiency ratings than the individual cooling units discussed earlier. It is important that the most efficient model be selected. For example, the 3-ton unit being discussed in this section has its engineering data shown in Table 8.36. As the temperature goes up, the cooling Btu/hr provided goes down; at 85°F (29°C) it produces 38 600 Btu/hr (11 300 W) for cooling using 4200 W, while at 95°F (35°C) it produces 36 400 Btu/hr (10 655 W) using 4430 W. Its SEER (seasonal energy efficiency ratings) ranges from 9.19 to 6.73. The SEER most commonly used for comparison of cooling units is founded on its performance at a 95°F temperature; based on that, this unit's SEER would be 8.21.

8.12   EVAPORATIVE (SWAMP) COOLER DESIGN

An evaporative (swamp) cooler basically consists of a large fan that draws air through water-wetted pads and introduces the cooled air into the building. It converts dry outdoor air into moist, cool air by saturating the air with water vapor and decreasing the sensible air temperature. Outside air is cooled by up to 30°F (17°C) as it is drawn through the wet pads and blown into the building. In the process, large quantities of indoor air (an air change every 2 to 3 min) are exhausted. This process contrasts with vapor compression refrigeration systems (e.g., air conditioning systems), which reduce relative humidity as they recycle the air in the building.

Temperature Differential That Can Be Achieved

With most evaporative coolers, the dry bulb temperature of the incoming outdoor air can be reduced by 60 to 80% of the dry bulb–wet bulb temperature differential, so a low wet bulb temperature of outdoor air is required for the system to work effectively. Typically, indoor dry bulb temperatures are maintained between 3° and 6°F (1.7° and 3.3°C) above that temperature.

Example 8.4

In Denver, Colorado, summer design conditions are 93°F (34°C) dry bulb temperature and 59°F (15°C) wet bulb temperature. Approximate the dry bulb temperature that can be achieved by an evaporative cooler.

The dry bulb–wet bulb temperature differential is found by:

$$\Delta T = 93°F - 59°F = 34°F$$

The approximate reduction in outdoor air temperature is found by:

$$34°F \cdot 60\% = 20°F$$
$$34°F \cdot 80\% = 27°F$$

The range of dry bulb temperature that can be achieved by an evaporative cooler is found by:

$$93°F - 20°F = 73°F$$
$$93°F - 27°F = 66°F$$

The dry bulb temperature that can be achieved by an evaporative cooler in Denver, Colorado, is between 66° and 73°F (19° and 23°C).

Climate is a major consideration in the effectiveness of an evaporative cooler because it converts dry outdoor air into moist, cool, indoor air. The saturation process slightly increases the relative humidity of the air entering the indoor space. Thus, evaporative cooling is especially effective in hot, dry climates. The maximum outdoor wet bulb temperature for an evaporative cooler to work effectively is about 70°F (21°C).

Design wet bulb temperatures are considered too high in many cities in the United States, including Houston, Texas, where summer design conditions are 96°F (36°C) dry bulb temperature and 77°F (25°C) wet bulb temperature; New York, New York, 92°F (33°C) dry bulb and 74°F (24°C) wet bulb; Chicago, Illinois, 94°F (34°C) dry bulb and 75°F (24°C) wet bulb; and Miami, Florida, where summer conditions are 91°F (33°C) dry bulb and 77°F (25°C) wet bulb.

Sizing an Evaporative Cooler

There are several methods used to size the capacity of an evaporative cooler. Most methods are based on an *industry standard CFM rating*, which is assigned to the evaporative cooler based on airflow testing by the manufacturer. This rating describes the volumetric flow rate of an operating evaporative cooler based on an external back pressure.

An approximation method used to size the required capacity is to base the computation on 2 to 4 industry standard CFM per square foot of building floor area. Typically, two (2) industry standard CFM are used in moderate to moist climates and four (4) industry standard CFM are used in an arid climate.

The most appropriate method is based on 1000 industry standard CFM per ton of cooling (12 000 Btu/hr or 3500 W). This method requires computation of a building cooling load. An evaporative cooler should be selected that is rated for at least the industry standard CFM arrived at by these sizing calculations.

Example 8.5

A commercial building has a summer design load of 125 000 Btu/hr. Determine the required size of evaporative cooler in industry standard CFM.

125 000 Btu/hr at 12 000 Btu/hr per ton of cooling
= 10.4 tons of cooling

10.4 tons of cooling · 1000 CFM/ton of cooling
= 10 400 industry standard CFM

An evaporative cooler that provides at least 10 400 industry standard CFM should be selected. An option would be to go with two coolers.

Exact matching of an evaporative cooler to the building cooling load is much less critical than it is with a vapor compression refrigeration system (e.g., air conditioning system). Most evaporative coolers have 2 or 3 fan speeds, and actually cool more efficiently at lower speeds, so oversizing is not a real concern. This is because fan efficiency is typically higher at slower blower speeds and there is higher saturation effectiveness as the wetted-pad temperature drops closer to the wet bulb temperature of the ambient air at lower air velocities.

A thermostat can be used to automatically control an evaporative cooler. In large installations, multiple coolers can more effectively distribute cool air. Coolers interconnected with forced-air ductwork should have dampers to isolate them from other system components (e.g., furnace). The damper may be a barometric, motorized, or manual type.

Because the evaporative cooler blower will be drawing considerable quantities of outdoor air, it should be located at least 10 ft (3 m) away from plumbing vents, combustion appliance flues, clothes dryer vents, or exhaust fan vents. Animals, chemicals, fuels, and solvents should also be kept away from the cooler inlet.

TABLE 8.41   EQUIVALENT RECTANGULAR DUCT DIMENSIONS—CUSTOMARY (U.S) UNITS.

Duct Diameter, in.	Rectangular Size, in.	Aspect Ratio														
		1.00	1.25	1.50	1.75	2.00	2.25	2.50	2.75	3.00	3.50	4.00	5.00	6.00	7.00	8.00
6	Width	—	6													
	Height	—	5													
7	Width	6	8													
	Height	6	6													
8	Width	7	9	9	11											
	Height	7	7	6	6											
9	Width	8	9	11	11	12	14									
	Height	8	7	7	6	6	6									
10	Width	9	10	12	12	14	14	15	17							
	Height	9	8	8	7	7	6	6	6							
11	Width	10	11	12	14	14	16	18	17	18	21					
	Height	10	9	8	8	7	7	7	6	6	6					
12	Width	11	13	14	14	16	16	18	19	21	21	24				
	Height	11	10	9	8	8	7	7	7	7	6	6				
13	Width	12	14	15	16	18	18	20	19	21	25	24	30			
	Height	12	11	10	9	9	8	8	7	7	7	6	6			
14	Width	13	14	17	18	18	20	20	22	24	25	28	30	36		
	Height	13	11	11	10	9	9	8	8	8	7	7	6	6		
15	Width	14	15	17	18	20	20	23	25	24	28	28	35	36	42	
	Height	14	12	11	10	10	9	9	9	8	8	7	7	6	6	
16	Width	15	16	18	19	20	23	23	25	27	28	32	35	42	42	48
	Height	15	13	12	11	10	10	9	9	9	8	8	7	7	6	6
17	Width	16	18	20	21	22	25	25	28	27	32	32	35	42	49	48
	Height	16	14	13	12	11	11	10	10	9	9	8	7	7	7	6
18	Width	16	19	21	23	24	25	28	28	30	32	36	40	42	49	56
	Height	16	15	14	13	12	11	11	10	10	9	9	8	7	7	7
19	Width	17	20	21	23	24	27	28	30	30	35	36	40	48	49	56
	Height	17	16	14	13	12	12	11	11	10	10	9	8	8	7	7
20	Width	18	20	23	25	26	27	30	30	33	35	40	45	48	56	56
	Height	18	16	15	14	13	12	12	11	11	10	10	9	8	8	7
21	Width	19	21	24	26	28	29	30	33	33	39	40	45	54	56	64
	Height	19	17	16	15	14	13	12	12	11	11	10	9	9	8	8
22	Width	20	23	26	26	28	32	33	36	36	39	44	50	54	56	64
	Height	20	18	17	15	14	14	13	13	12	11	11	10	9	8	8
23	Width	21	24	26	28	30	32	35	36	39	42	44	50	54	63	64
	Height	21	19	17	16	15	14	14	13	13	12	11	10	9	9	8
24	Width	22	25	27	30	32	34	35	39	39	42	48	55	60	63	72
	Height	22	20	18	17	16	15	14	14	13	12	12	11	10	9	9
25	Width	23	25	29	30	32	36	38	39	42	46	48	55	60	70	72
	Height	23	20	19	17	16	16	15	14	14	13	12	11	10	10	9
26	Width	24	26	30	32	34	36	38	41	42	46	52	55	66	70	72
	Height	24	21	20	18	17	16	15	15	14	13	13	11	11	10	9
27	Width	25	28	30	33	36	38	40	41	45	49	52	60	66	70	80
	Height	25	22	20	19	18	17	16	15	15	14	13	12	11	10	10
28	Width	26	29	32	35	36	38	43	44	45	49	56	60	66	77	80
	Height	26	23	21	20	18	17	17	16	15	14	14	12	11	11	10
29	Width	27	30	33	35	38	41	43	44	48	53	56	65	72	77	88
	Height	27	24	22	20	19	18	17	16	16	15	14	13	12	11	11
30	Width	27	31	35	37	40	43	45	47	48	53	60	65	72	77	88
	Height	27	25	23	21	20	19	18	17	16	15	15	13	12	11	11
31	Width	28	31	35	39	40	43	45	50	51	56	60	70	78	84	88
	Height	28	25	23	22	20	19	18	18	17	16	15	14	13	12	11
32	Width	29	33	36	39	42	45	48	50	54	56	60	70	78	84	96
	Height	29	26	24	22	21	20	19	18	18	16	15	14	13	12	12
33	Width	30	34	38	40	44	47	50	52	54	60	64	75	78	91	96
	Height	30	27	25	23	22	21	20	19	18	17	16	15	13	13	12
34	Width	31	35	39	42	44	47	50	52	57	60	64	75	84	91	96
	Height	31	28	26	24	22	21	20	19	19	17	16	15	14	13	12
35	Width	32	36	39	42	46	50	53	55	57	63	68	75	84	91	104
	Height	32	29	26	24	23	22	21	20	19	18	17	15	14	13	13
36	Width	33	36	41	44	48	50	53	55	60	63	68	80	90	98	104
	Height	33	29	27	25	24	22	21	20	20	18	17	16	15	14	13
38	Width	35	39	44	47	50	54	58	61	63	67	72	85	96	105	112
	Height	35	31	29	27	25	24	23	22	21	19	18	17	16	15	14

*Shaded area not recommended.

Duct Diameter, in.	Rectangular Size, in.	Aspect Ratio														
		1.00	1.25	1.50	1.75	2.00	2.25	2.50	2.75	3.00	3.50	4.00	5.00	6.00	7.00	8.00
40	Width	37	41	45	49	52	56	60	63	66	70	76	90	96	105	120
	Height	37	33	30	28	26	25	24	23	22	20	19	18	16	15	15
42	Width	38	43	48	51	56	59	63	66	69	74	80	90	102	112	120
	Height	38	34	32	29	28	26	25	24	23	21	20	18	17	16	15
44	Width	40	45	50	54	58	61	65	69	72	81	84	95	108	119	128
	Height	40	36	33	31	29	27	26	25	24	23	21	19	18	17	16
46	Width	42	48	53	56	60	65	68	72	75	84	88	100	114	126	136
	Height	42	38	35	32	30	29	27	26	25	24	22	20	19	18	17
48	Width	44	49	54	60	62	68	70	74	78	88	92	105	120	126	136
	Height	44	39	36	34	31	30	28	27	26	25	23	21	20	18	17
50	Width	46	51	57	61	66	70	75	77	81	91	96	110	120	133	144
	Height	46	41	38	35	33	31	30	28	27	26	24	22	20	19	18
52	Width	48	54	59	63	68	72	78	83	84	95	100	115	126	140	152
	Height	48	43	39	36	34	32	31	30	28	27	25	23	21	20	19
54	Width	49	55	62	67	70	77	80	85	90	98	104	120	132	147	160
	Height	49	44	41	38	35	34	32	31	30	28	26	24	22	21	20
56	Width	51	58	63	68	74	79	83	88	93	102	108	125	138	147	160
	Height	51	46	42	39	37	35	33	32	31	29	27	25	23	21	20
58	Width	53	60	66	70	76	81	85	91	96	105	112	130	144	154	168
	Height	53	48	44	40	38	36	34	33	32	30	28	26	24	22	21
60	Width	55	61	68	74	78	83	90	94	99	109	116	130	144	161	
	Height	55	49	45	42	39	37	36	34	33	31	29	26	24	23	
62	Width	57	64	71	75	82	88	93	96	102	112	120	135	150	168	
	Height	57	51	47	43	41	39	37	35	34	32	30	27	25	24	
64	Width	59	65	72	79	84	90	95	99	105	116	124	140	156		
	Height	59	52	48	45	42	40	38	36	35	33	31	28	26		
66	Width	60	68	75	81	86	92	98	105	108	119	128	145	162		
	Height	60	54	50	46	43	41	39	38	36	34	32	29	27		
68	Width	62	70	77	82	90	95	100	107	111	123	132	150	168		
	Height	62	56	51	47	45	42	40	39	37	35	33	30	28		
70	Width	64	71	80	86	92	99	105	110	114	126	136	155			
	Height	64	57	53	49	46	44	42	40	38	36	34	31			
72	Width	66	74	81	88	94	101	108	113	117	130	140	160			
	Height	66	59	54	50	47	45	43	41	39	37	35	32			
74	Width	68	76	84	91	98	104	110	116	123	133	144	165			
	Height	68	61	56	52	49	46	44	42	41	38	36	33			
76	Width	70	78	86	93	100	106	113	118	126	137	148	165			
	Height	70	62	57	53	50	47	45	43	42	39	37	33			
78	Width	71	80	89	95	102	110	115	121	129	140	152				
	Height	71	64	59	54	51	49	46	44	43	40	38				
80	Width	73	83	90	98	104	113	118	124	132	144	156				
	Height	73	66	60	56	52	50	47	45	44	41	39				
82	Width	75	84	93	100	108	115	123	129	135	147	160				
	Height	75	67	62	57	54	51	49	47	45	42	40				
84	Width	77	86	95	103	110	117	125	132	138	151	164				
	Height	77	69	63	59	55	52	50	48	46	43	41				
86	Width	79	88	98	105	112	119	128	135	141	154	168				
	Height	79	70	65	60	56	53	51	49	47	44	42				
88	Width	80	90	99	107	116	124	130	138	144	158					
	Height	80	72	66	61	58	55	52	50	48	45					
90	Width	82	93	102	110	118	126	133	140	147	161					
	Height	82	74	68	63	59	56	53	51	49	46					
92	Width	84	94	104	112	120	128	138	143	150	165					
	Height	84	75	69	64	60	57	55	52	50	47					
94	Width	86	96	107	116	124	131	140	146	153	168					
	Height	86	77	71	66	62	58	56	53	51	48					
96	Width	88	99	108	117	126	135	143	151	159						
	Height	88	79	72	67	63	60	57	55	53						
98	Width	90	100	111	119	128	137	145	154	162						
	Height	90	80	74	68	64	61	58	56	54						
100	Width	91	103	113	123	132	140	148	157	165						
	Height	91	82	75	70	66	62	59	57	55						
102	Width	93	105	116	124	134	142	153	160	168						
	Height	93	84	77	71	67	63	61	58	56						
104	Width	95	106	117	128	136	146	155	162							
	Height	95	85	78	73	68	65	62	59							

'Shaded area not recommended.

Duct Diameter, in.	Rectangular Size, in.	Aspect Ratio														
		1.00	1.25	1.50	1.75	2.00	2.25	2.50	2.75	3.00	3.50	4.00	5.00	6.00	7.00	8.00
106	Width	97	109	120	130	140	149	158	165							
	Height	97	87	80	74	70	66	63	60							
108	Width	99	110	122	131	142	151	160	168							
	Height	99	88	81	75	71	67	64	61							
110	Width	101	113	125	135	144	153	163								
	Height	101	90	83	77	72	68	65								
112	Width	102	115	126	137	146	158	165								
	Height	102	92	84	78	73	70	66								
114	Width	104	116	129	140	150	160									
	Height	104	93	86	80	75	71									
116	Width	106	119	131	142	152	162									
	Height	106	95	87	81	76	72									
118	Width	108	121	134	144	154	164									
	Height	108	97	89	82	77	73									
120	Width	110	123	135	147	158										
	Height	110	98	90	84	79										

*Shaded area not recommended.

SI UNITS

Circular Equivalents of Rectangular Duct for Equal Friction and Capacity[a]

Lgth Adj.[b]	Length of One Side of Rectangular Duct (a), mm																			
	100	125	150	175	200	225	250	275	300	350	400	450	500	550	600	650	700	750	800	900
100	109																			
125	122	137																		
150	133	150	164																	
175	143	161	177	191																
200	152	172	189	204	219															
225	161	181	200	216	232	246														
250	169	190	210	228	244	259	273													
275	176	199	220	238	256	272	287	301												
300	183	207	229	248	266	283	299	314	328											
350	195	222	245	267	286	305	322	339	354	383										
400	207	235	260	283	305	325	343	361	378	409	437									
450	217	247	274	299	321	343	363	382	400	433	464	492								
500	227	258	287	313	337	360	381	401	420	455	488	518	547							
550	236	269	299	326	352	375	398	419	439	477	511	543	573	601						
600	245	279	310	339	365	390	414	436	457	496	533	567	598	628	656					
650	253	289	321	351	378	404	429	452	474	515	553	589	622	653	683	711				
700	261	298	331	362	391	418	443	467	490	533	573	610	644	677	708	737	765			
750	268	306	341	373	402	430	457	482	506	550	592	630	666	700	732	763	792	820		
800	275	314	350	383	414	442	470	496	520	567	609	649	687	722	755	787	818	847	875	
900	289	330	367	402	435	465	494	522	548	597	643	686	726	763	799	833	866	897	927	984
1000	301	344	384	420	454	486	517	546	574	626	674	719	762	802	840	876	911	944	976	1037
1100	313	358	399	437	473	506	538	569	598	652	703	751	795	838	878	916	953	988	1022	1086
1200	324	370	413	453	490	525	558	590	620	677	731	780	827	872	914	954	993	1030	1066	1133
1300	334	382	426	468	506	543	577	610	642	701	757	808	857	904	948	990	1031	1069	1107	1177
1400	344	394	439	482	522	559	595	629	662	724	781	835	886	934	980	1024	1066	1107	1146	1220
1500	353	404	452	495	536	575	612	648	681	745	805	860	913	963	1011	1057	1100	1143	1183	1260
1600	362	415	463	508	551	591	629	665	700	766	827	885	939	991	1041	1088	1133	1177	1219	1298
1700	371	425	475	521	564	605	644	682	718	785	849	908	964	1018	1069	1118	1164	1209	1253	1335
1800	379	434	485	533	577	619	660	698	735	804	869	930	988	1043	1096	1146	1195	1241	1286	1371
1900	387	444	496	544	590	663	674	713	751	823	889	952	1012	1068	1122	1174	1224	1271	1318	1405
2000	395	453	506	555	602	646	688	728	767	840	908	973	1034	1092	1147	1200	1252	1301	1348	1438
2100	402	461	516	566	614	659	702	743	782	857	927	993	1055	1115	1172	1226	1279	1329	1378	1470
2200	410	470	525	577	625	671	715	757	797	874	945	1013	1076	1137	1195	1251	1305	1356	1406	1501
2300	417	478	534	587	636	683	728	771	812	890	963	1031	1097	1159	1218	1275	1330	1383	1434	1532
2400	424	486	543	597	647	695	740	784	826	905	980	1050	1116	1180	1241	1299	1355	1409	1461	1561
2500	430	494	552	606	658	706	753	797	840	920	996	1068	1136	1200	1262	1322	1379	1434	1488	1589
2600	437	501	560	616	668	717	764	810	853	935	1012	1085	1154	1220	1283	1344	1402	1459	1513	1617
2700	443	509	569	625	678	728	776	822	866	950	1028	1102	1173	1240	1304	1366	1425	1483	1538	1644
2800	450	516	577	634	688	738	787	834	879	964	1043	1119	1190	1259	1324	1387	1447	1506	1562	1670
2900	456	523	585	643	697	749	798	845	891	977	1058	1135	1208	1277	1344	1408	1469	1529	1586	1696

Lgth Adj.[b]	Length of One Side of Rectangular Duct (a), mm																			
	1000	1100	1200	1300	1400	1500	1600	1700	1800	1900	2000	2100	2200	2300	2400	2500	2600	2700	2800	2900
1000	1093																			
1100	1146	1202																		
1200	1196	1256	1312																	
1300	1244	1306	1365	1421																
1400	1289	1354	1416	1475	1530															
1500	1332	1400	1464	1526	1584	1640														
1600	1373	1444	1511	1574	1635	1693	1749													
1700	1413	1486	1555	1621	1684	1745	1803	1858												
1800	1451	1527	1598	1667	1732	1794	1854	1912	1968											
1900	1488	1566	1640	1710	1778	1842	1904	1964	2021	2077										
2000	1523	1604	1680	1753	1822	1889	1952	2014	2073	2131	2186									
2100	1558	1640	1719	1793	1865	1933	1999	2063	2124	2183	2240	2296								
2200	1591	1676	1756	1833	1906	1977	2044	2110	2173	2233	2292	2350	2405							
2300	1623	1710	1793	1871	1947	2019	2088	2155	2220	2283	2343	2402	2459	2514						
2400	1655	1744	1828	1909	1986	2060	2131	2200	2266	2330	2393	2453	2511	2568	2624					
2500	1685	1776	1862	1945	2024	2100	2173	2243	2311	2377	2441	2502	2562	2621	2678	2733				
2600	1715	1808	1896	1980	2061	2139	2213	2285	2355	2422	2487	2551	2612	2672	2730	2787	2842			
2700	1744	1839	1929	2015	2097	2177	2253	2327	2398	2466	2533	2598	2661	2722	2782	2840	2896	2952		
2800	1772	1869	1961	2048	2133	2214	2292	2367	2439	2510	2578	2644	2708	2771	2832	2891	2949	3006	3061	
2900	1800	1898	1992	2081	2167	2250	2329	2406	2480	2552	2621	2689	2755	2819	2881	2941	3001	3058	3115	3170

[a] Table based on $D_e = 1.30(ab)^{0.625}/(a+b)^{0.25}$.
[b] Length of adjacent side of rectangular duct (b), mm.

TABLE 8.42   SUGGESTED BONNET AND RETURN PRESSURE.

CFM	TYPE A	TYPE B	TYPE C
1000	0.10	0.15	0.20
1500	.12	.17	.22
2000	.13	.18	.24
2500	.14	.20	.27
3000	.16	.22	.29
3500	.17	.24	.31
4000	.18	.26	.33
4500	0.20	0.28	0.36
5000	.21	.29	.38
5500	.22	.31	.40
6000	.24	.33	.42
6500	.25	.34	.45
7000	.27	.36	.47
7500	.28	.38	.49
8000	0.29	0.39	0.51
8500	.30	.41	.53
9000	.32	.43	.56
9500	.34	.45	.58
10,000	.34	.47	.60
10,500	.37	.48	.60
11,000	.39	.50	.60
11,500	0.40	0.50	0.60
12,000	.40	.50	.60
12,500	.40	.50	.60
13,000	.40	.50	.60
14,000	.40	.50	.60

CFM = Maximum cfm in any one duct.
Type A = Systems for quiet operation in residences, churches, concert halls, broadcasting studios, funeral homes, etc.
Type B = Systems for schools, theaters, public buildings, etc.
Type C = Systems for industrial buildings.

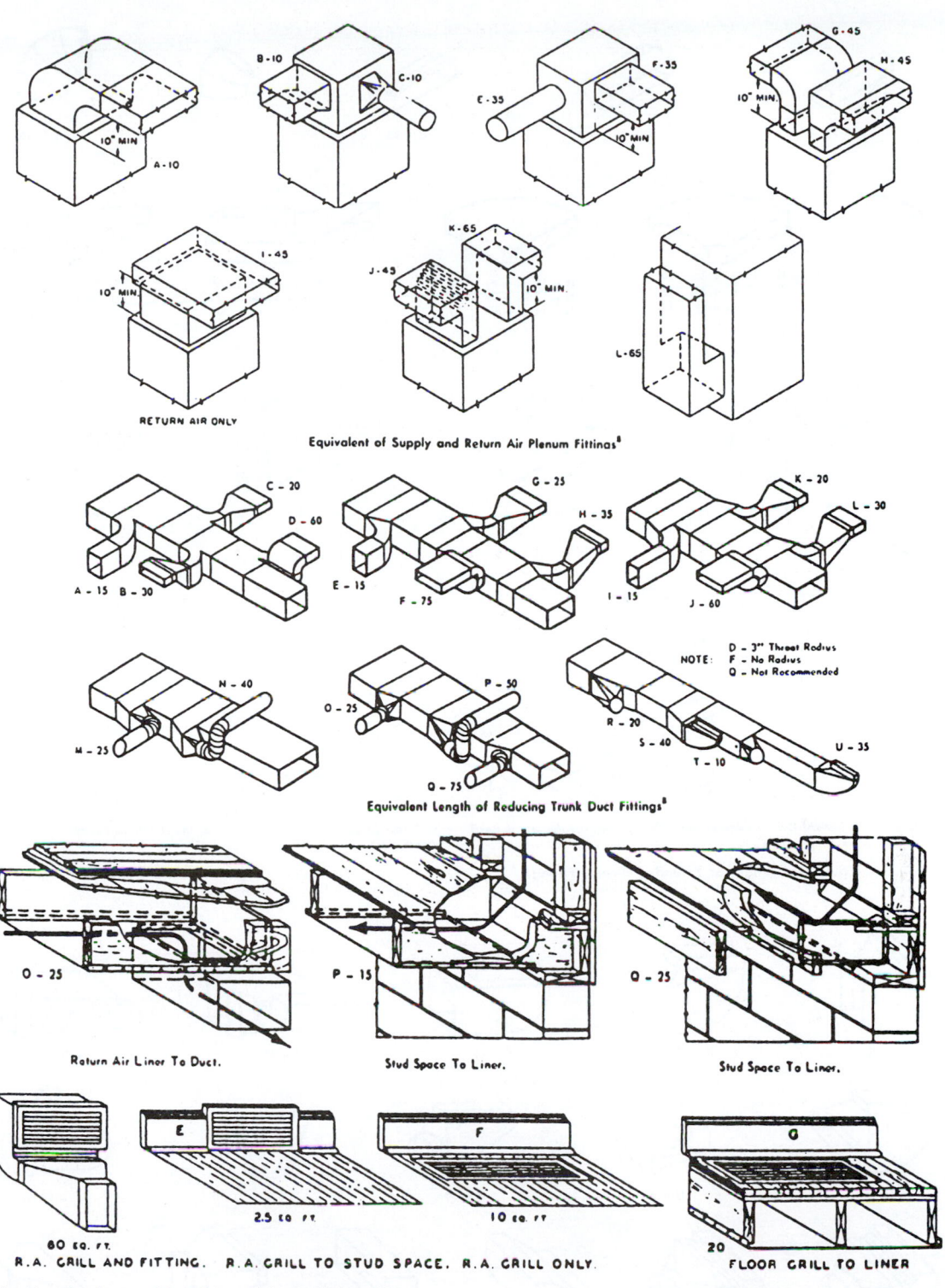

Equivalent of Supply and Return Air Plenum Fittings[a]

NOTE: D – 3" Throat Radius
F – No Radius
Q – Not Recommended

Equivalent Length of Reducing Trunk Duct Fittings[a]

Return Air Liner To Duct.   Stud Space To Liner.   Stud Space To Liner.

R.A. GRILL AND FITTING.   R.A. GRILL TO STUD SPACE.   R.A. GRILL ONLY.   FLOOR GRILL TO LINER

[a]Inside radius for A and B = 3 in., and for F and G = 5 in.

Equivalent Length of Angles and Elbows for Individual and Branch Ducts[a,b]

FIGURE 8.15 Equivalent length of fittings.

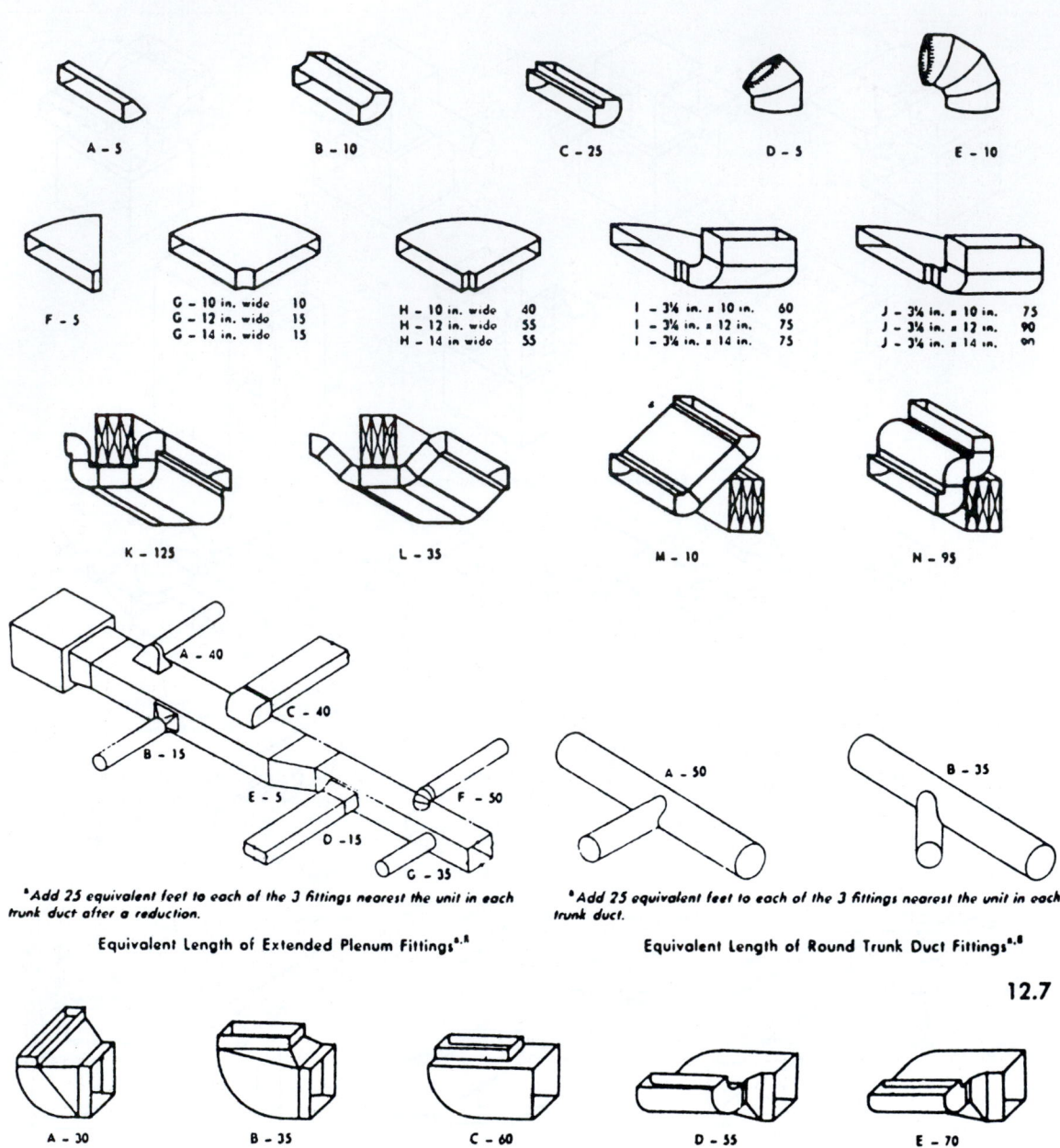

A – 5

B – 10

C – 25

D – 5

E – 10

F – 5

G – 10 in. wide   10
G – 12 in. wide   15
G – 14 in. wide   15

H – 10 in. wide   40
H – 12 in. wide   55
H – 14 in. wide   55

I – 3¼ in. x 10 in.   60
I – 3¼ in. x 12 in.   75
I – 3¼ in. x 14 in.   75

J – 3¼ in. x 10 in.   75
J – 3¼ in. x 12 in.   90
J – 3¼ in. x 14 in.   90

K – 125

L – 35

M – 10

N – 95

A – 40

C – 40

B – 15

E – 5

F – 50

D – 15

G – 35

A – 50

B – 35

[a] Add 25 equivalent feet to each of the 3 fittings nearest the unit in each trunk duct after a reduction.

Equivalent Length of Extended Plenum Fittings[a,b]

[a] Add 25 equivalent feet to each of the 3 fittings nearest the unit in each trunk duct.

Equivalent Length of Round Trunk Duct Fittings[a,b]

12.7

A – 30

B – 35

C – 60

D – 55

E – 70

F – 45

G – 30

H – 50

I – 5

J – 15

K – 30

L – 30

M – 5

N – 15

O – 15

P – 5

[a] These values may also be used for floor diffuser boxes.

FIGURE 8.15   (Continued)

306

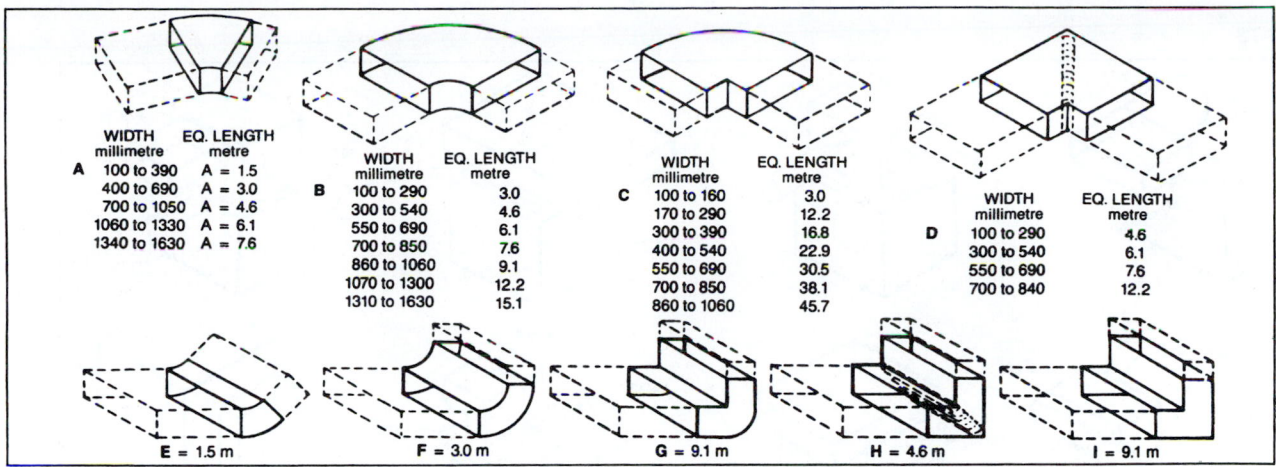

Equivalent Length of Angles and Elbows for Trunk Ducts (ACCA 1984)

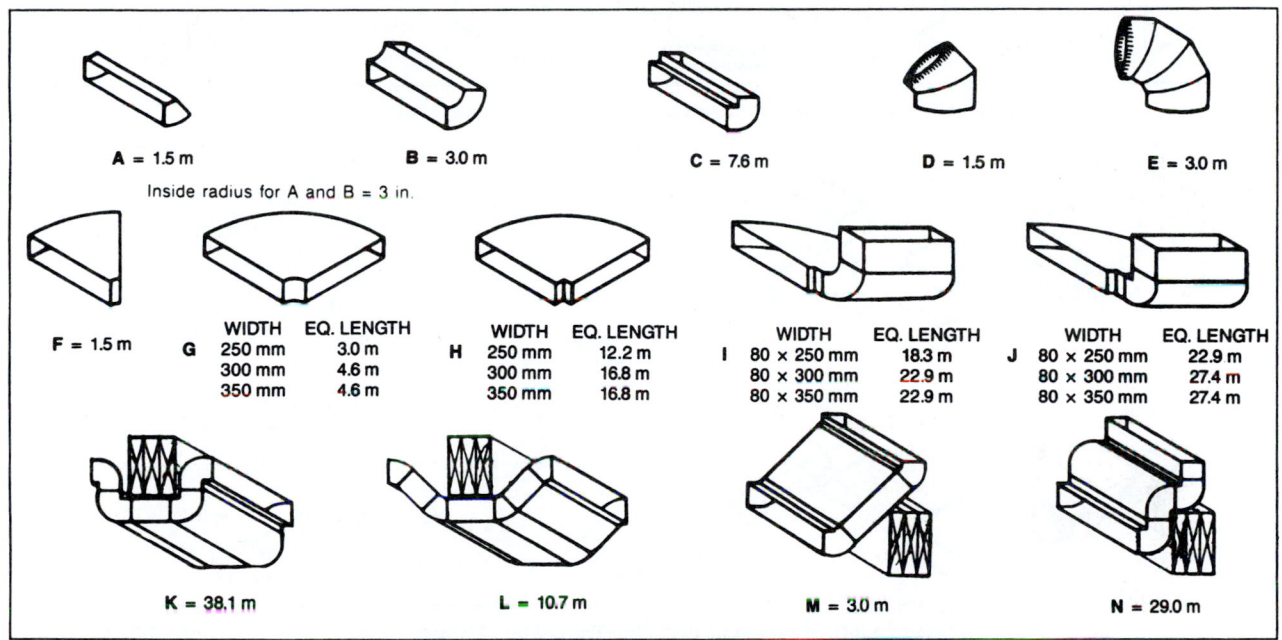

Equivalent Length of Angles and Elbows for Branch Ducts (ACCA 1984)

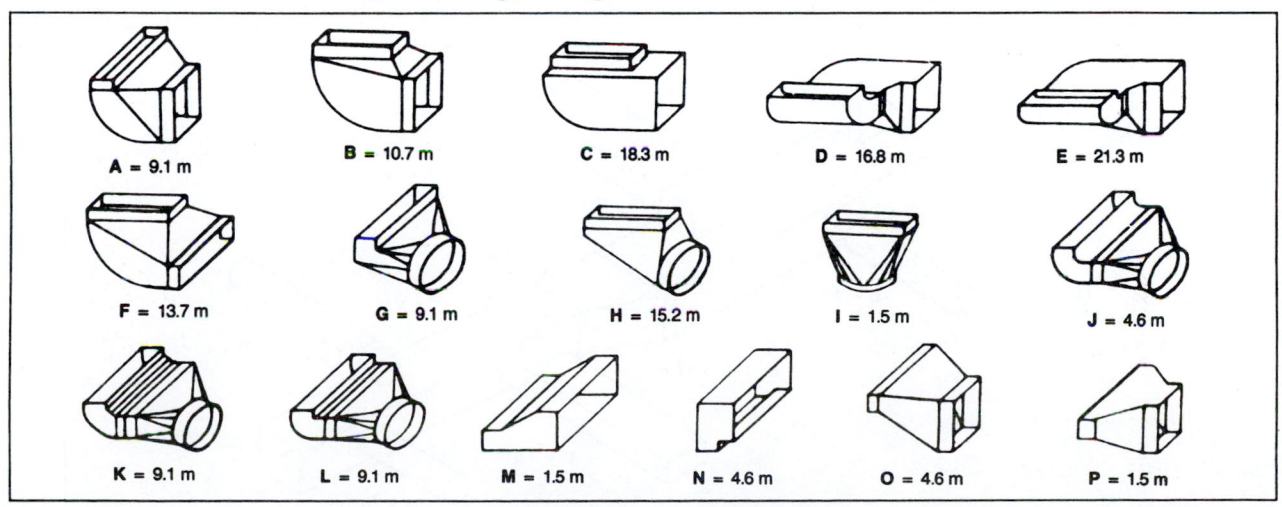

Equivalent Length of Boot Fittings

FIGURE 8.15   *(Continued)*

SI UNITS

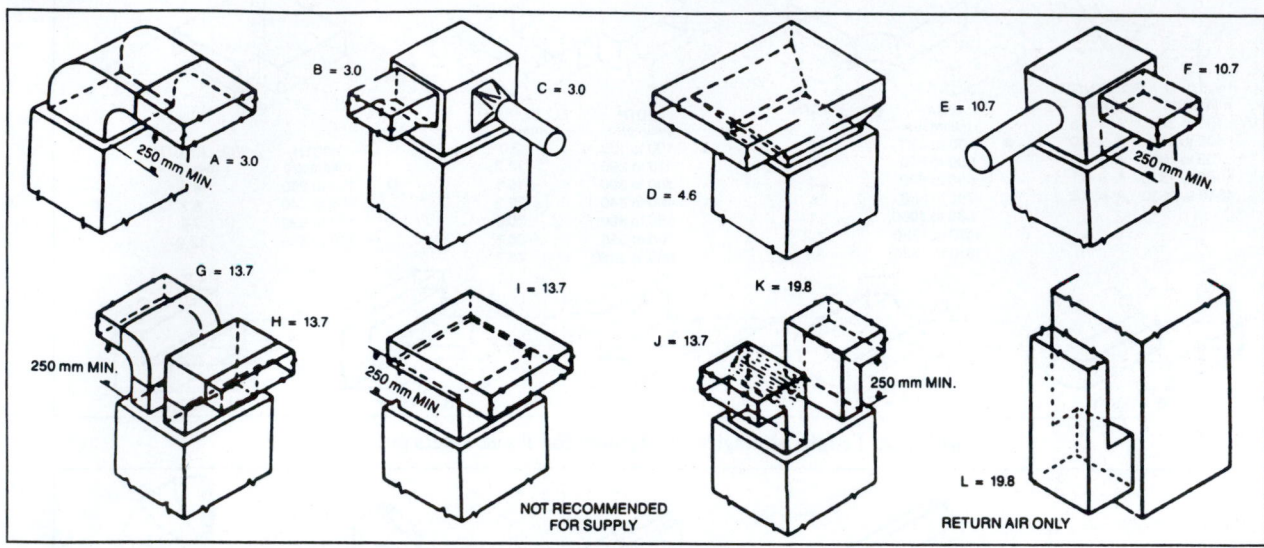

Equivalent Length in Metres of Supply and Return Air Plenum Fittings

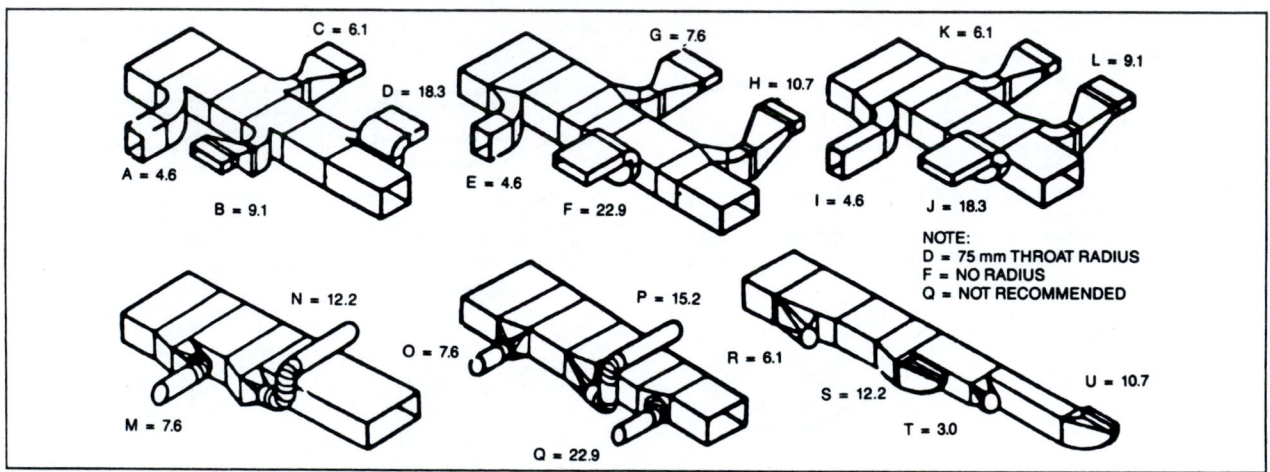

Equivalent Length in Metres of Reducing Trunk Duct Fittings

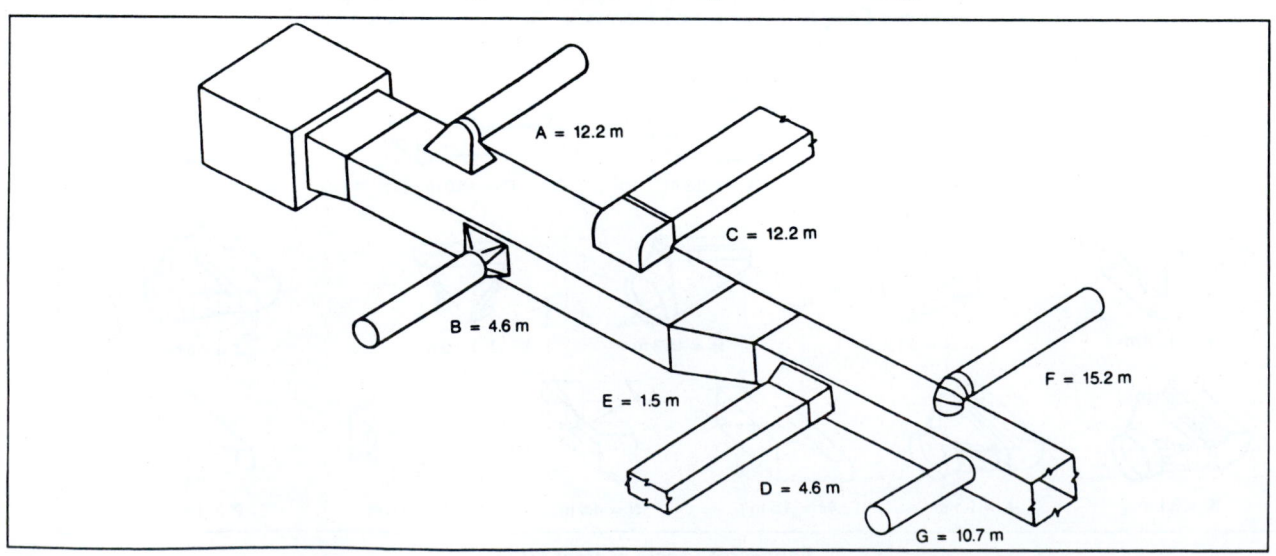

FIGURE 8.15 *(Continued)*

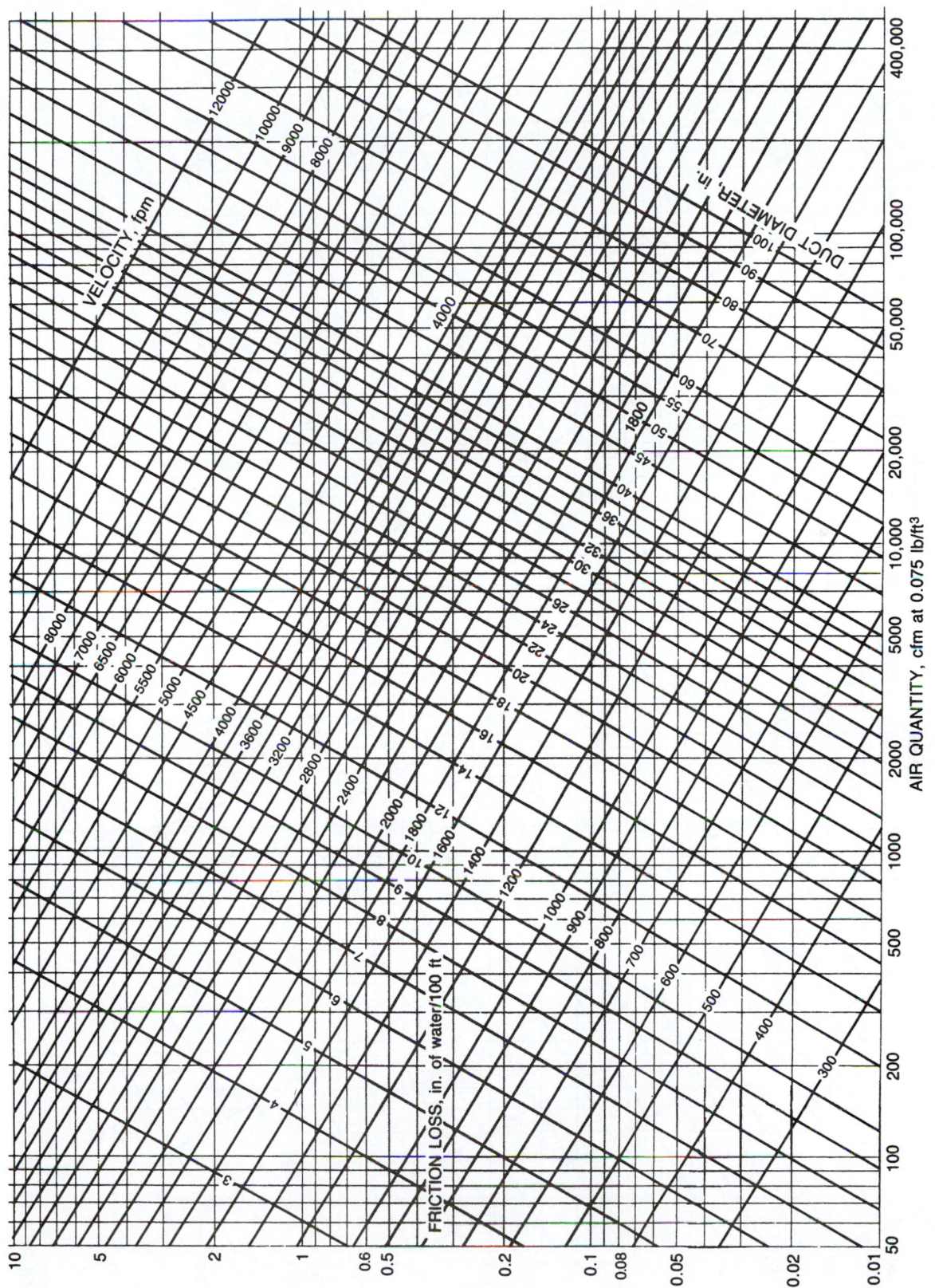

Reprinted with permission from ASHRAE, *Fundamentals Handbook*, 1993.

FIGURE 8.16 Friction loss in duct.

FIGURE 8.16 (Continued)

STUDY QUESTIONS

8-1. What are the advantages of a forced-air system as compared to a hot water system?

8-2. What types of materials are most commonly used for forced-air ducts?

8-3. What can be done to ensure that the system will be as quiet as possible?

8-4. What is the purpose of putting dampers in the ductwork?

8-5. What is the purpose of putting turning vanes in the ductwork and where are they used?

8-6. When is insulation around the ductwork recommended?

8-7. When heating is the predominant function of the forced-air system, what are the recommended locations for the supply and return?

8-8. Describe the layout of an extended plenum ductwork arrangement.

8-9. Describe the layout of a reducing plenum ductwork arrangement.

8-10. Describe the layout of an individual supply ductwork arrangement.

8-11. Describe the layout of a perimeter loop ductwork arrangement.

8-12. Describe the single-zone and multizone constant air volume systems. Explain the differences between the systems.

8-13. How does the equivalent diameter of a duct relate to a rectangular duct section?

8-14. Which duct system is more common, the low-velocity or high-velocity system?

8-15. Why should air velocity in the ducts be kept below certain limits? Give two reasons.

8-16. Why must automatic dampers be designed to operate so they close gradually?

8-17. What term is commonly used to refer to the cooling capacity of a cooling unit, and how does it relate to Btu/hr?

8-18. What does the term *SEER* mean, and why is it important to check the *SEER* of the cooling unit?

8-19. What type of equipment may be used when both heating and air conditioning are required?

8-20. What factors determine where the supplies and returns are located?

Design Exercises

8-21. Determine the perimeter length (in inches) and aspect ratio of the following ducts:

   a. 12 in × 12 in

   b. 18 in × 12 in

   c. 24 in × 12 in

   d. 36 in × 12 in

   e. 48 in × 12 in

8-22. For the duct sizes in the previous exercise, which duct section uses the least material?

8-23. Determine the perimeter length (in inches) and aspect ratio of the following ducts:

   a. 8 in × 8 in

   b. 16 in × 8 in

   c. 24 in × 8 in

   d. 32 in × 8 in

   e. 40 in × 8 in

8-24. For the duct sizes in the previous exercise, which duct section uses the least material?

8-25. Determine the perimeter length (in mm) and aspect ratio of the following ducts:

   a. 200 mm × 200 mm

   b. 400 mm × 200 mm

   c. 600 mm × 200 mm

   d. 800 mm × 200 mm

   e. 1000 mm × 200 mm

8-26. For the duct sizes in the previous exercise, which duct section uses the least material?

8-27. Determine the equivalent diameter of the following ducts:

   a. 12 in × 12 in

   b. 18 in × 12 in

   c. 24 in × 12 in

   d. 30 in × 12 in

   e. 48 in × 12 in

8-28. Determine the equivalent diameter of the following ducts:

   a. 8 in × 8 in

   b. 16 in × 8 in

   c. 24 in × 8 in

   d. 32 in × 8 in

   e. 40 in × 8 in

8-29. The minimum required equivalent diameter of a duct is 16 in but is too deep for the ceiling clearance allowed by the architect's design. The vertical clearance available for ductwork is 12 in. Identify a cost-effective rectangular duct section that meets these clearance and equivalent diameter requirements. Assume the duct is not lined.

8-30. The minimum required equivalent diameter of a duct is 30 in but is too deep for the ceiling clearance allowed by the architect's design. The vertical clearance available for ductwork is 20 in. Identify a cost-effective rectangular duct section that meets these clearance and equivalent diameter requirements. Assume the duct is not lined.

8-31. The minimum required equivalent diameter of a duct is 34 in but is too deep for the ceiling clearance allowed by the architect's design. The vertical clearance available for ductwork is 24 in. Identify a cost-effective rectangular duct section that meets these clearance and equivalent diameter requirements. Assume the duct is not lined.

8-32. The minimum required equivalent diameter of a duct is 500 mm but is too deep for the ceiling clearance allowed by the architect's design. The vertical clearance available for ductwork in the space above a ceiling is 350 mm. Identify a cost-effective rectangular duct section that meets these clearance and equivalent diameter requirements. Assume the duct is not lined.

8-33. The minimum required equivalent diameter of a duct is 900 mm but is too deep for the ceiling clearance allowed by the architect's design. The vertical clearance available for ductwork is 400 mm. Identify a cost-effective rectangular duct section that meets these clearance and equivalent diameter requirements. Assume the duct is not lined.

8-34. A 12 in diameter galvanized steel (smooth) straight duct section conveys air at 1000 ft³/min. It is 100 ft long.

a. Determine the pressure loss, in inches of w.c.

b. Determine the average velocity of air flowing through the duct.

c. Identify a cost-effective rectangular duct section with an equivalent pressure loss that has a side depth of 8 in.

8-35. A 24 in diameter galvanized steel (smooth) straight duct section conveys air at 5000 ft³/min. It is 50 ft long.

a. Determine the pressure loss, in inches of w.c.

b. Determine the average velocity of air flowing through the duct.

c. Identify a cost-effective rectangular duct section with an equivalent pressure loss that has a side depth of 16 in.

8-36. A 20 in diameter galvanized steel (smooth) straight duct section conveys air at 3000 ft³/min. It is 45 ft long.

a. Determine the pressure loss, in inches of w.c.

b. Determine the average velocity of air flowing through the duct.

c. Identify a cost-effective rectangular duct section with an equivalent pressure loss that has a side depth of 12 in.

8-37. A 34 in diameter galvanized steel (smooth) straight duct section conveys air at 10 000 ft³/min. It is 42 ft long.

a. Determine the pressure loss, in inches of w.c.

b. Determine the average velocity of air flowing through the duct.

c. Identify a cost-effective rectangular duct section with an equivalent pressure loss that has a side depth of 20 in.

8-38. A 500 mm diameter galvanized steel (smooth) straight duct section conveys air at 2000 L/s. It is 21 m long.

a. Determine the pressure loss, in Pa/m.

b. Determine the average velocity of air flowing through the duct.

c. Identify a rectangular duct section with an equivalent pressure loss that has a side depth of 300 mm.

8-39. A 250 mm diameter galvanized steel (smooth) straight duct section conveys air at 500 L/s. It is 21 m long.

a. Determine the pressure loss, in Pa/m.

b. Determine the average velocity of air flowing through the duct.

c. Identify a cost-effective rectangular duct section with an equivalent pressure loss that has a depth of 150 mm.

8-40. A branch duct will convey air at a volumetric flow rate of 500 ft³/min. The pressure loss available for this duct is 0.1 in w.c./100 ft. The duct is not lined.

a. Identify the minimum diameter of a round duct.

b. Identify a cost-effective rectangular duct section with an equivalent pressure loss that has a side depth of 8 in.

8-41. A branch duct will convey air at a volumetric flow rate of 5000 ft³/min. The pressure loss available for this duct is 0.1 in w.c./100 ft. The duct is not lined.

a. Identify the minimum diameter of a round duct.

b. Identify a cost-effective rectangular duct section with an equivalent pressure loss that has a side depth of 20 in.

8-42. A branch duct will convey air at a volumetric flow rate of 8000 ft³/min. The pressure loss available for this duct is 0.1 in w.c./100 ft. The duct is not lined.

a. Identify the minimum diameter of a round duct.

b. Identify a cost-effective rectangular duct section with an equivalent pressure loss that has a side depth of 20 in.

8-43. A branch duct will convey air at a volumetric flow rate of 8000 L/s. The pressure loss available for this duct is 10 Pa/m. The duct is not lined.

a. Identify the minimum diameter of a round duct.

b. Identify a cost-effective rectangular duct section with an equivalent pressure loss that has a side depth of 400 mm.

8-44. A branch duct will convey air at a volumetric flow rate of 1000 L/s. The pressure loss available for this duct is 10 Pa/m. The duct is not lined.

a. Identify the minimum diameter of a round duct.

b. Identify a cost-effective rectangular duct section with an equivalent pressure loss that has a side depth of 200 mm.

8-45. Design a forced hot air system for the residence in Appendix D based on the heat loss calculations for the geographic location where you reside.

8-46. Design a forced-air heating and cooling system for the residence in Appendix D based on the geographic location where you reside.

8-47. Design a forced-air heating and cooling system for one of the top-floor apartments and one of the lower-floor apartments in Appendix A based on the geographic location where you reside.

HVAC WATER (HYDRONIC) DISTRIBUTION SYSTEMS

9.1 SYSTEMS

Basic Systems

A *hydronic system* is an all-water HVAC distribution system that uses water as the heat transfer medium in heating and cooling systems. In a basic hydronic heating system, water is heated in a boiler by natural gas, oil, electricity, propane, or solid fuel. The heated water is then circulated to a heat transfer device where it emits heat before returning to the boiler for reheating. In large-scale commercial buildings, a hydronic system may include both a chilled and a heated water loop, to provide both space heating and cooling. Chillers, air-cooled and water-cooled condensing units (i.e., cooling towers), and heat pumps are used as means to provide water cooling, while boilers heat water. The focus of this chapter is heating systems.

In a basic hydronic system, when heat is desired, a circulating pump is automatically turned on and the hot water is circulated through an arrangement of pipes, passing through any of a variety of convector or coil unit types. As it circulates, it gives off the heat, primarily through the convector or coil units; then the pipes return the water to the boiler to be reheated and circulated again. Most of the convector or coil units used can be regulated slightly by opening or closing adjustable valves or dampers. A compression tank is included to accommodate the varying pressure in the closed system, because water expands as it is heated.

Piping Arrangements

The four different hot water piping arrangements commonly used in residential and light commercial installations are the series loop, one-pipe and two-pipe, and radiant arrangements. Three-pipe and four-pipe arrangements are used in commercial installations involving simultaneous heating and cooling.

Series Loop Arrangement

Most commonly used for space heating residences and small buildings, convector or coil units in the *series loop arrangement* are fed by a single pipe that goes through a convector and makes a loop around the building, or one portion of the building (Figures 9.1 and 9.2). The pipe acts as combination supply and return, with the water getting cooler as it progresses through the loop. Water temperature in the supply pipe becomes progressively cooler as it passes through the successive

heating units. These units must be selected accordingly. This means that larger convector or coil units are required to obtain the same amount of heat at the end of the loop as compared to convectors at the beginning because the water is cooler. Also, the longer the run of piping and the more convectors it serves, the cooler the water will become. For more uniform heating, the building may be broken into zones (Figure 9.3), each with its own series loop arrangement activated by its individual thermostat. It is economical to install, but any control of individual convectors is minimal. Only heating is supplied with this arrangement.

One-Pipe Arrangement

The *one-pipe arrangement* has a single pipe going around the building or a zone of the building. (See Figures 9.4 and 9.5.) A portion of the hot water is diverted through the convector or coil unit, where it gives off heat to the room. A special pipe fitting, called a *diverting* or *distributor tee*, is installed in the piping to force the water to divert out of the main pipe through a branch containing the heating device, then back to the main piping loop. Upon entering the supply pipe, it will slightly reduce the temperature of the water in the supply pipe. Water temperature in the supply pipe becomes progressively cooler as it passes through the successive heating units. These units must be selected accordingly. The primary advantage of this arrangement over the series loop is that each individual convector or coil unit may be controlled by a manual or thermostatically controlled valve.

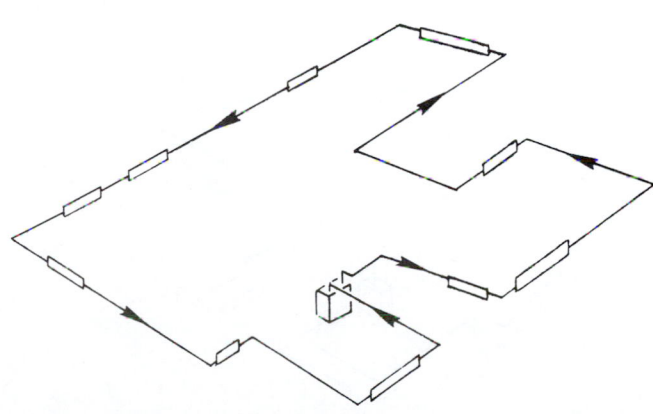

FIGURE 9.1 Series loop isometric.

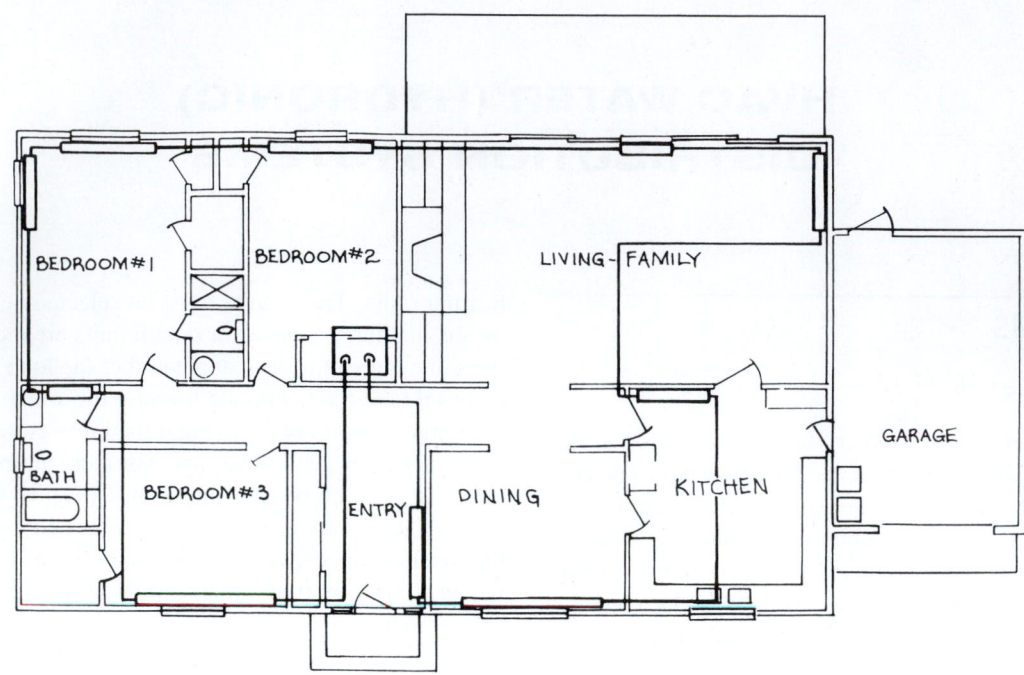

FIGURE 9.2 Series loop plan.

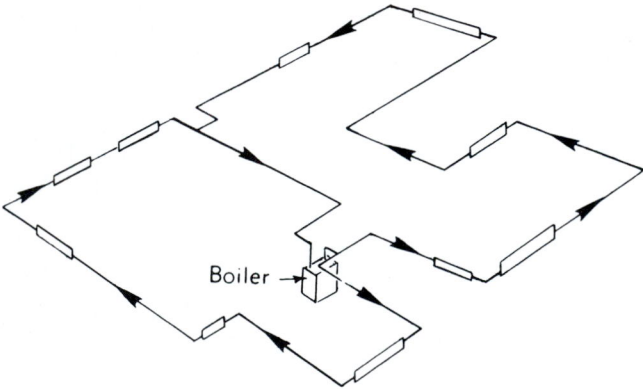

FIGURE 9.3 Series loop (zoned) isometric.

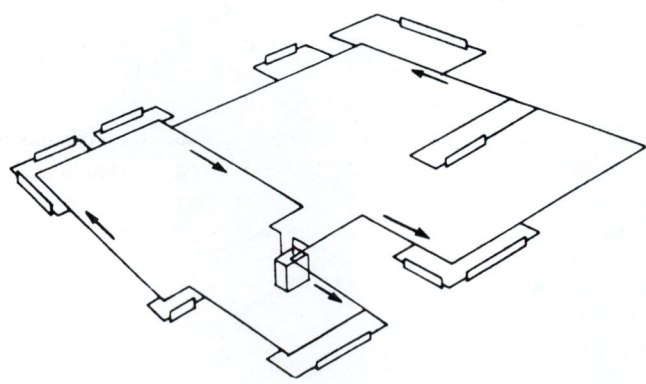

FIGURE 9.4 One-pipe (zoned) isometric.

The convector or coil units may be placed above the pipe (upfeed) or below it (downfeed). The upfeed is more effective because the water tends to be diverted more easily in that direction. This arrangement is more expensive than the series loop, as it requires additional piping, fittings, and valves. It may also be zoned for larger buildings where the temperature drop over a long pipe run would be excessive.

Two-Pipe Arrangement

The *two-pipe arrangement* has a supply pipe to the convector or coil units and a return pipe back to the chiller or boiler. To accomplish this, one pipe is the supply pipe, which then empties into another pipe used for return. This type of arrangement provides the water at as high a temperature as possible. The return may be classified as *reverse return* (Figure 9.6) or *direct return* (Figure 9.7). The reverse return results in a more uniform flow of water because the supply and return are of equal length, resulting in equal friction losses. The direct return requires slightly less piping. Individual convector or coil unit control is available by the installation of valves. This is the best arrangement available and the most expensive, so it is used in commercial installations. It would be unnecessary to use this arrangement in a residence.

Three- and Four-Pipe Arrangements

Simultaneous heating and cooling is necessary in a building when its thermal zones have separate heating and cooling loads occurring at the same time, so there is a need for hot water in

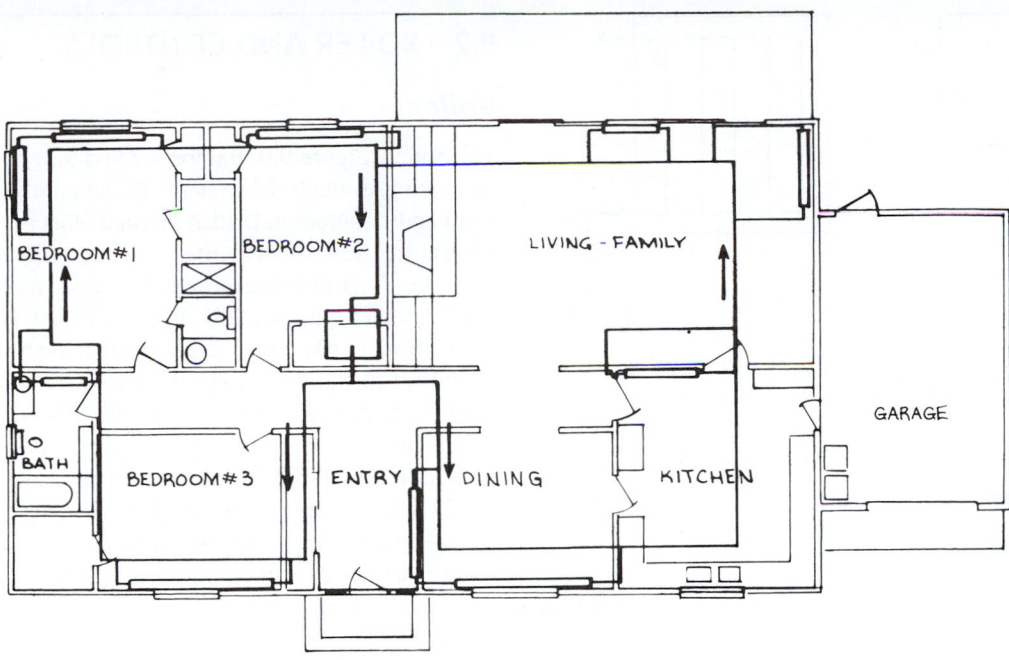

FIGURE 9.5 One-pipe plan.

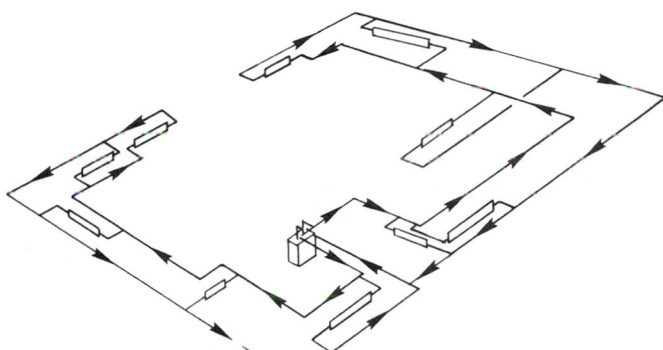

FIGURE 9.6 Two-pipe (zoned) reverse return.

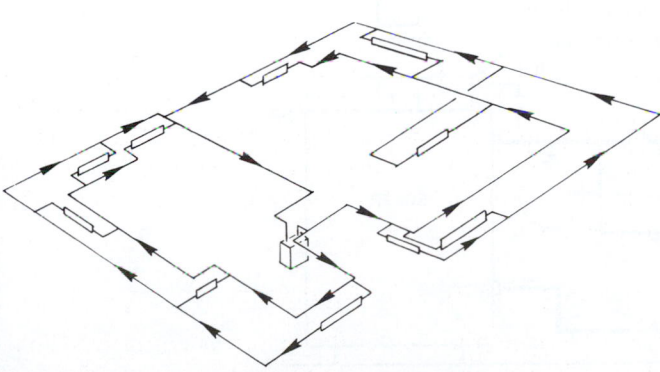

FIGURE 9.7 Two-pipe (zoned) direct return.

one zone and chilled water in another. Simultaneous heating and cooling cannot be delivered with the one-pipe or two-pipe arrangements. With these systems, a seasonal *changeover* (e.g., system change from cooling to heating in fall, and back in spring) is required in most climates, which presents overheating or overcooling potential for some spaces between seasons.

In large installations with heating and cooling capabilities, three or four pipes are used to keep the supply of hot water separate from chilled water. The *four-pipe arrangement* has two supply pipes (hot water supply and chilled water supply) and two return pipes (hot water return and chilled water return), each set carrying hot water or chilled water to separate coils in a fan-coil unit (or coils in an air-handling unit). As a result, it is the most complex arrangement with the highest initial cost. This configuration is like two separate two-pipe arrangements: one for heating and one for cooling. Simultaneous heating and cooling is possible with this arrangement. The four-pipe arrangement thus offers superior control because the terminal unit in any space can provide maximum heating or cooling at any time throughout the year. The *three-pipe arrangement* has separate supply pipes that convey hot water and chilled water to separate coils, but has a single return pipe. Hot water and chilled water are mixed in a single return, which makes this arrangement very inefficient. Seasonal changeover is not required in either arrangement.

Radiant

Radiant heating systems involve supplying heat directly to the floor or to radiant panels in the floor, wall, or ceiling of a building. The systems depend largely on radiant heat transfer: the delivery

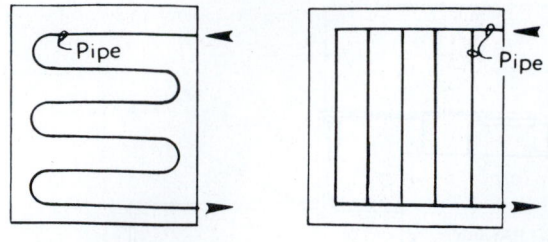

FIGURE 9.8   Radiant panels.

of heat directly from a hot surface of a heat source in a straight-line path to the people and objects in the room via infrared radiation. Infrared radiation only warms the people and objects in its path. Thermal shadowing is a concern, where radiant heat does not strike shielded objects (i.e., feet under a table). See Figure 9.8.

Piping and Fittings

In residential installations, the piping used for hot water heating is usually copper, but steel or thermoplastic pipe are sometimes used. Soldered copper pipe is the traditional piping material because it is lightweight and easy to work with. The types of copper piping used for residential hot water heating systems are Type L pipe and tubing and Type M pipe. Cross-linked polyethylene (PEX) and high-density polyethylene (HDPE) are common thermoplastic piping materials. The most commonly used fittings for the system are the same as those used in plumbing systems. In commercial HVAC installations, Schedule 40 steel pipe and L- or K-type copper are commonly used; thermoplastic pipe is generally not used. Piping materials, and piping and pumping design methods are discussed in Chapters 12 and 13.

9.2   BOILER AND CONTROLS

Boiler

The boiler (Figure 9.9, Photos 9.1 and 9.2) heats the water for circulation through the system. Boilers are generally rated on input and/or output in British thermal units per hour (Btu/hr) or kilowatts (1 kW = 3413 Btu/h). Boiler output should be adequate to offset building design heat loss, piping losses (if the pipes run through unheated space), plus any additional heating needs. Boiler efficiency increases when the boiler runs for long periods of time, so the unit selected should not be oversized or it will run intermittently and thus be less efficient. Hot water boilers may use oil, natural gas, propane gas, coal, or electricity as fuel. In residences, oil and natural gas boilers are most commonly used. Boilers were introduced in Chapter 6.

For a hot water heating system, the temperature that the water is heated to in the boiler is of prime importance because it has a direct relationship to the amount of heat that the radiation or convector units will give off. Typically, the boiler aquastat is set at about 180°F (82°C), but water temperatures as high as 220°F (105°C) are possible with pressures above atmospheric pressure. The aquastat that controls the water temperature is located somewhere on the boiler (usually in plain sight, sometimes behind a small sheet metal cover plate), and should be inspected after installation by the designer.

Thermostat

The *thermostat*, which controls the air temperature in the space, is placed in the building. With hot water heat, there will be one thermostat for each zone. The temperature desired is set on the

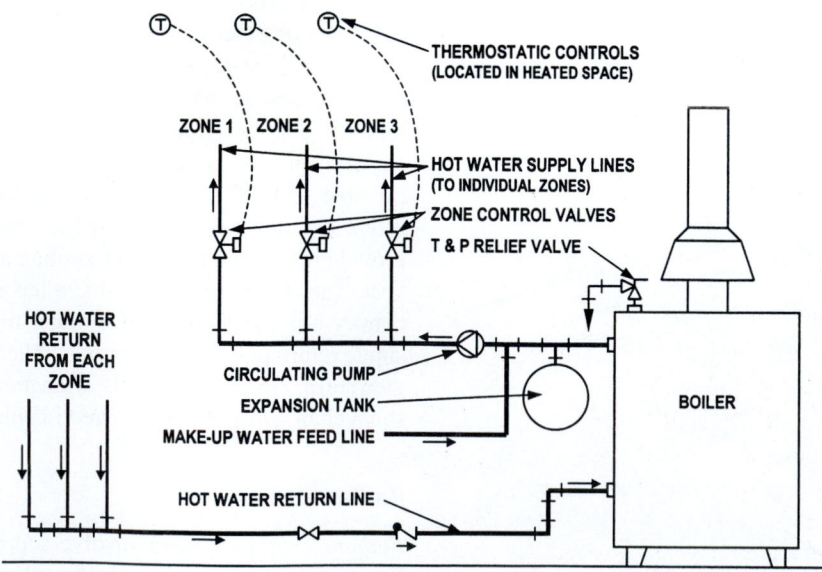

FIGURE 9.9   A typical residential boiler installation with three zones.

PHOTO 9.1 A commercial hot water boiler. (Used with permission of ABC)

PHOTO 9.2 A residential hot water boiler. (Courtesy of DOE/NREL)

thermostat, and when the temperature in the room falls below the desired temperature, the thermostat turns on the circulating pump. When the desired temperature is reached, the thermostat will turn the circulator (pump) off.

Expansion Tank

The *expansion tank* in a hydronic system (also called a *compression tank*) allows for the expansion of the water in the system as it is heated. It is located above or alongside the boiler.

Automatic Fill Valve

Hydronic systems are connected to a water supply (e.g., the public water supply). An *automatic fill valve* regulates the amount of water in the system and also prevents backflow of system water (and any water treatment chemicals) into the water supply. When the pressure in the system drops below about 12 psi (82.7 kPa), this valve opens, allowing city water into the

PHOTO 9.3 Pressure and temperature gauges on a hot water boiler. A small circulating pump used to circulate hot water.

system. A check valve automatically closes as the pressure increases. This maintains a minimum water level in the system.

Circulating Pump

Hot water is circulated from the boiler through the pipes and back to the boiler by a *circulator* or *circulating pump*. The circulator (see Photo 9.3) provides fast distribution of hot water through the system, thus delivering heat as quickly as possible. The circulating pump size is based on the delivery of water required, in gallons per minute (gal/min) or "gpm" (L/s), and the amount of friction head allowed. Most commercial and some residential installations have more than one circulator to serve separate zones (e.g., domestic hot water).

Flow Control Valve

The *flow control valve* or *zone valve* opens flow of hot water or closes off flow when the pump stops. Rather than have separate pumps for different zones, one circulator may serve several zones, with each zone regulated by a zone valve controlled by its own thermostat. A closed zone valve will prevent hot water from flowing through the system by gravity. Gravity flow of the water would cause the temperature in the room to continue to rise. Most residential zone valves are electrically powered on low voltage (typically 24 V AC).

Safety Relief Valve

When the pressure in the system exceeds about 30 psi (207 kPa), a spring-loaded *temperature/pressure relief valve* opens, permitting some of the water to bleed out of the system, which allows the pressure to drop. The valve should be located where the discharge will not cause any damage. The relief valve usually has a manual operating handle to allow testing and the flushing of contaminants (such as grit) that may cause the valve to leak under otherwise normal operating conditions.

Air Elimination Devices

All hydronic systems must have a means to eliminate air from the system. Air causes system noise in addition to interrupting heat transfer as the system fluids circulate throughout the system. Further, unless reduced below an acceptable level, the oxygen found within water will cause corrosion, which can create rust and scale buildup in the system piping. These particles can become loose, reduce flow, clog the system, and damage pump seals and other system components. Various devices are used to remove or eliminate air.

9.3 HOT WATER HEATING DISTRIBUTION DEVICES

Heat Distribution Device

A *heat distribution device, heat emitter,* or *terminal unit* must be used to distribute the heat efficiently from the hot water to the space being heated. Common terminal units for heating used in hydronic systems are convection heaters, baseboard convectors, convector radiators, radiators, fan-coil units, and unit ventilators. Cooling can be provided by fan-coil units, unit ventilators, and chilled beams. These were introduced in Chapter 7. These devices typically are classified as radiant, convector, or fan-coil, according to the way in which they transfer the heat.

Convector Units

Convector units have the heat transfer surface enclosed in a cabinet or some other enclosure. Transfer of heat occurs primarily through convection as air flows through the enclosure and past the heat transfer surface. For institutional and commercial projects, the convectors used may have several rows of large finned-tube radiators in the cabinet. (See Photos 9.4 and 9.5.) Chilled beams are a type of convector unit that can be installed in or hung from a ceiling to introduce cooling directly to the room air.

Fan-Coil Units

Fan-coil units (including unit ventilators) use a fan to distribute the air through the space. The unit usually consists of a heating

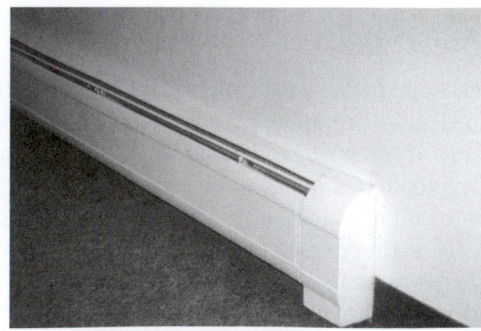

PHOTO 9.4 A baseboard convector typically used in residential installations. The position of the damper will control flow of air from natural convection and thus heat released by the unit. (Used with permission of ABC)

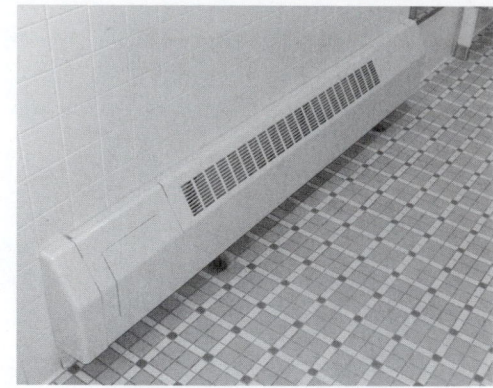

PHOTO 9.5 A baseboard convector typically used in commercial applications. (Used with permission of ABC)

coil and/or cooling coil and a fan enclosed in a cabinet. The heat is supplied by hot water. Often this type of unit is suspended from the ceiling in locations such as warehouses, storage areas, or large rooms, especially those with high ceilings. Such a unit is sized according to heat output. Layout of these units must provide adequate coverage of the space with warm air. This type of unit is also effective when the room is deep with a relatively small outside wall area for a convector or radiation unit. A typical situation would be a motel room, where the unit heater is placed under the glass area in the exterior wall and the air is fan blown through the room. Unit heaters may be recessed, surface mounted, or suspended from the ceiling or wall, or recessed or surface mounted on the floor.

Radiant Heating Units

Radiators have a heat transfer surface exposed to view within the room or space so that heat is transferred by radiation to the surrounding objects and by natural convection to the surrounding air. Technically these devices have a heat transfer liquid (e.g., steam or hot water) that circulates through exposed pipes, which may have fins or another means of increasing surface area. Traditional cast iron radiators are still used in many systems and are often an efficient and cost-effective option. Contemporary radiators are available in baseboard, wall-hung, room-divider, panel-type, and towel-drying styles that integrate well into the interior architecture while using minimal space. Because some radiator units transfer heat by both convection and radiation, they are often called *convector-radiators.* They work well in countering the radiant and convective heat loss near large window areas. (See Photos 9.6 through 9.8.)

Radiant Floor

In the *radiant floor* system, hot water (or a food-grade antifreeze mixture) is circulated through pipes or tubing that is embedded in a concrete floor or a thin concrete mixture on top of a wood-framed floor. Tubing can also be placed into a formed aluminum

PHOTO 9.6   A convector-radiator. (Used with permission of ABC)

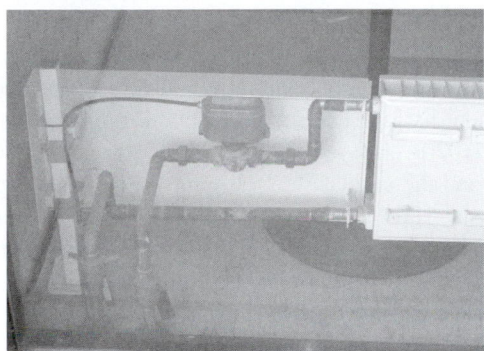

PHOTO 9.7   A rear view of the control valve and piping configuration serving a convector-radiator. (Used with permission of ABC)

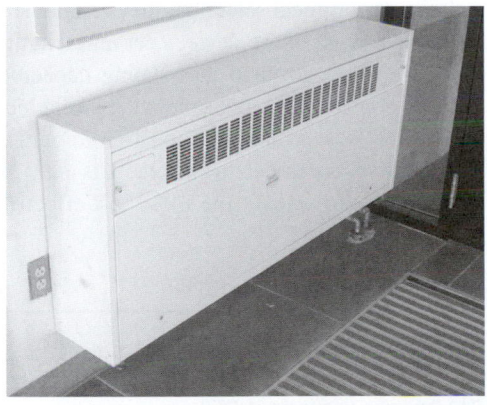

PHOTO 9.8   A cabinet heater. (Used with permission of ABC)

section that fastens to the underside of the wood subfloor. The formed aluminum section serves as fins that help distribute the heat uniformly over the floor surface. The tubing is laid out in an exaggerated S-pattern, with many variations. (See Figure 9.8 and Photo 9.9.) It is laid between 6 and 18 in (150 and 450 mm) on center. A 12 in on-center pattern is common.

Radiant floor heating systems typically use water temperatures of 85° to 140°F (30° to 60°C). Because of the large area of

PHOTO 9.9   A radiant floor system. (Courtesy of DOE/NREL)

this type of radiant floor system, the floor only needs to be heated a few degrees above the desired room temperature. To maintain comfort on most radiant floor systems, the floor surface temperature should not exceed 85°F (29°C). Zoning of radiant floors is usually done with advanced manifold modules that allow the water temperature to be varied in different zones. This provides flexibility for maintaining different temperatures in different rooms and for allowing differential heat delivery to spaces with and without solar gain. Because of the long lag time with concrete slab radiant floor heating systems, standard setback thermostats usually are not effective, although setback thermostats that have a built-in anticipation feature may work well for this application.

Radiant Ceilings and Walls

In the *radiant ceiling* or *wall* system, tubes are attached to the framing and plastered over drywall. The basic disadvantage with this type of system is that although it provides uniform heat over an entire room, heat is not uniformly lost; most of it is lost at exterior walls, usually at the doors and windows.

Radiant Panels

Wall- and ceiling-mounted *radiant panels* are usually made of aluminum and can be heated with either electricity or with tubing that carries hot water. The majority of commercially available radiant panels for homes are electrically heated. See Photos 9.10 and 9.11.

PHOTO 9.10   In-wall radiant panels. (Courtesy of DOE/NREL)

PHOTO 9.11 In-wall radiant panel controls (under construction). (Courtesy of DOE/NREL)

Device Ratings

All radiant, convector, coil units, and unit heaters are rated in terms of the amount of heat they will emit in an hour; the capacity ratings are given in Btu/hr (or kW) for small devices and in thousands of Btu/hr (MBH) for larger devices. Heating capacities vary, depending on the type of pipe, the size of the finned tube, the number of fins per foot of radiation, and the temperature of the water. At a water temperature of 180°F (82°C), residential/light commercial convectors can output 400 to 1200 Btu/hr per foot (384 to 1150 W per m), while large commercial-size convectors can deliver over 4000 Btu/hr per linear foot (3844 W per m). Typical cabinet units are about 1 to 3 ft (3 to 3 m) long with outputs ranging from about 12 000 to 80 000 Btu/hr (3.5 to 23.5 kW). Radiant floor, wall, and ceiling systems can produce about 30 Btu/hr per square foot (95 W/m²). The manufacturer's specifications (or technical data report) should be reviewed to ascertain the heating capacity of a particular unit.

Baseboard radiation units may be cast iron radiation units or finned tube convectors. A typical manufacturer's specification rating is shown in Table 9.1. The various types of elements available from this manufacturer are shown in the left column and the heat ratings for the various water temperatures on the

right. Notice that the heating capacity of the unit increases as the water temperature increases. Comparing the copper-aluminum elements, the capacity for the 2¾ in · 5 in · 0.020 · 40/ft—1¼ in tube element (the second entry on left) at 180°F is 850 Btu/hr, while at 200°F the capacity is 1030 Btu/hr. The capacity is increased by 180 Btu/hr simply by increasing the temperature of water that will flow through the element. This 180 Btu/hr increase represents a 21% gain in the heating capacity at no increase in the cost of boiler or heating device, only an increase in the temperature of the water. The effects of the fins and tube on the heating capacity of an element can be seen by comparing the first and third listings under the copper-aluminum elements. The smaller top element (2¾ in · 3¾ in · 0.011 · 50/ft—1 in tube) has a capacity of 840 Btu/hr at 180°F, while the third listing (2¾ in · 5 in · 0.020 · 50/ft—1¼ in tube) has a capacity of 910 Btu/hr. This is an increase of 70 Btu/hr or about 8.33%. The copper-aluminum elements may be compared with the steel elements by checking the second copper-aluminum listing against the second steel element listing. The second copper-aluminum listing (2¾ in · 5 in · 0.020 · 40/ft—1¼ in tube) has a capacity of 850 Btu/ hr at 180°F, and the similar steel element listing has a capacity of 710 Btu/hr at 180°F. The copper-aluminum element capacity is 140 Btu/hr, or 19.7% higher than that of the steel element.

As initial selections of heating devices are made, the various ratings must be checked. It may be necessary to use different sizes of heating devices in different rooms (usually depending on the amount of wall space available for mounting the devices), but once a basic decision is made as to type of material (steel or copper-aluminum), this will usually be used throughout.

To make an economical selection, the designer should also consider the relative costs of the devices per unit length (ft or m) and then compare these costs to the heating capacities of the units. Prices vary considerably, but unless the steel elements, installed, cost about 20% less than the copper-aluminum elements, they will not be as economical to install. It is part of the designer's responsibility to specify the most cost-effective device.

TABLE 9.1   FINNED TUBE RATINGS.

	Hot Water Output* 1 gal., (3.8 L) flow rate (for 5 gal. (19 L) flow rate use factor 1.067)								
Copper-Aluminum Elements	220°F	105°C	210°F	99°C	200°F	93°C	190°F	88°C	180°F	82°C
2¾″ × 3¾″ × .011 × 50/ft. 1″ tube	1240	1190	1120	1076	1020	980	930	890	840	800
2¾″ × 5″ × .020 × 40/ft. 1¼″ tube	1250	1200	1120	1076	1030	990	940	900	850	815
2¾″ × 5″ × .020 × 50/ft. 1¼″ tube	1340	1285	1200	1150	1100	1050	1000	960	910	870
Steel Elements									
2¾″ × 5 × 24 ga. × 40/ft. 1″ tube (IPS)	1020	980	920	880	840	800	770	735	690	660
2¾″ × 5 × 24 ga. × 40/ft. 1¼″ tube (IPS)	1040	1000	940	900	860	825	780	745	710	680

*Based on 65° (18°C) entering air.
°F values are per foot
°C values are per meter

9.4 BASIC DISTRIBUTION SYSTEM DESIGN

The following design approach will work for most systems (except radiant panels).

1. Determine the heat loss of each room and list them (Table 9.2).

   Heat losses for this residence are determined using the method outlined in Chapter 4.

2. Next, determine the pipe arrangement that will be used.

   For this design, a series loop system has been selected.

3. Using a floor plan, locate the approximate position of the heating devices in each room.

   The heating devices have been located on the plan in Figure 9.10.

4. Locate the boiler on the plan (Figure 9.11).

TABLE 9.2 ZONE HEAT LOSS TOTALS.

Trunk	32,480	BTUH	Zone 2	
Zone 1			Bedroom 3	3483
Entry		4061	Bath	1092
Dining		3056	Bedroom 1	4326
Kitchen		4160	Bedroom 2	2772
Living		9327	Int. Bath	203
		20,604		11,876
SI UNITS				
Trunk	9,546		Zone 2	
Zone 1			Bedroom 3	1013
Entry		1287	Bath	322
Dining		898	Bedroom 1	1267
Kitchen		1205	Bedroom 2	807
Living		2690	Int. Bath	57
		6080		3466

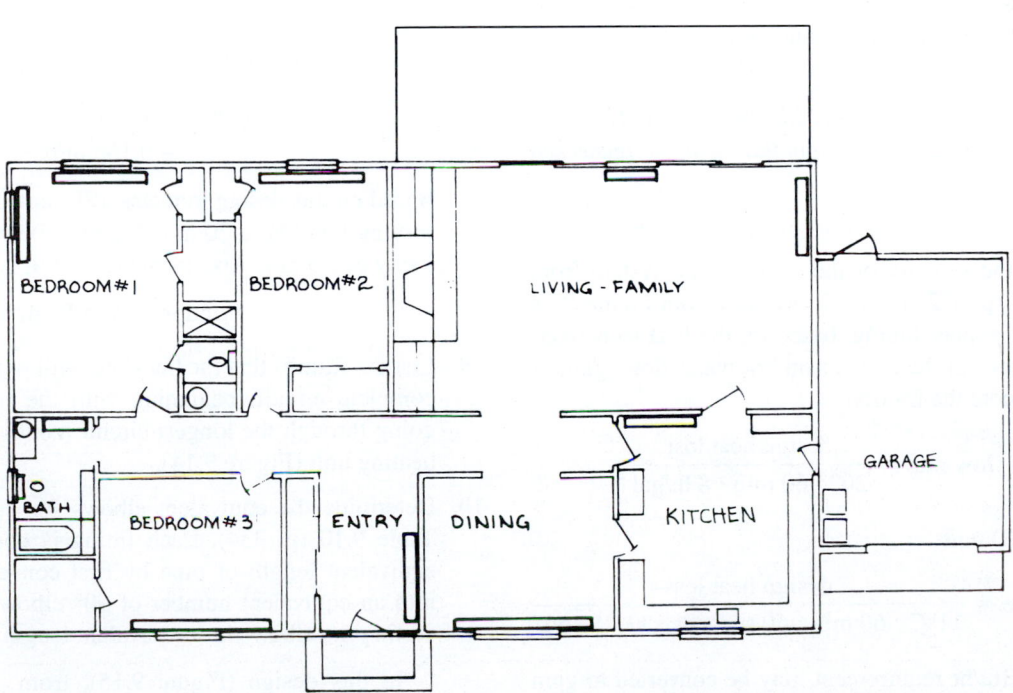

FIGURE 9.10 Series loop system.

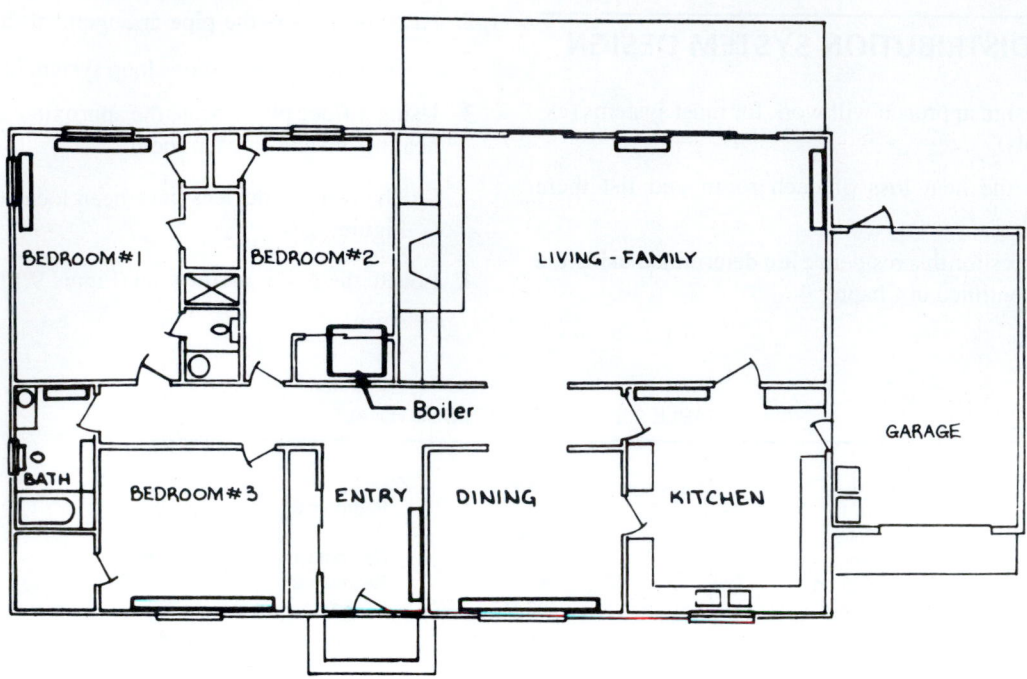

FIGURE 9.11 Locate boiler on plan.

5. Determine how many zones will be used in the design.

In this design, two zones will be used, as shown in Figure 9.12.

6. Determine the actual length of the longest zone (circuit).

From Figure 9.13, the actual length of the longest zone is taken as 145 lineal ft (44.3 m).

7. Assume an average pipe size for the system. *This pipe size is a preliminary selection that will be rechecked later.*

Assume ¾ in (20 mm) copper tubing for this design.

8. Determine the velocity of the water in the system from Figure 9.17 (p. 332) for steel pipe and from Figure 9.18 (p. 333) for copper tubing, based on the heat to be conveyed per hour in the zone or on hot water flow (gpm or L/s). Also, note the friction.

$$°F \text{ flow rate} = \frac{\text{design heat lost}}{20° \cdot 60 \text{ min} \cdot 8 \text{ lb/gal}}$$

In metric (SI) units:

$$°C \text{ flow rate} = \frac{\text{design heat loss}}{11°C \cdot 60 \text{ min} \cdot 60 \text{ seconds} \cdot 13.75 \text{ kg/L}}$$

Note: The Btu/hr requirement may be converted to gpm by dividing Btu/hr by 9600 (based on a 20°F temperature drop). Zone 1 requires 20 604 Btu/hr or 2.1 gpm.

In metric (SI) units, the watts requirement may be converted to L/s by dividing watts by 38 000 (based on an 11°C temperature drop). Zone 1 requires 6080 W or 0.16 L/s.

Based on this design (Figure 9.17), with the largest zone serving 20 604 Btu/hr, a ¾ in Type L pipe would have a velocity of 1.5 feet per second (fps). Also from the chart, it is determined that

$$\text{Friction} = 1.4 \text{ ft per 100 ft}$$
$$= 0.168 \text{ in/ft}$$

Based on this design in metric (SI) units, the largest zone serving 6080 W, a 20 mm Type L pipe would have a velocity of 0.5 m/s. Also from the chart it is determined that

$$\text{Friction} = 185 \text{ Pa/m}$$

9. List the fittings that the hot water will pass through in the complete circuit, beginning with the heating unit and going through the longest circuit (zone) and back to the heating unit (Figure 9.14).

10. Determine the equivalent elbows for each fitting from Table 9.10 (p. 334). Each fitting is converted into the equivalent length of pipe by first converting the fitting into an equivalent number of 90° elbows and then converting the elbows into equivalent length.

In this design (Figure 9.15), from Table 9.10, one boiler is equal to four 90° copper elbows or three 90° iron elbows. A check valve is equal to twenty 90° copper

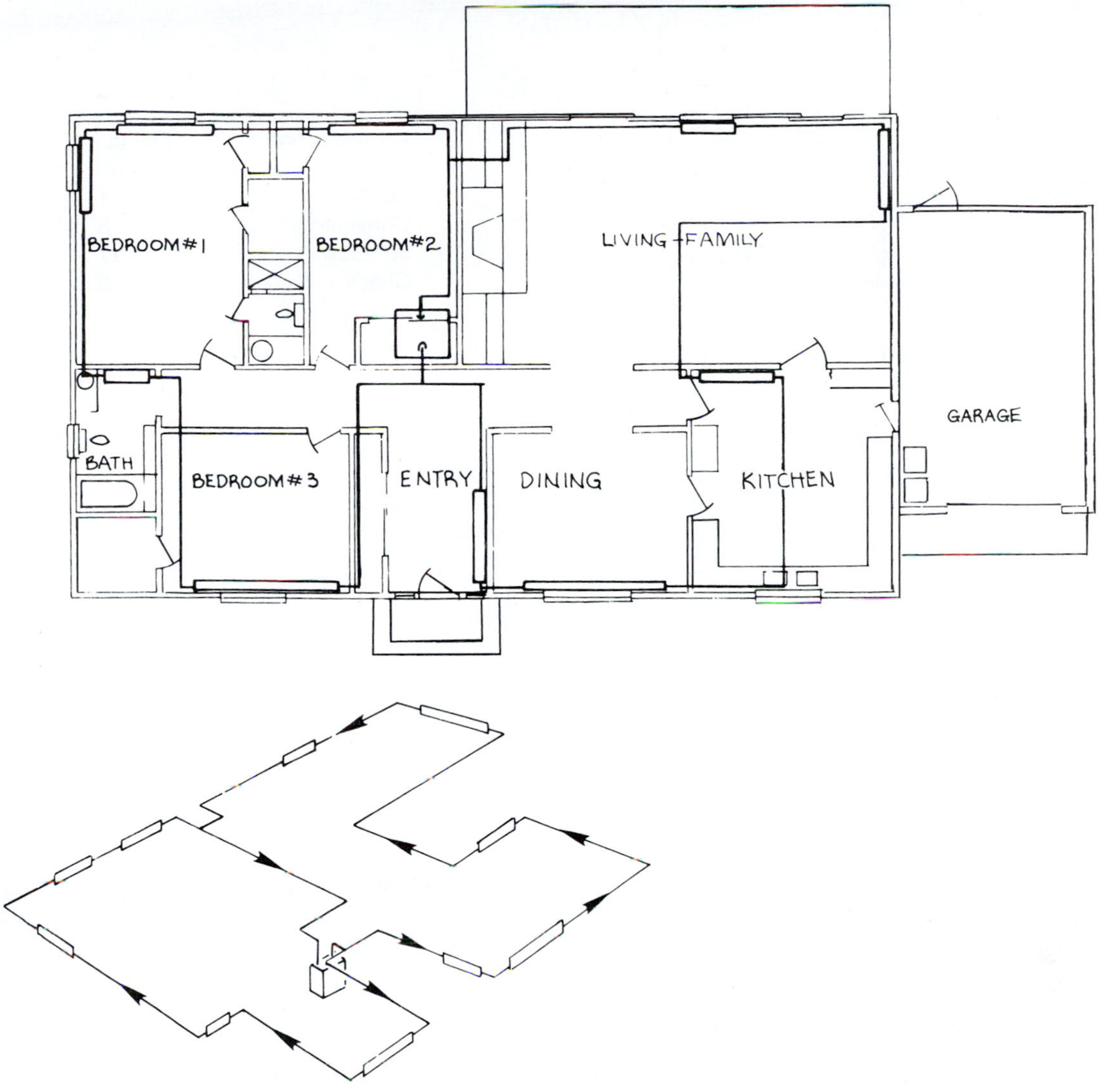

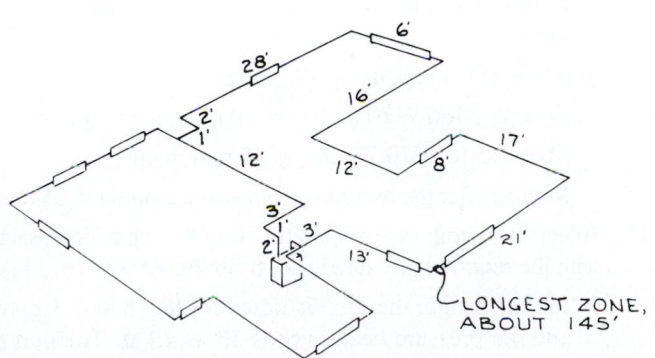

FIGURE 9.12   Determine zones.

FIGURE 9.13   Determine actual length of zones.

elbows or fifteen 90° iron elbows. The equivalent length will be found in Step 11.

11. Determine the total equivalent length of the fittings from Table 9.10. This table is based on pipe size and velocity. The pipe was tentatively selected in Step 7 and the velocity noted in Step 8.

   In this design, a ¾ in pipe and 1.5 fps have been selected. From Figure 9.11, there is an equivalent length of 1.8 ft of pipe for every elbow. The equivalent length of the fittings is 46 ft (81 elbows · 1.8).

   In this design in metric (SI) units, a 20 mm pipe and 0.5 m/s have been selected. From Figure 9.15, there is an equivalent length of 0.6 m of pipe for every elbow. The equivalent length of the fittings is 48.6 m (81 elbows · 0.6).

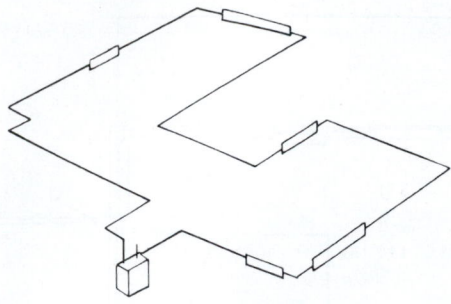

FIGURE 9.14 Fittings.

Longest Zone	Fittings No.
Boiler	1
Convectors	5
90° Elbows	17
Check Valves	2

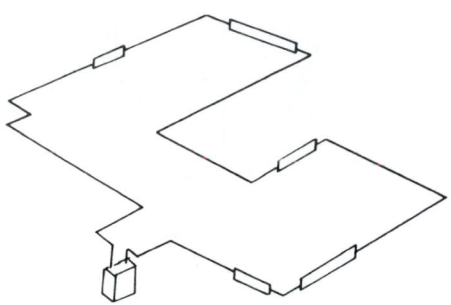

FIGURE 9.15 Equivalent elbows.

Longest Zone Fittings

	No.	Equiv Elbows
Boiler	1	4
Convectors	5	20
90° Elbows	17	17
Check Valves	2	40
		81

12. Add the length of the longest circuit (zone), Step 6, to the total equivalent length of fittings for the circuit, Step 11.

   In this design, the longest circuit is 145 lineal ft (Step 6), and the equivalent length of fittings is 146 lineal ft (Step 11), for a total equivalent length of 291 lineal ft.

   In this design in metric (SI) units, the longest circuit is 44.2 m (Step 6), and the equivalent length of fittings is 48.6 m (Step 11), for a total equivalent length of 92.8 m.

13. Determine the pressure head, using the friction head selected, Step 8, and the total equivalent length of pipe, Step 12, and Table 9.11 (p. 334).

   In this design, the friction head is about 1.4 ft per 100 ft (Step 8) and the total equivalent length is 291 lineal ft (Step 12). The pressure head required is determined by multiplying the friction loss per 100 ft times the equivalent length of the pipe. Note that the loss is *per 100 ft,* and the equivalent length must be adjusted to the number of 100 ft required (291 lf = 2.91).

   1.4 ft per 100 ft · 2.91 = 4.1 ft pressure head

   In this design in metric (SI) units, the friction head is about 195 Pa/m (Step 8) and the total equivalent length is 92.8 m (Step 12). The pressure head required

is determined by multiplying the friction loss per meter times the equivalent length of the pipe.

$$185 \text{ Pa/m} \cdot 92.8 = 17\ 186 \text{ Pa} = 17.2 \text{ kPa}$$
$$17.2 \text{ kPa} \div 9.8 \text{ kPa/m} = 1.76 \text{ m pressure head}$$

14. Select the rest of the pipe sizes required for the system, using the friction head selected and the heat it must supply, from Figure 9.18.

   Using 1.4 ft per 100 ft:

   Zone 2: 11 876 Btu/hr (1.1 gpm) = ¾ in pipe required

   Main: 32 480 Btu/hr (2.9 gpm) = 1 in main (Figure 9.18)

   Return: After the two zones join, same as main = 1 in pipe

   In metric (SI) units, using 185 Pa/m:

   Zone 2: 3466 W (0.09 l/s) = 20 mm pipe required

   Main: 9546 W (0.25 L/s) = 25 mm main

   Return: after the two zones join, same as main = 25 mm

15. Select the pump size required, using the gpm necessary with the required pressure head, from Figure 9.19 (p. 334).

   In this design, the gpm is noted in Step 8 as 2.1 gpm and the pressure head in Step 13 as 4.1 ft. The pump chart is shown in Figure 9.19. Find the gpm along the bottom and move vertically; now find the pressure

TABLE 9.3   EXTERIOR WALL AVAILABLE.

	BTUH Req.	L.F. of Ext. Wall	SI UNITS		W Req.	L.M. of Ext. Wall
Entry	4061	—	Entry		1287	—
Dining	3056	12	Dining		898	3.6
Kitchen	4160	—	Kitchen		1205	—
Living Room	9327	10	Living Room		2690	3.0

head along the left and move horizontally until they meet. On this chart, an A pump would provide a pressure head of about 6.0 ft. (Do not use this chart for actual design; secure the manufacturer's performance ratings.)

In this design in metric (SI) units, the L/s is noted in Step 8 as 0.16 L/s, and the pressure head in Step 13 as 1.72 m. The pump chart is shown in Figure 9.19. Find the L/s along the top and move vertically; now find the pressure head along the right and move horizontally until they meet. On this chart, an A pump would provide a pressure head of about 1.8 m. (Do not use this chart for actual design; secure the manufacturer's performance ratings.)

16. Find the temperature drop for the total circuit based on the total friction head in millinches (friction times total equivalent length) times the design temperature drop for the system divided by the pressure head (pressure head in millinches equals pressure head in feet times 12 000).

$$\frac{\frac{\text{Friction head}}{(\text{millinches})} \cdot \frac{\text{Total equivalent}}{\text{length (ft)}} \cdot \frac{\text{Design}}{\text{temperature}}}{\text{Pressure head (ft)} \cdot 12\ 000}$$

In this design, the friction head is 168 millinches, the total equivalent length is 291 ft, the design temperature drop is 20°F, and the pressure head of the pump is 6.0 ft.

$$\frac{168 \cdot 291\ \text{ft} \cdot 20°F}{6.0\ \text{ft} \cdot 12\ 000} = 13.58\ (\text{use } 14.0)°F\ \text{temperature drop in circuit}$$

In metric (SI) units:

$$\frac{\frac{\text{Friction head}}{(\text{Pa/m})} \cdot \frac{\text{Total equivalent}}{\text{length (m)}} \cdot \frac{\text{Design}}{\text{temperature}}}{\text{Pressure head (m)} \cdot \text{Pa/m}}$$

In this design, the friction head is 195 Pa/m, the total equivalent length is 48.6 m, the design temperature drop is 11°C, and the pressure head is 1.8 m.

$$\frac{185 \cdot 92.8 \cdot 11°C}{1.8 \cdot 9800} = 10.7\ (\text{use } 11°C)$$

17. The next phase of the design is to determine how many lineal feet (meters) of exterior wall are available for radiation units of some type (Table 9.3). In a room with little or no exterior wall available, it may be necessary to use interior wall space.

18. Calculate the temperature drop for each radiation unit served through the circuit, and record it in the tabulated form (Table 9.4) under "System Temperature Drop." The formula used to calculate the temperature drop for each radiation unit(s) is

$$\frac{\text{Radiation unit(s) heating capacity}}{\text{Circuit (zone) capacity}} \cdot \frac{\text{Temperature drop}}{\text{in total system}}$$

19. Calculate the temperature in the circuit main, in a series loop, or after the water returns from the radiation unit(s) served. Tabulate this temperature in the main in the form under "Enter." *This is the entering temperature for each of the radiation units* (Table 9.5).

Note: If designing a series loop system, go directly to Step 21a.

TABLE 9.4   CALCULATE TEMPERATURE DROP.

			SI UNITS		
ENTRY	$\frac{4061}{20,604} \times 14 = 2.8°F$		ENTRY	$\frac{1287}{6080} \times 11 = 2.3°C$	
DINING	$\frac{3056}{20,604} \times 14 = 2.1°F$		DINING	$\frac{898}{6080} \times 11 = 1.6°C$	
KITCHEN	$\frac{4160}{20,604} \times 14 = 2.8°F$		KITCHEN	$\frac{1205}{6080} \times 11 = 2.2°C$	
LIVING ROOM	$\frac{9327}{20,604} \times 14 = 6.3°F$		LIVING ROOM	$\frac{2690}{6080} \times 11 = 4.9°C$	

TABLE 9.5   CALCULATE TEMPERATURE IN MAIN.

	ENTER TEMP °F	TEMP LOSS IN CONVECTOR °F
ENTRY	200.0	2.8
DINING	197.2	2.1
KITCHEN	195.1	2.8
LIVING ROOM	192.3	6.3

SI UNITS

	ENTER TEMP °C	TEMP LOSS IN CONVECTOR °C
ENTRY	93°C	2.3
DINING	90	1.6
KITCHEN	89.1	2.2
LIVING ROOM	86.9	4.9

20. *Note:* This step is used only in the one-pipe system. For all other systems, go to Step 21.

Calculate the temperature drop of the water that is diverted through the radiation unit(s) and through the fitting used. The amount of water diverted depends on the size of the fitting used and location of the main above or below the radiation unit; Figure 9.16 is used for this percentage of diversion.

$$\text{Radiation unit(s) temp. drop} = \frac{\text{Temp. drop in circuit (zone)}}{\text{Percent diversion}} \cdot 100$$

Note: The 100 is a constant and changes the percentage to a decimal equivalent.

Tabulate the temperature drop for the radiation unit(s) as shown in Table 9.6.

21. Calculate the average temperature.

a. *Series loop only.* Calculate the average temperature in the radiation unit(s) by subtracting one-half of the temperature drop from the entering temperature of the circuit main. Tabulate the average temperature in the form under "Average Radiation Temperature."

b. *All except series loop.* Calculate the average temperature in the radiation unit(s) by subtracting one-half of the temperature drop that occurs in the radiation units (from Step 16) from the entering temperature of the circuit main. Tabulate the average temperature in the form.

22. Next, the heat emission rate of radiation required is determined by dividing the heat loss for each room by the available length of exterior wall.

TABLE 9.6   CALCULATE AVERAGE TEMPERATURE DROP.

	ENTER	LEAVE	AVE.
ENTRY	200.0	197.2	198.6
DINING	197.2	195.1	196.2
KITCHEN	195.1	192.5	193.8
LIVING ROOM (3.2°F loss)	192.5	189.3	190.9
LIVING ROOM (3.1°F loss)	189.3	186.2	187.7

SI UNITS

	ENTER	LEAVE	AVE.
ENTRY	93.0	90.7	91.8
DINING	90.7	89.1	89.9
KITCHEN	89.1	86.9	88.0
LIVING ROOM (2.5°C)	86.9	84.4	85.6
LIVING ROOM (2.4°C)	84.4	82	83.2

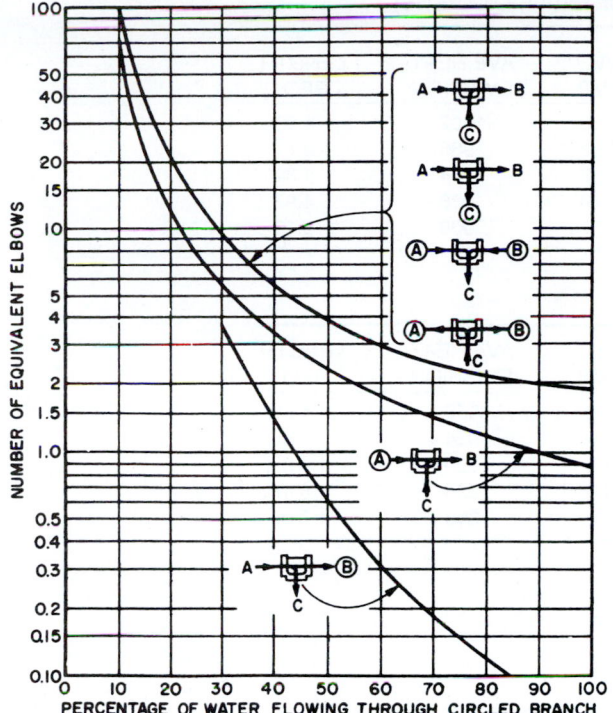

Notes: 1. The chart is based on straight tees, that is, branches A, B, and C are the same size.

2. Head loss in desired circuit is obtained by selecting proper curve according to illustrations, determining the flow at the circled branch, and multiplying the head loss for the same size elbow at the flow rate in the circled branch by the equivalent elbows indicated.

3. When the size of an outlet is reduced the equivalent elbows shown in the chart do not apply. The maximum loss for any circuit for any flow will not exceed 2 elbow equivalents at the maximum flow (gpm) occurring in any branch of the tee.

4. The top curve of the chart is the average of 4 curves, one for each of the tee circuits illustrated.

Elbow Equivalents of Tees at
Various Flow Conditions[1,4]

Reprinted with permission from ASHRAE, *Fundamentals Handbook*, 1993.

FIGURE 9.16 Elbow equivalents of tees.

23. Select the type of radiation unit(s) to be used. 2¾ in · 3¾ in · 0.11 · 50/ft in tube (Table 9.7).

24. Determine the length or amount of radiation unit(s) required to evenly heat each room by dividing the heat loss of the room by the heat emission rate of the radiation unit(s) selected. Tabulate the lengths on the form (Table 9.8).

25. Size the compression tank by determining the entire volume of water in the system and then finding the amount of expansion based on a temperature rise of 150°F (from 50°F entering to 200°F heated) times (0.004 lt − 0.0466) the net coefficient of expansion of the water. The compression tank must be sized to accommodate this water expansion. The volume of water in the system is determined from the figures in Table 9.9.

9.5 BOILER SELECTION

In residences, the boiler that provides hot water for heating is commonly located in the basement or crawl space or in a first-floor utility room, but it may also be in the attic. Boilers are most commonly fired by natural gas or oil, and occasionally by electricity, coal, or propane.

The boiler selected must be large enough to supply all of the heat loss in the spaces as calculated. The boiler must also have enough additional capacity to compensate for heat loss that will occur in the pipes that circulate the hot water through the system (called *pipe loss*). In addition, an additional reserve capacity is required so that the boiler can provide heat quickly to warm up the space. This *pick-up allowance* also provides some extra heating capacity when it is necessary to increase the temperature of a space several degrees, not merely to maintain a temperature. For example, a retail store that closes at 9 pm may turn its thermostat down to 60°F before closing. In the morning, as the thermostat is raised to 70°F, it is necessary to have additional capacity in the boiler to provide this extra heat as quickly as possible. Most buildings have automatic thermostatic controls that are used to regulate the temperatures throughout the day.

TABLE 9.7 SELECT RADIATION UNITS(S).

Copper-Aluminum Elements	Hot Water Output* 1 gal., (3.8 L) flow rate (for 5 gal., (19 L) flow rate use factor 1.067)									
	220°F	105°C	210°F	99°C	200°F	93°C	190°F	88°C	180°F	82°C
2¾" × 3¾" × .011 × 50/ft. 1" tube	1240	1190	1120	1076	1020	980	930	890	840	800
2¾" × 5" × .020 × 40/ft. 1¼" tube	1250	1200	1120	1076	1030	990	940	900	850	815
2¾" × 5" × .020 × 50/ft. 1¼" tube	1340	1285	1200	1150	1100	1050	1000	960	910	870

TABLE 9.8   DETERMINE LENGTHS OF RADIATION UNIT(S).

	BTUH REQ.	L.F. OF EXT.WALL	AVE,. TEMP	AVE.HEAT EMISSION	LENGTH REQ.
ENTRY	4061	—	198.6	1007	4.03'
DINING	3056	12	196.2	986	3.1'
KITCHEN	4160	—	193.8	964	4.32'
LIVING ROOM	4664	10	190.9	938	4.97'
LIVING ROOM	4663	—	187.7	909	5.13'

SI UNITS

	WATTS REQ.	L.F. OF EXT.WALL	AVE,. TEMP	AVE.HEAT EMISSION	LENGTH REQ.
ENTRY	1287	—	91.8	958	1.4 m
DINING	898	12	89.9	924	1.0 m
KITCHEN	1205	—	88.0	890	1.4 m
LIVING ROOM	1345	10	84.4	836	1.6 m
LIVING ROOM	1345	—	83.2	813	1.7 m

Note:The lengths vary slightly due to rounding off.

TABLE 9.9   VOLUME OF WATER IN STANDARD PIPE OR TUBE.

Volume of Water in Standard Pipe and Tube

Nominal Pipe Size In.	Standard Steel Pipe			Type L Copper Tube	
	Schedule No.	Inside Dia In.	Gallons per Lin Ft	Inside Dia In.	Gallons per Lin Ft
⅜	—	—	—	0.430	0.0075
½	40	0.622	0.0157	0.545	0.0121
⅝	—	—	—	0.666	0.0181
¾	40	0.824	0.0277	0.785	0.0251
1	40	1.049	0.0449	1.025	0.0429
1¼	40	1.380	0.0779	1.265	0.0653
1½	40	1.610	0.106	1.505	0.0924
2	40	2.067	0.174	1.985	0.161
2½	40	2.469	0.249	2.465	0.248
3	40	3.068	0.384	2.945	0.354
3½	40	3.548	0.514	3.425	0.479
4	40	4.026	0.661	3.905	0.622
5	40	5.047	1.04	4.875	0.970
6	40	6.065	1.50	5.845	1.39
8	30	8.071	2.66	7.725	2.43
10	30	10.136	4.19	9.625	3.78
12	30	12.090	5.96	11.565	5.46

Reprinted with permission from ASHRAE, *Applications Handbook,* 1991.

Typically, the designer allows about 33⅓% for the heat loss in the pipes and an additional 15% to 20% as a pick-up allowance, in addition to the calculated heat loss of the building. For example, in the residence being designed:

Calculated heat loss	32 480 Btu/hr
Piping loss (33⅓%)	+10 825
	43 305
Pick-up allowance (20%)	+ 8660
Total	51 965 Btu/hr
Calculated heat loss	9546 W
Piping loss (33⅓%)	+ 3182
	12 728
Pick-up allowance (20%)	2545
Total	15 273 W

Spaces that are heated intermittently, such as churches or auditoriums, will need a much larger pick-up allowance, perhaps as much as 50%. This is because the temperature will often be allowed to drop quite low (perhaps 50°F) when the space is not in use, and then, when the heat is raised to 70°F, it must heat up quickly.

Although most designers simply allow a certain percentage for piping loss, it can be calculated—but only once the piping layout is finalized and the installation decided upon and specified. From the piping layout, it would be necessary to determine the length of pipe and the various temperatures of the air through which the pipe travels. Where the pipe travels through an unheated basement or attic, the heat loss would be much greater than the heat loss of the same pipe traveling through a heated space.

Once the piping loss and the pick-up allowance have been taken into account, the boiler size may be selected. The boiler selected should be as close to the total Btu/hr as possible to provide the most efficient operation. Keep in mind that the outside design temperature used will not actually occur very often, that the boiler will be operating at less than full capacity most of the time, and that the pickup allowance also gives a little extra capacity. Also, oversizing the boiler will probably be of little value during any unusually long cold spells with temperatures well below the outside design temperature unless the radiation or convector units selected are able to put out additional heat.

In large projects (institutional and commercial), quite often two boilers are used. Because boilers are more efficient the longer they operate, the first boiler would be designed to satisfy about 60% of the heat required, and the second would supply the balance. In this way, the first unit would come on and supply the heat required for about 75% of the days that heat is needed. Only during colder weather would the second unit turn on. The more efficient operation of the smaller first unit (when compared with the operation of a single boiler) pays for the added cost of a second unit. Also, this provides a backup unit in case one of the boilers requires repairs.

9.6 INSTALLED AND EXISTING SYSTEMS

No matter how carefully a system is designed and laid out, there is the possibility that at least a portion of the building is not being heated satisfactorily. This, of course, is also true of many existing systems. Quite often the problems in these systems can be corrected easily and relatively inexpensively after a careful analysis of the system.

First, carefully take notes as to exactly what the problem is, what time of day it is most noticeable, what the outside temperature is during the time that sufficient heat is not being properly distributed, and whether the boiler is running continuously or intermittently.

Sample complaint: Temperature in house is 60° to 62°F during night; it rose to the desired temperature during the day

Heating equipment: Hot water, working satisfactorily

Outside temperature: 10°F

Wind: Light

Boiler water temperature: 180°F

In this example, note first that the general complaint is that the temperature of the entire house dropped to about 60° to 62°F during the night. During the day, it warmed up to the 70°F reading desired. The outside temperature that night reached a low of about −10°F, well below the design temperature of 0°F used for Albany, New York (the location of the residence).

The first observation is that the outside temperature during the night was about 10°F below the outside design temperature and the inside temperature registers about 8°F below the thermostat setting. Thus, the problem is how to get more heat out of the baseboard convector radiators installed in the house. The first possibility is to check the temperature to which the boiler is heating the hot water. Each boiler has a thermostat that sets the highest temperature to which it can heat the water, and then the boiler will shut off while the circulating pumps continue to push hot water through the system.

It is most important to know whether the boiler was operating continuously or intermittently. Quite often, the occupant confuses the fact that the pipes were hot and the hot water was being circulated with actual boiler operation. It may be necessary to obtain more information about the boiler operation before the system is corrected. If the boiler is operating continuously, it indicates that no matter how hard the boiler works, it cannot heat the water to the thermostat setting. When the boiler operates intermittently, it means that the boiler has heated the water to the thermostat setting and is waiting for cooler water to circulate to it before starting again. Intermittent operation is most commonly found and will be discussed first.

Typically, the boiler thermostat is set at 170° or 180°F during installation. Keeping in mind that the amount of heat delivered by the convectors will increase with hotter water (Table 9.1), the solution may be to increase the thermostat setting of the boiler. The hot water, in a closed heating system such as this, can be heated as high as 220°F. Generally, for every increase of 10°F in water temperature, the heat output of the radiation or convector units is increased about 10%. Intermittent operation is the most common problem and, fortunately, this is the most common solution.

If the temperature *is* set high and the boiler is operating intermittently, the next solution is to provide additional heat in each room by some means. In most rooms, the length of finned tube radiators may be increased, thus increasing the amount of heat put into the room. Many rooms have convector covers all along a wall, but check inside the cover; a great deal of that length may be plain tube and not finned tube. In such a case, a plumber can add some finned tube element. It may be necessary to add extra lengths of finned tube on an interior wall or to put a convector or unit heater up in a wall where baseboard

space is not available. Any revision or addition of this type must be carefully planned and may be costly.

9.7 SYSTEM INSTALLATION

The piping required for the hot water heating system will run in the walls, floors, and ceiling spaces. The heating devices may require recessed wall spaces or extra structural support if hung from the ceiling. The heating unit is often located in a basement or mechanical room. Radiant heating in the floor slab requires that the tubing be installed before the slab is placed. When located in the ceiling or floor, tubing is attached to the framing. All tubing and pipes should be inspected for leakage under an air pressure test before it is covered.

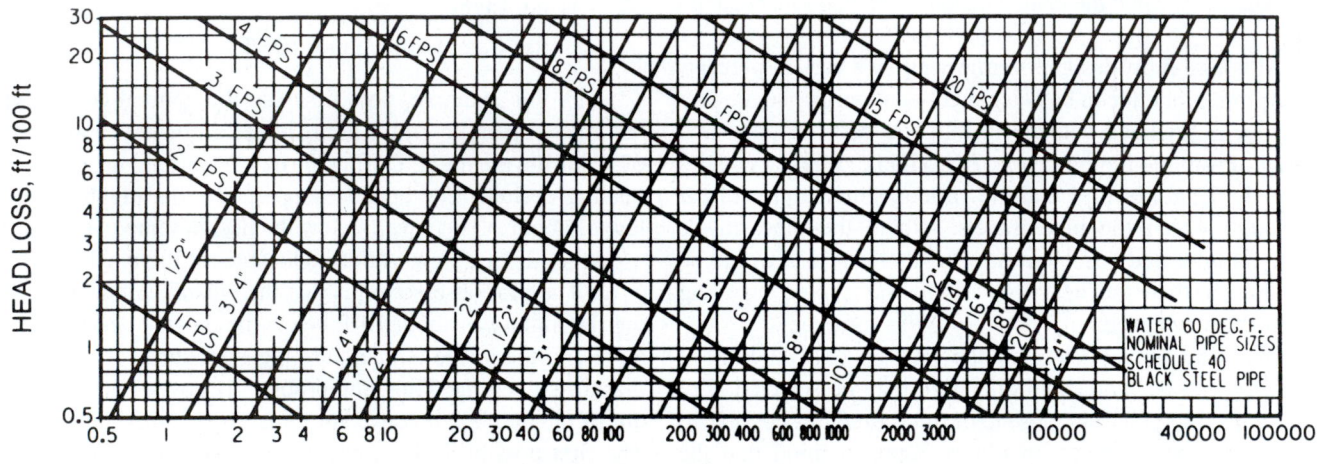

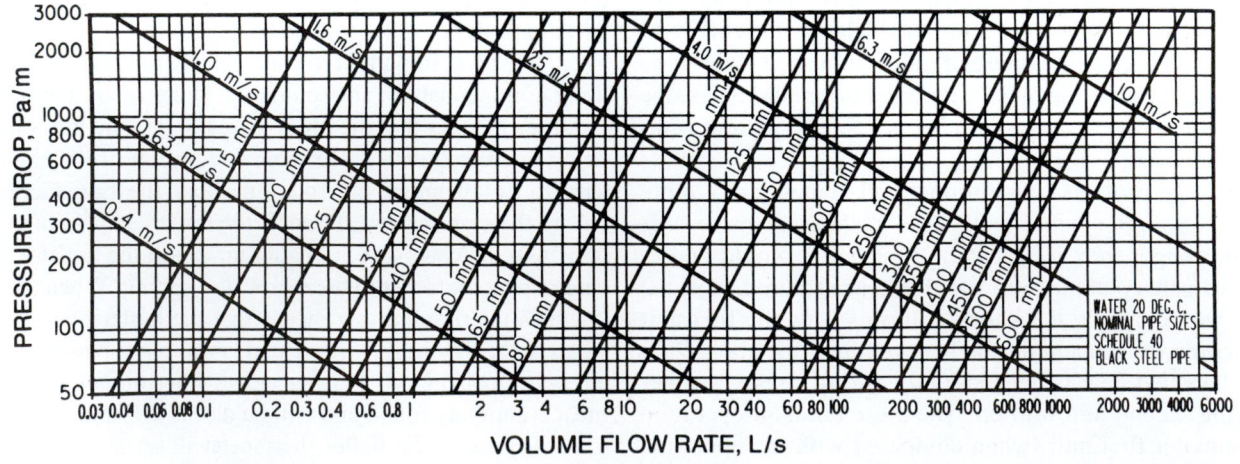

Reprinted with permission form ASHRAE, *Fundamentals Handbook,* 1993.

FIGURE 9.17 Friction loss for water in commercial steel pipe.

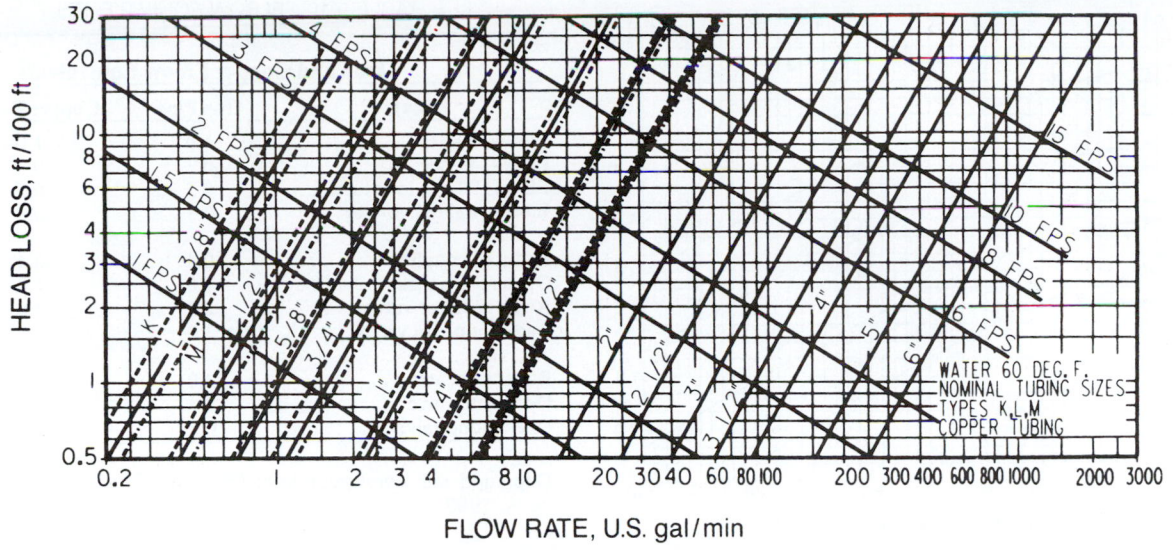

Reprinted with permission from ASHRAE, *Fundamentals Handbook,* 1993.

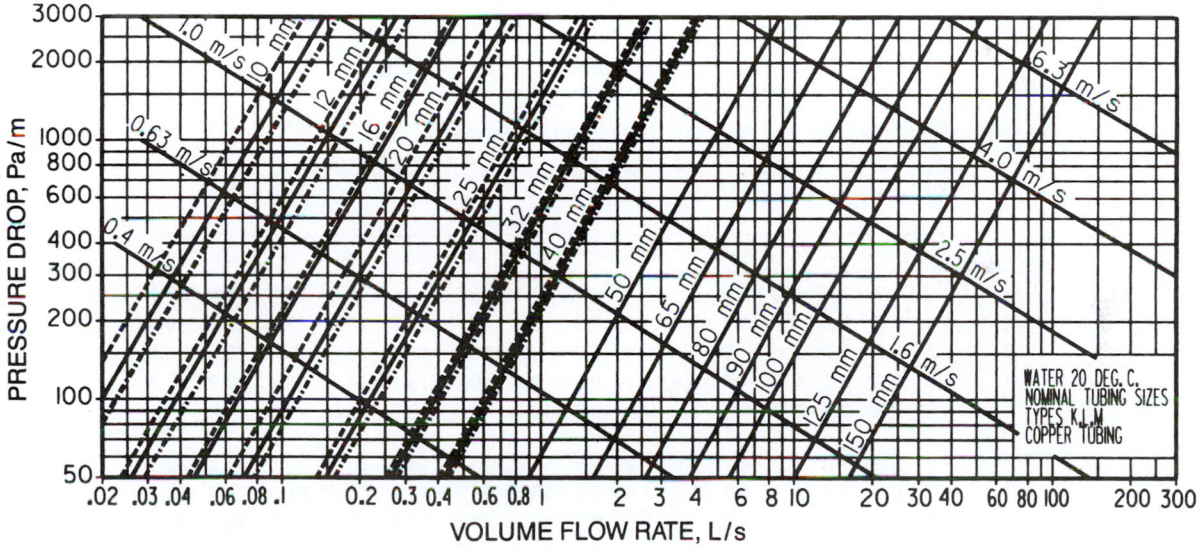

Reprinted with permission form ASHRAE, *Fundamentals Handbook,* 1993.

FIGURE 9.18   Friction loss for water in copper tubing.

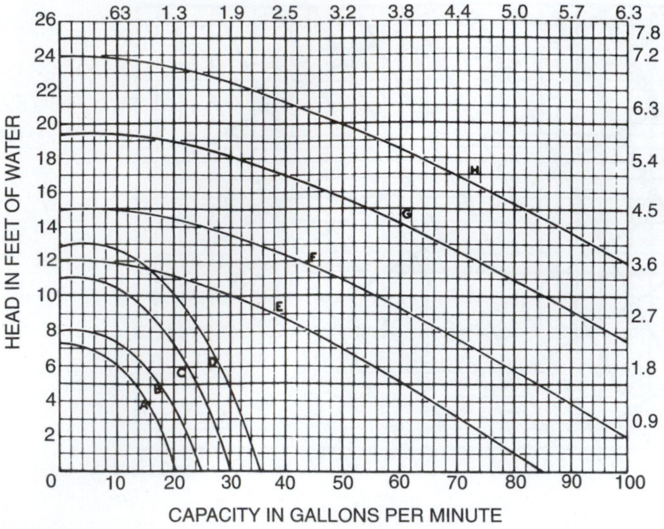

TABLE 9.10   ELBOW EQUIVALENTS.

HEAD IN FEET OF WATER

CAPACITY IN GALLONS PER MINUTE

Note: Metrics added by author.

FIGURE 9.19   Pump performance.

Iron and Copper Elbow Equivalents

Fitting	Iron Pipe	Copper Tubing
Elbow, 90-deg	1.0	1.0
Elbow, 45-deg	0.7	0.7
Elbow, 90-deg long turn. . .	0.5	0.5
Elbow, welded, 90-deg	0.5	0.5
Reduced coupling	0.4	0.4
Open return bend.	1.0	1.0
Angle radiator valve.	2.0	3.0
Radiator or convector	3.0	4.0
Boiler or heater	3.0	4.0
Open gate valve	0.5	0.7
Open globe valve	12.0	17.0

Reprinted with permission from ASHRAE, *Fundamentals Handbook,* 1993.

TABLE 9.11   EQUIVALENT LENGTH OF PIPE.

Reprinted with permission from ASHRAE, *Fundamentals Handbook,* 1993.

Equivalent Length of Pipe for 90-Deg Elbows

	Pipe Size														
Vel. Fps.	½	¾	1	1¼	1½	2	2½	3	3½	4	5	6	8	10	12
1	1.2	1.7	2.2	3.0	3.5	4.5	5.4	6.7	7.7	8.6	10.5	12.2	15.4	18.7	22.2
2	1.4	1.9	2.5	3.3	3.9	5.1	6.0	7.5	8.6	9.5	11.7	13.7	17.3	20.8	24.8
3	1.5	2.0	2.7	3.6	4.2	5.4	6.4	8.0	9.2	10.2	12.5	14.6	18.4	22.3	26.5
4	1.5	2.1	2.8	3.7	4.4	5.6	6.7	8.3	9.6	10.6	13.1	15.2	19.2	23.2	27.6
5	1.6	2.2	2.9	3.9	4.5	5.9	7.0	8.7	10.0	11.1	13.6	15.8	19.8	24.2	28.8
6	1.7	2.3	3.0	4.0	4.7	6.0	7.2	8.9	10.3	11.4	14.0	16.3	20.5	24.9	29.6
7	1.7	2.3	3.0	4.1	4.8	6.2	7.4	9.1	10.5	11.7	14.3	16.7	21.0	25.5	30.3
8	1.7	2.4	3.1	4.2	4.9	6.3	7.5	9.3	10.8	11.9	14.6	17.1	21.5	26.1	31.0
9	1.8	2.4	3.2	4.3	5.0	6.4	7.7	9.5	11.0	12.2	14.9	17.4	21.9	26.6	31.6
10	1.8	2.5	3.2	4.3	5.1	6.5	7.8	9.7	11.2	12.4	15.2	17.7	22.2	27.0	32.0

Note: 120 millinches per foot equals 1 foot per 100 feet.

Equivalent Length in Metres of Pipe for 90° Elbows

Velocity, m/s	Pipe Size, mm													
	15	20	25	32	40	50	65	90	100	125	150	200	250	300
0.33	0.4	0.5	0.7	0.9	1.1	1.4	1.6	2.0	2.6	3.2	3.7	4.7	5.7	6.8
0.67	0.4	0.6	0.8	1.0	1.2	1.5	1.8	2.3	2.9	3.6	4.2	5.3	6.3	7.6
1.00	0.5	0.6	0.8	1.1	1.3	1.6	1.9	2.5	3.1	3.8	4.5	5.6	6.8	8.0
1.33	0.5	0.6	0.8	1.1	1.3	1.7	2.0	2.5	3.2	4.0	4.6	5.8	7.1	8.4
1.67	0.5	0.7	0.9	1.2	1.4	1.8	2.1	2.6	3.4	4.1	4.8	6.0	7.4	8.8
2.00	0.5	0.7	0.9	1.2	1.4	1.8	2.2	2.7	3.5	4.3	5.0	6.2	7.6	9.0
2.35	0.5	0.7	0.9	1.2	1.5	1.9	2.2	2.8	3.6	4.4	5.1	6.4	7.8	9.2
2.67	0.5	0.7	0.9	1.3	1.5	1.9	2.3	2.8	3.6	4.5	5.2	6.5	8.0	9.4
3.00	0.5	0.7	0.9	1.3	1.5	1.9	2.3	2.9	3.7	4.5	5.3	6.7	8.1	9.6
3.33	0.5	0.8	0.9	1.3	1.5	1.9	2.4	3.0	3.8	4.6	5.4	6.8	8.2	9.8

STUDY QUESTIONS

9-1. What are the four different hot water piping arrangements commonly used in residential and light commercial installations? Describe each.

9-2. What are the differences between series loop and one-pipe hot water systems? What are the advantages of each?

9-3. Discuss the two-pipe system, how it works, and its advantages and disadvantages.

9-4. Why are multiple heating circuits (zones) often used?

9-5. How does a radiant hot water heating panel work?

9-6. Describe the following components of a boiler configuration in a hydronic system:

    a. Expansion tank

    b. Circulating pump

    c. Flow control valve

    d. Safety relief valve

    e. Air elimination device

9-7. Describe the following heat distribution device or terminal unit available to distribute heat to a space:

    a. Convector unit

    b. Fan coil unit

    c. Radiator

    d. Radiant floor unit

    e. Radiant wall unit

9-8. What is the one variable that affects the amount of heat given out by a finned tube convector?

9-9. What type of system is most commonly used in a residence?

Design Exercises

9-10. A 2¼ in × 5 in × 0.011–1-in copper-aluminum fin tube baseboard convector (from Table 9.1) has an output of 840 Btu/hr per foot of tube at a fluid temperature of 180°F. A room in a residence has a heating load of 7450 Btu/hr. Determine the length of convector tube (in feet) needed to meet the load.

9-11. A 2¾ in × 5 in × 0.020–1¼-in copper-aluminum fin tube baseboard convector (from Table 9.1) has an output of 850 Btu/hr per foot of tube at a fluid temperature of 180°F. A classroom in an elementary school has a heating load of 13 550 Btu/hr. Determine the length of convector tube (in feet) needed to meet the load.

9-12. A 2¾ in × 5 in × 24 gauge–1¼-in steel tube convector (from Table 9.1) has an output of 710 Btu/hr per foot of tube at a fluid temperature of 180°F. A classroom in an elementary school has a heating load of 17 850 Btu/hr. Determine the length of convector tube (in feet) needed to meet the load.

9-13. A 2¾ in × 5 in × 24 gauge–1¼-in steel tube convector (from Table 9.1) has an output of 860 Btu/hr per foot of tube at a fluid temperature of 200°F. A classroom in an elementary school has a heating load of 17 850 Btu/hr.

    a. Determine the length of convector tube (in feet) needed to meet the load.

    b. At a flow rate of 1 gal/min (8.3 lb/min), determine the temperature of the water leaving the convector, if the water enters the convector at a temperature of 200°F.

9-14. Design a series loop system, two zones, for the residence in Appendix D, based on the heat loss calculations for the geographic location where you reside.

9-15. Design a series loop system, one zone, for the apartments in Appendix A, based on the heat loss calculations for the geographic location where you reside.

HVAC ELECTRIC HEATING SYSTEMS

10.1 SYSTEMS

An electric resistance heating system consists of individual electric heaters or cable that uses resistance to electrical current flow to generate heat. The advantages of an electric resistance heating system include low installation cost, individual room control, quiet operation, and cleanliness. In addition, when cables or panels are used, there are no exposed heating units.

Electric resistance heating converts nearly 100% of the energy in the electricity to heat. However, most electricity is produced from oil, gas, or coal generators that convert only about 30% of the fuel's energy into electricity. Because of electricity generation and transmission losses, electric heat is often more expensive than heat produced in the home or business using combustion appliances (e.g., natural gas, propane, and oil furnaces).

Types of electric resistance heating systems are introduced below.

Baseboard

Electric *resistance baseboard units* have electric heating elements encased in metal pipes, which are enclosed in a metal case. The pipes, surrounded by aluminum fins to aid heat transfer, run the length of the cabinet. As air within the unit is warmed, it rises into the room, and cooler room air is drawn into the bottom of the unit. Heat transfer is mainly by natural convection, but some heat is also radiated from the pipe, fins, and housing.

This system offers individual room control, but furniture arrangement and draperies must not interfere with the operation of the units by blocking the natural flow of air. This system is economical for a builder to install and is especially popular in low-cost housing and housing built for speculation, although the cost of operation is generally higher than most other fuels.

Resistance Cable

Electric *resistance heating cable* is stapled to the drywall in a grid pattern and covered with plaster or gypsumboard. Individual thermostats control the heat in each room. However, because the cable is usually installed on the ceiling, the disadvantages of warm air rising and cold feet must be considered.

Construction drawings are not usually needed for this type of heating system. Instead, the amount of heat required is noted for each building space. The cable system allows complete freedom in furniture and drapery placement. It does create, however, a thermal shadow effect when furniture shadows radiant heat transfer.

Radiant Panels and Mats

Prefabricated *electric radiant ceiling* and *wall panels* have electric resistance wires, typically copper or nichrome, wrapped in a water-resistant plastic. Panels are usually about 1 in thick (25 mm) and range in size from 1 ft × 2 ft to 4 ft × 8 ft (about 300 mm × 600 mm to 1.2 m × 2.4 m). Electric radiant underfloor heating consists of electrical wire coils contained in plastic mesh-like mats that are placed on the subfloor below a floor covering such as tile, carpet, vinyl, or hardwood flooring. Mats are typically about $\frac{1}{8}$ in (3 mm) thick.

Electric radiant systems are typically powered by a 120 V, 208 V or 240 V electrical current. Panels typically operate at a temperature range of 150° to 170°F (65° to 77°C), with output based on physical size. Output of a floor mat system is typically about 50 Btu/hr per ft^2 (15 W/ft^2 or 160 W/m^2).

The panel or mat system has the same basic advantages and disadvantages as a resistance cable system, including flexibility of furniture and drapery arrangement. Another advantage is that radiant panels and mats can rapidly heat when turned on and quickly cool when turned off (within a few minutes), allowing for near-instantaneous space temperature control. Panels can be used to provide spot, supplementary, or zoned heating, or can be used as the primary source of heating for an entire building.

Unit Heaters

There are a variety of electric unit heaters available that may be used to supplement other heat sources in a space or to completely heat a room. In residences, such a unit heater is quite often installed in the bathroom, often to supplement other heat sources, when a higher temperature is necessary for occupants to feel comfortable when washing and after bathing. Unit heaters are also commonly installed in spaces where the heat is only used periodically, such as a basement, workshop, warehouse (Photo 10.1), or garage work area.

So that the warm air will be quickly spread throughout the area to be heated, unit heaters are equipped with a fan. It

PHOTO 10.1 Ceiling-mounted electric unit heater.

is preferable that the fan switch be the type that will not start until after the unit comes on and the air has warmed to a preset temperature; in this way, cold air is not pushed around the room.

Units can be mounted from the ceiling or high on a wall. One manufacturer makes these units available in the range of 2.5 to 50 kW (8500 to 210 000 Btu/hr). Additionally, units are available for use in walls (recessed into the wall) and in ceilings (recessed into the ceiling). They are also available in a cabinet that is located along a wall, or in an under-cabinet toe space.

Electric Furnaces

Electric furnaces operate much like fuel-fired furnaces. Heated air is delivered throughout the building through supply ducts and returned to the furnace through return ducts. Fans in electric furnaces move air over a group of three to seven electric resistance coils, called *elements,* each of which are typically rated at 5 kW (17 065 Btu/hr). The furnace's heating elements activate in stages to avoid overloading the building's electrical system. A built-in thermostat-like device called a *limit controller* prevents overheating. This limit controller may shut the furnace off if the blower fails or if a dirty filter is blocking the airflow. Electric furnaces tend to be more expensive to operate than other electric resistance systems because of their duct heat losses and the extra energy required to distribute the heated air. Electric duct heaters are also available. One manufacturer offers 1700 Btu/hr to 3.4 MMBtu/hr (0.5 to 1000 kW) capacities.

Rooftop Units

A rooftop unit (RTU) is a self-contained unit, installed directly above the conditioned space, in a weatherproof enclosure with supply and return ductwork that passes through the roof or that passes horizontally through an outside wall. These can be an electric cooling RTU with optional electric heat or an electric heat pump RTU. One manufacturer offers RTUs rated from 2 to 25 tons of (electric) cooling and up to 273 000 Btu/hr (80 kW) electric heating.

Infrared Heaters

An electric *infrared heater* is a resistance heating device that creates a higher surface temperature that transfers energy to a surface at a lower temperature through thermal radiation. Infrared energy passes through the air and heats the object or person. Electric infrared heaters generate heat by electrical current flowing through a high-resistant wire, a graphite ribbon, or a film element. Theoretically, an infrared heater is 100% efficient because it converts nearly all electrical energy into heat and emits its heat as infrared radiation, but some energy is lost from conduction or convection. Because they are directional, electric infrared heaters focus heat where desired and work well for spot heating without heating an entire area. Banks of several heaters combine to heat larger areas.

Because of high temperatures (>2000°F/1100°C), infrared heating units are installed high on the wall or on the ceiling, safely away from children or furnishings. There are several sizes and wattages of radiant heaters available that can be used in homes, apartments, grocery stores, outdoor restaurants, ski shops, offices, schools, and medical facilities. Overhead infrared heaters are suspended like fluorescent lighting fixtures but can be used in many positions. One manufacturer offers 6500 to 44 000 Btu/hr (2 to 13.5 kW) capacities. Lay-in infrared ceiling panels fit in a suspended, drop ceiling, with each 2 ft × 4 ft panel emitting 800 to 2400 Btu/hr (250 to 750 W).

Packaged Terminal Units

A *packaged terminal unit* is a through-the-wall, self-contained, air-cooled direct expansion (DX) cooling unit (e.g., with integral compressor, evaporator, and condenser), an optional heating unit, and an optional minimum ventilation intake. It offers heating and cooling capabilities in one compact package. One manufacturer offers *packaged terminal air conditioners (PTAC)* with heating unit and *packaged terminal heat pumps (PTHP)* with electric (or hydronic) heat in 208/230 V or 265/277 V models and nominal cooling capacities of 7000, 9000, 12 000, and 15 000 Btu/hr. Because these units are compact, they are often used in hotels, motels, nursing homes, apartments, condominiums, and small offices.

System Combinations

Quite often, in order to provide the best results, the designer may incorporate more than one type of system. For example, a building may be predominantly heated with baseboard units, but in areas where there is little free wall space, such as a kitchen, it may be desirable to use a unit heater in the ceiling, the wall, or under the cabinets. Another possibility for the kitchen might be to use resistance cables or panels in the ceiling. Bathrooms may be heated with unit heaters to provide added comfort.

Codes and Installation

An electrical heating system must be installed by a licensed electrical contractor. ASHRAE publishes guides that discuss the use of the units for proper heating conditions and the installation, but the *National Electric Code* (NEC) governs the installation of any electrical system.

10.2 BASEBOARD SYSTEM DESIGN

A baseboard electric system transfers heat to the space primarily by convection. It consists of baseboard units that may be mounted on the wall or recessed into the wall. A wall-mounted unit is shown in Photo 10.2. In selecting a baseboard unit, the designer must consider the direction in which the air will be discharged from the baseboard unit (Figure 10.1).

Baseboard units are rated in watts (W) and Btu/hr (or Btuh). These ratings, which may vary for different manufacturers, are given *for the length specified* and not per lineal foot. For example, in Table 10.1, the very first listing specifies a 36 in (3 ft or 0.9 m) length and a rating of 500 W and 1707 Btu/hr for the 3 ft (0.9 m) length. The manufacturer often has several models, sizes, and ratings available, and the ratings will probably vary with other manufacturers. In Tables 10.2 and 10.3, two separate series are listed; both have the same out-

side dimensions and the *B* series has a rating about 100% higher for each length than the *A* series. In addition, many manufacturers make larger units, with higher ratings, which are normally used in commercial, industrial, and institutional buildings. A 3 ft (0.9 m) length of this style may range from 750 to 2250 W.

In designing an electric baseboard system, it may be necessary to use the larger units in some areas in order to provide the Btu/hr required.

TABLE 10.1   BASEBOARD RATINGS.

Length	Watts	Btuh
36″ (3′-0″) (0.9 m)	500	1,707
52″ (4′-4″) (1.3 m)	750	2,560
68″ (5′-8″) (1.7 m)	1,000	3,413
100″ (8′-4″) (2.5 m)	1,500	5,120

TABLE 10.2   HIGH OUTPUT BASEBOARD RATINGS.

Length	Watts	Btuh
36″ (3′-0″) (0.9 m)	750	2,560
48″ (4′-0″) (1.2 m)	1,000	3,413
60″ (5′-0″) (1.8 m)	1,250	4,269
72″ (6′-0″) (1.8 m)	1,500	5,120
96″ (8′-0″) (2.4 m)	2,000	6,830

A Series

TABLE 10.3   HIGH OUTPUT BASEBOARD RATINGS.

Length	Watts	Btuh
36″ (3′-0″) (0.9 m)	1,500	5,120
48″ (4′-0″) (1.2 m)	2,000	6,830
60″ (5′-0″) (1.5 m)	2,500	8,538
72″ (6′-0″) (1.8 m)	3,000	10,245
96″ (8′-0″) (2.4 m)	4,000	13,660

B Series

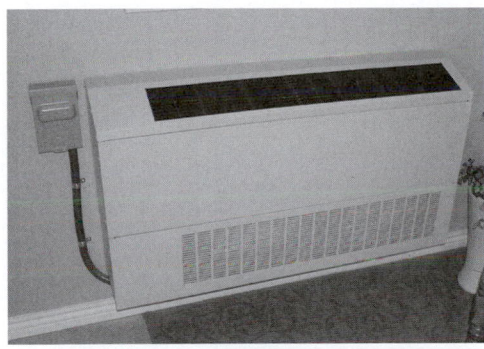

PHOTO 10.2   Wall-mounted unit.

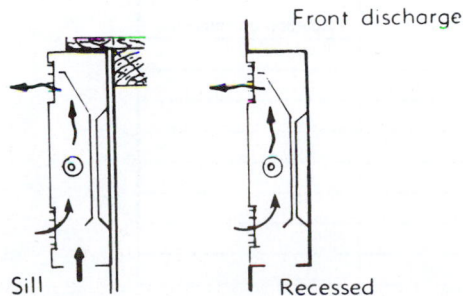
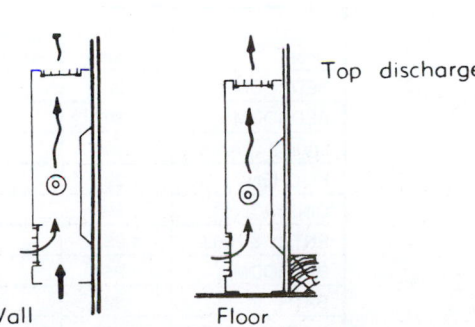

FIGURE 10.1   Baseboard air discharge.

10.3 BASEBOARD SYSTEM DESIGN EXAMPLE

1. Using the heating load computation method outlined in Chapter 4, compute heat loss in each space (see Table 10.4).

2. List the lineal feet of exterior wall that are available for baseboard convector units for each separate space, and tabulate them as shown (Table 10.5).

3. Using the manufacturer's ratings (Tables 10.1 and 10.2 are typical), determine the length and corresponding rating that will be used to provide heat to each space. Some walls, such as those in the living room of this design, will require two or more lengths of baseboard. All that is important is that the Btu/hr of heat loss calculated for each room is put back into that room.

   In some rooms, such as a kitchen, where there is little exterior wall space, it may be necessary to put a unit on an interior wall, or a larger unit that has a higher capacity may be used.

4. Tabulate the baseboard convector units selected for each room; list their lengths, wattage ratings, and Btu/hr ratings.

Based on the heat loss calculations, the units have been selected and tabulated as shown (Table 10.5).

Note: Because exterior wall space for radiant baseboard units is likely to be limited in some areas, it may be desirable to use other units or devices to supplement or take the place of the radiant baseboards. Kitchens and bathrooms quite often have unit heaters recessed in the wall. Kitchen heat may also be supplemented by a "kickspace" heater, which is placed in the kickspace under the kitchen cabinets.

10.4 RESISTANCE CABLE SYSTEM DESIGN

The amount of heat given off by cables will vary with the amount of heating cable used in the installation.

The electric cable used for ceiling installations comes in rolls; it is stapled to the ceiling and then covered with gypsum- or plasterboard in accordance with the manufacturer's specifications and code requirements. Basically, the cable must not be installed within 6 in (150 mm) of any wall, within 8 in (200 mm) of the edge of any junction box or outlet, or within 2 in (50 mm) of any recessed lighting fixtures.

TABLE 10.4   ROOM HEAT LOSS.

Room	Heat loss (Btuh)	Exterior wall (l.f.)	Watts (Btuh ÷3.413)	Radiation unit selected
BEDROOM 1	4,326			
BEDROOM 2	2,772			
LIVING	4,664			
KITCHEN	4,663			
DINING	3,056			
ENTRY & HALL	4,061			
BEDROOM 3	3,483			
BATH	1,092			
INTERIOR BATH	203			

SI UNITS

Room	Heat loss (Watts)	Exterior wall (m)		Radiation unit selected
BEDROOM 1	1,324			
BEDROOM 2	807			
LIVING	2,690			
KITCHEN	1,205			
DINING	898			
ENTRY & HALL	1,287			
BEDROOM 3	956			
BATH	322			
INTERIOR BATH	57			

Note: Watts vary from English design solution due to rounding off throughout the design.

TABLE 10.5   SELECT AND TABULATE BASEBOARD UNITS.

Room	Heat loss (Btuh)	Exterior wall (l.f.)	Watts (Btu ÷ 3.413)	Radiation unit selected
BEDROOM 1	4,326	30.0	1,268	2-750W, 4'-4"
BEDROOM 2	2,772	11.0	882	1-1,000W, 5'-8"
LIVING	9,327	23.3	2,968	3-1000W, 5'-8"
KITCHEN	4,160	—	1,324	2-750W, 4'-4"
DINING	3,056	14.0	973	2-500W, 3'
ENTRY & HALL	4,061	—	1,292	2-750W, 4'-4"
BEDROOM 3	3,483	14.0	1,094	1-1000W,5'-8"
BATH	1,092	—	348	1-500W, 3'
INTERIOR BATH	203	—	65	USE UNIT HEATER

SI UNITS

Room	Heat loss (Watts)	Exterior wall (m)	Radiation unit selected
BEDROOM 1	1,324	9.1	2-750W
BEDROOM 2	807	3.3	1-1,000W
LIVING	2,690	7.1	2-1000W, 1-750W
KITCHEN	1,205	—	1-500W, 1-750W
DINING	898	4.2	2-500W
ENTRY & HALL	1,287	—	2-750W
BEDROOM 3	956	4.2	2-500W
BATH	322	—	1-500W
INTERIOR BATH	57	—	USE UNIT HEATER

Note: Watts vary from English design solution due to rounding off throughout the design.

TABLE 10.6   TYPICAL CABLE RATINGS.

Btuh	Watts	Length Ft.	m
1365	400	145	44.2
2047	600	218	66.5
2730	800	292	89.1
3413	1000	362	110.4
4095	1200	436	133.0
5461	1600	582	177.5
6143	1800	654	199.5
6826	2000	728	222.0
7509	2200	800	244.0
8533	2500	910	277.6
10,239	3000	1090	332.5
11,287	3600	1310	399.6
15,700	4600	1672	510.0

PHOTO 10.3   Electric heating panel for wet plaster ceilings.

Cable assemblies are usually rated at 2.75 W per lineal foot (0.84 m), with generally available ratings from 400- to 5000-W lengths in 200-W increments, but the manufacturer's specifications should be checked to determine what is available. A typical list of available lengths, watts, and Btu/hr from one manufacturer is shown in Table 10.6. The cables have insulated coverings that are resistant to moderate temperatures, water absorption, and the effects of aging and chemical reactions (concrete, plaster, and so on); a polyvinyl chloride covering with a nylon jacket is commonly used. Each separate cable has an individual thermostat, providing flexible control throughout the building.

A typical cable installation in a plastered ceiling is shown in Photo 10.3, and installation details are shown in Figure 10.2. The space between the rows of heating cable is generally limited to a minimum of 1.5 in (37 mm), and some manufacturers recommend a 2 in (50 mm) minimum spacing when drywall construction is used. Another limitation on the spacing of the cable is a 2.5 in (62 mm) clearance required between cables under each joist (Figure 10.3), and a review of the layout in Photo 10.3 shows that the cable is installed parallel (in the same direction) as the joists.

To be certain the required quantity of heat is obtained, it is sometimes desirable to specify the maximum spacing of heat cable allowed in a room. This maximum spacing may be determined by using the formula:

$$s = 12\,(A_n/C)$$

where

s = cable spacing (in)
A_n = net available area for heat cables (ft^2)
C = length of a cable required to deliver required Btu/hr (ft)
12 = constant (used to convert ft to in)

In metric (SI) units, the maximum spacing may be determined by using the formula:

$$s = 1000\,(A_n/C)$$

where

s = cable spacing (mm)
A_n = net available area (m^2)
C = length of cable required to deliver required watts (m)
1000 = constant (used to convert m to mm)

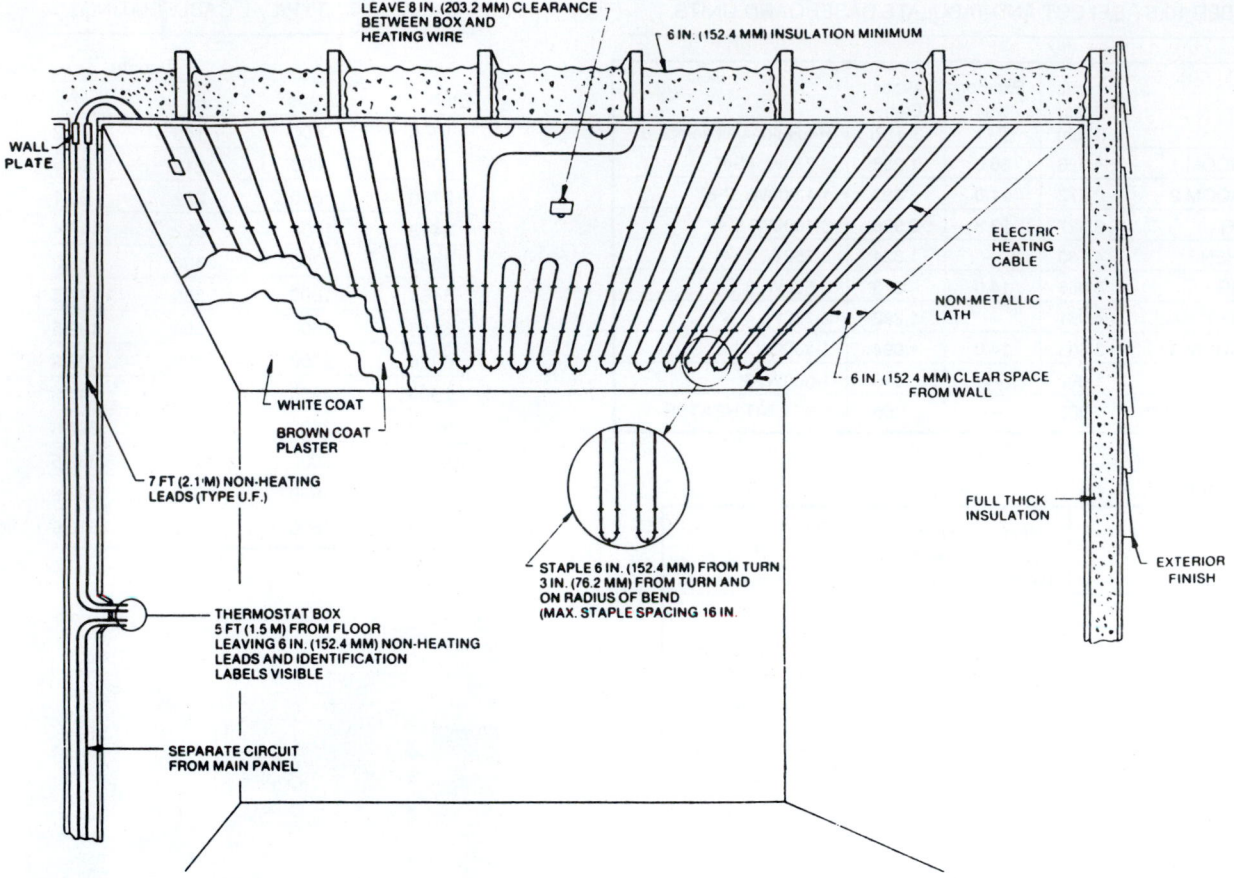

Reprinted with permission from ASHRAE, *HVAC Handbook*, 1993

FIGURE 10.2 Electric heating cable installation.

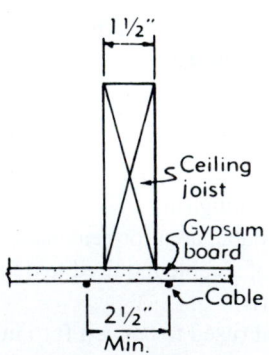

FIGURE 10.3 Cable detail.

The net available area for heating cables is equal to the total ceiling area minus any area in which cable cannot be placed (because of borders, ceiling obstructions, cabinets, and similar items). Although a small lighting fixture may be neglected, if the ceiling has several, their area should be deducted.

Example 10.1

Given: Assume that Bedroom 1 illustrated in Figure 10.4 is about 18 ft-3 in by 12 ft-3 in and requires 4326 Btu/hr. From the available cable lengths given in Table 10.6, a 1600-W (5461-Btu/hr), 582-ft cable is selected. Assume one ceiling

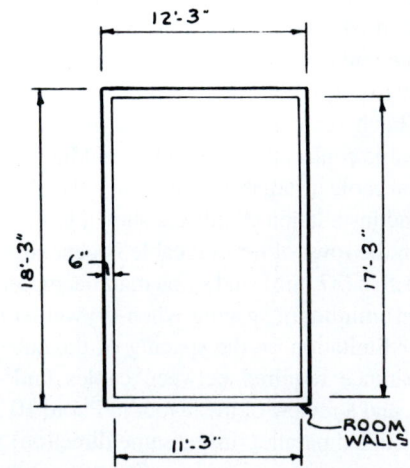

FIGURE 10.4 Cable area calculations.

fixture, which is neglected in the calculation, and a 6-in border required between the cable and the intersection of wall.

Determine the net area available for heating cables and the maximum cable spacing.

$$A_n = 18 \text{ ft-3 in} \cdot 12 \text{ ft-3 in} - (6 \text{ in around the} \\ \text{perimeter of room}) \text{ (See Figure 10.4.)}$$

$$17.25 \text{ ft} \cdot 11.25 \text{ ft} = 194 \text{ sq ft}$$

$$s = 12 \, (194 \text{ ft}^2/582 \text{ ft}) = 4 \text{ in maximum} \\ \text{spacing}$$

Bedroom 1 (Figure 10.4) requires a maximum cable spacing of 4 in.

Given: In metric (SI) units, assume that Bedroom 1 illustrated in Figure 10.4 is about 5.57 m by 3.74 m and requires 1324 W. From the available cable lengths in Table 10.6, a 1600 W 178 m cable is selected. Assume one ceiling fixture, which is neglected in the calculation, and a 0.15 m border required between the cable and the intersection of the wall.

$$A_n = 5.57 \text{ m} \cdot 3.74 \text{ m} - (0.15 \text{ m around the} \\ \text{perimeter of the room})$$

$$5.27 \text{ m} \cdot 3.44 \text{ m} = 18.13 \text{ m}^2$$

$$s = 1000 \, (18.13 \text{ m}^2/177.5 \text{ m}) = 102 \text{ mm} \\ \text{maximum spacing}$$

Bedroom 1 requires a maximum cable spacing of 102 mm.

10.5 RESISTANCE CABLE DESIGN EXAMPLE

1. Determine the heat loss of each room and tabulate.

The heat loss has already been calculated for the residence being designed. Figures are tabulated by room as shown in Table 10.7.

TABLE 10.7 ROOM HEAT LOSS.

Room	Heat loss (Btuh)	Cable selected (Btuh)	Cable length (ft.)	Ceiling space available (ft.)	Ceiling area (s.f.)	Maximum spacing (in.)
BEDROOM 1	4,326					
BEDROOM 2	2,772					
LIVING	9,327					
KITCHEN	4,160					
DINING	3,056					
ENTRY & HALL	4,061					
BEDROOM 3	3,483					
BATHROOM	1,092					
INTERIOR BATH	203					

SI UNITS

Room	Heat loss	Cable selected (Watts)	Cable length (m)	Ceiling space available (sq m)	Ceiling area (sq m)	Maximum spacing (mm)
BEDROOM 1	1,324					
BEDROOM 2	807					
LIVING	2,690					
KITCHEN	1,205					
DINING	898					
ENTRY & HALL	1,287					
BEDROOM 3	956					
BATHROOM	322					
INTERIOR BATH	57					

TABLE 10.8   SELECT AND TABULATE CABLE LENGTHS.

Room	Heat loss (Btuh)	Cable selected (Btuh)	Cable length (ft.)	Ceiling space available (ft.)	Ceiling area (s.f.)	Maximum spacing (in.)
BEDROOM 1	4,326	5,461	582			
BEDROOM 2	2,772	2,730	292			
LIVING	9,327	10,239	1,090			
KITCHEN	4,160	4,095	436			
DINING	3,056	3,413	362			
ENTRY & HALL	4,061	4,095	436			
BEDROOM 3	3,483	4,095	436			
BATHROOM	1,092	1,365	145			
INTERIOR BATH	203	USE ELECTRIC UNIT HEATER				

SI UNITS

Room	Heat loss	Cable selected (Watts)	Cable length (m)	Ceiling space available (sq m)	Ceiling area (sq m)	Maximum spacing (mm)
BEDROOM 1	1,324	1,600	177.5			
BEDROOM 2	807	1,000	110.4			
LIVING	2,690	3,000	332.5			
KITCHEN	1,205	1,200	133.0			
DINING	898	1,000	110.4			
ENTRY & HALL	1,287	1,600	177.5			
BEDROOM 3	956	1,000	110.4			
BATHROOM	322	400	44.2			
INTERIOR BATH	57	USE UNIT HEATER				

2. Select the cable required to provide the heat required for each room. Do not use a cable that will provide less heat than calculated.

From Table 10.6, select the required cable for each room and tabulate each cable rating (Btu/hr, watts, length) as shown in Table 10.8.

3. Calculate the net ceiling area for each room.

Calculate the net ceiling area for each room based on a 6-in border. In the kitchen, deduct for the wall cabinets and the fluorescent light. Tabulate the net ceiling area for each room as shown in Table 10.9 (p. 345).

4. Calculate the maximum cable spacing that can be used in each space.

Using the formula $s = 12 (A_n/C)$, calculate the cable spacing for each room and tabulate the information as shown in Table 10.10 (p. 346).

This information is often given on the drawings in tabulated form, similar to that shown in Table 10.10. The location of each thermostat must be shown on the drawing and may be put on the general construction (architectural) drawings, but often the information is put on the electrical drawings, as the electrician will locate and install the thermostat and tie the cable circuit into the power panel that serves it.

10.6   RADIANT PANEL SYSTEM DESIGN

The heating rates for radiant ceiling panels are generally given in watts per panel, Btu/hr per panel, or both. As shown in Table 10.11 (p. 347), the ratings for a 2 ft · 4 ft panel may vary from about 500 to 750 W per panel, depending on the panel selected and the manufacturer. Because 1 kW = 3413 Btu/hr, 500 W = 1707 Btu/hr, and 750 W = 2560 Btu/hr. It is important that the designer get accurate ratings by checking the engineering specifications for the type of panel that will actually be used on the project.

10.7   RADIANT PANEL DESIGN EXAMPLE

To design ceiling panels for the residence for which the heat loss was calculated in Chapter 4.

1. The first step in the design is to tabulate the Btu/hr for each space, as shown in Table 10.12 (p. 347).

2. List the square feet of ceiling area available for use in each space (Table 10.12).

Any space that does not have sufficient ceiling area for the panels must have supplemental heat provided by

TABLE 10.9   CALCULATE NET CEILING AREAS.

Room	Heat loss (Btuh)	Cable selected (Btuh)	Cable length (ft.)	Ceiling space available (ft.)	Ceiling area (s.f.)	Maximum spacing (in.)
BEDROOM 1	4,326	5,461	582	17.25 × 11.25	194	
BEDROOM 2	2,772	2,730	292	14.3 × 11	157	
LIVING	9,327	10,239	1,090	17 × 27.6	470	
KITCHEN	4,160	4,095	436	15 × 12	160	
DINING	3,056	3,413	362	13 × 11.5	150	
ENTRY & HALL	4,061	4,095	436	7 × 11.5 + 3 × 30	170	
BEDROOM 3	3,483	4,095	436	13 × 11.5	150	
BATHROOM	1,092	1,365	145	5 × 9	45	
INTERIOR BATH	203	USE ELECTRIC UNIT HEATER				

SI UNITS

Room	Heat loss	Cable selected (Watts)	Cable length (m)	Ceiling space available (sq m)	Ceiling area (sq m)	Maximum spacing (mm)
BEDROOM 1	1,324	1,600	177.5	5.27 × 3.44	18.13	
BEDROOM 2	807	1,000	110.4	4.4 × 3.4	14.96	
LIVING	2,690	3,000	332.5	5.2 × 8.4	43.68	
KITCHEN	1,205	1,200	133.0	4.6 × 3.7	17.02	
DINING	898	1,000	110.4	4.0 × 3.5	14.0	
ENTRY & HALL	1,287	1,600	177.5	2.1 × 3.5 + 0.9 × 9.1	15.54	
BEDROOM 3	956	1,000	110.4	4.0 × 3.5	14.0	
BATHROOM	322	400	44.2	1.5 × 2.7	4.05	
INTERIOR BATH	57	USE UNIT HEATER				

adding wall panels or electric baseboard unit heaters. Another solution would be to use another type of system, such baseboard or unit heaters, and not put any ceiling panels in the space.

It may be decided to put unit heaters in certain spaces where it is desirable to have a fan circulate the air through the space. A bathroom is a typical example of where a unit heater might be used for this reason.

3. Select the heating capacity for the space (Table 10.13).
4. Determine the number of ceiling panels required for each space (Table 10.14).

The biggest disadvantage of ceiling panels is that there is a tendency for the heat to stay high in the space (keep in mind that warm air rises), requiring higher temperature settings to achieve comfort. Also, this type of heating will not heat "hidden" air spaces, such as under a table or desk; these will feel quite cool.

But from the builder's standpoint, it is an acceptable system because of its low initial cost, individual room control, and "invisibility" (there are no portions of the system exposed except the thermostat).

Ceiling panels may be installed so that the heat is effectively put into the space at the place where most of it is lost—at the windows. In general, best results are obtained with this type of unit when the rooms are generally small (such as in rental apartments) and when the glass area is limited.

10.8   SYSTEM INSTALLATION

Baseboard units that are recessed into the wall will have to be framed into the wall. Otherwise, they simply need the conductors that bring them the power and a thermostat control.

Unit heaters may be quite large, and may also be recessed into a wall. If hung from the ceiling, the structure will need to be designed to carry the extra weight. Wiring to bring power to them and a thermostat must be considered.

Resistance cable may be installed in concrete slabs and on walls, but is most commonly placed in ceilings (Figure 10.2). During construction, care must be taken not to damage the cable.

Prefabricated wall and ceiling panels do not require special system installation except for the wiring and thermostat.

TABLE 10.10 CALCULATE MAXIMUM CABLE SPACING.

Room	Heat loss (Btuh)	Cable selected (Btuh)	Cable length (ft.)	Ceiling space available (ft.)	Ceiling area (s.f.)	Maximum spacing (in.)
BEDROOM 1	4,326					
BEDROOM 2	2,772					
LIVING	9,327					
KITCHEN	4,160					
DINING	3,056					
ENTRY & HALL	4,061					
BEDROOM 3	3,483					
BATHROOM	1,092					
INTERIOR BATH	203					

SI UNITS

Room	Heat loss	Cable selected (Watts)	Cable length (m)	Ceiling space available (sq m)	Ceiling area (sq m)	Maximum spacing (mm)
BEDROOM 1	1,324					
BEDROOM 2	807					
LIVING	2,690					
KITCHEN	1,205					
DINING	898					
ENTRY & HALL	1,287					
BEDROOM 3	956					
BATHROOM	322					
INTERIOR BATH	57					

STUDY QUESTIONS

10-1. Discuss the advantages and disadvantages of electric heating systems.

10-2. What is the primary reason that electric heating systems are used?

10-3. What are the types of electric resistance heating systems?

10-4. What model code typically governs the installation of electric resistance heating systems?

Design Exercises

10-5. Determine the output of a 1.5 kW electric resistance heater (100% efficient), in Btu/hr.

10-6. A packaged terminal air conditioning unit is specified as "One-ton PTAC with 3.5 kW Electric Heater." Determine the output of the heater (100% efficient), in Btu/hr.

10-7. A packaged terminal air conditioning unit is specified as "One-ton PTAC with 5.0 kW Electric Heater." Determine the output of the heater (100% efficient), in Btu/hr.

10-8. An electric resistance duct heater is rated at 21.5 kW. Determine the output of the heater (100% efficient), in Btu/hr.

10-9. One manufacturer's brand of cabinet heaters is rated from 2 kW to 32 kW. Determine the range of heater output for this brand (100% efficient), in Btu/hr.

10-10. Electric resistance baseboard convectors (from Table 10.1) are under consideration for heating a vacation home. The bedroom has a design heating load of 4000 Btu/hr. Identify the length of unit (in feet) needed to meet the load.

10-11. Series A high-output, electric resistance baseboard convectors (from Table 10.2) are under consideration for heating a vacation home. The bedroom has a design heating load of 4800 Btu/hr. Identify the length of unit (in feet) needed to meet the load.

10-12. Series B high-output, electric resistance baseboard convectors (from Table 10.3) are under consideration for heating a vacation home. The bedroom has a design

TABLE 10.11   RADIANT CEILING PANEL RATINGS.

Watts	Btuh	Size
500	1707	2'-0"x4'-0"
750	2560	2'-0"x4'-0"
560	1911	2'-0"x4'-0"
700	3019	2'-0"x5'-0"
500	1707	2'-0"x3'-0"
750	2560	2'-0"x3'-0"
1000	3413	2'-0"x3'-0"

TABLE 10.12   TABULATE BTU/HR.

Room	Heat loss (Btuh)	Net ceiling area (s.f.)	Watts (Btuh ÷ 3413)	No. of panels
BEDROOM 1	4,326	194		
BEDROOM 2	2,772	157		
LIVING	9,327	470		
KITCHEN	4,160	160		
DINING	3,056	150		
ENTRY & HALL	4,061	170		
BEDROOM 3	3,483	150		
BATH	1,092	45		
INTERIOR BATH	203	14		

SI UNITS

Room	Heat loss (Watts)	Net ceiling area (sq m)	No. of panels
BEDROOM 1	1,324	59	
BEDROOM 2	807	48	
LIVING	2,690	124	
KITCHEN	1,194	49	
DINING	898	46	
ENTRY & HALL	1,287	52	
BEDROOM 3	956	46	
BATH	322	14	UNIT HEATER
INTERIOR BATH	57	4	UNIT HEATER

heating load of 11 000 Btu/hr. Identify the length of unit (in feet) needed to meet the load.

10-13. Electric baseboard heaters will be used to heat a room with a design heating load of 5050 Btu/hr. Identify the length of each of the following units (in feet) required to meet the load.

a. Standard electric resistance baseboard convectors (from Table 10.1)

b. Series A high-output, electric resistance baseboard convectors (from Table 10.2)

c. Series B high-output, electric resistance baseboard convectors (from Table 10.3)

10-14. Electric baseboard heaters will be used to heat a room with a design heating load of 3500 Btu/hr. Identify the

length of each of the following units (in feet) required to meet the load.

a. Standard electric resistance baseboard convectors (from Table 10.1)

b. Series A high-output, electric resistance baseboard convectors (from Table 10.2)

c. Series B high-output, electric resistance baseboard convectors (from Table 10.3)

10-15. Electric baseboard cable imbedded in a ceiling will be used to heat a room with a design heating load of 5050 Btu/hr. Identify the length of cable (in feet) needed to meet the load.

10-16. Electric baseboard cable imbedded in a ceiling will be used to heat a room with a design heating load of

TABLE 10.13   TABULATE WATTS.

Room	Heat loss (Btuh)	Net ceiling area (s.f.)	Watts (Btuh ÷ 3413)	No. of panels
BEDROOM 1	4,326	194	1,268	
BEDROOM 2	2,772	157	882	
LIVING	9,327	470	2,968	
KITCHEN	4,160	160	1,324	
DINING	3,056	150	973	
ENTRY & HALL	4,061	170	1,292	
BEDROOM 3	3,483	150	1,094	
BATH	1,092	45	348	
INTERIOR BATH	203	14	65	

TABLE 10.14   TABULATE NUMBER OF PANELS.

Room	Heat loss (Btuh)	Net ceiling area (s.f.)	Watts (Btuh ÷ 3413)	No. of panels
BEDROOM 1	4,326	194	1,268	1-500W, 1-750W
BEDROOM 2	2,772	157	882	1-1000W
LIVING	9,327	470	2,968	1-750W, 2-1000W
KITCHEN	4,160	160	1,324	1-500W, 1-750W
DINING	3,056	150	973	2-500W
ENTRY & HALL	4,061	170	1,292	1-500W, 1-750W
BEDROOM 3	3,483	150	1,094	2-500W
BATH	1,092	45	348	UNIT HEATER
INTERIOR BATH	203	14	65	UNIT HEATER

SI UNITS

Room	Heat loss (Watts)	Net ceiling area (sq m)	No. of panels
BEDROOM 1	1,324	59	2-750W
BEDROOM 2	807	48	1-1000W
LIVING	2,690	124	1-750W, 2-1000W
KITCHEN	1,194	49	1-500W, 1-750W
DINING	898	46	2-500W
ENTRY & HALL	1,287	52	2-750W
BEDROOM 3	956	46	2-500W
BATH	322	14	UNIT HEATER
INTERIOR BATH	57	4	UNIT HEATER

4000 Btu/hr. Identify the length of cable (in feet) needed to meet the load.

10-17. Design an electric baseboard heating system for the residence in Appendix D, based on the heat loss calculations for the geographic location where you reside.

10-18. Design a resistance cable heating system for the residence in Appendix D, based on the heat loss calculations for the geographic location where you reside.

10-19. Design an electric baseboard heating system for a typical apartment (not the top floor) in Appendix A, based on the heat loss calculations for the geographic location where you reside.

10-20. Design a resistance cable heating system for the residence in Appendix D, based on the heat loss calculations for the geographic location where you reside.

SOLAR THERMAL SYSTEMS IN BUILDINGS

11.1 PRINCIPLES OF SOLAR ENERGY

Solar energy is a radiant energy source that causes natural processes such as photosynthesis, thermal heating, and wind. The earth receives about one-half of one-billionth of the sun's energy output. Yet, if less than 0.1% of this solar radiation was converted to usable energy, it would meet the energy demand of the entire world. Solar radiation can be collected, stored, and used in buildings and building systems to provide heat and power. Solar thermal systems that are used for space heating and cooling buildings and heating domestic water will be covered in this chapter. Solar radiation can also be converted to electricity and used to power building systems (this topic is covered in Chapter 25).

Solar Radiation

Solar radiation is emitted from the sun and travels through space in a straight-line path. It consists of radiation of varying wavelengths known as the solar spectrum. Visible light (0.4 to 0.7 μm) accounts for about 43% of the energy emitted from the sun even though it is a relatively small segment of the wavelengths emitted. About 52% is from the infrared band of the spectrum (0.7 to 2.5 μm). Slightly less than 5% is ultraviolet radiation (0.3 to 0.4 μm). Most of the ultraviolet component of solar radiation is reflected back into space at the outer atmosphere (the ozone layer). Thus, the sun's energy striking the earth's surface is mostly in the form of infrared and visible light.

Insolation (*incident solar radiation*) or *solar irradiance* is a measure of the amount of solar radiation (all wavelengths) received on a specific surface area in a specific time. It is commonly expressed as average irradiance expressed in units of Btu per hour per square foot (Btu/hr · ft^2) or watts per square meter (W/m^2). The quantity of extraterrestrial insolation (solar radiation outside the earth's atmosphere) and proportion of the different wavelengths in the solar spectrum remains relatively constant. Insolation received on a surface is largest when the surface directly faces the sun because there is no reflection in space.

The *solar constant* is the average quantity of solar radiation per unit area that is received at the outer edge of the earth's atmosphere at the mean earth/sun distance. The solar constant is 433.3 Btu/ft^2 · hr (1366 W/m^2). It is not a measure of solar energy that actually reaches the earth's surface, because up to 35% of the solar radiation intercepted by the earth's atmosphere and surface is reflected back into space. Additionally, atmospheric water vapor and gases (e.g., constituents of air and pollution) absorb another 15%. Thus, on average only about 35 to 40% of the solar radiation striking the outer atmosphere actually reaches the earth's surface. Weather phenomenon (e.g., clouds) reflects more sunlight into space. Such climatic irregularities make the amount of solar energy available at a building site inconsistent and unpredictable.

Earth's Rotation and Seasonal Tilt

In 1543, the Polish astronomer Nicolaus Copernicus (1473–1543), in his book *De Revolutionibus Orbium Coelestium,* revolutionized astronomy by professing that the earth revolves annually around the sun in a heliocentric or sun-centered system. He also theorized that the earth's rotation caused day and night. As the earth orbits the sun, it rotates about its own axis at a rate of one revolution every 24 hours. In addition, there is a slight wobble as it rotates. In relation to its plane of orbit, the earth's axis is at an inclination (tilted at) of 23.45°. With respect to the sun, this inclination varies from +23.45° to −23.45° or a total variation of 46.9° over the course of a full year. It is the inclination of the earth and its orbit around the sun that affect the hours of sunlight received at a particular location on the earth's surface.

The daily rotation of the earth and the seasonal tilt of the earth's axis affect the hours of daily and seasonal sunlight at a particular location on the earth's surface. See Photo 11.1. They also affect the length of atmosphere through which the sun's rays must pass before striking a particular location on the earth's surface. At either end of daylight hours, the sun is lower in the sky, so solar radiation must pass through more atmosphere. These phenomena cause the amount of solar radiation available to vary by building site (latitude), season, and time of day.

All locations on earth receive sunlight during some part of the year. However, the number of hours of sunlight received at a location varies over the year. Solstices and equinoxes are astronomical events that describe the annual positions of the sun. See Figure 11.1.

Solstices
Solstices occur twice a year at two specific moments in time (not a whole day) when the tilt of the earth's axis is most oriented toward or away from the sun. On the *summer solstice* (between June 20 and 23), the earth's north pole is fully tilted toward the sun, the northern hemisphere receives the maximum number of hours of daily sunlight, and the sun is highest in the

PHOTO 11.1 The daily rotation of the earth and the seasonal tilt of the earth's axis affect the hours of daily and seasonal sunlight as evidenced by the changing shadow on a sundial. (Used with permission from ABC)

sky. On the *winter solstice* (December 21 or 22), the earth's north pole is fully tilted away from the sun, the northern hemisphere receives the minimum number of hours of daily sunlight, and the sun is lowest in the sky.

Equinoxes
An *equinox* takes place at two specific moments each year when the center of the sun can be observed to be directly overhead at the earth's equator. The tilt of the

earth's axis is oriented neither towards nor away from the sun. On the *vernal equinox* (about March 20 or 21) and *autumnal equinox* (about September 22 or 23), the earth receives about 12 hours of sunlight.

In the northern hemisphere during the days between the vernal equinox and autumnal equinox (summer months), the sun's higher arc of travel between sunrise and sunset causes days to have more hours of sunlight. In contrast, during winter days between the autumnal equinox and vernal equinox, the sun's arc of travel is lower in the sky and there are fewer hours of sunlight. On the vernal and autumnal equinoxes, hours of sunlight are nearly 12 hours at all locations. Of course, the reverse is true in the southern hemisphere.

Solar Altitude and Azimuth

As the sun appears to sweep across the sky, its location in the sky changes. It is higher in the sky in the summer and lower in the winter. The direction that the sun rises and sets also change each day. The sun is, however, always due south at solar noon. The solar altitude and azimuth angles define the sun's location in the sky.

The *solar altitude angle (β)* is the angle measured vertically between the sun's rays and the horizontal plane on the earth's surface. See Figure 11.2. It is an angular vertical measure of how high the sun is in the sky. At sunset/sunrise, the solar altitude is zero (0°) because it is at the horizon. The sun directly overhead (in the center of the sky) has a solar altitude of 90°.

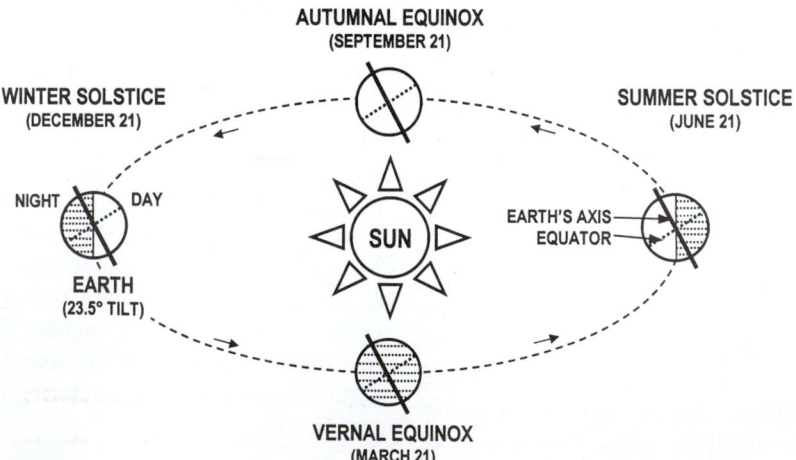

FIGURE 11.1 The daily rotation of the earth and the seasonal tilt of the earth's axis affect the hours of daily and seasonal sunlight at a particular location on the earth's surface. On the summer solstice (~June 21), the northern hemisphere receives the maximum number of hours of daily sunlight and the sun is highest in the sky. On the winter solstice (~December 21), the northern hemisphere receives the minimum number of hours of daily sunlight and the sun is lowest in the sky. The reverse is true in the southern hemisphere. At the vernal equinox (~March 21) and autumnal equinox (~September 21), the earth receives about 12 hours of sunlight.

TABLE 11.1 HOURLY SOLAR ANGLES AND SOLAR INTENSITY ON A CLEAR DAY FOR VARIOUS SOUTH-FACING SURFACES ON OR ABOUT THE ANNUAL EQUINOXES AND SOLSTICES AT 40° NORTH LATITUDE.

Day	Solar Time AM	Solar Time PM	Solar Position Altitude	Solar Position Azimuth	Total Insolation on Surfaces, Btu/hr·ft²						
					Normal	Horiz.	South-Facing Surface (Angle from Horizontal)				
							30°	40°	50°	60°	90°
December 21 Winter Solstice	8	4	5.5	53.0	89	14	39	45	50	54	56
	9	3	14.0	41.9	217	65	135	152	164	171	163
	10	2	20.7	29.4	261	107	200	221	235	242	221
	11	1	25.0	15.2	280	134	239	262	276	283	252
	Solar Noon		26.6	0.0	285	143	253	275	290	296	263
	Full-Day Totals				1978	782	1480	1634	1740	1796	1646
March 21 Vernal Equinox	7	5	11.4	80.2	171	46	55	55	54	51	35
	8	4	22.5	69.6	250	114	140	141	138	131	98
	9	3	32.8	57.3	282	173	215	217	213	202	138
	10	2	41.6	41.9	297	218	273	276	271	258	176
	11	1	47.7	22.6	305	247	310	313	307	293	200
	Solar Noon		50.0	0.0	307	257	322	326	320	305	208
	Full-Day Totals				2916	1852	2308	2330	3384	2174	1484
June 21 Summer Solstice	5	7	4.2	117.3	22	4	3	3	2	2	1
	6	6	14.8	108.4	155	60	30	18	17	16	10
	7	5	26.0	99.7	216	123	92	77	59	41	14
	8	4	37.4	90.7	246	182	159	142	121	97	16
	9	3	48.8	80.2	263	233	219	202	179	151	47
	10	2	59.8	65.8	272	272	266	248	224	194	74
	11	1	69.3	41.9	277	296	296	278	253	221	92
	Solar Noon		73.5	0.0	279	304	306	289	263	230	98
	Full-Day Totals				3180	2648	2434	2224	1974	1794	610
September 21 Autumnal Equinox	7	5	11.4	80.2	149	43	51	51	49	47	32
	8	4	22.5	69.6	230	109	133	134	131	124	84
	9	3	32.8	57	263	167	206	208	203	193	132
	10	2	41.6	41.9	280	211	262	265	260	247	168
	11	1	47.7	22.6	287	239	289	301	295	281	192
	Solar Noon		50.0	0.0	290	249	310	313	307	292	200
	Full-Day Totals				2708	1788	2210	2228	2182	2074	1416

The *solar azimuth angle* (ψ) is the angle measured horizontally (along the horizon) between the due-south direction line and the horizontal projection of the sun's rays. It is a horizontal angular measure of how far away the sun is from the south. True south is used—not "magnetic" south.

Table 11.1 shows the hourly position of the sun at different angles for south-facing surfaces (angle from horizontal) on annual equinoxes and solstices at 40° north latitude. Times are based on solar time, not local (clock) time. These angles vary by time of day, day of year, and latitude.

Types of Solar Radiation

Solar radiation can be in the form of direct rays that pass through space or the atmosphere without being affected, or it can be absorbed or scattered by the gases that make up the atmosphere. Forms of solar radiation include the following:

Direct (beam) radiation arrives in a direct line from the sun and passes through the atmosphere without being affected. On a clear day, about 85% of the solar radiation striking the earth is high-energy, direct radiation. In

TABLE 11.1 HOURLY SOLAR ANGLES AND SOLAR INTENSITY ON A CLEAR DAY FOR VARIOUS SOUTH-FACING SURFACES ON OR ABOUT ANNUAL EQUINOXES AND SOLSTICES AT 40° NORTH LATITUDE. (CONTINUED)

Day	Solar Time AM	Solar Time PM	Solar Position Altitude	Solar Position Azimuth	Total Insolation on Surfaces, W/m² Normal	Horiz.	South-Facing Surface (Angle from Horizontal) 30°	40°	50°	60°	90°
December 21 Winter Solstice	8	4	5.5	53.0	281	44	123	142	158	170	177
	9	3	14.0	41.9	684	205	426	479	517	539	514
	10	2	20.7	29.4	823	337	631	697	741	763	697
	11	1	25.0	15.2	883	423	754	826	870	892	795
	Solar Noon		26.6	0.0	899	451	798	867	914	933	829
	Full-Day Totals				6237	2466	4666	5152	5486	5663	5190
March 21 Vernal Equinox	7	5	11.4	80.2	539	145	173	173	170	161	110
	8	4	22.5	69.6	788	359	441	445	435	413	309
	9	3	32.8	57.3	889	545	678	684	672	637	435
	10	2	41.6	41.9	936	687	861	870	854	813	555
	11	1	47.7	22.6	962	779	977	987	968	924	631
	Solar Noon		50.0	0.0	968	810	1015	1028	1009	962	656
	Full-Day Totals				9194	5839	7277	7346	10670	6855	4679
June 21 Summer Solstice	5	7	4.2	117.3	69	13	9	9	6	6	3
	6	6	14.8	108.4	489	189	95	57	54	50	32
	7	5	26.0	99.7	681	388	290	243	186	129	44
	8	4	37.4	90.7	776	574	501	448	382	306	50
	9	3	48.8	80.2	829	735	691	637	564	476	148
	10	2	59.8	65.8	858	858	839	782	706	612	233
	11	1	69.3	41.9	873	933	933	877	798	697	290
	Solar Noon		73.5	0.0	880	959	965	911	829	725	309
	Full-Day Totals				10 027	8349	7674	7012	6224	5656	1923
September 21 Autumnal Equinox	7	5	11.4	80.2	470	136	161	161	154	148	101
	8	4	22.5	69.6	725	344	419	423	413	391	265
	9	3	32.8	57	829	527	650	656	640	609	416
	10	2	41.6	41.9	883	665	826	836	820	779	530
	11	1	47.7	22.6	905	754	911	949	930	886	605
	Solar Noon		50.0	0.0	914	785	977	987	968	921	631
	Full-Day Totals				8538	5638	6968	7025	6880	6539	4465

contrast, on overcast days the sun is obscured by clouds and the direct beam radiation is zero or nearly zero.

Diffuse radiation is scattered from the direct beam by molecules, water vapor, and particulate matter (e.g., dust, smoke, pollen) in the atmosphere. Because it comes from all regions of the sky, it is also referred to as *sky radiation*. The portion of total solar radiation that is diffuse is typically about 10% to 20% for clear skies. On cloudy days, most or all of the solar radiation is diffuse.

Reflected radiation is solar energy reflected by the surface in front of the receiving surface. It also contributes to the total solar radiation received. It can be in the form of diffuse or direct radiation. Without diffuse and reflected radiation, visible light could not enter windows in walls facing the opposite side of the sun.

Global (total) solar radiation striking a surface can be composed of direct (beam) radiation and diffuse radiation. It is the sum of the direct beam, diffuse, and

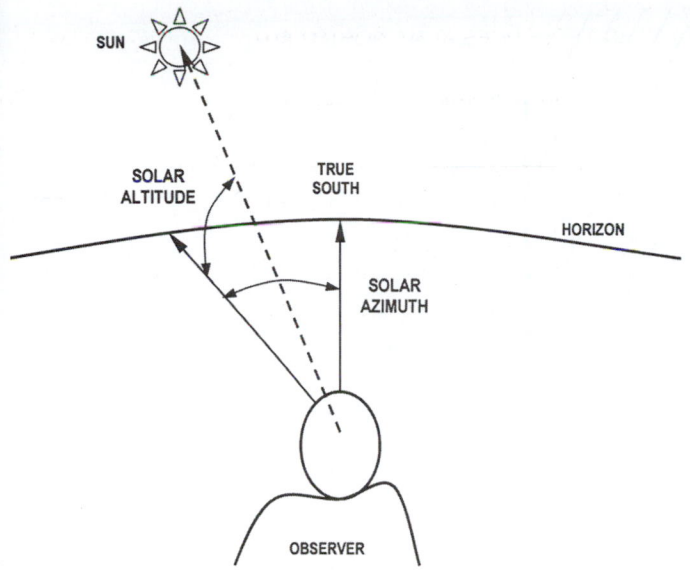

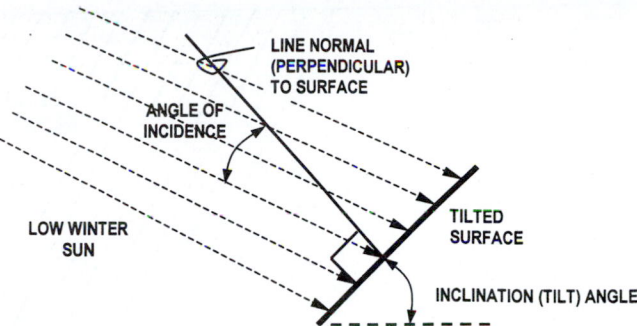

FIGURE 11.3 The angle created by incoming radiation and a line perpendicular to an intercepting surface is called the *angle of incidence*.

FIGURE 11.2 Solar altitude (β) is the angle measured vertically between the sun's rays and the horizontal plane on the earth's surface. It is an angular measure of how high the sun is in the sky. Solar azimuth (ψ) is the angle measured vertically between the due-south direction line and the horizontal projection of the sun's rays.

ground-reflected radiation arriving at the surface. Table 11.2 (See pp. 386–387) provides clear day daily solar insolation for selected cities in the United States.

Direct solar radiation contains more energy than diffuse solar radiation, because it travels unimpeded through the atmosphere. As a result, locations that enjoy clear, sunny days throughout the year have the best solar resources.

Behavior of Solar Radiation

Solar radiation travels in a straight-line path. The angle created by incoming radiation and a line perpendicular to an intercepting surface is called the *angle of incidence*. (See Figure 11.3.) A surface in a perpendicular position with respect to the incoming solar radiation (a surface said to be normal) receives the maximum density of solar radiation. A deviation from normal (away from perpendicular) reduces the radiation density and thus the amount of energy received by the surface. This is illustrated in Figure 11.4 and Table 11.3, which show that an increase in the angle of incidence results in the radiation being spread over a greater area, thereby causing a reduction in intercepted radiation.

Table 11.1 provides data on solar intensity during a clear day for various south-facing surfaces on annual equinoxes and solstices at 40° north latitude. Solar intensity is a measure of direct solar radiation intercepted on a receiving surface. These values vary significantly by surface tilt, time of day, and season of the year.

As introduced in Chapter 2, when radiation strikes a surface it can be reflected, transmitted, or absorbed, depending on

the texture, color, and transparency of the surface. An *opaque material* reflects and absorbs radiation, but does not transmit radiation. A *translucent material* reflects, absorbs, and transmits radiation. A smooth, polished opaque surface reflects solar radiation uniformly at an angle equal to the angle of incidence, while an irregular opaque surface such as concrete reflects solar radiation in a scattered manner.

An opaque material with a rough black surface absorbs nearly all wavelengths in the visible spectrum, while a *transparent material* such as window glass allows nearly all the radiation to pass through it with comparatively little reflection or absorption, and without deflecting it from its parallel lines of travel. Translucent materials also transmit radiation but can scatter the rays as they pass. It should be noted that relatively few materials are excellent reflectors, transmitters, or absorbers of solar radiation.

Solar radiation is converted from solar energy to thermal (heat) energy when it strikes the surface of a material and the material absorbs it. The differences in temperatures of a white surface and a black surface exposed the same direct sunlight serves as a simple display of this conversion: the temperature of the black surface is higher because it is absorbing more solar energy.

The transparent nature of glass varies with size of wavelength of radiation. Glass transmits nearly all of the solar radiation that it receives but is less transparent to most infrared radiation. The temperature buildup in a closed car on a sunny day is evidence of this characteristic. Solar energy passing through windows is absorbed by interior materials (e.g., the dashboard and seats) and is converted to thermal energy; the interior materials heat up and reradiates the energy in the form of infrared energy, which is unable to pass back through the glass to the outdoors. The thermal energy becomes trapped inside the car, causing the car interior to heat up. This phenomenon is known as the *greenhouse effect*.

A surface emits thermal radiation to surrounding surfaces when it is at a higher temperature than the surrounding surfaces.

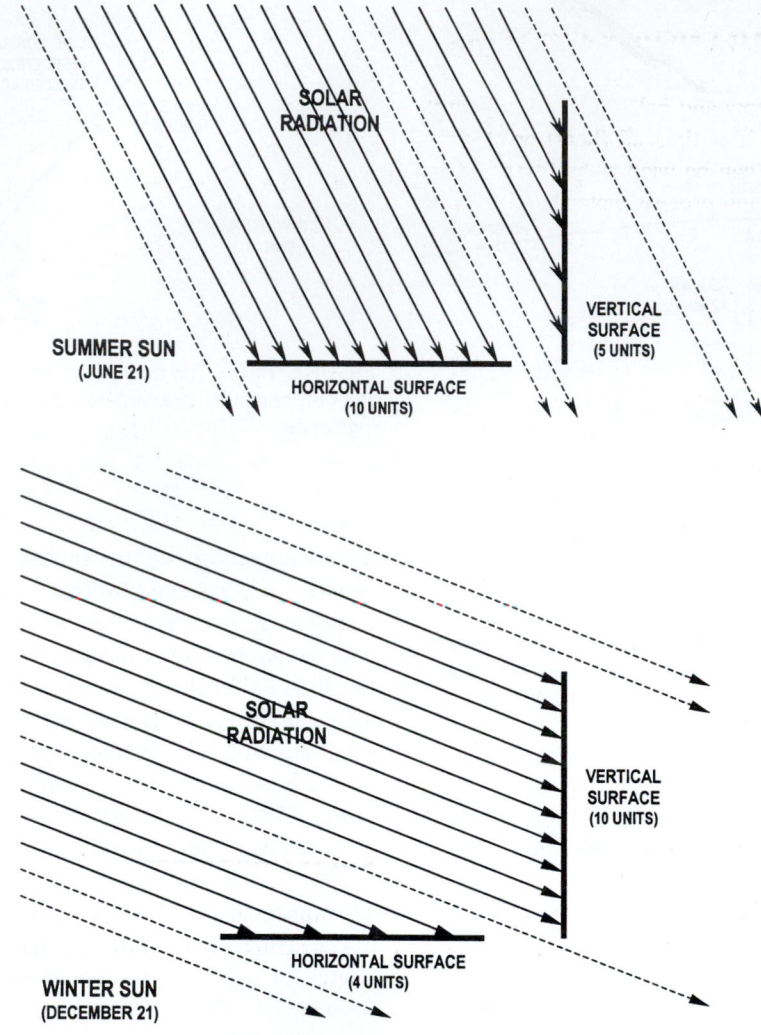

FIGURE 11.4 The altitude of the sun at solar noon changes with the seasons. In the middle latitudes, it is higher in the sky during the summer months and lower in the winter months. During the winter, the low mid-day sun results in a vertical (wall) surface receiving more solar energy than a horizontal (roof) surface. The reverse is true in the summer months.

TABLE 11.3   THE FRACTION OF RADIATION INTERCEPTED
WITH CHANGE IN INCIDENT ANGLE.

Angle of Incidence	Fraction of Intercepted Radiation
0	1.00
5	0.99
10	0.98
15	0.97
20	0.94
25	0.91
30	0.87
40	0.77
50	0.64
60	0.50
70	0.34
80	0.17
90	0.00

In buildings, this radiation is in the form of infrared radiation. Some surface finishes give off radiation better than others. *Emissivity* is a term that describes the extent to which a material can emit radiant energy. The emissivity of a surface depends on the temperature of the material and nature of its surface. A surface with a low emissivity is a poor emitter of thermal radiation. Most polished metal surfaces (e.g., aluminum foil) tend to be poor emitters (and poor absorbers) of thermal energy.

Specially designed *selective surfaces* are good absorbers and poor emitters of radiant energy and can be used to effectively capture solar energy without reradiating too much infrared energy. This type of surface works effectively on the absorption surface of a solar collector because it absorbs solar energy well, but does not release thermal energy back outdoors.

11.2 BUILDING SOLAR ENERGY SYSTEMS

A solar energy system intercepts and collects energy from the sun, stores it, and distributes it as thermal (heat) or electrical energy. Solar energy systems can be used in buildings to provide or supplement domestic and process water heating, space heating, and space cooling.

Basic Types of Solar Systems

There are three basic types of solar systems used in buildings:

- An *active solar system* uses mechanical devices such as fans or pumps powered by an external energy source (e.g., electricity) in addition to solar energy to collect, store, and distribute thermal (heat) energy from the sun. Active solar systems are commonly used for space heating building interiors and heating domestic and process water.

- A *passive solar system* is integrated into the building elements (e.g., windows) and materials (e.g., tile-covered concrete floors). This type of system does not rely on mechanical equipment to collect, store, and distribute thermal (heat) energy from the sun. Passive solar systems are typically used for space heating building interiors, but can also be used for domestic and process water heating.

- A *photovoltaic solar system* is a system of semiconductor solar cells that convert solar radiation directly into electricity. Because of the relatively low density of solar radiation and high manufacturing costs of solar cells, photovoltaic solar systems are only used on buildings at remote sites, where utility power is not readily available. These systems are introduced in Chapter 25.

Solar Savings

The intent behind incorporating a solar energy heating system into a building is to reduce conventional energy use and costs. The *solar fraction (SF)* is the ratio of the solar energy contribution to a solar-heated building or system and the potential energy consumed by the same building or system if a solar system were not installed. It is a useful way to compare a solar-heated building or system to a similar nonsolar but energy-efficient building or system. For example, if a passive solar system saved 60% of the energy consumed in comparison to a similar building without the solar system, the SF would be 0.60, a 60% savings.

The SF of a particular system is dependent on several factors (i.e., building load, the collection and storage sizes, the operation, and the climate). For example, a solar-thermal water heating system installed in a Denver, Colorado, single-family house might have an SF of 0.70, while in Buffalo, New York, which has a much colder and cloudier climate, an SF of 40% may be achieved. Generally, between 40 and 70% of the water heating demand can be cost-effectively achieved by a residential solar hot water system, resulting in an SF of 0.40 to 0.70. The SFs on commercial solar hot water systems are usually between 0.15 and 0.50. A well-designed, passive solar system in a home can have an SF of up to 0.80, even in a cold climate.

11.3 ACTIVE SOLAR HEATING SYSTEMS

An *active solar system* uses electric pumps and valves or fans and dampers to circulate water, antifreeze solutions, air or other heat-transfer fluids through the collectors to collect, store, and distribute thermal (heat) energy from the sun. Active solar systems can be used to space heat building interiors or heat domestic/process water. An active solar system relies on a *solar collector* to capture solar heat and convert it to thermal (heat) energy and a *thermal storage medium* to store heat. A *heat transfer medium* such as air, water, or some other fluid conveys heat from the collector to thermal storage. If the solar system cannot fully meet the space or hot water heating demand, an *auxiliary* or *back-up* system provides the additional supplemental heat.

Solar Collectors

A *solar collector* captures radiant energy from the sun. There are basically three types of solar collectors: flat plate, evacuated tube, and concentrating. These are discussed below.

Flat-Plate Collectors

A *flat-plate collector* is an insulated, weatherproofed box containing a dark absorber plate under one or more glazings (e.g., glass or plastic) that cover the front opening in the box. Transparent or translucent glazing allows solar radiation to be admitted into the collector body but reduces the amount of heat that can escape. Solar radiation strikes the absorber plate, which heats up, converting solar energy into thermal (heat) energy. Heat is transferred to an air or liquid heat transfer medium moving through the collector. Absorber plates are often made of metal, usually copper or aluminum, because a metal is a good conductor of heat. Copper is more expensive, but is a better conductor and is less prone to corrosion than aluminum. Absorber plates are commonly painted black or coated with a selective surface material. Examples of these collectors are shown in Photos 11.2 through 11.7 and Figure 11.5. The sides and bottom of the collector body are insulated, further minimizing heat loss.

Flat-plate collectors fall into two basic categories based on the heat transfer medium used: liquid and air. Generally, flat-plate liquid collectors heat distribution fluids to between 90° and 120°F (32° and 49°C).

PHOTO 11.2   A cross section of a liquid flat plate collector.

PHOTO 11.3   Two roof-mounted flat-plate collectors for domestic water heating at this Habitat for Humanity house in Denver, Colorado. (Courtesy of DOE/NREL)

PHOTO 11.4   A commercial active solar installation with almost 4000 flat-plate collectors. (Used with permission from ABC)

PHOTO 11.5   A liquid flat-plate collector system on a residence as part of an active solar heating and domestic water heating system. (Used with permission from ABC)

PHOTO 11.6   Two roof-mounted flat-plate collectors used in an active solar domestic water heating system. (Used with permission from ABC)

PHOTO 11.7   Homes with south-facing roof-mounted solar collectors. (Used with permission from ABC)

In a *liquid flat-plate collector,* solar energy heats a liquid as it flows through tubes in or adjacent to the absorber plate. The tubes are attached to the absorber plate so the heat absorbed by the absorber plate is easily conducted to the heat transfer liquid. Tubes are typically routed in parallel, using inlet and outlet headers to which the system piping is connected.

Liquid flat-plate collectors can be either glazed or unglazed. Glazed liquid collectors are used for domestic/process water and space heating systems. They use a glazing material, usually a sheet of glass, to cover the absorber plate to reduce heat loss from the collector. This delivers a higher temperature heat transfer medium but increases collector cost. Unglazed liquid collectors have the absorber plate exposed to the outdoor air. They are commonly used to heat water for outdoor swimming pools. Because these collectors only need to deliver moderate temperatures, they can use less expensive materials such as

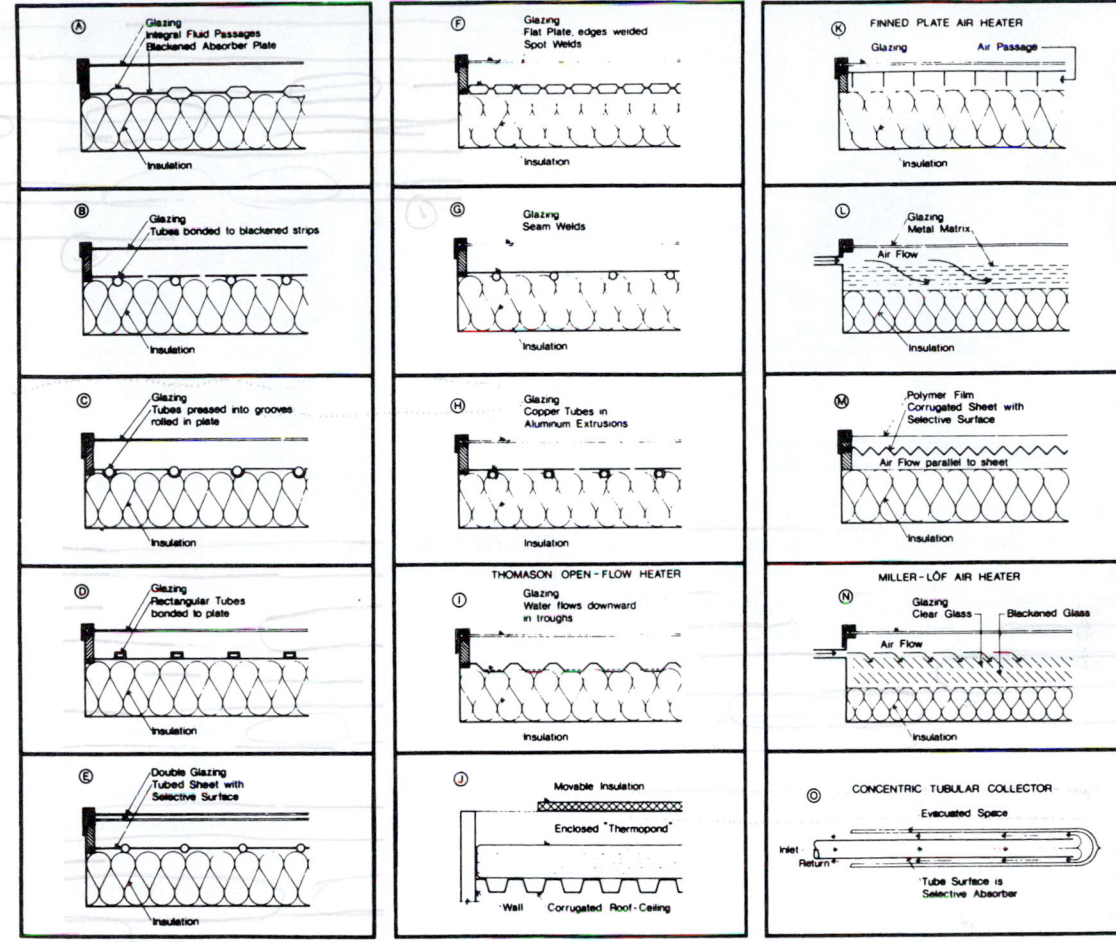

FIGURE 11.5 Cross-sectional variations of active solar collectors.

plastic or rubber and do not require freeze proofing because swimming pools are generally used only in warm weather.

Air flat-plate collectors are simple collectors used primarily for space heating. The absorber plate in an air collector is constructed of metal or nonmetallic materials. Because air conducts heat much less easily than does a liquid, less heat is transferred between the air and the absorber than in a liquid collector. Fins or corrugations built into the absorber material can increase air turbulence and improve heat transfer. However, this design modification increases a fans power requirements and thus increases costs of operation. Air typically moves in a cavity behind the absorber plate and the back of the collector body.

Evacuated-Tube Collectors

An *evacuated-tube collector* is made of rows of parallel tubes that consist of an outer glass tube and an inner tube-shaped absorber, which is covered with a selective surface coating that absorbs solar energy but impedes radiation heat loss. The air is withdrawn or evacuated from the space between the tubes to form

a near vacuum, which eliminates conductive and convective heat loss. Some radiant heat loss still occurs but this loss is small compared with the amount of heat transferred to the liquid in the absorber tube. They can operate at higher temperatures, in the range of 170° to 350°F (77° to 177°C). An example of an evacuated tube collector is shown in Photo 11.8.

There are two basic types of evacuated-tube collectors.

Direct-Flow Evacuated-Tube Collector The *direct-flow evacuated-tube collector* has a loop of pipe that runs through each of the evacuated tubes that make up the collector. The heat transfer fluid travels through the loop and comes into direct contact with the absorbing surface. As this fluid is pumped through the collector tubes, it is heated by sunlight.

Evacuated-Tube Heat-Pipe Collectors An *evacuated-tube heat-pipe collector* uses a heat pipe installed within each evacuated tube as a method of capturing and transferring heat. A basic *heat pipe* is a closed vessel consisting of a capillary wick structure and a small amount of fluid. A heat pipe uses an evaporating–condensing cycle, which accepts heat from an

PHOTO 11.8   An evacuated heat-pipe solar collector. Each evacuated glass tube holds a small-diameter copper heat pipe containing a small amount of water that is superheated by the sun. Heat exchangers at the top of the collector transfer heat energy to pipes containing an antifreeze solution, which is then routed to large hot water storage tanks in the basement. (Courtesy of DOE/NREL)

external source (the sun), uses this heat to evaporate the liquid and capture latent heat, and then releases the latent heat by condensation at a heat sink region. This process is repeated continuously by returning the condensed fluid back to the collector evaporation zone.

A typical evacuated-tube heat-pipe collector consists of a set of glass tubes, each containing a sealed heat pipe with a black copper fin that fills the tube and serves as the absorber. Extending beyond the top of each tube is a metal tip attached to the sealed pipe called the condenser. These tubes are mounted, the metal tips up into a manifold, which serves as a heat exchanger. As sunlight strikes the fin, the fluid is heated, some of the fluid evaporates, and hot vapor rises to the top of the pipe. Water, or glycol, flows through the manifold and picks up the heat from the tubes. The heated liquid circulates through another heat exchanger and gives off its heat to water that is stored in a solar storage tank. The condensation zone must be at a higher level than the evaporation zone. As a result, a heat-pipe collector must be mounted with a minimum tilt angle of around 25° to allow the internal fluid of the heat pipe to return to the hot absorber.

Reflectors placed behind the evacuated tubes can help focus additional solar radiation on the collector. An evacuated-tube collector is used in applications that require moderate to high water temperatures. This type of collector is more efficient than a flat-plate collector: it performs well in both direct and diffuse solar radiation. Second, the circular shape of the evacuated tube keeps the tube in direct sunlight most of the day. Although evacuated-tube collectors achieve higher efficiencies than flat-plate collectors, they are also more expensive.

PHOTO 11.9   A close-up view of a concentrating collector. (Used with permission from ABC)

Concentrating Collectors

A *concentrating collector* for residential and light commercial applications is typically constructed of parabolic troughs with mirrored surfaces that reflect and concentrate the solar energy on an absorber tube, called a receiver, which contains a heat transfer liquid. Solar radiation collected over a large area is reflected onto a smaller absorber area to achieve high temperatures. The intensity of the solar energy is concentrated so much that the receiver can attain temperatures of several hundred degrees. A view of a concentrating collector is shown in Photo 11.9.

If they are to perform effectively, concentrating collector systems use a tracking mechanism to move the collectors during the day to keep them focused on the sun. *Single-axis trackers,* used in most residential installations, move east to west. *Dual-axis trackers* move east and west and north and south. Because tracking systems are fairly expensive and require frequent maintenance, concentrating collector systems are used mostly in commercial applications. Concentrating collectors can only focus direct solar radiation, so performance is reduced significantly on hazy or cloudy days. As a result, concentrating collector systems are most practical in areas of high insolation such as those close to the equator and in the southwest desert region of the United States.

Transpired Collectors

A *transpired collector* is a dark, perforated metal wall installed on the south-facing side of a building, which creates approximately a 6 in (150 mm) cavity between it and the building's structural wall. Air is used as the heat transfer medium. Solar radiation strikes the metal and heats it. Fans mounted at the top of the wall pull outside air through the transpired collector's perforations, and the thermal energy collected by the wall is

PHOTO 11.10 A transpired collector on the entire south (left) side of the Federal Express building in Englewood, Colorado. Solar energy absorbed by the wall provides heated ventilation air to the warehouse. (Courtesy of DOE/NREL)

PHOTO 11.11 A close-up of the perforated transpired collector material. Outside air is drawn through the perforations and moves behind the metal material, which heats the air before it is drawn into an air-handling unit where it is further heated, if necessary. Lack of a need for glass on this system reduces collector costs. (Courtesy of DOE/NREL)

transferred to the air passing through the holes. The fans then distribute the heated air into the building through ducts mounted near the ceiling. Examples of transpired collectors are shown in Photos 11.10 and 11.11.

On a sunny winter day, the collector can generate air temperatures up to 40°F (21°C) higher than the outdoor air temperature. Because these collectors require no glazing or insulation, they are inexpensive to manufacture.

Collector Orientation and Tilt

To optimize collection of solar energy, fixed solar collectors in the northern hemisphere are mounted so they face south. In the southern hemisphere, they are oriented to face north. A south-facing orientation, within 30° east or west of true south, will provide around 90% of the maximum static solar collection potential. The optimum directional orientation will depend on site-specific factors that may determine where a collector can be located and on local features such as trees, hills, or other buildings that may shade the collector during certain times of the day.

The most effective tilt angle at which to mount collectors depends on latitude of the installation and type of heating required (i.e., year-round versus seasonal heating). For year-round applications such as generating domestic and process hot water, collectors should be tilted at an angle relative to the horizontal plane in a range equal to the latitude plus or minus 15°. For winter space heating in latitudes above about 30°, the collector should be situated near vertical. For summer heating (e.g., heating swimming pool water), the collector should be situated near horizontal.

Thermal Storage

The intermittent nature of solar radiation as an energy source means that the collection system must be increased substantially so additional energy can be stored to meet demands when solar energy is unavailable or when availability is insufficient. Thermal storage is typically achieved by a liquid heat storage medium such as water or an antifreeze solution; or by a large solid mass such as concrete or crushed rock.

In addition to water and rock thermal storage systems, more compact *phase-change thermal storage* mediums can be used. These rely on latent heat of fusion of a phase-change material (PCM) to store energy. The phase-change characteristics of animal fats and eutectic salts (salts that melt at low temperatures) can also be used.

Types of Active Heating Systems

There are two basic types of active solar heating systems, categorized by type of fluid heated in the collectors: liquid or air. Liquid systems use water or an antifreeze solution as the heat transfer medium in a hydronic collector. Air systems use air as the heat transfer medium in a collector designed to heat air. Both of these systems collect and absorb solar radiation, transferring the solar heat directly to the interior space or to a storage system, from which the heat is distributed.

Liquid Systems

Liquid active solar heating systems can be used for domestic/process water heating and central space heating. In the collector of a liquid system, a heat transfer fluid such as water, antifreeze (usually nontoxic propylene glycol), oil, or other type of liquid is pumped through a collector and absorbs the captured solar heat. An electronic controller operates a circulating pump that moves the fluid through the collector. The liquid flows rapidly through the collectors, so its temperature increases only 10° to 20°F (5.6° to 11°C). Higher temperature differences mean greater piping and collector heat losses and lower efficiency. The liquid flows to either a storage tank or a heat exchanger for immediate use. System components include piping, pumps, valves, an expansion tank, a heat exchanger, a storage tank, and controls. A view of a liquid system is shown in Figure 11.6. There are three types of liquid systems:

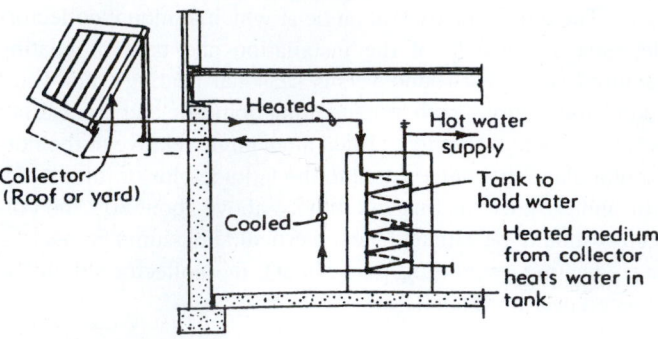

FIGURE 11.6 Liquid active solar heating systems heat water or an antifreeze solution when a hydronic collector absorbs solar heat. A storage tank holds water.

Open-Loop Liquid Systems An *open-loop active solar system* uses pumps to circulate domestic or process water through the collectors. This design is efficient and lowers operating costs, but is not appropriate if your water is hard or acidic because scale and corrosion quickly disable the system.

A *recirculating system* is a specific type of open-loop active solar system that provides freeze protection by circulating warm water from the solar storage tank through the collectors and exposed piping when temperatures approach freezing. Activating the freeze protection mode wastes electricity and stored heat, so these systems should only be considered where mild freezes occur only once or twice a year at most (e.g., Florida) or in nonfreezing climates (e.g., Hawaii, Mexico). Additionally, in a power outage during freezing weather, the pump will not work and the system can freeze. A freeze valve can be installed to provide additional protection in the event the pump does not operate. In freezing weather, the valve trickles warmer water through the collector to prevent freezing.

Closed-Loop Liquid Systems A *closed-loop active solar system* pumps an antifreeze or oil heat transfer medium (e.g., glycol–water solution) through the collectors. A heat exchanger transfers the captured solar heat from the heat transfer fluid to the domestic/process water that is then stored in the solar storage tank(s). The heat exchanger is located exterior to or within the tank. Double-walled heat exchangers prevent contamination of domestic/process water. Most codes require double walls when the heat transfer medium is anything but domestic water.

Closed-loop glycol systems are used in geographic regions subject to extended freezing temperatures because they offer good freeze protection. However, glycol antifreeze and oil systems are a bit more expensive to purchase, install and operate due to the added heat exchanger and cost of antifreeze replacement. The antifreeze glycol–water solution must be tested each year and replaced every 3 to 5 years, depending on glycol quality and system temperatures.

Drainback Liquid Systems *Drainback active solar systems* use water as the heat transfer fluid in the collector loop. A pump circulates the water through the collectors. The water drains by gravity to the solar storage tank and heat exchanger; there are no valves to fail. When the pumps are off, the collectors are empty, which assures freeze protection and allows the system to turn off if the water in the storage tank becomes too hot.

Liquid systems store captured solar heat in storage tanks or in the concrete mass of a radiant slab system. Most liquid solar systems use a tank for thermal storage. Many commercially available residential systems use a converted domestic electric water heater or a solar storage tank plumbed in series with a conventional water heater. Solar storage tanks can be constructed of stainless steel, fiberglass, concrete, or high-temperature polymers. Specialty or custom tanks built on-site may be necessary in systems with very large storage requirements. It is often more cost-effective to use several smaller tanks rather than one large one. Tank insulation is necessary to prevent excessive heat loss. A solar domestic hot water system with a natural gas backup is shown in Photo 11.12.

Most storage tanks for domestic hot water installations require 1.5 to 2 gal (3.8 to 7.6 L) of water for each ft^2 (0.093 m^2) of collector area. For domestic water heating systems with small storage requirements (e.g., a residence), standard water heaters are a good option because they are designed to meet pressure vessel requirements, are lined to inhibit corrosion, and connect to pipes and fittings easily.

Solar storage tanks in residential installations are typically 50, 60, 80, or 120 gal (190, 230, 300, or 460 L) capacity. A 50 to 60 gal (190 to 230 L) tank is sufficient for 1 or 2 occupants; an 80 gal (300 L) tank is adequate for 3 or 4 occupants, and a 120 gal (460 L) tank is appropriate for 4 to 6 occupants. For large commercial and process water heating systems, and space heating systems, a larger tank or several tanks are required. A solar storage tank for space heating an average-size residence in a moderate heating climate is typically about 750 to 1000 gal (2850 to 3800 L) or more.

With domestic/process water heating installations, hot water is drawn directly from the tank. If supplemental heating is required, water from the solar storage tank is introduced into

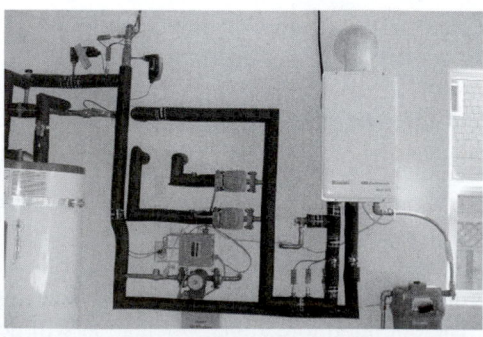

PHOTO 11.12 A solar domestic hot water system with a natural gas backup. This project is a demonstration project showing all components, except the flat-plate solar collectors mounted on the roof and associated piping. (Courtesy of DOE/NREL)

a conventional water heating system (e.g., water heater or boiler) where it is heated to the appropriate temperature. The solar-heated water can also serve as preheated water.

In space heating installations, heated water from the solar storage tank is distributed to the indoor space by various methods such as through a radiant floor, through hydronic (hot water) coils, convectors, or radiators, or with a central forced-air system (e.g., furnace or air-handling unit).

A radiant floor system circulates solar-heated liquid through pipes embedded in a concrete floor slab or located below the subfloor. The heated floor then radiates heat to the indoor space. This type of distribution system was introduced in Chapter 9. Radiant floor heating is ideal for liquid solar systems because radiant heating performs well at relatively low-grade temperatures. A well-designed system may not need a separate heat storage tank, except for temperature control. A conventional boiler or domestic water heater can supply backup heat.

To function well, hydronic (hot water) convectors and radiators require a water temperature in the range of 140° and 180°F (60° and 82°C). Generally, flat-plate liquid collectors heat distribution fluids to between 90° and 120°F (32° and 49°C), so these distribution systems require that either the surface area of the baseboard or radiators be larger, that the distribution liquid be heated further (with the backup system), and/or that a medium-temperature solar collector, such as an evacuated-tube collector, be used.

A liquid system can be integrated into a forced-air heating system such as a furnace or air-handling unit. A common method is to place a liquid-to-air heat exchanger in the return air duct just before the furnace or air-handling unit. The heated water from the solar storage tank is pumped into the heat exchanger. Air returning from the indoor space is heated as it passes over the solar-heated liquid in the heat exchanger. Supplemental heat is supplied as necessary by an electric heating element, furnace, or air-handling unit. The heat exchanger must be large enough to transfer sufficient heat to the air, especially when the heat exchanger liquid is at 90°F (32°C).

Example 11.1

A small car wash has a hot water load of 1100 gal/day at a 50°F temperature rise. A solar system at the location of the building can collect, store, and deliver about 1130 Btu/day per square foot of collector area.

a. Approximate the collector area needed to meet 50% of the hot water load.

$$(1100 \text{ gal/day})(8.33 \text{ Btu/lb} \cdot °F)(50°F) \cdot 50\% = 229\ 075 \text{ Btu/day}$$
$$(229\ 075 \text{ Btu/day})/(1130 \text{ Btu/day}) = 203 \text{ ft}^2$$

b. Approximate the required storage tank capacity based on 1.8 gal of water for each ft² of collector area.

$$203 \text{ ft}^2 \cdot 1.8 \text{ gal/ft}^2 = 365 \text{ gal}$$

Air Systems

An *active air solar heating system* uses air as the heat transfer medium for absorbing and transferring solar energy. Solar air collectors can directly heat individual rooms or be integrated into a central heating system. Air collectors produce heat earlier and later in the day than liquid systems. Therefore, air systems may produce more usable energy over a heating season than a liquid system of the same size. Also, unlike liquid systems, air systems do not freeze, and minor leaks in the collector or distribution ducts will not cause problems. Types of air systems include the following:

Conventional Air Solar Systems *Conventional air solar systems* use flat-plate air collectors and a large bin of rocks for thermal storage. They generally can be integrated into a central heating system, which serves to provide supplemental or backup heating. Small systems simply discharge air directly into the living space.

In a typical design, a blower circulates warm air from the air collectors through a large duct that carries the heated air through the rocks, which absorb heat. A large duct at the opposite side of the bin returns the air to the collectors for reheating. When the building needs heat, air is drawn from return ducts in the building through the bin in a reverse direction and is forced into supply ducts for heating individual spaces. The rocks serve as a storage medium and the rock surfaces serve as a heat exchanger. Bin temperatures can reach 140°F (60°C). If the air in the bin is too cool, a supplemental heating system (e.g., a furnace) heats the air leaving the bin to the desired temperature before distributing it.

A rock bin requires ½ to 1 ft³ (0.014 to 0.028 m³) of volume for every ft² of collector area. The heat capacity of rock is about one-third that of water, which means rock storage needs to be about 3 to 4 times the volume of a liquid system. Rocks of uniform size, typically ¾ to 1½ in (19 to 38 mm), are used. They must be clean, dry, without any dirt or gravel, and kept dry inside the bin to prevent moisture problems (e.g., mold, mildew). The bin is typically located in an underfloor crawl space or basement, because warm air naturally rises to the indoor space, but it can be located outdoors, above or below ground. A bin can be built of masonry, concrete, or wood. It must be tightly constructed to prevent air leaks and moisture intrusion, and well insulated.

Heat in an air system can also be stored in a water tank. An air-to-water heat exchanger located exterior to the tank is required in this design. A blower circulates solar-heated air through the heat exchanger and back to the collectors. A thermostatic controller operates the blower fan. The heat is then distributed from the storage in the same way as in a liquid system.

The electrical energy required to power fairly large fans reduces the overall efficiency of an air system. The volume of storage reduces the cost-effectiveness of this type of system. As a result, solar systems that use air as the storage medium are typically not used.

PHOTO 11.13   7800 ft² (725 m²) of transpired solar collector area (the upper dark wall) used as a ventilation air preheating system on a helicopter maintenance hangar at Fort Carson U.S. Army Base, south of Colorado Springs, Colorado. (Courtesy of DOE/NREL)

PHOTO 11.15   A thermosiphon air panel discharges air directly into the living space. (Courtesy of DOE/NREL)

PHOTO 11.14   A transpired solar collector installed on the south wall of a home improvement center. (Courtesy of DOE/NREL)

Transpired Air Systems   *Transpired air systems* are designed to trap solar energy to preheat outdoor ventilation air. The building's ventilation fan draws outdoor air through a transpired air collector. The intake air is heated as it passes through perforations in the heated metal and through the space between the facade and the south-facing wall of the building. The heated air then enters the building. Examples of this system are shown in Photos 11.13 and 11.14.

This contemporary technology is ideally suited for buildings with at least moderate ventilation requirements in sunny locations with long heating seasons. Most industrial and commercial buildings require large quantities of ventilation air to maintain a healthy indoor environment. In many regions, this ventilation air needs to be heated to provide a comfortable work environment. During the cooling season, ventilation could be brought in from another location outside of the collector. A transpired air system captures between 60 and 75% of the available solar energy, making it one of the most efficient solar collectors designed to date.

Solar Room Air Heaters   A *room air solar heating unit* is a basic collector that is installed on a roof or an exterior (south-facing) wall for heating one or more rooms. The collector is an airtight and insulated metal or wood box with an opening covered with glazing

in front and a black metal plate under the glazing in the bottom of the box for absorbing heat. Solar radiation is admitted into the box and heats the plate, which heats the air in the collector. A thermostatically controlled, electrically powered fan or blower draws air from the room, pushes it through the collector, and blows it back into the room(s). Roof-mounted collectors require ducts for supplying air from the room(s) to the collector and for distribution of the warm air into the room(s). Wall-mounted collectors are placed directly on a south-facing wall. Holes are cut through the wall for the collector air inlet and outlets. Simple window box units fit in an existing window opening. These units only provide a small amount of heat, because the collector area is relatively small. An example of this system is shown in Photo 11.15.

Room air solar heating units can be active (using a fan) or passive. In passive types, airflow is driven by natural convection: room air enters the bottom of the collector, rises as it is heated, and enters the room. A baffle or damper keeps the room air from flowing back into the panel (reverse thermosiphoning) when the sun is not shining.

A room air solar heating unit is easier and less expensive to install than a central solar heating system. They do not have a dedicated storage system or extensive ductwork. Thermal mass in the floors, walls, and furniture will absorb some solar heat, which will help keep the room warm for a few hours after sundown. Masonry walls and tile floors will provide more thermal mass, and thus provide heat for longer periods into the evening hours. Manufactured room air solar heating units for on-site installation are available.

Pumps and Fans

In residential active solar applications, temperature sensors and automatic controllers activate the circulating pumps or fans. Pumps and fans have fairly low power requirements. Typically, power is 120 V alternating current (AC). In small water heating installations, some suppliers offer direct current (DC) pumps powered by small photovoltaic (PV) panels. PV panels convert solar radiation into DC electricity. Although they initially cost more to install, they are very simple and require little maintenance. Such systems cost nothing to operate and continue to function during power outages.

11.4 PASSIVE SOLAR HEATING SYSTEMS

Introduction

A *passive solar heating system* is integrated into building elements (e.g., windows) and materials (e.g., tile-covered concrete floors) and does not rely on mechanical equipment to collect, store, and distribute thermal (heat) energy from the sun. Collection of solar energy and thermal energy storage and movement occur naturally through radiation, conduction, and natural convection. Some passive systems use a small amount of conventional energy to control dampers, shutters, night insulation, and other devices that enhance solar energy collection, storage, and use, and reduce undesirable heat loss. A true passive solar system does not use auxiliary energy.

The goal of all passive solar heating systems is to capture solar energy within the building envelope and naturally release that heat during periods when the sun is not shining (e.g., during nighttime hours and during periods of cloudy weather). Ideally, it has little to no operating costs, has low maintenance costs, and emits no greenhouse gases in operation. To accomplish these objectives, a passive solar system must be optimized to deliver the best performance and economic payback, and

must maintain indoor thermal comfort (e.g., limit daily indoor temperature swings and overheating).

The key to designing a passive solar building is to take advantage of local climate. Important design considerations include window location and glazing type, insulation, air sealing, thermal mass, shading, and usually an auxiliary heating system.

The five necessary elements of a passive solar design include the following (see Figure 11.7).

Collector Area/Aperture

The *collector area* or *aperture* is the glass/window area through which sunlight enters the building. In the northern hemisphere, a south-facing exposure of transparent material (e.g., glass, plastic) allows solar energy to enter the building envelope.

Absorber

The *absorber* is the darkened surface of the storage element that sits in the direct path of sunlight. It could be a masonry wall, floor, or partition, or a water container. Sunlight strikes the surface and is absorbed as heat.

Thermal Mass

Thermal mass includes the massive materials that retain or store the heat produced by solar energy. The difference between the absorber and thermal mass, although they often form the same wall or floor, is that the absorber is an exposed surface whereas storage is the material below or behind that surface. These materials in the building interior must absorb solar energy for later use and keep the space comfortable but not overheated.

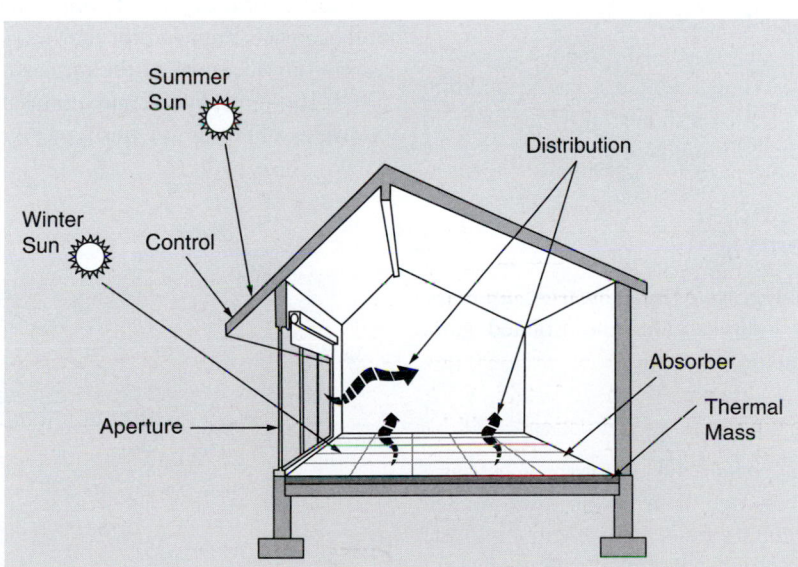

FIGURE 11.7 The five necessary elements of a passive solar system are aperture (glass area), thermal mass (to store energy collected), absorber material (to enhance heat transfer from sunlight to storage), distribution (of heat), and control (shading). (Source: U.S. DOE)

PHOTO 11.16 Insulating shutters on a passive solar system enhance the thermal insulating capability and reduce heat loss through the building envelope. (Used with permission from ABC)

Distribution

Distribution is the method by which solar heat circulates from the collection and storage points to different areas of the house. A passive design will naturally use the three natural heat transfer modes (e.g., conduction, convection, and radiation) exclusively. In a hybrid installation, heat is mechanically moved with fans, ducts, and/or blowers to help with the distribution of heat through the house.

Control

Controls are integrated within the system to naturally manage heat gain to prevent underheating or overheating. Roof overhangs and awnings shade the collector area during summer months. Operable blinds, shades, louvers, vents, and dampers can be used to moderate heat gain, as shown in Photo 11.16. In a hybrid installation, an electronic sensing device (e.g., thermostat) can be used to signal a fan to operate.

Passive solar heating is the most cost-effective means of providing heat to buildings. When integrated in the building design, it adds little to the initial overall cost of a building and reduces heating costs once the building is in operation.

Fundamental Passive Solar Space Heating Systems

There are three fundamental passive solar systems used for space heating: direct gain, indirect gain, and isolated gain. These types are introduced in the following.

Direct Gain

In a *direct-gain passive solar system,* the indoor space acts as a solar collector, heat absorber, and distribution system. See Photo 11.17. South-facing glass admits solar energy into the building interior where it directly heats (radiant energy absorption) or indirectly heats (through convection) thermal mass in the building such as concrete or masonry floors and walls. The floors and walls acting as thermal mass are incorporated as functional parts of the building and temper the intensity of heat

PHOTO 11.17 A direct-gain passive solar system. (Courtesy of DOE/NREL)

during the day. At night, the heated thermal mass radiates heat into the indoor space. It is also possible to use some other form of thermal storage such as water containers or PCMs inside the building to store heat. See Figure 11.8.

Sufficient thermal mass is required to prevent large temperature fluctuations in indoor air. Overheating of the building interior can result in insufficient or poorly designed thermal mass. Thermal mass in floors and walls should be kept as bare as functionally and aesthetically possible. Wall-to-wall carpeting and large wall hangings should be avoided. Typically, for every ft^2 of south-facing glass, 1 ft^3 of concrete or masonry is required for thermal mass (1 m^2 per 5 to 10 m^2). Maximum thickness of the mass should be limited to 3 to 4 in (75 to 100 mm). When accounting for minimal wall and floor coverings and furniture, this typically equates to about 5 to 10 ft^2

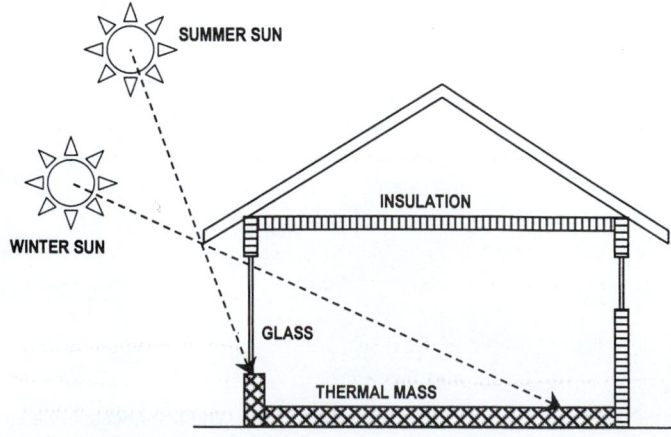

FIGURE 11.8 In a *direct-gain* passive solar system, the indoor space acts as a solar collector, heat absorber, and distribution system.

per ft^2 (5 to 10 m^2 per m^2) of south-facing glass, depending on whether the sunlight strikes the surface directly. The simplest rule of thumb is that thermal mass area should have an area of 5 to 10 times the (uncovered) surface area of the direct-gain collector (glass) area.

Thermal mass should be relatively thin, no more than about 4 in (100 mm) thick. Thermal masses with large exposed areas and those in direct sunlight for at least part of the day perform best. Medium-to-dark colors should be used on surfaces of thermal mass elements that will be in direct sunlight. Thermal mass not in contact with sunlight can be any color. Lightweight elements (e.g., drywall walls and ceilings) can be any color. Covering the glazing with tight-fitting, moveable insulation panels during dark, cloudy periods and nighttime hours will greatly enhance performance of a direct-gain system.

Typically, with adequate thermal mass, south-facing glass areas should be limited to 15 to 20% of the floor area (e.g., 150 to 200 ft^2 for a 1000 ft^2 floor area). This should be based on the net glass or glazing area. Note that most windows have a net glass area that is 75 to 85% of the overall window area.

Example 11.2

A single-story home has a floor area of 2000 ft^2. It will be heated with a direct gain passive solar system with south-facing collector (glass) area sized at 15% of the floor area. The windows will have a glass area that measures 2 ft-6in wide by 6 ft-4 in high. Thermal mass will be based on 5 ft^2 of mass surface area per ft^2 of collector (glass) area.

a. Determine the direct-gain collector (glass) area.

$$2000 \text{ ft}^2 \cdot 15\% = 300 \text{ ft}^2$$

b. Determine the number of windows required.

$$2.5 \text{ ft} \cdot 6.33 \text{ ft} = 15.83 \text{ ft}^2 \text{ window area}$$
$$(300 \text{ ft}^2)/(15.83 \text{ ft}^2) = 19 \text{ windows}$$

c. Determine the surface area of thermal mass.

$$5 \text{ ft}^2/\text{ft}^2 \text{ of collector area} \cdot 300 \text{ ft}^2 = 1500 \text{ ft}^2$$

Most building designers unconsciously use direct gain for space heating with placement of south-facing windows but forget thermal mass. This can result in severe overheating. Unless good thermal mass strategies are implemented, the area of south-facing glass should be limited to about 5 to 7% of the floor area. A larger area of south-facing glass without adequate thermal mass leads to overheating.

A *sun-tempered building* is a type of direct-gain system in which the building envelope is well insulated, is elongated in an east–west direction, and has a large fraction of the windows on the south side. It has little added thermal mass beyond what is already in the building (i.e., just framing, wall board, and so forth). Again, the south-facing window area should be limited to about 5 to 7% of the total floor area to prevent overheating.

PHOTO 11.18 A passive solar residence with direct gain (right) and sunspace systems (left). Note that shades are used for control to prevent overheating and eliminate glare. (Used with permission from ABC)

PHOTO 11.19 A passive solar residence with direct gain (small windows), thermal storage wall (large glazing on lower ground level) and attic sunspace systems (on roof). (Courtesy of DOE/NREL)

Indirect Gain

In an *indirect-gain passive solar system,* the thermal mass (concrete, masonry, or water tanks) is located directly behind glass and in front of the heated indoor space. The position of the mass prevents sunlight from entering the indoor space and obstructs the view through the glass. There are two types of indirect gain systems: thermal storage wall systems and roof pond systems.

Thermal Storage Walls In a *thermal storage wall* system a massive wall is located directly behind south-facing glass. See Photos 11.20 and 11.21. It typically consists of an 8 to 18 in (200 to 450 mm) thick masonry wall coated with a dark, heat-absorbing finish (or a selective surface) and covered with a single or double layer of glass. The glass is typically placed from ¾ in to 2 in from the wall to create a small airspace. In some designs, the mass is located 1 to 2 ft (0.6 m) away from the glass, but the space is still not useable. The thermal mass absorbs the solar radiation that strikes it and stores it for nighttime use. An *unvented storage wall* captures solar energy on the exterior surface, heats up, and conducts heat to the interior surface, where it radiates from the interior wall surface to the indoor space later in the day. A *water wall* is a type of thermal mass that consists of tanks or tubes of water used as thermal mass. See Figure 11.9.

PHOTO 11.20   A close-up view of the thermal storage wall.
(Used with permission from ABC)

PHOTO 11.21   The trombe wall entails the lower obscured (but translucent) glass sections. It is exposed to direct sunlight (unshaded) during winter months and is shaded by the overhangs during the summer months. Heat from the sun is trapped between the transluscent panes of glass and black selective coating on a masonry wall. The wall stores solar heat for later release into the building. (Courtesy of DOE/NREL)

A *trombe wall,* named after Felix Trombe, a French engineer who popularized it's use in the 1960s, is commonly called a *vented thermal storage wall.* It has operable vents near the ceiling and floor levels of the thermal mass wall that allow room air to flow through them by natural convection. These vents permit indoor air to be drawn through the lower vent, then into the space between the glass and wall to get solar heated, increasing its temperature and causing it to rise, and then through the top (ceiling) vent back into the indoor space.

Vents should be closed at night so radiant heat from the interior surface of the storage wall heats the indoor space. If vents are left open at night (or on cloudy days), a reversal of convective airflow will occur, wasting heat by dissipating it outdoors. See Figure 11.10. Generally, vents are closed during summer months when heat gain is not needed. Vented thermal storage walls have proven somewhat ineffective, mostly because they deliver too much heat during the day.

Although the position of a thermal storage wall prevents daytime overheating of the indoor space, a well-insulated home should be limited to approximately 0.2 to 0.3 ft^2 of thermal mass wall surface per ft^2 of floor area being heated (0.2 to 0.3 m^2 per m^2 of floor area), depending upon climate. A water wall should have about 0.15 to 0.2 ft^2 of water wall surface per ft^2 (0.15 to 0.2 m^2 per m^2) of floor area.

The thickness of a thermal storage wall should be approximately 10 to 14 in (250 to 350 mm) for brick, 12 to 18 in (300 to 450 mm) for concrete, 8 to 12 in (200 to 300 mm) for earth/adobe, and at least 6 in (150 mm) for water. These thicknesses delay movement of heat such that indoor surface temperatures peak during late evening hours. Heat will take about 8 to 10 hr to reach the interior of the building (heat travels through a concrete wall at a rate of about one inch per hour).

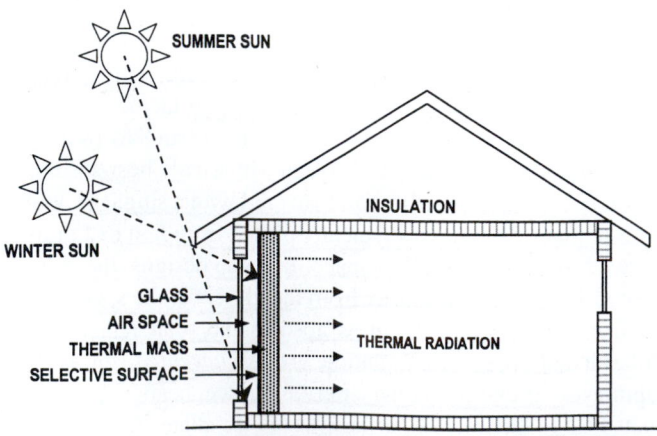

FIGURE 11.9   An *unvented thermal storage wall* is an indirect-gain passive solar system that captures solar energy on the exterior surface of the wall, heats up, and conducts heat to the interior surface, where it radiates from the interior wall surface to the indoor space.

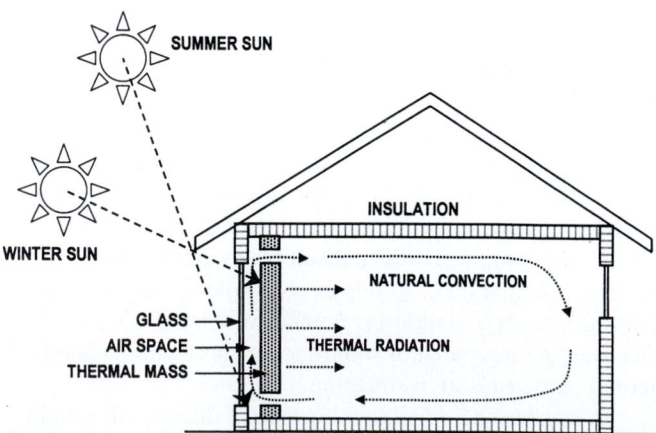

FIGURE 11.10   A *vented thermal storage wall* or *trombe wall* is an indirect-gain passive solar system that has operable vents at the top and bottom of the thermal storage wall that permit indoor air to be drawn through the lower vent by natural convection into the cavity between the glass and wall, to get solar heated, increasing its temperature and causing it to rise, and to pass through the top vent back into the indoor space.

Nighttime thermal losses through the thermal mass can still be significant in cold climates. Covering the glazing with tight-fitting, moveable insulation panels during dark, cloudy periods and nighttime hours will enhance performance of a thermal storage system. A selective surface (high-absorbing/low-emitting surface) applied to the exterior surface of the thermal storage wall will typically achieve a similar improvement in performance without the need for daily installation and removal of insulating panels.

Photos 11.22 through 11.32 show types of thermal storage wall systems.

Roof Pond System A *roof pond passive solar system* typically holds 6 to 12 in (150 to 300 mm) of water on a flat roof. Water is stored in large plastic or fiberglass containers that can be left unglazed or covered by glazing. Solar radiation heats the water, which acts as a thermal storage medium. At night or during cloudy weather, the containers can be covered with insulating panels. The indoor space below the roof pond is heated by thermal energy emitted by the roof pond storage above. These systems require good drainage systems, movable insulation, and an enhanced structural system to support a 35 to 70 lb/ft^2 (1.7 to 3.3 kN/m^2) dead load. Roof pond systems perform better for cooling in hot, low humidity climates. See Figure 11.11.

PHOTO 11.24 Double-glass windows covering thermal storage wall on the south side of the residence. (Used with permission from ABC)

PHOTO 11.25 An interior view of the direct-gain glass and unvented thermal storage wall. (Used with permission from ABC)

PHOTO 11.22 A simple passive solar residence incorporating direct-gain and unvented thermal storage wall systems. (Used with permission from ABC)

PHOTO 11.26 A direct gain and unvented thermal storage wall system at the National Renewable Energy Laboratory (NREL) Visitor Center, Golden, Colorado. (Used with permission from ABC)

PHOTO 11.23 An unvented thermal storage wall under construction. The thermal storage wall is constructed of cast-in-place concrete and painted black to improve absorption of sunlight. Note the shadow line indicating shading of the wall. Photograph was taken at solar noon in early May. (Used with permission from ABC)

PHOTO 11.27 The articulated thermal storage wall at the NREL Visitor Center, Golden, Colorado. (Courtesy of DOE/NREL)

PHOTO 11.28 Construction of the fully grouted, reinforced concrete masonry articulated thermal storage wall at the NREL Visitor Center, Golden, Colorado. (Courtesy of DOE/NREL)

PHOTO 11.29 A water wall and stone fireplace used as a thermal storage wall in a classroom in a nature center facility. Fiberglass tubes contain about 75 gal/285 L (~625 lb/280 kg) of light blue-colored water. Floors are colored concrete, which also serve as thermal mass. (Used with permission from ABC)

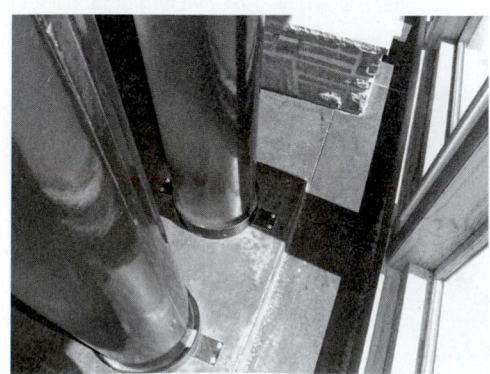

PHOTO 11.30 Fiberglass tubes are anchored to the concrete floor. (Used with permission from ABC)

PHOTO 11.31 Fiberglass tubes are capped at top and secured at ceiling to prevent tipping. (Used with permission from ABC)

PHOTO 11.32 Exterior view of glass of water wall. (Used with permission from ABC)

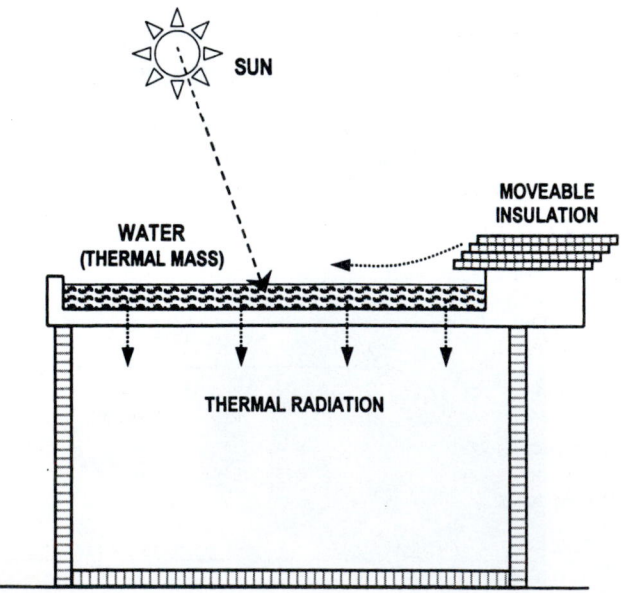

FIGURE 11.11 A *roof pond* is an indirect-gain passive solar system that has water stored on the roof in large plastic or fiberglass containers covered by glazing. The indoor space below the roof pond is heated by thermal energy emitted by the roof pond storage above.

Isolated Gain

In an *isolated gain passive solar system,* the components (e.g., collector and thermal storage) are isolated from the indoor area of the building. A sunspace (e.g., sunroom or greenhouse) and a convective loop system are examples.

Sunspace System Sunspaces are essentially attached green-houses that make use of a combination of direct-gain and indirect-gain system characteristics. South-facing glass collects solar energy as in a direct-gain system. A thermal mass wall situated on the back of the sunspace adjacent to the living space will function like an indirect-gain thermal mass wall. Solar energy entering the sunspace is retained in the thermal mass. Solar heat is conveyed into the building by conduction through the shared mass wall in the rear of the sunspace and by vents (like an unvented thermal storage wall) or through openings in the wall that permit airflow from the sunspace to the indoor space by convection (like a vented thermal storage wall). See Figure 11.12. Examples of sunspaces are shown in Photos 11.33 through 11.35.

A sunspace with a masonry thermal wall will need approximately 0.3 ft² of thermal mass wall surface per ft² of floor area being heated (0.3 m² per m² of floor area), depending on climate. Wall thicknesses should be similar to a thermal storage wall. If a water wall is used between the sunspace and living space, use about 0.20 ft² of thermal mass wall surface per ft² of floor area being heated (0.2 m² per m² of floor area). In most climates, a ventilation system is required in summer months to prevent overheating. Generally, vast overhead (horizontal) and east- and west-facing glass areas should not be used in a sunspace without special precautions for summer overheating such as using heat-reflecting glass and providing summer-shading system areas.

Thermosiphon Systems A *thermosiphon system* uses a south-facing air collector to collect solar radiation, which then allows

PHOTO 11.33 A commercially available attached sunspace. (Used with permission from ABC)

PHOTO 11.34 Thermal mass in this passive solar sunspace absorbs solar energy during the day without overheating the sunspace area receiving the direct sunlight. During the evening, thermal mass releases energy reducing the daily temperature swing (fluctuation) in the area. (Courtesy of DOE/NREL)

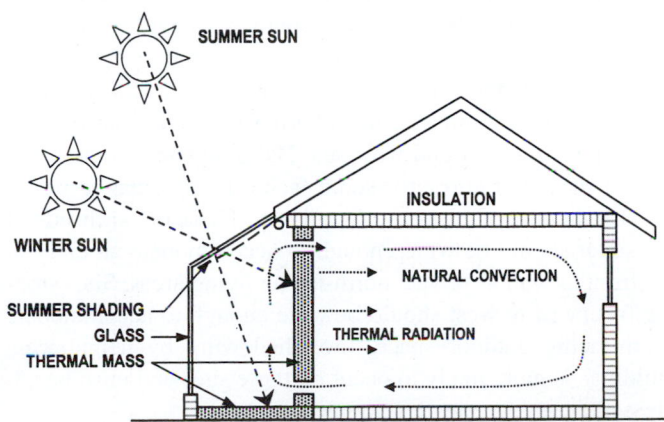

FIGURE 11.12 A *sunspace* is an isolated-gain passive solar system that has south-facing glass that collects solar energy and a thermal mass wall situated on the back of the sunspace that heats up and conducts heat to the interior surface, where it radiates from the interior wall surface to the indoor space.

PHOTO 11.35 The interior of a sunspace integrated in a new home. Insufficient thermal mass and the upper glass will create overheating and drastic temperature swings (fluctuations). Shading of upper glass and venting will be necessary. (Used with permission from ABC)

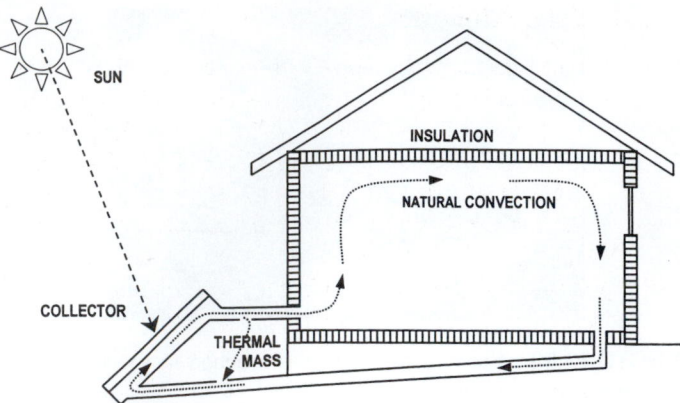

FIGURE 11.13 A *thermosiphon* is an isolated-gain passive solar system that uses a south-facing air collector to collect solar radiation, which then allows heated air to flow into a storage bin through natural convection.

heated air to flow into a storage bin through natural convection. This system is a variation on the active solar system air collector except that there is no fan or blower forcing airflow. Convective air collectors must be located lower than the storage area so that the heated air generated in the collector naturally rises into the storage area and is replaced by return air from the lower cooler section of the storage area. Heat can be released from the storage area either by opening vents in the rock storage or by conduction if the storage bin is built into the building. See Figure 11.13.

Example 11.3

A residence in a cold climate has a heating load of 900 000 Btu/day on a cold day during the middle of the heating season. A direct-gain passive solar heating system at the location of the residence can collect, store, and deliver about 710 Btu/day per square foot of window area. Approximate the south-facing window area needed to meet 50% of this heating load.

$$[(900\,000 \text{ Btu/day})/(710 \text{ Btu/day} \cdot \text{day})] \cdot 50\% = 634 \text{ ft}^2$$

Passive Solar Water Heaters

Passive systems can also be designed to heat or preheat domestic or process water. There are two basic passive solar systems used for heating water: batch heaters and thermosiphon systems. These are discussed as follows:

Batch Water Heaters A *batch water heater,* also known as a *breadbox heater,* is a simple passive water heating system consisting of one or more storage tanks placed in an insulated box that has a glazed side facing the sun. Batch heaters are inexpensive and have few components, resulting in less maintenance and fewer failures. A batch heater is mounted on the ground or on the roof. Some batch heaters use selective surfaces on the tank(s).

In climates where freezing occurs, batch heaters must either be protected from freezing or drained for the winter. In well-designed systems, the most vulnerable components for freezing are the pipes, if located in uninsulated areas that lead to the solar water heater. In mild climates, if these pipes are well insulated, the warmth from the tank will prevent freezing. Certified systems clearly state the temperature level that can cause damage.

Thermosiphon Water Heating Systems A *thermosiphon water heating system* relies on warm water rising by natural convection to circulate water through the collectors and storage tank. For natural convection to occur, the storage tank must be located above the collector. As water in the collector is heated by solar radiation, it becomes less dense and rises naturally through pipes into the storage tank. This forces cooler water in the bottom of the tank to flow through pipes to the bottom of the collector, causing circulation throughout the system. The storage tank is attached to the top of the collector so that thermosiphoning can occur. These systems are reliable and relatively inexpensive, but require careful planning in new construction because the water tanks are bulky.

Other Passive Solar Space Heating Strategies

Several fundamental strategies must be integrated in design of any passive solar system. These include the following:

Site Considerations Solar access is the term used to describe the amount or number of hours of unobstructed, direct sunlight that reaches a building site or the building itself. Unobstructed access to the sun is necessary for optimum performance of active and passive solar energy systems. The desired amount of solar access varies with climate. Generally, the building's south elevation (north elevation in the southern hemisphere) should receive direct sunlight between the hours of about 9 AM and 3 PM (solar time, not local time) during the heating season.

Sites sloping from north to south are ideal because they receive good access to the sun with minimum potential for overshadowing by neighboring buildings or natural obstructions. In summer months, neighboring buildings can provide protection from east and west sun. Building sites on the south side of the street allow for south-facing living areas to be located at the rear of the house for privacy. Those on north side of the street should be wide enough to accommodate an entry at the front as well as private north-facing living areas. Sites sloping from east to west should be wide enough to accommodate south-facing outdoor space. Overshadowing by neighboring buildings is more likely to occur on these sites and must be addressed during the site development process.

A south-facing slope (north facing in the southern hemisphere) increases the potential for access to the southern sun and is ideal for higher housing densities. A north-facing slope increases the potential for overshadowing. Sites with views to the south (north in the southern hemisphere) are ideal, because south is the best direction to locate windows and living areas. If the view is to the north (south in the southern hemisphere), it is necessary to avoid large areas of glass to minimize winter heat loss. West- or east-facing glass areas will cause overheating in summer if not properly shaded, so large areas of glass should be avoided.

Building Orientation   With optimum orientation, it is much easier to design a building to incorporate solar features, such as passive space heating and cooling and daylighting. Ideally, the building should be oriented towards true south. However, an orientation within 30° east or west of true south will provide about 90% of the maximum static solar collection potential. The optimum directional orientation is dependent on site-specific factors that may determine where a solar installation can be located and on local landscape features such as trees, hills, or other buildings that may shade the building during certain times of the day.

Building Shape   Except in extreme climates and latitudes, the building shape should be elongated along an east–west axis to increase south-facing wall area and reduce east- and west-facing wall area. This configuration allows for maximum solar glazing (windows) to the south, and is advantageous for summer cooling because it minimizes east and west exposure to morning and afternoon summer sunlight.

Well-Insulated/Tight Building Envelope   The building envelope should be well insulated and sealed adequately to reduce transmission and infiltration losses. A reduced heating load results in a smaller solar collector (glass) area, which reduces the potential for overheating in mild weather, and cuts solar system cost. Passive solar only becomes cost-effective when the heating load is minimized below standard practices. These concepts were addressed in Chapters 3 and 4.

Interior Spaces   Habitable interior spaces requiring the most light and heat should be located along the south face of the building. Less used spaces (e.g., closets, storage areas, garages) should be located on the north wall where they can act as a buffer between high-use living space and the cold north side.

Window Placement   In general, some window/glass area is desirable in each room of a building to allow sunlight into these spaces for daylighting and, in buildings with operable windows, to provide for natural ventilation capabilities. However, the apparent path followed by the sun influences effective area and placement of windows and glass. It is necessary for a designer to address these influences, especially in buildings with light loads (e.g., residences). Nighttime window insulation (R-9) should be used because it improves the performance of the solar system by up to 40%. These concepts are addressed in more detail later in this chapter.

Shading   Exterior shading techniques (e.g., trees, overhangs) should be used to prevent summer sunlight from entering the interior. These concepts are addressed later in this chapter.

Passive Solar Windows

The apparent path the sun follows with respect to a fixed point on the earth's surface varies daily with season and latitude. In the northern hemisphere at latitudes where most of the population of the United States and Canada resides (30° to 60° NL, the middle latitudes), the sun is high in the sky in the summer months and low in the sky in the winter months. Sunrise and sunset also vary with season. In the summer, the sun rises roughly on the northeastern horizon, is south and almost directly overhead at solar noon, and it sets in the northwest. In the early winter, the sun rises at the southeastern horizon, is relatively low in the southern sky at solar noon, and sets more or less in the southwest. As a result, there are many more hours of sunlight during summer months in comparison to winter months.

East/West-Facing Glass   In the winter months at middle latitudes, there is only a modest amount of intercepted solar radiation on east- and west-facing glass because the sun is in the southern sky for most of the day and the sunlight glances off east- and west-facing vertical surfaces. In summer months, however, solar radiation is nearly normal to east- and west-facing vertical surfaces during midmorning and late afternoon hours. Thus, east- and west-facing vertical glass will have significant solar gains in summer months, which can overheat the building and severely affect occupant comfort and cooling costs, especially in sunny climates (e.g., the southwestern United States).

Although the design preference on the east and west sides may be to simply eliminate glass entirely, some glass area is desirable for daylighting and ventilation. It is suggested that glass areas be kept to 2 to 5% of the floor area being served, with the east side having slightly more window area. When it is necessary to take advantage of a west-facing view, care should be used to ensure that the windows are shaded externally.

North-Facing Glass   In the northern hemisphere at middle latitudes, north-facing vertical surfaces remain in the shade during the colder months of the year. Thus, they are losing energy. In the summer months when solar gain is not wanted, the glass on north-facing vertical surfaces receives small amounts of solar energy, mainly in the early morning and early evening hours. Again, the design preference may be to simply eliminate north-facing glass, but daylighting and ventilation requirements need to be achieved. It is suggested that glass areas be kept to 2 to 5% of the floor area being served, the less the better.

Horizontal Glass (Skylights)   During summer months in the northern hemisphere at middle latitudes, the sun is very nearly overhead at midday hours, and horizontally positioned or slightly sloped glass (e.g., skylights) will have significant solar gains. This can overheat the building and severely affect occupant comfort and cooling costs. In the winter, the sun is low in the sky and any sunlight glances off horizontally positioned glass surfaces; there is little solar gain. The design preference should be to meet daylighting requirements only. It is suggested that glass areas be kept to 2 to 5% of the floor area being served, the less the better.

South-Facing Glass   In a properly orientated home in the middle latitudes of the northern hemisphere, south-facing walls receive the most benefit from the sun. In the winter months, the sun's apparent path situates it relatively low in the southern portion of the sky from midmorning to midafternoon (about 9 AM and 3 PM solar time). Solar radiation is nearly normal to the building's south walls. Reflection from the ground (e.g., snow, ice) increases solar gain. In the winter, south-facing vertical glass areas are annual net energy gainers in all but extremely cold climates. In the summer when the sun is directly overhead, solar radiation strikes the building's south walls at a shallow angle. At such a shallow angle, it mostly reflects off of the south-facing vertical glass surface, minimizing solar gain.

Although most window/glass area should be on south-facing vertical walls, the recommended area is based on the amount of thermal mass to absorb, store, and distribute heat. Too much glass or too little thermal mass will result in overheating during mild temperatures in the fall and spring. Design strategies for south-facing glass used as a collector are discussed later.

Selecting Window Properties

When selecting the thermal performance characteristics for glass, there are two important properties: solar heat gain coefficient and U-factor. The *solar heat gain coefficient (SHGC)* is the fraction of solar heat that is transmitted through the glazing and that becomes heat. By comparison, the *visible transmittance (VT)* refers to the percentage of the visible spectrum (380 to 720 nm) that is transmitted through the glazing. When daylighting of a space is desirable, high VT glazing is a smart choice. As introduced in Chapter 2, the *overall coefficient of heat transmission (U-factor)* is a measure of how easily heat travels through a material. The lower the U-factor, the lower the amount of heat transfer through the window.

Table 11.4 lists performance measures for different types of windows. VT is important for natural lighting. It makes

economic sense to select a glass with a high U-factor (high R-value) to limit the amount of heat transfer through the window. Typically, however, a glass with a low U-factor (high R-value) also rejects solar gains (has a low SHGC), which defeats the purpose of solar glass. Simply, it is difficult to obtain a window with a low U-factor and a high SHGC. Therefore, selecting the thermal performance characteristics of glass is related to the orientation of the window and how it must perform (e.g., solar or nonsolar). The visible light-to-solar gain ratio (L/SG) is the ratio between SHGC and VT. The L/SG provides a gauge of the relative efficiency of different glass types in transmitting daylight while blocking heat gains. The higher the ratio number, the brighter the room is without adding excessive amounts of heat.

South-facing windows designed to admit solar heat during the heating season should admit the maximum amount of solar radiation. High SHGCs on south windows will not typically result in indoor overheating problems during the cooling season because of the lower solar radiation levels on south-facing windows, especially those with overhangs.

On the other hand, skylights and east- and west-facing glass typically necessitate lower SHGCs because they transmit the most solar heat during summer months. The SHGC makes little difference on north-facing glass.

The general rules of thumb are as follows:

- South-facing glass should have a high SHGC (greater than 0.80) but only if the glass is properly shaded during the summer months.
- Skylights and east- and west-facing glass should have a low SHGC (less than 0.40).
- North-facing glass should have a low U-factor.

Passive Solar Collector Area (Aperture)

Optimization of the size of the solar collector (aperture) area is based on climate, system type, and building load. It is difficult to establish exactly. A myriad of software packages are available to fine-tune a passive solar system. The preliminary design

TABLE 11.4   PERFORMANCE MEASURES FOR DIFFERENT TYPES OF WINDOWS.

Window and Glazing Types	Center-of-Glass Thermal Properties		Solar Heat Gain Coefficient (SHGC)		Visible Transmittance (VT)		Visible Light-to-Solar Gain Ratio (L/SG)	
	R-Value	U-Factor	Total Window	Center of Glass	Total Window	Center of Glass	Total Window	Center of Glass
Single glazed, clear	0.91	1.10	0.79	0.86	0.69	0.90	0.87	1.04
Double glazed, clear	2.04	0.49	0.58	0.76	0.57	0.81	0.98	1.07
Double glazed, bronze	2.04	0.49	0.48	0.62	0.43	0.61	0.89	0.98
Double glazed, spectrally selective	2.04	0.49	0.31	0.41	0.51	0.72	1.65	1.75
Double glazed, low-e (e = 0.10)	3.13	0.32	0.26	0.32	0.31	0.44	1.19	1.38
Triple glazed, low-e (e = 0.15)	3.70	0.27	0.37	0.49	0.48	0.68	1.29	1.39

values in Table 11.5 (See pp. 388–392) can be used to approximate the SF during preliminary design.

These preliminary design values are based upon a building heat loss of the following:

Heating Degree-Days (°F)	Btu/(°F·day·ft²)
Less than 1000	9
1000 to 3000	8
3000 to 5000	7
5000 to 7000	6
Greater than 7000	5

If the heat loss is different than recommended, then the glazing area should be adjusted proportionately.

Example 11.4

Approximate the passive solar glazing area requirements and solar savings fraction for a 2000 ft² residence in Denver, Colorado.

a. Find the minimum and maximum solar collector (aperture) area.

From Table 11.5, the minimum and maximum solar collector (aperture) areas are 12% and 23% respectively:

$$2000 \text{ ft}^2 \cdot 12\% = 240 \text{ ft}^2$$
$$2000 \text{ ft}^2 \cdot 23\% = 460 \text{ ft}^2$$

b. Estimate the minimum and maximum length of a south-facing glass needed, based upon a glazing height of 6 ft-8 in (6.67 ft) for these collector (aperture) areas.

$$\text{length} = \text{area/height} = 240 \text{ ft}^2/6.67 \text{ ft} = 36 \text{ ft long}$$
$$\text{length} = \text{area/height} = 460 \text{ ft}^2/6.67 \text{ ft} = 69 \text{ ft long}$$

c. Estimate the solar savings fraction without night insulation for the minimum and maximum solar collector (aperture) areas.

From Table 11.5, the SF ranges from:

27% to 43%

d. Estimate the solar savings fraction with night insulation for the minimum and maximum solar collector (aperture) areas.

From Table 11.5, the SF ranges from:

47% to 74%.

Example 11.5

a. Approximate the passive solar glazing area requirements for a 16 ft × 20 ft south-facing family room in a residence in Denver, Colorado.

From Table 11.5, the minimum and maximum solar collector (aperture) areas are 12% and 23% respectively:

$$(16 \text{ ft} \times 20 \text{ ft}) \cdot 12\% = 38.4 \text{ ft}^2 \text{ glass area}$$
$$(16 \text{ ft} \times 20 \text{ ft}) \cdot 23\% = 73.6 \text{ ft}^2 \text{ glass area}$$

b. Estimate the number of 30 in × 80 in glazing units that must be placed on the south wall.

Number of glazing units at 12% SF: ft²/
[(30 in × 80 in)/144 in²/ft²] = 2.3 ≈ 2 or 3 units

Number of glazing units at 23% SF: 73.6 ft²/
[(30 in × 80 in)/144 in²/ft²] = 4.4 ≈ 4 or 5 units

Passive Solar Heating System Performance

As introduced earlier in this chapter, the solar fraction (SF) of a specific solar system is the actual amount of useful solar energy received divided by the total energy required to heat the building without the solar contribution. On a passive solar system, the SF is dependent on several factors (i.e., building load, collector area, thermal storage configuration and sizes, operation, and climate). As the SF increases, the building relies more on solar gains. A system with a large area of collector aperture (glass) and properly sized thermal mass yields a higher SF. However, attempting to achieve a high SF by adding collector area without having sufficient thermal mass will result in overheating on clear days, especially at both ends of the heating season when outdoor air temperature is moderate and the space heating load is small.

In extremely cold weather when the space heating load is at its maximum, all of the heat collected by a passive solar heating system is used for space heating. However, in mild weather, only a portion of the system is used because only a small amount of space heating is needed. This means a fraction of the total area of collector aperture has a greater seasonal output *per unit area* than the total collector aperture area. This difference in performance is because a fraction of the collector aperture generates useful energy during most of the season, whereas a fraction of the collector aperture generates useful energy only during a few weeks a year (when the weather is extremely cold). This difference in performance is evident in Table 11.6, which shows that as a system's SF increases, the annual solar gain to accomplish the SF decreases. As a result, it makes little economic sense to strive to achieve an SF = 1.0. Also, the increase in performance with use of night insulation is evident on reviewing this table. On a thermal storage wall, use of a selective surface will see a similar increase in performance.

Thermal Storage and Distribution

A *thermal mass* is a material with the capacity to absorb and store thermal energy and later release significant amounts of heat. It typically includes materials such as concrete, masonry, and earth. Water is also a widely available thermal storage

TABLE 11.6 PERFORMANCE OF A DIRECT-GAIN PASSIVE SOLAR SYSTEM AT A LOCATION IN DENVER/BOULDER, COLORADO, BASED ON NATURAL GAS AT $0.80/CCF AND 83 000 BTU/CCF; AND ELECTRICITY AT $0.085/kWh.

| Solar Fraction | Annual Solar Gain | Annual Energy Cost Savings ($/ft² per year) | | | |
| | | Natural Gas | | Electric Resistance | Geothermal Heat Pump |
SF(%)	Btu/ft²·yr	80% Efficiency	92% Efficiency	100% Efficiency	COP ≈ 3.0
Direct Gain (No Insulation)					
0.10	67 379	$0.81	$0.71	$1.68	$0.56
0.20	48 128	$0.58	$0.50	$1.20	$0.40
0.30	25 869	$0.31	$0.27	$0.64	$0.22
0.40	6618	$0.08	$0.07	$0.16	$0.06
Direct Gain with R-9 Night Insulation from 5 PM to 8 AM					
0.10	131 149	$1.58	$1.37	$3.27	$1.09
0.20	117 914	$1.42	$1.24	$2.94	$0.98
0.30	104 678	$1.26	$1.10	$2.61	$0.87
0.40	93 850	$1.13	$0.98	$2.34	$0.78
0.50	77 005	$0.93	$0.81	$1.92	$0.64
0.60	57 152	$0.69	$0.60	$1.42	$0.48
0.70	39 104	$0.47	$0.41	$0.97	$0.33
0.80	22 259	$0.27	$0.23	$0.55	$0.19
0.90	7821	$0.09	$0.08	$0.19	$0.07

material. Buildings constructed of high thermal mass have a distinctive thermal advantage because they absorb energy slowly and hold it for longer periods of time than do less massive materials, which allows energy to be stored and moderates indoor temperature fluctuations (temperature swings).

The R_t-values and U-factors used to describe heat transfer have only a modest effect on buildings with large amounts of thermal mass. R-values and U-factors alone are unsatisfactory as the lone measure used in describing the heat transfer properties of construction assemblies with significant amounts of thermal mass. Typically, specific heat capacity (described earlier in this chapter) is used to better characterize wall construction with thermal mass. Building materials such as concrete, masonry, ceramic tile, gypsum wallboard, and earth have relatively high heat capacities and are said to have high thermal mass. Lightweight materials such as timber, insulation, and engineered wood have relatively low heat capacities and have low thermal mass.

A material with a high thermal mass is generally not a good thermal insulator. In buildings, thermal mass should not be used as a substitute for insulation. These materials perform different functions. Insulation impedes heat flow into or out of the building envelope while thermal mass stores and releases heat inside the building. Thermal mass should be used in conjunction with insulation. The thermal mass should be located so that it is exposed to the building interior. Insulation should be located on the outside of the thermal mass to limit heat transfer

to the outdoors. Table 11.7 lists the thermal storage properties of common construction materials.

There are two basic thermal storage strategies using thermal mass. *Direct* thermal storage materials (e.g., masonry or tiles) are placed to be directly in the sunlight (for at least 4 hr) so that direct (beam) radiation enters them quickly. *Diffuse* thermal storage materials are placed throughout the building. They can absorb heat by radiation, the reflectance of sunlight as

TABLE 11.7 THERMAL STORAGE PROPERTIES OF COMMON CONSTRUCTION MATERIALS.

Material	Density lb/ft³	Heat Capacity Btu/in·ft²·°F
Poured Concrete	120 to 150	2.0 to 2.5
Clay Masonry		
Molded brick	120 to 130	2.0 to 2.2
Extruded brick	125 to 135	2.1 to 2.3
Pavers	130 to 135	2.2 to 2.3
Concrete Masonry		
Block	80 to 140	1.3 to 2.3
Brick	115 to 140	1.9 to 2.3
Pavers	130 to 150	2.2 to 2.5
Gypsum Wallboard	50	0.83

Source: National Concrete Masonry Association.

it bounces around a room, and air heated elsewhere in the building (e.g., from a sunspace).

Masonry or Concrete Floors Flooring of concrete, tile, brick, or stone materials, exposed to direct sunlight, is probably the most common form of thermal storage selected for passive solar buildings. Masonry materials have high thermal capacity and their natural dark color aids in the absorption of sunlight. They also provide an attractive and durable floor surface, are widely available, and are readily accepted by contractors and building occupants. Masonry's effectiveness can be inhibited if occupants place furniture and carpets over the floors. Thus, masonry floors should be used only in the areas where direct heat gain and storage is required.

Masonry or Concrete Walls Many buildings, especially low-rise commercial buildings, are constructed with concrete or masonry walls that can provide excellent thermal mass to absorb excess solar heat and stabilize indoor temperatures. In most climates masonry walls are most energy efficient when they are insulated on the outside of the building, which allows them to absorb excess heat within the building, without wicking it away to the outside. However, there are barriers to using this technique. It is not common practice for contractors, and it may seem redundant to cover up an existing excellent weather surface. Insulated masonry also adds extra width to a wall, making it difficult to finish at the edges of windows, roofs, and doors.

Fortunately, new technologies have lowered the cost and increased the options for insulated masonry. Various foam insulations are available in panels that can be adhered directly to the masonry surface and then protected with a troweled or sprayed-on weathering skin; masonry-insulated structural panels are also available. Manufacturers are also developing self-insulating masonry materials that both increase the thermal capacity of the building and slow the flow of heat through the walls.

Double Gypsum Board The thermal capacity of a building can be increased by simply increasing the thickness of the gypsum board used on interior wall surfaces of the building or by using thicker gypsum board products. Increasing the thickness of all of the wall surfaces can raise the thermal capacity of the building for little additional material cost and practically no labor cost. It has the added benefits of increasing the fire safety and acoustic privacy of interior spaces. This diffuse thermal mass approach depends on effective convective airflows because room air is the heat transfer medium. To effectively charge the walls, temperatures within the space must be allowed to fluctuate a little more than standard design assumptions, on the order of 5°F above and below the thermostat setting.

Water Storage Containers Water, stored in plastic, fiberglass, or glass-lined steel containers, is not only the lowest cost, widely available thermal storage material, but it also has the highest thermal energy storage capability, about twice that of common masonry materials. Water also has the advantage that convection currents distribute heat more evenly throughout the medium. Passive solar designers have experimented with a wide variety of water storage containers built primarily into walls. Creative solutions include enclosing water containers in seating boxes under south windows or using water as an indoor feature such as a large tropical aquarium, pond, or pool.

Thermal Storage Wall A thermal storage wall is a south-facing masonry wall covered with glass. The glass and airspace keep the heat from radiating back to the outside. Heat is transferred by conduction as the masonry surface warms up and is slowly delivered to the building some hours later.

Thermal storage walls can provide carefully controlled solar heat to a space without the use of windows and direct sunlight, thus avoiding potential problems from glare and overheating, if thermal storage is inadequate. The masonry wall is part of the building's structural system, effectively lowering costs. The inside, or discharge, surface of the thermal storage wall can be painted white to enhance lighting efficiency within the space. However, the outside large dark walls sheathed in glass must be carefully designed for both proper performance and aesthetics.

Phase-Change Materials *Phase-change materials (PCMs)* store energy while maintaining constant temperatures, using chemical bonds to store and release latent heat. PCMs include solid–liquid Glauber's salt, paraffin wax, and solid–solid linear crystalline alkyl hydrocarbons. Although these compounds are fairly inexpensive, the packaging and processing necessary to get consistent and reliable performance from them is complicated and costly.

Phase-change materials use chemical bonds to store and release heat. The thermal energy transfer occurs when a material changes in state or phase, from a solid to a liquid, or from a liquid to a solid. The types include the following:

Solid–Liquid PCMs Solid–liquid PCMs perform like conventional storage materials; their temperature rises as they absorb solar heat. Unlike conventional (sensible) storage materials, when they reach the temperature at which they change phase (their melting point), they absorb large amounts of heat without getting hotter. When the ambient temperature in the space around the PCM material drops, the PCM solidifies, releasing its stored latent heat. PCMs absorb and emit heat while maintaining a nearly constant temperature. Within the human comfort range of 68° to 86°F (20° to 30°C), latent thermal storage materials are very effective. They store 5 to 14 times more heat per unit volume than sensible storage materials such as water, masonry, or rock.

Solid–Solid PCMs Solid–solid PCMs absorb and release heat in the same manner as solid–liquid PCMs. These materials do not change into a liquid state under normal conditions. They merely soften or harden. Relatively few of the solid–solid PCMs that have been identified are suitable for thermal storage applications.

Liquid–Gas PCMs Liquid–gas PCMs are not yet practical for use as thermal storage. Although they have a high heat of transformation, the increase in volume during the phase change from liquid to gas makes their use impractical. The PCM applications described next are with liquid–solid materials.

TABLE 11.8   THERMAL STORAGE PROPERTIES OF PHASE-CHANGE MATERIALS.

Material	Density lb/ft^3	Heat of Fusion Btu/lb	Melting Temperature °F	Heat Storage Capacity at a 50°F Temperature Difference	
				Btu/lb	Btu/ft^3
Phase-Change Materials					
Glauber's salt, $Na_2SO_4 \cdot 10 H_2O$	91	108	88 to 90	108*	9828*
Hypo, $Na_2S_2O_3 \cdot 5 H_2O$	104	90	118 to 120	90*	9360*
Calcium chloride, $CaCl_2 \cdot 6 H_2O$	102	75	84 to 102	75*	3825*
Paraffin	51	75	112	75*	7650*
Construction Materials (for Comparison)					
Water	1.0	—	—	50	3120
Rock	0.2	—	—	10	1000

*Neglects sensible storage.

The main application for PCMs is when space restrictions limit larger thermal storage units in direct-gain or sunspace passive solar systems. Although PCMs can store 5 to 14 times more heat per unit volume than traditional materials, they are usually more expensive than conventional heat storage materials. Phase-change drywall or wallboard is an exciting type of building-integrated heat storage material that has only been produced for research. This type of gypsum wallboard incorporates phase-change materials within its structure to moderate the thermal environment within the building. Table 11.8 lists properties of phase-change materials.

Use of Thermal Mass

Appropriate use of thermal mass throughout a building can make an enormous difference to comfort and heating and cooling costs. A building with high thermal mass uses less energy than a similar building with low thermal mass because of the reduced heat transfer through the massive elements and the tempered temperature swings. During summer, thermal mass absorbs heat, keeping the house comfortable. In winter, the same thermal mass can store the heat from the sun or heaters to release it at night, helping the home stay warm. Thermal mass is particularly advantageous where there is a large difference between day and night outdoor temperatures. Providing supplementary materials dedicated exclusively for thermal mass can be costly unless it is integrated in the building structure (e.g., use of thick gypsum board or high-density concrete floor systems). The most cost-effective method usually is to take advantage of thermal mass in the building structure.

To adequately temper indoor temperature swings, thermal mass should be exposed to the interior air and insulated from outdoor temperature variations, especially in extreme climates. For thermal mass to be effective in buildings, air must circulate freely through the building to carry the heat from the thermal mass to the spaces where it is needed. In large commercial buildings, fans and ducts are used. In homes, natural convection will usually circulate the air sufficiently but doors must be left open for this approach to work. In passive solar buildings, concrete, masonry, ceramic tile, earth, or even water-filled walls or phase-change materials can provide a method of storing the solar energy.

Table 11.9 lists the thermal mass required in a direct-gain system. Mass requirements should be increased by 4 if mass is not in direct sunlight.

Example 11.6

A 12 ft by 24 ft family room will be heated with a direct-gain system having a solar savings fraction of 50%. The collector (glass) area will be about 58 ft^2. Determine the mass requirements for this system.

a. For a 4 in thick concrete floor.

From Table 11.9, at an SF of 50% and for a 4 in thick concrete slab, 3.1 ft^2 of concrete per ft^2 of solar collection (glazing) area are required.

58 ft^2 collector (glass) area · 3.1 ft^2

= 180 ft^2 of concrete surface in direct sunlight

(720 ft^2 of concrete surface if not in direct sunlight)

b. For water storage.

From Table 11.9, at an SF of 50% and for water, 3.6 gal or 0.48 ft^3 of water container per ft^2 of solar collection (glazing) area are required.

58 ft^2 collector (glass) area · 3.6 gal = 209 gal of water

(836 gal of water if not in direct sunlight)

or

58 ft^2 collector (glass) area · 0.48 ft^3 = 28 ft^3 of water

(112 ft^3 of water if not in direct sunlight)

TABLE 11.9   THERMAL MASS REQUIRED IN A DIRECT-GAIN SYSTEM. INCREASE THE MASS REQUIREMENT BY FOUR IF MASS IS NOT IN DIRECT SUNLIGHT.

Solar Savings Fraction (SF) %	Recommended Amounts of Thermal Storage Per ft^2 of Solar Collection (Glazing) Area					
	Water*			Concrete**		
	lb	gal	ft^3	lb	ft^3	ft^2 at 4 in thickness
30	18	2.2	0.29	90	0.63	1.9
40	24	2.9	0.39	120	0.83	2.5
50	30	3.6	0.48	150	1.04	3.1
60	36	4.3	0.57	180	1.25	3.8
70	42	5.0	0.67	210	1.46	4.4
80	48	5.8	0.78	240	1.67	5.0
90	54	6.5	0.87	270	1.88	5.6

*Water weighs 62.4 lb/ft^3.
**Standard concrete weighs about 144 lb/ft^3.

Thermal Lag

Buildings with high thermal mass will experience a significant delay or lag in peak heating and cooling loads. Buildings with low thermal mass will delay peak loads only slightly. This delay is known as thermal lag. *Thermal lag* is the time required for a material to reach a new constant rate of heat gain or loss. Correct use of thermal mass can delay heat flow through the building envelope by as much as 10 to 12 hr, producing a warmer building at night in winter and a cooler building during the day in summer. Buildings with low thermal mass will experience a thermal lag of only a few hours. This effect is evidenced during the summer in an unconditioned light frame residence when the midafternoon's heat peaks indoors about dinnertime.

A greater thermal lag is an indication that the peak load for both heating and cooling are reduced. It results in less energy consumption and a decrease in required capacity of heating and cooling equipment, resulting in significant immediate cost savings. Thermal mass can be used to shift energy demand to off-peak time periods when cooling and heating loads are less and when utility rates are lower. This shift of energy demand is especially helpful in commercial buildings that have high occupancies during the periods of peak heating and cooling loads.

An improper application of thermal mass can have a huge influence on energy conservation and comfort. It can make worse the temperature extremes of a climate. It can radiate heat all night during a summer heat wave, making a space unbearably hot, or it can absorb all the heat produced in the winter, making a space intolerably cold. To be effective, thermal mass must be integrated into the building with sound design techniques.

In passive solar systems, the thermal mass must be exposed directly to winter sunlight if it is to be the most effective. Thermal mass exposed to sunlight is 3 to 4 times more effective in absorbing the sun's heat than thermal mass in the shade. Thus, positioning the thermal mass next to the solar glazing is crucial. Designers must plan for furniture, throw rugs, and artwork that can shade thermal mass, reducing its effectiveness.

In residential construction, integration of thermal mass typically includes the use of materials such as concrete, tile, brick, adobe, gypsum wallboard, stone, or masonry floors. Thick gypsum wallboard can be used throughout the spaces that are thermally linked to solar windows or clerestories. This dispersed thermal mass approach relies mostly on convection currents to transfer the solar heat gain to the wall surfaces. The thin concrete topping often placed between floors for fire and acoustic separation of multifamily apartments and condominiums can be made from high-density concrete to create thermal mass. However, this floor surface should not be carpeted but should be finished with materials that absorb heat such as tile, stone, or brick masonry. Resilient floor tiles may work if they do not insulate (e.g., cork) and are not too reflective.

Poured concrete or the precast slabs and shear walls used in commercial buildings provide high thermal mass. A 4 in (100 mm) slab thickness is the most effective, because there is little improvement in performance beyond that thickness. A 3 in (75 mm) thick slab provides over 90% of the performance of a 4 in (100 mm) thick slab. High-density concrete provides more thermal mass than low-density concrete because it has a higher specific heat capacity.

Improving airflow across the surface is usually the most cost-effective means of using thermal mass. Maximizing the exposed surface area can increase the amount of heat transferred to and from walls and floor slabs. The corrugated profile of steel decking sections used in concrete composite floors can increase surface area by 15% to 50%.

The Passive Solar Absorber

A *solar absorber* is the component of a solar system that absorbs solar radiation and converts it to heat. It is a surface treatment (e.g., paint, coating, film) that has high absorptivity (α) at wavelengths corresponding to that of the solar spectrum. In a passive solar system, sunlight passing through glass is absorbed by the solar absorber and then stored in thermal mass. The outside surface of the mass wall should have an absorptivity greater than 0.90. A flat, dark-colored surface is a good absorber. Painting the wall with black or dark-green absorptive-type paint will help the wall to absorb the sun's heat.

A *selective surface* is a unique type of solar absorber material (e.g., black chrome, black nickel, or black crystal) that is highly absorbent in the solar spectrum but also highly reflective in the infrared, resulting in a surface where absorptivity is high and emissivity is low. Unlike flat black painted surfaces, selective surfaces are able to restrict the energy lost through radiation, thus making them more efficient. A selective surface must be protected, so it is used on thermal storage walls and the absorber plates of solar collectors.

On a thermal storage wall, a selective surface is a copper foil (for high conductivity) with glue on the back, a chrome coating (for low emissivity at subvisible wavelengths), and a further coating of copper oxide (which is black and provides the high absorptivity at visible wavelengths). It should be applied to achieve 100% adhesion to the thermal mass wall.

A selective surface applied to a thermal storage wall improves its performance by reducing the amount of infrared energy radiated back through the glass. Use of selective surface will increase the efficiency of the wall by approximately 30% in Denver and up to 60% in more northerly climates. A selective surface source is MTI Solar Inc., 220 Churchill Ave. Somerset, NJ 08873; phone: (732) 246-1000.

Solar Control

Sun control of glass/window areas is an important design consideration in passive solar design. Unprotected (e.g., poorly shaded) glass is the largest source of heat gain in a well-insulated home, which can lead to severe overheating during the summer months.

Shading   *External shading devices* (e.g., overhangs, awnings, exterior shutters, trees, trellises, and solar screens) can control the entry of solar heat into the window/glass itself by providing internal shade. *Internal shading devices* (e.g., roller shades, curtains, blinds, indoor shutters) can control solar heat into the building interior somewhat by reflecting solar energy back through the window, but they also let in some of the undesirable heat that is trapped inside the window. As a result, exterior shading devices are more effective than internal devices at blocking solar heat because energy is reflected or blocked before it enters the building envelope.

Shading requirements vary by climate and house orientation. In climates where winter heating is necessary, shading devices should allow full winter sun to penetrate but keep out summer sun. This is most simply achieved on south-facing walls. East- and west-facing windows require different shading solutions to south-facing windows. In climates where no heating is required, shading of the whole home and outdoor spaces will improve comfort and save energy.

There are some disadvantages to shading devices; for example, shutters, solar screens, curtains, and blinds tend to make rooms dark. Exterior shading devices may create problems with the building's aesthetics and are sometimes expensive to build. Additionally, it is impractical to construct roof overhangs to effectively shade east- and west-facing windows.

Overhangs   *Overhangs* are shading devices integrated into the building envelope that block the high summer sun, but allow the lower winter sun to strike the building walls, windows, and doors. Overhangs can be solid or louvered and can be fixed, operable, and/or removable. They can support vegetation or combine all of these aspects. Shutters, eaves, trellises, light shelves, solar screens, and awnings can be designed to serve the same purpose as an overhang. See Photos 11.36 and 11.37.

Overhangs are most effective for south-facing elements (in the northern hemisphere) during the middle of the day. They are also particularly effective for exposures oriented within 30° of true south (true north in the southern hemisphere); beyond this orientation, effectiveness starts to decrease significantly. Overhangs usually only affect the amount of direct solar radiation that strikes a surface.

As shading devices, overhangs have limitations. First, they do not directly reduce diffuse and reflected radiation gains, which can be significant in some climates. Overhangs cannot

PHOTO 11.36   An overhang providing summer shade. (Courtesy of DOE/NREL)

PHOTO 11.37   A view of a commercial building under construction with overhangs over windows that provide summer shading. (Courtesy of DOE/NREL)

adequately shade east- and west-facing windows. A roof overhang on a two-story building will not shade a first-story window.

Fixed shading devices will always be a design compromise, because the sun's angle is the same in spring and autumn. Solar gain may be needed through a window in March, but not in September when the air temperature is warmer. As a result, there is as yet no universal formula for sizing overhangs. Although one overhang formula works well for some building sites, it can be completely unsuitable for others.

A properly sized overhang will adequately provide summer shade without affecting winter sunlight. The overhang must extend long enough to block much of the summer sun, yet short enough to allow the winter sun to be received by the solar glazing (e.g., south-facing window). To determine the angle based on the sun's angle at the winter solstice and summer solstice, the following methods may be used:

- To get full shade for 3 weeks in summer and full sun for 3 weeks in winter, use a summer ray that is about 2.5° less than at summer solstice and a winter ray that is about 2.5° more than at winter solstice.
- To get full shade for about 6 weeks in summer and full sun for 6 weeks in winter, use a summer ray that is about 5° less than at summer solstice and a winter ray that is about 5° more than at winter solstice.

Awnings An *awning* is a type of overhang that works by blocking high-angle sunlight like a visor on a baseball cap. On buildings, awnings can cover individual windows or sections of outside walls. They are most effective on the south side of the building. Awnings come in a variety of shapes, sizes, and colors to match many building designs. Some awnings stay in a fixed position, while others are rolled up in the winter to allow low-angle sun to reach the building. Awnings are an effective method for adding shading to an existing building. A disadvantage of an awning is that it tends to hang low over the window, which blocks the view from the top portion of the window.

Solar Screens Exterior *solar screens* are a special type of screen mesh (similar to an insect screen) that is installed on the exterior of a window to limit direct solar radiation entering the window. These devices are often called *sunscreens, shade cloths,* or *solar shields.* Solar screens are made from lightweight and durable aluminum or plastic mesh. Unlike insect screens, solar screens are specially made to block out a portion of the solar energy. Usually, this is between 50 and 90% of the energy striking the outside of the window. The term *shading coefficient* describes the amount of solar energy that penetrates the screen. A lower shading coefficient corresponds to less solar energy getting through. The view through a solar screen is obscured somewhat.

Trees Properly positioned trees offer excellent natural shading because they cast shade over the exterior walls and roof.

They also will shade driveways, sidewalks, and patios that can reflect heat to the building. Because full-size trees provide more shade than miniature trees, a site plan should preserve as many existing trees as possible. New trees should be planted immediately after construction.

Deciduous trees offer canopies that are broad and dense only during cooling months. When the leaves fall in the winter, deciduous trees allow solar heat to reach the building. (Evergreens work well for north and northwest areas of the site.) The closer a tree is to the building and the faster it grows, the more hours of shade it will provide. To be effective, trees should be planted between 10 and 20 feet from the building.

Trees provide an additional cooling benefit. To keep themselves cool, trees pump water from the ground into their leaves. As this water evaporates from the surface of the leaves, it cools the tree. This evaporative cooling effect cools the surrounding area. Shrubs offer less shading, but they have several other advantages. They usually cost less, reach mature size more quickly, and require less space. Shrubs can shade walls and windows without blocking roof-mounted solar panels.

Trellises *Trellises* are permanent structures that partially shade the outside of a building. Clinging vines growing over the trellis add more shade and evaporative cooling. A special trellis to shade air conditioners, heat pumps, and evaporative coolers improves the equipment's performance. Be sure not to restrict airflow to the equipment.

Fast growing vines create shade quickly, while trees can take years to provide useful shade. Deciduous vines lose their leaves in winter, allowing the solar energy to strike the building. Trellises and climbing plants are a design solution that's attractive and flexible.

Window Shades Interior window shades, such as roller shades, blinds, and drapes, can reduce heat gain. However, shading devices are not as effective in blocking sunlight, as are exterior shading devices. The percentages of radiant energy that different types of internal shading devices transmit, reflect, or absorb are provided in Table 11.10. Interior shades and blinds work fairly well, with curtains only marginally effective.

Interior shades work by reflecting sunlight back out the window before it can turn into heat; blocking the movement of hot air from the area around the window into the room; and by

TABLE 11.10 THE PERCENTAGES OF RADIANT ENERGY THAT DIFFERENT TYPES OF INTERNAL SHADING DEVICES TRANSMIT, REFLECT, OR ABSORB.

Internal Shading Device	Transmittance	Reflection	Absorption
Roller shades	Up to 25	15 to 80	20 to 65
Vertical blinds	0	23	77
Venetian blinds	5	40 to 60	35 to 55

insulating the interior space from the hot surfaces and radiant temperature effect of the window glass and frame. Ideally, interior shades should

- Have a light-colored surface on the side that faces the window
- Fit tightly to prevent air movement into the room
- Be made of an insulating material
- Cover the entire window, including glass and frame

11.5 PASSIVE SOLAR SPACE COOLING

Several strategies can be employed to passively cool a building interior. These strategies work well in buildings with low to moderate heat gains such as dwellings and light commercial buildings with low occupant densities and little heat gain from lighting, appliances, and equipment. These strategies include the following.

Ventilation

A primary strategy for buildings without mechanical cooling in hot, dry climates is to employ natural ventilation. Prevailing summer winds are region and site specific. The designer should ascertain the direction of summer prevailing winds at the site and place operable windows toward that exposure (the windward side). Operable windows should also be placed on the opposite side of the building (the leeward side) to exhaust warm air and create cross-ventilation. If a room can have windows on only one side, two widely spaced windows should be used.

Casement windows offer the best airflow in comparison to single- or double-hung windows because they open fully and can trap the wind. Awning or hopper windows can also be fully opened. Awning windows offer the best rain protection and perform better than single- or double-hung windows, but airflow is directed to the ceiling. Insect screens on window openings decrease the velocity of slow breezes more than stronger breezes (i.e., about a 60% decrease at 1.5 mph and 28% decrease at 6 mph). Screening a porch with large openings will not reduce air speeds as much as screening the windows.

Wing walls are vertical solid panels that can be placed alongside of the windows perpendicular to the wall. A wing wall can be used to channel and funnel prevailing winds toward open windows. Properly positioned wing walls on the windward side of the building can be used to accelerate the natural wind speed from pressure differences created by the wing wall.

A *windcatcher* is an architectural device that has been used for many centuries to create natural ventilation in buildings. A windcatcher consists of a tall tower that is integrated into the building structure and opens into the building interior. It is capped and has several directional ports (openings) at the top. Windcatchers come in various designs (i.e., unidirectional, bidirectional, and multidirectional), depending on the direction of the summer prevailing winds. By closing all but the one facing away from the incoming wind, air is drawn upwards. This generates significant ventilation within the structure.

Thermal Mass Cooling

Buildings with large amounts of thermal mass can coast through warm periods of the day and remain relatively comfortable in hot, dry weather as long as nighttime temperatures fall at least 20°F (11°C) below the desired indoor temperature. High mass buildings can be cooled with night ventilation through open windows providing that interior furnishings and finishes (e.g., light furniture, drapes, carpeting, and so on) are minimized in the building. A high mass building should be kept closed (e.g., windows and doors closed) during the day and opened at night. During periods of the cooling season without the 20°F (11°C) nighttime temperature drop, mechanical ventilation (night flush ventilation) will generally be required. This concept was discussed in Chapter 7.

Thermal Chimney

A *thermal chimney* makes use of the buoyancy effect of air and the resulting natural convective currents to carry warm air out of a building. By creating a warm or hot zone with exterior inlet and exhaust outlets, air can be drawn into the building, ventilating the structure as the warm air is exhausted. Sunspaces can be designed to perform this function. The excessive heat generated in a south-facing sunspace during the summer can be vented at the top. By connecting lower vents to the living space coupled with open windows on the north side, air is drawn through the living space to be exhausted through the sunspace upper vents.

Thermal chimneys can be constructed in a narrow configuration (like a chimney) with a black metal absorber on the inside behind a glazed cover that can reach high temperatures and be insulated from the building. The chimney must terminate above the roof. A rotating metal scoop at the top that opens opposite to the wind direction allows heated air to exhaust without being overcome by the prevailing wind. The thermal chimney strategy can be integrated into open stairwells and atria.

Passive Downdraft Evaporative Cooling Tower

A *passive downdraft evaporative cooling (PDEC) tower* uses evaporation of moisture on top of the tower to contribute to indoor cooling. Water is introduced at the top of a tower by using evaporative cooling pads or by spraying water. Evaporation

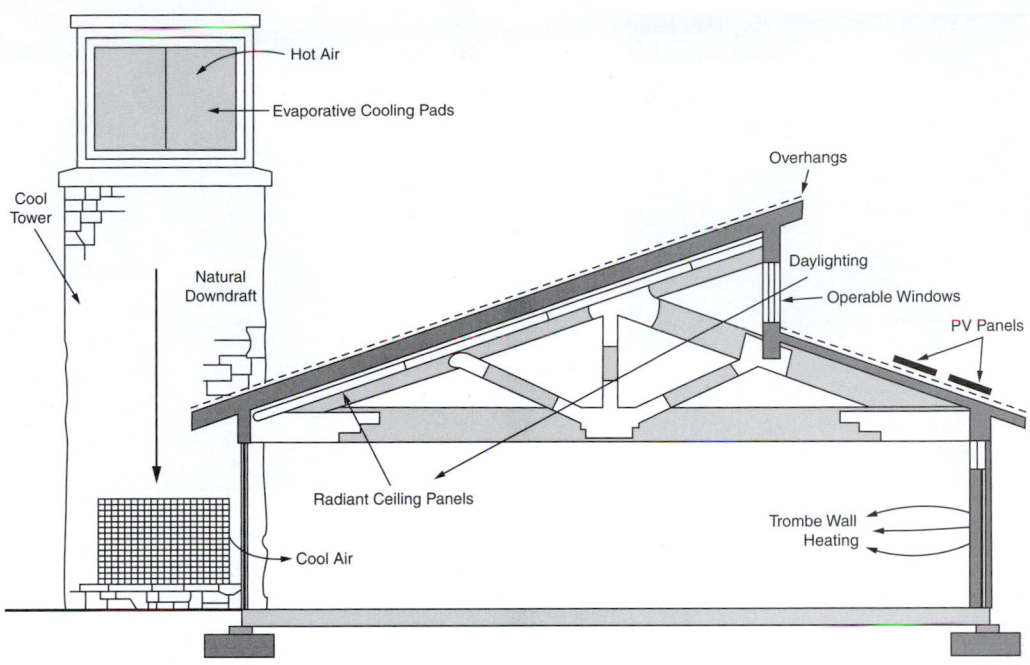

FIGURE 11.14 A cross section of the passive solar heated and cooled Zion National Park Visitor Center, in Springdale, Utah. (Courtesy of DOE/NREL/NPS)

cools the incoming air, causing a downdraft of cool air that is introduced into the building interior. Airflow can be increased by using a solar chimney on the opposite side of the building to help in venting hot air to the outside.

Earth cooling tubes, also known as *ground-coupled heat exchangers,* use the earth's nearly constant subterranean temperature to cool air that is drawn through tubes placed underground. They are often a viable and economical alternative to conventional heating, cooling, or heat pump systems as there are no compressors, chemicals, or burners and only blowers are required to move the air. Large (6 to 24 in/150 to 600 mm) plastic-coated metal pipes or plastic pipes serve as ducts that carry the air. They must be coated with an antimicrobial treatment and have a condensate drain to remove condensation to impede the growth of molds and bacteria within the tubes.

These strategies passively cool a building interior but do not dehumidify the indoor air. They work best in arid and semi-arid climates. In humid climates, mechanical refrigeration is typically necessary.

11.6   A MODEL SOLAR BUILDING

The U.S. National Park Service's (NPS) Zion National Park Visitor Center is an example of a 7600 ft^2 (706 m^2) visitor center and 1100 ft^2 (102 m^2) comfort station that effectively uses passive solar heating and cooling. These buildings are shown in Figure 11.14 and Photos 11.38 through 11.46.

Integrated into the buildings are thermal storage (Trombe) walls and a direct-gain system with thermal mass for passive solar heating and PDEC towers for natural ventilation cooling. Energy-efficient lighting with advanced controls and daylighting strategies (see Chapter 20) are also incorporated into the design. A roof-mounted photovoltaic solar electric system (see Chapter 25) provides up to 7.2 kW of electrical power, which reduces the amount of power purchased from the utility by about 30%. Annual savings at this facility are about $14 000 per year, from a 74.4% reduction in energy use, and about 310 000 lb of CO_2 emissions (2006 records).

PHOTO 11.38 A view of the sustainable Visitor Center at Zion National Park, Springdale, Utah. (Courtesy of DOE/NREL)

PHOTO 11.39 The south exposure of the Visitor Center at Zion National Park provides a well-suited location for trombe walls that provide most of the building's heat. In addition, a 7.2 kW photovoltaic array provides a significant portion of the electricity needed by the building and serves as an uninterrupted power supply for use during power outages. (Courtesy of DOE/NREL)

PHOTO 11.42 A close-up photograph of the PDEC tower of the Zion Visitor Center. (Courtesy of DOE/NREL)

PHOTO 11.40 The trombe wall, clerestory windows, and passive downdraft evaporative cooling (PDEC) tower of the Zion Visitor Center. (Courtesy of DOE/NREL)

PHOTO 11.43 The interior vents and thermal mass of one of the PDEC towers of the Zion Visitor Center. (Courtesy of DOE/NREL)

PHOTO 11.41 The trombe wall and clerestory windows of the Zion Visitor Center. (Courtesy of DOE/NREL)

PHOTO 11.44 Clerestory windows at the Zion Visitor Center provide daylighting and ventilation. (Courtesy of DOE/NREL)

PHOTO 11.45 Rooftop photovoltaic (PV) solar collectors at the Zion Visitor Center provides electricity. PV technology is covered in Chapter 25. (Courtesy of DOE/NREL)

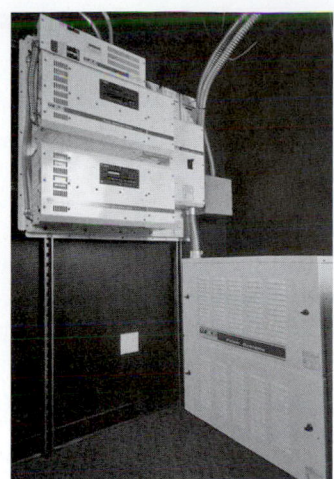

PHOTO 11.46 The PV solar system supplies power to inverters and battery storage that provide an uninterruptible power supply to the Zion Visitors Center. PV technology is covered in Chapter 25. (Courtesy of DOE/NREL)

STUDY QUESTIONS

11-1. What is insolation?

11-2. Explain the solar constant.

11-3. Describe the solar altitude and solar azimuth.

11-4. Describe the following forms of solar radiation and explain how they differ:

    a. Direct (beam) radiation

    b. Diffuse radiation

    c. Reflected radiation

11-5. With respect to a building, what is the greenhouse effect?

11-6. Describe the three basic types of solar systems used in buildings.

11-7. What is the solar fraction (SF)?

11-8. Identify the components of an active solar system. Explain each.

11-9. Describe the operation and differences of the following solar collectors:

    a. Flat-plate collector

    b. Evacuated-tube collector

    c. Concentrating collector

    d. Transpired collector

11-10. What are the three basic types of heat transfer mediums on active solar systems?

11-11. What is the function of a storage medium in the solar energy system?

11-12. What are the types of active liquid solar systems and how do they function?

11-13. What are the types of active air solar systems and how do they function?

11-14. Describe the five necessary elements of a building passive solar design.

11-15. Explain the types of passive solar space heating systems and how they function:

    a. Direct gain

    b. Thermal storage wall (vented and unvented)

    c. Roof pond

    d. Sunspace

    e. Thermosiphon

11-16. Describe the differences between sun-tempered and direct-gain solar systems.

11-17. What is the maximum percentage of south-facing window area that can be used to prevent overheating in a sun-tempered building?

11-18. What are the types of passive solar water heating systems and how do they function?

11-19. Why is supplementary heat often required in solar heating systems?

11-20. Explain the solar heat gain coefficient (SHGC).

11-21. Describe the recommended window performance (i.e., SHGC and U-factor) characteristics for glass used in buildings within 30° to 60° north latitudes for the following glass surfaces:

    a. South facing

    b. North facing

c. East facing

d. West facing

e. Horizontal or nearly horizontal (i.e., skylights)

11-22. Identify and describe strategies that can be employed to passively cool a building interior.

Design Exercises

11-23. Optimum collector tilt is tied to the heating application and collector location. Determine the optimum collector tilt for Boulder, Colorado (40° north latitude) for the following locations:

a. Year-round heating applications (i.e., generating domestic and process hot water)

b. Winter space heating

c. Summer heating (e.g., heating swimming pool water)

11-24. Optimum collector tilt is tied to the heating application and collector location. Determine the optimum collector tilt for Boulder, Colorado (40° north latitude) for the following applications:

a. Year-round heating (i.e., generating domestic and process hot water)

b. Winter space heating

c. Summer heating (e.g., heating swimming pool water)

11-25. Optimum collector tilt is tied to the heating application and collector location. Determine the optimum collector tilt for Ely, Minnesota (48° north latitude) for the following applications:

a. Year-round heating (i.e., generating domestic and process hot water)

b. Winter space heating

c. Summer heating (e.g., heating swimming pool water)

11-26. Determine the optimum collector tilt for year-round solar heating (i.e., generating domestic and process hot water) at the following locations:

a. Fargo, North Dakota (47° north latitude)

b. Columbus, Ohio (40° north latitude)

c. Savannah, Georgia (32° north latitude)

d. Miami, Florida (26° north latitude)

e. Lihue, Hawaii (22° north latitude)

11-27. For year-round heating applications (i.e., generating domestic and process hot water), solar collectors should be tilted at an angle relative to the horizontal plane that is approximately equal to the latitude angle. Determine the closest architectural roof slope (i.e., $\frac{4}{12}$, $\frac{6}{12}$, and so forth) related for the optimum collector tilt for year-round heating at the following locations:

a. Fargo, North Dakota (47° north latitude)

b. Columbus, Ohio (40° north latitude)

c. Savannah, Georgia (32° north latitude)

d. Miami, Florida (26° north latitude)

e. Lihue, Hawaii (22° north latitude)

11-28. A residence in a temperate climate has a heating load of 720 000 Btu/day on a cold day during the heating season. A solar heating system at the location of the residence can collect, store, and deliver about 800 Btu/day per square foot of collector area. Approximate the collector area needed to fully meet this heating load on a sunny day.

11-29. A residence in a temperate climate has a heating load of 720 000 Btu/day on a cold day during the heating season. A solar heating system at the location of the residence can collect, store, and deliver about 800 Btu/day per square foot of collector area. Approximate the collector area needed to meet 50% of the heating load.

11-30. A small residence has a domestic hot water load of 105 gal/day at a 90°F temperature rise. A solar system at the location of the residence can collect, store, and deliver about 1350 Btu/day per square foot of collector area.

a. Approximate the collector area needed to fully meet this hot water load.

b. Approximate the required storage tank capacity based on 1.5 gal of water for each ft^2 of collector area.

11-31. A small residence has a domestic hot water load of 90 gal/day at an 80°F temperature rise. A solar system at the location of the residence can collect, store, and deliver about 1350 Btu/day per square foot of collector area.

a. Approximate the collector area needed to fully meet this hot water load.

b. Approximate the required storage tank capacity based on 1.5 gal of water for each ft^2 of collector area.

11-32. Approximate the passive solar glazing area requirements and solar savings fraction for a 2000 ft^2 residence at the location where you reside.

a. Find the minimum and maximum solar collector (aperture) area.

b. Estimate the solar savings fraction without night insulation for the minimum and maximum solar collector (aperture) areas.

c. Estimate the solar savings fraction with night insulation for the minimum and maximum solar collector (aperture) areas.

11-33. Approximate the passive solar glazing area requirements and solar savings fraction for a 2000 ft^2 residence in Pittsburgh, Pennsylvania.

a. Find the minimum and maximum solar collector (aperture) area.

b. Estimate the solar savings fraction without night insulation for the minimum and maximum solar collector (aperture) areas.

c. Estimate the solar savings fraction with night insulation for the minimum and maximum solar collector (aperture) areas.

11-34. Approximate the passive solar glazing area requirements and solar savings fraction for a 2000 ft² residence in Seattle, Washington.

a. Find the minimum and maximum solar collector (aperture) area.

b. Estimate the solar savings fraction without night insulation for the minimum and maximum solar collector (aperture) areas.

c. Estimate the solar savings fraction with night insulation for the minimum and maximum solar collector (aperture) areas.

11-35. Approximate the passive solar glazing area requirements and solar savings fraction for a 2000 ft² residence in Phoenix, Arizona.

a. Find the minimum and maximum solar collector (aperture) area.

b. Estimate the solar savings fraction without night insulation for the minimum and maximum solar collector (aperture) areas.

c. Estimate the solar savings fraction with night insulation for the minimum and maximum solar collector (aperture) areas.

11-36. Approximate the passive solar glazing area requirements and solar savings fraction for a 2000 ft² residence in New York City, New York.

a. Find the minimum and maximum solar collector (aperture) area.

b. Estimate the solar savings fraction without night insulation for the minimum and maximum solar collector (aperture) areas.

c. Estimate the solar savings fraction with night insulation for the minimum and maximum solar collector (aperture) areas.

11-37. A 14 ft by 20 ft master bedroom will be heated with a sun-tempered system. Determine the following:

a. The maximum percentage of south-facing glass area that can be used to prevent overheating

b. The maximum area (ft²) of south-facing glass area that can be used to prevent overheating

11-38. A 4.5 m by 6.75 m master bedroom will be heated with a sun-tempered system. Determine the following:

a. The maximum percentage of south-facing glass area that can be used to prevent overheating

b. The maximum area (m²) of south-facing glass area that can be used to prevent overheating

11-39. A 14 ft by 20 ft master bedroom will be heated with a direct-gain system having a solar savings fraction of 70%. The collector (glass) area will be about 60 ft². Determine the mass requirements for this system.

a. For a 4 in thick concrete floor

b. For water storage

11-40. A 20 ft by 32 ft room will be heated with a direct-gain system having a solar savings fraction of 40%. The collector (glass) area will be about 60 ft². Determine the mass requirements for this system.

a. For a 4 in thick concrete floor

b. For water storage

11-41. A 6 m by 10 m room will be heated with a direct-gain system having a solar savings fraction of 40%. The collector (glass) area will be about 6.5 m². Determine the mass requirements for this system.

a. For a 100 mm thick concrete floor

b. For water storage

11-42. A 14 ft by 20 ft master bedroom will be heated with a thermal storage wall. It should have approximately 0.2 to 0.3 ft² of thermal mass wall surface per ft² of floor area being heated. Determine the following:

a. The maximum percentage of south-facing glass area

b. The maximum area (ft²) of south-facing glass area

11-43. A 4.5 m by 6.25 m master bedroom will be heated with a thermal storage wall. It should have approximately 0.2 to 0.3 m² of thermal mass wall surface per m² of floor area being heated. Determine the following:

a. The maximum percentage of south-facing glass area

b. The maximum area (m²) of south-facing glass area

TABLE 11.2 DAILY SOLAR INSOLATION FOR SELECTED CITIES IN THE UNITED STATES. FROM U.S. DEPARTMENT OF ENERGY.

State	City	Daily Solar Insolation					
		Btu/ft²·day			kWh/m²·day		
		High	Low	Average	High	Low	Average
AK	Fairbanks	1865	674	1268	5.87	2.12	3.99
	Matanuska	1665	553	1128	5.24	1.74	3.55
AL	Montgomery	1490	1071	1344	4.69	3.37	4.23
AR	Bethel	1999	753	1211	6.29	2.37	3.81
	Little Rock	1681	1233	1490	5.29	3.88	4.69
AZ	Tucson	2358	1910	2088	7.42	6.01	6.57
	Page	2320	1795	2021	7.30	5.65	6.36
	Phoenix	2266	1837	2091	7.13	5.78	6.58
CA	Santa Maria	2072	1722	1888	6.52	5.42	5.94
	Riverside	2018	1700	1865	6.35	5.35	5.87
	Davis	1935	1052	1621	6.09	3.31	5.10
	Fresno	1967	1087	1710	6.19	3.42	5.38
	Los Angeles	1951	1598	1786	6.14	5.03	5.62
	Soda Springs	2056	1398	1780	6.47	4.40	5.60
	La Jolla	1665	1363	1516	5.24	4.29	4.77
	Inyokern	2765	2183	2434	8.70	6.87	7.66
CO	Grandby	2374	1637	1808	7.47	5.15	5.69
	Grand Lake	1862	1131	1614	5.86	3.56	5.08
	Grand Junction	2015	1662	1859	6.34	5.23	5.85
	Boulder	1818	1411	1548	5.72	4.44	4.87
DC	Washington	1490	1071	1344	4.69	3.37	4.23
FL	Apalachicola	1900	1563	1745	5.98	4.92	5.49
	Miami	1989	1605	1786	6.26	5.05	5.62
	Gainesville	1846	1497	1675	5.81	4.71	5.27
	Tampa	1958	1672	1802	6.16	5.26	5.67
GA	Atlanta	1640	1300	1506	5.16	4.09	4.74
	Griffin	1719	1354	1586	5.41	4.26	4.99
HI	Honolulu	2132	1776	1913	6.71	5.59	6.02
IA	Ames	1525	1185	1398	4.80	3.73	4.40
ID	Boise	1853	1058	1563	5.83	3.33	4.92
	Twin Falls	1722	1087	1494	5.42	3.42	4.70
IL	Chicago	1297	467	998	4.08	1.47	3.14
IN	Indianapolis	1595	810	1338	5.02	2.55	4.21
KN	Manhattan	1614	1150	1452	5.08	3.62	4.57
	Dodge City	1316	1678	1840	4.14	5.28	5.79
KY	Lexington	1897	1144	1570	5.97	3.60	4.94
LA	Lake Charles	1821	1363	1567	5.73	4.29	4.93
	New Orleans	1815	1154	1563	5.71	3.63	4.92
	Shreveport	1586	1230	1471	4.99	3.87	4.63
MA	East Wareham	1424	972	1268	4.48	3.06	3.99
	Boston	1357	950	1220	4.27	2.99	3.84
	Blue Hill	1392	1058	1287	4.38	3.33	4.05
	Natick	1468	982	1303	4.62	3.09	4.10
	Lynn	1462	740	1204	4.60	2.33	3.79
MD	Silver Hill	1497	1220	1420	4.71	3.84	4.47
ME	Caribou	1786	817	1332	5.62	2.57	4.19
	Portland	1662	1131	1433	5.23	3.56	4.51
MI	Sault Saint Marie	1535	740	1335	4.83	2.33	4.20
	East Lansing	1497	858	1271	4.71	2.70	4.00
MN	Saint Cloud	1726	1122	1440	5.43	3.53	4.53

State	City	Daily Solar Insolation					
		Btu/ft^2·day			kWh/m^2·day		
		High	Low	Average	High	Low	Average
MO	Columbia	1748	1262	1503	5.50	3.97	4.73
	St. Louis	1548	1030	1392	4.87	3.24	4.38
MS	Meridian	1544	1157	1408	4.86	3.64	4.43
MT	Glasgow	1897	1300	1637	5.97	4.09	5.15
	Great Falls	1811	1163	1567	5.70	3.66	4.93
	Summit	1643	750	1268	5.17	2.36	3.99
NM	Albuquerque	2275	1973	2151	7.16	6.21	6.77
NE	Lincoln	1716	1392	1522	5.40	4.38	4.79
	Omaha	1678	1354	1557	5.28	4.26	4.90
NC	Cape Hatteras	1846	1490	1687	5.81	4.69	5.31
	Greensboro	1605	1271	1497	5.05	4.00	4.71
ND	Bismarck	1741	1262	1592	5.48	3.97	5.01
NJ	Sea Brook	1513	1017	1338	4.76	3.20	4.21
NV	Las Vegas	2266	1856	2037	7.13	5.84	6.41
	Ely	2059	1745	1900	6.48	5.49	5.98
NY	Binghamton	1249	515	1004	3.93	1.62	3.16
	Ithaca	1452	728	1204	4.57	2.29	3.79
	Schenectady	1246	804	1128	3.92	2.53	3.55
	Rochester	1341	502	1052	4.22	1.58	3.31
	New York City	1579	963	1297	4.97	3.03	4.08
OH	Columbus	1672	845	1319	5.26	2.66	4.15
	Cleveland	1522	855	1252	4.79	2.69	3.94
OK	Stillwater	1754	1341	1586	5.52	4.22	4.99
	Oklahoma City	1989	1583	1776	6.26	4.98	5.59
OR	Astoria	1513	632	1182	4.76	1.99	3.72
	Corvallis	1815	604	1281	5.71	1.90	4.03
	Medford	1856	642	1433	5.84	2.02	4.51
PA	Pittsburgh	1332	461	1042	4.19	1.45	3.28
	State College	1411	887	1243	4.44	2.79	3.91
RI	Newport	1490	1138	1344	4.69	3.58	4.23
SC	Charleston	1818	1344	1608	5.72	4.23	5.06
SD	Rapid City	1878	1449	1662	5.91	4.56	5.23
TN	Nashville	1652	998	1414	5.20	3.14	4.45
	Oak Ridge	1608	1023	1389	5.06	3.22	4.37
TX	San Antonio	1869	1478	1684	5.88	4.65	5.30
	Brownsville	1745	1405	1563	5.49	4.42	4.92
	El Paso	2358	1865	2136	7.42	5.87	6.72
	Midland	2012	1662	1853	6.33	5.23	5.83
	Fort Worth	1907	1525	1726	6.00	4.80	5.43
UT	Salt Lake City	1935	1201	1672	6.09	3.78	5.26
	Flaming Gorge	2107	1741	1853	6.63	5.48	5.83
VA	Richmond	1430	1071	1312	4.50	3.37	4.13
WA	Seattle	1535	508	1134	4.83	1.60	3.57
	Richland	1948	639	1411	6.13	2.01	4.44
	Pullman	1929	922	1503	6.07	2.90	4.73
	Spokane	1757	369	1424	5.53	1.16	4.48
	Prosser	1973	972	1598	6.21	3.06	5.03
WI	Madison	1541	1042	1363	4.85	3.28	4.29
WV	Charleston	1309	785	1160	4.12	2.47	3.65
WY	Lander	2164	1748	1926	6.81	5.50	6.06

TABLE 11.5 PRELIMINARY DESIGN VALUES FOR PASSIVE SOLAR COLLECTOR SIZE.

	Percentage of Solar Glass to Floor Area (sca)			Solar Savings Fraction (SF)					
				No Night Insulation			R-9 Night Insulation		
	Minimum	Average	Maximum	Minimum	Average	Maximum	Minimum	Average	Maximum
City	1	2	3	1	2	3	4	5	6
Alabama									
Birmingham	9	13.5	18	22	29.5	37	34	46	58
Mobile	6	9	12	26	35	44	34	47	60
Montgomery	7	11	15	24	32.5	41	34	46.5	59
Arizona									
Phoenix	6	9	12	37	48.5	60	48	61.5	75
Prescott	10	15	20	29	38.5	48	44	58	72
Tucson	6	9	12	35	46	57	45	59	73
Winslow	12	18	24	30	38.5	47	48	61	74
Yuma	4	6.5	9	43	54.5	66	51	64.5	78
Arkansas									
Fort Smith	10	15	20	24	31.5	39	38	51	64
Little Rock	10	14.5	19	23	30.5	38	37	49.5	62
California									
Bakersfield	8	11.5	15	31	40.5	50	42	54.5	67
Daggett	7	11	15	35	45.5	56	46	59.5	73
Fresno	9	13	17	29	37.5	46	41	53	65
Long Beach	5	7.5	10	35	46.5	58	44	58	72
Los Angeles	5	7	9	36	47	58	44	58	72
Mount Shasta	11	16	21	24	31	38	42	54.5	67
Needles	6	9	12	39	50	61	49	62.5	76
Oakland	7	11	15	35	45	55	46	59	72
Red Bluff	9	13.5	18	29	37.5	46	41	53	65
Sacramento	9	13.5	18	29	38	47	41	53.5	66
San Diego	4	6.5	9	37	49	61	46	60	74
San Francisco	6	9.5	13	34	44	54	45	58	71
Santa Maria	5	8	11	31	42	53	42	55.5	69
Colorado									
Colorado Springs	12	18	24	27	34.5	42	47	60.5	74
Denver	12	17.5	23	27	35	43	47	60.5	74
Eagle	14	21.5	29	25	30	35	53	65	77
Grand Junction	13	20	27	29	36	43	50	63	76
Pueblo	11	17	23	29	37	45	48	61.5	75
Connecticut									
Hartford	17	26	35	14	16.5	19	40	52	64
Delaware									
Wilmington	15	22	29	19	24.5	30	39	51	63
Washington	12	17.5	23	18	23	28	37	49	61
Florida									
Apalachicola	5	7.5	10	28	37.5	47	36	48.5	61
Daytona	5	6.5	8	30	40.5	51	36	49.5	63
Jacksonville	5	7	9	27	37	47	35	48.5	62
Miami	1	1.5	2	27	37.5	48	31	42.5	54
Orlando	3	4.5	6	30	41	52	37	50	63
Tallahassee	5	8	11	26	35.5	45	35	47.5	60
Tampa	93	49.5	6	30	41	52	36	49.5	63
West Palm Beach	1	2	3	30	40.5	51	34	46.5	59
Georgia									
Atlanta	8	12.5	17	22	29	36	34	46	58
Augusta	8	12	16	24	32	40	35	47.5	60
Macon	7	11	15	25	33	41	35	47	59
Savannah	6	9.5	13	25	34	43	35	47.5	60
Idaho									
Boise	14	21	28	27	32.5	38	48	59.5	71
Lewiston	15	22	29	22	25.5	29	44	54.5	65
Pocatello	13	19.5	26	25	30	35	51	62.5	74

	Percentage of Solar Glass to Floor Area (sca)			Solar Savings Fraction (SF)					
				No Night Insulation			R-9 Night Insulation		
	Minimum	Average	Maximum	Minimum	Average	Maximum	Minimum	Average	Maximum
City	1	2	3	1	2	3	4	5	6
Illinois									
Chicago	17	26	35	17	20	23	43	55	67
Moline	20	29.5	39	17	19.5	22	46	58	70
Springfield	15	22.5	30	19	23.5	28	42	54.5	67
Indiana									
Evansville	14	20.5	27	19	24	29	37	49	61
Fort Wayne	16	24.5	33	13	15	17	37	48.5	60
Indianapolis	14	21	28	15	18	21	37	48.5	60
South Bend	18	26.5	35	12	13.5	15	39	50	61
Iowa									
Burlington	18	27	36	20	23.5	27	47	59	71
Des Moines	21	32	43	19	22	25	50	62.5	75
Mason City	22	33	44	18	18.5	19	56	67.5	79
Sioux City	23	34.5	46	20	22	24	53	64.5	76
Kansas									
Dodge City	12	17.5	23	27	34.5	42	46	59.5	73
Goodland	13	20	27	26	32.5	39	47	60.5	74
Topeka	14	21	28	24	29.5	35	45	58	71
Wichita	14	21	28	26	33.5	41	45	58.5	72
Kentucky									
Lexington	13	20	27	17	21.5	26	35	46.5	58
Louisville	13	20	27	18	22.5	27	35	47	59
Louisiana									
Baton Rouge	6	9	12	26	34.5	43	34	46.5	59
Lake Charles	6	8.5	11	24	32.5	41	32	44.5	57
New Orleans	5	8	11	27	36.5	46	35	48	61
Shreveport	8	11.5	15	26	34.5	43	36	48.5	61
Maine									
Caribou	25	37.5	50	—	—	—	53	63.5	74
Portland	17	25.5	34	14	15.5	17	45	57	69
Maryland									
Baltimore	14	20.5	27	19	24.5	30	38	50	62
Massachusetts									
Boston	15	22	29	17	21	25	40	52	64
Michigan									
Alpena	21	31.5	42	—	—	—	47	58	69
Detroit	17	25.5	34	13	15	17	39	50	61
Flint	15	23	31	11	11.5	12	40	51	62
Grand Rapids	19	28.5	38	12	12.5	13	39	50	61
Sault Ste. Marie	25	37.5	50	—	—	—	50	60	70
Traverse City	18	27	36	—	—	—	42	52	62
Minnesota									
Duluth	25	37.5	50	—	—	—	50	60	70
International Falls	25	37.5	50	—	—	—	47	56.5	66
Minneapolis-St. Paul	25	37.5	50	—	—	—	55	65.5	76
Rochester	24	36.5	49	—	—	—	54	65	76
Mississippi									
Jackson	8	11.5	15	23	31	39	34	46	58
Meridian	8	11.5	15	23	31	39	34	46	58
Missouri									
Columbia	13	19.5	26	20	25	30	41	53.5	66
Kansas City	14	21.5	29	22	27	32	44	57	70
Springfield	13	19.5	26	22	28	34	40	52.5	65

City	Percentage of Solar Glass to Floor Area (sca)			Solar Savings Fraction (SF)					
				No Night Insulation			R-9 Night Insulation		
	Minimum	Average	Maximum	Minimum	Average	Maximum	Minimum	Average	Maximum
	1	2	3	1	2	3	4	5	6
Montana									
Billings	16	24	32	24	27.5	31	53	64.5	76
Cut Bank	24	36.5	49	22	22.5	23	62	71.5	81
Dillon	16	24	32	24	28	32	54	65.5	77
Glasgow	25	37.5	50	—	—	—	55	65	75
Great Falls	18	27.5	37	23	25.5	28	56	66.5	77
Helena	20	29.5	39	21	23	25	55	66	77
Lewistown	19	28.5	38	21	23	25	54	65	76
Miles City	23	35	47	21	22	23	60	70	80
Missoula	18	27	36	15	15.5	16	47	57.5	68
Nebraska									
Grand Island	18	27	36	24	28.5	33	51	63.5	76
North Omaha	20	30	40	21	25	29	51	63.5	76
North Platte	17	25.5	34	25	30.5	36	50	63	76
Scottsbluff	16	23.5	31	24	30	36	49	63.5	78
Nevada									
Elko	12	18.5	25	27	33	39	52	64	76
Ely	12	17.5	23	27	34	41	50	63.5	77
Las Vegas	9	13.5	18	35	45.5	56	48	61.5	75
Lovelock	13	19	25	12	30	48	53	65.5	78
Reno	11	16.5	22	31	39.5	48	49	62.5	76
Tonopah	11	17	23	31	39.5	48	51	64	77
Winnemucca	13	19.5	26	28	35	42	49	62	75
New Hampshire									
Concord	17	25.5	34	13	14	15	45	56.5	68
New Jersey									
Newark	13	19	25	19	24	29	39	51.5	64
New Mexico									
Albuquerque	11	16.5	22	29	38	47	46	59.5	73
Clayton	10	15	20	28	36.5	45	45	61	77
Farmington	12	18	24	29	37	45	49	62.5	76
Los Alamos	11	16.5	22	25	32.5	40	44	58	72
Roswell	10	11.5	13	30	39.5	49	45	59	73
Truth or Consequences	9	13	17	32	41.5	51	46	59.5	73
Tucumcari	10	15	20	30	39	48	45	59	73
Zuni	11	16	21	27	35	43	45	59	73
New York									
Albany	21	31	41	13	14	15	43	54.5	66
Binghamton	15	22.5	30	—	—	—	35	45.5	56
Buffalo	19	28	37	—	—	—	36	46.5	57
Massena	25	37.5	50	—	—	—	50	60.5	71
New York (Central Park)	15	22.5	30	16	20.5	25	36	47.5	59
Rochester	18	27.5	37	—	—	—	37	47.5	58
Syracuse	19	28.5	38	—	—	—	37	48	59
North Carolina									
Asheville	10	15	20	21	28	35	36	48.5	61
Cape Hatteras	9	13	17	24	32	40	36	48	60
Charlotte	8	12.5	17	23	30.5	38	36	48	60
Greensboro	10	15	20	23	30	37	37	50	63
Raleigh-Durham	9	14	19	22	29.5	37	36	48.5	61
North Dakota									
Bismarck	25	37.5	50	—	—	—	56	66.5	77
Fargo	25	37.5	50	—	—	—	51	61.5	72
Minot	25	37.5	50	—	—	—	52	62	72

	Percentage of Solar Glass to Floor Area (sca)			Solar Savings Fraction (SF)					
				No Night Insulation			R-9 Night Insulation		
	Minimum	Average	Maximum	Minimum	Average	Maximum	Minimum	Average	Maximum
City	1	2	3	1	2	3	4	5	6
Ohio									
Akron-Canton	15	23	31	12	14	16	35	46	57
Cincinnati	12	18	24	15	19	23	35	46	57
Cleveland	15	23	31	11	12.5	14	34	44.5	55
Columbus	13	20.5	28	13	15.5	18	35	46	57
Dayton	14	21	28	14	17	20	36	47.5	59
Toledo	17	25.5	34	13	15	17	38	49.5	61
Youngstown	16	24	32	—	—	—	34	44	54
Oklahoma									
Oklahoma City	11	16.5	22	25	33	41	43	55	67
Tulsa	11	16.5	22	24	31	38	40	52.5	65
Oregon									
Astoria	9	14	19	21	27.5	34	37	48.5	60
Burns	13	19	25	23	27.5	32	47	59	71
Medford	12	18	24	21	26.5	32	38	49	60
North Bend	9	13	17	25	33.5	42	38	51	64
Pendleton	14	20.5	27	22	26	30	43	53.5	64
Portland	13	19.5	26	21	26	31	38	49	60
Redmond	13	20	27	26	32	38	47	59	71
Salem	12	18	24	21	26.5	32	37	48	59
Pennsylvania									
Allentown	15	22	29	16	20	24	39	51	63
Erie	17	25.5	34	—	—	—	35	45	55
Harrisburg	13	19.5	26	17	21.5	26	38	50	62
Philadelphia	15	22	29	19	24	29	38	50	62
Pittsburgh	14	21	28	12	14	16	33	44	55
Wilkes-Barre-Scranton	16	14	12	13	15.5	18	37	48.5	60
Rhode Island									
Providence	15	22.5	30	17	20.5	24	40	52	64
South Carolina									
Charleston	7	10.5	14	25	33	41	34	46.5	59
Columbia	8	12.5	17	25	33	41	36	48.5	61
Greenville-Spartanburg	8	12.5	17	23	30.5	38	36	48	60
South Dakota									
Huron	25	37.5	50	—	—	—	58	68.5	79
Pierre	22	32.5	43	21	22	23	58	69	80
Rapid City	15	22.5	30	23	27.5	32	51	63.5	76
Sioux Falls	22	33.5	45	18	18.5	19	57	68	79
Tennessee									
Chattanooga	9	14	19	19	25.5	32	33	44.5	56
Knoxville	9	13.5	18	20	26.5	33	33	44.5	56
Memphis	9	14	19	22	29	36	36	48	60
Nashville	10	15.5	21	19	24.5	30	33	44	55
Texas									
Abilene	9	13.5	18	29	38	47	41	54.5	68
Amarillo	11	16.5	22	29	37.5	46	45	58.5	72
Austin	6	9.5	13	27	36.5	46	37	50	63
Brownsville	3	4.5	6	27	36.5	46	12	34.5	57
Corpus Christi	5	7	9	29	39	49	36	49.5	63
Dallas	8	12.5	17	27	35.5	44	38	51	64
Del Rio	6	9	12	30	40	50	39	52.5	66
El Paso	9	13	17	32	42.5	53	45	58.5	72
Forth Worth	9	13	17	26	35	44	38	51	64
Houston	6	8.5	11	25	34	43	34	46.5	59
Laredo	5	7	9	31	41.5	52	39	51.5	64
Lubbock	9	14	19	30	39.5	49	44	58	72

City	Percentage of Solar Glass to Floor Area (sca)			Solar Savings Fraction (SF)					
				No Night Insulation			R-9 Night Insulation		
	Minimum	Average	Maximum	Minimum	Average	Maximum	Minimum	Average	Maximum
	1	2	3	1	2	3	4	5	6
Texas (continued)									
Lufkin	7	10.5	14	26	34.5	43	35	48	61
Midland-Odessa	9	13.5	18	32	42	52	44	58	72
Port Arthur	6	8.5	11	26	35	44	34	47	60
San Angelo	8	11.5	15	29	38.5	48	40	53.5	67
San Antonio	6	9	12	28	38	48	38	51	64
Sherman	10	15	20	25	33	41	38	51	64
Waco	8	11.5	15	27	36	45	38	51	64
Wichita Falls	10	15	20	27	36	45	41	54	67
Utah									
Bryce Canyon	13	19	25	26	32.5	39	52	65	78
Cedar City	12	18	24	29	36	43	48	61.5	75
Salt Lake City	13	19.5	26	27	33	39	48	60	72
Vermont									
Burlington	22	32.5	43	—	—	—	46	57	68
Virginia									
Norfolk	9	14	19	23	30.5	38	37	49.5	62
Richmond	11	16.5	22	21	27.5	34	37	49	61
Roanoke	11	17	23	21	27.5	34	37	49	61
Washington									
Olympia	12	17.5	23	20	24.5	29	38	53.5	69
Seattle-Tacoma	11	16.5	22	21	25.5	30	39	49	59
Spokane	20	29.5	39	20	22	24	48	58	68
Yakima	18	27	36	24	27.5	31	49	59.5	70
West Virginia									
Charleston	13	19	25	16	20	24	32	44.5	57
Huntington	13	19	25	17	22	27	34	45.5	57
Wisconsin									
Eau Claire	25	37.5	50	—	—	—	53	64	75
Green Bay	23	34.5	46	—	—	—	53	64	75
La Crosse	21	32	43	—	—	—	52	63.5	75
Madison	20	30	40	15	16	17	51	62.5	74
Milwaukee	18	26.5	35	15	16.5	18	48	59.5	71
Wyoming									
Casper	13	19.5	26	27	33	39	53	65.5	78
Cheyenne	11	16	21	25	32	39	47	60.5	74
Rock Springs	14	21	28	26	32	38	54	66.5	79
Sheridan	16	23.5	31	22	26	30	52	63.5	75
Canada									
Alberta									
Edmonton	25	37.5	50	—	—	—	54	63	72
Suffield	25	37.5	50	28	29	30	67	76	85
British Columbia									
Nanaimo	13	19.5	26	26	30.5	35	45	55.5	66
Vancouver	13	19.5	26	20	24	28	40	50	60
Manitoba									
Winnipeg	25	37.5	50	—	—	—	54	64	74
Nova Scotia									
Dartmouth	14	21	28	17	20.5	24	45	57.5	70
Ontario									
Moosonee	25	37.5	50	—	—	—	48	57.5	67
Ottawa	25	37.5	50	—	—	—	59	69.5	80
Toronto	18	27	36	17	20	23	44	56	68
Quebec									
Normandin	25	37.5	50	—	—	—	54	64	74

PLUMBING FUNDAMENTALS

12.1 BUILDING PLUMBING SYSTEMS

Tidbits from Plumbing History

Nearly 4000 years ago, the ancient Greeks had hot and cold water systems in buildings. The Minoan Palace of Knossos on the isle of Crete had terra cotta (baked clay) piping laid beneath the palace floor. These pipes provided water for fountains and faucets of marble, gold, and silver that offered hot and cold running water. Drainage systems emptied into large sewers constructed of stone. Surprisingly, although hot and cold water systems were in place, for the Spartan warrior it was unmanly to use hot water.

The first storm sewers of Rome were built about 2800 years ago. Over 2000 years ago, the Romans had in place highly developed community plumbing system in which water was conveyed over many miles by large aqueducts. Water was then distributed to residences in lead pipes. By the 4th century C.E., Rome had 11 public baths, over 1300 public fountains and cisterns, and over 850 private baths.

The Roman *plumber* was an artisan who worked with lead. Both male and female plumbers soldered, installed, and repaired roofs, gutters, sewers, drains, and every part of the plumbing supply, waste, and storm drainage systems. The term *plumbing* is derived from the Latin word *plumbum* for lead (Pb). Historians theorize that lead leaching into drinking water from water supply pipes and lead from other sources poisoned the Roman aristocracy, contributing to the decline of the Roman Empire.

King Minos of Crete owned the world's first flushing water closet with a wooden seat and a small reservoir of water, over 2800 years ago. In the Far East, archaeologists in China recently uncovered an antique water closet in the tomb of a king of the Western Han Dynasty (206 B.C.E. to 24 C.E.). It was complete with running water, a stone seat, and a comfortable armrest.

The decline of the Roman Empire and an outbreak of deadly bubonic plague that killed an estimated one-third of the European population during the Middle Ages resulted in the decline of public baths and fountains. The period from 500 to 1500 C.E. was a dark age in terms of human hygiene; community plumbing became almost nonexistent. At the end of the Middle Ages, London's first water system was rebuilt around 1500. It consisted partly of the rehabilitated Roman system with the remainder patterned off of the Roman's design.

Pumping devices have been an important way of moving fluids for thousands of years. The ancient Egyptians invented water wheels with buckets mounted on them to move water for irrigation. Over 2000 years ago, Archimedes, a Greek mathematician, invented a screw pump made of a screw rotating in a cylinder (now known as an Archimedes screw). This type of pump was used to drain and irrigate the Nile Valley.

The beginnings of modern plumbing began in the early 1800s, when steam engines became capable of supplying water under pressure and inexpensive cast iron pipes could be produced to carry it. Still it was considered unhealthy to bathe. In 1835, the Common Council of Philadelphia nearly banned wintertime bathing (the ordinance failed by two votes). Ten years later, Boston prohibited bathing except on specific medical advice.

Finally, it was through observation of several cholera epidemics in the mid-1800s that epidemiologists finally recognized the link between sanitation and public health. This discovery provided the thrust for modern water and sewage systems. In 1848, England passed the national Public Health Act, which later became a model plumbing code for the world to follow. It mandated some type of sanitary disposal in every residence such as a flushing toilet, a privy, or an ash pit.

In America, like Europe, colonial hygiene and sanitation were poor. Colonial bathing consisted of infrequent baths in ponds or streams. New World settlers emulated the Native Americans' discharge of waste and refuse in running water, open fields, shrubs, or forests. As in Europe, colonials living in town would empty their chamber pots by tossing excrement out the front door or window onto the street. As early as 1700, local ordinances were passed to prevent people from throwing waste in a public street. Eventually, use of the privy or outhouse slowly became accepted.

Drinking water in colonial America came from streams, rivers, and wells. It was commonly believed at the time that foul-tasting mineral water had medicinal value. Around the time of the American Revolution, Dr. Benjamin Rush, a signer of the Declaration of Independence and surgeon general under George Washington, had the bad fortune of having a well with horribly tasting water at the site of his Pennsylvania home. Townspeople rushed to his well to get drinking water in hopes that its medicinal value would cure ailments. Unfortunately, when Dr. Rush's well dried out from overuse, it was discovered too late that the well was geologically connected underground to the doctor's privy.

Boston and later New York built the country's first water-works to provide water for firefighting and domestic use about 1700. The wooden pipe system, laid under roads, provided water at street pumps or hydrants. Water pipes were made of bored-out logs. Wooden pipes were common until the early 1800s, when the increased pressure required to pump water into rapidly expanding streets began to split the pipes. In 1804, Philadelphia earned the distinction as the first city in the world to adopt cast iron pipe for its water mains. Chicago is credited with having the first comprehensive sewerage project in the United States, designed in 1885.

Inside running water and toilets were not common in the U.S. home until well into the mid-1900s. The Census of 1910 indicated that only about 10% of American homes had inside running water. Farms during that time relied on well water, with many powered by hand pumps and windmills.

Modern Plumbing Systems

Modern cities have sophisticated water delivery and waste-water treatment systems. In buildings, the plumbing system performs two primary functions: water supply and waste disposal. A complete plumbing arrangement consists of a water supply system, a sanitary drainage system, and a wastewater treatment system.

The *water supply system* consists of the piping and fittings that supply hot and cold water from the building water supply to the fixtures, such as lavatories, bathtubs, water closets, dishwashers, clothes washers, and sinks. Water supply design is introduced in Chapter 13.

The waste disposal system consists of the piping and fittings required to take that water supplied to the fixtures out of the building and into the sewer line or disposal field. This system is typically referred to as a *sanitary drainage system* or *drain, waste and vent (DWV) system*. Design of the sanitary drainage system is introduced in Chapter 14.

Because of environmental concerns, *wastewater treatment* is also an important component of waste disposal from building plumbing systems. Although most buildings rely upon district or community water treatment plants to dispose of their sewage, some buildings and facilities operate their own operations. These are generally known as septic or *on-site sewage treatment (OSST)* systems. OSST systems are introduced in Chapter 15.

Essentially, a plumbing system is a network of pipes, fittings, and valves that carry and control flow of supply water and wastewater to and from points of use known as fixtures. *Fixtures* are components, receptacles, or pieces of equipment that use water and dispose of wastewater at the point of water use. *Piping* is a series of hollow channels that carry water to and wastewater from plumbing fixtures. *Fittings* are used to connect lengths of pipe in the piping network. *Valves* are used to regulate or control flow of water. Types of plumbing pipes, fittings,

valves, and fixtures are discussed in this chapter because they relate to the water supply and sanitary drainage systems.

12.2 WATER: THE SUBSTANCE

Any study of a plumbing system must begin with the substance it carries, water. *Water* is the name given to the liquid compound H_2O. A molecule of water is composed of one oxygen atom and two hydrogen atoms. In a pure state, it is tasteless and odorless. The physical properties of water are provided in Tables 12.1 and 12.2.

Under standard atmospheric pressure (14.696 psi, 101.04 kPa), the boiling point temperature of water is 212°F (100°C). The temperature at which water boils decreases with lower atmospheric or system pressure and increases at higher pressures. Thus, the temperature at which water boils decreases with elevation increase. For example, at standard atmospheric conditions at an elevation of 5000 ft (1524 m) above sea level, water boils at 202.4°F (94.7°C). It boils at 193.2°F (89.6°C) at 10 000 ft (3048 m) above sea level. The freezing point of water is 32°F (0°C).

Fundamental Units

Several fundamental units describe the properties and behavior of water in building plumbing systems. Customary and SI unit conversions for water based on weight, pressure, flow, and volume are provided in Table 12.3.

The following are definitions of the fundamental units.

Specific Weight (Density)

Specific weight (w) or *density* is weight per unit volume. Water density varies with temperature; it is most dense at 39°F (4°C). Below this temperature, crystals begin to form, increasing its volume and therefore decreasing its density. Water attains a specific weight of 62.42 lb/ft^3 (1.00 kg/L) at a temperature of 39°F (4°C). Above and below 39°F, water is less dense; for example, the specific weight of water at 80°F (27°C) is 62.2 lb/ft^3 (0.996 kg/L). A specific weight of 62.4 lb/ft^3 (1.00 kg/L) is commonly used for liquid water in engineering computations.

Specific Gravity

The specific gravity (s.g.) of a fluid or solid is the ratio of the specific weight of the fluid or solid to the specific weight of water at a temperature of 39°F (4°C), the temperature at which water is most dense (62.42 lb/ft^3 or 1.00 kg/L). It is a comparison of its weight with the weight of an equal volume of water. Materials with a specific gravity less than 1.0 are less dense than water (e.g., oil) and will float on pure water; substances with a specific gravity more than 1.0 are denser than water and will sink. The specific gravity of water is assumed to be 1.0 at common plumbing system temperatures.

TABLE 12.1   PHYSICAL PROPERTIES OF LIQUID WATER.

Temperature		Specific Volume		Specific Weight		Specific Gravity	Weight	
°F	°C	ft³/lb	m³/kg	lb/ft³	kg/m³	—	lb/gallon	kg/L
32	0	0.01602	0.001000	62.41	999.7	1.000	8.344	1.000
40	4.4	0.01602	0.001000	62.42	999.9	1.000	8.345	1.000
50	10.0	0.01602	0.001000	62.41	999.7	1.000	8.343	1.000
60	15.6	0.01603	0.001001	62.37	999.1	0.999	8.338	0.999
70	21.1	0.01605	0.001002	62.31	998.0	0.998	8.329	0.998
80	26.7	0.01607	0.001003	62.22	996.6	0.997	8.318	0.997
90	32.2	0.01610	0.001005	62.12	995.0	0.995	8.304	0.995
100	37.8	0.01613	0.001007	62.00	993.1	0.993	8.288	0.993
110	43.3	0.01617	0.001009	61.86	990.9	0.991	8.270	0.991
120	48.9	0.01620	0.001012	61.71	988.5	0.989	8.250	0.989
130	54.4	0.01625	0.001014	61.55	985.9	0.986	8.228	0.986
140	60.0	0.01629	0.001017	61.38	983.1	0.983	8.205	0.983
150	65.6	0.01634	0.001020	61.19	980.1	0.980	8.180	0.980
160	71.1	0.01640	0.001024	60.99	977.0	0.977	8.154	0.977
170	76.7	0.01645	0.001027	60.79	973.7	0.974	8.126	0.974
180	82.2	0.01651	0.001031	60.57	970.2	0.970	8.097	0.970
190	87.8	0.01657	0.001035	60.34	966.6	0.967	8.067	0.967
200	93.3	0.01664	0.001039	60.11	962.8	0.963	8.035	0.963
210	98.9	0.01671	0.001043	59.86	958.9	0.959	8.002	0.959
212	100.0	0.01672	0.001044	59.81	958.1	0.958	7.996	0.958

TABLE 12.2   BOILING POINTS, BAROMETRIC READINGS, AND ATMOSPHERIC PRESSURES OF WATER AT VARIOUS ALTITUDES.

Altitude (Above Sea Level)		Barometer Reading		Atmospheric Pressure		Boiling Point	
Ft	m	in Hg	mm Hg	psi	kPa	°F	°C
−1000	−305	31.0	788	15.2	104.8	213.8	101.0
−500	−152	30.5	775	15.0	103.4	212.9	100.5
0	0	29.9	760	14.7	101.4	212.0	100.0
500	152	29.4	747	14.4	99.3	211.1	99.5
1000	305	28.9	734	14.2	97.9	210.2	99.0
1500	457	28.3	719	13.9	95.8	209.3	98.5
2000	610	27.8	706	13.7	94.5	208.4	98.0
2500	762	27.3	694	13.4	92.4	207.4	97.4
3000	914	26.8	681	13.2	91.0	206.5	96.9
3500	1067	26.3	668	12.9	88.9	205.6	96.4
4000	1219	25.8	655	12.7	87.6	204.7	95.9
4500	1372	25.4	645	12.4	85.5	203.8	95.4
5000	1524	24.9	633	12.2	84.1	202.9	94.9
5500	1676	24.4	620	12.0	82.7	201.9	94.4
6000	1829	24.0	610	11.8	81.4	201.0	93.9
6500	1981	23.5	597	11.5	79.3	200.1	93.4
7000	2134	23.1	587	11.3	77.9	199.2	92.9
7500	2286	22.7	577	11.1	76.5	198.3	92.4
8000	2438	22.2	564	10.9	75.2	197.4	91.9
8500	2591	21.8	554	10.7	73.8	196.5	91.4
9000	2743	21.4	544	10.5	72.4	195.5	90.8
9500	2896	21.0	533	10.3	71.0	194.6	90.3
10 000	3048	20.6	523	10.1	69.6	193.7	89.8
15 000	4572	16.9	429	8.3	57.2	184.0	84.4

TABLE 12.3   CONVERSIONS FOR WATER BASED ON WEIGHT, PRESSURE, FLOW, AND VOLUME.

Weights

1 U.S. gallon of water	8.3357 pounds
1 cu. foot of water	62.3554 pounds
1 imperial gallon of water	10.0 pounds
1 liter of water	2.2 pounds

Volume

1728 cubic inches	1 cubic foot
231 cubic inches	1 gallon
27 cubic feet	1 cubic yard
1 cubic foot	7.48052 gallons
1 cubic yard	202 gallons
16 drams	1 ounce
1 pint	16 ounces
2 pints	1 quart
32 ounces	1 quart
8 pints	1 gallon
4 quarts	1 gallon
1 gallon	3.785 liters
1 gallon	0.00379 cubic meters
1 gallon	0.833 imperial gallon
1000 liters	1 cubic meter
1 liter	0.2642 gallon
27.154 gallons	1 acre inch
325 851 gallons	1 acre foot
1 000 000 gallons	3.0689 acre feet
1 acre foot	43 560 cubic feet

Volume Formulas

Volume of a cube	WLH (width · length · height)
Volume of a pyramid	$1/3 LA_{base}$
Volume of a sphere	$5236 D^3$
Volume of a cylinder	$\pi LD^2/4$

Flows

1 gallon/minute	0.002228 cubic foot/second
1 gallon/minute	0.13368 cubic foot/minute
1 gallon/minute	8.0208 cubic foot/hour
1 gallon/minute	0.06309 liters/second
1 gallon/minute	3.78533 liters/minute
1 gallon/minute	0.0044192-acre feet/24 hours
1 cubic foot per second	448.83 gpm
1 liter per second	15.85 gpm
1 acre inch per hour	452.57 gpm
1 acre foot per day	226.3 gpm
1 000 000 gallons per day	694.4 gpm
1 cubic foot per second	0.992 acre inches/hour

Pressure

1 atmosphere	33.94 feet of water @ 62°F
1 atmosphere	14.6963 pounds/square inch
1 pound per square inch	2.31 feet of head
1 pound per square inch	27.7612 inches of water
1 foot of head	0.433 pounds/square inch
1 inch of water column	0.0360 pounds/square inch
1 kilogram/sq. centimeter	14.22 pounds/square inch
1 foot of water	62.3554 pounds/square foot

Example 12.1

Raw sewage is found to weigh 60.5 lb/ft³. Determine its specific gravity.

$$s.g. = 60.5 \text{ lb/ft}^3/62.42 \text{ lb/ft}^3 = 0.969$$

Volume

Volume (V) is the amount of space occupied by a substance. Water volume is typically expressed in cubic inches (in³) or cubic feet (ft³) in the customary system, and in cubic meters (m³) or liters (L) in the SI system. In plumbing system design, volume is commonly expressed in gallons (g or gal). There are 7.48 gallons in a cubic foot (ft³). A gallon is approximately 3.8 L.

Volumetric Flow Rate

Volumetric flow rate (Q), frequently called the *flow rate,* is the volume of a substance that passes a point in a system per unit of time. Flow rate is usually expressed in liters per second (L/s), liters per minute (L/min), or cubic meters per second (m³/s) in the SI system. In the customary system, volumetric flow rate is expressed in cubic feet per second (cfs or ft³/s), cubic feet per minute (cfm or ft³/min), gal per second (gps or g/s), and gal per minute (gpm or g/min).

Volumetric flow rate (Q) may be determined with volume (V) and time:

$$Q = V/time$$

Example 12.2

a. Determine the volumetric flow rate, in gpm, for water flowing out of a faucet based on 2 gal in 23 s.

$$Q = V/time = 2 \text{ gal}/(23 \text{ s} \cdot (1 \text{ min}/60 \text{ s})) = 5.2 \text{ gpm}$$

b. Determine the volumetric flow rate, in L/s, for water flowing out of a faucet based on 6 L in 8 s.

$$Q = V/time = 6 \text{ L}/(8 \text{ s} \cdot (1 \text{ min}/60 \text{ s})) = 45 \text{ L/s}$$

A faucet supplying water at a volumetric flow rate of 5 gpm will fill a 5 gal bucket in exactly 1 min or a 1 gal bucket in 12.5 s (one-fifth of a minute). In plumbing system design, volumetric flow rate is found by multiplying the area of the inside diameter of the pipe carrying the water by the average velocity of the flowing water.

Velocity

Velocity is the rate of linear motion of a substance in one direction. The magnitude of velocity, known as *speed,* is usually expressed in terms of distance covered per unit of time. In the customary system of weights and measures, velocity is expressed in inches per second (in/s) or feet per second (ft/s). In the international system of measure (the SI system), velocity is expressed in meters per second (m/s).

In a fluidic system such as a plumbing system, water velocity is expressed as an *average velocity* because water molecules each have different speeds and directions of travel; that is, water molecules flowing in the center of a pipe tend to travel faster than water molecules at or near the inner wall of the pipe.

Average velocity (v) of a fluid (such as water) flowing through a pipe may be found by the following equations based upon average volumetric flow rate (Q) and cross-sectional area (A) or inside diameter (D_i). Units must be consistent in these equations (e.g., volume, area, and diameter must be expressed in units of in, ft, m, and so on).

$$v = Q/A = 4Q/\pi D_i^2$$

The following equation, in customary units, is useful in plumbing system design. It may be used to find the average velocity (v) of a fluid flowing through a pipe, in ft/s, based on the volumetric flow rate (Q), in gpm, and an inside diameter (D_i) of the pipe, in inches:

$$v = 0.409Q/D_i^2$$

Example 12.3

Determine the average velocity for water flow in a pipe under the following conditions:

a. A ¾ in diameter, Type L copper tube (0.875 in outside diameter and 0.785 in inside diameter) carrying water at a volumetric flow rate of 10 gpm.

$$v = 0.409Q/D_i^2 = (0.409 \cdot 10 \text{ gpm})/(0.785 \text{ in})^2 = 6.6 \text{ ft/s}$$

b. A 2 in diameter, Schedule 40 chlorinated polyvinyl chloride (CPVC) pipe (2.375 in outside diameter and 2.047 in inside diameter) carrying cold water at a volumetric flow rate of 40 gpm.

$$v = 0.409Q/D_i^2 = (0.409 \cdot 40 \text{ gpm})/(2.047 \text{ in})^2 = 3.9 \text{ ft/s}$$

The following equation, in SI (metric) units, may be used to find the average velocity (v) of a fluid flowing through a pipe, in m/s, based on the volumetric flow rate (Q), in L/min, and an inside diameter (D_i) of the pipe, in mm:

$$v = 21.22Q/D_i^2$$

Example 12.4

Determine the average velocity for water flow in a 20 mm diameter copper tube (21.4 mm outside diameter and 19.9 mm

inside diameter) carrying water at a volumetric flow rate of 40 L/min.

$$v = 21.22Q/D_i^2 = (21.22 \cdot 40 \text{ L/min})/(19.9 \text{ mm})^2 = 2.14 \text{ m/s}$$

Pressure

Pressure (P) is the force per unit area exerted by liquid or gas on a surface such as the sidewall of a container or pipe. In the customary system of measure, pressure is expressed in pounds per square inch (lb/in² or psi) or pounds per square foot (lb/ft² or psf). In the international system (SI), pressure is expressed in Newton per square meter or the Pascal (N/m² or Pa). Although units of lb/in² are dimensionally correct, the acronym "psi" will be used for pounds per square inch of gauge pressure because it is universally accepted in the plumbing industry. The acronym "psia" will be used for absolute pressure.

Standard atmospheric pressure (P_s) is the typical barometric pressure of air at sea level and 70°F (21°C). It is equal to 14.696 psia (101 325 Pa). Atmospheric pressure varies with weather conditions and elevation. In Denver, Colorado, atmospheric pressure is about 11.8 psia because Denver is about a mile above sea level; it is above about 20% of the earth's atmosphere.

Gauge pressure (P_g) is the pressure of a fluid (gas or liquid) *excluding* pressure exerted by the atmosphere. Pressure can be expressed in terms of absolute and gauge pressure: *Absolute pressure (P_a)* is the pressure of a fluid (gas or liquid) *including* pressure exerted by the atmosphere:

$$P_g + P_s = P_a$$

Consider the following example, which illustrates the difference between absolute and gauge pressures.

Example 12.5

a. At sea level, atmospheric pressure is 14.7 psia (101 325 Pa). A pressure gauge placed at the bottom of an 8 ft (2.45 m) deep tank filled with water measures a water pressure at the tank bottom of 3.5 psi (24 130 Pa). Determine the absolute and gauge pressure.

Gauge pressure at the bottom of the tank is 3.5 psi (24 130 Pa).

Absolute pressure at the bottom of the tank is 18.2 psia, as found by:

$$P_g + P_s = P_a$$
$$3.5 \text{ psi} + 14.7 \text{ psi} = 18.2 \text{ psia}$$
$$(24\ 130 \text{ Pa} + 101\ 325 \text{ Pa} = 125\ 455 \text{ Pa})$$

b. At sea level, atmospheric pressure is 14.7 psia (101 325 Pa). A bicycle tire (at sea level) is inflated to 50 psi (344 737 Pa). Determine the gauge and absolute pressures in the inflated tire.

The inflated tire has a gauge pressure of 50 psi (344 737 Pa).

The inflated tire has an absolute pressure of 64.7 psia (446 062 Pa) because:

$$P_g + P_s = P_a$$
$$50 \text{ psi} + 14.7 \text{ psia} = 64.7 \text{ psia}$$
$$(344\ 737 \text{ Pa} + 101\ 325 \text{ Pa} = 446\ 062 \text{ Pa})$$

c. In Denver, atmospheric pressure is about 11.8 psi (81 358 Pa). A bicycle tire (in Denver) is inflated to 50 psi (344 737 Pa). Determine the gauge and absolute pressures in the inflated tire.

The inflated tire has a gauge pressure of 50 psi (344 737 Pa).

The inflated tire has an absolute pressure of 61.8 psia (426 095 Pa) because:

$$P_g + P_s = P_a$$
$$50 \text{ psi} + 11.8 \text{ psia} = 61.8 \text{ psi}$$
$$(344\ 737 \text{ Pa} + 81\ 358 \text{ Pa} = 426\ 095 \text{ Pa})$$

A gauge is frequently used to record the pressure difference between the system and the atmospheric pressure. Normally, if pressure in a system is below atmospheric pressure, it is called *vacuum pressure* or a *suction pressure*. It is expressed as a negative gauge pressure.

Saturation vapor pressure is the pressure that water vapor molecules exert when the air is fully saturated at a given temperature. Saturation vapor pressure is directly proportional to the temperature: it increases with rising temperature and falls with decreasing temperature.

In plumbing systems there are three additional classifications of pressure: *Static pressure* is the pressure that exists without any flow. It is the pressure available at a location in the system. *Residual pressure* is the pressure available at a fixture or outlet during a period of maximum demand. It is the pressure that exists after pressure losses from friction from water flow, elevation change, and other pressure losses in the system are subtracted. *Discharge pressure* is the pressure of the water at the point of discharge, such as at the mouth of a showerhead, faucet, or hose bibb.

The constituent gases that make up a mixture of gases such as air each exert a *partial pressure* that contributes to the total pressure exerted by the gas mixture. For example, atmospheric air consists of about 75% nitrogen, by weight, so 75% of the total pressure exerted by atmospheric air is from the nitrogen constituent. Thus, under standard conditions (14.696 psia), the partial pressure of nitrogen is 11 psia, 75% of the total pressure. The partial pressure of a constituent gas in a mixture of gases equals the pressure it would exert if it occupied the same volume alone at the same temperature.

Water vapor pressure is the pressure that the water vapor molecules alone exert in air. It is based on the amount of water vapor that exists in the air. Like all constituent gases in air, water vapor exerts a pressure, which is known as vapor pressure.

Water vapor pressure is the contribution of water vapor to the total pressure exerted by a gas. In buildings, this gas is atmospheric air.

A difference in pressure in a system is required for a fluid to flow; for example, a pump increases the pressure of the liquid passing through it, thereby causing flow. Flow will always be from a high-pressure region to a low-pressure region. Pressure difference is the driving force behind water flow. An increase in pressure difference will increase flow.

12.3 WATER SUPPLY

Water Sources

A supply of good water is more important to human survival than food. *Potable* is clean water that is suitable for human drinking. It must be available for drinking, cooking, and cleaning. *Nonpotable* water may be used for flushing water closets (toilets), irrigating grass and gardens, washing cars, and for any use other than drinking, cooking, or cleaning. An abundant supply of potable water that is easily distributed is vital to a prosperous economy.

Rain and snowmelt are the sources of most of the water available for our use. When it rains or a snowfield melts, water flows into streams and rivers or soaks into the ground. By definition, *surface water* is the rain that runs off the surface of the ground into streams, rivers, and lakes. *Groundwater* is water found below the surface of the earth. It is water that has percolated through porous soil until it reaches an impervious stratum, upon which it collects.

Surface Water

Surface water readily provides much of the water needed by cities, counties, large industry, and others. However, this source is dependent on recurring rain. During a long period of drought, the flow of water may be significantly reduced. *Reservoirs* hold surface water during periods of high runoff and release water during periods of low runoff. Surface water is typically treated to provide the potable water required. Where nonpotable water may be used, no treatment of the water is necessary.

Surface water can be collected in a storage tank called a *cistern*. A cistern can fill with rainwater as it drains from the roof of a building or a more elaborate collection system. Collected water is then pumped into the supply line of the building for use. (See Figure 12.1.) The need for water is so important on certain islands that the government has covered part of the land surface (usually the side of a mountain or a hill facing the direction from which the rains usually come) with a plastic film so rain can be collected and stored for later use.

In remote buildings without a collection system, a tank truck can deliver water and fill a cistern periodically. Some volunteer fire departments partially fund their operations by delivering water to remote residences that rely on a cistern for their supply of domestic water. This is common in mountain homes that do not have access to public water service or a well.

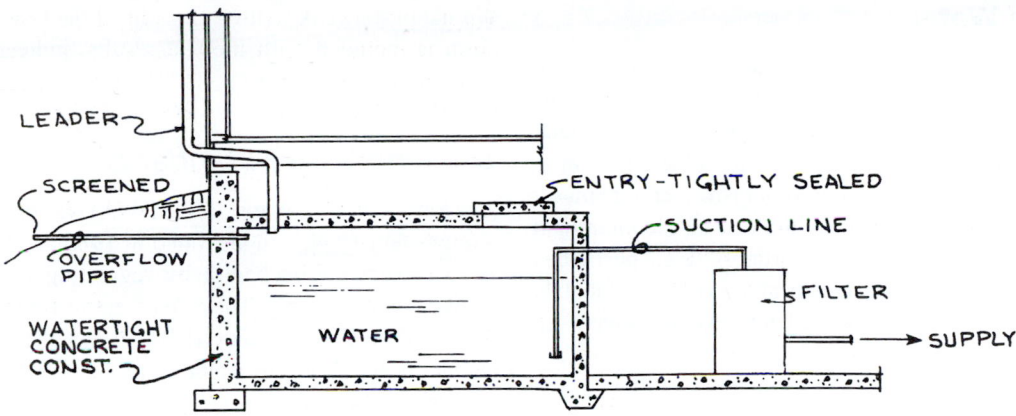

FIGURE 12.1 A cistern is a reservoir that holds drinking water.

Groundwater

Groundwater seeps through the soil and is trapped on impervious stratum, a layer of soil or rock that water cannot pass through. The water collects in pores of permeable stratum; a layer of porous earth that water can pass through such as sands, gravels, limestone, or basalt. Saturated permeable stratum capable of providing a usable supply of water is known as an *aquifer.*

Groundwater can be captured at many layers below a building site—that is, there may be several aquifers at different depths. Very deep-lying groundwater can remain undisturbed for thousands or millions of years. However, most groundwater lies at shallower depths. An aquifer can serve as an underground reservoir with almost unlimited capacity. However, if water is removed from the aquifer at too rapid a rate, the water level may drop so drastically that groundwater can no longer be reached. (See Figure 12.2.)

The level of groundwater is referred to as the *water table.* The distance from the ground surface to the water table is referred to as the *water table depth.* Depth of a shallow water table can vary considerably. Generally, shallow water table depth will vary with amount of rain. During a long dry spell, the water table depth will usually drop and during the rainy season it will likely rise. Depth of a water table can also change artificially by seepage from a nearby stream, lake, reservoir, or irrigation ditch.

Because the water table is formed by an accumulation of water over impervious stratum, the availability of water follows the irregular path of the stratum. At one location the water table may be close to the surface while dropping well below the surface nearby. The underground supply of water flows approximately horizontally. If it reaches a low spot in the ground surface, it may outflow as a spring or seep out creating a swampy area. If the flowing underground water becomes confined between impervious strata, significant pressure may be built up. Under this condition, if the water pressure is released by drilling through the top stratum or through a natural opening

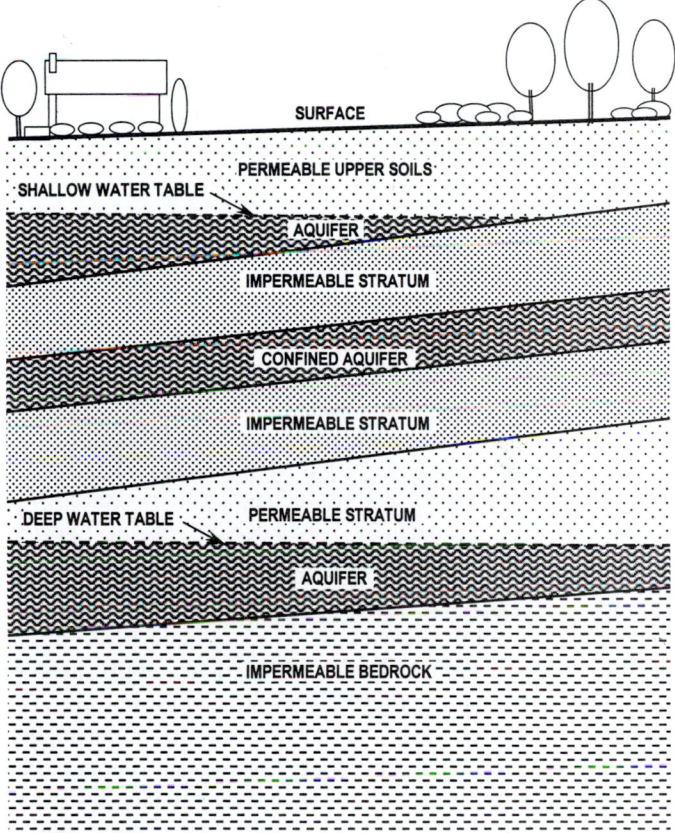

FIGURE 12.2 Groundwater can be captured at many levels below a building site. A saturated permeable stratum capable of holding a usable supply of water is known as an aquifer.

in the stratum, the water will be forced upward creating an *artesian well.*

Groundwater may require treatment to be potable, but often it does not. When treatment is required, it is generally less treatment than is required when making surface water potable.

Impurities in Water

All water sources contain some impurities. It is the type and amount of these impurities that affect water quality and suitability for a particular use. Whenever animal or human fecal material connects with a water source, it is possible one or more pathogenic (disease-causing) microorganisms could invade the water source. Chemicals and toxins can also contaminate the water source such as agricultural runoff (e.g., pesticides, herbicides, and so forth) and industrial runoff (e.g., metals, mine tailings, and so on). Some toxic bacteria can spawn algae in warm, shallow water and turn it green.

Surface water generally contains larger quantities of turbidity (cloudiness) and bacteria than groundwater. Groundwater generally contains higher concentrations of dissolved chemicals. Seawater contains high concentrations of dissolved chemicals and some microscopic organisms. As surface water runs over the ground, it may pick up various organic substances. These include algae, fungi, bacteria, vegetable matter, animal decay and wastes, fertilizer, garbage wastes, and sewage. This contaminated runoff returns to the river, lake, or reservoir. Unless treated, these waters remain tainted and are not potable.

As groundwater percolates down through the soil, it dissolves minerals such as calcium, iron, silica, sulfates, fluorides, and nitrates, and may also entrap gases such as sulfide, sulfur dioxide, and carbon dioxide. It may also pick up contamination from public or private underground garbage and sewage wastes. Generally, as it percolates, it will filter out any organic matter that may have been accumulated at the surface or in the ground.

Characteristics of water sources will vary greatly. Impurities in the surface water and groundwater may be harmful, of no importance, or even beneficial to a person's health. To determine what is in the water, it must be tested.

Water Testing

All potable water supplies should be tested and certified before being put in use and periodically recertified during use. The governmental entity (e.g., special district, city, municipality, and so forth) that controls the supply of water to a community regularly tests its water to be certain it is potable. Testing of private water supplies, such as wells and streams, is the responsibility of the property owner or user. The responsible governmental health entity often refers the users of private water supplies to independent testing laboratories.

The test for potable water provides a chemical analysis of the water, indicating the parts per million (ppm) of each chemical found in the water. A separate test is made for bacteriological quality of the water, providing an estimate of the density of bacteria in the water supply. Of particular concern in this test is the presence of any coliform organisms, which indicate that the water supply may be contaminated with human or animal wastes (perhaps seepage from a nearby septic tank field or animal pasture). A written analysis of the test or a standardized form is included with the test results, indicating whether the water is potable or not.

Drinking Water Standards

The Safe Drinking Water Act (SDWA) was originally passed by Congress in 1974, and amended in 1986 and 1996. Its purpose is to protect public health by regulating the nation's public drinking water supply. The SDWA authorizes the U.S. Environmental Protection Agency (EPA) to set national health-based standards for drinking water to protect against both naturally occurring and human-made contaminants that may be found in drinking water.

In the United States, the *National Primary Drinking Water Regulations* are legally enforceable mandatory standards that apply to public water systems and that protect public health by limiting the levels of contaminants in drinking water. A partial list of contaminants in drinking water, maximum acceptable levels, common sources of contaminants, and potential health effects from ingestion of water is provided in Table 12.4. These standards apply to public water systems that provide water for human consumption through at least 15 service connections, or regularly serve at least 25 individuals. Public water systems include municipal water companies, homeowner associations, schools, businesses, campgrounds, and shopping malls.

The *National Secondary Drinking Water Regulations* are guidelines regulating contaminants that may cause undesirable cosmetic effects (such as skin or tooth discoloration), aesthetic effects (e.g., taste, odor, or color) and other characteristics (e.g., corrosivity, pH) in drinking water. These secondary contaminants include metals (e.g., copper, aluminum, iron, manganese, silver, and zinc) and other chemicals (e.g., chloride, fluoride, sulfate, and so on). Individual states may choose to adopt these secondary standards as enforceable standards.

Water Treatment

Water quality and taste vary considerably from place to place, depending on the water source of the area, the chemical and bacteria contents of the water, and the amount and type of treatment given the water before it is put into the system.

Potable water can have an objectionable odor and taste and even be cloudy and slightly muddied or colored in appearance. Although the odor, taste, or appearance of potable water may not mean you want to drink it, it is still safe to drink. Several methods are used to improve water quality and taste:

- Problems with undesirable taste and odor are overcome by use of filtration equipment or by aeration of the water.

- Bacteria are destroyed by the addition of a few parts per million of chlorine. The taste of chlorine is then removed with sodium sulfite.

TABLE 12.4 A PARTIAL LIST OF CONTAMINANTS IN DRINKING WATER, MAXIMUM ACCEPTABLE LEVELS, COMMON SOURCES OF CONTAMINANTS, AND POTENTIAL HEALTH EFFECTS FROM INGESTION OF WATER (FROM NATIONAL PRIMARY DRINKING WATER REGULATIONS, U.S. EPA).

Contaminant Name	Maximum Acceptable Level	Common Sources	Health Effects
Arsenic	0.010 mg/L	Natural deposits, smelters, glass, electronics wastes, orchards	Skin damage or problems with circulatory systems, and may have increased risk of getting cancer
Asbestos	7 MFL (million fibers/liter)	Natural deposits, asbestos, cement in water systems	Increased risk of developing benign intestinal polyps
Barium	2 mg/L	Natural deposits, pigments, epoxy sealants, spent coal	Increase in blood pressure
Bacteria/coliform	n/a	Animal and human waste, septic fields, sewage, farming	Not a health threat in itself; it is used to indicate whether other potentially harmful bacteria may be present
Cadmium	0.005 mg/L	Natural deposits, galvanized pipe corrosion, batteries, paints	Kidney damage
Chlordane	0.002 mg/L	Leaching from soil, treatment for termites	Liver or nervous system problems; increased risk of cancer
Chlorine	4 mg/L	Water disinfection	Eye and nose irritation; stomach discomfort
Chromium	0.1 mg/L	Natural deposits, mining, electroplating, pigments	Liver, kidney, circulatory disorders
Copper	1.3 mg/L	Natural/industrial deposits, wood preservatives, plumbing	Gastrointestinal illness (e.g., diarrhea, vomiting, cramps)
Cryptosporidium	Zero	Animal or human waste, contaminated food products	Gastrointestinal illness (e.g., diarrhea, vomiting, cramps)
E. coli (bacteria)	Zero	Naturally occurring, human or animal wastes	Gastrointestinal disorders (often severe)
Fluoride	4 mg/L	Natural deposits, fertilizer, aluminum industries, water additive	Bone disease (pain and tenderness of the bones); children may get mottled teeth
Giardia	Zero	Naturally occurring, human or animal wastes	Gastrointestinal illness (e.g., diarrhea, vomiting, cramps)
Hydrogen sulfide	n/a	Natural deposits	Rotten egg taste and odor
Iron	0.3 mg/L	Natural deposits	Staining of laundry, plumbing, appliances
Mercury	0.002 mg/L	Crop runoff, natural deposits, batteries, electrical switches	Kidney, nervous system disorders
Microbiological contaminants	—	Animal and human waste, septic fields, sewage, farming	Various illnesses
Nitrate/nitrite	10 mg/L	Animal waste, fertilizer, natural deposits, septic tanks, sewage	Infants below the age of six months could become seriously ill; shortness of breath and blue-baby syndrome
Radium	5 pCi/L	Natural deposits	Increased risk of cancer
Radon		Natural deposits	Increased risk of cancer
Selenium	0.05 mg/L	Natural deposits, mining, smelting, coal/oil combustion	Liver damage
Total dissolved solids	500 mg/L	Erosion of naturally occurring mineral deposits	Gastrointestinal irritation
Toxaphene	0.003 mg/L	Insecticide formerly used on cattle, cotton, soybeans	Increased risk of cancer
Trihalomethanes (TTHM) (chlorination by-products)	0.08 mg/L	By-product of chlorination in drinking water	Liver, kidney, or central nervous system problems; increased risk of cancer
Turbidity	n/a	Soil runoff	A measure of cloudiness of water; higher turbidity levels are associated with disease-causing microorganisms, which cause symptoms such as nausea, cramps, diarrhea
Uranium	0.03 mg/L	Natural occurring	Kidney disorders, cancer
Viruses	—	Animal and human waste, septic fields, sewage, farming	Gastrointestinal illness (e.g., diarrhea, vomiting, cramps)
Volatile organic compounds (VOCs)	Varies	Varies	Risk varies with compound: nervous system, reproductive system, circulatory system, kidney, spleen, and liver problems including cancer

- Suspended organic matter that supports bacterial life and suspended mineral matter are removed by the addition of a flocculating and precipitating agent, such as alum, before settling or filtration.

- Excessive hardness, which renders the water unsuitable for many industrial purposes, is reduced by the addition of slaked, or hydrated, lime or by an ion-exchange process.

In addition to treating water for quality and taste, artificial *fluoridation* of public water is done in many U.S. communities. It is an established method of reducing tooth decay in children.

In many regions, the need for additional potable water supply has forced the development of processing water. In the *desalination process,* saline (salt) is removed from water (e.g., seawater) thereby making the water potable. This process is used in areas where seawater is the only source of water available or where groundwater is high in saline. Desalination plants provide much of the potable water on islands such as the Bahamas, Malta, and Catalina and in arid countries such as Kuwait, United Arab Emirates, and Qatar. Although the desalination process is successful, the cost of treating seawater is much higher than that for treating high salinity in fresh water.

Additionally, desalination is used when river water is overused for irrigation; salt that dissolves from soil and rocks becomes concentrated in the water that returns to the river, causing an increase in the salinity of the water in the river. High salinity makes desalination of river water necessary. For example, a desalination plant near the U.S.–Mexico border removes salt from the Colorado River and enables the United States to provide Mexico with usable water.

Water Use

Traditionally, water use rates are described in units of *gallons per capita per day (gpcd)* or *liters per capita per day (Lpcd).* Of the potable water supplied by public water systems, only a small portion is actually used for drinking.

The United States uses more water than other industrialized countries, even those that are equally well developed. Significant amounts of water are used for lawn and garden sprinkling, automobile washing, and kitchen and laundry appliances. According to an American Water Works Association study on residential end uses of water in the United States, daily indoor per capita water use in a typical single family home in the United States is 69.3 gal (262.4 L). See Table 12.5 for a breakdown of end use. The average daily domestic demands in commercial/industrial settings range between 20 and 35 gal per day (gpd) per employee.

The amount of water we use in residences varies by time of day:

- Lowest rate of use—11:30 PM to 5:00 AM
- Sharp rise/high use—5:00 AM to noon (peak use from 7:00 AM to 8:00 AM)

TABLE 12.5   RESIDENTIAL END USES OF WATER IN THE UNITED STATES BY CATEGORY OF USE.

Category of End Use	Percentage of Household per Capita Water Use	Household per Capita Water Use	
		gpcd (gal/person · day)	Lpcd (L/person · day)
Water closets (toilets)	26.7%	18.5	70.0
Clothes washers	21.7%	15.0	56.8
Showers	16.8%	11.6	43.9
Faucets	15.7%	10.9	41.3
Leaks	12.7%	9.5	36.0
Baths	1.7%	1.2	4.5
Dishwashers	1.4%	1.0	3.8
Other indoor domestic uses	2.2%	1.6	6.1
Total	100%	69.3	262.4

- Moderate use—noon to 5:00 PM (low around 3:00 PM)
- Increasing evening use—5:00 to 11:00 PM (second minor peak, 6:00 to 8:00 PM)

Significant differences in use exist, depending upon where water is drawn from. A typical family of four living in a dwelling connected to a community water system uses about 350 gal (1325 L) per day in and around their home. In contrast, a typical household that gets its water from a private well or cistern uses about 200 gal (757 L) for a family of four. Additionally, in U.S. cities, an estimated 35 to 50 gal (132 to 189 L) of water per person per day is used for public activities such as firefighting, street washing, and park maintenance, but this does not fully account for these considerable differences in use.

Commercial and industrial businesses also place heavy demands on public water supplies in developed countries. In most water supply systems, the predominant number of user connections is residential, but the few connections to nonresidential customers typically accounts for a significant fraction of the system-wide water use.

Water Conservation

Potable water is not an infinite resource. As population and demand for water grows, it is in short supply and is becoming more expensive, particularly in arid and semiarid climates and in all regions during periods of long drought. Efforts are underway to cut back on its use. The introduction of running water and waste systems in buildings is a rather new experience, occurring only in the last hundred years. During this time, we have progressed from taking a bath once a week (the Saturday night bath) to bathing once or twice daily.

Water Conservation

Conservation methods, such as flow restrictors on all water outlets (such as sinks and showers), can reduce the amount of water used by 50%, depending on the type installed. For example, an older showerhead may have a water flow of 5 to 7 gpm (19 to 26.6 L/min), so just a 5-min shower consumes 25 gal (95 L) of water. Reduced-flow showerheads with 1.8 to 2.5 gpm (6.8 to 9.5 L/min) flow are now commonplace. Reduced flow not only conserves water but also energy used to heat water.

Similar savings are realized by using the modern water-saving toilets. Originally, the water tank used to flush a toilet bowl was mounted high on the wall so the water would gain enough velocity to flush away the waste. This could be accomplished with between 1 and 1½ gal (3.8 to 5.7 L) of water. Over the years, it became fashionable to put the tank lower, and finally units were designed that were single low-profile units. All of these improvements meant that more water volume was necessary to wash away the waste, and each flush used from 5 to 7 gal per flush (gpf, 19 to 26 Lpf) of water.

U.S. Energy and Water Conservation Standards require toilets manufactured after January 1, 1994, use no more than 1.6 gpf (6 Lpf). In addition, there are a number of devices available for use in existing water closet tanks to cut the use of water by 50% and more. Additionally, urinals must operate at 1 gpf or less, whereas the historic (pre-1994) flush rate for a urinal was 3 gpf (11 Lpf). Showerheads must operate at or below 2.5 gpm (9.5 L/min). Kitchen faucets are limited to 2.2 gpm (8.4 L/min). These installations save significant amounts of potable water. In addition, they greatly reduce the amount of water that must be treated by sewage treatment plants and reduce the need for additional treatment plants.

Use of Untreated Water

Presently, potable water is used for many functions that could be done with nonpotable water. As the potable water supply becomes more valuable and costs increase, more communities will require the use of *nonpotable* or *untreated* water wherever possible. These communities are installing separate water mains to provide nonpotable water to homes and businesses to preserve their supply of potable water. Water from nonpotable supplies is typically used for landscape irrigation. In some communities, the cost of potable water is so high that many of the residents have put in shallow wells that provide them with irrigation water for lawns and gardens and for washing cars.

Use of Gray Water

Another approach to conserving potable water is a water reuse system. This system, known as a *gray-water* system, involves the processing of household wastewater for reuse. In the design of the gray-water system, the water from the bathtub or shower and the washing machine is run into a collection tank instead of going into the sewer lines. From the collection tank, the water

is filtered and chlorinated and then reused as water to flush the toilets. This water reuse system cuts water consumption by about one-half. The potable water system is kept completely separate from the reuse portion of the system to prevent cross-contamination. Soiled water from water closets goes directly into the sewer. These are discussed in Chapter 15.

In another experiment, all of the household water, except that from the garbage disposal and water closets, is processed for multiple reuses in the system. This results in savings of up to 70% of overall household water consumption. In addition to the reduced amount of water required, savings from such a system result in the following:

1. Smaller community or private sewer systems
2. Smaller community treatment plants required to treat sewage
3. Smaller community treatment plants required to treat supply water (when required)

12.4 WATER SUPPLY SYSTEMS

In large metropolitan areas, most of the drinking water originates from a surface source such as a lake, stream, river, or reservoir. In rural areas, people are more likely to drink groundwater that was pumped from a well. The design of any building water supply system begins with an evaluation of the system from which the water will be obtained. Basically, water is available through systems that serve a community or through private systems.

Community Systems

Community water supply systems are public or private entities that install and provide a central supply of water to a neighborhood, city or special district. They are government owned, as in most cities, or privately owned, such as in a rural housing development where the builder or real estate developer develops the water system. The water for these systems may have been obtained from any of the water sources discussed previously. Quite often water is drawn from more than one source. For example, part of the water may be taken from a river or reservoir, and it may be supplemented by deep wells.

Before proceeding with the design of the water supply, the following information should be obtained and evaluated:

1. What is the exact location of the water main (pipe) in relation to the construction site?
2. If the main is on the other side of the street from the construction site, what procedures must be followed to get permission (in writing) to cut through the street, set up barricades, and patch the street? If available, obtain the specifications (written requirements) concerning the cutting and patching of the street.
3. What permits are required from local authorities, how much do the permits cost, and who will inspect the work and when?

4. If the water main does not run to the construction site, can it be extended from its present location to the property? Who pays for the extension? How long will it take?

5. Is there a charge to connect (tap) onto the community system? Many communities charge a fee just to tap, and the charge is often high (e.g., thousands or tens of thousands of dollars).

6. What is the water pressure in the main? Plumbing fixtures are manufactured to operate efficiently with water pressures from about 30 to 60 psi (200 to 400 kPa). Pressure outside this range may result in poor operation.

   a. A storage tank and pump may be required to raise the pressure if the pressure is too low (below 30 psi or about 200 kPa for a residence). Such a system is often used on commercial and industrial projects where the pressure requirements may have to be quite high to meet the water demands.

   b. Water pressure that is too high (above 80 psi or about 551 kPa for a residence) will probably require a pressure-reducing valve in the system to decrease the pressure to an acceptable level.

Low pressure may cause certain fixtures to operate at a low flow rate, especially showers, flush valve water closets (toilets), and garden hoses. Rapid wearing out of the washers and valves, and noises in the piping, is typical of problems associated with high pressure. The required water pressure at various fixtures and the water pressure from the main to the fixtures are discussed in Chapter 13.

7. What is the cost of the water? Typically, a water meter is installed, either out near the road or somewhere in the project, and there is a charge for the water used. After determining what the charges are, a cost analysis may show that it is cheaper to put in a private system. Some areas do not allow private systems for potable water, but quite often it will be desirable to put in a well to provide nonpotable water for irrigating the lawn and garden and for washing the car. Where costs for potable water are extremely high, it may be feasible to use separate potable and nonpotable water supply systems within the project. This consideration is especially true on industrial and commercial projects.

Water meters are required in all community systems that charge for water usage or in systems where water consumption must be monitored. The water meter measures the amount of water that passes through it, and then the user is billed for that amount. In cold-climate areas, the meter is usually located where it is least likely to freeze, typically in a basement or underground below the frost line. The amount of water is transmitted to a recording device outside the building so it can be read at any time. In warm-climate regions, the water supply line is near the surface of the ground, and the meter is often located in a small box near the street.

Private Systems

Approximately 42 million people in the United States obtain water from their own private drinking water supplies. Private systems may also use any of the water sources discussed. Most private water is drawn from groundwater through wells (U.S. Geological Survey, 1995).

Large industrial and commercial projects may draw all of their supply from one source (e.g., a stream) or they may draw part of their supply from one source (such as a stream) and supplement the supply with another source (e.g., a well). Such systems often include treatment plants, water storage towers, and sometimes reservoirs to store the water.

Small private systems, such as those used for residences, usually rely on a single source of water to supply potable water to the system. Installing a well is the most commonly used method of obtaining a source of water. Springs may be used when available. Most private water systems use wells to tap the underground water source.

Many industries and businesses that draw their water from community systems have alternative private systems that can be put into operation in case of a water shortage from breakdown of the community system or a prolonged drought.

Experts (usually consulting mechanical engineers, geological engineers, or water supply and treatment specialists) should be consulted early in the planning for any large project requiring its own private water system. Such specialists can make tests, interpret what the tests mean to the project, and make recommendations as to the quality and amount of water available. Permits from state or local governmental entities are also required.

Well Systems

A modern well system consists of a well, a motor-driven pump, and a storage tank. In most systems, the pump draws water from the well where it is stored in a storage tank. A switch activated by water pressure controls the pump. As water is consumed in the building, it is drawn from the storage tank and the pressure in the tank decreases. When tank pressure drops to a preset cut-in pressure level, the switch activates the well pump. Pumped water replaces the water drawn from the tank. The pump is switched off when the tank pressure rises to a preset cut-out pressure.

Types of Wells

Wells are classified according to their depth and the method used to construct the well:

Classification	Depth	Construction Method
Shallow	Less than 25 ft (7.6 m) in depth	Dug, driven, and shallow bored
Deep	25 ft (7.6 m) or more in depth	Drilled and bored

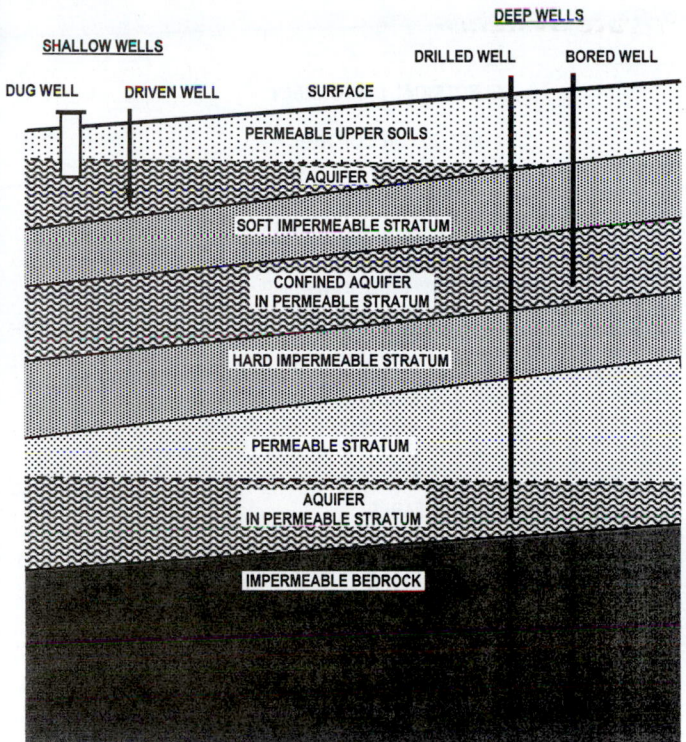

FIGURE 12.3 There are three common types of water wells: dug, driven, and bored/drilled.

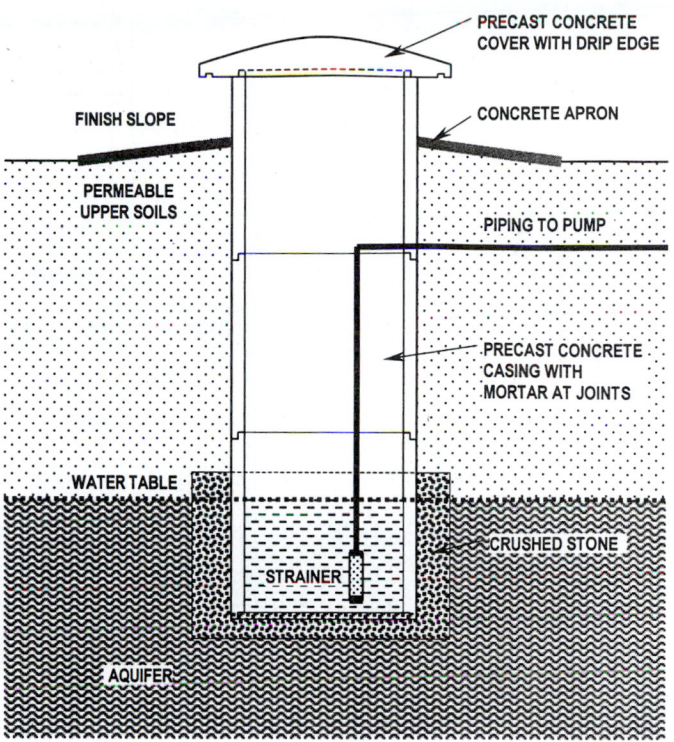

FIGURE 12.4 A dug well is shallow. It is excavated at surface level. The top of a dug well extends above grade. A concrete cap has a drip edge to prevent contamination.

The depth of the well is determined by the depth of the water table and the amount of water that can be pumped from that depth. Where the water table is high, it may be necessary to go less than 25 ft (7.6 m) into the earth. Where a suitable water table is deep, it is not unusual for a well to be 1000 ft (305 m) or more deep.

There are three common types of water wells: dug, driven, and bored/drilled. (See Figure 12.3.) Today, the drilled well is the most common type for private water supply.

Dug Wells Dug wells are shallow wells, generally not more than 25 ft (7.6 m) deep, and typically 3 to 6 ft (1 to 2 m) in diameter. They are typically made by excavation with a backhoe or excavator but can be hand shoveled, as this was the method typically used for hundreds of years. (See Figure 12.4.)

The excavation of a dug well is lined with rock, masonry, cast concrete, prefabricated concrete pipe, ceramic material, or another substance. Water enters the well through joints in the top 20 ft or so of steel or concrete casing. To minimize the chances of surface contamination, the well should have a watertight top and walls. The top should be either above the ground or sloped so that surface water will run away from it and not over it. The watertight walls should extend at least 10 ft (3 m) into the earth.

Generally, groundwater will flow into the well through the bottom of the well and the water in the well will rise to about the level of the water table. Some wells also allow water to seep through the walls by using porous construction materials near the bottom of the wall. This porous construction may be perforated seepage pits or concrete masonry placed without mortar.

The placing of washed gravel in the bottom of the dug well, and on the sides of the well when porous walls are used, will reduce the sand particles or discoloration in the well water. Washed gravel is gravel (rounded river stone) that has been put through a wash (water sprayed over the stone) to remove much of the sand or clay from the stone. To further protect the water from possible contamination, a tight seal is required around the suction line where it passes through the wall.

Because they are shallow, dug wells are vulnerable to contamination from surface pollution sources such as cattle manure, fertilizer, trash, and so forth. Surface flooding from rivers and streams can also carry pollutants into the well. Because of this susceptibility, dug wells are typically no longer permitted to provide potable water in developed areas. They can be used for irrigation and industrial applications.

Driven Wells Driven wells, also referred to as *sand point wells*, consist of lengths of 1¼ to 2 in (32 to 50 mm) diameter pipe that is driven into the ground. In driving this type of well, a sharp well point and drive cap are attached to a pipe. An impact-loading device such as a small pile driver is used to drive the assembly into the earth by striking the drive cap. The *drive cap* allows the assembly to be driven into the earth without the pipe end being damaged. The assembly is driven into the ground until the well point extends below the water table. As the well point is driven,

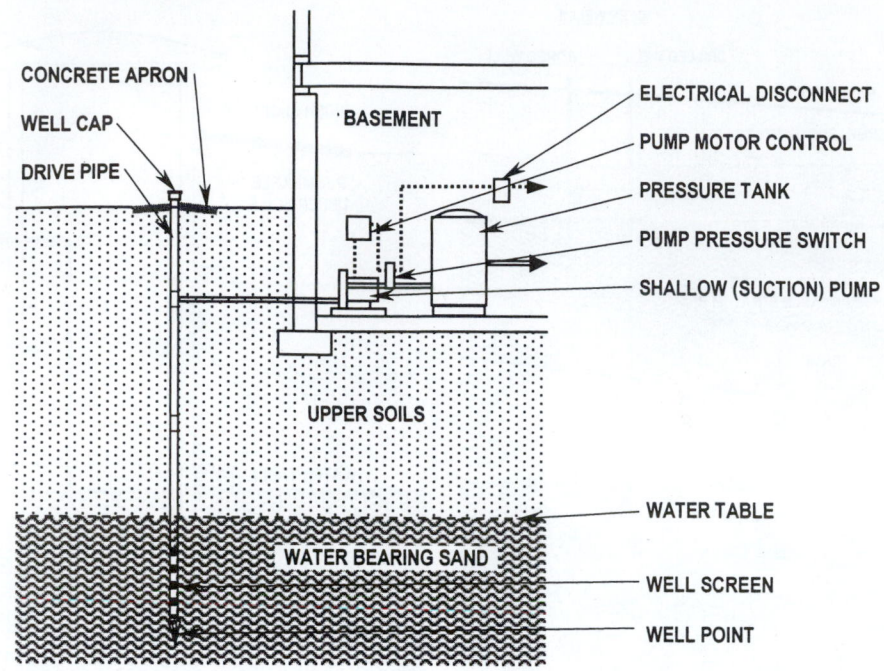

FIGURE 12.5   A driven well.

additional lengths of pipe may be attached (usually 5 ft [1.5 m] lengths are used) to the assembly by the use of a coupling. The *well point* is a pointed perforated pipe or a pipe with a pointed well screen that allows water to be sucked up the pipe to the surface by a shallow well pump. (See Figures 12.5 and 12.6.)

Drilled/Bored Wells   Drilling or boring methods are used for deep wells. A well-drilling rig is used to create the well hole. *Drilled wells* have the holes formed by using rotary bits. *Bored wells* have the holes formed by using an auger and covered with a casing. Only the drilling method is effective in cutting through hard rock. Drilled wells typically have holes 200 feet or more in depth. Shallow wells may have to be bored or drilled if it is necessary to pass through rock to reach the water table. The *well shaft*, or *borehole*, is lined with a solid pipe that seals out contaminants and stabilizes the hole. (See Figure 12.7.)

During drilling or boring a hole, a pipe-like *casing* is lowered into the hole. This pipe is usually 4 to 6 in (100 to 150 mm) in diameter in sections with threaded or welded joints that must be watertight. Steel and thermoplastic pipe are the most practical casing materials for wells. Brass, copper, and stainless steel pipe and tubing have been used where soil or water corrosion seriously limited the estimated life of steel pipe. Thermoplastic (e.g., styrene rubber [SR]) and fiberglass pipe have been used for casing rotary-drilled and straight percussion drilled wells. The casing prevents collapse of the well hole and the entrance of contaminants. It also allows placement of a pump or pumping equipment.

The *wellhead* is the top of the well, the part that rises above the surface of the soil. Usually, at least 12 in of casing

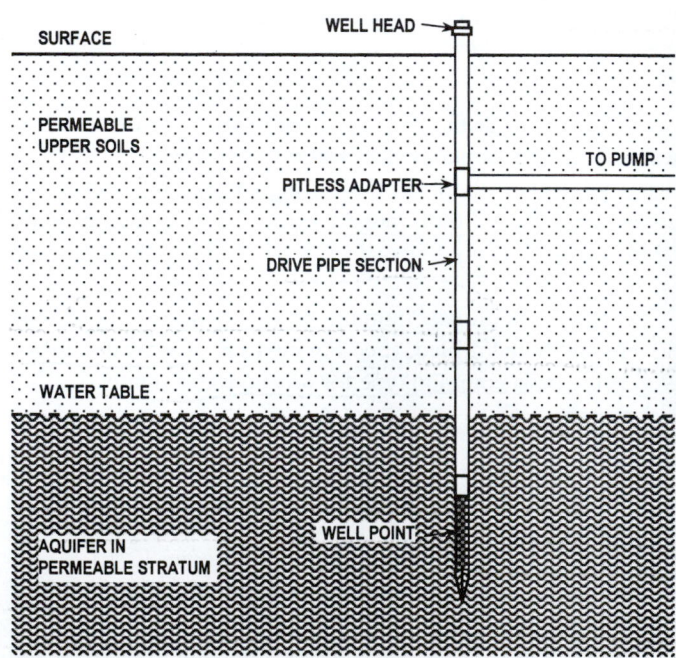

FIGURE 12.6   Driven wells consist of lengths of pipe attached one on top of another and driven into the ground. A sharp well point and drive cap are attached to a pipe.

must extend above the soil line, which is capped, then stabilized with a concrete slab. The slab slopes outward, and extends at least 2 ft on all sides. The space between the casing and the sides of the hole provides a direct channel for surface water (and pollutants) to reach the water table. To seal off that

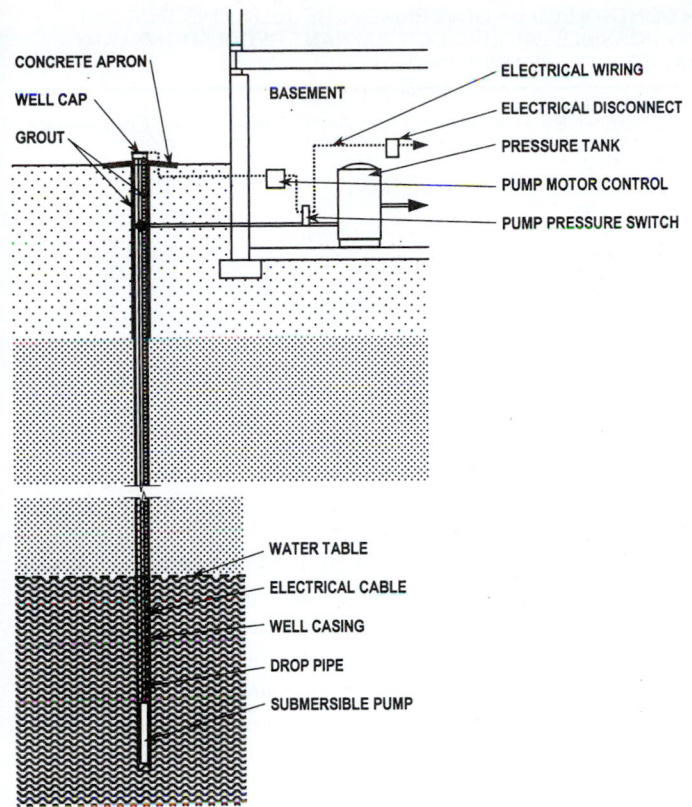

FIGURE 12.7 Drilled wells typically have holes 200 ft or more in depth. A submersible pump is dropped into the well casing to pump water up the well.

channel the space between the casing and soil is filled with a mixture of gravel and cement called grout. To protect the inside of the casing, the driller installs a tight-fitting *well cap*. To further protect against surface drainage and contamination, a *concrete apron*, sloping away from the well, is generally cast around the casing at the ground surface.

Well Pumps

Pumps are used to bring well water to the surface. Well pumps are referred to as *shallow well* and *deep well*, depending on the type and depth of well. *Pumping level*, expressed in feet or meters, is the vertical distance between the pump and the lowest water level, taking into account level draw down by pumping and lower levels during dry seasons. Most well pumps are powered on 120 or 240 V AC single-phase electricity. They require a motor control box, which includes motor starter, relay, thermal overload, and capacitors.

There are two general types of well pumps: *submersible well pumps* and *jet pumps*.

Well Jet Pumps The *well jet pump* combines centrifugal and ejection pumping. In addition to a motor, impeller, and diffuser, the jet pump includes a jet (ejector) assembly that consists of a nozzle and venturi tube. Jet pumps are self-priming, but prim-

ing (manually filling with water) prior to initial use is required for the pump to operate.

Shallow well jet pumps are used for wells with a pumping level up to 25 ft (7.6 m) deep, which is the pump's suction lift limit. It has no working parts submerged in water. The jet assembly is located on the suction side of the pump impeller. Water is supplied to the nozzle under pressure from the pump. As the drive water moves through the nozzle and venturi, a partial vacuum is created, drawing more water from the well up the suction pipe. A portion of the water is directed to the discharge outlet for the distribution system; the rest is recirculated to the ejector.

Deep well jet pumps can be used for wells with a pumping level up to 120 ft (37 m) deep. It works the same as the shallow well type but with the jet assembly located in the well 10 to 20 ft (3 to 9 m) below water table level. A two-pipe system uses a drive pipe (pressure pipe) to push pressurized water from the pump down into the ejector. Well water is drawn up through the foot valve by suction at the ejector, and with the drive water flows up a return pipe (suction pipe) to the pump. A portion of the water is then recirculated back to the ejector and the rest flows to the distribution system. Deep well jet pumps come in single- or multistage models (i.e., two or more impeller and diffuser). The convertible jet pump allows for conversion to either shallow well or deep well operation.

Submersible Well Pumps *Submersible well pumps* are centrifugal pumps designed to operate submersed in water near the bottom end of the well shaft. It is typically used in wells with a pumping level of at least 75 ft (23 m) deep. The pump is submerged into the well water, usually to about 20 ft (6 m) from the bottom of the well. It may be set hundreds of feet beneath the water in a well. This system usually alleviates pump-freezing concerns by placing the pump deep inside the well.

Centrifugal pumps work by using their rotation of impellers to push water outward and then upward through the well shaft. A small electric motor, called a *driver*, is installed in the well shaft, usually below the pump itself. An electric cable is attached to the pump's motor. Electric cable runs through the length of the shaft and into the pump house or building. When the pump is activated, the motor rotates the impellers and pushes water up out of the well. Because the diameter of wells is restrictive, the impellers have to be stacked on top of each other to exert enough pressure to force the water up through the pipe; for this reason, submersible well pumps are very long. A standard 4 in (100 mm) submersible pump measures 24 to 48 in (0.6 to 1.2 m) in length.

Well Tanks

Water drawn from a well is pumped into a storage tank where it is stored for use by building occupants. Elevated storage tanks are discussed later in this chapter. A typical utility-powered well system consists of a pump delivering water into a pressure tank.

A *pressure tank* is a type of closed storage container designed to store water under pressure. In a well system, a pressure tank is used to hold water under pressure after it is pumped

TABLE 12.6   WELL LOCATION AND CONSTRUCTION ARE OFTEN CONTROLLED BY GOVERNMENT REGULATIONS THAT SET MINIMUM DISTANCES BETWEEN THE WELL AND ANY POSSIBLE GROUND CONTAMINANT. REGULATIONS VARY WITH BUILDING CODE. A TYPICAL SET OF MINIMUM DISTANCES IS PROVIDED.

Possible Ground Contaminant	Distances Between the Well and Any Possible Ground Contaminant	
	ft	m
Road or highway boundary, property line	25	12
Building sewer line	50	15
Distribution box	50	15
Grave or cemetery, termite-treated building slab, unplugged abandoned wells, existing operating well, barnyards, agricultural fields, home gardens	100	15
Drain field	100	30
Seepage pit	100	30
Landfill or lagoon, road salt stock piles, storage tank for petroleum, petroleum products or chemicals, storage areas for manure, commercial fertilizers or chemicals	Varies 200 to 1000	Varies 61 to 305

to ensure steady water pressure in the building. This type of tank is divided into two internal compartments by a flexible diaphragm or bladder.

In the diaphragm pressure tank, the diaphragm separates the water storage section from a pressurized air chamber. In the bladder design, a balloon-like bladder holds pressurized air. The pressurized chamber is charged with air, which applies a force against the water, inducing and maintaining a consistent pressure in the stored water. The bladder or diaphragm separate the pressurized air from the water pumped into the tank.

A pressure tank stores the energy the pump has produced in the form of pressure. Air is compressed in the pressurized air chamber. When water is drawn from a plumbing fixture, the pressure in the tank is released in the form of water flow. After about a third to half of the tank capacity is drained, a device called a *pressure switch* turns the pump on to restore the pressure in the tank. When the pressure in the tank reaches an upper limit, usually about 50 psi (340 kPa), the pump cycles off. The pressure is stored in the tank until it is needed again.

A pressure tank will hold storage water on reserve under pressure so small demands do not require the pump to switch on. It extends the time between pumping cycles and therefore prolongs the life of the pump and motor. The tank must be protected from freezing by locating it in a pump house or basement, or by burying it in a pump pit below the frost line.

Well Design and Installation

The maximum flow rate of water drawn from a well is referred to as the *yield* or capacity of the well; it is expressed in gpm or L/m. A well yielding 10 gpm provides five times the water, as does a well yielding 2 gpm. A well testing company certifies well yield.

Water demand is the amount of water required to meet the demands of the building served by the well system. For example, a general average for each member of a household, for all purposes, including kitchen, laundry, bath, and toilet (but not including irrigation water), is about 75 gal (284 L) per day per person.

Once demand has been calculated, it can be compared to well yield to determine storage tank size and whether additional wells will be needed to provide the required water for the project. Methods of estimating water demand are introduced in Chapter 13.

Well location and construction are often controlled by government regulations that set minimum distances between the well and any possible ground contaminant. See Table 12.6. Various authorities and government regulations require different minimum distances, and there is no single set of standards used. Additionally, when certain types of construction methods are used, these regulations may even require that licensed well drillers install the well. It is important that local regulations be observed for each project.

Possible well contamination from an underground flow of contaminants through rock formations that allow free-flowing groundwater to travel long distances is always possible, especially through strata of eroded limestone. Frequent testing of water quality is required wherever there is a possibility of such contamination. For the well contaminated in this manner, three methods are used to eliminate the problem: water treatment, relocation of the well, and elimination of the source of the contamination.

Before planning the well, local conditions should be analyzed to provide background information. For example, existing local wells should be checked for depth and yield of water. This information can be obtained from local well drillers and government agencies and, if possible, verified by testing existing or just-completed wells.

All wells should be tested to determine the yield and the quality of the water. It is important that this be done at an early stage in the design so that the size of the water storage tank can be determined and so that any required water treatment equipment can be designed and space allowed in the design of the project to locate the tank and equipment. Where insufficient information on well yields is available, and especially where large projects will require substantial water supplies, it may be necessary to have test holes made so that the yield can be checked.

When a large supply of water is required for the continuous operation of the project, it may be necessary to put in other wells to be certain that the water yield will be sufficient to meet the projected demand. For example, if one well provides adequate water, it may be a good investment to have a second well put in to act as a backup in case the first well should fail. A backup well is usually not required for residences. However, for industries or businesses that require large amounts of water such as a car wash, farm, or apartment complex, it is a wise investment.

When more than one well is required, the wells must be spaced so that the use of one well will not lower the water table in the other well. In general, deep wells must be 500 to 1000 ft (150 to 300 m) apart, while shallow wells must be 20 to 100 ft (6 to 30 m) apart. Because of geological variables, the minimum distance between wells can be determined only by testing.

Water Towers and Elevated Storage Tanks

Water towers used in community systems and *elevated water storage tanks* used in private systems carry a reserve capacity of water. They serve many additional purposes, including what follows:

- To introduce pressure to the water supply system
- To equalize supply and demand over periods of high consumption
- To supply water during equipment failure or maintenance
- To supply water for firefighting demand

A water tower must be tall enough to deliver adequate pressure to all of the houses and businesses in the area of the tower. Each foot of water height provides 0.433 psi (pounds per square inch) of pressure. (This is discussed in Chapter 13.) A typical community water supply maintains pressures between 50 and 100 psi (344 and 688 kPa), whereas plumbing fixtures require 8 psi (55 kPa) to 30 psi (206 kPa). An example of a private water tower is shown in Photo 12.1.

Water in a water tower tank must be 100 to 200 ft (30 to 60 m) above the highest plumbing fixture being served. Therefore water towers are typically located on high ground, and they are tall enough to provide the necessary pressure. In mountainous regions, a ground-level water storage tank or reservoir located on the highest hill in the area can sometimes substitute for a water tower.

The capacity of a community water tower for even a small community is quite large; normally it will hold 1 000 000 gal (3 800 000 L) or more. In comparison, a typical in-ground residential swimming pool might hold 20 000 gal (76 000 L). The tank of the water tower is typically sized to hold about two days of water supply. If the pumps fail during a power failure or are down for maintenance, stored water provides an adequate supply under pressure. The extra supply also serves as a reserve for the high demand for water during firefighting situations.

PHOTO 12.1 A private water storage tank. (Used with permission of ABC)

Elevated water storage tanks serve buildings that are too tall to rely on street water pressure. Water is pumped to a storage tank located on top of the building. An elevated storage tank that is 30 to 35 ft (10 to 12 m) above the highest plumbing fixture being served is generally required. Elevated water storage tanks are sized to hold one to two days of water supply plus a reserve for firefighting. An alternative to an elevated storage tank in tall buildings is a pressurized tank—a storage tank that is pressurized to the appropriate pressure.

A big benefit of a water tower and elevated water storage tanks is that water pumps are sized for average rather than peak demand. During periods of high demand, water flows from the tank to the consumer while during periods of low demand the tank refills. For example, water consumption for a pumping station averaging 750 gpm (2850 L/min) is equivalent to 1 080 000 gal/day (4 100 000 L/day). During a period of high demand such as from 7 to 8 AM, water consumption may peak at 3000 gpm (11 400 L/min) and water is removed from the tank at a greater rate than the pumping station is filling the tank. When demand drops off, say at midnight, the tank is refilled. Even though water demand peaks at 3000 gpm (11 400 L/min), the smaller 750 gpm (2850 L/min) pumping capacity is sufficient. A 3000 gpm (11 400 L/min) pumping capacity is not needed.

12.5 PIPING MATERIALS

Pipe is a round, hollow channel used to transport liquids such as water or solid–liquid mixtures such as wastewater from one point to the next. In a building plumbing system, pipes transport hot and cold water and remove liquid and solid wastes. Piping in buildings is also used in transporting natural and liquefied petroleum gases, fuel oil, compressed air, refrigerants, and irrigation water.

Water pipe generally falls into one of two categories: *pressure pipe,* which delivers supply water; and *drain, waste, and vent (DWV) pipe,* which carries waste and soil water away. Both categories are sold in metal and plastic; however, metal (copper in plumbing systems) dominates the pressure category. Plastic and cast iron are the most common piping material for DWV.

Pressure pipe must be heavy enough to hold continuous pressure without rupture, and all connections must be leak proof. This pipe tends to be of a smaller diameter, and it must be made of material that will not react with the chemicals or minerals in the water. DWV pipe provides a channel for waste materials to flow freely away from the fixtures and the building by the force of gravity. It is typically lighter weight with thinner walls than pressure pipe, and joints do not need to be as tightly sealed because there is no pressure exerted on them. DWV pipe is generally larger in size than pressure pipe to allow for free gravity flow, and it must not react to common chemicals that might be poured down a drain. In both pressure and DWV piping, fitting design and joining techniques must be compatible with the pressures and temperatures encountered when the pipe is placed in service.

Piping Materials

Many types of piping materials most commonly used in building plumbing systems as described in the following sections. Photos 12.2 to 12.7 show examples of piping materials. Table 12.7 provides a comparison of common pipe materials.

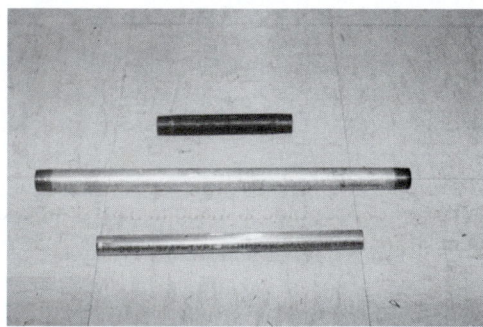

PHOTO 12.2   Piping materials. (Used with permission of ABC)

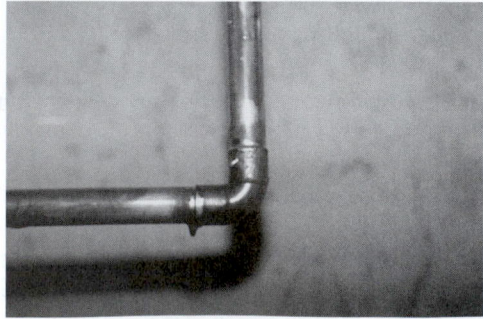

PHOTO 12.3   A sweat soldered copper elbow. (Used with permission of ABC)

PHOTO 12.4   Various copper fittings. (Used with permission of ABC)

PHOTO 12.5   Cast iron pipes and elbow connected with compression fittings. (Used with permission of ABC)

PHOTO 12.6   ABS pipe connected with a solvent welded connection. (Used with permission of ABC)

PHOTO 12.7   Various sanitary drainage fittings. (Used with permission of ABC)

TABLE 12.7   A COMPARISON OF COMMON PIPE MATERIALS.

Type of Pipe	Ease of Working	Water Flow Effectiveness	Common Fittings Used	Manner Usually Stocked	Principal Uses	Comments
Copper— hard temper	Easier to work with than steel	High	Solder connections	20 ft rigid lengths cut to size wanted	Water supply	Most common pipe used in building plumbing systems
Copper— soft temper	Easier to work with than hard copper—bends readily by using a bending tool	High	Solder connections, flare fittings	Common wall thickness: K (thick wall) L (medium wall) coils (soft)	Water supply	Good for remodeling
Black or galvanized steel	Must be threaded; more difficult to cut; measurements for jobs must be exact	Lower than copper because of inner wall roughness	Screw connections	Rigid lengths up to 21 ft; usually cut to size needed	Generally found in older buildings	Recommended if lines are in a location subject to impact
Thermoplastic pipe	Easily cut with a saw or knife	High	Cement-solvent welded; insert couplings; and clamps	Rigid, semirigid— 10 ft, 20 ft lengths; some pipes are flexible	Good alternative to metal pipe	Lightest, most flexible, and most brittle of all pipe materials

Copper Pipe and Tubing

Copper tubing is traditionally the most popular water supply pipe material. It is also used in water space heating (hydronic) systems, air conditioning and refrigeration systems, sanitary drainage, and natural gas and liquid petroleum gas piping. The thin walls of copper tubing are usually soldered to fittings. This allows the pipes and fittings to be set into place before the joints are connected with solder. This advantage generally allows faster installation of copper pipe in comparison to treaded steel or brass.

The types of copper tubing available are K, L, and M, with K having the thickest walls, then L, and finally M, with the thinnest walls of this group. DWV copper tubing is used for drainage, waste, and vent piping. Types K and L are preferred for pressure applications. Type M and DWV are used for low- and no-pressure applications. Types of commercially available copper tubing are summarized in Table 12.8. They are described in the following:

Type K   Type K copper tube is available as either rigid (hard temper) or flexible (soft temper). Type K is used primarily for underground water service in water supply systems. It is available in the following nominal diameters: ⅜, ½, ¾, 1, 1¼, 1½, 2, 2½, 3, 3½, 4, 5, 6, and 8 in.

Soft temper tubing 1 in and smaller is usually available in coils 60 or 100 ft (18.3 or 30.5 m) long, while 1¼ and 1½ in tubing is available in 40- or 60-ft (12.2 or 18.3 m) coils. Hard temper is available in 12- and 20-ft (3.7 and 6.1 m) straight lengths. Type K copper tubing is color coded in green for quick visual identification.

Type L   Type L copper tube is also available in either hard or soft temper and in coils (soft temper only) and straight lengths much like Type K. It is available in the following nominal diameters: ⅜, ½, ¾, 1, 1¼, 1½, 2, 2½, 3, 3½, 4, 5, 6, 8, and 10 in. The soft temper tubing is often used as replacement plumbing because the flexibility of the tube allows easier installation. Hard temper tubing is often used for new installations, particularly in commercial work. Type L copper tubing is color coded blue. This type of tubing is most popular for use in water supply systems.

Type M   Type M copper tube is made in hard temper only and is available in straight lengths of 12 and 20 ft (3.7 and 6.1 m). It is used for branch supplies where water pressure is not too great, but it is not used for risers and mains. It is available in the following nominal diameters: ½, ¾, 1, 1¼, 1½, 2, 2½, 3, 3½, 4, 5, 6, 8, and 10 in. It is also used for chilled water systems, exposed lines in hot water heating systems, and drainage piping. Type M copper tubing is color coded red.

DWV   DWV copper tube is the thinnest copper tube and is used in nonpressure applications. It is made in hard temper only and is obtainable in straight lengths of 20 ft (6.1 m). It is available in the following nominal diameters: 1¼, 1½, 2, 2½, 3, 3½, 4, 5, 6, 8, and 10 in.

The diameter of copper pipe is expressed in nominal size. The actual size is ⅛ in larger than the nominal size expressed (e.g., a 1-in copper pipe has an actual outside diameter of 1⅛ in). Regardless of type, the outside diameter does not vary for a specific nominal diameter. Inside diameter will vary with wall thickness. The inside diameter of a thin-wall pipe will be

TABLE 12.8   TYPES OF COMMERCIALLY AVAILABLE COPPER TUBING.

Tube Type	Color Code	Typical Application or Use	Nominal Sizes (in)		Drawn (Hard)	Annealed (Soft)
Type K	Green	Domestic water Service/distribution Fire protection Solar Fuel/fuel oil HVAC Snow melting	**Straight**			
			¼ in to 8 in		20 ft	20 ft
			10 in		18 ft	18 ft
			12 in		12 ft	12 ft
			Coils			
			¼ in to 1 in		—	60 ft
					—	100 ft
			1¼ in and 1½ in		—	60 ft
			2 in		—	40 ft
					—	45 ft
Type L	Blue	Domestic water Service/distribution Fire protection Solar Fuel/fuel oil LP gas Snow melting	**Straight:**			
			¼ in to 8 in		20 ft	20 ft
			12 in		18 ft	18 ft
			Coils			
			¼ in to 1 in		—	60 ft
					—	100 ft
			1¼ in to 1½ in		—	60 ft
			2 in		—	40 ft
					—	45 ft
Type M	Red	Domestic water Service/distribution Solar Fuel/fuel oil HVAC Snow melting	**Straight**			
			¼ in to 12 in		20 ft	N/A
DWV	Yellow	Drain, waste, vent Solar HVAC	**Straight**			
			1¼ inch to 8 inch		20 ft	N/A
ACR	Blue	Air conditioning Refrigeration Natural gas LP gas	**Straight:**			
			⅜ in to 4⅛ in		20 ft	N/A
			Coils:			
			⅛ into 1⅝ in		—	50
OXY, MED OXY/MED OXY/ACR ACR/MED	(K) Green (L) Blue	Medical gas	**Straight**			
			¼ in to 8 in		20 ft	N/A
Type G	Yellow	Natural gas LP gas	**Straight**			
			⅜ in to 1⅛ in		12 ft	12 ft
					20 ft	20 ft
			Coils			
			⅜ in to ⅞ in		—	60 ft
					—	100 ft

greater than the inside diameter of a thick-wall pipe. Weights and dimensions of copper tubing are provided in Table 12.9. Weights and dimensions of copper and brass pipes are provided in Table 12.10.

Compared with iron or steel pipe, copper pipe has the advantage of not rusting and of being highly resistant to any accumulation of scale (particles) in the pipe. Copper tubing has a lower friction loss than wrought iron or steel, providing an additional advantage. Also, the outside dimensions of the fittings are smaller, which makes a neater, better-looking job. With wrought iron and steel pipe, the larger outside dimensions of the fittings sometimes require that wider walls be used in the building.

Copper piping should not be installed if it will carry water having a pH of 6.8 or less, as this could cause copper to corrode from the acidic nature of the water at this pH. The majority of public utilities supply water at a pH between 7.2 and 8.0. Private well water systems often have a pH below 6.8. When this it the case, it is suggested that a treatment system be installed to make the water less acidic.

Brass Pipe

Red brass piping, consisting of approximately 85% copper and 15% zinc, is used as water supply piping. The pipe is threaded for fitting connections, but this requires thicker walls to

TABLE 12.9   WEIGHTS AND DIMENSIONS OF COPPER TUBING.

Nominal Size (in)	Outside Diameter (in)	Type K		Type L		Type M	
		Inside Diameter (in)	Weight per ft (lb)	Inside Diameter (in)	Weight per ft (lb)	Inside Diameter (in)	Weight per ft (in)
⅛	0.250	0.186	0.085	0.200	0.068	0.200	0.068
¼	0.375	0.311	0.134	0.315	0.126	0.325	0.106
⅜	0.500	0.402	0.269	0.430	0.198	0.450	0.144
½	0.625	0.527	0.344	0.545	0.284	0.569	0.203
⅝	0.750	0.652	0.418	0.666	0.362	0.690	0.263
¾	0.875	0.745	0.641	0.785	0.454	0.811	0.328
1	1.125	0.995	0.839	1.025	0.653	1.055	0.464
1¼	1.375	1.245	1.04	1.265	0.882	1.291	0.681
1½	1.625	1.481	1.36	1.505	1.14	1.571	0.940
2	2.125	1.959	2.06	1.985	1.75	2.009	1.46
2½	2.625	2.435	2.92	2.465	2.48	2.495	2.03
3	3.125	2.907	4.00	2.945	3.33	2.981	2.68
3½	3.625	3.385	5.12	3.425	4.29	3.459	3.58
4	4.125	3.857	6.51	3.905	5.38	3.935	4.66
5	5.125	4.805	9.67	4.875	7.61	4.907	6.66
6	6.125	5.741	13.87	5.845	10.20	5.881	8.91
8	8.125	7.583	25.90	7.725	10.29	7.785	16.46

TABLE 12.10   WEIGHTS AND DIMENSIONS OF COPPER AND BRASS PIPES.

Nominal Size (in)	Outside Diameter (in)	Inside Diameter (in)	Weight per ft (in lb)		
			67% Copper	85% Copper	100% Copper
⅛	0.405	0.281	0.246	0.253	0.259
¼	0.540	0.375	0.437	0.450	0.460
⅜	0.675	0.494	0.612	0.630	0.643
½	0.840	0.625	0.911	0.938	0.957
¾	1.050	0.822	1.24	1.27	1.30
1	1.315	1.062	1.74	1.79	1.83
1¼	1.660	1.368	2.56	2.63	2.69
1½	1.900	1.600	3.04	3.13	3.20
2	2.375	2.062	4.02	4.14	4.23
2½	2.875	2.500	5.83	6.00	6.14
3	3.500	3.062	8.31	8.56	8.75
3½	4.000	3.500	10.85	11.17	11.41
4	4.500	4.000	12.29	12.66	12.94
4½	5.000	4.500	13.74	14.15	14.46
5	5.563	5.063	15.40	15.85	16.21
6	6.625	6.125	18.44	18.99	19.41
7	7.625	7.062	23.92	24.63	25.17
8	8.625	8.000	30.05	30.95	31.63

accommodate the threading, making installation and handling more difficult than for copper. In addition, its relatively higher total cost, installed on the job, limits its usage. Brass piping has seen limited use in new construction.

Steel and Iron Pipe

Steel pipe is available in the following nominal diameters: ⅜, ½, ¾, 1, 1¼, 1½, 2, 2½, 3, 3½, 4, 5, 6, 8, 10, and 12 in. It is typically sold in lengths of 21 ft. When steel pipe is forged, a black oxide scale forms on its surface that gives it a dull black finish, and as a result it is called *black pipe*. Because steel is subject to rust and corrosion, the pipe manufacturer also coats it with protective oil. Black pipe is most commonly used for natural gas supply lines and fire suppression sprinkler system lines.

Galvanized steel pipe is covered with a protective coating of zinc that greatly reduces its tendency to corrode and thus

TABLE 12.11   WEIGHTS AND DIMENSIONS OF STANDARD WEIGHT SCHEDULE 40 STEEL PIPE.

Nominal Size (in)	Outside Diameter (in)	Inside Diameter (in)	Wall Thickness (in)	Weight per Foot Plain Ends Pounds	Inside Cross-Sectional Area (in²)	gpm at 10 ft/s Velocity
⅛	0.405	0.269	0.068	0.244	0.057	1.8
¼	0.540	0.364	0.088	0.424	0.104	3.2
⅜	0.675	0.493	0.091	0.567	0.191	6.0
½	0.840	0.622	0.109	0.852	0.304	9.5
¾	1.050	0.824	0.113	1.130	0.533	16.6
1	1.315	1.049	0.133	1.678	0.864	26.9
1¼	1.660	1.380	0.140	2.272	1.495	46.6
1½	1.900	1.610	0.145	2.717	2.036	63.5
2	2.375	2.067	0.154	3.652	3.355	105.0
2½	2.875	2.469	0.203	5.793	4.788	149.0
3	3.500	3.068	0.216	7.575	7.393	230.0
3½	4.000	3.548	0.226	9.109	9.886	308.0
4	4.500	4.026	0.237	10.79	12.73	397.0
5	5.563	5.047	0.258	14.62	20.01	623.0
6	6.625	6.065	0.280	18.97	28.89	900.0

extends its life expectancy. It is moderately corrosion resistant and suitable for mildly acid water. It was commonly used for water supply, waste, and vent lines in plumbing systems through the early 1950s. It is not frequently used for water supply lines today because the minerals in the water react with the galvanizing material and form scale, which builds up over time and will eventually clog the pipe.

Weights and dimensions of standard weight steel pipe are provided in Tables 12.11, 12.12, 12.13, and 12.14.

Steel pipe is typically cut and threaded to fit the job. Fittings for this type of pipe are of malleable (soft) cast iron. They connect by screwing onto the threaded pipe, after applying a small amount of pipe joint compound on the threads. Larger diameter pipe is typically welded rather than threaded.

Concerns arise as galvanized steel pipe ages: corrosion of the inner surface of the pipe restricts water flow; development of rust flakes loosen, collect, and restrict water flow in fittings and valves; and leaks form from corrosion. Therefore, galvanized steel pipe is not used extensively in water supply systems. Steel pipe is connected to its fittings with threaded connections. Steel pipe also has a higher friction loss than copper.

Lightweight wrought-iron pipe, designated Standard (or Schedule 40), is the type most commonly used for water supply and fire suppression sprinkler systems. The most commonly used wrought-iron pipe is galvanized. The zinc-galvanized coating adds extra corrosion resistance. Occasionally, it is used as the service main from the community main to the riser.

TABLE 12.12   WEIGHTS AND DIMENSIONS OF EXTRA STRONG (XH) WEIGHT SCHEDULE 80 STEEL PIPE.

Nominal Size (in)	Outside Diameter (in)	Inside Diameter (in)	Wall Thickness (in)	Weight per Foot Plain Ends Pounds	Inside Cross-Sectional Area (in²)	gpm at 10 ft/s Velocity
⅛	0.405	0.215	0.095	0.314	0.036	1.1
¼	0.540	0.302	0.119	0.535	0.072	2.2
⅜	0.675	0.423	0.126	0.738	0.141	4.4
½	0.840	0.546	0.147	1.087	0.234	7.3
¾	1.050	0.742	0.154	1.473	0.433	13.5
1	1.315	0.957	0.179	2.171	0.719	22.4
1¼	1.660	1.278	0.191	2.996	1.283	40.0
1½	1.900	1.500	0.200	3.631	1.767	55.0
2	2.375	1.939	0.218	5.022	2.953	92.0
2½	2.875	2.323	0.276	7.661	4.238	132.0
3	3.500	2.90	0.300	10.25	6.605	206.0
3½	4.000	3.364	0.318	12.51	8.888	277.0
4	4.500	3.826	0.337	14.98	11.50	358.0
5	5.563	4.813	0.375	20.78	18.19	567.0
6	6.625	5.761	0.432	28.57	26.07	812.0

TABLE 12.13   WEIGHTS AND DIMENSIONS OF SCHEDULE 160 STEEL PIPE.

Nominal Size (in)	Outside Diameter (in)	Inside Diameter (in)	Wall Thickness (in)	Weight per Foot Plain Ends Pounds	Inside Cross-Sectional Area (in²)	gpm at 10 ft/s Velocity
½	0.840	0.466	0.187	1.310	0.171	5.3
¾	1.050	0.614	0.218	1.940	0.296	9.2
1	1.315	0.815	0.250	2.850	0.522	16.3
1¼	1.660	1.160	0.250	3.764	1.057	32.9
1½	1.900	1.338	0.281	4.862	1.406	43.8
2	2.375	1.689	0.343	7.450	2.240	69.8
2½	2.875	2.125	0.375	10.01	3.547	111.0
3	3.500	2.626	0.437	14.30	5.416	169.0
4	4.500	3.438	0.531	22.60	9.28	289.0
5	5.563	4.313	0.625	32.96	14.61	455.0
6	6.625	5.189	0.718	45.30	21.15	659.0

TABLE 12.14   WEIGHTS AND DIMENSIONS OF DOUBLE EXTRA STRONG (XXH) WEIGHT STEEL PIPE.

Nominal Size (in)	Outside Diameter (in)	Inside Diameter (in)	Wall Thickness (in)	Weight per Foot Plain Ends Pounds	Inside Cross-Sectional Area (in²)	gpm at 10 ft/s Velocity
½	0.840	0.252	0.294	1.714	0.050	1.6
¾	1.050	0.434	0.308	2.440	0.148	4.6
1	1.315	0.599	0.358	3.659	0.282	8.8
1¼	1.660	0.896	0.382	5.214	0.630	19.6
1½	1.900	1.100	0.400	6.408	0.950	29.6
2	2.375	1.503	0.436	9.029	1.774	55.3
2½	2.875	1.771	0.552	13.70	2.464	76.8
3	3.500	2.300	0.600	18.58	4.155	129.0
3½	4.000	2.728	0.636	22.85	5.845	182.0
4	4.500	3.152	0.674	27.54	7.803	243.0
5	5.563	4.063	0.750	38.55	12.97	404.0
6	6.625	4.897	0.864	53.16	18.84	587.0

Wrought-iron pipe is threaded for connection to the fittings, and it can be identified by a red spiral stripe on the pipe. The higher cost of wrought-iron pipe limits its increased use. Wrought-iron pipe also has a higher friction loss than copper. Wrought-iron pipe used in buildings is available in the following nominal diameters: ⅜, ½, ¾, 1, 1¼, 1½, 2, 2½, 3, 3½, 4, 5, 6, 8, 10, and 12 in.

Cast iron pipe is commonly used in gravity building and storm drain/sewer systems. Cast iron pipes and fittings are limited to gravity pressure systems. It is available in two grades: Service (SV) for above-grade installations; and, Extra Heavy (XH) for applications below grade. Cast iron pipes are available in 5 and 10 ft lengths with the following nominal diameters: 2, 3, 4, 5, 6, 8, 10, 12, and 15 in. Cast iron pipes and fittings are connected using two methods: hub (female end) and spigot (male end) that are joined by sliding the spigot into the hub; and the no-hub connection that is connected with a rubber gasket and screw-type clamp that is similar to a hose clamp. Dimensions of cast iron pipe are provided in Table 12.15.

Thermoplastic Pipe

Thermoplastic pipe, sometimes referred to simply as *plastic pipe,* is used for water supply systems because its economy and ease of installation make it popular, especially on projects such as low-cost housing or apartments where cost economy is important. It is important to check the plumbing code in force in your locale because some areas still do not allow the use of plastic pipe for water supply systems.

A variety of thermoplastics are used for pipe and fittings in building plumbing systems. Types of thermoplastic materials and their uses are summarized in Table 12.16.

Acrylonitrile butadiene styrene (ABS) thermoplastic pipe is typically black in color. It is generally approved for use in DWV applications. It is available in two grades: Schedule 40 and Service. It is available in straight lengths in the following nominal diameters: 1, 1¼, 1½, 2, 2½, 3, 3½, 4, 5, 6, 8, and 10 in. Solvent-cement welding is used to join ABS pipe and fittings. Dimensions of ABS pipe for drainage, waste, and vent systems are provided in Table 12.17.

TABLE 12.15   DIMENSIONS OF CAST IRON PIPE.

Nominal Diameter (in)	CLASS A 100 Foot Head 43 Pounds Pressure			CLASS B 200 Foot Head 86 Pounds Pressure		
	Outside Diameter (in)	Wall Thickness (in)	Inside Diameter (in)	Outside Diameter (in)	Wall Thickness (in)	Inside Diameter (in)
3	3.80	0.39	3.02	3.96	0.42	3.12
4	4.80	0.42	3.96	5.00	0.45	4.10
6	6.90	0.44	6.02	7.10	0.48	6.14
8	9.05	0.46	8.13	9.05	0.51	8.03
10	11.10	0.50	10.10	11.10	0.57	9.96
12	13.20	0.54	12.12	13.20	0.62	11.96
14	15.30	0.57	14.16	15.30	0.66	13.98
16	17.40	0.60	16.20	17.40	0.70	16.00
18	19.50	0.64	18.22	19.50	0.75	18.00
20	21.60	0.67	20.26	21.60	0.80	20.00
24	25.80	0.76	24.28	25.80	0.89	24.02
30	31.74	0.88	29.98	32.00	1.03	29.94
36	37.96	0.99	35.98	38.30	1.15	36.00
42	44.20	1.10	42.00	44.50	1.28	41.94
48	50.50	1.26	47.98	50.80	1.42	47.96
54	56.66	1.35	53.96	57.10	1.55	54.00
60	62.80	1.39	60.02	63.40	1.67	60.06
72	75.34	1.62	72.10	76.00	1.95	72.10
84	87.54	1.72	84.10	88.54	2.22	84.10

Nominal Diameter (in)	CLASS C 300 Foot Head 130 Pounds Pressure			CLASS D 400 Foot Head 173 Pounds Pressure		
	Outside Diameter (in)	Wall Thickness (in)	Inside Diameter (in)	Outside Diameter (in)	Wall Thickness (in)	Inside Diameter (in)
3	3.96	0.45	3.06	3.96	0.48	3.00
4	5.00	0.48	4.04	5.00	0.52	3.96
6	7.10	0.51	6.08	7.10	0.55	6.00
8	9.30	0.56	8.18	9.30	0.60	8.10
10	11.40	0.62	10.16	11.40	0.68	10.04
12	13.50	0.68	12.14	13.50	0.75	12.00
14	15.65	0.74	14.17	15.65	0.82	14.01
16	17.80	0.80	16.20	17.80	0.89	16.02
18	19.92	0.87	18.18	19.92	0.96	18.00
20	22.06	0.92	20.22	22.06	1.03	20.00
24	26.32	1.04	24.22	26.32	1.16	24.00
30	32.40	1.20	30.00	32.74	1.37	30.00
36	38.70	1.36	39.98	39.16	1.58	36.00
42	45.10	1.54	42.02	45.58	1.78	42.02
48	51.40	1.71	47.98	51.98	1.96	48.06
54	57.80	1.90	54.00	58.40	2.23	53.94
60	64.20	2.00	60.20	64.82	2.38	60.06
72	76.88	2.39	72.10	—	—	—
84	87.54	1.72	84.10	88.54	2.22	84.10

Polybutylene (PB) pipe is a flexible (coils) thermoplastic pipe generally approved for use in potable hot and cold water supply applications. Because of several lawsuits tied to this type of pipe, it is no longer recommended for use in building plumbing systems.

Interior PB pipe is easily recognized by its gray color. Underground service laterals are typically blue in color.

PB is available in copper tube size (CTS) and iron pipe size (IPS). PB pipe cannot be solvent-cement welded, so special fittings are used: a brass, copper, or acetyl plastic insert fitting that slides into the pipe and a crimp ring around the outside of the pipe; a compression fitting with a nut, ring, and cone; and an instant connect fitting that involves sliding the pipe into the fitting and rotating the

TABLE 12.16   TYPES OF THERMOPLASTIC PIPE MATERIALS.

Plastic	Characteristics	Joins By	Colors	Uses
ABS (acrylonitrile-butadiene-styrene)	Rigid	Solvent welding	Black	DWV, sewer, and drain pipe; tubular parts
CPVC (chlorinated polyvinyl chloride)	Rigid, heat resistant	Solvent welding	Beige	Hot and cold water supply tubes, indoors and buried
PB (polybutylene)	Flexible, heat resistant	Mechanical couplings	Gray, beige or blue	Hot and cold water supply tubes, indoors and buried; riser tubes
PE (polyethylene)	Flexible, low cost	Clamped couplings	Black or milky white	Cold water only, outdoor piping, buried
PEX (cross-linked polyethylene)	Flexible, heat resistant	Mechanical couplings	Colored	Hot and cold water supply tubes, indoors; riser tubes
PP (polypropylene)	Semirigid with high heat and chemical resistance	Slip-jam-nut couplings	Beige	Tubular drainage products for fixtures
PVC (polyvinyl chloride)	Rigid with high chemical resistance	Solvent welding	White, gray, beige, and many others	DWV, sewer, and drain pipe; cold water buried pressure pipe; tubular goods
S or RS (styrene or rubber-styrene)	Rigid, low cost	Solvent welding	Black or milky white	Drain pipe outdoors and buried

TABLE 12.17   DIMENSIONS OF ACRYLONITRILE-BUTADIENE-STYRENE (ABS) PLASTIC PIPE FOR DRAINAGE, WASTE, AND VENT SYSTEMS.

Nominal Pipe Size (in)	Outside Diameter (in)	Average Wall Thickness (in)	Inside Diameter (in)	Average Length (ft)
1¼	1.660	0.140	1.380	20
1½	1.900	0.145	1.610	10, 20
2	2.375	0.154	2.067	10
3	3.500	0.216	3.068	10, 20
4	4.500	0.237	4.026	10, 20
6	6.625	0.280	6.065	20

fitting, which causes the fitting and pipe to press together. It is available for water distribution applications in the following nominal diameters: ½, ¾, 1, 1¼, 1½, and 2 in.

Polyethylene (PE) is a flexible (coils) thermoplastic pipe. Black PE pipe is used for buried cold building water supply and irrigation (yard) piping. PE pipe is also approved for use in piping for natural gas and liquefied petroleum gas (LPG) applications, but only when it is directly buried and outside the building foundation. Fusion (melt) welding and compression and flanged connections are used to join PE pipe and fittings carrying gas. PE pipe is available for water distribution applications in the following nominal diameters: ½, ¾, 1, 1¼, 1½, and 2 in IPS. PE fittings are typically copper alloy or plastic barbed insert.

Cross-linked polyethylene (PEX) is a specific type of medium- or high-density polyethylene with individual molecules linking one polymer chain to another. This type of bond makes PEX stronger and more stable than PE with respect to temperature extremes, chemical

attack, and creep deformation. In contrast to metal pipes, it is freeze-break resistant. As a result, PEX plastic pipe is ideally suited for interior potable cold and hot water plumbing applications. PEX tubing has been in use successfully in Europe for plumbing, radiant heating, and snow melt applications since the 1960s. PEX is commonly available in ½, ¾, 1, 1½, and 2 in outside diameter CTS and is packaged in coils or 20 ft straight lengths. See Table 12.18. Some tubing is color coded for easy identification of hot (red) and cold (blue) lines. PEX fittings are generally made of brass, copper, and engineered plastic barbed insert fittings specifically designed for PEX.

Polyvinyl chloride (PVC) is a rigid thermoplastic pipe generally approved for use in pressure applications such as cold water supply applications outside the building (e.g., the building service and in DWV and irrigation piping). It is generally white or gray in color, but can be other colors. PVC is typically rated at 73°F (23°C) and 100 psi (690 kPa), so it is not suitable for potable hot

TABLE 12.18 DIMENSIONS OF CROSS-LINKED POLYETHYLENE (PEX) TUBING.

Nominal Pipe Size (in)	Outside Diameter (in)	Wall Thickness (in)	Inside Diameter (in)
¼	0.375	0.065	0.250
⅜	0.500	0.075	0.350
½	0.625	0.075	0.475
⅝	0.750	0.088	0.574
¾	0.875	0.102	0.671
1	1.125	0.132	0.863
1¼	1.375	0.161	1.053
1½	1.625	0.191	1.243
2	2.125	0.248	1.629

water distribution. It is available in straight lengths in the following nominal diameters: 1, 1¼, 1½, 2, 2½, 3, 3½, 4, 5, and 6 in. Solvent-cement welding and threaded or flanged connections are used to join PVC pipe and fittings. Dimensions of PVC pipe for drainage, waste, and vent systems are provided in Table 12.19.

Chlorinated polyvinyl chloride (CPVC) is a rigid thermoplastic pipe generally approved for use in potable hot and cold water supply, fire suppression sprinkler systems in residences, and in process piping. CPVC is rated at 180°F (82°C) and 100 psi (690 kPa), making it suitable for potable hot water distribution. Because of its excellent chemical resistance, it can also be used in sanitary drainage applications. CPVC tubing and fittings are beige or tan in color. CPVC pipe is available as Schedule 40 and Schedule 80 in straight lengths in the nominal diameters from ½ to 12 in. Dimensions of Schedule 40 and Schedule 80 CPVC pipe are provided in Table 12.19. CPVC is also available CTS, which is designed for use in hot and cold water distribution systems in buildings. Dimensions of CTS-CPVC plumbing pipe are provided in Table 12.20. CPVC pipe with a standard dimension ratio (SDR) of 13.5 is used for fire sprinkler piping (see Table 12.21). Solvent-cement welding and threaded and flanged connections are used to join CPVC pipe and fittings.

Styrene rubber (SR) is a rigid thermoplastic pipe that is generally approved for use in septic tanks, drain fields, and storm sewers. It is available in straight lengths in the following nominal diameters: 3, 3½, 4, 5, 6, and 8 in.

Polypropylene (PP) is a thermoplastic pipe material that is typically used in chemical waste lines. It can also be used for hot and cold water applications. It is rarely used in building plumbing systems, likely because it is joined by heat fusion.

Polyvinylidene fluoride (PVDF) is an extremely expensive thermoplastic pipe that is used in ultrapure water systems and industrial applications (e.g., pharmaceutical industry). It is joined by heat fusion and, for smaller pipe sizes, by mechanical joining techniques. PVDF is not used in building plumbing systems.

Reinforced thermosetting plastic pipe is a thermoplastic resin used in combination with reinforcement and fillers. The most commonly used reinforced thermosetting plastic pipe products are based on polyester or epoxy resins. The reinforcement may be organic (e.g., synthetic fiber)

TABLE 12.19 DIMENSIONS OF POLYVINYL CHLORIDE (PVC) AND CHLORINATED POLYVINYL CHLORIDE (CPVC) PIPE.

Diameter (in)	Outside Diameter (ins)	Schedule 40			Schedule 80		
		Wall Thickness (in)	Inside Diameter (in)	Wall Pressure 73°F (psi)	Wall Thickness (in)	Inside Diameter (in)	Wall Pressure 73°F (psi)
½	0.840	0.109	0.602	600	0.147	0.526	850
¾	1.050	0.113	0.804	480	0.154	0.722	690
1	1.315	0.133	1.029	450	0.179	0.936	630
1¼	1.660	0.141	1.360	370	0.191	1.255	520
1½	1.900	0.145	1.590	330	0.200	1.476	470
2	2.375	0.154	2.047	280	0.218	1.913	400
2½	2.875	0.203	2.445	300	0.276	2.290	420
3	3.500	0.216	3.042	260	0.300	2.864	370
4	4.500	0.237	3.998	220	0.337	3.786	320
6	6.625	0.280	6.031	180	0.432	5.709	280
8	8.625	0.322	7.941	160	0.500	7.565	250
10	10.750	0.365	9.976	140	0.593	9.493	230
12	12.750	0.406	11.888	130	0.687	11.294	230
14	14.000	0.438	13.072	130	0.750	12.412	220
16	16.000	0.500	14.936	130	0.843	14.224	220
18	18.000	0.562	16.809	130	0.937	16.014	220
20	20.000	0.593	18.743	120	1.031	17.814	220
24	24.000	0.687	22.544	120	1.218	21.418	210

TABLE 12.20   DIMENSIONS OF COPPER TUBE SIZE CHLORINATED POLYVINYL CHLORIDE (CTS-CPVC) PLUMBING PIPE.

Nominal Pipe Size (in)	Outside Diameter (in)	Inside Diameter (in)	Wall Thickness (in)	Wall Pressure 73°F (psi)	Wall Pressure 180°F (psi)
½	0.625	0.469	0.068	400	100
¾	0.875	0.695	0.080	400	100
1	1.125	0.901	0.102	400	100
1¼	1.375	1.105	0.125	400	100
1½	1.625	1.309	0.148	400	100
2	2.125	1.716	0.193	400	100

TABLE 12.21   DIMENSIONS OF CHLORINATED POLYVINYL CHLORIDE (CPVC) WITH A STANDARD DIMENSION RATIO (SDR) OF 13.5. USED FOR FIRE SPRINKLER PIPE.

Nominal Pipe Size (in)	Outside Diameter (in)	Wall Thickness (in)	Inside Diameter (in)
¾	1.050	0.078	0.874
1	1.315	0.097	1.101
1¼	1.660	0.123	1.394
1½	1.900	0.141	1.598
2	2.375	0.176	2.003
2½	2.875	0.213	2.423
3	3.500	0.259	2.950

or inorganic (e.g., glass fiber). Glass fiber is the most common reinforcement used in this type of pipe. Reinforced thermosetting plastic pipe will typically consist of 15 to 70% glass fiber, 0 to 50% filler (e.g., sand), and 30 to 75% thermosetting resin.

Reinforced thermosetting plastic pipe is produced as a standard product with a full line of fittings. It is especially suited to applications such as water distribution, sewage, and effluent disposal exposed to highly corrosive conditions. Pipes made from conventional polyester range in diameter from 1 in to 16 ft (25 mm to 4.9 m) or more. Because of high cost, reinforced thermosetting plastic-based pipes constitute only a small proportion of the total plastic piping in use.

Thermoplastic piping materials are available in two sizes: pipe sizes and tubing sizes. Although both are sized nominally according to inside diameter, pipes go by IPS and tubes go by CTS. Pipes and tubes and their fittings, even in the same designated sizes, should not be interchanged. With plastic piping, you can choose from a wide selection of materials.

Many plumbing subcontractors, engineers, and architects still prefer copper for the water supply systems. This dates back to very early concerns about possible toxicity (poisoning) resulting from the use of plastic pipe; this concern has since been proved groundless. Another concern is thermal expansion: plastics expand several times more than metal from temperature increases. Additionally, some thermoplastics degrade in sunlight so they must be protected (e.g., painted or covered).

In recent years, there have been reports of premature deterioration and catastrophic failure of ABS and PB pipe in building plumbing systems. Several class-action lawsuits worth nearly a billion dollars have been filed. Problems range from the development of leaks at the fittings, to cracking, splitting, and deterioration of the pipe material. Reportedly, failure caused by many factors including the nature of the material, flaws in manufacture of fittings and pipe, and improper installation. Although the early pipe and fitting failures were attributed mostly to defective materials, the industry is placing blame for current problems on poor installation.

PEX tubing is an approved material in the current model-plumbing codes. It is replacing copper as the material of choice in residential water supply systems. In some jurisdictions using older versions of these codes may not have amended the code to include PEX tubing.

Composite Pipe

Composite pipe is a flexible pipe material that is constructed of an aluminum tube laminated between two layers of polyethylene thermoplastic. It is available in ⅜, ½, ⅝, ¾, and 1 in nominal diameter coils ranging from 100 to 1000 ft (30 to 30 m). Fittings are joined to the pipe with a compression or crimped connection and to fixtures and other fittings with a threaded connection. Branches can extend from a main manifold and extend uninterrupted to the plumbing fixture (e.g., sink, lavatory, bathtub, and so on). It is available in two types:

PE-AL-PE Pipe   *PE-AL-PE pipe* is an aluminum (AL) tube laminated between two layers of PE plastic. It carries long-term pressure and temperature ratings of 200 psi at 73°F, and 160 psi at 140°F, which is approved for use in cold water and compressed air applications. PE-AL-PE pipe is coded dark blue in color.

PEX-AL-PEX Pipe   *PEX-AL-PEX pipe* is an aluminum (AL) tube laminated between two layers of temperature-resistant, PEX plastic. Cross-linking of PE means that the molecular chains are

TABLE 12.22   DIMENSIONS OF AND BENDING REQUIREMENTS FOR PEX-AL-PEX COMPOSITE PIPE.

Nominal Pipe Size		Outside Diameter		Inside Diameter		Minimum Bend Radius	
in	mm	in	mm	in	mm	in	mm
⅜	9	0.500	12.7	0.364	9.2	3.0	76
½	12	0.625	16.5	0.485	12.3	3.8	97
¾	30	0.875	22.2	0.681	17.3	7.0	178
1	25	1.125	28.6	0.875	22.2	9.0	229

linked into a three-dimensional network that makes PEX remarkably durable within a wide range of temperatures, pressures, and chemicals. It is color coded orange, light blue, or black. Black is used in exposed installations. It is approved for use in cold and hot water and high-pressure applications and can also be used in radiant floor heating systems. It has a long-term pressure rating of 125 psi at 180°F. PEX-AL-PEX pipe is more costly than PE-AL-PE pipe and tends to be used in hot water applications only.

Composite pipe is extremely light; a 1000 ft (300 m) coil weighs about 40 lb (178 N). Dimensions of and bending requirements for PEX-AL-PEX composite pipe are provided in Table 12.22. As a flexible pipe, minimum radius requirements limit the minimum size of bend based upon the diameter of the pipe. Ease of handling and installation makes this type of pipe a cost-effective alternative to copper.

Clay and Concrete Pipe

Clay pipe is made from vitrified clay. *Concrete pipe* is cast from concrete. These pipes are traditionally used for sewage, industrial waste, storm water, and drain field applications. Concrete pipe is also used as large water supply pipe. These materials are not commonly used in building plumbing systems.

Pipe is normally supplied in three end styles: PE or plain end; BE, or beveled end for welding; or T&C for threaded and supplied with one coupling per length. Steel pipe can be cut to any length and sold threaded both ends (TBE) or threaded on one end only (TOE). Copper and thermoplastic pipe are sold PE only.

Tubing and Pipe Sizes

Historically, pipe size was based on the inside diameter of the pipe that was characteristic of the period, which was cast iron. For example, a half-inch cast iron pipe had an inside diameter (ID) that was exactly one-half inch. The thickness of its wall determined the outside diameter. Later, the standard was changed so that pipe size related to a specific outside diameter to ensure that all pipes and fittings would fit together for a specific size.

Pipe is thick walled and available in standard *iron pipe size (IPS)*. IPS remains the standard by which pipe size is measured. With materials other than iron, the wall thickness of pipe is different. Consequently, inside diameters of pipes of different materials vary for a specific pipe size. Thus, a half-inch pipe is neither a half-inch on the inside nor the outside, but it is still called a half-inch pipe based on the nominal diameter. Under the IPS designation, female fittings are identified by FIP and male fittings are MIP. The terms *nominal pipe size (NPS)* and IPS are interchangeable and refer to the nominal diameter of the pipe, not the actual diameter.

Pipe is distinguished from *tubing* by the standard by which it is measured. When copper tubing was developed, the walls were much thinner than cast iron or steel. Because of copper's unique characteristics, it was not necessary that it be made in IPS sizes. A new standard called *copper tube size (CTS)* was developed. The actual size of CTS is much closer to its nominal size than that of pipe.

The standard has evolved so that any product made in IPS size is called pipe and any product made in CTS size is called tubing, without regard to any differences in material or manufacturing process. CPVC is an exception, being called pipe but sold in CTS.

Pipe is available in a number of different thicknesses or schedules. The American Society for Testing and Materials (ASTM) establishes the standards by which they are graded. The ASTM has assigned standards to each schedule of pipe made, and those standards dictate their use.

Pipe Pressure Rating

With the exception of sewer and drainage pipe, all pipe is pressure rated. There are several different methods of determining pressure ratings:

The *schedule number* is obtained from the expression $1000 \times P/S$, where P is the service pressure and S is the allowable stress, both being expressed in the same units. For example, on types of steel pipe with IPS sizes thru 12 in, wall thickness is assigned schedule numbers from Schedule 10 (S.10) thru Schedule 160 (S.160), which represent approximate values for 1000 times the pressure–stress ratios.

The *SDR* is calculated by dividing the outside diameter of the pipe by its wall thickness. Pipe with an SDR of 13.5 has an outside diameter that is 13.5 times thicker than the wall thickness.

The *pressure-level rating* provides the pressure rating of the pipe at a given temperature. Pipes are available commercially at many pressure ratings, and the most popular of these are 50, 100, and 125 psi (340, 690, and 860 kPa); 160, 200, 250, and 315 psi (1.1, 1.4, 1.7, and 2.2 MPa).

Weights designations are used for steel and iron pipe: standard wall (Std), extra strong wall (XS), and double extra strong wall (XXS). These last two designations are sometimes referred to as extra heavy wall (XH) and double extra heavy wall (XXH), respectively. Wrought-iron pipe is referred to as Std, XS, and XXS and not by schedule numbers.

12.6 FITTINGS AND VALVES

Fittings

A variety of *fittings* must be used to connect pipe lengths and make all the pipe turns, branch lines, couplings that join the straight runs, and stops at the end of the runs. Fittings for steel and wrought-iron pipe are made of malleable iron and cast iron. The fittings for plastic, copper, and brass pipe are made of the same materials as the pipe being connected.

Elbows

Elbows, usually at 45° and 90°, are angular fittings used to change the direction of a supply pipe. On a sanitary drainage system, a *sanitary bend* makes a more gradual turn to prevent blockage.

Tees

Tees are used in a supply system when a line must branch off at a straight run. A *reducing tee* allows different pipe sizes to be joined together in a supply system. *Sanitary T* and *sanitary Y* are tee-like fittings used in sanitary drainage systems that make a more gradual turn to prevent blockage. A sanitary Y can accept two or three branches before combining flow into one pipe.

Couplings

Couplings are used to join straight runs of pipe. A *union* joins straight runs of pipe but also allows the pipes to be more easily disconnected when future piping revisions are expected or equipment needs to be replaced.

A *reducer* is a straight fitting used to decrease the diameter in a pipe in a water supply system. An *increaser* is a straight fitting used to increase the diameter in a pipe in a sanitary drainage system.

Adapters

Adapters are used in a supply system where threaded pipe is being connected to copper or thermoplastic. Adapters have one threaded end to accommodate threaded pipe.

Joining Pipes and Fittings

Pipes and fittings are joined through a number of techniques. Pipes and fittings can be joined mechanically. *Threaded joints, insert fittings* with *crimped connections* or *clamped connec-*tions; *hub and spigot;* and *flared (metal to metal) joints* are popular mechanical joining techniques. Fire suppression sprinkler pipes are frequently joined using a *grooved Victaulic* fitting.

A *compression fitting* is a type of connection for tubing or pipe where a nut, and then a sleeve or ferrule, is placed over a copper or plastic tube, and is compressed tightly around the tube as the nut is tightened, forming a positive grip and seal without soldering.

Soldering, brazing, and welding are ways of joining metal surfaces. Soldering and brazing are methods of joining two or more metal surfaces by melting nonferrous filler metal with a melting temperature well below the metals to be joined. The melted filler metal distributes itself between the surfaces to be bonded by capillary action. *Soldering* involves melting solder to a temperature below 840°F (449°C), usually in the range of 350° to 550°F (177° to 288°C). *Brazing* involves melting the metal filler above 430°C (800°F), usually in the range of 1100° to 1500°F (593° to 816°C), but still below the melting temperature of the metals to be joined. Soldered joints are used when the service temperature does not exceed 205°F (96°C). Brazed joints offer greater strength and should be used where operating temperatures are up to 400°F (204°C). *Welding* typically involves joining two or more pieces of metal by the application of heat. Unlike soldering or brazing, welding involves a partial melting of the surfaces of the metals to be joined. It offers the greatest physical strength.

Solvent-cementing and fusion welding can join some plastic pipe materials. *Solvent cementing* involves coating the plastic surfaces with a prime coat and a solvent cement coat before they are joined. The cement cures joining the surfaces in a manner similar to the cementing technique used to attach the pieces of a plastic model airplane. *Fusion welding* involves heating the surfaces until they melt, allowing them to be joined.

Valves

Valves are used to control flow of the water throughout the system. Proper location of valves simplifies repairs to the system, fixtures, or equipment being serviced. Valves also regulate flow to deliver the appropriate quantity of water and reduce water consumption. In building plumbing systems, there are usually valves at risers (vertical pipe serving the building), branches (horizontal pipe serving the fixtures), and pipes to individual fixtures or equipment. The inner workings of most valves are generally accessible for repairs.

Valves generally fall into four categories: gate, globe, check and angle. These are described below.

Gate Valves

The *gate valve* is a manual valve that has a wedge-shaped leaf that, when closed, seals tightly against two metal seats that are set at slight angles. (See Figure 12.8.) This type of valve is usually used where the flow of the water is left either completely opened or closed for most of the time. Because the flow of

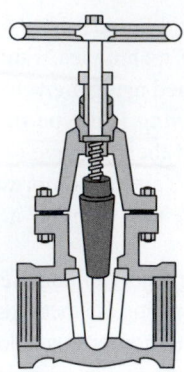

FIGURE 12.8   A cross-section of a gate valve.

water passes straight through the valve, there is very little water pressure lost to friction. The gate valve is not used to regulate flow of water. Instead, it is used to shut off the flow of water such as to fixtures and equipment when repairs or replacement must be made.

Globe Valves

The *globe valve* is a manual, compression-type valve, commonly used where there is occasional or periodic use, such as lavatories (faucets) and hose connections (called *hose bibbs*). This type of valve regulates the flow of water. Design of the globe valve is such that the water passing through is forced to make two 90° turns, which greatly increases the friction loss in this valve compared with that in a gate valve. (See Figure 12.9.)

Angle Valves

The *angle valve* is a manual valve similar in operation to the globe valve, utilizing the same principle of compressing a washer against a metal seat to cut the flow of water. It is commonly used for outside hose bibbs. (See Figure 12.10.) The angle valve has a much higher friction loss than the gate valve and about half the friction loss of the globe valve.

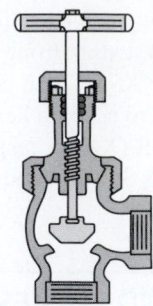

FIGURE 12.10   A cross-section of an angle valve.

Check Valves

The *check valve* opens to allow the flow of water in the direction desired and prevents flow in the other direction. There are two types of check valves, swing types and spring types. In the *swing check valve* design, the pressure of the water forces the valve gate to swing open, but once the flow stops, gravity causes the gate to fall closed, preventing a reversal of the flow. This type of valve must be mounted vertically or horizontally to work properly. (See Figure 12.11.)

In contrast, the gate in a *spring check valve* is spring loaded. Water pressure forces the gate open much like the swing type, but when the flow stops, a spring (not gravity) forces the gate closed. This enables the valve to be mounted in any position and at any angle. This valve is used in such places as the water feed line to a boiler (heating unit) where the water from the boiler might pollute the system if it backed up.

All valves are modifications of these fundamental types. Valves can also be categorized by their function. There are several types of special valves that justify a description.

Ball Valves

A *ball valve* is a manual valve that has a ball with a hole through it that is mounted between two seats. When the ball hole is in line with the valve openings, full flow of water occurs. A 90° rotation of the ball causes the valve to be fully closed. Ball valves are available in both on/off shutoff control and controlled-flow designs. Controlled-flow ball valves are designed to regulate the flow of water.

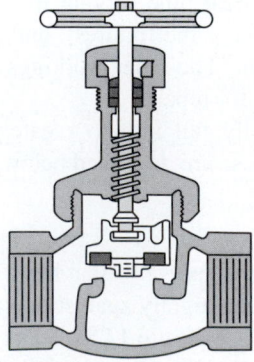

FIGURE 12.9   A cross-section of a globe valve.

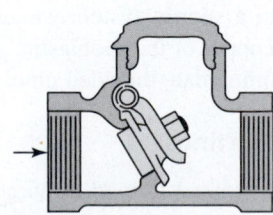

FIGURE 12.11   A cross-section of a check valve.

Metered Valves

Metered valves are designed to automatically discharge for a specific length of time and thus deliver a fixed quantity of water before closing off flow. They operate by pushing down or against the valve handle. They are used on lavatories in public restrooms such as in transportation terminals, restaurants, and convention halls to ensure that water is shut off after a short period of time. A *flushometer valve* is a metered valve that discharges a predetermined quantity of water to fixtures for flushing purposes (e.g., water closets and urinals) and is closed by direct water pressures.

Flow Control Valves

A *flow control valve* automatically adjusts the rate of water flow to a predetermined flow rate as pressure in the system varies. They can be used to limit flow at a fixture outlet thereby holding demand to a required minimum.

Thermostatic Valves

A *thermostatic valve,* frequently called a *tempering valve* or *mixing valve,* is an automatic valve thermostatically blends hot and cold water to desired temperatures and to prevent scalding.

Temperature-Pressure Relief Valves

A *temperature-pressure relief (T/P) valve* is a safety valve designed to limit pressure of a liquid vapor or gas. These valves are specified such that the valve remains closed at normal operating pressures yet it is allowed to open to release excessive pressure. They are commonly found as a safety feature on water heaters and boilers.

Pressure-Reducing Valves

A *pressure-reducing valve* is an adjustable valve designed to reduce pressure to a specific setting. These valves are commonly used in building plumbing systems where street water pressure is excessive and needs to be reduced before being sent to plumbing fixtures. A pressure-reducing valve is shown in Photo 12.8.

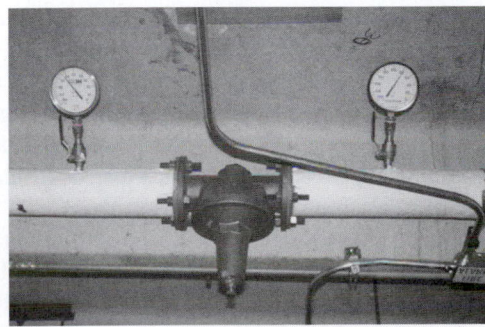

PHOTO 12.8　A pressure-reducing valve. (Used with permission of ABC)

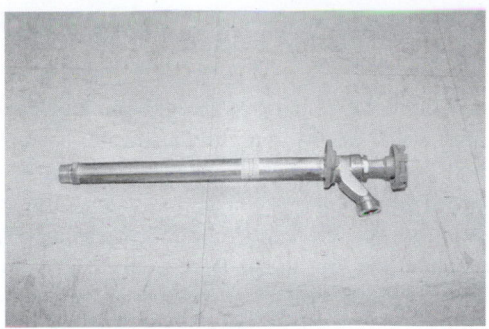

PHOTO 12.9　A freezeless hose bibb or sill cock. (Used with permission of ABC)

Hose Bibbs

A *hose bibb,* sometimes called a *sill cock,* is a valve designed to accept the threaded connection of a hose. A *freezeless hose bibb* has a long body that when placed in an exterior wall, cuts off the water supply near the interior wall surface. This allows water near the exterior wall surface to drain out when the valve is closed to avoid freezing of water and valve damage in severe winter temperatures. *Secured hose bibbs* require a specially designed knob to open the valve, which prevents use by the general public. A hose bibb is shown in Photo 12.9.

Flushometer

A *flushometer* is a valve-like device designed to supply a fixed quantity of water for flushing toilets and urinals. When operated, it automatically shuts off after a measured amount of water flow in order to conserve water. It uses pressure from the water supply system rather than the force of gravity to discharge water.

Sensor-Operated Valves

Modern urinals and water closets (toilets) use a *sensor-operated valve* that automatically flushes the fixture when a user departs. The unit uses an infrared proximity sensor to detect a user approaching the fixture, then waits until the user departs. A solenoid is used to actuate the flush. Typically, a batter contained within the unit powers the sensor circuit.

Valves referred to as *standard weight* are designed to withstand pressures up to 125 psi (860 kPa). High-pressure valves are also available. Most small valves have bronze bodies, while large valves (2 in [50 mm] and larger) have iron bodies with noncorrosive moving parts and seats that must be replaced periodically. They are available threaded or soldered to match the pipe or tubing used.

Valves must be installed in the appropriate direction of flow. An arrow cast in the body of the valve usually indicates direction of flow. Some valves are better than others in regulating flow. Gate valves and ball valves undergo excessive wear (from cavitation) when they are partially closed. Globe valves are designed to more easily and effectively regulate flow.

12.7 PLUMBING FIXTURES

A *plumbing fixture* is an approved receptacle, device, or appliance that uses water and discharges wastewater such as a water closet, urinal, faucet, shower, dishwasher, drinking fountain, hose connection, hose bibb, water heater, water softener, underground sprinkler, hot tub, spa, and clothes washer. They must be made of dense, durable, nonabsorbent materials with smooth, impermeable surfaces. Plumbing fixtures are the only part of the plumbing system that the owners or occupants of the building will see regularly, because most of the plumbing piping is concealed in walls and floors.

The designer of the plumbing system, the architect, the engineer, the plumber, and/or the owner may select plumbing fixtures. It is important that the designer of the plumbing system know what fixtures will be used (and even the manufacturer and model number, if possible) in order to do an accurate job in designing the system. All fixtures should be carefully selected as they will be in use for years, perhaps for the life of the building.

The available sizes for each fixture should be carefully checked in relation to the amount of space available. Most manufacturers supply catalogs that show the dimensions of the fixtures they supply. Whoever selects the fixtures should check with the local supplier to be certain that those chosen are readily available; if not, they may have to be ordered far in advance of the time they are required for installation. Most of the fixtures are available in white or colors, so the color must also be selected.

Plumbing fixtures are classified according to their use. Groups of two or more like fixtures that are served by a common drainage branch are known as a *group of fixtures*. Types of plumbing fixtures and related design concerns are as follows.

Water Closets

A *water closet* is a plumbing fixture that serves as an indoor receptacle and removal system for human waste. Although this fixture is commonly called a *toilet* or *commode,* the building code specifically refers to it as a water closet. Water closets are typically made of solid vitrified china cast with an integral (built-in) trap. They are also available in stainless steel that is typically specified for high-vandalism installations such as at highway rest stops, outdoor recreation areas, jails, and detention centers. Examples of water closets are shown in Photos 12.10 and 12.11.

In North America, water closets are available as *single-flush,* flush tank, or flush valve fixtures. Present requirements limit average water consumption to 1.6 gal (6.0 L) per flush. These are known as *ultra-low flush (ULF)* water closets. Infrared and ultrasonic sensors can be built into the flush valve to automatically flush and avoid nonflushing or double flushing.

PHOTO 12.10   A flush tank water closet. (Used with permission of ABC)

PHOTO 12.11   A wall-mounted, flush valve water closet. Note the wall cleanout cover at the floor line to the right of the water closet. (Used with permission of ABC)

A *flush tank* water closet has a water tank as part of the fixture. (See Figure 12.12.) As the handle or button on a water closet is pushed, it lifts the valve in the tank, releasing the water to flush out the bowl. Then, when the handle is released, the valve drops and the tank fills through a tube attached to the bottom of the tank. This type of water closet cannot be effectively flushed again until the tank is refilled. Foam lining can be installed in the tank to minimize condensation on the outside of a toilet tank by insulating the cold water in the tank from warm, humid air.

Flush tank models range from those having the tank as a separate unit set on the closet bowl to those having a low tank silhouette with the tank cast as an integral part of the water closet. Generally, clients prefer this low-slung appearance but it is considerably more expensive.

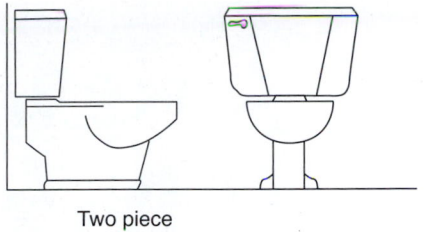
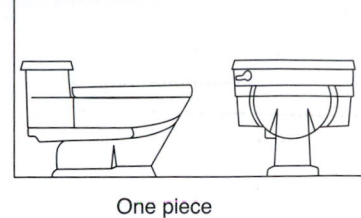

Two piece | One piece

FIGURE 12.12 Flush tank water closets.

Floor mounted | Wall mounted

FIGURE 12.13 Flush valve water closets.

Flush valve water closets have no tank to supply water. Instead, when the handle is pushed, the water to flush the bowl comes directly from the water supply system at a high rate of flow. When used, it is important that the water supply system be designed to supply the high flow required. Although most of the fixtures operate effectively at a pressure of 20 psi (140 kPa), the manufacturer's specifications should be confirmed because higher pressure is often required.

Water closets may be floor or wall mounted. The floor-mounted fixture is much less expensive in terms of initial cost, but the wall-mounted fixture allows easier and generally more effective cleaning of the floor. It is acceptable for most residential applications. Wall-mounted fixtures are considered desirable for public use, and some codes even require their use in public places. When wall-mounted fixtures are used in wood stud walls, a wider wall will be required than is sometimes used with floor-mounted fixtures.

The *dual-flush water closet,* a technology first developed in the early 1980s, takes water conservation one step further by using 1.6 gal (6.0 L) of water to flush solid waste but only 0.8 gal (3.0 L) to flush liquid waste. Although this technology is mandated in some countries (e.g., Australia and Singapore), it is optional in North America.

The National Association of Home Builders Research Center (NAHB) completed performance tests on 49 popular toilet models. One element of this study provided a relative rating called the *Flush Performance Index (FPI).* FPI ratings ranged from 0 to 82, with lower numbers being better. The FPI for each toilet was calculated from the ability of a toilet to flush varying amounts of sponges and paper. The study was limited: Only two specimens of each toilet model were tested, and the sponges and paper used in the testing may not be a suitable substitute for measuring real-world toilet performance. Thus, the testing results do not serve as an exact measure of how specific toilets perform. The development of the measure is indicative of the demand for a functioning toilet.

Urinals

Urinals are plumbing fixtures that are commonly used in public restrooms where it is desirable to reduce possible contamination of the water closet seats. They are commonly available in

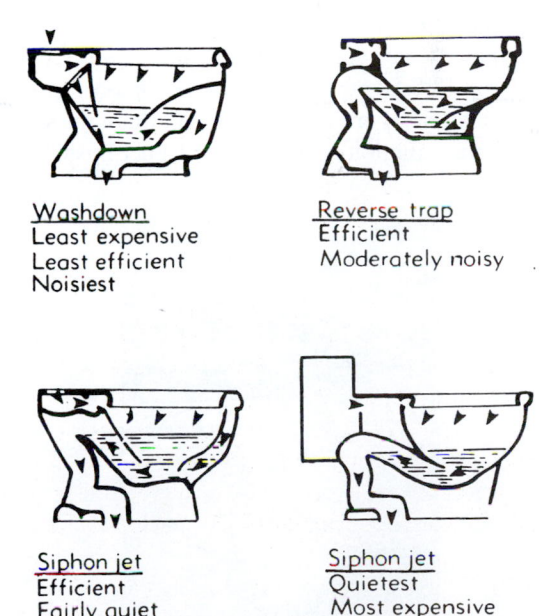

Washdown
Least expensive
Least efficient
Noisiest

Reverse trap
Efficient
Moderately noisy

Siphon jet
Efficient
Fairly quiet

Siphon jet
Quietest
Most expensive

FIGURE 12.14 Types of flushing actions used in a water closet. Water flows into the bowl from the bowl rim. This raises the water level in the bowl to fill the gooseneck pathway. As water fills the gooseneck, the water and waste remaining in the bowl is sucked up and into the gooseneck by a siphoning action.

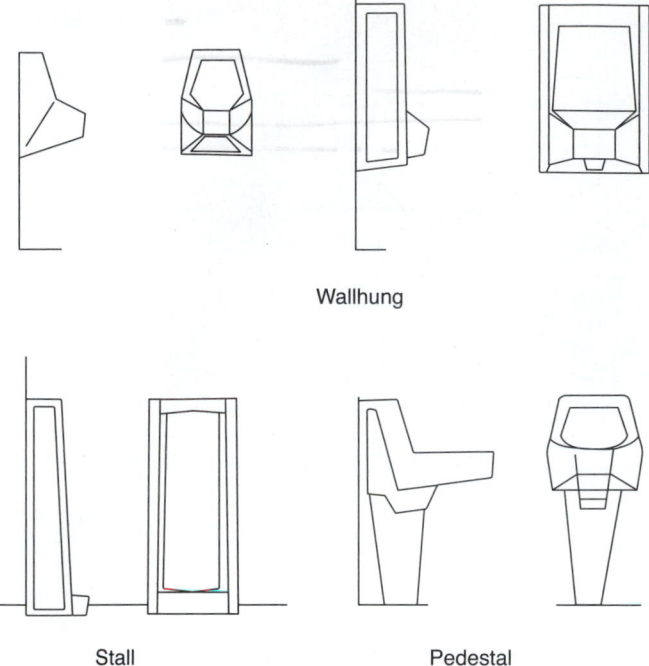

Wallhung

Stall   Pedestal

FIGURE 12.15   Types of urinals.

vitreous china and sometimes in enameled iron. They are also available in stainless steel for high-vandalism installations. Floor and trough-type urinals are no longer allowed in new construction. Examples of urinals are shown in Figure 12.15 and Photos 12.12 and 12.13.

Urinals are available as flush tank or flush valve fixtures. Present requirements typically limit average water consumption to 1.0 gal (3.8 L) per flush. These are known as the *ULF* urinals. Special metal urinals with straight drain lines limit average water consumption to 0.5 gal (1.9 L) per flush.

PHOTO 12.13   A group of urinals separated by partitions. (Used with permission of ABC)

Waterless Urinals

A *waterless urinal* is a urinal that is specifically engineered to eliminate potable water consumption for urinal flushing. It looks very much like a conventional urinal except the flush valve and piping that is normally positioned above the fixture is omitted. See Photo 12.14. Conventional urinals use about 3.5 (13 L) of water per flush and modern water-saving urinals use about 0.5 to 1 gal (2 to 4 L) of water per flush. The chief benefit of waterless urinals is that they do not use water. In office buildings and schools, waterless urinals can save up to 25 000 gal (100 000 L) of potable water per year per fixture, saving water and sewer costs and reducing the burden on the municipal sewage and sewage treatment system.

A waterless urinal fixture blocks odors and gasses by a means other than a traditional trap. Each manufacturer constructs its waterless urinal fixture differently. The most popular

PHOTO 12.12   A wall-mounted, flush valve urinal. (Used with permission of ABC)

PHOTO 12.14   A waterless urinal, which represents the most water-efficient urinal option because they provide first-cost savings (e.g., eliminating the need to provide a water line and flush valve) and less maintenance (e.g., leaks, valve repairs, and water overflows) over the conventional urinals. (Courtesy of NREL/DOE)

types use a removable cartridge that needs to be replaced on a regular maintenance schedule, or a liquid sealant that must be regularly flushed and refilled periodically, usually by the house-keeping staff. With most designs, urine flows by gravity off the smooth surface of the urinal into a trapped liquid with a lighter-than-water specific density. The liquid floats, allowing the urine to flow through it to the drain. The liquid remains and serves to trap the odor. One manufacturer recommends cartridge replacement after an estimated 12 000 to 15 000 uses, which on average will be every 3 to 4 months depending on usage.

Waterless urinals are easier to install than conventional urinals in both new construction and retrofit applications. No water connection and no flush mechanisms must be installed, saving construction time and costs. Waterless urinal systems also avoid the need for frost- and vandal-prone plumbing accessories. They avoid flooding when the drain is blocked from scale or vandalism. Most problems with the waterless urinals arise from faulty installation or improper maintenance.

Waterless urinals are much more hygienic than conventional flush urinal systems. The dry surface of a waterless urinal does not allow bacteria to survive. In contrast, the conventional flush urinal discharges an unavoidable bacterial spray with each flush, which then falls on nearby wall and floor surfaces and forms new bacterial growth. Additionally, the dry seal of a waterless urinal is more sanitary than the pool of diluted urine that seals the conventional urinal drain, which also eliminates urinal odor by isolating urine from the restroom atmosphere.

Bidets

Bidets are personal hygiene plumbing fixtures used for genital and perineal cleanliness. It is typically used after using the water closet. Equipped with valves for hot and cold water, the inside walls of the bowl are washed the same way as a standard toilet. The bidet is not designed or intended to carry away solid human waste. It is installed alongside the water closet. The user sits on the fixture facing the wall (and the water controls) and is cleansed by a rinsing spray. Bidets are available in vitreous china. Some bidets have a warm air dryer that is used to blow dry the genital and perineal area after washing. The bidet is used extensively in Europe and South America and is enjoying increased usage in Canada and the United States. (See Figure 12.16.)

Bathtubs

Bathtubs are plumbing fixtures used for bathing. See Photos 12.15 through 12.17. They are available in enameled iron, cast

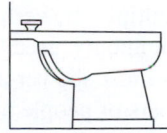 

FIGURE 12.16 A bidet.

PHOTO 12.15 A luxurious bathtub with power jets in a residential bathroom. (Used with permission of ABC)

PHOTO 12.16 A bathtub for a master bathroom. (Used with permission of ABC)

PHOTO 12.17 An enameled iron bathtub stored before installation. (Used with permission of ABC)

iron, or fiberglass. Tubs are available in a variety of sizes, the most common being 30 or 32 in (760 or 810 mm) wide; 12, 14, or 16 in (300, 350 or 400 mm) high; and 4 to 6 ft (1.2 to 1.8 m) long. *Whirlpool bathtubs* are fitted with jets that propel a current of warm water in a swirling motion.

Enameled iron tubs are formed of steel that is clad with a porcelain enamel finish. They are generally available in lengths of 4½ and 5 ft (1.37 and 1.53 m); widths of 30 to 31 in (760 to 785 mm); and typical depths of 15 to 15½ in (375 to 387 mm).

The most commonly available length of fiberglass bathtubs is 5 ft (1.5 m), and it takes 34 to 36 in (865 to 915 mm) of width to install. Generally, the fiberglass units are cast in a single piece. Many include three walls (eliminating the need for a ceramic tile tub surround). It is this single-piece feature, with no cracks or sharp corners to clean, which makes the fiberglass tub so popular with clients.

The size of the fiberglass unit typically makes it almost impossible to fit it through the standard bathroom door; it must therefore be ordered and delivered early to be set in place before walls and doors are finished. In wood-frame buildings, these units are usually delivered to the job and put in place before the plaster or gypsum board is put on the walls or the doors installed.

When selecting fiberglass tubs, be certain to specify only manufacturers who are widely known and respected, with long experience in the plumbing fixture field. Off-brands often give unsatisfactory results in that the fiberglass "gives" as it is stepped on, making a slight noise. In addition, some may be far more susceptible to scratching and damage.

Bathtub fittings may be installed on only one end of a tub, and the end at which they are placed designates the tub. As you face the tub, if the fittings are placed on the left, it is called a *left-handed* tub, and if placed on the right, it is *right-handed*.

Showers

A *showerhead* is an overhead nozzle that sprays water down on the bather. Shower fittings may be placed over bathtubs instead of having a separate shower space; this is commonly done in residences, apartments, and motels. However, it is important that when a showerhead is used with a bathtub fixture, the walls be constructed of an impervious material such as ceramic tile. See Photos 12.18 through 12.21.

Present requirements for average water consumption by a showerhead are that flow rates not exceed 2.5 gpm (9.5 L/min). These are known as *low-flow showerheads*. A *handshower* is a showerhead attached to the end of a flexible hose, which the bather can hold during bathing or showering.

Shower surrounds cover the walls that enclose a shower stall. A *shower enclosure* consists of glass panels, either framed or frameless, used to enclose bathtubs, shower modules, shower receptors, and custom-tiled showering spaces. A *receptor* or *shower pan* is a shallow basin used to catch and contain water in the bottom of a showering space. They are available in units of porcelain enameled steel, fiberglass, tile, terrazzo, marble, cement, or molded compositions. Special

PHOTO 12.18   A tub faucet with tub/shower control. (Used with permission of ABC)

PHOTO 12.19   A showerhead. (Used with permission of ABC)

PHOTO 12.20   A roughed-in plastic shower pan. (Used with permission of ABC)

shower surrounds available include corner units and gang head units. A *gang head shower* has multiple showerheads extending from the top of a post. It is commonly used in institutions, schools, factories where workers must shower after work, and other locations where large numbers of people must shower.

Shower surrounds and receptors of tile, concrete, or marble may be built to any desired size or shape. Typically lead or plastic sheets are site-formed into shower pans on custom-made

PHOTO 12.21 A three-quarter bathroom (lavatory, water closet, and shower) with glass doors on the shower stall. (Used with permission of ABC)

showers. Preformed shower stall surrounds are most commonly available in sizes of 30 in by 30 in (760 mm by 760 mm) and 30 in by 36 in (760 mm by 915 mm). Other sizes may be ordered. Steel shower surrounds are usually available in sizes of 30 in by 30 in (760 mm by 760 mm) and 30 in by 36 in (760 mm by 915 mm). Special sizes may also be ordered. Fiberglass shower surrounds are commonly available in sizes of 36 in by 36 in (915 mm by 915 mm) and 36 in by 48 in (915 mm by 1220 mm).

Code generally sets a minimum shower size (except as permitted herein) of at least 1024 in^2 (0.66 m^2) of interior cross-sectional space with a minimum interior dimension of 30 in (760 mm). The only exception is a prefabricated one-piece

shower designed to accommodate a 32 in by 32 in (800 mm by 800 mm) roughed-in opening, provided it has at least 900 in^2 (.56 m^2) of interior area.

Lavatories

A *lavatory* is a bathroom basin or sink used for personal hygiene. Lavatories are generally available in vitreous china or enameled iron, or they may be cast in plastic or a plastic compound with the basin an integral part of the countertop. They are also available in stainless steel for high-vandalism applications. See Figure 12.17 and Photos 12.22 through 12.24.

Present requirements for nonmetered lavatory faucets limit the average water consumption to 2.2 gpm (8.4 L/min). Metered lavatory faucets are designed to shut off after a short period of time. They are used in public restrooms such as in transportation terminals, restaurants, and convention halls to ensure that water is shut off and not flowing freely. Metered faucets used on lavatories should not deliver more than 0.25 gal (1.0 L) per use. Infrared and ultrasonic sensors can be installed to operate faucets and limit waste.

Lavatories are available in a large variety of sizes and the shapes are usually square, rectangular, round, or oval. The lavatory may be wall-hung, set on legs, set on a stand, or built into a cabinet. Lavatory styles are usually classified as flush-mount, self-rimming, undercounter, integral, or as units that can be wall-hung or supported on legs. *Self-rimming lavatories* have a finished rim that is placed directly over the countertop

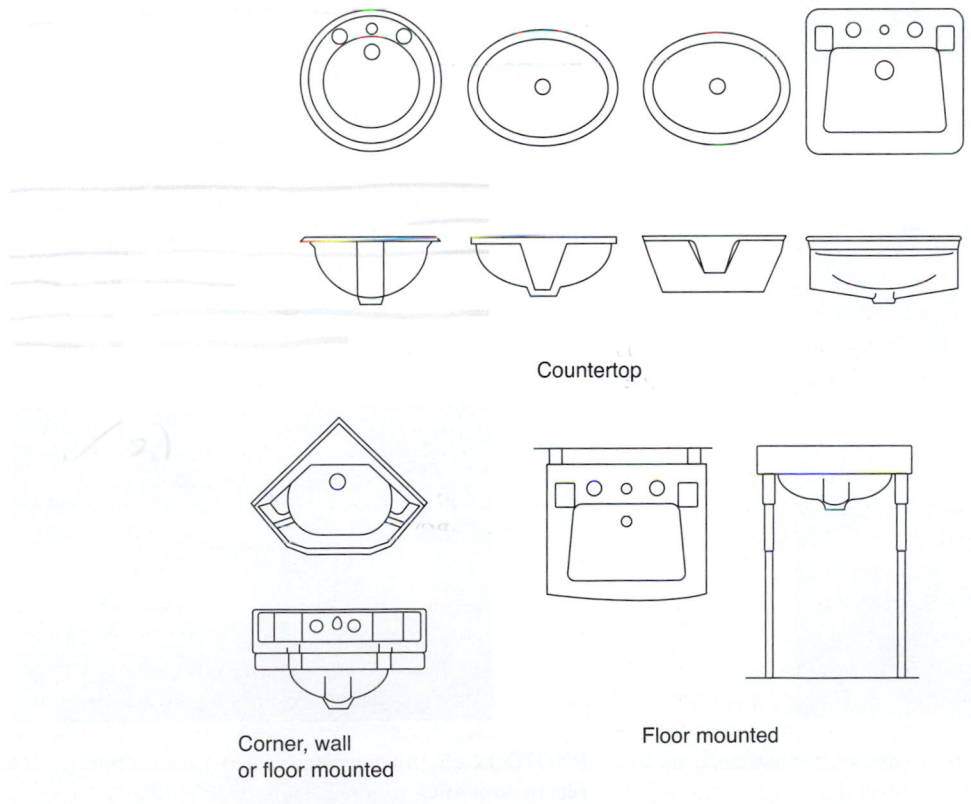

Countertop

Corner, wall or floor mounted

Floor mounted

FIGURE 12.17 Types of lavatories.

PHOTO 12.22 Luxurious lavatories in a residential bathroom. (Used with permission of ABC)

PHOTO 12.23 A pedestal lavatory. (Used with permission of ABC)

PHOTO 12.24 A cast, vitrified china wall-mounted lavatory in a public restroom. (Used with permission of ABC)

opening. *Undercounter lavatories* are an installation in which a lavatory (or sink) is attached to the underside of a countertop. *Pedestal lavatories* have a basin that is supported primarily by a freestanding pedestal leg.

Special fittings for lavatories include the following: foot controls (often used in institutions such as hospitals and nursing homes); self-closing faucets, which are commonly used in public facilities (especially on hot water faucets) to conserve water; and automatic "no-touch" flow, which operates automatically when a sensor recognizes that hands are positioned under the faucet. Residential lavatories have a *lift rod* that opens the pop-up drain when the lift rod is depressed. When rod is lifted, the drain closes so the lavatory will retain water.

Sinks

Kitchen sinks are most commonly made of enameled cast iron or stainless steel. Sinks are usually available in a single- or a double-bowl arrangement; some even have a third bowl, which is much smaller. A *waste disposal* is typically connected to one of the sink drains. Kitchen sinks are generally flush-mounted into a plastic laminate or into a composition plastic counter. Present water conserving requirements for residential kitchen sink faucets limits the average water consumption to 2.5 gpm (9.5 L/min). A common sink width for the kitchen is 30 in.

A *utility* or *service sink* has a deep, fixed basin that is supplied with hot and cold water and is used for rinsing mops and disposing cleaning water. They are often called *slop sinks* or *mop sinks*. These sinks are made of enameled cast iron or vitreous china. Most service sinks have high backs, and there may be two or as many as three bowl compartments. Other sinks commonly used are laundry trays, pantry sinks, bar sinks, and surgeon's sinks. Service sinks are generally available in enameled iron or in stainless steel. A *floor-mount sink* is installed into the center of a concave floor to dispose of water. The dome strainer and grate provide a convenient drain and catch basin for general cleaning and maintenance tasks. Examples of types of sinks are shown in Photos 12.25 through 12.28.

PHOTO 12.25 An enameled cast iron kitchen sink. (Used with permission of ABC)

PHOTO 12.26 A stainless steel kitchen sink. (Used with permission of ABC)

PHOTO 12.27 A stainless wet bar with sink. (Used with permission of ABC)

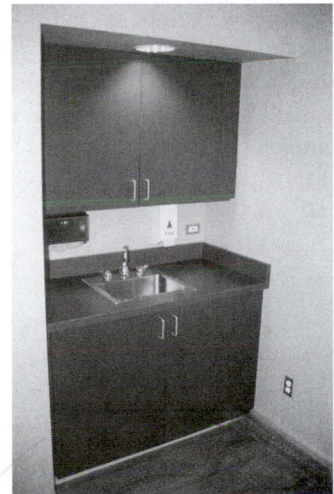

PHOTO 12.28 A stainless steel service sink. (Used with permission of ABC)

Laundry Tubs and Trays

Laundry tubs, sometimes called *trays,* are a large deep sink used in laundry rooms. They are usually available in a single- or a double-bowl arrangement. Laundry tubs are typically floor or wall-mounted units available in low-cost plastic, enameled iron, or stainless steel. A laundry tray is shown in Photo 12.29.

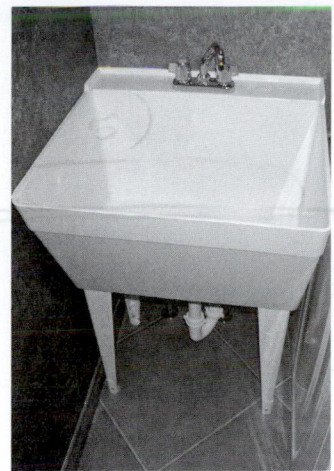

PHOTO 12.29 A laundry tray. (Used with permission of ABC)

Drinking Fountains and Water Coolers

Drinking fountains offer users a limitless supply of drinking water at any location where water and sanitary drainage are readily available. *Water coolers* can deliver 8 gal/hr (30 L/hr) or more of chilled drinking water. They require connections to power, water, and drainage. Drinking fountains and water coolers are available in wall-mounted and floor units. Drinking fountains and water coolers should not be installed in public restrooms. (See Photo 12.30.)

Other Fixtures

Emergency fixtures include eye-face washes, drench showers, decontamination units, portables, and accessories designed for use wherever hazardous substances are present. Other types of fixtures include baptisteries, ornamental ponds, fountains, and aquariums. An emergency drench shower at a university laboratory is shown in Photo 12.31.

PHOTO 12.30 A wall-mounted drinking fountain unit. Chilled water is provided to the unit. (Used with permission of ABC)

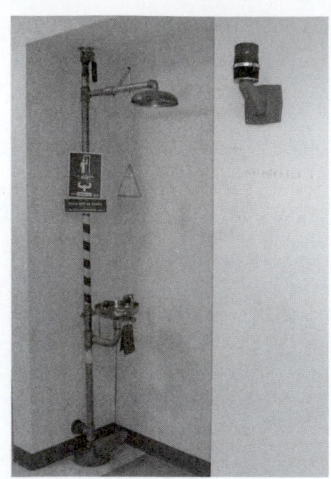

PHOTO 12.31   An emergency drench shower at a university laboratory. (Used with permission of ABC)

"Approved" Fixtures

Federal law mandates that all plumbing fixtures meet or exceed the minimum Energy Policy Act (EPACT) requirements based on maximum flow.

Toilets	1.6 gpf (6.1 Lpf)
Urinals	1.0 gpf (3.8 Lpf)
Showerheads	2.5 gpm (9.5 L/min)
Faucets	2.2 gpm (8.4 L/min)
Metering faucets	0.25 gal (1 L) per cycle

Applicable state and local codes should be checked prior to design as some of them have "approved fixture" lists; some code officials have not approved the waterless urinal and low-flush toilet technologies.

Basic Design Considerations for Restrooms

A *restroom* is a personal hygiene facility provided to allow use of a water closet by members of the public, or by patrons or customers. In other parts of the world, the restroom is known as a *washroom* (Canada), *public toilet* (Great Britain, Australia, and Hong Kong) or *comfort station* (Africa, Middle East, and Southeast Asia).

A properly designed restroom improves the experience of both users and those who maintain the facilities. Restrooms with multiple water closets should have partitioned stalls (required by code). Wall partitions between urinals (mandated by code) provide privacy and prevent splash from spreading. Automated "touch-free" fixtures and dispensers (e.g., door openers, toilet and urinal sensing flush valve, faucets, hand dryers, and liquid soap, paper towel, and toilet paper dispensers) reduce the spread of disease and cut cost by controlling product usage. Because these devices are occasionally prone to failure, they should have the capability to operate manually. Consideration should be given to

mirror placement, because improperly positioned mirrors can provide a reflected view to areas where privacy was intended.

Public restrooms can present accessibility challenges for people with disabilities. As a result, building regulations in the United States establish requirements to allow accessibility for occupants and provide for use of public restroom facilities. With the passage of the Americans with Disabilities Act (ADA) in 1991, public restroom facilities in the United States must be designed to accommodate people with disabilities. New commercial construction in the United States is required to comply with Title III of the *ADA Standards for Accessible Design* as enforced by the U.S. Department of Justice. New multifamily dwellings must meet the requirements of the Fair Housing Amendment Act (FHAA) as enforced by the U.S. Department of Housing and Urban Development.

Designers of restroom spaces should always consult local building codes for requirements specific to the building site. Typical accessibility requirements for restrooms are as follows:

a. An accessible entry on an accessible route.

b. A 60 in (1500 mm) diameter clear space within the restroom. Only one door, in any position, can infringe upon this clear space by no more than 12 in (300 mm).

c. At least one men's and one women's accessible water closet stall door with a 32 in (800 mm) minimum.

d. Grab bars behind (rear) and on one side of each accessible water closet that are mounted 33 to 36 in (840 to 910 mm) from the centerline of the grab bar to the finished floor. Side grab bars shall be 42 in (1050 mm) long and mounted 12 in (300 mm) from the rear wall, and rear grab bars should be a minimum of 36 in long (900 mm) and mounted a maximum of 6 in (150 mm) from the side wall. The diameter of grab bars should be between 1¼ to 1½ in (30 to 40 mm) with 1½ inch (40 mm) clearance from the wall.

PHOTO 12.32   An accessible water closet with rear and side grab bars. (Used with permission of ABC)

TABLE 12.23   AN EXAMPLE OF MINIMUM NUMBER OF PLUMBING FIXTURES. THE DESIGNATION 1:100 INDICATES 1 FIXTURE TYPE PER 100 OCCUPANTS. NUMBER OF OCCUPANTS IS TYPICALLY DETERMINED BY FLOOR AREA. VARIES WITH VARIOUS MODEL CODES AND LOCAL REGULATIONS.

| Water Closets | | Urinals | Lavatories | | Bathtubs or Showers | Drinking Fountains |
Male	Female	Male	Male	Female		
1: 1 to 100 2: 101 to 200 3: 201 to 400	3: 1 to 50 4: 51 to 100 8: 201 to 200 11: 201 to 400	1: 1 to 100 2: 101 to 200 3: 201 to 400 4: 401 to 600	1: 1 to 200 2: 201 to 400 3: 401 to 750	1: 1 to 200 2: 201 to 400 3: 401 to 750	—	1 per 150
Over 400, add one fixture for each additional 500 male occupants and one fixture for each additional female occupant		Over 600, add one fixture for each additional 300 male occupants	Over 750, add one fixture for each additional 500 occupants			One per occupied floor

e. A 48 in (1200 mm) clear space in front, and 32 in (800 mm) clear space on one side, of each accessible water closet.

f. Top of each accessible water closet seat measured between 17 to 19 in (425 to 475 mm) from floor.

g. At least one accessible lavatory with 29 in (725 mm) clearance underneath, bladed valve handles, and insulated hot water drain pipes.

h. At least one accessible mirror bottom mounted 40 in (1000 mm) maximum above floor.

i. Dispensers (at least one towel, sanitary napkin, seat cover, soap) mounted with highest operable part 40 in (1000 mm) maximum above floor.

Minimum Number of Fixtures

Codes generally set the minimum number of fixtures that must be installed on a project according to the type of occupancy and, in some cases, the number of occupants. When designing any residential, commercial, industrial, or institutional project, this minimum plumbing facility chart must be followed. Requirements vary by occupancy number and type as dictated by local code. For example, single- and multiple-dwelling units and apartments must have one water closet, one lavatory, and one bathtub or shower per dwelling or apartment unit. Commercial, industrial, or institutional projects have requirements tied to occupant load (number of occupants) in a space.

Table 12.23 shows a characteristic set of requirements for number of fixtures per number of occupants served in offices and public buildings. Table 12.24 provides minimum toilet requirements in kindergarten through 12th-grade schools based on population of each gender at the school site. Of course, these requirements vary by local code.

For example, in applying the requirements in Table 12.23, a three-story office building serving 250 occupants (125 male and 125 female) must have two water closets, two urinals, and one lavatory for males and eight water closets and one lavatory for females. Three drinking fountains are required (one per occupied floor). More fixtures may be required to effectively serve occupants in some facility layouts.

TABLE 12.24   AN EXAMPLE OF MINIMUM TOILET REQUIREMENTS IN KINDERGARTEN THROUGH 12TH-GRADE SCHOOLS BASED ON POPULATION OF EACH GENDER AT THE SCHOOL SITE. VARIES BY SCHOOL DISTRICT.

	Male	Female
Kindergarten (toilets to be within kindergarten complex)	1 toilet serves 1 to 20 2 toilets serve 21 to 50 Over 50, add 1 toilet for every 50 people	1 toilet serves 1 to 20 2 toilets serve 21 to 50 Over 50, add 1 toilet for every 50 people
Elementary	1 urinal for 75+ 1 toilet for 30	1 toilet for 25
Secondary	1 urinal for 35+ 1 toilet for 40	1 toilet for 30
Staff	1 toilet serves 1 to 15 2 toilets serve 16 to 35 3 toilets serve 36 to 55 Over 55, add 1 toilet for every 40 men Provide 1 urinal for every 50 men	1 toilet serves 1 to 15 2 toilets serve 16 to 35 3 toilets serve 36 to 55 Over 55, add

12.8   CODES AND STANDARDS

Building codes in the United States began as fire regulations written and enacted by several large cities during the 19th century. These regulations have evolved into a code that contains standards and specifications for materials, construction methods, structural strength, fire resistance, accessibility, egress (exiting), ventilation, lighting, energy conservation, and other considerations. There are many codes and standards that govern building design and engineering. An architect and engineer must be familiar with and maintain a working-level understanding of current codes and standards.

Standards

A *standard* is a set of specifications written by a professional organization or group of professionals that seek to standardize materials, components, equipment, or methods of construction/ operation. Many organizations develop technical standards, specifications, and design techniques that govern the design and construction of buildings and building systems. Many organizations exist that write standards for the plumbing industry, including those that follow:

- *American Gas Association, Inc. (AGA),* a national organization that develops standards, tests, and qualifies products used in gas lines and gas appliance installations.

- *American National Standards Institute (ANSI),* a private, nonprofit organization that coordinates the work between standards writing groups in the United States (e.g., International Standard Organization, American Society of Mechanical Engineers, American Society for Testing of Materials, and so on).

- *American Society for Testing of Materials (ASTM),* an international standards-writing organization that develops voluntary standards for materials, products, systems, and services.

- *American Society of Mechanical Engineers (ASME),* a national organization that develops standards for plumbing materials and products.

- *American Society of Sanitary Engineering, Inc. (ASSE),* a national organization that develops standards and qualifies products for plumbing and sanitary installations.

- *American Water Works Association (AWWA),* a national organization that promotes public health through improvement of the quality of water. Develops standards for drinking water, valves, fittings, and other equipment.

- *Canadian Gas Association, Inc. (CGA),* a Canadian association that develops standards, tests, and qualifies products used in gas lines and gas appliance installations.

- *International Standard Organization (ISO),* a worldwide standard coordinating organization that offers

internationally recognized certification for manufacturers that comply with high standards of quality control.

- *Mechanical Standardization Society of the Valve and Fittings Industry, Inc. (MSS),* a nonprofit technical association consisting of a group of manufacturers that develops technical codes and standards for the valve and fitting industry.

- *National Sanitation Foundation, Inc. (NSF),* a nonprofit organization known for its role in developing standards for equipment, products, and services. Many standards govern drinking water treatment chemicals and plumbing system components.

- *Underwriter's Laboratory, Inc. (UL),* a nonprofit organization that tests and qualifies valve and fitting products under UL standards.

Building Codes

By definition, a *building code* is a local ordinance (a law) that establishes the *minimum* requirements for design, construction, use, renovation, alteration, and demolition of a building and its systems. The intent of a building code is to ensure health, safety, and welfare of the building occupants. A *model building code* is a collection of standards and specifications written and compiled by group of professionals and made available for adoption by state and local jurisdictions.

A model building code does not become a law until it is formally adopted as a public ordinance by a local governmental entity such as a city or county or until it is enacted into public law by the state government. A few states have adopted a uniform statewide building code, but most states legally assign code adoption to counties and municipalities within the state. A municipality or county may write its own building code, but typically it relies on adoption of model codes as the base of its building code. Amendments to the model code are made to address local issues.

Model building codes and technical standards are formally revised periodically, usually every three to six years, to remain current with advancements and new practices in industry. Each time a new model code or standard is revised, it needs to be adopted into law by the governmental entity. As a result, different code editions may be in effect in neighboring communities. Design professionals must work hard to keep abreast of revisions in the code and standards.

The model code organizations in the United States created the International Code Council (ICC) to oversee the development of a single set of model codes to replace the corresponding model code of each organization. Initial efforts have focused on a common code format, with the ultimate goal of a single national model code referred to as the International Codes.

A typical set of plumbing standards extracted from a local code is provided in Figure 12.18 (See pp. 437–439). These standards are in addition to regulations in a model building code (e.g., International Plumbing Code) adopted by that municipality.

Construction Drawings

A complete set of construction drawings and specifications of the building plumbing system is needed to convey design information to the contractor. Plumbing construction drawings show the layout and design of a plumbing installation. Common plumbing symbols used on construction drawings are shown in Figure 12.19 (See pp. 439–440). The following construction drawings and details are generally required:

1. Water supply and distribution plan showing piping sizes, valves, water heater details, and temperature-pressure relief valves with discharge pipe.

2. Potable water system riser diagram showing piping sizes and provisions for protection of potable water supply.

3. Plumbing piping plan showing layout, pitch of drainage lines, cleanouts, size of traps, and riser diagram.

4. Sanitary drainage and vent system riser diagram showing drainage fixture units (DFU), sizes, and vent termination details through the roof.

5. Piping support and installation schedule.

6. Storm drainage details, including rain gutter or roof drain sizes and downspout/leader sizes.

7. Other special details (e.g., health care fixtures).

Specifications should contain piping and material fixture specifications, including:

1. The occupant load used to determine the number of required plumbing fixtures

2. Number and distribution based on the use group

3. Separate facilities for each gender

4. Accessible plumbing facilities and details

Administration of the Code

Where codes are in force, there will be a building department or department of building within the local governmental entity (e.g., city, county, and so forth). The governmental building department issues permits for the construction, addition, alteration, repair, occupancy, use, and maintenance of all buildings, structures, or utilities within its jurisdiction.

A *building inspector* is a representative of a governmental entity who performs the local administration and enforcement of the codes. The inspector first reviews the proposed construction documents (drawings and specifications) for compliance with the applicable codes, regulations, and ordinances. After approval of drawings and specifications, the inspector then checks for compliance during construction. A single building inspector may inspect all phases of construction or there may be a separate inspector for the building structure, plumbing, HVAC, electrical installations, and fire protection, health, and safety systems.

Plumbing inspections are mandatory on the local level in those municipalities that have implemented inspection requirements in their plumbing code. Although it varies by local jurisdiction, the typical residential or commercial code-compliance inspections occur in three phases.

1. *Underground Inspection*   The inspector reviews the sewer and water services coming from the city mains to the property. The inspector verifies that acceptable materials were used in the construction of the water and sewer services. The burial depth of the pipes is also measured to ensure that pipes have adequate protection from freezing. Therefore, the inspection must be made while all piping is exposed; that is, it must be completed before any plumbing trenches are backfilled.

2. *Rough-In Inspection*   This is an inspection of the interior drainage, waste vent, and water supply piping. Verification of the adequacy of the plumbing materials used in the project is made, including an examination of the sizing of water, waste, and vent piping, the grade of drainage piping, and the quality of connections between pipe and fittings. This inspection must be made while all piping is exposed before walls and ceilings are built. Frequently, a pressure test of the piping system is required.

3. *Final Inspection*   This is the inspection of the final setting of fixtures (bathtub, water closet, lavatory, kitchen sink, and so on.). The inspector looks for proper fixture setting and alignment, proper caulking around fixtures, acceptable shower valves, and so forth.

(404-408)

STUDY QUESTIONS

12-1. What is the difference between potable and nonpotable water? For what purposes may each be used?

12-2. What sources of water supply may be available to a city?

12-3. Why should any source of water be tested before the water is used?

12-4. What is the basic difference between a community and a private water supply system, and what are the advantages and disadvantages of each?

12-5. When a project (building) being designed is going to connect to a community water supply system, what information about the system must be obtained?

12-6. How are wells classified, and what methods of construction may be used for each type?

12-7. Show with a sketch how a well may be protected from surface water contaminants.

12-8. How are wells protected from possible contamination from sewage disposal fields?

12-9. What two types of well pumps are used, and what are the limitations of each?

12-10. What materials are most commonly used for the pipes and tubing in a building water supply system?

12-11. What materials are most commonly used for the pipes and tubing in a building sanitary sewage system?

12-12. Identify and describe the function of five common types of plumbing fixtures.

12-13. What is a waterless urinal? Why is its use advantageous in place of a conventional urinal?

12-14. How and where are the following valves used in the plumbing system?

a. Globe

b. Gate

c. Angle

d. Check

e. Thermostatic

f. Metered

12-15. Describe the differences between the three types of code-compliance inspections.

Design Exercises

12-16. The specific weight of water is 62.4 lb/ft^3.

a. Assuming 7.48 gal/ft^3, determine the weight of 1 gal of water.

b. Determine the weight of 1 L of water, in N.

12-17. A tank contains 500 gal of water. Determine the volume of water in:

a. Square feet

b. Liters

12-18. A pipe carries water at a flow rate of 10 gpm. Determine its volumetric flow rate in:

a. ft^3/s

b. L/min

12-19. A ¾ in diameter copper pipe (0.785 in inside diameter) carries water at a flow rate of 10 gpm. Determine the average velocity of the fluid.

12-20. A 20 mm diameter copper pipe (19.94 mm inside diameter) carries water at a flow rate of 40 L/s. Determine the average velocity of the fluid.

12-21. A pipe carries water at a flow rate of 25 gpm. Determine the minimum inside diameter so that the average velocity of the fluid does not exceed:

a. 10 ft/s

b. 8 ft/s

c. 4 ft/s

12-22. A pipe carries water at a flow rate of 90 L/s. Determine the minimum inside diameter so that the average velocity of the fluid does not exceed:

a. 3 m/s

b. 2 m/s

c. 1.4 m/s

12-23. Atmospheric pressure is about 14.7 psia. Water pressure at the base of a water tower is 110 psig. Determine the following:

a. Gauge pressure at the base of the tower

b. Absolute pressure at the base of the tower

12-24. Atmospheric pressure is about 80 kPa (absolute). Water pressure at the base of a water tower is 600 kPa (gauge). Determine the following:

a. Gauge pressure at the base of the tower

b. Absolute pressure at the base of the tower

12-25. Determine the volume of water contained in a ½ in diameter (nominal) copper tube per foot of length. Use actual pipe dimensions.

12-26. Determine the volume of water contained in a 1 in diameter (nominal) copper tube per foot of length. Use actual pipe dimensions.

12-27. Determine the volume of water contained in a 2 in diameter (nominal) standard weight Schedule 40 steel pipe per foot of length. Use actual pipe dimensions.

12-28. Determine the volume of water contained in a ½ in diameter (nominal) cross-linked polyethylene (PEX) tube per foot of length. Use actual pipe dimensions.

12-29. Determine the volume of water contained in a 1 in diameter (nominal) chlorinated polyvinyl chloride (CPVC) pipe with a standard dimension ratio (SDR) of 13.5 per foot of length. Use actual pipe dimensions.

12-30. Using a container with a known volume (e.g., 1 gal, 5 gal, and so on) and a watch with a second hand, determine the flow rate of the plumbing fixtures in your home or apartment (urinals, water closet, and clothes washer excluded). Record the flow rate for each fixture in gpm.

12-31. Using a container with a known volume (e.g., 1 L, 2 L, and so forth) and a watch with a second hand, determine the flow rate of the plumbing fixtures in your home or apartment (urinals, water closet, and clothes washer excluded). Record the flow rate for each fixture in L/min.

12-32. Using a copy of the local plumbing code, determine the minimum number of plumbing fixtures required for an auditorium planned for a total of 5500 people (assume 50% of each gender).

12-33. Using a copy of the local plumbing code, determine the minimum number of plumbing fixtures required for a restaurant planned for 250 seats (50% of each gender).

12-34. Using a copy of the local plumbing code, determine the minimum number of plumbing fixtures required for a higher education school building planned to accommodate 750 people (50% of each gender).

No person shall occupy, own, and allow to be occupied, or let to another for occupancy, any dwelling that does not comply with the requirements in this chapter.

(a) Every plumbing fixture and water and waste pipe in a dwelling shall be installed as provided in the city plumbing code and maintained in sanitary and sound condition, free from all sewage leaks and obstructions, and free from potable water leaks that are a constant flow of water. It is a specific defense to a charge of plumbing code non-conformity that the installation was in conformance with the plumbing code in effect at the time of installation and the work was done with a permit and approved, but this specific defense shall not apply to any requirements specifically listed in this section.

(b) Every dwelling unit shall contain a kitchen sink in sound condition and properly connected to an approved water and sewer system.

(1) A kitchen sink shall be of seamless construction and impervious to water and grease. The internal surfaces shall be smooth with rounded internal angles and corners, easily cleanable, free from cracks or breaks that leak or that could cut or injure a person, and impervious to water and grease.

(2) A kitchen sink shall be no smaller than twenty inches by sixteen inches, with a minimum uniform depth of six inches and a maximum uniform depth of twenty inches. Stone, plastic, and concrete laundry tubs, lavatory basins, or bathtubs are not acceptable substitutes for required kitchen sinks.

(c) Every dwelling shall contain a room completely enclosed by partitions, doors, or opaque windows from floor to ceiling and wall-to-wall that is equipped with a flush water closet in sound condition and properly connected to an approved water and sewer system.

(1) Every flush water closet shall have an integral water-seal trap.

(2) Water closets shall have smooth, impervious, easily cleanable surfaces that are free from cracks or breaks, that leak or that could cut or injure a person, and from makeshift repairs and shall be equipped with seats and flush tank covers constructed of smooth materials that are impervious to water and are free of cracks and breaks that could cut or injure a person.

(d) Every dwelling shall contain a lavatory basin in sound condition and properly connected to an approved water and sewer system and located in the same room as the flush water closet or as near to that room as practicable. If a dwelling contains a flush water closet in more than one room, it shall also contain a lavatory basin in each room with the flush water closet or as near to each such room as practicable.

(1) The lavatory basin shall be designed, intended, and located for use exclusively for ablutionary purposes.

(2) Lavatory basin surfaces shall be smooth, easily cleanable, impervious to water and grease, and free from cracks and breaks that leak or that could cut or injure a person. Stone, plastic, and concrete laundry tubs, sinks used for kitchen purposes, and bathtubs are not acceptable substitutes for lavatory purposes.

(e) Every dwelling shall contain, within a room completely enclosed by partitions, doors, or opaque windows from floor to ceiling and wall-to-wall, a

FIGURE 12.18 A typical set of plumbing standards extracted from local code. These standards are in addition to the regulations in a model-building code (e.g., International Plumbing Code) adopted by that municipality.

bathtub or shower in sound condition and properly connected to an approved water and sewer system.

(1) Every bathtub shall have a smooth, impervious, and easily cleanable inner surface free from makeshift repairs and free from cracks and breaks that leak or that could cut or injure a person.

(2) Every shower compartment or cabinet shall have a base with a leak-proof receptor that is made of materials such as precast stone, cement aggregates, plastic, preformed metals or materials of similar characteristics and whose pitch is sufficient to drain completely. The interior walls and ceiling surfaces of the shower cabinet or compartment shall be made of smooth, non-absorbent materials free of sharp edges that could cut or injure a person. Finishes of walls and ceilings that peel readily are not acceptable. The interior of every shower cabinet or compartment shall be watertight, maintained in sound condition, and easily cleanable. Repairs shall be required if more than 2 square feet of compartment wall is no longer waterproof or more than four linear feet of caulking has failed, or if the leak is causing an unsafe electrical condition.

(3) Built-in bathtubs with overhead showers shall have waterproof joints between the tub and adjacent walls and waterproof walls. Repairs shall be required if more than 2 square feet of compartment wall is no longer waterproof or more than four linear feet of caulking has failed, or if the leak is causing an unsafe electrical condition.

(f) Every dwelling shall have at least one flush water closet, lavatory basin, and bathtub or shower for each eight occupants thereof or for every family related by marriage, adoption, or blood plus two roomers residing therein. To meet the requirement of this subsection for occupants not related by blood, marriage, or adoption, in dwellings with more than one sleeping room, all facilities shall be accessible to the occupants without passing through any sleeping room. Under no circumstances shall occupants be required to pass through another dwelling unit to reach facilities, and all facilities shall be in the same building and located so that the occupants are not required to go outside the building to reach facilities.

(g) Every kitchen sink, lavatory basin, bathtub, or shower required under this section shall be connected to both hot and cold water lines as provided in the city plumbing code.

(h) Shared toilet and bath facilities shall be located on the same or adjacent floor as the rooming unit that they serve.

(i) Plumbing fixtures, except those having integral traps, shall be separately trapped by a water seal trap that is located as near the fixture outlet as possible and readily accessible for inspection. But one trap may be installed for a set of not more than three single compartment sinks or laundry trays or three lavatory basins immediately adjacent to each other in the same room, if the waste outlets are not more than thirty inches apart and if the trap is centrally located for the set of sinks.

(j) All exterior openings into the interior of the building, including without limitation those in crawl spaces, provided for the passage of piping shall be properly sealed with snug fitting collars of metal or other rat-proof material securely fastened into place.

Water Supply and Distribution Standards

(a) Potable water shall be provided for all dwelling units. When the premises are connected to a private water supply system, the private system shall be tested and approved as a sanitary source of water supply by the County Health

FIGURE 12.18 *(Continued)*

Department. The city manager may require additional testing of a private or public water supply, if there is a reason to believe contamination has occurred.

(b) Potable and non-potable water supplies shall be distributed through systems entirely independent of each other. There shall be no actual cross-connections between such supplies.

(c) Materials for water distributing pipes and tubing shall be brass, copper, cast iron, wrought iron, open-hearth iron, steel, or other approved material with fittings meeting the requirements of the city plumbing code.

(d) Water pressure shall at all times be adequate to permit a reasonable and proper flow of water into all plumbing fixtures.

FIGURE 12.18 *(Continued)*

PLUMBING ACRONYMS AND ABBREVIATIONS

ABS	Acrylonitrile Butadiene Styrene		LF	Low flush
ADA	American Disability Act		MGH	Male garden hose
AGA	American Gas Association		MHT	Male hose thread
AL	Aluminum		Mil	1/1000 inch
ANSI	American National Standards Institute		MIP	Male iron pipe
ASME	American Society of Mechanical Engineers		MIP	Male iron pipe
ASSE	American Society of Sanitary Engineering		MIPS	Male iron pipe size
ASTM	American Society for Testing of Materials		MPT	Male pipe thread
AWWA	American Water Works Association		MSS	Mechanical Standardization Society of the Valve and Fittings Industry
BC	Brushed Chrome			
BN	Brushed Nickel		Nom	Nominal
BOD	Biochemical oxygen demand		NPS	National pipe straight threads standard
Brs	Brass		NPSH	Net positive suction head
BSHW	Building service hot water		NPT	National pipe tapered threads standard
CGA	Canadian Gas Association		NSF	National Sanitation Foundation
CP	Chrome plated		OD	Outside diameter
CP	Polished Chrome		OSST	On-site sewage treatment
CPVC	Chlorinated Polyvinyl Chloride		Oz	Ounce
CTS	Copper tube size		PB	Polybutylene
CU	Copper		PB	Polished brass
CW	Cold water		PB	Polished brass
CWT	Copper water tube		PE	Polyethylene
DC	Die cast		PEX	Cross-linked Polyethylene
DFU	Drainage fixture unit		Pln	Plain
DHW	Domestic hot water		POC	Point of connection
Dia or Φ	Diameter		PP	Polypropylene
DWV	Drain, waste and vent		PR	Pressure regulator
Fct	Faucet		PRV	Pressure reducing valve
FGH	Female garden hose		psi	Pounds per square inch
FHR	First hour rating		Pt	Point
FHT	Female hose thread		PVC	Polyvinyl Chloride
FIP	Female iron pipe		PVDF	Polyvinylidene Fluoride
FIPS	Female iron pipe size		SDR	Standard dimension ratio
FM	Factory Mutual		SF	Satin finish
FPT	Female pipe thread		SN	Satin Nickel
Ga	Gauge		SP	Stand pipe
Gal	Gallon		SR	Styrene rubber
GPF	Gallons per flush		SS	Stainless steel
GPM	Gallons per minute		SWT	Sweat
Hd	Head		T/P	Temperature-pressure
Hdl	Handle		THD	Thread
Hgt	Height		UL	Underwriter's Laboratory
HP	Horsepower		ULF	Ultra-low flush
HW	Hot water		UNC	Unified National Coarse Thread
HWR	Hot water recirculating		UNF	Unified National Fine Thread
IAPMO	International Association of Plumbing and Mechanical Officials		VTR	Vent through roof
			WC	Water closet
ID	Inside diameter		WH	White
IPS	Iron pipe size		WSFU	Water supply fixture unit

FIGURE 12.19 Plumbing symbols used on construction drawings.

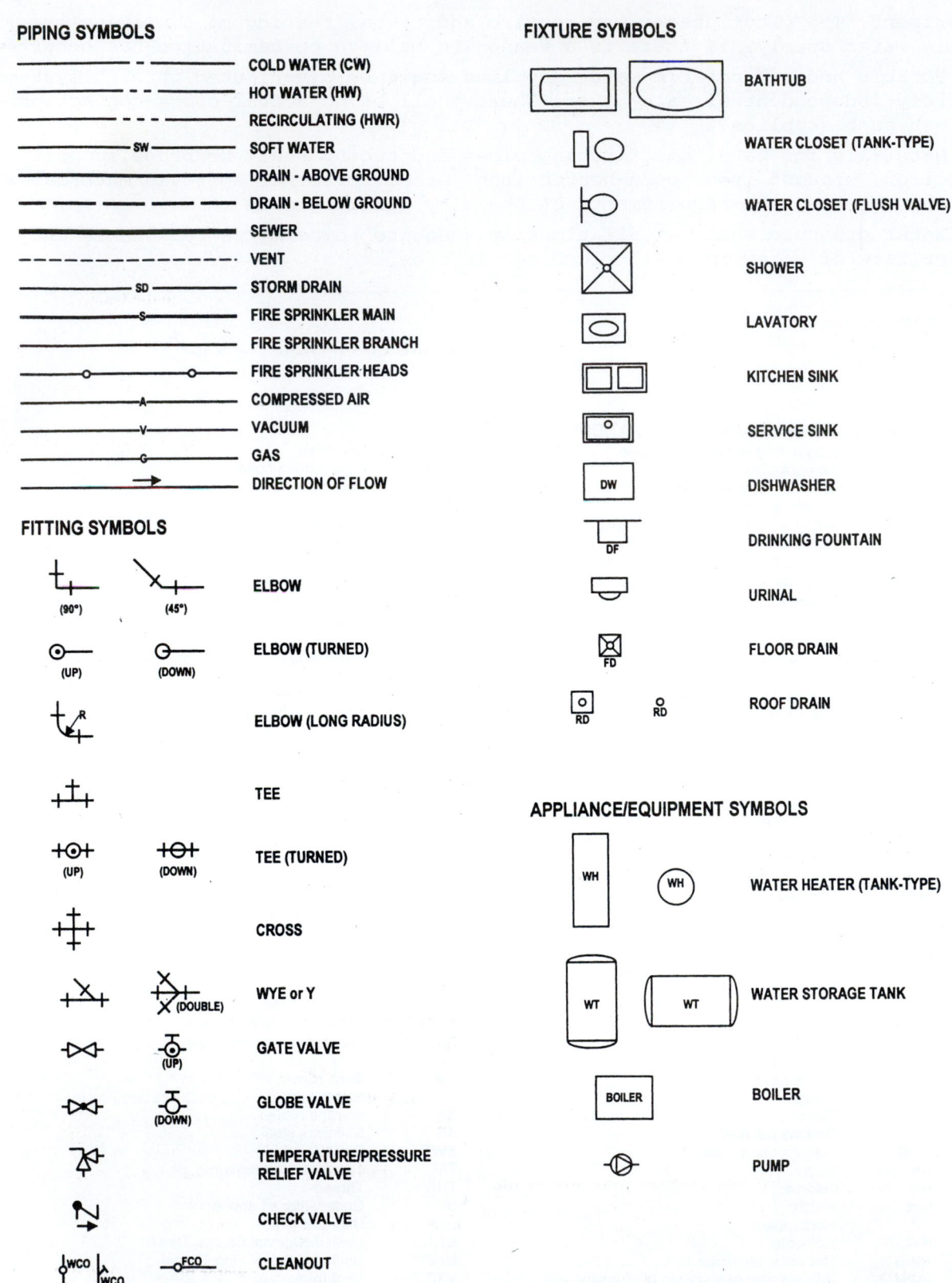

PIPING SYMBOLS

COLD WATER (CW)
HOT WATER (HW)
RECIRCULATING (HWR)
SOFT WATER
DRAIN - ABOVE GROUND
DRAIN - BELOW GROUND
SEWER
VENT
STORM DRAIN
FIRE SPRINKLER MAIN
FIRE SPRINKLER BRANCH
FIRE SPRINKLER HEADS
COMPRESSED AIR
VACUUM
GAS
DIRECTION OF FLOW

FITTING SYMBOLS

ELBOW
(90°) (45°)
ELBOW (TURNED)
(UP) (DOWN)
ELBOW (LONG RADIUS)
TEE
TEE (TURNED)
(UP) (DOWN)
CROSS
WYE or Y
(DOUBLE)
GATE VALVE
(UP)
GLOBE VALVE
(DOWN)
TEMPERATURE/PRESSURE
RELIEF VALVE
CHECK VALVE
CLEANOUT
(WALL CLEANOUT) (FLOOR CLEANOUT)

FIXTURE SYMBOLS

BATHTUB
WATER CLOSET (TANK-TYPE)
WATER CLOSET (FLUSH VALVE)
SHOWER
LAVATORY
KITCHEN SINK
SERVICE SINK
DISHWASHER
DRINKING FOUNTAIN
URINAL
FLOOR DRAIN
ROOF DRAIN

APPLIANCE/EQUIPMENT SYMBOLS

WATER HEATER (TANK-TYPE)
WATER STORAGE TANK
BOILER
PUMP

FIGURE 12.19 *(Continued)*

BUILDING WATER SUPPLY
SYSTEMS

13.1 THE BUILDING WATER SUPPLY SYSTEM

Main Parts of a Water Supply System

Plumbing codes require that a potable water supply be adequately furnished to all plumbing fixtures. The water supply system in a building carries cold and hot water through distribution pipes and delivers it to the plumbing fixtures. Schematic drawings of conventional residential and commercial systems are shown in Figures 13.1 and 13.2. The water service line carries water from a district supply pipe to the building. The main parts of a typical water supply system include the following.

Building Supply

The *building supply* or *water service* is a large water supply pipe that carries potable water from the district or city water system or other water source to the building.

Water Meter

A *water meter* is required by most district water supply systems to measure and record the amount of water used. It may be placed in a meter box located in the ground near the street or inside the building.

Building Main

The *building main* is a large pipe that serves as the principal artery of the water supply system. It carries water through the building to the furthest riser. The building main is typically run (located) in a basement, in a ceiling, in a crawl space, or below the concrete floor slab.

Riser

A *riser* is a water supply pipe that extends vertically in the building at least one story and carries water to fixture branches. It is typically connected to the building main and runs vertically in the walls or pipe chases.

Fixture Branch

A *fixture branch* is a water supply pipe that runs from the riser or main to the fixture being connected. In a water supply system, it is any part of a piping system other than a riser or main pipe. Fixture branch pipes supply the individual plumbing fixtures. A fixture branch is usually run in the floor or in the wall behind the fixtures.

Fixture Connection

A *fixture connection* runs from the fixture branch to the fixture, the terminal point of use in a plumbing system. A shut-off valve is typically located in the hot and cold water supply at the fixture connection.

General Water Distribution System Layout

The water service pipe is an underground pipe that is typically called a *lateral*. It extends from the underground street main that is part of a district or city water system, and delivers pressurized potable water to a building plumbing system. The water service lateral is connected to a water meter that measures consumption. The water meter is typically located in an underground curb box located in the building's front yard or is located in the building interior, in which case it is connected to a remote readout on the exterior of the building, which allows easy access for meter readings. If the building plumbing system is served by a well, a water meter is not needed unless monitoring of consumption is required. A water service shut-off valve is typically located at the meter location.

As the building supply piping enters the building, pipes split off to supply water to hose bibbs, the irrigation system, and any industrial process equipment using water that does not need to be heated. A building shut-off valve is typically located at this location. If pressure available from the water service is too high, a pressure-reducing valve or pressure-reducing arrangement drops the water pressure to an acceptable level.

If water softening or treatment is desired in the building, water is passed through a softening or treatment device before the water is distributed to the building. Water softening or treatment is done after irrigation water and/or water for process equipment is removed. Once the water supply main passes through the softening or treatment device (if used), the main pipe splits to provide water to a water heater or water heating system that generates hot water before it is distributed.

A fixture connection links the hot or cold water branch to the fixture, the terminal point of use in a plumbing system. A shut-off valve is typically located in the hot and cold water supply at the fixture connection. Air chambers are installed as

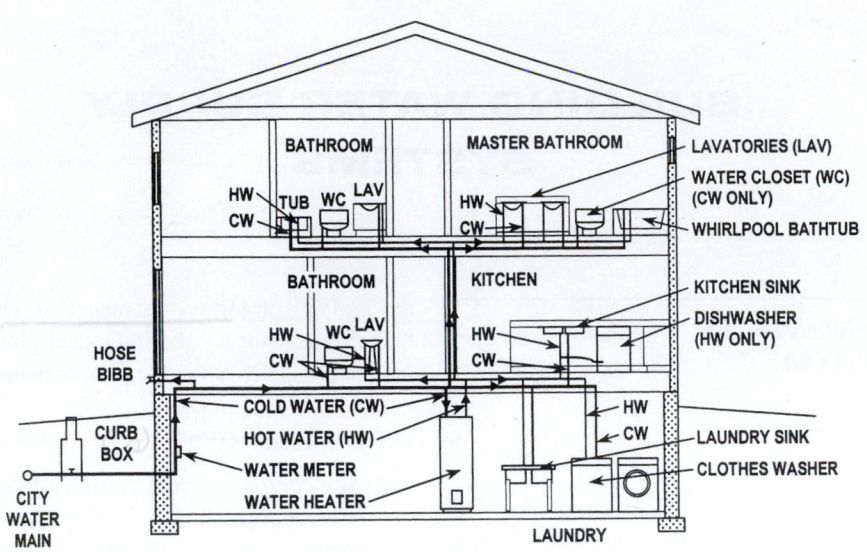

FIGURE 13.1   Basic parts of a residential rigid-pipe distribution configuration. Branch pipes follow a straight-line path.

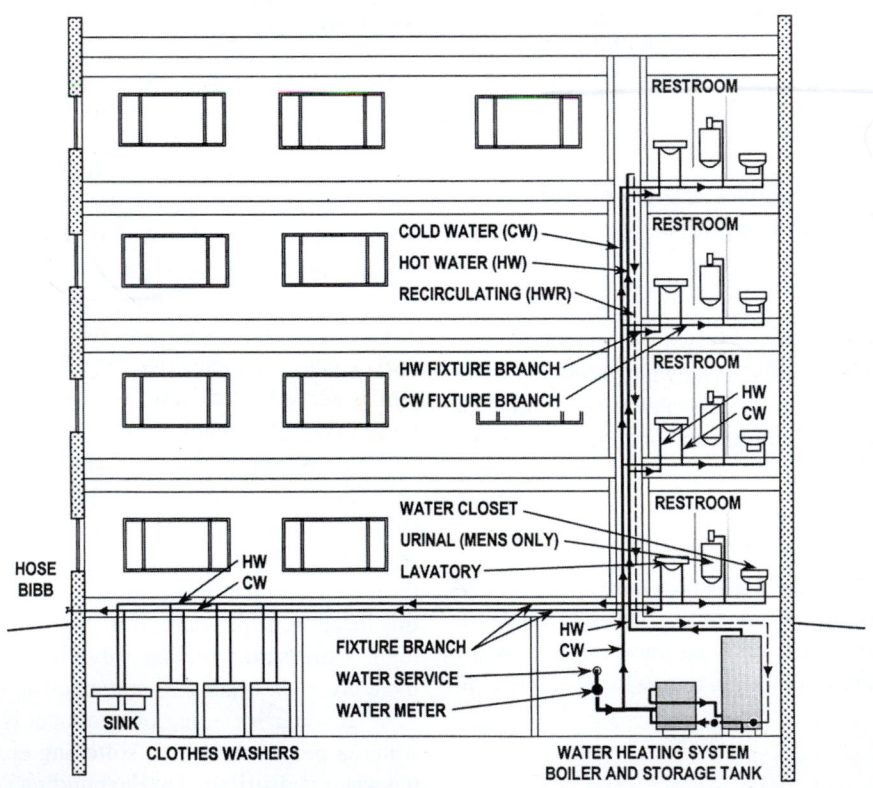

FIGURE 13.2   Basic parts of a commercial rigid-pipe distribution configuration. Note the common plumbing chase.

close as possible to the fixture valves or faucet and at the end of long runs of pipe to reduce water hammer.

In a conventional rigid-pipe water distribution method, fixture branches extend from a riser or main to the individual fixture being connected. A fixture branch is usually run in the floor or in the wall behind the fixtures. In the innovative homerun water distribution method, individual branches begin at a main manifold located in a utility room or basement, usually near the water service. Individual hot and cold water branches extend uninterrupted to each plumbing fixture or a fixture group. The distribution configuration methods are described in the following.

Rigid-Pipe Distribution Configuration

In the conventional *rigid-pipe distribution configuration,* the hot and cold water distribution pipes are installed parallel to one another as they convey hot and cold water to risers and branch pipes. Running pipes parallel with building walls and floors arrange pipes in an organized manner. Hot and cold pipes should be spaced at least 6 in (150 mm) apart or have insulation placed between them to prevent heat interchange. In top-quality work, both pipes should be insulated so that hot water pipes minimize heat loss and cold water pipes prevent surface condensation. Schematics of this system are shown in Figures 13.1 and 13.2. Photos 13.1 through 13.17 show components of the rigid-pipe water distribution system.

A branch supplying water to two or more fixtures is called a *zone.* A zone can supply one or many fixtures on one floor or on a few floors. Fixtures are typically located in clusters called *groups.* For example, in a commercial building or school, restrooms for men and women are grouped together with fixtures arranged against a common plumbing chase; that is, fixtures back up on both sides of walls enclosing the

PHOTO 13.3 A residential service line with an interior water meter, shut-off valve (above meter), and pressure-reducing valve (below meter). (Used with permission of ABC)

PHOTO 13.1 Rough-in of a residential service line passing through a foundation wall. (Used with permission of ABC)

PHOTO 13.4 A wall-mounted exterior water meter. (Used with permission of ABC)

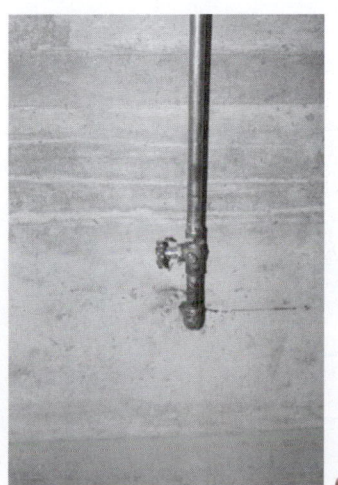

PHOTO 13.2 A residential service line with a shut-off valve. (Used with permission of ABC)

PHOTO 13.5 An underground curb box ready for water meter installation. (Used with permission of ABC)

PHOTO 13.6 Hot and cold water supply fixture branches. (Used with permission of ABC)

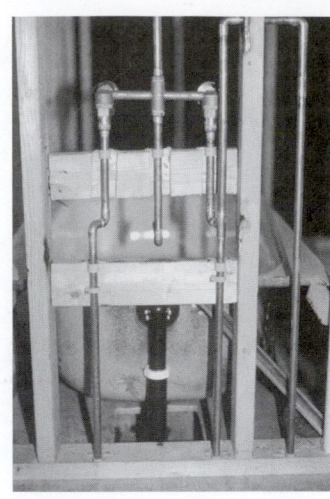

PHOTO 13.7 Fixture branches connecting to a bathtub. (Used with permission of ABC)

PHOTO 13.8 A front side view of branches connecting to a bathtub and showerhead. (Used with permission of ABC)

PHOTO 13.9 A front side view of branches connecting to a bathtub and showerhead control valve. (Used with permission of ABC)

PHOTO 13.10 The rough-in connection for a showerhead. (Used with permission of ABC)

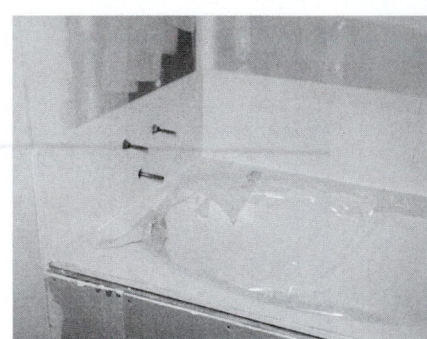

PHOTO 13.11 A rough-in of fixture branches for a bathtub (no showerhead). (Used with permission of ABC)

PHOTO 13.12 A rough-in of fixture branches for a lavatory and water closet. (Used with permission of ABC)

PHOTO 13.13 A rough-in of fixture branches and drain for a laundry connection. (Used with permission of ABC)

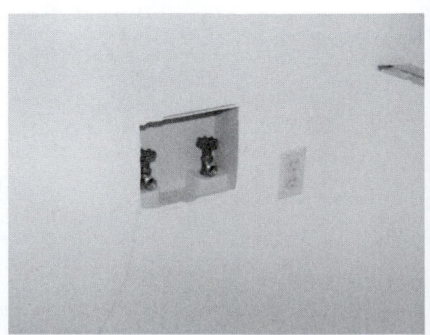

PHOTO 13.14 A finished laundry connection. (Used with permission of ABC)

PHOTO 13.15 A rough-in of fixture branches for a group of lavatories in a public restroom. (Used with permission of ABC)

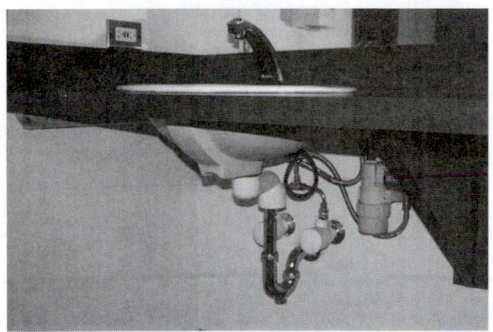

PHOTO 13.16 Hot and cold water supply fixture connections. (Used with permission of ABC)

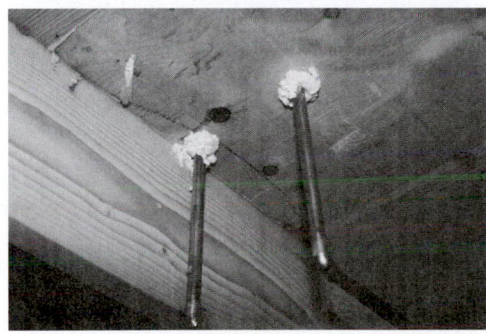

PHOTO 13.17 Sealed openings to prevent flanking of sound from a bathroom above. (Used with permission of ABC)

plumbing chase. Water fountains are also located in the area of the restrooms. Likewise, a set of showerheads is located on a common chase or wall. A group arrangement allows several fixtures to be served by the same water main or branch. For efficiency and convenience, risers and chases should be located near plumbing groups.

In wood and light-gauge steel frame construction, holes are drilled or cut to allow the passage of the pipes. Holes are located in the center of the framing members to avoid nails or screws penetrating the pipe and to minimize structural damage in load-bearing members. Metal plates cover pipes in wood

framing when the pipe surface comes too close to the surface of the framing and there is a possibility of nail or screw penetration. There are times when the width of a wall needs to be increased to allow for pipes running horizontally to pass by drainage pipes (or other pipes) running vertically. These walls of increased thickness are called *plumbing walls*.

The spaces in the open web of wood or steel truss construction (e.g., wood trusses and steel joists) make it easier to run piping through these structural members without affecting their structural integrity. It is sometimes difficult to feed long lengths of pipe through closely spaced trusses or open web joists after they are installed. A point of difficulty also exists where pipe encounters some other large pipe or ductwork going in the opposite direction. This necessitates coordination with the subcontractors erecting the trusses and installing the heating, air conditioning, or ventilating ductwork.

On projects with concrete walls and ceilings, it is typically necessary to provide sleeves (holes) in the concrete for the pipes to pass through to get from space to space. It will also be necessary to provide inserts and hangers to support the pipes. Pipes placed below poured concrete slabs and in concrete–steel composite decking must be placed before the concrete is poured. Typically, both the water supply and drainage pipes are laid out next to each other, because they go to the same areas of the project. Many times, they are planned so they will come up in a partition, so accurate placement of piping is crucial. Tub, shower, and water closet piping will need to be placed in the exact location where the fixture is to go. All piping must be carefully located and the system checked for leaks before the concrete is poured because any relocation or repairs of pipes would be costly.

In multistory buildings, risers are pipes that carry water vertically through walls or through enclosures called chases. A *chase* is a vertical opening through a floor or several floors that is enclosed with walls between floors. A chase can enclose piping only or it can enclose electrical wiring and/or mechanical system ducting and/or pipes that run vertically from floor to floor through the building.

Pipe tunnels may be used on large projects to provide concealed space for the passage of mechanicals at ground level and from building to building. Hangers from the top or side of the tunnel are used to support the pipes. Access may be from either end of the tunnel, or access floors may be provided.

Readily accessible valves used to close off the water supply to a fixture, appliance, or system are called *shut-off valves*. A shut-off valve is required on the discharge side of the water meter. When more than one building is served by a single water service line, a shut-off valve must be installed at each building. A readily accessible shut-off valve must be installed on the cold water supply line feeding a water heater in the proximity of the appliance. In multifamily installations, single or multiple shut-off valves must be installed in each dwelling unit so water to the unit can be shut off without interrupting the supply to other units. Valves must be accessible in the dwelling unit they control.

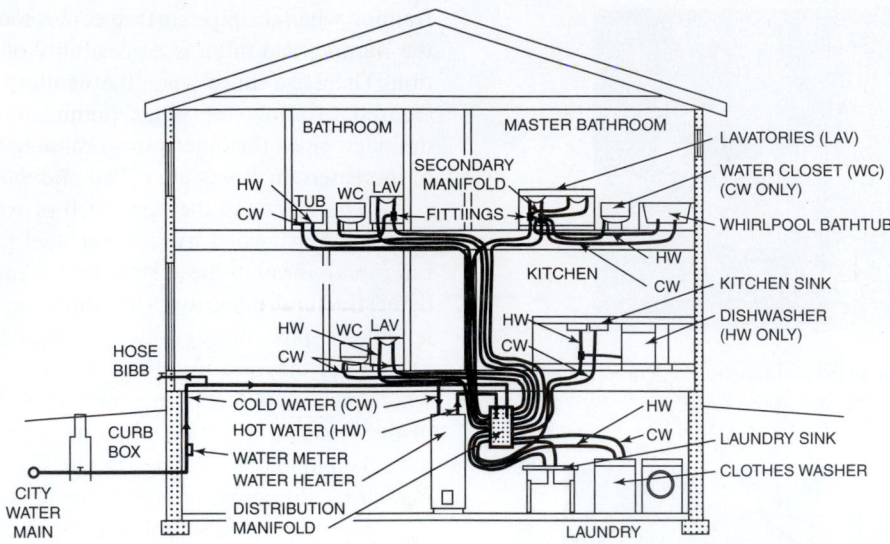

FIGURE 13.3 Basic parts of a homerun (manifold) distribution configuration. Hot and cold water lines dedicated to each fixture originate at individual ports in the manifold and extend to the individual fixture.

Homerun (Manifold) Distribution Configuration

A *homerun* or *manifold distribution configuration* consists of a plastic or metal plumbing manifold and flexible plastic piping. A schematic of this system is shown in Figure 13.3. The manifold serves as a common location from which all the plumbing fixtures are supplied. Manifolds have two separate chambers: the cold water chamber is supplied from the main water supply line and the hot water chamber is fed from the water heater. A water line dedicated to each fixture originates at a port in the manifold and extends to the individual fixture, so fewer fittings are required. Some manifolds offer shut-off valves so each fixture can be turned off individually at the manifold. In a hybrid configuration, termination or secondary manifolds may feed the plumbing requirements for a room or set of rooms and reduce the number of fittings required in the plumbing system. Photos 13.18 through 13.22 show components of the homerun (manifold) water distribution system.

Homerun configurations typically use cross-linked polyethylene (PEX) or composite PEX-AL-PEX piping, which is suitable for cold and hot water use. Because of the flexibility of PEX, there is less of a need for piping tees and elbows. With one fitting needed at the manifold and a second transition fitting at the fixture, a single run is made from the manifold to the fixture. As a result, homerun configurations can be installed more quickly than rigid plumbing. Fittings and couplings are available for unique situations, such as creating changes in direction that are tighter than the minimum bend radius allowable for the piping. The homerun configuration requires much more pipe than the rigid configuration, but the plastic pipe used is much less expensive than the metal (copper) pipe used in the

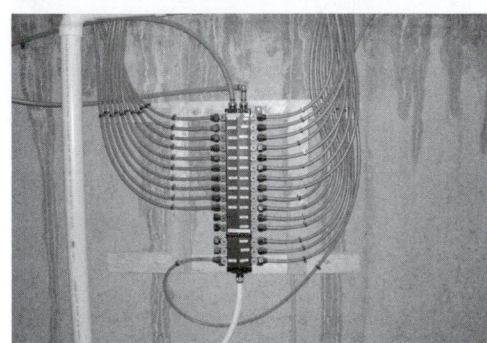

PHOTO 13.18 The manifold of a homerun water supply distribution configuration. Water enters the manifold at the bottom. The pipes at the top of the manifold serve and return from the water heater. Flexible pipes (cold water on the right and hot water on the left) run to individual fixtures. (Used with permission of ABC)

PHOTO 13.19 A clearly marked set of hot water (left) and cold water (right) ports on a manifold. (Used with permission of ABC)

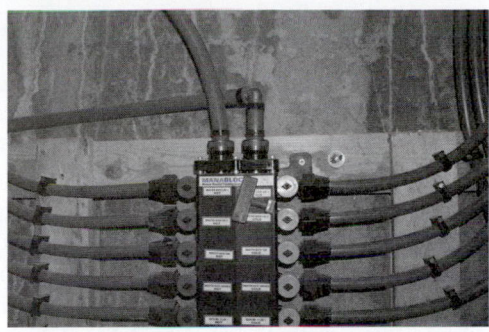

PHOTO 13.20 A close-up of the manifold of a homerun water supply distribution configuration shown in the previous photograph. A water line dedicated to each fixture originates at a port in the manifold and extends to the individual fixture. A valve at each port can shut off water to the individual fixture. (Used with permission of ABC)

PHOTO 13.21 Water lines dedicated to each fixture run from the manifold to the individual fixture. Because of the flexibility of PEX pipe, fittings are not needed except at the start (manifold port) and end (fitting at fixture). (Used with permission of ABC)

PHOTO 13.22 The backside of the terminating fitting at the end of runs of PEX pipe in the homerun water supply distribution configuration. (Used with permission of ABC)

rigid-pipe configuration. As a result, the home run configuration is typically cheaper.

Homerun configurations equalize pressure, which allows several fixtures to be operated simultaneously without drastic changes in pressure or temperature. Because each fixture has its own supply line, the size of the pipe can be adjusted for a fixture's specific use. In addition, for some fixtures, PEX piping can be downsized (usually ⅛ in diameter smaller) in comparison to a conventional rigid-piping configuration. A smaller diameter means that hot water arrives at fixtures faster, and less hot water is left standing in pipes after a fixture is operated.

Upfeed and Downfeed Distribution

13-3

Two basic types of water supply distribution systems are used in buildings: the upfeed (or upflow) system and the downfeed (or downflow) systems. See Figures 13.4 and 13.5. Variations of these distribution systems are described in the paragraphs that follow.

In a conventional upfeed system, water pressure from the water supply main is relied on to drive water flow through the system. Water pressure in building water supply mains typically ranges from 40 to 80 psi (275 to 550 kPa), with 80 psi (550 kPa) considered the upper limit for most systems plumbed with metal pipe and 40 psi the upper limit for plastic pipe. (Note: psi is an abbreviation for lb/in^2.) This available pressure places limits on how far water can be driven upward in a plumbing system. Part of the available pressure is expended in friction losses as the water passes through the meter and the various pipes and fittings; and part of the pressure is expended to overcome gravity, which is the pressure required to push the weight of water upward vertically (up the riser). Additionally, there must be sufficient pressure left at the remote fixture to drive flow of water through the fixture.

It takes 0.433 psi to push water up 1 ft vertically or, in the SI (metric) system, 9.8 kPa to push water up 1 m vertically. Conversely, a 1.0 psi pressure can push water upward 2.31 ft vertically or, in the SI (metric) system, 1 kPa to push water up 0.144 m vertically. (This concept is further discussed later in this chapter.) Pushing water up 20 ft (6.1 m) vertically requires a pressure at the base of the riser of at least 8.68 psi (42 kPa), because 20 · 0.433 psi = 8.68 psi.

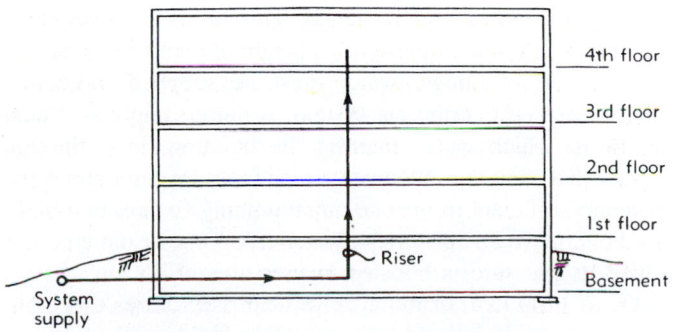

FIGURE 13.4 An upfeed plumbing system supply relies on street pressure to drive water flow.

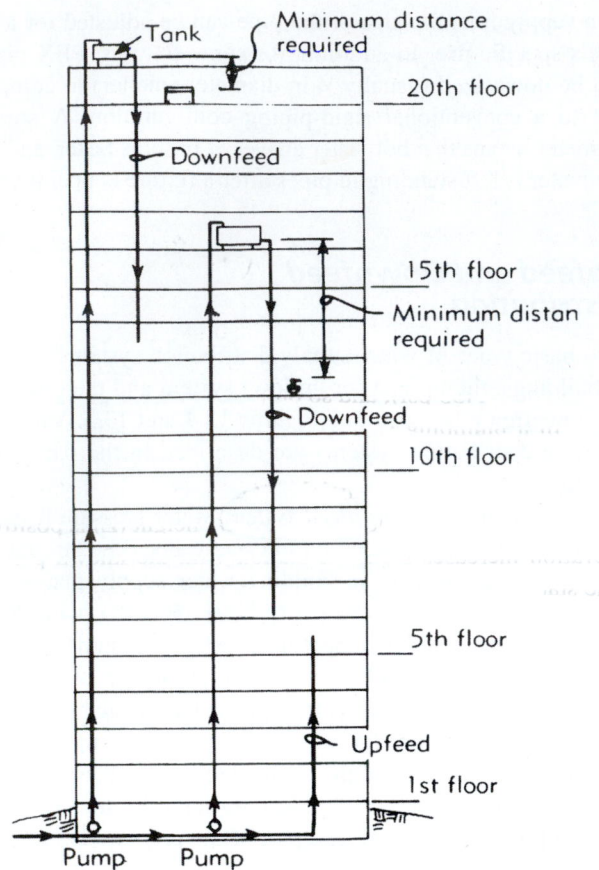

FIGURE 13.5 A downfeed plumbing system supply is needed to provide adequate pressure on upper floors. It is usually combined with an upfeed system that provides water to the lower floors.

If a pressure of 80 psi (550 kPa) is available at the base of a plumbing system riser, the maximum height water can rise vertically is 185 ft (56.5 m), because 80 psi · 2.31 ft = 185 ft. Depending on the exact floor-to-floor height, a pressure of 80 psi (550 kPa) would drive water about 15 stories, but only if there were no friction losses or fixture operation pressures to consider. When friction loss and fixture operation are taken into account, practical design limitations typically establish about 60 ft (18 m) as the standard maximum height and 40 ft (12 m) as the preferred maximum height. This limits a conventional upfeed system to buildings with a height of about 5 stories.

In tall buildings, water must be supplied through a *pumped upfeed* distribution system. A pumped upfeed system is one in which water entering the building flows through pumps that maintain adequate water pressure throughout the structure sufficient to operate any plumbing fixture. In a high-rise building (e.g., 50 stories), water enters one or more pumps where its pressure is boosted to pressures of 150 to 250 psi (1000 to 1700 kPa) or more. A vertical riser carries this high-pressure water to fixtures at the top of the building. Such a pressure in the distribution system is too great to use in plumbing fixtures (e.g., lavatories and water closets). For this reason, at

several zones water is removed from the vertical riser, reduced in pressure at *pressure-reducing stations* and distributed to the fixtures in that zone. The pressure-reducing stations, which are located about every 10 floors, monitor and adjust for any variation in pressure. This ensures that water available to plumbing fixtures is always kept under a constant pressure.

In buildings that cannot be adequately serviced to the top floor by an upfeed system, water is pumped to elevated storage tanks in, or on, the building, and the water is fed down into the building by gravity. This gravity system, fed from the upper stories to the lower, is called a *downfeed* distribution system. Water entering the building flows through pumps that develop sufficient water pressure to drive water to storage tanks serving zones of about 10 floors each. To develop adequate pressure, the storage tanks are placed above the zones that they serve.

13.2 WATER PRESSURE CONSIDERATIONS

Hydrostatic Pressure

Fluid (gas or liquid) molecules tend to seek equilibrium (a stability of forces). When forces acting on a fluid are unequal, molecules in the fluid move in the direction of the resultant forces. Therefore, an elementary property of any fluid at rest (not flowing) is that the force exerted on any molecule within the fluid is the same in all directions.

A *hydrostatic force* is a force exerted by the weight of the fluid against the walls of a vessel containing the fluid. *Hydrostatic pressure*, the hydrostatic force per unit area, is perpendicular to the interior walls at every point. If the pressure were not perpendicular, an unbalanced force component would exist and the fluid would flow.

If gravity is the only force acting on a fluid (e.g., the water in a gravity plumbing system), the hydrostatic pressure at any point in the system is directly proportional to the weight of a vertical column of that water. Additionally, the pressure is directly proportional to the depth below the surface and is independent of the size or shape of the container. For example: the hydrostatic pressure at the bottom of a 6 ft (2 m) high pipe that is filled with water is the same as the hydrostatic pressure at the bottom of a tank or pool that is 6 ft (2 m) deep. A 12 ft (4 m) pipe that is filled with water, and slanted so that the top is only 6 ft (2 m) above the bottom (measured vertically), will have the same hydrostatic pressure exerted at the bottom of the 6 ft (2 m) vertical pipe even though the distance along the 12 ft (4 m) pipe is much longer.

Water Pressure

Water pressure difference is the driving force behind fluid flow. Water pressure available at the water service is lost as water flows through the piping of a plumbing system. This pressure loss or pressure drop in a plumbing system is from friction loss as the water moves through the system and pressure loss as

TABLE 13.1   REQUIRED RESIDUAL PRESSURE FOR DIFFERENT FIXTURE TYPES. RESIDUAL PRESSURE IS THE PRESSURE IN THE PIPE AT THE ENTRANCE TO THE PARTICULAR FIXTURE.

Fixture	Required Residual Pressure	
	psi	kPa
Ordinary basin faucet	8	55
Self-closing basin faucet	12	80
Sink faucet—⅜ in	10	70
Sink faucet—½ in	5	35
Bathtub faucet	5	35
Laundry tub cock —½ in	5	35
Shower	12	80
Water closet	15	100
Urinal (lush valve)	15	100
Garden hose, 50 ft, and hose bibb/sill cock	30	200

water is forced to a higher elevation (e.g., from the basement to an upper story).

Water pressure available at the water service is considered acceptable in the range of 40 to 80 psi (275 to 550 kPa) or greater in mountainous regions. In most residential and commercial systems, the upper limit is 80 psi (550 kPa). Some systems with thermoplastic supply piping set the upper pressure limit much lower, usually about 40 psi (275 kPa). When water service pressure is deemed too great, a pressure reducer is used to limit pressure and reduce potential for leaks in the thermoplastic supply piping.

An insufficient pressure at a plumbing fixture results in low flow of water at that fixture. An excessive pressure at a fixture may cause disturbingly high flow, will waste water, and may cause damage to or premature deterioration of the fixture.

Residual water pressure is the pressure available at the outlet, just before a fixture. It affects water output of a fixture. The residual pressure requirement at the many types of plumbing fixtures varies. Code specifies that the highest (most remote outlet) fixture have a minimum pressure of 8 psi (55 kPa) for flush tanks and 15 psi (103 kPa) for fixtures with flushometer valves. Table 13.1 provides recommended residual pressure for different plumbing fixture types.

Pressure Difference

When forces acting on a fluid are unequal, molecules in the fluid move in the direction of the least pressure. Fluid flow is caused by a pressure difference in the fluid. A fluid will always flow from a higher pressure region to a lower pressure region. A pressure difference must exist at a plumbing fixture to cause water to flow—that is, water pressure at the fixture must be at a higher level than atmospheric pressure for water to flow from the fixture. *Pressure difference (ΔP)* is the driving force of fluid flow.

Pressure Difference from Elevation Change (Static Head)

In the building plumbing supply system, water is the fluid under consideration. Water has a maximum specific weight of 62.4 lb/ft^3. So at its maximum weight, a 1 ft by 1 ft by 1 ft cube of water exerts a maximum force of 62.4 lb at its base, which equates to a pressure of 62.4 lb/ft^2 or a pressure of 0.433 psi at the base of the cube (62.4 lb/ft^2 divided by 144 in^2/ft^2 = 0.433 lb/in^2). Therefore, a 1 ft high column of water creates a pressure of 0.433 psi at its base; a 2 ft high column exerts a pressure of 0.866 psi at its base (2 · 0.433 psi); a 3 ft column, 12.99 psi at its base (3 · 0.433 psi); and so on.

In a plumbing supply system, pressure difference from elevation change or simply *static head (ΔP$_{static}$)* is found by multiplying the vertical height (Z), in feet, by the factor of 0.433 psi/ft. By convention, the vertical height (Z) is positive if elevation increases from the station with the known pressure (the station is higher than the station with the known pressure) and negative if elevation decreases.

$$\text{static head, in psi, } \Delta P_{static} = -0.433Z$$

In the SI (metric) units, where vertical height (Z) is measured in meters:

$$\text{static head, in kPa, } \Delta P_{static} = -9.8Z$$

Pressure difference is negative (a loss) if the elevation change from the known pressure is upward (a positive Z) and positive if elevation change is downward (a negative Z).

Pressure change in a plumbing system from elevation change may be computed by multiplying the vertical height of the fixture outlet to the street main (a known pressure) by the pressure the water creates per foot. Conversely, a 2.31 ft of elevation change results in a pressure difference of 1 psi and a 0.102 m change results in a pressure difference of 1 kPa.

Example 13.1

A plumbing fixture outlet is 24 ft above the water service line. Pressure available at the water service is 45 psi.

a. Determine the change in pressure from elevation.

$$\Delta P_{static} = -0.433Z = -0.433(24)$$
$$= -10.4 \text{ psi (a 10.4 psi loss)}$$

Change in elevation is upward from the known pressure, so ΔP is negative.

b. For pressure available at fixture:

45 psi + (−10.4 psi) = 34.6 psi available at fixture outlet

Pressure Losses from Friction

Pressure losses from friction, *friction head (ΔP$_{friction}$)*, are more difficult to compute, as they are related to flow rate (gpm,

FRICTION LOSS IN SMOOTH PIPE

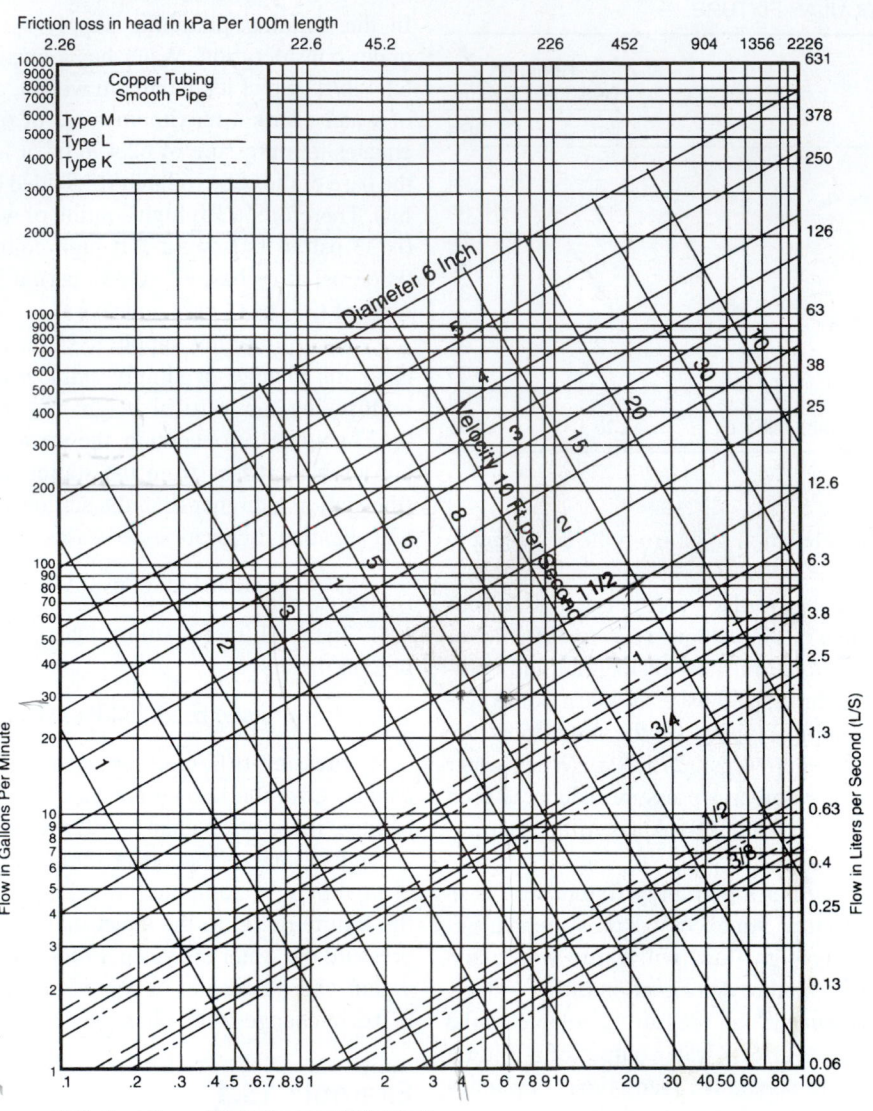

1. This chart applies to smooth new copper tubing with recessed (streamlined) soldered joints and to the actual sizes of types indicated on the diagram.

Note: Metric added by author

FIGURE 13.6 Pressure loss from friction caused by water flow in smooth pipe (copper tubing), based on flow rate and pipe diameter.

L/min or L/s), fluid velocity (ft/s or m/s), pipe diameter, pipe material and surface roughness, pipe length, and number of fittings and valves. Experimentation has led to a variety of pressure drop charts for pipes of many different materials. Pressure drop charts are provided in Figures 13.6 through 13.8. A review of the pressure drop chart in Figure 13.6 for water in smooth pipe shows that pressure drop is related to water flow rate, velocity, and nominal pipe diameter.

The pressure drop chart in Figure 13.6 applies to smooth pipe, such as copper and plastic pipe and tubing. Pipe diameters from ⅜ to 1 in are also based Types K, L, and M copper tubing sizes; refer to the legend in the upper left hand corner of the chart. Beyond 1 in diameter tubing, pressure drop disparity with difference in wall thickness is negligible.

Figure 13.7 applies to friction loss in rough pipe, such as steel and iron pipe. Roughness of the inner walls of pipe

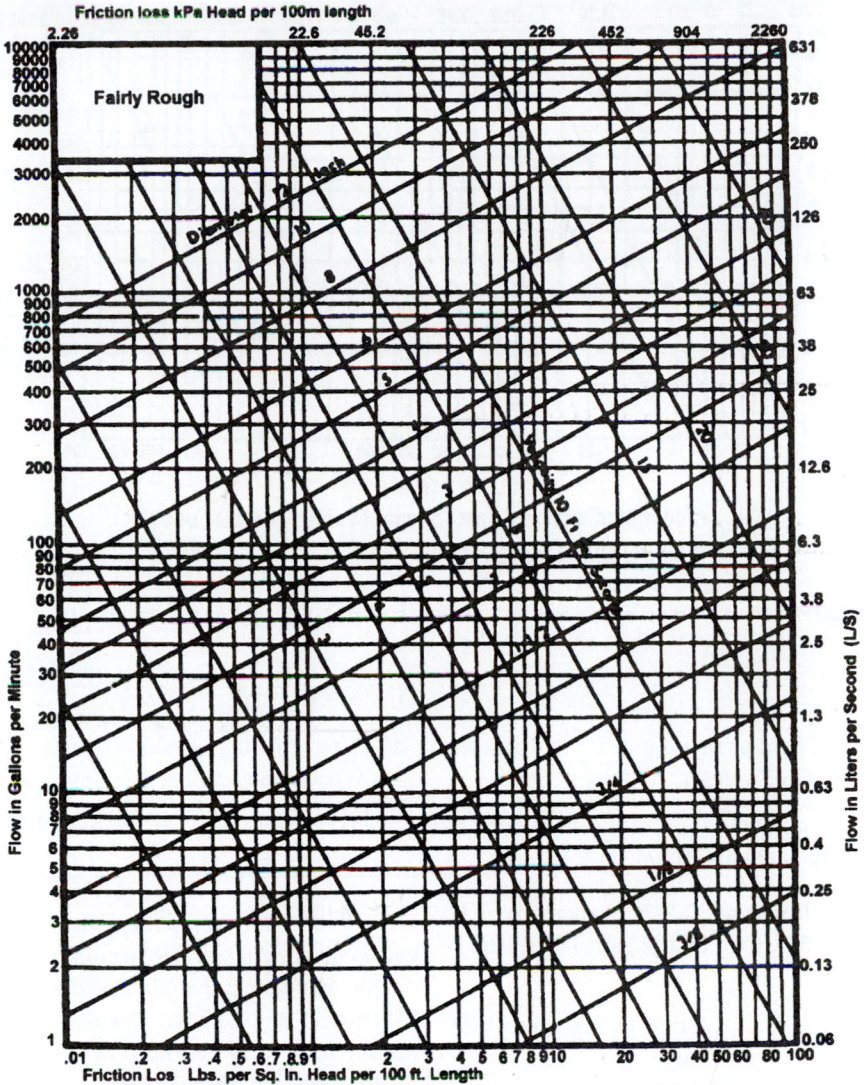

Friction loss kPa Head per 100m length

Flow in Gallons per Minute

Fairly Rough

Flow in Liters per Second (L/S)

Friction Los Lbs. per Sq. In. Head per 100 ft. Length

1. **This chart applies to fairly rough pipe and to actual diameters which in general will be less than the actual diameters of the new pipe of the same kind.**

** **Metric added by author.**

FIGURE 13.7 Pressure loss from friction caused by water flow in rough pipe (steel or iron), based on flow rate and pipe diameter.

influences friction loss. Pressure drop for a particular rough pipe diameter is greater than in a smooth pipe having the same inside diameter.

The pressure drop chart in Figure 13.8 applies to pressure drop as water flows through a water meter. Water meter design typically reduces pressure significantly. It should not be neglected in pressure drop computations.

Pressure drop charts have many lines and numbers; use them with care and review the information on the chart before using it. Along the left and right is the volumetric flow rate, and along the bottom and top is the friction loss in the pipe. On the charts provided, pressure drop from friction is expressed in psi per 100 ft. The heavy, solid lines sloping diagonally to the left

represent the nominal diameters of pipe; the lines running perpendicular (at a 90° angle) to the pipe diameter lines represent the velocity of the water in a pipe of a specific nominal diameter.

Example 13.2

a. Determine the pressure drop across a 1 in diameter Type L copper pipe that is 20 ft long and is carrying water at 20 gpm.

From Figure 13.6, at 20 gpm a 1 in diameter Type L copper pipe (center line) has a pressure drop of about 10 psi per 100 ft. Thus:

$$\Delta P_{friction} = (10 \text{ psi}/100 \text{ ft}) \cdot 20 \text{ ft} = 2 \text{ psi loss or } -2 \text{ psi}$$

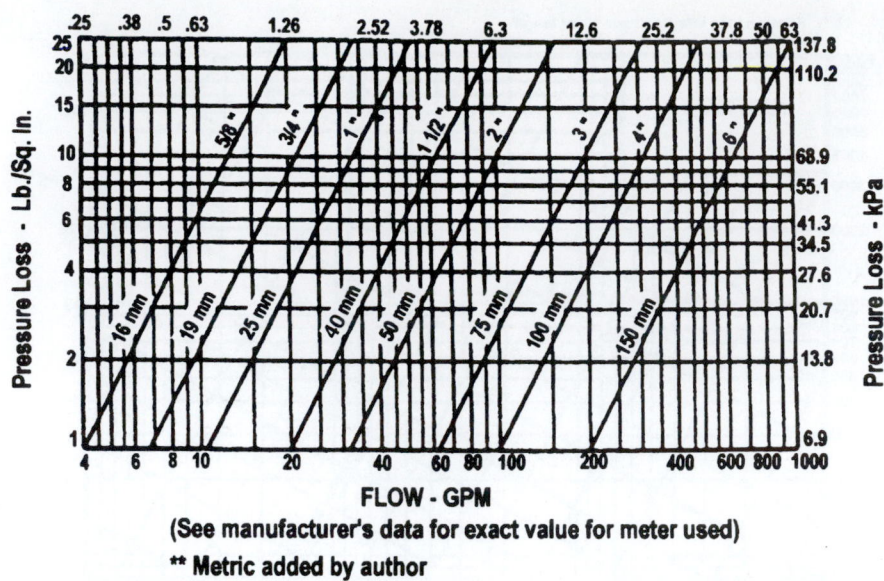

FIGURE 13.8   Pressure loss in a water meter, based on flow rate and meter size.

b. Determine the velocity of water flowing in a 1 in diameter Type L copper pipe carrying water at 20 gpm.

From Figure 13.6, at 20 gpm and for 1 in diameter Type L copper pipe, the velocity is about 7.8 ft/s.

c. Determine the pressure drop across a 4 in diameter Type K copper pipe that is 236.5 ft long and is carrying water at 400 gpm.

From Figure 13.6, at 400 gpm a 4 in pipe has a pressure drop of about 3.5 psi per 100 ft. Thus:

$$\Delta P_{friction} = (3.5 \text{ psi}/100 \text{ ft}) \cdot 236.5 \text{ ft}$$
$$= 8.3 \text{ psi loss or } -8.3 \text{ psi}$$

d. Determine the velocity of water flowing in a 4 in diameter Type K copper pipe carrying water at 20 gpm.

From Figure 13.6, at 400 gpm and for 4 in diameter Type L copper pipe, the velocity is about 10 ft/s.

In addition to pipe length, friction loss from fittings and valves must be taken into account. Analysis is usually founded on the equivalent length based on type of valve or fitting. Tables 13.2 and 13.3 indicate the equivalent length, in terms of friction losses, for various types and sizes of valves and fittings for tubing (smooth pipe) and threaded pipe. Tables 13.2a and 13.3a are in customary units and Tables 13.2b and Tables 13.3b are in modified metric (SI) units.

TABLE 13.2A   EQUIVALENT LENGTH IN FEET OF TUBING FOR FRICTION LOSS IN VALVES AND FITTINGS (TUBING).

Fitting or Valve	Equivalent Feet of Pipe for Various Nominal Diameters							
	½ in	¾ in	1 in	1¼ in	1½ in	2 in	2½ in	3 in
45″ elbow (wrought)	0.5	0.5	1.0	1.0	2.0	2.0	3.0	4.0
90″ elbow (wrought)	0.5	1.0	1.0	2.0	2.0	2.0	2.0	3.0
Tee, straight run (wrought)	0.5	0.5	0.5	0.5	1.0	1.0	2.0	—
Tee, branch (wrought)	1.0	2.0	3.0	4.0	5.0	7.0	9.0	—
45″ elbow (cast)	0.5	1.0	2.0	2.0	3.0	5.0	8.0	11.0
90″ elbow (cast)	1.0	2.0	4.0	5.0	8.0	11.0	14.0	18.0
Tee, straight run (cast)	0.5	0.5	0.5	1.0	1.0	2.0	2.0	2.0
Tee, branch (cast)	2.0	3.0	5.0	7.0	9.0	12.0	16.0	20.0
Compression stop	13.0	21.0	30.0	—	—	—	—	—
Globe valve	—	—	—	53.0	66.0	90.0	—	—
Gate valve	—	—	1.0	1.0	2.0	2.0	2.0	2.0

TABLE 13.2B   EQUIVALENT LENGTH IN METERS OF TUBING FOR FRICTION LOSS IN VALVES AND FITTINGS (TUBING).

Fitting or Valve	Equivalent Meters of Pipe for Various Nominal Diameters							
	13 mm	18 mm	25 mm	32 mm	40 mm	50 mm	65 mm	80 mm
45″ elbow (wrought)	0.2	0.2	0.3	0.3	0.6	0.6	0.9	1.2
90″ elbow (wrought)	0.2	0.3	0.3	0.6	0.6	0.6	0.6	0.9
Tee, straight run (wrought)	0.2	0.2	0.2	0.2	0.3	0.3	0.6	—
Tee, branch (wrought)	0.3	0.6	0.9	1.2	1.5	2.1	2.8	—
45″ elbow (cast)	0.2	0.3	0.6	0.6	0.9	1.5	2.4	3.4
90″ elbow (cast)	0.3	0.6	1.2	1.5	2.4	3.4	4.3	5.5
Tee, straight run (cast)	0.2	0.2	0.2	0.3	0.3	0.6	0.6	0.6
Tee, branch (cast)	0.6	0.9	1.5	2.1	2.8	3.7	4.9	6.1
Compression stop	4.0	6.4	9.2	—	—	—	—	—
Globe valve	—	—	—	16.2	20.1	27.5	—	—
Gate valve	—	—	0.3	0.3	0.6	0.6	0.6	0.6

TABLE 13.3A   EQUIVALENT LENGTH IN FEET OF THREADED PIPE FOR FRICTION LOSS IN VALVES AND FITTINGS (THREADED PIPE).

Fitting or Valve	Equivalent Feet of Pipe for Various Nominal Diameters							
	½ in	¾ in	1 in	1¼ in	1½ in	2 in	2½ in	3 in
45″ elbow	1.2	1.5	1.8	2.4	3.0	4.0	5.0	6.0
90″ elbow	2.0	2.5	3.0	4.0	5.0	7.0	8.0	10.0
Tee, straight run	0.6	0.8	0.9	1.2	1.5	2.0	2.5	3.0
Tee, branch	3.0	4.0	5.0	6.0	7.0	10.0	12.0	15.0
Gate valve	0.4	0.5	0.6	0.8	1.0	1.3	1.6	2.0
Balancing valve	0.8	1.1	1.5	1.9	2.2	3.0	3.7	4.5
Plug-type cock	0.8	1.1	1.5	1.9	2.2	3.0	3.7	4.5
Check valve, swing	5.6	8.4	11.2	14.0	16.8	22.4	28.0	33.6
Globe valve	15.0	20.0	25.0	35.0	45.0	55.0	65.0	80.0
Angle valve	8.0	12.0	15.0	18.0	22.0	28.0	34.0	40.0

TABLE 13.3B   EQUIVALENT LENGTH IN FEET OF THREADED PIPE FOR FRICTION LOSS IN VALVES AND FITTINGS (THREADED PIPE).

Fitting or Valve	Equivalent Meters of Pipe for Various Nominal Diameters							
	13 mm	18 mm	25 mm	32 mm	40 mm	50 mm	65 mm	80 mm
45″ elbow	0.4	0.5	0.6	0.8	0.9	1.2	1.5	1.8
90″ elbow	0.6	0.8	0.9	1.2	1.5	2.1	2.4	3.1
Tee, straight run	0.2	0.2	0.3	0.4	0.5	0.6	0.8	0.9
Tee, branch	0.9	1.2	1.5	1.8	2.1	3.1	3.7	4.6
Gate valve	0.1	0.2	0.2	0.2	0.3	0.4	0.5	0.6
Balancing valve	0.2	0.4	0.5	0.6	0.7	0.9	1.1	1.4
Plug-type cock	0.2	0.4	0.5	0.6	0.7	0.9	1.1	1.4
Check valve, swing	1.7	2.6	3.4	4.3	5.1	6.8	8.6	10.0
Globe valve	4.6	6.1	7.6	10.7	13.7	16.8	19.8	24.4
Angle valve	2.4	3.7	4.6	5.5	6.7	8.6	10.4	12.2

Example 13.3

a. Determine the equivalent length of six 90° elbows, three straight-run tees (all wrought), and two gate valves for 1 in diameter copper pipe:

6	90° elbows (1.0 ft equivalent length)	6 · 1.0 ft = 6.0 ft
3	tees—straight run (0.5 ft equivalent length)	3 · 0.5 ft = 1.5 ft
2	gate valves (1.0 ft equivalent length)	2 · 1.0 ft = 2.0 ft
	Total equivalent length:	9.5 ft

b. Determine the pressure drop from friction across a plumbing line containing 45 ft of 1 in diameter Type L copper pipe with fittings and valves noted above. Assume the system is carrying water at a volumetric flow rate of 20 gpm.

Total equivalent length of pipe, valves, and fittings:

$$45 \text{ ft} + 9.5 \text{ ft} = 54.5 \text{ ft}$$

From Figure 13.6, at 20 gpm a 1 in pipe has a pressure drop of about 10 psi per 100 ft. Thus:

$$\Delta P_{friction} = (10 \text{ psi per } 100 \text{ ft}) \cdot 54.5 \text{ ft}$$
$$= 5.45 \text{ psi loss or } -5.45 \text{ psi}$$

c. Determine the pressure available at the fixture outlet when water is flowing at 20 gpm. Assume the pressure available at the water service is 45 psi; the water meter has a pressure drop of 12 psi when water is flowing at 20 gpm; and the plumbing line serves a fixture outlet that is 16 ft above the water service line.

$\Delta P_{static} = 0.433(Z) = 0.433(-16) =$	−6.93 psi
$\Delta P_{friction}$ (from b. above) =	−5.45 psi
Pressure drop through meter =	−12.0 psi
Total change:	−24.38 psi

d. For residual pressure available before fixture:

$$45 \text{ psi} + (-24.38 \text{ psi}) = 20.62 \text{ psi}$$
available at fixture connection.

As a minimum, pressure drop analysis should be performed to ensure that pressure available at the most remote outlet (farthest and highest fixture) is adequate. This proceeds once the layout of the plumbing system has been performed. Thus, the constant velocity method may be used in initial sizing.

13.3 WATER SUPPLY DESIGN CONCERNS

Water Velocity

Noise, erosion of inner pipe walls and valves, and economy of installation, operation, and maintenance dictate the minimum and maximum water velocity in a plumbing system; as a result, these have a bearing on pipe diameter. If pipe diameters are small, cost is low but noise, erosion (from high velocities), and pumping costs (from high-pressure losses) are high. In contrast, large diameter pipes reduce noise erosion and pumping costs, but result in high installation costs. An intermediate pipe diameter is desirable.

Typically, plumbing codes set velocity limits in water supply piping. Maximum water velocities in plumbing water supply piping are usually limited to a range of 5 to 10 ft/s (1.5 to 3 m/s). Maximum velocities of up to 15 ft/s (4.5 m/s) are allowed for equipment feed lines in mechanical rooms (e.g., boiler feed lines) where noise is less of a concern. The maximum safe velocity for thermoplastic pipe is about 5 ft/s (1.5 m/s).

Cavitation

Cavitation is a physical phenomenon that occurs in a liquid when it experiences a drastic drop in pressure that causes the liquid to vaporize into small vapor bubbles. Vaporization is a problem because the liquid being vaporized expands greatly; for example, 1 ft³ (or 1 m³) of water at room temperature becomes 1600 ft³ (or 1600 m³) of vapor at the same temperature. As the low pressure returns to normal pressure levels, these bubbles implode as the vapor changes phase back to a liquid and thus drastically decreases its volume. This implosion causes noise and high levels of erosion where the imploding bubbles contact the walls of a pipe, fitting, pump, or valve. The noise that develops sounds similar to gravel flowing through the system in the area where the cavitation is developing. Over time, the erosion results in excessive wear; this eventually manifests itself as pinhole leaking.

Valves can develop cavitation when they are partially closed and flow is restricted. The result is noise and possible damage from erosion. Cavitation can also develop in a pump, which is noisy and can adversely affect pump performance by causing violent and damaging vibration and a sharp drop in discharge pressure. It occurs if the pumped liquid on the suction side of the pump drops below its vapor pressure. Eliminating this potential for cavitation is necessary because a cavitating pump can be completely damaged in a few hours of operation.

Cross-Connections

A *cross-connection* is an unsatisfactory connection or arrangement of piping that can cause nonpotable water to enter the potable water system. A cross-connection can cause used or contaminated water to mix with the water supply. It is an unsanitary and potentially hazardous condition. For example, a garden hose with one end immersed in a bucket of soapy water or in a swimming pool are possible backflow conditions. The American Water Works Association (AWWA) estimates over 100 000 cross-connections occur each day—half of these from garden hoses.

Most plumbing fixtures are designed to prevent a cross-connection. A gap exists between the faucet and the rim of the bowl in lavatories, sinks, and tubs to create a separation and avert a cross-connection. An *air gap* is the vertical distance through open air between an opening in a fixture or faucet conveying potable water to the flood level rim of a tank or fixture. As a general rule, the minimum air gaps for cross-connection protection for fixtures against one wall are as follows:

Lavatory	1 in (25 mm)
Sink	1½ in (38 mm)
Laundry tray	1½ in (38 mm)
Bathtub	2 in (51 mm)

Air gaps must be increased if the lavatory, sink, or tray is against two walls instead of one. Building codes will cite the minimum air gap based on fixture type; this is typically two times the inner diameter of the pipe ($2D_{inside}$) serving the fixture.

Backflow

Backflow is a type of cross-connection that occurs when contaminated water or some other liquid or substance unintentionally flows backwards into distribution pipes containing potable water. Simply, it is water flowing in the opposite direction from normal flow. Backflow can allow contaminants to enter the potable drinking water system through cross-connections. A backflow can be a serious plumbing problem that causes illness and even death. There are over 10 000 reported backflow situations that develop in the United States each year.

Backpressure or *back siphoning* is backflow caused by a negative pressure (vacuum) in a potable water system. A downstream pressure that is greater than the supply pressure causes *backpressure backflow*. It can result from a reduction in the supply pressure, an increase in downstream pressure, or a combination of both, such as from firefighting efforts, a water main break, consumer high-side pressure (pumped), or when a fire hydrant is opened for testing. Any buildings near such a break or unusual fire hydrant use will experience lower water pressure.

A backflow can be avoided by the use of proper protection devices. Backflow prevention protects the potable water system from minor, moderate, and severe hazards. A *backflow prevention device,* often called a *vacuum breaker,* is a device or plumbing assembly that when properly installed in a plumbing system prevents backflow. Different types of backflow prevention assemblies are required depending on the extent of the hazard. A high hazard exists when there is danger that backflow could create a health threat. Examples of this classification include lawn irrigation systems with chemical injection, hospitals, and manufacturing plants where dyes or chemicals are mixed. Moderate hazards occur when there is no health threat, but backflow could cause discolored, smelly, or objectionable water. Retail stores and office buildings are examples of this classification.

The *atmospheric vacuum breaker (AVB),* the most common type, consists of a body, a check valve-like member (to prevent backflow), and an atmospheric opening. The AVB is

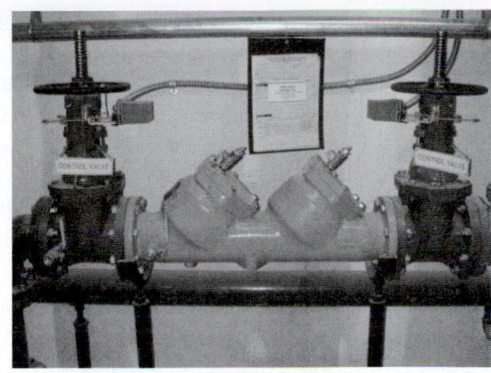

PHOTO 13.23   A double check assembly (DCA) is a backflow prevention device that consists of two check valves assembled in series usually with a ball valve or gate valve installed at each end for isolation and testing. The DCA shown is for the water supply to a fire sprinkler system. (Used with permission of ABC)

not a testable device. A *pressure vacuum breaker (PVB)* is a type of backflow prevention device used to keep nonpotable (or contaminated) water from entering the water supply. A PVB is similar to an AVB, except that the PVB contains a spring-loaded poppet. This makes it acceptable for applications that are high hazard or where valves are downstream. PVB devices are generally required on small (residential size) irrigation (sprinkler) systems to keep water contaminated with pesticides and fertilizers from reentering the building's plumbing system through the irrigation system. Generally, the PVB must be installed on the main line leading to the control valves. It must also be installed above ground and it must be 6 in (150 mm) higher than the highest sprinkler head or drip emitter controlled by any of the valves.

A *double check assembly (DCA)* or *double check valve* is a backflow prevention device assembly that consists of two check valves assembled in series usually with a ball valve or gate valve installed at each end for isolation and testing. See Photo 13.23. Test cocks (small valves) are in place to attach test equipment for evaluating whether the double check assembly is functional. The DCA is suitable for prevention of back pressure and back siphonage, but is not suitable for high hazard applications. A DCA is commonly used on fire sprinkler and boiler systems.

Water Hammer

A large pressure develops when fluid moving through a pipe is suddenly stopped. In a plumbing supply system, the sudden closing of a valve will cause fast-flowing water to stop quickly, resulting in a large increase in pressure that is known as *water hammer*. For example, in a pipe with water flowing at 10 ft/s (3 m/s), the maximum theoretical pressure that will develop if a valve is instantly closed is 635 psi (4.35 MPa). Although the elasticity of the pipe material reduces this theoretical pressure to some extent, a large pressure and force still develops against the valve and inner walls of the pipe.

As a minimum, water hammer produces a force that makes pipes rattle with banging or thumping sounds as they expand and contract from exposure to an increase in water pressure. This sound is frequently heard as a loud thump when the automatic shut-off valve on an older appliance such as a dishwasher or clothes washer rapidly shuts off the supply of water. In extreme cases when the water flow rate is high and the valve is closed very rapidly, water hammer can cause a valve to rupture and the pipe walls to burst.

To avert water hammer damage to a buildings plumbing system, air chambers or water hammer arrestors are used in the supply branches serving each fixture. These devices use trapped air to cushion the hydraulic shock.

Air Chambers

Air chambers are 15 in to 5 ft long pipes or pipe-like devices. They are installed vertically above the fixture water connection and are concealed in the wall. Air is trapped within the air chamber. The trapped air is compressible, which cushions the pressure surge as the valve is closed and absorbs the hydraulic shock.

Water Hammer Arrestors

Water hammer arrestors are patented devices that absorb hydraulic shock. (See Figure 13.9 and Photo 13.24). Such devices, when installed, must be accessible for maintenance. One type should be placed at the end of the branch line between the last two fixtures served. Additional arrestors should be placed at the midpoint of a run longer than 20 ft.

Thermal Expansion

No matter what type of piping material is used in the water system, some expansion in the pipe will occur. This expansion must be considered in the design of the system. The amount of expansion will depend on the type of piping material and the

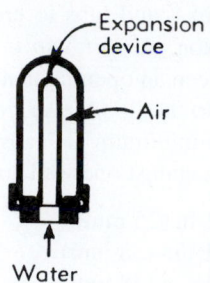

FIGURE 13.9   A cross-section view of a water hammer arrestor, a patented device that absorbs hydraulic shock.

PHOTO 13.24   Water hammer arrestors (two) above shut-off valves at a clothes washing machine connection. (Used with permission of ABC)

range of temperatures that the pipe will be subjected. Expansion of copper tubing is slightly greater than one inch in a 100 ft length in a 100°F temperature change. Plastic pipe will expand even more; up to 10 inches in a 100 ft length in a 100°F temperature change. See Table 13.4.

Piping for commercial hot water may have to withstand a temperature range from about 68°F (20°C), the average indoor

TABLE 13.4   DIMENSIONAL CHANGE FROM THERMAL EXPANSION OF SELECTED PIPE MATERIALS.

Pipe Material	Coefficient of Linear Thermal Expansion		Dimensional Change	
	in/in · °F	m/m · °C	Per 100 ft in 100°F Change — in	Per 50 m in 50°C Change — mm
Cast iron	0.0000057	0.000010	0.7	26
Stainless steel (12% chromium)	0.0000058	0.000010	0.7	26
Wrought iron/steel	0.0000074	0.000013	0.9	33
Monel (nickel/copper alloy)	0.0000077	0.000014	0.9	35
Copper	0.0000095	0.000017	1.1	43
Brass	0.0000096	0.000017	1.2	43
Chlorinated polyvinyl chloride (CPVC)	0.0000500	0.000090	6.0	225
Acrylonitrile-butadiene-styrene (ABS)	0.0000530	0.000095	6.4	239
Polyvinyl chloride (PVC)	0.0000560	0.000101	6.7	252
Polybutylene (PB)	0.0000833	0.000150	10.0	375
Cross-linked polyethylene (PEX)	0.0000833	0.000150	10.0	375

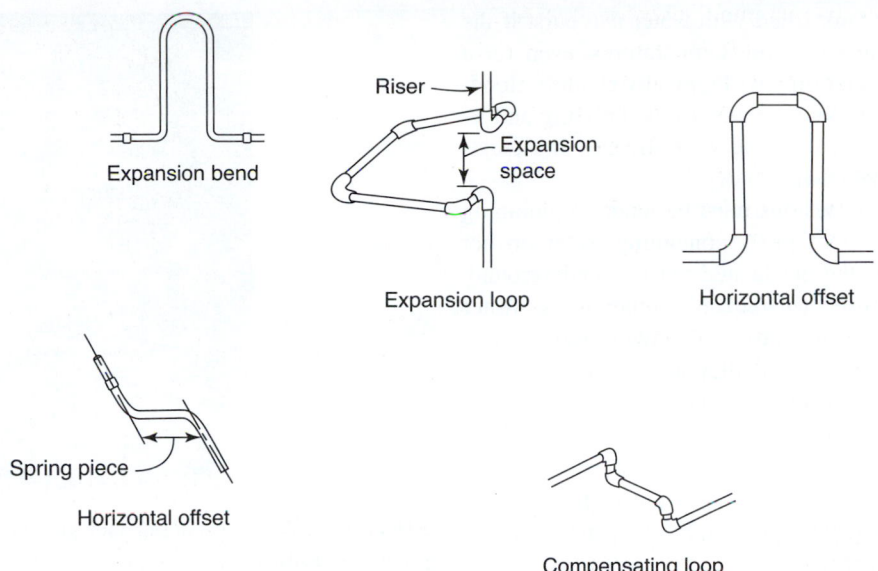

FIGURE 13.10 Types of expansion bends and loops. Thermal expansion in the pipes must be accommodated to minimize damage from thermal movement.

temperature, to over 180°F (82°C), the temperature of the sanitary rinse water. In a residence, the upper limit for hot water pipes is usually 125°F (52°C). Cold water piping will be subjected to a much smaller temperature range, usually with a low of 35°F (2°C) and a high of about 80°F (27°C). These ranges will vary, sometimes considerably, on projects and must be checked.

Thermal expansion in the pipes of a plumbing system must be accommodated to minimize damage from thermal movement. Expansion from temperature increases can push a pipe through a wall or cause it to burst. There are two methods in common use for providing for expansion in pipelines: expansion bends and expansion joints.

Expansion bends and *expansion loops* make use of pipe fabricated with U-shaped or circular bends. (see Figure 13.10). The increase in the length of pipe from thermal expansion is accommodated by flexing or springing of the bends or loops. A connection of four elbows and three short sections of straight pipe, connected in the form of a U-shape between the long lengths of pipe, can also be used to accommodate thermal expansion.

Expansion joints in common use include the slip expansion joint and the corrugated expansion joint. The *slip expansion joint* consists of a slip pipe and a flange, which is bolted to an adjoining pipe. The slip pipe fits into the main body of the joint, which is fastened to the end of the other adjoining pipe. When piping expands with temperature change, the slip pipe slides into the joint body. To prevent leakage between the slip pipe and the joint body, packing is used around the outside of the slip pipe and the slip pipe moves within the packing. A *corrugated expansion joint* consists of a flexible corrugated section. The corrugated, accordion-like section is able to absorb a certain amount of end movement of the pipe.

Viscosity

As water flows through a pipe, its viscosity (thickness) decreases with temperature decrease. Water at 40°F (4°C) is twice as viscous as water at 90°F (32°C) and four times as much at 170°F (77°C). As a result, pumping energy and cost are higher when water temperatures are lower.

Volume Change with Temperature Change

Water is the only substance that can exist as a solid, liquid, and gas at ordinary temperatures. Like most substances, water expands when it is heated. Unlike most substances, the volume of water increases when it freezes. Below 39°F (4°C), water molecules begin to align themselves into the crystal structure of ice, causing its volume to increase slightly. Under normal conditions, water freezes at 32°F (0°C). It expands significantly when the molecules of water are pushed farther apart than they were in a liquid state.

Freezing

A phase change from liquid (water) to solid (ice) results in about a 10% increase in volume.

Example 13.4

Water in a 50-gal water heater cools to a temperature below 32°F (0°C) and freezes fully. Determine the volumetric change, in gallons, as a result of freezing.

$$\Delta V = 10\% \cdot 50\,\text{gal} = 5.0\,\text{gal increase in volume}$$

A pipe, fitting, or tank filled with water can burst if the water is exposed to below-freezing temperatures, even for a short period of time. A burst pipe will typically result in flooding that can cause catastrophic damage to the building and its contents. Flooding may occur as the water freezes and bursts the pipe, or after the frozen water thaws.

In most climates, provisions must be made in plumbing system design to ensure that pipes containing water do not freeze. Pipes must be located in a heated space or underground, where they are not exposed to freezing temperatures. Water supply lines located in unheated attic and crawl spaces are susceptible to freezing, particularly if they are located next to a crawl space or attic space vent, or make-up air (outdoor air) inlet. Water supply pipes should be located in interior plumbing walls when possible. Water pipes located inside exterior walls must be placed on the heated side of the wall insulation. Underground lines must be adequately insulated by soil and should be placed well below the frost line.

Expanding Water

Liquid water expands above 39°F (4°C). Expansion is about 4.37% from 40°F (4.4°C) to 212°F (100°C). This volumetric change from expansion (ΔV) equates to about 0.0254% per °F (0.0457% per °C).

Example 13.5

Water in a 50-gal water heater is heated from a temperature of 50°F (10°C) to 125°F (51.7°C). Determine the volumetric change, in gallons.

$$\Delta V = 0.0254\% \text{ per } °F(125°F - 50°F)$$
$$= 1.91\% \text{ increase in volume}$$
$$1.91\% \cdot 50 \text{ gal} = 0.96 \text{ gal}$$

In any piping system, provisions for expansion and contraction of liquid water must be considered. In an *open plumbing system,* such as a building water supply system, pressure buildup from thermal expansion of water is released each time a faucet or valve opens. Excess pressure is released into the surrounding air.

In a *closed plumbing system,* where water is contained fully within the system (e.g., a hydronic heating system), provisions are necessary. A temperature-pressure relief (T/P) valve is a safety valve installed in a system that remains closed at normal operating pressures yet is permitted to open to release excessive pressure. They are commonly found as a safety feature on water heaters and boilers. *Expansion tanks,* installed in a closed system, provide additional volume in the closed system for expansion of water from temperature increase. See Photos 13.25 and 13.26.

Steam

Under standard atmospheric pressure (14.696 psi, 101.04 kPa), the boiling point temperature of water is 212°F (100°C). As

PHOTO 13.25 An expansion tank for potable water. (Used with permission of ABC)

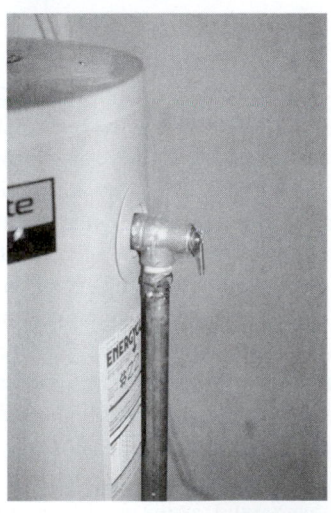

PHOTO 13.26 A temperature-pressure relief valve on a storage tank water heater. (Used with permission of ABC)

discussed in Chapter 12, the boiling point temperature varies with atmospheric or system pressure. Water boils and is converted to steam at its boiling point temperature. This phase change from liquid (water) to water vapor (steam) results in a 160 000% increase in volume (an increase of 1600 times).

Although steam is not found in building plumbing systems under ordinary conditions, a faulty water heating appliance such as a boiler or domestic water heater may result in the development of steam (e.g., a water heater with defective thermostatic controls), which can have catastrophic results. Reports of malfunctioning boilers exploding and water heaters rocketing through the roof of a building are not exaggerated. Provisions must be made in a system to release steam and reduce the buildup of pressure.

A T/P valve is a safety device that prevents the buildup of pressure in a closed vessel such as a tank, water heater, or boiler. It is designed to open to release excessive pressure if such a malfunction occurs. A T/P valve is installed in any system or component that heats water.

Aging

As pipes in a plumbing system are used, their inner walls become increasingly rough. The effects of aging in a plumbing system are related to piping material, quality of water (e.g., hard versus soft), and water temperature. Buildup from calcium deposits (especially in high-temperature hard water) and corrosion (especially in ferrous pipe materials) reduces the inside opening in the pipe, which restricts flow.

Over several decades of use, aging galvanized steel and iron pipe can result in a capacity loss of up to 80%. Copper piping experiences about half the capacity loss from aging as do steel and iron pipes. Plastic pipes experience little loss of capacity as they age.

Pipe Insulation

Pipe insulation is applied to the outer walls of piping to reduce heat loss from the pipe or prevent condensation on the outside pipe walls. Foam and covered fiberglass insulation are common pipe insulation materials. Hot water supply lines, especially on hot water recirculating systems, should be insulated to reduce heat loss and save energy. In most commercial buildings, heat loss increases the cooling load and the costs of air conditioning. In residential applications in temperate to cold climates, insulating hot water lines is less cost-effective because heat lost from hot water pipes contributes to space heating during the heating season; that is, the heat lost is extremely small. Building codes generally require pipe insulation to be ½ to 1½ in thick in commercial installations, depending on pipe diameter, and ½ in thick in residential installations. When the pipe is exposed, insulation is usually cased with a protective plastic sheathing. See Photo 13.27.

Under high humidity conditions it is necessary to insulate cold water lines to keep condensation from forming. Cold

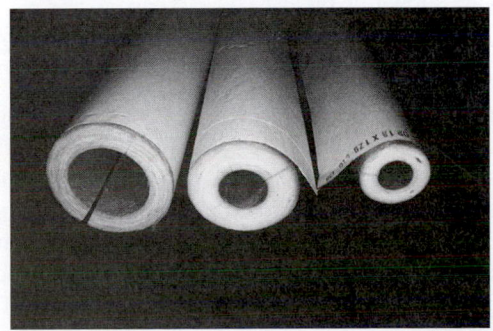

PHOTO 13.27 Pipe insulation with protective sheathing. (Used with permission of ABC)

water in a pipe cools the outside pipe walls, which lowers the temperature of the air surrounding the pipe. If the temperature of air surrounding the pipe drops below the dew point temperature, moisture condenses out of the air on to the pipe surface. Dripping water from condensation can cause cosmetic or, over time, even structural damage. The potential for forming condensation increases when water is supplied from an underground water source where water temperatures are significantly lower than indoor temperatures and flow is great, and when water flows through a pipe located in a cool unheated space (e.g., a crawl space or garage).

Testing

The water supply system should be tested for leaks before it is covered with finish materials to determine if it is watertight. Tests commonly run on water systems require that it be watertight under a hydrostatic water pressure of 125 psi for a minimum of 1 hr. Any leaks that occur should be repaired with the joint compound originally used.

Leaks

Surprisingly, plumbing leaks contribute significantly to water consumption in operating plumbing systems. Leaks account for about 12.7% of household per capita water use in a typical U.S. home (AWWA). A leak of just one drop per second will waste about 2700 gal (10 200 L) of water a year. Leaks not only waste money and water, they can cause damage to walls, flooring, ceilings, furniture, and electrical systems. Leaking pipes also create an environment for mold and mildew to thrive. Water damage claims may result in a building owner's insurance premiums being raised or nonrenewal of policies. Leaks can develop from substandard piping, improper use of materials, poor workmanship, and improper design.

Pinhole leaks in copper plumbing affects property owners throughout the United States and elsewhere in the world. Although much research has been conducted around the issue, no definitive cause has been determined to date. Potential causes include the following: galvanic (dissimilar metal) corrosion; changes in water chemistry and chemical water compositions; the improper addition of chemicals during the water treatment process; aggressive water, which includes factors such as pH content, presence of chlorides, metal ions, and dissolved gases in the water; and excessive water velocity in water pipes (>4 ft/s), which may erode or work on weak points in the materials.

13.4 WATER SUPPLY PIPE DESIGN METHODS

Flow Rates

Flow rate will vary by type of fixture and water pressure at the fixture. Each fixture in a plumbing system is designed to operate at a specific flow rate, expressed in gallons per minute

TABLE 13.5   APPROXIMATE FLOW RATES AND TYPICAL CONSUMPTION BY TYPE OF PLUMBING FIXTURE AND LOCATION.
FLOW RATE VARIES SIGNIFICANTLY BY RESIDUAL PRESSURE (WATER PRESSURE BEFORE THE FIXTURE).

Approximate Flow Rates by Fixture Type

Bathtub faucet	4 to 8 gpm	15 to 30 L/min	0.25 to 0.5 L/s
Clothes washer (laundry)	3 to 5 gpm	11 to 19 L/min	0.18 to 0.32 L/s
Dishwasher (domestic)	3 to 5 gpm	11 to 19 L/min	0.18 to 0.32 L/s
Drinking fountain	0.75 gpm	3 L/min	0.05 L/s
Hose bibb (½ in)	3 to 12 gpm	11 to 27 L/min	0.18 to 0.45 L/s
Kitchen sink	2 to 5 gpm	8 to 57 L/min	0.13 to 0.95 L/s
Lavatory faucet	2 to 3 gpm	8 to 11 L/min	0.13 to 0.18 L/s
Lawn sprinkler (domestic)	3 to 7 gpm	11 to 27 L/min	0.18 to 0.45 L/s
Showerhead (low flow design)	2 to 3 gpm	8 to 11 L/min	0.13 to 0.18 L/s
Showerhead (standard flow design)	3 to 6 gpm	11 to 23 L/min	0.18 to 0.38 L/sec
Toilet, flush tank	3 to 5 gpm	11 to 19 L/min	0.18 to 0.32 L/s
Toilet or urinal, 1 in flush valve (15 to 30 psi)	27 to 35 gpm	103 to 133 L/min	1.72 to 2.22 L/s

Typical Consumption for Urban or Suburban Residence

Per person per day, for all purposes	30 to 60 gal	114 to 228 L	—
Per shower, per use	25 to 60 gal	95 to 225 L	—
Fill bathtub, per fill	35 gal	133 L	—
Fill sink, per fill	1 to 2 gal	4 to 8 L	—
Flush toilet, per flush	1.6 to 7 gal	6 to 27 L	—
Irrigate with ¼ in of water, per 1000 ft^2 of lawn	160 gal	600 L	—
Dishwasher, per load	10 to 20 gal	38 to 76 L	—
Washing machine, per load	30 to 50 gal	114 to 450 L	—
Domestic water softener regeneration cycle	50 to 150 gal	190 to 570 L	—

Typical Consumption for Rural Residence

Per person per day, for all purposes	60 gal	228 L	—
Per horse, dry cow, or beef animal per day	12 gal	46 L	—
Per milking cow per day	35 gal	133 L	—
Per hog per day	4 gal	15 L	—
Per sheep per day	2 gal	8 L	—
Per 100 chickens per day	6 gal	23 L	—

(gpm), liters per second (L/s), or liters per minute (L/min) of flow. Residual water pressure (pressure available at the outlet before a fixture) affects the flow rate of a fixture. A higher residual pressure results in a greater flow rate and thus more water consumption. Approximate flow rates and typical consumption by type of plumbing fixture and location are shown in Table 13.5.

Water Consumption

Water use in many homes is lowest from about midnight to 5 AM, averaging less than one gallon per person per hour. Use climbs sharply in the morning around 6 AM, to about 3 gallons per person per hour. During the day, water use drops off moderately and rises again in the early evening hours. Weekly peak flows may occur in some homes on weekends, especially when all adults work during the week. In U.S. homes, average water use is approximately 45 gallons per person per day, but may range from 30 to 60 gallons or more.

Peak flows at retail stores and other businesses typically occur during business hours. Peak flow occurs during meal times at restaurants. Rental properties, resorts, and commercial establishments in tourist areas typically have flow rates that vary by season. Approximate water requirements for selected buildings are shown in Table 13.6.

Water Demand

The instantaneous peak demand for water in a pipe serving a number of plumbing fixtures or serving an entire building is referred to as the design load. The *design load* is the maximum probable or peak instantaneous demand for domestic water by a group of fixtures. The design load is typically expressed in gpm, L/min or L/s.

The design load of a pipe serving a group of plumbing fixtures or for the total fixtures installed on a project will depend not only on the number and type of fixtures installed but on the operation of the fixtures. Some fixtures may, at times, have continuous flow (e.g., faucets, hose bibbs, and showerheads), while

TABLE 13.6   APPROXIMATE WATER REQUIREMENTS FOR SELECTED BUILDINGS.

Occupancy Type	Typical Flow Rate and Demand Load	Number of Fixtures						
		0 to 50	51 to 100	101 to 200	201 to 400	401 to 800	801 to 1200	Over 1200
Apartments[a]	Flow rate, gpm/fixture	.50	.35	.30	.28	.25	.24	.24
	Minimum capacity, gpm	16	30	40	65	120	210	300
	Maximum capacity, gpm	25	35	60	115	200	290	—
Hospitals[a,b]	Flow rate, gpm/fixture	1.0	.80	.60	.50	.40	.40	.40
	Minimum capacity, gpm	25	55	85	125	210	330	500
	Maximum capacity, gpm	50	80	120	200	320	480	—
Hotels[a,b]	Flow rate, gpm/fixture	.65	.55	.45	.35	027	.25	.20
	Minimum capacity, gpm	25	35	60	100	150	225	300
	Maximum capacity, gpm	33	55	90	140	210	300	—
Mercantile[a,c]	Flow rate, gpm/fixture	1.3	.75	.70	.60	.55	.50	.50
	Minimum capacity, gpm	40	70	80	150	250	460	620
	Maximum capacity, gpm	65	75	140	240	440	600	—
Office[a,c]	Flow rate, gpm/fixture	1.1	.70	.60	.50	.37	.30	.27
	Minimum capacity, gpm	35	60	80	140	210	320	380
	Maximum capacity, gpm	55	70	120	200	300	360	—
Schools[a]	Flow rate, gpm/fixture	1.0	.60	.50	.40	.40	.40	.40
	Minimum capacity, gpm	20	50	70	110	180	340	500
	Maximum capacity, gpm	50	60	100	160	320	480	—

[a] Values are based on an equal number of men and women occupants. If the majority of occupants are women, increase the capacity by 15%.
[b] Where laundry is operated in connection with the building, increase the capacity by 10%.
[c] These values do not include water for special process work. The extra amount should be determined and added to the total capacity.

other fixtures have an intermittent flow (e.g., water closets and urinals). Nevertheless, it would be highly unlikely that every sink, dishwasher, water closet, bathtub, shower, clothes washer, and garden hose in a building would be all operating at one time. So, simply totaling fixture flow rates for all fixtures in an entire building distribution system would give the total demand for water usage only if all fixtures were used at one time. In most instances, totaling fixture requirements provides a very high estimate that results in overdesign of the piping.

Method 1: Simple Empirical Design Method

Pipe sizes for the water supply system of a single-family house and similar simple structures can be determined on the basis of experience and pertinent code requirements. Detailed analysis is not necessary in the design of simple systems. The fixture fed by the branch will influence branch pipe diameter. Pipe diameter is determined by the pipe size serving the fixture.

In the empirical design method, piping is sized with rules of thumb based on observation and experience. For example, the mains that serve fixture branches can be sized as follows:

- Up to three ½ inch branches can be served by a ¾ inch main
- Up to three ¾ inch branches or up to six ½ inch branches can be served by a 1 inch main
- Up to five ¾ inch branches or up to ten ½ inch branches can be served by a 1¼ inch main

Branch pipes can be sized from minimum branch requirements cited in the building code, such as those provided in Table 13.7.

The empirical approach is used in design of plumbing systems for residences and similar buildings with simple plumbing systems. Typically, a qualified plumber does design during rough-in of the piping. This approach can lead to system problems in complex piping arrangements.

Fixture Units

A method of estimating the design load for a group of plumbing fixtures is typically based on a quantity called the fixture unit. The *fixture unit* is an arbitrarily chosen measure that allows all types of plumbing fixtures to be expressed in common terms; that is, a fixture having twice the instantaneous flow rate of a second fixture would have a fixture unit value twice as large. The sole purpose of the fixture unit concept is to make it possible to calculate the design load on a system composed of different types of fixtures, each having different flow rates.

Fixture unit values are assigned to the different types of plumbing fixtures. The total number of fixture units is then used to establish the maximum probable water supply load and drainage load. (Drainage fixture unit load is discussed in Chapter 14.) Dr. Roy Hunter of the National Bureau of Standards (now the National Institute of Standards and Technology, NIST) developed this method over a half century ago and it still serves as the basis for estimating the design load of a plumbing system.

TABLE 13.7 SUPPLY, BRANCH SIZE, AND WATER SUPPLY FIXTURE UNIT LOAD REQUIREMENTS OF COMMON PLUMBING FIXTURES.

Fixture Type	Occupancy	Supply Control	Minimum Branch Pipe Size		Water Supply Fixture Unit Load (WSFU)		
			in	mm	Total[a]	Cold Water[b]	Hot Water[b]
Bathtub	Private	Faucet	½	15	2	1½	1½
Bathtub	Public	Faucet	½	15	4	3	3
Bidet	Private	Faucet	⅜	10	1	¾	¾
Clothes washer	Private	Faucet	½	15	2	1½	1½
Clothes washer	Public	Faucet	½	15	4	3	3
Dishwasher	Domestic	Faucet	½	15	1	—	1
Drinking fountain	Public	Shut off	⅜	10	0.5	0.5	—
Hose bibb	Private	Valve	½	15	2.5	2.5	—
	Public	Valve	½	15	2.5	2.5	
Kitchen sink	Private	Faucet	½	15	2	1½	1½
Kitchen sink	Public	Faucet	¾	20	4	3	3
Laundry trays	Private	Faucet	½	15	3	2¼	2¼
Lavatory	Private	Faucet	⅜	10	1	¾	¾
Lavatory	Public	Faucet	⅜	10	2	1½	1½
Service sink	Public	Faucet	½	15	3	2¼	2¼
Showerhead[c]	Private	Mixing valve	½	15	2	1½	1½
Showerhead[c]	Public	Mixing valve	½	15	4	3	3
Urinal	Public	Flush valve	1	25	10	10	—
Urinal	Private	Flush valve	¾	20	5	5	
Urinal	Public	Flush tank	½	15	3	3	
Water closet	Private	Flush valve	1	25	6	6	—
Water closet	Private	Flush tank	⅜	10	3	3	—
Water closet	Public	Flush valve	1	25	10	10	
Water closet	Public	Flush tank	⅜	10	5	5	
Tap size							
⅜ in dia	Private	—	⅜	10	1	¾	¾
	Public	—	⅜	10	2	1½	1½
½ in dia	Private		½	15	2	1½	1½
	Public		½	15	4	3	3
¾ in dia	Private		¾	20	3	2¼	2¼
	Public		¾	20	6	4½	4½
1 in dia	Private		1	25	6	4½	4½
	Public		1	25	10	7½	7½

[a] The listed WSFU values represent their total load on the cold water service.

[b] Cold water and hot water branches serving one fixture shall each be taken as three-quarters (¾) of the listed total WSFU value of the fixture.

[c] A showerhead over a bathtub does not increase the fixture value.

The Hunter method assigns a water supply fixture unit to each fixture. The *water supply fixture unit (WSFU)* is a probability factor that represents each fixture connected to the water supply system and used to determine the total use of water within a given system. Table 13.7 provides typical WSFU load requirements for common plumbing fixtures. The total WSFU quantity listed in the table relates to the WSFU load for that fixture. For example, a kitchen sink in a private residence (2 WSFU) has twice the flow rate in comparison to a lavatory in a private residence (1 WSFU).

Most plumbing fixtures are connected to both cold water and hot water branches. Water is typically drawn from both the cold water and hot water supply lines. The fixture unit value for fixtures having both cold and hot water connections are taken as three-quarters (¾) of the listed total value of the fixture. Thus, when calculating the peak demand flow rate for the hot or cold water distribution system, if both hot and cold branches supply a fixture (which, except for urinals, water closets, dishwashers, and hose bibs, is nearly always the case), then the WSFU for the fixture is reduced by a factor of 0.75. These reduced WSFU values are provided in Table 13.7 as cold water and hot water values.

A standard one-bath home with kitchen sink, dishwasher, water heater, clothes washer, flush tank toilet, lavatory, tub/shower

TABLE 13.8   WATER SUPPLY FIXTURE UNIT (WSFU) LOAD AND RELATED DESIGN LOAD, IN GPM, L/MIN, AND L/S, BASED ON HUNTER'S WORK ON INSTANTANEOUS DOMESTIC WATER DEMAND.

Water Supply Fixture Unit Load (WSFU)	System with Predominantly Flush Tanks			System with Predominantly Flush Valves			Water Supply Fixture Unit Load (WSFU)
	gpm	L/min	L/s	gpm	L/min	L/s	
6	5.0	19	0.3	27.0	103	1.7	6
8	6.5	25	0.4	—	—	—	8
10	8.0	30	0.5	—	—	—	10
12	9.2	35	0.6	28.6	109	1.8	12
14	10.4	40	0.7	30.2	115	1.9	14
15	11.0	42	0.7	31.0	118	2.0	15
16	11.6	44	0.7	31.8	121	2.0	16
18	12.8	49	0.8	33.4	127	2.1	18
20	14.0	53	0.9	35	133	2.2	20
25	17.0	65	1.1	38	144	2.4	25
30	20.0	76	1.3	41	156	2.6	30
35	22.5	86	1.4	44	167	2.8	35
40	25	95	1.6	47	179	3.0	40
45	27	103	1.7	49	186	3.1	45
50	29	110	1.8	51	194	3.2	50
60	33	125	2.1	55	209	3.5	60
70	36	137	2.3	59	223	3.7	70
80	39	148	2.5	62	236	3.9	80
90	41	156	2.6	65	247	4.1	90
100	44	167	2.8	68	258	4.3	100
120	49	186	3.1	74	281	4.7	120
140	53	201	3.4	78	296	4.9	140
160	57	217	3.6	83	315	5.3	160
180	61	232	3.9	87	331	5.5	180
200	65	247	4.1	91	346	5.8	200
225	70	266	4.4	95	361	6.0	225
250	75	285	4.8	100	380	6.3	250
275	80	304	5.1	105	399	6.7	275
300	85	323	5.4	110	418	7.0	300
400	105	399	6.7	125	475	7.9	400
500	125	475	7.9	140	532	8.9	500
750	170	646	10.8	175	665	11.1	750
1000	210	798	13.3	218	828	13.8	1000
1250	240	912	15.2	240	912	15.2	1250
1500	270	1026	17.1	270	1026	17.1	1500
1750	300	1140	19.0	300	1140	19.0	1750
2000	325	1235	20.6	325	1235	20.6	2000
2250	348	1322	22.0	350	1330	22.2	2250
2500	380	1444	24.1	380	1444	24.1	2500
2750	402	1528	25.5	405	1539	25.7	2750
3000	435	1653	27.6	435	1653	27.6	3000
4000	525	1995	33.3	525	1995	33.3	4000
5000	600	2280	38.0	600	2280	38.0	5000
6000	650	2470	41.2	650	2470	41.2	6000
7000	700	2660	44.3	700	2660	44.3	7000
8000	730	2774	46.2	730	2774	46.2	8000
9000	760	2888	48.1	760	2888	48.1	9000
10 000	790	3002	50.0	790	3002	50.0	10 000

combo, and two hose bibbs would be counted as 18 WSFU. Most standard two-bath homes would be counted as about 24 WSFU. Most standard three-bath homes would be counted as 34 WSFU.

Hunter developed a curve that establishes the flow rate for any given water supply fixture unit value. The design load, in gpm or L/min, is determined based on the number of

WSFU using the Hunter method. See Table 13.8. The design load is different, depending on whether the system consists of water closets and urinals in the system are predominantly flush tank or flush valve. Systems that consist of water closets and urinals with flush valves have a higher flow rate and thus a higher design load below about 1000 WSFU.

TABLE 13.9   WATER SUPPLY FIXTURE UNIT (WSFU) LOAD AND RECOMMENDED DESIGN LOAD, IN GPM, L/MIN, AND L/S, FOR SELECTED BUILDING TYPES.

Water Supply Fixture Unit Load (WSFU)	Hotels and Motels			Office Buildings			Institutional Buildings (e.g., Hospitals)			Water Supply Fixture Unit Load (WSFU)
	gpm	L/min	L/s	gpm	L/min	L/s	gpm	L/min	L/s	
200	82	312	5.2	73	277	4.6	95	361	6.0	200
250	85	323	5.4	76	289	4.8	98	372	6.2	250
300	90	342	5.7	80	304	5.1	100	380	6.3	300
400	95	361	6.0	85	323	5.4	110	418	7.0	400
500	98	372	6.2	90	342	5.7	115	437	7.3	500
600	100	380	6.3	100	380	6.3	120	456	7.6	600
900	125	475	7.9	120	456	7.6	135	513	8.6	900
1200	145	551	9.2	135	513	8.6	155	589	9.8	1200
1500	165	627	10.5	150	570	9.5	185	703	11.7	1500
2000	200	760	12.7	190	722	12.0	200	760	12.7	2000
2500	240	912	15.2	220	836	13.9	260	988	16.5	2500
5000	400	1520	25.3	350	1330	22.2	430	1634	27.2	5000

Table 13.9 provides WSFU loads and the recommended design load, in gpm, L/min, and L/s, for selected building types. It is based on a modification of Hunter's method. It is modified from the original Hunter method because the original research assumed only two types of buildings: one that used tank-type water closets and one that used flush valve water closets. It did not account for variations in operation of the different types of buildings. A sports arena will have a higher demand on a water distribution system than an office building, a school, or a residence. Additionally, the original fixture unit design concept was based on water use over a half century ago, when fixtures discharged at a much higher flow rate. New research is suggesting changes in WSFU rates for many fixtures and different categories of buildings (i.e., one- and two-family dwellings, multifamily dwellings, high-use assembly buildings, and other buildings or commercial buildings). The new WSFU method simply uses the adjusted WSFU values but still relies upon the Hunter curve.

Not all systems can be sized using the WSFU method. Like all sizing methods, there are restrictions regarding high and low limits of some water distribution systems. For example, the WSFU method will result in an undersized water distribution system for sports stadiums and arenas because of high peak demands. In these cases, the pipe size is selected and the pressure loss in the piping system is evaluated. Any sizing method needs a common sense approach for establishing the flow rate in the piping system.

Method 2: WSFU Design Table Method

In residential and small commercial buildings, WSFU design tables can be used to establish meter and distribution pipe size based on the total demand in WSFUs and the supply pressure (the available static pressure after static head loss). Table 13.10 represents of WSFU tables used to size building supply and branch lines, and meter and service lines.

Meter and distribution pipe can be sized using the following methods:

1. Obtain minimum service water pressure for the location of construction. Usually this is available through the municipal water department.

2. Compute the total WSFUs, including proposed and projected future plumbing fixtures.

3. Calculate the maximum developed length of water piping: the actual length of pipe between the source of supply and the most remote fixture plus the developed length of fittings. Developed length can be approximated by multiplying the actual length to the most remote fixture by 1.2 to compensate for loss of meter and fittings.

4. Compute the static head (the pressure loss from elevation change) and subtract it from the service water pressure. Static head (ΔP_{static}) is found by multiplying the vertical height (Z), in feet or meters:

$$\text{static head, in psi, } \Delta P_{static} = -0.433Z$$

$$\text{static head, in kPa, } \Delta P_{static} = -9.8Z$$

5. Use Table 13.10 to determine the meter and distribution pipe sizing based on the total demand in WSFUs, maximum developed length of water piping, and the supply pressure (the available static pressure after static head loss).

Example 13.6

Using the WSFU design table method, determine the minimum meter and distribution pipe sizes. Assume the following:

- Minimum service water pressure for the location is 65 psi.

- Total WSFUs is 28.

TABLE 13.10   MAXIMUM WATER SUPPLY FIXTURE UNITS (WFSU) FOR A SUPPLY PRESSURE (THE AVAILABLE STATIC PRESSURE AFTER HEAD LOSS) AND SPECIFIC PIPE SIZE.

Supply Pressure	Meter and Street Service Pipes (in)	Building Supply and Branch Pipes (in)	Maximum Developed Length (ft)								
			40	60	80	100	150	200	300	400	500
30 to 45 psi	¾	½	6	5	4	3	2	1	1	0	0
	¾	¾	16	16	14	12	9	6	5	4	4
	¾	1	29	25	23	21	17	15	12	10	8
	1	1	36	31	27	25	20	17	13	12	10
	1	1¼	54	47	42	38	32	28	23	19	17
	1	1½	85	84	79	65	56	48	38	32	28
45 to 60 psi	¾	½	7	7	6	5	4	3	2	1	1
	¾	¾	20	20	19	17	14	11	8	6	5
	¾	1	39	39	36	33	28	23	19	17	14
	1	1	39	39	39	36	30	25	20	18	15
	1	1¼	78	78	76	67	52	44	36	30	27
	1	1½	85	85	85	85	85	85	67	55	49
Over 60 psi	¾	½	7	7	7	6	5	4	3	2	1
	¾	¾	20	20	20	20	17	13	10	8	7
	¾	1	39	39	39	39	35	30	24	21	17
	1	1	39	39	39	39	38	32	26	22	18
	1	1¼	78	78	78	78	74	62	47	39	31
	1	1½	85	85	85	85	85	85	85	81	64

- Actual length of pipe between the source of supply and the most remote fixture is 83 ft.
- Elevation above the source of supply is 23 ft.

The maximum developed length of water piping is approximated by: 83 ft · 1.2 = 99.6 ft ≈ 100 ft

The pressure loss from static head, in psi, is found by:

$$\Delta P_{static} = -0.433Z = -0.433(23 \text{ ft})$$
$$= -9.96 \text{ psi} \approx -10 \text{ psi}$$

Supply pressure is found by: 65 psi − 10 psi = 55 psi

For the maximum developed length of 100 ft and a supply pressure in the range of 45 to 60 psi, at 27 WSFU (<33 WSFU in 100 ft column):

- The meter and street service pipe size: ¾ in
- The building supply and branch pipe size: 1 in

Method 3: Velocity Design Method

The *velocity design method* entails selecting the smallest pipe diameter without exceeding a preestablished maximum velocity for the design load in the pipe. It is typically used accurately in a downfeed system and works well in preliminary design of a plumbing system provided the system layout is reasonably symmetrical. This method does require an investigation of pressure loss to ensure that residual pressure at the most remote fixture is adequate.

The velocity design method involves computing the number of WSFUs served by a pipe, converting total WSFUs to a design load (in gpm or L/min), and then sizing the pipe based on a maximum velocity. Maximum velocities in plumbing systems typically range from 5 to 10 ft/s (1.5 to 3.1 m/s). Pipe sizes are calculated at strategic points in the system (e.g., wherever the WSFU served and, therefore, the design load change).

The procedure is outlined below:

1. Sum the total number of WSFUs for hot water and cold water. (See Table 13.7.)

2. Determine maximum probable demand in gpm. (See Table 13.8 or Table 13.9.)

3. Based on the maximum desired velocity (e.g., 8 ft/s or 2.4 m/s) and design load (Q), solve for the minimum required diameter ($D_{i\text{-min}}$):

   In customary units, minimum required diameter ($D_{i\text{-min}}$) of the pipe, in inches, is based on the maximum desired velocity (v) of a fluid flowing through a pipe, in ft/s, and the volumetric flow rate (Q), in gpm:

   $$D_{i\text{-min}} = \sqrt{0.409Q/v}$$

   In SI (metric) units, minimum required diameter ($D_{i\text{-min}}$) of the pipe, in mm, is based on the maximum desired velocity (v) of a fluid flowing through a pipe, in m/s, and the volumetric flow rate (Q), in L/min:

   $$D_{i\text{-min}} = \sqrt{21.22Q/v}$$

4. Select a pipe size for the appropriate pipe material (from design tables such as Tables 12.9 through 12.15, 12.17 through 12.22 in Chapter 12) with an inside diameter equal to or greater than the minimum required diameter, $D_{i\text{-min}}$.

Example 13.7

Using the velocity design method, determine the minimum required size of hot and cold water supply pipes serving two apartments, each containing a kitchen sink and a bathroom group as noted in the listing that follows. Use a maximum velocity of 6 ft/s, because of noise limitations. Assume a system with predominantly flush tanks and Type L copper tube (Table 12.9 in Chapter 12).

From Table 13.7, the WSFU loads for hot and cold supply pipes are determined:

	Cold	Hot
2 bath tubs (1½ WSFU each for hot and cold)	3	3
2 water closets—flush tank (3 WSFU for cold only)	6	—
2 lavatories (¾ WSFU each for hot and cold)	1.5	1.5
2 kitchen sinks (1½ WSFU each for hot and cold)	3	3
Total WSFU:	13.5	7.5

For cold water supply pipe: 13.5 WSFU = 10 gpm (Table 13.8, for system with predominantly flush tanks)

$$D_{i\text{-min}} = \sqrt{0.409Q/v} = \sqrt{0.409(10 \text{ gpm})/(6 \text{ ft/s})}$$
$$= 0.826 \text{ in diameter}$$

From Table 12.9 in Chapter 12, a 1 in diameter Type L copper tube with an inside diameter of 1.025 in is acceptable.

For hot water supply pipe: 7.5 WSFU = 6 gpm (Table 13.8, for system with predominant flush tanks)

$$D_{i\text{-min}} = \sqrt{0.409Q/v} = \sqrt{0.409(6 \text{ gpm})/(6 \text{ ft/s})}$$
$$= 0.640 \text{ in diameter}$$

From Table 12.9 in Chapter 12, a ¾ in diameter Type L copper tube with an inside diameter of 0.785 in is acceptable.

Example 13.8

Using the velocity design method, determine the minimum required size of cold water supply pipe serving two apartments with design load of 10 gpm and a system with predominant flush tanks and Type L copper tube (same as previous exercise). Use a maximum velocity of 8 ft/s (rather than the 6 ft/s used in previous example).

$$D_{i\text{-min}} = \sqrt{0.409Q/v} = \sqrt{0.409(10 \text{ gpm})/(8 \text{ ft/s})}$$
$$= 0.715 \text{ in diameter}$$

From Table 12.9 in Chapter 12, a ¾ in diameter Type L copper tube with an inside diameter of 0.785 in is acceptable.

In comparing the cold water supply pipe sizes determined in the previous two examples, it is evident that the velocity limitation influences pipe size. Selection of a pipe size is a balance between economy (a smaller pipe diameter is less expensive) and noise and erosion (a smaller pipe diameter is noisier and will wear faster).

Example 13.9

Using the velocity design method, determine the minimum required size of hot and cold water supply pipes serving 16 apartments, each containing a kitchen sink and a bathroom group as noted in the listing that follows. Use a maximum velocity of 8 ft/s. Assume a system with predominant flush tanks and Type L copper tube (Table 12.9 in Chapter 12).

	Cold	Hot
16 bath tubs (1½ WSFU each for hot and cold)	24	24
16 water closets—flush tank (3 WSFU for cold only)	48	—
16 lavatories (¾ WSFU each for hot and cold)	12	12
16 kitchen sinks (1½ WSFU each for hot and cold)	24	24
Total WSFU:	108	60

For cold water supply pipe: 108 WSFU = 46 gpm (Table 13.8, for a system with predominant flush tanks)

$$D_{i\text{-min}} = \sqrt{0.409Q/v} = \sqrt{0.409(46 \text{ gpm})/(8 \text{ ft/s})}$$
$$= 1.53 \text{ in diameter}$$

A 2 in diameter Type L copper tube with an inside diameter of 1.985 in (from Table 12.9 in Chapter 12) is acceptable.

For hot water supply pipe: 60 WSFU = 33 gpm (Table 13.8, for a system with predominant flush tanks)

$$D_{i\text{-min}} = \sqrt{0.409Q/v} = \sqrt{0.409(33 \text{ gpm})/(8 \text{ ft/s})}$$
$$= 1.30 \text{ in diameter}$$

A 1½ in diameter type L copper tube with an inside diameter of 1.505 in (from Table 12.9 in Chapter 12) is acceptable.

In Example 13.9, the designer should select a 2 in diameter copper tube for the cold water pipe based on the 8 ft/s velocity limitation. If this velocity limitation is arbitrary (e.g., not precisely required by design constraints or code requirements), the designer may chose to select a 1½ in diameter tube. The 1½ in diameter tube has an inside diameter slightly smaller than the minimum required diameter ($D_{i\text{-min}}$). As shown in the next example, if the smaller pipe is selected, the water velocity will exceed 8 ft/s slightly. The smaller pipe may not present a concern unless design or code velocity limits are exceeded.

Example 13.10

Determine the water velocity in the following pipes based on a flow rate of 46 gpm.

a. A 2 in diameter Type L copper tube (inside diameter of 1.985 in)

$$v = 0.409Q/D_i^2 = 0.409(46 \text{ gpm})/(1.985 \text{ in})^2 = 4.8 \text{ ft/s}$$

b. A 1½ in diameter Type L copper tube (inside diameter of 1.505 in)

$$v = 0.409Q/D_i^2 = 0.409(46 \text{ gpm})/(1.505 \text{ in})^2 = 8.3 \text{ ft/s}$$

Example 13.11

A large 10-story office building has a demand load of 60 WSFU for cold water and 25 WSFU for hot water at each floor. An additional supply of 18 gpm of cold water is required in the basement for makeup water for mechanical equipment and a landscaping sprinkler system. The system will be plumbed with Type L copper tube (Table 12.9 in Chapter 12).

a. Use the velocity design method to determine the minimum required size of hot and cold water supply pipes serving the building, based on a maximum velocity of 8 ft/s.

For cold water supply pipe:

$$\text{Total cold water demand load} = 60 \text{ WSFU} \cdot 10 \text{ floors}$$
$$= 600 \text{ WSFU}$$

$$600 \text{ WSFU} = 100 \text{ gpm (Table 13.9)}$$
$$100 \text{ gpm} + 18 \text{ gpm} = 118 \text{ gpm}$$

$$D_{i\text{-min}} = \sqrt{0.409 Q / v} = \sqrt{0.409(118 \text{ gpm})/(8 \text{ ft/s})}$$
$$= 2.46 \text{ in diameter}$$

A 2½ in diameter Type L copper tube with an inside diameter of 2.465 in (from Table 12.9 in Chapter 12) is acceptable.

For hot water supply pipe:

$$\text{Total hot water demand load} = 25 \text{ WSFU} \cdot 10 \text{ floors}$$
$$= 250 \text{ WSFUs}$$

$$250 \text{ WSFU} = 76 \text{ gpm (Table 13.9)}$$

$$D_{i\text{-min}} = \sqrt{0.409 Q / v} = \sqrt{0.409(76 \text{ gpm})/(8 \text{ ft/s})}$$
$$= 1.971 \text{ in diameter}$$

A 2 in diameter Type L copper tube with an inside diameter of 1.985 in (from Table 12.9 in Chapter 12) is acceptable.

b. Use the velocity design method to determine the minimum required size of the main supply pipe serving the building, based on a maximum velocity of 8 ft/s.

$$\text{Total demand load for hot and cold} = 600 \text{ WSFU}$$
$$+ 250 \text{ WSFU} = 850 \text{ WSFU}$$

$$850 \text{ WSFU} = 117 \text{ gpm (interpolated from Table 13.9)}$$
$$117 \text{ gpm} + 18 \text{ gpm} = 135 \text{ gpm}$$

$$D_{i\text{-min}} = \sqrt{0.409 Q / v} = \sqrt{0.409(135 \text{ gpm})/(8 \text{ ft/s})}$$
$$= 2.627 \text{ in diameter}$$

A 3 in diameter Type L copper tube with an inside diameter of 2.945 in (from Table 12.9 in Chapter 12) is acceptable.

Table 13.11 indicates the maximum demand at common maximum supply velocities for common Type L copper tube (pipe) sizes.

Method 4: Equal Friction Design Method

A more accurate approach to sizing the pipe diameter in a complex network of pipes is the *equal friction design method*. In this design method, it is necessary to determine the total pressure drop required between the water service and the fixture and equate this to a pressure drop per 100 ft over the equivalent length of pipe. Pressure available at the water service minus pressure head and desired pressure at a fixture determines the desired pressure drop. This method is more complex and more accurate, but usually requires several iterations before pipe diameters are selected.

Sizing the piping using the equal friction design method requires several iterations. Often, it is a matter of trial and error, even for experienced engineers; the process involves first selecting a pipe size for the building main, which runs from the water system to the riser(s), and then determining the friction loss for the pipe used from the charts in Figures 13.6 and 13.7. The chart used will depend on the type of pipe roughness (material).

To make a pipe selection for a specific condition using the equal friction design method:

1. Find the volumetric flow rate along the side of the chart.

2. Move horizontally across the chart to the pipe diameters and, for specific nominal diameters, note associated pressure drops and velocities.

TABLE 13.11 TYPE L COPPER TUBE (PIPE) DIMENSIONS AND MAXIMUM DEMAND IN GPM AT COMMON MAXIMUM SUPPLY VELOCITIES.

Nominal Pipe Size	Maximum Demand, in gpm, at Common Maximum Supply Velocities						
	4 ft/s	5 ft/s	6 ft/s	7 ft/s	8 ft/s	9 ft/s	10 ft/s
½	2.9	3.6	4.4	5.1	5.8	6.5	7.3
¾	6.0	7.5	9.0	10.5	12.1	13.6	15.1
1	10.3	12.8	15.4	18.0	20.6	23.1	25.7
1¼	15.7	19.6	23.5	27.4	31.3	35.2	39.1
1½	22.2	27.7	33.2	38.8	44.3	49.8	55.4
2	38.5	48.2	57.8	67.4	77.1	86.7	96.3
2½	59.4	74.3	89.1	104.0	118.9	133.7	148.6
3	84.8	106.0	127.2	148.4	169.6	190.8	212.1
3½	114.7	143.4	172.1	200.8	229.4	258.1	286.8
4	149.1	186.4	223.7	261.0	298.3	335.6	372.8

3. Select a pipe diameter having the desired pressure drop (including fittings) without exceeding the velocity limitation requirements.

Example 13.12

a. Select a nominal pipe diameter of copper pipe (Type L) with a volumetric flow rate of 100 gpm. The desired pressure drop from friction loss is 16 psi in 48 ft. Use a 50% increase in pressure drop from fitting losses.

From Figure 13.6, these nominal pipe diameters of copper pipe have the associated pressure drops and velocities. Pressure drop and velocity for each pipe size were found at the intersection of the line representing nominal pipe size and the line representing a flow rate of 100 gpm.

4 in pipe	0.27 psi per 100 ft	2.6 ft/s
3½ in pipe	0.53 psi per 100 ft	3.5 ft/s
3 in pipe	1.05 psi per 100 ft	4.7 ft/s
2½ in pipe	2.70 psi per 100 ft	6.7 ft/s
2 in pipe	7.80 psi per 100 ft	10.4 ft/s
1½ in pipe	29.0 psi per 100 ft	18.1 ft/s

Pressure drop for the 48 ft long pipe for each pipe size is computed as follows:

For 4 in pipe $\Delta P_{friction}$ = 0.27 psi per 100 ft · 48 ft · 150%
= 0.19 psi

For 3½ in pipe $\Delta P_{friction}$ = 0.53 psi per 100 ft · 48 ft · 150%
= 0.38 psi

For 3 in pipe $\Delta P_{friction}$ = 1.05 psi per 100 ft · 48 ft · 150%
= 0.76 psi

For 2½ in pipe $\Delta P_{friction}$ = 2.70 psi per 100 ft · 48 ft · 150%
= 1.94 psi

For 2 in pipe $\Delta P_{friction}$ = 7.80 psi per 100 ft · 48 ft · 150%
= 5.62 psi

For 1½ in pipe $\Delta P_{friction}$ = 29.0 psi per 100 ft · 48 ft · 150%
= 20.9 psi

A pipe with a 2 in nominal diameter should be selected because it has a pressure drop from friction loss that is closest to 6 psi.

b. Select a nominal pipe diameter of copper pipe (Type L) with a volumetric flow rate of 100 gpm. Limit velocity of water flow to a maximum of 8 ft/s.

Based on the information shown above, a pipe with a 2½ in nominal diameter should be selected because its velocity does not exceed 8 ft/s.

c. The total water pressure available at one end of a 48 ft long pipe is 50 psi. The pipe will have a volumetric flow rate of 100 gpm and a vertical elevation increase of 24 ft. Limit velocity of water flow to a maximum of 8 ft/s. Use a 50% increase in pressure drop from fitting losses. Select a nominal pipe diameter of copper pipe (Type L) based on total pressure drop (including static and friction losses) so that the minimum pressure available at the fixture is 12 psi.

Total pressure available: 50 psi

Pressure loss from elevation change:

$$\Delta P_{static} = -0.433Z = -0.433(24) = -10.4 \text{ psi (a loss)}$$

Desired residual pressure: 12 psi

Pressure available for friction is found by:

$$\Delta P_{friction} = 50 \text{ psi} - 10.4 \text{ psi} - 12 \text{ psi} = 27.6 \text{ psi}$$

A friction pressure loss in the pipe of 27.6 psi is the best selection. From the solution in a. above, a 2 in diameter pipe provides a pressure drop including fittings of 5.62 psi (much too small) and a 1½ in diameter pipe of 20.9 psi (slightly too large). The 1½ in diameter pipe would likely be selected initially but the water velocity, at 18.1 ft/s for this pipe size, exceeds the 8 ft/s significantly.

With the velocity constraint, a 2½ in diameter pipe is the only alternative, at a friction pressure drop of 1.94 psi. The remaining pressure of 37.66 psi (27.6 psi – 10.4 psi – 1.94 psi = 37.66 psi) is higher than the minimum pressure required at the fixture of 12 psi. It could be deemed acceptable or a valve could be installed in the system to reduce the pressure further.

d. Using the velocity design method, select a nominal pipe diameter of copper pipe (Type L) based on the design criteria in c.

$$D_{i\text{-min}} = \sqrt{0.409Q/v} = \sqrt{0.409(100 \text{ gpm})/(8 \text{ ft/s})}$$
$$= 2.26 \text{ in}$$

A 2½ in diameter (Type L) copper pipe with an inside diameter of 2.465 in is acceptable.

The velocity limitation dictated pipe diameter in the previous example. Had the pipe been 480 ft long (instead of 48 ft long) with the same pressure drop from fitting losses, friction pressure loss would have been 19.4 psi (ten times greater). The friction pressure loss in the longer pipe is much closer than the optimal pressure loss of 27.6 psi. Frequently, velocity requirements dictate pipe diameter in short lengths of pipe such as in the previous example.

13.5   WATER TEMPERATURES

Cold Water

The temperature of water available in the water supply system will vary. Water from shallow (just below frost line) underground water service lines enters the building at a temperature that fluctuates with climate, season, and location of the water service line. In a moderate climate, it may range from 35°F (2°C) in midwinter to 70°F (21°C) or more in midsummer. In hot climate regions, water service temperature can reach temperatures above 90°F (32°C) or more in the summer. In harsh desert climates, water must actually be chilled for drinking.

Water pumped from deep wells is generally at a fairly constant temperature of 50° to 55°F (10° to 13°C) year-round. Water that remains standing in an underground storage pipe or in an interior water supply line for a long period of time will equilibrate to the temperature of the surrounding space (e.g., room or crawl space temperature).

Chilled Water

The varying temperature of water extracted from cold water supply lines is normally acceptable. In some instances *chilled water* is desired, such as from a drinking fountain. In this case, a *chiller* is used to cool water to a temperature of about 50°F (10°C) before it is used. A chiller is a vapor compression refrigeration system that cools water much like a refrigerator.

Heated Water

In modern buildings, hot water is desired for bathing, cleaning, washing, and other associated purposes. By definition, *hot water* is potable water that is heated to at least 120°F. Heated water below 120°F (49°C) is typically called *tempered water*. Hot water used for household functions such as bathing, dishwashing, and clothes washing is referred to as *domestic hot water (DHW)*. In commercial installations, hot water used in nondomestic applications is referred to as *building service hot water (BSHW)*. The temperature of heated water available in the water supply system will vary depending upon need.

Typical temperatures of heated water in buildings are provided in Table 13.12. Generally, a lower hot water temperature is more energy efficient and safer (less likely to scald). A higher water temperature (180°F [82°C]) achieves sanitation that is required in commercial (e.g., restaurants) and institutional (e.g., hospital and health center) dishwashing final rinse and laundering operations.

The hot water distribution temperature must be higher than 120°F (49°C) because of the concern over *Legionella pneumophila* (Legionnaires' disease). This bacterium, which can cause serious illness when inhaled, can grow in domestic hot water systems at temperatures of 115°F (46°C) or less. Bacteria colonies have been found in system components, such as showerheads and faucet aerators, and in uncirculated sections of storage-type water heaters. A water temperature of approximately

TABLE 13.12   TYPICAL TEMPERATURES OF HEATED WATER IN BUILDINGS. COMPILED FROM VARIOUS GOVERNMENTAL STANDARDS.

Type	Typical Temperatures	
Lavatory (hand washing, shaving)	105°F	41°C
General Bathing (showering and bathing)	115°F	46°C
Residential dishwashing and laundry (hot)	140°F	60°C
Commercial/institutional laundry	160°F	71°C
Commercial/institutional dishwashing—final rinse	180°F	82°C
Sanitizing of nursing utensils	180°F	82°C

TABLE 13.13   LENGTH OF TIME FOR HOT WATER TO CAUSE SCALDING.

Temperature of Water		
°F	°C	Time to Cause Scalding Injury to Skin
125	52	1½ to 2 min
130	54	About 30 s
135	57	About 10 s
140	60	Less than 5 s
145	63	Less than 3 s
150	66	About 1½ s
155	68	About 1 s

140°F is suggested to reduce the potential of growth of this bacterium. This higher temperature, however, increases the possibility of scalding. Periodic supervised flushing of fixture heads with 170°F water is recommended in hospitals and health care facilities because the already weakened patients are generally more susceptible to infection.

Water temperatures above 110°F (45°C) can be painfully hot to the touch. Exposure to water temperatures of 120°F (49°C) and above will cause scalding (burning) of the skin. See Table 13.13. The elderly, young children, and incapacitated individuals are most susceptible to scalding because their reaction time is impeded. In institutional applications (e.g., hospitals and long-term health care centers), hot water available at lavatories, baths, and showers is limited by health regulations to temperatures of about 105°F (41°C). As a result of an unfortunate scalding accident that resulted in a several million-dollar settlement in the early 1990s, water heater manufacturers have lowered the preset temperature on storage tank water heaters from 140°F (60°C) to about 120°F (49°C).

Tempered Water

Tempered water is a blend of hot and cold water that is mixed at a thermostatic (mixing) valve. It is used in applications requiring low-temperature hot water. Typical applications include shower rooms and group showers, domestic water for small buildings, and tempered water for light industrial processes (e.g., photographic laboratories). The temperature requirement for tempered water typically ranges from 90° to 110°F (32° to 43°C). Usually the mixing valve is set at the desired temperature and delivers water within ±3°F (±1.7°C) that temperature setting.

An *antiscald shower valve* is a type of thermostatic valve required in residential showers. This valve can be preset to any temperature between 60° and 120°F (15.5° and 49°C) with flow rates between 0.5 gpm and 10 gpm. It compensates for both temperature and pressure changes in the supply lines and protects against water supply failure. Under minor temperature and pressure variations, the delivery temperature of an antiscald shower valve will hold within ±3°F (±1.7°C). If temperature conditions vary greatly (a change of ±25°F [±14°C]), delivery temperature will vary.

TABLE 13.14   TYPICAL HOT WATER USE IN A U.S. HOUSEHOLD
            BY PERCENTAGE. UNITED STATES DEPARTMENT
            OF ENERGY.

Use	Percentage
Shower	37
Clothes washing	26
Dishwashing	14
Bath	12
Sinks	11

13.6   DOMESTIC WATER HEATING

Consumption

The U.S. Department of Energy reports that the average U.S. household consumes about 60 gal (228 L) of hot water per day, about 15 gal/person (57 L/person) per day. An average European uses daily 36 L (10 gal) of 60°C hot water per day. Energy consumption for heating domestic water is the second largest energy user (after space heating) in the residential sector in industrialized countries, accounting for about 14% of the energy used in a typical household. Typical hot water usage in a U.S. household by percentage is listed in Table 13.14.

Types of Water Heaters

A *water heater* is an appliance or system that heats water for domestic or building service hot water use. There are three types of water heaters: storage tank water heaters, instantaneous on-demand (tankless) water heaters, and circulating water heaters. Different types of water heaters are shown in Photos 13.28 through 13.36.

PHOTO 13.29   A gas connection to a residential storage tank water heater. Note the drain valve on the right side of the tank. (Used with permission of ABC)

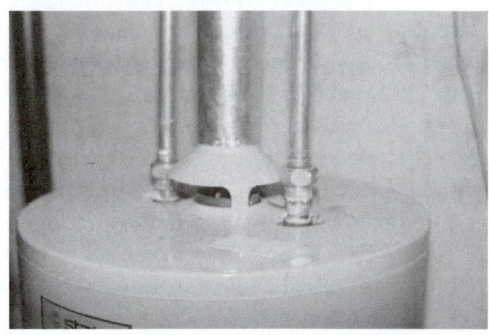

PHOTO 13.30   Plumbing connections to a residential storage tank water heater. Dielectric unions are used to minimize a galvanic reaction between the copper pipes and the water heater. (Used with permission of ABC)

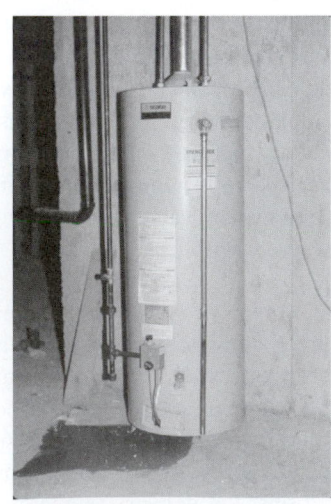

PHOTO 13.28   A residential storage tank water heater. (Used with permission of ABC)

PHOTO 13.31   A short storage tank water heater. (Used with permission of ABC)

PHOTO 13.32 A commercial, gas-fired storage tank water heater. (Used with permission of ABC)

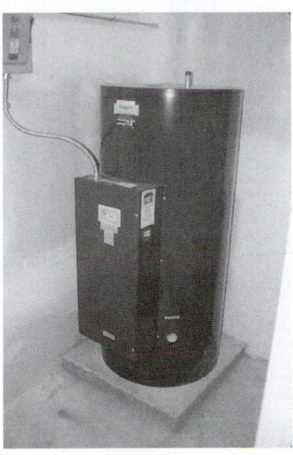

PHOTO 13.33 A commercial, electric storage tank water heater. Note the presence of an electrical connection and lack of presence of a vent. (Used with permission of ABC)

PHOTO 13.34 Two ultraefficient, gas-fired, power-vented, storage tank water heaters connected in series. Note the drain pans at the base of each unit that are designed to drain water into a floor drain. (Used with permission of ABC)

PHOTO 13.35 A gas-fired instantaneous water heater. (Used with permission of ABC)

PHOTO 13.36 A water heating boiler that provides hot water to a commercial building. (Used with permission of ABC)

Storage Tank Water Heaters

A *storage tank water heater* consists of a storage tank and a heating medium. Typically, storage tank sizes include 30, 40, 50, 60, 65, 75, 80, 100, and 120-gal (120- to 400-L) capacity. They are available in natural gas, propane (LP), fuel oil, and electric models. Storage tank water heaters remain the most popular type for residential heating needs in the United States. An average-size, three-bedroom residence uses a 50-gal (200-L) water heater.

Electric water heaters have coil-like elements that extend into the tank, which heat water as electric current passes through the elements. See Figure 13.11. Gas and fuel oil fired water heaters have burners located on the bottom of the tank and a vent that passes up through the center of the tank. The vent tube extracts residual heat from the combustion products before the gases are exhausted to the top of the tank into a vent that carries them outdoors. See Figure 13.12.

A storage tank water heater stores water for use on demand. Street pressure drives water flow. As hot water is drawn

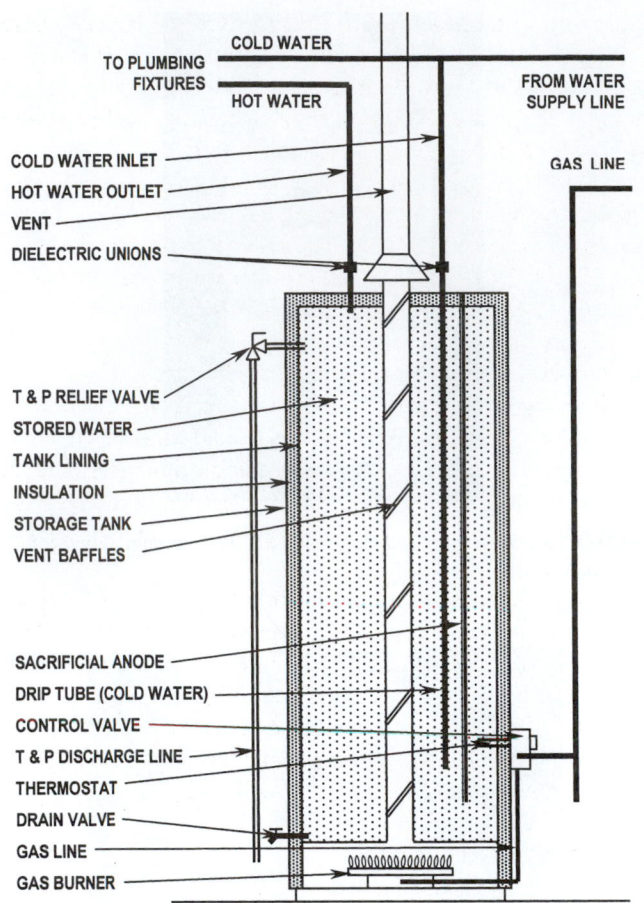

FIGURE 13.11   A gas-fired storage tank water heater.

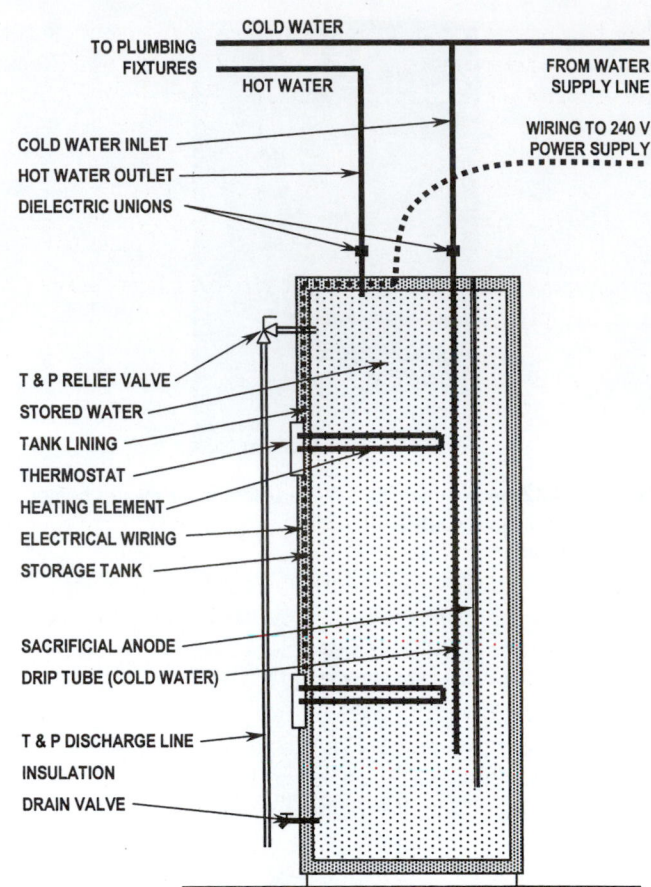

FIGURE 13.12   An electric storage tank water heater.

from the tank, it is replaced by water from the water service. This cools the water contained in the tank.

A thermostat that extends into the tank serves as a control of the heating medium: it switches the heating medium on when the temperature of the water in the tank has cooled below a set-point temperature and switches the heater off when the water in the tank reaches the preset temperature setting. The storage tank is insulated to reduce standby losses.

All storage tank water heaters are required to have T/P relief valves. Separate valves may be used, or a combination T/P relief valve may be installed. Both temperature relief valves and the combination T/P relief valves must be installed so that the temperature-sensing element is located in the top 6 in of the tank.

Types of storage tank water heaters include the following:

- *Residential storage tank water heaters* are designed for the residential market, but can be appropriate for many small commercial facilities. These water heaters, which are available with tank sizes up to 120 gal and gas inputs up to 75 000 Btu/hr, are manufactured in large quantities. As a result, they are relatively inexpensive and widely available.

- *Commercial storage tank water heaters* are similar to residential models except that they are available with much higher gas input ratings (1 000 000 Btu/hr or more) and larger storage tanks (up to 250 gal). They also feature larger pipe connections, more rugged controls, and a few features that are only rarely found in residential heaters, such as flue dampers and electronic ignition.

- *Ultraefficient water heaters* use power burners and enhanced heat exchangers to force hot combustion gasses into chambers and tubes that are submerged in the stored water. Ultraefficient water heaters are vented with plastic pipes that go directly through an outside wall. Because they draw combustion air directly from outside, through one of those pipes, their combustion processes are sealed off from the occupied space.

Storage tank water heaters can also be classified on their method of venting: nonventing (electric), naturally aspirated, power vented, and sealed combustion. *Naturally aspirated water heaters* have a flue that runs vertically through the center of the tank. Products of combustion are vented by relying on the natural buoyancy of the flue gases. Care should always be taken to ensure that naturally aspirated water heaters cannot backdraft because of strong winds at the chimney or because of high exhaust airflows in the building. *Power-vented water heaters* use a fan to exhaust flue gases. *Sealed-combustion water heaters* have supply air and exhaust air connections to the outdoors. These units do not require any indoor air to operate. There are two types of sealed-combustion water heaters. Some units have a flue up the center of the tank (similar to the power-vented water heaters). The second type of unit has a heat exchanger that wraps around the tank.

Instantaneous (Tankless) Water Heaters

Instantaneous water heaters, sometimes called *tankless water heaters* or *demand water heaters,* supply hot water on demand. They do not rely on a standby storage in a tank to artificially boost their capacity. Instead, they have a heating device that is activated by the flow of water when a hot water valve is opened. Once activated, the heater delivers a constant supply of hot water. The output of the heater, however, limits the rate of the heated water flow. They are more efficient because they do not have any tank standby losses. (See Figure 13.13.)

Instantaneous water heaters are available in natural gas, LP, and electric models. They come in a variety of sizes for different applications, such as a whole-house water heater, a hot water source for a remote bathroom or hot tub, or as a boiler to provide hot water for a home heating system. They can also be used as a booster heater for dishwashers and washing machines.

The largest gas-fired units can provide all the hot water needs of a household, and are installed centrally. Gas-fired models have a higher hot water output than electric models. As with many storage tank water heaters, even the largest whole-house tankless gas models may not supply enough hot water for simultaneous, multiple uses of hot water (i.e., showers and laundry). Large users of hot water, such as the clothes washer and dishwasher, need to be operated separately.

Some types of instantaneous water heaters are thermostatically controlled. They can modulate (vary) their output temperature according to the water flow rate and the inlet water temperature. This is useful when using a solar water heater for preheating the inlet water.

Instantaneous *point-of-use water heaters* fit compactly under a fixture (e.g., sink or lavatory) and heat water at the location where it is used. This device ensures that hot water is available without delay. Because only a cold water line needs to be brought to the fixture, it also saves on piping installation costs. This type of heater typically is electrically powered and is available in a 1 kW, 120 V, single-phase through 32 kW, three-phase power design.

Instantaneous water heaters are rated in gpm of hot water output at a specified temperature rise (temperature increase of incoming water). Residential units are available up to 5 gpm capacities. An average-size residence will typically demand 2 to 4 gpm (7.6 to 15.2 L/min). Smaller, ¼ to 2 gpm (1 to 8 L/min) point-of-demand units can be installed near demand points, such as under a sink or lavatory.

Circulating Water Heaters

Circulating water heaters consist of a separate storage tank that stores water heated by a heat exchanger. The heat exchanger may be a separate unit that is heated by stream or hot water from a boiler or may be contained in a boiler. The boiler may be designed to exclusively heat water for DHW or BSHW use or may be used for other purposes.

Tankless Coil and Indirect Water Heaters

A building's space heating boiler can also be used to heat water. Two types of water heaters that use this system are tankless coil and indirect. No separate storage tank is needed in the *tankless coil water heater* because water is heated directly inside the boiler in a hydronic (hot water) heating system. The water flows through a heat exchanger in the boiler whenever a hot water faucet is opened. During colder months, the tankless coil works well because the heating system is used regularly. However, the system is less efficient during warmer months and in warmer climates when the boiler is used less frequently or not at all; that is, the boiler must operate when hot water is needed.

A separate storage tank is required with an indirect water heater. Like the tankless coil, the *indirect water heater* circulates water through a heat exchanger in the boiler. But this heated water then flows to an insulated storage tank. Because the boiler does not need to operate as frequently, this system is more efficient than the tankless coil. In fact, when an indirect

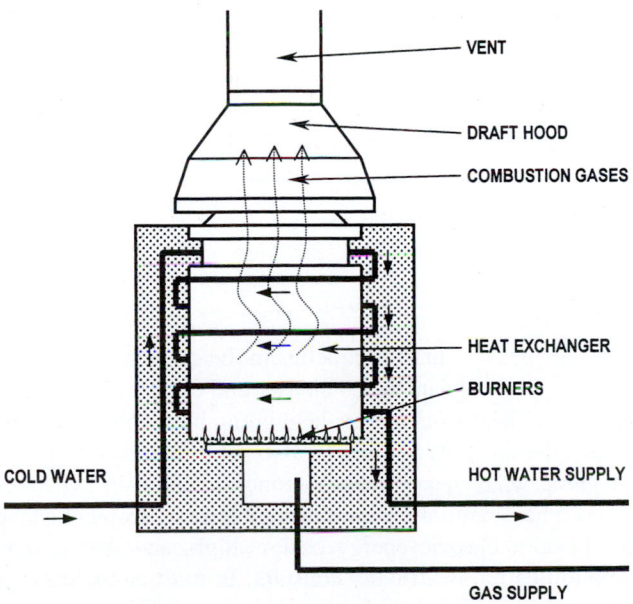

VENT

DRAFT HOOD

COMBUSTION GASES

HEAT EXCHANGER

BURNERS

COLD WATER

HOT WATER SUPPLY

GAS SUPPLY

FIGURE 13.13 An instantaneous (tankless) water heater.

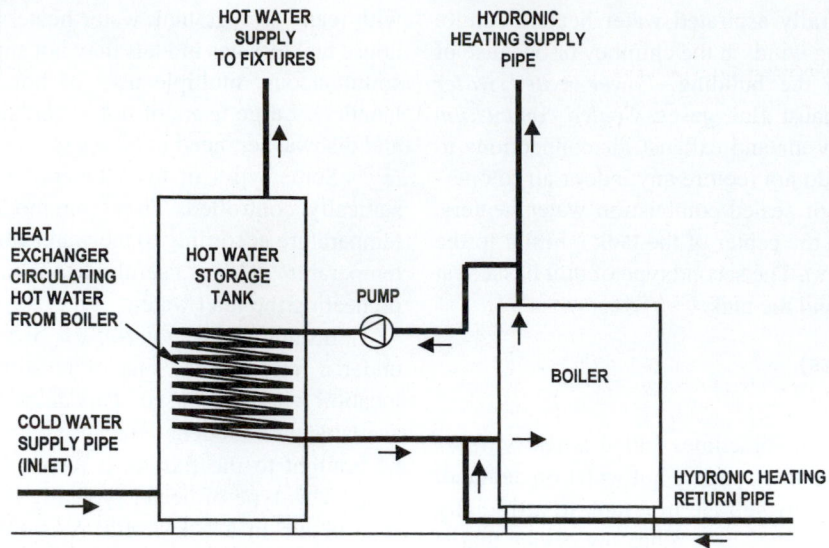

FIGURE 13.14 An indirect water heating system.

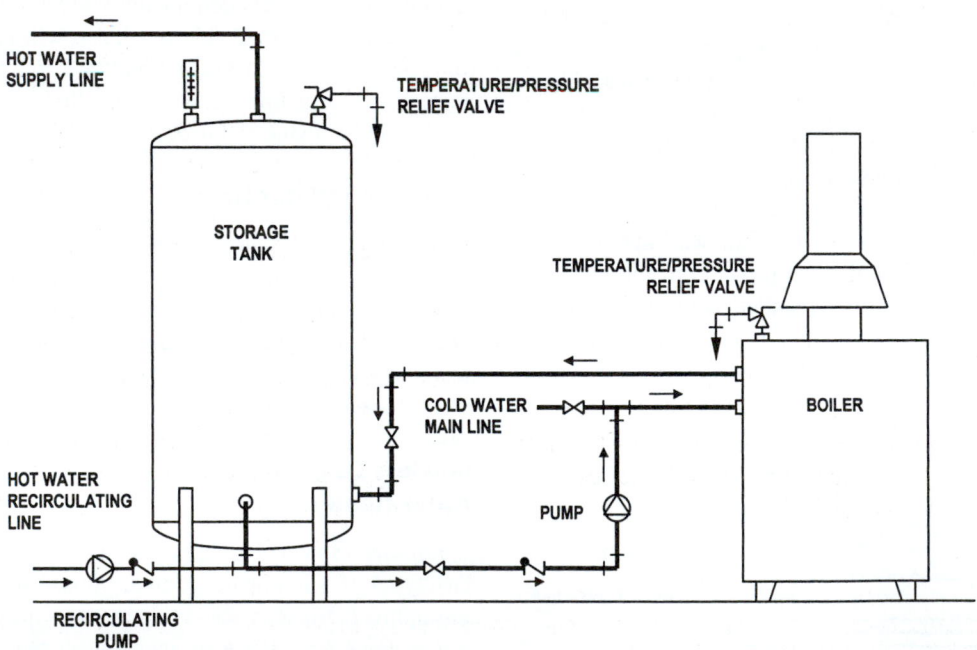

FIGURE 13.15 An indirect water heating system for a large building.

water heater is used with a highly efficient boiler, the combination may provide one of the least expensive methods of water heating. See Figures 13.14 and 13.15.

Heat Pump Water Heaters

Heat pump water heaters extract energy from outdoor air and use it to produce hot water very efficiently. Heat pump water heaters use an electric motor to run a compressor. The compressor draws a gaseous refrigerant through an evaporator, raising its pressure until it liquefies in the condenser. This heat pumping process heats the condenser and cools the evaporator. In removing heat from air, the heat pump both cools and dehumidifies the air, thus helping to also meet cooling needs. The heat pump water heater makes economic sense in hot humid regions where natural gas in unavailable (e.g., tropical islands like Hawaii), electric energy cost is high, and the need for dehumidification is virtually constant. In most cases, the extra costs of heat pump water heaters over standard electric water heaters are paid back in a few years.

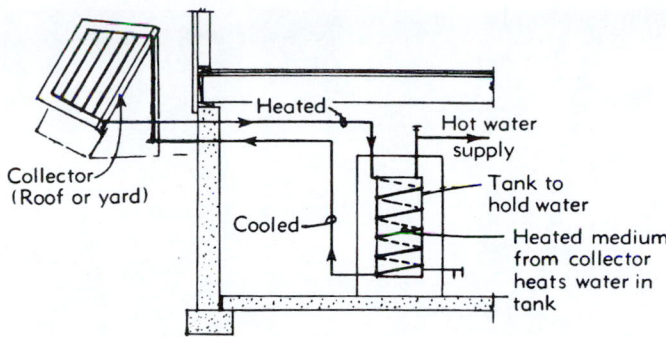

Tank—This tank could be used to just hold water heated by the collector medium or it could be a direct hot water heater used to supplement the collector during times when the collector cannot provide all of the hot water required.

FIGURE 13.16 A solar water heating system.

Solar Water Heaters

A *solar water heater* typically includes collectors mounted on the roof or in a clear area of the yard, a separate storage tank near the conventional heater in the home, connecting pipes, and an electronic controller. Throughout the year, the solar system preheats the water before it reaches the conventional water heater. Solar water heating systems can reduce the annual fuel cost of supplying hot water to a residence by more than half. There is, however, the added cost of the solar system. A schematic of a solar water heating system is shown in Figure 13.16. Solar water heating systems are discussed in Chapter 11.

Desuperheaters

A *desuperheater* is an attachment to an air conditioner or heat pump that allows waste heat from that device to assist in heating domestic water. In hot climates, a desuperheater can provide most of a home's hot water needs.

Energy Efficiency

The *energy factor (EF)* is a standardized measure used to express the efficiency of residential heaters. It was developed by the U.S. Department of Energy. It represents the amount of heat required to warm up hot water for a typical home in a year, divided by the amount of energy input into the heater to warm that water. A water heater with an EF of 0.60 transfers 60% of the energy it consumes to the hot water; the remaining energy is vented outdoors as waste heat. The higher the EF, the more energy efficient the water heater and the less energy consumed.

By varying the tank insulation, burner design, and a few other features, manufacturers make residential heaters available in energy factors ranging from 0.43 to greater than 0.95. Naturally aspirated water heaters units have the lowest EF, typically below 0.60. Many units also use a standing pilot light for ignition that further degrades seasonal performance. Power-vented water heaters have lower standby losses and therefore a higher EFs that range from 0.58 to 0.63. Sealed-combustion water heaters with the flue in the center of the tank will have an EF as high as 0.65. Those with a heat exchanger that wraps around the tank can have an EF over 0.80.

Electric resistance water heaters have an EF ranging from 0.7 and 0.95; gas water heaters from 0.5 to 0.6, with a few high-efficiency models ranging around 0.8; ultraefficient water heaters have thermal efficiencies of 94%, an EF of 0.94; oil water heaters from 0.7 and 0.85; and heat pump water heaters from 1.5 to 2.0. An instantaneous (tankless) water heater with an electronic ignition will have an EF above 0.90. Instantaneous water heaters will always have a higher EF in comparison to storage tank water heaters because they do not have tank standby losses. Characteristics of selected gas-fired tank-type and instantaneous (tankless) water heaters are provided in Table 13.15.

Unlike residential heaters, commercial heaters are not rated for overall efficiency. Instead, they are rated for thermal efficiency, which represents the portion of input gas energy that goes toward heating the water drawn from the tank. With few

TABLE 13.15   CHARACTERISTICS OF SELECTED GAS-FIRED TANK-TYPE AND INSTANTANEOUS (TANKLESS) WATER HEATERS. COMPILED FROM INDUSTRY SOURCES.

Water Heater Features	Water Heater Type			
	Storage Tank Type		Instantaneous (Tankless)	
Tank capacity	100 gal	50 gal	No tank	No tank
First hour rating	135 gal/hr	74 gal/hr	180 gal/hr	240 gal/hr
Btu input	75 000	40 000	37 000 to 165 000	20 000 to 185 000
Thermal efficiency	76%	76%	83%	85%
Energy factor	0.48 Gas	0.54 Gas	0.81 Gas	0.81 Gas
	0.44 LP gas	0.53 LP gas	0.84 LP gas	0.84 LP gas
Dimensions	Height: 64 in	Height: 57 in	Height: 24.5 in	Height: 24 in
	Diameter: 26.25 in	Diameter: 19.75 in	Width: 16.5 in	Width: 16.5 in
Wall hanging	No	No	Yes (optional)	Yes
Life expectancy	10 to 15 years	10 to 15 years	15 to 20 years	15 to 20 years

exceptions, commercial heaters are available with thermal efficiencies ranging from 78 to 80%. These heaters also must meet standards for standby loss, which represents the portion of the stored energy that is lost when the burners are not operating. Standby losses are typically not published but can be obtained from the manufacturer.

The instantaneous energy efficiency of a heat pump water heater system depends on incoming water temperature, intake air temperature, the heat transfer characteristics of the heat pump, and various conductive and convective losses throughout the system. Additionally, although in most circumstances the hot water output is useful throughout the year, the cold air output may not be. Accordingly, there is no simple index that accounts for both outputs and describes overall heat pump water heater efficiency. As a result, the heat pump water heater industry relies on two indexes of energy efficiency: coefficient of performance (COP), which is favored by manufacturers of larger heat pump water heater systems. The EF of a typical heat pump water heater will be about 2.2.

The COP is a measure of the instantaneous energy output of a system in comparison with its instantaneous energy input. Standby losses and the interaction of changing water and air temperatures are not reflected in measurements of COP. Accordingly, the COP of a standard hot water system is close to 1, and the COP of a typical heat pump water heater may be 3. Buyers of commercial systems should be aware that COP quoted by manufacturers sometimes reflects the combination of the production of cold air and hot water in relation to energy input. This is helpful if full use is made of the cold air, but not otherwise.

Hot Water Recirculating Systems

In many medium and large commercial buildings and even some residences, a *hot water recirculating system* continuously circulates hot water from the water-heating unit through the hot water supply piping and back to the water heater through hot water recirculating piping. This ensures that hot water is always available at the taps, thus avoiding the need to run water for a long time to obtain water at the desired temperature. See Photo 13.37 and Figures 13.17 and 13.18.

There are three types of hot water recirculating strategies as follows.

Continuous Recirculating

In *continuous recirculating,* water is constantly recirculated from the water heater through the piping. This type of system is best suited for buildings having round-the-clock occupancy. Hospitals and multiunit residential buildings typically require continuous hot water recirculating.

Timed Recirculating

Timed recirculating involves use of an electronic or electromechanical timer to shut off circulation of hot water when the

PHOTO 13.37   Properly marked domestic cold water, hot water, and hot water recirculating lines. (Used with permission of ABC)

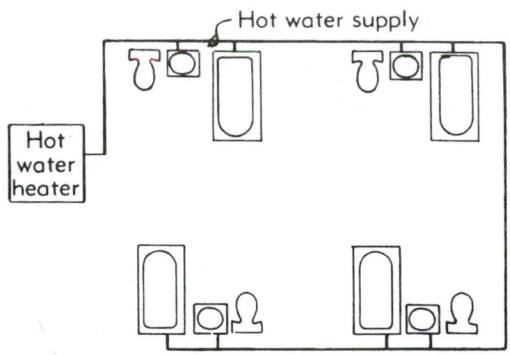

FIGURE 13.17   A noncirculating hot water line conveys hot water to the individual fixtures. For hot water to be received at the most remote fixture, stagnant water in the hot water supply line must first be evacuated.

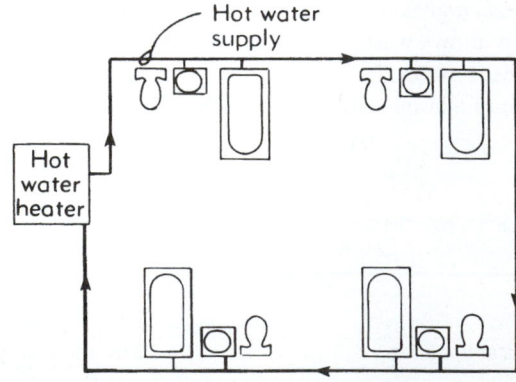

FIGURE 13.18   A hot water recirculating system continuously circulates hot water through the hot water supply piping, making hot water available at the fixtures without evacuating stagnant water.

building is not occupied. This shutdown would apply to commercial, institutional, and industrial buildings that are closed at night and/or on weekends. Nighttime housekeeping and maintenance operations should not be neglected when considering shutdown periods.

Thermostatically Controlled Recirculating

Thermostatically controlled recirculating relies on a sensor located at a remote location in the recirculating line, which senses water temperature and activates the recirculating pump when water temperature drops below a predetermined setting.

Small- to medium-size buildings typically have a single recirculating loop, a recirculating pump, controls, and a water heating system with storage tank. In large buildings, several recirculating loops may be necessary. Each loop will typically have its own piping and controls. One recirculating pump will usually serve all loops with balancing valves controlling flow to a specific loop. The pump requires a minimal horsepower drive in the range of $1/2$ or $1/16$ hp (just enough to keep the water moving). To reduce heat loss, the hot water recirculating piping must be insulated.

13.7   DETERMINING WATER HEATER SIZE

Design Load

Hot water use is not spread out over the entire day. Instead, domestic use of hot water tends to peak in the morning hours and again in the early evening (supper time). As a result, a water heating system must be designed to have sufficient capacity to provide hot water during periods of peak use. An accepted method of determining the maximum demand on a water heater is to determine peak usage during a particular time of day. For single- and two-family dwelling units, this peak occurs during the morning or early evening. Refer to Table 13.16 for typical household demands by use.

TABLE 13.16   TYPICAL HOUSEHOLD HOT WATER DEMANDS BASED ON USE. UNITED STATES DEPARTMENT OF ENERGY.

Household Hot Water Use	Hot Water (per Use)	
	gal	L
Bath	20	76
Shower	20	76
Shaving	2	8
Face/hands washing	4	15
Teeth brushing	2	8
Hair shampoo in sink	4	15
Food preparation	5	19
Automatic dishwasher	14	53
Hand dish washing of family meal—major clean up	14	53
Hand dish washing of some dishes—minor clean up	4	15
Automatic clothes washer	32	122

Example 13.13

Estimate the peak hot water demand of a household that has a morning routine of two showers, three hands/face washings, four teeth brushings, food preparation, and minor dishwashing.

From Table 13.16:

2	showers (20 gallons)	$2 \cdot 20 =$	40
3	hands/face washings (4 gallons)	$3 \cdot 4 =$	12
4	teeth brushings (2 gallons)	$4 \cdot 2 =$	8
1	food preparation (5 gallons)	$1 \cdot 5 =$	5
1	minor dish washing (4 gallons)	$1 \cdot 4 =$	4
	total peak hot water demand:		69 gallons

Typical hot water demands for various types of nonresidential buildings are provided in Table 13.17. An accepted method of determining the maximum demand on a water heater is to determine peak usage.

Example 13.14

Estimate the hot water demand of a junior high school serving 850 students and 65 staff.

From Table 13.17:

1.5 gpm/student · 850 students = 1275 gpm

Sizing Storage Tank Water Heaters

Residential water heaters are typically sized based on their first hour rating (FHR). This rating is contained on the U.S. Department of Energy Guide label found on all water heaters. The FHR relates to the gallons of hot water available for one hour of peak demand. It is the maximum output of the water heater over an hour, in gallons of hot water in a 100°F temperature rise. The FHR is the sum of the "standby" hot water found in the tank plus the capacity of the heater to heat water during that first hour. Typically, standby water is taken as 70% of the tank capacity.

FHR (in gallons) = 70% of tank capacity (gallons) + recovery rate (gallons per hour)

Tank capacities are usually available on the water heater nameplate. Recovery rate is indicated less frequently, but can be approximated from the heater input rating. *Recovery rate* is the quantity of water that the burner or element can heat to a 100°F (38°C) increase in one hour. This is referred to as a 100° temperature rise, which means that the water temperature is increased 100°F. For example, if a burner can take 40 gallons of 40°F (4.4°C) water and raise its temperature to 140°F (60°C) in one hour, that burner/heater has a 40 gal recovery rate.

TABLE 13.17   TYPICAL HOT WATER DEMANDS FOR VARIOUS TYPES OF BUILDINGS.

	Hot Water Demand					
	Maximum per Hour		Maximum per Day		Average per Day	
Type of Building	gal	L	gal	L	gal	L
Food service establishments, maximum	1.5	5.7	11.0	5.7	2.4	9.1
Type A: full meal restaurants and cafeterias	0.7	2.6	6.0	22.7	0.7	2.6
Type B: drive-ins, grilles, sandwich shops						
Apartment houses, per unit						
Number of apartments						
20 or less	12.0	45.5	80.0	303	42.0	159
50	10.0	37.9	73.0	277	40.0	152
75	8.5	32.2	66.0	250	38.0	144
100	7.0	26.5	60.0	227	37.0	140
200 or more	5.0	19	50.0	195	35.0	133
Men's dormitories, per student	3.8	14.4	22.0	83.4	13.1	49.7
Women's dormitories, per student	5.0	19	26.5	100	12.3	46.6
Motels, per unit number of motel units[a]						
20 or less	6.0	22.7	35.0	133	20.0	75.8
60	5.0	19.7	25.0	95	14.0	53.1
100 or more	4.0	15.2	15.0	57	10.0	37.9
Nursing homes, per bed	4.5	17.1	30.0	114	18.4	69.7
Office buildings, per person	0.4	1.5	2.0	7.6	1.0	3.8
Elementary schools, per student	0.6	2.3	1.5	5.7	0.6	2.3
Junior and senior high schools, per student[b]	1.5	3.8	3.6	13.6	1.8	6.8

[a] Interpolate for intermediate values.
[b] Per day of operation.

Average gas-fired residential and light commercial heater sizes fall in the range of 30 000 to 75 000 Btu/hr, which equates to a recovery rate of about 27 to 68 gallons per hour. For a natural gas-fired water heaters, multiplying the burner input rating by 0.0009 can approximate recovery rate (e.g., A 35 000 Btu/hr burner has a recovery rate of $35\,000 \cdot 0.0009 = 31.5$ gallons). Water heaters powered by electricity will have recovery rates of about one gallon/hour per 250 watts. A common residential electric water heater is rated at 3000 to 4500 watts, which would be a recovery rate of about 12 to 18 gallon per hour. The recovery rate of a gas-fired water heater is typically about double the rate of electric-powered units having the same storage tank size. To compensate and provide a suitable FHR, electric water heater storage tanks are typically larger.

In sizing a storage tank-type water heater, the FHR of a water heater must meet or exceed the peak hot water demand.

Table 13.18 is a guide to selecting the minimum recommended gas-fired storage tank water heater size, in gal. Table 13.19 provides specifications for selected storage tank water heaters.

Sizing Instantaneous Water Heaters

Instantaneous water heaters are selected based on the amount of hot water needed to meet the design load (peak instantaneous demand), in gpm or L/min, at a specific water temperature rise and other criteria. Table 13.20 provides typical specifications for a selected residential gas-fired instantaneous (tankless) water heater.

The design load (the flow rate of the instantaneous water heater) is determined by adding flow rates of fixtures used simul-

TABLE 13.18   GUIDE TO SELECTING THE MINIMUM GAS-FIRED STORAGE TANK WATER HEATER SIZE, IN GAL. ADD 20 GAL FOR EACH ADDITIONAL BEDROOM. MOVE TO THE NEXT LARGEST VALUE FOR HIGH DEMAND.

Number of bathrooms	1 to 1.5			2 to 2.5				3 to 3.5			
Number of bedrooms	1	2	3	2	3	4	5	3	4	5	6
Water heater tank size, gal	20	30	40	50	50	50	66	50	66	66	80
First hour capacity, gal/hr	35	45	55	70	70	70	86	70	86	86	100
W tank size, L	76	113	151	189	189	189	249	189	249	249	302
First hour capacity, L/hr	132	170	208	265	265	265	325	265	325	325	378

TABLE 13.19 SPECIFICATIONS FOR SELECTED TANK-TYPE WATER HEATERS. COMPILED FROM SEVERAL INDUSTRY SOURCES.

Gas Fired

Storage Tank Capacity		First Hour Rating		
gal	L	gal	L	Energy Factor
30	113	67	220	0.62
40	151	75	246	0.65
40	151	79	259	0.64
40	151	76	249	0.64
40	151	72	236	0.64
40	151	67	220	0.64
48	181	105	344	0.65
50	189	84	276	0.64
75	284	145	476	0.64
65	246	129	423	0.62
73	276	130	426	0.60

Oil Fired

Storage Tank Capacity		First Hour Rating		
gal	L	gal	L	Energy Factor
32	121	132	433	0.68
32	121	130	426	0.66
30	113	150	492	0.62
30	113	150	492	0.62

Electric

Storage Tank Capacity		First Hour Rating		
gal	L	gal	L	Energy Factor
30	113	42	138	0.94
30	113	39	128	0.94
30	113	35	115	0.94
40	151	50	164	0.95
40	151	52	171	0.94
40	151	48	157	0.94
40	151	45	148	0.94
50	189	58	190	0.95
50	189	61	200	0.94
50	189	58	190	0.94
66	249	76	249	0.95
85	321	91	298	0.92
80	302	89	292	0.92

taneously. The following assumptions on water flow for various residential fixtures may be used to determine the size of unit:

- Faucets: 0.75 gpm (2.84 L/min) to 2.5 gpm (9.46 L/min)
- Low-flow showerheads: 1.2 gpm (4.54 L/min) to 2 gpm (7.57 L/min)
- Older standard showerheads: 2.5 gpm (9.46 L/min) to 3.5 gpm (13.25 L/min)
- Clothes washers and dishwashers: 1 gpm (3.79 L/min) to 2 gpm (7.57 L/min)

Most instantaneous water heaters are rated for flow rates at a variety of inlet water temperatures or temperature rises. A good assumption in design of instantaneous water heaters is that that the incoming potable water temperature is no warmer than 50°F (10°C). A temperature of 35° to 40°F (2° to 4°C) should be used in cold climates. Water must typically be heated to 120°F (49°C) for most residential uses. To determine the required temperature rise needed, subtract the incoming water temperature from the desired output temperature. The needed temperature rise is typically at least 70°F (39°C). Table 13.21 provides test results of a single instantaneous water heater.

Example 13.15

Size an instantaneous water heater to meet the following conditions:

- One hot water faucet open with a flow rate of 0.75 gpm (2.84 L/min)
- One person showering using a showerhead with a flow rate of 2.5 gpm (9.46 L/min)
- An inlet water temperature is 40°F (4°C)
- Water must be heated to 120°F (49°C)

The design load is found by:

$$0.75 \text{ gal} + 2.5 \text{ gal} = 3.25 \text{ gal}$$

The required temperature rise is found by:

$$120°F - 50°F = 70°F$$

An instantaneous water heater with a flow rate of no less than 3.25 gpm (12.3 L/min) at a temperature rise of 80°F (44°C) should be selected.

Faster flow rates or cooler inlet (water supply) temperatures will reduce the water temperature available. Using low-flow showerheads and water-conserving faucets are a good idea with instantaneous water heaters.

Sizing a Large Multifamily Water Heating System

As with one- and two-family water heaters, a hot water system must be designed to provide a sufficient supply of hot water for use by building occupants during peak periods of use. In larger buildings, it is less likely that a large share of plumbing fixtures will be in use at a given time. In large multifamily dwellings, such as apartments or condominiums with 10 or more units, hot water demand tends to peak between 6 and 9 AM and again between 5 and 8 PM.

An approximation method used to determine hot water demand in large multifamily buildings (10 or more dwelling

TABLE 13.20 SPECIFICATIONS FOR SELECTED GAS-FIRED INSTANTANEOUS (TANKLESS) WATER HEATERS IN CUSTOMARY (U.S.) UNITS. COMPILED FROM INDUSTRY SOURCES.

Weight		55 lb/121 kg		50 lb/110 kg		50 lb/110 kg		36 lb/80 kg		50 lb/110 kg	
Maximum/Minimum Gas Rate (Input)		237 to 19 MBtu/hr 69 to 65 kW		199 to 15 MBtu/hr 678 to 51 kW		180 to 15 MBtu/hr 614 to 51 kW		150 to 19 MBtu/hr 512 to 65 kW		150 to 15 MBtu/hr 512 to 51 kW	
Hot Water Capacity		0.6 to 9.8 gal 2.3 to 37.1 L		0.6 to 9.4 gal 2.3 to 35.6 L		0.6 to 7.5 gal 2.3 to 28.4 L		0.6 to 6.3 gal 2.3 to 23.8 L		0.6 to 5.0 gal 2.3 to 18.9 L	
Energy Factor		.82		.82		.82		.82		.82	
Temperature Rise		gal/min at Δ°F	L/min at Δ°C	gal/min at Δ°F	L/min at Δ°C	gal/min at Δ°F	L/min at Δ°C	gal/min at Δ°F	L/min at Δ°C	gal/min at Δ°F	L/min at Δ°C
Δ°F	Δ°C										
20	11	9.8	37.1	9.4	35.6	7.5	28.4	6.3	23.8	5.0	18.9
25	14	9.8	37.1	9.4	35.6	7.5	28.4	6.3	23.8	5.0	18.9
30	17	9.8	37.1	9.4	35.6	7.5	28.4	6.3	23.8	5.0	18.9
35	19	9.8	37.1	9.4	35.6	7.5	28.4	6.3	23.8	5.0	18.9
40	22	9.8	37.1	8.3	31.4	7.5	28.4	6.2	23.5	5.0	18.9
45	25	8.8	33.3	7.4	28.0	6.7	25.4	5.5	20.8	5.0	18.9
50	28	7.9	29.9	6.6	25.0	6.0	22.7	5.0	18.9	5.0	18.9
55	31	7.2	27.3	6.0	22.7	5.5	20.8	4.5	17.0	4.6	17.4
60	33	6.6	25.0	5.5	20.8	5.0	18.9	4.2	15.9	4.2	15.9
65	36	6.1	23.1	5.1	19.3	4.6	17.4	3.8	14.4	3.9	14.8
70	39	5.7	21.6	4.7	17.8	4.3	16.3	3.6	13.6	3.6	13.6
75	42	5.3	20.1	4.4	16.7	4.0	15.1	3.3	12.5	3.4	12.9
80	44	4.9	18.5	4.2	15.9	3.8	14.4	3.1	11.7	3.2	12.1
85	47	4.7	17.8	3.9	14.8	3.5	13.2	2.9	11.0	3.0	11.4
90	50	4.4	16.7	3.7	14.0	3.3	12.5	2.8	10.6	2.8	10.6
95	53	4.2	15.9	3.5	13.2	3.2	12.1	2.6	9.8	2.7	10.2
100	56	4.0	15.1	3.3	12.5	3.0	11.4	2.5	9.5	2.5	9.5
105	58	3.8	14.4	3.2	12.1	2.9	11.0	2.4	9.1	2.4	9.1
110	61	3.6	13.6	3.0	11.4	2.7	10.2	2.3	8.7	2.3	8.7
115	64	3.4	12.9	2.9	11.0	2.6	9.8	2.2	8.3	2.2	8.3
120	67	3.3	12.5	2.8	10.6	2.5	9.5	2.1	7.9	2.1	7.9
125	69	3.2	12.1	2.7	10.2	2.4	9.1	2.0	7.6	2.0	7.6
130	72	3.0	11.4	2.6	9.8	2.3	8.7	1.9	7.2	1.9	7.2
135	75	2.9	11.0	2.5	9.5	2.2	8.3	1.8	6.8	1.9	7.2
140	78	2.8	10.6	2.4	9.1	2.1	7.9	1.8	6.8	1.8	6.8

TABLE 13.21 TEST RESULTS OF FLOW RATES AT VARIOUS RISES IN TEMPERATURE FOR A SELECTED RESIDENTIAL GAS-FIRED INSTANTANEOUS (TANKLESS) WATER HEATER BASED ON AN INLET TEMPERATURE OF 66°F (19°C). WITH LOWER SUPPLY TEMPERATURE, THE FLOW RATE WOULD BE LOWER AT THE SAME DELIVERY TEMPERATURES. THE HEATER MODULATES DELIVERY OF HEAT FROM 19 000 TO 180 000 BTU/HR (5.6 TO 53 kJ/s) IN 13 INCREMENTS ACCORDING TO INLET WATER TEMPERATURE AND HOT WATER DEMAND.

Delivery Temperature	Temperature Rise	Flow Rate gpm (L/s)
180°F (82°C)	114°F (46°C)	2.5 (9.5)
167°F (75°C)	101°F (38°C)	2.8 (10.6)
140°F (60°C)	74°F (23°C)	4.2 (15.9)
130°F (54°C)	64°F (18°C)	4.7 (17.8)
120°F (49°C)	54°F (12°C)	6.2 (23.5)
116°F (47°C)	50°F (10°C)	6.5 (24.6)
112°F (44°C)	46°F (8°C)	6.7 (25.4)

units) is based on the demand unit (DU). See Table 13.22. For apartments or condominiums, a DU is counted for each bathroom and clothes washing machine served by the water heating system. For apartments or condominiums with 10 or more units:

Maximum probable demand (MPD) of hot water, in gal/hr,
$$MPD = 350 + 11 \, (DU)$$

This expression assumes 11 gal/hr for each demand unit plus a 350 gal/hr reserve capacity.

Example 13.16

Estimate the peak hot water demand of a 24-unit condominium having a single water heating system. Each unit has two bathrooms, and there are four clothes washing machines in the common area.

24 condominiums (each having two bathrooms) = 48 DU

4 clothes washers = 4 DU

Total DU: 52 DU

$$MPD = 350 + 11(DU) = 350 + 11(52) = 922 \text{ gal/hr}$$

A hot water boiler is typically sized to heat water for buildings containing 10 or more dwelling units. It is customa~~ry~~ [obscured] indrical storage

to include a storage task with the boiler to act as a reserve for times when instantaneous demand exceeds boiler capacity. Size of the storage tank is again based on the number of DUs served by the water heating system. See Table 13.23.

Minimum storage tank capacity, in gal, $STC = DU \cdot \text{gal/DU}$

GAS FIRED STORAGE TANK

Example 13.17

Calculate the minimum storage tank size for a water heating system for the 24-unit condominium described in the previous example. Select a tank size, based on a height of 7 ft (to allow ceiling clearance).

From the previous example, the total DU = 52 DU

TABLE 13.22 WATER HEATING DEMAND UNITS (DU) VERSUS STORAGE TANK CAPACITY PER DEMAND UNIT.

Number of Demand Units (DU)	Gallons per Demand Unit	Liters per Demand Unit
0 to 10	30	114
11 to 25	25	95
26 to 50	20	76
Over 50	15	57

TABLE 13.23A CAPACITIES OF CYLINDRICAL TANKS, IN GALLONS. CAPACITIES PROVIDED ARE TANK SIZE EXCLUDING INSULATION THICKNESS.

Length or Height (ft)	Diameter (in)												
	36	42	48	54	60	66	72	78	84	90	96	102	108
1	53	72	94	120	145	180	210	250	290	330	375	425	475
2	106	144	188	240	200	360	420	500	580	660	750	850	950
3	159	216	282	360	435	540	630	750	870	990	1125	1275	1425
4	212	288	376	480	580	720	840	1000	1160	1320	1500	1700	1900
5	265	360	470	600	725	900	1050	1250	1450	1650	1875	2125	2375
6	318	432	564	720	870	1080	1260	1500	1740	1980	2250	2550	2850
7	371	504	658	840	1015	1260	1470	1750	2030	2310	2625	2975	3325
8	424	578	752	960	1160	1440	1680	2000	2320	2640	3000	3400	3800
9	477	648	846	1080	1305	1620	1890	2250	2610	2970	3375	3825	4275
10	530	720	940	1200	1450	1800	2100	2500	2900	3300	3750	4250	4750
11	583	792	1034	1320	1595	1980	2310	2750	3190	3630	4125	4675	5225
12	636	804	1128	1440	1740	2160	2520	3000	3480	3960	4500	5100	5700
13	689	936	1222	1560	1885	2340	2730	3250	3770	4280	4875	5525	6175
14	742	1008	1316	1680	2030	2520	2940	3500	4060	4620	5250	5950	6650
15	795	1080	1410	1800	2175	2700	3150	3750	4350	4950	5625	6375	7125
16	848	1152	1504	1920	2320	2880	3360	4000	4640	5280	6000	6800	7800

TABLE 13.23B CAPACITIES OF CYLINDRICAL TANKS, IN LITERS. CAPACITIES PROVIDED ARE INTERIOR TANK SIZE EXCLUDING INSULATION THICKNESS.

Length or Height (m)	Diameter (m)										
	1.0	1.2	1.4	1.6	1.8	2.0	2.2	2.4	2.6	2.8	3.0
0.5	393	565	770	1005	1272	1571	1901	2262	2655	3079	3534
1.0	785	1131	1539	2011	2545	3142	3801	4524	5309	6158	7069
1.5	1178	1696	2309	3016	3817	4712	5702	6786	7964	9236	10 603
2.0	1571	2262	3079	4021	5089	6283	7603	9048	10 619	12 315	14 137
2.5	1963	2827	3848	5027	6362	7854	9503	11 310	13 273	15 394	17 671
3.0	2356	3393	4618	6032	7634	9425	11 404	13 572	15 928	18 473	21 206
3.5	2749	3958	5388	7037	8906	10 996	13 305	15 834	18 583	21 551	24 740
4.0	3142	4524	6158	8042	10 179	12 566	15 205	18 096	21 237	24 630	28 274
4.5	3534	5089	6927	9048	11 451	14 137	17 106	20 358	23 892	27 709	31 809
5.0	3927	5655	7697	10 053	12 723	15 708	19 007	22 619	26 546	30 788	35 343

TABLE 13.24   CAPACITIES OF SELECTED GAS-FIRED WATER HEATING BOILERS.

Input (Btu/hr)	Output (Btu/hr)	Water Heating Rate							
		gal/hr Output for Temperature Rise					L/hr Output for Temperature Rise		
		60°F	70°F	80°F	90°F	100°F	20°C	30°C	40°C
100 000	80 000	160	137	120	107	96	534	422	349
150 000	120 000	240	206	180	160	144	800	633	523
200 000	160 000	320	274	240	213	192	1067	844	698
250 000	200 000	400	343	300	267	240	1334	1055	872
300 000	240 000	480	411	360	320	288	1601	1266	1047
350 000	280 000	560	480	420	373	336	1868	1477	1221
400 000	320 000	640	549	480	427	384	2135	1688	1396
450 000	360 000	720	617	540	480	432	2401	1899	1570
500 000	400 000	800	686	600	533	480	2668	2110	1745
550 000	440 000	880	754	660	587	528	2935	2321	1919
600 000	480 000	960	823	720	640	576	3202	2532	2094
650 000	520 000	1040	891	780	693	624	3469	2743	2268
700 000	560 000	1120	960	840	747	672	3736	2954	2442
750 000	600 000	1200	1029	900	800	720	4002	3165	2617
800 000	640 000	1280	1097	960	853	768	4269	3376	2791
850 000	680 000	1360	1166	1020	907	816	4536	3587	2966
900000	720000	1440	1234	1080	960	864	4803	3798	3140
950 000	760 000	1520	1303	1140	1013	912	5070	4009	3315
1 000 000	800 000	1600	1371	1200	1067	960	5336	4220	3489
1 125 000	900 000	1800	1543	1350	1200	1080	6004	4747	3925
1 250 000	1 000 000	2000	1714	1500	1333	1200	6671	5274	4362
1 375 000	1 100 000	2200	1886	1650	1467	1320	7338	5802	4798
1 500 000	1 200 000	2400	2057	1800	1600	1440	8005	6329	5234
1 625 000	1 300 000	2600	2229	1950	1733	1560	8672	6857	5670
1 750 000	1 400 000	2800	2400	2100	1867	1680	9339	7384	6106
1 875 000	1 500 000	3000	2571	2250	2000	1800	10 006	7912	6542
2 000 000	1 600 000	3200	2743	2400	2133	1920	10 673	8439	6978
2 250 000	1 800 000	3600	3086	2700	2400	2160	12 007	9494	7851
2 500 000	2 000 000	4000	3429	3000	2667	2400	13 341	10 549	8723
2 750 000	2 200 000	4400	3771	3300	2933	2640	14 675	11 604	9595
3 000 000	2 400 000	4800	4114	3600	3200	2880	16 009	12 659	10 468
3 250 000	2 600 000	5200	4457	3900	3467	3120	17 344	13 713	11 340
3 500 000	2 800 000	5600	4800	4200	3733	3360	18 678	14 768	12 212
3 750 000	3 000 000	6000	5143	4500	4000	3600	20 012	15 823	13 085
4 000 000	3 200 000	6400	5486	4800	4267	3840	21 346	16 878	13 957
4 250 000	3 400 000	6800	5829	5100	4533	4080	22 680	17 933	14 829
4 500 000	3 600 000	7200	6171	5400	4800	4320	24 014	18 988	15 702
4 750 000	3 800 000	7600	6514	5700	5067	4560	25 348	20 043	16 574
5 000 000	4 000 000	8000	6857	6000	5333	4800	26 682	21 098	17 446
5 250 000	4 200 000	8400	7200	6300	5600	5040	28 016	22 153	18 318
5 500 000	4 400 000	8800	7543	6600	5867	5280	29 351	23 207	19 191
5 750 000	4 600 000	9200	7886	6900	6133	5520	30 685	24 262	20 063
6 000 000	4 800 000	9600	8229	7200	6400	5760	32 019	25 317	20 935
6 250 000	5 000 000	10 000	8571	7500	6667	6000	33 353	26 372	21 808
6 500 000	5 200 000	10400	8914	7800	6933	6240	34 687	27 427	22 680
6 750 000	5 400 000	10800	9257	8100	7200	6480	36 021	28 482	23 552
7 000 000	5 600 000	11 200	9600	8400	7467	6720	37 355	29 537	24 425
7 250 000	5 800 000	11600	9943	8700	7733	6960	38 689	30 592	25 297
7 500 000	6 000 000	12 000	10 286	9000	8000	7200	40 024	31 647	26 169

From Table 13.22, the 52 DU should be computed at 15 gal per DU:

$$STC = DU \cdot gal/DU = 52 \cdot 15 = 780 \text{ gal minimum}$$

From Table 13.23, for a minimum capacity of 780 gal and length (height) of 7 ft:

A 54 in diameter by 7 ft-840-gal tank is acceptable.

Table 13.24 provides capacities of selected gas-fired water heating boilers. To meet the 922 gal/hr hour demand (from Example 13.17), a gas-fired boiler with a capacity of 900 000 Btu/hr input (960 gal/hr) at a 90°F temperature rise could be selected.

An alternate solution would be to presume that some of the 840 gal storage tank capacity (from Example 13.17) could be shared over the 3-hr peak demand period: 840 gal/3 hr =

280 gal/hr. Therefore, the boiler capacity can be decreased, based on 922 gal/hr hour demand minus 280 gal/hr or 642 gal/hr. At a 90°F temperature rise, a gas-fired boiler with a capacity of 650 000 Btu/hr input (693 gal/hr) would be selected. The alternate solution is a more cost-effective solution because of lower initial cost and lower operating costs. It is, however, a less conservative approach to design.

13.8 WATER SUPPLY SYSTEM DESIGN EXAMPLE

The following examples convey a typical design approach for the apartment building shown in Appendix A. Both downfeed and upfeed systems for the apartment building are covered.

Demand Load Example

1. The first step in determining the demand load is to list the plumbing fixtures required on the project. The following information is gathered:

   4-story apartment building

   3 in (75 mm) service main

   Street main pressure: 50 psi (345 kPa)

   Fixtures per floor:

   2 flush valve water closets (toilets)

   2 tubs with showerheads

   2 lavatories (bathroom sinks)

   2 kitchen sinks

   4 hose bibbs on the first floor (⅜ in [10 mm] supply, general use)

   It should be noted at this point that the apartment building being sized for a water system has a repetitive floor plan for each floor. This allows the entire apartment to be serviced from a single water pipe (riser) going up the building (Figures 13.19 through 13.22).

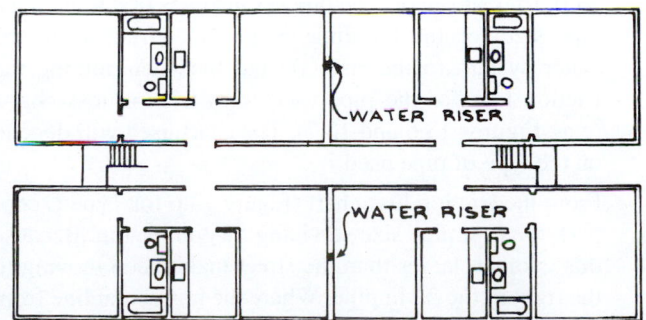

FIGURE 13.19 Location of water risers in apartment building design example are required on both sides of the hallway.

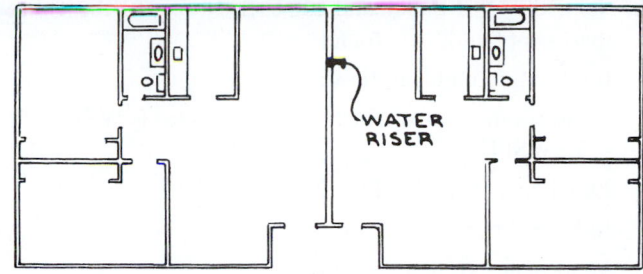

FIGURE 13.20 Water riser location in apartment building design example.

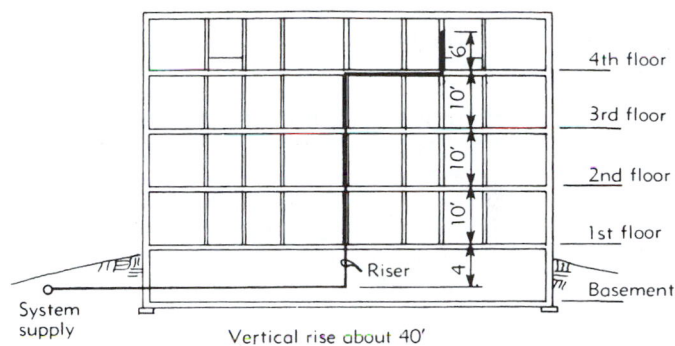

FIGURE 13.21 Elevation of water riser in apartment building design example.

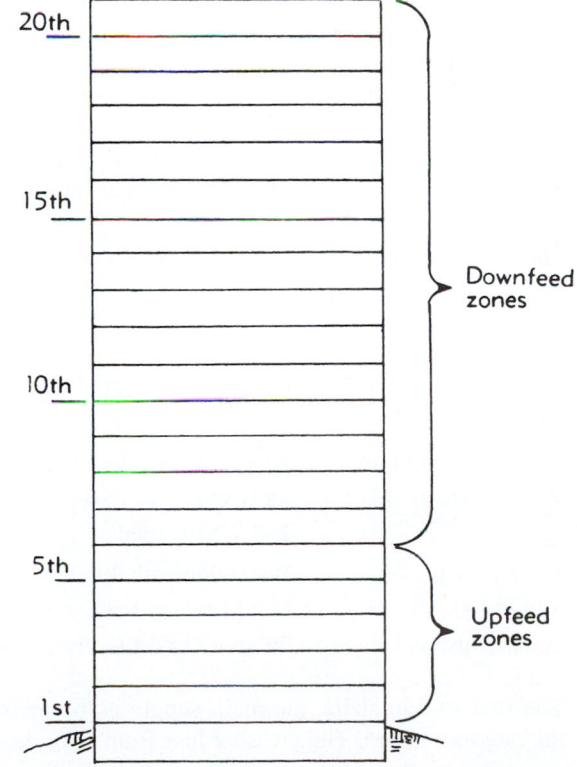

FIGURE 13.22 Water zoning in apartment building design example is comprised of upfeed and downfeed zones.

2. Next, the WSFU demand for each floor and the entire apartment building is found:

Total fixture units per floor:

2 bathroom groups, flush valve, private, 8 WSFU · 2 = 16 WSFU

2 kitchen sinks, 2 WSFU · 2 = 4 WSFU

20 WSFU per floor

Apartment fixture units on riser:

20 WSFU (per floor) · 4 (floors) = 80 WSFU

Additional fixture units:

2 hose bibbs (first floor) 2 WSFU · 2 = 4 WSFU

Total fixture units on this riser:

80 WSFU (apartments) + 4 WSFU (hose bibbs) = 84 WSFU

3. The demand for water in gallons per minute can now be determined by using the Table 13.9 for various WSFU loads.

$$84 \text{ WSFU} = 40 \text{ gpm}$$
$$84 \text{ WSFU} = 2.52 \text{ L/s}$$

Upfeed System Design Example

Water supply pipes must be of sufficient size to provide adequate pressure to all fixtures in the system at a reasonable cost. Selection of economical sizes for the piping and the meter is based on the total demand for water (see earlier Demand Load Example); this process must take into account the available water pressure (street main pressure), the pressure loss from static head (pressure loss to raise the water up the pipe), the pressure loss from friction in piping and fittings ($\Delta P_{friction}$), and the pressure required to operate the fixture requiring the most pressure on the top floor.

The main pressure in the street must be adequate to supply:

required fixture flow pressure + static head (ΔP_{static})
+ friction in pipe and meter ($\Delta P_{friction}$)

The accumulated information to this point (from above) is as follows:

4-story apartment building	20 WSFU per floor (per riser)
3 in (75 mm) service main	84 WSFU (per riser), 168 WSFU total
Street main pressure: 50 psi (345 kPa)	Water demand: 40 gpm (2.52 L/s) per riser
Predominate flush valve	59 gpm (3.72 L/s) for main

1. The first step in sizing the main supply is to determine the pressure losses. The pressure loss from static head is computed by multiplying 0.434 psi per ft (9.8 kPa per m) times the vertical distance of the riser from the service main into the building to the top branch water line (Figure 13.21):

Static head loss, $\Delta P_{static} = 0.434 \text{ psi} \cdot 40 \text{ ft} = 17.36 \text{ psi}$

In metric (SI) units:

Static head loss, $\Delta P_{static} = 9.8 \text{ kPa} \cdot 12.2 \text{ m} = 120 \text{ kPa}$

2. The residual pressure required at the most remote fixture on the top floor (e.g., the fixture requiring the most pressure) must be determined. The design should also leave some excess pressure so that more than one fixture can be operated properly at the same time:

Water closet: 15 psi (from Table 13.1)

In metric (SI) units:

Water closet: 100 kPa (from Table 13.1)

3. The pressure loss from friction in the main and riser pipes, the fittings, and the meter must be found next. Add the static head and the pressure required for the most remote fixture, then subtract the total from the available street main pressure to find the pressure left, which is the maximum amount that can be lost to friction:

Street main pressure:	50.00 psi
Static head + fixture pressure (17.36 + 15)	−32.36 psi
Pressure left to overcome friction:	17.64 psi (residual pressure available)

In metric (SI) units:

Street main pressure	345 kPa
Static head + fixture pressure (120 + 100)	−220 kPa
Pressure left to overcome friction	125 kPa (residual pressure available)

Residual pressure available for the water closet is acceptable.

4. Sizing the piping using the equal friction design method (see Method 4: Equal Friction Design Method) is a matter of trial and error. The process involves first selecting a pipe size for the building main, which runs from the water system to the riser(s), and then determining the friction loss for the pipe used from friction loss charts (e.g., Figures 13.6 and 13.7). The chart used will depend on the type of pipe used.

From the friction loss chart (Figure 13.6 for Type L copper), the first pipe size servicing 59 gpm is 4 in. Because this is much larger than the street main, keep moving to the right to the 2½ in pipe. Where the horizontal line from 59 gpm touches the line representing 2½ in, draw a vertical line to the bottom (or top), and the friction loss reads about 1.26 psi per 100 ft. For the other pipes, the following friction readings are taken:

$$2\tfrac{1}{2} \text{ in pipe} = 1.26 \text{ psi per } 100 \text{ ft}$$
$$2 \text{ in pipe} = 3.20 \text{ psi per } 100 \text{ ft}$$
$$1\tfrac{1}{2} \text{ in pipe} = 13.0 \text{ psi per } 100 \text{ ft}$$
$$1\tfrac{1}{4} \text{ in pipe} = 28.0 \text{ psi per } 100 \text{ ft}$$
$$1 \text{ in pipe} = 74.0 \text{ psi per } 100 \text{ ft}$$

In metric (SI) units from the friction loss chart (Figure 13.7 for Type L copper), going from the left side to the right, the first pipe size servicing 3.72 L/s is 100 mm. Because this is much larger than the street main, keep moving to the right to a smaller pipe (65 mm). Where the horizontal line from 3.72 L/s touches the line representing 65 mm, draw a vertical line to the top and the friction loss reads about 28.4 kPa per 100 m. For the other pipes, the following friction readings are taken:

$$65 \text{ mm pipe} = 28.4 \text{ kPa per } 100 \text{ m}$$
$$50 \text{ mm pipe} = 72.3 \text{ kPa per } 100 \text{ m}$$
$$40 \text{ mm pipe} = 293.8 \text{ kPa per } 100 \text{ m}$$
$$32 \text{ mm pipe} = 632.8 \text{ kPa per } 100 \text{ m}$$
$$25 \text{ mm pipe} = 167.2 \text{ kPa per } 100 \text{ m}$$

Because all friction losses are given per 100 ft (100 m) of pipe length, the distance from the street main (or point of service, such as a storage tank) to the base of the riser must be checked on the plot plan. Referring to Appendix A, the horizontal building main distance is estimated at 60 ft. So, for example, the friction loss in the building main pipe will be 60/100, 3/5, 0.6, or 60% of the values recorded. Friction losses for 60 ft of pipe are as follows:

$$2\tfrac{1}{2} \text{ in pipe} = 1.26 \text{ psi per } 100 \text{ ft} = 1.26 \text{ psi } (0.6) = 0.76 \text{ psi}$$
$$2 \text{ in pipe} = 3.20 \text{ psi per } 100 \text{ ft} = 3.20 \text{ psi } (0.6) = 1.92 \text{ psi}$$
$$1\tfrac{1}{2} \text{ in pipe} = 13.0 \text{ psi per } 100 \text{ ft} = 13.0 \text{ psi } (0.6) = 7.8 \text{ psi}$$
$$1\tfrac{1}{4} \text{ in pipe} = 28 \text{ psi per } 100 \text{ ft} = 28 \text{ psi } (0.6) = 16.8 \text{ psi}$$
$$1 \text{ in pipe} = 74 \text{ psi per } 100 \text{ ft} = 74 \text{ psi } (0.6) = 44.4 \text{ psi}$$

By observation, the 1 in pipe is eliminated because its friction loss is slightly greater than the pressure left in the calculations (17.64 psi). The 1¼ in, 1½ in, and 2½ in pipe are all possibilities at this point.

In metric (SI) units, referring to Appendix A, the horizontal building main distance is estimated at 18.3 m. So, for this example, the friction loss in the building main pipe will be 18.3/100 or 18.3% of the values recorded. Friction losses for 18.3 m of pipe are as follows:

$$65 \text{ mm pipe} = 28.4 \text{ kPa per } 100 \text{ m} = 28.4 \,(0.183)$$
$$= 5.2 \text{ kPa}$$
$$50 \text{ mm pipe} = 72.3 \text{ kPa per } 100 \text{ m} = 72.3 \,(0.183)$$
$$= 13.2 \text{ kPa}$$
$$40 \text{ mm pipe} = 293.8 \text{ kPa per } 100 \text{ m} = 293.8 \,(0.183)$$
$$= 53.8 \text{ kPa}$$

$$32 \text{ mm pipe} = 632.8 \text{ kPa per } 100 \text{ m} = 632.8 \,(0.183)$$
$$= 115.8 \text{ kPa}$$
$$25 \text{ mm pipe} = 167.2 \text{ kPa per } 100 \text{ m} = 167.2 \,(0.183)$$
$$= 306 \text{ kPa}$$

By observation, the 25 mm pipe is eliminated because its friction loss is higher than the pressure left for friction loss in the calculations (125 kPa). All of the other pipes are still possibilities at this time.

5. The next step is to determine the pressure loss as the water passes through the meter (if one is required; if not, this step is eliminated). The chart in Figure 13.8 gives the pressure loss for various meter sizes with the flow in gpm along the bottom and the pressure loss on the left. When the main size is 1½ in or greater, the most common main-meter selection will have the meter one size smaller than the main.

For this problem, locate the gpm along the bottom—in this case, 59 gpm—and go vertically to the first diagonal line (3 in meter); at this point, move horizontally to the pressure loss reading (0.8 psi). Referring to Figure 13.8, the pressure loss for a 2 in pipe meter is about 3.8 psi.

$$3 \text{ in meter} = 0.8 \text{ psi}$$
$$2 \text{ in meter} = 3.8 \text{ psi}$$
$$1\tfrac{1}{2} \text{ in meter} = 9.7 \text{ psi}$$

In metric (SI) units:

$$75 \text{ mm meter} = 5.56 \text{ kPa}$$
$$50 \text{ mm meter} = 26.2 \text{ kPa}$$
$$40 \text{ mm meter} = 66.9 \text{ kPa}$$

Totaling the pressure for main and meter:

$$2\tfrac{1}{2} \text{ in pipe and 2 in meter} = 0.76 + 3.8 = 4.56 \text{ psi}$$
$$2\tfrac{1}{2} \text{ in pipe and } 1\tfrac{1}{2} \text{ in meter} = 0.76 + 9.7 = 10.46 \text{ psi}$$
$$2 \text{ in pipe and 2 in meter} = 1.92 + 3.8 = 5.72 \text{ psi}$$
$$2 \text{ in pipe and } 1\tfrac{1}{2} \text{ in meter} = 1.92 + 9.7 = 11.62 \text{ psi}$$
$$1\tfrac{1}{2} \text{ in pipe and } 1\tfrac{1}{2} \text{ in meter} = 16.8 + 9.7 = 26.5 \text{ psi}$$

In metric (SI) units, totaling the pressure for main and meter:

$$65 \text{ mm pipe and 50 mm meter} = 5.2 + 26.2 = 31.4 \text{ kPa}$$
$$65 \text{ mm pipe and 40 mm meter} = 5.2 + 66.9 = 72.1 \text{ kPa}$$
$$50 \text{ mm pipe and 50 mm meter} = 13.2 + 26.2 = 39.4 \text{ kPa}$$
$$50 \text{ mm pipe and 40 mm meter} = 13.2 + 66.9 = 80.1 \text{ kPa}$$
$$40 \text{ mm pipe and 40 mm meter} = 293.8 + 66.9 = 360.7 \text{ kPa}$$

Referring back, there is a pressure of 17.64 psi left to overcome pressure losses from friction in the main, meter, and riser. The totals above are pressure losses for the main and meter; the riser pressure loss must still be calculated.

Because the most economical solution involves the use of the smallest pipe sizes, which will provide an adequate flow of water throughout the system, at this point a preliminary selection of 2½ in pipe and 1½ in meter is made. This means that of the 17.64 psi that are left to overcome friction, 10.46 psi will be used in the 2½ in pipe and the 1½ in meter. This leaves 7.18 psi as the maximum friction loss in the riser.

> 17.64 psi available for friction loss
>
> −10.46 psi friction loss in 2½ in main and 1½ in meter
>
> 7.18 psi available for riser friction loss

Using a 2 in main and 1½ in meter would leave (17.64 − 11.62) = 6.02 psi for riser friction loss. Using the 2½ in main and 2 in meter would leave (17.64 − 4.56) = 13.08.

In metric (SI) units:

> 125 kPa available friction loss
>
> −72.1 kPa friction loss in 65 mm pipe and 40 mm meter
>
> 52.9 kPa available for riser friction loss

Using a 50 mm main and a 40 mm meter would leave (125 − 80.1) = 44.9 kPa for friction loss. Using the 65 mm main and a 50 mm meter would leave (125 − 31.4) = 93.6 kPa.

6. The next step is to select the possible pipe sizes for the riser. Using the friction-loss pipe chart (Figure 13.6), find the 40 gpm (2.52 L/s) flow on the side; the following friction-loss readings are then taken:

> 2 in pipe = 1.6 psi per 100 ft
>
> 1½ in pipe = 6.0 psi per 100 ft
>
> 1¼ in pipe = 13 psi per 100 ft
>
> 1 in pipe = 34 psi per 100 ft

In metric (SI) units:

> 50 mm pipe = 36.2 kPa per 100 m
>
> 40 mm pipe = 135.6 kPa per 100 m
>
> 32 mm pipe = 293.8 kPa per 100 m
>
> 25 mm pipe = 768.4 kPa per 100 m

Once again, note that the friction losses are given per 100 ft or 100 m of pipe length. This means that the vertical distance from the main to the highest and most remote fixture must be estimated. On some projects, this may mean a meeting with the architect to ascertain floor-to-floor height.

For this apartment building, with floor-to-floor heights of approximately 10 ft-0 in (3.0 m), the overall riser height will be approximately 40 ft (12.2 m). In such piping there is an additional friction loss for fittings (such as valves and tees). When the entire system has been designed and all pipes sized and fittings located, it is possible to determine the friction loss for the fittings. An example of how to calculate fittings is provided in Pressure Losses from Friction. Because, during this stage of design, the fittings have not been determined, it is common practice to allow an additional 50% of the piping length to account for the friction

loss in the fittings. In this case, the pipe length is 40 ft and, adding 50% for friction loss in the fittings, the total equivalent length of piping is 40 + 20 = 60 ft.

The friction loss for riser pipes and fittings is calculated next. Using the friction loss for various pipe sizes per 100 ft (previously found), convert the friction loss to the equivalent length. In this case, 60 ft or 60/100, 0.60, or 60% of the values recorded.

> 2 in pipe = 1.6 psi per 100 ft = 1.6 psi (0.60) = 0.96 psi
>
> 1½ in pipe = 6.0 psi per 100 ft = 6.0 psi (0.60) = 3.60 psi
>
> 1¼ in pipe = 13 psi per 100 ft = 13 psi (0.60) = 20.4 psi

In metric (SI) units:

> 50 mm pipe = 36.2 kPa per 100 m = 36.2 (.183) = 6.62 kPa
>
> 40 mm pipe = 135.6 kPa per 100 m = 135.6 (.183) = 24.8 kPa
>
> 32 mm pipe = 293.8 kPa per 100 m = 293.8 (.183) = 53.8 kPa
>
> 25 mm pipe = 768.4 kPa per 100 m = 768.4 (.183) = 140.62 kPa

These friction losses are now compared with the preliminary main size selections, and a final selection for main and riser is made. The choices are as follows:

a. Using a 2½ in main and 1½ in meter that has 10.46 psi friction loss, leaving 7.18 psi for riser friction loss, the choice is limited to 1½ in and 2 in. riser pipe.

b. Using a 2 in main and 1½ in meter that has 11.62 psi friction loss, leaving 6.02 psi for riser friction loss, the riser choices are only the 1½ in, and 2 in. pipe. The final selection may also depend on whether there is any possibility of adding any length to the riser (by adding more stories to the building, making it taller).

The final selection would depend on other variables; in particular, whether the main will serve any other building and whether there is any chance of further development on the site (perhaps another apartment building). There is no single correct answer; in this case, the solution that seems most economical is:

> 2½ in main and 1½ in meter with 1½ in riser
>
> 50 mm main and 40 mm meter with a 40 mm riser

At this point, the information on the upfeed system should be tabulated. The required information includes the fixture units per floor, the accumulated fixture units (which are the units that the riser must serve at any given point), the demand flow and the pipe size. The tabulated information for this problem is shown in Tables 13.25 through 13.27.

TABLE 13.25   UPFEED DATA.

Floor	Accumulated w.s.f.u.	Demand load (gpm)	Friction loss (psi per 100[1])	Pipe size
		WATER SUPPLY		
		UPFEED ZONE, APT. BLDG. PROJ. #20–1		
1	84	40		
2	60	33		
3	40	25		
4	20	14		

Floor	Accumulated w.s.f.u.	Demand load (L/s)	Friction loss (kPa per 100 m)	Pipe size (mm)
		SI UNITS		
		WATER SUPPLY		
		UPFEED ZONE, APT. BLDG. PROJ. #20–1		
1	84	2.52		
2	60	2.08		
3	40	1.58		
4	20	0.88		

TABLE 13.26   UPFEED DATA.

Floor	Accumulated w.s.f.u.	Demand load (gpm)	Friction loss (psi per 100[1])	Pipe size
		WATER SUPPLY		
		UPFEED ZONE, APT. BLDG. PROJ. #20–1		
1	84	40	1.97	1½"
2	60	33		1¼"
3	40	25		1¼"
4	20	14		1"

Floor	Accumulated w.s.f.u.	Demand load (L/s)	Friction loss (kPa per 100 m)	Pipe size
		SI UNITS		
		WATER SUPPLY		
		UPFEED ZONE, APT. BLDG. PROJ. #20–1		
1	84	2.52	289	40
2	60	2.08	289	32
3	40	1.58	289	32
4	20	0.88	289	25

TABLE 13.27   TABULATED UPFEED DATA (PREDOMINATELY FLUSH VALUE).

Floor	Accumulated w.s.f.u.	Demand load (gpm)	Friction loss (psi per 100[1])	Pipe size
		WATER SUPPLY		
		UPFEED ZONE, FLOORS 1-5 PROJ. #20–2		
1	235	97		
2	188	89		
3	141	78		
4	94	66		
4	47	50		

Floor	Accumulated w.s.f.u.	Demand load (L/s)	Friction loss (kPa per 100 m)	Pipe size
		SI UNITS		
		WATER SUPPLY		
		UPFEED ZONE, FLOORS 1-5 PROJ. #20–2		
1	235	6.2		
2	188	5.6		
3	141	4.9		
4	94	4.2		
5	47	3.2		

Now the riser for the upper floors can be sized. First, the accumulated fixture units are computed:

2nd floor   20 WSFU   60 accumulated WSFU
3rd floor   30 WSFU   40 accumulated WSFU
4th floor   20 WSFU   20 accumulated WSFU

Next, the demand load is determined for each floor by using Table 13.9 and the accumulated fixture units for each floor (use line 2, predominant flush tank):

2nd floor   60 accumulated WSFU   33 gpm
3rd floor   40 accumulated WSFU   25 gpm
4th floor   20 accumulated WSFU   14 gpm

The riser pipe sizes are now selected by using the demand load (gpm) for each floor, the pressure available for friction loss in the riser, and the chart in Figure 13.6.

2nd floor   33 gpm   1½ in pipe
3rd floor   25 gpm   1½ in pipe
4th floor   14 gpm   1 in pipe

In metric (SI) units:

2nd floor   2.08 kPa   32 mm pipe
3rd floor   1.58 kPa   32 mm pipe
4th floor   0.88 kPa   25 mm pipe

Once all of this information is tabulated as shown in Table 13.26, the design is complete.

Downfeed System Design Example

For all tall buildings, a downfeed system must be used (Figure 13.5).

1. The first step in the design is to list all the information about the project:

20-story public building
Floors 1–18:

50 psi at community main
Floors 19 and 20:

3 flush valve water closets	3 flush tank water closets
2 urinals, flush valve	2 urinals, flush tank
2 lavatories	2 lavatories
1 service sink	1 service sink

Floor-to-floor height: 11 ft (3.4 m)

Distance from community main to riser: 40 ft (12.2 m)

2. To determine the main, meter, and riser sizes for the system, find the WSFU per floor and the total WSFU in the building.

For Floors 1–18:

3 flush valve water closets	10 WSFU each = 30
2 urinals, flush valve (¾)	5 WSFU each = 10
2 lavatories	2 WSFU each = 4
1 service sink	3 WSFU each = 3
	47 WSFU per floor

47 WSFU (per floor) · 18 · floors = 846 WSFU
for Floors 1–18

For Floors 19 and 20:

3 flush tank water closets	5 WSFU each = 15
2 urinals, flush tank	3 WSFU each = 6
2 lavatories	2 WSFU each = 4
1 service sink	3 WSFU each = 3
	28 WSFU per floor

28 WSFU (per floor) · 2 (floors) = 56 WSFU
for Floors 19 and 20

Total fixture units:

846 WSFU (Floors 1–18) + 56 WSFU (Floors 19 and 20) = 902 WSFU total

3. The next step is to determine the demand load for water in gpm (from Table 13.9).

The system is predominantly flush valve, and the approximate demand load is:

$$902 \text{ WSFU} = 201 \text{ gpm}$$
$$902 \text{ WSFU} = 12.68 \text{ L/s}$$

4. Because the bottom floors of the building will be serviced by upfeed zones (Figure 13.5 and Figure 13.22) and

the upper floors by downfeed zones, a preliminary decision must be made on how many floors each zone will supply and the number of zones required.

The upfeed zone is limited to a practical maximum height of approximately 60 ft (20 m). With a floor-to-floor height given as 11 ft (3.4 m), the upfeed system will serve five to six floors (stories), depending on the depth of the water main (a deep main would limit the zone to five floors). Because on this project the main is shallow (about 4 ft [1.2 m] below ground level), it will be possible to have an upfeed zone of six floors, if necessary.

5. Downfeed zones also have maximum heights (distances they can serve) without using special pressure-reducing and control valves. In this design, the flush tank water closet on the top floor requires a water pressure of 15 psi to operate (Table 13.1). So, if 15 psi is required and the water pressure increases 0.434 psi per foot, then 15 psi · 0.434 psi = 34.6 ft of vertical distance would be required between the tank and the top fixture to provide the pressure required. (This distance would be even greater if flush valve water closets were used on the top floors, as they may require more pressure than the flush tank. In addition to the 34.6 ft of distance required to operate the top fixture, a small additional height would be required to overcome the friction loss in the pipes and fittings (an additional 5 ft is usually sufficient unless very small risers are used).

In metric units for this design, the flush tank water closet on the top floor requires a water pressure of 100 kPa to operate (Table 13.1). So, if 100 kPa is required and the water pressure increases 9.8 kPa per meter, then 100 kPa · 9.8 kPa = 10.2 m of vertical distance that would be required between the tank and the top of the fixture to provide the pressure required. To overcome the friction loss in the pipes and fittings will require an additional height of about 2 m.

6. The zoning for the downfeed systems is determined next. Because the water pressure is increasing as it goes down the downfeed riser, the vertical distance that a zone can service is limited to the maximum amount of pressure that can be put on the lowest fixture and still have it operate properly. This maximum pressure varies with the type and manufacturer of the fixtures being used; the allowable pressure may range from about 45 to 60 psi (275 to 415 kPa).

In this design, a pressure of 55 psi will be assumed to be the preferred maximum. Thus, the downfeed zone can be 55 psi divided by 0.434 psi per foot, or about 126 ft. Because there will be some pressure loss from friction in the riser pipe, the zone may be slightly longer computed.

In review, then, the upfeed portion of the system can easily service the bottom five floors. This leaves 15 floors to be serviced by the downfeed system. With a floor-to-floor height of 11 ft, this is a vertical distance of about

165 ft, which cannot be served by one downfeed zone. Two downfeed zones will easily provide the water service required to the 15 floors that the upfeed zone (Figure 13.22) does not serve.

Note: In reviewing the schematic downfeed water supply system (Figure 13.5), note that the house tank used to provide water for Zone 2 must be placed high enough above the fixtures it is to supply that there will be sufficient pressure for the fixtures to operate efficiently. In this case, 20 psi will be required to operate the flush valve water closets, calling for a height of 20 psi $\cdot$ 0.434 psi per ft = 46 ft, plus a short distance to overcome friction in the riser (totaling about 50 to 52 ft, depending on the final riser size selected).

In metric units for this design, 380 kPa will be assumed to be the preferred maximum. This means that the downfeed zone can be 380 kPa divided by 9.8 kPa per meter, or about 38.8 m. As there will be some pressure loss from friction in the riser pipe, the zone may be slightly longer if required.

In review, then, the upper portion of the system can easily service the bottom five floors. This leaves 15 floors to be serviced by the downfeed system. With a floor-to-floor height of 3.4 m, this is a vertical distance of about 51 m, which cannot be served by one downfeed zone. Two downfeed zones will easily provide the water service required.

Note: In reviewing the schematic downfeed water supply system (Figure 13.5), note that the house tank used to provide water for Zone 2 must be placed high enough about the fixtures it is to supply to provide sufficient pressure for the fixtures to operate efficiently. In this design 140 kPa will be required to operate the flush valve water closets, calling for a height of (140 kPa divided by the static head of 9.8 kPa per meter) 14.3 m plus a short distance to overcome friction in the riser (totaling about 16 m), depending on the final riser size selected.

7. The accumulated information for the upfeed zone (Zone 1) is now tabulated (Table 13.27), with each floor listed along with its fixture units, the accumulated fixture units (remember that the water at the bottom floor must service the floors above it, so the fixture units are the accumulated total of what must be serviced), the demand flow (gpm) for the accumulated fixture units, and the pipe size selected.

8. Static head loss is determined:

From Figure 13.5, the vertical height of the riser from the top of the main to the top of the upfeed zone is about 55 ft.

Static head loss, $\Delta P_{static} = 0.434$ psi $\cdot$ 55 ft = 23.87 psi

Pressure required to operate fixture at the top (5th) floor of the upfeed zone

Flush valve water closet: 20 psi

Pressure left to overcome friction:

$$55 \text{ psi} - (23.87 \text{ psi} + 20 \text{ psi})$$
$$= 55 \text{ psi} - 43.87 \text{ psi} = 11.13 \text{ psi}$$

In metric (SI) units, from Figure 13.5, the vertical height of the riser from the top of the main to the top of the upfeed zone is about 16.8 m.

Static head loss, $\Delta P_{static} = 9.8$ kPa $\cdot$ 16.8 m = 138 kPa

Pressure to operate fixture at the top floor of the upfeed zone:

Flush valve water closet: 138 kPa

Pressure remaining to overcome friction:

$$380 - (165 + 138) = 380 - 303 = 77 \text{ kPa}$$

9. At this point, the service main should be selected for 902 accumulated water supply fixture units with a demand load of 201 gpm. Referring to Figure 13.6, the preliminary main pipe selections are:

5 in main = 0.35 psi per 100 ft
4 in main = 1.1 psi per 100 ft
3½ in main = 2.1 psi per 100 ft
3 in main = 3.9 psi per 100 ft
2½ in main = 9.9 psi per 100 ft

In metric (SI) units:

125 mm = 7.9 kPa per 100 m
100 mm = 24.9 kPa per 100 m
75 mm = 88.1 kPa per 100 m
65 mm = 224 kPa per 100 m

Because the main distance is given as 40 ft, the friction loss for each pipe would be:

5 in main   = 0.35 psi per 100 ft = 0.35 (0.4) = 0.14 psi
4 in main   = 1.1 psi per 100 ft   = 1.1 (0.4)   = 0.44 psi
3½ in main = 2.1 psi per 100 ft   = 2.1 (0.4)   = 0.84 psi
3 in main   = 3.9 psi per 100 ft   = 3.9 (0.4)   = 1.56 psi
2½ in main = 9.9 psi per 100 ft   = 9.9 (0.4)   = 3.96 psi

In metric (SI) units, because the main distance is given as 12.2 m, the friction loss for each pipe size would be:

125 mm main = 7.9 kPa per 100 m = 7.9 (.122) = .96 kPa
100 mm main = 24.9 kPa per 100 m = 24.9 (.122) = 3.0 kPa
75 mm main = 88.1 kPa per 100 m = 88.1 (.122) = 10.7 kPa
65 mm main = 224 kPa per 100 m = 224 (.122) = 27.3 kPa

10. The friction loss in the meter (if required) is checked next (Figure 13.8).

4 in meter, 201 gpm: 4.1 psi (a 3 in meter requires a 5 in pipe)

3 in meter, 201 gpm: 10.1 psi (a 3 in meter requires a 4 in pipe)

In metric (SI) units:

100 mm meter, 12.68 L/s: 28.3 kPa (a 100 mm meter requires a 125 mm pipe)

75 mm meter, 12.68 L/s: 69.6 kPa (a 75 mm meter requires a 100 mm pipe)

The 3 in main will not work, and the 4 in and 5 in mains are next considered. (It should be noted here that 3½ in pipe is not readily available in many areas, and in such a situation, the 4 in pipe would be used.) At this point, a preliminary selection of a 5 in main and 4 in meter is made and tabulated.

Pressure available for friction loss	11.13 psi
Friction loss in 5 in main and 4 in meter (0.14 psi + 4.1 psi)	−4.28 psi
Pressure available for riser friction loss	6.85 psi

In metric (SI) units, the 65 mm main will not work. The 100 mm and 125 mm are considered next. At this point, a preliminary selection of a 125 mm main and a 100 mm meter is made and tabulated.

Pressure available for friction loss	74 kPa
Friction loss in 125 mm main and 100 mm meter (0.96 + 28.3)	29.3 kPa
Pressure available for riser friction loss	44.7 kPa

11. Before the riser selections can be made, the water supply fixture units and demand load (in gpm) for this upfeed zone must be calculated and put in the tabulated form.

Fixture units, Floors 1–5, were previously calculated at 47 WSFUs per floor:

$$47 \text{ WSFUs (per floor)} \cdot 5 \text{ (floors)}$$
$$= 235 \text{ (accumulated) WSFUs}$$

Demand load (Table 13.9) predominant flush valve: 97 gpm.

Demand load (Table 13.9) predominant flush value: 6.2 L/s

Available pressure left to overcome friction loss will be converted into loss per 100 ft for equivalent length:

$6.85 \cdot 100 = 685$ ft
$55 + 22.5 = 77.5 = 8.84$ psi loss per 100 ft (maximum)

In metric (SI) units:

$$44.7 \cdot 100 = 4470 = 177 \text{ kPa}$$
$$16.8 + 8.4 = 25.2 \text{ m}$$

12. Riser pipe selection is as follows:

2 in pipe = 7.7 psi per 100 ft
2½ in pipe = 2.9 psi per 100 ft
3 in pipe = 1.3 psi per 100 ft

In metric (SI) units:

50 mm pipe = 173.5 kPa per 100 m
65 mm pipe = 65.5 kPa per 100 m
75 mm pipe = 29.4 kPa per 100 m

The size of the riser may be decreased in the upper floors, as the gpm decreases, as long as the friction loss for each segment does not increase.

While the 2 in riser could be used, the 2½ in pipe is selected to allow some extra pressure at the top fixtures. Now, complete the tabulations for the upfeed zone. (Later, once the design is fully laid out, the actual equivalent length would be rechecked.)

In metric (SI) units, while the 50 mm riser could be used, the 65 mm pipe is selected to allow some extra pressure at the top fixtures.

Once all of this information is tabulated, the design is complete.

13.9 PUMPS

A *pump* is a mechanical device used to move a fluid by converting mechanical energy to pressure energy called *head*. In buildings, pumps are used to circulate or pump water in domestic hot and cold water systems, hydronic (hot water) and chilled water distribution systems, feed water systems for boilers and chillers, condensate return systems, well systems, wastewater treatment systems, sump installations, and other process piping systems.

Types of Pumps

There are basically two types of pumps used in building systems: positive displacement pumps and centrifugal pumps. Types of pumps are discussed in the sections that follow. Centrifugal pumps are most common in building systems and are discussed in more detail.

Positive Displacement Pumps

A *positive displacement pump* has an expanding cavity on the suction side of the pump and a decreasing cavity on the discharge side. An example of a positive displacement pump is the human heart. Liquid is allowed to flow into the pump as the cavity on the suction side expands. The liquid is forced out of the discharge as the cavity collapses. This principle applies to all types of positive displacement pumps including piston, rotary lobe, gear within a gear, diaphragm, or screw pumps.

Centrifugal Pumps

A centrifugal pump is made up of an outer *casing* that has a rotating wheel-like component called an *impeller* inside a stationary cavity created by the casing and called the *volute*. Fluid is drawn into the inlet port at the center of the rotating impeller. As the impeller spins, vanes on the impeller force the fluid to rotate, thrusting it outward radially and thus moving the fluid from inlet to outlet under its own momentum. As it does this, it creates a vacuum, which draws more fluid into the inlet.

Manufacturers will offer different size impellers that fit inside a particular casing to offer many different pump capabilities. Thus, the impeller can be changed out to change pump flow and pressure characteristics. This reduces manufacturing and storage costs associated with pump casings. However, a pump's efficiency typically decreases as impeller size is decreased for a particular pump casing.

The number of impellers determines the number of *stages* of a centrifugal pump. A *single-stage pump* has just one impeller and is better for low head service. A *two-stage pump* has two impellers mounted in series for medium head service. A *multistage pump* has three or more impellers mounted in series for high head service such as in deep well pumps.

Centrifugal pumps can operate at different speeds. The faster the impeller rotates, the faster the fluid movement and the stronger its force.

Pump Drives

Pumps are driven by a *drive,* usually an electric motor in building plumbing systems. A fuel-powered engine can also drive a pump, but this type of drive is typically reserved for temporary or emergency applications. The pump drive may be *close coupled* such that it is on the same shaft as the pump impeller; that is, the drive and pump are directly connected. This design approach is less costly and is acceptable on small pumps where vibration and noise are not problems. *Flexible coupled* pump-drive configurations minimize vibration and wear between the pump and pump drive, especially for large pumps. Pumps used in residential and small commercial installations are typically small enough that they can be close coupled. Typically large pumps used in building plumbing and heating and cooling systems need to be flexible coupled.

The speed of a pump is measured in revolutions per minute (rpm). *Constant speed drives* are drives designed to operate at a specific speed (e.g., 1750, 3500 rpm). These drives are usually sized to handle the largest loads so they are typically oversized for normal operating loads. They are economical in terms of initial cost but not economical in lifetime operating cost.

There are two types of speed control in pumps: multispeed drives and variable speed drives. *Multispeed drives* have separate speed settings (e.g., high, medium, and low) so they can be adjusted to control speed and, thus, pump flow rate. *Variable-speed drives* provide speed control over a continuous range. Variable speed drives control pump speed by changing the speed of the driver and thus flow rate. These drives are useful in meeting the needs of installations with highly variable loads, In such cases, the use of multiple pumps (e.g., two or more pumps), multispeed drives, or variable-speed drives often improve system performance over the range of operating conditions. They are the most expensive type of drive in initial cost but the least costly to operate.

Pump-drive installations need to be secured well for vibration and sound control. A heavy foundation or base is required for large pumps. The weight of a foundation should be 3 to 5 times the combined weight of the pump and drive. Isolation, by vibration damping, is another solution for vibration and sound control.

Pumping Configurations

For some pumping installations it may be necessary to use multiple pumps to meet design requirements. In systems with highly variable loads, pumps that are sized to handle the largest loads may be oversized for normal operating loads. In such cases, the use of multiple pumps improves system performance over the range of operating conditions. Multiple pumps can be staged so both pumps operate only during periods of full demand. A single pump operates when demand is low. Multiple pumps can be configured in parallel or series.

Pumping in Parallel

Parallel pumping entails installing two pumps side by side in a piping system. When installing two identical pumps in parallel, the combined flow rate will be less than double. Pumps with different flow rates can be installed in parallel and configured such that the small pump operates during periods of low to average demand while the larger pump operates during periods of high demand. One major benefit of parallel pumping is the high level of standby capacity provided by single-pump operation. When one pump is out of operation for maintenance or repair, the other pump continues to pump water through the installation.

Pumping in Series

Series pumping involves installing two pumps one in line with the other in a single pipe in a piping system. Pumps in series double the head at the same flow condition point. One pump discharge is piped into the suction of the second pump, producing twice the head capability of each pump separately. The second pump must be capable of operating at the higher suction pressure that is produced by pump number one. This mode of operation is a very cost-effective way of overcoming high discharge heads when the flow requirement remains the same.

Pump Performance

The fundamental performance considerations of centrifugal pumps are capacity, total dynamic head, brake horsepower, pump efficiency, and pump speed. These characteristics and related terms are discussed in the following sections.

Pump Capacity

Pump capacity (Q_{pump}) is the flow rate at which liquid is moved or pushed by a pump to the desired point in the system. It is commonly measured in either gallons per minute (gpm), liters per minute (L/min), or cubic meters per hour (m³/hr). Pump capacity depends on pressure, temperature, and viscosity and specific gravity of the liquid being pumped; on the size and speed of the impeller; and the size and shape of the cavities between the impeller vanes.

For a pump with a specific impeller design and that is running at a certain speed, the basic factors that affect flow rate are the pressures at the pump inlet and outlet (i.e., pressure difference between the pump inlet and outlet). The larger the pressure difference a specific pump must overcome, the lower the flow rate of the liquid. Furthermore, the bigger the pressure difference a specific pump must overcome, the greater the horsepower required to drive the pump. Conversely, a small pressure difference means a higher flow rate and a smaller horsepower requirement.

A centrifugal pump is not limited to a single flow rate at a given speed. Its flow rate depends on the amount of head it encounters in the piping, so it operates at the point on the performance curve where its total dynamic head matches the resistance in the piping (i.e., the sum of the static head and friction head). As a general rule, an increase in flow rate causes a decrease in the total dynamic head the pump produces. Conversely, a decrease in flow rate increases the total dynamic head the pump produces.

Total Dynamic Head

Total dynamic head (ΔP_{TDH}) is a pressure difference, expressed in feet (meters), that represents the measurement of the height of a liquid column that the pump can generate from the kinetic energy imparted to the liquid. An illustration to clarify this unit of measure would be a pump shooting a stream of water straight up into the air; the maximum height attained by the water stream would be the total dynamic head produced by the pump.

To pump water at a specific flow rate, the total dynamic head generated by a pump must overcome the head of the system through which the pump is pushing water. Thus, total dynamic head developed by a pump must overcome a combination of the static head and friction head of the piping system:

$$\text{total dynamic head } (\Delta P_{TDH}) = \text{static head } (\Delta P_{static}) + \text{friction head } (\Delta P_{friction})$$

Static head (ΔP_{static}) is the actual vertical distance measured from the water level in the reservoir from which the pump draws the fluid to the highest point in the discharge piping. Friction head ($\Delta P_{friction}$) is the additional head created in the discharge system from resistance to flow within its piping components (e.g., friction from pipes and fittings). Computation of

static head and friction head was introduced previously in Section 13.3.

Head is a term that can have units of a length (feet or meters) and pressure (e.g., psi, Pa). The principal reason for using head in units of height (instead of units of pressure) to express a centrifugal pump's pumping capability is that the pressure from a pump will change if the specific gravity (weight) of the liquid changes, but head will not change. Because most centrifugal pumps can move various fluids, with different specific gravities, it is simpler to express the pump's head in feet (or meters). In systems with water, this change in specific gravity is generally not a consideration because most pump performance criteria is based on the specific weight of water.

Power and Hydraulic Efficiency

The work performed by a pump is related to the total dynamic head and weight of the liquid pumped in a specific time period. *Brake horsepower (BHP)* is the actual horsepower delivered to the pump shaft under stated operating conditions of the pump where horsepower is 550 foot/pounds per second. Pump output is the *water horsepower (WHP)* delivered by the pump. The brake horsepower of a pump is greater than the water horsepower because of the mechanical and hydraulic losses incurred in the pump. These are determined by manufacturer's testing of the pump under a specific flow rate and head.

The *water efficiency (η_{water})* of a pump (sometimes called the *hydraulic efficiency*) is expressed as a percentage of hydraulic horsepower to BHP over the recommended operating range of the indicated impeller size(s). It describes the change of centrifugal force (expressed as the velocity of the fluid) into pressure energy. The relationship between BHP, WHP, and η_{water} of a pump is:

$$BHP = WHP/\eta_{water}$$

Water efficiency of a specific pump/impeller configuration will vary with flow rate and head, but some configurations are better than others at a specific flow rate and head.

A pump operating at high efficiency uses less energy to operate than one operating at low efficiency. The *best efficiency point (BEP)* is the operating condition at which a pump most efficiently converts shaft power to flow. This is also the point where there is no radial deflection of the shaft caused by unequal hydraulic forces acting on the impeller. Operation of a pump well outside the BEP results in unstable conditions like vibration, erosion, and cavitation, which result in premature bearing and mechanical seal failures. An industry design standard used to approximate the lowest performance level of a pump is an efficiency of 55%. If possible, it is best to avoid operating any pump at efficiencies of less than 55%.

BHP, the power delivered to the pump shaft under stated operating conditions of the pump, is found by the following expression:

$$BHP = (Q \cdot \Delta P_{TDH} \cdot \text{s.g.})/(3960 \cdot \eta_{water})$$

As introduced in Chapter 12, the specific gravity (s.g.) of a fluid or solid is the ratio of the specific weight of the fluid or solid to the specific weight of water at a temperature of 39°F (4°C), the temperature at which water is most dense. The specific gravity of water is 1.0 at common plumbing system temperatures. The expression above implies that, under specific conditions, a fluid heavier than water (a fluid with a specific gravity greater than one) will require more horsepower to pump. Viscosity of a fluid (i.e., fluid thickness) will also affect BHP; a higher viscosity fluid requires increased BPH.

Net Positive Suction Head

Pumps are designed to pump only liquids, not vapors. Vaporization of the liquid being pumped should not occur at any condition of operation. As discussed earlier, vaporization results in cavitation, which is undesirable because it can obstruct the pump, impair performance and flow capacity, and damage the impeller and other pump components. To prevent cavitation, a pump always needs to have a sufficient amount of suction head present to prevent this vaporization at the lowest pressure point in the pump.

The inlet of a pump sucks liquid into the pump. The term used to describe this suction pressure of a pump is *net positive suction head (NPSH)*. It is expressed in height, as feet or meters. The minimum head required to prevent cavitation with a specific liquid at a specific flow rate is called *net positive suction head required (NPSHr)*. Pump manufacturers determine the NPSHr of a specific impeller design by conducting a test using water as the pumped fluid. NPSHr varies with speed and capacity of a specific pump. It increases as capacity increases because as the velocity of a liquid increases, the pressure (or head) decreases. The NPSHr shown on pump performance curves is for fresh water at 68°F (20°C), which will not relate directly to another type of fluid or combination of fluids being pumped.

In a pumping installation, the difference between the actual head of the liquid available (as measured at the pump's suction inlet) and the vapor pressure of that liquid is called the *net positive suction head available (NPSHa)*. This is the amount of NPSH available to the pump from the suction line. It is defined as atmospheric pressure + gauge pressure + static pressure − vapor pressure − friction loss in the suction piping.

NPSHa is directly related to *system* design, while NPSHr is directly related to *pump* design. NPSHr can be roughly thought of as the maximum height that a pump can be located above the reservoir from which it is pumping (assuming negligible vapor pressure and friction loss in the suction piping). Technically, for a pump to operate properly, the system's NPSHa must always be greater than the pump's NPSHr. So, if the system's NPSHa falls below the pump's NPSHr, then suction pressure is too great and the water being pumped will spontaneously vaporize in the pump casing, causing cavitation.

It is prudent that a margin of safety be provided between the pump's NPSHr cited by the manufacturer and the system's NPSHa at the operating conditions. It is normal practice to have at least 2 ft (0.6 m), and in some cases more, of extra NPSHa existing at the suction inlet to avoid cavitation problems under demands outside design conditions. Thus, excessively long runs of suction piping that potentially cause the system's NPSHa to be greater than the pump's NPSHr should be avoided in system design.

Pump Speed

Pump speed (N) is tied to the rotational speed of the drive (motor). It is expressed in revolutions per minute (rpm).

Pump Affinity Laws

Pump *affinity laws* are scientific relationships that describe changes in pump capacity, total dynamic head, and BHP when a change is made to pump speed, impeller diameter, or both. The affinity laws are only valid under conditions of constant efficiency. According to affinity laws:

Capacity (Q_{pump}) changes in direct proportion to impeller diameter (D) ratio, where the subscript 1 refers to the original impeller diameter and subscript 2 refers to the new diameter:

$$Q_{pump\,2} = Q_{pump\,1}(D_2/D_1)$$

Capacity (Q_{pump}) changes in direct proportion to the ratio of pump impeller speeds (N), where speed is in rpm, the subscript 1 refers to the original speed, and subscript 2 refers to the new speed:

$$Q_{pump\,2} = Q_{pump\,2}(N_2/N_1)$$

From these relationships it can be observed that capacity of a pump can be increased by increasing pump impeller diameter or pump speed. Manufacturers make several impeller sizes that fit in one size of pump casing, making it easier to change to another impeller size. Variable-speed drives (i.e., motors) control flow by changing pump speed.

Total dynamic head (ΔP_{TDH}) changes in direct proportion to the square of impeller diameter (D) ratio, where the subscript 1 refers to the original impeller diameter, and subscript 2 refers to the new diameter:

$$\Delta P_{TDH\,2} = \Delta P_{TDH\,1}(D_2/D_1)^2$$

Total dynamic head (ΔP_{TDH}) changes in direct proportion to the square of the ratio of pump impeller speeds (N), where the subscript 1 refers to the original speed, and subscript 2 refers to the new speed:

$$\Delta P_{TDH\,2} = \Delta P_{TDH\,1}(N_2/N_1)^2$$

Increasing impeller diameter or speed will increase a pump's total dynamic head. Because of the square function of the expression, even slight changes in impeller size or speed produce large changes in total dynamic head.

BHP changes in direct proportion to the cube of the impeller diameter ratio (D_2/D_1), where the subscript 1 refers to the original impeller diameter, and subscript 2 refers to the new diameter:

$$BHP_2 = BHP_1(D_2/D_1)^3$$

BHP changes in direct proportion to the cube of the speed ratio (N_2/N_1), where N_1 refers to the original speed and N_2 refers to the new speed:

$$BHP_2 = BHP_1(N_2/N_1)^3$$

Because of the cubic function of the expression, increasing impeller diameter or speed will significantly increase the BHP required to drive the pump. Slight changes in impeller size or speed result in a need for larger drive size and power consumed.

Capacity (Q), total dynamic head (ΔP_{TDH}), and BHP can be found by these equations if changes are made to both impeller diameter (D) and pump speed (N), where the subscript 1 refers to the original impeller diameter and speed, and subscript 2 refers to the new diameter and speed:

$$Q_{pump\,2} = Q_{pump\,1}[(D_2N_2)/(D_1N_1)]$$
$$\Delta P_{TDH\,2} = \Delta P_{TDH\,1}[(D_2N_2)/(D_1N_1)]^2$$
$$BHP_2 = BHP_1[(D_2N_2)/(D_1N_1)]^3$$

Example 13.18

A pump with a 6-in diameter impeller is operating at 1750 rpm and delivering 25 gpm with a total dynamic head of 40 ft. The pump's BHP is 0.5 hp under these conditions. Assume constant efficiency.

a. Approximate the pump's capacity if impeller size is decreased to 5 in.

$$Q_{pump\,2} = Q_{pump\,1}(D_2/D_1) = 25\ gpm\ (5\ in/6\ in)$$
$$= 20.8\ gpm$$

b. Approximate the pump's capacity if pump speed is increased to 3500 rpm.

$$Q_{pump\,2} = Q_{pump\,2}(N_2/N_1) = 25\ gpm\ (3500/1750\ rpm)$$
$$= 50\ gpm$$

c. Approximate the pump's total dynamic head if impeller size is decreased to 5 in.

$$\Delta P_{TDH\,2} = \Delta P_{TDH\,1}(D_2/D_1)^2 = 40\ ft\ (5\ in/6\ in)^2$$
$$= 27.8\ ft$$

d. Approximate the pump's total dynamic head if pump speed is increased to 3500 rpm.

$$\Delta P_{TDH\,2} = \Delta P_{TDH\,1}(N_2/N_1)^2 = 40\ ft\ (3500/1750)^2$$
$$= 160\ ft$$

e. Approximate BHP if impeller size is decreased to 5 in.

$$BHP_2 = BHP_1(D_2/D_1)^3 = 0.5\ hp\ (5\ in/6\ in)^3$$
$$= 0.29\ hp$$

f. Approximate BHP if pump speed is increased to 3500 rpm.

$$BHP_2 = BHP_1(N_2/N_1)^3 = 0.5\ hp\ (3500\ rpm/1750\ rpm)^3$$
$$= 4.0\ hp$$

Example 13.19

A pump with a 6-in diameter impeller is operating at 1750 rpm and delivering 25 gpm with a total dynamic head of 40 ft. The pump's BHP is 0.5 hp under these conditions. Approximate capacity (Q), total dynamic head (ΔP_{TDH}), and BHP if impeller size is increased to 8 in and pump speed is increased to 3500 rpm. Assume constant efficiency.

Capacity (Q) is found by:

$$Q_{pump\,2} = 25\ gpm\ [(8\ in \cdot 3500\ rpm)/(6\ in \cdot 1750\ rpm)]$$
$$= 66.7\ gpm$$

Capacity is found by:

$$\Delta P_{TDH\,2} = 40\ ft\ [(8\ in \cdot 3500\ rpm)/(6\ in \cdot 1750\ rpm)]^2$$
$$= 284\ ft$$

BHP is found by:

$$BHP_2 = 0.5\ hp\ [(8\ in \cdot 3500\ rpm)/(6\ in \cdot 1750\ rpm)]^3$$
$$= 9.5\ hp$$

Pump Performance Curves

A *pump performance curve* is a graph that shows the flow rate that a specific pump model and impeller size is capable of pumping over a range of pressure differences. The pump curve also shows pump speed (in rpm) and other information such as pump size and type, and impeller size. The pump performance curve is produced by a pump manufacturer from actual tests performed on a single model of pump. Manufacturers usually incorporate several performance curves on a single chart, with each curve representing a different impeller size.

A chart of pump performance curves for a specific pump model is shown in Figure 13.23. This chart shows several performance curves, impeller sizes, and drive horsepower ratings. If a pump is available with only one size of impeller, there will be just a single performance curve on the entire chart. Similarly, if a pump is only available with one drive (motor) size, the chart will not have separate horsepower lines.

On the chart in Figure 13.23, pump capacity with the range of flow rates is shown along the bottom of the chart. The left side of the chart indicates the amount of total dynamic head a pump is capable of generating. The bold curved lines represent

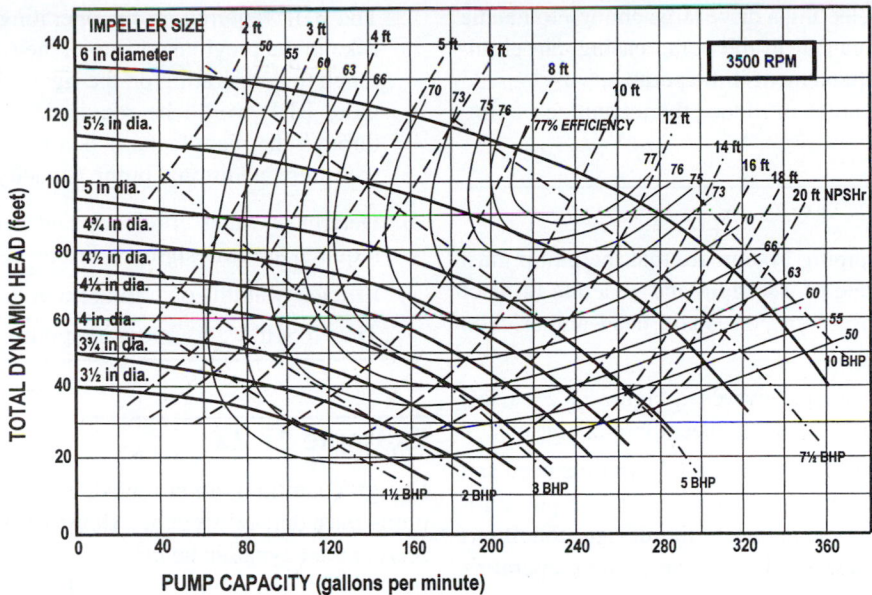

FIGURE 13.23 Pump performance curves for a specific pump model having several impeller sizes.

the performance curves for this pump model for impeller sizes available. The dash-dot lines represent BHP ratings available for this pump. The top right of the chart shows pump speed; in the chart provided, this is 3500 rpm.

Curved lines representing efficiency are also shown on the chart. Efficiency varies with pump impeller size and operating conditions. The BEP is the point on each of the performance curves where the pump operates at highest efficiency. For example, in Figure 13.23, the BEP for a pump with the 6-in diameter impeller occurs at a pump capacity of 240 gpm and a total dynamic head of 103 ft. All points on a performance curve to the right or left of the BEP have a lower efficiency.

Using Pump Performance Curves

Two variables affect pump performance, drive horsepower, and impeller size: total dynamic head and pump capacity. These hydraulic design conditions are specific to a particular piping installation and are used to determine the required size of a pump.

The first step in selecting pump size is computing the system head, the sum of the static head and friction head of the piping system (as introduced in Section 3.2), and establishing the required flow rate of the system. A pump overcomes system head by producing total dynamic head, so system head relates to total dynamic head. The required system flow rate is tied to the desired pump capacity.

The intersecting point on a pump performance curve where the desired pump capacity and total dynamic head match the installation's requirement is known as the *operating point* (sometimes called the *duty point*). The operating point will always fall on the pump performance curve of a specific impeller.

Selection of a pump impeller size and drive horsepower are based on this operating point.

In most cases when selecting a pump impeller size, desired pump capacity and total dynamic head needed do not intersect a performance curve exactly; that is, the intersecting point falls between two impeller curves. In this case, the larger impeller is selected. However, this moves the operating point to the performance curve of the selected impeller. Movement is up and to the right because system head increases with an increase in flow rate. To control the flow rate of a centrifugal pump, it is normally necessary to further restrict flow through the discharge pipeline with a valve. By partially closing the valve, friction head (resistance to flow) in the system is increased and flow in the discharge piping is reduced to the desired flow rate.

When selecting a centrifugal pump and impeller size, the pump efficiency under hydraulic design conditions (i.e., capacity required and total dynamic head) must be taken into consideration. Theoretically, the BEP is the ideal spot for a pump to operate, so when evaluating several different pumps or impeller diameters for a specific pump model, the pump that best matches the design conditions (i.e., capacity and head) and that comes closest to the BEP is hypothetically the best choice. Practically, selecting a pump with the flow and head requirements that are slightly to the left of the BEP makes good engineering sense. This offers the capability to make adjustments if necessary to flow and head conditions for the system to operate properly while providing a high level of efficient operation. For optimal performance, it is desirable that pumps operate within a range of 70 to 115% of BEP.

With small pumps, manufacturers typically match a drive to the pump. For large pumps, the designer must select the driver based on the BHP related to the selected pump model and impeller diameter. Selecting a drive large enough to handle the largest impeller that can fit in the pump casing can eliminate the need for drive replacement if the pump.

The following examples introduce the technique of simple pump selection.

Example 13.20

Design conditions for a piping system are that the pump must deliver water at a flow rate of 180 gpm and generate 60 ft of total dynamic head to overcome the static head and friction head of the piping system.

a. Based on pump performance curves in Figure 13.23, size a pump impeller and drive that can meet these conditions.

At the intersecting point of 60 ft of total dynamic head and a flow rate of 180 gpm, the pump performance curve for the 4¾-in diameter impeller meets the design conditions exactly (and conveniently). This is the pump's operating point. A 4¾-in diameter impeller is selected.

The BHP at the operating point is midway between 3 BHP and 5 BHP. A 4 hp drive (motor) would be adequate and could be selected, if one was available. However, the 4 hp drive would have a tough time operating on the right side of the curve because more brake horsepower is required. A 5 hp drive available through the manufacturer is a better selection because, in the future, additional pump capacity may be needed.

b. Determine water efficiency and net positive suction head required at the design conditions.

Efficiency at this operating point is about 70%.

The NPSHr at this operating point is about 9 ft.

Example 13.21

Design conditions for a piping system are that the pump must deliver water at a flow rate of 180 gpm and generate 80 ft of total dynamic head to overcome the static head and friction head of the piping system.

a. Based on pump performance curves in Figure 13.23, size a pump impeller and drive that can meet these conditions.

The intersecting point of 80 ft of total dynamic head and a flow rate of 180 gpm falls between the pump performance curves for the 5-in and 5½-in diameter impellers. The larger 5½-in diameter impeller is selected.

The larger impeller will have a flow rate much higher than the required 180 gpm if left unthrottled. A valve must be installed in the discharge piping to increase head and reduce flow. To achieve a flow rate of 180 gpm, the valve will need to increase the head to about 92 ft (the intersecting point of the 180 gpm line and the 5½-in diameter impeller performance curve). This is the pump's operating point.

The BHP required at this operating point is about 6 BHP. Like in the previous example, the 6 hp pump would have a tough time operating on the right side of the curve because more BHP would be required. A 7½ hp drive available through the manufacturer is a better selection because, in the future, additional pump capacity may be needed.

b. Determine water efficiency and net positive suction head required at the design conditions.

Efficiency at this operating point is about 74.5%.

The NPSHr at this operating point is about 7.3 ft.

Example 13.22

Design conditions for a renovated piping system are that the pump must deliver water at a flow rate of 275 gpm and generate 33 ft of total dynamic head to overcome the static head and friction head of the piping system. The existing pump has a 5-in diameter impeller and a 5 hp drive. Based on the pump performance curve in Figure 13.23, evaluate the existing pump's water efficiency.

The operating point of the pump will be the intersecting point of a 275 gpm pump capacity and 33 ft of total dynamic head, which falls directly on the 5-in diameter impeller's performance curve. Water efficiency at this operating point is about 50%. It is best to avoid operating any pump at efficiencies of less than 55%, so a different model pump should be considered.

13.10 WATER SOFTENING

Hard Water

The presence of excess calcium (Ca) and/or magnesium (Mg) mineral ions in water results in the water being referred to as *hard water*. Calcium and magnesium minerals generally are measured in grains of hardness per gallons of water (GPG), parts per million (ppm), or milligrams per liter (mg/L). *Soft water* has a hardness of 3.5 GPG (60 ppm/60 mg/L) or less. Hard water has a hardness of 7 GPG (121 ppm/121 mg/L) or more. *Softened water* is any water that is treated to reduce hardness. Water hardness can be tested by independent water testing laboratories and some governmental health departments. Most public community water systems record the hardness of the water they supply.

Mineral ions in hard water react with metal components and chemical agents, which causes two main problems in plumbing systems: *scale,* a hard, crusty deposit visible on taps, shower fittings, toilets, and water-using appliances (e.g., boilers, water heaters), which clogs pipes, fixtures, and appliances, reducing equipment efficiency and increasing energy costs; and, *scum,* which reduces the cleaning effectiveness of detergents used for laundry, dish washing, and bathing.

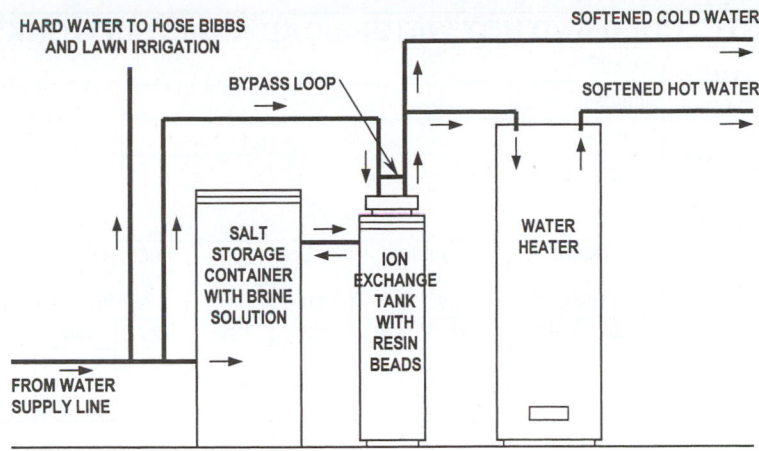

FIGURE 13.24   The layout of an ion-exchange water softener.

Water Softeners

Water softening is a process that reduces or removes calcium and magnesium ions from hard water. There are many processes used in to soften water. The most used softening methods in residential installations are ion exchange and reverse osmosis. They are introduced in the sections that follow. The industrial methods used are very complex and include electrodialysis reversal (EDR), flashpoint distillation, thin film evaporation, and lime-soda ash softening.

Ion-Exchange Water Softener

An *ion-exchange water softener* operates by passing hard water through porous resin beads that have soft sodium/potassium ions attached to them. The ions contributing to hardness are replaced or exchanged with other ions that do not contribute to hardness, thereby softening the water. After softening a large quantity of hard water the beads become saturated with calcium and magnesium ions. The resin beads then need to be regenerated by flushing a brine solution (salt–water mix) through the beads. The sodium ions in the brine recharge the beads so they will again be able to exchange the hard ions for the soft ions. Frequency of regeneration depends on the hardness of the water, the amount of water used, the size of the softener, and the capacity of resins to remove hardness. It is done manually, automatically with a timer, or founded on a demand-initiated control based on quantity of water used. (See Figure 13.24.) The ion-exchange softening process adds sodium (salt) to the water, which could be troublesome for individuals on sodium-restricted diets.

Reverse Osmosis Water Softener

A *reverse osmosis water softener* works by forcing water under pressure against a semipermeable membrane, where water molecules form a barrier that allows other water molecules to pass through while excluding most suspended and dissolved materials, including ions contributing to hardness. It is called "reverse osmosis" because mechanical pressure is used to force water flow in a direction that is the reverse of natural osmosis (e.g., from a dilute to a concentrated solution). These softeners typically incorporate an activated carbon filter, which can provide added treatment for the volatile organic compounds (VOCs) not treated by the membrane itself. As a result, they are often called "water treatment devices." In residential applications, maintenance of a reverse osmosis softener involves the replacement of a membrane cartridge every two or three years, and replacement of the carbon filter cartridges once or twice per year.

DOMESTIC COLD AND HOT WATER LOAD ANALYSIS WORKSHEET

ADVANCED BUILDING CONSULTANTS
12450 WEST BELMONT AVENUE

LITTLETON, CO 80127-1515

303-933-1300

Project:									No.:	
Address:										
Designer:									Date:	
Checked by:									Date:	

No.	Description of Fixture Type or Description of Zone	Fixture Load (WSFU)		Total Load (WSFU)		Demand Load (gpm or L/s)		Friction Loss (psi/100 ft or kPa/100 m)		Minimum Pipe Diameter (in or mm)		Notes
		CW	HW	CW	HW	CW	HW	CW	HW	CW	HW	

PAGE ____OF___

FIGURE 13.25   An example of a load analysis worksheet for domestic water computations.

STUDY QUESTIONS

13-1. Identify and describe the six main parts of a building water supply system.

13-2. What is residual water pressure?

13-3. Identify and describe the two basic types of water supply distribution systems used in buildings, and where each will be used.

13-4. What are the factors that dictate the minimum and maximum water velocity in a plumbing system?

13-5. What is cavitation and where might it occur?

13-6. What is a cross-connection and where might it occur?

13-7. What is water hammer, and how can it be reduced in the system?

13-8. How can thermal expansion damage a pipe?

13-9. What types of pipe materials expand the most with temperature increase?

13-10. What are the two basic types of water supply systems used?

13-11. How can freezing water damage a pipe and how does damage occur?

13-12. How can heated water damage a pipe and how does damage occur?

13-13. What effect does aging have on different types of pipe?

13-14. Why is pipe insulation necessary and where is it used?

13-15. During what hours does water usage peak in a home?

13-16. What is a fixture unit and where is it used?

13-17. What is meant, in water supply systems, by static head and friction head?

13-18. Describe the terms *cold water, chilled water, heated water,* and *tempered water*.

13-19. What is the difference between a storage tank water heater and an instantaneous water heater?

13-20. Why is more than one storage tank water heater sometimes used in larger residences?

13-21. What is the difference between recirculating and non-circulating hot water systems, and where would the circulating system most likely be used?

13-22. Describe the operation of the two types of pumps commonly used in buildings. Which type of pump is most common?

13-23. With respect to a pump, what is pump capacity?

13-24. With respect to a pump, what is total dynamic head?

13-25. With respect to a pump, what is brake horsepower?

13-26. With respect to a pump, what is net positive suction head?

13-27. With respect to a pump, what is a pump performance curve?

13-28. What are parallel and series pumping?

13-29. What is the difference between hard and soft water?

13-30. What are the two main problems with hard water in plumbing systems?

13-31. What are the two types of water softening methods and how do they function?

Design Exercises

13-32. Using the velocity design method, determine the minimum required size of hot and cold water supply pipes serving an apartment containing a kitchen sink, dishwasher, lavatory, bathtub, and tank-type water closet. Use a maximum velocity of 8 ft/s. Base pipe size on Type L copper tube (Table 12.9 in Chapter 12).

13-33. Using the velocity design method, determine the minimum required size of hot and cold water supply pipes serving 8 apartments, each containing a kitchen sink, dishwasher, lavatory, bathtub, and tank-type water closet. Use a maximum velocity of 8 ft/s. Base pipe size on Type L copper tube (Table 12.9 in Chapter 12).

13-34. A plumbing fixture outlet is 18 ft above the water service line. Pressure available at the water service is 80 psi. Determine the change in pressure from elevation.

13-35. A plumbing fixture outlet is 120 ft above the water service line. Pressure available at the water service is 80 psi. Determine the change in pressure from elevation.

13-36. Determine the pressure drop across a ¾-in diameter Type L copper pipe that is 60 ft long and is carrying water at 12 gpm.

13-37. Determine the pressure drop across a 3-in diameter Type L copper pipe that is 80 ft long and is carrying water at 120 gpm.

13-38. A plumbing branch is composed of 68 ft of 1-in diameter Type L copper pipe, three 90° elbows, and a gate valve. The elevation increase is 12 ft. Determine the total pressure drop across the plumbing branch if it is carrying water at 45 gpm.

13-39. Estimate the peak hot water demand of a household that has a morning routine of three showers, two hands/face washings, five teeth brushings, food preparation, and automatic dishwashing.

13-40. Determine the demand load of water required for the apartment building in Appendix B.

13-41. Size an instantaneous water heater to meet the following conditions:

- Two hot water faucets open with a flow rate of 0.75 gpm
- Two persons showering using showerheads with a flow rate of 2.0 and 2.5 gpm

- An inlet water temperature is 40°F (4°C)
- Water must be heated to 120°F (49°C)

13-42. Estimate the peak hot water demand of an 18-unit condominium having a single water heating system. Each unit has two full bathrooms and a clothes washing machine. Calculate the minimum storage tank size for this water heating system.

13-43. Design an upfeed water supply system for the apartment building in Appendix B. Use Type L copper pipe for this design problem. Assume each apartment has its own storage tank water heater.

13-44. Determine the hot water consumption and the equipment sizes required for the building in Appendix B.

13-45. Determine the hot water consumption and the equipment sizes required for a 150-room motel based on the following information:

- Hot water required: 30 gal per day per person
- Occupancy rate: 2.4 persons per room, 100% occupied
- Storage capacity (tank): 60% of total daily use (of which 75% is usable)
- Maximum hourly demand: ⅛ of total daily use
- Peak demand time: 3 hr

13-46. Design the upfeed hot water supply system required for the apartment in Appendix B. Use copper pipe and tubing.

13-47. A pump with an 8-in diameter impeller is operating at 1750 rpm and delivering 40 gpm with a total dynamic head of 55 ft. The pump's brake horsepower (BHP) is 1.2 hp under these conditions. Assume constant efficiency.

a. Approximate the pump's capacity if impeller size is increased to 10 in.

b. Approximate the pump's capacity if pump speed is increased to 3450 rpm.

c. Approximate the pump's total dynamic head if impeller size is increased to 10 in.

d. Approximate the pump's total dynamic if pump speed is increased to 3450 rpm.

e. Approximate BHP if impeller size is increased to 10 in.

f. Approximate BHP if pump speed is increased to 3450 rpm.

13-48. A pump with an 8-in diameter impeller is operating at 1750 rpm and delivering 40 gpm with a total dynamic

head of 55 ft. The pump's BHP is 1.2 hp under these conditions. Approximate the capacity (Q), total dynamic head (ΔP_{TDH}), and BHP if impeller size is increased to 10 in and pump speed is increased to 3450 rpm.

13-49. A pump with a 150-mm diameter impeller is operating at 1750 rpm and delivering 3.2 L/s with a total dynamic head of 16 meters. The pump's BHP is 3.4 kW (0.45 hp) under these conditions. Assume constant efficiency.

a. Approximate the pump's capacity if impeller size is increased to 200 mm.

b. Approximate the pump's capacity if pump speed is increased to 3450 rpm.

c. Approximate the pump's total dynamic head if impeller size is increased to 200 mm.

d. Approximate the pump's total dynamic if pump speed is increased to 3450 rpm.

e. Approximate BHP if impeller size is increased to 200 mm.

f. Approximate BHP if pump speed is increased to 3450 rpm.

13-50. A pump with a 150-mm diameter impeller is operating at 1750 rpm and delivering 3.2 L/s with a total dynamic head of 16 meters. The pump's BHP is 3.4 kW (0.45 hp) under these conditions. Approximate the capacity (Q), total dynamic head (ΔP_{TDH}), and BHP if impeller size is increased to 200 mm and pump speed is increased to 3450 rpm.

13-51. Design conditions for a piping system are that the pump must deliver water at a flow rate of 180 gpm and generate 50 ft of total dynamic head to overcome the static head and friction head of the piping system.

a. Based on pump performance curves in Figure 13.23, size a pump impeller and drive that can meet these conditions.

b. Find water efficiency and net positive suction head required at the design conditions.

13-52. Design conditions for a piping system are that the pump must deliver water at a flow rate of 120 gpm and generate 100 ft of total dynamic head to overcome the static head and friction head of the piping system.

a. Based on pump performance curves in Figure 13.23, size a pump impeller and drive that can meet these conditions.

b. Find water efficiency and net positive suction head required at the design conditions.

SANITARY DRAINAGE SYSTEMS

In Chapter 13, building hot and cold water supply systems were described with pipes to provide sufficient running water to all of the fixtures throughout a building or project. Following the flow of the water through the plumbing system, the next step will be to design a system to dispose of the sanitary waste and wastewater.

Wastewater, sometimes referred to as *sewage,* is used water. It comes from almost all sections of the building, including bathrooms, kitchens, and laundry areas, and in commercial projects, equipment being serviced. Because organic waste in wastewater tends to decompose quickly, one of the primary objectives of the sanitary drainage system is to dispose of decaying wastes rapidly, before they cause objectionable odors or become hazardous to health. Wastewater treatment is covered in Chapter 15.

14.1 SANITARY DRAINAGE SYSTEM

Conventional Sanitary Drainage and Vent System

A *sanitary drainage and vent system,* sometimes referred to as the *drain, waste, and vent (DWV) system,* is a network of pipes that remove wastewater from a building. In this section the terminology and function of each of the parts are explained, but first a discussion of system operation is needed. (See Figure 14.1.)

In typical plumbing system operation, the *sanitary drainage* side of the system consists of traps at each fixture, and fixture branch, stack, and drain pipes that carry wastewater away from the plumbing fixtures and out of the building. Water transports wastes into the sanitary drainage piping and out of the building sewer line, leading to a community wastewater treatment plant or to a private sewage treatment system. Gravity is the driving force behind wastewater flow, so the sanitary drainage system is known as a *gravity system.*

The *vent system* side of the system introduces and circulates air in the system to maintain atmospheric pressure in the drain lines and ensure adequate gravity flow of wastewater. Venting prevents a negative pressure (suction) in the system that could suck water from fixture traps and allow sewer gases to infiltrate the building. The vent system also exhausts sewer gases to the outdoors.

The chief components of a sanitary drainage and vent system are described as follows.

Traps

A *trap* is a U-shaped pipe that catches and holds a small quantity of wastewater that is poured down a fixture drain. The trapped water prevents gases resulting from wastewater decomposition from entering the building through the drain pipes and the fixture. (See Figure 14.2.) Traps are made of copper, plastic, steel, wrought iron, or brass, with plastic most commonly used. The most acceptable type of trap is called a *P-trap* (see Figure 14.3; see also Photos 14.1 and 14.2). S-traps and U-traps can easily be siphoned, so they are prohibited by the building code. An *integral trap* is built in as part of the fixture. The integral trap in a vitreous china water closet is cast as part of the fixture.

In most instances, a trap is installed immediately downstream of the fixture, as close to the fixture as possible, usually within 2 ft (0.6 m) of it, unless the fixture is designed with an integral trap (e.g., a water closet). Because the trap may occasionally need to be cleaned, access is necessary. A removable plug in the trap may provide access or the trap may have screwed connections on each end for easy removal.

In locations where a fixture is infrequently used, water in the trap may evaporate and, with the water seal not working, gases may back up from the sewer and drainage pipes through the fixture and into the building. Floor drains, which are used to take away the water after washing floors or which may be used only in case of equipment malfunctions or repairs, present the most serious possibility of losing their water seal. When floor drains are connected to the drainage system, the possibility of a serious gas problem exists. The designer of the system can avoid such a situation by not tying the floor drain into the drainage system. Instead, the floor drains could be tied into a drywell, from which there will be no gases.

The water seal in a trap may be broken if there is a great deal of vacuum pressure in the pipes. A vent system is attached to the sanitary drainage system to reduce the vacuum and to equalize the pressure throughout the system.

Historically, a *building trap* was located at the end of the building drain (inside the building and just before it connected to the sewer line). It was theorized that this trap would act as a seal to keep gases from entering the building's sanitary drainage system from the sewer line. On the other hand, a building trap may impede the flow of wastes in the system. For this reason, codes disallow use of a building trap except in special installations.

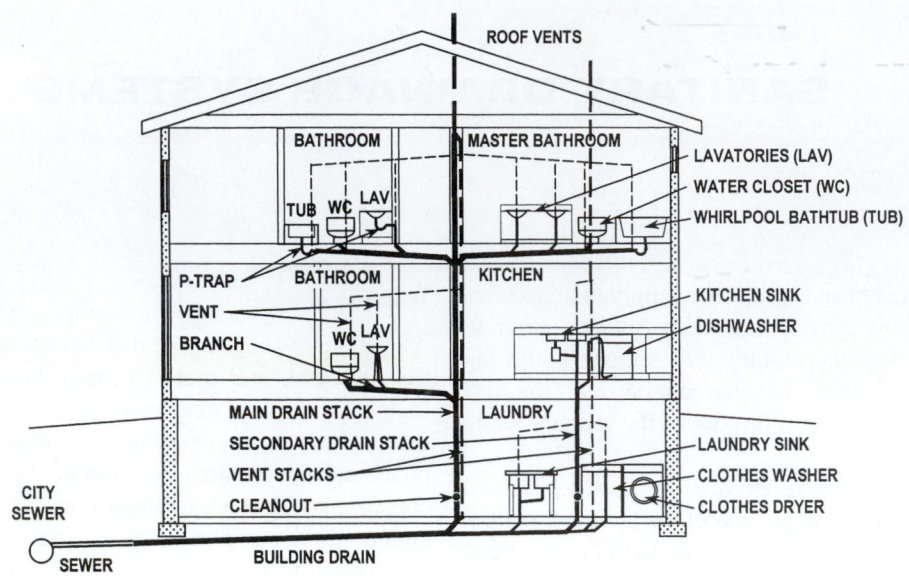

FIGURE 14.1A   Basic parts of residential sanitary drainage and vent plumbing systems.

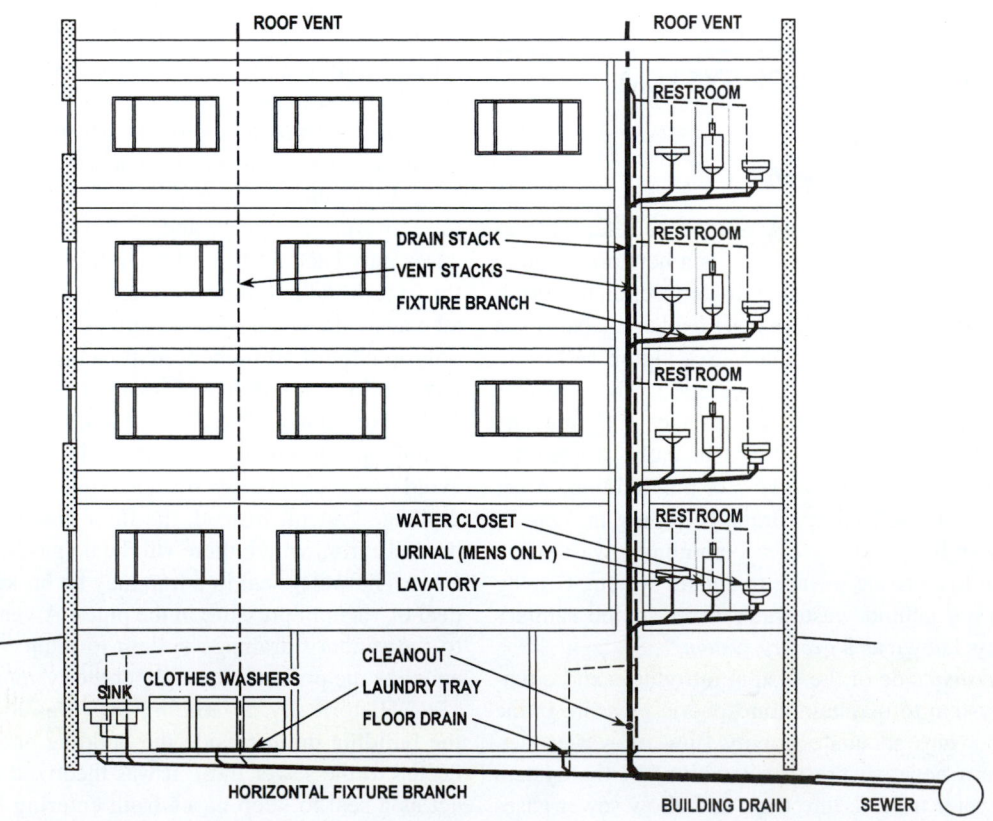

FIGURE 14.1B   Basic parts of commercial sanitary drainage and vent plumbing systems.

FIGURE 14.2 An integral trap in a water closet.

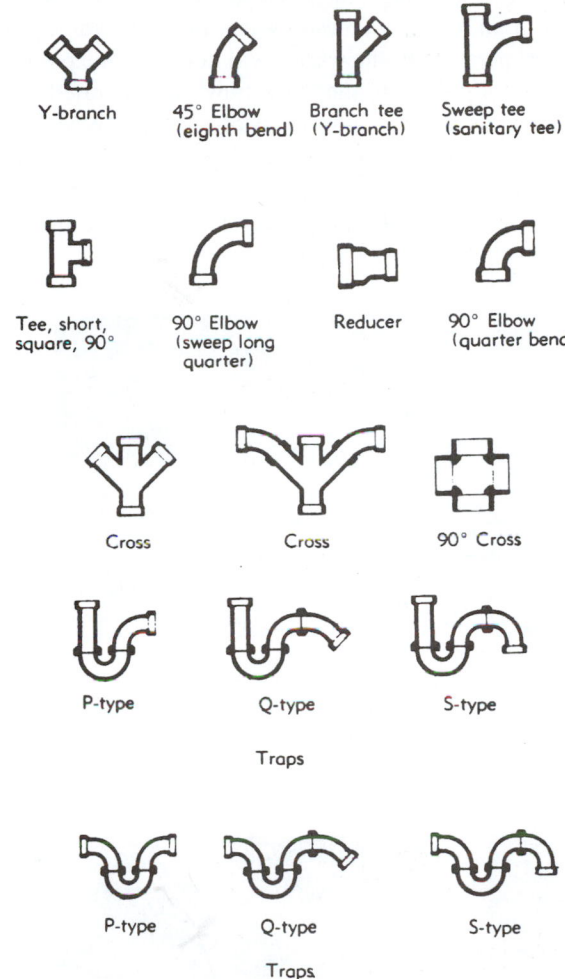

FIGURE 14.3 Drainage fittings and traps. Q-traps (sometimes called a ¾ S-trap) and S-traps are generally not permitted.

PHOTO 14.1 A P-trap. (Used with permission of ABC)

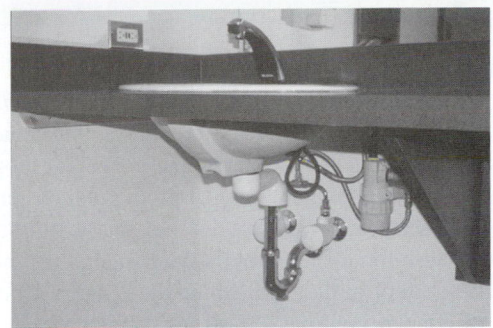

PHOTO 14.2 A P-trap under a bathroom lavatory. (Used with permission of ABC)

Interceptors

Many substances (e.g., grease, fat, oil, hair, sand, clay, wax, or debris) are accidentally or intentionally placed into a building drain, potentially creating blockages that can cause backups and overflows, or contaminating wastewater, which makes treatment difficult and more costly. *Interceptors* are passive devices designed into a plumbing system that trap, separate, and retain these toxic or undesirable substances from wastewater before it is discharged into the sewer line.

Grease can solidify and coat the inner walls of pipes, creating a stoppage. Restaurants, cafeterias, and other commercial food establishments with cooking facilities must have a *grease interceptor* or *grease trap* that receives wastewater from sources such as sinks, dishwashers, floor drains, and washing area drains before draining to the municipal sewer system. Manufacturing plants, vehicle service facilities, car washes, and other similar establishments must have an *oil-sand interceptor* to separate and remove floatable material (oils) and settleable materials (sands and metals) from wastewater before it is discharged to the municipal sewer system. Barber shops, beauty salons, pet grooming facilities, and any other establishments that discharge hair and/or other fibrous materials in wastewater must have a *hair interceptor.*

Typically, interceptors are sized for at least a 30-min peak wastewater flow detention time from all contributory sources. A grease interceptor for a restaurant can have a capacity of 1000 gal (3800 L) or more. Interceptors should be located as close to the discharging fixture as possible; they sometimes also serve as the trap, with some exceptions. Grease traps must be located at least 10 ft (3 m) from hot water faucets. All hot water must cool to 120°F before entering the grease trap.

An interceptor must be readily accessible for periodic cleaning, inspection, and testing. Wastes captured in an interceptor must be disposed of following health standards. Precious metals (e.g., from polishing jewelry in manufacturing plants) can be recovered.

Fixture Branches

Each plumbing fixture is connected horizontally to the sanitary drainage system by a drain line called a *fixture branch*. Beginning with the fixture farthest from the stack, the branch must slope ⅛ to ½ in per ft (10.4 to 41.6 mm per meter) for proper flow of wastes through the branch. Branch piping, which serves urinals, water closets, showers, or tubs, is usually run under the floor. When these fixtures are not on the branch, the piping may be run in the floor or in the wall behind the fixtures. Branch piping may be copper, approved plastic, galvanized steel, or cast iron. (See Photos 14.3 through 14.10.)

Stacks

The fixture branches feed into a vertical pipe referred to as a *stack*. When the wastewater that the stack will carry includes human waste from water closets (or from fixtures that have similar functions), the stack is referred to as a *soil stack*. When the stack will carry all wastes except human waste, it is referred to as a *waste stack*. Soil and waste stacks may be copper, plastic, galvanized steel, or cast iron. These stacks service the fixture branches beginning at the top branch and go vertically downward to the building drain.

In larger buildings, the point where the stack ties into the building drain rests on a masonry pier or steel post so that the downward pressure of the wastes will not cause the piping system to sag. In addition, the stack must be supported at 10-ft (3-m) intervals to limit movement of the pipe. When a stack length is greater than 80 ft (24.4 m), horizontal offsets are used to reduce free-fall velocity and air turbulence. Connections to fixture branches and the building drain should be angled 45° or more to allow the smooth flow of wastewater.

PHOTO 14.3 Fixture branches for a kitchen sink and dishwasher tied into a waste stack (vertical pipe). Note the metal plates secured to the wood studs that are covering the pipes to protect them from screws when the wall surface is finished with drywall. (Used with permission of ABC)

PHOTO 14.4 An unfinished plumbing wall with cast iron drain and vent lines. Copper fixture branches extend from the drain pipes and will be connected to trap arms that will extend from the wall. (Used with permission of ABC)

PHOTO 14.5 A floor-mounted water closet flange. (Used with permission of ABC)

PHOTO 14.6 A flange and bracket connection to hold a wall-mounted water closet and connect it to a fixture branch. (Used with permission of ABC)

PHOTO 14.7 A drain stack passes through the wall above and picks up additional drain branches before transitioning to a short horizontal section and back to a vertical section to accommodate changes in wall alignment. The vent connecting to the horizontal section is under construction and not fully connected. (Used with permission of ABC)

PHOTO 14.8 Horizontal branches connecting to a horizontal line. (Used with permission of ABC)

PHOTO 14.9 Horizontal branches connecting to a horizontal line, which then connects to a stack. A floor cleanout is located at the base of the stack. (Used with permission of ABC)

PHOTO 14.10 Placement of drain lines (tall pipes), water closet flanges, and floor drain (far right) before a concrete floor is cast. Note that the pipes are covered with an expansion material to isolate them from the concrete slab. This will prevent damage to the pipes that could be caused by movement of the concrete floor slab. (Used with permission of ABC)

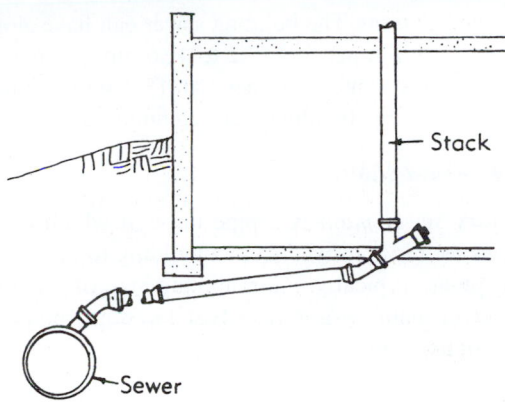

FIGURE 14.4 A drain stack connected to an underfloor building drain. Note the floor cleanout at the base of the stack. The building drain connects to sewer just outside the building foundation. The building sewer line connects to the large city sewer.

Most designers try to layout plumbing fixtures to line up vertically floor after floor so that a minimum number of stacks will be required. Many times, a central core of a multistory building will be used as a plumbing core, and a pipe chase, a space that is left to put the pipes in, runs from the first floor to the roof of the building. (See Figure 14.4.)

Building Drains

The soil or waste stacks feed into a main horizontal pipe referred to as the *building drain*. By definition, the building drain extends to a point 2 to 5 ft (0.6 to 1.5 m) outside the foundation wall of the building. The building drain slopes ¹⁄₁₆ to ½ in per foot (5.2 to 41.6 mm per meter) as it feeds the wastewater into the building sewer outside the building. Slopes of ⅛ to ¼ in per foot (10.4 to 20.8 mm per meter) are common in most buildings.

Location of the building drain in the building depends primarily on the elevation of the community sewer line. Ideally, all of the plumbing wastes of the building will flow into the sewer (whether it is a community or a private sanitary system) by gravity. The drain is typically placed below the first floor or below the basement floor. If the height of the sewer requires the drain to be placed above the lowest fixtures, it will be necessary for the low fixtures to drain into a sump pit. When the level in the sump pit rises to a certain point, an automatic float or control will activate a pump that raises the wastewater out of the pit and into the building drain.

Building drains are usually made of approved plastics, copper (for above the floor), or extra-heavy cast iron (for below the floor) pipe.

Building Sewer

The *building sewer* is an extension of the building drain that carries wastewater from the building drain to a community sanitary sewer main or an individual on-site sewage treatment (OSST) system. In community sanitary wastewater systems, the building sewer may also be known as a *house* or *building connection,* or *sanitary sewer lateral.* The building sewer can have slopes that range from ⅟₁₆ to ½ in per foot (5.2 to 41.6 mm per meter). The extremely shallow slope of ⅟₁₆ per foot (5.2 mm per meter) is only common in large buildings serving hundreds of fixtures.

Sanitary Sewer Main

The *sanitary sewer main* is a pipe through which the wastewater flows as it is conveyed from a building to the wastewater treatment plant. Typically, the minimum size of a community sanitary sewer main for a gravity-based system should be 8 in (200) mm in diameter.

Cleanouts

Provisions must be made to allow cleaning of the sanitary drainage system. *Cleanouts* are screw-type fittings with a cap that can unscrew to allow access to the inside of the sanitary drain pipes. A cleanout should not have a plumbing fixture installed in it or be used as a floor drain. *Floor cleanouts (FCO)* are found in horizontally positioned building drain or sewer lines that are installed in the floor or in the ground. *Wall cleanouts (WCO)* are placed in vertically positioned stacks. All cleanouts in vertical stacks should be located no higher than 48 in (1.2 m) above the floor.

Cleanouts are generally required:

- At the base of soil and waste stacks
- At the upper end of building drains
- At each change of direction of the horizontal building drainage system greater than 60°; the total of the fittings between cleanouts shall not exceed 120°
- At the junction between the building drain and building sewer (usually 2 to 4 ft away from the building foundation)

In addition:

- Cleanouts should be no more than 50 ft apart, including the developed length of the cleanout pipe, in horizontal drainage lines of 4 in or less size.
- Cleanouts should be no more than 100 ft apart, including the developed length of the cleanout pipe, in horizontal drainage lines of sizes over 4 to 10 in.
- Cleanouts should not be more than 150 ft apart, including the developed length of the cleanout pipe, in horizontal drainage lines exceeding sizes of 10 in.

Cleanout size is related to pipe size: 1½ and 2 in diameter pipe have a 1½ in cleanout; 2½ and 3 in diameter pipe have a 2½ in cleanout; and 4 in diameter and larger pipe have a 3½ in cleanout. (See Photos 14.11 and 14.12.)

Venting

Vents are pipes that introduce sufficient air into the drainage system to reduce air turbulence (from siphoning or back pressure)

PHOTO 14.11   A close-up of a wall cleanout. (Used with permission of ABC)

PHOTO 14.12   A cleanout fitting allows easy access to the drain system. (Used with permission of ABC)

PHOTO 14.13   A vent connection and penetration through the roof decking. (Used with permission of ABC)

and to release sewer gases to the outside. (See Photo 14.13.) The prime purpose of venting is to protect the trap seal. If traps did not exist in a drainage system, a venting could be eliminated. Without a vent, as water drains from a fixture, the moving wastewater tends to siphon water from the trap of another fixture as it falls through the drain pipes. As a result, vents must serve the various fixtures, or groups of fixtures, as well as the rest of the drainage system. Vent piping may be copper, plastic, cast iron, or steel. Types of venting methods are as follows.

Individual Vents The *individual venting* technique is defined as the installation of a vent pipe for every trap or trapped fixture. It is the easiest method of ensuring the preservation of a trap seal but the most costly because of the number of vent pipes required in the venting system. An individual vent must be located in close proximity to the trap to properly vent it. A more effective way of reducing the cost of venting has been in the combining of vents into a system. This would include common venting, circuit venting, wet venting, combination drain and vent, waste stack venting, and single stack systems. Venting methods are described below.

Common Vents The *common venting* method serves two fixtures located on the same floor; it is essentially an individual vent that serves no more than two traps or trapped fixtures. This type of vent must be located close to the traps it vents to properly vent it. When the fixture connects at different levels, the drainage pipe between the two traps must be increased to compensate for the combined water and airflow.

Wet Vents The *wet venting* method uses a single vent pipe to provide venting for all of the fixtures of one or two bathroom groups (e.g., a water closet, lavatory, shower, bathtub, and bidet) that are located on the same floor. The vent pipe for the lavatory typically serves as the vent for the other fixtures in the bathroom. Plumbing codes used to require the water closet to be the last fixture in line on a wet vent system. However, recent tests provided evidence that the order of the fixtures does not influence the overall performance of the wet vent system. The most recent standard permits the fixtures to be located in any order when connecting to the system.

Circuit Vents A *circuit venting* system is a horizontal venting pipe serving up to eight fixtures. Each fixture must be connected to a single horizontal drain in this technique. The vent connection is made between the two upstream fixtures—that is, those fixtures connected to the horizontal drain pipe that are the farthest away from the vent stack. In this system, all of connections and the main piping must remain in the horizontal orientation. Vertical drops are generally not permitted.

Combination Drain and Vent A *combination drain and vent* system allows the distance from trap to vent to be extended infinitely, provided the drain stays in the horizontal orientation and there is a vent somewhere within the horizontal branch. It is based on oversizing the horizontal drain, so there is an increased likelihood of stoppage in the drain line. This is the most popular method of venting a floor drain or venting island fixtures. A combination drain and vent is a marginally effective venting method.

Relief Vent A *relief vent* is a continuous pipe of lesser or equal diameter running parallel and alongside the soil and waste stack

in a multistory plumbing system. It is used to equalize air pressure within the stack.

Vent stack configurations are shown in Figures 14.5 through 14.9. Codes limit the distance between the trap outlet and the vent to ensure proper venting. These distances depend on the venting technique and size of the drain and lines.

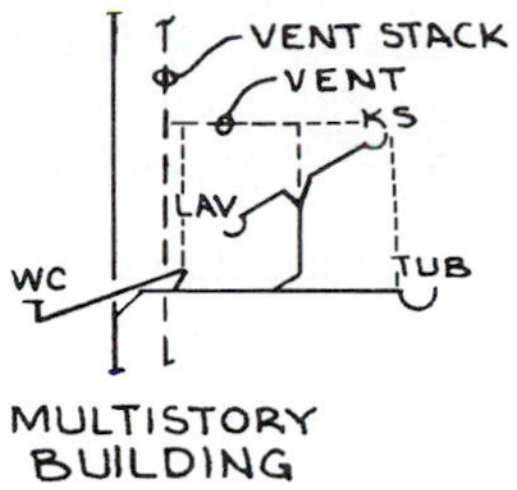

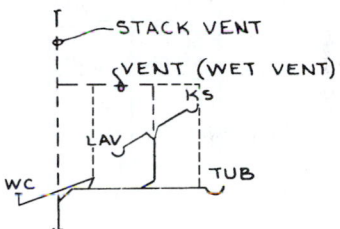

FIGURE 14.5 A vent-to-vent stack configuration.

FIGURE 14.6 A vent-to-stack vent is typically used on upper floors. Fixtures are individually vented.

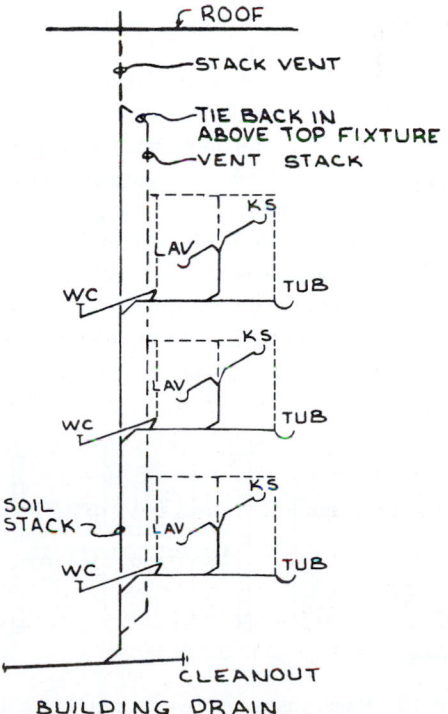

FIGURE 14.7 A stack-to-stack vent is typically used in multistory buildings.

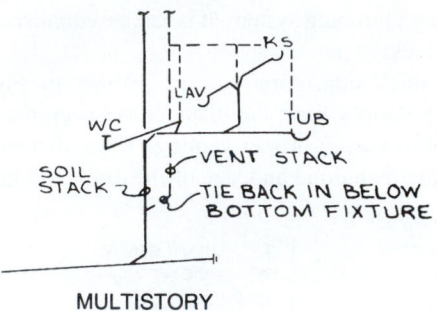

FIGURE 14.8   A vent-to-soil-stack connection.

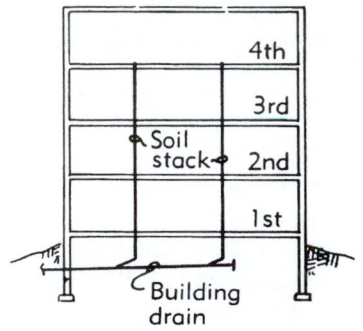

FIGURE 14.9   Multiple stacks are required in most multistory buildings. These stacks connect to the building drain below the bottom floor.

A *vent stack* extends vertically through the building and up through the roof to the exterior of the building. Vents from a fixture or group of fixtures ties in with the main vent stack, which extends to the exterior. It must extend beyond the roof at least 6 in (152 mm) and terminate to open air well beyond attic vents, windows, doors, or intake air vents. A vent stack is used in multistory buildings where a pipe is required to provide the flow of air throughout the drainage system. The vent stack can also begin at the soil or waste pipe, just below the lowest horizontal connection, and may go through the roof or connect back into the soil or waste pipe not less than 6 in (150 mm) above the top of the highest fixture. See Figure 14.10.

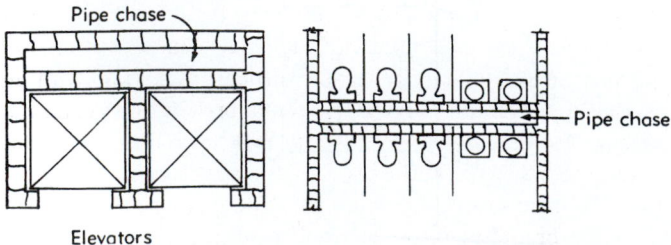

FIGURE 14.10   Pipe chases run from floor to floor to allow soil stacks and vents to pass vertically between floors. Chases are typically located alongside elevator hoist ways and common plumbing walls.

Air Admittance Valves

An *air admittance valve (AAV)* is a pressure-activated, one-way mechanical venting port used to eliminate the need for expensive venting and roof penetrations (See Photo 14.14.) Wastewater discharges cause the AAV to open, allowing air to circulate in the vent system. When there is no discharge, the valve remains closed, preventing the escape of sewer gas and maintaining the trap seal. Individual or branch-type air admittance valves may be used for venting individual, branch, and circuited fixtures. AAVs are not permitted for venting combination drain and vent systems and wet vented systems.

AAVs are typically made from polyvinyl chloride (PVC) plastic materials with ethylene propylene diene monomer (EPDM) rubber valve diaphragms. Valves come in two sizes: one for fixture venting and a larger size for system venting. The valves fit standard diameter pipes, ranging from 1¼ to 4 in. Screening protects the valves from foreign objects and vermin. Using AAVs can significantly reduce the amount of venting materials needed for a plumbing system, increase plumbing labor efficiency, allow greater flexibility in the layout of fixtures, and reduce long-term maintenance problems where conventional vents penetrate the roof surface.

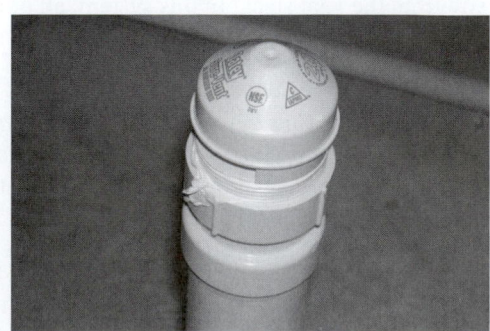

PHOTO 14.14   An air admittance valve (AAV) is a pressure-activated, one-way mechanical venting port used to eliminate the need for venting through roof penetrations. (Used with permission of ABC)

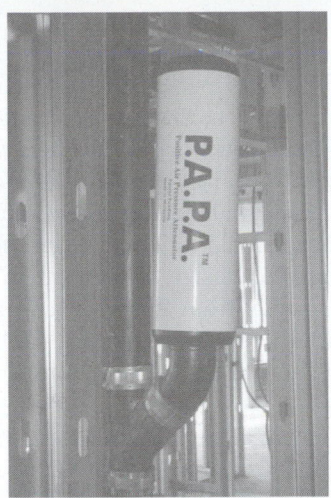

PHOTO 14.15 A positive air pressure attenuator (PAPA) is a product developed to protect buildings of 10 or more stories against the unwanted positive pressures (i.e., back pressure/positive transients) generated in the DWV system. PAPAs are installed at the base of the soil and waste stack and at various floor intervals, depending on the height of the building. (Used with permission of ABC)

Positive Air Pressure Attenuator

A *positive air pressure attenuator (PAPA)* is a product developed to protect buildings of 10 or more stories against the unwanted positive pressures (i.e., back pressure/positive transients) generated in the DWV system. PAPAs are installed at the base of the soil and waste stack and at various floor intervals, depending on the height of the building (See Photo 14.15.) The unsteady nature of the water flows cause pressure fluctuations (known as pressure transients), which can compromise water trap seals and provide a path for sewer gases to enter the habitable space. A PAPA/AAV system counters the tendency for the loss of trap water seals resulting from positive pressure pulses in a soil and waste stack. The PAPA/AAV system may be used in sanitary plumbing systems as an alternative to relief venting, eliminating the need for a continuous parallel relief vent pipe. It is a viable option to the Sovent® system.

Sovent® Drain and Vent System

Multistory buildings traditionally rely on a complex drain and vent system with two stacks that run vertically from floor to floor and vents and branches to every fixture. In high-rise buildings, if it works successfully, a drain/vent scheme with a single stack and branches without vents is an effective substitute for the traditional two-pipe, drain and vent system.

The *Sovent® system* is a system that combines the drain stack, branches, and vents into one pipe system by using patented Sovent® fittings. Fritz Sommer of Switzerland, whose work was mainly driven by a need for resource-conserving construction techniques, developed and patented the Sovent® fittings in the 1950s.

The system consists of four components: vertical stack piping, horizontal branches to the fixtures, aerator fitting, and de-aerator fittings. These components work together to collect wastes from the plumbing fixtures and transport them down a stack to the building drain.

The system works on the principle that wastewater flowing down a vertical pipe tends to cling to the interior wall surface and continue downward in a swirling motion. As the wastewater travels down the walls of the pipe, the pipe center remains open and serves as an airway. The airway provides venting so there is a balance of pressures within the drainage system. It eliminates the need for a separate venting system. However, if the fall rate of wastewater is uncontrolled, the falling water will increase speed and meet air resistance, which will flatten out the falling waste until it blocks the stack. This downward moving blockage can throw off the pressure balance in the system and suck water out of fixture traps. Specially designed fittings are placed in the vertical stack at each floor to eliminate speed buildup and blockage, thereby maintaining the airway and allowing for good drainage.

Horizontal branches and branch runouts connect to the plumbing fixture and transport the wastes to a specially designed stack. Generally, vents to individual fixtures are not required if fixture placement is near the stack; for example, a 4-in soil/waste line may be run horizontally out to 27 ft from the Sovent® stack without the use of traditional venting methods.

The *Sovent® stack* is a vertical pipe that conveys wastes from the upper levels of a building to the base of the stack. The stack begins just above the bottom-most de-aerator fitting (to be described later) and continues to just above the highest fixture connection. The main difference between the specially designed stack and the traditional waste and vent stack is the Sovent® stack will remain one size throughout its entire length. It is not permitted to change diameter because it functions for both drainage and venting purposes. Sovent® stack size is based on the total number of drainage fixture units that connect to that stack. The stack will penetrate the roof to the atmosphere much like traditional vent systems.

The *Sovent® aerator fitting* is made of two separate chambers. The first chamber, called an *offset chamber,* allows falling waste from the upper floors to enter the chamber and pass around the horizontal branch inlets. This offset reduces the falling waste's velocity, eliminating blockage before it is allowed to form. The second chamber, named the *mixing chamber,* is fully separated from vertical stack flow with an internal separation baffle. As horizontal branch flows enter the aerator fitting, it must transition to a vertical flow, smoothly uniting with any vertical stack flow that may exist. Aerator fittings can have several branch inlets. The mixing chamber provides the branches with sufficient air circulation to balance any pressure fluctuations that may occur. A second internal baffle in the mixing chamber is located perpendicular to the separation baffle to prevent crossflow from opposing branch inlets on that floor.

A *Sovent® de-aerator fitting* must be located at the base of each Sovent® stack and at any horizontal stack offset. This fitting is designed to effectively deal with pressure fluctuations that occur when vertical falling wastes suddenly turn horizontal.

Sewage Ejection

For the most part, sanitary drainage systems rely on the force of gravity to create flow to discharge wastewater. In some building installations, however, a fixture or group of fixtures must to be installed below the level of the nearest available sewer line. In these cases, wastewater must be lifted to the level of the main drain or sewer by a pumping system called a *sewage ejector*. Typically, a sewage ejector can pump solids from 2 to 4 in (50 to 100 mm) in size or grinds solid wastes before passing them through the ejector. Photo 14.16 shows an installed swage ejector.

A *sewage ejector system* consists of the sump basin, a motor-pump assembly, and a system of automatic electrical controls. Wastewater from the sanitary pipes flow by gravity into the sump basin, a pit that collects wastewater. As the wastewater level rises, it triggers a float switch that activates the pump. The pump then lifts the wastewater through a check valve and discharge line into a typical building drain line, where it gravity flows into the building sewer. It operates much like a sump pump.

The check valve in the discharge line prevents backflow. Without it, the pump will cycle continuously. A vent pipe connects to the sump basin to relieve the suction created by the pump. A high water alarm is generally added to the system, to warn of pump failure or backup to prevent flooding. Basins are typically fabricated of fiberglass, cast iron, or high-density polyethylene thermoplastic; they are typically set in a hole in a concrete floor slab.

A single ejector pump is installed in a small system, such as a single-family residence or small commercial building. Larger commercial and industrial installations require two pumps to ensure continued operation if one pump fails. The additional pump also provides extra capacity in times of extra heavy loads.

PHOTO 14.16   A sewage ejector. (Used with permission of ABC)

The size and capacity of a sewage ejector system is determined by the application. The manufacturers' literature specifies the capacity of the pump and the maximum size of wastewater solids that can be handled by a particular pump. Attempting to eject solid matter that exceeds this rated size or materials that expand in water have the potential to clog the system.

Typically, residential ejector systems must have the capacity of ejecting solids up to 2 in (50 mm) in size. Depending on pump impeller design, a 4-in pump will normally handle spherical solids from 2 to 3 in and typically range in motor size from $\frac{1}{3}$ to 2 horsepower. Additionally, these systems are generally rated to a maximum temperature of 180°F (82°F).

Sizing commercial or industrial installations involves application of complex formulas. In each case, the design must consider total dynamic head, the highest vertical point, and the size of the basin provided. The farther the distance the waste must be lifted, the more powerful the pump must be to do the job. Regardless of peak flow requirement for a given application, the pump must always be able to provide a minimum velocity of 2 ft per second through the line. Typically, in residential and small commercial applications, the water supply fixture unit (WSFU) load can be used to estimate usage demands of plumbing fixtures served. Larger capacity systems are required in motels, apartment complexes, and large office buildings because of higher peak demands. The installation must conform to the local building code.

14.2   DRAIN AND VENT PIPE DESIGN

Drainage Fixture Units

The draining rate for plumbing fixtures is based upon the *drainage fixture unit (DFU)*. Refer to Table 14.1. Similar to the water supply fixture unit introduced in Chapter 13, the DFU is an arbitrarily chosen measure that allows all of types of plumbing fixtures to be expressed in common terms; that is, a fixture having twice the instantaneous drainage flow rate of a second fixture would have a fixture unit value twice as large. The WSFU and DFU may differ slightly for a single fixture, because the rates of filling and draining are different.

Design Approach

The approach used to size drain and vent lines relies on tabular information found in code. Table 14.2 indicates the maximum load in DFU and maximum pipe length for a given pipe diameter. The minimum pipe diameter is based on the total connected DFU. In the case of vent lines, maximum developed length for a given pipe is also a criterion. Developed length is the "centerline" length of the lines, excluding traps and trap arms. It is important to ensure that a larger pipe diameter does not flow into a pipe having a smaller diameter.

TABLE 14.1 DRAINAGE FIXTURE UNITS (DFU) AND MINIMUM TRAP SIZE FOR SELECTED PLUMBING FIXTURES.

Type of Fixture or Group of Fixtures	Minimum Trap Size		DFU
	in	mm	
Automatic dishwasher			3
Bathtub group (water closet, lavatory, and bathtub or shower stall)	1½	38	6
Bathtub	1½	38	2
Bidet	1¼	32	1
Clothes washer	1½	38	2
Combination sink/tray with waste disposal (grinder)	2½	65	4
Combination sink/tray with one 1½ in (38 mm) trap	1½	38	2
Combination sink/tray with separate 1½ in (38 mm) trap	1½	38	3
Dental unit or cuspidor	1¼	32	1
Dental lavatory	1¼	32	1
Drinking fountain	1¼	32	½
Dishwasher (domestic)	1½	38	2
Dishwasher (commercial)	2	50	3
Floor drain with 2 in (50 mm) waste	2	50	3
Kitchen sink (domestic) with 1½ in (38 mm) trap	1½	38	2
Kitchen sink (domestic) with waste disposal (grinder)	1½	38	2
Kitchen sink (domestic) with waste disposal and dishwasher, 1½ in (38 mm) trap	1½	38	3
Kitchen sink (domestic) with dishwasher, 1½ in (38 mm) trap	1½	38	3
Lavatory with 1¼ in (32 mm) waste	1¼	32	1
Laundry tray (1 or 2 compartments)	1½	38	2
Shower stall (domestic)	2	50	2
Showers per head (group)	2	50	2
Sink (surgeon's)	1½	38	3
Sink (flushing rim with valve)	3	75	6
Sink (service-trap standard)	3	75	3
Sink (pot, scullery)	2	50	2
Sink per faucet (circular or multiple wash)	1½	38	2
Urinal (siphon jet blowout)	—	—	4
Urinal	—	—	6
Urinal (waterless)	—	—	0.5
Water closet (private)	—	—	4
Water closet (general use)	—	—	6
Waste disposal (commercial)	2	50	3
Waste disposal (domestic)	1½	38	2
Fixture not listed—trap size of 1¼ in (32 mm) or less	1¼	32	1
Fixture not listed—trap size of 1½ in (38 mm) or less	1½	38	2
Fixture not listed—trap size of 2 in (50 mm) or less	2	50	3
Fixture not listed—trap size of 2½ in (65 mm) or less	2½	65	4
Fixture not listed—trap size of 3 in (75 mm) or less	3	75	5
Fixture not listed—trap size of 4 in (100 mm) or less	4	100	6

Traps and trap arms are sized based on a specific type of fixture. Refer to Table 14.1 for minimum trap sizes. Some fixtures such as urinals and water closets have integral traps built into the fixture so trap size does not need to be specified.

Example 14.1

The following number and type of plumbing fixtures serve two apartment units: two bathtubs, two water closets, two lavatories, and two kitchen sinks. Assume the horizontal fixture branch serving these fixtures flows into the waste stack. Assume the vent stack extends through the roof and is 22 ft long. Determine the minimum pipe diameter required for the horizontal fixture branch, waste stack, and vent stack.

From Table 14.1, the DFU values for the plumbing fixtures are extracted. DFU are then totaled:

2	Bathtubs (2 DFU each)	4
2	Water closets—flush tank (4 DFU each)	8
2	Lavatories (1 DFU each)	2
2	Kitchen sinks (2 DFU each)	4
	Total DFU:	18

For the horizontal fixture branch, from Table 14.2, a 3-in diameter pipe is selected. A 3-in diameter pipe used as a horizontal fixture branch can serve up to 20 DFU.

TABLE 14.2   MAXIMUM DRAINAGE FIXTURE UNITS (DFU) THAT MAY BE CONNECTED TO HORIZONTAL FIXTURE BRANCH AND STACK BASED ON PIPE SIZE.

Nominal Pipe Diameter		Maximum DFU that May Be Connected			
				Stacks with More Than Three Branch Intervals	
in	mm	Any Horizontal Fixture Branch[a]	One Stack of Three Branch Intervals or Less	Total for Stack	Total at One Branch Interval
1½	38	3	4	8	2
2	50	6	10	24	6
2½	65	12	20	42	9
3	75	20[b]	48[b]	72[b]	20[b]
4	100	160	240	500	90
5	125	360	540	1100	200
6	150	620	960	1900	350
8	200	1400	220	3600	600
10	250	2500	3800	5600	1000
12	300	3900	6000	8400	1500

[a] Does not include branches of the building drain.

[b] Not more than 2 water closets or bathroom groups within branch interval nor more than 6 water closets or bathroom groups on the stack.

NOTE: A *branch interval* is a length of soil or waste stack receiving horizontal branches from one floor and corresponding to a one-story (at least 8 ft/2.6 m) height.

TABLE 14.3   MAXIMUM DRAINAGE FIXTURE UNITS (DFU) THAT MAY BE CONNECTED TO A BUILDING DRAIN OR BUILDING SEWER BASED ON PIPE SIZE WHEN THE BUILDING DRAIN AND SEWER SERVES ONE BUILDING.

Nominal Pipe Diameter		Maximum DFU that May Be Connected			
		Slope			
in	mm	1/16 in/ft (5.2 mm/m)	1/8 in/ft (10.4 mm/m)	1/4 in/ft (20.8 mm/m)	1/2 in/ft (41.6 mm/m)
2	50	—		21	26
2½	65	—		24	31
3	75	—		42	50
4	100	—	180	216	250
5	125	—	390	480	575
6	150	—	700	840	1000
8	200	1400	1600	1920	2300
10	250	2500	2900	3500	4200
12	300	2900	4600	5600	6700
15	400	7000	8300	10 000	12 000

For the waste stack, from Table 14.2, a 2½ in diameter pipe can be selected but the 3 in diameter horizontal fixture branch would then flow into a smaller pipe. A 3 in diameter waste stack is a prudent choice.

For the vent stack, from Table 14.4a, a 1½ in diameter pipe is selected, based on a 3 in diameter soil and waste stack and a capacity of up to 30 DFU.

Example 14.2

The following number and type of plumbing fixtures serve six apartment units with two apartments on each floor: six bathtubs, six water closets, six lavatories, and six kitchen sinks. Assume horizontal fixture branches serving these fixtures flow into the waste stack at three locations (three intervals), two apartments

per interval. Assume the building drain is sloped at ¼ in per ft and the vent stack extends through the roof and is 42 ft long. Determine the minimum pipe diameter required for the horizontal fixture branches, waste stack, building drain, and main vent stack.

6	Bath tubs (2 DFU each)	12
6	Water closets—flush tank (4 DFU each)	24
6	Lavatories (1 DFU each)	6
6	Kitchen sinks (2 DFU each)	12
	Total DFU:	54

For the horizontal fixture branch, from Table 14.2, a 3 in diameter pipe is selected. A 3 in diameter pipe used as a horizontal fixture branch can serve up to 20 DFU. Two apartment units have two bathtubs, two water closets, two lavatories, and two kitchen sinks—a total of 18 DFU (see Example 14.2).

TABLE 14.4A REQUIRED VENT DIAMETER AND MAXIMUM LENGTH OF VENT BASED ON SOIL AND WASTE STACK SIZE AND CONNECTED DRAINAGE FIXTURE UNITS (DFU).

Soil or Waste Stack Size (in)	Connected DFU on Stack	Required Vent Diameter (in)								
		1¼	1½	2	2½	3	4	5	6	7
		Maximum Developed Length of Vent (ft)								
1½	8	50	150							
2	12	30	75	200						
2	20	26	50	150						
2½	42		30	100	300					
3	10		30	100	300	600				
3	30			60	200	500				
3	60			50	80	400				
4	100			35	100	260	1000			
4	200			30	90	250	900			
4	500			20	70	180	700			
5	200				35	80	350	1000		
5	500				30	70	300	900		
5	1100				20	50	200	700		
6	350				25	50	200	400	1300	
6	620				15	30	125	300	1100	
6	960					24	100	250	1000	
6	1900					20	70	200	700	
8	600						50	150	500	1300
8	1400						40	100	400	1200
8	2200						30	80	350	1100
8	3600						25	60	250	800
10	1000							75	125	1000
10	2500							50	100	500
10	3800							30	800	350
10	5600							25	600	250

TABLE 14.4B REQUIRED VENT DIAMETER AND MAXIMUM LENGTH OF VENT BASED ON SOIL AND WASTE STACK SIZE AND CONNECTED DRAINAGE FIXTURE UNITS (DFU), FOR SI (METRIC) UNITS.

Soil or Waste Stack Size (mm)	Maximum Connected DFU on Stack	Required Vent Diameter (mm)								
		32	38	50	65	80	100	125	150	200
		Maximum Length of Vent (m)								
38	8	15	45							
50	12	9	23	61						
50	20	8	15	45						
65	42		9	31	92					
80	10		9	31	31	183				
80	30			18	61	153				
80	60			15	24	122				
100	100			11	31	79	305			
100	200			9	27	76	275			
100	500			6	21	55	214			
125	200				11	24	107	305		
125	500				9	21	92	275		
125	1100				6	15	61	214		
150	350				8	15	61	122	396	
150	620				5	9	38	92	336	
150	960					7	31	76	305	
150	1900					6	21	61	214	
200	600						15	45	153	396
200	1400						12	31	122	366
200	2200						9	24	107	336
200	3600						8	18	76	244
250	1000							23	38	305
250	2500							15	31	153
250	3800							9	24	107
250	5600							8	18	76

For the waste stack, from Table 14.2, a 4 in diameter waste stack is selected. A 3 in diameter pipe used as a waste stack can serve up to 240 DFU.

For the building drain, from Table 14.3, a 4 in diameter pipe is required. A 4 in diameter pipe used as a building drain can serve up to 216 DFU at a slope of ¼ in/ft.

For the vent stack, from Table 14.4a, a 2½ in diameter pipe is selected, based on a 4 in diameter soil and waste stack, a capacity of up to 100 DFU, and a developed length of 42 ft.

14.3   SYSTEM INSTALLATION

On a small project, the drainage piping typically varies in size from 1½ to 4 in. It can be much larger in large hotels, apartments, and office buildings. This larger size of pipe often requires special provisions in wall width or furred-out areas.

Poured concrete slabs will require that the plumbing layout be carefully considered. The pipes need to be placed in the ground before the slab is poured, so accurate placement is crucial. Typically, both the water supply and drainage pipes are laid out next to each other, as they go to the same areas of the building. String is usually stretched over the slab area to mark where the pipes should be located. Many times, they are planned so they will come up in a wall. However, the tub, shower, and water closet piping will need to be placed in the exact location where the fixture is to go. All piping must be carefully located and the system checked for leaks before the concrete is poured because any relocation or repairs of pipes would be costly.

On larger projects with concrete walls and ceilings, it is usually necessary to provide sleeves (holes) in the concrete for the pipes to pass through to get from space to space. It will also be necessary to provide inserts and hangers to support the pipes.

The open spaces provided in truss-type construction make it easier to run piping through to the desired location. The only points of difficulty would be where it needs to pass by ductwork or some other large pipe that is going in the opposite direction. This will require coordination with the contractor installing any heating, air conditioning, or ventilating ductwork.

In wood frame construction, the holes are sometimes drilled to allow the passage of the pipes. These should be at the middle of any load-bearing wood members so that a minimum of structural damage is done. There are times when the width of a wall needs to be increased to allow for pipes running horizontally to pass by drainage pipes (or other pipes) running vertically.

Pipe chases run from floor to floor to allow stacks and vents to pass vertically between floors. A view of the interior of a plumbing (wall) chase is shown in Photo 14.17. Chases are typically located alongside elevator hoist ways and common plumbing walls. A drain stack is shown in Photo 14.18. *Pipe tunnels* (Figure 14.11) may be used on large projects to provide concealed space for the passage of mechanicals at ground level and from building to building. *Hangers* from the top or side of

PHOTO 14.17   A view of the interior of a plumbing (wall) chase. DWV lines are cast iron (black). (Used with permission of ABC)

PHOTO 14.18   A drain stack interval in a high-rise condominium building. (Used with permission of ABC)

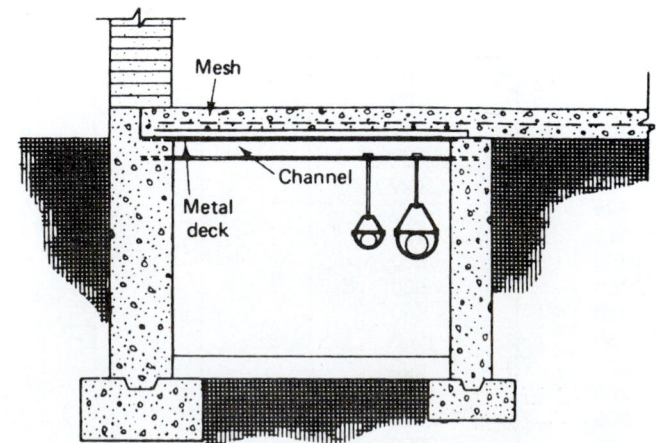

FIGURE 14.11   In large installations with multiple buildings, pipe tunnels house pipes. They allow easy access.

the tunnel are used to support the pipes. Access may be from either end of the tunnel or access floors may be provided.

14.4 SANITARY DRAINAGE AND VENT SYSTEM DESIGN EXAMPLE

The following example conveys a typical design approach for the four-story apartment building used in the water supply design example in Chapter 13. A review of the drawings for the apartments (in Appendix A) shows that each floor contains two bathroom groups and two kitchen sinks. Assume the project will be connected to a community sewer. In this project, it is assumed that there will be two plumbing stacks, one for each tier of apartments (See Figure 14.12).

1. The first step is to sketch an isometric drawing of the drainage piping. The stack is located on the plan as shown in Figure 14.12. The vertical stack is then sketched as shown in Figure 14.13, and beginning at the top floor, the fixture branch and the connection at each fixture. Next, the other three floors are sketched as shown in Figure 14.14. The vent stack is added from a point just below the bottom fixture branch to a point above the top fixture and the building drain (See Figure 14.15).

2. Next, the minimum trap size for each fixture is selected from Table 14.1 and the trap size noted on the schematic pipe layout, as shown in Figure 14.16. Assume the kitchen sink has a small P.O. (plumbing outlet) plug.

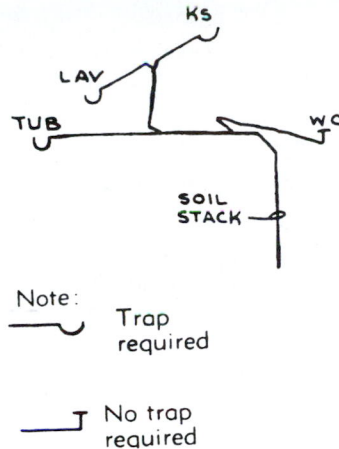

FIGURE 14.13 Trap arms from the kitchen sink (ks), bathroom lavatory (lav), and bathtub (tub) connect to a fixture branch, which then connects to a soil stack. An integral trap is found in a water closet (wc), so a trap does not need to be constructed in the piping.

3. The fixture branch is the first drainage pipe to be sized. The first portion of the branch to be sized is from the fixture farthest from the stack to the next fixture, as shown in Figure 14.17. Begin by determining the fixture units that this short piece of pipe will serve from Figure 14.17. In this problem, the branch serves a bathtub with a fixture-unit value of 2, based on Table 14.1.

4. Next, select the branch size to serve the bathtub from Table 14.2. In this case, the horizontal fixture branch size is being

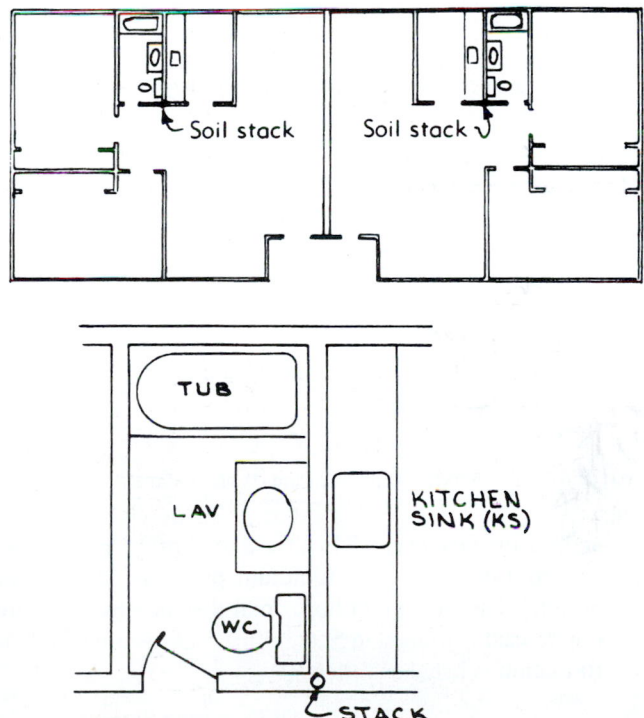

FIGURE 14.12 Soil and vent stacks run vertically from floor to floor and are located near plumbing fixtures.

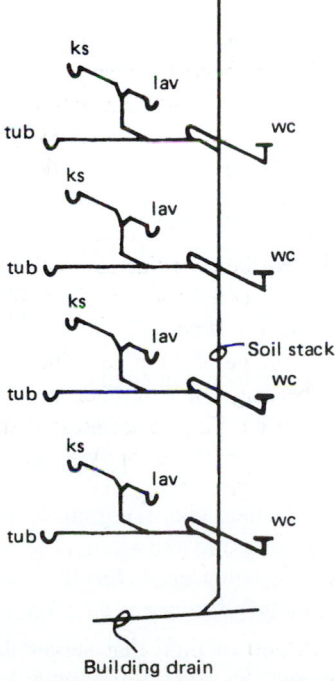

FIGURE 14.14 Fixture branches connect to a soil stack, which then connects to the building drain. Vents are not shown for clarity.

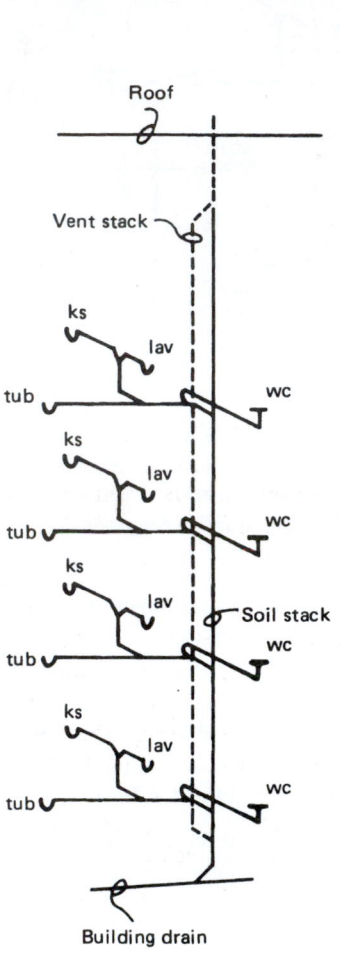

FIGURE 14.15 A vent stack parallels the soil stack and extends to through the roof. Individual vents are not shown for clarity.

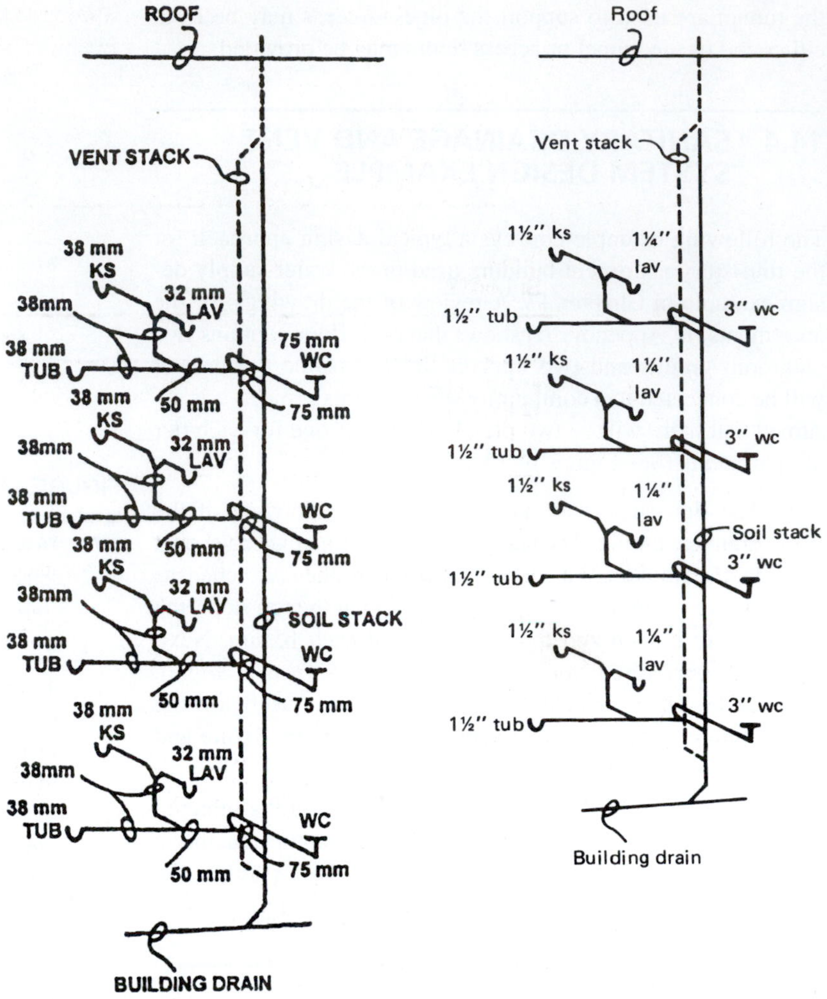

FIGURE 14.16 Trap sizes are noted on the drawing. Individual vents are not shown for clarity.

selected, so go to that column (Table 14.2). Go down the column until the fixture-unit number is the same as or more than the amount being served—in this case, 3 DFU, which is greater than the 2 DFU value of the bathtub. Now, move horizontally to the left and select the minimum pipe size required. In this case, a 1½ in or 38 (40) mm. Now, note the information accumulated in tabular form as shown:

	DFU	Pipe Size
No. 1 tub	2	1½ in (38 or 40 mm)

For all fixtures, there is a maximum horizontal distance between the fixture trap and a vent. The distance depends on the pipe size being used. In this case, the plumbing code states that this distance is 5 ft (1.5 m).

5. DFU. The section of pipe that serves the lavatory and kitchen sink must be sized. From Table 14.1, the number of fixture units for the kitchen sink being served is 2 DFU

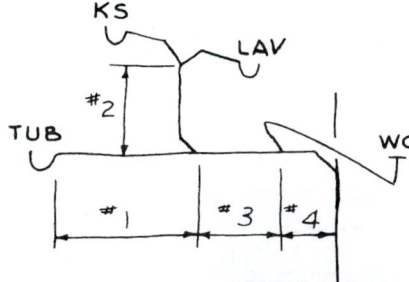

FIGURE 14.17 A fixture branch for the design example.

and for the lavatory, 1 DFU, for a total of 3 DFU. Referring to Table 14.2, the minimum pipe size is 1½ in or 38 mm. The maximum horizontal distance from fixture trap to vent is limited to 5 ft (1.5 m). The accumulated information is listed as follows.

Total DFU

	DFU	Pipe Size
No. 2 sink/lav. to branch	3	1½ in (38 or 40 mm)

TABLE 14.5   HORIZONTAL BRANCH TABULATION.

	d.f.u.	Pipe	Dist. to Vent
#1 Tub to Lav	2	1½″	5′-0″
#2 KS, Lav to Tub	3	1½″	5′-0″
#3 KS, Lav, Tub to WC	5	2″	8′-0″
#4 KS, Lav, Tub, WC to Stack	8	3″	10′-0″

SI UNITS

		Pipe	Dist. to Vent
#1 Tub to Lav	2	38 mm	1.5 m
#2 KS, Lav to Tub	3	38 mm	1.5 m
#3 KS, Lav to Tub to WC	5	50 mm	2.4 m
#4 KS, Lav, Tub, WC to Stack	8	75 mm	3.0 m

6. The section of branch marked No 3 serves the tub, kitchen sink, and lavatory. From Table 14.1:

Bathtub	2 DFU
Kitchen sink	2 DFU
Lavatory	1 DFU
Total	5 DFU

From Table 14.2 the pipe size is 2 in or 50 mm. Add this information to the tabular form.

	DFU	Pipe Size
No. 3 sink, lav., & tub	5	2 in (50 mm)

7. Now, the last section (No. 4) is calculated and tabulated. At this point, the branch services all of the fixtures, which are a fixture group and a kitchen sink. From Table 14.1, the fixture units are:

1 bathroom group	6 DFU
1 kitchen sink	2 DFU
Total	8 DFU

From Table 14.2, the pipe size is 2½ in or 65 mm, but since the pipe cannot be smaller than the largest fixture trap, a 3-in or 75-mm pipe is required (water closet, 3-in or 75-mm trap).

	DFU	Pipe Size
No. 4 bathroom and sink	8	3 in (75 mm)

8. With all branch sizes tabulated (Table 14.5), transfer the sizes to the isometric sketch, as shown in Figure 14.18. Because each floor is exactly the same, all of the branch sizes are the same, and all floors are noted on the isometric drawing. If any of the branches were different, the same procedure would be followed to determine the pipe sizes in these different branches.

9. The stack size is selected next. Because the stack handles human waste, it is a *soil* stack. The stack must be selected and sized for the total fixture units that it must handle. As noted on the sketch in Figure 14.19, each floor has a total of 8 DFU, for an overall total of 32 DFU.

Based on 32 DFU for the entire stack, the minimum stack size from Table 14.2 would be 2½ in, except that the stack

SI UNITS

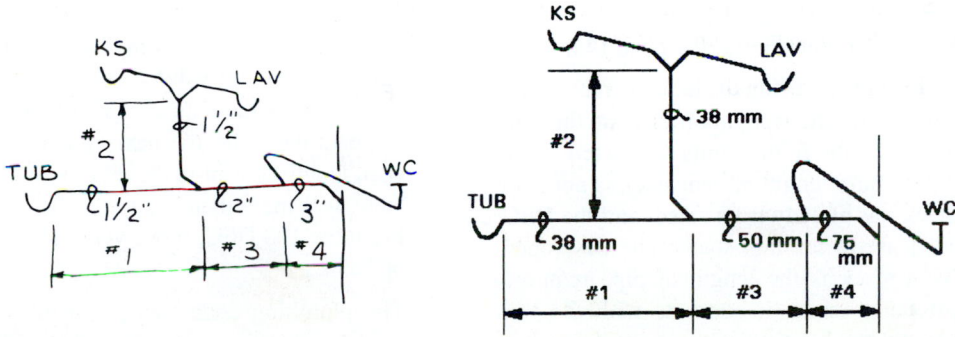

FIGURE 14.18   A horizontal fixture branch for the design example.

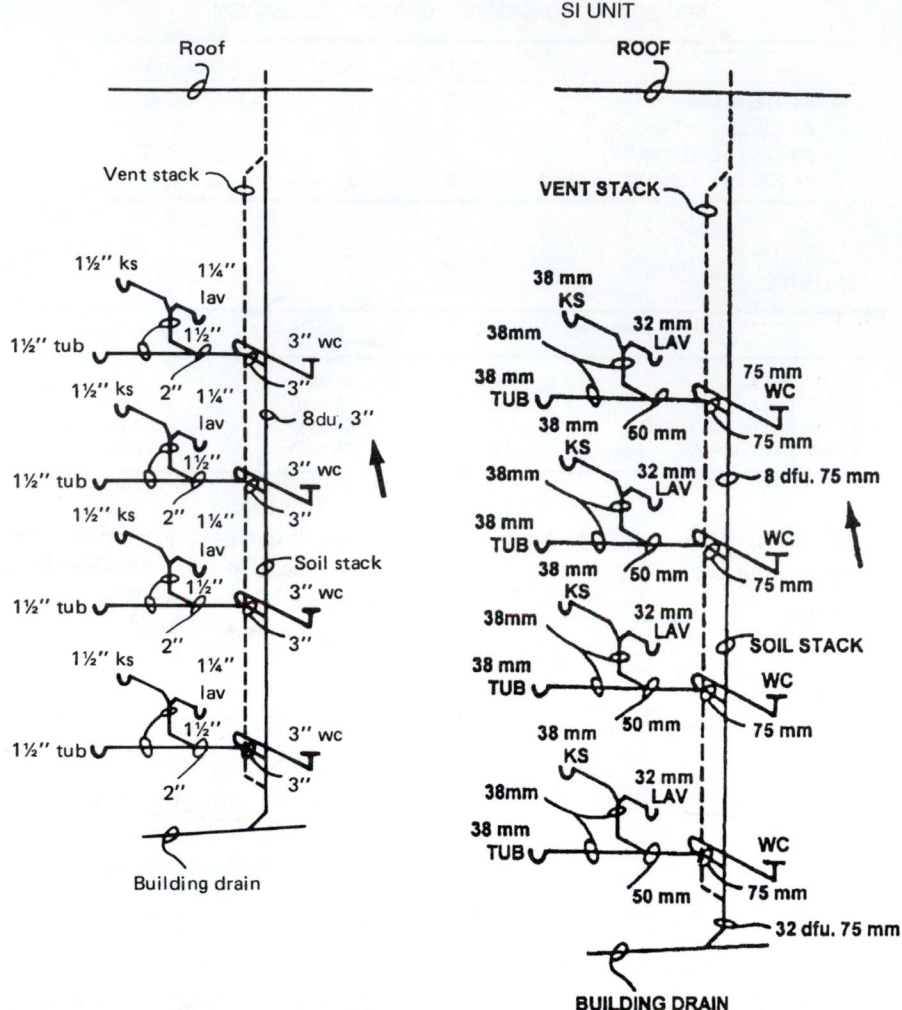

FIGURE 14.19 Assigned drainage fixture units for the design example. Individual vents are not shown for clarity.

must be at least as large as the fixture branch: in this design, it would be 3 in. Note the stack size on the sketch as shown in Figure 14.20 and tabulated in Table 14.6.

In metric (SI) units, based on 32 DFU, for the entire stack, the minimum size from the chart would be 65 mm, except that the stack must be at least as large as the fixture branch—in this design, 75 mm. Note the stack size on the sketch as shown in Figure 14.20, and tabulated in Table 14.6.

10. The vent stack is sized next, using the table in Table 14.4. It is necessary to know the maximum size of the soil stack (left column) and the fixture units connected to the vent stack; the maximum length of vent stack is noted in feet. The soil stack has already been sized, and the DFU have been totaled as 32 for this stack. The developed length of the vent stack is the length of pipe required from the lowest point where it connects with the soil stack to the point where it terminates outside the building, as illustrated in Figure 14.21.

The next step is to determine the size and location of the main (vent) stack. The code requires that every building "shall have at least one main stack." This main stack must be sized to handle the total fixture units on the system.

In this exercise, the developed length is about 52 ft. To size the stack use Table 14.4, find the soil stack size (4 in) and then move horizontally to the right to the fixture-unit column. There are three listings for a 4-in stack, all with different fixture-unit values. In this case, there are 64 DFU, which means the listing of 100 DFU must be used. Now move to the right; the next number represents the developed length of vent in the design; in this case, 52 ft. In the table, it is more than 35, so the 100 DFU column is used. From the 100 DFU, move vertically up to read a vent size of 2½ in.

The plumbing code states that the main vent and vent stack shall be sized in accordance with the table "and be not less than 3 in in diameter." In addition, it states that

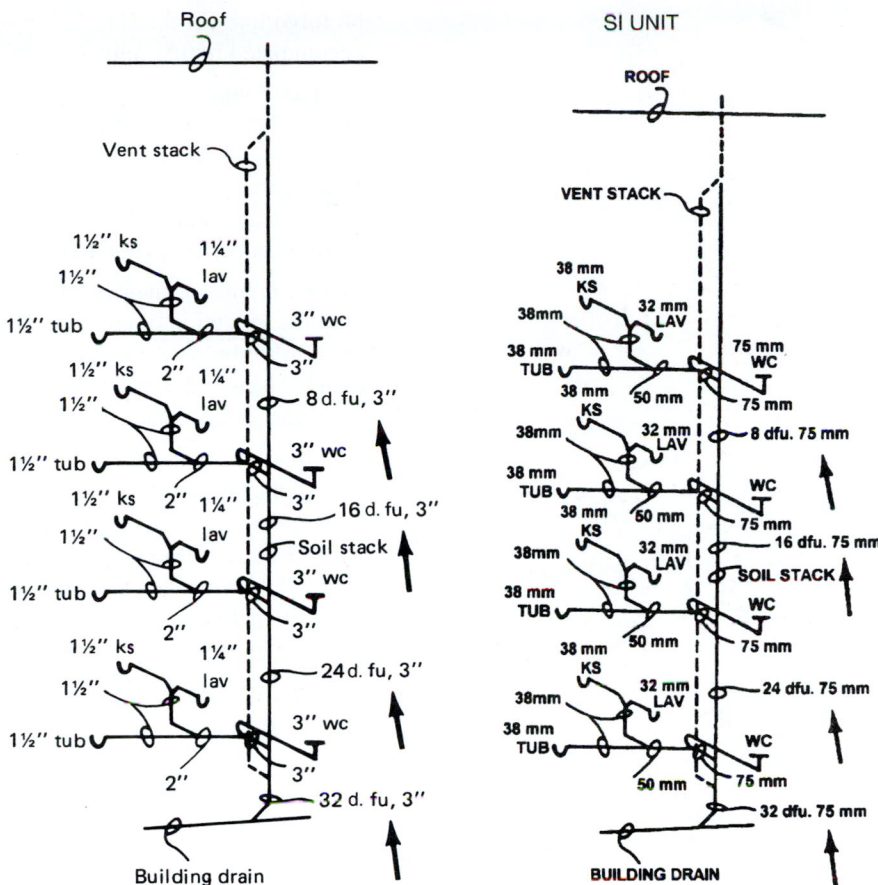

FIGURE 14.20 Pipe sizes for the design example. Individual vents are not shown for clarity.

TABLE 14.6   TABULATED STACK SIZE.

STACK SIZE			
Floor	fu	Accum. d.f.u.	Pipe size
4	8	8	3″ (75 mm)
3	8	16	3″ (75 mm)
2	8	24	3″ (75 mm)
1	8	32	3″ (75 mm)

the size of the vent stack must not be reduced all the way through the roof. So in this design, the minimum vent stack size of 3 in is selected.

In metric (SI) units, the developed length is about 15.9 m. To size the stack, find the soil stack size (100 mm) and move horizontally to the right to check the fixture unit column. At this point, note that there are three listings for a 100 mm soil stack, all with different fixture-unit values. In this case there are 64 DFU, which means the listing of 100 DFU must be used. Now move to the right; the next number represents the developed length of vent in the design—in this case, 15.9 m. In the table it is more than

11 m, so the 30 m column is used. From the 30 m, move vertically up to read a vent size of 65 mm.

The plumbing code states that the main vent and vent stack shall be sized in accordance with the table "and be not less than 75 mm in diameter." In addition, it states that the size of the vent stack must not be reduced all the way through the roof. So in this design, the minimum vent stack size of 75 mm is selected.

Code requirements also stipulate that the vent stack must terminate no less than 6 in (150 mm) above the roof (Figure 14.22), and if the roof is to be used for other than weather protection (for example, as a terrace or balcony), the vent stack must run at least 5 ft (1.5 m) above the roof. Most codes require at least a 3-in (75-mm) vent through the roof.

11. The next step in the design is to size the building drain. The building drain must be sized for the total amount of DFU connected to the point at which the main vent is connected.

In this exercise, there are two stacks serving the building, which means that the two stacks must be connected to the building drain. A sketch similar to Figure 14.23 helps to prevent any confusion or the possibility that a stack may

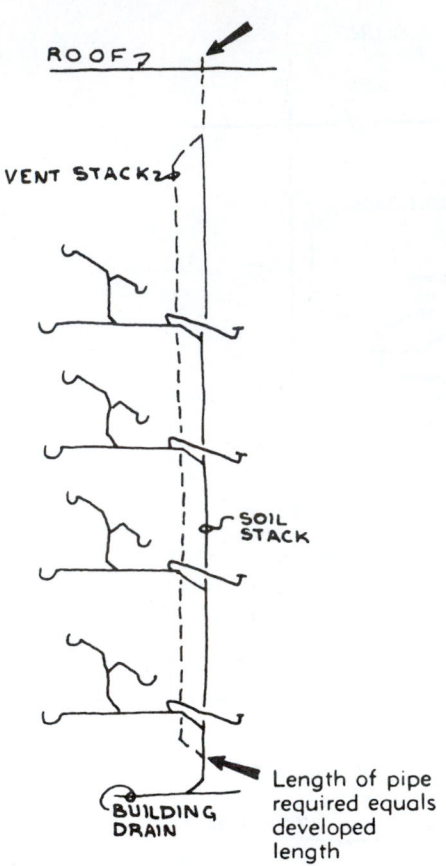

FIGURE 14.21 The developed length of a vent is computed to its terminating point at the roof.

be forgotten. On the sketch, and in tabular form, list the accumulated fixture units.

Next, the slope of the drain must be determined and noted on the sketch; in this case, a ⅛-in slope per ft (10.4 mm per m) is selected. Building drain sizes are found in Table 14.3. In this design, the main vent is considered to be the end vent, as noted in Figure 14.24, so the entire building drain must be sized for 64 DFU.

12. The plumbing drainage system has now been designed, with the exception of checking the location of the vent in relation to the fixtures. As the fixture branch sizes were being selected, the maximum distance from the trap to a vent was also tabulated (Table 14.5).

In reviewing the fixture branch and trap sizes in Table 14.5, each of the tub, lavatory, and sink traps can be no more than 5 ft (1.5 m) from a vent, and the water closet no more than 10 ft (3 m). A review of the floor plan finds that the total width of the bathroom is 7 ft-6 in (2.25 m). This means that the vent could be located in the wall near the lavatory and be within the limits of all listed horizontal distances.

If fixtures are so spread out that one or more fixtures are beyond the maximum horizontal distance, there are two options:

a. Increase the size of the horizontal fixture branch, which automatically increases the allowable horizontal distance. For example, change length 1 from 1½ in (38 mm) to 2 in (50 mm), and it

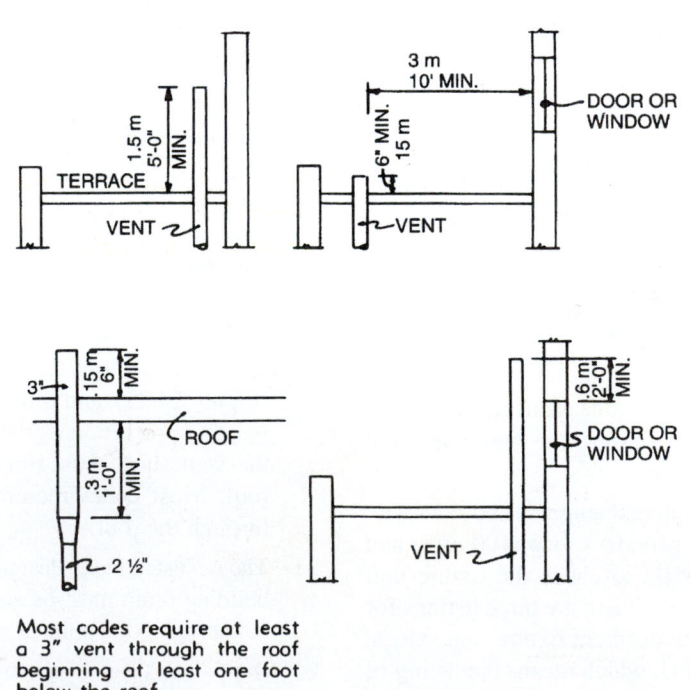

Most codes require at least a 3" vent through the roof beginning at least one foot below the roof.

FIGURE 14.22 Vents must terminate away from walls, windows, and doors.

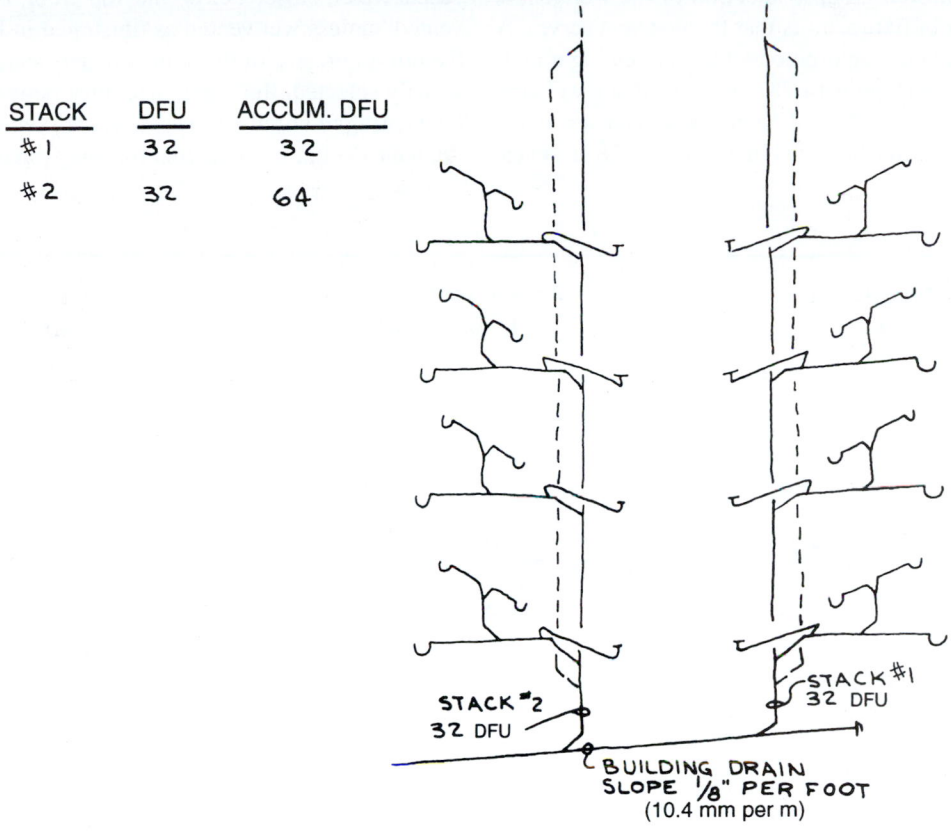

STACK	DFU	ACCUM. DFU
#1	32	32
#2	32	64

FIGURE 14.23 Sketch of building drain for the design example.

changes the horizontal distance from 5 ft to 8 ft (1.5 m to 2.4 m). Many times this is the most economical solution.

b. Add more vents. Instead of trying to service a group of fixtures with one vent, perhaps a vent could be added to service any fixtures that are beyond the allowable distance. A variety of venting solutions for various groups of fixtures is shown in Figs. 14.25 through 14.28.

In this example, a single vent off the top of the lavatory–kitchen sink fixtures (referred to as a *wet vent* and discussed in the following paragraph) will be used.

The size of the wet vent is selected from Tables 14.4 and Figures 14.25 through 14.28. But a little explanation is needed. First, a review of the sketch in Figure 14.25 shows that the portion of the vent referred to as a *wet vent* is actually a part of the drainage system for the lavatory and the kitchen sink. Because it acts as both a *vent* and a *drainage* pipe, it is referred to as a *wet vent*.

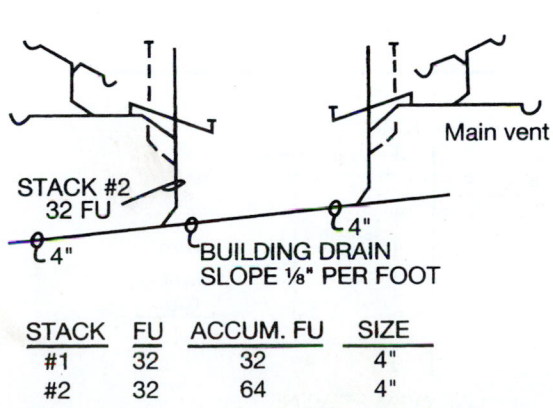

S1 UNITS

STACK	FU	ACCUM. FU	SIZE
#1	32	32	4"
#2	32	64	4"

STACK	FU	ACCUM. FU	SIZE
#1	32	32	100 mm
#2	32	64	100 mm

FIGURE 14.24 Tabulation and sketches for building drain size.

For single bathroom groups, selection of the wet vent is based on the number of fixture units that the wet vent serves. A careful review of the code requirements for wet venting a multistory bathroom group (Figure 14.28) indicates that a wet vent and its extension must be 2 in diameter and can serve the kitchen sink, lavatory, and bathtub. In Figure 14.28 it states,

"each water closet below the top floor is individually back vented" unless wet vented as illustrated in Figure 14.26. When the minimum size of the vent is larger than the waste sizes previously selected, the larger size must be used. Note the size of the venting required on the schematic sketch, and the plumbing drainage design is completed for this project.

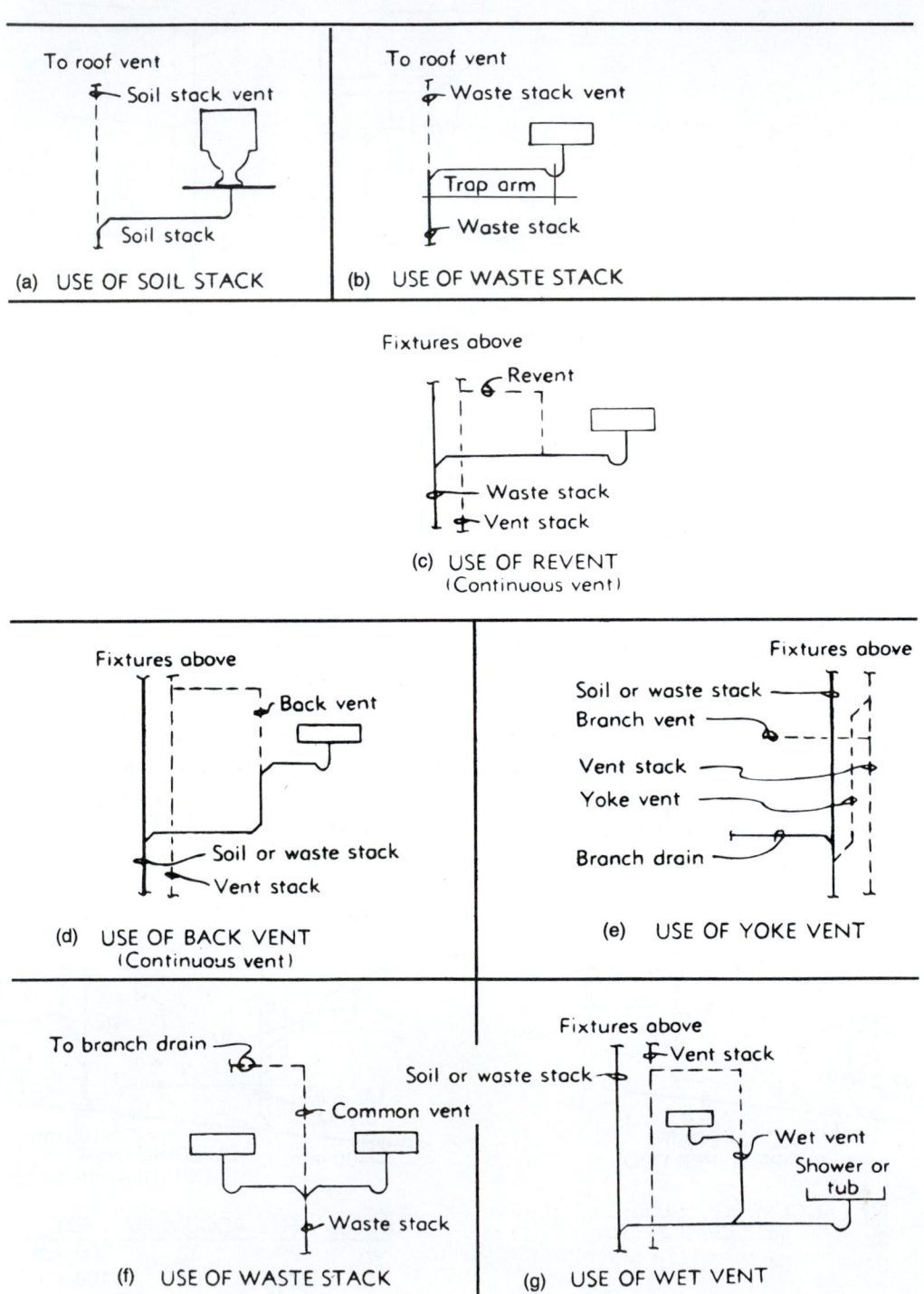

FIGURE 14.25 Venting options.

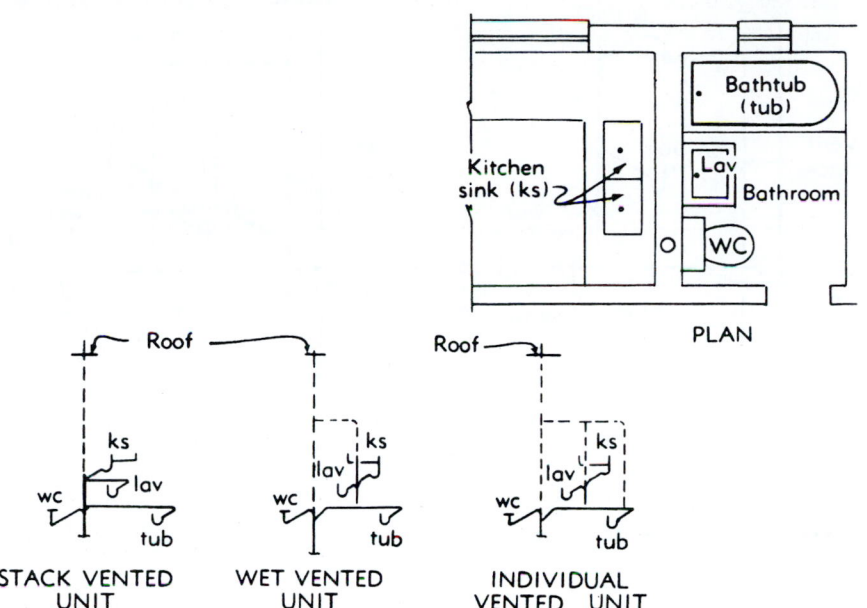

FIGURE 14.26 Vent for a single-family dwelling for the design example.

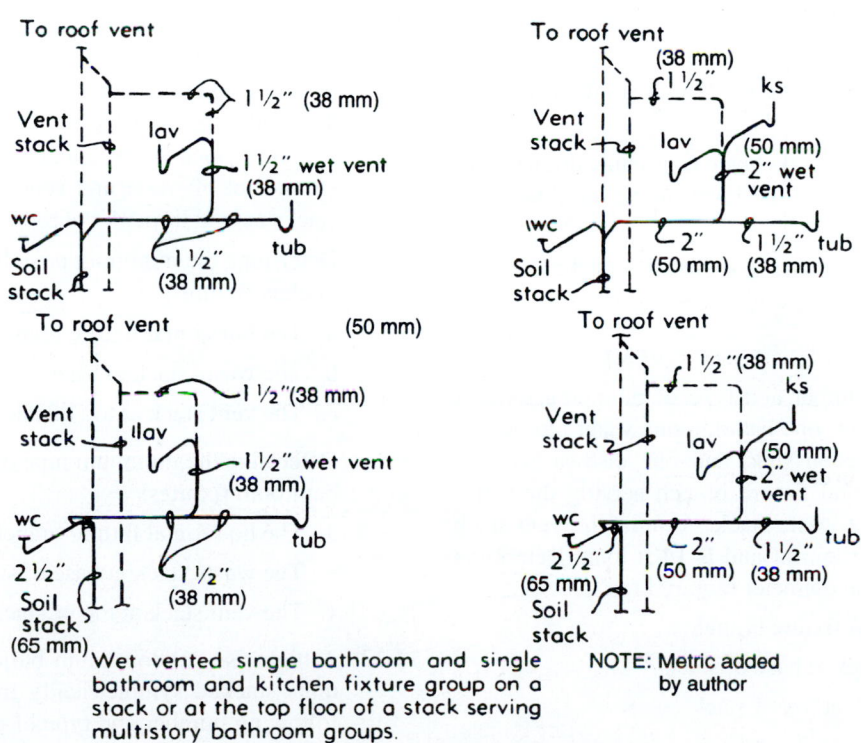

Wet vented single bathroom and single bathroom and kitchen fixture group on a stack or at the top floor of a stack serving multistory bathroom groups.

NOTE: Metric added by author

FIGURE 14.27 Wet venting at the top floor for the design example.

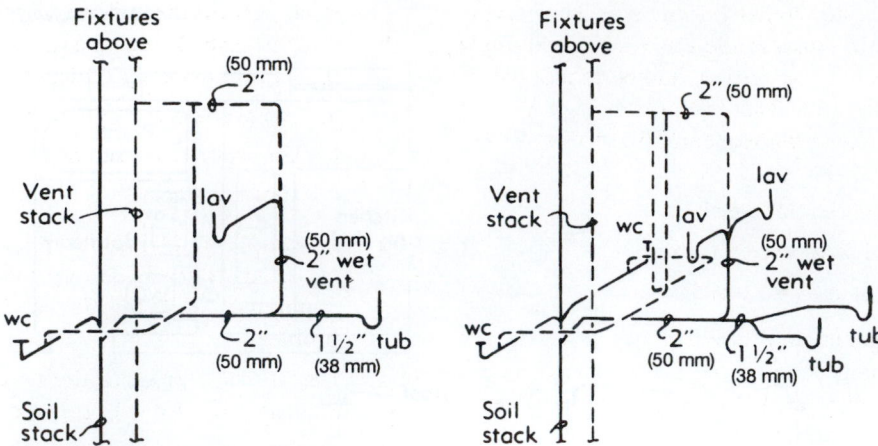

FIGURE 14.28   Wet venting below the top floor for the design example.

STUDY QUESTIONS

14-1. What is a trap, where is it located, and how does it work?

14-2. Why are vents required on the waste system? Where are they located in reference to the fixture?

14-3. What is a wet vent, and how does it differ from other types of vents?

14-4. What is the difference between a stack vent and a vent stack? Using a sketch, show the location of a stack vent and a vent stack in a multistory design.

14-5. What is the difference between a soil stack and a waste stack?

14-6. Sketch and locate the house (building) drain and the sewer.

14-7. What provisions must be made to provide drainage for fixtures located below the level of the building drain and the sewer?

14-8. Why does the designer make sketches of the drainage piping?

Design Exercises

14-9. The following number and type of plumbing fixtures serve an apartment: one bathtub, one water closet, one lavatory, one kitchen sink, and one dishwasher. Assume the horizontal fixture branch serving these fixtures flows into a waste stack. Assume the vent stack extends through the roof and is 14 ft long. Determine the minimum pipe diameter required for:

a. The horizontal fixture branch

b. The waste stack at base of stack

c. The vent stack at top of stack

14-10. The following number and type of plumbing fixtures serve a condominium: two bathtubs, two water closets,

two lavatories, one kitchen sink, and one dishwasher. Assume the horizontal fixture branch serving these fixtures flows into a waste stack. Assume the vent stack extends through the roof and is 18 ft long. Determine the minimum pipe diameter required for:

a. The horizontal fixture branch

b. The waste stack at base of stack

c. The vent stack at top of stack

14-11. The following number and type of plumbing fixtures serve a residence: two bathtubs, three water closets, three lavatories, one kitchen sink, and one dishwasher. The bathroom fixtures are served by one set of waste and vent stacks. The kitchen fixtures are served by a second set of waste and vent stacks. Assume the vent stacks extend through the roof and are 20 ft long.

Determine the minimum pipe diameter required for the kitchen fixtures:

a. The horizontal fixture branch

b. The waste stack at base of stack

c. The vent stack at top of stack

Determine the minimum pipe diameter required for the bathroom fixtures:

d. The horizontal fixture branch

e. The waste stack at base of stack

f. The vent stack at top of stack

14-12. A high-rise condominium building has condominium units stacked symmetrically from floor to floor. The following number and type of plumbing fixtures serve each condominium: one bathtub, two water closets, two lavatories, one kitchen sink, and one dishwasher.

Assume that the horizontal fixture branches serving these fixtures flow into a single waste stack and single vent stack. Each stack serves seven condominium units. Assume the vent stack extends through the roof and is 78 ft long. Determine the minimum pipe diameter required for:

a. The horizontal fixture branch

b. The waste stack at base of stack

c. The vent stack at top of stack

14-13. A high-rise condominium building has condominium units stacked symmetrically from floor to floor. The following number and type of plumbing fixtures serve each condominium: one bathtub, two water closets, two lavatories, one kitchen sink, and one dishwasher. Assume that the horizontal fixture branches serving these fixtures flow into a single waste stack and single vent stack. Each stack serves ten condominium units. Assume the vent stack extends through the roof and is 112 ft long. Determine the minimum pipe diameter required for:

a. The horizontal fixture branch

b. The waste stack at base of stack

c. The vent stack at top of stack

14-14. Design the plumbing drainage for the apartment building in Appendix B. Use acrylonitrile-butadiene-styrene copolymer (ABS) pipe and tubing for this design problem.

14-15. Design the plumbing drainage for the residence in Appendix C. Use ABS pipe for this design problem.

14-16. Design the plumbing drainage for the residence in Appendix D. Use ABS pipe for this design problem.

DRAIN, WASTE AND VENT LOAD ANALYSIS WORKSHEET

ADVANCED BUILDING CONSULTANTS, LLC
12450 WEST BELMONT AVENUE
LITTLETON, CO 80127 -1515
303-933-1300

Project:		No.:
Address:		
Designer:		Date:
Checked by:		Date:

no.	fixture type or description	trap arm		drain		vent		notes
		load (DFU)	size (dia)	load (DFU)	size (dia)	load (DFU)	size (dia)	

PAGE ____ OF ____

FIGURE 14.29   An example of a load analysis worksheet for sanitary drain waste and vent computations.

WASTEWATER TREATMENT AND DISPOSAL SYSTEMS

15.1 WASTEWATER TREATMENT STANDARDS

History

Before about 1850, human and liquid industrial wastes were typically dumped in the street or alley or conveyed to the nearest body of water without treatment. Groundwater and other sources for drinking and bathing were regularly contaminated with raw sewage. Epidemics of cholera, typhoid, dysentery, and other waterborne diseases killed thousands. Outbreaks of these diseases were especially devastating in densely populated areas. Bathing was not accepted as a common practice and was viewed as a health hazard.

About 1854, the connection between cholera and sewage-contaminated water was first discovered. Better attempts were made to treat and dispose of sewage separately from drinking water. However, until the latter part of the 1800s, most U.S. communities still allowed discharges of untreated or inadequately treated wastewater from homes and industries through combined storm and sanitary sewers into lakes, bays, oceans, streams, and rivers. Treatment standards varied from municipality to municipality. As population increased, damage to the environment and risks to public health reached dangerous levels.

Because of the vast land and water resources available for dumping wastes and wastewater, treatment of wastewater in the United States received little attention until the early 1900s. By this time health problems associated with improper disposal of waste and wastewater were causing real problems, especially in larger cities. Cities were having trouble obtaining areas for disposal, which led to more intensive methods of treatment.

In the 1970s, the U.S. Congress passed the Clean Water Act. This legislation led to the establishment of national water quality standards and limits for the discharge of pollutants. In recent years, amendments to this legislation have transferred implementation of pollution control to individual state governments. However, a few U.S. communities were still permitted to release untreated or inadequately treated wastewater into streams and rivers through the late 1900s.

Today, governmental leaders and health authorities are responsible for ensuring that state standards for wastewater treatment and water quality are met consistently—not only at inspection time, but always to protect public health and the environment. In some cases, treatment plant operators and community leaders may be held personally liable for noncompliance. Homeowners also are personally liable for malfunctioning on-site systems. On-site systems must be properly operated and maintained to protect groundwater and other drinking water sources, as well as the health of family and neighbors.

Wastewater

As described in the previous chapter, wastewater (sewage) is "used" water. It contains the waste products, excrement, or other discharge from the bodies of human beings or animals, and other noxious or poisonous substances that are harmful to the public health, or to animal or aquatic life, or to the use of water for domestic water supply or for recreation. Wastewater is a combination of the liquid and water-carried wastes from residences, commercial buildings, industrial plants, and institutions, together with any groundwater, surface water, and storm water that has infiltrated the public sewage system. Although the terms *wastewater* and *sewage* are used interchangeably, sewage technically contains waste products or excrement from human beings or animals and wastewater may not. Buildings nearly always generate sewage, and thus wastewater.

Wastewater from residences, apartments, motels, office buildings, and other similar types of buildings is referred to as *domestic wastewater*. There are two types of domestic wastewater: gray water and black water.

Gray water is wastewater that typically contains the residues of washing processes. It is generated in the bathtub, shower, sink, lavatory, and clothes washing machine. Gray water accounts for about two-thirds of the wastewater produced in a typical residence. *Black water* is wastewater that contains fecal matter and urine. It is produced in water closets (toilets), urinals, and bidets. Black water and gray water have different characteristics, but both contain pollutants and disease-causing agents that require treatment. Some areas in the United States permit the use of innovative *gray water reuse systems* that safely recycle household gray water for reuse in toilets or for irrigation to conserve water and reduce the flow to treatment systems. These systems are discussed later in this chapter.

Commercial wastewater is nontoxic, nonhazardous wastewater from commercial and institutional food service operations and beauty salons. It is usually similar in composition to domestic wastewater, but may occasionally have one or more of its constituents exceed typical domestic ranges. *Industrial wastewater* is process and nonprocess wastewater from manufacturing, commercial, laboratory, and mining operations, including the runoff from areas that receive pollutants associated with industrial or commercial storage, handling, or processing. Industrial wastewater requires special handling, and such treatment is not discussed in detail here. It is important to note that such wastes should not be put into a community wastewater system without first receiving approval from the governing authorities.

Because of the variety of characteristics tied to commercial and industrial wastewater, communities need to assess each source individually to ensure that adequate treatment is provided. For example, public restrooms may generate wastewater with some characteristics similar to domestic sewage, but usually at higher volumes and at different peak hours. The volume and pattern of wastewater flows from rental properties, hotels, and recreation areas often vary seasonally as well. Laundries differ from many other nonresidential sources because they produce high volumes of wastewater containing lint fibers. Restaurants typically generate a lot of oil and grease. It may be necessary to provide pretreatment of oil and grease from restaurants or to collect it prior to treatment, for example, by adding grease traps to septic tanks.

Industrial wastewater also may require additional treatment steps. For example, storm water should be collected separately to prevent the flooding of treatment plants during wet weather. Screens often remove trash and other large solids from storm sewers. In addition, many industries produce wastewater high in chemical and biological pollutants that can overburden on-site and community systems. Additionally, with dairy farms and breweries, communities may require that these types of nonresidential sources provide their own treatment or preliminary treatment to protect community systems and public health.

Wastewater Constituents

Wastewater is mostly water by weight. Wastewater released by residences, businesses, and industries is approximately 99.94% water. Only about 0.06% of the wastewater is dissolved and suspended solid material. Other matter makes up only a small portion of wastewater, but can be present in large enough quantities to endanger public health and the environment. Practically anything that can be flushed down a toilet, drain, or sewer can be found in wastewater. Even household wastewater contains many potential pollutants. As a result, wastewater treatment is as important to public health as drinking water treatment.

The wastewater constituents of most concern are those that have the potential to cause disease or detrimental environmental effects. A wide range of unhealthy constituents, with widely ranging problems, may be found in the wastewater of residences and commercial and industrial facilities. These include the following:

Organisms Many different types of organisms live in wastewater and some are essential contributors to treatment. Bacteria, protozoa, and worms work to break down certain carbon-based (organic) pollutants in wastewater. Through this process, organisms turn wastes into carbon dioxide, water, or new cell growth. Bacteria and other microorganisms are particularly plentiful in wastewater and accomplish most of the treatment. Most wastewater treatment systems are designed to rely in large part on these biological processes.

Pathogens Many disease-causing viruses, parasites, and bacteria are present in wastewater. These pathogens often originate from people and animals that are infected with or are carriers of a disease. Gray water and black water can contain enough pathogens to pose a sufficient risk to public health from sources in communities that include residences, hospitals, schools, farms, and food processing plants. In addition, municipal drinking water sources are vulnerable to health risks from wastewater pathogens if they are contaminated.

Some illnesses from wastewater-related sources are quite common. Gastroenteritis (from parasitic protozoa *Giardia lambia* and *Cryptosporidium*) is not unusual in the United States. Other diseases transmitted in wastewater include hepatitis A, typhoid, polio, cholera, and dysentery. Outbreaks of these diseases can occur when drinking water from wells is polluted by wastewater or there is exposure from recreational activities in polluted waters. Animals and insects that come in contact with wastewater can spread some illnesses.

Organic Matter Organic materials that originate from plants, animals, or synthetic organic compounds are found everywhere in the environment. They enter wastewater in the form of human wastes, paper products, detergents, cosmetics, foods, and other agricultural, commercial, and industrial sources. *Biodegradable* waste is easily broken down by aerobic bacteria, which are natural organisms. Waste that cannot be broken down by other living organisms is *nonbiodegradable*. Many organics are biodegradable (e.g., proteins, carbohydrates, or fats).

Large quantities of biodegradable materials are dangerous, because the bacteria consume large amounts of dissolved oxygen in water, which robs other aquatic life of the dissolved oxygen they need to live. *Biochemical* or *biological oxygen demand (BOD)* is a measurement procedure used to assess how fast biological organisms are depleting dissolved oxygen in a body of water, as measured in milligrams per liter (mg/L) over a 5-day period. High BOD levels indicate that much of the available dissolved oxygen has been consumed by aerobic bacteria. Pristine waterways have a 5-day BOD level of less than 1 mg/L, while moderately polluted waters are typically in the range of 2 to 8 mg/L. Effectively treated municipal sewage has a 5-day BOD level of 20 mg/L or less. Untreated sewage varies, but is typically around 200 mg/L in the United States.

Many organic compounds developed for agriculture and industry are nonbiodegradable. In addition, certain synthetic organics are highly toxic. Pesticides and herbicides are frequently disposed of improperly or are leached from soil and carried by storm water. In receiving waters, they kill or contaminate fish, making them unfit for human consumption. They also can damage processes in treatment plants.

Oil and Grease   Bacteria do not quickly break down fatty organic materials from animals, vegetables, and petroleum. When large amounts of oils and greases are discharged to receiving waters from community systems, they increase BOD and may float to the surface and solidify, causing aesthetically unpleasing conditions. Oils and greases also can trap trash, plants, and other materials, causing foul odors, attracting flies and mosquitoes. On-site sewage treatment systems can be harmed by too much oil and grease, which can clog on-site system drainage field pipes and soils, adding to the risk of system failure. Excessive grease also adds to the septic tank scum layer, causing more frequent tank pumping to be required. Petroleum-based oils used for vehicle motors are considered hazardous waste and should be collected and disposed of separately from wastewater.

Inorganics   Inorganic minerals, metals, and compounds, such as sodium, potassium, calcium, magnesium, cadmium, copper, lead, nickel, and zinc, are common in wastewater from both residential and nonresidential sources. They can originate from a variety of sources in the community, including industrial and commercial sources, storm water, and inflow and infiltration from cracked pipes and leaky manhole covers. Large amounts of many inorganic substances can contaminate soil and water. Some are toxic to animals and humans and may accumulate in the environment. For this reason, extra treatment steps are often required to remove inorganic materials from industrial wastewater sources. For example, heavy metals that are discharged with many types of industrial wastewaters are difficult to remove by conventional treatment methods. Poisonings from heavy metals in drinking water are possible. In receiving waters, they kill or contaminate aquatic life.

Nutrients   Wastewater often contains large amounts of the nutrients nitrogen and phosphorus, which promote plant growth. Organisms only require small amounts of nutrients in biological treatment, so there normally is excess available in treated wastewater. In severe cases, excessive nutrients in receiving waters cause algae and other plants to grow quickly, depleting oxygen in the water. Deprived of oxygen, fish and other aquatic life die. Nutrients from wastewater have also linked to ocean "red tides" that poison fish and cause illness in humans.

Solids   Solid materials in wastewater can consist of organic and/or inorganic materials and organisms. The solids must be significantly reduced by treatment or they can increase BOD when discharged to receiving waters and provide places for microorganisms to escape, reducing the effectiveness of disinfection system. Suspended solids can also clog soil absorption fields in on-site sewage treatment systems. Small particles of certain wastewater materials can dissolve like salt in water. Micro-

organisms in wastewater consume some dissolved materials, but others, such as heavy metals, are difficult to remove by conventional treatment. Excessive amounts of dissolved solids in wastewater can have adverse effects on the environment.

Gases   Many gases in wastewater can cause odors or are dangerous. For example, methane gas, a by-product of anaerobic biological treatment, is highly combustible. The gases hydrogen sulfide and ammonia can be toxic and pose asphyxiation hazards. Ammonia as a dissolved gas in wastewater is dangerous to fish. Many gases emit odors, which can be a serious nuisance and, in severe cases, lower property values and affect the local economy. Unless effectively contained or minimized by design and location, wastewater odors can affect quality of life. Special precautions need to be taken near septic tanks, manholes, treatment plants, and other areas where wastewater gases can collect.

Other Important Wastewater Characteristics

In addition to the many substances found in wastewater, there are other characteristics that system designers and operators use to evaluate wastewater. For example, the color, odor, and turbidity of wastewater give clues about the amount and type of pollutants present and the treatment necessary. The following are some other important wastewater characteristics that can affect public health and the environment, as well as the design, cost, and effectiveness of treatment.

Temperature   The best temperatures for wastewater treatment range from 77° to 95°F. In general, biological treatment activity accelerates in warm temperatures and slows in cool temperatures, but extreme hot or cold can stop treatment processes altogether. Therefore, some systems are less effective during cold weather and some may not be appropriate for very cold climates. Wastewater temperature also affects receiving waters. Hot water, for example, which is a by-product of many manufacturing processes, can be a pollutant. When discharged in large quantities, it can raise the temperature of receiving streams locally and disrupt the natural balance of aquatic life.

pH   The acidity or alkalinity of wastewater affects both treatment and the environment. Low pH indicates increasing acidity, while a high pH indicates increasing alkalinity (a pH of 7 is neutral). The pH of wastewater needs to remain between 6 and 9 to protect organisms. Acids and other substances that alter pH can inactivate treatment processes when they enter wastewater from industrial or commercial sources.

Flow   Whether a system serves a single residence or an entire community, it must be able to handle fluctuations in the quantity and quality of wastewater it receives to ensure proper continuous treatment. Systems that are under-designed or that are overloaded may fail to provide treatment and allow the release of pollutants to the environment.

To design systems that are both as safe and as cost-effective as possible, engineers must estimate the average and maximum (peak) amount of flows generated by various sources. Because extreme fluctuations in flow can occur, estimates are based on observations of the minimum and maximum amounts of water used on an hourly, daily, weekly, and seasonal basis. The possibility of instantaneous peak flow events that result from fixtures being used at once is also taken into account. The number of possible users or units and the number, type, and efficiency of water-using fixtures and appliances at the source are considered in design. Estimating flow volumes for centralized treatment systems is a complicated task, especially when designing a new treatment plant in a community where one has never existed previously.

15.2   COMMUNITY WASTEWATER TREATMENT AND DISPOSAL

Wastewater Removal

A *community wastewater treatment and disposal system,* a network of pipes that transport wastewater to treatment plants where it is treated and released to the environment, serves most buildings in the United States. At a typical wastewater treatment plant, several million gallons of wastewater are treated each day, about 50 to 100 gallons for every person using the system.

In these systems, wastewater is carried from the building to the community treatment facility through below-ground piping systems that are generally classified according to the type of wastewater flowing through them. Sanitary sewage is typically separated from storm water (from rain and snowmelt) by a separate piping arrangement. In less modern systems, known as combined systems, the system carries both sanitary water and storm water.

The building drain in a building sanitary drainage system is connected to a *building sewer* about 2 to 5 ft (0.6 to 1.5 m) outside of the building. The building sewer connects the building sanitary drainage system to the community sanitary sewer main located several feet below the surface of a street or alley. Sewer lines must be sloped to permit gravity flow of wastewater at a velocity of at least 1.5 ft/s (0.46 m/s), because at lower velocities the solid waste tends to settle in the pipe.

Storm water mains are similar in design to sanitary sewers except that they have a much larger diameter. Certain types of sewers, such as inverted siphons and pipes from pumping stations, flow under pressure, and are thus called *force mains.* Urban sewer mains generally discharge into *interceptor sewers* that join to form a large *trunk line,* which discharges wastewater into the community sewage treatment plant where it is treated.

After being treated at the community wastewater treatment plant, the treated wastewater is usually released into a lake or stream, where it flows toward the ocean. Reconditioned wastewater will generally be used again and again along the way for irrigation, by industry, and as drinking water, or it will evaporate into the atmosphere and return again as rain as part of the hydrological cycle. Wastewater treatment plants operate at a critical point in the water cycle to protect against excessive pollution.

Sewage Treatment and Disposal

Sewage treatment is a multistage process designed to restore the quality of wastewater before it reenters a body of water such as a stream, river, or lake. The objective is to reduce or entirely remove organic matter, solids, nutrients, disease-causing organisms, and other pollutants from wastewater.

Community wastewater treatment plants are operated by a municipality or special district and serve a community. Processes involved in large wastewater treatment plants are usually classified as being part of preliminary, primary, secondary, or tertiary treatment.

Preliminary Wastewater Treatment

Preliminary treatment to screen out, grind up, or separate debris is the first step in wastewater treatment. Sticks, rags, large food particles, sand, gravel, toys, and so on are removed at this stage to protect the pumping and other equipment in the treatment plant. Treatment equipment such as bar screens, comminutors (a large version of a garbage disposal), and grit chambers are used as the wastewater first enters a treatment plant. The collected debris is usually disposed of in a landfill.

Primary Wastewater Treatment

Primary treatment is the second step in wastewater treatment. It separates suspended solids and greases from wastewater. Wastewater is held in a quiet tank for several hours, allowing the solid particles to settle to the bottom and the greases to float to the top. The solids drawn off the bottom and skimmed off the top receive further treatment as sludge. The clarified wastewater then flows on to the next stage of wastewater treatment. Clarifiers and septic tanks are usually used to provide primary treatment.

Secondary Wastewater Treatment

Secondary treatment is a biological treatment process to remove dissolved organic matter from wastewater. Sewage microorganisms are cultivated and added to the wastewater. The microorganisms absorb organic matter from wastewater as their food supply. Three approaches are used to accomplish secondary treatment: fixed film, suspended film, and lagoon systems.

- *Fixed film systems* grow microorganisms on substrates such as rocks, sand, or plastic. The wastewater is spread over the substrate, allowing the wastewater to flow past the film of microorganisms fixed to the substrate. As organic matter and nutrients are absorbed from the wastewater, the film of microorganisms grows and thickens. Trickling filters, rotating biological contactors, and sand filters are examples of fixed film systems.

- *Suspended film systems* stir and suspend microorganisms in wastewater. As the microorganisms absorb organic matter and nutrients from the wastewater, they grow in size and number. After the microorganisms have been suspended in the wastewater for several hours, they are settled out as sludge. Some of the sludge is pumped back into the incoming wastewater to provide "seed" microorganisms. The remainder is wasted and sent on to a sludge treatment process. Activated sludge, extended aeration, oxidation ditch, and sequential batch reactor systems are all examples of suspended film systems.

- *Lagoon systems* are shallow basins that hold the wastewater for several months to allow for the natural degradation of sewage. These systems take advantage of natural aeration and microorganisms in the wastewater to recondition wastewater.

Final Treatment (Disinfection)

Final treatment involves *disinfection;* the removal of disease-causing organisms from wastewater. Treated wastewater can be disinfected by adding chemicals to the water such as chlorine, bromine, iodine and ozone, or by exposing it to ultraviolet light.

Chlorination has become the standard method of disinfection because it remains in the water after the treatment. Although chlorine introduces an unnatural taste and smell to the water, it does remove or mask tastes and odors caused by other potential ingredients. High levels of chlorine may be harmful to aquatic life in receiving streams. Treatment systems often add a chlorine-neutralizing chemical to the treated wastewater before stream discharge.

Other treatments include the following: *aeration,* the exposure of treated water to air, which removes odors and improves taste; *corrosion removal,* to balance the pH (acidity) of the treated water to prevent corrosion damage to pipes, remove odors, and improve taste; and to soften the water. The *softening process* removes calcium and magnesium that have dissolved in water, making the hard water soft. Hard water leaves deposits on plumbing fixtures and does not allow soap to clean as effectively.

15.3   ON-SITE INDIVIDUAL SEWAGE TREATMENT

On-Site Sewage Treatment

Individual sewage treatment systems are used in areas not served by a community wastewater treatment plant. In the

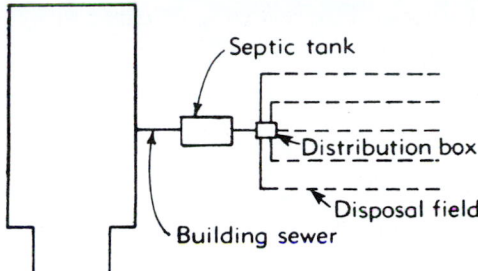

FIGURE 15.1   An on-site sewage treatment (OSST) system for a residence.

United States, about one-fifth of the building stock relies on on-site individual sewage treatment systems to treat wastewater.

On-site sewage treatment (OSST) systems, traditionally called *septic systems,* usually consist of the building sewer, which leads from the building into a septic tank and then into a distribution box that feeds the fluid (effluent) into a drainage field or disposal field. Complex systems may contain a mound drain field system, an aerobic treatment unit, a gray water system tank, a grease interceptor, a dosing tank, and a solids or effluent pump. An OSST system can be installed beyond the building sewer on land of the owner or on another nearby site to which the owner has the legal right to install a system. OSST systems treat wastewater from rural and suburban homes, mobile home developments, apartments, schools, retail facilities, and businesses that do not have access to a community wastewater treatment and disposal system. (See Figure 15.1.)

An OSST system consists of a primary treatment component, such as a *septic tank,* and a disposal component, which is typically the *drainage field.* Household and human wastes flow in a pipe from the building's sanitary drainage system to a septic tank. Inside the septic tank, anaerobic and aerobic bacteria convert the wastes into minerals, gases, and a liquid waste called the *effluent. Clarified effluent* leaves the septic tank and flows in a pipe to a drainage field. If the drainage field piping is higher in elevation than the septic tank outlet, a pump tank and pump are required. The effluent is then distributed as evenly as possible throughout the drainage field where it percolates through the soil. Gravity is typically the driving force behind wastewater and effluent flow.

City and county health departments issue permits for the construction and alteration of OSST systems. These are based on an evaluation of the soil at the site. Consulting geological engineers are hired to conduct extensive soil evaluations for a particular site. State and federal soil scientists can provide assistance to health departments, agriculturalists, and others with local soil problems.

Primary Treatment Equipment

Wastewater from a building is first treated in primary treatment equipment such as tanks or filters. In the primary treatment

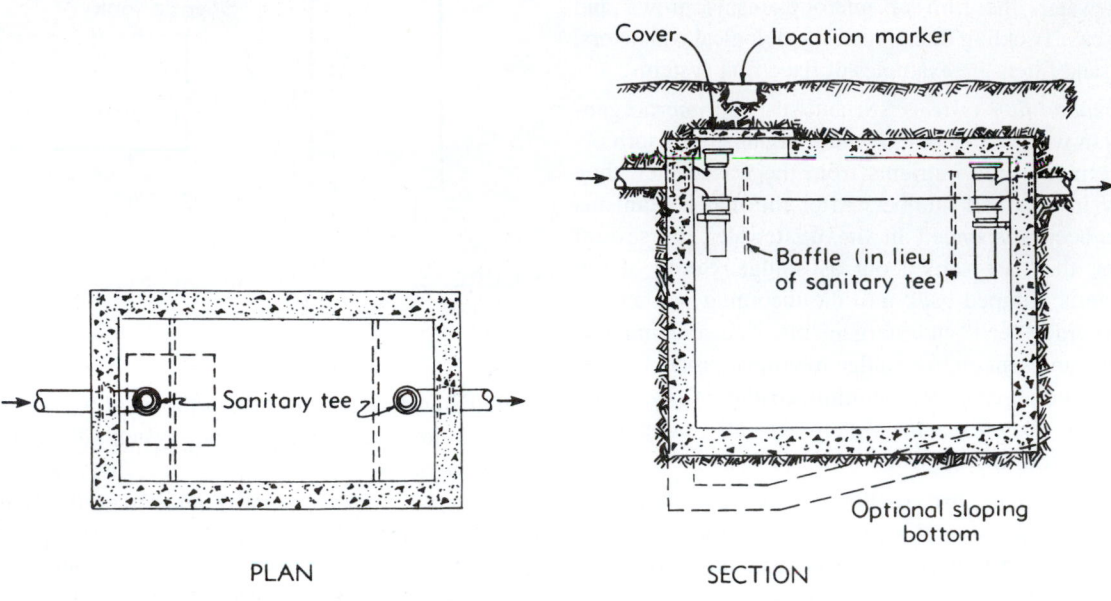

FIGURE 15.2 A rectangular septic tank is a watertight, covered container designed to settle out and hold solid wastes and partially treat wastewater with beneficial bacteria.

process, anaerobic digestion and settlement of solids in wastewater takes place. Types of primary treatment equipment used in OSST systems include the following.

Septic Tank

The *septic tank* is a watertight, covered container designed to settle out and hold solid wastes and partially treat wastewater with beneficial bacteria. It represents a relatively low-cost, low-maintenance method for primary treatment of wastewater. Septic tanks are constructed of concrete, metal, fiberglass, or plastic (fiberglass and polyethylene) and are commonly placed underground with the top surface covered with grass. An access cover built into the top of the tank allows periodic inspection and removal of sludge and scum that collects in the tank.

A septic tank can be thought of as a separating device. It allows heavier solids to settle to the bottom of the tank and lighter particles such as grease and soap float to the top of the tank. The lighter particles form a layer known as the *scum*. Aerobic and anaerobic bacteria digest up to 50% of the wastewater solids that enter the septic tank. The remaining solids accumulate as *sludge* in the bottom of the tank. The bacteria require at least 24 hr for proper digestion of solids. The greater the surface area of the liquid in the tank, the more solids the bacteria can digest, thus improving the clarity (quality) of effluent discharged to the drainage field. (See Figures 15.2 and 15.3 and Photos 15.1 through 15.6.)

A typical septic tank is divided into two compartments to enhance the treatment process. The first compartment is usually larger than the second. For larger flows, two or more noncompartmented tanks can be arranged in series. A septic tank is equipped with baffles to prevent the scum and sludge from leaving the tank. The baffles allow only the liquid effluent between the scum and sludge to leave the tank and enter the drainage field.

The size of the septic tank is based on the amount of wastewater flowing into it from the building or group of buildings it serves. A septic tank must permit detention of incoming sewage for a minimum of 2 days based on maximum daily flow. For a home, this flow is based on the number of bedrooms,

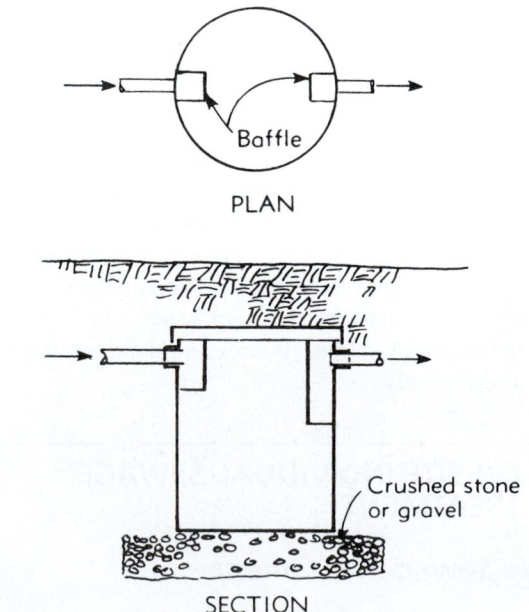

FIGURE 15.3 A steel septic tank.

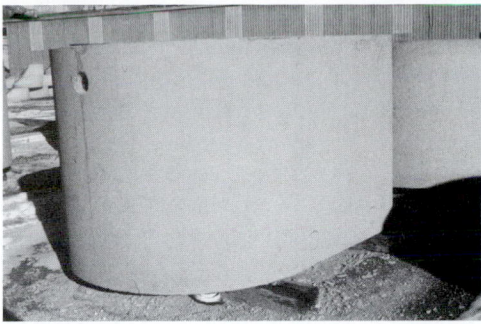

PHOTO 15.1 A precast concrete septic tank at the manufacturing plant. (Used with permission of ABC)

PHOTO 15.4 Concrete baffles used to separate compartments in a septic tank. (Used with permission of ABC)

PHOTO 15.2 Lid of a two-compartment septic tank. Note the access openings and crane "picks" for lifting the lid. (Used with permission of ABC)

PHOTO 15.5 A 1250 gallon (4725 L) septic tank in hole before hole is backfilled. The three covers allow access to each compartment. Tank will be covered with soil. (Used with permission of ABC)

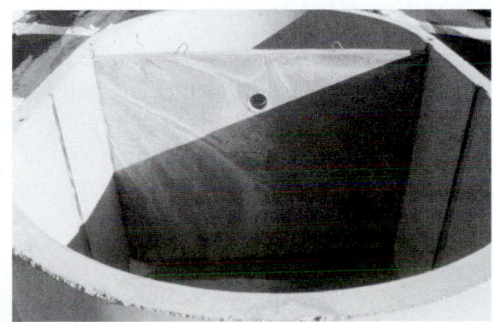

PHOTO 15.3 The interior of a three-compartment septic tank. (Used with permission of ABC)

PHOTO 15.6 A polyethylene plastic septic tank.

533

the square footage of the home, and the use of water-saving fixtures. For commercial installations, the flow is usually based on the number of persons visiting or working in the facility and the processes used.

A septic tank should be inspected annually by removing the cover. Sludge and scum that collects in the septic tank must be removed (pumped out) regularly. In a residence with a properly sized septic tank, this must be done about every 3 to 5 years. Kitchen garbage disposals and similar devices are strongly discouraged in a conventional septic system because they introduce an extremely high amount of solids to a system. Additionally, chemical household cleaners and too-frequent pumping of the septic tank destroy the beneficial bacteria and the ability of the septic tank to properly treat wastewater.

Aerobic Tank

Aerobic tanks are a substitute for a septic tank. They consist of a trash tank, an aeration chamber, and a settling chamber. Some systems require a trash tank to be installed external to and in front of the unit. Premanufactured aerobic tanks use wastewater treatment processes similar to municipal wastewater treatment processes. The clarified effluent is then usually discharged into a drainage field.

Aerobic tanks work more effectively than septic tanks and thus can be smaller in size. To remain effective, aerobic treatment components require regular maintenance and continuous monitoring. Additionally, an aerobic tank must be used on a regular basis to maintain the aerobic unit's microbe digestion process. Treatment bacteria survive on a constant flow so aerobic systems should not be used for weekend vacation homes and similar nonregular uses.

Pump Tank

A *pump tank* is a watertight container used to temporarily store clarified effluent before it flows into a drainage field. Wastewater is first treated in an aerobic or septic tank. The effluent then flows by gravity into the pump tank. When the level of stored effluent reaches a preset elevation, a float switch turns on the pump. The pump discharges the effluent to the drainage field several times a day. Excess effluent produced during periods of high generation of wastewater can be stored in the pump tank temporarily and then discharged in the drainage field during low generation periods. A remote alarm sounds if the effluent level in the tank becomes too high. Pump tank materials are typically concrete; plastic (fiberglass and polyethylene) tanks are also used.

Sand Filters

A *sand filter* is a lined, impermeable container containing a bed of granular material that provides additional treatment of effluent as it flows from the primary treatment tank to the drainage field. They are usually placed underground with the top surface covered with grass. At sites that have near-surface bedrock or a high water table, sand filters are usually constructed with above-ground concrete walls.

There are two types of sand filters: the intermittent sand filter and the recirculating intermittent sand filter. In an *intermittent sand filter,* the effluent is dispersed throughout the upper portion of the granular bed through perforated pipes. The effluent then flows by gravity through the granular material until it reaches another series of perforated pipes, where it flows to a pump tank. The effluent is then pumped from the pump tank to the drainage field. A *recirculating intermittent sand filter* requires a recirculating pump tank. The pump is used to mix filtrate with incoming septic tank effluent. The effluent is circulated several times through the sand filter media before it is pumped to the drainage field.

Trash/Grease Tank

A *trash tank* is occasionally used in conjunction with an aerobic tank. The trash tank removes materials that treatment microorganisms are unable to degrade. Several types of aerobic tanks used for small wastewater flows have the trash tank enclosed inside the unit. Grease tanks are used with septic and aerobic tanks, usually in commercial applications.

Cesspool

A *cesspool* is a covered underground container that receives untreated sewage directly from a building and discharges it into soil. Openings in the cesspool walls allow untreated sewage to pass through and seep into the surrounding soil. Because of health concerns tied to the discharge of raw sewage, use of a cesspool is considered unacceptable today in most applications in developed countries.

High-Level Alarms

A *high-level alarm* is used to alert the homeowner or building operator if liquid inside a tank reaches a level that is higher than it would be if the pump were operating normally or if the liquid inflow is greater than the maximum pumping capacity of the pump. High-level alarms are used in sump, sewage, or effluent installations. These alarms can either be purchased separately or as part of a preassembled package.

Secondary Treatment and Disposal Equipment

A *drainage field* provides secondary treatment and is the final disposal location for clarified effluent from wastewater. A piping network carries effluent from a septic or pump tank to the drainage field for further treatment within the soil or disbursement into the air. Types of drainage fields are described below.

Absorption Drainage Field

An *absorption drainage field* consists of rows (called *lines*) of underground pipes through which the clarified effluent passes. Perforated pipes are laid in shallow underground beds or narrowly cut trenches filled with gravel, then are typically covered with grass. Effluent from a septic or pump tank is discharged into

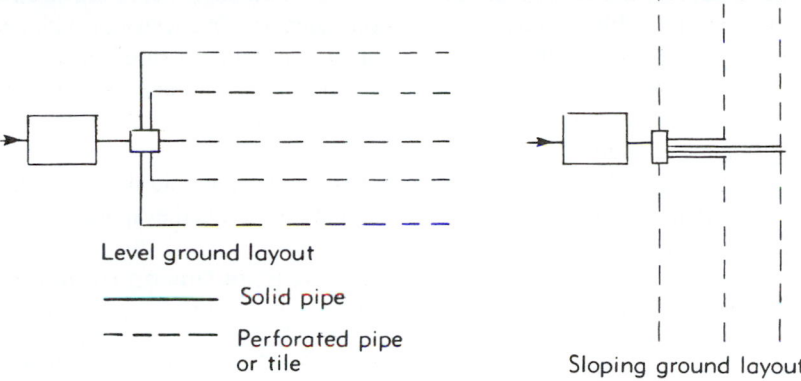

Level ground layout

———————— Solid pipe

– – – – Perforated pipe
or tile

Sloping ground layout

FIGURE 15.4 An absorption drainage field consists of rows (called *lines*) of underground pipes through which the clarified effluent is distributed. Effluent seeps into and percolates through the soil.

the drainage field, where it is distributed to the soil bed or trench and percolates through the soil. Absorption drainage field soils must have good absorption and filtration qualities. Dense clay soils and gravel soils do not absorb or filter the effluent properly and therefore are considered unacceptable soils. (See Figure 15.4 and Photo 15.7.)

A *distribution box* receives the effluent from the septic tank and distributes it equally to each individual line of the drainage field. The box is connected to the septic tank with a watertight sewer line. Perforated pipe (pipe with holes), frequently called *drain tiles* because they used to be made of clay tile pipe, disperse the clarified effluent throughout the drainage field. The effluent then gravitates into gravel or other media covered with a geotextile fabric and loamy soil. Vegetation covering the soil absorption system uses the water and nutrients to grow. The quantity of effluent flowing to the drainage field and the percolation rate determines the drainage field size.

An *absorption drainage trench* is an absorption field that consists of one or more individual trenches containing the drainage pipes. Trenches typically have a maximum width of

3 ft with a minimum 6 ft between trenches. An *absorption drainage bed* is an open drainage field without trenches containing the drainage pipes. A drainage field may take any of a number of shapes, depending on the contours (slope) of the ground, the size of the lot and the location of any well or stream on the property.

Gravel-Less Drain Fields

A *gravel-less drain field* distributes effluent into the soils through *gravel-less drain pipe* instead of gravel. As a result, these systems are also called *no gravel* or *no rock drain field systems*. Like a conventional OSST system, gravel-less pipe requires only a septic tank to pretreat the wastewater. When installed, the pipe is not surrounded by gravel or rock. These systems can be installed with small equipment and in hand-dug trenches in areas with steep slopes where conventional gravel systems would not be possible. This pipe takes many forms, including geotextile fabric-wrapped pipe and open-bottomed chambers. (see Photo 15.8).

Fabric-wrapped drain field systems use proprietary, large-diameter, corrugated, and perforated polyethylene tubing covered with permeable nylon geotextile filter-like fabric to distribute water around the pipe. The pipe is available in 8 and

PHOTO 15.7 An absorption drainage trench. Trench will be covered with soil. (Used with permission of ABC)

PHOTO 15.8 Gravel-less drain pipe during construction. Pipe will be covered with soil. (Used with permission of ABC)

10 in (200 and 250 mm) diameters and 10 and 20 ft (3 and 6 m) lengths. Effluent can circulate behind the filter fabric, within the pipe corrugations, allowing it to discharge into the soil from all sides of the pipe. A variant of the fabric-wrapped pipe is to produce it from permeable synthetic materials (i.e., expanded polystyrene foam chips) that strain the effluent.

The pipe is laid in a trench that is backfilled with native soil. It requires a well-aerated soil; it cannot be installed in clayey soils. The area of fabric in contact with the soil provides the surface for the septic tank effluent to infiltrate the soil. The pipe is placed in a 12 to 24 in (300 to 600 mm) wide trench. Multiple trenches are connected with a solid 4 in diameter pipe. The pipe is flexible, which enables it to be placed in curved trenches excavated to a specified elevation on a sloping site. Each trench must also have a cleanout/inspection port to allow sludge to be pumped out and air to enter the pipe.

A *chamber-type drain field system* consists of several arch-shaped, open-bottomed plastic chamber segments connected to form the underground drain field network. Chamber sections are placed in an open trench and backfilled with native soil. Distribution pipes hung from the inner chamber walls distribute and disperse the effluent to the open-bottomed trenches, where it infiltrates the soil. These systems are also called *leaching chambers, galleys,* and *flow diffusers.*

A typical leaching chamber consists of several high-density polyethylene injection-molded arch-shaped chamber segments. Chamber segments are available in a variety of shapes and sizes. A typical chamber has an average inside width of 16 to 40 in (0.4 to 1.0 m) and an overall length of 6 to 8 ft (1.8 to 2.4 m). The chamber segments are usually about 1 ft (300 mm) high, with wide slotted sidewalls. Because leaching chamber systems can be installed without heavy equipment, they are easy to install and repair.

Chamber-type drain field systems can provide more efficient storage than conventional gravel systems because the empty chamber provides additional storage volume to handle peak effluent loads. The additional storage volume may reduce overall drain field costs and the number of trees that must be removed from the drain field area. Maintenance requirements are comparable to those of aggregate drain field systems.

Evapotranspiration Bed or Trench Drainage Field

An *evapotranspiration (ET) bed* or *evapotranspiration trench* drainage field treats wastewater by evaporating the water from the soil and by transpiring the water into the air through plants and grasses. Vegetation covering the soil prospers on nutrients introduced by the effluent. An ET drainage field typically consists of two sets of perforated drainage pipes and gravel, soil lining (if needed), and topsoil backfill for vegetation growth. The surface area of the ET drainage field is divided into two bed sections. By periodically alternating flow to a section of bed, the other section of bed is periodically rested.

ET drainage fields are used at sites with impermeable soils such as dense clays or with very permeable soils such as sandy gravel or fractured limestone. Heavy elastomeric (rubbery plastic) membrane sheeting is commonly used for liners in a lined ET system. Impermeable dense clay soils do not need to be lined and are less expensive than lined ET drainage fields. ET drainage fields do not perform well in wet regions where rainfall rates exceed evaporation rates.

Low-Pressure Dosing Drainage Field

Low-pressure dosing (LPD) drainage fields typically consist of narrowly cut 6 to 12 in wide trenches, containing small-diameter PVC dispersion pipes. LPD systems also include a pump tank and an electrical control system. As effluent flows from the septic tank into the pump tank, the water level rises inside the tank. When a fixed water level is reached in the pump tank, the pump introduces water into the drainage field. In the drainage field, the effluent is introduced into the drainage field through small holes in the drainage pipe. The soil provides additional wastewater treatment. Soil particles filter the effluent, and microbes in the soil kill the bacteria and pathogens.

Pumped effluent low-pressure dosing drainage fields are a basic LPD trench system intended for single-family homes and cannot be used for commercial or institutional facilities. They consist of a septic tank, pump tank, and drainage field. The pump periodically doses a measured quantity of effluent evenly within the drainage field. Use of a pump allows the drainage field to be located upslope from the septic tank. A 2% maximum slope (2 ft vertical rise in 100 ft vertical run) of natural grade is allowed with this type of system.

Absorption Mounds

Absorption mounds consist of septic tank(s), a pump tank, effluent pump and controls, and an above-grade drainage system. The drainage system consists of perforated drainage pipes laid in a deep layer of sand to sandy loam soil that is placed above the natural soils. The mound created by this deep layer soil is a topsoil cap and grass cover. Effluent from the pump tank flows through holes in the perforated drainage pipes and into the mound. The effluent moves downward through the mound by gravity flow and is partially treated by the time it reaches the natural soil.

Mounds are used at sites that have shallow soils. They typically require a minimum depth of 12 in of natural soil to any limiting layer such as bedrock or high water table. Natural surface slopes greater than 10% (10 ft vertical rise in 100 ft vertical run) are not suitable for absorption mounds. The main sizing standard for mound construction is the contact area of the fill sand with the existing soil. For slowly permeable clay and clay loam soils, a loading rate of from 0.20 to 0.25 gallons per day per square foot is usually satisfactory. This typically means that a large drainage area is required for absorption mounds, which can substantially increase land area requirements for this type of drainage field system.

Spray Distribution

Spray distribution systems spray the disinfected effluent onto the ground surface in a manner similar to a lawn irrigation system. They require an advanced treatment process that purifies the water, a pathogen-removing disinfecting system (chlorinating is the most common method), a pump tank, distribution pipe, and a disposal area containing the spray heads. The wastewater must be treated to secondary-quality effluent level, which usually requires an aerobic tank for primary treatment or an intermittent sand filter. Spray systems can be used in almost all soil conditions (subject to regulatory approval), but typically have higher cost and maintenance requirements compared with most other system types.

Leaching Chamber Drainage Field

Leaching chambers are proprietary, commercially produced plastic chambers pre-molded into a dome shape. They can be easily assembled and placed in trenches. Depending on site conditions, leaching chamber drainage fields can require less land area than other conventional absorption drainage fields.

Leaching chambers distribute effluent to the soil in a manner similar to conventional gravel-filled trench systems. Each chamber dome supports the soil above it while maintaining sufficient volume inside for effluent storage. The sides are louvered and the bottom is open to allow for passage of effluent into surrounding soils. Leaching chamber domes usually vary in width from 15 to 36 in.

Subsurface Drip Drainage Field

Subsurface drip systems consist of a septic tank, a pump tank, a filtering device, and a drip distribution system. An aerobic tank is typically used for the primary treatment process. When a septic tank is used, additional treatment in a sand filter is required. Treated wastewater is pumped through a filtering device that removes larger particles to reduce clogging of the drip emitters. The filtered effluent is distributed to tubing laid just below the ground surface. Effluent is introduced to the soil through emitters spaced along and inserted into the tubing wall, where bacteria and pathogens are removed. Emitters can become clogged and an ongoing maintenance contract is required. Subsurface drip systems can be installed on steep slopes and at sites containing dense clay soils and soils overlying shallow bedrock.

Seepage Pit

A *seepage pit* is a deep underground container that receives clarified effluent from a septic tank. It has openings in its walls that allow effluent to pass through and seep into the surrounding soil. (See Figure 15.5.) A seepage pit functions like an absorption drainage field except that effluent seeps through openings in the sidewalls of the pit rather than through the bottom of a trench or bed. A seepage pit tends to be deeper than an absorption bed or trench. It is designed on the basis of the sidewall area. It can serve as an alternative to a drainage field,

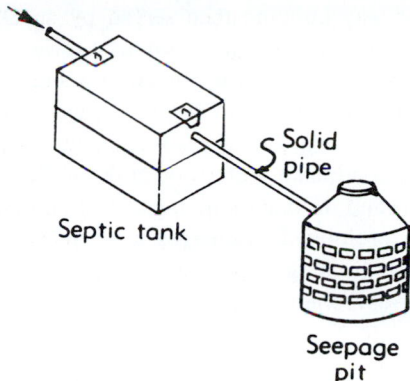

FIGURE 15.5 A septic tank to seepage pit system. The seepage pit has openings in its walls that allow effluent to pass through and seep into the surrounding soil.

but only when surrounding soil has a good percolation rate (usually below 30 min/in) and the water table is very deep.

A seepage pit can be a precast concrete or molded plastic unit. It can also be constructed of individual special concrete masonry units (CMU) laid with the cells (holes) placed horizontally to allow the effluent to seep into the ground. See Photo 15.9. The tapered cells of the block are set with the widest area to the outside to reduce the amount of loose material behind the lining that might fall into the pit. The bottom of the pit is lined with coarse gravel a minimum of 1 ft (0.3 m) deep before the block or concrete is placed. Between the block or concrete and the soil is a minimum of 6 in (0.15 m) of clean crushed stone or gravel. The top of the pit should have an opening with a watertight cover to provide access to the pit if necessary. The construction of the pit above the inlet pipe should be watertight.

General Regulations

Typically, an OSST system should not be installed, repaired, or rehabilitated where a community sanitary sewer system is available or where a local ordinance requires connection to a community system. An OSST system is generally not permitted when a building site is located within 200 ft from any community sewer. When a community sanitary sewer becomes available

PHOTO 15.9 A concrete seepage pit.

within 200 ft, any building then served by an OSST system should connect to the community sanitary sewer within a time frame or under conditions set by the governmental authority.

Generally, it is illegal to discharge any wastewater from OSST systems (except under a governmental permit) to any ditch, stream, pond, lake, waterway, or drain tile, or to the surface of the ground. In older buildings, existing systems having any of these prohibited discharges are typically required to eliminate these discharges by constructing a system in compliance with current requirements.

15.4 TESTING OF SOIL AND WATER FOR DRAINAGE

Soil Evaluation

Soil is an important part of an OSST system because it treats and disposes of the septic tank effluent. Soil for a drainage field must effectively filter effluent, and not allow inhibiting consolidation or swelling of the soil. For example, dense clay soils do not absorb or filter the effluent properly and are considered an unacceptable soil for an OSST system with an absorption drainage field.

A *soil evaluation* is an assessment of subsurface soil conditions at a specific site that is conducted under the supervision of a professional engineer or professional geologist. Consideration is given to the influence of the following: topography; drainage ways; terraces; floodplain; percentage of land slope; location of property lines; location of easements; buried utilities; existing and proposed tile lines; existing, proposed, and abandoned water wells; amount of available area for the installation of the system; evidence of unstable ground; alteration (cutting, filling, compacting) of existing soil profile; and soil factors determined from a soil analysis, percolation tests, and soil survey maps if available.

The evaluation involves boring test holes to a depth of 6 to 8 ft (depending on local regulations) at the site to examine the soil profile. The soil profile is examined to determine if there is existence of a limiting soil or a bedrock layer that will hinder proper absorption and filtering of wastewater in the planned area of the absorption system.

A minimum depth of unsaturated soil beneath the planned soil absorption system is required to restore wastewater before it reaches a limiting layer. A *limiting layer* may be an impervious soil layer, bedrock, or a high water table. If a limiting layer is within 5 to 7 ft of the surface (depending on local regulations), the site is not suitable for an absorption drainage field system. If a limiting layer is located within 2 ft (0.6 m) of the surface, the site is not suitable for a mound system. Other conditions that restrict use of a drainage field system include a ground slope of greater than 30% (30 ft drop in 100 ft, 30 m in 100 m), adverse effects on adjoining property, and a proposed location in a 100-year floodplain.

Soils can vary significantly from site to site and even at a single building site. An accurate assessment of soils at the building site is needed to ensure that an OSST system will not fail. Depth of the soil and how rapidly it will absorb water is used to determine the suitability of the disposal area for a septic system. This information, coupled with the estimated wastewater produced by the building or buildings, is used to determine the size of the soil absorption system.

Soil Percolation Test

A *soil percolation test* is a subsurface soil test at a depth of a proposed absorption system to determine the water absorption capability of the soil. The test identifies the soil percolation rate, the rate that water seeps through saturated soil. (See Figure 15.6.)

The percolation test is conducted by boring a series of test holes at the site under consideration and observing water seepage rates in holes placed throughout the intended area of the drainage field. Each hole is presoaked with water in an attempt to saturate the soil. The water seepage rate is then measured in each hole and reported as the number of minutes it takes an inch (min/in) of water to soak into the soil.

Certain soils do not absorb or filter the effluent properly and therefore are considered unacceptable soils. Gravels tend to pass water very quickly and will not do an adequate job of treatment before the wastewater reaches the limiting zone. Dense clay soils accept water too slowly and will have difficulty treating all the wastewater from a building, causing the system to become overloaded.

A soil percolation test is required for any system dependent on soil absorption for effluent disposal. Many local health departments use the soil percolation test to gather soils information before locating and sizing the drainage field system. Typically, this test must be conducted under the supervision of a professional engineer or professional geologist.

Percolation testing procedures are established by a governmental health entity, commonly a municipal, county, or state office of health. Figure 15.7 is a legal description of a typical

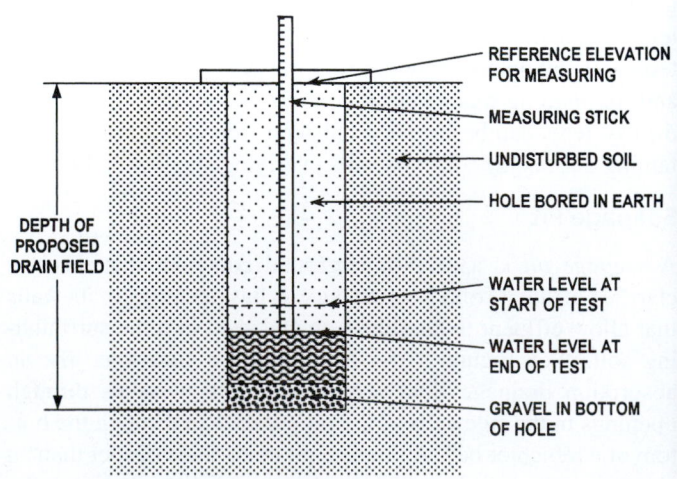

FIGURE 15.6 A trench for a percolation test.

1. **Number and location.** Six or more tests shall be made in separate test holes spaced uniformly over the proposed absorption area site.

2. **Results.** Percolation holes located within the proposed absorption area shall be used in the calculation of the arithmetic average percolation rate.

3. **Type of hole.** Holes having a uniform diameter of 6 to 10 inches shall be bored or dug as follows:

   a. To the depth of the proposed absorption area, where the limiting zone is 60 inches or more from the mineral soil surface.

   b. To a depth of 20 inches if the limiting zone is identified as seasonal high water table, whether perched or regional; rock formation; other stratum; or other soil condition which is so slowly permeable that it effectively limits downward passage of effluent, occurring at less than 60 inches from the mineral soil surface.

   c. To a depth 8 inches above the limiting zone or 20 inches, whichever is less, if the limiting zone is identified as rock with open joints or with fractures or solution channels, or as masses of loose rock fragments including gravel with insufficient fine soil to fill the voids between the fragments, occurring at less than 60 inches from the mineral soil surface.

4. **Preparation.** The bottom and sides of the hole shall be scarified with a knife blade or sharp-pointed instrument to completely remove any smeared soil surfaces and to provide a natural soil interface into which water may percolate. Loose material shall be removed from the hole. Two inches of coarse sand or fine gravel shall be placed in the bottom of the hole to protect the soil from scouring and clogging of the pores.

5. **Procedure for presoaking.** Holes shall be presoaked to approximate normal wet weather or in-use conditions in the soil, according to the following procedure:

   a. **Initial presoak.** Holes shall be filled with water to a minimum depth of 12 inches over the gravel and allowed to stand undisturbed for 8 to 24 hours prior to the percolation test.

   b. **Final presoak.** Immediately before the percolation test, water shall be placed in the hole to a minimum depth of 6 inches over the gravel and readjusted every 30 minutes for 1 hour.

6. **Determination of measurement interval.** The drop in the water level during the last 30 minutes of the final presoaking period shall be applied to the following standard to determine the time interval between readings for each percolation hole:

FIGURE 15.7   A typical soil percolation test procedure. Procedure will vary by governmental jurisdiction. Check with health authority having jurisdiction at the building site.

a. If water remains in the hole, the interval for readings during the percolation test shall be 30 minutes.

b. If no water remains in the hole, the interval for readings during the percolation test may be reduced to 10 minutes.

7. **Measurement.** After the final presoaking period, water in the hole shall again be adjusted to approximately 6 inches over the gravel and readjusted when necessary after each reading.

a. Measurement to the water level in the individual percolation holes shall be made from a fixed reference point and shall continue at the interval determined from paragraph (6) for each individual percolation hole until a minimum of eight readings are completed or until a stabilized rate of drop is obtained whichever occurs first. A stabilized rate of drop means a difference of 1/4 inch or less of drop between the highest and lowest readings of four consecutive readings.

b. The drop that occurs in the final period in percolation test holes, expressed, as minutes per inch, shall be used to calculate the arithmetic average percolation rate.

c. When the rate of drop in a percolation test is too slow to obtain a measurable rate, the rate of 240 minutes per inch shall be assigned to that hole for use in calculating the arithmetic average percolation rate. The absorption area may be placed over holes with no measurable rate when the average percolation rate for the proposed absorption area is within the limits established in Table A.

When a percolation test hole is dry at the end of a 10-minute testing interval, that hole may not be used in the calculation of the arithmetic average percolation rate. If 1/3 or more of the percolation test holes are dry at the end of a 10 min testing interval, the proposed absorption area may not be designed or installed over these holes unless the local agency determines that an anomaly caused the fast percolation rate and a retest of the area is within the acceptable percolation rate limits. If no anomaly is discovered, the local agency may accept the percolation test results from the remaining holes if the results are supplemented with the results of additional percolation testing conducted outside of the area in which the dry percolation holes were found.

FIGURE 15.7 *(Continued)*

procedure for testing soil percolation at a building site. Test standards vary by local requirements. For specifics, the designer should check with the governmental entity having jurisdiction at the building site.

BOD Test

The *BOD test* measures the amount of dissolved oxygen organisms are likely to need to degrade wastes in wastewater. This test is important for evaluating both how much treatment wastewater is likely to require and the potential effect that it can have on receiving waters.

To perform the test, wastewater samples are placed in BOD bottles and are diluted with specially prepared water containing dissolved oxygen. The dilution water is also seeded with bacteria when treated wastewater is being tested.

The amount of dissolved oxygen in the diluted sample is measured and the samples are then stored at a constant temperature of 68°F (20°C). Common incubation periods are 5, 7, or 20 days, with 5 days (or BOD_5) the most common. At the end of the incubation period, the dissolved oxygen is measured again. The amount that was used (expressed in milligrams per liter) is an indication of wastewater strength.

Coliform Test

The *coliform test* determines whether wastewater has been adequately treated and whether water quality is suitable for drinking and recreation. Because they are very abundant in human wastes, coliform bacteria are much easier to locate and identify in wastewater than viruses and other pathogens that cause severe diseases. Therefore, coliform bacteria are used as indicator organisms for the presence of other, more serious pathogens. Some coliform are found in soil, so tests for fecal coliform are considered the most reliable. However, tests for both total coliform and fecal coliform are commonly used.

There are two methods for determining the presence and density of coliform bacteria. The membrane filter (MF) technique provides a direct count of colonies trapped and then cultured. The multiple tube fermentation method provides an estimate of the most probable number (MPN) per 100 milliliters from the number of test tubes in which gas bubbles form after incubation.

15.5 DESIGN EXAMPLE OF AN INDIVIDUAL OSST SYSTEM

Design of a Septic Tank/Absorption Field

An individual OSST system must be designed to have a capacity capable of disposing of the sewage produced by the building or facility being served. Design is usually based on the maximum daily wastewater flow rate and average percolation rate.

Maximum daily wastewater flow rate (Q) is typically expressed in gallons per day or liters per day. In new construction, the anticipated maximum sewage flow rate is based on specific criteria and parameters outlined by the local public health authority, including number of bedrooms, bathrooms, occupants, and/or drainage fixture units. Tables 15.1 and 15.2 provide information on typical maximum daily wastewater flow rates. In existing buildings, the maximum daily sewage flow rate can be determined by measurement of existing conditions.

The septic tank is typically sized for detention of incoming sewage for a minimum of 2 days based on maximum daily flow. Typical allowable design capacities of septic tanks serving a residence are provided in Table 15.3. These capacities vary by governmental jurisdiction. Check with the health authority having jurisdiction at the building site.

Minimum drainage area requirements for septic tank effluent are based on type of system and *average percolation rate* (t_{perc}) of the soil in min/in (min/cm). The percolation rate is

TABLE 15.1 TYPICAL MAXIMUM DAILY WASTEWATER FLOW RATES USED IN ABSORPTION DRAINAGE FIELD DESIGN OF DWELLING UNITS. VARIES BY GOVERNMENTAL JURISDICTION. CHECK WITH HEALTH AUTHORITY HAVING JURISDICTION AT THE BUILDING SITE.

Use	Flow Rate gal/min	L/min	Load	
Adult or child	—	—	75 gal/day	284 L/day
Infant	—	—	100 gal/day	379 L/day
Clothes washer	5	19	30 to 50 gal/load	114 to 189 L/load
Dish washer	2	8	7 to 15 gal/load	26 to 57 L/load
Garbage disposal	3	11	4 to 6 gal/day	15 to 23 L/day
Kitchen sink	3	11	2 to 4 gal/use	8 to 15 L/use
Shower or bath tub	5	19	25 to 60 gal/use	95 to 227 L/use
Toilet (per flush)	3	11	4 to 7 gal/use	15 to 26 L/use
Lavatory	2	8	1 to 2 gal/use	4 to 8 L/use

TABLE 15.2   WASTEWATER FLOW RATES BY TYPE OF ESTABLISHMENT. VARIES BY GOVERNMENTAL JURISDICTION. CHECK WITH HEALTH AUTHORITY HAVING JURISDICTION AT THE BUILDING SITE.

Type of Establishment	Wastewater Flow Rate Per Day per Person, Unless Noted Otherwise	
	gal	L
Airports, per passenger	3 to 5	11 to 19
Assembly halls, per seat	2	8
Boarding schools	50	189
Boarding schools	100	378
Bowling alleys, per lane	75	284
Churches (large with kitchens), per sanctuary seat	5.7	22
Churches (small), per sanctuary seat	3 to 5	11 to 19
Country clubs, per locker	25	95
Dance halls	2	8
Day camps	25	95
Day workers at schools and offices	15	57
Drive-in theaters, per car space	5	19
Factories (per shift—exclusive of industrial wastes)	25	95
General hospitals	150	568
Hotels (with connecting baths)	50	189
Hotels (with private baths—2 persons per room)	100	378
Kitchen wastes from hotels, camps, boarding houses, and so on, serving three meals per day	10	38
Laundries (coin operated), per machine	400	1514
Luxury residences and estates	150	568
Marinas		
Flush toilets, per fixture per hour	36	136
Urinals, per fixture per hour	10	38
Wash basins, per fixture per hour	15	57
Showers, per fixture per hour	150	568
Motels with bath, toilet, and kitchen wastes, per bed space	60	227
Motels, per bed space	50	189
Nursing homes	75	284
Public institutions (other than hospitals)	100	378
Public picnic parks (bathhouse, showers, and flush toilets)	10	38
Public picnic parks (toilet wastes only)	5	19
Restaurants (toilet and kitchen wastes per unit of serving capacity	25	95
Rooming houses	40	151
Schools (toilet and lavatories only)	15	57
Schools (with above plus cafeteria and showers)	35	132
Schools (with above plus cafeteria)	25	95
Service stations		
First bay	1000	3780
Each additional bay	500	1892
Service stations, per vehicle served	10	38
Stores, per toilet room	400	1514
Subdivisions or individual homes	75	284
Swimming pools and beaches	10	38
Trailer parks or tourist camps (with built-in bath)	50	189
Trailer parks or tourist camps (with central bathhouse)	35	132
Work or construction camps	50	189

determined in a test at the site that is performed under the supervision of a professional engineer or professional geologist adhering to procedures established by the governmental health entity, commonly a municipal, county, or state health office (discussed previously in this chapter). Table 15.4 provides typical computation information.

Usually, the bottom of an absorption drainage field must be a minimum of 4 ft above the seasonal high water table and be a minimum of 4 ft above bedrock. The field must be covered with a 12 in minimum depth of ½ to 2 in gravel/rock. An absorption drainage field must be sloped in the direction of flow. Code typically limits the maximum slope to 6 in (150 mm) in 100 ft (30 m) so that the effluent will not simply flow to the end of the line and then back up. Code also limits any individual line to a length of 100 ft (30 m) and sets the minimum separation between lines at 6 ft (2 m).

TABLE 15.3 TYPICAL ALLOWABLE DESIGN CAPACITY OF A SEPTIC TANK. VARIES BY GOVERNMENTAL JURISDICTION. CHECK WITH HEALTH AUTHORITY HAVING JURISDICTION AT THE BUILDING SITE.

Single Family Dwelling			Multifamily Dwelling Units		Maximum Daily Sewage Flow Rate		Minimum Capacity of Septic Tank(s)	
Maximum Number of Bedrooms	Maximum Number of Bathrooms	Maximum Number of Occupants	Units	Maximum Drainage Fixture Units Served	gal/day	L/day	Gallons	Liters
3	2	6	—	20	500	1890	1000	3785
4	3	8	2	25	600	2270	1200	4540
5 or 6	4	10	3	33	750	2840	1500	5680
7 or 8	5	12	4	45	1000	3790	2000	7570
—	—	—	5	55	1125	4260	2250	8520
—	—	—	6	60	1250	4730	2500	9460
—	—	—	7	70	1375	5200	2750	10 410
—	—	—	8	80	1500	5680	3000	11 360
—	—	—	9	90	1625	6150	3250	12 300
—	—	—	10	100	1750	6620	3500	13 250
—	—	—	11	110	1875	7090	3750	14 200
—	—	—	12	120	2000	7560	4000	15 150
—	—	—	13	130	2125	8030	4250	16 100
—	—	—	14	140	2250	8500	4500	17 050
—	—	—	15	150	2375	8970	4750	18 000
—	—	—	16	160	2500	9440	5000	18 950
—	—	—	17	170	2625	9910	5250	19 900
—	—	—	18	180	2750	10 380	5500	20 850
—	—	—	19	190	2875	10 850	5750	21 800
—	—	—	20	200	3000	11 320	6000	2275

TABLE 15.4 MINIMUM DRAINAGE AREA REQUIREMENTS FOR SEPTIC TANK EFFLUENT. VARIES BY GOVERNMENTAL JURISDICTION. CHECK WITH HEALTH AUTHORITY HAVING JURISDICTION AT THE BUILDING SITE.

Average Percolation Rate (t_{perc})		Minimum Drainage Area Requirement [Multiply by 0.0245 to convert from area (ft^2) per gallon per day to area (m^2) per liter per day]	
		All OSST Systems, Except Elevated Sand Mounds and Subsurface Sand Filters	Elevated Sand Mound and Subsurface Sand Filter OSST Systems
min/in	min/cm	Area (ft^2) per Gallon per Day	Area (ft^2) per Gallon per Day
Less than 3.0[d]	Less than 1.2[d]	Unsuitable	Unsuitable
3 to 5[c]	1.2 to 2.0[c]	Unsuitable	1.50[ab]
6 to 15[c]	2.4 to 5.9[c]	1.19[b]	1.50[ab]
16 to 30[c]	6.3 to 11.8[c]	$(t_{perc} - 15) \cdot (0.040) + 1.19$[b]	1.50[ab]
31 to 45[c]	12.2 to 17.7[c]	$(t_{perc} - 30) \cdot (0.030) + 1.79$[b]	$(t_{perc} - 30) \cdot (0.026) + 1.50$[ab]
46 to 60[c]	18.1 to 23.6[c]	$(t_{perc} - 45) \cdot (0.028) + 2.24$[b]	$(t_{perc} - 45) \cdot (0.022) + 1.89$[a]
61 to 90[c]	24.0 to 35.4[c]	$(t_{perc} - 60) \cdot (0.023) + 2.66$[a]	$(t_{perc} - 60) \cdot (0.020) + 2.22$[a]
91 to 120[acd]	35.8 to 47.2[acd]	Unsuitable	$(t_{perc} - 90) \cdot (0.017) + 2.82$[a]
121 to 150[cd]	47.6 to 59.1[cd]	Unsuitable	$((t_{perc} - 120) \cdot (0.015) + 3.33)(1.05)$[a]
151 to 180[cd]	59.4 to 70.9[cd]	Unsuitable	$((t_{perc} - 150) \cdot (0.014) + 3.78)(1.10)$[a]
Greater than 181[cd]	Greater than 71.3[cd]	Unsuitable	Unsuitable

[a]Pressure dosing required.
[b]One-third reduction may be permitted with use of an aerobic tank.
[c]Experimental or alternate proposals may be considered.
[d]May be unsuitable.

Example 15.1

An OSST system consisting of a conventional septic tank and an absorption bed drainage field will serve a single-family residence with 4 bedrooms and 4 bathrooms. The maximum daily sewage flow rate is anticipated to be 720 gal/day.

a. Determine the required septic tank size.

From Table 15.3, a 1500-gal septic tank is required.

b. Determine the area required for an absorption bed drainage field based on a percolation rate of 12 min/in.

From Table 15.4, the required drainage area is 1.19 ft² per gallon per day. Required area of the drainage field is found by:

$$\text{Required area} = 1.19 \text{ ft}^2 \cdot 720 \text{ gal/day} = 857 \text{ ft}^2$$

c. Determine the minimum trench length. Assume a trench width of 3 ft. *100'*

$$\text{minimum trench length} = 857 \text{ ft}^2/3 \text{ ft} = 286 \text{ ft}$$

d. Determine the area required for an absorption bed drainage field based on a percolation rate of 50 min/in.

From Table 15.4, the required drainage area must be determined based on a specified formula:

$$(t_{perc} - 45) \cdot (0.028) + 2.24^b = (50 - 45) \cdot (0.028)$$
$$+ 2.24^b = 2.38 \text{ ft}^2 \text{ per gal/day}$$

Required area = 2.38 ft² per gal/day · 720 gal/day = 1714 ft²

e. Determine the minimum trench length. Assume a trench width of 3 ft.

$$\text{minimum trench length} = 1714 \text{ ft}^2/3 \text{ ft} = 595 \text{ ft}$$
571

f. Assume an elevated sand mound system will be used in lieu of an absorption drain field based on the same percolation rate of 50 min/in.

From Table 15.4, the required drainage area must be determined based on a specified formula:

$$(t_{perc} - 45) \cdot (0.022) + 1.89^a = (50 - 45) \cdot (0.022)$$
$$+ 1.89^a = 2.0 \text{ ft}^2 \text{ per gal/day}$$

Required area = 2.0 ft² per gal/day · 720 gal/day = 1440 ft²

Example 15.2

The four-story apartment building (in Appendix A) has 8 apartments, averaging 2.5 persons per apartment.

a. Determine the maximum daily sewage flow rate.

From Table 15.1, the load is 75 gal/day for adults/children, therefore:

$$75 \text{ gal/day} \cdot 2.5 \text{ persons/apartment} \cdot 8 \text{ apartments}$$
$$= 1500 \text{ gal/day}$$

b. Determine the required septic tank size.

From Table 15.3, a 3000 gal septic tank is required.

c. Determine the area required for an absorption bed drainage field based on a percolation rate of 15 min/in.

From Table 15.4, the required drainage area is 1.19 ft² per gallon per day. Required area of the drainage field is found by:

$$\text{Required area} = 1.19 \text{ ft}^2 \cdot 1500 \text{ gal/day} = 1785 \text{ ft}^2$$

d. Determine the minimum trench length. Assume a trench width of 3 ft.

$$\text{minimum trench length} = 1785 \text{ ft}^2/3 \text{ ft} = 595 \text{ ft}$$

Design of a Seepage Pit

The size of the seepage pit is typically based on the outside area of the walls of the seepage pit. Some designers exclude the open bottom area of the pit from the absorption area required to allow for a safety factor, while others calculate the total area available and then size the system to allow some safety factor.

Wall areas for seepage pits of various diameters are provided in Table 15.5. Seepage pits are sized from Table 15.6, which lists percolation rates and the minimum required absorption area required for 100 gal (378.5 L) of wastewater. (See Figure 15.8.)

When more than one seepage pit is used, the pipe from the septic tank must be laid out so that the effluent will be spread uniformly to the pits. To provide equal distribution, a distribution box with separate laterals (each lateral feeding no

TABLE 15.5   ABSORPTION AREAS IN FT² AND M² OF A SELECTED SEEPAGE PIT DESIGN. 6 IN (150 MM) TO DEPTH PROVIDED TO ACCOMMODATE THE COVER.

Diameter		Depth Customary Units (ft)					Depth Metric SI Units (m)				
ft	m	4	5	6	8	10	1.22	1.53	1.83	2.44	3.05
4	1.22	50.2	62.8	75.3	100.4	125.6	4.66	5.83	7.00	9.33	11.67
5	1.53	62.8	78.5	94.2	125.6	157.0	5.83	7.29	8.75	11.67	14.59
6	1.83	75.4	94.2	113.0	150.7	188.4	7.00	8.75	10.50	14.00	17.50
8	2.44	100.4	125.6	150.7	200.9	251.2	9.33	12.56	14.00	18.67	23.34

TABLE 15.6 TYPICAL REQUIRED ABSORPTION AREAS FOR A SEEPAGE PIT. THE EFFECTIVE ABSORPTION AREA LISTED IS FOR EACH GALLON (OR LITER) OF WASTEWATER PER DAY BASED ON PERCOLATION RATE OF SOIL.

Percolation Rate	Required Absorption Area of Wastewater per Day	
min/in (min/25 mm)	ft² per gal	m² per L
1	0.32	0.0079
2	0.40	0.0098
3	0.45	0.0111
5	0.56	0.0137
10	0.75	0.0185
15	0.96	0.0235
20	1.08	0.0264
25	1.39	0.0341
30	1.67	0.0410

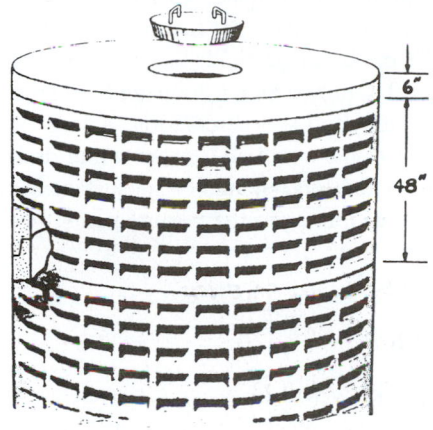

FIGURE 15.8 A seepage pit.

more than two pits) provides the best results. The distance between the outside walls of the pits should be a minimum of 3 pit diameters and not less than 10 ft (3 m).

Example 15.3

An OSST system serves a single-family residence. It has a 1500-gal septic tank and a maximum daily sewage flow rate of 720 gal/day.

a. Determine the required absorption area for a seepage pit based on a percolation rate of 12 min/in.

From Table 15.6, at a percolation rate of 12 min/in, the required area is between 0.75 and 0.96 ft² per gal per day. It is interpolated to be 0.834 ft² per gal. Therefore:

$$\text{required absorption area } 0.834 \text{ ft}^2 \text{ per gal} \cdot 720 \text{ gal} = 600.5 \text{ ft}^2$$

b. Determine the number of 8 ft diameter × 8 ft deep seepage pits that must be used in this design.

From Table 15.5, an 8 ft diameter × 8 ft seepage pit has an absorption area of 200.9 ft². Therefore:

$$\text{number of seepage pits} = 600.5 \text{ ft}^2/200.9 \text{ ft}^2 = 3.0 \text{ seepage pits}$$

In Example 15.3, three seepage pits are required. Other combinations may be used depending on the area required to place the pits, the sizes available locally, whether precast concrete is used, and the designer's own preference. (See Figure 15.9.)

Example 15.4

The four-story apartment building (in Appendix A) with 8 apartments, averaging 2.5 persons per apartment, has a maximum daily sewage flow rate of 5680 L/day (1500 gal/day). See Example 15.3.

a. Determine the required septic tank size.

From Table 15.3, an 11 360 L (3000 gal) septic tank is required.

b. Determine the required absorption area for a seepage pit based on a percolation rate of 5 min/25 mm (5 min/in).

From Table 15.6, at a percolation rate of 5 min/25 mm (5 min/in), the required area is 0.0137 m²/L (0.56 ft²/gal). Therefore:

$$\text{required absorption area } = 0.0137 \text{ m}^2/\text{L} \cdot 5680 \text{ L/day} = 77.8 \text{ m}^2 \text{ (840 ft}^2)$$

c. Determine the number of 1.83 m diameter × 2.44 m deep (6 ft diameter × 8 ft) seepage pits that must be used in this design.

From Table 15.5, a 1.83 m diameter × 2.44 m deep (6 ft diameter × 8 ft) seepage pit has an absorption area of 14.00 m² (150.7 ft²). Therefore:

$$\text{Number of seepage pits } = 77.8 \text{ m}^2/14.00 \text{ m}^2 = 5.6 \text{ or } 6 \text{ seepage pits}$$

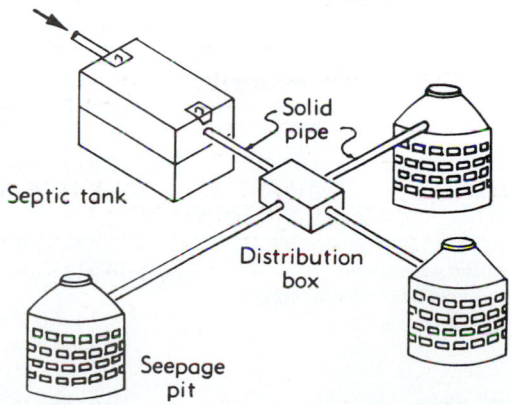

FIGURE 15.9 A septic tank to seepage pit system distributes clarified effluent flows through a distribution box to multiple seepage pits.

Avoiding Potential Design Problems

The uncertainty of exactly how much the soil will actually absorb is reflected in the values given in the various tables provided in OSST regulations. Even so, many designers prefer to oversize the OSST system slightly to allow for poor soil absorption and for future increased amounts of effluent, either because more people are using the facility than anticipated or because of an addition to the individual residence or an addition of various water-using fixtures that may not have been included in the original design.

Recent environmental regulations have led to increased land area requirements for the drainage field size of OSST systems. In some jurisdictions, additional land must be reserved for a future a replacement drainage field. With limited land available for development in many areas, the amount of land occupied by individual OSST systems, including mandatory spacing and setbacks, must be considered at early stages of development.

One of the most wasteful uses of a residential OSST system is the connection of a clothes washing machine. Many times, connecting a washing machine to an older system has resulted in more water flow than the soil can handle through the existing system. A clothes washer generates a significant amount of water, especially in large families. One of the simplest solutions is to spread washing out over several days, giving the soil a chance to absorb the water. Other solutions are to increase the size of the disposal field or add a seepage pit.

The connection of storm drainage (e.g., gutters, downspouts, and roof drains) to a drainage field may cause periodic overloads. When these are connected, the designer must increase the size of the system to accommodate the periodic additional flow. Most designers prefer to run such connections into drywells if no storm drain system is available.

15.6   OSST SYSTEM INSTALLATION

Location

Individual sewage disposal requires finding an appropriate location on the site to place all of the system components, especially the drain field, which takes a considerable amount of area. Typically local health agencies also sets minimum distances for the location of the various parts of the OSST system; for example, the septic tank must be a minimum of 50 ft (15 m) from any well or suction line.

Minimum clearance and setback distance requirements for OSST system components are shown in Table 15.7. These distances greatly reduce the danger of contaminating drinking water if leaks should occur in the tank or pipes (lines). Many local codes require longer distances and therefore regulations must be reviewed for a specific building site.

As illustrated in Exercises 15.1 and 15.2, high percolation rates (i.e., more minutes per inch) result in a larger area requirement for the drainage field; that is, a site with poor soil percolation will require a larger drainage field area in comparison to a site with good soil percolation. A large area drainage field may not fit on a small site. A substitute drainage system such as the elevated sand mound or subsurface sand filter system may reduce drainage area requirements, but will likely increase system costs.

Installation

Typically, a backhoe will be needed to excavate holes for the septic tank or seepage pits and any trenches for distribution piping and disposal fields. In a community OSST system, a trench must be dug from the building to the community line. The final elevation of the two must be checked to ascertain the sewage will flow properly.

TABLE 15.7   MINIMUM CLEARANCE AND SETBACK REQUIREMENTS FOR OSST SYSTEM COMPONENTS. VARIES BY GOVERNMENTAL JURISDICTION CHECK WITH HEALTH AUTHORITY HAVING JURISDICTION AT THE BUILDING SITE.

Minimum Horizontal Distance (Clearance) Required From	Building Sewer	Septic Tank	Drain Field	Seepage Pit or Cesspool
Buildings or structures[a]	2 ft (0.6 m)	5 ft (1.5 m)	8 ft (2.4 m)	8 ft (2.4 m)
Disposal field	—	5 ft (1.5 m)	4 ft (1.2 m)	5 ft (1.5 m)
Distribution box	—	—	5 ft (1.5 m)	5 ft (1.5 m)
On-site domestic water service line	1 ft (0.3 m)	5 ft (1.5 m)	5 ft (1.5 m)	5 ft (1.5 m)
Pressure public water main	10 ft (3.0 m)	10 ft (3.0 m)	10 ft (3.0 m)	10 ft (3.0 m)
Property line adjoining private property	—	5 ft (1.5 m)	5 ft (1.5 m)	8 ft (2.4 m)
Seepage pits or cesspools	—	5 ft (1.5 m)	5 ft (1.5 m)	12 ft (3.7 m)
Streams	50 ft (15.2 m)	50 ft (15.2 m)	50 ft (15.2 m)	100 ft (30.5 m)
Trees	—	10 ft (3.0 m)	—	10 ft (3.0 m)
Water supply wells	50 ft (15.2 m)	50 ft (15.2 m)	100 ft (30.5 m)	150 ft (45.7 m)

[a]When disposal fields and/or seepage pits are installed in sloping ground, the minimum horizontal distance between any part of the leaching system and ground surface should be 15 ft (4.6 m).

15.7 GRAY WATER REUSE SYSTEMS

Gray Water and Black Water

Gray water (also spelled *grey water*) is untreated wash water from bathtubs, showers, lavatory fixtures, and clothes washing machines. Although no longer potable, it is water that is relatively free of toilet wastes. Kitchen waste is usually not considered gray water because it contains excessive levels of grease, oils, and fats and waste disposal residue. Although gray water will contain bacterial contamination (e.g., total and fecal coliforms and heterotrophic plate count [HPC] bacteria) from normal bathing and washing of underwear and diapers, it does not contain bacterial contamination at the same level as does water with toilet wastes. About 50 to 80% of domestic wastewater is gray water.

Although gray water may contain grease, food particles, hair, and other impurities, it may still be suitable for irrigation and/or sewage processing (flushing toilets). An obvious use of gray water is that it may potentially replace treated tap water used for landscape irrigation. It may benefit plants because it often contains nutrients such as nitrogen or phosphorus. Gray water use also helps overtaxed municipal wastewater treatment plants because gray water use diminishes sewer flows, thereby lessening the need to overwork or expand such treatment facilities.

Black water contains toilet wastes (e.g., feces, cellulose from toilet paper, and nitrogen compounds) from water closets (toilets), urinals, and bidets. Black water may also carry hazardous chemicals from activities such as cleaning car parts, washing greasy/oily rags, or disposing of waste solutions from home photo labs or similar hobbyist activities. Compared to gray water, black water has a large amount of organic and pathogenic pollutants.

A significant difference between black water and gray water is in the rate of decay of the pollutants in each. The pollutants in gray water decompose rapidly. For example, fecal matter from underwear and diapers washed in a clothes washing machine breaks down, exposing potential pathogens to detergent. Conversely, black water contains a substantial quantity of organic compounds that have already been exposed to the digestive tract of the human body, so they do not rapidly further decompose when placed in water. Cellulose from toilet paper in black water also decays slowly.

It is estimated that about 35 gal of gray water is generated each day per person in new construction and approximately 46 gals each day per person in existing homes. Provided that it can be treated and reclaimed, gray water represents a large water conservation resource in buildings. It can result in a significant reduction in the need for potable water and a reduction in the amount of wastewater entering on-site sewage treatment systems or a community wastewater treatment system. For example, in a residence, reuse of gray water can amount to an annual savings of 30 000 to 50 000 gal per year. As a result, innovative systems that safely recycle gray water for landscape irrigation and/or water for toilet flushing are being installed.

Gray Water Reuse

Depending on health regulations, technology for gray water reuse can be as simple as saving rinse water from bathing and the clothes washer to rather complex treatment systems, such as one in which the gray water flows to an aerobic treatment unit, then to a recirculating filter. Although the level of contamination of gray water is low in comparison to black water, concerns about health and public safety need to be managed, especially those related to the potential for transmission of disease. As a general rule, gray water originating at a building is required to be contained and used within the property boundary. See Figures 15.10 and 15.11.

Gray water needs to be filtered to remove large waste particles before being used or treated. This primary treatment will reduce the solids in the wastewater. Once coarse filtered, gray water may be treated using a sand filter. The basic structure is a waterproof box filled with coarse sand laid over a gravel bed. Gray water flows in at the top and out the bottom. A number of commercial sand filters are available. Reed beds and sand filters treat the wastewater through filtration and some biological nutrient uptake. Gray water needs to be pretreated to allow removal of large particles, or else clogging will occur and the lifetime of the system will be reduced. Secondary treatment

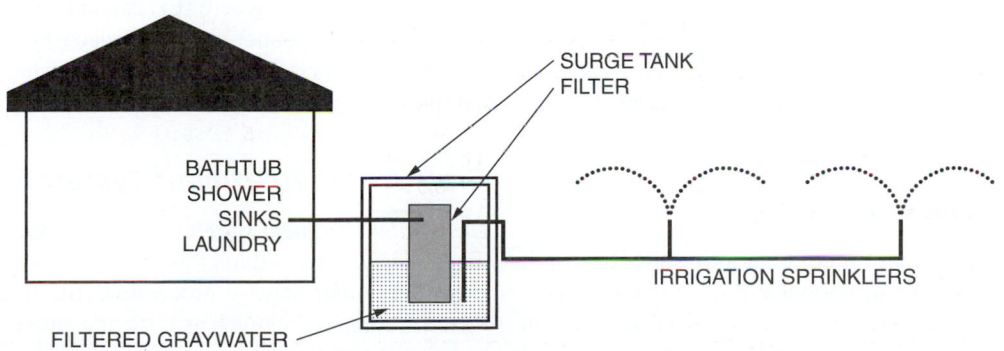

FIGURE 15.10 A simple filtered gray water system for irrigation.

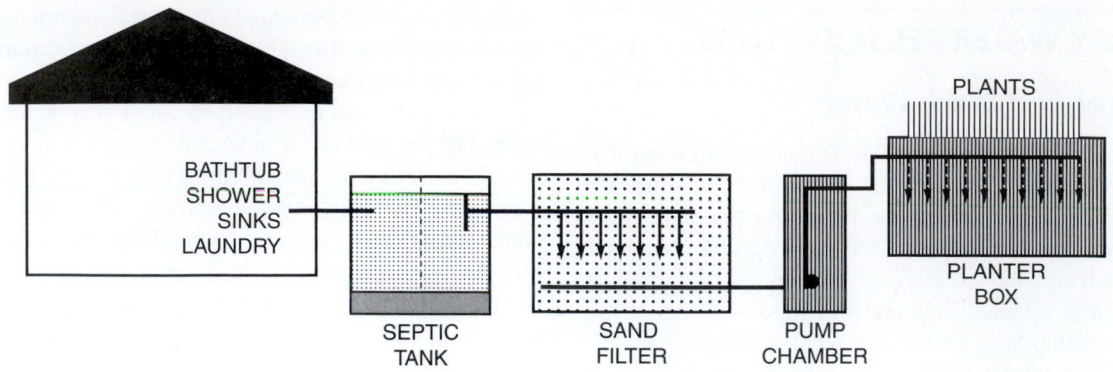

FIGURE 15.11 A gray water treatment (GTS) system that collects, stores, and treats gray water to a quality water standard for irrigation.

removes pollutants from the remaining liquid. Treated gray water has had most nutrients like nitrogen and phosphorus removed, so it is safer to use in large quantities. Disinfection is required for general reuse of gray water. All disinfection requires regular maintenance. Chlorine is most commonly used for disinfection, but it has been found to have adverse environmental impacts. Ultraviolet (UV) or ozone disinfection can be used in place of chlorination.

The amount and quality of gray water will to a certain extent determine how it can be reused. Gray water is commonly used for irrigation. Untreated gray water can only be reused for subsurface irrigation. Treated gray water can be reused for either subsurface or above-ground irrigation. Only treated and disinfected gray water should be used for above-ground irrigation because of the potential presence of pathogens. The health effect of use of untreated gray water directly on edible vegetable plants has not been fully researched. The use of gray water for irrigation of vegetable gardens should be avoided if vegetables will be eaten raw or lightly cooked.

Healthy soil is a complex system that requires moisture, organic material, microorganisms, and other life forms (e.g., worms, and so forth). By introducing large amounts of detergent into the soil, the microorganisms can be destroyed, which adversely affects the quality of soil. Gray water from the wash cycle of a clothes washer contains higher chemical concentrations from soap powders and soiled clothes, and is high in suspended solids, lint, turbidity, and oxygen demand; if applied untreated, it can damage the soil and lead to environmental damage. Water from the first and second rinse cycles contains pollutants, although these are greatly reduced.

Gray Water Diversion Devices

Gray water diversion devices do not treat gray water but use it directly as irrigation water. Gray water is piped to outdoor vegetation through subsurface irrigation lines. Human contact with gray water must be avoided. To prevent human exposure, a soil barrier of at least 4 in (100 mm) must separate the surface from the subsurface irrigation lines. There are two types of diversion devices: gravity diversion devices and pump diversion devices.

A *gravity diversion device* is a fitting configuration that diverts gray water from a plumbing fixture directly to an irrigation line. A Suldi Gray Water Diversion Valve is designed to fit under a laundry sink. A hose extends from the valve out through the wall of the building and is linked to a garden hose, which is then connected to a slow release soaker hose. It can then be easily turned off to allow soapy wash water to go down a conventional drain line and on to use the rinse water for irrigation.

A *pump diversion device* has a similar irrigation piping arrangement to the gravity diversion device, but includes a surge tank to temporarily store and limit the flow of gray water during sudden surges. The surge tank should be capable of momentarily storing a volume of water equivalent to the emptying discharge of fixtures connected to it. However, it should not serve as a storage tank. The surge tank should be fitted with a high water level alarm to warn of pump failure or system blockage.

In both systems an overflow connection is joined to the conventional drain line. The flow of gray water in a gray water diversion device is usually activated through a hand operated valve, tap, or switch. During wet or cold (below freezing) weather, the valve, tap, or switch directs gray water to a conventional drain line, where it is discharged to the sewer. A diversion device does not treat gray water but it is good practice to install a course screen to remove material that may clog pumps, block pipes, or pollute the soil.

Gray Water Treatment Systems

Gray water treatment systems (GTS) are more complex systems that collect, store, and treat gray water to a quality standard for irrigation and/or sewage processing (flushing toilets). A GTS includes components such as wetlands, intermittent sand filters, soil filters, gray water septic tanks, and aerated wastewater treatment systems.

A GTS requires separate black water and gray water waste lines. Separation of lines is achieved through the use of two independent plumbing arrangements. Plumbing configurations should not allow contamination by wastes from the toilets, cause undesirable odors, and result in aesthetic degradation of yards and gardens. Wastewater separation is relatively easy to accomplish in new construction, but can be difficult in retrofitted systems in existing dwellings. The treatment process varies according to how the gray water is used and includes settling of solids, floatation of lighter materials, anaerobic digestion in a septic tank, aeration, clarification, and finally disinfection.

Commercially available gray water treatment systems include the following:

- BioSeptic Pty Limited Aerated wastewater treatment systems (AWTS)
- Wattworks
- Graywater Saver
- Rain Reviva
- ECO Wastewater Recycling System
- Clivus Multrum Graywater Prefilter
- Gough Plastics Gray Water Reuse Unit

Gray Water Reuse Regulations

Regulations affecting reuse of gray water vary from state to state and from municipality to municipality within the same state. In many locations gray water reuse is not approved because it is assumed to carry pathogenic (disease-causing) bacteria and other hazardous pollutants. In jurisdictions where gray water systems are permitted, they are strictly regulated. In many cases, they must meet all design and construction standards for OSST (septic) systems, so disadvantages include cost for additional equipment and the requirement of gaining approval of a septic system.

As codes are revised, a septic tank may not be required for disposal of gray water only. An approved filter system may be used in place of the septic tank as long as no garbage disposal waste or waste from a toilet enters the gray water disposal system.

If garbage disposal waste or liquid waste from a composting *toilet* enters the gray water disposal system, the gray water must be treated before discharge. The conventional treatment method, a septic tank large enough to provide at least a two-day retention time, allows grease to cool, solidify, and float to the top. This is particularly important when kitchen wastes are part of the gray water flow. Garbage disposals are strongly discouraged as they stress the conventional septic system and provide an extremely high amount of solids to a system designed to handle only gray water. If a garbage disposal is installed, it should be plumbed as part of the black water system.

The State of Arizona has a progressive law on offering permits for gray water systems. It is a performance-based code that uses a three-tiered method of classifying systems:

Tier 1 Private residential systems that process less than 400 gal per day do not need a special permit if all the following conditions are met:

1. Human contact with gray water and soil irrigated by gray water is avoided.

2. Gray water originating from the residence is used and contained within the property boundary for household gardening, composting, lawn watering, or landscape irrigation.

3. Surface application of gray water is not used for irrigation of food plants, except for citrus and nut trees.

4. The gray water does not contain hazardous chemicals derived from activities such as cleaning car parts, washing greasy or oily rags, or disposing of waste solutions from home photo labs or similar hobbyist or home occupational activities.

5. The application of gray water is managed to minimize standing water on the surface.

6. The gray water system is constructed so that if blockage, plugging, or backup of the system occurs, gray water can be directed into the sewage collection system or on-site wastewater treatment and disposal system, as applicable. The gray water system may include a means of filtration to reduce plugging and extend system lifetime.

7. Any gray water storage tank is covered to restrict access and to eliminate habitat for mosquitoes or other vectors.

8. The gray water system is sited outside of a floodway.

9. The gray water system is operated to maintain a minimum vertical separation distance of at least 5 ft from the point of gray water application to the top of the seasonally high groundwater table.

10. For residences using an on-site wastewater treatment facility for black water treatment and disposal, the use of a gray water system does not change the design, capacity, or reserve area requirements for the on-site wastewater treatment facility at the residence, and ensures that the facility can handle the combined black water and gray water flow if the gray water system fails or is not fully used.

11. Any pressure piping used in a gray water system that may be susceptible to cross-connection with a potable water system clearly indicates that the piping does not carry potable water.

12. Gray water applied by surface irrigation does not contain water used to wash diapers or similarly soiled

or infectious garments unless the gray water is disinfected before irrigation

13. Surface irrigation by gray water is only by flood or drip irrigation.

Tier 2   Systems that process over 400 but less than 3000 gal a day or do not meet the list of conditions (e.g., commercial, multifamily, and institutional systems) require a standard permit.

Tier 3   Systems that process over 3000 gallons a day require a special permit and are considered on an individual basis.

In many states and local communities, regulators apply oversight to gray water systems based on their possible impacts.

15.8   ALTERNATIVE WASTEWATER TREATMENT SYSTEMS

Alternative wastewater treatment systems serve as an option for the sewage treatment systems described previously. They may be required because one of the systems described earlier cannot be used or because they are too costly (e.g., the geological conditions are not suitable). These systems do not use water to treat or transport human body wastes. If appropriately designed, they conserve water and avoid disposal of effluent and pollutants into waterways and the general environment.

Privy/Latrine

One of the oldest and most basic methods of waste disposal is the *pit privy* or *latrine,* a pit dug below an outhouse structure that collects human body wastes. Liquid wastes seep into the soil and percolate through the soil. Solid wastes remain and partially decompose so the pit becomes full over several years, depending on size and the number of users.

When the wastes in the pit reach a certain depth from the ground surface, the pit is cleaned out or the outhouse is moved to another location and the pit is covered with earth. Cleaning the pit is an unpleasant job and may result in exposure to fresh fecal material.

In long-term installations, a good arrangement is to plan for at least two pits. When the first pit is full, the outhouse structure is moved to the second pit and the first or is covered with earth and allowed to compost. When the second pit is full, the first pit is cleaned out and the slab and the outhouse structure are moved back over the first pit. The second pit is covered and allowed to compost.

If soil conditions are suitable and no other system is available, a privy can be a safe method of waste disposal. It is, however, inconvenient for users and often produces offensive odors. Important considerations are that the pit be designed so it will not pollute groundwater or permit access by insects or rodents. If wells or other supplies of water are nearby, there is a

risk of contamination. As a result, a privy may not be an approved alternative waste treatment system. This system should only be used for the disposal of human body waste in extremely remote locations or in undeveloped areas. In some cases, however, health authorities in undeveloped countries allow such installations; as a result, it is discussed.

Composting Toilets

A *composting toilet* is a self-contained waste treatment system that uses natural biological decomposition to convert toilet wastes into water vapor, carbon dioxide, and a stable compost-like end product. It consists of a toilet seat and cover over a riser that connects to a compartment or vault that receives, holds, and converts human toilet wastes and other composting materials. It is essentially an "in-house" outhouse. A venting system that extends up through the roof prevents sewer gases from entering the building.

The decomposition process in a composting toilet is achieved by aerobic (oxygen-using) bacteria and fungi. The complex population of microorganisms in the composting material makes conditions unfavorable for the growth of disease-causing organisms that can be present in human waste. Pathogenic organisms die off or are consumed by the composting organisms as long as the composting process is proceeding normally and has adequate time to work.

To produce a thoroughly decomposed compost product, three conditions are necessary:

1. Microorganisms that decompose the waste need oxygen to flourish, so the process must remain aerobic. Aerobic conditions are sustained by mixing the pile and by controlling moisture.

2. The compost must be maintained at the correct moisture content. If the compost becomes too dry, decomposition will not occur. If the compost is too wet, it will not remain aerobic and decomposition will cease. Humans excrete a much higher volume of liquid than solid each day. This excess liquid must be managed to ensure the composting waste does not become too wet. Excess liquid is managed either by evaporating it off using exhaust fans and heater units inside the compost chamber or by collecting it at the bottom of the unit where it must be disposed of in an acceptable manner.

3. Temperatures must be maintained above 60°F for composting to proceed effectively. At lower temperatures bacterial activity is inhibited and the composting process slows.

There are two types of composting toilets available for buildings.

Bi-Level Composting Toilet

Bi-level composting toilets are relatively large, two-story, watertight containers equipped with a chute that connects the

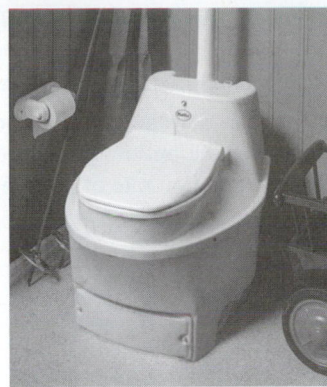

PHOTO 15.10 A self-contained composting toilet.

toilet receptacle to the composting unit located in the basement. The bottom of the composting unit often has an inclined floor where solid wastes decompose and slide to the lower end as new waste enters at the upper end. Excess liquid is drained to the lowest part of the tank, where it is either evaporated or collected and removed. Compared to self-contained composting toilets (described below), bi-level composting toilets have a large compost volume and long retention time. Thus, the composting process is more stable than in smaller units, is better able to cope with peak loads, and can withstand intermittent or seasonal use. This type of composting toilet is similar to an in-house outhouse. In most units, a power vent removes excess gases and discharges them outdoors. Compost generally needs to be emptied every few years. The best-known bi-level composting toilet is the Clivus Multrum™. (See Photo 15.10.)

Self-Contained Composting Toilet

The second type of composting toilet is a smaller unit in which the toilet receptacle and composting tank comprise a single self-contained unit located on the bathroom floor. These units

have traditionally been installed for intermittent use such as in vacation homes. They are similar in design to a port-a-potty. These units are marketed by BioLet USA, Inc. (the "BioLet"), Sun-Mar Corp. (the "Sun-Mar"), and SanCor Industries Ltd. (the "Envirolet").

Incinerating Toilet

Incinerating toilets are self-contained waterless systems that do not require being hooked up to a sewer system or an in-ground septic system (except to dispose of gray water). They rely on electric power or natural or propane gas to incinerate human waste to sterile clean ash. When properly installed, these systems are simple to use, safe, clean, and relatively easy to maintain.

These waterless systems look much like a standard household water closet. Between the gas and electric incinerating toilets, there are some mechanical and operational differences, but the overall treatment processes work the same. Both systems accept human waste, both solid and liquid, into a burn chamber. The burn chamber reaches temperatures of 970° to 1400°F (520° to 760°C) and reduces human waste into clean sterile ash.

Holding Tanks

Holding tanks are used for wastewater disposal when soil, slope, lot size, groundwater, or other features on the site render all other tank/drainage field solutions impossible to achieve. They are used as a method of last resort in vacation homes or other short-term use facilities. Holding tanks are rarely used in facilities that generate wastewater on a daily basis.

Holding tanks are typically sized to hold seven days of wastewater flow. When the tank has been filled to within 75% of its capacity, a visual and audible alarm is automatically triggered, which alerts the user that the tank must be promptly pumped. Because holding tanks might have to be pumped on a regular basis, they can be costly to maintain.

STUDY QUESTIONS

15-1. Describe gray water and black water.

15-2. Describe the constituents of wastewater.

15-3. Describe preliminary, primary, secondary, or tertiary wastewater treatment processes.

15-4. Describe where OSST systems are used.

15-5. Describe the components of a septic tank.

15-6. Describe the differences between a septic tank and aerobic tank.

15-7. Why do most codes set minimum distances between the parts of the sewage system and wells and streams?

15-8. How is the ability of the soil to absorb wastewater determined?

15-9. Why should the designer consider not connecting pipes carrying certain wastewater (such as water from dishwashing machines) to the private OSST system?

Design Exercises

15-10. Draw a sketch of the major parts of an individual OSST system (septic tank and absorption field) in relation to a residence. Show clearance and setback requirements.

15-11. Design a private sewage system (septic tank and disposal field) for the apartment building in Appendix B. In this design, use a percolation rate of 16 min.

15-12. Design a private sewage system (seepage pits) for the apartment building in Appendix B. In this design, use a percolation rate of 16 min. Assume the maximum water table elevation is 2 ft (0.6 m) below grade.

15-13. Design a private sewage system (septic tank and disposal field) for the building in Appendix C. In this design, use a percolation rate of 22 min.

15-14. Design a private sewage system (septic tank and disposal field) for the building in Appendix D. In this design, use a percolation rate of 32 min.

BUILDING STORM WATER DRAINAGE SYSTEMS

16.1 STORM WATER

Storm water is the result of rain, snowmelt, sleet, or hail. It is part of the natural hydrologic process. If not properly controlled, storm water runoff can result in moisture problems. Most storm water problems in buildings are caused by poor roof and surface drainage. These problems develop when storm water from rain and snowmelt does not adequately drain away from the building, and thus finds its way into the structure.

When storm water is allowed to pond next to the building foundation or when the underground water table rises such that it is near the basement or crawl space floor, water can be driven into the building area below grade. The driving force behind water flow is hydrostatic pressure: the same force that drives water down a stream or through a hole in a leaking bucket. Hydrostatic pressure can drive water into a building through construction joints such as the intersection of a concrete floor slab and foundation wall, through cracks that develop in the floor slab or foundation wall, and through porous materials such as concrete and brick mortar.

When it precipitates (i.e., rains, snows, sleets, or hails), storm water must be controlled. Runoff from roofs, courtyards, and paved areas (such as parking lots) must be carried away from the building and properly disposed. Storm water may be directed to drains in the building roofs, parking areas, courtyards and then be directed into:

- A private storm sewer or drywell
- A community storm sewer
- A creek, stream, lake, or pond

If the storm water is directed to a community storm sewer system, the only concern will be that the elevation of the storm sewer line is low enough that the private storm line can run into it. Many communities have storm sewer systems, which are also used to drain storm water from the streets and safely away. (See Photo 16.1.)

Reoccurrence Interval

Statistical techniques are used to estimate the probability of the occurrence of a given precipitation event (i.e., rain storm, snow storm, and so on). The *recurrence interval* is based on the probability that a given event will be equaled or exceeded in any given year. For example, assume there is a 1 in 50 chance that

PHOTO 16.1 A community storm drainage line before placement of backfill. (Used with permission of ABC)

TABLE 16.1 RECURRENCE INTERVALS AND PROBABILITIES OF OCCURRENCES.

Recurrence Interval, in Years (Average)	Probability of Occurrence in Any Given Year	Percentage Chance of Occurrence in Any Given Year
500	1 in 500	0.2
100	1 in 100	1
50	1 in 50	2
25	1 in 25	4
10	1 in 10	10
5	1 in 5	20
2	1 in 2	50

5.0 in of rain will fall in a certain area in a 24-hr period during any given year. Thus, a rainfall total of 5.0 in during a consecutive 24-hr period is said to have a 50-year recurrence interval. See Table 16.1. Storm water systems are typically designed to handle storm water from a specific precipitation event. Typically it is based upon a recurrence interval of 100 or 500 years.

16.2 STORM SEWERS

Private Storm Sewers

If the water is collected into a private storm sewer line, the line can discharge (outflow) into:

- A drywell or drainage field, which allows the water to be absorbed irectly into the ground (see Figures 16.1 through 16.3).

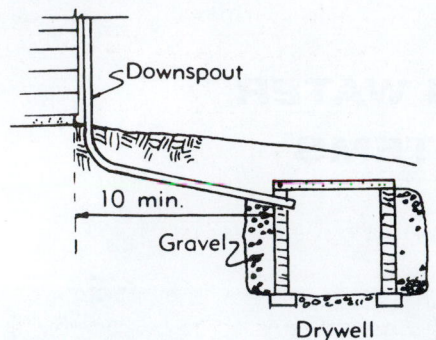

FIGURE 16.1 A downspout (or leader) discharging into a drywell.

- An area of low elevation on the site, which allows the water to percolate into the soil
- A nearby river, creek, or stream
- A public or private lake or pond

If the storm sewer line serves a large area, the force of the running water, after a heavy rain, may cause considerable damage where it runs out the end of the line. An engineer with experience in drainage should carefully consider potential for damage caused by outflow.

Running the storm sewer private line so that it outflows into a creek, stream, or lake may require a permit and may not be allowed in some areas. Usually, before approval, the design of the system must be analyzed to ascertain that there is no possibility of sewage wastes (chemical, human, or industrial wastes) getting into the storm sewer line and contaminating a lake, stream, or creek. Many times on large projects, the storm water is run into a pond so that the water will be available for nonpotable uses, such as for irrigating lawns and gardens or circulating in fountains.

Combined Community Sewers

In some cities, especially in older areas, storm sewers from the city streets and the private buildings, driveways, and parking areas all run into the sanitary sewer line. This system is called a *combined sewer.* It should be done only if no other solution is available and if the city allows storm lines to be tied into its sewage lines. This storm water creates a tremendous excess load for the city (county or municipality) sewage treatment plant. This unnecessary burden often requires that the sewage treatment plants built be much larger than they would be if only sewage were to be treated. In cities with such systems, the cost to separate the storm and sewage lines now would be prohibitive. (See Figure 16.4.)

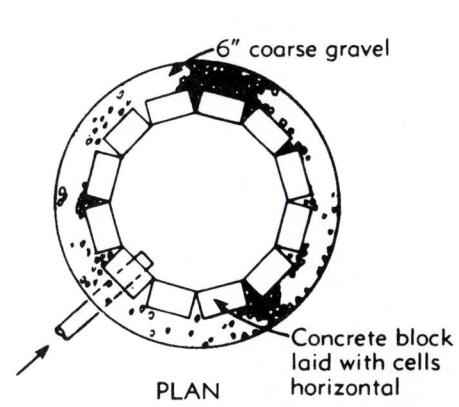

FIGURE 16.2 Construction of a drywell.

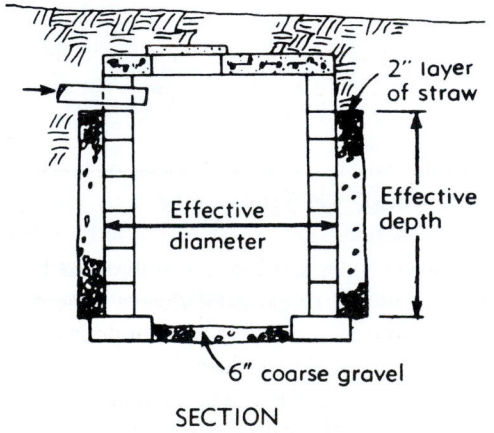

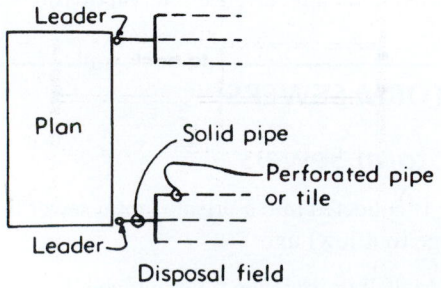

FIGURE 16.3 Leaders or downspouts discharging into disposal fields.

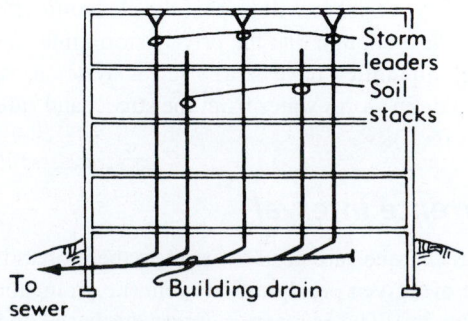

FIGURE 16.4 Soil stacks and storm leaders discharging into a combined (storm and waste) drain system.

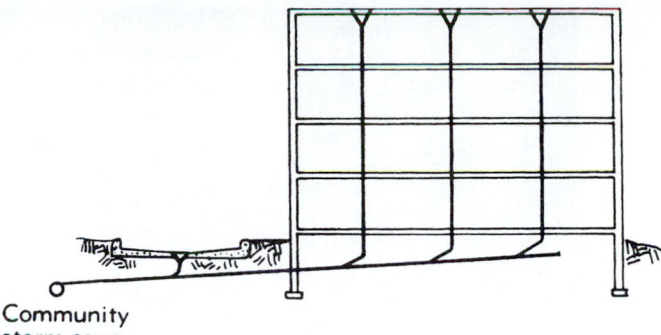

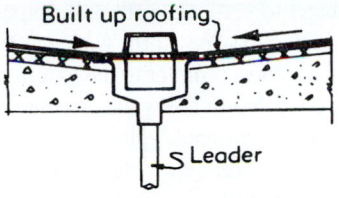

FIGURE 16.6   A cross-section of a roof drain.

FIGURE 16.5   Building and street drains discharging into a community storm sewer.

Separate Community Storm Sewers

In most instances, water pollution regulations require that there be separate storm and sewage lines, and that only storm water runoff be discharged to storm sewers. Such regulations typically require that only the following can go to the storm sewers:

- Storm water from the land, parking lots, and streets
- Roof drainage
- Water from subfloor and footing drains

None of the inlets to the storm sewer system should be used to dispose of any waste materials. Cities with only sewage lines may or may not allow storm water to be introduced into the sewage lines, and this should be carefully verified with the local authorities. (See Figure 16.5.)

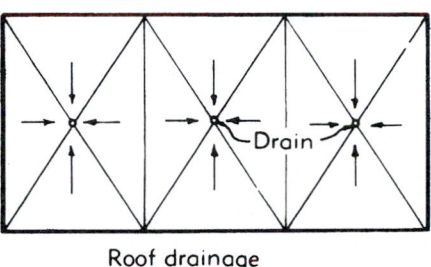

Roof drainage

FIGURE 16.7   A low-sloped (near flat) roof surface is sloped to interior roof drains.

16.3   ROOF DRAINAGE DESIGN

Storm water from the roof can be removed by any of three methods: a roof drain system, a gutter and downspout/leader system, and allowing the water to run off the building without drains or gutters. The third method is generally considered unacceptable unless provisions are made to ensure adequate drainage away from the building. In most instances, storm water collecting near the building foundation saturates the foundation soils. Most basement moisture problems are tied directly to poor drainage of roof and surface storm water. Foundation problems can also develop as a result of excess soil moisture.

Roof Drain System

A *roof drain system* consists of roof drains at regular intervals on the roof surface that collect storm water, transport it through pipes, and discharge it at ground level or into a storm sewer. A *roof drain* is a bowl-shaped collecting sump with a strainer on top that prevents leaves and other debris from entering the drain. It is designed for draining storm water from low-sloped (essentially flat) roofs. Low-sloped roof surfaces are typically pitched slightly toward the drain to prevent storm water from collecting on the roof surface. (See Figures 16.6 through 16.9.)

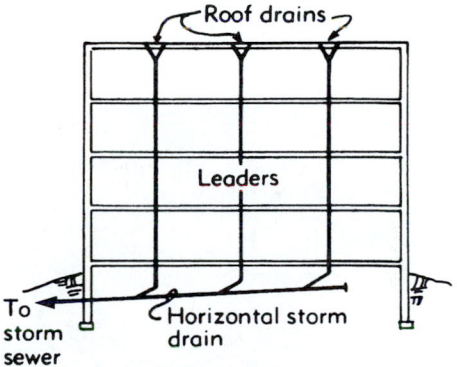

FIGURE 16.8   Roof drains collect and direct water into vertical storm drains (leaders). Collected water then drains into a horizontal storm drain, which discharges storm water into a private or community storm sewer.

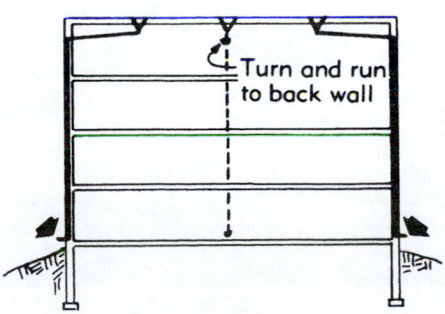

FIGURE 16.9   Leaders from roof drains can run through exterior walls and outflow at the building perimeter.

Drains connect to vertical storm drain pipes, called *leaders* or *conductors,* which carry the storm water away from the drain and into a horizontal storm drain or to the exterior of the building. Leaders may be concealed in the walls or columns if the sight of an exposed pipe is objectionable. However, once enclosed, it is more expensive to make repairs if necessary. A *downspout* is a vertical storm drain pipe that is secured to the building exterior.

If the leader/downspout runs to the outside of the building and empties, precautions must be taken so that the water will be directed away from the building. Immediately adjacent to the building, a concrete or plastic pad, called a *splashblock,* or other similar device is required so that the water coming from the end of the pipe will not strike the soil with such force that it will wash it away, causing soil erosion and permitting the possibility of wet foundation walls. Minor problems, such as staining the building, may also occur. However, undermining of the building structure or other elements (e.g., walks, stairs, driveways, and so forth) is possible in extreme cases. (See Photos 16.2 through 16.5.)

Once the water is directed away from the building, surrounding contours (slopes of grade) must keep the water flowing away from the building. Good drainage must prevent ponding at the building site and flow of excessive storm water to adjacent building sites.

In locations where the leaders/downspouts can be tied into the sewage system, the system is referred to as a *combined*

PHOTO 16.2 Interior roof drains on a low-sloped commercial roof system. (Used with permission of ABC)

PHOTO 16.3 Conductors (pipes) on the underside of the roof surface carry storm water away from the roof drain. (Used with permission of ABC)

PHOTO 16.4 A cleanout is typically provided at the base of a vertical conductor. (Used with permission of ABC)

PHOTO 16.5 The discharge of storm water at the building exterior. (Used with permission of ABC)

sewer. Because many locales do not allow the installation of combined sewers, be certain to check with local authorities. Figure 16.4 illustrates how the leader might be tied into the building drain. In this situation, most codes require a trap on the leader (storm sewer systems seldom require traps, except with combined sewers), and they may specify that the trap shall be a minimum of 10 ft (3 m) from any stack.

A roof drain system is commonly placed on low-sloped (flat) roofs. These roofs have at least a ¼ in/ft slope to ensure that the water will not pond on the roof after rain or snowmelt. It is detrimental to the roofing materials to have water pond on the roof surface, so the roof surface should not be flat.

With roof drains, the roof surface should be sloped toward the roof drains to be certain that all storm water is drained off the roof. Most new structures are designed to slope toward the roof drains. In older buildings with a flat roof deck, the required slope may be accomplished by using a layer of lightweight concrete or asphalt. When the deck is made of poured gypsum or concrete, the slope is cast as the deck is poured. A tapered insulation system may also be installed below the roof membrane to create the necessary slope.

Gutter and Downspout System

Storm water may be directed off the roof into gutters. A *gutter* is a horizontal trough or channel that runs along the eaves of a building to capture and divert rainwater or snowmelt from the roof and allow it to drain into a downspout or leader. The roof surface must be sloped to direct the flow of storm water toward the gutter. Gutters can be used on low-sloped roofs (roofs with a slope of less than 4 in 12). They are generally always incorporated in the storm drainage system of steep-sloped roofs (roofs with a slope of 4 in 12 or greater). (See Photos 16.6 through 16.8.)

Storm water collected by a gutter drains into a leader or downspout, a vertical pipe that carries storm water away from the roof gutter. The downspout may be tied to a storm sewer line (either community or private) or may empty outside the building onto a splashblock or some other means to disperse the water at the ground level. (See Figure 16.10.)

PHOTO 16.8 A residential roof drainage system consists of an ogee-shaped gutter to capture roof water (at roof overhand), a downspout (running vertically), and a downspout extension (at ground level), which is designed to discharge storm water well away from the building. (Used with permission of ABC)

PHOTO 16.6 An exterior roof drainage system. Storm water from the low-sloped roof drains through openings in a parapet wall into funnel-like conductor heads at the top of the vertical conductor or downspout. The vertical conductor has an open face to prevent blockage from ice or debris. (Used with permission of ABC)

PHOTO 16.7 A series of exterior roof drains. (Used with permission of ABC)

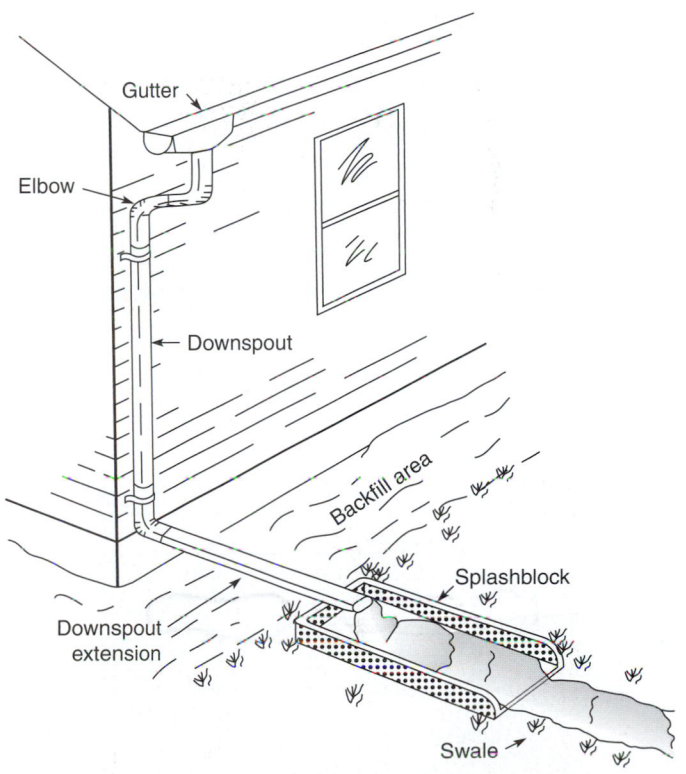

FIGURE 16.10 Components of a residential roof drainage system. (Courtesy of David Noe, 2007, *A Guide to Swelling Soil for Colorado Homebuyers and Homeowners,* second edition, Colorado Geological Survey Special Publication 43)

Downspout extensions are horizontally sloped pipes at the base of a downspout that extend the outflow of the downspout well beyond the building foundation. They should extend a minimum of 5 ft or more away from the building's foundation so roof drainage is beyond the area of construction excavation and backfill. This is especially important in newer homes where backfill soils are unsettled.

Another possible solution is to run the downspout into a small catch basin or disposal area filled with gravel. This may be effectively used on smaller buildings, such as residences, but it is important that the catch basin be located at least 10 ft (3 m) from the foundation walls to reduce any chance for wet walls from the water.

Materials

Gutters and downspouts/leaders are made of copper, galvanized steel, aluminum, or vinyl. Vinyl and aluminum gutters and downspouts can be custom-made in one piece without seams and are called *seamless*. Aluminum and steel gutters and downspouts come in many different colors that are ideal for matching the color of trim and cladding. Gutter colors are baked on at the factory or painted on-site. Vinyl is typically available in only a few colors (e.g., brown, black, ivory, or white). Copper comes unpainted and ultimately oxidizes to a green patina.

Most gutters come in several sizes and shapes called *profiles*. These include the U-shaped trough (a half round channel shape) and the K- or ogee-shaped configuration (a front that looks like the letter K). Common gutter profiles are available in the sizes shown in Table 16.2. The 5 in (125 mm) ogee-shape is most common on residences. Downspout choices are shown in Table 16.3. The 3 in × 4 in (75 mm × 100 mm) rectangular size is most common on residential installations. On a commercial project, the interior storm drainage piping typically varies in size from 3 in to 8 in

TABLE 16.2   COMMON GUTTER SIZES AND DIMENSIONS. ACTUAL SIZE VARIES BY MANUFACTURER.

Nominal Size		Actual Size					
		Top Width		Depth		Bottom Width	
in	mm	in	mm	in	mm	in	mm
K- or Ogee-Shaped Profile							
4	100	4	100	3⅛	79	2¾	70
5	125	5	125	3⅝	92	3⅜	60
6	150	6	150	4⅞	124	3⅞	86
7	175	7	175	5⅞	149	4⁹⁄₁₆	116
8	200	8	200	7¼	184	5⅜	137
U- or Half-Round Shaped Profile							
4	100	4	100	2	50	—	—
5	125	5	125	2½	6½	—	—
6	150	6	150	3	75	—	—
7	175	7	175	3½	82½	—	—
8	200	8	200	4	100	—	—

TABLE 16.3   COMMON DOWNSPOUT/LEADER SIZES AND DIMENSIONS. ACTUAL SIZE VARIES BY MANUFACTURER.

Nominal Size		Actual Size					
		Depth		Width		Diameter	
in × in	mm × mm	in	mm	in	mm	in	mm
Rectangular Profiles							
2 × 3	50 × 75	1¾	44	2¼	57	—	—
3 × 4	75 × 100	2¼	57	4¼	105	—	—
3 × 5	75 ×125	3¼	93	5	127	—	—
4 × 5	100 × 125	4	102	5	127	—	—
4 × 6	100 × 150	4	102	6	152	—	—
Plain and Corrugated Round Profiles							
3	75	—	—	—	—	3	75
4	100	—	—	—	—	4	100
5	125	—	—	—	—	5	125
6	150	—	—	—	—	6	150

(75 mm to 200 mm). The size of exterior piping and gutters may range to 24 in (600 mm) and larger.

The size of the drain pipe often requires special provisions in wall width or furred-out areas. In unfinished spaces (e.g., warehouses), the drain pipe may run exposed next to a column. In finished spaces (e.g., offices), the drain pipe may be enclosed in fireproofing that protects the column. Storm drains that serve the exterior areas of a building project are larger and require extensive planning because they often run under roads and buildings.

16.4   SURFACE DRAINAGE

When designing drainage for driveways, parking lots, and surrounding ground, the site plan of the project must be reviewed to determine the effect the existing and revised contours will have on the flow of the surface water. It is important that the flow of water be away from the building and not toward it. In housing developments and other large projects, the ground should be contoured so water will flow toward the storm sewer system, usually to a catch basin. A *catch basin* is an underground structure, usually at the curb line, with an open grate cover to collect storm water from streets and pathways and discharge it to a storm sewer system. (See Figure 16.11.)

The elevation of driveways and parking lots should also be reviewed. When storm water is simply being allowed to run off onto the surrounding ground, the driveway and any curbs should be constructed higher than the surrounding ground so that the water will run off and onto the ground. When a storm sewer system will be used to collect and carry away the water, the driveway and curbs may be set lower than the surrounding ground and should be generally sloped toward the catch basins.

On projects without a city storm sewer system, the water must be directed away from buildings, driveways, and parking lots. This is accomplished with a *swale*, a depression or ditch designed to move storm water away from the building. It typically contains flowing water only during and after a rainfall or snow melt, the intent being to move storm water away from the

building and allow it to percolate through the ground or discharge away from the building.

Regulations may require that storm water flow into detention or retention ponds. A *detention pond* is a human-made or natural pool or basin area used to detain and control flow of storm water runoff during heavy rainstorms. The pond fills with excess storm water runoff and impedes storm water flow. Detained storm water is released gradually into natural or human-made outlets at a rate not greater than the flow before the development of the property. If it drains into a combined sewer system, it must flow at a rate that reduces the chance of sewer surcharge (overflow). A *retention pond* is a human-made or natural pool or basin area used for permanent storage of storm water runoff. It is designed to collect and hold surface and subsurface water. (See Photo 16.9.)

Land development and construction activities can significantly alter natural drainage patterns and pollute storm water runoff. Runoff picks up pollutants as it flows over the ground or paved areas and carries these pollutants into the storm sewer system. Common sources of pollutants from construction sites include the following: sediments from soil erosion; construction materials and waste (e.g., paint, solvents, concrete, and drywall); landscaping runoff containing fertilizers and pesticides; and spilled oil, fuel, and other fluids from construction vehicles and heavy equipment.

Most municipalities are required by federal regulations to develop programs to control the discharge of pollutants to the storm drain system, including the discharge of pollutants from construction sites and areas of new development or significant redevelopment. As a result, development and construction projects may be subject to new requirements designed to improve storm water quality, such as expanded plan check and review, new contract specifications, and increased site inspection.

16.5   SUBSURFACE DRAINAGE SYSTEMS

Most groundwater infiltration problems are easily corrected by the roof and surface drainage measures outlined in the previous sections. In extreme cases when the building site slopes severely

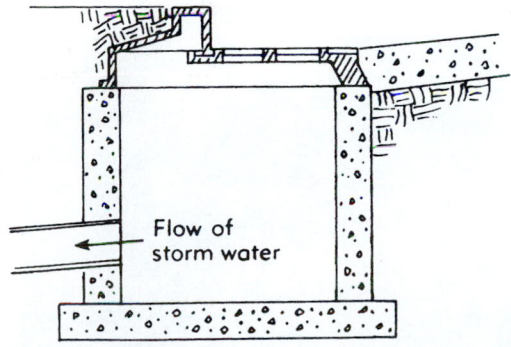

FIGURE 16.11   A catch basin. An underground structure, usually at the curb line, with an open grate cover to collect storm water from streets and pathways and discharge it to a storm sewer system.

Flow of storm water

toward the building or if the water table consistently rises above the basement or crawl space floor line, there is need for a more complex solution to these groundwater infiltration problems. Remedies include installation of the following systems.

Sump and Sump Pump

A pit or reservoir called a *sump* can be located in the basement or crawl space floor. It is used for collecting water that is discharged by a *sump pump.* Some building contractors find it cost-effective to install a sump in the basement slab in case a problem develops. They add a sump pump system only when a problem with subsurface water exists. A building can have one or more sump systems. (See Photos 16.10 through 16.14.)

Interior Perimeter (French) Drain

An *interior perimeter drain,* commonly called a *french drain,* is a subsurface drainage system consisting of a drainpipe or drain

tile placed in a trench below the basement slab and near the inside perimeter of foundation walls. Pipes from the interior perimeter typically drain into a sump (pit). Water is pumped from the sump and discharged to the exterior by the action of a sump pump. Some building contractors install an interior

PHOTO 16.12   An exterior sump discharge line. (Used with permission of ABC)

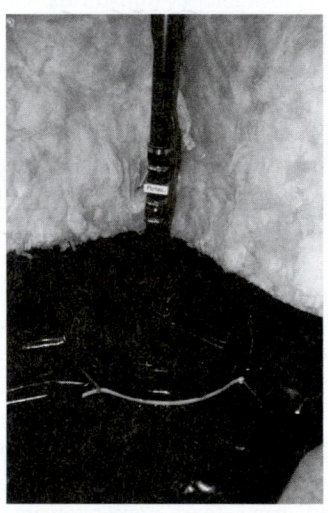

PHOTO 16.10   A sump in a crawl-space floor. (Used with permission of ABC)

PHOTO 16.13   An exterior perimeter drain placed against a foundation wall during construction of a home. (Used with permission of ABC)

PHOTO 16.11   A view into the sump in the previous photograph. A sump pump is situated in the bottom of the sump. (Used with permission of ABC)

PHOTO 16.14   A below-slab floor drain. (Used with permission of ABC)

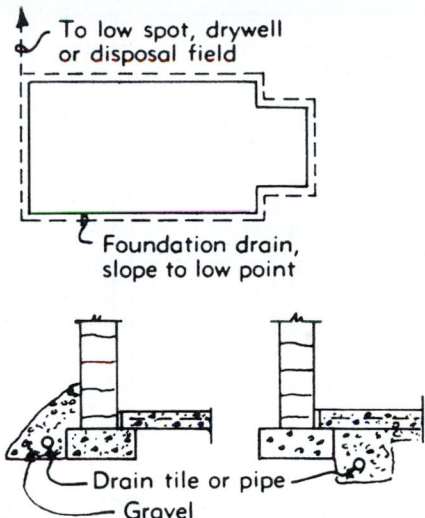

FIGURE 16.12 An interior perimeter drain consists of a drainpipe or drain tile placed in a trench below the basement slab and near the inside perimeter of foundation walls. An exterior perimeter drain is laid in a trench around the outside perimeter of the building foundation and backfilled with gravel.

perimeter drain and sump system in case a problem develops. (See Figure 16.12.)

Exterior Perimeter (Peripheral) Drain

An *exterior perimeter drain,* commonly called a *peripheral drain,* is a subsurface drainage system consisting of drainpipe or drain tile laid in a trench around the outside perimeter of the building foundation and backfilled with gravel. It discharges into a ditch, sump, or storm sewer.

Interceptor Drain

An *interceptor drain* is a subsurface drainage system involving a drainpipe or drain tile laid in a trench between the building and an uphill source of water. It intercepts the water and discharges it into a ditch or storm sewer away from the building. (See Figure 16.13.)

Drain pipes are 4 in (100 mm) diameter or larger pipes made of hard fibrous materials with holes, perforated thermoplastic pipe, or clay tile spaced about ¼ in (6 mm) apart with the upper half of the joint covered. The pipes are typically laid in a trench in a layer of gravel. They are covered with fibrous material and gravel and then soil is backfilled over the pipe or tile. The fibrous (fabric-like) material that covers the pipe is porous enough to allow water to penetrate yet impermeable enough to prevent soil and gravel from clogging the pipe. The drain must be installed at a slope so water flows into a sump, drywell, storm sewer, or low spot on the site. Whenever possible, this drain should not be connected to the sanitary sewer system.

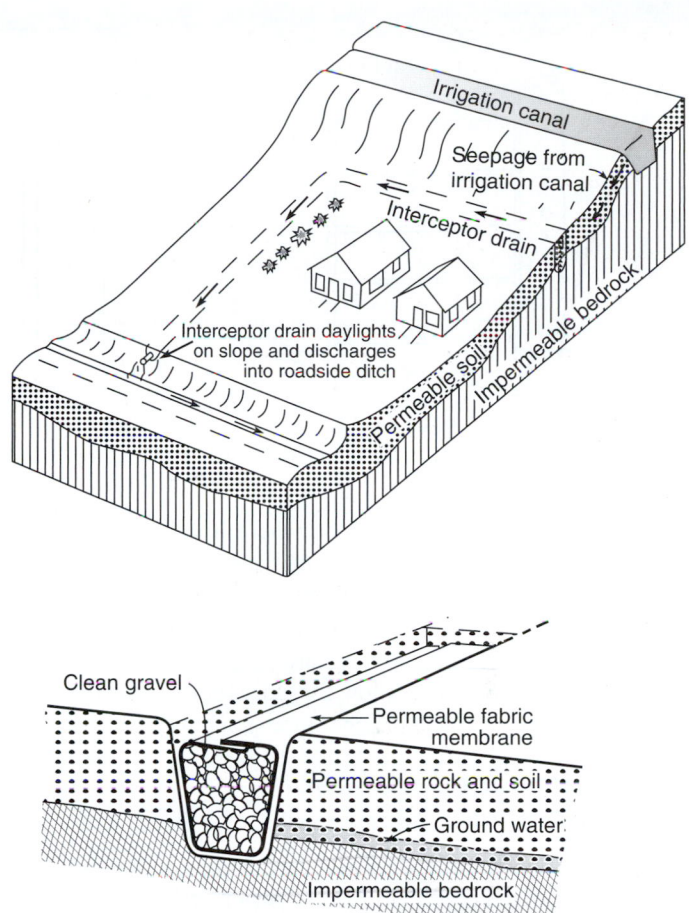

FIGURE 16.13 Components of an interceptor drainage system. (Courtesy of David Noe, 2007)

16.6 STORM WATER DRAINAGE SYSTEM INSTALLATION

Roof Drainage Installation Considerations

It is most important that the gutters be installed with a definite slope toward the leaders/downspouts. This reduces the potential for collection of water that may freeze or cause corrosion. It also minimizes the collection of debris (e.g., roofing aggregate, leaves, and so forth) in the gutter. In roof areas exposed to collecting leaves, snow, or ice, larger downspout diameters than those required minimize clogs.

Poured concrete slabs will require that the interior storm drainage layout be carefully considered. The pipes need to be placed in the ground before the slab is poured, so their accurate placement is crucial. All piping must be carefully located and the system checked for leaks before the concrete is poured because any relocation or repairs of pipes would be costly.

The open spaces provided in truss-type construction make it easier to run piping through to the desired location. The

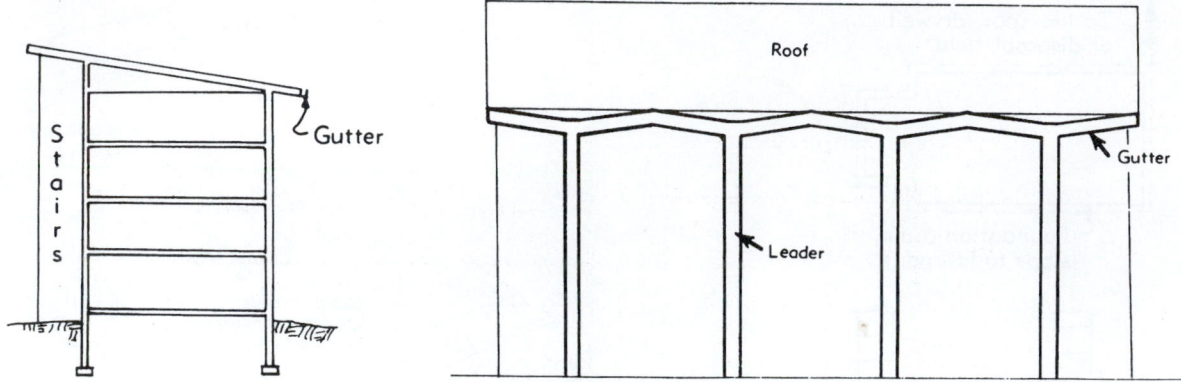

FIGURE 16.14   Sloped roof and gutters and leaders for the design example.

only points of difficulty would be where it needs to pass by ductwork or some other large pipe that is going in the opposite direction. This will require coordination with the contractor installing any heating, air conditioning, or ventilating ductwork. In wood frame construction, there are times when the width of a wall must be increased to allow for pipes running horizontally to pass by drainage pipes (or other pipes) running vertically.

Storm Drainage Installation Considerations

Pipe material for storm sewers may be the same as that used for the sanitary drainage system. Storm sewer systems, however, may include pipe of much larger sizes than are needed for sanitary sewers. Plain or reinforced concrete pipe (rather than clay, cast iron, or asbestos cement) is generally used for the larger lines. Also, it is not so important that the joints be watertight in storm sewer systems. In fact, the mortar is sometimes omitted from a portion of the joint and washed gravel is placed next to the opening; the storm drain thus also serves as an underdrain to pick up subsurface water.

Storm and sanitary systems may differ in the installation of the piping. Building storm drains should generally be graded at least ¼ in per foot whenever feasible. This amount of drop per foot provides an unobstructed and self-scouring flow. However, a greater drop per foot may be given because fixture traps that might lose their seals are not associated with storm sewer systems.

When a change of direction is necessary, long radius fittings are used and a cleanout need not be installed. This is especially true in and under buildings. But a manhole is used outside of buildings when a change of direction is necessary, or when two or more lines are connected together.

16.7   DESIGN EXAMPLE OF A STORM WATER DRAINAGE SYSTEM

The four-story apartment building (found in Appendix A) will be used in this design example. The surface roof is sloped to the rear of the structure (See Figure 16.14). A gutter that serves the entire 3200 ft² (300 m²) roof area will be installed along the edge of the rear roof overhang to collect storm water. The maximum rate of rainfall will be assumed to be 4 in/hr (100 mm/hr).

Gutters and downspouts are sized according to the area of roof they serve following the procedure outlined below.

Sizing Roof Gutters and Downspouts

1. The size of each downspout is based on the number of downspouts and the roof area each downspout will serve.

   In this design, four downspouts will be used, each serving an area of 800 ft² (3200 ft²/4 downspouts = 800 ft²). From Table 16.4a for a maximum rate of rainfall of 4 in/hr, each downspout must be at least of a 3 in diameter. If the downspouts are connected to a horizontal storm drain, the drain is then sized as discussed later in this section.

   In metric (SI) units in this design, four downspouts will be used, each serving an area of 75 m² (300 m²/4 downspouts = 75 m²). From Table 16.4b, for a maximum rate of rainfall of 100 mm/hr, each downspout would be 75 mm in diameter. If the downspouts are connected to a horizontal storm drain, the drain is then sized as discussed later in this chapter.

2. The gutter size depends on the area of the roof that each portion of the gutter serves and the slope of the gutter when it is installed.

   In this design, each of the four downspouts serves 800 ft². The layout would be such that the gutter would feed into the downspout from two sides, so that each portion of the gutter would serve 400 ft² of roof area. Assuming a ⅛ in slope per foot, from Table 16.5a for a maximum rate of rainfall of 4 in/hr, a 6 in diameter is selected for the gutters.

   In metric (SI) units, each of the four downspouts serves 75 m². The layout would be such that the gutter would feed into the downspout from two sides, so that each

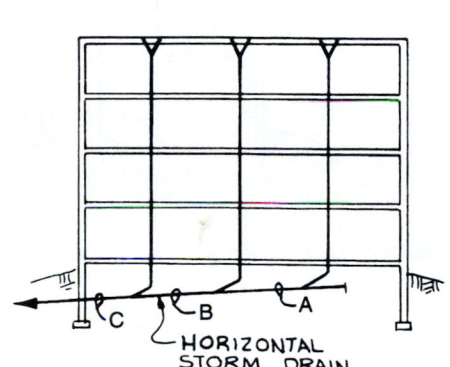

HORIZONTAL STORM DRAIN			
	s.f.	Accum. s.f.	Pipe size
A	1067	1067	
B	1067	2134	
C	1066	3200	

SI UNITS			
HORIZONTAL STORM DRAIN			
	sq m	Accum. sq m	Pipe size
A	100	100	
B	100	200	
C	100	300	

FIGURE 16.15 Horizontal storm drain sketches and computations for the design example.

portion of the gutter serves 37.5 m² of roof area. Assuming a 10.4 mm per m slope from Table 16.5b, a 150 mm diameter is selected for the gutters.

Sizing the Horizontal Storm Drain

Once the roof drains or gutters and downspouts have been selected and sized, the next step is to size the horizontal storm drain (if one is to be used). The horizontal storm drain data applies to any horizontal storm drain location (e.g., under the roof slab or below the floor slab).

1. The horizontal storm drain is sized from Table 16.5a; its size depends on the area being served and the slope

at which the pipe is installed. The pipe is typically increased in size as it collects the downspouts, so the first step will be to make a sketch of the system (Figure 16.15). Next, add the area that each downspout serves and the slope selected for the horizontal drain to the sketch.

In this design, three downspouts are used, serving each 1067 ft², and the slope is ¼ in per ft.

In metric (SI) units, three downspouts are used serving 100 m² each, and the slope is 20.9 mm per m.

2. With this information, the first length of drain (labeled "A" on Figure 16.16) is sized from Table 16.5 as a 3-in pipe. Be certain that the column for a ¼-in slope is used in this case. The next length of drain ("B") services 2134 ft²

HORIZONTAL STORM DRAIN			
	s.f.	Accum. s.f.	Pipe size
A	1067	1067	3″
B	1067	2134	4″
C	1066	3200	5″

Slope ¼″ per foot

SI UNITS			
HORIZONTAL STORM DRAIN			
	sq m	Accum. sq m	Pipe size
A	100	100	75 mm
B	100	200	100 mm
C	100	300	125 mm

Slope 20.8 mm per m

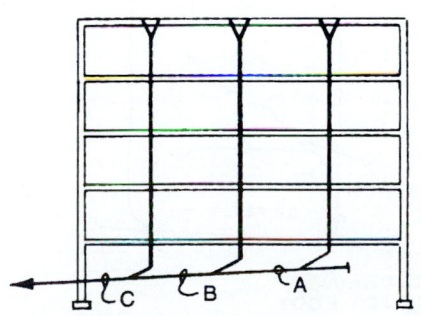

FIGURE 16.16 Horizontal storm drain sketches and sizes for the design example.

and is a 4-in pipe, while the last length ("C") serves 3200 ft^2 and is a 4-in pipe. The pipe sizes on the sketch are as shown in Figure 16.16.

In metric (SI) units, the length of drain labeled "A" in Figure 16.16 is taken from Table 16.5. Using a slope of 20.9 mm per meter, a 75-mm drain pipe is selected. The next length of drain, "B," services 200 m^2 and a 100-mm pipe is required. The last length, "C," serves 300 m^2 and a 125-mm pipe is required.

Sizing a Combined Sewer

If the roof downspouts/leaders are permitted by code to be connected to the building drain (not common) in a combined sewer (Figure 16.17), it will be necessary to convert the roof area into an equivalent number of drainage fixture units (DFU) so that the building drain can be sized to reflect the increased load. (The procedure used to size building drains for sewage waste was covered in Chapter 14). A typical procedure follows:

1. First, a schematic sketch of the stacks, downspouts, and building drain should be made so that the relationship of the stacks and the downspouts to the building drain can be envisioned.

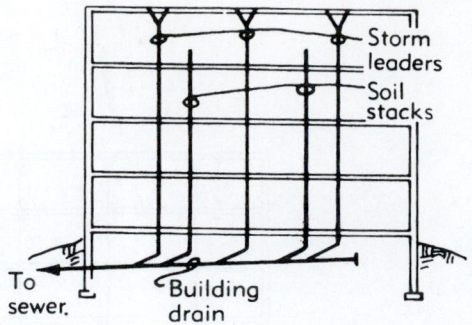

FIGURE 16.17   Combined sewer illustration for the design example.

In this example, the design will continue with the four-story apartment building (Appendix A). The sanitary waste stacks have been previously sized (Chapter 14), and a schematic of the design would look similar to Figure 16.18. The DFU total (from Chapter 14) of each of the waste stacks is added to the sketch, as is the slope of the building drain selected (⅛ in per foot or 10.4 mm per meter slope).

2. Next, the fixture units served by each stack must be converted into equivalent square feet or square meters. The

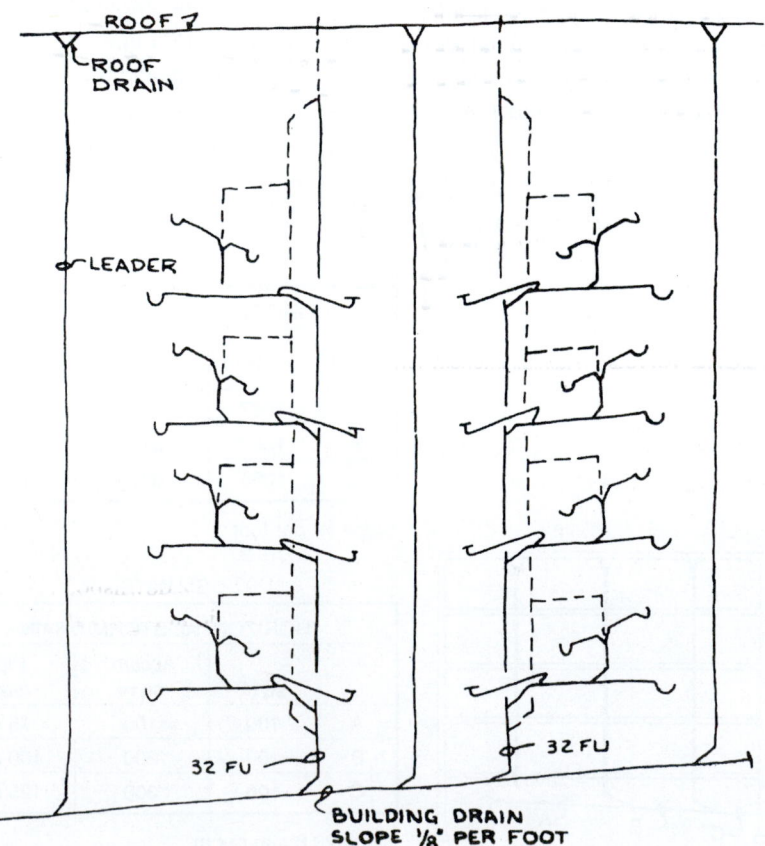

FIGURE 16.18   Combined sewer sketch for the design example.

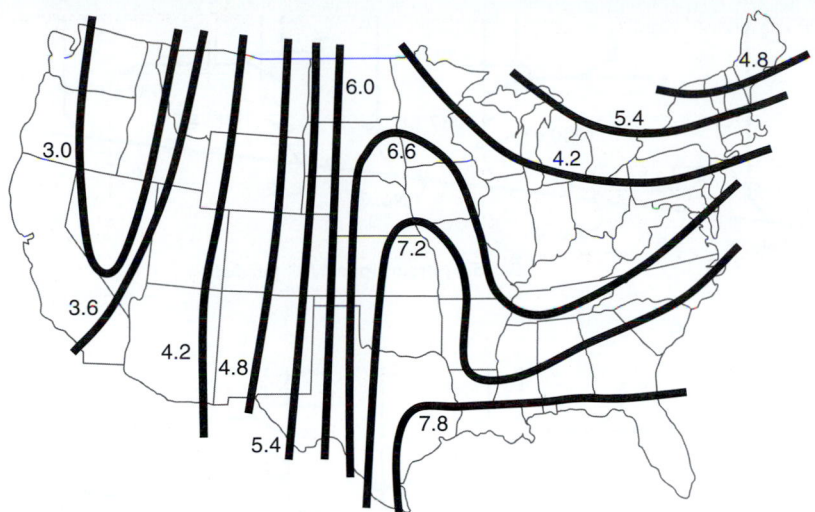

FIGURE 16.19A Rainfall intensity for 15 min of precipitation (in).

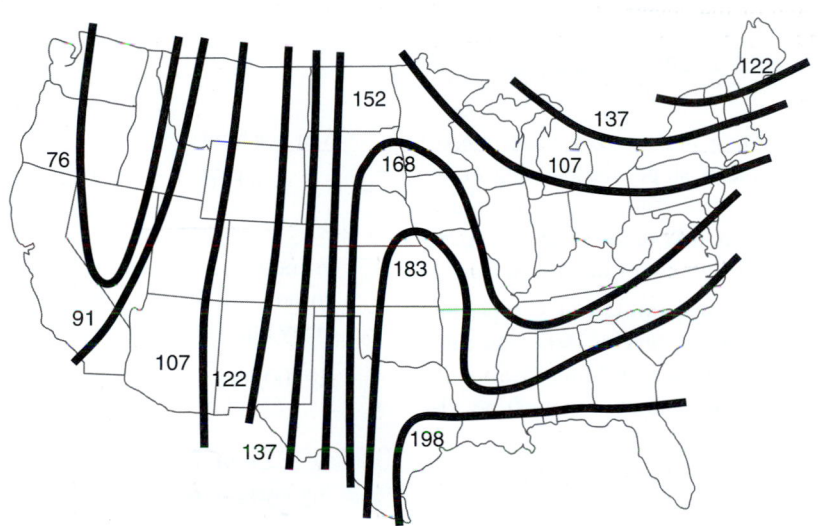

FIGURE 16.19B Rainfall intensity for 15 min of precipitation (mm).

code sets up a ratio of DFU and equivalent square feet or equivalent square meters. (Figure 16.19).

The equivalent area, based on code, for the first 256 DFU is 1000 ft². Any additional DFU are converted into equivalent area on the basis that 1 DFU equals 3.9 ft².

In this design, the 64 DFU would be the equivalent of 1000 ft².

In metric (SI) units, the equivalent area for the first 256 DFU is 93 m². Any additional DFU are converted into equivalent square meters on the basis that 1 DFU equals 0.36 m.

In this design in metric (SI) units, the 64 DFU would be the equivalent of 93 m².

3. The total equivalent area being served by the building drain is determined by adding the equivalent area to the roof area being collected.

The total equivalent area is 3200 ft² + (1000 ft² equivalent area · 2 downspouts) = 5200 ft². This is noted on the schematic in Figure 16.20.

From Table 16.5, the building drain size is determined. Based on a ⅛ in slope, the building drain must be 6 in to the main stack. The complete tabulation is shown in Table 16.8. A review of the sketch in Figure 16.20 indicates that there is less than 200 ft developed length.

In metric (SI) units, the total equivalent area is 300 m² + (93 m² equivalent area · 2 downspouts) = 486 m². This is noted on the schematic in Figure 16.20.

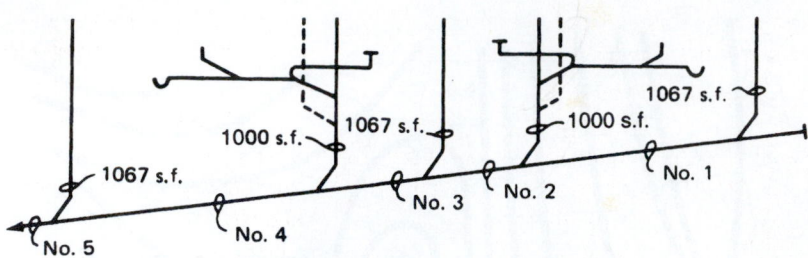

S1 UNITS

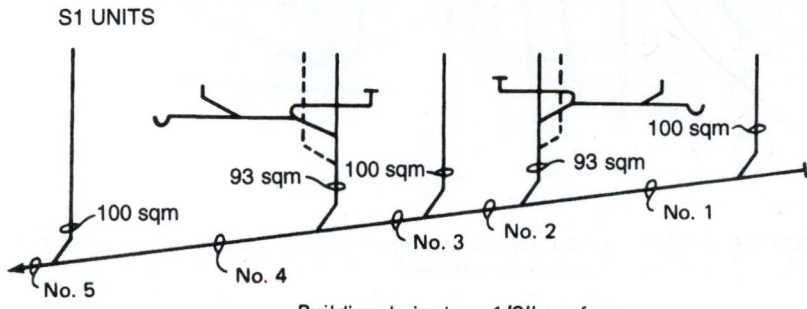

Building drain slope 1/8″ per foot

FIGURE 16.20 Leader and riser fixture sizes.

From Table 16.5, the building drain size is determined. Based on a 10.4 mm per meter slope, the building drain must be 150 mm to the main stack. The complete compilation is shown in Table 16.8. A review of the sketch in Figure 16.18 indicates there is less than 60 m of developed length.

4. The equivalent area is based on a rainfall of 4 in per hr. A check of the local weather service will indicate whether the rainfall in the proposed building location is more or less. If so, then the equivalent area must be adjusted proportionately.

Assume the rainfall rate in the proposed location for the apartment is 5 in per hr. How many equivalent area would the 64 DFU equal?

At 4 in per hour:

$$64 \text{ DFU} = 1000 \text{ equivalent ft}^2$$

5 in per hour:

$$64 \text{ DFU} = 1000 \text{ equivalent ft}^2 \cdot (5 \text{ in}/4 \text{ in})$$
$$= 1250 \text{ equivalent ft}^2$$

In metric (SI) units, assume the rainfall rate in the proposed location of the apartment is 125 mm per hour. How many equivalent feet would the 64 DFU equal?

At 100 mm per hr:

$$64 \text{ DFU} = 93 \text{ m}^2 \text{ equivalent area}$$

At 125 mm per hour:

$$63 \text{ DFU} = 93 \text{ m}^2 \text{ equivalent area} \cdot (125 \text{ mm}/100 \text{ mm})$$
$$= 116 \text{ m}^2 \text{ equivalent area}$$

STUDY QUESTIONS

16-1. What methods may be used to dispose of the water from roofs, courtyards, and parking lots?

16-2. What is a combined sewer, and under what conditions should this system of storm drainage be used?

16-3. What methods are commonly used for roof drainage?

16-4. When the roof drainage water is run into roof drains or gutters, what solutions may be used to disperse the water at the end of the downspout?

16-5. What methods are commonly used for subsurface drainage?

Design Exercises

16-6. Design the storm drainage system serving a 16 ft by 32 ft roof area with two downspouts. Assume a sloped roof with gutters (⅛ in per foot slope) and downspouts to the ground and a 4 in per hr rainfall.

16-7. Design the storm drainage system serving a 16 ft by 32 ft roof area with two downspouts. Assume a sloped roof with gutters (¼ in per foot slope) and downspouts to the ground and a 5 in per hr rainfall.

16-8. Design the storm drainage system serving a 64 ft by 84 ft roof area with four downspouts. Assume a sloped roof with gutters (⅛ in per foot slope) and downspouts to the ground and a 5 in per hr rainfall.

16-9. Design the storm drainage system serving a 16 ft by 32 ft roof area with two downspouts. Assume a sloped roof with gutters (20.8 mm per meter slope) and downspouts to the ground and a 100 mm per hr rainfall.

16-10. Design the storm drainage system for the apartment building in Appendix B. In this design, assume a sloped roof with gutters (¼ in per foot or 20.8 mm per meter slope) and downspouts to the ground and a 3 in (75 mm) per hr rainfall.

16-11. Design the storm drainage system for the building in Appendix B. Use three roof drains and size the down-spouts and horizontal storm drainage piping (½ in per foot or 41.6 mm per meter slope) to a community storm drainage system and a 5 in (125 mm) per hr rainfall.

16-12. Design the storm drainage system for the residence in Appendix C. Use gutters (⅛ in per foot or 10.4 mm per meter slope) and downspouts to the ground, and assume a 4 in (100 mm) per hr rainfall.

16-13. Design the storm drainage system for the residence in Appendix D. Use gutters (¼ in per foot or 20.8 mm per meter slope) and downspouts to the ground, and assume a 3 in (75 mm) per hr rainfall.

TABLE 16.4A   REQUIRED SIZE OF VERTICAL CONDUCTOR, LEADER, OR DOWNSPOUT, IN CUSTOMARY (US) UNITS. QUANTITIES IN MAIN BODY OF TABLE ARE MAXIMUM HORIZONTAL PROJECTED ROOF AREAS, IN FT^2.

Rainfall (in/hr)	Size of Drain or Downspout (in)[a]					
	2	3	4	5	6	8
1	2880	8800	18 400	34 600	54 000	116 000
2	1440	4400	9200	17 300	27 000	58 000
3	960	2930	6130	11 530	17 995	38 660
4	720	2200	4 600	8 650	13 500	29 000
5	575	1760	3680	6920	10 800	23 200
6	480	1470	3070	5765	9000	19 315
7	410	1260	2630	4945	7715	16 570
8	360	1100	2300	4325	6750	14 500
9	320	980	2045	3845	6000	12 890
10	290	880	1840	3460	5400	11 600
11	260	800	1675	3145	4910	10 545
12	240	730	1530	2880	4500	9660

[a]Round, square, or rectangular rainwater pipe may be used and is considered equivalent when enclosing a scribed circle equivalent to the downspout diameter.

TABLE 16.4B   REQUIRED SIZE OF VERTICAL CONDUCTOR, LEADER, OR DOWNSPOUT, IN METRIC (SI) UNITS. QUANTITIES IN MAIN BODY OF TABLE ARE MAXIMUM HORIZONTAL PROJECTED ROOF AREAS, IN M^2.

Rainfall (mm/hr)	Size of Drain or Downspout (mm)[a]					
	50	75	100	125	150	200
25	267	817	1709	3214	5016	10 776
50	133	408	854	1607	2508	5388
75	89	272	569	1071	1671	3591
100	66	204	427	803	1254	2694
125	53	163	341	642	1003	2155
150	44	136	285	535	836	1794
175	38	117	244	459	716	1539
200	33	102	213	401	627	1347
225	29	91	190	357	557	1197
250	26	81	170	321	501	1077
275	24	74	155	292	456	979
300	22	67	142	267	418	897

[a]Round, square, or rectangular rainwater pipe may be used and is considered equivalent when enclosing a scribed circle equivalent to the leader/downspout diameter.

TABLE 16.5A   SIZE OF HORIZONTAL RAINWATER PIPING BASED ON MAXIMUM RAINFALL AND ROOF AREA, IN CUSTOMARY (US) UNITS. QUANTITIES IN MAIN BODY OF TABLE ARE MAXIMUM HORIZONTAL PROJECTED ROOF AREAS, IN FT^2.

| | Pipe Size (in) | Maximum Rainfall (in/hr) | | | | |
		2	3	4	5	6
⅛ in/ft Slope	3	1644	1096	822	657	548
	4	3760	2506	1800	1504	1253
	5	6680	4453	3340	2672	2227
	6	10 700	7133	5350	4280	3566
	8	23 000	15 330	11 500	9200	7600
	10	41 400	27 600	20 700	16 580	13 800
	12	66 600	44 400	33 300	26 650	22 200
	15	109 000	72 800	59 500	47 600	39 650

| | Pipe Size (in) | Maximum Rainfall (in/hr) | | | | |
		2	3	4	5	6
¼ in/ft Slope	3	2320	1546	1160	928	773
	4	5300	3533	2650	2120	1766
	5	9440	6293	4720	3776	3146
	6	15 100	10 066	7550	6040	5033
	8	32 600	21 733	16 300	13 040	10 866
	10	58 400	38 950	29 200	23 350	19 450
	12	94 000	62 600	47 000	37 600	31 350
	15	168 000	112 000	84 000	67 250	56 000

| | Pipe Size (in) | Maximum Rainfall (in/hr) | | | | |
		2	3	4	5	6
½ in/ft Slope	3	3288	2295	1644	1310	1096
	4	7520	5010	3760	3010	2500
	5	13 360	8900	6680	5320	4450
	6	21 400	13 700	10 700	8580	7140
	8	46 000	30 650	23 000	18 400	15 320
	10	85 800	55 200	41 400	33 150	27 600
	12	133 200	88 800	66 600	53 200	44 400
	15	238 000	158 800	119 000	95 300	79 250

TABLE 16.5B SIZE OF HORIZONTAL RAINWATER PIPING BASED ON MAXIMUM RAINFALL AND ROOF AREA, IN METRIC (SI) UNITS. QUANTITIES IN MAIN BODY OF TABLE ARE MAXIMUM HORIZONTAL PROJECTED ROOF AREAS, IN M².

		Maximum Rainfall (mm/hr)				
	Pipe Size (mm)	50	75	100	125	150
10.4 mm/m Slope	75	152	101	76	61	50
	100	349	232	174	139	116
	125	620	413	310	248	206
	150	994	662	497	397	331
	200	2136	1424	1068	854	706
	250	3846	2564	1923	1540	1282
	300	6187	4124	3093	2475	2062
	375	10 126	6763	5527	4422	3683
		Maximum Rainfall (mm/hr)				
	Pipe Size (mm)	50	75	100	125	150
20.9 mm/m Slope	75	215	143	107	86	71
	100	492	328	246	197	164
	125	877	584	438	350	292
	150	1402	935	701	561	467
	200	3028	2019	1514	1211	1009
	250	5425	3618	2712	2169	1806
	300	8732	5815	4366	3493	2912
	375	15 607	10 404	7803	6247	5202
		Maximum Rainfall (mm/hr)				
	Pipe Size (mm)	50	75	100	125	150
41.7 mm/m Slope	75	305	213	152	121	101
	100	698	465	349	279	232
	125	1241	826	620	494	413
	150	1988	1272	994	797	663
	200	4274	2847	2136	1709	1423
	250	7692	5128	3846	3079	2564
	300	12 374	8249	6187	4942	4124
	375	22 110	14 752	11 055	8853	7362

TABLE 16.6A SIZE OF ROOF GUTTERS BASED ON MAXIMUM RAINFALL AND ROOF AREA, IN CUSTOMARY (US) UNITS.
QUANTITIES IN MAIN BODY OF TABLE ARE MAXIMUM HORIZONTAL PROJECTED ROOF AREAS, IN FT².

	Gutter Size (in)	Maximum Rainfall (in/hr)				
		2	3	4	5	6
¹⁄₁₆ in/ft Slope	3	340	226	170	136	113
	4	720	480	360	288	240
	5	1250	834	625	500	416
	6	1920	1280	960	768	640
	7	2760	1840	1380	1100	918
	8	3980	2655	1990	1590	1325
	10	7200	4800	3600	2880	2400

	Gutter Size (in)	Maximum Rainfall (in/hr)				
		2	3	4	5	6
⅛ in/ft Slope	3	480	320	240	192	160
	4	1020	681	510	408	340
	5	1760	1172	880	704	587
	6	2720	1815	1360	1085	905
	7	3900	2600	1950	1560	1300
	8	5600	3740	2800	2240	1870
	10	10 200	6800	5100	4080	3400

	Gutter Size (in)	Maximum Rainfall (in/hr)				
		2	3	4	5	6
¼ in/ft Slope	3	680	454	340	272	226
	4	1440	960	720	576	480
	5	2500	1668	1250	1000	834
	6	3840	2560	1920	1536	1280
	7	5520	3680	2760	2205	1840
	8	7960	5310	3980	3180	2655
	10	14 400	9600	7200	5750	4800

	Gutter Size (in)	Maximum Rainfall (in/hr)				
		2	3	4	5	6
½ in/ft Slope	3	960	640	480	384	320
	4	2040	1360	1020	816	680
	5	3540	2360	1770	1415	1180
	6	5540	3695	2770	2220	1850
	7	7800	5200	3900	3120	2600
	8	11 200	7460	5600	4480	3730
	10	20 000	13 330	10 000	8000	6660

TABLE 16.6B   SIZE OF ROOF GUTTERS BASED ON MAXIMUM RAINFALL AND ROOF AREA, IN METRIC (SI) UNITS. QUANTITIES IN MAIN BODY OF TABLE ARE MAXIMUM HORIZONTAL PROJECTED ROOF AREAS, IN M².

Gutter Size (mm)	Maximum Rainfall (mm/hr)				
	50	75	100	125	150
5.2 mm/m Slope					
75	31	21	15	12	10
100	66	44	33	26	22
125	116	77	58	46	38
150	178	119	89	71	59
175	256	170	128	102	85
200	369	246	184	147	123
250	668	445	334	267	223

Gutter Size (mm)	Maximum Rainfall (mm/hr)				
	50	75	100	125	150
10.4 mm/m Slope					
75	44	29	22	17	14
100	94	63	47	37	31
125	163	108	81	65	54
150	252	168	126	100	84
175	362	241	181	144	120
200	520	347	260	208	173
250	947	631	473	379	315

Gutter Size (mm)	Maximum Rainfall (mm/hr)				
	50	75	100	125	150
20.9 mm/m Slope					
75	63	42	31	25	21
100	133	89	66	53	44
125	232	155	116	92	77
150	356	237	178	142	118
175	512	341	256	204	170
200	739	493	369	295	246
250	133	891	668	534	445

Gutter Size (mm)	Maximum Rainfall (mm/hr)				
	50	75	100	125	150
41.7 mm/m Slope					
75	89	59	44	35	29
100	189	126	94	75	63
125	328	219	164	131	109
150	514	343	257	206	171
175	724	483	362	289	241
200	1040	693	520	416	346
250	1858	1238	929	743	618

TABLE 16.7   PRECIPITATION RATE IN INCHES/HOUR, BASED
ON 15 MINUTES OF PRECIPITATION, FOR
SELECTED CITIES IN CANADA.

Alberta

Calgary	36 mm
Edmonton	48 mm
Grand Prairie	40 mm
Jasper	24 mm

Atlantic Provinces

Battle Harbour	24 mm
Bonavista	44 mm
Cape Race	44 mm
Chatham	50 mm
Halifax	44 mm
Schefferville	30 mm
Yarmouth	48 mm

British Columbia

Kamloops	10 mm
Prince George	13 mm
Vancouver	20 mm
Victoria	16 mm

Manitoba

Brandon	62 mm
Grand Rapids	56 mm
The Pas	47 mm
Winnipeg	68 mm

Ontario

London	72 mm
Ottawa	64 mm
Sault Ste. Marie	48 mm
Sudbury	56 mm
Toronto	72 mm

Saskatchewan

Lloydminister	48 mm
Prince Albert	46 mm
Regina	57 mm
Saskaton	52 mm

Quebec

Chibougaman	52 mm
Montreal	64 mm
Rivierie-du-Loup	40 mm
Roberval	60 mm
Quebec	54 mm

TABLE 16.8   BUILDING DRAIN TABULATIONS.

SI UNITS

BUILDING DRAIN			
	s.f.	Accum. s.f.	Pipe size
No. 1	1067	1067	4″
No. 2	1000	2067	4″
No. 3	1067	3134	5″
No. 4	1000	4134	5″
No. 5	1086	5200	6″

Slope 1/8″ per foot

BUILDING DRAIN			
	sq m	Accum. sq m	Pipe size
No. 1	100	100	100 mm
No. 2	93	193	100 mm
No. 3	100	293	125 mm
No. 4	93	386	125 mm
No. 5	100	486	150 mm

Slope 10.4 mm per m

ELECTRICITY THEORY

17.1 INTRODUCTION

Electricity is a form of energy tied to the existence of electrical charge and, as a result, is related to magnetism. It plays a fundamental role in all the technologies we use today. Everyday work and play activities through manufacturing and scientific research use electricity as a source of energy. In this chapter, the theory of electricity, fundamental units, and costs are introduced. Devices, equipment, and materials used to distribute electricity from the power utility to points of use in the building and building electrical system design methods are discussed in Chapters 18 and 19.

The History of Electricity

The study and development of electricity occurred over many centuries. It has its roots about 600 B.C.E. when a Greek mathematician named Thales documented what eventually became known as static electricity. He recorded that after rubbing amber, a yellowish, translucent mineral, with a piece of wool or fur other light objects such as straw or feathers were attracted to the amber. For centuries this distinctive property was thought to be unique to amber.

There was little development in the understanding of electricity until about 1600 when English scientist William Gilbert described the electrification of many substances. He coined the term *electricity,* which is derived from the Latin term *electricus,* meaning to "produce from amber by friction." It has its roots in the Greek term *elektor,* which means, "beaming sun." Gradual improvements in the understanding of electricity have led to the invention of motors, generators, telephones, radio and television, and computers.

In 1660, a German experimenter named Otto von Guericke built the first electric generating machine. It was constructed of a ball of sulfur, rotated by a crank with one hand and rubbed with the other. Other experimenters recognized that other substances, such as copper, silver, and gold, did not attract anything. An Englishman, Stephen Gray, distinguished between materials that were conductors and nonconductors in 1729.

About 1746, Ewald Georg von Kleist, a German inventor, and Dutch physicist Pieter van Musschenbroek of the University of Leyden, working independently, invented an electrical storage device called a Leyden jar, a glass jar coated inside and outside with tin foil. Static electricity could be discharged by simultaneously touching the inner and outer foil layers. It demonstrated that electricity could be stored for future use.

In 1747, American inventor and statesman, Benjamin Franklin, suggested the existence of an electrical fluid and surmised that an electric charge was made up of two types of electric forces, an attractive force and a repulsive force. To identify these two forces, he gave the names positive and negative, which are still in use today. Franklin conducted his famous kite experiment in 1752. He flew a kite with a stiff wire pointing upward as a thunderstorm was about to break. He attached a metal key to the other end of the hemp string, and let it hang close to a Leyden jar. Rain moistened the string, which could then conduct electricity. Sparks jumped from the key to the jar. Although there was no lightning, there was enough electricity in the air for Franklin to prove that electricity and lightning are the same thing.

In 1786, an Italian anatomy professor, Luigi Galvani, observed that a discharge of static electricity made a dead frog's leg twitch. Ensuing experimentation produced what was a simple electron cell using the fluids of the leg as an electrolyte and the muscle as a circuit and indicator. Expanding on Galvani's findings, Alessandro Volta, another Italian, built the voltaic pile, an early type of electric cell or battery.

In 1820, H. C. Oersted, a Danish physicist, discovered that a magnetic field surrounds a current-carrying wire, by observing that electrical currents affected the needle on a compass. Within two years Andre Marie Ampere, a French mathematician, observed that a coil of wires acts like a magnet when electrical current is passed thorough it. Shortly thereafter, D. F. Arago invented the electromagnet and Joseph Henry, an American, demonstrated an electromagnetic device that was capable of lifting over a thousand pounds. Also as the result of the newly discovered electromagnet, Michael Faraday, an Englishman, developed a crude electric motor in 1831, but a practical motor was not developed until 1870. Both Faraday and Joseph Henry, working independently, invented the electric generator with which to power the motor.

In 1831, American Samuel Morse conceived the idea of sending coded messages over wires using the electromagnetic telegraph and a code of electrical impulses identified as dots and dashes that eventually became known as "Morse Code." The first message sent by the electric telegraph was "What hath God wrought," from the Supreme Court Room in the U.S. Capitol to the railway depot at Baltimore on May 24, 1844. Morse's electric telegraph is recognized as the first practical use of electricity and the first system of electrical communication.

Charles de Coulomb was the first person to measure the amount of electricity and magnetism generated in a circuit. G. S. Ohm, a German college teacher, formulated a law showing the relationship between volts, amps, and resistance. Henry and Ohm demonstrated that in a long electric line it was better to have relatively high voltage and low current. Additionally, J. P. Joule, G. R. Kirchhoff, and J. C. Maxwell also developed mathematical relationships and rules concerning electrical circuiting.

In the late 1800s, electric lighting was viewed as an ideal use of electrical energy. Although arc lights were invented and put to practical use for lighting streets by 1860, it was not until 1879 that a practical incandescent lamp was developed independently by Thomas Edison in America and Joseph Swan in England. Edison was the first to patent the commercially feasible incandescent lamp so he is recognized as the inventor. The development of electric lighting is covered in Chapter 20.

In 1882, the Edison Electric Light Company, later known as General Electric, successfully demonstrated the use of artificial lighting by powering incandescent streetlights and lamps in London and New York City. By the end of the 1880s, small electrical stations based on Edison's designs were in use a number of U.S. cities. However, each power station was able to power only a few city blocks. Edison's designs still serve as the basis of how we to distribute electricity from power stations with the exception that Edison's systems were direct current systems. Direct current systems had the problematic characteristic that current could not be economically transmitted over long distances.

American Nikola Tesla of Croatian decent, one of Edison's former employees and a rival of Edison at the end of the 19th century, is the inventor of 3-phase power distribution, the alternating current motor, wireless transmission. He began experimenting on generators in 1883, and discovered the rotating magnetic field. This phenomenon serves as the basic principle of the alternating current generator. Tesla then developed plans for an alternating current induction motor, which become the first step towards the successful utilization of alternating current.

In 1885, George Westinghouse, head of the Westinghouse Electric Company, bought the patent rights to Tesla's alternating current system. In America, in 1886 the first alternating current power station was placed in operation, but as no alternating current motor was available, the output of this station was limited to lighting. In 1888, the alternating current motor was introduced and ultimately became the most commonly used electric motor in buildings (e.g., for fans, air conditioners, and refrigerators). L. Caulard and J. D. Gibbs announced the first transformer in 1883. This allowed alternating current power to be generated at low voltage, then stepped up to high voltage for efficient transmission, and then stepped down to an even lower voltage for safety reasons.

Large-scale electric power distribution began on August 26, 1895, when water flowing over Niagara Falls was diverted through a pair of high-speed turbines that were coupled to two 5000-horsepower generators that powered nearby manufacturing plants. The following year a portion was transmitted 20 miles to the city of Buffalo, where it was used for powering lighting and streetcars. This project involved generators produced by Westinghouse and later by General Electric.

The Niagara project clearly demonstrated that large-scale generation and transmission of electricity was conceptually sound, technically feasible, and economically practical. Gradually, electrical power became commercialized in urban areas of the U.S. Gas lighting that had been used in streetlights was replaced by electric lights and overhead wires eventually connected homes to a large-scale power plant operated by privately owned electric companies.

By about 1930 most of the occupants of large cities in the United States had electricity, yet only 10% of the Americans who lived in rural areas had electricity. At this time, private electric utilities determined that it was too expensive to run long transmission lines to spaced farms. The Roosevelt administration believed that if private enterprise could not supply electric power to the people, then it was the duty of the government to do so. In 1935, the Rural Electric Administration (REA) was created to bring electricity to these rural areas. The REA helped to establish hundreds of electric cooperatives that served millions of rural households.

Reliance on electricity has grown significantly over the past decade in all countries. For example, in 2002, the U.S. Department of Energy reported that total U.S. net generation of electricity was 3811 billion kWh, with 50% produced by coal-fired plants, 20% from nuclear plants, 18% from gas plants, 7% from hydroelectric plants, 2% from petroleum-fired plants, and 2% from renewable power sources (e.g., wind, solar electric, and so on).

In less than a century, the developed world has become extremely dependent on electricity and problems have occurred because of this dependence. In 1965, 1977, and 2003, power failures blacked out much of the northeastern United States and Canada. In 1994 California enacted legislation intended to deregulate the electric power business in the state and establish a competitive market. It was heralded as a model for developed countries to follow. But, by January 2001, flaws in the California approach had become evident with the state's utilities driven to the brink of bankruptcy and Californians suffering electricity shortages and blackouts. The effects of shortages and blackouts experienced by the general public have underscored the significance of electricity in every day life.

17.2 ELECTRICAL THEORY

The Phenomenon of Electricity

Electricity is a physical phenomenon tied to the behavior of positively and negatively charged elementary particles of an atom. An introduction on the elementary particles of an atom is necessary to develop a sense of what electricity is and how it behaves.

Two theories exist: the classical theory and the modern theory. Both are briefly introduced in the sections that follow.

Classical Theory: Flow of Electrons

Chemical *elements* are the fundamental materials that make up matter; they are the building blocks of the universe. Hydrogen, oxygen, nitrogen, carbon, silicon, copper, iron, silver, and gold are among the more than 100 known elements. *Atoms* are the smallest unit of an element and are composed of several kinds of elementary particles, including protons, neutrons, and electrons. See Figure 17.1. The *proton* of an atom is positively charged (+), the *electron* is negatively charged (−), and the *neutron* is electrically neutral (o).

The Law of Charges states that opposite charges attract each other and like charges repel each other. Thus, negatively charged electrons are attracted to positively charged protons. Conversely, negatively charged electrons tend to repel one another.

Each atom of a single element has an equal number of protons and electrons: hydrogen has 1 proton, oxygen has 8, iron has 26, and so on. The number of electrons contained in an atom determines its chemical and electrical properties. When an atom has an equal number of electrons and protons, charges cancel and the atom is electrically neutral. It has no charge. An atom containing fewer electrons than protons is positively charged. On the other hand, an atom containing more electrons than protons is negatively charged. Charged atoms are called *ions*. A *positive ion* is a positively charged atom and a negatively charged atom is called a *negative ion*.

Protons and neutrons occupy the nucleus of an atom. Most of the mass of the atom resides in its nucleus. Although the like charges of protons tend to cause protons to repel each other, the gravitational attraction from their relatively large masses overcomes this effect. Electrons whirl in a high-velocity orbit around an atom's nucleus much like a satellite orbits the earth. An electron's high rate of speed causes it to attempt to escape the attraction of the proton. However, the strong attraction between the positively charged proton and negatively charged electron causes the electron to reside in a balanced orbit around the nucleus.

When in a balanced orbit, electrons move in spherical paths called *orbital shells* that surround the nucleus of an atom. Each orbital shell has a maximum capacity of electrons; that is, a maximum number of electrons can reside in a shell. The first orbital shell can contain up to 2 electrons; the second shell can contain up to 8; the third shell, up to 18; the forth shell, up to 32; and so on. See Figure 17.2. The outermost orbital shell of an atom is called the *valence shell*. Electrons contained in the valence shell are called *valence electrons*. The outermost shell can contain no more than half of its capacity or a maximum of 8 electrons before the next shell begins to fill. The number of valence electrons determines the electrical properties of a material.

In classical theory, electrical current is electron flow. Electrons in an orbital shell near the nucleus have a strong attraction to the protons in the nucleus and thus are difficult to free. Electrons in outer orbital shells experience a weaker attraction and are more easily freed. Energy can be added to an electron to move it to the next higher orbital shell. If sufficient additional energy is added, a valence electron can be forced out of the atom. Such an electron is said to be free. These free electrons make up electrical current flow.

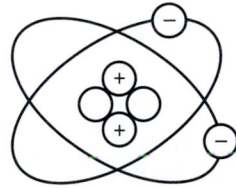

FIGURE 17.1 Atoms are the smallest unit of an element and are composed of several kinds of elementary particles called protons, neutrons, and electrons. The proton of an atom is positively charged (+), the electron is negatively charged (−), and the neutron is electrically neutral (o). Protons and neutrons occupy the nucleus, the center of the atom. Although the like charges of protons tend to cause protons to repel each other, the gravitational attraction from their relatively large masses overcomes this effect. Electrons whirl in a high-velocity orbit around an atom's nucleus. An electron's high rate of speed causes it to attempt to escape the attraction of the proton. However, the strong attraction between the positively charged proton and negatively charged electron causes the electron to reside in a balanced orbit around the nucleus.

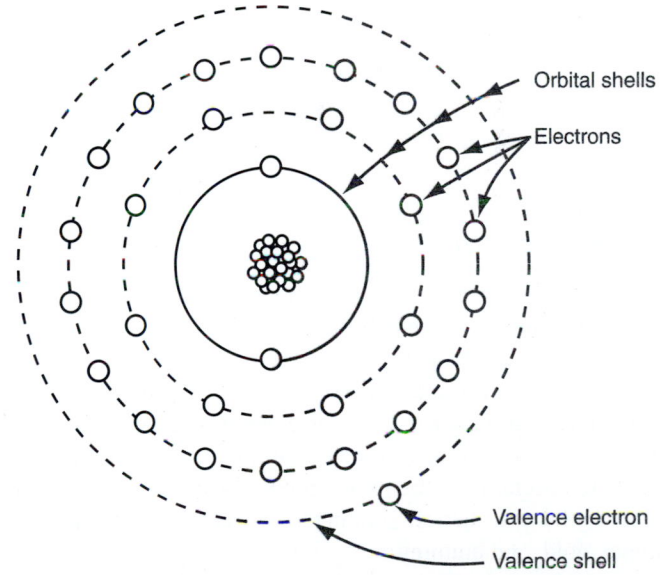

FIGURE 17.2 When in a balanced orbit, electrons move in spherical paths called *orbital shells* that surround the nucleus of an atom. Each orbital shell has a maximum capacity of electrons. The first orbital shell can contain up to 2 electrons; the second shell, up to 8; the third shell, up to 18; the forth shell, up to 32; and so on. The outermost orbital shell of an atom is called the *valence shell*. Electrons contained in the valence shell are called *valence electrons*. Shown is a molecule of copper. It has one valence electron, in this case, in the fourth shell. A single electron in the valence shell makes copper a good conductor of electricity.

Modern Theory: Flow of Charged Particles

In modern theory, electricity is tied to even smaller subatomic particles that possess either a positive or negative electromagnetic charge. Not all subatomic particles have a charge. It is only the subatomic *charged particles*, those with an electromagnetic charge, that are associated with electricity. As in the classical theory, charged particles with like charges repel one another and charged particles with unlike charges attract one another. The electromagnetic force between two charged particles is greater than the gravitational force between the two, so flow of electricity is the flow of charged subatomic particles caused by these repelling and attracting forces.

Electrical Current

In both classical and modern theories, electricity is the movement of subatomic particles (electrons or charged particles) that is attributable to the existence of a charge. The movement of these infinitesimal masses is caused by the attraction of masses with unlike charges and the repulsion of masses that have like. Masses that are oppositely charged, one positive and one negative, are attracted toward each other and masses that are similarly charged, either positive or both negative, have a repulsive force between them.

A flow of electric charge through a conductor is an *electrical current* or, simply *current*. When opposite charges are placed across a conductor, negatively charged subatomic particles move from the negative charge to the positive charge. Actual movement of a single subatomic particle is fairly slow, averaging about one-half inch per second. However, the chain-reaction effect of current flow occurs very rapidly, at about the speed of light (about 186 000 miles/s or 300 000 m/s).

Conductors, Insulators, and Semiconductors

A *conductor* carries electrical current without providing too much resistance to current flow. Some materials convey electricity better than others and are good conductors of electricity. In classical theory, the electrons of metal atoms migrate freely from atom to atom across the entire metal body while in modern theory it is charged particles that move freely through the conductor. Metallic elements are good conductors of electricity. Silver is the best natural conductor of electricity, followed by copper, gold, and aluminum.

Insulators are materials that resist the flow of electricity. They have electrons that tend to retain electrons on their original atoms, making it difficult for electrons to move and conduct electricity. Insulators are nonmetallic elements and compounds such as glass and other ceramic materials. Most ceramics such as glass, rubber, and plastics are good insulators. Even air is a good insulator at low to medium voltages.

Semiconductors are materials that are neither good conductors nor good insulators. They behave like good conductors at high temperature and insulators at low temperature. (Heat has the opposite effect on conductors.) At room temperature, the conductivity of some semiconductors falls somewhere between that of a good conductor and an insulator. Semiconductors are solid-state devices such as diodes, transistors, and integrated circuits, which are used in most electronic devices such as computers and sound systems. The two most common materials used in the production of electronic components are silicon and germanium. Of the two, silicon is used more often because of its ability to withstand heat.

Producing Current Flow

Electricity is the flow of current through a conductor. Current must be forced to flow in a conductor by the presence of a charge. There are six primary ways that current can be forced to move through a conductor:

Static electricity from friction: Simply rubbing two materials together produces a charge of static electricity. Heat energy caused by friction frees electrons near the surface of one material and they move to the other material. A comb becomes *electrostatically* charged when you run it through your hair. The soles of shoes become statically charged when they are rubbed across carpet. This charge is discharged when an electrical fixture or door handle is touched.

Thermoelectricity is electricity from heat. When two dissimilar metals are joined, a thermoelectric charge is created when the joined metals are heated. This device is called a *thermocouple*. Heat frees electrons in one metal and they transfer to the other metal creating the charge. When the materials cool, the charge dissipates.

Piezoelectricity is electricity from pressure. Certain crystalline materials produce a piezoelectric charge when a force deforms or strains the material. The pressure forces the electrons to one side of the material, causing it to be negatively charged while the side losing the electrons becomes positively charged. Strain gauges and phonograph needles apply this principle.

Electrochemistry is electricity from a chemical reaction. A galvanic reaction produces opposite electrical charges in two dissimilar metals when they are placed in certain chemical solutions. The ordinary flashlight and car battery produce electricity using this principle.

Photoelectricity is electricity from light. When small particles of light called photons strike a material, they release energy that can cause atoms to release electrons. When light strikes the surface of one of two plates that are joined together, energy from the light forces electrons to be released to the second plate. The plates build up opposite electrical charges.

Magnetoelectricity is electricity from magnetism. The force of a magnetic field can drive electron flow. When any good conductor such as a copper or aluminum wire

---

Symbols Used in Lighting and Electrical System Design

In electrical design calculations, the symbol E is commonly used for voltage and the symbol I for amperage. In lighting design, E is used as a symbol for illuminance and I is the symbol for luminous intensity. The reader is cautioned that the lighting and electrical design professions use common symbols with different meanings.

Water System/Electrical System Analogy

Forces that influence flow of current in an electrical system resemble those found in a water system. It is helpful to relate electron flow to water flow to assist in developing a good understanding of how the fundamental units of electricity relate. Voltage is similar in nature to water pressure. In a water system, a greater system pressure results in a greater flow of water. With electrical systems, voltage is the electrical pressure. A higher voltage level produces a higher level of electron flow. Voltage is the driving force of current flow. Amperage describes the rate of electrical current flow. In water systems, rate of flow is described in a flow rate, usually expressed in number of gallons of water per minute (gpm) that flow past a given point in a pipe. In an electrical system, amperage describes the number of electrons that flow past a given point. Pressure losses reduce pressure available at a given point in a water system. A greater resistance to water flow results when pipe diameter is decreased or pipe length is increased. In an electrical system, resistance to electron flow increases as conductor diameter decreases or length increases.

moves through a magnetic field, the force of the field causes free electrons to move in one direction across the conductor. Reversing the direction of conductor movement reverses direction of electron flow.

17.3   UNITS OF ELECTRICITY

Fundamental Units of Electricity

Units used to describe electricity are voltage, amperage and ohms. These are defined in the sections that follow.

Voltage

Voltage or *electromotive force (E or EMF)*   This is the driving force behind current flow. A difference in charge creates an electrical pressure, which moves current in one direction. The unit of electrical pressure is the *volt* (V). Voltage level governs the amount of electrical energy that will flow through a wire. A boost in voltage increases current flow and a drop in voltages reduces flow.

Amperage

Amperage or Inductive Flow (I)   The rate of current flow in a closed electrical system is measured in a unit called the *ampere,* frequently called the *amp.* An *ampere (A)* is related to the number of electrons flowing through a section of conductor (wire) over a period of time. It is equal to one Coulomb (6.280×10^{18} or 6 280 000 000 000 000 000 electrons) passing through one point in an electrical circuit in one second.

Resistance

Resistance (R)   The length of a conductor (wire), the diameter of the conductor, type of conductor material, and temperature of the conductor affect the resistance to flow of current. The unit used to measure electrical resistance is the *ohm (Ω).* One ohm is that resistance that allows one amp to flow when pushed by a pressure of one volt.

Ohm's Law

Current flow is caused by electromotive force or voltage. Amperage is the rate of current flow and may be referred to as in-

ductive flow. Resistance (R) refers to the ability of a conductor to resist current flow and is measured in ohms. Voltage (E), amperage (I), and resistance (Ω) in an active electrical circuit are related through *Ohm's Law:*

$$E = IR$$

Ohms' Law makes it possible to determine one of these values, if the other two are known. It will be used analytically in the discussion that follows later in this text.

Example 17.1

Small- and medium-gauge electrical conductors used as wiring in buildings are typically categorized by the American Wire Gauge (AWG). The AWG number (No.) of a conductor is inversely proportional to the cross-section diameter of the wire—that is, a smaller gauge number identifies a thicker wire. Approximate the resistance in 100 and 500 ft lengths for the following conductors.

a. A No. 12 AWG copper conductor (wire) with a resistance of 1.62 Ω/1000 ft.

For the 100 ft conductor:

$$R = 1.62 \ \Omega/1000 \ \text{ft} \cdot 100 \ \text{ft} = 0.162 \ \Omega$$

For the 500 ft conductor:

$$R = 1.62 \ \Omega/1000 \ \text{ft} \cdot 500 \ \text{ft} = 0.810 \ \Omega$$

b. A No. 10 AWG (thicker than No. 12 AWG) copper conductor with a resistance of 1.02 Ω/1000 ft.

For the 100 ft conductor:

$$R = 1.02 \ \Omega/1000 \ \text{ft} \cdot 100 \ \text{ft} = 0.102 \ \Omega$$

For the 500 ft conductor:

$$R = 1.02 \ \Omega/1000 \ \text{ft} \cdot 500 \ \text{ft} = 0.510 \ \Omega$$

As shown in Example 17.1, large-diameter conductors have smaller resistances than do small-diameter conductors. Also, conductors of shorter length produce less resistance to current flow. Conductors with less resistance allow electricity to flow more freely.

Power

Power is the rate at which work is accomplished; it is work or energy released divided by time. The unit of power measurement that most individuals are likely familiar with is horsepower. One horsepower is equivalent to 33 000 foot-pounds (ft-lb) of work per minute (550 ft-lb/s). This is the equivalent of lifting a one-ton weight at a rate of 6½ feet per minute. One horsepower is equivalent to 746 watts power.

The electrical unit of power is the watt. In theory, the watt can be related to other measures of power:

$$1 \ \text{horsepower (hp)} = 746 \ \text{watts}$$

$$1 \ \text{watt (W)} = 3.413 \ \text{Btu/hr}$$

$$1 \ \text{kilowatt (kW)} = 1000 \ \text{W}$$

$$1 \ \text{megawatt (MW)} = 1 \ 000 \ 000 \ \text{W}$$

On a direct current circuit, voltage (E) and amperage (I) are related to wattage through the *DC power equation,* also known as Joule's Law:

$$P = EI$$

The concept of power in an alternating current system is covered later in this chapter.

Example 17.2

A lamp is designed for use at 120 V and has a current draw of 0.5 A. Determine the power consumed.

The power consumed by the lamp is calculated as follows:

$$P = EI = 120 \ \text{V} \cdot 0.5 \ \text{A} = 60 \ \text{W}$$

Energy

If power used by an appliance is multiplied by the amount of time that the unit operates, the energy consumption value or amount of work accomplished is determined. The measurement of electrical energy consumption, the rate at which power is being consumed over a specified period of time in hours, results

---

Electrocution

Humans are conductors of electricity and have electrical resistance similar to any other material. When a person comes in contact with electricity, that person can feel the current flow through his or her body, ranging from faint tingling sensations to death.

The lowest level at which people can perceive electrical current is about 0.001 A (1 milliamp). Slightly above this level, a mild tingling sensation is felt. At currents higher than 0.05 A (50 milliamps), heat produced by electrical current is enough to burn human skin and tissue. At levels of current flow exceeding 0.1 A (100 milliamps), the heart stops. A person may survive an electrocution if his or her heart can be started again. Care should be exercised when working with electricity.

---

in watt-hours (Wh) of energy. The standard billing for energy consumption is the kilowatt-hour (kWh), which is equivalent to 1000 watt-hours.

$$1000 \ \text{watt-hours (Wh)} = \text{one kilowatt hour (kWh)}$$

Electrical energy consumption (q) in watt-hours may be determined by the following expression, where power (P) is expressed in watts and time (t) in hours of operation:

$$q = Pt$$

17.4   ELECTRICAL CIRCUITS

The Basic Electrical Circuit

An *electric circuit* is a continuous path along which an electric current can flow. A simple circuit is composed of a *power source* (e.g., battery or generator); the *load,* an electrical component or group of components that consume electricity (e.g., a lamp or appliance); and a set of *conductors* that carry current from the source to the load (e.g., wires). See Figure 17.3. If the circuit is broken at any point, current will not flow.

Closed Circuit

To keep current flowing in an electrical conductor, there must be a difference in charge between the ends of the conductor. A *closed circuit* is an uninterrupted path that allows a continuous flow of current through an electrical circuit. In a building electrical system, a circuit is closed when a switch is turned on, allowing current to flow uninterrupted and the lamp to light. The switch completes the conductor path, which allows current to flow from the power source through the conductor (a wire) to the lamp. To complete the circuit, a second conductor runs from the lamp back to the power source.

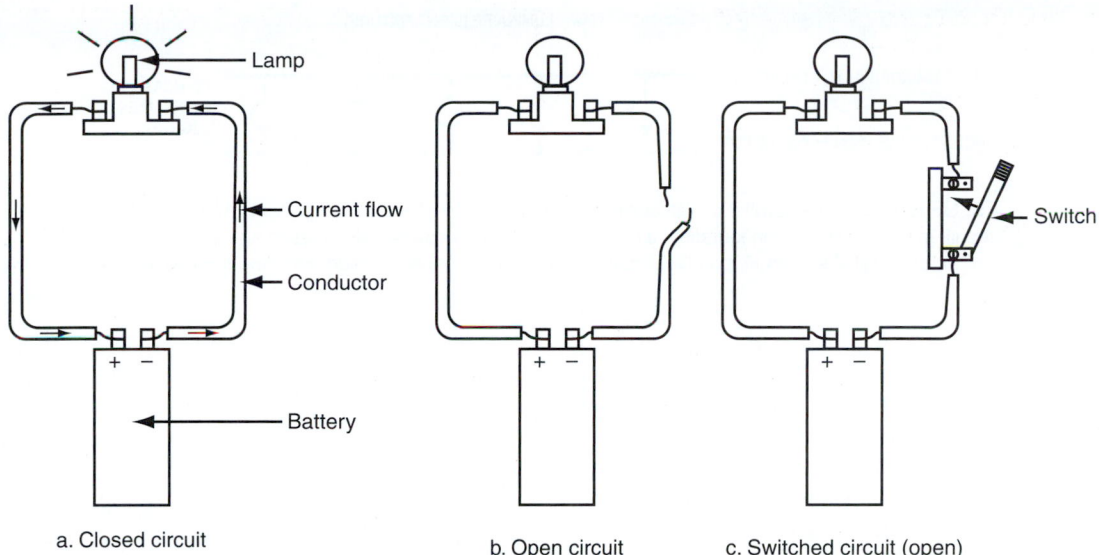

a. Closed circuit   b. Open circuit   c. Switched circuit (open)

FIGURE 17.3 A *closed circuit* is an uninterrupted path that allows a continuous flow of current through an electrical conductor. Shown in (a) is a closed lighting circuit in which current flows from the battery to the lamp and back to the battery. The resistance to current flow in the lamp's filament causes the lamp to illuminate. If the path of current flow in a circuit is interrupted or opened (turned off), an *open circuit* results. The broken wire in the second circuit (b) causes the circuit to open. A switch is installed in a circuit (c) to allow the circuit to open or close to control operation of the lamp.

Open Circuit

If the path of current flow is interrupted such as if the switch in a circuit is opened (turned off), an *open circuit* results. The switch breaks the conductor path, which prevents current from flowing from the power source through the conductor to the lamp. An open circuit prevents flow of current.

Short Circuit

If an inadvertent shortcut develops in a circuit that permits current flow through an unintentional path, a *short circuit* is created. A short circuit occurs when current leaks out of the intended conductor path such as out of a wire with damaged insulation. A short circuit is a dangerous condition especially if it incorrectly energizes the metal cabinet or housing of an appliance—a hazard if a person touches the appliance.

A circuit may also have a control device and/or a protective device, but these are optional. A *control device* either opens or closes the path of the circuit. Light switches, thermostats, and time clocks are examples of common control devices found in circuits. An *overcurrent protection device* is used to protect either the load and/or the conductors from excessive heat from high amperage conditions. Most protective devices open the circuit, thereby interrupting the path of current if excessive current is flowing in the circuit. Common examples of protective devices include fuses and circuit breakers.

Circuiting Configurations

There are two basic types of circuiting configurations used in electrical systems: series and parallel. These are shown in Figures 17.4 and 17.5.

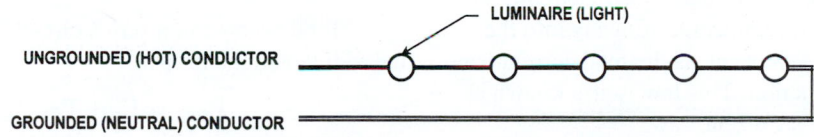

FIGURE 17.4 A series circuit is connected so that current passes through each component in the circuit without branching off to individual components in the circuit. Although a series circuit requires fewer connections, if one lamp fails the circuit becomes open and all lamps go out (like a string of low-cost Christmas lamps).

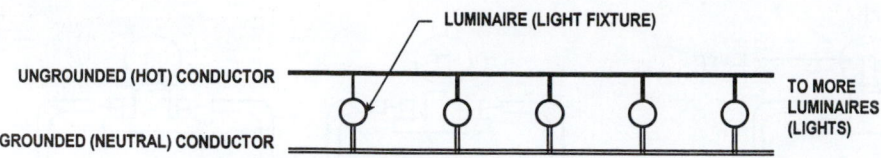

FIGURE 17.5 In a parallel circuit, current branches off to individual components in the circuit. In this circuiting configuration, if one lamp fails, the circuit remains closed and all other lamps remain lit. As a result, parallel circuiting is the most frequently used circuiting technique.

Series Circuits

A *series circuit* is connected so that current passes through each component in the circuit without branching off to individual components in the circuit. It is like water flowing down a river. The same quantity of water flows past each point in the river (unless of course water is added or removed by a connecting stream).

Parallel Circuits

In a *parallel circuit,* current branches off to individual components in the circuit. It is like water flowing down a river and being divided into several smaller streams. Smaller quantities of water flow in each stream but the total quantity of water flowing is the sum of the quantities flowing in the streams.

Although a series circuit requires fewer connections than a parallel circuit, if one lamp fails the circuit becomes open and all lamps go out (e.g., a string of low-cost Christmas lamps). In a parallel circuit, if one lamp fails the circuit remains closed and all other lamps remain lit. Therefore, most circuits in buildings are wired in parallel. Series circuiting is seldom used in permanent lighting installations in buildings.

Current and Voltage Laws

In 1857, German physicist Gustav Kirchhoff's established two laws known today as Kirchhoff's Laws. These laws state the general restrictions on the current and voltage in an electric circuit. These laws are paraphrased as follows:

Law No. 1:   The sum of the potential differences (voltages) in a complete circuit must be zero. This law is also known as Kirchhoff's Voltage Law.

Law No. 2:   At any specific instant at any junction (e.g., connection) in an electric circuit, the total current (amperage) flowing into the junction is the same as the total current leaving the junction. This law is also known as Kirchhoff's Current Law.

According to the Current Law, when a charge enters a junction, it has no place to go except to leave. By convention, currents flowing into a junction are assumed to be negative and currents flowing out of the junction positive, and by this law

the sum of these two quantities is equal to zero. So, no matter how many paths into and out of a single junction, all the current leaving that junction must equal the current arriving at that junction.

The Voltage Law states the relationship between voltage drops and voltage sources in a complete circuit. By convention, with voltage drops are assumed to be negative and voltage gains positive, and according to this law the sum of these two quantities in a complete electrical circuit is equal to zero. This means that the voltage drops around any closed loop in a circuit must equal the voltages applied.

Kirchhoff's Current and Voltage Laws indicate that the load(s) consumes all of the power provided from the source in an electric circuit and that energy and charge in an electric circuit are conserved. Simply, energy and charge cannot be destroyed. These principles are known as the Law of Conservation of Energy and Law of Conservation of Charge. These laws are useful in understanding the movement of electricity through an electric circuit. They serve as a prelude for the circuit analysis that follows and the discussion on voltage drop and power loss that is introduced in the next chapter.

Series and Parallel Circuits

Series and parallel circuits perform and function differently. Several principles apply to series and parallel circuits. These principles are outlined in the following sections.

Series Circuit Principles

Components in a series circuit share the same amperage, so the total amperage in a series circuit is equal to the sum of the individual amperages:

$$I_{total} = I_1 + I_2 + I_3 + \ldots + I_n$$

Total voltage in a series circuit is equal to the sum of the individual voltage drops:

$$E_{total} = E_1 + E_2 + E_3 + \ldots + E_n$$

Total resistance in a series circuit is equal to the sum of the individual resistances, making it greater than any of the individual resistances:

$$R_{total} = R_1 + R_2 + R_3 + \ldots + R_n$$

Parallel Circuit Principles

Components in a parallel circuit share the same voltage, so the total voltage in a parallel circuit is equal to the sum of the individual voltages:

$$E_{total} = E_1 = E_2 = E_3 = \ldots = E_n$$

Total amperage in a parallel circuit is equal to the sum of the individual branch amperages:

$$I_{total} = I_1 + I_2 + I_3 + \ldots + I_n$$

Total resistance in a parallel circuit is less than any of the individual resistances:

$$R_{total} = 1/(1/R_1 + 1/R_2 + 1/R_3 + \ldots + 1/R_n)$$

Example 17.3

Six lamps are wired in a circuit. Each lamp draws 0.75 A and has a resistance of 110 Ω. Neglecting the effects of the wiring, determine the following:

1. Total amperage in a series circuit

    $$I_{total} = I_1 + I_2 + I_3 + \ldots + I_n = (6 \cdot 0.7 \text{ A}) = 4.5 \text{ A}$$

2. Total amperage in a parallel circuit

    $$I_{total} = I_1 + I_2 + I_3 + \ldots + I_n = (6 \cdot 0.7 \text{ A}) = 4.5 \text{ A}$$

3. Total resistance in a series circuit

    $$R_{total} = R_1 + R_2 + R_3 + \ldots + R_n = (6 \cdot 110 \text{ Ω})$$
    $$= 660 \text{ Ω}$$

4. Total resistance in a parallel circuit.

    $$R_{total} = 1/(1/R_1 + 1/R_2 + 1/R_3 + \ldots + 1/R_n)$$
    $$= 1/(6 \cdot (1/110 \text{ Ω})) = 18.33 \text{ Ω}$$

17.5 THE RELATIONSHIP BETWEEN MAGNETISM AND ELECTRICAL CURRENT

Magnetism is a force of attraction between ferromagnetic metals such as iron, nickel and cobalt and a force of repulsion between diamagnetic materials such as antimony and bismuth. A *magnet* displays the properties of magnetism. Magnets occur naturally such as the mineral magnetite. Coiling insulated wire around an iron core and running current through the wire can produce an *electromagnet*, as shown in Figure 17.6.

A simple magnet has two poles: a north pole and a south pole. A free-hanging magnet within the earth's magnetic field will orient itself longitudinally between the earth's poles. The magnet's north pole will face the earth's north pole and the magnet's south pole will face south. Similar poles of different magnets repel each other and the dissimilar poles attract each other—that is, north and south poles attract one another. A

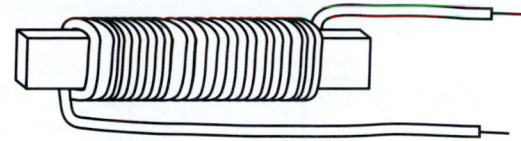

FIGURE 17.6 An electromagnet can be made by coiling insulated wire around an iron core and running current through the wire by connecting it to a battery.

magnetic field radiates out from the two poles of a single magnet or between the poles of two magnets. See Figure 17.7.

A strong link exists between electricity and a magnetic field. The force of a magnetic field can produce electrical current flow in a conductor. On the other hand, electrical current flow in a conductor produces a magnetic field. This relationship is introduced in the following paragraph.

When a conductor is moved through a magnetic field or a magnetic field is moved across a fixed conductor, a voltage is produced in the conductor. The voltage causes current to flow through the conductor. When this happens, current flow is *induced* in the conductor and the phenomenon is called *induction*. Moving the conductor in one direction across the magnetic field causes current to flow in one direction. Reversing direction of conductor movement reverses direction of current flow. When the conductor is no longer moved through a magnetic field, current flow stops. See Figure 17.8.

When a constant voltage is applied in a closed circuit, the voltage forces current to move in one direction through the conductor. As current flows in one direction, the magnetic fields of the electrons (or charged particles) align and combine to produce a strong magnetic field that extends around the conductor. Increasing voltage, and thus increasing current flow, produces a stronger magnetic field. Decreasing the voltage, and thus decreasing current flow, reduces the magnetic field.

When the circuit is opened, current flow through the conductor stops and the electrons (or charged particles) again move in random paths. Their magnetic fields cancel. With no current

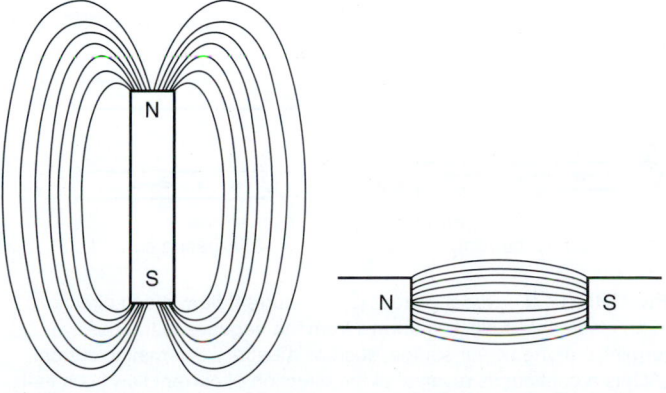

FIGURE 17.7 A magnetic field radiates out between the two poles of a single magnet or between the poles of two magnets.

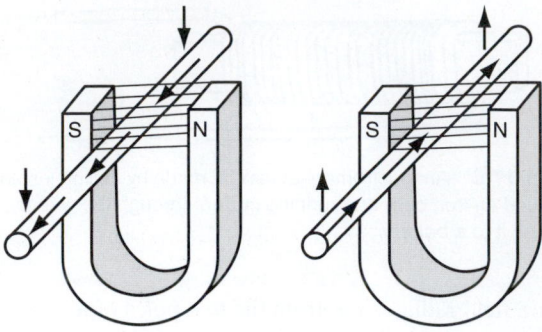

FIGURE 17.8   When a conductor is moved through a magnetic field, the force of the magnetic field generates a voltage in the conductor. This voltage causes current to flow through the conductor. When this happens, current flow is induced in the conductor. This phenomenon is called *induction*. Current flows in one direction when the conductor moves down through the magnetic field, and it reverses direction when the conductor moves up through the field. When the conductor is stationary, there is no current flow.

flow, there is no magnetic field surrounding the conductor. Finally, if the connections of the conductor to the power source were switched, the polarity of the circuit would change. Current would flow in the opposite direction and the polarity of the magnetic field around the conductor would reverse.

17.6   DIRECT AND ALTERNATING CURRENT

Direct Current

Direct current (DC) is current flow in one direction in an electrical circuit. See Figure 17.9. It is always from the negative to the positive terminals of the power source such as a battery. Flashlights and automobile electrical installations are designed to operate on a DC power. A graph of DC voltage versus time is shown in Figure 17.10.

Alternating Current

Alternating current (AC) is a continuous reversal of the direction of current flow such that at a point in time the current flow

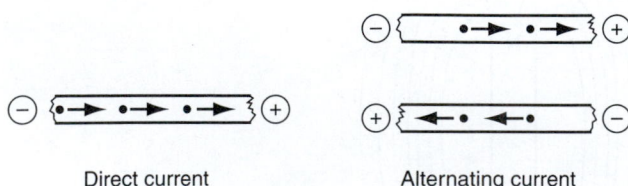

Direct current   Alternating current

FIGURE 17.9   Direct current (DC) is current flow in one direction in an electrical circuit. It is always from the negative to the positive terminals of the power source, such as a battery. Alternating current (AC) is a continuous reversal of the direction of current flow such that at one time current flow is in one direction and another time it is in the reverse direction. Direction of current flow reverses as the polarity of the power source in the circuit reverses.

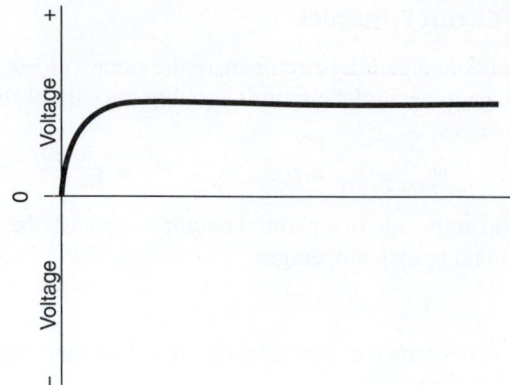

FIGURE 17.10   When a direct-current circuit is closed, the voltage in the circuit climbs rapidly to a constant voltage, which produces a steady flow of current in one direction.

is in one direction and at another point in time it is in the reverse direction. Direction of current flow reverses as the polarity of the power source in the circuit reverses. A graph of DC voltage versus time is shown in Figure 17.11. The change from one direction to the next and back again is called a *cycle*. In a cycle, rapidly changing attracting and repelling forces exist in the circuit so flow of current oscillates back and forth. *Frequency* is the term that describes cycles per second. It is expressed in *hertz (Hz)*.

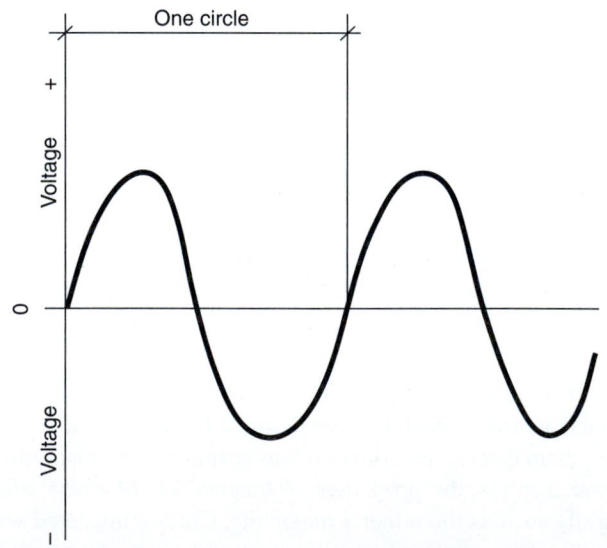

FIGURE 17.11   When an AC circuit is closed, current flows and voltage in the circuit climbs and falls rapidly. When plotted on a graph in terms of voltage and time, the curve of voltage is in the form of a sine wave. As shown, the voltage in a circuit goes from zero to the maximum voltage and back to zero every half cycle and then polarity instantaneously reverses as voltage goes from zero to maximum voltage and back to zero for the second half cycle. The continuous change in voltage and reversal of polarity produces a continuous increase, decrease, and reversal of the direction of current flow.

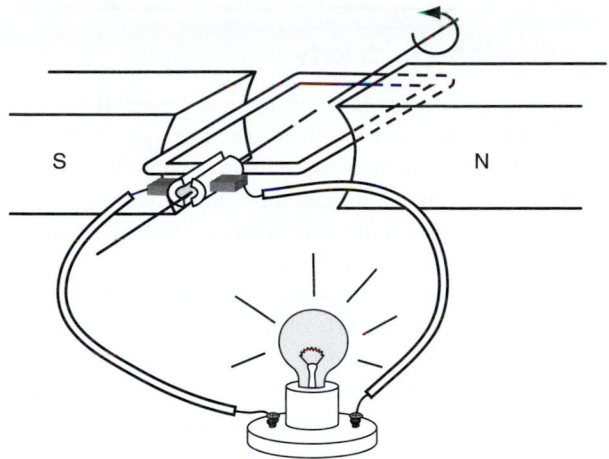

FIGURE 17.12 Shown is a simple AC generator. When the conductor loop rotates through the magnetic field between the poles of a stationary magnet, current is induced in the windings. As long as the loop is rotating, current will flow. If rotation stops, so does current flow.

A *waveform* is a representation of how AC varies with time. The most common AC waveform is the sine wave, which derives its name for the reason that the current or voltage varies with the sine of the elapsed time. Other common AC waveforms are the square wave, the ramp, the sawtooth wave, and the triangular wave, which may be produced by certain types of electronic oscillators, uninterruptible power supplies operating from a battery, and some audio amplifiers. Current produced at a power plant has a sine waveform.

Power generated in the United States and Canada is produced at 60 Hz; that is, 60 times every second power is positive, 60 times it is negative, and 120 times every second there is no voltage in the circuit. In countries in Europe and the Middle East, power generated at a frequency of 50 Hz is more popular. Naval and aviation equipment (and some large mainframe computer equipment) use 400/415 Hz power. However, 400/415 Hz power is not likely to be encountered within facilities.

Frequency affects the physical size and weight of motors and transformers. Higher frequencies mean lighter motors and transformers. A higher frequency means more wear on a motor because it spins faster. The military operates motors and transformers at several hundred hertz, where the trade-off for weight and wear is acceptable. Power for residential, commercial, and industrial applications is generated at 50 Hz or 60 Hz, which is a balance between cost of appliance weight and wear.

Voltage is the driving force of current flow. In an AC circuit voltage is never constant but is changing rapidly. Voltage from an AC power source goes from zero to the maximum voltage and back to zero every half cycle; as polarity instantaneously reverses, voltage goes from zero to maximum voltage and back to zero for the second half of the cycle. Change in voltage is typically in the form of a sine wave. This fluctuating "push" and "pull" of voltage results in an oscillating current flow. See Figures 17.12 and 17.13.

Voltage produced by an AC power source is specified by its *average* voltage. For example, on the 120 V, single-phase, AC system (the voltage available at most household receptacles), voltage reaches a maximum of about 170 V but averages 120 V. See Figure 17.14. The area below the sine wave determines the average voltage, so the maximum voltage on a 120 V AC circuit is 169.71 V (120 V multiplied by the square root of 2).

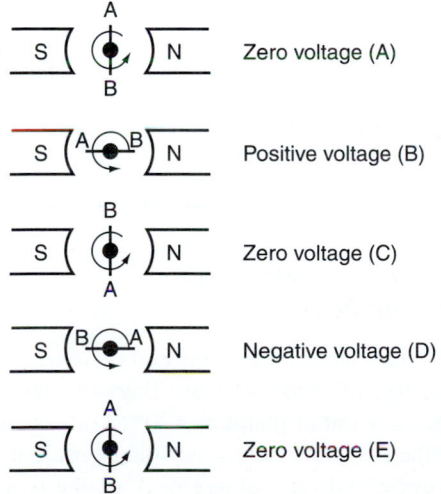

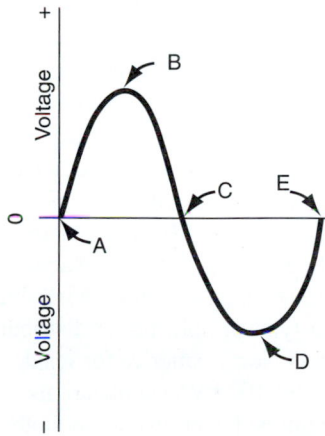

FIGURE 17.13 An *AC generator* produces alternating current that keeps reversing its direction of flow. When the conductor windings move through the magnetic field between the poles of the stationary magnet, current is induced in the windings. During one-half of the armature rotation, voltage changes from zero to maximum voltage, then back to zero and current is pushed in one direction. During the second half of the rotation, voltage forces current in the opposite direction because the generator windings are moving through the magnetic field in the opposite direction.

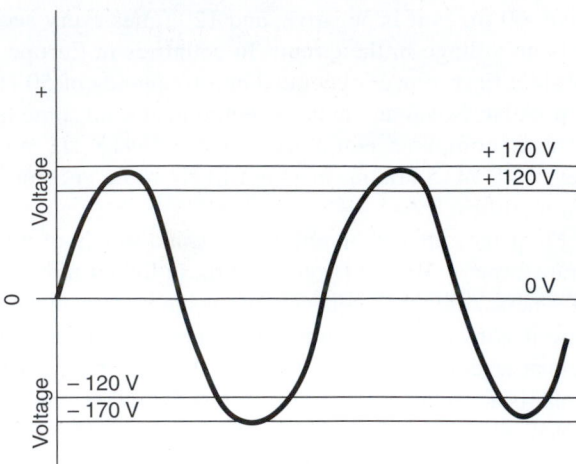

FIGURE 17.14 Voltage produced by an AC power source is usually specified by its average voltage. In the 120 V, single-phase, AC system voltage available at most home receptacles, voltage reaches a maximum of about 170 V but averages about 120 V.

Because of the voltage fluctuation, there is a fluctuation in current flow. An incandescent lamp on an AC circuit fluctuates in light intensity 120 times a second; that is, for an instant it glows brightly and then dims slightly each time the voltage changes from zero to maximum. One cannot see the light intensity fluctuate because it is beyond the flicker threshold of the human eye. (Frequencies lower than 20 Hz tend to result in noticeable flicker of incandescent lamps.) This fluctuation in voltage does create vibration that can be felt in motors, which presents significant concerns during the operation of large motors, where it can result in excessive noise and even catastrophic failure.

Single-Phase Alternating Current Power

A *single-phase (1Φ) alternating current* distribution system refers to a system in which all the voltages of the supply vary in unison. A basic system typically has two conductors: one is neutral and the other carries current (the *hot* or *live* conductor).

A variant of the single-phase AC circuit is the *split-phase alternating current distribution system,* which is where there are two voltage sources, 180° phase shifted from each other, which power two series-connected loads. A three-wire single-phase distribution system is a type of split-phase distribution system that is commonly used in North America for residential and light commercial (up to about 100 kVA) applications.

Single-phase power requires fewer wires on both the electrical distribution system and the customer's system (in comparison to the three-phase distribution system discussed in the next section), which results in lower installed costs. It is most often used in homes, small businesses, and farms. In large commercial buildings and industrial locations where large motors are used, single-phase power is not generally adequate.

Generating Electricity

An *electric generator* is a mechanical device that converts mechanical energy into electrical energy. An armature, a shaft with conductor windings (coils) wrapped around an iron core, is rotated through a stationary magnetic field. Movement of the conductor windings through the field produces current flow.

An *AC generator* produces alternating current that keeps reversing its direction of flow. (See Figure 17.13.) When the conductor windings move through the magnetic field between the poles of the stationary magnet, current is induced in the windings. During one-half of the rotation of the armature, current is pushed in one direction. During this half rotation, voltage changes from zero to maximum voltage then back to zero. During the second half of the rotation, voltage pushes current in the opposite direction because the windings are moving through the magnetic field in the opposite direction. Again, voltage changes from zero to maximum voltage then back to zero. This alternating change in voltage and thus current direction is why the system is called *alternating current*. Most power generated in the United States and Canada is produced with the generator armature rotating 60 times a second (60 Hertz); therefore, 60 times every second current flow is in one direction, 60 times every second it is in the other direction, and 120 times every second there is no current flow.

A rotary engine called a *turbine* is connected to the generator and drives the rotation of the armature shaft. In a steam turbine, high-pressure steam moves through the turbine, driving rotation of discs attached to the turbine shaft. Steam is produced by heating water by burning coal, oil, or natural gas, or with heat created by a nuclear reaction. In a hydroelectric plant, falling water strikes the blades or buckets of a turbine, causing the turbine shaft to rotate.

Three-Phase Alternating Current Power

A *three-phase (3Φ) alternating current* distribution system consists of three separate lines of single-phase power with each line out of phase by 120° (⅓ of a cycle). The voltage peak of the first phase leads the voltage peak of the second phase by ⅓ cycle and the voltage peak of the second phase leads the voltage peak of the third phase by ⅓ cycle and so on. See Figure 17.15. Basic three-phase circuits typically have three current-carrying (hot or live) conductors plus one grounded (neutral) conductor.

A distinct advantage of three-phase AC power is in its use with motors. Each of the three separate phases peak at one

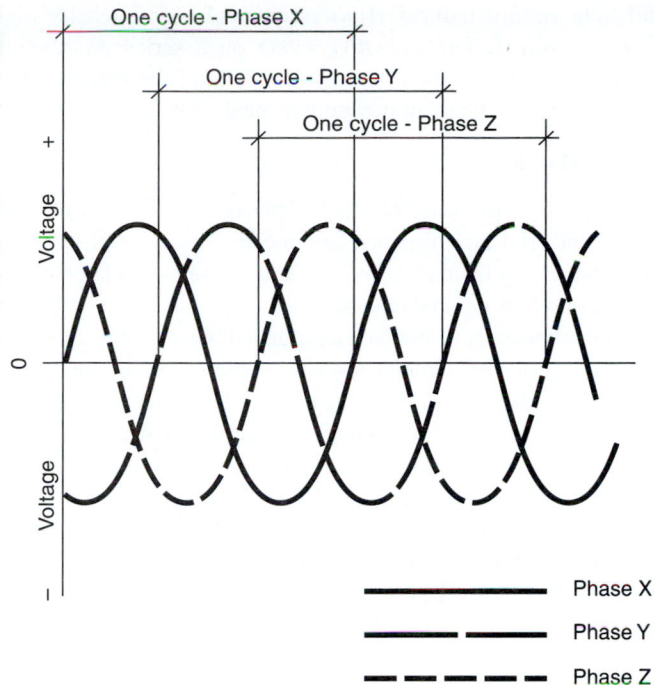

FIGURE 17.15 Three-phase alternating current consists of three separate phases of single-phase power with each out of phase by ⅓ of a cycle. In theory, three generators are needed to produce three-phase power with each generator rotating one-third out of phase. In practice, a single generator with three separate sets of windings is used.

third of a cycle apart, which attenuates the thrust of the individual voltage peak—that is, over a full cycle there are three voltage peaks instead of one. Rotation of a motor shaft is caused by the expanding and collapsing magnetic field acting against a rotor located on the shaft. These impulses of magnetic force push the rotor about the motor shaft, thereby causing the motor shaft to rotate. Although multiple magnetic windings can be used, each time the AC voltage alternates from zero to peak voltage it creates an impulse of force. On large single-phase motors, the oscillating impulses create huge vibrations that are potentially catastrophic. Three-phase alternating current provides three peaks instead of one on the three-phase cycle. Three-phase alternating current power is used to run motors that drive large machinery and large air-conditioning blowers.

Power with multiple phases is referred to as *polyphase alternating current*. Theoretically, more phases results in less motor vibration, but more power lines and more turnings on motors are required for more phases. The three-phase power standard is an economical choice made by utility companies and manufacturers of motors.

Power companies in the United States generally only generate and transmit three-phase AC power. Single-phase power for commercial building or residential use is obtained from one leg of the three phase power lines. Typically, three-phase power is used in buildings where the load exceeds 50 kVA or where it is needed to power three-phase equipment.

17.7 TRANSFORMING VOLTAGE AND CURRENT

In an AC circuit the strength of the voltage and the polarity of the power source are always alternating, so rate and direction of current flow are always changing. As current flow increases in the circuit, the magnetic field surrounding the conductor increases, and as current flow decreases, the field decreases. Fluctuation in rate of current flow causes the magnetic field surrounding the energized conductor to expand and collapse. If a second stationary conductor is positioned within this expanding and collapsing magnetic field, the force of the field induces a voltage across the second conductor. This voltage drives current flow in the second conductor.

The second conductor does not have to touch the energized conductor nor does it have to move. Instead, the expanding and collapsing magnetic field of the energized conductor induces a voltage in the second conductor. Current induced in the second conductor speeds up as the magnetic field expands and slows down as it collapses. Likewise, as current flow in the first conductor changes direction, so does current flow in the stationary conductor. Although current flow in both conductors is not exactly synchronized (there is some delay in the response of the current to change in voltage), the properties of the magnetic forces created by the energized conductor directly affect current flow in the second conductor.

Transformers

A *transformer* is an electrical device that transfers an alternating current and voltage from one circuit to another using the induction phenomenon. The device is used in a circuit to change voltage, current, phase, and other electrical characteristics. As shown in Figure 17.16, a simple transformer consists of sets of wire coils or windings around an iron core. The expanding and collapsing magnetic field in the *primary winding* of the transformer induces voltage in the *secondary winding*. The primary winding receives

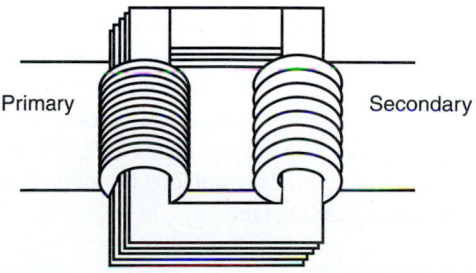

FIGURE 17.16 A simple single-phase transformer consists of sets of wire coils or windings around an iron core. The expanding and collapsing magnetic field in the *primary windings* of the transformer induces voltage in the *secondary windings*. The primary windings receive energy from an AC source. Power transfer is accomplished completely by the expanding and collapsing magnetic field. The windings are not connected to each other in any way.

energy from an AC source. Power transfer is accomplished completely by the expanding and collapsing magnetic field—that is, the windings are not connected to each other in any way.

Transformers serve as an efficient way of converting power at a primary voltage and amperage to the equivalent power at a different secondary voltage and amperage. In a simple transformer composed of two coils, the ratio of the alternating current output voltage to the AC input voltage is approximately equal to the ratio of the number of turns in the secondary coil to the number of turns in the primary coil. Thus, the theoretical relationship between primary (E_p) and secondary (E_s) voltages is proportional to the number of windings in the primary (N_p) and secondary (N_s) windings is expressed as:

$$E_p/N_p \approx E_s/N_s$$

Primary voltage can be stepped up to a higher secondary voltage or stepped down to a lower secondary voltage based on the ratio of number of windings:

$$E_s \approx E_p(N_s/N_p)$$

Example 17.4

A 225 kVA transformer located outside a building is used to step down the voltage for the building. It is connected to a 7200 V AC power source. The ratio of the number of primary windings to the number of secondary windings on the transformer is 30 to 1.

a. Approximate the voltage provided to the building.

$$E_s \approx E_p (N_s/N_p) = 7200 \text{ V } (1/30) \approx 240 \text{ V}$$

b. Approximate the current available in the building, in amps.

By rearranging terms in the equation of power, P = EI:

$$I = P/E = 225\ 000 \text{ VA}/240 \text{ V} = 937.5 \text{ A}$$

Alternating current is used exclusively in the United States in residential and commercial wiring because it provides greater flexibility in voltage selection and simplicity of equipment design. The principle reason is that when AC is used, the voltage can be increased (stepped up) or decreased (stepped down) very easily and efficiently through the use of transformers. A transformer is about 95% efficient in making the conversion. By comparison, an electrical DC generator—the only means of changing DC voltage—is about 75% efficient.

Types of commercially available building transformers are discussed in the next chapter.

17.8 IMPEDANCE AND THE POWER FACTOR

Inductors

An *inductor* is a coil of wire that creates an electromagnetic field. On AC circuits, inductive loads are created as current flows through coils or windings found in motors, transformers, and light fixture ballasts (fluorescent and high-intensity discharge fixtures). The *inductive effect* on a series AC circuit causes the phase of the current to lag behind the phase of the voltage—that is, peak amperage lags peak voltage.

Capacitors

A *capacitor* is composed of metal plates separated by air or a dielectric material such as paper, ceramic, or mica. Capacitors store electrical energy in an electrostatic field and release it later, much like your body stores and releases static electricity as you rub your feet across the carpet and touch a grounded object. The *capacitive effect* on a series AC circuit causes the phase of the current to lead the phase of the voltage—that is, peak voltage lags peak current.

Although there are no inductive and capacitive effects on a DC circuit, current flow on an AC circuit is impeded by inductance and capacitance. *Impedance (Z)* is a measure of resistance to current flow on an AC circuit due to the combined effect of resistance, inductance and capacitance. Impedance is measured in ohms (Ω). *Ohm's Law for AC circuits* is:

$$E = IZ$$

Example 17.5

The impedance values of incandescent lamps with three wattage ratings are noted below. With a voltage of 120 V on an AC circuit, determine the impedance of each lamp.

$$E = IZ, \text{ therefore: } I = E/Z$$

a. 100 W (121 Ω)

$$I = E/Z = 120 \text{ V}/121 \ \Omega = 0.992 \text{ A}$$

b. 75 W (161 Ω)

$$I = E/Z = 120 \text{ V}/161 \ \Omega = 0.745 \text{ A}$$

c. 40 W (300 Ω)

$$I = E/Z = 120 \text{ V}/300 \ \Omega = 0.400 \text{ A}$$

Power Factor

DC and AC circuits perform differently with respect to power use. On a DC circuit, the product of measured voltage and measured amperage equals wattage (VA = W). In contrast, on most AC circuits the computed volt-amperage is different than power consumed (wattage); that is, the product of the measured voltage and amperage (V · A) does not equal wattage (VA ≠ W). This phenomenon is directly related to the inductive effects in circuits powering motors, transformers, and magnetic ballasts as described earlier Inductors section. The three components of AC power are:

Real power is the "working power" that performs useful effort in a circuit (e.g., creating heat, light, and motion); it is expressed in watts (W) or kilowatts (kW). Inductance on an AC circuit can result in periodic reversals of the direction of energy flow. Real power is the net result after discounting these periodic reversals. It is that portion of the power flow, averaged over a complete cycle of the AC waveform, that results in a net transfer of energy in one direction. In a circuit, a wattmeter reads real power. The utility company charges its customer based on real power.

Reactive power is the power that generates the magnetic field required for inductive devices to operate. It dissipates no energy in the load but which returns to the source on each alternating current cycle; it is expressed in units called volt-amps-reactive (VAR) or kilovolt-amperes-reactive (kVAR), rather than watts. Reactive power required by inductive loads increases the amount of apparent power in a distribution system.

The *apparent power* is the "power available to use." It is expressed in volt-amperes (VA) or kilovolt-ampere (kVA), because it is the simple product of voltage and current. Apparent power is the "total" power required by an inductive device that is a composite (vector sum) of the real power and reactive power. It can be computed as the product of measurements made with a voltmeter and ammeter in an AC circuit.

The *power factor (PF or cosφ)* for a single-phase circuit is the ratio between real power and apparent power in a circuit:

$$PF = (\text{real power/apparent power})$$
$$PF = \text{watts/(volt} \cdot \text{amps)} = W/VA$$

The *power factor* is a number between 0 and 1 (frequently expressed as a percentage, e.g., a 0.7 PF = 70% PF). In a single-phase circuit or balanced three-phase circuit, the power factor can be measured with the wattmeter-ammeter-voltmeter method, where the measured power in watts is divided by the product of measured voltage and current.

Equipment specifications provided by the manufacturer and motor nameplates often designate the power factor. Where the AC waveform is pure (purely sinusoidal), true power and apparent power can be though of as forming the adjacent and hypotenuse sides of a right triangle, so the power factor ratio is also equal to the cosine of that *phase angle (φ)*. For this reason, equipment nameplates and manufacturer information will frequently abbreviate the power factor as *cosφ*.

Example 17.6

A circuit consumes 3000 W of real power when the apparent power is 3600 VA.

a. Determine the power factor.

$$PF = (\text{real power/apparent power})$$
$$= 3000 \text{ kWh}/3600 \text{ kVA} = 0.833$$

b. Determine phase angle φ.

$$\varphi = \cos^{-1} 0.833 = 33.6°$$

By algebraic manipulation, PF = (real power/apparent power) can be converted to:

$$\text{real power} = \text{apparent power} \cdot PF$$
$$W = V \cdot A \cdot PF$$

In Example 17.6, the power factor is less than 1, which means that apparent power and real power are not equal (VA ≠ W). When an AC circuit is purely resistive, it draws current (amperage) exactly synchronized and proportional to voltage—that is, voltage and current peaks occur simultaneously. As a result, the apparent power and real power are equal (VA ≠ W) and the power factor is 1.0 (or 100%). If the circuit in the previous example had been purely resistive (PF = 1.0), a full 3600 W would be delivered to the load, rather than the 3000 W. A poor power factor makes for an inefficient delivery of power.

On an AC circuit, voltage (E), amperage (I), and the power factor (PF) are applied to determine power (P) through the *AC power equation:*

$$\text{apparent power, } P_A = EI$$
$$\text{real power, } P_R = EI(PF)$$

Example 17.7

An AC circuit is powering an electric heater (i.e., pure resistance, PF = 1.0). Assume the voltage is 240 V and a current draw of 10 A. Compute the apparent power and real power.

$$\text{apparent power, } P_A = EI = 240 \text{ V} \cdot 10 \text{ A}$$
$$= 2400 \text{ W}$$
$$\text{real power, } P_R = EI(PF) = 240 \text{ V} \cdot 10 \text{ A} \cdot 1.0$$
$$= 2400 \text{ W}$$

Example 17.8

An AC circuit is powering a motor (i.e., inductive load, PF < 1.0). Assume the voltage is 240 V and a current draw of 10 A. Compute the real power, assuming a power factor of 0.833.

$$\text{apparent power, } P_A = EI = 240 \text{ V} \cdot 10 \text{ A}$$
$$= 2400 \text{ W}$$
$$\text{real power, } P_R = EI(PF) = 240 \text{ V} \cdot 10 \text{ V} \cdot 0.833$$
$$= 1999 \text{ W}$$

In buildings, circuits are hardly ever purely resistive, so the PF is generally less than 1.0. Circuits serving inductive loads tend to draw current with a phase shift (i.e., time lag)—that is, voltage and current peaks occur out of sync. This phase shift causes voltage to sometimes be positive when current is negative or current to be positive when voltage is negative. In instants where power is negative, a motor is actually behaving like a generator feeding power back into the system. As a result, it takes more current to deliver a fixed amount of power when current is shifted out of phase with voltage. Real power and apparent power are not equal. Inductive circuits have a power factor of less than 1.0 (PF < 1.0). As the amount of inductive load in a circuit or system increases, the PF decreases. The main contributor to a low PF is a motor operated at less than full load.

The power factor measures how effectively the total delivered power is being used. A motor running with a 0.7 PF indicates that only 70% of the current provided by the electrical utility is being used to produce useful work. In contrast, a motor running with a 0.85 PF will use 85% of the power. A higher PF at the motor means more of the supplied energy will be transferred into useable work by the electrical system. In contrast, a low PF means the current flowing through electrical system components is higher than necessary to do the required work—an inefficient use of power.

A low PF is caused by an electrical circuit having a greater amount of inductive load than capacitive load. Lightly loaded induction motors (e.g., power saws, conveyors, compressors, and grinders) or varying-load inductive equipment (e.g., HVAC systems, machining equipment, welders, and presses) are examples of equipment that can have a low PF. An inexpensive fluorescent fixture with magnetic ballast can have a PF as low as 0.50. In contrast, resistance-type appliances (e.g., water heaters, oven/range elements, electric resistance heaters, and incandescent lamps) have a PF of very nearly 1.0.

In buildings, circuits containing inductive loads (e.g., circuits feeding equipment with motors, transformers or light fixture ballasts for fluorescent and high intensity discharge luminaires) have PFs that range from 0.30 to over 0.90. Effects of the individual building circuits merge at the building electrical service, yielding a system PF. Industrial and commercial electric customers operate with system PFs between 0.75 and 0.98. A PF for a building system below about 0.95 is considered low.

The PF is an important consideration in design of an AC circuit or system, because when a PF is less than 1.0, the circuit's wiring has to carry more current than would be necessary to deliver the same amount of real power. For example, consider a building drawing 200 A at 480 V. The system equipment (i.e., conductors, transformers, switchgear) must be rated at 96 kVA (because 200 A · 480 V = 96 kVA). But if the PF of the loads is only 0.7, then only 67.2 kVA of real power is being consumed (because 200 A · 480 V · 0.7 = 67.2 kVA).

Power companies supply customers with apparent power measured in volt-amperes (VA), but charge them for real power, expressed in watts (W) consumed. So, from the perspective of the power company, a low overall PF means the power company must provide more generating capacity than actually is required. Additionally, all power company system equipment (e.g., conductors, transformers, and generators) must be increased in size to handle the "additional" reactive power caused by inductive loads. This results in additional costs for the power company. The goal of a power company is a PF of 1.0 or a *unity*

power factor. As a result, power companies require their customers, especially those with large loads, to maintain their PFs above a specified amount (usually 0.90 to 0.95 or higher). Power companies measure reactive power used by high-demand customers and charge higher rates accordingly.

PF Correction

Commercial or industrial operations rarely achieve a PF of 1.0 because inductive loads from motors, transformers and light fixture ballasts are common. This inductive effect causes the phase of the current to lag behind the phase of the voltage. A greater lag in current results in a poorer power factor. The PF can be improved by adding capacitance to the circuit. A capacitor has the effect of causing the phase of the voltage to lag the phase of the current and thus reduces the lagging effect caused by the inductive loads. Capacitors or synchronous motors are frequently used to counter the effect of inductive motors and transformers in industrial applications.

Some consumers install PF correction devices (e.g., a capacitor) to cut down on higher costs associated with a low PF. For example, modern fluorescent fixtures have a PF correction device that improves the PF to 0.90. A PF factor in an AC circuit can be corrected (to a value close to 1.0) by adding a parallel reactance opposite the effect of the load's reactance. Most commonly the load's reactance is inductive in nature, so a parallel capacitance can be added to correct the poor PF. For example, industrial plants tend to have a lagging PF, where the current lags the voltage (like an inductor). This is primarily the result of having several electric induction motors in operation. Some industrial sites will have large banks of capacitors, called *power factor correction capacitors*, specifically for the purpose of correcting the PF back toward 1 to save on power company charges. PF correction capacitors are not perfect. They only provide a PF close to 1.0 when the power supply is at or near its full rated load.

The main advantages of the PF correction are as follows:

1. A high PF reduces the load currents, resulting in a considerable saving in hardware costs (i.e., conductors, switchgear, substation transformers, and so on).

2. Power companies typically impose low power factor penalties, so by correcting the PF, this penalty can be avoided.

3. The electrical load on the power company is reduced, which allows the power company to supply the surplus power to other consumers without increasing its generation capacity.

17.9 COST OF ELECTRICAL ENERGY AND POWER

A utility company will charge its customers for the electrical energy consumed and, for all but small users (e.g., most residential customers), the rate at which energy is consumed.

Conversion Formulas

Design of a building electrical installation involves computations with power (P), expressed in watts (W); voltage (E), in volts (V); amperage (I), in amps (A); and volt-amps (VA), in volts and amperes. In AC systems, the power factor (PF) is used. A summary of conversion formulas for direct current and alternating current systems is provided in Table 17.1. These formulas are useful in determining loads in circuiting design.

TABLE 17.1   A SUMMARY OF CONVERSION FORMULAS FOR DIRECT CURRENT AND ALTERNATING CURRENT SYSTEMS, WHERE POWER (P) IS EXPRESSED IN WATTS (W), VOLTAGE (E) IN VOLTS (V), AMPERAGE (I) IN AMPS (A), AND VOLT-AMPS (VA) IS IN VOLT-AMPERES. WHEN USED IN THESE FORMULAS, POWER FACTOR (PF) SHOULD BE EXPRESSED AS A DECIMAL. FOR TWO-PHASE, THREE-WIRE ALTERNATING CURRENT SYSTEMS, SUBSTITUTE 1.41 FOR 1.73. FOR TWO-PHASE, FOUR-WIRE ALTERNATING CURRENT SYSTEMS, SUBSTITUTE 2.00 FOR 1.73.

To Find	Direct Current (DC) Systems	Alternating Current (AC) Systems	
		Single Phase	Three Phase
DC power (W)	$P = E \cdot I$	—	—
AC real power (W)	—	$EI(PF)$	$1.73 \cdot EI(PF)$
AC apparent power (VA)	—	(VA)	$1.73 \cdot (VA)$
Amperes (I)	$I = P/E$	$I = P/E \cdot PF$	$I = P/1.73 \cdot E \cdot PF$
	—	$I = (VA)/E$	$I = (VA)/1.73 \cdot E$
Power factor (PF)	—	$PF = P/E \cdot I$ or $PF = P/(VA)$	$PF = P/1.73 \cdot E \cdot I$ or $PF = P/1.73(VA)$

Note: VA is the rating of the appliance or piece of equipment in volt-amperes (VA).

These charges make up the largest part of a typical electric bill and will be discussed in the following sections.

Energy Charge

The *energy charge* is simply the cost of electrical energy consumed ($\$_{energy}$). This may be computed by the following equation, where energy consumption (q) is expressed in kilowatt-hours and unit cost of electricity ($/kWh) is expressed in dollars per kilowatt-hours:

$$\$_{energy} = q \cdot \$/kWh$$

The energy charge is based on energy consumed by the customer during a billing period, say once a month or every 30 days. Energy consumed is determined by reading the electric meter.

Example 17.9

A 60 W lamp remains lighted for 24 hr a day for 30 days. Determine the electrical energy consumed over this period. Calculate the energy charge for the billing period at a rate of $0.1172/kWh.

$q = Pt$
$= (60 \text{ W}) \cdot (24 \text{ hr/day} \cdot 30 \text{ days/billing period})$
$= 43\,200 \text{ Wh}$
$= 43\,200 \text{ Wh} (1 \text{ kW}/1000 \text{ W}) = 43.2 \text{ kWh}$

$\$_{energy} = q \cdot \$/kWh$
$= 43.2 \text{ kWh/billing period } \$0.1172/kWh$
$= \$5.06/billing period$

Example 17.10

A large residence consumes 1155 kWh of electrical energy over a billing period. Determine the total charge for the billing period based on the rate schedule provided in Table 17.2.

From Table 17.2, the following charges are found:

Service Charge . . . .$5.16 Energy Charge
First 1000 kWh of billing period . . . .$0.117 per kWh
Next 2000 kWh of billing period . . . .$0.109 per kWh
The total charge is found by:

Total charge = $5.16 + ($0.117/kWh · 1000 kWh)
             + ($0.109/kWh · 155 kWh)
             = $5.16 + $117.00 + $15.90 = $139.06

Power "Demand" Charge

Commercial and industrial consumers (and some all-electric residential users) are assessed an additional charge based on the highest rate that energy is consumed. *Maximum demand* is the user's highest rate at which energy is consumed in kilowatts

TABLE 17.2 A SAMPLE RESIDENTIAL RATE SCHEDULE.

Public Utility Company, Inc.
General Residential Rate Schedule (Nondemand)

Service Charge

Nondemand $5.16

Energy Charge

First 1000 kWh of billing period .... $0.117 per kWh
Next 2000 kWh of billing period .... $0.109 per kWh
Additional kWh of billing period .... $0.102 per kWh

TABLE 17.3 A SAMPLE RATE SCHEDULE FOR COMMERCIAL AND INDUSTRIAL OPERATIONS.

Public Utility Company, Inc.
General Commercial and Industrial Rate Schedule (Demand)

Service Charge

Single phase .... $ 6.10
Three phase $10.36

Energy Charge

First 1000 kWh of month .... $0.0976 per kWh
Next 2000 kWh of month .... $0.0885 per kWh
Next 5000 kWh of month .... $0.0821 per kWh
Next 10 000 kWh of month .... $0.0712 per kWh
Additional kWh of month .... $0.0657 per kWh

Demand Charge

First 25 kW of month .... $8.85 per kW per month
Next 475 kW of month .... $7.95 per kW per month
Next 1000 kW of month .... $7.50 per kW per month
Additional kW of month .... $7.20 per kW per month
Demand charge is based upon greatest billing demand month for
that month and eleven successive months.

(kW) over a small time interval (usually 15 min but sometimes 30 or 60 min) that is measured by the electric meter during a billing period. A *demand charge* is the billing fee related to maximum demand. Depending on the billing rate, a high demand charge may remain at that rate for 12 months even though the demand for succeeding months is significantly lower.

Take, for example, a small manufacturing plant that typically consumes power at a demand rate of 400 kW during the day and 200 kW at night. The demand charge would be based on maximum demand, which in this case is 400 kW. Of course maximum demand will change day-to-day and month-to-month. Now, assume that because of an irregularity in plant operation, demand for power increases triples to 1200 kW over a 15-min period on one day during the billing period and the demand for succeeding periods returns to 400 kW. The demand charge would now be based on 1200 kW. On some billing rate schedules, the demand charge will remain at the rate for 1200 kW for 12months even though this increase in demand occurred only once for a 15-min period. As shown in the following two examples, a one-time peak in demand can be very costly.

Example 17.11

A small manufacturing plant on a single three-phase service consumes 104 000 kWh of electrical energy over a monthly billing period. Peak demand is measured at 400 kW during the day and 200 kW at night. Determine the total monthly charge based on the rate schedule provided in Table 17.3.

From Table 17.3, the following billing rates are found:
Service Charge for three-phase service $10.36
Energy Charge
First 1000 kWh of month .... $0.0976 per kWh
Next 2000 kWh of month .... $0.0885 per kWh
Next 5000 kWh of month .... $0.0821 per kWh
Next 10 000 kWh of month .... $0.0712 per kWh
Additional kWh of month .... $0.0657 per kWh
Demand Charge
First 25 kW of month .... $8.85 per kW per month
Next 475 kW of month .... $7.95 per kW per month
Next 1000 kW of month .... $7.50 per kW per month

The total charge is computed by:
Service Charge for three-phase service = $10.36
Energy Charge for 104 000 kWh:
First 1000 kWh: (1000 kWh · $0.0976/kWh) = $ 97.60
Next 2000 kWh: (2000 kWh · $0.0885/kWh) = $ 177.00
Next 5000 kWh: (5000 kWh · $0.0821/kWh) = $ 410.50
Next 10 000 kWh: (10 000 kWh · $0.0712/kWh) = $ 712.00
Additional kWh: (remaining 86 000 kWh ·
                         $0.0657/kWh) = $5650.20
Energy Charge Subtotal: $7047.30

Demand Charge
First 25 kW: (25 kW · $8.85/kW) = $ 221.25
Next 475 kW: (remaining 375 kW · $7.95/kW) = $2981.25
Demand Charge Subtotal: $3202.50

Total Monthly Charge:
$10.36 + $7047.30 + $3202.50 = $10 260.16

Example 17.12

Much like in Example 17.11, a small manufacturing plant on a single three-phase service consumes 126 000 kWh of electrical energy over a monthly billing period. Peak demand is typically measured at 400 kW during the day and 200 kW at night. However, because of an abnormality in plant operation, demand for power increases to 1200 kW over a 15-min period on one day during the month. Determine the total monthly charge based on the rate schedule provided in Table 17.3.

From Table 17.3, the following billing rates are found:

Service Charge for three-phase service $10.36

Energy Charge

First 1000 kWh of month . . . . $0.0976 per kWh

Next 2000 kWh of month . . . . $0.0885 per kWh

Next 5000 kWh of month . . . . $0.0821 per kWh

Next 10 000 kWh of month . . . . $0.0712 per kWh

Additional kWh of month . . . . $0.0657 per kWh

Demand Charge

First 25 kW of month . . . . $8.85 per kW per month

Next 475 kW of month . . . . $7.95 per kW per month

Next 1000 kW of month . . . . $7.50 per kW per month

The total charge is computed by:

Service Charge for three-phase service = $10.36

Energy Charge for 104 000 kWh:

First 1000 kWh: (1000 kWh · $0.0976/kWh) = $ 97.60

Next 2000 kWh: (2000 kWh · $0.0885/kWh) = $ 177.00

Next 5000 kWh: (5000 kWh · $0.0821/kWh) = $ 410.50

Next 10 000 kWh: (10 000 kWh · $0.0712/kWh) = $ 712.00

Additional kWh: (remaining 86 000 kWh ·

              $0.0657/kWh) = $5650.20

Energy Charge Subtotal: $7047.30

Demand Charge:

First 25 kW: (25 kW · $8.85/kW) = $ 221.25

Next 475 kW: (475 kW · $7.95/kW) = $3776.25

Next 1000 kW: (remaining 700 kW · $7.50/kW) = $5250.00

Demand Charge Subtotal: $9247.50

Total Monthly Charge:

$10.36 + $7047.30 + $9247.50 = $16 305.16

Demand Limiting and Load Shedding

As shown in Examples 17.11 and 17.12, a one-time peak in maximum demand cost the manufacturer an extra demand charge of $6045/month. On billing rate schedules where maximum demand is carried for 12 months, this equates to a $72 540 annual increase! The additional demand charge appears excessive but is rationalized because utility companies are legally charged with the task of providing power regardless of demand. So, a utility company must have the extra generating capacity to produce electricity during periods of peak demand, say during the afternoon of the hottest day of the year, even if this extra capacity is used only a few hours per year. Simply, the demand charge is needed for the utility to recover its investment in extra generating, transmission, and distribution capacity that is required to meet infrequent peak electrical demands.

It is more profitable for the utility company and thus more economical for the consumer to keep demand uniform. When demand is uniform, the utility more effectively utilizes its generating, transmission, and distribution system. The demand penalty imposed by the utility to keep demand uniform encourages the consumer to level off demand peaks.

Leveling demand is achieved many ways. *Demand limiting* is accomplished by disconnecting loads that are not needed during periods of high demand. *Load shedding* is a method by which nonessential equipment and appliances are deliberately switched off to maintain a uniform load and thus limit demand. *Load shifting* moves nonessential loads to periods of low demand. In small installations, load shedding and shifting may be accomplished manually by scheduling the startup and shutdown of a piece of equipment during certain periods of the day when known peaks occur. In sophisticated commercial and industrial installations, load shedding and shifting are accomplished with an energy management system that monitors demand and sheds and shifts loads in a prioritized manner.

Figure 17.17 is a hypothetical graph of the energy use in a retail store. Peak use occurs as the store opens for operation. If some HVAC and domestic hot water consumption could be shifted to before 9 am, peak use could be dropped to about 5kW. In this case, load shifting can result in an almost 30% decrease in demand.

Energy storage and alternate sources of energy can also be used to reduce demand peaks, a technique that is called *peak shaving*. For example, in an ice storage system, ice is made by a building's air conditioning system at night when demands are low. The stored ice is then melted to chill water for cooling the building when air conditioning demand peaks the afternoon. Ice storage has the added benefit of requiring reduced air conditioning capacity. A cogeneration system that produces electricity from waste process heating or by consuming fossil fuels can be used to limit demand.

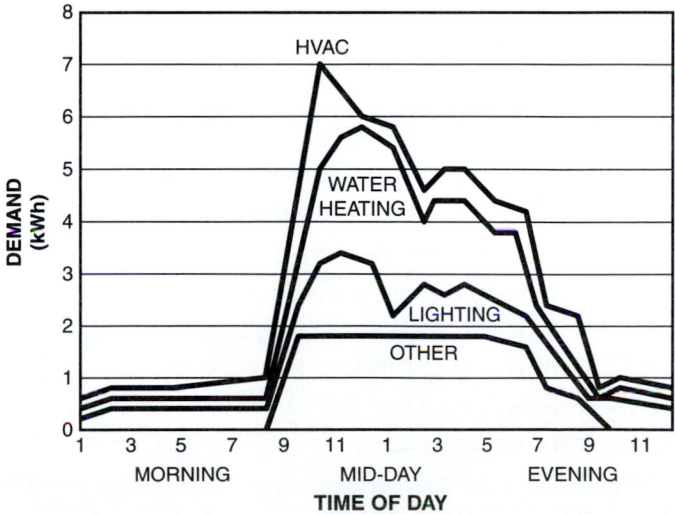

FIGURE 17.17 A graph of the energy use in a retail store. Peak use occurs as the store opens for operation. If some HVAC and domestic hot water consumption could be shifted to before 9 AM, peak usage could be dropped to about 5 kW.

Time-of-Use Rates

The *time-of-use (TOU) rate* rewards the user for reducing power consumption during periods when electrical demand is highest and a lower rate for the remainder of the year. In most climates, this peak period occurs during the midday and early evening hours during the cooling season, because the weather is hottest and power consumption is greatest. For example, one utility charges its peak TOU rate between 1 PM and 7 PM daily, June through September. See Table 17.4.

Additional Charges

Additional surcharges for service and fuel adjustment may also be assessed. The *service* or *billing charge* covers the cost of metering and bill collecting activities such as meter reading and preparing and mailing billing statements. This charge is assessed even if no electricity was consumed. The *fuel adjustment*

TABLE 17.4   AN EXAMPLE OF A TIME-OF-USE (TOU) RATE.

Public Utility Company, Inc.
Residential Time-of-Use (TOU) Rates

Period	Definition	Price per kWh
Summer peak	1 to 7 PM, June 1 to Sept 30	23.081¢ kWh—Single family 21.972¢ kWh—Multifamily
Summer off-peak	7:01 to 12:59 PM, June 1 to Sept 30	7.578¢ kWh—Single family 8.079¢ kWh—Multifamily
All other times	24 hr a day, Oct 1 to May 31	7.097¢ kWh—Single family 7.817¢ kWh—Multifamily

charge reflects periodic changes in the cost of purchasing, delivering, handling, and storing raw fuel (e.g., coal, natural gas) that is used to produce electricity. It is typically applied to each kilowatt consumed.

STUDY QUESTIONS

17-1. What is an insulator?

17-2. What is a conductor?

17-3. What is a semiconductor?

17-4. What are voltage, amperage, and resistance?

17-5. What is the difference between power and energy?

17-6. What are the three components of an electric circuit?

17-7. What are open, closed, and short circuits?

17-8. Why is overcurrent protection used on an electrical circuit?

17-9. Why is a control device used on an electrical circuit?

17-10. What are the differences between series and parallel circuits?

17-11. How do magnetism and electricity relate to each other?

17-12. How do DC and AC differ?

17-13. How do single-phase AC and three-phase AC differ?

17-14. What is a transformer and where is it used?

17-15. What is the power factor?

17-16. What is a power factor correction?

17-17. What is the reason for a utility company charging for demand of electricity?

17-18. What is demand limiting?

17-19. What is load shedding?

17-20. What is the concept of time of use (TOU) rates and why are they needed?

Design Exercises:

17-21. A No. 12 AWG copper conductor has a resistance of 0.162 Ω/100 ft. A No. 10 AWG (thicker than No. 12)

conductor has a resistance of 0.102 Ω/100 ft. Approximate the resistance in each conductor for total lengths of 10, 25, and 50 ft.

17-22. A No. 10 AWG (thicker than a No. 12) copper conductor has a resistance of 0.531 Ω/100 m. Approximate the resistance in lengths of 10, 25, and 50 meters.

17-23. An appliance is designed for use at 120 V and has a current draw of 15 A. Determine the power consumed in watts.

17-24. An appliance is designed for use at 36 V and has a current draw of 1.5 A. Determine the power consumed in watts.

17-25. A circuit consumes 3200 W of real power when the apparent power is 3800 VA.

   a. Determine the power factor.

   b. Determine phase angle φ.

17-26. A circuit consumes 4800 W of real power when the apparent power is 5600 VA.

   a. Determine the power factor.

   b. Determine phase angle φ.

17-27. An AC circuit is powering an electric heater (i.e., pure resistance, PF = 1.0). Assume the voltage is 120 V and the current draw is 10 A. Compute the apparent power and real power.

17-28. An AC circuit is powering an electric heater (i.e., pure resistance, PF = 1.0). Assume the voltage is 240 V and the current draw is 15 A. Compute the apparent power and real power.

17-29. An AC circuit is powering a motor (i.e., inductive load, PF < 1.0). Assume the voltage is 240 V and the current draw is 15 A. Compute the real power, assuming a power factor of 0.8.

17-30. An AC circuit is powering a motor (i.e., inductive load, PF < 1.0). Assume the voltage is 240 V and the current draw is 30 A. Compute the real power, assuming a power factor of 0.75.

17-31. An AC circuit serves inductive and resistive loads. Assume the voltage is 240 V and the current draw is 20 A. Compute the real power, assuming a power factor of 0.9.

17-32. A 25 kVA closet transformer located inside a building is used to step down the voltage for the building. It is connected to a primary 7200 V AC power source. The ratio of the number of primary windings to secondary windings on the transformer is 30 to 1.

　a. Approximate the secondary voltage provided.

　b. Approximate the current available, in amps.

17-33. A 20 kVA closet transformer located inside a building is used to step down the voltage for the building. It is connected to a primary 4800 V AC power source. The ratio of the number of primary windings to the number of secondary windings on the transformer is 20 to 1 and 40 to 1.

　a. Approximate the secondary voltage provided.

　b. Approximate the current available, in amps.

17-34. A 100 W incandescent lamp remains lit for 24 hr a day during a 30-day billing period.

　a. Determine the energy consumed over this period.

　b. Calculate the utility energy charges for this period at a rate of $0.12/kWh.

17-35. A 23 W compact fluorescent lamp (equivalent to a 100 W incandescent lamp) remains lit for 24 hr a day during a 30-day billing period.

　a. Determine the energy consumed over this period.

　b. Calculate the utility energy charges for this period at a rate of $0.12/kWh.

17-36. Ten 100 W incandescent lamps remain lit for 24 hr a day for a one-year period.

　a. Determine the energy consumed over this period.

　b. Calculate the utility energy charges for this period at a rate of $0.12/kWh.

17-37. A 23 W compact fluorescent lamp (equivalent to a 100 W incandescent lamp) remains lit for 12 hr a day for a one-year period.

　a. Determine the energy consumed over this period.

　b. Calculate the utility energy charges for this period at a rate of $0.12/kWh.

17-38. A circulating pump consumes 450 W for 24 hr a day.

　a. Determine the energy consumed for a one-year period.

　b. Calculate the utility energy charges for this period at a rate of $0.12/kWh.

　c. Determine the energy saved for a one-year period, if the pump is switched off 8 hr a day.

　d. Calculate the energy cost savings for this period at a rate of $0.12/kWh, if the pump is switched off 8 hr a day.

17-39. Equipment and appliances shown in the following table (see next page) will be used on a 120/240 V AC household circuit. Complete the table. Base analysis on a 120 V circuit and an energy charge of $0.15/kWh.

Appliance or Piece of Equipment	Power Requirement W or VA	Average Daily Operation Hours	Current Requirement at 120 V Amp (A)	Monthly Bill (30 days)	
				Energy Use kWh	Energy Cost ($/month)
Heating/Cooling Equipment					
Air conditioner, window type	1800	3			
Air conditioner, household	6000	3			
Furnace, electric	30 000	4			
Furnace, gas	1200	4			
Heater, baseboard electric	1500	4			
Heater, portable electric	1500	2			
Water heater, electric	4500	2			
General Appliances					
Computer	150	4			
Hair dryer	1500	0.1			
Hot tub (360 gal)	1500	4			
Radio	8	4			
Stereo	250	2			
Television, large	300	4			
VCR	75	2			
Waterbed heater	300	12			
Kitchen Appliances					
Clothes washer, household size	1200	0.5			
Range/oven unit	12 000	0.1			
Refrigerator, household size	600	3			
Household Appliances					
Clothes dryer, household size	6000	0.5			
Iron	600	0.5			
Sewing machine	200	1			
Vacuum cleaner	600	0.5			
Lighting					
Fluorescent bulb (18 W[a])	20	2			
Incandescent (60 W)	60	2			
Fluorescent (32 W[a])	37	2			
Incandescent (100-watt)	100	2			
Fluorescent (18 W[a])	20	24			
Incandescent (60 W)	60	24			
Fluorescent (32 W[a])	37	24			
Incandescent (100 W)	100	24			
Incandescent (500 W)	500	24			
Metal halide (175 W[a])	205	24			

[a]Does not include power consumed by required ballast.

BUILDING ELECTRICAL
MATERIALS AND EQUIPMENT

18.1 POWER GENERATION AND TRANSMISSION

Nearly all of the electrical energy consumed in the United States and Canada is generated, transmitted, and distributed by large public or privately owned utility companies. Individual residential and commercial consumers purchase electricity from these utility companies, either directly or through a cooperative association that purchases it from the utility. Some large manufacturers serve as small producers of electrical energy by operating a small power generating station or by generating electrical energy as part of their manufacturing processes. In a few instances, a homeowner with a bank of photoelectric solar cells or a wind generator serves as a small power producer, selling excess electrical energy back through utility lines owned by the utility company.

In most cases, electricity is produced at a utility-owned power station. A *power station* is an industrial facility that houses equipment to generate electrical energy. A *generator* is a mechanical device that converts mechanical energy into electrical energy. A generator rotates an *armature,* a shaft with conductor windings wrapped around an iron core, through a stationary magnetic field, to produce current flow. A rotary engine called a *turbine* is connected to the generator and drives the rotation of the armature shaft. In a steam turbine, high-pressure steam moves through the turbine, driving rotation of discs attached to the turbine shaft. Steam is produced by heating water by burning coal, oil, or natural gas, or with heat created by a nuclear reaction. In a hydroelectric plant, falling water strikes the blades or buckets of a turbine, causing the turbine shaft to rotate. The efficiency of conversion of heat to

mechanical energy is about 40% for a fossil-fuel plant and about 30% for a nuclear plant. A large power station produces more than 1000 megawatts (MW) of electrical power. One megawatt equals 1 000 000 watts of power. In the United States, a city with a population of 1 000 000 people needs about 3000 to 4000 MW of power.

A schematic of a power transmission and distribution system is shown in Figure 18.1. Electricity is conveyed through a transmission system of overhead metal cables supported on high towers. Public utility companies and most small power producers are interconnected in a power grid. These companies buy and sell electricity from each other and from small producers. The most economical power stations connected to the grid typically generate power continuously, unless off line for maintenance or repairs. During periods of heavy demand, less-economical power stations are brought on-line. If one generating station is off-line or fails, another is brought on line to take over the load.

At the power station, electricity is usually generated in the range of 2.4 kV to 13.2 kV. For economical transmission over long power lines, a step-up transformer steps up voltage to above 39 kV, usually between 115 kV to 765 kV. Ultrahigh-voltage transmission is now being explored for equipment and lines rated for 1100 kV. Power transmission at high voltages requires less amperage and subsequently there are lower power losses in the transmission lines.

To avoid the danger of high-voltage power transmission through populated areas, power substations located near housing developments step down the transmission voltages. *Power substations* are small facilities in fenced yards that contain transformers, switches, and other electrical equipment that reduce transmission voltages to safer distribution levels. Generally,

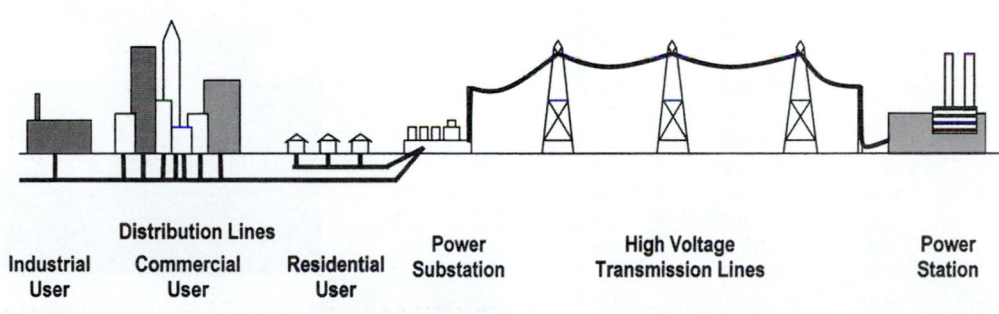

| Industrial User | Distribution Lines
Commercial User | Residential User | Power Substation | High Voltage Transmission Lines | Power Station |

FIGURE 18.1   Power transmission and distribution.

distribution line voltages are generally 4160 V, 4800 V, 6900 V or 13 200 V, although higher voltages can be made available if loads are large enough.

Overhead, underground, and underwater distribution lines carry electrical power from the substation to transformers mounted on a pole or resting on the ground near the building property line. Distribution line voltages are still too high for most typical uses, such as in homes, businesses, and small industrial users, so a distribution transformer further steps down the voltage to safer, more-useable levels for residential, commercial, and industrial uses such as 120/240 V single phase, 120/208 V three phase, or 480 Y/277 V three phase in the United States and Canada. In other countries, voltage may be stepped down to different levels; in Europe, for example, 400 Y/230 V supplies both industrial and residential loads.

Power utilities generally generate, transmit, and distribute three-phase AC power. Single-phase AC power is obtained from one phase of a set of three-phase lines. Utilities use the highest transmission and distribution voltages consistent with safe and economical use. It is very economical to carry power at high voltage because it requires smaller conductors. However, high voltages introduce safety clearance hazards, so high voltages are reserved for use in unpopulated areas or special applications. Table 18.1 presents the relationship between generated and transmitted voltages and the associated current for a 2000 kVA load. It shows that, for transmission of a specific quantity of power, higher voltages reduce current (amperage) and thus smaller power lines are needed.

TABLE 18.1   THE RELATIONSHIP BETWEEN THE GENERATED, TRANSMISSION, AND DISTRIBUTION VOLTAGES AND ASSOCIATED CURRENT (AMPERAGE) FOR A 2000 KVA LOAD AND A POWER FACTOR OF 1.0. HIGHER VOLTAGE MEANS LOWER AMPERAGE (AND SMALLER CONDUCTOR SIZE) FOR A SPECIFIC LOAD. IT IS MORE ECONOMICAL, BUT MORE DANGEROUS, TO TRANSMIT POWER AT HIGH VOLTAGE.

| | Common Voltages | | Current Requirement for 2000 kVA Load* |
Application	Kilovolts (kV)	Volts (V)	Amperes (A)
Generated voltages	13.2	13 200	151.5
	2.4	2400	833.3
Transmission voltages	765	765 000	2.6
	500	500 000	4.0
	230	230 000	8.7
	138	138 000	14.5
	115	115 000	17.4
Distribution voltages	34.5	34 500	58.0
	13.2	13 200	151.5
	6.9	6900	289.9
	4.8	4800	415.7
	4.16	4160	4808.8

* $I = P/E$; for example: 2 000 000 VA/13 200 V = 151.5 A.

18.2   BUILDING ELECTRICAL SERVICE EQUIPMENT

Service Entrance Conductors

Underground or overhead *service entrance conductors* carry power from the transformer through a metering device to the building's service disconnects. Overhead wires extend from a pole-mounted distribution transformer to the building's service entrance and are generically referred to as the *service drop*. Underground (buried) service entrance conductors are typically called the *service lateral*. Overhead service entrance conductors extending from pole-mounted transformers are shown in Photo 18.1.

Service entrance conductors must be insulated except the neutral, which may be bare in overhead installations. Single-phase service drops will have either 2 or 3 wires while three phase service drops will have either 3 or 4 wires. Typically, service entrance conductors are provided by the utility. Each utility has standards on wire size and minimum clearances.

Service Entrance

The *service entrance* includes the components that connect the utility-supplied wiring (the service lateral or service drop) to the service disconnect, excluding the utility's metering equipment. *Service entrance equipment* receives the service entrance conductors. The service equipment includes a method of measuring power (metering equipment), a method of cutting off power (main disconnect or switch gear), and overcurrent protection devices (circuit breakers or fuses) that protect the service entrance conductors. Overcurrent protection devices are discussed later in this chapter.

The *electric meter* is an instrument that is used by the utility company to measure and record electrical energy consumed. In building services rated up to about 400 A, a feed-through

PHOTO 18.1   Overhead service entrance conductors extending from pole-mounted transformers. (Used with permission of ABC)

meter is used. Essentially, this meter is a small electric motor with a speed that is proportional to the power consumed. As current flows through transformer-like coils in the meter, a disc rotates and pointers on dials revolve to record the amount of power used in kilowatt-hours (kWh). Services rated above about 400 A use a current-transformer (C/T) to meter use. C/T devices measure amperage through each ungrounded (hot) conductor. C/T metering devices are housed in an enclosure called a C/T cabinet that is part of the service entrance equipment.

A *service disconnect* is a required part of the service entrance equipment that allows electrical service from the utility company to be switched off so that power is disconnected to the building installation. It is a set of fuses or a circuit breaker that protects the service entrance conductors. Service entrances and equipment configurations are shown in Figures 18.2 through 18.4.

Switchboards

A *switchboard* is a large cabinet or assembly of metal cabinets in which is connected disconnecting switches, overcorrect protection devices (fuses or circuit breakers), other protective devices, and instruments designed to divide large amounts of electrical current into smaller amounts of current used by electrical equipment. It contains one or more devices that can be used to manually and automatically disconnect a circuit from its power source. Switchboards are the highest capacity components in building distribution and protection devices. They are typically floor mounted, rated for current levels of 1200 to 6000 amperes (A), and voltages below 600 volts (V). Access to a switchboard is usually through the front and rear of the switchboard cabinet. An example of a switchboard is shown in Photo 18.2.

Panelboards

A *panelboard* is one or more metal cabinets that serve as a single unit, including buses, automatic overcurrent protection devices (fuses or circuit breakers). It is equipped with or without switches for the control of light, heat, and power circuits. By definition, a panelboard is a cabinet or cutout enclosure placed in or against a wall or partition that is accessible from the front. A panelboard may be referred to in the trade as a *power panel, load center, distribution center,* or *main power panel.* By definition, a *load center* is a panelboard containing a preassembled

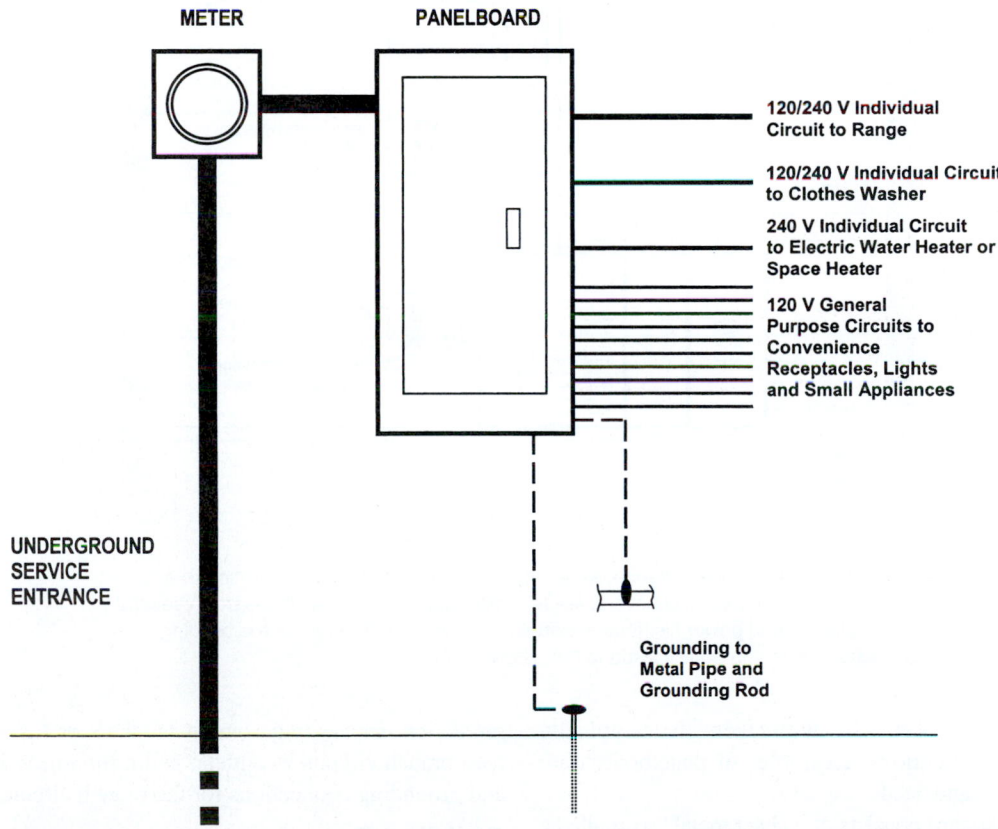

FIGURE 18.2 A typical service entrance/panelboard configuration in a simple residential building electrical system consists of several circuits extending from the panelboard to outlets throughout a building, much like branches extend from a tree's trunk. The main disconnect (contained behind the door of the panelboard) is not shown.

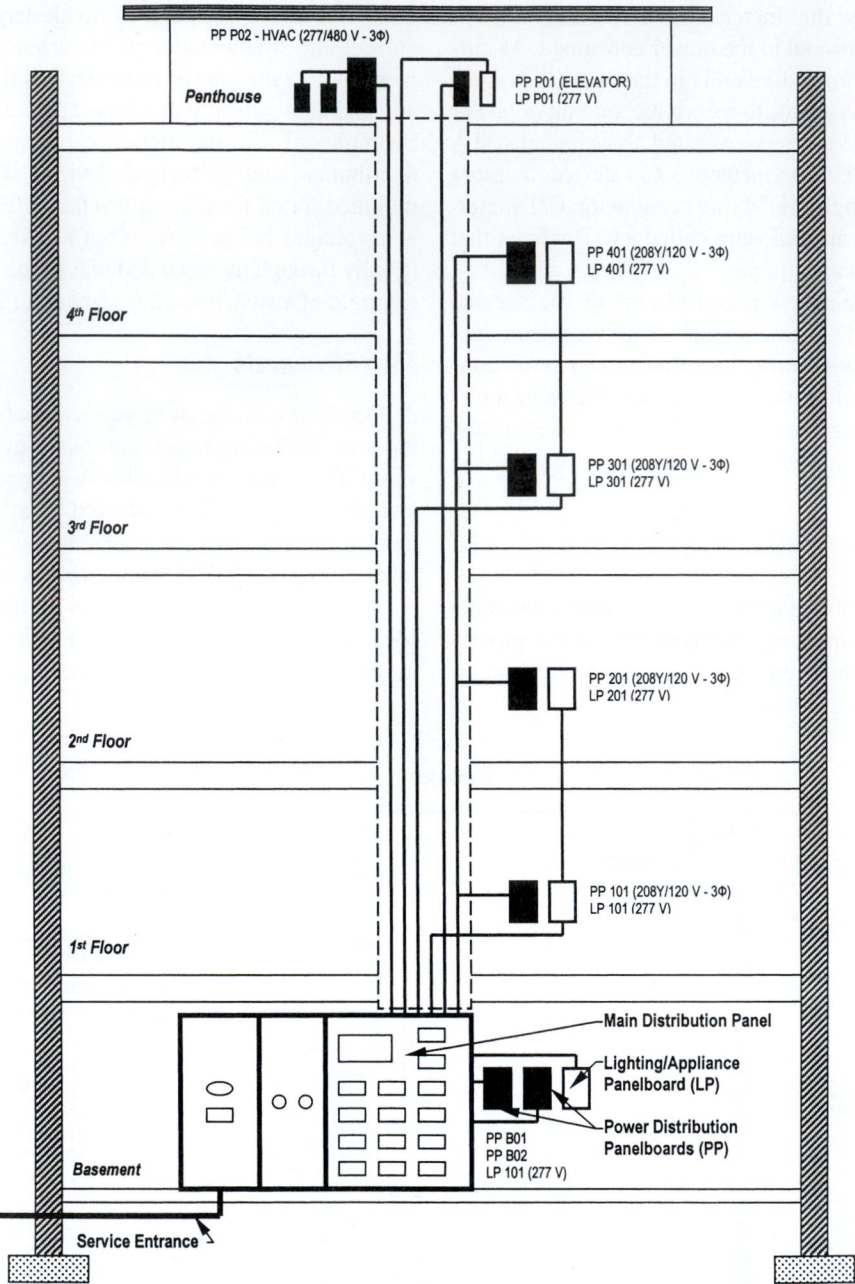

FIGURE 18.3 A service entrance, switchboard, and distribution configuration in a commercial building consists of several feeders extending from the switchboard to several lighting/appliance and power distribution panelboards located throughout the building. Circuits extend from the panelboards to the different outlets.

disconnect and the necessary circuit breakers. It is typically used in residential applications. Examples of panelboards are shown in Photos 18.3 and 18.4.

A typical panelboard consists of a sheet metal box, called a *cabinet,* and a *cover* that encloses and conceals the panelboard interior to limit access to power distribution components. The cabinet and cover enclose and protect the interior section containing *vertical buses,* which are used to distribute power; *overcurrent*

protection devices (e.g., circuit breakers or fuses) to protect and feed branch circuits to outlets; and *termination buses* for neutral and grounding connections for the branch circuits.

Building Transformers

Transformers are used in transmitting and distributing power from the power plant to a substation. The operation of a large commercial installation depends on power distribution that, in

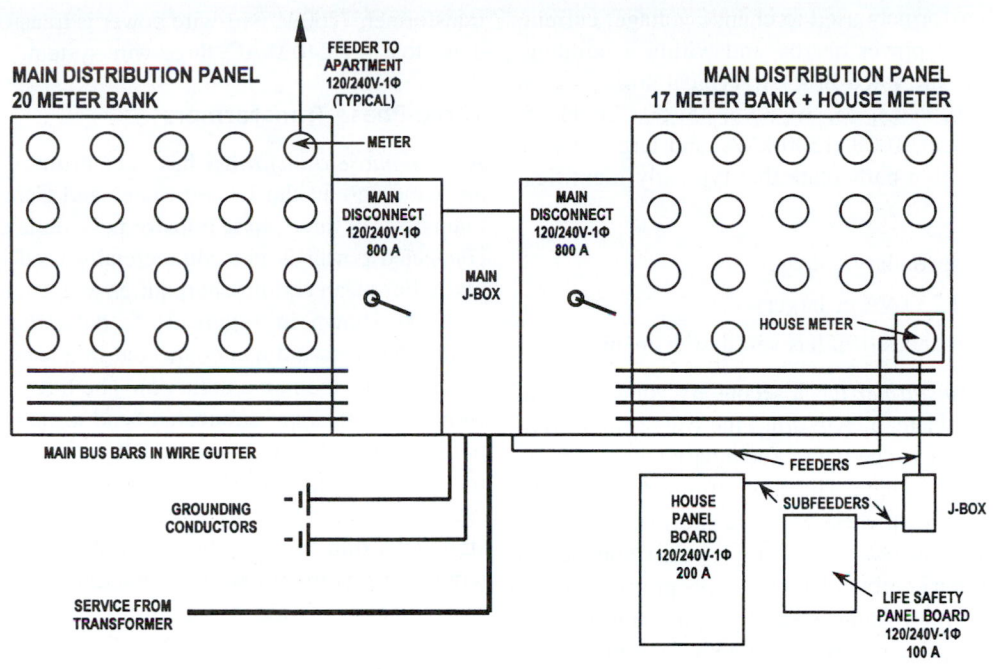

SERVICE ENTRANCE AND MAIN DISTRIBUTION EQUIPMENT

FIGURE 18.4 Service entrance and distribution panels serving a multifamily dwelling consists of several feeders extending from the main distribution panels to panelboards located at each apartment unit. A house panel serves outlets in common spaces (e.g., halls, laundry room, mechanical room, and so on). A life safety panelboard serves emergency lighting and smoke detectors. Circuits extend from the panelboards to the different outlets.

PHOTO 18.2 A switchboard divides large amounts of current into smaller amounts used by individual pieces of electrical equipment (e.g., air-handling units, pump motors, and so on) or panelboards. (Used with permission of ABC)

PHOTO 18.3 Power distribution and lighting panelboards serving one floor of a commercial office building. (Used with permission of ABC)

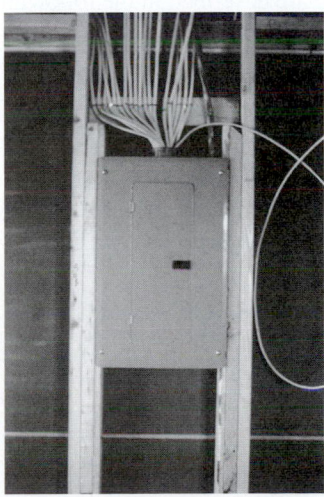

PHOTO 18.4 A panelboard serving a single-family residence. The service entrance conductors (dark, thick sheathed conductors) and branch circuit wiring (light colored) extending from the panel are evident. (Used with permission of ABC)

turn, depends on transformers used to change voltage, current, and phase of electrical power nearby and within a building. Building transformers are rated in kVA. Typical sizes used in buildings include 3, 6, 9, 15, 25, 30, 37.5, 45, 50, 75, 112.5, 150, 225, 300, 500, 750, 1000, 1500 kVA, and larger. Every transformer comes with a nameplate that typically identifies:

- Rated kVA

- Primary and secondary voltage

- Impedance (if 25 kVA or larger)

- Required clearances (if it has ventilating openings)

A *step-down transformer* has a secondary voltage that is less than its primary voltage. It *steps down* the voltage applied to it. A *step-up transformer* is one with a secondary voltage that is greater than its primary voltage. This kind of transformer *steps up* the voltage applied to it. Step-down transformers are typically used in buildings to reduce building system voltages to useable levels. Common commercially available primary to secondary voltage transformer voltages are shown in Table 18.2.

Excessive heat prematurely deteriorates a transformer. Operation at only 20°F (11°C) above the transformer rating will cut transformer life by half. Heat is caused by internal losses from loading, high ambient air temperature, and in exterior locations, solar radiation. Small transformers are typically air cooled by ventilation. Larger transformers are liquid cooled. *A ventilated dry-type transformer* has its core and coils in a gaseous or dry compound. Dry-type distribution transformers are usually found inside larger commercial/industrial facilities and are generally owned by the facility. A *liquid-immersion transformer* has its core and coils immersed in an insulating liquid. In building installations, a transformer can be single phase or three phase. These are described in the following sections.

Single-Phase Transformers

A *single-phase transformer* has a single primary winding and a single secondary winding. The 7200/240/120 V AC, single-phase, three-wire transformer is used in most residential and small commercial applications where 120 V and 240 V are required. In this

transformer, 7200 V, two-wire power is transformed and stepped down to a 120/240 V AC, three-wire system.

Three-Phase Transformers

A *three-phase transformer* has three primary and three secondary windings. In the United States and Canada, there are two main types of three phase transformers: delta and wye. The delta connected power is not commercially used in Europe. Three-phase European equipment requires wye-connected power.

As shown in Figure 18.5, the *delta-connected transformer* has its windings connected in a series in the form of a triangle, thus the name delta (Δ). The three independent transformer windings are connected head to toe. There is no single point common to all phases. A delta-connected transformer has only a single voltage level available: the phase-to-phase voltages. Other voltages can be obtained only by using step-up or step-down transformers. For example, on a 120/240 V Δ system, the midpoint of one phase winding is grounded to provide 120 V between Phase A and ground and Phase C and ground. There are 240 V between the windings of each phase—that is, between Phases A and B, Phases A and C, or Phases B and C. Between Phase B and ground there is 208 V available.

A *wye-connected transformer (Y)* has three independent transformer windings connected at a common point, called a *neutral* or *star point,* as shown in Figure 18.6. The physical arrangement of windings resembles a Y. The angular displacement between each winding of the Y is 120°. The center point is

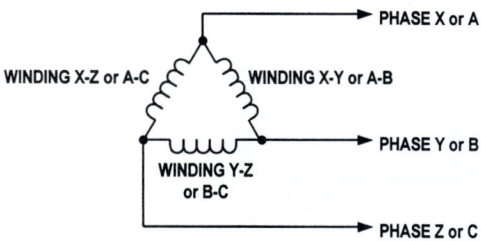

FIGURE 18.5 A delta-connected transformer has its windings connected in series in the form of a triangle. The physical arrangement of windings resembles the Greek letter Delta (Δ).

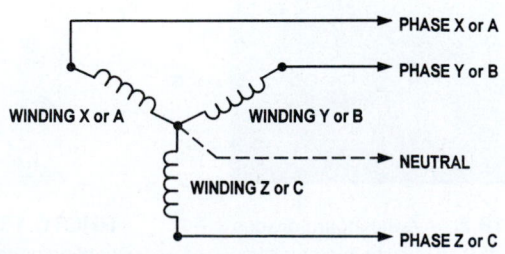

FIGURE 18.6 A wye-connected transformer (Y) has three independent transformer windings that are connected at a common point, called a *neutral* or *star point.* The physical arrangement of windings resembles a Y.

TABLE 18.2   COMMON COMMERCIALLY AVAILABLE PRIMARY TO SECONDARY VOLTAGE TRANSFORMERS.

Primary Voltages	Secondary Voltages Available
2400	
2400/4260Y	
4160	
4800	120/240
4160/7200Y	240/480
6900/11950Y	277
7200/12420Y	480
7620/13200Y	
8320/14400Y	

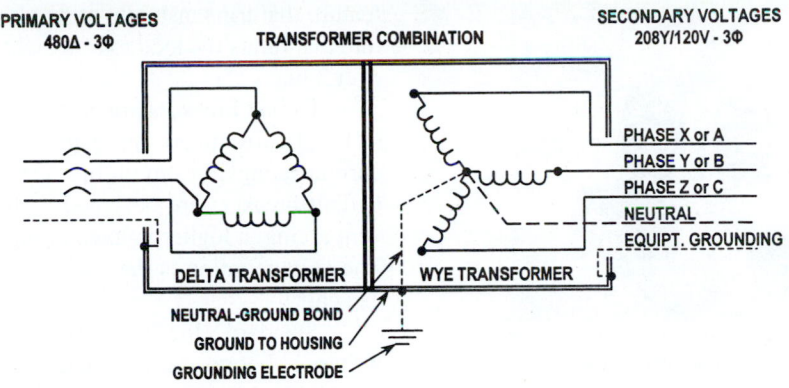

FIGURE 18.7 A delta-wye (Δ-Y) transformer combination. Additional transformer combinations, such as wye-wye (Y-Y), delta-delta (Δ-Δ), and wye-delta (Y-Δ) are available.

the common return point for the neutral conductor. Connecting one terminal from three equal voltage transformer windings together to make a common terminal forms the wye-connected system. The Y in 480 Y/277 V relates to a wye (Y) winding configuration of the secondary windings.

A delta-wye (Δ-Y) transformer is an electrical device that converts three-phase electric power without a neutral wire into three-phase power with a neutral wire. A Δ-Y combination is shown in Figure 18.7. Additional transformer combinations, such as wye-wye (Y-Y), delta-delta (Δ-Δ), and wye-delta (Y-Δ) are available.

Historically, standardized system voltages related to the use of delta transformers were based on multiples of 120 V (e.g., 120 V, 240 V, 480 V, 7200 V). Voltages that have the 1.732 as a multiplier are wye connections (e.g., 208 V, 4160 V, 12.47 kV). New distribution transformers are very efficient, with losses of less than 0.25% in large units. Most large facility distribution transformers convert at least 95% of input power into useable output power. Smaller closet transformers have efficiencies of 98% or above. Examples of building transformers are shown in Photos 18.5 and 18.6.

PHOTO 18.6 A 75 kW, 480 V delta to 208 Y/120 V closet transformer during construction. It provides three-phase and single-phase power to a portion of a college building. (Used with permission of ABC)

PHOTO 18.5 A pad-mounted building transformer serving a commercial office building. (Used with permission of ABC)

18.3 OVERCURRENT PROTECTION: FUSES AND CIRCUIT BREAKERS

An *overcurrent protection (OCP) device* safeguards the building service or an individual circuit from excessive current flows. It protects the circuit components from severe overheating when current flowing through the circuit reaches an amperage that will cause an excessive or dangerous temperature rise in conductors. Fuses and circuit breakers serve as automatic overcurrent protection devices. They are designed to open a circuit if the amount of current, in amps, that flows through the circuit exceeds the OCP device rating. Most OCP devices respond to both short-circuit or ground-fault current values in

PHOTO 18.7 Various sizes of circuit breakers. (Used with permission of ABC)

addition to overload conditions. Types of overcurrent protection devices are introduced in the following sections.

Circuit Breakers

A *circuit breaker* is an overcurrent protection device that serves two purposes: It acts as a switch that can be opened and closed manually, and most importantly, it automatically "trips off," which opens the circuit when current flowing through it exceeds the circuit rating. This action instantaneously interrupts current flow. Once it trips, it can be reset (closed like a switch) and will continue to allow electricity flow as long as the current flowing through it does not exceed the circuit rating. Various types of circuit breakers are shown in Photo 18.7.

The most popular circuit breaker is the *thermal-magnetic* type. It consists of a bimetallic strip that bends when it is heated by power loss created when current flows through it. When current flow is excessive, the circuit breaker heats up, bends, and trips a release that opens contacts and interrupts current flow. A magnetic device also can trip the release instantly if a short circuit develops. When tested in open air, the thermal-magnetic circuit breaker can carry a load of about 10% above its rating. It can carry up to 50% above its rating for a minute and up to three times its rating for a few seconds. The ability to handle a load substantially above its rating for a short interval avoids the annoyance of tripping a breaker with motor startup. It is still safe because the heat from power loss takes some time to create damage. Other characteristics include the current-limiting feature that trips instantly at its rating and the 100% rated feature that limits the load to 80%, when the load is continuous for over 3 hr.

Circuit breakers are classified by a voltage rating in volts (V) and current-carrying capacity in amperage (A). Standard current ratings for circuit breakers are provided in Table 18.3. Circuit breakers are designed with specific clearances to prevent arcing; a higher voltage rating requires a larger clearance. The voltage rating is the maximum voltage the circuit breaker can carry.

Standard circuit breaker ratings recognized by Underwriters Laboratories, Inc. (UL), for alternating current include 120 V, 120/240 V, 277 V, 277/480 V, 480 V, and 600 V. The UL standard for direct current includes 125 V, 250 V, and 600 V. The voltage of the electrical circuit or system being protected by the circuit breaker must not exceed the circuit breaker voltage rating. Circuit breakers protect the ungrounded (hot) conductors in a circuit. They are identified as *single pole (SP)* if protecting a single ungrounded conductor; *two pole (2P)* when protecting two ungrounded conductors such as on a 240 V circuit; and *three pole (3P)* when protecting three ungrounded conductors on a three-phase circuit.

Fuses

A *fuse* is an overcurrent protection device that consists of a strip of metal with a low melting temperature. Under normal operation, electricity flows through the metal strip. However, when its current rating is exceeded, the metal strip heats up and melts and the circuit is opened, thereby interrupting current flow. In this case, it is said that the fuse has "blown." A new fuse of the same rating must replace a blown fuse, which is discarded. Most fuses are rated for 250 V or 600 V although 125 V and 300 V are available. Standard current ratings for fuses are provided in Table 18.3

There are three basic types of fuses used in building electrical systems: plug, cartridge, and time delay fuses. Types of fuses are shown in Photos 18.8 and 18.9. These are discussed as follows:

> *Plug fuses* screw into sockets much like a lamp. They are rated from 5 A to 30 A. An *Edison-base plug fuse* has threads similar to an ordinary incandescent lamp base.

TABLE 18.3   STANDARD UL CURRENT RATINGS CIRCUIT BREAKERS AND FUSES.

Overcurrent Protection Device	Rating (Amperes)
Circuit breakers	15 A, 20 A, 30 A, 40 A, 50 A, 60 A, 70 A, 80 A, 90 A, 100 A, 110 A, 125 A, 150 A, 200 A, 225 A, 250 A, 300 A, 350 A, 400 A, 450 A, 500 A, 600 A, 700 A, 800 A, 1000 A, 1200 A, 1600 A, 2000 A, 2500 A, 3000 A, 4000 A, 5000 A, and 6000 A
Fuses	1 A, 3 A, 6 A, 10 A, 15 A, 20 A, 30 A, 40 A, 50 A, 60 A, 70 A, 80 A, 90 A, 100 A, 110 A, 125 A, 150 A, 200 A, 225 A, 300 A, 350 A, 400 A, 450 A, 500 A, 600 A, 700 A, 800 A, 1000 A, 1200 A, 1600 A, 2000 A, 2500 A, 3000 A, 4000 A, 5000 A, and 6000

PHOTO 18.8 Edison screw-type fuses (round shapes) in a circa 1950 lighting subpanelboard with switches (light-colored rectangular shapes). (Used with permission of ABC)

PHOTO 18.9 Three cartridge fuses installed to protect a three-phase circuit. (Used with permission of ABC)

Edison plug fuses are interchangeable regardless of rating, allowing a 30 A fuse to improperly protect a 20 A circuit. This interchange ability makes Edison fuses unsafe, so they are no longer allowed in new and retrofit installations. *Type S plug fuses* have bases of different sizes that match the fuse rating to the circuit rating.

Cartridge fuses are cylindrical in shape and available in two types. The ferrule-contact type have round copper contacts at their end and are rated up to 60 A. The knife-blade type has flat blades sticking out at each end. Type S fuses and cartridge fuses are allowed in new and retrofit installations.

Time delay fuses can handle an overload for fraction of a second without blowing. They are desirable on circuits serving electric motors such as air conditioners and machinery because motors draw much more current at startup than during normal operation.

Circuit breakers can be more convenient than fuses because they are easily reset rather than having to be replaced. However, circuit breakers deteriorate with age and tend to lose some degree of sensitivity each time they trip. They must be replaced periodically to ensure maximum overcurrent protection. Overcurrent protection is typically installed in

TABLE 18.4   COMMON PANELBOARD AND LOAD CENTER RATINGS AND CAPACITIES.

Single-Phase, Three-Wire Panelboards and Load Centers

Frame Size	Disconnect Rating	Capacity
Amperes	Amperes	Maximum Number of Circuit Breakers
50	50	8, 10, 12
100	100	6, 8, 10, 12, 14, 16, 18, 20
125	125	6, 8, 10, 12, 14, 16, 18, 20, 24
150	150	20, 24, 28, 30
200	200	20, 24, 28, 30, 36, 40

Three-Phase, Four-Wire Panelboards

Frame Size	Disconnect Rating	Capacity
Amperes	Amperes	Maximum Number of Poles
100	100	16, 18, 20, 22, 24, 26, 28, 30
125	125	20, 22, 24
225	225	22, 24, 26, 28, 30, 32, 34, 36, 38, 40, 42
400	400	30, 42

an enclosure such as a switchboard, panelboard, motor control center, or an individual enclosure. Common panelboard and load center ratings and OCP device capacities are provided in Table 18.4.

Adequate overcurrent protection is the single-most important safety feature in any electrical circuit as it safeguards conductors and equipment. Although an OCP device safeguards against overheating and fire, it does not protect an occupant from electrical shock.

OCP Device Ratings

OPDs have two current ratings: *overcurrent* and *amperes interrupting current*.

Overcurrent Rating

The *overcurrent rating* of an OCP device is the highest amperage it can carry continuously without exceeding a specific temperature limit (e.g., without overheating). If the current (amperage) flowing through the protection device exceeds the device setting for a significant period, the OCP device will open. The overcurrent rating of an OCP is listed in amperes, such as 15 A, 20 A, or 30 A. The amperage carried by the electrical circuit or system protected by an OCP device must not exceed the maximum current rating of the circuit breaker.

Interrupting Rating

OCP devices must have an interrupting rating sufficient for the maximum possible fault-current (short-circuit). If the OCP is not rated to interrupt at the available fault-current, it could explode while attempting to clear the fault and/or the downstream equipment could suffer serious damage, causing possible hazards to occupants and property. The *amperes interruption current (AIC)* rating for circuit breakers is 5000 A and 10 000 A for fuses. Circuit breakers and fuses typically have an AIC rating of 10 000 A.

18.4 UTILIZATION EQUIPMENT AND DEVICES

Utilization equipment is a broad category of electrical or electronic machine or instrument designed to perform a specific mechanical, chemical, heating, or lighting function through the use of electrical energy. Electric motors, air conditioning, refrigeration and heating units, signs, industrial machinery, cranes, hoists, elevators, and escalators fit in the category of utilization equipment.

An *appliance* is an end-use piece of utilization equipment designed to perform a specific function such as cooking, cleaning, cooling, or heating. Electric ranges, refrigerators, clothes washers and dryers, freezers, blenders, toasters, and hair dryers are appliances. *Fixed appliances* are permanently attached installations such as a built-in electric cook top or oven. *Stationary appliances* are situated and used at a specific location but can be moved to another outlet such as a refrigerator, clothes washer, or clothes dryer. *Portable appliances* are appliances that can be easily carried or moved such as a hair dryer or toaster.

An electrical *device* is a component in an electrical system that is designed to carry but not use electricity. This includes components such as switches, receptacles, and relays.

Outlets, Receptacles, and Plugs

In an electrical system, an *outlet* is the location in a branch circuit where electricity is used. For example, a *lighting outlet* is the location in a branch circuit where conductors provide power to a light fixture. A *receptacle* is a female connecting device with slotted contacts. It is installed at an outlet or on equipment, where it is intended to easily establish an electrical connection with an inserted plug. A *plug* is a male connecting device that has two or more prongs that are inserted into a receptacle to connect to an electrical circuit. A plug is typically connected to a flexible cord that is attached to a portable appliance, light, or equipment. Receptacles and plugs offer a simple way to attach or detach an appliance or piece of equipment to/from an electrical outlet.

Tables 18.5, 18.6, and 18.7 provide information on different types and configurations of receptacles and plugs. According

TABLE 18.5   CONFIGURATION FOR NONLOCKING RECEPTACLES AND PLUGS.

System	Voltage Rating	Number/Type of Conducting Poles			Receptacle (R) and Plug (P) Ratings and NEMA Designations				
		Ungrounded	Grounded (Neutral)	Grounding	15 A	20 A	30 A	50 A	60 A
Two pole, two wire	125	1	1	—	1-15R				
					1-15P				
	250	2	—	—		2-20R	2-30R		
						2-20P	2-30P		
Two pole, three wire, grounding	125	1	1	1	5-15R	5-20R	5-30R	5-50R	
					5-15P	5-20P	5-30P	5-50P	
	250	2	—	1	6-15R	6-20R	6-30R	6-50R	
					6-15P	6-20P	6-30P	6-50P	
Three pole, three wire	277	1	1	1	7-15R	7-20R	7-30R	7-50R	
					7-15P	7-20P	7-30P	7-50P	
	125/250, 1Φ	2	1	—		10-20R	10-30R	10-50R	
						10-20P	10-30P	10-50P	
	250Δ, 3Φ	3	—	—	11-15R	11-20R	11-30R	11-50R	
					11-15P	11-20P	11-30P	11-50P	
Three pole, three wire, grounding	125/250, 1Φ	2	1	1	14-15R	14-20R	14-30R	14-50R	14-60R
					14-15P	14-20P	14-30P	14-50P	14-60P
	250Δ, 3Φ	3	—	1	15-15R	15-20R	15-30R	15-50R	15-60R
					15-15P	15-20P	15-30P	15-50P	15-60P
Four pole, four wire	208Y/120, 3Φ	3	1	0	18-15R	18-20R	18-30R	18-50R	18-60R
					18-15P	18-20P	18-30P	18-50P	18-60P

TABLE 18.6 RATING, APPLICATION, AND CONFIGURATIONS FOR COMMON TYPES OF RECEPTACLES AND PLUGS USED PRINCIPALLY IN RESIDENTIAL AND LIGHT COMMERCIAL APPLICATIONS.

NEMA Designation		Rating	Application	Configuration	
Receptacle	Plug			Receptacle	Plug
1-15R	1-15P	15 A, 125 V two pole, three wire	For replacement only. Former standard, wall-mounted receptacle for residential, commercial, and industrial applications		
5-15R	5-15P	15 A, 125 V Grounding two pole, three wire	Standard, wall-mounted receptacle for residential, commercial, and industrial applications		
5-20R	5-20P	20 A, 125 V Grounding two pole, three wire	For heavy-duty appliances such as room air conditioners and portable shop tools in residential, commercial, and industrial applications		
10-30R	10-30P	30 A, 125/250 V three pole, three wire	Typically for replacement only or nongrounding applications. Formerly for clothes dryers and other heavy-duty equipment and appliances		
14-30R	14-30P	30 A, 125/250 V Grounding three pole, four wire	Clothes dryers in residences and heavy-duty equipment and appliances in commercial/industrial applications		
10-50R	10-50P	50 A, 125/250 V three pole, three wire	Typically for replacement only or nongrounding applications. Formerly for ranges in residences and other heavy-duty equipment and appliances		
14-50R	14-50P	50 A, 125/250 V Grounding three pole, four wire	Ranges in residences and heavy-duty equipment and appliances in commercial/industrial applications		

to National Electrical Manufacturing Association (NEMA) standards, the different types of receptacles and plugs are identified by a specific designation. A standard, wall-mounted, 125 V/15 A convenience receptacle outlet is shown in Photo 18.10. The NEMA designation ties the number and configuration of the slots in the receptacle and the matching slots in a plug to the voltage and amperage permitted. For example, the 5-15R and 5-15P designation is associated with the matching receptacle (R) and plug (P) that are rated at 15 amperes and 125 V and have equipment-grounding capability. These devices are the matching receptacles and plugs commonly associated with residential and commercial wall-mounted outlets. Different receptacles and plug combinations are generally tied to a specific application. For example, the 5-15R receptacle discussed earlier serves as a common outlet for portable lights and small appliances, while the 14-50R and 14-50R receptacle/plug configuration is typically used to connect a kitchen range or other heavy-duty electrical equipment to an outlet.

The screw terminals on a receptacle are color coded: brass-colored screws are for connection to the ungrounded conductor, silver-colored screws are for the grounded or neutral conductor, and green-colored screws are for the equipment grounding conductor. Receptacles marked "CO/ALR" can be connected to copper, aluminum, or copper-clad aluminum conductors. Those marked "CU/AL" were formerly allowed for use with copper or aluminum conductors, but can only be connected to copper conductors. They are no longer deemed acceptable for use with aluminum conductors. (See Wire and Cable sections later in this chapter.)

Switches

A simple *switch* is a device placed between two or more electrical conductors in a circuit to safely and intentionally open or close the circuit or to redirect the path of current in a circuit. Contacts in a switch open (switches off) a circuit, close

TABLE 18.7   RATING, APPLICATION, AND CONFIGURATIONS FOR COMMON TYPES OF RECEPTACLES AND PLUGS USED IN SPECIAL COMMERCIAL AND INDUSTRIAL APPLICATIONS.

NEMA Designation		Rating	Application	Configuration	
Receptacle	Plug			Receptacle	Plug
5-30R	5-30P	30 A, 250 V Grounding two pole, three wire	For office copy machines, commercial air conditioners, and other heavy equipment in commercial and industrial applications		
5-50R	5-50P	50 A, 125 V Grounding two pole, three wire	For production copy machines, commercial air conditioners, and other heavy equipment in commercial and industrial applications		
6-15R	6-15P	15 A, 250 V Grounding two pole, three wire	For heavy-duty appliances such as room air conditioners and portable shop tools in commercial and industrial applications		
6-20R	6-20P	20 A, 250 V Grounding two pole, three wire	For heavy-duty appliances such as room air conditioners and ovens in commercial and industrial applications		
6-30R	6-30P	30 A, 250 V Grounding two pole, three wire	For heavy-duty appliances such as large air conditioners and ovens in commercial and industrial applications		
6-50R	6-50P	50 A, 250 V Grounding two pole, three wire	For production copy machines, commercial air conditioners, and other heavy equipment in commercial and industrial applications		
7-15R	7-15P	15 A, 277 V Grounding two pole, three wire	For heavy commercial and industrial lighting		
14-60R	14-60P	60 A, 125/250 V Grounding three pole, four wire	For heavy-duty commercial and industrial applications and house trailers		
18-20R	18-20P	20 A, 250 V–3Φ four pole, four wire, three phase	For three-phase motors in commercial and industrial air conditioners and equipment		
18-60R	18-60P	20 A, 250 V–3Φ four pole, four wire, three phase	For heavy-duty, three-phase motors in commercial and industrial air conditioners and equipment		

(switches on) a circuit, or divert current from one conductor path to another. Switches are rated by purpose, voltage, and amperage, and are classified for AC or DC.

Safety switches are used in building electrical systems. They are designed to reduce the possibility of contact with bare electrical conductors and have current interrupting capability. Types of switches are shown in Photos 18.11 through 18.16.

There are two categories of safety switches: heavy duty and general duty.

Heavy-duty safety switches are designed for heavy industry, commercial, and institutional applications where safety, performance, and continuity of service are required. These are enclosed and may be fused or unfused.

PHOTO 18.10 A standard, wall-mounted, 125V/15A receptacle outlet. (Used with permission of ABC)

PHOTO 18.11 General duty wall-mounted switch. (Used with permission of ABC)

PHOTO 18.12 Heavy-duty switch. (Used with permission of ABC)

PHOTO 18.13 Weatherproof switch. (Used with permission of ABC)

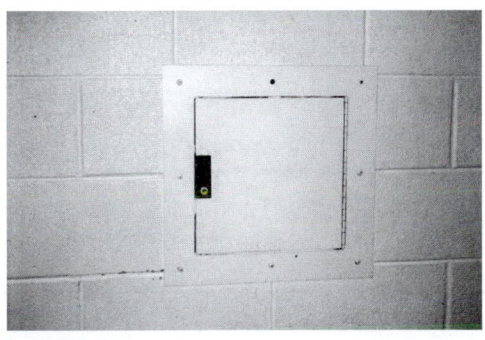

PHOTO 18.14 Locked lighting switch panel. (Used with permission of ABC)

PHOTO 18.15 Keyed switch. (Used with permission of ABC)

PHOTO 18.16 Emergency safety cut-off switch. Used to shut down equipment in an emergency (e.g., personal accident, fire, and so on). (Used with permission of ABC)

They are used extensively as motor circuit switches, service entrance switches, and feeder disconnects, as well as for industrial furnaces, capacitors, transformers, and welders. Ratings up to 1200 A, 600 V are available.

General duty safety switches are intended for industrial, general commercial and residential loads where economy is important and requirements are less stringent. They are used on lighting, heating, appliance, and intermittent motor loads. Ratings up to 600 V are available.

In switching terminology, the term *pole* refers to the number of conductors the switch is opening and closing; the term *throw* refers to the number of operations a switch can perform. A *single-pole, single-throw (SPST) switch* is a simple on/off switch. It opens or closes a single ungrounded conductor in a circuit. It is the most commonly used switch found in buildings, where it is typically used to control a lighting installation from a single location.

A *single-pole, double-throw (SPDT) switch* diverts current from one conductor path to another. A special type of SPDT switch is known as a *three-way switch (S_3)*, which allows the control of an installation from two locations (for instance, turning a light on or off from either end of a flight of stairs). A *double-pole, single-throw (DPST) switch* opens or closes two conductors in a circuit. It is equivalent to two SPST switches controlled by a single mechanism. It can be used to switch off the ungrounded and grounded conductors in a single lighting circuit that is serving a paint spray booth containing explosive vapors. *Double-pole, double-throw (DPDT)* and *three-pole, single-throw switches* are also available.

Switching Configurations

Switches provide control from one or more points in a circuit. A single-pole, single-throw switch (S) is used to provide control from one point by opening or closing the ungrounded conductor in the circuit. See Figures 18.8 through 18.10. Three-way (S_3) and four-way (S_4) switches are used when multiple control points are needed. A schematic of circuiting configurations and required switches are shown in Figures 18.8 through 18.14. Common switching configurations are described in the following:

Control from one point:	One single-pole, single-throw switch (S) is required. See Figures 18.8 through 18.10.
Control from two points:	Two three-way switches (S_3) are required. See Figures 18.11 and 18.12.
Control from three points:	Two three-way switches (S_3) and a four-way switch (S_4) are required. The four-way switch must be wired between the three-way switches. See Figure 18.13.
Control from four or more points:	Two three-way switches (S_3) and two or more four-way switches (S_4) are required. The four-way switches must be wired between the three-way switches. See Figure 18.14.

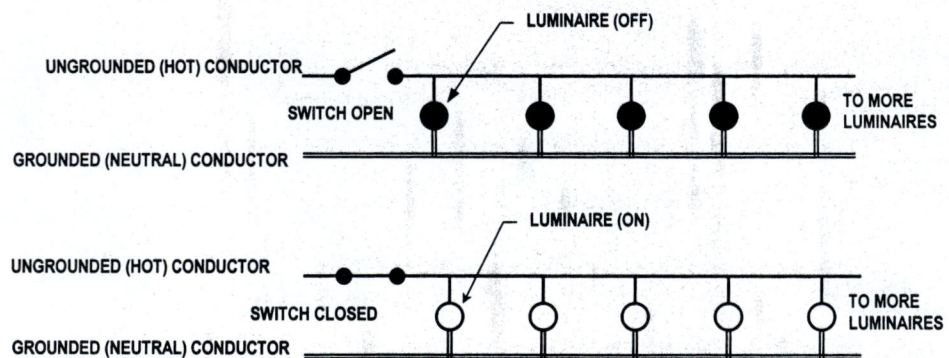

FIGURE 18.8 A single-pole, single throw switch (S) is used to provide control from one point by opening or closing the ungrounded (hot) conductor in the circuit. In the first figure, the switch is open and the lights in the circuit are off. In the second figure, the switch is closed, the circuit is complete, and the lights in the circuit are on.

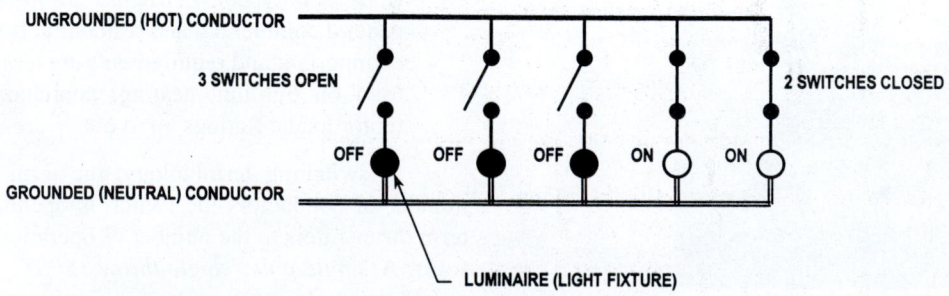

FIGURE 18.9 Several single-pole, single-throw switches can be used to control individual lights in a circuit.

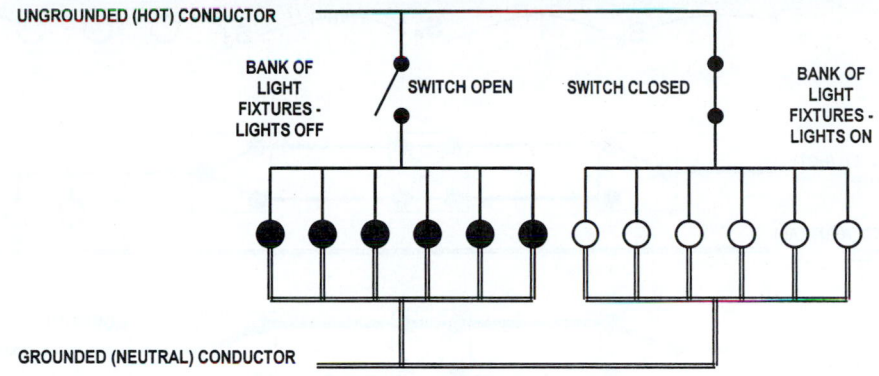

FIGURE 18.10 Single-pole, single-throw switches can be used to control a bank (group) of lights in a circuit.

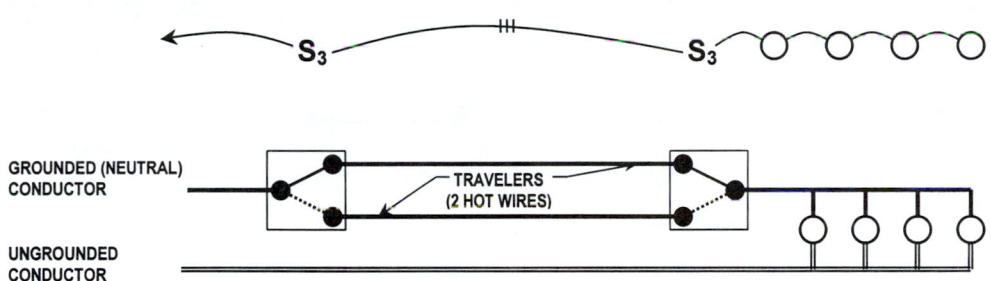

FIGURE 18.11 Shown are a one-line schematic drawing of electrical symbols and the related switching configuration for switching at two locations. The dashed line at each switch shows the new configuration of the circuit when the switch is flipped. Two three-way (S_3) switches are used to control from two points. Note that two ungrounded conductors called *travelers* are used between the switches. As configured, the top traveler carries current. The bottom traveler will carry current only when the switches are tripped. The arrow on the wiring (to the left of the first three-way switch) indicates that the circuit originates at a panelboard.

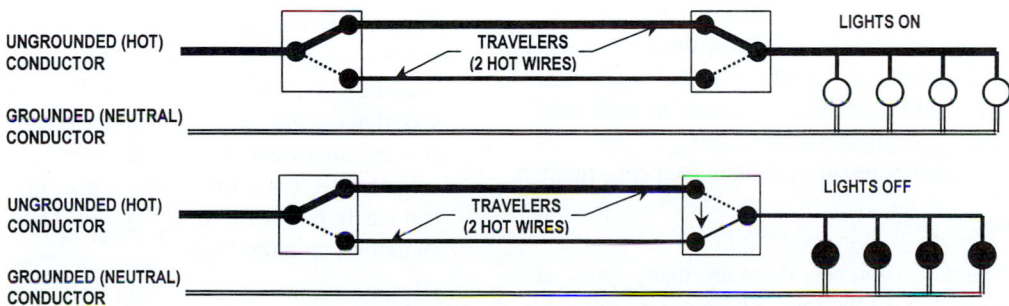

FIGURE 18.12 Current flowing through the circuit with two three-way switches can be interrupted by flipping a single switch. In the top figure, current flows through both switches, the circuit is complete, and the lights are on. In the bottom figure, current flows through the first switch but is interrupted at the second (right) switch, the circuit is open, and the lights are switched off.

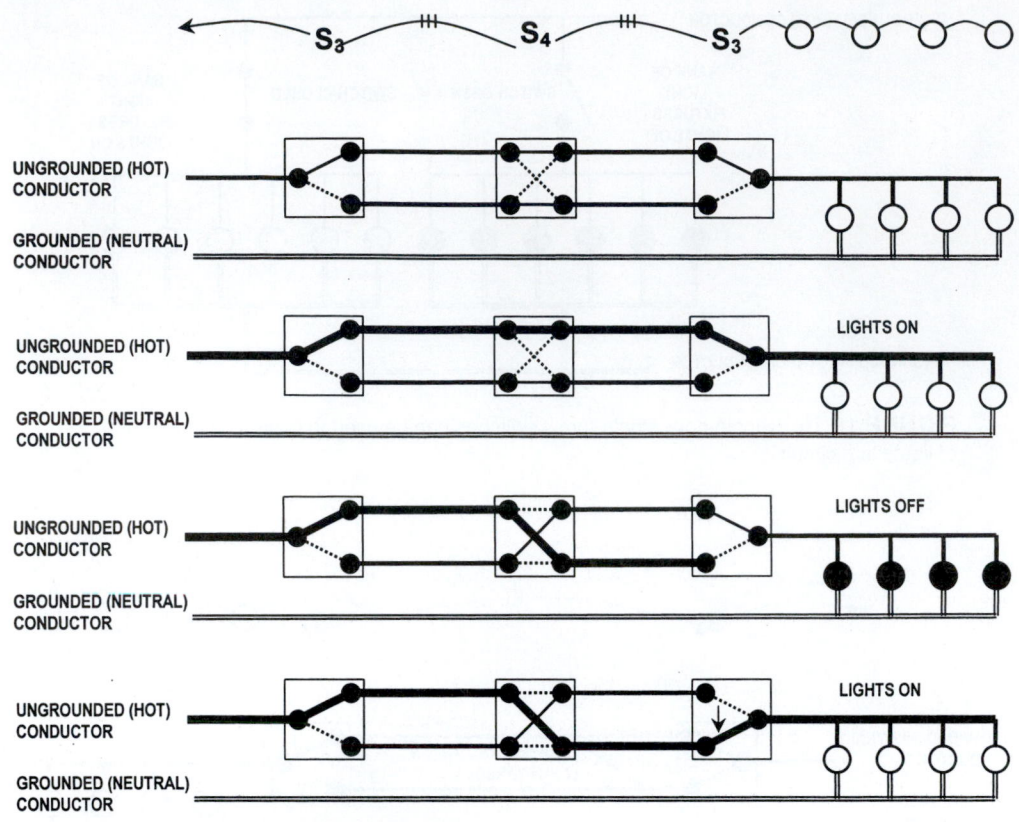

FIGURE 18.13 Shown is a one-line schematic drawing of electrical symbols and the related switching configuration for switching at three locations. The dashed line at each switch shows the new configuration of the circuit when the switch is flipped. Two three-way (S₃) switches and a four-way switch (S₄) are used to control from three points. Not that two ungrounded conductors called travelers are used between the switches. In this switching configuration, current flows through the switches, the circuit is complete, and the lights are on. When any switch is flipped, current flow is interrupted and the lights go off.

Single-pole, single-throw (S), three-way (S₃) and four-way (S₄) switches generally control operation of lighting installations. Dimmer switches (sometimes identified as S_D) can used to adjust brightness of a lighting installation by adjusting the current flowing through the circuit. Key-operated switches (S_K) require a key to operate the switch. These are used where operation of a lighting installation or other piece of equipment must be further controlled by limiting operation to personnel holding a key such as in lighting in a high school gymnasium or large meeting room and equipment in a manufacturing plant.

Specialty Switches

In addition to the standard switches, there are many types of switches that perform special functions. These are as follows:

Automatic switches deactivate a circuit after a preset time period has lapsed. They are available as a twist-turn device where the operator determines the operating time interval by how far the switch is twisted or as an electronic device that looks like a normal on/off switch but is designed to automatically switch off after a preset time.

A *dimmer switch (S_D)* is a device in the electrical circuit for varying power to a circuit. Dimmers are usually included in a lighting installation to vary the intensity of light emitted by the lights.

Time clocks can be used to control the time period that a piece of equipment or a lighting installation operates. Traditionally, a time clock is an electrical-mechanical device that controls time of operation by pin placement on a moving time wheel that repeats a daily cycle as the wheel rotates. *Electronic timers* allow greater flexibility as they can easily be set for 7-day cycles. They do require relay switching on large loads.

Photocell controls sense light and open or close a circuit with the presence of light. They can be used to control night lighting in lieu of a time clock. Recent advancements in photocell technology allow them to be used effectively to control illumination levels in spaces that have daylight available. As illumination levels from daylighting increases, the photocell dims the lights.

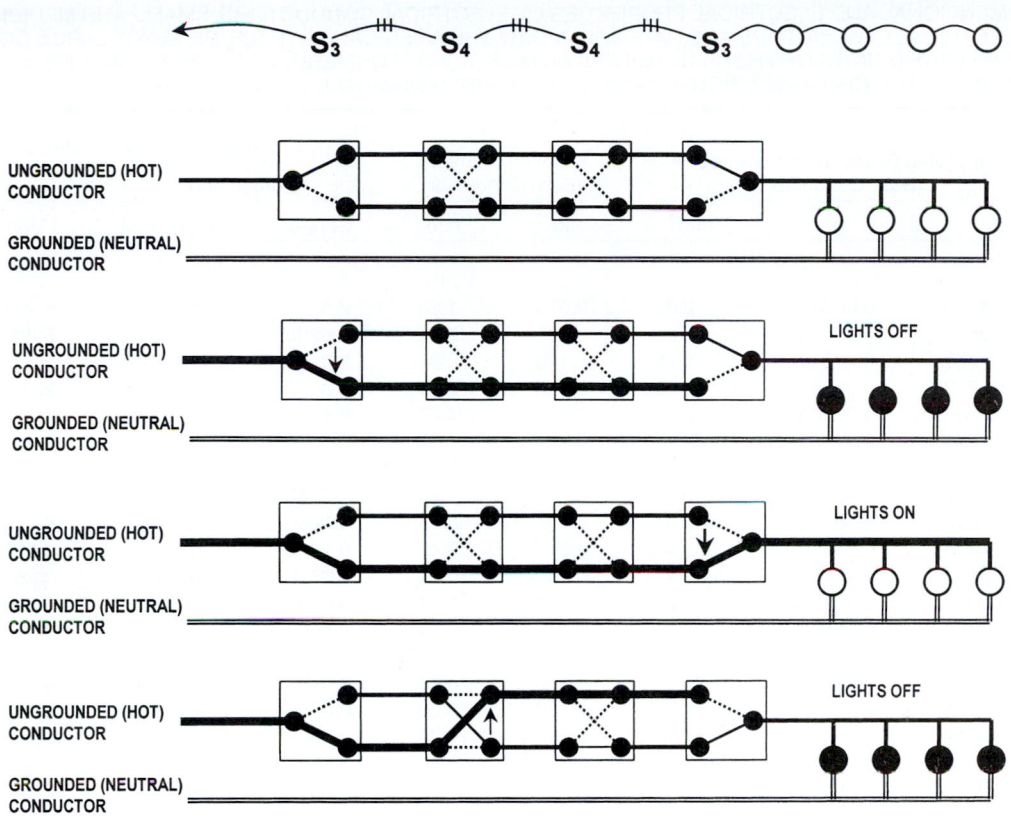

FIGURE 18.14 Shown is a one-line schematic drawing of electrical symbols and the related switching configuration for switching at four locations. The dashed line at each switch shows the new configuration of the circuit when the switch is flipped. Two three-way (S_3) switches and two four-way switches (S_4) are used to control from four points. Additional four-way switches (S_4) may be added to the circuit to control from additional points. In this switching configuration, current flows through the switches, the circuit is complete, and the lights are on. When any switch is flipped, current flow is interrupted and the lights go off.

Occupancy sensors control a lighting or equipment installation by sensing occupants in a space. *Infrared sensors* respond to the motion of an infrared heat source, such as a person or animal. The field of view of the detector's lens is divided into several zones. Motion is detected when the heat source moves from one zone to another. The sensor must have a direct line of sight to the infrared heat source to detect motion. Relatively small movements, such as typing on a keyboard, may not be sufficient to trigger the sensor.

Ultrasonic sensors emit a high-frequency sound that is in the range of 25 to 40 kilohertz and well above the capacity of normal human hearing. Objects moving in the space shift the frequency of the returning signal, which in turn is detected by the sensor. Ultrasonic sensors do not require a direct line of sight to the occupant. They usually can detect small movements, but extraneous signals from wind-blown curtains or people walking by in nearby spaces can trigger the sensor. Both infrared and ultrasonic sensors work well in interior spaces such as classrooms

and offices. Occupancy sensors are usually mounted on the wall or ceiling.

18.5 CONDUCTORS

Wire

An *electrical conductor* is any material that conducts electrical current. A wire is a common electrical conductor. Most conductors used in building applications are classified according to a wire gauge standard and on the cross-sectional area of the wire in units called circular mils. A *mil* is equal to 1/1000 inch, so one *circular mil (cmil)* is equal to the cross-sectional area of a 0.001 in diameter circle.

Common conductor sizes are provided in Table 18.8. Small- and medium-gauge electrical conductors are categorized by the *American Wire Gauge (AWG)*, a standardized wire gauge system used predominantly in the United States since 1857 for round, solid, electrically conducting wire. The

TABLE 18.8   DIMENSIONAL AND ELECTRICAL PROPERTIES OF ELECTRICAL CONDUCTORS. SMALL- AND MEDIUM-GAUGE ELECTRICAL CONDUCTORS ARE CATEGORIZED BY THE AMERICAN WIRE GAUGE (AWG). LARGE CONDUCTORS ARE SIZED IN UNITS OF THOUSAND CIRCULAR MILS (KCMIL, FORMERLY MCM). ONE CIRCULAR MIL (CMIL) IS EQUAL TO THE CROSS-SECTIONAL AREA OF A 0.001 INCH DIAMETER CIRCLE.

Conductor (Wire) Size	Nominal Cross-Sectional Area (Bare Conductor)			Nominal Diameter (Bare Conductor)		DC Resistance per Conductor at 77°F (25°C)			
						Ohms/1000 ft		Ohms/1000 m	
	cmil	in^2	mm^2	in	mm	Copper	Aluminum	Copper	Aluminum
AWG									
14	4110	0.00323	2.08	0.073	1.85	2.5700	4.2200	8.4219	13.8289
12	6530	0.00513	3.31	0.092	2.34	1.6200	2.6600	5.3087	8.7168
10	10 380	0.00815	5.26	0.116	2.95	1.0200	1.6700	3.3425	5.4726
8	16 510	0.01296	8.36	0.146	3.71	0.6410	1.0500	2.1006	3.4409
6	26 240	0.02060	13.29	0.184	4.67	0.4030	0.6610	1.3206	2.1661
4	41 740	0.03277	21.14	0.232	5.89	0.2530	0.4150	0.8291	1.3600
3	52 620	0.04131	26.65	0.260	6.60	0.2010	0.3300	0.6587	1.0814
2	66 360	0.05209	33.61	0.292	7.42	0.1590	0.2610	0.5210	0.8553
1	83 690	0.06570	42.38	0.332	8.43	0.1260	0.2070	0.4129	0.6783
0 or 1/0	105 600	0.08290	53.48	0.373	9.47	0.1000	0.1640	0.3277	0.5374
00 or 2/0	133 100	0.10448	67.41	0.419	10.64	0.0795	0.1300	0.2605	0.4260
000 or 3/0	167 800	0.13172	84.98	0.470	11.94	0.0630	0.1030	0.2065	0.3375
0000 or 4/0	211 600	0.16611	107.16	0.528	13.41	0.0500	0.0820	0.1639	0.2687
kcmil									
250	250 000	0.19625	126.61	0.575	14.61	0.0431	0.0708	0.1412	0.2320
300	300 000	0.23550	151.94	0.630	15.00	0.0360	0.0590	0.1180	0.1933
350	350 000	0.27475	177.26	0.681	17.30	0.0308	0.0505	0.1009	0.1655
400	400 000	0.31400	202.58	0.728	18.49	0.0270	0.0442	0.0885	0.1448
500	500 000	0.39250	253.23	0.813	20.65	0.0216	0.0354	0.0708	0.1160
600	600 000	0.47100	303.87	0.893	22.68	0.0180	0.0295	0.0590	0.0967
700	700 000	0.54950	354.52	0.964	24.49	0.0144	0.0236	0.0472	0.0773
750	750 000	0.58875	379.84	0.998	25.35	0.0108	0.0177	0.0354	0.0580
800	800 000	0.62800	405.16	1.030	26.16	0.0104	0.0170	0.0340	0.0557
900	900 000	0.70650	455.81	1.090	27.69	0.0095	0.0156	0.0311	0.0511
1000	1 000 000	0.78500	506.45	1.150	29.21	0.0086	0.0142	0.0283	0.0465
1500	1 500 000	1.17750	759.68	1.410	35.81	0.0072	0.0118	0.0236	0.0387
2000	2 000 000	1.57000	1012.90	1.630	41.40	0.0054	0.0089	0.0177	0.0290

AWG gauge number is inversely proportional to the cross-section diameter of the wire—that is, a smaller number identifies a thicker wire. AWG gauges range from No. 36 AWG through No. 0000 AWG. A No. 10 AWG is thicker than a No. 12; a No. 12 AWG is thicker than a No. 14 AWG; and so on. Large AWG wire sizes are expressed as a zero or multiple zeros from No. 0 AWG through No. 0000 AWG (known as "ought" through "four-ought"). As a rough rule of thumb, if the diameter of a wire is doubled, the AWG will decrease by about 6 (e.g., No. 1 AWG wire is about twice the diameter of No. 6 AWG wire).

Large conductors are sized in units of *thousand circular mils (kcmil,* formerly *MCM). A circular mil* is a unit of circular cross-sectional area, equal to the area of a circle with a diameter of one mil. A mil is one thousandth of an inch. Wire sizes in the kcmil system increase as the numbers get larger, which is exactly opposite from the AWG system. A 250 kcmil conductor contains one-half the cross-sectional area of a 500 kcmil conductor. Large conductors range from 250 kcmil to the 2000 kcmil size.

Electrical conductors are either solid or stranded. *Solid conductors* are a single solid length of conductor called a *wire. Stranded conductors* consist of smaller wire strands. The choice between solid and stranded depends on the need for flexibility in handling and working with the conductor. Stranded conductors conduct current more efficiently because electrons flow more easily near the surface of a conductor, in contrast to the center; this is known as the *skin effect.* The AWG No. and kcmil gauge of a stranded wire is determined by the total cross-sectional area of the conductor strands. Because there are small gaps between strands, a stranded wire will always have a slightly larger overall diameter than a solid wire of the same gauge.

In building electrical systems, conductor sizes No. 14 AWG through about No. 8 AWG are typically solid conductors. On occasion, stranded conductors are used in these sizes because stranded wiring is easier to snake through a conduit. Larger conductors are typically stranded. The number of strands of wire is standardized; sizes No. 18 AWG through No. 2 AWG are 7-strand cables; AWG No. 1 AWG through No. 4/0 AWG are 19-strand cables; and so on.

Buses

A *bus*, sometimes called a *bus bar,* is an electrical conductor (usually copper or aluminum) that serves as a common connection for two or more electrical circuits. Buses are typically solid bars used for power distribution. They are commonly found in panelboards, switchboards, and other power distribution equipment. Busbars are either flat strips or hollow tubes as these shapes allow heat to dissipate more efficiently because of their high surface area to cross-sectional area ratio. The skin effect generally makes hollow or flat shapes prevalent in higher current applications. Table 18.9 provides specifications on selected commercially available bus bars.

Conductor Material

Scientifically, silver is the best electrical conductor material (other than a superconductor material) because it has the least resistance of common materials. It works so well as a conductor that several hundred pounds of silver wiring was used in the first

TABLE 18.9   PROPERTIES AND APPROXIMATE AMPACITIES OF SELECTED COMMERCIALLY AVAILABLE BUS BAR CROSS-SECTIONS. COMPILED FROM INDUSTRY SOURCES.

Dimensions (in)	Cross-Section Area		Temperature Rise					
	in²	Circular mils (kcmil)	Copper			Aluminum		
			86°F Rise	122°F Rise	149°F Rise	86°F Rise	122°F Rise	149°F Rise
			30°C Rise	50°C Rise	65°C Rise	30°C Rise	50°C Rise	65°C Rise
⅛ × ½	0.0625	79.6	153	205	235	92	123	141
⅛ × ¾	0.0938	119	215	285	325	129	171	195
⅛ × 1	0.125	159	270	360	415	162	216	249
⅛ × 1½	0.188	239	385	510	590	231	306	354
⅛ × 2	0.250	318	495	660	760	297	396	456
⅛ × 2½	0.312	397	600	800	920	360	480	552
⅛ × 3	0.375	477	710	940	1100	426	564	660
⅛ × 3½	0.438	558	810	1100	1250	486	660	750
⅛ × 4	0.500	636	910	1200	1400	546	720	840
¼ × ½	0.125	159	240	315	360	144	189	216
¼ × ¾	0.188	239	320	425	490	192	255	294
¼ × 1	0.250	318	400	530	620	240	318	372
¼ × 1½	0.375	477	560	740	860	336	444	516
¼ × 2	0.500	637	710	940	1100	426	564	660
¼ × 2½	0.625	796	850	1150	1300	510	690	780
¼ × 3	0.750	955	990	1300	1550	594	780	930
¼ × 3½	0.875	1110	1150	1500	1750	690	900	1050
¼ × 4	1.00	1270	1250	1700	1950	750	1020	1170
¼ × 5	1.25	1590	1500	2000	2350	900	1200	1410
¼ × 6	1.50	1910	1750	2350	2700	1050	1410	1620
¼ × 8	2.00	2550	2250	3000	3450	1350	1800	2070
¼ × 10	2.50	3180	2700	3600	4200	1620	2160	2520
¼ × 12	3.00	3820	3150	4200	4900	1890	2520	2940
½ × 1	0.500	637	620	820	940	372	492	564
½ × 1½	0.750	955	830	1100	1250	498	660	750
½ × 2	1.00	1270	1000	1350	1550	600	810	930
½ × 2½	1.25	1590	1200	1600	1850	720	960	1110
½ × 3	1.50	1910	1400	1850	2150	840	1110	1290
½ × 3½	1.75	2230	1550	2100	2400	930	1260	1440
½ × 4	2.00	2550	1700	2300	2650	1020	1380	1590
½ × 5	2.50	3180	2050	2750	3150	1230	1650	1890
½ × 6	3.00	3820	2400	3150	3650	1440	1890	2190
½ × 8	4.00	5090	3000	4000	4600	1800	2400	2760
½ × 10	5.00	6360	3600	4800	5500	2160	2880	3300
½ × 12	6.00	7640	4200	5600	6400	2520	3360	3840
¾ × 4	3.00	3820	2050	2750	3150	1230	1650	1890
¾ × 5	3.75	4770	2400	3250	3750	1440	1950	2250
¾ × 6	4.50	5730	2800	3750	4300	1680	2250	2580
¾ × 8	6.00	7640	3500	4700	5400	2100	2820	3240
¾ × 10	7.50	9550	4200	5600	6500	2520	3360	3900
¾ × 12	9.00	11 500	4900	6500	7500	2940	3900	4500

mainframe computer. Gold is also an excellent conductor. Both materials, however, are too costly for building installations.

Traditionally, copper and aluminum conductors are used in building conductor (wiring) installations as a compromise between good conductivity and economy. Copper clad aluminum wire was introduced as an alternative to bare aluminum. The wire core is aluminum with a thin coating of copper that is metallurgically bonded to the core. It looks like copper but functions like aluminum.

As a conductor material, aluminum has significant weight and cost advantages over copper. Copper does, however, conduct electricity better than aluminum, so an aluminum wiring installation requires a thicker gauge. Also, extra care in splicing and terminating aluminum wire is important.

An increased fire-hazard potential exists with old-technology aluminum wiring that was manufactured and used before about 1972. Concerns typically are tied to the use of solid, small-gauge aluminum conductors (No. 10 AWG and No. 12 AWG) that are commonly used on the 15 A and 20 A circuits serving receptacle outlets, light fixtures, and small appliances. As a result of these concerns, the industry typically reserves use of aluminum conductors for large gauge (No. 6 AWG and above) installations such as branch circuits feeding large appliances (e.g., kitchen ranges).

Conductor Insulation

Conductors are covered with *insulation* that provides electrical isolation and some physical protection of the conductor material. It prevents loss of power and the danger of short circuits and ground faults. The type of insulation protecting a conductor determines the environment in which it can be used safely. Wires used indoors are subjected to less exposure to the elements than those designed for outdoor use. Outdoor wiring is exposed to water and ultraviolet light, so the insulation is designed to withstand these elements. Insulation on wires buried in the ground must also be able to withstand the damp, corrosive environment of the soil. For special applications, a *jacket* is applied over the insulation. A jacket provides the necessary chemical, physical, or thermal protection required by the application.

Most electrical conductors available for use in buildings can be operated at voltages up to 600 V, although some power cord and fixture wire are rated at 300 V maximum. At voltages above 600 V, conductors must be better insulated and shielded to eliminate electromagnetic interferences, which makes building system voltages above 600 V expensive.

Most of the electrical wires used in buildings today have insulation coverings made of plastic, which offers a long-lasting life. Many older wires used cloth or rubber insulation, but these materials are not used much any more because they deteriorate over time. There are variations in the capabilities of different insulations to withstand heat generated, so type of insulation material will have an effect on the maximum temperature that a specific conductor can withstand without premature deterioration. Common temperature ratings include 140°F (60°C), 167°F (75°C), and 194°F (90°C).

Insulation is also rated for dry, damp, or wet locations. A *dry* location typically applies to conductors that are enclosed in a building and remain dry except during initial construction. A *damp* location includes wiring in above-ground, protected outdoor locations such as in covered open decks and canopies. A *wet* location is when the conductor is exposed to weather (i.e., buried conductors and conductors encased in concrete in contact with the ground) or exposed to severe moisture conditions such as where the conductor is subjected to applications involving water (i.e., cleaning or washing processes such as a car wash).

There are various letter designations and classifications for different types of insulation on electrical wires. These letter designations indicate the type of material the insulation is made of and in what type of environment it can safely be used in without deteriorating. Table 18.10 provides a description of common conductor insulation designations used in building systems.

Tables 18.11 through 18.13 contain technical information on common types of conductors used in building electrical systems. Figure 18.15 shows types of conductors.

Cable

A *cable* contains more than one conductor bundled together in a factory assembly of wires. An outer *sheathing* encases and protects the conductors, simplifying installation of multiple wiring. Common designations of sheathed cables used in building systems are provided in Table 18.14.

Tables 18.15 through 18.21 contain technical information on common types of cable used in building electrical systems. Figure 18.15 illustrates types of cable. Types of sheathed cables include the following.

Nonmetallic-Sheathed Cable

Nonmetallic-sheathed cable is classified as type NM or NMC and is commonly called by its trade name, Romex. See Photo 18.17. NM consists of two or more insulated conductors enclosed within a moisture-resistant, flame-retardant outer sheathing or jacket that is very flexible. NMC has conductors encased in the sheathing. NM is reserved for use in dry, indoor applications. NMC can be used in dry and damp applications but not wet and exposed conditions. Type NM and Type NMC cables can typically be used only in one- and two-family dwellings, and in multifamily dwellings permitted to be of Type III, IV, and V construction.

Underground Feeder Cable

Underground feeder (UF) cables are flame retardant and moisture, fungus, and corrosion resistant. UF cable is available in No. 14 AWG copper and No. 12 aluminum AWG through No. 4/0 AWG. It looks much like NM or NMC except that the sheathing fully encases the insulation-covered conductors. UF cable is used in direct-burial applications as a feeder or branch

TABLE 18.10 CONDUCTOR INSULATION AND MATERIAL DESIGNATIONS AND MARKINGS. COMPILED FROM INDUSTRY SOURCES.

Insulation Designations

Designation	Description
A	Glass fiber or similar insulation material (formerly asbestos, which is obsolete)
FEP	Fluorinated ethylene propylene insulation
L	Lead sheathed
N	Nylon jacketed
PF	Perfluoroalkooxy insulation
R	Thermoset insulation (e.g., rubber or synthetic rubber)
S	Silicone (thermoset) insulation
T	Thermoplastic insulation
X	Cross-linked synthetic polymer (plastic) insulation
Z	Modified tetrafluoroethylene insulation
−2	194°F (90°C) and wet or dry
H[a]	Heat resistant: temperature rated at 167°F (75°C)
HH	Extra heat resistant: temperature rated at 194°F (90°C)
W[b]	Moisture resistant

Common Insulation Material Markings[c]

Marking	Description
RHH	Thermoset (e.g., rubber or synthetic rubber). 194°F (90°C) temperature rating. Extra heat resistant. Not moisture resistant (not suitable for conduits exposed to weather). Not sunlight resistant.
RHW	Thermoset (e.g., rubber or synthetic rubber). 167°F (75°C) temperature rating. Moisture and heat resistant. Not sunlight resistant.
RHW-2	Thermoset (e.g., rubber or synthetic rubber). 194°F (90°C) temperature rating. Extra heat resistant. Not sunlight resistant.
THHN	Thermoplastic. 194°F (90°C) temperature rating. Extra heat resistant. Not moisture resistant (not suitable for conduits exposed to weather). Nylon jacket over insulation (slides through conduit easier). Not sunlight resistant.
THW	Thermoplastic. 167°F (75°C) temperature rating. Extra heat resistant. Wet and dry rating. Not sunlight resistant.
THW-2	Thermoplastic. 194°F (90°C) temperature rating. Extra heat resistant. Wet and dry rating. Not sunlight resistant.
TWHN	Thermoplastic. 167°F (75°C) temperature rating. Moisture and heat resistant. Nylon jacket over insulation (slides through conduit easier). Not sunlight resistant.
TWHN-2	Thermoplastic. 194°F (90°C) temperature rating. Moisture and extra heat resistant. Nylon jacket over insulation (slides through conduit easier). Not sunlight resistant.
XHHW-2	Thermoset (e.g., cross-linked synthetic polymer). 194°F (90°C) temperature rating. Extra heat resistant. Wet and dry rating.

Conductor Material and Temperature Rating

Conductor Material	Temperature Rating of Conductor		Insulation Designation
Copper (CU)	140°F	60°C	TW, UF
	167°F	75°C	FEPW, RH, RHW, THHW, THW, XHHW, USE, ZW
	194°F	90°C	TA. TBS, SA, SIS, FEP, FEPB, MI, RHH, RHW-2, THHN, THHW, THW-2, THWN-2, USE-2, XHH, XHHW, XHHW-2, ZW-2
Aluminum (AL)	140°F	60°C	TW, UF
	167°F	75°C	RH, RHW, THHW, THW, THWN, XHHW, USE
	194°F	90°C	TA, TBS, SA, SIS, RHH, RHW-2, THHN, THHW, THW-2, THWN-2, USE-2, XHH, XHHW, XHHW-2, ZW-2

[a]Lack of "H" indicates 140°F (60°C) temperature rating.
[b]Lack of "W" indicates no moisture resistance.
[c]Double or triple markings indicate the conductor or cable has the same properties of those markings.

TABLE 18.11 PROPERTIES OF COPPER AND ALUMINUM CONDUCTORS WITH THHN AND THWN-2 INSULATION THAT ARE USED IN CONDUIT AND CABLE TRAYS FOR SERVICES, FEEDERS, AND BRANCH CIRCUITS IN COMMERCIAL OR INDUSTRIAL APPLICATIONS. COMPILED FROM INDUSTRY SOURCES.

THHN or THWN-2 Copper Conductor

Size (AWG or kcmil)	Number of Strands	Nominal Outside Diameter (mils)		Approx. Net Weight per 1000 ft (lb)		Ampacity		
		Solid	Stranded	Solid	Stranded	60°C 140°F	75°C 167°F	90°C 194°F
14	1, 19	102	109	15	16	15	15	15
12	1, 19	119	128	23	24	20	20	20
10	1, 19	150	161	37	38	30	30	30
8	19	—	213	—	62	40	50	55
6	19	—	249	—	95	55	65	75
4	19	—	318	—	152	70	85	95
3	19	—	346	—	188	85	100	110
2	19	—	378	—	234	95	115	130
1	19	—	435	—	299	110	130	150
1/0	19	—	474	—	371	125	150	170
2/0	19	—	518	—	461	145	175	195
3/0	19	—	568	—	574	165	200	225
4/0	19	—	624	—	717	195	230	260
250	37	—	694	—	850	215	255	290
300	37	—	747	—	1011	240	285	320
350	37	—	797	—	1173	260	310	350
400	37	—	842	—	1333	280	335	380
500	37	—	926	—	1653	320	380	430
600	61	—	1024	—	1985	355	420	475
750	61	—	1126	—	2462	400	475	535
1000	61	—	1275	—	3254	455	545	615

THHN or THWN-2 Aluminum Conductor

Size (AWG or kcmil)	Number of Strands	Nominal Outside Diameter (mils)		Approx. Net Weight per 1000 ft (lb)		Ampacity		
		Solid	Stranded	Solid	Stranded	60°C 140°F	75°C 167°F	90°C 194°F
6	7	—	239	—	38	40	50	60
4	7	—	305	—	62	55	65	75
2	7	—	360	—	90	75	90	100
1	18	—	413	—	116	85	100	115
1/0	18	—	450	—	141	100	120	135
2/0	18	—	490	—	171	115	135	150
3/0	18	—	537	—	209	130	155	175
4/0	18	—	589	—	257	150	180	205
250	35	—	656	—	310	170	205	230
300	35	—	706	—	364	190	230	255
350	35	—	752	—	417	210	250	280
400	35	—	795	—	470	225	270	305
500	35	—	872	—	575	260	310	350
600	58	—	971	—	698	285	340	385
700	58	—	1035	—	802	310	375	420
750	58	—	1066	—	855	320	385	435
1000	58	—	1218	—	1115	375	445	500

TABLE 18.12 PROPERTIES OF COPPER AND ALUMINUM CONDUCTORS WITH XHHW-2 INSULATION THAT ARE USED IN CONDUIT OR OTHER RECOGNIZED RACEWAYS FOR SERVICES, FEEDERS, AND BRANCH CIRCUIT WIRING. COMPILED FROM INDUSTRY SOURCES.

XHHW-2 Copper Conductor *type of conductor*

Size (AWG or kcmil)	Number of Strands	Nominal Outside Diameter (mils) Solid	Nominal Outside Diameter (mils) Stranded	Approx. Net Weight per 1000 ft (lb) Solid	Approx. Net Weight per 1000 ft (lb) Stranded	Ampacity 60°C 140°F	Ampacity 75°C 167°F	Ampacity 90°C 194°F
14	7	—	130	—	18	15	15	15
12	1	—	141	—	25	20	20	20
10	7	—	171	—	40	30	30	30
8	7	—	232	—	66	40	50	55
6	7	—	267	—	99	55	65	75
4	7	—	314	—	149	70	85	95
2	7	—	370	—	230	95	115	130
1	19	—	434	—	292	110	130	150
1/0	19	—	473	—	363	125	150	170
2/0	19	—	517	—	452	145	175	195
3/0	19	—	567	—	565	165	200	225
4/0	19	—	623	—	705	195	230	260
250	37	—	691	—	835	215	255	290
300	37	—	744	—	995	240	285	320
350	37	—	794	—	1155	260	310	350
400	37	—	839	—	1314	280	335	380
500	37	—	923	—	1633	320	380	430
600	61	—	1029	—	1966	355	420	475
700	61	—	1098	—	2283	385	460	520
750	61	—	1131	—	2441	400	475	535
1000	61	—	1280	—	3230	455	545	615

XHHW-2 Aluminum Conductor

Size (AWG or kcmil)	Number of Strands	Nominal Outside Diameter (mils) Solid	Nominal Outside Diameter (mils) Stranded	Approx. Net Weight per 1000 ft (lb) Solid	Approx. Net Weight per 1000 ft (lb) Stranded	Ampacity 60°C 140°F	Ampacity 75°C 167°F	Ampacity 90°C 194°F
8	7	—	227	—	30	30	40	45
6	7	—	262	—	42	40	50	60
4	7	—	306	—	58	55	65	75
2	7	—	361	—	86	75	90	100
1	18	—	412	—	110	85	100	115
1/0	18	—	449	—	134	100	120	135
2/0	18	—	489	—	163	115	135	150
3/0	18	—	536	—	200	130	155	175
4/0	18	—	588	—	247	150	180	205
250	35	—	653	—	296	170	205	230
300	35	—	703	—	349	190	230	255
350	35	—	749	—	401	210	250	280
400	35	—	792	—	452	225	270	305
500	35	—	869	—	556	260	310	350
600	58	—	976	—	679	285	340	385
700	58	—	1040	—	782	310	375	420
750	58	—	1071	—	833	320	385	435
1000	58	—	1223	—	1090	375	445	500

TABLE 18.13 PROPERTIES OF COPPER AND ALUMINUM CONDUCTORS WITH RHH OR RHW-2 OR USE-2 INSULATION THAT ARE SUITABLE FOR USE AS UNDERGROUND SERVICE ENTRANCE CABLE FOR DIRECT BURIAL. COMPILED FROM INDUSTRY SOURCES.

RHH or RHW-2 or USE-2 Copper Conductor

Size (AWG or kcmil)	Number of Strands	Nominal Outside Diameter (mils)		Approx. Net Weight per 1000 ft (lb)		Ampacity		
		Solid	Stranded	Solid	Stranded	60°C 140°F	75°C 167°F	90°C 194°F
14	7	—	160	—	21	15	15	15
12	7	—	177	—	30	20	20	20
10	7	—	201	—	44	30	30	30
8	7	—	262	—	72	40	50	55
6	7	—	297	—	106	55	65	75
4	7	—	344	—	156	70	85	95
2	7	—	400	—	238	95	115	130
1	19	—	484	—	309	110	130	150
1/0	19	—	523	—	381	125	150	170
2/0	19	—	567	—	472	145	175	195
3/0	19	—	617	—	586	165	200	225
4/0	19	—	673	—	729	195	230	260
250	37	—	751	—	867	215	255	290
300	37	—	804	—	1029	240	285	320
350	37	—	854	—	1191	260	310	350
400	37	—	899	—	1352	280	335	380
500	37	—	983	—	1674	320	380	430
600	61	—	1089	—	2012	355	420	475
700	61	—	1158	—	2332	385	460	520
750	61	—	1191	—	2492	400	475	535
800	61	—	1223	—	2652	410	490	555
900	61	—	1283	—	2970	435	520	585
1000	61	—	1340	—	3288	455	545	615

RHH or RHW-2 or USE-2 Aluminum Conductors

Size (AWG or kcmil)	Number of Strands	Nominal Outside Diameter (mils)		Approx. Net Weight per 1000 ft (lb)		Ampacity		
		Solid	Stranded	Solid	Stranded	60°C 140°F	75°C 167°F	90°C 194°F
8	7	—	257	—	36	30	40	45
6	7	—	292	—	49	40	50	60
4	7	—	336	—	65	55	65	75
2	7	—	391	—	94	75	90	100
1	18	—	462	—	126	85	100	115
1/0	18	—	499	—	151	100	120	135
2/0	18	—	539	—	182	115	135	150
3/0	18	—	586	—	221	130	155	175
4/0	18	—	638	—	269	150	180	205
250	35	—	713	—	326	170	205	230
300	35	—	763	—	381	190	230	255
350	35	—	809	—	435	210	250	280
400	35	—	852	—	488	225	270	305
500	35	—	929	—	595	260	310	350
700	58	—	1100	—	829	310	375	420
750	58	—	1131	—	881	320	385	435
1000	58	—	1283	—	1145	375	445	500

Common Conductors

THHN or THWN-2 conductor

RHH or RHW-2 conductor

Common Cables

Nonmetallic-sheathed (NM) cable

Armored (AC) cable

Underground feeder (UF) cable

Service entrance (SE) cable

Service entrance (SER) cable

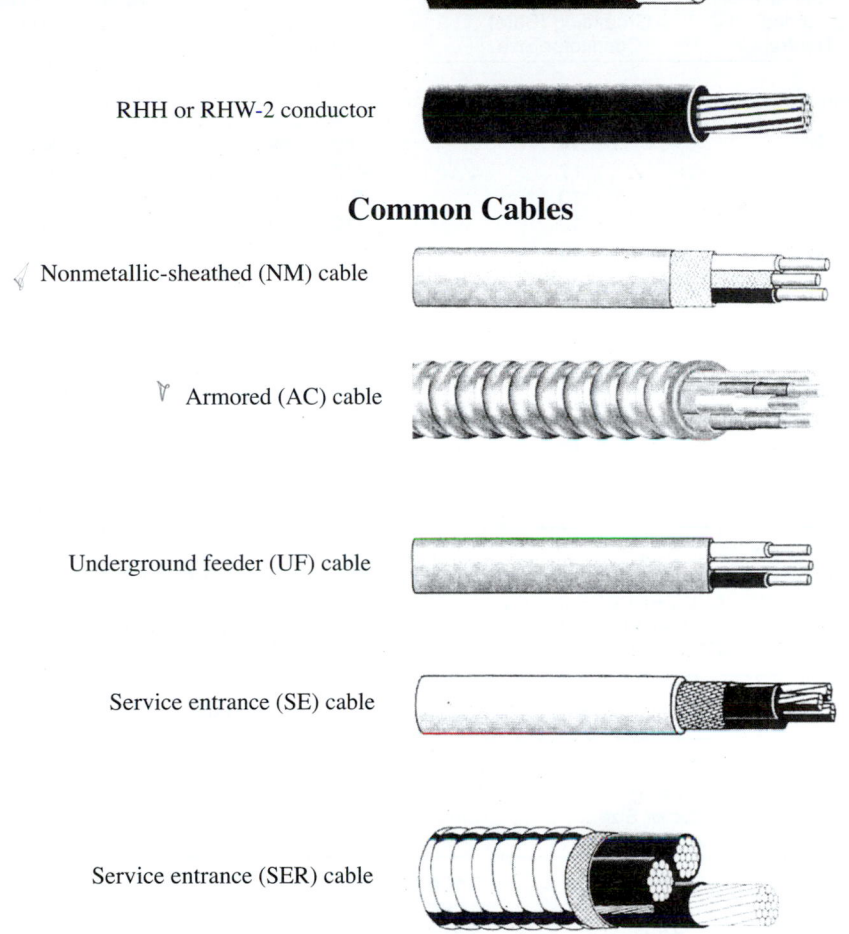

FIGURE 18.15 Common types of conductors and cable used in building electrical systems.

TABLE 18.14   COMMON CABLE MARKINGS. COMPILED FROM NUMEROUS INDUSTRY SOURCES.

Marking	Description
ACHH	Armored cable with conductors having thermoset insulation. 194°F (90°C) temperature rating.
ACTH	Armored cable with conductors having thermoplastic insulation. 167°F (75°C) temperature rating.
ACTHH	Armored cable with conductors having thermoplastic insulation. 194°F (90°C) temperature rating.
NM	Non-metallic sheathed cable. 140°F (60°C) temperature rating. Approved for use in dry, interior residential wiring if properly installed inside framed walls, floors, and ceilings. Not sunlight resistant.
SE	Service entrance.
UF	Underground feeder. 140°F (60°C) temperature rating. Generally not sunlight resistant, unless marked.
USE	Underground service entrance. 167°F (75°C) temperature rating, wet insulation rating, heat and moisture resistant. Sunlight resistant.
USE-2	Underground service entrance. 194°F (90°C) temperature rating, wet insulation rating. Heat and moisture resistant. Sunlight resistant.

TABLE 18.15 PROPERTIES OF COPPER CONDUCTORS IN NM-B (NONMETALLIC-SHEATHED) CABLE THAT IS USED IN RESIDENTIAL WIRING SUCH AS BRANCH CIRCUITS FOR OUTLETS, SWITCHES, AND OTHER LOADS. NM-B CABLING CAN ONLY BE USED FOR BOTH EXPOSED AND CONCEALED WORK IN NORMALLY DRY LOCATIONS AT TEMPERATURES NOT TO EXCEED 90°C (WITH AMPACITY LIMITED TO THAT FOR 60°C CONDUCTORS). COMPILED FROM INDUSTRY SOURCES.

NM-B Cable with Type THHN/THWN Copper Conductors

Conductor Size (AWG or kcmil)	Number of Ungrounded and Neutral	Grounding (Bare) Conductor Size	Nominal Size (mil)	Approx. Net Weight per 1000 ft (lb)	Ampacity
Three Conductors Plus Grounding Conductor					
14	2	14	360 × 162	58	15
12	2	12	410 × 179	83	20
10	2	10	494 × 210	126	30
8	2	10	612 × 269	187	40
6	2	10	683 × 304	256	55
Four Conductors Plus Grounding Conductor					
14	3	14	307	75	15
12	3	12	347	109	20
10	3	10	422	167	30
8	3	10	565	254	40
6	3	10	650	357	55
4	3	8	892	593	70
2	3	8	1034	856	95

TABLE 18.16 PROPERTIES OF COPPER AND ALUMINUM CONDUCTORS IN AC (ARMORED) CABLE THAT IS SUITABLE FOR USE IN BRANCH CIRCUITS AND FEEDERS IN BOTH EXPOSED AND CONCEALED WORK IN COMMERCIAL, INDUSTRIAL, INSTITUTIONAL, AND MULTIRESIDENTIAL APPLICATIONS WHERE IT IS NOT SUBJECT TO PHYSICAL DAMAGE. COMPILED FROM NUMEROUS INDUSTRY SOURCES.

AC Cable with Type THHN/THWN Copper Conductors

Conductor Size (AWG or kcmil)	Number of Ungrounded and Neutral Conductors	Aluminum Bonding Conductor Size (AWG)	Nominal Outside Diameter (mils)	Approx. Net Weight per 1000 ft (lb)	Ampacity 60°C 140°F	75°C 167°F	90°C 194°F
14	2	16	0.464	84	15	15	15
	3	16	0.485	104	15	15	15
	4	16	0.517	124	15	15	15
12	2	16	0.498	105	20	20	20
	3	16	0.521	133	20	20	20
	4	16	0.557	162	20	20	20
10	2	16	0.560	142	30	30	30
	3	16	0.588	184	30	30	30
	4	16	0.632	229	30	30	30

TABLE 18.17 PROPERTIES OF COPPER CONDUCTORS IN UF (UNDERGROUND FEEDER) CABLE THAT IS GENERALLY USED AS A FEEDER TO OUTSIDE POST LAMPS, PUMPS, AND OTHER LOADS OR APPARATUS FED FROM A DISTRIBUTION POINT IN AN EXISTING BUILDING. COMPILED FROM NUMEROUS INDUSTRY SOURCES.

UF-B Cable with Type THHN/THWN Copper Conductors

Conductor Size (AWG or kcmil)	Number of Ungrounded and Neutral Conductors	Grounding (Bare) Conductor Size	Nominal Size (mil)	Approx. Net Weight per 1000 ft (lb)	Ampacity
14	2	14	581 × 168	97	15
12	2	12	626 × 183	131	20
10	2	10	727 × 215	194	30
6	2	10	1223 × 361	480	55
8	2	10	1059 × 319	345	70

TABLE 18.18 PROPERTIES OF COPPER AND ALUMINUM CONDUCTORS IN SERVICE ENTRANCE (SE) CABLE IS USED TO CONVEY POWER FROM THE SERVICE DROP TO THE METER BASE AND FROM THE METER BASE TO THE DISTRIBUTION PANELBOARD. COMPILED FROM NUMEROUS INDUSTRY SOURCES.

SE Cable with Type XHHW-2 Copper Conductor

Conductor Size (AWG or kcmil)	Stranding of Conductors		Nominal Size (mil)	Approx. Net Weight per 1000 ft (lb)	Ampacity		
	Ungrounded and Neutral	Grounding (Bare)			60°C	75°C	90°C
					140°F	167°F	194°F
10-10-10	1	12	453 × 295	133	30	30	30
8-8-8	7	8	627 × 400	219	40	50	55
6-6-6	7	12	698 × 435	316	55	65	75
4-4-4	7	12	815 × 506	471	70	85	95
3-3-3	7	12	883 × 548	583	85	100	110
2-2-2	7	15	944 × 578	718	95	115	130
1-1-1	19	14	1093 × 664	904	110	130	150
1/0-1/0-1/0	19	18	1171 × 703	1122	125	150	170
2/0-2/0-2/0	19	18	1275 × 763	1378	145	175	195
3/0-3/0-3/0	19	14	1421 × 858	1711	165	200	225
4/0-4/0-4/0	19	18	1533 × 914	2145	195	230	260

SE Cable with Type XHHW-2 Aluminum Conductors

Conductor Size (AWG or kcmil)	Stranding of Conductors		Nominal Size (mil)	Approx. Net Weight per 1000 ft (lb)	Ampacity		
	Ungrounded and Neutral	Grounding (Bare)			60°C	75°C	90°C
					140°F	167°F	194°F
8-8-8	1	—	556	106	30	40	45
6-6-6	7	—	650	150	40	50	60
4-4-4	7	—	745	203	55	65	75
3-3-3	7	—	799	241	65	75	85
2-2-2	7	—	864	290	75	90	100
1-1-1	18	—	974	361	85	100	115
1/0-1/0-1/0	18	—	1054	435	100	120	135
2/0-2/0-2/0	18	—	1140	527	115	135	150
3/0-3/0-3/0	18	—	1242	641	130	155	175
4/0-4/0-4/0	18	—	1354	784	150	180	205

circuit provided it is protected by an overcurrent protection device (fuse or circuit breaker) before if leaves the panelboard.

Service Entrance Cable

There are several conductors that can be used specifically for underground and overhead service entrances. *Service entrance (SE) cable* is suitable for exposed above-grade conditions. *Underground service entrance (USE) cable* is used in underground service applications. A conductor marked with only type USE or USE-2 may not be installed in conduit inside buildings because it does not have the necessary flame retardant. In many cases, USE cable is accompanied by the RHW-type marking and USE-2 cables are dual marked with RHW-2. These dual-marked cables are suitable for use as exposed single-conductor cables and as cables inside conduits in buildings.

Armored Cable

Armored cable, classified as either AC or ACT, is sometimes referred to by the trade name, BX cable. This cable consists of two to four copper conductors between 14 AWG and 1 AWG in size that are enclosed within a flexible spiral-shaped metallic enclosure. Armored cable is classified as type ACT if the conductor insulation is thermoplastic and AC if the insulation is rubber. AC cable contains a 16 AWG bonding strip, which is in constant contact with the metal armor, allowing the armor-bonding strip combination to act as an equipment ground. Armored (AC) installed in a steel-framed partition is shown in Photo 18.18.

Metal-Clad Cable

Metal-clad (MC) cable is similar to armored cable except it is not limited to the number sizes (from 18 AWG to 2000 kcmil) of conductors it can carry. The conductors in MC cable may be copper, aluminum, or copper-clad aluminum. The metal armor may be a smooth tube, corrugated tube, or interlocked metal armor. MC cable does not contain a bonding strip like AC cable, and the armor cannot be used by itself as an equipment ground. However, it supplements the internal grounding

TABLE 18.19 PROPERTIES OF COPPER AND ALUMINUM CONDUCTORS IN SERVICE ENTRANCE (SER) CABLE PRIMARILY USED AS PANEL FEEDER IN MULTIPLE-UNIT DWELLINGS. COMPILED FROM NUMEROUS INDUSTRY SOURCES.

SER Cable with Type XHHW-2 Copper Conductors

Conductor Size (AWG or kcmil)	Stranding of Conductors		Nominal Size (mil)	Approx. Net Weight per 1000 ft (lb)	Ampacity		
	Ungrounded and Neutral	Grounding (Bare)			60°C 140°F	75°C 167°F	90°C 194°F

Three Conductors (No Grounding Conductor)

8-8-8	7	—	586	236	40	50	55
6-6-6	7	—	669	342	55	65	75
4-4-4	7	—	771	499	70	85	95
3-3-3	7	—	829	612	85	100	110
2-2-2	7	—	896	753	95	115	130
1-1-1	19	—	1024	948	110	130	150
1/0-1/0-1/0	19	—	1110	1169	125	150	170
2/0-2/0-2/0	19	—	1203	1445	145	175	195
3/0-3/0-3/0	19	—	1313	1793	165	200	225
4/0-4/0-4/0	19	—	1434	2228	195	230	260

Four Conductors (No Grounding Conductor)

8-8-8-8	7	7	645	291	40	50	55
6-6-6-6	7	7	738	428	55	65	75
4-4-4-6	7	7	852	586	70	85	95
3-3-3-5	7	7	917	720	85	100	110
2-2-2-4	7	7	992	889	95	115	130
1-1-1-3	19	7	1134	1119	110	130	150
1/0-1/0-1/0-2	19	7	1231	1384	125	150	170
2/0-2/0-2/0-1	19	19	1335	1715	145	175	195
3/0-3/0-3/0-1/0	19	19	1458	2131	165	200	225
4/0-4/0-4/0-2/0	19	19	1593	2653	195	230	260

SER Cable with Type XHHW-2 Aluminum Conductors

Conductor Size (AWG or kcmil)	Stranding of Conductors		Nominal Size (mil)	Approx. Net Weight per 1000 ft (lb)	Ampacity		
	Ungrounded and Neutral	Grounding (Bare)			60°C 140°F	75°C 167°F	90°C 194°F

Three Conductors (No Grounding Conductor)

8-8-8	1	—	556	117	30	40	45
6-6-6	7	—	650	167	40	50	60
4-4-4	7	—	745	222	55	65	75
3-3-3	7	—	799	263	65	75	85
2-2-2	7	—	864	313	75	90	100
1-1-1	18	—	974	392	85	100	115
1/0-1/0-1/0	18	—	1054	470	100	120	135
2/0-2/0-2/0	18	—	1140	565	115	135	150
3/0-3/0-3/0	18	—	1242	684	130	155	175
4/0-4/0-4/0	18	—	1354	832	150	180	205

Four Conductors (Including Grounding Conductor)

8-8-8-8	1	1	612	136	30	40	45
6-6-6-6	7	7	717	196	40	50	60
4-4-4-6	7	7	823	252	55	65	75
3-3-3-5	7	7	883	299	65	75	85
2-2-2-4	7	7	956	359	75	90	100
1-1-1-3	18	7	1079	449	85	100	115
1/0-1/0-1/0-2	18	1	1168	540	100	120	135
2/0-2/0-2/0-1	18	1	1264	653	115	135	150
3/0-3/0-3/0-1/0	18	1	1378	793	130	155	175
4/0-4/0-4/0-2/0	18	1	1503	968	150	180	205

TABLE 18.20 PROPERTIES OF COPPER CONDUCTORS IN MC (METAL CLAD) CABLE SUITABLE FOR USE AS FOLLOWS: BRANCH, FEEDER, AND SERVICE POWER DISTRIBUTION IN COMMERCIAL, INDUSTRIAL, MULTIRESIDENTIAL BUILDINGS, THEATERS, AND PLACES OF ASSEMBLY; INSTALLATION IN CABLE TRAY AND APPROVED RACEWAYS. COMPILED FROM NUMEROUS INDUSTRY SOURCES.

MC Cable with Type THHN/THWN Copper Conductors

Conductor Size (AWG or kcmil)	Number of Ungrounded and Neutral Conductors	Stranding	Grounding Conductor Size and Stranding	Overall Outside Diameter (in)	Approx. Net Weight per 1000 ft (lb)	Ampacity		
						60°C 140°F	75°C 167°F	90°C 194°F
14	2	Solid	14 solid	0.439	81	15	15	15
	3	Solid	14 solid	0.464	99	15	15	15
	4	Solid	14 solid	0.494	118	15	15	15
12	2	Solid	12 solid	0.475	109	20	20	20
	3	Solid	12 solid	0.505	136	20	20	20
	4	Solid	12 solid	0.539	163	20	20	20
10	2	Solid	10 solid	0.542	158	30	30	30
	3	Solid	10 solid	0.580	199	30	30	30
	4	Solid	10 solid	0.623	241	30	30	30
12	2	19	12/19 strand	0.494	114	20	20	20
	3	19	12/19 strand	0.527	142	20	20	20
	4	19	12/19 strand	0.564	170	20	20	20
10	2	19	10/19 strand	0.566	165	30	30	30
	3	19	10/19 strand	0.607	208	30	30	30
	4	19	10/19 strand	0.653	251	30	30	30
8	2	19	10/19 strand	0.644	232	40	50	55
	3	19	10/19 strand	0.678	299	40	50	55
	4	19	10/19 strand	0.732	369	40	50	55
6	2	19	8/19 strand	0.716	331	55	65	75
	3	19	8/19 strand	0.756	431	55	65	75
	4	19	8/19 strand	0.819	535	55	65	75
4	2	19	8/19 strand	0.905	471	70	85	95
	3	19	8/19 strand	0.986	635	70	85	95
	4	19	4/19 strand	1.077	800	70	85	95
3	2	19	6/19 strand	0.965	585	85	100	110
	3	19	6/19 strand	1.053	786	85	100	110
	4	19	6/19 strand	1.152	988	85	100	110
2	2	19	6/19 strand	1.034	685	95	115	130
	3	19	6/19 strand	1.130	934	95	115	130
	4	19	2/19 strand	1.239	1183	95	115	130
1	3	19	6/19 strand	1.203	1138	110	130	150
	4	19	6/19 strand	1.351	1468	110	130	150
1/0	3	19	6/19 strand	1.246	1350	125	150	170
	4	19	6/19 strand	1.367	1749	125	150	170
2/0	3	19	6/19 strand	1.339	1633	145	175	195
	4	19	6/19 strand	1.491	2195	145	175	195
3/0	3	19	4/19 strand	1.449	2040	165	200	225
	4	19	4/19 strand	1.614	2724	165	200	225
4/0	3	19	4/19 strand	1.590	2552	195	230	260
	4	19	4/19 strand	1.749	3322	195	230	260
250	3	37	4/19 strand	1.737	2977	215	255	290
	4	37	4/19 strand	1.913	3891	215	255	290
350	3	37	3/19 strand	1.960	4026	260	310	350
	4	37	3/19 strand	2.162	5280	260	310	350
500	3	37	2/19 strand	2.238	5572	320	380	430
	4	37	2/19 strand	2.477	7331	320	380	430

TABLE 18.21 PROPERTIES OF ALUMINUM CONDUCTORS IN MC (METAL CLAD) CABLE SUITABLE FOR USE AS FOLLOWS: BRANCH, FEEDER, AND SERVICE POWER DISTRIBUTION IN COMMERCIAL, INDUSTRIAL, MULTIRESIDENTIAL BUILDINGS, THEATERS, AND PLACES OF ASSEMBLY; INSTALLATION IN CABLE TRAY AND APPROVED RACEWAYS. COMPILED FROM NUMEROUS INDUSTRY SOURCES.

MC Cable with Type XHHW-2 Aluminum Conductors

Conductor Sizes (AWG/kcmil)	Subassembly (in)	Overall Nominal Diameter (in)	Approx. Net Weight Per 1000 ft (lb)	Ampacity (Wet or Dry)	
				75°C 140°F	90°C 167°F
Three Conductors Plus Grounding Conductor					
6-6-6-6	0.17	0.78	239	50	60
4-4-4-6	0.17	0.88	299	65	75
2-2-2-4	0.78	1.00	412	90	100
1-1-1-4	0.89	1.11	498	100	115
1/0-1/0-1/0-4	0.97	1.19	581	120	135
2/0-2/0-2/0-4	1.06	1.27	682	135	150
3/0-3/0-3/0-4	1.16	1.38	807	155	175
4/0-4/0-4/0-2	1.27	1.51	1048	180	205
250-250-250-2	1.41	1.65	1222	205	230
350-350-350-1	1.62	1.86	1592	250	280
500-500-500-1	1.88	2.12	2106	310	350
600-600-600-1/0	2.11	2.35	2540	340	385
750-750-750-1/0	2.31	2.55	3043	385	435
Four Conductors Plus Grounding Conductor					
2-2-2-2-4	0.87	1.09	515	72	80
1-1-1-1-4	0.99	1.21	631	80	92
1/0-1/0-1/0-1/0-4	1.08	1.30	740	96	108
2/0-2/0-2/0-2/0-4	1.18	1.40	872	108	120
3/0-3/0-3/0-3/0-4	1.29	1.53	1109	124	140
4/0-4/0-4/0-4/0-2	1.42	1.66	1341	144	164
250-250-250-250-1	1.58	1.81	1584	164	184
350-350-350-350-1/0	1.81	2.05	2077	200	224
500-500-500-500-2/0	2.10	2.34	2788	248	280
600-600-600-600-2/0	2.36	2.59	3340	272	308
750-750-750-750-3/0	2.59	2.82	4054	308	348

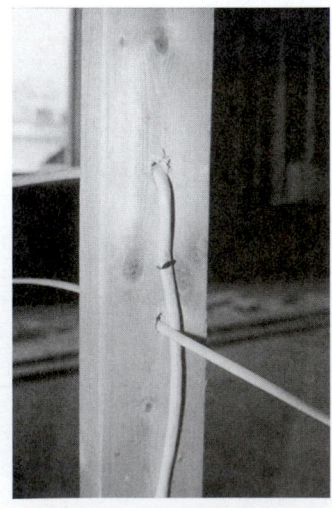

PHOTO 18.17 Nonmetallic (NM) sheathed cabling. NM cabling can pass through studs in residential construction. NM cabling running parallel with framing members must be secured (stapled). (Used with permission of ABC)

PHOTO 18.18 Armored (AC) in a steel-framed partition. (Used with permission of ABC)

conductor. MC cable can be used in many locations not allowed with AC cable.

Flat Conductor Cable

Flat conductor cable (FCC) is a wiring system composed of very thin cable with three or more conductors and special connectors and terminals. It is designed to rest between the topside of a smooth continuous subfloor and carpet squares. It can serve general purpose and appliance circuits up to 20 A and individual circuits up to 30 A, with a system voltage between the ungrounded conductors not exceeding 300 V.

Thermostat Cable

Thermostat cable is used in applications with voltages less than 30 V, such as wiring to doorbells, chimes, and thermostats. It generally contains No. 16 AWG or No. 18 AWG conductors that are bundled in a thin plastic sheathing.

Cords

Cords are made of stranded conductors within a flexible insulated sheathing material. They are designed for flexibility and bending. Cords are generally manufactured of a lighter gauge (e.g., No. AWG 18 or No. AWG 16) stranded conductors. They are designed for use on power tools, large stationary equipment, or detachable computer power cords. *Extension cords* are made of stranded wires because they require flexibility, allowing the cord to be bent and twisted without stressing the conductors.

Concealed Knob-and-Tube Wiring

Concealed knob-and-tube (K&T) wiring consists of an old style wiring technique using insulated conductors strung between glass or porcelain knobs and tubes. In this wiring method, the ungrounded (hot) wire is run along one side of the joist/stud bay and the neutral is run along the other. To secure it to the wood, the wire is wrapped around ceramic *knobs* spaced every 18 in or so. To penetrate a joist/stud and prevent abrasion, the wire is separated from the wood joist/stud by a ceramic *tube*.

K&T wiring is installed in walls or ceilings so it is concealed from view when finish materials such as plaster is applied. Loose or blown-in insulation in framing cavities can encase the knob-and-tub conductors, causing heat build-up in walls or ceilings with insulation. Therefore, concealed knob-and-tube wiring is not permitted in framing cavities where insulation presents this problem.

K&T wiring was the general wiring method until the 1930s, when it was replaced by armored cable. By 1907, the NEC began to recognize the inherent problems with concealment of this type of wiring. It began to require placement of the wires in dry areas only and a separation distance of 5 in between the wires. It was still used in homes through the 1950s. Although it is generally considered inferior to present-day

wiring, permanent concealed knob-and-tube wiring is still allowed in original installations. However, many insurance companies require a certificate of inspection from a licensed electrician before they will insure a building with K&T wiring.

A form of K&T wiring is still used to provide temporary lighting at the construction site, at roadside stands, and at carnival tents. Some restaurants and amusement rides use it for aesthetic appeal. It is not allowed in commercial garages, motion picture studios, theaters, and other hazardous locations.

Conductor Power Loss

Heat generated by current flow through a conductor results in a loss of power. This lost power is referred to as *power loss* or *line loss*. Power loss (P_{loss}) in a conductor can be computed with amperage (I) or voltage (V) and resistance (R) by the following formula:

$$P_{loss} = I^2R = V^2/R$$

Power loss is converted directly to heat. Power loss is equivalent to heat produced. The relationship between power and heat is 1 W = 3.413 Btu/hr. Heat produced (q) for a known power loss (P_{loss}) can be computed by the following formula:

$$q = 3.413 \, P_{loss}$$

Example 18.1

A copper conductor (wire) used in a 20 amp common household circuit has a resistance of 1.62 Ω/1000 ft (from Chapter 17). The total conductor length is 80 ft.

a. Determine the power loss in the conductors, when the circuit is fully loaded.

The resistance in the conductors is found by:

$$R = (1.62 \, \Omega/1000 \text{ ft}) \cdot 80 \text{ ft} = 0.130 \, \Omega$$

The power loss in the conductors is found by:

$$P_{loss} = I^2R = (20 \text{ A})^2 \cdot 0.130 \, \Omega = 52.0 \text{ W}$$

b. Determine the heat produced by the conductors.

The heat produced by the circuit is found by:

$$q = 3.413 \, P_{loss} = 3.413 \text{ Btu/hr/W} \cdot 52.0 \text{ W}$$
$$= 177.5 \text{ Btu/hr}$$

The I^2 characteristic of the power loss equation means that heat produced increases as the square of the current increases—that is, doubling current increases power loss and heat generation by four times, tripling current provides an increase of nine times, and so on. Excessive heat from power loss causes premature deterioration of the conductor insulation, so by limiting power loss to acceptable levels the safety of a wiring system is improved.

Example 18.2

A set of copper conductors on a branch circuit serve a tank pump motor rated at 20 A. The pump is 50 ft from a 120 V power source. Compare the line losses of No. 14, No. 12, and No. 10 AWG copper conductors. A No. 14 AWG conductor has a resistance of 2.57 Ω/1000 ft, a No. 12 AWG conductor has a resistance of 1.62 Ω/1000 ft, and a No. 10 AWG conductor has a resistance of 1.02 Ω/1000 ft.

Conductors are 50 ft (one way), so the total conductor length is: 50 ft $\cdot$ 2 = 100 ft.

$$P = I^2R = (20\ A)^2 \cdot (2.57\ \Omega/1000\ ft) \cdot 100\ ft$$

$$= 102.8\ W\text{—No. 14 AWG conductor}$$

$$= (20\ A)^2 \cdot (1.62\ \Omega/1000\ ft) \cdot 100\ ft$$

$$= 64.8\ W\text{—No. 12 AWG conductor}$$

$$= (20\ A)^2 \cdot (1.02\ \Omega/1000\ ft) \cdot 100\ ft$$

$$= 40.8\ W\text{—No. 10 AWG conductor}$$

Power consumed by the pump is:

$$P = EI = 120\ V \cdot 20\ A = 2400\ W$$

Percentage of power loss is:

102.8 W/2400 W = 4.28% line loss—No. 14 AWG conductor

64.8 W/24000 W = 2.70% line loss—No. 12 AWG conductor

40.8 W/24000 W = 1.70% line loss—No. 10 AWG conductor

In Example 18.2, if the upper limit of power loss is 3%, then the No. 14 AWG copper conductor is unacceptable for use in this branch circuit because its line losses are too great. No. 12 AWG conductors are acceptable. A No. 14 AWG conductor is smaller in size than a No. 12 AWG conductor. The No. 14 AWG conductor has a larger resistance to current flow. This added resistance leads to an unacceptable amount of power loss, which will result in excessive heating. Such excessive heating may lead to failure of the insulation surrounding the conductor and the potential for short-circuiting and fire.

The indirect method of determining power loss that is used in Example 18.2 is intended to provide a detailed overview of the concept of voltage drop and how it relates to power loss. Voltage drop can be easily computed by using Ohm's Law: E = IR. Percentage of voltage drop is simply the relationship between voltage drop and system voltage. Line voltage is system voltage less the voltage drop.

Example 18.3

A No. 12 AWG copper conductor with a two-way length of 250 ft is carrying a load of 20 A on a 277 V, two-wire circuit. Approximate the voltage drop in the conductor, the percentage of voltage drop, and the line voltage.

From Table 18.8, a No. 12 AWG conductor has a resistance of 1.62 Ω/1000 ft. Voltage drop is found by:

$$E_{drop} = IR = 20\ A \cdot 1.62\ \Omega/1000\ ft \cdot 250\ ft = 8.1\ V$$

Percentage of voltage drop in this circuit is found by:

$$8.1\ V/277\ V = 0.029 = 2.9\%$$

Line voltage is found by:

$$277\ V - 8.1\ V = 268.9\ V$$

Conductor Ampacity

Electrical current flowing through a circuit produces heat from the resistance of the conductor material to current flow. In building wiring systems, some heat is permitted as part of design. Excessive heating is considered undesirable and unsafe because it will prematurely degrade conductor insulation, resulting in the danger of short circuits and ground faults. As a result, a type of conductor is rated based upon the highest amperage (current) it can carry safely without overheating and damaging its insulation.

A conductor's *ampacity* is the maximum current (in amperes) it can carry continuously without exceeding the temperature limitations of the insulation and sheathing material. Simply, it is a conductor's maximum current-carrying capacity. Ampacity is based on the following:

- Wire thickness (thicker wires have larger cross-sectional areas and can carry more electrical current without overheating)

- Type of conductor material (at a specific current and conductor size, aluminum produces more heat than copper)

- Insulation and sheathing type (some insulation materials handle heat better than others)

- Number of conductors bundled in the sheathing or in proximity of one another (more conductors concentrate heat in an area)

- Temperature and exposure of the conductor (e.g., buried, in free air, in attic, in crawl space, and so forth).

Ampacity tables specified in the National Electrical Code, Canadian Electrical Code, and technical literature from professional organizations (e.g., Insulated Cable Engineers Association) are basic approximations with a substantial safety margin. They are sufficient for most installations. There are instances where the application of the ampacity tables including the safety margins is insufficient, requiring engineers, installers, and inspectors to perform actual calculations.

Tables 18.11 through 18.13 and 18.15 through 18.21 contain ampacities for various conductors, conductor insulations, and sheathings. Ampacities provided in these tables are values based on a normal operating temperature of 86°F (30°C). Computation methods are beyond the scope of this text.

Voltage Drops in Conductors

Because of power losses, voltage is reduced in a closed circuit—that is, voltage across two conductors is lower at the usage end than at the power supply end of the circuit. Appliances and equipment work inefficiently on voltages lower than the voltage for which they were designed. In heating devices, the heat output varies with the square of the voltage applied to the device. A 10% drop in the voltage results in a 19% decrease in the heat output. Under the same conditions, the input current (amperage) at rated mechanical load to a motor would increase about 11% and the heating of the motor conductors would increase about 23%, a hazardous condition. Therefore, in addition to ampacity, voltage drop is a design concern with long conductor runs.

Voltage drop is directly proportional to power loss. This is true because current flow (I) through a conductor will not change. And, with the power equation $P = EI$, because power is lost (wattage available is less), the voltage (E) must drop. Refer to Kirchhoff's Current and Voltage Laws introduced in Chapter 17. This is demonstrated in Figure 18.16, where voltage available to a 20 A load at an outlet 50 ft away (100 ft two way wire length) from the panelboard is 116.8 V. This voltage is acceptable because it is in the range of 110 V to 130 V. However, for the 250 ft length (five times the length), voltage available at the end usage point is 103.8 V, an unacceptable voltage.

Consider the power losses in a No. 12 AWG conductor in Example 18.3 (see Figure 18.16): A 100 ft length of this No. 12 AWG conductor was found to have a power loss of about 65 W when carrying a current of 20 A. Power available at the outlet is:

$$2400 \text{ W} - 65 \text{ W} = 2335 \text{ W}$$

With the power equation ($P = EI$) introduced in Chapter 17 and a current flow of 20 A, voltage available at the point of usage is:

$$E = P/I = 2335 \text{ W}/20 \text{ A} = 116.8 \text{ V}$$

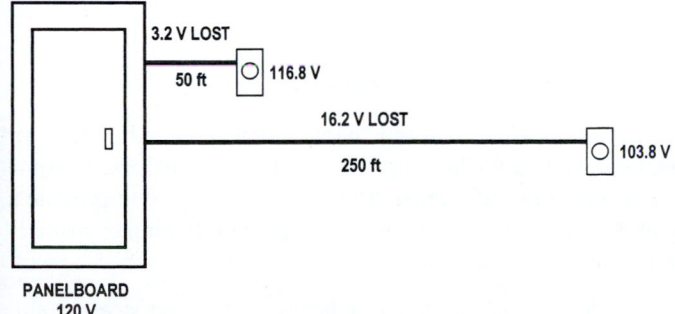

FIGURE 18.16 Voltage available to a 20 A load at an outlet 50 ft away (100 ft two-way wire length) from the panelboard is 116.8 V, which is acceptable because it is in the range of 110 V to 130 V. Assuming a 250 ft length (five times the length), voltage available at the end usage point is 103.8 V, an unacceptable voltage.

The slight drop in voltage will not greatly affect most appliances, because they are usually designed to operate in the range of 110 V to 130 V. However, assuming a 500 ft length (five times the original 100 ft length), power loss is approximately $5 \cdot 65 \text{ W} = 325 \text{ W}$. Power and voltage available at the end usage point of the 500 ft run is:

$$2400 \text{ W} - 325 \text{ W} = 2075 \text{ W}$$

$$2075 \text{ W}/20 \text{ A} = 103.75 \text{ V}$$

Such a low voltage at an outlet is unacceptable on a 120 V system, because it is outside the acceptable 110 V to 130 V range. This type of voltage drop will result in an unsatisfactory operation of electrical equipment. Motors and similar equipment operate at only a fraction of their rated capacity, leading to lower output, overloading, hotter operation, and reduced life. Excessive voltage drop in a circuit can cause lights to flicker or burn dimly and heaters to heat less effectively. Similarly, electronic devices, such as video displays, televisions, and radios, will not operate at this low voltage (e.g., generally a TV picture gets smaller as voltage drops). Thus, with long conductor runs, resistance to current flow causes an excessive voltage drop. The conductor size must be increased to limit the voltage drop to acceptable levels.

The previous analysis approach does yield somewhat inaccurate results because the analysis did not account for the voltage drop at the end of each 100 ft segment (or shorter segments) of the 500 ft conductor. It does demonstrate the effect of a voltage drop in long lengths of conductor. All wiring sections and the entire system should be considered for voltage drop concerns.

18.6 ENCLOSURES AND RACEWAYS

Enclosures

Enclosures are electrical boxes and cabinets made of metal (e.g., steel, galvanized steel, aluminum, and so on) or non-metallic (plastic) materials that provide protection for conductors, connections, controls, and other electrical equipment. They protect the wiring, devices, and equipment from damage and deterioration from accidental contact, wear, corrosive atmospheric exposure, and sunlight. In cases where the enclosed conductors, connections, controls, and electrical equipment are faulty, enclosures contain the arc and flaming that result, thereby confining damage from fire.

Electrical Boxes

Electrical boxes are metal and non-metallic (plastic) enclosures that hold devices such as switches or outlets and safely permit wiring connections. Boxes are available in four primary shapes: square, rectangular, octagonal, or round. Boxes can be joined (ganged) together or are manufactured to accommodate multiple outlets or switches. These are known as two-gang, three-gang or four gang boxes. *Junction boxes* (J-boxes) are a special type of electrical box used to enclose conductor connections. Connections are commonly called *junctions* in the trade. A *pull box* is a type of junction box that allows access to a raceway for snaking conductors through the raceway. *Knockouts* in most boxes and other enclosures can be easily removed to allow wiring to enter the box. Common sizes of electrical boxes are provided in Table 18.22. Examples of electrical boxes are shown in Photos 18.19 and 18.20.

   Cover plates for various single or combinations of switches, convenience outlets, blank covers, and dimmers are required. Plastic is the most common material used, although

PHOTO 18.19   Switches in a four-gang nonmetallic electrical box. Nonmetallic (NM) sheathed cabling is entering the box. (Used with permission of ABC)

TABLE 18.22   COMMON SIZES OF ELECTRICAL BOXES. COMPILED FROM INDUSTRY SOURCES.

Enclosure/Box Type	Enclosure/Box Dimensions				
	Depth in	Length in	Height in	Number of Gangs	Capacity in³
Round or octagonal (lighting)	1¼	4	4	—	12.5
	1½	4	4	—	15.5
	2⅛	4	4	—	21.5
Square (devices)	1¼	4	4	1	18
	1½	4	4	1	21
	1¼	4¹¹⁄₁₆	4¹¹⁄₁₆	1	25.5
	1½	4¹¹⁄₁₆	4¹¹⁄₁₆	1	29.5
	1½	4¹¹⁄₁₆	4¹¹⁄₁₆	—	29.5
	2⅛	4¹¹⁄₁₆	4¹¹⁄₁₆	1	42
	2⅛	4¹¹⁄₁₆	4¹¹⁄₁₆	—	44
	2⅛	4¹¹⁄₁₆	4¹¹⁄₁₆	—	44
	2⅛	4¹¹⁄₁₆	4¹¹⁄₁₆	—	44
	2⅛	4¹¹⁄₁₆	4¹¹⁄₁₆	—	44
Rectangular (devices)	1½	2	3	1	7.5
	1½	2⅛	4	1	10.3
	1⅝	3½	4½	1	22.5
	1⅝	7¹⁄₁₆	4½	2	45
	1⅝	8⅞	4½	3	58
	1⅝	10¹¹⁄₁₆	4½	4	70
	1⅝	12½	4½	5	85
	1⅝	14⁵⁄₁₆	4½	6	95
	1⅝	16⅛	4½	7	105
	1⅝	17¹⁵⁄₁₆	4½	8	115
	1⅝	19¾	4½	9	125
	1⅝	21⁹⁄₁₆	4½	10	135
	2	2	3	1	10
	2¼	2	3	1	10.5
	2½	2	3	1	12.5
	2¾	2	3	1	14
	2½	3½	4½	1	35.5
	2½	7¹⁄₁₆	4½	2	71
	2½	8⅞	4½	3	90
	2½	10¹¹⁄₁₆	4½	4	110
	2½	12½	4½	5	132
	2½	14⁵⁄₁₆	4½	6	150
	2½	16⅛	4½	7	168
	2½	17¹⁵⁄₁₆	4½	8	186
	2½	19¾	4½	9	204
	2½	21⁹⁄₁₆	4½	10	222

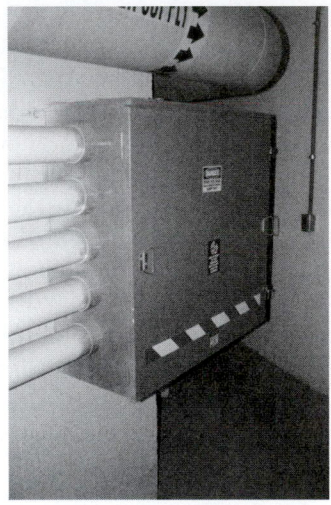

PHOTO 18.20   A large junction box. (Used with permission of ABC)

stainless steel, ceramic, or plain galvanized steel is also available. Covers are attached with screws to the switch or receptacles. Where lighting fixtures are attached directly to the box, no cover plate is required, but all other boxes should be finished with a cover plate.

Conduit and Other Raceways

A *raceway* is as an enclosed channel such as a conduit, tube, or gutter designed for holding wires, cables, or busbars. In some cases, metal raceways that are fully bonded serve to provide a path for the flow of fault current to ground. Examples of raceway materials or systems follow:

   Rigid metal conduit is a heavy galvanized steel or aluminum tube that looks like the galvanized steel pipe used for plumbing applications except it is much smoother and is labeled with a *UL Listed* stamp or label. It has threaded connections much like pipe. It can be used in contact with earth and embedded in concrete. See Photo 18.21.

PHOTO 18.21 Four electrical boxes connected to metal conduit in a steel-framed partition. (Used with permission of ABC)

Intermediate metal conduit (IMC) is a galvanized steel or aluminum tube that has a thinner wall than rigid metal conduit. It has threaded connections much like pipe or threadless connectors and couplings. It can be used in contact with earth and embedded in concrete.

Electrical metallic tubing (EMT) is a thin-walled galvanized steel or aluminum tube in nominal diameters up to 4 in. Unlike conduit, EMT cannot be threaded. EMT is labeled with a *UL Listed* stamp or label. It is joined with threadless compression couplings, couplings that are crimped into the tubing with a special indenting tool, or fittings that are tightened with screws.

Electrical nonmetallic tubing (ENT) is a flame-retardant corrugated plastic tube that is semiflexible such that it can be bent by hand. ENT must be used behind fire-rated finish materials. It is permitted in damp locations and can be set in concrete above ground. Plastic connectors snap together. It can also be solvent welded to rigid polyvinyl chloride (PVC) conduit. It can be used in any building, but special precautions are required when used more than three floors above grade.

Rigid nonmetallic conduit is a thin-walled pipe of PVC. It is joined with fittings that are solvent welded.

Flexible metal conduit is similar to armored cable, but it is installed without cables or wiring in it; wiring must be pulled.

Liquid tight flexible metal conduit is similar to flexible metal conduit, but it is covered with a plastic, watertight jacket that is sunlight resistant. Unlike flexible metal conduit, the special connectors are watertight.

Liquid tight flexible nonmetallic conduit is a flexible plastic conduit used in a manner similar to flexible metal conduit. It cannot be used over 6 ft in length.

Cellular concrete floor raceways are hollow voids in floors made of precast concrete slabs (core slabs) found in certain precast concrete buildings.

Sizes of common electrical conduits are provided in Table 18.23.

The use of rigid conduit, EMC, EMT, and ENT involves cutting, bending, and in some cases, threading of lengths. Cutting is typically done with a saw. Bends in a conduit must be made so that it will not be damaged and that the internal diameter of the conduit will not be effectively reduced. Bends can be made at the factory or in the field with a manual bender, called a *hickey,* or a *hot box heater* for rigid nonmetallic (plastic) conduit.

A run of conduit should be as straight and direct as possible. All raceways must be installed as a complete system before any conductors are pulled into them. Straps or hangers must support conduit throughout the entire run. The number of supports needed and spacing of supports depends on the type of conduit being used. The number of conductors you can have in a conduit is based on the size of the conduit, the type of conductor insulation, and the size of the conductors. When several conductors are fed into a conduit, they should be kept parallel, straight, and free from kinks and bends.

Busways

A *busway* is of a standardized, factory-assembled enclosure consists of outer duct-like housing, bus bars, and insulators. Busway systems are typically used in service equipment or as feeders because these systems are designed to carry large amounts of current.

Two general types of busway systems are available: feeder and plug-in. A *feeder busway* is used to deliver large amounts of power with low voltage drop. It is available in sizes from 600 A to several thousand amps. A *plug-in busway* is used to provide power tap-offs at multiple points. It is available in 100 A to 3000 A sizes. Plug-in busways make power distribution flexible. Power is available anywhere along the busway without necessitating a main service interruption. Thus, equipment (e.g., for manufacturing and machining) can be relocated and placed in operation simply, economically, and quickly, merely by inserting a plug-in unit into the nearest busway tap-off point. Connections may be made at any point along the busway. Rigid and flexible conduit or flexible bus drop cable extends from the busway to the piece of equipment.

As a result of initial cost, busways are only used in large commercial and industrial installations (e.g., manufacturing plants) where large loads are concentrated (about 600 A and higher), where a large number of taps are required, or where superior fire and mechanical injury protection is desirable. Busways can also be salvaged and moved more easily and cost-effectively than ordinary conductors.

Wireways

Wire gutters or *wireways* are sheet metal or nonmetallic, flame-resistant plastic troughs that serve as a housing that encloses and protects conductors. Access to the enclosure interior is through a hinged door or removable cover. Wire gutters and wireways typically carry large conductors. Photo 18.22 shows conductors in an overhead wire tray.

TABLE 18.23   COMMON SIZES OF ELECTRICAL CONDUIT. COMPILED FROM NUMEROUS INDUSTRY SOURCES.

Trade Size	Electrical Metallic Tubing (EMT)				Rigid Conduit (Metal)				Intermediate Metal Conduit (IMC)			
	Outside Diameter	Inside Diameter	Wall Thickness	Inside Area	Outside Diameter	Inside Diameter	Wall Thickness	Inside Area	Outside Diameter	Inside Diameter	Wall Thickness	Inside Area
in	in	in	in	in²	in	in	in	in²	in	in	in	in²
½	0.706	0.622	0.042	0.30	0.840	0.632	0.104	0.31	0.815	0.675	0.070	0.36
¾	0.922	0.824	0.049	0.53	1.050	0.836	0.107	0.55	1.029	0.879	0.075	0.61
1	1.163	1.049	0.057	0.86	1.315	1.060	0.126	0.88	1.290	1.120	0.085	0.99
1¼	1.510	1.380	0.065	1.50	1.660	1.394	0.133	1.53	1.638	1.468	0.085	1.69
1½	1.740	1.610	0.065	2.04	1.900	1.624	0.138	2.07	1.883	1.703	0.090	2.28
2	2.197	2.067	0.065	3.36	2.375	2.083	0.146	3.41	2.360	2.170	0.095	3.70
2½	2.875	2.731	0.072	5.86	2.875	2.489	0.193	4.87	2.857	2.597	0.130	5.30
3	3.500	3.356	0.072	8.85	3.500	3.090	0.205	7.50	3.476	3.216	0.130	8.12
3½	4.000	3.834	0.083	11.54	4.000	3.570	0.215	10.01	3.971	3.711	0.130	10.82
4	4.500	4.334	0.083	14.75	4.500	4.050	0.225	12.88	4.466	4.206	0.130	13.89
5	—	—	—	—	5.563	5.073	0.245	20.21	—	—	—	—
6	—	—	—	—	6.625	6.093	0.266	29.16	—	—	—	—

Trade Size	PVC Schedule 40 Conduit				PVC Schedule 80 Conduit			
	Outside Diameter	Inside Diameter	Wall Thickness	Inside Area	Outside Diameter	Inside Diameter	Wall Thickness	Inside Area
in	in	in	in	in²	in	in	in	in²
½	0.840	0.622	0.109	0.30	0.840	0.546	0.147	0.23
¾	1.050	0.824	0.113	0.53	1.050	0.742	0.154	0.43
1	1.315	1.049	0.133	0.86	1.315	0.957	0.179	0.72
1¼	1.660	1.380	0.140	1.50	1.660	1.278	0.191	1.28
1½	1.900	1.610	0.145	2.04	1.900	1.500	0.200	1.77
2	2.375	2.067	0.154	3.36	2.375	1.939	0.218	2.95
2½	2.875	2.469	0.203	4.79	2.875	2.323	0.276	4.24
3	3.500	3.068	0.216	7.39	3.500	2.900	0.300	6.61
3½	4.000	3.548	0.226	9.89	4.000	3.364	0.318	8.89
4	4.500	4.026	0.237	12.73	4.500	3.826	0.337	11.50
5	5.563	5.047	0.258	20.01	5.563	4.813	0.375	18.19
6	6.625	6.065	0.280	28.89	6.625	5.761	0.432	26.07

PHOTO 18.22 Conductors in an overhead wire tray. (Used with permission of ABC)

18.7   ELECTRIC MOTORS

In homes, electric motors are found in refrigerators, freezers, dishwashers, kitchen sink waste disposal, portable kitchen appliances, exhaust and ventilation fans, clothes washers and dryers, furnaces, air conditioners, and paddle fans. In commercial and industrial buildings, electric motors drive exhaust/ventilation fans, air-handling unit blowers, refrigeration equipment, air and fluid compressors and pumps, heating system circulators, and manufacturing equipment and machinery. The electric motor serves us well beyond the limits of what one human being can produce.

Motors are designed to operate on voltages slightly less than the building system voltage, because voltage available at the motor outlet is less than the building system voltage because of a voltage drop on the branch circuit conductors. For example, a motor may be designed for operation at a line voltage of 230 V rather than a building system voltage of 240 V.

Most AC 60 Hz electric motors used in North America are designed to operate at theoretical 1800 or 3600 revolutions per minute (rpm). When delivering its rated horsepower, the actual speed of a motor will be somewhat less, about 1725 and 3450 rpm, respectively. Other motors are designed to operate at lower speeds of slightly less than 900 and 1200 rpm. Motors designed to operate at slower speeds are larger, operate quieter, and last longer because they encounter less friction.

When an appliance or piece of equipment must operate at a slower speed, this is accomplished by using a gear or pulley/belt system to gear down or gear up to the required rpm. For example, a clothes dryer tumbler is driven by a motor turning at 1800 (actually about 1750) rpm that is geared down 25 times so the tumbler revolves at a rate of 72 (actually about 70) rpm.

Motor Ratings

Electric motors are rated in *horsepower (hp)*. One horsepower is equivalent to 33 000 foot-pounds (ft-lb) of work per minute (550 ft-lb/s). This is about ⅛ the power an adult can produce continuously. In theory, one horsepower is equivalent to 746 W. However, because of losses from heat and friction, an electric motor delivers less work than the theoretical equivalent; more power is consumed than is produced. For example, a 1 hp motor draws about 1400 W, much more than the theoretical equivalent of 746 W.

At startup, an electric motor (especially a single-phase motor) consumes substantially more electrical power than when it is operating at its rated speed and load. This is known as inrush current and is discussed later in this chapter (see the section titled Inrush Current Protection). A 230 V, 1 hp motor that normally draws about 1000 W (4.3 A) under full load may draw 6000 W (26.1 A) or more at startup. Once operating at its design speed, electrical power and thus amperage drawn by the motor is related to the load it is delivering; that is, a 230 V 1 hp motor delivering ½ hp will consume about 600 W (2.6 A) and the same motor delivering 1 hp will consume about 1000 W (4.3 A). As is evident when comparing power requirements for half and full loading, the electric motor operates more efficiently at its full horsepower rating.

Electric motors have a unique advantage over fuel-powered motors in that they can deliver more horsepower than their rating. It does require more power; that is, the same 230 V 1 hp motor can deliver 1½ hp, but it will draw more power (about 1700 W) and amperage (7.4 A). The ability to handle an overload temporarily allows the motor to overcome an obstacle much like a table saw works harder to cut through a tough knot that it encounters while cutting through a piece of wood.

Continuous overloading, however, causes a decrease in efficiency and overheating, which greatly reduces the life of the motor.

An electric motor's *locked rotor amperage* (LRA) is the highest amperage that a motor pulls. The LRA occurs the moment the motor is first switched on but before the motor rotor begins to turn. Once the motor rotor begins to move, the amperage drops until the motor reaches its operating speed and amperage. A motor operating at its rated speed, voltage, and horsepower draws current at its *full load amperage* (FLA) or *running load amperage* (RLA). An electric motor's LRA is usually four to six times greater than the FLA/RLA. Conductors serving an electric motor are generally sized to the LRA. Approximate full load ratings for selected motors, in amperes, based on horsepower, phase, and line voltage are provided in Table 18.24.

Most electric motors are designed to run at 50% to 100% of rated load, with maximum efficiency typically near 75% of rated load. For example, a 10-horsepower (hp) motor has an acceptable load range of 5 to 10 hp; peak efficiency is at 7.5 hp. A motor's efficiency tends to decrease dramatically below about 50% load.

A motor is considered underloaded when it is in the range where efficiency drops significantly with decreasing load. Overloaded motors can overheat and lose efficiency. Frequently motors are oversized because they must accommodate peak conditions (e.g., a pumping system that must satisfy occasionally high demands). If equipment with motors operates for extended periods under a 50% load, modifications should be considered. Options available to meet variable

TABLE 18.24 APPROXIMATE FULL LOAD RATING FOR SELECTED MOTORS, IN AMPERES, BASED ON HORSEPOWER, PHASE, AND LINE VOLTAGE.

Horsepower	Single Phase			Three Phase		
	120 V	208 V	230 V	208 V	240 V	460 V
⅓	7.0	3.8	3.5	2.1	2.0	1.0
½	8.4	4.6	4.2	2.6	2.4	1.2
¾	9.6	5.3	4.8	3.0	2.8	1.4
1	11.8	6.5	5.9	3.7	3.4	1.7
1.5	15.6	9.1	8.3	5.0	4.8	2.4
2	20.8	11.4	10.4	6.2	6.0	3.0
3	—	18.3	15.6	9.8	9.6	4.8
5	—	25.0	23.0	14.5	14.2	7.1
10	—	—	—	29.0	28.0	14.0
15	—	—	—	42.0	40.0	20.0
20	—	—	—	52.0	50.0	25.0
25	—	—	—	—	60.0	30.0
30	—	—	—	—	72.0	36.0
40	—	—	—	—	104.0	52.0
50	—	—	—	—	120.0	60.0
60	—	—	—	—	146.0	73.0
75	—	—	—	—	184.0	92.0
100	—	—	—	—	230.0	115.0
125	—	—	—	—	290.0	145.0
150	—	—	—	—	330.0	165.0

loads include two-speed motors, adjustable speed drives, and load management strategies that maintain loads within an acceptable range.

A motor's *service factor* indicates the percentage a motor can be continuously overloaded without developing damage related to overheating. For example, a 10-hp motor with a 1.15 service factor can handle an 11.5-hp load for short periods of time without incurring significant damage. Although many motors have service factors of 1.15, running the motor continuously above rated load reduces efficiency and motor life. Motors over 1 hp generally have service factors up to 1.20. *Fractional motors,* those rated less than 1 hp, can have service factors approaching 1.50. Electric motors operate most efficiently at their rated horsepower. A proper match between an electric motor rating in horsepower and its load in horsepower is good design practice.

Motors are designed to operate at an ambient (surrounding air) temperature of 40°C (104°F). A motor operating in a location above this temperature should be cooled with a blower or fan or not used at its rated horsepower. Motors are typically rated for use at an altitude of up to 3300 ft (1000 m) above sea level. The cooling effect of less-dense air found at high altitudes is less than air at or near sea level. A motor operated at its full-rated horsepower at high altitudes will overheat, so compensation is necessary.

Types of Motors

The following is a description of the types of motors.

Universal Motor

The universal motor is a fractional horsepower (less than one horsepower) motor designed to operate on both AC and DC power. Its rate of rotation varies considerably with load. It operates at high speeds under light load and low speeds with heavy load. The universal motor can free idle at up to 20 000 rpm. Universal motors are used on appliances such as blenders and vacuum cleaners and power tools such as routers and electric drills.

Split-Phase Motor

The split-phase motor operates on single-phase AC only. The motor windings are configured so that single-phase AC power is split into two phases that are ½ out of phase. This type of motor starts slowly with low torque so it is not capable of starting heavy loads. The split-phase motor is available in sizes up to ⅓ horsepower.

Capacitor Motor

This type of motor operates on single-phase AC only. Capacitor-start motors have a capacitor that stores and discharges energy to help start the motor rotor. Capacitor-run motors have one or more capacitors to help start and run the motor. The capacitor motor is more efficient and has a better starting torque than the split-phase motor.

Induction Motors

These motors use electromagnetic induction to cause the motor rotor to turn. Repulsion-start induction motors are capable of handling heavy starting loads.

Three-Phase Motors

Large motors operate more efficiently on three-phase AC power. Electric motors rated up to 7½ hp can operate on a single-phase AC system. However, a single-phase motor rated at 7½ hp draws a large amount of instantaneous current at startup (up to 200 A on a 240 V single-phase circuit). As a result, conductor size is large and costly. Three-phase motors also vibrate less than single-phase motors, so they are lighter and less costly.

Dual-Voltage Motors

Motors above ¼ hp are designed to operate on one of two different line voltages such as 115 V or 230 V. Larger motors operate more efficiently on a higher line voltage because of lower I^2R losses. So, if a higher line voltage is available, it is used.

The specifications of an electric motor are typically listed on the nameplate. Information provided may include:

Manufacturer's type and frame designation

Horsepower output

Power factor

Maximum ambient temperature for which motor is designed

Insulation system designation

Revolutions per minute at rated load

Frequency

Number of phases

Rated load current

Voltage

Motor Controllers

A *motor controller* is a switching device designed to start, stop, and protect the motor. A controller might also be called on to provide functions such as reversing, jogging (repeated starting and stopping), plugging (rapid stopping by momentarily reversing the polarity of the motor), operating at several speeds, or at reduced levels of current and motor torque.

Industrial plants and manufacturing facilities have hundreds of motors all requiring specialized control and protection. Rather than scatter them throughout the facility, they are grouped together into *motor control centers (MCC)*. An MCC is a centrally located, sheet metal, cabinet-like enclosure that houses starters and controls that control and protect several motors. The front panel of an MCC contains operator controls and gauges. The interior of the MCC contains plug-in units such as starters, controls, and specialized units.

A *variable-frequency drive (VFD)* is a solid-state electronic power conversion device used for controlling the rotational speed of an AC electric motor by controlling the frequency of the electrical power supplied to the motor. The motor used in a VFD system is typically a three-phase induction motor, but some types of single-phase motors can be used. Induction motors are suitable for most purposes and are generally the most economical choice. When a VFD starts a motor, it initially applies a low frequency (typically 2 Hz or less) and voltage to the motor, which avoids the high current surge that occurs when a motor is typically started by simply turning on a switch. The stopping sequence is just the opposite as the starting sequence.

Pumps, fans, conveyors, air compressors, and chillers operate more efficiently when the speed and torque of their motor can be varied. The savings potential for centrifugal pumps and fans is the largest because the theoretical input power varies with the cube of fan/pump speed and volume. For example, a fan operating at half speed will require only about 13% of its full-speed power. Losses in the VFD will reduce savings somewhat, but the savings are still significant.

18.8   OCCUPANT PROTECTION

Need for Occupant Protection

In the United States, hundreds of people are accidentally electrocuted each year. Electrocution occurs when a small amount of electrical current flows through the heart for 1 to 3 s. The amount of 0.006 to 0.2 A (6 to 200 milliamps, or mA) of current flowing through the heart disrupts the normal coordination of heart muscles. These muscles lose their vital rhythm and begin to fibrillate. Death soon follows. To provide an example of how small an amount of current it takes to kill: a 15 W nightlight on a 120 V circuit draws about 13 mA, enough amperage to cause electrocution.

Tamper-Resistant Receptacles

According to U.S. Consumer Product Safety Commission data, approximately 2400 children suffer electrical injuries each year from incidents involving electrical outlets or receptacles. *Tamper-resistant receptacles* have built-in shutter systems that prevent foreign objects from touching electrically live components when these are inserted into the slots. The shutters protect against electrical burns without impairing normal plug insertion, removal, or function. Tamper-resistant receptacles have emerged because of the tendency of children to unwittingly endanger themselves by inserting keys, pins, paper clips, or other items into unprotected receptacles. Products such as plastic plug-in inserts and wall plates with contact shutters are available for tamper resistance, but do not meet the requirements of a tamper-resistant receptacle.

There are several methods to achieve tamper-resistance operation, the most common being the use of a spring-loaded shutter mechanism. Inside the face of the receptacle is a spring-loaded thermoplastic safety shutter. Under normal, unused conditions, the shutters are closed and both contact openings are covered. Upon insertion of a grounded or ungrounded plug, the blades of the plug simultaneously compress the shutters against the spring. The simultaneous force allows the shutters to slide and open up access to the receptacle contacts. The plug becomes fully inserted and securely fits into the receptacle. When the plug is removed, the shutters instantly close, covering the contact openings. When a foreign object, such as a paper clip or small bladed screwdriver, is inserted into only one of the openings, the safety shutter will not allow access to the live contact. For these receptacles, the safety mechanism covers only the line and load contacts, and not the ground. Because they accept a two-pronged plug, these receptacles can be used with standard household and office appliances like light fixtures, clocks, and radios.

Ground Fault Interruption

A *ground fault* is the unintentional flow of electrical current between a power source, such as an ungrounded (hot) wire, and a grounded surface. A ground fault occurs when electrical current leaks or escapes to ground. When a hot bare conductor inside an appliance inadvertently touches the metal housing, the housing may become charged with electricity. If a person touches the faulty appliance and at the same time touches a grounded metal object, such as a water faucet or metal sink, the person will receive a shock because the person's body serves as an inadvertent path to ground.

A ground fault is a danger to humans because only a very low current is required to affect human heart rhythm. The commonly cited threshold for inducing life-threatening ventricular fibrillation (abnormal irregular heart rhythm that causes cardiac arrest) in an average adult male is 100 mA (0.0001 A). The 120 V circuit typically found in building electrical systems can easily drive this amount of current through a total resistance of 1000 Ω, which is in the range of the resistance of a human body with skin-to-conductor contact. Current levels this small are very small in comparison to current at a normal load and thus are not detected by fuses and circuit breakers. As a result, circuits where ground faults are a concern that require specialized protection.

The U.S. Consumer Product Safety Commission reports that, in the United States, there are over 300 accidental electrocution deaths each year that are caused by ground faults. Additionally, there are several thousand more individuals who suffer injury or burns from severe electrical shock. The Commission advises that a majority of these injuries or fatalities could have been prevented by use of an inexpensive device called a ground fault circuit interrupter.

A *ground fault circuit interrupter (GFCI)* is an electrical device that detects an extremely low leak (6 mA) of electrical current (called *ground faults*) and acts quickly to shut off power. It is designed to protect the user of an electrical appliance much like a circuit breaker or fuse safeguards the wiring in an electrical system. A GFCI continuously monitors the current drawn through the ungrounded (hot) and neutral conductors of an electrical circuit. When a leakage to ground that exceeds 6 mA is detected, the GFCI instantaneously switches off power to the branch circuit or appliance, thereby protecting a person from the dangerous effects of electrical shock. Power is shut off in less than a heartbeat (about 1/40th of a second), which is hopefully before serious or fatal shock occurs.

Three types of GFCIs are commonly available for use in a building.

Receptacle Outlet Type

This type is generally used in place of standard duplex convenience outlets that are commonly found throughout the house. A GFCI convenience outlet fits into the standard electrical outlet box and protects the user against ground faults whenever an electrical appliance is plugged into the outlet. Most convenience outlet-type GFCIs can be installed so that they will also protect other electrical outlets further downstream in the branch circuit. Many people are unaware of the presence of GFCI protection in a home, yet they are familiar with the unique-looking TEST and RESET buttons found on the GFCI-protected convenience outlets in modern bathrooms, kitchens, and outdoor outlets. A receptacle outlet-type ground fault circuit interrupter is shown in Photo 18.23.

Circuit Breaker Type

A GFCI circuit breaker can be installed in the panelboard in buildings equipped with circuit breakers. The GFCI circuit breaker gives protection to the entire branch circuit. The wiring and each outlet that is served by the branch circuit is protected by the GFCI breaker. By providing overcurrent protection as a circuit breaker and serving to provide GFCI protection, the GFCI circuit breaker serves a dual purpose: it will interrupt power in the event of a ground fault and it will trip when a short circuit or an power overload occurs.

Portable Type

Where permanent GFCIs are not possible or practical, portable GFCIs may be used. One type contains the GFCI circuitry in a plastic enclosure with plug blades in the back and convenience outlet slots in the front. It can be plugged into a convenience outlet so an electrical appliance plugged into the GFCI is protected. Another type of portable GFCI is one that is part of an extension cord, such as those required on new-model hair dryers.

One- and two-pole GFCI circuit breakers are available. One-pole GFCI breakers can be installed on the 120 V portion of a 120/240 V AC, 1Φ-3W system that is typically found in residential and small commercial installations. Two-pole GFCI breakers are typically used in commercial and industrial applications. They can be installed on a 120/240 V AC, 3Φ-4W system, the 120/240 V portion of a 120/240 V AC, 3Φ-4W system, and two ungrounded phases and the grounded phase of a 120/208 V AC, 3Φ-4W system.

Arc Fault Protection

In the United States, there are 42 900 fires per year from electrical equipment according to National Fire Protection Association (NFPA) data. These fires cause 370 fatalities and over $615 million in property damage annually. Of these fires, 15 200 are from fixed wiring, 7800 are from faulty cords and plugs, and 8400 are from problems in lamp and light fixtures.

In a 1987 study, the U.S. Consumer Product Safety Commission found that fires are located in every area of residential dwellings: bedrooms, living rooms, kitchens, closet/storage areas, garages, bathrooms, laundries, halls, and dining rooms (listed in decreasing order of occurrence). It has been estimated that arcing and sparking problems in building wiring are tied to more than 40 000 home fires each year. These fires claim over 350 lives and injure about 1400 victims annually.

An *arc fault* is an unintentional electrical discharge (an electrical arc) characterized by low and erratic current. Arcing generates high-intensity heat and expels burning particles, which can easily ignite combustible materials. Arc faults are caused by the breakdown of the protective insulation that surrounds household wiring. These breakdowns occur naturally as the wiring ages and can be exacerbated by dust, settling and shifting of a home's foundation, or by rodents. New wiring is also at risk from drywall staples, picture-hanging nails, or any sharp object that could nick the wire. Current residential breakers only detect and react to power overloads, not arc faults.

PHOTO 18.23 Receptacle outlet-type ground fault circuit interrupter. (Used with permission of ABC)

Types of arc faults common in building wiring are the following:

- Parallel arcing faults result from direct contact of two wires of opposite polarity. Examples of this type of fault include appliance or extension cords that are frayed or ruptured; staples or other fasteners that pierce or pinch insulation on construction wire and appliance or extension cords; and wire or cord insulation that has cracked from age, heat, corrosion, or bending stress.

- Ground arcing faults are arcs between a single conductor and ground, such as in the cases of wire or cords that touch vibrating metal; in appliances, wall plugs or switches where the internal wires were not installed properly; and where connections became loose.

- Series arcing faults occur across the break of a single conductor—for example, in the case of an electrical wire cut by a nail or screw used to mount a wall hanging.

A safety device, called an *arc fault circuit interrupter (AFCI)*, provides enhanced protection from fires resulting from arc faults. This device uses electronics to recognize an arc fault and interrupts the circuit when the fault occurs. Essentially, an AFCI continuously monitors the current and voltage characteristics in a circuit, senses variations in these characteristics, and automatically opens the circuit (trips the AFCI breaker) when arc fault characteristics are detected.

An AFCI detects low-level arc faults that traditional overcurrent protective devices (fuses and circuit breakers) cannot detect. Traditional overcurrent protection perceives a low-level arc fault as a normal load unless the current flow exceeds its rating. As a result, fuses and circuit breakers do not respond to early arcing and sparking conditions in building wiring. By the time a fuse or circuit breaker opens a circuit to stop these conditions, a fire may already have begun.

The basic application of an AFCI is protection of 15 A and 20 A branch circuits in single- and multifamily residential occupancies. They are available as circuit breakers with built-in AFCI features that combine traditional thermal-magnetic overcurrent protection with the ability to detect and interrupt electrical arcs. There are four basic types of AFCIs.

Circuit Breaker Type

A branch/feeder AFCI breaker with protection provided to branch-circuit wiring in the form of a circuit breaker.

Convenience Outlet Type

An outlet AFCI for protecting connected cord sets and power-supply cords in the form of an outlet receptacle.

Portable Type

A portable AFCI for protecting connected cord sets and power-supply cords that can be moved from outlet to outlet.

Cord-Mounted type

A cord-mounted AFCI for protecting the power-supply cord connected to it (in the form of an attachment plug on a power-supply cord).

Nuisance Tripping

Because GFCIs and AFCIs are extremely sensitive, they have a tendency to trip frequently. This repeated tripping is referred to as *nuisance tripping*, as the general public sees it only as a nuisance. Appliances that are beginning to fail can cause nuisance tripping. They should be repaired or replaced. Lightning can also cause nuisance tripping. For this reason, it is not a recommended practice to connect essential equipment and appliances containing perishable products (e.g., refrigerators, freezers) into an outlet with GFCI or AFCI protection.

Extremely Low-Frequency Electromagnetic Fields

Extremely low-frequency electromagnetic fields (EMF) are silent, invisible magnetic fields produced any time electricity runs through a wire, an appliance, or piece of equipment. In buildings, higher levels of EMF can cause computer monitor interference and raise potential health concerns. High levels of EMF produce *electromagnetic interference (EMI)*, which reveals itself as visible screen jitter in video displays, humming

TABLE 18.25 A COMPARISON OF TYPES OF PROTECTION DEVICES.

Type of Protection Device	Type of Protection Devices				
	Fuse	Circuit Breaker	Ground Fault Circuit Interruption	Arc Fault Circuit Interruption	Surge Protection
Thermal protection	Yes	Yes	Yes	Yes	
Overload protection	Yes	Yes	Yes	Yes	
Short circuit protection		Yes	Yes	Yes	
Ground fault protection			Yes	Possibly*	
Arc fault protection				Yes	
Surge protection					Yes

*May become available on some devices.

in telephone/audio equipment, and data errors in magnetic media or digital signals. In the United States, EMF is measured in units called *milligauss (mG)*. In most of the world and the scientific community, magnetic fields are measured in units of *microtesla (µT)*. The relationship between these two units is $1 \, \mu T = 10 \, mG$.

EMF measurement is performed with an instrument called an ELF (extremely-low frequency) meter. The strength of an EMF weakens rapidly with the distance away from the electrical source, so these EMFs decrease rapidly with distance away from the source. Measurements must be done in multiple locations over a considerable period of time, because there are large variations in fields over distance and time. Field surveys are required that provide precise data taken at numerous points at multiple elevations (e.g., floors above and below the suspected source).

EMF Exposure

Most Americans get their greatest exposure to EMFs from household appliances or business equipment, not power lines. EMF from power transmission lines varies greatly with distance away from the lines, voltage, and design. Below a high-voltage (115 to 765 kV) power transmission line, EMF can approach 1.0 mG (0.01 µT); 30 ft (10 m) from a medium voltage (12 kV) distribution line, EMF will be in the range of 2.0 to 10 mG (0.2 to 1.0 µT). Electric fields from power lines have little ability to penetrate the building envelope.

Household appliances that have the highest EMF are those with high currents or high-speed electric motors (e.g., vacuum cleaners, microwave ovens, electric washing machines, dishwashers, blenders, can openers, electric shavers). Contacting appliances like blow dryers (1400 mG/140 µT), can openers (4000 mG/400 µT), and electric shavers (1600 mG/160 µT) or equipment like power drills (500 mG/50 µT) expose one to high levels of EMF. The usual background levels of exposure in the typical household are 1.5 to 2.0 mG (0.15 to 0.2 µT).

Occupants in offices, schools, and other settings can be exposed to EMF levels ranging between 2 to 4 mG (0.2 to 0.4 µT). Power passing through electrical circuits in some commercial buildings can create areas with relatively high magnetic fields (above 100 mG/10 µT), typically near transformer vaults, network protectors, secondary feeders, switchboards, distribution busways, and electrical rooms. EMF exposure can also come from video displays, computers, networks, electronic instruments, audio/video equipment, and magnetic media. The sensitivity of most computer monitors to magnetic field interference is commonly found in environments with magnetic field strengths in the range of 5 to 7 mG (0.5 to 0.7 µT). Some models of cardiac pacemakers have been shown to be susceptible to EMF interference as low as 1000 mG (100 µT).

Wiring errors in conduits, busways, and other distribution circuits can cause elevated levels of EMF in buildings. In a properly wired circuit, current goes out and returns along the same circuit path (e.g., through the same cable, conduit, or busway). This causes a natural canceling effect of magnetic fields from electric currents in each conductor, unless circuiting is done improperly. In addition, mistakes in wiring can cause current to take an alternative (and improper) neutral path back to ground. These wiring errors are known as *net-current* conditions. They produce elevated EMF levels in building areas, through which circuit wiring is routed, which will expose large sections of the building to high EMF levels.

Currently, there are no Federal standards for EMF levels. Expert opinions vary on recommendations for safe human exposure limits, ranging from 2.5 to 100 mG (0.25 to 10 µT). For equipment affected by EMF, the threshold for EMF interference depends on many variables but has commonly been estimated at between 5 to 10 mG (0.5 to 1.0 µT).

EMF Health Studies

There have been hundreds of studies that examined the relationship between human health and EMF. A study of child leukemia deaths in Denver and a later study in Sweden of nearly one-half million people indicate that children living near power lines had a higher rate of leukemia than other children did. These results prompted further investigation. Most epidemiological studies show little evidence of a relationship to exposure to power lines or electrical occupations (e.g., welders, electricians) and an increase in cancer. The International Agency for Research on Cancer classified EMF as "possibly carcinogenic" to humans, requiring further research. Although far from being conclusive, many studies suggest an association between exposure to electromagnetic fields and the development of certain health problems (e.g., sleep rhythm disorders, leukemia in children) but not cancer.

EMF Mitigation

Several methods are successful in reducing EMF exposure. First, the source generating EMF or the work area or equipment affected by high EMF can be relocated away from the field, if practical. If relocation is not possible or economical, shielding of magnetic fields can be employed to reduce EMF levels.

Passive shielding can be accomplished by using a conductive sheet material, similar to a window screen, in front of the appliance or equipment. The shield must be connected to electrical ground for maximum efficiency. With extremely high levels, active shielding can be used to eliminate high levels. *Active shielding* uses a system that senses the existence of a magnetic field in the building area to be shielded and generates a current on additional conductors such that it reduces (cancels the effect of) the magnetic field.

Magnetic fields caused by net-current wiring problems are very difficult to shield using traditional shielding materials and methods. This is mainly because of the large area that can be affected. In these cases, high EMF levels are best reduced by correction of wiring errors.

18.9 EQUIPMENT PROTECTION

Ground Fault Protection of Equipment

Most relatively new commercial buildings in the United States are served with 480/277 V electrical service (e.g., versus 240/120 V and 208/120 V; the electrical services commonly used in the past). Although the higher voltage provides increased capacity, it can result in an undesirable side effect tied to a combination of the higher voltage and the characteristic spacing between energized parts and grounded metal in electrical distribution equipment. Problems can be attributed to arcing to ground at current levels well below standard load current so that if it happens it does not blow a fuse or trip a circuit breaker. Simply put, the small spacing between electrical components and high-voltage power creates a condition in which arcing faults can develop and persist fairly easily, resulting in melting of metal components, burning of insulation, and explosive arcing.

Ground fault protection (GFP) is designed to detect and rapidly interrupt low-level equipment ground faults. A GFP operates on the same principle as GFCI protection, by monitoring the current drawn through the conductors of an electrical circuit and quickly shutting off power when current is not equal. A special type of sensing transformer, called a *current transformer (CT)*, encircles the conductors in the switchboard or panelboard and produces a low-current output signal if all current flowing to the load does not return to the source through the phase or neutral conductors. The low-current output is sensed by a ground fault relay, which has adjustable sensitivity and time delay settings. When a signal above the sensitivity setting persists beyond the time delay setting, the relay operates to open the switch or circuit breaker.

Typical ground fault relays have pickup settings that are adjustable from 100 to 1200 A and time delays that are adjustable from instantaneous tripping to 0.5 s. Ground fault protection is generally required on low-voltage systems with solidly grounded neutrals rated 1000 A or more when the voltage from any phase to ground is over 150 V. In most buildings, GFP is applied only at the main service disconnect because it is the only circuit rated over 1000 A.

Surge Protection

A *power surge* is a sudden increase in electrical current or voltage that is very short in duration. This increase may be caused by a lightning strike or from a sudden power spike caused by a problem in the utility transmission and generating system. The switching of fluorescent light fixtures or other heavy equipment can also cause a power surge. A power surge can damage sensitive electronic equipment such as computers, fax machines, televisions, stereos, VCRs, and electronic phone systems. Even where no obvious damage has occurred, the life of electronic equipment can be significantly reduced following a power surge. Failure of equipment may not happen immediately, but may take months to happen.

A *surge protection device (SPD)* is an electrical device that prevents power surges from reaching electric and electronic equipment or other device. SPDs work by instantaneously limiting the transient voltage from a power surge to a level that is safe for the equipment they protect by diverting the large surge current safely to ground. The SPD diverts the surge by allowing the current to flow past rather than through the protected equipment.

Ordinary surge protectors simply divert harmful surge current from the hot line to the neutral and ground wires, but this will *increase* the likelihood of damage to audio, video, and computer data lines, because these lines use the power line ground wire circuit for their reference voltage. More sophisticated surge protection technology is designed so that unwanted voltages are not passed through to protected equipment via power, telephone, telemetry, or other cables. They prevent excessive energy from reaching sensitive parts of equipment or systems by disconnecting signal lines and diverting surges to earth. Once the surge current has subsided, the advanced SPD automatically restores normal operation and resets to a state ready to receive the next surge. Types of SPD include the following.

Metal Oxide Varistor (MOV)

A semiconductor device used mainly on AC power applications that has surge ratings ranging from a few hundred to many thousands of amps.

Gas Discharge Tubes (GDT)

A primary surge protection component with a surge rating of several thousand amps that is comprised of a sealed tube containing a special gas that breaks down at a given voltage rise time.

Zener Diode

A secondary surge protection component used for accurate clamping of surge voltages. It has a quicker response time than the GDTs and provides a more accurate clamping voltage than MOVs.

Hybrid Circuit

A circuit comprising different types of surge protection component, taking advantage of each component's strengths. For example, a hybrid may combine the high surge capability of a GDT with the accurate voltage clamping of a surge diode.

An ideal SPD operates rapidly and is capable of carrying large currents (amperage) for short periods while limiting the voltage across or the current through protected equipment to levels below which damage can take place. Maintenance-free and self-resetting devices are normally preferred where interruptions to service should be avoided. The *surge rating* is a measure of the level of surge that an SPD can withstand. Manufacturers specify the surge ratings of their SPDs.

Inrush Current Protection

Inrush current or *input surge current* refers to the peak instantaneous current, measured in amperes (A), drawn by an electrical appliance or piece of equipment (e.g., power supplies, AC motors, lamps, and lighting ballasts) when it is first turned on. Current can be as high as 100 times the normal steady state current. On a normal 60 hertz cycle, this momentary surge of current usually lasts from less than ½ cycle to a few cycles depending upon the piece of equipment being powered. For example, AC electric motors can draw several times their normal full-load current when first energized; incandescent lamps have high inrush currents until their filaments warm up and their resistance increases; and transformers can have an inrush current up to 50 times larger than normal rated current flow when first energized. Usually, an inrush current is not dangerous but can be a nuisance because the high surge of current can blow fuses and trip circuit breakers, thereby shutting the down the circuit.

Selection of overcurrent protection devices such as fuses and circuit breakers is more complex when high inrush currents are anticipated. Ideally, overcurrent protection must react quickly to an overload or short-circuit, but must not interrupt the circuit when the momentary inrush current flows. Typically, time delay fuses or circuit breakers can be used to avoid nuisance tripping from inrush current. Because inrush current is short in duration, the time delay prevents such events.

Inrush current can also be reduced by an *inrush current limiter*, which protects circuits that encounter high inrush loads up to 30 times the rated current for one-half cycle at 60 Hz and eliminate nuisance tripping. Although many techniques are used to limit inrush current, *negative temperature coefficient (NTC) thermistors* are the most commonly used inrush current limiter for power supplies and motors. A *thermistor* is a thermally sensitive resistor with a resistance that changes significantly and predictably as a result of temperature changes. When equipment is first turned on, the NTC thermistor is cool and its resistance is high, which limits current flow. Resistance decreases rapidly as the thermistor self-heats, which then allows near-full current to flow (at $<1\ \Omega$). A disadvantage is that when the NTC thermistor is switched off, it initially retains heat, so it offers little resistance and cannot limit the inrush current until heat is dissipated. Cool-down time varies according to the particular device, its mounting method, and ambient temperature. The typical cool-down time is about 1 min. NTC resistors are rated according to their resistance at room temperature.

Lightning Protection

More than 100 lightning flashes occur every second in the atmosphere, which equates to about 8.6 million per day and over 3 billion annually. About 20% of these flashes actually strike the ground.

Lightning discharges have been measured from 2000 A to more than 200 000 A, with rise times to peak current of less than 10 microseconds (10 millionths of a second). As a result, lightning causes more damage, injuries, and deaths each year than tornadoes, hurricanes, or floods. In the United States, about 250 people are killed annually by direct lightning strikes and another 500 will die in lightning-caused fires. Additionally, at least 1500 persons are injured by lightning each year. Many of the deaths and injuries would not have occurred if proper lightning protection equipment was installed. Yet, the Lightning Protection Institute reports that 95% of today's homes are not protected against lightning surges.

The idea of protecting buildings and other structures from the effects of direct lightning strikes by the use of protective conductors was first suggested by Benjamin Franklin. The theory presented has not changed. Simply put, lightning protection must provide a direct path for the lightning bolt to follow to ground and it must prevent destruction, injury, or death as the current travels that path.

Vertical lightning rods, sometimes called *air terminals*, are placed at the top of the protected structure. These lightning rods are equally spaced a maximum of 20 ft apart along the high points on the structure. Rods are also placed within 2 ft of the building ends and on the top of cupolas, chimneys, antenna mast, and towers. A network of conductive copper or aluminum tape (thick, flat cable) bonds the rods to grounding electrodes that are driven into the earth. Metal roof vents in proximity of the lightning rods are also bonded to the conductive tape.

The system of vertical rods, tape, and ground electrodes is designed to present a low impedance (resistance) path to the lightning current that diverts current away from the structural parts of the building. Modern buildings of reinforced concrete or brick-clad steel frames may use the structural steelwork as part or the entire conductor network. There is also a wide variety of other structural metal in buildings, which may be used as part of the protection network or as air terminators (e.g., window cleaning rails).

A lightning strike does not have to directly contact the building to cause damage to the building's electrical system. A lightning strike on nearby power lines can induce high voltages that damage the building's wiring system and electronic equipment within the building. Although proper grounding of the building's electrical system can limit damage to the building's wiring, unprotected electronic equipment is usually damaged. A surge protection device can minimize the damaging effects of a lightning strike (see the earlier section on SPDs). Additionally, power passing through electrical circuits can generate relatively high magnetic fields that damage electronic equipment. Shielding the wiring can minimize damage (see the earlier section on EMF).

Equipment for Hazardous Locations

In a hazardous location, such as a gasoline station, paint-spray booth, or factory, explosion-proof or explosion-resistant equipment and wiring must be used. A *hazardous location* is a location in which fire or explosion hazards may exist because of the presence of flammable gases or vapors, flammable liquids, combustible dust, or easily ignited fibers under normal operation or abnormal operating conditions. Types of hazardous locations are grouped by class and division or by zone. Buildings or building spaces that are identified as hazardous locations must meet regulations specific to the group it falls into.

Technically, the definitions of explosion-proof and explosion-resistant differ. Explosion-proof receptacles, switches, enclosures, fixtures, equipment are specially designed to withstand an explosion that may occur within it and of preventing the ignition of a specified gas or vapor surrounding the enclosure. Conduit that is used in an explosion-proof location must be sealed so that gas may not move from one enclosure to another. In contrast, explosion-resistant equipment is designed to prevent ignition of an explosive or flammable material. A Nationally Recognized Testing Laboratory (NRTL), such as UL, accomplishes testing of explosion-proof or explosion-resistant equipment and wiring components.

STUDY QUESTIONS

18-1. What are power generation, transmission, and distribution?

18-2. What is the service entrance of a building electrical system and what are its components?

18-3. What are the functions and differences of switchboards and panelboards?

18-4. What is the difference between a wye-connected and delta-connected transformer?

18-5. What is the difference between a step-down and step-up transformer?

18-6. What is a circuit breaker and where is it used in a building electrical system?

18-7. What is a fuse and where is it used in a building electrical system?

18-8. What types of fuses are used in building electrical systems?

18-9. What is the difference between a receptacle and plug-in building electrical systems?

18-10. Distinguish between the terms *appliance* and *device*. Give examples of each.

18-11. What types of switching configuration is used to control a lighting installation from two points? What types of switches are used and how are switches arranged in the circuit?

18-12. What types of switching configuration is used to control a lighting installation from three points? What types of switches are used and how are switches arranged in the circuit?

18-13. What types of switching configuration is used to control a lighting installation from four points? What types of switches are used and how are switches arranged in the circuit?

18-14. Describe series and parallel circuit. Which is used in building electrical systems?

18-15. What are the types of specialty switches?

18-16. What are types of conductor (wire) materials and where are they used in building electrical systems?

18-17. What are types of conductor insulation materials and where are they used in building electrical systems?

18-18. What is a cable and where is it used in building electrical systems?

18-19. What is a cord and where is it used in building electrical systems?

18-20. What is a bus (bar)?

18-21. What are power loss and voltage drop and how are they related?

18-22. What is ampacity?

18-23. What factors is the ampacity of a conductor based on?

18-24. What are types of raceways used in building electrical systems? Explain each.

18-25. What is a busway and where is it used in building electrical systems?

18-26. How are electric motors rated?

18-27. With regard to electric motors, what are LRA, RLA, and FLA?

18-28. What are the types of motors used in building electrical systems?

18-29. What does a motor service factor indicate?

18-30. What is a motor controller?

18-31. What is a variable-frequency drive (VFD) and what is its advantage?

18-32. What is a tamper-resistant receptacle?

18-33. What is a GFCI, how does it function, and where is it used?

18-34. What is an AFCI, how does it function, and where is it used?

18-35. What is a surge protection device (SPD), how does it function, and where is it used?

18-36. What is GFP, how does it function, and where is it used?

18-37. What is lightning protection, how does it function and where is it used?

18-38. What is EMF and why is it a potential concern in buildings?

18-39. What is an inrush current limiter, and where is it used?

18-40. In a single panelboard, what is the maximum number of overcurrent protection devices that may be used for protecting lighting/appliance branch circuits?

18-41. It is referred to in the trade by several names: power panel, load center, distribution center, or main power panel. Code refers to it by a single name. Identify this name.

Design Exercises

18-42. With respect to conductor insulation, interpret the following insulation designations:

a. T

b. N

c. H

d. HH

e. W

18-43. With respect to conductors, interpret the following designations:

a. THHN

b. THWN

c. XHHW

d. AL

e. CU

18-44. With respect to cable, interpret the following designations:

a. NM

b. ACTH

c. UF

d. USE

18-45. With respect to conductor insulation, describe the following location ratings:

a. Dry

b. Damp

c. Wet

18-46. A single-phase, three-wire panelboard must feed 30 circuits. From tables provided in this chapter, identify the minimum frame size required.

18-47. A single-phase, three-wire panelboard must feed 36 circuits. From tables provided in this chapter, identify the minimum frame size required.

18-48. A three-phase, four-wire panelboard must feed 42 circuits. From tables provided in this chapter, identify the minimum frame size required.

18-49. A three-phase, four-wire panelboard must feed 42 circuits. From tables provided in this chapter, identify the frame sizes available to meet this requirement.

18-50. From tables provided in this chapter, identify the NEMA designation for the following wall-mounted, grounding-type devices used in a residence that are rated at 15 A, 125 V, and serve as a connection method for a two-pole, three-wire circuit:

a. Receptacle

b. Plug

18-51. From tables provided in this chapter, identify the NEMA designation for the following heavy-duty grounding-type devices that are rated at 20 A, 125 V, and serve as a connection method for a two-pole, three-wire circuit:

a. Receptacle

b. Plug

18-52. From tables provided in this chapter, identify the NEMA designation for the following grounding-type devices used for clothes dryers:

a. Receptacle

b. Plug

18-53. From tables provided in this chapter, identify the NEMA designation for the following grounding-type devices used for a kitchen range requiring a 50 A, 125/250 V rating:

a. Receptacle

b. Plug

18-54. From tables provided in this chapter, identify the NEMA designation for the following grounding-type devices used for copy machines and air conditioners requiring a 50 A, 125 V rating:

a. Receptacle

b. Plug

18-55. From tables provided in this chapter, identify the NEMA designation for the following grounding-type devices requiring a 20 A, 250 V four-pole, four-wire, three-phase rating:

a. Receptacle

b. Plug

18-56. From tables provided in this chapter, identify the NEMA designation for the following grounding-type devices requiring a 20 A, 250 V four-pole, four-wire, three-phase rating:

a. Receptacle

b. Plug

18-57. From tables provided in this chapter, identify the ampacity of the following conductors:

a. No. 8 AWG copper conductor with THHN insulation and a temperature rating of 60°C/140°F

b. No. 8 AWG copper conductor with THHN insulation and a temperature rating of 75°C/167°F

   c. No. 8 AWG copper conductor with THHN insulation and a temperature rating of 90°C/194°F

   d. No. 8 AWG aluminum conductor with THHN insulation and a temperature rating of 75°C/167°F

18-58. From tables provided in this chapter, identify the ampacity of the following conductors:

   a. No. 4/0 AWG aluminum conductor with THHN insulation and a temperature rating of 60°C/140°F

   b. No. 4/0 AWG aluminum conductor with THHN insulation and a temperature rating of 75°C/167°F

   c. No. 4/0 AWG aluminum conductor with THHN insulation and a temperature rating of 90°C/194°F

18-59. From tables provided in this chapter, identify the ampacity of the following conductors:

   a. No. 500 kcmil aluminum conductor with XHHW-2 insulation and a temperature rating of 60°C/140°F

   b. No. 500 kcmil aluminum conductor with XHHW-2 insulation and a temperature rating of 75°C/167°F

   c. No. 500 kcmil aluminum conductor with XHHW-2 insulation and a temperature rating of 90°C/194°F

18-60. Nonmetallic (NM-B) cable is commonly used in single-family residential installations. From tables provided in this chapter, identify the ampacity of the following conductors:

   a. NM-B cable with three No. 14 AWG copper conductors having THHN or THWN insulation

   b. NM-B cable with three No. 12 AWG copper conductors having THHN or THWN insulation

   c. NM-B cable with three No. 10 AWG copper conductors having THHN or THWN insulation

18-61. Armored cable (AC) is commonly used in commercial, industrial, institutional, and multiresidential installations. From tables provided in this chapter, identify the ampacity of the following conductors:

   a. AC cable with three No. 14 AWG copper conductors having THHN or THWN insulation

   b. AC cable with three No. 12 AWG copper conductors having THHN or THWN insulation

   c. AC cable with three No. 10 AWG copper conductors having THHN or THWN insulation

18-62. From tables provided in this chapter, approximate the full load rating for a 10 hp, 208 V, three-phase electric motor, in amperes.

18-63. From tables provided in this chapter, approximate the full load rating for a 1½ hp, 120 V, single-phase electric motor, in amperes.

18-64. From tables provided in this chapter, approximate the full load rating for a 100 hp, 460 V, three-phase electric motor, in amperes.

BUILDING ELECTRICAL DESIGN PRINCIPLES

19.1 ELECTRICAL CODES, LICENSES, AND PERMITS

The Electrical Code

Design of an electrical system involves applying the standards written into the building code. An electrical code specifies the minimum provisions necessary for protecting people and property from the improper use of electricity and electrical equipment. It applies to both the manufacture and installation of electrical equipment.

National Electrical Code

The *National Electrical Code (NEC)* is a set of specifications and standards in the form of a model code that can be adopted into local law by the local governmental entity. Most municipalities and counties in the United States require that residential and commercial electrical wiring conform to the NEC. In some jurisdictions, certain NEC requirements are superseded by local requirements.

The NEC was established in 1897 through the combined efforts of insurance, electrical, architectural, and other interested groups. In 1911, the *National Fire Protection Association (NFPA)* became the sponsor of the NEC and continues to act in this capacity. The NEC is revised every three years to ensure that only the latest safety methods and procedures are used in electrical installations. In the United States, a new edition of the NEC does not become a legal regulation until it is adopted into law by the local governmental entity. Each new edition must be enacted into law after a formal review of the new release. Generally, review of the new edition takes time and effort, so it is possible that a city ordinance may still require use of an earlier edition even though a more recent edition has been released.

Canadian Electrical Code

The *Canadian Standards Association (CSA)* is the body that publishes and administers the *Canadian Electrical Code (CEC)*. The *Canadian Electrical Code Committee* develops the Canadian Electrical Code. This committee's members represent regulatory authorities, manufacturers, installers, consumer groups, and the Canadian Electrical Association (utility industry advocate). The CEC establishes safety requirements for electrical work and the installation of electrical equipment operating, or intended to operate, at all voltages in electrical installations for buildings, structures, and premises across Canada. It is similar but not identical to the National Electrical Code; there are differences.

The CEC is revised every four years. When a new edition of the CEC is published, Canada's provinces and territories act to adopt it into legislation, usually with local revisions.

As the Code is revised and a new edition is enacted into local law, previously acceptable regulations may become outdated; that is, what was formerly an acceptable electrical installation may no longer be acceptable. The newly enacted Code generally applies to new installations only. Existing electrical installations that do not meet the current, more stringent regulations still remain permissible as long as they are not regarded as a hazard.

A *grandfather clause* allows outdated electrical installations to remain in use if they are not regarded as hazardous. An outdated installation must be changed only if it is part of a renovation project. For example, at the time a home was built in 1975, NEC requirements for ground fault circuit interruption (GFCI) protection of convenience receptacles in the kitchen did not exist and GFCI protection was likely not installed. The existing non-GFCI receptacles in the kitchen can remain in use without GFCI protection even though GFCI protection of certain kitchen receptacles was introduced in the 1987 edition of the NEC. If the kitchen was remodeled today, GFCI protection must be added to these outlets.

Manufacturing Standards

Today, all electrical equipment, appliances, and devices should meet specific safety standards based on regular product testing. An *approved* product meets minimum safety standards as determined by extensive testing by an independent testing company or organization.

Underwriters Laboratory, Inc. (UL), is a not-for-profit product safety testing and certification organization. It is the leading third-party certification organization in the United States and the largest in North America. It has been evaluating products in the interest of public safety since 1894. UL is a leading developer of safety standards. The UL Standards for Safety are designed to be compatible with nationally recognized installation, building, and safety codes. *ETL Testing Laboratories, Inc.,* of Cortland, New York, is another internationally recognized, fully independent testing company. In Canada, the *CSA* develops industrial standards. All electrical equipment for sale in Canada must bear a CSA product certification mark.

A UL or ETL Mark on a product (CSA Mark in Canada) indicates that the device or piece of equipment has met minimum safety standards as determined by extensive testing. Testing of the product is done at the laboratory and by unannounced visits to the manufacturing plant to check ongoing production to ensure the product continues to meet applicable standards. These marks, sometimes known as *labels,* are found on electrical devices, fixtures, appliances, panelboards, and thousands of other products.

The *National Electrical Manufacturing Association (NEMA)* is the leading trade association in the United States representing the interests of electrical manufacturers. NEMA's member companies manufacture products used in the generation, transmission and distribution, control, and end use of electricity. NEMA has been developing standards for the electrical manufacturing industry since 1926 and is today one of the leading standards development organizations in the world. As such, it contributes to an orderly marketplace and helps ensure the public safety.

The *Council for Harmonization of Electrotechnical Standards of the Nations of the Americas* (founded in 1992) is known in most other countries in North and South America as *Consejo de Armonizacion de Normas Electrotecnicas de las Naciones de las Americas (CANENA).* The purpose of CANENA is to facilitate and promote the development of harmonized electrotechnical codes and standards and uniform conformity assessment methods. CANENA is not a standards developing organization. It is an industry-driven organization that encourages the reduction of nontariff trade barriers between the member countries. The ultimate objective of CANENA is to have one standard for a product for all of the western hemisphere, submit the product to a conformance assessment-testing laboratory in one of the countries, and upon successful completion of the testing, receive listings in all countries.

Licensing

Most municipalities have ordinances (local laws) that require that any person who wishes to engage in the business of installing electrical systems must be licensed (usually by the state or province). This means that the person must have a minimum number of years of experience working with a licensed electrician and must pass a written test that deals with the electrical code being used and with methods of installation.

By requiring a license, it is assured that the electrician knows, at a minimum, the code requirements and the installation procedures. There are areas where no laws require that only licensed electricians may install electrical systems and, in effect, there is no protection for the consumer against an unskilled electrician. It is good practice to insist on licensed electricians for all installations.

In some states, electrical contractors must register with the state electrical board. Documentation of current workers' compensation and unemployment insurance coverage is usually required. The individual must sign an acknowledgment of responsibility form to register as an electrical contractor.

Permits

Most municipalities require that a permit be issued before any electrical installations may be made on a project. A complete electrical construction drawing may also be required for review and approval by a plans examiner before installation begins. This is typically the case on large projects. Other municipalities may not require drawings at all. In general, most municipalities that require electrical permits also require licensed electricians.

Municipalities that require a permit will have electrical inspectors who check the project during regularly scheduled visits. Typically, they will inspect the installation after the rough wiring is in but before it is concealed behind construction materials. When all of the fixtures and devices are installed and wired back to the panel and the service and meter installed, the final inspection is completed. On large projects, it may be necessary for several rough-in and final inspections as electrical work may be done in phases. For example, conduit encased in concrete may have to be inspected before the concrete is poured, and conduit to be built into masonry walls will have to be inspected before the walls are begun.

The installer and designer should become aware of when inspections are required and what will be inspected. Also, it is important that close coordination and cooperation be maintained with the inspector, because the inspector could slow down the progress of the work if the inspection is not made promptly. Whenever possible, the electrical inspector will need to know as early as possible when inspections will be scheduled.

19.2 ELECTRICAL CONSTRUCTION DRAWINGS

Electrical construction drawings show the layout and design of an electrical installation. A complete set of construction drawings and specifications of the building electrical system is needed to convey design information to the contractor. The following construction drawings and details are generally required:

1. Complete plans and specifications of all electrical work
2. Labeling criteria of all electrical equipment
3. Lighting floor plan(s) including electrical circuits indicating conduit and wiring sizes
4. Power floor plans including electrical circuits indicating conduit and wiring sizes, equipment, and disconnect switches
5. Exit sign/means of egress lighting location and power supply

6. Panelboard schedule

7. Lighting fixture schedule

8. Symbol schedule and diagrams

Specifications and drawings should include requirements for:

1. Raceway and conduit with fittings

2. Wire and cable

3. Electrical boxes, fittings, and installation

4. Electrical connections

5. Electrical wiring devices

6. Circuit and motor disconnects

7. Hangers and supporting devices

8. Electrical identification

9. Service entrance and details

10. Overcurrent protection

11. Switchboards

12. Grounding

13. Transformers

14. Panelboards

15. Motor control centers

16. Lighting fixtures

Symbols are used on construction drawings to represent lighting and power distribution components and equipment. Figure 19.1 shows commonly used symbols. No specific symbol set is universally accepted by the industry. The symbols provided are typically modified by the engineering consulting firm and used on their drawings.

Many manufacturing companies have their own set that may be distinctly different than the symbols provided. For example, the symbol S_D may mean a door-controlled switch but might denote a dimmer switch on another set of drawings. The legend of symbols is an important part of the construction drawings. It should always be reviewed to ascertain the meaning of an electrical symbol on a specific drawing.

Symbols are arranged on the drawings in the form of a one-line diagram. One line represents the conductors extending between two symbols, regardless of the number of conductors. Generally, the single line without any additional identification indicates a two-wire set of conductors—either an ungrounded (hot) and grounded conductor or two ungrounded conductors. A single line with three slashes indicates a three-wire set of conductors and so on.

ABBREVIATIONS AND ACRONYMS

1Φ	Single phase
3Φ	Three phase
A	Ampere
A	Glass fiber or similar insulation material
AC	Alternating current
AFCI	Arc fault circuit interrupter
AFI	Arc fault interrupter
AH	Ampere-hour
AIC	Amperes interruption current rating
AL	Aluminum
ANSI	American National Standards Institute
AWG	American wire gauge
C	Capacitance
CB	Circuit breaker
CSA	Canadian Standards Association
CU	Copper
CV	Constant voltage
CVT	Constant voltage transformer
DC	Direct current
DPDT	Double-pole, double throw (switch)
DPST	Double-pole, single throw (switch)
E	Voltage
EMF	Electromotive force
EMI	Electromagnetic interference
F	Farad (measure of capacitance)
FEP	Fluorinated ethylene propylene insulation material
GFCI	Ground fault circuit interrupter
GFI	Ground fault interrupter
GND	Ground
H	Heat resistant insulation material
HH	Extra heat resistant insulation material
hp	Horsepower
Hz	Hertz (frequency in cycles per second)
k	Kilo (1000 units)
kA	Kilo ampere
kcmil	Thousand circular mils
kV	Kilovolts
kVA	Kilovolt-amp
kW	Kilowatts
kWh	Kilowatt-hours
L	Inductance
L	Lead sheathed
M	Mega
MCM	Thousand circular mils
MW	Megawatt
MWh	Megawatt-hour
N	Nylon jacketed insulation material
NC	Normally closed (switch/relay)
NEC	National Electrical Code
NEMA	National Electrical Manufacturers Association
NFPA	National Fire Protection Association
NO	Normally open (switch/relay)
NSF	National Science Foundation
OSHA	Occupational Safety and Health Administration
OVP	Over voltage protection
PB	Panelboard
PCB	Printed circuit board
PF	Power factor
PF	Perfluoroalkooxy insulation material
PWR	Power
R	Recessed
R	Resistance
R	Thermoset insulation material (e.g., synthetic rubber)
RPM	Revolutions per minute
S	Silicone (thermoset) insulation
SPDT	Single pole double throw (switch)
SPST	Single pole single throw (switch)
T	Thermoplastic insulation
UL	Underwriters Laboratories. Inc.
V	Volt
VA	Volt ampere
VAC	Volts alternating current
VAR	Volt amperes reactive
VDC	Volts direct current
W	Moisture resistant
Wh	Watt hour
X	Cross-linked synthetic polymer insulation material
Z	Impedance
Z	Modified tetrafluoroethylene insulation material

FIGURE 19.1 Common electrical abbreviations, acronyms, and symbols. Symbols will vary with set of drawings.

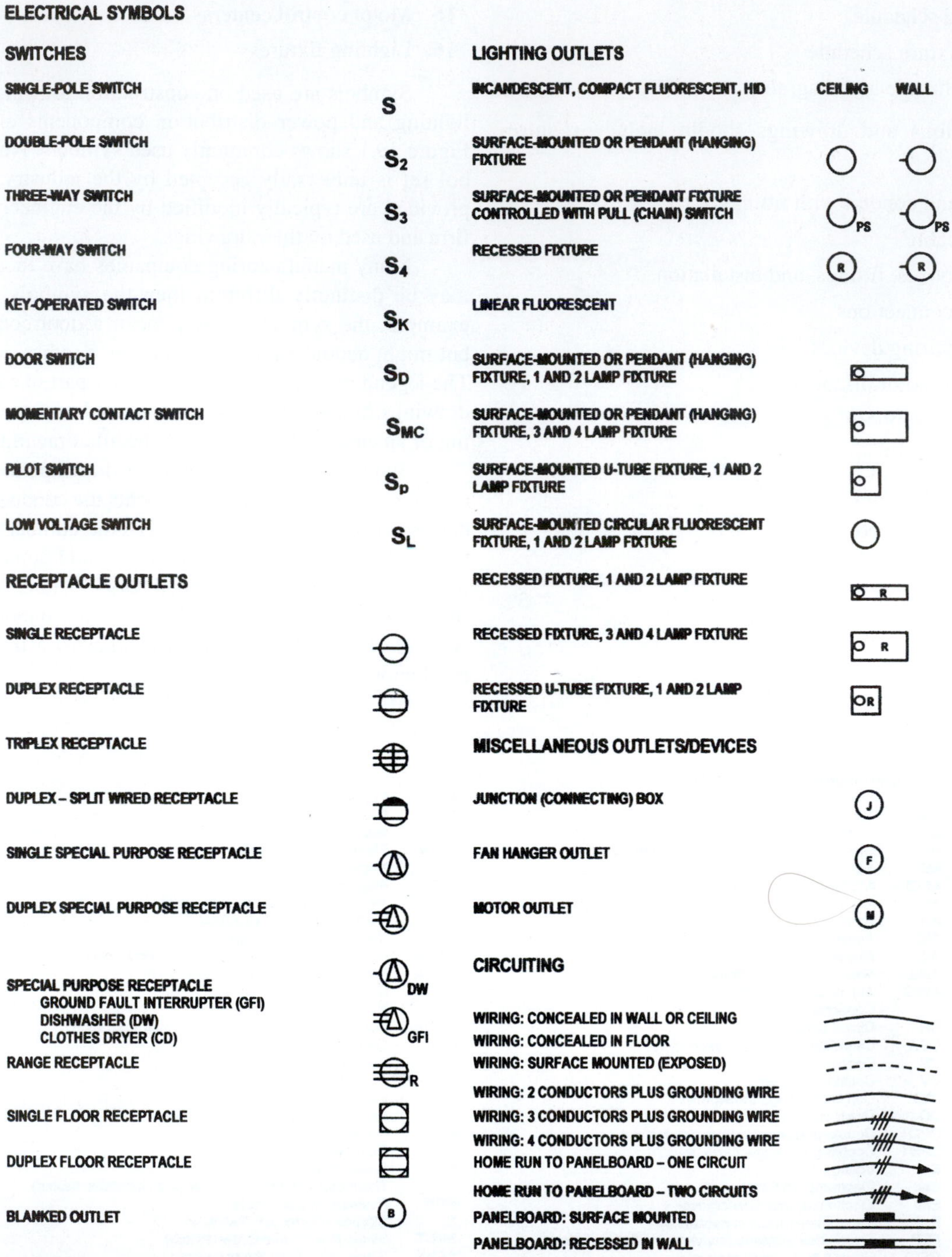

ELECTRICAL SYMBOLS

SWITCHES

SINGLE-POLE SWITCH	S
DOUBLE-POLE SWITCH	S_2
THREE-WAY SWITCH	S_3
FOUR-WAY SWITCH	S_4
KEY-OPERATED SWITCH	S_K
DOOR SWITCH	S_D
MOMENTARY CONTACT SWITCH	S_{MC}
PILOT SWITCH	S_p
LOW VOLTAGE SWITCH	S_L

RECEPTACLE OUTLETS

SINGLE RECEPTACLE

DUPLEX RECEPTACLE

TRIPLEX RECEPTACLE

DUPLEX – SPLIT WIRED RECEPTACLE

SINGLE SPECIAL PURPOSE RECEPTACLE

DUPLEX SPECIAL PURPOSE RECEPTACLE

SPECIAL PURPOSE RECEPTACLE
    GROUND FAULT INTERRUPTER (GFI)
    DISHWASHER (DW)
    CLOTHES DRYER (CD)

RANGE RECEPTACLE

SINGLE FLOOR RECEPTACLE

DUPLEX FLOOR RECEPTACLE

BLANKED OUTLET

LIGHTING OUTLETS

INCANDESCENT, COMPACT FLUORESCENT, HID	CEILING	WALL
SURFACE-MOUNTED OR PENDANT (HANGING) FIXTURE		
SURFACE-MOUNTED OR PENDANT FIXTURE CONTROLLED WITH PULL (CHAIN) SWITCH		
RECESSED FIXTURE		
LINEAR FLUORESCENT		
SURFACE-MOUNTED OR PENDANT (HANGING) FIXTURE, 1 AND 2 LAMP FIXTURE		
SURFACE-MOUNTED OR PENDANT (HANGING) FIXTURE, 3 AND 4 LAMP FIXTURE		
SURFACE-MOUNTED U-TUBE FIXTURE, 1 AND 2 LAMP FIXTURE		
SURFACE-MOUNTED CIRCULAR FLUORESCENT FIXTURE, 1 AND 2 LAMP FIXTURE		
RECESSED FIXTURE, 1 AND 2 LAMP FIXTURE		
RECESSED FIXTURE, 3 AND 4 LAMP FIXTURE		
RECESSED U-TUBE FIXTURE, 1 AND 2 LAMP FIXTURE		

MISCELLANEOUS OUTLETS/DEVICES

JUNCTION (CONNECTING) BOX

FAN HANGER OUTLET

MOTOR OUTLET

CIRCUITING

WIRING: CONCEALED IN WALL OR CEILING
WIRING: CONCEALED IN FLOOR
WIRING: SURFACE MOUNTED (EXPOSED)
WIRING: 2 CONDUCTORS PLUS GROUNDING WIRE
WIRING: 3 CONDUCTORS PLUS GROUNDING WIRE
WIRING: 4 CONDUCTORS PLUS GROUNDING WIRE
HOME RUN TO PANELBOARD – ONE CIRCUIT
HOME RUN TO PANELBOARD – TWO CIRCUITS
PANELBOARD: SURFACE MOUNTED
PANELBOARD: RECESSED IN WALL

FIGURE 19.1 *(Continued)*

19.3 BUILDING SYSTEM VOLTAGES

Supply Voltages

Power is delivered by the utility company to the user at *supply voltages*. Supply voltage is expressed as a nominal voltage because it varies slightly. During normal conditions, supply voltages can vary from about 90 to 105% of nominal voltage. Variations from nominal voltages are caused by a number of reasons, including load variation and changes in conditions at the utility power system. Additionally, transient voltages, caused by phenomena such as lightning strikes, some types of faults, and the switching of some types of user loads, cause variations in voltage available to the user.

System Voltages

The principle voltages available in a building are called the *system voltages*. *Medium and high voltage systems* carry voltages above 600 V may be used in special cases such as the 2400/4160 V three-phase system found in industrial and commercial installations such as for large signage, sports lighting in stadiums, and services for large manufacturing plants and skyscrapers. There are, however, drawbacks to voltages higher than 600 V because significant and costly special precautions such as heavy insulation and conductor shielding are needed. As a result, *low voltage systems* that carry voltages less than 600 V are typically used in buildings. System voltage is expressed as a nominal voltage because it varies slightly for the reasons mentioned earlier.

Design of a building's electrical system begins with establishing the desired building system voltage. A higher voltage means that a circuit can carry more current. A 208 V circuit can carry 1.73 times the current of a 120 V circuit (208 V/120 V = 1.73); a 240 V circuit can carry twice the current of a 120 V circuit; a 277 V circuit can carry 2.31 times the current of a 120 V circuit; and so on. Thus, higher voltage means smaller conductor sizes. The savings for larger conductors (feeders) of moderate length can be quite significant. Higher voltage, however, is more dangerous.

There are a numerous system voltage levels and combinations used in buildings throughout the world. Availability of a particular system voltage is dependent on utility lines and equipment at or near the building site. In the United States and Canada, several system voltage configurations are in use. However, a particular system voltage may not be available at a specific building site.

Circuit Wiring

Before entering into a detailed discussion on system voltages, it is necessary to introduce types of wiring found in a circuit. A minimum of two types of conductors is required to deliver alternating current in a building electrical system: the ungrounded conductor and the neutral conductor. A third conductor, called a grounding conductor, is added to most circuits. See Figure 19.2.

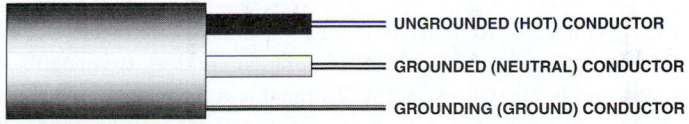

UNGROUNDED (HOT) CONDUCTOR

GROUNDED (NEUTRAL) CONDUCTOR

GROUNDING (GROUND) CONDUCTOR

FIGURE 19.2 Two types of conductors are required to deliver alternating current to the building system: the ungrounded conductor and the grounded conductor. A third conductor, called a *grounding conductor,* is added to most circuits. The ungrounded conductor is the current-carrying conductor in an alternating current system. A grounded or neutral conductor is required to complete the circuit by connecting the ungrounded conductor to ground. A third conductor, the grounding conductor, provides additional protection.

Each conductor plays an important role in how electricity is distributed to and circuited in a building. The different function of each conductor is introduced here as a prelude to understanding building system voltages. A more detailed discussion of these conductors is introduced later in this chapter.

Ungrounded Conductor

The *ungrounded conductor* is the initial current-carrying conductor in an AC system. The ungrounded conductor is frequently known as the *hot* or *live conductor* because it feeds current to the circuit. When an ungrounded conductor is grounded (connected to ground), a closed circuit in single-phase results. This is the type of circuit used to power small appliances (e.g., toaster, portable microwave oven, and so forth), small pieces of equipment (e.g., computer, electric drill, and so on) and lighting (e.g., desk lamps, office lighting, and so forth). When two associated ungrounded conductors are connected in a single circuit, a higher voltage is delivered.

Grounded/Neutral Conductor

A *grounded (neutral) conductor* is required to complete a single-phase circuit by connecting the ungrounded (hot) conductor to ground. The neutral conductor is a grounded conductor that serves more than one circuit. It carries the unbalanced load between two ungrounded (hot) conductors. Both conductors complete the circuit(s) by connecting it to ground and, as a result, are treated as current carrying conductors.

Grounding Conductor

A third conductor known as the *grounding* conductor provides supplementary but important grounding protection. The grounding conductor is not normally a current-carrying conductor, but is energized only on a temporary, emergency basis when there is a fault between an ungrounded (hot) conductor and any metal associated to the electrical equipment. Confusion often exists between the definitions of the grounded conductor and the grounding conductor because the grounding conductor is commonly referred to as a "ground." It is more correctly called a "grounding" conductor.

In a simple single-phase circuit, the ungrounded conductor provides power to the load and the grounded conductor provides a path from the load back to the power source, which completes the circuit. Voltage in the circuit is equal to the voltage on the ungrounded conductor (e.g., 120 V on a 120 V circuit). When two ungrounded (hot) conductors in a single-phase circuit are connected in a circuit, voltage in the circuit is double the voltage available on each ungrounded conductor (e.g., It is 240 V between two ungrounded conductors that are 120 V each).

A three-phase circuit requires three ungrounded conductors that are one-third out of phase of each other. In the case where two ungrounded conductors from a three-phase wye-connected transformer are connected, voltage in the circuit is

TABLE 19.1   A COMPARISON OF THE NUMBER OF CONDUCTORS AND VOLTAGES BETWEEN CONDUCTORS ON COMMON BUILDING SYSTEM VOLTAGES.

Building System Voltage	Number of Conductors			Voltage Between Conductors	
	Ungrounded (Hot) Conductors	Grounded (Neutral) Conductors	Grounding (Ground) Conductors	Voltage Between One Grounded Conductor and One Ungrounded Conductor	Voltage Between Two Ungrounded Conductors
120 V, single phase, two wire	1	1	1	120	—
120/240 V, single phase, three wire	2	1	1	120	240
208 Y/120 V, three phase, four wire	3	1	1	120	208
480 Y/277 V, three phase, four wire	3	1	1	277	480
600 Y/346 V, three phase, four wire	3	1	1	346	600
240 Δ/120 V, three phase, four wire	3	1	1	X phase: 120 Y phase: 120 Z phase: 208	240

1.732 times the voltage available on each ungrounded conductor (e.g., 208 V between two ungrounded conductors that are 120 V each but one-third out of phase). The factor 1.732 is the square root of 3.

Common Building System Voltages

A comparison of the number of conductors and voltages between conductors on common building system voltages is found in Table 19.1. The following is a description of the common building system voltages used in the United States and Canada.

120 Volt, Alternating Current, Single-Phase, Two-Wire System (120 V AC, 1Φ-2W)

The 120 V AC, 1Φ-2W system is the most basic system voltage used. It was used in the first electrical services to buildings; however, nearly all have since been upgraded. Today, this system is used to serve outbuildings and farm buildings because its use is limited to buildings with loads up to 6000 VA (50 A). The service entrance provided to the service equipment (switchboard or panelboard) is by two conductors: one ungrounded (hot) conductor carrying 120 V and one neutral conductor. Voltage measured between the ungrounded (hot) and neutral conductors is 120 V.

Before about 1945, only two conductors were used to feed electrical energy beyond the panelboard to the 120 V branch circuits in the building; this configuration relates to the two-slotted convenience outlets found in older buildings. Later, the grounding conductor became a standard part of branch circuits. It relates to the extra slot found in the three-slotted convenience outlets in newer homes. Today, three conductors are used to deliver electrical energy to the branch circuits in the

building: one ungrounded (hot) conductor carrying 120 V, one neutral conductor, and one grounding conductor.

120/240 Volt, Alternating Current, Single-Phase, Three-Wire System (120/240 V AC, 1Φ-3W)

The 120/240 V AC, 1Φ-3W is the most common residential electrical service in use today. It is also used on a limited basis in light commercial buildings such as small office buildings, churches, and retail shops and stores.

The service entrance conductors feeding the service equipment (switchboard or panelboard) are three conductors: two ungrounded (live) conductors, each carrying 120 V and one neutral conductor. See Figure 19.3. The ungrounded (hot) conductors are known as the A and B legs. At the switchboard or panelboard, the grounding conductor is added. The availability of 120 V or 240 V leads to a number of circuit or feeder arrangements that can supply the following loads:

- 120 V, single-phase, two-wire branch circuit
- 240 V single-phase, two-wire branch circuit
- 120/240 V single-phase, three-wire feeder or branch circuit

On a 120/240 V system, a 120 V branch circuit provides electrical energy to convenience (receptacle) outlets, small appliances, and light fixtures. A 240 V branch circuit serves large appliances and equipment such as electric-resistance baseboard heaters, water heaters, and air conditioning equipment. A 120/240 V branch circuit provides both 120 V and 240 V to an appliance such as a range and clothes dryer; controls, and light fixtures. Typically, small motors run on 120 V and heating elements operate on 240 V. A grounding conductor runs continuously through all branch circuits and serves as a safety circuit in case of a short-circuit.

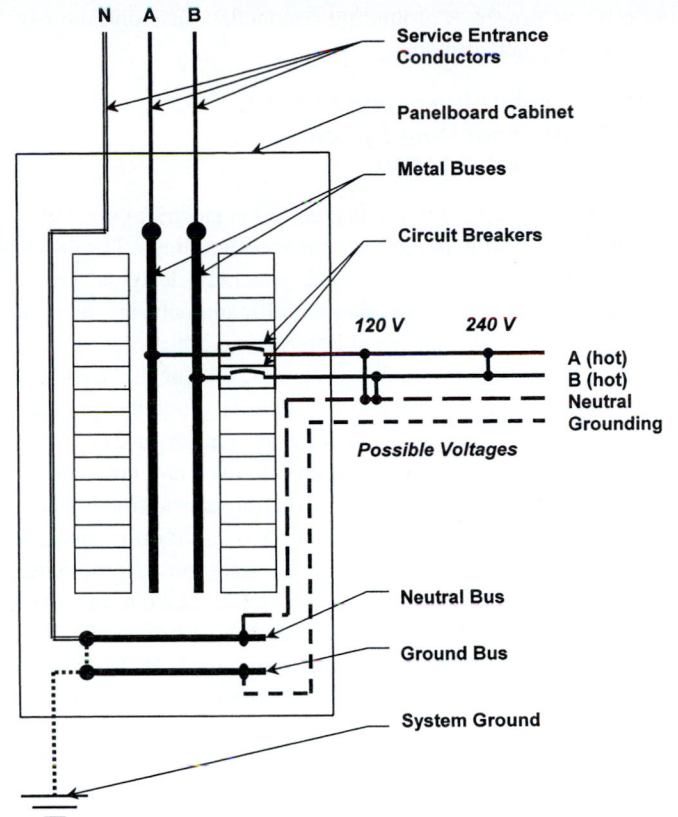

FIGURE 19.3 Shown is a schematic of a 120/240 V, alternating current, single-phase, three-wire system. The service entrance conductors include ungrounded (live) conductors A and B and one grounded conductor (N). At the panelboard, a grounding conductor that runs continuously through all branch circuits is added. The availability of 120 V or 240 V leads to a number of circuit or feeder arrangements that can supply the following loads: 120 V, single-phase, two-wire branch circuit; 240 V single-phase, two-wire branch circuit; and 120/240 V single-phase, three-wire feeder or branch circuit. Main disconnect is not shown for clarity.

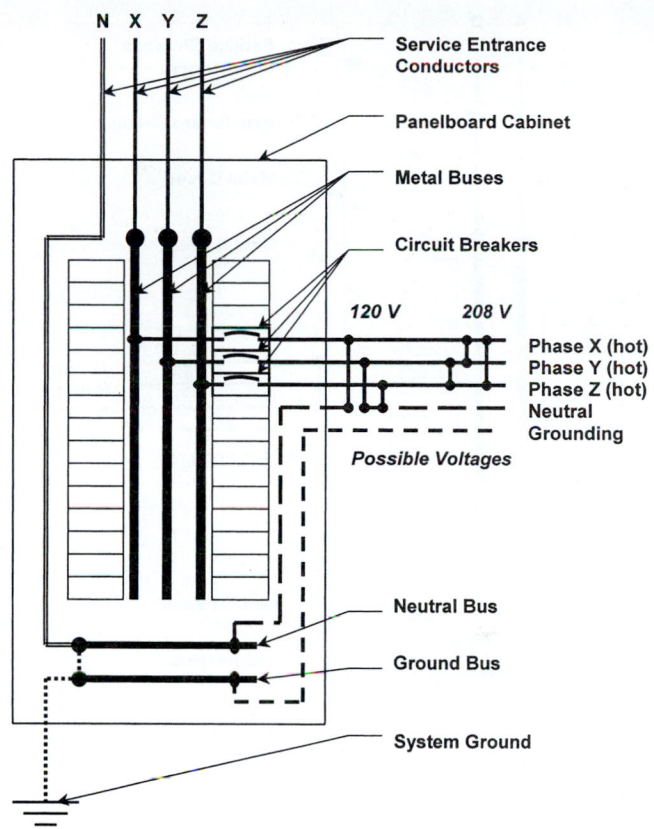

FIGURE 19.4 Shown above is a schematic of a 208 Y/120 V, alternating current, three-phase, four-wire system. The service entrance conductors include ungrounded (live) conductors A, B, and C, and one grounded conductor (N). At the panelboard, a grounding conductor that runs continuously through all branch circuits is added. Voltage measured across the connection of any single ungrounded conductor (X, Y, or X phase) and the grounded conductor provides 120 V single-phase power. Voltage across any two ungrounded conductor (X, Y, or Z phase) is 208 V single-phase power, because each phase is one-third out of phase. Main disconnect is not shown for clarity.

208 Y/120 Volt, Alternating Current, Three-Phase, Four-Wire System (208 Y/120 V AC, 3Φ-4W)

The 208 Y/120 V AC, 3Φ-4W is an older electrical service found in small commercial buildings (e.g., office buildings and schools) and high-rise buildings where three-phase motors (motors above about ½ horsepower) and equipment such as large air conditioners are used. It is not used very often in industry because a 480 V system is more economical for large motor loads.

The Y in 208 Y/120 V relates to a wye (Y) winding configuration of the secondary windings of the transformer producing three-phase current. Service entrance provided to the service equipment (switchboard or panelboard) is by four conductors: three ungrounded conductors, each at 120 V and one-third out of phase, and one neutral conductor. See Figure 19.4.

The ungrounded (hot) conductors are known as the X, Y, and Z legs or phases. Voltage measured across the connection of any single ungrounded conductor (X, Y, and Z phase) and the neutral conductor provides 120 V single-phase power. Voltage across any two ungrounded conductor (X, Y, and Z phase) is 208 V single-phase power because each phase is one-third out of phase. Equipment designed to operate on three-phase power, such as a three-phase motor, is connected to the three ungrounded conductors (X, Y, and Z phases). At the panelboard, a grounding conductor is added and it runs continuously through all branch circuits. The availability of 120 V or 208 V in single- or three-phase leads to a number of circuit or feeder arrangements that can supply the following loads:

- 120 V, single-phase, two-wire branch circuit
- 208 V single-phase, two-wire branch circuit

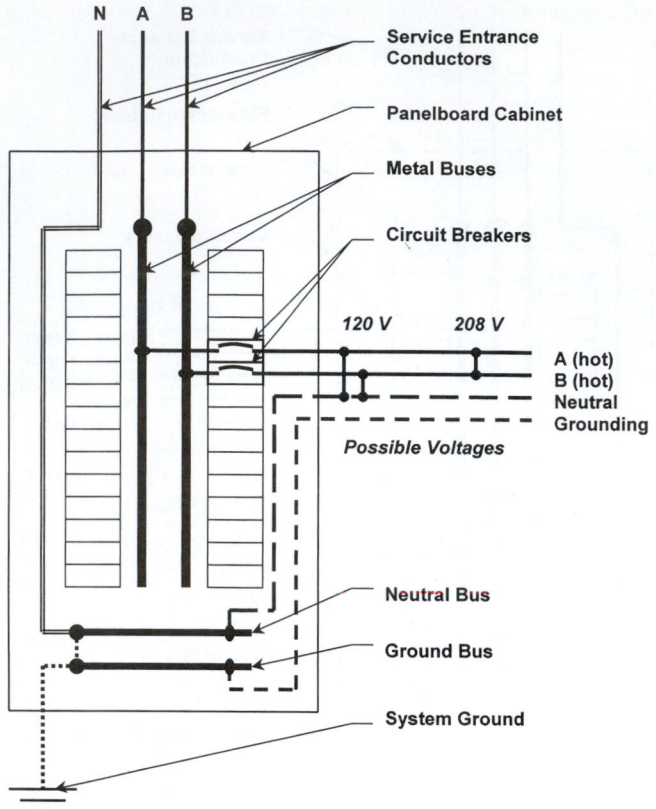

FIGURE 19.5 Shown is a schematic of a 208 Y/120 V, alternating current, single-phase, three-wire system. The service entrance conductors include ungrounded (live) conductors A and B, and one grounded conductor (N). At the panelboard, a grounding conductor that runs continuously through all branch circuits is added. The availability of 120 V or 208 V leads to a number of circuit or feeder arrangements that can supply the following loads: 120 V, single-phase, two-wire branch circuit; 240 V single-phase, two-wire branch circuit; and 120/208 V single-phase, three-wire feeder or branch circuit. This system is extracted from a 208 Y/120 V, alternating current, three-phase, four-wire system. Main disconnect is not shown for clarity.

- 208 V three-phase, three-wire branch circuit
- 120/208 V three-phase, three-wire feeder or branch circuit

As shown in Figure 19.5, this system is occasionally broken down within the building to provide 120/208 V on a single-phase, three-wire system. In some cases where 208 Y/120 V is the only system voltage available, it may serve residential customers. The 120/208 V, single-phase, three-wire system works much like a 120/240 V, single-phase system, except 208 V is available instead of 240 V, about 87% of the voltage and thus the power. Distribution is by three conductors and a grounding conductor. In this configuration, voltage measured across the connection of any single ungrounded (hot) conductor and the neutral conductor provides 120 V single-phase power. Voltage across any two ungrounded conductor is 208 V

single-phase power. A grounding conductor runs continuously through all branch circuits.

480 Y/277 Volt, Alternating Current, Three-Phase, Four-Wire System (480 Y/277 V AC, 3Φ-4W)

The 480Y/277 V AC, 3Φ-4W is a common electrical service in most modern medium to large commercial buildings. The 480 V three-phase power is used to power specially designed heavy machinery (e.g., at machine shops and manufacturing plants). High-voltage, 277 V fluorescent lighting and other single-phase devices have also been developed specifically for use with this system. Large retail shopping malls, schools, grocery supermarkets, and office buildings may use this system for its 277 V fluorescent lighting capabilities, where fixtures are not located closer than 3 ft away from windows, platforms, and fire escapes.

The Y in 480 Y/277 V relates to a wye-winding configuration of the secondary windings of the transformer producing three-phase current. In this system, service entrance to the service equipment (switchboard or panelboard) is provided by four conductors: three ungrounded conductors, each at 277 V and one-third out of phase, and one neutral conductor. Small transformers located in electrical rooms or closets in the building step down the voltage from 480 V to 120 V for small equipment and convenience outlets.

At the panelboard or switchboard, a grounding conductor is added. The ungrounded (hot) conductors are known as the X, Y, and Z phases. Voltage measured across the connection of any single ungrounded conductor (X, Y, and Z phase) and the neutral conductor provides 277 V single-phase power. Voltage across any two ungrounded conductor (X, Y, and Z phase) is 480 V single-phase power. See Figure 19.6. Three-phase motors and other equipment that operate on three-phase power are connected to the three ungrounded conductors (X, Y, and Z phases). The availability of 277 V or 480 V in single or three -phase offers the following circuit or feeder configurations:

- 277 V, single-phase, two-wire branch circuit
- 480 V single-phase, two-wire branch circuit
- 480 V three-phase, three-wire branch circuit
- 277/480 V three-phase, three-wire feeder or branch circuit

When compared with the 208 Y/120 V system, the 480 Y/277 V system has economic advantages from the standpoint of equipment and conductors. Because a given conductor can carry more than twice the VA load at 480 V than at 208 V, the savings in wire size for feeders can be quite significant with the 480 Y/277 V system. Additionally, the smaller current at 480 V for any supply transformer capacity permits the use of protective devices with both smaller frame size and interrupting rating. Both of these factors permit significant savings.

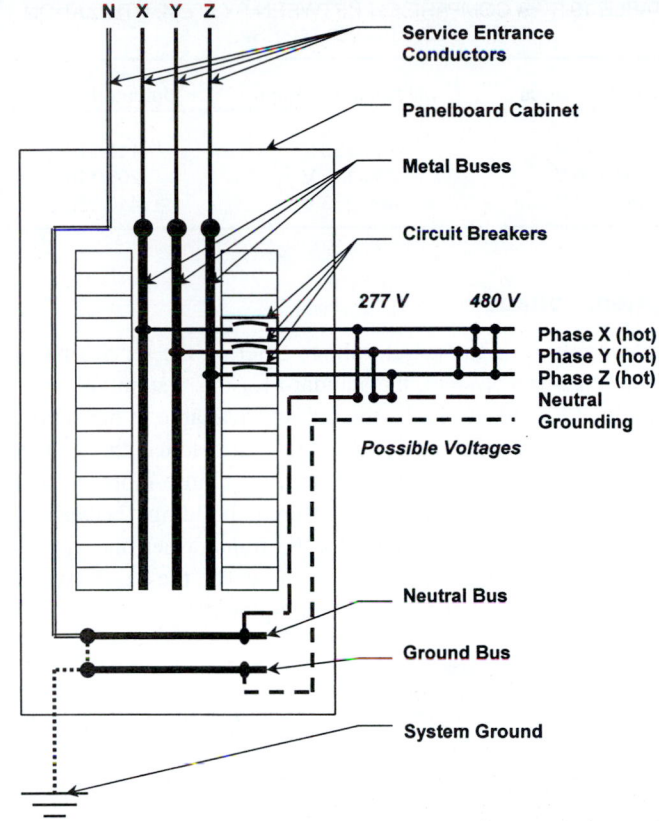

FIGURE 19.6 Shown is a schematic of a 480 Y/277 V, alternating current, three-phase, four-wire system. In this system, service entrance conductors include three ungrounded conductors, each at 277 V and one-third out of phase, and one grounded conductor. At the panelboard, a grounding conductor is added. The ungrounded (live) conductors are known as the X, Y, and Z phases. Voltage measured across the connection of any single ungrounded conductor (X, Y, and Z phase) and the grounded conductor provides 277 V single-phase power. Voltage across any two ungrounded conductor (X, Y, and Z phase) is 480 V single-phase power. Three-phase motors and other equipment that operate on three-phase power are connected to the three ungrounded conductors (X, Y, and Z phases). Main disconnect is not shown for clarity.

600 Y/346 Volt, Alternating Current, Three-Phase, Four-Wire System (600 Y/346 V AC, 3Φ-4W)

The 600Y/346 V AC, 3Φ-4W is a less common electrical service in large commercial and industrial buildings that is used to power specially designed heavy machinery. This system is designed like the 480 Y/277 V AC, 3Φ-4W described earlier, except that 600 V and 346 V are available in the circuit or feeder configurations.

The 600 Y/346 V system has additional economic advantages from the standpoint of equipment and conductor sizing in comparison to the 208 Y/120 V and 480 Y/277 V systems. However, the 550 V or 575 V equipment used on the 600Y/346 V system is not as readily available as the 460 V equipment used on the 480 Y/277 V system. As a result, this electrical service is

used primarily in industries where the 600 Y/346 V systems are a tradition.

240 Δ/120 Volt, Alternating Current, Three-Phase, Four-Wire System (240 Δ/120 V AC, 3Φ-4W)

The 240 Δ/120 AC, 3Φ-4W is another fairly common electrical service found in commercial and industrial buildings where three-phase motors (motors above about ½ horsepower) and equipment such as large air conditioners are used. On this three-phase, four-wire, delta-connected (Δ) system, the mid-point of one phase winding is neutral to provide 120 V between Phase X and ground and Phase Z and ground only. There are 240 V between the windings of each phase—that is, between Phases X and Y, Phases X and Z, or Phases Y and Z. Between Phase Y and ground there is 208 V available. Phase Y is known as the *high leg* or *wild leg* because it has a higher voltage to ground than the other legs. See Figure 19.7.

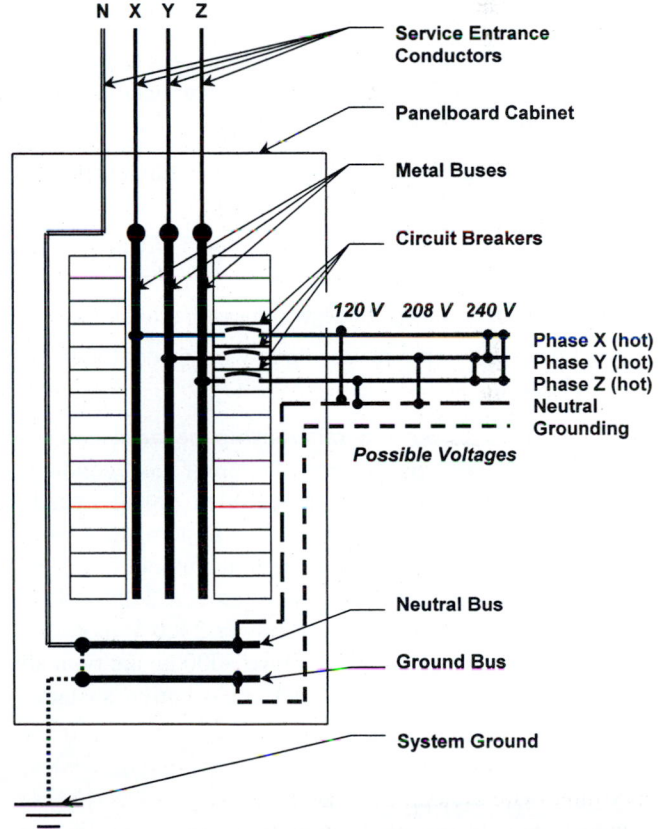

FIGURE 19.7 Shown is a schematic of a 240 Δ/120 V, alternating current, three-phase, four-wire system. In this three-phase, four-wire, delta-connected (Δ) system, the midpoint of one phase winding is grounded to provide 120 V between Phase X and ground and Phase Z and ground. There are 240 V between the windings of each phase—that is, between Phases X and Y, Phases X and Z, or Phases Y and Z. Between Phase Y and ground there is 208 V available. The Phase Y leg is known as the high leg or wild leg because it has a higher voltage to ground than the other legs. Main disconnect is not shown for clarity.

TABLE 19.2 THE RELATIONSHIP BETWEEN THE BUILDING
SYSTEM VOLTAGE, CURRENT (AMPERAGE), AND
REQUIRED CONDUCTOR SIZE FOR A 20 KVA LOAD
AND A POWER FACTOR OF 1.0. HIGHER VOLTAGE
MEANS LOWER AMPERAGE (AND SMALLER
CONDUCTOR SIZE) FOR A SPECIFIC LOAD.
ALTHOUGH IT IS LESS SAFE TO TRANSMIT POWER
AT HIGH VOLTAGE, IT IS MORE ECONOMICAL
BECAUSE OF SMALLER WIRE SIZES REQUIRED
AT HIGHER VOLTAGES.

Common Building System Voltages	Current Requirement for 20 kVA Load*	Required Conductor Size (AWG)
Volts (V)	Amperes (A)	Copper—75°C (167°F) Rating
480	41.7	No. 8
277	72.2	No. 4
240	83.3	No. 4
208	96.2	No. 3
120	166.7	No. 4/0

*I = P/E; for example: 20 000 VA/480 V = 41.7 A.

The availability of 120 V, 208 V, and 240 V in single or three phase offers a number of circuit or feeder options that can supply the following loads:

- 120 V, single-phase, two-wire branch circuit (Phase X and ground and Phase Z and ground)
- 208 V single-phase, two-wire branch circuit (Phase Y and ground)
- 240 V three-phase, three-wire branch circuit
- 120/240 V three-phase, three-wire feeder, or branch circuit

Table 19.2 shows the relationship between common building system voltage, current (amperage) and conductor size. These are the common building system voltages used in the United States and Canada but other system voltages are used for special applications. Large manufacturing plants use 2400 V or 4160 V distribution systems. A 2400 V distribution system is typically used to power 2300 V motors up to about 2000 hp. Motors at and above 3000 hp are typically 4000 V motors and thus a 4160 V distribution system is the standard.

System, Utilization, and Maximum Voltages

A building system voltage may be specified as 120/240 V but is sometimes referred to as 110/220 V, 115/230 V or 125/250 V. The difference in the cited voltages has to do with variations in how the voltage is being defined. There are three ways that a voltage is defined: system voltage, utilization voltage and maximum voltage. A comparison is provided in Table 19.3. System, utilization and outdated voltages are defined as follows.

TABLE 19.3 A COMPARISON BETWEEN SYSTEM, UTILIZATION, AND OUTDATED VOLTAGES.

System Voltages	Utilization Voltages	Outdated Voltages
120/240 V	115/230 V	110/220 V
208/120 V	200/115 V	208/110 V
480/277 V	460/265 V	440/255 V

System Voltage

System voltage is the target voltage entering the service panel. On a 120/240 V system, the standard for the system voltage is actually 120/240 V; that is, the voltage available at the service equipment is approximately 120/240 V. In practice, this voltage is sometimes a little less and sometimes a little more. System voltage will vary slightly for different buildings because of variations of voltage available at the transformer and voltage drop in the service conductors. It is, however, the target voltage distributed to a building's service equipment.

Utilization Voltage

A voltage drop occurs as current flows from the service equipment through the branch circuit conductors to the outlet (point of use in the building). Because of these voltage drops, voltage available at an outlet in the building is less than the system voltage. The *utilization voltage* accounts for anticipated voltage drops on branch circuit conductors. On a 120/240 V system, approximately 115/230 V is available at the outlet of the branch circuit and not the 120/240 V available at the service equipment. Appliances and equipment connected to the 120/240 V building system voltages are designed for 115 V, 230 V, or 115/230 V. (The commonly cited 110/220 V standard is obsolete and no longer specified.) Again, there are slight variations in utilization voltage. Measured voltage at an outlet or connection is called the *line voltage*.

Maximum Voltage

Wiring devices such as switches, receptacles, relays and conductors, and electrical equipment are manufactured to endure voltages slightly higher than the utilization voltage. The highest voltage to which a wiring device can be exposed is known as the *maximum voltage*. For example, a 5-15R duplex receptacle that is the wall-mounted receptacle common in most homes and offices is designed to handle a maximum voltage of 125 V, but is intended for use on a 120 V circuit, where the line voltage is likely about 115 V.

19.4 GROUNDED AND UNGROUNDED CONDUCTORS

The grounded and ungrounded (neutral) conductors are necessary to complete an electrical circuit. They provide a continuous path from the load to ground. Both conductors are connected to the earth (to ground), but there are differences in the location and function of each of these conductors. These differences are addressed in the following sections.

Grounded Conductor

In single-phase branch circuits (beyond the service equipment and any feeders and originating at the panelboard), a *grounded conductor* serves as the grounded leg of the circuit. It completes the circuit by connecting the ungrounded (hot) conductor to ground. Thus, in circuit design, the grounded conductor is considered to be a current carrying conductor because it serves as a return path back to the circuit's power source.

On a two-wire branch circuit (e.g., a 120 V circuit with one ungrounded conductor and one grounded conductor), the grounded conductor carries current equal to the load. For example, if the ungrounded (hot) conductor is feeding a load of 12 A, then the grounded conductor carries 12 A to ground. As a result, the grounded conductor must be sized at the same ampacity as the ungrounded (hot) conductor.

Neutral Conductor

A *neutral conductor* performs the function of a grounded conductor for at least two ungrounded (hot) conductors that have sources from different voltage phases, such as on a multiwire branch circuit, multiwire feeder, and the electrical service. The conductors served by a neutral must measure voltage between the ungrounded conductors and be protected by a double- or triple-pole breaker or set of fuses. A neutral conductor cannot by definition serve a single 120 V circuit because it has only one ungrounded conductor; a neutral conductor is a grounded conductor that is shared between two or more ungrounded conductors. Thus, a neutral conductor is frequently called a *shared neutral* or *common neutral*.

In a 120/240 V three-wire service entrance, feeder, or branch circuit, there is a neutral conductor and two ungrounded (hot) conductors, each with 120 V to ground. A neutral conductor is also present in 208 Y/120 V and 480 Y/277 V services, feeders, and circuits.

Grounded and neutral conductors should always be insulated because they can carry current. They should be continuous because they provide a path to complete the circuit or circuits. The neutral conductor is grounded at the transformer and at the service equipment (switchboard or panelboard). In these cases it is connected to ground much like the grounding conductor. At subpanelboards served by feeders, the equipment grounding conductor and the neutral conductor remain separate. This way, each circuit extending from the subpanelboard has a separate equipment grounding conductor and the neutral conductor.

Grounded and neutral conductors should be wired so a circuit breaker, fuse, or switch does not interrupt them. A switch or overcurrent device located in the grounded or neutral conductor allows the remainder (and much of) the circuit

wiring to remain electrically hot if it opens the circuit. Consider the following example: A branch circuit serves several light fixtures with a switch connected to the grounded conductor instead of the ungrounded (hot) conductor. When the switch is closed (on), the circuit is complete and the lamps light; when the switch is open (off), the lamps are off. It appears the circuit is controlled and wired properly, but there are concerns. With the switch open (off), the circuit wiring and light fixtures are still hot; that is, power runs through the ungrounded wiring to the lamps and through the grounded wiring to the open switch where it is interrupted. The lamps in the light fixtures do not light but the circuit wiring is hot back to the open switch. Similarly, a circuit breaker or fuse located on the grounded conductor may open the circuit because of a short-circuit or current overload, but the circuit wiring will remain hot. A hot circuit is potentially hazardous condition because it unnecessarily exposes occupants to hot wiring.

Interruption of the grounded or neutral conductor with a circuit breaker, fuse, or switch is considered poor practice in most instances. (There are unique cases when the grounded and ungrounded conductors are controlled with a switch.) When switching control is desired in a circuit, the ungrounded (hot) wire should be the conductor that is switched. Also, the circuit should be protected with a circuit breaker or fuse in the ungrounded (hot) wire so, if a short circuit or current overload occurs, the entire circuit is interrupted. In both cases, power is interrupted near the source (e.g., at the panelboard) so that circuit wiring does not remain hot.

Load Balancing

In multiwire branch circuits, feeders, and services (e.g., a 120/240 V or 480 Y/277 V circuits with two or more ungrounded conductors and one neutral conductor), the common neutral conductor carries current only when dissimilar loads exist on the ungrounded (hot) conductors in the circuit. For example, in a 120/240 V single-phase, three-wire circuit, if one ungrounded (hot) conductor feeds a 16 A load on one circuit and the other feeds a 4 A load on the other circuit, the common neutral conductor carries 12 A to ground. This is the difference between the two loads.

A common neutral conductor carries only the unbalanced load. In multiwire circuits, feeders, and services the neutral load can be quite high if loads on different circuits are uneven from occupant use of electrical equipment or poor design. High neutral loads have been known to overheat neutral conductors and distribution transformers without tripping overcurrent protection. For example, in office buildings, computer use is very sporadic (unlike lighting, which is fairly constant), which causes disproportionate loading on individual circuits and a heavy load on the neutral conductor. Under extreme conditions of use, this unbalanced load can be excessive and unsafe.

Designing circuits to avoid or to accommodate the high neutral currents can avert the conditions created by unbalanced loading on a common neutral. Typically, the steps are balancing the circuit loads, treating the neutral as a current carrying conductor, and properly sizing the common neutral conductor in multiwire branch circuits, feeders, and services.

Load balancing is the practice of dividing loads as evenly as possible between the ungrounded conductors on a multiwire circuit, feeder, or service. A well-designed neutral is balanced so that under load little or no current flows through the neutral conductor. Similarly, the designer should attempt to balance the loads in feeders and building services. This is accomplished by dividing loads between the A and B legs or X, Y, and Z legs; that is, on a three-wire, single-phase system the circuit loads should be evenly shared (as possible) between the A and B hot conductors, and on a four-wire, three-phase system the circuit loads should be evenly divided between the X, Y, and Z hot conductors. On a multiwire circuit, each ungrounded conductor should carry loads equally.

Load balancing of 240 V circuits on a 120/240 V three-wire, single-phase system is easily accomplished because the load is the same on each phase conductor serving the circuit. Two-pole circuit breakers connect to the panelboard so that one breaker connects to the Phase A bus and the other breaker to the Phase B bus. Balancing 120 V circuits on a 120/240 V three-wire, single-phase system is more difficult. Loads must be evenly divided between hot conductors. These breakers or fuses serving ungrounded conductors must be connected to the hot buses in the panelboard so that loads on each bus tend to equate with loads on the other bus.

On a three-phase system, the neutral conductor is common to all phases. Three-phase motors and other equipment operating on three-phase systems are connected to the three ungrounded (hot) conductors and a common neutral. Balancing is easily accomplished because the common neutral conductor is shared and carries no load. Three-pole circuit breakers connect to the panelboard such that one breaker connects to the Phase X bus, a second connects to the Phase Y bus, and the third breaker connects to the Phase Z bus. A two-wire circuit on a three-phase system (e.g., 277 V circuit) is more difficult because it requires that loads be evenly divided between phase conductors.

The basic way to attempt to balance the loads at a panelboard is to connect an equal number of branch circuits to each ungrounded bus, but this only accomplishes good balancing when all circuits are equally loaded. On lighting panelboard, this is accomplished by connecting equal loads to each circuit. When circuit or feeder loads are not equal, circuits must be divided so that the total load on each phase conductor is equal. This requires the designer to identify the load on each circuit. In most cases the load on a circuit is not known exactly so the designer predicts an anticipated load and uses it to balance the neutral load.

Because a neutral conductor will typically not carry current equivalent to the ungrounded conductors in the multiwire circuit, feeder, or service it serves, it is generally not required to be sized at the same ampacity rating as the ungrounded (hot) conductor. Code prescribes methods to derate (reduce) the size to an acceptable size. Discretion must be exercised by the designer in this instance because it is possible to undersize the neutral, particularly if loading of a panelboard is changed or expanded as a result of future renovation. As a current-carrying conductor, an undersized neutral will overheat, producing a hazardous condition, and of course, this condition can be avoided by using a larger neutral conductor.

The neutral conductor is sized to carry the maximum unbalanced current in the circuit (for instance, the largest load between the neutral and any one ungrounded phase conductor). On large circuits, feeders, and the service, the first 200 A of neutral current is computed at 100%. After the first 200 A, additional resistive loads on the neutral can be reduced by a demand factor of 70%, but all inductive neutral current must be computed at 100% (no demand factor). Additionally, for cooking equipment or clothes dryers, the feeder neutral load should be computed at 70% of the demand load.

Example 19.1

The neutral load on a 120/240 V, three-wire feeder circuit carries an unbalanced load on the ungrounded conductors of 68 A on Phase A and 58 A on Phase B. Determine the grounded neutral conductor load.

$$\text{Unbalanced load} = 68\,\text{A} - 58\,\text{A} = 10\,\text{A}$$

The grounded neutral conductor load is 10 A for these conditions.

19.5 SYSTEM AND CIRCUIT GROUNDING

Grounding

In an electrical system, *grounding* is required to protect building occupants and electrical equipment. Grounding an electrical system begins with a *ground,* an electrode in direct contact with the earth itself. The grounding conductor is a continuous conductor that connects the ground to the neutral bus bar and the grounding conductor bus bar in the service equipment/main panelboard. The grounding conductor does not normally carry current. Instead, it links ground to the metal frames or housings of appliances and motors and the metal boxes containing outlets and switches. If needed, the grounding conductor can safely carry current to ground in the event of a lightning strike or in cases of damage or defect in the circuiting, appliances, devices, or equipment.

Grounding of an electrical branch circuit enables current to take an alternate path back to the overcurrent protection device if an electrical device or appliance short-circuits. It requires an additional, supplemental wire, called the *grounding conductor,* which is connected to the appliance cabinet or housing and provides an additional grounding path, in addition to the grounded conductor.

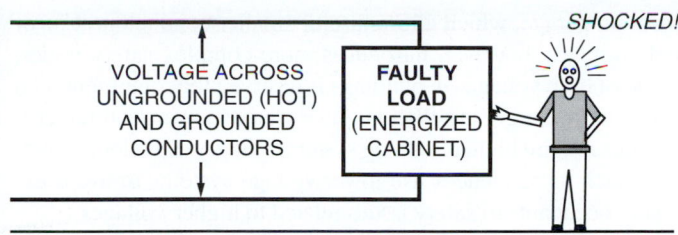

FIGURE 19.8 When the ungrounded (hot) conductor contacts the cabinet of an electrical device or appliance, the cabinet becomes "hot." If a person contacts the cabinet, a ground fault occurs and the person feels a shock because current flows through the person to ground. This dangerous condition is typically referred to as a *ground fault*.

Without this additional grounding path, current could flow through a user that was touching a faulty appliance.

For example, assume a damaged ungrounded (hot) conductor of an appliance (i.e., refrigerator) has contacted the appliance's metal housing, making the housing hot. See Figure 19.8. Without a grounding conductor connecting the appliance housing to ground, the appliance housing would remain hot. This would expose the user to a hazardous condition and the possibility of electrical shock if the user touched the live housing. As shown in Figure 19.9, with a grounding conductor connecting the appliance housing to ground, a faulty but complete circuit is created and the circuit breaker or fuse protecting the circuit would open, thereby shutting down power to the faulty appliance and eliminating occupant hazard. Essentially, the grounding conductor provides an added safety feature to the wiring system.

Grounding Requirements

In building electrical systems, there are two types of grounding: system grounding and equipment grounding. These are discussed as follows.

System Grounding

System grounding is that part of a building electrical system that provides protection against electrical shock, lightning, and fires. A lightning strike near the building or a high-voltage transmission line contacting the service entrance conductors can

introduce high voltage to a building electrical system. A properly grounded electrical system reduces danger and minimizes damage to the wiring and appliances from such an occurrence.

System grounding relates to bonding (connecting) all building electrical system components at the service entrance equipment at the neutral bus of the main panelboard or switchboard. A grounding system must be connected to some or all of the following elements if available on the building premises:

- An underground metal water (not gas) pipe in direct contact with the earth for no less than 10 ft (3 m); the metal building frame where it is effectively grounded

- An electrode made of at least 20 ft (6 m) of electrically conductive steel reinforcing bars (No. 4 AWG or greater) or bare copper wire no smaller than No. 2 AWG that is encased in at least 2 in of concrete that is part of a foundation or footing in direct contact with the earth

- An electrode made of a steel or iron plate that is at least ¼ in thick or copper plate that is at least 0.06 in thick with at least 2 ft^2 (0.2 m^2) of the plate surface in contact with exterior soil

- An electrode made of a grounding ring of bare copper wire no smaller than No. 2 AWG that encircles the building at a depth no less than 2.5 ft (0.75 m) below grade

- The structural metal frame of the building where the frame is effectively grounded

In systems where only a connection to an underground metal water pipe in direct contact with the earth is the only means of grounding, a supplemental electrode is required. A metal pipe, rod, or plate driven or placed into the earth are acceptable as a supplemental electrode.

Equipment Grounding

Equipment grounding refers to a grounding conductor or grounding path that connects the noncurrent-carrying metal components of equipment. This may be accomplished by installing an additional grounding conductor in all circuits or by permanently bonding (joining) metal components such as metallic conduit in a circuit to form a good conductive path. Equipment grounding extends from the outlets to the neutral bus bar at the service entrance equipment.

The *equipment-grounding conductor* is a bare conductor or a green-colored, insulated conductor that connects (bonds) the outlet boxes, metallic raceways, other enclosures and frames on motors, appliances, and other electrical equipment. All receptacles must be of the grounding type (with supplementary grounding slot) and must be connected to the equipment-grounding conductor. If properly bonded, a metal raceway (conduit) or armored cable system can serve as a means of equipment grounding so an equipment-grounding conductor is not actually needed.

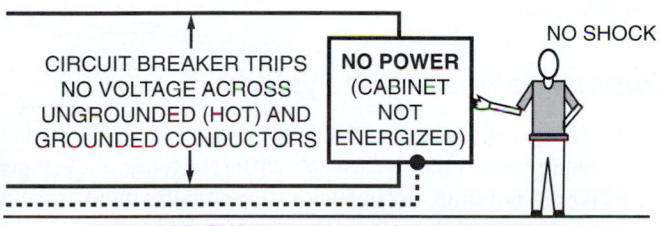

FIGURE 19.9 Grounding of an electrical branch circuit enables current to take an alternate path back to the overcurrent protection device (i.e., circuit breaker or fuse) if an electrical device or appliance short circuits. This condition causes the overcurrent protection to shut off the circuit so the unsafe condition does not exist.

Equipment-grounding and system-grounding electrodes must be bonded. *Bonding* is accomplished by installing an additional grounding conductor or by permanently joining metal components in a circuit. When joining components to form a good bond, special connections called *bonding jumpers* may be required to ensure a good connection between the metal components. Bonding jumpers are required in instances when flexible conduit is used.

Tables available in Code indicate the minimum size equipment-grounding conductors for grounding raceways and equipment. They are used to size the equipment-grounding conductor. Equipment-grounding conductors should be routed in the same raceway, cable, or cord as the circuit conductors. Where more than one circuit is installed in a single raceway, one equipment-grounding conductor can be installed in the raceway, but it must be sized for the largest overcurrent protection device serving conductors in that raceway. The equipment-grounding conductor is never required to be larger than the circuit conductors. When conductors are run in parallel in more than one raceway, the equipment-grounding conductor should also be run in parallel.

Double Insulation

Double insulation of an appliance or power tool protects the user from electric shock by creating a nonconducting barrier between the user and the electric components inside the appliance or tool. An appliance or tool that is double insulated has two levels of insulating materials between the electrical parts of the appliance and any parts on the outside that can be touched by the user. If the first layer of insulation fails, the second layer provides protection. Small appliances and power tools with double insulation are not required to have a grounding conductor—that is, they are allowed to have a two-prong plug.

The primary difference between an appliance with a three-prong plug and an appliance with a two-prong plug is the appliance casing. If an appliance casing is a good conductor (e.g., made of metal), then it must have a grounding conductor and three-prong plug. Many of newer household electrical appliances and tools are double insulated. Examples include coffee makers, blow dryers, electric drills, and other similar small power tools and appliances.

19.6   THE BUILDING ELECTRICAL SYSTEM

There are many possible electrical systems that can be used to distribute power in a building. Typically, system design begins with selection of a building system voltage, which is dependent on sizes and types of the connected loads, utilities near the building, local codes and ordinances, economics, and safety.

It is more economical to distribute power at high voltage. Amperage determines conductor size, and higher voltages mean lower amperage, which allows use of smaller conductors. Use of high voltages, however, introduces more complex safety issues, so use of high voltages in buildings is reserved for equipment with heavy loads. Larger installations have greater power requirements and thus require higher building system voltages. Smaller installations such as residences use lower voltage systems to avoid exposing occupants to safety issues related to higher voltages.

Typical types of building electrical systems are introduced in the following sections.

Residential Systems

Normally, a 120/240 V, three-wire, single-phase service entrance serves a residence. In some cases where it is the only service available, a 120/208 V, three-wire, single-phase service entrance is used. Power from a ground- or pole-mounted utility transformer located outside the building is brought to the building service equipment through underground or overhead service entrance lines.

From a single panelboard rated from 100 to 200 A or more, power is distributed throughout the residence through branch circuits. These branch circuits originate at the panelboard, are protected by overcurrent protection in the panelboard, and terminate at outlets serving appliances, equipment, and lighting. In large residences, a set of feeders may extend from the main panelboard and bring power to one or more subpanelboards located at a remote area of the building. Branch circuits originating at the subpanelboard feed outlets in these outlying areas.

Multifamily Dwellings

In multifamily dwelling units (e.g., apartments, condominiums), power is brought from a utility transformer to the building service equipment. It is then divided at a main distribution panel, passes through individual meters, and is distributed to the individual dwelling units through feeders. Each dwelling unit is served by a separate panelboard located in the dwelling unit. Branch circuits extend from a panelboard to feed outlets within the unit. A *house panelboard* has branch circuits that serve common areas (e.g., corridors, laundry rooms, lobbies, and so forth).

Photos 19.1 through 19.22 show examples of residential electrical installations.

Commercial/Industrial Systems

Large commercial and industrial facilities have large and varied power requirements that necessitate different types of systems for different building occupancies. Essentially there are so many variations that there is no standard type of system.

In a typical large building, electrical power is provided to a transformer located outside the building or it enters a transformer vault located at the service level in the building. A *transformer vault* is a basement- or ground-level structure or

PHOTO 19.1 Underground electrical distribution lines serving a residential development being buried in a trench. (Used with permission of ABC)

PHOTO 19.2 Underground distribution lines readied for connection to exterior transformer. (Used with permission of ABC)

PHOTO 19.3 A power transformer (right) steps down distribution voltage to building system voltage (120/240 V). The power pedestal (at center right) is where the connection is made before being fed to the residence through the service entrance conductors. (Used with permission of ABC)

PHOTO 19.4 Power pedestals prepared for installation. (Used with permission of ABC)

PHOTO 19.5 Underground service entrance for a multifamily dwelling unit. (Used with permission of ABC)

PHOTO 19.6 A 120/240 V–150 A, underground service entrance for a single-family residence. (Used with permission of ABC)

PHOTO 19.7 A service entrance with the electrical meter and main disconnect (below meter and above circuit breakers). Circuit breakers are properly marked on the panelboard face. (Used with permission of ABC)

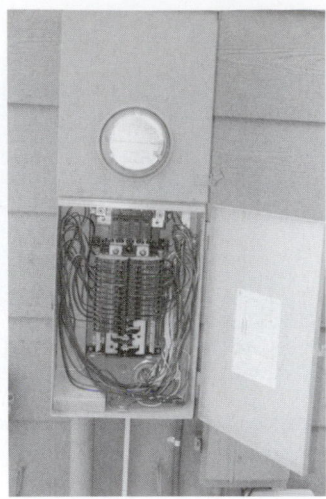

PHOTO 19.8 Panelboard with face removed. (Used with permission of ABC)

PHOTO 19.9 A close-up view of the inside of the panelboard. Conductors from the meter connect to the two lugs in the top of the photo. Power is fed through two bus bars connected to the lugs and then through each circuit breaker and the dark-colored ungrounded conductors (wires). On the right side of the panelboard, the white grounded (neutral) conductors and bare grounding conductors connect to the neutral/grounding bus. (Used with permission of ABC)

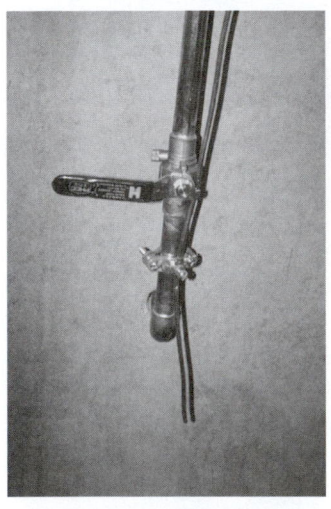

PHOTO 19.10 A main ground extends from the panelboard to the water service pipe and to a grounding rod (not shown). (Used with permission of ABC)

PHOTO 19.11 Branch circuit cables shown before installation of the panelboard. (Used with permission of ABC)

PHOTO 19.12 Branch circuit cables extend through the inside of the wall (behind the panelboard). (Used with permission of ABC)

PHOTO 19.13 Branch circuit conductors are fed to outlets. (Used with permission of ABC)

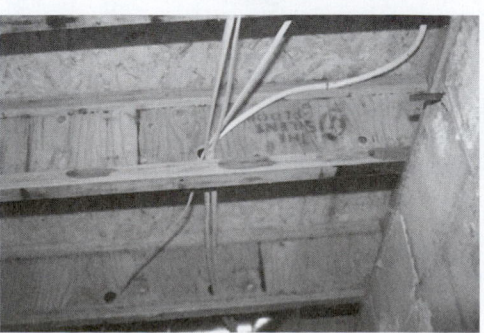

PHOTO 19.14 Circuit cables can pass through floor and ceiling joists. (Used with permission of ABC)

PHOTO 19.15 Rough wiring of branch circuits extends through framing to outlets. (Used with permission of ABC)

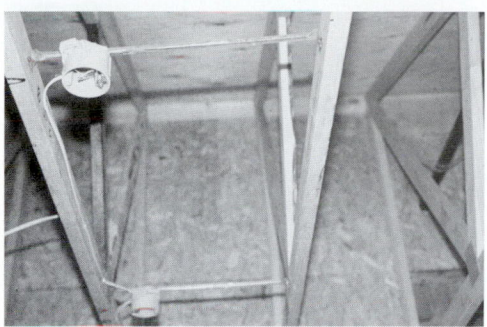

PHOTO 19.16 Roughed-in lighting box. (Used with permission of ABC)

PHOTO 19.17 Roughed-in recessed light fixture. (Used with permission of ABC)

PHOTO 19.18 Exposed lighting fixture. (Used with permission of ABC)

PHOTO 19.19 Roughed-in receptacle box. (Used with permission of ABC)

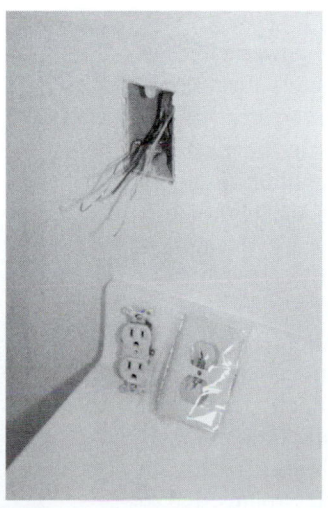

PHOTO 19.20 Exposed box in finished drywall with unattached receptacle and faceplate. (Used with permission of ABC)

PHOTO 19.21 Covered convenience receptacle. (Used with permission of ABC)

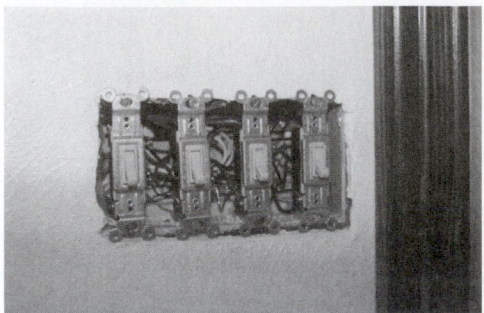

PHOTO 19.22 Ganged switches with faceplate removed. (Used with permission of ABC)

PHOTO 19.23 Switchboard for a manufacturing facility. (Used with permission of ABC)

room in which power transformers, network protectors, voltage regulators, circuit breakers, meters, and so on are housed. Three-phase power normally enters the building with three un-grounded (hot) conductor lines sharing a common neutral conductor. A transformer steps down power to voltages below 600 V (e.g., 480 Y/277 V, three phase). At the stepped-down level, the electrical power passes out of the transformer to the main switchboard serving the building.

The switchboard is located in a separate room, near the transformer or adjacent to the transformer vault. At the switchboard, power passes through meters and circuit breakers or fuses as it is divided into primary circuits and is then carried by feeder busways or sets of large conductors into the different parts of the building. Several feeder circuits provide power to power distribution panelboards throughout the building, where it is used to run large motors powering mechanical equipment (e.g., HVAC fan motors and plumbing pumps) and elevators. Others feeder circuits feed lighting and appliance panelboards where the power is divided to smaller circuits serving lighting and convenience outlets.

Primary feeders are separated into high-power and low-power bus risers, busways, or sets of conductors that supply different system voltages to the many levels or areas of the building. As the bus riser or busway extends to the electrical room of each floor or area, a feeder circuit is split off to supply the high-power panelboard (480/277 V). In the high-power panel, additional branch circuits are divided out. For example, a three-phase circuit may be established to drive air-handling units in the mechanical room. Single-phase, 277 V circuits provide power to overhead fluorescent fixtures. Another feeder circuit is taken off of the high-power panel by a local three-phase step-down transformer every three floors, which steps down voltage to a 208/120 V for low-power circuits. Low power is then distributed from a panelboard through branch circuits to outlets in the individual spaces (e.g., offices, classrooms, and so on). Closet transformers may be located throughout the building to accommodate localized power requirements. Photos 19.23 through 19.38 show various components of commercial and industrial systems.

PHOTO 19.24 Service entrance conductors and feeder conductors encased in conduit and a wire gutter. Note the grounding wire (bottom wire). (Used with permission of ABC)

PHOTO 19.25 Ground wire connection to grounding rod that extends into earth on other side of foundation wall. (Used with permission of ABC)

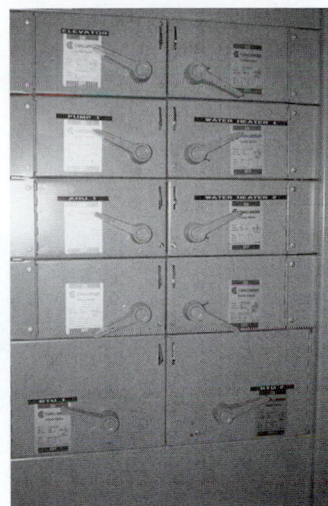

PHOTO 19.26 Distribution panels with fused disconnects. (Used with permission of ABC)

PHOTO 19.27 Open distribution panels exposing two sets of fuses. (Used with permission of ABC)

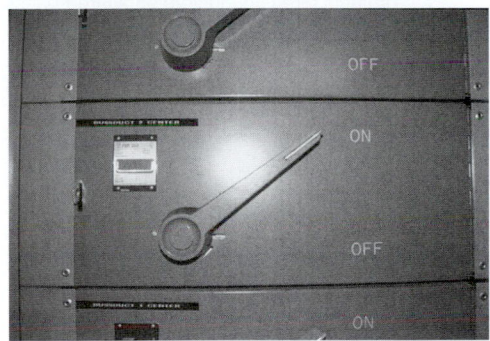

PHOTO 19.28 A close-up of a single distribution panel with fused disconnect. (Used with permission of ABC)

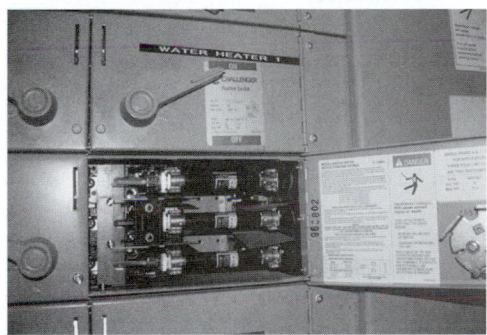

PHOTO 19.29 Open distribution panel with three cartridge fuses protecting a three-phase feeder (three ungrounded conductors). (Used with permission of ABC)

PHOTO 19.30 A 480 Y/277 V–400 A, three-phase panelboard providing three-phase power to equipment (e.g., milling machines, lathes, machining centers, and so on) or 277 V single-phase power for lighting. (Used with permission of ABC)

PHOTO 19.31 Three feeders from the main distribution panel (see Photo 19.26) connected to lugs in the panelboard (see Photo 19.30). The lugs connect to the three bus bars to which circuit breakers are connected. The fourth conductor (left side of photograph) is the grounding conductor, which connects the panelboard ground to the main distribution panel near the service entrance. (Used with permission of ABC)

PHOTO 19.32 The panelboard with circuit breakers installed. (Used with permission of ABC)

PHOTO 19.33 A close-up view of the three bus bars. (Used with permission of ABC)

PHOTO 19.34 An individual three-phase breaker protects three conductors on a three-phase circuit. (Used with permission of ABC)

PHOTO 19.35 Conductors for each circuit extend through conduits from the main panelboard (under construction). (Used with permission of ABC)

PHOTO 19.36 Circuits to equipment are controlled by heavy-duty switches. (Used with permission of ABC)

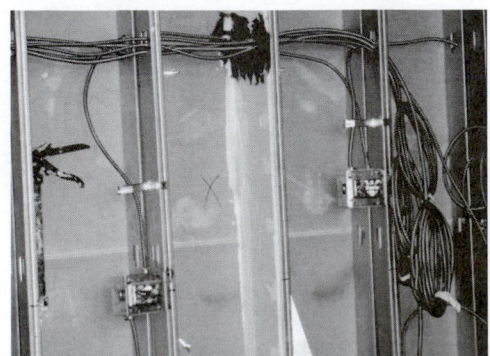

PHOTO 19.37 Armored cable carrying circuiting from panelboard to electrical boxes; a commercial installation. (Used with permission of ABC)

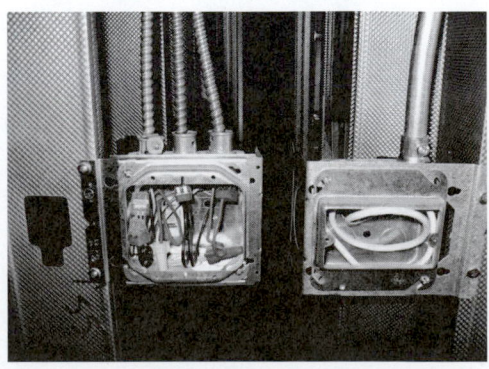

PHOTO 19.38 Electrical boxes receiving conductor-carrying armored cable. (Used with permission of ABC)

19.7 CONDUCTOR REQUIREMENTS

Conductor Materials

Copper and aluminum are the most common conductor materials used in building electrical wiring, although other materials can be used. As a general rule, solid copper conductors are used in small conductor sizes (up to about 8 AWG) because safety issues associated with aluminum are avoided and weight and cost are not significantly affected. Medium- and large-gauge stranded aluminum cable (above No. 8 AWG) is safely used on circuits as long as the terminals or connectors on the circuit are rated CU-AL (copper-aluminum) and an antioxidant paste is used on properly tightened connections.

Stranded aluminum conductors are widely used on larger (above 30 A) circuits serving large motors, equipment, and appliances such as clothes dryers, kitchen ranges, or central air conditioners. They are also used as service entrance and feeder conductors that carry electrical energy from the transformer to the building service equipment (the switchboard or panelboard). Use of UL-listed 15 A and 20 A switches, outlets, and other devices marked CO/ALR is presently accepted.

Table 19.4 shows the normal range of applications for commonly used conductor sizes.

Conductor Insulation

Conductors are covered with insulation to provide electrical isolation and physical protection of the conductor material. The type of insulation material determines the environment in which a wire or cable can be used safely. Wires used indoors are subjected to less exposure to the elements than those designed for outdoor use. Outdoor wiring is exposed to water and sunlight, so the insulation is designed to withstand these elements. Insulation on wires buried in the ground must also be able to withstand the damp, corrosive environment of the soil. Insulation used to cover electrical conductors are designated with a set of letters that describes the properties of the insulation. These designations were described in Chapter 18.

Conductor Ampacity Requirements

Tables 18.14 through 18.16 and 18.18 through 18.24 contain ampacities for various conductors, conductor insulations, and sheathings. Ampacities provided in these tables are values based on a normal operating temperature of 86°F (30°C). Ampacity values for each conductor size are for different equipment terminal (where connections of wiring are made) temperatures. Heat generated at the equipment terminals can damage the conductors if it is not properly dissipated. Unless equipment terminals are marked otherwise, circuit conductors are to be sized according to 140°F (60°C) for equipment rated 100 A and less, and equipment rated over 100 A must be sized to 167°F (75°C).

Constant exposure to high temperatures increases the deterioration rate of insulation, so a conductor exposed to higher temperatures will have a lower ampacity than one that is operating in the normal temperature range. Conversely, exposure to lower temperatures extends insulation life and increases a conductor's ampacity. Therefore, in building electrical systems, the ampacity of a conductor may need to be adjusted with correction factors when conditions related to the temperature of the surroundings and the number of conductors in a raceway (e.g., conduit or cable) fall outside normal operating ranges. Correction factors typically applied are addressed in the following sections.

Temperature Correction Factor

Ambient temperature is the temperature of a surrounding medium (e.g., air, soil). In the case of electrical wiring, it is the temperature of the medium surrounding the conductor. Ambient temperature can affect allowable current-carrying capacity of a conductor. As ambient temperature rises, less current-generated heat is needed to reach the temperature rating of the insulation. Therefore, ampacity is governed by contribution of ambient heat.

The *ambient temperature rating* of a conductor refers to the normal temperature range in the environment in which that conductor is to be used (e.g., the temperature of the surrounding

TABLE 19.4   COMMON USES OF CONDUCTORS IN BUILDINGS BY CONDUCTOR SIZE.

Conductor Gauge	Common Uses
No. 20 AWG and smaller	Electronic circuits and phone extensions
No. 16 to 18 AWG	Light-gauge extension cords, lamp cords, door chime wiring, small appliance cords
No. 12 to 14 AWG	Normal 15 A and 20 A branch circuits serving small appliances, convenience (receptacle) outlets, and luminaires
No. 4 to 10 AWG	Larger branch circuits at 30 A and above serving electrical appliances such as electric water heaters, clothes dryers, air conditioning equipment, and water pumps
No. 2 to 4/0 AWG	Residential and light commercial service entrance conductors and feeders to panelboards
250 kcmil and larger	Heavy commercial and industrial service entrance conductors, large feeders to closet transformers, and panelboards

air, water, or earth). Conductor ampacity is adjusted for changes in ambient temperature, including temperatures below 78°F (26°C) and above 86°F (30°C). Most electrical and mechanical rooms and attic spaces where electrical equipment is installed will exceed 86°F (30°C) under typical operating conditions.

A *temperature correction factor (F_t)* for conductors is applied based on the ambient temperature of the conductor. For example, the temperature correction factor for THHN, THWN, and XHHW (75°C, 167°F) insulation in an ambient temperature range of 123° to 131°F (51° to 55°C) is 0.67. This means a conductor operating in this ambient temperature range will have an ampacity (allowable current capacity) that is 67% of one in the normal operating temperature range—that is, a larger conductor is required at higher temperatures. Conversely, the temperature correction factor in an ambient temperature range of 70° to 77°F (21° to −25°C) is 1.05. Refer to the local code for other correction factors and temperature ranges.

Bundling Correction Factor

When several current-carrying conductors are contained in a raceway or cable, the temperature of the conductors will increase under normal loading conditions. The additional current-generated heat causes a conductor surrounded by several other current-carrying conductors to exceed the temperature rating of the insulation more easily. As a result, ampacity of a conductor grouped with several other conductors is influenced by contribution of ambient heat. Simply, when several (more than three) current-carrying conductors are added to a raceway or bundled in a cable, ampacity of a current-carrying conductor is decreased to compensate for the extra heat.

A *bundling correction factor (F_N)* must be applied for four or more conductors in a raceway or cable installed in the same raceway or conduit or any bundled cables that are more than 24 in (0.63 m) long. For raceways and cables with 4 through 6 current-carrying conductors, the bundling correction factor is 0.80; for 7 through 9 conductors, it is 0.70; and for 10 through 20, it is 0.50. Refer to the local code for correction factors for additional conductors.

The number of current-carrying conductors includes any ungrounded conductor or grounded conductor.

A shared neutral (see definition in the Neutral Conductor section above: a single 120 V circuit is not served by a neutral conductor) that is not current carrying, and is not counted. However, on four-wire, three-phase wye-connected systems the shared neutral must be counted as a current-carrying conductor. Equipment-grounding conductors are not current-carrying conductors and are not counted.

To account for ambient temperatures outside of this normal range, ampacity for a conductor in the normal operating temperature range (the base ampacity) is adjusted with reduction factors discussed in Example 19.3. These factors are applied such that the ampacity (I_ampacity) of a conductor at a specific operating temperature is the ampacity (I_normal) at normal operating temperatures multiplied by any applicable reduction factors for ambient temperature (F_t) and conductor bundling (F_N):

$$I_{ampacity} = I_{normal}(F_t)(F_N)$$

Example 19.2

Determine the ampacity of a No. 8 AWG copper conductor with THHN insulation that will be used in a 120 V, two-wire circuit in an environment with an average ambient air temperature of no greater than 86°F (30°C).

No corrections must be made for temperature or more than three conductors in a raceway, so $I_{ampacity} = I_{normal}$

Circuit conductors are sized according to 140°F (60°C) for equipment rated 100 A and less.

From Table 18.11 in Chapter 18, the ampacity of this conductor is 40 A.

Example 19.3

Determine the ampacity of a No. 8 AWG copper conductor with THHN insulation that will be used in a 120 V, two-wire circuit in an industrial environment with an average ambient air temperature of no greater than 125°F (51.7°C). Three similar circuits will be carried in a conduit. Assume that a shared neutral is not used.

From the main body of the text, the temperature correction factor for THHN insulation in an ambient temperature range of 123° to 131°F (51° to 55°C) is 0.67. From the main body of the text, the temperature correction factor for more than 6 conductors in a raceway is 0.80. As in Example 19.2, circuit conductors are sized according to 140°F (60°C) for equipment rated 100 A and less. From Table 18.11 in Chapter 18, the ampacity of this conductor is 40 A.

$$I_{ampacity} = I_{normal}(F_t)(F_N) = 40\ A\ (0.67)(0.80) = 21.4\ A$$

The ampacity of this conductor under these conditions is 21.4 A.

Example 19.4

Determine the ampacity of a No. 8 AWG copper conductor with THHN insulation that will be used in a 120 V, two-wire circuit in an industrial environment with an average ambient air temperature of no greater than 125°F (51.7°C). Three similar circuits will be carried in a conduit. Assume that a shared neutral is not used. Equipment terminal connections are rated at 167°F (75°C).

From the main body of the text, the temperature correction factor for THHN insulation in an ambient temperature range of 123° to 131°F (51° to 55°C) is 0.67. From the main body of the text, the temperature correction factor for more than 6 conductors in a raceway is 0.80. Equipment terminal connections are rated at 167°F (75°C), so from Table 18.11 in Chapter 18, the ampacity of this conductor is 50 A.

$$I_{ampacity} = I_{normal} (F_t)(F_N) = 50 \text{ A } (0.67)(0.80) = 26.8 \text{ A}$$

The ampacity of this conductor under these conditions is 26.8 A.

Conductor Voltage Drop Requirements

In addition to ampacity requirements, branch circuits and feeders should be analyzed for voltage drop because of the adverse effect it can have on performance and operating life of appliances and equipment. Although not required, branch circuits are typically designed to ensure that voltage drop at full load does not exceed 3% from the panelboard to the farthest outlet. Feeders are typically designed to ensure that voltage drop at full load does not exceed 2%. Total voltage drop in the feeders and branch circuits should not exceed 5%. For circuits feeding critical and sensitive equipment (e.g., medical, testing, and other high-fidelity electronic equipment), tighter limits are recommended: no more than a 1% voltage drop on branch circuits and a total voltage drop of 2% for feeders and branch circuits. When the feeder and branch circuit conductors for long circuits are sized on this basis, conductor sizes are increased beyond that required for ampacity.

The basic formula for determining voltage drop (E_{drop}) in a two-wire AC circuit or three-wire AC single-phase circuit with a balanced load at 100% power factor (neglecting reactance) is based on the one-way circuit length (L), in feet or meters; conductor resistance (R), in $\Omega/1000$ ft or $\Omega/1000$ m; and the circuit load (I) in amperes:

$$E_{drop} = 2 \text{ LRI}/1000$$

The percentage of voltage drop is determined by the ratio of voltage drop and system voltage.

Example 19.5

A No. 8 copper conductor with THHN insulation has a one-way length of 200 ft. It is carrying a load of 16 A on a 120 V, two wire circuit. Approximate the voltage drop and percentage of voltage drop in the circuit.

From Table 18.11 in Chapter 18, a No. 8 AWG conductor has a resistance of 0.6410 $\Omega/1000$ ft. Voltage drop is found by:

$$\begin{aligned} E_{drop} &= (2 \text{ LRI}) \\ &= (2 \cdot 200 \text{ ft} \cdot 0.6410 \ \Omega/1000 \text{ ft} \cdot 16 \text{ A}) = 4.1 \text{ V} \end{aligned}$$

The percentage of voltage drop is determined by:

$$4.1 \text{ V}/120 \text{ V} = 0.034 = 3.4\%$$

In Example 19.5, voltage drop is excessive (above 3%). Conductor size would need to be increased.

The following equations can be used to *approximate* the maximum one-way distance (L) that a set of conductors can run, in feet. It is based on voltage (E), amperage (I), conductor size in circular mils (cmil), and voltage drop (E_{drop}) expressed as a coefficient (i.e., 2% voltage drop = 0.02, 3% = 0.03, and so on). These equations are an approximation based on the basic voltage drop formula introduced earlier. Accuracy is limited to sizes up to 4/0 conductors.

For single-phase circuits, the equations are:

Copper conductors: $L = E(E_{drop})(cmil)/21.6 \text{ I}$
Aluminum conductors: $L = E(E_{drop})(cmil)/36 \text{ I}$

In SI (metric) computations, substitute the constant 70.8 for 21.6 and 118 for 36 to find length (L) in meters.

For three-phase circuits, equations are:

Copper conductors: $L = E(E_{drop})(cmil)/18.7 \text{ I}$
Aluminum conductors: $L = E(E_{drop})(cmil)/31.2 \text{ I}$

In SI (metric) computations, substitute the constant 61.3 for 18.7 and 102.2 for 31.2 to find length (L) in meters.

Example 19.6

Approximate the maximum distance two No. 10 AWG conductors can carry a current of 20 A on a 120 V, single-phase circuit. Use a maximum voltage drop of 3%.

No. 10 AWG conductor has a cross-sectional area of 10 380 cmils (from Chapter 18).

For copper conductors:

$$\begin{aligned} L &= E(E_{drop})(cmil)/21.6 \text{ I} \\ &= (120 \text{ V})(0.03)(10 \ 380 \text{ cmils})/(21.6)(20 \text{ A}) = 86.5 \text{ ft} \end{aligned}$$

For aluminum conductors:

$$\begin{aligned} L &= E(E_{drop})(cmil)/36 \text{ I} \\ &= (120 \text{ V})(0.03)(10 \ 380 \text{ cmils})/(36)(20 \text{ A}) = 51.9 \text{ ft} \end{aligned}$$

The terms in Example 19.6 equations can be rearranged to create an equation to determine *minimum conductor size* in circular mils (cmil):

For single-phase circuits, the equations are:

Copper conductors: $cmil = 21.6 \text{ IL}/E(E_{drop})$
Aluminum conductors: $cmil = 36 \text{ IL}/E(E_{drop})$

For three-phase circuits, the equations are:

Copper conductors: $cmil = 18.7 \text{ IL}/E(E_{drop})$
Aluminum conductors: $cmil = 31.2 \text{ IL}/E(E_{drop})$

Example 19.7

A set of conductors on a branch circuit serve a swimming pool pump motor rated at 20 A. The pump is 250 ft from the 120 V power source. Determine the required size in cmil and AWG for copper and aluminum conductors at a maximum voltage drop of 3%.

For copper conductors:

$$\text{cmil} = 21.6 \, IL/E(E_{drop})$$
$$= (21.6 \cdot 20 \, A \cdot 250 \, ft)/(120 \, V \cdot 0.03) = 30 \, 000 \, \text{cmil}$$

A No. 4 AWG copper conductor is required (41 740 cmils)

For aluminum conductors:

$$\text{cmil} = 36 \, ID/E(E_{drop})$$
$$= (36 \cdot 20 \, A \cdot 250 \, ft)/(120 \, V \cdot 0.03) = 50 \, 000 \, \text{cmil}$$

A No. 2 AWG aluminum conductor is required (66 360 cmils).

Table 19.5 provides ampacities of common insulated conductors of different lengths for 120 V, single-phase circuits

TABLE 19.5   AMPACITIES OF INSULATED CONDUCTORS OF DIFFERENT LENGTHS FOR 120 V, SINGLE-PHASE CIRCUITS FOR NO MORE THAN THREE CONDUCTORS IN A RACEWAY, CABLE OR EARTH BURIAL BASED ON AN AMBIENT TEMPERATURE OF 86°F (30°C) FOR RHH, RHW-2, THHN, THHW, THW-2, THWN-2, USE-2, XHH, XHHW, AND XHHW-2 INSULATIONS. VALUES IN ITALICS ARE AMPACITIES GOVERNED BY VOLTAGE DROP FOR THE ONE-WAY DISTANCE PROVIDED, BASED ON A MAXIMUM 3% VOLTAGE DROP (AS COMPUTED USING THE TECHNIQUE IN EXAMPLE 19.7).

Wire Gauge/Size	Ampacity (U.S. Customary Units)											
	One-Way Distance (ft)											
	Copper						Aluminum					
AWG	50 ft	100 ft	200 ft	300 ft	400 ft	500 ft	50 ft	100 ft	200 ft	300 ft	400 ft	500 ft
14	15	7	3	2	2	1	—	—	—	—	—	—
12	20	11	5	4	3	2	15	7	3	2	2	1
10	30	17	9	6	4	3	20	10	5	3	3	2
8	40	28	14	9	7	6	30	17	8	6	4	3
6	50	44	22	15	11	9	40	26	13	9	7	5
4	85	70	35	23	17	14	50	42	21	14	10	8
3	100	88	44	29	22	18	75	53	26	18	13	11
2	115	111	55	37	28	22	90	66	33	22	17	13
1	130	130	70	46	35	28	100	84	42	28	21	17
0 or 1/0	150	150	88	59	44	35	120	106	53	35	26	21
00 or 2/0	175	175	111	74	55	44	135	133	67	44	33	27
000 or 3/0	200	200	140	93	70	56	155	155	84	56	42	34
0000 or 4/0	230	230	176	118	88	71	180	180	106	71	53	42

Wire Gauge/Size	Ampacity (SI Metric Units)											
	One-Way Distance (m)											
	Copper						Aluminum					
AWG	15 m	30 m	60 m	90 m	120 m	150 m	15 m	30 m	60 m	90 m	120 m	150 m
14	15	7	3	2	2	1	—	—	—	—	—	—
12	20	11	6	4	3	2	15	7	3	2	2	1
10	30	18	9	6	4	4	20	10	5	3	3	2
8	40	28	14	9	7	6	30	17	8	6	4	3
6	50	44	22	15	11	9	40	26	13	9	7	5
4	85	71	35	24	18	14	50	42	21	14	11	8
3	100	89	45	30	22	18	75	53	27	18	13	11
2	115	112	56	37	28	22	90	67	33	22	17	13
1	130	130	71	47	35	28	100	84	42	28	21	17
0 or 1/0	150	150	89	60	45	36	120	106	53	35	27	21
00 or 2/0	175	175	113	75	56	45	135	134	67	45	34	27
000 or 3/0	200	200	142	95	71	57	155	155	85	56	42	34
0000 or 4/0	230	230	179	120	90	72	180	180	107	71	53	43

for no more than three conductors in a raceway, cable, or earth burial for ambient temperatures in the normal operating range. Values in italics were governed by voltage drop for the one-way distance provided, based on a maximum 3% voltage drop.

Insulation Color Coding and Identification Markings

The insulation on small- and medium-size conductors is color coded for identification. Larger conductors requiring color identification are marked at the terminal ends with a hand-painted stripe or colored tape wrapped around the conductor insulation. Multiple ungrounded conductors in a raceway or wire gutter must be indicated with a colored stripe. The insulation of grounded and neutral conductors at No. 6 AWG or less in size must be color coded with a continuous white or natural gray color. The grounding (ground) conductor insulation must be color coded green, green with one or more yellow stripes, or may be a bare conductor on small conductors in cables.

Code no longer requires specific colors for color coding of ungrounded (hot) conductors, except the Phase Y on delta-connected, three-phase systems such as the 240 Δ/120 AC, 3Φ-4W system discussed earlier. On delta-connected systems, Phase Y has a higher voltage to ground (e.g., 208 V on the 240 Δ/120 AC, 3Φ-4W system) than the other legs (120 V on the 240 Δ/120 AC, 3Φ-4W system). Code requires that this phase be marked by the color orange, with the intent of preventing the connection of single-phase loads to this phase and avoiding the resulting equipment damage. Otherwise, ungrounded conductors may be any color, except white, gray, and green.

The North American standard for color coding is black or any color, except white, gray, and green (ungrounded/hot); white (grounded/neutral); and green (ground). The commonly used but not mandatory color sequence of conductors serving single-phase circuiting is:

Two wires:	grounded: white	ungrounded: black
Three wires:	grounded: white	ungrounded: black and red
Four wires:	grounded: white	ungrounded: black, red, and blue
Five wires:	grounded: white	ungrounded: black, red, blue, and yellow

On three-phase circuits, the color sequence of conductors tends to be:

Four wires:	grounded: white	ungrounded: brown, orange, and yellow

Internationally, the standard is brown (ungrounded/hot), blue (grounded/neutral), and green with yellow strip (ground). Historically, different color codes have been used. On three-phase circuits, the color sequence of conductors tends to be:

first-phase conductor (brown), second-phase conductor (black), and third-phase conductor (gray).

In large commercial and industrial facilities where several system voltages are available, it is useful (and safe) to mark or label conductors and equipment for identification in addition to color coding it. Identification markings typically indicate system voltage and phase. Markings must be durable so they withstand the environment. A sample color and marking code for conductors is shown in Table 19.6.

19.8   CABLE, RACEWAY, AND ENCLOSURE REQUIREMENTS

Cable and Raceway Requirements

All building wiring must be enclosed in a cable, conduit, wireway, or raceway. During installation, conductors are snaked through conduit or tubing, are laid in a wireway, or are contained in cables and secured to structural framing. Care must be exercised in placing conductors, as conductors and insulation can be easily damaged. For example, if a conductor is pulled through a tight conduit, it can stretch. Deformation caused by stretching reduces the cross-sectional area of the wire, thereby reducing its ampacity. This creates an unsafe condition because the conductor can overheat.

Conductors that are run through a raceway must have sufficient open air space to prevent overheating. The number of current-carrying conductors that can run through a raceway is limited by code. A current-carrying conductor found in a raceway is any ungrounded conductor or grounded conductor. Equipment grounding (bare or green colored) and shared neutral conductors are not current carrying, and are not counted.

Table 19.7 indicates the *recommended* maximum number of conductors allowed in conduit or tubing for conductors with THWN or THHN insulation. Similar tables exist in the electrical code for this and other insulation types. These recommended values are less than those typically established as maximum values in the electrical code. They are recommended maximums because they allow for less cumbersome installation (e.g., pulling conductors through conduit) and are based on industry experience. They allow for future expansion (adding conductors).

Example 19.8

Recommend the maximum number of No. 12 AWG THWN conductors that should be placed in a ¾ EMT conduit, based on information in Table 19.7.

From Table 19.7, 8 conductors maximum.

TABLE 19.6   A SYSTEM FOR COLOR CODING AND IDENTIFICATION MARKINGS FOR CONDUCTORS.

Voltage System	Color and Identification Marking Code
120 V AC single phase, two wire	• Grounded neutral, white (first or only neutral in raceway, box, auxiliary gutter, or other types of enclosures) • Grounded neutral, white with black stripe running entire length of insulation (when neutral is installed in raceway, box, auxiliary gutter, or other types of enclosures with another neutral) • Grounding conductor, green, green with one or more yellow stripes, green tape, or bare • Ungrounded conductor, black with "120V-1PH" marking
120/240 V AC single phase, three wire	• Grounded neutral, white (first or only neutral in raceway, box, auxiliary gutter, or other types of enclosures) • Grounded neutral, white with brown stripe running entire length of insulation (when neutral is installed in raceway, box, auxiliary gutter, or other types of enclosures with another neutral) • Grounding conductor, green, green with one or more yellow stripes, green tape, or bare • Ungrounded conductor, black with "240/120V-1PH-A" marking • Ungrounded conductor, red with "240/120V-1PH-B" marking
208 Y/120 V AC three phase, four wire	• Grounded neutral, white (first or only neutral in raceway, box, auxiliary gutter, or other types of enclosures) • Grounded neutral, white with red stripe running entire length of insulation (when neutral is installed in raceway, box, auxiliary gutter, or other types of enclosures with another neutral) • Grounding conductor, green, green with one or more yellow stripes, green tape, or bare • Phase X (ungrounded) conductor, black with "208Y/120V-3PH-X" marking • Phase Y (ungrounded) conductor, red with "208Y/120V-3PH-Y" marking • Phase Z (ungrounded) conductor, blue with "208Y/120V-3PH-Z" marking
480 Y/277 V AC three phase, four wire	• Grounded neutral, white (first or only neutral in raceway, box, auxiliary gutter, or other types of enclosures) • Grounded neutral, white with yellow stripe running entire length of insulation (when neutral is installed in raceway, box, auxiliary gutter, or other types of enclosures with another neutral) • Grounding conductor, green, green with one or more yellow stripes, green tape, or bare • Phase X (ungrounded) conductor, brown with "480Y/277V-3PH-X" marking • Phase Y (ungrounded) conductor, orange with "480Y/277V-3PH-Y" marking • Phase Z (ungrounded) conductor, yellow with "480Y/277V-3PH-Z" marking
240 Δ/120 V AC three phase, three wire	• Grounding conductor, green, green with one or more yellow stripes, green tape, or bare • Phase X (ungrounded) conductor, black with "240VD-3PH-X" marking • Phase Y (ungrounded) conductor, black with "240VD-3PH-Y" marking • Phase Z (ungrounded) conductor, black with "240VD-3PH-Z" marking
480 Δ V AC three phase, three wire	• Grounding conductor, green, green with one or more yellow stripes, green tape, or bare • Phase X (ungrounded) conductor, brown with "480VD-3PH-X" marking • Phase Y (ungrounded) conductor, orange with "480VD-3PH-Y" marking • Phase Z (ungrounded) conductor, yellow with "480VD-3PH-Z" marking

Example 19.9

Recommend the minimum EMT conduit size for three No. 4 AWG THWN current-carrying conductors, based on information in Table 19.7.

From Table 19.7, a 1¼ in conduit is recommended.

Rigid conduit and tubing (e.g., rigid metal and nonmetallic conduit, IMC, EMT, ENT) are favorite raceway materials used to protect conductors in all types of buildings. Because it is available in straight lengths, it must be bent to accommodate changes in direction. To prevent stretching of conductors as they are pulled through the conduit or tube, bends must be made no smaller than the minimum radius specified by code, about 6 to 8 times the conduit diameter. Typically, conduit and tubing can have up to four 90° bends or the equivalent (360° total) in one run. In runs requiring more bends, a pull box is added in the run to assist in pulling conductors and to allow access. Bends can be made at the factory or at the job site for small to medium conduits.

Flexible conduit (e.g., flexible metal conduit, liquid tight flexible metal, and nonmetallic conduit) offers the advantage of easier installation and can be salvaged easily when circuits are rearranged. It is used where mechanical protection is needed and easy relocation of equipment is desired. It can have up to four 90° bends or the equivalent (360° total) in one run. Again in runs requiring more bends, a pull box is added.

Conduit should be supported to prevent wearing away against structure and to avoid stressing its end fittings. Rigid conduit and tubing must be supported within 3 ft of a box or other connection and at intervals of 10 ft (about 3 m). If it is continuous, metal conduit can be used as the grounding conductor. Flexible conduit must be supported every 4 to 6 in (1.35 m) and within 12 in (300 mm) of a box, except up to 36 in (about 1 m) is allowed at usage points where flexibility is

TABLE 19.7 RECOMMENDED MAXIMUM NUMBER OF CONDUCTORS IN METAL (EMT) AND PLASTIC (PVC) CONDUIT FOR CONDUCTORS WITH THWN, THHN, AND THW-2 INSULATION. VALUES SUGGESTED IN THIS TABLE ARE LESS THAN THOSE TYPICALLY ESTABLISHED AS MAXIMUM VALUES IN THE ELECTRICAL CODE. THE RECOMMENDED MAXIMUM VALUES PROVIDED ALLOW FOR LESS CUMBERSOME INSTALLATION (E.G., PULLING CONDUCTORS THROUGH CONDUIT).

		Wire Size (THWN, THHN) Conductor Size																				
		AWG									kcmil											
Trade Size		14	12	10	8	6	4	3	2	1	1/0	2/0	3/0	4/0	250	300	350	400	500	600	700	750
Metal (EMT)	½	6	5	3	2	1	1	1	1	1	1	—	—	—	—	—	—	—	—	—	—	—
	¾	11	8	5	3	2	1	1	1	1	1	1	1	1	—	—	—	—	—	—	—	—
	1	18	13	8	5	4	2	2	2	1	1	1	1	1	1	1	1	—	—	—	—	—
	1¼	31	23	14	8	6	4	3	3	2	2	1	1	1	1	1	1	1	1	1	1	—
	1½	42	31	19	11	8	5	4	4	3	2	2	1	1	1	1	1	1	1	1	1	1
	2	69	51	32	18	13	8	7	6	4	4	3	3	2	2	1	1	1	1	1	1	1
	2½	121	88	56	32	23	14	12	10	8	6	5	4	4	3	3	2	2	2	1	1	1
	3	182	113	84	48	35	22	18	15	11	10	8	7	6	5	4	3	3	3	2	2	2
	3½	238	174	110	63	46	28	24	20	15	13	10	9	7	6	5	5	4	3	3	2	2
	4	304	222	140	81	58	36	30	26	19	16	13	11	9	8	7	6	5	4	4	3	3
Plastic (PVC Schedule No. 80)	½	5	3	2	1	1	1	1	1	1	—	—	—	—	—	—	—	—	—	—	—	—
	¾	9	6	4	2	2	1	1	1	1	1	1	1	—	—	—	—	—	—	—	—	—
	1	14	10	7	4	3	2	2	1	1	1	1	1	1	1	—	—	—	—	—	—	—
	1¼	26	19	12	7	5	3	3	2	2	1	1	1	1	1	1	1	1	—	—	—	—
	1½	35	26	16	9	7	4	4	3	2	2	2	1	1	1	1	1	1	1	1	1	1
	2	59	43	27	16	11	7	6	5	4	3	3	2	2	2	1	1	1	1	1	1	1
	2½	85	62	39	23	16	10	9	7	5	5	4	3	3	2	2	2	2	1	1	1	1
	3	133	97	61	35	26	16	13	11	8	7	6	5	4	3	3	3	2	2	2	1	1
	3½	179	131	82	48	34	21	18	15	11	9	8	7	5	4	4	3	3	3	2	2	2
	4	232	169	107	62	45	27	23	20	15	12	10	9	7	6	5	4	4	3	3	2	2
	5	368	269	169	98	71	43	37	31	23	19	16	13	11	9	8	7	6	5	4	4	4
	6	528	385	243	140	101	62	53	44	33	28	23	19	16	13	11	10	9	7	6	5	5

required (e.g., a pump motor or air conditioner condensing unit) and 6 ft (about 2 m) between a recessed light fixture and a box.

If it is continuous and properly bonded, IMC and EMT can be used as the grounding conductor. Rigid nonmetallic conduit and tubing (ENT) cannot be used as the grounding conductor. With a few exceptions, a grounding conductor is required for flexible conduit. The grounding conductor can be bare or insulated.

Rigid nonmetallic conduit and tubing is manufactured from thermoplastics that have very high rates of thermal expansion (over 6 in for a 100°F temperature change per 100 ft/over 300 mm in a 50°C temperature change per 30 m). Provisions must be made if it is exposed to significant temperature changes.

Nonmetallic-sheathed cable (NM and NMC) is permitted in single- and multifamily dwelling units and some other buildings. It cannot be used underground, in buildings that are more than three stories above grade or in commercial garages, motion picture studios, theaters, places of assembly, elevator hoist ways, and other corrosive or hazardous locations. NM and NMC must be supported every 4'-6" (1.35 m) and within 12 in (300 mm) of a box.

Armored cable (AC) is for use in dry, indoor applications. It is not allowed in commercial garages, motion picture studios, theaters, places of assembly, elevator hoistways, and other corrosive or hazardous locations. Bends in armored cable are limited to no less than five times the diameter. It must be supported every 4 to 6 in (1.35 m) and within 12 in (300 mm) of a box, except up to 24 in (600 mm) is allowed at points of usage where flexibility is required (i.e., at a pump motor). AC cannot be embedded in masonry, concrete, or plaster.

Conduit and cable should be installed away from locations where building occupants might use it as a handhold or footstep. Drain holes should be provided at the lowest point in a conduit run. Drilling and culling burrs should be carefully removed to prevent damage to conductor insulation.

Box/Enclosure Requirements

All electrical connections must be made in a protective enclosure such as a panelboard, junction, or device box, fixture, or appliance. Every switch, outlet, and connection must be contained in an electrical box and every lighting fixture must be mounted on a box. All wiring must begin or terminate at a panelboard or in a box that is housing a switch, luminaire connection, receptacle, conductor junction, or some similar terminal fitting. Junction boxes must be mounted so that the blank cover plate is visible and readily accessible. All electrical boxes must be adequately secured to the building structure.

There are no requirements that specify that a certain type of electrical box be installed for a specific purpose. The typical trade practice is to install octagonal and round boxes for lighting outlets and to install rectangular and square boxes for switches and receptacle outlets. Round boxes are normally installed overhead for lighting installations. However, almost any box can be used for any purpose.

Conductors in an electrical box must have sufficient open air space to prevent overheating. Thus, the more conductors and/or the conductor size, the bigger the box must be. The capacity of a box, in cubic inches, is determined by its length, width, and depth. An increase in one or more of these dimensions increases box capacity. Some boxes are designed so they can be ganged together to increase box capacity.

The size and number of conductors to be installed in a box influences selection of type of box and box size. By convention, a conductor that runs through the box is counted as one conductor and each conductor that terminates in the box counts as one. Fixture wires and conductors that do not leave the box, such as an internal grounding wire, are not counted. One conductor is deducted for one or more grounding conductors that enter the box.

Table 19.8 recommends the maximum number of conductors allowed to enter electrical boxes of selected sizes. Similar

TABLE 19.8 RECOMMENDED MAXIMUM NUMBER OF CONDUCTORS TO ENTER ELECTRICAL BOXES OF SELECTED SIZES. SIMILAR TABLES EXIST IN THE ELECTRICAL CODE. VALUES SUGGESTED IN THIS TABLE ARE LESS THAN THOSE TYPICALLY ESTABLISHED AS MAXIMUM VALUES IN THE ELECTRICAL CODE. THE RECOMMENDED MAXIMUM VALUES PROVIDED ALLOW FOR LESS CUMBERSOME CONNECTIONS WITHIN THE ENCLOSURE/BOX.

	Enclosure/Box Dimensions			Recommended Maximum Number of Conductors Entering an Enclosure/Box by Gauge (AWG)				
	Length	Height	Depth					
Enclosure/Box Type	in	in	in	No. 14	No. 12	No. 10	No. 8	No. 6
Round or Octagonal (Lighting Boxes)	4	4	1¼	5	4	4	3	2
	4	4	1½	6	6	5	4	2
	4	4	2⅛	9	8	7	6	3
Square (Device Boxes)	4	4	1¼	7	6	6	5	3
	4	4	1½	8	7	7	6	3
	4 11/16	4 11/16	1¼	10	9	8	7	4
	4 11/16	4 11/16	1½	12	10	9	8	5
	4 11/16	4 11/16	2⅛	18	16	14	12	7
Rectangular (Device Boxes)	2	3	2¾	1	14	6	5	2
	2	3	1½	3	3	2	2	—
	2⅛	4	1½	4	4	3	3	2
	2	3	2	4	4	3	3	2
	2	3	2¼	4	4	3	3	2
	2⅛	4	1⅞	5	5	4	3	2
	2	3	2½	5	4	4	3	2
	2⅛	4	2⅛	6	5	5	4	2
	2	3	3½	7	6	6	5	3
	3½	4½	1⅝	9	8	7	6	4
	3½	4½	2½	14	13	11	9	6
	7 1/16	4½	1⅝	18	16	14	12	7
	8⅞	4½	1⅝	23	21	19	15	9
	10 11/16	4½	1⅝	28	25	22	19	11
	7 1/16	4½	2½	28	25	23	19	11
	12½	4½	1⅝	34	30	27	23	14
	8⅞	4½	2½	36	32	29	24	14
	14 5/16	4½	1⅝	38	34	30	25	15
	16⅛	4½	1⅝	42	37	34	28	17
	10 11/16	4½	2½	44	39	35	29	18
	17 15/16	4½	1⅝	46	41	37	31	18
	19¾	4½	1⅝	50	44	40	33	20
	12½	4½	2½	53	47	42	35	21
	21 9/16	4½	1⅝	54	48	43	36	22
	14 5/16	4½	2½	60	53	48	40	24
	16⅛	4½	2½	67	60	54	45	27
	17 15/16	4½	2½	74	66	60	50	30
	19¾	4½	2½	82	73	65	54	33
	21 9/16	4½	2½	89	79	71	59	36

tables exist in the electrical code. These recommended values are less than those typically established as maximum values in the electrical code. They are recommended maximums because they allow for less cumbersome installation (e.g., making connections) and are based on industry experience. They allow for future expansion (adding connections).

Example 19.10

Recommend the maximum number of No. 12 AWG THWN conductors that can enter a 4 in × 4 in × 1½ in square electrical box, based on information in Table 19.8.

> From Table 19.8, 10 No. 12 AWG THWN conductors are the minimum number recommended.

Example 19.11

Recommend the minimum square box size for six No. 8 AWG THWN conductors entering the box, based on information in Table 19.8.

> From Table 19.8, a 4 in × 4 in × 1½ in square box is the minimum recommended.

19.9   BRANCH CIRCUIT REQUIREMENTS

Branch Circuiting

A building electrical system consists of several circuits that extend out from the switchboard or panelboard. A *branch circuit* is that portion of a building wiring system that extends beyond the final overcurrent protection device that is protecting a circuit. It provides power from a circuit breaker or fuse in the panelboard to single or multiple points of use called outlets. An *outlet* is a point in a wiring system where current is taken to supply an appliance, piece of equipment, or lighting installation. A branch circuit is composed of an overcurrent protection device (fuse or circuit breaker), wiring, and one or more outlets.

Types of Branch Circuits

Four primary types of branch circuits are recognized for general use.

Individual Branch Circuit

This type of branch circuit serves only one receptacle or piece of equipment such as for a range, clothes dryer, large copy machine, or other piece of machinery. These circuits usually lead directly from the distribution panel to the appliance and do not serve any other electrical devices. The individual branch circuit is sometimes known in the trade as a *dedicated* or *special purpose circuit*.

General Purpose Branch Circuit

A branch circuit supplies two or more outlets for lighting and appliances. This type of circuit may be referred to as a *lighting* circuit; this is a carryover from the days when electricity was first used in buildings and its predominant purpose was lighting. There are usually a number of general purpose branch circuits supplying lights and outlets in different rooms around a residence or commercial or industrial building.

Appliance Branch Circuit

This is the type of branch circuit that supplies energy to one or more outlets to which appliances are to be connected. They supply fixed electric equipment such as refrigerators, washers, and other large appliances and electrical devices. Appliance branch circuits do not supply lighting fixtures. Appliance branch circuits cannot exceed 20 A.

Multiwire Branch Circuit

A branch circuit consisting of two or more ungrounded (hot) conductors having a voltage between them and a common neutral (grounded) conductor that is shared between the ungrounded conductors such as in a 120/240 V three-wire circuit. In this circuit, all conductors must originate from the same panelboard.

Other branch circuits specific to a particular occupancy may also be required. For example, the following branches are required in health care facilities such as hospitals, nursing homes, and dental facilities.

Life Safety Branch Circuit

An emergency system of feeders and branch circuits that provides adequate power to patients and personnel. It must automatically connect to an alternate power source such as a generator when the normal power source is interrupted.

Critical Branch Circuit

An emergency system of feeders and branch circuits intended to provide power to task illumination, special power circuits, and selected receptacles serving areas and functions related to patient care. It must automatically connect to an alternate power source such as a generator when the normal power source is interrupted.

Split Wiring Receptacles

Split wired duplex receptacles are fed with a 120/240 V circuit having two ungrounded (hot) conductors, a grounded (neutral) conductor, and a grounding conductor. One ungrounded (hot) conductor feeds power to the upper outlet and the other ungrounded (hot) conductor feeds the lower outlet. The grounded (neutral) conductor is shared between both circuits. Split wiring allows power to be drawn from two separate circuits on one duplex receptacle.

Branch Circuit Rating and Loads

The *branch circuit rating* is determined by the rating of the overcurrent protection device (fuse or circuit breaker) used to protect the wiring in the circuit from excessive current flow. The rating of the overcurrent protection device is related to the connected load or loads being fed by the branch circuit. Simply put, the wiring in the circuit must safely deliver current to the connected load and the overcurrent protection device protects this wiring, so the circuit rating matches the rating of the overcurrent protection device.

The *connected load* on a branch circuit is the sum of all loads connected in a circuit. It is found by totaling the connected volt-amp (VA) load at each outlet connected to the circuit. The *volt-amp (VA)* is similar in nature to the watt, except that it describes the instantaneous current rather than power (wattage).

As discussed in Chapter 17, design of a building electrical installation involves computations with power (P), expressed in watts (W); voltage (E), in volts (V); amperage (I), in amps (A); and volt-amps (VA) in volts and amperes. A summary of conversion formulas for direct current and alternating current systems is provided in Table 19.9. These formulas are useful in determining loads in circuiting design.

Example 19.12

For a 120 V, 15 A, two-wire branch circuit, determine the maximum load that can be connected to this circuit.

From Table 19.9, $P = E \cdot I$

$$P = E \cdot I = 120 \text{ V} \cdot 20 \text{ A} = 2400 \text{ VA}$$

All electric-resistant appliance or pieces of equipment such as an oven, water heater, and space heater (no blower) and lights generally have power (wattage) ratings equal to their connected load (VA rating). For example, a 4500 W water heater has a 4500 VA rating; ten 150 W lamps have a connected load of 1500 VA; and an oven rated at 12 kW has a connected load of 12 000 VA.

In contrast, all electromechanical (having a motor) appliances or pieces of equipment generally will have a VA rating larger than the power (W) rating. At startup, an electric motor consumes substantially more electrical power than when it is operating at its rated speed and load. For example, a piece of equipment containing a 230 V, 1 hp motor that normally draws about 1000 W (4.3 A) under full load can draw 6000 VA (26.1 A) at startup. Because voltage remains constant, this motor's starting amperage is much greater than amperage required when it is at full operation. Thus, this device may be rated at 6000 VA even though it only consumes 1000 W when operating at full speed. This initial power surge is evident by the temporary dimming of lights at the instant a heavy-duty, motor-driven appliance such as a vacuum cleaner or table saw is switched on.

Design of a branch circuit involves sizing conductors and overcurrent protection (fuse or circuit breaker) that match the circuit rating. Volt-amperage (VA) loads and corresponding amperage (A) loads on single-phase and three-phase circuits for various system voltages are provided in Tables 19.10 and 19.11. Maximum connected loads, voltages, and requirements for common circuit ratings are provided in Tables 19.12 through 19.16. Approximate full load ratings (amperes) for single-phase and three-phase motors based on motor horsepower are provided in Table 19.17.

Check with manufacturer for actual specifications. Manufacturers' data should always be used in computations. Requirements for common branch circuits are introduced in the following.

General Purpose Circuits

General purpose circuits feed more than one outlet for lighting or other purpose. According to requirements, the rating of general

---

TABLE 19.9 A SUMMARY OF CONVERSION FORMULAS FOR DIRECT CURRENT AND ALTERNATING CURRENT SYSTEMS WHERE POWER (P) IS EXPRESSED IN WATTS (W); VOLTAGE (E) IN VOLTS (V); AMPERAGE (I) IN AMPS (A); AND VOLT-AMPS (VA) IS IN VOLT-AMPERES. WHEN USED IN THESE FORMULAS, POWER FACTOR (PF) SHOULD BE EXPRESSED AS A DECIMAL. FOR TWO-PHASE, THREE-WIRE ALTERNATING CURRENT SYSTEMS, SUBSTITUTE 1.41 FOR 1.73. FOR TWO-PHASE, FOUR-WIRE ALTERNATING CURRENT SYSTEMS, SUBSTITUTE 2.00 FOR 1.73.

To Find	Direct Current (DC) Systems	Alternating Current (AC) Systems	
		Single Phase	Three Phase
DC power (watts)	$P = E \cdot I$	—	—
AC real power (watts)	—	EI(PF)	$1.73 \cdot EI(PF)$
AC apparent power (VA)	—	(VA)	$1.73 \cdot (VA)$
Amperes (I)	$I = P/E$	$I = P/E \cdot PF$	$I = P/1.73 \cdot E \cdot PF$
	—	$I = (VA)/E$	$I = (VA)/1.73 \cdot E$
Power factor (PF)	—	$PF = P/E \cdot I$	$PF = P/1.73 \cdot E \cdot I$
		or	or
		$PF = P/(VA)$	$PF = P/1.73(VA)$

Note: VA is the rating of the appliance or piece of equipment in volt-amperes (VA).

TABLE 19.10 VOLT-AMPERAGE (VA) LOADS AND CORRESPONDING AMPERAGE (A) LOADS ON SINGLE-PHASE AND THREE-PHASE CIRCUITS FOR VARIOUS SYSTEM VOLTAGES BASED UPON A POWER FACTOR OF 1.0.

Load	Amperage (A)						Load
	Single-Phase Circuits				Three-Phase Circuits (Balanced Load)		
Volt-Amperage (VA)	120 V	240 V	277 V	480 V	240 V	480 V	Volt-Amperage (VA)
100	0.83	0.42	0.36	0.21	0.24	0.12	100
200	1.67	0.83	0.72	0.42	0.48	0.24	200
250	2.08	1.04	0.90	0.52	0.60	0.30	250
300	2.50	1.25	1.08	0.63	0.72	0.36	300
400	3.33	1.67	1.44	0.83	0.96	0.48	400
500	4.17	2.08	1.81	1.04	1.20	0.60	500
600	5.00	2.50	2.17	1.25	1.44	0.72	600
700	5.83	2.92	2.53	1.46	1.68	0.84	700
800	6.67	3.33	2.89	1.67	1.92	0.96	800
900	7.50	3.75	3.25	1.88	2.17	1.08	900
1000	8.33	4.17	3.61	2.08	2.41	1.20	1000
1100	9.17	4.58	3.97	2.29	2.65	1.32	1100
1200	10.00	5.00	4.33	2.50	2.89	1.44	1200
1300	10.83	5.42	4.69	2.71	3.13	1.56	1300
1400	11.67	5.83	5.05	2.92	3.37	1.68	1400
1500	12.50	6.25	5.42	3.13	3.61	1.80	1500
1600	13.33	6.67	5.78	3.33	3.85	1.92	1600
1700	14.17	7.08	6.14	3.54	4.09	2.04	1700
1800	15.00	7.50	6.50	3.75	4.33	2.17	1800
1900	15.83	7.92	6.86	3.96	4.57	2.29	1900
2000	16.67	8.33	7.22	4.17	4.81	2.41	2000
2100	17.50	8.75	7.58	4.38	5.05	2.53	2100
2200	18.33	9.17	7.94	4.58	5.29	2.65	2200
2300	19.17	9.58	8.30	4.79	5.53	2.77	2300
2400	20.00	10.00	8.66	5.00	5.77	2.89	2400
2500	20.83	10.42	9.03	5.21	6.01	3.01	2500
2600	21.67	10.83	9.39	5.42	6.25	3.13	2600
2700	22.50	11.25	9.75	5.63	6.50	3.25	2700
2800	23.33	11.67	10.11	5.83	6.74	3.37	2800
2900	24.17	12.08	10.47	6.04	6.98	3.49	2900
3000	25.00	12.50	10.83	6.25	7.22	3.61	3000
3250	27.08	13.54	11.73	6.77	7.82	3.91	3250
3500	29.17	14.58	12.64	7.29	8.42	4.21	3500
3750	31.25	15.63	13.54	7.81	9.02	4.51	3750
4000	33.33	16.67	14.44	8.33	9.62	4.81	4000
4250	35.42	17.71	15.34	8.85	10.22	5.11	4250
4500	37.50	18.75	16.25	9.38	10.83	5.41	4500
4750	39.58	19.79	17.15	9.90	11.43	5.71	4750
5000	41.67	20.83	18.05	10.42	12.03	6.01	5000
6000	50.00	25.00	21.66	12.50	14.43	7.22	6000
7000	58.33	29.17	25.27	14.58	16.84	8.42	7000
7500	62.50	31.25	27.08	15.63	18.04	9.02	7500
8000	66.67	33.33	28.88	16.67	19.25	9.62	8000
9000	75.00	37.50	32.49	18.75	21.65	10.83	9000
10 000	83.33	41.67	36.10	20.83	24.06	12.03	10 000

TABLE 19.11   MAXIMUM CONNECTED LOADS, VOLTAGES, AND REQUIREMENTS FOR COMMON 120 V RESIDENTIAL CIRCUIT RATINGS.

120 V Circuit Rating	Maximum (100%) Connected Load VA	Maximum Continuous Load (80% of Connected Load) VA	50% of Maximum Load VA	Typical Conductors (Excluding Grounding) AWG	Overcurrent Protection A	Recommended Uses
15 A	1800	1440	900	2-No. 14 CU or 2-No. 12 CU	15	Lighting and individual circuits.
20 A	2400	1920	1200	2-No. 12 CU	20	Convenience receptacles, small appliance, and individual circuits.
30 A	3600	2880	1800	2-No. 10 CU	30	Clothes dryer.
40 A	4800	3840	2400	2-No. 8 CU	40	Small electric range or electric range top.
50 A	6000	4800	3000	2-No. 6 AL	50	Typical electric range or electric range top.

TABLE 19.12   STANDARD CONNECTED LOADS, VOLTAGES AND REQUIREMENTS FOR KITCHEN APPLIANCES AND EQUIPMENT.

Appliance/Equipment	Common Connected Load, VA	Connected Voltage, V	Conductors (Excluding Grounding), AWG	Overcurrent Protection, A	Standard NEMA Device	Recommendations
Range/oven unit (household size)	12 000	120/240	3-No. 6 AL	50	14-50R	Individual circuit.
Range/oven unit (apartment size)	8000	120/240	3-No. 6 AL	40	14-50R	Individual circuit.
Cooktop (large, separate)	8000	120/240	3-No. 6 AL	40	14-50R	Individual circuit.
Cooktop (medium size)	6000	120/240	3-No. 6 AL	50	14-50R	Individual circuit.
Oven (built-in, separate)	4500	120/240	3-No. 10	30	14-30R	Individual circuit.
Microwave oven (fixed)	1200	120	2-No. 12	20	5-15R	Individual circuit.
Microwave oven (counter top)	800	120	2-No. 12	20	5-15R	Part of kitchen appliance circuit.
Dishwasher	1000	120	2-No. 12	20	5-15R	Individual circuit.
Waste disposal (sink)	1000	120	2-No. 12	20	5-15R	Individual circuit.
Freezer	400	120	2-No. 12	20	5-15R	Individual circuit.
Exhaust hood (kitchen)	500	120	2-No. 12	20	5-15R	Individual circuit.
Trash compactor	1200	120	2-No. 12	20	5-15R	Individual circuit.
Refrigerator(household size)	600	120	2-No. 12	20	5-15R	Part of kitchen appliance purpose circuit.
Refrigerator(apartment size)	400	120	2-No. 12	20	5-15R	Part of kitchen appliance circuit.
Coffee maker (10 cup)	1500	120	2-No. 12	20	5-15R	Part of kitchen appliance circuit.

TABLE 19.13 STANDARD CONNECTED LOADS, VOLTAGES, AND REQUIREMENTS FOR ENTERTAINMENT EQUIPMENT AND LIGHTING IN HABITABLE SPACES.

Equipment	Common Connected Load, VA	Connected Voltage, V	Conductors (Excluding Grounding), AWG	Overcurrent Protection, A	Standard NEMA Device	Recommendations
Television (19 inch)	150	120	2-No. 12	20	5-15R	Part of general purpose circuit. Surge protection is recommended.
Television (46 in)	300	120	2-No. 12	20	5-15R	Part of general purpose circuit. Surge protection is recommended.
VCR/DVD player	10-75	120	2-No. 12	20	5-15R	Part of general purpose circuit. Surge protection is recommended.
Stereo (large rack type)	100	120	2-No. 12	20	5-15R	Part of general purpose circuit. Surge protection is recommended.
Incandescent light luminaire 1-100 W lamp (fixed)	100	120	2-No. 12	20	Lighting outlet	Part of general purpose circuit.
Incandescent light luminaire 1-75 W lamp (fixed)	75	120	2-No. 12	20	Lighting outlet	Part of general purpose circuit.
Chandelier light luminaire 6-60 W lamps (fixed)	360	120	2-No. 12	20	Lighting outlet	Part of general purpose circuit.
Fluorescent light luminaire 2-40 W lamps with ballast	76	120 or 277	2-No. 12	20	Lighting outlet	Part of general purpose circuit.
Fluorescent light luminaire 4-40 W lamps with ballast	152	120 or 277	2-No. 12	20	Lighting outlet	Part of general purpose circuit.
Metal halide light luminaire 175 W lamps with ballast	205	120	2-No. 12	20	Lighting outlet	Part of general purpose circuit.
Metal halide light luminaire 400 W lamps with ballast	435	120	2-No. 12	20	Lighting outlet	Part of general purpose circuit.

TABLE 19.14 STANDARD CONNECTED LOADS, VOLTAGES, AND REQUIREMENTS FOR APPLIANCES AND EQUIPMENT IN LAUNDRY SPACES.

Appliance/Equipment	Common Connected Load, VA	Connected Voltage, V	Conductors (Excluding Grounding), AWG	Overcurrent Protection, A	Standard NEMA Device	Recommendations
Clothes washer (household)	1200	120	2-No. 12	20	5-15R	Individual circuit.
Clothes washer (apartment)	800	120	2-No. 12	20	5-15R	Individual circuit.
Clothes dryer (electric) (household)	6000	120/240	3-No. 10	30	14-30R	Individual circuit.
Clothes dryer (electric) (apartment)	4000	120/240	3-No. 10	30	14-30R	Individual circuit.
Ironing center	1200	120	2-No. 12	20	5-15R	Individual circuit.

TABLE 19.15 STANDARD CONNECTED LOADS, VOLTAGES, AND REQUIREMENTS FOR APPLIANCES AND EQUIPMENT IN OFFICE SPACES.

Appliance/Equipment	Common Connected Load, VA	Connected Voltage, V	Conductors (Excluding Grounding), AWG	Overcurrent Protection, A	Standard NEMA Device	Recommendations
Personal computer	600	120	2-No. 12	20	5-15R	Part of general purpose circuit. Surge protection is recommended.
Computer printer (desk size)	800	120	2-No. 12	20	5-15R	Part of general purpose circuit. Surge protection is recommended.
Facsimile (fax) machine	250	120	2-No. 12	20	5-15R	Part of general purpose circuit. Surge protection is recommended.
Copy machine (45 copies/min)	5800	120/240	3-No. 10	30	14-30R	Individual circuit. Surge protection is recommended.
Cash register	100	120	2-No. 12	20	5-15R	Part of general purpose circuit. Surge protection is recommended.
Coffee maker (10 cup)	1500	120	2-No. 12	20	5-15R	Part of general purpose circuit. Surge protection is recommended.
Water cooler (8 gal/hr)	700	120	2-No. 12	20	5-15R	Part of general purpose circuit.
Microwave oven (countertop)	800	120	2-No. 12	20	5-15R	Part of kitchen appliance purpose circuit.
Cold food/beverage dispenser	1800	120	2-No. 12	20	5-15R	Individual circuit.

TABLE 19.16 STANDARD CONNECTED LOADS, VOLTAGES, AND REQUIREMENTS FOR MISCELLANEOUS EQUIPMENT.

Appliance/Equipment	Common Connected Load, VA	Connected Voltage, V	Conductors (Excluding Grounding), AWG	Overcurrent Protection, A	Standard NEMA Device	Recommendations
Air conditioner, central 2 ton (apartment size)	3000	120/240 or 120/208	3-No. 10	30	Outlet	Individual circuit.
Air conditioner, central 5 ton (household size)	6000 to 12 000	120/240 or 120/208	3-No. 6 AL	50	Outlet	Individual circuit.
Air conditioner, window 1½ ton (large room size)	1800	120	2-No. 12	20	5-15R	Individual circuit.
Air conditioner, window 1 ton (room size)	1200	120	2-No. 12	20	5-15R	Part of general purpose circuit.
Bathroom ventilator	500	120	2-No. 12	20	Outlet	Individual circuit.
Bathroom ventilator (with heat lamp)	1000	120	2-No. 12	20	Outlet	Individual circuit.
Fan (attic ventilation)	800	120	2-No. 12	20	Outlet	Individual circuit.
Fan (whole-house ventilation)	800	120	2-No. 12	20	Outlet	Individual circuit.
Furnace (gas, ½ hp blower)	1200	120	2-No. 12	20	Outlet	Individual circuit.
Humidifier (in duct)	450	120	2-No. 12	20	5-15R	Part of general purpose circuit.
Spa (electrically heated)	12 000	120/240	3-No. 6 AL	50	14-50R	Individual circuit. GFCI protection generally required.
Swimming pool heater (electrically heated)	12 000	120/240	3-No. 6 AL	50	14-50R	Individual circuit. GFCI protection generally required.
Central vacuum system	1200	120	2-No. 12	20	5-15R	Part of general purpose circuit.
Water heater (40 to 50 gal)	3000	240 or 208	2-No. 10	30	Outlet	Individual circuit.

TABLE 19.17 APPROXIMATE FULL LOAD RATINGS (AMPERES) FOR SINGLE- PHASE AND THREE-PHASE MOTORS BASED ON MOTOR HORSEPOWER. VALUES WILL VARY WITH UNIT. CHECK WITH MANUFACTURER FOR ACTUAL SPECIFICATIONS.

| Motor Horsepower | Line Voltage | | | | | Minimum Transformer Rating KVA |
	115 V	208 V	230 V	460 V	575 V	
Single-Phase Motors						
⅙	4.4	2.4	2.2	—	—	0.53
¼	5.8	3.2	2.9	—	—	0.70
⅓	7.2	4.0	3.6	—	—	0.87
½	9.8	5.4	4.9	—	—	1.18
¾	13.8	7.6	6.9	—	—	1.66
1	16	8.8	8	—	—	1.92
1½	20	11	10	—	—	2.40
2	24	13.2	12	—	—	2.88
3	34	18.7	17	—	—	4.10
5	56	30.8	28	—	—	6.72
7½	80	44	40	—	—	9.60
10	100	55	50	—	—	12.0
Three-Phase Motors						
½	—	2.2	2.0	1.0	0.8	0.9
¾	—	3.1	2.8	1.4	1.1	1.2
1	—	4.0	3.6	1.8	1.4	1.5
1½	—	5.7	5.2	2.6	2.1	2.1
2	—	7.5	6.8	3.4	2.7	2.7
3	—	10.7	9.6	4.8	3.9	3.8
5	—	16.7	15.2	7.6	6.1	6.3
7½	—	24	22	11	9	9.2
10	—	31	28	14	11	11.2
15	—	46	42	21	17	16.6
20	—	59	54	27	22	21.6
25	—	75	68	34	27	26.6
30	—	88	80	40	32	32.4
40	—	114	104	52	41	43.2
50	—	143	130	65	52	52
60	—	170	154	77	62	64
75	—	211	192	96	77	80
100	—	273	248	124	99	103
125	—	342	312	156	125	130
150	—	396	360	180	144	150
200	—	528	480	240	192	200

Note: If motors are started more than once per hour, add 20% additional KVA.

purpose branch circuits must be 15 A, 20 A, 30 A, 40 A, or 50 A. Maximum connected loads, voltages, and other requirements for general purpose circuit ratings are shown in Table 19.11.

General purpose circuits typically provide power to convenience receptacles and lighting outlets with ratings of 20 A and 15 A because of their ease of running and pulling the slender conductors required (No. 12 AWG and No. 14 AWG copper, respectively). Some electricians use 15 A circuits for lighting installations (No. 14 AWG copper conductors) and 20 A circuits for receptacles (No. 12 AWG copper conductors).

General purpose circuits are typically limited according to what will be connected to them:

1. When a general purpose circuit feeds fixed appliances and luminaires or portable appliances, the total of the fixed appliances should be no more than 50% of the branch circuit rating. Assuming a 15 A, 120 V branch circuit, it would have a maximum rating of 15 A · 120 V = 1800 VA (refer to Chapter 18 for formula explanation). In this case, the fixed appliances would be limited to 900 VA, leaving the other 900 VA available to supply luminaires or portable appliances also served by the branch circuit. Refer to Table 19.11.

2. A 20 A, 120 V branch circuit would have a theoretical maximum of 2400 VA (20 A · 120 V = 2400 VA), but it is common practice to limit the connected load to 80% of the circuit rating (e.g., 20 A · 120 V · 80% = 1920 VA). Refer to Table 19.11.

3. When the load on the circuit will be a continuous operating load (e.g., for store lights), the total load should not exceed 80% of the circuit rating. The lighting load must include

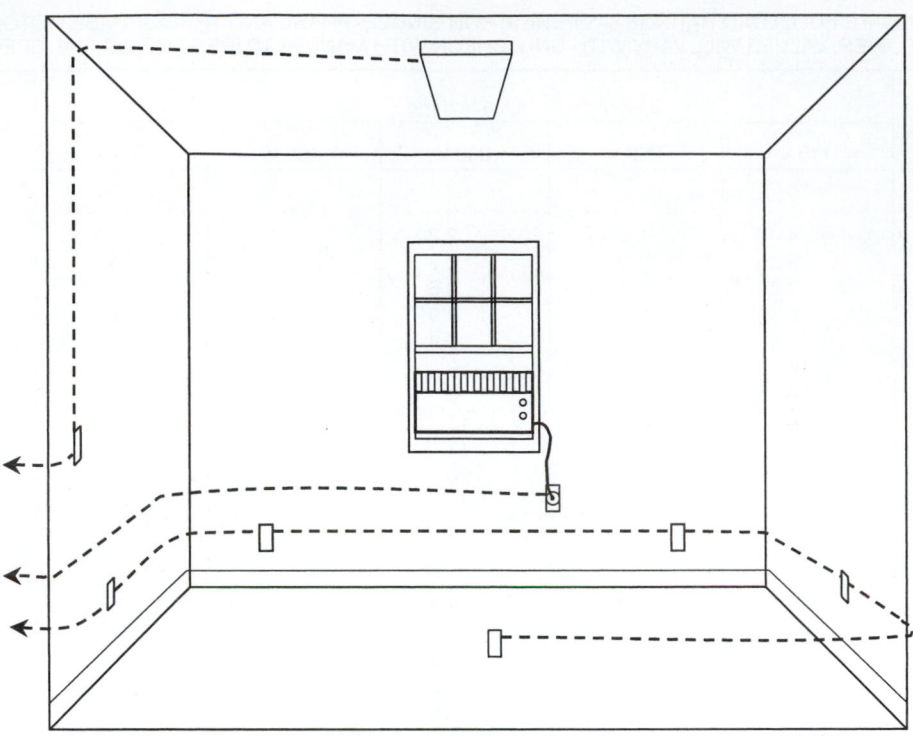

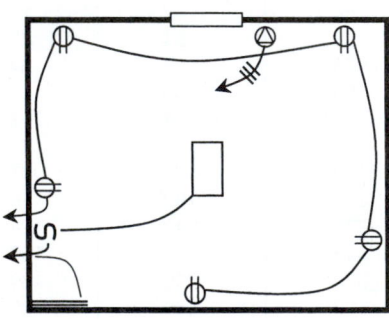

FIGURE 19.10   Shown is a cut-away pictorial view of a room with separate circuits for lighting, convenience receptacles, and window air conditioner. Wiring is shown as a dashed line because it is hidden from view (within the walls). The related one-line schematic drawing of electrical symbols is also shown. The arrows indicate that each circuit originates at a panelboard.

any ballasts, transformers, or autotransformers, which are part of the lighting system. Because a 15 A branch has a full rating of 1800 VA, the limit would be 80%, or 12 A and 1440 VA (e.g., 15 A · 120 V · 80% = 1440 VA). A 20 A, 2400 VA branch would be limited to 16 A and 1920 VA of connected load. Again, refer to Table 19.11.

4. When portable appliances will be used on a general purpose circuit, the limit for any one portable appliance is 80% of the branch circuit rating.

5. In commercial applications, convenience receptacles are computed at a load of 1.5 A (180 VA) per receptacle and are limited to 80% of the rating. This limits a branch circuit serving only receptacles to its rating divided by 1½ A. For example, a 15 A circuit is limited to a maximum of 8 outlets and a 20 A circuit to 10 outlets.

Example 19.13

A 120 V, 20 A, two-wire branch circuit feeds incandescent lighting that stays on for 3 hr or more. Determine the maximum *continuous* load permitted on this branch circuit.

From Table 19.9, P = E · I

$$P = E \cdot I = 120 \text{ V} \cdot 20 \text{ A} \cdot 1.0 = 2400 \text{ VA}$$

SCHEMATIC LAYOUT

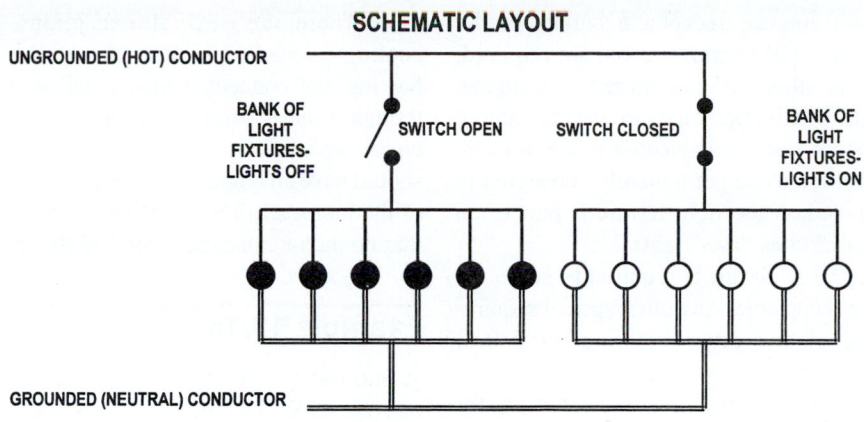

FLOOR PLAN LAYOUT

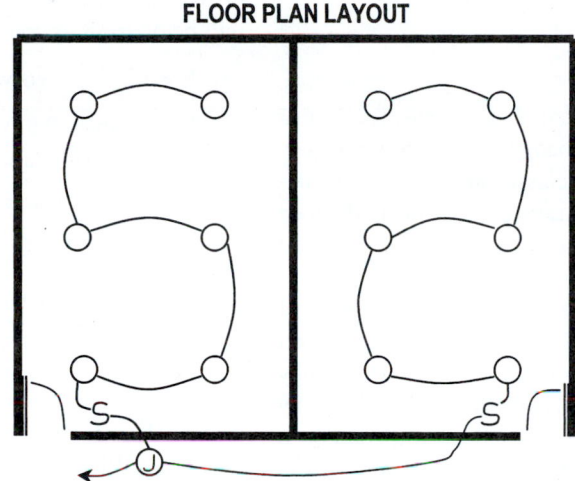

FIGURE 19.11 Shown are the switching configuration and related one-line schematic drawing for a lighting circuit serving two rooms. Wiring from each room is connected at a junction (J) box were the circuit returns to a panelboard.

Individual Circuits

These circuits provide power to a single outlet such as a receptacle serving a range, clothes dryer, or copy machine. Although there are no size limitations for an individual circuit rating, appliances and equipment rated at above 25 A must be placed on a separate individual circuit because of the 50% maximum single load limitation in general purpose circuiting. Usually connected loads above 20 A are placed on an individual circuit. It is good practice to provide individual circuits for loads above 1500 W. Motors above ⅛ hp should also be placed on an individual circuit. Generally, individual circuits are required for the following appliances and equipment:

- Kitchen range (both stand-alone and counter-mounted units)
- Oven
- Microwave (built-in)
- Waste disposal
- Dishwasher
- Clothes washer
- Clothes dryer
- Electric water heater
- Furnace
- Boiler circulating pump motor (large commercial and industrial)
- HVAC air-handling unit
- Large machinery (e.g., table saw, lathe, milling machine, machining center, elevator)
- Large equipment (e.g., large copy machines, compressors, HVAC blowers)

Appliance Circuits

These circuits serve two or more outlets to which only appliances are connected. In dwelling units, two or more 20 A small appliance circuits for convenience receptacle outlets in the kitchen, dining room, pantry, and breakfast room are required. These are in addition to the other outlets required (e.g., range, waste disposal, and so on). Small appliance circuits can supply any refrigerators and freezers, but not an electric range or oven. A light fixture is not permitted to be permanently connected to a small appliance circuit, unless the light fixture is part of an appliance (e.g., an oven or clothes dryer light).

A minimum of one 20 A circuit is required to serve the laundry room convenience receptacle. Any other special requirements, such as an electric clothes dryer requiring 120/240 V, must also be provided.

Tables 19.12 through 19.17 provide recommendations for circuiting common appliances and pieces of equipment. Manufacturers' data should always be used in computations.

Continuous Loads

A *continuous load* is a connected load that operates for 3 hr or more at any time. Many electrical loads fit within this category such as circuits serving office and classroom lighting installations. When determining a circuit rating, most loads deemed continuous must have a circuit rating calculated at 125% of the circuit's connected load.

The intent behind the 125% factor is from the inability of the overcurrent protection device to handle a continuous load without overheating; that is, most circuit breakers trip if they carry their rated load for any significant time period. In many installations, it is good practice for *all* connected loads not to exceed 80% of the individual circuit rating.

Example 19.14

A lighting installation containing 16 to 100 W incandescent luminaires on one circuit will be operated approximately 10 hr per workday. Determine the circuit rating.

The 1600 W connected load is a continuous load because it operates for 3 hr or more. The continuous load multiplier (125%) must be applied:

$$1600 \text{ VA} \cdot 1.25 = 2000 \text{ VA}$$

The circuit rating is found by algebraically manipulating the power equation, P = IE:

$$I = P/E = 2000 \text{ VA}/120 = 16.7 \text{ A}$$

A 20 A circuit breaker is the closest circuit breaker available for overcurrent protection (see Chapter 18 for sizes available). The rating for this circuit is 20 A.

Another way of looking at this is that the connected load on a continuous-load circuit should not exceed 80% of the individual circuit rating. Thus, the maximum continuous load on a 20 A circuit is 16 A (80% of 20 is 16).

There are several exceptions to application of the continuous-load multiplier. On circuits involving electric space heating, the connected load is taken at 100% of the load even though it may operate for more than 3 hr at a time. Any branch circuit serving a single motor or a device containing a motor should have an ampacity (amperage rating) of not less than 125% of the motor's full load current rating (this is the same as saying that no motor can exceed 80% of the branch circuit rating).

Example 19.15

A fluorescent lighting fixture has a ballast with a nameplate rating of 185 W. Determine the number of such fixtures that can be connected to a two-wire circuit protected by a 20 A circuit breaker if the fixtures are used in an office building where they will be on for 8 hr continuously each day.

a. For a 277 V, two-wire lighting system:

From Table 19.9, P = E · I. The continuous-load multiplier (80%) must be applied.

$$P = E \cdot I = 277 \text{ V} \cdot 20 \text{ A} \cdot 80\% = 4432 \text{ VA}$$
$$4432 \text{ VA}/185 \text{ VA} = 23.96 = 24 \text{ fixtures}$$

b. For a 120 V, two-wire lighting system:

From Table 19.9, P = E · I · PF. The continuous-load multiplier (80%) must be applied.

$$P = E \cdot I = 120 \text{ V} \cdot 20 \text{ A} \cdot 80\% = 1920 \text{ VA}$$
$$1920 \text{ VA}/185 \text{ VA} = 10.4 = 10 \text{ fixtures}$$

Branch Circuit Conductor Size

On a branch circuit, conductor size is tied to circuit rating. Generally, ungrounded (hot) and grounded (neutral) conductors in the circuit must be sized so that conductor ampacity is at least the branch circuit rating. The ampacity of a conductor can be larger than the circuit rating but not smaller. For example, the No. 8 AWG copper conductor with THW insulation that will be used in a 120 V, two-wire circuit in an environment with an average ambient air temperature of no greater than 86°F (30°C) has an ampacity of 50 A. It can safely carry up to 50 A without overheating.

A 50 A overcurrent protection device (fuse or circuit breaker) would adequately protect a circuit wired with this conductor size, because the overcurrent protection device would disconnect if current drawn exceeded 50 A. A 40 A overcurrent protection device would also protect a circuit wired with the No. 8 AWG copper conductors because the 50 A ampacity of the conductor would not be exceeded. However, a 60 A overcurrent protection device would be too large and unsafe. It would not adequately protect the circuit, because it will allow current to flow that exceeds the 50 A ampacity of the conductor.

Example 19.16

A clothes dryer rated at 5 kW is connected to a 120/240 V, single-phase three-wire, individual branch circuit. Determine the following:

a. The connected full load, in amps.

From Table 19.9, $P = E \cdot I$. Therefore, $I = P/E$.

$$I = P/E = 5000 \text{ VA}/240 \text{ V} = 20.8 \text{ A} = 21 \text{ A}$$

b. The minimum size circuit rating (of the overcurrent device) and the minimum copper wire size based on THWN-2 insulation at a temperature rating 167°F/75°C (see Chapter 18). Neglect any correction factors for temperature and conductor bundling.

From Table 18.3, the minimum size overcurrent device that can be used on this circuit is 25 A.

From Table 18.11, an AWG No. 10 copper conductor rated at 30 A is the minimum required.

The designer would likely select a 30 A circuit rating, and thus a fuse or circuit breaker rated at 30 A.

c. The rating of the receptacle and plug used to connect the clothes dryer.

From Table 18.6 based on a 30 A circuit rating, a three-pole, four-wire 14-30R receptacle and 14-30P plug rated at 30 A, 125/250 V.

Residential Branch Circuit Wiring

Type NMB cable (e.g., Romex®) is the most widely used wiring method in residential dwellings. NM cable must have 194°F (90°C) conductor insulation rating, which is designated by a "B" on the cable sheath. Typically, AWG No. 12, and AWG No. 14 are used for receptacle and lighting circuits; AWG No. 10/2 is commonly used for electric water heaters; AWG No. 10/3 with ground for electrical dryers and cooktops; and AWG No. 8/3 with ground or AWG No. 6/3 with ground for ranges and wall-mounted ovens. Type SER or other four-wire cable is used for electrical ranges, cooktops, wall ovens, and clothes dryers.

19.10 DEVICE AND EQUIPMENT REQUIREMENTS

Requirements for Switches and Receptacles

Switches must be selected to match the load they control. Large lighting installations that require many switches may have the switches contained within a panelboard-like enclosure called a lighting control panel. Receptacles must be selected to match the appliance or equipment they serve.

Ordinary convenience receptacles and switches are generally wall mounted. There are no specific mounting height requirements for wall switches and receptacles. Switches are normally mounted approximately 48 in (1.2 m) above finished floor (AFF), unless otherwise specified. Convenience receptacles are normally mounted approximately 16 in AFF (400 mm), unless otherwise specified. Convenience receptacles in bathrooms and restrooms are normally mounted approximately 44 in AFF (1.1 m).

Switches are typically oriented so they trip off in a downward orientation (e.g., switch down). Receptacles are customarily installed with the grounding slot oriented downward, unless the receptacle is controlled by a switch, in which case the receptacle is frequently aligned with the grounding slot upward. Preference of grounding slot orientation (e.g., grounding slot up or down) appears to be a cosmetic decision because there are no regulations that mandate orientation of the grounding slot.

It is recommended that receptacles be installed with the grounding slot oriented upward in contrast to historical practice. This recommendation is made to improve safety. With the grounding pin of a plug installed so it is oriented downward as is common practice, if a metal object (e.g., a tool or kitchen utensil) falls and strikes a grounded plug, the first blades the object could contact are the ungrounded (hot) and grounded conductors. A contacting condition would result in a short-circuit and arcing. By orienting the grounding slot upward, the grounding pin would be the first blade to contact the object, which would not create a short-circuit.

Overcurrent Protection (Circuit Breakers and Fuses) Requirements

An overcurrent protection (OCP) device, a fuse, or circuit breaker serves to limit current levels in a conductor by interrupting power when current limitations are exceeded. It prevents excessive heat from damaging conductors and related equipment. Therefore, the overcurrent device must be matched to the conductor and equipment so that the current-carrying capacity of the conductor and equipment are not exceeded.

The current carried (amperage) by the electrical circuit or system protected by an OCP device must not exceed the maximum current rating of the circuit breaker. Additionally, conductors must be protected in accordance with their ampacity. For example, a circuit with a 20 A rating should have a 20 A fuse or circuit breaker protecting it and the ampacity of the conductors in the circuit must have an ampacity (after corrections) of at least 20 A.

The voltage rating of a fuse or circuit breaker must be equal to or greater than the voltage of the circuit in which the fuse is applied. For power systems of 600 V or less, fuses of a higher voltage rating can be applied on circuits of a lower system voltage. For example, a fuse rated at 600 V can be used on a 480 V system.

Additionally, the amperes interruption current (AIC) rating for circuit breakers should be at least 5000 A and 10 000 A for fuses.

The fuse or circuit breaker must be installed at a location in the circuit where the conductors receive power—that is, generally at the panelboard where the circuit originates. The OCP device must protect the ungrounded conductors in a circuit to ensure that power to the circuit is interrupted by the OCP device where the circuit originates (generally the panelboard). The neutral (grounded) and grounding conductors are not protected by overcurrent protection.

Feeder Requirements

A *feeder* is a set of conductors that carry a comparatively large amount of power from the service equipment to a second panelboard, called a *subpanelboard,* where branch circuits further distribute the power. For example, a feeder may originate at the main panelboard and feed a lighting subpanelboard that further divides power to branch circuiting for lighting. In an apartment building electrical system, several individual feeders will run from an individual meters in at a central meter bank to individual apartment panelboards where current is ultimately distributed to the apartment unit by branch circuits.

Feeders must be designed to provide sufficient power to the branch circuits they supply so feeder conductor size is based on the maximum load to be supplied by the feeder. Feeders should be capable of carrying the amount of current required by the load, plus any current that may be required in the future.

It is not likely that all connected loads on a feeder will be in operation at a specific time. Thus, feeder conductors do not need to be sized to carry the total connected load served by the feeder. A variety of demand factors reduce the connected load to the computed load. These demand factors are too numerous to mention here. The reader is referred to local Code requirements. Additional capacity may be warranted for future expansion.

Switchboard and Panelboard Requirements

Switchboards and panelboards can be used as distribution equipment, at a point downstream from the service entrance equipment. By definition, panelboards feeding lighting and convenience receptacles and having at least 10% of the circuits rated at 30 A or less are identified as *lighting and appliance panelboards. Power distribution panelboards* feed other panelboards (called subpanelboards), motors, and transformers, but not circuits powering lights and convenience receptacles. In a single panelboard, not more than 42 overcurrent protection devices may be used for protecting lighting and appliance branch circuits.

Switchboards and panelboards used as service equipment should have a rating not less than the minimum allowable service capacity of the computed load. Panelboards used as subpanelboards should have a rating not less than the minimum feeder capacity of the computed load. A variety of demand factors reduce the connected load to the computed load. These demand factors are too numerous to mention here. The reader is referred to local Code requirements. Additional capacity may be warranted for future expansion.

When locating overcurrent protection in a panelboard, it is important to balance the anticipated load so that both bus bars are carrying a similar load. The loads should be balanced on phase bus, as discussed earlier. Similarly loaded circuits should be shared between bars so one bus bar is not overloaded. Two- and three-pole beakers (e.g., for a kitchen range or three-phase motor) automatically share the load on each phase. When possible, single-phase circuits should be distributed such that there anticipated loads served by that circuit are equal on each phase bus.

In service equipment panelboards, the neutral and equipment grounding conductors are bonded (connected) together. In subpanelboards, the neutral is isolated from ground; that is, there are separate neutral/grounded and grounding buses. Neutral and grounded conductors serving branch circuits originating at a subpanelboard originate at the neutral bus and grounding conductors originate at the grounding bus.

The minimum headroom requirement of working spaces about service equipment is 6.5 ft (2 m). In general, a fuse or circuit breaker must be installed at a location in the circuit where the conductors receive power. Generally, this location is in the panelboard or load center before the conductors leave to convey current to the outlets in the circuit.

Example 19.17

A feeder on a 120/240 V, three-wire circuit consisting of aluminum conductors with THHN insulation will serve a continuous load on a panelboard with 75°C/167°F terminals. The continuous load is computed to be 115 A.

a. Size the panelboard served by this feeder, based on single-phase panels in Chapter 18, Table 18.4.

The feeder must be sized not less than 125% of 115 A:

$$115 \text{ A} \times 125\% = 144 \text{ A}$$

The panelboard served by this feeder must be sized no less than 144 A. From Table 18.4, a 150 A panelboard and disconnect are selected.

b. Select a feeder conductor (THHN, 75°C/167°F).

A feeder conductor must be sized no less than 125% of the continuous load. Initially, it appears that a feeder conductor should be selected such that it is sized no less than 125% of the continuous load (115 A × 125% = 144 A). However, because a 150 A overcurrent protection device (disconnect) must protect the conductor, the ampacity of the conductor must be sized no less than the rating of this device. Therefore, conductor size must be increased to 350 kcmil, which has an ampacity of 250 A.

Service Entrance Equipment Requirements

Service equipment must be large enough to supply the computed load of the building or area of the building being served. It is calculated using code requirements and utility regulations. In most commercial and industrial installations, several disconnects may be used. Commonly, a maximum of six service disconnects per service are allowed. So, in large installations where more than six switches or circuit breakers could be used as disconnects, a single main service disconnect must be provided to disconnect power to the building. In very large buildings, there can be several service entrances.

The most common sizes of residential service equipment are 100, 125, 150, 175, and 200 A. RHW, THWN, THHN, XHHW, and USE aluminum conductors are commonly used. Minimum conductor sizes for service entrance conductors are provided in Table 19.18. Larger service equipment is used in commercial and industrial applications.

In single-family residences and multifamily dwellings, the main service panelboard can be mounted either outside or inside the dwelling as near as possible to the point of entrance of the service conductors to the building. All service equipment and electrical panels shall have a clear area 30 in (0.75 m) wide and 36 in (0.9 m) deep in front. This clear area must extend from floor to ceiling with no intrusions from other obstructions (e.g., equipment, cabinets, counters, and appliances). Panels are not allowed in clothes closets or bathrooms.

Generally, all electrical service entrances rated at 250 V and over or exceeding 250 A must be located in a separate room used for no other purpose, except that telecommunications equipment may terminate in that room. Where a separate electrical room is required, it should have a 1 HR fire rating unless the building and room is sprinklered. Boiler and furnace rooms are not acceptable for service entrance or main distribution equipment. No pipes containing a liquid should enter that room unless the pipe terminates in that room and is intended for use in that room (i.e., sprinkler system).

TABLE 19.18 MINIMUM SIZE CONDUCTOR FOR SERVICE ENTRANCE CONDUCTORS FOR RHW, THWN, THHN, XHHW, AND USE CONDUCTORS.

| Service Rating | Minimum for Service Entrance Conductor Size | |
	Aluminum and Copper-Clad Aluminum	Copper
Amps	AWG	AWG
100	2	4
125	1/0	2
150	2/0	1
175	3/0	1/0
200	4/0	2/0

It is necessary to maintain safe working clearances from overhead electrical conductors to reduce risk of accidental contact. The required clearance can be found in the Code. Typically, minimum vertical clearances of 18 ft (5.5 m) above roadways, 12 ft above driveways, and 10 ft above sidewalks are the required minimum. An 8 ft (2.4 m) clearance is required above low-sloped (less than 4 in 12 slope) rooflines. A 3 ft (0.9 m) clearance is required for steep-sloped roofs.

It is also necessary to bury underground conductors sufficiently below grade to reduce the hazard of unintentional contact. The required depth of burial can be found in the Code. Minimum earth cover varies from 6 to 24 in (150 to 600 mm), depending on whether the conductors are in a cable or protected by a conduit, or covered with soil or below a concrete walk or street.

The maximum single-span distance that utilities will run overhead service drop conductors to the point of service entrance varies, but typically it is 100 to 125 ft (30.5 to 38.1 m). Building heights, large conductors, or the necessity for street, driveway, or sidewalk crossings may reduce maximum permissible spans.

Transformer Requirements

Transformers may be located in a building to step up or step down the building system voltage. Transformer combinations, such as wye-wye (Y-Y), delta-delta (Δ-Δ), delta-wye (Δ-Y), and wye-delta (Y-Δ) are available for use in buildings. The first symbol (Y or Δ) indicates the configuration of the primary windings and the second the configuration of the secondary windings. The delta-wye (Δ-Y) is the most commonly found transformer combination. The reasons for choosing a Y or Δ configuration for transformer winding connections are the same as for any other three-phase application: Y connections provide the opportunity for multiple voltages, while Δ connections enjoy a higher level of reliability (if one winding fails to open, the other two can still maintain full line voltages to the load). For example, when there is no need for a neutral conductor in the secondary power system, Δ-Δ connection schemes are preferred because of the inherent reliability of the Δ configuration.

A 480 VΔ primary, 208 VY/120 V secondary, three-phase transformer is a popular unit used in large commercial buildings and industrial facilities. From a four-wire 480/277 V supply, 277 V lighting and 480 V heavy equipment can be powered before being stepped down to a 208 Y/120 V, three-phase, four-wire system for convenience receptacles and light-duty equipment and appliances. Standard voltages for common commercially available primary to secondary voltage transformers are provided in Table 18.2 in Chapter 18. Approximate full loads (amperes) for single-phase and three-phase transformers are provided in Table 19.19.

Equipment designed to operate from delta-connected power, such as air conditioners or motors, can also operate from wye-connected power, because the phase-to-phase voltages are available in both systems. However, equipment that requires

TABLE 19.19   APPROXIMATE FULL LOADS (AMPERES) FOR SINGLE-PHASE AND THREE-PHASE TRANSFORMERS. VALUES WILL VARY WITH UNIT. CHECK WITH MANUFACTURER FOR ACTUAL SPECIFICATIONS.

				Single Phase			
Transformer Rating				Full Load Current (Amperes)			
KVA	120 V	208 V	240 V	277 V	380V	480 V	600 V
0.050	0.4	0.2	0.2	0.2	0.1	0.1	0.1
0.100	0.8	0.5	0.4	0.3	0.2	0.2	0.2
0.150	1.2	0.7	0.6	0.5	0.4	0.3	0.3
¼	2.0	1.2	1.0	0.9	0.6	0.5	0.4
½	4.2	2.4	2.1	1.8	1.3	1.0	0.8
¾	6.3	3.6	3.1	2.7	2.0	1.6	1.3
1	8.3	4.8	4.2	3.6	2.6	2.1	1.7
1½	12.5	7.2	6.2	5.4	3.9	3.1	2.5
2	16.7	9.6	8.3	7.2	5.2	4.2	3.3
3	25	14.4	12.5	10.8	7.9	6.2	5
5	41	24	20.8	18	13.1	10.4	8.3
7½	62	36	31	27	19.7	15.6	12.5
10	83	48	41	36	26	20.9	16.7
15	125	72	62	54	39	31	25
25	208	120	104	90	65	52	41
37½	312	180	156	135	98	78	62
50	416	240	208	180	131	104	83
75	625	360	312	270	197	156	125
100	833	480	416	361	263	208	166
167	1391	802	695	602	439	347	278

				Three Phase			
Transformer Rating				Full Load Current (Amperes)			
KVA	120 V	208 V	240 V	277 V	380 V	480 V	600 V
3	—	8.3	7.2	—	—	3.6	2.9
6	—	16.6	14.4	—	—	7.2	5.8
9	—	25	21.6	—	—	10.8	8.6
15	—	41	36	—	—	18	14.4
22	—	62	54	—	—	27	21.6
30	—	83	72	—	—	36	28
45	—	124	108	—	—	54	43
75	—	208	180	—	—	90	72
112½	—	312	270	—	—	135	108
150	—	416	360	—	—	180	144
225	—	624	541	—	—	270	216
300	—	832	721	—	—	360	268
500	—	1387	1202	—	—	601	481
750	—	2084	1806	—	—	903	723

*For motor service factor greater than 1.0, increase full load amps proportionally (e.g., service factor is 1.18, increase amperage values by 18%).

wye-connected power cannot operate from a delta-connected source because the phase to neutral voltages are not available. A special isolation transformer, designed to convert delta to wye, is needed in this case.

An overcurrent protection device, typically rated at 150 to 200% of the rated primary current of the transformer, must protect the primary circuit on a transformer, which ensures adequate short-circuit protection for primary conductors and the transformer but less opportunity for nuisance tripping. The secondary overcurrent device is set at 125% of full load.

Primary and secondary conductors are sized to carry 100% of the ampere rating of the overcurrent protection. The equipment-grounding conductor size is based on the ampacity of the phase conductors. System grounding of a transformer is required to remove static electricity and if a short-circuit develops, such as if the transformer windings come in contact with the enclosure. It is based on the ampacity of the phase conductors on the secondary side of the transformer.

Transformers located within buildings generally must be located in transformer closets (rooms) or vaults that have adequate fire ratings. They must be readily accessible to qualified personnel for inspection and maintenance. Small, dry-type transformers can be located in the open on walls, columns, or structures, or above suspended ceilings or in hollow spaces of buildings if they are readily accessible. Generally, transformers rated at 112½ kVA or less must have a separation from combustible materials of at least 12 in and those rated above 112½ kVA must be located in an approved transformer room.

Transformer vaults and rooms must be constructed of walls and ceilings that are structurally adequate and that have at least a 3 hr fire rating (i.e., 6 in thick reinforced concrete), unless the vault or room has an automatic fire protection system (e.g., sprinklers), in which case the rating must be 1 hr. The vault door must be tight fitting with a minimum fire-resistance rating of 3 hr [450.43(A)]. This minimum fire resistance (for the vault and the door) drops to 1 hr, where an automatic sprinkler system protects the vault. Vault doors must swing out, be equipped with panic bars or pressure plates so the door can open from inside under simple pressure, and be provided with locks that are accessible only to qualified persons. The vault cannot be used as a storage room. The aim of a transformer vault is keeping the transformer(s) cool and away from building occupants. Nothing unrelated to the transformers can go in a vault—that is, it cannot be used as a telecommunications closet or house a water heater. All transformer installations must provide enough ventilation so the transformer does not overheat.

19.11 OCCUPANT PROTECTION REQUIREMENTS

Tamper-Resistant Receptacle Requirements

The NEC has required tamper-resistant receptacles be installed in all 125-V, 15- and 20-A electrical receptacles in hospital pediatric areas for nearly three decades. Recently, the NEC introduced requirements for all 125-V, 15- and 20-A electrical receptacles in new residential construction to be tamper-resistant receptacles. The move comes in an effort to better protect small children from suffering electrical burns when they accidentally insert items into conventional outlets. Products such as plastic plug-in inserts and wall plates with contact shutters are available for tamper resistance, but do not meet these requirements. There are many other settings where children may be at risk, including day care centers, children's play areas, elementary and nursery schools, doctor's offices and lobbies, and retail establishments featuring children's attire or toys. Tamper-resistant receptacles should be considered in these areas. All tamper-resistant receptacles must have either the words "tamper resistant" or the letters "TR" (minimum 3/16 in or 5 mm high) on the device as a clear indication that this is a tamper-resistant receptacle.

Ground Fault Circuit Protection Requirements

The NEC introduced requirements for use of GFCI in residences in 1973 when it required GFCI protection of outdoor convenience receptacles within 6½ ft (2 m) of grade level. In residential installations, GFCI protection is required at all 125 V, single-phase 15 A and 20 A outlets in the locations listed below. The requirements do not apply to equipment rated at 240 V (e.g., baseboard heater, room air conditioners, welding receptacles, or other outlets that are not rated at 125 V). The NEC introduced these requirements beginning in the years listed in parentheses:

- Outdoor convenience outlets within 6½ ft (2 m) of grade level (1973 edition) and revised to include all exterior convenience outlets (1993 edition)
- Bathroom convenience outlets (1975 edition)
- Readily accessible convenience outlets in garages (1978 edition), except where not readily accessible such as outlets dedicated to an overhead door opener or freezer
- Convenience outlets within 6 ft (1.8 m) of kitchen sink (1987 edition) and revised to include all kitchen convenience outlets that serve countertops, including islands (1996 edition) but not those serving fixed kitchen appliances (e.g. range and oven) and the refrigerator or freezer
- Convenience outlets in unfinished basements and crawl spaces, except laundry (1990 edition)
- Convenience outlets within 6 ft (1.8 m) of laundry, utility room, or wet bar sink (1993 edition)

Other locations requiring GFCI protection on 120 V, single-phase 15 A and 20 A outlets include:

- Commercial kitchens
- Restrooms in commercial, industrial, and in any other nondwelling type buildings
- Receptacles with grade-level access and at rooftop locations
- Commercial garages
- Elevator pits
- Agricultural buildings
- Aircraft hangers
- Wet locations in health care facilities
- Boathouses, marinas, and boatyards
- Receptacle outlets on roofs (except dwelling units)
- Circuits to resistance (impedance) heating units such swimming pool, spa, hot tub, de-icing and snow-melting heaters;
- Receptacle and lighting outlets near swimming pools, spas, hot tubs, and fountains

- Receptacles in temporary locations (i.e., on construction sites, carnivals, circuses, and fairs)

It is common practice to provide GFCI protection for all other wet locations and outdoor receptacles even though it may not be required.

Some types of equipment (e.g., appliances with large motors) have inherent leakage current levels that exceed the standard trip setting of a GFCI. This will cause nuisance tripping of GFCI devices. For this reason, the Code usually exempts from the requirements for GFCI protection a receptacle dedicated to a single appliance, which is not accessible for other uses. This equipment must rely on a solid grounding connection to provide adequate protection.

GFCI circuit breakers may be added in panelboards of older buildings to replace ordinary circuit breakers. For homes protected by fuses, the convenience receptacle or portable-type GFCI protection is highly recommended. GFCI protection should also be used whenever operating electrically powered garden equipment (mower, hedge trimmer, edger, and so on). Circuit breakers are used as switches.

Arc Fault Circuit Protection Requirements

The NEC requires AFCIs for bedroom circuits in new residential construction (starting in January 2002). Future editions of the Code will likely expand coverage to include commercial and industrial applications such as use in fire station sleeping areas, military housing, hospitals, outpatient clinics, rest homes, retirement homes, and in other locations where extension cords or cord-connected equipment may be used and where the general occupancy may be at risk from arc faults.

Older homes with ordinary circuit breakers may benefit from the added protection against arcing faults that can occur in aging wiring systems. Buildings wired with solid aluminum conductors used in the late 1960s and early 1970s are prone to poor connections. Such building electrical systems can also be retrofitted with AFCI protection to identify faulty connections.

19.12 ELECTRICAL SYSTEM DESIGN

Preliminary Design Guidelines

Ideally, the electrical designer should be involved in the design of the project from the very beginning. It would be best, in many situations, for the electrical designer to be involved in the selection of the site for the project. On a large project, it may be necessary to extend high-voltage lines to the project site and the owner may have to pay part of the cost. The electrical designer is the person who could best discuss the electrical component of project planning with the power company. All of the utilities (whether sewer, water storm sewer, natural gas lines,

and so forth) must be checked to determine if they are near the property and whether they can be brought to the property economically. Such information is needed early in the design stage.

Before actually beginning the design layout of the project, the designer will need to accumulate certain information:

1. Determine whether electrical service is available at the site, and what type of system voltage is available (e.g., 120/240 V AC, 1Φ-3W, 208 Y/120 V AC, 3Φ-4W, and so forth). If service is not available, arrangements must be made with the power company to extend service to the building site. Large projects may require more power or a different system voltage than the existing service can supply. Coordination with the power company is desirable as early in the design stage as possible. Costs that must be paid by the owner should be thoroughly discussed, written, and received by the owner.

2. Obtain a list from the owner of all the types and locations of equipment and appliances to be used in the building that will require electricity. Although the electrical designer may be aware of electrical requirements for much of the equipment, it may be necessary to find the manufacturer's specifications for certain equipment (e.g., motor sizes, power and system voltage required).

3. Work with the architectural designer to best locate all of the electrical equipment and appliances on the floor plan. On commercial projects, this sometimes takes several meetings with the architects, engineers, consultants, owner's representatives, and manufacturer's representatives. There are times when the type of equipment used and its location must be approved by governmental agencies.

4. Review with the architect where the basic mechanical equipment (e.g., HVAC and plumbing), the service entrance equipment, the power and lighting panels, and the conduit or cable will be located.

5. Discuss with the owners any future plans for changing or expanding the facilities (e.g., remodeling, constructing additions or other buildings, future equipment requirements) and anything else that could potentially affect the size and location of the electrical service. Many times, the service entrance must be sized to anticipate future expansion as well as present building plans. Once the basic information has been gathered, the designer can begin to design the system itself.

Basic Design Considerations

In electrical design there are numerous possible solutions. Experience guides the designer to a solution that best suits the need of the building occupant. The designer achieves a good solution by:

- Observing and evaluating existing installations and adapting them to meet the project requirements

- Applying electrical systems theory
- Applying Code requirements.

Design of an electrical system begins with the layout of all outlets or outlet connections. See Figure 19.1. Symbols are used to indicate the approximate location of convenience receptacles, lighting outlets, and special purpose outlets. At first, these symbols are located without consideration of circuit design. Connecting the symbols with wiring circuits comes after all outlet locations are identified.

Design Guidelines for Common Spaces

Convenience Outlets and Switches

1. The number and type of lighting outlets should be fitted to the various seeing tasks. Lighting outlets should be located to meet the desired lighting effects and fixtures to be used. Refer to Chapter 18.

2. All convenience receptacles on 15 A and 20 A general purpose circuits should be of the grounding type, minimizing the hazard of shock from short circuits.

3. GFCI protection should be provided on convenience receptacles where required by local code (e.g., where the occupant is exposed to water).

4. AFCI protection should be used on convenience receptacles where required by local code (e.g., in bedroom circuits).

5. All rooms that have more than one entrance should be equipped with multiple-switch controls (e.g., two-way or three-way switching) at each principal entrance. Principal entrances are those commonly used for entry to the room when going from a normally lighted to an unlighted condition. If this recommendation would result in the placing of switches controlling the same lighting installation within 8 ft (2.5 m) of each other, multiple-switch controls may not be required.

6. Wall switches should be located at the lockset or latch side of doors or at the traffic side of arches, and within the room or area where the lighting outlets are located.

7. Convenience receptacles in living rooms, bedrooms, dining areas, and other habitable spaces should be placed so that no point along the floor line in any usable wall space is more than 6 ft (1.8 m) from an outlet in that space. Any wall 2 ft (0.6 m) or more in length must have a convenience receptacle. Preferably, convenience receptacles should be located near the ends of a wall space, rather than near the center of the wall, to reduce the likelihood of being concealed behind large pieces of furniture. Outlets should not be placed above electrical baseboards, hot air registers, and hot water or steam registers. The intent is to eliminate cords having to pass over hot or conductive surfaces wherever possible.

Building Exterior

1. One or more lighting outlets should be located at or near all exterior entrances. Outlets should be switched or automatically controlled.

2. For each single-family dwelling, at least one duplex receptacle shall be installed outdoors to be readily available from ground level. Weatherproof convenience receptacles should be provided on exterior walls for outside work. GFCI protection is required for outdoor receptacles.

3. One or more outlets may be required for exterior equipment (e.g., swimming pool pump, well pump, and so on).

Common Areas and Living Rooms

1. Outlets for ambient and task lighting should be provided. General illumination outlets should be wall-switch controlled.

2. Convenience receptacles in living areas should be placed so that no point along the floor line in any usable wall space is more than 6 ft (0.6 m) from a receptacle outlet in that space.

3. One or more receptacles for entertainment equipment should be provided at bookcases, shelves, or other suitable locations.

4. When general illumination is to be provided from portable lamps, then at least two separate wall-switched plug-in positions should be provided. These can be provided with two switched regular duplex receptacles or one switched plug-in position in each of two split-receptacle outlets.

5. An outlet for a fireplace may be required.

6. A smoke detector/alarm on a 120 V circuit should be considered.

Food Preparation/Kitchen/ Cooking Areas

1. Lighting design should provide for ambient and local/task illumination of the work areas, sink, range, counters, and tables. Lighting outlets should be switch controlled.

2. Special purpose receptacles should be provided for all fixed appliances (e.g., range, built-in microwave, exhaust hood, dishwasher, trash disposal unit, waste disposal, and so on).

3. At least two 20 A small appliance circuits for kitchen countertops should be provided. These circuits are in addition to those required for refrigerators, ranges, microwaves, lighting, and so forth. Outlets on these circuits should serve only the kitchen, pantry, and/or dining room areas. The following convenience receptacles should be connected to small appliance circuits:

   - One receptacle for each 2 linear ft (0.6 m) of work-surface face.

- At least one receptacle to serve each separate work surface. Any counterspace wider than 12 in should have a convenience receptacle.

4. Convenience receptacles serving countertop areas (e.g., except behind refrigerator and those serving fixed appliances) should be GFCI protected. It is recommended that receptacles serving countertop areas be split wired.

5. A smoke detector/alarm on a 120 V circuit should be considered.

Sleeping Areas/Bedrooms

1. General illumination should be provided from either ceiling or wall outlets, controlled by one or more wall switches.

2. A convenience receptacle should be placed on each side and within 6 ft (1.8 m) of the centerline of each probable individual bed location. Preferably, convenience receptacles should be located near the ends of a wall space, rather than near the center of the wall, to reduce the likelihood of being concealed behind large pieces of furniture.

3. All 120 V branch circuits that supply outlets in dwelling unit bedrooms must be protected by an AFCI device.

4. A smoke detector/alarm must be provided on a 120 V circuit and should be AFCI protected.

5. In master bedrooms, outlets should be considered for a television and entertainment equipment (e.g., DVD/VCR player).

6. In master bedrooms, an outlet for a fireplace may be required.

Bathrooms/Restrooms

1. Lighting sources at the mirror should be capable of illuminating both sides of the face.

2. At least one GFCI-protected receptacle within 3 ft of the outside edge of each lavatory basin should be provided. A receptacle that is a part of a bathroom lighting fixture is not typically suitable for this purpose. Placing each bathroom on a separate circuit should be considered because of the heavy use and demands of these receptacles. The circuit serving bathrooms should have no other outlets (e.g., it cannot supply power to an outside receptacle or a garage receptacle). Where the 20 A circuit supplies only a single bathroom, it can supply power to outlets for other equipment within the same bathroom (i.e., lighting outlets or an exhaust fan).

3. A wall-switched or timer-operated, built-in ventilating fan capable of providing a minimum of 8 to 10 air changes per hour per water closet (50 cfm/water closet) should be provided where no natural ventilation through windows is included.

4. Wall switches should be located so as not to be readily accessible while standing in the tub or shower stall.

Laundry Areas

1. Outlets for fixed lights should be installed to provide illumination of work areas, such as laundry tubs, sorting tables, washing, ironing, and drying centers. Lighting outlets should be wall-switched controlled.

2. In the laundry area, one 20 A receptacle for the clothes washer and a special receptacle for the clothes dryer are required. Outlets for other workstations (e.g., sewing, ironing, repairing, and so on) should be provided.

3. One outlet and one switch for a ventilation fan should be provided.

4. A smoke detector/alarm on a 120 V circuit should be considered.

Halls/Corridors

1. Ceiling fixtures should be installed for proper illumination of the entire area with particular attention paid to irregularly shaped spaces.

2. Convenience receptacles in hallways within a dwelling unit should be placed so that no point in the hallway shall be more than 10 ft (1.0 m) from a duplex receptacle as measured by the shortest path that the supply cord of an appliance connected to the receptacle would follow without passing through an opening fitted with a door (the "vacuum-cleaner" rule). Each hall over 25 ft^2 (2.3 m^2) in floor space should have at least one receptacle.

3. In entrance foyers, convenience receptacles should be placed so that no point along the floor line in any usable wall space is more than 10 ft (3.1 m) from a receptacle in that space.

Stairways

1. Fixed wall or ceiling lighting outlets should be installed to provide adequate illumination of each stair flight.

2. Outlets should be so arranged that the stair system can be fully illuminated from either floor.

3. A smoke detector/alarm on a 120 V circuit should be considered at the top of the stairs.

Utility Rooms

1. Lighting outlets should be placed to illuminate the furnace/boiler area and work area. At least one lighting outlet should be wall-switch controlled.

2. Convenience receptacles should be provided.

3. Outlets should be provided for each piece of mechanical equipment requiring electrical connections such as the boiler, chiller, furnace, water pump, or compressor.

4. A special purpose outlet may be required for an electric-resistance water heater, and/or an electric-resistance furnace.

Shops/Garages

1. Lighting outlets should be placed to illuminate the work areas. Task lighting should be provided above workbenches. Lighting outlet should be wall-switch controlled.

2. At least one duplex receptacle should be provided for each space in a garage or carport.

3. Convenience outlets should be provided at workbenches. In garages or near water, these outlets should be GFCI protected.

4. Outlets should be provided for automatic overhear (garage) door operators in the ceiling above each bay.

5. Special purpose outlets should be provided for special equipment such as air compressors, welding equipment, tire changer, dust collection equipment, machining equipment (e.g., table saws, drill presses, milling machines, lathes, machining centers), and so on. Check with owner on equipment voltage requirements, load, and placement.

6. A smoke detector/alarm on a 120 V circuit should be considered.

Storage Rooms/Closets

1. Generally, one lighting outlet for each large closet or storage room should be provided. Where shelving or other conditions make the installation of lights within a closet ineffective or unsafe, convenience receptacles in the adjoining space should be so located as to provide light within the closet.

2. Wall switches or automatic door switches are preferred, but pull switches are acceptable.

Electrical/Telecommunications Closets

1. Lighting outlets should be placed to illuminate the area.

2. A minimum of two dedicated convenience receptacles on separate circuits is required. Additional duplex convenience receptacles should be placed at 6 ft (1.8 m) intervals around the perimeter.

3. Rooms should be located away from sources of electromagnetic interference (e.g., transformers, motors, x-ray equipment, induction heaters, arc welders, radios, radar systems, and so forth).

4. Emergency power should be considered and supplied.

5. A smoke detector/alarm on a 120 V circuit should be considered.

19.13   DESIGN EXAMPLE

An example of system design of a 61 ft-6 in by 36 ft, single-story, single-family residence is used in the explanation. This residence contains general lighting (lights and receptacle outlets), equipment, and appliances, including what follows: water heater (3800 VA), clothes dryer (4400 VA), dishwasher (1000 VA), range (11 700 VA), waste disposal (1000 VA), air conditioner (6300 VA), and garage door opener (1000 VA). Loads used are from manufacturers' data.

Circuit Design

1. Locate the convenience receptacles on the floor plan with symbols used to represent each. Receptacles should be located conveniently following the guidelines provided. All switching, sizing of conductors, and circuit layout will be done later. GFCI- and AFCI-protected receptacles should be located following the requirements specified. See Figure 19.12.

2. Locate all appliance and equipment outlets on the floor plan using the appropriate symbols for the various receptacles required. See Figure 19.12.

   This process calls for coordination with the architectural designer because each appliance and piece of equipment must be known to locate the outlets properly. Many times, the architectural plans do not convey each appliance and piece of equipment that must be connected. For example, a kitchen waste disposal must be connected with a switch on the wall, yet the waste disposal and switch are seldom shown on the floor plan. The electrical designer should request that the architectural designer list all appliances and equipment in writing or on the drawing for complete coordination.

   The electrical designer should also check the number of appliances or equipment required. For example, two furnaces may be used in large residences and both will need connections. Also, in commercial buildings, typically more than one heating and/or cooling unit may be used.

   The electrical designer will need to know the specified voltage and amperage requirements for equipment and appliances. A list of typical appliances that require electrical connections and their requirements are listed in Tables 19.12 through 19.17. These requirements will vary among different manufacturers. Manufacturer nameplate ratings should be confirmed and used in computations.

3. Locate the lighting fixtures, using the appropriate symbols to represent each. See Figure 19.13.

   The electrical designer should make a list of types of luminaires and the load requirements. This list will be used later when grouping circuits. It may also be included in the specifications or on the drawings as a *luminaire (light fixture) schedule*. The fixture schedule will be used by the electrical estimator when determining the cost to be charged for the work; next by the electrical purchasing agent when the fixtures and materials are ordered for the project; and then by the electrician who actually

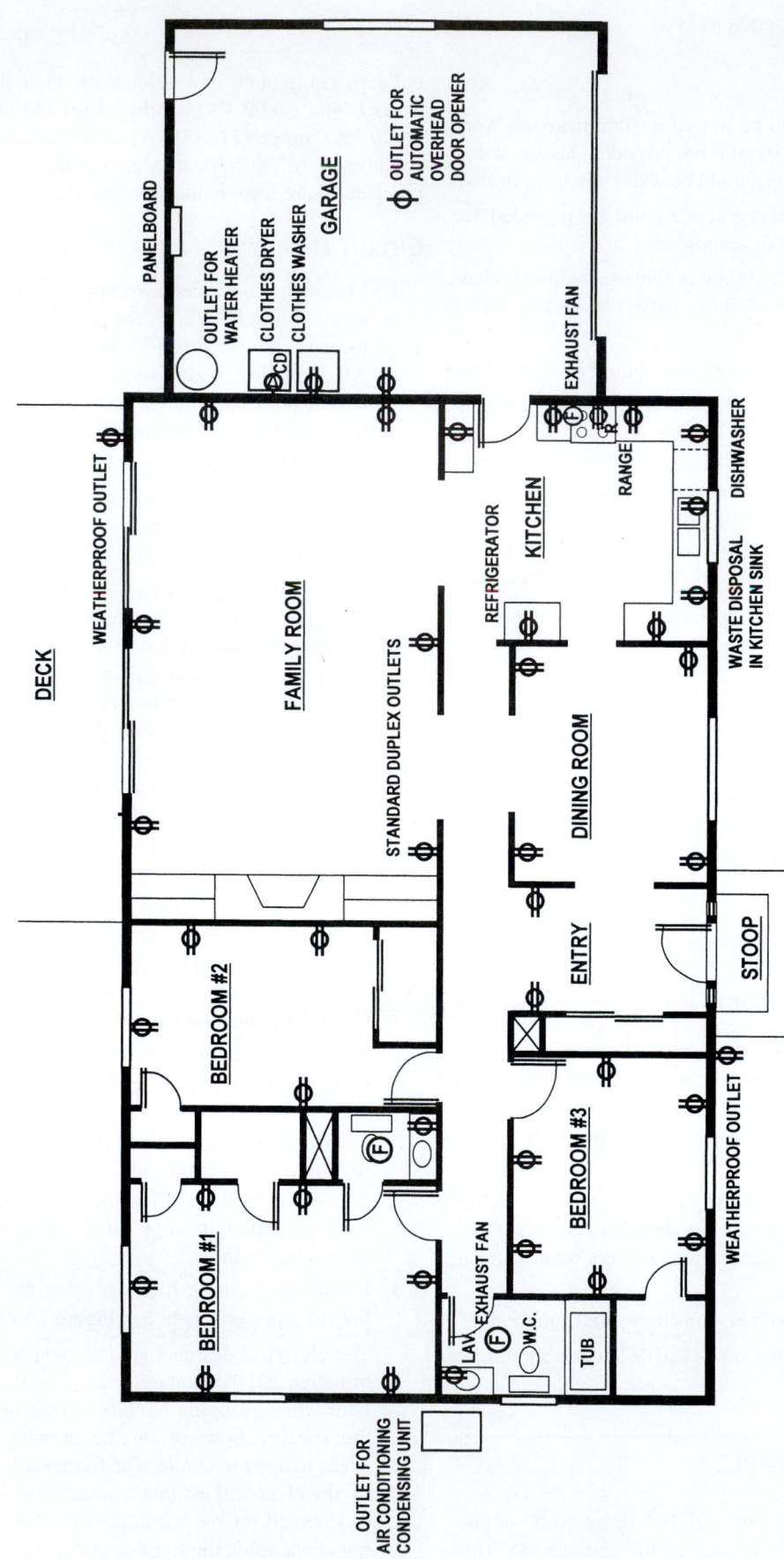

FIGURE 19.12 Layout of locations of convenience receptacles and other outlets for design example.

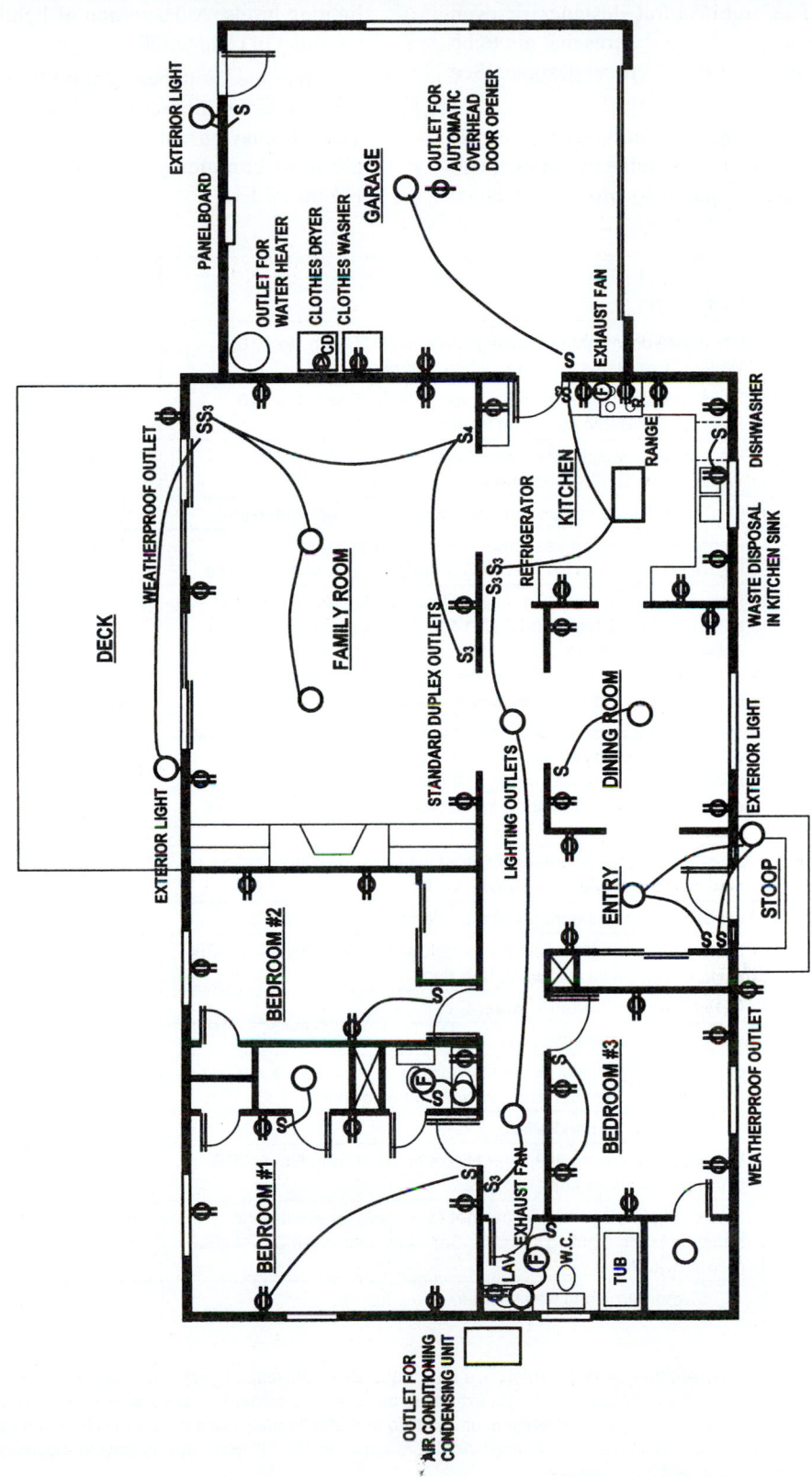

FIGURE 19.13 Layout of lighting outlets and related switches for lights and outlets for design example.

installs the work. The architectural designer or even the building owner may decide the fixtures that are to be used, often with assistance of the electrical designer. See Figure 19.14.

On commercial, industrial, and institutional projects, the electrical designer will need to determine the number and types of lighting fixtures required to provide adequate lighting levels. A discussion of lighting system design is provided in Chapter 20.

4. Lay out the switches required to control the lights, appliances, equipment, and any desired receptacles. The discussion of switches outlines where they are most commonly used and the symbols used. See Figure 19.13.

Fixture Designation	Fixture Type	Description	Lamp Type	Mounting
A 75W R30	A	Recessed downlight or similar Halo's H7UICT, with No. 310W White Coilex Baffle cone and white trim ring.	75W R30	Ceiling
A1 75W R30	A1	Recessed downlight or similar Halo's H7UICT, with No. 310 Black Coilex Baffle cone and white trim ring.	75 W R30	Ceiling
B 75W R30	B	Recessed downlight or similar Halo's H7T, with No. 310W White Coilex Baffle cone and white trim ring.	75W R30	Ceiling
	C	Owner furnished hung chandelier/ceiling mounted light fixture.		Ceiling
D 90W PAR38	D	Recessed downlight designed for sloped ceilings, or similar Halo's H47ICT, with No. 14199 Coilex Baffle and white trim ring.	90W PAR38	Sloped ceiling
E1 75W	E1	Recessed combination light fixture/exhaust fan unit similar to Nutone's "QuieTTest" Model QT140L. Fan to be capable of delivering a minimum of 150 cfm and must have a sound rating of 2.0 sones or better. Provide new exhaust ductwork to exterior in accordance with manufacturer's specifications.	75W	Ceiling
E2 75W	E2	Recessed exhaust fan unit similar to Nutone's "QuieTTest" Model QT200. Fan to be capable of delivering a minimum of 150 cfm and must have a sound rating of 2.0 sones or better. Provide new exhaust ductwork to exterior in accordance with manufacturer's specifications.	75W	Ceiling
F 40RS	F	Surface mounted fluorescent fixture with prismatic lens, 48" long, sim. to KLP's Model MWB140A.	40RS	Ceiling
	G	Owner furnished exterior wall mounted fixture.		Wall
H 20W 12V MR16	H	Recessed downlight or similar to Halo's H1499T, with No. 1493P white Coilex baffle and white trim ring.	20W 12V MR16	Ceiling
L	L	Owner furnished exterior lantern.		Ceiling/wall
M 40W	M	Bare bulb porcelain fixtures.	40W	Ceiling
N	N	Owner furnished incandescent fixture to be mounted to wall over mirror or to mirror face, as directed by owner.		Wall/Mirror
P	P	Owner furnished paddle fan.		Ceiling
S 60W A19	S	Recessed downlight or similar Halo's H7UICT, with No. 170PS Showerlite Albalite Lens.	60W A19	Ceiling
T 10H2 Halogen	T	Low voltage halogen undercabinet task lights, or similar and equal to Ardee Lighting's Diskus Series mini-halogen downlights, Model DULWF10.	10H2 Halogen	Under cabinet
W	W	Owner furnished wall sconce.		Wall

Light Fixture Notes

1. Light fixture products specified above are for contractor reference only. Equivalent fixtures, subject to owner and architect review and approval may be accepted. Contractor to submit cuts on all light fixtures to architect and owner for approval prior to ordering any light fixtures.
2. Fixtures to be recessed in direct contact with insulated ceilings or installed in damp locations are to be rated for such applications.
3. Contractor to coordinate with owner regarding method of installation of rough wiring for "J" type fixtures (owner's cabinetry subcontractor may design method of wiring concealment into casework).
4. Contractor to review exterior lighting control concept with owner prior installation of any materials.

FIGURE 19.14 An example of a light fixture (luminaire) schedule.

5. Locate the panelboard in a convenient location. The location must be accessible (e.g., not in a closet or storage room, unless dedicated). Follow the requirements outlined earlier. See Figure 19.12.

6. Layout circuiting for large appliances and equipment served by an individual branch circuit. Examples of individual branch circuits would be circuits from the panelboard to a dishwasher (120 V), electric clothes dryer (120/240 V), an electric oven or range (120/240 V), or an electric water heater (240 V).

7. Layout circuiting for lighting and convenience receptacles on general purpose (lighting) branch circuits. Usually 15 A and 20 A general purpose branch circuits are used for convenience receptacles, luminaires, and small appliances. See Figure 19.15.

The designer must comply with code requirements. In practice, a good designer tends to be a little more conservative, generally limiting a 15 A branch circuit to 1000 to 1200 VA and a 20 A branch circuit to 1300 to 1600 VA. Convenience receptacles are generally limited to about 6 to 8 on a circuit. This allows the circuit to take additional loads, such as when higher wattage lamps are used to replace those originally installed and calculated.

Because more and more small appliances and equipment are being purchased and connected to receptacles, the designer must anticipate future requirements. Such a layout allows for the extension of a circuit if it is necessary to add a light or a convenience receptacle instead of adding a whole new circuit from the panelboard. If it is possible that the occupant will desire to install individual air conditioners, an individual branch circuit may be desired.

8. Lay out the panel circuits, either on the drawing or in a table as shown in Figure 19.16. In large designs, with more than one panel, this provides the electrician with a schedule of what circuits will be served from what box. Although a panelboard layout is not often done for a residence, it is helpful to both the electrician and the designer if one is included. For commercial projects, a panel layout is almost always included.

Load Computations

Load computations involve computing the *demand load* for a building system or a distribution system extending from a panelboard. This load includes the total of all general lighting, appliance, and equipment loads in the building. The demand load allowed by the code takes into account that all of the electrical connections will not be in use at one time. As a result, the demand load is not a total of all connected loads, but rather a fraction of the connected loads. Code prescribes methods of computing the demand load, which trims down the total connected load to a safe, more reasonable level.

Demand load computation and panelboard layout worksheets such as the sheets provided in Figure 19.17 assist the designer in these computations. General considerations in demand load computations follow.

1. Compute the *general lighting load*. This is calculated for all types of occupancies based on the unit load given in the table (in watts) times the square footage of the building. (For this exercise it is based upon Code specified 3 VA/ft² for residences). The floor area is determined using the outside dimensions of the building involved and the number of stories. For dwellings, do not include any open porches, garages, or carports. Any unfinished or unused spaces do not have to be included in the square footage *unless* they are adaptable for future use.

For the (hypothetical) residential system example, the outside dimensions (excluding garage) of the garage are:

61 ft 6 in · 36 ft, there are 2214 ft² of floor area

The minimum general lighting load, based on the specified 3 VA/ft² for residences:

3 VA/ft² × 2214 ft² = 6642 VA

2. Compute the *appliance and laundry circuit load*. Code requires at least two 20 A appliance branch circuits for the kitchen. The load is based on 1500 VA (from Code) for each appliance branch circuit in the kitchen. In addition, one 20 A circuit is required for laundry room appliances.

For the residential system example, this results in a total of three 20 A branch circuits for appliances:

Appliance and laundry load = 3 circuits × 1500 VA
= 4500 VA

3. Subtotal the general lighting, appliance, and laundry branch circuit loads.

For the residential system example:

General lighting	6642 VA
Appliance and laundry circuits	4500 VA
Subtotal	11 142 VA

4. The demand load allowed by the Code takes into account that all of the electrical connections will not be in use at one time. Although there are limits to this reduction for certain types of occupancies, in a dwelling the first 3000 VA are taken as 100%, and from 3000 to 120 000 VA, only 35% of the load is calculated.

For the residential system example, the load subtotal is 11 142 VA, so:

First 3000 VA at 100%	3000 VA
Remaining 8142 VA at 35%	2850 VA
Total demand load	5850 VA

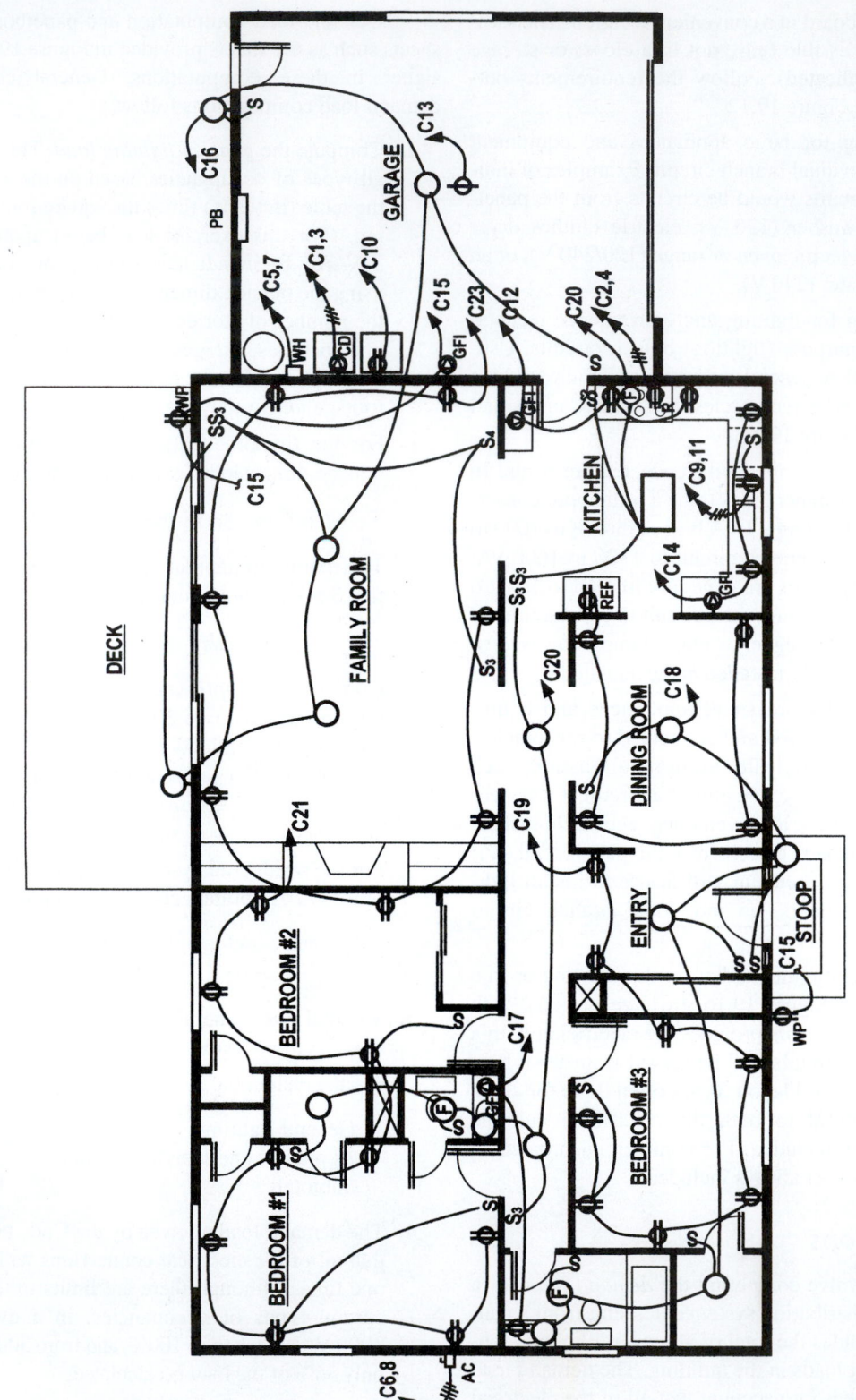

FIGURE 19.15 Circuit layout and numbering for design example.

The loads of all other appliances and equipment (motors) must be added to this demand load to determine the total service load on the system.

5. To determine the *appliance and equipment load,* all appliances and equipment that will not be on the lines discussed above must be listed along with their electrical requirements. Although typical ratings are given in Tables 19.12 through 19.16, nameplate ratings from manufacturers' data should be used in the design.

For the residential system example, the following is a list of fixed appliances and equipment and their ratings from manufacturer's data:

Water heater	3800 VA
Clothes dryer	4400 VA
Dishwasher	1000 VA

Range	11 700 VA
Waste disposal	1000 VA
Air conditioner	6300 VA
Garage door opener	1000 VA

The demand load for an *electric range,* consisting of an oven and a cooktop unit, is taken from manufacturer's data.

For the residential system example, an electric range with a rating of 11.7 kW, the demand load would be 8 kW (or 8000 VA):

$$\text{Electric range demand load} = 8000 \text{ VA}$$

The demand load for the *clothes dryer* is the total amount of power required according to the manufacturer's data.

150A, 24 POLE PANELBOARD LAYOUT
120/240V, 1Φ, 3-WIRE

CIRCUIT NO.	CIRCUIT DESCRIPTION	CONNECTED LOAD VA	BREAKER A	HOT LEGS A	B	BREAKER A	CONNECTED LOAD VA	CIRCUIT DESCRIPTION	CIRCUIT NO.
1	CLOTHES DRYER	2200	30			50	5850	RANGE	2
3	CLOTHES DRYER	2200	30			50	5850	RANGE	4
5	WATER HEATER	1900	30			40	3150	AIR CONDITIONER	6
7	WATER HEATER	1900	30			40	3150	AIR CONDITIONER	8
9	DISHWASHER	1000	20			20	1000	LAUNDRY	10
11	WASTE DISPOSAL	1000	20			20	1500	SMALL APPLIANCE	12
13	GARAGE DOOR OPENER	1000	20			20	1500	SMALL APPLIANCE	14
15	GFCI RECEPTACLES	1500	20			20	800	LIGHTS	16
17	GFCI RECEPTACLES	1500	20			20	800	LIGHTS	18
19	RECEPTACLES	1600	20			20	800	LIGHTS	20
21	RECEPTACLES	1600	20					SPARE	22
23	RECEPTACLES	1600	20					SPARE	24

WIRING LAYOUT

C1,3 ← 3 - No. 10 AWG	CLOTHES DRYER	RANGE	3 - No. 6 AWG → C2,4
C5,7 ← 3 - No. 10 AWG	WATER HEATER	AIR CONDITIONER	3 - No. 6 AWG → C6,8
C9 ← 2 - No. 12 AWG	DISHWASHER	LAUNDRY	2 - No. 12 AWG → C10
C11 ← 2 - No. 12 AWG	WASTE DISPOSAL	SMALL APPLIANCE	2 - No. 12 AWG → C12
C13 ← 2 - No. 12 AWG	GARAGE DOOR OPENER	SMALL APPLIANCE	2 - No. 12 AWG → C14
C15 ← 2 - No. 12 AWG	RECEPTACLES	LIGHTS	2 - No. 12 AWG → C16
C17 ← 2 - No. 12 AWG	RECEPTACLES	LIGHTS	2 - No. 12 AWG → C18
C19 ← 2 - No. 12 AWG	RECEPTACLES	LIGHTS	2 - No. 12 AWG → C20
C21 ← 2 - No. 12 AWG	RECEPTACLES	SPARE	
C23 ← 2 - No. 12 AWG	RECEPTACLES	SPARE	

FIGURE 19.16 Panelboard and wiring layout and numbering for design example.

PANELBOARD LAYOUT
120/240 V, 1Φ, 3-Wire

CIRCUIT NO.	CIRCUIT DESCRIPTION	CONNECTED LOAD VA	BREAKER A	HOT LEGS A B	BREAKER A	CONNECTED LOAD VA	CIRCUIT DESCRIPTION	CIRCUIT NO.
1								2
3								4
5								6
7								8
9								10
11								12
13								14
15								16
17								18
19								20
21								22
23								24
25								26
27								28
29								30
31								32
33								34
35								3
37								38
39								40
41								42

PANELBOARD LAYOUT
208Y/120 V or 480Y/277 V, 3Φ, 4-Wire

CIRCUIT NO.	CIRCUIT DESCRIPTION	CONNECTED LOAD VA	BREAKER A	HOT LEGS A B C	BREAKER A	CONNECTED LOAD VA	CIRCUIT DESCRIPTION	CIRCUIT NO.
1								4
2								5
3								6
7								8
8								10
9								12
13								16
14								17
15								18
19								22
20								23
21								24
25								28
26								29
27								30
31								34
32								35
33								36
37								40
38								41
39								42

FIGURE 19.17 Examples of load computation and panelboard layout worksheets.

DWELLING UNIT CIRCUITING/PANELBOARD RATING

ADVANCED BUILDING CONSULTANTS
12450 WEST BELMONT AVENUE
LITTLETON, CO 80127 -1515
303-933-1300

Project:		No.:
Address:		
Designer:		Date:
Checked by:		Date:

1. LOADS FOR DWELLING UNIT

General lighting load: _____ ft² of floor area @ _____ VA per ft²: _____ VA
Electric range/cooktop(s): _____ watts: _____ VA
Electric resistance heating or cooling: _____ kW: _____ VA
Dishwasher: _____ VA
Other: _____ VA

2. REQUIRED (MINIMUM NUMBER) BRANCH CIRCUITS IN DWELLING UNIT

General lighting load (lights/outlets): _____ VA/120V = _____ amps: ___ - _____ amp circuits
Small appliance: minimum of two 20 amp circuits required (NEC Section 220-4(b)):   ___ - _____ amp circuits
Electric range: _____ VA @ _____ % (NEC Table 220-19, column C)
           _____ VA/240V* = _____ amps: ___ - _____ amp circuits
Electric space heating or cooling: _____ VA/240V* = _____ amps: ___ - _____ amp circuits
Dishwasher: ___ - _____ amp circuits
Waste disposal: ___ - _____ amp circuits
Clothes washer: ___ - _____ amp circuits
Clothes dryer: ___ - _____ amp circuits
Other: ___ - _____ amp circuits
Other: ___ - _____ amp circuits

3. MINIMUM SERVICE OR FEEDER CIRCUIT RATING FOR DWELLING UNIT

Computed Load - Ungrounded (hot) Legs

General lighting load: _____ VA
Small appliance load - _____ @ 1500VA each (two required):   _____ VA
           _____ VA Total

Application of Demand Factor for General Lighting and Small Appliance Loads (NEC Table 220-11):
      First 3000VA @ 100%:   _____ VA
      3001 - 120000VA @ 35% - _____ VA @ 35%:   _____ VA
      Remainder @ 25%- _____ VA @ 25%   _____ VA
Electric range load - _____ VA @ _____ % (from above):   _____ VA
Electric space heating or cooling- _____ VA @ 100% (NEC Section 220-15):   _____ VA
Other: _____ VA
Other: _____ VA
Other: _____ VA
           Computed total load carried by ungrounded (hot) legs: _____ VA
For 120V/240V*, 3-wire system:_____ VA/240V* = _____ **amps per ungrounded (hot) leg (minimum)**

Computed Load - Neutral Leg

Lighting and Small Appliance Load (after application of Demand Factor):   _____ VA
Electric range load - _____ VA @ _____ % (above) @ 70% (NEC Section 220-20):   _____ VA
Electric space heating or cooling - _____ VA (on 240V* heating - no neutral used):   _____ VA
Other: _____ VA
Other: _____ VA

           Computed total load carried neutral leg: _____ VA

For 120V/240V*, 3-wire system: VA/240V* =   **amps for neutral leg (minimum)**

CONDUCTOR SPECIFICATIONS:	
Ungrounded Conductor Size:	
Grounded Conductor Size:	
Grounding Conductor Size:	

EQUIPMENT SPECIFICATIONS:

* use 208V on a 120V/208V, 3-wire system.

FIGURE 19.17 Examples of load computation and panelboard layout worksheets. (*Continued*)

For the residential system example, the full 4400 VA must be used in the calculation:

$$\text{Clothes dryer demand load} = 4400 \text{ VA}$$

The demand for *fixed appliances* (other than the range, clothes dryer, and air conditioning and space heating equipment) is taken as 100% of the total amount required, *except* that when there are four or more of these fixed appliances (other than those omitted), the demand load can be taken as 75% of the fixed appliance load.

For the residential system example, there are three fixed appliances: the water heater at 3800 VA, the dishwasher at 1000 VA, and the waste disposal at 1000 VA. The total of the ratings is 4800 VA, also the demand load.

$$\text{Fixed appliances demand load} = 5800 \text{ VA}$$

Motors, such as those used in central air conditioners, have their demand loads calculated as 125% of the motor rating.

For the residential system example, the air conditioner is rated at 6300 VA and the garage door opener is at 1000 VA. The total demand load will be 7300 VA × 125% = 9125 VA.   6300 + 1000

$$\text{Motor (air conditioner/opener) demand load} = 9125 \text{ VA}$$

The total demand load for all of the lighting and appliances is then tabulated.

For the residential system example:

General lighting, appliances, and laundry	5850 VA
Electric range	8000 VA
Clothes dryer	4400 VA
Fixed appliances	5800 VA
Motor (air conditioner/opener)	9125 VA
Total demand load	33 175 VA

Service Entrance Design

The service entrance conductors and equipment are designed based on the computed total demand load.

1. The minimum service entrance size is found by dividing the demand load for the building by the voltage serving the building. Most commonly, 240 V service is used.

   For the residential system example, the total demand load of 33 175 VA is divided by the 240 V service for a minimum service entrance of 143 A:

$$\begin{aligned}\text{minimum service entrance} &= 33\ 175 \text{ VA}/240 \text{ V}\\ &= 138.2 \text{ or } 138 \text{ A}\end{aligned}$$

   The 138 A computed demand load would be rounded up to the nearest commercially available panelboard rating. From Table 18.4 (Chapter 18), a panelboard with a 150 A
   ↑
   603P

rating would be selected. A 200 A panelboard would be a better choice if expansion for additional future circuits for appliances and other equipment was anticipated. Assume a 150 A panel is deemed adequate.

2. The next step is sizing the service entrance conductors and any conductors between the service equipment and the branch overcurrent device (circuit breaker or fuse). The feeder size is based on the computed demand load. The size is then selected from Table 18.12 (in Chapter 18).

   For the residential system example, using XHHW (75°C, 167°F) aluminum service entrance conductors and a 150 A load, as calculated:

$$\text{150 A feeder demand load, 3/0 AWG, XHHW, aluminum}$$

3. The size of the *neutral conductor* may be determined as 70% of the demand load calculated for the range plus all other demand loads on the system.

   For the residential system example, the neutral feeder demand load would be:

Range load (8000 W × 70%)	5600 VA
All other demand loads	25 175 VA
Neutral demand load	30 778 VA

$$\begin{aligned}\text{Neutral net computed demand load} &= 30\ 778 \text{ VA}/240 \text{ V}\\ &= 128.2 \text{ A}\end{aligned}$$

   Select the size of the neutral conductor for a 128 A neutral feeder demand load from Table 18.12 (in Chapter 18):

$$\text{2/0 AWG, XHHW, aluminum conductor}$$

   For the residential system example, the three service conductors are:

   Two No. 3/0 AWG and one 2/0 AWG aluminum conductors

Circuit Design

Circuit design involves ascertaining the number and rating of circuits needed in the panelboard. It generally involves the following steps:

1. Determine the minimum number of lighting circuits by dividing the general lighting load by the voltage, finding the amperage required and dividing the amperage into circuits.

   For the residential system example, the general lighting load was calculated above as 6642 VA and the voltage used for the lighting is 120 V. Therefore:

$$6642 \text{ VA}/120 \text{ V} = 56 \text{ A}$$

   Because the general purpose branch circuit size is limited to 80% of the rating, four 20 A branch circuits for a total of 64 A, or five 15 A branches for a total of 60 A, are needed.

For 20 A circuits: 56 A/(20 A · 80%) = 3.5
$$= 4 \text{ circuits}$$

For 15 A circuits: 56 A/(15 A · 80%) = 4.7
$$= 5 \text{ circuits}$$

This is the minimum number of branch circuits required to serve the residence. In laying out the circuits, almost all designs will have more circuits than the minimum number required. This is because most designers will limit each circuit to fewer receptacles, lights, or a combination of receptacles and lights.

2. Lay out and number all branch circuits on the drawing. Bear in mind that, in most cases, all of these general use receptacles and all lighting will use 120 V service. In large commercial applications, 277 V may be used for interior lighting. Note that each circuit is numbered.

For the residential system example, there are a total of 14 circuits.

Panelboard Selection

Select the panelboard based on the number of circuits and the required amperage. Be certain that all the pole space is not taken up so there is room for expansion. Keep in mind that each 120 V circuit takes up one pole (for a one-pole circuit breaker), while each 240 V or 120/240 V circuit takes up two poles.

For the residential system example, there are:

14-120-V circuits	14 poles
4-240-V circuits (4 × 2 poles)	8 poles
Total poles required	22 poles

A 150 A, 24-pole panelboard should be selected (from Table 18.4, Chapter 18).

STUDY QUESTIONS

19-1. What electrical code is universally used in the United States?

19-2. Describe the following branch circuits and where each is used:
   a. Individual
   b. General purpose
   c. Appliance

19-3. Describe a split wired circuit.

19-4. How many circuits are required to accommodate kitchen appliances in an average size home?

19-5. When selecting the panelboard size, what considerations for the future should be taken into account?

19-6. How is the general lighting load for a building determined?

19-7. How is the minimum service entrance determined?

19-8. In selecting the service entrance size, what should be considered?

19-9. On a 120/240 V, single-phase, three-wire system, identify the following:
   a. Voltage between one grounded conductor and one ungrounded conductor
   b. Voltage between two ungrounded conductors

19-10. On a 208 Y/120 V, three-phase, four-wire system, identify the following:
   a. Voltage between one grounded conductor and one ungrounded conductor
   b. Voltage between two ungrounded conductors

19-11. On a 240 Δ/120 V, three-phase, four-wire system, identify the following:
   a. Voltage between one grounded conductor and one ungrounded conductor
   b. Voltage between two ungrounded conductors

19-12. On a 480 Y/277 V, three-phase, four-wire system, identify the following:
   a. Voltage between one grounded conductor and one ungrounded conductor
   b. Voltage between two ungrounded conductors

19-13. Identify five types of equipment that may require voltages higher than 120 V AC.

19-14. Identify three types of equipment that may require three-phase power.

19-15. Identify the type of cable permitted in single- and multifamily dwelling units but that cannot be used underground, nor in buildings that are more than three stories above grade, nor in commercial garages, motion picture studios, theaters, places of assembly, elevator hoist ways, and other corrosive or hazardous locations.

19-16. Identify the type of cable permitted in direct-burial applications such as a feeder or branch circuit provided it is protected by an overcurrent protection device (fuse or circuit breaker) before if leaves the panelboard.

Design Exercises

19-17. Determine the ampacity of a No. 12 AWG copper conductor with THHN insulation that will be used in a 120 V, two-wire circuit in an environment with an average ambient air temperature of no greater than 86°F (30°C). Neglect temperature and bundling correction factors.

19-18. Determine the ampacity of a No. 14 AWG copper conductor with THHN insulation that will be used in a 120 V, two-wire circuit in an environment with an average ambient air temperature of no greater than 86°F (30°C). Neglect temperature and bundling correction factors.

19-19. Determine the ampacity of a No. 12 AWG copper conductor with THHN insulation that will be used in a 277 V, two-wire circuit in an environment with an average ambient air temperature of no greater than 86°F (30°C). Neglect temperature and bundling correction factors.

19-20. Determine the ampacity of a No. 4 AWG copper conductor with THHN insulation that will be used in a 120 V, two-wire circuit in an environment with an average ambient air temperature of no greater than 86°F (30°C). Neglect temperature and bundling correction factors.

19-21. Determine the ampacity of a No. 1/0 AWG copper conductor with underground feeder (UF) insulation that will be used in a 120 V, two-wire circuit in an environment with an average ambient air temperature of no greater than 86°F (30°C). Neglect temperature and bundling correction factors.

19-22. Determine the maximum one-way distance two conductors can carry a current of 20 A on a 120 V, single-phase circuit based on a maximum voltage drop of 3%. Base your analysis on the following conductor sizes:

a. No. 14 AWG—copper

b. No. 14 AWG—aluminum

c. No. 12 AWG—copper

d. No. 12 AWG—aluminum

e. No. 8 AWG—copper

f. No. 8 AWG—aluminum

19-23. Determine the maximum distance two conductors can carry a current of 20 A on a 277 V, single-phase circuit based on a maximum voltage drop of 3%. Base your analysis on the following conductor sizes:

a. No. 14 AWG—copper

b. No. 14 AWG—aluminum

c. No. 12 AWG—copper

d. No. 12 AWG—aluminum

e. No. 8 AWG—copper

f. No. 8 AWG—aluminum

19-24. A set of conductors will feed power to a well station with pump having a load of 30 A. The building is 150 ft from the 120 V power source. Determine the required wire size for copper and aluminum conductors at a maximum voltage drop of 3%.

19-25. A set of conductors will feed power to an outbuilding with equipment having a total load of 12 000 VA. The building is 200 ft from the 120 V power source. Determine the required wire size for copper and aluminum conductors at a maximum voltage drop of 3%.

19-26. Determine the circuit rating for the following appliances or equipment on a 120/240 V circuit:

a. Household range

b. Trash compactor

c. Household clothes washer

d. Household clothes dryer (electric)

e. Central air conditioner (5 ton)

f. Eight luminaires with 4-32 W fluorescent lamps per luminaire (138 W including ballast)

19-27. Determine the circuit rating for the following appliances or equipment on a 277 V circuit:

a. Eight luminaires with 4-32 W fluorescent lamps per luminaire (138 W including ballast)

b. Sixteen luminaires with 4-32 W fluorescent lamps per luminaire (138 W including ballast)

c. Twenty luminaires with 4-32 W fluorescent lamps per luminaire (138 W including ballast)

19-28. Design the electrical service entrance system for the residence in Appendix D. Base your analysis on:

a. Water heater 3800 W

b. Clothes dryer 4400 W

c. Dishwasher 1000 W

d. Range 11 700 W

e. Air conditioner 9000 W

19-29. Design the electrical service entrance system for one of the apartments in Appendix A. Calculate the total load and service for the 38 ft × 28 ft (11.6 m × 8.55 m) apartment building. Base your analysis on:

a. Air conditioner, 2000 W

b. Electric range, 10 500 W

c. Water heater, 3500 W

d. Dishwasher, 1000 W

e. No clothes dryer

CHAPTER TWENTY

LIGHT AND ARCHITECTURAL
LIGHTING SYSTEMS

Lighting accounts for 20 to 25% of the electricity consumed in the United States and Canada and almost 20% of total global electricity consumption. Power generated for lighting creates pollution and greenhouse gas emissions, but it is crucial to the performance of everyday work and play activities. The fundamental reasons for providing light in a space are to make the objects in the space visible and to conduct activities that must take place in the space. Good architectural lighting provides the right quantity of light, with excellent color rendition and minimal glare. Quality lighting has been shown to improve productivity and enhance worker satisfaction.

Part of the light required in a building interior may come from natural lighting. Windows and skylights allow sunlight to come into the interior space. This natural light may be sufficient to provide all the light required during the day. Full reliance on natural lighting is complicated. In the northern hemisphere, as the sun moves, it shines first on the east portion of the building, and it moves toward the west as the day progresses. This means that windows facing the east are subjected to the strong rays of the sun in the morning, but must count on reflected light later in the day. Windows facing north get no direct sun, while windows facing south get a considerable amount. And on cloudy days, no direct sunlight is available. In effect, although light from windows and skylights can be counted on to provide some general light, natural light does not provide a controlled light source by which to perform activities. Therefore, in most projects, at least some electric lighting must be introduced to provide a constant, controlled amount of light so that the space can be used whenever necessary, whether sunny or cloudy, day or night.

This chapter introduces basic lighting terminology and the procedures of simple lighting design. This presentation highlights material that acquaints the technician with basic design principles.

20.1   A CHRONICLE OF ARTIFICIAL LIGHTING

Historical Perspective

Great significance has been placed on light throughout history. The foot-candle, a common measure of illuminance, is the oldest physical unit still in common usage. Acceptance of this unit has survived while use of other common units has become obsolete. Even in the Old Testament of the Bible, the Book of Genesis (1:3) begins with, "And God said, 'Let there be light,' and there was light."

For centuries, wood, candle wax, whale oil, coal oil, coal gas, and kerosene were used to illuminate building interiors. During the end of the Industrial Revolution (mid-1800s), much effort went into developing electricity and an electric lamp. However, well into the first part of the 20th century, artificial lighting in buildings continued to come from natural lighting and flame sources.

On December 21, 1879, Thomas Alva Edison announced the successful development of an incandescent lamp with a baked carbonized cotton thread filament. However, he was not the first to develop a working incandescent lamp. Sir Joseph Swan demonstrated use of a carbon-filament lamp on February 5 of that same year. Swan's lamp was similar to one demonstrated by German chemist Herman Sprengel in 1865. Edison, however, was the first to develop a commercially feasible electric light, a low-cost lamp that could remain lit for a long period of time. When speaking about his success, Edison said, "Electricity will make lighting so cheap that only the rich will be able to afford candles." At the same time, gas suppliers that provided natural gas for street and building lighting dismissed the effects of the development of Edison's electric light as a temporary passing craze. Flame sources continued to light the world until the equipment to produce and transmit electricity was commercially developed.

In 1882, the Edison Electric Light Company successfully demonstrated the use of artificial lighting by powering incandescent streetlights and lamps in approximately 30 buildings in part of London beginning on January 12 and later by illuminating parts of New York City beginning on September 4. Serving as a model for future utilities, the company later supplied power to Manhattan. The company eventually became known as the Consolidated Edison Company, the utility company that presently provides power to New York City. On August 26, 1895, the Niagara Falls Power Company became the first commercial utility to produce and transmit hydroelectric power. Eventually, communities switched from flame sources to electricity for lighting buildings and streets, principally because of safety and cost-effectiveness.

As electric lighting took a foothold, several advancements in lighting took place. In 1939, General Electric introduced fluorescent lighting, an electric light source that is more efficient than the incandescent lamp. The first high-pressure discharge lamp developed was the mercury vapor lamp, known by the bluish-greenish street lamps of the middle few decades of the 20th century. Other forms of gaseous discharge lighting were also developed about that time. After World War II, much of the developed world became dependent on artificial lighting.

Professional Organizations

In the past several decades, professional organizations developed technical standards, specifications and design techniques that govern the design and construction of building lighting and electrical systems. Some of the organizations related to building lighting and electrical systems include Illuminating Engineering Society of North America (IESNA), International Association of Lighting Designers (LALD), National Electrical Manufacturers Association (NEMA), and the American Lighting Association (ALA).

Standards and Codes

The *National Electrical Code (NEC)* is a model code that specifies the minimum provisions necessary for protecting people and property from the use of electricity and electrical equipment. It applies to both the manufacture and installation of electrical equipment. Most municipalities and counties require that residential and commercial electrical wiring conform to the NEC. In some jurisdictions, certain NEC requirements are superseded by local requirements. Lighting installations powered by electricity must comply with this code.

ASHRAE/IESNA 90.1 (American Society of Heating, Refrigeration and Air Conditioning Engineers, Inc./Illuminating Engineering Society of North America, Inc., Standard 90.1) addresses energy consumption in commercial and high-rise residential buildings. It sets minimum performance standards for building systems and components that have an impact on building energy consumption, including the building envelope, heating, ventilation and air-conditioning (HVAC) systems, and lighting. For lighting systems, the standard defines a number of basic requirements, including minimum number of lighting controls, ballast performance, and limits on installed lighting power. ASHRAE/IESNA 90.1 is incorporated in most codes.

20.2 ELEMENTS OF SEEING

Design of a good lighting system involves application of a blend between scientific principles, artistic skill, and design experience. *Optics* is that branch of physics that relates to the properties of light and the function of vision. It involves a study of the human visual system and how it interacts with light. The basic principles of vision and light as related to building lighting systems will be introduced. These principles serve as an introduction to design of building lighting systems.

The Visual System

The *visual system* of a human is composed of the eye, optic nerve, and certain parts of the brain. The eye is the organ that allows a human to sense light and produce electrical impulses. These electrical impulses are sent through the optic nerve to the brain. The brain is that part of the visual system where the impulses are processed. Each element of the visual system is needed to transform light stimuli into nerve excitations that allow a human to view an object.

The Eye

The eye functions much like a simple, very crude camera. Rays of light pass through the transparent *cornea* of the eye and through an opening called the *pupil*. As shown in Figure 20.1, the *iris* surrounds the pupil and adjusts for the amount of light available. It opens and closes to control the quantity of light the interior of the eye receives, much like the aperture on a camera opens and closes to limit the light to which the film is exposed. Behind the iris is the lens. The *lens* is a transparent ellipsoidal medium that changes thickness, allowing it to bend and focus the rays of light entering the interior of the eye. The lens thickens to focus rays of light from objects nearby and narrows to focus light from distant objects. The lens concentrates the rays of light on the retina, a membrane on the back of the eye.

The retina is composed of nerve cells with photoreceptors that are shaped like rods and cones. The rod-shaped photoreceptors sense extremely low levels of light and provide efficient vision in dim light. These photoreceptors do not discern color well, so dimly lit objects are perceived as being uncolored—that is, seen in shades of gray. The photoreceptors that are shaped like cones provide color vision and respond best to bright light. Higher levels of light are required for the eye to discern color. As rays of light strike the photoreceptors of the retina, they produce electrical impulses that travel from the nerve cells on the retina through the optic nerve to the brain. These impulses are processed in the brain and give the perception of seeing.

There are three different types of cone-shaped photoreceptors in the retina. Each type responds to one of the primary

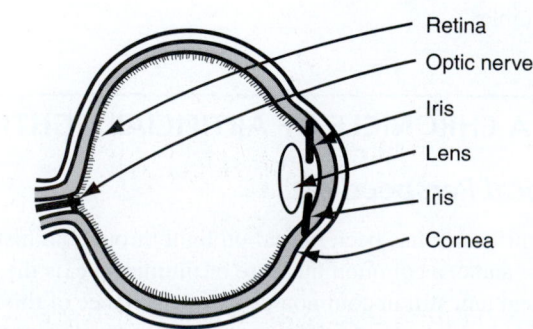

FIGURE 20.1 A cross-section view of the human eye.

colors of light: red, green, and blue. In individuals with normal vision, the electrical impulses from each cone blend together to create the sensation of other colors. However, about 5% of the population, mostly males, has defective color vision.

Color blindness is the inability to distinguish colors. The most common form of color blindness is found in those individuals who have difficulty distinguishing red from green. Individuals who are completely red–green color-blind see yellows and blues normally, but have trouble differentiating reds and greens. They tend to see reds and greens as yellow. Individuals who are totally color-blind see only black, white, and shades of gray.

20.3 LIGHT

Electromagnetic Radiation

Light is that form of electromagnetic radiation that allows the eye to see. In fundamental theory, *electromagnetic radiation* is energy radiated in the form of a wave caused by an electric field interacting with a magnetic field. It is the result of the acceleration of charged particles. Contemporary theory suggests that electromagnetic radiation also behaves as a group of particles called *photons*. Nevertheless, light tends to travel in a straight-line path unless influenced by a gravitational, magnetic, or some other force. Unimpeded, light travels at the speed of 186 000 miles per second (300 000 000 meters per second) in air or in a vacuum.

Electromagnetic radiation is categorized by wavelength and frequency. *Wavelength* (λ) is measured as the distance from one peak of one wave to the next wave. It is expressed in meters or nanometers (1 nm = 0.000 000 001 meter or one billionth of a meter). One inch contains about 25.4 million nanometers. *Frequency* is the number of wave cycles per second. It is expressed in units of hertz (Hz). TV and radio waves are several meters long with frequencies ranging from 10 kilohertz to 300 000 megahertz. In contrast, visible light waves are only about 0.000 000 5 m in length.

There is a very broad spectrum of radiant, electromagnetic energy, of which visible light is but one narrow band. As outlined in Figure 20.2, types of electromagnetic radiation are TV and radio waves, microwaves, infrared radiation, visible light, ultraviolet radiation, x-rays, gamma radiation, and cosmic radiation.

Wavelength and frequency of electromagnetic radiation are dependent on the source emitting the radiation. Large antennas such as those found on the tops of buildings and mountains must be used to generate radio and TV waves. Gamma radiation, the shortest and most powerful form of electromagnetic radiation, result from changes within the nucleus of the atom.

The sun and electric lamps are light sources because they transform energy to the visible electromagnetic wavelengths that can be perceived by the human eye. Most light sources emit electromagnetic radiation composed of different wavelengths of light. Sunlight striking the earth's outer atmosphere is made up of ultraviolet (about 5%), visible light (about 45%),

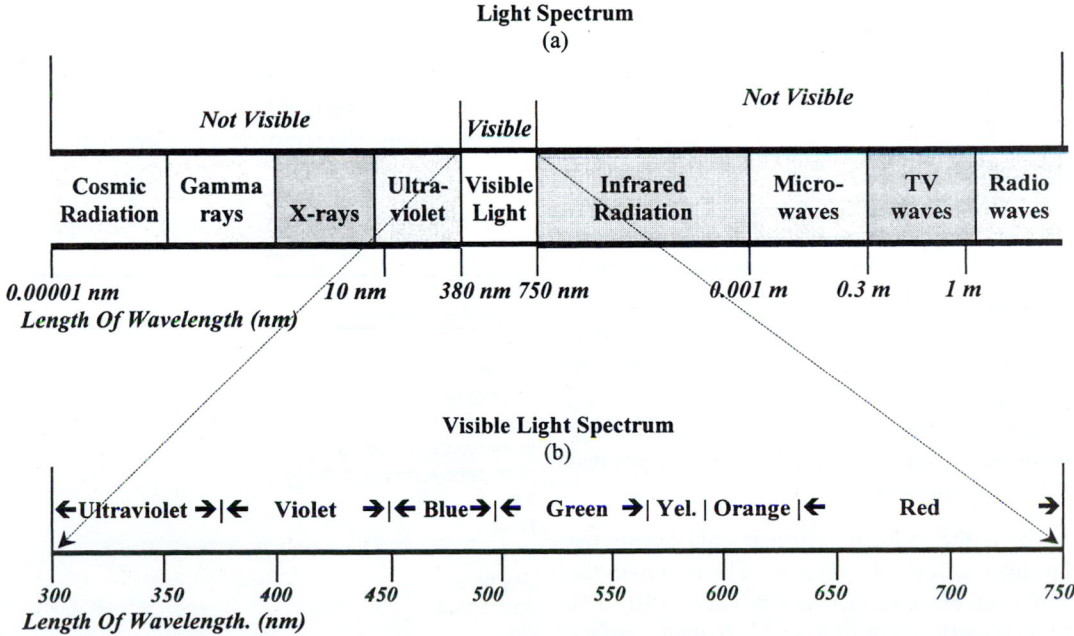

FIGURE 20.2 Classifications of electromagnetic radiation spectrum are grouped by wavelength. Shown are the (a) full light spectrum and the (b) visible light spectrum that is part of the light spectrum. One nanometer (nm) is 1/1 000 000 000 meter (m).

TABLE 20.1 WAVELENGTH RANGES OF VISIBLE LIGHT BY COLOR. VISIBLE LIGHT IS THE WAVELENGTHS OF ELECTROMAGNETIC RADIATION SPECTRUM TO WHICH THE HUMAN EYE IS SENSITIVE.

Color	Wavelength (nm)
Violet	380 to 450
Blue	450 to 490
Green	490 to 570
Yellow	570 to 590
Orange	590 to 630
Red	630 to 770

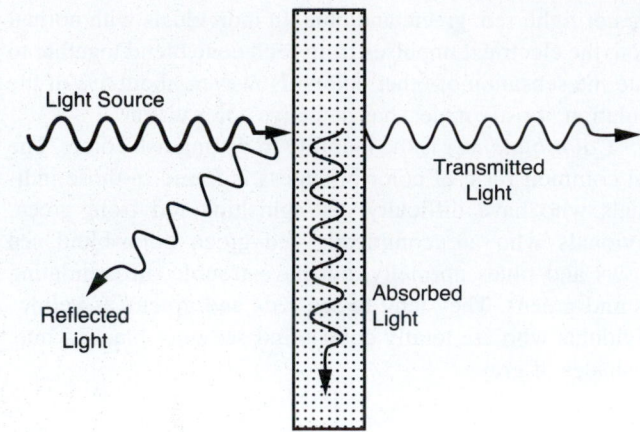

FIGURE 20.3 Light striking an object will be reflected, absorbed, or transmitted.

and infrared (about 50%) radiation. Ordinary incandescent lamps also emit radiation of different wavelengths in varying magnitudes, principally in the visible light and infrared sectors of the electromagnetic spectrum.

Visible Light

Visible light is that part of electromagnetic radiation spectrum capable of exciting the retina and ultimately producing a visual sensation; it is the wavelengths of electromagnetic radiation to which the human eye is sensitive. Wavelengths of visible light range from about 380 nm to about 750 nm. Beyond this range is darkness. Although the eye may be exposed to many other wavelengths of radiant energy, they are not capable of initiating responses in the eye. Much like a human ear cannot hear the high-frequency sound of a dog whistle, a human eye cannot see radiation that exists outside of the visible light portion of the electromagnetic spectrum. See Table 20.1.

Behavior of Light

When light strikes a surface, the surface is illuminated. *Illuminance* is the amount of light incident on (striking) a surface. Although an object may be illuminated, the eye cannot see it without visible light leaving (reflecting off) the object in the direction of the viewer. *Luminance* is the amount of light leaving an object. It is how bright an object appears. The terms *illuminance* and *luminance* are further defined later in this chapter. The visual system perceives luminance, not illuminance; that is, the eye sees visible light leaving the object, not the light arriving at (incident on) the object.

When a surface is illuminated, the illuminance striking the surface can be reflected from, absorbed by, and transmitted through the body (see Figure 20.3).

Reflectance (ρ) is the ratio of reflected light versus the light striking the surface (illuminance). For example, the reflectance of a dull black surface may be about 0.10 (10% is reflected) while reflectance of a polished, white surface may be 0.85 (85% is reflected). *Specular reflection* occurs when light is reflected off a polished or mirror-like surface—that is, the reflected image is maintained. See

Table 20.2. *Diffuse reflection* results when reflected light is scattered after striking the surface. See Figure 20.4.

Transmittance (τ) is the ratio of light transmitted through the body versus the light illuminating the surface. When light strikes ordinary window glass at an angle perpendicular to the surfaces, the transmittance is about 0.90—that is 90% of the light striking the glass is transmitted. A *transparent body* transmits light through it without distorting the image. A *translucent body* transmits light but obscures the image because the light is scattered. Ordinary window glass is an example of a transparent medium while frosted glass is translucent.

Absorptance (α) is the ratio of the light absorbed versus the light striking the surface. For example, the absorptance of a dark black surface may be about 0.90 (90% is absorbed) while absorptance of a white surface may be 0.05 (% is absorbed). Absorbed light manifests itself as energy by raising the temperature of the body receiving the light.

TABLE 20.2 REFLECTANCES (ρ) OF COMMON COLORED SURFACES

Color of Surface	Reflectance (ρ)
White	0.80 to 0.85
Light gray	0.45 to 0.70
Dark gray	0.20 to 0.25
Ivory (yellowish white)	0.60 to 0.70
Buff (light yellowish brown)	0.40 to 0.70
Tan	0.30 to 0.50
Brown	0.20 to 0.40
Pink	0.50 to 0.70
Sky blue	0.35 to 0.40
Azure blue	0.50 to 0.60
Red	0.20 to 0.40
Green	0.25 to 0.50
Olive	0.20 to 0.30
Black	0.10 to 0.00

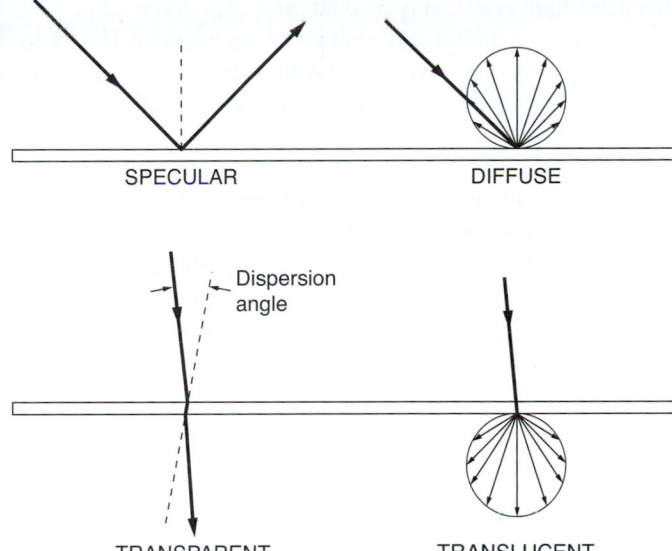

FIGURE 20.4 Reflected light can be specular or diffuse. Light that is transmitted through the glass will behave in a transparent or translucent manner.

When a *translucent* or *transparent* body is illuminated, the illuminance striking the surface will be reflected, absorbed, and transmitted. Therefore:

Reflectance (ρ) + Absorbtance (α) + Transmittance (τ) = 1.0

When light strikes an *opaque* surface, the illuminance will be reflected and absorbed but not transmitted. Therefore:

Reflectance (ρ) + Absorbtance (α) = 1.0

The type of light striking the surface, the properties of the surface receiving it, and the angle at which the light is received dictates the amount of reflectance, transmittance, and absorptance. A shinny or polished surface will tend to have higher reflectance; a light-colored surface will have greater reflectance than a dark surface; and an increase in incident angle (measured from perpendicular to the surface) will tend to increase reflection.

A light source can produce diffuse or direct light. *Diffuse light* is light that is widely spread or scattered. It is the type of light available on a hazy day. *Direct light* is a strong, directional type of light. It is the type available on a cloudless, sunny day where shadows are well defined.

20.4   THE COLOR OF LIGHT

Color Perception

Color perception is the ability to distinguish and interpret different wavelengths of visible light. Artificial and natural (daylight) light sources are composed of many wavelengths having varying magnitudes. An opaque object will selectively absorb different proportions of each of these wavelengths and reflect the remainder toward the viewer. So, an object that absorbs all of the longer wavelengths and reflects the shorter wavelengths will be perceived as violet or blue. The perceived color of an object is dependent on the wavelengths emitted by the light source, the wavelengths illuminating the object, and the selective reflectance of the object's surface.

A light source producing a blend of wavelengths that are evenly distributed across the light spectrum is perceived by the human eye as white or normal light. This light is actually made up of all wavelengths of visible light. It contains the individual red, green, and blue wavelengths that make up the *primary colors of light*. The primary colors of light are additive; they can be used in various combinations to produce any other color by adding color. When they are combined together, primary colors of light produce white. The *colors of pigments* are magenta (purplish red), cyan (greenish blue), and yellow. They are opposite of the colors of light. Colors of pigments are subtractive; they can be used in various combinations to produce any other color by subtracting a color. When the colors of pigment are combined together, they produce black.

The perceived color of an object depends on the dominant wavelengths that luminate from it. In normal light, a surface that reflects all wavelengths appears bright white while a surface that absorbs all wavelengths appears black. A colored surface selectively reflects only the wavelengths that make up its own color. A blue surface reflects only those wavelengths that are predominantly in the blue portion of the electromagnetic spectrum. A yellow object reflects only yellow wavelengths. A red object reflects only red. Any other wavelengths are either absorbed or transmitted.

A *chromatic* light source emits a fairly even distribution of all wavelengths of light. Most light sources are not fully chromatic. A *monochromatic* light source produces visible light in a very small range of wavelengths: a red light emits predominantly red wavelengths; a green light gives off predominantly green wavelengths; and so on. Type of wavelengths emitted by a light source will influence color. A yellow light renders even the most vibrant colors in shades of yellow and orange. So, an object seen as white in normal light, when illuminated by a yellow light source, will be perceived as a yellow.

When a monochromatic light source illuminates a surface, only the color emitted by the light source will luminate. All other colors are seen as shades of gray. For example, a multicolored object when illuminated by a yellow light will be perceived as shades of yellow or gray. Those colors that do not contain some yellow will not reflect yellow light. They are seen as gray. This principle is evident when observing objects illuminated by yellow lamps commonly used to light street intersections at night. These monochromatic lamps emit most of their energy as wavelengths in the yellow portion of the electromagnetic radiation spectrum. The yellow light makes white and surfaces containing some yellow (yellows, oranges, and browns) appear yellow. Surfaces of other colors appear gray.

Color Quality

A light source that produces a blend of wavelengths evenly distributed across the light spectrum is perceived as white light. White light renders all colors—that is, a multicolored surface illuminated by a white light will render all colors evenly. Generally, a good quality light source renders all colors uniformly. A light source of poor color quality renders the different colors unevenly.

Light sources do not produce an even distribution of wavelengths across the light spectrum. Some light sources render all colors better than others. Ordinary incandescent lamps found in homes are generally strong in yellow, orange, and red wavelengths and somewhat deficient in green, blue, and violet wavelengths. The color emitted by fluorescent lamps used for general lighting in most offices and classrooms typically varies by type of lamp, but they are generally strong in red, green, and blue wavelengths. High-intensity lamps commonly used to light remote street intersections emit mostly yellow wavelengths. Even sunlight may pass through a reflective-coated window glass that removes some of the wavelengths in the light spectrum.

Two methods are used to rate the quality of color emitted by a light source: the color temperature and color rendering index. A third technique involves review of the spectral power distribution curve. This is a method of appraising the colors emitted by a light source by evaluating the power emitted in small wavelength groups across the light spectrum. These methods are introduced later in the chapter.

Visual Acuity

Visual acuity is the ability to distinguish fine details. It is the "keenness" of vision that is necessary to perform tasks such as reading, writing, drafting, sewing, or surgery. Technically, visual acuity is measured as the ability to identify black symbols (letters) on a white background at a specific distance as the size of the symbols is varied. A person's visual acuity depends on the function of the eye (the retina) and the interpretative ability of the brain. Acuity of vision is also influenced by many external factors, such as black–white (or color) contrast between an object and its background, size of the object being viewed, the amount of light leaving the object in the direction of the viewer (luminance), time available for seeing, and variations in the seeing capacity of the viewer (e.g., age, disability, and so forth).

The inability to distinguish the details of an object becomes more pronounced as the object decreases in size as seen by the viewer. In Figure 20.5, the smaller text is more difficult to see when viewed at a typical reading distance. All of the text would be difficult to see if it was viewed from across the room.

Contrast between the object being viewed and the background influences the eye's ability to distinguish fine detail. As shown in Figures 20.5, text is more difficult to see as contrast between the text and the background decreases. Black text on a white background is easy to see because black ($\rho = 0.10$) and white ($\rho = 0.80$) have vastly different reflectances. On the other hand, light gray text ($\rho = 0.60$) on a dark gray background ($\rho = 0.40$) have similar reflectances so the object is harder to see. A similar effect is experienced with color contrast (e.g., blue text on a purple background is more difficult to see than blue text on a yellow background).

The speed and accuracy with which a task is performed is related to the amount of light available; more light means higher seeing accuracy and efficiency. Surgery, sewing, and drafting require higher lighting levels than reading and writing because there is more detail required. The visual system tends to deteriorate with age because of yellowing of the lens and clouding of the fluid in the eye. Such deterioration necessitates additional illumination for an older person to perform a specific seeing task.

20.5 CHARACTERISTICS OF ARTIFICIAL LIGHTING

Lamps, Luminaires, and Light Fixtures

In the lighting industry, a *lamp* is a device that generates light. Although frequently called a *light bulb* by the layperson, the term *lamp* (rather than *bulb*) is used in the lighting industry to avoid confusion when describing a light source. Simply, a bulb is the glass portion of a lamp that encloses and protects the working parts of the light source, whereas a lamp is a source of light.

A variety of lamps are available commercially. Lamps fit in two broad categories: incandescent and gaseous discharge. Gaseous discharge lamps can be further divided into two subcategories: low-pressure lamps and high-pressure or high-intensity discharge (HID) lamps. The group of light sources that are low-pressure discharge sources consists of the fluorescent lamps and the low-pressure sodium lamps. High-pressure or HID lamps include mercury vapor, metal halide, and high-pressure sodium lamps. Lamps are rated by efficacy, life and color temperature, and rendition.

In buildings, a *luminaire* is a complete lighting unit, which consists of a lamp (or lamps), lamp socket(s), any lenses, refractors, or louvers, any ballast (or ballasts), and the housing. Luminaires include fluorescent troffers, downlights, spot or accent lighting, task lighting, and outdoor area and flood lighting. A *light fixture* is the luminaire less the lamp(s). It includes the structural parts of a luminaire including any lenses, mounting supports, wiring, and ballasts, but does not include the lamps. In street and road lighting, the term *lantern* is also customarily used to describe a luminaire.

Light Output

Luminous flux is the measure of the perceived power of light expressed in *lumens (lm)*. It is adjusted from the total power of radiant energy emitted by a light source to correct for differences in sensitivity of the human eye to different wavelengths

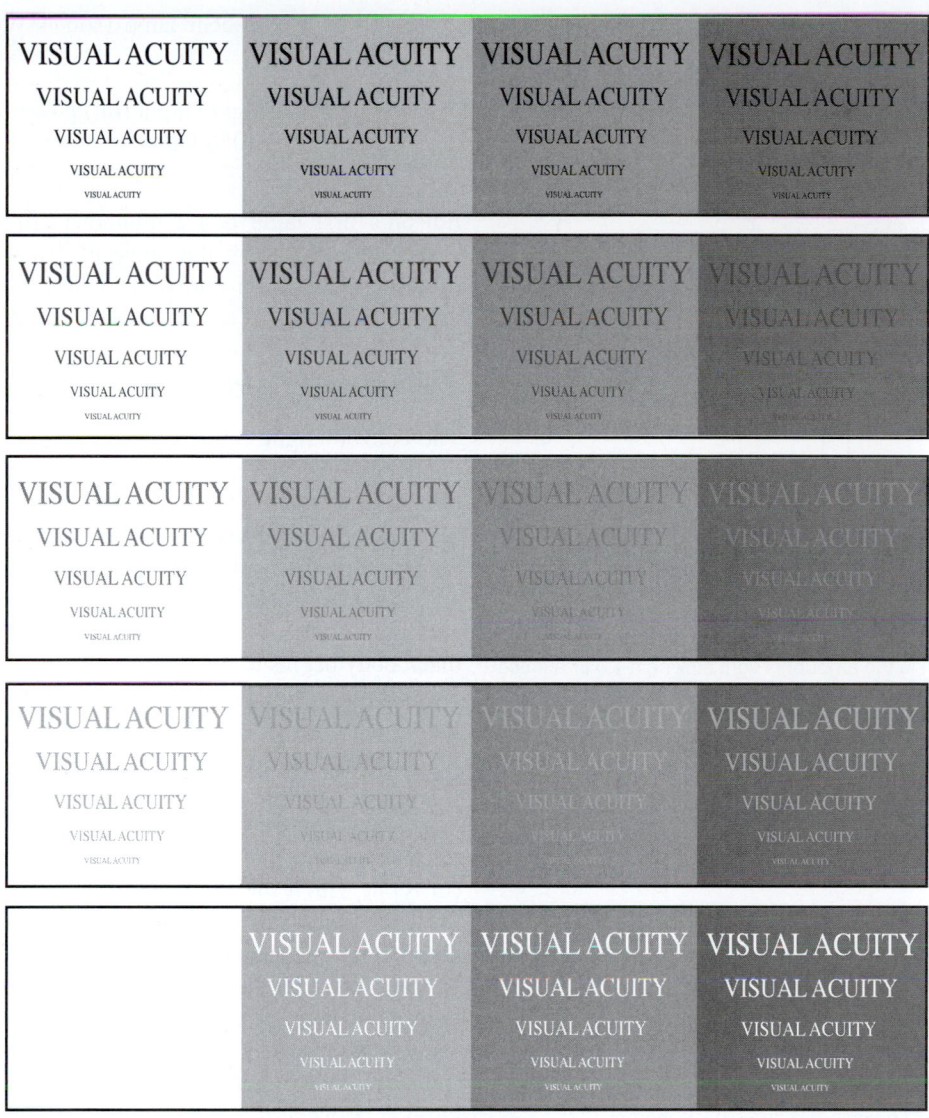

FIGURE 20.5 Contrast between an object and its background and size of the object influence visual acuity.

of light. Technically, a lumen is defined as the quantity of light given out through a steradian by a source of one candela of intensity radiating equally in all directions. The steradian (sr) is the SI unit of solid angular measure, where there are 4π or about 12.5664 steradians in a complete sphere—about 7.96% of the area of the sphere. The lumen is most commonly used for measurement of total output of a light source. A 100 W soft white incandescent lamp and a 23 W compact fluorescent lamp emit roughly 1500 lumens.

A *candela (cd)* is the SI unit of luminous intensity—that is, the power emitted by a light source in a particular direction. One candela is one lumen per steradian. It differs from the lumen in that it is a measure of light flux in a specific direction. Technically, a candela is the luminous intensity, in a given direction, of a source that emits a monochromatic radiation

with a frequency of 540×10^{12} hertz and that has a radiant intensity in that direction of 1/683 watt per steradia, which describes how to produce a light source that emits one candela. Originally, the unit of luminous intensity was expressed in terms of units called *candles*. As a result, the candela is sometimes improperly called "candlepower," which was formerly defined as the light emitted from a wax candle.

The candela is roughly the amount of light generated by a single ordinary dinner candle. As shown in Figure 20.6, a lumen is roughly equal to the amount of candlelight striking a curved 1.0 ft² (or 1 m²) surface that at all points is held 1 ft (or 1 m) away from the flame of a single candle emitting one candela of illumination. Another way to understand the terms *lumens* and *candela* is to imagine that a uniform point source of one candela (e.g., a candle) is at the center of a sphere that has

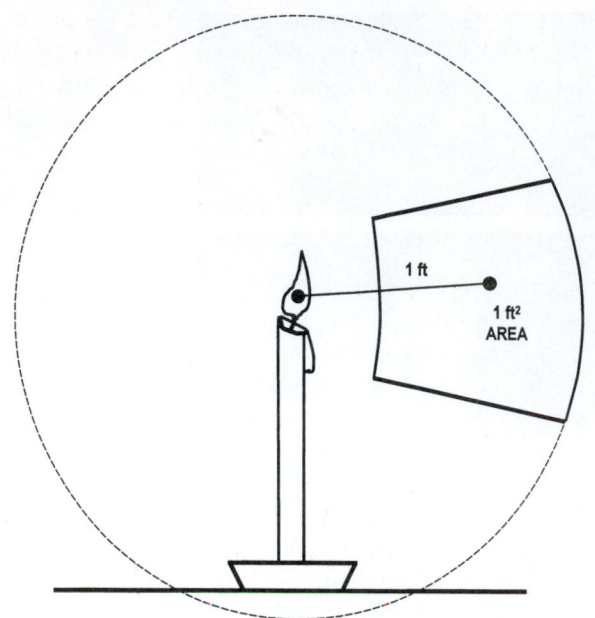

FIGURE 20.6 A lumen is the amount of light striking a 1.0 ft² surface that at all points is held one foot away from the flame of a standard candle (roughly one candela of intensity).

a radius of 1 ft (or m). If the sphere had an opening in the surface that is a 1-ft² (or 1-m²) area, the quantity of light that passes through the opening is a lumen. The sphere has a total surface area of 12.5664 ft² (or 12.5664 m²). Because, by definition, a lumen flows to each square foot (or square meter) of the spherical surface area, a uniform point source of one candela produces 12.57 lumens. An ordinary wax dinner candle (roughly one candela) emits about 12 to 13 lumens.

Luminous Efficacy

Luminous efficacy is the ratio of the light output of a light source (a specific lamp) to the electrical energy consumed (including the ballast if applicable) to produce that light source. It is expressed in units of lumens per watt of electric power (lm/W) and is thus typically referred to as LPW.

$$LPW = \text{light output (lm)/power input (W)}$$

Luminous efficacy is a good measure of the efficiency of a *bare* lamp. It will decrease over the life of the lamp because lamp output declines over time.

Example 20.1

Determine the luminous efficacy of the following lamps:

a. A 40 W incandescent lamp that emits 505 lm of light

$$LPW = \text{light output (lm)/power input (W)}$$
$$= 505 \text{ lm}/40 \text{ W} = 12.6 \text{ lm/W}$$

b. A 40 W fluorescent lamp (and 4 W ballast) that emits 3050 lm of light

$$LPW = \text{light output (lm)/power input (W)}$$
$$= 3050 \text{ lm}/44 \text{ W} = 69.3 \text{ lm/W}$$

Edison's first lamp had a luminous efficiency of 1.4 lm/W. Today's lamps fall in the range of 10 (incandescent) to 180 (low-pressure sodium) lm/W. The theoretical maximum for luminous efficiency of a white light is 225 lm/W.

Correlated Color Temperature

The color of light emitted by a light source correlates to the temperature of the emitting source. In physics, a black body is a theoretically ideal body that perfectly radiates energy. As a black body is heated well above room temperature, it begins to incandesce—that is, it emits visible light as a result of heating. As a black body gets hotter, it turns red, orange, yellow, white, and finally blue. At a temperature of about 1000°F (800 K), a black body begins to glow with red light; at 5000°F (3000 K) it emits bright yellow light; at 8500°F (5000 K) it gives off white light; and at about 14 000°F (8000 K) the black body emits a soft blue light.

The color of a lamp can be rated by its *correlated color temperature (CCT)* in degrees Kelvin (K), the SI unit of temperature. The CCT of a light source is the temperature (in K) at which the heated black body matches the color of the light source in question. A color temperature less than about 3500 K provides *warm lighting* by emitting mostly yellow and red wavelengths. A color temperature greater than about 5000 K provides *cool lighting* because it emits light predominantly in the white and blue wavelengths. Neutral *daylighting* occurs when the color temperature of the lamp is between temperatures of 4500 and 6500 K, with the lower side of the range warmer and the upper side cooler. By comparison, a candle flame is about 1800 K and noon sunlight is about 5500 K. A warm light is produced by a light source that has a lower temperature. Typical color temperatures for commercially available lights are 2800 K (incandescent), 3000 K (halogen), 4100 K (cool white fluorescent), and 5000 K (daylight-simulating fluorescent). Table 20.3 contains approximate color temperatures of various theoretical (black body), natural, and commercially available sources of light.

Technically, color temperature only applies directly to light sources that incandesce (emit light from heating) such as the ordinary incandescent lamp. Gaseous discharge lamps such as fluorescent, mercury vapor, metal halide, and high- and low-pressure sodium lamps do not incandesce. They are, however, rated by their *apparent* or *correlated color temperature*. Much of the work in the lighting industry has been directed toward developing cost-efficient lamps that emit an even distribution of radiant power.

TABLE 20.3 THE APPROXIMATE COLOR TEMPERATURE, IN DEGREES KELVIN (K), OF VARIOUS THEORETICAL (BLACK BODY), NATURAL, AND ARTIFICIAL LIGHT SOURCES.

Light Source	Color Temperature (K)
Dull red glow (black body)	800
Low-pressure sodium lamp*	1750
Candle flame/sunlight at sunrise	1800
High-pressure sodium lamp*	2100
Tungsten filament incandescent lamp	2600 to 3000
Bright yellow glow (black body)	3000
Fluorescent, mercury vapor, metal halide lamps*	3000 to 5000
"Warm white" fluorescent	3000
"White" fluorescent	3500
"Cool white" fluorescent	4100
"Daylight" incandescent lamp	4700
White glow (black body)	5000
Noon sunlight	5500
LED (light-emitting diode)	6000
"Daylight" fluorescent	7000
Overcast sky	7000
Light blue glow (black body)	8000
Clear blue sky	10 000
Brilliant blue glow (black body)	60 000

*Apparent or correlated color temperature.

Color Rendering Index

The *color rendering index (CRI)* is a method of numerically comparing the color distribution of a light source to a reference lamp. The reference lamp emits light over the full visible light spectrum at the same color temperature as the light source being tested. Light emitted by the test lamp is compared to light emitted by the reference lamp at eight specific colors. A deficiency in lumination of one or more of these specific colors decreases the CRI of the test lamp.

Approximate CRI values for different commercially available lamps are provided in Table 20.4. In theory; a lamp with a higher CRI more closely emulates the reference lamp and provides a better color rendition. A lamp with good color distribution

TABLE 20.4 APPROXIMATE COLOR RENDERING INDEX (CRI) FOR DIFFERENT COMMERCIALLY AVAILABLE LAMPS.

Lamp Type	Approximate CRI
Natural light	100
Incandescent	95 to 99
Fluorescent	22 to 92
Mercury vapor	15 to 50
Metal halide	65 to 75
High-pressure sodium	20 to 25*
Low-pressure sodium	0

*One color-enhancing type has achieved a CRI of 70.

has a CRI of at least 80. However, the principal drawback of the CRI is that it does not specify deficiencies in a specific color; for example, a lamp may have a relatively high CRI but be missing a color. Also, it is meaningless to compare the CRI of lamps with significantly different color temperatures.

Spectral Power Distribution Data

A *spectral power distribution curve* is a graphic presentation of the quantities of light emitted by a lamp by wavelength component. These curves are a plot of radiant power versus specific wavelength across the light spectrum. They provide a visual fingerprint of the color characteristics of the light emitted by the source over the visible part of the spectrum. Spectral power distribution curves for specific lamps are available from the manufacturer. Figure 20.7 shows spectral power distribution curves of various commercially available lamps.

A spectral power distribution curve of an ordinary household incandescent lamp will typically show a large increase in radiant power from 380 nm to 780 nm; that is, light emitted by the source is weak in violet wavelengths (380 to 450 nm), it increases through blue, green, yellow, and orange wavelengths, and very strong in red wavelengths (630 to 750 nm). This type of lamp illuminates yellow-, orange-, and red-colored objects very well. Violets and blues are not illuminated as well. In contrast, the monochromatic low-pressure sodium lamps that are frequently used to illuminate street intersections typically have a spectral power distribution curve that is concentrated between 589 and 589.6 nm, the yellow wavelengths. Light is not emitted at other wavelengths. These lamps produce a spectral power distribution curve that peaks between 589 and 589.6 nm and is essentially nonexistent at all other wavelengths. This lamp renders all surfaces in an orange-yellow version of blacks and white.

Lamp Life

The *rated life* of a lamp is its median life expectancy, expressed in hours. It is the cumulative time that lapses before 50% of a representative group of lamps has failed and the other lamps in the group continue operating. For example, a 60 W soft white incandescent lamp can be expected, on average, to burn for 1000 hr. Based on continuous testing of lamps in laboratories, the 1000-hr rating is the point in time when 50% of the test samples have burned out and 50% are still burning.

In other cases, rated life is difficult to determine because these lamp types will continue to emit light but get dimmer and dimmer with time—that is, they do not burn out. In such cases, the manufacturer normally arrives at the life by determining the most economical point for replacement in the life of the lamp. Information on rated lamp life is available from the lamp manufacturer.

Lamp life generally decreases each time a lamp is switched on and off. A continuously lit lamp will normally last

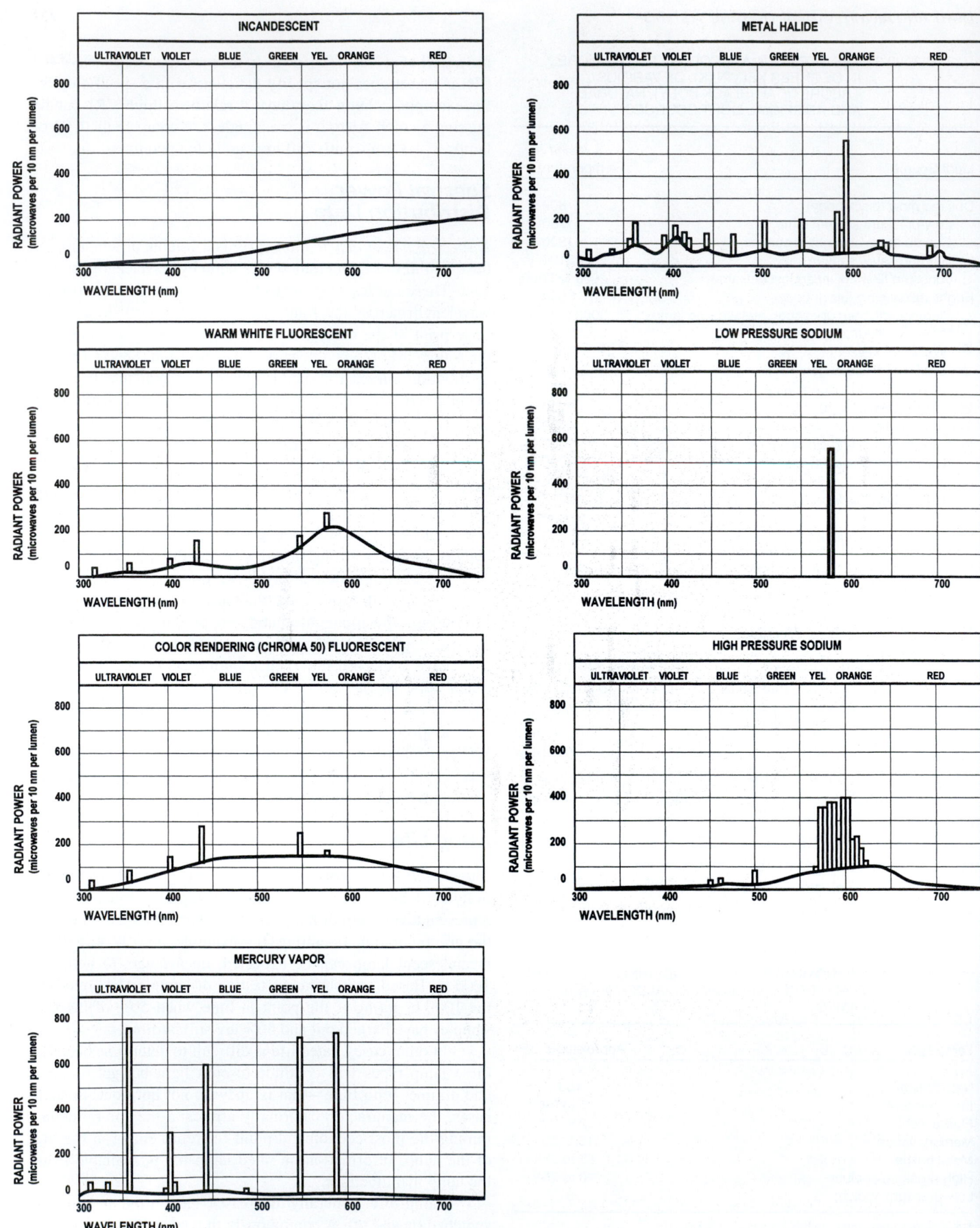

FIGURE 20.7 Spectral power distribution curves of various artificial light sources.

much longer than one that is switched on and off regularly. Because of energy costs, this reduction in life does *not* warrant leaving lights on in an unoccupied space. It does, however, influence lamp life test results. Testing is typically performed in 3-hr start intervals for fluorescent lamps and 10-hr start intervals for most other lamps.

Life of incandescent lamps will be affected by the voltage they operate at versus the voltage for which they were designed; a slightly lower voltage means increased life because of cooler operation but this reduces light (lumen) output. Life of fluorescent and HID is greatly affected by type of ballast. Lamp and ballast must be matched to ensure proper operation.

Example 20.2

A lamp manufacturer's specifications for a 40 W fluorescent lamp indicates an output of 2680 lm, a rated life of 20 000 hr and a cost of $6.00 per lamp. Data for a 150 W incandescent lamp indicates an output of 2850 lm, a rated life of 750 hr, and a cost of $2.00 per lamp. When in use, it is anticipated that these lamps will be operated 10 hr per day, 220 days per year.

a. For each lamp, approximate the time period that elapses before replacement is necessary.

150 W incandescent lamp:

750 hr/(220 days/year · 10 hr/day)
= 0.34 years (4.1 months)

40 W fluorescent lamp:

20 000 hr/(220 days/year · 10 hr/day)
= 9.1 years (110 months)

b. Approximate the cost of each type of lamp over a 10-year period, assuming use and lamp costs remain constant.

150 W incandescent lamp:

10 years/0.34 years per lamp = 29.4 lamps
29.4 lamps over 10 years @ $2.00 per lamp = $58.80

40 W fluorescent lamp:

10 years/9.1 years per lamp = 1.1 lamps
1.1 lamps over 10 years @ $6.00 per lamp = $6.60

Depreciation of Lamp Output

As lamps operate the luminous flux (lumen output) will decrease. As incandescent lamps operate their light output degrade to about 85% of lamp's initial lumen output. With fluorescent and HID lamps, lumen depreciation will vary by lamp type, size, and use. Degradation to 60 to 80% of the lamp's initial lumen output is common at the end of a fluorescent lamp's rated life. Degradation to 50 to 70% of the initial lumen output is typical at the end of an HID lamp's rated life.

Lumen depreciation will also be affected by dirt and dust in the environment: Lamps in soiled or dusty industrial and outdoor environments will become covered with dust or a soiled coating, so lamp output will depreciate further. Usually planned maintenance programs are established to replace lamps at the end of their useful life to maintain proper illumination levels. This process is generally called *relamping*.

20.6 TYPES OF ARTIFICIAL LIGHT SOURCES

The first decision in the design of a good lighting system is the choice of light source. Several light sources are available, each with a unique combination of properties and operating characteristics. Although there are hundreds of commercially available lamps, they can be categorized into three groups by their construction and operating characteristics: incandescent, fluorescent, and HID. HID lamps can be grouped into four major classes: high-pressure sodium, metal halide, mercury, and low-pressure sodium.

Incandescent Lamps

Incandescent lamps (I) emit visible light as a result of heating; they incandesce. They are the most familiar source of light and are known as a "light bulb" by the lay consumer. Incandescent lamps offer advantages of low lamp cost, reliability, familiarity, and good color rendition, but they have a short rated life and poor efficacy (LPW). They are ideal for applications needing excellent color rendition such as retail, furniture, clothing and grocery stores, hair and beauty salons, restaurants, and art studios.

The basic design of the incandescent lamp has essentially remained unchanged since Thomas Edison introduced a commercially successful lamp in the late 1890s. A thin filament is supported within an evacuated, gas-filled glass bulb. The filament is attached to conductors (wires) that are connected to a base. The base attaches to the glass bulb and is designed to screw or plug into a standard socket to connect the lamp to an electrical circuit.

When electricity flows through the filament of an incandescent lamp, the filament provides resistance against electrical current flow causing it to heat up and incandesce. The filament temperature rapidly rises until it reaches temperature equilibrium—that is, it loses heat at the same rate that heat is being generated. Design of incandescent lamps is a balance between efficacy and lamp life: The hotter the filament, the more efficient the lamp is in converting electricity into light and the whiter the light. Hotter operation does, however, reduce lamp life. As a result, incandescent lamps of the same wattage can have different outputs and rated lives.

Incandescent filaments can reach temperatures of between 4600 and 6000°F (2300 to 3000 K). At these temperatures, the electromagnetic radiation from the filament includes a significant amount of visible light, mainly in red, orange, and

TABLE 20.5   LAMP DESIGNATIONS OF SELECTED BULBS.

Lamp Designation	Description of Bulb Shape
A	Standard light bulb shape
ER	Ellipsoidal reflector
IR	Infrared reflecting
MR	Multifaceted reflector
PAR	Parabolic aluminized reflector
PS	Pear-shaped bulb
R	Reflector
T	Slender, tube-shaped bulb

Incandescent lamps have wattages of between 3 to 1000 W and voltages of 6 to 277 V; the 120, 125, and 130 V lamps are the most common. *Shape* refers to the lamp bulb profile, as shown in Table 20.5 and Figure 20.8. Lamps are also designated with a number digit that indicates tube diameter in eighths of an inch. An R30 is an internal reflector lamp that is 30/8ths or 3¾ inches in diameter whereas an A19 is an ordinary pear-shaped lamp that is 19/8ths of an inch or 2⅜ inches in diameter.

Threaded or pinned bases hold the lamp in the socket and attach the lamp to the electrical circuit: The *medium base* is the familiar threaded base found with ordinary lamps up to 300 W; the *candelabra base* is a smaller threaded base used in ornamental lighting such as chandeliers; and the *mogul base* is a larger threaded base found on lamps that are 300 W or greater.

Special use incandescent lamps are also available. *Long-life lamps* are used in applications where the difficulty or cost of changing lamps is prohibitive. However, a more economical solution may be to use lamps rated for a slightly higher voltage, but this reduces light (lumen) output. *Rough service lamps* are special versions of the standard lamps that are helpful for rough or vibration service. They are designed with extra filament supports to withstand bumps, shocks, and vibrations with some loss in lumen output. They are used in applications where there is excessive vibration such as portable lighting. In applications where lamps are broken fairly often, there are special *plastic-coated lamps* available. These lamps are not unbreakable but they break less frequently and, when they break, the shattered

yellow wavelengths. Most of the radiation emitted from an incandescent lamp is infrared radiation. Thus, efficacy of an incandescent lamp is relatively low. The lamp's warm yellow glow has been accepted as the standard for indoor lighting.

As an incandescent lamp burns, the filament evaporates and deposits on the inner surface of the glass bulb. Deterioration of the filament also occurs over time. These effects reduce light output to about 85% of lamp's initial output. Basic incandescent lamp design has been improved by using tungsten filaments, which evaporate more slowly than the carbon filament used in the early lamps. The filament can be operated at higher temperatures with chemically inert fill gases such as argon, krypton, or xenon.

Standard incandescent lamps come in an assortment of wattages, voltages, shapes, and types of bases.

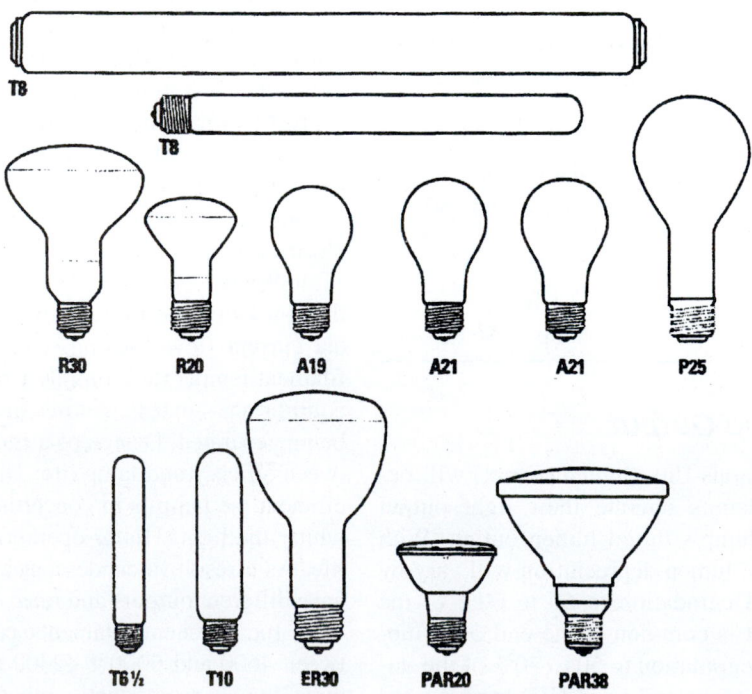

FIGURE 20.8   Selected incandescent lamp shapes.

TABLE 20.6   LAMP CHARACTERISTICS FOR SELECTED INCANDESCENT LAMPS.
COMPILED FROM VARIOUS INDUSTRY SOURCES.

Input	Bulb Shape	Base	Description	Rated Life	Output	Efficacy (LPW)	Color Rendition
W				hr	lm	lm/W	CRI
7.5	S11	Medium	White	1400	39	5.2	95+
10	S14	Medium	Clear	1500	77	7.7	95+
15	S15	Medium	Clear	3000	110	7.3	95+
25	A19	Medium	Clear	2500	215	8.6	95+
40	A19	Medium	Standard	1000	505	12.6	95+
	A19	Medium	Clear	1500	480	12.0	95+
50	A21	Medium	Inside frost	1000	875	17.5	95+
60	A19	Medium	Standard	1000	865	14.4	95+
	A19	Medium	Clear	1000	870	14.5	95+
	A19	Medium	Long life	1500	820	13.7	95+
75	A19	Medium	Standard	750	1190	15.9	95+
	A19	Medium	Clear	750	1220	16.3	95+
	A19	Medium	Long life	1125	1125	15.0	95+
100	A19	Medium	Standard	750	1730	17.3	95+
	A19	Medium	Clear	750	1750	17.5	95+
	A19	Medium	Long life	1125	1600	16.0	95+
150	A21	Medium	Standard	750	2850	19.0	95+
	A21	Medium	Clear	750	2850	19.0	95+
200	A21	Medium	Standard	750	3910	19.6	95+
	A21	Medium	Clear	750	4010	20.1	95+
300	PS25	Medium	Clear	750	6360	21.2	95+
500	PS35	Mogul	Clear	1000	10 850	21.7	95+
750	PS52	Mogul	Clear	1000	17 040	22.7	95+
1000	PS52	Mogul	Clear	1000	23 740	23.7	95+

glass is contained within the plastic coating. *Low-voltage lamps* are used in decorative and accent lighting applications that require good control or highlighting.

Lamp characteristics for selected incandescent lamps are provided in Table 20.6.

Tungsten-Halogen Lamps

Tungsten-halogen lamps, frequently called *halogen lamps*, are a smaller, brighter, and more expensive version of the incandescent lamp. These lamps contain high-pressure halogen gases such as iodine or bromine, which allow the tungsten filaments to be operated at higher temperatures and higher efficacies. A higher temperature chemical reaction involving tungsten and the halogen gas recycles evaporated particles of tungsten back onto the filament surface. This offers the advantages of better color rendition, more light output, and a longer life. Selected tungsten-halogen lamp shapes are shown in Figure 20.9.

Halogen lamps generate intense heat and require adequate clearance and good ventilation for heat dissipation. They are pressurized and can shatter if scratched or damaged during installation. Use of a protective cover/shield is needed to protect against minor bulb explosion. The tungsten-halogen lamp cannot be touched by hand without depositing residual skin oils that substantially reduce the life of the quartz glass bulb. Residual oil can be cleaned from the bulb before it is operated.

Gaseous Discharge Lamps

A gaseous discharge is electricity passing through a gas such that it causes the gas to arc. In nature, lightning is a gaseous discharge that produces an immense amount of visible light when electricity flows through the atmosphere. The bright flash produced by lightning is momentary and very intense, so it is not good for seeing.

A *gaseous discharge lamp* produces *continuous* light by passing electricity through a gas contained within the lamp. This electric discharge generates light directly or by producing ultraviolet radiation that causes a phosphor coating on the inside of the lamp to glow. So, unlike incandescent lamps, gaseous discharge lamps do not produce light by getting a filament hot. Instead, light is produced by a continuous gaseous discharge. Fluorescent, mercury vapor, metal halide and low- and high-pressure sodium lamps fit within the gaseous discharge classification.

All gaseous discharge lamps require an additional device called a ballast. A *ballast* is a voltage transformer and current-limiting device designed to start and properly control the flow of power to discharge light sources such as fluorescent and HID lamps. It does not produce light directly. Instead, it is needed to create the starting voltage kick that causes the gas to arc and then controls the electrical (voltage, current, and waveform) input to the lamp that is necessary for proper operation. As a

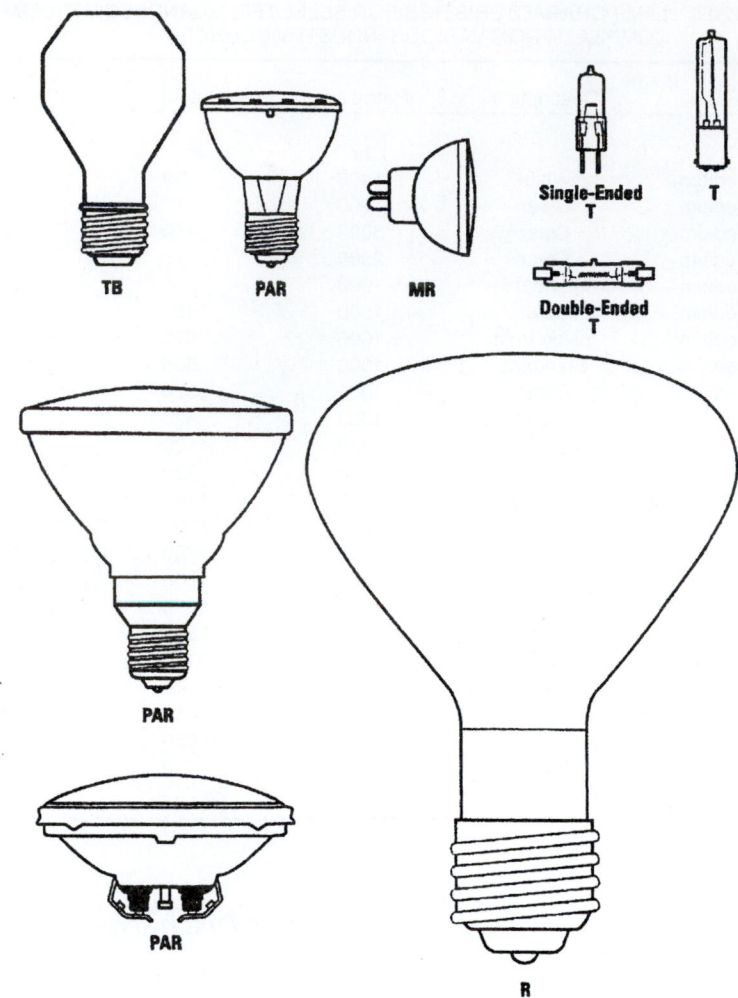

FIGURE 20.9 Selected tungsten-halogen lamp shapes.

transformer, the ballast dictates the voltage output to the lamp and thus determines the type of lamp that can be used in a luminaire. Ballasts can consume up to 20% of the electricity used to power a luminaire.

There are many types of ballasts. Magnetic or iron ballasts increase and control voltage. They are heavy and make noise (hum). Electronic ballasts use solid-state electronic components. They tend to operate more efficiently with less noise and are capable of dimming lamp (lumen) output. There are letter ratings used to rank the sound given off by ballasts: A is extremely quiet (e.g., libraries, churches); B is slightly noisy (e.g., classrooms); C is moderately noisy (e.g., workshops); and so on. Ballast noise is extremely annoying in quiet spaces so a quiet ballast should be selected.

Actual lamp output and life will vary depending on the characteristics of the ballast energizing the lamp. The *ballast factor (BF)* is the ratio of a lamp's rated output (lumens) when it is operated on a specific commercially available ballast as compared to light output (lumens) when operated on a refer-

ence ballast. It is the percentage of a lamp's rated output that can be expected when it is operated on a specific ballast; for example, a lamp energized by a ballast having a ballast factor of 0.90 will emit 90% of its rated lumen output whereas the same lamp being energized by a ballast with a ballast factor of 0.80 will emit only 80% of its rated lumen output. Ballasts and lamps should be matched so the luminaire can produce the largest light output.

Fluorescent Lamps

A *fluorescent lamp* is composed of a tubular glass bulb that is covered with a thin phosphor coating on its inside surface. The glass bulb is filled with a low-pressure gas containing mercury vapor. The base or bases at the end(s) of the lamp are designed to connect the lamp to an electrical circuit. *Cathodes* are filament-like coils at the end(s) of the bulb that act as terminals for the electric arc. When energized, ultraviolet radiation from the mercury arc causes the phosphor coating to fluoresce; that is, they emit visible light as a result of being irradiated by electro-magnetic radiation. A ballast is needed to control the electrical current that maintains the arc.

Linear fluorescent lamps (LFL) are commercially available in straight, U-shaped of circular tubes in a variety of sizes, wattages, voltages, colors, and types of bases. (See Figure 20.10.) A code composed of several digits is used to identify a specific lamp. For example, the standard lamp for general classroom and office lighting for several decades has been the F40T12ES: The digit F is for fluorescent; the digits 40 indicate *nominal* wattage; the digit T is for the tubular shape (e.g., U for U-shaped, C for circular shaped, and so forth); the digit 12 indicates tube diameter in eighths of an inch (e.g., 12 indicates 12/8's or 1.5 in); and ES stands for energy saving. Common lamp diameters include:

T12 lamp that measures 12/8's or 1½ inches in diameter

T10 lamp that measures 10/8's or 1¼ inches in diameter

T8 lamp that measures 8/8's or 1 inch in diameter

T6 lamp that measures 6/8's or ¾ inch in diameter

T5 lamp that measures 5/8's or ⅝ inch in diameter

T9 lamp that measures 9/8's or 1⅛ inches in diameter (circular lamps).

For the traditional T12 (1.5 in) tube, the wattage (except for newer energy-saving fluorescent lamp types) is approximately ⅚ of the length in inches; for example, a F40T12 tube is 48 inches long and ⅚ of 48 equals 40 W. For a smaller diameter T8 (1 inch) tube, the wattage is approximately ⅔ of lamp length

in inches; for example, a T8 tube that is 48 inches long uses ⅔ of 48 is 32 W.

T8 lamps (one-inch diameter) are rapidly replacing T12 lamps because these newer lamps have higher efficacy ratings. Depending on the ballast used, a T8 lamp delivers roughly the same lumen output for about 20% fewer watts. T8 lamps are manufactured with high color-rendering phosphors that provide a better color rendition than the older fluorescent sources. They have good lumen maintenance at 94%, only losing 6% of its light output in the first 40% of rated life.

T5 lamps have a ⅝ in diameter and miniature bi-pin bases, so they cannot be easily retrofitted into T12 and T8 fixtures. Physically, these lamps are intentionally not as long as T8 or T12 lamps because they are not intended for retrofit applications. The main characteristic that distinguishes T5 lamps from other types of fluorescent lamps is that they are designed to peak in their lumen output at a higher ambient temperature, which provides higher light output in applications with little or no air circulation. T5 lamps are available in the three standard fluorescent color temperatures: cool at 4100 K; warm at 3000 K; and neutral at 3500 K.

T5 lamps were developed in Europe and are only available in metric (SI) sizes, so they

High output (HO) and *very high output (VHO)* fluorescent lamps offer higher light (lumen) output in comparison to standard output fluorescent lamps, but these lamps consume more power and have significantly lower efficacies. They are used in relamping applications where improved light levels are required such as warehouses, factories, and school gymnasiums.

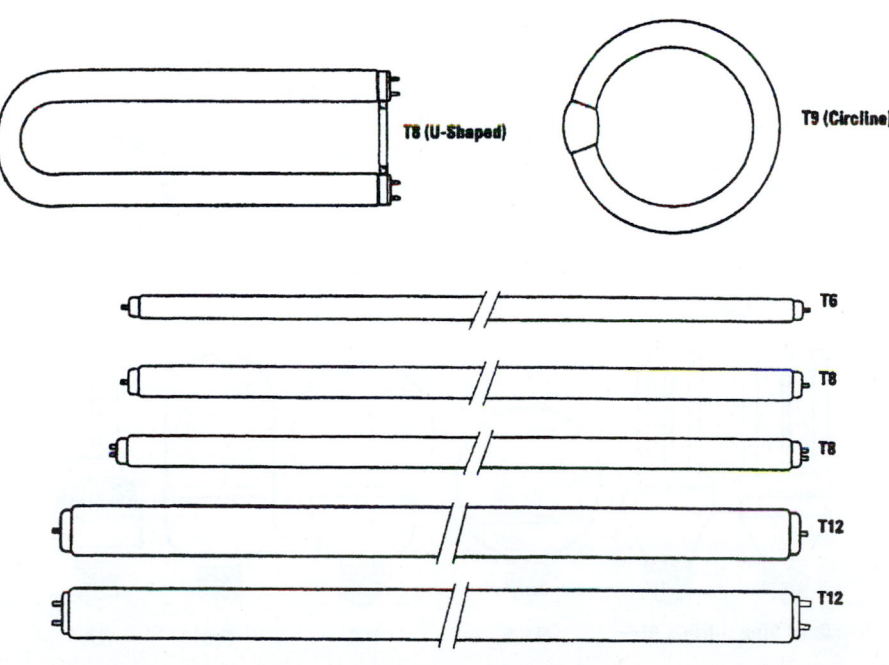

FIGURE 20.10   Selected linear fluorescent lamp shapes.

Compact fluorescent lamps (CFL) are miniaturized fluorescent lamps. Traditionally, CFLs are manufactured with an integral ballast and a standard screw base that can be installed in standard light fixtures in place of incandescent lamps. They are designed as a substitute for standard size incandescent lamps:

7 W CFLs replace 25 W incandescent lamps

11 W CFLs replace 40 W incandescent lamps

13 W CFLs replace 60 W incandescent lamps

19 W CFLs replace 75 W incandescent lamps

23 W CFLs replace 100 W incandescent lamps

Newer, innovative luminaire designs accommodate plug-in compact fluorescent lamps that have a two- or four-pin base. Early compact fluorescent lamps usually produce as little as 25% of their full light output when first started. Newer lamps have improved start-up output but still do not reach full-light output until they warm up for a few minutes. (See Figure 20.11 and Photo 20.1.)

Fluorescent light fixtures come in a variety of shapes, wattages, and voltages. Straight-, circular-, and U-shaped tubes are available, with the 48-in straight tube the most common. Fluorescent lamps are available in wattages of between 20 to 125 W and lengths of 6 to 96 in. The 120 V lamps are the most

PHOTO 20.1 Compact fluorescent lamps are ENERGY STAR®-compliant lighting, energy-efficient lamps. (Courtesy of NREL/DOE)

commonly used fluorescent lighting in residential and light commercial applications. Structures that have 277/480 V, three-phase power available such as retail shopping malls, schools, hospitals, grocery supermarkets, and office buildings typically use the 277 V component of this system for the fluorescent lighting where fixtures are not located closer than 3 ft away from windows, platforms, and fire escapes.

Fluorescent lamps come in a variety of colors or temperatures such as Daylight, Cool White, Soft White, Warm White, and Deluxe Warm White that cover the color temperature range

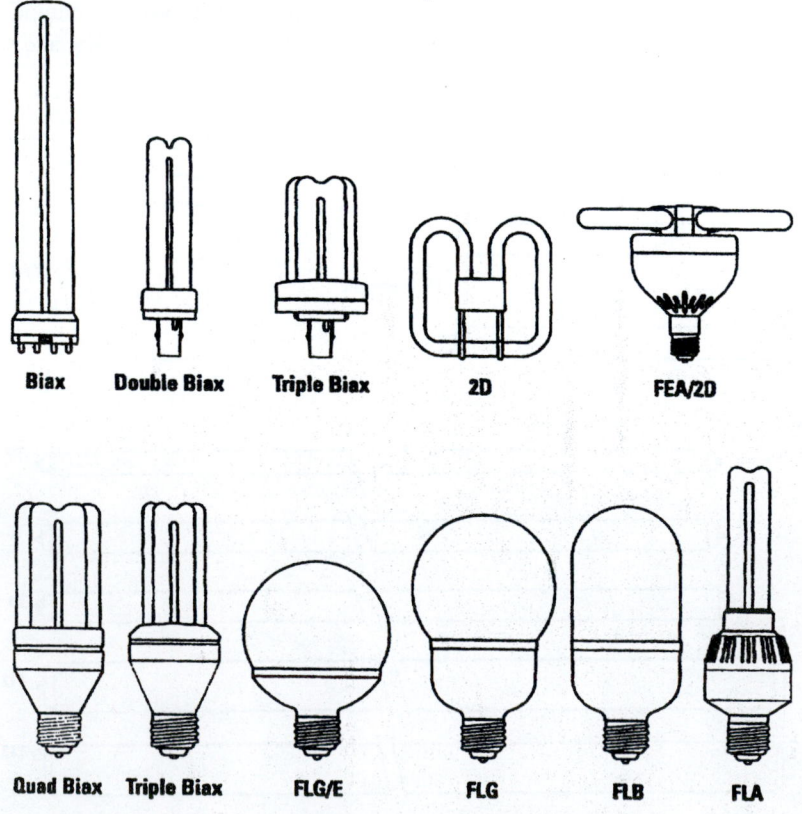

FIGURE 20.11 Selected compact fluorescent lamp shapes.

TABLE 20.7 LAMP CHARACTERISTICS FOR SELECTED COMPACT FLUORESCENT LAMPS. COMPILED FROM VARIOUS INDUSTRY SOURCES.

Input	Description or Lamp Designation	Rated Life	Output		Efficacy (LPW)		Color	
			Initial	*Mean*	Initial	*Mean*	Temperature	Rendition
W		hr	lm	*lm*	lm/W	*lm/W*	K	CRI
Rapid-Start, Four-Pin, Plug-In (Biax T5) Lamps								
18	F18BX/SPX30/RS	20 000	1250	*1130*	69.4	*62.8*	3000	82
18	F27BX/SPX41/RS	20 000	1250	*1130*	69.4	*62.8*	4100	82
27	F27BX/SPX30/RS	12 000	1800	*1620*	66.7	*60.0*	3000	82
27	F27BX/SPX41/RS	12 000	1800	*1620*	66.7	*60.0*	4100	82
39	F39BX/SPX30/RS	12 000	2850	*2510*	73.1	*64.4*	3000	82
39	F40/30BX/SPX41/RS	20 000	3150	*2840*	80.8	*72.8*	4100	82
50	F50BX/SPX30/RS	14 000	4000	*3400*	80.0	*68.0*	3000	82
50	F50BX/SPX41/RS	14 000	4000	*3400*	80.0	*68.0*	4100	82
Compact Fluorescent Ballasted, One-Piece, Screw-In, Lamps for Medium-Base Incandescent Sockets								
15	FLE15TBZ/SPX27EC	10 000	825	*700*	55.0	*46.7*	2700	82
20	FLE20TBZ/SPX27EC	10 000	1200	*1020*	60.0	*51.0*	2700	82
23	FLE23TBZ/SPX27EC	10 000	1520	*1290*	66.1	*56.1*	2700	82
28	FLE28TBZ/SPX27EC	10 000	1750	*1485*	62.5	*53.0*	2700	82
Compact Fluoescent Twin-Tube Lamps								
7	CFT7W	10 000	400	—	57.1	—	2700	82
9	CFT9W	10 000	580	—	64.4	—	2700	82
13	CFT13W	10 000	800	—	61.5	—	2700	82
Compact Fluorescent Twin-Tube Lamps								
18	FT18W	20 000	1250	—	69.4	—	3000	82
24	FT24W	12 000	1800	—	75.0	—	3000	82
36	FT36W	12 000	2900	—	80.6	—	3000	82
40	FT40W	20 000	3150	—	78.8	—	3000	82
55	FT55W	12 000	4800	—	87.3	—	3000	82

from 3000 to 6500 K. Color or temperature varies with the type of phosphor coating.

The phosphor blend in a fluorescent lamp determines the color temperature and color rendering of the light emitted by the lamp. Halophosphors are the most popular, least expensive, and lowest quality phosphors. They are used in standard cool white and warm white lamps. Rare earth phosphors are more expensive, but they produce light in the red, blue, and green wavelengths to which our eyes are most sensitive. They produce a higher color rendering light than halophosphors, are more efficient, and maintain their light output longer.

All fluorescent lamps are designed to operate at a specific air temperature, usually a surrounding air temperature of 77°F (25°C). Lamp performance is greatly affected by low temperature. Below about 50°F (10°C), most fluorescent lamps will not operate reliably. Energy saving (ES) fluorescent lamps will not function below about 60°F (15.5°C). Special lamps and ballasts are required under low-temperature conditions and are available for reliable use down to about −20°F (−29°C).

The primary drawbacks of fluorescent lamps in comparison to other types of light sources are that they are larger in size and general offer inferior color rendition. Their large size makes it difficult to control the light they emit and direct it to the location where it is needed. To overcome the size problem, there has been a recent increase in use of the physically smaller circular- and U-shaped tubes and compact fluorescent lamps. The color problem can be resolved with higher-priced, high color rendering lamps.

Although fluorescent lamps are more expensive than incandescent lamps, their higher efficacy and rated life makes them extremely cost-effective. Additionally, fluorescent lamps produce a diffuse light source that is relatively glare free and visually comfortable, unlike the concentrated light emitted by other commercially available light sources. As a result, these lamps have been the standard for general classroom and office lighting for several decades.

Lamp characteristics for selected fluorescent lamps are provided in Tables 20.7 through 20.9.

TABLE 20.8   LAMP CHARACTERISTICS FOR SELECTED T12 (1½ INCH DIAMETER), 48 INCH NOMINAL LENGTH,
FLUORESCENT LAMPS. COMPILED FROM VARIOUS INDUSTRY SOURCES.

Input	Description or Lamp Designation	Rated Life	Output		Efficacy (LPW)		Color	
			Initial	Mean	Initial	Mean	Temperature	Rendition
W		hr	Lumens	Lumens	lm/W	lm/W	K	CRI
Rapid-Start Lamps								
40	F40CW—Cool White	20 000	3050	2680	76.3	67.0	4150	62
40	F40WW—Warm White	20 000	3150	2770	78.8	69.3	3000	52
40	F40D—Daylight	20 000	2550	2240	63.8	56.0	6250	75
40	F40N—Natural	20 000	2100	1740	52.5	43.5	3700	90
40	F40C50—Chroma 50	20 000	2250	1870	56.3	46.8	5000	90
40	F40C75—Chroma 75	20 000	1950	1680	48.8	42.0	7500	92
Energy-Saving Rapid-Start Lamps								
34	F40CW—Cool White	20 000	2650	2280	77.9	67.1	4150	62
34	F40WW—Warm White	20 000	2750	2370	80.8	69.7	3000	52
34	F40D—Daylight	20 000	2225	1910	65.4	56.2	6250	75
34	F40C50—Chroma 50	20 000	2000	1720	58.8	50.6	5000	90
Color-Enhancing, Rapid-Start Lamps								
40	F40SPX30	20 000	3250	2960	81.3	74.0	3000	82
40	F40SPX35	20 000	3250	2960	81.3	74.0	3500	82
40	F40SPX41	20 000	3250	2960	81.3	74.0	4100	82
40	F40SPX50	20 000	3250	2820	81.3	70.5	5000	82
Energy-saving, Color-Enhancing, Rapid-Start Lamps								
34	F40SPX30	20 000	2850	2570	83.8	75.6	3000	82
34	F40SPX35	20 000	2850	2570	83.8	75.6	3500	82
34	F40SPX41	20 000	2850	2570	83.8	75.6	4100	82
34	F40SPX50	20 000	2700	2430	79.4	71.5	5000	82
High Output Lamps								
60	F48T12/CW/HO—Cool White	12 000	4050	3520	67.5	58.7	4150	62
60	F48T12/D/HO—Daylight	12 000	3400	2960	56.7	49.3	6250	75
60	F48T12/WW/HO—Warm White	12 000	4130	3590	68.8	59.8	3000	62
Energy Saving, Color-Enhancing, High-Output Lamps								
60	F48T12/SPX30/HO	12 000	4350	3920	72.5	65.3	3000	82
60	F48T12/SPX35/HO	12 000	4350	3920	72.5	65.3	3500	82
60	F48T12/SP30/HO	12 000	4250	3830	70.8	63.8	3000	70
60	F48T12/SP35/HO	12 000	4250	3830	70.8	63.8	3500	73
60	F48T12/SP41/HO	12 000	4250	3830	70.8	63.8	4100	70
Very High Output Lamps								
110	F48T12/CW/1500—Cool White	10 000	6200	4030	56.4	36.6	4150	62
110	F48T12/WW/HO—Warm White	10 000	6280	4080	57.1	37.1	3000	62

Fluorescent Luminaire Efficacy

The *luminaire efficacy rating (LER)* is the ratio of light (the luminous flux, in lumens) emitted by a *fluorescent* luminaire to the electrical energy consumed, including the ballast. It is the luminous efficacy of the luminaire as a whole, not just the bare lamp.

The LER is found by the following equation, where the luminaire efficiency (EFF), luminaire power input (watts), and ballast factor (BF) are from the fixture manufacturer's data and the total rated lamp lumens (TLL) is from the lamp manufacturer's data:

$$LER = (EFF \cdot TLL \cdot BF)/W$$

TABLE 20.9   LAMP CHARACTERISTICS FOR SELECTED T8 (1 INCH DIAMETER), 48 INCH NOMINAL LENGTH, TRAMLINE FLUORESCENT LAMPS. COMPILED FROM VARIOUS INDUSTRY SOURCES.

Input	Description or Lamp Designation	Rated Life	Output		Efficacy (LPW)		Color	
			Initial	*Mean*	Initial	*Mean*	Temperature	Rendition
W		hr	Lumens	*Lumens*	lm/W	*lm/W*	K	CRI
32	F32T8/SPX30	20 000	2950	*2650*	92.2	*82.8*	3000	84
32	F32T8/SPX35	20 000	2950	*2650*	92.2	*82.8*	3500	84
32	F32T8/SPX41	20 000	2950	*2650*	92.2	*82.8*	4100	80
32	F32T8/SPX50	20 000	2800	*2520*	87.5	*78.8*	5000	80
32	F32T8/SP30	20 000	2850	*2570*	89.1	*80.3*	3000	75
32	F32T8/SP35	20 000	2850	*2570*	89.1	*80.3*	3500	75
32	F32T8/SP41	20 000	2850	*2570*	89.1	*80.3*	4100	75
32	F32T8/SP65	20 000	2700	*2565*	84.4	*80.2*	6500	75
32	F32T8/SP50	20 000	2750	*2610*	85.9	*81.6*	5000	78
32	F32T8/SP30	20 000	2850	*2710*	89.1	*84.7*	3000	78
32	F32T8/SP35	20 000	2850	*2710*	89.1	*84.7*	3500	78
32	F32T8/SPX35	20 000	2950	*2800*	92.2	*87.7*	3500	86
32	F32T8/SPX41	20 000	2950	*2800*	92.2	*87.7*	4100	86
32	F32T8/SP41/C	20 000	2850	*2710*	89.1	*84.7*	4100	78
32	F32T8/SPX50	20 000	2800	*2660*	87.5	*83.1*	5000	86

Extra-Life

32	F32T8/XL/SP30ECO	24 000	2850	*2710*	89.1	*84.7*	3000	78
32	F32T8/XL/SP35ECO	24 000	2850	*2710*	89.1	*84.7*	3500	78
32	F32T8/XL/SP41ECO	24 000	2850	*2710*	89.1	*84.7*	4100	78
32	F32T8XL/SPX30ECO	24 000	2950	*2800*	92.2	*87.5*	3000	86
32	F32T8XL/SPX35ECO	24 000	2950	*2800*	92.2	*87.5*	3500	86
32	F32T8XL/SPX41ECO	24 000	2950	*2800*	92.2	*87.5*	4100	86
32	F32T8SXL/SP30ECO	30 000	2850	*2675*	89.1	*83.6*	3000	82
32	F32T8SXL/SP35ECO	30 000	2850	*2675*	89.1	*83.6*	3500	82
32	F32T8SXL/SP41ECO	30 000	2850	*2675*	89.1	*83.6*	4100	81

The LER is part of a voluntary program being implemented by the lighting industry. The LER rates the effectiveness of like luminaires in a manner similar to the EPA miles/gallon rating for automobiles. It is a good measure of the efficiency of *similar* fluorescent luminaires.

Example 20.3

Fixture manufacturer's data for a two-lamp fluorescent luminaire provides a luminaire efficiency of 0.60, a luminaire power input (watts) of 85 W, and ballast factor of 0.87. Data from the lamp manufacturer indicates that the fluorescent lamps under consideration will output 3050 lumens. Compute the LER.

$$LER = (EFF \cdot TLL \cdot BF)/W$$
$$LER = (0.6 \cdot 2 \cdot 3050 \text{ lm} \cdot 0.87)/85 \text{ W} = 37.5 \text{ lm/W}$$

Comparing the LER of two luminaires is not quite as simple as comparing the EPA mileage ratings of two different automobiles. The standard distinguishes between the major categories of fluorescent luminaires. Each rating contains a two-

letter code indicating source and product category, such as FL, FP, FW, and so on. The digit F stands for fluorescent. The second digit may include lensed (L), parabolic louvered (P), wraparound (W), strip (S), and industrial (I) luminaire categories. Only luminaires within a specific product category should be compared. By the nature of their different applications, a lensed fixture and a strip fixture (for example) would fall into different LER ranges.

High-Intensity Discharge Lamps

High-intensity discharge (HID) lamps produce a very bright light by discharging an arc when electrical current passes through a metal gas contained under high pressure in a glass bulb. HID lamps include mercury vapor, metal halide, and high-pressure sodium lamps. These lamps differ from fluorescent lamps in that their gas is under higher pressure, the lamp is physically smaller, and the emitted light is more concentrated.

HID lamps have an internal arc tube that contains the fill gas. A large outer glass bulb surrounds the arc tube to avoid shattering that could result from direct contact with water and other fluids. Use of a protective cover/shield is recommended to protect against minor bulb explosion. The following is a description of the common types of HID lamps:

Mercury vapor (MV) lamps were the first commercially available HID lamps. These lamps are constructed of an internal quartz tube enclosed in an outer glass envelope. A small amount of liquid mercury is sealed in an argon gas fill inside the quartz tube. After the warm-up period, the arc emits both visible and ultraviolet (UV) light. High-pressure MV lamps without color correction produce a blue-white light directly from their discharge arc. Phosphor coatings can be added to improve color rendition MV lamps are available in wattages ranging from 50 to 1000 W. They are approaching obsolescence, now that other HID light sources with better efficacy, more accurate color rendering, and improved lumen maintenance are becoming available at lower cost, especially in larger lamp sizes. As a result, mercury vapor–tungsten blended lamps are being introduced. (see Blended Lamps section, which follows).

Metal halide (MH) lamps are constructed in a manner similar to MV lamps except that in addition to the mercury and argon, various metal halides are included in the gas fill. Use of metal halides increases luminous efficiency and improves color rendition. No phosphor is needed to produce a cool white color, but some may be added to improve rendering of oranges and reds. MH lamps have all but replaced MV lamps because they offer much better efficacy and color rendition. They are available in wattages ranging from 32 to 1000 W. The small 32 W MH lamp produces 2500 lumens, about as much light as a 150 W incandescent or a 40 W (4 ft) fluorescent. This high intensity limits the range of application for metal halide lamps. MH units produce high levels of UV radiation that must be shielded by glass in the lamp or fixture.

High-pressure sodium (HPS) lamps contain an internal arc tube made of a translucent ceramic material rather than quartz glass because of the high temperature (2350°F/1300°C) at which it operates. The arc tube is enclosed in an outer glass envelope like other HID lamps. A small amount of solid metallic sodium and mercury is sealed in a xenon gas fill inside the ceramic arc tube. HPS lamps are available in wattages ranging from 18 to 1000 W and vary more widely than other HID lamps in their efficacy and color quality. Low-quality outdoor HPS lamps produce an orange-white light with a very high efficacy, but a CRI of 21. Genera purpose indoor HPS lamps offer a CRI of about 60 and color-enhancing "white" types have achieved a CRI of up to 80, but these lamps have a lower efficacy.

Low-pressure sodium (LPS) lamps are really a blend between HID and fluorescent technologies. These lamps are constructed of a large sodium-resistant glass tube containing sodium and a neon–argon gas mixture. Excess sodium is contained in the arc tube because the glass may absorb or react with some of the sodium. As the tube is rather large, it is bent into a tight U-shape and enclosed in an outer bulb. The inner surface of the outer bulb is coated with a material that reflects infrared radiation but passes visible light. A LPS discharge arc produces a monochromatic yellow light consisting almost entirely of orange-yellow (589.0 and 589.6 nm) wavelengths. The monochromatic yellow light creates a very poor color rendition with all surfaces illuminated in an orange-yellow version of blacks and white. Colors are dramatically changed: most reds appear bright orange and greens and blues appear dark.

With HID lamps, full light output does not occur immediately when power is applied. There is a start-up time delay called the *strike time* before the lamp reaches it peak output. During the start-up process, a low-pressure discharge is established in the gas. As the pressure increases, light starts being produced by the discharge through the high-pressure metal vapor. A strike time of one to several minutes is needed before the lamp reaches its peak output. With a short power interruption (e.g., accidental shutoff or temporary power outage), the discharge arc is extinguished. Full light output does not occur immediately even if the power outage was brief. A hot lamp cannot be restarted until it has cooled. A restart delay called the *restrike time* of one to several minutes is needed before the lamp reaches its peak output.

An HID lamp should not be operated if the outer glass envelope is cracked or broken. This condition is dangerous because the extremely hot arc tube can explode. Additionally, the mercury arc produces substantial amounts of short-wave UV radiation that is extremely hazardous. An undetected breakage will usually result in the lamp failing to emit light relatively quickly.

Unlike fluorescent lamps, HID lamps generate full light output at low air temperatures, although a special ballast may be needed. This advantage makes HID lamps more suitable than fluorescent lamps for outdoor use. With the exception of the MV lamp, HID lamps offer the best efficacy (LPW) and lowest operating cost of all types of electric lights. HID lamps can be used for commercial and industrial indoor lighting and exterior landscape, recreation/sports court, parking lot, and street lighting.

Selected HID lamp shapes are shown in Figure 20.12. Typical lamp characteristics for selected HID lamps are provided in Table 20.10.

Blended Lamps

Blended lamps combine the luminous efficiency of an HID lamp (e.g., MV) with the good color rendering capability of an incandescent lamp. The most common blended lamp has mercury

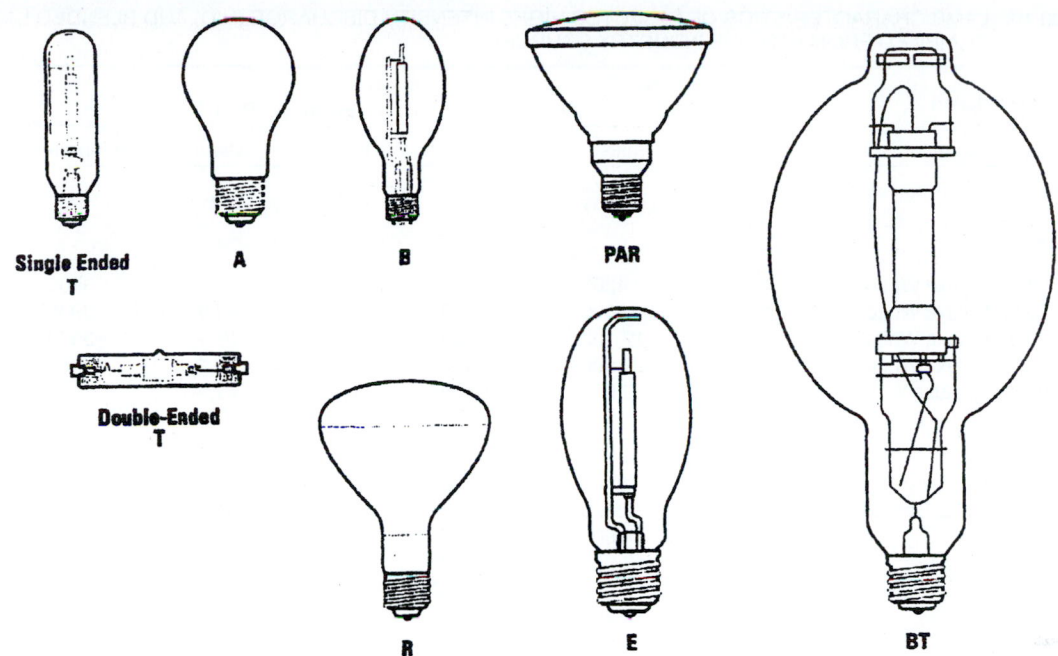

FIGURE 20.12 Selected high-intensity discharge (HID) lamp shapes.

vapor combined a with tungsten filament. The tungsten filament acts as an incandescent lamp, so a blended lamp requires no external ballast or other control devices. There is essentially no start time and thus total luminous flux is available immediately. Color rendition is fair to good: between 50 and 63. A key advantage of mercury–tungsten blended lamps is that they are directly interchangeable with incandescent lamps and have a longer lamp life. A disadvantage is that they cannot be dimmed. Lamp characteristics for selected mercury–tungsten blended lamps are provided in Table 20.10.

Solid-State Lighting

Solid-state lighting (SSL) refers to a type of lighting that uses *light-emitting diodes (LEDs), organic light-emitting diode lights (OLEDs),* or *polymer light-emitting diodes (PLEDs),* instead of traditional lighting sources that use an electrical filament or gas inside a glass bulb. A *light-emitting diode (LED)* is a semiconductor that consists of a chip of semiconducting material treated to create a structure with two electron-charged materials. When connected to a power source, electrons jump from one material to the other. As an electron jumps, it emits energy in the form of a photon (light). The color of light created depends on the amount of energy in that photon, which depends on the material used for the layers. See Photo 20.2.

An individual LED chip emits light in a specific wavelength, so the light source is near monochromatic. The specific wavelength or color emitted by an LED depends on the materials used to make the diode. Red LEDs are based on aluminum gallium arsenide (AlGaAs); blue from indium gallium nitride (InGaN); and green from aluminum gallium phosphide (AlGaP). White light can be achieved with LEDs in two main ways: (1) phosphor conversion, in which a blue or near-UV chip is coated with phosphor(s) to emit white light; and (2) RGB systems, in which light from multiple monochromatic LEDs (red, green, and blue) is mixed, resulting in white light. The phosphor conversion approach is most commonly based on coating a blue LED with yellow phosphor.

A single LED is about the size of a pea. It is powered by about 1 W. Typically, multiple LEDs are ganged together into an array that outputs more light. One manufacturer offers a single 10 W LED component that produces more light than a 100 W incandescent light bulb.

LEDs are vibration and shock resistant; they have no moving parts, no fragile glass, no mercury, no toxic gasses, and no filament. They are not subject to sudden failure or burnout and have an exceptionally long life (50 000+ to 100 000 hours). Like all lamps, they gradually degrade in performance over time. Also, an LED is low voltage and cool to the touch. As a result, SSL devices promise to replace conventional incandescent and fluorescent light sources. SSL technology is rapidly evolving to provide light sources for general illumination. This technology holds promise for lower energy consumption and reduced maintenance.

LEDs emit light in a specific focused direction, reducing the need for reflectors and diffusers that can trap light. This allows

TABLE 20.10 LAMP CHARACTERISTICS OF SELECTED HIGH0 INTENSITY DISCHARGE (HID) AND BLENDED LAMPS. COMPILED FROM VARIOUS INDUSTRY SOURCES.

Input	Description or Lamp Designation	Rated Life	Output		Efficacy (LPW)		Color	
			Initial	*Mean*	Initial	*Mean*	Temperature	Rendition
W		hours	Lumens	*Lumens*	lm/W	*lm/W*	K	CRI
Mercury Vapor Lamps								
100	HR100DX38 Deluxe White	24 000	4200	*3200*	42.0	*32.0*	3900	50
175	HR175DX39 Deluxe White	24 000	8600	*7200*	49.1	*41.1*	3900	50
250	HR250DX37 Deluxe White	24 000	12 100	*9800*	48.4	*39.2*	3900	50
400	HR400DX33 Deluxe White	24 000	22 500	*17 500*	56.3	*43.8*	3900	50
1000	HR1000DX34 Deluxe White	16 000	62 000	*47 700*	62.0	*47.7*	3900	50
Metal Halide								
100	MVR100/U/MED Clear	15 000	8100	*5750*	81.0	*57.5*	4000	75
	MVR100/C/U/MED Coated	15 000	7600	*4850*	76.0	*48.5*	4000	75
175	MVR175/U Clear	10 000	14 000	*10 350*	80.0	*59.1*	4000	65
	MVR175/C/U Diffuse	10 000	13 200	*9750*	75.4	*55.7*	3900	70
250	MVR250/U Clear	10 000	21 000	*17 000*	84.0	*68.0*	4200	65
	MVR250/C/U Diffuse	10 000	19 800	*16 000*	79.2	*64.0*	3900	70
400	MVR400/U Clear	20 000	36 000	*28 800*	90.0	*72.0*	4000	65
	MVR400/C/U Clear	20 000	33 900	*27 100*	84.8	*67.8*	3700	70
1000	MVR1000/U Clear	12 000	110 000	*88 000*	110.0	*88.0*	4000	65
	MVR1000/C/U Clear	12 000	105 000	*79 800*	105.0	*79.8*	3400	70
High-Pressure Sodium								
100	LU100 Clear	24 000	9500	*8550*	95.0	*85.5*	2000	22
150	LU150/MED Clear	24 000	16 000	*14 400*	106.7	*96.0*	2000	22
200	LU200 Clear	24 000	22 000	*19 800*	110.0	*99.0*	2100	22
250	LU250 Clear	24 000	28 000	*27 000*	112.0	*108.0*	2100	22
310	LU310 Clear	24 000	37 000	*33 300*	119.4	*107.4*	2100	22
400	LU400 Clear	24 000	51 000	*45 000*	127.5	*112.5*	2100	22
1000	LU1000 Clear	24 000	140 000	*126 000*	140.0	*126.0*	2100	22
Mercury–Tungsten Blended Lamps								
160	MVT 160	6000	2400	*1920*	15.0	*12.0*	3500+/−	63
160	MVT 160	6000	2500	*2000*	16.0	*12.8*	3500+/−	63
160	MVT 160	6000	2900	*3625*	18.0	*14.4*	3500+/−	63
160	MVT 160	6000	3000	*2400*	18.75	*15.0*	3500+/−	63
160	MVT 160	6000	3100	*2480*	19.5	*15.6*	3500+/−	63
250	MVT 250	6000	5600	*4480*	22.5	*18.0*	3500+/−	62
250	MVT 250	6000	5800	*4640*	23.2	*18.6*	3500+/−	62
500	MVT 500	6000	14 000	*11 200*	28.0	*22.4*	3500+/−	50
500	MVT 500	6000	14 200	*11 360*	28.5	*22.8*	3500+/−	50

them to be more efficient in throwing light, which makes them best suited for recessed downlights, kitchen undercabinet lighting, portable desk/task, and path and step light designs. Developments in LED technology and luminaire design are enabling significant energy savings in downlighting applications. LEDs are well suited for decorative and accent lighting such as that found in retail stores, theaters, and restaurants. The New Year's Eve Ball that is lowered in Times Square, New York, was fitted with more than 9500 LEDs to replace about 600 incandescent and halogen lamps. The ball is now twice as bright, has a palette of 16 million colors, and uses less than 20% of the energy.

Although a SSL presently costs more than incandescent, the energy savings and longevity of LED lighting make it a good substitute for incandescent and halogen. It is expected that the manufacturing costs of white LED lights will fall substantially in the future.

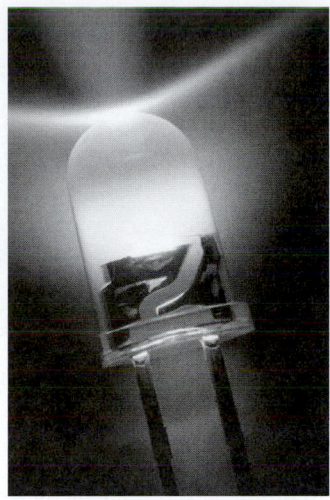

PHOTO 20.2   An LED lamp. (Courtesy of NREL/DOE)

PHOTO 20.3   Ambient lighting is used to provide uniform illumination. In this case, HID uplight luminaires are used to provide lighting in a gymnasium. (Used with permission of ABC)

Mercury Content of Lamps

Fluorescent and HID lamps have some mercury content and may need to be recycled or disposed of in hazardous waste landfills instead of municipal solid waste landfills under the U.S. Resource Conservation and Recovery Act. These lamps cannot be incinerated. The U.S. Environmental Protection Agency (U.S. EPA) and some states have relaxed regulations regarding the disposal of these lamps. In recent years, the major lamp manufacturers have introduced a number of low-mercury fluorescent and HID lamps, allowing them to be disposed of in municipal solid waste landfills.

20.7   FORMS OF ARCHITECTURAL LIGHTING

Architectural Lighting

Good architectural lighting is crucial to the performance of everyday activities and to the appreciation of the built environment. The basic functions that light performs in interior building spaces include the following:

Ambient lighting, sometimes called *general lighting,* provides uniform illumination throughout the space. It softens shadows on people's faces and provides the illumination for color and texture. Ambient lighting is the most essential form of lighting because it fills the volume of a room or space with uniform light. The amount and type of ambient lighting depends largely on the activities or tasks performed in the space and color of the walls, ceiling, and furniture. (See Photo 20.3.)

Task or *local lighting* is the illumination provided for a specific visual function, which is additional to and con-

trolled separately from the ambient lighting. Exacting tasks, such as sewing, drafting, assembly, and surgery, generally require this type of lighting. In open, modular office spaces, local lighting is used to provide additional illumination at workstations with ambient lighting providing lower levels in the surrounding areas. The color quality and intensity of local lighting can influence the effectiveness of the individual performing the task. (See Photo 20.4.)

Accent lighting is directional lighting used to emphasize a particular object or area. It is used to call attention to or orchestrate interest by emphasizing a particular architectural feature, piece of artwork, photograph, or plant. In open retail spaces, ambient lighting is typically provided at low levels and accent lighting is used to provide additional illumination to highlight merchandise displays. Accent lighting used to highlight a display is shown in

PHOTO 20.4   Local lighting used to highlight a task area. (Used with permission of ABC)

PHOTO 20.5   Accent lighting used to highlight a display. (Used with permission of ABC)

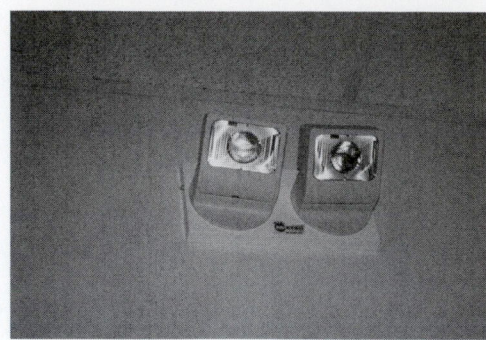

PHOTO 20.7   Emergency lighting installation. (Used with permission of ABC)

Photo 20.5. Care must be used to ensure that decorative light sources are not too bright so they dominate the space or the furniture, occupants, and activities within the space.

Decorative lighting is the light source that adds a quality of interest to the space. It combines with other types of lighting to give an overall "feel" to a room that serves little purpose other than to look attractive. This type of lighting may be a hanging crystal chandelier used as a main feature in an entry or dining space. (See Photo 20.6.) Novelty lights such as Christmas lights and colored rope lights are decorative.

A comparison of characteristics by lamp type is provided in Tables 20.11 and 20.12. Table 20.13 indicates the best application by lamp type.

Emergency and Safety Lighting

Emergency and safety lighting is a secondary lighting system that provides backup illumination when the power supply to the normal lighting system is interrupted or fails, such as in the case of a public utility power outage.

Emergency lighting is required in the critical care and emergency spaces found in hospitals, nursing homes, and police, fire protection, and crisis management areas. *Stand-by lighting* is part of the emergency lighting system. It enables normal activities to continue substantially unchanged. This type of lighting system is typically powered by an emergency generator or battery backup system. These systems must meet stringent regulations and be tested frequently to ensure proper operation during a power interruption. (See Photo 20.7.)

Safety lighting is the part of emergency lighting system that ensures the safety of people involved in a potentially hazardous process. *Escape lighting* provides illumination to ensure that an escape route can be effectively identified and used in the case of failure of the normal lighting system. *Exit lighting* is the part of the escape lighting system that includes illuminated signage used to provide clear directions for an emergency exit of building occupants. Exit signage is typically required above exit doorways and in egress corridors that serve as means of safe exit from the building interior.

20.8   LIGHTING INSTALLATIONS

Traditional Lighting Installations

There are many types of luminaires and lighting installations that can illuminate a building space or highlight a centerpiece. Common types of luminaires are listed in Table 20.14. Common lighting installations and types of luminaires are further defined in the following paragraphs.

A *pendent* is a luminaire that is hung with a cord, chain, or tube that enables it to be suspended from a ceiling or other support. It broadcasts light over the entire space. An *uplight* is a luminaire where a shielded light source directs its light to the ceiling, where it is reflected back to the space. A *downlight* is a luminaire that is usually attached to or recessed in the ceiling

PHOTO 20.6   Decorative lighting draws attention to the space. (Used with permission of ABC)

TABLE 20.11   A COMPARISON OF DIFFERENT LAMP TYPES.

Incandescent Lamps

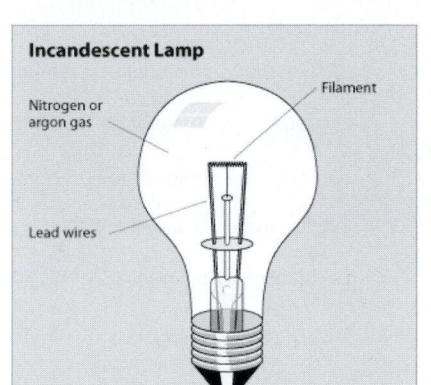

- Do not require a ballast
- Warm color appearance
- Low color temperature
- Excellent color rendering (CRI ~100)
- Compact light source
- Simple maintenance from screw-in Edison base
- Poor efficacy (lm/W)
- Shorter service life
- Filament is sensitive to vibrations and jarring
- Bulb can get very hot during operation
- Must be properly shielded because incandescent lamps can produce direct glare as a point source
- Line voltage variations can adversely affect light output and service life

Fluorescent (Linear and Compact) Lamps

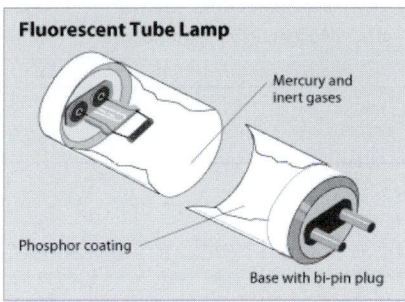

- Require a ballast (built into some compact fluorescent lamps)
- Lamp must be compatible with ballast
- Good range of color temperatures and color rendering capabilities
- Low surface brightness compared to point sources
- Cooler operation
- Good to high efficacy (lm/W)
- Ambient temperatures can affect light output and life
- Low temperatures can affect starting unless a "cold weather" ballast is specified

HID (i.e., Mercury Vapor, Metal Halide, High/Low-Pressure Sodium) Lamps

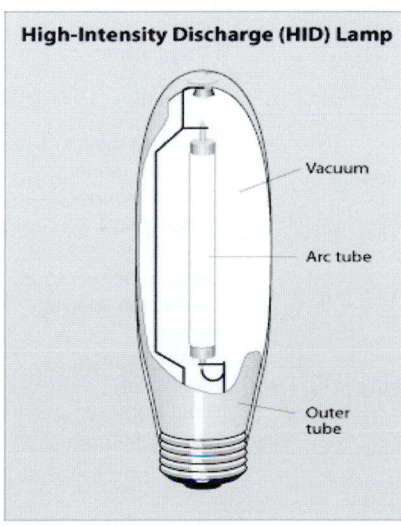

- Require a ballast
- Ambient temperature does not affect light output
- Low ambient temperatures require a special ballast
- Compact light source
- High efficacy (lm/W)
- Point light source
- Range of color temperatures and color rendering abilities depending on the lamp type
- Long service life

TABLE 20.12   A COMPARISON OF CHARACTERISTICS BY LAMP TYPE.

Lamp Type	Typical Efficacy (LPW)	Typical Rated Life	Typical Operating Life (2000 hr/yr)	Color Rendition	Color Temperature	Start Characteristics Start Time	Restart Time
	lm/W	hr	years	CRI	K	min	min
Standard incandescent	7 to 15	750 to 2000	0.4 to 1.0	99 Excellent	2300 to 3000	Immediate	Immediate
Halogen	12 to 25	2000 to 5000	1.0 to 2.5	99 Excellent	3000	Immediate	Immediate
Mercury–tungsten blended	15 to 33	6000	3.0	50 to 63 Good	3500+/−	Immediate	Immediate
Mercury vapor	25 to 55	12 000 to 29 000	6.0 to 14.5	15 to 50 Fair to good	3000 to 5000	5 to 7	3 to 6
Compact fluorescent	40 to 75	9000 to 10 000	4.5 to 5.0	73 to 85 Very good	2700 to 6500	Immediate	Immediate
Linear fluorescent	61 to 95	12 000 to 24 000	6.0 to 12.0	52–92 Good to very good	3000 to 6500	Immediate	Immediate
Metal halide	45 to 100	10 000 to 20 000	5.0 to 10.0	60–75 Good to very good	3000 to 5000	2 to 5	15 to 20 (Aging) 5 to −7 (New)
High-pressure sodium	45 to 110	10 000 to 29 000	5.0 to 14.5	2 to 25 Fair	2100	3 to 4	1
Low-pressure sodium	80 to 150	14 000 to 18 000	7.0 to 9.0	0 to 10 Poor	1750	5 to 10	Immediate
Light emitting diode	25 to 50	50 000 to 100 000	25+	82 Very good	2700 to 5000	Immediate	Immediate

TABLE 20.13   BEST APPLICATIONS BY LAMP TYPE.

Lamp Type	Application
Incandescent	Interior—ambient (small spaces), task, accent, decorative Exterior—ambient (small areas), task, accent, decorative
Halogen	Interior—task, accent, decorative
Compact fluorescent	Interior—ambient (small spaces), task, accent Exterior—ambient (small areas), task, accent, decorative
Fluorescent	Interior—ambient (small and large spaces), task, accent Low ceiling spaces (e.g., offices, classrooms, retail, supermarket, shops)
Mercury vapor	High ceiling spaces (e.g., warehouses, hangers)
Metal halide	Interior—ambient (medium and large spaces) Sporting arenas (e.g., gyms, hockey rinks, basketball courts) Low ceiling spaces (e.g., offices, classrooms, retail, supermarket, shops) High ceiling spaces (e.g., warehouses, hangers) General exterior (e.g., parking areas, tunnels, roadways)
High-pressure sodium	Interior—ambient (medium and large spaces) Low ceiling spaces (e.g., offices, classrooms, retail, supermarket, shops) High ceiling spaces (e.g., warehouses, hangers) General exterior (e.g., parking areas, tunnels, roadways)
Low-pressure sodium	General exterior (e.g., parking areas, tunnels, roadways)

TABLE 20.14   COMMON TYPES OF LUMINAIRES.

Indoor

Adjustable accent
Bath/vanity
Cabinet
Cable/rail
Chandelier
Cornice
Cove
Decorative
Demanding environment
Desk
Downlight
Exit/emergency
Floodlight
Floor Lamps
Fluorescent linear
Fluorescent troffer
Fluorescent wraparound
High bay
Industrial fluorescent
Inspection/task
Low bay
Pendent
Recessed
Sconce
Showcase/display
Sports
Spotlight
Table lamps

Track
Wallwasher
Valance
Workstation panel mount
Uplight
Undercabinet

Outdoor

Area post mount
Area wall mount
Canopy mount
Floodlighting
High mast
Landscape low mount
Landscape specialty
Pathway
Parking garage
Parking lot
Road work
Sign lighting
Sports
Street

Specialty

Entertainment/stage
Fiber optics
Light pipe
Linear (rope, tube, string)

and emits a concentrated light downward. A *high hat* is a type of downlight that is a recessed, canister-shaped luminaire with a shielded lamp that emits light downward. *Recessed* luminaires are mounted above the ceiling or behind a wall or other surface so that any visible projection of light is insignificant. *Sconces* are decorative, wall-mounted luminaires that provide ambient illumination. They can direct light upward, downward, or in all directions.

High bay luminaires are used in high-ceiling areas, 20 ft (6 m) or higher, that require uniform illumination. High bay fixtures installed in a grid pattern are typically used to provide ambient illumination in warehouses, gymnasiums, and retail superstores. These luminaires are high wattage (200 W or more) so they can deliver light over long distances. The fixture size is fairly large, about 2 ft (0.6 m) for a typical fixture. *Low bay* luminaires are more compact. They are designed for use in low- to medium-ceiling areas, 12 to 20 ft (3 to 6 m). They are typically used for general illumination in offices, retail spaces, and loading dock areas. Both high bay and low bay luminaires can be used in uplighting and downlighting applications.

Directional lighting is illumination where light received at the work plane or light illuminating an object is incident predominantly from a particular direction. The only light that is important is the light that falls where it is needed. Spot, flood, and track lighting are common types of directional lighting.

A *spotlight* is a luminaire that is designed to emit an intense, concentrated beam of light with usually no more than a 20° divergence (spread) from where it is directed. A *floodlight* is a luminaire that emits an intense light that is broader than a spotlight and that is capable of being pointed in any direction. *Track lighting* is a directional lighting installation where luminaires are attached to and are moveable along a metal track. A more subtle method of directional lighting involves the use of one of the many different recessed downlights.

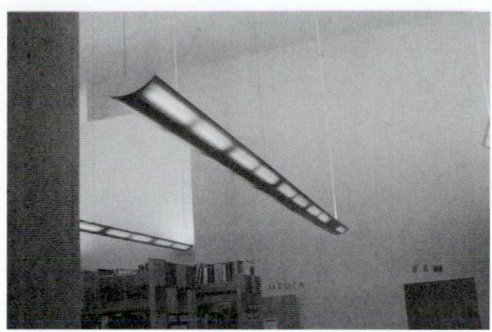

PHOTO 20.9   A linear fluorescent uplight used to provide ambient lighting in a library. (Used with permission of ABC)

PHOTO 20.10   An HID uplight luminaire. (Used with permission of ABC)

PHOTO 20.11   An HID uplight installation providing ambient lighting in a gymnasium. (Used with permission of ABC)

PHOTO 20.8   A decorative pendent light suspended from a ceiling, which provides ambient lighting. (Used with permission of ABC)

PHOTO 20.12   A recessed compact fluorescent luminaire. (Used with permission of ABC)

PHOTO 20.13   A recessed light fixture during construction. After it is secured to the joists, placement of the fixture can be adjusted (left to right). (Used with permission of ABC)

PHOTO 20.16   A low-bay HID luminaire. (Used with permission of ABC)

PHOTO 20.14   A pendent-type, recessed, compact fluorescent luminaire. (Used with permission of ABC)

PHOTO 20.17   A surface-mounted floodlight luminaire. (Used with permission of ABC)

PHOTO 20.18   A track lighting installation. (Used with permission of ABC)

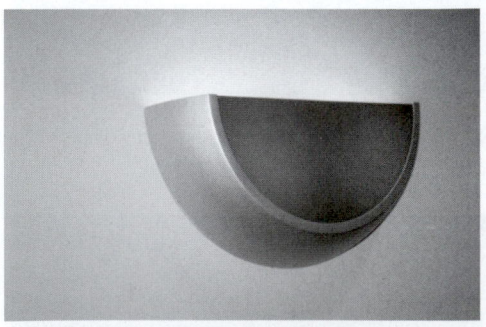

PHOTO 20.15   A decorative wall sconce. (Used with permission of ABC)

Diffused lighting is a lighting installation in which the light on the working plane or on an object is not incident predominantly from a particular direction. Diffused light is cast and disbursed over a large area. A *diffuser* is a lighting component such as a translucent glass refractor that redirects or scatters the light from a source. This type of lighting can be produced by a variety of *indirect lighting* techniques, where much of the light reaches the work plane after it has reflected off another surface such as a wall or ceiling. Diffused lighting is frequently used to provide soft, background lighting.

Cornice, cove, and valance lighting fit in a category of indirect, diffused lighting installations called *wallwashers*. *Cornice lighting* is a lighting installation where the light source is shielded by a panel that is parallel to the wall and attached to the ceiling; it distributes light over the wall. *Cove lighting* is an uplighting installation where the light source is shielded by a

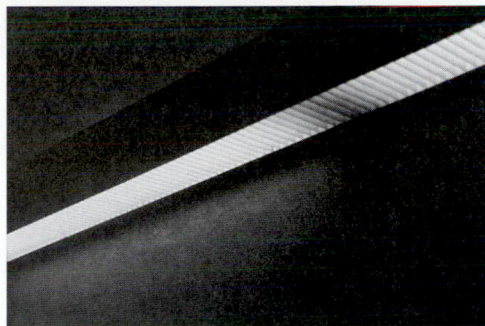

PHOTO 20.19 A valance lighting installation with linear fluorescent lamps. (Used with permission of ABC)

PHOTO 20.21 A linear fluorescent luminaire. Baffles create openings for emitting light from the fixture but cut down on glare. (Used with permission of ABC)

ledge or recess with light dispersed over the ceiling and upper wall. *Valance lighting* is a lighting installation where the light source is shielded by a panel that is parallel to the wall at the top of a window. (See Photo 20.19.).

Fluorescent lighting is widely used for general lighting installations in buildings. Fluorescent luminaires are available in recessed, surface-mounted, or pendent style fixture designs. Unlike the concentrated light emitted by other commercially available light sources, fluorescent lamps produce a diffuse light that is relatively glare free and visually comfortable. As a result, they have been and continue to be the standard for ambient illumination for classroom and office lighting installations.

Fluorescent luminaires can be *lensed* (have a flat lens to diffuse the light), *parabolic louvered* (have parabolic-shaped reflectors and open louvers to direct the light downward), or *wraparound* (have a lens that wraps around the lamps to diffuse and direct the light outward). A *troffer* is a linear luminaire constructed of an inverted metal trough that serves as a fixture for fluorescent lighting lamps. It is usually installed with the opening flush with the ceiling. Fluorescent luminaires can be incorporated in a *luminous ceiling* that is constructed of open fluorescent luminaires mounted above a translucent suspended ceiling. They can also be used in a *strip* installation consisting of rows of bare-bone fixtures with exposed lamps.

PHOTO 20.22 Recessed linear fluorescent luminaries with deep-cell parabolic louvers. Louvers create openings for emitting light and reflect light from the fixture, which reduces glare. (Used with permission of ABC)

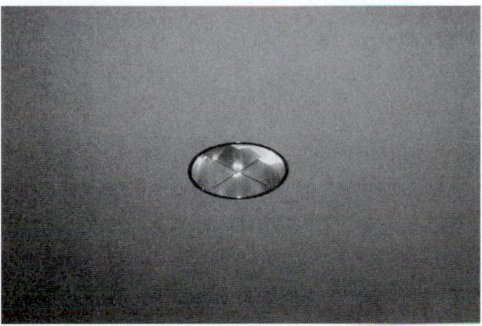

PHOTO 20.23 Louvers on this recessed compact fluorescent luminaire reduces glare. (Used with permission of ABC)

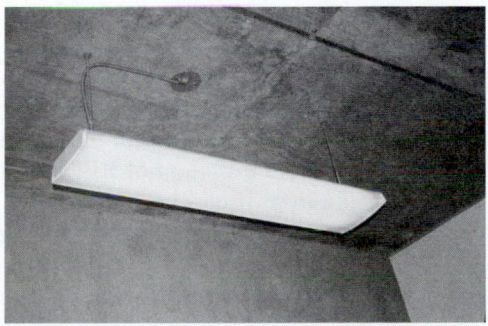

PHOTO 20.20 A linear fluorescent luminaire with prismatic wraparound lens that diffuses light. (Used with permission of ABC)

New lamp and luminaire designs offer excellent lighting solutions to specific lighting problems. Halogen and compact fluorescent lamps are small, offer low maintenance, and have excellent light and color qualities. Metal halide lamps have good color rendering qualities and work well in spaces with high ceilings. Because incandescent, halogen, and HID lamps produce a very intense light, a direct view of the lamp can be extremely irritating. Even with diffusion globes or lenses, the light can be very intense. HPS lamps are used when color rendition is not important such as in warehouses and factories. LPS lamps are rarely used in building interiors because they produce a monochromatic yellow light that renders even the most vibrant colors in shades of yellow.

Hazardous locations such as paint spray booths, flammable chemical areas, and grain mills, and corrosive environments such as in chemical and refinement plants, require luminaires that are approved for these types of use. Usually a testing agency such as Underwriters Laboratory, Inc. (UL), tests and lists approved fixtures for a specific use. Special fixtures and lamps are also required when a luminaire will be used in installations involving extreme temperatures.

With the drastic variations in wavelength composition of commercially available lamps, it is necessary that interior finishes (carpet, drapes, furniture) be selected by viewing them under light from the lamp that will be installed in the space. A slight change in lamp can cause a subtle difference in wavelengths emitted and in the shade of color seen. Investing in the proper lamp and fixture, one that delivers a specific type of light where it is needed, can reduce the amount of lighting required, and offset the cost of a more expensive luminaire.

Reflector Lamps

Reflector lamps are incandescent, compact fluorescent, or HID lamps with a built-in reflecting surface. Incandescent and HID versions are made from a single piece of blow-molded soft or hard glass. Compact fluorescent versions may be one piece or may be designed so that the inner lamp can be replaced.

An *elliptical reflector (ER) lamp* is an incandescent lamp with an elliptically shaped reflector. This shape produces a focal point directly in front of the lamp that reduces light absorption in some types of luminaries. It is particularly effective at increasing efficiency of baffled downlights.

A *parabolic aluminized reflector (PAR) lamp,* which may use an incandescent filament, a halogen filament tube, or HID arc tube, is a precision pressed-glass reflector lamp. PAR lamps rely on both the internal reflector and prisms in the lens for the control of the light beam.

Remote-Source Lighting Installations

Remote-source lighting (RSL) is an advanced lighting technology that transports light from a single source over a distance to one or more light outlets or emits light evenly along the way. It is typically more costly than traditional lighting installations, but can provide an aesthetic appeal that cannot be achieved by other means and therefore can be effective as accent or decorative

lighting. Additionally, health and safety benefits resulting from the RSL delivery system include a decrease in infrared and ultraviolet energy introduced into a space and improved safety in damp or explosive environments. Losses in the light distribution system generally make RSL less energy efficient than conventional systems. In many cases precision targeting of light can make up for those losses. Remote-source light is distributed using fiber optic or light pipe systems:

Fiber optic RSL systems consist of a light source; a set of reflectors, filters, and lenses to feed the light to the fiber optic cables; and a fixture to distribute the light at the point of illumination. The number of fixtures that can be fed from one light source depends on the intensity of the light source and the lighting requirements at the end of the run. For decorative applications where light distribution is not crucial, several hundred fixtures might be fed from one source.

Light pipe RSL systems feature a hollow tube with a reflective inner surface that directs light through the tube. The most common linings are prismatic films and mirrored surfaces. Most light pipe applications require side-emitting tubes that can carry artificial light or daylight. These feature a layer of translucent or transparent material surrounding a layer of prismatic light-reflecting film. The light is released evenly along the length of the tube. Mirrored surface tubes are most often used with daylight pipes that emit light at the end of the tube.

Most RSL systems use metal halide lamps or sunlight as the light source. Metal halide lamps provide high-quality light and high light intensities. The distribution system prevents radiation from these lamps from reaching the illuminated space. Filters can also be added to the light source to provide a variety of colorful effects. Use of sunlight reduces energy costs, improves productivity, and creates health benefits. The most common approach is the use of daylight pipes that carry light from the roof of a building into living or working spaces. The basic components include a clear plastic dome that sits on the roof and lets in sunlight; a reflective tube that carries light into the interior; and a light diffuser, which looks like a ceiling light fixture and distributes light around the room. Originally developed for residential applications, daylight pipes are also being used in commercial and industrial spaces. Use of lasers and LEDs appear to be promising light sources for the future of fiber optics.

The biggest light-source breakthrough for remote-source lighting could come from the evolution of LEDs. In the past, LEDs have been unable to provide high levels of brightness, but recent developments are changing the picture. The newest LEDs offer a brightness of 25 lumens per watt—far more than earlier versions and better than an incandescent lamp, but still short of fluorescent and metal halide sources. Laboratory prototypes have produced LEDs with efficiencies similar to those of fluorescent lamps.

RSL is more expensive than conventional lighting for most applications. It has aesthetic appeal in certain markets, such as sidewalk and building outline lighting; underwater lighting in pools, fountains, and aquariums; and special effects lighting, accent lighting, wall washing, and downlighting.

20.9 LIGHT DISTRIBUTION AND GLARE

Distribution of Light

One of the primary functions of a luminaire is to throw light only where it is needed. Light can be directed downward for maximum use or directed to the ceilings and upper walls and then reflected to all parts of a space. Table 20.15 shows the classification of light distribution produced by luminaires.

It is helpful for a designer to know how a specific luminaire broadcasts light. A *candlepower (candela) distribution curve* is a graphical representation that illustrates the luminous intensity around the cross- section of a lamp or luminaire. It is a curve on a polar graph that shows the relative luminous intensity, in candela, around the lamp or luminaire.

A candlepower distribution curve can be thought of as a cross-sectional map of intensity (candelas) measured at many different angles. It is a two-dimensional representation and therefore shows data for one plane only. If the distribution of the unit is symmetric, the curve in one plane is sufficient for all calculations. If asymmetric, such as with street lighting and fluorescent units, three or more planes are required. In general, a single vertical plane of photometry describes incandescent and HID reflector units. Fluorescent luminaires require a minimum of one plane along the lamp axis, one across the lamp axis, and one at a 45° angle.

Candlepower (candela) distribution curves of different types of downlighting luminaires are shown in Figures 20.13 and 20.14. The heavy dark line is a curve that represents a specific luminous intensity at various positions beneath the luminaire opening.

Figure 20.14 shows distribution curves for three different compact fluorescent luminaires. In this figure, it is evident that distribution is characteristic of a specific luminaire. A candlepower distribution curve for a specific luminaire is available from the luminaire manufacturer. This information is useful to the designer because it shows how much light is emitted in each direction and the relative proportions of downlighting and uplighting.

TABLE 20.15   THE LIGHT DISTRIBUTION PRODUCED BY LUMINAIRES IS CATEGORIZED AS SHOWN BELOW.

Luminaire Type	Distribution Produced by Luminaires
Direct	90 to 100% of the light is directed downward for maximum use
Indirect	90 to 100% of the light is directed to the ceilings and upper walls and is reflected to all parts of a room or 90 to 100% of the light is directed downward for maximum use
Semidirect	60 to 90% of the light is directed downward with the remainder directed upward
General diffuse or direct-indirect	Equal portions of the light are directed upward and downward
Highlighting	Beam projection distance and focusing ability characterize this luminaire

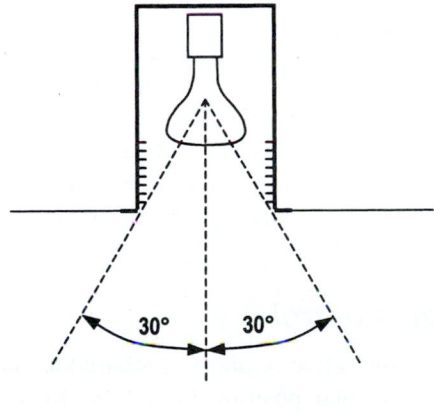

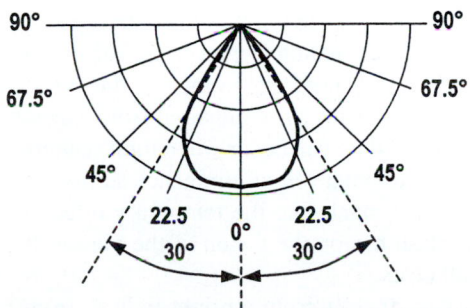

FIGURE 20.13 A candela (candlepower) distribution curve is a polar graph that represents the luminous intensity of a lamp or luminaire at measured horizontal distances and vertical angles. The heavy, dark line in the curve represents luminous intensity at various positions beneath the luminaire opening.

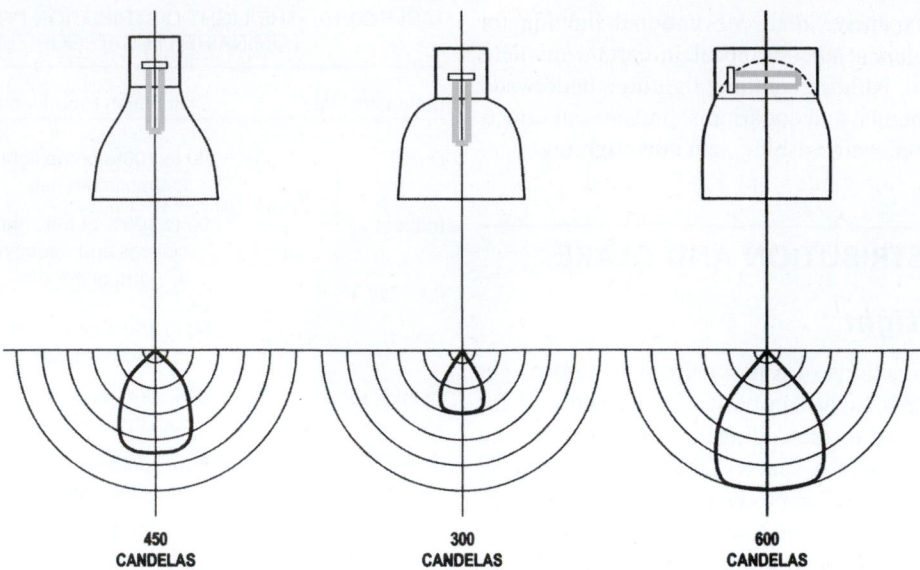

FIGURE 20.14 Shown are candlepower (candela) distribution curves for three different compact fluorescent luminaires. Candlepower distributions are one of the basic ways of comparing lighting system performance.

Glare

Glare is excessive brightness in the field of vision that causes discomfort or, in extreme cases, produces a disability from a temporary loss of vision. It is a visual sensation caused by luminance that is sufficiently greater than the luminance to which the eyes are adapted.

Glare can be an irritant of or impediment to vision. *Discomfort glare* causes visual discomfort without necessarily impairing vision. To prevent the design of lighting installations that produce discomfort, the visual comfort probability is calculated. The *visual comfort probability (VCP)* is a rating of a lighting installation expressed as a percentage of individuals who, when viewing from a specified location and in a specified direction, will be expected to find it acceptable in terms of discomfort glare. In most spaces, the VCP should exceed %; that is, at least 70% of a statistical group of individuals will rate glare levels acceptable in the space being illuminated.

Guidelines from the Illuminating Engineering Society of North America (IESNA) are used to compute the VCP. *Disability glare* occurs when visibility is impaired from excessive brightness. This type of glare can occur when light scatters in the eye in such a way as to put a veil of luminance across the retina. This type of glare is much like the temporary effect a close-up photographic flash has on the vision of the viewer. It should be avoided at all costs.

Glare can be caused directly from a bright light or from the reflection of light off a glossy surface. *Direct glare* occurs when excessive light enters the eye directly from a light source. It can occur even if the eye is not looking directly at the source. Problems with direct glare arise from improper placement of the light source, an exposed, bright light source, or an improperly installed or poorly designed light fixture. The lighting professional must anticipate the position of occupants in a space and then work to properly select and position light sources to avoid direct glare.

Reflected glare is the result of light entering the eye after reflecting off a glossy surface. This type of glare can occur if a light source is positioned in such a way that its reflected light creates a veil of luminance across the visual task. A common type of reflected glare is called a *veiling reflection*—the reflection on a glossy surface that impedes seeing such as the reflection on a computer screen or on a glossy sheet in a magazine. Veiling reflections occur in a work space when the light source is positioned above or slightly in front of the visual task. Instead, light sources should be positioned to the side of the visual task without casting shadows. A simple test for veiling reflections is to place a mirror on the visual task. If the image of any direct light source appears in the mirror, then veiling reflections are present and must be removed.

Glare Control

Preventing glare requires special attention to fixture design, selection, and position to prevent light from entering the glare zone. The *glare zone* is defined as a zone above a 45° angle from the fixture's vertical axis. When light is emitted into the glare zone, direct glare and reflected glare are likely, so light emitted in the glare zone can cause visual discomfort. Positioning and selecting light fixtures that do not throw light in the glare zone is a good way to reduce problems with glare.

In small spaces, indirect lighting techniques consisting of cornice, cove, and valance lighting installations diffuse light well and prevent glare. As a shielded light source, an uplight directs its light to the ceiling, where it is reflected back to the space. Uplights also prevent glare and work well in larger spaces such as gymnasiums and conference spaces. Properly designed downlights use baffles to prevent light from being thrown into the glare zone. Most linear fluorescent fixtures use either a lens or a louver to prevent direct viewing of the lamps so direct glare is prevented.

Lenses are typically made from clear UV-stabilized plastics. Clear lens types include prismatic, batwing, linear batwing, and polarized lenses. White translucent diffusers are much less efficient than clear lenses, and they result in relatively low visual comfort probability. New low-glare lens materials are available for retrofit and provide high visual comfort (a VCP of 80 or more) and high efficiency. Lenses are usually much less expensive but they provide less glare control than louvered fixtures.

Louvers are slats in a light fixture that create openings for emitting light. Louvers provide superior glare control and high visual comfort compared with lens-diffuser systems. The most common application of louvers is to eliminate the fixture glare reflected on computer screens. Deep-cell parabolic louvers (with 5 to 7 in cell apertures and depths of 2 to 4 in) provide a good balance between visual comfort and luminaire efficiency. Although small-cell parabolic louvers provide the highest level of visual comfort, they reduce luminaire efficiency to about 40%. A disadvantage of the deep-cell louver is that it adds 2 to 4 in to the overall depth of a luminaire.

Baffles are opaque or translucent elements that shield a light source from direct view. They are typically part of the light fixture. In the cross-section of the light fixture shown in Figure 20.13, baffles prevent the light source from throwing light beyond a 30° angle from the fixture's vertical axis. Baffles keep the light from being broadcast into the glare zone.

20.10 ILLUMINANCE AND LUMINANCE

As described previously, *illuminance* is the amount of light incident on (striking) a surface. *Luminance* is the amount of light leaving an object, thus relating to how bright an object appears to the human eye. Although an object may be illuminated, the eye cannot see it without visible light leaving (reflecting off) the object in the direction of the viewer. The luminance of a surface is equal to the reflected illuminance.

Luminance

Luminance is a luminous intensity of a surface in a given direction per unit of projected area of the surface. Luminance can be expressed in two ways: in candelas per unit area or in lumens per unit area. There are many different standard units of measurement. For example: *Candela per square inch (cd/in²),*

foot lambert (luminance of a surface emitting one lumen per square foot), and the *lambert* (luminance of a surface emitting one lumen per square cm). The relationships behind these units of measurement are:

$$1 \ cd/in^2 = 452 \ foot \ lamberts$$
$$1 \ lambert = 929 \ foot \ lamberts = 2.054 \ cd/in^2$$

It is not necessary to go beyond these brief definitions, because luminance is seldom used.

Illuminance

Illuminance it is the density of luminous flux striking a surface. It describes how bright a surface *may* appear to the human eye depending upon the surface's reflectance. The appropriate units of measurement for illuminance are the foot-candle and lux. The *foot-candle (fc)* is the unit of measure of illuminance (E), the intensity of light falling on a surface. It is equal to one lumen of light uniformly illuminating a surface over an area of one square foot:

$$1 \ fc = 1 \ lm/ft^2$$

The foot-candle, the oldest physical unit still in common use, was originally defined as the illuminance striking a surface held one foot away from a burning beeswax candle. Foot-candles incident on a surface are related to the lumens (lm) striking the surface and the area of the surface (A), in ft^2:

$$E = lm/A$$

The SI (metric) unit for illuminance is the *lux (lx),* which is one lumen of light uniformly illuminating a one square meter surface (lm/m^2). Lux incident on a surface are related to the lumens (lm) striking the surface and the area of the surface (A), in m^2: $E = lm/A$.

One foot-candle is roughly equal to 10 lux:

$$foot\text{-}candle \ (fc) = 10.764 \ lx$$
$$lux = 0.0929 \ fc$$

Typical values of illuminance for selected conditions are provided in Table 20.16. By comparison, natural sunlight illuminates a surface 100 000 times more than light cast by a full moon on clear night.

TABLE 20.16   TYPICAL VALUES OF ILLUMINANCE FOR SELECTED CONDITIONS.

Condition	Approximate Illuminance	
	Foot-Candles	Lux
Light cast by full moon on clear night	0.1	1
Street lighting	1	10
Work space lighting	10 to 100	100 to 1000
Lighting for surgery	1000	10 000
Natural sun light on clear day	6 000	60 000

Example 20.4

Assume that all of the light emitted by a 150 W incandescent lamp (2850 lm) illuminates a 100 ft^2 (9.29 m^2) surface.

a. Determine the average illuminance of the surface in foot-candles.

$$E = 2850 \text{ lm}/100 \text{ ft}^2 = 28.5 \text{ fc}$$

b. Determine the average illuminance of the surface in lux.

$$E = 2850 \text{ lm}/9.29 \text{ m}^2 = 307 \text{ lx}$$

In Example 20.4 (although highly theoretical because not all the light emitted from a lamp will strike a surface), 28.5 foot-candles (307 lx) of light illuminates the surface. This level of illuminance is bright enough for visual tasks of high contrast or large size such as reading newsprint or writing. It is the level of lighting commonly found in office and conference spaces, excluding local lighting used to light a task. In contrast, an illuminance of 300 fc (3000 lx) would be required to perform sewing and precision machining because these tasks involve working with low contrast and very small size. Even higher levels of illuminance, say above 750 fc (7500 lx), may be required for prolonged, visually demanding tasks such as transplant surgery or exacting inspection.

Illuminance at a specific location in a room or space can be measured with a portable, hand-held *photometer* that is commonly called a *light meter*. The IESNA prescribes exact methods of determining illuminance in its *Lighting Handbook*. Light meter readings at a specific location are generally performed by holding or placing the meter in a horizontal position on a desk or tabletop in the location where the task will be performed. In special instances such as warehouse shelving and library book stacks, the meter is held vertically. Care must be exercised to ensure that shadows from the meter user do not improperly influence results.

Average illuminance in a space can be determined by dividing the room into a grid pattern that does not correspond with the spacing of the light fixtures and averaging the readings. Because of site conditions and manufacturing tolerances, average measured values may vary by 20% or more from calculated design values.

Recommended Illuminance

Historically, the chief goal of architectural lighting was to introduce as much light as possible to a building interior; that is, more light was good. The result was large lighting installations that consumed a lot of energy. Current thinking is that too much light is excessive because it is not needed for less-difficult seeing tasks and is wasteful because of the higher investment, maintenance, and energy (electricity and cooling) costs; that is, just enough light is better. Simply, a task requiring observation of high contract and large objects requires less light than one involving small objects and low contrast.

Light Output Versus Light Intensity

Light output (measured in lumens) and light intensity (measured in foot-candles) are two entirely different concepts. A small battery-powered flashlight has a relatively small light output. Its beam, when focused closely onto a very small area, appears bright. It may provide more light on a concentrated area in comparison to a 100 W table lamp. The distribution of the light output determines how bright lighting is. Distribution of light depends on the characteristics of the source and the distance the light has to travel.

IESNA publishes recommended lighting levels for various tasks and spaces. Recommended levels of illuminance are published in the IESNA *Lighting Handbook*. IESNA divides these levels of illuminance into nine categories, each with three ranges of illuminance. For example, the three ranges for a conference room are cited as 20-30-50 fc (200-300-500 lx). The three ranges of illuminance are based on occupant age, room surface reflectances, task performance, and safety. Higher room surface reflectances (greater than 70%) or younger occupants (age under 40) require less lighting, so the lower value of the three ranges is used.

The illuminance categories are ranked in alphabetical order (A through I) by type of activity and seeing difficulty as outlined in the following:

- Categories A through C have ranges of illuminance from 2-3-5 to 10-15-20 fc (20-30-50 to 100-150-200 lx). They are for ambient lighting in public and working spaces, including corridors, lobbies, stairs, and conversation areas (e.g., restaurants and exhibition areas).
- Categories D through F have ranges of illuminance from 20-30-50 to 100-150-200 fc (200-300-500 to 1000-1500-2000 lx). They are for common visual tasks requiring more difficulty such as in classrooms, office and drafting areas, laboratories, and kitchen areas.
- Categories G through I have ranges of illuminance from 200-300-500 to 1000-1500-2000 fc (2000-3000-5000 to 10 000-15 000-20 000 lx). They are for prolonged, visually demanding tasks involving sewing, precision machining, and surgery.

Recommendations for levels of lighting in various spaces are provided in Table 20.17. In general, recommended lighting levels increase as the size and contrast of the visual task decrease. Thus, the recommended lighting level will be near the lower level of the range when the size and/or contrast of the visual task is large and will be near the upper level of the range when the size and/or contrast of the visual task is small. Governmental entities may specify levels of illumination.

TABLE 20.17   RECOMMENDATIONS FOR LEVELS OF LIGHTING IN VARIOUS SPACES.

Space or Area	Foot-Candles	Lux
Schools		
Corridors	20	200
Auditoriums	20 to 30	200 to 300
Study halls	70 to 80	700 to 800
Classrooms	70	700
Whiteboard area	150	1500
Drafting rooms	80 to 100	800 to 1000
Laboratories	80	800
Art rooms	100	1000
Sewing rooms	120 to 150	1200 to 1500
Shop areas	100	1000
Gymnasium	50	500
Stores		
Circulation areas	30	300
Merchandise areas	100 to 150	1000 to 1500
Showcases	150 to 200	1500 to 2000
Industrial		
General lighting	30 to 50	300 to 500
Laundries	30 to 50	300 to 500
Locker rooms	30	300
Machines, rough	40 to 50	400 to 500
Machines, fine	80 to 100	800 to 1000
Intricate work	150 to 400	1500 to 4000
Assembly areas	50 to 100	500 to 1000
Theaters		
Entrance, lobby	30 to 40	300 to 400
Foyer	30 to 40	300 to 400
Auditorium	10	100
Post Office		
Lobby	30	300
Mail sorting area	100	1000
Corridors and storage	30	300

Minimum levels of lighting (abridged) from Canada's Occupational Health and Safety Regulations are provided in Table 20.18.

20.11   PRINCIPLES OF LIGHTING DESIGN

Basic Design Approach

Proper design of a lighting installation requires use of two calculation procedures: computing illuminance from a point source and computing average illuminance levels. Calculation of illuminance at specific points in a lighting installation is done to evaluate lighting uniformity, especially when using luminaires where maximum spacing recommendations are not supplied or where task lighting levels must be confirmed.

Computations for Illuminance from a Single Point Source

Computing illuminance from a point is complex and generally involves computer analysis. Design is usually based on a designer's experience with different sources. Analysis is, however, founded on the principle that the intensity of light traveling away from a source decreases as it gets farther from the source; that is, a constant beam of light is spread over a larger area as the plane being illuminated is held farther away from the light. See Figure 20.15.

Technically, the *inverse square law* states that the illuminance (E) at a point on a plane perpendicular to the line joining the point and a source is inversely proportional to the square of the distance between the source and the plane (d), where the intensity of the source is expressed in luminous intensity (I):

$$E = I/d^2$$

This law is theoretical and generally does not apply at great distances because it is unrealistic to assume that all the light emitted by a distant source actually strikes the plane being illuminated. As shown in Example 20.5, illuminance decreases rapidly as the source is moved away from the plane being illuminated. This makes a strong case for placement of the luminaire as close to a task as is physically possible. It also suggests that task lighting rather than ambient lighting is a good choice when high illumination levels are desired.

Example 20.5

A 500-candela light source (a source that gives off 500 lm of luminous flux in one direction) emits a beam of light. Determine the illuminance (E), in foot-candles, on a surface held perpendicular to and in line with the light beam at the following distances:

a. at 1 ft

$$E = I/d^2 = 500 \text{ cd}/(1 \text{ ft})^2 = 500 \text{ fc}$$

b. at 10 ft

$$E = I/d^2 = 500 \text{ cd}/(10 \text{ ft})^2 = 5 \text{ fc}$$

c. at 100 ft

$$E = I/d^2 = 500 \text{ cd}/(100 \text{ ft})^2 = 0.05 \text{ fc}$$

Symbols Used in Lighting and Electrical System Design

In electrical design, the symbol E is commonly used for voltage and the symbol I for amperage. In lighting design, E is used as a symbol for illuminance and I is the symbol for luminous intensity. The reader is cautioned that the lighting and electrical design professions use common symbols with different meanings.

TABLE 20.18 MINIMUM LEVELS OF LIGHTING (ABRIDGED) FROM CANADA'S OCCUPATIONAL HEALTH AND SAFETY REGULATIONS.

Task Position or Area	Lux	Foot-Candles
Desk Work		
Task positions at which cartography, designing, drafting, plan-reading, or other very difficult visual tasks are performed	1000	100
Task positions at which business machines are operated or stenography, accounting, typing, filing, clerking, billing, continuous reading or writing, or other difficult visual tasks are performed	500	50
Video Display Terminal (VDT) Work		
Task positions at which data entry and retrieval work are performed intermittently	500	50
Task positions at which data entry work is performed exclusively	750	75
Air traffic controller areas	100	10
Telephone operator areas	300	30
Other Office Work		
Conference and interview rooms, file storage areas, switchboard or reception areas, or other areas where ordinary visual tasks are performed	300	30
Service Areas		
Stairways and corridors that are used frequently	100	10
Stairways and corridors that are used infrequently	50	5
Stairways that are used only in emergencies	30	3
Garages		
Main repair and maintenance areas (with some exceptions)	300	30
Main repair and maintenance areas used for repairing and maintaining cranes, bulldozers, and other major equipment	150	15
General work areas adjacent to a main repair and maintenance area	50	5
Fueling areas	150	15
Battery rooms	100	10
Laboratories		
Areas in which instruments are read and where errors in such reading may be hazardous to the health or safety of an employee	750	75
Areas in which a hazardous substance is handled	500	50
Areas in which laboratory work requiring close and prolonged attention is performed	500	50
Areas in which other laboratory work is performed	300	30
Loading Platforms, Storage Rooms, and Warehouses		
Areas in which packages are frequently checked or sorted	250	25
Areas in which packages are infrequently checked or sorted	75	7.5
Docks (indoor and outdoor), piers, and other locations where packages or containers are loaded or unloaded	150	15
Areas in which grain or granular material is loaded or unloaded in bulk	30	3
Areas in which goods are stored in bulk or where goods in storage are all of one kind	30	3
Areas where goods in storage are of different kinds	75	7.5
Machine and Woodworking Shops		
Areas in which medium or fine bench or machine work is performed	500	50
Areas in which rough bench or machine work is performed	300	30
Manufacturing and Processing Areas		
Major control rooms or rooms with dial displays	500	50
Areas in which a hazardous substance is processed, manufactured, or used in main work areas	500	50
Areas in which a hazardous substance is processed, manufactured, or used in surrounding areas	200	20
Areas in which substances that are not hazardous substances are processed, manufactured, or used, or where automatically controlled equipment operates in main work areas	100	10
Areas in which substances that are not hazardous substances are processed, manufactured, or used, or where automatically controlled equipment operates in surrounding areas	50	5

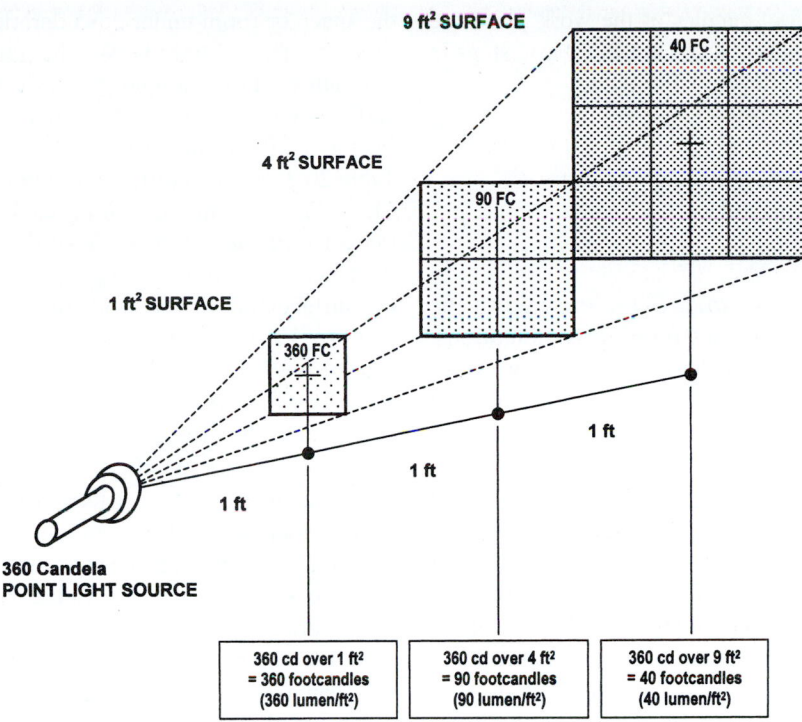

FIGURE 20.15 A beam of light spreads over a larger area as the plane (surface) being illuminated is held farther away from the light source. Illuminance (E) is the density of luminous flux, the intensity of light on a surface. The illuminance of a surface decreases as the surface gets farther from the light source.

A surface held perpendicular to a light source receives more light than a surface held at an angle to the source. Most times, the surface being illuminated is not perpendicular to the light source. The *cosine law of incidence* states that illuminance (E) at a point on a plane is proportional to the cosine of the angle of light incidence, where the angle of light incidence is measured between the direction of the incident light and the normal to the plane of the surface.

$$E = (I/d^2) \cos \theta$$

Example 20.6

A 500-candela light source (a source that gives off 500 lm of luminous flux in one direction) emits a beam of light to point on a drafting table. The source is held 2 ft above the table surface and the drafting surface is sloped at various angles measured from horizontal.

a. Determine the illuminance (E), in foot-candles, on a drafting surface at an angle of 10° from horizontal.

$$E = I/d^2 \cos \theta = [500 \text{ cd}/(2 \text{ ft})^2](\cos 10°) = 123 \text{ fc}$$

b. Determine the illuminance (E), in foot-candles, on a drafting surface at an angle of 45° from horizontal.

$$E = I/d^2 \cos \theta = [500 \text{ cd}/(2 \text{ ft})^2](\cos 45°) = 88 \text{ fc}$$

c. Determine the illuminance (E), in foot-candles, on a drafting surface at an angle of 60° from horizontal.

$$E = I/d^2 \cos \theta = [500 \text{ cd}/(2 \text{ ft})^2](\cos 60°) = 63 \text{ fc}$$

Average Illuminance Levels at the Work Plane

The designer is frequently interested in achieving a specific level of illumination at a reference work plane throughout the room or space; for example, 15 fc may be acceptable on the floor of a lobby, 30 fc on a table in a conference room, or 75 fc on tables in a drafting production area. This *target illumination* is actually the average illuminance at a reference work plane. The *work plane* is an area, usually horizontally positioned, at which work is performed and on which the illuminance is specified and measured. It may be a plane at desk or tabletop level (usually 28 to 30 in above the floor) or at counter or bench height (36 to 48 in above the floor).

Calculating Average Illuminance Levels

The *zonal cavity method* is the currently accepted method for calculating average illuminance levels for indoor areas unless the light distribution is extremely asymmetric. The basic principle behind this method is that foot-candles at the work plane are equal to flux over the work plane. There are four basic steps in any calculation of illuminance level:

1. Determine cavity ratio (CR)
2. Determine effective cavity reflectances
3. Select coefficient of utilization
4. Compute average illuminance level

In these computations, the *area (A)* is the area of the space to be illuminated. In ambient lighting computations, this is the floor area or the area of the work plane; for example, a 30 ft by 40 ft room that will be uniformly illuminated will have an area of 1200 ft^2. The illuminance (E) is the desired target illuminance or, in the case of an existing installation, the illumination at the work plane. In design, target values for illuminance are usually extracted from IESNA standards.

Lumen output (LM) is generally based on the number of lumens initially output by each lamp. However, lumen output of a lamp decreases over its life because of decrease in lamp output as the lamp ages (referred to as *lamp lumen depreciation*) and because the collection of dirt and dust on the luminaire (referred to as *luminaire surface depreciation*).

The *light loss factor (LLF)* is the product of all considered factors that contribute to a lighting installation's reduced light output over a period of time. It takes into account dirt accumulation on luminaires and room surfaces, lamp depreciation, maintenance procedures, and atmosphere conditions. In a lighting installation, the LLF accounts for the reduction of light output while in service. It allows for acceptable operation at less than initial conditions.

The *number of lamps in luminaire (n)* is the quantity of lamps found in the specified or existing luminaire. The *number of luminaires (#)* is the minimum required or the given number of luminaires in the space.

The *coefficient of utilization (CU)* is the ratio of the amount of light (lumens) illuminating the work plane to the output of the lamp(s) in a fixture. It is generally found in technical literature available through the light fixture manufacturer (similar to those shown in Figure 20.16). The CU is unique to a type of luminaire, the geometry of the space (room cavities), and the reflectances of the surfaces or cavities in the space.

A room is made up of three spaces called *cavities* (see Figure 20.17). The space between the luminaires and the work plane is referred to as the *room cavity;* the space between the work plane and the floor is called the *floor cavity;* and the space between the ceiling and the luminaires (if they are suspended) is defined as the *ceiling cavity.* Again, for our purposes, computing the room CR is sufficient. CR will vary based on the geometry of the space, as shown in Figure 20.18.

Identification of the CU begins with establishing the effective reflectance of the ceiling, room and floor cavities of the space or room under consideration. Reflectance (ρ) is the ratio of reflected light versus the light striking the surface (illuminance). For example, the reflectance of medium-to-dark surfaces tends to be in the range of 0.10 to 0.40 while reflectance of a white surface may be 0.85 (85% is reflected). Table 20.2 provides reflectances of common colored surfaces. The reflectance of the ceiling cavity (ρ_{cc}) is the average reflectance of the surfaces above the luminaire plane. The reflectance of the floor cavity (ρ_{fc}) is the average reflectance of the surfaces below the work plane. Reflectance of wall surfaces (ρ_w) is for the wall between the plane of the luminaires and the work plane.

The IESNA *Handbook* prescribes a detailed method of determining the effective reflectance of the ceiling, walls, and floor cavities for a specific space. In this method, effective cavity reflectances must be determined for the ceiling cavity and for the floor cavity. These are determined based on the cavity ratio and actual reflectance of ceiling, walls, and floor.

To keep things uncomplicated for the purposes of our study, reflectance values of these surfaces and cavities may be approximated based on good judgment. Keep in mind that dark wall hangings or widows (that reflect little) will result in reflectances that are much lower than the reflectance of the painted wall or ceiling surfaces. Therefore, although the anticipated finish of a wall may be off-white in color with a reflectance of 0.80 to 0.85 (from Table 20.2), windows in the wall will prompt the designer to choose a lower reflectance value; for example, an average reflectance of wall surfaces of 0.50.

In addition to reflectance of walls, ceilings, and floors, the geometry of the space will influence the CU of a luminaire. For example, luminaires close to the work plane will provide more illuminance of the work plane than the same fixtures held a great distance from the plane. Because most spaces are not simple geometric spaces, the zonal cavity method is used to determine the cavity ratio in a space. The *cavity ratio (CR)* is a number indicating cavity proportions from length, width, and height.

The CR is determined by either of the following formulas:

$$CR = [5\ MH(L + W)]/LW$$

or

$$CR = (2.5\ MHp)/A$$

where

MH: distance between the plane of the luminaires and reference work plane, in inches, feet, or meters.
L: length of the space, in feet or meters
W: width of the space, in feet or meters
p: perimeter of space, in feet or meters
A: area of the space to be illuminated, in square feet or square meters

The first formula works well for rectangular-shaped rooms. The second formula applies to odd-shaped rooms.

Surface Prismatic Wraparound Fluorescent Luminaire (Four Lamps)

	Spacing Criteria (CS)	Reflectance	Coefficient of Utilization (CU)											
4 Linear Fluorescent Lamps		Ceiling cavity (ρ_{cc})	80			50			30			10		
		Wall Surfaces (ρ_w)	50	30	10	50	30	10	50	30	10	50	30	10
	1.5⊥ 1.2‖	Cavity Ratio (CR)												
		10	.26	.21	.17	.24	.19	.16	.22	.18	.15	.21	.17	.17
		9	.29	.23	.19	.26	.22	.18	.25	.21	.18	.23	.20	.17
		8	.31	.27	.23	.29	.25	.21	.28	.24	.21	.26	.23	.20
		7	.35	.29	.26	.32	.28	.25	.30	.27	.24	.29	.26	.23
		6	.39	.33	.29	.36	.31	.29	.33	.29	.25	.31	.29	.26
		5	.44	.38	.34	.39	.35	.32	.37	.33	.30	.35	.31	.29
		4	.48	.44	.39	.44	.40	.37	.41	.38	.35	.39	.36	.33
		3	.54	.49	.46	.48	.46	.43	.46	.43	.40	.43	.40	.38
		2	.61	.56	.53	.54	.51	.48	.50	.58	.47	.47	.46	.44
		1	.67	.66	.63	.61	.59	.57	.56	.55	.53	.52	.51	.50
		0	.76	.76	.76	.67	.67	.67	.63	.63	.63	.57	.57	.57

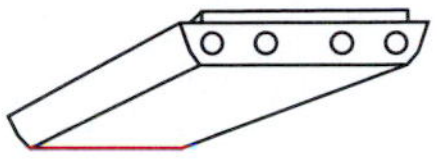

Surface Prismatic Wraparound Fluorescent Luminaire (Two Lamps)

	Spacing Criteria (CS)	Reflectance	Coefficient of Utilization (CU)											
2 Linear Fluorescent Lamps		Ceiling Cavity (ρ_{cc})	80			50			30			10		
		Wall surfaces (ρ_w)	50	30	10	50	30	10	50	30	10	50	30	10
	1.5⊥ 1.2‖	Cavity Ratio (CR)												
		10	.27	.22	.18	.25	.20	.17	.23	.19	.16	.22	.18	.18
		9	.30	.24	.20	.27	.23	.19	.26	.22	.19	.24	.21	.18
		8	.33	.28	.24	.30	.26	.22	.29	.25	.22	.27	.24	.21
		7	.37	.31	.27	.34	.29	.26	.32	.28	.25	.30	.27	.24
		6	.41	.35	.31	.38	.33	.30	.35	.31	.26	.33	.30	.27
		5	.46	.40	.36	.41	.37	.34	.39	.35	.32	.37	.33	.31
		4	.51	.46	.41	.46	.42	.39	.43	.40	.37	.41	.38	.35
		3	.57	.52	.48	.51	.48	.45	.48	.45	.42	.45	.42	.40
		2	.64	.59	.56	.57	.54	.51	.53	.61	.49	.49	.48	.46
		1	.71	.69	.66	.64	.62	.60	.59	.58	.56	.55	.54	.53
		0	.80	.80	.80	.71	.71	.71	.66	.66	.66	.60	.60	.60

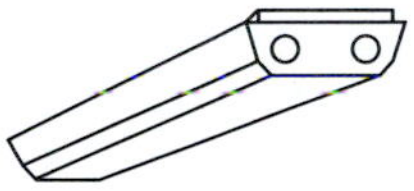

High Bay Intermediate-Distribution Reflector HID Luminaire

	Spacing Criteria (CS)	Reflectance	Coefficient of Utilization (CU)											
1 HID Lamp		Ceiling Cavity (ρ_{cc})	80			50			30			10		
		Wall Surfaces (ρ_w)	50	30	10	50	30	10	50	30	10	50	30	10
	1.0	Cavity Ratio (CR)												
		10	.38	.33	.29	.37	.32	.29	.36	.32	.29	.35	.31	.28
		9	.41	.36	.31	.40	.35	.32	.39	.35	.32	.38	.35	.32
		8	.45	.40	.36	.44	.39	.36	.43	.39	.35	.46	.42	.39
		7	.50	.44	.45	.52	.48	.45	.51	.47	.44	.50	.47	.44
		6	.54	.49	.45	.52	.46	.45	.51	.47	.44	.50	.47	.44
		5	.59	.54	.50	.57	.53	.50	.56	.52	.49	.54	.51	.48
		4	.65	.60	.56	.62	.58	.55	.60	.57	.54	.59	.56	.54
		3	.71	.66	.63	.67	.64	.61	.65	.62	.60	.63	.61	.59
		2	.77	.73	.70	.73	.70	.68	.70	.68	.66	.68	.66	.65
		1	.84	.81	.79	.79	.77	.76	.76	.74	.73	.73	.72	.71
		0	.91	.91	.91	.84	.84	.84	.81	.81	.81	.77	.77	.77

FIGURE 20.16 Coefficients of utilization for selected fixtures.

Sphere-Shaped Pendant Luminaire												
1 Incandescent or Compact Fluorescent Lamp	Spacing Criteria (CS)	Reflectance	Coefficient of Utilization (CU)									
		Ceiling Cavity (ρ_{cc})	80			50			30			10		
		Wall Surfaces (ρ_w)	50	30	10	50	30	10	50	30	10	50	30	10
		Cavity Ratio (CR)												
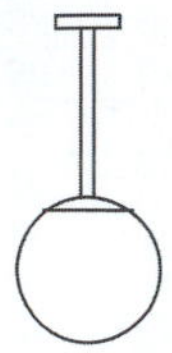		10	.23	.17	.13	.19	.14	.10	.16	.12	.09	.13	.09	.07
		9	.26	.19	.15	.20	.15	.12	.17	.13	.10	.14	.11	.08
		8	.29	.22	.17	.23	.17	.14	.19	.15	.12	.15	.12	.09
		7	.32	.25	.20	.25	.20	.16	.21	.16	.13	.17	.13	.11
		6	.36	.28	.23	.28	.23	.19	.23	.19	.16	.19	.15	.13
		5	.40	.33	.27	.32	.26	.22	.26	.22	.19	.21	.18	.15
		4	.46	.38	.33	.36	.30	.26	.30	.26	.22	.24	.21	.18
		3	.52	.45	.39	.41	.36	.31	.34	.30	.28	.27	.24	.22
		2	.61	.54	.49	.47	.43	.39	.39	.36	.33	.32	.29	.27
		1	.71	.67	.63	.56	.53	.50	.47	.45	.43	.39	.37	.35
		0	.87	.87	.87	.69	.69	.69	.59	.59	.59	.49	.49	.49

Open Reflector Recessed Downlight Luminaire (Vertically Mounted Lamp)												
1 Compact Fluorescent Lamp	Spacing Criteria (CS)	Reflectance	Coefficient of Utilization (CU)									
		Ceiling Cavity (ρ_{cc})	80			70			50			30		
		Wall Surfaces (ρ_w)	50	30	10	50	30	10	50	30	10	50	30	10
	0.8	Cavity Ratio (CR)												
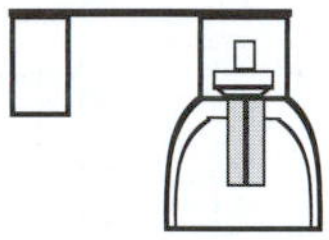		10	.42	.38	.36	.42	.38	.36	.41	.38	.36	.41	.38	.35
		9	.45	.41	.38	.45	.41	.38	.44	.41	.38	.44	.40	.38
		8	.48	.44	.42	.48	.44	.41	.47	.44	.41	.46	.43	.41
		7	.51	.48	.45	.51	.47	.45	.50	.47	.45	.49	.46	.44
		6	.55	.51	.49	.54	.51	.48	.53	.50	.48	.53	.50	.48
		5	.59	.55	.52	.58	.54	.52	.57	.54	.51	.56	.53	.51
		4	.62	.59	.56	.62	.59	.56	.60	.57	.55	.59	.57	.55
		3	.66	.63	.61	.65	.63	.60	.64	.61	.59	.62	.60	.58
		2	.71	.68	.66	.70	.67	.65	.67	.66	.64	.66	.64	.63
		1	.75	.74	.73	.74	.73	.71	.71	.70	.69	.69	.68	.67

Open Reflector Recessed Downlight Luminaire (Horizontally Mounted Lamp(s))												
1 or 2 Compact Fluorescent Lamp(s)	Spacing Criteria (CS)	Reflectance	Coefficient of Utilization (CU)									
		Ceiling Cavity (ρ_{cc})	80			70			50			30		
		Wall Surfaces (ρ_w)	50	30	10	50	30	10	50	30	10	50	30	10
	1.5	Cavity Ratio (CR)												
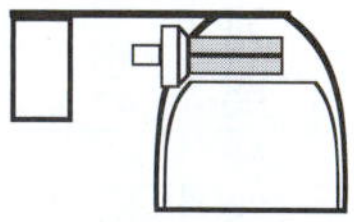		10	.31	.27	.24	.38	.27	.24	.30	.26	.24	.30	.26	.24
		9	.34	.30	.27	.41	.30	.27	.33	.29	.27	.33	.29	.27
		8	.38	.33	.30	.44	.33	.30	.37	.33	.30	.36	.32	.30
		7	.41	.37	.34	.47	.37	.34	.40	.36	.34	.39	.36	.33
		6	.45	.41	.38	.51	.41	.38	.44	.40	.38	.43	.40	.37
		5	.50	.45	.42	.54	.45	.42	.48	.44	.42	.47	.44	.41
		4	.54	.50	.47	.58	.50	.47	.52	.49	.46	.50	.48	.46
		3	.58	.55	.52	.61	.54	.52	.56	.53	.51	.54	.52	.50
		2	.63	.61	.58	.65	.60	.58	.60	.58	.56	.58	.57	.55
		1	.69	.67	.65	.69	.66	.64	.65	.63	.62	.62	.61	.60

FIGURE 20.16   (Continued)

Recessed Lay-in 2 ft × 4 ft Open Parabolic Troffer Luminaire (Two Fluorescent Lamp)									
2 Linear Tube Fluorescent Lamps	Spacing Criteria (CS)	Reflectance	Coefficient of Utilization (CU)						
		Ceiling Cavity (ρ_{cc})	80			70			50	
		Wall Surfaces (ρ_w)	70	50	30	70	50	30	50	30
		Cavity Ratio (CR)								
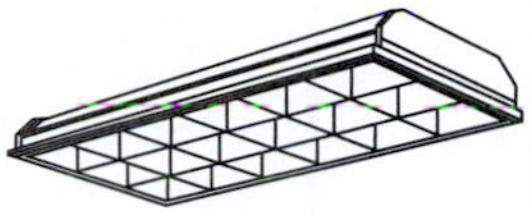	1.4⊥ 1.2∥	10	.41	.32	.26	.40	.30	.25	.29	.25
		9	.44	.34	.28	.42	.34	.28	.33	.28
		8	.47	.36	.30	.46	.36	.30	.35	.29
		7	.51	.40	.34	.50	.40	.34	.39	.34
		6	.55	.46	.39	.54	.45	.39	.42	.38
		5	.59	.51	.44	.58	.50	.44	.47	.42
		4	.65	.56	.50	.64	.56	.50	.54	.48
		3	.70	.64	.57	.69	.63	.56	.60	.56
		2	.78	.71	.66	.76	.70	.67	.68	.65
		1	.84	.81	.79	.82	.80	.78	.77	.75

Recessed Lay-in 2 ft × 4 ft Open Parabolic Troffer Luminaire (Three Fluorescent Lamp)									
3 Linear Tube Fluorescent Lamps	Spacing Criteria (CS)	Reflectance	Coefficient of Utilization (CU)						
		Ceiling Cavity (ρ_{cc})	80			70			50	
		Wall Surfaces (ρ_w)	70	50	30	70	50	30	50	30
		Cavity Ratio (CR)								
	1.4⊥ 1.2∥	10	.39	.29	.25	.38	.28	.23	.28	.23
		9	.40	.32	.27	.40	.32	.27	.30	.26
		8	.44	.34	.29	.42	.34	.28	.34	.28
		7	.47	.39	.33	.46	.38	.33	.36	.32
		6	.51	.42	.36	.50	.41	.36	.40	.35
		5	.56	.46	.41	.54	.46	.40	.45	.40
		4	.59	.53	.46	.58	.52	.46	.50	.46
		3	.65	.58	.54	.64	.57	.53	.56	.52
		2	.70	.66	.61	.68	.65	.60	.63	.59
		1	.77	.75	.71	.75	.72	.70	.69	.68

Recessed Lay-in 2 ft × 4 ft Open Parabolic Troffer Luminaire (Four Fluorescent Lamp)									
4 Linear Tube Fluorescent Lamps	Spacing Criteria (CS)	Reflectance	Coefficient of Utilization (CU)						
		Ceiling Cavity (ρ_{cc})	80			70			50	
		Wall Surfaces (ρ_w)	70	50	30	70	50	30	50	30
		Cavity Ratio (CR)								
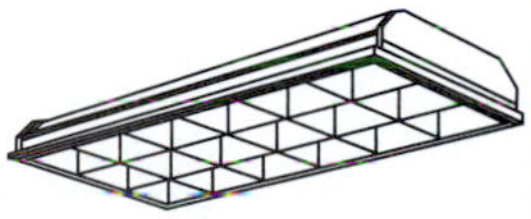	1.4⊥ 1.2∥	10	.35	.28	.23	.34	.28	.23	.27	.23
		9	.38	.29	.26	.36	.29	.25	.28	.25
		8	.40	.33	.28	.40	.32	.28	.32	.27
		7	.44	.35	.30	.42	.34	.30	.34	.29
		6	.46	.39	.34	.46	.39	.34	.38	.34
		5	.51	.42	.39	.48	.42	.38	.41	.38
		4	.55	.47	.42	.53	.46	.42	.46	.41
		3	.58	.54	.48	.57	.53	.48	.51	.47
		2	.64	.59	.56	.61	.58	.56	.56	.54
		1	.68	.67	.65	.67	.65	.64	.63	.60

Recessed Lay-in 2 ft × 4 ft Lensed Troffer Luminaire (Four Fluorescent Lamp)									
4 Linear Tube Fluorescent Lamps	Spacing Criteria (CS)	Reflectance	Coefficient of Utilization (CU)						
		Ceiling Cavity (ρ_{cc})	80			70			50	
		Wall Surfaces (ρ_w)	70	50	30	70	50	30	50	30
		Cavity Ratio (CR)								
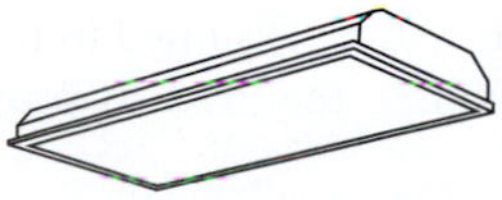	1.4⊥ 1.2∥	10	.46	.35	.29	.44	.35	.29	.34	.28
		9	.49	.38	.31	.47	.37	.31	.36	.31
		8	.52	.41	.34	.51	.41	.34	.40	.34
		7	.56	.45	.38	.54	.44	.38	.43	.37
		6	.60	.49	.42	.58	.49	.42	.47	.41
		5	.65	.55	.48	.63	.54	.47	.52	.46
		4	.70	.61	.54	.68	.60	.53	.57	.52
		3	.76	.68	.61	.74	.66	.61	.64	.59
		2	.83	.76	.71	.81	.75	.70	.72	.68
		1	.90	.86	.83	.88	.85	.82	.81	.79

FIGURE 20.16 (Continued)

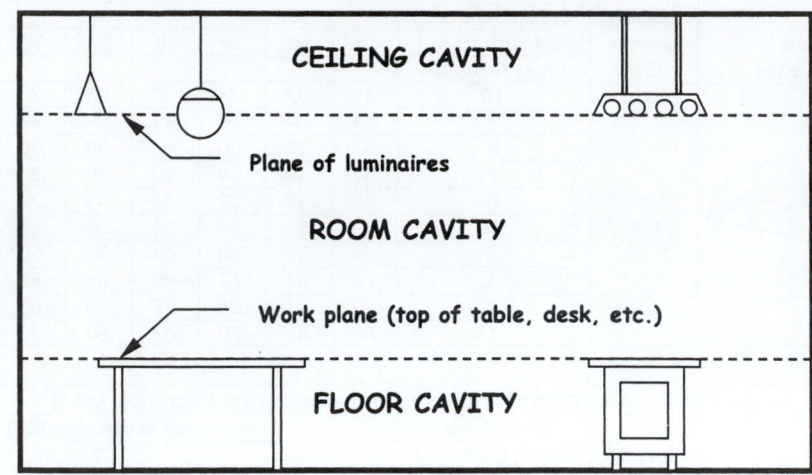

FIGURE 20.17 The space between the luminaires and the work plane is referred to as the *room cavity*; the space between the work plane and the floor is called the *floor cavity*; and the space between the ceiling and the luminaires (if they are suspended) is defined as the *ceiling cavity*.

With reflectances of the ceiling, room, and floor cavities approximated and the room CR determined, the CU is found in technical literature specific to the light fixture. The basic formulas used in average illuminance computations are shown below. Typically formula number 1 is used in design of a lighting installation.

1. Number of luminaires (#) required in a space:

$$\# = (E \cdot A)/(n \cdot LM \cdot CU \cdot LLF)$$

2. Average illuminance in a space (foot-candles in service):

$$E = (\# \cdot n \cdot LM \cdot CU \cdot LLF)/A$$

3. Minimum required area per luminaire to maintain the desired illuminance:

$$A = (n \cdot LM \cdot CU \cdot LLF)/E$$

where

A:	area of the space to be illuminated, in ft^2 or m^2
CU:	coefficient of utilization of luminaire
E:	illuminance, in foot-candles or lux
#:	number of luminaires (light fixtures)
n:	number of lamps in luminaire
LLF:	light loss factor
LM:	lamp output, in lumens

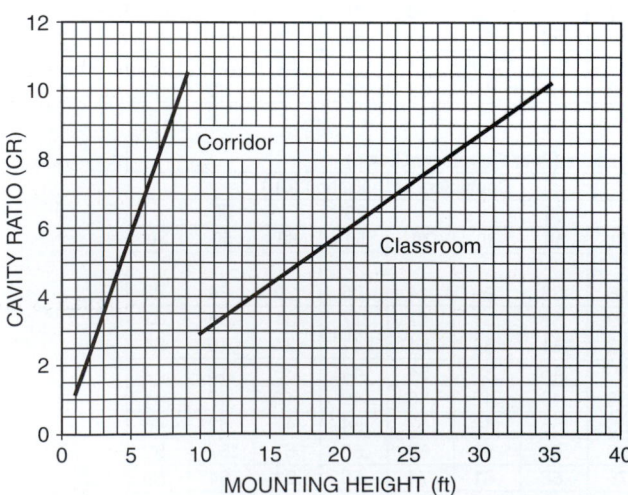

FIGURE 20.18 Relationship between cavity ratio (CR) and mounting height (MH) in a 30 ft by 40 ft classroom and a 30 ft by 5 ft corridor.

Example 20.7

A 30 ft by 40 ft classroom with 9-ft high ceilings will have an ambient lighting target illuminance of 60 fc at a work plane that is 30 in above the floor. The walls and ceiling will be off-white in color such that the anticipated ceiling cavity reflectance is 0.80 and the average wall reflectance with artwork is about 0.50. The space will be illuminated with surface-mounted, prismatic wraparound luminaires with four fluorescent lamps as shown in Figure 20.16. The initial output of the lamps is 2850 lm. The LLF will be assumed to be 0.65. Determine the number of luminaires required to provide uniform illumination in the space.

The CR must be computed before finding the CU, where the mounting height is found by: 9 ft − 2.5 ft = 6.5 ft:

$$CR = [5\ MH(L + W)]/LW$$
$$= [5 \cdot 6.5\ ft\ (40\ ft + 30\ ft)]/(40\ ft \cdot 30\ ft) = 1.896 \approx 2$$

From Figure 20.16, for reflectances of $\rho_{cc} = 80\%$ and $\rho_w = 50\%$ and a CR of 2, the CU for a four-lamp prismatic wraparound luminaire is found to be 0.61.

With all the values known, formula 1 is used to determine the number of luminaires (#) required in a space:

$$\# = (E \cdot A)/(n \cdot LM \cdot CU \cdot LLF)$$
$$= (60 \text{ fc} \cdot 1200 \text{ ft}^2)/(4 \text{ lamps} \cdot 2850 \text{ lm} \cdot 0.61 \cdot 0.65)$$
$$= 15.93 \approx 16 \text{ luminaires}$$

Example 20.8

A 100-ft by 140-ft gymnasium with 27-ft high ceilings will have luminaires for ambient lighting hung 24-ft above the floor. The walls vary in color and the ceiling cavity is off-white such that the anticipated ceiling cavity reflectance is 0.80 and the average wall reflectance is about 0.30. The space will be illuminated with high-bay, intermediate-distribution, ventilated reflector luminaires as shown in Figure 20.16. The target illuminance is 50 fc at the floor plane. The light loss factor will be assumed to be 0.60. Determine the number of luminaires required to provide uniform illumination in the space for the following lamp sizes:

a. 400 W clear metal halide lamps with an initial output of 36 000 lm

The cavity ratio must be computed before finding the CU where the mounting height is 24 ft because the luminaires are hung 24 ft above the work plane:

$$CR = [5 \text{ MH}(L + W)]/LW$$
$$= [5 \cdot 24 \text{ ft}(140 \text{ ft} + 100 \text{ ft})]/(140 \text{ ft} \cdot 100 \text{ ft})$$
$$= 2.06 = 2$$

From Figure 20.16, for reflectances of $\rho_{cc} = 80\%$ and $\rho_w = 30\%$ and a cavity ratio of 2, the CU for an intermediate-distribution, ventilated reflector with a clear HID lamp is found to be 0.73. With all the values known, formula 1 is used to determine the number of luminaires (#) required in a space:

$$\# = (E \cdot A)/(n \cdot LM \cdot CU \cdot LLF)$$
$$= (50 \text{ fc} \cdot 14 000 \text{ ft}^2)/(1 \text{ lamp} \cdot 36 000 \text{ lm} \cdot 0.73 \cdot 0.60)$$
$$= 44.4 = 45 \text{ lm}$$

b. 1000 W clear metal halide lamps with an initial output of 110 000 lm

The CR of 2 and CU of 0.73 remain the same, only lamp output changes.

$$\# = (E \cdot A)/(n \cdot LM \cdot CU \cdot LLF)$$
$$= (50 \text{ fc} \cdot 14 000 \text{ ft}^2)/(1 \text{ lamp} \cdot 110 000 \text{ lm} \cdot 0.73 \cdot 0.60)$$
$$= 14.5 = 15 \text{ lm}$$

In Example 20.8, it was found that with higher output lamps, the number of required luminaires decreased. In theory this holds true for *average* illuminance but not *uniform* illuminance. Simply, fewer luminaires can result in spotty lighting. The trade-off is that an installation with fewer luminaires is more cost-effective initially (lower number of fixtures) and over the life of the installation (higher wattage lamps tend to be more efficient).

The number of luminaires computed in Example 20.8a is likely too many because of higher installation costs. The number computed in the second part of the Example 20.5b is too few because of concerns with inconsistent (spotty) lighting. A likely choice for the space is about 25-1000 W luminaires, but this will lead to higher illumination levels than the target levels. See Example 20.9.

Example 20.9

Compute the average illuminance in the space described in Example 20.8. Illumination will be provided by 25 high-bay, intermediate-distribution, reflector luminaires (as shown in Figure 20.16) with 1000 W clear metal halide lamps having an initial output of 110 000 lm.

The CR of 2 and CU of 0.73 remain the same as in Example 20.8.

Average illuminance in a space (foot-candles in service) is found by:

$$E = (\# \cdot n \cdot LM \cdot CU \cdot LLF)/A$$
$$E = (1 \text{ lamp} \cdot 25 \text{ lm} \cdot 110 000 \text{ lm} \cdot 0.73 \cdot 0.60)/14 000 \text{ ft}^2$$
$$= 86 \text{ fc}$$

Example 20.10

A 100 ft by 140 ft conference center lobby area will have luminaires for ambient lighting hung 48 ft above the floor. The ceiling cavity reflectance is 0.80 and the average wall reflectance is about 0.30. The space will be illuminated with high-bay, intermediate-distribution, reflector luminaires as shown in Figure 20.16. 400 W clear metal halide lamps with an initial output of 36 000 lm will be used. The target illuminance is 50 fc at the floor plane. The LLF will be assumed to be 0.60. Determine the number of luminaires required to provide uniform illumination in the space.

The CR must be computed before finding the CU, where the mounting height is 48 ft because the luminaires are hung 48 ft above the work plane:

$$CR = [5 \text{ MH}(L + W)]/LW$$
$$= [5 \cdot 12 \text{ ft}(140 \text{ ft} + 100 \text{ ft})]/(140 \text{ ft} \cdot 100 \text{ ft})$$
$$= 1.03 \approx 1$$

From Figure 20.16, for reflectances of $\rho_{cc} = 80\%$ and $\rho_w = 30\%$ and a CR of 1, the CU for an intermediate-distribution, ventilated reflector with a clear HID lamp is found to be 0.81. With all the values known, formula 1 is used to determine the number of luminaires (#) required in a space:

$$\begin{aligned}\# &= (E \cdot A)/(n \cdot LM \cdot CU \cdot LLF) \\ &= (50 \text{ fc} \cdot 14\ 000 \text{ ft}^2)/(1 \text{ lamp} \cdot 36\ 000 \text{ lm} \cdot 0.81 \cdot 0.60) \\ &= 40 \text{ luminaires}\end{aligned}$$

When comparing the results in Example 20.8b with results in Example 20.10, it becomes evident that higher ceilings result in the need for more luminaires to achieve the same levels of illumination.

Luminaire Spacing

Once the number of luminaires (#) required for uniform illumination has been determined, the luminaires must be arranged. Incandescent and HID luminaires generally cast light evenly in all directions, which permits uniform spacing in even rows and columns, although even rows and offset columns may be used. Tubular fluorescent luminaires generally require spacing that is greater perpendicular to the axis of the lamps versus parallel with the lamps.

A large spacing between luminaires can result in inconsistent (spotty) illumination, particularly at small mounting heights. Spaces with a small quantity of luminaires or a small mounting height generally require more fixtures to ensure uniform illumination. Additionally, some light fixtures throw light better laterally than others. For example, a pendent light will disburse light well in a lateral direction while a reflector downlight throws most of its light downward. To ensure uniform illumination of the work plane, luminaires cannot be spaced too far apart.

Spacing Criterion

The *spacing criterion (CS)* ratio is an approximate maximum spacing-to-mounting height ratio required to ensure uniform illumination on the work plane. It is used as a conservative guide to determine maximum center-to-center luminaire spacing. It takes into account the direct component of illumination only and ignores the indirect component of light, which can contribute significantly to the uniformity. However, used within its limits, the spacing criterion can be a valuable design tool.

To use the spacing criterion, multiply the mounting height (luminaire to work plane) by the spacing criteria number to find the luminaire spacing:

$$S \le (CS \cdot MH)$$

where

    S: luminaire spacing, in feet or meters
    MH: mounting height, in feet or meters
    CS: spacing criteria, from manufacturer's literature

For most incandescent and HID luminaires, *spacing (S)* is the center-to-center luminaire spacing, in feet or meters, between two successive luminaires in a lighting installation.

Example 20.11

The 1000 W reflector luminaires specified in Example 20.8b have a spacing coefficient of 1.0. The luminaires will be mounted 24 ft above the floor. The work plane is the floor. Determine the maximum center-to-center spacing of the luminaires.

$$S = CS \cdot MH = 1.0 \cdot 24 \text{ ft} = 24 \text{ ft}$$

With average illuminance and spacing requirements known, the number of the luminaires in a simple rectangular space can be *approximated* by the following expression:

$$\# \cdot A/S^2$$

where

    A: area of the space to be illuminated, in square feet or meters
    S: luminaire spacing, in feet or meters
    #: number of luminaires (light fixtures)

For fluorescent fixtures longer than 4 ft, spacing is measured 2 ft from the end of the fixture. *Mounting height (MH)* is the distance between the plane of the luminaires and reference work plane, in feet or meters. Fluorescent luminaires and some others distribute light asymmetrically; that is, they throw light farther perpendicular ($\perp$) to the length of the tube rather than parallel ($\parallel$). As a result, they will have two spacing coefficients that are expressed together such as 1.5/1.2; an SC of 1.5 in the perpendicular ($\perp$) direction and a 1.2 for the parallel ($\parallel$) direction.

Example 20.12

The 48-in fluorescent luminaires specified in Example 20.7 have spacing coefficients of 1.5/1.2. They are mounted 9 ft above the floor. The work plane is 30 in from the floor. Determine the maximum center-to-center spacing of the luminaires and arrange the 16 fixtures according to the spacing requirements.

$$MH = 9 \text{ ft} - 2.5 \text{ ft} = 6.5 \text{ ft and } S \le (CS \cdot MH)$$

Perpendicular ($\perp$) to the length of the tube:

$$S_\perp = CS \cdot MH = 1.5 \cdot 6.5 \text{ ft} = 9.75 \text{ ft} = 9 \text{ ft-9 in}$$

Parallel ($\parallel$) to the length of the tube:

$$S_\parallel = CS \cdot MH = 1.2 \cdot 6.5 \text{ ft} = 7.80 \text{ ft} = 7 \text{ ft-9}\tfrac{5}{8} \text{ in}$$

The arrangement of luminaires is shown in Figure 20.19.

General lighting installations in offices and other similar spaces should not have light fixtures with spacing criteria ratios that exceed approximately 1.5. A spacing criteria of greater than 1.5 results in luminaires that emit more light at high vertical angles, which may produce a glare. On the other hand, luminaires with spacing ratios less than about 1.0 offer less

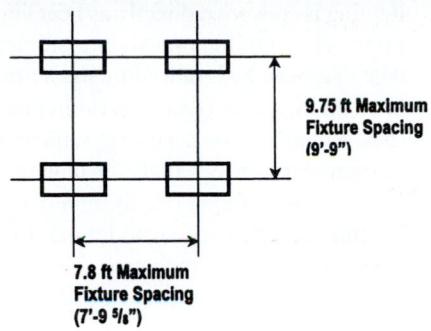

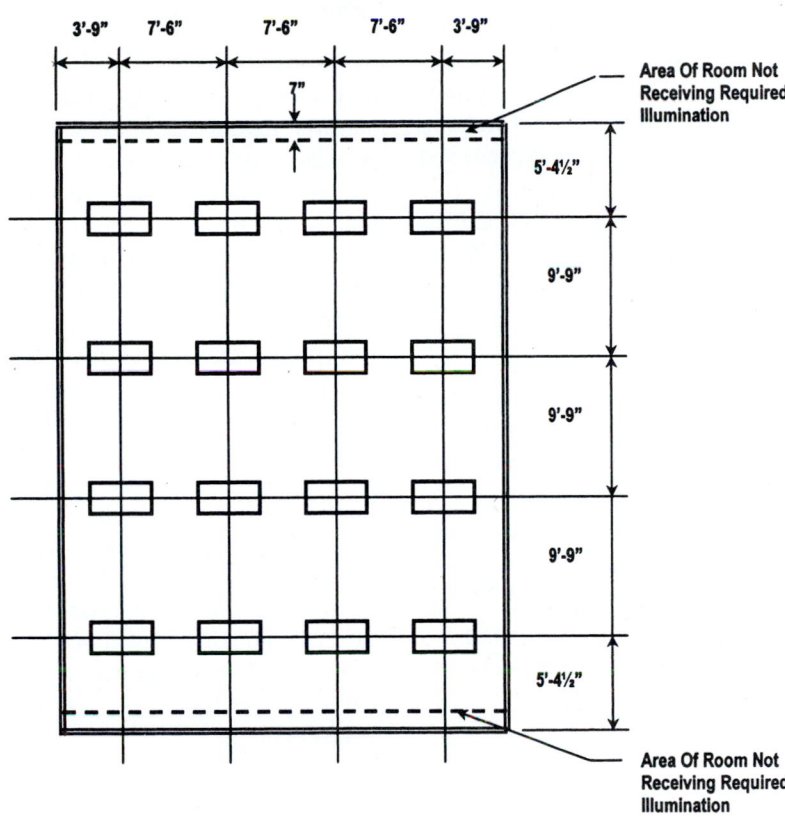

FIGURE 20.19   Arrangement of luminaires from Example 20.7.

vertical distribution but can produce spotty lighting conditions. With spacing criteria ratios of less than 1.0, perimeter walls will be in shadow unless the last row of fixtures is close by. This shadow condition can happen with parabolic troffers, also. The designer must pay attention to the perimeter spacing of these fixtures or room lighting will take on a spotty appearance.

Power Density and Power Allowance

Unit power density (UPD) of a lighting installation is the power consumed for illumination (W) divided by the area (ft^2 or m^2) served by the lighting installation. It is expressed in W/ft^2 or

W/m^2. It is frequently used as a measure of installed building lighting efficiency; a lower UPD indicates a higher efficiency. Typical power densities in new lighting installations are:

- General fluorescent lighting systems, such as in an office building, have lighting loads of 1.5 to 2.0 W per square foot

- A fluorescent lighting system with both task and ambient lighting has a lighting load of 1.2 to 4.0 W per square foot

- An HID lighting system has a load of 1.0 to 2.0 W per square foot

Example 20.13

In Example 20.7 it was found that illumination of a 30 ft by 40 ft classroom requires 16 fluorescent luminaires containing four lamps (32 W per lamp) per light fixture. Assume each luminaire consumes 144 W including the ballast. Determine the power density of this lighting installation.

$$\text{Power density} = (16 \text{ luminaires} \cdot 144 \text{ W/luminaire})/$$
$$(30 \text{ ft} \cdot 40 \text{ ft}) = 1.9 \text{ W/ft}^2$$

Energy standards such as ASHRAE/IESNA 90.1 provide a *power allowance* (W/ft^2 or W/m^2) that limits the unit power density for lighting installations in various space and function types. Selected power allowances (in W/ft^2) for various spaces are provided in Table 20.19. According to the standard, power allowances are adjusted for type and characteristics of lighting controls (e.g., occupancy sensors versus manual switching) and use of daylighting in a complex computation method. This standard deals only in power allowances; it does not address adequacy or quality of a lighting installation. It is up to the designer to use the allowed power to produce a pleasant and effective visual environment.

Reducing the Cooling Load

The intent behind the power allowance standard is to stimulate use of more efficient lighting, but it also reduces a building's cooling load by reducing waste heat. All of the energy consumed by lighting is converted directly as heat into the space. Therefore, any improvement of lighting system efficiency reduces the amount of heat that must be removed by the air cooling system. This results in air cooling energy savings during the operation of the building. In new construction, an energy-efficient lighting design can result in significant savings in the installed and operation costs of cooling systems. A rough rule of thumb is that every 2.0 to 3.0 kW of lighting load requires an additional kW of cooling. Therefore, in theory, every 1 kW reduction in lighting often results in a 1.3 to 1.5 kW decrease in total electrical energy consumption.

Benefits of Energy Efficient Lighting

Lighting accounts for 20 to 25% of the electricity consumed in the United States and Canada. Commercial, institutional, and industrial lighting (e.g., for schools, industries, stores, offices, and warehouses) represents from 80 to 90% of the total lighting electricity consumption. If energy efficient lighting were used in all places that it was profitable, the electricity required for lighting would be cut by at least 50%. Frequently, a blended lighting installation with a mix of efficient lighting (e.g., HID) can be used to provide efficient ambient lighting and combined with lamps for good color rendition (e.g., incandescent) to provide for a more efficient system. (See Photo 20.24.).

A 50% reduction would free about 20 billion dollars annually for investment elsewhere in our economy because less power plants would need to operated and built to meet demands. Environmentally, it would reduce annual carbon dioxide emissions by about 230 million tons, the equivalent of eliminating emissions from about 42 million cars. It would also reduce sulfur dioxide emissions by 1.7 million tons and nitrogen oxide emissions by about 1 million tons. Other forms of pollution (ash, scrubber waste, acidic drainage and waste from coal mining, radioactive waste, and natural gas leakage) would also be reduced.

TABLE 20.19   SELECTED POWER ALLOWANCES EXPRESSED AS THE MAXIMUM UNIT POWER DENSITY FOR VARIOUS BUILDING TYPES AND SPACES. COMPILED FROM VARIOUS GOVERNMENTAL SOURCES.

Building Type and Space	Power Allowances	
	W/ft^2	W/m^2
Building Type		
Offices, range depending on the floor area	1.50 to 1.90	0.14 to 0.18
Retail, range depending on the floor area	2.10 to 3.30	0.20 to 0.31
Schools, range depending on the floor area and type of school	1.50 to 2.40	0.14 to 0.22
Room/Space Type		
Corridors	0.8	0.01
Office, open	1.3	0.13
Office, enclosed	1.5	0.15
Restroom	1.0	0.10
Corridor	0.7	0.07
Fast food restaurants	1.3	0.12
Leisure dining restaurants	2.5	0.23
Hospital emergency room	2.3	0.21

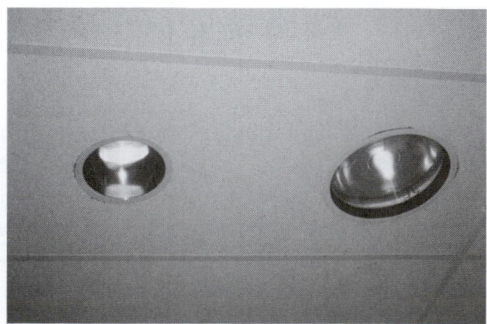

PHOTO 20.24   A blended lighting installation can be used to provide efficient ambient lighting with good color rendition. An incandescent luminaire (right) offers good color rendering characteristics and an HID luminaire (left) provides highly efficient lighting, which also serve to accent the space. (Used with permission of ABC)

20.12 LIGHTING DESIGN PRACTICES AND CONSIDERATIONS

Historical Design Practices

Historically, it was common practice to have ambient lighting provide all the lighting in a space. Banks of light fixtures were routinely used to uniformly light classrooms, offices, manufacturing work spaces, and retail stores. Uniform banks of light fixtures provided light in all areas of the space, which offered flexibility in situating and repositioning the workstations in the space (e.g., desks in an open office could be moved with little concern over inadequate lighting).

Uniform ambient lighting is uninteresting to the occupant and very wasteful because of high investment in rows and rows of light fixtures. It has been found that ambient lighting is more efficient and psychologically effective when used as a soft, background light combined with additional local lighting at the workstation and accent lighting to feature artwork or a display.

Architectural Lighting Design

Architectural lighting design is both a science and an art, and there is no single correct lighting solution for all situations. A good designer studies and evaluates existing lighting installations and blends knowledge gained from these observations with use of scientific principles in creating new designs. The scientific principles introduced in this chapter are relied on to formulate proper levels of illuminance and to make the lighting installation efficient. The creative side of lighting design helps make the space aesthetically interesting and psychologically comfortable. Because lighting is an integral part of the building, the lighting designer must work closely with the architect to achieve a lighting solution that blends with and becomes a part of the architectural design. Photos 20.25 and 20.26 show the contrast between a traditional classroom lighting installation and a modern installation, which offers a more comfortable seeing environment because there is less glare.

PHOTO 20.26 A classroom space lit with uplights, a more comfortable seeing environment because there is less glare. (Courtesy of NREL/DOE)

The main goals in designing an aesthetically pleasing architectural lighting system are to:

- Provide a sufficient quantity of light to meet the seeing needs of the occupants that is not excessive
- Provide the appropriate quality (color rendition) of light
- Provide a balance and variety to add interest to the space being illuminated

The designer should not depend on a single source of light for a space. Instead, the designer should use a variety of types of light sources and have the ability to balance them to suit the occupants' needs. This balance is usually accomplished by layering. *Layered light* is a nonuniform, balanced use of all types of architectural lighting. It tends to create a composition quality that best suits good atmosphere within most spaces by blending color rendering and other properties of the different light sources.

Ambient Lighting Ambient lighting is the most essential form of lighting because it is the basic background of light for almost every room. In a residence, this is especially true in the kitchen, living room, and family room. Ambient lighting can come from several sources. Fluorescent, halogen, and incandescent recessed lights in the ceiling cast light directly downward and outward. Halogen floor lamps shine their bright light up at the ceiling, which then reflects a diffuse light throughout the room. Wallwashers cast their light against the wall, which in turn provide ambient light without creating a glare. Portable table lamps with translucent lampshades cast soft light in a room. In many applications, it is desirable to have the major ambient light source dimmable.

Task/Local Lighting Task or local lighting illuminates a specific visual function and can influence the effectiveness of the individual performing the task. Good task lighting makes work easier on the eyes, prevents headaches, makes

PHOTO 20.25 A conventionally lit classroom space. (Courtesy of NREL/DOE)

cooking more enjoyable, and allows the individual to better concentrate on the job at hand. Task lights focus light in a particular direction or area, without casting any glare. Halogen and high-intensity desk lamps, table lamps with opaque lampshades, and spotlights with reflective coatings are examples of task lighting applications.

Accent Lighting   Accent lighting is used to add drama and emphasize a particular object or area. Low-voltage halogen spotlights, especially track lighting, are excellent applications of accent lighting. Picture lights focus attention on wall paintings of interest, up-lights, or cans of light on the floor shining upward, illuminate plants and sculptures. Strip lights can highlight a ceiling, a mantel or stairs. Entryways, foyers, vestibules, high-use corridors, dining rooms, living rooms, and other areas of special architectural interest benefit most from accent lighting.

Decorative Lighting   Decorative lighting aesthetically adds interest to the space by blending with other types of lighting to give an overall "feel" to a space. Track, halogen, and strip lights can be used in a decorative manner. (See Photos 20.27 through 20.30.) Decorative, colored sconces may also be used. Although decorative lighting may provide some ambient lighting, it serves principally as an eye-catching attraction. Use decorative lighting sparingly.

The following are some of the design considerations associated with lighting installations in common spaces.

Entrances   The entrance or foyer is the transitional space from public to private and exterior to interior. It offers the first impression of a home or building. Lighting in homes and commercial establishments should draw attention to the building entry. Ambient illumination is also important to make guests/occupants feel welcome and comfortable.

PHOTO 20.27   A bank of recessed luminaries provides ambient and task lighting in a kitchen area. (Used with permission of ABC)

PHOTO 20.28   Ambient lighting provided by decorative ceiling and wall-mounted fixtures. (Used with permission of ABC)

PHOTO 20.29   Halogen accent lighting used as task lighting over a conference table. (Used with permission of ABC)

PHOTO 20.30   Decorative halogen accent lighting can add interest to a space but can also cause glare. These eye-level accent lights were intended to provide accent and task lighting, but they adversely affect visual comfort from glare at eye level. (Used with permission of ABC)

Offices/Studies   Reading and writing require quality lighting that limits eye fatigue caused from glare. Often, this space can take advantage of a beautiful vista out the window and good natural ambient light daylight. At night, an ambient light source should be provided to compensate for daylight. Computers require special attention. Soft indirect lighting is enough illumination and should be properly located so as not cause patterns on the screen.

Restrooms/Bathrooms In restrooms and bathrooms, the best light washes the face from all directions, softening shadows. There should be sufficient light to see detail for grooming but not so much to cause glare and discomfort. Quality lighting can be accomplished within any design style because there is great variety in well-engineered fixtures.

Dining Spaces Lighting should make the food, the table setting, and the people look attractive, so excellent color rendition is a must. Ambient lighting should be low to make the occupants sitting at the dining tables feel isolated. A center-pendent luminaire works well if not too obtrusive. A chandelier is very appropriate in formal dining, particularly in residences. The light level from the center-pendent luminaire or chandelier should be indirect so it does not draw attention from other, more important views. Ambient lighting should be low to draw attention away from the occupants sitting at dining tables. A balance of accent lighting showing off artwork, a special sideboard, and the table centerpiece is very effective and desirable.

Kitchen Spaces In the home, kitchen lighting has become very important because this space has become a central focal point. As a gathering point and a place of special work, lighting must be a well-conceived part of the overall design. Inappropriate lighting can make rich materials look dark and cause glare off shiny counters. Ambient and well-placed task lighting are the successful solution and the ability to control light levels is essential in the open-plan house. In commercial kitchens, adequate, diffused lighting with good color rendition is a must.

Exit Illumination and Marking

Building codes generally require that building exits be illuminated any time that the building is occupied, with light having an intensity of not less than one foot-candle at the floor/walking surface level. Exit illumination is not required in individual dwelling units, guest rooms, and sleeping rooms. In auditoriums, theaters, and concert or opera halls, the illumination at the floor/walking level may be reduced during performances to not less than 0.2 foot-candles. Required exit illumination should be so arranged that the failure of any single lighting unit (e.g., burning out of an electric bulb) will not leave any area in darkness.

Exit signage, like that shown in Photo 20.31, should safely mark exits or direct occupants to an exit where the route is not directly apparent. When two or more exits from a story are required by code (most commercial buildings), exit signs are required at stair enclosure doors, horizontal exits, and other required exits from the story. Main entry exit doors that can be clearly recognized as exits need no exit sign posted. Additionally,

PHOTO 20.31 Exit lighting. (Used with permission of ABC)

dwellings, lodging houses, congregate residences accommodating ten occupants or less, and individual units of hotels and apartment houses are not required to be posted with exit signs.

Each exit sign must have the word "EXIT" in plainly legible letters not less than 6 in high nor less than ¾ in wide. Any door, passage, or stairway that is neither an exit nor a means of exit and that may be mistaken for an exit shall be identified by a sign reading, "NOT AN EXIT," or a similar phrase. An exit sign shall be distinctive in color and must offer contrast with decorations, interior finish, or other signs in the space. Decorations, furnishings, protrusions, or equipment cannot impair visibility of an exit sign.

Directional exit signs guide occupants to an exit where the route is not directly apparent. A sign reading "EXIT" or a similar phrase, with an arrow indicating the direction, must be located at locations where the route to the nearest exit is not apparent.

20.13 LIGHTING SYSTEM CONTROLS

A variety of switches and controls are used to manage operation of lighting installations. A *switch* is an electrical device that opens an electrical circuit, thereby shutting the lights off or closing the circuit to energize the lighting installation. A switch or group of switches can control the lighting installation from one or multiple locations. Switches are covered in Chapter 18 and electrical circuiting of switching is covered in Chapter 19.

Following are some of the common types of switches used in lighting installations:

- *Single-pole, single throw (SPST) switches* are a simple on/off switch that controls a lighting installation from one location.
- *Three-way switches (S_3)* and *four-way switches (S_4)* can be circuited to control a lighting installation from two or more locations. Two three-way switches are required to control a lighting installation from two locations. Two three-way switches and one or more four-way switch are required to control a lighting installation from three or more locations.

- *Automatic switches* deactivate the lighting circuit after a preset time period has lapsed.

- *Dimmer switches* can be used to vary the luminous flux (lumen output) from lamps in a lighting installation. A wall-mounted dimmer switch is shown in Photo 20.32. Incandescent installations can be easily dimmed with electronic dimmers or variable transformers. Fluorescent and HID lamps require special dimming ballasts and controls. Frequently, variable transformers are used on large fluorescent and HID installations, but do not provide full-range dimming. Dimming of HID lamps often results in a color shift.

- *Keyed switches* limit access to switches for lights and equipment to authorized personnel. They are available with SPST, S_3, and S_4 capabilities. A wall-mounted, ganged (multiple) key switch installation is shown in Photo 20.33.

- *Door switches* can activate a lighting circuit when a door is open or closed (e.g., similar to a refrigerator light).

- *Time clocks* can be used to control the time period that a lighting installation operates. Traditionally, time clocks are an electrical-mechanical device that controls lighting operation by pin placement on a moving time wheel that repeats a daily cycle as the wheel rotates.

- *Electronic timers* automatically control operation of a lighting installation with electronic components that are wired into the circuitry. They allow great flexibility as they can easily be set for daily and weekly cycles. Electronic timers do require relay switching on larger lighting loads.

- *Photocell controls* sense light and can be used to control night lighting in lieu of a time clock or timer. Recent advancements in photocell technology allow them to be used effectively to control illumination levels in spaces that have daylight available: as daylighting increases the photocell dims the lights.

- *Occupancy sensors* control a lighting installation by sensing occupants in a space. They work well in interior spaces such as classrooms and offices. Occupancy sensors are usually mounted on the wall or ceiling.

A central *lighting control system* offers building operators full control of multiple lighting installations from a single location. Central control systems control the lighting installations in a building along with controlling HVAC equipment, fire protection and security systems, and other devices. They are usually integrated in a building energy management system (BEMS), which was discussed in Chapter 7. An advantage of a central control system is that light circuits can be controlled in a remote location that, on some systems, can be controlled remotely off-site.

PHOTO 20.32 A wall-mounted dimmer switch used to control several banks of track lights. (Used with permission of ABC)

PHOTO 20.33 Ganged (multiple) key switched. Key switches are often used in lighting installations to provide better control of lighting (accidental or disruptive switch off). (Used with permission of ABC)

20.14 DAYLIGHTING PRINCIPLES

Fundamental Principles

Daylighting is the efficient and effective use of direct, diffuse, or reflected sunlight to provide full or supplemental illumination for building interiors during hours of sunlight. Sunlight has excellent color rendition and brilliance. It diminishes the need for artificial light in buildings and thus saves lighting costs. In some buildings, an overall saving of up to 80% in lighting energy costs can be achieved during sunlight hours.

Effective use of daylighting will reduce electrical costs (both energy and demand costs) and operating and maintenance costs (lamp life is extended). Daylighting in buildings also typically decreases space heating and cooling costs. Because daylight produces less heat per unit of illumination than artificial lights, daylighting reduces the cooling load when it replaces artificial lighting within the space being cooled. As part of a passive solar heating system, sunlight can also provide supplementary building heat.

Daylighting generally improves occupant comfort by providing a more pleasant, naturally lit indoor environment.

The human eye tends to adapt more easily to daylight. Additionally, the glazed (clear or translucent panels of plastic or glass) surfaces needed for daylighting give the occupants a sense of contact with the outdoors. Various studies suggest that daylighting increases worker and student productivity, and reduces absenteeism. Daylighting has also been shown to contribute to higher sales in retail stores. In factories, daylighting can reduce the loss of worker productivity during power failures.

The major disadvantage of daylighting is the unpredictable availability of sunlight. The varying intensity of sunlight with passing clouds, cloud cover on a rainy day, and shading by natural features (e.g., hills, vegetation, and so forth) and by human-made objects (e.g., tall buildings and upper stories) can affect the availability of adequate daylighting. The changing position of the sun with respect to the space being illuminated must also be addressed.

If not designed properly, oversized glazed areas that allow daylight into a building may contribute to higher building heat losses in the winter and undesirable heat gain in the summer. These additional heating and cooling costs diminish and may fully offset savings from reduced artificial lighting costs. Occupant comfort can also be adversely affected by uneven heating or cooling and glare from direct sunlight. Additionally, sunlight can fade interior finishes and furnishings.

Daylight

Daylight is a combination of direct, reflected, and diffuse sunlight.

- *Direct light* travels in a straight-line path from the sun and tends to be more intense than diffuse and reflected sunlight. Glare from direct sunlight can cause discomfort when the occupant is watching television, working at a computer, or reading.

- *Reflected light* strikes a surface, such as a sidewalk or automobile, and reflects off the surface in another direction. Because the receiving surface absorbs some sunlight before reflecting the light, the reflected light is less intense than direct sunlight. Glare from reflected sunlight can cause occupant discomfort.

- *Diffuse light* is light that has been reflected or refracted by clouds, glazing, or other objects. It accounts for most of the daylight received at the earth's surface on a cloudy day. It is less intense than direct and reflected light and typically results in less glare.

Glazings

Typical performance measures for the center of glass for selected types of glazings are provided in Table 20.20. A description of these measures follows.

- The *overall coefficient of heat transfer (U)* is a measure of how easily heat travels through an assembly of materials: the lower the U factor, the lower the rate of heat transfer through the glazing and the more efficient the glazing. The U factor has units of $Btu/hr \cdot ft^2 \cdot °F(W/m \cdot °C)$. Thermal insulating ability is also measured by the *thermal resistance (R):* a higher R factor indicates a better insulating performance. The R factor has units of $hr \cdot ft^2 \cdot °F/Btu(m \cdot °C/W)$. The R factor is the inverse of the U factor (e.g., R = 1/U and U = 1/R; a U factor of 0.5 $Btu/hr \cdot ft^2 \cdot °F$ is the same as an R factor of 2.0 $hr \cdot ft^2 \cdot °F/Btu$). These factors are discussed in Chapter 4.

  The *overall* or whole-window U factor of a window or skylight depends on the type of glazing, frame materials and size, glazing coatings, and type of gas between the panes. The overall U factor should be used because energy-efficient glazings can be compromised with poor frame designs. R factors for common glazing materials range from 0.9 to 3.0 (U factors from 1.1 to 0.3), but some highly energy-efficient exceptions also exist. Experimental super window glazings have a center-of-glass R factor of 8 to 10 (U factor of about 0.1), but have an overall window R factor of only about 4 to 5 (U factor of about 0.25 to 0.2), because of edge and frame losses.

- The *solar heat gain coefficient (SHGC)* is the fraction of solar heat that is transmitted through the glazing and ultimately becomes heat. This includes both directly transmitted and absorbed solar radiation. The lower the

TABLE 20.20 TYPICAL CENTER-OF-GLASS PERFORMANCE MEASURES FOR SELECTED TYPES OF GLAZINGS. TOTAL WINDOW VALUES ARE SIGNIFICANTLY DIFFERENT. COMPILED FROM VARIOUS INDUSTRY SOURCES.

Glazing Types	Overall Coefficient of Heat Transfer (U Factor)	Solar Heat Gain Coefficient (SHGC)	Visible Transmittance (VT)	Light-to-Solar Gain Ratio (LSG)
Single glazed, clear	1.10	0.86	0.88	1.04
Double glazed, clear	0.49	0.76	0.81	1.07
Double glazed, bronze	0.49	0.62	0.61	0.98
Double glazed, spectrally selective	0.49	0.32	0.44	1.38
Double glazed, low-e (e = 0.10)	0.32	0.60	0.77	1.28
Triple glazed, low-e (e = 0.15)	0.27	0.49	0.68	1.39

SHGC, the less solar heat is transmitted through the glazing and the greater its shading ability. In general, south-facing windows in buildings designed for passive solar heating should have windows with a high SHGC to allow in beneficial solar heat gain in the winter. East- and west-facing windows that receive undesirable direct sunlight in mornings and afternoons should have lower SHGC assemblies.

- *Visible transmittance (VT)* is the percentage of visible light (light in the 380 to 720 nm range) that is transmitted through the glazing. When daylight in a space is desirable, glazing is a logical choice. However, low VT glazing such as bronze, gray, or reflective-film windows are more logical for office buildings or where reducing interior glare is desirable. A typical clear, single-pane window has a VT of about 0.88, meaning it transmits 88% of the visible light.

- The *light-to-solar gain ratio (LSG)* is the ratio between SHGC and VT of a single glazing. It provides a gauge of the relative efficiency of different glazing types in transmitting daylight while blocking heat gains. The higher the LSG ratio the brighter the room is without adding unnecessary amounts of solar heat.

Types of Glazing Materials

Historically, only *glass* was used as a glazing material. Advantages of glass include long life, high light transmission, hardness, and stiffness. The main disadvantage of glass is brittleness and the safety hazard that falling broken glass creates. Glass can be made more resistant to breakage by increasing its thickness, by heat treatment (tempered glass), and by combining it with reinforcing materials such as wire or a plastic film. All safety improvements for glass add cost and typically add weight.

During the past two decades, plastic glazings have become an alternative to glass. They are much lighter in weight and are resistant to shattering, so they pose less of a safety hazard. The plastics commonly used for glazing are acrylics and polycarbonates. Polycarbonates are stronger, but acrylics are more resistant to degradation from the UV component of sunlight. All plastics deteriorate in strength and light transmission over several years. Horizontal-laying plastic glazing material buckles as it ages. The main causes of deterioration of plastics are UV light, heat, and oxidation. The service life of plastic glazing can be extended greatly with additives such as UV inhibitors.

Plastics can be reinforced with fibers of various materials, including glass, to increase strength and service life. The fibers cause some light loss and diffuse light, which is useful in most daylighting applications. Additionally, nonreinforced plastics are easily molded; an entire skylight assembly can be formed from a single plastic sheet.

Glass and plastic can be combined in a composite to minimize the limitations of each. Glass is used for the outer sheet, where it can provide considerable protection to the plastic, while the inner plastic sheet protects against glass breakage. Ordinary window glass strongly absorbs the damaging UV portion of sunlight, so a plastic material will survive longer if it is installed inside glass.

Today, several types of advanced glazing systems are available to help control heat loss or gain. The advanced glazings include double- and triple-pane windows with such coatings as low-emissivity (low-e), spectrally selective, heat-absorbing (tinted), or reflective; gas-filled windows; and windows incorporating combinations of these options. These were introduced in Chapter 3. Typical center-of-glass performance measures for selected types of glazings are provided in Table 20.20.

Daylighting Strategies

Daylighting systems must be designed to provide illumination in areas where it is most needed (e.g., northern spaces, internal spaces, and ground levels of tall buildings) and to do so without adversely affecting occupant comfort (e.g., limiting glare and excessively spotty lighting). A basic daylighting strategy is to increase the number and size of glazed areas in the walls or ceilings of a building, but simply using large glazed openings does not ensure good daylighting. The following are common daylighting strategies.

Windows

The main functions of windows are to bring in daylight and fresh air (if windows are operable) and introduce a view of the outdoors. Windows located in walls of a building are the most common method of introducing daylight into the interior building spaces. *Fenestration* is a design term that describes window size, arrangement, and glazing type (the glass and coatings used in the window). Fenestration affects daylighting, passive solar heating, space cooling, and natural ventilation.

A *clerestory window system* is a fenestration arrangement in an upper story wall that extends above one roof surface; it introduces daylight into the ceiling area of the space. An *atrium* is an interior courtyard covered with glazing. Rooms and spaces adjoining the atrium receive daylight entering through the glazed roof. Clerestories and atriums are effective daylighting strategies because they introduce daylighting deep into a structure.

A fenestration arrangement of several smaller windows can provide uniform daylight illumination if properly distributed in the space being illuminated. A large window area tends to overlight and produce spotty lighting. Good distribution is difficult to achieve in large deep rooms or in interior spaces without outside walls. Position and orientation of a window system will also affect overall performance.

In cold climates in the northern hemisphere, south-facing, vertical windows work very effectively by providing good daylight and good solar gain control all year round. South-facing windows limit solar gains in the summer because the steep angle of the summer sun with respect to the glass surface results in considerable reflection off the glass surface and minimal solar gain. Properly sized overhangs can further prevent overheating in the summer by shading windows from the high summer sun.

In the winter, the low position of the sun produces a shallow angle and a heat gain that contributes to space heating. In fact, south-facing insulating glass (e.g., double glazing) generally results in a net energy gain over the winter season in all but severe climates. Provisions may need to be made to control solar glare from the low winter sun. Additionally, a slanted south-facing window surface should be avoided because it gains heat all year long and is particularly poor at limiting solar heat gains in the summer.

In cold climates in the northern hemisphere, window areas facing east and west should be minimized. In late spring through early fall, the midmorning eastern sun and afternoon western sun are relatively low in the sky. East- and west-facing windows are exposed to considerable direct sunlight during these times of the day. Windows facing in these directions should only be used when no other method of introducing daylight is possible and where control of direct sunlight and glare is achievable. Although overhangs are impractical for east- and west-facing windows, vertical shading can be used. Additionally, vegetation (e.g., trees and shrubs) can be strategically located to shade window areas facing east and west.

Although north-facing windows in the northern hemisphere do provide good quality daylighting, excessive heat loss and mean radiant temperature in the winter is a concern in cold climates. In hot climates, north-facing windows can provide good daylighting without heating the building.

Skylights

A *skylight* is a transparent panel located in a roof opening that allows direct and diffuse sunlight into the building. A *roof monitor* is a type of skylight system that is a raised, typically triangular shaped, extension of a roof and that has at least one glazed surface. Roof monitors and skylights can provide large quantities of reflected and diffuse light into the interior of the building.

An arrangement of several smaller skylights provides illumination that is much more uniform than daylighting with a single large skylight. Thus, it is better to use a larger number of smaller skylights, rather than one or a few large skylights. Installations with large skylights commonly suffer from excess brightness directly below the skylight, accompanied by gloomy dark areas surrounding the skylight. Repetitive layout of small skylights provides uniform illumination within the space. Additionally, less alteration of the roof structure is needed with smaller skylights.

Direct sunlight through skylights is not suitable for illumination because it is too intense and it forms localized bright spots. Diffusion corrects or reduces these problems by distributing sunlight in a fairly uniform pattern. It also minimizes changes in illumination caused the motion of the sun.

Skylights are effective for many industrialized and maintenance operations. Warehousing can be a favorable application. They can be used to provide a sense of natural ambience, which is valuable in applications such as restaurants, transportation centers, and other public areas. Skylights can also be effective for retailing because sunlight has excellent color rendition and brilliance.

A major disadvantage of skylights is the large fluctuations in illumination intensity caused by movement of clouds across the sun. Skylights are less likely to be satisfactory where reading tasks occur, as in offices, drafting areas, and reading rooms. Broad fluctuations of sunlight intensity are more noticeable in applications that require concentration on text. In addition, daylighting makes it more difficult to avoid veiling reflections, which are a problem especially with reading tasks (e.g., reading paperwork and a computer screen).

Skylights must be located where the sun can shine on them directly. When clouds pass in front of the sun, they cause abrupt changes in illumination level that reduce illumination from daylight by a factor of five to ten. Therefore, skylights are not valuable in regions that regularly have heavy cloud cover for a large fraction of the time. An exception is in a mild climate where the structure can accommodate a large area of skylights. Additionally, skylights do not produce a useful amount of daylight if it is shaded by adjacent structures or vegetation.

Skylights are commonly made from glass, glass composites, plastics, and plastic composites. All these materials can be treated to maintain visible light transmission and reduce cooling load, by adding tints and coatings that absorb light. All glazing materials can be provided with diffusing properties.

Heat loss through horizontal or steeply slanted glazing is two to three times higher than when it is installed vertically (e.g., a window). In cold climates, use of three or four panes (layers) of glazing material is economical. This relatively low thermal resistance causes skylights to condense moisture more easily than vertical windows. Skylight design should include gutters to capture condensation that flows off the interior surface of the glazing. Gutters should be large enough to hold the condensation until it can evaporate back into the interior space.

PHOTO 20.34 Skylights on a building roof. (Used with permission of ABC)

PHOTO 20.35 Exterior windows designed to introduce daylight in a gymnasium. (Used with permission of ABC)

PHOTO 20.38 Stepped windows provide soft, natural lighting to the open office area below. (Courtesy of NREL/DOE)

PHOTO 20.36 Naturally lit high school corridor area. (Used with permission of ABC)

PHOTO 20.39 Stepped windows and light shelves throw natural light deep into the building. (Courtesy of NREL/DOE)

PHOTO 20.37 Naturally lit bank lobby area. (Used with permission of ABC)

PHOTO 20.40 Side lights and a roof monitor in the center of the ceiling throws natural light evenly. (Courtesy of NREL/DOE)

Reflective Light Shelves

A *light shelf* is a passive (nonmechanical) architectural element or mechanism that allows sunlight to enter deep into a building. They can be a separate element or mechanism or can be an integral part of the building structure. They may be located on the interior or exterior of the building envelope. Exterior light shelves may also function as overhangs that shade the window from the high summer sun. Typical installations usually make the light shelf a unique architectural feature. Examples include light shelves suspended by stainless steel cables, fabric light shelves stretched over metal tubes, and assemblies of commercially available components created by ceiling manufacturers.

Properly designed light shelves allow daylight to penetrate the interior of the space up to 2½ times the distance between the floor and the top of the window. Advanced light shelves increase this distance up to four times.

Light Pipes

Most commercially available *light pipes* consist of an exterior, roof-mounted transparent dome, a reflecting metal pipe, and a diffuser for installation at the ceiling level of the space. See Figure 20.20. In effect, a light pipe is a small skylight with an integral reflective enclosure. The pipe may be rigid or flexible. Flexible pipes are easier to install, but they suffer more light loss from increased light absorption from reflection and scatter on the inside surface of the pipe wall. Light pipes are available commercially. Table 20.21 provides data on the output of a commercially available light pipe.

When a building has an attic space, installing skylights in the roof above the attic is difficult unless the attic is empty. Light pipes can pass through attic spaces much more easily.

A *sun tracking light pipe* is a type of light pipe that has a movable mirror or light-refracting system and that can be used to align the incoming sunlight with the axis of the light pipe, thereby minimizing reflection losses. Sun trackers are available commercially. Their main limitation is that they lose effectiveness if the sky does not remain clear. The system is designed to collect light from the sun, which is a point source. The light-reflecting apparatus gets in the way of the whole sky when the sun is obscured. Another disadvantage is the need for occasional maintenance.

Fabric Roof Membranes

Translucent *fabric roof membranes* can be used to introduce daylight through the roof itself. On clear days, these roof membranes virtually eliminate the need for indoor artificial lighting. The main use for fabric structures has been for sports stadiums, shopping malls, harbors, and airports (Denver International Airport).

Fabric structures may be divided into two categories: tension and air supported. Air-supported structures use a mem-

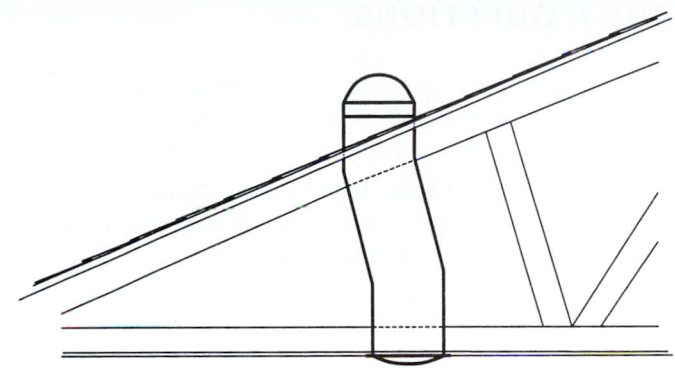

FIGURE 20.20   A sketch of a light pipe.

brane, supported by air pressure, to act as the roof and walls of the structure. Tension structures use a membrane, supported by a cable, to act as the roof of a structure.

Polyvinyl chloride-polyester and Teflon-coated woven fiberglass fabrics are commercially available fabric roof membranes that have been used successfully. These serve as the roof membrane. A second inner fabric liner is needed for insulation, sound control, and protection against condensation.

How Much Area

Outdoor sunlight from a clear sky produces an illuminance of about 6000 fc (60 000 lx). When illuminance with an intensity of 6000 fc is cast over an area that is one hundred times greater, it results in an illuminance of 60 fc (600 lx), an acceptable intensity for general lighting tasks. So, in theory, clear sky sunlight can be distributed to adequately light an area that is one hundred times greater than the area of captured sunlight. If one assumes that one-quarter of the sunlight entering a daylighting system can effectively be used for daylighting, then an aperture of sunlight of about 4% is required—roughly 4 ft² of window or skylight area for every 100 ft² of floor area.

TABLE 20.21   REPORTED OUTPUT OF A COMMERCIALLY AVAILABLE LIGHT PIPE. COMPILED FROM VARIOUS INDUSTRY SOURCES.

Size		Full Sun Summer Approx. 10500 fc		Overcast Summer Approx. 4500 fc		Clear Sky Winter Approx. 2000 fc		Approximate Area that Can Be Lit to Normal Daylight Level	
in	mm	fc	lx	fc	lx	fc	lx	ft²	m²
8	200	30	300	22	220	10	100	80	7.5
13	330	63	630	45	450	24	240	150	14
18	450	95	950	68	680	35	350	230	22
21	530	122	1220	87	870	43	430	320	30
24	600	154	1540	110	1100	54	540	430	40
30	750	215	2150	148	1480	62	620	530	50
36	900	258	2580	172	1720	70	700	650	60
40	1000	310	3100	211	2110	76	760	750	70
By Comparison									
100 W lamp		17	170	17	170	17	170	–	–

STUDY QUESTIONS

20-1. A light source emits light with dominant wavelengths in the range of 650 to 690 nm. What is the principal color of light emitted by the source?

20-2. A light source emits light with dominant wavelengths in the range of 470 to 515 nm. What is the principal color of light emitted by the source?

20-3. A light source emits light with dominant wavelengths in the range of 589.0 to 589.6 nm. What is the principal color of light emitted by the source?

20-4. Noon sunlight strikes a stained glass window. After it passes through the glass, the transmitted light appears blue from the inside of the building.

 a. What size wavelengths of light are transmitted through the glass?

 b. What size wavelengths of light did not pass through the glass?

20-5. Noon sunlight strikes a stained glass window. After it passes through the glass, the transmitted light appears red from the inside of the building.

 a. What size wavelengths of light are transmitted through the glass?

 b. What size wavelengths of light did not pass through the glass?

20-6. Light from a chromatic light source strikes a surface. Twenty percent is reflected and 30% is absorbed.

 a. What percentage of the light incident on the surface is transmitted?

 b. What are the reflectance, absorptance, and transmittance values?

 c. In your opinion, is the surface receiving the light or dark in color?

 d. Is the surface translucent or opaque?

20-7. Light from a chromatic light source strikes a surface. Ten percent of the light is absorbed and 40% is reflected.

 a. What percentage of the light incident on the surface is transmitted?

 b. What are the reflectance, absorptance, and transmittance values?

 c. In your opinion, is the surface receiving the light or dark in color?

 d. Is the surface translucent or opaque?

20-8. Light from a chromatic light source strikes a surface. Seventy percent of the light is reflected and 30% is absorbed.

 a. What percentage of the light incident on the surface is transmitted?

 b. What are the reflectance, absorptance, and transmittance values?

 c. In your opinion, is the surface receiving the light or dark in color?

 d. Is the surface translucent or opaque?

20-9. Light from a chromatic light source strikes a surface. Seventy-three percent of the light is reflected and 27% is absorbed.

 a. What percentage of the light incident on the surface is transmitted?

 b. What are the reflectance, absorptance, and transmittance values?

 c. Is the surface light or dark in color?

 d. Is the surface translucent or opaque?

20-10. Two coins are dropped on the sidewalk at night. One coin falls under a street lamp and the other coin falls a distance away from the lamp.

 a. What must happen for the coin that fell under the street lamp to be seen?

 b. What must happen for the coin that fell a distance away from street lamp to be seen?

 c. What factors influence whether these coins will be seen?

20-11. Identify and describe the types of glazings used for daylighting.

20-12. Identify and describe the types of daylighting strategies.

Design Exercises

20-13. Determine the luminous efficacy of the following lamps:

 a. A 150 W incandescent lamp that emits 2850 lm

 b. A 34 W linear fluorescent lamp that consumes 36 W, including the ballast and emits 3400 lm

20-14. Determine the luminous efficacy of the following lamps:

 a. A 75 W incandescent lamp that emits 1170 lm

 b. A 40 W linear fluorescent lamp that consumes 44 W, including the ballast and emits 3400 lm

 c. A 175 W metal halide lamp that consumes 205 W, including the ballast and emits 16 600 lm

20-15. Determine the luminous efficacy of the following lamps having comparable illumination outputs:

 a. A 300 W incandescent lamp that emits 5820 lm

 b. Two 32 W linear fluorescent lamps that emit 2950 lm per lamp in a luminaire that consumes 70 W including the ballasts

 c. A 70 W metal halide lamp that consumes 88 W, including the ballast and emits 5500 lm

20-16. Fixture manufacturer's data for a four-lamp linear fluorescent luminaire provides a luminaire efficiency of 0.62, a luminaire power input (watts) of 162 W, and ballast factor of 0.87 .Data from the lamp manufacturer indicates that the fluorescent lamps under consideration will output 3050 lumens. Compute the luminaire efficacy rating (LER).

20-17. Fixture manufacturer's data for a two-lamp linear fluorescent luminaire provides a luminaire efficiency of 0.68, a luminaire power input (watts) of 70 W, and ballast factor of 0.95. Data from the lamp manufacturer indicates that the fluorescent lamps under consideration will output 2750 lumens. Compute the LER.

20-18. Lamp manufacturer's specifications for a 40 W fluorescent lamp indicate an output of 2680 lumens, a rated life of 20 000 Hr, and a cost of $5.20 per lamp. Data for a 150 W incandescent lamp indicates an output of 2850 lumens, a rated life of 750 hr, and a cost of $2.30 per lamp. When in use, it is anticipated that these lamps will be operated 14 hr per day, 280 days per year.

   a. For each lamp, approximate the time period that elapses before replacement is necessary.

   b. Approximate the energy and replacement costs of each type of lamp over a 10-year period. Assume use and lamp costs remain constant.

20-19. You are considering replacing a 75 W incandescent lamp that costs $1.75/lamp with a compact fluorescent lamp. The incandescent lamp outputs 1190 lumens and has a rated life of 750 hr. It is anticipated that the lamp will operate 8 hr per day, 350 days per year.

   a. Identify a ballasted, one-piece, screw-in, compact fluorescent lamp that can replace the incandescent lamp without reducing light output. Base the comparison on mean lumen output of the compact fluorescent lamp. A slight increase in output is acceptable.

   b. Approximate the cost of each type of lamp over a 10-year period, assuming use and lamp costs remain constant. Assume the cost of fluorescent lamp is $5.00 per lamp.

20-20. Assume that half of the light emitted by a 100 W incandescent lamp (1750 lumens) illuminates a 100 ft^2 surface.

   a. Determine the average illuminance of the surface in foot-candles.

   b. Determine the average illuminance of the surface in lux.

   c. What would the average illuminance be if only half of the light illuminated the surface?

20-21. Assume that 40% of the light emitted by a 1000 W metal halide lamp (110 000 lumens) illuminates a surface with the area of a 32 ft diameter.

   a. Determine the average illuminance of the surface in foot-candles.

   b. Determine the average illuminance of the surface in lux.

   c. What would the average illuminance be if only half of the light illuminated the surface?

20-22. A 1000 candela light source emits a beam of light. Determine the luminous intensity, in foot-candles, on a surface held perpendicular to and in line with the light beam at distances in 1 ft intervals in the range of 1 to 10 ft.

20-23. A 1000 candela light source emits a beam of light. Determine the luminous intensity, in foot-candles, on a surface held perpendicular to and in line with the light beam at distances in 2 ft intervals in the range of 2 to 50 ft.

20-24. A 1000 candela light source emits a beam of light. The source is held exactly 4 ft above a moveable surface. The surface can be sloped at angles in 10° intervals in the range of 0° to 90°, measured from horizontal. Determine the luminous intensity, in foot-candles, on the surface at each angle.

20-25. A 2000 candela light source emits a beam of light. The source is held 4 ft above the surface. The surface can be sloped at various angles. Determine the luminous intensity, in foot-candles, on the surface if it is held at 10°, 20°, and 30° angles measured from horizontal.

20-26. A 30 ft by 40 ft classroom with 8 ft high ceilings will have an ambient lighting target illuminance of 70 fc at a work plane that is 28 in above the floor. It is anticipated that the ceiling reflectance is 0.50 and the average wall reflectance is about 0.3. The space will be illuminated with surface-mounted, prismatic wraparound luminaires with four lamps, as shown in Figure 20.16. The initial output of the fluorescent lamps is 2850 lumen. The light loss factor will be assumed to be 0.70. Neglecting the spacing criteria, determine the minimum number of luminaires required to provide uniform illumination in the space.

20-27. A 30 ft by 40 ft classroom with 8 ft high ceilings will have an ambient lighting target illuminance of 70 fc at a work plane that is 28 in above the floor. It is anticipated that the ceiling reflectance is 0.50 and the average wall reflectance is about 0.3. The space will be illuminated with recessed lay-in 2 ft × 4 ft open parabolic troffer luminaires with four lamps, as shown in Figure 20.16. The initial output of the fluorescent lamps is 2850 lumen. The light loss factor will be assumed to be 0.70. Neglecting the spacing criteria, determine the minimum number of luminaires required to provide uniform illumination in the space.

20-28. A 30 ft by 40 ft classroom with 8 ft high ceilings will have an ambient lighting target illuminance of 70 fc at a work plane that is 28 inches above the floor. It is anticipated that the ceiling reflectance is 0.50 and the average wall reflectance is about 0.3. The space will be illuminated with recessed lay-in 2 ft × 4 ft open parabolic troffer luminaires with three lamps, as shown in Figure 20.16. The initial output of the fluorescent lamps is 2850 lumen. The light loss factor will be assumed to be 0.70. Neglecting the spacing criteria, determine the minimum number of luminaires required to provide uniform illumination in the space.

20-29. A 14 ft by 18 ft kitchen with 8 ft high ceilings will have an ambient lighting target illuminance of 50 fc at cabinet countertops (the work plane) that are 36 in above the floor. It is anticipated that the ceiling reflectance is 0.70 and the average wall reflectance is about 0.1. The space will be illuminated with open reflector recessed downlight luminaire with one vertically mounted lamp, as shown in Figure 20.16. The initial output of the 39 W compact fluorescent lamps is 2850 lumen. The light loss factor will be assumed to be 0.70. Neglecting the spacing criteria, determine the minimum number of luminaires required to provide uniform illumination in the space.

20-30. A 14 ft by 18 ft kitchen with 10 ft high ceilings will have an ambient lighting target illuminance of 50 fc at cabinet countertops (the work plane) that are 36 in above the floor. It is anticipated that the ceiling reflectance is 0.70 and the average wall reflectance is about 0.1. The space will be illuminated with open reflector recessed downlight luminaire with one vertically mounted lamp, as shown in Figure 20.16. The initial output of the 39 W compact fluorescent lamps is 2850 lumen. The light loss factor will be assumed to be 0.70. Neglecting the spacing criteria, determine the minimum number of luminaires required to provide uniform illumination in the space.

20-31. A 100 ft by 200 ft warehouse with 18 ft high ceilings will have luminaires for ambient lighting hung 15.5 ft above the floor. It is anticipated that the ceiling reflectance is 0.70 and the average wall reflectance is about 0.10. The space will be illuminated with intermediate-distribution, reflector luminaires, as shown in Figure 20.16 with 1000 W clear metal halide lamps with an initial output of 110 000 lumen. The target illuminance is 30 fc at the floor plane. The light loss factor will be assumed to be 0.65. Neglecting the spacing criteria, determine the number of luminaires required to provide uniform illumination in the space.

20-32. A 160 ft by 200 ft practice space for a hockey team will have luminaires for ambient lighting hung 24 ft above ice level. It is anticipated that the ceiling reflectance is 0.80 and the average wall reflectance is about 0.50. The space will be illuminated with intermediate-distribution, ventilated reflector luminaires, as shown in Figure 20.16 with 1000 W clear metal halide lamps with an initial output of 110 000 lm. The target illuminance is 50 fc at the floor plane. The light loss factor will be assumed to be 0.60.

   a. Neglecting the spacing criteria (initially), determine the number of luminaires required to provide uniform illumination in the space.

   b. Determine the maximum center-to-center spacing of the luminaires.

20-33. Prismatic-wraparound, fluorescent luminaires have spacing coefficients of 1.6/1.2. They are mounted 8 ft above the floor. The work plane is 36 in from the floor. Determine the maximum center-to-center spacing of the luminaires:

   a. Perpendicular to the length of the tube

   b. Parallel to the length of the tube

20-34. Parabolic-louvered fluorescent luminaires have spacing coefficients of 1.4/1.1. They are mounted 10 ft above the floor. The work plane is 28 in from the floor. Determine the maximum center-to-center spacing of the luminaires:

   a. Perpendicular to the length of the tub

   b. Parallel to the length of the tube

20-35. Illumination of a 100 ft by 140 ft gymnasium requires 45-400 W metal halide luminaires. Assume each luminaire consumes 435 W, including the ballast. Determine the power density of this lighting installation.

20-36. Illumination of a 100 ft by 140 ft conference center lobby requires 40-400 W metal halide luminaires. Assume each luminaire consumes 435 W including the ballast. Determine the power density of this lighting installation.

20-37. Illumination of a 24 ft by 28 ft classroom requires 28 fluorescent luminaires containing two lamps (32 W per lamp) per light fixture. Assume each luminaire consumes 76 W, including the ballast. The energy code limits power density to 2.0 W/ft² in this application. Is this lighting installation acceptable?

LIFE SAFETY SYSTEMS
IN BUILDINGS

21.1   FIRE IN BUILDINGS

History of Firefighting

The Roman emperor Augustus is credited with instituting a corps of firefighting *vigiles* (watchmen) in 24 B.C.E. Regulations for checking for and preventing fires were developed. In this era most cities had watchmen who sounded an alarm at signs of fire. The principal piece of firefighting equipment in ancient Rome and into early modern times was the bucket, passed from hand-to-hand to deliver water to the fire. Another important firefighting tool was the ax, used to remove the fuel and prevent the spread of fire as well as to make openings that would allow heat and smoke to escape a burning building. In major blazes, long hooks with ropes were used to pull down buildings in the path of an approaching fire to create firebreaks.

Following the Great Fire of London in 1666, insurance companies formed fire brigades. The government was not involved until 1865, when these brigades became London's Metropolitan Fire Brigade. The first modern standards for the operation of a fire department were not established until 1830, in Edinburgh, Scotland. These standards explained, for the first time, what was expected of a good fire department.

After a major fire in Boston in 1631, the first fire regulation in America was established. In 1648 in New Amsterdam (now New York) fire wardens were appointed, thereby establishing the beginnings of the first public fire department in North America. Up until a hundred years ago, church bells were the only means to alarm citizens of a fire. The most common method of fighting fires after the church bells rang was the bucket brigade: a line of volunteers passing buckets and using water supplies from private wells to quench a fire.

Modern fire departments are a fairly recent phenomenon, developing over the past two centuries. Their personnel are either volunteer (nonsalaried) or career (salaried). Typically, volunteer firefighters are found mainly in smaller communities, career firefighters in cities. The modern fire department with salaried personnel and standardized equipment became an integral part of municipal administration only in late 1800s.

Noteworthy Building
Fire Catastrophes

Several noteworthy fire-related catastrophes have led to sweeping changes in building codes and revised techniques used to prevent and fight fires in buildings. These events include the following catastrophes.

Iroquois Theatre Fire

On December 30th, 1903, fire broke out at the Iroquois Theater, Chicago, Illinois, when an arc light ignited a velvet curtain. At the time of the fire, approximately 1900 people filled the theater to standing room only capacity. The fire resulted in over 600 deaths and was the deadliest blaze in Chicago history.

When the fire first broke out at the Iroquois Theater, the orchestra continued to play and the audience was told to remain calm and that everything was under control. The fire rapidly erupted into an uncontrolled blaze. Many occupants, still in their seats, died of smoke inhalation. Most occupants made there way to the 27 exits, where they found several of them blocked with wrought iron gates. Some of the gates were locked. Other gates were unlocked but latched and required operation of a latch that was unfamiliar to most theater occupants. Some gates opened inwards. Occupants in front of the unopened gates were trampled and crushed against the doors by a surge of panicking occupants. Trampled bodies were piled ten high in the stairwell area where exits from the balcony met the exit from the main floor. A public inquiry revealed that most injuries occurred within 15 min of the start of the fire, which was put out by the fire department within a half hour. A large fraction of the injuries were caused by being crushed. As a result of these investigations, the fire code was changed to require theater doors to open outward and to have fire exits clearly marked. Theaters were also required to have employees practice fire drills.

Triangle Shirtwaist Company Fire

On March 25, 1911, a fire broke out at the Triangle Shirtwaist Company factory on the eighth, ninth, and tenth floors of the Asch Building in the lower Manhattan garment district of New York City. Although the fire lasted less than 30 min, 146 of the 500 employees at the factory were killed.

The factory workers, mostly young female immigrants from Europe who worked long hours for low wages, died because of inadequate safety precautions and lack of fire escapes. To keep the employees working at their sewing machines, doors leading to the exits were locked once the workday began. When the fire rapidly engulfed the factory, panicked workers rushed to the stairs, the freight elevator, and the fire escape.

Nearly all workers on the eighth and tenth floors escaped. Most workers on the ninth floor died because they were unable to force open the locked exit door. The rear fire escape collapsed, killing many workers and eliminating an escape route for others still trapped. Several workers tried to slide down elevator cables but lost their grip. Observers of the fire witnessed many frantic leaps by workers from the ninth floor windows. Others were simply burned to death.

The Asch Building itself was constructed of modern construction techniques and was classified as fireproof. An investigation concluded that the fire was caused by combustible shirtwaists and fabric scraps that littered the floors. This tragedy eventually led to the introduction of fire-prevention legislation, factory inspections, liability insurance, and better working conditions for employees.

Coconut Grove Nightclub Fire

A fire struck the Cocoanut Grove nightclub in Boston, Massachusetts, on November 28, 1942. On the night of the fire, the nightclub had approximately 1000 occupants, many of whom were military personnel preparing to go overseas for World War II. Almost half (492) of the occupants were killed, and many more were seriously injured, in less than 10 min.

Combustible contents such as satin drapes, plastic upholstery, and paper decorations spread thick smoke and fire rapidly. One exit door, equipped with panic hardware, was chained shut. The two revolving doors at the main entrance had bodies stacked up to five deep after the fire was brought under control. Authorities estimated that possibly 300 of those killed could have been saved had the doors swung outward. The "safe" capacity of the structure had also been exceeded.

The Coconut Grove fire prompted major efforts in the field of fire prevention and control for nightclubs and other related places of assembly. Immediate steps were taken to provide for emergency lighting and occupant capacity limitations in places of assembly. Exit lights were also mandated as a result of the concern generated by this fire.

World Trade Center Attack

A terrorist attack catastrophically destroyed the twin towers of World Trade Center in New York City. The two towers were unable to endure the effects of a direct hit by two hijacked commercial jetliners on the morning of September 11, 2001. Shortly after the attack, both towers collapsed, killing nearly 3000 people.

Although the towers were designed to withstand being struck by an aircraft, the resultant explosions and fires weakened the structure of the building, collapsing the upper floors and creating too much load for the lower floors to bear. Once one story collapsed, all floors above began to fall. The huge mass of falling upper structure gained momentum, crushing the structurally intact floors below and resulting in catastrophic failure of the entire structure. This tragedy will unquestionably

have a long-term effect on building codes, fire prevention, evacuation plans, and firefighting tactics in skyscrapers. It has resulted in more stringent emergency evacuation procedures and improved safety regulations for all high-rise commercial and residential buildings.

The Station Nightclub Fire

On February 20, 2003, a one-story nightclub called The Station, in West Warwick, Rhode Island, was engulfed in flames within 3 min after an on-stage pyrotechnics display spread to highly combustible soundproofing foam. This led to 100 fire-related deaths and 180 injuries in a few minutes. Fire officials had inspected the nightclub two months prior to the tragedy, as part of a reapplication for a liquor license. After the fire, they reported that no permit had been granted to use the pyrotechnics. Although fire officials reported that the club was below its occupancy limit of 300, films of the event show that the club was crowded at the time of the fire. Eyewitnesses claimed that patrons were late in reacting to the fire because they believed the flames spreading along the walls of the stage were part of the pyrotechnics display. Panic ensued when thick black smoke began spreading across the ceiling.

Although several of the deaths were the result of burns and smoke inhalation, many were determined to have resulted from occupants getting trapped at crowded exits. Computer simulations conducted for a National Institute of Standards and Technology (NIST) investigation concluded that a sprinkler system would have contained the fire enough to give the occupants time to exit the building safely. An automatic fire sprinkler system was not required in the structure because of its age and small size. The concerns that grew from this fire were the lack of automatic sprinkler protection and the use of highly combustible sound-deadening foam. The rapid spread of this fire in this incident makes a case for regulations for automatic sprinkler protection in small commercial and residential buildings.

Fire

Fire is a combustion reaction that requires oxygen (air), heat, and a fuel. Typically, a spark or flame ignites the fire, beginning the combustion reaction. In order for combustion to continue, there must be sufficient heat given off by the reaction and a proper blend of oxygen and fuel. The rate at which a fire burns is dependent on the composition of the fuel, the surface area of the fuel, the rate at which fuel absorbs heat, and the amount of oxygen that is present.

A fuel must be in a gaseous state for combustion to occur. Heat from ignition and later heat generated by the flames of the fire cause solid and liquid fuel to decompose into volatile gases. These volatile gases enter the flame, mix with oxygen in the surrounding air, ignite, burn to create heat, causing more fuel to decompose and make additional gas that enters the flame. This chain reaction continues as long as there is the proper blend of oxygen (air), heat, and a fuel.

Combustible gases (e.g., natural gas, propane, and so on) mix easily with air and will burn continuously as long as the proper air–gas blend is present.

Different fuels ignite at different temperatures.

Piloted Ignition Temperature

The *piloted ignition temperature* of a fuel is the temperature at which a fire can start when a flame or spark begins the combustion reaction. The fuel is hot enough that it releases sufficient flammable gases for combustion to occur, but a catalyst is needed to begin ignition. A large mass requires a greater rate of heating to reach the piloted ignition temperature than a small mass (e.g., igniting a large log opposed to a stick).

Autoignition Temperature

The *autoignition temperature,* sometimes called the *spontaneous ignition temperature,* is the lowest temperature at which a combustible material ignites in air without a spark or flame. Some materials do not ignite spontaneously because they break down into other substances at high temperatures and never achieve a spontaneous ignition temperature. Spontaneous combustion often occurs in piles of oily rags, green hay, dust, leaves, or coal; it can constitute a serious fire hazard.

An uncontrolled fire can engulf an enclosed building space very rapidly. A wastepaper basket full of combustible paper can turn into an uncontrolled blaze in less than a minute. An ignited upholstered chair will fill a room with black smoke in 90 s. Temperatures from combustion of ordinary building materials in an enclosed space can exceed 1200°F (650°C) in a matter of minutes and 1600°F (870°C) in a half hour. Burning flammable liquids (e.g., gasoline, jet fuel, and so forth) can cause temperatures in an enclosed space to exceed 2000°F (1100°C).

Progression of Fire

There are four stages in the progression of a fire: ignition, flame spread, flashover, and consumption. The first stage of any fire begins with the *ignition* of a fuel source. Ignition requires the proper blend of oxygen (air), heat, and fuel.

The second stage is *flame spread,* which is characteristic of rapid crawling tongues of fire that lick across the surface of walls, ceilings, floors and supporting timbers. The nature and combustibility of the material govern the speed and intensity of flame spread.

As the fire intensifies, the heated material releases large volumes of volatile gases into the air. When the mixture of gases and air reach critical proportions, the material ignites in a great ball of fire called the *flashover* stage. Flashover instantly consumes the surrounding oxygen and can raise the premise temperature to exceed 1500°F (816°C). During the flashover stage, the fire might reach explosive proportions.

The final stage in the burning sequence is the fiery *consumption* of the material itself as it burns to ash. The rate of destruction depends on the amount of oxygen-rich air reaching the burning area and the combustibility of the fully ignited material.

Classifications of Fires

Generally, fires are classified into four groups by type of fuel:

Group A: Ordinary combustibles (e.g., wood, paper, plastics, trash, grass, and so on)

Group B: Flammable liquids (e.g., gasoline, oil, grease, acetone, and so on)

Group C: Electrical equipment (e.g., any electrical wiring, connection, equipment, and so on)

Group D: combustible metals (e.g., potassium, sodium, aluminum, magnesium, and so on)

Extinguishing a Fire

Building fires typically begin with the ignition of building contents (e.g., a smoldering cigarette sets fire to upholstered chair or mattress). If the flames are not extinguished quickly (while the fire is in the content phase), the fire will expand throughout the structure. Fire will spread throughout concealed spaces and cavities in walls, floors, crawl space, and attic, and eventually to the outside of the building.

Once ignited, fires become self-sustaining as the increase in temperature heats the fuel above its flash point. Fires must be extinguished by eliminating at least one of the constituents in the chemical reaction: fuel, oxygen, or heat energy. Taking away the fuel, cutting the oxygen supply, and lowering the temperature of the burning mass and surroundings are effective methods.

Extinguishing a building fire is more complex than quenching a content fire. The spreading flames that are sometimes concealed must be located and disrupted in addition to extinguishing the original content fire. To accomplish this effectively, the firefighter must know the various ways fire can spread throughout a building structure.

Performance of Materials in a Fire

Building materials exposed to the high temperatures in a fire can fail rapidly. Structural collapse from high temperature is a real safety concern in buildings, as evidenced by the collapse of the World Trade Center towers after the terrorist attacks and resulting fires on September of 11, 2001.

The materials most commonly used in building structure assemblies are steel, wood, brick, and concrete. Their performance in a fire varies significantly:

Steel

Steel is a noncombustible material, yet it displays a significant loss in strength at high temperatures. Structural steel loses about half of its strength at a temperature of about 950°F (510°C). At temperatures of about 1350°F (730°C), steel loses about 90% of its strength. As a result, structural steel is typically protected from fire by an insulating layer of fire-resisting material.

A fire-resisting material limits temperature rise of steel in a fire to keep it from losing strength. Some light materials such as gypsum plaster and wallboard are effective as fire protection of steel. Stronger materials such as concrete or masonry may also contribute to the load-carrying capacity of the assembly, thus extending in some cases the fires endurance.

Lumber and Timber

Wood is a good insulator, but when it is exposed to fire at temperatures as low as 300°F (150°C), it will burn until it is destroyed. In a fire, wood loses strength by charring. The reduction in effective cross-sectional area is dependent on the number of faces exposed to the fire.

The penetration of surface charring of the wood surface in a fire is fairly consistent with time. It is estimated that the depth of char in wood surfaces exposed to the standard endurance test temperature is about 1.5 in (37.5 mm) per hour (about 1/40th of an inch per minute).

Fired Clay Masonry

Brick and other fired clay products are vitrified in a kiln (oven) at high temperatures during their manufacture. As a result, fired clay masonry units are relatively stable in a fire endurance test. Brick also displays reasonably good thermal performance. One of the more significant factors in the fire endurance of hollow brick masonry is the amount of solid material in the wall thickness.

Hollow clay masonry units having thin face shells and webs are subject to stresses resulting from unevenly distributed thermal expansion. The tendency to spalling and shattering has been observed in hollow clay tiles.

Concrete

Concrete, which is similar to brick in thermal performance, loses strength gradually during exposure to high temperatures. It retains about half its original strength at 950°F (510°C) and one-third of its original strength at about 1300°F (700°C). This loss in strength is irreversible because it is from the deterioration of the cement binder and, in some cases, degradation of the aggregate.

The fire endurance properties of concrete depend on the type of aggregate, the proportions of the concrete mix, and moisture content at the time of fire exposure. Wide variations in performance are possible. Concretes composed of limestone aggregate display generally favorable performance in fire, whereas some quartz and granite aggregates used in concrete have a tendency to spall when exposed to high temperatures.

There are two factors to be taken into account in assessing the fire endurance of reinforced concrete. One is the thickness of concrete required to limit the temperature rise on the unexposed surface to 250°F for the period desired; the other is the cover required to keep the temperature of the reinforced steel below that at which it will lose its effective strength. Prestressed concrete requires greater thickness of cover to the reinforcement than regular reinforced concrete because a lower temperature will release the prestress and bring about collapse of the assembly.

Like clay masonry, hollow concrete masonry units (CMU) have face shells and webs that are subject to temperature variations and uneven thermal expansion stresses. With CMU, however, the face shells and webs are thicker and thus the spalling and shattering observed in hollow clay tiles is much less apparent with CMU. Additionally, CMU cells are frequently filled with grout, a cementitious material that increases the apparent thickness of the concrete, making it perform like a thick concrete member.

With most materials, an assembly of small members exposed to high temperatures in a fire is more vulnerable than an assembly of large members.

Building Construction Types

There are five fundamental categories of building construction in the United States known as *types of building construction,* as summarized in Table 21.1. Each type of building construction has fire-resistive strengths and weaknesses—that is, some types burn much more readily than others. The five building construction types are arranged in the form of a scale based on the amount of combustible material used in their construction. For example, a Type I fire-resistive building has the least amount of combustible material in its structure, whereas a Type V wood-frame building has the most combustibles in it.

In addition to the relative combustibility of the materials in the five types of building construction, unique fire spread problems are inherent in each type. These recurring fire spread hazards increase firefighting problems. The following are chronic problems that allow fire to develop in each one of the five basic types of building construction.

Fire-Resistive (Type I) Construction

Fire-resistive (Type I) construction, with its concrete and protected steel walls, floors, and structural framework, was initially intended to confine a fire by its method of construction—that is, by containing the fire with noncombustible wall, ceiling, and floor assemblies so it is confined to one floor or one space on a floor. However, fire does spread several floors in a modem fire-resistive building through two paths: through ductwork in the central heating, ventilating, and air conditioning (HVAC) system and by flames extending vertically from window to window.

A system of HVAC ducts can spread fire and smoke throughout a building that is fire-resistive construction. Air ducts delivering air to interior spaces in a central HVAC system

TABLE 21.1 TYPES OF BUILDING CONSTRUCTION THAT SERVE AS THE FUNDAMENTAL CATEGORIES OF BUILDING CONSTRUCTION IN THE UNITED STATES.

Type	Category Name	Combustibility	Description of Structure
I	Fire-Resistive	Least combustible	Noncombustible wall, ceiling, and floor assemblies; concrete, masonry, and protected steel walls, floors, and structural framework. Roof covering is noncombustible.
II	Noncombustible		Noncombustible steel or concrete walls, floors, and structural framework. Roof covering is combustible.
III	Ordinary		Noncombustible masonry-bearing walls, but the floors, structural framework, and roof can be made of wood or another combustible material.
IV	Heavy-timber		Structure consists of large solid wood timbers.
V	Wood-frame	Most combustible	Interior framing and exterior walls are constructed of slender repetitive wood studs, joists, rafters, and trusses that burn very rapidly.

go through walls, floors, partitions, and ceilings. They penetrate fire barriers and fire separations. Fire or hot gases in a room near a fresh air intake or return air duct will be sucked into the duct system and be blown throughout the structure if the system continues to operate. Fire can spread to other areas of the building. Deadly smoke can also be distributed throughout the building. Therefore, the first action taken in a burning fire-resistive building should be to shut down the HVAC air system.

The vertical spread of flames from windows below to windows above is another way fire spreads throughout a Type I building. Flames erupting out of a heat-shattered window can break or melt glass in a window directly above. Once the window above is open, flames can enter and ignite combustible ceiling tile, wall hangings, or furnishings. Even if the windows do not melt or break from heat, concealed cavities between the exterior wall and the end of the floor slab can allow vertical spread of fire and smoke from floor to floor above and near a window.

Noncombustible (Type II) Construction

Noncombustible (Type II) construction is also built of noncombustible steel or concrete walls, floors, and structural framework; however, the roof covering is combustible, which can burn and spread fire. The roof covering of a Type II building can be constructed of a combustible built-up roof covering, a layered asphalt and felt paper covering, or an ethylene propylene diene monomer (EPDM) or polyvinylchloride (PVC) thermoplastic membrane. Combustible foams may be used as thermal insulation. When a fire occurs inside a Type II building, flames can rise to the underside of the steel roof deck, conduct heat through the metal, and ignite the combustible roof covering. The asphalt, felt paper, and foam insulation may burn and spread fire along the roof covering.

Ordinary (Type III) Construction

Ordinary (Type III) construction is built of noncombustible masonry-bearing walls, but the floors, structural framework, and

roof can be made of wood or another combustible material. The major recurring fire spread problem with Type III construction is concealed spaces and penetration. These small voids, cavities, and openings through which smoke and fire can spread are found behind the partition walls, floors, and ceilings. Wood studs, floor joists, and suspended ceilings create concealed spaces. Penetrations are created by small openings for utilities. These small openings around pipes and wires allow fire to spread into concealed spaces. Flames can spread vertically several stories or horizontally to adjoining occupancies through concealed spaces. Fire spreads inside concealed spaces of a Type III building by convection, the transfer of heat by motion of a liquid or gas. Heated fire gases and flames in a concealed space can travel upwards several floors and break out in an attic space, engulfing the entire building envelope.

Heavy-Timber (Type IV) Construction

Heavy-timber (Type IV) construction is built of a structure that consists of large timbers. In this type of construction, a wood column cannot be less than 8 in thick in any dimension and a wood beam cannot be less than 6 in thick. The floor and roof decking can be thick wood planks. Exposed timber beams, columns, and decks, if ignited in a fire, create large radiated heat waves after the windows break during a blaze. If a fire in a heavy-timber building is not extinguished by the initial firefighting attack, a tremendous fire with flames shooting out of the windows will spread fire to adjoining buildings by radiated heat. A fully involved type IV building requires large water supply sources to protect nearby buildings.

Wood-Frame (Type V) Construction

Wood-frame (Type V) construction is the most combustible of the five types of building construction. A wood-frame building is the only one of the five types of construction that has combustible exterior walls. The interior framing and exterior walls are typically constructed of slender repetitive wood studs, joists, rafters, and trusses that burn very rapidly. Flames can spread out a window and then along the outside wood walls in addition to the interior fire spread. A Type five building is rapidly engulfed in flame and is therefore reserved for small structures with small occupancies.

Fire Damage in Buildings

Although heat alone can prove deadly to occupants, toxic gases in smoke cause the majority of deaths and injuries. About half of all fatalities from fires are from carbon monoxide poisoning, and more than a third are from cardiopulmonary complications.

On average, there are about 2.1 million fires reported annually in the United States. Losses from all natural disasters combined (e.g., floods, hurricanes, tornadoes, and earthquakes) average a fraction of the annual direct dollar loss from fire. Fires in U.S. homes have taken a high toll of life and property each year: over 5000 deaths. Additionally, there are an average of 28 300 reported civilian injuries and an average of 54 500 firefighter injuries annually. Direct property loss from building fires averages about $10 billion dollars every year.

Fire is one of the greatest fears of any homeowner, business owner, or director of an institution. Although the prime concern is always loss of lives in a fire, more than half of all businesses never reopen after the devastating effect of a fire.

The United States has the sixth highest fire death rate among all industrialized countries. According to the National Fire Protection Association (NFPA), 75 to 80% of all deaths by fire happen in dwellings. More than half of these deaths occurred in buildings without smoke alarms. The threat of a fire destroying lives and property can be reduced tremendously by proper installation of fire detection, alarm, and suppression equipment. In residences, automatic sprinkler systems cut the chances of dying in a fire by more than half. When combined with smoke alarms, they cut the chances of dying in a fire by more than 80%, relative to having neither. Sprinklers also cut the average property loss in a fire by one-half to two-thirds.

21.2 PASSIVE FIRE PROTECTION

Passive fire protection in buildings involves constructing walls, floors, ceilings, beams, columns, and shaft enclosures so they can resist, control, and contain the damaging effects of a fire. It is intended to entail the following:

- Provide structural and thermal integrity of floor, wall, and ceiling assemblies during a fire for a specified time period

- Compartmentalize a room or space to control the fire spread

- Provide exiting systems for occupants to safely and rapidly evacuate the building

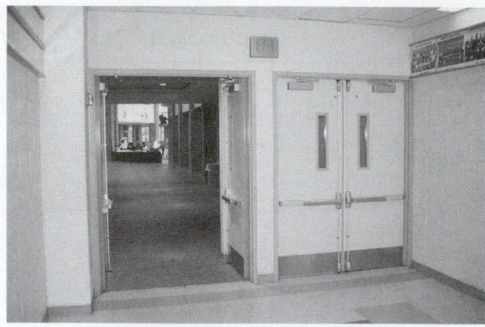

PHOTO 21.1   Fire separation doors are used to compartmentalize the building. (Used with permission of ABC)

PHOTO 21.2   Exit signage for occupant egress. (Used with permission of ABC)

If well designed and maintained properly, passive fire protection systems are extremely effective in protecting building occupants and controlling the spread of fire. These systems require periodic inspection and necessary maintenance. Breaches in structural and thermal integrity caused by renovation can lead to lack of proper protection in a fire emergency situation. Examples of passive fire protection measures that are evident in most public buildings are shown in Photos 21.1 and 21.2.

Fire-Resistive Construction

A principal objective of *fire-resistive construction* is to use materials and construction assemblies that contain the fire in a small area and confine the fire in the room or area for a specific period of time. Fire-resistive construction provides protection for a specific time period so building occupants can be made aware of the fire, the occupants can be evacuated from the building, and firefighters can fight the fire.

A good example of a building that was suitably compartmented against spread of fire is the Empire State Building. In

1945 an aircraft hit the 78th and 79th stories and a severe fire involving large quantities of fuel broke out. Despite the severity of the fire, there were no casualties among the many occupants of the floors both above and below the fire area.

A factor that plays a great role in reducing the overall fire risk in a building is the extent to which fire-resisting construction is used to divide a building into fire-resisting compartments that will contain a fire and prevent its propagation to neighboring compartments. *Compartmentalizing* means separating a building into compartments so that if there is a fire, the fire damage is confined to certain a room or certain section of the building only. This requires fire separation barriers on walls, floors, and ceilings for each zone of the building that serves as fire compartment.

Fire walls are fire-rated walls that form a required barrier to restrict the spread of fire throughout the building. They serve as a means of dividing a large structure into compartments. Firewalls are normally built of brick, concrete, or masonry. Typically, a firewall must extend from the foundation and intersect a noncombustible roof surface or extend beyond the roof by a specified vertical distance, usually 32 in (813 mm). Openings such as door and window openings are restricted in size and must contain an approved glass (e.g., wire glass) or fire door.

A *fire separation* is similar to a fire wall except that it does not extend from the foundation to the roof assembly. It is used to divide different occupancies in a building (e.g., a garage from a residence) or enclose exit corridors and stairs. A *shaft wall* is a protective fire-rated enclosure around an elevator hoist way or mechanical chase. Depending upon construction type (e.g., protected steel, unprotected steel), type of occupancy and size of building, a fire separation or shaft wall must meet a specific fire resistive rating.

A *firestop* is a specific construction technique consisting of all materials that fill the opening around penetrating items such as cables, cable trays, conduits, ducts, and pipes and their means of support through the wall or floor to prevents the spread of fire. Firestops are needed to compartmentalize a fire. All penetrations through fire separations must be sealed with an approved firestop material that meets the requirements of applicable fire and building codes. The integrity of these barriers must be maintained to provide smoke and flame containment.

Fire-Protective Materials

Several site-applied fire-protective coverings, insulations, and coatings can be used to insulate structural members from the effects of high temperatures generated in fire. Gypsum wallboard is a fire-protective covering that consists of approximately 21% water chemically bonded to calcium sulfate. In a fire, a large amount of energy is released to evaporate water in the gypsum material, giving it excellent fire-protective qualities. Insulating materials, include rockwool (a fibrous insulation made from volcanic rock) and vermiculite (a natural insulating material), also functions well. The performance of these insulations is dependent on the thickness: a thicker material provides greater fire resistance. Concrete and masonry also serve well as fire-protective coverings.

An *intumescent material* swells, enlarges, inflates, and expands when exposed to heat. Fire-protective intumescent coatings are applied like paint to structural steel members at a thickness that ranges from 0.03 to 0.4 in (0.8 to 10 mm). These intumescent coatings expand approximately 15 to 30 times their volume when exposed to high temperatures in a fire, and thus provide a good fire-protective barrier. Most intumescent coatings generate an ash-like char layer during their expansion process. As the fire exposure continues, the char layer erodes, exposing the remaining intumescent coating, which chars again. Because of their paint-like qualities and the ability to overcoat an intumescent layer with a decorative sealing coat, intumescent coatings have seen increased use in recent years. Intumescent materials perform well as a firestop to sealing penetrations through fire separations.

Fire Doors and Windows

Fire doors are typically of steel or solid wood construction and are installed with specially tested components including closers, latching hardware, and fire-rated glass lites (windows). It is essential that every part of the door and frame assembly contribute the required level of performance. Intumescent seals are fitted to the edge of the door leaf or in the frame reveal. An intumescent seal expands in fire to seal the gap between the edge of the leaf and the frame, as a result preventing the passage of smoke and flame. As shown in Figure 21.1, fire door assemblies are labeled indicating their fire rating.

Fire-resistant glass can be classified in two categories: insulating and transmitting glass. Fire-resistant *heat-transmitting glass* contains flames and inflammable gas for a short period of time, but does not prevent the transmission of heat to the other

FIGURE 21.1 A typical label indicating the fire rating of a door.

side of the glazing (e.g., wired glass, reinforced laminated glass). Fire-resistant *insulating glass* contains flames and inflammable gas for a longer period of time and prevents not only the transmission of flames and smoke but also of heat to the other side of glazing.

Fire and Smoke Dampers

Another common way for fire to spread from one compartment to another is through the HVAC ductwork. *Fire dampers* automatically close to obstruct smoke and fire from a building blaze. Fire dampers are installed in the plane of the firewall to protect these openings. Upon detection of heat, the fusible link (available in 165°F, 212°F, and 285°F) melts, closing the fire damper blades and blocking the flame from penetrating the partition into the adjoining compartment.

Smoke dampers close upon detection of smoke, preventing the circulation of air and smoke through a duct or a ventilation opening. They can be part of an engineered smoke control system designed to control smoke migration using walls and floors as barriers to create pressure differences. Pressurizing the areas surrounding the fire prevents the spread of smoke into other areas. They are controlled by a smoke or heat detector signal that is part of a fire alarm control system.

Fire and Smoke Ratings

Several fire and smoke ratings are used to classify the behavior and performance in a fire. Common ratings include the following

Fire-Resistance Ratings

A *fire-resistance rating*, expressed in hours or minutes, is a measure of fire endurance, the elapsed time during which a material or assembly continues to exhibit fire resistance under specified conditions. It is assigned to building assemblies (walls, columns, girders, beams, and composite assemblies for ceilings, floors, and roofs) based on results from laboratory testing that determine their ability to withstand the effects of a fire for a period of time. An assembly meeting the 1 hr exposure in the standard fire test receives a 1 hr rating; an assembly meeting the 2 hr exposure receives a 2 hr rating, and so forth. Fire-resistance ratings of selected construction assemblies are summarized in Table 21.2.

The fire-resistance rating is determined in a standard fire endurance test such as the method specified in *ASTM E 119—Standard Methods of Fire Tests of Building Construction and Material* ASTME: American Society for Testing and Materials. This test method evaluates how long a construction assembly will contain a fire and how long it will retain its structural integrity during a predetermined fire exposure. Other tests include *ASTM E 152—Fire Tests of Door Assemblies* and *ASTM E 163—Fire Tests of Window Assemblies*.

The ASTM E 119 test of floor and roof assemblies is conducted using a furnace with horizontal dimensions of approximately 13 ft by 17 ft (4 × 5). The assembly is installed on top of the furnace and loaded to its design capacity. The furnace temperature is regulated along a standard time–temperature curve. Fire temperatures start at room temperature and then rise to 1000°F (540°C) at 5 min; 1300°F (705°C) at 10 min; 1700°F (925°C) at 1 hr; 1850°F (1010°C) at 2 hr; and 2000°F (1093°C) at 4 hr. The test is terminated and the rating time is established when:

- Hot gases passing through the assembly ignite cotton waste.
- Thermocouples on top of the assembly show a temperature rise averaging 250°F (140°C).
- A single rise of 325°F (180°C) is achieved.
- The assembly collapses.

Horizontal assemblies such as floors, ceilings, and roofs are tested for fire exposure from the underside only. This is because a fire in the compartment below presents the most severe threat. For this reason, the fire-resistance rating is required from the underside of the assembly only. The E-119 test of walls and doors is similar to the floor and roof test. A vertical furnace is used. For load-bearing walls, the test requires the maximum load permitted by design standards to be superimposed on the assembly. Loading during the test is critical as it affects the capacity of the wall assembly to remain in place and serve its purpose in preventing fire spread. The fire-resistance rating of load-bearing wall assemblies is typically lower than that of a similarly designed nonload-bearing assembly.

Partitions (interior non-load-bearing walls) required to have a fire-resistance rating are rated equally from each side because a fire could develop on either side of the fire separation. If they are not symmetrical, the fire-resistance rating of the assembly is determined based on testing from the weakest side.

Exterior walls only require a rating for fire exposure from within a building. This is for the reason that fire exposure from the exterior of a building is not likely to be as severe as that from a fire in an interior room or compartment. Because this rating is required from the inside only, exterior wall assemblies do not have to be symmetrical in design.

Another time measured in the test is the finish or membrane rating, which is a measure of ceiling performance. Thermocouples are installed on the lowest face of the structural members and the finish rating time is obtained using the same temperature rise criteria described earlier. A fire-resistive assembly can be negated by poor construction or with a small hole cut through the assembly after the building is occupied (e.g., by a cable, duct, or electrical equipment installer). A small hole, although it may seem insignificant, breaches the fire resistance of the assembly and causes the space to no longer be confined. In a fire, the hole can allow extreme heat to pass through the assembly to another space, which limits confinement of the fire. Unfortunately, as building ages, many unprotected openings exist and often are not repaired by the installer; they are not routinely inspected and repaired.

TABLE 21.2   FIRE-RESISTANCE RATINGS OF SELECTED CONSTRUCTION ASSEMBLIES. COMPILED FROM VARIOUS INDUSTRY SOURCES.

Fire-Resistive Rating	Basic Description of Construction Assembly*
	Walls/Partitions
1 hr	One layer of ⅝ in, Type X gypsum wallboard appropriately nailed or screwed on each side of wood studs spaced at 16 in O.C.
	One layer of ⅝ in, Type X gypsum wallboard appropriately screwed on each side of metal studs spaced at 24 in O.C.
	4 in thick, cored brick units and hollow tile units.
	4 in thick, solid and hollow concrete masonry units, Type S or N concrete.
	4 in thick, solid concrete and concrete panels.
2 hr	Two layers of ⅝ in, Type X gypsum wallboard appropriately nailed or screwed on each side of wood studs spaced at 16 in O.C.
	Two layers of ⅝ in, Type X gypsum wallboard appropriately screwed on each side of metal studs spaced at 24 in O.C.
	6 in thick, solid concrete and concrete panels.
3 hr	8 inch thick, clay or shale, hollow brick (75% or more solid).
	6 in thick, solid and hollow concrete masonry units, Type S or N concrete.
	8 in thick, solid concrete and concrete panels.
4 hr	8 in thick, clay or shale, solid brick.
	12 in thick, clay or shale, hollow brick (64% solid).
	8 in thick, hollow brick (60% solid), fully filled with loose insulation.
	8 in thick, solid and hollow concrete masonry units, Type S or N concrete.
	8 in thick, solid concrete and concrete panels.
	Floors/Ceilings
1 hr	One layer of ⅝ in, Type X gypsum wallboard appropriately nailed or screwed on bottom side of wood joists spaced at 16 in O.C. A 1 in thick nominal solid wood subfloor and finish floor appropriately nailed or screwed to topside of joists.
	One layer of ½ in, Type X gypsum wallboard appropriately nailed or screwed furring channels spaced at 24 in O.C. secured to bottom side of wood joists spaced at 16 in O.C. A 1 in thick nominal solid wood subfloor and finish floor appropriately nailed or screwed to topside of joists.
2 hr	One layer of ½ in, Type X gypsum wallboard appropriately screwed to furring channels wire tied to the bottom chord of steel joists. Topside of steel joist is covered with a 4 in thick concrete slab on 28 gauge steel decking.
	One layer of ½ in, Type X gypsum wallboard appropriately screwed to furring channels secured to the bottom web of a reinforced concrete joist or double tee web (at least 10 in deep).
	Steel Columns
1 hr	Two layers of gypsum wallboard: Base layer of ½ in gypsum wallboard wire tied to column at 15 in O.C. Face layer of ½ in gypsum wallboard adhered to base layer with a laminating adhesive over entire surface.
2 hr	Two layers of gypsum wallboard: Base layer of ½ in Type X gypsum wallboard screwed to metal studs secured to column outer flange surfaces and side layer attached to ends of flanges. Face layer of ½ in Type X gypsum wallboard appropriately screwed to base layer.
3 hr	Three layers of gypsum wallboard: Base layer of ⅝ in Type X gypsum wallboard screwed to metal studs secured to column outer flange surfaces and side layer attached to ends of flanges. Two face layers of ⅝ in Type X gypsum wallboard appropriately screwed to base layer.
	Doors (includes fire door, frame, hardware, and other accessories)
3 hr (A) 1½ hr (B) ¾ hr (C) ¾ hr or 20 min	**Steel doors** Rating varies with door type and size, frame type, type and size of lite (glazing), use if intumescent seals, and hardware.
90 min 60 min 45 min 20 min	**Wood doors** Rating varies with door type and size, frame type, type and size of lite (glazing), use if intumescent seals, and hardware.

*Some construction specifications are neglected for clarity. Check with the local building code for precise specifications of construction assembly.

Flame-Spread Ratings

Another common measure used to evaluate the performance of a material in a fire is its flame spread. The *flame-spread rating (FSR)* describes the surface-burning characteristics of a building material. The most widely accepted flame-spread classification system is specified in the NFPA Life Safety Code, NFPA No. 101. The NFPA Life Safety Code primarily applies this FSR classification to interior wall and ceiling finish materials. Roof coverings must meet a different set of criteria, as discussed later.

The FSR is expressed as a number on a continuous scale where inorganic reinforced cement board is 0 and red oak is 100. The scale is divided into three classes. The NFPA Life Safety Code groups the following classes in accordance with their flame-spread and smoke development based on Classes A, B, and C. Some older model building codes (e.g., International Conference of Building Officials, Inc. [ICBO], Uniform Building Code and Building Officials and Code Administrators International, Inc. [BOCA], National Building Code) refer to the three categories as Class I, II, and III, respectively.

- Class A or I   Flame spread 0–25   Good resistance to flame spread
- Class B or II   Flame spread 26–75   Fair resistance to flame spread
- Class C or III   Flame-spread 76–200   Poor resistance to flame spread

Flame-spread ratings and classifications for common materials are found in Table 21.3. In general, inorganic materials such as brick or tile are Class A or I materials. Reconstituted wood materials such as plywood, particleboard, and hardboard are Class C or III. Although different species of wood differ in their surface-burning (flame-spread) rates, most wood products have a flame-spread rating less than 200 and are considered Class C or III material. A few species have a flame-spread index slightly less than 75 and qualify as Class B or II materials. Flame-spread ratings and classifications for common materials are provided in Tables 21.3 and 21.4.

The FSR informs designers how likely a fire is to move from its point of origin, and how fast. For instance, if the wall coverings and furnishings in a room are all Class A or I surfaces, a wastebasket fire will probably burn itself out without doing much more than scorching the surroundings. In contrast, if one or more of the surfaces has a high flame-spread rating, the fire will spread rapidly, consuming the entire room and its contents. Once the room and the furnishings are totally consumed, their flame-spread ratings no longer affect the fire's progress. Instead, it becomes a matter of the fire resistance of the walls, the floor, the ceiling, and the amount of combustible material in the room.

The FSR is specific to a particular surface and substrate. Other aspects of the wall or ceiling structure do not affect these surface-burning characteristics. However, painting the surface changes the surface-burning characteristics. For this reason, special care must be exercised in determining the type and thickness of paint applied to Class A or I surfaces because the FSR can be changed.

TABLE 21.3 FLAME-SPREAD RATINGS AND CLASSIFICATIONS FOR COMMON MATERIALS. COMPILED FROM VARIOUS INDUSTRY SOURCES.

Material/Species	Flame Spread Rating	Class
Brick	0	A or I
Fiber-cement exterior materials	0	A or I
Gypsum sheathing	15 to 20	A or I
Gypsum wallboard	10 to 15	A or I
Inorganic reinforced cement board	0	A or I
Plywood, fire-retardant-treated construction	0 to 25	A or I
Cedar, Western Red	69	B or II
Hemlock, West Coast	73	B or II
Spruce, Engelmann	55	B or II
Birch, Yellow	80	C or III
Douglas fir	90	C or III
Fiberboard, medium density	167	C or III
Hardboard siding panels	<200	C or III
Idaho White Pine	82	C or III
Maple	104	C or III
Masonite	<200	C or III
Oak, Red or White	100	C or III
Oriented strand board (OSB)	150	C or III
Particleboard	116 to 178	C or III
Pine, Lodge Pole	98	C or III
Pine, Ponderosa	115	C or III
Plywood, oak	125 to 185	C or III
Plywood, pine	120 to 140	C or III
Wood structural panels	76 to 200	C or III

TABLE 21.4 FLAME-SPREAD CLASSIFICATIONS FOR COMMON SIDING AND SHEATHING MATERIALS. COMPILED FROM VARIOUS INDUSTRY SOURCES.

Material	Flame-Spread Class	Typical Use
Cement fiber panel	A or I	Siding
Cement fiberboard	A or I	Siding
1 in gypsum sheetrock	A or I	Sheathing
⅝ in Type X exterior gypsum wallboard	A or I	Sheathing
T1-11 plywood panel	C or III	Siding
Hardboard board	C or III	Siding
Hardboard plank	C or III	Siding
1 in log veneer (pine)	C or III	Siding
1 in oriented strand board (OSB)	C or III	Sheathing

Class A-B-C Roof Coverings

Class A, B, or C roof coverings are sometimes confused with the Class A-B-C/I-II-III flame-spread categories mentioned previously. The tendency is to assume that a Class A roof system has a Class A flame-spread rating, and so on, but there is no correlation. Roof coverings must meet a different set of test criteria. The roof-covering classification test does not produce a flame-spread rating. Instead, it is a pass–fail test under which a product either passes the criteria as a Class A, B, or C roof-covering system or it does not.

It is an entirely different test from the FSR test (e.g., it includes weather exposure determined in a rain penetration test).

The highest classification for a roof covering is Class A and Class C is the lowest. Note that a Class C roof system is considered fire resistant while an FSR Class C (or III) building material is not. Nonclassified roof systems have no fire rating.

Smoke Developed Rating

The *smoke developed rating* is a single-number classification of a building material as determined by an ASTM E 84 test of its surface-burning characteristics. It is expressed as a ratio of the smoke emitted by a burning material to the smoke emitted by the red oak standard material.

21.3 ACTIVE FIRE PROTECTION AND SUPPRESSION

Active Fire Protection

Active fire protection systems include standpipe, sprinkler, and spray systems designed to extinguish the fire outright or control the fire by delaying its damaging effects. Types of firefighting media include water, foams, inert gases, and chemical powders.

Active fire protection systems are extremely effective in containing and fighting a fire if they are designed and maintained so they work properly. These systems require regular inspection, testing, and maintenance. Poor maintenance leads to a false sense of security and lack of proper protection when the system is needed under an emergency situation.

Standpipe Systems

A *standpipe system* is an internal piping network connected to fire-hose stations that are used to rapidly suppress a fire. Firefighters can use hoses connected to the standpipe system or connect their hoses to valve outlets near the fire. Firefighters have great difficulty fighting fires from the ground when flames and smoke are visible above the fourth floor of a building. So, standpipe systems are mandated in buildings where it may be difficult for the fire department to adequately pump water on the fire (e.g., in buildings that are over six stories or 75 ft in height). A standpipe system also provides water that trained occupants or employees can manually discharge through hoses until the fire department arrives.

Piping in a standpipe system runs vertically (up and down) and horizontally (side to side) throughout the building.

The standpipes running vertically are usually called *risers*. The risers are usually located in the staircase enclosures or in the hallways in the building. This piping system supplies water to every floor in the building.

At selected locations in the building, the standpipe is connected to a *hose*. The hose is usually stored on a quick-release rack called a *hose reel*. Fire-hose and reel stations are strategically positioned throughout the building. Gate valves control these connections. The occupant or firefighter must manually open the gate valve to open flow to the hose. A *nozzle* is attached at the end of the hose. The nozzle is used to direct the stream of water from the hose.

The hose and nozzle must be easy to reach at all times. The hose outlets are located so that every part of the building may be reached with a hose stream. The maximum length of a single hose line is 125 ft. Sometimes the hoses are installed in cabinets. If the hose is installed in a cabinet, it must be labeled "FIRE HOSE." When the hose outlets are not easy to see, signs should be posted telling where the hose outlets are located. (See Photos 21.3 and 21.4.)

PHOTO 21.3 A properly marked fire hose cabinet (closed). (Used with permission of ABC)

PHOTO 21.4 A typical fire hose outlet and release rack connected to a standpipe. (Used with permission of ABC)

A standpipe system may be combined with an automatic fire protection system. For example, a standpipe system and a sprinkler system may be installed in the same building. The standpipe and the sprinkler systems may even share the same water supply and riser piping. The top of the standpipe riser extends up onto the roof. Hose connections are attached to the top of the standpipe riser. These three connections make up the roof manifold. The roof manifold is used when extinguishing fires on the roof. It is also used when testing the water flow in the standpipe.

Types of Standpipe Systems

Several types of standpipe systems are approved for use. The four common types are briefly described in the following.

Wet Standpipe This system always has water in the piping. The water in the system is always under pressure. In some cases a fire pump may be used to increase the water pressure. The wet pipe system is the most commonly used standpipe system. It is used in heated buildings where there is no danger of the water in the piping freezing. Any part of the standpipe system that is exposed to freezing temperatures should be insulated. It is very important that the water in the piping does not freeze. Frozen water may prevent a standpipe system from working.

Dry Standpipe with an Automatic Dry Pipe Valve This system is usually supplied by a public water main. Under normal conditions there is no water in the piping. Instead, there is air under pressure in the piping. A dry pipe valve is installed to prevent water from entering the standpipe system. The dry pipe valve is designed to open when there is drop of air pressure in the standpipe. When a hose is opened it causes a drop in air pressure in the standpipe system. Then the dry pipe valve automatically lets water flow into the standpipe. A control valve is installed at the automatic water supply connection. This valve should be kept open at all times to supply the standpipe system. This system is usually installed in a building that is not heated.

Dry Standpipe with a Manual Control Valve This system is supplied by a public water main. Under normal conditions this system has no water in the piping. The water is not allowed into the standpipe until a control valve is manually operated. The control valve remains closed until a fire occurs. This system is usually used in a building that is not heated.

Dry Standpipe with No Permanent Water Supply Under normal conditions, this system has no water in the piping. Water is pumped into the standpipe system by the local fire department. The water is pumped in through the Siamese connection (see Siamese connection as follows). This system cannot be used unless water is supplied by the fire department. A sign must be attached to each of the hose outlets. It should read "Dry Standpipe for Fire Department Use Only." This system is usually used in a building that is not heated.

Classes of Standpipes Systems

Standpipe systems are classified depending on who is expected to use the system. The three classes are briefly described in the following.

Class I This system is designed for use by professional firefighters. For example, fire department personnel use the system. The fire hoses in these systems are 2½ inches in diameter. The large hose diameter makes it difficult to control the stream of water from the hose.

Class II This system is designed for use by the occupants of a building. The hose and nozzle are connected to the standpipe. They are ready to be used by occupants in case of a fire. The hose is 1½ inches in diameter. The hose stream is easier to control than the Class I hose.

Class III This system may be used by either professional firefighters or by occupants of the building. The hosing may be adjusted to either 1½ or 2½ inches in diameter. Attaching special reducing valves to the hose line does this.

Automatic Fire Protection Systems

Studies by various federal and state agencies and private organizations suggest that installation of automatic fire suppression systems could save hundreds of lives annually, prevent thousands of injuries, and eliminate hundreds of millions of dollars in property losses each year.

Fire suppression systems are intended to extinguish or control a fire. These include automatic water sprinkler systems and systems that use a gas agent or foam to eliminate oxygen and suffocate the fire. *Smoke control systems* are designed to limit the spread of smoke to maintain passable occupant egress routes for a given period of time and to aid firefighters in fighting the fire.

An *automatic fire protection system* provides a warning to occupants of the building, notifies emergency personnel responding to the alarm, and activates fire suppression systems to reduce the growth rate of a fire or the movement of smoke. Typically, smoke detectors sense the presence of fire in the building. The fire control panel then sounds an alarm, shuts down air-handling equipment, disconnects power from the protected equipment, and then releases agent into the protected area.

Conventional Automatic Sprinkler Systems

An *automatic sprinkler system* consists of the sprinkler heads and a network of pipes placed in a horizontal pattern near the ceiling and is designed to automatically dispense water on a fire. Examples of sprinkler system piping components are shown in Photos 21.5 through 21.10. The system is designed to extinguish the fire entirely, or to prevent the spread of the fire.

A conventional sprinkler system is fitted with automatic devices designed to release water on a fire. These devices are called *sprinkler heads*. A rise to a predetermined temperature causes the sprinkler head to open. Water is then discharged in the form of spray. When the sprinkler heads open, they are said to have fused. The sprinkler heads are fitted at standard intervals on the piping. If more than one head opens, the area sprayed by each overlaps that of the sprinkler head next to it.

An *approved automatic sprinkler system* is installed in accordance with fire or building codes. It uses the proper automatic sprinkler heads for the structure's occupancy and construction, has an adequate and reliable supply of water, has been tested and shown to be in working order, and has been found acceptable to the appropriate governmental authority.

PHOTO 21.7 A cross main (large pipe) connected to a sprinkler branch. (Used with permission of ABC)

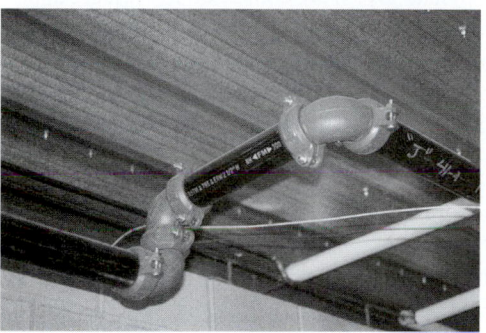

PHOTO 21.8 Victaulic elbows used to connect a cross main. (Used with permission of ABC)

PHOTO 21.5 Wet-pipe sprinkler control with pressurized water in the pipe and mains. (Used with permission of ABC)

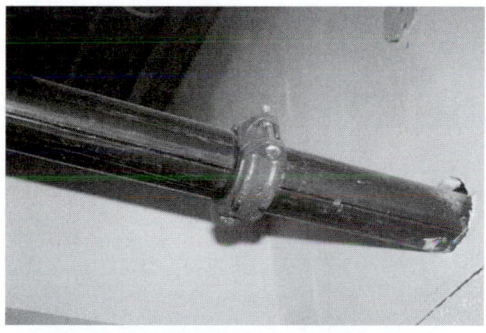

PHOTO 21.9 A Victaulic coupling is a quick method of joining sprinkler pipes. (Used with permission of ABC)

PHOTO 21.6 Sprinkler pipe precut, numbered, and ready for installation. (Used with permission of ABC)

PHOTO 21.10 A Victaulic tee coupling. (Used with permission of ABC)

Types of Conventional Automatic Sprinkler Systems

There are several types of automatic sprinkler fire suppression systems. The following are the conventional types.

Wet-Pipe Automatic Sprinkler Systems *Wet-pipe automatic sprinkler systems* have pressurized water in the pipe and mains. Water is released when the sprinkler head is activated. Because of the potential for freezing, this system is suitable for buildings where the indoor ambient temperature is not lower than about 40°F (5°C). Wet-pipe sprinkler systems are the most common in use today. In wet systems exposed to freezing temperatures, pipes containing an antifreeze solution of water–glycerin or water–propylene glycol are connected to a water supply. The antifreeze solution, followed by water, discharges from sprinkler heads opened by a fire. This type of system is used in locations subject to freezing. Use of antifreeze solutions is limited because they are costly and are difficult to maintain. Figure 21.2 shows a schematic drawing of a wet-pipe sprinkler system.

Dry-Pipe Automatic Sprinkler Systems *Dry-pipe automatic sprinkler systems* have pipes filled with compressed air or nitrogen. The pressure in these lines is slightly above the water pressure, and this pressure difference is what keeps the water out of the sprinkler lines. When a sprinkler head is activated, the air will begin to be released and the air pressure will drop.

As air pressure drops, water will begin to advance throughout the lines and flow through the activated head(s). The dry-pipe type is typically used in unheated buildings where there is danger that the water in the pipes would freeze and burst the pipes.

Preaction Automatic Sprinkler Systems *Preaction automatic sprinkler systems* are similar to dry-pipe except that the water first fills the pipe as an alarm is set off, providing an opportunity to extinguish the fire manually before the sprinklers open. Water is stopped at feeders (in the walls before the pipes supplying the sprinkler heads) by a valve. This valve is electronically activated by a heat-detecting device within the area, and a signal is sent to the valve and the valve opens. Water will then flow to all heads, but will only discharge through the activated heads. If there is an accidental break of a sprinkler line, water will not immediately discharge because the valve is holding back the water flow and not the sprinkler heads (unlike the wet-pipe or dry-pipe systems). The preaction sprinkler system is often used where the use of sprinklers could cause extensive material or equipment damage, such as in retail stores and computer areas.

Deluge Automatic Sprinkler Systems *Deluge automatic sprinkler systems* allow all sprinkler heads to go off at the same time. This system is very similar to the preaction system, except all sprinkler heads are open. Once a heat-detecting device activates the valve, water will flow from all heads within the area. Deluge systems are generally installed in hazardous areas where

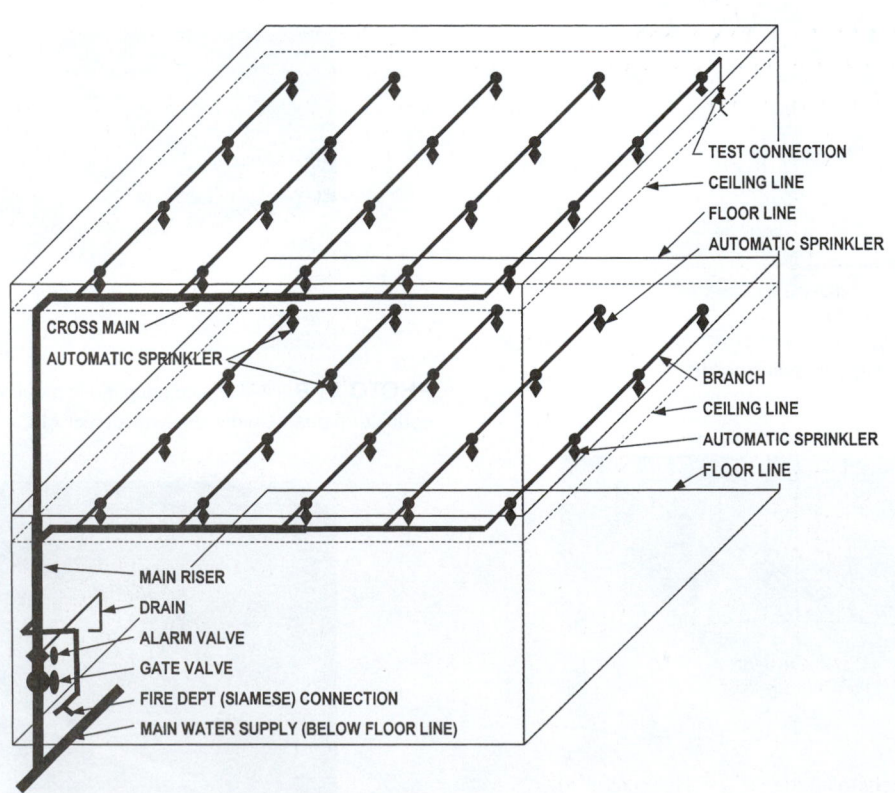

FIGURE 21.2 A drawing of a wet-pipe sprinkler system.

extremely rapid fire spread is anticipated and that requires immediate application of water.

Automatic Sprinklers

Automatic sprinklers are devices that open automatically to discharge water when an excessive temperature is detected. Each sprinkler is typically individually heat activated. When the heat of a fire raises the sprinkler temperature to its operating point (e.g., 165°F/75°C), a solder link will melt or a liquid-filled glass bulb will shatter to open that single sprinkler. Once the sprinkler is open, water from the sprinkler pipes flows directly over the source of the heat, as shown in Figure 21.3. The system works immediately upon sensing excessive temperature to prevent the fast-developing fires of intense heat, which are capable of trapping and killing building occupants.

Sprinkler heads are made of metal. They are screwed into the piping at standard intervals. Water contained within the system is prevented from leaving the sprinkler head by an arrangement of levers and links. The levers and links are soldered together on the sprinkler head. The solder is a metal alloy with a fixed melting point. Other types of sprinkler heads use a quartz bulb that expands and breaks under heat. A third type uses a solid chemical held in a cylinder that is broken by heat action. The latest type of sprinkler head is called the "cycling sprinkler." This sprinkler head cycles water on and off depending on the temperature. When the disc reaches a temperature of 165°F, the valve opens, permitting water to flow. When the disc temperature cools, the valve closes to shut off the water. Sprinkler head installations are shown in Photos 21.11 through 21.15.

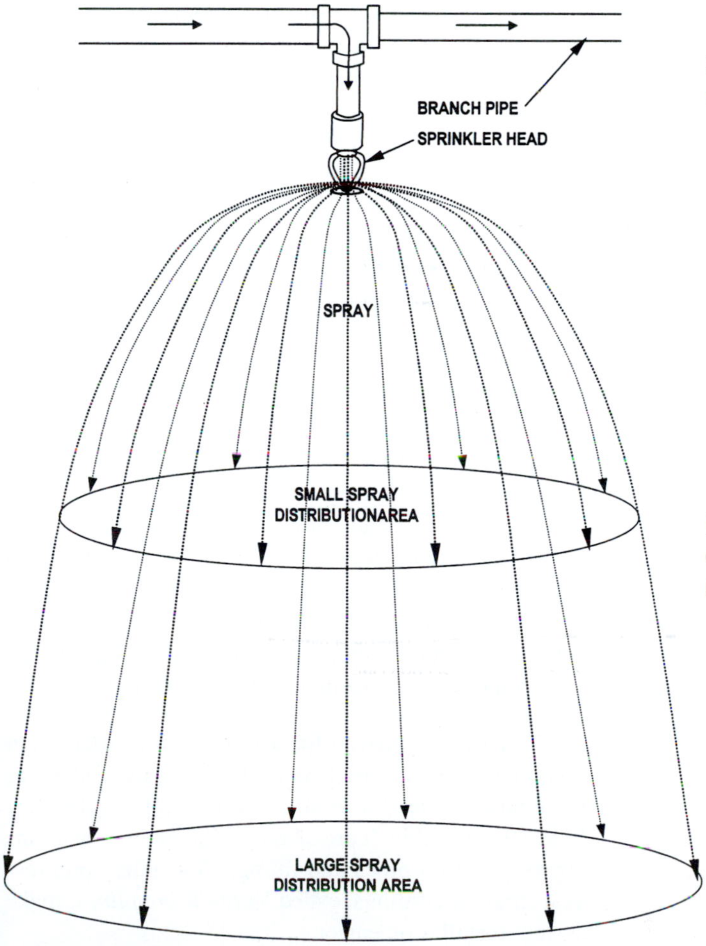

PHOTO 21.11   A sprinkler head connected to a branch before being covered with a suspended ceiling. (Used with permission of ABC)

PHOTO 21.12   A covered sprinkler head in a finished suspended ceiling. The cover is used to conceal the sprinkler. It should not be painted. (Used with permission of ABC)

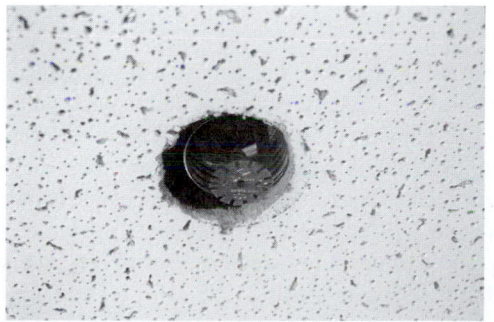

PHOTO 21.13   An exposed sprinkler head in a finished suspended ceiling (cover removed). (Used with permission of ABC)

BRANCH PIPE

SPRINKLER HEAD

SPRAY

SMALL SPRAY DISTRIBUTION AREA

LARGE SPRAY DISTRIBUTION AREA

FIGURE 21.3   Water discharging from a sprinkler head sprays over a wide area. Distribution of the spray depends on spinkler head height and design.

PHOTO 21.14   A quartz-bulb pendent sprinkler head exposed in a finished suspended ceiling. At a specific temperature, the bulb breaks releasing water. (Used with permission of ABC)

PHOTO 21.15   A fusible link pendent sprinkler head exposed in a finished ceiling. In this design, the water is prevented from leaving the sprinkler head by an arrangement of levers and links. The levers and links are soldered together on the sprinkler head. The solder is a metal alloy with a fixed melting point. (Used with permission of ABC)

Quartz-bulb sprinkler heads have a colored, heat-sensitive glycerin solution inside the bulb that will expand with temperature increase, causing the bulb to shatter and the sprinkler head to activate. Sprinklers with red (or orange) glass bulbs have a light hazard classification (i.e., residences, schools, offices, churches) and will activate at a ceiling temperature of 155°F (68°C), plus or minus a few degrees. Sprinklers with green (or yellow) bulbs will activate at 212°F (100°C) and are designed for ordinary hazards (i.e., laundries, manufacturing plants, and warehouses). Blue solutions in the bulb activate at around 286°F (141°C) and are designed for high hazards (i.e., aircraft

hangers and occupancies dealing with flammable liquids or gases or explosives). Mauve or purple solutions in the bulb will activate at 340°F (171°C) and are rarely used; they are indicative of a very high hazard sprinkler classification.

Some sprinkler heads are designed for special situations. Sprinkler heads exposed to corrosive conditions are often covered with a protective coat of wax or lead. Corrosive vapors are likely to make automatic sprinklers inoperative or slow down the speed of operation. They can also seriously block the spray nozzles in the sprinkler heads. They can damage, weaken, or destroy the delicate parts of the sprinkler heads. In most cases such corrosive action takes place over a long time so sprinkler heads must be carefully monitored for signs of corrosion. Care must be taken to ensure that the protective coating is not damaged when handling or replacing the heads.

Sprinkler Operation and Layout

Sprinkler systems are installed in accordance with building and fire codes. In conventional systems, individual fire sprinklers are spaced throughout the ceiling of a building at predetermined intervals or positions. Most sprinklers have a ½ in discharge opening. Each sprinkler can cover approximately 100 ft².

Sprinkler heads are usually positioned upright, pendent (pointing downward), or sidewall (pointing sideways). Most commonly used is the pendent type, which hangs below the pipes. This allows the piping to be concealed above the suspended ceilings with only the pendent head showing. Upright heads sit on top of the pipe and the entire system is exposed to view. They are commonly used in warehouses and in retrofitting older buildings. Sidewall sprinklers are often used in small rooms where they can throw a spray of water across the entire room. In this manner, only one sprinkler is needed in the room.

Once activated, conventional sprinklers continue to run until the main valve is manually closed. This can result in excessive water damage, one of the major causes of property loss in fires. Flow control sprinklers close automatically once the ceiling temperatures are reduced. Many of these sprinkler heads are reactivated if the temperature goes back up. They are not designed for use in dry-pipe systems. Regardless of the type of sprinkler head, they must be replaced after they have been activated.

A network of pipe delivers water to the sprinkler heads. Types of pipe approved for use include the following:

- *Steel pipe* are approved for use in all fire suppression sprinkler applications. It is available in the following nominal diameters: ⅜, ½, ¾, 1, 1¼, 1½, 2, 2½, 3, 3½, 4, 5, 6, 8, 10, and 12 in. Threaded and flanged connections are used to join pipes and fittings. Specialty compression strap-type fittings, called *Victaulic couplings,* make system installation easier.

- *Copper tubing* is the most popular water supply pipe material, but it is used less frequently in fire sprinkler systems. The thin walls of copper tubing are usually soldered to fittings. It is available in the following nominal diameters: ⅜, ½, ¾, 1, 1¼, 1½, 2, 2½, 3, 3½, 4, 5, 6, and 8 in.

- *Chlorinated polyvinyl chloride (CPVC)* is a rigid plastic pipe generally approved for use in fire suppression sprinkler systems in residential and many light commercial applications. It is available in straight lengths in the following nominal diameters: 1, 1¼, 1½, 2, 2½, 3, 3½, 4, 5, 6, 8, and 10 in. Solvent-cement welding, threaded, and flanged connections are used to join CPVC pipe and fittings. CPVC pipe and fittings for fire sprinkler systems are orange in color.

Alarm and Control Valves

An alarm valve and gate valve serve to control flow in the network of sprinkler system. The *alarm valve* initiates an alarm signal when water flows through the sprinkler system. The *gate valve* opens or closes flow to the system. It should be maintained in the open position at all times. Large buildings must be protected by multiple sprinkler systems, each having an alarm valve and gate valve.

The alarm valve and gate valve must be functioning in order for the system to work. Operation of the control valves can be accomplished by the automatic fire alarm system by provision of valve tamper switches that will initiate a trouble or supervisory signal at the fire alarm control panel to indicate a closed valve. If this type of monitoring cannot be provided, the valves should be locked in the open position.

Primary Sources of Water Supply

Water for automatic sprinkler and standpipe systems must be available in sufficient quantity and at sufficient volume and pressure at all times to ensure reliable operation in the event of fire. Potential sources of water supply for sprinkler and standpipe systems include a direct connection to the public water system, gravity tank, pressure tank, or automatic fire pump. Where a municipal water supply cannot meet flow and pressure requirements, a reservoir or storage tank must be provided to provide a secure water supply. Approximate water application rates for various manual and automatic suppression methods are provided in Table 21.5.

A *gravity storage tank* may be located on the top of a building or on a tall tower. The water in the tank is distributed throughout the sprinkler or standpipe system by the force of gravity. *Pressure tanks* are often used where there is enough water from a supply source, but the water pressure is too low or in tall buildings that need the extra water pressure to supply the highest line of sprinklers or the highest line of hoses.

A *fire pump* is a part of a fire sprinkler system's water supply in high-rise installations where the local public water system cannot provide sufficient pressure, where systems require high pressure at the fire sprinkler in order to flow a large volume of water (large warehouse or manufacturing plant), and

TABLE 21.5   APPROXIMATE WATER APPLICATION RATES FOR VARIOUS MANUAL AND AUTOMATIC SUPPRESSION METHODS. COMPILED FROM VARIOUS INDUSTRY SOURCES.

Delivery Method	Approximate Water Application Rate	
	gal/min	L/min
Portable fire extinguisher	2.5	10
Occupant use fire hose	100	380
Residential automatic sprinkler (system dependent)	10 to 20	38 to 75
Commercial automatic sprinkler (system dependent)	25 to 72	95 to 260
Fire department single 1.5 in (38 mm) hose	100	380
Fire department double 1.5 in (38 mm) hose	200	760
Fire department single 2.5 in (64 mm) hose	250	950
Fire department double 2.5 in (64 mm) hose	500	1900

where water is from a storage tank. It is designed to pump water into the fire suppression system under high pressure. The pump intake is either connected to the public water supply piping or a static water source (e.g., tank, reservoir, lake). The pump provides water flow at a higher pressure to the sprinkler system risers and hose standpipes. The fire pump starts when the pressure in the fire sprinkler system drops below a pressure threshold, such as when the sprinkler system pressure drops significantly while one or more fire sprinklers are open releasing water.

Supplementary Sources of Water Supply

Supplementary sources of water supply for sprinkler and standpipe systems include manually or automatically activated fire pumps or Siamese connections. Tanks used to provide the required primary water supply to a standpipe system may also be used as a supply for an automatic sprinkler system.

A *Siamese connection* is a Y-shaped inlet connection for fire department use in supplementing or supplying water for standpipes and sprinkler systems. If a firefighting unit arriving at a fire finds that the sprinkler or standpipe systems is not receiving sufficient water and pressure, a fire (truck) pumper is connected to the sprinkler system to supply additional water. Siamese connections should be provided for all sprinkler and standpipe systems in commercial buildings. Types of fire department hookups are shown in Photos 21.16 through 21.18.

Metal signs should be fastened to or above the Siamese connection indicating the area protected. Where the building has two or more frontages, additional metal signs should be

PHOTO 21.16 A fire hydrant for fire department hookup. (Used with permission of ABC)

PHOTO 21.17 A Siamese connection for fire department hookup to automatic sprinkler system. These connections are a standard part of automatic sprinkler and standpipe systems. Firefighters connect a hose to the connection to introduce water to a standpipe or, in this case, a sprinkler system. (Used with permission of ABC)

PHOTO 21.18 Automatic sprinkler system and standpipe connections for fire department hookup. (Used with permission of ABC)

installed indicating the location of the Siamese connection. The installation and construction of Siamese connections should be the same as required for fire standpipe systems, except that the caps of each automatic sprinkler Siamese connection should be painted green. The entire Siamese connection of a nonautomatic sprinkler system should be painted aluminum. The caps of each Siamese connection used for combination standpipe and sprinkler systems should be painted yellow and signs should be provided.

When a fire suppression system is connected to a potable water system (in most cases, the introduction of any nonpotable water creates a potential contamination problem). In the aftermath of a fire department pumper truck connecting to the system, the piping must be cleaned and disinfected.

Backflow and Backflow Prevention

Poor water quality is found in most fire sprinkler systems because of stagnation of water and contamination from cross-connections that typically occur in a fire. For example, most fire sprinkler systems are constructed with piping material that is not approved for potable water. Stagnant water in piping is subject to contamination from corrosion, which leaves a considerable amount of particulate matter and metals (e.g., zinc, cadmium, iron, copper, and lead) deposited in the water. Corrosion inhibitors and antifreeze solutions are toxic to humans and animals. Microorganisms in stagnant water in piping create taste and odor problems and can be hazardous to drink. A fire system connection to a supplemental source of water may pump a noxious or toxic substance into the fire sprinkler piping system. A loss of pressure on the potable water supply main during firefighting can allow tainted water from the fire system to enter the potable supply.

Fire sprinkler systems must have backflow prevention capabilities installed to protect the public or private potable water distribution systems from backflow of water. The backflow prevention method required is dependent upon the hazard level or classification of the fire protection system. Typically, a *double check valve assembly (DCVA)* or a *reduced pressure backflow prevention assembly (RBPA)* is required in residential, commercial, and industrial installations. Sprinkler systems using secondary sources of water or that have chemicals added should be RBPA systems, while other systems can use the DCVA method. The backflow prevention assembly should be properly maintained, inspected, and tested annually by a licensed fire sprinkler contractor. Concepts of backflow and backflow prevention are related to a building's plumbing system and were introduced in Chapter 13.

Alternative Fire Suppression Systems

Conventional sprinklers demand high water supply rates and are associated with fixed large diameter pipe networks around the area to be protected. The necessity for large amounts of water has some inherent disadvantages: it damages most of the building's contents and interior finishes; flammable oils tend to float on the water's surface and continue to burn; it conducts electricity; and if it vaporizes into steam, it may be harmful to the firefighters. Other methods may be considered when these disadvantages are of major concern. These alternative methods include the following.

Water Mist Automatic Sprinkler Systems *Water mist automatic sprinkler systems* rely upon a fine spray of water to suppress a fire. A typical system consists of cylinders of water under pressure, heat/smoke detectors, and discharge nozzles connected to a network of pipes. The mist, with its small droplets of water, is very efficient in absorbing a large amount of heat as the droplets contact the fire and is converted to steam. This conversion to a vapor causes an expansion of over 1600 times its original volume, thereby displacing oxygen and disrupting the combustion process. The small droplets affect heat transfer and tend to wet and cool combustibles much like a conventional sprinkler system.

Water mist systems must produce a directional mist or fog of fine water drops through a nozzle. The optimum water droplet size ranges from 0.003 to 0.005 in (80 to 200 μm), although larger droplet sizes can be used. The nozzle design must produce a small droplet with an orifice sufficiently large to avoid clogging from suspended particulates that may be present in the water stream. The droplets must be small enough to penetrate all areas behind obstructions, yet large enough to penetrate to the surface of the combusting fuel.

Like conventional sprinkler systems, there are requirements on the placement of nozzles and the height of the area to be protected. Unlike conventional systems, a number of additional requirements are mandatory, such as the installation of strainers or filters, the use of corrosion-resistant piping materials, and the need for a separate detection system for activation.

The ability of small droplets to absorb heat and create steam at the fire source is key to the effectiveness of fine mist systems in controlling and extinguishing fires. This process makes water mist systems more effective than a conventional automatic sprinkler system. But like conventional systems, water mist systems cannot be used on water-reactive materials.

When compared to conventional automatic sprinkler systems, the amount of water required in a water mist system is significantly less. Only very small amounts of water and air are needed to extinguish even the most intense fires (including gasoline fires) because a small amount of water is effectively spread into fine water droplets covering a large area.

The principal advantages of a water mist system are that the small water droplets are not harmful to occupants, they are effective on flammable liquid fires, and they have minimal clean-up problems. This system is only permitted in a small number of applications. Approved applications include general machinery spaces and gas turbine enclosures. Testing continues on residential occupancies, flammable liquid storage facilities, and electrical equipment spaces with encouraging results.

Clean Agent Gas Fire Suppression Systems *Clean agent gas fire suppression systems* discharge as a gas on the surface of combusting materials. A typical system consists of cylinders of a liquid agent under high pressure, heat/smoke detectors, and discharge nozzles connected to a network of pipes. See Figure 21.4. Large amounts of heat energy are absorbed from the surface of the burning material, lowering the surface temperature below the ignition point.

Clean agent gases can be released in a building space without leaving residue. When released, they extinguish the fire rapidly but do little harm to building occupants, firefighters, interior contents, and equipment. Typical installations often protected by clean agent gas fire suppression include art and historical collections, computer rooms, data processing centers,

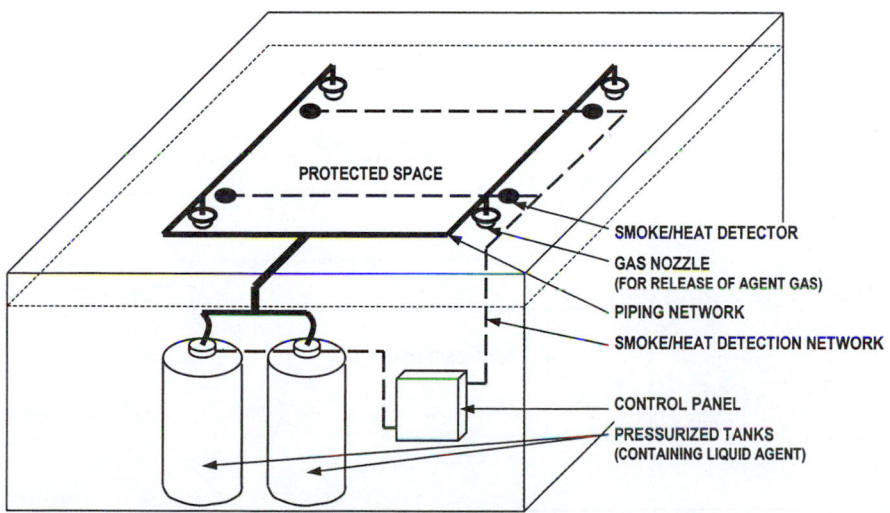

FIGURE 21.4 A sketch of a clean agent gas fire suppression system.

telecommunications centers, telephone switching rooms, electrical switchgear and transformer closets, vaults, tape storage areas, and raised floor spaces.

Halogenated hydrocarbons, known as *halons,* have been used in clean agent gas fire suppression systems for decades. The most commonly used agent was Halon 1301, an inert gas. However, because of its damaging effect on the earth's ozone layer, production of Halon 1301 ceased on January 1, 1994. Halon alternatives include FM-200® (heptafluoropropane), Inergen® (a mixture of three gases: approximately 52% nitrogen, 40% argon, and 8% carbon) and FE-13® (trifluromethane). These are environmentally friendly with no ozone depletion, no global warming potential, and zero atmospheric lifetime.

Carbon Dioxide (CO₂) Fire Suppression Systems *Carbon dioxide (CO₂) fire suppression systems* discharge a CO_2 gas that extinguishes fire by displacing oxygen or taking oxygen away from the fire. CO_2 is also very cold as it comes out of the extinguisher, so it also cools the fuel. A typical system consists of cylinders of liquid CO_2 under high pressure, heat/smoke detectors, and discharge nozzles connected to a network of pipes. CO_2 systems have been an industry standard for many decades and are still the preferred agent in many confined space or tank applications.

The principal problem with CO_2 is that it must be used in fairly high concentrations and because high CO_2 concentrations deplete much of the oxygen in a space, this type of system cannot be used with occupants or other living beings present. As a result, it is usually used in confined areas such as mechanical chases, unventilated areas, and display cases. There are several common application systems that utilize CO_2 to extinguish fires in engine compartments, dip tanks, and quench tanks and operations where spilled fuel is a possibility.

Foam Fire Suppression Systems *Foam fire suppression systems* discharge a high volume of gas-filled bubbles that rapidly fill a space. Foam masses are lighter than water and flammable liquids, and they may be either air or chemical gas bubbles. They float on the surface of burning liquids to deplete oxygen and smother the fire.

A typical system consists of foam generators, heat/smoke detectors, and a blower system that distributes the foam (see Figure 21.5). Other systems discharge foam in a manner similar to a conventional sprinkler system. Foam is very effective on flammable liquid fires and most popular in areas where flammable fuel is likely to be, such as airplane or jet hangars.

Automatic Sprinkler Testing and Maintenance

After installation of an active fire protection system is completed and approved, fire codes generally require a certificate from the contractor indicating that the system was installed in compliance with the code and that all of the required tests were performed. The contractor issuing the certificate must be licensed in installation of fire protection systems. Systems are typically hydrostatically tested at a pressure of at least 200 psi (1380 kPa), which is well above street pressure.

Active fire protection systems must be well maintained to ensure reliability. In particular, systems using water and water-based foam are prone to rust deposits that can obstruct sprinkler heads and spray nozzles. Periodic testing and maintenance is essential to ensuring that a fire protection system will work as intended in a fire situation. (See Photo 21.19.) Procedures should be in place to ensure regular maintenance and testing of systems. Maintenance contracts are often placed with a fire protection system contractor. The building operator or manager should keep records of these activities.

Influence on Building Design

The presence of an automatic fire sprinkler system and/or smoke control installation influences the allowable size and

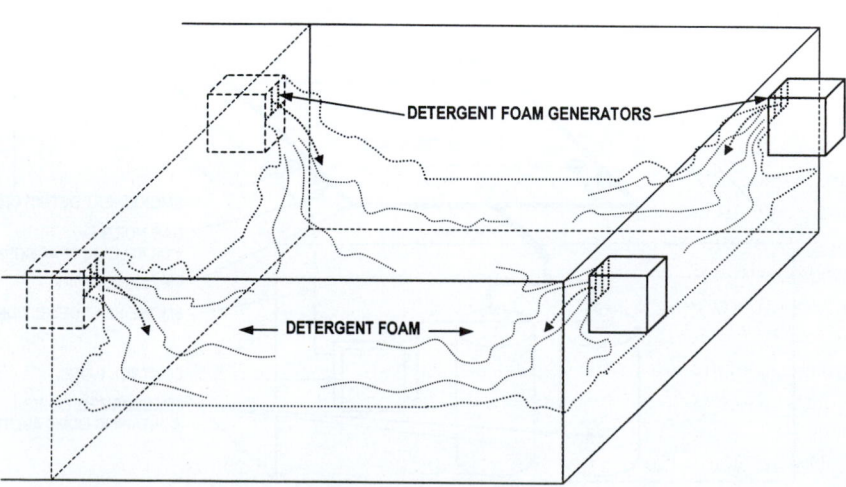

FIGURE 21.5 A sketch of a foam fire suppression system.

PHOTO 21.19 A pressure test of a sprinkler system. (Used with permission of ABC)

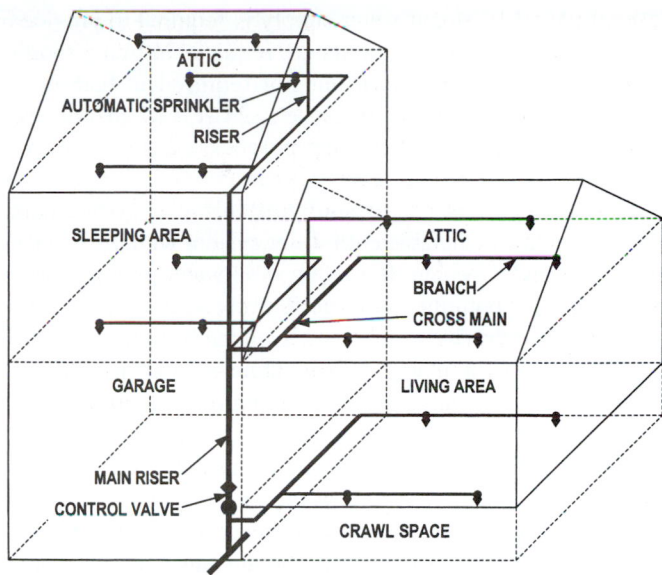

FIGURE 21.6 A drawing of a residential sprinkler system.

configuration of a building because it extends the distance an occupant can travel when exiting the building.

Typically, the maximum distance of travel from any point to an exterior exit door, horizontal exit, exit passageway, or an enclosed stairway in a facility, building, or structure not equipped with an automatic fire sprinkler system should not exceed 150 ft. This distance is extended 200 ft in a facility, building, or structure equipped with an automatic fire sprinkler system installation throughout.

In a single-story factory, warehouse, or airplane hangar, the exit travel distance may typically be increased to 400 ft if the building is equipped with an automatic fire sprinkler system installation throughout and provided with an approved smoke control installation. In an open parking garage, the exit travel distance may be increased to 250 ft with these installations.

Automatic Sprinkler Systems in Residences

In the United States, almost 80% of all fire fatalities and more than 70% of all fire injuries result from fires starting in residential living rooms, bedrooms, or kitchen areas. Yet, automatic sprinklers are not commonplace in existing single-family residences except at many remote building sites where there is no access to an adequate water supply. A move is underway to require automatic sprinklers in new home construction. This movement is based on a ten-year study from 1986 to 1996 in Scottsdale, Arizona, where all new single-family residences were required to be fitted with automatic sprinklers. Sprinklers now protect almost half of the dwellings in this community. The efforts in Scottsdale resulted in no fire deaths, an 80% reduction of fire injuries and property damage, and a 95% reduction in water usage for fire control. Cost of the sprinkler systems was less than ½% of the cost of construction. Homes with residential sprinkler systems received an average discount on insurance of about 10%. As a result of these findings, many communities in the United States have adopted codes or introduced incentive programs for sprinkler installation in one- and two-family dwellings.

Automatic sprinkler systems installed in single-family residences are simplified versions of the systems introduced earlier, usually a variation of the wet-pipe system. A schematic drawing of a residential sprinkler system is shown in Figure 21.6. Chlorinated polyvinyl chloride (CPVC) pipe being installed as part of a sprinkler system for a dwelling unit in a high-rise condominium building is shown in Photo 21.20. Specially designed sprinkler heads spray water on the walls as well as the floor. They are sensitive to both smoldering and rapidly developing fires. Most codes related to systems in residences require automatic sprinklers in all habitable rooms, including the kitchen but excluding small bathrooms and closets. Garages, underfloor crawl spaces, and attic spaces must also be sprinklered.

A residential sprinkler system typically uses a ½ in (12.7 mm) orifice, standard sprinkler, with a maximum of 256 ft² (23.8 m²) coverage, and a 25 gpm (94.6 L/m) flow rate. If the system is not supplied by an adequate public water source, a

PHOTO 21.20 Chlorinated polyvinyl chloride (CPVC) pipe being installed as sprinkler pipe. It will be serving a dwelling unit in a high-rise condominium building. (Used with permission of ABC)

250 gal (946.3 L) stored water supply is required to provide a 10 min water supply. Sprinklers are required in living rooms, bedrooms, or kitchen areas, but not required in bathrooms 40 ft² (3.7 m²) or less, small closets, 24 ft² (2.2 m²) or less, attics not used as a living space, porches, carports, garages, and foyers.

A *multipurpose residential fire sprinkler system* combines the domestic potable cold water system with the residential fire sprinkler system. It uses the cold water piping to serve as a supply for both the domestic fixtures (i.e. sinks, showers, and so on) and the fire sprinklers. Given the likelihood for a reduced amount of pipe and fittings, there is a potential for reduced system cost. Piping products commonly accepted for use in these systems match those used in standard plumbing systems, including copper, CPVC, and cross-linked polyethylene (PEX). A minimum ¾-in water meter will be required to ensure adequate flow. The two most remote sprinklers in the structure must be designed to operate simultaneously, which usually adds a flow requirement of 16 to 20 gal/min (65 to 75 L/min). Because multipurpose systems are potable water systems, they eliminate the need for cross-connection control—that is, backflow prevention is not required.

Portable Fire Extinguishers

Portable fire extinguishers can be used to put out most fires in their early stages. They are classified according to their ability to handle specific classes and sizes of fires. Not all fuels are the same, and if a fire extinguisher is used on the wrong type of fuel, it can make matters worse. Labels on extinguishers indicate the class and relative size of fire that they can be expected to handle. A clearly marked fire extinguisher in a cabinet is shown in Photo 21.21.

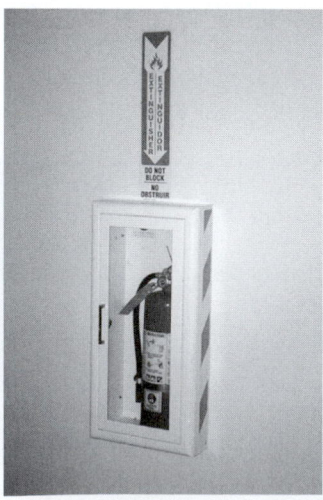

PHOTO 21.21   A clearly marked fire extinguisher in a cabinet. (Used with permission of ABC)

Types of Portable Fire Extinguishers

There are four classifications of extinguishers.

Class A Extinguishers   *Class A extinguishers* are suitable for use on fires in ordinary combustibles such as wood, paper, rubber, trash, and many plastics, where a quenching-cooling effect is required. The numeral indicates the relative fire extinguishing effectiveness of each unit. Class A extinguishers are rated from 1-A to 40-A. (A discussion on the numbering system follows.) Extinguishers rated for Class A hazards are water, foam, and multipurpose dry chemical types.

Class B Extinguishers   *Class B extinguishers* are suitable for use on fires in flammable liquids, gases, and greases, where an oxygen-exclusion or flame-interruption effect is essential. Class B extinguishers are rated from 1-B to 640-B. (A discussion follows.) Extinguishers rated for Class B hazards are foam, Halon alternative, and CO_2 and multipurpose dry chemical.

Class C Extinguishers   *Class C extinguishers* are suitable for use on fires involving energized electrical equipment and wiring where the dielectric conductivity of the extinguishing agent is of importance. For example, water-solution extinguishers cannot be used on electrical fires because water conducts electricity and the operator could receive a shock from energized electrical equipment via the water.

Class D Extinguishers   *Class D extinguishers* are suitable for use on fires in combustible metals such as magnesium, titanium, zirconium, sodium, and potassium. No numeral is used for Class D extinguishers; the relative effectiveness of these extinguishers for use on specific combustible metal fires is detailed on the extinguisher nameplate.

In addition to the letter designations, Class A and B type extinguishers also carry a numerical rating code. The numbers indicate the level of effectiveness in extinguishing fires, with 10 rated 10 times more effective than 1. A 1-A fire requires 1¼ gal (5 L) of water to extinguish. A 2-A fire needs 2½ gal of water (10 L) or twice that of the 1-A fire. So an extinguisher rated 5-A will put out a fire five times as large as one rated 1-A. For Class B extinguishers, the numerical codes are even more complicated, and generally this type of information is of most use to professional firefighters.

Fire extinguishers may contain mixtures of water, but they are also available with gases or dry chemicals. Some of the common types of fire extinguishers are as follows.

Air-Pressurized Water Fire Extinguishers

Air-pressurized water (APW) extinguishers are large, silver tanks filled about two-thirds water, and then pressurized with air. An APW is a giant squirt gun that stands about 2 ft tall and weighs approximately 25 lbs when full. They are designed for Class A fires only (solid combustible materials that are not metals such as wood, paper, cloth, trash, and plastics).

Carbon Dioxide Fire Extinguishers

Carbon dioxide (CO_2) fire extinguishers are filled with non-flammable carbon dioxide gas under extreme pressure. They are designed for Class B and C fires only (flammable liquid and electrical). CO_2 is a gas that extinguishes fire by displacing oxygen. CO_2 is also very cold as it comes out of the extinguisher, so it also cools the fuel. CO_2 may be ineffective at extinguishing Class A fires because they may not be able to displace enough oxygen to successfully put the fire out. Burning materials may smolder and reignite. CO_2 extinguishers are used in laboratories, mechanical rooms, kitchens, and flammable liquid storage areas.

CO_2 extinguishers are filled with nonflammable CO_2 gas under extreme pressure. A CO_2 extinguisher can be recognized by its hard horn and lack of pressure gauge. CO_2 cylinders are red and range in size from 5 lbs to 100 lbs or larger. In the larger sizes, the hard horn will be located on the end of a long, flexible hose.

Dry Chemical Fire Extinguishers

Dry chemical fire extinguishers put out fire by coating the fuel with a thin layer of dust, separating the fuel from oxygen in the air. The powder also works to interrupt the chemical reaction of fire, so these extinguishers are extremely effective at putting out fire. Dry chemical extinguishers come in a variety of types. "ABC" indicates that they are designed to extinguish Class A, B, and C fires; "BC" indicates that they are designed to extinguish Class B and C fires.

Class ABC fire extinguishers are considered multipurpose. They are usually filled with ammonium phosphate. They are not ideal for electrical fires because they leave a hard residue that is difficult to remove. Class ABC fire extinguishers are found in a variety of locations. New buildings will have them located in public hallways. They may also be found in laboratories, mechanical rooms, break rooms, chemical storage areas, offices, and vehicles. ABC extinguishers are red and typically range in size from 5 lbs to 20 lbs. Class BC extinguishers may be located in places such as commercial kitchens or areas with flammable liquids.

Fire Extinguisher Location

Portable fire extinguishers must be strategically situated. They must be located so the travel distance is not more than 75 ft for Class A and Class D hazard areas, and not more than 50 ft for Class B hazard areas. Extinguishers must be located close to the likely hazards, but not so close that they would be damaged/isolated by the fire. If possible, they should be located along normal paths of egress from the building. Where highly combustible material is stored in small rooms or enclosed space, extinguishers should be located outside the door and never inside where they might become inaccessible.

Extinguishers must not be blocked or hidden by stock, material, or machines. They should be located or hung where they will not be damaged by trucks, cranes, and harmful operations, or corroded by chemical processes, and where they will not obstruct aisles or injure passers-by. All extinguisher locations should be made conspicuous. For example, if an extinguisher is hung on a large column or post, a distinguishing red band can be painted around the post. Also, large signs can be posted directing attention to extinguishers. Extinguishers should be kept clean and should not be painted in any way that could camouflage them or obscure labels and markings.

Smoke Control Systems

Smoke can cause significant damage to interior spaces and equipment. For example, smoke can corrode electronic equipment, such as printed circuit boards found in computers. By the time heat accumulates at the ceiling and activates heat detectors or sprinkler heads, smoke could have already done damage. As a result, smoke detectors are recommended in spaces containing electronic equipment such as computer rooms and sound booths.

A *smoke control system* is an engineered system that uses mechanical fans to produce airflows and pressure differences across smoke barriers to limit and direct smoke movement. It is the part of a fire protection system that manages and directs smoke to protect building occupants and property (both the building and its contents). This system can also be used to assist firefighting activities.

A water flow indicator from the automatic sprinkler system or an area smoke detector automatically actuates smoke control systems. In-duct smoke detectors can also be used but only in conjunction with water flow indicators and/or area smoke detectors. They should not be the sole means of activation because they often do not react quickly enough to fires in building spaces.

The concept of zoned smoke control requires a building to be divided into a number of smoke control zones. Physical barriers such as walls, floors, ceilings, and doors separate the zones from each other. In a fire scenario, the spread of smoke from the zone of fire origin (called the smoke zone) to adjacent zones is limited by pressure differences and airflows. Pressure differences across the barriers of the zone of origin can be accomplished by supplying outside air to the other zones, by venting the smoke zone, or by a combination of both.

In large buildings, each floor is typically a separate zone. However, the system can also be designed so that a floor can have a number of zones, or a single zone can consist of more than one level.

21.4 FIRE DETECTION AND ALARM SYSTEMS

Fire Alarm Systems

Fire alarm systems detect products of combustion, such as smoke (aerosol particulate), heat, and light, and provide early occupant notification to allow the safe egress of the occupants.

These systems use various methods to detect the products of combustion, including various heat and smoke detection techniques. They use audible and visual alarms to alert and warn building occupants of a fire.

In a residence, a fire alarm system may be composed of a few stand-alone units. Fire alarm systems in large buildings are more sophisticated; they include individual components such as smoke or heat detectors, control panels, fire command centers, communication centers, and alarm horns or speakers.

Smoke Alarms

A *smoke alarm* is a fire-safety device that detects the products of combustion and gives off an audible and/or visual warning to building occupants. It is a smoke detector and alarm in one unit. They typically use an audible alarm signal to alert and warn building occupants of a fire. Smoke alarms are typically used in single-family homes, multifamily dwelling. and in some instances, light commercial applications. Ionization and photoelectric are the types of commercially available smoke detectors used in these applications. (These are described later.) Smoke alarms are powered by battery or are hard-wired into the building electrical system with a battery backup. In new construction, hard-wired units are typically required in common spaces on each floor and in bedrooms. (See Photos 21.22 and 21.23.)

PHOTO 21.22 A smoke detector in a finished ceiling of a commercial building. (Used with permission of ABC)

PHOTO 21.23 A hard-wired smoke alarm in an unfinished ceiling of a residence. (Used with permission of ABC)

Smoke and Heat Detectors

A *smoke detector* is a sensing device that identifies products of combustion in air. *Heat detectors* are a sensing device that recognizes a high temperature or a rapid increase in temperature. In large buildings, smoke and/or heat detectors are placed in building spaces and in the ductwork of air-handling units. They are then connected to a fire alarm control panel, which is designed to recognize the signal of the detector and alert building occupants of the fire through a separate alarm system. Conventional detectors typically use a signal that is electronically sent to a fire alarm/control panel, where the signal is interpreted. The detector either reports an alarm condition or a normal condition.

Smoke and heat detectors serve as the first line of defense against fire in buildings. They provide an early warning to building occupants, allowing them valuable time to escape a fire. They alert sleeping occupants who would otherwise have been overcome by smoke and toxic gases in their sleep. A description of various detectors follows.

Fixed-Temperature Heat Detectors

Fixed-temperature heat detectors signal an alarm after the temperature at the detector reaches a set value. A number of fixed-temperature designs are available, including what follows:

- *Fusible alloy* detectors that employ metal alloys designed to rapidly melt at a desired temperature.
- *Bimetallic* detectors use sensing elements made of two strips of different metals, each having different thermal expansion coefficients. When the metallic strips are heated, the sensing element deflects, closing the contact and initiating an alarm.
- *Electrical conductivity* detectors have a sensing element in which resistance varies as a function of temperature.
- *Heat-sensitive cable* detectors consist of two current-carrying wires separated by heat-sensitive insulation that softens at the rated temperature, thus allowing the wires to make electrical contact.
- *Liquid expansion* detectors have a sensing element comprising a liquid that expands with an increase in temperature.

Fixed-temperature detectors are more suitable for property protection rather than life safety applications. This is because, in many cases, spaces will be untenable from smoke prior the ceiling temperature reaching the alarm threshold. This depends on the ceiling height, room configuration, and other factors.

Rate-of-Rise Heat Detectors

Rate-of-rise heat detectors signal an alarm when the temperature at the detector increases at a rate exceeding a preset value.

- *Pneumatic rate-of-rise* detectors have an air chamber with a diaphragm enclosing a portion of the chamber.

Electrical contacts are attached to the diaphragm. The chamber is also provided with an air vent. As the detector is exposed to heat, air within the chamber expands. The vent is provided to allow the normally expanding air to escape. The vent is sized such that if air within the chamber expands very quickly, it cannot vent enough air to equalize pressures within the chamber. The diaphragm expands to compensate for the increased pressures. As the diaphragm expands, it completes a circuit within the detector and initiates an alarm. Some designs use a tube connected to the diaphragm.

- *Electrical conductivity rate-of-rise* detectors have a sensing element that changes its resistance with a change in temperature. It is connected to control equipment and sends an alarm when the rate of temperature increase exceeds a preset value.

These detectors respond to temperature changes. If the temperature changes slowly over time, this type of detector will not detect a fire. Rate-of-rise detectors can, in many cases, detect fires more quickly than fixed-temperature detectors, because of the length of time it takes for temperatures at the detector to reach that of fixed-temperature detectors.

Flame Detectors

Flame detectors optically sense high levels of either infrared (IR) radiation or ultraviolet (UV) radiation. Combination UV/IR detectors are also commercially available. A flame emits UV and IR radiation in wavelength ranges that are not emitted by sunlight or artificial lights (e.g., fluorescent, incandescent, metal halide, and so on). These detectors sense specific wavelength ranges of UV or IR radiation and send an alarm signal. Although radiation travels in a straight-line path, either reflector will detect reflected UV or IR radiation off of wall, floor, and ceiling surfaces.

Ionization Smoke Detectors

Ionization smoke detectors are designed with a sensing chamber that has a radioactive element. The chamber has two charged electrodes. The radioactive element charges the air within the chamber, which creates an electrical current between the electrodes. Smoke particles entering the sensing chamber change the electrical balance of the air. The greater the amount of smoke, the higher the electrical imbalance. When combustion particles enter the smoke alarm, they obstruct the flow of the current.

Ionization smoke detectors respond first to fast-flaming fires because these fires produce invisible particles that less than 1 micron in size. They respond less quickly to smoldering fires that tend to generate larger particles. A flaming fire consumes combustibles extremely fast, spreads rapidly, and generates considerable heat with little smoke. Ionization alarms are best suited for rooms that contain highly combustible materials.

These types of material include cooking fat/grease, flammable liquids, newspaper, paint, and cleaning solutions.

Photoelectric Smoke Detectors

Photoelectric smoke detectors use a light scattering or light obscuration principle. They contain a light emitting diode (LED) that is adjusted to direct a narrow IR light across the unit's detection chamber. When smoke particles enter this chamber, they interfere with the beam and scatter the light. A photodiode monitors the amount of light scattered within the chamber. When a preset level of light strikes the photodiode, the alarm is activated.

Photoelectric smoke detectors respond first to slow smoldering fires. A smoldering fire generates large amounts of thick, black smoke with little heat and may smolder for hours before bursting into flames. Because photoelectric smoke detectors function on visible smoke, they are more suitable for detecting large smoke particles produced by smoldering fires.

Photoelectric models are best suited for living rooms, bedrooms, and kitchens. This is because these rooms often contain large pieces of furniture, such as sofas, chairs, mattresses, and countertops, which will burn slowly and create more smoldering smoke than flames. These smoke detectors are also less prone to nuisance alarms in the kitchen area than ionization models.

Air-Sampling Smoke Detectors

Air-sampling smoke detectors use a similar approach to light obscuration detectors, however, a laser or xenon tube is typically used as a light source. Unlike conventional smoke detectors that passively wait for smoke to reach them, air samples are continuously drawn through a sampling pipe network to the detection chamber. A filter assembly screens out large airborne dust particles. Once inside the detector, the samples are exposed to a light beam that bounces light off small particles released by the combusting materials in the protected area. The scattered light is converted to an electronic signal and passed to the control system.

Air-sampling technology has revolutionized the fire alarm industry and allows for extremely early warning in the event of a fire. Systems can be designed to be 1000 times more sensitive than standard ionization and photoelectric detection systems, yet with proper installation, will not false alarm. This technology is used in applications where downtime is a critical factor.

Heat and smoke detectors are located on ceilings and walls in the spaces they protect. Placement and spacing is determined by ceiling configuration and projections that impede flow of smoke. Other factors include airflow rate, elevation, humidity, and temperature.

Although not a substitute for room-mounted units, heat and smoke detectors are also required in air-handling units. When mounted in ducts, they sense smoke in the air stream. These detectors shut down HVAC equipment to prevent the spread of smoke and heat, and/or to close fire dampers in a zoned system.

PHOTO 21.24   A manual pull station. (Used with permission of ABC)

PHOTO 21.25   A fire alarm with emergency voice communication and warning (strobe) light capabilities. (Used with permission of ABC)

Manual Pull Stations

Manual pull stations are lever-like devices mounted on a wall or pole in strategic places in the building and that are connected to a building fire alarm control panel or directly to the municipal or district fire alarm system. When the pull station lever is pulled, an alarm is sounded. Some models are contained behind a glass cover that must be broken before the device can be operated. The need to break the glass cover reduces the frequency of false alarms. Manual pull stations are available commercially in both indoor and weatherproof outdoor models. An example of a manual pull station is shown in Photo 21.24.

Alarms

The most common method of alerting occupants during a fire emergency is an audible evacuation signal delivered through bells, horns, chimes, buzzers, and sirens. Strobe lights are also used in combination with the audible signal to ensure that hearing-impaired occupants recognize the need to evacuate. Newer systems use voice commands to direct the occupants to evacuate the building.

Most building occupants demonstrate complacent behaviors when audible warning signals are sounded. They either pay no attention to the warning or rationalize that the warning is not influencing their safety. They take too much time to evacuate the building without realizing how they badly they are placing themselves at risk. Studies have shown that people react better to voice instructions as compared to loud sounds or tones. Thus, the use of voice command fire alarm systems has become more commonplace in buildings.

Emergency Voice Communication Systems

An *emergency voice communication system* provides preprogrammed recorded messages that offer direction, instructions,

and a calming voice in an emergency situation. Initial voice messages are preprogrammed and are automatically transmitted to speakers located throughout the building. Once the fire department and other emergency personnel assess the situation, live instructions may be delivered to certain zones or floors of the building to instruct occupants, relocate occupants to aid fire department operations, or merely reassure remaining occupants of their safety during an incident.

Emergency voice communication systems can also be provided with a two-way communication subsystem. This subsystem enables responding firefighters and other emergency personnel to communicate throughout the facility using dedicated two-way communication phone lines with portable handsets or emergency communication telephone stations provided in remote locations. Although often provided together, this is an independent system and is not always required.

Voice fire alarm systems have been required in high-rise buildings for many years. Fire situations in high-rise buildings are treated differently from fires in typical low-rise buildings. Occupants in the area of the fire incident and other exposed fire areas must be notified and directed to relocate to a safe area. The area affected typically includes the floor where the emergency situation is occurring, the floor above, and the floor below. The actual bounds of the area evacuated are based on the overall fire protection program, arrangement of fire areas, and the facility emergency plan. Occupants in other areas of the building are informed of fire conditions by automatic means, advised of the situation, and instructed to wait in their current location until receiving further instructions. A fire alarm with emergency voice communication and warning (strobe) light capabilities is shown in Photo 21.25.

Fire Detection and Alarm Systems

In medium to large buildings and building complexes, a fire detection and alarm system includes all or some of the following:

- A system control unit
- A primary or main electrical power supply

- A secondary (stand-by) power supply, usually batteries or an emergency generator
- Alarm-initiating devices such as automatic fire detectors, manual pull stations, and/or sprinkler system flow devices, connected to initiating circuits of the system control unit
- Alarm-indicating devices, such as bells or lights, connected to initiating circuits of the system control unit
- Ancillary controls such as ventilation shutdown functions, connected to output circuits of the system control unit
- Remote alarm indication to an external response location, such as the fire department
- Control circuits to activate a fire protection system or smoke control system

The *fire system control unit* serves as the center of the fire alarm system. The fire alarm control panel is the central part of a fire detection/alarm network in schools, municipal buildings, nursing homes, hospitals, apartment buildings, warehouses, office buildings, retail malls, and department stores. In large buildings, there are several fire system control units.

A *fire command center* is a remote panel or set of panels connected to the fire system control panels. The command center contains the following:

- Voice fire alarm system panes and controls
- Fire department two-way communication service panels
- Telephone for outside communications
- Sprinkler valve and water flow status panels
- Smoke management controls
- Elevator location status panels and annunciators
- Fire pump status panels
- Emergency generator status panels

A fire detection and alarm system are shown in Photos 21.26 through 21.28. The fire command center is typically located at an entrance of the building in a space with a minimum 1-hr fire resistance-rated construction. The location is determined by the jurisdiction having authority (e.g., municipality or special fire district) based on how emergency equipment will first arrive at the site. This location serves as the command center for interior firefighting actions during a fire incident. Large building facilities may have more than one fire command center.

Intelligent fire alarm systems provide useful information on the type of fire and the specific location instead of simply sensing the potential of a fire. These systems provide a unique point identifier that can pinpoint the exact location of the detector sensing a fire. Each detector or alarm is provided with an individual address that can be programmed to give the operator or responding personnel useful information. A detector can be addressed to allow the system to identify a room or specific location

PHOTO 21.26 A fire command center. (Used with permission of ABC)

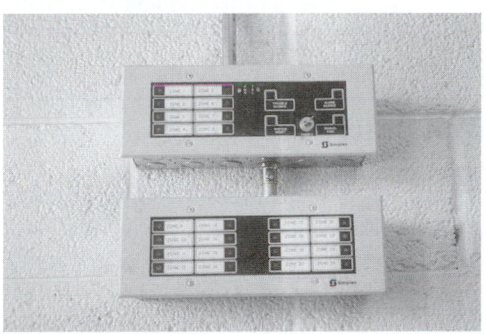

PHOTO 21.27 A fire status panel. (Used with permission of ABC)

PHOTO 21.28 Low-voltage wiring for networking a fire alarm and detection system. (Used with permission of ABC)

where the detector is installed. If the detector sends an alarm signal, the system displays a message that informs the operator of its specific location. This information can be useful and thus provide intelligent information.

Operation of the fire alarm control system begins with the smoke or heat detector sensing a fire and sending a signal to the control panel. At the control panel, the signal is processed. Depending on smoke levels and the preprogrammed alarm levels, the appropriate output signals are generated. The first stage of the three-staged alarm levels (alert) may simply indicate that the system has detected something out of the ordinary that should be investigated. The second stage (action) indicates that a potential fire exists and that emergency procedures should begin. The third stage (fire) signifies an actual fire condition. Once emergency personnel arrive at the scene, they take control of the situation at the fire command center.

Fire alarm control systems can initiate fire suppression by closing of fire doors and dampers that may be otherwise held open to allow normal building function. Where permitted, a fire alarm control system may be combined with other automated building operations (e.g., energy management) and/or security stations. A security system can work together with a fire alarm system to enhance the fire alarm system goals. For example, a facility that has a security system is well covered with closed-circuit television surveillance equipment, which can be utilized during a fire incident to visually assess the situation (e.g., determine whether the incident is a small trash can fire or a room is totally engulfed in fire).

19.5   BUILDING SECURITY

Except in a relatively few instances, building security has traditionally received little consideration during the design stage. However, today we face a new set of challenges associated with security threats. Crimes against businesses and increasing threats of violence, terrorist attacks, bombings, and shootouts that have resulted in tens of billions of dollars of costs annually have increased interest in building security significantly. Airport security has increased in an attempt to prevent future acts of terrorism and, for the most part, to limit what can be taken on an aircraft and potentially used as a weapon. The utility and chemical industries have implemented extra security procedures to provide better protection and help prevent contamination and loss of power. Most industries have taken additional steps to strengthen themselves against potential terrorist/criminal actions.

A building security system can be thought of as a life-safety system. It can assist in the safety of building occupant and business personnel because it minimizes prohibited entry by unauthorized persons such as disgruntled former employees, terrorists, or common criminals. It also prevents damage that can result in building occupant injuries.

Building security must be approached in building design, selection of materials used in the building, occupant control, and surveillance and alarm systems.

Initial Building Design

Security issues in a building should be considered in the early planning and architectural design stages. This will simplify the ensuing security system design, decrease security costs, and reduce the potential for crimes to occur in the building or building complex. All aspects of the design should be considered including access routes, landscaping, signage, lighting, materials, colors, entryways, and interior layouts.

Designers of a building should pay particular attention to the locations of doors, windows, loading docks, and money-handling rooms. These areas should all be easy to view from other surrounding areas and be well lit (as should elevators and stairways). Ledges and exterior ornamentation that might allow people to climb up the side of the building should be avoided. One of the best investments in terms of security is the provision of high levels of illumination throughout the project.

Material Selection

Doors, windows, and their hardware should be carefully selected to be certain that they will discourage intrusion. Windows should be designed so that the glass cannot be removed from the outside, and the hardware should be such that the window cannot be opened from the outside with a plastic card or wire. Doors in question should be metal, and locks should be selected so they cannot be opened with a credit card. Also be certain that the door's hinge pins cannot be taken out from the outside of the building and the door removed.

Many building codes require that all doors that lead into stairways and exits be operable from both directions in case of fire or emergency. When such a door is not to be used except in an emergency, it should be wired to an alarm system that activates when the door is opened; large signs should be placed on such doors warning of the alarm system setup.

Buildings that are more prone to security problems because of the location or the type of occupancy should be built of materials that prevent easy entrance. For example, exterior wood frame walls are much easier to "open up" than cast-in-place concrete, CMU, or brick walls. The roof assembly should also be carefully selected. It is easier to cut or pry an opening in a plywood roof than a concrete or gypsum roof. Roof scuttles (hatches), stair access, ventilation openings, and other equipment on the roof should also be carefully chosen.

Building Security Systems

For most projects, an electromechanical security system provides economical and effective security. Such systems have been used extensively in residences and in commercial, industrial, and institutional buildings.

A basic *electronic security system* consists of a control panel, keypad, digital communicator, backup battery, transformer, sensing devices, and a horn, buzzer, or siren. Various sensing devices such as door/window contacts, motion detectors, glass break sensors, smoke detectors, heat sensors, temperature sensors, carbon monoxide sensors, flood/water sensors, panic/hold-up buttons, and medical panic pendents are used, depending upon the size and complexity of the installation. The system might also use closed circuit television cameras and monitors.

The setup of an electronic security system is quite simple, especially when it is installed in new construction. In the simplest system, an electric circuit to all the doors and windows selected and to the alarm connects to the monitoring system. Once the system is programmed, if the circuit is broken at any of the contact points (at a door or window), the alarm goes off. There are many types of installations and equipment available, depending on the equipment manufacturer and the desired security.

Perimeter Protection

The first level of protection for any security system is perimeter protection. Perimeter protection includes such areas as doors and windows. It may also include closed circuit TV video surveillance (discussed later).

Alarm contacts, either the plunger or the magnetic type, can be installed on doors and windows to protect against unauthorized entry. The *plunger-type contact* is installed so that when the door or window is closed, a plunger is pushed in, and when the door or window is opened, the plunger pushes out, setting off the alarm. The *magnetic-type contact* consists of two contacts that are surface mounted, so that when the door or window is closed they make contact with each other, and when the door or window is opened the magnets are separated, setting off the alarm.

Interior Protection

Interior protection provides a backup to perimeter protection. This includes:

- *Infrared (IR) motion detectors* pick up body heat from any living being within the protected areas. IR detectors can be strategically placed inside a space to provide this kind of protection. In homes, pet immune motion detectors will allow a large animal to move past them without activating the alarm.

- *Glass break detectors* are digital microphones that have a range of up to 25 ft (8 m) and can recognize the sound associated with breaking glass. Glass break detectors can be installed near any window or pane of glass that can be broken by forced entry.

- *Floor mat detectors* will activate the alarm if stepped on. Common concealed locations for the floor mat entry detector include under the rugs, inside all exterior doors, and just below windows.

- *Emergency immediate-response keys and pendents* are used to summon aid in an emergency. When pressed for 2 s, the emergency key or pendent will send an alarm to the central monitoring station for immediate notification of the proper authorities.

Video and Audio Surveillance

Closed circuit television (CCTV) video surveillance systems can allow a small number of security staff to monitor indoor areas of a building and the building site. A few basic applications include loading docks, parking areas, areas of high shoplifting potential, schools, casinos, and banking institutions.

Surveillance cameras serve as a way to monitor events and to provide a deterrent. On remote control surveillance cameras, pan, tilt, and zoom capabilities can be controlled from a central location. The signal is transmitted through a local area network or via the Internet. Examples are shown in Photos 21.29 and 21.30. The most effective surveillance systems are those that use others monitoring devices such as motion sensors and audible alarms to notify personnel when a problem occurs.

Surveillance systems can record activities. Recorder playback functions can be used to review a past incident and assist police in catching the perpetrators of a crime. Digital video recording (DVR) equipment stores data that can be digitally enhanced and analyzed.

Control Panels and Centers

In residential installations, the control panel used to activate and deactivate a detector circuit is powered by the regular household power (120 V) available, with a stand-by battery power source to take over in case of a power failure. Many of the units have lights that indicate when the circuit is on. In addition, they may have a test alarm button so that the entire system can be tested.

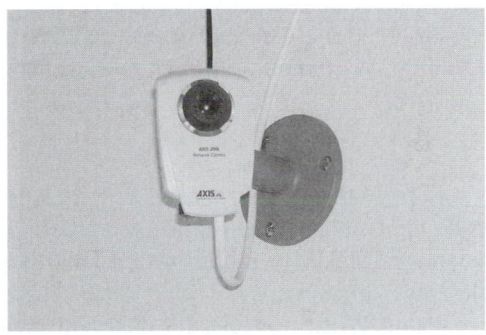

PHOTO 21.29 An indoor camera used for remote surveillance via a local area network or the Internet. (Used with permission of ABC)

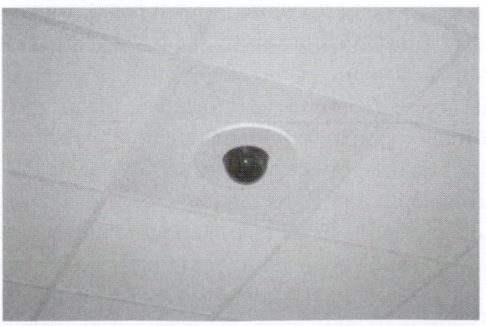

PHOTO 21.30 A remotely controlled closed circuit television (CCTV) video surveillance camera. (Used with permission of ABC)

To allow the occupant to leave and re-enter without sounding the alarm, an exit/entry control panel is installed. It includes an outside key-operated switch and an inside switch at the door. This type of control panel will sound the alarm if any attempt is made to disconnect the outside wall plate. Indicator lights on the wall plate turn on when the system is activated. Some exterior control panels are simply activated and de-activated by an outside key-operated switch, with none of the protective features mentioned.

When the outside control panel is used, a push-button interior alarm switch is installed. One inside alarm switch may be installed for the entire system, or several may be installed to operate individual doors and windows. The latter allows the opening of a door or window without having to de-activate the entire system.

Also available is a control panel with a time-delay exit and entry feature. This control delays the operation of the system, allowing the occupant to leave or re-enter the building without setting off the alarm. With this control panel, an outside key-operated switch is necessary. Control panels should be mounted where they will be readily accessible to the adult occupants but not to any children or visitors (e.g., in a closet, laundry area, bookcase, or cabinet).

In commercial installations, a main control panel performs more complex functions. It consists of a keypad, video console, backup battery, and transformer. Remote sensing devices are connected to the main panel. The system typically includes CCTV cameras and monitors. On large installations, remote control panels are tactically positioned at outlying locations. The main control panel and remote control panels are connected so they can communicate with one another.

Alarms

The alarm set off by the various systems may be a bell inside and/or outside the building, a bell inside and a horn outside, a horn inside and/or outside, or a horn with a light (beacon). In many locales, the alarm system can be connected directly to the fire or police station. When available, this is the best arrangement.

Generally, the next best arrangement is to have an inside bell or horn with an exterior horn and a light. The more noise the systems make, the more likely it is that neighbors will hear it and call police—and the more likely it is that whoever breaks in will leave without taking anything. To be effective, this type of system must depend on the cooperation of neighbors, so be certain that they understand what the alarm means and that they should call the police to report any attempted break-in.

More elaborate electronic security systems utilizing ultrasonic sound, photoelectric and microwave devices, audio-activated devices, and closed-circuit television are also available. Many of the systems are used in combination to achieve maximum control. Often, the integrated system is set up with a number of alarms that may warn of smoke, fire, and intruders. Obviously, the larger the building complex is, the more money should be available for security systems.

Electronic Access Control Systems

Many large companies use *electronic access control systems* to control employee entrance by identifying an authorized individual and allowing that person access to a restricted area. Access control can deny an unauthorized employee or outsider access into a restricted building or space. It also allows management to increase employee productivity by preventing unrestricted traffic to different areas of the building. It can also track employee movement through a building or room. The owner can recall this information at a later time as the system can store this information in a database.

A commonly used access control method requires insertion of a coded electronic *cardkey* (similar to an encoded credit card) into a wall-mounted receptacle that decodes the key and activates the door lock only for the proper key. This type of lock is much more secure than the standard cylinder lock used on most doors. A wall or door mounted *keypad* can also allow the employee access after the individual has properly entered the keypad code. These control devices can operate independently or be connected to a central control panel. Codes can be updated or changed regularly to purge access for former employees.

Another control method known as *photo identification* requires the employee to punch a coded number on a keyboard. Within seconds, a picture of the person who is assigned that code is flashed on the screen for the guard to check. It operates by the same principle as the instant replay technique used on television and uses the same equipment.

Biometric identification, the latest technology, is the automatic assessment and recognition of a unique body feature (e.g., fingerprint, eye and face recognition) or personal action (e.g., voice recognition). With biometric identification, a person may place a hand on a hand plate; the device then measures such things as finger size, fingertip angle, and skin translucency, and compares this information either with an identification card that is magnetically coated or with information that has been placed in the memory of the computer.

Security Personnel

Security guards are personnel that provide security in many buildings by serving as a physical presence that tends to deter crime and improper activities. Guards are positioned at strategic points in the facility so they have an overlapping view to monitor activities. They might sit at a control desk near the main entrance to the building and monitor or control access. The central console of surveillance equipment might be used to assist the guard in monitoring events. Guards might also be used for occupant or vehicular traffic control in emergency situations.

In most instances, the training level and competency of a security guard should not be compared with that of a policeman. Most security guards do not have the training that policemen receive. Many guards hired range from those too inexperienced to do the job properly (e.g., college students who do homework on the job) to those who are not physically capable of handling the job (e.g., some semiretired personnel).

Security guard jobs are generally low paying. One survey has shown over a third of those guards interviewed indicated that the guard job was the best job available to them and that they had been unemployed before they took the guard job. It is unfortunate that those persons hired to protect the building and materials are frequently the lowest paid personnel in many companies, perhaps making them more susceptible to stealing or taking bribes.

Neither guards nor electronic devices by themselves are the complete answer to building security. The selective use of electronics, tied into central control points, reduces the need for security guards. Using this approach, adequate money should be available to hire and train qualified guards.

Emergency Power Systems

Lengthy or recurring power outages are an inconvenience because lighting, space heating, and air conditioning and alarm systems no longer operate. In the business world, computers, office machines, and Universal Product Code (UPC) scanners do not function, which will result in financial loss. In the health care industry, power outages can result in loss of life.

Emergency power is electricity that is generated locally on a limited basis for the purpose of supplying electricity to critical devices during a general power outage. Emergency power can be provided by a backup battery system or emergency stand-by generator system. Typically only small devices such as emergency lighting and alarm and communication panels operate on battery backup systems.

Emergency power systems consist of two major parts: a *generator* that produces electricity and an *engine* that drives the generator. The generator can be switched on manually or automatically.

Generator engines can be powered by gasoline, diesel fuel, liquid petroleum gas (propane), or natural gas. Selection of a fuel is related to availability of the fuel in a power outage (e.g., natural gas versus diesel fuel). Effective storage of gasoline and diesel fuels requires a storage tank and the periodic addition of a chemical stabilizer. Propane requires a storage tank, either above or below ground. Natural gas is delivered by pipeline and may not be available in some emergencies (e.g., earthquake). The decision to select a gas or fuel is based on whether a particular fuel type is already being stored or used on site for another use. Concerns with engine noise dictate the location of the generator.

On large systems, an *automatic transfer switch (ATS)* monitors utility power from the utility line. When utility power fails, the backup system will start within 15 to 30 s and automatically transfers to generator power. Once utility power returns, the ATS monitors that the power is stable and transfers back to utility power when it is acceptable.

Emergency generators are designed to produce either single- or three-phase power. Three-phase generators produce

PHOTO 21.31 An emergency generator. (Used with permission of ABC)

PHOTO 21.32 An automatic transfer switch for an emergency generator system. (Used with permission of ABC)

120/208 or 277/480 V and single-phase generators are 120 or 120/240 V. One manufacturer makes generators available in 12, 17, 25, 40, 50, 55, 60, 75, and 100 kW capacities. Smaller systems are available for residential installations. Examples of an emergency generator and an automatic transfer switch for an emergency generator system are shown in Photos 21.31 and 21.32.

Emergency Lighting

Emergency lights enable building occupants to safely escape the building in the event of a power failure. Equipped with backup batteries, these lights are capable of powering exit and emergency lights for more than an hour. Emergency lighting is a critical element of a safety system in the event of a power failure. Emergency illumination is typically required in all buildings where the exiting system serves an occupant load of 100 or more. (See Photo 21.33.)

Emergency Action Planning

Emergencies and other threats to life and property can be caused by both human-made and natural events (e.g., fire,

PHOTO 21.33 Emergency lighting with a remote battery backup for power outages. (Used with permission of ABC)

hurricanes, tornadoes, blizzards, floods, earthquakes, tsunamis, toxic chemical and nuclear releases, explosions, bomb threats, acts of terrorism, and so on). They can occur rapidly and without warning. Emergency plans and procedures must be established to protect building occupants and employees from serious injury or loss of life and property from further damage in the event of an actual or potential major disaster.

A *building emergency action (BEA) plan* provides for immediate, positive, and orderly action to safeguard life and property in the event of any emergency or disaster (except enemy attack). The BEA plan should address emergencies that the employer may reasonably expect in the workplace. It establish policies, procedures, and an organizational structure for response to emergencies and identifies the roles played by various personnel.

An *emergency management team* is formed to address ongoing emergency planning and coordination efforts, including issues regarding road closings, access to buildings, emergency evacuation procedures, providing updated security information to employees, and so forth. This team formulates the BEA plan. During an emergency, the designated leader of this team puts into service immediate actions and ensures that response efforts are coordinated with the various groups responsible for public and employee safety. This team directs emergency actions from the *emergency operations center,* a designated central communication room. An *incident response team* with medical, firefighting, and/or hazardous material handling personnel tend to the incident and report back to the emergency operations center, where the emergency management team takes further action, if necessary.

A *building evacuation plan* is a central part of the BEA plan that ensures orderly evacuation of building occupants by establishing emergency escape procedures and escape route assignments. A decision to evacuate is made by the designated leader of the emergency management team based typically on the worst-case scenario of the incident. Consideration for evacuation is given to the specific threat (e.g., fire, bomb threat, explosion, hazardous material incident, and so on), its extenuating circumstances (e.g., time of day, likelihood of further damage, and so forth), and the recommendation of local governmental officials.

Typically, every building or building area (in large buildings) has an *evacuation team* to assist in emergency evacuations. Members of the team include *floor monitors* and *stairway monitors,* who act in response to the specific emergency and coordinate safe evacuation. Floor monitors check rooms and other enclosed spaces for occupants who may be trapped or otherwise unable to evacuate the area. Stairway monitors direct occupant egress on stairways. Monitors (and fellow employees) also assist handicapped occupants who may need extra assistance. Once evacuation is accomplished, the monitors verify that all employees are in the safe areas. Normally, one monitor for every 20 occupants provides sufficient supervision.

At the time of an emergency, monitors and other employees should know what type of evacuation is necessary and what their role is in carrying out the plan. In cases where there is a serious emergency inside the building (e.g., uncontrolled fire, indoor toxic chemical and nuclear release, explosion, bomb threat, and so on), total and immediate evacuation of all occupants may be necessary. In some cases, only those employees in the immediate area of the emergency are directed to evacuate or move to a safe area. In other cases, it may be necessary to direct occupants to a safer area of the building (e.g., tornado, hurricane, outdoor toxic chemical and nuclear releases, acts of terrorism, and so forth).

In some emergencies, an evacuation of nonessential occupants with a delayed evacuation of others may be needed for continued operation of building operations. *Essential personnel* are employees who have been selected in advance to remain behind to care for essential operations. They perform these tasks until evacuation becomes absolutely necessary. Essential operations may include monitoring of power supplies, water supplies, and other essential services that cannot be shut down for every minor emergency. Essential operations may also include tending to chemical or manufacturing processes that must be shut down in stages or where specific employees must be present to assure that safe shut down procedures are accomplished (e.g., shut down of a nuclear plant).

The use of emergency floor plans that clearly show emergency escape routes should be included in the BEA plan. These floor plans and signage should be displayed in corridors and stairways of the building to educate and guide building occupants. The designation of *refuge* or *safe areas* for evacuation should be determined in advance and identified in the emergency floor plan. In a building that is compartmentalized into fire zones, the refuge area may be within the same building but in a different zone from where the incident occurs. Exterior refuge or safe areas include parking lots, open fields, or streets that are located away from the site of the emergency and that provide sufficient space to accommodate occupants. Occupants should be instructed to move away from the exit doors of the building, and to avoid congregating close to the building, where they may hinder emergency operations.

Before implementing the BEA plan and on a re-occurring basis, a sufficient number of personnel to assist in safe and orderly emergency evacuation of occupants (e.g., employees, visitors, and so on) should be designated and trained. This training generally involves a review of the BEA plan, technical training in equipment use for emergency responders, and evacuation drills. In large facilities (e.g., manufacturing plants, hospitals, airports, and so forth), full-scale training exercises should take place regularly. Essentially, employees must know what is expected of them in a potential emergency situation to ensure their safety from fire or some other emergency.

STUDY QUESTIONS

21-1. What three elements must exist for a combustion reaction (fire) to occur?

21-2. Describe the difference between piloted ignition temperature and autoignition temperature.

21-3. Describe the four stages in the progression of a fire.

21-4. Describe the four classifications of fire by type of fuel.

21-5. Describe the five types of building construction.

21-6. What is compartmentalizing and why is it necessary in fighting a fire?

21-7. What is an intumescent material? Where is it used?

21-8. What is a fire-resistant rating?

21-9. How are fire-resistance ratings established?

19-10. What is a flame-spread rating?

21-11. How are flame-spread ratings established?

21-12. What are passive fire protection and active fire protection? How are they different?

21-13. What is a standpipe system?

21-14. What are the four types of conventional automatic fire sprinkler systems and how do they function?

21-15. What are the four types of alternative fire suppression systems and how do they function?

21-16. What are the four classes of fire extinguishers and where are they used?

21-17. What is a smoke control system and why is it needed?

21-18. What is a smoke detector?

21-19. What is a heat detector?

21-20. Describe the types of heat and smoke detectors.

21-21. Describe a fire alarm control system.

21-22. Why has building security become increasingly important?

21-23. Identify and describe things that can be implemented in building design that improve security.

21-24. What types of electronic devices are used to provide building security?

21-25. What types of electronic access control devices are used to provide building security?

21-26. What is a building emergency action (BEA) plan?

21-27. What is a building evacuation plan?

21-28. What are the roles of the members of the emergency management team?

21-29. What are the roles of the members of the incident response team?

21-30. What are the roles of the members of the building evacuation team?

Design Exercises

21-31. A 184 ft by 242 ft room in a warehouse will be protected by individual fire sprinklers spaced throughout the ceiling. Each sprinkler head covers a 10 ft by 10 ft area and will discharge approximately 40 gal/min (gpm). Sprinkler heads are located in straight rows spaced at 10 ft on center. For preliminary design, approximate the following:

a. The minimum number of sprinkler heads required

b. The minimum number of rows of sprinkler heads, if rows run transversely (widthwise) through the building

c. The minimum number of rows of sprinkler heads, if rows run longitudinally (lengthwise) through the building

d. The flow rate of the system, in gpm, based on the number of sprinkler heads

21-32. A 184 ft by 242 ft room in a warehouse will be protected by individual fire sprinklers spaced throughout the ceiling. Each sprinkler head covers a 12 ft by 12 ft area and will discharge approximately 50 gal/min (gpm). Sprinkler heads are located in straight rows spaced at 12 ft on center. For preliminary design, approximate the following:

a. The minimum number of sprinkler heads required

b. The minimum number of rows of sprinkler heads, if rows run transversely (widthwise) through the building

c. The minimum number of rows of sprinkler heads, if rows run longitudinally (lengthwise) through the building

d. The flow rate of the system, in gpm, based on the number of sprinkler heads

21-33. A 62 m by 80 m room in a warehouse will be protected by individual fire sprinklers spaced throughout the ceiling. Each sprinkler head covers a 4 m by 4 m area and will discharge approximately 180 liters/minute (L/min). Sprinkler heads are located in straight rows spaced at 4 m on center. For preliminary design, approximate the following:

  a. The minimum number of sprinkler heads required

  b. The minimum number of rows of sprinkler heads, if rows run transversely (widthwise) through the building

  c. The minimum number of rows of sprinkler heads, if rows run longitudinally (lengthwise) through the building

  d. The flow rate of the system, in L/min, based on the number of sprinkler heads

21-34. Draw a sketch of a fire sprinkler system for the residence in Appendix A.

21-35. Draw a sketch of a fire sprinkler system for the apartment building in Appendix B.

21-36. Make a visit to a residence with a fire sprinkler system. Make a sketch of the system and identify chief components of the system. Write out control strategies used.

21-37. Make a visit to a commercial building with a fire sprinkler system. Make a sketch and describe chief components of the system. Describe control strategies used.

21-38. Make a visit to a high-rise (over 75 ft) office building with a fire sprinkler system. Make a sketch and describe chief components of the system. Describe control strategies used.

21-39. Visit a building manager and discuss the building's emergency action plan.

BUILDING TELECOMMUNICATION SYSTEMS

22.1 TELECOMMUNICATION SYSTEMS

Historical Perspective

Methods of communicating over long distances have evolved over many millennia. Although carrier pigeons were used to convey messages from about 700 B.C.E., the first long-distance communication systems were based on signals of sound and light (e.g. drums and horns, smoke signals and beacon fires). Signal fires alerted the British of the arrival of the Spanish Armada in 1588 C.E. The Chinese used rockets as signals to warn of an imminent attack on the Great Wall. Native Americans communicated by covering and uncovering a bonfire with a blanket to produce smoke signals or by beating drums. The British Navy sent signals at night by raising and lowering a lantern, which coincidentally was the same way Paul Revere was signaled with news of the arrival of the British. In instances when clear vision was difficult (e.g., fog), bells or whistles and fired weapons sent signals. Until almost 1800, traditional long-distance communication was by horse-mounted dispatch riders.

In 1793 Frenchman Claude Chappe developed an optical telegraph (semaphore) system of stations built on rooftops or towers that were visible from a great distance. Each semaphore station consisted of a column-like tower with a moveable beam. Attached to the beam were two moveable arms. The beam and arms were swiveled with ropes, conveying different signal patterns representing upper- and lowercase letters, punctuation marks, and numbers. A set of patterns was translated into words by an observer at another station, who then sent it on to the next station. This system allowed the French to send a concise message over 100 miles (160 km) in less than 5 min as long as visibility was good. Swede A. N. Edelcrantz developed another type of optical telegraph system with ten collapsible iron shutters, which when placed in various positions formed combinations of numbers that were translated into letters, words, or phrases. Crude semaphore systems were also used in Boston, New York City, and San Francisco at that time.

Communications by sending electrical signals over wires came only after the demonstration of electromagnetism by Danish physicist Christian Oersted in 1820 and electrical flow by Michael Faraday and others before him. In 1830, American Joseph Henry transmitted the first practical electrical signal by sending electricity through a long set of wires to produce electromagnetism that was used to ring a bell. The next year, in 1831, American, Samuel Morse patented the first functional electrical communication system: the electric telegraph with its system of electrical impulses identified as dots and dashes that eventually became known as Morse Code.

The first message sent by electric telegraph was "What hath God wrought," from the Supreme Court Room in the U.S. Capitol to the railway depot at Baltimore on May 24, 1844. Three decades later, in 1861, there were over 2000 telegraph offices in operation across America and the East and West coasts were connected. Six years later, the first transatlantic cable was laid, connecting England and the United States. The telegraph flourished as a method of long-distance communication throughout the world.

On March 10, 1876, in Boston, Massachusetts, Alexander Graham Bell invented an electrical speech machine that transmitted voice over wires and became known as the telephone. Bell's assistant, Thomas Watson, fashioned the device from a funnel, a cup of acid, and copper wire attached to a wooden stand. "Mr. Watson, come here, I want you!" were the first words accidentally spoken into the new invention. Four years later in 1880, the first telephone company, American Bell, was formed and over 30 000 phones were in use. Within 40 years (about 1920), over ten million American Bell System telephones were in service.

In 1895, Italian inventor Gugliemo Marconi demonstrated the first radio transmission that was received out of a line of sight (about 2 miles) on the grounds of his family estate in Italy. Six years later in Newfoundland, Canada, Marconi's radio received a weak signal that was sent across the Atlantic Ocean by one of his associates in Cornwall, England. The signal was an "S" sent in Morse Code format, "dot, dot, dot." It demonstrated that radio waves could bounce off the upper atmosphere. The first true radio message was sent a year later. Less than 50 years after the telephone was invented, transatlantic communications from New York to London became operational with signals transmitted by radio waves.

In 1865, Italian physicist Giovanni Caselli invented a pantelegraph for transmitting pictures, the first commercial fax system. On May 19, 1924, the first transmission of pictures over telephone wires was publicly demonstrated. On January 23, 1926, John Logie Baird of Scotland gave the first public demonstration of a mechanical television with images of living human faces, not just outlines or silhouettes. It was with this use of radio waves that transmission of pictures took a major step toward the television we use today.

New engineering and scientific discoveries continued during the last half of the twentieth century with the gradual introduction of automatic switching devices, Teletype machines, transatlantic cables, microwave and fiber optic technologies, communications satellites (e.g., Telstar I in 1962), personal computers, fax machines, wireless (cellular) phone service, and the Internet. Today's telecommunications industry includes simple voice telephone calls, fax transmissions, video conferencing, cable TV, access to the Internet, high-speed data communications, satellite communications, and surfing the World Wide Web.

Fundamentals of Telecommunications Systems

By industry definition, telecommunication is the transmission, emission, or reception of signs, signals, writing, images, sounds, or information of any nature by wire, radio, optical, or other electromagnetic systems. A *telecommunication system* uses electricity, light (visible and infrared), or radio waves to transmit signals that carry voice and data transmissions.

Telecommunication systems function when a *transmitter* converts sound waves (e.g., those created when a person speaks into a telephone mouthpiece) or data into signals, which travel along wires or through the air before reaching their destination. When a *receiver* intercepts the signals, they are converted back into useful data or sound waves that become distinguishable by the human ear and recognized by brain. A *transceiver* is a telecommunications device that functions as a transmitter and receiver.

Historically, telecommunications systems such as a telephone system have used analog transmission. Modern systems use digital transmission technology. The following is a description of these transmission formats:

Analog transmission in an electronic network is the conversion of useful sound or data into electrical impulses. It is capable of transmitting both voice and nonvoice messages (e.g., telex, telegrams, data). However, nonvoice transmissions are bulky when transmitted in an analog format, so they cannot be transmitted rapidly.

Digital transmission in an electronic network involves a transmission of a signal that varies in voltage to represent one of two separate states (e.g., on and off or 0 and 1). In an optical network, digital signaling can involve either pulsating (on and off) light or a variation in the intensity of the light signal. Digital transmission over radio systems (microwave, cellular, or satellite) can be accomplished by varying the amplitude of the wave. Digital transmission technology offers a rapid method of voice and nonvoice transmission.

In telecommunications systems, *bandwidth* is the range between the highest and lowest frequencies of transmission, measured in *hertz (Hz),* cycles per second. Bandwidth varies with the type and method of transmission. It is a measure of the information capacity.

Telecommunication Networks

A *telecommunications network* is a collection of communication equipment and devices that are interconnected so they can communicate in order to share data, hardware, and software or perform an electronic function. The network includes a series of connecting points called *nodes* (e.g., a telecommunication terminal such as a telephone receiver or computer) that are interconnected with cables (wiring). Networks can also interconnect with other networks and contain subnetworks.

In design and layout of communication networks, the term *topology* describes the configuration of a network, including its nodes, connecting cables and equipment. It describes the manner in which the cable is run to individual workstations on the network. As shown in Figure 22.1, there are three basic network topologies: the bus, the star, and the ring. Acronyms and abbreviations used in the telecommunications industry are shown in Table 22.1.

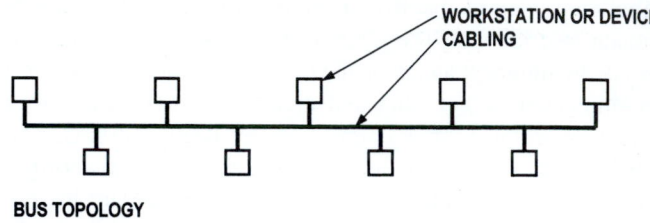

BUS TOPOLOGY

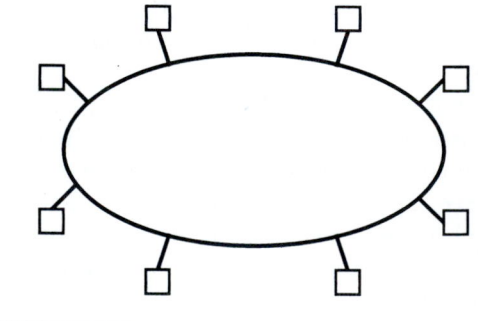

RING TOPOLOGY

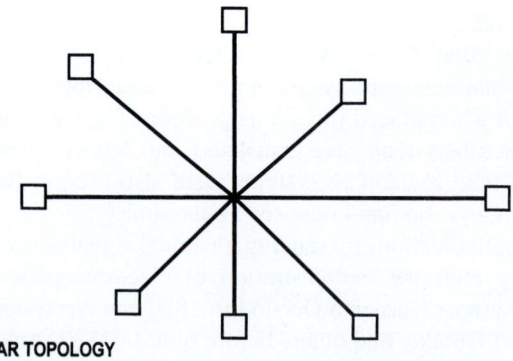

STAR TOPOLOGY

FIGURE 22.1 The basic network topologies used in building telecommunication systems: the bus, the ring, and the star.

TABLE 22.1   ACRONYMS AND ABBREVIATIONS USED IN THE TELECOMMUNICATIONS INDUSTRY.

ACR	Attenuation to cross-talk ratio		m	Meter
ANSI	American National Standards Institute		MAC	Media access control (layer)
ASTM	American Society for Testing and Materials		MAC(s)	Moves, adds, and changes
ATM	Asynchronous transfer mode		MAU	Medium attachment unit
AUI	Attachment unit interface		Mbs	Megabits per second
AWG	American Wire Gauge		MC	Main cross-connect
BER	Bit error rate		MDF	Main distribution frame
BICSI	Building Industry Consulting Service International		MHz	Megahertz
CCITT	International Telegraph and Telephone Consultative Committee		mm	Millimeter
			NBC	National Building Code
COAX	Coaxial cable		NEC	National Electrical Code
COSAC	Canadian Open Systems Application Criteria		NEMA	National Electrical Manufacturers Association
CSA	Canadian Standards Association		NeXT	Near-end cross-talk
CSMA/CD	Carrier sense multiple access/collision detection		NI	Network interface
EF	Entrance facility		NIR	Near-end cross-talk to insertion loss ratio
EIA	Electronic Industries Association		NIST	National Institute of Standards and Technology
EMI	Electromagnetic interference		nm	Nanometer
EMI	Electrical metallic tubing		NRZ	Nonreturn to zero
EP	Entrance point		OSI	Open systems interconnection
ER	Equipment room		PBX	Private branch exchange
Ethernet	Precursor to, and almost identical with, the IEEE802.3 standard		PVC	Polyvinyl chloride
			PWA	Provisioned work area
FDDI	Fiber distributed data interface		RCDD	Registered communications distribution designer
FIPS PUB	Federal Information Processing Standard Publication		RFI	Radio frequency interference
FTE	Field test equipment		ROI	Return on investment
HC	Horizontal cross-connect		SQL	Structured query language
HVAC	Heating, ventilation, and air conditioning		STP	Shielded twisted pair
Hz	Hertz		TBITS	Treasury Board Information Technology Standard
IC	Intermediate cross-connect		TC	Telecommunications closet
IDC	Insulation displacement contact		TIA	Telecommunications Industry Association
IEC	International Electro-Technical Commission		TO	Telecommunications outlet
IEEE	Institute of Electrical and Electronics Engineers		TP/PMD	Twisted pair/physical media dependent
ISDN	Integrated services digital network		TR	Token Ring
ISO	International Organization for Standardization		TSB	Telecommunications System Bulletin
ITU	International Telecommunications Union—Telecommunications Standardization Section		UTP	Unshielded twisted pair
			UL	Underwriters Laboratories, Inc.
kHz	Kilohertz		WA	Work area
km	Kilometer		WAN	Wide area network
LAN	Local area network		X	Cross-connect
LED	Light emitting diode			

A *bus topology* connects each workstation (node) to a single cable trunk. All signals are broadcast to all workstations. Each computer checks the address on the signal as it passes along the bus. If the signal's address matches that of the computer, the computer processes the signal. If the address does not match, the computer takes no action and the signal travels down the bus to the next computer. Next, in a *star topology*, all workstations (nodes) are connected to a central unit called a *hub*. Home runs are cables that extend from the hub to the terminal without splicing or other connections. This configuration allows cables to have a direct link between entrance facilities/ equipment room equipment, telecommunications closet devices, and workstation equipment (e.g., computers, printers, telephone receiver, and so on). Third, a network that is wired in the *ring topology* connects workstation equipment and devices in a point-to-point serial manner in an unbroken circular configuration.

Not all networks are the same. The various types provide different services, use different technology, have different resources and require users to follow different procedures. Networks can be distinguished in terms of spatial distance between nodes such as *local area networks (LAN)*, *metropolitan area networks (MAW)*, and *wide area networks (WAN)*. Large telephone networks and networks using their infrastructure (such as the Internet) have sharing and exchange arrangements with other companies so that large WANs are created. In building telecommunication systems, LANs are used. LANs connect computers and hardware such as printers located relatively close together and sharing resources, equipment, and files. Types of LANs include the Ethernet, ARCnet, and Token Ring, each having their own method of transmitting data.

The transmitting medium used in networks can be copper wire, glass, or plastic (fiber optic cable), and air (microwave and radio wave). A signal sent through a telecommunications network can be sent through any or all of these media.

Transmission Media

Cable is the most common medium through which voice and data usually move from one network device to another. It serves as the pipeline of a telecommunication system. There are several types of cable in use, including copper wire, coaxial cable, and optical fibers. Copper wiring used in building telecommunication transmission is being replaced by optical fibers because they have much greater signal capacity. *Wireless* transmission capabilities are also used in buildings and are replacing the need for hard-wired direct connections.

Connectors are the devices that connect cable to the network device (e.g., computer, printer, entertainment center, and so forth). Connectors may come with the equipment purchased or it may be necessary to purchase them individually. Connections on a cable system tend to be the weakest element in any network, so they must be made properly.

Types of transmission media include the following.

Copper Wiring

Historically, copper wiring has been the principal telecommunications transmission medium. It consists of one or more pairs of solid copper wires. Bundles of pairs of twisted insulated copper wires form the majority of the telephone lines in the United States and elsewhere. See Photo 22.1.

Twisted pair cable consists of pairs of copper wires that are twisted to certain specifications. Each pair is twisted with a specified number of twists per inch to help eliminate interference from adjacent pairs and other electrical devices; the tighter the twisting, the higher the supported transmission rate but the greater the cost. Each signal on a twisted pair requires both wires. Because some telephone sets or desktop locations require multiple connections, a twisted pair is sometimes installed in two or more pairs, all within a single cable. Typically, twisted pair cable has four pairs of wires inside the jacket. Twisted pair comes with each pair uniquely color coded when it is packaged in multiple pairs.

Twisted pair wiring is available in shielded and unshielded versions. *Unshielded twisted pair (UTP) wiring* consists of multiple pairs of twisted insulated copper conductors bound in a single sheath. It is unshielded from electromagnetic waves and therefore is sensitive to electrical interference. UTP wiring is adequate for basic voice, fax, or data communications. For some applications, a twisted pair is enclosed in a shield and is known as *shielded twisted pair (STP) wiring*. An outer covering or shield is added to the ordinary twisted pair wires; the shield functions as a ground. STP is suitable for environments with electrical interference; however, the extra shielding can make the cables quite bulky. Thus, the more common type of wire used is not shielded. STP is commonly used in Token Ring networks and UTP in Ethernet networks, where it is referred to as 10baseT.

The quality of UTP will vary from telephone grade to extremely high-speed cable. The Telecommunications Industry Association (TIA), an offshoot of the Electronic Industries

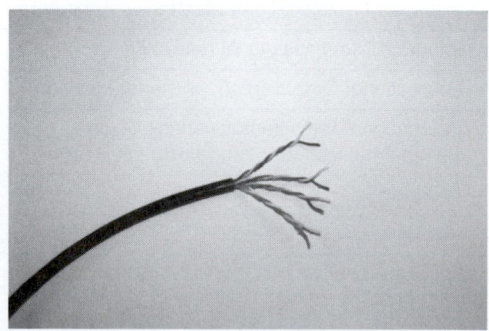

PHOTO 22.1   Twisted pair copper cable. (Used with permission of ABC)

TABLE 22.2   ANSI/TIA/EIA STANDARD 568 CATEGORIES (CAT) OF TWISTED PAIR CABLING.

Category	Maximum Data Rate	Usual Application
CAT 1	Less than 1 Mbps	Used for basic voice telecommunications (e.g., telephone, intercom) and limited power circuit cables (e.g., alarm and doorbell wiring). Not suitable for networking applications.
CAT 2	4 Mbps	Mainly used in the IBM cabling system for Token Ring networks.
CAT 3	16 Mbps	Used in low-speed data applications, primarily for telephone wiring and 10BaseT Ethernet.
CAT 4	20 Mbps	Rarely used. Primarily for Token-Ring networks.
CAT 5	100 Mbps 1000 Mbps (4 pair)	Provides optimal performance for all data and phone systems and has become the standard for high-speed applications. Will run with peak accuracy, efficiency, and throughput. Primarily for 10BaseT Ethernet, and 100BaseT Ethernet.
CAT 5e	100 Mbps	
CAT 6	Up to 250 Mbps	Extremely fast broadband applications.
CAT 7	Up to 600 Mbps	Super-fast broadband applications.

Association (EIA), has established TIA/EIA standards of UTP and rated categories of wire as shown in Table 22.2.

In addition to the EIA/TIA standards, the U.S. standard for wire conductor size applied to copper electrical power and telephone wiring is *American Wire Gauge (AWG)*. The gauge refers to wire thickness: the higher the gauge number, the thinner the wire. Wiring used for typical household power circuiting is AWG 14 or 12, greater on circuits serving large equipment. Telecommunications wire is thinner, typically AWG 22, 24, or 26. Because thicker wire carries current more efficiently (because it has less electrical resistance over a specific length), a thicker wire is more efficient for longer distances. For this reason, where extended distance is required, a company installing a network might prefer telephone wire with the lower gauge, thicker wire of AWG 22. 22 AWG wire is typically used in telephone and UTP wire.

RJ45 connectors are the standard female connectors used in a telecommunication system for UTP cable. A slot allows the RJ-45 to be inserted only one way. RJ stands for *registered jack,* implying that the connector follows a standard borrowed from the telephone industry. The RJ45 is an eight-pin connector used for data transmission or networking and some business telephones. Pins are numbered 1 through 8 with a locking clip at the top. Telephone connectors, referred to as RJ11 or RJ12, have four or six pins, respectively.

Coaxial Cable

Coaxial cable has two conductors: an inner solid wire surrounded by an outer braided metal sheath. The conductors both run concentrically along the same axis; thus the name *coaxial (COAX)*. Insulation separates the two concentric conductors, and a hard casing protects the entire cable. Several coaxial cables can be arranged in bundles protected by an outer sheathing, called a *jacket*.

The primary types of coaxial cabling are as follows:

Thin coaxial cable is also referred to as *thinnet*. Thinnet is about ¼ inch (8 mm) in diameter and is very flexible. It looks like regular TV cable. The *10Base2* designation refers to specifications for thin coaxial cable. The 2 refers to the *approximate* maximum segment length being 200 m (654 ft), but the maximum practical segment length is actually 185 m (605 ft).

Thick coaxial cable is referred to as *thicknet*. *10Base5* refers to the specifications for thick coaxial cable. The 5 refers to the maximum segment length being 500 m (1635 ft). Thick coaxial cable has an extra protective plastic cover that helps keep moisture away from the center conductor. This makes thick coaxial a better choice when running longer lengths in a linear network. A disadvantage of thick coaxial is that it does not bend easily and is difficult to install. Thicknet is not commonly used except as a backbone within and between buildings.

Triax cable is a type of coax cable with an additional outer copper braid insulated from signal carrying conductors. It has a core conductor and two concentric conductive shields.

Twin axial cable (Twinax) is a type of communication transmission cable consisting of two center conductors surrounded by an insulating spacer, which in turn is surrounded by a tubular outer conductor (usually a braid, foil, or both). The entire assembly is then covered with an insulating and protective outer layer. It is similar to coaxial cable except that there are two conductors at the center.

Common types of coaxial cable are shown in Table 22.3.

Coaxial cable is very effective at carrying many analog signals at high frequencies. In contrast to twisted pair wires, coaxial has a much higher bandwidth to carry more data, and

TABLE 22.3   COMMON TYPES OF AXIAL CABLE.

Cable Type	Illustration	Description
Coaxial cable		Coaxial cables are constructed with an inner conductor surrounded by a dielectric, which is enclosed by an outer conductor that also acts as a shield. A protective jacket covers the outer conductor and also acts as insulation.
Dual-shielded coaxial cable		Dual-shielded coaxial cables have two outer conductors, or shields, enclosing the dielectric. Dual shielding is needed for strength and abrasion resistance. Offers a decrease in attenuation and the possibility of unwanted external signals.
Twin axial cable		Twin axial cable is composed of two insulated single conductor cables or hook-up wires twisted together, having a common shield and protective jacket.
Triaxial cable		Triaxial cable is coaxial cable with one inner conductor and two shields all separated with dielectric material. Triaxial cable signals may be transported by both the inner conductor and the inner shield while the outer shield is at ground potential.

offers greater protection from noise and interference. Although coaxial cabling is difficult to install, it is highly resistant to signal interference. In addition, it can support greater cable lengths between network devices than twisted pair copper cable.

High-capacity coaxial cable is widely used in cable television systems because it is capable of carrying many TV and radio signals simultaneously. Coaxial cable is used by telephone and cable television companies from the central office to the user. It is also widely installed for use in business and school LANs. Gradually, existing coaxial lines are being replaced by optical fibers.

The most common type of connector used with coaxial cables is the *Bayonet Neil-Concelman (BNC) connector*. A BNC male connector has a pin that connects to the primary conducting (core) wire and then is locked in place with an outer ring that turns into locked position. Different types of adapters are available for BNC connectors, including a T-connector, barrel connector, and terminator.

Optical Fibers

Optical fibers are long, thin strands of very pure silicon glass or plastic about the diameter of a human hair. A single optical fiber consists of three elements: a *core,* the thin glass center of the fiber where the light travels; *cladding,* the outer material surrounding the core that reflects the light back into the core; and a *buffer coating,* a plastic coating that protects the fiber from damage and moisture. Each strand can pass a signal in only one direction, so fiber optic cable on a network typically consists of at least two strands: one for sending and one for receiving.

Hundreds or thousands of optical fibers are arranged in bundles called *optical cables*. The cable's outer sheathing, called a *jacket,* protects these bundles. Optical fibers come in two types: *single-mode fibers* that are used to transmit one

signal per fiber (used in telephones and cable TV); and *multimode fibers* that are used to transmit many signals per fiber (used in computer networks, local area networks).

The most common connectors used with fiber optic cable are the ST and SC connectors. The *ST connector* is barrel shaped, similar to a BNC connector. The *SC connector* has a squared face and is easier to connect in a confined space.

Optical fiber carries much more information than copper wire and is in general not subject to electromagnetic interference. This makes it ideal for environments that contain a large amount of electrical interference. This characteristic has made it the standard transmission medium for connecting networks between buildings.

Fiber optics refers to the technology in which communication signals in the form of modulated light beams are transmitted over a glass fiber transmission medium. The light in a single optical fiber travels through the core by reflecting from the mirror-like cladding, a physical principle called *total internal reflection*. Light reflects from the cladding no matter what angle the fiber itself is bent. Because the cladding does not absorb any light from the core, the light wave can travel great distances. Light is generated by a laser or a light-emitting diode (LED). Lasers have more power than LEDs, but vary light output more with changes in temperature and are more expensive.

A *fiber optic relay system* transmits and receives a light signal that is transmitted through an optical fiber. An *optical transmitter* produces and encodes the light signal that is sent through the optical fiber. An *optical receiver* that decodes the signal receives the light signal. The receiver uses a *photocell* or *photodiode* to detect the light signal, decodes it, and sends an electrical signal to a computer, TV, or telephone. Over long distances, an *optical regenerator* is needed to boost the light signal. One or more optical regenerators may be spliced along a long cable to amplify the degraded light signal.

TABLE 22.4   COMMON TYPES OF ETHERNET CABLE IN USE.

Cable Type	Specification	Maximum Distances	
		Meters	Feet
Unshielded twisted pair	10BaseT	100	325
Thin coaxial	10Base2	185	600
Thick coaxial	10Base5	500	1635
Fiberoptic	10BaseF	2000	6540
Unshielded twisted pair	100BaseT	100	325
Unshielded twisted pair	100BaseTX	220	719

The most common wavelengths of light signals in a fiber optic system are 850 nm, 1300 nm, and 1550 nm, which are all wavelengths within the nonvisible, infrared area of the electromagnetic light spectrum. Single-mode fibers have small cores (about 0.00035 in or 9 μm diameter) while multimode fibers have larger cores (about 0.0025 in or 62.5 μm diameter). Optical fibers made from plastic require a much larger core (0.04 in or 1 mm diameter) and transmit visible red light (wavelength = 650 nm) from LEDs.

As light passes through the optical fiber, the light signal degrades over its length. Degradation is principally caused by impurities in the glass or plastic. The extent that the signal degrades depends on the purity of the medium and the wavelength of the transmitted light (e.g., higher wavelengths tend to have less degradation). The finest optical fibers offer signal degradation of less than 10% per km at wavelengths 1550 nm.

Fiber optic cable has the ability to transmit signals over much longer distances than coaxial and twisted pair cabling and can carry information at much greater speeds. This capacity broadens communication possibilities to include services such as video conferencing and interactive services. The cost of fiber optic cabling is comparable to copper cabling; however, it is more difficult to install and modify. *10BaseF* refers to the specifications for fiber optic cable carrying Ethernet signals. Common types of Ethernet cable in use are shown in Table 22.4.

Wireless

Wireless is a term used to describe telecommunications in which electromagnetic waves (instead of some form of wire) carry the signal. Wireless communications can take several forms: microwave, synchronous satellites, low-earth-orbit satellites, cellular, and personal communications service (PCS). In every case, a wireless system eliminates the need for a complex hard-wired infrastructure. *Fixed wireless* is the operation of wireless devices or systems in homes and offices, and in particular, equipment connected to the Internet by the use of specialized modems. A fixed wireless network enables users to establish and maintain a wireless connection throughout or between buildings, without the limitations of wires or cables.

There are two types of wireless networks: peer-to-peer and access point or base station. A *peer-to-peer wireless network*

consists of a number of computers, each equipped with a wireless networking interface card. Each computer can communicate directly with all of the other wireless-enabled computers and equipment (e.g., printers). An *access point* or *base station wireless network* has a computer or receiver that serves as the point at which the network is accessed. It acts like a hub, which provides connectivity for the wireless equipment.

Two modes of transmission are used in fixed wireless systems in buildings: infrared and radio frequency. *Infrared (IR) wireless* is the use of technology in devices or systems that convey data through infrared radiation. *Radio frequency (RF) wireless* transmits data through radio wavelengths. Infrared radiation is electromagnetic energy at wavelengths somewhat longer than those of visible red light. Radio wavelengths are much longer than infrared wavelengths. The shortest wavelength IR borders visible red in the electromagnetic radiation spectrum; the longest wavelength IR borders radio waves. Both infrared and radio wavelengths are invisible to the unaided eye.

IR wireless is used for short- and medium-range communications. Some systems operate in a *line-of-sight mode,* which means that there must be a visually unobstructed straight-line path through space between the transmitter (source) and receiver (destination). Other systems operate in *diffuse mode,* where the system can function when the source and destination are not directly visible to each other. However, IR wireless cannot pass through walls, so a link is not possible between different rooms in a building or between different buildings (unless they have facing windows). Despite these limitations, most IR wireless systems offer a level of security comparable to that of hard-wired systems. It is difficult, for example, to eavesdrop on a well-engineered, line-of-sight, IR communications link.

IR wireless technology can be used in home entertainment control units; robot control systems; medium-range, line-of-sight communications (e.g., cordless microphones, headsets, modems, printers, and other peripherals).

RF wireless technology uses radio waves to send and receive information, similar to a garage door opener, baby monitor, walkie-talkie, or portable phone. It can transmit data through walls and between nearby buildings. This characteristic offers flexible linking capability between communication devices. However, communication on RF wireless is less private; it is much easier to eavesdrop on a RF communications link in comparison to IR wireless and hard-wired system technologies.

Wi-Fi (derived from the term *wireless fidelity*) is the popular expression used to describe high-frequency *wireless local area network (WLAN)* technology. *Wireless hotspots* provide Internet access using wireless network devices installed in public locations. By installing an inexpensive PC card, a laptop computer can send and receive data at a very high speed, to any other computer in range. Wi-Fi technology can be set up for use either free or with a paid subscription in public places such as airports, hotels, coffee shops, civic plazas, conference centers, school buildings, and libraries where the user can access e-mail or the Internet without being directly connected with wiring to a local network. It can serve as a LAN in a building that has not been pre-equipped with cable. Wi-Fi technology can be used at home where a computer can be connected to the Internet anywhere in the home without being wired. As a result, Wi-Fi is the pre-eminent technology for building general purpose wireless networks.

Wi-Fi technology can be used for both data and voice (e.g., telephone) transmission. It is rapidly gaining acceptance in the business world as an alternative to a wired LAN. However, unless adequately protected with security safeguards (e.g., firewalls and encryption techniques), Wi-Fi technology can be susceptible to access from the outside by unauthorized users who simply access the Internet for free or pirate company secrets.

The transmission media chosen for a network is related to the topology, protocol, and size of the network. In some cases, a network will use only one type of cable, while other networks will use a variety of cable types, and others will rely upon wireless technology.

Electromagnetic Interference

Electrical current flow in power lines generates an electromagnetic field that surrounds the electrical conductor. Electrical equipment, especially large motors, generators, induction heaters, arc welders, x-ray equipment, and radio frequency, microwave, or radar sources, also produce a powerful electromagnetic field. The ballasts of fluorescent and high-intensity discharge (HID) fixtures also produce a significant electromagnetic field.

A telecommunication cable placed within an electromagnetic field will have its telecommunication signal affected. This is known as *electromagnetic interference*. Because of potential for electromagnetic interference, voice and data telecommunications cabling should not be run adjacent and parallel to power (electrical) cabling unless the cables are shielded and grounded. For low-voltage telecommunication cables, a minimum 5-in (125 mm) distance is needed from any fluorescent lighting fixture or power line over 2000 volt-amperes (VA) and up to 24 in from any power line over 5000 VA. In general, telecommunications cabling is routed separately, or several feet away from power cabling. For similar reasons, telecommunications cabling must be routed away from electrical equipment.

22.2 STRUCTURED BUILDING TELECOMMUNICATION SYSTEMS

Wiring and Cabling Standards

Prior to 1991, the manufacturers of electronics equipment controlled the specifications of telecommunications cabling. End-users were frequently confused by manufacturers' conflicting claims concerning transmission performance and were forced to pay high installation and administration costs for proprietary

systems. The communications industry recognized the need to define a cost-effective, efficient cabling system that would support the widest possible range of applications and equipment.

The EIA, TIA, and a large consortium of leading telecommunications companies worked cooperatively to create the American National Standards Institute (ANSI)/TIA/EIA-568-1991 *Commercial Building Telecommunications Wiring Standard.* Additional standards documents covering pathways and spaces, administration, cables, and connecting hardware were subsequently released. The ANSI/TIA/EIA-568-1991 was revised in 1995, and is now referred to as ANSI/TIA/EIA-568-A *Commercial Building Telecommunications Cabling Standard.* The goal of these standards is to define structured cabling a telecommunications cabling system that can support virtually any voice, imaging, or data application that an end-user chooses.

As the acceptance of standards-compliant structured cabling has grown, the price of installed networking equipment has dropped and performance has exponentially increased. The physical layer has evolved into an affordable bandwidth-rich business resource.

Telecommunication Cabling and Pathways

Telecommunication cabling is the medium through which voice and data move from one telecommunication device to another. Cabling physically carries electrical or optical signals to and from devices and equipment in a telecommunication system. Cabling media typically used include UTP and STP copper wire, coaxial cable, and optical fibers. Wireless technology can also be used.

A *pathway* is a passageway, and thus a path, for cable to travel when interconnecting devices, components, and equipment in a telecommunication system. Pathways are typically a raceway, a channel, or trough designed to hold wires and cables (e.g., conduit, cable trough, cellular floor, electrical metallic tubing, sleeves, slots, underfloor raceways, surface raceways, lighting fixture raceways, wireways, busways, auxiliary gutters, and ventilated flexible cableways). Raceways may be metallic or nonmetallic and may totally or partially enclose the cabling.

Pathways can carry existing cable and that easily allow additional cabling to be installed to accommodate the addition of equipment or upgrades in technology. In a building telecommunications system, pathways typically run between the building entrance facilities/equipment rooms, telecommunication closets, and the work area where telecommunication equipment and devices are used by building occupants.

A *backbone* is a generic term used to describe a main pathway or cabling media that interconnects a number of telecommunication devices. A backbone is used to connect networks in a building or in separate buildings. Fiber optic cable is typically used for this type of backbone. *Drop cables* may be attached from the backbone to connect individual workstations. Common types of backbone cabling are provided in Table 22.5.

TABLE 22.5   COMMON TYPES OF BACKBONE CABLING IN USE.

Cabling Types	Maximum Backbone Distances	
	Meters	Feet
Twisted Copper Wire		
100 ohm UTP (24 or 22 AWG) for voice transmission	800	2625
150 ohm STP for data transmission	90	295
Optical Fiber Cables		
Multimode optical fiber (62.5/125 μm) for voice/data transmission	2000	6560
Single-mode optical fiber (8.3/125 μm) for voice/data transmission	3000	9840

Structured Cabling Systems

A *structured cabling system* is the cabling, devices, and equipment that integrate the voice, data, video, and electronic management systems of a building (e.g., safety alarms, security access, energy management and control systems, and so on). Design and installation of structured cabling systems adheres to national and international standards.

In commercial buildings, structured telecommunications cabling systems include seven subsystems. Figure 22.2 is a schematic of a structured telecommunications cabling system. These subsystems are described in the following sections.

Interbuilding Backbone

The *interbuilding backbone* is the cabling and pathways *outside* of the building, including the cables carrying local exchange carrier (LEC) services (e.g., outside telephone company), Internet service provider services, and private branch exchange (PBX) telecommunication cable (e.g., private phone network between buildings at a school campus or business park). Simply, the interbuilding backbone caries telecommunication services to the building.

Building Entrance Facilities

The *building entrance facility* is an entrance to the building for both public and private network service cables. It includes the cables, connecting hardware, protection devices, and other equipment needed to connect the interbuilding backbone cabling to the backbone cabling in the building. See Photos 22.2 and 22.3.

PHOTO 22.2   Internet service building entrance. (Used with permission of ABC)

PHOTO 22.3   Local exchange carrier (LEC) and private branch exchange (PBX) building entrance. (Used with permission of ABC)

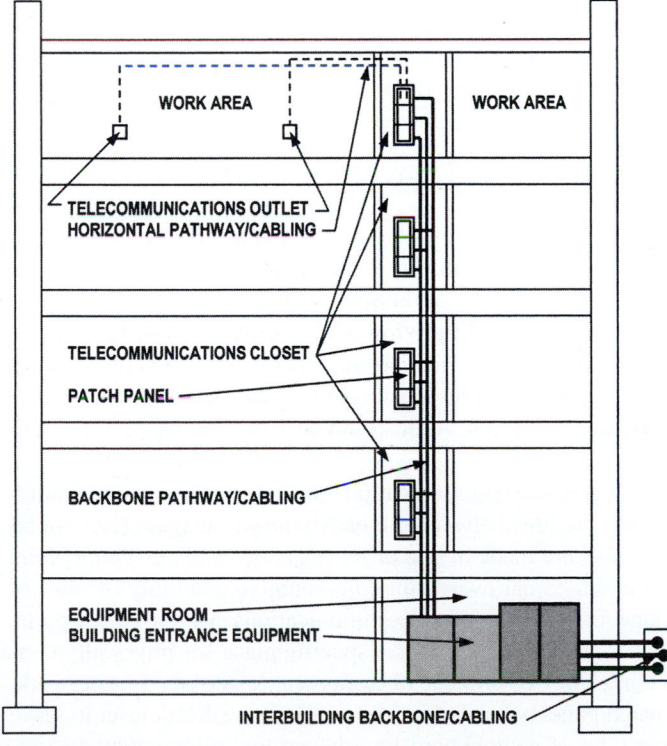

FIGURE 22.2   A schematic of a structured telecommunications cabling system.

In buildings with a finished floor area larger than 20 000 ft² (1870 m²), a secured (locked), dedicated, enclosed room is recommended for the building entrance. An industry standard is to allow 1 ft² (0.1 m²) of ¾-in (20-mm) plywood wall-mount area for each 200 ft² (19 m²) area of finished floor area. The plywood allows mounting capabilities for equipment and panels. In large buildings, rack-mounted and freestanding frames may also be required to support entrance equipment within the build entrance facilities.

Telecommunications Equipment Room

A *telecommunications equipment room* is a centralized space for housing main telecommunications equipment. It is a large, dedicated, centralized room that provides a controlled environment to house equipment, connecting hardware, splice closures, grounding and bonding facilities, and protection apparatus. Equipment rooms typically accommodate equipment of higher complexity than telecommunications closets (see below); however, any or all of the functions of a telecommunications closet may be performed in an equipment room (see patch panels). A telecommunications equipment room serves a building or multiple buildings in a campus or business park environment. See Photos 22.4 and 22.5.

The building entrance facilities should be located adjacent to or contained within the equipment room to allow shared air conditioning, security, fire control, lighting, and limited access. An industry standard is to allow 0.75 ft² (0.07 m²) of equipment room floor area for each 100 ft² (9 m²) of user workstation area, or about 1 to 2 ft² (0.1 to 0.2 m²) of equipment room floor area per workstation, with a minimum floor area of 150 ft² (14 m²). At least two walls should be covered with 8 ft (2.6 m) high, ¾ in (20 mm) thick, fire-rated plywood to attach equipment.

A secured (locked), dedicated equipment room is recommended. Doors providing access to an equipment room should be at least 36 in (900 mm) wide by 8 ft (2.45 m) high. Piping, ductwork, mechanical equipment, or electrical wiring should not enter the equipment room. The room should not serve as an unrelated storage room (e.g., for storage of paper and cleaning supplies).

PHOTO 22.4 Telecommunications room and equipment, including server, router and switches. (Used with permission of ABC)

PHOTO 22.5 Electrical panelboard providing filtered "clean" power to telecommunications room equipment. Battery-backup provides an interruptible power supply. Note the fire alarm and manual pull station nearby. (Used with permission of ABC)

The equipment room should be located away from sources of electromagnetic interference (e.g., transformers, motors, x-ray equipment, induction heaters, arc welders, radios, radar systems, and so on). Air conditioning should be provided (24 hours/day, 365 days/year, at temperatures of 64° to 75°F, with 30% to 55% relative humidity). General lighting (50 footcandles @ 3 ft above floor) and a minimum of two dedicated 15 A, 120 V ac duplex convenience receptacles on separate circuits should be provided. Additional duplex convenience receptacles should be placed at 6 ft (2 m) intervals around the perimeter. Emergency power should be considered and supplied.

Telecommunications Closet

A *telecommunications closet* is a dedicated room on each floor in a building that houses intermediate voice and data telecommunications equipment and related cable connections. A large building will have several telecommunications closets, and more than one on a floor. The telecommunications closet should be located in a space that is central to the work areas it serves. A telecommunications closet is shown in Photo 22.6.

Each telecommunications closet serves as a location where junctions between the backbone pathway and horizontal pathways are made at one or more patch panels. A *patch panel* is a mounted hardware unit containing an assembly of rows of connecting locations in a communications system, called ports. A *port* is receptacle that is a specific place for physically connecting a device or piece of equipment to another. In a network, a patch panel is located in a telecommunications closet to serve as a type of switchboard-like device that allows network circuiting arrangements and rearrangements by simply plugging and unplugging a patch cord. A *patch cord* is a type of jumper

PHOTO 22.6   Telecommunications closet. (Used with permission of ABC)

cable that is used to create a connection from one port in a patch panel to another port. (See Photos 22.7 and 22.8.)

A locked, dedicated telecommunications closet is recommended. The recommended closet size is 10 ft × 12 ft (3 m × 4 m), about 120 ft^2 (11 m^2) for each 10 000 ft^2 (940 m^2) useable floor area served. More than one telecommunications closet per floor is required if the distance to a work area exceeds 300 ft (90 m), or if the floor area served exceeds 10 000 ft^2 (940 m^2).

Two walls in a telecommunications closet should be covered with 8 ft (2.6 m) high, ¾ in (20 mm) thick, fire-rated plywood to attach equipment (e.g., patch panels). Power, lighting, and air conditioning requirements for a telecommunications closet are the same as for a telecommunications equipment room (see previous section). (See Photo 22.9.)

PHOTO 22.7   Patch panels and patch cables. (Used with permission of ABC)

PHOTO 22.8   Telephone patch panel and cables. (Used with permission of ABC)

When possible, telecommunications closets should be stacked vertically above each other on each floor. An industry standard is to provide at least three 4-in (100 mm) diameter sleeves (a stub of conduit through the floor) per 50 000 ft^2 (4676 m^2) of finished floor area served. An equivalent 4 in × 12 in (100 mm × 300 mm) slot may be used in lieu of three sleeves. If closets are not vertically aligned, then 4 in (100 mm) diameter horizontal conduit runs are required, with no more than two 90° bends between pull points. When there are multiple telecommunications closets on a single floor, it is recommended that these multiple closets be interconnected with at least one 3-in (75 mm) diameter conduit or an equivalent pathway. To prevent the spread of fire, provisions for a firestop are required in every opening that penetrates the telecommunications closet compartment (e.g., walls and floors).

Backbone Pathway

Within a building telecommunications system, the *backbone pathway* connects the entrance facilities/equipment room to the telecommunications closets for cabling that interconnects equipment and devices in these spaces. It contains several backbone

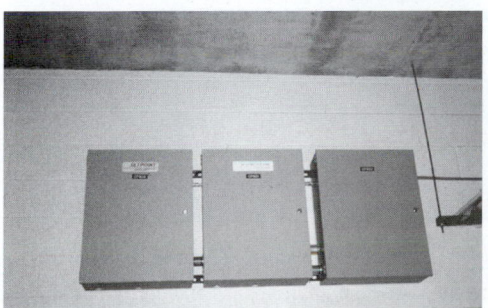

PHOTO 22.9   Wall-mounted telecommunications panels. (Used with permission of ABC)

(main) cables that carry the heaviest telecommunications traffic throughout the building. It is usually a vertical arrangement that connects floors in a multistory building. However, the same function may be served by a lateral backbone for horizontal distribution in a large building with spacious floors.

A building's backbone pathway consists of the backbone cables, intermediate and main cross-connects, mechanical terminations, and patch cords used for backbone-to-backbone cross-connection, connections between floors (risers), and cables between an equipment room and building cable entrance facilities.

The backbone pathway can hold any type or combination of transmission media, but cabling typically includes UTP, STP, and optical fiber cable. Backbone cabling distances are dependent on the type of system, data speed, and the manufacturer's specifications for the system electronics and the associated components used (e.g., adapters, line drivers, and so on). All cables in the backbone pathway are typically strung in a star topology. This configuration allows modifications to be made without the hassle of having to pull new cables. (See Photos 22.10, 22.11, and 22.12.)

Horizontal Pathways

Horizontal pathways connect the backbone cabling entering the telecommunications closet with the terminal equipment in the work area (e.g., computers, data terminals, telephones, and so on). Horizontal pathways can include underfloor ducts embedded in concrete decks or slabs, modular/cellular (raised) floors, underground trench ducts, and raceways (e.g., conduits, cable trays, recessed molding). A raised floor system is shown in Photos 22.13 and 22.14.

The most commonly used horizontal pathway consists of cable bundles run from the telecommunications closet along J-hooks or cable trays suspended above a plenum ceiling. Once a work area is reached, the cabling fans out and individual cable drops through interior walls, support columns, or chases, eventually terminating at a telecommunications outlet.

The *horizontal cabling* system extends from the work area (workstation) outlet to the telecommunications closet and consists of horizontal cabling, telecommunications outlet, table terminations, and cross-connections. Types of media used for horizontal cabling include UTP, STP, and optical cable, each extending from the telecommunications closet to the work area at a maximum distance of 294 ft (90 m).

PHOTO 22.10   Horizontal backbone cabling in a wire tray. (Used with permission of ABC)

PHOTO 22.12   Horizontal backbone cabling entering a firewall. The openings are sealed with an approved firestopping material. (Used with permission of ABC)

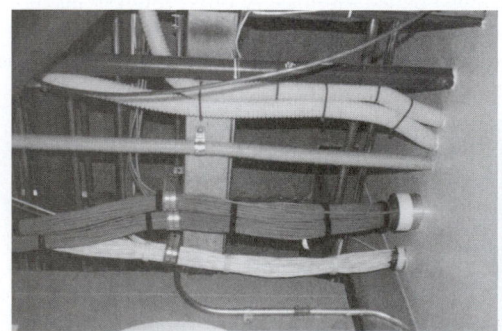

PHOTO 22.11   Bundled horizontal backbone cabling. CAT 3 PBX telephone cable, CAT 5e data transmission cable, interduct sheathing containing fiberoptic cable, and two bundled interduct sheathings containing fiberoptic cable. (From rear to front of photograph). (Used with permission of ABC)

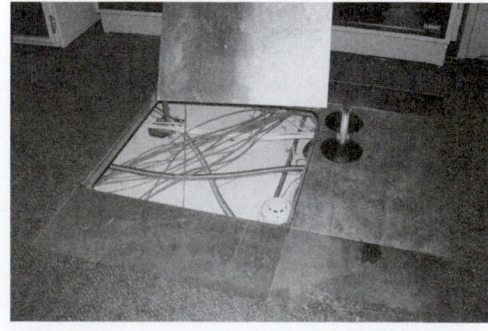

PHOTO 22.13   Raised floor system (open) containing cabling. (Used with permission of ABC)

PHOTO 22.14 Raised floor system (closed). (Used with permission of ABC)

PHOTO 22.16 A work area above a raised floor system. (Used with permission of ABC)

All cables in horizontal pathways should be strung in a star topology so cables directly link the telecommunications closet with each telecommunications outlet. Again, this arrangement allows alterations to be made without the hassle of having to pull new cables.

An industry standard is to size horizontal pathways by providing 1 in² (645 mm²) of cross-section area for every 100 ft² (9.3 m²) of workspace area being served. Easy access to the horizontal cabling is desirable. A pull box, splice box or pulling point is required for any pathway where there are more than two 90° bends, a 180° reverse bend or length more than 100 feet.

PHOTO 22.17 A workstation. (Used with permission of ABC)

Work Area

The *work area* is the space containing workstation (terminal) equipment and components. The workstation components include equipment and devices (e.g., telephones, personal computers, graphic or video terminals, fax machines, modems) and terminal patch cables (e.g., modular cords, PC adapter cables, fiber jumpers, and so forth) that connect work area equipment to the network. Work area wiring is designed to be relatively simple to interconnect so that modifications and additions can be easily accomplished. The work area can also be served by a wireless access point. (See Photos 22.15 through 22.19.)

PHOTO 22.18 A workstation connection with telecommunications and power outlets. (Used with permission of ABC)

A typical *telecommunications outlet* is made in a 2 in × 4 in (50 mm × 100 mm) electrical box with horizontal cabling terminating at a connector on the faceplate covering the box. It is necessary to consider the number and type of devices to be connected when selecting the outlet capabilities and capacities. Industry practice is to provide a minimum of two telecommunications outlet/connectors at each work area. At areas where telecommunications use is anticipated to be heavier than normal (e.g., reception areas, secretarial areas, and control desk areas), additional outlets will be required. Patch cables connect work area equipment and devices to a telecommunications outlet. A maximum length of 33 ft (10 m) is typically allowed for work area patch cables.

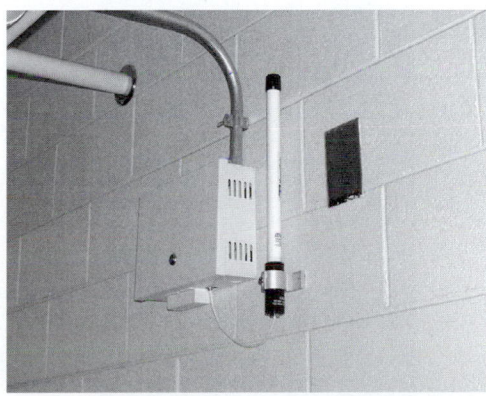

PHOTO 22.15 A wireless access point (station). (Used with permission of ABC)

PHOTO 22.19   A telecommunications outlet with coded female connectors. (Used with permission of ABC)

22.3   ADVANCED WIRING SYSTEMS FOR HOMES

Advanced Home Wiring Systems

An *advanced home wiring system* allows a homeowner to integrate the control and management of the following subsystems:

- Communication subsystem (e.g., intercom, phone, message recording, fax and e-mail)

- Entertainment subsystem (e.g., whole-house stereo, VCR, cable, digital and satellite television, and home theater system)

- Home office subsystem (e.g., computers, printers and scanners)

- Environmental control/energy management subsystem (e.g., control of HVAC equipment, water heater, lighting and other appliances)

- Security/property protection subsystem (e.g., video surveillance with closed circuit TV, control entry gates and garage doors, control lawn irrigation)

Wiring System Components

An advanced home wiring system is typically consists of three main components.

Service Center

The *service center,* sometimes called the *central hub* or *distribution center,* is the central part of the system that accepts incoming services and distributes services throughout the home. It is where all external communication services enter the residence, including cable TV, telephone, digital satellite transmission, and Internet service. It serves as a central hub

that distributes these services to locations throughout the house in a way similar to how an electrical panelboard distributes and controls flow of electricity.

A single metal box typically serves as the distribution center. It is a stand-alone piece of equipment that contains distribution devices. It provides universal access to various networking elements within the home as well as connection to service providers. The service center must allow the wiring system to be customized periodically as the network evolves to accommodate future technologies. The distribution center must be located in a place that is readily accessible to cabling maintenance. Home telecommunication components and wiring are shown in Photos 22.20 through 22.22.

Universal Multiuse Outlets

The telecommunications outlets in a room determine the services that are available in that room. Universal multiuse outlets can be customized to a consumer's specific needs based on the services that are desired in each room (e.g., cable, Internet

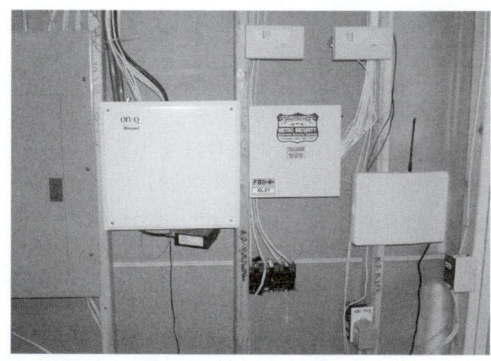

PHOTO 22.20   An advanced home wiring system service center with security and wireless capabilities. (Used with permission of ABC)

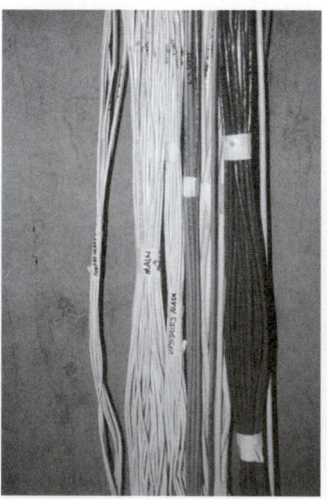

PHOTO 22.21   Home telecommunication wiring. (Used with permission of ABC)

PHOTO 22.22 A residential computerized control panel with IPOD capabilities. (Used with permission of ABC)

access, telephone, and so on). These outlets are designed to support a full range of communication technologies with a variety of flexible configurations, including voice, data, and video. Even after the wiring system installation is complete, outlets can be changed to meet the changing needs of the homeowner.

High-Performance Cabling

Several types of high-performance cable are used, including CAT 5 UTP copper wire and coaxial cable. Wireless transmission technologies can also be used. In the future, fiber optic cable may eventually become a medium of choice for some audio/video applications.

Pathways for cabling run between the service center and the outlets. For new construction, cable is run in concealed pathways such as ducts and conduits. Provisions should also be made for conduit for cable pathways to allow for pulling additional wiring in the future. For retrofits of an existing structure, wiring can be concealed in attics or crawl spaces wherever possible. For exposed retrofit cabling, the cables should be enclosed in a protective surface-mount raceway.

STUDY QUESTIONS

22-1. Describe the analog and digital transmission formats.

22-2. Sketch the three types of wiring topologies used in a telecommunications network.

22-3. Identify and describe the types of transmission media used in a telecommunications network.

22-4. Identify and describe the main subsystems of a structured cabling system for a building telecommunications network.

22-5. Describe UTP and STP wiring.

22-6. Describe the types of coaxial cable.

22-7. Describe the function of a fiber optic system.

22-8. Describe types of wireless communications used in buildings.

22-9. Describe the following components:

    a. Pathway

    b. Drop cables

    c. Patch panel

    d. Patch cord

    e. Telecommunications outlet

22-10. How does electrical equipment (e.g., motors, generators, and so on) affect telecommunication signals?

22-11. How does fluorescent and HID lighting affect telecommunication signals?

22-12. Identify and describe the components of an advanced home wiring system.

Design Exercises

22-13. Make a visit to a residence with an advanced home wiring system. Make a sketch of the system and identify chief components of and their locations in the system.

22-14. Make a visit to a commercial building with a telecommunications network. Make a sketch of the telecommunications network and identify chief components and their locations in the system. Describe the location, physical size, and function of each component.

ACOUSTICAL CONTROL SYSTEMS IN BUILDINGS

23.1 FUNDAMENTALS OF SOUND

Acoustics

Acoustics is the science of sound, including the generation, transmission, and effects of sound waves. This chapter introduces the elements of acoustical design to limit noise issues in buildings. The control of acoustical problems is of prime concern, and the best results are achieved by anticipating the problems before they occur.

The Human Ear

The human ear converts sound energy to mechanical energy to a nerve impulse that is transmitted to the brain. It allows us to perceive the pitch of sounds by detection of the wave's frequencies, the loudness of sound by detection of the wave's amplitude, and the resonance of the sound by the detection of the various frequencies that make up a complex sound wave.

The human ear consists of three fundamental parts: the outer ear, the middle ear, and the inner ear. Each part serves a specific purpose in the tasks of detecting and interpreting sound.

The outer ear collects and channels the incoming pressure waves that make up sound through the ear canal. The middle ear is an air-filled cavity that consists of an eardrum and three tiny, interconnected bones.

The eardrum is the receiver of the incoming pressure waves of sound. It is a very durable and tightly stretched membrane that vibrates as the oscillating pressure waves make contact with it. The high-pressure component of the sound wave forces the eardrum inward and the low-pressure component allows the eardrum to expand outward, thus vibrating the eardrum at the same frequency as the sound wave.

Movement of the eardrum sets the three interconnected bones into motion at the same frequency as the sound wave. These bones act as levers to amplify the vibrations of the sound wave and ultimately transform these vibrations into a mechanical compression wave in the inner ear.

The inner ear is a fluid-filled spiral shell with groups of the delicate hair sensors on the membrane that respond to different frequencies transmitted down the spiral. They convert movements to nerve impulses, which are transmitted by the auditory nerve to the hearing center of the brain where the sound is interpreted. These hair sensors are one of the few cell types in the human body that do not regenerate and are irreversibly damaged by large noise doses.

Sound

Sound is defined as a rapidly varying pressure wave within a fluid medium such as air or water that is capable of being detected by the human ear. Sound is produced when the air is disturbed in some way (e.g., by speaking, by clapping hands, or by a vibrating object). These extremely small, rapid oscillations in air pressure occur both above and below atmospheric pressure. When someone speaks, their vocal cords vibrate, which creates extremely small, rapid oscillations in air pressure that travel to the ears of the listener, much as waves travel across a pond when a stone is dropped in the water. The waves in a pond are two-dimensional, while the waves of sound emanate in three dimensions. A speaker cone from a stereo system serves as a good illustration of the vibration. It is possible to see the movement of a base speaker cone, providing it is producing very low-frequency sound. As the cone moves forward, the air immediately in front of the cone is compressed, causing a slight increase in air pressure. It then moves back past its resting position, which causes a reduction in air pressure (below atmospheric pressure). Sound is made up of waves of alternating high and low pressures radiated away from the source.

Sound *frequency* is the number of sound wave crests per unit time that pass a fixed location; it is a measure of the tone or pitch of a sound. The wavelength, or distance between wave crests, is related to frequency: lower frequencies have longer wavelengths. With a high-pitched sound, such as a squeak or shrill (higher frequency sound), the waves are closer together; with a low-pitched sound, such as the pounding of a bass drum (lower frequency sound), the waves are farther apart.

Sound that is audible by a normal human ear is in a frequency range of 20 and 20 000 Hertz (Hz)—that is, the waves occur between 20 and 20 000 times per second. Nearly all spoken sound is in the range of 600 and 4000 Hz. Frequencies of common sounds are provided in Table 23.1. In air, all frequencies of sound travel at the same speed. However, when bending waves travel through a flexible structure, low frequencies travel faster than high frequencies. The human ear does not respond to all frequencies equally. Instead, the ear acts like a filter favoring certain frequencies. It is more sensitive to sounds from 1000 to 5000 Hz, and most sensitive to sounds of about 4000 Hz.

TABLE 23.1 FREQUENCIES OF COMMON SOUNDS.

Common Sound	Frequency
Female voice range	600 to 800 Hz
Fluorescent fixture "humming sound" range	550 to 650 Hz
Musical "A" tone	440 Hz
Musical high "C" tone	4186 Hz
Pedal of pipe organ	10 to 20 Hz
Span of human hearing	20 to 20 000 Hz
Span of AM radio	200 to 5000 Hz

The *threshold of hearing* is the minimum sound pressure difference level at which a person can hear a specified frequency of sound. It is commonly accepted that the threshold of human hearing for a 1000 Hz sound wave is about 0.000 000 003 lb/in² (0.00002 Pa or 20 μPa). More intense sounds are caused by a much greater variation in pressure. By comparison, standard atmospheric pressure (atmospheric pressure at sea level) is about 14.7 lb/in² (101 325 Pa).

Intensity of Sound

The *decibel (dB)* is a measure of the intensity of sound. The decibel scale runs from the faintest sound the human ear can detect, which is labeled 0 dB, to over 180 dB, the noise of a rocket at the pad during launch. A chart showing relative sound levels in dB is shown in Table 23.2. From this chart, observe that a level below 40 dB is reasonably quiet, while noise levels of 70 dB and up become increasingly noticeable.

Decibels are measured logarithmically. A logarithmic scale is based on powers or multiples of 10. This means that as decibel intensity increases by 10 units, the increase is 10 times the lower value. Thus, a 40 dB (e.g., quiet music) sound has 10 times the sound intensity of 30 dB (e.g., a whisper); 60 dB (e.g., normal conversation) is 1000 times more intense than 30 dB; and, 90 dB (e.g., running lawnmower or noisy street) is 1 000 000 times more intense than 30 dB.

A *bel* is another unit for comparing intensities of sound: 1 bel equals 10 decibels. Sound intensity falls on the bel scale from 0 bel (threshold of hearing) to 13 bel (the threshold of pain). The origin of the decibel is the bel, which was used to express the transmission losses of long telephone lines. It is named for Alexander Graham Bell. (Note that one "l" is omitted from the name of the unit.) The bel, being too large for practical use, was later changed to decibel. Many air conditioning manufacturers still use the bel sound rating in their product literature. For example, 7.6 bels has been established as a standard for air conditioner sound performance. A rating of 7.2 or lower is very quiet and above 7.8 is usually unacceptable.

Loudness is the human perception of the strength of a sound. The perceived loudness of a noise does not necessarily correlate with its sound intensity level. Apparent change in loudness varies greatly with actual change in sound level. A change of 1 dB is not distinguishable by a human ear; a change of 3 dB is noticeable by a careful listener, a change of 5 dB is typically noticeable, and a change of 10 dB makes a significant difference in perception of sound intensity. A change from 50 to 60 dB makes the noise seem twice as loud while a change from 50 to 40 dB makes it seem half as loud. However, the difference in sound intensity (pressure difference) in these examples is by a factor of 10. Refer to Table 23.3.

The *A-weighted decibel (dBA)* is a measure of sound intensity level that is adjusted to match the acuity of the human ear, which does not respond equally to all sound frequencies. The human ear has a tendency to magnify sounds having frequencies in the range from 1000 Hz to 5000 Hz—that is, sounds within these ranges seem louder to the human ear. Sounds outside of this range are perceived as being greatly diminished. By using an electronic filter, a sound level meter will read a numerical value proportional to the human perception of the strength of sound. The value measured is expressed in dBA.

Sound level for airborne sound, unless specified otherwise, is the A-weighted (dBA) sound level. There are also B-weighted (dBB) and C-weighted decibel (dBC) scales, but these are used infrequently.

Behavior of Sound

Sound moves in a straight-line path unless it is interrupted. When sound waves strike an object, they can be reflected, absorbed, or transmitted. Reflection and absorption are dependent on the wavelength of the sound. The percentage of the sound transmitted through an obstacle depends on how much sound is reflected and how much is absorbed. See Figure 23.1. High-frequency sounds tend to pass through thin materials and curve more easily around barriers. Low-frequency sounds are easily reflected by relatively thin materials and do not bend around most barriers. Thus, it is more difficult to isolate high-frequency sounds.

Speed of sound is nearly independent of frequency and atmospheric pressure, but the resultant sound velocity may be substantially altered by wind velocity. The formula for the speed of sound (c) in air, in ft/s, is as follows, where T is in degrees Fahrenheit:

$$c = 49\sqrt{(460 + T)}$$

In SI (metric) units, the formula for the speed of sound (c) in air, in m/s, is as follows, where T is in degrees Celsius:

$$c = 20\sqrt{(273 + T)}$$

The speed of sound in air at a temperature of 0°F is 1051 ft/s and about 98 ft/s faster at 90°F.

Sound level decreases as the distance from a sound source increases. The inverse square law quantifies this decrease in sound level; that is, sound energy decrease is proportional to the square of the distance increase. For example, if the listening distance from a sound source is doubled (increased by a factor of 2), the direct sound energy is decreased by a factor

TABLE 23.2   SOUND POWER AND INTENSITY OF SOUND FROM DIFFERENT SOURCES.

	Source	Sound Power (W)	Noise Level dBA
Deafening	Saturn rocket (close exposure)	100 000 000	200
		10 000 000	190
		1 000 000	180
	Jet engine (close exposure)	100 000	170
	Immediate loss of hearing	10 000	160
	Inside jet engine test cell	1000	150
	Jet plane at takeoff at 75 ft (25 m)	100	140
	Rock concert (maximum) Symphonic orchestra (loud) Artillery fire Machine gun	10	130
	Human pain threshold Siren at 100 ft (30 m) Thunder (close) Sonic boom	1	120
Extremely Loud	Large aircraft 150 ft overhead Busy woodworking shop Accelerating motorcycle Loud radio	0.1	110
	Large air compressor Air chisel Magnetic drill press High-pressure gas leak Automobile on highway	0.01	100
	Permissible sound exposure up to 8 hr (OSHA) Heavy diesel vehicle Heavy city traffic Lawn mower Unmuffled cars, trucks, motorcycles at 50 ft (15 m) Jackhammer, bulldozer at 50 ft (15 m)	0.001	90
Loud	Average manufacturing plant Noisy office Kitchen blender Alarm clock	0.0001	80
	Average street noise; car at 50 ft (15 m) Freight train at 100 ft (30 m) Vacuum cleaner at 10 ft (3 m) Forceful conversational voice	0.00001	70
Moderate	Large department store Busy restaurant Average office Noisy home Normal conversation from 3 ft (1 m)	0.000001	60
	Room with window air conditioner Quiet restaurant	0.0000001	50
Quiet	Quiet conversation Private office Quiet radio	0.00000001	40
	Room in a quiet residence Broadcast studio	0.000000001	30
	Very soft whisper Empty auditorium Noise from rustling leaves	0.0000000001	20
Very Faint	Quietest audible sound (under normal conditions) Human breathing Human heartbeat	0.00000000001	10
	Soundproof room Quietest audible (under laboratory conditions) Threshold of hearing	0.000000000001	0

TABLE 23.3   THE EFFECTS OF INCREASING SOUND.

Decibel Increase	Effect
Δ1 dBA	Not distinguishable by a human
Δ3 dBA	Barely perceptible by a careful listener
Δ5 dBA	Clearly noticeable
Δ10 dBA	Perceived as twice as loud; ten times the intensity of 1 dBA
Δ15 dBA	Big noticeable change
Δ20 dBA	Perceived as three times as loud; a hundred times the intensity of 1 dBA

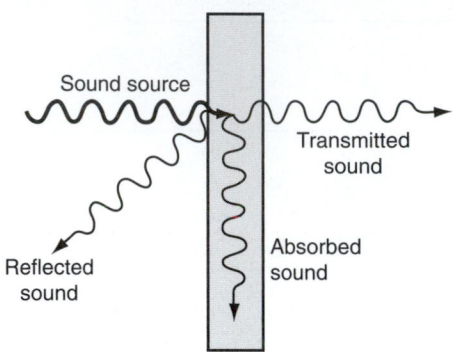

FIGURE 23.1   Sound waves striking an object will be reflected, absorbed, or transmitted.

of $4(2^2 = 4)$. This translates to a 6 dBA reduction in sound intensity.

The surrounding environment, especially close to the ground/floor, has a great effect on sound received at a distant location. Ground/floor reflection affects sound levels more than a few feet away (e.g., distances greater than the height of the sound source or the receiver above the ground).

Noise

By definition, *noise* is any disagreeable or undesirable sound that interferes with rest, sleep, mental concentration, or speech communication. Simply, noise is sound that is not part of what a person is trying to hear.

Ambient noise is the all-inclusive noise associated with a given environment at a specified time. It is usually a composite of sound from many sources and from many directions in which no particular sound is dominant. *Background noise* is the total noise from all sources other than a particular sound that is of interest (e.g., other than the sound being measured or other than the speech or music being listened to). *White noise* is undistinguishable background noise.

Exposure to Excessive Noise

Unprotected exposure to sound pressure levels above 100 dBA will cause permanent hearing loss. Hearing protection should be used when exposed to noise levels above 85 dBA (about the sound level of a lawn mower when it is being pushed over a grassy surface), especially when prolonged exposure (more than a fraction of an hour) is anticipated. Damage to hearing from loud noise is cumulative and irreversible. Exposure to high noise levels is also one of the main causes of tinnitus, ringing in the ear or noise sensed in the head.

Noise dose is the exposure to noise expressed as a percentage of the allowable daily exposure. For the Occupational Safety and Health Administration (OSHA), a 100% noise dose would equal an 8-hr exposure to a continuous 90 dBA noise. A 50% dose would equal an 8-hr exposure to an 85 dBA noise or a 2 hr exposure to a 95 dBA noise.

The permissible exposure limit (PEL) is published and enforced by OSHA as a legal standard (OSHA Hearing Conservation Standard 29 CFR 1910.95). The PEL refers to levels of exposure and conditions under which it is believed that nearly all healthy workers may be repeatedly exposed, day after day without adverse effects. Currently, the OSHA PEL for noise is 90 dBA as an 8-hr time-weighted average (TWA). Exposures at and above this level are considered hazardous. Maximum permissible exposures limit based upon Occupational Safety and Health Act (OSHA) regulations are provided in Table 23.4.

In cases where an employee is exposed to varying noise levels during the day, OSHA states that the daily total of the ratios of actual exposure to allowable exposure should not exceed 1. For example, a worker exposed to a noise level of 90 dBA for 3 hrs. (8 hr are allowable, as shown in Table 23.4), 92 dBA for 1 hr (6 hr are allowable), and 100 dBA for 1 hr (2 hr are allowable) exceeds the limit because:

$$3 \text{ hr}/8 \text{ hr} + 1 \text{ hr}/6 \text{ hr} + 1 \text{ hr}/2 \text{ hr}$$
$$= 1.04, \text{ which is greater than } 1$$

Manufacturing plants and construction sites with noise levels above 90 dB are required by OSHA to minimize the working time in those areas or the use of personal hearing protection devices (e.g., ear muffs or plugs). Installing sound-controlled rooms will maximize the productive time in an area.

Reverberant Sound

Sound waves in a room typically travel at about 1130 ft/s (345 m/s). The sound coming directly from a source in a room will generally reach a listener after a time of from 0.01 to 0.20 s. Sound waves from a sound source can take a direct path to a listener's ear or can reflect off surfaces (i.e., walls, ceiling or floor surfaces, and people) before being heard. The reflected sound waves arrive at the listener slightly later than direct sound because they travel a longer distance to the ear.

Reflected sound is generally weaker because only a fraction of the direct sound is reflected as the receiving surfaces will absorb some of the sound energy received. Reflected sound is noticeable by the listener. Within about 0.05 s after the arrival of the direct sound, a series of fairly distinct reflections from various reflecting surfaces, known as *early reflections,*

TABLE 23.4 MAXIMUM PERMISSIBLE EXPOSURES LIMIT BASED ON OCCUPATIONAL SAFETY AND HEALTH ACT (OSHA) REGULATIONS.

Duration Per Day (hr)	Permissible Exposure Limit (PEL) (dBA)	Example of Sound Level (dBA)
8	90	Jackhammer or bulldozer at 50 ft (15 m)
6	92	
4	95	
3	97	Large air compressor Air chisel
2	100	Magnetic drill press
1½	102	
1	105	
½	110	Busy woodworking shop
¼ or less	115	

reach the listener. They typically arrive from the nearest side wall or from the ceiling for listeners in the center of the space. They tend to reinforce the direct sound. After about 0.05 s, additional reflections merge into what are called *successive reflections*. These reflections are typically of lower amplitude and very closely spaced in time.

Reverberant sound, referred to as *reverberation,* is made up of successive reflections from the surfaces in an enclosed space (i.e., a room) after the direct sound has ceased. When sound is produced in a space, a large number of reverberations build up and then slowly decay as the sound energy is absorbed by the walls and air. Reverberation accompanies any sound emerged in an enclosed environment. Reverberation is most noticeable when the sound source rapidly stops but the reflections continue, decreasing in intensity until they can no longer be heard. Reverberation in an enclosed space dies away with time as the sound energy is absorbed by recurring interactions with the surfaces of the room.

Reverberant sound is different than an echo or a series of echoes. An *echo* is a distinct, delayed reflection of a direct sound with a delay more than 0.1 s. A reverberation is perceived when the reflected sound wave reaches an ear in less than 0.1 s after the original direct sound wave. Because of the short delay, humans do not completely discern each reflection as a reproduction of the original sound but do hear the effect that the entire series of reflected sound waves makes.

Reverberation Time

Reverberation time (RT$_{60}$) is the time required for reflections of direct sound intensity (sound pressure level) to decay to 1/1 000 000th (60 dB) below the level of the direct sound. Over 100 years ago, a Harvard physics professor named Wallace Sabine developed what has become known as Sabine's formula.

Sabine's formula for reverberation time (seconds) is based on room volume (V) in ft^3 or m^3, the individual surface area of each material in the room (S), in ft^2 or m^2, and the absorption coefficient (ac) of each individual surface at specific frequency. The constant 0.049 may be substituted with 0.16 for metric (SI) analysis. The formula is:

$$RT_{60} = 0.049 \, V/\Sigma S(ac)$$
$$= 0.049 \, V/[S_1(ac_1) + S_2(ac_2) + S_3(ac_3) + \ldots]$$

Absorption coefficients (ac) represent the fraction of sound energy (not sound level decibels), the material will absorb as a decimal from 0 to 1, as determined through laboratory measurements. Absorption coefficients for various building and insulation materials vary with frequency of sound. See Table 23.5.

Example 23.1

A 48 ft by 144 ft church sanctuary with 24 ft high ceilings has walls with an absorption coefficient (1000 Hz) of 0.09, ceiling with an absorption coefficient of 0.20, and seating (unoccupied) with an absorption coefficient of 0.35. Approximate the reverberation time.

Volume of the space is: 48 ft · 144 ft · 24 ft = 165 888 ft^3

Ceiling and floor each have an area of:

$$48 \text{ ft} \cdot 144 \text{ ft} = 6912 \text{ ft}^2$$

Walls have a total area of:

$$24 \text{ ft} (48 \text{ ft} + 48 \text{ ft} + 144 \text{ ft} + 144 \text{ ft}) = 9216 \text{ ft}^2$$

$$RT_{60} = 0.049 \, V/\Sigma S(ac)$$
$$= 0.049 \cdot 165 \, 888 \text{ ft}^3/[(9216 \text{ ft}^2 \cdot 0.09) + (6912 \text{ ft}^2 \cdot 0.20) + (6912 \text{ ft}^2 \cdot 0.35)]$$
$$= 1.76 \text{ s}$$

TABLE 23.5   ABSORPTION COEFFICIENTS FOR VARIOUS BUILDING AND INSULATION MATERIALS VARY WITH
FREQUENCY OF SOUND.

Material	Frequency					
	125 Hz	250 Hz	500 Hz	1000 Hz	2000 Hz	4000 Hz
Acoustic ceiling tiles	0.70	0.66	0.72	0.92	0.88	0.75
Brick wall, painted	0.012	0.013	0.017	0.020	0.023	0.025
Carpet on concrete	0.09	0.08	0.21	0.26	0.27	0.37
Carpet on foam	0.08	0.24	0.57	0.69	0.71	0.73
Cast-in-place concrete, unpainted	0.010	0.012	0.016	0.019	0.023	0.035
Chalkboard	0.01	0.01	0.01	0.01	0.02	0.02
Concrete (block) masonry, painted	0.10	0.05	0.06	0.07	0.09	0.08
Draperies	0.05	0.12	0.35	0.45	0.40	0.44
Drywall (gypsum wallboard)	0.25	0.15	0.08	0.06	0.04	0.04
Glass	0.35	0.25	0.18	0.12	0.07	0.04
Plaster, rough surface	0.039	0.056	0.061	0.089	0.054	0.070
Resilient tile flooring	0.02	0.03	0.03	0.03	0.03	0.02
Wood, plywood panel (⅜ in/9 mm)	0.28	0.22	0.17	0.09	0.10	0.11
People (adults)	0.25	0.35	0.42	0.46	0.50	0.50
Wooden pews (fully occupied)	0.57	0.61	0.75	0.86	0.91	0.86

Approximate the reverberation time if the church sanctuary is occupied (floor absorption coefficient of 0.55).

$$RT_{60} = 0.049 \ V/\Sigma S(ac)$$
$$= 0.049 \cdot 165\,888 \text{ ft}^3/[(9216 \text{ ft}^2 \cdot 0.09)$$
$$+ (6912 \text{ ft}^2 \cdot 0.20) + (6912 \text{ ft}^2 \cdot 0.55)]$$
$$= 1.35 \text{ s}$$

Reverberation time is defined for wide band of frequencies. In a complete analysis, the calculation shown in Example 23.1 should be performed for each frequency band. Absorption coefficients can vary considerably at different frequencies, which affects the reverberation time. For a rough estimate for rooms used for speaking (i.e., classrooms), the reverberation time can be approximated by using the frequency band representative of speech (i.e., 1000 Hz). Reverberation time is often calculated with the room unoccupied. Because people and their clothing provide additional sound absorption, an unoccupied room is the worst-case scenario, although not an unreasonable one, as occupancy varies.

Optimum Reverberation Times

The *optimal reverberation time* is the most desirable reverberation time for a specific room volume, room type (i.e., auditorium, church sanctuary, classroom), use (i.e., speech versus music), and sound frequency. In large enclosed spaces (i.e., cathedrals, gymnasia, indoor swimming pools) reverberation time is long and can clearly be heard. If reverberation time is excessive, it makes the reflected sounds run together, creating garbled (difficult to decipher) sound. Rooms for speech require a shorter reverberation time than for music as a longer reverberation time can make it difficult to understand speech. If

reverberation time from syllable to syllable overlaps, it may make it difficult to identify the word.

The optimum reverberation time depends upon the intended use of the space. The preferred ranges of reverberation time at midfrequency (average of reverberation at 500 and 1000 Hz) for a variety of spaces are shown in Table 23.6. These ranges are based on the experience of listeners with normal hearing in finished spaces. Satisfactory listening conditions can be achieved that have different reverberation times within the preferred range provided other acoustical needs are achieved. In general, large rooms ($>500\,000$ ft^3) should be nearer the upper limit of the reverberation time range; smaller rooms ($>50\,000$ ft^3) of the same type should be nearer the lower limit.

Reverberation time is primarily controlled by change the reflectivity (absorption coefficient) of the surfaces in the space and, if possible, adjusting the size of the space. The *absorption*

TABLE 23.6   PREFERRED RANGES OF REVERBERATION TIME AT MID FREQUENCY (AVERAGE OF REVERBERATION AT 500 AND 1000 HZ) FOR LISTENERS WITH NORMAL HEARING SELECTED SPACES.

Space	Reverberation Time (s)	
	Lower Limit	Upper Limit
Auditorium	1.4	2.0
Church	1.2	1.8
Cinema (theatre)	0.8	1.3
Classroom	0.6	1.0
Conference room	0.6	1.0
Opera theatre	1.4	1.8
Orchestra concert hall	1.0	2.6
Theatre	1.0	1.4

coefficients (ac) of the surfaces of the room determine how much sound energy is lost in each reflection. Highly reflective materials (low absorption coefficients) will increase the reverberation time, while absorptive materials (low absorption coefficients) reduce the reverberation time. Similarly, people and furnishings absorb energy, reducing the reverberation time. Additionally, large, spacious spaces ($>100\,000$ ft^3) tend to have longer reverberation times because the sound waves travel a longer distance between reflections. A designer can increase reverberation time by increasing volume in a space (i.e., raising the ceiling).

Vibration

Vibration is an oscillating movement of a body. When a body moves back and forth about a stationary position, it can be said to vibrate. Vibration can cause unwanted noise. Examples of unwanted vibration noise are the movement of a building near a railway line when a train passes or the vibration of the floor is caused by operation of a mechanical device or appliance (e.g., washing machine, clothes dryer, or unbalanced fan).

Vibration can be reduced with *vibration isolators,* a system of connected springs and masses for damping. The concept of vibration isolation is demonstrated with a weight suspended from a rubber band. If the band is moved up and down very slowly, the suspended weight will move by the same amount. At resonance the weight will move much more and possibly in the opposite direction. As the frequency of vertical movement is further increased, however, the weight will become almost stationary. Springs are more often used in compression than in tension. Although vibration can be reduced with vibration isolators, sometimes it is at the risk of increased machinery vibration and premature wear. See Section 23.3.

Sound Transmission in Buildings

Sound may be transmitted through the air or through the structure. *Airborne sound is* transmitted through air or gas as a medium rather than a liquid or solid. *Structure-borne sound is* transmitted through a construction assembly (e.g., wall, floor, and ceiling).

Planning is required to reduce the transmission of sound from the exterior to the interior and from room to room. Basically, in each space it is necessary to keep the necessary sounds in and undesirable sounds out.

Sound is not selective; it passes (transmits) through most building materials and any openings, no matter how small. As sound passes through building materials (walls, floors, and ceilings), its intensity is reduced. The amount of reduction is dependent on the types of materials and the construction used. Sound passes more readily through lighter, more porous materials than through heavy, dense, massive materials and assemblies of materials. The use of sound-absorbent materials on walls, ceilings, and floors will increase the transmission of sound through the surface; that is, it will make the room in which the sound originates quieter, but part of the sound will travel to the surrounding rooms. See Section 23.3.

Construction Noise

Governmental entities typically establish regulations that control unreasonable, unnecessary, and excessive noise. Construction noise can be a nuisance to nearby businesses and residences. As a result, construction projects, especially in residential zones, are subject to the permissible noise levels. A typical noise abatement ordinance limits permissible noise levels to 80 dBA during the day (e.g., between 7:00 AM and 7:00 PM) and 70 dBA at night (e.g., between 7:00 PM and 7:00 AM).

23.2   SOUND RATINGS

Sound control is an important consideration in the design of a project. The design of a building must ensure that all spaces will be insulated from unreasonable outside sources of noise. Sound within a space must also be controlled.

Sound isolation between rooms and floor levels and noise reduction within a room are measured by several single number rating systems. A single number rating offers an engineering measure of performance between wall, ceiling, and floor assemblies, and within the space.

Transmission/Isolation Ratings

Sound rating systems that evaluate the transmission of sound through a construction assembly (e.g., partitions, walls, ceilings, and floors) are as follows.

Sound Transmission Class

The *sound transmission class (STC)* is a single-number rating system that rates the ability of a wall or other construction to block sound transmission. It is used to compare the airborne sound-isolating characteristics of construction assemblies used to separate occupied spaces. STC values are provided in Table 23.7. The STC value can be determined through testing of the assembly in the field, in which case the reported value is referred to as the *field sound transmission class (FSTC).*

Essentially, the STC is a decibel measure of the difference between the sound energy striking the panel or construction on one side and the sound energy transmitted from the other side. This includes sound from all angles of direction, and from low and high sound frequencies. The higher the STC rating, the better the construction assembly is in reducing sound transmission. Table 23.8 shows the effectiveness of STC ratings in muting common sounds.

A designer uses the STC as measure of performance when selecting a surrounding wall and floor–ceiling assemblies. Minimum or desired STC ratings are specified in company standards

TABLE 23.7 SOUND TRANSMISSION CLASS (STC) RATINGS FOR SELECTED CONSTRUCTION ASSEMBLIES, COMPILED FROM VARIOUS INDUSTRY AND GOVERNMENT SOURCES.

Description of Construction Assembly	Typical Sound Transmission Class (STC)
Framed Wall Assemblies	
Framed Wall/No Insulation: Studs 16 in on center (OC), single layer ½ in gypsumboard on two sides, openings not sealed.	30
Framed Wall/No Insulation: Studs 16 in OC, single layer ½ in gypsumboard on two sides, openings sealed.	35
Framed Wall/No Insulation: Studs 16 in OC, single layer ½ in gypsumboard on one side, two layers of ½ in gypsumboard on opposite side, openings sealed.	37
Framed Wall/Insulation: Single steel studs 16 in OC, single layer ½ in Type X gypsumboard on one side, single layer ⅝ in Type X gypsumboard on opposite side, 1½ in sound absorbing insulation in stud cavities.	38
Framed Wall/Insulation: Single wood studs 16 in OC, single layer ⅝ in Type X gypsumboard each side, stud cavities filled with sound-absorbing insulation.	42
Framed Wall/Insulation: Single steel studs 16 in OC single layer ½ in Type X gypsumboard each side 1½ in sound-absorbing insulation in stud cavities.	43
Framed Wall/Insulation: Single steel studs 16 in OC, single layer ½ in Type X gypsumboard on one side, double layer ½ in Type X gypsumboard on opposite side, 1½ in sound-absorbing insulation in stud cavities.	45
Framed Wall/Insulation: Single steel studs 16 in OC, single layer ½ in Type X gypsumboard on one side, single layer ⅝ in Type X gypsumboard on opposite side, stud cavities filled with sound-absorbing insulation.	47
Framed Wall/Insulation: Single wood studs 16 in OC, single layer ⅝ in Type X gypsumboard on one side, double layer ½ in Type X gypsumboard on opposite side, stud cavities filled with sound-absorbing insulation.	47
Framed Wall/Insulation: Single wood studs 24 in OC, single layer ½ in type X gypsum board each side, 1½ in sound absorbing insulation in stud cavities.	47
Framed Wall/Insulation: Single wood studs 16 in OC, single layer ½ in Type X gypsumboard on one side, single layer ⅝ in gypsumboard on opposite side, stud cavities filled with sound-absorbing insulation.	48
Framed Wall/Insulation: Single wood studs 24 in OC, single layer ½ in Type X gypsumboard on resilient channels on one side, single layer ½ in Type X gypsumboard on opposite side, 1½ in sound-absorbing insulation in stud cavities.	52
Framed Wall/Insulation: Single steel studs 16 in OC, single layer ⅝ in Type X gypsumboard on one side, single layer ½ in Type X gypsumboard mounted on resilient channels on opposite side, stud cavities filled with sound-absorbing insulation.	53
Framed Wall/Insulation: Single wood studs 24 in OC, single layer ½ in Type X gypsumboard on one side, single layer ⅝ in Type X gypsumboard on opposite side, stud cavities filled with sound-absorbing insulation.	54
Framed Staggered Double Wall/Insulation: Single layer ½ in Type X gypsumboard on each side of single wood studs 24 in OC, 1½ in sound-absorbing insulation in stud cavities, 1 in air gap, single wood studs 24 in OC, single layer ½ in Type X gypsumboard on one side, single layer ⅝ in Type X gypsumboard on opposite side, stud cavities filled with sound-absorbing insulation.	61
Framed "Soundproofed" Wall: Proprietary design.	72
Masonry Wall Assemblies	
Concrete Masonry Unit (CMU) Wall: 4 in nominal thickness, standard or heavyweight CMU, unpainted surface.	40
CMU Wall: 4 in nominal thickness, standard or heavyweight CMU, painted surface.	41
CMU Wall: 8 in nominal thickness, lightweight CMU, unpainted surface.	43
CMU Wall: 8 in nominal thickness, lightweight CMU, painted surface.	44
CMU Wall: 8 in nominal thickness, lightweight CMU with ½ in gypsumboard attached directly to CMU surface on one side of wall.	45
CMU Wall: 8 in nominal thickness, standard or heavyweight CMU, unpainted surface.	45
Brick Wall: 4 in nominal thickness brick.	45
CMU Wall: 12 in nominal thickness, lightweight CMU, unpainted surface.	46

Description of Construction Assembly	Typical Sound Transmission Class (STC)
Masonry Wall Assemblies (*continued*)	
CMU Wall: 8 in nominal thickness, standard or heavyweight CMU with ½ in gypsumboard attached directly to CMU surface on one side of wall.	47
CMU Wall: 12 in nominal thickness, standard or heavyweight CMU, unpainted surface.	48
CMU Wall: 8 in nominal thickness, fully grouted, lightweight CMU.	48
CMU Wall: 8 in nominal thickness, fully grouted, standard or heavyweight CMU.	49
Brick Cavity Wall: Two wythes of 4 in nominal thickness brick with 2 in wide hollow cavity.	50
CMU Wall: 8 in nominal thickness, standard or heavyweight CMU with plaster applied directly to CMU surface on one side of wall.	51
CMU/Brick Composite Wall: 8 in nominal thickness. 4 in standard or heavyweight CMU and 4 in brick, unpainted surface.	51
Brick Cavity Wall: Two wythes of 4 in nominal thickness brick with 2 in wide grouted cavity.	51
CMU Wall: 8 in nominal thickness, light eight CMU with ½ in gypsumboard attached to resilient furring channels on one side of wall.	52
Brick Wall: Two wythes of 4 in nominal thickness brick.	52
CMU/Brick Composite Wall: 8 in nominal thickness. 4 in standard or heavyweight CMU and 4 in brick with plaster applied directly to CMU surface on one side of wall.	53
CMU/Brick Cavity Wall: 10 in nominal thickness. 4 in standard or heavyweight CMU, 2 in hollow cavity, and 4 in brick, unpainted surface.	54
CMU Wall: 8 in nominal thickness, lightweight CMU with ½ in gypsumboard attached to resilient furring channels on one side of wall.	54
CMU Wall: 8 in nominal thickness, lightweight CMU with ½ in gypsumboard attached to resilient furring channels on both sides of wall.	54
CMU/Brick Composite Wall: 8 in nominal thickness. 4 in standard or heavyweight CMU and 4 in brick with ½ in gypsumboard attached to resilient furring channels on one side of wall.	56
CMU/Brick Cavity Wall: 10 in nominal thickness. 4 in standard or heavyweight CMU, 2 in hollow cavity, and 4 in brick with plaster applied directly to CMU surface on one side of wall.	57
CMU/Brick Composite Wall: 10 in nominal thickness. 4 in standard or heavy eight CMU, 2 in hollow cavity, and 4 in brick with ½ in gypsumboard attached to resilient furring channels on one side of wall.	59
CMU Wall: 8 in nominal thickness, standard or heavyweight CMU with ½ in gypsumboard attached to resilient furring channels on both sides of wall.	60
Glass and Glazings	
Single glass, single strength (³⁄₃₂ in thick) float glass.	26
Insulating glass, two ⅛ in glass panes with ½ in air space.	28
Single glass, double strength (⅛ in thick) float glass.	29
Single glass, ¼ in thick float glass.	29
Single glass, ⅜ in thick float glass.	30
Insulating glass, two ⅛ in glass panes with ¾ in air space.	32
Insulating glass, two ¼ in glass panes with ½ in air space.	32
Single glass, ½ in thick float glass.	33
Single laminated glass, approx. ¼ in total thickness. 2 panes of glass and 0.054 in thick plastic interlayer.	33
Single laminated glass, approx. ½ in total thickness. 2 panes of glass and 0.054 in thick plastic interlayer.	36

TABLE 23.7   SOUND TRANSMISSION CLASS (STC) RATINGS FOR SELECTED CONSTRUCTION ASSEMBLIES. COMPILED FROM VARIOUS INDUSTRY AND GOVERNMENT SOURCES. (CONTINUED)

Description of Construction Assembly	Typical Sound Transmission Class (STC)
Glass and Glazings (*continued*)	
Insulating glass, two ⅛ in glass panes with 2 in air space.	38
Single laminated glass, approx. ¾ in total thickness. 2 panes of glass and 0.054 in thick plastic interlayer.	39
Insulating glass, two ⅛ in glass panes with 4 in air space.	42
Floor Assemblies	
Floor panel, single 2 in × 10 in floor joists 16 in OC, ½ in waferboard subfloor, ½ in particleboard main floor, carpet, pad, single layer ⅝ in Type X gypsumboard mounted on resilient channels, 1 in sound-absorbing insulation sprayed in joist cavities.	45
Floor panel, single 2 in × 10 in floor joists 16 in OC, ½ in waferboard subfloor, ½ in particleboard main floor, carpet, pad, single layer ⅝ in Type X gypsumboard mounted on resilient channels, 2 in sound-absorbing insulation sprayed in joist cavities.	54
Hollow Core Slab: 4 in thick hollow core slab.	44
Hollow Core Slab: 6 in thick hollow core slab.	45
Hollow Core Slab: 8 in thick hollow core slab.	47
Hollow Core Slab: 10 in thick hollow core slab.	50
Hollow Core Slab Floor: 8 in thick hollow core slab with 2 in thick concrete topping.	53
Doors	
Hollow core, wood door. 1⅜ in thick, no gaskets or weather stripping.	15
Hollow core, wood door. 1⅜ in thick, sealed with gaskets or weather stripping.	25
Insulated metal door, sealed with gaskets or weather stripping.	27
Solid core, wood door, sealed with gaskets or weather stripping.	27
"Soundproof" door. Proprietary design.	70
Floor Assemblies	
Floor panel, single 2 in × 10 in floor joists 16 in OC, ½ in waferboard subfloor, ½ in particleboard main floor, carpet, pad, single layer ⅝ in Type X gypsumboard mounted on resilient channels, 1 in sound-absorbing insulation sprayed in joist cavities.	45
Floor panel, single 2 in × 10 in floor joists 16 in OC, ½ in waferboard subfloor, ½ in particleboard main floor, carpet, pad, single layer ⅝ in Type X gypsumboard mounted on resilient channels, 2 in sound-absorbing insulation sprayed in joist cavities.	54
Hollow Core Slab: 4 in thick hollow core slab.	44
Hollow Core Slab: 6 in thick hollow core slab.	45
Hollow Core Slab: 8 in thick hollow core slab.	47
Hollow Core Slab: 10 in thick hollow core slab.	50
Hollow Core Slab Floor: 8 in thick hollow core slab with 2 in thick concrete topping.	53

or local building codes. The U.S. Department of Housing (HUD) sound transmission class requirements for multifamily housing projects are provided in Table 23.9. Recommended STC requirements for construction assemblies (e.g., walls, floors, and so on) in schools and dwellings are provided in Table 23.10.

The STC value can be determined through testing of the assembly once it is constructed on the building site, in which case the reported value is referred to as the *field sound transmission class (FSTC)*. The required FSTC value is generally permitted to be slightly less than the specified STC.

TABLE 23.8   EFFECTIVENESS OF STC RATINGS IN MUTING COMMON SOUNDS.

STC	Noise Heard through Floor or Wall	Effectiveness of STC in Muting this Type of Sound
30	Normal speech audible as a murmur but not decipherable. Loud speech understood fairly well; decipherable.	Suppresses normal neighborhood noises.
35	Loud speech audible but unintelligible; not decipherable.	Suppresses noises near a busy road or highway.
40	Loud speech audible as a murmur; not decipherable.	Suppresses average street noise; car at 50 ft (15 m).
45	Loud speech barely audible.	Suppresses loud music from neighboring apartment.
50	Loud speech not audible.	Suppresses loud nearby train or airport sounds.

TABLE 23.9   U.S. DEPARTMENT OF HOUSING (HUD) REQUIREMENTS FOR SOUND TRANSMISSION CLASS (STC) IN MULTIFAMILY HOUSING.

Location	Minimum STC Requirement[d]
Wall Assemblies	
Living unit to living unit, corridor or public space, average noise[b]	45
Living unit to public space and service areas, high noise[c]	50
Floor-Ceiling Assemblies	
Floor–ceiling separating living units from other living units, public space[a], or service areas[b]	45
Floor–ceiling separating living units from public space and service areas (high noise), including corridor floors over living units[c]	50

[a]These values assume floors in corridor are carpeted: otherwise increase STC by 5.
[b]Public space of average noise includes lobbies, storage rooms, stairways, and so on.
[c]Areas of high noise include boiler rooms, mechanical equipment rooms, elevator shafts, laundries, incinerator shafts, garages, and most commercial uses.
[d]Increase STC by 5 when over or under mechanical equipment that operates at high noise levels.

TABLE 23.10   RECOMMENDED SOUND REQUIREMENTS FOR CONSTRUCTION ASSEMBLIES (E.G., WALLS, FLOORS, AND SO ON) BETWEEN ROOM AND ADJACENT SPACE IN SCHOOLS AND DWELLINGS.

Room	Adjacent Space	Recommended Sound Requirement	
		Sound Transmission Class (STC)	Impact Insulation Class (IIC)
Schools			
Classroom	Classroom	37	47
Classroom	Corridor	42	—
Music room	Classroom	52	55
Music room	Theater	52	62
Office	Office	33	47
Theater	Classroom	52	—
Theater	Music rehearsal	57	—
Dwellings			
Bedroom	Bedroom	44	52
Bedroom	Mechanical room	52	—
Kitchen	Family room	52	52
Living room	Bedroom	46	57
Living room	Kitchen	52	52

Transmission Loss

Sound *transmission loss (TL)* is a measure of sound insulation provided by a structural configuration. Transmission loss performance of the construction assembly is tested over a range of different frequencies. The results of these tests, at each frequency, are averaged together for a transmission loss rating. This does not take into account that some people are more sensitive to noise in some of the frequency ranges than in others. As a result, it is possible that the construction assembly does not provide the transmission loss desired at the proper frequencies.

Noise Isolation Class

The noise isolation class (NIC) is a single number rating used to measure noise reduction between two areas or rooms. It provides an evaluation of the sound isolation between two enclosed spaces that are acoustically connected by one or more paths.

Placement of openings that provide a flanking path in a wall or ceiling greatly influences sound transmission in a building. Poorly placed openings can significantly decrease the sound insulation of gypsumboard walls. For example, research has shown that a decrease of up to six STC points occurs with poorly placed electrical boxes, when compared to walls with no boxes.

TABLE 23.11   IMPACT INSULATION CLASS (IIC) RATINGS FOR SELECTED FLOOR–CEILING ASSEMBLIES.
COMPILED FROM VARIOUS INDUSTRY AND GOVERNMENT SOURCES.

Floor—Ceiling Assembly	Impact Insulation Class (IIC)
Hollow Core Slab: 4 in thick slab, no floor finish.	20
Hollow Core Slab: 6 in thick slab, no floor finish.	22
Hollow Core Slab: 8 in thick slab, no floor finish.	27
Hollow Core Slab: 10 in thick slab, no floor finish.	31
Cast-in-Place Concrete Slab: 4 in thick solid slab, no floor finish.	20
Cast-in-Place Concrete Slab: 6 in thick ribbed slab (over decking), no floor finish.	21
Cast-in-Place Concrete Slab: 6 in thick solid slab, no floor finish.	27
Cast-in-Place Concrete Slab: 6 in thick ribbed slab (over decking) with two layers of ½ in gypsumboard suspended from resilient metal channels spaced at 16 in OC, no floor finish.	36
Cast-in-Place Concrete Slab: 8 in thick solid slab, no floor finish.	39
Wood—Solid Framing: 2 ×10 solid wood joists at 16 in OC, ⅝ in thick OSB subfloor, ⅝ in thick gypsumboard secured to bottom edge of joist. Cavity empty (no sound insulation).	28
Wood—Solid Framing: 2 × 10 solid wood joists at 16 in OC, ⅝ in thick OSB subfloor, ⅝ in thick gypsumboard secured to solid wood furring at 24 in OC secured to bottom edge of joist. Cavity empty (no sound insulation).	32
Wood—Solid Framing: 2 × 10 solid wood joists at 16 in OC, ⅝ in thick OSB subfloor, ⅝ in thick gypsumboard secured to resilient channels at 24 in OC secured to bottom edge of joist. Cavity empty (no sound insulation).	37
Wood—Insulated Solid Framing: 2 × 10 solid wood joists at 16 in OC, ⅝ in thick OSB subfloor, ⅝ in thick gypsumboard secured to resilient channels at 24 in OC secured to bottom edge of joist. 6 in thickness of sound insulation.	44
Wood—Insulated I-Joist Framing: 2 × 10 solid wood joists at 16 in OC, ⅝ in thick OSB subfloor, ⅝ in thick gypsumboard secured to resilient channels at 24 in OC secured to bottom edge of joist. 6 in thickness of sound insulation.	44
Wood—Insulated Solid Framing: 2 × 10 solid wood joists at 16 in OC, double layer of ⅝ in thick OSB subfloor, ⅝ in thick gypsumboard secured to resilient channels at 24 in OC secured to bottom edge of joist. 6 in thickness of sound insulation.	47
Wood—Insulated Solid Framing: 2 × 10 solid wood joists at 16 in C, double layer of ⅝ in thick OSB subfloor, double layer of ⅝ in thick gypsumboard secured to resilient channels at 24 in OC secured to bottom edge of joist. 6 in thickness of sound insulation.	52
Steel—Insulated Light Gauge Framing: 10 in deep, 16 gauge steel joists at 16 in OC, ⅝ in thick OSB subfloor with 1 in thick concrete topping, ⅝ in thick gypsumboard secured to resilient channels at 24 in OC secured to bottom edge of joist. Cavity empty (no sound insulation).	24
Steel—Insulated Light Gauge Framing: 10 in deep, 16 gauge steel joists at 16 in OC, ⅝ in thick OSB subfloor with 1 in thick concrete topping, ⅝ in thick gypsumboard secured to resilient channels at 24 in OC secured to bottom edge of joist. 6 in thickness of sound insulation.	28

Impact Ratings

An *impact sound* is produced by the collision of two solid objects. Typical sources in a building are footsteps and objects dropped or thrown onto an interior surface (e.g., wall, floor, or ceiling). Impact sound ratings include the following.

Impact Insulation Class

The *impact insulation class (IIC)* is a number rating, based on standardized test performance, for evaluating the effectiveness of assemblies in isolating impact sound transmission. The IIC is the current rating system used to rate the ability of floor–ceiling assemblies to resist the transmission of impact sound. IIC ratings are expressed in decibels and are more closely related to the STC ratings than the INR system for airborne sound transmission introduced next. As a result, the IIC rating system is used today. Recommended IIC requirements for construction assemblies (e.g., walls, floors, and so on) in schools and dwellings are provided in Table 23.10. IIC ratings for selected floor–ceiling assemblies are provided in Table 23.11.

Impact Noise Rating

The *impact noise rating (INR)* is an obsolete rating system that rates the ability of floor–ceiling assemblies to resist the transmission of impact sound. Ratings are based on plus (+) and minus (−) ratings from a standard curve, with a plus rating indicating a better-than-standard performance and a minus rating indicating a less-than-standard performance. The standard (0) is based on average background noises that might exist in typical, moderately quiet, suburban apartments. An INR of +5 indicates that the assembly averages 5 dBA better than the standard. Where louder background noises are found, as in urban areas, a minus INR (to about −5) rating may be used without any detrimental effects. In quieter areas, an INR rating of +5 to +10 would be used because not as much background noise is available to mask the sound. The disadvantages of this rating system are its lack of relationship to the STC ratings and the negative values, which tend to cause confusion in the use of INR.

The IIC rating of any construction assembly will be about 51 dBA higher than the same assembly's INR rating; therefore, an IIC rating of 61 would be comparable to an INR of +10. It should be noted that deviations can occur, and individual test data should be reviewed.

Absorption Ratings

Sound absorption is a property of materials that causes a reduction in the amount of sound reflected from a surface. High absorbing surfaces in a room will reduce sound intensity in that room by not reflecting all of the sound energy striking the room's surfaces. Common absorption ratings are the following.

Absorption Coefficient

As introduced earlier in the chapter, the *absorption coefficient (a)* is a measure of the sound-absorbing ability of a surface. It is defined as the fraction of incident sound absorbed or otherwise not reflected by a surface. Unless otherwise specified, a diffuse sound field is assumed. The values of the sound-absorption coefficient usually range from about 0.01 for marble slate to almost 1.0 for long absorbing wedges often used in anechoic rooms (e.g., soundproof recording or TV studios). Absorption coefficients vary with frequency of sound (see Table 23.5). A designer must select materials that perform best at the sound frequencies anticipated in the space.

Noise Reduction Coefficient

The *noise reduction coefficient (NRC)* is also used to measure the ability of a material to absorb sound rather than reflect it. The material absorbs part of the sound striking the surface of the material and part of it is transmitted through to the other side of the material.

The NRC is established by averaging the absorption coefficients of the middle frequencies (250, 500, 1000, and 2000 Hz), expressed to the nearest multiple of 0.05. It can theoretically range from 0.0 to 1.0, with a perfectly reflective material rated at 1. If a material has an absorption rating of 0.25, the material will absorb 25% of any sound that strikes it and will reflect the remaining 75% of the sound. A difference of 0.10 in the NRC rating is seldom detectable in a completed installation.

NRCs for selected materials and finishes are provided in Table 23.12. The recommended NRC for surfaces in most rooms is between 0.60 and 0.75. Special consideration (NRC over 0.75) should be given to computer and accounting rooms, classrooms, kitchens, and any room where excessive noise is generated.

Ceiling Attenuation Class

The *ceiling attenuation class (CAC)*, previously referred to as the ceiling sound transmission class (CSTC), rates a ceiling's efficiency as a barrier to airborne sound transmission between adjacent closed offices. The CAC is a single-number rating. An acoustical unit with a high CAC may have a low NRC.

Equipment Noise

Noise from rooftop or mechanical room equipment can affect both occupants inside a building and nearby property owners. Typically, a building specification will require that mechanical equipment noise be less than a specific level in occupied areas of the building based upon the following ratings.

Room Criteria

The *room criteria (RC)* are a single-number noise rating system used to diagnose and rate the HVAC noise exposure in a room. RC ratings contain both a numerical value and a letter rating related to sound quality. The RC numerical rating is simply the arithmetic average of the sound pressure level in the 500, 1000, and 2000 Hz octave bands, which is the speech interference level (SIL). These SIL frequencies affect speech communication privacy and impairment.

Studies show that an RC between 35 and 45 will usually provide speech privacy in open-plan offices, while a value below 35 does not. Above RC-45, the sound is likely to interfere with speech communication. The four-letter designations currently in use are rumble (R), hiss (H), acoustically induced vibration (V), and neutral (N). These letters are determined by analyzing the low- and high-frequency sound introduced into the room.

A typically specified value in private offices is RC-35N. An RC-25N to RC-30N is desirable in classrooms and lecture halls. A much higher RC-50 value is acceptable in kitchens, shops, gymnasiums, and warehouses where noise is less of a concern.

This RC rating system is more effective than the NC system (introduced next). It does a better job in rating noise with strong low-frequency content.

Noise Criteria

The *noise criteria (NC)* is a single-number noise rating system used to rate continuous noise in a room from all types of HVAC

TABLE 23.12 NOISE REDUCTION COEFFICIENTS (NRC) FOR SELECTED MATERIALS AND FINISHES. COMPILED FROM VARIOUS INDUSTRY AND GOVERNMENT SOURCES.

Material Surface	Typical Noise Reduction Coefficient (NRC)
Walls–Ceilings	
CMU wall, heavy aggregate, unpainted	0.27
CMU wall, heavy aggregate, spray painted two coats	0.22
CMU wall, heavy aggregate, brush painted two coats, oil or latex	0.10
CMU wall, heavy aggregate, spray painted two coats, cement base	0.02
CMU wall, lightweight aggregate, unpainted	0.45
CMU wall, lightweight aggregate, spray painted two coats, cement base	0.04
CMU wall, light weight aggregate, spray painted two coats	0.36
CMU wall, lightweight aggregate, brush painted two coats, oil or latex	0.16
Brick wall, unpainted	0.05
Brick wall, painted	0.02
Concrete wall, unpainted	0.02
Plaster, acoustical	0.21
Wood paneling	0.06
Plaster, gypsum, or lime, rough finish on lath	0.05
Plaster, gypsum, or lime, smooth finish on tile or brick	0.04
Acoustical ceiling tile	0.55 to 0.85
Floors	
Carpet, heavy, pad, or foam underlay	0.55
Carpet, heavy, on concrete	0.45
Wood	0.03
Concrete or terrazzo	0.02 to 0.04
Glass	0.02 to 0.12
Marble or glazed tile	0.01
Linoleum, asphalt, rubber, or cork tile on concrete	0.03 to 0.08
Fabrics	
Light, 10 oz. per sq. yd. hung straight	0.20
Medium, 14 oz. per sq. yd. draped to half area	0.57
Heavy, 18 oz. per sq. yd. draped to half area	0.63
Furniture	
Bed	0.80
Sofa	0.85
Chairs, wood veneer seat and back	0.50
Wood pews	0.40
Chairs, metal or wood	0.20
Leather covered upholstered chair	0.50
Cloth covered upholstered chair	0.70

TABLE 23.13   NOISE CRITERIA RATING SYSTEM FOR HVAC EQUIPMENT NC 30 IS TYPICAL FOR OFFICE.

Noise Criteria	Communication Environment	Typical Occupancy
Below NC 25	Extremely quiet environment; suppressed speech is quite audible; suitable for acute pickup of all sounds	Broadcasting studios, concert halls, music rooms
NC 30	Very quiet office; suitable for large conferences; telephone use satisfactory	Residences, theaters, libraries, executive offices, directors' rooms
NC 35	Quiet office; satisfactory for conference at a 15 ft (5 m) table; normal voice 10 to 30 ft (3 to 9 m); telephone use satisfactory	Private offices, schools, hotel rooms, courtrooms, churches, hospital rooms
NC 40	Satisfactory for conferences at a 6 to 8 ft (2 to 2.5 m) table; normal voice 6 to 12 ft (2 to 4 m); telephone use satisfactory	General office, labs, dining rooms
NC 45	Satisfactory for conferences at a 4 to 5 ft (1.5 m) table; normal voice 3 to 6 ft (1 to 2 m); raised voice 6 to 12 ft (2 to 4 m); telephone use occasionally difficult	Retail stores, cafeterias, lobby areas, large drafting and engineering offices, reception areas
Above NC 50	Unsatisfactory for conference of more than two or three persons; normal voice 1 to 2 ft (0.3 to 0.6 m); raised voice 3 to 6 ft (1 to 2 m); telephone use slightly difficult	Computer rooms, stenographic pools, print machine rooms, process areas

equipment (e.g., fans, mixing boxes, diffusers). HVAC air terminal products (e.g., diffusers, grilles, and so forth) are commonly specified and reported in NC, rather than RC. NC ratings have been common in specifications for a number of years, with the most common requirements being between NC-30 through 45. Although NC offers a great improvement over previous single-number ratings (e.g., sones, bels, and dBA requirements), it gives little indication of the quality of the sound. (See Table 23.13.)

A typical NC rating for classrooms, private offices, and conference rooms is NC-30. In a gymnasium, where a higher noise level is more tolerable, a typical rating is NC-45.

23.3   ACOUSTICAL DESIGN IN BUILDINGS

Included in this section is a discussion of the various architectural and construction elements that must be considered to effectively control sound. Some of these elements fall outside of the designer's principal areas of concern and responsibility, but the designer should be aware of them and be certain that they are not overlooked.

Generally, the designer would consider the following elements:

- Surrounding environment
- Arrangement and layout of rooms
- Shape of rooms
- Reflecting surfaces
- Isolation of vibration

- Isolation of impact
- Isolation of sound
- Background noise

Surrounding Environment

The surrounding environment should be carefully reviewed before a building site is actually purchased and, once purchased, before planning room arrangements and wall materials. In addition, the zoning should be carefully checked to determine what types of businesses (and accompanying noise sources) might move into the surrounding area. The location for a building should be chosen with the use of the building in mind. For example, a nursing home should not be placed under the approach pattern of an airport.

Once a site has been selected, exterior noise problems can be reduced by orienting the building on the site to reduce direct sound transmission and reflective sound from surrounding buildings (Figure 23.2) and equipment. Another method of sound reduction is to shield the building from major noise sources using other buildings, barrier walls, and natural topography and vegetation (e.g., berms and trees). Any barriers used such as walls or buildings should be as close to the noise source and as high as possible.

In addition, the general layout and design must consider existing or possible future noises from surrounding areas. If there is an existing noise source (e.g., noisy equipment used in an existing building), the general layout of the proposed building should place rooms that require quiet away from the existing noise and from noisy rooms. In the new building, perhaps

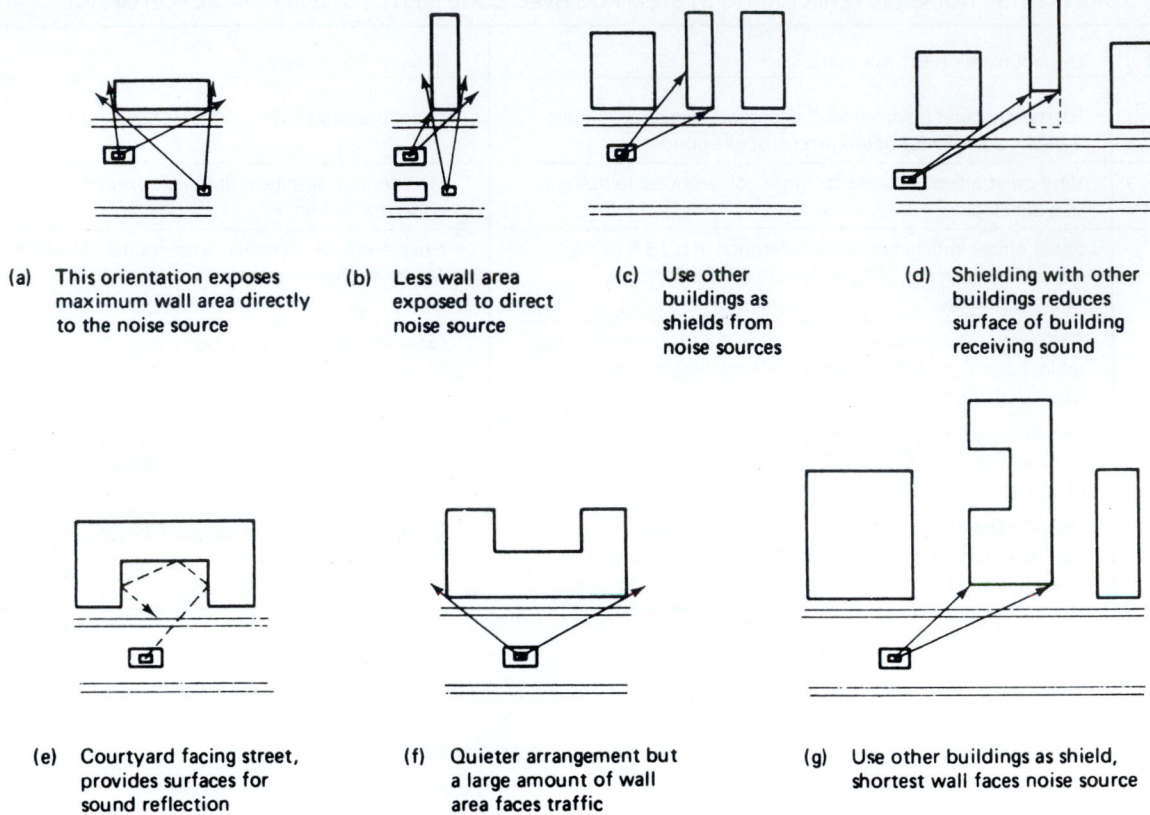

(a) This orientation exposes maximum wall area directly to the noise source

(b) Less wall area exposed to direct noise source

(c) Use other buildings as shields from noise sources

(d) Shielding with other buildings reduces surface of building receiving sound

(e) Courtyard facing street, provides surfaces for sound reflection

(f) Quieter arrangement but a large amount of wall area faces traffic

(g) Use other buildings as shield, shortest wall faces noise source

FIGURE 23.2 Orienting the building on the site is necessary to reduce direct sound transmission and reflective sound from surrounding buildings.

the rooms for mechanical equipment may be placed closer to the noise source as they will not be affected by it. Figure 23.3 shows a suggested schematic solution to such a design problem.

Quite frequently, the site itself and the project being designed necessitate that the quiet areas must be located adjacent to a noise source. Noise can be reduced by placing solid walls on the side of the building facing the noise source and placing the windows on the side facing quieter areas (see Figure 23.4). Still, the designer must be careful that, in placement of win-

dows, doors, and other openings, noise does not enter through the windows by bouncing off an adjacent wall (see Figure 23.5).

Arrangement and Layout of Rooms

Noisy rooms should be separated from quiet rooms by as great a distance as possible. The building layout should be designed so that rooms that are not as susceptible to noise (e.g., closets

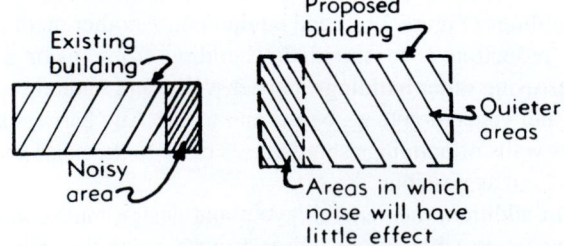

FIGURE 23.3 The general layout of the proposed building should place rooms that require quiet away from the existing noise and from noisy rooms.

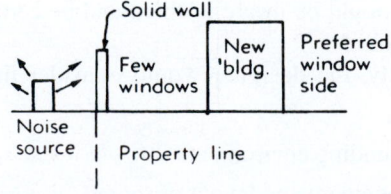

FIGURE 23.4 Placing massive solid walls on the side of the building facing the noise source and placing the windows on the side facing quieter areas can reduce noise.

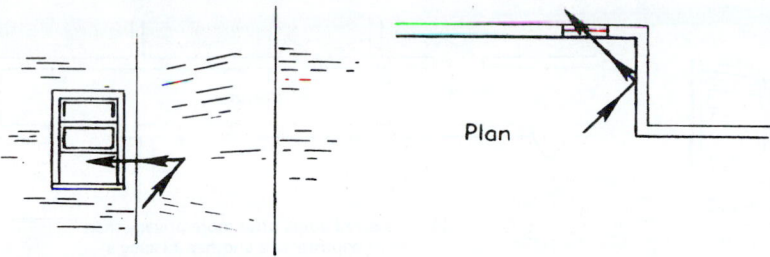

FIGURE 23.5 Windows, doors and other openings, must be placed so noise does not enter through the windows by bouncing off an adjacent wall.

and corridors) act as buffers between those areas that contain noise sources and those that require quiet (see Figure 23.6). The rooms from which the noise will originate should be located wherever noise from exterior sources may be expected. Quiet areas should be located as far as possible from exterior noise sources. For example, a conference room should not be placed next to manufacturing areas or even noisy business machine areas or secretarial pools. Sound travel between rooms must also be controlled by avoiding air paths in the placement and design of doors and windows (see Figure 23.7).

Shape of Rooms

The room proportions used will affect the sound reflection within the space. Room shapes to be avoided are long, narrow rooms or corridors with high ceilings and rooms that are nearly

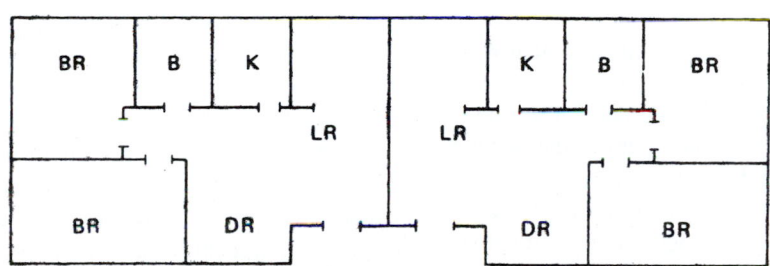

(a) By mirroring (flopping over) the plan an approximately equal noise level on each side of the wall for both apartments is obtained.

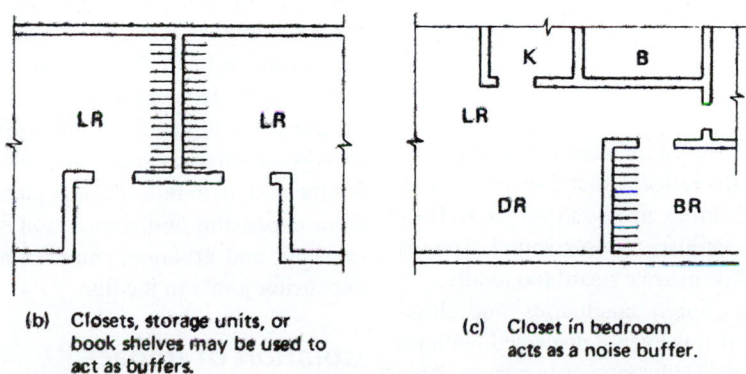

(b) Closets, storage units, or book shelves may be used to act as buffers.

(c) Closet in bedroom acts as a noise buffer.

FIGURE 23.6 The building layout should be designed so that rooms that are not as susceptible to noise (e.g., closets and corridors) act as buffers between those areas that contain noise sources and those that require quiet.

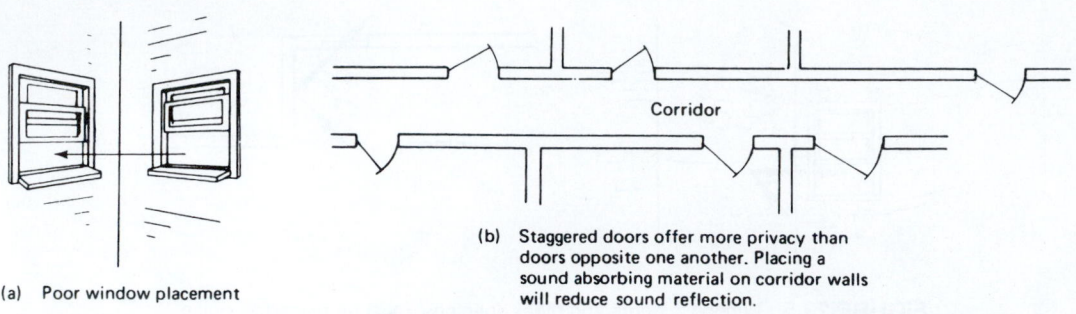

(a)   Poor window placement

(b)   Staggered doors offer more privacy than
doors opposite one another. Placing a
sound absorbing material on corridor walls
will reduce sound reflection.

FIGURE 23.7   Sound travel between rooms must also be controlled by avoiding flanking air paths in the placement and design of doors and windows.

cubical. Each of these room proportions will cause excessive reverberation (sound reflection or echoing).

Ceilings that are domed or vaulted (or any concave surfaces) tend to focus the sound, causing it to be distorted. Large auditoriums with low ceilings also create a situation in which it is difficult for some of the audience to hear. Concave surfaces and large flat surfaces may be broken with splayed areas in order to diffuse the reflection of sound and to direct the sound as desired.

Absorbing Surfaces

Sound is absorbed when it encounters a material that will convert some or all of it into heat, or that allows it to pass through not to return. All walls, floors, ceilings, and furnishings have sound-absorbing characteristics that control the amount of sound they reflect and absorb. If the surfaces in a room tend to be highly reflective, the room will seem loud, and it may have a slight echo. An example of a room with highly reflective surfaces is one with plaster ceilings and walls and a terrazzo floor.

The highly reflective qualities of a room can be controlled by using materials with a high NRC (e.g., carpeting, heavy drapes, and upholstered furniture). Acoustical tile is commonly used when a surface material with high absorption characteristics is desired. Acoustical tile is most frequently used on ceilings, but is also an economical wall covering.

Focusing Surfaces

Reflected sound waves can be focused intentionally or inadvertently. A flat surface (i.e., a wall) reflects sound better than an irregular surface, which tends to break up sound waves. Reflection from large concave surfaces also focuses sound. Sounds near the focus of a curved surface may be heard too loudly.

In auditoriums, theatres, church sanctuaries, and classrooms, sound is more pleasing if it is evenly dispersed, with no prominent echoes, no significant dead spots, or live spots. This even dispersion is usually achieved by avoiding any focusing surfaces and avoiding large flat areas that reflect sound into the listing area. Rough walls tend to diffuse sound, reflecting it in a variety of directions. This allows a spectator to perceive sounds from every part of the room, making it seem lively and full. For this reason, auditorium and concert hall designers prefer construction materials that are rough rather than smooth. Sometimes it is desirable to add some antifocusing surfaces or devices called *acoustical baffles* and *banners,* which are used to absorb reflected noise (reverberated sound) bouncing off reflective walls, floors, or ceilings.

Isolation of Vibration

Because the structural elements of the building readily transmit sound throughout the building, noisy or vibrating equipment should not be attached to the structure. Equipment within the building, such as heating units, pumps, and motors, should be mounted on commercially available resilient machinery mounts and bases. Soft, resilient subfloor materials are often used under the wearing surface of the floors and under the equipment to reduce the transfer of vibration from the equipment to the structure.

Equipment located in the basement of a building may be placed on resilient mounts and a concrete pad that is isolated, by expansion joints, from the rest of the floor. This is an example of discontinuous construction, which is very effective when it can be used.

Vibration noises from equipment are also transmitted through pipes, ducts, or other similar conductors. Providing a break where they leave the equipment may reduce the transmission of noise from the equipment. Ducts and certain types of pipes may be isolated with flexible couplings and resilient connections (see Figure 23.8). Water pipes should be provided with expansion valves, expansion tanks, and air chambers throughout the system to reduce water hammering and knocking. Noises from expansion and contraction of the water pipes (usually a creaking and groaning sound) can be controlled by installing expansion joints in the line.

Isolation of Impact

Impact sound is that sound caused when one object strikes another. Typical examples of impact sound are footsteps, falling objects (such as shoes, toys, and machine parts), and hammering. When a structure is rigid and continuous, the sound easily

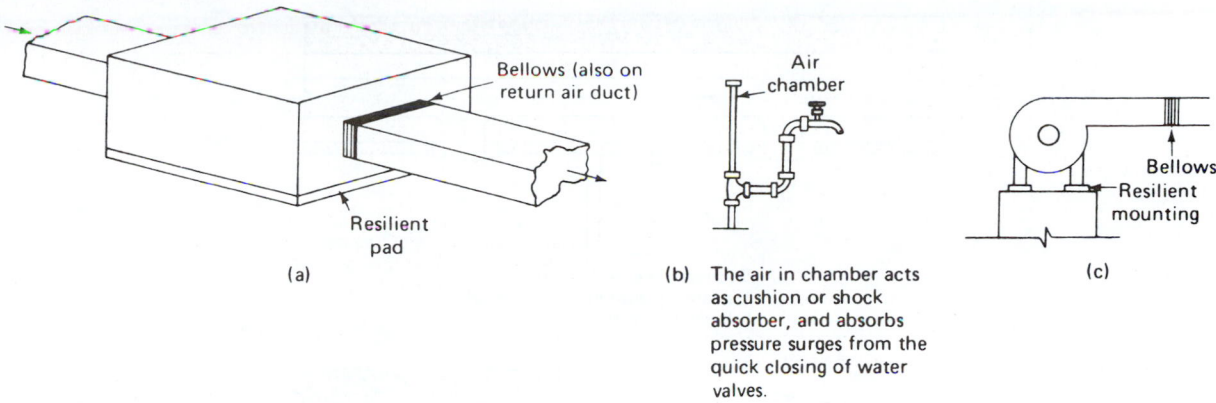

FIGURE 23.8 Vibration noises from equipment can be prevented.

travels through it. Impact noise is controlled most effectively by using absorptive materials (such as carpeting), by isolating the noise sources (using discontinuous construction (Figure 23.9), and by reducing flanking (Figure 23.10). The introduction of a masking or background sound should also be considered, but its effectiveness is limited when the sounds to be covered over are loud.

Isolation may be accomplished by separating the surface that will receive the impact from the structure supporting it. Floors may be isolated from the structure (Figure 23.11) by using resilient subflooring and underlayment materials. Walls

may be isolated from the structure by mounting the finish materials on resilient channels. The ceilings and walls in the surrounding spaces may also be mounted on resilient channels. Carpeting with a thick pad below it is one of the best absorptive materials in terms of reducing impact noises.

Isolation of Sound

The amount of airborne sound transmitted between rooms will depend on the materials and methods used in the construction. Sound transmission through walls depends on the mass (or unit

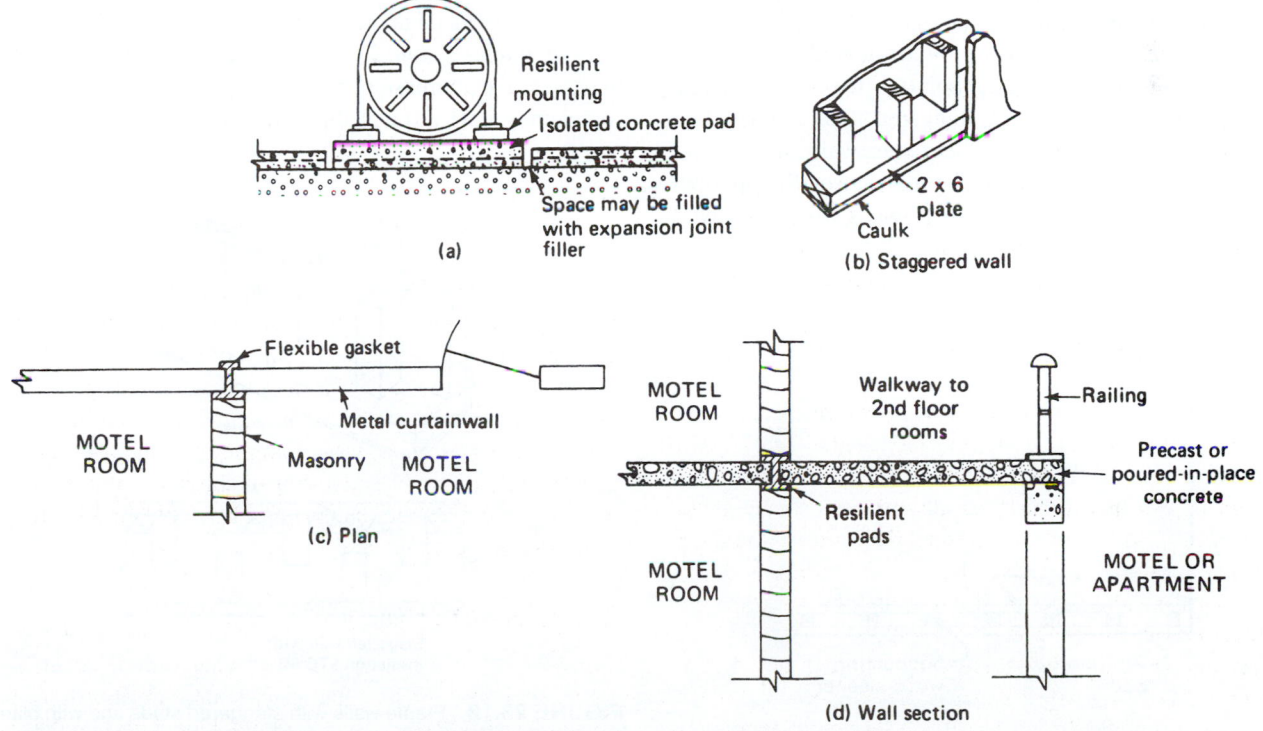

FIGURE 23.9 A variety of methods can be used to isolate sound transmission through solid structural assemblies.

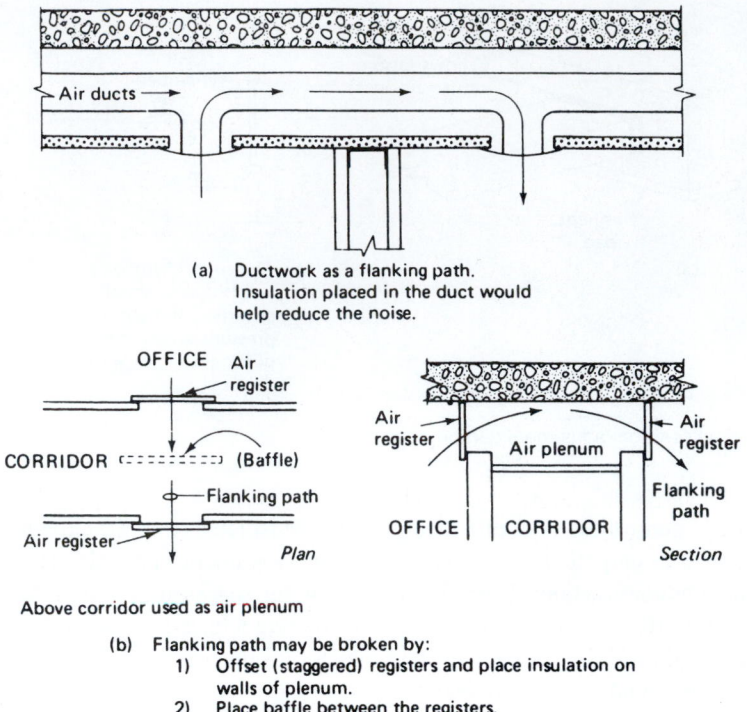

Air ducts
(a) Ductwork as a flanking path.
Insulation placed in the duct would help reduce the noise.
OFFICE
Air register
CORRIDOR
(Baffle)
Flanking path
Air register
Plan
Air register
Air plenum
Air register
Flanking path
OFFICE
CORRIDOR
Section
Above corridor used as air plenum
(b) Flanking path may be broken by:
1) Offset (staggered) registers and place insulation on walls of plenum.
2) Place baffle between the registers.

FIGURE 23.10 Eliminating flanking paths can reduce airborne sound transmitted between rooms.

weight) of the walls and on their inelasticity. Lead, as a dense material, works well in isolating sound. Massive, thick walls provide excellent sound barriers, but economically there is a point of diminishing returns. As the mass is doubled, the transmission of sound will be reduced by about 6 dBA. Therefore, mass alone is not an economical solution for isolating sound.

Building frame walls with staggered studs and with blanket sound insulation between the outer wall surfaces (see Figure 23.12) is a good method of sound control. Slightly more effective is the building of two separate walls with no structural connections between them (see Figure 23.13). Another control method involves the use of resilient channels to mount the wall-covering surface, thus dissipating the sound energy in the channels. These methods also have their practical limitations. For a specific mass, as the air space doubles, the transmission of sound is reduced about 5 dBA.

A *flanking sound path* is a sound transmission path that bypasses a transmission barrier. It is a route through which sound may easily travel through a wall or ceiling assembly.

Room
Floor
Resilient pad
Supporting wood sleepers

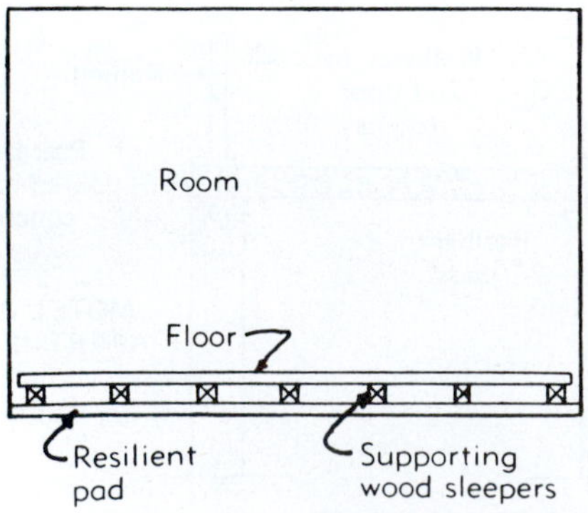

FIGURE 23.11 Floors can be isolated from the structure by using resilient subflooring and underlayment materials.

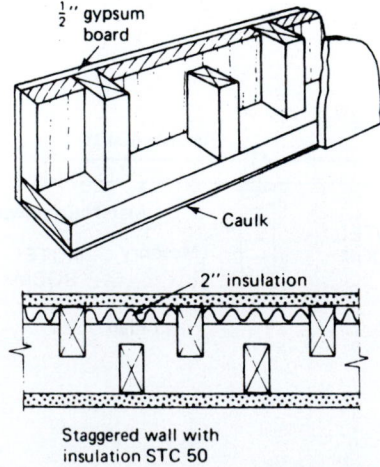

½" gypsum board
Caulk
2" insulation
Staggered wall with insulation STC 50

FIGURE 23.12 Frame walls with staggered studs and with blanket sound insulation between the outer wall surfaces is a good method of sound control.

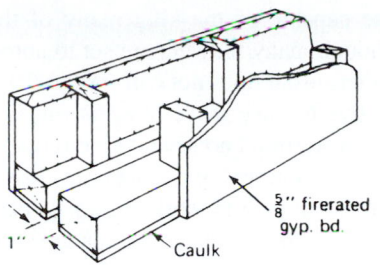

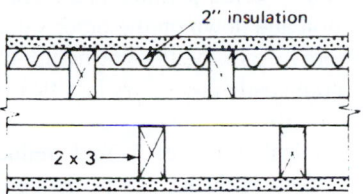

Double wall
with insulation
STC 51

FIGURE 23.13 Frame walls having two separate walls with no structural connections between them and with blanket sound insulation is an excellent method of sound control.

Openings in doors, windows, and electrical outlets create good flanking sound paths. Rigidly connected electrical conduit, pipes, and air ducts are additional examples of flanking paths. Airborne sound passing through a flanking path can significantly diminish an assembly's sound isolation capability. For example, a ¼-in perimeter crack surrounding a 96 ft^2 partition system represents an approximate 1-ft^2 opening. In terms of sound rating, this untreated perimeter crack will reduce the overall performance from 53 STC to 29 STC. Examples of construction techniques are shown in Photos 23.1 through 23.3.

The weak point of sound isolation in the building envelope is frequently the windows and doors. Double glazing will afford noticeably better protection than single glazing. In areas of high external noise it might be preferable to have double windows with a large air gap (1 to 4 in [25 to 100 mm]) and acoustic absorbent material on the perimeter reveal around that gap. Doors also typically represent a serious weakness in a good wall. Insulation ratings for solid-core doors vary from about STC-15 for a swinging door with clearance all around to STC-27 for the same door completely sealed by gaskets or weather stripping all around. A more typical installation, with stops on three sides but no gaskets, is about STC-20.

In commercial buildings, a common sound isolation problem exists where wall partitions terminate at the suspended

PHOTO 23.2 Slotted acoustical concrete masonry units dampen sound, thus improving the noise reduction attributes of an interior space. Acoustical units are often used in schools, industrial plants, churches, and other facilities to improve internal acoustics. (Used with permission of ABC)

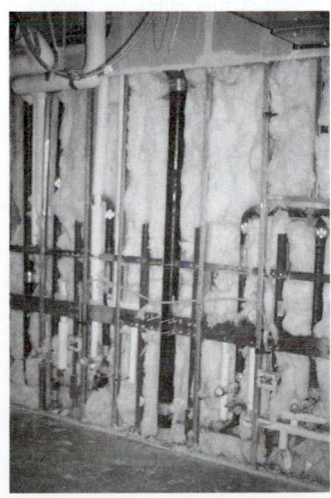

PHOTO 23.1 Acoustical insulation in a plumbing wall. (Used with permission of ABC)

PHOTO 23.3 Sealed openings to prevent flanking of sound from a bathroom above. (Used with permission of ABC)

ceiling level. With such a system, sound can be transmitted up through the suspended ceiling into the plenum space, through the plenum above the partition and down on the other side. Extending the wall to the bottom side of the floor structure eliminates this problem, but is more costly and offers less flexibility with future renovations.

The practical, economical limit of sound reduction through walls is about 50 dBA. With the use of background masking noises, this reduction is adequate for most conditions. At the same time, great care should be taken to eliminate as much sound transmission as possible through flanking paths, air ducts, soffits, and items placed back to back, such as medicine cabinets, electrical outlets, and lighting fixtures. Sound transmission through doors and door arrangements and through windows and window arrangements must also be carefully considered in achieving noise control.

Sound Masking

Sound masking is the addition of natural or artificial sound of a different frequency into an environment to mask or cover-up noise. Sound masking reduces or eliminates awareness of existing noise in a specific area. It can make a work environment more comfortable and more productive by creating speech privacy and making workers less distracted.

An effective method of sound masking is to introduce background sound within the room. *White noise*, a type of background noise that is produced by combining sounds of all different frequencies together into an unintelligible sound, may be introduced into the space. Because white noise contains all frequencies, it is frequently used to mask other sounds. Figure 23.14 illustrates a typical example of the effect of background sound in masking undesirable noise.

Masking sounds include piped-in music, radios, or even the hum-drum white noise made by heating and air-conditioning systems in the building. Some manufacturers provide white noise sound machines and white noise generators. The blowers on most heating and cooking units can provide a background noise capable of masking many of the noises in a building. Although many blowers are set to automatically shut off when the thermostat does not call for heat or cooling, it is sometimes effective to have these blowers run all of the time to provide both background noise and air circulation.

Many stores and restaurants use piped-in music to mask the noise of people working, traffic passing, and multiple conversations. Background sounds are often introduced into very quiet "dead" rooms (e.g., sleeping areas in nursing or rest homes or in residences on busy streets) where any noises (i.e., coughing) would be disturbing. It also provides some privacy for speaking and telephone communications.

Acceptable levels of background noise range from 25 to 35 dBA in bedrooms; 30 to 40 dBA in living rooms, offices, and conference rooms; and 35 to 45 dBA in large offices, reception areas, and secretarial areas. When acceptable levels of background sound is combined with effective transmission ratings for walls and floors, it is possible to tolerate relatively high surrounding noise levels.

Active Noise Control

Active noise control (ANC), also known as *active noise reduction (ANR)* or *antinoise,* is a method of reducing unwanted sound through sound energy cancellation. In its simplest form, a one-dimensional ANC system uses a microphone connected to electronic circuitry that recognizes the unwanted sound and a speaker that generates an antinoise sound wave resulting in destructive interference, which cancels out the noise. A three-dimensional ANC system requires many microphones and speakers. Each of the speakers tends to interfere with nearby speakers, reducing the system's overall performance, particularly when there are multiple listeners. Another type of noise-canceling technology involves treating glass with small squares of piezoelectric material that measures and countervibrates the glass to cancel out sound transmission entirely.

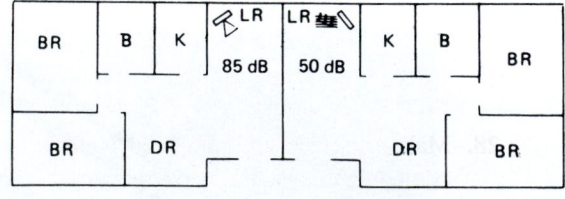

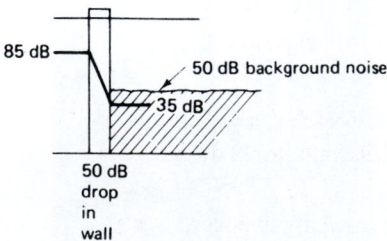

Sound inaudible — the level of the sound
must be brought at least 10 dB below
the background noise to make it inaudible.

FIGURE 23.14 An effective method of reducing the awareness of unwanted sound is to use background noise within the room; a measure called sound masking.

STUDY QUESTIONS

23-1. What is sound?

23-2. What units are used to identify the intensity of sound?

23-3. What is the difference between sound and noise?

23-4. What is sound transmission?

23-5. What is the difference between airborne sound and structure-borne sound?

23-6. If the sound level changes from 42 to 45 dBA, how noticeable would it be to a person in the same room? What if the sound level changes from 42 to 62 dBA?

23-7. What is the STC?

23-8. What is the IIC?

23-9. What is the NRC?

23-10. How can the surrounding environment affect sound control?

23-11. What methods might be used to control and isolate vibration noises in buildings?

23-12. How can building orientation affect the sound level inside the building?

23-13. What is a flanking sound path, and how may this problem be avoided?

23-14. Define background noise.

23-15. How is background noise used in controlling objectionable sound?

23-16. How can background noises be introduced into a space?

23-17. Why should the surrounding environment be checked before the building is designed?

23-18. How can the surrounding noise environment affect the design of a building?

Design Exercises

23-19. Over the workday, a worker is exposed to a noise level of 90 dBA for 4 hr, 92 dBA for 2.5 hr, and 100 dBA for 1.5 hr.

   a. Find the daily exposure.

   b. Does this exposure exceed the OSHA permissible exposure limit (PEL)?

23-20. Over the workday, a worker is exposed to a noise level of 90 dBA for 4 hr, 92 dBA for 3.5 hr, and 100 dBA for 0.5 hr.

   a. Find the daily exposure.

   b. Does this exposure exceed the OSHA permissible exposure limit (PEL)?

23-21. A 64 ft by 220 ft church sanctuary with 22 ft high ceilings has walls with an absorption coefficient (1000 Hz) of 0.09, ceiling with an absorption coefficient of 0.24, and seating (unoccupied) with an absorption coefficient of 0.35.

   a. Approximate the reverberation time.

   b. Approximate the reverberation time if the church sanctuary is occupied (floor absorption coefficient of 0.60).

23-22. A 32 ft by 84 ft museum hall with 20 ft high ceilings has walls with an absorption coefficient (1000 Hz) of 0.09, ceiling with an absorption coefficient of 0.20, and seating (unoccupied) with an absorption coefficient of 0.35.

   a. Approximate the reverberation time.

   b. Approximate the reverberation time if the hall is occupied (floor absorption coefficient of 0.45).

23-23. A 16 m by 48 m concert hall with 8 m high ceilings has walls with an absorption coefficient (1000 Hz) of 0.09, ceiling with an absorption coefficient of 0.25, and seating (unoccupied) with an absorption coefficient of 0.35.

   a. Approximate the reverberation time.

   b. Approximate the reverberation time if the hall is occupied (floor absorption coefficient of 0.55).

23-24. A wall separates a space with noise levels of about 70 dBA and another room with sound levels of about 30 dBA. Recommend an STC for the wall.

23-25. A wall separates a space with noise levels of about 80 dBA and another room with sound levels of about 40 dBA. Recommend an STC for the wall.

23-26. A wall separates a space with noise levels of about 85 dBA and another room with sound levels of about 45 dBA. Recommend an STC for the wall.

23-27. For the apartment building in Appendix B, identify U.S. Department of Housing (HUD) requirements for sound transmission class (STC). Use Table 23.9. Identify construction assemblies that meet these requirements.

23-28. Make a visit to a commercial building. Identify chief components, systems, and strategies used for acoustical control.

23-29. Make a visit to a theatre. Identify chief components, systems, and strategies used for acoustical control.

CHAPTER TWENTY-FOUR

BUILDING CONVEYING
SYSTEMS

Building conveying systems mechanically move occupants and goods. In most buildings, these systems include passenger and freight elevators, dumbwaiters, escalators, moving ramps and walkways, and lifts for people and wheelchairs. Business and industrial operations may require specially designed material-handling equipment (i.e., conveyors, chutes, and pneumatic tube systems), hoists, cranes, and scaffolding. Facilities on large sites may have monorails and other types of people movers.

Manually operated elevators were first used for lifting freight in warehouses and manufacturing plants as early as the 1600s. They began as simple rope or chain hoists that moved an open platform. In 1852, Elisha Otis introduced the safety elevator, which prevented the fall of the elevator cab if the cable broke. Otis's safety device consisted of a knurled roller located below the elevator platform and a governor device that monitored descending speed. The safety device locked when the elevator descended at a higher than normal speed. It was this safety feature that made the elevator a safe conveying system for building occupants, which then made skyscrapers achievable.

Building conveying systems consume a significant fraction of the total energy used in tall buildings. In low-rise and midrise buildings, elevator energy use is lower, but still substantial. Estimates are that elevators in North American buildings account for about 5% of the building's overall energy use (American Academy for an Energy-Efficient Economy). They also contribute significantly to the cost of the building operation. As a result, building conveying systems are an important consideration during initial design. Occupant-conveying systems are really a subsystem of the occupant traffic system, and must be considered along with corridors, stairs, and exits as part of the overall occupant traffic system.

In the United States, the manufacture of elevators, escalators, and other vertical transportation devices is governed by the *ASME A17.1—Safety Code for Elevators and Escalators.* Most states and local jurisdictions have adapted at least a portion of this code as a referenced standard for elevators and escalators.

24.1 ELEVATORS

Elevator Technologies

An *elevator* is a conveying device used to move people or freight vertically, usually between floors of a building. Photo 24.1 shows an elevator car with hoistway doors open. The two basic types of elevator technologies are hydraulic and traction. Type and design speeds for elevators are compared in Table 24.1.

Hydraulic Elevators

Hydraulic elevators use a fluid-driven hydraulic jack to lift the elevator car. Basic components of a hydraulic elevator system are shown in Figure 24.1. It consists of a *hydraulic jack* (cylinder and plunger); a *pump*, powered by an electric motor, that increases the pressure in the hydraulic fluid; a *control valve* between the cylinder and reservoir controls the pressure in the jack; and a fluid reservoir (tank). It operates when the pump draws oil from the reservoir, pressurizes it, pushing the oil through the oil line to the jack, and driving the elevator car upward. A release of fluid through the control valve and back to the reservoir decreases oil pressure, which allows the plunger and connected elevator car to move downward. Photo 24.2 shows some of the components of a hydraulic elevator.

On a *conventional (holed) hydraulic elevator*, an in-ground hydraulic jack lifts the elevator car. A long plunger requires a deep hole below the bottom landing. The hole is usually drilled into the ground and cased with a plastic or metal casing before the building is erected. See Figures 24.1 and 24.2a. The hole must be about as deep as the desired vertical movement of the elevator car. A *telescopic hydraulic elevator* has a telescoping plunger consisting of concentric tubes that slide within one another, allowing a shallow hole below the lowest floor. Conventional (holed) hydraulic elevators are the most balanced type of hydraulic elevator configuration because the lifting point on the bottom of the elevator car is centered.

PHOTO 24.1 First-floor entrances to two elevator cars (doors closed). Note the hall lanterns between the doors (above sign). The hall station is below the sign.

TABLE 24.1   TYPE AND DESIGN SPEEDS FOR ELEVATORS.

Type	Residential	Commercial	Speed	ft/min	m/min
Hydraulic	6 floors or less	3 floors or less 6 floors or less	Low	100 150	30 45
Traction—geared	18 floors or less	5 floors or less 9 floors or less 18 floors or less	Moderate	200 350 500	60 105 135
Traction—gearless	Over 18 floors	15 floors or less 15 to 25 floors above 25 floors	High	500 700 1000	150 210 305
	-	above 75 stories	Very high	3500	1080

Holeless hydraulic elevators have one or two jacks situated beside the rails that lift the platform. See Figure 24.2c. Because they do not require holes to be dug for the hydraulic jack(s), they are referred to as "holeless." The dual or twin jack configuration can have two (front and rear) entrances, while the single jack configuration can only have one (front) entrance. *Roped hydraulic elevators* use a combination of both ropes and hydraulic power to raise and lower cars. See Figure 24.2d. They typically consist of a cantilevered car that is lifted by ropes that pass over a sheave (pulley) fastened to the top of a hydraulic plunger. As the plunger rises, so does the elevator car. Single rope configuration cannot have rear entrances.

Traction Elevators

Traction elevators have a drive machine with an electric motor and pulley-like (grooved) *drive sheave* that holds cables that move the elevator car up or down. See Figures 24.1 and 24.2b. Steel cables, called *hoisting ropes,* support the elevator and counterweight during normal operation. There are typically

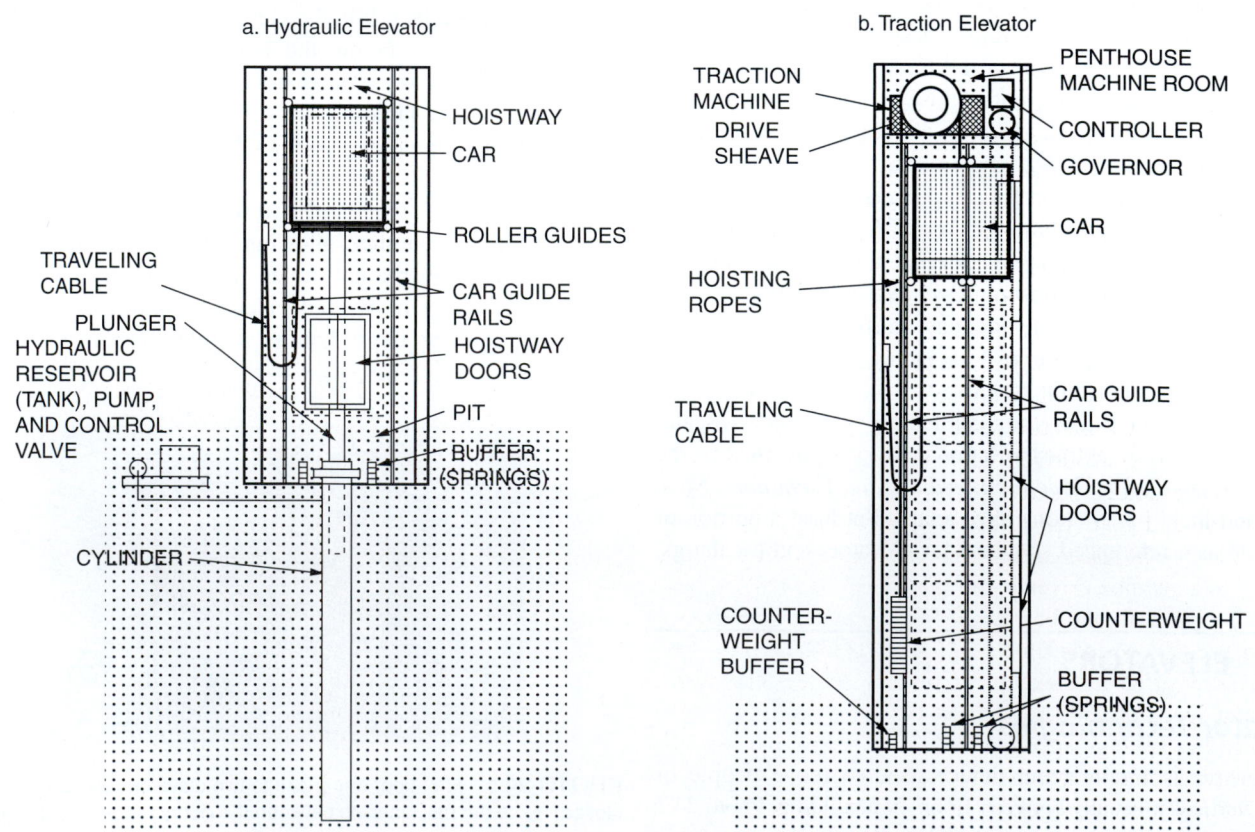

FIGURE 24.1 Basic types of elevators.

PHOTO 24.2 Hydraulic elevators use a fluid-driven hydraulic jack to lift the elevator car. Shown is the bottom side of a conventional (holed) hydraulic elevator. The elevator car is made of the cab, which carries the passengers and the platform, which consists of structural framework. The hydraulic plunger connects to the center beam of the car platform. Behind and alongside the cylinder (to left) is the traveling cable. Guide rails are evident on the left and right side of the hoistway walls. (Used with permission of ABC)

three to eight cables for each elevator. Most traction elevators generally use wire ropes that are ¼ to 1¼ inches (6 to 32 mm) in diameter and are composed of multistrand soft steel wire wound around a hemp or polymeric core. Traditionally, the 8 × 19 wire rope (eight strands with 19 wires per strand) pattern was used, but this is being replaced with modern patterns. Wire ropes must match the drive sheave perfectly to ensure that they have a long life. The *counterweight* is a set of steel or iron plates fastened to one end of the hoisting rope that counterbalances the car. It is weighted to be equal to the car's dead weight plus 40 to 50% of car load capacity. The weight of the car and counterweight presses ropes into grooves on a drive sheave. The friction between the hoisting ropes and the drive sheave is used to move the elevator car with the cable.

A *traction machine* is an electric machine in which the friction between the hoist ropes and the drive sheave is used to move the elevator car with the cable. Traction machines are driven by AC or DC electric motors. See Photo 24.3. *Geared traction machines* are driven by low-speed (low rpm), high-torque electric motors. The electric motor drives a reduction unit of the worm and gear type to mechanically control movement of elevator cars by rolling steel hoist ropes over a drive sheave attached to a gearbox driven by a high-speed motor. These machines are generally the best option for basement or overhead traction used for speeds up to 500 ft/min (152 m/min). One manufacturer offers an elevator for skyscrapers that moves at a maximum speed of 3540 ft/min (1080 m/min). *Gearless traction machines* are high-speed, direct-drive electric motors. The drive sheave is directly attached to the end of motor.

The *machine room-less (MRL) elevator* uses a gearless traction machine that is mounted within the hoistway itself and is attached directly to the top of the car. It employs a smaller

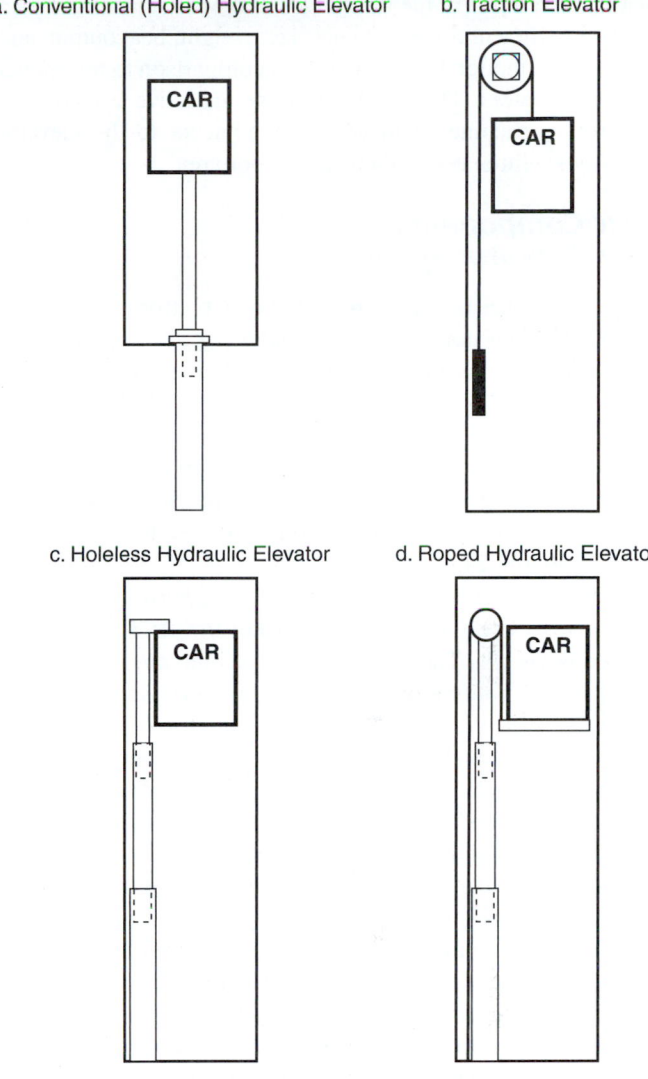

FIGURE 24.2 Variations of basic types of elevators.

PHOTO 24.3 Traction elevators have a drive machine with an electric motor and pulley-like (grooved) drive sheave that holds cables that move the elevator car up or down. Shown is a traction machine. (Used with permission of ABC)

sheave. Developed in the mid-1990s, the MRL machine uses a small and efficient drive that cuts size, weight, heat output, and energy consumption by about 50% in comparison to traditional traction elevators. These machines are attractive because the need for a machine room above or adjacent to the elevator hoistway is eliminated, which saves floor area.

Basic Components of an Elevator System

An elevator consists of a car mounted on a platform that is connected to the elevator drive. The elevator *car,* also called a cage, is the load-carrying unit, including the frame, enclosure, and car door. The *platform* is a flat, relatively horizontal framework to which a car is mounted and on which passengers stand or the load is placed. The *cab* is a decorative room in which occupants ride in a passenger elevator. These are shown in Photo 24.2. The basic components of the two types of elevator systems are shown in Figure 24.1.

The elevator car moves vertically in a *hoistway,* which is the shaft-like space in which the elevator travels. It is enclosed by fireproof walls. The hoistway includes the *pit,* which is the space at the bottom of the hoistway under the car, or a *penthouse,* which is the space between the top of the elevator hoistway and the underside of the roof. The elevator car stops at the *landing,* which is the portion of a floor, balcony, or platform used to receive and discharge passengers or freight. *Hoistway doors* at the landings provide an opening in the hoistway to allow passengers or freight to access the elevator car at a landing. These doors remain closed when the elevator car is not present at the landing, which ensures occupant safety and maintains the fire enclosure. A *blind hoistway* has no hoistway door openings or landings on the lower part of a hoistway. It allows the elevator to service only the upper floors of the hoistway. Most building codes require an access door every three floors for rescue purposes. A hoistway and components within a hoistway are shown in Photos 24.4 and 24.5.

PHOTO 24.4   An open elevator hoistway serving two stories. (Used with permission of ABC)

PHOTO 24.5   Hoistway doors as viewed from inside of hoistway. Guide rails are evident on the left and right side of the hoistway walls. (Used with permission of ABC)

A *drive unit* is an assembly of an electric motor, brake, and power transmission or hydraulic system that supplies the power for movement of a car. The elevator car slides on steel T-, round, or formed sections installed vertically in the hoistway to guide and direct the car and the counterweight. A *buffer* is an energy absorber located at the bottom of the hoistway to soften the force with which a car runs into the pit during an emergency. Buffers can be large springs or an oil dampener-spring combination.

Car operating station is a panel mounted in the car that contains: the car operating controls, call register buttons, door open and close, alarm emergency stop, key switches, and other buttons that are required for operation. The *hall station* is a control panel located outside the elevator doorway in the corridor that houses the call button. Nearby, is the hall lantern, a corridor mounted signal light indicating than an elevator car is approaching that landing and the direction in which the car is to travel. A *traveling cable* is a set of electric conductors that provide an electrical connection between the car and outlet in the hoistway or machine room.

The modern *elevator controller* is a microprocessor-based system that directs starting, acceleration, deceleration, and stopping of the elevator cab. It is designed to minimize average passenger wait time. Multiple elevators have an integrated control system. The controller automatically compares the location of all the cars in the elevator bank and sends the car to the nearest one. It is typically programmed to respond differently at different times of the day. For example, in office buildings during morning hours, all unassigned cars go to the ground floor, while during late afternoon hours, cars go to upper floors. Modern elevators have smart traffic pattern recognition capabilities that allow them to "learn" and respond to daily and weekly elevator use patterns. See Photo 24.6.

Elevators have several safety devices. A *brake* is a spring-loaded clamping mechanism that works to prevent car movement when it is at rest or when no power is supplied to the hoist motor. A *governor* is a speed-monitoring device on traction

PHOTO 24.6 An elevator controller. Note that Floor 13 is "missing."

elevators that triggers the safety when the elevator moves too quickly. *Door interlock* mechanisms prevent the operation of the elevator unless the hoistway door is locked in the closed position. The door interlock also prevents the opening of the hoistway doors from the landing side unless the elevator is in the landing zone and either stopped or coming to a stop.

Emergency power operation allows cars to return to a predetermined landing in the event of a power failure. These systems typically operate on generator power. All cars have an *emergency exit,* which is a removable panel. Removable only from the top of the car, the emergency exit permits passengers to be evacuated from the elevator during an emergency. The *emergency stop switch* is a hand-operated switch in the car push button station that, when thrown to the off position, stops the elevator and prohibits its running. All elevators are required to have communication connection to an outside 24 hr emergency service, automatic recall capability in a fire emergency, and special access for firefighters' use in a fire.

Elevators can be placed in the fire/emergency service mode by operating a keyed switch. This operation mode removes all cars from normal use, sends them to a designated landing, and permits special operation for firefighters or emergency personnel.

Classifications of Elevators

Elevators fall into basic classifications based on use.

Passenger Elevators

Passenger elevators are designed to carry people and small packages. They typically have capacities from 1500 to 5000 lb (680 to 2300 kg), in 500 lb (230 kg) increments. Passenger elevators are operated by the passenger and have attractive interior finishes. They can also be used to move freight. Standard dimension and reactions of passenger elevators are provided in Table 24.2. Recommended capacity and design strategies for

passenger elevators in various occupancies are shown in Tables 24.3, 24.4, and 24.5.

Freight Elevators

Freight elevators are used to carry material, goods, equipment, and vehicles, rather than people. In many freight elevator installations, passengers often accompany the freight, so they are governed by the same safety requirements. Freight elevators are typically capable of carrying heavier loads than a passenger elevator, generally from 5000 to 10 000 lb (2300 to 4500 kg), but can be up to 13 tons/26 000 lb (11 700 kg) weight capacity. Elevators designed to only carry freight (not passengers) are required to post a written notice in the car that the use by passengers is prohibited. Freight elevators may have manually operated doors, and often have sturdy interior finishes to prevent damage during loading and unloading.

Dumbwaiters

A *dumbwaiter* is a small freight elevator used to transport lightweight freight such as food, laundry, books, records, and other small items. Passengers are not permitted on dumbwaiters. Dumbwaiters are generally driven by a small electric motor with a counterweight or may be hand operated using a roped pulley. They are generally limited to a capacity of about 750 lb (340 kg).

Manlifts

A *manlift* is an elevator installed in a variety of structures and locations to provide vertical transportation of authorized personnel and their tools and equipment only. These elevators are typically installed in structures such as grain elevators, radio antennas and bridge towers, underground facilities, dams, power plants, and similar structures. Typically available in 300 lb (140 kg), 500 lb (230 kg), 650 lb (300 kg) and 1000 lb (467 kg) capacities.

Elevator Design Criteria

Because of accessibility regulations, passenger elevators are often a building code requirement in new buildings with multiple floors. Model building codes require compliance with the American Society of Mechanical Engineers (ASME) standards for the installation, maintenance, and inspection of elevators.

One of the first elements to identify is the extent to which vertical transport will be provided within the building. High-rise buildings require a set of elevators. Selecting the technology to be used in new elevator installations depends on many parameters. Traction elevator motor size and power consumption is significantly lower than hydraulic elevators, but elevator and building costs are higher. Traction elevators are much quicker than hydraulic elevators. On the other hand, hydraulic elevators have lower installed costs. Installing hydraulic jacks becomes impractical for tall hoistways because of the height limitations of

TABLE 24.2   STANDARD DIMENSIONS AND REACTIONS OF PASSENGER ELEVATORS.

Capacity			Speed (rpm)		Entrance Opening				Car	
					Width		Height		Internal Dimensions	External Dimensions
Persons	lb	kg	ft/min	m/min	in	mm	in	mm	in (mm)	in (mm)
8	1200	550	200 300 350	60 90 105	32	800	84	2100	56 × 41 (1400 × 1030)	57 × 48 (1445 × 1200)
10	1500	680	200 300 350	60 90 105	32	800	84	2100	56 × 50 (1400 × 1250)	57 × 56 (1445 × 1420)
11	1650	750	200 300 350	60 90 105	32	800	84	2100	56 × 54 (1400 × 1350)	58 × 61 (1445 × 1520)
13	1950	900	200 300 350 400 500	60 90 105 120 150	36	900	84	2100	64 × 54 (1600 × 1350)	66 × 61 (1645 × 1520)
15	2250	1000	200 300 350 400 500 600	60 90 105 120 150 180	36	900	84	2100	64 × 62 (1600 × 1550)	66 × 69 (1645 × 1720)
17	2550	1150	200 300 350	60 90 105	40	1000	84	2100	72 × 60 (1800 × 1500)	74 × 67 (1850 × 1675)
			400 500 600	120 150 180	44	1100	84	2100		
20	3000	1350	200 300 350	60 90 105	40	1000	84	2100	72 × 69 (1800 × 1730)	74 × 76 (1850 × 1905)
			400 500 600	120 150 180	44	1100	84	2100		
24	3600	1650	200 300 350	60 90 105	40	1000	84	2100	80 × 72 (2000 × 1800)	82 × 79 (2050 × 1975)
			400 500 600	120 150 180	44	1100	84	2100		

the plunger, so these elevators become more costly beyond about 60 ft (20 m). Thus, hydraulic elevators are quite common in low- and medium-rise buildings (2 to 5 stories). For high-rise buildings (>75 ft/25 m) traction elevators must be used.

Design criteria differ depending on the building type (i.e., hotel, apartment, office). For office buildings, one elevator group can generally serve all floors in buildings up to 15 to 20 floors, depending on the building population. When there are more than 20 floors, single grouping is not efficient and would normally result in long travel times and congestion in the elevator lobbies during peak periods. The passenger elevators for buildings with more than 20 floors (up to about 35 floors) should be separated into low rise or *local* service and high rise or *express* service. Elevators in the low-rise group should serve the lower portion of the building while elevators in the high-rise group travel directly from the main stop to the upper portions of the

TABLE 24.3   RECOMMENDED CAPACITY AND DESIGN STRATEGIES FOR PASSENGER ELEVATORS IN VARIOUS OCCUPANCIES.

Building Type	Capacity Pounds	Size Width	Depth	Design Rules of Thumb
Office buildings	3500	6 ft-8 in	5 ft-5 in	• One elevator per 45 000 usable ft^2 • Number in a single group should not exceed eight • No single group should serve more than 16 levels • A separate service elevator should be considered above 4 floors • Additional elevators for cafeterias, central supplies, and so on
Apartment buildings	2500	6 ft-8 in	4 ft-3 in	• One elevator for every 60 to 75 rooms • Do not exceed 150 ft from farthest room • One service elevator to move furniture
Hotels/motels/dorms	3500	6 ft-8 in	5 ft-5 in	
Service	4500	5 ft-4 in	8 ft-5 in	• Requirements vary by facility

TABLE 24.4   AN EXAMPLE OF ELEVATOR LOADING CAPACITIES FOR PASSENGER AND FREIGHT ELEVATORS.

Rated Load (lb)	Inside Net Platform Area (ft^2)	Maximum Number of Passengers	Rated Load (lb)	Inside Net Platform Area (ft^2)	Maximum Number of Passengers
500	7.0	3	5000	50.0	33
600	8.3	4	6000	57.7	40
700	9.6	5	7000	65.3	47
1000	13.25	7	8000	72.9	53
1200	15.6	8	9000	80.5	60
1500	18.9	10	10 000	88.0	67
1800	22.1	12	12 000	103.0	80
2000	24.2	13	15 000	125.1	100
2500	29.1	17	18 000	146.9	120
3000	33.7	20	20 000	161.2	133
3500	38.0	23	25 000	196.5	167
4000	42.2	27	30 000	231.0	200

TABLE 24.5   RECOMMENDED PERFORMANCE OF ELEVATORS.

Type of Building	Average Waiting Interval	Percentage of Total Population Handled over Peak (5 min period)
Office buildings	25 to 30 s	12 to 15%
Apartment buildings	50 to 80 s	5 to 8%
Dormitories	50 to 70 s	10 to 15%
Hotels	40 to 70 s	10 to 15%

building. In large buildings, groups of local and express (if necessary) elevators are located in the service core and other elements (i.e., stairs, mechanical and electrical chases) are designed around the core. See Figures 24.3 and 24.4.

Elevator hoistways are sized according to car shapes and sizes and door sizes, with consideration given to space requirements for guide rails and brackets, counterweight systems, running clearances, and ancillary equipment. Sufficient space should be provided around cars and elevator counterweights to minimize buffeting and airborne noise during operation.

24.2   ESCALATORS

An *escalator* is a power-driven, continuously moving stairway system used for transporting people. They can move in a linear or spiral (curved) manner. *Spiral* escalators are designed to match the curve of a building. They are used to move pedestrian traffic in places where elevators would be impractical or less efficient. Analysis typically shows that 15 to 20 elevators are needed to move the occupant capacity of an escalator system. Various types of escalators are shown in Photo 24.7.

Escalators can be placed in the same physical space as stairs. They have the capacity to move large numbers of people. In contrast to the elevator, escalators have no waiting interval. As a result, they are typically used in department stores, shopping malls, sporting arenas, stadiums, airports, convention centers, hotels, subways, office complexes, and public buildings. Factors that affect escalator design include physical requirements (vertical and horizontal distance to be spanned), location, traffic patterns, safety considerations, and aesthetic preferences.

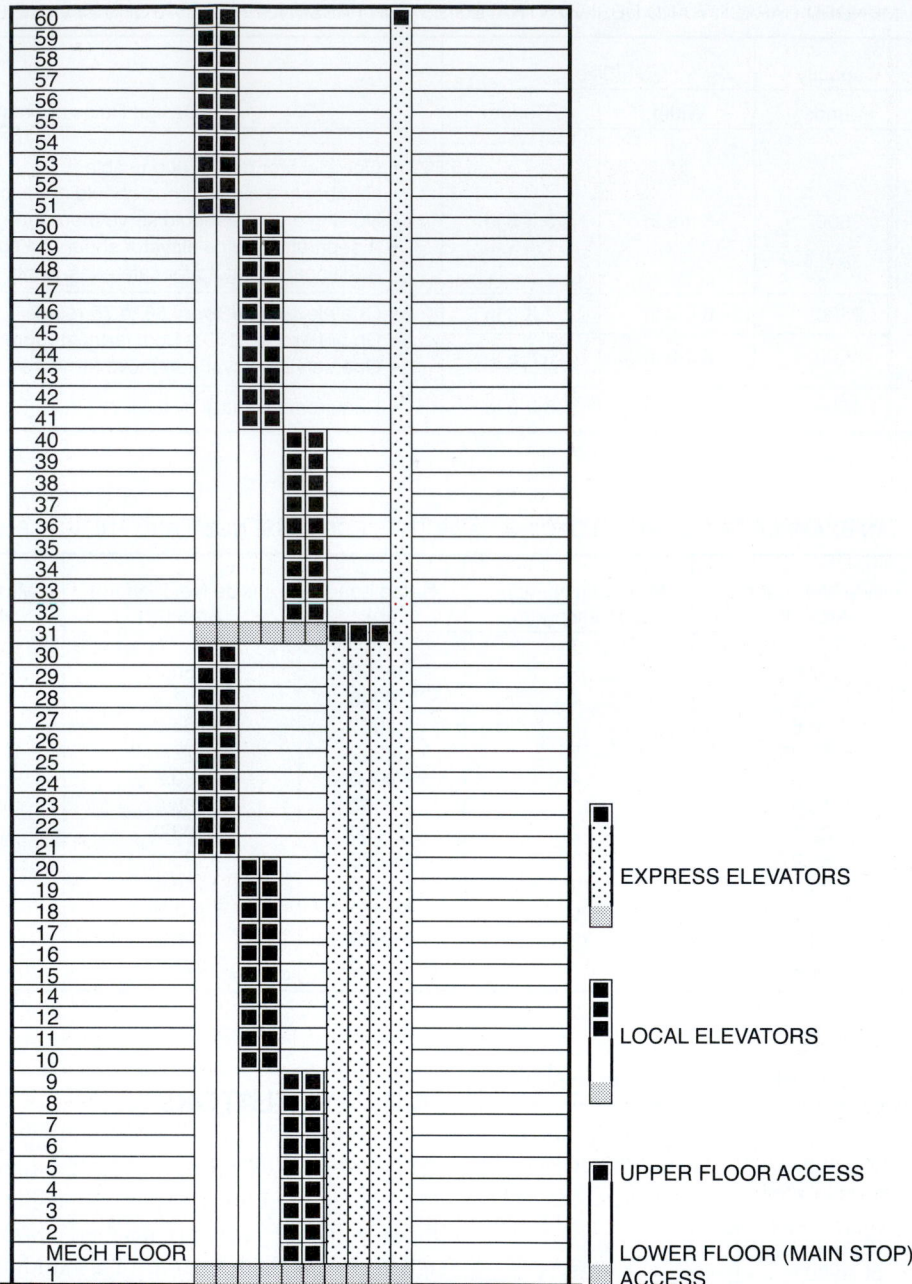

FIGURE 24.3 Elevators in a skyscraper are separated into low-rise or local-service elevators and high-rise or express-service elevators. Elevators in the low-rise group should serve the lower portion of the building while elevators in the high-rise group travel directly from the main stop to the upper portions of the building.

Standard dimensions and design capacities of escalators are listed in Tables 24.6 and 24.7.

Basic Components of an Escalator System

Although expensive and large, escalators are basic machines. Escalator components are shown in Figure 24.5. The escalator *drive unit* is a machine that drives the escalator. It is comprised of an electric motor, decelerator, electromagnetic brake, V belt, sprocket, and other components. It is powered by constant-speed AC electric motor. The *sprocket drive* is comprised of wheels installed at top and bottom (each end) to drive the steps (pallets). The top sprocket drives the moving steps, while the bottom sprocket turns the steps. The *steps* (pallets) are made from one-piece, die-cast aluminum or steel. They serve as the

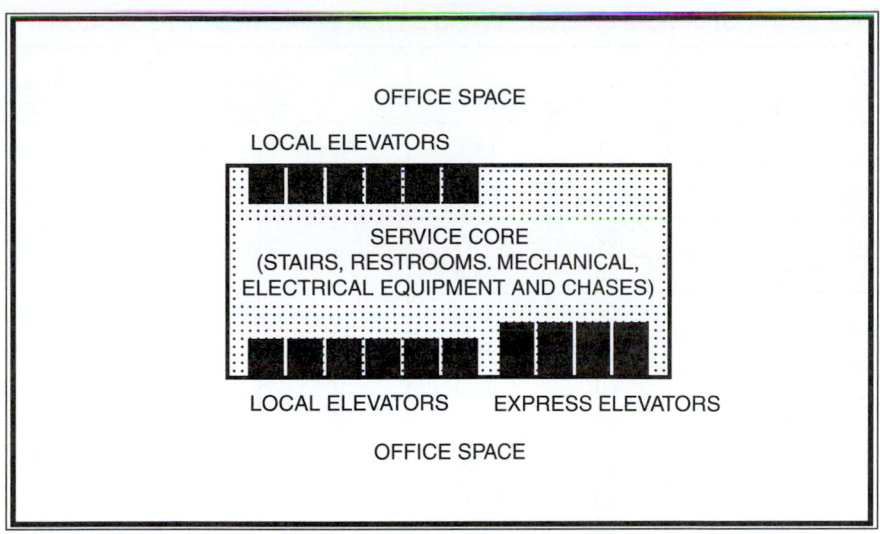

FIGURE 24.4   Elevators in a skyscraper are located in a service core.

PHOTO 24.7   Various types of escalators.

moving platform on which an escalator passenger rides. Individual steps move up or down on tracks, which keep the topside of the steps (treads) horizontal. Steps are attached to a continuously circulating belt or pallet system. The *balustrade* is the side of an escalator system. It extends above the steps and includes skirt panels, interior panels, decks, and handrails.

A *moving handrail* provides a handhold that riders use for balance and safety on their ride up or down. The handrail is powered by the same system that powers the steps. It moves along the top of the balustrade in synchronization with the steps.

The *truss* is an assembly of structural steel that serves to support the escalator load. Ends of the truss are attached to top and bottom landing platforms. The machinery of an escalator is hidden beneath its steps and within the truss. At the top of the escalator, housed in the truss, is an electric motor that runs the drive gears. There are two drive gears on either side at the top and two return gears on either side at the bottom. These gears have chains that loop around the gears and run down each side of the escalator. Connected to each step, these chains help the steps make their way up, or down, the escalator.

Escalators are powered by constant-speed alternating current motors and are designed to move at approximately 1 to 2 ft (0.3 to 0.6 m) per second. A single passenger (24 in/ 600 mm) escalator traveling at 1.5 feet (0.45 m) per second can move about 170 persons per 5-min period. The carrying capacity of an escalator must match the expected peak traffic demand. The maximum *angle of inclination* of an escalator, measured

TABLE 24.6   BASIC DESIGN CAPACITIES OF ESCALATORS.

Size	Width (between Balustrade Panels)	Single-Step Capacity	Applications
Small	24 in (600 mm)	One passenger	Low-volume use, uppermost levels of department stores, when space is limited
Medium	32 in (800 mm)	One passenger and one package or one piece of luggage	Shopping malls, department stores, smaller airports
Large	40 in (1000 mm)	Two passengers	Metro transit systems, larger airports, train stations, large retail use

TABLE 24.7   STANDARD DIMENSIONS AND DESIGN CAPACITIES OF ESCALATORS.

Type	Capacity	Step Width		Speed		Angle of Inclination	Horizontal Steps
	Persons/hr	in	mm	ft/min	m/s	Degrees	Number
Commercial escalator	4500	24	600	100	0.50	30°	2
	6750	32	800				
	9000	40	1000				
	4500	24	600				3
	6750	32	800				
	9000	40	1000				
	4500	24	600			35°	2
	6750	32	800				
	9000	40	1000				
	4500	24	600				3
	6750	32	800				
	9000	40	1000				
Slimline escalator	4500	24	600	100	0.50	30°	2
	6750	32	800				
	9000	40	1000				
	4500	24	600				3
	6750	32	800				
	900	40	1000				
	4500	24	600			35°	2
	6750	32	800				
	9000	40	1000				3
	4500	24	600				
	6750	32	800				
	9000	40	1000				
Heavy-duty escalator	4500	24	600	100	0.50	30°	3
	6750	32	800				
	9000	40	1000				
Multilevel escalator	4500	24	600	100	0.50	30°	2
	6750	32	800				
	9000	40	1000				

from the horizontal floor level, is typically 30°. A standard total rise for a commercial escalator can be up to about 60 ft (18 m). Transit escalators can have a total rise up to 164 ft (50.0 m).

Escalator Arrangement

A *single escalator* takes passengers up one floor; that is, it travels up but there is no downward travel. A *single noncontinuous arrangement* is a set of interrupted escalators, all traveling up from floor to floor; it has no downward travel. It requires a passenger traveling multiple floors to get off, walk a distance to the other side of the escalator system to get on the next escalator before traveling to the next upper floor. A *single continuous arrangement* is a set of up-only escalators that zigzag back and forth as they move floor to floor; a passenger traveling multiple floors gets off one escalator, takes a few steps, and gets on the next escalator to travel to the next upper floor. See Figure 24.6. Single escalators are energy efficient because they only travel upward, saving the energy consumed to take passengers downward. They are popular in many areas of the world but not in the United States. Components of escalators are shown in Photos 24.8 through 24.11.

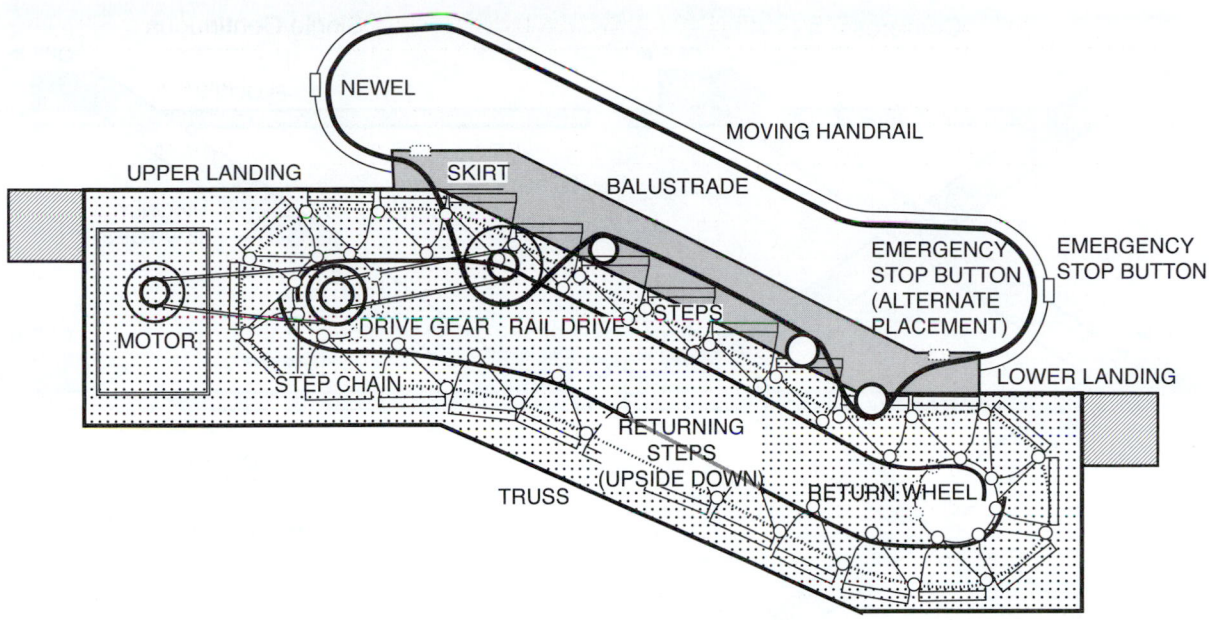

FIGURE 24.5 Components of an escalator.

PHOTO 24.8 A continuous parallel escalator system.

PHOTO 24.10 An upper landing of an escalator. Note how the stairs fold in at the top of the escalator.

PHOTO 24.9 A lower landing of an escalator. Note how the stairs raise at the beginning of the escalator. An emergency stop switch extends from the right handrail. (Used with permission of ABC)

PHOTO 24.11 An emergency stop switch on an escalator. (Used with permission of ABC)

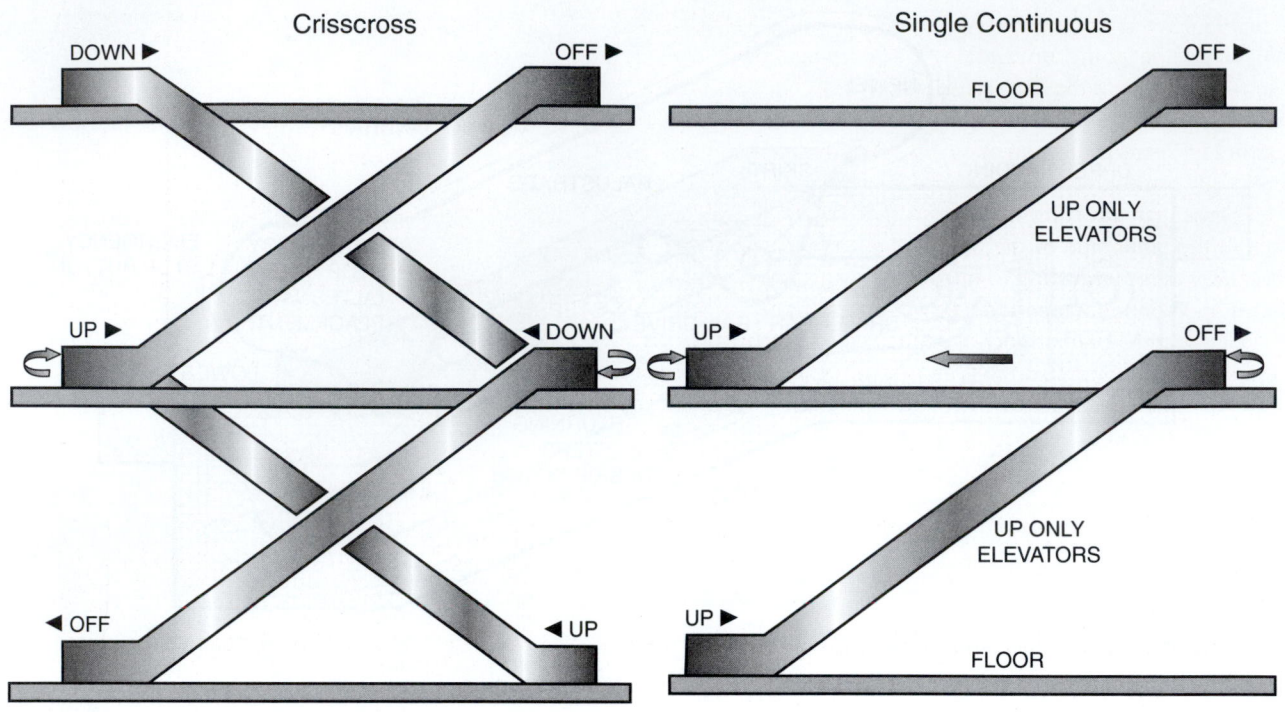

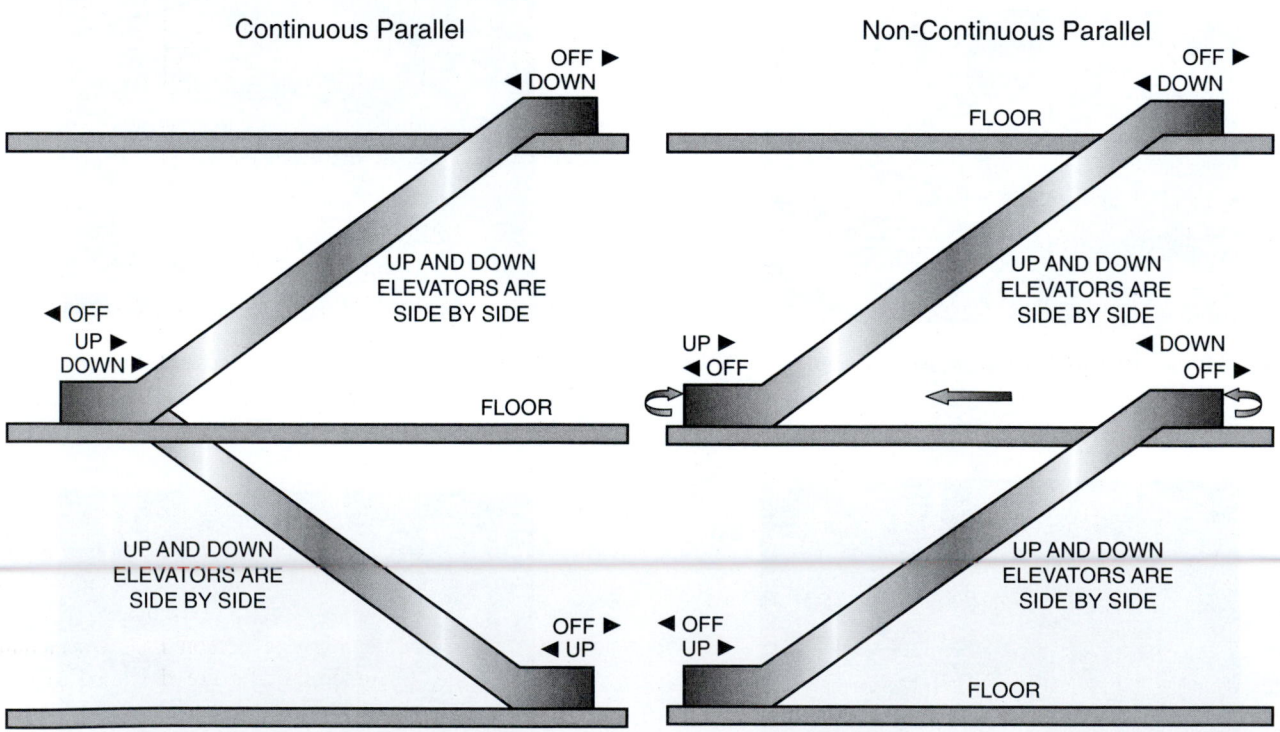

FIGURE 24.6 Common escalator arrangements.

In the United States, escalators are typically installed in pairs with one going up and the other going down. The arrangements used are the *crisscross, continuous parallel,* and *noncontinuous parallel* systems. See Figure 24.6. Travel on the noncontinuous parallel arrangement is less efficient than the crisscross and continuous parallel arrangements because it requires a passenger traveling multiple floors to get off, walk a distance to the other side of the escalator system to get on the next escalator before traveling to the next upper floor. Although crisscross and continuous parallel arrangements move passengers efficiently, the noncontinuous parallel arrangement is preferred in shopping centers and malls. It makes shoppers get off and walk to the next escalator, forcing them to peruse merchandise as they walk on between escalators.

24.3 WALKWAYS AND RAMPS

A *moving walkway* is a power-driven, continuous, slow-moving conveyor belt that transports people horizontally. They are also called *moving sidewalk, moving pavement, walkalator,* and *travelator.* An *inclined moving walkway,* also called a *moving ramp* or *power ramp,* is a moving walkway that transports people on an incline, up to a 12° angle of inclination. Moving walkways are more accessible to those in wheelchairs, as they are usually thought easier to use than getting in and out of small elevators. Components of moving walkways are shown in Photos 24.12 through 24.14.

There are two types of walkway technologies: pallet type and moving belt. The *pallet-type walkway* is a continuous series of flat metal plates, called *pallets,* that are joined together to form a walkway. Usually there is a metal or rubber surface (extra traction). *Moving belt walkway systems* are comprised of a mesh metal or rubber belt with a rubber walking surface that move over metal rollers. Once on the walkway, riders can stand or walk. Some riders complain that the rollers below the belt tend to cause a "bouncy" feel.

Walkways are typically installed in pairs, with one for each direction of travel. Walkways typically operate at 90 to

PHOTO 24.13   A (horizontal) moving walkway. (Used with permission of ABC)

PHOTO 24.14   A pallet-type walkway is a continuous series of flat metal plates, called *pallets,* that are joined together to form a walkway. (Used with permission of ABC)

120 fpm (27 to 37 m/min) and are up to 500 ft (~150 m) long. Capacities and design strategies for moving walks and moving ramps are shown in Table 24.8.

24.4 OTHER SYSTEMS

Lifts

A *wheelchair lift* is a powered device designed to raise a wheelchair or scooter and its occupant to overcome a step or similar vertical barrier, usually 6 ft (1.8 m) or less. They often are designed to accommodate just one person in a wheelchair or scooter at a time. Commercial lifts are designed to raise a wheelchair or scooter and its occupant up to one story (about 12 ft/4 m). See Photo 14.15. *Platform lifts* supply access to decks, porches, stages, and elevated surfaces. A *stair lift* will carry a user safely up stairs. To use a stair lift, the user sits on the lift's seat; the seat will then transfer the user up or downstairs via a staircase-mounted track. Types of accessibility lifts are introduced in Table 24.9. A wheelchair lift in a residential garage is shown in Photo 24.15.

PHOTO 24.12   An inclined moving walkway. (Used with permission of ABC)

TABLE 24.8 CAPACITIES AND DESIGN STRATEGIES FOR MOVING WALKS AND MOVING RAMPS.

Capacity	Persons	One Person		Two Persons		Three Persons
Approximate width	ft-in	4 ft-0 in		5 ft-4 in		6 ft-8 in
	mm	1244 mm		1650 mm		2056 mm
Speed (horizontal)	ft/min	90	120	90	120	90
	m/min	27	36	27	36	27
Capacity	Occupant/hr	6000	8000	9000	12 000	12 000
Inclination angle	Degrees	0° to 12°	0° to 12°	0° to 12°	0° to 12°	0° to 3°

TABLE 24.9 TYPES OF ACCESSIBILITY LIFTS.

Vertical platform lift	This minielevator travels in a straight vertical path and contains a one- or two-passenger cab with 42 in high walls. Its maximum travel height is 14 ft, with up to three stops, and is generally housed in a hoistway.
Inclined platform lift	A folding platform with 6 in to 8 in safety walls and wraparound safety arms that travels along the incline of an existing stairway. Designed for one passenger, the number of stops and maximum travel length is limited by practicality.
Inclined stairway chairlift	A one-passenger folding seat that travels along the incline of an existing stairway.
Emergency evacuation device	Transportation device with rubber-tracked wheelbase that transports persons with disabilities down stairways and out of buildings that may be under threat of fire, hurricane, earthquake, or bombing.
Portable wheelchair lift	A device that transports persons with disabilities up and down stairs when a permanent lift is not available.
Limited use/limited application (lula) elevator	Small-scale fully automatic elevators with shallow pits that can be utilized in limited use applications.

A *car lift* is installed in small parking garages where ramps are not feasible. The platforms are raised and lowered hydraulically and are connected to steel chain gears. In addition to the vertical motion, the platforms can rotate about its vertical axis (up to 180°) to ease driver access and/or accommodate building plans. In selecting lift equipment, the building designer typically relies on specifications available from equipment suppliers.

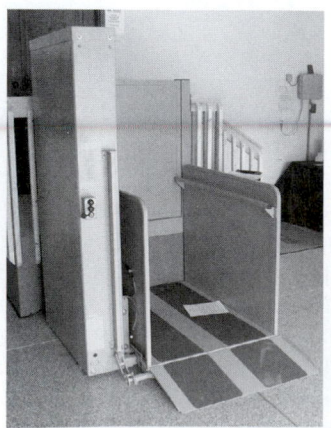

PHOTO 24.15 A wheelchair lift in a residential garage. (Used with permission of ABC)

People Movers

An *automated people mover (APM)* is a fully automated, grade-separated mass transit system. An APM system typically serves relatively small facilities such as airports, downtown districts, or theme parks, but is sometimes applied to considerably more complex automated systems. It may use technologies such as monorail, duorail, automated guideway transit, or magnetically levitating (maglev) method. Propulsion may involve conventional on-board electric motors, linear motors, or cable traction. APMs are common at large airports in the United States.

Material-Handling Equipment

Material-handling equipment is a mechanical device used to move and store materials and goods. This equipment consists of trolleys, conveyors, forklifts, automated storage/retrieval systems, cargo and baggage handlers, carousels, rail-guided vehicles, automated guided vehicles, intelligent flexible modular conveyors, pick-and-place units, and overhead hoists and cranes. In selecting material-handling equipment, the building designer typically relies on specifications available from equipment suppliers. The conveyor belt of an automatic book-sorting system at a public library is shown in Photo 24.16.

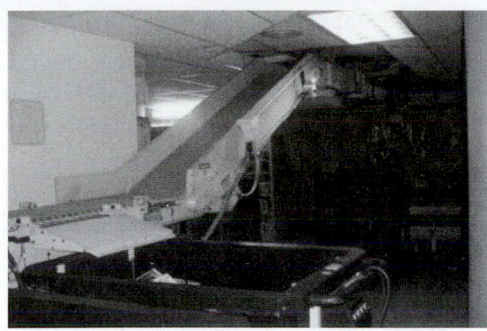

PHOTO 24.16 The conveyor belt of an automatic book-sorting system at a public library. (Used with permission of ABC)

Paternoster

A *paternoster* is a special type of elevator consisting of a constantly moving chain of boxes. A similar concept moves only a small platform, which the rider mounts while using a handhold and was once seen in multistory industrial plants. Passengers can step on or off at any floor they like. Today, the installation of new paternosters is no longer allowed because of to their inherent danger.

24.5 SAFETY

Accidents

Elevators, escalators, and walkways are potential sources of injuries and deaths to the general public. In the United States, there are over 660 000 elevators and 30 000 escalators making an estimated 230 000 000 trips per day. Most elevator accidents and deaths involve falls into elevator shafts (including where an elevator door opened and there was no elevator car) and getting caught in the elevator door or between the elevator and door or shaft. The most frequent escalator accidents are as a result of passenger actions: loose clothing (including shoelaces); fingers becoming entrapped when people fall; childrens fingers or feet becoming caught; and footwear being caught. Several instances of multiple injuries were caused when an escalator suddenly sped up or reversed its direction of movement. The U.S. Consumer Product Safety Commission reports that about 6000 people/year make emergency room visits after accidents on escalators. About 20% of these accidents involved having hands, feet, or shoes trapped. Soft-soled shoes (e.g. Crocs® and sandals) have led to many serious injuries. Malfunctioning elevators and escalators during maintenance are also a cause of deaths or injuries.

Licensing

In most locations, licensing is part of performing work on elevators, escalators, dumbwaiters, stairlifts, wheelchair lifts, construction manlifts, material lifts, grain elevators, and any other type of conveying system. Licenses are issued through city, county, or state authorities. Exceptions include some types of work on residential systems and some types of electrical work that do not involve working in, under, or on top of the elevator car. Any person, firm, or company that performs work from paying customers must be a licensed elevator contractor. In addition, any person wishing to work on conveyances must be a licensed elevator mechanic employed by a licensed elevator contractor.

STUDY QUESTIONS

24-1. Describe the hydraulic elevator.

24-2. Describe the traction elevator.

24-3. Explain the differences between the conventional (holed) hydraulic elevator, the telescopic hydraulic elevator, and the holeless hydraulic elevator.

24-4. Describe and distinguish between the following components of an elevator system:
   a. Car
   b. Platform
   c. Cab

24-5. Describe the following components of an elevator system:
   a. Pit
   b. Penthouse
   c. Landing

24-6. Describe the following components of an elevator system:
   a. Hoistway
   b. Blind hoistway
   c. Hoistway doors

24-7. Describe the following components of an elevator system:
   a. Elevator controller
   b. Car operating station
   c. Hall station
   d. Traveling cable

24-8. Describe the following components of an elevator system:
   a. Brake
   b. Governor
   c. Drive unit
   d. Buffer

24-9. Describe the classifications of elevators based on use.

24-10. Describe an escalator.

24-11. Describe the following components of an escalator system:

  a. Drive unit

  b. Sprocket drive

  c. Steps

  d. Balustrade

  e. Moving handrail

  f. Truss

  g. Emergency stop button

24-12. Describe the differences between the crisscross, single continuous, continuous parallel, and noncontinuous parallel escalator systems.

24-13. What is the typical inclination angle of an escalator?

24-14. Describe a moving walkway.

24-15. What are the two types of walkway technologies?

24-16. What is a wheelchair lift?

24-17. What is the difference between a platform lift and a stair lift?

24-18. Describe a car lift.

24-19. What is a paternoster?

Design Exercises

24-20. Using a stopwatch, time an elevator in a building. Determine the following:

  a. Average time required for the door to close once a car operating station button (i.e., floor button) is pushed

  b. Average time required for the door to fully open once the car stops at the floor

  c. Average time the door remains open at a floor (from the time it is fully open to the time it begins to close after a button is pushed)

  d. The average time it took to go up one story, excluding wait time (start the stopwatch as soon as the elevator starts to move, and stop it as soon as the elevator stops at the floor)

  e. The average time it took to go up one story, including wait time

  f. The average time it takes to walk up one story by using the stairs

24-21. Using a stopwatch, time an elevator in a high-rise building (>5 stories). Determine the following:

  a. The average time it took to go up one story, excluding wait time (start the stopwatch as soon as the elevator starts to move, and stop it as soon as the elevator stops at the floor)

  b. The approximate number of adults who can travel from floor to floor in 5 min, if the elevator is fully used

24-22. Using a stopwatch, time an escalator in a building. Determine the following:

  a. The average time it took to go up one story (start the stopwatch as soon as you step onto the escalator, and stop it as soon as you step off)

  b. The approximate number of adults who can travel from floor to floor in 5 min, if the escalator is fully used

EMERGING SUSTAINABLE TECHNOLOGIES

25.1 SUSTAINABLE BUILDINGS

Sustainability

The earth's resources are limited and world population is increasing rapidly. Our world is consuming natural resources more rapidly than preceding generations, which is affecting our ability to provide for future generations. Excessive human impact on our planet is likely the single greatest long-term problem facing humankind.

Sustainability is our ability to meet current needs without harming the environmental, economic, and societal systems on which future generations will rely for meeting their needs. It does not mean we must surrender our lifestyle by living in the cold and dark; nor does it mean building a hut in the woods where local lumber can be used. It simply means using resources wisely. Embracing values of sustainability ensures that decisions made and actions taken today do not hinder the existence of future generations.

In the United States, buildings consume almost 40% of the total energy, over 65% of the electricity, over 10% of the water. About 42% of energy used in buildings is used for comfort heating, cooling, ventilation, and refrigeration; lighting accounts for about 18%; and water heating about 10%. Buildings in the United States account for about 10% of global energy use. Such energy consumption makes buildings good candidates for sustainability efforts.

Sustainable Buildings

In the building construction industry, terms like *sustainable architecture, sustainable design, green design, high-performance design,* and *ecological design* are used to describe the process of designing and constructing sustainable buildings. Sustainable building design creates a balance between human needs for shelter and the natural and societal environments. It minimizes adverse environmental and cultural effects. The ideal approach is to build only when necessary; to use natural sustainable materials, collected on site if possible; to limit waste generated from construction and manage waste during building operation; and rely on renewable energy sources (e.g., solar or wind) to condition and power the building.

A *sustainable building,* also known as a *green building,* is a healthier and more resource-efficient structure that is designed, built, operated, renovated, reused, and eventually dismantled/demolished in a sustainable manner. It is designed to meet specific goals such as protecting occupant health; improving employee productivity; using energy, water, and other resources efficiently; and reducing the overall impact to the environment. Design and construction of a sustainable building should:

- Strive for design of compact and efficient interior spaces so overall building size and materials needed for construction and operation are minimized (i.e., a "compact design is better")
- Incorporate energy-conserving methods and strategies that complement the local climate in design and construction of the building envelope (e.g., solar orientation, superinsulation, and tightness)
- Use locally produced (indigenous), environmentally preferable building materials effectively
- Use renewable materials and components that can be reused or recycled without difficulty
- Avoid use of energy-intensive, environmentally damaging, waste-producing, and/or hazardous materials
- Use life-cycle analysis in decision making for materials and construction techniques
- Use simple, efficient, and "right-sized" mechanical and electrical (e.g., heating, cooling, plumbing, lighting, and so on) technologies
- Use renewable energy sources (e.g., solar, wind) where effective
- Strive for minimal environmental disruption, resource consumption, and material waste, and identify opportunities for reuse and recycling of construction debris
- Incorporate provisions for trouble-free recycling of wastes generated by occupants
- Plan for future expansion and/or use by other occupancies
- Plan for disassembly (deconstruction) or adaptation that maximizes recovery of materials and minimizes demolition waste (as an alternative to demolition)

A green building may cost more up front, but reduced operating costs over its life result in lower operating costs over the life of the building. An integrated systems approach is used to ensure that the building is designed as one system rather than a collection of stand-alone systems.

Sustainable Building Programs

Most municipal, state, and federal governments recognize the need for limiting energy and resource use. Several energy and resource efficiency programs have been established to drive society to pursue the concept of sustainability (e.g., ENERGY STAR® Program). A major driver in building sustainability is the U.S. Green Building Council (USGBC), a coalition of leaders in the building industry working to promote buildings that are environmentally responsible and sustainable, profitable, and healthy places to work and live.

The USGBC's *Leadership in Energy and Environment Design (LEED) Rating System* is a voluntary, consensus-based national standard for developing high-performance, sustainable buildings. It provides a complete framework for assessing building performance and meeting sustainability goals. The LEED rating system is a method for certifying a green building. It is not a punch list of how to design and build a green building. In LEED (Version 2.2), for new construction and major renovations for commercial buildings there are 69 possible points; buildings can qualify for four levels of certification:

- Certified: 26–32 points
- Silver: 33–38 points
- Gold: 39–51 points
- Platinum: 52–69 points

The LEED rating system addresses six major areas as follows:

- Sustainable sites (up to 14 points)
- Water efficiency (up to 5 points)
- Energy and atmosphere (up to 17 points)
- Materials and resources (up to 13 points)
- Indoor environmental quality (up to 15 points)
- Innovation and design process (up to 5 points)

USGBC members continue to expand and adapt the LEED system. The USGBC Web site, http://www.usgbc.org/, contains a wealth of information on green building and links to pertinent public and private sector Web sites. Green building professionals can become LEED accredited through the LEED Accredited Professional Exam. This accreditation enables an individual to facilitate the rating of buildings with the various LEED systems.

Sustainability and the "P-Green" Building

In recent years, many designers and constructors, along with manufacturers and other businesses, have jumped on the sustainability bandwagon by making green products and green buildings available. In many instances, this move is motivated by the good intentions of helping the environment, and in other cases, it is driven by marketing studies that show the general public embraces green practices. Some buildings and products are being marketed as being green even though they only partially fulfill the standards of sustainability or do not satisfy them at all. This makes these buildings and products *partly green (P-green)*. In the *P-green building,* the designer or builder attempts to incorporate some sustainability strategies (with good intentions or for marketing purposes) yet the overall building or many of its systems are not really green at all.

The P-green building may feature high-efficiency systems for heating, cooling, ventilating, and lighting, yet have indoor spaces that are quite large and opulent requiring these systems to consume significant amounts of energy. Such a building requires more resources to build and maintain, and more energy to operate in comparison to a basic, simpler building design. The P-green building may include a handful of inconsequential green features (such as imported bamboo flooring or nontoxic carpet) and neglect other, more important, significant green strategies. Perfect examples of P-green buildings in the residential market are the well-insulated McMansions that are popping up everywhere. Their high-efficiency heating, cooling, and lighting systems are coupled with lavish design features that include huge kitchens and voluminous rooms with three times the necessary floor area. In the commercial building market, an add-on solar system or wind turbine can be installed and the entire building and its construction methods are incorrectly promoted as sustainable, environmentally friendly construction. These are classic examples of P-green buildings—they are too large, poorly designed, use too much energy, and consume too many resources than a more simply designed structure.

Sometimes, an effort is made to unethically camouflage the P-green building and it is highly advertised to the uninformed consumer as being green when this is a huge embellishment. The P-green disguise is intended to throw off the consumer who desires a building that is sustainable. A product or building is not green because the manufacturer, designer, or builder says it is green. Caution should be exercised by the architecture, engineering, and construction professional to avoid the P-green trap. A question should always be asked: Is the product used or building designed green or P-green?

Material in this Chapter

Several topics related to sustainability in building electrical and mechanical systems have been introduced where applicable and relevant in other chapters in this text. Some of these topics and the related chapters include the following:

- Indoor air quality (Chapter 3)
- CO_2 ventilation control (Chapter 6)
- Displacement ventilation (Chapter 6)
- District heating and cooling (Chapter 6)
- Heat recovery (Chapter 6)
- Indirect evaporative cooling (Chapter 6)
- Night precooling/ventilation (Chapter 6)
- Thermal (ice) storage (Chapter 6)

- Chilled beams (Chapter 7)
- Water loop heat pump systems (Chapter 7)
- Solar thermal systems (Chapter 11)
- Low-flow and waterless plumbing fixtures (Chapter 12)
- Point-of-use domestic water heating (Chapter 13)
- Graywater reuse systems (Chapter 15)
- Compact fluorescent and LED lighting (Chapter 20)
- Daylighting (Chapter 20)

In this chapter, selected emerging sustainable technologies are introduced. Although many of these technologies have been used for many years (and in some cases, centuries and millennia), these technologies are being modernized and implemented to address the sustainability issue.

25.2   COMBINED HEAT AND POWER (COGENERATION) SYSTEMS

Conventional power plants release heat created as a by-product of electricity generation into the environment through cooling towers, flue gas, or by other means. *Cogeneration,* also known as *combined heat and power (CHP),* is the use of a heat engine (or other means) to simultaneously generate both electricity and heat. Its principal purpose is to produce electricity but, as a by-product, the heat produced is used for heating water, space heating, or industrial process heating. Conventional cogeneration is not a new technology. In 1882, Thomas Edison's Pearl Street Station in New York City, the world's first commercial power plant, produced electricity and thermal energy, which was used to warm neighboring buildings. The combined efficiency of Edison's Pearl Street Station plant was about 50%. In modern CHP systems, the efficiency of energy conversion increases to over 80% as compared to an average of 30 to 35% for conventional fossil fuel fired electricity generation systems. Cogeneration is most efficient when the heat can be used on-site or very close to it because overall efficiency is reduced when the heat must be transported through costly insulated pipes. Cogeneration units are generally classified by the type of prime mover (i.e. drive system), generator, and fuel used.

Conventional Cogeneration

The basic elements of conventional cogeneration system are the prime mover (engine), generator, heat recovery system, exhaust system, controls, and acoustic enclosure. The generator is driven by the engine, and the useful heat is recovered from the engine exhaust and cooling systems. On-site cogeneration systems use many engine technologies, including internal combustion reciprocating engine based cogeneration systems, microturbine based cogeneration systems, fuel cell based cogeneration systems, and reciprocating external combustion Stirling engine based cogeneration systems.

In commercial and institutional building installations, reciprocating engines are the most common and most efficient prime mover used in commercial cogeneration systems. Internal combustion reciprocating engines are based on the spark ignition or the compression-ignition cycle. In the *spark ignition (Otto cycle) engine,* the mixture of air and fuel is compressed in each cylinder before ignition is caused by an externally supplied spark. These engines are mainly four-stroke direct-injection engines fitted with a turbocharger and intercooler. They run on gasoline, natural gas, or liquid propane gas (LPG). The *compression-ignition (Diesel) engine* involves only the compression of air in the cylinder; the fuel is introduced into the cylinder towards the end of the compression stroke. The spontaneous ignition is thus caused by the high temperature of the compressed mixture. Diesel engines run on diesel fuel or heavy oil; as a result, they are primarily used for large-scale cogeneration.

In both systems, the engine drives a generator to produce electricity. Thermal energy is captured through heat recovery from the exhaust gas, engine oil, and cooling water, which can then be used for space or water. Not all of the heat produced in an internal combustion engine based cogeneration system can be captured in on-site electric generation, because some of the heat energy is lost as low-temperature heat within the exhaust gases and as radiation and convection losses from the engine and generator. Typical engine efficiencies are between 20 and 30%. By recovering heat from the cooling systems and exhaust, approximately 75 to 90% of the energy derived from the fuel is utilized to produce both electricity and useful heat.

Stirling Engine Cogeneration

A *Stirling engine* is a closed-cycle, piston-driven, external heat engine with a gaseous working fluid that under cooling, compression, heating, and expansion drives a piston. The working fluid is permanently sealed within the engine's system so that no gas enters or leaves the engine. No valves are required, unlike other types of piston engines. Piston movement is accomplished by moving the gas back and forth between hot and cold heat exchangers, often with a regenerator between the heater and cooler. Because of its high efficiency, quiet operation, and ease with which it can use waste heat, the Stirling engine is currently being researched as a prime driver (engine) in residential cogeneration systems.

Fuel Cell Cogeneration

A *fuel cell* is an electrochemical conversion device that directly converts a fuel into electrical energy. A hydrogen-oxygen fuel cell consumes hydrogen and oxygen and produces water as the principal by-product. In conventional commercial fuel cell systems, the chemical energy typically comes from hydrogen contained in natural gas, which is made from hydrogen and carbon. When hydrogen is separated from carbon and fed into a fuel cell, it combines with oxygen to produce water, electricity, and heat. The carbon is released as carbon dioxide, although in much smaller quantities. A typical single cell delivers up to 1 volt (V). To get sufficient power, a fuel cell stack is made of several single cells connected in series. A typical fuel cell has an efficiency of about 40%, which means that 40% of the energy content of the hydrogen is converted into electrical energy and the remaining 60% is converted into heat. A truly sustainable system involves using solar (photovoltaic) panels to power an electrolyzer, which makes hydrogen through electrolysis of water and can be used to fuel a fuel cell.

Fuel cells have been proven very effective as power sources in remote locations (e.g., spacecraft, remote weather stations, large parks, rural buildings). This technology has been used over the last three decades by National Aeronautics and Space Administration (NASA) to provide reliable power for space applications. In buildings, fuel cells are an emerging technology that has the potential for environmentally friendly power generation and cogeneration. A fuel cell cogeneration system can generate a constant supply of electric power (selling excess power back to the grid when it is not consumed) and produce hot water (or hot air) from the waste heat. Fuel cells in buildings typically use natural gas and release fewer environmentally harmful emissions than those produced by a combustion cogeneration plant.

Fuel cell cogeneration based systems have potential in residential and small-scale commercial applications because of the ability to produce electricity at relatively high efficiency with significantly less greenhouse gas emissions (compared to conventional power plants). When used for cogeneration application, fuel cells are expected to achieve electrical efficiencies between 30 and 60%, with an overall efficiency of 70 to 90%. Recent technological advancements are making fuel cells more affordable. One manufacturer is currently producing fuel cells with output power ranging from 0.5 to 100 kilowatts (kW).

Building CHP Systems

Building cogeneration or CHP systems produce electrical power for local buildings, and use the heat from that production to also provide heat to the buildings (often through underground steam or hot water piping systems). Because the heat is being productively used, rather than being discharged at a distant power plant, and because the electricity is being locally produced, rather than being subjected to the losses inherent in power transmission, a building CHP is 2 to 3 times the efficiency of power delivered over the grid. CHPs can be operated continuously or only during peak load hours to reduce peak demand charges.

Historically, CHP systems have been most used at hospital and university facilities, where a large number of buildings can be efficiently served. But new systems are bringing this same technology into other buildings, such as residences. Today, CHP technologies are commercially available over a wide range of sizes from 2 kW to more than 10 MW. Sizes of conventional CHP installations in buildings typically fall in

PHOTO 25.1 A residential-scale micro cogeneration (home heating and power) system that can produce about 1.2 kW of electric power and over 11 000 Btu/hr (3.26 kW) of heat. (Courtesy of DOE/NREL)

ranges as follows: household or single family ($<$5 kW), multifamily (5 to 30 kW), commercial (5 to 100 kW), and institutional (20 to 100 kW).

By definition, a *micro-cogeneration* installation is usually less than 5 kW and is typically installed in a house or small building. A *mini-cogeneration* installation is usually more than 5 kW and less than 500 kW and is typically installed in commercial or institutional buildings. A residential-scale micro-cogeneration is shown in Photo 25.1. Available building CHP systems can be fired on a broad variety of fuels. Reciprocating spark ignition engines are commonly used because of their suitability for most building CHP applications. Packaged internal combustion engine CHP systems of 50 to 100 kW capacities are currently in use in the commercial sector. One manufacturer's CHP unit suitable for a commercial building produces 60 kW electric and 138 kW (470 000 Btu/hr) thermal energy while achieving a total energy efficiency of 83.1%. Such heat recovery system produces up to 320°F (160°C) hot water. These systems are suitable for multifamily residential buildings and commercial applications (i.e., hotels, recreation centers, institutional buildings, and hospitals).

CHP units are currently being used to meet the heating load requirement of a building or facility, in addition to backup or peak shaving applications. CHP systems can also be connected to an absorption chiller that provides heating and cooling for the central heating, ventilation, and air conditioning (HVAC) system. The absorption chiller, which is powered by thermal energy, replaces a traditional electric chiller. CHP systems can also heat domestic water for use in the building. From a cooling standpoint, absorption chillers can be fueled by low-grade waste heat recovered from on-site electricity generation. Absorption chillers use environmentally kind refrigerants and absorbents instead of polluting chlorofluorocarbons (common in electrical chillers). They are available in capacities ranging from 100 to 1500 tons. Although absorption chillers have a low coefficient of performance compared to electrical chillers, they are extremely cost-effective and efficient when used with a CHP system because very little electricity is used to power the chiller.

Residential scale CHP units produce electricity and heat water for domestic hot water and space heating applications. They are more widely used in Europe in comparison to the United States. Reciprocating internal combustion engines used for single-family residential CHP applications are typically less than 5 kW. Typically, household CHPs consist of an internal combustion engine that burns natural gas to generate electricity. The heat produced is used either to create hot air (as with a traditional warm air furnace) or hot water (for radiator or radiant floor heated homes). A heat exchanger feeds any captured heat to a furnace, which then distributes the hot air. One manufacturer's compact household CHP unit produces 1 kW electric and 3.25 kW (11 000 Btu/hr) thermal while achieving a total energy efficiency of 85.5%. It operates using a four-stroke, water-cooled, single-cylinder natural gas or LPG fueled engine.

25.3   GEOTHERMAL ENERGY SYSTEMS

Geothermal energy, derived from the words *geo* (meaning "earth") and *thermal* (meaning "heat"), is energy derived from the heat in the earth's interior. Geothermal energy comes from the process of decay of radioactive elements within the earth's crust. It also comes from the capability of the earth to serve as an enormous heat sink; the earth absorbs and stores heat energy from the sun. Although it is not truly a renewable resource, geothermal energy is an enormous, underused heat and power resource.

Direct Use of Geothermal Energy

One method used to extract thermal energy from the earth is referred to as *hydrothermal.* Hydrothermal energy is manifested in as hot springs and geysers or through dry steam. Hydrothermal reservoirs of low- to moderate-temperature water in the range of 68° to 302°F (20° to 150°C) can provide direct heat for residential, industrial, and commercial uses. Hot water near the earth's surface has already been used to heat buildings, grow plants in greenhouses, dehydrate vegetables, heat water for fish farming, and pasteurize milk. In Iceland, geothermal energy is used extensively for district heating applications that use networks of piped hot water to heat buildings in whole communities. A geothermal hot spring is shown in Photo 25.2.

In the United States most near-surface hydrothermal reservoirs are located in the western states, such as Idaho, Colorado, Wyoming, Alaska, and Hawaii. Some cities pipe the hot water under roads and sidewalks to melt snow. Today, there are four district heating systems in Boise, Idaho, that provide heating to over 5 million square feet of residential, business, and governmental space. In Iceland, the high concentration of volcanoes and geothermal energy are used for heating and production of electricity. In Iceland, there are five major geothermal power plants that produce about 26% (2006) of the country's electricity. In addition, geothermal heating meets the heating and hot water requirements for around 87% of the nation's housing. Geothermal energy is so inexpensive that some roads in Reykjavík and Akureyri are heated in the wintertime. More than half of the energy consumed in Iceland comes from geothermal sources.

Engineers are developing technologies that will probe more than 10 miles below the earth's surface in search of geothermal energy. These deep wells are drilled into underground reservoirs to tap steam and very hot water that drive turbines that drive electricity generators. Technology is also being developed to drill into hot, dry rock, inject cold water down one well, circulate it through the hot, fractured rock, and draw off the heated water from another well.

PHOTO 25.2   A geothermal hot spring. (Courtesy of DOE/NREL)

Geothermal Heat Pump Systems

A *geothermal heat pump (GHP),* also known as a *ground source heat pump (GSHP), earth-coupled heat pump,* or *geoexchange system,* is an electrically powered heat pump system that consists of pipes buried in the shallow ground near the building, a heat pump, and ductwork in the building. Because the upper 10 ft (3 m) of earth's surface maintains a nearly constant temperature that is typically between 50 and 60°F (10° and 16°C) in the United States and ranging between 41° and 54°F (5° and 12°C) across Canada, it can be used a heat source or heat sink that can provide heating and cooling for both residential and commercial buildings.

As introduced in Chapter 6, a *heat pump* is a type of heating and/or cooling device that draws heat into a building from outside of the building during the heating season, and if designed with cooling capability, the heat pump expels heat from the building to the outside during the cooling season. A heat pump drives heat from a low temperature region to a high temperature region using the vapor-compression refrigeration process.

The fundamental components of a geothermal heat pump system include the following:

- **Ground loop**   A system of fluid-filled plastic pipes buried in the ground, or submersed in a body of water, near the building.

- **Heat pump**   A device that removes heat from the fluid in the pipes, concentrates it, and transfers it to the building as heat. For cooling, this process is reversed: heat is transferred from the building; it is carried in the fluid to the pipes, where it is transferred into the earth.

- **Air delivery system**   Conventional ductwork attached to the heat pump and used to distribute heated or cooled air throughout the building.

Engineering specifications for several residential and light commercial geothermal heat pumps are provided in Table 25.1. Application temperature limits of water for an air series geothermal heat pump are found in Table 25.2. Examples of geothermal heat pumps are shown in Photos 25.3 through 25.5.

In a geothermal heat pump system, loops of pipe are buried vertically or horizontally in the ground, or submersed in a pond or lake. Water or an anti-freeze solution is circulated through these pipes and heat is extracted from the earth. The loops are connected to a heat pump, which extracts the captured heat. The geothermal heat pump removes heat from the fluid in the earth connection, concentrates it, and then transfers it to the building. Because the earth is a heat sink, it can be used to provide cooling, in which case the process is reversed. Standard

TABLE 25.1   ENGINEERING SPECIFICATIONS FOR SEVERAL RESIDENTIAL AND LIGHT COMMERCIAL GEOTHERMAL HEAT PUMPS.

Cooling Capacity		Heating Capacity		Airflow		Water Flow		EER	COP	Power
Ton	kW	Btu/h	kW	ft³/min	m³/h	gal/min	m³/h			
0.8	2.9	6142	1.8	225	382	2.6	0.6	15.7	3.1	220 to 240 V/50 A/1φ 208 to 230 V/60 A/1φ
1.0	3.5	8189	2.4	301	511	3.5	0.8	15.7	3.2	220 to 240 V/50 A/1φ 208 to 230 V/60 A/1φ
1.3	4.5	10 236	3.0	449	763	4.0	0.9	18.4	3.2	220 to 240 V/50 A/1φ 208 to 230 V/60 A/1φ
1.5	5.4	13 307	3.9	526	893	5.3	1.2	16.4	3.0	220 to 240 V/50 A/1φ 208 to 230 V/60 A/1φ
1.8	6.2	14 330	4.2	600	1019	6.6	1.5	15.7	3.0	220 to 240 V/50 A/1φ 208 to 230 V/60 A/1φ
2.3	8.2	19 790	5.8	750	1274	8.4	1.9	15.4	3.2	220 to 240 V/50 A/1φ 208 to 230 V/60 A/1φ
2.7	9.6	23 202	6.8	900	1530	10.6	2.4	16.4	3.5	380 to 415 V/50 A/3φ 380 V/60 A/3φ 460 V/60 A/3φ
3.0	10.7	25 931	7.6	1049	1782	12.3	2.8	15.4	3.0	380 to 415 V/50 A/3φ 380 V/60 A/3φ 460 V/60 A/3φ
3.6	12.6	30 367	8.9	1349	2293	15.0	3.4	17.7	3.6	380 to 415 V/50 A/3φ 380 V/60 A/3φ 460 V/60 A/3φ
4.4	15.4	37 532	11.0	1500	2549	16.7	3.8	15.7	3.0	380 to 415 V/50 A/3φ 380 V/60 A/3φ 460 V/60 A/3φ
5.3	18.5	44 356	13.0	1574	2675	18.9	4.3	17.4	3.3	380 to 415 V/50 A/3φ 380 V/60 A/3φ 460 V/60 A/3φ

TABLE 25.2   APPLICATION TEMPERATURE LIMITS OF WATER FOR AN AIR SERIES GEOTHERMAL HEAT PUMP.

Temperatures		Cooling		Heating	
		°C	°F	°C	°F
Entering air temperature	Minimum	10	50.0	10	50.0
	Normal (Dry Bulb/Wet Bulb [DB/WB])	27/19	80.6/66.2	20	68.0
	Maximum (DB/WB)	38/28	96.8/82.4	30	86.0
Ambient air temperature	Minimum	10	50.0	10	50.0
	Normal	27	80.6	20	68.0
	Maximum	38	96.8	20	68.0
Entering water temperature	Minimum	20	68.0	10	50.0
	Normal	30	86.0	20	68.0
	Maximum	43	109.4	32	89.6

PHOTO 25.3   A 1.4 ton geothermal heat pump unit, part of the geothermal heat pump system. (Courtesy of DOE/NREL)

PHOTO 25.4   A residential-scale 3-ton geothermal heat pump unit, part of the geothermal system. (Courtesy of DOE/NREL)

PHOTO 25.5   A medium-scale 16-ton geothermal heat pump unit serving Cavett Elementary School in Lincoln, Nebraska. (Courtesy of DOE/NREL)

ductwork is normally used to distribute heated or cooled air from the geothermal heat pump throughout the building.

In winter, heat from the comparatively warmer ground is carried through a liquid-to-air geothermal heat pump and introduced into the house as heated air. In summer, hot air from the house is pulled through the liquid-to-air geothermal heat pump, and heat is discharged and deposited from the pipes into the relatively cooler ground. Heat removed during the summer can also be used to heat water. It works by concentrating naturally existing heat in the earth, rather than by producing heat through combustion of fossil fuels.

Types of GHP Ground Loops

Geothermal heat pump systems are classified by the type of ground loop that they use. The type chosen depends on the available land areas and the soil and rock type at the installation site. (See Figure 25.1.) Types of systems in use include the following.

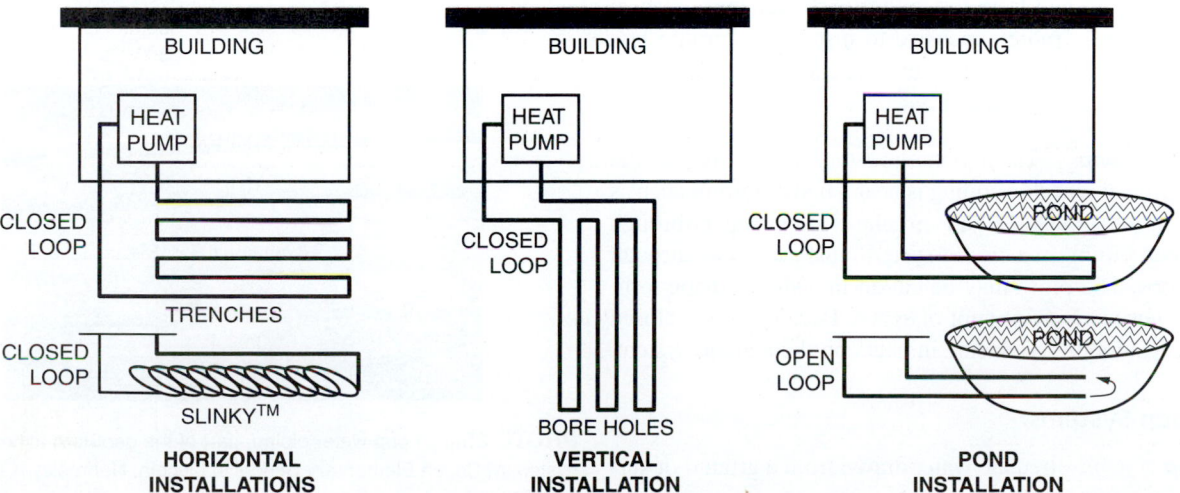

FIGURE 25.1   Geothermal heat pump systems are classified based on the type of installation of the ground loop.

Closed-Loop Systems

Closed-loop geothermal heat pump systems circulate a solution of water and antifreeze through a series of sealed loops of piping. These loops serve as a heat exchange medium between the fluid and the earth. The loops can be installed in the ground horizontally or vertically, or they can be placed in a body of water, such as a pond.

During the winter, the fluid collects heat from the earth and carries it through the system and into the building. During the summer, the system reverses itself to cool the building by pulling heat from the building, carrying it through the system, and placing it in the ground. This process creates free hot water in the summer and delivers substantial hot water savings in the winter.

Horizontal Ground Closed Loops A *horizontal ground closed-loop system* will use a number of horizontal trenches. Piping system designs range from a single pipe, to multiple pipes arranged vertically in a narrow trench, to multiple pipes in a wider trench. When sufficient land area suitable for drilling or trenching is available, horizontal loops are typically the most economically viable.

Workers use trenchers or backhoes to dig the trenches 3 to 6 ft below the ground in which they lay a series of parallel plastic pipes. They backfill the trench, taking care not to allow sharp rocks or debris to damage the pipes. Fluid runs through the pipe in a closed system. A typical horizontal loop will be 400 to 700 ft (122 to 213 m) long for each ton of heating and cooling.

Vertical Ground Closed Loops A *vertical ground closed-loop system* is used where there is little yard space, when surface rocks make digging impractical, or when you want to disrupt the landscape as little as possible. Vertical holes 150 to 500 ft (46 to 152 m) deep—much like wells—are bored in the ground, and a single loop of pipe with a U-bend at the bottom is inserted before the hole is backfilled. Each vertical pipe is then connected to a horizontal underground pipe that carries fluid in a closed system to and from the indoor exchange unit. Vertical loops are generally more expensive to install, but require less piping than horizontal loops because the earth's temperature is more stable farther below the surface.

Pond Closed Loops The *pond closed-loop system* may be the most economical when a building is near a body of water such as a shallow pond or lake. Fluid circulates underwater through polyethylene piping in a closed system, just as it does through ground loops. The pipes may be coiled in a slinky shape to fit more of it into a given amount of space. Because it is a closed system, it results in no adverse impacts on the aquatic system.

Open-Loop Systems

Open-loop systems circulate water drawn from a ground or surface water source (e.g., pond, lake). Once the heat has been transferred into or out of the water, the water is returned to a well or surface discharge (instead of being recirculated through the system).

Open-loop systems operate on the same principle as closed-loop systems and can be installed where an adequate supply of suitable water is available and open discharge is feasible. Benefits similar to the closed-loop system are obtained.

Standing Column Well System

Standing column wells, also called *turbulent wells* or *Energy Wells*™, have become an established technology in some regions, especially the northeastern United States. Standing wells are typically 6 inches in diameter and may be as deep as 1500 ft (450 m). Temperate water from the bottom of the well is withdrawn, circulated through the heat pump's heat exchanger, and returned to the top of the water column in the same well. Usually, the well also serves to provide potable water. However, groundwater must be plentiful for a standing well system to operate effectively. If the standing well is installed where the water table is too deep, pumping would be prohibitively costly. Under normal circumstances, the water diverted for building (potable) use is replaced by constant-temperature groundwater, which makes the system act like a true open-loop system. If the well water temperature climbs too high or drops too low, water can be "bled" from the system to allow groundwater to restore the well water temperature to the normal operating range. Permitting conditions for discharging the bleed water vary from locality to locality, but are eased by the fact that the quantities are small and the water is never treated with chemicals.

Photos of various geothermal loop installations during construction are shown in Photos 25.6 through 25.12. Ground loops are easily installed during rough grading of the building site and construction of the building. The addition of a geothermal heat pump ground loop system to an existing building will require disruption of the building site, which will typically increase overall installation costs.

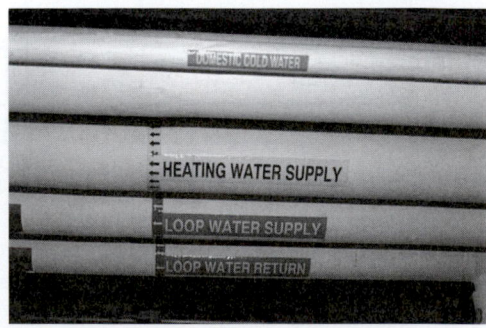

PHOTO 25.6 Loop water piping, part of the geothermal heat pump system at Cavett Elementary School in Lincoln, Nebraska. (Courtesy of DOE/NREL)

PHOTO 25.7 A 200 ft (61 m) loops as shipped from the manufacturer.
(Courtesy of DOE/NREL)

PHOTO 25.10 A commercial scale pump that circulates water from a geothermal heat pump to the ground source heat exchangers. (Courtesy of DOE/NREL)

PHOTO 25.8 Placement of a Slinky® loop as part of a horizontal ground closed-loop system. (Courtesy of DOE/NREL)

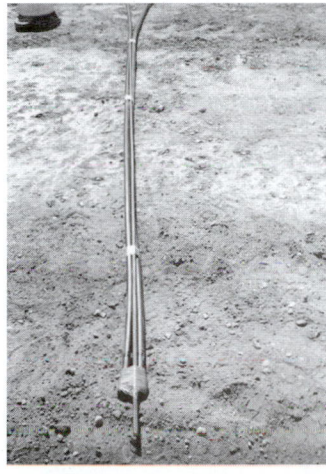

PHOTO 25.11 A 200 ft (61 m) loop being pressure tested. Note the 10-ft piece of rebar that is duct taped to the bottom of the loop. This rebar will help keep the loop straight as it is inserted into the hole. It also adds weight to help keep the loop in the hole. (Courtesy of DOE/NREL)

PHOTO 25.9 Drilling boreholes for a vertical ground closed-loop system. (Courtesy of DOE/NREL)

PHOTO 25.12 Vertical loops are installed and backfilled with coarse sand to a level of 30 ft (10 m), and then a 20% solid bentonite grout is pumped down to fill the remainder of the hole. (Courtesy of DOE/NREL)

GHP Site Evaluation

Because shallow ground temperatures are relatively constant throughout the United States and Canada, GHPs can be effectively used almost anywhere. However, the specific geological, hydrological, and physical characteristics of the site will help determine the best type of ground loop for a specific site.

Geology

Factors such as the composition and properties of soil and rock (which can affect heat transfer rates) require consideration when designing a ground loop. For example, soil with good heat transfer properties requires less piping to gather a certain amount of heat than soil with poor heat transfer properties. The amount of soil available contributes to system design as well. GHP system suppliers in areas with extensive hard rock or soil too shallow to trench may install vertical ground loops instead of horizontal loops.

Hydrology

Ground or surface water availability also plays a part in deciding what type of ground loop to use. Depending on factors such as depth, volume, and water quality, bodies of surface water can be used as a source of water for an open-loop system, or as a repository for coils of piping in a closed-loop system. Groundwater can also be used as a source for open-loop systems, provided the water quality is suitable and all groundwater discharge regulations are met.

Land Availability

The amount and layout of the land, the landscaping, and the location of underground utilities or sprinkler systems also contribute to system design. Horizontal ground loops (generally the most economical) are typically used for newly constructed buildings with sufficient land. Vertical installations or more compact horizontal (e.g., Slinky™) installations are often used for existing buildings because they minimize the disturbance to the landscape.

Land Area Requirements

The typical land area required for a groundwater system is usually not a crucial design factor. An estimate based on a 6 m radius per well, including the presence of injection wells, can be calculated. The typical land area for vertical closed-loop systems can be based on an average borehole depth of 91 m and a spacing of 5 m between the boreholes. Common land areas required for vertical systems can vary significantly but are usually around 5 to 10 m²/kW. Horizontal systems require the most land area. The amount of land required varies with the layout and piping arrangement required to minimize the pumping power. Typical values for land area for horizontal systems are as shown in Table 25.3.

TABLE 25.3   TYPICAL LAND AREA REQUIRED FOR GEOTHERMAL HEAT PUMP HORIZONTAL LOOP.

Configuration	Northern Climate Regions of North America		Southern Climate Regions of North America	
	ft²/(MBtu/hr)	m²/kW	ft²/(MBtu/hr)	m²/kW
One pipe	249	79	249	79
Two pipe	167	53	293	93
Four pipe	126	40	208	66
Six pipe	126	40	208	66

GHP System Efficiency

The efficiency of a heat pump varies as the temperatures and flows of the liquid and air pumped through the heat pump change. Manufacturers publish the ratings of their heat pump on the basis of a specific set of standard conditions called the ISO 13256-1 rating. The rating for a closed loop system is called the *ground loop heat pump (GLHP)* rating; and, the rating for an open loop or groundwater system is called the *groundwater heat pump (GWHP)* rating. In the heating season, a geothermal heat pump supplies 3 to 5 units of useable heat for every unit of electrical energy required to operate the system. This equates to 3 to 5 kilowatt-hours (kWh) of free energy for every one kWh of electrical energy consumed. In other words, a ground-source heat pump is 300% to 500% efficient.

25.4   BIOMASS

Introduction

Biomass is organic material from plant and animal growth. It includes wood, agricultural crops (e.g., corn, soybeans, sunflowers, and sugarcane), energy crops, and other organic by-products from a variety of agricultural processes (e.g., straw, stalks, and manure). Biomass is sunlight stored in the form of organic compounds that are formed in growing plant life through the process of photosynthesis, which takes energy from the sun and converts carbon dioxide into carbohydrates (sugars, starches, and cellulose). From the time when early humans began burning wood in campfires for heat, biomass has been a source of energy.

Today, most biomass is converted to energy by burning it for commercial and residential space heating and for industrial process heat and steam (known as *bioheat*). Biomass energy resources can be burned to create steam for generating electricity (called *biopower*). Additionally, biomass products can be refined into alternative fuels (*biofuel*), or into a variety of bio-based products (*bioproducts*) that replace petroleum-based goods (e.g., antifreeze, plastics, artificial sweeteners, toothpaste gel, lubricants, textile fibers, adhesives/glues, and soy-based inks).

The actual source of biomass differs depending on location, availability, and end use. Biomass can come from:

- Harvested timber from local forests
- Wood wastes from timber thinning (i.e., wood collected in forest fire mitigation efforts)
- Residue from paper mills, woodworking shops, and forest operations (e.g., sawdust, shavings, wood chips, and recycled untreated wood)
- Agricultural and animal wastes (e.g., manure and litter)
- Food processing (e.g., nutshells, olive pits)
- Garbage from paper, plant, or animal products (e.g., food scraps, lawn clippings, leaves, wood-based construction debris), but not made out of glass, plastic, and metals
- Intentionally grown energy crops (e.g., fast-growing native trees and grasses) and agricultural crops

Biomass is presently the leading source of renewable energy in the United States. In 2003 it accounted for 47% of all renewable energy or almost 4% of the total energy produced in the United States. Biomass currently provides about 2% of U.S. electricity, about 1% of the fuel used in cars and trucks, and some of the heat and steam used by homes, businesses, and schools. Industrial consumption of biomass by the pulp and paper industry makes up about half of the biomass energy consumed in the United States.

Biomass represents an enormous quantity of renewable energy in the United States, thereby increasing the potential for national energy independence. Additionally, biomass energy has the possibility of revitalizing rural economies and reducing pollution. It is regarded as green source of energy for several reasons. Biomass fuels can be infinitely renewable if resources such as energy crops and forests are properly managed. Additionally, biomass is neutral in terms of carbon dioxide (CO_2) emissions. Burning of biomass generates about the same amount of carbon dioxide as fossil fuels, but every time a new plant grows, carbon dioxide is actually removed from the atmosphere. As long as plants continue to be replenished for use as biomass energy, the net emission of carbon dioxide is zero unlike fossil fuels.

Types of Biomass Fuels

Several sources of biomass energy exist.

Solid Fuels

Current biomass consumption in the United States is dominated largely by wood and wood residue. Domestic and industrial heating with wood is the most efficient and competitive way of using biomass for energy. Common types of biomass fuels used in buildings include timber (logs) and manufactured logs, wood pellets, briquettes, and corn. An example of wood residue used

PHOTO 25.13 Wood residues from pulp and paper manufacturing, lumber mills, and other industrial wood users are used for producing biomass electricity. (Courtesy of DOE/NREL)

as a biomass fuel is shown in Photo 25.13. Manufactured fuels are made from wood wastes, wastepaper, cardboard, and agricultural residues or forest industry by-products.

Buildings, which are generally heated with a furnace or boiler that consumes fossil fuels (e.g., natural gas or fuel oil) as the primary fuel source, can use biomass for heating. For example, in Vermont, about two dozen schools heat with clean, efficient wood chip heating systems and these systems are being included in new school construction. Many other types of solid biomass fuels can be used as an alternative to fossil fuels. For example, corn is currently utilized for feed in the agricultural industry. It is readily available, flows smoothly in augur systems, is inexpensive when compared to fossil fuel, is renewable, does not contaminate the ground or groundwater, is nonexplosive, and is abundant.

Pellets, briquettes, and manufactured logs are different forms of dense fuels. These biomass fuels are made from wood wastes, wastepaper, and cardboard and agricultural residues. The first steps in densification are reducing the biomass feedstock to particles ¼ in or less in diameter and drying the feedstock to a moisture content of 10 to 15%. Mechanical compression or extrusion then forms the material into a product that has less than one-third of the feedstock's original volume.

Liquid Fuels

Alternative liquid transportation fuels (e.g., ethanol, biodiesel) can be produced from agricultural crops, mainly corn, soybeans, and sunflowers, instead of petroleum. Ethanol and biodiesel can be blended with or directly substituted for gasoline and diesel. *Ethanol* is an alcohol that is made from any biomass high in carbohydrates (starches, sugars, or celluloses), typically corn, through a fermenting process similar to brewing beer. Ethanol is presently used as a fuel additive to cut down a vehicle's carbon monoxide and other smog-causing emissions. Flexible-fuel vehicles, which run on mixtures of gasoline and up to 85% ethanol, are now available. *Biodiesel* is made by combining alcohol (usually methanol) with vegetable oil, animal fat, or recycled cooking greases. It can be used as an additive to reduce vehicle emissions (typically 20%) or in its pure form as a renewable alternative fuel for diesel engines. *Black liquor* is a recycled by-product formed during the pulping of wood in the papermaking industry that can be used as a source of energy. It is burned in a boiler to produce electricity and steam.

Gaseous Fuels

Methane gas is a combustible gas produced by the anaerobic, or oxygen-free, digestion of vegetable and/or animal wastes. Methane gas can be used to generate power. Within the agricultural sector, there is growing interest in using farm wastes to produce methane. Anaerobic methane digesters can offer three key benefits to farms: power production along with odor reduction and improved nutrient management.

When biomass is heated in the absence of oxygen it gasifies to a mixture of carbon monoxide and hydrogen, which is known as *synthesis gas* or *syngas*. Syngas burns more efficiently

and cleanly than the solid biomass from which it was made. Like natural gas, syngas can be burned to drive gas turbines, a more efficient electrical generation technology than steam boiler–turbine systems to which solid biomass and fossil fuels are limited. In a variety of conversion processes, syngas can be converted to liquid fuels known as *synthol*, which can be used as a transportation fuel.

The energy content of biomass fuels varies significantly with the biomass product. The average energy content of different conventional and biomass resources is provided in Table 25.4.

Types of Biomass Technologies

Several technologies exist to extract energy from raw biomass, which include the following.

Direct Combustion

The *direct combustion* process makes use of a furnace or boiler to convert biomass fuel into hot air, hot water, steam, and electricity for commercial or industrial uses. Steam drives a turbine, which turns a generator that produces electricity. In some

TABLE 25.4 AVERAGE ENERGY CONTENT OF DIFFERENT CONVENTIONAL AND BIOMASS RESOURCES. VALUES ARE APPROXIMATE.

Energy Source	Unit	Energy Content per Unit	
		Btu	kJ
Conventional Source			
Electricity	Kilowatt-hour	3413	3601
Natural gas	Cubic foot	1008 to 1034	1063 to 1091
Natural gas	Therm	100 000	105 500
Liquefied petroleum gas (LPG)	Gallon	95 475	100 726
Crude oil	Gallon	138 095	145 690
Crude oil	Barrel	5 800 000	6 119 000
Kerosene or light distillate oil	Gallon	135 000	142 425
Diesel fuel	Gallon	138 690	146 318
Residential fuel oil	Gallon	149 690	157 923
Gasoline	Gallon	125 000	131 875
Coal	Pound	8100 to 13 000	8546 to 13 715
Coal (subbituminous)	Pound	8000 to 12 000	8440 to 12 660
Biomass Source			
Biodiesel	Gallon	74 000 to 82 000	78 070 to 86 510
Biogas (anaerobic digestion)	Cubic foot	600	633
Ethanol	Gallon	84 400	89 042
Gasohol (10% ethanol, 90% gasoline)	Gallon	120 900	127 550
Methanol	Gallon	62 800	66 254
Mill waste (lower quality), 30 to 40% mc	Pound	4500	4748
Municipal solid waste (MSW), 30 to 40% mc	Pound	4500	4748
Paper, 15% mc	Pound	7500	7913
Shelled corn	Pound	7000 to 10 000	7385 to 10 550
Shelled corn, 15% mc	Pound	7000	7390
Straw, stalks, and stubble, 15% mc	Pound	7500	7910
Wood, 30% mc	Pound	5200	5490
Wood, 20% mc	Pound	6250	6540
Wood, 8% mc	Pound	7500	7910

mc: Moisture content.

industries, the steam from the power plant is also used for manufacturing processes or to heat buildings. Biomass such as wood logs, pellets, manufactured logs, and corn can also be burned directly in a fireplace or stove to heat a building or heat water. District energy systems produce heat at a central source and distribute it through pipes to several buildings. Utility-size, coal-fired power plants can use biomass as a supplementary energy source in high-efficiency boilers, which significantly reduces sulfur dioxide emissions.

Anaerobic Digestion

Anaerobic digestion is a biochemical process in which groups of bacteria, working symbiotically, break down complex organic wastes in animal manure and food-processing residue to produce biogas, mainly a methane and carbon dioxide mix. Controlled anaerobic digestion requires an airtight chamber, called a *digester,* which must maintain a temperature of at least 68°F (20°C). In landfills, wells can be drilled to release the methane from the decaying organic matter. Then pipes from each well carry the gas to a central point where it is filtered and cleaned before burning. Biogas is burned in a boiler to produce steam for electricity generation or for industrial processes.

Gasification

Biomass *gasification* is a thermochemical process that converts biomass into a combustible gas called *producer gas* that contains carbon monoxide, hydrogen, water vapor, carbon dioxide, tar vapor, and ash particles. This biogas has a low to medium heating value, depending on the process used. This biogas can be used for heating or to generate electricity.

Pyrolysis

In the *pyrolysis* process, very small, low-moisture particles of biomass fuel are rapidly heated to temperatures in the range of 840° to 1020°F (450° to 550°C) in the absence of oxygen, resulting in liquid pyrolysis oil, which can be used as a synthetic fuel oil. Although these oils can be used for heating or to generate electricity, they tend to be more valuable as chemicals for use in making other bioproducts.

Biofuels

Production of *biofuels* such as ethanol (alcohol), biodiesel, and black liquor involve different processes. Fermentation is the biochemical process that converts sugars (predominantly from corn) into ethanol. Biodiesel production is a chemical conversion process that converts oilseed crops into biodiesel fuel. Production of methanol (wood alcohol) involves a thermochemical conversion process that converts wood and agricultural residues into methanol. Although these fuels can be used for heating or to generate electricity, they tend to be more valuable as alternative fuels.

PHOTO 25.14 A small-scale modular biopower system. It is comparable to diesel/gas engine-generator technology except that wood chips are used as fuel. The wood chips are converted to a fuel gas (biomass gasification) and this gas fuels the engine. (Courtesy of DOE/NREL)

PHOTO 25.15 Biomass heat exchanger furnace can burn corn husks, wood residue, or other biomass fuels to produce warm air for space heating. (Courtesy of DOE/NREL)

Municipal Solid Waste

A *municipal solid waste (MSW)* program makes use of trash as a source of energy by either burning MSW in a waste-to-energy plant or by capturing biogas. In waste-to-energy plants, trash is burned to produce steam that can be used as process heat, to space heat buildings, or generate electricity. In landfills, biomass rots and releases methane gas, called *landfill gas.* Some landfills have a system that collects the methane gas so that it can be used as a fuel source.

Examples of biomass installations are shown in Photos 25.14 and 25.15.

Building Heating with Biomass

Biomass heating is a proven technology for space heating and domestic water heating in buildings. Timber, woodchips, pellets, and corn are popular biomass fuels that can replace conventional fuels (i.e., oil, gas, and coal) in buildings. The most significant form of biomass heating in the United States is wood burning. *Timber* can be harvested from a forest. Waste wood (i.e., waste timber from logging operations, beetle-killed trees) can be used. *Woodchips* typically come from a lumber mill or are forest residue. Large stationary chippers used at sawmills and trailer-mounted, whole-tree chippers used in the forest mechanically reduce logs, whole trees, or rejected lumber to chips of relatively uniform size, usually ³⁄₁₆ to 1 in (5 to 25 mm) thick. Woodchips typically have a moisture content of 30% or less. *Pellets* are made of compacted sawdust, wood bark, or a number of other mill or agricultural by-product materials. Wood pellets typically have a moisture content of 8 to 10%, which means they are more efficient and produce more heat per unit weight than woodchips. Some furnaces, boilers, and stoves can also accommodate other biomass fuels (i.e., corn, dried cherry pits, rice husks, nutshells, orchard prunings, and soybeans), but this tends to be driven by availability of indigenous fuels. For large-scale installations, woodchips are more cost-effective than pellets because woodchips are easier to handle in bulk. Because of convenience, pellets tend to be a more suitable fuel for smaller residential installations even though they are costly. Pellet stoves have lower smoke emissions and higher steady state combustion efficiencies.

Most biomass-fueled central heating systems in large commercial buildings create either hot water or steam in a boiler, which is then delivered through a conventionally piped hot water or steam distribution system. Some systems use a furnace that creates warm air that is delivered through a conventional duct distribution system. A multifaceted approach would be a biomass-fueled CHP plant, which can meet power needs while also producing heat. They have fully automated thermostatic control, precise automatic fuel feed, and a precisely controlled air supply ensuring optimum air-fuel ratios. These controls enable modern biomass heating systems to run at very high efficiencies, usually 85 to 90%.

Storage bins are required to store biomass fuel before it is burned. A *day bin* is a solid fuel storage bin that holds enough fuel to last approximately one day (or even a weekend). On larger commercial systems (i.e., school), a tractor with front-end bucket moves the fuel from a storage pile or bin to the day bin. This process is performed by an operator, once or twice daily. A *metering bin* is a small bin in the fuel feed stream just upstream of the boiler or furnace that allows a precise feed rate, or metering, of the fuel to the fire. An automated *fuel-handling system* moves the fuel (no operator labor) from the day bin to the metering bin to the combustion chamber of the boiler or furnace. An *injection auger,* which is a principal part of the fuel-handling system, moves a measured amount of solid fuel into the combustion chamber. Typically, the system is controlled with an on/off fuel feed that delivers fuel to the grates on an intermittent basis in response to water or air temperature and building heating load variations. Some systems use a *modulating fuel feed,* which adjusts the feed rate up or down in response to changes in the heating load. On large systems, a *char reinjector* collects unburned char and injects it back into the combustion chamber,

both to capture its energy through recombustion. As the fuel burns, ash collects under the grates of the combustion chamber. An automated or manually operated *ashing auger* discharges ash as the fire burns. Commercial boiler and furnace systems are rated by a thousand Btu per hour (MBH) (kW) output.

A residential installation consists of a stove, furnace, or small boiler. Freestanding stoves and fireplace inserts are designed to sit within the room and will heat a large space. A biomass-burning furnace heats air that is distributed through a conventional duct system. A biomass-burning boiler heats water that is distributed through a conventional pipe system. Other than burning cordwood, pellet fuel is a common biomass fuel for residential installations. It is typically sold in 40 lb (15 kg) bags. It is rated by ash content, with premium pellets having low ash content. Low ash content of pellet and other biomass fuels means less cleaning. Pellet fuel produced about one-third the particulate emissions of a conventional wood-burning stove. Matching the appropriate fuel to the stove is crucial for proper operation and efficient burning. Pellet fuel is placed in a hopper (weekly), which is connected to an electrically powered auger or feeding system. Electricity powers blowers that distribute warmed air into the space. Modern pellet stoves have automatic ignition, use a programmable thermostat and have self-cleaning capabilities, which automatically discharges ash as the fire burns and only requires emptying of the ash pan about once a week. Residential stoves and furnaces are rated by Btu/hr (W) output. With stoves, higher output relates to a larger space the unit can heat.

25.5 PHOTOVOLTAIC (SOLAR ELECTRICITY) POWER

PV Cells

A *photovoltaic (PV) cell* is a power-generating device that produces direct current electricity when it is exposed to light. PV cells are constructed of a semiconducting material, usually silicon, the same material used to make computer chips. When a cell absorbs light, electrons in the semiconducting material are dislodged, producing a flow of electrical current. This electricity can be used to power electrical appliances and equipment. The simplest PV cells power watches and calculators, while more complex PV systems can be used to power buildings. Figure 25.2 shows a schematic drawing of a simple photovoltaic circuit.

A typical PV cell consists of a weatherproof glass cover, an antireflective layer to keep light from reflecting away from the cell, a top metallic grid that serves as a contact to allow the electrons to enter a circuit, and a back contact layer to allow them to complete the circuit. Sandwiched between these contacts are the semiconductor layers where electron (electrical) flow develops. These layers are made of various thin-film materials. Presently, crystalline silicon and amorphous silicon are the PV materials used most often in residential applications.

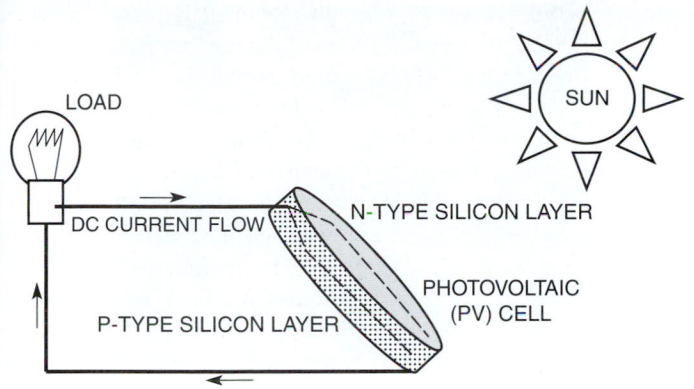

FIGURE 25.2 A schematic drawing of a simple photovoltaic circuit.

PHOTO 25.17 A PV system integrated into a horizontal roof consisting of 372 panels at the Williams Building in downtown Boston, Massachusetts. (Courtesy of DOE/NREL)

Solar radiation is composed of photons (particles) that contain energy. When a PV cell absorbs a photon, the energy of the absorbed photon is transferred to an electron in the semiconductor material that makes up the PV cell. This supplementary energy causes the electron to jump to a higher orbit and jump from atom to atom. A permanent electric field built into the cell forces electron movement in a particular direction: out of the cell, through an electrical circuit, and back to the other side of the cell. Electron flow is electricity, which can be used to drive electrical appliances and equipment.

Building-integrated PV (BIPV) materials are a special type of PV cell that is manufactured with the dual purpose of generating electricity and serving as a construction material. BIPV materials can replace customary building components such curtain walls, skylights, atrium roofs, awnings, roof tiles and shingles, and windows.

Examples of PV and BIPV installations are shown in Photos 25.16 through 25.24. Many of these photos show how BIPV can be incorporated within the design strategy of the building.

PHOTO 25.18 A 7-kW photovoltaic awning is situated above the eighth-floor windows of the 26-story University Center Tower on the Texas Medical Center campus in Houston, Texas. (Courtesy of DOE/NREL)

PHOTO 25.16 A 7.2 kW PV rooftop array at the Visitor Center at Zion National Park, Springdale, Utah, provides electricity. It provides about 12% of the Center's power needs. This Center was introduced at the end of Chapter 11. (Courtesy of DOE/NREL)

PHOTO 25.19 A 540-kW PV system provides shaded parking for about 1000 cars in a desert of scorching blacktop at the Cal Expo in Sacramento, California. (Courtesy of DOE/NREL)

PHOTO 25.20 Building-integrated PV (BIPV) at the Alfred A. Arraj U.S. Federal Courthouse, Denver, Colorado. (Courtesy of DOE/NREL)

PHOTO 25.21 BIPV shingles serve to substitute for composition shingles. (Courtesy of DOE/NREL)

PHOTO 25.22 BIPV shingles at the BigHorn Home Improvement Center retail complex in Silverthorne, Colorado, with a design capacity of 8 kW. (Courtesy of DOE/NREL)

PHOTO 25.23 BIPV panels. (Courtesy of DOE/NREL)

PHOTO 25.24 Installing flexible PV roofing shingles, rated at 17 W each. (Courtesy of DOE/NREL)

PV Cell Conversion Efficiency

The *conversion efficiency* of a PV cell is the proportion of solar radiation that the cell converts to electrical energy. Because of the low power density of sunlight, high efficiency is very important in making PV energy competitive with more traditional sources of energy (e.g., fossil fuels). Commercially available PV cells have a conversion efficiency of 8 to 20%. Experimental PV cells have efficiencies greater than 23%. By comparison, the earliest PV devices converted only about 1% of solar radiation into electric energy.

Commercially available PV cells measure between ½ to 4 in (12 to 100 mm) square and produces a maximum of about 1 to 2 W of power in direct sunlight. Output and dimensions of selected commercially available PV cells are provided in Table 25.5.

PV Cell Configurations

Cells are grouped together in a *PV module* that creates a power source capable of lighting a small electric lamp. A PV module can be made in a variety of shapes. To generate power required in a building, PV modules must be grouped together in a *PV array* that when combined with other components make up a *PV system*. Most building-scale systems involve fixed arrays

TABLE 25.5 OUTPUT AND DIMENSIONS OF SELECTED COMMERCIALLY AVAILABLE PHOTOVOLTAIC CELLS. COMPILED FROM VARIOUS MANUFACTURERS INFORMATION.

Current (A)		Power[a] (W)		Dimensions		
Maximum Output	Nominal Output	Maximum Output	Nominal Output	Metric (SI) Units mm × mm	Customary (U.S.) Units in × in	Shape
0.525	0.500	0.30	0.29	62.5 × 32	2.5 × 1.25	Square
3.5	3.4	2.03	1.97	103 × 103	4.0 × 4.0	Square
4.7	4.5	2.73	2.73	125 × 125	5.0 × 5.0	Square
0.075	0.065	0.04	0.04	26 × 10	1.0 × 0.4	Rectangular
0.30	0.25	0.17	0.15	17 × 62.5	0.7 × 2.5	Rectangular
0.310	0.300	0.18	0.17	62.5 × 17	2.5 × 0.7	Rectangular
0.42	0.40	0.2	0.23	25 × 62.5	1.0 × 2.5	Rectangular
1.4	1.2	0.81	0.70	62.5 × 62.5	2.5 × 2.5	Rectangular
2.5	2.3	1.45	1.33	125 × 62.5	5.0 × 2.5	Rectangular
3.7	3.5	2.15	2.03	103 × 103	4.0 × 4.0	Psuedosquare
4.8	4.5	2.78	2.62	125 × 125	5.0 × 5.0	Pseudosquare

[a]Based on 0.58 V.

mounted on rooftops to minimize installation and maintenance costs. Tracking systems that keep the PV array directed toward the sun are also available. See Figure 25.3.

PV Cell Position

Theoretically, a true south-facing orientation offers the maximum collection potential. Southeast- or southwest-facing arrays are also adequate and may even be advantageous depending on the power requirements and time of loads on the building power system. The most effective tilt angle for fixed PV arrays depends on latitude. The PV array should be tilted at an angle in a range equal to the latitude plus or minus 15° relative to the horizontal plane.

PV systems should be positioned to maximize exposure to the sun, especially during the prime generation time between 10:00 AM and 3:00 PM. Even small shadows falling on a module will dramatically reduce the amount of power generated. The best position and orientation can also depend on site-specific factors such as local features such as trees, hills, or other buildings that may shade the PV array during certain times of the day.

Roof surfaces are ideally suited for PV integration in a building because there is usually little shadowing at roof heights and a roof usually provides a large surface for integration. PV modules provide the same architectural appearance as tinted glass so they can be situated on a facade or in spandrel panel areas where they look like tinted windows. BIPV systems are also suitable.

Photovoltaic Systems

A *photovoltaic (PV) system* collects solar energy and converts it to electricity. A basic PV system incorporates the following components:

- An array of solar cells that converts sunlight into DC electricity
- An inverter that changes DC electricity into AC electricity
- A connection to the utility grid for additional power or a bank of batteries to store collected electricity

There are several generic types of PV systems.

Off-Grid PV Systems

Stand-alone, off-grid, or *autonomous PV systems* produce power independently of the utility grid. In remote locations, stand-alone systems can be more cost-effective than extending a power line to the electricity grid (a cost that can range from $15 000 to $50 000 per mile). In some locations as near as one-quarter mile from utility power lines, stand-alone PV energy systems can be

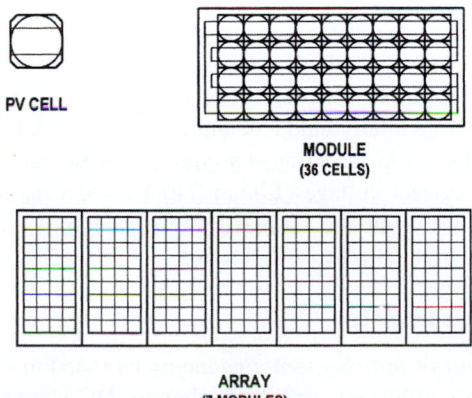

FIGURE 25.3 Photovoltaic (PV) cells are grouped together in a PV module. To generate power required in a building, PV modules must be grouped together in a PV array that when combined with other components make up a PV system.

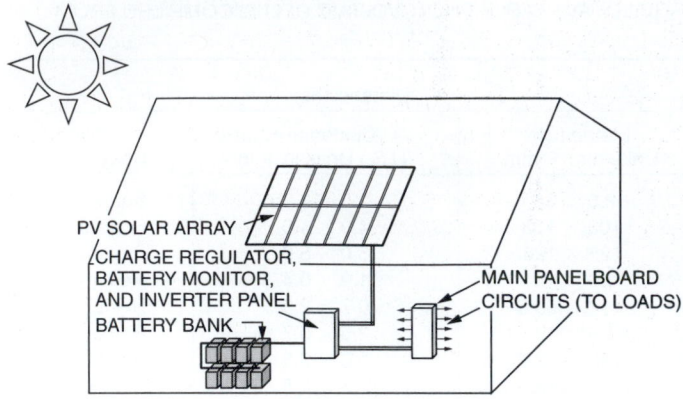

FIGURE 25.4   A schematic of a stand-alone, off-grid photovoltaic (PV) system.

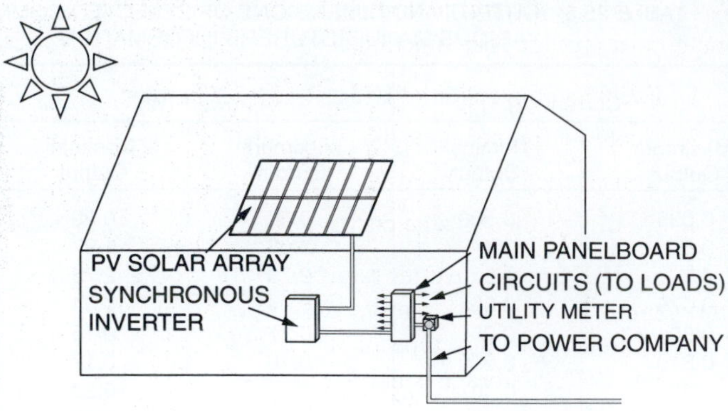

FIGURE 25.5   A schematic of a grid-connected photovoltaic (PV) system.

more cost-effective than extending power lines. They are especially appropriate for remote, environmentally sensitive areas, such as national parks, cabins, and remote homes. A schematic of a stand-alone, off-grid PV system is shown in Figure 25.4.

In rural areas, small stand-alone PV systems power farm lighting, fence chargers, and water pumps, which provide water for livestock. *Direct-coupled systems* need no electrical storage because they operate only during daylight hours, but most systems rely on battery storage. Some systems, called *hybrid systems,* combine solar power with additional power sources (e.g., wind turbine or diesel generator).

Battery systems can supply the owner with reserve power whenever energy demand exceeds that delivered by the PV system. This reserve power is needed during calm spells, but in situations where the storage capacity is taxed beyond its limits, a backup system, such as a portable gasoline or diesel generator, may be necessary. By combining two or more sources of energy, the size of energy storage can be decreased.

Grid-Connected PV Systems

Grid-connected PV systems interface with the utility grid. They supply surplus power back through the grid to the utility and obtain power from the utility grid when the building system's power supply is low. These systems eliminate the need for battery storage, although arranging for grid interconnection with the utility company can be complicated.

In this system, one side of a synchronous inverter is typically connected to the DC power source (e.g., PV array) and the other side is connected through a meter to the panelboard. This configuration enables the utility to measure the amount of power produced. Some utilities only use a single meter: the meter turns backwards when the system is producing more power than it is consuming. This measuring technique is called *net metering.* By law, public utilities allow net metering, which allows the PV system operator to sell excess power back to the

utility, but this is generally at a significantly reduced fee than what the utility charges. Although a grid-connected system does not constitute true storage, it provides power on demand at any time, in any amount. The possibilities for interconnection should be investigated early in the process of exploring the potential for a PV system. See Figure 25.5.

Inverters

An *inverter* transforms DC power to AC power to supply electricity to AC equipment and appliances and/or to feed excess power into the utility lines or to battery storage. When a DC power source (e.g., PV array, wind turbine) generates electricity, the inverter powers any appliances in use. If the generated power exceeds the load from the appliances, the inverter feeds excess power into the utility line or charges batteries. If the generated power cannot meet the load, power is drawn into the house from the utility line or from the batteries.

Types of inverters include the following.

Static Inverter

A *static, stand-alone,* or *battery-charging inverter* is used with an independent power system to draw power from battery storage. They operate totally independent of a utility grid. Inverters that convert DC to AC are typically equipped with features that produce the correct voltage (120 or 240 V) and constant frequency (60 cycles in the United States) of electricity, even when the electrical charge is fluctuating.

Synchronous Inverter

A *synchronous* or *utility-interactive inverter* is used in systems connected to a utility power line. It changes DC power to AC power and synchronizes its characteristics with those in the utility grid as it feds the power into the utility lines. This type of inverter produces AC electricity in synchronization with the power line and is of a quality acceptable to the utility company.

A synchronous inverter and related power conditioning equipment must provide four functions:

1. Efficient conversion of DC power to AC power
2. Accurate frequency of the AC cycle
3. Consistency of voltage
4. Quality of the AC sine curve

A synchronous inverter requires utility power to operate and cannot be used in a remote site, away from utility lines. Power output from this type of inverter is sold to the utility company. Battery storage is not used.

Multifunction Inverter

A *multifunction inverter* performs a dual purpose. It can operate as stand-alone inverter and serve as a synchronous inverter at the same time. Power output is sold to the utility company or fed to battery storage. The battery storage configuration is typically for emergency backup power.

Table 25.6 provides sample specifications of selected static and synchronous inverters. Inverters manufactured in the United States are typically designed for an input of 12, 24, 32, or 48 V direct current (V DC) and produce 120 V, 60-cycle

TABLE 25.6   SPECIFICATIONS OF SELECTED STATIC AND SYNCHRONOUS INVERTERS.

DC Input Volts	AC Volts/ Hertz	Continuous Watts	Surge Watts	Battery Charger Amps	On-Grid Capabilities
Static (Modified Sine Wave) Inverters					
12	120/60	1500	3300	0 to 70	No
12	120/60	1500	3300	0 to 70	No
12	120/60	2400	6200	0 to 120	No
24	120/60	1500	4800	0 to 35	No
24	120/60	2400	8500	0 to 70	No
24	120/60	3600	12 000	0 to 70	No
48	120/60	2500	6200	0 to 35	No
Synchronous (True Sine Wave) Inverters					
12	120/60	2500	5000	0 to 130	Yes
12	120/60	2500	7200	0 to 150	Yes
24	120/60	2500	6000	0 to 65	Yes
24	120/60	4000	9300	0 to 150	Yes
48	120/60	4000	9300	0 to 60	Yes
48	120/60	4000	—	n/a	Yes
48	240/60	4000	—	n/a	Yes
48	120/60	5500	9300	0 to 70	Yes
48	120/60	5500	—	n/a	Yes
48	240/60	5500	—	n/a	Yes
100	120/60	1000	—	n/a	Yes
125	240/60	1000	—	n/a	Yes
125	240/60	1500	—	n/a	Yes
125	240/60	2000	—	n/a	Yes
125	240/60	2500	—	n/a	Yes
600	240/60	2500	—	n/a	Yes

alternating current (V AC). Some inverters can be connected in series to produce 240 V AC. A second option is to use a step-up transformer, which increases the voltage from 120 V AC to 240 V AC. These manufacturers also produce 220 V, 50 Hz models for application in countries outside of North America. Popular sizes provide between 50 and 5000 W of AC power. Small inverters can power individual appliances; a large inverter can handle the load for all the appliances in a house or small commercial building.

Inverters have two output ratings: continuous and surge (or peak intermittent). The *continuous rating* is the maximum amount of power available on an uninterrupted basis. This rating relates to the continuous load an inverter can handle. The *surge rating* relates to the momentary surge of power (e.g., from startup of motors) the inverter can handle without shutting off to protect itself from overload. An inverter must be capable of powering every device connected to it that will be operated at the same time.

Inverter efficiency ratings are typically based on a partial power output point where maximum efficiency is achieved. Some inverters operate very efficiently at one-quarter or one-half of their rated capacity. With some inverters, efficiency may drop off on either side of this output. Consequently, it is prudent to match inverter capacity with the load as closely as possible. Quasi or modified sine-wave inverters typically have a higher operating efficiency than pure sine-wave inverters. The efficiency of modern inverters is typically in the range of 85 to 95%. This means that between 5 and 15% of the power converted is lost.

Some inverters draw power even when no appliances are operating. This idle or standby power drain can be as low as a few watts to as high as 10% of the maximum continuous power rating. More efficient inverters have a load-sensing feature that senses when a device is using the inverter and turns off the inverter when all loads are shut down. It uses only a small amount of electricity in standby mode.

Inverter Interference

Inverters may cause interference with electronic signals (e.g., radio and television reception, noise on telephones and audio equipment). Locating the inverter very close to the batteries, twisting together cables that connect the inverter to the battery, and locating the inverter away from appliances that are susceptible to interference can minimize interference.

Inverter Configuration

Synchronous inverters are also designed to disconnect from the grid during a utility grid power failure, so that utility company workers can safely make repairs on the lines. Installation and operation of a synchronous inverter must comply with safety requirements, local electrical codes, and utility regulations. Most utilities will also require inverters approved by the Underwriters Laboratory (UL), or some other nationally recognized testing organization, to make sure that the equipment meets their safety requirements. The building owner interested in

using a grid-connected inverter should first consult with experienced designers and installers of these systems, as well as with local utility officials.

Battery Storage

In a PV system, energy may not be used at the time it is produced but may be required at night or on cloudy days. A battery storage or backup system will be necessary if tapping into the utility grid is not an option and for which there are power demands other than during sunlight hours. A PV system with batteries operates by connecting the PV modules to a bank of batteries, and the batteries to the load. During sunny periods, the PV modules charge the batteries. During all hours, the bank of batteries supplies power to the load. A *charge regulator* or *charge controller* keeps batteries charged properly. Mainly, it operates by monitoring the charge to the batteries, shutting off electrical flow when the batteries are fully charged. It helps prolong battery life by protecting them from overcharging and excessive discharging.

PHOTO 25.25   PV inverter (left) and battery storage at the Manzanita Indian Reservation near Boulevard, California. (Courtesy of DOE/NREL)

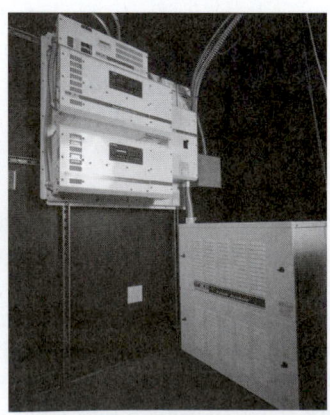

PHOTO 25.26   The PV solar system supplies power to inverters and battery storage that provide an uninterruptible power supply to the Zion Visitors Center. (Courtesy of DOE/NREL)

Batteries

Batteries can store and deliver only DC power. Unless an inverter is used to convert DC to AC, only DC appliances can be operated from the stored power of batteries. The battery voltage must be the same as the voltage needed to run the appliance. Standard battery voltage is 6 or 12 V. For an appliance requiring 24 V, two 12-V or four 6-V batteries connected in series are required. For a 120-V application, a series of ten 12-V batteries is needed. When wired in series, batteries are called a *battery bank*. The required size of the battery bank will depend on the storage capacity required, the maximum discharge rate at any time, the maximum charge rate, and the temperature of the batteries.

The least costly batteries for PV applications are deep-cycle, heavy-duty, industrial-type *lead-acid batteries,* such as those used in golf carts and forklifts designed for high reliability and long life. They can be fully charged and discharged, while standard lead-acid batteries (e.g., automobile type) cannot. Lead-acid batteries get their name from the materials they are made from: a number of lead plates submerged in an electrolyte, a mixture of sulfuric acid and water. Modern lead acid batteries have improved safety over traditional liquid acid batteries by containing the hydrogen produced during charging and by preventing the liquid acid from spilling.

Battery Ratings

A battery is rated by capacity and discharge rate. A battery's *storage capacity* is rated in *amp-hours,* the amount of electrical energy it can hold and a measure of its ability to deliver a specific amperage for a certain number of hours. For example, for a rating of 200 amp-hours, 10 amps can be delivered for 20 hour. If the system voltage is 120 V, this equates to 1200 W, because by the power equation power = voltage × amperage (120 V · 20 A = 2400 W). Table 25.7 lists specifications for selected, commercially available deep-cycle, maintenance-free, lead-acid batteries.

The *discharge rate* of a battery, expressed in *amp-hours,* is the amount of time a battery is designed to discharge its full capacity. For example, a 200-amp hour battery with a 20-hr discharge rate is designed to deliver 10 amps. This battery would be completely drained, if every hour it delivered 10 amp-hours so that in 20 hr it will have delivered 200 amp-hours.

If a battery is discharged faster than its discharge rate, its capacity will be less; if it is discharged slower, its capacity will be greater. If this same 200-amp-hour battery were discharged at a rate of 40 amps instead of its specified discharge rate of 20 amps, its capacity would drop to much less then 200 amp-hour. In contrast, charging a battery too fast (with too many amps) causes excessive heat to build up in the battery, which will shorten its life. Normally, a battery can be safely charged with a current (amps) equal to one-tenth or less of its amp-hour capacity. Industrial "rapid charge" batteries can be charged at a faster rate. The slower a battery is charged, the greater the charging efficiency.

A battery bank in a PV system should have sufficient amp-hour capacity to supply needed power during the longest expected period of cloudy weather (and no generated power)

TABLE 25.7 SPECIFICATIONS FOR SELECTED, COMMERCIALLY AVAILABLE DEEP CYCLE, MAINTENANCE FREE, LEAD-ACID BATTERIES.

	Overall Dimensions				Nominal Capacity-Ampere Hours at Rate Listed							
Volts	Length in (mm)	Width in (mm)	Height in (mm)	Unit Weight lbs (kg)	1 hr	2 hr	4 hr	8 hr	24 hr	48 hr	72 hr	120 hr
2	10.22 (260)	6.77 (172)	8.93 (227)	51 (23.2)	252	318	330	360	414	438	456	474
2	10.21 (260)	6.60 (168)	8.93 (227)	57 (25.9)	312	396	408	444	504	561	540	582
2	12.90 (328)	6.75 (171)	8.96 (228)	62 (28.2)	330	420	432	474	534	570	591	612
2	12.01 (305)	6.60 (168)	8.93 (227)	66 (30.0)	390	492	510	558	624	672	696	720
2	12.90 (328)	6.75 (171)	8.96 (228)	70 (31.8)	408	516	528	582	648	708	732	756
6	10.22 (260)	6.77 (172)	8.92 (227)	51 (23.2)	84	106	110	120	138	146	151	158
6	10.21 (259)	6.60 (168)	8.92 (227)	57 (25.9)	104	132	136	148	168	180	187	194
6	12.90 (328)	6.75 (171)	8.96 (228)	62 (28.2)	110	140	144	158	178	190	196	204
6	12.01 (305)	6.60 (168)	8.93 (227)	66 (30.0)	130	164	170	186	208	224	232	240
6	12.90 (328)	6.75 (171)	8.96 (228)	70 (31.8)	136	172	176	194	216	236	244	252
6	10.27 (261)	7.12 (181)	10.24 (260)	67 (30.4)	143	180	185	204	224	246	256	263
12	7.71 (196)	5.18 (132)	6.89 (175)	25 (11.4)	21	27	28	30	34	36	37	38
12	7.71 (196)	5.18 (132)	8.05 (204)	30 (13.6)	26	33	34	36	42	43	43	45
12	8.99 (228)	5.45 (138)	8.82 (224)	36 (16.4)	31	39	40	43	49	52	54	55
12	8.99 (228)	5.45 (138)	8.82 (224)	40 (18.2)	36	45	46	49	56	60	62	63
12	10.22 (260)	6.60 (168)	8.93 (227)	51 (23.2)	42	53	55	60	69	73	76	79
12	10.22 (260)	6.60 (168)	8.93 (227)	57 (25.9)	52	66	68	74	84	90	94	97
12	12.90 (328)	6.75 (172)	8.96 (228)	62 (28.2)	55	70	72	79	89	95	98	102
12	12.01 (305)	6.60 (168)	8.93 (227)	66 (30.0)	65	82	85	93	104	112	116	120
12	12.01 (305)	6.60 (168)	8.93 (227)	66 (30.0)	65	82	85	93	104	112	116	120
12	12.90 (328)	6.75 (172)	8.96 (228)	70 (31.8)	68	86	88	97	108	118	122	126
12	20.76 (527)	8.70 (221)	9.76 (248)	138 (62.7)	136	172	176	194	212	235	244	253
12	20.76 (527)	10.89 (277)	9.65 (245)	165 (75)	165	209	214	236	258	285	295	305

without being discharged more than 80%. If there is another source of power, such as a standby generator or utility grid connection, the battery bank does not have to be sized for such worst-case weather conditions.

A well-ventilated storage room is needed to accommodate the battery bank. Provisions for access should be provided because batteries must be routinely inspected for fluid level and corrosion. If allowed to accumulate, the hydrogen gas produced by some batteries can explode so ventilation is required. Ambient temperature in battery storage must be considered in design because the amp-hour capacity of a battery decreases with battery temperature. The stated amp-hour capacity of a battery is typically based on 80°F. When the same battery is 40°F, its capacity is only 75% of its rating and at 0°F, its capacity is only 50% of its rating. If batteries will be located in an unheated area, the lowest temperature should be considered in the sizing of the storage system.

Battery conversion efficiency is approximately 60% to 80%. Properly maintained batteries last for about 5 to 8 years, after which their capacity is significantly reduced. Overcharging and excessive discharging will cause electrolyte loss and will cause damage or destroy the battery plates.

Substantial research on battery technology has taken place since 1990, with the initiation of battery research for electric vehicles. Much of this research has focused on developing batteries that can be rapidly charged and are lighter in weight. Many new types of batteries are being developed. Two examples of near-future alternatives to lead-acid batteries are nickel-cadmium batteries and nickel-iron batteries. Both types generally provide good low-temperature performance and long life, but they are still more expensive than lead-acid batteries.

Heat Storage

When heat is the desired end product, hot water is an alternate way to store energy. In this system, electricity is sent to resistance heaters immersed in water. Electric resistance heaters can be DC or AC powered with unregulated voltage and frequency levels. Thus, the buyer has considerable flexibility in choosing a machine without the need for additional complex and expensive control or conditioning devices. The conversion efficiency of a resistance heater is nearly 100%, and heat loss is minimized if the water storage tank is well insulated. Resistance heaters can also be used directly to heat air, as with baseboard electric home heaters.

PV System Sizing

The size of a PV system needed to meet a desired electrical output depends on access to solar radiation and the efficiency of the PV cells. A typical size for a net-metered system is between 2 and 10 kW. Energy-efficient appliances and lighting should be used to reduce the electrical load and thus the overall system requirements. Once a PV system is designed and installed, the modular nature of PV allows more modules to be added if needed. Table 25.8 provides PV cell output for different system efficiencies for the average daily solar insolation at the city

TABLE 25.8   PHOTOVOLTAIC CELL OR PHOTOVOLTAIC SYSTEM OUTPUT FOR DIFFERENT SYSTEM EFFICIENCIES FOR THE AVERAGE DAILY SOLAR INSOLATION AT THE CITY NOTED.

State	City	Average Daily Solar Insolation (Clear Day) Btu/ft^2 · day	Photovoltaic Cell Output Based on Photovoltaic System Efficiency Watts per Day per ft^2 Wh/ft^2 · day						Photovoltaic Array Size Based on Photovoltaic System Efficiency Area (ft^2) to Produce 1 kWh per Day ft^2/kWh · day					
Photovoltaic System Efficiency			**10%**	**12%**	**14%**	**16%**	**18%**	**20%**	**10%**	**12%**	**14%**	**16%**	**18%**	**20%**
AK	Fairbanks	1268	37.2	44.6	52.0	59.4	67.0	74.0	27	22	19	17	15	14
	Matanuska	1128	33.1	39.7	46.3	52.8	60.0	66.0	30	25	22	19	17	15
AL	Montgomery	1344	39.4	47.3	55.1	63.0	71.0	79.0	25	21	18	16	14	13
AR	Bethel	1211	35.5	42.6	49.7	56.8	64.0	71.0	28	23	20	18	16	14
	Little Rock	1490	43.7	52.4	61.1	69.8	79.0	87.0	23	19	16	14	13	12
AZ	Tucson	2088	61.2	73.4	85.6	97.8	110.0	122.0	16	14	12	10	9	8
	Page	2021	59.2	71.1	82.9	94.8	107.0	118.0	17	14	12	11	9	9
	Phoenix	2091	61.3	73.5	85.8	98.0	110.0	123.0	16	14	12	10	9	8
CA	Santa Maria	1888	55.3	66.4	77.4	88.6	100.0	111.0	18	15	13	11	10	9
	Riverside	1865	54.6	65.6	76.5	87.4	98.0	109.0	18	15	13	11	10	9
	Davis	1621	47.5	57.0	66.5	76.0	86.0	95.0	21	18	15	13	12	11
	Fresno	1710	50.1	60.1	70.1	80.2	90.0	100.0	20	17	14	12	11	10
	Los Angeles	1786	52.3	62.8	73.3	83.8	94.0	105.0	19	16	14	12	11	10
	Soda Springs	1780	52.2	62.6	73.0	83.4	94.0	104.0	19	16	14	12	11	10
	La Jolla	1516	44.4	53.3	62.2	71.0	80.0	89.0	23	19	16	14	13	12
	Inyokern	2434	71.3	85.6	99.8	114.0	128.0	143.0	14	12	10	9	8	7
CO	Grandby	1808	53.0	63.6	74.2	84.8	95.0	106.0	19	16	13	12	11	10
	Grand Lake	1614	47.3	56.7	66.2	75.6	85.0	95.0	21	18	15	13	12	11
	Grand Junction	1859	54.5	65.4	76.3	87.2	98.0	109.0	18	15	13	11	10	9
	Boulder/Denver	1548	45.4	54.4	63.5	72.6	82.0	91.0	22	18	16	14	12	11
DC	Washington	1344	39.4	47.3	55.1	63.0	71.0	79.0	25	21	18	16	14	13
FL	Apalachicola	1745	51.1	61.4	71.6	81.8	92.0	102.0	20	16	14	12	11	10
	Miami	1786	52.3	62.8	73.3	83.8	94.0	105.0	19	16	14	12	11	10
	Gainesville	1675	49.1	58.9	68.7	78.6	88.0	98.0	20	17	15	13	11	10
	Tampa	1802	52.8	63.4	73.9	84.4	95.0	106.0	19	16	14	12	11	10
GA	Atlanta	1506	44.1	53.0	61.8	70.6	79.0	88.0	23	19	16	14	13	12
	Griffin	1586	46.5	55.8	65.1	74.4	84.0	93.0	22	18	15	13	12	11
HI	Honolulu	1913	56.1	67.3	78.5	89.6	101.0	112.0	18	15	13	11	10	9
IA	Ames	1398	41.0	49.2	57.3	65.6	74.0	82.0	24	20	17	15	13	12
ID	Boise	1563	45.8	55.0	64.1	73.2	82.0	92.0	22	18	16	14	12	11
	Twin Falls	1494	43.8	52.5	61.3	70.0	79.0	88.0	23	19	16	14	13	12
IL	Chicago	998	29.2	35.1	40.9	46.8	53.0	58.0	34	28	24	21	19	17
IN	Indianapolis	1338	39.2	47.0	54.9	62.8	71.0	78.0	26	21	18	16	14	13
KN	Manhattan	1452	42.5	51.1	59.6	68.0	77.0	85.0	24	20	17	15	13	12
	Dodge City	1840	53.9	64.7	75.5	86.2	97.0	108.0	19	15	13	12	11	10
KY	Lexington	1570	46.0	55.2	64.4	73.6	83.0	92.0	22	18	16	14	12	11
LA	Lake Charles	1567	45.9	55.1	64.3	73.4	83.0	92.0	22	18	16	14	12	11
	New Orleans	1563	45.8	55.0	64.1	73.2	82.0	92.0	22	18	16	14	12	11
	Shreveport	1471	43.1	51.7	60.3	69.0	78.0	86.0	23	19	17	14	13	12
MA	East Wareham	1268	37.2	44.6	52.0	59.4	67.0	74.0	27	22	19	17	15	14
	Boston	1220	35.7	42.9	50.0	57.2	64.0	71.0	28	23	20	17	16	14
	Blue Hill	1287	37.7	45.3	52.8	60.4	68.0	75.0	27	22	19	17	15	14
	Natick	1303	38.2	45.8	53.4	61.0	69.0	76.0	26	22	19	16	14	13
	Lynn	1204	35.3	42.3	49.4	56.4	64.0	71.0	28	24	20	18	16	14
MD	Silver Hill	1420	41.6	49.9	58.2	66.6	75.0	83.0	24	20	17	15	13	12
ME	Caribou	1332	39.0	46.8	54.6	62.4	70.0	78.0	26	21	18	16	14	13
	Portland	1433	42.0	50.4	58.8	67.2	76.0	84.0	24	20	17	15	13	12
MI	Sault Saint Marie	1335	39.1	46.9	54.8	62.6	70.0	78.0	26	21	18	16	14	13
	East Lansing	1271	37.2	44.7	52.1	59.6	67.0	74.0	27	22	19	17	15	14
MN	Saint Cloud	1440	42.2	50.6	59.1	67.6	76.0	84.0	24	20	17	15	13	12
MO	Columbia	1503	44.0	52.8	61.7	70.4	79.0	88.0	23	19	16	14	13	12
	St. Louis	1392	40.8	48.9	57.1	65.2	73.0	82.0	25	20	18	15	14	13
MS	Meridian	1408	41.3	49.5	57.8	66.0	74.0	83.0	24	20	17	15	13	12
MT	Glasgow	1637	48.0	57.6	67.1	76.8	86.0	96.0	21	17	15	13	12	11
	Great Falls	1567	45.9	55.1	64.3	73.4	83.0	92.0	22	18	16	14	12	11
	Summit	1268	37.2	44.6	52.0	59.4	67.0	74.0	27	22	19	17	15	14

TABLE 25.8   PHOTOVOLTAIC CELL OR PHOTOVOLTAIC SYSTEM OUTPUT FOR DIFFERENT SYSTEM EFFICIENCIES FOR THE AVERAGE DAILY SOLAR INSOLATION AT THE CITY NOTED. (CONTINUED)

State	City	Average Daily Solar Insolation (Clear Day) Btu/ft² · day	Photovoltaic Cell Output Based on Photovoltaic System Efficiency Watts per Day per ft² Wh/ft² · day						Photovoltaic Array Size Based on Photovoltaic System Efficiency Area (ft²) to Produce 1 kWh per Day ft²/kWh · day					
Photovoltaic System Efficiency			10%	12%	14%	16%	18%	20%	10%	12%	14%	16%	18%	20%
NM	Albuquerque	2151	63.0	75.6	88.2	101.0	113.0	126.0	16	13	11	10	9	8
NE	Lincoln	1522	44.6	53.5	62.4	71.4	80.0	89.0	22	19	16	14	12	11
	Omaha	1557	45.6	54.7	63.9	73.0	82.0	91.0	22	18	16	14	12	11
NC	Cape Hatteras	1687	49.4	59.3	69.2	79.0	89.0	99.0	20	17	14	13	11	10
	Greensboro	1497	43.9	52.6	61.4	70.2	79.0	88.0	23	19	16	14	13	12
ND	Bismarck	1592	46.6	56.0	65.3	74.6	84.0	93.0	21	18	15	13	12	11
NJ	Sea Brook	1338	39.2	47.0	54.9	62.8	71.0	78.0	26	21	18	16	14	13
NV	Las Vegas	2037	59.7	71.6	83.6	95.4	107.0	119.0	17	14	12	10	9	9
	Ely	1900	55.7	66.8	77.9	89.0	100.0	111.0	18	15	13	11	10	9
NY	Binghamton	1004	29.4	35.3	41.2	47.0	53.0	59.0	34	28	24	21	19	17
	Ithaca	1204	35.3	42.3	49.4	56.4	64.0	71.0	28	24	20	18	16	14
	Schenectady	1128	33.1	39.7	46.3	52.8	60.0	66.0	30	25	22	19	17	15
	Rochester	1052	30.8	37.0	43.2	49.4	55.0	62.0	32	27	23	20	18	16
	New York City	1297	38.0	45.6	53.2	60.8	68.0	76.0	26	22	19	16	14	13
OH	Columbus	1319	38.6	46.4	54.1	61.8	69.0	77.0	26	22	18	16	14	13
	Cleveland	1252	36.7	44.0	51.4	58.6	66.0	73.0	27	23	19	17	15	14
OK	Stillwater	1586	46.5	55.8	65.1	74.4	84.0	93.0	22	18	15	13	12	11
	Oklahoma City	1776	52.0	62.4	72.9	83.2	94.0	104.0	19	16	14	12	11	10
OR	Astoria	1182	34.6	41.6	48.5	55.4	62.0	69.0	29	24	21	18	16	15
	Corvallis	1281	37.5	45.0	52.5	60.0	68.0	75.0	27	22	19	17	15	14
	Medford	1433	42.0	50.4	58.8	67.2	76.0	84.0	24	20	17	15	13	12
PA	Pittsburgh	1042	30.5	36.6	42.7	48.8	55.0	61.0	33	27	23	20	18	17
	State College	1243	36.4	43.7	51.0	58.2	66.0	73.0	27	23	20	17	15	14
RI	Newport	1344	39.4	47.3	55.1	63.0	71.0	79.0	25	21	18	16	14	13
SC	Charleston	1608	47.1	56.5	66.0	75.4	85.0	94.0	21	18	15	13	12	11
SD	Rapid City	1662	48.7	58.4	68.2	78.0	88.0	97.0	21	17	15	13	12	11
TN	Nashville	1414	41.4	49.7	58.0	66.2	75.0	83.0	24	20	17	15	13	12
	Oak Ridge	1389	40.7	48.8	57.0	65.2	73.0	81.0	25	20	18	15	14	13
TX	San Antonio	1684	49.3	59.2	69.1	79.0	89.0	99.0	20	17	14	13	11	10
	Brownsville	1563	45.8	55.0	64.1	73.2	82.0	92.0	22	18	16	14	12	11
	El Paso	2136	62.6	75.1	87.6	100.0	113.0	125.0	16	13	11	10	9	8
	Midland	1853	54.3	65.2	76.0	86.8	98.0	109.0	18	15	13	12	10	9
	Fort Worth	1726	50.6	60.7	70.8	81.0	91.0	101.0	20	16	14	12	11	10
UT	Salt Lake City	1672	49.0	58.8	68.6	78.4	88.0	98.0	20	17	15	13	11	10
	Flaming Gorge	1853	54.3	65.2	76.0	86.8	98.0	109.0	18	15	13	12	10	9
VA	Richmond	1312	38.4	46.1	53.8	61.6	69.0	77.0	26	22	19	16	14	13
WA	Seattle	1134	33.2	39.9	46.5	53.2	60.0	66.0	30	25	22	19	17	15
	Richland	1411	41.3	49.6	57.9	66.2	74.0	83.0	24	20	17	15	13	12
	Pullman	1503	44.0	52.8	61.7	70.4	79.0	88.0	23	19	16	14	13	12
	Spokane	1424	41.7	50.1	58.4	66.8	75.0	83.0	24	20	17	15	13	12
	Prosser	1598	46.8	56.2	65.5	75.0	84.0	94.0	21	18	15	13	12	11
WI	Madison	1363	39.9	47.9	55.9	63.8	72.0	80.0	25	21	18	16	14	13
WV	Charleston	1160	34.0	40.8	47.6	54.4	61.0	68.0	29	25	21	18	16	15
WY	Lander	1926	56.4	67.7	79.0	90.2	102.0	113.0	18	15	13	11	10	9

noted. It also provides PV system output for different system efficiencies for the average daily solar insolation at the city noted. These values can be used to approximate the area of PV cells needed to meet a specific demand.

Systems size and cost can vary dramatically depending on the amount of power required and the solar insolation available. PV installations have functioned well but costs are high. This generally excludes PV as a power source, except in remote locations. Constant developments in PV materials with improved efficiencies will likely transform this technology into a cost-effective power source in the near future.

Example 25.1

A photovoltaic system is being considered for a building with a load of 8 kW. The PV system will be approximately 16% efficient. Approximate the PV array size (in ft^2) required to meet this load in Chicago, Illinois.

From Table 25.8, a 21 ft^2 of PV array will produce one kWh per day at a PV system efficiency of 16%.

$$8 \text{ kW} \cdot 21 \text{ ft}^2/\text{kW} = 168 \text{ ft}^2$$

Example 25.2

A photovoltaic system is being considered for a building with a load of 8 kW. The PV system will be approximately 16% efficient. Approximate the PV array size (in ft^2) required to meet this load in Albuquerque, New Mexico.

From Table 25.8, a 9 ft^2 of PV array will produce one kWh per day at a PV system efficiency of 16%.

$$8 \text{ kW} \cdot 9 \text{ ft}^2/\text{kW} = 72 \text{ ft}^2$$

Conducting a Power Load Analysis

The first step in the process of investigating a PV system for a home or small business is to calculate the power load. A thorough examination of electricity needs of the building helps determine:

- The size of the system needed
- How energy needs fluctuate throughout the day and over the year
- Measures that can be taken to reduce electricity use and increase the efficiency

Conducting a power load analysis involves recording wattage and average daily use of all lights and power-consuming equipment and appliances (e.g., refrigerators, lights, televisions, and power tools). Some electrical equipment use electricity all the time (e.g., clock, smoke detector), while others use power intermittently (e.g., night lights, appliances, power tools). Power-consuming equipment that use electricity intermittently are referred to as *selectable loads*. By using selectable loads only when extra power is available, a smaller wind energy system can be selected. *Phantom loads* come from power-consuming equipment with continuous energy consumption that is not readily apparent (e.g., radio, TV, appliance control/indicator, power strip) and must be recorded.

To determine total energy consumption, wattage of lights and power-consuming equipment and appliances is multiplied by the number of hours they is used each day (seasonal variations should be accounted for). For selectable loads, record the time(s) of day the load exists. Generally, power use data can be found on a label, nameplate, or cord tag attached to the appliance. For appliances that do not furnish

PHOTO 25.27 A PV array at the Needles general store and campground in Canyonlands National Park, Utah. Output varies with snow cover. (Courtesy of DOE/NREL)

the wattage on a label, it can be calculated by multiplying the amperes times the volts. Table 25.9 provides information on selectable and phantom loads.

25.6 WIND ENERGY SYSTEMS

The Wind

Wind is a form of solar energy. Solar energy absorbed by the ground or water is transferred to air, where it causes differences in air temperature, density, and pressure that create forces that push air around. Wind is air moving from an area of high pressure to an area of low pressure. Wind results chiefly from natural convection driven by uneven heating of the earth's surface and to a certain extent from the rotation of the earth. Wind patterns are affected by the rotation of the planet, weather patterns, and surface terrain.

Even though only a few percent of the energy coming from the sun is converted into wind, wind energy has enormous potential as a source of energy. The United States Department of Energy (DOE) estimates that the wind energy potential in the United States is over 10 000 billion kWh/year, which is over three times the electricity generated annually. Installed wind energy generating capacity presently accounts for less than 1% of generated electricity.

Wind Speed

Wind is measured and recorded based upon its velocity and direction. *Wind speed* is the velocity of the wind in relation to the

TABLE 25.9 TYPICAL SELECTABLE AND PHANTOM LOADS. MANUFACTURERS INFORMATION SHOULD BE USED WHEN AVAILABLE.

Appliance or Piece of Equipment	Power Requirement (W or VA)		
	Connected (Surge) Load	Actual Load	Phantom Load
General Appliances			
Computer	150	50	5
Computer printer	150	120	5
Hair dryer	1500	1000	—
Hot tub (360 gal)	1500	1500	5
Phone, portable	120	20	4
Phone answering machine	120	10	4
Power tool (hand-held)	1200	800	—
Radio	8	8	3
Stereo	50	30	4
Television, 23 in	150	125	4
Satellite TV system	120	60	25
VCR/DVD player	60	50	8
Waterbed heater	300	Up to 300	8
Battery charger	120	10	2
Kitchen Appliances			
Blender	350	350	—
Cooktop/oven unit, household size	12 000	Up to 6000	8
Dishwasher, household size	1200	1200	8
Food processor	400	400	—
Microwave	1000	1000	—
Refrigerator, household size	600	600	8
Household Appliances			
Clothes dryer, household size	6000	6000 600 (motor only)	8
Clothes washer, household size	1200	800	8
Iron	600	Up to 600	—
Sewing machine	200	Up to 150	—
Vacuum cleaner	700	Up to 700	—
Lights			
60 W incandescent	60	60	—
75 W incandescent	75	75	—
100 W incandescent	100	100	—
13 W compact fluorescent	13	13	—
23 W compact fluorescent	23	23	—
32 W linear fluorescent	40	40	7
40 W linear fluorescent	48	48	7
100 W high-intensity discharge	120	120	7
175 W high-intensity discharge	205	205	7
250 W high-intensity discharge	280	280	7
400 W high-intensity discharge	435	435	7
1000 W high-intensity discharge	1050	1050	7
7 W nightlight	7	7	7
Heating/Cooling Equipment			
Air conditioner, window type	1800	Up to 1800	8
Air conditioner, household	6000	Up to 6000	8
Furnace, electric	30 000	Up to 30 000	8
Furnace, gas	1200	850	8
Heater, baseboard electric	1500	Up to 1500	5
Heater, portable electric	1500	Up to 1500	5
Water heater, electric	4500	4500	—

ground. Wind speed can be measured and recorded in units of miles per hour (mph), nautical miles per hour (knots), kilometers per hour (km/hr) and meters per second (m/s). The relationships between these units are:

$$1 \text{ mph} = 0.447 \text{ m/s}$$
$$= 1.609 \text{ km/h}$$
$$= 0.889 \text{ knots}$$

$$1 \text{ m/s} = 2.237 \text{ mph}$$
$$= 3.6 \text{ km/h}$$
$$= 1.944 \text{ knots}$$

$$1 \text{ knot} = 0.5144 \text{ m/s}$$
$$= 1.852 \text{ km/h}$$
$$= 1.125 \text{ mph}$$

Wind speeds are usually measured and recorded as 10-min averages. The relationship between wind speed and conditions at sea and on land is provided in Table 25.10. A *wind gust* is the speed reached during a sudden, brief increase in the strength of the wind. The duration of a gust is usually less than 20 s.

Wind Direction

Wind direction is the most commonly recorded of all weather elements in historical nautical logbooks. The convention of describing wind direction from which it is coming (rather than the direction it is blowing) was established during the 1500s. Historically, wind direction was expressed using the direction of a 32-point compass, in 11.25° increments expressed as an abbreviated direction (e.g., N, NbE, NNE, NEbN, NE, NEbE, ENE, EbN, E, and so on). See Table 25.11. The path the wind takes can be expressed as a direction or an azimuth. Wind direction is variable and changes with weather and season.

Wind azimuth is a bearing from true north in degrees clockwise in the direction from which the wind is blowing. It is expressed as an angle measured clockwise from north. An azimuth of 0° is north, 90° is east, 180° is south, and 270° is west, and so forth. Winds are also named after the direction from which they blow; for example, a west wind blows from the west toward the east. So, if the wind direction is 45°, the winds are coming out of the northeast and blowing towards the southwest. It would be called a northeasterly wind.

TABLE 25.10   THE RELATIONSHIP BETWEEN WIND SPEED AND CONDITIONS AT SEA AND ON LAND.

Wind Speed						
mph	m/s	km/hr	knots	Name	Conditions at Sea	Conditions on Land
<1	<0.5	<2	<1	Calm	Water like a mirror.	Smoke rises vertically.
1 to 4	0.5 to 1.8	1 to 5	1 to 3	Light air	Ripples only.	Smoke drifts and leaves rustle.
5 to 7	1.8 to 3.1	6 to 11	4 to 6	Light breeze	Small wavelets (6 in/0.2 m). Crests have a glassy appearance.	Wind felt on face.
8 to 11	3.6 to 4.9	12 to 19	7 to 10	Gentle breeze	Large wavelets (2 ft/0.6 m), crests begin to break.	Flags extended, leaves move.
12 to 18	5.4 to 8.0	20 to 29	11 to 16	Moderate breeze	Small waves (3 ft/1 m), some whitecaps.	Dust and small branches move.
19 to 24	8.5 to 10.7	30 to 39	17 to 21	Fresh breeze	Moderate waves (6 ft/1.8 m), many whitecaps.	Small trees begin to sway.
25 to 31	11.2 to 13.9	40 to 50	22 to 27	Strong breeze	Large waves (10 ft/3 m), some spray.	Large branches move, wires whistle, umbrellas are difficult to control.
32 to 38	14.3 to 17.0	51 to 61	28 to 33	Near gale	Mounting sea (13 ft/4 m) with foam blown in streaks downwind.	Whole trees in motion, inconvenience in walking.
39 to 46	17.4 to 20.6	62 to 74	34 to 40	Gale	Moderately high waves (18 ft/5.5 m), crests break into spindrift.	Difficult to walk against wind. Twigs and small branches blown off trees.
47 to 54	21.0-24.1	76 to 87	41 to 47	Strong gale	High waves (23 ft/7 m), dense foam, visibility affected.	Minor structural damage may occur (shingles blown off roofs).
55 to 63	24.6 to 28.2	88 to 102	48 to 55	Storm	Very high waves (30 ft/9 m), heavy sea roll, visibility impaired. Surface generally white.	Trees uprooted, structural damage likely.
64 to 73	28.6 to 32.6	103 to 118	56 to 63	Violent storm	Extremely high waves (36 ft/11 m), visibility poor.	Widespread damage to structures.
74+	33.1+	119+	64+	Hurricane	Enormously high waves (46 ft/14 m), air filled with foam and spray, visibility bad.	Severe structural damage to buildings, widespread devastation.

TABLE 25.11   WIND DIRECTION IS EXPRESSED USING A 32-POINT COMPASS OR BY THE WIND AZIMUTH. WIND DIRECTION IS THE DIRECTION THE WIND IS BLOWING FROM.

Point	Direction	Azimuth		Point	Direction	Azimuth	
0	N	0°-0′	0.0°	16	S	180°-0′	180.0°
1	N by E	11°-15′	11.25°	17	S by W	191°-15′	191.25°
2	NNE	22°-30′	22.5°	18	SSW	202°-30′	202.5°
3	NE by N	33°-45′	33.75°	19	SW by S	213°-45′	213.75°
4	NE	45°-0′	45.0°	20	SW	225°-0′	225.0°
5	NE by E	56°-15′	56.25°	21	SW by W	236°-15′	236.25°
6	ENE	67°-30′	67.5°	22	WSW	247°-30′	247.5°
7	E by N	78°-45′	78.75°	23	W by S	258°-45′	258.75°
8	E	90°-0′	90.0°	24	W	270°-0′	270.0°
9	E by S	101°-15′	101.25°	25	W by N	281°-15′	281.25°
10	ESE	112°-30′	112.5°	26	WNW	292°-30′	292.5°
11	SE by E	123°-45′	123.75°	27	NW by W	303°-45′	303.75°
12	SE	135°-0′	135.0°	28	NW	315°-0′	315.0°
13	SE by S	146°-15′	146.25°	29	NW by N	326°-15′	326.25°
14	SSE	157°-30′	157.5°	30	NNW	337°-30′	337.5°
15	S by E	168°-45′	168.75°	31	N by W	348°-45′	348.75°

In describing the wind, it is also helpful to easily express the sides with respect to the wind direction. By definition, *windward* is the direction from which the wind blows. *Leeward* is the direction in which the wind is blowing.

A *wind vane, windsock,* or *wind sensor* is used to detect wind direction. Wind speed is measured with an *anemometer* or *wind sensor*. See Photo 25.28. A cup anemometer has a vertical axis and three cups that catch the wind and cause it to rotate about its axis. Anemometers may also be fitted with propellers. The number of revolutions per minute is registered electronically and converted to a velocity. Ultrasonic sensors detect wind speed and direction. The phase shifting of sound and laser anemometers detect light reflected from the air molecules. The advantage of these anemometers is that they are less sensitive to icing. In practice, however, cup anemometers tend to be the most commonly used type. An anemometer used in design of a wind energy system must read accurately, with a with a measurement error of no more than 1%. Recorders store historical data using mechanical or electronic counters, paper strip charts, magnetic tapes, or electronic data chips or cards. Electronic recorders and other storage devices can be easily interfaced with a personal computer for data recording and analysis.

PHOTO 25.28   A cup anemometer (left) and wind vane (right) measures wind speed and direction. (Courtesy of DOE/NREL)

Fluctuating Wind Speed and Direction

Wind speed and direction are always fluctuating. Tables 25.12A and 25.12B show variations at a site in Boulder, Colorado. This fluctuation depends both on weather conditions and local surface

TABLE 25.12A   WIND DIRECTION FREQUENCY (PERCENTAGE OF TIME) AT THE BOULDER, COLORADO, WIND MONITORING SITE BETWEEN JANUARY 1, 1997, THROUGH DECEMBER 31, 2003.

Sensor	Height		Direction															
	m	ft	N	NNE	NE	ENE	E	ESE	SE	SSE	S	SSW	SW	WSW	W	WNW	NW	NNW
WD10	10	33	8.5	5.3	4.0	3.2	3.1	3.5	4.8	5.4	5.3	5.1	5.4	6.2	10.8	15.3	7.9	6.3
WD20	20	66	7.4	5.1	4.0	3.2	3.0	3.4	4.8	5.5	5.8	5.0	4.6	4.6	10.2	17.0	8.8	7.6
WD50	50	164	7.3	5.4	4.3	3.4	3.1	3.6	4.9	6.1	6.5	4.5	3.6	4.0	9.5	15.7	9.0	9.1
WD80	80	262	10.0	6.1	4.3	3.3	3.4	3.7	5.1	6.8	6.9	4.3	3.6	3.8	9.4	13.9	7.7	7.8

39.9136° North latitude, 105.2469° west longitude, elevation (ft): 6085.

TABLE 25.12B   WIND DIRECTION FREQUENCY (HOURS/YEAR) AT THE BOULDER COUNTY WIND MONITORING SITE BETWEEN JANUARY 1, 1997, THROUGH DECEMBER 31, 2003.

Sensor	Height		Direction															
	m	ft	N	NNE	NE	ENE	E	ESE	SE	SSE	S	SSW	SW	WSW	W	WNW	NW	NNW
WD10	10	33	745	464	350	280	272	307	420	473	464	447	473	543	946	1340	692	552
WD20	20	66	648	447	350	280	263	298	420	482	508	438	403	403	894	1489	771	666
WD50	50	164	639	473	377	298	272	315	429	534	569	394	315	350	832	1375	788	797
WD80	80	262	876	534	377	289	298	324	447	596	604	377	315	333	823	1218	675	683

39.9136° North latitude, 105.2469° west longitude, elevation (ft): 6085.

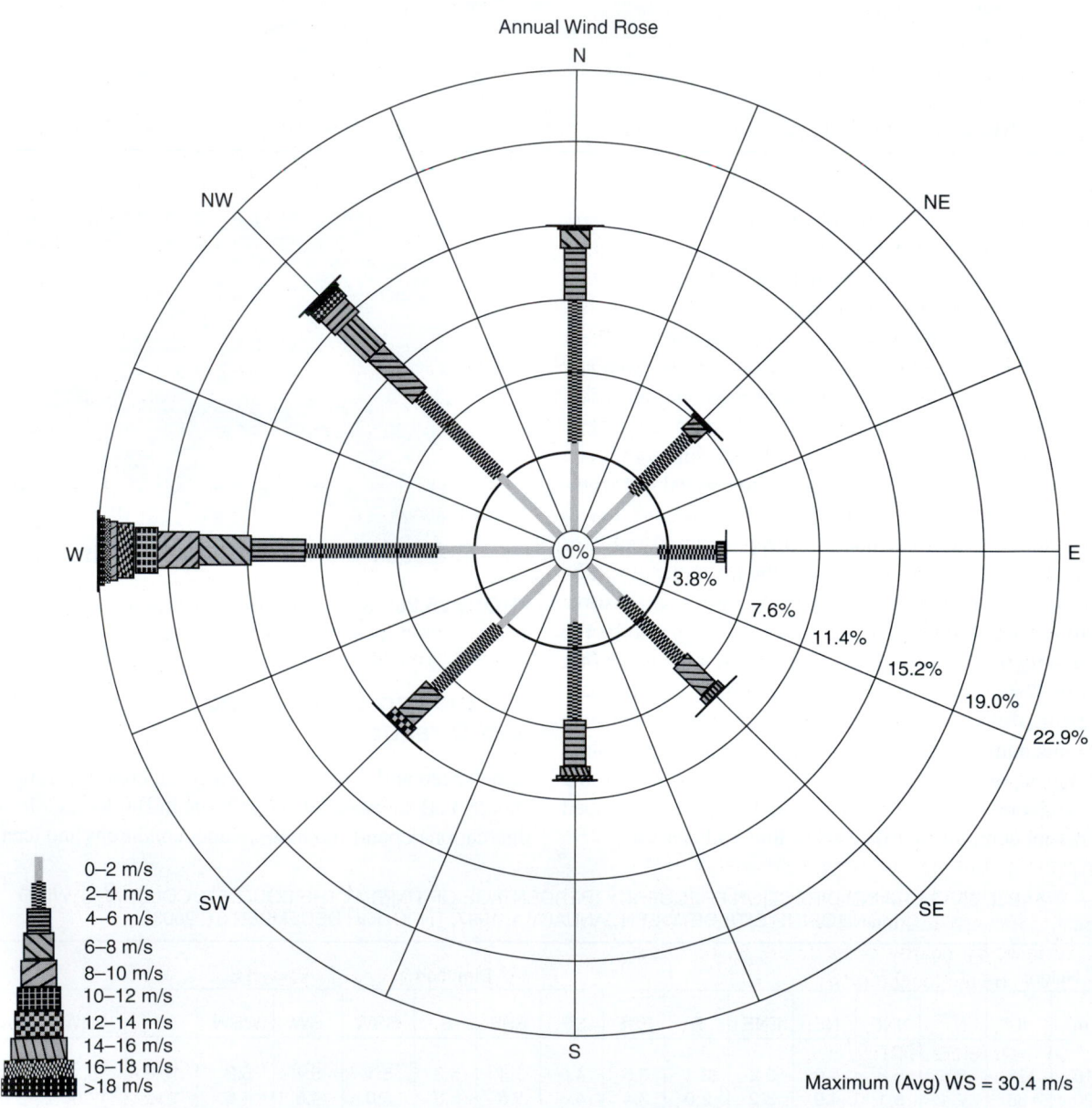

FIGURE 25.6   A wind rose is a polar bar graph that illustrates the distribution of winds at a specific location, showing wind strength and frequency by direction over a period of time. It is a diagram showing the amount of wind energy coming from different directions. Bars extending radially from the center of the diagram represent wind coming from a specific direction.

surroundings. It is normally windier during the day than at night because of larger temperature differences that exist during the day. The wind tends to change direction more frequently and be more turbulent during the daytime hours.

A *wind rose* is a polar bar graph that illustrates the distribution of winds at a specific location, showing wind strength and frequency by direction over a period of time. It is a diagram showing the amount of wind energy coming from different directions. See Figure 25.6. Bars extending radially from the center of the diagram represent wind coming from a specific direction. Usually, sets of 8 or 16 bars extend in the 8 or 16 compass directions (e.g., N, NNE, NE, and so on). An 8-direction rose is divided into 45° angle directions; a 16-direction rose is divided into 22.5° angle directions. Twelve directions (every 30°) is the standard used by the European Wind Atlas. A wind rose can illustrate the wind distribution over a month, a year, or some other period. The Wind Energy Resource Atlas of the United States offers information on wind strength in a general geographical area at http://rredc.nrel.gov/wind/pubs/atlas/maps.html.

On a wind rose, the length of each bar segment is proportional to the percentage of winds in that speed range, blowing from that particular direction. For example, if the bar pointing to the east is twice as long as the bar pointing to the north, then winds blew to the east twice as frequently as they did to the north over that period of time. Each bar is further divided into segments of different thickness, which represent wind speed ranges in that direction. A thin solid line extending from the center might indicate winds between 0 and 10 mph, a thicker line for winds between 5 and 10 mph, a thicker line to indicate speeds between 10 and 15 mph, and so on.

A wind rose provides historical information on the relative wind speeds in different directions at a specific location over a specific time period. Information varies by location and by time period. Wind patterns will vary seasonally and even from year to year, so a wind rose developed with information collected over many years is more reliable. Wind roses from nearby areas can be reasonably similar and it may be acceptable to interpolate between them, but only if there are no significant differences in topography, elevation, or surface roughness.

The historical information found on a wind rose is useful in locating a wind turbine site. It ensures adequate exposure to prevailing winds without exposure to obstacles. However, wind speeds are significantly influenced by the roughness of the surrounding surface, by nearby obstacles (e.g., buildings, trees, hills, other), and by the local topography.

Effects of Surface Roughness and Elevation

The roughness of the surface over which the wind blows affects its speed. Uneven surfaces, such as areas with trees and buildings, create more friction and turbulence than smooth surfaces

TABLE 25.13   COEFFICIENT (P) FOR DIFFERENT TYPES OF GROUND COVER.

Ground Cover	Coefficient (p)
A smooth surface (body of water)	0.10
Open field with short grass or uncultivated ground	0.16
High grass or low row crops	0.18
Tall row crops or low woods	0.20
Housing developments, small towns	0.28
Suburban area (clustered homes and small buildings)	0.30
Woods or forest with many trees	0.30
Urban area (clustered large buildings)	0.40

such as lakes or open land. Greater friction means reduced wind speeds near the ground. In contrast, higher elevations generally yield greater wind speeds.

The approximate increase of speed with height for different surfaces can be calculated from the following equation, where v_1 is the known (reference) wind speed at height h_1 above ground, v_2 is the speed at a second height h_2, and p is the coefficient determining the wind change (see Table 25.13):

$$v_2 = v_1(h_2/h_1)^p$$

Within dense vegetation (e.g., forest, orchard, mature cane field), the effective ground level is established at the height of the vegetation (e.g., top of tree canopy). Below this level there is little wind.

Example 25.3

The known mean wind velocity is 14.7 mph at a site at a height of 33 ft (about 10 m). Approximate the wind velocity at a height of 100 ft (about 30 m) for the following sites:

a. Site located in the suburbs (p = 0.30)

$$v_2 = v_1(h_2/h_1)^n = 14.7 \text{ mph } (100 \text{ ft}/33 \text{ ft})^{0.30} = 20.5 \text{ mph}$$

b. Site located in an open field with short grass (p = 0.16)

$$v_2 = v_1(h_2/h_1)^n = 14.7 \text{ mph } (100 \text{ ft}/33 \text{ ft})^{0.16} = 17.6 \text{ mph}$$

c. Site on island surrounded by open water (p = 0.10)

$$v_2 = v_1(h_2/h_1)^n = 14.7 \text{ mph } (100 \text{ ft}/33 \text{ ft})^{0.10} = 16.4 \text{ mph}$$

Wind Energy Systems

Introduction

The most familiar wind energy system is the Dutch windmill that was developed centuries ago and used in Europe for grinding grain and pumping water. In the United States, the fan-type water pumper is still seen throughout the countryside at remote watering wells in the western and Plains states. Today, the basic

Horizontal Axis Wind Turbines (HAWT)

Farm-type Windmill   Small-scale Wind Turbine   Large-scale Wind Turbine

Vertical Axis Wind Turbines (WAWT)

Savonius   Darrieus   H-Darrieus

PHOTO 25.29 Types of wind machines. (Photos courtesy of NREL/DOE)

wind energy conversion device is the wind turbine. A *wind turbine*, with its sleek aerodynamic blades and tall tower, collects kinetic energy from the wind and converts it to electricity. Although various designs and configurations exist, these turbines are generally grouped into two types: vertical-axis turbines and horizontal-axis turbines. Small turbines are sometimes used in connection with diesel generators, batteries, and photovoltaic systems. These systems are called *hybrid wind systems* and are typically used in remote, off-grid locations, where a connection to the utility grid is not available. Types of wind machines are shown in Photo 25.29 and are introduced as follows.

Horizontal-Axis Wind Turbines A *horizontal-axis wind turbine (HAWT)* has a rotor with an axis of rotation that is horizontal with respect to the ground and parallel to the wind stream. An HAWT must be oriented with respect to wind direction. Some HAWTs are designed to operate in an upwind mode, with the blades upwind of the tower. In this case, a *tail vane* is usually

used to keep the blades facing into the wind. Other HAWT designs operate in a downwind mode so that the wind passes the tower before striking the blades. The machine rotor naturally tracks the wind without a tail vane. Examples of HAWTs are shown in Photos 25.30 through 25.32. Very large (utility size) wind turbines use a motor-driven mechanism that rotates the turbine in response to a wind direction sensor mounted on the tower. See Figure 25.7.

Vertical-Axis Wind Turbines A *vertical-axis wind turbine (VAWT)* has blades with an axis of rotation that is vertical with respect to the ground and roughly perpendicular to the wind stream. Although vertical axis wind turbines have existed for centuries, they are not as common as HAWTs. The basic VAWT designs are the Darrieus, Giromill, and Savonius. See Figure 25.7.

A *Darrieus* VAWT has a rotor section that has multiple curved blades that look like an eggbeater. It is named after France inventor Georges Darrieus, who submitted a patent in 1931. Lift

PHOTO 25.30 A Bergey BWC EXCEL 10 kW wind turbine. It is designed to supply most of the electricity for an average total electric home in areas with an average wind speed of 12 mph. (Courtesy of DOE/NREL)

PHOTO 25.32 A 20-kW Jacobs wind turbine on a freestanding tower. (Courtesy of DOE/NREL)

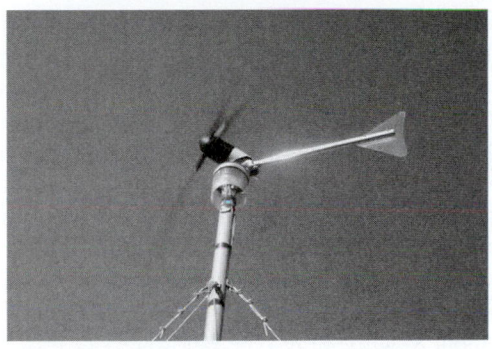

PHOTO 25.31 Furling of the wind turbine to prevent damage in very high winds. (Courtesy of DOE/NREL)

forces generated from a set of airfoil blades power it. It uses an elliptical blade that has a fixed angle of attack of 0° so it is not self-starting. Each blade experiences maximum lift twice per revolution, making for a sinusoidal, high-torque output. The *Giromill* is similar to the Darrieus VAWT except that it has straight vertical axis blades.

A *Savonius* VAWT has a rotor section that is S-shaped when viewed from above. A well-designed Savonius turbine can have a tip speed ratio of slightly greater than 1. It rotates relatively slowly, but yields a high torque. The simplest Savonius rotor consists of two oil-drum halves facing in opposite directions. This design has been used for pumping applications in undeveloped areas. Most Savonius rotors are only capable of achieving 15% efficiency.

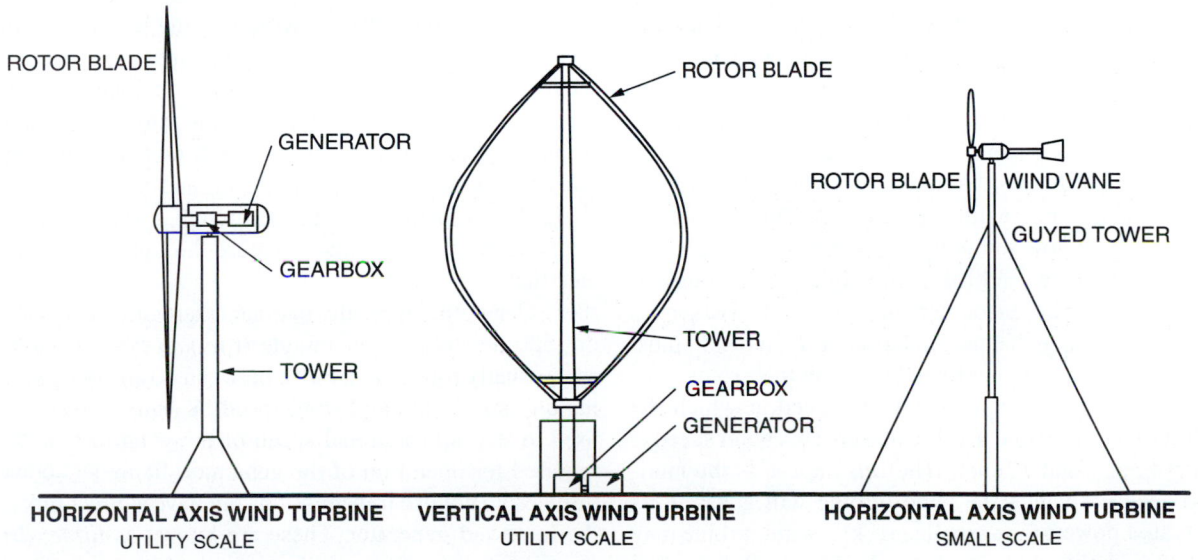

FIGURE 25.7 Wind turbines are generally grouped into two types: vertical-axis turbines and horizontal-axis turbines.

A VAWT does not need to be oriented into the wind direction. Because the shaft is vertical, the transmission and generator can be mounted at ground level, allowing easier servicing and a lighter weight, lower cost tower. VAWTs are very difficult to mount on a high tower to capture higher level winds. Because of this limitation, they are forced to accept the lower, more turbulent winds at surface levels. For these reasons, VAWT designs are not as efficient at collecting energy from the wind.

Wind Turbine Components

The basic components of a wind turbine include a blade or rotor, which converts the energy in the wind to rotational shaft energy; a drive train, usually including a gearbox and a generator; a tower that supports the rotor and drive train; and other equipment, including controls, electrical cables, ground support equipment, and interconnection equipment.

Rotor The *rotor* is the component of a wind turbine that collects energy from the wind. It typically consists of two or more wood, fiberglass or metal blades that rotate about an axis (horizontal or vertical) at a rate determined by the wind speed and the shape of the blades. The blades are attached to the rotor hub, which in turn is attached to the *nacelle,* which includes the gearbox, low- and high-speed shafts, generator, controller, and brake. A cover protects the components inside the nacelle.

Rotor blade designs operate on either the principle of drag or lift. In *drag* design, the wind pushes the rotor blades out of the way. Drag powered wind turbines have slower rotational speeds and high torque capabilities. They are useful for the pumping, sawing, or grinding. The *lift* design is aeronautically designed as an airfoil or wing that makes use of the principle that enables airplanes, kites, and birds to fly. When air flows through the blade, a wind speed and pressure differential is created between the two surfaces of the blade. The pressure at one surface is greater, which acts to lift the blade. On a wind turbine, the rotor blades are attached to a central axis, which causes the lift to be translated into a rotational motion. Lift-powered wind turbines generate much higher rotational speeds than drag types and are better suited for generation of electricity.

The number of blades that make up a rotor affects wind turbine performance. For a lift-type rotor to function effectively, the wind must flow smoothly over the blades. To prevent turbulence, spacing between blades should be great enough so that one blade does not encounter the disturbed, weaker air flow caused by the blade that passed before it. As a result, most wind turbines have only two or three blades on their rotors.

Cut-in speed is the minimum wind speed at which the blades will turn and generate useable power. This wind speed is typically between 7 and 10 mph. The *rated speed* is the minimum wind speed at which the wind turbine will generate its designated rated power; for example, a 5 kW wind turbine may not generate kW until wind speeds reach 25 mph. Rated speed for most wind turbines is in the range of 25 to 35 mph.

Between cut-in and rated speeds, the power output of a wind turbine varies with wind speed. Power output increases as wind speed increases. At lower wind speeds, the power output drops off sharply. The output of a wind turbine will level off above the rated speed because power output above the rated wind speed is mechanically or electrically maintained at a constant level, allowing more stable system control. Most manufacturers provide graphs, called *power curves,* that identify wind turbine output versus wind speed.

With extreme wind speeds, a wind turbine can become damaged or even self-destruct from high centrifugal forces, so most turbines automatically stop power generation and shut down at high wind speeds. The *cut-out speed* is the wind speed at which shutdown occurs. Cut-out speed is typically between 45 and 80 mph. Shutdown can be by twisting (pitching) the blades to spill the wind; through the use of spring-loaded spoilers that create drag; through activation of an automatic brake by a wind speed sensor; or by a spring-loaded device that rotates the blades away from the wind stream. Generally, wind turbine operation resumes when wind speed drops back to an acceptable level.

Tip speed ratio is the relationship between rotational speed of the tips of the rotor blade to the wind speed. The faster the rotation of the wind turbine rotor at a specific wind speed, the higher the tip speed ratio. A cup anemometer used for measuring wind speed is a drag design wind turbine. The velocity of the cups is essentially the same as or less than the wind speed. The ends of the cups will never go faster than the wind, so the maximum tip speed ratio is 1. A tip speed ratio below 1 indicates a drag, while a tip speed ratio above 1 requires lift. Typically, lift-type wind turbines have maximum tip speed ratios of about 8, while drag-type ratios are 1 or less. The generation of electricity requires high rotational speeds, so the lift-type wind turbine is better suited for this application.

Generator An electric generator is an electromechanical device that converts mechanical energy into electrical energy. A wind turbine *generator* converts the rotational motion of the rotor into electricity. The wind turbine rotor drives the rotation of an armature, which consists of conductor windings (wire coils) wrapped around an iron core and attached to the rotor shaft. The armature is rotated through a stationary magnetic field and movement of the armature through the field produces electricity.

Generators typically operate at a rotational speed of 1200 or 1500 revolutions per minute (rpm). A typical wind turbine rotor usually rotates in a range between 40 and 60 rpm, depending on rotor design and wind speed. A *gearbox transmission* is used to step up rotational speed of the wind turbine to speeds required for operation of the generator. Some DC-power wind turbines do not use transmissions but have a direct link between the rotor and generator. These are known as *direct drive systems.* Without a transmission, wind turbine complexity and maintenance requirements are reduced, but a much larger generator is required to deliver the same power output as other AC-type wind turbines.

Batteries and Inverters   Wind turbine generators produce DC power. A DC generator that is part of a wind turbine is used in battery-charging applications and for operating appliances and machinery designed specifically to operate on DC power. Battery systems are used to store energy. They require DC power and usually are configured at voltages of between 12 V and 120 V. Most home and office appliances and equipment operate on 120 V or 240 V, 60 cycle AC. DC generators can be configured to produce AC power with the use of an *inverter,* a device that converts DC to AC. Batteries and inverters were introduced in the previous section of this chapter (Photovoltaic [Solar Electricity] Power).

Tower   A *tower* is a structure that elevates the wind turbine so it can reach stronger winds at higher elevations and to provide blade-to-ground clearance. Towers can be constructed of a simple pipe or tube, a wooden pole, or a space/lattice frame of tubes, rods, and angle iron. A *freestanding tower* is built on a stable and substantial foundation and stands in place without the use of guy wires. It is typically erected in sections with a crane. A *guyed tower* is held in place with guy wires anchored to the ground on three or four sides of the tower. These towers can be erected by tilting them into position. Guyed towers generally cost less than freestanding towers, but require more land area to anchor the guy wires. Thin cylindrical or lattice frame towers are normally preferred over space frame towers because they minimize turbulence.

A tower must be strong enough to support the wind turbine and to sustain vibration, wind loading, and the overall weather elements for the lifetime of the wind turbine. Ideally, a tower is as high as is economically possible within zoning restrictions. Economic feasibility is evaluated based on the cost of taller towers versus the value of the resulting increase in energy generated. Studies have shown that the added cost of increasing tower height is often justified by the added power generated from the stronger winds available at higher elevations.

A general rule of thumb for proper and efficient operation of a wind turbine is that the bottom of the turbine's blades should be at least 10 ft (3 m) above the top of anything within 300 ft (about 100 m). This distance is adequate enough to reduce wind turbulence. Typically, the tower should be located at least 150 ft (about 50 m) from a residence. This distance moves the wind turbine sufficiently away from the home to limit noise concerns. Homes sitting on a one-acre parcel could probably accommodate such a small wind turbine, depending on local zoning restrictions.

Wind Turbine Size

Wind energy conversion systems come in many sizes, from very small micro systems, which can be mounted on a pole, to 1.5 (MW), utility-scale turbines that can supply energy to the electrical grid. Most stand-alone systems fall into one of three categories:

- Micro systems (100 W or less)
- Mini systems (100 W to 10 kW)
- Small systems (10 kW to 50 kW)

Single small turbines, below 50 kW, are used for individual homes, businesses, or small facilities, on grid or off grid. Common size 10 kW wind turbine has a blade diameter of about 7 to 25 ft (2 to 7.5 m) and usually requires a tower about 100 ft (30 m) tall. It can provide most of the electricity for an average total electric home at a moderate wind site. Table 25.14 provides specifications for selected modern commercially available high-speed (propeller-type) wind turbines.

Utility-scale turbines range in size from 50 kW to as large as several megawatts. Larger turbines are grouped together into wind farms, which provide bulk power to the electrical grid. These huge machines require high-wind resources because they must compete with conventional generation (coal, natural gas, oil, and nuclear) at the wholesale level.

A commercial, utility-size HAWT stands 200 ft or more in height and its blades can span 200 ft or more. HAWTs are the most common type in use today. Small HAWTs that are intended for residential or small business use are much smaller. Most have rotor diameters of 8 m or less and would be mounted on towers of 40 m in height or less.

Wind Turbine Location

Siting refers to the practice of selecting ideal sites or locations for wind turbines. The main criterion for a good wind power site is average wind speed. Wind speeds are generally higher at the highest elevation of the building site. An ideal location to site a wind turbine is to place it on top of hill, mountain, or ridge. In particular, it is always an advantage to have as wide a view as possible in the prevailing wind direction in the area.

A wind turbine should be located at least 30 ft above any physical wind barriers (e.g., trees, buildings, or bluffs) within 300 ft to avoid air turbulence. Most small wind turbine manufacturers recommend mounting turbines at least 65 ft high, but particular site conditions should be the primary factor when determining tower height. Towers from 80 to 120 ft tall may be optimal.

Trees and human-made structures (e.g., buildings, bridges, towers) are the most common obstacles to wind in the vicinity of a potential wind turbine site. They act to disturb the air both upwind and downwind of the obstruction by reducing wind speed and increasing turbulence. Mainly they will decrease the wind speed downstream from the obstacle.

Zoning restrictions on the site may limit the allowable tower height, requiring the seeking of a conditional use permit or variance from your city or county planning department. Local authorities have the discretion to issue zoning waivers for small turbines where appropriate.

TABLE 25.14 SPECIFICATIONS OF SELECTED MODERN COMMERCIALLY AVAILABLE HIGH-SPEED (PROPELLER) WIND TURBINES.

Specifications	10 kW Wind Turbine	7.5 kW Wind Turbine	3.2 kW Wind Turbine	1 kW Wind Turbine
Turbine type	3 Blade upwind	3 Blade upwind	2 Blade upwind	3 Blade upwind
Blade type	Fiber-reinforced plastic blades	Fiber-reinforced plastic blades	Fiber-reinforced plastic blades	Fiber-reinforced plastic blades
Rotor diameter	23 ft (7.0 m)	23.0 ft (7.0 m)	16 ft (4.9 m)	8.2 ft (2.5 m)
Rated power output	10 kW (Grid and pumping) 7.5 kW (battery charging)	7.5 kW	3.2 kW	1 kW
Output form	3 Phase AC, variable frequency (regulated 48 to 240 V DC or 240 V AC, 1Ø, 60 Hz with synchronous inverter)	3 Phase AC, variable frequency (regulated 48 to 240 V DC or 240 V AC, 1Ø, 60 Hz with synchronous inverter)	3 Phase AC, 12, 24, 36 or 48 V DC	24 V DC nominal
Gear	Direct drive	Direct drive	Direct drive	Direct drive
Temperature range	−40 to +140°F (−40 to +60°C)	−40 to +140°F (−40 to +60°C)	−40 to +140°F (−40 to +60°C)	−40 to +140°F (−40 to +60°C)
Generator	Permanent magnet alternator	Permanent magnet alternator	Permanent magnet alternator	Permanent magnet alternator
Start-up wind speed	7.5 mph (3.4 m/s)	7.5 mph (3.4 m/s)	7.5 mph (3.4 m/s)	6.7 mph (3 m/s) ()
Cut-in wind speed	7 mph (3.1 m/s)	7.0 mph (3.1 m/s)	7.0 mph (3.1 m/s)	5.6 mph (2.5 m/s)
Rated wind speed	31 mph (13.8 m/s)	29 mph (13.0 m/s)	22 mph (9.8 m/s)	24.6 mph (11 m/s)
Over speed protection	Furling	Furling	Furling	Furling
Cut-out wind speed	None	None	None	None
Furling wind speed	35 mph (15.6 m/s)	35 mph (15.6 m/s)	35 mph (15.6 m/s)	29 mph (13 m/s)
Maximum design wind speed	120 mph (54 m/s)	120 mph (54 m/s)	120 mph (54 m/s)	120 mph (54 m/s)

A wind turbine extracts energy from the wind. The wind leaving the turbine must have a lower energy content than the wind arriving in front of the turbine. The wake of a wind turbine is known as a *shadow*. A wind turbine will always cast a wind shade in the downwind direction. Thus, a wind turbine upwind from a second wind turbine serves as an obstruction by reducing wind speed and increasing turbulence. Multiple turbines are usually spaced about 7 rotor diameters apart in the prevailing wind direction, and about 4 rotor diameters apart in the direction perpendicular to the prevailing winds.

Wind Energy Conversion Systems

There are two generic types of *wind energy conversion systems (WECS)*; the stand-alone and grid-connected WECS. See Figures 25.8 and 25.9.

Stand-Alone Systems *Stand-alone off-grid systems* produce power independently of the utility grid. In remote locations, stand-alone systems can be more cost-effective than extending a power line to the electricity grid (the cost of which can ran range from $15 000 to $50 000 per mile). In some locations as near as one-quarter mile from utility power lines, stand-alone WECS can be more cost-effective than extending power lines. They are especially appropriate for remote, environmentally sensitive areas, such as national parks, cabins, and remote homes.

In rural areas, small stand-alone WECS power farm lighting, fence chargers, and water pumps, which provide water for livestock. *Direct-coupled systems* need no electrical storage because they operate only during daylight hours, but most systems rely on battery storage. Some systems, called *hybrid systems,* combine wind power with additional power sources (e.g., wind turbine or diesel generator).

Battery systems can supply the owner with reserve power whenever energy demand exceeds that delivered by the WECS.

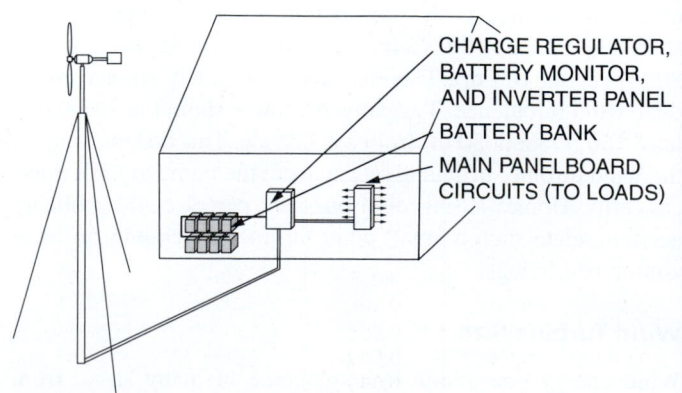

FIGURE 25.8 A schematic of a stand-alone, off-grid wind energy conversion system (WECS).

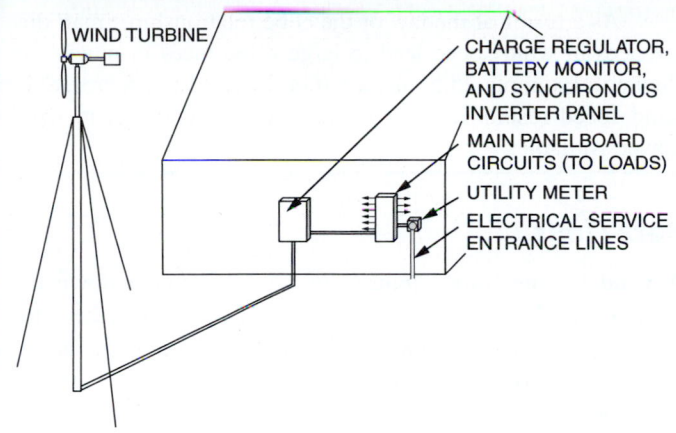

FIGURE 25.9 A schematic of a grid-connected (net-metered) wind energy conversion system (WECS).

This reserve power is needed during calm spells, but in situations where the storage capacity is taxed beyond its limits, a backup system, such as a portable gasoline or diesel generator, may be necessary. By combining two or more sources of energy, the size of energy storage can be decreased.

Grid-Connected Systems *Grid-connected systems* interface with the utility grid. They supply surplus power back through the grid to the utility and obtain power from the utility grid when the building system's power supply is low. These systems eliminate the need for battery storage, although arranging for grid interconnection with the utility company can be complicated.

In this system, one side of a synchronous inverter is typically connected to the DC power source (e.g., WECS) and the other side is connected through a meter to the panelboard. This configuration enables the utility to measure the amount of power produced. Some utilities only use a single meter: the meter turns backwards when the system is producing more power than it is consuming. This measuring technique is called *net metering*. By law, public utilities allow net metering, which allows the WECS operator to sell excess power back to the utility, but this is generally at a significantly reduced fee than what the utility charges. Although a grid-connected system does not constitute true storage, it provides power on demand at any time, in any amount. The possibilities for interconnection should be investigated early in the process of exploring the potential for a WECS.

Theoretical Wind Power

The power available in wind is related to energy in moving air. The kinetic energy (KE) of a moving body is proportional to its mass (or weight). Wind is a moving air mass and it therefore has kinetic energy. The fundamental formula for kinetic energy is $KE = \frac{1}{2}mv^2$, where m is mass and v is wind velocity. Mass of a moving fluid is the fluid's density (kinetic) and area (A) and the velocity (v) of the fluid: $m = \rho Av$. Substituting ρAv for mass yields the power in the wind.

Theoretical wind power (WP$_{theoretical}$) is a measure of the energy available in the wind over a period of time, so the kinetic energy equation can be converted into a flow equation, based on the power in the swept area (A) of the wind turbine rotor, and mass density (ρ) of the air:

$$WP_{theoretical} = \frac{1}{2}\rho Av^3$$

This equation can be used to find the raw power of the wind. It is impossible to extract all of the wind's available power. Actual power extracted is significantly less because of wind turbine and other system efficiency losses. These will be addressed under Wind Turbine Power later in this chapter.

Example 25.4

Approximate the theoretical power in wind moving at a speed of 6 m/s with an air temperature of 4°C striking a wind turbine with a 3 m diameter rotor.

From Table 25.15, mass density (ρ) of air at a temperature of 4°C is 1.277 kg/m³.

TABLE 25.15   MASS DENSITIES AND CORRECTION FACTORS OF DRY AIR AT DIFFERENT ALTITUDES AND TEMPERATURES.

Altitude Correction				Temperature Correction				
Altitude Above Mean Sea Level		Mass Density of Dry Air		Altitude Correction Factor	Air Temperature		Mass Density of Dry Air		Temperature Correction Factor
ft	m	lb/ft³	kg/m³	(F$_{alt}$)	°F	°C	lb/ft³	kg/m³	(F$_{temp}$)
0	0	0.076	1.224	1.00	−50	−46	0.097	1.560	1.27
1000	305	0.073	1.176	0.96	−25	−32	0.092	1.469	1.20
2000	610	0.071	1.143	0.93	0	−18	0.087	1.388	1.13
3000	915	0.068	1.095	0.89	20	−7	0.083	1.330	1.09
4000	1220	0.066	1.063	0.87	32	0	0.081	1.296	1.06
5000	1524	0.064	1.031	0.84	40	4	0.080	1.277	1.04
6000	1829	0.061	0.982	0.80	50	10	0.078	1.275	1.02
7000	2134	0.059	0.950	0.78	60	16	0.076	1.224	1.00
8000	2439	0.057	0.918	0.75	70	21	0.075	1.203	0.98
9000	2744	0.055	0.886	0.72	80	27	0.074	1.179	0.96
10 000	3049	0.053	0.854	0.70	100	38	0.071	1.137	0.93

The swept area (A) of the wind turbine rotor is found by:

$$A = \pi D^2/4 = \pi (3 \text{ m})^2/4 = 7.07 \text{ m}^2$$

The theoretical power in wind is found by:

$$WP_{theoretical} = \frac{1}{2} \rho A v^3 = \frac{1}{2}(1.277 \text{ kg/m}^3)(7.07 \text{ m}^2)(6 \text{ m/s})^3$$
$$= 975 \text{ W}$$

Wind Power Density

Theoretical wind power density (WPD_{theoretical}) is the unit power (W/m^2) in moving air at a specific speed (v), in mph or m/s. It can be computed by $WPD_{theoretical} = \frac{1}{2} \rho v^3$. Based on an air temperature of 60°F (16°C) and density of dry air at mean sea level (0.076 lb/ft³ and 1.224 kg/m³), the theoretical wind power density can be found by the following simplified equations. Mass densities and correction factors of dry air at different altitudes and temperatures are provided in Table 25.15.

$$WPD_{theoretical} \text{ in W/m}^2 = 0.054 \text{ } v^3 \text{ where v is in mph}$$
$$= 0.602 \text{ } v^3 \text{ where v is in m/s}$$

Again, this equation is based on the raw power of the wind. It is impossible to extract all of the wind's power. Actual power extracted will be much less. It does, however, serve to provide a basis on which to examine the affect of wind speed on wind power.

From this equation, it can be discerned that wind power is related to the cube (third power) of wind speed, which is known as the *law of the cube*. Based on this law, if wind speed is doubled, power in the wind increases by a factor of 8 ($2^3 = 8$); if wind speed is tripled, wind power increases by a factor of 27 ($3^3 = 27$). Even small differences in wind speed lead to large differences in power; for example, a 10% increase in wind speed produces a 33% increase in wind power ($1.1^3 = 1.33$). The effect of this relationship is demonstrated in Example 25.5.

Example 25.5

Approximate the theoretical power density in wind with an air temperature of 60°F moving at speeds of 8 mph, 10 mph, 15 mph, 20 mph, and 25 mph.

The theoretical wind power density at the different speeds is found by the following:

At 8 mph: $WPD_{theoretical} = 0.054 \text{ } v^3 = 0.054(8 \text{ mph})^3$
$$= 28 \text{ W/m}^2$$

At 10 mph: $WPD_{theoretical} = 0.054 \text{ } v^3 = 0.054(10 \text{ mph})^3$
$$= 54 \text{ W/m}^2$$

At 15 mph: $WPD_{theoretical} = 0.054 \text{ } v^3 = 0.054(15 \text{ mph})^3$
$$= 182 \text{ W/m}^2$$

At 20 mph: $WPD_{theoretical} = 0.054 \text{ } v^3 = 0.054(20 \text{ mph})^3$
$$= 432 \text{ W/m}^2$$

At 25 mph: $WPD_{theoretical} = 0.054 \text{ } v^3 = 0.054(25 \text{ mph})^3$
$$= 844 \text{ W/m}^2$$

As a result of the law of the cube relationship, small differences in wind speed lead to large differences in power. As shown in Example 25.5, just a 5 mph increase (or decrease) in wind speed results in a large increase (or decrease) in wind power available.

Betz Limit

A wind turbine rotor captures energy by slowing down the wind as air flows over the rotor blades and through the rotor area. A wind turbine cannot extract all of the energy because it would stop airflow. Some flow is required to carry the air past the wind turbine rotor; otherwise, the rotor itself would stop.

The maximum aerodynamic efficiency of a wind turbine rotor is 0.593 (or 59.3%). This value is known as the *Betz limit*. It means that, theoretically, the maximum amount of energy that can be extracted by a wind turbine rotor is 59.3%. In practice, the *coefficient of power* (C_p) of a wind turbine rotor is between 10% and 45%. Ranges for common types of rotors are:

High speed (propeller):	0.30 to 0.45
Darrieus:	0.25 to 0.35
Multiblade farm type:	0.20 to 0.25
Savonius:	0.10 to 0.15

The aerodynamic efficiency of a wind turbine rotor is normally illustrated in *coefficient of power curves*. As shown in Figure 25.10, drag-type rotors (e.g., Savonius rotor and multiblade farm type) have lower tip speed ratios and lower power coefficients when compared to the lift type (e.g., high-speed propeller and Darrieus). The flat curve from propeller type rotor indicates that the rotor is able to maintain high efficiency over a long range of rotor rpm while the sharp curve for the Savonius rotor, multiblade farm type, and Darrieus rotor suggests it experiences a drastic efficiency drop when the rotor rpm moves away from the narrow optimum range.

A complete wind energy system, including rotor, gearbox transmission, generator, storage, and other devices, which all have efficiencies losses, will produce between 5 and 30% of the energy available in the wind. The actual energy produced is dictated by system design.

Wind Turbine Power

The power available in the wind is related to the speed of the wind and the density of the air. The theoretical wind power density equation introduced above relates to the power in a free-flowing stream of air at standard conditions (i.e., based on the density of air at mean sea level and at a temperature of 60°F [16°C]). Air density, and thus power density, must be adjusted for air temperature and altitude. The air temperature correction factor (F_{temp}) and altitude correction factor (F_{alt}) are provided in Table 25.15. Additionally, there are efficiency losses associated with the turbine generator and gearbox

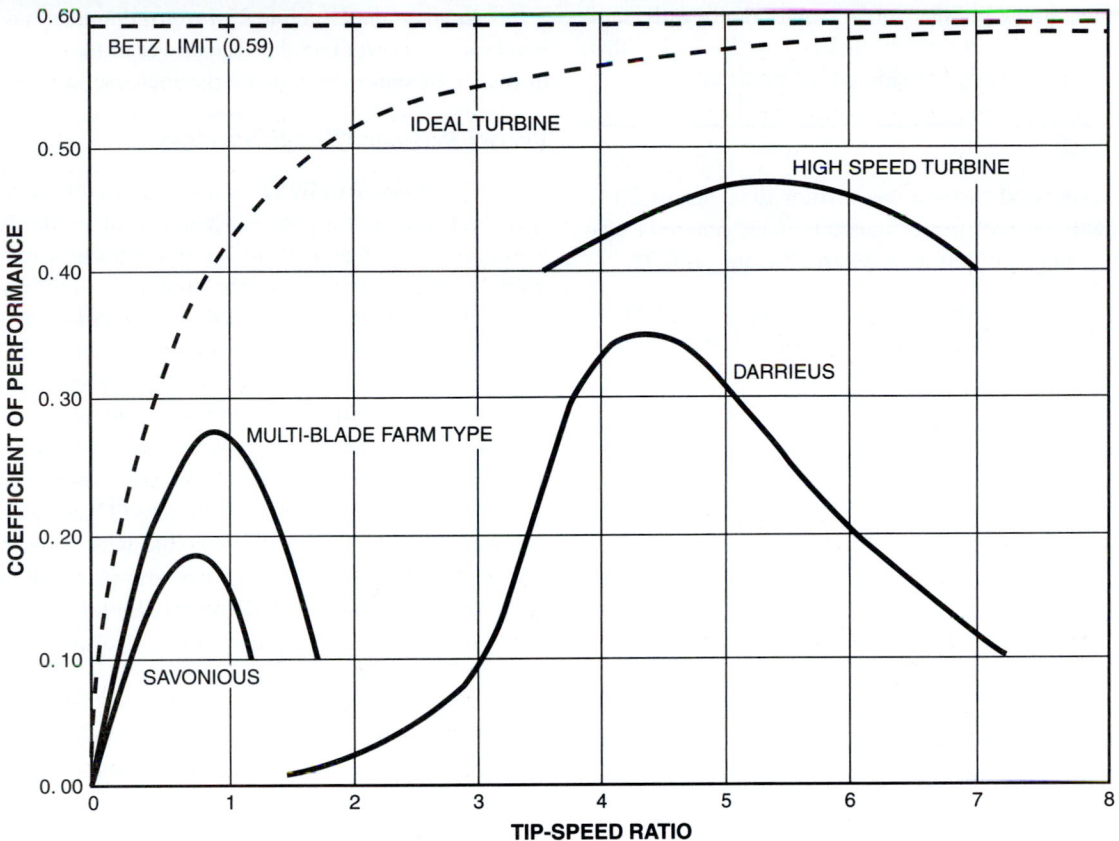

FIGURE 25.10 According to the Betz limit, the maximum aerodynamic efficiency of a wind turbine rotor is 0.593 (or 59.3%). This means that, theoretically, the maximum amount of energy that can be extracted by a wind turbine rotor is 59.3%. In practice, the coefficient of power (C_p) of a wind turbine rotor is between about 10% and 45%, as illustrated in these coefficient of power curves.

transmission. Accounting for these factors, *wind turbine power density (WTPD)* is found by:

$$WTPD \text{ in } W/m^2 = 0.005 \, AF_{alt}F_{temp}C_p\eta_{gen}\eta_{gb}v^3$$
$$\text{where a is in ft}^2 \text{ and v is in mph}$$

$$= 0.602 \, AF_{alt}F_{temp}C_p\eta_{gen}\eta_{gb}v^3$$
$$\text{where a is in m}^2 \text{ and v is in m/s}$$

For horizontal-axis wind turbines with a rotor diameter (D), this can be further converted to *wind turbine power (WTP)*, the power extracted from the wind by a wind turbine of specific design and size:

$$WTP \text{ in } W = 0.00392 \, F_{alt}F_{temp}C_p\eta_{gen}\eta_{gb}v^3D^2$$
$$\text{where D is in ft and v is in mph}$$

$$= 0.4728 \, F_{alt}F_{temp}C_p\eta_{gen}\eta_{gb}v^3D^2$$
$$\text{where D is in m and v is in m/s}$$

and where

WTP: wind turbine power (W)

WTPD: wind turbine power density (in W/m²)

A: wind turbine rotor swept area (ft² or m²)

D: wind turbine rotor diameter (ft or m)

C_p: coefficient of performance

high-speed (propeller):	0.30 to 0.45
Darrieus:	0.25 to 0.35
multiblade farm type:	0.20 to 0.25
Savonius:	0.10 to 0.15

F_{temp}: air temperature correction factor (see Table 25.15)

F_{alt}: altitude correction factor (see Table 25.15)

v: wind speed (mph or m/s)

η_{gen}: generator efficiency (0.80 for high-efficiency units with a permanent magnet generator or grid-connected induction generator)

η_{gb}: gearbox transmission efficiency (0.95 for high-efficiency transmissions)

These equations are based on the density of *dry* air. Because air does contain some moisture, air density is slightly higher. Depending on air temperature and pressure (altitude), the density of atmospheric "moist" air can be up to 4% greater (at full saturation or 100% relative humidity) than dry air under the same conditions. Atmospheric "moist" air is rarely fully saturated and it typically has a density that is 1 to 2% higher than dry air and it is thus usually considered insignificant.

Example 25.6

A horizontal axis wind turbine has a rotor diameter of 20 ft. The turbine coefficient of performance is 0.35, generator efficiency is 0.76, and gearbox transmission efficiency is 0.92.

a. Determine the wind turbine power at sea level at wind at speeds of 10 mph, 15 mph, and 20 mph, and an air temperature of 50°F.

$$WTP = 0.00392\, F_{alt}F_{temp}C_p\eta_{gen}\eta_{gb}v^3D^2$$

At 10 mph: WTP = 0.00392(1.00)(1.02)(0.35)(0.76)
$$(0.92)(10\text{ mph})^3(20\text{ ft})^2 = 391\text{ W}$$

At 15 mph: WTP = 0.00392(1.00)(1.02)(0.35)(0.76)
$$(0.92)(15\text{ mph})^3(20\text{ ft})^2 = 1321\text{ W}$$

At 20 mph: WTP = 0.00392(1.00)(1.02)(0.35)(0.76)
$$(0.92)(20\text{ mph})^3(20\text{ ft})^2 = 3131\text{ W}$$

b. Determine the wind turbine power at an elevation of 6000 ft above sea level (Denver, Colorado) at wind at a speed of 15 mph and an air temperature of 50°F.

At 15 mph: WTP = 0.00392(0.80)(1.02)(0.35)(0.76)
$$(0.92)(15\text{ mph})^3(20\text{ ft})^2 = 1057\text{ W}$$

Wind speed is always fluctuating, so power available in the wind is always changing. Because of the law of the cube, much more power is available in winds above the mean (average) wind speed than below the mean speed. This characteristic of wind was discussed earlier and is evident in Example 25.6 when comparing wind turbine power at different wind speeds. Therefore, use of mean (average) wind speed to determine monthly or average wind energy available at a site is inaccurate.

Example 25.7

A horizontal axis wind turbine has a turbine coefficient of performance of 0.35, generator efficiency of 0.76, and gearbox transmission efficiency of 0.92. Determine the wind turbine power for rotor diameters of 10 ft, 12 ft, and 14 ft at a wind speed of 20 mph.

$$WTP = 0.00392\, F_{alt}F_{temp}C_p\eta_{gen}\eta_{gb}v^3D^2$$

For 10 ft rotor: WTP = 0.00392(1.00)(1.02)(0.35)(0.76)
$$(0.92)(20\text{ mph})^3(10\text{ ft})^2 = 783\text{ W}$$

For 12 ft rotor: WTP = 0.00392(1.00)(1.02)(0.35)(0.76)
$$(0.92)(20\text{ mph})^3(12\text{ ft})^2 = 1127\text{ W}$$

For 14 ft rotor: WTP = 0.00392(1.00)(1.02)(0.35)(0.76)
$$(0.92)(20\text{ mph})^3(14\text{ ft})^2 = 1534\text{ W}$$

As shown in Example 25.7, rotor diameter has an effect on wind turbine power based on the power of two. A slight increase in rotor diameter results in a large increase in power generated.

Wind Power Classification

Wind power density (WPD) is a useful way to assess the wind energy available at a potential wind turbine site. Expressed in watts per square meter (W/m^2), it indicates how much energy is available at the site for conversion by a wind turbine.

Estimates of wind resource at a potential site are expressed in *wind power classes* ranging from Class 1 (weakest) to Class 7 (strongest), with each class representing a range of mean wind power density (W/m^2) at equivalent mean speed at specified heights above the ground. Estimates of wind resource at potential sites in the United States are shown in Figure 25.11. See also Table 25.16. Areas designated Class 4 or greater are suitable sites. Class 3 may be suitable with sophisticated wind turbine technology. Class 2 areas are marginal and Class 1 areas are unsuitable for wind energy development. Wind power classifications for two standard wind measurement heights are listed in Table 25.16.

Approximate power output (W) for selected blade sizes (ft and m diameter) and wind speeds (mph) and at sea level based on 35% of the energy available in the wind is provided in Table 25.17. This data can be used in preliminary design to approximate power output based on average wind speed and turbine size.

Wind System Sizing

Sizing a WECS is similar to sizing a photovoltaic (PV) system (see earlier section, Photovoltaic [Solar Electricity] Power). A 10-kw wind turbine can generate about 10 000 kWh annually at a site with wind speeds averaging 12 mph, or about enough to power a typical household. The size of a WECS needed to meet a desired electrical output depends on wind energy available at the site and the efficiency of the WECS. A typical size for a net-metered system is at least 5 kW. Energy-efficient appliances and lighting should be used to reduce the electrical load and thus the overall system requirements.

WECS size and cost can vary dramatically depending on the amount of power required and wind energy available. WECS installations have functioned well, but costs are high. This generally excludes WECS as a power source, except in remote locations. Constant developments in WECS technology with improved efficiencies will likely transform this technology into a cost-effective power source in the near future.

Sizing Grid-Connected Systems

The *generating capacity* of a wind turbine for a particular installation depends on the maximum overall load and on the wind conditions at the site. It is impracticable to presume that all energy needs can be met economically by wind energy

UNITED STATES ANNUAL AVERAGE WIND POWER

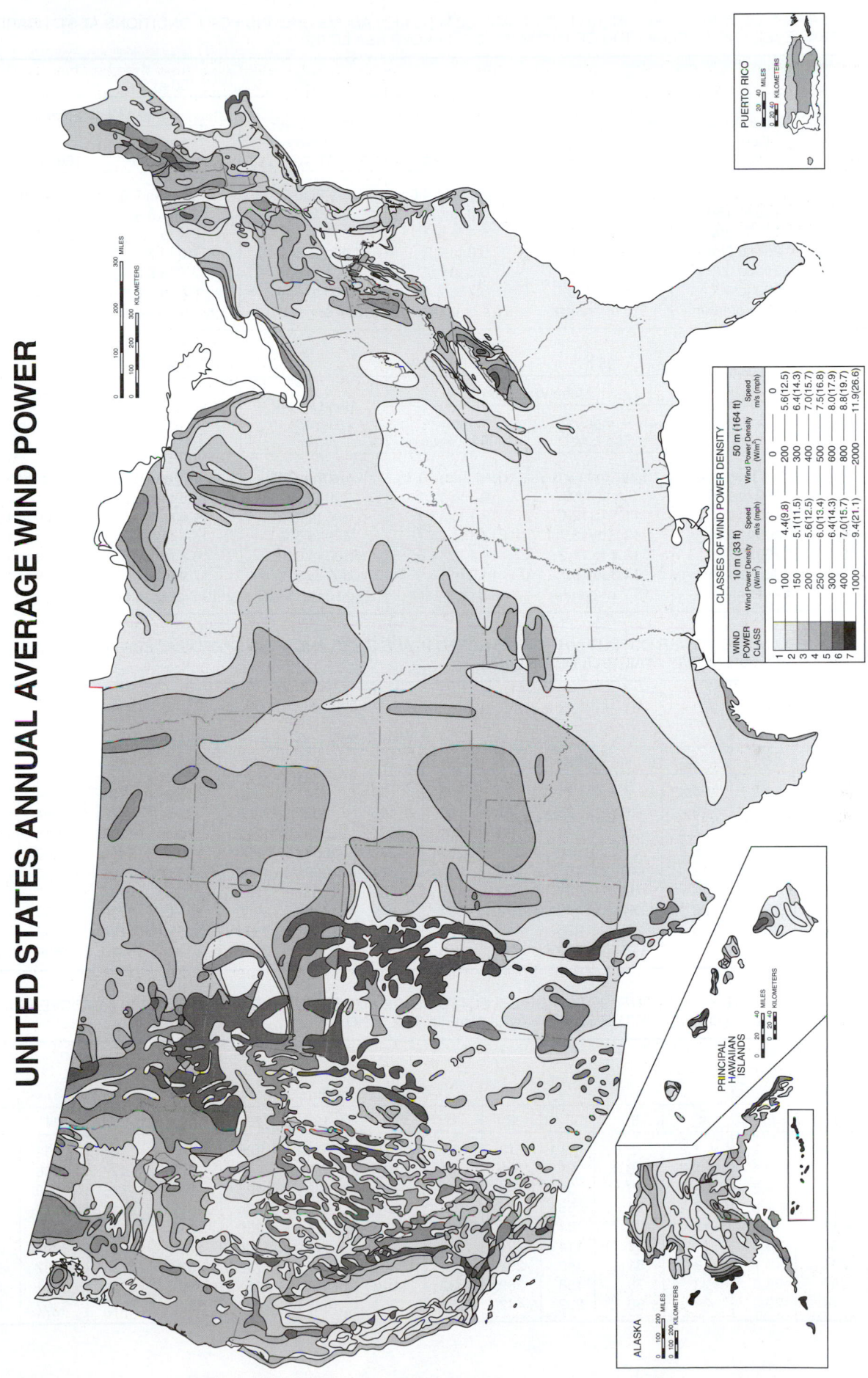

WIND POWER CLASS	10 m (33 ft)		50 m (164 ft)	
	Wind Power Density (W/m²)	Speed m/s (mph)	Wind Power Density (W/m²)	Speed m/s (mph)
1	0	0	0	0
	100	4.4(9.8)	200	5.6(12.5)
2	150	5.1(11.5)	300	6.4(14.3)
3	200	5.6(12.5)	400	7.0(15.7)
4	250	6.0(13.4)	500	7.5(16.8)
5	300	6.4(14.3)	600	8.0(17.9)
6	400	7.0(15.7)	800	8.8(19.7)
7	1000	9.4(21.1)	2000	11.9(26.6)

CLASSES OF WIND POWER DENSITY

RIDGE CREST ESTIMATES LOCAL RELIEF > 1000 FT‡

PUERTO RICO

MILES

KILOMETERS

ALASKA

PRINCIPAL HAWAIIAN ISLANDS

FIGURE 25.11 Estimates of wind resource at potential sites, expressed in wind power classes ranging from Class 1 (weakest) to Class 7 (strongest); Class 1 (white) through Class 7 (black).

TABLE 25.16 CLASSES OF WIND POWER AT 10 M (33 FT) AND 50 M (164 FT) ABOVE GROUND FOR CONDITIONS AT STANDARD SEA LEVEL AND AN ELEVATION OF 1667 M (5000 FT) ABOVE SEA LEVEL.

Wind Power Class	10 m			50 m		
	Wind Power Density W/m²	Mean Wind Speed		Wind Power Density W/m²	Mean Wind Speed	
		m/s Sea Level	m/s 1667 m Elevation		m/s Sea Level	m/s 1667 m Elevation
1	Less than 100	Less than 4.4	Less than 4.6	Less than 200	Less than 5.6	Less than 5.9
2	100 to 150	4.4 to 5.1	4.6 to 5.4	200 to 300	5.6 to 6.4	5.9 to 6.7
3	150 to 200	5.1 to 5.6	5.4 to 5.9	300 to 400	6.4 to 7.0	6.7 to 7.4
4	200 to 250	5.6 to 6.0	5.9 to 6.3	400 to 500	7.0 to 7.5	7.4 to 7.9
5	250 to 300	6.0 to 6.4	6.3 to 6.7	500 to 600	7.5 to 8.0	7.9 to 8.4
6	300 to 400	6.4 to 7.0	6.7 to 7.4	600 to 800	8.0 to 8.8	8.4 to 9.2
7	400 or greater	7.0 or greater	7.4 or greater	800 or greater	8.8 or greater	9.2 or greater

Wind Power Class	33 ft			164 ft		
	Wind Power Density W/m²	Mean Wind Speed		Wind Power Density W/m²	Mean Wind Speed	
		mph Sea Level	mph 5000 ft Elevation		mph Sea Level	mph 5000 ft Elevation
1	Less than 100	Less than 9.8	Less than 10.3	Less than 200	Less than 12.5	Less than 13.1
2	100 to 150	9.8 to 11.5	10.3 to 12.1	200 to 300	12.5 to 14.3	13.1 to 15.0
3	150 to 200	11.5 to 12.5	12.1 to 13.1	300 to 400	14.3 to 15.7	15.0 to 16.5
4	200 to 250	12.5 to 13.4	13.1 to 14.1	400 to 500	15.7 to 16.8	16.5 to 17.6
5	250 to 300	13.4 to 14.3	14.1 to 14.8	500 to 600	16.8 to 17.9	17.6 to 18.8
6	300 to 400	14.3 to 15.7	14.8 to 16.5	600 to 800	17.9 to 19.7	18.8 to 20.7
7	400 or greater	15.7 or greater	16.5 or greater	800 or greater	19.7 or greater	20.7 or greater

TABLE 25.17A APPROXIMATE POWER OUTPUT (W) FOR SELECTED BLADE SIZES AND WIND SPEEDS AT SEA LEVEL BASED ON 35% OF THE ENERGY AVAILABLE IN THE WIND.

Blade Size				Wind Speed (mph)											
Diameter		Swept Area													
ft	m	ft²	m²	4	6	8	10	12	14	16	18	20	22	24	26
4	1.2	12.6	1.2	1.4	4.6	11	22	38	60	90	128	176	234	304	386
6	1.8	28.3	2.6	3.1	10	25	49	85	136	202	288	396	526	684	868
8	2.4	50.2	4.7	5.5	19	45	88	151	241	359	512	703	934	1215	1542
10	3.0	78.5	7.3	8.6	29	70	137	237	377	564	800	1099	1460	1900	2410
12	3.7	113	10.5	12	42	101	198	341	524	809	1152	1582	2102	2735	3469
14	4.3	154	14.3	17	57	137	270	465	739	1103	1571	2156	2864	3727	4728
16	4.9	201	18.7	22	74	179	352	607	965	1439	2050	2814	3739	4864	6171
18	5.5	254	23.6	28	94	226	444	767	1219	1819	2591	3556	4724	6146	7798
20	6.1	314	29.2	35	116	279	533	948	1507	2248	3203	4396	5840	7599	9640

TABLE 25.17B APPROXIMATE POWER OUTPUT (W) FOR SELECTED BLADE SIZES AND WIND SPEEDS AT 5000 FT ABOVE SEA LEVEL BASED ON 35% OF THE ENERGY AVAILABLE IN THE WIND.

Blade Size				Wind Speed (mph)											
Diameter		Swept Area													
ft	m	ft²	m²	4	6	8	10	12	14	16	18	20	22	24	26
4	1.2	12.6	1.2	1.2	3.8	9.1	18	32	50	75	106	146	194	252	320
6	1.8	28.3	2.6	2.6	8.3	21	41	71	113	168	239	329	437	568	720
8	2.4	50.2	4.7	4.6	16	37	73	125	200	298	425	583	775	1008	1280
10	3.0	78.5	7.3	7.1	24	58	114	197	313	468	664	912	1212	1577	2000
12	3.7	113	10.5	10	35	84	164	283	435	671	956	1313	1745	2270	2879
14	4.3	154	14.3	14	47	114	224	386	613	915	1304	1789	2377	3093	3924
16	4.9	201	18.7	18	61	149	292	504	801	1194	1702	2336	3103	4037	5122
18	5.5	254	23.6	23	78	188	369	637	1012	1510	2151	2951	3921	5101	6472
20	6.1	314	29.2	29	96	232	442	787	1251	1866	2658	3649	4847	6307	8001

alone. As a general rule, a grid-connected wind system should be sized to supply 25% to 75% of the building's electrical energy requirements. Most residential applications require a wind turbine capacity of between 1 and 10 kW. Farm use requires 10 to 50 kW, and commercial/small industrial uses typically require 20 kW or larger.

The variability of your energy consumption and the amount of money you are willing to spend on a wind system should also guide your selection. For example, a user whose consumption is erratic or concentrated during short periods of the day should size a wind turbine differently than a user with a fairly constant energy demand. In the former case, wind turbine size should be a function of off-peak or average energy demand.

The synchronous inverter to convert from DC to AC power should be sized to match the maximum wind energy produced by the wind turbine. Future additional loads should also be considered. It is generally less costly to purchase an inverter with larger input and output ratings than is needed at present, if expansion is likely in the future.

Sizing of Stand-Alone, Off-Grid Systems

Sizing stand-alone, off-grid systems is substantially different than sizing a wind system for utility interconnection because remote systems must be designed to supply the entire electrical demand. Before one can size the components of a remote system, one must determine the load requirements of the site. This means quantifying the power demand on a daily and seasonal basis. The goal is to compare the amount of energy needed at different times of the day and year to when it is available on average from the wind. After taking into account the wind's intermittence, you can determine the size and type of energy storage or other energy sources needed to meet your total demand.

Battery storage should be sized large enough to handle at least three windless days. Backup generators are often included in remote electrical systems as a supplemental backup. They help to power large, infrequently used loads and to preserve the life of the batteries by minimizing the number of times they are completely discharged. Also, they run most efficiently at full load. For these reasons, generators are often sized larger than the average expected load of the system so that they can also charge the batteries at the same time, keeping run time and fuel consumption to a minimum. Many wind system dealers are familiar with remote system designs and can assist you in selecting an optimum, cost-effective system.

Wind Turbine Economics

Generally, annual average wind speeds of about 12 mph (about 6 m/s) are cost-effective for grid-connected applications. Annual average wind speeds of 7 to 9 mph (3 to 4 m/s) may be cost-effective for nonconnected electrical and mechanical applications such as battery charging and pumping water. Wind resources exceeding these speeds are available in many parts of the world.

25.7 HYDROPOWER SYSTEMS

Basic Systems

Hydropower involves hydraulic systems that use the energy in flowing or falling water to produce mechanical energy (motion), which can then be converted to electricity. Most hydroelectricity produced in the United States is from *large-scale hydropower projects* that generate more than 30 MW. By comparison, 1 MW of electricity can power about 500 American households. *Small-scale hydropower systems* are those that generate between kW to 30 MW of electricity. Hydropower systems that generate up to 100 kW of electricity are called *micro hydropower (microhydro) systems*. *Pico hydropower systems* generate less than 5 kW of electricity. Most of the systems used by home and small business owners would qualify as micro or pico hydropower systems. A 5 kW system can generally provide enough power for a moderately sized residence. A crude pico hydropower system is shown in Photo 25.33.

In the United States today, hydropower projects provide 81% of the nation's renewable electricity generation and about 10% of the nation's total electricity. According to the National Hydropower Association, that is sufficient to power 37.8 million homes.

Although there are several ways to harness the moving water, most large hydropower systems use storage reservoirs. Water from these reservoirs flow through a pressurized pipeline called a *penstock* that delivers it to a waterwheel or turbine. An *impoundment hydropower system* uses a dam to store river water in a reservoir. The water may be released either to meet changing electricity needs or to maintain a constant reservoir level. When the demand for electricity is low, a *pumped storage hydropower system* stores energy by pumping water from a lower reservoir to an upper reservoir. During periods of high electrical demand, the water is released back to the lower reservoir to generate electricity. These are typically used on large- or medium-scale hydropower projects.

PHOTO 25.33 A crude pico hydropower system in Vietnam.

Run-of-the-River Systems

Run-of-the-river or *diversion hydropower systems* do not require large storage reservoirs, but use a portion of a river's water, which is diverted to a channel, pipeline, or pressurized penstock. The moving water rotates the wheel or turbine, which spins a shaft. The motion of the shaft can be used for mechanical processes, such as pumping water, or it can be used to power an alternator or generator to generate electricity. Small run-of-the-river hydropower systems consist of these basic components (see Figure 25.12):

- *Water conveyance* a channel, pipeline, or pressurized pipeline (penstock) that delivers the water
- *Turbine* or *waterwheel* transforms the energy of flowing water into rotational energy
- *Alternator* or *generator* transforms the rotational energy into electricity
- *Regulator* controls the generator
- *Wiring* delivers the electricity

Many systems use an inverter to convert the low-voltage DC electricity produced by the system into 120 or 240 V AC. Some systems use batteries to store the electricity generated by the system. However, because hydro resources tend to be more seasonal than wind or solar resources, batteries may not always be practical for hydropower systems.

Waterwheels

The *waterwheel* is the oldest hydropower system component. Waterwheels are still available, but they are not very practical for generating electricity because of their slow speed and bulky structure.

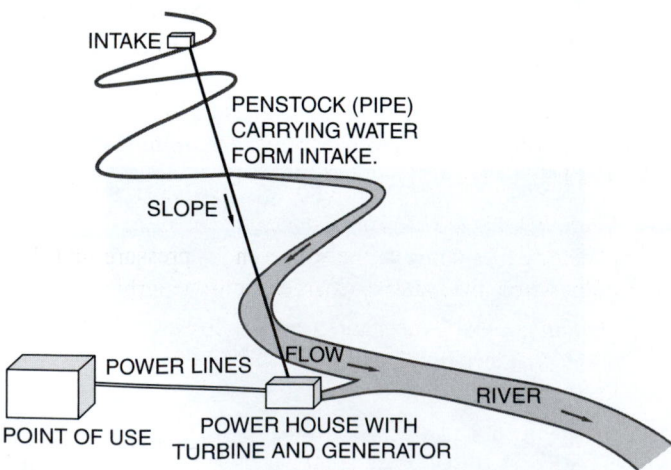

FIGURE 25.12 A schematic of a run-of-the-river hydropower system.

Turbines

Turbines have hydraulically designed blades that are fixed to a shaft (much like a waterwheel). They are designed so that when moving water strikes the surfaces of the turbine blades, they spin a shaft. The rotating part of a turbine is called the *runner*. Turbines are more efficient and compact in relation to waterwheels; they have fewer gears and require less material for construction. There are two general classes of turbines: impulse and reaction.

Impulse Turbines   *Impulse turbines* have the least complex design and are most commonly used for high head pico hydropower systems. They rely on the velocity of water to move the turbine runner. The most common types of impulse turbines include:

Pelton Wheel   The *Pelton wheel* uses the concept of jet force to create energy. Water is funneled into a pressurized pipeline with a narrow nozzle at one end. The water sprays out of the nozzle in a jet, striking the double-cupped buckets attached to the wheel. The impact of the jet spray on the curved buckets creates a force that rotates the wheel at high-efficiency rates of 70 to 90%. Pelton wheel turbines are available in various sizes and operate best under low-flow and high-head conditions.

Turgo Impulse Wheel   The *Turgo impulse wheel* is an advanced version of the Pelton wheel. It uses the same jet spray concept, but the Turgo jet, which is half the size of the Pelton, is angled so that the spray hits three buckets at once. As a result, the Turgo wheel moves twice as fast. It is also less bulky, needs few or no gears, and is generally trouble free. The Turgo can operate under low-flow conditions but requires a medium or high head.

Jack Rabbit   The *Jack Rabbit*, referred to by its trade name as the *Aquair UW Submersible Hydro Generator*, is a drop-in-the-creek turbine that can generate power from a stream with as little as 13 in of water and no head. Output from the Jack Rabbit is a maximum of 100 W, so daily output averages 1.5 to 2.4 kWh, depending on the site.

Reaction Turbines   *Reaction turbines* are highly efficient and are generally used in large-scale hydropower applications. They depend on pressure rather than velocity to produce energy. All blades of the reaction turbine maintain constant contact with the water. Because of their complexity and high cost, they are not usually used on pico hydropower projects. An exception is the *propeller turbine,* which works much like a boat's propeller and typically has three to six usually fixed blades set at different angles aligned on the runner. The propeller turbine comes in many different designs. The bulb, tubular, and Kaplan tubular are variations of the propeller turbine. The Kaplan turbine, which is a highly adaptable propeller system, can be used for pico hydropower sites.

Conventional Pumps

Conventional pumps can be used as substitutes for hydraulic turbines because, when the action of a pump is reversed, a pump operates like a turbine. Because pumps are mass produced, they are readily available and less expensive than turbines. However, adequate pump performance at a pico hydropower site requires fairly constant head and flow. Pumps are also less efficient and more prone to damage. Because of these disadvantages, pumps are rarely used for pico hydropower applications.

Channels, Storage, and Filters Before water enters the turbine or waterwheel, it is first funneled through a series of components that control its flow and filter out debris. These components are the headrace, forebay, and water conveyance (channel, pipeline, or penstock).

The *headrace* is a waterway running parallel to the water source. A headrace is sometimes necessary for hydropower systems when insufficient head is provided. It often is constructed of cement or masonry. The headrace leads to the *forebay,* which also is made of concrete or masonry. It functions as a settling pond for large debris that would otherwise flow into the system and damage the turbine.

Water from the forebay is fed through the *trash rack,* a grill that removes additional debris. The filtered water then enters through the controlled gates of the spillway into the water conveyance, which funnels water directly to the turbine or waterwheel. These channels, pipelines, or penstocks can be constructed from plastic pipe, cement, steel, and even wood. They often are held in place above ground by support piers and anchors.

Dams or Diversion Structures Dams or diversion structures are rarely used in pico hydropower projects. They are an added expense and require professional assistance from a civil engineer. In addition, dams increase the potential for environmental and maintenance problems.

Hydraulics

An introduction to basic hydraulics of water flow was provided in Chapters 12 and 13. A brief review of these principles as related to hydropower systems is provided here. Table 25.18 provides common conversion factors for hydraulic systems.

Head is related to the pressure at the base of a vertical column of water. It is usually measured in feet, meters, or units of pressure such as lb/in^2 or Pa. The relationship between feet of head and lb/in^2 is:

$$1 \text{ ft}_{head} = 0.433 \text{ lb/in}^2$$
$$1 \text{ lb/in}^2 = 2.1 \text{ ft}_{head}$$

Most small hydropower sites are categorized as low head or high head. A *high head system* has a change in elevation of more than 10 m (3 m). A *low head system* refers to a change in elevation of less than 10 m (3 m). A high head system requires less water to produce a given amount of power, so smaller, less expensive equipment is required. A vertical drop of less than 3 ft (1 m) will probably make a small-scale hydropower system unfeasible. However, for extremely small power generation amounts, a flowing stream with as little as 12 in (300 mm) of water can support a submersible turbine, like the type used originally to power scientific instruments towed behind oil exploration ships.

Head also is a function of the characteristics of the channel or pipe through which it flows. When determining head, the difference between gross head and net head must be considered.

Static head is the vertical distance between the top of the penstock that conveys the water under pressure and the point where the water discharges from the turbine.

Dynamic head is the static head less losses from friction and turbulence in the piping; thus, it is a fraction of static head.

Static head losses depend on the type, diameter, and length of the penstock piping, and the number of bends or elbows. Dynamic head can be used to approximate power availability and determine general feasibility, but dynamic head

TABLE 25.18 COMMON CONVERSION FACTORS FOR HYDRAULIC SYSTEMS.

Length

1 inch (in) = 25.4 millimeters (mm)
1 foot (ft) = 0.3048 meters (m)
1 meter (m) = 3.28 feet (ft)

Volume

1 gallon = 3.785 liters (L)
1 cubic foot (ft^3) = 7.48 gallons (gal)
1 cubic foot (ft^3) = 0.028 cubic meters (cm)
1 cubic foot (ft^3) = 28.31 liters (L)
1 cubic meter (m^3) = 35.3 cubic foot (ft^3)

Weight

1 pound (lb) = 0.454 kilograms (kg)
1 kilogram (kg) = 2.205 pounds (lb)
1 cubic foot (ft^3) of water = 62.4 pounds (lb)

Volumetric Flow Rate

1 cubic foot per second (ft^3/s) = 448.8 gallons per minute (gpm)
1 cubic foot per second (ft^3/s) = 1698.7 liters per minute (L/min)
1 cubic meter per second (m^3/s) = 15 842 gallons per minute (gpm)

Pressure

1 pound per square inch (psi) of pressure = 2.31 feet (head) of water

Power

1 kilowatt (kW) = 1.34 horsepower (hp)
1 horsepower (hp) = 746 watts (W)

must be used to calculate the actual power available. Standard engineering books and other references cover methods used to calculate head loss.

Approximating Static Head

Static head of a hydropower site is determined by finding the vertical distance between the point where water will enter the penstock and the point where water will discharge from the turbine. It can be roughly approximated from the vertical drop shown on a U.S. Geological Survey map. For an elevation drop of several hundred feet, another technique is to use an altimeter. A fair approximation technique is the hose-tube measurement method described below. If preliminary measurements using any of these techniques appear promising, a more accurate professional survey of the site should be made to determine the exact static head.

The *hose-tube measurement method* can be used to establish a rough estimate of the static head (vertical distance) of a small hydropower site. It involves taking measurements across the stream from the penstock intake (upstream location) to the turbine outflow (downstream location). A hose-tube measurement requires a 20 to 30 ft (6 to 9 m) length of small-diameter garden hose or other flexible tubing, a funnel at one end of the hose, and a measuring stick or tape. The measurement is made by stretching the funnel end of the hose up the stream to the point that is the most practical elevation for the penstock intake. The upstream end of the hose (with the funnel) is held underwater as near the surface as possible. Once the funnel is underwater and water flows through the hose, the downstream end of the hose (end without funnel) is lifted until water stops flowing from it. The static head for that section of stream is determined by measuring the vertical distance between the end of the tube and the surface of the water. Hose-length measurements continue until the future location of the turbine is reached. The sum of these measurements gives a rough approximation of the static head for the site.

Volumetric Flow Rate

The quantity of water flowing past a point in the stream per unit time is called the *volumetric flow rate*. Flow rate is measured in gallons per minute (gpm), cubic feet per second (ft³/s), or liters per second (L/s). A stream's flow rate can be obtained from local offices of the U.S. Geological Survey, the U.S. Army Corps of Engineers, the U.S. Department of Agriculture, city or county engineer, local water supply district, or flood control authorities. A common method for measuring flow on very small streams is the bucket measurement method. Generally, unless a storage reservoir will be used, the lowest average flow of the year should be used as the basis of the system design.

Determining Flow

The *bucket measurement method* involves damming the stream with logs or boards to divert the stream flow into a bucket or container. The rate that the container fills is the flow rate. For example, a 5 gal (or 20 L) bucket that fills in one minute is a flow rate of 5 gpm (or 20 L/m).

The *cross-sectional flow method* is a more accurate method used to approximate volumetric flow at the narrowest point in the stream. Needed to perform this measurement are an assistant, a tape measure, a yardstick or calibrated measuring rod, a weighted float (i.e., a partially filled plastic milk jug), a stopwatch, and graph paper. Begin by calculating the cross-sectional area of the streambed during the time of lowest water flow. Select a stretch of the stream with the straightest channel and most uniform depth and width as possible, then measure the width and at depth of the stream at equal increments across the stream that are perpendicular to water flow. Determine the cross-sectional area of each section of stream and sum the section areas to find the total cross-sectional area.

Next, establish a distance downstream of where measurements of the stream depth were made. Determine the velocity of each section of stream by recording the time it takes a weighted float to travel the specific distance downstream. The float should not hit or drag on the bottom of the stream. Repeat this procedure several times to get an average velocity for each section of stream. Then, divide the distance between the two points by the float time in seconds to determine the velocity for each section of stream, in ft/s (or m/s). Multiply the average velocity of each section by the cross-sectional area for the same section of stream to determine the flow rate for that section. Add the measured flow rates for each section to determine the measured flow rate.

Finally, multiply the measured flow rate by a factor that accounts for the roughness of the stream channel (0.8 for a sandy stream bed, 0.7 for a bed with small- to medium-sized stones, and 0.6 for a bed with many large stones). The result will be a fair estimate of the volumetric flow rate, in ft³/s (or m³/s), at the time of measurement.

Repeat the procedure several times during the low flow season to more accurately estimate the average low water flow. To make cross-sectional area measurements easier, measure the water depth above or below the water level when you first measured the stream, and calculate the area of greater or less water, and add or subtract this from the benchmark area. Alternatively, install a gauge (made from a calibrated rod or post) on the bank to read the water depth and recalculate the cross-sectional area of the stream. You will need to repeat the flow velocity procedure each time.

Volumetric flow data may correlate well with long-term precipitation data or reported flow data from nearby rivers, to get an estimate of long-term, seasonal low, high, and average flows for your stream. Remember that, no matter what the volume of the flow is at any one time, you may be able to legally divert only a certain amount or percentage of the flow. Also try to determine if there are any plans for development or changes in land use upstream from your site. Activities such as logging upstream from your site can greatly alter stream flows and water quality.

Estimating Power Output

The power available at any instant in a stream is primarily a product of the volumetric flow rate and head. Turbine and generator efficiencies also affect power output. These characteristics depend on operating conditions (head and flow). The higher the head the better, because less water is necessary to produce a given amount of power, and smaller, more efficient, and less costly turbines and piping can be used. A high volumetric flow rate can compensate for low head, but a larger, and more costly turbine will be necessary. Additionally, low-head, low-speed waterwheels are less efficient than high-head, high-speed turbines. The overall efficiency of a system will range between 40% and 70%. A well-designed system will achieve an efficiency of 50 to 55%.

The power output (watts) for a small hydropower system can be approximated based on dynamic head (ft_{head}); the vertical distance available after subtracting losses from pipe friction and flow rate (gpm); and a 55% efficiency:

$$\text{power output (W)} = (ft_{head} \times gpm)]/10$$

Small-Scale Turbines and Generators

Commercially available turbines and generators are usually sold as a package. Only a few companies make pico hydropower turbines. Most are high-head turbines. Low-head, low-flow turbines are difficult to find, and may have to be custom-made. Pico hydropower turbines can produce anywhere from 20 to 2500 W. Table 25.19 has the specifications on output of a pico hydroelectric system, in watts.

One supplier sells hydropower generators for flows 200 to 600 gpm that produce from 400 W to 2 kW. They generate 120, 240, or 480 V wild (unregulated) AC, which is then stepped down with a transformer and rectifier. Another supplier sells turbines with one, two, or four nozzles, depending on water flow and power requirements. (PVC penstock with one shut-off valve on two-nozzle machines and 3 cutoff valves on four-nozzle machines included.) These turbines can be fitted with nozzles up to ½ inches in diameter. Each hydropower system is custom-built to match site specifications based on head, flow,

pipe size and length, electrical transmission line length, and battery voltage. For a static head range of 20 to 600 ft and volumetric flow rates of 4 to 450 gpm, they produce a maximum of 700 W (at 12 V), a maximum of 1400 W (at 24 V), and a maximum of 2500 W (at 48 V).

System Performance

Generally, low-head, low-speed water wheels are less efficient than high-head, high-speed turbines. The overall efficiency of a system will range between 40% and 70%. A well-designed system will achieve an average efficiency of 55%. Turbine manufacturers provide a close estimate of potential power output for their turbine for the head and flow conditions at a site.

Legal Restrictions and Requirements

Use, access to, control, or diversion of water flows is highly regulated in most regions. In most states, a separate water right to produce power is needed, even if one already has a water right for any other use. The same is true for any physical alteration of a stream channel or bank that may affect water quality or wildlife habitat, regardless of whether the stream is on private property. There are many local, state, and federal regulations that govern, or will effect, the construction and operation of any hydropower system. The larger the system, the more complicated, drawn out, and expensive the permitting and approval process will be. Penalties for not having permits or necessary approvals can be harsh.

When planning a hydropower system in the U.S., the county engineer should be contacted initially. He or she will be the most informed about what restrictions govern the development and/or control of water resources in the area. The two primary federal agencies that influence hydropower permits are the Federal Energy Regulatory Commission (FERC) and the U.S. Army Corps of Engineers. FERC is responsible for licensing all nonfederal government hydroelectric projects under its jurisdiction. A hydroelectric project is within the jurisdiction of FERC if any of the following conditions apply:

- The project is on a navigable waterway
- The project will affect interstate commerce (i.e., if the system is to be connected to a regional electric transmission grid)
- The project uses federal land
- The project will use surplus water or waterpower from a federal dam

If the project falls under the jurisdiction of FERC, it is necessary to apply for a license or exemption from FERC. The FERC application process will require contacting and consulting other federal, state, and local government agencies, and providing evidence that this was accomplished. If the project involves a discharge of dredge or fill material into a watercourse or wetland, a permit from the Army Corps of Engineers may be needed.

TABLE 25.19 OUTPUT OF A TYPICAL PICO HYDROELECTRIC SYSTEM, IN WATTS (W).

Volumetric Flow Rate, gpm	Dynamic Head, ft					
	25	50	75	100	200	300
3	—	—	—	—	30	70
6	—	—	25	35	100	150
10	—	35	60	80	180	275
15	20	60	95	130	260	400
20	30	80	130	200	400	550
30	50	125	210	290	580	850
50	115	230	350	500	950	1400
100	200	425	625	850	1500	—
200	—	520	850	1300	—	—

It must be determined whether, and to what extent, water can be diverted from the stream channel, and what restrictions apply to construction and operation of the system. Each state controls water rights, and the project may require a separate water right to produce power, even if the owner already has a water right for any other use.

Other federal government agencies that may require permits include: the U.S. Fish and Wildlife Service; the Federal Aviation Administration (if a power line will be constructed near an airport); and the U.S. Forest Service or Bureau of Land Management, if the project will use land administered by these agencies.

Selling Power to a Utility

The Public Utility Regulatory Policies Act (PURPA) of 1978 requires electric utilities to purchase power from independent power producers if certain conditions are met. You will need to contact your local utility and/or public utility commission to determine what these technical and operating requirements are and the price that the utility will pay you for the electricity you generate. You may also need a license from FERC. The utility will require that you insure the system. The interconnect requirements and insurance premiums may cost more than what you earn from selling the power. An alternative to selling power is "net metering or billing," where your system offsets the amount of power you purchase from a utility. Many states in the United States have net metering provisions; however, you will still have to negotiate with the utility concerning their interconnection requirements.

Other Considerations

Many other factors will determine whether developing the site is practical. Penstock routing and placement is important. You will need to inspect and clean the penstock intake regularly. Freezing weather, livestock, and vandals can damage exposed piping, but burying it may not be practical or cost-effective. The piping must have adequate support to keep it from breaking apart or moving under the weight and pressure of the water. The turbine/generator should be above the stream's flood stage. A power line from the generator could be expensive.

25.8   RAINWATER HARVESTING

Rainwater

Rainwater can provide clean, safe, and reliable water so long as the collection system is properly constructed and maintained, and the water is treated appropriately for its intended use. It is one of the purest sources of water available, nearly always exceeding the quality of ground or surface water. Additionally, rainwater is soft, which can significantly lower the quantity of detergents and soaps needed for cleaning.

Rainwater Collection/Harvesting

Rainwater collection/harvesting systems intercept and collect storm water runoff and detain or retain it for later use. Harvested water can be used for toilet flushing, car washing, indoor plant watering, pet and livestock watering or washing, and lawn/garden irrigation.

Rainwater Harvesting Systems

Rainwater harvesting systems (RWHS) are made of a catchment area, conveyance devices (e.g., gutters, conductors, and downspouts), filters, storage tanks, and distribution systems. Rainwater is collected in the *catchment area*, which is usually a rooftop, patio, terrace, driveway, or other impermeable surface. Buildings and landscapes can be designed to maximize the amount of catchment area, thereby increasing rainwater-harvesting capabilities.

Roof catchment areas are made of a variety of materials including wood, ceramic, cementitious, and metal. Wood shakes, wood shingles, and asphalt composition shingles are more likely than other materials to support the growth of mold, algae, bacteria, and moss, which can contaminate water supplies. Metal roofs are relatively smooth and are therefore less susceptible to contamination. Stainless steel is an exceptionally inert metal that works well, but it is expensive. Aluminum is also relatively inert, but the health effects of ingesting small amounts of aluminum are somewhat controversial, with the debate surrounding a possible link to the development of Alzheimer's disease. More commonly used is corrugated steel roofing, which is constructed of cold-formed mild steel sheets protected with a coating of hot-dipped or electrolytic galvanized (zinc) coating or paint. Zinc has a low toxicity. Another good roof product is galvanized steel with a baked-enamel, certified lead-free finish. Some paints may cause some risk to health and/or may impart an objectionable taste to the rainwater. To be most effective, the roof should be fully exposed and away from overhanging tree branches, which can result in extra contamination.

Storage systems make use of above-ground containers (e.g., barrel or fiberglass tank) known as *cisterns*. They are the single largest investment for most rainwater harvesting systems. Most cisterns are cylindrical. In cold climates, the cistern must be protected from freezing by burying it underground or incorporating it into a basement. In northern or arid climates where rainwater is the only water source, it may be necessary to significantly oversize the cistern to provide carryover capacity during a significant portion of the season when rain does not fall. The use of two or more cisterns allows servicing one of the units without losing the operation of the system. An example of rainwater cisterns are shown in Photo 25.34.

Cisterns can be made of galvanized steel, concrete, wood, fiberglass, or plastic (polyethylene). They have a lid to reduce evaporation, thwart mosquito breeding, impede animal infestation (insects, rodents, and amphibians), prevent contamination, and protect child safety. Cisterns should be made of an opaque

PHOTO 25.34   Rainwater harvesting cisterns. (Courtesy of DOE/NREL)

material to prevent light transmission and inhibit photosynthesis and algae and bacterial growth. Annual cleaning of a tank may be necessary in a potable system.

Rainwater captured in the catchment area can be conveyed to the cistern through gutters and downspouts. A *gutter* is a horizontal trough or channel that runs along the eaves of a building roof to capture and divert rainwater. Water collected by a gutter drains into a *downspout,* which is a vertical pipe that carries water away from the roof gutter. A *conductor* is a horizontal pipe that carries water. Gutters and downspouts are made of copper, galvanized steel, aluminum, or vinyl. Bamboo can be used. Vinyl and aluminum gutters and downspouts can be custom-made in one piece without seams and are called *seamless.* Seamless gutters are preferred because they have few joints. Aluminum and steel gutters and downspouts come in many different colors that are ideal for matching the color of trim and cladding. Gutter colors are baked on at the factory or painted on site. Gutters need to be supported so they do not sag or fall off when loaded with water. Dirt, debris, and other materials from the roof surface may contaminate rainwater. It is necessary to filter and screen out contaminants before they enter the cistern.

Pipes from the storage tank channel water to the point of use (e.g., an irrigation system feeding vegetation located at low points in the yard). An overflow pipe drains extra rain that falls after the cistern is full. Water flow is by gravity-flow conveyance. Where gravity flow is not possible, a pump and pressure tank are required for water distribution.

Quality of Harvested Water

Rainwater captured in a sanitized system is good for drinking. However, infrequency of rainfall results in buildup of contaminants on the catchment area surface and in the conveyance devices. Contaminants on the catchment area surface can include pollen, dust, mold, bacteria, algae, bird and animal droppings, dead insects, smoky condensate residues from heaters, atmospheric pollutants, and chemicals from painted or waterproofed surfaces or made from fibrous cement. Rainwater will wash

many types of contaminants off that surface and into the storage tank. After a dry spell, these impurities may occur in high concentrations in rooftop runoff.

A *first-flush washing* prevents the initial flow of rainwater with pollen, mold spores, and other contaminants from draining into the storage tank, thereby preventing large concentrations of surface contamination from being collected in the storage tank. Such devices include tipping buckets that dump when water reaches a certain level. In addition, there are containers with a ball that floats with the rising water to close off an opening after a certain amount of inflow. Water is then diverted to the storage container. First-flush washing is not needed for water used solely for irrigation purposes.

Dirty storage containers may become a health hazard or a breeding ground for mosquitoes and other pests (e.g., insects, lizards, and other small animals, which may enter the tank). Filtering rainwater is necessary to impede flow of debris and reduce sediment buildup in the irrigation system.

Because of concerns of microbial contamination of harvested rainwater, it is not recommended as a direct source of drinking water for humans. Any harvested rainwater must be purified into a potable state before human consumption. To become potable, rainwater must go through several purification steps: screening, settling, filtering, and disinfecting. Screening prevents leaves and other debris from entering the storage tank. Settling in the storage tank helps remove fine particles of dirt and dust by allowing them to settle to the bottom of the tank. Filtering can remove sediment and contaminants, and trap particulate matter depending on what types of filters are used. Disinfecting with chlorine, ozone, or ultraviolet light kills microorganisms. Properly designed, constructed, and maintained purification systems that include disinfection steps have been successfully used for private domestic water supplies.

Design of a Rainwater Harvesting System

Design of a rainwater harvesting system should take into account the intended use (potable or nonpotable) of the system, water consumption rates, local rainfall data and weather patterns, and the size and type catchment area and cistern.

Water Consumption Rate

The design of a rainwater harvesting system must take into account how much water is needed and when. For systems relying only on harvested rainwater, water consumption includes toilet flushing, bathing, clothes washing, dishwashing, outdoor watering, and other water use by occupants. As cited earlier, Americans use almost 100 gal (378.5 L) of drinking water per person per day. Approximate flow rates and typical consumption by type of plumbing fixture and location are provided in Table 25.20. With prudent use, low-flow, water-conserving plumbing fixtures and little outdoor water use, residential consumption can be reduced to a range from 40 to 70 gal (160 to 280 L) per person per day.

TABLE 25.20   APPROXIMATE FLOW RATES AND TYPICAL CONSUMPTION BY TYPE OF PLUMBING FIXTURE AND LOCATION. FLOW RATE VARIES SIGNIFICANTLY BY RESIDUAL PRESSURE (WATER PRESSURE BEFORE THE FIXTURE).

Approximate Flow Rates by Fixture Type

Bathtub faucet	4 to 8 gpm	15 to 30 L/min
Clothes washer (laundry)	3 to 5 gpm	11 to 19 L/min
Dishwasher (domestic)	3 to 5 gpm	11 to 19 L/min
Drinking fountain	0.75 gpm	3 L/min
Hose bibb (½ in)	3 to 12 gpm	11 to 27 L/min
Kitchen sink	2 to 5 gpm	8 to 57 L/min
Lavatory faucet	2 to 3 gpm	8 to 11 L/min
Lawn sprinkler (domestic)	3 to 7 gpm	11 to 27 L/min
Showerhead (low flow design)	2 to 3 gpm	8 to 11 L/min
Showerhead (standard flow design)	3 to 6 gpm	11 to 23 L/min
Toilet, flush tank	3 to 5 gpm	11 to 19 L/min
Toilet or urinal, 1 in flush valve (15 to 30 psi)	27 to 35 gpm	103 to 133 L/min

Typical Consumption for Urban or Suburban Residence

Per person per day, for all purposes	30 to 60 gal	114 to 228 L
Per shower, per use	25 to 60 gal	95 to 225 L
Fill bathtub, per fill	35 gal	133 L
Fill sink, per fill	1 to 2 gal	4 to 8 L
Flush toilet, per flush	1.6 to 7 gal	6 to 27 L
Irrigate with ¼ in of water, per 1000 ft^2 of lawn	160 gal	600 L
Dishwasher, per load	10 to 20 gal	38 to 76 L
Washing machine, per load	30 to 50 gal	114 to 450 L
Domestic water softener regeneration cycle	50 to 150 gal	190 to 570 L

Typical Consumption for Rural Residence

Per person per day, for all purposes	60 gal	228 L
Per horse, dry cow, or beef animal per day	12 gal	46 L
Per milking cow per day	35 gal	133 L
Per hog per day	4 gal	15 L
Per sheep per day	2 gal	8 L
Per 100 chickens per day	6 gal	23 L

Consumption rates for irrigation systems vary significantly based on type and maturity of vegetation. In semiarid climates, turf grass requires irrigation rates ranging from 5 in/ft^2 (26 gal/ft^2) for Buffalo grass to 24 in/ft^2 (125 gal/ft^2) for Kentucky bluegrass over the growing season.

Amount of Rainfall

The amount of rainfall in a particular period is very unpredictable, especially in arid and semiarid climates where annual precipitation is less than 20 in (500 mm). Even in wet climates, rain does not fall evenly throughout the year. Some months are drier than others. Rainfall also varies with slight changes in location, so that data collected at a weather station even a few miles (3 km) away may be misleading. Table 25.21 provides normal monthly precipitation averages for several U.S. cities based on recorded 30-year average from 1961 to 1990. This data can be used as an approximation. More accurate data may be available through the state meteorologist or climatologist office.

Rainfall intensity is the precipitation rate at which rain falls over a short duration of time (usually based on 15 min of precipitation). The units for rainfall intensity are in/hr and mm/hr. The map in Figure 25.13 shows rainfall intensity rates in the United States. Table 25.22 shows rainfall intensity rates, in mm/hr, based on 15 min of precipitation, for selected cities in Canada.

Rainwater Collected

The *catchment area (A)* is computed using the footprint of the building plus the length of the overhangs. One inch of rain typically produces about 0.60 gal/ft^2 of catchment area. Not all the rain that falls can actually be collected. A small amount of rain (usually 0.03 to 0.1 in/0.75 to 2.5 mm) is needed to dampen the roof, gutter, and downspout surfaces before water begins to actually flow. Some water will be lost through splash and first-flush washing. The efficiency is usually assumed to be between 75% (rough concrete) and 90% (metal), depending on system design and capacity.

To approximate the rainwater collected (R), multiply 0.60 gal/ft^2 of catchment area by the catchment area (A), the average rainfall (r) for the period under consideration, and the system efficiency (η), expressed as a coefficient:

$$R = (0.60 \cdot A \cdot r \cdot \eta)$$

TABLE 25.21   NORMAL MONTHLY PRECIPITATION, INCHES, BASED ON 30-YEAR AVERAGE (1961 TO 1990).

Location	State	Jan	Feb	Mar	Apr	May	Jun	Jul	Aug	Sep	Oct	Nov	Dec	Annual
Birmingham AP	AL	5.10	4.72	6.19	4.96	4.85	3.73	5.25	3.59	3.93	2.81	4.33	5.12	54.58
Huntsville	AL	5.17	4.87	6.62	4.93	5.08	4.13	4.85	3.47	4.08	3.25	4.86	5.87	57.18
Mobile	AL	4.76	5.46	6.41	4.48	5.74	5.04	6.85	6.96	5.91	2.94	4.10	5.31	63.96
Montgomery	AL	4.68	5.48	6.26	4.49	3.92	3.90	5.19	3.69	4.09	2.45	4.08	5.20	53.43
Anchorage	AK	0.79	0.78	0.69	0.67	0.73	1.14	1.71	2.44	2.70	2.03	1.11	1.12	15.91
Annette	AK	10.07	8.79	7.89	7.79	6.03	4.67	4.27	6.15	9.28	15.47	11.67	11.20	103.28
Barrow	AK	0.17	0.15	0.17	0.20	0.16	0.28	0.94	0.96	0.60	0.45	0.25	0.16	4.49
Bethel	AK	0.58	0.43	0.59	0.70	0.78	1.44	1.98	2.91	2.04	1.45	1.07	1.02	14.99
Bettles	AK	0.69	0.64	0.68	0.64	0.61	1.44	1.94	2.38	1.72	1.20	0.90	0.90	13.74
Big Delta	AK	0.33	0.28	0.25	0.29	0.92	2.56	2.72	1.92	1.06	0.71	0.52	0.40	11.96
Cold Bay	AK	2.84	2.27	2.16	1.97	2.29	2.10	2.52	3.24	4.41	4.34	4.19	3.67	36.00
Fairbanks	AK	0.47	0.40	0.37	0.32	0.61	1.37	1.87	1.96	0.95	0.90	0.80	0.85	10.87
Gulkana	AK	0.45	0.53	0.36	0.24	0.61	1.60	1.78	1.50	1.38	0.89	0.61	0.92	10.87
Homer	AK	2.40	2.13	1.70	1.28	1.15	1.03	1.49	2.24	3.29	3.24	2.62	2.82	25.39
Juneau	AK	4.54	3.75	3.28	2.77	3.42	3.15	4.16	5.32	6.73	7.84	4.91	4.44	54.31
King Salmon	AK	1.05	0.81	1.07	1.13	1.34	1.58	2.23	2.95	2.74	2.07	1.48	1.37	19.82
Kodiak	AK	7.38	5.28	4.63	4.20	5.52	4.78	3.70	5.15	6.99	7.18	5.96	6.81	67.58
Kotzebue	AK	0.43	0.32	0.35	0.37	0.33	0.52	1.46	1.78	1.58	0.73	0.59	0.52	8.98
Mcgrath	AK	0.75	0.66	0.80	0.82	0.87	1.47	2.05	2.47	1.98	1.40	1.29	1.40	15.96
Nome	AK	0.79	0.60	0.54	0.68	0.62	1.12	2.17	2.71	2.43	1.35	1.04	0.83	14.88
St. Paul Island	AK	1.85	1.33	1.22	1.34	1.28	1.23	1.90	2.75	2.59	2.81	2.80	2.22	23.32
Talkeetna	AK	1.30	1.44	1.46	1.59	1.67	2.63	3.63	4.52	4.23	3.10	1.73	1.91	29.21
Unalakleet	AK	0.69	0.51	0.70	0.63	0.87	1.35	2.28	3.04	2.28	1.24	1.09	0.91	15.59
Valdez	AK	5.60	5.13	4.70	3.16	3.83	3.08	3.84	5.96	8.37	8.07	5.50	6.80	64.04
Yakutat	AK	12.18	10.67	10.72	9.92	9.66	7.30	8.18	11.54	18.65	22.97	14.52	14.94	151.25
Flagstaff	AZ	2.04	2.09	2.55	1.48	0.72	0.40	2.78	2.75	2.03	1.61	1.95	2.40	22.80
Phoenix	AZ	0.67	0.68	0.88	0.22	0.12	0.13	0.83	0.96	0.86	0.65	0.66	1.00	7.66
Tucson	AZ	0.87	0.70	0.72	0.30	0.18	0.20	2.37	2.19	1.67	1.06	0.67	1.07	12.00
Winslow	AZ	0.45	0.52	0.55	0.26	0.31	0.31	1.20	1.39	0.91	0.91	0.57	0.66	8.04
Yuma	AZ	0.35	0.22	0.21	0.14	0.04	0.02	0.26	0.64	0.31	0.29	0.24	0.45	3.17
Fort Smith	AR	1.90	2.60	3.95	3.97	5.24	3.39	2.99	2.92	3.24	3.68	3.99	3.03	40.90
Little Rock	AR	3.91	4.36	5.31	6.21	7.02	7.84	8.19	8.06	7.41	6.30	5.21	4.28	74.10
North Little Rock	AR	3.21	3.53	4.97	5.08	5.24	3.26	3.31	3.10	4.08	3.62	5.16	4.69	49.25
Bakersfield	CA	0.86	1.06	1.04	0.57	0.20	0.10	0.01	0.09	0.17	0.29	0.70	0.63	5.72
Bishop	CA	1.11	0.95	0.39	0.26	0.29	0.18	0.23	0.18	0.24	0.13	0.57	0.84	5.37
Blue Canyon	CA	0.00	0.00	0.00	0.00	0.00	0.00	0.00	0.00	0.00	0.00	0.00	0.00	0.00
Eureka	CA	6.00	4.73	5.32	2.88	1.44	0.51	0.13	0.48	0.89	2.67	6.44	6.04	37.53
Fresno	CA	1.96	1.80	1.89	0.97	0.30	0.08	0.01	0.03	0.24	0.53	1.37	1.42	10.60
Long Beach	CA	2.47	2.47	1.96	0.68	0.17	0.04	0.01	0.11	0.29	0.27	1.65	1.68	11.80
Los Angeles AP	CA	2.40	2.51	1.98	0.72	0.14	0.03	0.01	0.15	0.31	0.34	1.76	1.66	12.01
Los Angeles C.O.	CA	2.92	3.07	2.61	1.03	0.19	0.03	0.01	0.14	0.45	0.31	1.98	2.03	14.77
Mount Shasta	CA	6.31	5.16	5.12	2.63	1.54	0.82	0.35	0.48	0.82	2.40	5.78	5.61	37.02
Redding	CA	6.06	4.45	4.38	2.08	1.27	0.56	0.17	0.46	0.91	2.24	5.21	5.51	33.30
Sacramento	CA	3.73	2.87	2.57	1.16	0.27	0.12	0.05	0.07	0.37	1.08	2.72	2.51	17.52
San Diego	CA	1.80	1.53	1.77	0.79	0.19	0.07	0.02	0.10	0.24	0.37	1.45	1.57	9.90
San Francisco AP	CA	4.35	3.17	3.06	1.37	0.19	0.11	0.03	0.05	0.20	1.22	2.86	3.09	19.70
San Francisco C.O.	CA	4.06	2.95	3.07	1.29	0.25	0.15	0.04	0.07	0.26	1.26	3.21	3.10	19.71
Santa Barbara	CA	3.21	3.61	2.84	1.06	0.16	0.04	0.02	0.11	0.45	0.52	2.03	2.20	16.25
Santa Maria	CA	2.16	2.62	2.27	0.99	0.20	0.03	0.01	0.05	0.30	0.49	1.46	1.78	12.36
Stockton	CA	2.84	1.97	2.17	1.08	0.26	0.08	0.06	0.07	0.35	0.79	2.17	2.11	13.95
Alamosa	CO	0.26	0.29	0.45	0.49	0.64	0.67	1.19	1.12	0.89	0.70	0.43	0.44	7.57
Colorado Springs	CO	0.29	0.40	0.94	1.19	2.15	2.25	2.90	3.02	1.33	0.84	0.47	0.46	16.24
Denver	CO	0.50	0.57	1.28	1.71	2.40	1.79	1.91	1.51	1.24	0.98	0.87	0.64	15.40
Grand Junction	CO	0.56	0.48	0.90	0.75	0.87	0.50	0.65	0.81	0.82	0.98	0.71	0.61	8.64
Pueblo	CO	0.32	0.31	0.78	0.88	1.25	1.25	2.09	1.99	0.90	0.57	0.43	0.42	11.19
Bridgeport	CT	3.24	3.01	3.75	3.75	3.93	3.46	3.78	3.25	3.07	3.11	3.81	3.50	41.66
Hartford	CT	3.41	3.23	3.63	3.85	4.12	3.75	3.19	3.65	3.79	3.57	4.04	3.91	44.14
Wilmington	DE	3.03	2.91	3.43	3.39	3.84	3.55	4.23	3.40	3.43	2.88	3.27	3.48	40.84
Washington Dulles AP	D.C.	2.70	2.81	3.17	3.11	4.02	3.92	3.49	3.94	3.36	3.20	3.30	3.22	40.24

(continues)

Location	State	Jan	Feb	Mar	Apr	May	Jun	Jul	Aug	Sep	Oct	Nov	Dec	Annual
Washington Nat'l AP	D.C.	2.72	2.71	3.17	2.71	3.66	3.38	3.80	3.91	3.31	3.02	3.12	3.12	38.63
Apalachicola	FL	3.90	3.79	4.25	2.72	2.67	4.55	7.35	7.50	7.54	3.40	3.20	4.08	54.95
Daytona Beach	FL	2.75	3.11	2.90	2.23	3.45	5.99	5.40	6.16	6.34	4.13	2.84	2.59	47.89
Fort Myers	FL	1.84	2.23	3.07	1.06	3.87	9.52	8.26	9.66	7.82	2.94	1.57	1.53	53.37
Gainesville	FL	3.35	4.21	3.65	2.64	3.76	6.77	6.80	8.01	5.27	1.82	2.26	3.27	51.81
Jacksonville	FL	3.31	3.93	3.68	2.77	3.55	5.69	5.60	7.93	7.05	2.90	2.19	2.72	51.32
Key West	FL	2.01	1.80	1.71	1.75	3.46	5.09	3.61	5.03	5.85	4.42	2.84	2.02	39.59
Miami	FL	2.01	2.08	2.39	2.85	6.21	9.33	5.70	7.58	7.63	5.64	2.66	1.83	55.91
Orlando	FL	2.30	3.02	3.21	1.80	3.55	7.32	7.25	6.78	6.01	2.42	2.30	2.15	48.11
Pensacola	FL	4.68	5.40	5.63	3.77	4.20	6.40	7.42	7.39	5.32	4.21	3.54	4.29	62.25
Tallahassee	FL	4.77	5.56	6.21	3.74	4.75	6.93	8.82	7.53	5.58	2.92	3.87	5.03	65.71
Tampa	FL	1.99	3.08	3.01	1.15	3.10	5.48	6.58	7.61	5.98	2.02	1.77	2.15	43.92
Vero Beach	FL	2.14	2.91	3.09	1.90	4.36	6.46	6.09	6.10	7.15	5.52	3.27	2.17	51.16
West Palm Beach	FL	2.80	2.69	3.66	2.91	6.13	8.09	6.14	6.02	8.53	6.60	4.69	2.49	60.75
Athens	GA	4.60	4.42	5.46	3.99	4.37	3.93	4.88	3.70	3.36	3.28	3.66	4.09	49.74
Atlanta	GA	4.75	4.81	5.77	4.26	4.29	3.56	5.01	3.66	3.42	3.05	3.86	4.33	50.77
Augusta	GA	4.05	4.27	4.65	3.31	3.77	4.13	4.24	4.50	3.02	2.84	2.48	3.40	44.66
Columbus	GA	4.59	4.85	5.77	4.30	4.17	4.07	5.54	3.73	3.23	2.22	3.56	4.97	51.00
Macon	GA	4.56	4.74	4.79	3.46	3.57	3.58	4.30	3.63	2.78	2.18	2.73	4.31	44.63
Savannah	GA	3.59	3.22	3.78	3.03	4.09	5.66	6.38	7.46	4.47	2.39	2.19	2.96	49.22
Hilo	HI	9.88	10.29	13.92	15.26	9.91	6.20	9.71	9.34	8.53	9.60	14.51	12.04	129.19
Honolulu	HI	3.55	2.21	2.20	1.54	1.13	0.50	0.59	0.44	0.78	2.28	3.00	3.80	22.02
Kahului	HI	4.14	2.87	2.72	1.84	0.77	0.27	0.38	0.49	0.35	1.23	2.59	3.27	20.92
Lihue	HI	5.89	3.33	4.17	3.50	3.15	1.69	2.13	1.76	2.37	4.41	5.45	5.15	43.00
Boise	ID	1.45	1.07	1.29	1.24	1.08	0.81	0.35	0.43	0.80	0.75	1.48	1.36	12.11
Lewiston	ID	1.28	0.89	1.09	1.13	1.31	1.25	0.67	0.78	0.78	0.90	1.15	1.20	12.43
Pocatello	ID	1.04	0.92	1.26	1.20	1.35	1.02	0.65	0.67	0.85	0.91	1.16	1.11	12.14
Cairo	IL	0.00	0.00	0.00	0.00	0.00	0.00	0.00	0.00	0.00	0.00	0.00	0.00	0.00
Chicago	IL	1.53	1.36	2.69	3.64	3.32	3.78	3.66	4.22	3.82	2.41	2.92	2.47	35.82
Moline	IL	1.54	1.23	2.98	3.90	4.30	4.27	4.95	4.22	4.02	2.93	2.51	2.23	39.08
Peoria	IL	1.51	1.42	2.91	3.77	3.70	3.99	4.20	3.10	3.87	2.65	2.69	2.44	36.25
Rockford	IL	1.28	1.14	2.46	3.65	3.66	4.52	4.12	4.15	3.80	2.88	2.57	2.05	36.28
Springfield	IL	1.51	1.77	3.24	3.68	3.62	3.43	3.52	3.29	3.33	2.60	2.53	2.73	35.25
Evansville	IN	2.66	3.12	4.71	4.02	4.75	3.49	4.04	3.11	2.97	2.87	3.73	3.67	43.14
Fort Wayne	IN	1.87	1.91	2.90	3.38	3.44	3.59	3.45	3.37	2.67	2.49	2.79	2.89	34.75
Indianapolis	IN	2.32	2.46	3.79	3.70	4.00	3.49	4.47	3.64	2.87	2.63	3.23	3.34	39.94
South Bend	IN	2.23	1.90	3.10	3.82	3.22	4.11	3.82	3.67	3.62	3.08	3.27	3.30	39.14
Des Moines	IA	0.96	1.11	2.33	3.36	3.66	4.46	3.78	4.20	3.53	2.62	1.79	1.32	33.12
Dubuque	IA	1.26	1.32	2.89	3.72	4.26	4.13	4.02	4.69	4.67	2.73	2.71	1.96	38.36
Sioux City	IA	0.55	0.71	1.96	2.34	3.67	3.71	3.27	2.97	2.88	1.94	1.08	0.78	25.86
Waterloo	IA	0.80	1.08	2.30	3.30	4.08	4.47	4.83	3.64	3.51	2.57	1.82	1.30	33.70
Concordia	KS	0.58	0.75	2.20	2.31	4.29	4.49	3.65	3.54	3.01	1.99	1.13	0.84	28.78
Dodge City	KS	0.49	0.62	1.56	2.05	3.03	3.10	3.24	2.73	1.91	1.28	0.83	0.65	21.49
Goodland	KS	0.41	0.39	1.18	1.30	3.49	3.19	2.87	1.80	1.57	0.90	0.69	0.41	18.20
Topeka	KS	0.95	1.04	2.46	3.08	4.45	5.54	3.59	3.89	3.81	3.06	1.93	1.43	35.23
Wichita	KS	0.79	0.96	2.43	2.38	3.81	4.31	3.13	3.02	3.49	2.22	1.59	1.20	29.33
Greater Cincinnati AP	KY	2.59	2.69	4.24	3.75	4.28	3.84	4.24	3.35	2.88	2.86	3.46	3.15	41.33
Jackson	KY	3.76	3.82	4.77	3.95	4.63	4.25	5.14	3.91	3.66	3.20	4.20	4.38	49.67
Lexington	KY	2.86	3.21	4.40	3.88	4.47	3.66	5.00	3.93	3.20	2.57	3.39	3.98	44.55
Louisville	KY	2.86	3.30	4.66	4.23	4.62	3.46	4.51	3.54	3.16	2.71	3.70	3.64	44.39
Paducah	KY	3.27	3.90	4.92	5.01	4.94	4.05	4.19	3.34	3.69	3.00	4.32	4.68	49.31
Baton Rouge	LA	4.91	5.52	4.81	5.37	4.89	4.48	6.74	6.00	4.85	3.48	4.31	5.53	60.89
Lake Charles	LA	4.52	3.59	3.29	3.33	5.67	4.96	5.20	5.33	5.69	3.95	4.26	5.05	54.84
New Orleans	LA	5.05	6.01	4.90	4.50	4.56	5.84	6.12	6.17	5.51	3.05	4.42	5.75	61.88
Shreveport	LA	3.88	3.92	3.59	3.75	5.18	4.29	3.67	2.43	3.12	3.73	4.45	4.10	46.11
Caribou	ME	2.42	1.92	2.43	2.45	3.07	2.91	4.01	4.07	3.45	3.10	3.55	3.22	36.60
Portland	ME	3.53	3.33	3.67	4.08	3.62	3.44	3.09	2.87	3.09	3.90	5.17	4.55	44.34
Baltimore	MD	3.05	3.12	3.38	3.09	3.72	3.67	3.69	3.92	3.41	2.98	3.32	3.41	40.76
Blue Hill	MA	4.15	4.31	4.41	4.05	3.80	3.43	3.49	3.92	3.82	3.94	4.92	4.71	48.95
Boston	MA	3.59	3.62	3.69	3.60	3.25	3.09	2.84	3.24	3.06	3.30	4.22	4.01	41.51

TABLE 25.21   NORMAL MONTHLY PRECIPITATION, INCHES, BASED ON 30-YEAR AVERAGE (1961 TO 1990). (CONTINUED)

Location	State	Jan	Feb	Mar	Apr	May	Jun	Jul	Aug	Sep	Oct	Nov	Dec	Annual
Worcester	MA	3.68	3.46	3.95	3.91	4.33	3.88	3.85	3.82	4.01	4.32	4.49	4.05	47.75
Alpena	MI	1.64	1.29	2.11	2.25	2.74	3.04	2.92	3.40	3.11	2.10	2.20	2.03	28.83
Detroit	MI	1.76	1.74	2.55	2.95	2.92	3.61	3.18	3.43	2.89	2.10	2.67	2.82	32.62
Flint	MI	1.39	1.28	2.16	2.94	2.65	3.21	2.71	3.49	3.56	2.18	2.60	2.11	30.28
Grand Rapids	MI	1.83	1.42	2.63	3.37	3.13	3.68	3.19	3.57	4.24	2.81	3.32	2.85	36.04
Houghton Lake	MI	1.50	1.16	2.02	2.22	2.57	3.02	2.58	3.37	3.41	2.18	2.27	1.95	28.25
Lansing	MI	1.49	1.36	2.30	2.81	2.61	3.71	2.52	3.20	3.56	2.10	2.63	2.33	30.62
Marquette	MI	2.17	1.73	2.77	2.64	3.03	3.48	2.88	3.41	4.08	3.61	2.89	2.61	35.30
Muskegon	MI	2.34	1.49	2.51	2.90	2.60	2.35	2.10	3.41	3.88	2.80	3.15	3.03	32.56
Sault Ste. Marie	MI	2.42	1.74	2.30	2.35	2.71	3.14	2.71	3.61	3.69	3.23	3.45	2.88	34.23
Duluth	MN	1.22	0.80	1.91	2.25	3.03	3.82	3.61	3.99	3.84	2.49	1.80	1.24	30.00
International Falls	MN	0.88	0.63	1.06	1.58	2.47	3.93	3.59	3.11	3.15	1.97	1.15	0.84	24.36
Minneapolis-St. Paul	MN	0.95	0.88	1.94	2.42	3.39	4.05	3.53	3.62	2.72	2.19	1.55	1.08	28.32
Rochester	MN	0.78	0.74	1.78	2.73	3.40	3.72	4.20	3.88	3.47	2.32	1.61	1.03	29.66
Saint Cloud	MN	0.74	0.63	1.41	2.35	3.16	4.60	3.11	3.96	3.16	2.21	1.27	0.83	27.43
Jackson	MS	5.24	4.70	5.82	5.57	5.05	3.18	4.51	3.77	3.55	3.26	4.81	5.91	55.37
Meridian	MS	5.15	5.43	6.75	5.46	4.42	3.63	5.15	3.58	3.52	3.06	4.49	6.07	56.71
Tupelo	MS	4.89	4.72	6.07	5.25	5.72	3.84	4.30	3.05	3.60	3.42	4.85	6.16	55.87
Columbia	MO	1.45	1.84	3.17	3.83	5.01	4.32	3.67	3.28	3.86	3.22	2.93	2.47	39.05
Kansas City	MO	1.09	1.10	2.51	3.12	5.04	4.72	4.38	4.01	4.86	3.29	1.92	1.58	37.62
St. Louis	MO	1.81	2.12	3.58	3.50	3.97	3.72	3.85	2.85	3.12	2.68	3.28	3.03	37.51
Springfield	MO	1.79	2.17	3.89	4.18	4.38	5.09	2.92	3.51	4.62	3.58	3.75	3.16	43.04
Billings	MT	0.90	0.64	1.16	1.74	2.57	1.99	0.94	1.01	1.36	1.14	0.84	0.79	15.08
Glasgow	MT	0.37	0.27	0.41	0.69	1.77	2.11	1.72	1.35	1.00	0.61	0.28	0.38	10.96
Great Falls	MT	0.91	0.57	1.10	1.41	2.52	2.39	1.24	1.54	1.24	0.78	0.66	0.85	15.21
Helena	MT	0.63	0.41	0.73	0.97	1.78	1.87	1.10	1.29	1.15	0.60	0.48	0.59	11.60
Kalispell	MT	1.53	1.10	1.02	1.10	1.87	2.21	1.12	1.40	1.26	0.87	1.30	1.73	16.51
Missoula	MT	1.24	0.79	0.97	0.96	1.78	1.78	0.91	1.20	1.12	0.74	0.81	1.16	13.46
Grand Island	NE	0.46	0.72	1.89	2.50	3.82	3.91	2.83	2.82	2.85	1.35	1.04	0.71	24.90
Lincoln	NE	0.54	0.72	2.09	2.76	3.90	3.89	3.20	3.41	3.48	2.12	1.27	0.88	28.26
Norfolk	NE	0.52	0.77	1.86	2.29	3.68	4.46	3.22	2.55	2.45	1.60	1.01	0.74	25.15
North Platte	NE	0.36	0.43	1.20	1.99	3.43	3.37	3.06	1.74	1.61	0.98	0.66	0.47	19.30
Omaha Eppley AP	NE	0.74	0.77	2.04	2.66	4.52	3.87	3.51	3.24	3.72	2.28	1.49	1.02	29.86
Omaha (North)	NE	0.65	0.78	2.13	2.74	4.36	3.90	3.27	3.22	3.65	2.41	1.35	0.93	29.39
Scottsbluff	NE	0.50	0.47	1.09	1.58	2.77	2.64	2.06	1.07	1.10	0.81	0.62	0.56	15.27
Valentine	NE	0.29	0.43	1.04	1.67	3.16	2.89	3.06	2.28	1.53	0.91	0.60	0.37	18.23
Elko	NV	0.98	0.80	0.96	0.82	1.00	0.91	0.33	0.65	0.62	0.65	1.11	1.10	9.93
Ely	NV	0.70	0.65	0.96	1.00	1.15	0.88	0.69	0.83	1.01	0.89	0.67	0.70	10.13
Las Vegas	NV	0.48	0.48	0.42	0.21	0.28	0.12	0.35	0.49	0.28	0.21	0.43	0.38	4.13
Reno	NV	1.07	0.99	0.71	0.38	0.69	0.46	0.28	0.32	0.39	0.38	0.87	0.99	7.53
Winnemucca	NV	0.74	0.62	0.78	0.84	0.83	0.86	0.27	0.45	0.40	0.62	0.94	0.88	8.23
Concord	NH	2.51	2.53	2.72	2.91	3.14	3.15	3.23	3.32	2.81	3.23	3.66	3.16	36.37
Mt. Washington	NH	7.94	8.56	8.97	8.17	7.51	7.82	7.08	8.24	7.38	7.19	10.38	9.72	98.96
Atlantic City AP	NJ	3.46	3.06	3.62	3.56	3.33	2.64	3.83	4.14	2.93	2.82	3.58	3.32	40.29
Newark	NJ	3.39	3.04	3.87	3.84	4.13	3.22	4.50	3.91	3.66	3.05	3.91	3.45	43.97
Albuquerque	NM	0.44	0.46	0.54	0.52	0.50	0.59	1.37	1.64	1.00	0.89	0.43	0.50	8.88
Clayton	NM	0.24	0.31	0.55	0.94	1.99	2.27	2.70	2.61	1.77	0.90	0.52	0.29	15.09
Roswell	NM	0.35	0.46	0.33	0.46	1.04	1.61	1.71	2.58	2.02	1.05	0.52	0.45	12.58
Albany	NY	2.36	2.27	2.93	2.99	3.41	3.62	3.18	3.47	2.95	2.83	3.23	2.93	36.17
Binghamton	NY	2.40	2.33	2.82	3.13	3.36	3.60	3.50	3.36	3.32	2.89	3.28	3.00	36.99
Buffalo	NY	2.70	2.31	2.68	2.87	3.14	3.55	3.08	4.17	3.49	3.09	3.83	3.67	38.58
Islip	NY	3.69	3.48	4.10	4.23	3.94	3.82	3.46	4.04	3.48	3.55	4.23	4.05	46.07
New York (C. Park)	NY	3.42	3.27	4.08	4.20	4.42	3.67	4.35	4.01	3.89	3.56	4.47	3.91	47.25
New York (JFK AP)	NY	3.17	3.02	3.59	3.90	3.80	3.65	3.80	3.41	3.30	2.88	3.65	3.42	41.59
New York (Laguardia AP)	NY	3.04	2.86	3.60	3.79	3.82	3.57	4.07	3.75	3.39	3.02	3.81	3.40	42.12
Rochester	NY	2.08	2.10	2.28	2.61	2.72	3.00	2.71	3.40	2.97	2.44	2.92	2.73	31.96
Syracuse	NY	2.34	2.15	2.77	3.33	3.28	3.79	3.81	3.51	3.79	3.24	3.72	3.20	38.93
Asheville	NC	3.25	3.91	4.63	3.36	4.43	4.23	4.52	4.69	3.87	3.59	3.59	3.52	47.59
Cape Hatteras	NC	5.30	4.12	4.29	3.53	4.00	4.11	4.98	6.00	5.27	4.98	4.97	4.54	56.09

(continues)

TABLE 25.21   NORMAL MONTHLY PRECIPITATION, INCHES, BASED ON 30-YEAR AVERAGE (1961 TO 1990). (CONTINUED)

Location	State	Jan	Feb	Mar	Apr	May	Jun	Jul	Aug	Sep	Oct	Nov	Dec	Annual
Charlotte	NC	3.71	3.84	4.43	2.68	3.82	3.39	3.92	3.73	3.50	3.36	3.23	3.48	43.09
Grnsboro-Wnstn-Salem	NC	3.17	3.32	3.72	2.84	4.02	3.81	4.51	3.88	3.52	3.50	2.97	3.36	42.62
Raleigh	NC	3.48	3.69	3.77	2.59	3.92	3.68	4.01	4.02	3.19	2.86	2.98	3.24	41.43
Wilmington	NC	3.87	3.70	3.88	2.87	4.43	5.98	8.13	6.94	5.04	2.69	3.11	3.63	54.27
Bismarck	ND	0.45	0.43	0.77	1.67	2.18	2.72	2.14	1.72	1.49	0.90	0.49	0.51	15.47
Fargo	ND	0.67	0.45	1.06	1.82	2.45	2.82	2.70	2.43	1.99	1.68	0.73	0.65	19.45
Williston	ND	0.53	0.42	0.69	1.28	1.99	2.28	2.10	1.25	1.33	0.77	0.45	0.58	13.67
Akron	OH	2.16	2.23	3.33	3.16	3.73	3.18	4.08	3.32	3.32	2.35	3.01	2.95	36.82
Cleveland	OH	2.04	2.19	2.91	3.14	3.49	3.70	3.52	3.40	3.44	2.54	3.17	3.09	36.63
Columbus	OH	2.18	2.24	3.27	3.21	3.93	4.04	4.31	3.72	2.96	2.15	3.22	2.86	38.09
Dayton	OH	2.13	2.17	3.42	3.46	3.88	3.82	3.54	3.20	2.54	2.48	3.07	2.93	36.64
Mansfield	OH	1.98	2.02	3.30	3.64	4.35	3.95	4.04	4.08	3.38	2.34	3.51	3.07	39.66
Toledo	OH	1.75	1.73	2.66	2.96	2.91	3.75	3.27	3.25	2.85	2.10	2.81	2.93	32.97
Youngstown	OH	2.13	2.03	3.11	3.06	3.52	3.94	4.07	3.32	3.48	2.62	3.11	2.93	37.32
Oklahoma City	OK	1.13	1.56	2.71	2.77	5.22	4.31	2.61	2.60	3.84	3.23	1.98	1.40	33.36
Tulsa	OK	1.54	1.97	3.46	3.72	5.60	4.44	3.09	3.12	4.70	3.66	3.13	2.16	40.59
Astoria	OR	10.00	7.59	7.07	4.60	3.02	2.40	1.15	1.33	2.91	5.73	10.05	10.55	66.40
Burns	OR	0.99	0.76	1.01	0.65	0.98	0.83	0.40	0.66	0.56	0.72	1.25	1.15	9.96
Eugene	OR	7.91	5.64	5.52	3.11	2.16	1.43	0.51	1.08	1.67	3.41	8.32	8.61	49.37
Medford	OR	2.69	1.93	1.82	1.16	1.00	0.58	0.26	0.52	0.86	1.49	3.23	3.32	18.86
Pendleton	OR	1.51	1.14	1.16	1.04	0.99	0.64	0.35	0.53	0.59	0.86	1.58	1.63	12.02
Portland	OR	5.35	3.85	3.56	2.39	2.06	1.48	0.63	1.09	1.75	2.67	5.34	6.13	36.30
Salem	OR	5.92	4.50	4.17	2.42	1.88	1.34	0.56	0.76	1.55	2.98	6.28	6.80	39.16
Sexton Summit	OR	5.60	4.07	3.91	2.19	1.53	0.86	0.35	0.77	1.24	2.98	6.28	5.90	35.68
Allentown	PA	3.16	2.95	3.28	3.52	4.20	3.75	4.14	4.28	3.93	2.94	3.88	3.49	43.52
Erie	PA	2.22	2.28	3.00	3.24	3.44	4.09	3.43	4.06	4.39	3.77	4.02	3.59	41.53
Harrisburg	PA	2.84	2.93	3.28	3.24	4.26	3.85	3.59	3.31	3.51	2.93	3.52	3.24	40.50
Middletown/Harrisburg AP	PA	2.84	2.93	3.28	3.24	4.26	3.85	3.59	3.31	3.51	2.93	3.52	3.24	40.50
Philadelphia	PA	3.21	2.79	3.46	3.62	3.75	3.74	4.28	3.80	3.42	2.62	3.34	3.38	41.41
Pittsburgh	PA	2.54	2.39	3.41	3.15	3.59	3.71	3.75	3.21	2.97	2.36	2.85	2.92	36.85
Avoca	PA	2.10	2.15	2.55	2.97	3.65	3.98	3.79	3.32	3.31	2.79	3.06	2.51	36.18
Williamsport	PA	2.54	2.76	3.19	3.23	3.86	4.32	3.98	3.39	3.39	3.30	3.73	3.03	40.72
Providence	RI	3.88	3.61	4.05	4.11	3.76	3.33	3.18	3.63	3.48	3.69	4.43	4.38	45.53
Charleston AP	SC	3.45	3.30	4.34	2.67	4.01	6.43	6.84	7.22	4.73	2.90	2.49	3.15	51.53
Charleston C.O.	SC	3.36	3.06	4.30	2.44	3.53	5.83	6.05	7.31	4.67	2.78	2.29	2.90	48.52
Columbia	SC	4.42	4.12	4.82	3.28	3.68	4.80	5.50	6.09	3.67	3.04	2.90	3.59	49.91
Greenville-Spartanburg AP	SC	4.10	4.41	5.39	3.86	4.42	4.77	4.63	3.95	3.96	3.99	3.65	4.14	51.27
Aberdeen	SD	0.37	0.47	1.34	1.95	2.41	3.15	2.75	2.13	1.86	1.12	0.59	0.41	18.55
Huron	SD	0.41	0.68	1.66	2.09	2.87	3.35	2.67	1.97	1.72	1.47	0.72	0.47	20.08
Rapid City	SD	0.39	0.52	1.03	1.89	2.68	3.06	2.04	1.67	1.23	1.10	0.56	0.47	16.64
Sioux Falls	SD	0.51	0.64	1.64	2.52	3.03	3.40	2.68	2.85	3.02	1.78	1.09	0.70	23.86
Bristol-Jhnsn Cty-Kngsprt	TN	3.23	3.44	3.70	3.30	3.84	3.54	4.32	3.17	3.26	2.59	2.94	3.39	40.72
Chattanooga	TN	4.89	4.81	6.03	4.31	4.37	3.52	4.85	3.53	4.15	3.22	4.61	5.17	53.46
Knoxville	TN	4.17	4.06	5.09	3.72	4.13	3.97	4.67	3.13	3.07	2.84	3.75	4.54	47.14
Memphis	TN	3.73	4.35	5.41	5.46	4.98	3.57	3.79	3.43	3.53	3.01	5.10	5.74	52.10
Nashville	TN	3.58	3.81	4.85	4.37	4.88	3.57	3.97	3.46	3.46	2.62	4.12	4.61	47.30
Oak Ridge	TN	4.57	4.34	5.68	4.08	4.68	4.34	5.45	3.70	3.86	3.18	4.59	5.30	53.77
Abilene	TX	1.03	1.16	1.36	1.90	2.97	2.86	2.09	2.80	3.21	2.51	1.48	1.03	24.40
Amarillo	TX	0.50	0.61	0.96	0.99	2.48	3.70	2.62	3.22	1.99	1.37	0.69	0.43	19.56
Austin	TX	1.71	2.17	1.87	2.56	4.78	3.72	2.04	2.05	3.30	3.43	2.37	1.88	31.88
Brownsville	TX	1.56	1.06	0.53	1.56	2.94	2.73	1.90	2.77	6.00	2.80	1.51	1.25	26.61
Corpus Christi	TX	1.71	1.96	0.94	1.72	3.33	3.38	2.39	3.31	5.52	3.02	1.59	1.26	30.13
Dallas-Fort Worth	TX	1.83	2.18	2.77	3.50	4.88	2.98	2.31	2.21	3.39	3.52	2.29	1.84	33.70
Del Rio	TX	0.56	0.95	0.69	1.98	2.03	2.11	1.85	1.47	2.83	2.24	0.92	0.61	18.24
El Paso	TX	0.40	0.41	0.29	0.20	0.25	0.67	1.54	1.58	1.70	0.76	0.44	0.57	8.81
Galveston	TX	3.26	2.26	2.23	2.43	3.59	4.44	3.96	4.47	5.93	2.84	3.37	3.50	42.28
Houston	TX	3.29	2.96	2.92	3.21	5.24	4.96	3.60	3.49	4.89	4.27	3.79	3.45	46.07
Lubbock	TX	0.39	0.68	0.89	0.97	2.35	2.75	2.37	2.51	2.60	1.86	0.75	0.53	18.65
Midland-Odessa	TX	0.40	0.62	0.58	0.83	1.98	1.55	1.70	1.69	2.62	1.74	0.69	0.56	14.96
Port Arthur	TX	4.77	3.38	3.24	3.51	5.71	5.59	5.38	5.34	6.31	4.29	4.85	4.81	57.18

Location	State	Jan	Feb	Mar	Apr	May	Jun	Jul	Aug	Sep	Oct	Nov	Dec	Annual
San Angelo	TX	0.80	1.07	0.91	1.67	3.00	2.33	1.06	1.93	3.41	2.40	1.08	0.79	20.45
San Antonio	TX	1.71	1.81	1.52	2.50	4.22	3.81	2.16	2.54	3.41	3.17	2.62	1.51	30.98
Victoria	TX	2.16	2.00	1.55	2.41	4.50	4.89	3.34	3.01	5.60	3.46	2.45	2.04	37.41
Waco	TX	1.65	2.09	2.33	3.19	4.58	3.28	1.99	1.68	3.52	3.36	2.43	1.86	31.96
Wichita Falls	TX	1.04	1.46	2.21	3.01	4.07	3.52	1.72	2.48	3.82	2.74	1.54	1.29	28.90
Milford	UT	0.63	0.71	1.17	1.03	0.74	0.48	0.78	0.94	0.96	0.84	0.78	0.70	9.76
Salt Lake City	UT	1.11	1.23	1.91	2.12	1.80	0.93	0.81	0.86	1.28	1.44	1.29	1.40	16.18
Burlington	VT	1.82	1.63	2.23	2.76	3.12	3.47	3.65	4.06	3.30	2.88	3.13	2.42	34.47
Lynchburg	VA	2.86	3.04	3.47	3.09	3.91	3.45	4.16	3.59	3.24	3.70	3.14	3.23	40.88
Norfolk	VA	3.78	3.47	3.70	3.06	3.81	3.82	5.06	4.81	3.90	3.15	2.85	3.23	44.64
Richmond	VA	3.24	3.16	3.61	2.96	3.84	3.62	5.03	4.40	3.34	3.53	3.17	3.26	43.16
Roanoke	VA	2.62	3.04	3.48	3.25	3.98	3.19	3.91	4.15	3.50	3.85	3.19	2.97	41.13
Wallops Island	VA	3.38	3.34	3.67	2.67	3.60	3.34	3.94	3.73	3.28	2.83	2.81	3.34	39.93
Olympia	WA	8.01	5.77	4.95	3.29	2.09	1.63	0.82	1.29	2.26	4.31	8.05	8.12	50.59
Quillayute	WA	14.37	12.59	11.48	7.51	5.40	3.12	2.57	2.54	4.88	10.52	14.74	15.46	105.18
Seattle C.O.	WA	5.35	4.03	3.77	2.51	1.84	1.59	0.85	1.22	1.94	3.25	5.65	6.00	38.00
Seattle Sea-Tac AP	WA	5.38	3.99	3.54	2.33	1.70	1.50	0.76	1.14	1.88	3.23	5.83	5.91	37.19
Spokane	WA	1.98	1.49	1.49	1.18	1.41	1.26	0.67	0.72	0.73	0.99	2.15	2.42	16.49
Yakima	WA	1.21	0.74	0.67	0.50	0.45	0.53	0.16	0.40	0.40	0.47	1.03	1.41	7.97
Beckley	WV	2.92	2.94	3.40	3.43	3.98	3.84	4.70	3.38	3.33	2.89	2.99	3.23	41.03
Charleston	WV	2.91	3.04	3.63	3.31	3.94	3.59	4.99	4.01	3.24	2.89	3.59	3.39	42.53
Elkins	WV	3.08	3.00	3.83	3.82	4.12	4.46	4.53	4.35	3.76	3.08	3.33	3.48	44.84
Huntington	WV	2.83	2.90	3.68	3.43	4.24	3.51	4.65	3.83	2.93	2.83	3.30	3.36	41.49
Green Bay	WI	1.15	1.03	2.05	2.40	2.82	3.39	3.10	3.50	3.47	2.23	2.16	1.53	28.83
La Crosse	WI	0.93	0.90	1.98	2.88	3.26	3.90	3.79	3.92	3.79	2.20	1.73	1.27	30.55
Madison	WI	1.07	1.08	2.17	2.86	3.14	3.66	3.39	4.04	3.37	2.17	2.09	1.84	30.88
Milwaukee	WI	1.60	1.45	2.67	3.50	2.84	3.24	3.47	3.53	3.38	2.41	2.51	2.33	32.93
Casper	WY	0.55	0.60	0.95	1.56	2.13	1.46	1.26	0.67	0.94	0.97	0.77	0.66	12.52
Cheyenne	WY	0.40	0.39	1.03	1.37	2.39	2.08	2.09	1.69	1.27	0.74	0.53	0.42	14.40
Lander	WY	0.48	0.57	1.15	2.08	2.32	1.46	0.81	0.53	1.10	1.13	0.80	0.58	13.01
Sheridan	WY	0.73	0.64	0.97	1.72	2.39	2.25	0.88	0.82	1.37	1.18	0.83	0.70	14.48

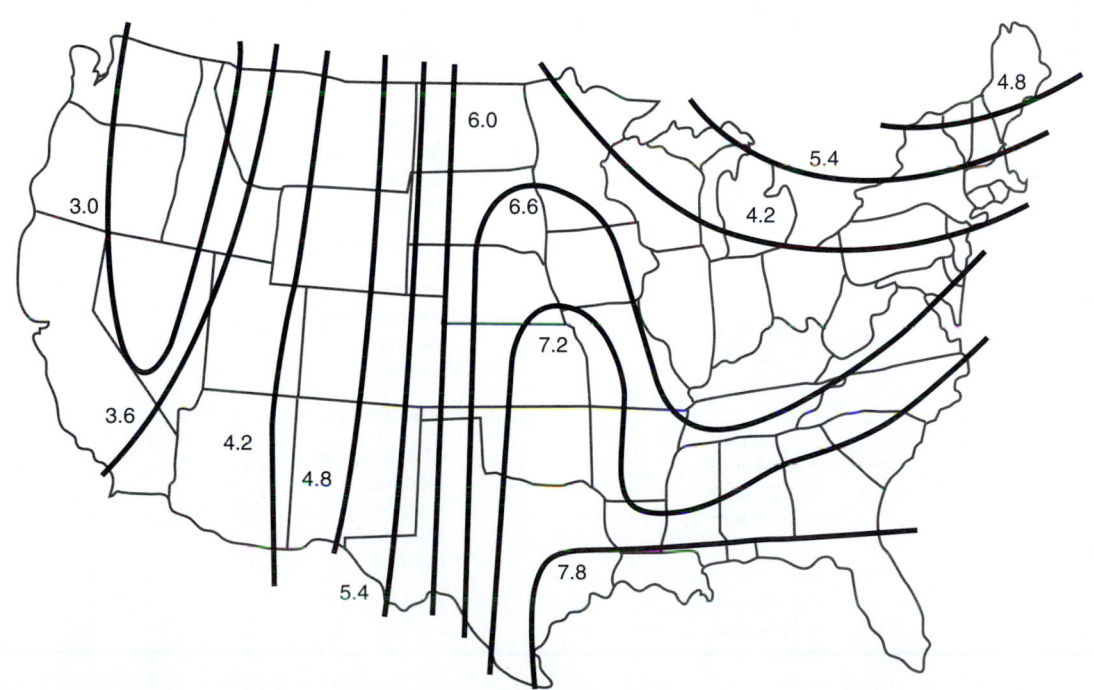

FIGURE 25.13A   Rainfall intensity for 15 min of precipitation (in).

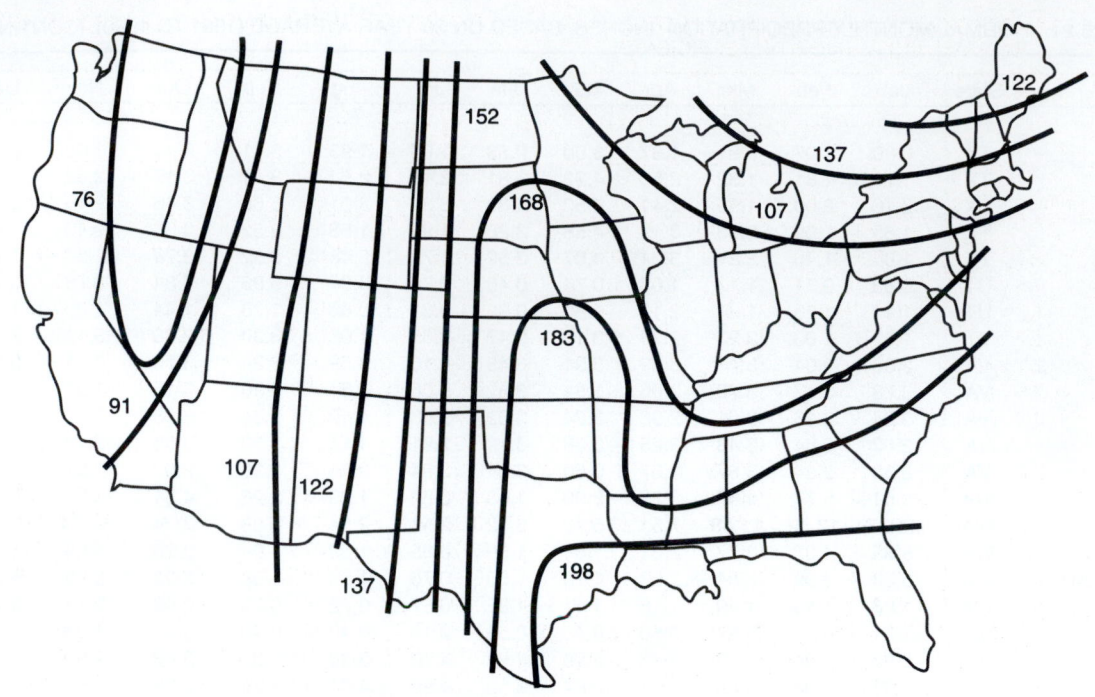

FIGURE 25.13B Rainfall intensity for 15 min of precipitation (mm).

TABLE 25.22   RAINFALL INTENSITY RATES, IN MM/HR, BASED ON 15 MINUTES OF PRECIPITATION, FOR SELECTED CITIES IN CANADA.

Alberta		**The Pas**	47 mm
		Winnipeg	68 mm
Calgary	36 mm		
Edmonton	48 mm	**Ontario**	
Grand Prairie	40 mm		
Jasper	24 mm	London	72 mm
		Ottawa	64 mm
Atlantic Provinces		Sault Ste. Marie	48 mm
		Sudbury	56 mm
Battle Harbour	24 mm	Toronto	72 mm
Bonavista	44 mm		
Cape Race	44 mm	**Saskatchewan**	
Chatham	50 mm		
Halifax	44 mm	Lloydminister	48 mm
Schefferville	30 mm	Prince Albert	46 mm
Yarmouth	48 mm	Regina	57 mm
		Saskaton	52 mm
British Columbia			
		Quebec	
Kamloops	10 mm		
Prince George	13 mm	Chibougaman	52 mm
Vancouver	20 mm	Montreal	64 mm
Victoria	16 mm	Rivierie-du-Loup	40 mm
		Roberval	60 mm
Manitoba		Quebec	54 mm
Brandon	62 mm		
Grand Rapids	56 mm		

TABLE 25.23 APPROXIMATE RAINWATER COLLECTED, IN GAL/DAY PER FT^2 OF CATCHMENT AREA, BASED ON ANNUAL RAINFALL (IN) AND SYSTEM EFFICIENCY.

| Rainfall | | Approximate Rainwater Collected, in gal/day per ft^2 of Catchment Area | | | | | | | |
| | | System Efficiency (η) | | | | | | | |
(in/year)	(in/month)	60	65	70	75	80	85	90	95
5	0.4	0.005	0.005	0.006	0.006	0.007	0.007	0.007	0.008
10	0.8	0.010	0.011	0.012	0.012	0.013	0.014	0.015	0.016
15	1.3	0.015	0.016	0.017	0.018	0.020	0.021	0.022	0.023
20	1.7	0.020	0.021	0.023	0.025	0.026	0.028	0.030	0.031
25	2.1	0.025	0.027	0.029	0.031	0.033	0.035	0.037	0.039
30	2.5	0.030	0.032	0.035	0.037	0.039	0.042	0.044	0.047
35	2.9	0.035	0.037	0.040	0.043	0.046	0.049	0.052	0.055
40	3.3	0.039	0.043	0.046	0.049	0.053	0.056	0.059	0.062
45	3.8	0.044	0.048	0.052	0.055	0.059	0.063	0.067	0.070
50	4.2	0.049	0.053	0.058	0.062	0.066	0.070	0.074	0.078
55	4.6	0.054	0.059	0.063	0.068	0.072	0.077	0.081	0.086
60	5.0	0.059	0.064	0.069	0.074	0.079	0.084	0.089	0.094
65	5.4	0.064	0.069	0.075	0.080	0.085	0.091	0.096	0.102
70	5.8	0.069	0.075	0.081	0.086	0.092	0.098	0.104	0.109
75	6.3	0.074	0.080	0.086	0.092	0.099	0.105	0.111	0.117
80	6.7	0.079	0.085	0.092	0.099	0.105	0.112	0.118	0.125
85	7.1	0.084	0.091	0.098	0.105	0.112	0.119	0.126	0.133
90	7.5	0.089	0.096	0.104	0.111	0.118	0.126	0.133	0.141
95	7.9	0.094	0.102	0.109	0.117	0.125	0.133	0.141	0.148
100	8.3	0.099	0.107	0.115	0.123	0.132	0.140	0.148	0.156
105	8.8	0.104	0.112	0.121	0.129	0.138	0.147	0.155	0.164
110	9.2	0.108	0.118	0.127	0.136	0.145	0.154	0.163	0.172
115	9.6	0.113	0.123	0.132	0.142	0.151	0.161	0.170	0.180
120	10.0	0.118	0.128	0.138	0.148	0.158	0.168	0.178	0.187
125	10.4	0.123	0.134	0.144	0.154	0.164	0.175	0.185	0.195
130	10.8	0.128	0.139	0.150	0.160	0.171	0.182	0.192	0.203
135	11.3	0.133	0.144	0.155	0.166	0.178	0.189	0.200	0.211
140	11.7	0.138	0.150	0.161	0.173	0.184	0.196	0.207	0.219
145	12.1	0.143	0.155	0.167	0.179	0.191	0.203	0.215	0.226
150	12.5	0.148	0.160	0.173	0.185	0.197	0.210	0.222	0.234

To determine daily rainwater collected, divide the rainwater collected (R) by the number of days (D) in period under consideration (e.g., 365 days for a year):

$$R = (0.60 \cdot A \cdot r \cdot \eta)/D$$

If rainwater is the sole source of water or for rainwater harvesting under drought conditions, it is wise to assume that the rainwater collected (R) will be about half of normal conditions:

$$R = (0.60 \cdot A \cdot r \cdot \eta)/2$$

Table 25.23 provides the approximate rainwater collected, in gal/day per ft^2 of catchment area, based on annual rainfall (in) and system efficiency.

Required Catchment Area

The minimum required catchment area (A_{min}) to meet a specific daily rate of consumption is based on average rainfall (r) in period under consideration, the system efficiency (η), expressed as a coefficient, and the number of days (D) in period under consideration (e.g., 365 days for a year):

$$A_{min} = (Q_{total} \cdot D)/(0.60 \cdot r \cdot \eta)$$

Cistern Size

The cistern must store enough water storage for reasonably expected dry periods. The required cistern storage capacity (ST) can be approximated by the daily consumption rate (Q_{total}), in gal/day or L/day, and the longest average dry spell ($D_{dry\ spell}$), in days:

$$ST = Q_{total} \cdot D_{dry\ spell}$$

This simple method assumes sufficient rainfall and catchment area, and is therefore only applicable in areas where this is the situation. If rainwater is the sole source of water or for rainwater harvesting under drought conditions, the cistern ST should be doubled:

$$ST = 2 \cdot Q_{total} \cdot D_{dry\ spell}$$

Table 25.24 provides volume capacities of cylindrical tanks.

TABLE 25.24A   CAPACITIES OF CYLINDRICAL TANKS, IN GALLONS. CAPACITIES PROVIDED ARE TANK SIZE.

Length or Height (ft)	Diameter (in)												
	36	42	48	54	60	66	72	78	84	90	96	102	108
1	53	72	94	120	145	180	210	250	290	330	375	425	475
2	106	144	188	240	200	360	420	500	580	660	750	850	950
3	159	216	282	360	435	540	630	750	870	990	1125	1275	1425
4	212	288	376	480	580	720	840	1000	1160	1320	1500	1700	1900
5	265	360	470	600	725	900	1050	1250	1450	1650	1875	2125	2375
6	318	432	564	720	870	1080	1260	1500	1740	1980	2250	2550	2850
7	371	504	658	840	1015	1260	1470	1750	2030	2310	2625	2975	3325
8	424	578	752	960	1160	1440	1680	2000	2320	2640	3000	3400	3800
9	477	648	846	1080	1305	1620	1890	2250	2610	2970	3375	3825	4275
10	530	720	940	1200	1450	1800	2100	2500	2900	3300	3750	4250	4750
11	583	792	1034	1320	1595	1980	2310	2750	3190	3630	4125	4675	5225
12	636	804	1128	1440	1740	2160	2520	3000	3480	3960	4500	5100	5700
13	689	936	1222	1560	1885	2340	2730	3250	3770	4280	4875	5525	6175
14	742	1008	1316	1680	2030	2520	2940	3500	4060	4620	5250	5950	6650
15	795	1080	1410	1800	2175	2700	3150	3750	4350	4950	5625	6375	7125
16	848	1152	1504	1920	2320	2880	3360	4000	4640	5280	6000	6800	7800

TABLE 25.24B   CAPACITIES OF CYLINDRICAL TANKS, IN LITERS. CAPACITIES PROVIDED ARE INTERIOR TANK SIZE.

Length or Height (m)	Diameter (m)										
	1.0	1.2	1.4	1.6	1.8	2.0	2.2	2.4	2.6	2.8	3.0
0.5	393	565	770	1005	1272	1571	1901	2262	2655	3079	3534
1.0	785	1131	1539	2011	2545	3142	3801	4524	5309	6158	7069
1.5	1178	1696	2309	3016	3817	4712	5702	6786	7964	9236	10 603
2.0	1571	2262	3079	4021	5089	6283	7603	9048	10 619	12 315	14 137
2.5	1963	2827	3848	5027	6362	7854	9503	11 310	13 273	15 394	17 671
3.0	2356	3393	4618	6032	7634	9425	11 404	13 572	15 928	18 473	21 206
3.5	2749	3958	5388	7037	8906	10 996	13 305	15 834	18 583	21 551	24 740
4.0	3142	4524	6158	8042	10 179	12 566	15 205	18 096	21 237	24 630	28 274
4.5	3534	5089	6927	9048	11 451	14 137	17 106	20 358	23 892	27 709	31 809
5.0	3927	5655	7697	10 053	12 723	15 708	19 007	22 619	26 546	30 788	35 343

Example 25.8

A home in Denver, Colorado, has a catchment area equivalent to the 2400 ft^2 roof footprint (including the roof overhangs). The home will have 4 occupants and a water consumption rate of 55 gal per person per day.

a. Approximate the average annual rainfall.

Denver, Colorado, receives an average rainfall of 15.40 in/year (from Table 25.21).

b. Approximate the rainwater collected over a year. Assume an efficiency of 85%.

$$R = (0.60 \cdot A \cdot r \cdot \eta) = 0.60 \text{ gal/ft}^2 \cdot 2400 \text{ ft}^2 \cdot 15.40 \text{ in/year} \cdot 0.85 = 18\,850 \text{ gal/year}$$

c. Approximate the daily rainwater collected.

$$18\,850 \text{ gal/year}/365 \text{ days} = 51.6 \text{ gal/day (less than demand for one occupant)}$$

d. Approximate the required cistern storage capacity. Assume the longest average dry spell will be 4 weeks.

$$ST = Q_{total} \cdot D_{dry\ spell} = 220 \text{ gal/day} \cdot 4 \text{ weeks} \cdot 7 \text{ day/week} = 6160 \text{ gal}$$

Clearly, the 2400 ft^2 catchment area for the home in Denver, Colorado (from Example 25.8), is undersized. It meets less than the demand for one occupant and about one-fourth of the total consumption requirement. With this system, rainwater can only be used for supplementing consumption.

Example 25.9

Approximate the minimum required catchment area for the home in Denver, Colorado, described in the previous example.

$$Q_{total} = 55 \text{ gal/person-day} \cdot 4 \text{ persons} = 220 \text{ gal/day}$$

$$A_{min} = (Q_{total} \cdot D)/(0.60 \cdot r \cdot \eta)$$

$$A_{min} = (220 \text{ gal/day} \cdot 365 \text{ days})/(0.60 \text{ gal/ft}^2 \cdot 15.40 \text{ in/year} \cdot 0.85) = 10\,224 \text{ ft}^2$$

Example 25.10

A home in Atlanta, Georgia, has a catchment area equivalent to the 2400 ft^2 roof footprint (including the roof overhangs). The home will have 4 occupants and a water consumption rate of 55 gal per person per day.

a. Approximate the average annual rainfall.

Atlanta, Georgia, receives an average rainfall of 50.77 in/year (from Table 25.21).

b. Approximate the rainwater collected over a year. Assume an efficiency of 85%.

$$R = (0.60 \cdot A \cdot r \cdot \eta) = 0.60 \text{ gal/ft}^2 \cdot 2400 \text{ ft}^2 \cdot 50.77 \text{ in/year} \cdot 0.85 = 62\,142 \text{ gal/year}$$

c. Approximate the daily rainwater collected.

62 142 gal/year/365 days = 170 gal/day (demand for four occupants)

d. Approximate the minimum required catchment area.

$$Q_{total} = 55 \text{ gal/person-day} \cdot 4 \text{ persons} = 220 \text{ gal/day}$$

$$A_{min} = (Q_{total} \cdot D)/(0.60 \cdot r \cdot \eta) = (220 \text{ gal/day} \cdot 365 \text{ days})/(0.60 \text{ gal/ft}^2 \cdot 50.77 \text{ in/year} \cdot 0.85) = 3101 \text{ ft}^2$$

e. Approximate the required cistern storage capacity. Assume the longest average dry spell will be 2 weeks.

$$Q_{total} = 55 \text{ gal/person-day} \cdot 4 \text{ persons} = 220 \text{ gal/day}$$

$$ST = Q_{total} \cdot D_{dry\ spell} = 220 \text{ gal/day} \cdot 2 \text{ weeks} \cdot 7 \text{ day/week} = 3080 \text{ gal}$$

Gutter/Downspout Size

Rainwater captured in the catchment area can be conveyed to the cistern through gutters and downspouts. Most gutters come in several sizes and shapes called *profiles*. These include a U-shaped trough (a half-round channel shape) and a K- or ogee-shaped configuration (a front that looks like the letter *K*). Common gutter profiles are available in the sizes shown in Table 25.25. The 5 in (125 mm) ogee-shape is most common on residences. Downspout choices are shown in Table 25.26. The 3 in × 4 in (75 mm × 100 mm) rectangular size is most common on residential installations.

Gutter and downspout size depends on the catchment area of the roof that is served by each portion of the gutter or downspout, the slope of the gutter when it is installed, and

TABLE 25.25   COMMON GUTTER SIZES AND DIMENSIONS. ACTUAL SIZE VARIES BY MANUFACTURER.

Nominal Size		Actual Size					
		Top Width		Depth		Bottom Width	
in	mm	in	mm	in	mm	in	mm
K- or Ogee-Shaped Profile							
4	100	4	100	3⅛	79	2¼	70
5	125	5	125	3⅝	92	3⅜	60
6	150	6	150	4⅞	124	3⅞	86
7	175	7	175	5⅞	149	4⁹⁄₁₆	116
8	200	8	200	7¼	184	5⅜	137
U- or Half-Round Shaped Profile							
4	100	4	100	2	50	—	—
5	125	5	125	2½	6½	—	—
6	150	6	150	3	75	—	—
7	175	7	175	3½	82½	—	—
8	200	8	200	4	100	—	—

TABLE 25.26   COMMON DOWNSPOUT/LEADER SIZES AND DIMENSIONS. ACTUAL SIZE VARIES BY MANUFACTURER.

Nominal Size		Actual Size					
		Depth		Width		Diameter	
in × in	mm × mm	in	mm	in	mm	in	Mm
Rectangular Profiles							
2 × 3	50 × 75	1¾	44	2¼	57	—	—
3 × 4	75 × 100	2¼	57	4¼	105	—	—
3 × 5	75 × 125	3¼	93	5	127	—	—
4 × 5	100 × 125	4	102	5	127	—	—
4 × 6	100 × 150	4	102	6	152	—	—
Plain and Corrugated Round Profiles							
3	75	—	—	—	—	3	75
4	100	—	—	—	—	4	100
5	125	—	—	—	—	5	125
6	150	—	—	—	—	6	150

rainfall intensity. Table 25.27 gives the recommended size of roof gutters based on maximum rainfall intensity and roof area served. Table 25.28 provides the recommended size of vertical downspout piping based on maximum rainfall intensity and roof area served. Table 25.29 provides the recommended size of horizontal rainwater piping based on maximum rainfall and roof area, in customary (U.S.) units.

Example 25.11

The home in Denver, Colorado, has roof footprint that measures 32 ft by 75 ft. Gutters to harvest rainwater will be installed on the 75 ft sides of the roof. Each gutter will have a ¹⁄₁₆-in/ft slope and two downspouts located at the gutter ends.

a. Determine the required gutter size.

The catchment area served by the gutter is equivalent to one-fourth of 2400 ft², or 600 ft². The rainfall intensity (from map above) is 4.8 in/hr in Denver, Colorado, so assume 5 in/hr.

From Table 25.27A for a rainfall intensity of 5 in/hr and a ¹⁄₁₆-in/ft gutter slope, a 6 in gutter is selected, which can handle a catchment area of up to 768 ft².

b. Determine the required downspout size.

The catchment area served by one downspout is equivalent to one-fourth of 2400 ft², or 600 ft². From Table 25.28A for a rainfall intensity of 5 in/hr, a 3 in downspout is selected, which can handle a catchment area of up to 1760 ft².

TABLE 25.27A SIZE OF ROOF GUTTERS BASED ON MAXIMUM RAINFALL INTENSITY AND CATCHMENT AREA, IN CUSTOMARY (U.S.) UNITS. QUANTITIES IN MAIN BODY OF TABLE ARE MAXIMUM HORIZONTAL CATCHMENT AREAS, IN FT^2.

	Gutter Size (in)	Maximum Rainfall (in/hr)				
		2	3	4	5	6
1/16 in/ft Slope	3	340	226	170	136	113
	4	720	480	360	288	240
	5	1250	834	625	500	416
	6	1920	1280	960	768	640
	7	2760	1840	1380	1100	918
	8	3980	2655	1990	1590	1325
	10	7200	4800	3600	2880	2400
	Gutter Size (in)	Maximum Rainfall (in/hr)				
		2	3	4	5	6
1/8 in/ft Slope	3	480	320	240	192	160
	4	1020	681	510	408	340
	5	1760	1172	880	704	587
	6	2720	1815	1360	1085	905
	7	3900	2600	1950	1560	1300
	8	5600	3740	2800	2240	1870
	10	10 200	6800	5100	4080	3400
	Gutter Size (in)	Maximum Rainfall (in/hr)				
		2	3	4	5	6
1/4 in/ft Slope	3	680	454	340	272	226
	4	1440	960	720	576	480
	5	2500	1668	1250	1000	834
	6	3840	2560	1920	1536	1280
	7	5520	3680	2760	2205	1840
	8	7960	5310	3980	3180	2655
	10	14 400	9600	7200	5750	4800
	Gutter Size (in)	Maximum Rainfall (in/hr)				
		2	3	4	5	6
1/2 in/ft Slope	3	960	640	480	384	320
	4	2040	1360	1020	816	680
	5	3540	2360	1770	1415	1180
	6	5540	3695	2770	2220	1850
	7	7800	5200	3900	3120	2600
	8	11 200	7460	5600	4480	3730
	10	20 000	13 330	10 000	8000	6660

TABLE 25.27B SIZE OF ROOF GUTTERS BASED ON MAXIMUM RAINFALL INTENSITY AND CATCHMENT AREA, IN METRIC (SI) UNITS. QUANTITIES IN MAIN BODY OF TABLE ARE MAXIMUM HORIZONTAL CATCHMENT AREAS, IN M^2.

	Gutter Size (mm)	Maximum Rainfall (mm/hr)				
		50	75	100	125	150
5.2 mm/m Slope	75	31	21	15	12	10
	100	66	44	33	26	22
	125	116	77	58	46	38
	150	178	119	89	71	59
	175	256	170	128	102	85
	200	369	246	184	147	123
	250	668	445	334	267	223
	Gutter Size (mm)	Maximum Rainfall (mm/hr)				
		50	75	100	125	150
10.4 mm/m Slope	75	44	29	22	17	14
	100	94	63	47	37	31
	125	163	108	81	65	54
	150	252	168	126	100	84
	175	362	241	181	144	120
	200	520	347	260	208	173
	250	947	631	473	379	315
	Gutter Size (mm)	Maximum Rainfall (mm/hr)				
		50	75	100	125	150
20.9 mm/m Slope	75	63	42	31	25	21
	100	133	89	66	53	44
	125	232	155	116	92	77
	150	356	237	178	142	118
	175	512	341	256	204	170
	200	739	493	369	295	246
	250	133	891	668	534	445
	Gutter Size (mm)	Maximum Rainfall (mm/hr)				
		50	75	100	125	150
41.7 mm/m Slope	75	89	59	44	35	29
	100	189	126	94	75	63
	125	328	219	164	131	109
	150	514	343	257	206	171
	175	724	483	362	289	241
	200	1040	693	520	416	346
	250	1858	1238	929	743	618

TABLE 25.28A REQUIRED SIZE OF VERTICAL DOWNSPOUT, IN CUSTOMARY (U.S.) UNITS. QUANTITIES IN MAIN BODY OF TABLE ARE MAXIMUM HORIZONTAL CATCHMENT AREAS, IN FT^2.

Rainfall (in/hr)	Size of Drain or Downspout (in)*					
	2	3	4	5	6	8
1	2880	8800	18 400	34 600	54 000	116 000
2	1440	4400	9200	17 300	27 000	58 000
3	960	2930	6130	11 530	17 995	38 660
4	720	2200	4 600	8 650	13 500	29 000
5	575	1760	3680	6920	10 800	23 200
6	480	1470	3070	5765	9000	19 315
7	410	1260	2630	4945	7715	16 570
8	360	1100	2300	4325	6750	14 500
9	320	980	2045	3845	6000	12 890
10	290	880	1840	3460	5400	11 600
11	260	800	1675	3145	4910	10 545
12	240	730	1530	2880	4500	9660

*Round, square, or rectangular rainwater pipe may be used and is considered equivalent when enclosing a scribed circle equivalent to the downspout diameter.

TABLE 25.28B REQUIRED SIZE OF VERTICAL DOWNSPOUT, IN METRIC (SI) UNITS. QUANTITIES IN MAIN BODY OF TABLE ARE MAXIMUM HORIZONTAL CATCHMENT AREAS, IN M^2.

Rainfall (mm/hr)	Size of Drain or Downspout (mm)					
	50	75	100	125	150	200
25	267	817	1709	3214	5016	10 776
50	133	408	854	1607	2508	5388
75	89	272	569	1071	1671	3591
100	66	204	427	803	1254	2694
125	53	163	341	642	1003	2155
150	44	136	285	535	836	1794
175	38	117	244	459	716	1539
200	33	102	213	401	627	1347
225	29	91	190	357	557	1197
250	26	81	170	321	501	1077
275	24	74	155	292	456	979
300	22	67	142	267	418	897

*Round, square, or rectangular rainwater pipe may be used and is considered equivalent when enclosing a scribed circle equivalent to the leader/downspout diameter.

TABLE 25.29A SIZE OF HORIZONTAL RAINWATER PIPING BASED ON MAXIMUM RAINFALL AND ROOF AREA, IN CUSTOMARY (U.S.) UNITS. QUANTITIES IN MAIN BODY OF TABLE ARE MAXIMUM HORIZONTAL CATCHMENT AREAS, IN FT².

	Pipe Size (in)	Maximum Rainfall (in/hr)				
		2	3	4	5	6
⅛ in/ft Slope	3	1644	1096	822	657	548
	4	3760	2506	1800	1504	1253
	5	6680	4453	3340	2672	2227
	6	10 700	7133	5350	4280	3566
	8	23 000	15 330	11 500	9200	7600
	10	41 400	27 600	20 700	16 580	13 800
	12	66 600	44 400	33 300	26 650	22 200
	15	109 000	72 800	59 500	47 600	39 650

	Pipe Size (in)	Maximum Rainfall (in/hr)				
		2	3	4	5	6
¼ in/ft Slope	3	2320	1546	1160	928	773
	4	5300	3533	2650	2120	1766
	5	9440	6293	4720	3776	3146
	6	15 100	10 066	7550	6040	5033
	8	32 600	21 733	16 300	13 040	10 866
	10	58 400	38 950	29 200	23 350	19 450
	12	94 000	62 600	47 000	37 600	31 350
	15	168 000	112 000	84 000	67 250	56 000

	Pipe Size (in)	Maximum Rainfall (in/hr)				
		2	3	4	5	6
½ in/ft Slope	3	3288	2295	1644	1310	1096
	4	7520	5010	3760	3010	2500
	5	13 360	8900	6680	5320	4450
	6	21 400	13 700	10 700	8580	7140
	8	46 000	30 650	23 000	18 400	15 320
	10	85 800	55 200	41 400	33 150	27 600
	12	133 200	88 800	66 600	53 200	44 400
	15	238 000	158 800	119 000	95 300	79 250

TABLE 25.29B SIZE OF HORIZONTAL RAINWATER PIPING BASED ON MAXIMUM RAINFALL AND CATCHMENT AREA, IN METRIC (SI) UNITS. QUANTITIES IN MAIN BODY OF TABLE ARE MAXIMUM HORIZONTAL CATCHMENT AREAS, IN M².

	Pipe Size (mm)	Maximum Rainfall (mm/hr)				
		50	75	100	125	150
10.4 mm/m Slope	75	152	101	76	61	50
	100	349	232	174	139	116
	125	620	413	310	248	206
	150	994	662	497	397	331
	200	2136	1424	1068	854	706
	250	3846	2564	1923	1540	1282
	300	6187	4124	3093	2475	2062
	375	10 126	6763	5527	4422	3683

	Pipe Size (mm)	Maximum Rainfall (mm/hr)				
		50	75	100	125	150
20.9 mm/m Slope	75	215	143	107	86	71
	100	492	328	246	197	164
	125	877	584	438	350	292
	150	1402	935	701	561	467
	200	3028	2019	1514	1211	1009
	250	5425	3618	2712	2169	1806
	300	8732	5815	4366	3493	2912
	375	15 607	10 404	7803	6247	5202

	Pipe Size (mm)	Maximum Rainfall (mm/hr)				
		50	75	100	125	150
41.7 mm/m Slope	75	305	213	152	121	101
	100	698	465	349	279	232
	125	1241	826	620	494	413
	150	1988	1272	994	797	663
	200	4274	2847	2136	1709	1423
	250	7692	5128	3846	3079	2564
	300	12 374	8249	6187	4942	4124
	375	22 110	14 752	11 055	8853	7362

STUDY QUESTIONS

25-1. Explain the concept of sustainability.

25-2. Describe a sustainable building.

25-3. What is the LEED program?

25-4. What is a CHP system and why is it efficient?

25-5. Identify and describe the three types of cogeneration systems.

25-6. Explain geothermal energy.

25-7. Describe direct use of geothermal energy.

25-8. Describe a geothermal heat pump.

25-9. Explain the direct use of geothermal energy and a geothermal heat pump.

25-10. Explain the fundamental components of a heat pump system:

    a. Ground loop

    b. Heat pump

    c. Air delivery system

25-11. Describe the types of heat pump loops.

25-12. Explain the characteristics of a geothermal site that require consideration.

25-13. What is biomass?

25-14. Describe the types of biomass energy that can be used in buildings.

25-15. Describe the types of biomass technologies that can be used in buildings.

25-16. Describe photovoltaic (PV) power and where it can be used in buildings.

25-17. Describe the following PV systems and how they differ:

    a. Stand alone off grid

    b. Grid connected

25-18. Describe the components of in a PV system.

    a. PV cells

    b. Inverters

    c. Batteries

    d. Charge controllers

25-19. What is a wind energy conversion system?

25-20. Explain the difference between a wind turbine and a windmill.

25-21. How does elevation of a wind turbine affect turbine output?

25-22. Describe the following wind systems:

    a. Stand alone off grid

    b. Grid connected

25-23. What is Betz's limit?

25-24. What is a hydropower system?

25-25. Explain the components of a run-of-the-river hydropower system.

25-26. Describe a rainwater collection/harvesting systems.

25-27. With respect to a rainwater harvesting system, explain first-flush washing.

Design Exercises

Information for Design Exercises 25.28 and 25.29.

Selecting the best-suited HVAC system depends on factors such as: cost and availability of the energy source; appliance or system efficiency; cost to purchase, install, and maintain the appliance or system; and environmental impacts associated with the fuel. Energy prices vary because of type of fuel, market pressures, supply availability, production capacity, weather, season, and politics. They tend to vary somewhat by geographic location. A common way to compare heating energy costs at a geographical location is by determining the delivered heating cost, expressed in consistent units (i.e., cost per million Btu, $/MMBtu), based on the cost of the fuel ($/unit), heating value of the fuel (HV), and seasonal efficiency of the system or appliance (η_{hs}), expressed in decimal form (i.e., 80% = 0.80).

Delivered Heating Cost ($/MMBtu)
$$= (\$/\text{unit} \cdot 1\,000\,000 \text{ Btu/MMBtu})/(\eta_{hs} \cdot \text{HV})$$

25-28. Compute the delivered heating cost ($/MMBtu) for a wood-fired heating unit based on a heating value of 21 000 000 Btu/cord (pine) and a cost of $225/cord:

    a. For a fireplace with glass doors with a seasonal efficiency of 15%.

    b. For a modern, furnace, or boiler with a seasonal efficiency of 80%.

25-29. Compute the delivered heating cost ($/MMBtu) for the following fuels burned in a modern biomass heating unit (i.e., furnace or boiler) with a seasonal efficiency of 75%:

    a. Wood pellets based on a heating value of 8000 Btu/lb and a cost of $0.10/lb.

    b. Shelled corn (15% moisture content based on a heating value of 7000 Btu/lb) and a cost of $0.04/lb. (Shelled corn has had kernels removed from cob. Cost is based on $2.0/bushel. A typical bushel of corn weighs 56 lb/25.4 kg and contains 72 800 kernels.)

    c. BioHeat (5 to 20% biodiesel with heating oil) based on a heating value of 126 000 Btu/gal and a cost of $2.90/gal.

25-30. A photovoltaic (PV) system is being considered for a building with a load of 8 kW. The PV system will be approximately 16% efficient. Approximate the PV array size (in ft^2) required to meet this load at the geographical location where you reside.

25-31. A PV system is being considered for a building with a load of 10 kW. The PV system will be approximately 14% efficient. Approximate the PV array size (in ft^2) required to meet this load at the following locations:

   a. Seattle, Washington

   b. Phoenix, Arizona

   c. Los Angeles, California

   d. Miami, Florida

   e. Saint Louis, Missouri

   f. Pittsburgh, Pennsylvania

   g. Lincoln, Nebraska

   h. New York City, New York

   i. Boston, Massachusetts

25-32. A PV system is being considered for a building with a load of 8 kW. The PV system will be approximately 16% efficient. Approximate the PV array size (in ft^2) required to meet this load at the following locations:

   a. Seattle, Washington

   b. Phoenix, Arizona

   c. Los Angeles, California

   d. Miami, Florida

   e. Saint Louis, Missouri

   f. Pittsburgh, Pennsylvania

   g. Lincoln, Nebraska

   h. New York City, New York

   i. Boston, Massachusetts

25-33. A PV system is being considered for a building with a load of 6 kW. The PV system will be approximately 18% efficient. Approximate the PV array size (in ft^2) required to meet this load at the following locations:

   a. Seattle, Washington

   b. Phoenix, Arizona

   c. Los Angeles, California

   d. Miami, Florida

   e. Saint Louis, Missouri

   f. Pittsburgh, Pennsylvania

   g. Lincoln, Nebraska

   h. New York City, New York

   i. Boston, Massachusetts

25-34. The known mean wind velocity is 13.2 mph at a site at a height of 33 ft (about 10 m). Approximate the wind velocity at a height of 100 ft (about 30 m) for the following sites:

   a. Site located in suburban area (clustered homes and small buildings)

   b. Site located in urban area (clustered large buildings)

   c. Site on island surrounded by open water

25-35. The known mean wind velocity is 15.3 mph at a site at a height of 33 ft (about 10 m). Approximate the wind velocity at a height of 100 ft (about 30 m) for the following sites:

   a. Site located in forest with many trees

   b. Site located in an open field with short grass

   c. Site on island surrounded by open water

25-36. Approximate the theoretical power in wind moving at a speed of 7 m/s with an air temperature of 4°C striking a wind turbine with the following rotor sizes at a site at sea level:

   a. 2 m

   b. 4 m

   c. 6 m

   d. 8 m

   e. 10 m

25-37. Approximate the theoretical power in wind moving at a speed of 14 mph with an air temperature of 40°F striking wind turbine with the following rotor sizes at a site at sea level:

   a. 10 ft

   b. 15 ft

   c. 20 ft

   d. 25 ft

   e. 30 ft

25-38. Approximate the theoretical power in the wind at a site at sea level with an air temperature of 50°F striking a wind turbine with a 15 ft rotor size at the following wind velocities:

   a. 10 mph

   b. 15 mph

   c. 20 mph

   d. 25 mph

   e. 30 mph

25-39. A horizontal axis wind turbine has a rotor diameter of 24 ft. The turbine coefficient of performance is 0.37, generator efficiency is 0.78, and gearbox transmission efficiency is 0.94. Air is at a temperature of 60°F. Determine the wind turbine power at sea level at wind at speeds of 10 mph, 20 mph, and 30 mph.

25-40. A horizontal axis wind turbine has a rotor diameter of 24 ft. The turbine coefficient of performance is 0.37, generator efficiency is 0.80, and gearbox transmission efficiency is 0.94. Air is at a temperature of 60°F. Determine the wind turbine power at an elevation of 5000 ft above sea level at wind at speeds of 10 mph, 20 mph, and 30 mph.

25-41. Approximate the power output of a small hydropower system based upon a dynamic head of 80 ft and the following flow rates:

    a. 50 gallons per minute

    b. 100 gallons per minute

    c. 200 gallons per minute

25-42. A home has a catchment area equivalent to the 2200 ft^2 roof footprint (including the roof overhangs). The home will have 4 occupants and a water consumption rate of 60 gal per person per day. Average annual rainfall is 60 in/year.

    a. Approximate the rainwater collected over a year. Assume an efficiency of 85%.

    b. Approximate the daily rainwater collected.

    c. Approximate the minimum required catchment area.

    d. Approximate the required cistern storage capacity. Assume the longest average dry spell will be 2 weeks.

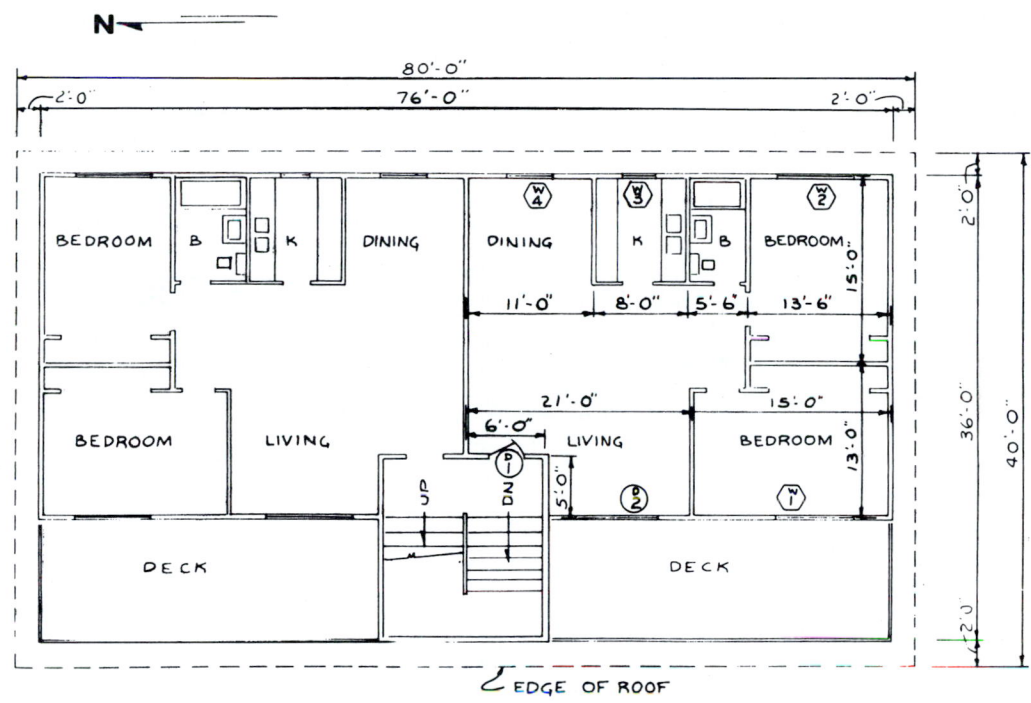

TYPICAL FLOOR PLAN

Design data

Apartment building, 4 story

Floor to floor height: 10'-0"

3-in service main

Street main pressure: 50 psi

2 exterior hose bibs

Street main to riser distance: 60 ft

Building drain slope, ½ in per foot.

Flat roof

Occupancy: 2.5 persons per apartment

Heat loss values (U)

    walls 0.077 Btu/hr • ft^2 • °F

    roof 0.025 Btu/hr • ft^2 • °F

    windows 0.29 Btu/hr • ft^2 • °F

    doors 0.27 Btu/hr • ft^2 • °F

Infiltration: Average (ACH=0.5)

D1-1¾ in steel door, insulated core, steel fasteners, no thermal break

D2- Vinyl with thermal break, metal, double glazing, ¼-in air space.

Windows–same construction as D2.

SCHEDULE		
NO.	SIZE	TYPE
D1	3'-0" × 6'-8"	—
D2	6'-0" × 6'-8"	ALUM.-SLIDING
W1	5'-0" × 4'-0"	—
W2	5'-0" × 3'-0"	—
W3	3'-0" × 4'-0"	—
W4	4'-0" × 4'-0"	—

Electric

Air Conditioning, 2000 watts

Electric Range, 10 500 watts

Lighting 3 watts per sq ft

Water heater, 3500 watts

Dishwasher, 1000 watts

No clothes dryer

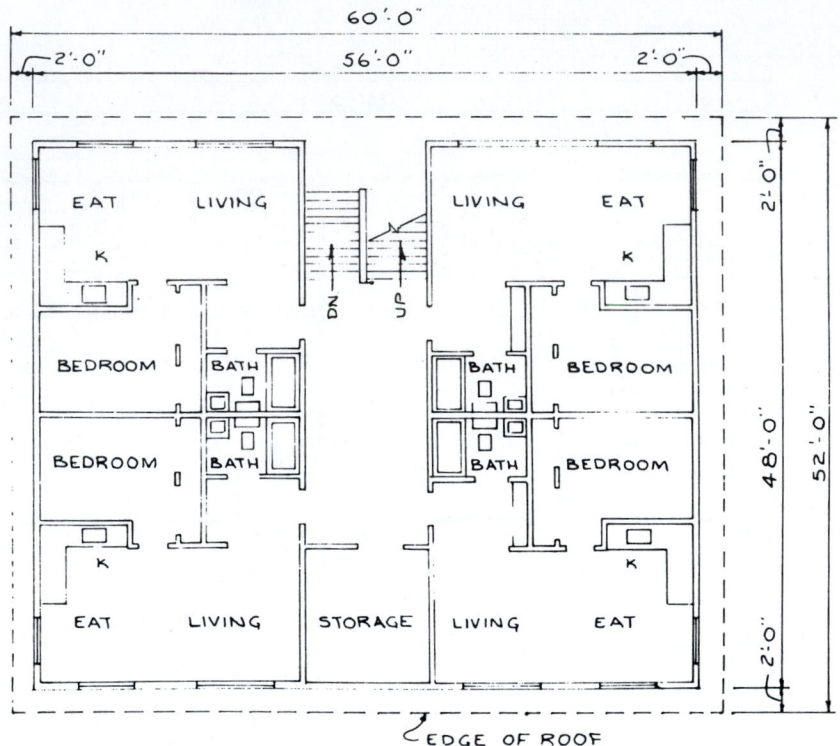

TYPICAL FLOOR PLAN

Design data
Apartment building, 3 story
Floor to floor height: 9'-0"
3-in service main
Street main pressure: 55 psi
3 exterior hose bibs
Street main to riser distance: 85 ft
Flat roof
Occupancy: 1.5 persons per apartment
Food disposal units in each apartment

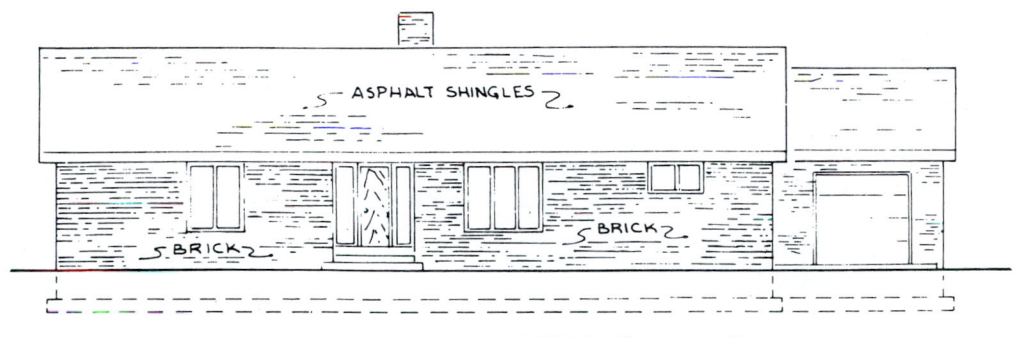

NORTHEAST ELEVATION

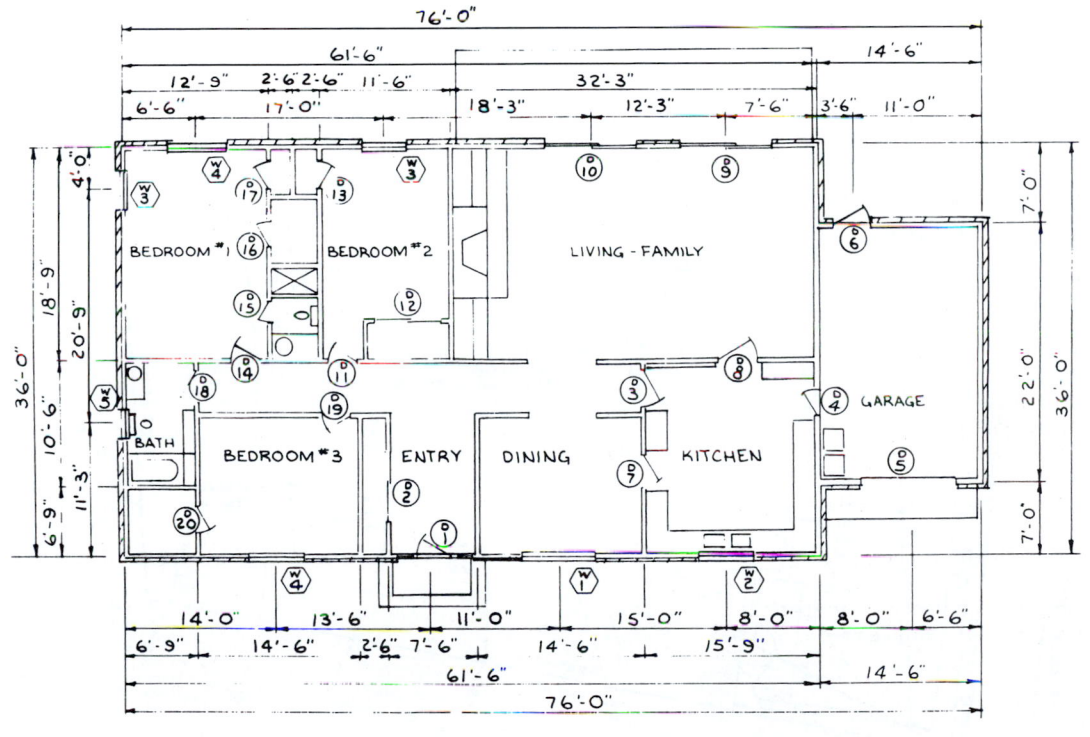

FLOOR PLAN

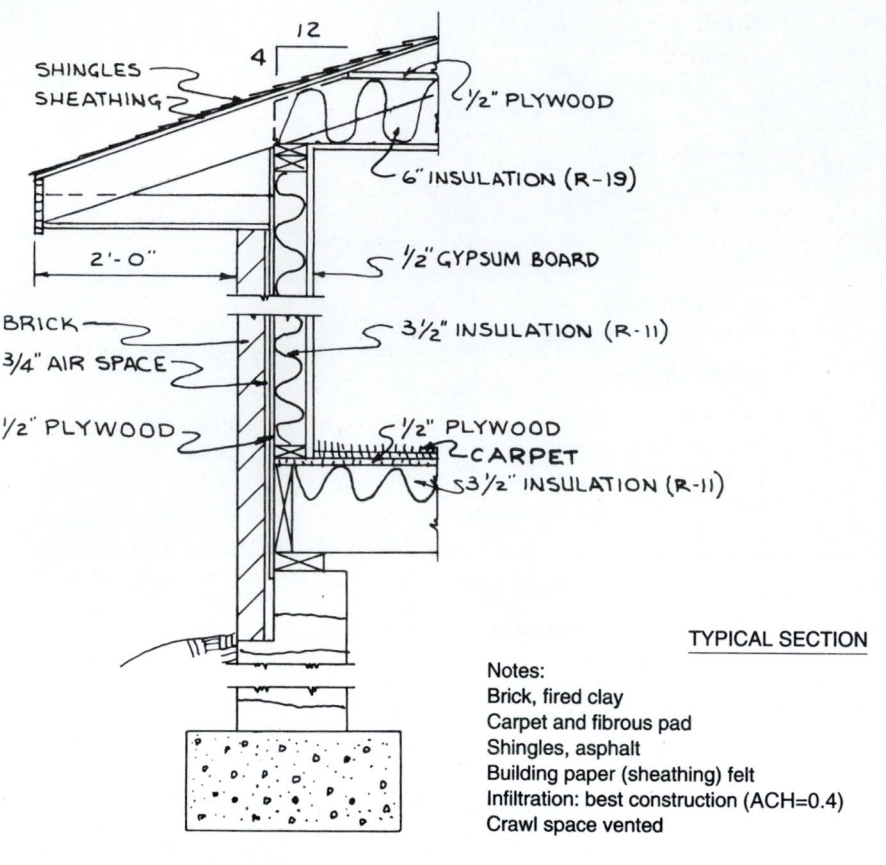

SHINGLES
SHEATHING
12
4
½" PLYWOOD
6" INSULATION (R-19)
½" GYPSUM BOARD
2'-0"
BRICK
3½" INSULATION (R-11)
3/4" AIR SPACE
½" PLYWOOD
½" PLYWOOD
CARPET
3½" INSULATION (R-11)

TYPICAL SECTION

Notes:
Brick, fired clay
Carpet and fibrous pad
Shingles, asphalt
Building paper (sheathing) felt
Infiltration: best construction (ACH=0.4)
Crawl space vented

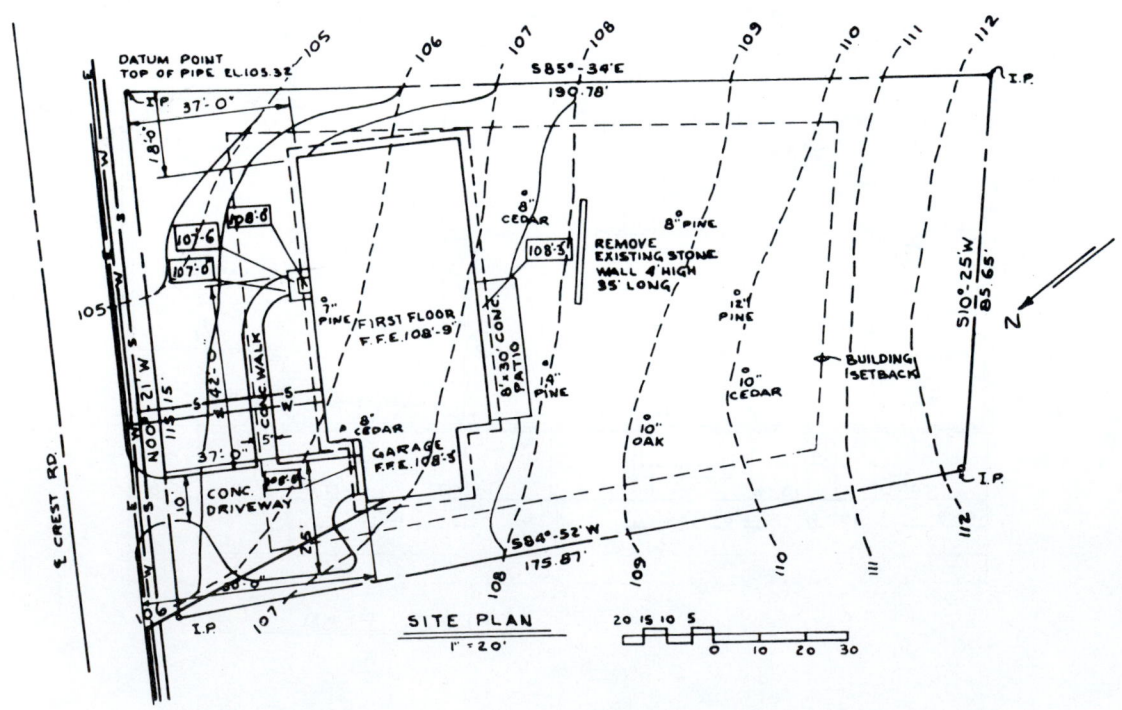

SITE PLAN
1" = 20'

WINDOW SCHEDULE

NO.	SIZE	TYPE
1	6'-0" × 5'-0"	CASEMENT
2	4'-0" × 3'-0"	"
3	3'-0" × 4'-0"	"
4	4'-0" × 4'-0"	"
5	2'-0" × 2'-0"	"

INSULATING GLASS

DOOR SCHEDULE

NO.	SIZE	REMARKS
1, 4, 6	3'-0" × 6'-8"	
9, 10	6'-0" × 6'-8"	
5	7'-0" × 8'-0"	
2, 12	2-2'-6" × 6'-8"	
3, 7, 8, 11 12, 14, 19	2'-8" × 6'-8"	
13, 15, 16 17, 18, 20	2'-4" × 6'-8"	

Design data
Occupancy: 2 adults, 4 children
Design temperature:
    heating: 10°F, 70°F
    cooling: 95°F, medium
Draperies, except kitchen
Roof, dark

Windows—double glazing, ½ in air space, operable, metal edge of glass, aluminum with thermal break.

DOORS—

D1—1¾ in steel door, mineral wool core, steel stiffeners, no thermal break

D4—1⅜ in hollow core

D9 & D10—double glazing, ½ in air space, metal edge of glass, double door, aluminum with thermal break

Entry glass—double glazing, ¼ in air space, metal edge of glass

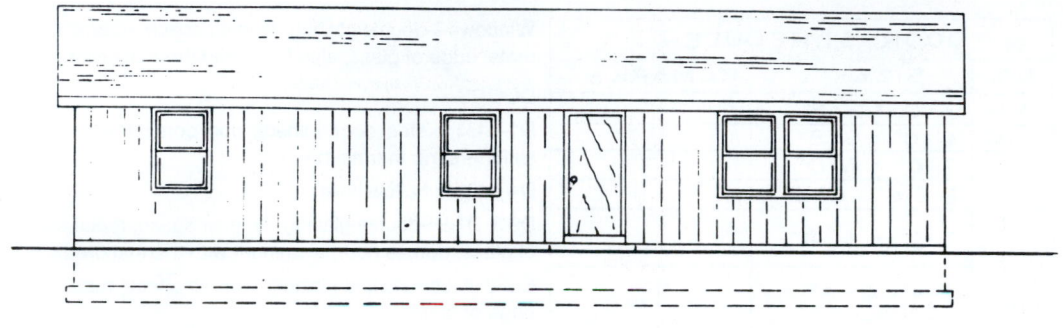

FRONT ELEVATION

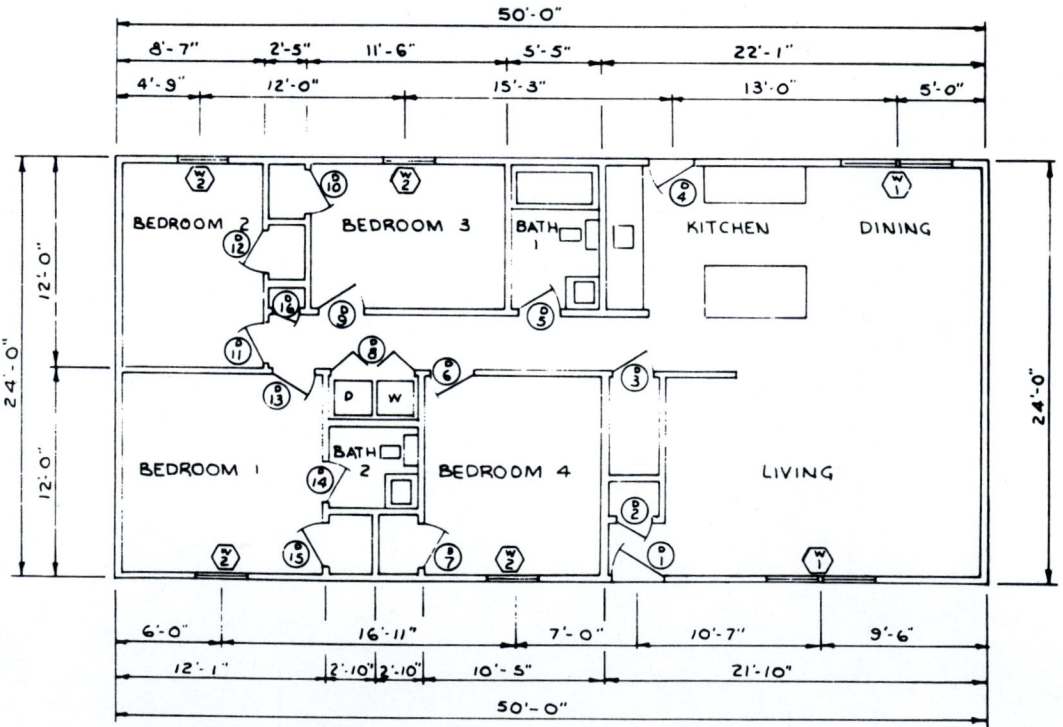

FLOOR PLAN

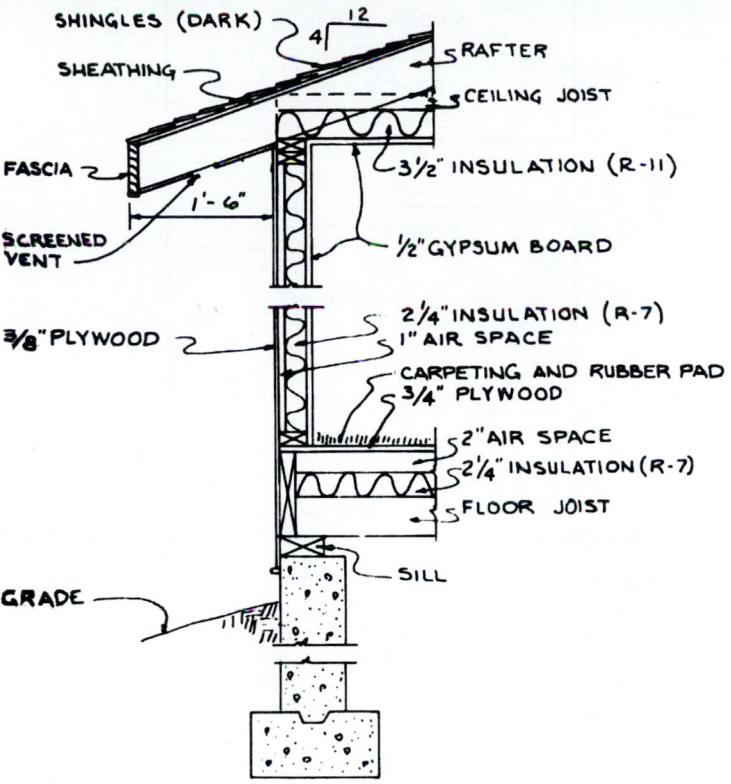

SHINGLES (DARK)

SHEATHING

RAFTER

CEILING JOIST

FASCIA

3½" INSULATION (R-11)

SCREENED VENT

1'-6"

½" GYPSUM BOARD

3/8" PLYWOOD

2¼" INSULATION (R-7)

1" AIR SPACE

CARPETING AND RUBBER PAD

3/4" PLYWOOD

2" AIR SPACE

2¼" INSULATION (R-7)

FLOOR JOIST

SILL

GRADE

TYPICAL WALL SECTION

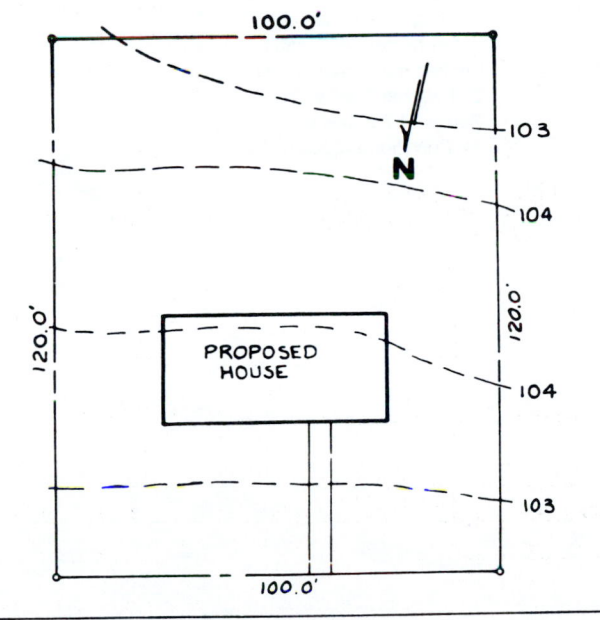

100.0'

103

N

104

120.0'

120.0'

PROPOSED HOUSE

104

103

100.0'

BELL VIEW RD.

SITE PLAN

WINDOW SCHEDULE		
NO.	SIZE	TYPE
1	2-3'-6" × 4'-4"	
2	3'-6" × 4'-4"	

DOOR SCHEDULE		
NO.	SIZE	REMARKS
1,4	3'-0" × 6'-8"	
6,9,11,13	2'-6" × 6'-8"	
2,3	2'-0" × 6'-8"	
16	1'-6" × 6'-8"	
8	5'-0" × 6'-8"	BIFOLD
5,7,10, 12,14,15	2'-4" × 6'-8"	

Design data
Location: Richmond, Va
Occupancy: 2 adults, 3 children
Food disposal
No window shading
Roof, dark
Infiltration, average
Windows—Aluminum with thermal break, operable, metal,
        double glazing, ¼ in air space.
D1—1¾ in solid core flush doors
D2—1¾ in panel door with 1⅛ in panels

Electrical
  Lighting 3 watts per foot
  Water heater 3800 watts
  Clothes dryer, 4400 watts
  Dishwasher 1000 watts
  Range 11 700 watts
  Air Conditioner 2000 watts

ABBREVIATIONS, SI UNITS, AND CONVERSIONS

TYPICAL ABBREVIATIONS

British thermal units	Btu
British thermal units per hour	Btu/hr
cubic feet	ft^3
cubic feet per minute	ft^3/min
cubic meters	m^3
feet	ft
gallons	gal
gallons per minute	gpm
grams	g
inches	in
kilograms	kg
kilopascals	kPa
kilowatts	kW
kilowatt-hours	kWh
liters	L
liters per second	L/s
meters	m
millimeters	mm
square feet	ft^2
square inches	in^2
square meters	m^2
watts	W
watts per square meters	W/m^2

CONVERSION FACTORS

Multiply	By	To Get
Btu/hr	0.2928	watts
Btu/hr/ft^2	3.152	watts per square meter
Btu/hr/ft^2	5.673	watts per Kelvin per square meter
cubic feet	0.028	cubic meters
cubic feet	28.32	liters
cubic feet per minute	0.472	liters per second
cubic meters	35.32	cubic feet
feet	0.305	meters
feet	304.8	millimeters
gallons	3.785	liters
gallons per minute	0.06308	liters per second
inches	25.4	millimeters
liters	0.2642	gallons
liters per second	2.119	cubic feet per minute
liters per second	951	gallons per hour
liters per second	15.85	gallons per minute
meters	3.281	feet
millimeters	0.039	inches
pounds of force per square inch	6.895	kilopascals
square feet	0.0929	square meters
square inches	645.2	square millimeters
square meters	10.76	square feet
watts	3.1412	British thermal units per hour
watts per square meter	0.317	British thermal units per hour